Problems Plus in IIT Mathematics

Asit Das Gupta, PhD
Formerly University Professor of Mathematics
Department of Mathematics
Science College, Patna

Bharati Bhawan
PUBLISHERS & DISTRIBUTORS

Published by

BHARATI BHAWAN (Publishers & Distributors)

www.bharatibhawan.in email: editorial@bbpd.in

4271/3 Ansari Road, Daryaganj, NEW DELHI 110 002, Phone: 23286557
A-61 B/2 Sector 63, NOIDA 201 307, Phone: 4757400
Thakurbari Road, PATNA 800 003, Phone: 2670325
10 Raja Subodh Mallick Square, KOLKATA 700 013, Phone: 22250651
No. 98 Sirsi Circle, Mysore Road, BENGALURU 560 018, Phone: 26740560

First edition 1998
Seventh edition 2005
Revised print 2010
2025 print

Problems Plus in IIT Mathematics
Printed at Goyal Offset Works Pvt. Ltd., Haryana

Preface

Problems Plus in IIT Mathematics is meant for students who aspire to take the entrance examinations of different technical institutions, especially IIT. In writing this book, I have drawn heavily from my experience in teaching students preparing for JEE (IIT). Many students do well at the board examinations and in their coaching courses, and yet falter at JEE (IIT). The reasons for this are:

- They are unable to apply their knowledge to new situations.
- They often cannot detect the area(s) of mathematics on which a problem is based. At JEE (IIT), the student's ability to apply the concepts learnt to new situations is tested. A number of concepts from different areas of mathematics are mixed, creating a new situation.
- They fail to complete the paper within the given time.
- They are often unable to detect the 'form' of the problem. A problem may be set in any one of the following forms:

  1. Direct 2. Indirect 3. Inverted

  To illustrate the different forms, we take three problems from coordinate geometry.

  1. Prove that the points (1, –3), (–2, 1) and $\left(\frac{1}{4}, -2\right)$ are collinear. *[Direct form]*
  2. Find the value of a such that the point $(a, -2)$ is on the diameter of the circle $x^2 + y^2 + 4x - 2y - 20 = 0$ through the point (1, –3). *[Indirect form]*
  3. Find the point whose ordinate is –2 and which is collinear with the points (1, –3) and (–2, 1). *[Inverted form]*

  Basically, the three problems test the same knowledge, namely the condition of collinearity of three points.

  You know that three points (x_r, y_r); $r = 1, 2, 3$ are collinear if the area of the triangle with the points as its vertices is zero, i.e.,

$$\Delta = \frac{1}{2}\begin{vmatrix} x_1 & y_1 & 1 \\ x_2 & y_2 & 1 \\ x_3 & y_3 & 1 \end{vmatrix} = 0.$$

  Problem 1 is easily solved by proving that $\frac{1}{2}\begin{vmatrix} 1 & -3 & 1 \\ -2 & 1 & 1 \\ \frac{1}{4} & -2 & 1 \end{vmatrix} = 0.$

  For Problem 2, we have to find the third point which is the centre of the circle [or the other end of the diameter, because (1, –3) is on the circle]. So, the knowledge of finding the centre from the equation of a circle is also required. Here, centre = (–2, 1) and (1, –3) satisfies the equation of the circle. So we have to find the value of a such that the points $(a,-2)$, $(-2,1)$ and $(1,-3)$ are collinear, i.e., $\frac{1}{2}\begin{vmatrix} a & -2 & 1 \\ -2 & 1 & 1 \\ 1 & -3 & 1 \end{vmatrix} = 0,$

which gives $a = \frac{1}{4}$.

For Problem 3, we take any point $(a,-2)$. So, we have to find the unknown a by using collinearity of the three points $(1,-3)$, $(-2,1)$ and $(a,-2)$. Many students find it difficult to visualise the 'basics' of a problem in the fashion shown above.

The aim of this book is to provide sufficient practice in a methodical manner to help students overcome these difficulties. It is assumed that they have already gone through standard textbooks of the +2 level and know the different mathematical concepts at this level. They know how to solve problems up to a certain level of difficulty, and are now ready to tackle problems which require a bit more application or in some cases, additional knowledge. The collection of these problems has, therefore, been dubbed *Problems Plus.* The salient features of the book are:

- At the beginning of every chapter, a summary of facts, formulae and working techniques are given, with examples wherever necessary. A number of finer points, tricks, techniques, etc., have been given and marked with the icon TIP
- A large number of problems of higher-difficulty level have been worked out. These include many problems already set in competitive examinations, and which illustrate an important point or technique, and many newly framed ones. Since basic knowledge of the chapter is assumed, the *Selected Solved Examples* section starts with problems with slightly higher level of difficulty.
- A large number of the problems in the exercises are newly framed. They have been put in short-answer, long-answer and objective forms. Problems have been given in direct, indirect and inverted forms.
- At the end of each chapter there is a time-bound chapter test. Students should try to finish the test within the time indicated.
- At the end of the book, seven practice tests, based on the knowledge of all the chapters, are given.
- This is followed by *Miscellaneous Exercises*. This will help the reader in gaining confidence in solving problems without prior knowledge of the chapter(s) to which the problems belong.

The book contains about 1000 worked-out problems and about 4000 problems given as exercise. After going through this book, I am certain that the readers will find themselves more comfortable and confident in facing new problems. Any suggestion to improve the book will be thankfully received.

Asit Das Gupta

Note to the Third Edition

I am grateful to the students and teachers, particularly Dr S S Chawla of FIITJEE for sending invaluable suggestions. In this edition a number of new problems have been added, especially in Miscellaneous Exercises.

Asit Das Gupta

Contents

Vectors

Probability

Coordinate Geometry of Three Dimensions

□

Algebra

1. Progressions, Related Inequalities and Series

Recap of Facts and Formulae

1. General forms of progressions

(a) An arithmetic progression is known by its first term a and the common difference d. So while solving a problem on AP, we have to take the first term and the common difference. If given then take them as they are given, otherwise assume the first term = a and the common difference = d.

- General form of an AP is $a, a + d, a + 2d, a + 3d, \dots$.
- If three or four numbers in AP are to be considered, they can be taken conveniently as $a - d, a, a + d$ or $a - 3d, a - d, a + d, a + 3d$.

(b) A geometric progression is known by its first term a and the common ratio r. So while solving a problem on GP we have to take the first term and the common ratio. If given then take them as they are given, otherwise assume the first term = a and the common ratio = r.

- General form of a GP is $a, ar, ar^2, ar^3, \dots$.
- If three or four numbers in GP are to be considered, they can be taken conveniently as

$$\frac{a}{r}, a, ar \quad \text{or} \quad \frac{a}{r^3}, \frac{a}{r}, ar, ar^3.$$

(c) A harmonic progression is known by the AP formed by the reciprocals of its terms.

- General form of an HP is

$$\frac{1}{a}, \frac{1}{a+d}, \frac{1}{a+2d}, \frac{1}{a+3d}, \dots$$

- If three or four numbers in HP are to be considered, they can be taken as

$$\frac{1}{a-d}, \frac{1}{a}, \frac{1}{a+d}$$

or $\dfrac{1}{a-3d}, \dfrac{1}{a-d}, \dfrac{1}{a+d}, \dfrac{1}{a+3d}$ respectively.

2. Classification of a given progression

When we are required to verify whether a given sequence is in AP, GP or HP we proceed as follows:

(here t_n = nth term)

- If $t_n - t_{n-1}$ = constant, for all $n \geq 2$ then the sequence is in AP.
- If $\dfrac{t_n}{t_{n-1}}$ = constant, for all $n \geq 2$ then the sequence is in GP.
- If $\dfrac{1}{t_n} - \dfrac{1}{t_{n-1}}$ = constant, for all $n \geq 2$ then the sequence is in HP.

3. Relations between variables and facts about AP

- nth term = $t_n = a + (n-1)d$.
- Sum to n terms = $S_n = \dfrac{n}{2}\{2a + (n-1)d\} = \dfrac{n}{2}\{a + t_n\}$.
- By knowing the sum to n terms of an AP we can know the nth term using $t_n = S_n - S_{n-1}$.
- The terms of an AP are either increasing (when $d > 0$) or decreasing (when $d < 0$), i.e.,

$$t_n > t_{n-1} \quad \text{or} \quad t_n < t_{n-1} \text{ for all } n \geq 2.$$

- In the AP $a_1, a_2, a_3, \dots, a_n$ we have the following:

$a_2 - a_1 = a_3 - a_2 = \dots = a_n - a_{n-1}$ = common difference

$a_1 + a_n = a_2 + a_{n-1} = \dots$ = constant

$a_1 + k, a_2 + k, a_3 + k, \dots, a_n + k$ are also in AP for $k > 0$ or $k < 0$

$a_1 \times k, a_2 \times k, a_3 \times k, \dots, a_n \times k$ are also in AP for any k

$a_1, a_3, a_5, \dots$ are also in AP

$a_p, a_{p+q}, a_{p+2q}, \dots$ are also in AP.

- The arithmetic mean A of two numbers a, b is

$$A = \frac{a+b}{2}.$$

- The arithmetic mean A of n numbers $a_1, a_2, a_3, \dots, a_n$ is

$$A = \frac{a_1 + a_2 + a_3 + \dots + a_n}{n}.$$

- If n numbers are inserted between a and b such that $a, A_1, A_2, \ldots, A_n, b$ are in AP then

$$A_r = a + (r+1-1)\frac{b-a}{n+1}$$

and $A_1 + A_2 + A_3 + \ldots + A_n = n \times \dfrac{a+b}{2}$.

4. Relations between variables and facts about GP

- nth term $= t_n = ar^{n-1}$.
- Sum of n terms $= S_n = \dfrac{a(1-r^n)}{1-r}$.
- Sum of infinite terms $= S = \dfrac{a}{1-r}$ provided $|r| < 1$.
- By knowing the sum to n terms of a GP we can know the nth term using $t_n = S_n - S_{n-1}$.
- In a GP of positive terms the terms are either increasing (when $r > 1$) or decreasing (when $r < 1$).
- In a GP, $a_1, a_2, a_3, \ldots, a_n$ we have the following:

$$\frac{a_2}{a_1} = \frac{a_3}{a_2} = \ldots = \frac{a_n}{a_{n-1}} = \text{common ratio}$$

$$a_1 a_n = a_2 a_{n-1} = \ldots = \text{constant}$$

$a_1 \cdot k, a_2 \cdot k, a_3 \cdot k, \ldots, a_n \cdot k$ are also in GP, $(k \neq 0)$

$a_1, a_3, a_5, \ldots$ are also in GP

$a_p, a_{p+q}, a_{p+2q}, \ldots$ are also in GP.

- The geometric mean G of two numbers a and b is $G = \sqrt{ab}$.
- The geometric mean G of n numbers $a_1, a_2, a_3, \ldots, a_n$ is

$$G = \sqrt[n]{a_1 \cdot a_2 \cdot a_3 \ldots a_n}.$$

- If n numbers are set between a and b such that $a, G_1, G_2, \ldots, G_n, b$ are in GP then

$$G_m = ar^m \text{ where } r = \left(\frac{b}{a}\right)^{\frac{1}{n+1}}$$

and $G_1 \cdot G_2 \cdot G_3 \cdot \ldots \cdot G_n = (\sqrt{ab})^n$.

5. Relations between variables and facts about HP

- nth term $= t_n = \dfrac{1}{n\text{th term of corresponding AP}}$.
- If $a_1, a_2, a_3, \ldots, a_n$ are in HP then

$$\frac{1}{a_2} - \frac{1}{a_1} = \frac{1}{a_3} - \frac{1}{a_2} = \ldots = \frac{1}{a_n} - \frac{1}{a_{n-1}} = \text{constant.}$$

- The harmonic mean of two numbers a and b is

$$H = \frac{2ab}{a+b}.$$

- The harmonic mean of n nonzero numbers $a_1, a_2, a_3, \ldots, a_n$ is

$$H = \frac{1}{\dfrac{\frac{1}{a_1} + \frac{1}{a_2} + \ldots + \frac{1}{a_n}}{n}}$$

- If n numbers are set between a and b such that $a, H_1, H_2, \ldots, H_n, b$ are in HP then

$$H_m = \frac{1}{\frac{1}{a} + md} \text{ where } d = \frac{\frac{1}{b} - \frac{1}{a}}{n+1}.$$

6. Relations between AM, GM and HM —establishing inequalities of positive numbers

- For two numbers, $AH = G^2$, i.e., A, G, H are in GP.
- For two positive numbers, $A \geq G \geq H$, equality holding for equal numbers.
- For n nonzero positive numbers, $A \geq G \geq H$, equality holding when all the numbers are equal.

7. Summation of series of natural numbers.

- $\sum_{n=1}^{n} n^0 = 1^0 + 2^0 + 3^0 + \ldots + n^0 = \sum_{n=1}^{n} 1 = n$
- $\sum_{n=1}^{n} n = 1 + 2 + 3 + \ldots + n = \dfrac{n(n+1)}{2}$
- $\sum_{n=1}^{n} n^2 = 1^2 + 2^2 + 3^2 + \ldots + n^2 = \dfrac{n(n+1)(2n+1)}{6}$
- $\sum_{n=1}^{n} n^3 = 1^3 + 2^3 + 3^3 + \ldots + n^3 = \left\{\dfrac{n(n+1)}{2}\right\}^2 = \left(\sum_{n=1}^{n} n\right)^2$
- $\sum_{n=1}^{n} n^4$ can be found by using the identity

$$k^5 - (k-1)^5 = 5k^4 - 10k^3 + 10k^2 - 5k + 1$$

which gives

$$n^5 = \sum_{k=1}^{n} \{k^5 - (k-1)^5\}$$

$$= 5\sum_{k=1}^{n} k^4 - 10\sum_{k=1}^{n} k^3 + 10\sum_{k=1}^{n} k^2 - 5\sum_{k=1}^{n} k + \sum_{k=1}^{n} 1.$$

- $(a_1 + a_2 + a_3 + \ldots + a_n)^2 = \sum_{i=1}^{n} a_i^2 + 2\sum_{i<j=1}^{n} a_i a_j$,

i.e., (sum of numbers)^2

= sum of squares of numbers + 2 (sum of products of numbers taking two at a time)

TIP This formula is useful to find the sum of products of numbers taking two at a time if the sum of the numbers and the sum of their squares can be obtained.

8. Summation of related series

- If the nth term of a series is of the form

$$t_n = an^3 + bn^2 + cn + d + kp^n$$

then $S_n = \Sigma t_n = a\Sigma n^3 + b\Sigma n^2 + c\Sigma n + d\Sigma 1 + k\Sigma p^n$

can be obtained by using the formulae for Σn^3, etc., the series Σp^n being in GP.

- Series of following forms admit of summation by using the above method (known by Σ method):

(i) $a \cdot b + (a+d)(b+c) + (a+2d)(b+2c) + \ldots$

which is obtained by the sum of the products of corresponding terms of two (or more) APs.

For example $1 \cdot 4 + 3 \cdot 7 + 5 \cdot 10 + \ldots$

in which the first factors 1, 3, 5, ... as well as the second factors 4, 7, 10, ... are in AP.

(ii) $a, a+b, a+2b+c, a+3b+3c, a+4b+6c, \ldots$

in which the differences of consecutive terms are $b, b+c, b+2c, b+3c, \ldots$ and these are in AP.

(iii) $a, a+b, a+b+br, a+b+br+br^2, \ldots$

in which the differences of consecutive terms are $b, br, br^2, \ldots$ and these are in GP.

- The Σ method is not applicable in a series whose nth term t_n cannot be written as sum of positive integral powers of n and powers of constant, i.e., $an^3 + bn^2 + cn + d + kp^n$. If t_n contains a term of the form $n \cdot p^n$ or $n^2 p^n$, etc., the Σ method cannot be applied.

In series like $a \cdot b + (a+d)br + (a+2d)br^2 + \ldots$, which is obtained by the sum of the products of corresponding terms of an AP and a GP, we get a term as mentioned above. Such series are known as arithmetico-geometric series. The method of solving such series is as follows:

$$S_n = ab + (a+d)br + (a+2d)br^2 + \ldots + (a+\overline{n-1}\,d)br^{n-1}$$

$$r \cdot S_n = abr + (a+d)br^2 + \ldots + (a+\overline{n-2}\,d)br^{n-1} + (a+\overline{n-1}\,d)br^n$$

Subtracting,

$$(1-r)S_n = ab + dbr + dbr^2 + \ldots + dbr^{n-1} - (a+\overline{n-1}\,d)br^n$$

The series on the right, from the second term to the last but one term, is a GP which can be added.

- Some series have their terms in none of the above forms, but by rearranging the terms we get a sum of two or more series which can be summed up.

For example $1^2 - 2 + 3^2 - 4 + \ldots$ to 20 terms, can be rearranged in two series as follows:

$$(1^2 + 3^2 + 5^2 + \ldots \text{ to 10 terms}) - (2 + 4 + 6 + \ldots \text{ to 10 terms})$$

and each of the series can be easily summed.

9. Method of difference in summation of series

- If the nth term of a series can be written as

$$t_n = f(n) - f(n \pm 1)$$

then $S_n = \sum_{n=1}^{n} t_n = \sum_{n=1}^{n} f(n) - \sum_{n=1}^{n} f(n \pm 1)$

$= \{f(1) + f(2) + \ldots + f(n)\} - \{f(2) + f(3) + \ldots + f(n+1)\}$ when '+' sign is taken

or $\{f(1) + f(2) + \ldots + f(n)\} - \{f(0) + f(1) + \ldots + f(n-1)\}$ when '–' sign is taken

$= f(1) - f(n+1)$ when '+' sign is taken

or $f(n) - f(0)$ when '–' sign is taken

- If the series be such that the rth term and $(r+1)$th term can be related as follows then also the series can be added.

(i) $kt_r = f(r)t_r - f(r+1)t_{r+1}$

Then

$$\sum_{r=1}^{n} kt_r = \sum_{r=1}^{n} f(r)t_r - \sum_{r=1}^{n} f(r+1)t_{r+1}$$

or $kS_n = \{f(1) \cdot t_1 + f(2) \cdot t_2 + \ldots + f(n) \cdot t_n\} - \{f(2) \cdot t_2 + f(3) \cdot t_3 + \ldots + f(n+1)t_{n+1}\}$

or $S_n = \frac{1}{k}\{f(1)t_1 - f(n+1)t_{n+1}\}$.

(ii) $kt_r = f(r+1)t_{r+1} - f(r)t_r$

Then

$$\sum_{r=1}^{n} kt_r = \sum_{r=1}^{n} f(r+1)\,t_{r+1} - \sum_{r=1}^{n} f(r)t_r$$

or $kS_n = \{f(2)t_2 + f(3)t_3 + \ldots + f(n+1)t_{n+1}\} - \{f(1) \cdot t_1 + f(2) \cdot t_2 + \ldots + f(n)t_n\}$

or $S_n = \frac{1}{k}\{(n+1)t_{n+1} - f(1) \cdot t_1\}$.

Selected Solved Examples

1. Prove that square roots of three unequal prime numbers cannot be three terms of an AP.

If possible let $\sqrt{p}, \sqrt{q}, \sqrt{r}$ be three terms of an AP. $a, a+d, a+2d, \ldots$ where $p \neq q \neq r$ and they are prime numbers.

Let them be the λth, μth and νth terms respectively.

$\therefore \quad \sqrt{p} = a + (\lambda - 1)d$

$\sqrt{q} = a + (\mu - 1)d$

$\sqrt{r} = a + (\nu - 1)d$

$\therefore \quad \sqrt{p} - \sqrt{q} = (\lambda - \mu)d$

and $\sqrt{q} - \sqrt{r} = (\mu - \nu)d$

$$\therefore \quad \frac{\sqrt{p} - \sqrt{q}}{\sqrt{q} - \sqrt{r}} = \frac{\lambda - \mu}{\mu - \nu}$$

$$\text{or} \quad \frac{(\sqrt{p} - \sqrt{q})(\sqrt{q} + \sqrt{r})}{(\sqrt{q} - \sqrt{r})(\sqrt{q} + \sqrt{r})} = \frac{\lambda - \mu}{\mu - \nu}$$

$$\text{or} \quad \sqrt{pq} + \sqrt{pr} - q - \sqrt{qr} = \frac{\lambda - \mu}{\mu - \nu} \cdot (q - r)$$

$$\text{or} \quad \sqrt{pq} + \sqrt{pr} - \sqrt{qr} = q + \frac{\lambda - \mu}{\mu - \nu}(q - r)$$

$$= \text{rational number.}$$

As p, q, r are unequal primes, $\sqrt{pq}, \sqrt{pr}$ and $\sqrt{qr}$ are unequal pure irrational numbers. Hence, LHS is irrational. But irrational $\neq$ rational.

Hence, the problem.

2. If 3, $3\log_y x$, $3\log_z y$, $7\log_x z$ be in AP, show that $x^{18} = y^{21} = z^{28}$.

Let $\log x = a$, $\log y = b$, $\log z = c$. From the question,

$$3, 3\frac{\log x}{\log y}, 3\frac{\log y}{\log z}, 7\frac{\log z}{\log x} \text{ are in AP}$$

$$\text{or} \quad 3, 3\frac{a}{b}, 3\frac{b}{c}, 7\frac{c}{a} \text{ are in AP.}$$

$$\therefore \quad \frac{3a}{b} = 3 + k, \frac{3b}{c} = 3 + 2k, \frac{7c}{a} = 3 + 3k$$

where k is the common difference.

Multiplying these,

$$(3 + k)(3 + 2k)(3 + 3k) = \frac{3a}{b} \cdot \frac{3b}{c} \cdot \frac{7c}{a} = 63$$

$\therefore \quad 2k^3 + 11k^2 + 18k - 12 = 0$

or $\quad 2k^3 - k^2 + 12k^2 + 18k - 12 = 0$

or $\quad k^2(2k - 1) + 6(2k^2 + 3k - 2) = 0$

or $\quad k^2(2k - 1) + 6(2k - 1)(k + 2) = 0$

or $\quad (2k - 1)(k^2 + 6k + 12) = 0$

$\Rightarrow \quad k = \frac{1}{2}$; other values are imaginary.

$$\therefore \quad \frac{3a}{b} = 3 + \frac{1}{2}, \frac{3b}{c} = 3 + 2 \cdot \frac{1}{2}, \frac{7c}{a} = 3 + 3 \cdot \frac{1}{2}$$

$$\text{or} \quad \frac{a}{b} = \frac{7}{6}, \frac{b}{c} = \frac{4}{3}, \frac{c}{a} = \frac{9}{14}$$

$$\therefore \quad \frac{a}{7} = \frac{b}{6}, \frac{b}{4} = \frac{c}{3}, \frac{a}{14} = \frac{c}{9}$$

$$\therefore \quad \frac{a}{14} = \frac{b}{12} = \frac{c}{9};$$

multiplying by 14×18, $18a = 21b = 28c$

or $\quad 18\log x = 21\log y = 28\log z$

$\therefore \quad x^{18} = y^{21} = z^{28}$.

3. If x, y, z are real numbers satisfying the equation $25(9x^2 + y^2) + 9z^2 - 15(5xy + yz + 3zx) = 0$ then prove that x, y, z are in AP.

We have

$$(15x)^2 + (5y)^2 + (3z)^2 - (15x)(5y) - (5y)(3z) - (3z)(15x) = 0$$

$$\text{or} \quad (15x - 5y)^2 + (5y - 3z)^2 + (3z - 15x)^2 = 0$$

$$[\because a^2 + b^2 + c^2 - bc - ca - ab = \frac{1}{2}\{(a - b)^2 + (b - c)^2 + (c - a)^2\}]$$

$\Rightarrow \quad 15x - 5y = 0, 5y - 3z = 0, 3z - 15x = 0$

$\Rightarrow \quad 15x = 5y = 3z \quad \Rightarrow \quad \frac{x}{1} = \frac{y}{3} = \frac{z}{5} (= k)$

$\therefore \quad x = k, y = 3k, z = 5k.$

So x, y, z are in AP.

4. For what values of the parameter a are there real values of x such that $5^{1+x} + 5^{1-x}$, $\frac{a}{2}$, $25^x + 25^{-x}$ are three consecutive terms of an AP?

We should have

$$2 \cdot \frac{a}{2} = (5^{1+x} + 5^{1-x}) + (25^x + 25^{-x})$$

$$\text{or} \quad a = 5(5^x + 5^{-x}) + (5^{2x} + 5^{-2x})$$

$$= 5\{(5^{x/2} - 5^{-x/2})^2 + 2 \cdot 5^{x/2} \cdot 5^{-x/2}\} + \{(5^x - 5^{-x})^2 + 2 \cdot 5^x \cdot 5^{-x}\}$$

$$= 5\{(5^{x/2} - 5^{-x/2})^2 + 2\} + (5^x - 5^{-x})^2 + 2$$

$\therefore \quad a \geq 5 \cdot 2 + 2 = 12.$

5. Let $a_1, a_2, a_3, \dots, a_n$ be in AP where $a_1 = 0$ and common difference $\neq 0$. Show that

$$\frac{a_3}{a_2}+\frac{a_4}{a_3}+\frac{a_5}{a_4}+\dots+\frac{a_n}{a_{n-1}}-a_2\left(\frac{1}{a_2}+\frac{1}{a_3}+\dots+\frac{1}{a_{n-2}}\right)=\frac{a_{n-1}}{a_2}+\frac{a_2}{a_{n-1}}.$$

Here $t_1 = a_1 = 0$;

common difference $= a_2 - a_1 = a_2 - 0 = a_2$.

$$\text{LHS} = \frac{a_3 - a_2}{a_2}+\frac{a_4 - a_2}{a_3}+\frac{a_5 - a_2}{a_4}+\dots+\frac{a_{n-1}-a_2}{a_{n-2}}+\frac{a_n}{a_{n-1}}$$

$$= \frac{a_2}{a_2}+\frac{a_3}{a_3}+\frac{a_4}{a_4}+\dots+\frac{a_{n-2}}{a_{n-2}}+\frac{a_n}{a_{n-1}}$$

$\{\because t_n - \text{common diff.} = t_{n-1}\}$

$$= (n-3)+\frac{a_n}{a_{n-1}} = (n-3)+\frac{a_1+(n-1)a_2}{a_1+(n-2)a_2}$$

$\{\because a_2 = \text{common difference}\}$

$$= (n-3)+\frac{n-1}{n-2} \quad \{\because a_1 = 0\}$$

$$\text{RHS} = \frac{a_1+(n-2)a_2}{a_2}+\frac{a_2}{a_1+(n-2)a_2}$$

$$= (n-2)+\frac{1}{n-2} = (n-3)+1+\frac{1}{n-2}$$

$$= (n-3)+\frac{n-1}{n-2}$$

$\therefore$ LHS = RHS.

6. Prove that $(2n)^2$ where $n \in N$ can be expressed as the sum of n terms of a series of integers in AP while $(2n+1)^2$ can be expressed as $s_n + 1$ where s_n is the sum of an AP of n terms.

$(2n)^2 = 4n^2 = 4(1+3+5+\dots$ to n terms)

$= 4 + 12 + 20 + \dots$ to n terms,

which is in AP whose common diff. = 8.

$$(2n+1)^2 = 4n^2+4n+1 = 4n(n+1)+1 = 8\cdot\frac{n(n+1)}{2}+1$$

$$= 8\{1+2+3+\dots+n\}+1$$

$$= \{8+16+24+\dots+8n\}+1$$

$$= s_n + 1$$

where s_n is the sum of n terms of the AP 8, 16, 24, ...

7. Let $a_1, a_2, a_3, \dots, a_n, \dots$, be an AP and

$$S_k = a_{(k-1)n+1}+a_{(k-1)n+2}+a_{(k-1)n+3}+\dots+a_{kn}.$$

Prove that $S_1, S_2, S_3, \dots$ are in AP whose common difference

$= n^2 \times$ (the common difference of the given AP).

Here $S_k = a_{(k-1)n+1}+a_{(k-1)n+2}+\dots+a_{(k-1)n+n}$

$= [a_1+(k-1)nd]+[a_1+\{(k-1)n+1\}d]+\dots$

$+ [a_1+\{(k-1)n+(n-1)\}d]$,

where d = common diff. of the given AP.

$= na_1+(k-1)nd\cdot n+\{1+2+3+\dots+(n-1)\}d$

$$= na_1+\frac{(n-1)n}{2}d+(k-1)dn^2$$

$$\therefore \quad S_{k+1} = na_1+\frac{(n-1)n}{2}d+(k+1-1)dn^2$$

$\therefore \quad S_{k+1}-S_k = dn^2$ for all $k = 1, 2, 3, \dots$

$\therefore \quad S_2 - S_1 = S_3 - S_2 = \dots;\ \therefore\ S_1, S_2, S_3, \dots$ are in AP.

The common difference is dn^2, i.e., $n^2 \times d$, i.e., $n^2 \times$ {common difference of the given AP}.

8. The sum of r terms of an AP is denoted by S_r and $\frac{S_m}{S_n}=\frac{m^2}{n^2}$. Find the ratio of the mth term and the nth term of the AP.

Let the first term = a, the common difference = d.

$$\therefore \quad S_m = \frac{m}{2}[2a+(m-1)d]$$

and $S_n = \frac{n}{2}[2a+(n-1)d]$

$$\therefore \quad \frac{S_m}{S_n} = \frac{\frac{m}{2}\{2a+(m-1)d\}}{\frac{n}{2}\{2a+(n-1)d\}}$$

or $\frac{m^2}{n^2} = \frac{m}{n}\cdot\frac{2a+(m-1)d}{2a+(n-1)d}$

or $\frac{2a+(m-1)d}{2a+(n-1)d} = \frac{m}{n}$. This is an identity in m and n.

So, putting $2m-1$ for m and $2n-1$ for n we get

$$\frac{2a+(2m-1-1)d}{2a+(2n-1-1)d} = \frac{2m-1}{2n-1}$$

or $\frac{a+(m-1)d}{a+(n-1)d} = \frac{2m-1}{2n-1}$

or $\frac{m\text{th term}}{n\text{th term}} = \frac{2m-1}{2n-1}$.

9. Let r be the common ratio of the GP $a_1, a_2, a_3, \dots, a_n$. Show that

$$\frac{1}{a_1^m+a_2^m}+\frac{1}{a_2^m+a_3^m}+\dots+\frac{1}{a_{n-1}^m+a_n^m}=\frac{1-r^{m(1-n)}}{a_1^m(r^m-r^{-m})}.$$

$$\text{LHS} = \frac{1}{a_1^m + (a_1 r)^m} + \frac{1}{a_2^m + (a_2 r)^m} + \dots + \frac{1}{a_{n-1}^m + (a_{n-1} \cdot r)^m}$$

$$= \frac{1}{1 + r^m}\left\{\frac{1}{a_1^m} + \frac{1}{a_2^m} + \dots + \frac{1}{a_{n-1}^m}\right\}$$

As $a_1, a_2, a_3, \dots$ form a GP, their reciprocals

$$\frac{1}{a_1}, \frac{1}{a_2}, \frac{1}{a_3}, \dots$$

are also in GP whose common ratio $= \frac{1}{r}$.

Again, their powers $\left(\frac{1}{a_1}\right)^m, \left(\frac{1}{a_2}\right)^m, \left(\frac{1}{a_3}\right)^m, \dots$ are also in GP whose common ratio will be $\left(\frac{1}{r}\right)^m$.

$$\therefore \quad \text{LHS} = \frac{1}{1+r^m} \cdot \frac{\frac{1}{a_1^m}\left\{1 - \left(\frac{1}{r^m}\right)^{n-1}\right\}}{1 - \frac{1}{r^m}}$$

$$= \frac{1 - r^{-mn+m}}{a_1^m (1 + r^m)(1 - r^{-m})}$$

$$= \frac{1 - r^{m(1-n)}}{a_1^m (r^m - r^{-m})}.$$

10. The sum of squares of three distinct real numbers which are in GP is S^2. If their sum is $\alpha \cdot S$, show that $\alpha^2 \in \left(\frac{1}{3}, 1\right) \cup (1, 3)$.

Three numbers in GP are $\frac{a}{r}, a, ar$. As they are distinct, $r \neq 1, -1$.

From the question, $\frac{a^2}{r^2} + a^2 + a^2 r^2 = S^2$ ... (1)

and $\frac{a}{r} + a + ar = \alpha \cdot S$ ... (2)

Let $r + \frac{1}{r} = x$. So $x \neq 2, -2$ as $r \neq 1, -1$.

(1) $\Rightarrow \quad a^2\left(r^2 + \frac{1}{r^2}\right) + a^2 = S^2$

or $\quad a^2\left\{\left(r + \frac{1}{r}\right)^2 - 2\right\} + a^2 = S^2$

$\therefore \quad x^2 - 2 + 1 = \frac{S^2}{a^2}$

or $\quad x^2 - 1 = \frac{S^2}{a^2}$ ... (3)

(2) $\Rightarrow \quad a\left(r + \frac{1}{r}\right) + a = \alpha \cdot S$

or $\quad ax + a = \alpha \cdot S$

or $\quad x + 1 = \alpha \cdot \frac{S}{a}$ ... (4)

(3) and (4) $\Rightarrow \quad \frac{(x+1)^2}{x^2 - 1} = \alpha^2$

or $\quad \frac{x+1}{x-1} = \alpha^2$ ... (5)

$\therefore \quad x = \frac{\alpha^2 + 1}{\alpha^2 - 1}$

or $\quad r + \frac{1}{r} = \frac{\alpha^2 + 1}{\alpha^2 - 1}$

or $\quad \frac{r^2 + 1}{r} = \frac{\alpha^2 + 1}{\alpha^2 - 1}$

or $\quad (\alpha^2 - 1)r^2 - (\alpha^2 + 1)r + \alpha^2 - 1 = 0$

As r is real, $D \geq 0$

or $\quad (\alpha^2 + 1)^2 - 4(\alpha^2 - 1)^2 \geq 0$

or $\quad \{\alpha^2 + 1 + 2(\alpha^2 - 1)\}\{\alpha^2 + 1 - 2(\alpha^2 - 1)\} \geq 0$

or $\quad (3\alpha^2 - 1)(3 - \alpha^2) \geq 0$

$\therefore$ either $3\alpha^2 - 1 \geq 0$ and $3 - \alpha^2 \geq 0$ | or $3\alpha^2 - 1 \leq 0$ and $3 - \alpha^2 \leq 0$

$\therefore$ either $\alpha^2 \geq \frac{1}{3}$ and $\alpha^2 \leq 3$ | $\therefore$ $\alpha^2 \leq \frac{1}{3}$ and $\alpha^2 \geq 3$ (contradictory)

$\therefore \quad \alpha^2 \geq \frac{1}{3}$ and $\alpha^2 \leq 3$

From (5), $\alpha^2 \neq 3, \frac{1}{3}$ because $x \neq 2, -2$

Also $\alpha^2 \neq 1$ for otherwise $\frac{x+1}{x-1} = 1$, which implies $1 = -1$.

$\therefore \quad \frac{1}{3} < \alpha^2 < 1$ and $1 < \alpha^2 < 3$.

So $\alpha^2 \in \left(\frac{1}{3}, 1\right) \cup (1, 3)$.

11. Find the sum of all the proper divisors of $N = 2^p \cdot 3^q \cdot 5^r$ where p, q, r are positive integers. Also, find the sum of all the proper divisors that are multiples of 3.

[**Note** If we multiply the three geometric series

$$a^0 + a^1 + a^2 + a^3 + \dots + a^m,$$

$$b^0 + b^1 + b^2 + \dots + b^n$$

and $c^0 + c^1 + c^2 + \dots + c^p$

then we get the sum of all the terms of the form

$a^i b^j c^k$ where $i = 0, 1, 2, \ldots m$; $j = 0, 1, 2, \ldots n$ and $k = 0, 1, 2, \ldots p$, i.e., terms like

$a^0b^0c^0, a^0b^1c^0, a^0b^0c^1, a^0b^1c^1, a^0b^0c^2, a^0b^2c^0,$

$a^0b^1c^2, a^0b^2c^1, a^0b^0c^3, a^0b^3c^0, \ldots$

$a^1b^0c^0, a^1b^1c^0, a^1b^0c^1, \ldots$

$a^2b^0c^0, a^2b^1c^0, a^2b^0c^1, \ldots$

....

$a^m\ b^n\ c^p.$

Clearly each of the term is a divisor of $a^m b^n c^p$ of which $a^0b^0c^0$ and $a^m b^n c^p$ are not proper divisors.]

The sum of all the divisors of $2^p \cdot 3^q \cdot 5^r$

$$= (2^0 + 2^1 + 2^2 + 2^3 + \ldots + 2^p) \times (3^0 + 3^1 + 3^2 + 3^3 + \ldots + 3^q) \times (5^0 + 5^1 + 5^2 + 5^3 + \ldots + 5^r)$$

$$= \frac{2^{p+1} - 1}{2 - 1} \cdot \frac{3^{q+1} - 1}{3 - 1} \cdot \frac{5^{r+1} - 1}{5 - 1}$$

$$= \frac{1}{8}(2^{p+1} - 1)(3^{q+1} - 1)(5^{r+1} - 1)$$

But 1, i.e., $2^0 \cdot 3^0 \cdot 5^0$ and the number, i.e., $2^p \cdot 3^q \cdot 5^r$ are not proper divisors.

$\therefore$ the required sum

$$= \frac{1}{8}(2^{p+1} - 1)(3^{q+1} - 1)(5^{r+1} - 1) - 1 - 2^p \cdot 3^q \cdot 5^r.$$

The sum of all the divisors of $2^p \cdot 3^q \cdot 5^r$ that are multiples of 3

$$= (2^0 + 2^1 + 2^2 + \ldots + 2^p)(3^1 + 3^2 + 3^3 + \ldots + 3^q) \times (5^0 + 5^1 + 5^2 + \ldots + 5^r)$$

$$= \frac{2^{p+1} - 1}{2 - 1} \cdot \frac{3(3^q - 1)}{3 - 1} \cdot \frac{5^{r+1} - 1}{5 - 1}.$$

But $2^p \cdot 3^q \cdot 5^r$ is not a proper divisor of the given number.

$\therefore$ the required sum

$$= (2^{p+1} - 1)\frac{3}{2}(3^q - 1)\frac{1}{4}(5^{r+1} - 1) - 2^p \cdot 3^q \cdot 5^r$$

$$= \frac{3}{8}(2^{p+1} - 1)(3^q - 1)(5^{r+1} - 1) - 2^p \cdot 3^q \cdot 5^r.$$

12. Two geometric series have the same first term 1. The common ratio of the first series is greater than that of the second series and both are positive. If S_n and s_n denote sum of n terms of the two series respectively then prove that $\dfrac{S_n}{S_{n+1}} < \dfrac{s_n}{s_{n+1}}$.

Clearly S_n, S_{n+1}, s_n and s_{n+1} are all positive.

Let a, b be their common ratios respectively. So $a > b > 0$.

$$s_n \cdot S_{n+1} - s_{n+1} \cdot S_n$$

$$= (1 + b + b^2 + \ldots + b^{n-1})(1 + a + a^2 + \ldots + a^n) - (1 + b + b^2 + \ldots + b^n)(1 + a + a^2 + \ldots + a^{n-1})$$

$$= (1 + b + b^2 + \ldots + b^{n-1})(1 + a + a^2 + \ldots + a^{n-1}) + a^n(1 + b + b^2 + \ldots + b^{n-1}) - (1 + b + b^2 + \ldots + b^{n-1}) \times (1 + a + a^2 + \ldots + a^{n-1}) - b^n(1 + a + a^2 + \ldots + a^{n-1})$$

$$= a^n(1 + b + b^2 + \ldots + b^{n-1}) - b^n(1 + a + a^2 + \ldots + a^{n-1})$$

$$= (a^n - b^n) + ab(a^{n-1} - b^{n-1}) + a^2b^2(a^{n-2} - b^{n-2}) + \ldots + a^{n-1}b^{n-1}(a - b),$$

each term of which are positive because $a > b > 0$ and so $a^n > b^n$ for all $n \in N$.

$\therefore \quad s_n \cdot S_{n+1} - s_{n+1} \cdot S_n > 0$

or $\quad s_{n+1} \cdot S_n < s_n \cdot S_{n+1}$

$$\therefore \quad \frac{S_n}{S_{n+1}} < \frac{s_n}{s_{n+1}}.$$

13. If $S_1, S_2, S_3, \ldots, S_n$ are the sums of infinite geometric series whose first terms are 1, 2, 3, ..., n and whose common ratios are $\dfrac{1}{2}, \dfrac{1}{3}, \dfrac{1}{4}, \ldots, \dfrac{1}{n+1}$ respectively then find the value of $\sum\limits_{r=1}^{2n-1} S_r^2$.

The rth series has the first term $a = r$,

common ratio $= \dfrac{1}{r+1}$, sum $= S_r$

$$\therefore \quad S_r = \frac{r}{1 - \dfrac{1}{r+1}} = \frac{r(r+1)}{r} = r + 1.$$

So, $\quad S_r^2 = (r + 1)^2$

$$\therefore \quad \sum_{r=1}^{2n-1} S_r^2 = \sum_{r=1}^{2n-1} (r + 1)^2 = 2^2 + 3^2 + 4^2 + \ldots + (2n)^2$$

$$= \{1^2 + 2^2 + 3^2 + \ldots + (2n)^2\} - 1$$

$$= \frac{2n(2n + 1)(4n + 1)}{6} - 1$$

$$= \frac{1}{3}n(2n + 1)(4n + 1) - 1.$$

14. If α is a root of the equation $x^2 - 20x + 64 = 0$ and

$$\alpha = e^{(1 + |\cos\theta| + \cos^2\theta + |\cos^3\theta| + \cos^4\theta + \ldots \text{ to } \infty)\log_e 4}$$

then find $\theta \in (0, \pi)$.

Let $|\cos\theta| = t$. Then

$$\alpha = e^{(1 + t + t^2 + t^3 + \ldots \text{ to } \infty)\log_e 4} \quad \{\because t^2 = |t|^2\}$$

$$\therefore \quad \alpha = e^{\frac{1}{1-t}\log_e 4}$$

$\{\because t = |\cos\theta| < 1$ in the interval $(0, \pi)\}$

$$= e^{\log_e 4^{\frac{1}{1-t}}} = 4^{\frac{1}{1-t}}.$$

But roots of $x^2 - 20x + 64 = 0$

or $(x-4)(x-16) = 0$ are 4, 16.

$$\therefore \quad 4^{\frac{1}{1-t}} = 4, 4^2 \quad \Rightarrow \quad \frac{1}{1-t} = 1, 2, \text{ i.e., } t = 0, \frac{1}{2}$$

$$\therefore \quad |\cos\theta| = 0, \frac{1}{2}; \quad \therefore \quad \cos\theta = 0, \frac{1}{2}, -\frac{1}{2}$$

$$\therefore \quad \theta = \frac{\pi}{2}, \frac{\pi}{3}, \pi - \frac{\pi}{3}; \quad \text{i.e.,} \quad \frac{\pi}{3}, \frac{\pi}{2}, \frac{2\pi}{3}.$$

15. Prove that the natural number 777...7 consisting of 77 sevens is a composite number divisible by a multiple of 7.

$$777\ldots7 = 7 \times 111\ldots1$$
$$= 7 \times \{10^{76} + 10^{75} + 10^{74} + \ldots + 10 + 1\}$$
$$= 7 \times \frac{10^{77} - 1}{10 - 1} = 7 \times \frac{10^{77} - 1}{10^7 - 1} \times \frac{10^7 - 1}{10 - 1}$$
$$= 7 \times \frac{(10^7)^{11} - 1}{(10^7) - 1} \times \frac{10^7 - 1}{10 - 1}$$
$$= 7 \times (1 + 10^7 + 10^{14} + \ldots + 10^{70}) \times (1 + 10 + 10^2 + \ldots 10^6)$$
$$= 7 \times \text{integer} \times \text{integer}.$$

Hence the problem.

16. Given a GP and an AP with positive terms $a, a_1, a_2, a_3, \ldots, a_n, \ldots$ and $b, b_1, b_2, b_3, \ldots, b_n, \ldots$ respectively where the common ratio of the GP $\neq 1$. Prove that there exists a positive real number x such that $\log_x a_n - b_n = \log_x a - b$ for all $n \in N$. Also find that x.

Let the common ratio of the GP be r and the common difference of the AP be d. As the progressions have positive terms,

$$a > 0, r > 0, a_n > 0, b > 0, b_n > 0, d > 0 \text{ for all } n.$$

Now, $a_n = ar^n$ and $b_n = b + nd$

$$\therefore \quad \frac{a_n}{a} = r^n \quad \text{and} \quad b_n - b = nd$$

$$\therefore \quad \log_x \frac{a_n}{a} = \log_x r^n$$

$$\therefore \quad \log_x \frac{a_n}{a} = b_n - b \quad \text{if} \quad \log_x r^n = nd$$

$$\Rightarrow \quad n\log_x r = nd \quad \Rightarrow \quad \log_x r = d$$

$$\Rightarrow \quad x^d = r \quad \Rightarrow \quad x = r^{\frac{1}{d}} > 0$$

$$\therefore \quad \log_x a_n - b_n = \log_x a - b \quad \text{where} \quad x = r^{\frac{1}{d}} > 0.$$

17. If $\frac{1}{a} + \frac{1}{c} + \frac{1}{a-b} + \frac{1}{c-b} = 0$, prove that a, b, c are in HP or $a, \frac{b}{2}, c$ are in AP.

Here $\left(\frac{1}{a} + \frac{1}{c-b}\right) + \left(\frac{1}{c} + \frac{1}{a-b}\right) = 0$

or $\frac{a+c-b}{a(c-b)} + \frac{c+a-b}{c(a-b)} = 0$

or $(a+c-b)\left\{\frac{1}{a(c-b)} + \frac{1}{c(a-b)}\right\} = 0$

or $(a+c-b)\frac{c(a-b) + a(c-b)}{ac(c-b)(a-b)} = 0$

$\therefore$ either $a + c - b = 0$ ... (1)

or $c(a-b) + a(c-b) = 0$ ... (2)

(1) $\Rightarrow$ $b = a + c$

$\Rightarrow$ $\frac{b}{2} = \frac{a+c}{2}$

$\Rightarrow$ $a, \frac{b}{2}, c$ are in AP.

(2) $\Rightarrow$ $2ac = bc + ab$

$\Rightarrow$ $b = \frac{2ac}{a+c}$ $\Rightarrow$ b is the HM between a, c

$\therefore$ a, b, c are in HP.

18. If the $(m+1)$th, $(n+1)$th and $(r+1)$th terms of an AP are in GP and m, n, r are in HP, show that the ratio of the common difference to the first term of the AP is $\frac{-2}{n}$.

Let the AP be $a, a+d, a+2d, \ldots$

Then $t_{m+1} = a + md$, $t_{n+1} = a + nd$, $t_{r+1} = a + rd$

which are in GP.

$$\therefore \quad (a + nd)^2 = (a + md)(a + rd)$$

or $a^2 + 2nad + n^2d^2 = a^2 + (m+r)ad + mrd^2$

or $\{2n - (m+r)\}ad = (mr - n^2)d^2$

$\therefore \quad \dfrac{d}{a} = \dfrac{2n - (m+r)}{mr - n^2}$

$\therefore \quad \dfrac{d}{a} = \dfrac{2n^2 - n(m+r)}{n(mr - n^2)}$

$= \dfrac{2n^2 - 2mr}{n(mr - n^2)} \left\{\because m, n, r \text{ are in HP} \Rightarrow n = \dfrac{2mr}{m+r}\right\}$

$= \dfrac{2(n^2 - mr)}{-n(n^2 - mr)}$

$= -\dfrac{2}{n}.$

19. An AP, a GP and an HP have *a* and *b* for their first two terms, show that their (*n* + 2)th terms will be in GP if $\dfrac{b^{2n+2} - a^{2n+2}}{ba(b^{2n} - a^{2n})} = \dfrac{n+1}{n}$.

Clearly, the common difference of the AP $= b - a$, the common ratio of the GP $= \dfrac{b}{a}$, and the common difference of the corresponding AP of the HP $= \dfrac{1}{b} - \dfrac{1}{a}$.

$\therefore \quad (n+2)$th term of the AP $= t_{n+2} = a + (n+1)(b-a)$

$(n+2)$th term of the GP $= t'_{n+2} = a \cdot \left(\dfrac{b}{a}\right)^{n+1}$

$(n+2)$th term of the HP $= t''_{n+2}$

$= \dfrac{1}{(n+2)\text{th term of the corresponding AP}}$

$= \dfrac{1}{\dfrac{1}{a} + (n+1)\left(\dfrac{1}{b} - \dfrac{1}{a}\right)}.$

Now, $t_{n+2}, t'_{n+2}, t''_{n+2}$ are in GP if

$(t'_{n+2})^2 = t_{n+2} \cdot t''_{n+2}$

$\Rightarrow \quad \left\{a \cdot \left(\dfrac{b}{a}\right)^{n+1}\right\}^2 = \dfrac{a + (n+1)(b-a)}{\dfrac{1}{a} + (n+1)\left(\dfrac{1}{b} - \dfrac{1}{a}\right)}$

or $\dfrac{b^{2n+2}}{a^{2n}} = \dfrac{ab\{(n+1)b - na\}}{b + (n+1)(a-b)}$

or $\dfrac{b^{2n+1}}{a^{2n+1}} = \dfrac{(n+1)b - na}{(n+1)a - nb}$

or $(n+1)ab^{2n+1} - nb^{2n+2} = (n+1)a^{2n+1} \cdot b - na^{2n+2}$

or $(n+1)\{ab^{2n+1} - a^{2n+1}b\} = n(b^{2n+2} - a^{2n+2})$

or $(n+1)ab(b^{2n} - a^{2n}) = n(b^{2n+2} - a^{2n+2})$;

$\therefore \quad \dfrac{n+1}{n} = \dfrac{b^{2n+2} - a^{2n+2}}{ab(b^{2n} - a^{2n})}.$

20. If *P* be the *r*th term when *n* arithmetic means are inserted between *x* and *y* and *Q* be the *r*th term when *n* harmonic means are inserted between *x* and *y* then show that $\dfrac{P}{x} + \dfrac{y}{Q}$ is independent of *n* and *r*.

Let $x, A_1, A_2, A_3, \ldots, A_n, y$ be in AP ... (1)

and $x, H_1, H_2, H_3, \ldots, H_n, y$ be in HP ... (2)

According to the question, $P = A_{r-1}$ and $Q = H_{r-1}$.

Now, from (1), $y = (n+2)$th term $= x + (n+1)d$

$\therefore \quad d = \dfrac{y - x}{n+1}$

$\therefore \quad P = A_{r-1} = r$th term

$= x + (r-1)d = x + (r-1) \cdot \dfrac{y-x}{n+1}$

$\therefore \quad \dfrac{P}{x} = 1 + (r-1) \cdot \dfrac{y-x}{x(n+1)}$

Now, from (2), $\dfrac{1}{x}, \dfrac{1}{H_1}, \dfrac{1}{H_2}, \ldots, \dfrac{1}{H_n}, \dfrac{1}{y}$ are in AP.

$\therefore \quad \dfrac{1}{y} = (n+2)$th term

$= \dfrac{1}{x} + (n+1)d'$

$\therefore \quad d' = \dfrac{\dfrac{1}{y} - \dfrac{1}{x}}{n+1}$

$\therefore \quad \dfrac{1}{Q} = \dfrac{1}{H_{r-1}} = r$th term

$= \dfrac{1}{x} + (r-1)d' = \dfrac{1}{x} + (r-1) \cdot \dfrac{\dfrac{1}{y} - \dfrac{1}{x}}{n+1}$

$\therefore \quad \dfrac{y}{Q} = \dfrac{y}{x} + (r-1) \cdot \dfrac{1 - \dfrac{y}{x}}{n+1}$

$\therefore \quad \dfrac{P}{x} + \dfrac{y}{Q} = 1 + (r-1) \cdot \dfrac{y-x}{x(n+1)} + \dfrac{y}{x} + (r-1) \cdot \dfrac{1 - \dfrac{y}{x}}{n+1}$

$= 1 + \dfrac{r-1}{n+1} \cdot \dfrac{y-x}{x} + \dfrac{y}{x} + \dfrac{r-1}{n+1} \cdot \dfrac{x-y}{x}$

$= 1 + \dfrac{y}{x} =$ independent of n and r.

21. Take two sets of m and n parallel lines, lines of one set cutting those of the other. Thus $m \times n$ small rhombuses are formed. Two rhombuses are called neighbouring rhombuses if they have a common side. In each of these rhombuses a natural number is put such that the number written in a rhombus is the AM of the numbers written in all its neighbouring members. Show that all the numbers put in the rhombuses are equal.

There are two possibilities:

(i) all the numbers are equal

(ii) at least two numbers are different

If all the numbers are equal, there is nothing to prove.

If at least two numbers are different, there will be the smallest number. Select the rhombus with the smallest number p in it. As p is the AM of the numbers in the neighbouring rhombuses, p must be greater than at least one of the numbers in the neighbouring rhombuses.

($\because$ AM of two numbers a, b is $\frac{a+b}{2}$ which is bigger than one but smaller than the other; AM of three numbers a, b, c is $\frac{a+b+c}{3}$ which cannot be smaller than a, b as well as c, etc.). In that case p is not the least among the numbers put in all the rhombuses.

Thus the possibility (ii) is absurd.

Hence, all the numbers are equal.

22. A traveller starts from a certain place on a certain day and travels 1 km on the first day and on subsequent days, he travels 2 km more than the previous day. After 3 days, a second traveller sets out from the same place and on his first day he travels 12 km and on subsequent days he travels 1 km more than the previous day. On how many days will the second traveller be ahead of the first?

The first traveller travels on different days as follows (in km) 1, 1 + 2, 1 + 2 + 2, ...

After 3 days the first traveller is already ahead by (1 + 3 + 5) km, i.e., 9 km.

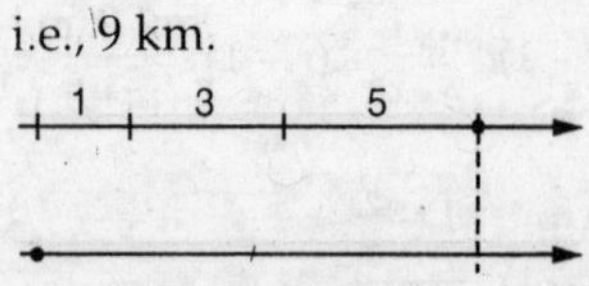

The second traveller travels on different days as follows: 0, 0, 0, 12, 13, 14, ...

After n days from the day the second traveller starts, the distance covered by the first

$$= 1 + 3 + 5 + (7 + 9 + \dots \text{ to } n \text{ terms})$$

$$= 1 + 3 + 5 + \dots \text{ to } (n+3) \text{ terms}$$

$$= (n+3)^2$$

and the distance covered by the second

$$= 12 + 13 + 14 + \dots \text{ to } n \text{ terms}$$

$$= \frac{n}{2}\{24 + (n-1)\cdot 1\} = \frac{n(n+23)}{2}.$$

The second traveller is ahead of the first on the nth day (after the second sets off) if

$$\frac{n(n+23)}{2} > (n+3)^2$$

or $\quad n^2 + 23n > 2(n^2 + 6n + 9)$

or $\quad n^2 - 11n + 18 < 0$

or $\quad (n-2)(n-9) < 0.$

So $\quad n - 2 > 0$ and $n - 9 < 0$ ... (1)

or $\quad n - 2 < 0$ and $n - 9 > 0$ ... (2)

(1) $\Rightarrow$ $n > 2$ and $n < 9$

(2) $\Rightarrow$ $n < 2$ and $n > 9$ (absurd)

Thus, from the beginning of the 3rd day to the end of the 9th day the second traveller is ahead of the first.

So, the second is ahead of the first on the 3rd, 4th, 5th, ..., 9th days (after the second sets off).

Hence, the required number of days = 7.

23. If $0 < r < 1$ and m is a positive integer, show that

$$\mathbf{(2m+1)r^m(1-r) < 1 - r^{2m+1}.}$$

$$\frac{1 - r^{2m+1}}{1 - r} = 1 + r + r^2 + \dots + r^{2m}$$

$$= (1 + r^{2m}) + (r + r^{2m-1}) + \dots$$

$$+ (r^{m-1} + r^{m+1}) + r^m \quad \dots (1)$$

making pairs of equidistant terms from the beginning and the end.

Now, for unequal positive numbers $A > G$

$$\therefore \quad \frac{1 + r^{2m}}{2} > \sqrt{1 \cdot r^{2m}}, \quad \frac{r + r^{2m-1}}{2} > \sqrt{r \cdot r^{2m-1}}, \dots,$$

$$\frac{r^{m-1} + r^{m+1}}{2} > \sqrt{r^{m-1} \cdot r^{m+1}}$$

$$\Rightarrow \quad 1 + r^{2m} > 2r^m, \ r + r^{2m-1} > 2r^m, \dots, r^{m-1} + r^{m+1} > 2r^m$$

$\therefore$ using (1),

$$\frac{1 - r^{2m+1}}{1 - r} > (2r^m + 2r^m + \dots \text{ to } m \text{ terms}) + r^m$$

$$= (2m+1)r^m$$

$$\therefore \quad 1 - r^{2m+1} > (2m+1)r^m \cdot (1 - r), \text{ because } 1 - r > 0.$$

24. If x, y, z are positive and $x+y+z=1$, prove that $\left(\frac{1}{x}-1\right)\left(\frac{1}{y}-1\right)\left(\frac{1}{z}-1\right)\geq 8.$

Here $\left(\frac{1}{x}-1\right)\left(\frac{1}{y}-1\right)\left(\frac{1}{z}-1\right)$

$$=\left(\frac{x+y+z}{x}-1\right)\left(\frac{x+y+z}{y}-1\right)\left(\frac{x+y+z}{z}-1\right),$$

because $x+y+z=1$

$$=\frac{(y+z)(z+x)(x+y)}{xyz}$$

Now, for positive numbers $A \geq G$.

$\therefore$ for y, z we get $\frac{y+z}{2}\geq\sqrt{yz}$

or $y+z\geq 2\sqrt{yz}$ ... (1)

For z, x we get $\frac{z+x}{2}\geq\sqrt{zx}$

or $z+x\geq 2\sqrt{zx}$ ... (2)

For x, y we get $\frac{x+y}{2}\geq\sqrt{xy}$

or $x+y\geq 2\sqrt{xy}$ ... (3)

Multiplying (1), (2) and (3) we get

$$(y+z)(z+x)(x+y)\geq 8\sqrt{yz\cdot zx\cdot xy}=8xyz$$

or $\frac{(y+z)(z+x)(x+y)}{xyz}\geq 8$, for $xyz>0$

$\therefore$ $\left(\frac{1}{x}-1\right)\left(\frac{1}{y}-1\right)\left(\frac{1}{z}-1\right)\geq 8.$

25. If $a>0, b>0$ and $a+b=1$, show that $\left(a+\frac{1}{a}\right)^2+\left(b+\frac{1}{b}\right)^2\geq 12.5.$

$\because$ $a+b=1$ and $a>0, b>0$, we take

$a=\cos^2\theta, b=\sin^2\theta.$

Then $\left(a+\frac{1}{a}\right)^2+\left(b+\frac{1}{b}\right)^2=a^2+\frac{1}{a^2}+b^2+\frac{1}{b^2}+4$

$$=\cos^4\theta+\sec^4\theta+\sin^4\theta+\operatorname{cosec}^4\theta+4$$

$$=(\cos^4\theta+\sin^4\theta)+(1+\tan^2\theta)^2+(1+\cot^2\theta)^2+4$$

$$=(\cos^4\theta+\sin^4\theta)+(\tan^4\theta+\cot^4\theta)$$

$$+2(\tan^2\theta+\cot^2\theta)+6 \quad \dots (1)$$

Now, $\cos^4\theta+\sin^4\theta=(\cos^2\theta+\sin^2\theta)^2-2\cos^2\theta\cdot\sin^2\theta$

$$=1-\frac{1}{2}\cdot\sin^2 2\theta\geq 1-\frac{1}{2}\cdot 1=\frac{1}{2}$$

because the maximum value of $\sin^2 2\theta=1$

$\because$ $A\geq G$, $\frac{\tan^4\theta+\cot^4\theta}{2}\geq\sqrt{\tan^4\theta\cdot\cot^4\theta}=1$

$\therefore$ $\tan^4\theta+\cot^4\theta\geq 2$

$\because$ $A\geq G$, $\frac{\tan^2\theta+\cot^2\theta}{2}\geq\sqrt{\tan^2\theta\cdot\cot^2\theta}=1$

$\therefore$ $\tan^2\theta+\cot^2\theta\geq 2$

$\therefore$ from (1),

$$\left(a+\frac{1}{a}\right)^2+\left(b+\frac{1}{b}\right)^2\geq\frac{1}{2}+2+2\cdot 2+6=12.5.$$

26. If a, b, c are unequal positive numbers in HP then prove that $\frac{a+b}{2a-b}+\frac{c+b}{2c-b}>4.$

$\because$ a, b, c are in HP, $\frac{2}{b}=\frac{1}{a}+\frac{1}{c}$ ... (1)

$$\text{LHS}=\frac{\frac{1}{b}+\frac{1}{a}}{\frac{2}{b}-\frac{1}{a}}+\frac{\frac{1}{b}+\frac{1}{c}}{\frac{2}{b}-\frac{1}{c}}=\frac{\frac{1}{b}+\frac{1}{a}}{\frac{1}{c}}+\frac{\frac{1}{b}+\frac{1}{c}}{\frac{1}{a}}, \text{ using (1)}$$

$$=\frac{c}{b}+\frac{c}{a}+\frac{a}{b}+\frac{a}{c}=\frac{a+c}{b}+\frac{a}{c}+\frac{c}{a}$$

Now $A>G$ $\Rightarrow$ $\frac{\frac{a}{c}+\frac{c}{a}}{2}>\sqrt{\frac{a}{c}\cdot\frac{c}{a}}$

or $\frac{a}{c}+\frac{c}{a}>2.$

$A>H$ $\Rightarrow$ $\frac{a+c}{2}>\frac{2ac}{a+c}=b$, using (1)

or $\frac{a+c}{b}>2.$

$\therefore$ $\text{LHS}=\frac{a+c}{b}+\left(\frac{a}{c}+\frac{c}{a}\right)>2+2=4.$

27. Show that $a^2(1+b^2)+b^2(1+c^2)+c^2(1+a^2)\geq 6abc.$

$\text{LHS}=(a^2+b^2+c^2)+(a^2b^2+b^2c^2+c^2a^2)$

$A\geq G$ $\Rightarrow$ $\frac{a^2+b^2+c^2}{3}\geq\sqrt[3]{a^2b^2c^2}$

or $a^2+b^2+c^2\geq 3a^{2/3}b^{2/3}c^{2/3}$

$A\geq G\Rightarrow$ $\frac{a^2b^2+b^2c^2+c^2a^2}{3}\geq\sqrt[3]{a^2b^2\cdot b^2c^2\cdot c^2a^2}$

or $a^2b^2+b^2c^2+c^2a^2\geq 3a^{4/3}b^{4/3}c^{4/3}$

$\therefore$ $\text{LHS}\geq 3\{a^{2/3}b^{2/3}c^{2/3}+a^{4/3}b^{4/3}c^{4/3}\}$

$$\geq 3\cdot 2\sqrt{a^{2/3}\cdot b^{2/3}\cdot c^{2/3}\cdot a^{4/3}\cdot b^{4/3}\cdot c^{4/3}}, \text{ for } A\geq G$$

$\therefore$ $\text{LHS}\geq 6\sqrt{a^2b^2c^2}=6abc.$

28. If $a > 0, b > 0, c > 0$ prove that $\frac{1}{s-a}+\frac{1}{s-b}+\frac{1}{s-c}>\frac{9}{2s}$ where $s = a + b + c$.

For three positive numbers x, y, z we have

$$A=\frac{x+y+z}{3},\ H=\frac{1}{\frac{\frac{1}{x}+\frac{1}{y}+\frac{1}{z}}{3}} \text{ and } A > H.$$

Here $s-a, s-b, s-c$ are positive.

$$\therefore \quad \frac{s-a+s-b+s-c}{3}>\frac{1}{\frac{\frac{1}{s-a}+\frac{1}{s-b}+\frac{1}{s-c}}{3}}$$

$$\text{or} \quad \frac{3s-(a+b+c)}{3}>\frac{3}{\frac{1}{s-a}+\frac{1}{s-b}+\frac{1}{s-c}}$$

$$\text{or} \quad \frac{2s}{3}>\frac{3}{\frac{1}{s-a}+\frac{1}{s-b}+\frac{1}{s-c}}$$

$$\text{or} \quad \frac{1}{s-a}+\frac{1}{s-b}+\frac{1}{s-c}>\frac{9}{2s}.$$

29. If a, b, x, y are positive rational numbers such that $\frac{1}{x}+\frac{1}{y}=1$ then prove that $\frac{a^x}{x}+\frac{b^y}{y}\ge ab$.

Let LCM of denominators x and y be k.

Then kx, ky are positive integers.

Consider the positive numbers:

$a^x, a^x, \ldots ky$ times and $b^y, b^y, \ldots kx$ times

For all these numbers,

$$A=\frac{\{a^x+a^x+\ldots ky \text{ times}\}+\{b^y+b^y+\ldots kx \text{ times}\}}{kx+ky}$$

$$=\frac{kya^x+kxb^y}{kx+ky}=\frac{ya^x+xb^y}{x+y}$$

$$G=\{(a^x\cdot a^x\cdot\ldots ky \text{ times})\times(b^y\cdot b^y\cdot\ldots kx \text{ times})\}^{\frac{1}{kx+ky}}$$

$$=\{(a^x)^{ky}\cdot(b^y)^{kx}\}^{\frac{1}{kx+ky}}$$

$$=(ab)^{\frac{kxy}{kx+ky}}=(ab)^{\frac{xy}{x+y}}$$

$$\text{But } A\ge G \Rightarrow \frac{ya^x+xb^y}{x+y}\ge(ab)^{\frac{xy}{x+y}} \quad \ldots (1)$$

As $\frac{1}{x}+\frac{1}{y}=1,\ \frac{x+y}{xy}=1$, i.e., $x+y=xy$.

$\therefore$ (1) becomes $\frac{ya^x+xb^y}{xy}\ge ab$

or $\frac{a^x}{x}+\frac{b^y}{y}\ge ab.$

30. If $a_i > 0; i = 1, 2, 3, \ldots n$ and $(n-1)s = a_1+a_2+a_3+\ldots+a_n$, prove that $a_1\cdot a_2\cdot a_3\cdot\ldots a_n\ge(n-1)^n\cdot(s-a_1)(s-a_2)\ldots(s-a_n)$.

Consider the $(n-1)$ positive numbers:

$s-a_2, s-a_3, \ldots, s-a_n$

For these numbers,

$$A=\frac{(s-a_2)+(s-a_3)+\ldots+(s-a_n)}{n-1}$$

$$=\frac{(n-1)s-(a_2+a_3+\ldots+a_n)}{n-1}=\frac{a_1}{n-1}$$

$$G=\{(s-a_2)(s-a_3)\ldots(s-a_n)\}^{\frac{1}{n-1}}$$

But $A\ge G$

$$\therefore \quad \frac{a_1}{n-1}\ge\{(s-a_2)(s-a_3)\ldots(s-a_n)\}^{\frac{1}{n-1}}$$

Similarly, $\frac{a_2}{n-1}\ge\{(s-a_1)(s-a_3)\ldots(s-a_n)\}^{\frac{1}{n-1}}$, etc.

Multiplying these n inequalities,

$$\frac{a_1\cdot a_2\cdot\ldots a_n}{(n-1)^n}\ge(s-a_1)((s-a_2)\ldots(s-a_n)$$

because each of $s-a_1$, etc., appear $(n-1)$ times on the right-hand side.

$$\therefore \quad a_1\cdot a_2\cdot\ldots a_n\ge(n-1)^n(s-a_1)(s-a_2)\ldots(s-a_n).$$

31. If $2a+b+3c=1$ and $a>0,\ b>0,\ c>0$, find the greatest value of $a^4b^2c^2$ and obtain the corresponding values of a, b, c.

Consider the positive numbers

$$\frac{2a}{4},\frac{2a}{4},\frac{2a}{4},\frac{2a}{4},\frac{b}{2},\frac{b}{2},\frac{3c}{2},\frac{3c}{2}.$$

{As there is a^4, take 4 equal parts of $2a$; as there is b^2, take 2 equal parts of b; as there is c^2, take 2 equal parts of $3c$.}

For the numbers,

$$A=\frac{\frac{2a}{4}+\frac{2a}{4}+\frac{2a}{4}+\frac{2a}{4}+\frac{b}{2}+\frac{b}{2}+\frac{3c}{2}+\frac{3c}{2}}{4+2+2}$$

$$=\frac{2a+b+3c}{8}=\frac{1}{8} \quad (\because 2a+b+3c=1)$$

$$G=\left(\frac{2a}{4}\cdot\frac{2a}{4}\cdot\frac{2a}{4}\cdot\frac{2a}{4}\cdot\frac{b}{2}\cdot\frac{b}{2}\cdot\frac{3c}{2}\cdot\frac{3c}{2}\right)^{\frac{1}{8}}$$

$$= \left(\frac{1}{2^4}\cdot\frac{1}{2^2}\cdot\frac{1}{2^2}\cdot 3^2a^4b^2c^2\right)^{\frac{1}{8}}$$

$\therefore \quad A \ge G \Rightarrow \dfrac{1}{8} \ge \left(\dfrac{3^2}{2^8}a^4b^2c^2\right)^{\frac{1}{8}}$

or $\quad \dfrac{1}{8^8} \ge \dfrac{3^2}{2^8}a^4b^2c^2$

or $\quad \dfrac{2^8}{3^2\cdot 8^8} \ge a^4b^2c^2$

or $\quad \dfrac{1}{9\cdot 4^8} \ge a^4b^2c^2.$

So the greatest value of $a^4b^2c^2 = \dfrac{1}{9\cdot 4^8}$.

When the equality holds, the greatest value takes place. We know $A = G$ when the numbers are all equal,

i.e., $\dfrac{2a}{4} = \dfrac{b}{2} = \dfrac{3c}{2} \quad \Rightarrow \quad a = b = 3c$

$\therefore \quad \dfrac{a}{3} = \dfrac{b}{3} = \dfrac{c}{1} = k$

$\therefore \quad a = 3k, b = 3k, c = k$

$\therefore \quad 2a + b + 3c = 1 \quad \Rightarrow \quad 6k + 3k + 3k = 1$

$\therefore \quad k = \dfrac{1}{12}$

$\therefore \quad a = \dfrac{3}{12}, \; b = \dfrac{3}{12}, \; c = \dfrac{1}{12},$

i.e., $a = \dfrac{1}{4}, b = \dfrac{1}{4}, c = \dfrac{1}{12}$.

32. Given the perimeter of a triangle, prove that the triangle of the greatest area is equilateral.

Let a, b, c be the sides and $a + b + c = 2s$ (given).

Let the area of the triangle be Δ. Then

$$\Delta^2 = s(s-a)(s-b)(s-c)$$

$\therefore$ Δ is the greatest when $(s-a)(s-b)(s-c)$ is the greatest, because s is constant. In a triangle, sum of two sides is greater than the third. So $s - a$, etc., are positive.

$\therefore \quad A \ge G$

$\Rightarrow \quad \dfrac{(s-a)+(s-b)+(s-c)}{3} \ge \sqrt[3]{(s-a)(s-b)(s-c)}$

$\therefore \quad \dfrac{3s-(a+b+c)}{3} \ge \sqrt[3]{(s-a)(s-b)(s-c)}$

or $\quad \left(\dfrac{s}{3}\right)^3 \ge (s-a)(s-b)(s-c)$

$\therefore \quad \dfrac{s^4}{27} \ge s(s-a)(s-b)(s-c)$

$\therefore \quad \dfrac{s^4}{27} \ge \Delta^2; \qquad \therefore \quad \dfrac{s^2}{3\sqrt{3}} \ge \Delta.$

$\therefore$ Δ has the maximum value $\dfrac{s^2}{3\sqrt{3}}$ and it takes place when $A = G$.

But $A = G$ when the numbers are equal,

i.e., $s-a, s-b, s-c$ are equal.

But $s - a = s - b = s - c \quad \Rightarrow \quad a = b = c.$

$\therefore$ Δ is maximum when $a = b = c$, i.e., the triangle is equilateral.

33. Sum the series
$1\cdot n + 2(n-1) + 3(n-2) + 4(n-3) + \ldots + (n-1)\cdot 2 + n\cdot 1.$

Also, find the coefficient of x^{n-1} in the expansion of $(1 + 2x + 3x^2 + \ldots + nx^{n-1})^2$.

Here $t_r = r(n-r+1)$ because rth term of 1, 2, 3, ... is r and rth term of $n, n-1, n-2, \ldots$ is $n-(r-1)$.

$$\therefore \quad \text{sum} = \sum_{r=1}^{n} t_r = \sum_{r=1}^{n} r(n-r+1)$$

$$= (n+1)\sum_{r=1}^{n} r - \sum_{r=1}^{n} r^2$$

$$= (n+1)(1+2+3+\ldots+n) - (1^2+2^2+3^2+\ldots+n^2)$$

$$= (n+1)\cdot\frac{n(n+1)}{2} - \frac{n(n+1)(2n+1)}{6}$$

$$= \frac{n(n+1)}{6}\{3(n+1)-(2n+1)\}$$

$$\therefore \quad \text{sum} = \frac{n(n+1)(n+2)}{6} \qquad \ldots (1)$$

Now, $(1 + 2x + 3x^2 + \ldots + nx^{n-1})^2$

$$= (1 + 2x + 3x^2 + \ldots + nx^{n-1}) \times (1 + 2x + 3x^2 + \ldots + nx^{n-1})$$

$\therefore$ coefficient of x^{n-1} in $(1 + 2x + 3x^2 + \ldots + nx^{n-1})^2$

$$= 1\cdot n + 2\cdot(n-1) + 3(n-2) + \ldots + n\cdot 1$$

$$= \frac{n(n+1)(n+2)}{6}, \text{ from (1).}$$

34. Prove that the sum of products of r consecutive integers at a time, taking from the first n natural numbers ($n > r$) is equal to $\dfrac{(n+1)!}{(r+1)(n-r)!}$.

Let $t_k = k(k+1)(k+2)\ldots(k+r-1) \qquad \ldots (1)$

Clearly $1 \le k \le n-r+1$ because there are n natural numbers.

Also $t_{k+1} = (k+1)(k+2)(k+3)\ \dots\ (k+r)$

$\therefore \quad (k+r)t_k = k\, t_{k+1}$

or $\quad (k-1)t_k - k\, t_{k+1} = -(r+1)t_k$

Putting $k = 1, 2, 3 \dots (n-r)$ we get

$$0 \cdot t_1 - 1 \cdot t_2 = -(r+1)t_1$$
$$1 \cdot t_2 - 2 \cdot t_3 = -(r+1)t_2$$
$$2 \cdot t_3 - 3 \cdot t_4 = -(r+1)t_3$$
$$\dots\dots\dots\dots\dots\dots$$
$$(n-r-1)t_{n-r} - (n-r)t_{n-r+1} = -(r+1)t_{n-r}$$

Adding these,

$$0 - (n-r)t_{n-r+1} = -(r+1)(t_1 + t_2 + t_3 + \dots + t_{n-r})$$

$$\therefore \quad t_1 + t_2 + t_3 + \dots + t_{n-r} = \frac{n-r}{r+1}\, t_{n-r+1}$$

$$\therefore \quad t_1 + t_2 + t_3 + \dots + t_{n-r} + t_{n-r+1}$$

$$= \left(\frac{n-r}{r+1} + 1\right) t_{n-r+1}$$

$$= \frac{n+1}{r+1}\, t_{n-r+1}$$

$$= \frac{n+1}{r+1}(n-r+1)(n-r+2)\dots n \quad \{\text{using (1)}\}$$

$$= \frac{(n+1)!}{(r+1)\,(n-r)!}.$$

Particular Results

$1 \cdot 2 + 2 \cdot 3 + 3 \cdot 4 + \dots + (n-1)n$

$$= \frac{(n+1)!}{(2+1)\,(n-2)!} = \frac{(n+1)n(n-1)}{3} \quad [\text{Here } r = 2]$$

$1 \cdot 2 \cdot 3 + 2 \cdot 3 \cdot 4 + \dots + (n-2)(n-1)n$

$$= \frac{(n+1)!}{(3+1)\,(n-3)!}$$

$$= \frac{(n+1)n(n-1)(n-2)}{4} \quad [\text{Here } r = 3]$$

$1 \cdot 2 \cdot 3 \cdot 4 + 2 \cdot 3 \cdot 4 \cdot 5 + 3 \cdot 4 \cdot 5 \cdot 6 + \dots$

$$+ (n-3)(n-2)(n-1)n$$

$$= \frac{(n+1)!}{(4+1)\,(n-4)!}$$

$$= \frac{(n+1)n(n-1)(n-2)(n-3)}{5}. \quad [\text{Here } r = 4]$$

35. Find the nth term and the sum to n terms of the series: $2 + 5 + 12 + 31 + 86 + \dots$

The series is

$$2 + (2+3) + (2+3+7) + (2+3+7+19) + (2+3+7+19+55) + \dots$$

Let the series in the mth bracket have the sum t_m.

Then $t_m = 2 + 3 + 7 + 19 + 55 + \dots$ to m terms

$\therefore \quad t_m - 2 = (3) + (3+4) + (3+4+12) + (3+4+12+36) + \dots$ to $(m-1)$ brackets.

Let the series in the rth bracket have the sum T_r.

Then $T_r = 3 + 4 + 12 + 36 + \dots$ to r terms

$$= 3 + \{4 + 12 + 36 + \dots \text{ to } (r-1) \text{ terms}\}$$

$$= 3 + \frac{4(3^{r-1} - 1)}{3-1}$$

$$= 3 + 2(3^{r-1} - 1)$$

$$= 1 + 2 \cdot 3^{r-1}.$$

So, $\displaystyle t_m - 2 = \sum_{r=1}^{m-1} T_r = \sum_{r=1}^{m-1} (1 + 2 \cdot 3^{r-1})$

$$= (m-1) + 2\sum_{r=1}^{m-1} 3^{r-1}$$

$$= m - 1 + 2(1 + 3 + 3^2 + \dots + 3^{m-2})$$

$$= m - 1 + 2 \cdot \frac{3^{m-1} - 1}{3-1}$$

$$= m - 1 + 3^{m-1} - 1$$

$$= m - 2 + 3^{m-1}$$

$\therefore \quad t_m = m + 3^{m-1}$; so nth term $= n + 3^{n-1}$.

$\therefore \quad$ the required sum of the series

$$= \sum_{m=1}^{n} t_m = \sum_{m=1}^{n} (m + 3^{m-1})$$

$$= (1 + 2 + 3 + \dots + n) + (1 + 3 + 3^2 + \dots + 3^{n-1})$$

$$= \frac{n(n+1)}{2} + \frac{3^n - 1}{3-1}$$

$$= \frac{1}{2}\{n(n+1) + 3^n - 1\}.$$

36. If $f(x) = x - 2x^2 + 3x^3 - \dots + (-1)^{n-1} \cdot nx^n$ where $x \in R$ then find $f'\left(\frac{1}{2}\right)$.

Here $f'(x) = 1 - 2^2 \cdot x + 3^2x^2 - 4^2x^3 + \dots$

$$+ (-1)^{n-1} \cdot n^2x^{n-1}$$

$$\therefore \quad f'\left(\frac{1}{2}\right) = 1^2 - 2^2 \cdot \frac{1}{2} + 3^2 \cdot \frac{1}{2^2} - 4^2 \cdot \frac{1}{2^3} + \ldots + (-1)^{n-1} \cdot n^2 \cdot \frac{1}{2^{n-1}}$$

$$\therefore \quad -\frac{1}{2} f'\left(\frac{1}{2}\right) = -1^2 \cdot \frac{1}{2} + 2^2 \cdot \frac{1}{2^2} - 3^2 \cdot \frac{1}{2^3} + \ldots - (-1)^{n-1} \cdot n^2 \cdot \frac{1}{2^n}$$

Subtracting,

$$\frac{3}{2} f'\left(\frac{1}{2}\right) = \left\{1 - 3 \cdot \frac{1}{2} + 5 \cdot \frac{1}{2^2} - 7 \cdot \frac{1}{2^3} + \ldots \text{ to } n \text{ terms}\right\} + (-1)^{n-1} \cdot n^2 \cdot \frac{1}{2^n}$$

Let $S_n = 1 - 3 \cdot \frac{1}{2} + 5 \cdot \frac{1}{2^2} - 7 \cdot \frac{1}{2^3} + \ldots + (-1)^{n-1} \frac{2n-1}{2^{n-1}}$

$$-\frac{1}{2} S_n = -1 \cdot \frac{1}{2} + 3 \cdot \frac{1}{2^2} - 5 \cdot \frac{1}{2^3} + \ldots + (-1)^n \frac{2n-1}{2^n}.$$

Subtracting,

$$\frac{3}{2} S_n = \left\{1 - 2 \cdot \frac{1}{2} + 2 \cdot \frac{1}{2^2} - 2 \cdot \frac{1}{2^3} + \ldots \text{ to } n \text{ terms}\right\} - (-1)^n \frac{2n-1}{2^n}$$

$$\therefore \quad \frac{3}{2} S_n = \left(1 - 1 + \frac{1}{2} - \frac{1}{2^2} + \ldots \text{ to } n \text{ terms}\right) - (-1)^n \cdot \frac{2n-1}{2^n}$$

$$= \frac{1}{2} - \frac{1}{2^2} + \ldots \text{ to } (n-2) \text{ terms} - (-1)^n \cdot \frac{2n-1}{2^n}$$

$$= \frac{\frac{1}{2}\left\{1 - \left(-\frac{1}{2}\right)^{n-2}\right\}}{1 - \left(-\frac{1}{2}\right)} - (-1)^n \cdot \frac{2n-1}{2^n}$$

$$\therefore \quad S_n = \frac{2}{9}\left\{1 - \left(-\frac{1}{2}\right)^{n-2}\right\} - (-1)^n \cdot \frac{2n-1}{3 \cdot 2^{n-1}}$$

Thus $\frac{3}{2} f'\left(\frac{1}{2}\right) = \frac{2}{9}\left\{1 - \left(-\frac{1}{2}\right)^{n-2}\right\} - (-1)^n \cdot \frac{2n-1}{3 \cdot 2^{n-1}} + (-1)^{n-1} \cdot \frac{n^2}{2^n}$

$$= \frac{2}{9}\left\{1 - \left(-\frac{1}{2}\right)^{n-2}\right\} + (-1)^{n-1} \cdot \frac{3n^2 + 4n - 2}{3 \cdot 2^n}$$

$$\therefore \quad f'\left(\frac{1}{2}\right) = \frac{4}{27}\left\{1 - \left(-\frac{1}{2}\right)^{n-2}\right\} + (-1)^{n-1} \cdot \frac{3n^2 + 4n - 2}{9 \cdot 2^{n-1}}.$$

37. Sum to n terms:
$$\mathbf{1 + 2^2 - 3^3 + 4 + 5^2 - 6^3 + 7 + 8^2 - 9^3 + \ldots}.$$

Regrouping the terms

the series $= (1 + 4 + 7 + \ldots) + (2^2 + 5^2 + 8^2 + \ldots) - (3^3 + 6^3 + 9^3 + \ldots)$

The number of terms in each group will depend on n.

If $n = 3m$, i.e., n is *divisible by* 3

Sum of the first group $= 1 + 4 + 7 + \ldots$ to m terms

$$= \frac{m}{2}\{2 \cdot 1 + (m-1)3\} = \frac{m(3m-1)}{2} = \frac{n}{6}(n-1). \quad \ldots (1)$$

Sum of the second group $= 2^2 + 5^2 + 8^2 + \ldots$ to m terms

$$= \sum_{r=1}^{m} \{2 + (r-1)3\}^2 = \sum_{r=1}^{m} (3r-1)^2 = \sum_{r=1}^{m} (9r^2 - 6r + 1)$$

$$= 9 \sum_{r=1}^{m} r^2 - 6 \sum_{r=1}^{m} r + \sum_{r=1}^{m} 1$$

$$= 9 \cdot \frac{m(m+1)(2m+1)}{6} - 6 \cdot \frac{m(m+1)}{2} + m$$

$$= 3m(m+1)\left\{\frac{1}{2}(2m+1) - 1\right\} + m$$

$$= \frac{3m(m+1)(2m-1)}{2} + m \quad \ldots (2)$$

$$= \frac{n\left(\frac{n}{3}+1\right)\left(\frac{2n}{3}-1\right)}{2} + \frac{n}{3} = \frac{n(n+3)(2n-3)}{18} + \frac{n}{3}.$$

Sum of the third group $= 3^3 + 6^3 + 9^3 + \ldots$ to m terms

$$= 3^3(1^3 + 2^3 + 3^3 + \ldots + m^3)$$

$$= 27\left\{\frac{m(m+1)}{2}\right\}^2 \quad \ldots (3)$$

$$= \frac{27}{4}\left\{\frac{n}{3}\left(\frac{n}{3}+1\right)\right\}^2$$

$$= \frac{27}{4} \cdot \frac{1}{81} n^2(n+3)^2 = \frac{n^2(n+3)^2}{12}$$

$\therefore$ the sum

$$= \frac{n(n-1)}{6} + \frac{n(n+3)(2n-3)}{18} + \frac{n}{3} - \frac{n^2(n+3)^2}{12}$$

when n is divisible by 3.

If $n = 3m + 1$, i.e., *when n is divided by 3, the remainder is* 1.

In this case the first group will have $m + 1$ terms while others will have m terms each.

$\therefore$ in (1) we shall get

$$\frac{(m+1)\{3(m+1)-1\}}{2}, \text{ i.e., } \frac{(m+1)(3m+2)}{2}$$

$\therefore$ the sum

$$= \frac{(m+1)(3m+2)}{2} + \frac{3m(m+1)(2m-1)}{2} + m - 27\left\{\frac{m(m+1)}{2}\right\}^2$$

$$= \frac{\left(\frac{n-1}{3}+1\right)(n-1+2)}{2} + \frac{(n-1)\left(\frac{n-1}{3}+1\right)\left(2\cdot\frac{n-1}{3}-1\right)}{2} + \frac{n-1}{3} - \frac{27}{4}\left\{\frac{n-1}{3}\left(\frac{n-1}{3}+1\right)\right\}^2$$

$$= \frac{(n+2)(n+1)}{6} + \frac{(n-1)(n+2)(2n-5)}{18} + \frac{n-1}{3} - \frac{1}{12}\{(n-1)(n+2)\}^2,$$

when n is of the form $3m+1$.

If $n = 3m+2$, i.e., *when n is divided by 3, the remainder is 2*

In this case the first and the second group will have $(m+1)$ terms each while the third will have m terms.

$\therefore$ in (1) we shall get

$$\frac{(m+1)\{3(m+1)-1\}}{2}, \text{ i.e., } \frac{(m+1)(3m+2)}{2},$$

in (2) we shall get $\dfrac{3(m+1)(m+2)(2m+1)}{2} + (m+1)$

$$\therefore \text{ the sum} = \frac{(m+1)(3m+2)}{2} + \frac{3(m+1)(m+2)(2m+1)}{2} + (m+1) - 27\cdot\left\{\frac{m(m+1)}{2}\right\}^2 \quad \{\text{using (3)}\}$$

$$= \frac{\left(\frac{n-2}{3}+1\right)n}{2} + \frac{3\left(\frac{n-2}{3}+1\right)\left(\frac{n-2}{3}+2\right)\left(2\cdot\frac{n-2}{3}+1\right)}{2} + \left(\frac{n-2}{3}+1\right) - \frac{27}{4}\cdot\left\{\frac{n-2}{3}\left(\frac{n-2}{3}+1\right)\right\}^2$$

$$= \frac{(n+1)n}{6} + \frac{(n+1)(n+4)(2n-1)}{18} + \frac{n+1}{3} - \frac{1}{12}\{(n-2)(n+1)\}^2,$$

when n is of the form $3m+2$.

38. Sum the series to n terms

$0.7 + 7.7 + 0.77 + 77.7 + 0.777 + 777.7 + 0.7777 + \dots$.

$$\text{Sum} = (0.7 + 0.77 + 0.777 + \dots) + (7.7 + 77.7 + 777.7 + \dots)$$

$$= \frac{7}{9}(0.9 + 0.99 + 0.999 + \dots) + \frac{7}{90}(99 + 999 + 9999 + \dots)$$

$$= \frac{7}{9}\left\{\left(1-\frac{1}{10}\right)+\left(1-\frac{1}{10^2}\right)+\left(1-\frac{1}{10^3}\right)+\dots\right\} + \frac{7}{90}\{(10^2-1)+(10^3-1)+(10^4-1)+\dots\}$$

$$= \frac{7}{9}\left\{(1+1+\dots)-\left(\frac{1}{10}+\frac{1}{10^2}+\frac{1}{10^3}+\dots\right)\right\} + \frac{7}{90}\{(10^2+10^3+10^4+\dots)-(1+1+\dots)\}.$$

If *n is even*, i.e., $n = 2m$ then

$$\text{sum} = \frac{7}{9}\left\{(1+1+\dots \text{ to } m \text{ terms}) - \left(\frac{1}{10}+\frac{1}{10^2}+\dots \text{ to } m \text{ terms}\right)\right\} + \frac{7}{90}\{(10^2+10^3+10^4+\dots \text{ to } m \text{ terms}) - (1+1+\dots \text{ to } m \text{ terms})\}$$

$$= \frac{7}{9}\left\{m - \frac{\frac{1}{10}\left(1-\frac{1}{10^m}\right)}{1-\frac{1}{10}}\right\} + \frac{7}{90}\left\{\frac{10^2(10^m-1)}{10-1} - m\right\}$$

$$= \frac{7}{9}\left\{m - \frac{1}{9}\left(1-\frac{1}{10^m}\right)\right\} + \frac{7}{90}\left\{\frac{100}{9}(10^m-1) - m\right\}$$

$$\therefore \text{ sum} = \frac{7}{9}\left\{\frac{n}{2} - \frac{1}{9}\left(1-\frac{1}{10^{n/2}}\right)\right\} + \frac{7}{90}\left\{\frac{100}{9}(10^{n/2}-1) - \frac{n}{2}\right\}, \text{ when } n \text{ is even.}$$

If *n is odd*, i.e., $n = 2m+1$ then

$$\text{sum} = \frac{7}{9}\left\{(1+1+\dots \text{to } (m+1) \text{ terms}) - \left(\frac{1}{10}+\frac{1}{10^2}+\dots \text{ to } (m+1) \text{ terms}\right)\right\} + \frac{7}{90}\left\{(10^2+10^3+\dots \text{ to } m \text{ terms}) - (1+1+\dots \text{ to } m \text{ terms})\right\}$$

$$= \frac{7}{9}\left\{m+1 - \frac{1}{9}\left(1-\frac{1}{10^{m+1}}\right)\right\} + \frac{7}{90}\left\{\frac{100}{9}(10^m-1) - m\right\}$$

$$=\frac{7}{9}\left\{\frac{n-1}{2}+1-\frac{1}{9}\left(1-\frac{1}{10^{(n+1)/2}}\right)\right\}$$

$$+\frac{7}{90}\left\{\frac{100}{9}\left(10^{\frac{n-1}{2}}-1\right)-\frac{n-1}{2}\right\}$$

$$=\frac{7}{9}\left\{\frac{n+1}{2}-\frac{1}{9}+\frac{1}{9\cdot 10^{(n+1)/2}}\right\}$$

$$+\frac{7}{90}\left\{\frac{100}{9}\left(10^{\frac{n-1}{2}}-1\right)-\frac{n-1}{2}\right\},$$

when n is odd.

39. Prove that

$$\frac{3}{1^2}+\frac{5}{1^2+2^2}+\frac{7}{1^2+2^2+3^2}+\frac{9}{1^2+2^2+3^2+4^2}+\ldots$$

$$\textbf{to } n \textbf{ terms} = \frac{6n}{n+1}.$$

Also find the sum to infinite terms of the series on the left-hand side.

$$t_n=\frac{3+(n-1)2}{1^2+2^2+3^2+\ldots+n^2}=\frac{2n+1}{n(n+1)(2n+1)/6}$$

$$=\frac{6}{n(n+1)}=6\left\{\frac{1}{n}-\frac{1}{n+1}\right\}$$

$$\therefore \quad t_1=6\left(1-\frac{1}{2}\right)$$

$$t_2=6\left(\frac{1}{2}-\frac{1}{3}\right)$$

$$t_3=6\left(\frac{1}{3}-\frac{1}{4}\right)$$

....

$$t_n=6\left(\frac{1}{n}-\frac{1}{n+1}\right)$$

Adding, LHS $=6\left(1-\dfrac{1}{n+1}\right)=\dfrac{6n}{n+1}.$

Thus $S_n=6\left(1-\dfrac{1}{n+1}\right)$

$\therefore$ sum of infinite terms

$$=\lim_{n\to\infty}S_n=\lim_{n\to\infty}6\left(1-\frac{1}{n+1}\right)=6.$$

40. If $a_1, a_2, a_3, \ldots, a_n$ are in AP where $a_i \neq k\pi$ for all i, prove that

$$\textbf{cosec } a_1\cdot \textbf{cosec } a_2+\textbf{cosec } a_2\cdot\textbf{cosec } a_3+\ldots$$

$$+\textbf{cosec } a_{n-1}\cdot\textbf{cosec } a_n=\frac{\cot a_1-\cot a_n}{\sin(a_2-a_1)}.$$

$$\text{LHS}=\frac{1}{\sin a_1\cdot\sin a_2}+\frac{1}{\sin a_2\cdot\sin a_3}+\ldots$$

$$+\frac{1}{\sin a_{n-1}\cdot\sin a_n}$$

$$t_r=\frac{\sin(a_{r+1}-a_r)}{\sin a_r\sin a_{r+1}}\times\frac{1}{\sin(a_{r+1}-a_r)}$$

$$=\frac{\sin a_{r+1}\cdot\cos a_r-\cos a_{r+1}\cdot\sin a_r}{\sin a_r\cdot\sin a_{r+1}}\cdot\frac{1}{\sin d}$$

where common difference $= d$

$$=(\cot a_r-\cot a_{r+1})\cdot\frac{1}{\sin d}$$

$$\therefore \quad t_1=(\cot a_1-\cot a_2)\cdot\frac{1}{\sin d}$$

$$t_2=(\cot a_2-\cot a_3)\cdot\frac{1}{\sin d}$$

$$t_3=(\cot a_3-\cot a_4)\cdot\frac{1}{\sin d}$$

....

$$t_{n-1}=(\cot a_{n-1}-\cot a_n)\cdot\frac{1}{\sin d}$$

Adding, LHS $=(\cot a_1-\cot a_n)\cdot\dfrac{1}{\sin d}$

$$=\frac{\cot a_1-\cot a_n}{\sin d}=\frac{\cot a_1-\cot a_n}{\sin(a_2-a_1)}.$$

41. Find the sum of n terms of the series

$$\cot^{-1}3+\cot^{-1}7+\cot^{-1}13+\cot^{-1}21+\ldots.$$

Consider the sequence 3, 7, 13, 21, ...,

i.e., $3+(3+4)+(3+4+6)+(3+4+6+8)+\ldots$

$\therefore$ nth term of the sequence

$= 3+4+6+8+\ldots$ to n terms

$= 1+(2+4+6+8+\ldots$ to n terms)

$= 1+2(1+2+3+4+\ldots$ to n terms)

$$=1+2\cdot\frac{n(n+1)}{2}=n^2+n+1.$$

For the given series,

$$t_n=\cot^{-1}(n^2+n+1)=\tan^{-1}\frac{1}{n^2+n+1}$$

$$=\tan^{-1}\frac{1}{1+n(n+1)}=\tan^{-1}\frac{(n+1)-n}{1+n(n+1)}$$

$$=\tan^{-1}(n+1)-\tan^{-1}n$$

$\therefore \quad t_1=\tan^{-1}2-\tan^{-1}1,$ putting $n=1$

$t_2=\tan^{-1}3-\tan^{-1}2,$ putting $n=2$

$t_3 = \tan^{-1}4 - \tan^{-1}3$, putting $n = 3$

....

$t_n = \tan^{-1}(n+1) - \tan^{-1}n$

Adding, LHS $= \tan^{-1}(n+1) - \tan^{-1}1$

$$= \tan^{-1}\frac{(n+1)-1}{1+(n+1)\cdot 1} = \tan^{-1}\frac{n}{n+2}.$$

42. Sum to *n* terms

$$\frac{1}{1\cdot 3} + \frac{2}{1\cdot 3\cdot 5} + \frac{3}{1\cdot 3\cdot 5\cdot 7} + \frac{4}{1\cdot 3\cdot 5\cdot 7\cdot 9} + \ldots.$$

Here $t_n = \dfrac{n}{1\cdot 3\cdot 5 \ldots (2n+1)}$

because there are two factors in the denominator of the first

$$= \frac{1}{2}\cdot\frac{(2n+1)-1}{1\cdot 3\cdot 5\cdot \ldots (2n+1)}$$

$$= \frac{1}{2}\left[\frac{1}{1\cdot 3\cdot 5\cdot \ldots (2n-1)} - \frac{1}{1\cdot 3\cdot 5\cdot \ldots (2n+1)}\right]$$

$$= \frac{1}{2}[T_{n-1} - T_n]$$

where $T_n = \dfrac{1}{1\cdot 3\cdot 5\cdot \ldots (2n+1)}$

$\therefore \quad 2t_n = T_{n-1} - T_n;$

putting $n = 2, 3, 4, \ldots, n$ we get

$2t_2 = T_1 - T_2$

$2t_3 = T_2 - T_3$

$2t_4 = T_3 - T_4$

....

$2t_n = T_{n-1} - T_n$

Adding, $2(S_n - t_1) = T_1 - T_n$

or $\quad 2S_n = 2t_1 + T_1 - T_n$

$$\therefore \quad 2S_n = 2\cdot\frac{1}{1\cdot 3} + \frac{1}{1\cdot 3} - \frac{1}{1\cdot 3\cdot 5\cdot \ldots (2n+1)}$$

$$= 1 - \frac{1}{1\cdot 3\cdot 5\cdot \ldots (2n+1)}$$

$$\therefore \quad S_n = \frac{1}{2}\left[1 - \frac{1}{1\cdot 3\cdot 5 \ldots (2n+1)}\right].$$

43. Sum the series to *n* terms

$$\frac{1}{x+1} + \frac{2x}{(x+1)(x+2)} + \frac{3x^2}{(x+1)(x+2)(x+3)} + \ldots$$

Here $t_n = \dfrac{nx^{n-1}}{(x+1)(x+2)(x+3)\ldots(x+n)}$

$$= \frac{x^{n-1}(x+n) - x^n}{(x+1)(x+2)(x+3)\ldots(x+n)}$$

$$= \frac{x^{n-1}}{(x+1)(x+2)\ldots(x+n-1)} - \frac{x^n}{(x+1)(x+2)\ldots(x+n)}$$

$$= \frac{x}{n-1}\cdot\frac{(n-1)x^{n-2}}{(x+1)(x+2)\ldots(x+n-1)} - \frac{x}{n}\cdot\frac{nx^{n-1}}{(x+1)(x+2)\ldots(x+n)}$$

$$\therefore \quad t_n = \frac{x}{n-1}\cdot t_{n-1} - \frac{x}{n}\cdot t_n;$$

putting $n = 2, 3, \ldots$ we get

$t_2 = \frac{x}{1}\cdot t_1 - \frac{x}{2}\cdot t_2$

$t_3 = \frac{x}{2}\cdot t_2 - \frac{x}{3}\cdot t_3$

$t_4 = \frac{x}{3}\cdot t_3 - \frac{x}{4}\cdot t_4$

....

$t_n = \frac{x}{n-1}\cdot t_{n-1} - \frac{x}{n}\cdot t_n$

Adding, $S_n - t_1 = \frac{x}{1}\cdot t_1 - \frac{x}{n}\cdot t_n \quad (\because \quad S_n = t_1 + t_2 + \ldots + t_n)$

or $\quad S_n = (x+1)t_1 - \frac{x}{n}\cdot t_n$

$$= (x+1)\cdot\frac{1}{x+1} - \frac{x}{n}\cdot\frac{nx^{n-1}}{(x+1)(x+2)\ldots(x+n)}$$

$$= 1 - \frac{x^n}{(x+1)(x+2)\ldots(x+n)}.$$

44. Sum to *n* terms

$$\frac{x}{1-x^2} + \frac{x^2}{1-x^4} + \frac{x^4}{1-x^8} + \ldots$$

Also find the sum to infinite terms when $|x| < 1$.

Here $t_n = \dfrac{x^{2^{n-1}}}{1-x^{2^n}} = \dfrac{(1+x^{2^{n-1}})-1}{(1-x^{2^{n-1}})(1+x^{2^{n-1}})}$

$$= \frac{1}{1-x^{2^{n-1}}} - \frac{1}{1-x^{2^n}};$$

putting $n = 1, 2, 3, \ldots$ we get

$$t_1 = \frac{1}{1-x} - \frac{1}{1-x^2}$$

$$t_2 = \frac{1}{1-x^2} - \frac{1}{1-x^4}$$

$$t_3 = \frac{1}{1-x^4} - \frac{1}{1-x^8}$$

....

$$t_n = \frac{1}{1-x^{2^{n-1}}} - \frac{1}{1-x^{2^n}}$$

Adding,

$$\text{sum} = \frac{1}{1-x} - \frac{1}{1-x^{2^n}} = \frac{x - x^{2^n}}{(1-x)(1-x^{2^n})}$$

$$= \frac{x(1-x^{2^n-1})}{(1-x)(1-x^{2^n})}$$

$\therefore$ sum to infinite terms

$$= \lim_{n\to\infty} \frac{x(1-x^{2^n-1})}{(1-x)(1-x^{2^n})} = \frac{x}{1-x}$$

for $x^{2^n} \to 0$ as $n \to \infty$ when $|x| < 1$.

45. Find the sum of

$$\frac{1}{1+x} + \frac{2x}{1+x^2} + \frac{4x^3}{1+x^4} + \ldots + \frac{2^n x^{2^n-1}}{1+x^{2^n}}.$$

Let $\quad f(x) = \log(1+x) + \log(1+x^2) + \log(1+x^4) + \ldots$

$$+ \log(1+x^{2^n}) \quad \ldots (1)$$

Then $f'(x) = \dfrac{1}{1+x} + \dfrac{2x}{1+x^2} + \dfrac{4x^3}{1+x^4} + \ldots$

$$+ \frac{2^n x^{2^n-1}}{1+x^{2^n}} \quad \ldots (2)$$

Now, $f(x) = \log\{(1+x)(1+x^2)(1+x^4)\ldots(1+x^{2^n})\}$

$$= \log\left\{\frac{1}{1-x}(1-x)(1+x)(1+x^2)(1+x^4)\ldots(1+x^{2^n})\right\}$$

$$= \log\left\{\frac{1}{1-x}(1-x^2)(1+x^2)(1+x^4)\ldots(1+x^{2^n})\right\}$$

....

$$= \log\left\{\frac{1}{1-x}(1-x^{2^{n+1}})\right\}$$

$$= \log(1-x^{2^{n+1}}) - \log(1-x)$$

$$\therefore \quad f'(x) = \frac{-2^{n+1}\cdot x^{2^{n+1}-1}}{1-x^{2^{n+1}}} + \frac{1}{1-x}$$

$$= \frac{1}{1-x} + \frac{2^{n+1}\cdot x^{2^{n+1}-1}}{x^{2^{n+1}}-1}.$$

From (2), this value of $f'(x)$ is the required sum.

Alternatively

If S be the sum then consider $S - \dfrac{1}{1-x}$.

Now $S - \dfrac{1}{1-x} = \dfrac{1}{1+x} - \dfrac{1}{1-x}$

$$+ \left(\frac{2x}{1+x^2} + \frac{4x^3}{1+x^4} + \ldots + \frac{2^n x^{2^n-1}}{1+x^{2^n}}\right)$$

$$= -\frac{2x}{1-x^2} + \frac{2x}{1+x^2}$$

$$+ \left(\frac{4x^3}{1+x^4} + \frac{8x^4}{1+x^8} + \ldots + \frac{2^n x^{2^n-1}}{1+x^{2^n}}\right)$$

$$= -\frac{4x^3}{1-x^4} + \frac{4x^3}{1+x^4} + \left(\frac{8x^4}{1+x^8} + \ldots + \frac{2^n x^{2^n-1}}{1+x^{2^n}}\right)$$

....

$$= -\frac{2^{n+1}\cdot x^{2^{n+1}-1}}{1-x^{2^{n+1}}}$$

$$\therefore \quad S = \frac{1}{1-x} + \frac{2^{n+1}\cdot x^{2^{n+1}-1}}{x^{2^{n+1}}-1}.$$

46. If there be m quantities in a GP whose common ratio is r and S_n denotes the sum of the first n terms of the GP, prove that the sum of their products (two by two) is $\dfrac{r}{r+1} S_m \cdot S_{m-1}$.

Let $S_n = a + ar + ar^2 + \ldots + ar^{n-1} = \dfrac{a(1-r^n)}{1-r}$;

$$\therefore \quad S_m = \frac{a(1-r^m)}{1-r}, \quad S_{m-1} = \frac{a(1-r^{m-1})}{1-r}.$$

Let the sum of the products of $a, ar, ar^2, \ldots, ar^{m-1}$ taken two at a time be S.

Now $(a_1 + a_2 + a_3 + \ldots + a_m)^2$

$$= a_1^2 + a_2^2 + a_3^2 + \ldots + a_m^2 + 2S$$

$\therefore \quad (a + ar + ar^2 + \dots + ar^{m-1})^2$

$= a^2 + a^2r^2 + a^2r^4 + \dots + a^2r^{2(m-1)} + 2S$

or $\left\{\frac{a(1-r^m)}{1-r}\right\}^2 = \frac{a^2(1-r^{2m})}{1-r^2} + 2S$

$\therefore \quad 2S = \frac{a^2(1-r^m)^2}{(1-r)^2} - \frac{a^2(1-r^{2m})}{1-r^2}$

$= \frac{a^2\{(1+r)(1-r^m)^2 - (1-r)(1-r^{2m})\}}{(1-r)^2(1+r)}$

$$= a^2 \cdot \frac{\left\{\begin{array}{l}1 - 2r^m + r^{2m} + r - 2r^{m+1} + r^{2m+1} \\ \qquad - 1 + r^{2m} + r - r^{2m+1}\end{array}\right\}}{(1-r)^2 \cdot (1+r)}$$

$= a^2 \cdot \frac{2(r^{2m} - r^{m+1} - r^m + r)}{(1-r)^2 \cdot (1+r)}$.

$\therefore \quad S = a^2 \cdot \frac{r^m(r^m - 1) - r(r^m - 1)}{(1-r)^2(1+r)}$

$= a^2 \cdot \frac{(r^m - 1)(r^m - r)}{(1-r)^2(1+r)}$

$= \frac{a(r^m - 1)}{1-r} \cdot \frac{a(r^m - r)}{1-r} \cdot \frac{1}{1+r}$

$= \frac{a(1-r^m)}{1-r} \cdot \frac{a(1-r^{m-1})}{1-r} \cdot \frac{r}{1+r}$

$= \frac{r}{1+r} S_m \cdot S_{m-1}$.

47. Find the coefficient of x^{98} in the continued product $(x+1)(x+2)(x+3) \dots (x+100)$.

We know that

$(x + a_1)(x + a_2)(x + a_3) \dots (x + a_n)$

$= x^n + (\Sigma a_1)x^{n-1} + (\Sigma a_1a_2)x^{n-2}$

$\qquad + (\Sigma a_1a_2a_3)x^{n-3} + \dots + (a_1a_2a_3 \dots a_n)$

$\therefore$ the coefficient of x^{98} in the product of 100 linear factors will be sum of the products of 1, 2, 3, ..., 100 taking two at a time.

Now $(1 + 2 + 3 + \dots + 100)^2 = 1^2 + 2^2 + 3^2 + \dots + 100^2$

$\qquad + 2 \times$ (required sum of products)

$\therefore$ the required coefficient

$= \frac{1}{2}[(1 + 2 + 3 + \dots + 100)^2$

$\qquad - (1^2 + 2^2 + 3^2 + \dots + 100^2)]$

$= \frac{1}{2}\left[\left(\frac{100 \times 101}{2}\right)^2 - \frac{100 \times 101 \times 201}{6}\right] = 12582075$.

48. Show that the sum of the cubes of the first n terms of an AP is divisible by the sum of the terms of the AP.

Let the AP be $a, a + d, a + 2d, \dots, a + (n-1)d$

whose sum $= S_n = \frac{n}{2}\{2a + (n-1)d\}$.

Sum of cubes of th above terms

$= a^3 + (a+d)^3 + (a+2d)^3 + \dots + \{a + (n-1)d\}^3$

$= \sum_{r=0}^{n-1} (a + rd)^3 = \sum_{r=0}^{n-1} (a^3 + 3a^2dr + 3ad^2r^2 + d^3r^3)$

$= a^3 \sum_{r=0}^{n-1} 1 + 3a^2d \sum_{r=0}^{n-1} r + 3ad^2 \sum_{r=0}^{n-1} r^2 + d^3 \sum_{r=0}^{n-1} r^3$

$= a^3n + 3a^2d \cdot \frac{(n-1)n}{2} + 3ad^2 \frac{(n-1)n(2n-1)}{6}$

$\qquad + d^3 \cdot \left\{\frac{(n-1)n}{2}\right\}^2$

$= \frac{n}{2}\left[2a^3 + 3a^2d(n-1) + ad^2(n-1)2n - 1 + \frac{d^3}{2}n(n-1)^2\right]$

$= \frac{n}{2}\left[a^2\{2a + (n-1)d\} + 2a^2d(n-1) + ad^2\{(n-1)^2 + n(n-1)\} + \frac{1}{2}d^3n(n-1)^2\right]$

$= \frac{n}{2}\left[a^2\{2a + (n-1)d\} + ad(n-1)\{2a + d(n-1)\} + ad^2n(n-1) + \frac{1}{2}d^3n(n-1)^2\right]$

$= \frac{n}{2}\left[a^2\{2a + (n-1)d\} + ad(n-1)\{2a + d(n-1)\} + \frac{1}{2}d^2n(n-1)\{2a + (n-1)d\}\right]$

$= \frac{n}{2}\{2a + (n-1)d\}\left[a^2 + (n-1)ad + \frac{1}{2}n(n-1)d^2\right]$

$= S_n \cdot \left[a^2 + (n-1)ad + \frac{1}{2}n(n-1)d^2\right]$

$\therefore$ the sum of the cubes is divisible by the sum of the numbers in AP.

Exercises

1. If in an AP $mt_m = nt_n$ then show that $t_{m+n} = 0$.

2. Prove that the numbers $\sqrt{2}, \sqrt{3}, \sqrt{5}$ cannot be three terms (not necessarily consecutive) of an AP.

3. If a, b, c be any three unequal terms of an AP, show that $\frac{b-c}{a-b}$ must be a rational number.

4. Show that there can be infinite number of sets of four distinct real numbers in AP which are such that the square of the last term is equal to the sum of the squares of the first three terms.

5. If $\log_l x, \log_m x, \log_n x$ are in AP where $x \neq 1$ and x is positive, prove that $n^2 = (ln)^{\log_l m}$.

6. If $\log_3 2, \log_3 (2^x - 5)$ and $\log_3 \left(2^x - \frac{7}{2}\right)$ are in AP, find the value of x.

7. Each of two triplets of numbers $\log a, \log b, \log c$ and $\log a - \log 2b, \log 2b - \log 3c, \log 3c - \log a$ is in AP. Prove that a, b, c can be the lengths of the sides of a triangle. Also find $a : b : c$.

8. Find three-digit numbers that are divisible by 5 as well as 9 and whose consecutive digits are in AP.

9. 1000 ! is divisible by 10^n. Find the largest positive integral value of n.

10. Find the number of common terms in the following sequences:

(i) 3, 7, 11, ... to 100 terms and 2, 5, 8, ... to 100 terms.

(ii) 1, 5, 9, ... to 100 terms and 4, 7, 10, ... to 60 terms.

11. Find four numbers in AP whose sum is 50 and in which the greatest number is four times the least.

12. The sum of four whole numbers in AP is 24 and their product is 945; find the numbers.

13. Divide 20 in four parts which are in AP such that the product of the first and the fourth is to the product of the second and the third = 2 : 3.

14. There are two sets of numbers, each having 3 terms in AP and the sum of each set is 15. The common difference of the first set is greater by 1 than the common difference of the second set and the product of the first set is to the product of the second set as 7 to 8. Find the numbers.

15. Find four numbers between 4 and 40 so that the six numbers are consecutive terms of an AP.

16. The sum of three numbers which are consecutive terms of an AP is 2 and the sum of their squares is $\frac{14}{9}$. Find the numbers.

17. The sum of three numbers which are consecutive terms of an AP is 3 and the sum of their cubes is 4. Find the numbers.

18. If $a_1, a_2, a_3, \ldots, a_n$ are in AP, prove that

(i) $a_1 + a_n = a_r + a_{n-r+1}$

(ii) $\frac{1}{a_1 a_2} + \frac{1}{a_2 a_3} + \frac{1}{a_3 a_4} + \ldots + \frac{1}{a_{n-1} a_n} = \frac{n-1}{a_1 a_n}$

(iii) $\frac{1}{\sqrt{a_1} + \sqrt{a_2}} + \frac{1}{\sqrt{a_2} + \sqrt{a_3}} + \ldots + \frac{1}{\sqrt{a_{n-1}} + \sqrt{a_n}}$
$= \frac{n-1}{\sqrt{a_1} + \sqrt{a_n}}$ provided $a_i > 0$ for all i.

19. If $a_1, a_2, a_3, \ldots, a_n$ are in AP, prove that

$$\frac{1}{a_1 a_n} + \frac{1}{a_2 a_{n-1}} + \frac{1}{a_3 a_{n-2}} + \ldots + \frac{1}{a_n a_1}$$
$$= \frac{2}{a_1 + a_n}\left(\frac{1}{a_1} + \frac{1}{a_2} + \frac{1}{a_3} + \ldots + \frac{1}{a_n}\right).$$

20. If $a_1, a_2, a_3, \ldots, a_{2n}$ be in AP, show that

$$a_1^2 - a_2^2 + a_3^2 - a_4^2 + \ldots + a_{2n-1}^2 - a_{2n}^2$$
$$= \frac{n}{2n-1}(a_1^2 - a_{2n}^2).$$

21. If the nonzero numbers $a_1, a_2, a_3, \ldots, a_n, \ldots$ are in AP, prove that

$$\frac{1}{a_1 a_2 a_3} + \frac{1}{a_2 a_3 a_4} + \ldots + \frac{1}{a_{n-2} a_{n-1} a_n}$$
$$= \frac{1}{2(a_2 - a_1)}\left(\frac{1}{a_1 a_2} - \frac{1}{a_{n-1} a_n}\right).$$

22. If a, b, c are in AP, prove that

(i) $b + c, c + a, a + b$ are also in AP

(ii) $a^2(b + c), b^2(c + a), c^2(a + b)$ are also in AP.

23. If a^2, b^2, c^2 are in AP, prove that

(i) $\frac{1}{b+c}, \frac{1}{c+a}, \frac{1}{a+b}$ are also in AP

(ii) $\frac{a}{b+c}, \frac{b}{c+a}, \frac{c}{a+b}$ are also in AP.

24. If x, y, z are in AP, prove that

$\frac{1}{\sqrt{y} + \sqrt{z}}, \frac{1}{\sqrt{z} + \sqrt{x}}, \frac{1}{\sqrt{x} + \sqrt{y}}$ will be in AP.

25. If $S_n = an^2 + bn$, verify that the series Σt_n is arithmetic where $S_n = \sum_{n=1}^{n} t_n$.

26. If $S_n = an^2 + bn + c$ where $c \neq 0$ and S_n denotes the sum to n terms of a series, verify whether the series is arithmetic.

27. The sum of n terms of a series is $2n^2 + 3n$. Is the series arithmetic? If so, find it.

28. Find the sum of all numbers from 100 to 550 which are divisible by 9?

29. Find the sum of all positive integers less than 200 that are

(i) divisible by 3 and 5

(ii) divisible by 3 but not by 5

(iii) divisible by 3 or 5.

30. Find the sum of all natural numbers $n \in [100, 300]$ that are neither divisible by 4 nor by 6.

31. Find the sum of all 3-digit natural numbers which are of the form $3m + 2$, $m \in N$, i.e., leaves the remainder 2 when divided by 3.

32. Sum the series to n terms:

$$\frac{1}{1+\sqrt{x}} + \frac{1}{1-x} + \frac{1}{1-\sqrt{x}} + \dots .$$

33. How many terms of $1 + 3 + 5 + 7 + \dots$ amount to 1234321?

34. What is the maximum sum of the series

$$20 + 19\frac{1}{3} + 18\frac{2}{3} + \dots ?$$

35. Find the sum of the first 19 terms of the AP $a_1, a_2, a_3, \dots$ if it is known that $a_1 + a_8 + a_{12} + a_{19} = 224$.

36. Find the sum of the first hundred numbers appearing in both the progressions 17, 21, 25, ... and 16, 21, 26, ...

37. If in an AP, $S_p = S_q$ then prove that $S_{p+q} = 0$ where S_n denotes the sum of the first n terms.

38. If in an AP, $S_p = a$, $S_q = b$ and $S_r = c$ then prove that

$$\frac{a}{p}(q-r) + \frac{b}{q}(r-p) + \frac{c}{r}(p-q) = 0$$

where S_n = sum to the first n terms.

39. If $a_1, a_2, a_3, \dots, a_n, \dots$ are in AP and S_n denotes $a_1 + a_2 + a_3 + \dots + a_n$ then prove that $S_{3n} = 3(S_{2n} - S_n)$.

40. If in an AP, $t_m = \frac{1}{n}$ and $t_n = \frac{1}{m}$ then prove that

$S_{mn} = \frac{1}{2}(mn + 1)$ where t_r denotes the rth term and S_r denotes the sum to the first r terms.

41. In an AP whose first term is a, if the sum of the first p terms is zero, show that the sum of the next q terms is $-\frac{a(p+q)}{p-1} \cdot q$.

42. In an AP of natural numbers, the sum of the first 9 terms is greater than 200 but less than 220. Find the progression if the second term is 12.

43. If in an AP, the sum of m terms is equal to the sum of the next n terms as well as the sum of the next p terms then prove that

$$(m+n)\left(\frac{1}{m} - \frac{1}{p}\right) = (m+p)\left(\frac{1}{m} - \frac{1}{n}\right).$$

44. If $S_1, S_2, S_3, \dots, S_p$ are the sums of n terms of p arithmetic series whose first terms are 1, 2, 3, ..., p and whose common differences are 1, 3, 5, ... respectively then prove that $\sum_{i=1}^{p} S_i = \frac{np}{2}(np + 1)$.

45. If S_1 be the sum of $(2n + 1)$ terms of an AP and S_2 be the sum of its odd terms then prove that

$$\frac{S_1}{S_2} = \frac{2n+1}{n+1}.$$

46. If for an AP of odd number of terms, the sum of all the terms is $\frac{15}{8}$ times the sum of the terms in odd places then find the number of terms in the AP.

47. The number of terms in an AP is even. The sum of the odd terms is 24 while that of the even terms is 30. If the last term exceeds the first by 10.5 then find the number of terms.

48. The ratio of the sums of n terms of two APs is $(3n - 13) : (5n + 21)$. Find the ratio of the 24th terms of the two progressions.

49. Find the AP in which the ratio of the sum to n terms to the sum of succeeding n terms is independent of n.

50. If x, $2x + 2$, $3x + 3$ are the first three terms of a GP, find the fifth term of the sequence.

51. In a GP, $t_4 = 27$ and $t_7 = 729$. find t_{11} where t_n denotes the nth term.

52. In a GP, $t_{p+q} = a$ and $t_{p-q} = b$. Prove that $t_p = \sqrt{ab}$.

53. If the pth, qth and rth terms of a GP are respectively x, y and z then prove that $x^{q-r} \cdot y^{r-p} \cdot z^{p-q} = 1$.

54. If a be the first term, b be the nth term and P be the product of n terms of a GP then prove that $P^2 = (ab)^n$.

55. If a, b, c are in GP, prove that

$$(a^2 - b^2)(b^2 + c^2) = (b^2 - c^2)(a^2 + b^2).$$

56. If a, b, c, d are in GP then prove that

(i) $(b-c)^2 + (c-a)^2 + (d-b)^2 = (a-d)^2$

(ii) $a + b$, $b + c$, $c + d$ are in GP

(iii) $a(b-c)^3 = d(a-b)^3$

(iv) $ax^3 + bx^2 + cx + d$ has a factor $ax^2 + c$.

57. Prove that a, b, c, d are in GP

$\Leftrightarrow a^2 + b^2 + c^2,\ ab + bc + cd,\ b^2 + c^2 + d^2$ are in GP.

58. Let $a_1, a_2, a_3, \dots, a_n$ be in GP whose common ratio is r. Show that

$$\sum_{k=1}^{n-1} \frac{1}{a_k^2 - a_{k+1}^2} = \frac{1 - r^{2(n-1)}}{a_1^2 \cdot r^{2(n-2)} \cdot (1 - r^2)^2}.$$

59. Does there exist a GP containing 27, 8 and 12 as three of its terms? If it exists, how many such progressions are possible?

60. Prove that no GP can have three of its terms (not necessarily consecutive) as three consecutive nonzero integers.

61. The product of three numbers in GP is 216 and their sum is 19. Find the numbers.

62. Divide 63 into three parts that are in GP and the product of the first and the second part is $\frac{3}{4}$ times of the third part.

63. The first term of a GP is unity. For what value of the common ratio of the progression the sum of 4 times the second term and 5 times the third term will be minimum?

64. For the series Σt_n, $S_n = \sum_{n=1}^{n} t_n = 2t_n - 1$. Is the series geometric? If so, find the sum to n terms of the series.

65. If S_1, S_2, S_3 are the sums to $n, 2n, 3n$ terms respectively for a GP then prove that

(i) $S_1, S_2 - S_1, S_3 - S_2$ are in GP

(ii) $S_1^2 + S_2^2 = S_1(S_2 + S_3)$

66. If the sum of n terms of a GP is S, their product is P and the sum of their reciprocals is R then prove that $P^2 = (S/R)^n$.

67. Find $\sum_{n=1}^{n} u_n$ if $u_n = \sum_{n=0}^{n} \frac{1}{2^n}$.

68. Prove that

$$(666 \ldots \text{to } n \text{ digits})^2 + (888 \ldots \text{to } n \text{ digits}) = (444 \ldots \text{to } 2n \text{ digits}).$$

69. Prove that each number is the square of an odd integer in the sequence 49, 4489, 444889, ... in which every number is made by inserting 48 in the middle of the previous number as indicated.

70. In a GP the third term exceeds the second by 6 and the second term exceeds the first by 9. Find the sum of the first 10 terms.

71. $a_1, a_2, a_3, \ldots$ are in GP. If $a_n > a_m$ when $n > m$ and $a_1 + a_n = 66$ while $a_2 \cdot a_{n-1} = 128$ and $\sum_{i=1}^{n} a_i = 126$, find the value of n.

72. The sum of the first three terms of a GP is 13 and the sum of their squares is 91. Find the common ratio of the GP of real terms.

73. Express the ratio of the sum of the squares of the odd number of terms of a GP to the sum of those terms as a polynomial of the common ratio of the GP.

74. Find the sum to $2n$ terms of the series whose every even term is a times the term before it and every odd term is c times the term before it, the first term being unity.

75. If $1 + 3 + 9 + 27 + \ldots$ to n terms exceeds 1000, find the minimum value of n.

76. It is known by Fermat's theorem that $n^p - n = M(p)$ = a multiple of p, if p is a prime number and n is prime to p. If n is a prime number which divides neither a, b nor $a + b$, prove that

$$a^{n-2} \cdot b - a^{n-3} \cdot b^2 + a^{n-4} \cdot b^3 - \ldots + ab^{n-2} = 1 + (\text{a multiple of } n).$$

77. Find the sum of the proper divisors of the following:

(i) $2^{10} \cdot 5^7 \cdot 11^8$ (ii) $6^a \cdot 10^b \cdot 15^c$

where a, b, c are positive integers.

78. Find the sum of all the proper divisors of $36^m \cdot 10^n$ that are divisible by 5 where m, n are positive integers.

79. There are twos, threes and fives each being 8 in number. Find the sum of the products of these numbers taking at least one 2, 3 and 5.

80. The sum of an infinite GP is 32 and the sum of the first two terms is 24; find the series.

81. Prove that in an infinite GP whose common ratio r has its absolute value less than 1, the ratio of any term to the sum of all the succeeding terms is $\frac{1-r}{r}$.

82. The first term of an infinitely decreasing GP is unity and its sum is S. Find the sum of the squares of the terms of the progression.

83. $u_1 + u_2 + u_3 + \ldots$ is a geometric series and $\sum_{i=1}^{\infty} u_i = 3$ while $\sum_{i=1}^{\infty} u_i^2 = 3$. Find u_n.

84. The sum of an infinite GP is 2 and sum of the cubes of the terms of the GP is 24. Find the series.

85. The ratio of the sum of the cubes of the decreasing terms of an infinite GP to the sum of the squares of the terms of the GP is 12 : 13. Find the progression if the sum of the first two terms is $\frac{4}{3}$.

86. If $S_n = 1 + r^n + r^{2n} + r^{3n} + \ldots$ to ∞ and $s_n = 1 - r^n + r^{2n} - r^{3n} + \ldots$ to ∞ then prove that $S_n + s_n = 2S_{2n}$.

87. If $x = \sum_{n=0}^{\infty} a^n$, $y = \sum_{n=0}^{\infty} b^n$ where $|a| < 1$, $|b| < 1$ then prove that

$$\sum_{n=0}^{\infty} (ab)^n = \frac{xy}{x + y - 1}.$$

88. If $\exp\{(\sin^2 x + \sin^4 x + \sin^6 x + \dots \text{ to } \infty) \ln 2\}$ satisfies the equation $x^2 - 9x + 8 = 0$, find the value of $\dfrac{\cos x}{\sin x + \cos x}$, $0 \le x < \dfrac{\pi}{2}$.

89. If $S_1, S_2, S_3, \dots, S_n$ are the sums of n infinite geometric series whose first terms are $1, 2, 3, \dots, n$ respectively and their common ratios are $\dfrac{1}{2}, \dfrac{1}{3}, \dfrac{1}{4}, \dots, \dfrac{1}{n+1}$ respectively then

(i) prove that $S_1 + S_2 + S_3 + \dots + S_n = \dfrac{1}{2} n(n+3)$

(ii) find $\sum_{n=1}^{n} s_n$ where $s_n = S_1 + S_2 + S_3 + \dots + S_n$

(iii) find $\sum_{i<j=1}^{n} S_i S_j$.

90. If a, b, c are in GP, prove that $\log a^n, \log b^n, \log c^n$ are in AP.

91. If a, b, c are in AP and x, y, z are in GP, prove that $x^{b-c} \cdot y^{c-a} \cdot z^{a-b} = 1$.

92. If a, b, c are in AP and a, b, d are in GP, prove that $a, a-b, d-c$ are in GP.

93. If $a, b, c, \dots$ are in GP and $a^{\frac{1}{x}} = b^{\frac{1}{y}} = c^{\frac{1}{z}} = \dots$ then prove that $x, y, z, \dots$ are in AP.

94. If a, b, c be in GP and $\log_c a, \log_b c, \log_a b$ be in AP, prove that the common difference of the AP is $\dfrac{3}{2}$.

95. If a, x, b as well as c, x, d are in GP while a^2, y, b^2 as well as c^2, y, d^2 are in AP then prove that

$a^n + b^n = c^n + d^n$ where n is an even integer or

$a^n + b^n + c^n + d^n = 0$ where n is an odd integer.

96. The nth term of an AP and a GP are a_n and b_n respectively. Prove that $a_n < b_n$ for $n \ge 3$ if $a_1 = b_1, a_2 = b_2, a_1 \ne a_2$ and $a_n > 0$ for all n.

97. If the pth, qth and the rth terms of both an AP and a GP be respectively a, b and c then prove that $a^{b-c} \cdot b^{c-a} \cdot c^{a-b} = 1$.

98. If the pth, qth, rth and sth terms of an AP be in GP then prove that $p-q, q-r, r-s$ are in GP.

99. The second, third and sixth terms of an AP are consecutive terms of a GP. Find the common ratio of the GP.

100. Find an AP of distinct terms whose first term is unity such that the second, tenth and thirty-fourth terms form a GP.

101. There are four numbers of which the first three are in GP and the last three are in AP whose common difference is 6. If the first number and the last number are equal, find the numbers.

102. There are four numbers of which the first three are in GP whose common ratio is $\dfrac{1}{2}$ and the last three are in AP. If the last number is 2 less than the first, find the four numbers.

103. Prove that three unequal numbers cannot be in GP when each number increased (or decreased) by the same quantity are in AP.

104. The sum of three numbers in AP is 15. If 1, 4, 19 are added to them respectively, the resulting series is in GP. Find the numbers.

105. The sum of three numbers in GP is 14. If the first two terms are each increased by 1 and the third is decreased by 1, the resulting numbers are in AP. Find the numbers.

106. The sum of three numbers in GP is 70. If the two extremes be multiplied each by 4 and the mean by 5, the products are in AP. Find the numbers.

107. Three nonzero real numbers are in AP and their squares, in the same order, are in GP. Find all possible values of the common ratio of the GP.

108. Find three numbers a, b, c between 2 and 18 such that their sum is 25 while the numbers $2, a, b$ are in AP and $b, c, 18$ are in GP.

109. Four numbers are such that the first three are in GP and the last three are in AP. The sum of the extreme numbers is 32 and that of the middle numbers is 24. Find the numbers.

110. The first term of an AP is the same as that of a GP. The common difference of the AP and the common ratio of the GP are each equal to 2. If the sum of the first five terms of each series be the same, find the sixth term of each series.

111. The sum of the first ten terms of an AP of integers is 155 and the sum of the first two terms of a GP of integers is 9. The first term of the AP is equal to the common ratio of the GP and the first term of the GP is equal to the common difference of the AP. Find the two progressions.

112. Find a three-digit number whose digits are in AP and the three-digit number which is greater by 400, has the digits in GP.

113. Three successive terms of a GP are the consecutive digits of a three-digit number. If 792 is subtracted from the number we get a number with the digits reversed. Also a number smaller than the original number by 400 has its digits in AP. Find the number.

114. If the numbers $\dfrac{1}{k+2}, \dfrac{1}{k^2}$ and $\dfrac{1}{4k+5}$ are three consecutive terms of an HP then find k.

115. If in an HP, $t_m = n$ and $t_n = m$ then prove that $t_{m+n} = \dfrac{mn}{m+n}$.

116. The sum of three consecutive terms in HP is 37 and the sum of their reciprocals is $\frac{1}{4}$. Find the numbers.

117. If the pth, qth and the rth terms of an HP are respectively a, b and c then prove that
$$(q-r)bc + (r-p)ca + (p-q)ab = 0.$$

118. If $4x^2 + 9y^2 + 16z^2 - 6xy - 12yz - 8zx = 0$ then prove that x, y, z will be in HP.

119. If a, b, c are in HP then prove that

(i) $\dfrac{b+a}{b-a} + \dfrac{b+c}{b-c} = 2$

(ii) $\dfrac{1}{b-a} + \dfrac{1}{b-c} = \dfrac{1}{a} + \dfrac{1}{c}$

(iii) $b^2(a-c)^2 = 2\{c^2(b-a)^2 + a^2(c-b)^2\}$

(iv) $\log(c+a) + \log(a-2b+c) = 2\log(a-c)$

120. If a, b, c, d are in HP, prove that $ab + bc + cd = 3ad$.

121. If $x_1, x_2, x_3, \ldots, x_n$ are in HP, prove that
$$x_1x_2 + x_2x_3 + x_3x_4 + \ldots + x_{n-1}x_n = (n-1)x_1x_n.$$

122. If $b+c, c+a, a+b$ are in HP then prove that $\dfrac{a}{b+c}, \dfrac{b}{c+a}, \dfrac{c}{a+b}$ are in AP.

123. If $\frac{1}{2}(x+y), y, \frac{1}{2}(y+z)$ are in HP then prove that x, y, z are in GP.

124. If a, b, c, d be four numbers of which the first three are in AP and the last three are in HP then prove that $ad = bc$.

125. If $\dfrac{a-x}{px} = \dfrac{a-y}{qy} = \dfrac{a-z}{rz}$ and p, q, r are in AP then prove that x, y, z are in HP.

126. If $a^x = b^y = c^z = d^w$ and a, b, c, d are in GP then prove that x, y, z, w are in HP.

127. If three numbers are in GP then prove that the three numbers obtained by adding the middle one to each of the numbers will be in HP.

128. A GP and an HP have the same pth, qth, rth terms a, b, c respectively. Prove that
$$a(b-c)\log a + b(c-a)\log b + c(a-b)\log c = 0.$$

129. a, b, c are in AP; b, c, d are in GP and c, d, e are in HP. Prove that a, c, e are in GP.

130. a, b, c are in AP; x, y, z are in HP and ax, by, cz are in GP then prove that $\dfrac{x}{z} + \dfrac{z}{x} = \dfrac{a}{c} + \dfrac{c}{a}$.

131. a, b, x are AP; a, b, y are in GP and a, b, z are in HP, then prove that $4z(x-y)(y-z) = y(x-z)^2$.

132. If a, b, c are in AP and a^2, b^2, c^2 in HP then prove that either $a = b = c$ or $-\dfrac{a}{2}, b, c$ are in GP or $-\dfrac{c}{2}, b, a$ are in GP.

133. If $x, 1, z$ are in AP and $x, 2, z$ are in GP then prove that $x, 4, z$ will be in HP.

134. Let a, b, c, d, e be five real numbers such that a, b, c are in AP; b, c, d are in GP and c, d, e are in HP. If $a = 2$ and $e = 18$, find all possible values of b, c, d.

135. The value of xyz is $\frac{15}{2}$ if a, x, y, z, b are in AP, while xyz is $\frac{18}{5}$ if a, x, y, z, b are in HP. If a and b are positive integers find them.

136. An AP and an HP have the same first term, the same last term and the same number of terms. Prove that the product of the rth term from the beginning of the first sequence and the rth term from the end of the second sequence is independent of r.

137. If $a, ma, a_1, a_2, a_3, \ldots$ are in AP; $a, ma, g_1, g_2, g_3, \ldots$ are in GP and $a, ma, h_1, h_2, h_3, \ldots$ are in HP then prove that a_n, g_n, h_n will be in GP if
$$(m^{2n+2} - 1) : (m^{2n} - 1) = m(n+1) : n.$$

138. If l, m, n are three positive integers in GP, prove that the first term of an AP whose lth, mth and nth terms are in HP is to the common difference as $m+1$ to 1.

139. Three unequal numbers are in HP and their squares are in AP. Show that the numbers are in the ratio $1-\sqrt{3} : -2 : 1+\sqrt{3}$ or $1+\sqrt{3} : -2 : 1-\sqrt{3}$.

140. Insert 12 arithmetic means between -3.5 and -42.5.

141. Insert 6 geometric means between 27 and $\frac{1}{81}$.

142. Find the 5th harmonic mean when n harmonic means are inserted between 1 and 2.

143. If n AMs are inserted between two numbers, prove that the sum of the means equidistant from the beginning and the end is constant.

144. Prove that the product of n geometric means between two quantities is equal to the nth power of a geometric mean of those two quantities.

145. If n arithmetic means are inserted between 1 and 31 such that the 7th mean : the $(n-1)$th mean $= 5 : 9$ then find n.

146. Between two numbers whose sum is $2\frac{1}{6}$, an even number of arithmetic means are inserted. The sum of these means exceeds their number by unity. How many means are there?

147. If the AM between m and n and the geometric mean between a and b are each equal to $\frac{ma+nb}{m+n}$, find m and n in terms of a and b.

148. Prove that $\frac{a^{n+1}+b^{n+1}}{a^n+b^n}$ is the GM or HM between nonzero numbers a and b according as $n=-\frac{1}{2}$ or -1.

149. If n is a root of the equation

$$x^2(1-ac)-x(a^2+c^2)-(1+ac)=0$$

and if n harmonic means are inserted between a and c, show that the difference between the first and the last mean is equal to $ac(a-c)$.

150. If the HM of $y-x$ and $y-z$ is $2(y-a)$ then prove that $x-a, y-a, z-a$ are in GP.

151. If n harmonic means $H_1, H_2, H_3, \ldots, H_n$ are inserted between a and b then prove that

$$\frac{H_1+a}{H_1-a}+\frac{H_n+b}{H_n-b}=2n.$$

152. If A and H be respectively the AM and HM of a and b then prove that

$$\frac{a-A}{a-H}\times\frac{b-A}{b-H}=\frac{A}{H}.$$

153. If 9 AMs and again 9 HMs be inserted between 2 and 3 then prove that $A+\frac{6}{H}=5$, A is any AM and H the corresponding HM.

154. If two arithmetic means A_1, A_2, two geometric means G_1, G_2 and two harmonic means H_1, H_2 be inserted between two given quantities a and b then prove that $\frac{G_1\cdot G_2}{H_1\cdot H_2}=\frac{A_1+A_2}{H_1+H_2}=\frac{(2a+b)(a+2b)}{9ab}$.

155. If $(2n-1)$ arithmetic, geometric and harmonic means be inserted between two numbers, prove that the nth means are in geometrical progression.

156. If b is the GM between a, c and x, y are the AMs of a, b and b, c respectively then prove that

$$\frac{a}{x}+\frac{c}{y}=2 \text{ and } \frac{1}{x}+\frac{1}{y}=\frac{2}{b}.$$

157. If one arithmetic mean A and two geometric means p and q be inserted between two given quantities, show that $\frac{p^2}{q}+\frac{q^2}{p}=2A$.

158. If one geometric mean G and two arithmetic means p and q be inserted between two given quantities, prove that $G^2=(2p-q)(2q-p)$.

159. If b, a, c are in AP and b, G_1, G_2, c are in GP then prove that $G_1^3+G_2^3=2abc$.

160. If a_1, a_2 be two AMs; g_1, g_2 be two GMs and h_1, h_2 be two HMs between x and y then prove that

$$a_1h_2=a_2h_1=g_1g_2=xy.$$

161. If the arithmetic mean between a and b is twice as great as their geometric mean, show that

$$a:b=(2+\sqrt{3}):(2-\sqrt{3}).$$

162. If the arithmetic mean between a and b is m times their harmonic mean, prove that

$$a:b=(\sqrt{m}+\sqrt{m-1}):(\sqrt{m}-\sqrt{m-1}).$$

163. If the GM between a and b be m times their HM then prove that

$$a:b=(m+\sqrt{m^2-1}):(m-\sqrt{m^2-1}).$$

164. If the HM between two quantities is to their GM as 12 to 13, prove that the quantities are in the ratio of 4 to 9.

165. If the AM of two numbers exceeds their GM by 2 and the GM exceeds their HM by $1\frac{3}{5}$ then find the numbers.

166. The HM of two numbers is 4, their AM, A and GM, G satisfy the relation $2A+G^2=27$. Find the numbers.

167. The value of $x+y+z$ is 15 if a, x, y, z, b are in AP while the value of $\frac{1}{x}+\frac{1}{y}+\frac{1}{z}$ is $\frac{5}{3}$ if a, x, y, z, b are in HP. Find a and b.

168. The AM, GM and HM of 25 and n are three terms of the sequence $25, 26, 27, \ldots, n-1, n$. Then find the smallest possible value of n.

169. Can the sides of a quadrilateral (not necessarily taken in order) be in AP in which a circle can be inscribed?

170. The internal angles of a polygon are in AP. The smallest angle is $120°$ and the common difference is $5°$. Find the number of sides of the polygon and the sum of internal angles.

171. There are 25 trees at equal distances of 5 metres in a line with a well, the distance of the well from the nearest tree being 10 metres. A gardener waters all the trees separately starting from the well and he returns to the well after watering each tree to get water for the next. Find the total distance the gardener will cover in order to water all the trees.

172. A gentleman deposits one paise in a bank on the first of March. In the subsequent days he deposits two paise more than his balance with the bank till the previous day. What will be his balance at the end of the month?

173. There are n stones in a line, the distance between the consecutive stones being 1 m, 3 m, 5 m, etc. There is a basket fixed at the first stone. A man starting from the basket collects all the stones in the basket one by one. Find the total distance described by him in collecting all the stones.

174. If you have the choice of a salary of Rs 100 per day for a month of 31 days or Re 1 for the first day, Rs 2 for the second day, Rs 4 for the third day and each day thereafter for the rest of the month your salary would be doubled, what choice would you make for better salary? Justify your answer.

175. A square is given. By joining the middle points of its sides a second square is constructed. Again joining the middle points of the sides of the second square a third square is constructed and so on. Prove that the area of the given square is equal to the sum of the areas of the subsequent squares.

176. Balls are arranged in rows to form an equilateral triangle. The first row consists of one ball, the second of 2 balls, the third of 3 balls and so on. If 669 more balls are added then all the balls can be arranged in the shape of a square and each of its sides then contains 8 balls less than each side of the triangle did. Determine the initial number of balls.

177. Two trains A and B start from the same station at the same time. A covers half the distance from the station to another station with a velocity x and the other half with a velocity y. The train B covers the whole distance with a velocity $\frac{x+y}{2}$. Which train will cross the second station earlier?

178. Two chords AB and CD of a circle cut at O such that $OA, OC, \frac{1}{2} \cdot OB$ and OD are respectively four consecutive terms of an AP. Prove that $OB = CD$.

179. A 5-km-long road is marked by road marks starting from the beginning and having marks at distances 1 m, 3 m, 6 m, 10 m, … . The marks are, in general, yellow in colour but those at k metres away from the beginning, where k is a multiple of 5, are white. How many marks of the road are white, the mark at the beginning being red?

180. A man agrees to lend Rs 10,000 to a friend under the following terms of payment:
At the end of every 5 years he will have to pay, as interest, an amount equal to the principal together with a simple interest of 5% per annum on the principal for the whole period of the loan. What will be the total money that the friend will have to pay for the first time by the end of 50 years?

181. A man arranges to pay off a debt of Rs 3,600 in 40 annual instalments which form an AP. When 30 of the instalments are paid, he dies leaving behind one-third of his debt unpaid; find the value of the first instalment.

182. After striking the floor, a certain ball rebounds $\frac{4}{5}$th of its height from which it has fallen. Find the total distance that it travels before coming to rest if it is gently dropped from a height of 120 metres.

183. The sides of a right-angled triangle are in AP. Prove that the sides are in the ratio 3 : 4 : 5.

184. If the sides of a right-angled triangle are in GP then find the tangents of the acute angles of the triangle.

185. An odd number of stones are placed at equal intervals of 10 metres in a row on the field. All stones are to be gathered at the place of the stone in the middle. If a man can carry only one stone at a time and starts from one of the end stones, he has to cover 3 km for the work. How many stones were there?

186. A, B, C have together Rs 5,700 and the amount of money possessed by them form a GP. If B had Rs 150 more, the amount would form an AP. Find the amount they possess.

187. A piece of a paper in the shape of a square is cut into 20 parts by cutting parallel to a side. The breadths of successive parts are in AP and the breadth of the first part is $\frac{1}{10}$th of the side. If every second piece is coloured red, what fraction of the piece of paper is coloured red?

188. If $x > 0$, prove that the least value of $x + \frac{1}{x}$ is 2.

189. If $0 < x < \frac{\pi}{2}$, prove that $\tan x + \cot x \geq 2$.

190. If $a > 1, b > 1$ prove that $\log_b a + \log_a b \geq 2$.

191. For what real values of a can the equation $2^x + 2^{-x} = a$ have real solutions for x?

192. If $\log_2 x + \log_2 y \geq 6$, prove that the smallest possible value of $x + y$ is 16.

193. If $x > 1$, find the least value of
$2\log_{100} x - \log_x 0.0001$.

194. If $x > 0, y > 0, z > 0$ prove that
$(x+y)(y+z)(z+x) \geq 8xyz$.

195. If a, b, c are positive fractions and $a+b+c=1$, prove that $\left(1+\frac{1}{a}\right)\left(1+\frac{1}{b}\right)\left(1+\frac{1}{c}\right) \geq 64$.

196. If $a^2+b^2=1$, prove that $\left(a^2-\frac{1}{a^2}\right)^2+\left(b^2-\frac{1}{b^2}\right)^2 \geq \frac{9}{2}$.

197. If $a > b > 0$ and $n \in N$, prove that
$$a^n - b^n \geq n(a-b)(ab)^{\frac{n-1}{2}}.$$

198. If bc, ca, ab are in AP where $a>0, b>0, c>0$ then prove that
(i) $a^n + c^n \geq 2b^n$ (ii) $a^3 + c^3 \geq 2b^3$.

199. If a, b, c, d are positive numbers in HP then prove that $a + d > b + c$.

200. If a, b, c are in AP or GP or HP where $a > 0, b > 0, c > 0$ then prove that

$b^2 >$ or $=$ or $< ac$.

201. If $a > 0, b > 0, c > 0$, prove that

$$(a+b+c)\left(\frac{1}{a}+\frac{1}{b}+\frac{1}{c}\right) \geq 9.$$

202. If $a > 0, b > 0, c > 0$, prove that

$$\frac{a}{b+c}+\frac{b}{c+a}+\frac{c}{a+b} \geq \frac{3}{2}.$$

203. Prove that for positive quantities a, b, c we have

$$\frac{bc}{b+c}+\frac{ca}{c+a}+\frac{ab}{a+b} \leq \frac{a+b+c}{2}.$$

204. If $x > 0, y > 0, z > 0$, prove that

$$x^{\log y - \log z} + y^{\log z - \log x} + z^{\log x - \log y} \geq 3.$$

205. If $s = a + b + c + d$ where a, b, c, d are positive unequal numbers then prove that

$$(s-a)(s-b)(s-c)(s-d) > 81abcd.$$

206. If a, b, c are natural numbers, prove that

$$\left(\frac{a^2+b^2+c^2}{a+b+c}\right)^{a+b+c} \geq a^a b^b c^c \geq \left(\frac{a+b+c}{3}\right)^{a+b+c}$$

207. Prove that $1^1 \cdot 2^2 \cdot 3^3 \ldots n^n \leq \left(\frac{2n+1}{3}\right)^{\frac{n(n+1)}{2}}$

208. Prove that

(i) $\left(\frac{n+1}{2}\right)^n \geq n\,!$

(ii) ${}^{2n+x}C_n \cdot {}^{2n-x}C_n \leq ({}^{2n}C_n)^2$.

209. Prove that the product of the first n odd natural numbers cannot exceed n^n.

210. If s be the sum of n positive unequal quantities $a, b, c, \ldots$ then prove that

$$\frac{s}{s-a}+\frac{s}{s-b}+\frac{s}{s-c}+\ldots > \frac{n^2}{n-1}.$$

211. If $a_i > 0, i = 1, 2, 3, \ldots, n$ then prove that

$$\frac{a_1}{a_2}+\frac{a_2}{a_3}+\frac{a_3}{a_4}+\ldots+\frac{a_{n-1}}{a_n}+\frac{a_n}{a_1} \geq n.$$

212. If $a \neq b$ and a, b, x, y are positive rationals such that $x + y = 1$ then prove that $a^x b^y < ax + by$.

213. If x, y, z be three sides of a triangle, prove that

$$xyz \geq (y+z-x)(z+x-y)(x+y-z).$$

214. In a triangle ABC, prove that

(i) $(a+b+c)^3 \geq 27(b+c-a)(c+a-b)(a+b-c)$

(ii) $\left(1+\frac{b-c}{a}\right)^a \cdot \left(1+\frac{c-a}{b}\right)^b \cdot \left(1+\frac{a-b}{c}\right)^c \leq 1.$

215. If a, b, c are three positive quantities whose sum is equal to their product then prove that $a + b + c \geq 3\sqrt{3}$. Also prove that if $a + b + c = 3\sqrt{3}$ then $a = b = c$.

216. If $a + 2b + 3c = 1$ and $a > 0, b > 0, c > 0$ then find the greatest value of a^3b^2c.

217. If $4a + 3b + 2c = 5$ and $a > 0, b > 0, c > 0$ then for what values of a, b, c the value of $a^2b^3c^4$ will be the greatest?

218. If $a + b + c = 1$ and $a > 0, b > 0, c > 0$ then prove that

$$ab^2c^3 \leq \frac{1}{432}.$$

219. Find the greatest value of the product of three positive numbers if the sum of their products taking two at a time is 12.

220. Sum to n terms: $1 \cdot 2 \cdot 3 + 2 \cdot 3 \cdot 4 + 3 \cdot 4 \cdot 5 + \ldots$

221. Sum to n terms: $1 \cdot 3^2 + 3 \cdot 5^2 + 5 \cdot 7^2 + \ldots$

222. Sum the series:

$$1 \cdot (n-1) + 2(n-2) + 3(n-3) + \ldots + (n-1) \cdot 1.$$

Hence find the coefficient of x^n in the expansion of

$$(1 + x + 2x^2 + 3x^3 + \ldots + nx^n)^2.$$

223. Sum the series

$$(1 \cdot 2) + (1 \cdot 2 + 2 \cdot 3) + (1 \cdot 2 + 2 \cdot 3 + 3 \cdot 4) + \ldots$$

to 12 brackets.

224. If $a_1, a_2, a_3, \ldots, a_n$ be in AP whose common difference is d then prove that

$$\sum_{i=1}^{n} a_i a_{i+1} = n\left\{a_1^2 + na_1 d + \frac{n^2-1}{3} d^2\right\}$$

225. Find the sum:

$$1 \cdot 2 \cdot 3 \cdot 4 \cdot 5 + 2 \cdot 3 \cdot 4 \cdot 5 \cdot 6 + 3 \cdot 4 \cdot 5 \cdot 6 \cdot 7 + \ldots + (n-4)(n-3)(n-2)(n-1)n$$

226. Find the *n*th term and the sum to n terms:

(i) $1 + 9 + 24 + 46 + 75 + \ldots$

(ii) $3 + 6 + 11 + 18 + \ldots$

227. If $t_n = \sum_1^n n$, find $S_n = \sum_1^n t_n$.

228. Find the sum to n terms:

$$1^2 + 3^2 + 6^2 + 10^2 + 15^2 + \ldots$$

229. Find the sum to n terms

(i) $2 + 5 + 14 + 41 + \ldots$

(ii) $1 + \left(1 + \frac{1}{2} + \frac{1}{2^2}\right) + \left(1 + \frac{1}{2} + \frac{1}{2^2} + \frac{1}{2^3} + \frac{1}{2^4}\right) + \ldots$

230. Sum to infinite terms:

$$1 + (1+a)r + (1+a+a^2)r^2 + (1+a+a^2+a^3)r^3 + \ldots$$

231. Find the *n*th term and sum to n terms:

$$1 + 5 + 19 + 49 + 101 + 181 + 295 + \ldots$$

232. Sum to 16 terms:

$$\frac{1^3}{1}+\frac{1^3+2^3}{1+3}+\frac{1^3+2^3+3^3}{1+3+5}+\dots$$

233. Sum to n terms:

$$\frac{1^2}{1}+\frac{1^2+2^2}{1+2}+\frac{1^2+2^2+3^2}{1+2+3}+\dots$$

234. Evaluate $\sum_{1}^{10}(2^{n-1}+8n-3)$.

235. Find Σn^3 where $100 < n^3 < 10{,}000$.

236. Evaluate $\sum_{1}^{n}\left\{\sum_{1}^{n}(2^n+3n)\right\}$.

237. Sum: $1\cdot 1+2\cdot 3+4\cdot 5+8\cdot 7+\dots$ to n terms.

238. Find the constant term in the product $f(x)\cdot g(x)$ of the functions:

$$f(x)=1+2x+3x^2+\dots+(n+1)x^n$$

$$g(x)=1+\frac{1}{3x}+\frac{1}{9x^2}+\dots+\frac{1}{3^n\cdot x^n},\ x\neq 0$$

239. Sum: $1-\frac{4}{2}+\frac{7}{2^2}-\frac{10}{2^3}+\dots$ to n terms.

240. Sum to infinite terms:

$$1\cdot 2+2\cdot 3x+3\cdot 4\cdot x^2+\dots,\ (|x|<1).$$

241. Evaluate:

(i) $\sum_{r=1}^{\infty} r^2x^{r-1},\ (|x|<1)$ (ii) $\sum_{n=1}^{15}(2n+1)2^n$

242. If $f(x)=x+3x^2+5x^4+7x^8+\dots$ to n terms then find the value of $f'(1)$.

243. If $g(x)=x+3x^3+5x^7+7x^{15}+\dots$ to n terms then find $\int_0^1 g(x)dx$.

244. Show that

$$\frac{n-1}{n+1}+3\left(\frac{n-1}{n+1}\right)^2+5\left(\frac{n-1}{n+1}\right)^3+\dots \text{ to } \infty=\sum_{r=1}^{n-1} r.$$

245. Find the sum of n terms of the series:

$$1^3+3\cdot 2^2+3^3+3\cdot 4^2+5^3+3\cdot 6^2+\dots$$

246. Sum to n terms (brackets):

$$\left(x^2+\frac{1}{x^2}+2\right)+\left(x^4+\frac{2}{x^4}+5\right)+\left(x^6+\frac{3}{x^6}+8\right)+\dots$$

247. Sum to infinite terms:

$$\frac{1}{5}-\frac{2}{7^2}+\frac{3}{5^3}-\frac{4}{7^4}+\dots$$

248. If $f(x)=x-2x^2+3x^3-\dots+(-1)^{n-1}nx^n$ where $x\in R$ find $f'(1)$.

249. Sum to n terms:

(i) $0.4+0.44+0.444+\dots$

(ii) $5+555+55555+\dots$

250. The natural numbers have been grouped as follows:

$(1); (2, 3); (4, 5, 6); (7, 8, 9, 10); \dots$

Prove that the sum of the numbers in the nth group is $\frac{n(n^2+1)}{2}$.

251. Prove that the sum of the numbers in any group in the following is the square of an odd integer:

$(1); (1, 3, 5); (1, 3, 5, 7, 9); \dots$

252. Find the sum to n terms. Also find the sum to infinite terms.

(i) $1+\frac{1}{1+2}+\frac{1}{1+2+3}+\dots$

(ii) $\frac{1}{1\cdot 2\cdot 3}+\frac{1}{2\cdot 3\cdot 4}+\frac{1}{3\cdot 4\cdot 5}+\dots$

(iii) $\frac{1}{1+1^2+1^4}+\frac{2}{1+2^2+2^4}+\frac{3}{1+3^2+3^4}+\dots+\frac{n}{1+n^2+n^4}$

253. Sum the series to n terms:

$$\frac{1}{(1+x)(1+2x)}+\frac{1}{(1+2x)(1+3x)}+\frac{1}{(1+3x)(1+4x)}+\dots$$

254. Sum to n terms:

$$\frac{1}{1\cdot 2\cdot 3\cdot 4}+\frac{1}{2\cdot 3\cdot 4\cdot 5}+\frac{1}{3\cdot 4\cdot 5\cdot 6}+\dots$$

255. Let t_r denote the rth term in the series

$$\frac{0!}{5!}+\frac{1!}{6!}+\frac{2!}{7!}+\frac{3!}{8!}+\dots.$$

Prove that $4t_{r+1}=rt_r-(r+1)t_{r+1}$ and hence find the sum to n terms of the series.

256. Sum to infinite terms if $|x|<1$:

$$\frac{1}{(1-x)(1-x^3)}+\frac{x^2}{(1-x^3)(1-x^5)}+\frac{x^4}{(1-x^5)(1-x^7)}+\dots$$

257. Evaluate $\sum_{n=1}^{n}\frac{n^2+n-1}{(n+2)!}$.

258. Evaluate $\sum_{n=1}^{n} n\,(n!)$.

259. If $a_1, a_2, a_3, \dots, a_n$ are in AP and $a_i\neq(2k-1)\frac{\pi}{2}$ for all i, find the sum

$$\sec a_1\cdot\sec a_2+\sec a_2\cdot\sec a_3+\sec a_3\cdot\sec a_4+\dots+\sec a_{n-1}\cdot\sec a_n.$$

260. Prove that

$$\tan^{-1}\frac{d}{1+a_1a_2}+\tan^{-1}\frac{d}{1+a_2a_3}+\ldots+\tan^{-1}\frac{d}{1+a_{n-1}a_n}=\tan^{-1}\frac{(n-1)d}{1+a_1a_n}$$

where d is the common difference of the AP $a_1, a_2, a_3, \ldots$.

261. Sum the series:

$\cot^{-1}2+\cot^{-1}8+\cot^{-1}18+\cot^{-1}32+\ldots$ to n terms.

262. Sum to n terms: $\frac{1}{1\cdot 2}+\frac{2}{1\cdot 2\cdot 3}+\frac{3}{1\cdot 2\cdot 3\cdot 4}+\ldots$

263. Find the sum:

$$\frac{1}{x+1}+\frac{2}{x^2+1}+\frac{4}{x^4+1}+\ldots+\frac{2^n}{x^{2^n}+1}$$

264. Find the sum of the product of the integers 1, 2, 3, ..., n taken two at a time and show that it is equal to half the excess of the sum of the cubes of the given integers over the sum of their squares.

265. Find the sum of the products of the first n odd natural numbers taking two at a time.

266. Find the sum of the products of $1, \frac{1}{2}, \frac{1}{2^2}, \ldots, \frac{1}{2^{10}}$ taking two at a time.

267. Find the sum of the products taken two together from the numbers 1, 4, 7, ..., $(3n-2)$.

268. Show that the sum of the products of every pair of the squares of the first n natural numbers is

$$\frac{1}{360}n(n^2-1)(4n^2-1)(5n+6).$$

269. Show that the sum of the products of the numbers $a, a-b, a-2b, \ldots, a-(n-1)b$ taking two at a time vanishes when n is of the form $3m^2-1$ and $2a=(3m-2)(m+1)b$.

270. Find the coefficient of (i) x^{49}, (ii) x^{48} in the product

$$(x+1)(x+3)(x+5)\ldots(x+99).$$

271. (i) Find the coefficient of x^{99} in the product $(2x-1)(2x-3)(2x-5)\ldots(2x-199)$.

(ii) Find the coefficient of x^n in the continued product $(x+1)(2x+1)(4x+1)\ldots(2^nx+1)$.

272. Prove that the sum of the cubes of any number of consecutive natural numbers (not beginning with 1) is divisible by the sum of the natural numbers.

273. In an AP of $2n$ terms the middle pair of terms are $\alpha+\beta$ and $\alpha-\beta$. Show that the sum of the cubes of the terms of the AP is $2n\alpha\{\alpha^2+(4n^2-1)\beta^2\}$.

274. Prove that in the series

$ab+(a+x)(b+x)+(a+2x)(b+2x)+\ldots$ to $2n$ terms, the excess of the sum of the last n terms over the sum of the first n terms is to the excess of the last term over the first as n^2 to $2n-1$.

275. Prove that

$$(1+x^{-1})(1+x^{-2})(1+x^{-4})(1+x^{-8})\ldots(1+x^{-2^n})=\frac{x}{x-1}\left(1-\frac{1}{x^{2^{n+1}}}\right).$$

Hence find the product

$$(1+3^{-1})(1+3^{-2})(1+3^{-4})(1+3^{-8})\ldots \text{ to } \infty.$$

Objective Questions

Fill in the blanks.

276. If $x^2, x, -8$ are in AP then $x=$ ______

277. (i) If the sum to n terms of a series be $5n^2+2n$ then the second terms = ______

(ii) The sum of n terms of an AP is n^2. Then the common difference = ______

278. If a series of terms in AP be collected into groups of n terms and the terms of each group be added together, the resultant series is an AP. Then

$$\frac{\text{the common difference of the resulting series}}{\text{the common difference of the original series}} = \_\_\_\_\_\_ .$$

279. The value of the product $\sqrt{3}\cdot\sqrt[4]{9}\cdot\sqrt[8]{27}\cdot\sqrt[16]{81}\ldots$ to infinite terms is equal to ______ .

280. For two unequal numbers, AM = 4 and GM = 2. Then HM = ______ .

281. 50^2-13^2 can be expressed as the sum of ______ consecutive odd integers.

282. The sum of the series $(n-1)^3+(n-2)^3+(n-3)^3+\ldots+2^3$ is ______ .

283. The sum of the first n terms of the series $1^2+2\cdot 2^2+3^2+2\cdot 4^2+5^2+2\cdot 6^2+\ldots$ is $\frac{n(n+1)^2}{2}$ when n is even. When n is odd, the sum is ______ .

284. The coefficient of x^{99} in the polynomial $(x-1)(x-2)(x-3)\ldots(x-100)$ is equal to ______ .

285. The sums of n terms of two AP series are in the ratio of $(n+1):(n+3)$. The ratio of their sixth terms is ______ .

286. Let the HM and GM of two positive numbers be in the ratio 4 : 5. Then the ratio of the two numbers is ______ .

287. In a GP of even number of distinct terms, the sum of all terms is 5 times the sum of odd terms. Then the common ratio of the GP is ______ .

288. The value of $\sum_{r=1}^{n}\int_0^{\pi/2}(r+\sin\theta)^2\cos\theta\,d\theta=$ ______ .

289. The greatest positive term in the sequence $\frac{2}{5}, \frac{12}{23}, \frac{3}{4}, \ldots$ has the value ______ .

290. $A_1, A_2, A_3, \ldots$ are points in the first quadrant on the parabola $y^2 = 4x$. If the x-coordinates of the points be in HP, the first two being $1, \frac{1}{2}$ then the y-coordinate of the point A_n is ______ .

Choose the correct option(s).

291. If a, b, c are in GP then $\frac{1}{c}, \frac{1}{b}, \frac{1}{a}$ are in

(a) AP (b) GP
(c) HP (d) none of these

292. If a, b, c are in GP then a^2, b^2, c^2 are in

(a) AP (b) GP
(c) HP (d) none of these

293. If $\frac{a-b}{b-c} = \frac{a}{c}$ then a, b, c are in

(a) AP (b) GP
(c) HP (d) none of these

294. If a, b, c are in GP then $\log_a \rho, \log_b \rho, \log_c \rho$ are in

(a) AP (b) GP
(c) HP (d) none of these

295. If $\log a, \log b, \log c$ are in AP then a, b, c are in

(a) AP (b) GP
(c) HP (d) none of these

296. If $l = 1 + a + a^2 + \ldots$ to ∞, $m = 1 + b + b^2 + \ldots$ to ∞, $n = 1 + c + c^2 + \ldots$ to ∞
where $|a| < 1, |b| < 1, |c| < 1$
and a, b, c are in AP then l, m, n are in

(a) AP (b) GP
(c) HP (d) none of these

297. If a, b, c are in AP then $\frac{a}{bc}, \frac{1}{c}, \frac{2}{b}$ are in

(a) AP (b) GP
(c) HP (d) none of these

298. If $\frac{1}{b-a} + \frac{1}{b-c} = \frac{1}{a} + \frac{1}{c}$ and $b \neq a + c$ then a, b, c are in

(a) AP (b) HP
(c) GP (d) HP and GP both

299. If a, b, c, d and p are distinct real numbers such that $(a^2 + b^2 + c^2)p^2 - 2(ab + bc + cd)p + (b^2 + c^2 + d^2) \leq 0$ then a, b, c, d are in

(a) AP (b) GP (c) HP
(d) $ab = cd$ (e) none of these

300. $\frac{1}{\log_3 2}, \frac{1}{\log_6 2}, \frac{1}{\log_{12} 2}$ are in

(a) AP (b) GP
(c) HP (d) none of these

301. If $a_n = \int_0^{\pi} \frac{\sin(2n-1)x}{\sin x} dx$ then which of the following is true about the progression $a_1, a_2, a_3, \ldots, a_n$?

(a) AP only (b) AP as well as GP
(c) AP, GP and HP (d) GP only

302. In an AP, $S_p = q$ and $S_q = p$ then S_{p+q} is equal to

(a) 0 (b) $-(p+q)$
(c) $p+q$ (d) pq

303. Which term of the sequence $2, 1, 2^{-1}, 4^{-1}, 8^{-1}, \ldots$ is $\frac{1}{128}$?

(a) eighth (b) tenth
(c) ninth (d) seventh

304. Which term of the sequence

$$25 + 22\frac{3}{4} + 20\frac{1}{2} + 18\frac{1}{4} + \ldots$$

is numerically smallest?

(a) thirteenth (b) twelfth
(c) fourteenth (d) eleventh

305. The middle term of the progression
2, 5, 8, 11, … , 92 is

(a) 47 (b) 48 (c) 46 (d) 49

306. The third term of a GP is 4. Then the product of the first 5 terms is

(a) 4^3 (b) 4^5
(c) 4^4 (d) none of these

307. In a GP of positive terms, any term is equal to the sum of the next two terms. Which of the following is the common ratio?

(a) 1 (b) $\frac{\sqrt{5}-1}{2}$ (c) $\frac{1-\sqrt{5}}{2}$ (d) $\frac{2}{\sqrt{5}}$

308. If $(1-p)(1 + 2x + 4x^2 + 8x^3 + 16x^4 + 32x^5) = 1 - p^6$, $(p \neq 1)$, then a value of $\frac{p}{x}$ is

(a) $\frac{1}{2}$ (b) 2 (c) $\frac{1}{4}$ (d) 4

309. If the sum of the series $1 + \frac{3}{x} + \frac{9}{x^2} + \frac{27}{x^3} + \ldots$ to ∞ is a finite number then

(a) $x < 3$ (b) $x > \frac{1}{3}$ (c) $x < \frac{1}{3}$ (d) $x > 3$

310. $x^{\frac{1}{2}} \cdot x^{\frac{1}{4}} \cdot x^{\frac{1}{8}} \cdot x^{\frac{1}{16}} \cdot \ldots$ to ∞ is equal to

(a) 0 (b) 1 (c) x (d) ∞

311. In a GP of $2n$ terms whose common ratio is r, the ratio of the sum of odd terms to that of even term is

(a) $\frac{1}{r}$ (b) r (c) n (d) nr

312. Let S_n denote the sum of the first n terms of an AP. If $S_{2n} = 3S_n$ then the ratio $S_{3n} : S_n$ is equal to

(a) 4 (b) 6 (c) 8 (d) 10

313. Sum of the first n terms of the series

$\frac{1}{2} + \frac{3}{4} + \frac{7}{8} + \frac{15}{16} + \dots$ is equal to

(a) $2^n - n - 1$ (b) $1 - 2^{-n}$

(c) $n + 2^{-n} - 1$ (d) $2^n - 1$

314. For $0 < \phi < \frac{\pi}{2}$ if

$$x = \sum_{n=0}^{\infty} \cos^{2n}\phi,\ y = \sum_{n=0}^{\infty} \sin^{2n}\phi,\ z = \sum_{n=0}^{\infty} \cos^{2n}\phi \cdot \sin^{2n}\phi$$

then

(a) $xyz = xz + y$ (b) $xyz = xy + z$

(c) $xyz = x + y + z$ (d) $xyz = yz + x$

315. The sum of infinite series

$\frac{1}{1\cdot 4} + \frac{1}{4\cdot 7} + \frac{1}{7\cdot 10} + \dots$ to ∞ is

(a) $\frac{1}{3}$ (b) 3 (c) $\frac{1}{4}$ (d) ∞

316. If the first and $(2n-1)$th terms of an AP, a GP and an HP are equal and their nth terms are a, b and c respectively then

(a) $a = b = c$ (b) $a \ge b \ge c$

(c) $a + c = b$ (d) $ac - b^2 = 0$

317. $\frac{a^n + b^n}{a^{n-1} + b^{n-1}}$ is the AM between a and b if n is

(a) 0 (b) $\frac{1}{2}$ (c) 1 (d) -1 (e) $-\frac{1}{2}$

318. If H be the harmonic mean between P and Q then the value of $\frac{H}{P} + \frac{H}{Q}$ is

(a) 2 (b) $\frac{PQ}{P+Q}$

(c) $\frac{P+Q}{PQ}$ (d) none of these

319. If $x > 0$ then the minimum value of $x^3 + \frac{1}{x^3}$ is

(a) 3 (b) 0 (c) 2 (d) $\frac{1}{2}$ (e) 1

320. The product of n positive numbers is unity. Then their sum is

(a) a positive integer (b) divisible by n

(c) equal to $n + \frac{1}{n}$ (d) never less than n

321. If $\sum_{k=1}^{n}\left(\sum_{m=1}^{k} m^2\right) = an^4 + bn^3 + cn^2 + dn + e$ then

(a) $a = \frac{1}{12}$ (b) $b = \frac{1}{6}$

(c) $d = \frac{1}{6}$ (d) $e = 0$

State whether the statements are true or false.

322. The value of $\log_5 3 + \log_7 5 + \log_9 7 \ge \frac{3}{\sqrt[3]{2}}$.

323. If x, y, z are any three distinct positive numbers then $x^3 + y^3 + z^3 > 3xyz$.

324. If a, b, c are in HP, the value of $\left(\frac{1}{b} + \frac{1}{c} - \frac{1}{a}\right)\left(\frac{1}{c} + \frac{1}{a} - \frac{1}{b}\right)$ is $\frac{2a-b}{ab^2}$.

325. For $x = \log_5 2$, the number $\log_{10}(2^x - 1)$ is the AM of $\log_{10} 2$ and $\log_{10}(2^x + 3)$.

326. $\tan\frac{\pi}{9} + \tan\frac{5\pi}{18}$ is the AM of $\tan\frac{\pi}{9}$ and $\tan\frac{7\pi}{18}$.

327. The sum of

$\frac{1}{10} + \frac{23}{10^3} + \frac{23}{10^5} + \frac{23}{10^7} + \dots$

as a recurring decimal fraction is $0.1\dot{2}\dot{3}$.

Answers

6. 3 7. 9 : 6 : 4
8. 135, 630, 765 9. 249
10. (i) 25 (ii) 15 11. 5, 10, 15, 20
12. 3, 5, 7, 9 13. 2, 4, 6, 8
14. 3, 5, 7 and 4, 5, 6 or 21, 5, −11 and 22, 5, −12
15. $11\frac{1}{5}, 18\frac{2}{5}, 25\frac{3}{5}, 32\frac{4}{5}$ 16. $\frac{1}{3}, \frac{2}{3}, 1$
17. $\frac{\sqrt{6}-1}{\sqrt{6}}, 1, \frac{\sqrt{6}+1}{\sqrt{6}}$ 27. yes, $t_1 = 5$ and $d = 4$
28. 16425 29. (i) 1365 (ii) 5268 (iii) 9168
30. 26634 31. 164850 32. $\frac{n\{2 + (n-3)\sqrt{x}\}}{2(1-x)}$
33. 1111 34. 310 35. 1064 36. 101100
42. $t_1 = 8, d = 4$ 46. 15 47. 8 48. 1 : 2
49. $k, 3k, 5k, \dots$ for all k 50. $\frac{-81}{4}$ 51. 3^{10}
59. yes, innumerable 61. 4, 6, 9
62. 3, 12, 48 63. $\frac{-2}{5}$ 64. yes, $2^n - 1$

67. $2^{-n}+2n-1$ **70.** $-\dfrac{58025}{729}$

71. 6 **72.** 3 or $\dfrac{1}{3}$

73. $a\{1-r+r^2-r^3+\ldots+r^{2n}\}$

74. $\dfrac{(1+a)(c^n a^n-1)}{ca-1}$ **75.** 7

77. (i) $\dfrac{1}{40}(2^{11}-1)(5^8-1)(11^9-1)-2^{10}\cdot 5^7\cdot 11^8-1$

(ii) $\dfrac{1}{8}(2^{a+b+1}-1)(3^{c+a+1}-1)(5^{b+c+1}-1)$ $-2^{a+b}\cdot 3^{c+a}\cdot 5^{b+c}-1$

78. $\dfrac{5}{8}(2^{2m+n+1}-1)(3^{2m+1}-1)(5^n-1)$ $-2^{2m+n}\cdot 3^{2m}\cdot 5^n$

79. $\dfrac{15}{4}(2^8-1)(3^8-1)(5^8-1)$

80. $16+8+4+2+\ldots$ or $48-24+12-6+\ldots$

82. $\dfrac{S^2}{2S-1}$ **83.** $\dfrac{3}{2^n}$ **84.** $3-\dfrac{3}{2}+\dfrac{3}{4}-\ldots$

85. $1, \dfrac{1}{3}, \dfrac{1}{9}, \ldots$ **88.** $1, \dfrac{1}{2}(\sqrt{3}-1)$

89. (ii) $\dfrac{n(n+1)(n+5)}{6}$

(iii) $\dfrac{1}{2}\left\{\dfrac{n^2(n+3)^2}{4}-\dfrac{(n+1)(n+2)(2n+3)}{6}+1\right\}$

99. 3 **100.** $1, \dfrac{4}{3}, \dfrac{5}{3}, \dfrac{6}{3}, \ldots$

101. 8, – 4, 2, 8 **102.** $2, 1, \dfrac{1}{2}, 0$

104. 2, 5, 8, or 26, 5, –16 **105.** 2, 4, 8 or 8, 4, 2

106. 40, 20, 10 **107.** 1 **108.** 5, 8, 12

109. 2, 6, 18, 30 or 32, 16, 8, 0

110. $10\dfrac{10}{13}, 24\dfrac{8}{13}$ resp. **111.** 2, 5, 8, ... and 3, 6, 12, ...

112. 531 **113.** 931 **114.** $\dfrac{7}{2}, -1$

116. 15, 12, 10 **134.** $b=4, -2$; $c=6, -6$; $d=9, -18$

135. 1, 3 or 3,1 **140.** –6.5, –9.5, –12.5, ... , –39.5

141. $9, 3, 1, \dfrac{1}{3}, \dfrac{1}{9}, \dfrac{1}{27}$ **142.** $\dfrac{2(n+1)}{2n-3}$

145. 14 **146.** 12

147. $m=\dfrac{2b\sqrt{a}}{\sqrt{a}+\sqrt{b}}, n=\dfrac{2a\sqrt{b}}{\sqrt{a}+\sqrt{b}}$ **165.** 16, 4

166. 6, 3 **167.** 9, 1 **168.** 225 **169.** yes

170. 9, 1260° **171.** 3370 m **172.** $(3\cdot 2^{30}-2)$ paise

173. $\dfrac{(n-1)n(2n-1)}{3}$ metres **174.** second choice

176. 1540 **177.** B **179.** 39

180. Rs 2,37,500 **181.** Rs 51

182. 1080 metres **184.** $\sqrt{\dfrac{1+\sqrt{5}}{2}}, \sqrt{\dfrac{2}{1+\sqrt{5}}}$

185. 25 **186.** Rs 2700, Rs 1800, Rs 1200 or reversely

187. $\dfrac{9}{19}$th part **191.** $a\ge 2$ **193.** 4 **216.** $\dfrac{1}{5184}$

217. $a=\dfrac{5}{18}, b=\dfrac{5}{9}, c=\dfrac{10}{9}$ **219.** 8

220. $\dfrac{1}{4}n(n+1)(n+2)(n+3)$ **221.** $\dfrac{1}{3}n(6n^3+16n^2+9n-4)$

222. $\dfrac{n(n^2-1)}{6}, \dfrac{n(n^2+11)}{6}$ **223.** 2730

225. $\dfrac{1}{6}n(n^2-1)(n-2)(n-3)(n-4)$

226. (i) $\dfrac{7}{2}n^2-\dfrac{5}{2}n, \dfrac{1}{6}n(n+1)(7n-4)$

(ii) $n^2+2, \dfrac{1}{6}n(2n^2+3n+13)$

227. $\dfrac{1}{6}n(n+1)(n+2)$

228. $\dfrac{1}{20}n^5+\dfrac{1}{4}n^2(n+1)^2-\dfrac{1}{24}n(n+1)(2n+1)$ $+\dfrac{1}{8}n(n+1)-\dfrac{1}{20}n$

229. (i) $\dfrac{1}{4}(3^{n+1}-3)+\dfrac{n}{2}$ (ii) $2n-\dfrac{4}{3}\left\{1-\dfrac{1}{2^{2n}}\right\}$

230. $\dfrac{1}{(1-r)(1-ar)}$

231. $n^3-n^2+1, \dfrac{n(n^2-1)(3n+2)}{12}+n$ **232.** 446

233. $\dfrac{1}{3}n(n+2)$ **234.** 1433

235. 53261 **236.** $2^{n+2}-4-2n+\dfrac{n(n+1)(n+2)}{2}$

237. $3-2^{n+1}+(2n-1)2^n$ **238.** $\dfrac{9}{4}\left(1-\dfrac{1}{3^{n+1}}\right)-\dfrac{n+1}{2\cdot 3^n}$

239. $\dfrac{2}{3}\left[\left(-\dfrac{1}{2}\right)^{n-1}-(3n-2)\left(-\dfrac{1}{2}\right)^n\right]$ **240.** $\dfrac{2}{(1-x)^3}$

241. (i) $\dfrac{1+x}{(1-x)^3}$ (ii) $29\cdot 2^{16}+2$

242. $(n-1)2^{n+1}-2^n+3$ **243.** $3-\dfrac{1}{2^{n-2}}-\dfrac{2n-1}{2^n}$

245. $\dfrac{n^2(n+1)^2}{4}-\dfrac{n^2(n+2)^2}{8}+\dfrac{n(n+2)(n+1)}{2}$, when n is even

$\dfrac{n^2(n+1)^2}{4}-\dfrac{n^2-1}{8}+\dfrac{n(n^2-1)}{2}$, when n is odd

246. $\dfrac{(1-x^{2n})(x^{2n}-x^{2n+2}-1)}{x^{2n-2}(1-x^2)^2}+\dfrac{n}{x^{2n}(1-x^2)}+\dfrac{n}{2}(3n+1)$

247. $\dfrac{211}{1152}$

248. $-\dfrac{n(n+1)}{2}$ if n is even, $\dfrac{n(n+1)}{2}$ if n is odd

249. (i) $\dfrac{4}{81}\left\{9n-\dfrac{10^n-1}{10^n}\right\}$ (ii) $\dfrac{50}{891}(10^{2n}-1)-\dfrac{5}{9}n$

252. (i) $\dfrac{2n}{n+1}$, 2 (ii) $\dfrac{n(n+3)}{4(n+1)(n+2)}$, $\dfrac{1}{4}$

(iii) $\dfrac{n(n+1)}{2(n^2+n+1)}$, $\dfrac{1}{2}$

253. $\dfrac{n}{(1+x)\{1+(n+1)x\}}$

254. $\dfrac{1}{18}-\dfrac{1}{3(n+1)(n+2)(n+3)}$

255. $\dfrac{1}{4\,(4\,!)}-\dfrac{n\,!}{4\,\{(n+4)\,!\}}$ 256. $\dfrac{1}{(1+x)(1-x)^2}$

257. $\dfrac{1}{2}-\dfrac{n+1}{(n+2)\,!}$ 258. $(n+1)\,!-1$

259. $\dfrac{\sin\{(n-1)(a_2-a_1)\}}{\sin\alpha}\sec a_n\cdot\sec a_1$

261. $\tan^{-1}\dfrac{n}{n+1}$ 262. $1-\dfrac{1}{(n+1)\,!}$

263. $\dfrac{1}{x-1}-\dfrac{2^{n+1}}{x^{2^{n+1}}-1}$ 265. $\dfrac{n}{6}(3n^3-4n^2+1)$

266. $\dfrac{1}{3}-\dfrac{1}{2^{11}}+\dfrac{1}{3}\cdot\dfrac{1}{2^{21}}$ 267. $\dfrac{1}{8}n(n-1)(9n^2-9n-2)$

270. (i) 2500 (ii) 3041675

271. (i) $-2^{99}\cdot 100^2$ (ii) $2^{\frac{n^2+n+2}{2}}\cdot\left(1-\dfrac{1}{2^{n+1}}\right)$

275. $\dfrac{3}{2}$	276. 4, −2	277. (i) 17 (ii) 2	
278. $n^2:1$	279. 9	280. 1	281. 37

282. $\dfrac{1}{4}(n^2-n+2)(n^2-n-2)$ 283. $\dfrac{(n+1)n^2}{2}$

284. −5050	285. 6 : 7	286. 4 : 1 or 1 : 4	287. 4
288. $\dfrac{1}{3}n(n^2+3n+3)$	289. 6	290. $\dfrac{2}{\sqrt{n}}$	
291. (b)	292. (b)	293. (c)	294. (c)
295. (b)	296. (c)	297. (d)	298. (b)
299. (b)	300. (a)	301. (c)	302. (b)
303. (c)	304. (b)	305. (a)	306. (b)
307. (b)	308. (b)	309. (d)	310. (c)
311. (a)	312. (b)	313. (c)	314. (b)
315. (a)	316. (d)	317. (c)	318. (a)
319. (c)	320. (d)	321. (a), (c), (d)	
322. true	323. true	324. false	325. false
326. true	327. true		

□

Chapter Test

Time: 120 minutes

1. In the following, each question has one or more than one correct answers. Indicate the correct answer(s).

(a) If $a > b > 0$ and AM of a, b is twice the GM of the numbers then which of the following is the value of $a : b$?

A. $(2 + \sqrt{3}) : (2 - \sqrt{3})$ B. $1 : (7 - 4\sqrt{3})$

C. $2 : \sqrt{3}$ D. $(7 + 4\sqrt{3}) : 1$

(b) If $a_1 = 0$ and $a_1, a_2, a_3, \dots, a_n$ are real numbers such that $|a_i| = |a_{i-1} + 1|$ for all i then the AM of the numbers $a_1, a_2, \dots, a_n$ has the value x where

A. $x < -1$ B. $x < -\frac{1}{2}$ C. $x \geq -\frac{1}{2}$ D. $x = -\frac{1}{2}$

(c) If in an AP, $t_p = q$ and $t_q = p$ then t_{p+q} is equal to

A. 0 B. pq C. $p + q$ D. $-(p + q)$

2. In the following, each statement is incomplete. Fill in the blank so that each resulting statement is correct.

(a) $\frac{1}{\sqrt{2}+\sqrt{5}} + \frac{1}{\sqrt{5}+2\sqrt{2}} + \frac{1}{2\sqrt{2}+\sqrt{11}} + \frac{1}{\sqrt{11}+\sqrt{14}} + \dots$ to n terms = ______

(b) If $a_n = (0 \cdot 2)^{\log_{\sqrt{5}}\left(\frac{1}{4}+\frac{1}{8}+\frac{1}{16}+\dots+\frac{1}{4 \cdot 2^{n-1}}\right)}$ then $\lim\limits_{n \to \infty} a_n =$ ______

(c) If $x, 2y, 3z$ are in AP where the distinct numbers x, y, z are in GP then the common ratio of the GP = ______

3. If $a_n^{-1} = \int_0^{\pi/2} \frac{\sin^2 nx}{\sin^2 x}\, dx$, prove that $a_1, a_2, a_3, \dots$ are in HP.

4. It is known that $\frac{1}{1^2} + \frac{1}{2^2} + \frac{1}{3^2} + \dots$ to $\infty = \frac{\pi^2}{6}$. Find the sum of the series $\frac{1}{1^2} + \frac{1}{3^2} + \frac{1}{5^2} + \dots$ to ∞.

5. If a_n denotes sum of n terms of the HP $\frac{1}{2}, \frac{1}{3}, \frac{1}{4}, \dots$ then prove that a_n cannot be equal to any integer for $n \in N$. Use summation of series.

6. Prove that there is no GP with integral common ratio whose three consecutive terms are sides of a scalene triangle.

7. Let $a_1, a_2, a_3, \dots$ be in AP. If $S_n = \sum_{n=1}^{n} a_n$ and $\frac{S_{mn}}{S_n}$ is independent of n, prove that the common difference is twice the first term.

8. Prove that m^r where $m \in N, r \in N, r \geq 2$ can be expressed as sum of m consecutive odd integers. Find the first odd integer in case of 8^5.

9. Prove that the difference between the two natural numbers 111 ... to $2n$ ones and 222 ... to n twos is an odd perfect square number.

10. If the AM of $a_1, a_2, a_3, \dots, a_n$ is a and $b_1, b_2, b_3, \dots, b_n$ have the AM b and $a_i + b_i = 1$ for $i = 1, 2, 3, \dots, n$ then prove that $\sum_{i=1}^{n} (a_i - a)^2 + \sum_{i=1}^{n} a_i b_i = nab$.

11. If $f(x) = 3x^2 - 2\left(\frac{1}{d} - \frac{1}{a}\right)x + \left\{\frac{1}{a^2} + \frac{1}{d^2} + 2\left(\frac{1}{b^2} + \frac{1}{c^2}\right)\right\} - \frac{2}{abcd}(ab + cd + ad)$ and $f(\alpha) \le 0$ for some $\alpha \in R$ where a, b, c, d are nonzero real numbers then prove that a, b, c, d are in HP.

12. If $\sum_{r=1}^{n} \frac{4r+1}{(4r-1)^2 \cdot (4r+3)^2} > \frac{1}{75}$ then find the least possible value of $n \in N$.

13. Let f be a real-valued function such that $f(x + y) = f(x) + f(y)$ for all rational x, y. Find $f(x)$ if $\sum_{n=1}^{5} \sum_{r=1}^{n} rf(r) = 315$.

14. Let the positive numbers $a_1, a_2, a_3, \ldots$ be in GP. Let A_n, G_n and H_n be the AM, GM and HM of $a_1, a_2, a_3, \ldots, a_n$ respectively. Prove that

(i) The GM of $G_1, G_2, G_3, \ldots, G_n$ = the GM of $A_1, A_2, A_3, \ldots A_n, H_1, H_2, H_3, \ldots H_n$

(ii) the GM of $G_1, G_2, G_3, \ldots, G_n$ = the GM of $\sqrt{a_1 A_n}$ and $\sqrt{a_1 H_n}$.

□

Answers

1. (a) A, B, D (b) C (c) A

2. (a) $\frac{n}{\sqrt{3n+2} + \sqrt{2}}$ (b) 4 (c) $\frac{1}{3}$

4. $\frac{\pi^2}{8}$

8. 4089

12. 4

13. $3x$

□

2. Determinants and Cramer's Rule

Recap of Facts and Formulae

1. Minors and cofactors of elements of a determinant

- In determinants of order 3 or more, every element has a minor and a cofactor.
- In the determinant

$$\begin{vmatrix} a & b & c \\ d & e & f \\ g & h & i \end{vmatrix} \text{ of the third order,}$$

the minor of $a = \begin{vmatrix} e & f \\ h & i \end{vmatrix}$

which is a determinant of the second order and is obtained by removing all elements of the determinant that fall on the horizontal and vertical lines drawn through a. Similarly,

the minor of $e = \begin{vmatrix} a & c \\ g & i \end{vmatrix}$.

In the determinant

$$\begin{vmatrix} a_1 & b_1 & c_1 & d_1 \\ a_2 & b_2 & c_2 & d_2 \\ a_3 & b_3 & c_3 & d_3 \\ a_4 & b_4 & c_4 & d_4 \end{vmatrix} \text{ of the fourth order,}$$

the minor of $a_1 = \begin{vmatrix} b_2 & c_2 & d_2 \\ b_3 & c_3 & d_3 \\ b_4 & c_4 & d_4 \end{vmatrix}$

which is a determinant of the third order and it is expanded as in case of determinants of third order.

- The cofactor of an element $= (-1)^{r+c} \times$ minor, where r and c denote the number of the row and the column respectively to which the element belongs.

In the determinant

$$\begin{vmatrix} a & b & c \\ d & e & f \\ g & h & i \end{vmatrix}, \text{ the cofactor of } a = (-1)^{1+1} \cdot \begin{vmatrix} e & f \\ h & i \end{vmatrix};$$

the cofactor of $f = (-1)^{2+3} \cdot \begin{vmatrix} a & b \\ g & h \end{vmatrix}$.

The cofactor of an element is denoted by the corresponding capital letters.

$G =$ cofactor of $g = (-1)^{3+1} \cdot \begin{vmatrix} b & c \\ e & f \end{vmatrix}$.

Similarly, we get the cofactor of an element in case of determinants of fourth or higher order.

2. Expansion of a determinant

- $\begin{vmatrix} a & b \\ c & d \end{vmatrix} = ad - bc$ (= value of the determinant).

- $$\Delta = \begin{vmatrix} a & b & c \\ d & e & f \\ g & h & i \end{vmatrix} = aA + bB + cC$$

$$= a\begin{vmatrix} e & f \\ h & i \end{vmatrix} - b\begin{vmatrix} d & f \\ g & i \end{vmatrix} + c\begin{vmatrix} d & e \\ g & h \end{vmatrix}$$

$$= a(ei - fh) - b(di - fg) + c(dh - eg)$$

Similarly,

$$\Delta = dD + eE + fF = -d\begin{vmatrix} b & c \\ h & i \end{vmatrix} + e\begin{vmatrix} a & c \\ g & i \end{vmatrix} - f\begin{vmatrix} a & b \\ g & h \end{vmatrix}$$

$$\Delta = cC + fF + iI = c\begin{vmatrix} d & e \\ g & h \end{vmatrix} - f\begin{vmatrix} a & b \\ g & h \end{vmatrix} + i\begin{vmatrix} a & b \\ d & e \end{vmatrix}, \text{ etc.}$$

Note A determinant having all elements in a row (or column) equal to 0, has the value zero.

- $$\Delta = \begin{vmatrix} a_1 & b_1 & c_1 & d_1 \\ a_2 & b_2 & c_2 & d_2 \\ a_3 & b_3 & c_3 & d_3 \\ a_4 & b_4 & c_4 & d_4 \end{vmatrix} = a_1A_1 + b_1B_1 + c_1C_1 + d_1D_1$$

$$\Delta = a_1\begin{vmatrix} b_2 & c_2 & d_2 \\ b_3 & c_3 & d_3 \\ b_4 & c_4 & d_4 \end{vmatrix} - b_1\begin{vmatrix} a_2 & c_2 & d_2 \\ a_3 & c_3 & d_3 \\ a_4 & c_4 & d_4 \end{vmatrix} + c_1\begin{vmatrix} a_2 & b_2 & d_2 \\ a_3 & b_3 & d_3 \\ a_4 & b_4 & d_4 \end{vmatrix} - d_1\begin{vmatrix} a_2 & b_2 & c_2 \\ a_3 & b_3 & c_3 \\ a_4 & b_4 & c_4 \end{vmatrix}$$

where each of the third order determinant is to be expanded.

3. Reduction and increase of order of a determinant

- If all the elements in a row (or a column), except one element, are zeros the determinant reduces to a determinant of an order less by one.

For example $\begin{vmatrix} 9 & 0 & 0 \\ -2 & 3 & 7 \\ 4 & 5 & 1 \end{vmatrix} = 9\begin{vmatrix} 3 & 7 \\ 5 & 1 \end{vmatrix}$

because cofactors of the other two elements in the first row cannot contribute to the value of the determinant. Similarly,

$$\begin{vmatrix} 4 & 7 & 0 & 5 \\ 8 & -5 & 2 & 3 \\ 6 & 0 & 0 & 0 \\ 1 & 2 & 3 & 4 \end{vmatrix} = 6\begin{vmatrix} 7 & 0 & 5 \\ -5 & 2 & 3 \\ 2 & 3 & 4 \end{vmatrix}.$$

- A determinant can be replaced by a determinant of a higher order by one.

For example $\begin{vmatrix} 2 & 3 \\ -5 & 8 \end{vmatrix} = \begin{vmatrix} 1 & 0 & 0 \\ 0 & 2 & 3 \\ 0 & -5 & 8 \end{vmatrix}$

or $\begin{vmatrix} 2 & 3 & 0 \\ -5 & 8 & 0 \\ 0 & 0 & 1 \end{vmatrix}$ or $\begin{vmatrix} 0 & 0 & 1 \\ 2 & 3 & 0 \\ -5 & 8 & 0 \end{vmatrix}$

or $\begin{vmatrix} 0 & 2 & 3 \\ 0 & -5 & 8 \\ 1 & 0 & 0 \end{vmatrix}$.

4. Properties of determinants

(i) A determinant can be transformed without changing value by any one of the following properties (these properties hold for determinants of all orders):

- Corresponding rows and columns can be interchanged. Symbolically, $\Delta = \Delta_{R \rightleftharpoons C}$.

For example $\begin{vmatrix} a_1 & b_1 & c_1 \\ a_2 & b_2 & c_2 \\ a_3 & b_3 & c_3 \end{vmatrix} = \begin{vmatrix} a_1 & a_2 & a_3 \\ b_1 & b_2 & b_3 \\ c_1 & c_2 & c_3 \end{vmatrix}$.

- Two consecutive rows (or columns) can be interchanged by multiplying the changed determinant by (–1). Symbolically,

$$\Delta = -\Delta_{R_i \rightleftharpoons R_{i+1}} \quad (\text{or } -\Delta_{C_j \rightleftharpoons C_{j+1}})$$

For example $\begin{vmatrix} a_1 & b_1 & c_1 \\ a_2 & b_2 & c_2 \\ a_3 & b_3 & c_3 \end{vmatrix} = (-1)\begin{vmatrix} a_2 & b_2 & c_2 \\ a_1 & b_1 & c_1 \\ a_3 & b_3 & c_3 \end{vmatrix}$

$$= (-1)^2\begin{vmatrix} a_2 & b_2 & c_2 \\ a_3 & b_3 & c_3 \\ a_1 & b_1 & c_1 \end{vmatrix}.$$

- If a factor is common in all elements of a row (or column), the factor can be taken outside the determinant. Symbolically, $\Delta = m \cdot \Delta_{R_i \to \frac{1}{m}R_i}$

For example $\begin{vmatrix} ma_1 & mb_1 & mc_1 \\ a_2 & b_2 & c_2 \\ a_3 & b_3 & c_3 \end{vmatrix} = m\begin{vmatrix} a_1 & b_1 & c_1 \\ a_2 & b_2 & c_2 \\ a_3 & b_3 & c_3 \end{vmatrix}$

which also implies

$$\begin{vmatrix} a_1 & b_1 & c_1 \\ a_2 & b_2 & c_2 \\ a_3 & b_3 & c_3 \end{vmatrix} = \frac{1}{m}\begin{vmatrix} ma_1 & mb_1 & mc_1 \\ a_2 & b_2 & c_2 \\ a_3 & b_3 & c_3 \end{vmatrix}$$

i.e., $\Delta = \frac{1}{m} \cdot \Delta_{R_i \to mR_i}$.

- m times of the elements of a row (or column) can be added to the corresponding elements of another row (or column). Symbolically,

$$\Delta = \Delta_{R_i \to R_i + mR_j} \quad (\text{or} \quad \Delta_{C_i \to C_i + mC_j})$$

For example

$$\begin{vmatrix} a_1 & b_1 & c_1 \\ a_2 & b_2 & c_2 \\ a_3 & b_3 & c_3 \end{vmatrix} = \begin{vmatrix} a_1 & b_1 & c_1 \\ a_2 & b_2 & c_2 \\ a_3 + ma_1 & b_3 + mb_1 & c_3 + mc_1 \end{vmatrix},$$

$(R_3 \to R_3 + mR_1)$

(ii) The value of a determinant reduces to zero in any one the following cases.

- The value is zero if all the elements in a row (or column) are zeros.

For example $\begin{vmatrix} 4 & 0 & 13 \\ -7 & 0 & 11 \\ 2 & 0 & 6 \end{vmatrix} = 0.$

- The value is zero if any two rows (or columns) are identical. Symbolically,

$$\Delta_{R_i \equiv R_j} = 0 \quad \text{or} \quad \Delta_{C_i \equiv C_j} = 0.$$

For example $\begin{vmatrix} -2 & 3 & 7 \\ 9 & 1 & 4 \\ -2 & 3 & 7 \end{vmatrix} = 0, \quad \begin{vmatrix} 5 & 5 & 2 \\ -1 & -1 & 4 \\ 7 & 7 & 11 \end{vmatrix} = 0.$

5. Properties of cofactors of a determinant

- The sum of the products of elements of a row (or column) with their corresponding cofactors is equal to the value of the determinant.

For example $aA + bB + cC = \Delta$

$aA + dD + gG = \Delta$

where $\Delta = \begin{vmatrix} a & b & c \\ d & e & f \\ g & h & i \end{vmatrix}$.

- The sum of the products of elements of a row (or column) with the cofactors of the corresponding elements of another row (or column) is equal to 0.

For example $aD + bE + cF = 0$

$cA + fD + iG = 0$

where $\Delta = \begin{vmatrix} a & b & c \\ d & e & f \\ g & h & i \end{vmatrix}$.

6. Sum of determinants, a determinant as sum of determinants

- Let $\Delta_1 = \begin{vmatrix} a_1 & b_1 & c_1 \\ a_2 & b_2 & c_2 \\ a_3 & b_3 & c_3 \end{vmatrix}$, $\Delta_2 = \begin{vmatrix} d_1 & b_1 & c_1 \\ d_2 & b_2 & c_2 \\ d_3 & b_3 & c_3 \end{vmatrix}$

be two third order determinants which have $3 - 1$ (one less than the common order) columns identical. Then their addition is also a determinant of the third order given by

$$\Delta_1 + \Delta_2 = \begin{vmatrix} a_1 + d_1 & b_1 & c_1 \\ a_2 + d_2 & b_2 & c_2 \\ a_3 + d_3 & b_3 & c_3 \end{vmatrix}.$$

Similarly, if $\Delta_3 = \begin{vmatrix} a_1 & b_1 & c_1 \\ d_2 & e_2 & f_2 \\ a_3 & b_3 & c_3 \end{vmatrix}$ then

$$\Delta_1 + \Delta_3 = \begin{vmatrix} a_1 & b_1 & c_1 \\ a_2 + d_2 & b_2 + e_2 & c_2 + f_2 \\ a_3 & b_3 & c_3 \end{vmatrix}.$$

But $\Delta_2 + \Delta_3$ cannot be found like this.

A determinant having two or more terms in the elements of a row (or column) can be written as the sum of two or more determinants as follows:

$$\begin{vmatrix} a_1 + b_1 & c_1 & d_1 \\ a_2 + b_2 & c_2 & d_2 \\ a_3 + b_3 & c_3 & d_3 \end{vmatrix} = \begin{vmatrix} a_1 & c_1 & d_1 \\ a_2 & c_2 & d_2 \\ a_3 & c_3 & d_3 \end{vmatrix} + \begin{vmatrix} b_1 & c_1 & d_1 \\ b_2 & c_2 & d_2 \\ b_3 & c_3 & d_3 \end{vmatrix},$$

decomposing along the first column.

$$\begin{vmatrix} a_1 & b_1 & c_1 \\ a_2 & b_2 & c_2 \\ d_1 + d_2 + d_3 & e_1 + e_2 + e_3 & f_1 + f_2 + f_3 \end{vmatrix}$$

$$= \begin{vmatrix} a_1 & b_1 & c_1 \\ a_2 & b_2 & c_2 \\ d_1 & e_1 & f_1 \end{vmatrix} + \begin{vmatrix} a_1 & b_1 & c_1 \\ a_2 & b_2 & c_2 \\ d_2 & e_2 & f_2 \end{vmatrix} + \begin{vmatrix} a_1 & b_1 & c_1 \\ a_2 & b_2 & c_2 \\ d_3 & e_3 & f_3 \end{vmatrix},$$

decomposing along the third row.

7. Product of two determinants

Two determinants of the same order can be multiplied and the product is also a determinant of the same order. The rule is as displayed by the illustrations.

If $\Delta_1 = \begin{vmatrix} a_1 & b_1 & c_1 \\ a_2 & b_2 & c_2 \\ a_3 & b_3 & c_3 \end{vmatrix}$ and $\Delta_2 = \begin{vmatrix} x_1 & y_1 & z_1 \\ x_2 & y_2 & z_2 \\ x_3 & y_3 & z_3 \end{vmatrix}$

then $\Delta_1 \cdot \Delta_2$ can be found in any one of the two ways given below.

(i) $\Delta_1 \cdot \Delta_2 = \begin{vmatrix} a_1 & b_1 & c_1 \\ a_2 & b_2 & c_2 \\ a_3 & b_3 & c_3 \end{vmatrix} \cdot \begin{vmatrix} x_1 & y_1 & z_1 \\ x_2 & y_2 & z_2 \\ x_3 & y_3 & z_3 \end{vmatrix}$

$$= \begin{vmatrix} a_1x_1 + b_1x_2 + c_1x_3 & a_1y_1 + b_1y_2 + c_1y_3 & a_1z_1 + b_1z_2 + c_1z_3 \\ a_2x_1 + b_2x_2 + c_2x_3 & a_2y_1 + b_2y_2 + c_2y_3 & a_2z_1 + b_2z_2 + c_2z_3 \\ a_3x_1 + b_3x_2 + c_3x_3 & a_3y_1 + b_3y_2 + c_3y_3 & a_3z_1 + b_3z_2 + c_3z_3 \end{vmatrix}$$

is obtained by using sum of products of elements in rows of the first and the columns of the second, pattern being as follows:

For the first row of the product, the elements in the different columns are shown below.

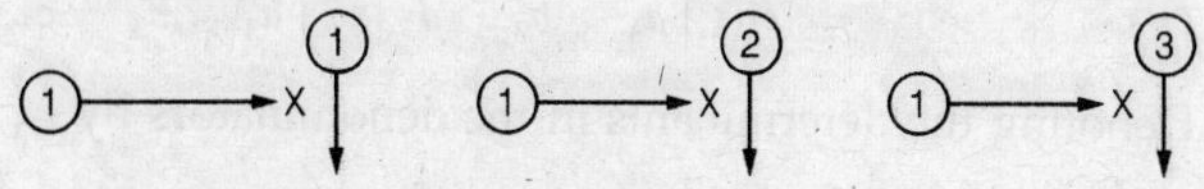

Similarly for other rows.

(ii) $\Delta_1 \cdot \Delta_2 = \begin{vmatrix} a_1 & b_1 & c_1 \\ a_2 & b_2 & c_2 \\ a_3 & b_3 & c_3 \end{vmatrix} \cdot \begin{vmatrix} x_1 & y_1 & z_1 \\ x_2 & y_2 & z_2 \\ x_3 & y_3 & z_3 \end{vmatrix}$

$$= \begin{vmatrix} a_1x_1 + b_1y_1 + c_1z_1 & a_1x_2 + b_1y_2 + c_1z_2 & a_1x_3 + b_1y_3 + c_1z_3 \\ a_2x_1 + b_2y_1 + c_2z_1 & a_2x_2 + b_2y_2 + c_2z_2 & a_2x_3 + b_2y_3 + c_2z_3 \\ a_3x_1 + b_3y_1 + c_3z_1 & a_3x_2 + b_3y_2 + c_3z_2 & a_3x_3 + b_3y_3 + c_3z_3 \end{vmatrix}$$

This is also obtained by interchanging rows and columns of the post-multiplier in (i) and then multiplying as in (i).

8. Derivative of a determinant

A determinant can be differentiated in any one of the two ways given below.

- Expand the determinant and then differentiate the expression.
- If $\Delta(x) = \begin{vmatrix} u(x) & v(x) & w(x) \\ p(x) & q(x) & r(x) \\ \lambda(x) & \mu(x) & \nu(x) \end{vmatrix}$ then

$$\Delta'(x) = \begin{vmatrix} u'(x) & v'(x) & w'(x) \\ p(x) & q(x) & r(x) \\ \lambda(x) & \mu(x) & \nu(x) \end{vmatrix} + \begin{vmatrix} u(x) & v(x) & w(x) \\ p'(x) & q'(x) & r'(x) \\ \lambda(x) & \mu(x) & \nu(x) \end{vmatrix} + \begin{vmatrix} u(x) & v(x) & w(x) \\ p(x) & q(x) & r(x) \\ \lambda'(x) & \mu'(x) & \nu'(x) \end{vmatrix}$$

This differentiation can also be done columnwise.

9. Cramer's Rule for solving simultaneous linear equations

The solution of the linear equations

$$\left.\begin{aligned} a_1x + b_1y + c_1z + d_1 = 0 \\ a_2x + b_2y + c_2z + d_2 = 0 \\ a_3x + b_3y + c_3z + d_3 = 0 \end{aligned}\right\} \quad \ldots \text{(A)}$$

is given by

$$\frac{x}{\begin{vmatrix} b_1 & c_1 & d_1 \\ b_2 & c_2 & d_2 \\ b_3 & c_3 & d_3 \end{vmatrix}} = \frac{-y}{\begin{vmatrix} a_1 & c_1 & d_1 \\ a_2 & c_2 & d_2 \\ a_3 & c_3 & d_3 \end{vmatrix}}$$

$$= \frac{z}{\begin{vmatrix} a_1 & b_1 & d_1 \\ a_2 & b_2 & d_2 \\ a_3 & b_3 & d_3 \end{vmatrix}} = \frac{-1}{\begin{vmatrix} a_1 & b_1 & c_1 \\ a_2 & b_2 & c_2 \\ a_3 & b_3 & c_3 \end{vmatrix}}$$

Denoting the determinants in the denominators by Δ_x, Δ_y, Δ_z and Δ respectively,

$$\frac{x}{\Delta_x} = \frac{-y}{\Delta_y} = \frac{z}{\Delta_z} = \frac{-1}{\Delta}; \qquad \therefore \; x = \frac{-\Delta_x}{\Delta},\; y = \frac{\Delta_y}{\Delta},\; z = \frac{-\Delta_z}{\Delta}$$

10. Number of solutions of a system of linear equations

For the above system of linear equations (A),

- $\Delta \neq 0 \;\Rightarrow$ the system has unique solution (*independent system*)
- $\Delta = 0$ but at least one of $\Delta_x, \Delta_y, \Delta_z \neq 0$
  $\Rightarrow$ the system has no solution (*unsolvable or inconsistent system*)
- $\Delta = 0 = \Delta_x = \Delta_y = \Delta_z \Rightarrow$ the system has infinite solution (*dependent system*)

11. Condition for consistency and existence of nontrivial solution

- The system of linear equations $a_rx + b_ry + c_r = 0$; $r = 1, 2, 3$ are consistent if

$$\begin{vmatrix} a_1 & b_1 & c_1 \\ a_2 & b_2 & c_2 \\ a_3 & b_3 & c_3 \end{vmatrix} = 0 \qquad \ldots (1)$$

- The system of homogeneous linear equations $a_rx + b_ry + c_rz = 0$; $r = 1, 2, 3$ will have nontrivial (at least one nonzero) solutions if the above condition (1) is satisfied.
- If $a_rx + b_ry + c_rz = 0$; $r = 1, 2, 3$ have a non-trivial solution then it will have infinite number of solutions.

Selected Solved Examples

1. If $|a_1| > |a_2| + |a_3|$, $|b_2| > |b_3| + |b_1|$, $|c_3| > |c_1| + |c_2|$ then prove that

$$\begin{vmatrix} a_1 & b_1 & c_1 \\ a_2 & b_2 & c_2 \\ a_3 & b_3 & c_3 \end{vmatrix} \neq 0.$$

Let $|a_1| = |a_2| + |a_3| + p$, $|b_2| = |b_3| + |b_1| + q$, $|c_3| = |c_1| + |c_2| + r$, where p, q, r are positive.

Now $\Delta = a_1 \begin{vmatrix} b_2 & c_2 \\ b_3 & c_3 \end{vmatrix} - b_1 \begin{vmatrix} a_2 & c_2 \\ a_3 & c_3 \end{vmatrix} + c_1 \begin{vmatrix} a_2 & b_2 \\ a_3 & b_3 \end{vmatrix}$

$$= a_1(b_2c_3 - b_3c_2) - b_1(a_2c_3 - a_3c_2) + c_1(a_2b_3 - a_3b_2)$$

$$= a_1b_2c_3 - a_1b_3c_2 - b_1a_2c_3 + b_1a_3c_2 + c_1a_2b_3 - c_1a_3b_2$$

[We know $|a \pm b \pm c| > |a| - |b| - |c|$, etc., and $|ab| = |a|\,|b|$. Use these inequalities.]

Now $|\Delta| = |a_1b_2c_3 - a_1b_3c_2 - b_1a_2c_3 + b_1a_3c_2 + c_1a_2b_3 - c_1a_3b_2|$

$$\geq |a_1b_2c_3| - |a_1b_3c_2| - |b_1a_2c_3| - |b_1a_3c_2| - |c_1a_2b_3| - |c_1a_3b_2|$$

$$\therefore \quad |\Delta| \geq |a_1|\,|b_2|\,|c_3| - |a_1|\,|b_3|\,|c_2| - |b_1|\,|a_2|\,|c_3| - |b_1|\,|a_3|\,|c_2| - |c_1|\,|a_2|\,|b_3| - |c_1|\,|a_3|\,|b_2|$$

Now, RHS of the above

$$= (|a_2| + |a_3| + p)(|b_3| + |b_1| + q) \times (|c_1| + |c_2| + r) - (|a_2| + |a_3| + p)\,|b_3|\,|c_2|$$
$$- |b_1|\,|a_2|(|c_1| + |c_2| + r) - |b_1|\,|a_3|\,|c_2|$$
$$- |c_1|\,|a_2|\,|b_3| - |c_1|\,|a_3|(|b_3| + |b_1| + q)$$

$\therefore$ RHS > 0 because all the negative terms cancel.

$\therefore$ $|\Delta| > 0$. Hence $\Delta \neq 0$.

2. Prove that $\begin{vmatrix} a^2 + 2a & 2a + 1 & 1 \\ 2a + 1 & a + 2 & 1 \\ 3 & 3 & 1 \end{vmatrix} > \text{ or } = \text{ or } < 0$ according as $a > 1$ or $a = 1$ or $a < 1$.

TIP: Sometimes a determinant has all its elements = 1 (or equal) in a row (or column) {here in the third column}. By using $R_1 \to R_1 - R_3$, etc. it is very easy to get at least two zeros in a particular row or column and this reduces the order of the determinant.

$$\Delta = \begin{vmatrix} a^2 + 2a - 3 & 2a - 2 & 0 \\ 2a - 2 & a - 1 & 0 \\ 3 & 3 & 1 \end{vmatrix}, \quad \begin{aligned} R_1 &\to R_1 - R_3 \\ R_2 &\to R_2 - R_3 \end{aligned}$$

$$= \begin{vmatrix} a^2 + 2a - 3 & 2(a - 1) \\ 2(a - 1) & a - 1 \end{vmatrix}$$

$$= \begin{vmatrix} (a+3)(a-1) & 2(a-1) \\ 2(a-1) & a-1 \end{vmatrix}$$

$$= (a-1)^2 \begin{vmatrix} a+3 & 2 \\ 2 & 1 \end{vmatrix}$$

$$= (a-1)^2 \cdot (a+3-4) = (a-1)^3.$$

$\therefore \quad \Delta > 0$ if $(a-1)^3 > 0$, i.e., $a > 1$

$\Delta = 0$ if $(a-1)^3 = 0$, i.e., $a = 1$

$\Delta < 0$ if $(a-1)^3 < 0$, i.e., $a < 1$.

Hence the problem.

3. Show that the roots of the equation $\begin{vmatrix} x & m & n & 1 \\ a & x & n & 1 \\ a & b & x & 1 \\ a & b & c & 1 \end{vmatrix} = 0$ **are independent of m and n.**

$$\Delta = \begin{vmatrix} x-a & m-x & 0 & 0 \\ 0 & x-b & n-x & 0 \\ 0 & 0 & x-c & 0 \\ a & b & c & 1 \end{vmatrix}, \quad \begin{matrix} R_1 \to R_1 - R_2 \\ R_2 \to R_2 - R_3 \\ R_3 \to R_3 - R_4 \end{matrix}$$

$$= \begin{vmatrix} x-a & m-x & 0 \\ 0 & x-b & n-x \\ 0 & 0 & x-c \end{vmatrix} = (x-a)(x-b)(x-c)$$

$\therefore \quad \Delta = 0 \quad \Rightarrow \quad (x-a)(x-b)(x-c) = 0$

whose roots are $x = a, b, c$ and they are independent of m, n.

4. Evaluate $\begin{vmatrix} \sqrt{13}+\sqrt{3} & 2\sqrt{5} & \sqrt{5} \\ \sqrt{15}+\sqrt{26} & 5 & \sqrt{10} \\ 3+\sqrt{65} & \sqrt{15} & 5 \end{vmatrix}$.

$$\Delta = \sqrt{5} \cdot \sqrt{5} \begin{vmatrix} \sqrt{13}+\sqrt{3} & 2 & 1 \\ \sqrt{15}+\sqrt{26} & \sqrt{5} & \sqrt{2} \\ 3+\sqrt{65} & \sqrt{3} & \sqrt{5} \end{vmatrix}$$

$$= 5 \begin{vmatrix} \sqrt{13}+\sqrt{3} & 2 & 1 \\ \sqrt{15}-\sqrt{6} & \sqrt{5}-2\sqrt{2} & 0 \\ 3-\sqrt{15} & \sqrt{3}-2\sqrt{5} & 0 \end{vmatrix}, \quad \begin{matrix} R_2 \to R_2 - \sqrt{2}R_1 \\ R_3 \to R_3 - \sqrt{5}R_1 \end{matrix}$$

$$= 5 \begin{vmatrix} \sqrt{15}-\sqrt{6} & \sqrt{5}-2\sqrt{2} \\ 3-\sqrt{15} & \sqrt{3}-2\sqrt{5} \end{vmatrix}$$

$$= 5\sqrt{3} \begin{vmatrix} \sqrt{5}-\sqrt{2} & \sqrt{5}-2\sqrt{2} \\ \sqrt{3}-\sqrt{5} & \sqrt{3}-2\sqrt{5} \end{vmatrix}$$

$$= 5\sqrt{3}\{(\sqrt{5}-\sqrt{2})(\sqrt{3}-2\sqrt{5}) - (\sqrt{3}-\sqrt{5})(\sqrt{5}-2\sqrt{2})\}$$

$$= 5\sqrt{3}\{\sqrt{15} - 10 - \sqrt{6} + 2\sqrt{10} - \sqrt{15} + 2\sqrt{6} + 5 - 2\sqrt{10}\}$$

$$= 5\sqrt{3}(\sqrt{6} - 5).$$

5. For a fixed $n \in N$, if

$$D = \begin{vmatrix} n! & (n+1)! & (n+2)! \\ (n+1)! & (n+2)! & (n+3)! \\ (n+2)! & (n+3)! & (n+4)! \end{vmatrix}$$

then $\left\{\dfrac{D}{(n!)^3} - 4\right\}$ is divisible by n.

Here

$$D = n! \cdot (n+1)! \,(n+2)! \begin{vmatrix} 1 & n+1 & (n+2)(n+1) \\ 1 & n+2 & (n+3)(n+2) \\ 1 & n+3 & (n+4)(n+3) \end{vmatrix}$$

$$= n! \cdot (n+1) \cdot (n+2)! \begin{vmatrix} 1 & n+1 & (n+2)(n+1) \\ 0 & 1 & (n+2)2 \\ 0 & 1 & (n+3)2 \end{vmatrix},$$

$$R_3 \to R_3 - R_2,\ R_2 \to R_2 - R_1$$

$$= n!\,(n+1)!\,(n+2)!\,[(n+3)2 - (n+2) \cdot 2]$$

$$= n!\,(n+1)!\,(n+2)! \cdot 2$$

$$\therefore \quad \frac{D}{(n!)^3} = 2(n+1)^2(n+2) = 2n^3 + 8n^2 + 10n + 4$$

$$\therefore \quad \frac{D}{(n!)^3} - 4 = 2n(n^2 + 4n + 5)$$

$\therefore \quad \left\{\dfrac{D}{(n!)^3} - 4\right\}$ is divisible by n.

6. If $a \neq p, b \neq q, c \neq r$ and $\begin{vmatrix} p & b & c \\ a & q & c \\ a & b & r \end{vmatrix} = 0$ then find the value of $\dfrac{p}{p-a} + \dfrac{q}{q-b} + \dfrac{r}{r-c}$.

$$\text{Here } \Delta = \begin{vmatrix} p & b & c \\ a-p & q-b & 0 \\ 0 & b-q & r-c \end{vmatrix}, \quad \begin{matrix} R_3 \to R_3 - R_2, \\ R_2 \to R_2 - R_1 \end{matrix}$$

$$= (p-a)(q-b)(r-c) \begin{vmatrix} \frac{p}{p-a} & \frac{b}{q-b} & \frac{c}{r-c} \\ -1 & 1 & 0 \\ 0 & -1 & 1 \end{vmatrix}$$

$$= (p-a)(q-b)(r-c) \times \begin{vmatrix} \frac{p}{p-a} + \frac{b}{q-b} + \frac{c}{r-c} & \frac{b}{q-b} & \frac{c}{r-c} \\ 0 & 1 & 0 \\ 0 & -1 & 1 \end{vmatrix}$$

$$C_1 \to C_1 + C_2 + C_3$$

$$\therefore \quad \Delta = (p-a)(q-b)(r-c)\left(\frac{p}{p-a} + \frac{b}{q-b} + \frac{c}{r-c}\right)$$

$= 0$, from the question.

But $a \neq p, b \neq q, c \neq r$. So $\dfrac{p}{p-a} + \dfrac{b}{q-b} + \dfrac{c}{r-c} = 0$

$$\therefore \quad \frac{p}{p-a}+\left(\frac{b}{q-b}+1\right)+\left(\frac{c}{r-c}+1\right)=1+1$$

$$\text{or} \quad \frac{p}{p-a}+\frac{q}{q-b}+\frac{r}{r-c}=2.$$

7. Let $a > 0, d > 0$. Find the value of

$$\begin{vmatrix} \frac{1}{a} & \frac{1}{a(a+d)} & \frac{1}{(a+d)(a+2d)} \\ \frac{1}{a+d} & \frac{1}{(a+d)(a+2d)} & \frac{1}{(a+2d)(a+3d)} \\ \frac{1}{a+2d} & \frac{1}{(a+2d)(a+3d)} & \frac{1}{(a+3d)(a+4d)} \end{vmatrix}.$$

$$\Delta = \frac{1}{a(a+d)^2(a+2d)^3(a+3d)^2(a+4d)} \times \begin{vmatrix} (a+d)(a+2d) & a+2d & a \\ (a+2d)(a+3d) & a+3d & a+d \\ (a+3d)(a+4d) & a+4d & a+2d \end{vmatrix}$$

$$= \frac{1}{a(a+d)^2(a+2d)^3(a+3d)^2(a+4d)} \times \begin{vmatrix} (a+d)(a+2d) & a+2d & a \\ (a+2d)2d & d & d \\ (a+3d)2d & d & d \end{vmatrix}, \begin{matrix} R_3 \to R_3 - R_2 \\ R_2 \to R_2 - R_1 \end{matrix}$$

$$= \frac{1}{a(a+d)^2(a+2d)^3(a+3d)^2(a+4d)} \times \begin{vmatrix} (a+d)(a+2d) & a+2d & -2d \\ (a+2d)2d & d & 0 \\ (a+3d)2d & d & 0 \end{vmatrix}, C_3 \to C_3 - C_2$$

$$= \frac{-2d}{a(a+d)^2(a+2d)^3(a+3d)^2(a+4d)} \begin{vmatrix} (a+2d)2d & d \\ (a+3d)2d & d \end{vmatrix}$$

$$= \frac{-4d^3}{a(a+d)^2(a+2d)^3(a+3d)^2(a+4d)} \begin{vmatrix} a+2d & 1 \\ a+3d & 1 \end{vmatrix}$$

$$= \frac{4d^4}{a(a+d)^2(a+2d)^3(a+3d)^2(a+4d)}.$$

8. Solve $\begin{vmatrix} a^2+\lambda & ab & ac \\ ab & b^2+\lambda & bc \\ ac & bc & c^2+\lambda \end{vmatrix} = 0, \lambda \in R.$

TIP Some determinants are such that there is a factor common between the elements in the first row and the first column but not necessarily at the junction of the row and column. Similar situation may exist for the other rows and columns. In the above problem those common factors are a, b, c respectively. In such problems common factors are taken out rowwise (or columnwise) and re-entered columnwise (or rowwise) respectively.

$$\text{Here } \Delta = abc \begin{vmatrix} a+\frac{\lambda}{a} & b & c \\ a & b+\frac{\lambda}{b} & c \\ a & b & c+\frac{\lambda}{c} \end{vmatrix}$$

$$= \begin{vmatrix} a^2+\lambda & b^2 & c^2 \\ a^2 & b^2+\lambda & c^2 \\ a^2 & b^2 & c^2+\lambda \end{vmatrix}$$

$$= \begin{vmatrix} \lambda+(a^2+b^2+c^2) & b^2 & c^2 \\ \lambda+(a^2+b^2+c^2) & b^2+\lambda & c^2 \\ \lambda+(a^2+b^2+c^2) & b^2 & c^2+\lambda \end{vmatrix}, \quad C_1 \to C_1 + C_2 + C_3$$

$$= \{\lambda+(a^2+b^2+c^2)\} \begin{vmatrix} 1 & b^2 & c^2 \\ 1 & b^2+\lambda & c^2 \\ 1 & b^2 & c^2+\lambda \end{vmatrix}$$

$$= \{\lambda+(a^2+b^2+c^2)\} \begin{vmatrix} 1 & b^2 & c^2 \\ 0 & \lambda & 0 \\ 0 & 0 & \lambda \end{vmatrix}, \quad R_2 \to R_2 - R_1, R_3 \to R_3 - R_1$$

$$= \lambda^2(\lambda+a^2+b^2+c^2)$$

$\therefore$ the equation is $\lambda^2(\lambda+a^2+b^2+c^2)=0$

or $\lambda = 0, -(a^2+b^2+c^2)$.

9. Prove that

$$\begin{vmatrix} (b+c)^2 & c^2 & b^2 \\ c^2 & (c+a)^2 & a^2 \\ b^2 & a^2 & (a+b)^2 \end{vmatrix} = 2(bc+ca+ab)^3.$$

$$\text{Here } \Delta = \frac{1}{a^2b^2c^2} \begin{vmatrix} a^2(b+c)^2 & a^2c^2 & a^2b^2 \\ c^2b^2 & b^2(c+a)^2 & a^2b^2 \\ b^2c^2 & a^2c^2 & c^2(a+b)^2 \end{vmatrix}.$$

Putting $bc = x$, $ca = y$, $ab = z$ we get

$$\Delta = \frac{1}{xyz} \begin{vmatrix} (y+z)^2 & y^2 & z^2 \\ x^2 & (z+x)^2 & z^2 \\ x^2 & y^2 & (x+y)^2 \end{vmatrix}$$

$$= \frac{1}{xyz} \begin{vmatrix} (x+y+z)(y+z-x) & (x+y+z)(y-z-x) & 0 \\ 0 & (x+y+z)(z+x-y) & (x+y+z)(z-x-y) \\ x^2 & y^2 & (x+y)^2 \end{vmatrix},$$

$$R_1 \to R_1 - R_2, \; R_2 \to R_2 - R_3$$

$$= \frac{(x+y+z)^2}{xyz}\begin{vmatrix} y+z-x & y-z-x & 0 \\ 0 & z+x-y & z-x-y \\ x^2 & y^2 & (x+y)^2 \end{vmatrix}$$

$$= \frac{(x+y+z)^2}{xyz}\begin{vmatrix} y+z-x & y-z-x & 2(x-y) \\ 0 & z+x-y & -2x \\ x^2 & y^2 & 2xy \end{vmatrix}, \quad C_3 \to C_3 - C_1 - C_2$$

$$= \frac{(x+y+z)^2}{xyz}\begin{vmatrix} y+z-x & 0 & -2y \\ 0 & z+x-y & -2x \\ x^2 & y^2 & 2xy \end{vmatrix}, \quad R_1 \to R_1 + R_2$$

$$= \frac{(x+y+z)^2}{xy \cdot xyz}\begin{vmatrix} x(y+z-x) & 0 & -2xy \\ 0 & y(z+x-y) & -2xy \\ x^2 & y^2 & 2xy \end{vmatrix}$$

$$= \frac{(x+y+z)^2}{xy \cdot xyz}\begin{vmatrix} x(y+z) & y^2 & 0 \\ x^2 & y(z+x) & 0 \\ x^2 & y^2 & 2xy \end{vmatrix}, \quad R_1 \to R_1 + R_3, R_2 \to R_2 + R_3$$

$$= \frac{(x+y+z)^2}{xy \cdot xyz} \cdot 2xy \begin{vmatrix} x(y+z) & y^2 \\ x^2 & y(z+x) \end{vmatrix}$$

$$= 2 \cdot \frac{(x+y+z)^2}{xyz} \cdot x \cdot y \begin{vmatrix} y+z & y \\ x & z+x \end{vmatrix}$$

$$= \frac{2}{z}(x+y+z)^2 \begin{vmatrix} x+y+z & x+y+z \\ x & z+x \end{vmatrix}, R_1 \to R_1 + R_2$$

$$= \frac{2}{z}(x+y+z)^2 \cdot (x+y+z)\begin{vmatrix} 1 & 1 \\ x & z+x \end{vmatrix}$$

$$= \frac{2}{z} \cdot (x+y+z)^3 \cdot z$$

$$= 2(x+y+z)^3 = 2(bc+ca+ab)^3.$$

10. Prove that

$$\begin{vmatrix} ax-by-cz & ay+bx & cx+az \\ ay+bx & by-cz-ax & bz+cy \\ cx+az & bz+cy & cz-ax-by \end{vmatrix} = (x^2+y^2+z^2)(a^2+b^2+c^2)(ax+by+cz).$$

$$\Delta = \frac{1}{a}\begin{vmatrix} a^2x-aby-caz & ay+bx & cx+az \\ a^2y+abx & by-cz-ax & bz+cy \\ acx+a^2z & bz+cy & cz-ax-by \end{vmatrix}$$

$$= \frac{1}{a}\begin{vmatrix} (a^2+b^2+c^2)x & ay+bx & cx+az \\ (a^2+b^2+c^2)y & by-cz-ax & bz+cy \\ (a^2+b^2+c^2)z & bz+cy & cz-ax-by \end{vmatrix}, \quad C_1 \to C_1 + bC_2 + cC_3$$

$$= (a^2+b^2+c^2)\frac{1}{a}\begin{vmatrix} x & ay+bx & cx+az \\ y & by-cz-ax & bz+cy \\ z & bz+cy & cz-ax-by \end{vmatrix}$$

$$= (a^2+b^2+c^2)\frac{1}{ax}\begin{vmatrix} x^2 & axy+bx^2 & cx^2+azx \\ y & by-cz-ax & bz+cy \\ z & bz+cy & cz-ax-by \end{vmatrix}$$

$$= (a^2+b^2+c^2)\frac{1}{ax}$$

$$\times \begin{vmatrix} x^2+y^2+z^2 & b(x^2+y^2+z^2) & c(x^2+y^2+z^2) \\ y & by-cz-ax & bz+cy \\ z & bz+cy & cz-ax-by \end{vmatrix}, \quad R_1 \to R_1 + yR_2 + zR_3$$

$$= (a^2+b^2+c^2)(x^2+y^2+z^2)\frac{1}{ax}$$

$$\times \begin{vmatrix} 1 & b & c \\ y & by-cz-ax & bz+cy \\ z & bz+cy & cz-ax-by \end{vmatrix}$$

$$= \frac{(a^2+b^2+c^2)(x^2+y^2+z^2)}{ax}$$

$$\times \begin{vmatrix} 1 & b & c \\ 0 & -cz-ax & bz \\ 0 & cy & -ax-by \end{vmatrix}, \quad R_2 \to R_2 - yR_1, \; R_3 \to R_3 - zR_1$$

$$= \frac{(a^2+b^2+c^2)(x^2+y^2+z^2)}{ax}$$

$$\times \begin{vmatrix} -cz-ax & bz \\ cy & -ax-by \end{vmatrix}$$

$$= \frac{(a^2+b^2+c^2)(x^2+y^2+z^2)}{ax}$$

$$\times \{(cz+ax)(ax+by) - bcyz\}$$

$$= \frac{(a^2+b^2+c^2)(x^2+y^2+z^2)}{ax}$$

$$\times \{acxz + bcyz + a^2x^2 + abxy - bcyz\}$$

$$= \frac{(a^2+b^2+c^2)(x^2+y^2+z^2)}{ax} ax(cz+ax+by)$$

$$= (x^2+y^2+z^2)(a^2+b^2+c^2)(ax+by+cz).$$

11. If $x_r \neq 0$; $r = 1, 2, 3$ then prove that

$$\begin{vmatrix} x_1+a_1b_1 & a_1b_2 & a_1b_3 \\ a_2b_1 & x_2+a_2b_2 & a_2b_3 \\ a_3b_1 & a_3b_2 & x_3+a_3b_3 \end{vmatrix} = x_1x_2x_3\left(1+\frac{a_1b_1}{x_1}+\frac{a_2b_2}{x_2}+\frac{a_3b_3}{x_3}\right).$$

$$\Delta = \frac{1}{b_1^2}\begin{vmatrix} x_1+a_1b_1 & a_1b_1b_2 & a_1b_1b_3 \\ a_2b_1 & b_1x_2+a_2b_1b_2 & a_2b_1b_3 \\ a_3b_1 & a_3b_1b_2 & b_1x_3+a_3b_1b_3 \end{vmatrix}$$

$$= \frac{1}{b_1^2}\begin{vmatrix} x_1+a_1b_1 & -b_2x_1 & -b_3x_1 \\ a_2b_1 & b_1x_2 & 0 \\ a_3b_1 & 0 & b_1x_3 \end{vmatrix}, \quad C_2 \to C_2 - b_2C_1, \; C_3 \to C_3 - b_3C_1$$

$$= \begin{vmatrix} x_1 + a_1b_1 & -b_2x_1 & -b_3x_1 \\ a_2 & x_2 & 0 \\ a_3 & 0 & x_3 \end{vmatrix}$$

$$= x_1x_2x_3 \begin{vmatrix} 1 + \frac{a_1b_1}{x_1} & -b_2 & -b_3 \\ \frac{a_2}{x_2} & 1 & 0 \\ \frac{a_3}{x_3} & 0 & 1 \end{vmatrix}$$

$$= x_1x_2x_3 \begin{vmatrix} 1 + \frac{a_1b_1}{x_1} + \frac{a_2b_2}{x_2} + \frac{a_3b_3}{x_3} & 0 & 0 \\ \frac{a_2}{x_2} & 1 & 0 \\ \frac{a_3}{x_3} & 0 & 1 \end{vmatrix},$$

$R_1 \to R_1 + b_2R_2 + b_3R_3$

$$= x_1x_2x_3 \left(1 + \frac{a_1b_1}{x_1} + \frac{a_2b_2}{x_2} + \frac{a_3b_3}{x_3}\right).$$

12. Prove that $\begin{vmatrix} \sin\alpha & \sin\beta & \sin\gamma \\ \cos\alpha & \cos\beta & \cos\gamma \\ \sin 2\alpha & \sin 2\beta & \sin 2\gamma \end{vmatrix}$

$$= 4\sin\frac{1}{2}(\beta-\gamma)\cdot\sin\frac{1}{2}(\gamma-\alpha)\cdot\sin\frac{1}{2}(\alpha-\beta)$$
$$\times\{\sin(\beta+\gamma)+\sin(\gamma+\alpha)+\sin(\alpha+\beta)\}.$$

$$\Delta = \frac{1}{\cos\gamma}\begin{vmatrix} \sin\alpha\cdot\cos\gamma & \sin\beta\cdot\cos\gamma & \sin\gamma\cdot\cos\gamma \\ \cos\alpha & \cos\beta & \cos\gamma \\ \sin 2\alpha & \sin 2\beta & \sin 2\gamma \end{vmatrix}$$

$$= \frac{1}{\cos\gamma} \times \begin{vmatrix} \sin(\alpha-\gamma) & \sin(\beta-\gamma) & 0 \\ \cos\alpha & \cos\beta & \cos\gamma \\ \sin 2\alpha - 2\cos\alpha\cdot\sin\gamma & \sin 2\beta - 2\cos\beta\sin\gamma & 0 \end{vmatrix},$$

$R_1 \to R_1 - \sin\gamma \times R_2,\ R_3 \to R_3 - 2\sin\gamma \times R_2$

$$= \frac{1}{\cos\gamma}(-\cos\gamma) \times \begin{vmatrix} \sin(\alpha-\gamma) & \sin(\beta-\gamma) \\ 2\cos\alpha\,(\sin\alpha-\sin\gamma) & 2\cos\beta\,(\sin\beta-\sin\gamma) \end{vmatrix}$$

$$= -2\begin{vmatrix} 2\sin\frac{\alpha-\gamma}{2}\cdot\cos\frac{\alpha-\gamma}{2} & 2\sin\frac{\beta-\gamma}{2}\cdot\cos\frac{\beta-\gamma}{2} \\ \cos\alpha\cdot 2\cos\frac{\alpha+\gamma}{2}\cdot\sin\frac{\alpha-\gamma}{2} & \cos\beta\cdot 2\cos\frac{\beta+\gamma}{2}\cdot\sin\frac{\beta-\gamma}{2} \end{vmatrix}$$

$$= -4\sin\frac{\alpha-\gamma}{2}\sin\frac{\beta-\gamma}{2} \times \begin{vmatrix} \cos\frac{\alpha-\gamma}{2} & \cos\frac{\beta-\gamma}{2} \\ 2\cos\alpha\cdot\cos\frac{\alpha+\gamma}{2} & 2\cos\beta\cdot\cos\frac{\beta+\gamma}{2} \end{vmatrix}.$$

$$= -4\sin\frac{\alpha-\gamma}{2}\cdot\sin\frac{\beta-\gamma}{2} \times \begin{vmatrix} \cos\frac{\alpha-\gamma}{2} & \cos\frac{\beta-\gamma}{2} \\ \cos\frac{\alpha-\gamma}{2} + \cos\frac{3\alpha+\gamma}{2} & \cos\frac{\beta-\gamma}{2} + \cos\frac{3\beta+\gamma}{2} \end{vmatrix}$$

$$= -4\sin\frac{\alpha-\gamma}{2}\sin\frac{\beta-\gamma}{2}\begin{vmatrix} \cos\frac{\alpha-\gamma}{2} & \cos\frac{\beta-\gamma}{2} \\ \cos\frac{3\alpha+\gamma}{2} & \cos\frac{3\beta+\gamma}{2} \end{vmatrix},$$

$R_2 \to R_2 - R_1$

$$= -4\sin\frac{\alpha-\gamma}{2}\sin\frac{\beta-\gamma}{2} \times \left\{\cos\frac{\alpha-\gamma}{2}\cdot\cos\frac{3\beta+\gamma}{2} - \cos\frac{\beta-\gamma}{2}\cdot\cos\frac{3\alpha+\gamma}{2}\right\}$$

$$= -2\sin\frac{\alpha-\gamma}{2}\sin\frac{\beta-\gamma}{2} \times \left\{\left(\cos\frac{\alpha-3\beta-2\gamma}{2} + \cos\frac{\alpha+3\beta}{2}\right) - \left(\cos\frac{3\alpha-\beta+2\gamma}{2} + \cos\frac{3\alpha+\beta}{2}\right)\right\}$$

$$= -2\sin\frac{\alpha-\gamma}{2}\sin\frac{\beta-\gamma}{2} \times \left\{\left(\cos\frac{\alpha-3\beta-2\gamma}{2} - \cos\frac{3\alpha-\beta+2\gamma}{2}\right) + \left(\cos\frac{\alpha+3\beta}{2} - \cos\frac{3\alpha+\beta}{2}\right)\right\}$$

$$= -2\sin\frac{\alpha-\gamma}{2}\cdot\sin\frac{\beta-\gamma}{2} \times \left\{2\sin(\alpha-\beta)\cdot\sin\frac{\alpha+\beta+2\gamma}{2} + 2\sin\frac{\alpha-\beta}{2}\sin(\alpha+\beta)\right\}$$

$$= -4\sin\frac{\alpha-\gamma}{2}\sin\frac{\beta-\gamma}{2}\sin\frac{\alpha-\beta}{2} \times \left\{2\cos\frac{\alpha-\beta}{2}\cdot\sin\frac{\alpha+\beta+2\gamma}{2} + \sin(\alpha+\beta)\right\}$$

$$= -4\sin\frac{\alpha-\gamma}{2}\sin\frac{\beta-\gamma}{2}\sin\frac{\alpha-\beta}{2}$$

$$\times\{\sin(\alpha+\gamma)+\sin(\beta+\gamma)+\sin(\alpha+\beta)\}$$

$$= 4\sin\frac{1}{2}(\beta-\gamma)\cdot\sin\frac{1}{2}(\gamma-\alpha)\sin\frac{1}{2}(\alpha-\beta)$$

$$\times\{\sin(\beta+\gamma)+\sin(\gamma+\alpha)+\sin(\alpha+\beta)\}.$$

13. Prove that the following determinant vanishes if any two of x, y, z are equal or $x + y + z = \pi/2$:

$$\begin{vmatrix} \sin x & \sin y & \sin z \\ \cos x & \cos y & \cos z \\ \cos^3 x & \cos^3 y & \cos^3 z \end{vmatrix}$$

Here $\Delta = \cos x\cos y\cos z \begin{vmatrix} \tan x & \tan y & \tan z \\ 1 & 1 & 1 \\ \cos^2 x & \cos^2 y & \cos^2 z \end{vmatrix}$

$$= \cos x\cos y\cos z$$

$$\times\begin{vmatrix} \tan x & \tan y-\tan x & \tan z-\tan y \\ 1 & 0 & 0 \\ \cos^2 x & \cos^2 y-\cos^2 x & \cos^2 z-\cos^2 y \end{vmatrix}$$

$$C_3\to C_3-C_2,\ C_2\to C_2-C_1$$

$$= -\cos x\cos y\cos z$$

$$\times\begin{vmatrix} \tan y-\tan x & \tan z-\tan y \\ \cos^2 y-\cos^2 x & \cos^2 z-\cos^2 y \end{vmatrix}$$

$$= -\cos x\cos y\cos z\begin{vmatrix} \dfrac{\sin(y-x)}{\cos x\cos y} & \dfrac{\sin(z-y)}{\cos y\cdot\cos z} \\ \sin^2 x-\sin^2 y & \sin^2 y-\sin^2 z \end{vmatrix}$$

$$= \begin{vmatrix} \cos z\cdot\sin(x-y) & \cos x\cdot\sin(y-z) \\ \sin(x+y)\cdot\sin(x-y) & \sin(y+z)\cdot\sin(y-z) \end{vmatrix} \quad \ldots (1)$$

$$= \sin(x-y)\sin(y-z)\begin{vmatrix} \cos z & \cos x \\ \sin(x+y) & \sin(y+z) \end{vmatrix}$$

$$= \sin(x-y)\sin(y-z)$$

$$\times\{\sin(y+z)\cos z-\sin(x+y)\cos x\}$$

$$= \frac{1}{2}\sin(x-y)\sin(y-z)[\{\sin(y+2z)+\sin y\}$$

$$-\{\sin(y+2x)+\sin y\}]$$

$$= \frac{1}{2}\sin(x-y)\sin(y-z)[\sin(y+2z)-\sin(y+2x)]$$

$$= \frac{1}{2}\sin(x-y)\sin(y-z)\cdot 2\cos(x+y+z)\cdot\sin(z-x)$$

$$= \sin(x-y)\sin(y-z)\sin(z-x)\cos(x+y+z).$$

Clearly, Δ is zero when any two of x, y, z are equal or $x+y+z = \dfrac{\pi}{2}$.

Note The conclusion for the sum can also be made at the stage (1).

14. Prove that in a $\triangle ABC$, $\begin{vmatrix} \sin 2A & \sin C & \sin B \\ \sin C & \sin 2B & \sin A \\ \sin B & \sin A & \sin 2C \end{vmatrix} = 0$.

[We know from properties of triangle in trigonometry that in a $\triangle ABC$,

$$\frac{a}{\sin A} = \frac{b}{\sin B} = \frac{c}{\sin C} = 2R$$

and $\cos A = \dfrac{b^2+c^2-a^2}{2bc}$, etc.]

$$\Delta = \begin{vmatrix} 2\sin A\cdot\cos A & \sin C & \sin B \\ \sin C & 2\sin B\cos B & \sin A \\ \sin B & \sin A & 2\sin C\cdot\cos C \end{vmatrix}$$

$$= \begin{vmatrix} 2\cdot\dfrac{a}{2R}\cdot\dfrac{b^2+c^2-a^2}{2bc} & \dfrac{c}{2R} & \dfrac{b}{2R} \\ \dfrac{c}{2R} & 2\cdot\dfrac{b}{2R}\cdot\dfrac{c^2+a^2-b^2}{2ca} & \dfrac{a}{2R} \\ \dfrac{b}{2R} & \dfrac{a}{2R} & 2\cdot\dfrac{c}{2R}\cdot\dfrac{a^2+b^2-c^2}{2ab} \end{vmatrix}$$

$$= \frac{1}{8R^3}\begin{vmatrix} \dfrac{a(b^2+c^2-a^2)}{bc} & c & b \\ c & \dfrac{b(c^2+a^2-b^2)}{ca} & a \\ b & a & \dfrac{c(a^2+b^2-c^2)}{ab} \end{vmatrix}$$

$$= \frac{1}{8R^3}\cdot\frac{1}{bc}\cdot\frac{1}{ca}\cdot\frac{1}{ab}$$

$$\times\begin{vmatrix} a(b^2+c^2-a^2) & bc^2 & b^2c \\ c^2a & b(c^2+a^2-b^2) & a^2c \\ ab^2 & a^2b & c(a^2+b^2-c^2) \end{vmatrix}$$

$$= \frac{1}{8R^3\cdot abc}\begin{vmatrix} b^2+c^2-a^2 & c^2 & b^2 \\ c^2 & c^2+a^2-b^2 & a^2 \\ b^2 & a^2 & a^2+b^2-c^2 \end{vmatrix}$$

$$= \frac{1}{8R^3abc}\begin{vmatrix} b^2-a^2 & b^2-a^2 & b^2-a^2 \\ c^2-b^2 & c^2-b^2 & c^2-b^2 \\ b^2 & a^2 & a^2+b^2-c^2 \end{vmatrix},$$

$$R_1\to R_1-R_2,\ R_2\to R_2-R_3$$

$$= \frac{(b^2-a^2)(c^2-b^2)}{8R^3abc}\begin{vmatrix} 1 & 1 & 1 \\ 1 & 1 & 1 \\ b^2 & a^2 & a^2+b^2-c^2 \end{vmatrix}$$

$= 0$ because $R_1 \equiv R_2$.

15. Prove that the value of

$$\begin{vmatrix} \cos(\theta+\phi) & -\sin(\theta+\phi) & \cos 2\phi \\ \sin\theta & \cos\theta & \sin\phi \\ -\cos\theta & \sin\theta & \lambda\cos\phi \end{vmatrix}$$

is independent of θ for all real λ and independent of θ and ϕ when $\lambda = -1$.

Here $\Delta = \dfrac{1}{\sin\phi\cdot\cos\phi}$

$$\times \begin{vmatrix} \cos(\theta+\phi) & -\sin(\theta+\phi) & \cos 2\phi \\ \sin\theta\cdot\sin\phi & \cos\theta\cdot\sin\phi & \sin^2\phi \\ -\cos\theta\cdot\cos\phi & \sin\theta\cdot\cos\phi & \lambda\cos^2\phi \end{vmatrix}$$

$$= \frac{1}{\sin\phi\cos\phi}\begin{vmatrix} 0 & 0 & \cos 2\phi + \sin^2\phi + \lambda\cos^2\phi \\ \sin\theta\cdot\sin\phi & \cos\theta\cdot\sin\phi & \sin^2\phi \\ -\cos\theta\cos\phi & \sin\theta\cdot\cos\phi & \lambda\cos^2\phi \end{vmatrix},$$

$R_1 \to R_1 + R_2 + R_3$

$$= \frac{\cos 2\phi + \sin^2\phi + \lambda\cos^2\phi}{\sin\phi\cdot\cos\phi} \times \begin{vmatrix} \sin\theta\cdot\sin\phi & \cos\theta\cdot\sin\phi \\ -\cos\theta\cdot\cos\phi & \sin\theta\cdot\cos\phi \end{vmatrix}$$

$$= (\cos^2\phi + \lambda\cos^2\phi)\begin{vmatrix} \sin\theta & \cos\theta \\ -\cos\theta & \sin\theta \end{vmatrix}$$

$$= (1+\lambda)\cos^2\phi$$

$\therefore$ Δ is independent of θ for all real values of λ.

But if $\lambda = -1$, $\Delta = 0$; so it is independent of θ and ϕ.

16. Prove that

$$\begin{vmatrix} -2a & a+b & a+c \\ b+a & -2b & b+c \\ c+a & c+b & -2c \end{vmatrix} = 4(b+c)(c+a)(a+b).$$

TIP When a determinant is to be proved equal to the product of linear factors mainly, we can adopt a new method which uses factor theorem of algebra.

Let the determinant be denoted by Δ. Clearly Δ is a function of a, b and c.

Now, putting $b = -c$ in the determinant Δ,

$$\Delta_{b=-c} = \begin{vmatrix} -2a & a-c & a+c \\ -c+a & 2c & 0 \\ c+a & 0 & -2c \end{vmatrix}$$

$$= \begin{vmatrix} c-a & -2c & a+c \\ -c+a & 2c & 0 \\ -c+a & 2c & -2c \end{vmatrix}, \quad C_1 \to C_1 + C_3,\ C_2 \to C_2 - C_3$$

$$= (-c+a)2c\begin{vmatrix} -1 & -1 & a+c \\ 1 & 1 & 0 \\ 1 & 1 & -2c \end{vmatrix} = 0, \text{ because } C_1 \equiv C_2.$$

Similarly, we can prove that $\Delta_{c=-a} = 0$, $\Delta_{a=-b} = 0$.

Hence, by factor theorem $(b+c)$, $(c+a)$ and $(a+b)$ are factors of the determinant.

$\therefore \quad \Delta = \lambda(b+c)(c+a)(a+b)$

where λ is any other possible factor of the determinant. As Δ is a determinant of the third order and each element is of the first degree in a, b, c we find that the highest degree terms in a, b, c are of the third degree.

Hence, Δ will have no other factor containing a, b, c.

So λ is a constant.

$$\text{Now,} \quad \begin{vmatrix} -2a & a+b & a+c \\ b+a & -2b & b+c \\ c+a & c+b & -2c \end{vmatrix} = \lambda(b+c)(c+a)(a+b)$$

where λ is a constant.

Putting $a = 0$, $b = 1$, $c = 2$ we get

$$\begin{vmatrix} 0 & 1 & 2 \\ 1 & -2 & 3 \\ 2 & 3 & -4 \end{vmatrix} = \lambda\cdot 3\cdot 2\cdot 1$$

$$\text{or} \quad \begin{vmatrix} 0 & 1 & 2 \\ 1 & -2 & 3 \\ 0 & 7 & -10 \end{vmatrix} = 6\lambda, \quad (R_3 \to R_3 - 2R_2)$$

$$\text{or} \quad -\begin{vmatrix} 1 & 2 \\ 7 & -10 \end{vmatrix} = 6\lambda$$

or $\quad 24 = 6\lambda; \quad \therefore \quad \lambda = 4.$

$\therefore \quad \Delta = 4(b+c)(c+a)(a+b).$

17. Show that $\begin{vmatrix} a_1+b_3 & a_1+b_2 & 1+a_1+b_1 \\ a_2+b_3 & 1+a_2+b_2 & a_2+b_1 \\ 1+a_3+b_3 & a_3+b_2 & a_3+b_1 \end{vmatrix}$

$$= \sum_{1\le i<j\le 3} (a_i - a_j)(b_i - b_j) - \sum_{i=1}^{3} (a_i + b_i) - 1.$$

$$\text{Here } \Delta = \begin{vmatrix} b_3-b_2 & -1+b_2-b_1 & 1+a_1+b_1 \\ -1+b_3-b_2 & 1+b_2-b_1 & a_2+b_1 \\ 1+b_3-b_2 & b_2-b_1 & a_3+b_1 \end{vmatrix},$$

$C_1 \to C_1 - C_2$, $C_2 \to C_2 - C_3$

$$= \begin{vmatrix} 0 & -1+b_2-b_1 & 1+a_1+b_1 \\ -1 & 1+b_2-b_1 & a_2+b_1 \\ 1 & b_2-b_1 & a_3+b_1 \end{vmatrix}$$

$$+ (b_3-b_2)\begin{vmatrix} 1 & -1+b_2-b_1 & 1+a_1+b_1 \\ 1 & 1+b_2-b_1 & a_2+b_1 \\ 1 & b_2-b_1 & a_3+b_1 \end{vmatrix}$$

$$= \begin{vmatrix} 0 & -1+b_2-b_1 & 1+a_1+b_1 \\ 0 & 1+2(b_2-b_1) & a_2+a_3+2b_1 \\ 1 & b_2-b_1 & a_3+b_1 \end{vmatrix} \quad (R_2 \to R_2+R_3)$$

$$+ (b_3-b_2) \begin{vmatrix} 1 & -1 & 1+a_1+b_1 \\ 1 & 1 & a_2+b_1 \\ 1 & 0 & a_3+b_1 \end{vmatrix}$$

$$\{C_2 \to C_2 - (b_2-b_1)C_1\}$$

$$= \begin{vmatrix} -1+b_2-b_1 & 1+a_1+b_1 \\ 1+2(b_2-b_1) & a_2+a_3+2b_1 \end{vmatrix}$$

$$+ (b_3-b_2) \begin{vmatrix} 0 & -1 & 1+a_1-a_3 \\ 0 & 1 & a_2-a_3 \\ 1 & 0 & a_3+b_1 \end{vmatrix},$$

$$R_1 \to R_1-R_3,\ R_2 \to R_2-R_3$$

$$= \begin{vmatrix} -1+(b_2-b_1) & 1+(a_1+b_1) \\ 3 & -2+(a_2-a_1)+(a_3-a_1) \end{vmatrix}$$

$$(R_2 \to R_2 - 2R_1)$$

$$+ (b_3-b_2) \begin{vmatrix} -1 & 1+a_1-a_3 \\ 1 & a_2-a_3 \end{vmatrix}$$

$$= 2-(a_2-a_1)-(a_3-a_1)-2(b_2-b_1)$$
$$+ (a_2-a_1)(b_2-b_1)+(a_3-a_1)(b_2-b_1)-3-3(a_1+b_1)$$
$$+ (b_3-b_2\{(a_3-a_2)-1+(a_3-a_1)\}$$
$$= (a_2-a_1)(b_2-b_1)+(a_3-a_1)(b_2-b_1)$$
$$+ (a_3-a_2)(b_3-b_2)+(a_3-a_1)(b_3-b_2)$$
$$+ \{-(a_2-a_1)-(a_3-a_1)-2(b_2-b_1)$$
$$- 3(a_1+b_1)-(b_3-b_2)\}+2-3$$
$$= (a_2-a_1)(b_2-b_1)+(a_3-a_2)(b_3-b_2)$$
$$+ (a_3-a_1)(b_2-b_1+b_3-b_2)$$
$$- \{(a_1+b_1)+(a_2+b_2)+(a_3+b_3)\}-1$$
$$= (a_1-a_2)(b_1-b_2)+(a_2-a_3)(b_2-b_3)$$
$$+ (a_3-a_1)(b_3-b_1)$$
$$- \{(a_1+b_1)+(a_2+b_2)+(a_3+b_3)\}-1$$
$$= \sum_{1 \le i < j \le 3} (a_i-a_j)(b_i-b_j) - \sum_{i=1}^{3} (a_i+b_i) - 1.$$

18. **$f(n)$, $g(n)$ and $h(n)$ are second degree polynomials in n having $n+2$ as a common factor, prove that $\sum_{r=1}^{n} \Delta_r$ is independent of n where**

$$\Delta_r = \begin{vmatrix} 2r+1 & 6n(n+2) & f(n) \\ 2^{r-1} & 3.2^{n+1}-6 & g(n) \\ r(n-r+1) & n(n+1)(n+2) & h(n) \end{vmatrix}.$$

TIP We have to find the sum of n determinants whose second and third columns are the same for all determinants. Hence, we use the sum of determinants.

$$\sum_{r=1}^{n} \Delta_r = \Delta_1 + \Delta_2 + \ldots + \Delta_n$$

$$= \begin{vmatrix} 2.1+1 & 6n(n+2) & f(n) \\ 2^0 & 3 \cdot 2^{n+1}-6 & g(n) \\ 1 \cdot (n+1-1) & n(n+1)(n+2) & h(n) \end{vmatrix}$$

$$+ \begin{vmatrix} 2.2+1 & 6n(n+2) & f(n) \\ 2^1 & 3.2^{n+1}-6 & g(n) \\ 2(n+1-2) & n(n+1)(n+2) & h(n) \end{vmatrix} + \ldots$$

to n determinants

$$= \begin{vmatrix} 2(1+2+3+\ldots+n)+(1+1+\ldots+1) & 6n(n+2) & f(n) \\ 2^0+2^1+2^2+\ldots+2^{n-1} & 3.2^{n+1}-6 & g(n) \\ 1 \cdot (n+1-1)+2(n+1-2)+\ldots+n(n+1-n) & n(n+1)(n+2) & h(n) \end{vmatrix}$$

$$= \begin{vmatrix} 2 \cdot \dfrac{n(n+1)}{2}+n & 6n(n+2) & f(n) \\ \dfrac{2^n-1}{2-1} & 3(2^{n+1}-2) & g(n) \\ (n+1)(1+2+\ldots+n)-(1^2+2^2+\ldots+n^2) & n(n+1)(n+2) & h(n) \end{vmatrix}$$

$$= \begin{vmatrix} n(n+2) & 6n(n+2) & f(n) \\ 2^n-1 & 6(2^n-1) & g(n) \\ (n+1)\dfrac{n(n+1)}{2}-\dfrac{n(n+1)(2n+1)}{6} & 6 \cdot \dfrac{n(n+1)(n+2)}{6} & h(n) \end{vmatrix}$$

$$= 6 \begin{vmatrix} n(n+2) & n(n+2) & f(n) \\ 2^n-1 & 2^n-1 & g(n) \\ \dfrac{n(n+1)}{6}(n+2) & \dfrac{n(n+1)(n+2)}{6} & h(n) \end{vmatrix}$$

$= 6 \times 0 \quad (\because \; C_1 \equiv C_2)$

$= 0;$

$\therefore \quad \sum_{r=1}^{n} \Delta_r$ is independent of n.

19. Prove that

$$\begin{vmatrix} 2 & \alpha+\beta+\gamma+\delta & \alpha\beta+\gamma\delta \\ \alpha+\beta+\gamma+\delta & 2(\alpha+\beta)(\gamma+\delta) & \alpha\beta(\gamma+\delta)+\gamma\delta(\alpha+\beta) \\ \alpha\beta+\gamma\delta & \alpha\beta(\gamma+\delta)+\gamma\delta(\alpha+\beta) & 2\alpha\beta\gamma\delta \end{vmatrix} = 0.$$

$$\Delta = \begin{vmatrix} 2 & \gamma+\delta-\alpha-\beta & \alpha\beta+\gamma\delta \\ \alpha+\beta+\gamma+\delta & (\alpha+\beta)(\gamma+\delta-\alpha-\beta) & \alpha\beta(\gamma+\delta)+\gamma\delta(\alpha+\beta) \\ \alpha\beta+\gamma\delta & \alpha\beta(\gamma+\delta-\alpha-\beta) & 2\alpha\beta\gamma\delta \end{vmatrix},$$

$$C_2 \to C_2 - (\alpha+\beta)\cdot C_1$$

$$= (\gamma+\delta-\alpha-\beta) \times \begin{vmatrix} 2 & 1 & \alpha\beta+\gamma\delta \\ \alpha+\beta+\gamma+\delta & \alpha+\beta & \alpha\beta(\gamma+\delta)+\gamma\delta(\alpha+\beta) \\ \alpha\beta+\gamma\delta & \alpha\beta & 2\alpha\beta\gamma\delta \end{vmatrix}$$

$$= (\gamma+\delta-\alpha-\beta) \begin{vmatrix} 1 & 1 & \alpha\beta+\gamma\delta \\ \gamma+\delta & \alpha+\beta & \alpha\beta(\gamma+\delta)+\gamma\delta(\alpha+\beta) \\ \gamma\delta & \alpha\beta & 2\alpha\beta\gamma\delta \end{vmatrix},$$

$$C_1 \to C_1 - C_2$$

$$= (\gamma+\delta-\alpha-\beta) \begin{vmatrix} 1 & 1 & 0 \\ \gamma+\delta & \alpha+\beta & 0 \\ \gamma\delta & \alpha\beta & 0 \end{vmatrix},$$

$$C_3 \to C_3 - \alpha\beta\cdot C_1 - \gamma\delta\cdot C_2$$

$= (\gamma+\delta-\alpha-\beta) \times 0 = 0.$

Second Method (using product of determinants)

TIP In this method if one can detect the determinants to be multiplied, working becomes very easy. This comes through experience.

Consider the following product of determinants.

$$\begin{vmatrix} 1 & 1 & 0 \\ \gamma+\delta & \alpha+\beta & 0 \\ \gamma\delta & \alpha\beta & 0 \end{vmatrix} \times \begin{vmatrix} 1 & \alpha+\beta & \alpha\beta \\ 1 & \gamma+\delta & \gamma\delta \\ 0 & 0 & 0 \end{vmatrix}$$

$$= \begin{vmatrix} 1.1+1.1+0.0 & 1\cdot(\alpha+\beta)+1\cdot(\gamma+\delta)+0\cdot 0 & 1\cdot\alpha\beta+1\cdot\gamma\delta+0\cdot 0 \\ (\gamma+\delta)\cdot 1+(\alpha+\beta)\cdot 1+0\cdot 0 & (\gamma+\delta)(\alpha+\beta)+(\alpha+\beta)(\gamma+\delta)+0\cdot 0 & (\gamma+\delta)\alpha\beta+(\alpha+\beta)\gamma\delta+0\cdot 0 \\ \gamma\delta\cdot 1+\alpha\beta\cdot 1+0\cdot 0 & \gamma\delta(\alpha+\beta)+\alpha\beta(\gamma+\delta)+0\cdot 0 & \gamma\delta\cdot\alpha\beta+\alpha\beta\cdot\gamma\delta+0\cdot 0 \end{vmatrix}$$

$$= \begin{vmatrix} 2 & \alpha+\beta+\gamma+\delta & \alpha\beta+\gamma\delta \\ \alpha+\beta+\gamma+\delta & 2(\alpha+\beta)(\gamma+\delta) & \alpha\beta(\gamma+\delta)+\gamma\delta(\alpha+\beta) \\ \alpha\beta+\gamma\delta & \alpha\beta(\gamma+\delta)+\gamma\delta(\alpha+\beta) & 2\alpha\beta\gamma\delta \end{vmatrix} = \Delta.$$

$$\text{But} \quad \begin{vmatrix} 1 & 1 & 0 \\ \gamma+\delta & \alpha+\beta & 0 \\ \gamma\delta & \alpha\beta & 0 \end{vmatrix} = 0, \quad \begin{vmatrix} 1 & \alpha+\beta & \alpha\beta \\ 1 & \gamma+\delta & \gamma\delta \\ 0 & 0 & 0 \end{vmatrix} = 0$$

$\therefore \quad 0 \times 0 = \Delta; \qquad \therefore \quad \Delta = 0.$

20. For all values of A, B, C and P, Q, R show that

$$\begin{vmatrix} \cos(A-P) & \cos(A-Q) & \cos(A-R) \\ \cos(B-P) & \cos(B-Q) & \cos(B-R) \\ \cos(C-P) & \cos(C-Q) & \cos(C-R) \end{vmatrix} = 0.$$

Breaking each element by using

$\cos(\alpha-\beta) = \cos\alpha\cdot\cos\beta + \sin\alpha\cdot\sin\beta$, we get

$$\Delta = \begin{vmatrix} \cos A\cdot\cos P+\sin A\cdot\sin P & \cos A\cdot\cos Q+\sin A\cdot\sin Q & \cos A\cdot\cos R+\sin A\cdot\sin R \\ \cos B\cdot\cos P+\sin B\cdot\sin P & \cos B\cdot\cos Q+\sin B\cdot\sin Q & \cos B\cdot\cos R+\sin B\cdot\sin R \\ \cos C\cdot\cos P+\sin C\cdot\sin P & \cos C\cdot\cos Q+\sin C\cdot\sin Q & \cos C\cdot\cos R+\sin C\cdot\sin R \end{vmatrix}$$

$$= \begin{vmatrix} \cos A\cdot\cos P & \cos A\cdot\cos Q+\sin A\cdot\sin Q & \cos A\cdot\cos R+\sin A\cdot\sin R \\ \cos B\cdot\cos P & \cos B\cdot\cos Q+\sin B\cdot\sin Q & \cos B\cdot\cos R+\sin B\cdot\sin R \\ \cos C\cdot\cos P & \cos C\cdot\cos Q+\sin C\cdot\sin Q & \cos C\cdot\cos R+\sin C\cdot\sin R \end{vmatrix}$$

$$+ \begin{vmatrix} \sin A\cdot\sin P & \cos A\cdot\cos Q+\sin A\cdot\sin Q & \cos A\cdot\cos R+\sin A\cdot\sin R \\ \sin B\cdot\sin P & \cos B\cdot\cos Q+\sin B\cdot\sin Q & \cos B\cdot\cos R+\sin B\cdot\sin R \\ \sin C\cdot\sin P & \cos C\cdot\cos Q+\sin C\cdot\sin Q & \cos C\cdot\cos R+\sin C\cdot\sin R \end{vmatrix}$$

$$= \cos P \begin{vmatrix} \cos A & \cos A \cdot \cos Q + \sin A \cdot \sin Q & \cos A \cdot \cos R + \sin A \cdot \sin R \\ \cos B & \cos B \cdot \cos Q + \sin B \cdot \sin Q & \cos B \cdot \cos R + \sin B \cdot \sin R \\ \cos C & \cos C \cdot \cos Q + \sin C \cdot \sin Q & \cos C \cdot \cos R + \sin C \cdot \sin R \end{vmatrix}$$

$$+ \sin P \begin{vmatrix} \sin A & \cos A \cdot \cos Q + \sin A \cdot \sin Q & \cos A \cdot \cos R + \sin A \cdot \sin R \\ \sin B & \cos B \cdot \cos Q + \sin B \cdot \sin Q & \cos B \cdot \cos R + \sin B \cdot \sin R \\ \sin C & \cos C \cdot \cos Q + \sin C \cdot \sin Q & \cos C \cdot \cos R + \sin C \cdot \sin R \end{vmatrix}$$

$$= \cos P \begin{vmatrix} \cos A & \sin A \cdot \sin Q & \sin A \cdot \sin R \\ \cos B & \sin B \cdot \sin Q & \sin B \cdot \sin R \\ \cos C & \sin C \cdot \sin Q & \sin C \cdot \sin R \end{vmatrix},$$

$$C_2 \to C_2 - \cos Q \cdot C_1,\ C_3 \to C_3 - \cos R \cdot C_1$$

$$+ \sin P \begin{vmatrix} \sin A & \cos A \cdot \cos Q & \cos A \cdot \cos R \\ \sin B & \cos B \cdot \cos Q & \cos B \cdot \cos R \\ \sin C & \cos C \cdot \cos Q & \cos C \cdot \cos R \end{vmatrix},$$

$$C_2 \to C_2 - \sin Q \cdot C_1,\ C_3 \to C_3 - \sin R \cdot C_1$$

$$= \cos P \cdot \sin Q \cdot \sin R \begin{vmatrix} \cos A & \sin A & \sin A \\ \cos B & \sin B & \sin B \\ \cos C & \sin C & \sin C \end{vmatrix}$$

$$+ \sin P \cdot \cos Q \cdot \cos R \begin{vmatrix} \sin A & \cos A & \cos A \\ \sin B & \cos B & \cos B \\ \sin C & \cos C & \cos C \end{vmatrix}$$

$= 0$ ($\because$ $C_2 \equiv C_3$ in each).

Second Method (using product of determinants)

Consider the following product of determinants.

$$\begin{vmatrix} \cos A & \sin A & 0 \\ \cos B & \sin B & 0 \\ \cos C & \sin C & 0 \end{vmatrix} \times \begin{vmatrix} \cos P & \cos Q & \cos R \\ \sin P & \sin Q & \sin R \\ 0 & 0 & 0 \end{vmatrix}$$

$$= \begin{vmatrix} \cos A \cdot \cos P + \sin A \cdot \sin P & \cos A \cdot \cos Q + \sin A \cdot \sin Q & \cos A \cos R + \sin A \sin R \\ \cos B \cdot \cos P + \sin B \cdot \sin P & \cos B \cdot \cos Q + \sin B \cdot \sin Q & \cos B \cos R + \sin B \sin R \\ \cos C \cdot \cos P + \sin C \cdot \sin P & \cos C \cdot \cos Q + \sin C \cdot \sin Q & \cos C \cos R + \sin C \sin R \end{vmatrix}$$

$$= \begin{vmatrix} \cos(A-P) & \cos(A-Q) & \cos(A-R) \\ \cos(B-P) & \cos(B-Q) & \cos(B-R) \\ \cos(C-P) & \cos(C-Q) & \cos(C-R) \end{vmatrix} = \Delta.$$

But

$$\begin{vmatrix} \cos A & \sin A & 0 \\ \cos B & \sin B & 0 \\ \cos C & \sin C & 0 \end{vmatrix} = 0, \quad \begin{vmatrix} \cos P & \cos Q & \cos R \\ \sin P & \sin Q & \sin R \\ 0 & 0 & 0 \end{vmatrix} = 0$$

$\therefore \quad 0 \times 0 = \Delta; \qquad \therefore \quad \Delta = 0.$

21. If α, β be the real roots of $ax^2 + bx + c = 0$, and $s_n = \alpha^n + \beta^n$ then prove that $as_n + bs_{n-1} + cs_{n-2} = 0$ for all $n \in N$. Hence or otherwise prove that

$$\begin{vmatrix} 3 & 1+s_1 & 1+s_2 \\ 1+s_1 & 1+s_2 & 1+s_3 \\ 1+s_2 & 1+s_3 & 1+s_4 \end{vmatrix} \geq 0 \text{ for all real } a, b, c.$$

Here $a\alpha^2 + b\alpha + c = 0$ and $a\beta^2 + b\beta + c = 0$;

also $\alpha + \beta = \dfrac{-b}{a}$, $\alpha\beta = \dfrac{c}{a}$.

Now, $as_n + bs_{n-1} + cs_{n-2}$

$$= a(\alpha^n + \beta^n) + b(\alpha^{n-1} + \beta^{n-1}) + c(\alpha^{n-2} + \beta^{n-2})$$

$$= \alpha^{n-2} \cdot (a\alpha^2 + b\alpha + c) + \beta^{n-2}(a\beta^2 + b\beta + c)$$

$$= \alpha^{n-2} \times 0 + \beta^{n-2} \times 0 = 0.$$

$$\therefore \text{ (A)} \begin{cases} as_2 + bs_1 + cs_0 = 0, \text{ where } s_0 = \alpha^0 + \beta^0 = 1 + 1 = 2 \\ as_3 + bs_2 + cs_1 = 0 \\ as_4 + bs_3 + cs_2 = 0 \end{cases}$$

Now,

$$\Delta = \frac{1}{c} \begin{vmatrix} 3c & c + cs_1 & c + cs_2 \\ 1+s_1 & 1+s_2 & 1+s_3 \\ 1+s_2 & 1+s_3 & 1+s_4 \end{vmatrix}$$

$$= \frac{1}{c} \begin{vmatrix} 3c + b(1+s_1) + a(1+s_2) & c + cs_1 + b(1+s_2) + a(1+s_3) & c + cs_2 + b(1+s_3) + a(1+s_4) \\ 1+s_1 & 1+s_2 & 1+s_3 \\ 1+s_2 & 1+s_3 & 1+s_4 \end{vmatrix},$$

$$R_1 \to R_1 + bR_2 + aR_3$$

$$= \frac{1}{c} \begin{vmatrix} a+b+c+(as_2 + bs_1 + 2c) & a+b+c+(as_3 + bs_2 + cs_1) & a+b+c+(as_4 + bs_3 + cs_2) \\ 1+s_1 & 1+s_2 & 1+s_3 \\ 1+s_2 & 1+s_3 & 1+s_4 \end{vmatrix}$$

$$= \frac{1}{c} \begin{vmatrix} a+b+c & a+b+c & a+b+c \\ 1+s_1 & 1+s_2 & 1+s_3 \\ 1+s_2 & 1+s_3 & 1+s_4 \end{vmatrix}, \text{ using (A)}$$

$$= \frac{a+b+c}{c}\begin{vmatrix} 1 & 1 & 1 \\ 1+s_1 & 1+s_2 & 1+s_3 \\ 1+s_2 & 1+s_3 & 1+s_4 \end{vmatrix}$$

$$= \frac{a+b+c}{c}\begin{vmatrix} 1 & 0 & 0 \\ 1+s_1 & s_2-s_1 & s_3-s_2 \\ 1+s_2 & s_3-s_2 & s_4-s_3 \end{vmatrix},$$

$$C_3 \to C_3 - C_2,\ C_2 \to C_2 - C_1$$

$$= \frac{a+b+c}{c}\begin{vmatrix} s_2-s_1 & s_3-s_2 \\ s_3-s_2 & s_4-s_3 \end{vmatrix}$$

$$= \frac{a+b+c}{c}\{s_2s_4 - s_3^2 + s_1s_3 - s_2^2 + s_2s_3 - s_1s_4\}$$

$$= \frac{a+b+c}{c}[\{(\alpha^2+\beta^2)(\alpha^4+\beta^4)-(\alpha^3+\beta^3)^2\}$$
$$+\{(\alpha+\beta)(\alpha^3+\beta^3)-(\alpha^2+\beta^2)^2\}$$
$$+\{(\alpha^2+\beta^2)(\alpha^3+\beta^3)-(\alpha+\beta)(\alpha^4+\beta^4)\}]$$

$$= \frac{a+b+c}{c}[\alpha^2\beta^2(\alpha-\beta)^2 + \alpha\beta(\alpha-\beta)^2$$
$$-\alpha\beta(\alpha+\beta)(\alpha-\beta)^2]$$

$$= \frac{a+b+c}{c}(\alpha-\beta)^2\,\alpha\beta[\alpha\beta+1-(\alpha+\beta)]$$

$$= \frac{a+b+c}{c}(\alpha-\beta)^2\cdot\frac{c}{a}\left[\frac{c}{a}+1+\frac{b}{a}\right]$$

$$= (\alpha-\beta)^2\cdot\frac{(a+b+c)^2}{a^2} \ge 0,$$

because α, β are given to be real.

Second Method (using product of determinants)

$$\Delta = \begin{vmatrix} 3 & 1+\alpha+\beta & 1+\alpha^2+\beta^2 \\ 1+\alpha+\beta & 1+\alpha^2+\beta^2 & 1+\alpha^3+\beta^3 \\ 1+\alpha^2+\beta^2 & 1+\alpha^3+\beta^3 & 1+\alpha^4+\beta^4 \end{vmatrix}$$

$$= \begin{vmatrix} 1 & 1 & 1 \\ 1 & \alpha & \beta \\ 1 & \alpha^2 & \beta^2 \end{vmatrix} \times \begin{vmatrix} 1 & 1 & 1 \\ 1 & \alpha & \alpha^2 \\ 1 & \beta & \beta^2 \end{vmatrix} = \begin{vmatrix} 1 & 1 & 1 \\ 1 & \alpha & \alpha^2 \\ 1 & \beta & \beta^2 \end{vmatrix}^2 \ge 0,$$

because α, β are real.

22. Without expanding a determinant at any stage, show that

$$\begin{vmatrix} x^2+x & x+1 & x-2 \\ 2x^2+3x-1 & 3x & 3x-3 \\ x^2+2x+3 & 2x-1 & 2x-1 \end{vmatrix} = xA+B$$

where A, B are determinants of the third order not involving x.

$$\Delta = \begin{vmatrix} x^2 & x+1 & x-2 \\ 2x^2 & 3x & 3x-3 \\ x^2 & 2x-1 & 2x-1 \end{vmatrix} + \begin{vmatrix} x & x+1 & x-2 \\ 3x-1 & 3x & 3x-3 \\ 2x+3 & 2x-1 & 2x-1 \end{vmatrix}$$

$$= \begin{vmatrix} x^2 & x+1 & x-2 \\ x^2 & 2x-1 & 2x-1 \\ x^2 & 2x-1 & 2x-1 \end{vmatrix} + \begin{vmatrix} 2 & 3 & x-2 \\ 2 & 3 & 3x-3 \\ 4 & 0 & 2x-1 \end{vmatrix}$$

$$(R_2 \to R_2 - R_1) \qquad (C_1 \to C_1 - C_3, C_2 \to C_2 - C_3)$$

$$= 0 + \begin{vmatrix} 2 & 3 & x \\ 2 & 3 & 3x \\ 4 & 0 & 2x \end{vmatrix} + \begin{vmatrix} 2 & 3 & -2 \\ 2 & 3 & -3 \\ 4 & 0 & -1 \end{vmatrix}$$

$$= x\begin{vmatrix} 2 & 3 & 1 \\ 2 & 3 & 3 \\ 4 & 0 & 2 \end{vmatrix} + \begin{vmatrix} 2 & 3 & -2 \\ 2 & 3 & -3 \\ 4 & 0 & -1 \end{vmatrix} = Ax+B,$$

where A, B are determinants of order 3 not involving x.

23. Without expanding at any stage, prove that

$$\begin{vmatrix} 0 & p-q & p-r \\ q-p & 0 & q-r \\ r-p & r-q & 0 \end{vmatrix} = 0.$$

$$\Delta = \begin{vmatrix} 0 & p-q & p-r \\ q-p & 0 & q-r \\ r-p & r-q & 0 \end{vmatrix}$$

$$= (-1)^3\begin{vmatrix} 0 & q-p & r-p \\ p-q & 0 & r-q \\ p-r & q-r & 0 \end{vmatrix}$$

$$= -\begin{vmatrix} 0 & p-q & p-r \\ q-p & 0 & q-r \\ r-p & r-q & 0 \end{vmatrix}, R \rightleftharpoons C$$

$\therefore \quad \Delta = -\Delta; \qquad \therefore \quad 2\Delta = 0; \qquad \therefore \quad \Delta = 0.$

24. Without expanding a determinant at any stage, show that the value of

$$\begin{vmatrix} 2 & 3+i & -1 \\ 3-i & 0 & -1+i \\ -1 & -1-i & 1 \end{vmatrix}$$ **is purely real.**

[Remember if $f(x+iy) = A+iB$ then $f(x-iy) = A-iB$.]

Let the value of the determinant be $A+iB$.

Now, $$\Delta = A+iB = \begin{vmatrix} 2 & 3+i & -1 \\ 3-i & 0 & -1+i \\ -1 & -1-i & 1 \end{vmatrix};$$

replacing i by $-i$, we get

$$A-iB = \begin{vmatrix} 2 & 3-i & -1 \\ 3+i & 0 & -1-i \\ -1 & -1+i & 1 \end{vmatrix}$$

$$= \begin{vmatrix} 2 & 3+i & -1 \\ 3-i & 0 & -1+i \\ -1 & -1-i & 1 \end{vmatrix}, \text{ by using } R \rightleftharpoons C$$

$$= A+iB$$

$\therefore \quad 2iB = 0; \qquad \therefore \quad B = 0.$

$\therefore \quad \Delta = A + i\cdot 0 = A =$ purely real.

25. If $f(x)=\begin{vmatrix}(1+x)^{a_1b_1} & (1+x)^{a_1b_2} & (1+x)^{a_1b_3}\\(1+x)^{a_2b_1} & (1+x)^{a_2b_2} & (1+x)^{a_2b_3}\\(1+x)^{a_3b_1} & (1+x)^{a_3b_2} & (1+x)^{a_3b_3}\end{vmatrix}$,

find the coefficient of x in the expansion of $f(x)$.

[Remember if $f(x)=A_0+A_1x+A_2x^2+A_3x^3+\ldots$

then $f'(x)=A_1+2A_2x+3A_3x^2+\ldots$

$f''(x)=2A_2+6A_3x+\ldots$

$\therefore \quad f'(0)=A_1=$ coefficient of x.

$f''(0)=2A_2=2\times$(coefficient of x^2), etc.]

Here $f'(x)$

$$=\begin{vmatrix}a_1b_1(1+x)^{a_1b_1-1} & a_1b_2(1+x)^{a_1b_2-1} & a_1b_3(1+x)^{a_1b_3-1}\\(1+x)^{a_2b_1} & (1+x)^{a_2b_2} & (1+x)^{a_2b_3}\\(1+x)^{a_3b_1} & (1+x)^{a_3b_2} & (1+x)^{a_3b_3}\end{vmatrix}$$

$$+\begin{vmatrix}(1+x)^{a_1b_1} & (1+x)^{a_1b_2} & (1+x)^{a_1b_3}\\a_2b_1(1+x)^{a_2b_1-1} & a_2b_2(1+x)^{a_2b_2-1} & a_2b_3(1+x)^{a_2b_3-1}\\(1+x)^{a_3b_1} & (1+x)^{a_3b_2} & (1+x)^{a_3b_3}\end{vmatrix}$$

$$+\begin{vmatrix}(1+x)^{a_1b_1} & (1+x)^{a_1b_2} & (1+x)^{a_1b_3}\\(1+x)^{a_2b_1} & (1+x)^{a_2b_2} & (1+x)^{a_2b_3}\\a_3b_1(1+x)^{a_3b_1-1} & a_3b_2(1+x)^{a_3b_2-1} & a_3b_3(1+x)^{a_3b_3-1}\end{vmatrix}$$

$$\therefore \quad f'(0)=\begin{vmatrix}a_1b_1 & a_1b_2 & a_1b_3\\1 & 1 & 1\\1 & 1 & 1\end{vmatrix}+\begin{vmatrix}1 & 1 & 1\\a_2b_1 & a_2b_2 & a_2b_3\\1 & 1 & 1\end{vmatrix}$$

$$+\begin{vmatrix}1 & 1 & 1\\1 & 1 & 1\\a_3b_1 & a_3b_2 & a_3b_3\end{vmatrix}=0+0+0=0$$

$\therefore \quad$ the coefficient of $x=f'(0)=0$.

26. Let α be a repeated root of the quadratic equation $f(x)=0$ and $A(x)$, $B(x)$ and $C(x)$ are polynomials of degree 3, 4 and 5 respectively. Show that the determinant

$$\begin{vmatrix}A(x) & B(x) & C(x)\\A(\alpha) & B(\alpha) & C(\alpha)\\A'(\alpha) & B'(\alpha) & C'(\alpha)\end{vmatrix} \text{ is divisible by } f(x)$$

where $A'(\alpha)=\left(\dfrac{dA}{dx}\right)_{x=\alpha}$, **etc.**

We know that α is a root of $f(x)=0$. This means $(x-\alpha)$ is a factor of $f(x)$. If α is a repeated root of $f(x)=0$ then $f(x)$ has a repeated factor $(x-\alpha)$, i.e., $(x-\alpha)^2$ is a factor of $f(x)$. But here $f(x)$ is quadratic.

$$\therefore \quad f(x)=\lambda(x-\alpha)^2 \qquad \ldots (1)$$

where λ is a constant.

Let $\Delta(x)=\begin{vmatrix}A(x) & B(x) & C(x)\\A(\alpha) & B(\alpha) & C(\alpha)\\A'(\alpha) & B'(\alpha) & C'(\alpha)\end{vmatrix}$

which is of the degree 5 at most and 3 at least.

Clearly, $\Delta(\alpha)=0 \qquad \ldots (2)$

Differentiating $\Delta(x)$ w.r.t. x,

$$\Delta'(x)=\begin{vmatrix}A'(x) & B'(x) & C'(x)\\A(\alpha) & B(\alpha) & C(\alpha)\\A'(\alpha) & B'(\alpha) & C'(\alpha)\end{vmatrix}+\begin{vmatrix}A(x) & B(x) & C(x)\\0 & 0 & 0\\A'(\alpha) & B'(\alpha) & C'(\alpha)\end{vmatrix}+\begin{vmatrix}A(x) & B(x) & C(x)\\A(\alpha) & B(\alpha) & C(\alpha)\\0 & 0 & 0\end{vmatrix}$$

because derivatives of constants = 0

$$=\begin{vmatrix}A'(x) & B'(x) & C'(x)\\A(\alpha) & B(\alpha) & C(\alpha)\\A'(\alpha) & B'(\alpha) & C'(\alpha)\end{vmatrix}$$

$$\therefore \quad \Delta'(\alpha)=\begin{vmatrix}A'(\alpha) & B'(\alpha) & C'(\alpha)\\A(\alpha) & B(\alpha) & C(\alpha)\\A'(\alpha) & B'(\alpha) & C'(\alpha)\end{vmatrix}=0 \qquad \ldots (3)$$

because $R_1\equiv R_3$.

We know that

$\phi(\alpha)=0 \Rightarrow (x-\alpha)$ is a factor of $\phi(x)$

and $\phi'(\alpha)=0 \Rightarrow (x-\alpha)^2$ is a factor of $\phi(x)$

$\therefore \quad$ from (2) and (3), $\Delta(x)$ has a factor $(x-\alpha)^2$.

$\therefore \quad \Delta(x)=(x-\alpha)^2\cdot F(x)$

$$=\frac{1}{\lambda}\lambda(x-\alpha)^2\cdot F(x)=\frac{1}{\lambda}f(x)\cdot F(x), \text{ using (1)}$$

$\therefore \quad \Delta(x)$ is divisible by $f(x)$.

27. Let $a_n=\displaystyle\int_0^{\pi/2}\frac{1-\cos 2nx}{1-\cos 2x}\,dx.$

Prove that a_{n+1} is the AM between a_n and a_{n+2}.

Hence or otherwise prove that $\begin{vmatrix}a_1 & a_2 & a_3\\a_4 & a_5 & a_6\\a_7 & a_8 & a_9\end{vmatrix}=0.$

Here a_n+a_{n+2}

$$=\int_0^{\pi/2}\frac{1-\cos 2nx}{1-\cos 2x}\,dx+\int_0^{\pi/2}\frac{1-\cos 2(n+2)x}{1-\cos 2x}\,dx$$

$$=\int_0^{\pi/2}\frac{2-\{\cos 2nx+\cos 2(n+2)x\}}{1-\cos 2x}\,dx$$

$$= \int_0^{\pi/2} \frac{2 - 2\cos 2(n+1)x \cdot \cos 2x}{1 - \cos 2x}\, dx.$$

Also $2 \cdot a_{n+1} = 2\int_0^{\pi/2} \frac{1 - \cos 2(n+1)x}{1 - \cos 2x}\, dx$

$\therefore \quad a_n + a_{n+2} - 2a_{n+1}$

$$= 2\int_0^{\pi/2} \frac{\{1 - \cos 2(n+1)x \cdot \cos 2x - 1 + \cos 2(n+1)x\}}{1 - \cos 2x}\, dx$$

$$= 2\int_0^{\pi/2} \frac{\cos 2(n+1)x \cdot \{1 - \cos 2x\}}{1 - \cos 2x}\, dx$$

$$= 2\int_0^{\pi/2} \cos 2(n+1)x\, dx$$

$$\because \quad a_n + a_{n+2} - 2a_{n+1} = 2\left[\frac{\sin 2(n+1)x}{2(n+1)}\right]_0^{\pi/2}$$

$$= \frac{1}{n+1}[0 - 0] = 0, \qquad \ldots (1)$$

for all n.

$$\therefore \quad a_{n+1} = \frac{a_n + a_{n+2}}{2}$$

$\therefore \quad a_{n+1}$ is the AM between a_n, a_{n+2}.

Now, $\Delta = \begin{vmatrix} a_1 & a_2 & a_3 \\ a_4 & a_5 & a_6 \\ a_7 & a_8 & a_9 \end{vmatrix} = \frac{1}{2}\begin{vmatrix} a_1 & 2a_2 & a_3 \\ a_4 & 2a_5 & a_6 \\ a_7 & 2a_8 & a_9 \end{vmatrix}$

$$= \frac{1}{2}\begin{vmatrix} a_1 & 2a_2 - (a_1 + a_3) & a_3 \\ a_4 & 2a_5 - (a_4 + a_6) & a_6 \\ a_7 & 2a_8 - (a_7 + a_9) & a_9 \end{vmatrix}, C_2 \to C_2 - C_1 - C_3$$

$$= \frac{1}{2}\begin{vmatrix} a_1 & 0 & a_3 \\ a_4 & 0 & a_6 \\ a_7 & 0 & a_9 \end{vmatrix}, \text{ using (1)}$$

$\therefore \quad \Delta = 0.$

28. Solve by Cramer's rule:

$$x + y + z = 1$$
$$ax + by + cz = k$$
$$a^2x + b^2y + c^2z = k^2$$

By transferring constants on the left, we have by Cramer's rule,

$$\frac{x}{\begin{vmatrix} 1 & 1 & -1 \\ b & c & -k \\ b^2 & c^2 & -k^2 \end{vmatrix}} = \frac{-y}{\begin{vmatrix} 1 & 1 & -1 \\ a & c & -k \\ a^2 & c^2 & -k^2 \end{vmatrix}}$$

$$= \frac{z}{\begin{vmatrix} 1 & 1 & -1 \\ a & b & -k \\ a^2 & b^2 & -k^2 \end{vmatrix}} = \frac{-1}{\begin{vmatrix} 1 & 1 & 1 \\ a & b & c \\ a^2 & b^2 & c^2 \end{vmatrix}}.$$

Now $\begin{vmatrix} 1 & 1 & 1 \\ a & b & c \\ a^2 & b^2 & c^2 \end{vmatrix} = \begin{vmatrix} 0 & 0 & 1 \\ a-b & b-c & c \\ a^2-b^2 & b^2-c^2 & c^2 \end{vmatrix}$,

$C_1 \to C_1 - C_2,\ C_2 \to C_2 - C_3$

$$= \begin{vmatrix} a-b & b-c \\ (a+b)(a-b) & (b+c)(b-c) \end{vmatrix}$$

$$= (a-b)(b-c)\begin{vmatrix} 1 & 1 \\ a+b & b+c \end{vmatrix}$$

$$= (a-b)(b-c)\{b + c - (a+b)\} = (a-b)(b-c)(c-a).$$

Similarly, $\begin{vmatrix} 1 & 1 & -1 \\ b & c & -k \\ b^2 & c^2 & -k^2 \end{vmatrix} = -\begin{vmatrix} 1 & 1 & 1 \\ b & c & k \\ b^2 & c^2 & k^2 \end{vmatrix}$

$$= -(b-c)(c-k)(k-b)$$

$$\begin{vmatrix} 1 & 1 & -1 \\ a & c & -k \\ a^2 & c^2 & -k^2 \end{vmatrix} = -\begin{vmatrix} 1 & 1 & 1 \\ a & c & k \\ a^2 & c^2 & k^2 \end{vmatrix}$$

$$= -(a-c)(c-k)(k-a)$$

$$\begin{vmatrix} 1 & 1 & -1 \\ a & b & -k \\ a^2 & b^2 & -k^2 \end{vmatrix} = -\begin{vmatrix} 1 & 1 & 1 \\ a & b & k \\ a^2 & b^2 & k^2 \end{vmatrix}$$

$$= -(a-b)(b-k)(k-a)$$

$$\therefore \quad x = -\frac{-(b-c)(c-k)(k-b)}{(a-b)(b-c)(c-a)} = -\frac{(b-c)(b-k)(c-k)}{(a-b)(b-c)(c-a)}$$

$$= -\frac{(b-k)(c-k)}{(a-b)(c-a)}$$

$$y = \frac{-(a-c)(c-k)(k-a)}{(a-b)(b-c)(c-a)} = -\frac{(c-a)(c-k)(a-k)}{(a-b)(b-c)(c-a)}$$

$$= -\frac{(c-k)(a-k)}{(a-b)(b-c)}$$

$$z = -\frac{-(a-b)(b-k)(k-a)}{(a-b)(b-c)(c-a)} = -\frac{(a-b)(a-k)(b-k)}{(a-b)(b-c)(c-a)}$$

$$= -\frac{(a-k)(b-k)}{(b-c)(c-a)}.$$

29. If x, y, z satisfy the equations

$$x\cos\alpha - y\sin\alpha + z = 1 + \cos\beta$$
$$x\sin\alpha + y\cos\alpha + z = 1 - \sin\beta$$
$$x\cos(\alpha + \beta) - y\sin(\alpha + \beta) + z = 2$$

then find the value of $x^2 + y^2 + z^2$.

By Cramer's rule,

$$\frac{x}{\begin{vmatrix} -\sin\alpha & 1 & -(1+\cos\beta) \\ \cos\alpha & 1 & -(1-\sin\beta) \\ -\sin(\alpha+\beta) & 1 & -2 \end{vmatrix}} = \frac{-y}{\begin{vmatrix} \cos\alpha & 1 & -(1+\cos\beta) \\ \sin\alpha & 1 & -(1-\sin\beta) \\ \cos(\alpha+\beta) & 1 & -2 \end{vmatrix}}$$

$$= \frac{z}{\begin{vmatrix} \cos\alpha & -\sin\alpha & -(1+\cos\beta) \\ \sin\alpha & \cos\alpha & -(1-\sin\beta) \\ \cos(\alpha+\beta) & -\sin(\alpha+\beta) & -2 \end{vmatrix}} = \frac{-1}{\begin{vmatrix} \cos\alpha & -\sin\alpha & 1 \\ \sin\alpha & \cos\alpha & 1 \\ \cos(\alpha+\beta) & -\sin(\alpha+\beta) & 1 \end{vmatrix}} \quad \ldots (1)$$

Denoting the determinants in the denominators by $\Delta_x, \Delta_y, \Delta_z, \Delta$ respectively, we get:

$$\Delta = \begin{vmatrix} \cos\alpha & -\sin\alpha & 1 \\ \sin\alpha & \cos\alpha & 1 \\ \cos(\alpha+\beta) & -\sin(\alpha+\beta) & 1 \end{vmatrix} = \begin{vmatrix} \cos\alpha & -\sin\alpha & 1 \\ \sin\alpha & \cos\alpha & 1 \\ 0 & 0 & 1-\cos\beta+\sin\beta \end{vmatrix},$$

$$R_3 \to R_3 - \cos\beta \times R_1 + \sin\beta \times R_2$$

$$= 1 - \cos\beta + \sin\beta.$$

$$\Delta_x = \begin{vmatrix} -\sin\alpha & 1 & -(1+\cos\beta) \\ \cos\alpha & 1 & -(1-\sin\beta) \\ -\sin(\alpha+\beta) & 1 & -2 \end{vmatrix} = \begin{vmatrix} -\sin\alpha & 1 & -\cos\beta \\ \cos\alpha & 1 & \sin\beta \\ -\sin(\alpha+\beta) & 1 & -1 \end{vmatrix}, C_3 \to C_3 + C_2$$

$$= \begin{vmatrix} -\sin\alpha & 1 & -\cos\beta \\ \cos\alpha & 1 & \sin\beta \\ 0 & 1-\cos\beta+\sin\beta & 0 \end{vmatrix}$$

$$= (1-\cos\beta+\sin\beta)\{-\cos(\alpha+\beta)\}.$$

$$\Delta_y = \begin{vmatrix} \cos\alpha & 1 & -(1+\cos\beta) \\ \sin\alpha & 1 & -(1-\sin\beta) \\ \cos(\alpha+\beta) & 1 & -2 \end{vmatrix} = \begin{vmatrix} \cos\alpha & 1 & -\cos\beta \\ \sin\alpha & 1 & \sin\beta \\ \cos(\alpha+\beta) & 1 & -1 \end{vmatrix}, C_3 \to C_3 + C_2$$

$$= \begin{vmatrix} \cos\alpha & 1 & -\cos\beta \\ \sin\alpha & 1 & \sin\beta \\ 0 & 1-\cos\beta+\sin\beta & 0 \end{vmatrix}$$

$$= (1-\cos\beta+\sin\beta)\cdot\sin(\alpha+\beta).$$

$$\Delta_z = \begin{vmatrix} \cos\alpha & -\sin\alpha & -(1+\cos\beta) \\ \sin\alpha & \cos\alpha & -(1-\sin\beta) \\ \cos(\alpha+\beta) & -\sin(\alpha+\beta) & -2 \end{vmatrix}$$

$$= \begin{vmatrix} \cos\alpha & -\sin\alpha & -(1+\cos\beta) \\ \sin\alpha & \cos\alpha & -(1-\sin\beta) \\ 0 & 0 & -2+\cos\beta(1+\cos\beta)-\sin\beta(1-\sin\beta) \end{vmatrix},$$

$$R_3 \to R_3 - \cos\beta \times R_1 + \sin\beta \times R_2$$

$$= -2 + \cos\beta(1+\cos\beta) - \sin\beta(1-\sin\beta)$$

$$= -1 + \cos\beta - \sin\beta = -(1 - \cos\beta + \sin\beta).$$

From (1), $\dfrac{x}{\Delta_x} = \dfrac{-y}{\Delta_y} = \dfrac{z}{\Delta_z} = \dfrac{-1}{\Delta} = \dfrac{\sqrt{x^2+y^2+z^2}}{\sqrt{\Delta_x^2+\Delta_y^2+\Delta_z^2}}$,

by ratio and proportion.

$$\therefore \quad x^2+y^2+z^2 = \frac{\Delta_x^2+\Delta_y^2+\Delta_z^2}{\Delta^2}$$

$$= \cos^2(\alpha+\beta) + \sin^2(\alpha+\beta) + 1 = 2.$$

30. For what value of λ the equations $2x+3y=8$, $7x-5y+3=0$ and $4x-6y+\lambda=0$ are consistent? Also find the solution of the system of equations for that value of λ.

Here the equations are linear. We have 3 equations in 2 unknowns.

$$\therefore \quad \text{they are consistent if } \begin{vmatrix} 2 & 3 & -8 \\ 7 & -5 & 3 \\ 4 & -6 & \lambda \end{vmatrix} = 0$$

or $\quad 2(-5\lambda+18) - 3(7\lambda-12) - 8(-42+20) = 0$

or $\quad -10\lambda + 36 - 21\lambda + 36 + 176 = 0$

or $\quad -31\lambda + 248 = 0; \qquad \therefore \quad \lambda = 8$

$\therefore$ for $\lambda = 8$ the system has a solution which can be obtained by solving any two of the three equations. Solving $2x+3y-8=0$

$7x-5y+3=0$ by Cramer's rule,

$$\frac{x}{\begin{vmatrix} 3 & -8 \\ -5 & 3 \end{vmatrix}} = \frac{-y}{\begin{vmatrix} 2 & -8 \\ 7 & 3 \end{vmatrix}} = \frac{1}{\begin{vmatrix} 2 & 3 \\ 7 & -5 \end{vmatrix}}$$

or $\quad \dfrac{x}{9-40} = \dfrac{-y}{6+56} = \dfrac{1}{-10-21}$

or $\quad \dfrac{x}{-31} = \dfrac{y}{-62} = \dfrac{1}{-31}; \qquad \therefore \quad x = 1, y = 2.$

31. Using condition of consistency of equations, prove that

$$\begin{vmatrix} ap & a & p \\ bq & b & q \\ cr & c & r \end{vmatrix} = 0 \text{ if } bc + qr = ca + rp = ab + pq = -1.$$

We have $bc + qr = -1$; multiplying by ap,

$$abcp + apqr + ap = 0 \quad \ldots (1)$$

$ca + rp = -1$; multiplying by bq,

$$abcq + bpqr + bq = 0 \quad \ldots (2)$$

$ab + pq = -1$; multiplying by cr,

$$abcr + cpqr + cr = 0 \quad \ldots (3)$$

(1), (2) and (3) may be considered as three equations in (abc) and (pqr).

Hence using consistency of equations, i.e., eliminating abc and pqr, we get

$$\begin{vmatrix} p & a & ap \\ q & b & bq \\ r & c & cr \end{vmatrix} = 0; \quad \therefore \quad \begin{vmatrix} ap & a & p \\ bq & b & q \\ cr & c & r \end{vmatrix} = 0.$$

32. Eliminate x, y from the following:

$$ax + a^2y + a^3 + 1 = 0$$
$$bx + b^2y + b^3 + 1 = 0$$
$$cx + c^2y + c^3 + 1 = 0$$

and prove that if no two of a, b, c are equal then $1 + abc = 0$.

Here we have three equations in two variables x, y. The condition for their consistency (eliminant of the equations) is

$$\begin{vmatrix} a & a^2 & a^3+1 \\ b & b^2 & b^3+1 \\ c & c^2 & c^3+1 \end{vmatrix} = 0$$

or $$\begin{vmatrix} a & a^2 & a^3 \\ b & b^2 & b^3 \\ c & c^2 & c^3 \end{vmatrix} + \begin{vmatrix} a & a^2 & 1 \\ b & b^2 & 1 \\ c & c^2 & 1 \end{vmatrix} = 0$$

or $$abc\begin{vmatrix} 1 & a & a^2 \\ 1 & b & b^2 \\ 1 & c & c^2 \end{vmatrix} + (-1)^2\begin{vmatrix} 1 & a & a^2 \\ 1 & b & b^2 \\ 1 & c & c^2 \end{vmatrix} = 0$$

or $$(abc+1)\begin{vmatrix} 1 & a & a^2 \\ 1 & b & b^2 \\ 1 & c & c^2 \end{vmatrix} = 0. \quad \ldots (1)$$

We know from circular determinants,

$$\begin{vmatrix} 1 & a & a^2 \\ 1 & b & b^2 \\ 1 & c & c^2 \end{vmatrix} = (a-b)(b-c)(c-a)$$

$\therefore$ from (1),

$$(1 + abc)(a-b)(b-c)(c-a) = 0;$$

$\because$ no two of a, b, c are equal,

$$a - b \neq 0, b - c \neq 0, c - a \neq 0. \quad \therefore \quad 1 + abc = 0.$$

33. If the system of following equations in x, y

$$ax + hy + g = 0$$
$$hx + by + f = 0$$
$$ax^2 + 2hxy + by^2 + 2gx + 2fy + c = t$$

admits of a solution then prove that

$$t = \begin{vmatrix} a & h & g \\ h & b & f \\ g & f & c \end{vmatrix} \div \begin{vmatrix} a & h \\ h & b \end{vmatrix}.$$

Clearly, $ax^2 + 2hxy + by^2 + 2gx + 2fy + c = t$

$\Rightarrow \quad x(ax + hy + g) + y(hx + by + f) + gx + fy + c = t$

The values of x, y that satisfy

$$ax + hy + g = 0 \quad \ldots (1)$$
$$hx + by + f = 0 \quad \ldots (2)$$

will satisfy the above equation if

$$gx + fy + c - t = 0 \quad \ldots (3)$$

The linear equations (1), (2) and (3) can admit of a solution if they are consistent. For this the condition is

$$\begin{vmatrix} a & h & g \\ h & b & f \\ g & f & c-t \end{vmatrix} = 0$$

or $$\begin{vmatrix} a & h & g \\ h & b & f \\ g & f & c \end{vmatrix} + \begin{vmatrix} a & h & 0 \\ h & b & 0 \\ g & f & -t \end{vmatrix} = 0$$

or $$\begin{vmatrix} a & h & g \\ h & b & f \\ g & f & c \end{vmatrix} - t\begin{vmatrix} a & h \\ h & b \end{vmatrix} = 0$$

$$\therefore \quad t = \begin{vmatrix} a & h & g \\ h & b & f \\ g & f & c \end{vmatrix} \div \begin{vmatrix} a & h \\ h & b \end{vmatrix}.$$

34. For what values of p and q the system of equations

$$2x + py + 6z = 8$$
$$x + 2y + qz = 5$$
$$x + y + 3z = 4$$

has (i) unique solution (ii) no solution (iii) infinite number of solutions?

Here the system of linear equations in x, y, z are

$$2x + py + 6z - 8 = 0$$
$$x + 2y + qz - 5 = 0$$
$$x + y + 3z - 4 = 0$$

$$\therefore \quad \Delta = \begin{vmatrix} 2 & p & 6 \\ 1 & 2 & q \\ 1 & 1 & 3 \end{vmatrix} = \begin{vmatrix} 2 & p-2 & 0 \\ 1 & 1 & q-3 \\ 1 & 0 & 0 \end{vmatrix},$$

$$C_2 \to C_2 - C_1, C_3 \to C_3 - 3 \times C_1$$

$$= \begin{vmatrix} p-2 & 0 \\ 1 & q-3 \end{vmatrix} = (p-2)(q-3)$$

$\therefore$ if $p \neq 2, q \neq 3$ then $\Delta \neq 0$

and so the system will have unique solution, i.e., the system will be independent/solvable/consistent.

If $p = 2$ or $q = 3$ then $\Delta = 0$

and so the system cannot have unique solution.

When $p = 2$,

$$\Delta_x = \begin{vmatrix} p & 6 & -8 \\ 2 & q & -5 \\ 1 & 3 & -4 \end{vmatrix} = \begin{vmatrix} 2 & 6 & -8 \\ 2 & q & -5 \\ 1 & 3 & -4 \end{vmatrix}$$

$$= 2\begin{vmatrix} 1 & 3 & -4 \\ 2 & q & -5 \\ 1 & 3 & -4 \end{vmatrix} = 0 \ (\because R_1 \equiv R_3)$$

$$\Delta_y = \begin{vmatrix} 2 & 6 & -8 \\ 1 & q & -5 \\ 1 & 3 & -4 \end{vmatrix} = 2\begin{vmatrix} 1 & 3 & -4 \\ 1 & q & -5 \\ 1 & 3 & -4 \end{vmatrix} = 0 \ (\because R_1 \equiv R_3)$$

$$\Delta_z = \begin{vmatrix} 2 & p & -8 \\ 1 & 2 & -5 \\ 1 & 1 & -4 \end{vmatrix} = \begin{vmatrix} 2 & 2 & -8 \\ 1 & 2 & -5 \\ 1 & 1 & -4 \end{vmatrix}$$

$$= 2\begin{vmatrix} 1 & 1 & -4 \\ 1 & 2 & -5 \\ 1 & 1 & -4 \end{vmatrix} = 0 \ (\because R_1 \equiv R_3)$$

$\therefore$ when $p = 2, \Delta = 0 = \Delta_x = \Delta_y = \Delta_z$.

$\therefore$ the system of equations will have infinite number of solutions (the system of equations will be dependent) for $p = 2$ and any real value of q.

When $q = 3$, $\Delta_x = \begin{vmatrix} p & 6 & -8 \\ 2 & q & -5 \\ 1 & 3 & -4 \end{vmatrix} = \begin{vmatrix} p & 6 & -8 \\ 2 & 3 & -5 \\ 1 & 3 & -4 \end{vmatrix}$

$$= \begin{vmatrix} p-2 & 0 & 0 \\ 2 & 3 & -5 \\ 1 & 3 & -4 \end{vmatrix}, R_1 \to R_1 - 2R_3$$

$$= (p-2)3$$

$\therefore$ if $p \neq 2$, $\Delta_x \neq 0$ and so the system of equations will have no solution, i.e., the system is unsolvable/ inconsistent when $q = 3$ but $p \neq 2$.

Thus we find that the system of equations will have

(i) unique solution if $p \neq 2$ and $q \neq 3$

(ii) no solution if $p \neq 2$ and $q = 3$

(iii) infinite number of solutions if $p = 2$.

35. Prove that the system of equations

$3x - y + 4z = 3, x + 2y - 3z + 2 = 0$ and $6x + 5y + \lambda z = -3$ has at least one solution for all $\lambda \in R$. Find the set of solutions when $\lambda = -5$.

Here the system of linear equations are

$$3x - y + 4z - 3 = 0 \quad \ldots (1)$$

$$x + 2y - 3z + 2 = 0 \quad \ldots (2)$$

$$6x + 5y + \lambda z + 3 = 0 \quad \ldots (3)$$

$$\therefore \quad \Delta = \begin{vmatrix} 3 & -1 & 4 \\ 1 & 2 & -3 \\ 6 & 5 & \lambda \end{vmatrix}$$

$$= 3(2\lambda + 15) + 1(\lambda + 18) + 4(5 - 12)$$

$$= 7\lambda + 35 = 7(\lambda + 5).$$

$\therefore$ if $\lambda \neq -5, \Delta \neq 0$ and then the system of equations will have unique solution.

If $\lambda = -5$, $\Delta = 0$ and so the system cannot have unique solution.

When $\lambda = -5$, the third equation becomes

$$6x + 5y - 5z + 3 = 0 \quad \ldots (4)$$

Obviously by adding 3 times of the second equation to the first we get,

$6x + 5y - 5z + 3 = 0$ which is the same as (4).

Hence, for $\lambda = -5$ the equation (3) is redundant. So we are left with only two equations in three unknowns x, y, z which indicates that the given system of equations are dependent and so the system will have infinite number of solutions.

Thus, for every real value of λ, the system has either unique solution or infinite number of solutions. So the system will have at least one solution for every value of λ.

When $\lambda = -5$, as above we are left with the equations

$$3x - y + 4z - 3 = 0$$
$$x + 2y - 3z + 2 = 0$$

Let $z = t$ satisfy both the equations. Then

$$3x - y = 3 - 4t \quad \ldots (5)$$

$$x + 2y = 3t - 2 \quad \ldots (6)$$

$2 \times (5) + (6) \quad \Rightarrow \quad 7x = 2(3 - 4t) + (3t - 2) = -5t + 4$

$$\therefore \quad x = \frac{1}{7}(4 - 5t)$$

$(5) - 3 \times (6) \quad \Rightarrow \quad -7y = 3 - 4t - 3(3t - 2) = -13t + 9$

$$\therefore \quad y = \frac{1}{7}(13t - 9)$$

$\therefore$ the required solution set when $\lambda = -5$ is

$$\left\{(x, y, z) \;\middle|\; x = \frac{4-5t}{7},\ y = \frac{13t-9}{7},\ z = t,\ t \in R\right\}.$$

36. For what value of λ the system of equations

$x - 2y + z = 0, kx - y + 2z = 0$ and $2x - y + z = 0$

has a nontrivial solution? Also find $x : y : z$ for that value of λ.

The system of linear homogeneous equations

$$x - 2y + z = 0$$
$$kx - y + 2z = 0$$
$$2x - y + z = 0$$

will have nontrivial solutions if

$$\Delta = \begin{vmatrix} 1 & -2 & 1 \\ k & -1 & 2 \\ 2 & -1 & 1 \end{vmatrix} = 0$$

or $1(-1+2) + 2(k-4) + 1(-k+2) = 0$

or $1 + 2k - 8 - k + 2 = 0; \quad \therefore \quad k = 5$

When $k = 5$, solving any two of the three equations, we get the nontrivial solutions. Taking

$$x - 2y + z = 0$$
$$5x - y + 2z = 0 \quad (\text{using } k = 5),$$

we get by cross-multiplication

$$\frac{x}{-4+1} = \frac{y}{5-2} = \frac{z}{-1+10}; \quad \therefore \quad \frac{x}{-3} = \frac{y}{3} = \frac{z}{9}$$

$$\therefore \quad \frac{x}{-1} = \frac{y}{1} = \frac{z}{3}; \quad \therefore \quad x : y : z = -1 : 1 : 3.$$

Note The set of nontrivial solutions

$$= \left\{ (x, y, z) \;\middle|\; x = -t, y = t, z = 3t \text{ and } t \in R \right\}.$$

37. Let λ, α be real. Find the set of all values of λ for which the system of equations

$$\lambda x + \sin\alpha \cdot y + \cos\alpha \cdot z = 0$$
$$x + \cos\alpha \cdot y + \sin\alpha \cdot z = 0$$
$$-x + \sin\alpha \cdot y - \cos\alpha \cdot z = 0$$

has nontrivial solution. If $\lambda = 1$, find all values of α.

The given system of linear homogeneous equations in x, y, z will have nontrivial solutions if

$$\Delta = \begin{vmatrix} \lambda & \sin\alpha & \cos\alpha \\ 1 & \cos\alpha & \sin\alpha \\ -1 & \sin\alpha & -\cos\alpha \end{vmatrix} = 0$$

or $\lambda(-\cos^2\alpha - \sin^2\alpha)$

$$-1(-\sin\alpha \cdot \cos\alpha - \sin\alpha \cdot \cos\alpha)$$
$$-1(\sin^2\alpha - \cos^2\alpha) = 0$$

or $-\lambda + \sin 2\alpha + \cos 2\alpha = 0$

or $\lambda = \cos 2\alpha + \sin 2\alpha$

Now, $\cos 2\alpha + \sin 2\alpha = \sqrt{2}\left(\frac{1}{\sqrt{2}}\cos 2\alpha + \frac{1}{\sqrt{2}}\sin 2\alpha\right)$

$$= \sqrt{2}\cos(2\alpha - \pi/4) \quad \dots (1)$$

As the maximum and minimum values of $\cos\left(2\alpha - \frac{\pi}{4}\right)$ are 1 and -1 respectively

$$-\sqrt{2} \le \cos 2\alpha + \sin 2\alpha \le \sqrt{2}; \text{ or } -\sqrt{2} \le \lambda \le \sqrt{2}.$$

When $\lambda = 1$, $\cos 2\alpha + \sin 2\alpha = 1$

$\therefore$ from (1), $\sqrt{2}\cos\left(2\alpha - \frac{\pi}{4}\right) = 1$

or $\cos\left(2\alpha - \frac{\pi}{4}\right) = \frac{1}{\sqrt{2}} = \cos\frac{\pi}{4}$

$\therefore \quad 2\alpha - \frac{\pi}{4} = 2n\pi \pm \frac{\pi}{4}$, i.e., $2\alpha = 2n\pi + \frac{\pi}{2}, 2n\pi$

$\therefore \quad \alpha = n\pi + \frac{\pi}{4}, n\pi$ where $n \in Z$, the set of integers.

Exercises

1. If $a \ne b \ne c$, prove that $\begin{vmatrix} 1 & 1 & 1 \\ a & b & c \\ a^3 & b^3 & c^3 \end{vmatrix}$ is divisible by $a + b + c$.

2. Prove that the value of $\begin{vmatrix} 1 & bc & a(b+c) \\ 1 & ca & b(c+a) \\ 1 & ab & c(a+b) \end{vmatrix}$ is independent of a, b and c.

3. Prove that $\begin{vmatrix} 1 & bc & bc(b+c) \\ 1 & ca & ca(c+a) \\ 1 & ab & ab(a+b) \end{vmatrix} = 0$.

4. Factorize $\begin{vmatrix} 1 & b+c & b^2+c^2 \\ 1 & c+a & c^2+a^2 \\ 1 & a+b & a^2+b^2 \end{vmatrix}$.

5. Prove that
$$\begin{vmatrix} b^2c^2 + a^2d^2 & bc + ad & 1 \\ c^2a^2 + b^2d^2 & ca + bd & 1 \\ a^2b^2 + c^2d^2 & ab + cd & 1 \end{vmatrix} = (a-b)(b-c)(c-a)(a-d)(b-d)(c-d).$$

6. If a, b, c are positive numbers which are the pth, qth and rth terms respectively of a GP then prove that
$$\begin{vmatrix} \log a & p & 1 \\ \log b & q & 1 \\ \log c & r & 1 \end{vmatrix} = 0.$$

7. Prove that $\begin{vmatrix} a^3 & 3a^2 & 3a & 1 \\ a^2 & a^2 + 2a & 2a + 1 & 1 \\ a & 2a + 1 & a + 2 & 1 \\ 1 & 3 & 3 & 1 \end{vmatrix} = (a-1)^6$.

8. Prove that $\begin{vmatrix} w^{3n} & w^{n} & w^{2n} \\ w^{n} & w^{2n} & w^{3n} \\ w^{2n} & w^{3n} & w^{n} \end{vmatrix} = 0$

where $w = \sqrt[3]{1}$ and n is an integer.

9. Prove that

$$\begin{vmatrix} a & b & c \\ a(a^2+1) & b(b^2+1) & c(c^2+1) \\ a+1 & b+1 & c+1 \end{vmatrix}$$

$$= (a-b)(b-c)(c-a)(a+b+c).$$

10. Prove that

$$\begin{vmatrix} a & b+c & a^2 \\ b & c+a & b^2 \\ c & a+b & c^2 \end{vmatrix} = -(a-b)(b-c)(c-a)(a+b+c).$$

11. Prove that $\begin{vmatrix} x+a & b & c \\ a & x+b & c \\ a & b & x+c \end{vmatrix} = x^2(x+a+b+c).$

12. Prove that $\begin{vmatrix} a & a+b & a+2b \\ a+2b & a & a+b \\ a+b & a+2b & a \end{vmatrix} = 9b^2(a+b).$

13. Factorize $\begin{vmatrix} a-b-c & 2a & 2a \\ 2b & b-c-a & 2b \\ 2c & 2c & c-a-b \end{vmatrix}.$

14. Solve for x: $\begin{vmatrix} 15-2x & 11 & 10 \\ 11-3x & 17 & 16 \\ 7-x & 14 & 13 \end{vmatrix} = 0.$

15. If a, b and c are rational, prove that

$$\frac{1}{2}\begin{vmatrix} a+b+2c & a & b \\ c & b+c+2a & b \\ c & a & c+a+2b \end{vmatrix}$$

is a perfect cube.

16. If $a+b+c = s$ then prove that

$$\begin{vmatrix} s+c & a & b \\ c & s+a & b \\ c & a & s+b \end{vmatrix} = 2s^3.$$

17. If $a+b+c > 0$ then prove that $\begin{vmatrix} a+b & b & c \\ b+c & c & a \\ c+a & a & b \end{vmatrix} \le 0$

for all real a, b and c.

18. Prove that

$$\begin{vmatrix} a & c & b \\ b & a & c \\ c & b & a \end{vmatrix} = (a+b+c)(a+bw+cw^2)(a+bw^2+cw)$$

where w is a nonreal cube root of unity.

19. Evaluate $\begin{vmatrix} \tan\pi/3 & \sin\pi/3 & \cos\pi/3 \\ \tan\pi/6 & \sin\pi/6 & \cos\pi/6 \\ \tan\pi/4 & \sin\pi/4 & \cos\pi/4 \end{vmatrix}.$

20. If $a+b+c=0$ and $\begin{vmatrix} a-x & c & b \\ c & b-x & a \\ b & a & c-x \end{vmatrix} = 0,$

prove that either $x = 0$ or $x = \pm\sqrt{\frac{3}{2}(a^2+b^2+c^2)}$.

21. Factorize $\begin{vmatrix} x & y & z \\ x^2 & y^2 & z^2 \\ yz & zx & xy \end{vmatrix}.$

22. Prove that

$$\begin{vmatrix} 1+a^2-b^2 & 2ab & -2b \\ 2ab & 1-a^2+b^2 & 2a \\ 2b & -2a & 1-a^2-b^2 \end{vmatrix}$$

is a perfect cube for all rationals a and b.

23. Factorize $\begin{vmatrix} x^2 & (y-z)^2 & yz \\ y^2 & (z-x)^2 & zx \\ z^2 & (x-y)^2 & xy \end{vmatrix}.$

24. Prove that $\begin{vmatrix} bc & bc'+b'c & b'c' \\ ca & ca'+c'a & c'a' \\ ab & ab'+a'b & a'b' \end{vmatrix}$

$$= (ab'-a'b)(bc'-b'c)(ca'-c'a).$$

25. Prove that

$$\begin{vmatrix} 1+a & 1 & 1 \\ 1 & 1+b & 1 \\ 1 & 1 & 1+c \end{vmatrix} = abc\left(1+\frac{1}{a}+\frac{1}{b}+\frac{1}{c}\right).$$

26. Prove that

$$\begin{vmatrix} \dfrac{1}{(a-a_1)^2} & \dfrac{1}{a-a_1} & \dfrac{1}{a_1} \\ \dfrac{1}{(a-a_2)^2} & \dfrac{1}{a-a_2} & \dfrac{1}{a_2} \\ \dfrac{1}{(a-a_3)^2} & \dfrac{1}{a-a_3} & \dfrac{1}{a_3} \end{vmatrix}$$

$$= \frac{-a^2(a_1-a_2)(a_2-a_3)(a_3-a_1)}{a_1a_2a_3(a-a_1)^2(a-a_2)^2(a-a_3)^2}.$$

27. Prove that

$$\begin{vmatrix} bc-a^2 & ca-b^2 & ab-c^2 \\ -bc+ca+ab & bc-ca+ab & bc+ca-ab \\ (a+b)(a+c) & (b+c)(b+a) & (c+a)(c+b) \end{vmatrix}$$

$$= 3(b-c)(c-a)(a-b)(a+b+c)(bc+ca+ab).$$

28. Prove that

$$\begin{vmatrix} (b+c)^2 & a^2 & a^2 \\ b^2 & (c+a)^2 & b^2 \\ c^2 & c^2 & (a+b)^2 \end{vmatrix} = 2abc(a+b+c)^3.$$

29. Evaluate $\begin{vmatrix} \dfrac{1}{a+1} & \dfrac{1}{a+2} & \dfrac{1}{a+3} \\ \dfrac{1}{a+2} & \dfrac{1}{a+3} & \dfrac{1}{a+4} \\ \dfrac{1}{a+3} & \dfrac{1}{a+4} & \dfrac{1}{a+5} \end{vmatrix}$

where $a \notin \{-1, -2, -3, -4, -5\}$.

30. Solve for $x \in R$:

$$\begin{vmatrix} (x+a)(x-a) & (x+b)(x-b) & (x+c)(x-c) \\ (x-a)^3 & (x-b)^3 & (x-c)^3 \\ (x+a)^3 & (x+b)^3 & (x+c)^3 \end{vmatrix} = 0,$$

a, b and c being distinct real numbers.

31. Prove that

$$\begin{vmatrix} x^2 & (x-1)^2 & (x-2)^2 \\ (x-1)^2 & (x-2)^2 & (x-3)^2 \\ (x-2)^2 & (x-3)^2 & (x-4)^2 \end{vmatrix}$$ is independent of x.

32. Prove that $\begin{vmatrix} x+1 & 3x+2 & 5x+4 \\ x & 3x+1 & 5x+3 \\ x^2 & (3x+1)^2 & (5x+3)^2 \end{vmatrix} \geq 0$

for all $x \geq -\frac{1}{2}$.

33. Prove that the determinant

$$\begin{vmatrix} a & b & ax+by \\ b & c & bx+cy \\ ax+by & bx+cy & 0 \end{vmatrix} = 0$$

if a, b, c be in GP or $ax^2 + 2bxy + cy^2 = 0$.

34. Prove that for all real x, y, z the determinant

$$\begin{vmatrix} 1 & z & -y \\ -z & 1 & x \\ y & -x & 1 \end{vmatrix} \geq 1.$$

35. If in a triangle, s denotes the semiperimeter and a, b, c denote the lengths of sides then prove that

$$\begin{vmatrix} a^2 & (s-a)^2 & (s-a)^2 \\ (s-b)^2 & b^2 & (s-b)^2 \\ (s-c)^2 & (s-c)^2 & c^2 \end{vmatrix} = 2s^3(s-a)(s-b)(s-c).$$

36. Prove that $\begin{vmatrix} b+c & a & a \\ b & c+a & b \\ c & c & a+b \end{vmatrix} = 4abc.$

37. Prove that

$$\begin{vmatrix} b^2+c^2 & ab & ac \\ ab & c^2+a^2 & bc \\ ca & cb & a^2+b^2 \end{vmatrix}$$

is always positive for all real a, b and c.

38. Evaluate $\begin{vmatrix} a^2+1 & ab & ac \\ ab & b^2+1 & bc \\ ac & bc & c^2+1 \end{vmatrix}$.

39. Prove that

$$\begin{vmatrix} (a+b)^2 & ca & bc \\ ca & (b+c)^2 & ab \\ bc & ab & (c+a)^2 \end{vmatrix} = 2abc(a+b+c)^3.$$

40. Prove that

$$\begin{vmatrix} -bc & b^2+bc & c^2+bc \\ a^2+ac & -ac & c^2+ac \\ a^2+ab & b^2+ab & -ab \end{vmatrix} \geq 27a^2b^2c^2$$

where a, b and c are of the same sign.

41. Prove that

$$\begin{vmatrix} (a-x)^2 & (a-y)^2 & (a-z)^2 \\ (b-x)^2 & (b-y)^2 & (b-z)^2 \\ (c-x)^2 & (c-y)^2 & (c-z)^2 \end{vmatrix}$$
$$= 2(b-c)(c-a)(a-b)(y-z)(z-x)(x-y).$$

42. Prove that

$$\begin{vmatrix} (b+x)(c+x) & (c+x)(a+x) & (a+x)(b+x) \\ (b+y)(c+y) & (c+y)(a+y) & (a+y)(b+y) \\ (b+z)(c+z) & (c+z)(a+z) & (a+z)(b+z) \end{vmatrix}$$
$$= (b-c)(c-a)(a-b)(y-z)(z-x)(x-y).$$

43. Prove that $\begin{vmatrix} 1 & a & a^2-bc \\ 1 & b & b^2-ca \\ 1 & c & c^2-ab \end{vmatrix} = 0.$

44. Prove that

$$\begin{vmatrix} zx+xy & yz & yz \\ zx & xy+yz & zx \\ xy & xy & yz+zx \end{vmatrix} \geq 0 \text{ for } x, y, z \in R.$$

45. Prove that

$$\begin{vmatrix} x-3 & 2x+1 & 2 \\ 3x+2 & x+2 & 1 \\ 5x+1 & 5x+4 & 5 \end{vmatrix}$$ is independent of x.

46. If $D_r = \begin{vmatrix} 1 & n & n \\ 2r & 2n^2 & n(n+1) \\ 6r^2 & 6n^3 & n(n+1)(2n+1) \end{vmatrix}$,

prove that $\sum_{r=1}^{n} D_r = 0$.

47. If x, y, z are unequal and $\begin{vmatrix} x & x^2 & 1+x^3 \\ y & y^2 & 1+y^3 \\ z & z^2 & 1+z^3 \end{vmatrix} = 0$,

prove that $1 + xyz = 0$.

48. Let $\Delta_a = \begin{vmatrix} a-1 & n & 6 \\ (a-1)^2 & 2n^2 & 4n-2 \\ (a-1)^3 & 3n^3 & 3n^2-3n \end{vmatrix}$.

Show that $\sum_{a=1}^{n} \Delta_a =$ constant.

49. If $D_r = \begin{vmatrix} 2^{r-1} & 2 \cdot 3^{r-1} & 3 \cdot 4^{r-1} \\ 1 & 2 & 3 \\ 2^n & 3^n+1 & 4^n+2 \end{vmatrix}$

then find $\sum_{r=1}^{n} D_r$.

50. Evaluate $\sum_{n=1}^{N} u_n$ if $u_n = \begin{vmatrix} n & 1 & 5 \\ n^2 & 2N+1 & 2N+1 \\ n^3 & 3N^2 & 3N \end{vmatrix}$.

51. If $a \neq b \neq c$ and $\begin{vmatrix} a & a^3 & a^4-1 \\ b & b^3 & b^4-1 \\ c & c^3 & c^4-1 \end{vmatrix} = 0$

then prove that $abc(ab+bc+ca) = a+b+c$.

52. Prove that $\begin{vmatrix} a & b-c & c+b \\ a+c & b & c-a \\ a-b & b+a & c \end{vmatrix}$

and $a+b+c$ have the same sign for all real values of a, b and c.

53. Evaluate $\begin{vmatrix} {}^xC_1 & {}^xC_2 & {}^xC_3 \\ {}^yC_1 & {}^yC_2 & {}^yC_3 \\ {}^zC_1 & {}^zC_2 & {}^zC_3 \end{vmatrix}$.

54. Show that

$$\begin{vmatrix} {}^xC_r & {}^xC_{r+1} & {}^xC_{r+2} \\ {}^yC_r & {}^yC_{r+1} & {}^yC_{r+2} \\ {}^zC_r & {}^zC_{r+1} & {}^zC_{r+2} \end{vmatrix} = \begin{vmatrix} {}^xC_r & {}^{x+1}C_{r+1} & {}^{x+2}C_{r+2} \\ {}^yC_r & {}^{y+1}C_{r+1} & {}^{y+2}C_{r+2} \\ {}^zC_r & {}^{z+1}C_{r+1} & {}^{z+2}C_{r+2} \end{vmatrix}.$$

55. If $e^{i\theta} = \cos\theta + i\sin\theta$, prove that

$$\begin{vmatrix} 1 & e^{i\pi/3} & e^{i\pi/4} \\ e^{-i\pi/3} & 1 & e^{i2\pi/3} \\ e^{-i\pi/4} & e^{-i2\pi/3} & 1 \end{vmatrix} = -(2+\sqrt{2}).$$

56. If $a_r = (\cos 2r\pi + i\sin 2r\pi)^{\frac{1}{9}}$ then prove that

$$\begin{vmatrix} a_1 & a_2 & a_3 \\ a_4 & a_5 & a_6 \\ a_7 & a_8 & a_9 \end{vmatrix} = 0.$$

57. Without breaking determinants at any stage,

prove that $\begin{vmatrix} bc & a & a^2 \\ ca & b & b^2 \\ ab & c & c^2 \end{vmatrix} = \begin{vmatrix} 1 & a^2 & a^3 \\ 1 & b^2 & b^3 \\ 1 & c^2 & c^3 \end{vmatrix}$.

58. Without breaking determinants at any stage,

prove that $\begin{vmatrix} x & y & z \\ x^2 & y^2 & z^2 \\ yz & zx & xy \end{vmatrix} = \begin{vmatrix} 1 & 1 & 1 \\ x^2 & y^2 & z^2 \\ x^3 & y^3 & z^3 \end{vmatrix}$.

59. Prove that $\begin{vmatrix} 1 & \cos\alpha - \sin\alpha & \cos\alpha + \sin\alpha \\ 1 & \cos\beta - \sin\beta & \cos\beta + \sin\beta \\ 1 & \cos\gamma - \sin\gamma & \cos\gamma + \sin\gamma \end{vmatrix}$

$$= 2\begin{vmatrix} 1 & 1 & 1 \\ \cos\alpha & \cos\beta & \cos\gamma \\ \sin\alpha & \sin\beta & \sin\gamma \end{vmatrix}.$$

60. Prove without expanding that

$$\begin{vmatrix} b+c & c+a & a+b \\ q+r & r+p & p+q \\ y+z & z+x & x+y \end{vmatrix} = 2\begin{vmatrix} a & b & c \\ p & q & r \\ x & y & z \end{vmatrix}.$$

61. Without expanding the determinants, prove that

$$\begin{vmatrix} 103 & 115 & 114 \\ 111 & 108 & 106 \\ 104 & 113 & 116 \end{vmatrix} + \begin{vmatrix} 113 & 116 & 104 \\ 108 & 106 & 111 \\ 115 & 114 & 103 \end{vmatrix} = 0.$$

62. Without evaluating the determinants, prove that

$$\begin{vmatrix} x_2-x_1 & y_2-y_1 \\ x_3-x_1 & y_3-y_1 \end{vmatrix} = \begin{vmatrix} x_1 & y_1 & 1 \\ x_2 & y_2 & 1 \\ x_3 & y_3 & 1 \end{vmatrix}.$$

63. Without breaking the determinant at any stage, prove that

$$\begin{vmatrix} 0 & b & -c \\ -b & 0 & a \\ c & -a & 0 \end{vmatrix} = 0.$$

64. Without breaking the determinant at any stage, prove that

$$\begin{vmatrix} -5 & 3+5i & \frac{3}{2}-4i \\ 3-5i & 8 & 4+5i \\ \frac{3}{2}+4i & 4-5i & 9 \end{vmatrix}$$

has a purely real value.

65. If A, B, C, etc., be the corresponding cofactors of a, b, c, etc., in the determinant $\begin{vmatrix} a & b & c \\ d & e & f \\ g & h & i \end{vmatrix}$

then prove that $\begin{vmatrix} A & B & C \\ D & E & F \\ G & H & I \end{vmatrix} = \begin{vmatrix} a & b & c \\ d & e & f \\ g & h & i \end{vmatrix}^2$.

66. Prove that

$$\begin{vmatrix} 2bc-a^2 & c^2 & b^2 \\ c^2 & 2ca-b^2 & a^2 \\ b^2 & a^2 & 2ab-c^2 \end{vmatrix} = \begin{vmatrix} a & b & c \\ b & c & a \\ c & a & b \end{vmatrix}^2.$$

67. Using the determinant $\begin{vmatrix} x & y & z \\ y & z & x \\ z & x & y \end{vmatrix}$, prove that

$$\begin{vmatrix} yz-x^2 & zx-y^2 & xy-z^2 \\ zx-y^2 & xy-z^2 & yz-x^2 \\ xy-z^2 & yz-x^2 & zx-y^2 \end{vmatrix} = \begin{vmatrix} r^2 & u^2 & u^2 \\ u^2 & r^2 & u^2 \\ u^2 & u^2 & r^2 \end{vmatrix}$$

where $r^2 = x^2+y^2+z^2$ and $u^2 = yz+zx+xy$.

68. Prove that $\begin{vmatrix} ax+a'x' & ay+a'y' & az+a'z' \\ bx+b'x' & by+b'y' & bz+b'z' \\ cx+c'x' & cy+c'y' & cz+c'z' \end{vmatrix} = 0$.

69. Prove that $\begin{vmatrix} \lambda & c & -b \\ -c & \lambda & a \\ b & -a & \lambda \end{vmatrix} = 0$

is a cubic equation in λ whose one root is real and two others are nonreal complex.

Hence or otherwise find the roots of the equation

$$\begin{vmatrix} a^2+\lambda^2 & ab+c\lambda & ca-b\lambda \\ ab-c\lambda & b^2+\lambda^2 & bc+a\lambda \\ ca+b\lambda & bc-a\lambda & c^2+\lambda^2 \end{vmatrix}=0.$$

70. If $\begin{vmatrix} 1 & x & x^2 \\ x & x^2 & 1 \\ x^2 & 1 & x \end{vmatrix}=3$, find the value of

$$\begin{vmatrix} x^3-1 & 0 & x-x^4 \\ 0 & x-x^4 & x^3-1 \\ x-x^4 & x^3-1 & 0 \end{vmatrix}.$$

71. Factorize

$$\begin{vmatrix} 3 & a+b+c & a^3+b^3+c^3 \\ a+b+c & a^2+b^2+c^2 & a^4+b^4+c^4 \\ a^2+b^2+c^2 & a^3+b^3+c^3 & a^5+b^5+c^5 \end{vmatrix}.$$

72. If $l_r^2+m_r^2+n_r^2=1;\ r=1,2,3$ and

$l_rl_s+m_rm_s+n_rn_s=0;\ r\neq s,\ r=1,2,3;\ s=1,2,3$

then find the value of

$$\begin{vmatrix} l_1 & m_1 & n_1 \\ l_2 & m_2 & n_2 \\ l_3 & m_3 & n_3 \end{vmatrix}.$$

73. Find all the values of θ such that

$$\begin{vmatrix} \sin 3\theta & -1 & 1 \\ \cos 2\theta & 4 & 3 \\ 2 & 7 & 7 \end{vmatrix}=0.$$

74. Prove that $\begin{vmatrix} \sin^3\alpha & \sin^3\beta & \sin^3\gamma \\ \sin\alpha & \sin\beta & \sin\gamma \\ \cos\alpha & \cos\beta & \cos\gamma \end{vmatrix}$

$=-\sin(\beta-\gamma)\sin(\gamma-\alpha)\sin(\alpha-\beta)\sin(\alpha+\beta+\gamma).$

75. Prove that in a ΔABC,

$$\begin{vmatrix} \sin^2 A & \sin A\cdot\cos A & \cos^2 A \\ \sin^2 B & \sin B\cdot\cos B & \cos^2 B \\ \sin^2 C & \sin C\cdot\cos C & \cos^2 C \end{vmatrix}$$

$=-\sin(B-C)\cdot\sin(C-A)\cdot\sin(A-B).$

76. Prove that in a ΔABC, $\begin{vmatrix} \sin^2 A & \cot A & 1 \\ \sin^2 B & \cot B & 1 \\ \sin^2 C & \cot C & 1 \end{vmatrix}=0.$

77. Prove that

$$\begin{vmatrix} \operatorname{cosec} x & 1 & 0 \\ 1 & 2\operatorname{cosec} x & 1 \\ 0 & 1 & 2\operatorname{cosec} x \end{vmatrix}\geq 1$$

for all $x\in(0,\pi)$.

78. If A, B and C are real angles, prove that

$$\begin{vmatrix} 1 & \cos(B-A) & \cos(C-A) \\ \cos(A-B) & 1 & \cos(C-B) \\ \cos(A-C) & \cos(B-C) & 1 \end{vmatrix}=0.$$

79. Prove that the value of the following determinant is independent of n.

$$\begin{vmatrix} 1 & a & a^2 \\ \cos nx & \cos(n-1)x & \cos(n-2)x \\ \sin nx & \sin(n-1)x & \sin(n-2)x \end{vmatrix}.$$

80. Evaluate $\begin{vmatrix} 1 & a & b \\ a & 1 & c \\ b & c & 1 \end{vmatrix}$.

Hence, prove that the determinant vanishes if $a=\cos x$, $b=\cos y$, $c=\cos z$ where $x+y+z=0$.

81. Prove that for all α, β the determinant

$$\begin{vmatrix} \cos\alpha & -\sin\alpha & \cos\beta \\ \sin\alpha & \cos\alpha & \sin\beta \\ -\cos(\alpha+\beta) & \sin(\alpha+\beta) & -\cos 2\beta \end{vmatrix}=0.$$

82. Prove that $\begin{vmatrix} \cos(\alpha-\beta) & \cos(\beta-\gamma) & \cos(\gamma-\alpha) \\ \cos(\alpha+\beta) & \cos(\beta+\gamma) & \cos(\gamma+\alpha) \\ \sin(\alpha+\beta) & \sin(\beta+\gamma) & \sin(\gamma+\alpha) \end{vmatrix}$

$=2\sin(\gamma-\alpha)\cdot\sin(\gamma-\beta)\cdot\sin(\alpha-\beta).$

83. If $\begin{vmatrix} 1 & \cos x & 0 & 0 \\ \cos x & 1 & \cos\alpha & \cos\beta \\ 0 & \cos\alpha & 1 & \cos\gamma \\ 0 & \cos\beta & \cos\gamma & 1 \end{vmatrix}=0$, prove that

$\sin^2\gamma\cdot\sin^2 x=\cos^2\alpha+\cos^2\beta-2\cos\alpha\cdot\cos\beta\cdot\cos\gamma.$

84. Prove that in any triangle ABC,

$$\begin{vmatrix} \tan A & 1 & 1 \\ 1 & \tan B & 1 \\ 1 & 1 & \tan C \end{vmatrix}=2.$$

85. Prove that in ΔABC, $\begin{vmatrix} 1 & \cos A & c\sin A \\ \cos A & 1 & b\sin A \\ c\sin A & b\sin A & a^2 \end{vmatrix}=0.$

86. Let the three digit numbers $A28$, $3B9$ and $62C$ where A, B and C are integers between 0 and 9, be divisible by a fixed integer k. Show that the determinant $\begin{vmatrix} A & 3 & 6 \\ 8 & 9 & C \\ 2 & B & 2 \end{vmatrix}$ is also divisible by k.

87. If $\Delta(x)=\begin{vmatrix} [x] & |x| & 1 \\ [x]^2 & |x|^2 & 1 \\ 1 & 1 & x-[x] \end{vmatrix}$,

find $\Delta(x)$ when $x>0$, $x\notin N$ where $[x]$ is the greatest integral function.

88. Prove that

$$\begin{vmatrix} 1-x & a & a^2 \\ a & a^2-x & a^3 \\ a^2 & a^3 & a^4-x \end{vmatrix}=x^2(1+a^2+a^4)-x^3.$$

89. If $(x_1-x_2)^2+(y_1-y_2)^2=a^2$

$(x_2-x_3)^2+(y_2-y_3)^2=b^2$

$(x_3-x_1)^2+(y_3-y_1)^2=c^2$

then prove that $4\begin{vmatrix} x_1 & y_1 & 1 \\ x_2 & y_2 & 1 \\ x_3 & y_3 & 1 \end{vmatrix}^2$

$= (a+b+c)(b+c-a)(c+a-b)(a+b-c)$.

90. Prove that

$$\begin{vmatrix} x-x_1 & y-y_1 & z-z_1 \\ x_2-x_1 & y_2-y_1 & z_2-z_1 \\ x_3-x_1 & y_3-y_1 & z_3-z_1 \end{vmatrix} = \begin{vmatrix} x & y & z & 1 \\ x_1 & y_1 & z_1 & 1 \\ x_2 & y_2 & z_2 & 1 \\ x_3 & y_3 & z_3 & 1 \end{vmatrix}.$$

91. If $x+y+z=0$, prove that

$$\begin{vmatrix} xa & yb & zc \\ yc & za & xb \\ zb & xc & ya \end{vmatrix} = xyz \begin{vmatrix} a & b & c \\ c & a & b \\ b & c & a \end{vmatrix}.$$

92. If a, b, c are in GP or $(x-\alpha)$ is a factor of $ax^2+2bx+c$ then prove that

$$\begin{vmatrix} a & b & a\alpha+b \\ b & c & b\alpha+c \\ a\alpha+b & b\alpha+c & 0 \end{vmatrix} = 0.$$

93. If $f(x)$ is a polynomial of the second degree, prove that

$$\begin{vmatrix} 1 & a & \dfrac{f(a)}{x-a} \\ 1 & b & \dfrac{f(b)}{x-b} \\ 1 & c & \dfrac{f(c)}{x-c} \end{vmatrix} = \frac{f(x)}{(x-a)(x-b)(x-c)} \begin{vmatrix} 1 & a & a^2 \\ 1 & b & b^2 \\ 1 & c & c^2 \end{vmatrix}.$$

94. If $D(x) = \begin{vmatrix} 1+\sin^2 x & \sin^2 x & \sin^2 x \\ \cos^2 x & 1+\cos^2 x & \cos^2 x \\ 4\sin 2x & 4\sin 2x & 1+4\sin 2x \end{vmatrix}$

then prove that $-2 \le D(x) \le 6$ for all real x.

95. Prove that for any AP $a_1, a_2, a_3, \ldots$ the determinant

$$\begin{vmatrix} a_p+a_{p+m}+a_{p+2m} & 2a_p+3a_{p+m}+4a_{p+2m} & 4a_p+9a_{p+m}+16a_{p+2m} \\ a_q+a_{q+m}+a_{q+2m} & 2a_q+3a_{q+m}+4a_{q+2m} & 4a_q+9a_{q+m}+16a_{q+2m} \\ a_r+a_{r+m}+a_{r+2m} & 2a_r+3a_{r+m}+4a_{r+2m} & 4a_r+9a_{r+m}+16a_{r+2m} \end{vmatrix} = 0.$$

96. If the polynomial $f(x)$ is expressed as

$$\begin{vmatrix} (1+x)^a & (2+x)^b & 1 \\ 1 & (1+x)^a & (2+x)^b \\ (2+x)^b & 1 & (1+x)^a \end{vmatrix}$$

then find (i) the constant term of $f(x)$,
(ii) the coefficient of x in $f(x)$.

97. If $a_r = \int_0^{\pi/2} (r+\sin\theta)^2 \cos\theta\, d\theta$,

find the value of the determinant

$$\begin{vmatrix} a_k & k & k^2 \\ a_{k+1} & k+1 & (k+1)^2 \\ a_{k+2} & k+2 & (k+2)^2 \end{vmatrix}$$

98. If $f(x) = \begin{vmatrix} 2\cos^2 x & \sin 2x & \sin x \\ \sin 2x & 2\sin^2 x & -\cos x \\ \sin x & -\cos x & 0 \end{vmatrix}$,

find $\int_0^{\pi/2} \{f(x)+f'(x)\}\, dx$.

99. If the vectors $a_1\vec{i}+a_2\vec{j}+a_3\vec{k}$ and $b_1\vec{i}+b_2\vec{j}+b_3\vec{k}$ are orthogonal, prove that

$$\begin{vmatrix} a_1b_1 & 1+a_1b_2 & 1+a_1b_3 \\ 1+a_2b_1 & a_2b_2 & 1+a_2b_3 \\ 1+a_3b_1 & 1+a_3b_2 & a_3b_3 \end{vmatrix} = 2 + \sum_{1\le i\ne j\le 3} a_ib_j.$$

100. Use calculus to prove that

$$\begin{vmatrix} 1+x\sin\theta & \cos(x+\theta) & \sin(x+\theta) \\ 2+x\sin\phi & \cos(x+\phi) & \sin(x+\phi) \\ 3+x\sin\psi & \cos(x+\psi) & \sin(x+\psi) \end{vmatrix}$$

is independent of x.

Solve by Cramer's rule.

101. $2x+3y-3z=0$
$5x-2y+2z=19$
$x+7y-5z=5$

102. $x+2y+3z=6$
$2x+4y+z=17$
$3x+2y+9z=2$

103. For what $\lambda \in R$ the system of equations
$2x-y+3=0$
$x+\lambda y+7=0$
$3x+2y-2=0$
is consistent? Solve the equations for that value of λ.

104. If a, b and c are three consecutive terms of a decreasing AP, prove that the following system of equations in x, y cannot be consistent

$\frac{x}{a}+\frac{y}{b}+\frac{1}{c}=0$, $ax+by+c=0$ and $x+y+1=0$.

105. If $c<1$ and the system of equations
$x+y-1=0$, $2x-y-c=0$ and $-bx+3by-c=0$
is consistent then find the possible real values of b.

106. If $x=cy+bz$, $y=az+cx$, $z=bx+ay$ where x, y, z are not all zeros, prove by using consistency of equations that
$a^2+b^2+c^2+2abc=1$.

107. Prove that at least two of the numbers a, b, c must have their absolute values equal in order that the system of following equations in x, y may be consistent.
$a^2x+b^2y+c^2=0$
$a^4x+b^4y+c^4=0$
$x+y+1=0$

108. Find the values of c for which the following system of equations in x, y is consistent.

$$2x + 3y = 3$$
$$(c+2)x + (c+4)y = c+6$$
$$(c+2)^2x + (c+4)^2y = (c+6)^2$$

Also solve the equations for those values of c.

109. Find the values of m and n for which the system of equations

$$x + y + z = 6$$
$$x + 2y + 3z = 10$$
$$x + 2y + mz = n$$

is (i) independent, (ii) dependent, (iii) inconsistent (i.e., unsolvable).

110. Show that the following system of equations is inconsistent for all real $k \neq 0$.

$$x + 2y + 3z = k$$
$$4x + 8y + 12z = 2k$$
$$9x - 7y + 16z = 5k$$

111. Prove that the following system of equations does not have real solutions.

$$2x^2 - 3y^2 + z^2 + 6 = 0$$
$$x^2 - 7y^2 - 2z^2 = 5$$
$$5x^2 - y^2 + 4z^2 + 23 = 0$$

112. Prove that the following system of equations in x, y, z has a unique solution for all α and β except $\beta = 2n\pi, 2n\pi - \dfrac{\pi}{2}$.

$$x\cos\alpha - y\sin\alpha + z = 3$$
$$x\sin\alpha + y\cos\alpha + z = 1$$
$$x\cos(\alpha+\beta) - y\sin(\alpha+\beta) + z = 2$$

113. For what value of $k \in R$ does the following system of equations admit a nontrivial solution?

$$x + ky + 3z = 0, \quad kx + 2y + 2z = 0$$
$$2x + 3y + 4z = 0$$

114. Consider the system of linear equations in x, y and z

$$\sin 3\theta \cdot x - y + z = 0$$
$$\cos 2\theta \cdot x + 4y + 3z = 0$$
$$2x + 7y + 7z = 0.$$

Find the value of θ for which this system has a nontrivial solution.

115. Show that there is no real λ for which he system has nontrivial solutions.

$$(\lambda+3)x + (\lambda+2)y + z = 0$$
$$3x + (\lambda+3)y + z = 0$$
$$2x + 3y + z = 0$$

116. For what value of k the equations

$$x + ky + 3z = 0, 3x + ky - 2z = 0, 2x + 3y - 4z = 0$$

possess a nontrivial solution over the set of rationals? For that value of k, find all solutions of the equations.

117. The system of equations

$$ax + by + cz = tx, \quad bx + cy + az = ty \text{ and}$$
$$cx + ay + bz = tz$$

has nontrivial solutions. Express the product of the possible values of the parameter t in the form of a determinant.

118. If the system of equations

$$ax + by + cz = 0, \quad bx + cy + az = 0, \quad cx + ay + bz = 0$$

has a nontrivial solution, prove that

$$x : y : z = 1 : 1 : 1 \text{ or } 1 : \omega : \omega^2 \text{ or } 1 : \omega^2 : \omega$$

where $\sqrt[3]{1} = 1, \omega, \omega^2$.

119. If a, b, c are three unequal real numbers, can the system

$$ax + by + cz = 0, \ a^2x + b^2y + c^2z = 0, \ x + y + z = 0$$

have nontrivial solutions? If no, why? If yes, find the nontrivial solutions.

120. Prove that the system of equations

$$(a-\lambda)x + by + cz = 0$$
$$bx + (c-\lambda)y + az = 0$$
$$cx + ay + (b-\lambda)z = 0$$

has nontrivial solutions for three values of λ in general. Also prove that the sum of the products of the three values taken two at a time cannot vanish unless $a = b = c$.

Objective Questions

Fill in the blanks.

121. The element in the ith row and the jth column of a determinant of the third order is equal to $a_i + j$. Then the value of the determinant is ______ .

122. With the nine numbers 1, 2, 3, ..., 9 all the possible determinants with nine different elements are constructed by arranging the numbers. The sum of all such determinants is ______ .

123. The value of the determinant

$$\begin{vmatrix} {}^{n-1}C_{r-1} & {}^{n-1}C_r & {}^{n-1}C_{r+1} \\ {}^{n-1}C_r & {}^{n-1}C_{r+1} & {}^{n-1}C_{r+2} \\ {}^nC_r & {}^nC_{r+1} & {}^nC_{r+2} \end{vmatrix} \text{ is ______ .}$$

124. The value of $\begin{vmatrix} 1! & 2! & 3! \\ 2! & 3! & 4! \\ 3! & 4! & 5! \end{vmatrix}$ is ______ .

125. Let $p\lambda^4 + q\lambda^3 + r\lambda^2 + s\lambda + t$

$$= \begin{vmatrix} \lambda^2 + 3\lambda & \lambda - 1 & \lambda + 3 \\ \lambda + 1 & -2\lambda & \lambda - 4 \\ \lambda - 3 & \lambda + 4 & 3\lambda \end{vmatrix}$$

be an identity in λ, where p, q, r, s and t are constants. Then the value of $t =$ ______ .

126. The value of $\begin{vmatrix} 4 & 9 & 7 \\ 3 & 5 & 7 \\ 5 & 4 & 5 \end{vmatrix}$ is ______.

127. The value of $\begin{vmatrix} 29 & 26 & 22 \\ 25 & 31 & 27 \\ 63 & 54 & 46 \end{vmatrix}$ is ______.

128. If $a+b+c$ is equal to n then the value of

$\begin{vmatrix} 1+a & b & c \\ a & 1+b & c \\ a & b & 1+c \end{vmatrix}$ in terms of n is ______.

129. The value of $\begin{vmatrix} b^2-ab & b-c & bc-ac \\ ab-a^2 & a-b & b^2-ab \\ bc-ac & c-a & ab-a^2 \end{vmatrix}$ is ______.

130. The ratio of the determinants

$\begin{vmatrix} 1 & a & a^2 \\ 1 & b & b^2 \\ 1 & c & c^2 \end{vmatrix}$ and $\begin{vmatrix} ab & a+b & 1 \\ bc & b+c & 1 \\ ca & c+a & 1 \end{vmatrix}$ is ______.

131. The value of $\begin{vmatrix} 1 & a & bc \\ 1 & b & ca \\ 1 & c & ab \end{vmatrix} - \begin{vmatrix} 1 & a & a^2 \\ 1 & b & b^2 \\ 1 & c & c^2 \end{vmatrix}$ is ______.

132. If $A+B+C=0$ then the value of

$\begin{vmatrix} 1 & \cos C & \cos B \\ \cos C & 1 & \cos A \\ \cos B & \cos A & 1 \end{vmatrix} =$ ______.

133. If $\begin{vmatrix} 1 & a^2 & a^4 \\ 1 & b^2 & b^4 \\ 1 & c^2 & c^4 \end{vmatrix} = \lambda \begin{vmatrix} 1 & 1 & 1 \\ a & b & c \\ a^2 & b^2 & c^2 \end{vmatrix}$ then $\lambda =$ ______.

Choose the correct option(s).

134. If $\lambda \in R$ and $\Delta = \begin{vmatrix} a & b \\ c & d \end{vmatrix}$ then $\lambda\Delta$ is equal to

(a) $\begin{vmatrix} \lambda a & \lambda b \\ \lambda c & \lambda d \end{vmatrix}$ (b) $\begin{vmatrix} \lambda a & b \\ c & d \end{vmatrix}$

(c) $\begin{vmatrix} \lambda a & b \\ \lambda c & d \end{vmatrix}$ (d) none of these

135. In a third order determinant, a_{ij} denotes the element in the ith row and the jth column.

If $a_{ij} = 0, \ i=j$
$\quad 1, \ i>j$
$\quad -1, \ i<j$

then the value of the determinant is

(a) 0 (b) 1

(c) –1 (d) none of these

136. If $A = \begin{vmatrix} a & d & l \\ b & e & m \\ c & f & n \end{vmatrix}$ and $B = \begin{vmatrix} l & m & n \\ a & b & c \\ d & e & f \end{vmatrix}$ then

(a) $A = B$ (b) $A + B = 0$

(c) $\frac{1}{A} + \frac{1}{B} = 0$ (d) none of these

137. If 7 and 2 are roots of the equation $\begin{vmatrix} x & 3 & 7 \\ 2 & x & 2 \\ 7 & 6 & x \end{vmatrix} = 0$ then the third root is

(a) –9 (b) 14

(c) $\frac{1}{2}$ (d) none of these

138. If $f(x,y) = \begin{vmatrix} 1 & 1 & 1 \\ 1 & 1+x & 1 \\ 1 & 1 & 1+y \end{vmatrix}$ then $f(x,y)$ is

(a) divisible by x (b) divisible by y

(c) divisible by x and y both

(d) divisible by neither x nor y

139. The value of $\begin{vmatrix} x+a & x+b & x+c \\ y+a & y+b & y+c \\ z+a & z+b & z+c \end{vmatrix}$ is

(a) $x+y+z+a+b+c$ (b) $xyz+abc$

(c) 0 (d) none of these

140. Which of the following is a factor of $\begin{vmatrix} a & b & c \\ b & c & a \\ c & a & b \end{vmatrix}$?

(a) $a+b+c$ (b) $a+bw+cw^2$

(c) $a+bw^2+cw$ (d) none of these

141. If x, y, z are positive, the value of the determinant

$\begin{vmatrix} 1 & \log_x y & \log_x z \\ \log_y x & 1 & \log_y z \\ \log_z x & \log_z y & 1 \end{vmatrix}$ is

(a) $\log_y x \cdot \log_z y \cdot \log_x z$ (b) 0

(c) 1 (d) none of these

142. The condition for the expression

$$ax^2 + 2hxy + by^2 + 2gx + 2fy + c$$

to be resolved into rational linear factors in the determinant form is

(a) $\begin{vmatrix} a & b & c \\ h & g & f \\ 1 & 1 & 1 \end{vmatrix} = 0$ (b) $\begin{vmatrix} a & h & f \\ h & b & g \\ f & g & c \end{vmatrix} = 0$

(c) $\begin{vmatrix} a & h & g \\ h & b & f \\ g & f & c \end{vmatrix} = 0$ (d) none of these

State whether the statements are true or false.

143. If $\begin{vmatrix} x_1 & y_1 & 1 \\ x_2 & y_2 & 1 \\ x_3 & y_3 & 1 \end{vmatrix} = \begin{vmatrix} a_1 & b_1 & 1 \\ a_2 & b_2 & 1 \\ a_3 & b_3 & 1 \end{vmatrix}$ then the two triangles whose vertices are $(x_1, y_1), (x_2, y_2), (x_3, y_3)$ and $(a_1, b_1), (a_2, b_2), (a_3, b_3)$ must be similar or congruent.

144. $\begin{vmatrix} x & 3x+2 & 2x-1 \\ 2x-1 & 4x & 3x+1 \\ 7x-2 & 17x+6 & 12x-1 \end{vmatrix} = 0$ is an identity in x.

145. $\begin{vmatrix} 1 & \alpha & \alpha^2 \\ \alpha & \alpha^2 & 1 \\ \alpha^2 & 1 & \alpha \end{vmatrix} = 0$ where $\alpha = \sqrt[3]{1}$.

146. $\begin{vmatrix} a & b & c \\ x & y & z \\ p & q & r \end{vmatrix} = \begin{vmatrix} y & b & q \\ x & a & p \\ z & c & r \end{vmatrix}$.

Answers

4. $(a-b)(b-c)(c-a)$ **13.** $(a+b+c)^3$

14. 4 **19.** $\frac{1}{2} - \sqrt{2} + \frac{1}{3}\sqrt{6}$

21. $(x-y)(y-z)(z-x)(xy+yz+zx)$

23. $(x+y+z)(x^2+y^2+z^2)(x-y)(y-z)(x-z)$

29. $\dfrac{4}{(a+1)(a+2)^2(a+3)^3(a+4)^2(a+5)}$

30. $0, \pm\sqrt{\frac{1}{3}(ab+bc+ca)}$ **38.** $1+a^2+b^2+c^2$

49. 0 **50.** 0

53. $\frac{xyz}{12}(x-y)(y-z)(z-x)$

69. $0, \pm i\sqrt{a^2+b^2+c^2}$ each root being repeated

70. 9 **71.** $(a+b+c)(a-b)^2(b-c)^2(c-a)^2$

72. ± 1 **73.** $n\pi, n\pi + (-1)^n \frac{\pi}{6}$ where $n \in \mathbb{Z}$

80. $1 + 2abc - (a^2+b^2+c^2)$

87. $([x]-x)\{[x](1-x^2) + x - 1 + x[x]^2\}$

96. (i) $2 - 3\cdot 2^b + 2^{3b}$ (ii) $3a - \frac{3}{2}(2a+b)2^b + 3b \cdot 2^{3b-1}$

97. $\frac{2}{3}$ **98.** $-\pi$

101. $x=3, y=6, z=8$ **102.** $x=1, y=4, z=-1$

103. $\lambda = \frac{-45}{13}; x = \frac{-4}{7}, y = \frac{13}{7}$ **105.** $b \in \left(-\frac{3}{4}, 3\right)$

108. $c = 0, -10; x=-3, y=3$ and $x = \frac{-1}{2}, y = \frac{4}{3}$ respectively

109. (i) all m, n except $m=3$ (ii) $m=3, n=10$ (iii) $m=3, n\neq 10$

113. $2, \frac{5}{4}$ **114.** $n\pi, n\pi + (-1)^n \frac{\pi}{6}, n \in \mathbb{Z}$

116. $k = \frac{33}{2}; x:y:z = -15:2:-6$

117. $\begin{vmatrix} a & b & c \\ b & c & a \\ c & a & b \end{vmatrix}$ **119.** no **121.** 0

122. 0 **123.** 0 **124.** 24 **125.** 0

126. 77 **127.** 132 **128.** $n+1$ **129.** 0

130. 1 : 1 **131.** 0 **132.** 0

133. $(a+b)(b+c)(c+a)$ **134.** (c) **135.** (a)

136. (a) **137.** (a) **138.** (a), (b), (c)

139. (c) **140.** (a), (b), (c) **141.** (b)

142. (c) **143.** false **144.** true

145. true **146.** true

Chapter Test

Time: 90 minutes

1. In the following, each question has one or more than one correct answers. Indicate the correct answer(s).

(a) $\begin{vmatrix} x+1 & x+2 & x+\lambda \\ x+2 & x+3 & x+\mu \\ x+3 & x+4 & x+\nu \end{vmatrix} = 0$, where λ, μ, ν are in AP, is

A. an equation whose all roots are real
B. an identity in x
C. an equation with only one real root
D. none of these

(b) If w is a nonreal cube root of unity then a root of the equation

$\begin{vmatrix} x+1 & w & w^2 \\ w & x+w^2 & 1 \\ w^2 & 1 & x+w \end{vmatrix} = 0$ is

A. $x=1$ B. $x=w$ C. $x=w^2$ D. $x=0$

2. In the following, each statement is incomplete. Fill in the blank so that each resulting statement is correct.

(a) The value of $\begin{vmatrix} (-1)^n a & (-1)^{n+1} b & (-1)^{n+1} c \\ a+1 & 1-b & 1+c \\ a-1 & b+1 & 1-c \end{vmatrix} + \begin{vmatrix} (-1)^{n+1} a & a+1 & a-1 \\ (-1)^n b & 1-b & b+1 \\ (-1)^{n+2} c & 1+c & 1-c \end{vmatrix}$

is equal to ______ .

(b) The sum of the roots of the equation $\begin{vmatrix} x & a & a \\ a & x & a \\ a & a & x \end{vmatrix} = 0$ in x is ______ .

3. If a, b, c are any three terms of an AP in order (not necessarily consecutive) whose common difference is positive then prove that

$$\begin{vmatrix} a & b & c \\ a^2 & b^2 & c^2 \\ 1 & 1 & 1 \end{vmatrix} > 0.$$

4. Without breaking a determinant at any stage, prove that

$$\begin{vmatrix} a-b+c & a+b-c & a-b-c \\ b-c+a & b+c-a & b-c-a \\ c-a+b & c+a-b & c-a-b \end{vmatrix} = 4\begin{vmatrix} a & c & b \\ b & a & c \\ c & b & a \end{vmatrix}.$$

5. Prove that $\begin{vmatrix} \sin(a+x) & \sin(a+y) & \sin(a+z) \\ \sin(b+x) & \sin(b+y) & \sin(b+z) \\ \sin(c+x) & \sin(c+y) & \sin(c+z) \end{vmatrix} = 0$

for all a, b and c and x, y, z.

6. If the number of positive integral solutions of $x+y+z=r$, $(r \geq 3)$ be denoted by P_r then

find the value of $\begin{vmatrix} P_r & P_{r+1} & P_{r+2} \\ P_{r+1} & P_{r+2} & P_{r+3} \\ P_{r+2} & P_{r+3} & P_{r+4} \end{vmatrix}$.

7. If a, b and c be the pth, qth and rth terms of an HP respectively, prove that the system of equations

$$bc \cdot x + ca \cdot y + ab \cdot z = 0$$
$$px + qy + rz = 0$$
$$x + y + z = 0$$

has infinite number of solutions.

8. Prove that the system of equations in x and y cannot have a solution for any α.

$$\cos^2\alpha \cdot x + \frac{1}{2}\sin 2\alpha \cdot y = \sin\alpha$$

$$\frac{1}{2}\sin 2\alpha \cdot x + \sin^2\alpha \cdot y + \cos\alpha = 0$$

$$\sin\alpha \cdot x - \cos\alpha \cdot y = 0.$$

9. Without breaking determinants at any stage, find $\lambda \in R$ such that the determinant

$$\begin{vmatrix} 4i & 8+i & 4+3i \\ -8+i & 16i & i \\ -4+\lambda i & i & 8i \end{vmatrix}$$

has purely imaginary value.

10. Let $L_1 \equiv 15x - 18y + 1$, $L_2 \equiv 12x + 10y - 3$ and $L_3 \equiv 6x + 66y - 11$. Prove that there are nonzero constants λ, μ, ν such that $\lambda L_1 + \mu L_2 + \nu L_3 \equiv 0$. Also find all possible real values of the triplet (λ, μ, ν).

□

Answers

1. (a) B (b) D

2. (a) 0 (b) 0

6. –1

9. 3

10. $(2k, -3k, k)$ where $k \in R$

□

3. Equations, Inequations and Expressions

Recap of Facts and Formulae

1. Identity

$f(x) = \phi(x)$ is an identity if $f(x)$ and $\phi(x)$ have the same value for every real value of x. An equation with arbitrary coefficients may be an identity under certain conditions.

- $ax^2 + bx + c = 0$ will be an identity (or can have more than two solutions) if coefficient of each power of x is separately zero, i.e., $a = 0, b = 0, c = 0$.
- $ax^3 + bx^2 + cx + d = 0$ will be an identity if
  $a = 0, b = 0, c = 0, d = 0$.

2. Polynomial equations and their solutions

If $f(x)$ is a function of x then $f(x) = 0$ is an equation in one unknown (or variable) and zeros of $f(x)$ or roots of $f(x) = 0$ are the values of x which make $f(x)$ equal to 0.

- If $f(x)$ is a polynomial of the first degree in x then the equation $f(x) = 0$ is of the first degree in one unknown.
  $ax + b = 0$ is an equation of the first degree in x.
  Its solution (or root) is found like this:
  $$ax = -b; \quad \therefore \quad x = \frac{-b}{a}.$$
  A first degree equation has only one solution.
- If $f(x)$ is a polynomial of the second degree in x then the equation $f(x) = 0$ is of the second degree (or quadratic equation) in one unknown.
  $ax^2 + bx + c = 0$ is an equation of the second degree in x where $a \neq 0$.
  The roots of $ax^2 + bx + c = 0$ are $x = \dfrac{-b \pm \sqrt{b^2 - 4ac}}{2a}$.
  The roots of $ax^2 + bx + c = 0$ can also be found by factorizing $ax^2 + bx + c$ and equating each factor to 0 separately.
  A second degree equation has two solutions (different or equal).
- If $f(x)$ is a polynomial of the degree three (or more) then the equation $f(x) = 0$ is cubic (or of higher degree) in one unknown.
  Such equations can be solved if $f(x)$ can be factorized in linear or quadratic factors.
  An equation of the nth degree has n solutions (different or equal).

For example Solve $x^3 + x^2 + x = 84$.

Here the equation is $x^3 + x^2 + x - 84 = 0$.

We have to factorize $x^3 + x^2 + x - 84$.

Trying $x = 1, -1, 2, -2$, etc., we get,

when $x = 4$, $x^3 + x^2 + x - 84 = 4^3 + 4^2 + 4 - 84 = 0$

$\therefore \quad (x - 4)$ is a factor of $x^3 + x^2 + x - 84$.

Dividing $x^3 + x^2 + x - 84$ by $x - 4$.

we get $x^2 + 5x + 21$.

$$\therefore \quad x^3 + x^2 + x - 84 = 0$$
$$\Rightarrow \quad (x - 4)(x^2 + 5x + 21) = 0$$
$$\therefore \quad x - 4 = 0 \quad \text{or} \quad x^2 + 5x + 21 = 0$$
$$\therefore \quad x = 4 \quad \text{or} \quad x = \frac{-5 \pm \sqrt{25 - 84}}{2}.$$

3. Exponential equations and their solutions

If the equation involves terms or factors of the type $a^{f(x)}$ or $\{\phi(x)\}^{f(x)}$, it will be an exponential equation.

- If the exponential equation is such that it can be put in the form
  $a^{f(x)} = a^{\psi(x)}$, $a \neq 1$ then $f(x) = \psi(x)$ will give the solution.

For example Solve $(2\sqrt{2})^{x^2} = 8^{3x}$.

$$\text{Here} \quad (2\sqrt{2})^{x^2} = \{(2\sqrt{2})^2\}^{3x}$$
$$\text{or} \quad (2\sqrt{2})^{x^2} = (2\sqrt{2})^{6x}$$
$$\therefore \quad x^2 = 6x; \quad \text{or} \quad x(x - 6) = 0;$$
$$\therefore \quad x = 0, 6.$$

- If the exponential equation cannot be put in the above form, select an exponential as y so that the equation changes into a polynomial equation in y.

- In the exponential a^x, a is greater than 0. So, no negative value of a^x is possible.

4. Logarithmic equations and their solutions

If the equation involves logarithm of some function of the unknown then it will be a logarithmic equation.

- If the logarithmic equation is such that it can be put in the form $\log_a f(x) = \log_a \phi(x)$ then $f(x) = \phi(x)$ will give the solution.
  Only those values of x from $f(x) = \phi(x)$ will give admissible solutions which make both $f(x)$ and $g(x)$ greater than 0.
- If the logarithmic equation cannot be put in the above form, select a logarithm as y so that the equation changes into a polynomial equation in y.
- If the bases of the logarithms are also functions of x, the admissible solutions must make the values of the bases greater than 0 but not equal to 1.

5. Equations involving modulus, greatest integer function, etc., and their solutions

- $f(x) = |x - a|$ is a piecewisely defined function whose definition is
  $$f(x) = x - a, \quad x \ge a$$
  $$\quad\;\; -(x - a), \quad x < a$$
- $f(x) = [x]$ is a piecewisely defined function whose definition is
  $f(x) = n,\ n \le x < n + 1$ where n is an integer.
- $f(x) = [x + n] = [x] + n$, where n is an integer.
- If the definition of the function is not uniform over R, the set R of real numbers should be divided into subsets according to the definitions of the function and the equation in the corresponding interval is to be solved. A solution will be admissible if it lies in the interval of definition of that equation.

6. Inequations and their solutions

(i) Laws of inequality are as follows:

- $a + b > a + c \;\Rightarrow\; b > c$
  and $a > b \;\Rightarrow\; a + c > b + c$.
- $a > b \;\Rightarrow\; ca > bc$ if $c > 0$
  $ca < bc$ if $c < 0$.
- $ab > ac \;\Rightarrow\; b > c$ if $a > 0$
  $b < c$ if $a < 0$.
- $a > b$ and $c > d \;\Rightarrow\; a + c > b + d$.
- $a > 0, b > 0 \;\Rightarrow\; a + b > 0$ and $ab > 0$
  $a < 0, b < 0 \;\Rightarrow\; a + b < 0$ and $ab > 0$
  $a > 0, b < 0 \;\Rightarrow\; ab < 0$
- $a^x > a^y \;\Rightarrow\; x > y$ if $a > 1$
  $x < y$ if $0 < a < 1$
- $\log_a x > \log_a y \;\Rightarrow\; x > y$ if $a > 1$
  $x < y$ if $0 < a < 1$

(ii) $f(x) > 0, f(x) < 0, f(x) \ge 0, f(x) \le 0$ are all inequations in one variable if they hold for all $x \in A \subset R$. But if they hold for all $x \in R$ then they are inequalities.

- $ax + b > 0$ is a linear inequation.
  For this inequation, $ax > -b$, we have
  $$x > \frac{-b}{a} \text{ if } a > 0 \text{ or } x < \frac{-b}{a} \text{ if } a < 0.$$
  The solution set is an infinite set.
- $ax^2 + bx + c > 0$ (or < 0 or ≥ 0 or ≤ 0) is a quadratic inequation.
  The solution of the inequation is the set of real values of x for which the inequality is true. The set can be obtained conveniently by sign-scheme.

(iii) **The sign-scheme for $ax^2 + bx + c, x \in R$**

It is as follows:

Let the roots of the corresponding equation $ax^2 + bx + c = 0$ be α, β.

- If α, β are real and unequal ($\alpha < \beta$) then

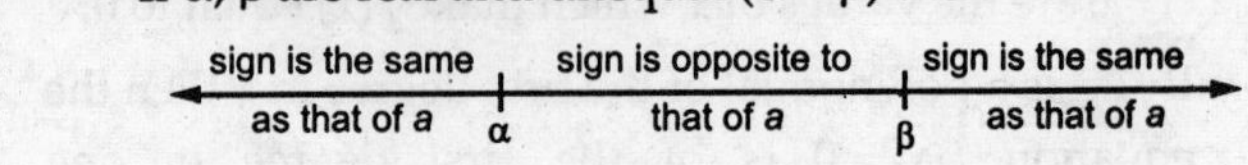

∴ if $a > 0$,

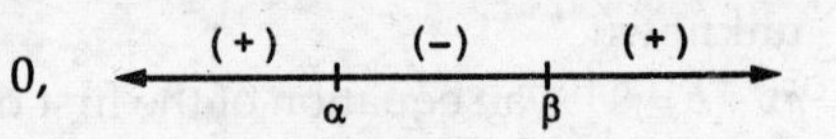

if $a < 0$.

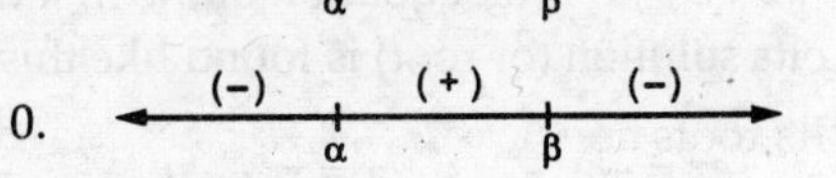

- If α, β are real and equal then

if $a > 0$,

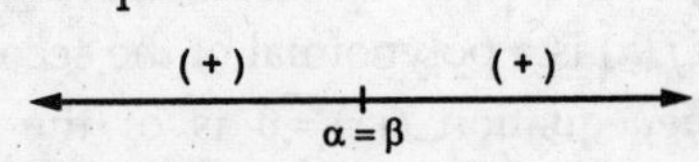

if $a < 0$,

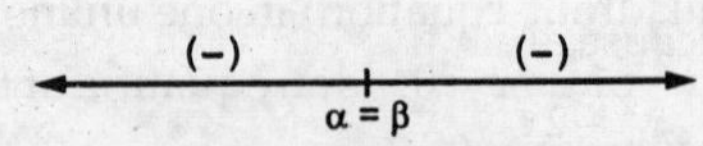

- If α, β are imaginary (nonreal complex) then

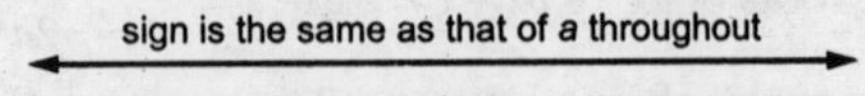

∴ if $a > 0$, the expression is always positive
if $a < 0$, the expression is always negative.

Note

(1) $|x| < a, (a > 0)$ holds when $-a < x < a$

(2) $|x| > a, (a > 0)$ holds when $x > a$ or $x < -a$

(3) $x^2 > a^2$ holds when $x > a$ or $x < -a$

(4) $x^2 < a^2$ holds when $-a < x < a$.

(iv) Positive definiteness and negative definiteness of a quadratic polynomial

- $ax^2 + bx + c > 0$ holds for all $x \in R$, i.e., $ax^2 + bx + c$ is positive definite, if
  $D < 0$ and $a > 0$ where $D = b^2 - 4ac$.
- $ax^2 + bx + c \geq 0$ holds for all $x \in R$, i.e., $ax^2 + bx + c$ is non-negative, if
  $D \leq 0$ and $a > 0$.
- $ax^2 + bx + c < 0$ holds for all $x \in R$, i.e., $ax^2 + bx + c$ is negative definite, if
  $D < 0$ and $a < 0$.

(v) The sign-scheme for $ax^3 + bx^2 + cx + d, x \in R$

It is as follows:

Let the roots of the corresponding equation $ax^3 + bx^2 + cx + d = 0$ be α, β, γ.

- If α, β, γ are real and unequal ($\alpha < \beta < \gamma$) then

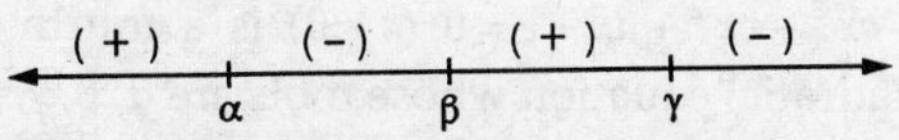

where a value between β and γ makes the expression positive;

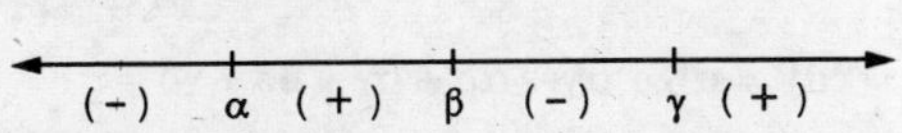

where a value between β and γ makes the expression negative.

For example Discuss the sign of $x^3 + 2x^2 - x - 2$, $x \in R$.

The corresponding equation is

$x^3 + 2x^2 - x - 2 = 0$

or $x^2(x + 2) - 1(x + 2) = 0$

or $(x + 2)(x^2 - 1) = 0$.

Its roots are $x = -2, 1, -1$.

$\therefore$ the sign-scheme for $x^3 + 2x^2 - x - 2, x \in R$ is as follows:

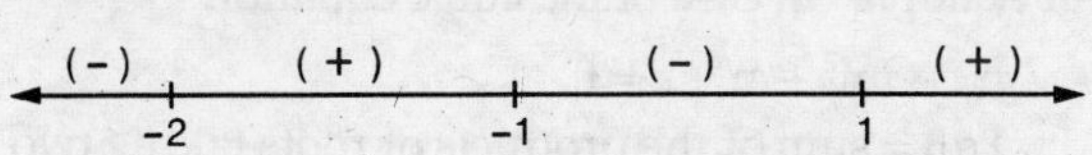

because at $x = 0 \in (-1, 1)$, the value is negative.

$\therefore$ $x^3 + 2x^2 - x - 2 > 0$ if $x \in (-2, -1) \cup (1, +\infty)$

$x^3 + 2x^2 - x - 2 < 0$ if $x \in (-\infty, -2) \cup (-1, 1)$

$x^3 + 2x^2 - x - 2 = 0$ if $x = -2, -1, 1$.

Note

(1) The sign-scheme for fourth or higher degree polynomials is also prepared as above by detecting the sign of the value of the polynomial for x belonging to an interval determined by two consecutive roots and then setting alternate signs in the other intervals.

(2) If two roots of $f(x) = 0$ for the inequation $f(x) > 0$ be real and equal to α then $(x - \alpha)^2$ is a factor of $f(x)$, which is positive for all real x except $x = \alpha$ where $f(x)$ is zero.

$\therefore$ the solution of $f(x) > 0$ will be the same as $\dfrac{f(x)}{(x-\alpha)^2} > 0$. So the omission of the factor $(x - \alpha)^2$ from $f(x)$ will not affect the solution of the inequation.

(3) If two roots of $f(x) = 0$ for the inequation be complex conjugate $\alpha \pm i\beta$ then $(x - \alpha)^2 + \beta^2$ is a positive factor of $f(x)$. So the omission of the factor $(x - \alpha)^2 + \beta^2$ from $f(x)$ will not affect the solution of the inequation.

(4) The solution of an inequation $f(x) > 0$ or < 0 or ≥ 0 or ≤ 0 is directly dependent on the solution of the corresponding equation $f(x) = 0$.

7. Quadratic equation and its roots

If $ax^2 + bx + c = 0, (a \neq 0)$ be a quadratic equation whose only two roots are α, β then

- roots $\alpha, \beta = \dfrac{-b \pm \sqrt{b^2 - 4ac}}{2a}$

  where $b^2 - 4ac$ is the discriminant D.

- The nature of the roots will be as follows:

  $D > 0 \Leftrightarrow$ roots are real and unequal (a, b, c being real)

  $D = 0 \Leftrightarrow$ roots are real and equal (a, b, c being real)

  $D < 0 \Leftrightarrow$ roots are nonreal conjugate complex (a, b, c being real)

  D is a perfect square $\Leftrightarrow$ roots are rational (a, b, c being rational)

  D is not a perfect square (but positive) $\Leftrightarrow$ roots are conjugate irrationals (a, b, c being rational)

Note

(1) If any of the coefficients a, b, c is nonreal complex and $p + iq$ is a complex root of $ax^2 + bx + c = 0$ then the other root need not be $p - iq$.

(2) If any of the coefficients a, b, c is irrational and $p + \sqrt{q}$ be an irrational root of $ax^2 + bx + c = 0$ then the other root need not be $p - \sqrt{q}$.

(3) The above notes (1), (2) hold for equations of higher degrees also.

- If $a + b + c = 0$ then the equation $ax^2 + bx + c = 0$ has a root $x = 1$.

- $\alpha + \beta = \dfrac{-b}{a}$, $\alpha\beta = \dfrac{c}{a}$.

- **Sign of real roots**

(a) When $D \geq 0$,

(i) both roots are positive if

$$\alpha + \beta = \frac{-b}{a} > 0, \alpha\beta = \frac{c}{a} > 0.$$

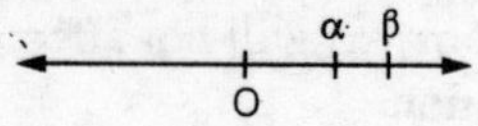

(ii) both roots are negative if

$$\alpha + \beta = \frac{-b}{a} < 0, \alpha\beta = \frac{c}{a} > 0.$$

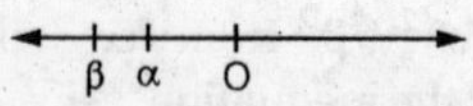

(b) When $D > 0$,

one root is positive and one root is negative if

$$\alpha\beta = \frac{c}{a} < 0.$$

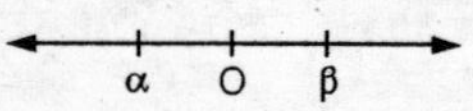

- **Location of real roots**

(a) When $D \geq 0$,

(i) both roots are greater than k

$$\Rightarrow \alpha - k > 0, \beta - k > 0$$

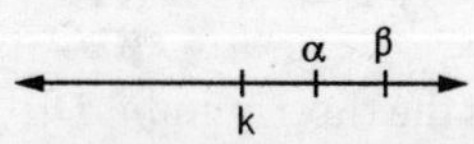

$$\Rightarrow (\alpha - k) + (\beta - k) > 0, (\alpha - k)(\beta - k) > 0$$

$$\Rightarrow \alpha + \beta - 2k > 0, \alpha\beta - k(\alpha + \beta) + k^2 > 0$$

(ii) both roots are less than k

$$\Rightarrow \alpha - k < 0, \beta - k < 0$$

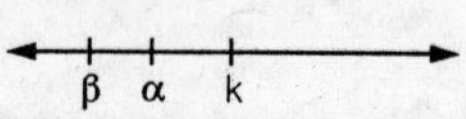

$$\Rightarrow (\alpha - k) + (\beta - k) < 0, (\alpha - k)(\beta - k) > 0$$

(iii) both roots lie between k and l $(k < l)$

$$\Rightarrow \alpha - k > 0, \beta - k > 0, \alpha - l < 0, \beta - l < 0$$

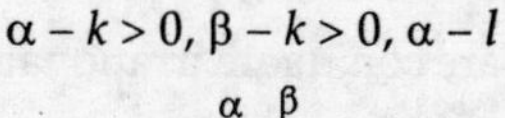

$$\Rightarrow \alpha - k + \beta - k > 0, \alpha - l + \beta - l < 0$$

$$(\alpha - k)(\beta - k) > 0, (\alpha - l)(\beta - l) > 0$$

(b) When $D > 0$,

(i) one root is less than k and the other greater than k

$$\Rightarrow \alpha - k < 0, \beta - k > 0$$

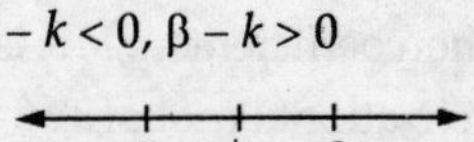

$$\Rightarrow (\alpha - k)(\beta - k) < 0$$

(ii) one root is less than k and the other greater than l $(k < l)$

$$\Rightarrow \alpha - k < 0, \beta - k > 0, \alpha - l < 0, \beta - l > 0$$

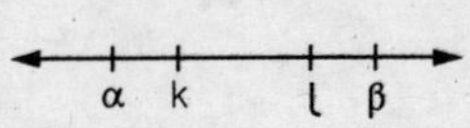

$$\Rightarrow (\alpha - k)(\beta - k) < 0, (\alpha - l)(\beta - l) < 0$$

8. Equations of higher degrees and their roots

$a_0x^n + a_1x^{n-1} + a_2x^{n-2} + \dots + a_n = 0$, where a_is are real $(a_0 \neq 0)$, is an nth degree polynomial equation in one variable x.

It has n roots (unequal or equal) which are either real or nonreal complex.

- $ax^3 + bx^2 + cx + d = 0, (a \neq 0)$ is a cubic equation whose roots are α, β, γ.

Then $\Sigma\alpha = \alpha + \beta + \gamma = \dfrac{-b}{a}$

$$\Sigma\alpha\beta = \alpha\beta + \beta\gamma + \gamma\alpha = \frac{c}{a} \qquad \alpha\beta\gamma = \frac{-d}{a}$$

These are obtained by equating coefficients of x^2, x and x^0 on both sides of

$$ax^3 + bx^2 + cx + d = a(x - \alpha)(x - \beta)(x - \gamma).$$

- $ax^4 + bx^3 + cx^2 + dx + e = 0, (a \neq 0)$ is a fourth degree (biquadratic) equation whose roots are $\alpha, \beta, \gamma, \delta$.

Then $\Sigma\alpha = \alpha + \beta + \gamma + \delta = \dfrac{-b}{a}$

$$\Sigma\alpha\beta = \alpha\beta + \alpha\gamma + \alpha\delta + \beta\gamma + \beta\delta + \gamma\delta = \frac{c}{a}$$

$$\Sigma\alpha\beta\gamma = \alpha\beta\gamma + \alpha\beta\delta + \alpha\gamma\delta + \beta\gamma\delta = \frac{-d}{a}$$

$$\alpha\beta\gamma\delta = \frac{e}{a}$$

Note

(1) The above results for an equation of the nth degree can be remembered as follows:

s_r = sum of the products of the roots taking r at a time

$$= (-1)^r \frac{\text{coefficient of the (highest} - r\text{th) power}}{\text{coefficient of the highest power}}$$

For example In case of the cubic equation

$$ax^3 + bx^2 + cx + d = 0,$$

$\Sigma\alpha\beta$ = sum of the products of roots taking two at a time

$$= (-1)^2 \cdot \frac{\text{coefficient of } x^{3-2}\text{, i.e., } x}{\text{coefficient of } x^3} = \frac{c}{a}$$

In case of the equation

$$ax^5 + bx^4 + cx^3 + dx^2 + ex + f = 0,$$

$\Sigma\alpha\beta\gamma$ = sum of the products of roots taking three at a time

$$= (-1)^3 \cdot \frac{\text{coefficient of } x^{5-3}\text{, i.e., } x^2}{\text{coefficient of } x^5} = -\frac{d}{a}.$$

(2) If the equation is of the third degree with real coefficients then

(i) roots are all real or

(ii) one root is real and two roots are nonreal complex which are conjugate to each other.

In case of (i) all roots are rational or one rational and two irrational conjugates.

9. Formation of equation when roots are known

- The quadratic equation with roots α, β is

  $(x-\alpha)(x-\beta)=0$, i.e., $x^2-(\alpha+\beta)x+\alpha\beta=0$.

- The cubic equation with roots α, β, γ is

  $(x-\alpha)(x-\beta)(x-\gamma)=0$, i.e.,

  $x^3-(\alpha+\beta+\gamma)x^2+(\alpha\beta+\beta\gamma+\gamma\alpha)x-\alpha\beta\gamma=0$.

- The nth degree equation with roots $\alpha_1, \alpha_2, \alpha_3, \ldots, \alpha_n$ is

  $(x-\alpha_1)(x-\alpha_2)(x-\alpha_3)\ldots(x-\alpha_n)=0$.

Note

(1) If one root of a quadratic equation with real coefficients is given to be $p+iq$ then the other root (another root) is also known to be $p-iq$. This is true for equations of higher degrees also.

(2) If one root of a quadratic equation with rational coefficients is given to be $p+\sqrt{q}$ then the other root (another root) is also known to be $p-\sqrt{q}$. This is true for equations of higher degrees also.

10. Common roots in two quadratic equations

- α is a common root in $a_1x^2+b_1x+c_1=0$ and $a_2x^2+b_2x+c_2=0$

$$\Rightarrow \quad a_1\alpha^2+b_1\alpha+c_1=0$$
$$a_2\alpha^2+b_2\alpha+c_2=0$$
$$\Rightarrow \quad \frac{\alpha^2}{b_1c_2-b_2c_1}=\frac{\alpha}{c_1a_2-c_2a_1}=\frac{1}{a_1b_2-a_2b_1}.$$

- If roots of $a_1x^2+b_1x+c_1=0$ are α, β and those of $a_2x^2+b_2x+c_2=0$ are α, γ (there being one common root) then

$$\alpha+\beta=-\frac{b_1}{a_1},\quad \alpha+\gamma=\frac{-b_2}{a_2},\quad \alpha\beta=\frac{c_1}{a_1},\quad \alpha\gamma=\frac{c_2}{a_2}.$$

- If two quadratic equations with real coefficients have a nonreal complex common root then both roots will be common, i.e., both the equations will be the same. So the coefficients of the corresponding powers of x will have proportional values.
- If two quadratic equations with rational coefficients have a common irrational root $p+\sqrt{q}$ then both roots will be common, i.e., no two different quadratic equations with rational coefficients can have a common irrational root $p+\sqrt{q}$.

11. Extreme values of $ax^2+bx+c, x \in R$

- If $ax^2+bx+c=y$ then $ax^2+bx+(c-y)=0$ will have real roots, i.e., $D \geq 0 \Rightarrow b^2-4a(c-y)\geq 0$;

  this is a linear inequation in y solving which we get the extreme values of y (i.e., ax^2+bx+c).

- If $\dfrac{a_1x^2+b_1x+c_1}{a_2x^2+b_2x+c_2}=y$

  then $(a_1-a_2y)x^2+(b_1-b_2y)x+(c_1-c_2y)=0$ will have real roots, i.e.,

  $D\geq 0 \Rightarrow (b_1-b_2y)^2-4(a_1-a_2y)(c_1-c_2y)\geq 0$;

  this is a quadratic inequation in y which can be solved by using sign-scheme. The solution of this inequation gives the extreme values of y.

12. Factors of polynomials

- $x-\alpha$ is a factor of the polynomial $f(x)$

  $\Leftrightarrow \alpha$ is a root of the equation $f(x)=0$, i.e., $f(\alpha)=0$.

- $y-mx$ is a factor of $f(x,y)$

  $\Leftrightarrow f(x, mx)=0$ or $f\left(\dfrac{y}{m}, y\right)=0$.

- ax^2+bx+c is a perfect square

  if $D=0$, i.e., $b^2-4ac=0$.

- If α, β are the roots of the corresponding equation then $ax^2+bx+c=a(x-\alpha)(x-\beta)$.
- $ax^2+2hxy+by^2+2gx+2fy+c$ can be resolved into linear factors $\Leftrightarrow abc+2fgh-af^2-bg^2-ch^2=0$.

13. Use of calculus in problems of equations and expressions

- If $f(\lambda)$ and $f(\mu)$ are of opposite signs then the equation $f(x)=0$ has a real root lying between the real numbers λ and μ.

  This is clear from the graph of the function $f(x)$.

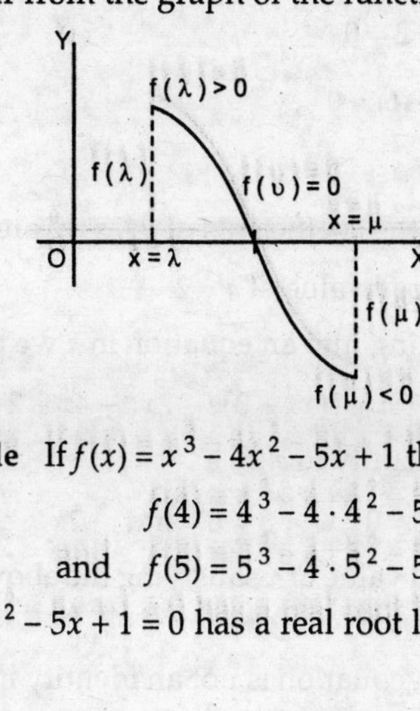

For example If $f(x)=x^3-4x^2-5x+1$ then

$$f(4)=4^3-4\cdot 4^2-5\cdot 4+1<0$$
and $$f(5)=5^3-4\cdot 5^2-5\cdot 5+1>0$$

$\therefore\ x^3-4x^2-5x+1=0$ has a real root lying between 4 and 5.

- If the equation $f(x) = 0$ has two real roots α and β then $f'(x) = 0$ will have a real root lying between α and β. (See *Rolle's theorem*)

For example The equation $2x^3 + x^2 - 2x - 1 = 0$ has two roots -1 and 1.

So, the derived equation $f'(x) = 6x^2 + 2x - 2 = 0$ has a root lying between -1 and 1.

- If α is a repeated root, i.e., two roots are α, α, of the equation $f(x) = 0$ then α will be a root of the derived equation $f'(x) = 0$ where $f'(x) = \dfrac{df}{dx}$.

The reason for this is as given below:

Let α be a repeated root of $f(x) = 0$.

Then two roots are α.

$\therefore \quad f(x) = (x-\alpha)(x-\alpha)\,\phi(x) = (x-\alpha)^2\,\phi(x)$

$\therefore \quad f'(x) = 2(x-\alpha)\,\phi(x) + (x-\alpha)^2\,\phi'(x)$

$\qquad = (x-\alpha)\{2\phi(x) + (x-\alpha)\,\phi'(x)\}$

$\therefore \quad f'(\alpha) = 0.$

- If α is a repeated root common in $f(x) = 0$ and $\phi(x) = 0$ then α is a common root in $f'(x) = 0$ and $\phi'(x) = 0$.
- $(x-\alpha)^2$ is a factor of $f(x)$

$\Rightarrow \quad x - \alpha$ is a factor of $f(x)$ as well as $f'(x)$.

Selected Solved Examples

1. For what values of a, the equation

$$(a^2 - a - 2)x^2 + (a^2 - 4)x + a^2 - 3a + 2 = 0$$

will have three solutions (more than two solutions)? Does there exist a value of x for which the above will become an identity in a?

We know $px^2 + qx + r = 0$ is an identity in x, i.e., the equation has three or more solutions, if $p = 0$, $q = 0$ and $r = 0$ simultaneously.

$\therefore$ the given quadratic equation will have three solutions if

$$a^2 - a - 2 = 0,\ a^2 - 4 = 0 \text{ and } a^2 - 3a + 2 = 0.$$

Now $a^2 - a - 2 = 0$

$\Rightarrow \quad (a-2)(a+1) = 0$

$\therefore \quad a = 2, -1$

Next $a^2 - 4 = 0$

$\Rightarrow \quad (a+2)(a-2) = 0$

$\therefore \quad a = 2, -2$

Also $a^2 - 3a + 2 = 0$

$\Rightarrow \quad (a-2)(a-1) = 0$

$\therefore \quad a = 2, 1$

$\therefore \quad$ for $a = 2$, the coefficients of x^2, x, x^0 are all zeros.

$\therefore \quad$ the required value of $a = 2$.

Now, rearranging into an equation in a we get

$$(x^2 + x + 1)a^2 - (x^2 + 3)a - 2x^2 - 4x + 2 = 0.$$

It will be an identity in a if

$$x^2 + x + 1 = 0,\ x^2 + 3 = 0 \text{ and } -2x^2 - 4x + 2 = 0.$$

There is no real value of x satisfying the above equations in x simultaneously.

$\therefore \quad$ the given equation is not an identity in a.

2. Obtain a quadratic equation in x and solve for it if

$$x = 1 + \cfrac{1}{3 + \cfrac{1}{2 + \cfrac{1}{3 + \cfrac{1}{2 + \dots}}}} \textbf{ to } \infty.$$

Here $x = 1 + \cfrac{1}{3 + \cfrac{1}{2 + (x-1)}}$

[because $x - 1 = \cfrac{1}{3 + \cfrac{1}{2 + \cfrac{1}{3 + \cfrac{1}{2 + \dots \text{ to } \infty}}}}$]

or $\quad x = 1 + \cfrac{1}{3 + \cfrac{1}{1+x}}$

or $\quad x = 1 + \cfrac{1}{\cfrac{4+3x}{1+x}}$

or $\quad x = 1 + \dfrac{1+x}{4+3x}$

or $\quad x = \dfrac{5+4x}{4+3x}$

or $\quad x(4+3x) = 5 + 4x$

or $\quad 3x^2 = 5$

$\therefore \quad x = \pm\sqrt{\dfrac{5}{3}}$.

But clearly x is positive; $\quad \therefore \quad x = \sqrt{\dfrac{5}{3}}$.

3. Solve for x: $(x^2 - 6x)^2 = 81 + 2(x-3)^2$.

Here, $x^2(x-6)^2 = 81 + 2(x-3)^2$.

Substitute $x - 3 = y$; then $x = y + 3$.

$\therefore$ the equation becomes,

$$(y+3)^2(y-3)^2 = 81 + 2y^2$$

or $(y^2-9)^2 = 81 + 2y^2$

or $y^4 - 18y^2 + 81 = 81 + 2y^2$

or $y^4 - 20y^2 = 0$, i.e., $y^2(y^2-20) = 0$

$\therefore$ $y^2 = 0, 20;$ $\therefore$ $y = 0, \pm\sqrt{20} = 0, \pm 2\sqrt{5}.$

$\therefore$ $x - 3 = 0, \pm 2\sqrt{5};$ $\therefore$ $x = 3, 3 \pm 2\sqrt{5}.$

[**Note** Roots of the equation are 3, 3, $3 \pm 2\sqrt{5}$.]

4. Solve for $x \in R$:

$$\sqrt{x+8+6\sqrt{x-1}} - \sqrt{x+3+4\sqrt{x-1}} = 1.$$

Here $\sqrt{x+8+6\sqrt{x-1}} - \sqrt{x+3+4\sqrt{x-1}} = 1$ ... (1)

$\therefore$ $(\sqrt{x+8+6\sqrt{x-1}} - \sqrt{x+3+4\sqrt{x-1}})$

$$\times (\sqrt{x+8+6\sqrt{x-1}} + \sqrt{x+3+4\sqrt{x-1}})$$

$$= \sqrt{x+8+6\sqrt{x-1}} + \sqrt{x+3+4\sqrt{x-1}}$$

$\therefore$ $x + 8 + 6\sqrt{x-1} - \{x + 3 + 4\sqrt{x-1}\}$

$$= \sqrt{x+8+6\sqrt{x-1}} + \sqrt{x+3+4\sqrt{x-1}}$$

or $\sqrt{x+8+6\sqrt{x-1}} + \sqrt{x+3+4\sqrt{x-1}} = 5 + 2\sqrt{x-1}$ ... (2)

(1) + (2) gives, $2\sqrt{x+8+6\sqrt{x-1}} = 6 + 2\sqrt{x-1}$

or $\sqrt{x+8+6\sqrt{x-1}} = 3 + \sqrt{x-1}$

$\therefore$ $\sqrt{x+8+6\sqrt{x-1}} - \sqrt{x-1} = 3$ ... (3)

$\therefore$ $(\sqrt{x+8+6\sqrt{x-1}} - \sqrt{x-1})$

$$\times (\sqrt{x+8+6\sqrt{x-1}} + \sqrt{x-1})$$

$$= 3(\sqrt{x+8+6\sqrt{x-1}} + \sqrt{x-1})$$

or $x + 8 + 6\sqrt{x-1} - (x-1)$

$$= 3(\sqrt{x+8+6\sqrt{x-1}} + \sqrt{x-1})$$

or $9 + 6\sqrt{x-1} = 3(\sqrt{x+8+6\sqrt{x-1}} + \sqrt{x-1})$

$\therefore$ $\sqrt{x+8+6\sqrt{x-1}} + \sqrt{x-1} = 3 + 2\sqrt{x-1}$ ... (4)

(4) – (3) gives, $2\sqrt{x-1} = 2\sqrt{x-1}$ which is an identity for all x.

But the given relation can hold for $x \geq 1$.

$\therefore$ solution-set $= R \cap [1, +\infty) = [1, +\infty)$.

5. Solve for x:

$$(x-1)^3 + (x-2)^3 + (x-3)^3 + (x-4)^3 + (x-5)^3 = 0.$$

Here AM of $\{x-1, x-5\}$ = AM of $\{x-2, x-4\} = x - 3$

Put $x - 3 = y$; then

$$x - 1 = y + 2, x - 2 = y + 1, x - 4 = y - 1, x - 5 = y - 2.$$

$\therefore$ the equation becomes,

$$(y+2)^3 + (y+1)^3 + y^3 + (y-1)^3 + (y-2)^3 = 0$$

or $\{(y+2)^3 + (y-2)^3\} + \{(y+1)^3 + (y-1)^3\} + y^3 = 0$

or $2(y^3 + 12y) + 2(y^3 + 3y) + y^3 = 0$

or $5y^3 + 30y = 0$

or $5y(y^2 + 6) = 0$

$\therefore$ $y = 0, \pm\sqrt{-6} = 0, \pm\sqrt{6}i.$ $\therefore$ $x = 3, 3 \pm \sqrt{6}i.$

6. Solve for x:

$$3x^3 = \{x^2 + \sqrt{18}\,x + \sqrt{32}\}\{x^2 - \sqrt{18}\,x - \sqrt{32}\} - 4x^2.$$

Here $3x^3 = \{x^2 + (\sqrt{18}x + \sqrt{32})\}\{x^2 - (\sqrt{18}x + \sqrt{32})\} - 4x^2$

$$= x^4 - (\sqrt{18}x + \sqrt{32})^2 - 4x^2$$

$$= x^4 - 2(3x+4)^2 - 4x^2$$

or $3x^3 + 4x^2 = x^4 - 2(3x+4)^2$

or $x^4 - 2(3x+4)^2 - x^2(3x+4) = 0.$

Put $3x + 4 = y$; then the equation becomes

$$x^4 - yx^2 - 2y^2 = 0$$

or $x^4 - 2yx^2 + yx^2 - 2y^2 = 0$

or $x^2(x^2 - 2y) + y(x^2 - 2y) = 0$

or $(x^2 - 2y)(x^2 + y) = 0$

$\therefore$ $x^2 - 2y = 0$ or $x^2 + y = 0.$

If $x^2 - 2y = 0$ then $x^2 - 2(3x+4) = 0$

or $x^2 - 6x - 8 = 0$

$\therefore$ $x = \dfrac{6 \pm \sqrt{36+32}}{2} = \dfrac{6 \pm \sqrt{68}}{2} = 3 \pm \sqrt{17}.$

If $x^2 + y = 0$ then $x^2 + 3x + 4 = 0$

$\therefore$ $x = \dfrac{-3 \pm \sqrt{9-16}}{2} = \dfrac{-3 \pm \sqrt{7}i}{2}.$

Thus $x = 3 \pm \sqrt{17}, \dfrac{-3 \pm \sqrt{7}i}{2}.$

7. If $\dfrac{x}{a}+\dfrac{y}{b}=1$ and $\dfrac{x^2}{a}+\dfrac{y^2}{b}=\dfrac{ab}{a+b}$, prove that

$$\frac{x^{n+1}}{a}+\frac{y^{n+1}}{b}=\left(\frac{ab}{a+b}\right)^n.$$

Here $\dfrac{x}{a}+\dfrac{y}{b}=1$ gives $\dfrac{y}{b}=1-\dfrac{x}{a}$

$$\therefore \quad y=\frac{b(a-x)}{a} \qquad \ldots (1)$$

Putting in the second equation,

$$\frac{x^2}{a}+\frac{1}{b}\left\{\frac{b(a-x)}{a}\right\}^2=\frac{ab}{a+b}$$

or $\dfrac{x^2}{a}+\dfrac{b(a-x)^2}{a^2}=\dfrac{ab}{a+b}$

or $ax^2+b(a-x)^2=\dfrac{a^3b}{a+b}$

or $(a+b)x^2-2abx+ba^2=\dfrac{a^3b}{a+b}$

or $(a+b)x^2-2abx+\left(ba^2-\dfrac{a^3b}{a+b}\right)=0$

or $(a+b)x^2-2abx+\dfrac{a^2b^2}{a+b}=0$

or $(a+b)^2x^2-2ab(a+b)x+(ab)^2=0$

or $\{(a+b)x-ab\}^2=0$

$\therefore \quad (a+b)x-ab=0; \quad \therefore \quad x=\dfrac{ab}{a+b}.$

Then from (1),

$$y=\frac{b}{a}\left\{a-\frac{ab}{a+b}\right\}=\frac{b}{a}\cdot\frac{a^2}{a+b}=\frac{ab}{a+b}$$

$$\therefore \quad \frac{x^{n+1}}{a}+\frac{y^{n+1}}{b}=\frac{1}{a}\cdot\left(\frac{ab}{a+b}\right)^{n+1}+\frac{1}{b}\cdot\left(\frac{ab}{a+b}\right)^{n+1}$$

$$=\left(\frac{ab}{a+b}\right)^n\cdot\left\{\frac{1}{a}\cdot\frac{ab}{a+b}+\frac{1}{b}\cdot\frac{ab}{a+b}\right\}$$

$$=\left(\frac{ab}{a+b}\right)^n\cdot\left(\frac{b}{a+b}+\frac{a}{a+b}\right)$$

$$=\left(\frac{ab}{a+b}\right)^n\cdot\frac{a+b}{a+b}=\left(\frac{ab}{a+b}\right)^n.$$

8. Solve $(18)^{8-4x}=(54\sqrt{2})^{3x-2}$.

Here $(3^2\cdot 2)^{8-4x}=(3^3\cdot 2\sqrt{2})^{3x-2}$

or $\{(3\sqrt{2})^2\}^{8-4x}=\{(3\sqrt{2})^3\}^{3x-2}$

or $(3\sqrt{2})^{16-8x}=(3\sqrt{2})^{9x-6}$

or $16-8x=9x-6$

$\therefore \quad 22=17x; \quad \therefore \quad x=\dfrac{22}{17}.$

9. Solve $2^{\sin^2 x}+5\cdot 2^{\cos^2 x}=7$.

Here $2^{\sin^2 x}+5\cdot 2^{1-\sin^2 x}=7.$

Putting $2^{\sin^2 x}=y$, the equation becomes

$$y+5\cdot\frac{2}{y}=7$$

or $y^2-7y+10=0$

or $(y-5)(y-2)=0$

$\therefore \quad y=5, 2$

$\therefore \quad 2^{\sin^2 x}=5, 2;$

but $2^{\sin^2 x}$ has the maximum value 2^1, i.e., 2.

$\therefore \quad 2^{\sin^2 x}\neq 5.$

Hence $2^{\sin^2 x}=2.$

$\therefore \quad \sin^2 x=1$

or $\sin x=\pm 1=\sin\left(\pm\dfrac{\pi}{2}\right)$

$\therefore \quad x=n\pi+(-1)^n\left(\pm\dfrac{\pi}{2}\right), n\in \mathbb{Z}$ (the set of integers).

10. Solve for x: $4^{x+1.5}+9^{x+0.5}=10\cdot 6^x$.

Here $2^{2(x+1.5)}+3^{2(x+0.5)}=10\cdot 3^x\cdot 2^x$

or $2^{2x+3}+3^{2x+1}=10\cdot 2^x\cdot 3^x$

or $8.2^{2x}+3.3^{2x}=10\cdot 2^x\cdot 3^x$

or $8(2^x)^2-10\cdot 2^x\cdot 3^x+3(3^x)^2=0$

or $8\left\{\left(\dfrac{2}{3}\right)^x\right\}^2-10\cdot\left(\dfrac{2}{3}\right)^x+3=0$

or $\left\{4\left(\dfrac{2}{3}\right)^x-3\right\}\left\{2\left(\dfrac{2}{3}\right)^x-1\right\}=0$

$\therefore \quad 4\left(\dfrac{2}{3}\right)^x=3 \quad$ or $\quad 2\left(\dfrac{2}{3}\right)^x=1$

$\therefore \quad \left(\dfrac{2}{3}\right)^x=\dfrac{3}{4} \quad$ or $\quad \left(\dfrac{2}{3}\right)^x=\dfrac{1}{2}$

$\therefore \quad x=\log_{\frac{2}{3}}\left(\dfrac{3}{4}\right) \quad$ or $\quad x=\log_{\frac{2}{3}}\left(\dfrac{1}{2}\right).$

11. Solve $(5+2\sqrt{6})^{x^2-3}+(5-2\sqrt{6})^{x^2-3}=10$.

Clearly $(5+2\sqrt{6})(5-2\sqrt{6})=5^2-(2\sqrt{6})^2=25-24=1$

$\therefore \quad 5-2\sqrt{6}=\dfrac{1}{5+2\sqrt{6}}$.

$\therefore$ the equation is

$$(5+2\sqrt{6})^{x^2-3}+\left(\frac{1}{5+2\sqrt{6}}\right)^{x^2-3}=10.$$

Writing $5+2\sqrt{6}=a,\ a^{x^2-3}+\left(\dfrac{1}{a}\right)^{x^2-3}=10$

or $\quad (a^{x^2-3})^2-10\cdot a^{x^2-3}+1=0$

$\therefore \quad a^{x^2-3}=\dfrac{10\pm\sqrt{100-4}}{2}=\dfrac{10\pm4\sqrt{6}}{2}=5\pm2\sqrt{6}$

$\therefore \quad (5+2\sqrt{6})^{x^2-3}=5+2\sqrt{6},\ 5-2\sqrt{6}$

$\qquad\qquad =(5+2\sqrt{6})^1,\ (5+2\sqrt{6})^{-1}$

$\Rightarrow \quad x^2-3=1,-1$

or $\quad x^2=4,2$

$\therefore \quad x=\pm2,\pm\sqrt{2}$.

12. Solve $(\sqrt{3}+1)^{2x}+(\sqrt{3}-1)^{2x}=2^{3x}$.

Here $(\sqrt{3}+1)^{2x}+(\sqrt{3}-1)^{2x}=(2^3)^x=8^x$

$\qquad =\{(2\sqrt{2})^2\}^x=(2\sqrt{2})^{2x} \quad \ldots (1)$

Clearly, $(\sqrt{3}+1)^2+(\sqrt{3}-1)^2=(2\sqrt{2})^2$

$$\therefore \quad \left(\frac{\sqrt{3}+1}{2\sqrt{2}}\right)^2+\left(\frac{\sqrt{3}-1}{2\sqrt{2}}\right)^2=1.$$

$\therefore$ if $\dfrac{\sqrt{3}+1}{2\sqrt{2}}=\cos\theta,\ \dfrac{\sqrt{3}-1}{2\sqrt{2}}=\sin\theta$

Now from (1), $\left(\dfrac{\sqrt{3}+1}{2\sqrt{2}}\right)^{2x}+\left(\dfrac{\sqrt{3}-1}{2\sqrt{2}}\right)^{2x}=1$

$\therefore \quad \cos^{2x}\theta+\sin^{2x}\theta=1 \quad \Rightarrow \quad x=1$

for $\cos^n\theta+\sin^n\theta=1$ holds only when $n=2$.

Hence, $x=1$.

13. If $a>1$, prove that the solutions of the following equation in x is independent of a.

$$(a+\sqrt{a^2-1})^{x^2-2x}+(a-\sqrt{a^2-1})^{x^2-2x}=2a$$

We know, $(a+\sqrt{a^2-1})(a-\sqrt{a^2-1})=a^2-(a^2-1)=1$.

$\therefore \quad a-\sqrt{a^2-1}=\dfrac{1}{a+\sqrt{a^2-1}}$

$$\therefore \quad (a+\sqrt{a^2-1})^{x^2-2x}+\left(\frac{1}{a+\sqrt{a^2-1}}\right)^{x^2-2x}=2a$$

or $\quad \left\{(a+\sqrt{a^2-1})^{x^2-2x}\right\}^2-2a(a+\sqrt{a^2-1})^{x^2-2x}+1=0$

$\therefore \quad (a+\sqrt{a^2-1})^{x^2-2x}=\dfrac{2a\pm\sqrt{4a^2-4}}{2}=a\pm\sqrt{a^2-1}$

$\qquad =a+\sqrt{a^2-1},\ (a+\sqrt{a^2-1})^{-1}$

$\therefore \quad x^2-2x=1,-1$

$\therefore \quad x^2-2x-1=0 \quad$ or $\quad x^2-2x+1=0$

$\therefore \quad x=\dfrac{2\pm\sqrt{4+4}}{2}=1\pm\sqrt{2}. \qquad \therefore \quad (x-1)^2=0; \quad \therefore \quad x=1$

$\therefore$ the solutions are $x=1,\ 1\pm\sqrt{2}$

which are independent of a ($a>1$ ensures $\sqrt{a^2-1}$ is real).

14. Solve $\log_{10}(2^x+1)+x=\log_{10}6+x\log_{10}5$.

Here $\log_{10}(2^x+1)=\log_{10}6+x(\log_{10}5-1)$

or $\quad \log_{10}(2^x+1)=\log_{10}6+x(\log_{10}5-\log_{10}10)$

or $\quad \log_{10}(2^x+1)=\log_{10}6+x\log_{10}\dfrac{1}{2}$

or $\quad \log_{10}(2^x+1)=\log_{10}6-\log_{10}2^x$

or $\quad \log_{10}(2^x+1)+\log_{10}2^x=\log_{10}6$

or $\quad \log_{10}\{2^x(2^x+1)\}=\log_{10}6$

or $\quad 2^x(2^x+1)=6$

or $\quad (2^x)^2+2^x-6=0$

or $\quad (2^x+3)(2^x-2)=0$

$\therefore \quad 2^x+3=0 \quad \ldots (1)$

or $\quad 2^x-2=0 \quad \ldots (2)$

As 2^x is always positive, (1) is not possible.

$\therefore$ from (2), $2^x=2 \quad \Rightarrow \quad x=1$.

15. Solve

$$\mathbf{\log_{2x+3}(6x^2+23x+21)=4-\log_{3x+7}(4x^2+12x+9).}$$

Here, $\log_{2x+3}(2x+3)(3x+7)=4-\log_{3x+7}(2x+3)^2$

or $\quad \log_{2x+3}(2x+3)+\log_{2x+3}(3x+7)$

$\qquad =4-2\log_{3x+7}(2x+3)$

or $\quad 1+\log_{2x+3}(3x+7)=4-2\log_{3x+7}(2x+3)$.

Let $\log_{2x+3}(3x+7)=y$; then $\log_{3x+7}(2x+3)=\dfrac{1}{y}$.

$\therefore$ the equation becomes $1+y=4-\dfrac{2}{y}$

or $\quad y^2=3y-2$

or $\quad y^2-3y+2=0$

or $(y-2)(y-1)=0;$ $\therefore$ $y=2, 1$

$\therefore$ $\log_{2x+3}(3x+7)=2, 1$

or $3x+7=(2x+3)^2, 2x+3.$

If $3x+7=(2x+3)^2$

then $3x+7=4x^2+12x+9$

or $4x^2+9x+2=0$

or $(4x+1)(x+2)=0;$ $\therefore$ $x=\frac{-1}{4}, -2.$

If $3x+7=2x+3$ then $x=-4.$

But when $x=-4, 2x+3=-8+3<0.$ But no base of logarithm can be negative. So, $x=-4$ is not a solution.

When $x=-2, 2x+3=-4+3<0.$ So $x=-2$ is also not a solution.

When $x=\frac{-1}{4}, 2x+3=\frac{-2}{4}+3>0$ and $\neq 1$

$3x+7=\frac{-3}{4}+7>0$ and $\neq 1$

$6x^2+23x+21=(2x+3)(3x+7)=(+ve)\times(+ve)$

$=(+ve)$

$4x^2+12x+9=(2x+3)^2>0.$

$\therefore$ $x=-\frac{1}{4}$ is the only solution.

16. Solve $25^{\log_{10}x}=5+4x^{\log_{10}5}$.

Here, $5^{2\log_{10}x}=5+4\times 5^{\log_5 x^{\log_{10}5}}$ $\{\because a=b^{\log_b a}\}$

$=5+4\times 5^{\log_{10}5\times\log_5 x}=5+4\times 5^{\log_{10}x}$

Note Our purpose was to write $x^{\log_{10}5}$ in terms of an exponential with the common base 5.

So, the equation becomes

$(5^{\log_{10}x})^2-4(5^{\log_{10}x})-5=0$

or $(5^{\log_{10}x}-5)(5^{\log_{10}x}+1)=0$

But $5^{\log_{10}x}+1\neq 0$ ($\because$ $5^{\log_{10}x}$ is positive)

$\therefore$ $5^{\log_{10}x}-5=0$

$\therefore$ $5^{\log_{10}x}=5;$ or $\log_{10}x=1$ $\therefore$ $x=10.$

17. Solve $|x^2+4x+3|+2x+5=0$.

Here $|x^2+4x+3|=-(2x+5)$ ... (1)

As $|y|\geq 0$, the above equation can have solution if $2x+5\leq 0$.

Also $|y|=a$ $\Rightarrow$ $y=\pm a$ where $a\geq 0$.

Thus, $x^2+4x+3=\mp(2x+5)$ where $2x+5\leq 0$.

Now, $x^2+4x+3=-(2x+5)$

$\Rightarrow$ $x^2+6x+8=0$

or $(x+4)(x+2)=0;$ $\therefore$ $x=-4, -2.$

Again $x^2+4x+3=2x+5$

$\Rightarrow$ $x^2+2x-2=0$

$\therefore$ $x=\frac{-2\pm\sqrt{4+8}}{2}=\frac{-2\pm 2\sqrt{3}}{2}=-1\pm\sqrt{3}.$

$\therefore$ $x=-4, -2, -1+\sqrt{3}, -1-\sqrt{3}.$

When $x=-4, 2x+5=-8+5<0$

$x=-2, 2x+5=-4+5>0$

$x=-1+\sqrt{3}, 2x+5=2(-1+\sqrt{3})+5=3+2\sqrt{3}>0$

$x=-1-\sqrt{3}, 2x+5=2(-1-\sqrt{3})+5=3-2\sqrt{3}<0$

As $2x+5\leq 0,$ the solutions are $x=-4, -1-\sqrt{3}.$

18. For $a<0$, determine all solutions of the equation $x^2-2a|x-a|-3a^2=0$.

Here $2a|x-a|=x^2-3a^2$

or $|x-a|=\frac{1}{2a}(x^2-3a^2).$

But $|x-a|$ is positive and a is negative.

So $x^2-3a^2\leq 0.$

$\therefore$ $x-a=\pm\frac{1}{2a}(x^2-3a^2)$, provided $x^2-3a^2\leq 0.$

Now, $x-a=\frac{1}{2a}(x^2-3a^2)$

$\Rightarrow$ $2ax-2a^2=x^2-3a^2$

or $x^2-2ax-a^2=0$

or $x=\frac{2a\pm\sqrt{4a^2+4a^2}}{2}=(1\pm\sqrt{2})a.$

Again, $x-a=-\frac{1}{2a}(x^2-3a^2)$

$\Rightarrow$ $2ax-2a^2=-x^2+3a^2$

or $x^2+2ax-5a^2=0$

$\therefore$ $x=\frac{-2a\pm\sqrt{4a^2+20a^2}}{2}$

or $x=\frac{-2a\pm 2\sqrt{6}a}{2}=(-1\pm\sqrt{6})a$

$\therefore$ we get $x=(1+\sqrt{2})a, (1-\sqrt{2})a, (-1+\sqrt{6})a, (-1-\sqrt{6})a.$

When $x=(1+\sqrt{2})a,$

$x^2-3a^2=(1+\sqrt{2})^2a^2-3a^2=2\sqrt{2}a^2>0.$

When $x=(1-\sqrt{2})a,$

$x^2-3a^2=(1-\sqrt{2})^2a^2-3a^2=-2\sqrt{2}a^2<0.$

When $x = (-1+\sqrt{6})a$,

$$x^2 - 3a^2 = (-1+\sqrt{6})^2 a^2 - 3a^2 = 4a^2 - 2\sqrt{6}a^2 < 0.$$

When $x = (-1-\sqrt{6})a$,

$$x^2 - 3a^2 = (-1-\sqrt{6})^2 a^2 - 3a^2 = 4a^2 + 2\sqrt{6}a^2 > 0.$$

As $x^2 - 3a^2 \le 0$, the solutions are $(1-\sqrt{2})a, (-1+\sqrt{6})a$.

19. Solve $(x+2)^2 - 3|x+2| + 2 = 0$.

We know $|x|^2 = (\pm x)^2 = x^2$

$\therefore \quad |x+2|^2 - 3|x+2| + 2 = 0$

or $\quad (|x+2| - 2)(|x+2| - 1) = 0$

$\therefore \quad |x+2| = 2, 1$

or $\quad x + 2 = \pm 2, \pm 1$

or $\quad x = \pm 2 - 2, \pm 1 - 2 = 0, -4, -1, -3.$

Note

If $|x| = 3, -1$ our conclusion is $x = \pm 3$ for $|x| \neq -1$ as $|x|$ is positive.

If $|x| = 1, 0$ our conclusion is $x = \pm 1, 0$.

If $|x| = -2, 0$ our conclusion is $x = 0$ for $|x| \neq -2$ as $|x|$ is positive.

If $|x| = -1, -2$ our conclusion is x has no value as $|x|$ is positive.

20. Solve $|x+1| + |x-1| = 2$.

We know $|x-a| = x - a$ if $x \ge a$

$\qquad\qquad\qquad\quad = a - x$ if $x < a$

So the modulus functions $|x+1|$ and $|x-1|$ are defined as follows:

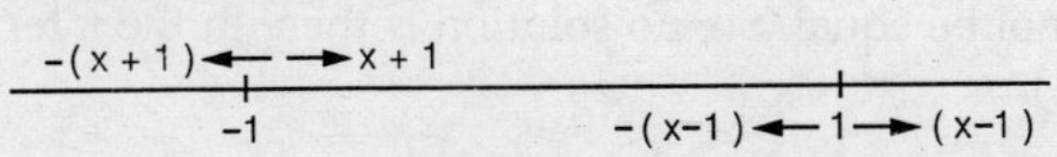

$\therefore \quad$ if $x < -1$, the equation is $-(x+1) - (x-1) = 2$

or $\quad -2x = 2; \quad \therefore \quad x = -1.$

But $x < -1$. So there is no solution.

If $-1 \le x < 1$, the equation is $x + 1 - (x-1) = 2$

or $\quad 2 = 2$ which is true for all x. But $-1 \le x < 1$.

$\therefore \quad$ we get $-1 \le x < 1$.

If $x \ge 1$, the equation is $x + 1 + x - 1 = 2$

or $\quad 2x = 2; \quad \therefore \quad x = 1$ which satisfies $x \ge 1$.

$\therefore \quad$ the required solution is $-1 \le x < 1$ or $x = 1$.

$\therefore \quad x \in [-1, 1]$.

Note

(1) In the above, for $-1 \le x < 1$ we got $2 = 2$ which is true for all x of the interval. Instead, if for $-1 \le x < 1$ we get $2 = 3$, our conclusion will be that x has no value in the interval $-1 \le x < 1$.

(2) If we have $|3-x|$, replace it by $|x-3|$ in the equation.

(3) If we have $|2x+1|$, replace it by $2\left|x+\frac{1}{2}\right|$ in the equation.

21. $|x^2 - 3x - 4| = 9 - |x^2 - 1|$.

Here $|(x-4)(x+1)| = 9 - |(x+1)(x-1)|$

or $\quad |x-4|\,|x+1| = 9 - |x+1|\,|x-1|.$

Here the modulus functions are defined as follows:

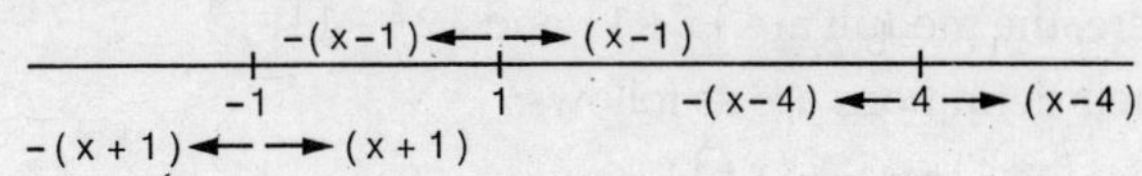

If $x < -1$, the equation is

$$\{-(x-4)\}\{-(x+1)\} = 9 - \{-(x+1)\}\{-(x-1)\}$$

or $\quad (x-4)(x+1) = 9 - (x+1)(x-1)$

or $\quad x^2 - 3x - 4 = 9 - (x^2 - 1)$

or $\quad 2x^2 - 3x - 14 = 0$

or $\quad 2x^2 - 7x + 4x - 14 = 0$

or $\quad x(2x-7) + 2(2x-7) = 0$

or $\quad (2x-7)(x+2) = 0; \quad \therefore \quad x = \frac{7}{2}, -2.$

But $x < -1$. So $x = -2$.

If $\quad -1 \le x < 1$, the equation is

$$\{-(x-4)\}(x+1) = 9 - (x+1)\{-(x-1)\}$$

or $\quad -(x-4)(x+1) = 9 + (x^2 - 1)$

or $\quad -x^2 + 3x + 4 = 8 + x^2$

or $\quad 2x^2 - 3x + 4 = 0$

or $\quad x = \dfrac{3 \pm \sqrt{9 - 32}}{4}$ which are not real.

So there is no solution in the interval.

If $1 \le x < 4$, the equation is

$$\{-(x-4)\}(x+1) = 9 - (x+1)(x-1)$$

or $\quad -(x^2 - 3x - 4) = 9 - (x^2 - 1)$

or $\quad 3x + 4 = 10; \quad \therefore \quad x = 2$ which satisfies $1 \le x < 4$.

If $x \ge 4$, the equation is

$$(x-4)(x+1) = 9 - (x+1)(x-1)$$

or $\quad x^2 - 3x - 4 = 9 - (x^2 - 1)$

or $\quad 2x^2 - 3x - 14 = 0$

or $\quad (2x-7)(x+2) = 0; \quad \therefore \quad x = \frac{7}{2}, -2.$

But $x \ge 4$; so x has no solution in this interval.

Hence, solutions are $x = -2, 2$.

Note Sometimes $ax^2 + bx + c$ in $|ax^2 + bx + c|$ cannot be factorized in real factors. It happens when $ax^2 + bx + c = 0$ has imaginary roots, i.e., $D < 0$. In this case, if $a > 0$, $ax^2 + bx + c$ will be always positive and so $|ax^2 + bx + c|$ can be replaced by $ax^2 + bx + c$. If $a < 0$, $ax^2 + bx + c$ will be always negative and so $|ax^2 + bx + c|$ will be replaced by $-(ax^2 + bx + c)$.

22. Solve $2^{|x+1|} - 2^x = |2^x - 1| + 1$.

Here, the moduli are $|x+1|$ and $|2^x - 1|$.
Their definitions are as follows:

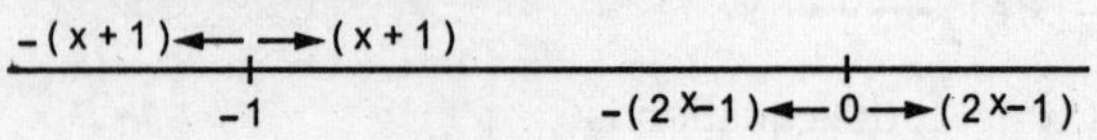

If $x < -1$, the equation becomes

$$2^{-(x+1)} - 2^x = -(2^x - 1) + 1$$

or $\quad 2^{-x-1} - 2^x = -2^x + 2$

or $\quad 2^{-x-1} = 2; \quad \therefore \quad -x-1 = 1$

$\therefore \quad x = -2$ which satisfies $x < -1$.

If $-1 \le x < 0$, the equation becomes

$$2^{x+1} - 2^x = -(2^x - 1) + 1$$

or $\quad 2^{x+1} = 2; \quad \therefore \quad x + 1 = 1$

$\therefore \quad x = 0$ which does not satisfy $-1 \le x < 0$.

If $x \ge 0$, the equation becomes

$$2^{x+1} - 2^x = 2^x - 1 + 1$$

or $\quad 2^{x+1} = 2 \cdot 2^x = 2^{x+1}$ which is true for all $x \ge 0$.

$\therefore \quad$ the solution is $x = \{-2\} \cup [0, +\infty)$.

23. Solve $\log_{|x+6|^2} 2 \times \log_2 (x^2 - x - 2) = 1$.

Here, $\log_{|x+6|^2} 2 \times \log_2(x^2 - x - 2) = 1$

or $\quad \log_{|x+6|^2}(x^2 - x - 2) = 1$

$$\{\because \log_n m \times \log_m p = \log_n p\}$$

$\Rightarrow \quad x^2 - x - 2 = |x+6|^2 \qquad \ldots (1)$

Clearly, for logarithmic functions to be defined

$$|x+6|^2 > 0; \quad |x+6|^2 \ne 1; \quad x^2 - x - 2 > 0$$

$\Rightarrow \quad x \ne -6; \quad |x+6| \ne 1; \quad (x-2)(x+1) > 0$

$\Rightarrow \quad x \ne -6; \quad x \ne -5, -7; \quad x < -1$ or $x > 2 \qquad \ldots (2)$

Now, (1) $\Rightarrow x^2 - x - 2 = (x+6)^2 \quad \{\because |x|^2 = x^2\}$

or $\quad x^2 - x - 2 = x^2 + 12x + 36$

or $\quad 13x = -38; \quad \therefore \quad x = \frac{-38}{13}$.

Clearly, $x = \frac{-38}{13}$ satisfies all the conditions of (2).

Hence, the solution is $x = \frac{-38}{13}$.

24. Find $x \in R$ satisfying the equation

$$|x-1|^{\log_3 x^2 - 2\log_x 9} = (x-1)^7.$$

If $x > 1$, $(x-1)^{\log_3 x^2 - 2\log_x 9} = (x-1)^7$

provided $x - 1 \ne 1$, i.e., $x \ne 2$

$\Rightarrow \quad \log_3 x^2 - 2\log_x 9 = 7$

or $\quad 2\log_3 x - 4\log_x 3 = 7$

or $\quad 2\log_3 x - 4 \cdot \frac{1}{\log_3 x} = 7$

or $\quad 2(\log_3 x)^2 - 7(\log_3 x) - 4 = 0$

or $\quad (2\log_3 x + 1)(\log_3 x - 4) = 0$

or $\quad \log_3 x = \frac{-1}{2}, 4$

$\Rightarrow \quad x = 3^{-1/2}, 3^4 = \frac{1}{\sqrt{3}}, 81$.

But $\frac{1}{\sqrt{3}} \ngtr 1$; so $x = 81$.

If $x = 2$, the equation is satisfied.

If $0 < x < 1$, $(1-x)^{\log_3 x^2 - 2\log_x 9} = -(1-x)^7$.

Here LHS is positive but the RHS is negative. So they cannot be equal. So, no solution is there in the interval (0, 1).

Clearly, $x \ne 1$ for then base $\ngtr 0$.

Hence, the solutions are $x = 81, 2$.

25. Let $\{x\}$ and $[x]$ denote the fractional and integral parts of a real number x respectively. Solve $4\{x\} = x + [x]$.

Here, $x = \{x\} + [x]$ according to the question.

Clearly, x is an integer or x is not an integer.

If x is an integer, let $x = n$. Then $[x] = n$, $\{x\} = 0$.

$\therefore \quad$ the equation $\quad 4\{x\} = x + [x] \qquad \ldots (1)$

gives $0 = n + n; \quad \therefore \quad n = 0; \quad \therefore \quad x = 0$.

If x is positive and not an integer, let $x = n + k$ where n is a positive integer and $0 < k < 1$

$\therefore \quad [x] = n, \{x\} = k$.

So, from (1), $4k = n + k + n$

or $3k = 2n$; $\therefore$ $k = \frac{2n}{3}$.

But $0 < k < 1$; $\therefore$ $0 < \frac{2n}{3} < 1$

or $0 < n < \frac{3}{2}$; $\therefore$ $n = 1$ and so $k = \frac{2}{3}$

$\therefore$ $x = n + k = 1 + \frac{2}{3} = \frac{5}{3}$.

If x is negative and not an integer, let $x = -n - k$, where n is any positive integer and $0 < k < 1$.

$\therefore$ $[x] = -n$, $\{x\} = -k$. So, from (1),

$4(-k) = -n - k + (-n)$

or $-4k = -2n - k$

or $3k = 2n$; $\therefore$ $k = \frac{2n}{3}$.

But $0 < k < 1$; $\therefore$ $0 < \frac{2n}{3} < 1$

or $0 < 2n < 3$; $\therefore$ $0 < n < \frac{3}{2}$. So $n = 1$.

$\therefore$ $x = -n - k = -1 - \frac{2}{3} = -\frac{5}{3}$.

Hence, the solutions are $x = 0, \pm\frac{5}{3}$.

Note

(1) In the above sum we have $[x]$, whose definition changes in the intervals as shown:

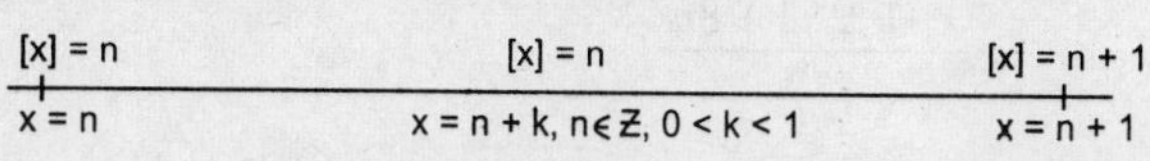

(2) If we have $[2x]$ instead of $[x]$, definition changes in the intervals as shown below:

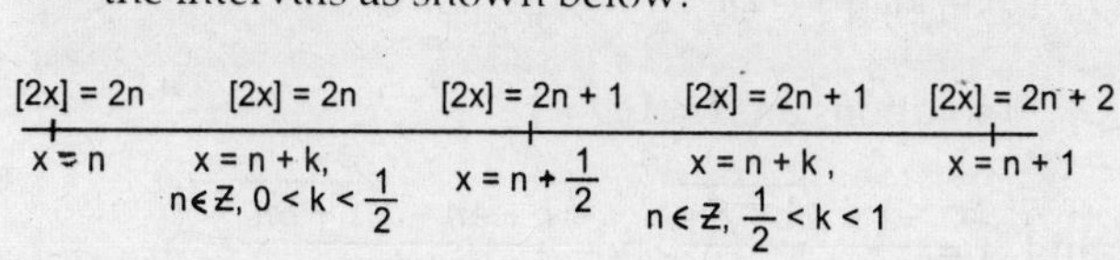

26. For $x \in R$, $\|x\|$ is defined as follows

$$\|x\| = \begin{cases} x + 1, & 0 \le x < 2 \\ |x - 4|, & 2 \le x \end{cases}$$

then solve the equation $\|x\|^2 + x = \|x\| + x^2$.

If $0 \le x < 2$, the equation is $(x + 1)^2 + x = x + 1 + x^2$

or $x^2 + 2x + 1 + x = x + 1 + x^2$

or $2x = 0$; $\therefore$ $x = 0$ which satisfies $0 \le x < 2$.

If $x \ge 2$, the equation is $|x - 4|^2 + x = |x - 4| + x^2$

or $(x - 4)^2 + x = |x - 4| + x^2$

or $x^2 - 8x + 16 + x = |x - 4| + x^2$

or $|x - 4| = 16 - 7x$

$\therefore$ $x - 4 = \pm(16 - 7x)$

provided $16 - 7x \ge 0$, i.e., $x \le \frac{16}{7}$.

When $x - 4 = 16 - 7x$ then $8x = 20$

or $x = \frac{5}{2}$ which does not satisfy $x \ge 2$ and $x \le \frac{16}{7}$.

When $x - 4 = -(16 - 7x)$ then $6x = 12$

or $x = 2$ which satisfies $x \ge 2$ and $x \le \frac{16}{7}$.

$\therefore$ the solutions are $x = 0, 2$.

27. Find all non-negative integral solutions of $y^2 + 6xy - 8x = 0$. Does it contain a positive integral solution which is a multiple of 3?

Here $8x - 6xy = y^2$; or $x(8 - 6y) = y^2$

$\therefore$ $x = \frac{y^2}{8 - 6y}$; but x is positive.

So non-negative integral values of y can be $y = 0, 1$ only.

When $y = 0, x = 0$.

When $y = 1, x = \frac{1^2}{8 - 6} = \frac{1}{2} \notin$ set of positive integers.

$\therefore$ only non-negative integral solution of x, y is $(0, 0)$.

Hence, the equation has only one non-negative integral solution. So it has no positive integral solutions which are multiples of 3.

28. Prove that $2x^4 + 1402 = y^4$ has no integral solution.

Let $x^2 = u, y^2 = v$. Then the equation is

$2u^2 = v^2 - 1402$.

As u, v are integers, $v^2 > 1402$.

But $37^2 = 1369, 38^2 = 1444$.

$\therefore$ minimum possible value of $v = 38$.

Also $2u^2$ is even; so $v^2 - 1402$ is even.

$\therefore$ v must be even.

Let $v = 38 + 2k$ where $k \in N$.

$\therefore$ $2u^2 = (38 + 2k)^2 - 1402 = 4k^2 + 152k + 42$

$\therefore$ $u^2 = 2k^2 + 76k + 21$

So $2k^2 + 76k + 21$ is a perfect square.

Hence $D = 0$

i.e., $76^2 - 4 \cdot 2 \cdot 21 = 0$ which is incorrect.

$\therefore$ u, v cannot have positive integral solutions.

So x, y cannot have integral solutions.

29. If x, y, z are positive integers and $x+y+z=240$ and $x^2+y^2=z^2$ then find the number of possible solutions.

As $x^2+y^2=z^2$, x, y and z are the sides of a right-angled triangle whose hypotenuse is z.

We know that $2n+1, 2n^2+2n$ and $2n^2+2n+1$ are sides of a right-angled triangle for all $n \in N$ because

$$(2n+1)^2+(2n^2+2n)^2=(2n^2+2n+1)^2.$$

Hence, the sides of a right-angled triangle are in the ratio $2n+1, 2n^2+2n, 2n^2+2n+1$.

So, taking $n=1, 2, 3, \ldots$ the solutions of $x^2+y^2=z^2$ are as follows:

when $n=1$, the solutions are (3, 4, 5)

when $n=2$, the solutions are (5, 12,13)

when $n=3$, the solutions are (7, 24, 25)

when $n=4$, the solutions are (9, 40, 41)

when $n=5$, the solutions are (11, 60, 61)

when $n=6$, the solutions are (13, 84, 85)

when $n=7$, the solutions are (15, 112, 113), etc.

So the solutions are either in the ratio 3, 4, 5 or 5, 12, 13 etc.

But $x+y+z=240$.

Now $3k+4k+5k=240$ if $k=20$.

So (60, 80, 100) is a solution.

$$5k+12k+13k=240 \text{ if } k=8.$$

So (40, 96, 104) is a solution.

$$7k+24k+25k=240 \text{ if } k=\frac{240}{56} \notin N.$$

So there is no solution proportional to (7, 24, 25).

$$9k+40k+41k=240 \text{ if } k=\frac{240}{90} \notin N.$$

So there is no solution proportional to (9, 40, 41).

$$11k+60k+61k=240 \text{ if } k=\frac{240}{132} \notin N.$$

So there is no solution proportional to (11, 60, 61).

$$13k+84k+85k=240 \text{ if } k=\frac{240}{182} \notin N.$$

So there is no solution proportional to (13, 84, 85).

$$15k+112k+113k=240 \text{ if } k=1.$$

So (15, 112, 113) is a solution.

$\therefore$ required solutions are $x=60, y=80, z=100$

$x=40, y=96, z=104$

$x=15, y=112, z=113$

$\therefore$ there are only 3 solutions.

30. Find $x>0, y>0$ satisfying the system of equations $x^{x+y}=y^n$ and $y^{x+y}=x^{2n}\cdot y^n$ where $n>0$.

Let $x^{x+y}=y^n$ ... (1)

$y^{x+y}=x^{2n}\cdot y^n$ ... (2)

Multiplying (1) and (2),

$$(xy)^{x+y}=x^{2n}\cdot y^{2n}=(xy)^{2n}$$

$\therefore$ $x+y=2n$ (provided $xy \neq 1$) or $xy=1$.

If $xy=1$, (1) becomes $x^{x+\frac{1}{x}}=x^{-n}$

$\Rightarrow$ $x=1$ or $x+\dfrac{1}{x}=-n$.

If $x=1$ then $y=\dfrac{1}{x}=1$.

If $x+\dfrac{1}{x}=-n<0$ then $x>0$ cannot hold.

If $xy \neq 1, x+y=2n$; then (1) gives $x^{2n}=y^n$

$\Rightarrow$ $\left(\dfrac{x^2}{y}\right)^n=1$

$\Rightarrow$ $n=0$ or $x^2=y$.

But $n>0$; so $x^2=y$.

$\therefore$ $x+y=2n$ gives $x+x^2=2n$

or $x^2+x-2n=0$

or $x=\dfrac{-1 \pm \sqrt{1+8n}}{2}$.

But $x>0$; so $x=\dfrac{-1+\sqrt{1+8n}}{2}$ and

$$y=2n-x=2n-\frac{-1+\sqrt{1+8n}}{2}$$

$\therefore$ solutions are $x=1, y=1$

and $x=\dfrac{-1+\sqrt{1+8n}}{2}, y=\dfrac{1+4n-\sqrt{1+8n}}{2}$.

31. Solve for x, y:

$$\log_{100}|x+y|=\frac{1}{2}$$

$$\log_{10} y-\log_{10}|x|=\log_{100} 4.$$

Here the two equations in x, y are

$\log_{100}|x+y|=\dfrac{1}{2}$ ... (1)

$\log_{10} y-\log_{10}|x|=\log_{100} 4.$ ... (2)

From (2), clearly $y>0$ for $\log_{10} y$ is real.

Also $x \neq 0, x+y \neq 0$.

From (1), $|x+y| = 100^{1/2} = 10$ ... (3)

the negative value being unacceptable as $|x+y| \geq 0$.

From (2), $\log_{10} \frac{y}{|x|} = \log_{100} 4 = \frac{\log_{10} 4}{\log_{10} 100}$

$$= \frac{2\log_{10} 2}{2\log_{10} 10} = \log_{10} 2$$

$\therefore \quad \frac{y}{|x|} = 2$, i.e., $2|x| = y$ ... (4)

Now, (3) $\Rightarrow$ $x + y = 10, -10.$

If $x + y = 10$, (4) gives $2|x| = 10 - x$

$\therefore \quad 2x = \pm(10 - x)$ provided $x \leq 10$.

When $2x = 10 - x$, $3x = 10$, i.e., $x = \frac{10}{3} < 10$.

When $2x = -(10 - x)$, $x = -10 < 10$

$\therefore \quad x = \frac{10}{3}, -10.$

$\therefore \quad y = 10 - x = 10 - \frac{10}{3}, 10 + 10 = \frac{20}{3}, 20.$

If $x + y = -10$, (4) gives $2|x| = -x - 10$

$\therefore \quad 2x = \pm(-x - 10)$

provided $-x - 10 \geq 0$, i.e., $x \leq -10$.

When $2x = -x - 10$, $3x = -10$, i.e., $x = \frac{-10}{3} > -10$.

But $x \leq -10$; so $x = \frac{-10}{3}$ is not a solution.

When $2x = -(-x - 10)$, $x = 10$. But $x \leq -10$.

So $x = 10$ is not a solution.

Thus, the solutions are $x = \frac{10}{3}$, $y = \frac{20}{3}$

$x = -10$, $y = 20$.

32. Find all values of θ in the interval $\left(-\frac{\pi}{2}, \frac{\pi}{2}\right)$ satisfying the equation $(1 - \tan\theta)(1 + \tan\theta)\sec^2\theta + 2^{\tan^2\theta} = 0$.

Here, $(1 - \tan^2\theta)(1 + \tan^2\theta) + 2^{\tan^2\theta} = 0.$

Put $\tan^2\theta = x$; then $(1 - x)(1 + x) + 2^x = 0$

or $1 - x^2 + 2^x = 0$

or $2^x = x^2 - 1$ ... (1)

By observation $x = 3$ is a root because

$2^x = 2^3 = 8$ and $x^2 - 1 = 3^2 - 1 = 8.$

$\therefore \quad x = 3$ is a solution. So $\tan^2\theta = 3$.

$\therefore \quad \tan\theta = \pm\sqrt{3}$; $\therefore \quad \theta = \frac{\pi}{3}, \frac{-\pi}{3}.$

Thus we find the equation has two solutions

$\theta = \frac{\pi}{3}, \frac{-\pi}{3}.$

Now, we want to find values of θ satisfying the given equation other than the values $\theta = \pm\frac{\pi}{3}$.

The solutions of (1) are given by the intersections of the graphs of $y = 2^x$ and $y = x^2 - 1$. The graphs are as follows:

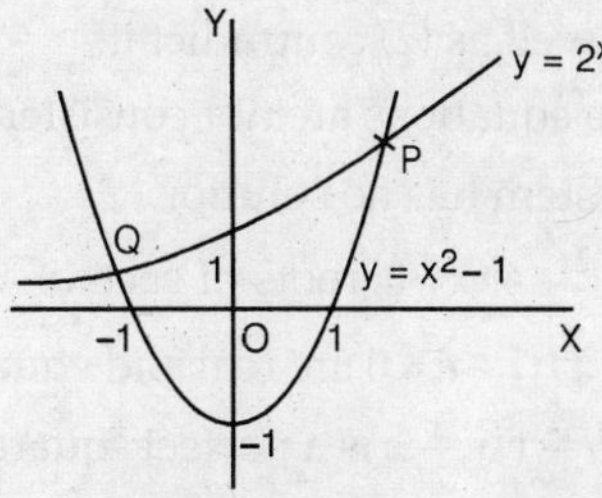

Clearly the curves $y = 2^x$ and $y = x^2 - 1$ intersect at two points only, one of which has positive x-coordinate.

As $x = \tan^2\theta$, x must be positive.

So x has only one admissible value 3.

Hence, the equation $2^x = x^2 - 1$ has no other solutions.

$\therefore$ the required values of θ are $\pm\frac{\pi}{3}$.

33. If $a, b, c \in R$ where $a \neq 0$ then for the system of quadratic equations in n variables $x_1, x_2, \ldots, x_n$

$$ax_1^2 + bx_1 + c = x_2$$
$$ax_2^2 + bx_2 + c = x_3$$
$$\ldots\ldots\ldots\ldots\ldots$$
$$ax_{n-1}^2 + bx_{n-1} + c = x_n$$
$$ax_n^2 + bx_n + c = x_1$$

(i) prove that the system has no solution if $(b-1)^2 < 4ac$

(ii) solve the equations when $(b-1)^2 = 4ac$

(iii) solve the equations when $(b-1)^2 > 4ac$.

Substituting

$x_2 = x_1 + X_1$, $x_3 = x_2 + X_2$, ... ,

$x_n = x_{n-1} + X_{n-1}$, $x_1 = x_n + X_n$

the equations become,

$$ax_1^2 + (b-1)x_1 + c = X_1$$
$$ax_2^2 + (b-1)x_2 + c = X_2$$
$$\ldots\ldots\ldots\ldots\ldots$$
$$ax_n^2 + (b-1)x_n + c = X_n.$$

(i) As $(b-1)^2 - 4ac < 0$, roots of each of

$ax_r^2 + (b-1)x_r + c = 0$ are imaginary.

If $a > 0$, each of $ax_r^2 + (b-1)x_r + c > 0$. So each of X_r is positive.

$\therefore \quad X_1 + X_2 + X_3 + \dots + X_n > 0 \quad \dots (1)$

If $a < 0$, each of $ax_r^2 + (b-1)x_r + c < 0$.

So each of X_r is negative.

So $X_1 + X_2 + \dots + X_n < 0 \quad \dots (2)$

But $X_1 + X_2 + X_3 + \dots X_n$

$= (x_2 - x_1) + (x_3 - x_2) + (x_4 - x_3) + \dots + (x_1 - x_n)$

$= 0$

$\therefore$ (1) as well as (2) contradict it.

Hence the equations are not consistent.

$\therefore$ the system has no solution.

(ii) As $(b-1)^2 - 4ac = 0$, roots of each of $ax_r^2 + (b-1)x_r + c = 0$ are real and equal and so each of $ax_r^2 + (b-1)x_r + c$ is a perfect square.

So each of X_r is greater than or equal to 0.

But as in (i), $X_1 + X_2 + \dots + X_n = 0$.

$\therefore \quad X_1 \ge 0, X_2 \ge 0, \dots, X_n \ge 0$

$\Rightarrow \quad X_1 = X_2 = X_3 = \dots = X_n = 0$

$\therefore \quad x_2 - x_1 = 0 = x_3 - x_2 = \dots = x_n - x_{n-1} = x_1 - x_n$

$\therefore \quad x_1 = x_2 = \dots = x_n.$

Also, putting $X_1 = 0$, $ax_1^2 + (b-1)x_1 + c = 0$

$$\therefore \quad x_1 = \frac{1 - b \pm \sqrt{(b-1)^2 - 4ac}}{2a} = \frac{1-b}{2a}$$

$$\therefore \quad x_1 = x_2 = \dots = x_n = \frac{1-b}{2a}.$$

(iii) As $(b-1)^2 - 4ac > 0$, roots of each of $ax_r^2 + (b-1)x_r + c = 0$ are real and unequal.

Let them be α, β.

If $a < 0$, for all $x_1 \in [\alpha, \beta]$,

$ax_1^2 + (b-1)x_1 + c \ge 0$, i.e., $X_1 \ge 0$.

Similarly for all $x_2 \in [\alpha, \beta]$,

$ax_2^2 + (b-1)x_2 + c \ge 0$, i.e., $X_2 \ge 0$, etc.

But $X_1 + X_2 + \dots + X_n = 0$

So $X_1 = X_2 = \dots = X_n = 0$

$\therefore \quad x_1 = x_2 = \dots = x_n.$

Then each $ax_r^2 + (b-1)x_r + c = 0$.

$$\text{So} \quad x_1 = x_2 = \dots = x_n = \frac{1 - b \pm \sqrt{(b-1)^2 - 4ac}}{2a}.$$

Also for all $x_1 \notin (\alpha, \beta)$,

$ax_1^2 + (b-1)x_1 + c \le 0$, i.e., $X_1 \le 0$.

But $X_1 + X_2 + \dots + X_n = 0$.

So $X_1 = X_2 = \dots = X_n = 0$;

$\therefore \quad x_1 = x_2 = \dots = x_n.$

Then each $ax_r^2 + (b-1)x_r + c = 0$;

$$\text{so} \quad x_1 = x_2 = \dots = x_n = \frac{1 - b \pm \sqrt{(b-1)^2 - 4ac}}{2a}.$$

Similarly, when $a > 0$, we get

$$x_1 = x_2 = \dots = x_n = \frac{1 - b \pm \sqrt{(b-1)^2 - 4ac}}{2a}.$$

34. Find the set of all real a such that $5a^2 - 3a - 2$, $a^2 + a - 2$ and $2a^2 + a - 1$ are the lengths of sides of a triangle.

As the sides have positive lengths and sum of two sides is greater than the third side,

$5a^2 - 3a - 2 > 0 \quad \dots (1)$

$a^2 + a - 2 > 0 \quad \dots (2)$

$2a^2 + a - 1 > 0 \quad \dots (3)$

$5a^2 - 3a - 2 + a^2 + a - 2 > 2a^2 + a - 1 \quad \dots (4)$

$5a^2 - 3a - 2 + 2a^2 + a - 1 > a^2 + a - 2 \quad \dots (5)$

$a^2 + a - 2 + 2a^2 + a - 1 > 5a^2 - 3a - 2 \quad \dots (6)$

The required set of real values of a is the set of solutions of a satisfying the inequations (1), (2), (3), (4), (5) and (6).

Solution of (1)

The corresponding equation is $5a^2 - 3a - 2 = 0$

or $(5a + 2)(a - 1) = 0$; $\therefore \quad a = 1, \dfrac{-2}{5}$

$\therefore$ the sign-scheme for the expression $5a^2 - 3a - 2$ is as follows:

(+) (−) (+)
$-\frac{2}{5}$ 1

$\therefore$ (1) is satisfied for $a < -\dfrac{2}{5}$ or $a > 1$.

Solution of (2)

The corresponding equation is $a^2 + a - 2 = 0$

or $(a - 1)(a + 2) = 0$; $\therefore \quad a = 1, -2$

$\therefore$ the sign-scheme for the expression $a^2 + a - 2$ is as follows:

(+) (−) (+)
−2 1

$\therefore$ (2) is satisfied for $a < -2$ or $a > 1$.

Solution of (3)

The corresponding equation is $2a^2 + a - 1 = 0$

or $(2a - 1)(a + 1) = 0$; $\therefore$

$\therefore$ the sign-scheme for the expression $2a^2 + a - 1$ is as follows:

(+) (–) (+) with points -1, $\frac{1}{2}$

$\therefore$ (3) is satisfied for $a < -1$ or $a > \frac{1}{2}$.

The set of values of a satisfying the inequations (1), (2), (3) is given by

$$a < -2 \quad \text{or} \quad a > 1. \quad \dots (7)$$

Now, (4) $\Rightarrow$ $4a^2 - 3a - 3 > 0$.

The corresponding equation is $4a^2 - 3a - 3 = 0$.

Its roots are $a = \frac{3 \pm \sqrt{9+48}}{8} = \frac{3 \pm \sqrt{57}}{8}$.

$\therefore$ as above we get

$$a < \frac{3-\sqrt{57}}{8} \quad \text{or} \quad a > \frac{3+\sqrt{57}}{8} \quad \dots (8)$$

(5) $\Rightarrow$ $6a^2 - 3a - 1 > 0$.

The corresponding equation is $6a^2 - 3a - 1 = 0$.

Its roots are $a = \frac{3 \pm \sqrt{9+24}}{12} = \frac{3 \pm \sqrt{33}}{12}$.

$\therefore$ as above we get

$$a < \frac{3-\sqrt{33}}{12} \quad \text{or} \quad a > \frac{3+\sqrt{33}}{12} \quad \dots (9)$$

(6) $\Rightarrow$ $2a^2 - 5a + 1 < 0$.

The corresponding equation is $2a^2 - 5a + 1 = 0$.

Its roots are $a = \frac{5 \pm \sqrt{25-8}}{4} = \frac{5 \pm \sqrt{17}}{4}$

$\therefore$ as above we get $\frac{5-\sqrt{17}}{4} < a < \frac{5+\sqrt{17}}{4}$ ... (10)

$\therefore$ values of a satisfying (7), (8), (9) and (10) are

$$\frac{3+\sqrt{57}}{8} < a < \frac{5+\sqrt{17}}{4}.$$

35. Solve in R: $(x+3)^5 - (x-1)^5 \geq 244$.

The AM of $x+3$, $x-1$ is $\frac{x+3+x-1}{2}$, i.e., $x+1$.

Put $x+1 = y$. Then $x+3 = y+2$, $x-1 = y-2$.

$\therefore$ the equation becomes $(y+2)^5 - (y-2)^5 \geq 244$

or $y^5 + {}^5C_1 \cdot y^4 \cdot 2 + {}^5C_2 \cdot y^3 \cdot 2^2 + {}^5C_3 \cdot y^2 \cdot 2^3 + {}^5C_4 \cdot y \cdot 2^4 + {}^5C_5 \cdot 2^5 - (y^5 - {}^5C_1 \cdot y^4 \cdot 2 + {}^5C_2 \cdot y^3 \cdot 2^2 - {}^5C_3 \cdot y^2 \cdot 2^3 + {}^5C_4 \cdot y \cdot 2^4 - {}^5C_5 \cdot 2^5) \geq 244$

or $2[{}^5C_1 \cdot y^4 \cdot 2 + {}^5C_3 \cdot y^2 \cdot 2^3 + {}^5C_5\, 2^5] \geq 244$

or $10y^4 + 80y^2 + 32 \geq 122$

or $y^4 + 8y^2 - 9 \geq 0$

or $(y^2+9)(y^2-1) \geq 0$

$\Rightarrow$ $y^2 - 1 \geq 0$ because $y^2 + 9 > 0$ for real y.

The corresponding equation $y^2 - 1 = 0$ has roots $1, -1$.

$\therefore$ the sign-scheme of $y^2 - 1$, $y \in R$ is as follows:

(+) (–) (+) with points -1, 1

$\therefore$ $y^2 - 1 \geq 0$ holds if $y \leq -1$ or $y \geq 1$.

Now $y \leq -1 \Rightarrow x + 1 \leq -1$ or $x \leq -2$

$y \geq 1 \Rightarrow x + 1 \geq 1$ or $x \geq 0$.

Hence, $x \leq -2$ or $x \geq 0$.

Therefore, the solution-set $= (-\infty, -2] \cup [0, +\infty)$.

36. Find all real values of x for which $\sqrt{\frac{(x+1)(x-3)}{x-2}}$ takes real values.

For real values of the expression we must have

$$\frac{(x+1)(x-3)}{x-2} \geq 0 \quad \text{and} \quad x - 2 \neq 0$$

or $(x+1)(x-3)(x-2) \geq 0$

(multiplying by $(x-2)^2$ which is greater than 0)

The corresponding equation is $(x+1)(x-3)(x-2) = 0$

$\therefore$ $x = -1, 3, 2$.

$\therefore$ the sign-scheme of the expression for real values of x is as follows:

(–) (+) (–) (+) with points -1, 2, 3

because for $x = 0 \in (-1, 2)$ the expression is positive.

$\therefore$ $(x+1)(x-3)(x-2) \geq 0$

$\Rightarrow$ $x \in [-1, 2]$ or $x \in [3, +\infty)$.

But $x \neq 2$. So the required set of values of x is $[-1, 2) \cup [3, +\infty)$.

37. Let $f(x) = \frac{2x}{2x^2+5x+2}$ and $g(x) = \frac{1}{x+1}$.

Find the set of real values of x for which $f(x) > g(x)$.

$$f(x) > g(x) \Rightarrow \frac{2x}{2x^2+5x+2} > \frac{1}{x+1}$$

where $x + 1 \neq 0$, $2x^2 + 5x + 2 \neq 0$

or $\dfrac{2x}{2x^2+5x+2} - \dfrac{1}{x+1} > 0$

or $\dfrac{2x}{(2x+1)(x+2)} - \dfrac{1}{x+1} > 0$

or $\dfrac{2x(x+1)-(2x^2+5x+2)}{(2x+1)(x+2)(x+1)} > 0$

or $\dfrac{-3x-2}{(2x+1)(x+2)(x+1)} > 0.$

Multiplying by $(2x+1)^2(x+2)^2(x+1)^2$ {which is positive},

$$(-3x-2)(2x+1)(x+2)(x+1) > 0 \qquad \dots (1)$$

Roots of the corresponding equation $= -\dfrac{2}{3}, \dfrac{-1}{2}, -2, -1.$

$\therefore$ the sign-scheme of the expression for real x is as follows:

(−) -2 (+) -1 (−) $-\frac{2}{3}$ (+) $-\frac{1}{2}$ (−)

because when $x = 0 \in \left(-\dfrac{1}{2}, +\infty\right)$ the expression is negative.

$\therefore$ (1) holds for $x \in (-2, -1), x \in \left(-\dfrac{2}{3}, -\dfrac{1}{2}\right)$

$\therefore$ the required set of real values of x is

$$(-2, -1) \cup \left(-\frac{2}{3}, -\frac{1}{2}\right).$$

38. Solve for $x \in R$: $1 \le \dfrac{3x^2-7x+8}{x^2+1} \le 2.$

As $x^2+1>0$, multiplying by x^2+1 we get

$$x^2+1 \le 3x^2-7x+8 \le 2(x^2+1).$$

Now, $x^2+1 \le 3x^2-7x+8$

$\Rightarrow \quad 2x^2-7x+7 \ge 0.$

The corresponding equation $2x^2-7x+7=0$ has

$$D = (-7)^2 - 4.2.7 < 0 \text{ and } a = 2 > 0.$$

$\therefore \quad 2x^2-7x+7 \ge 0$ for all $x \in R$.

Next, $3x^2-7x+8 \le 2(x^2+1)$

$\Rightarrow \quad x^2-7x+6 \le 0.$

The corresponding equation is $x^2-7x+6=0$

or $(x-6)(x-1)$; $\therefore \quad x = 6, 1.$

$\therefore$ the sign-scheme of $x^2-7x+6, x \in R$ is as follows:

(+) 1 (−) 6 (+)

$\therefore \quad x^2-7x+6 \le 0 \quad \Rightarrow \quad 1 \le x \le 6$

$\therefore$ the required solution set $= R \cap [1, 6] = [1, 6].$

39. Solve for $x \in R$: $4^{x^2} \le 3.2^{x^2+x} + 4^{x+1}.$

Here $2^{2x^2} \le 3.2^{x^2} \cdot 2^x + 4.2^{2x}$

or $(2^{x^2})^2 \le 3.2^{x^2} \cdot 2^x + 4 \cdot (2^x)^2$

or $(2^{x^2})^2 - 3.2^{x^2} \cdot 2^x - 4 \cdot (2^x)^2 \le 0$

or $(2^{x^2} - 4.2^x)(2^{x^2} + 2^x) \le 0 \qquad \dots (1)$

For all real x, $2^{x^2} + 2^x > 0$. So (1) will hold if

$2^{x^2} - 4.2^x \le 0$

or $2^{x^2} \le 2^{x+2}$

or $x^2 \le x+2$, because the base $2 > 1$

or $x^2 - x - 2 \le 0$; the corresponding equation is

$x^2 - x - 2 = 0$

$\therefore \quad (x-2)(x+1) = 0; \quad \therefore \quad x = 2, -1.$

$\therefore$ the sign-scheme for $x^2 - x - 2, x \in R$ is as follows:

(+) -1 (−) 2 (+)

$\therefore \quad x^2 - x - 2 \le 0 \quad \Rightarrow \quad -1 \le x \le 2.$

Hence the required solution-set $= [-1, 2].$

40. Solve $|x-1| + |x-2| + |x-3| \ge 6, x \in R.$

The definitions of the modulus functions are as follows:

$-(x-1) \leftarrow 1 \rightarrow (x-1)$; $\quad -(x-2) \leftarrow 2 \rightarrow (x-2)$; $\quad -(x-3) \leftarrow 3 \rightarrow (x-3)$

$\therefore$ if $x < 1$, the inequation becomes

$-(x-1)-(x-2)-(x-3) \ge 6$

or $-3x+6 \ge 6; \quad \therefore \quad x \le 0$

$\therefore$ x is such that $x<1$ and $x \le 0$. So $x \le 0$.

If $1 \le x < 2$, the inequation becomes

$(x-1)-(x-2)-(x-3) \ge 6$

or $-x+4 \ge 6; \quad \therefore \quad x \le -2.$

No such values satisfy $1 \le x < 2$.

If $2 \le x < 3$, the inequation becomes

$(x-1)+(x-2)-(x-3) \ge 6$

or $x \ge 6$. No such values satisfy $2 \le x < 3$.

If $x \ge 3$, the inequation becomes

$(x-1)+(x-2)+(x-3) \ge 6$

or $3x - 6 \geq 6$; or $3x \geq 12$

$\therefore$ $x \geq 4$ which satisfies $x \geq 3$.

$\therefore$ the values of x satisfy $x \leq 0$ or $x \geq 4$

$\therefore$ the required solution-set $= (-\infty, 0] \cup [4, +\infty)$.

41. Solve $||x - 2| - 1| \geq 3$.

If $|x - 2| > 1$ then $|x - 2| - 1 \geq 3$, (from the question)

$\therefore$ if $x < 2$ then $-(x - 2) > 1$ and $-(x - 2) - 1 \geq 3$

or $x < 1$ and $x \leq -2$.

So $x \leq -2$ which satisfies $x < 2$.

If $x \geq 2$ then $x - 2 > 1$ and $x - 2 - 1 \geq 3$

or $x > 3$ and $x \geq 6$.

So $x \geq 6$ which satisfies $x \geq 2$.

$\therefore$ the values of x are $x \leq -2$ or $x \geq 6$.

$\therefore$ the solution-set $= (-\infty, -2] \cup [6, +\infty)$.

42. If $x = \frac{4}{9}$ satisfies $\log_a(x^2 - x + 2) > \log_a(-x^2 + 2x + 3)$ then find all real solutions of the inequation.

Putting $x = \frac{4}{9}$ in the inequation,

$$\log_a\left\{\frac{16}{81} - \frac{4}{9} + 2\right\} > \log_a\left\{-\frac{16}{81} + \frac{8}{9} + 3\right\}$$

or $\log_a \frac{16 - 36 + 162}{81} > \log_a \frac{-16 + 72 + 243}{81}$

or $\log_a \frac{142}{81} > \log_a \frac{299}{81}$

As $\frac{142}{81} < \frac{299}{81}$ we can conclude $0 < a < 1$.

Thus we have

$$\log_a(x^2 - x + 2) > \log_a(-x^2 + 2x + 3)$$

where $0 < a < 1$.

$\therefore$ $x^2 - x + 2 < -x^2 + 2x + 3$

or $2x^2 - 3x - 1 < 0$.

The corresponding equation is $2x^2 - 3x - 1 = 0$

or $x = \frac{3 \pm \sqrt{9 + 8}}{4} = \frac{3 \pm \sqrt{17}}{4}$.

$\therefore$ the sign-scheme of the expression $2x^2 - 3x - 1$, $x \in R$ is as follows:

(+) $\frac{3 - \sqrt{17}}{4}$ (−) $\frac{3 + \sqrt{17}}{4}$ (+)

$\therefore$ $2x^2 - 3x - 1 < 0$

$\Rightarrow$ $\frac{3 - \sqrt{17}}{4} < x < \frac{3 + \sqrt{17}}{4}$ ... (1)

But, in order that the given logarithms may be defined we must have

$$x^2 - x + 2 > 0 \text{ and } -x^2 + 2x + 3 > 0.$$

The corresponding equation of $x^2 - x + 2 > 0$ is $x^2 - x + 2 = 0$ for which

$$D = 1 - 8 < 0 \text{ and coefficient of } x^2 = 1 > 0$$

$\therefore$ $x^2 - x + 2 > 0$ is true for all $x \in R$.

The corresponding equation of $-x^2 + 2x + 3 > 0$ is

$$-x^2 + 2x + 3 = 0$$

or $(x + 1)(3 - x) = 0$; $\therefore$ $x = -1, 3$.

The sign-scheme of $-x^2 + 2x + 3$, $x \in R$ is as follows:

(−) -1 (+) 3 (−)

$\therefore$ $-x^2 + 2x + 3 > 0$ $\Rightarrow$ $-1 < x < 3$ ... (2)

$\therefore$ the required values of x form the set of values of x common to (1) and (2).

$\therefore$ the solution-set $= \left(\frac{3 - \sqrt{17}}{4}, \frac{3 + \sqrt{17}}{4}\right)$.

43. Solve for $x \in R$: $\log_{3x + 5}(9x^2 + 8x + 8) > 2$.

For the logarithm to be defined

$$3x + 5 > 0, 3x + 5 \neq 1 \text{ and } 9x^2 + 8x + 8 > 0.$$

If $3x + 5 > 1$, the inequation gives

$$\log_{3x+5}(9x^2 + 8x + 8) > \log_{3x+5}(3x + 5)^2$$

or $9x^2 + 8x + 8 > (3x + 5)^2$

or $9x^2 + 8x + 8 > 9x^2 + 30x + 25$

or $22x + 17 < 0$; $\therefore$ $x < \frac{-17}{22}$.

But $3x + 5 > 1$ $\Rightarrow$ $x > \frac{-4}{3}$

and $9x^2 + 8x + 8 > 0$ holds for all $x \in R$ because

$$D = 8^2 - 4 \cdot 9 \cdot 8 < 0, a = 9 > 0.$$

$\therefore$ $-\frac{4}{3} < x < \frac{-17}{22}$.

If $0 < 3x + 5 < 1$, the inequation gives

$$\log_{3x+5}(9x^2 + 8x + 8) > \log_{3x+5}(3x + 5)^2$$

or $9x^2 + 8x + 8 < (3x + 5)^2$; as above $x > -\frac{17}{22}$.

But $0 < 3x + 5 < 1$ $\Rightarrow$ $-5 < 3x < -4$

or $\frac{-5}{3} < x < \frac{-4}{3}$.

There is no value of x satisfying

$$x > \frac{-17}{22} \text{ and } \frac{-5}{3} < x < \frac{-4}{3}.$$

Hence, the solution-set is $\left(\frac{-4}{3}, \frac{-17}{22}\right)$.

44. Solve for $x \in R$:

$$\log_{x-2} 6 + \log_{x+2} 6 > \log_{x-2} 6 \cdot \log_{x+2} 6.$$

The logarithms will be defined if

$$x - 2 > 0,\ x - 2 \neq 1,\ x + 2 > 0,\ x + 2 \neq 1.$$

Now, $\frac{\log 6}{\log(x-2)} + \frac{\log 6}{\log(x+2)} > \frac{\log 6}{\log(x-2)} \times \frac{\log 6}{\log(x+2)}$

or $\frac{\log(x+2) + \log(x-2)}{\log(x-2) \cdot \log(x+2)} > \frac{\log 6}{\log(x-2) \cdot \log(x+2)}$

or $\frac{\log(x^2-4) - \log 6}{\log(x-2) \cdot \log(x+2)} > 0$

or $\log_{10} \frac{x^2-4}{6} \cdot \log_{10}(x-2) \cdot \log_{10}(x+2) > 0,$

taking the base 10.

This is possible when any one of the following is true:

(i) $\log_{10} \frac{x^2-4}{6} > 0,\ \log_{10}(x-2) > 0,\ \log_{10}(x+2) > 0$

(ii) $\log_{10} \frac{x^2-4}{6} > 0,\ \log_{10}(x-2) < 0,\ \log_{10}(x+2) < 0$

(iii) $\log_{10} \frac{x^2-4}{6} < 0,\ \log_{10}(x-2) > 0,\ \log_{10}(x+2) < 0$

(iv) $\log_{10} \frac{x^2-4}{6} < 0,\ \log_{10}(x-2) < 0,\ \log_{10}(x+2) > 0$

(i) $\Rightarrow$ $\frac{x^2-4}{6} > 1,\ x - 2 > 1,\ x + 2 > 1$

$\Rightarrow$ $x^2 > 10,\ x > 3,\ x > -1$

$\Rightarrow$ $x > \sqrt{10}$, for $x - 2 > 0$ also.

Similarly

(ii) $\Rightarrow$ $x^2 > 10,\ x < 3,\ x < -1$

$\Rightarrow$ x has no value because $x - 2 > 0$ also.

(iii) $\Rightarrow$ $x^2 < 10,\ x > 3,\ x < -1$

$\Rightarrow$ x has no value.

(iv) $\Rightarrow$ $x^2 < 10,\ x < 3,\ x > -1$

$\Rightarrow$ $2 < x < 3$ because $x - 2 > 0$ also.

Hence the solution-set $= (2, 3) \cup (\sqrt{10}, +\infty)$.

45. If $(x^2 - 3x - 3)^{x+1} > 1$, find $x \in (-1, +\infty)$.

Clearly, $x^2 - 3x - 3 > 0$, because it is the base.

$\therefore$ by sign-scheme $\frac{3 - \sqrt{21}}{2} > x$ or $x > \frac{3 + \sqrt{21}}{2}$.

$\therefore$ taking logarithm,

$$\log_{10}(x^2 - 3x - 3)^{x+1} > \log_{10} 1 = 0$$

or $(x + 1)\log_{10}(x^2 - 3x - 3) > 0$.

Any one of the following is possible:

(i) $x + 1 > 0,\ \log_{10}(x^2 - 3x - 3) > 0$

(ii) $x + 1 < 0,\ \log_{10}(x^2 - 3x - 3) < 0$

(i) $\Rightarrow$ $x > -1,\ x^2 - 3x - 3 > 1$

Now, $x^2 - 3x - 3 > 1 \Rightarrow x^2 - 3x - 4 > 0$.

The corresponding equation is $x^2 - 3x - 4 = 0$

or $(x + 1)(x - 4) = 0;\ \therefore\ x = -1, 4$

$\therefore$ the sign-scheme of $x^2 - 3x - 4$, $x \in R$ is as follows:

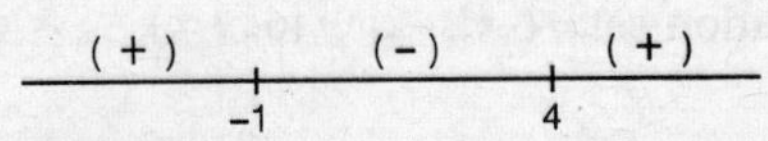

$\therefore$ $x^2 - 3x - 4 > 0 \Rightarrow x < -1$ or $x > 4$.

But $x > -1$. So $x > 4$.

Similarly, (ii) $\Rightarrow$ $x < -1$ and $-1 < x < 4$.

There is no such x.

$\therefore$ the solution-set $= (4, +\infty)$.

46. Let there be a quotient of two natural numbers in which the denominator is one less than the square of the numerator. If we add 2 to both numerator and denominator, the quotient will exceed $\frac{1}{3}$. If we subtract 3 from the numerator as well as the denominator, the quotient will lie between 0 and $\frac{1}{10}$. Determine the quotient.

Let the quotient be $\frac{p}{q}$ where $p \in N,\ q \in N$.

According to the question,

$$q = p^2 - 1 \quad \ldots (1)$$

$$\frac{p+2}{q+2} > \frac{1}{3} \quad \ldots (2)$$

$$0 < \frac{p-3}{q-3} < \frac{1}{10} \quad \ldots (3)$$

From (1) and (2), $\frac{p+2}{p^2 - 1 + 2} > \frac{1}{3}$

or $\frac{p+2}{p^2 + 1} > \frac{1}{3}$

or $\quad 3(p+2) > p^2+1 \quad \{\because p^2+1>0\}$

or $\quad p^2-3p-5<0;$

the corresponding equation is $p^2-3p-5=0$

$\therefore \quad p = \dfrac{3 \pm \sqrt{9+4\cdot 5}}{2} = \dfrac{3 \pm \sqrt{29}}{2}$

$\therefore$ the sign-scheme for $p^2-3p-5,\ p \in R$ is as follows:

(+) $\quad \dfrac{3-\sqrt{29}}{2}$ $\quad$ (−) $\quad \dfrac{3+\sqrt{29}}{2}$ $\quad$ (+)

$\therefore \quad p^2-3p-5<0 \quad \Rightarrow \quad \dfrac{3-\sqrt{29}}{2} < p < \dfrac{3+\sqrt{29}}{2}.$

$$\Rightarrow \quad -1.2 < p < 4.2$$

But p is a natural number. So possible values of $p = 1, 2, 3, 4$.

(3) $\Rightarrow \quad 0 < \dfrac{p-3}{p^2-1-3} < \dfrac{1}{10} \quad$ or $\quad 0 < \dfrac{p-3}{p^2-4} < \dfrac{1}{10}.$

When $p=1$, $\quad \dfrac{p-3}{p^2-4} = \dfrac{1-3}{1-4} = \dfrac{-2}{-3} = \dfrac{2}{3} \notin \left(0, \dfrac{1}{10}\right).$

When $p=2$, $\quad \dfrac{p-3}{p^2-4} = \dfrac{2-3}{4-4} \notin \left(0, \dfrac{1}{10}\right).$

When $p=3$, $\quad \dfrac{p-3}{p^2-4} = \dfrac{3-3}{9-4} = 0 \notin \left(0, \dfrac{1}{10}\right).$

When $p=4$, $\quad \dfrac{p-3}{p^2-4} = \dfrac{4-3}{16-4} = \dfrac{1}{12} \in \left(0, \dfrac{1}{10}\right).$

$\therefore$ the quotient $= \dfrac{p}{q} = \dfrac{p}{p^2-1} = \dfrac{4}{4^2-1} = \dfrac{4}{15}.$

47. Prove that the equation $x^2+2px+2q=0$ cannot have rational roots if p and q are odd integers.

Here, $D = (2p)^2 - 4\cdot 1\cdot 2q = 4(p^2-2q)$.

As p, q are odd integers, p^2 is odd while $2q$ is even. So $D \neq 0$.

$\therefore \quad p^2 < 2q \quad$ or $\quad p^2 > 2q$.

If $p^2 < 2q$ then D is negative and so the roots are imaginary and hence the roots are not rational.

If $p^2 > 2q$, let D be a perfect square if possible.

Then $p^2 - 2q$ = perfect square = odd − even = odd

$\therefore \quad p^2-2q$ must be a perfect square of an odd integer.

Let $p^2-2q = (2m-1)^2$. $\quad$ ... (1)

As p, q are odd, let $p = 2k-1$ and $q = 2h-1$.

(1) $\Rightarrow \quad (2k-1)^2 - 2(2h-1) = (2m-1)^2$

or $\quad (2k-1)^2 - (2m-1)^2 = 2(2h-1)$

or $\quad (2k-1+2m-1)(2k-1-2m+1) = 2(2h-1)$

or $\quad 2(k+m-1)2(k-m) = 2(2h-1)$

or $\quad 2(k+m-1)(k-m) = 2h-1$

$\therefore$ even integer = odd integer, which is absurd.

$\therefore$ D cannot be a perfect square when $p^2 > 2q$.

$\therefore$ under no condition, the equation has rational roots.

48. If the roots of the quadratic equation $x^2-ax+b=0$ are real and differ by a quantity less than 1, prove that $\dfrac{a^2-1}{4} < b < \dfrac{a^2}{4}$.

As the roots are real and unequal $a^2-4b>0$.

So $\quad b < \dfrac{a^2}{4}$ $\quad$ ... (1)

Also, from the question,

$|\alpha-\beta| < 1$ when α, β are roots.

Now $(\alpha-\beta)^2 = (\alpha+\beta)^2 - 4\alpha\beta = a^2-4b$

$\therefore \quad |\alpha-\beta|^2 = a^2-4b.$

But $|\alpha-\beta|<1$, i.e., $|\alpha-\beta|^2<1$.

So $\quad a^2-4b<1$

$\therefore \quad a^2-1<4b; \quad \therefore \quad \dfrac{a^2-1}{4} < b$ $\quad$ ... (2)

$\therefore$ from (1) and (2), $\dfrac{a^2-1}{4} < b < \dfrac{a^2}{4}.$

49. If $a<b<c<d$, prove that the equation $(x-a)(x-c)+k(x-b)(x-d)=0$ has real roots for all $k \in R$.

Here the equation is

$$x^2-(a+c)x+ac+k\{x^2-(b+d)x+bd\}=0$$

or $\quad (1+k)x^2 - \{a+c+k(b+d)\}x + ac + kbd = 0$ $\quad$ ... (1)

(1) will have real roots if

$$\{a+c+k(b+d)\}^2 - 4(1+k)(ac+kbd) \geq 0$$

or $\quad \{(b+d)^2-4bd\}k^2 + \{2(a+c)(b+d)-4ac-4bd\}k + (a+c)^2-4ac \geq 0$

or $\quad (b-d)^2k^2 + 2\{ab+ad+bc+cd-2ac-2bd\}k + (a-c)^2 \geq 0$ $\quad$ ... (2)

Now, the discriminant of the equation corresponding to (2) is

$D = 4\{ab+ad+bc+cd-2ac-2bd\}^2 - 4(b-d)^2(a-c)^2$

$= 4\{ab+ad+bc+cd-2ac-2bd+(b-d)(a-c)\}$

$\times \{ab+ad+bc+cd-2ac-2bd-(b-d)(a-c)\}$

$$= 4\{ab + ad + bc + cd - 2ac - 2bd + ab - bc - ad + cd\}$$

$$\times \{ab + ad + bc + cd - 2ac - 2bd - ab + bc + ad - cd\}$$

$$= 4 \times 2(ab + cd - ac - bd) \times 2(ad + bc - ac - bd)$$

$$= 16\{a(b - c) - d(b - c)\}\{d(a - b) - c(a - b)\}$$

$$= 16(b - c)(a - d)(a - b)(d - c)$$

$$= 16(-ve)(-ve)(-ve)(+ve) < 0,$$

because $a < b < c < d$.

Also $(b - d)^2 > 0$. So (2) is true for all $k \in R$.

Hence, the given equation has real roots.

50. Show that the equation

$$\frac{a}{x-1} + \frac{b}{x-2} + \frac{c}{x-3} + \frac{d}{x-4} = 1$$

cannot have imaginary roots if a, b, c, d are any four real numbers of the same sign.

If possible let the equation has imaginary roots $\alpha \pm i\beta$ ($\because$ imaginary roots occur in conjugate pairs)

$$\therefore \quad \frac{a}{\alpha + i\beta - 1} + \frac{b}{\alpha + i\beta - 2} + \frac{c}{\alpha + i\beta - 3} + \frac{d}{\alpha + i\beta - 4} = 1$$

or
$$\frac{a}{(\alpha - 1) + i\beta} + \frac{b}{(\alpha - 2) + i\beta} + \frac{c}{(\alpha - 3) + i\beta} + \frac{d}{(\alpha - 4) + i\beta} = 1 \quad \ldots (1)$$

and
$$\frac{a}{(\alpha - 1) - i\beta} + \frac{b}{(\alpha - 2) - i\beta} + \frac{c}{(\alpha - 3) - i\beta} + \frac{d}{(\alpha - 4) - i\beta} = 1 \quad \ldots (2)$$

From (2) – (1), we get

$$a\left[\frac{1}{(\alpha - 1) - i\beta} - \frac{1}{(\alpha - 1) + i\beta}\right] + b\left[\frac{1}{(\alpha - 2) - i\beta} - \frac{1}{(\alpha - 2) + i\beta}\right]$$
$$+ c\left[\frac{1}{(\alpha - 3) - i\beta} - \frac{1}{(\alpha - 3) + i\beta}\right] + d\left[\frac{1}{(\alpha - 4) - i\beta} - \frac{1}{(\alpha - 4) + i\beta}\right] = 0$$

or
$$a \cdot \frac{2i\beta}{(\alpha - 1)^2 + \beta^2} + b \cdot \frac{2i\beta}{(\alpha - 2)^2 + \beta^2} + c \cdot \frac{2i\beta}{(\alpha - 3)^2 + \beta^2} + d \cdot \frac{2i\beta}{(\alpha - 4)^2 + \beta^2} = 0.$$

Clearly $\beta \neq 0$ for then roots will be real.

$$\therefore \quad \frac{a}{(\alpha - 1)^2 + \beta^2} + \frac{b}{(\alpha - 2)^2 + \beta^2} + \frac{c}{(\alpha - 3)^2 + \beta^2} + \frac{d}{(\alpha - 4)^2 + \beta^2} = 0 \quad \ldots (3)$$

Each of the denominators is positive and the a, b, c, d have the same sign. So (3) cannot hold.

Hence, the equation cannot have imaginary roots.

51. If p be the first of n arithmetic means between two numbers and q be the first of n harmonic means between the same two numbers, prove that q cannot lie between $p \cdot \left(\frac{n+1}{n-1}\right)^2$ and p.

Let the numbers be x, y. Let the n AMs between them be $A_1, A_2, \ldots, A_n$.

Then $y = (n + 2)$th term $= x + (n + 1)d$,

d being the common difference.

$$\therefore \quad d = \frac{y - x}{n + 1}; \quad \therefore \quad A_1 = p = x + \frac{y - x}{n + 1}$$

$$\therefore \quad p = \frac{nx + y}{n + 1} \quad \ldots (1)$$

Similarly,
$$\frac{1}{q} = \frac{n \cdot \frac{1}{x} + \frac{1}{y}}{n + 1} = \frac{ny + x}{(n + 1)xy} \quad \ldots (2)$$

(1) $\Rightarrow$ $y = (n + 1)p - nx$; putting this in (2),

$$\frac{(n + 1)x}{q}\{(n + 1)p - nx\} = n\{(n + 1)p - nx\} + x$$

or
$$-\frac{n(n + 1)}{q}x^2 + \frac{(n + 1)^2 p}{q}x = n(n + 1)p + (1 - n^2)x$$

or
$$\frac{n(n + 1)}{q}x^2 - \left\{\frac{(n + 1)^2 p}{q} + (n^2 - 1)\right\}x + n(n + 1)p = 0$$

or
$$\frac{n}{q}x^2 - \left\{\frac{(n + 1)p}{q} + n - 1\right\}x + np = 0$$

or $nx^2 - \{(n + 1)p + (n - 1)q\}x + npq = 0$.

As x is real, $D \geq 0$

$$\therefore \quad \{(n + 1)p + (n - 1)q\}^2 - 4n \cdot npq \geq 0$$

or
$$(n - 1)^2 q^2 + \{2(n^2 - 1) - 4n^2\}pq + (n + 1)^2 p^2 \geq 0$$

or
$$q^2 - 2\frac{n^2 + 1}{(n - 1)^2}pq + \left(\frac{n + 1}{n - 1}\right)^2 p^2 \geq 0 \quad \ldots (3)$$

The corresponding equation in q is

$$q^2 - 2\frac{n^2 + 1}{(n - 1)^2}p \cdot q + \left(\frac{n + 1}{n - 1}\right)^2 p^2 = 0$$

$$\therefore \quad q = \frac{\left\{2 \cdot \frac{n^2+1}{(n-1)^2} p \pm \sqrt{4\left\{\frac{n^2+1}{(n-1)^2} p\right\}^2 - 4 \cdot \left(\frac{n+1}{n-1}\right)^2 p^2}\right\}}{2}$$

$$= \frac{n^2+1}{(n-1)^2} p \pm \frac{p}{n-1} \sqrt{\left(\frac{n^2+1}{n-1}\right)^2 - (n+1)^2}$$

$$= \frac{n^2+1}{(n-1)^2} p \pm \frac{p}{(n-1)^2} \sqrt{(n^2+1)^2 - (n^2-1)^2}$$

$$= \frac{p}{(n-1)^2} (n^2+1 \pm 2n) = p \left(\frac{n+1}{n-1}\right)^2, p.$$

$\therefore$ the sign-scheme for (3) is as follows:

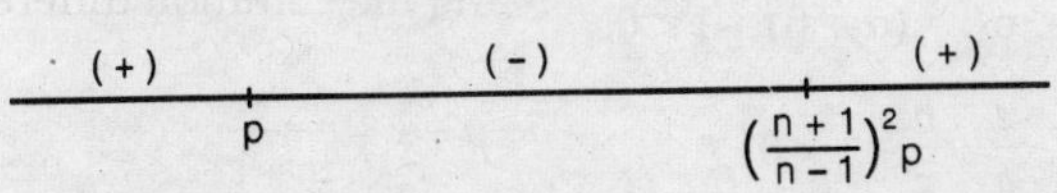

$\therefore$ (3) $\Rightarrow$ q cannot lie between p and $\left(\frac{n+1}{n-1}\right)^2 \cdot p$.

52. Let $a, b \in N$ and $a > 1$. Also p is a prime number. If $ax^2 + bx + c = p$ for two distinct integral values of x then prove that $ax^2 + bx + c \neq 2p$ for any integral value of x.

For the equation $ax^2 + bx + c - p = 0$, here we have

the product of roots $= \frac{c-p}{a}$ = integer ... (1)

and the sum of roots $= -\frac{b}{a}$ = integer ... (2)

If possible, let $ax^2 + bx + c = 2p$ for some integral x.

$\therefore$ the product of roots $= \frac{c-2p}{a} = \frac{c-p}{a} - \frac{p}{a}$

= integer − fraction {using (1) and p is prime}

$\neq$ integer

$\therefore$ at least one x satisfying $ax^2 + bx + c = 2p$ is not an integer.

Also sum of roots $= \frac{-b}{a}$ = integer {by (2)}.

But one of the roots is not an integer; so the other root also is not an integer.

$\therefore$ $ax^2 + bx + c \neq 2p$ for any integral value of x.

53. Find the values of the parameter $a \neq 0$ for which one of the roots of the quadratic equation $x^2 - x + 3a = 0$ is double of one of the roots of the equation

$$x^2 - x + a = 0.$$

Let α be a root of $x^2 - x + 3a = 0$.

$\therefore$ $\alpha^2 - \alpha + 3a = 0$ ... (1)

Then $\frac{\alpha}{2}$ will be a root of $x^2 - x + a = 0$

$\therefore$ $\frac{\alpha^2}{4} - \frac{\alpha}{2} + a = 0$

or $\alpha^2 - 2\alpha + 4a = 0$ ... (2)

(1) − (2) gives, $\alpha - a = 0$; $\therefore$ $\alpha = a$.

Putting in (1), $a^2 - a + 3a = 0$

or $a^2 + 2a = 0$

or $a(a+2) = 0$; $\therefore$ $a = 0, -2$.

But $a \neq 0$. So $a = -2$.

54. If a, b, c are in HP, find the relation between the roots of $ax^2 + bx + c = 0$.

Let α, β be the roots of the equation.

$\therefore$ $\alpha + \beta = \frac{-b}{a}, \alpha\beta = \frac{c}{a}$ ... (1)

As a, b, c are in HP, $b = \frac{2ac}{a+c}$

$\therefore$ $-a(\alpha + \beta) = \frac{2a \cdot a\alpha\beta}{a + a \cdot \alpha\beta}$ {using (1)}

or $-(\alpha + \beta) = \frac{2\alpha\beta}{1 + \alpha\beta}$

or $-(1 + \alpha\beta) = \frac{2\alpha\beta}{\alpha + \beta}$

$\therefore$ $-(1 + \alpha\beta)$ is the HM between α, β.

55. Let $\frac{\alpha}{\beta}, \frac{\beta}{\alpha}$ be two roots of $(x+1)^n + x^n + 1 = 0$ where α, β are roots of $x^2 + px + q = 0$. If α, β are also roots of the equation $x^{2n} + p^n x^n + q^n = 0$ then show that n must be an even integer when $p \neq 0$.

As $\frac{\alpha}{\beta}$ is a root of $(x+1)^n + x^n + 1 = 0$,

$$\left(\frac{\alpha}{\beta} + 1\right)^n + \left(\frac{\alpha}{\beta}\right)^n + 1 = 0$$

or $(\alpha + \beta)^n + \alpha^n + \beta^n = 0$. ... (1)

But α, β are roots of $x^2 + px + q = 0$

$\therefore$ $\alpha + \beta = -p, \alpha\beta = q$

$\therefore$ from (1), $(-p)^n + \alpha^n + \beta^n = 0$ ... (2)

As α, β are roots of $x^{2n} + p^n x^n + q^n = 0$, we get

$$\alpha^{2n} + p^n \alpha^n + q^n = 0$$

and $\beta^{2n} + p^n \beta^n + q^n = 0$.

Subtracting, $\alpha^{2n} - \beta^{2n} + p^n(\alpha^n - \beta^n) = 0$

or $(\alpha^n + \beta^n)(\alpha^n - \beta^n) + p^n(\alpha^n - \beta^n) = 0$

$\therefore$ $\alpha^n + \beta^n + p^n = 0$ ... (3)

$\{\because \alpha^n \neq \beta^n \text{ as } \alpha \neq \beta\}$

From (2) – (3), $(-p)^n - p^n = 0$

$\therefore$ $(-p)^n = p^n \Rightarrow n$ is an even integer.

56. If a root of $ax^2 + bx + c = 0$ is numerically equal but opposite in sign of a root of $a_1x^2 + b_1x + c_1 = 0$ then prove that the quadratic equation whose roots are the other roots of the two equations is

$$\frac{x^2}{\frac{b}{a} + \frac{b_1}{a_1}} + x + \frac{1}{\frac{b}{c} + \frac{b_1}{c_1}} = 0.$$

Let the roots of $ax^2 + bx + c = 0$ be α, β and the roots of the equation $a_1x^2 + b_1x + c_1 = 0$ be $-\alpha, \gamma$ (as per question)

$\therefore$ $\alpha + \beta = \frac{-b}{a}$ ... (1) $\alpha\beta = \frac{c}{a}$ ... (2)

$-\alpha + \gamma = \frac{-b_1}{a_1}$ ... (3) $-\alpha\gamma = \frac{c_1}{a_1}$ ... (4)

(1) + (3) gives $\beta + \gamma = -\left(\frac{b}{a} + \frac{b_1}{a_1}\right)$ ... (5)

(1) ÷ (2) gives $\frac{\alpha + \beta}{\alpha\beta} = -\frac{b}{c}$; $\therefore$ $\frac{1}{\beta} + \frac{1}{\alpha} = \frac{-b}{c}$

(3) ÷ (4) gives $\frac{-\alpha + \gamma}{-\alpha\gamma} = -\frac{b_1}{c_1}$; $\therefore$ $\frac{1}{\gamma} - \frac{1}{\alpha} = \frac{-b_1}{c_1}$.

Adding these, $\frac{1}{\beta} + \frac{1}{\gamma} = -\left(\frac{b}{c} + \frac{b_1}{c_1}\right)$ ... (6)

The quadratic equation whose roots are β, γ is

$$x^2 - (\beta + \gamma)x + \beta\gamma = 0$$

or $\frac{x^2}{-(\beta + \gamma)} + x + \frac{\beta\gamma}{-(\beta + \gamma)} = 0$ {dividing by $-(\beta + \gamma)$}

or $\frac{x^2}{\frac{b}{a} + \frac{b_1}{a_1}} + x + \frac{1}{-\left(\frac{1}{\beta} + \frac{1}{\gamma}\right)} = 0$ {using (5)}

or $\frac{x^2}{\frac{b}{a} + \frac{b_1}{a_1}} + x + \frac{1}{\frac{b}{c} + \frac{b_1}{c_1}} = 0$ {using (6)}.

57. Let a, b, c be real. If $ax^2 + bx + c = 0$ has two real roots α, β where $\alpha < -1$ and $\beta > 1$ then show that

$$1 + \frac{c}{a} + \left|\frac{b}{a}\right| < 0.$$

As α, β are real and unequal, $b^2 - 4ac > 0$.

Here $\alpha + \beta = -\frac{b}{a}$ and $\alpha\beta = \frac{c}{a}$.

Also $\alpha < -1, \beta > -1 \Rightarrow \alpha + 1 < 0, \beta + 1 > 0$ ... (1)

$\alpha < 1, \beta > 1 \Rightarrow \alpha - 1 < 0, \beta - 1 > 0$ ... (2)

From (1), $(\alpha + 1)(\beta + 1) < 0$

or $\alpha\beta + \alpha + \beta + 1 < 0$

$\Rightarrow$ $\frac{c}{a} - \frac{b}{a} + 1 < 0$ ... (3)

From (2), $(\alpha - 1)(\beta - 1) < 0$

or $\alpha\beta - (\alpha + \beta) + 1 < 0$

or $\frac{c}{a} + \frac{b}{a} + 1 < 0$ ... (4)

Combining (3) and (4), $1 + \frac{c}{a} \pm \frac{b}{a} < 0$

or $1 + \frac{c}{a} + \left|\frac{b}{a}\right| < 0$.

58. For what values of $m \in R$, both roots of the equation $x^2 - 6mx + 9m^2 - 2m + 2 = 0$ exceed 3?

Let α, β be the roots. As both roots exceed 3, they are real.

$\therefore$ $D \geq 0$ ($\because$ roots may be equal) and $\alpha > 3, \beta > 3$.

$D \geq 0 \Rightarrow 36m^2 - 4(9m^2 - 2m + 2) \geq 0$

or $8m - 8 \geq 0$; $\therefore$ $m \geq 1$ ... (1)

$\alpha > 3, \beta > 3 \Rightarrow \alpha - 3 > 0, \beta - 3 > 0$

$\Rightarrow (\alpha - 3) + (\beta - 3) > 0, (\alpha - 3)(\beta - 3) > 0$

$\Rightarrow \alpha + \beta - 6 > 0, \alpha\beta - 3(\alpha + \beta) + 9 > 0$

$\Rightarrow 6m - 6 > 0, 9m^2 - 2m + 2 - 3 \cdot 6m + 9 > 0$

$\Rightarrow m > 1, 9m^2 - 20m + 11 > 0$

$\Rightarrow m > 1, (9m - 11)(m - 1) > 0$

$\Rightarrow m > 1, m < 1$ or $m > \frac{11}{9}$ ... (2)

The values of m satisfying (1) and (2) at the same time are $m > \frac{11}{9}$.

59. Let a, b, c be positive integers. Consider the class of quadratic equations $ax^2 - bx + c = 0$ having two distinct real roots in the open interval (0, 1). Find the least positive integral value of a for such equations.

Here a, b, c are positive integers.

Let the distinct real roots be α, β. From the question,

$$0 < \alpha < 1, 0 < \beta < 1 \text{ and } b^2 - 4ac > 0$$

$\therefore$ $\alpha > 0, 1 - \alpha > 0$ as well as $\beta > 0, 1 - \beta > 0$

$$\therefore \quad 0 < \alpha(1-\alpha) = \alpha - \alpha^2 = \frac{1}{4} - \left(\alpha^2 - \alpha + \frac{1}{4}\right)$$

$$= \frac{1}{4} - \left(\alpha - \frac{1}{2}\right)^2 \le \frac{1}{4} \quad \ldots (1)$$

Similarly, $0 < \beta(1-\beta) \le \frac{1}{4}$. ... (2)

$\therefore$ multiplying (1) and (2), $0 < \alpha\beta(1-\alpha)(1-\beta) \le \left(\frac{1}{4}\right)^2$

or $\quad 0 < \frac{c}{a}\{1 - (\alpha + \beta) + \alpha\beta\} \le \frac{1}{16}$

or $\quad 0 < \frac{c}{a}\left\{1 - \frac{b}{a} + \frac{c}{a}\right\} \le \frac{1}{16}$

or $\quad 0 < c(a - b + c) \le \frac{a^2}{16}$.

As $c(a - b + c) > 0$ and it is an integer because a, b, c are all integers,

$$c(a - b + c) \ge 1, \text{ equality holding when } c = 1, a = b.$$

When $c = 1, a = b, b^2 - 4ac > 0$ implies $a^2 > 4a$.

$\therefore$ $a > 4$, as a is a positive integer.

If $1 < c(a - b + c) \le \frac{a^2}{16}$ then $a^2 > 16$

$\therefore$ a is at least 5.

But $b^2 - 4ac > 0$; so $b^2 > 4ac$.

As c is at least 1 and a is at least 5, $b^2 > 4 \cdot 5 \cdot 1 = 20$

$\therefore$ b is also at least 5.

Taking the least values of a, b, c we get the equation $5x^2 - 5x + 1 = 0$ whose roots are

$$\frac{5 \pm \sqrt{25 - 20}}{10}, \text{ i.e., } \frac{5 \pm \sqrt{5}}{10}$$

which lie between 0 and 1.

60. Find the values of a for which the inequation $x^2 + ax + a^2 + 6a < 0$ is satisfied for all $x \in (1, 2)$.

Let α, β be the roots of the corresponding equation

$$x^2 + ax + a^2 + 6a = 0 \quad \ldots (1)$$

As the coefficient of $x^2 = 1 > 0$, $x^2 + ax + a^2 + 6a < 0$ will be satisfied for all values of $x \in (\alpha, \beta)$ if α, β are real and unequal (let $\alpha < \beta$).

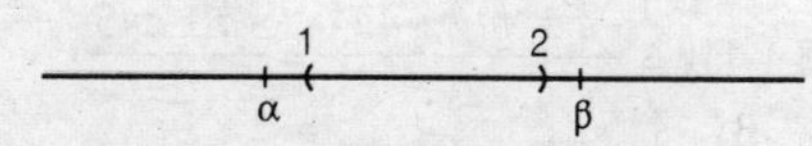

Hence the inequality will hold for all real $x \in (1, 2)$ if the interval (1, 2) is a subset of the interval (α, β).

Thus, for (1) we should have $D > 0$ and $\alpha < 1, \beta > 1$ as well as $\alpha < 2, \beta > 2$.

Now, $D > 0 \Rightarrow a^2 - 4(a^2 + 6a) > 0$

$\Rightarrow -3(a^2 + 8a) > 0$

$\Rightarrow a^2 + 8a < 0$

$\Rightarrow a(a + 8) < 0$

$\Rightarrow -8 < a < 0$ ... (2)

(using sign-scheme)

$\alpha < 1, \beta > 1 \Rightarrow \alpha - 1 < 0, \beta - 1 > 0$

$\Rightarrow (\alpha - 1)(\beta - 1) < 0$

$\Rightarrow \alpha\beta - (\alpha + \beta) + 1 < 0$

$\Rightarrow a^2 + 6a - (-a) + 1 < 0$

$\Rightarrow a^2 + 7a + 1 < 0.$

The corresponding equation has the roots

$$\frac{-7 \pm \sqrt{49 - 4}}{2}, \text{ i.e., } \frac{-7 \pm \sqrt{45}}{2}$$

$\therefore$ by sign-scheme, $\frac{-7 - \sqrt{45}}{2} < a < \frac{-7 + \sqrt{45}}{2}$ ... (3)

$\alpha < 2, \beta > 2 \Rightarrow \alpha - 2 < 0, \beta - 2 > 0$

$\Rightarrow (\alpha - 2)(\beta - 2) < 0$

$\Rightarrow \alpha\beta - 2(\alpha + \beta) + 4 < 0$

$\Rightarrow a^2 + 6a - 2(-a) + 4 < 0$

$\Rightarrow a^2 + 8a + 4 < 0.$

The corresponding equation has the roots

$$\frac{-8 \pm \sqrt{64 - 16}}{2}, \text{ i.e., } -4 \pm 2\sqrt{3}$$

$\therefore$ by sign-scheme $-4 - 2\sqrt{3} < a < -4 + 2\sqrt{3}$ ... (4)

Common values of a satisfying (2), (3) and (4) are

$$\frac{-7 - \sqrt{45}}{2} < a < -4 + 2\sqrt{3}.$$

61. Find all real values of the parameter a for which the equation $x(x+1)(x+a)(x+a+1) = a^2$ has four real roots.

Here the equation is

$$\{x(x + a + 1)\}\{(x + 1)(x + a)\} = a^2$$

or $\{x^2 + (a+1)x\}\{x^2 + (a+1)x + a\} = a^2$

$\therefore \quad \{x^2 + (a+1)x\}^2 + a\{x^2 + (a+1)x\} - a^2 = 0$

$\therefore \quad x^2 + (a+1)x = \dfrac{-a \pm \sqrt{a^2 + 4a^2}}{2} = \dfrac{-a \pm a\sqrt{5}}{2}.$

If $\quad x^2 + (a+1)x = \dfrac{-a + a\sqrt{5}}{2}$

then $x^2 + (a+1)x + \dfrac{a - a\sqrt{5}}{2} = 0$ ... (1)

If $\quad x^2 + (a+1)x = \dfrac{-a - a\sqrt{5}}{2}$

then $x^2 + (a+1)x + \dfrac{a + a\sqrt{5}}{2} = 0.$ ... (2)

If (1) and (2) have real roots then the given equation will have four real roots.

$\therefore$ the required conditions are as follows:

$D_1 = (a+1)^2 - 4 \cdot \dfrac{a - a\sqrt{5}}{2} \ge 0$

$\Rightarrow \quad a^2 + 2a + 1 - 2(a - a\sqrt{5}) \ge 0$

$\Rightarrow \quad a^2 + 2\sqrt{5}a + 1 \ge 0$

and $D_2 = (a+1)^2 - 4 \cdot \dfrac{a + a\sqrt{5}}{2} \ge 0$

$\Rightarrow \quad a^2 + 2a + 1 - 2(a + a\sqrt{5}) \ge 0$

$\Rightarrow \quad a^2 - 2\sqrt{5}a + 1 \ge 0$

Now, the required solution-set is the intersection of the solution-sets of

$a^2 + 2\sqrt{5}a + 1 \ge 0$ ... (3)

and $a^2 - 2\sqrt{5}a + 1 \ge 0$ ... (4)

For (3), the corresponding equation is

$a^2 + 2\sqrt{5}a + 1 = 0.$

Roots are $a = \dfrac{-2\sqrt{5} \pm \sqrt{20 - 4}}{2}$

$= -\sqrt{5} \pm 2$

$\therefore$ sign-scheme for $a^2 + 2\sqrt{5}a + 1$, $a \in R$ is as follows:

(+) $\quad -2-\sqrt{5} \quad$ (–) $\quad 2-\sqrt{5} \quad$ (+)

$\therefore \quad a \le -2 - \sqrt{5} \quad$ or $\quad a \ge 2 - \sqrt{5}$

For (4), the corresponding equation is $a^2 - 2\sqrt{5}a + 1 = 0.$

Roots are $a = \dfrac{2\sqrt{5} \pm \sqrt{20 - 4}}{2}$

$= \sqrt{5} \pm 2$

$\therefore$ sign-scheme for $a^2 - 2\sqrt{5}a + 1$, $a \in R$ is as follows:

(+) $\quad -2+\sqrt{5} \quad$ (–) $\quad 2+\sqrt{5} \quad$ (+)

$\therefore \quad a \le -2 + \sqrt{5} \quad$ or $\quad a \ge 2 + \sqrt{5}$

$\therefore$ taking the common values of a satisfying both (3) and (4),

$a \le -2 - \sqrt{5}$ or $-\sqrt{5} + 2 \le a \le \sqrt{5} - 2$ or $a \ge \sqrt{5} + 2$

$\therefore$ the solution-set

$= (-\infty, -2-\sqrt{5}] \cup [2-\sqrt{5}, -2+\sqrt{5}] \cup [2+\sqrt{5}, +\infty).$

62. Find all the integral values of a for which the quadratic equation $(x-a)(x-10) + 1 = 0$ has integral roots.

Here the equation is $x^2 - (10+a)x + 10a + 1 = 0$

$\therefore \quad x = \dfrac{10 + a \pm \sqrt{(10+a)^2 - 4(10a+1)}}{2}.$

To be integer, x must be real. So

$(10+a)^2 - 4(10a+1) \ge 0$

or $\quad a^2 - 20a + 96 \ge 0$

or $\quad (a-8)(a-12) \ge 0.$

Corresponding equation is $(a-8)(a-12) = 0$

$\therefore \quad a = 8, 12$

$\therefore$ sign-scheme for $a^2 - 20a + 96$, $a \in R$ is as follows:

(+) $\quad 8 \quad$ (–) $\quad 12 \quad$ (+)

$\therefore \quad a$ may be an integer ≤ 8 or ≥ 12.

Let $a = 12 + k$, k being a non-negative integer.

Then $D = a^2 - 20a + 96$

$= (12+k)^2 - 20(12+k) + 96 = k^2 + 4k$

D must be a perfect square. But $k^2 + 4k = (k+2)^2 - 4$ is a perfect square only when $k = 0$

(because the difference of two squares of non-negative integers can be 4 only if the integers are 0 and 2).

$\therefore$ when $a = 12$, x is an integer $\left(x = \dfrac{10+12}{2} = 11\right)$.

Let $a = 8 - p$, p being a non-negative integer.

Then $D = a^2 - 20a + 96$

$= (8-p)^2 - 20(8-p) + 96$

$= p^2 + 4p = (p+2)^2 - 4.$

This is a perfect square when $p = 0$

$\therefore$ when $a = 8$, x is an integer $\left(x = \dfrac{10+8}{2} = 9\right)$.

Thus $a = 12, 8.$

63. Find the parameter $a \in R$ such that $4^x - a \cdot 2^x - a + 3 \le 0$ for at least one $x \in R$.

Here $4^x + 3 - a(2^x + 1) \le 0$ ... (1)

As $4^x + 3$ and $2^x + 1$ are positive, (1) can hold for $a > 0$ only.

Now, $4^x - a \cdot 2^x - a + 3$

$$= (2^x)^2 - 2 \cdot \frac{a}{2} \cdot 2^x + \left(\frac{a}{2}\right)^2 - \left(\frac{a}{2}\right)^2 - a + 3$$

$$= \left(2^x - \frac{a}{2}\right)^2 - \left(\frac{a^2}{4} + a - 3\right) \le 0, \text{ (given)}$$

$$\therefore \quad \frac{a^2}{4} + a - 3 \ge \left(2^x - \frac{a}{2}\right)^2 \ge 0$$

or $\quad a^2 + 4a - 12 \ge 0.$

Corresponding equation is $a^2 + 4a - 12 = 0$

or $\quad (a + 6)(a - 2) = 0; \quad \therefore \quad a = -6, 2.$

$\therefore$ sign-scheme for $a^2 + 4a - 12$, $a \in R$ is as follows:

(+) –6 (–) 2 (+)

$\therefore \quad a^2 + 4a - 12 \ge 0$ when $a \le -6$ or $a \ge 2$.

But $a > 0$; so $a \ge 2$.

64. Find all values of the parameter a for which the inequality $a \cdot 9^x + 4(a-1)3^x + a > 1$ is satisfied for all real values of x.

Let $3^x = y$. Then the inequality becomes

$$ay^2 + 4(a-1)y + a > 1$$

or $\quad a\{y^2 + 4y + 1\} > 4y + 1$

or $\quad a\{(y+2)^2 - 3\} > 4y + 1 \qquad \ldots (1)$

As $y = 3^x > 0$, $4y + 1 > 1$ and minimum value of $(y+2)^2$ is greater than 4.

$\therefore \quad$ from (1), $a\{(y+2)^2 - 3\} > 4y + 1 > 1$

$\therefore \quad a > \dfrac{1}{(y+2)^2 - 3}$, because $(y+2)^2 - 3 > 4 - 3 > 0$.

Clearly, $0 < \dfrac{1}{(y+2)^2 - 3} < 1$, for all y.

Thus if $a \ge 1$, $a > \dfrac{1}{(y+2)^2 - 3}$ holds for all y and so for all real x.

Hence, $a \ge 1$.

65. Solve the equation $\sqrt{a(2^x - 2) + 1} = 1 - 2^x$ for every value of the parameter a.

Here $a(2^x - 2) + 1 = (1 - 2^x)^2$

or $\quad a \cdot 2^x + 1 - 2a = 1 - 2 \cdot 2^x + (2^x)^2$

or $\quad (2^x)^2 - (a+2) \cdot 2^x + 2a = 0$

$$\therefore \quad 2^x = \frac{a + 2 \pm \sqrt{(a+2)^2 - 4 \cdot 2a}}{2} = \frac{a + 2 \pm \sqrt{a^2 - 4a + 4}}{2}$$

$$= \frac{a + 2 \pm (a - 2)}{2} = a, 2$$

$\therefore \quad 2^x = a \quad$ or $\quad 2^x = 2.$

Clearly, $2^x = 2 \quad \Rightarrow \quad x = 1.$

If $a < 0$, $2^x = a$ gives no solution for $2^x > 0$ for all real x.

If $\quad a = 0, 2^x = 0 \quad \Rightarrow \quad x$ has no solution, for $2^x > 0$.

If $\quad a > 0, 2^x = a \quad \Rightarrow \quad x \log 2 = \log a$

or $\quad x = \dfrac{\log a}{\log 2} = \log_2 a.$

Thus we get $x = 1, \log_2 a$ where $a > 0$.

But $x = 1$ does not satisfy the given equation, because $\sqrt{1} = 1 - 2 = -1$ which is not true.

Now, from the question,

clearly, $2^x \le 1 \quad \Rightarrow \quad 2^x \le 2^0 \quad \Rightarrow \quad x \le 0$

and $\quad a(2^x - 2) + 1 \ge 0$

$\therefore \quad a(2^x - 2) \ge -1$

$\therefore \quad a(2 - 2^x) \le 1$

$\Rightarrow \quad a$ is a positive fraction or 1 because

$\qquad 2 - 2^x \ge 1 \ (\because x \le 0)$

Hence, $\quad x = \log_2 a \quad$ where $\quad 0 < a \le 1.$

66. Find all $a \in R$ for each of which $3 - |x - a| > x^2$ has at least one negative solution.

Here the inequation is $x^2 + |x - a| - 3 < 0 \qquad \ldots (1)$

This is a quadratic inequation where

coefficient of $x^2 = 1 > 0$.

If α, β $(\alpha < \beta)$ be the real and unequal roots of the corresponding equation then $x^2 + |x - a| - 3 < 0$ for all $x \in (\alpha, \beta)$.

$\therefore$ (1) will have at least one negative solution if there is at least one negative real number between α, β, i.e., at least one root of the corresponding equation is negative.

The corresponding equation of (1) is

$$x^2 + x - (a + 3) = 0 \qquad \ldots (2)$$

or $\quad x^2 - x + a - 3 = 0 \qquad \ldots (3)$

according as $x \ge a$ or $x < a$.

Both should have real and unequal roots. So

$$1 + 4(a + 3) > 0 \quad \text{and} \quad 1 - 4(a - 3) > 0$$

i.e., $a > \frac{-13}{4}$ and $a < \frac{13}{4}$.

Roots of (2) are $\frac{-1 \pm \sqrt{13 + 4a}}{2}$.

So it has one negative root.

Roots of (3) are $\frac{1 \pm \sqrt{13 - 4a}}{2}$. One of these will be negative provided $13 - 4a > 1$, i.e., $12 > 4a$, i.e., $a < 3$.

Thus, a should satisfy $-\frac{13}{4} < a, a < \frac{13}{4}$ and $a < 3$.

$\therefore \quad -\frac{13}{4} < a < 3$. Hence $a \in \left(-\frac{13}{4}, 3\right)$.

67. Find all values of the parameter $a \in R$ for which the following inequality is valid for all $x \in R$.
$1 + \log_5 (x^2 + 1) \geq \log_5 (ax^2 + 4x + a)$.

Here $1 \geq \log_5 (ax^2 + 4x + a) - \log_5 (x^2 + 1)$

or $\quad 1 \geq \log_5 \frac{ax^2 + 4x + a}{x^2 + 1}$

or $\quad \log_5 5 \geq \log_5 \frac{ax^2 + 4x + a}{x^2 + 1}$

$\therefore$ we have $5 \geq \frac{ax^2 + 4x + a}{x^2 + 1}$ ... (1)

Also, logarithms to be real,

$x^2 + 1 > 0$ and $ax^2 + 4x + a > 0$.

$x^2 + 1 > 0$ is true for all $x \in R$.

Now, $ax^2 + 4x + a > 0$ ... (2)

$\therefore$ the values of the parameter $a \in R$ should be such that both (1) and (2) hold for all $x \in R$.

Now, (2) holds for all $x \in R$ if

$D = 4^2 - 4 \cdot a \cdot a < 0$ and coefficient of $x^2 = a > 0$,

i.e., $16 - 4a^2 < 0, a > 0$

$\Rightarrow \quad a^2 > 4, a > 0 \quad \Rightarrow \quad a > 2$.

Now, from (1), $5(x^2 + 1) \geq ax^2 + 4x + a$ $\{\because\ x^2 + 1 > 0\}$

or $\quad (5 - a)x^2 - 4x + 5 - a \geq 0$.

This holds for all $x \in R$ if

$D = (-4)^2 - 4(5 - a)(5 - a) \leq 0$

and the coefficient of $x^2 = 5 - a > 0$,

i.e., $16 - 4(5 - a)^2 \leq 0, 5 - a > 0$

$\Rightarrow \quad (5 - a)^2 \geq 4, a < 5$

$\Rightarrow \quad \{5 - a \leq -2$ or $5 - a \geq 2\}, a < 5$

$\Rightarrow \quad \{a \geq 7$ or $a \leq 3\}, a < 5$

$\Rightarrow \quad a \leq 3$.

Hence, the required values of a satisfy $a > 2$ and $a \leq 3$.

$\therefore \quad$ the solution-set = (2, 3].

68. The real numbers x_1, x_2, x_3 satisfying the equation $x^3 - x^2 + \beta x + \gamma = 0$ are in AP. Find the intervals in which β and γ lie.

From the question, the real roots of $x^3 - x^2 + \beta x + \gamma = 0$ are x_1, x_2, x_3 and they are in AP.

As x_1, x_2, x_3 are in AP we take

$$x_1 = a - d, x_2 = a, x_3 = a + d.$$

Now, $x_1 + x_2 + x_3 = -\frac{-1}{1} = 1$

or $\quad a - d + a + a + d = 1; \quad \therefore \quad a = \frac{1}{3}$ ... (1)

$x_1x_2 + x_2x_3 + x_3x_1 = \frac{\beta}{1} = \beta$

or $\quad (a - d)a + a(a + d) + (a + d)(a - d) = \beta$ ... (2)

$x_1x_2x_3 = -\frac{\gamma}{1} = -\gamma$

or $\quad (a - d)a(a + d) = -\gamma$ ... (3)

From (1) and (2), we get $3a^2 - d^2 = \beta$

or $\quad 3 \cdot \frac{1}{9} - d^2 = \beta$; so, $\quad \beta = \frac{1}{3} - d^2 < \frac{1}{3}$.

From (1) and (3), we get $\frac{1}{3}\left(\frac{1}{9} - d^2\right) = -\gamma$

or $\quad \gamma = \frac{1}{3}\left(d^2 - \frac{1}{9}\right) > \frac{1}{3}\left(-\frac{1}{9}\right) = -\frac{1}{27}$

$\therefore \quad \beta \in \left(-\infty, \frac{1}{3}\right)$ and $\gamma \in \left(-\frac{1}{27}, +\infty\right)$.

69. If α, β, γ be roots of $2x^3 + x^2 - 7 = 0$ then find the value of $\Sigma\left(\frac{\alpha}{\beta} + \frac{\beta}{\alpha}\right)$.

We have, $\alpha + \beta + \gamma = \frac{-1}{2}$ ... (1)

$\alpha\beta + \beta\gamma + \gamma\alpha = \frac{0}{2} = 0$ ... (2)

$\alpha\beta\gamma = -\frac{-7}{2} = \frac{7}{2}$ ... (3)

Now, $(\alpha + \beta + \gamma)\left(\frac{1}{\alpha} + \frac{1}{\beta} + \frac{1}{\gamma}\right)$

$$= 3 + \frac{\alpha}{\beta} + \frac{\alpha}{\gamma} + \frac{\beta}{\alpha} + \frac{\beta}{\gamma} + \frac{\gamma}{\alpha} + \frac{\gamma}{\beta}$$

or $\quad (\alpha + \beta + \gamma) \frac{\beta\gamma + \gamma\alpha + \alpha\beta}{\alpha\beta\gamma}$

$$= 3 + \left(\frac{\alpha}{\beta} + \frac{\beta}{\alpha}\right) + \left(\frac{\gamma}{\alpha} + \frac{\alpha}{\gamma}\right) + \left(\frac{\beta}{\gamma} + \frac{\gamma}{\beta}\right)$$

$\therefore \quad -\frac{1}{2} \cdot \frac{0}{7/2} = 3 + \Sigma\left(\frac{\alpha}{\beta} + \frac{\beta}{\alpha}\right)$

$\therefore \quad \Sigma\left(\frac{\alpha}{\beta} + \frac{\beta}{\alpha}\right) = -3.$

70. If $f(x) = x^3 + bx^2 + cx + d$ and $f(0)$, $f(-1)$ are odd integers, prove that $f(x) = 0$ cannot have all integral roots.

$f(0) = d, f(-1) = -1 + b - c + d.$

From the question, $d = \text{odd}$

and $-1 + b - c + d = \text{odd}$

$\therefore \quad b - c = 1 + \text{odd} - d$

$= (1 + \text{odd}) - (\text{odd})$

$= \text{even} - \text{odd} = \text{odd}$ ... (1)

Thus, both d and $b - c$ are odd.

If possible let the three roots α, β, γ be all integers.

Now, $\alpha\beta\gamma = \frac{-d}{1} = -d = \text{negative odd integer}$ ... (2)

$\therefore$ α, β, γ are 3 integers whose product is odd.

So α, β, γ are all odd.

Again $\alpha + \beta + \gamma = -b$

and $\alpha\beta + \beta\gamma + \gamma\alpha = c$

Hence, b and c will be odd and so $b - c$ will be even which contradicts (1).

Hence, the three roots cannot be all integers.

71. If a, b, c be in GP then prove that the equations $ax^2 + 2bx + c = 0$ and $dx^2 + 2ex + f = 0$ have a common root if $\frac{d}{a}, \frac{e}{b}, \frac{f}{c}$ are in AP.

Here $b^2 = ac$. Let α be the common root. Then

$a\alpha^2 + 2b\alpha + c = 0$

and $d\alpha^2 + 2e\alpha + f = 0.$

$\therefore$ by cross-multiplication,

$$\frac{\alpha^2}{2(bf - ce)} = \frac{\alpha}{cd - af} = \frac{1}{2(ae - bd)}$$

$\therefore \quad (cd - af)^2 = 4(bf - ce)(ae - bd)$ ... (1)

$\{\because$ numerators $\alpha^2, \alpha, 1$ are in GP$\}$

Let $\frac{e}{b} - \frac{d}{a} = p$ and $\frac{f}{c} - \frac{e}{b} = q$ ... (2)

so $\quad \frac{f}{c} - \frac{d}{a} = p + q$ ... (3)

$\therefore \quad ae - bd = abp$ and $bf - ce = bcq$ [from (2)]

Putting in (1), $(cd - af)^2 = 4bcq \cdot abp$.

Dividing by a^2c^2, $\left(\frac{cd - af}{ac}\right)^2 = \frac{4bcq \cdot abp}{a^2c^2}$

or $\quad \left(\frac{d}{a} - \frac{f}{c}\right)^2 = 4\frac{b^2}{ac}pq = 4pq \quad \{\because b^2 = ac\}$

or $\quad (-p - q)^2 = 4pq$ [from (3)]

or $\quad (p + q)^2 - 4pq = 0$

or $\quad (p - q)^2 = 0 \quad \therefore \quad p = q$

$\therefore \quad \frac{e}{b} - \frac{d}{a} = p = q = \frac{f}{c} - \frac{e}{b}$

$\therefore \quad \frac{d}{a}, \frac{e}{b}, \frac{f}{c}$ are in AP.

72. If each pair of equations from $x^2 - b_r x + c_r = 0$; $r = 1, 2, 3$ have a common root, prove that $\Sigma b_1^2 + 4\Sigma c_1 = 2\Sigma b_1 b_2$.

As every pair has a common root, let

$x^2 - b_1 x + c_1 = 0$ have the roots α, β

$x^2 - b_2 x + c_2 = 0$ have the roots α, γ

$x^2 - b_3 x + c_3 = 0$ have the roots β, γ

$\therefore \quad \alpha + \beta = b_1$ ... (1)

$\alpha\beta = c_1$ ... (2)

$\alpha + \gamma = b_2$ ... (3)

$\alpha\gamma = c_2$ ... (4)

$\beta + \gamma = b_3$ ... (5)

$\beta\gamma = c_3$ ... (6)

Now, $\text{LHS} = b_1^2 + b_2^2 + b_3^2 + 4(c_1 + c_2 + c_3)$

$= (\alpha + \beta)^2 + (\alpha + \gamma)^2 + (\beta + \gamma)^2 + 4(\alpha\beta + \alpha\gamma + \beta\gamma)$

{using (1) through (6)}

$= 2(\alpha^2 + \beta^2 + \gamma^2) + 6(\alpha\beta + \alpha\gamma + \beta\gamma)$

$= 2\{\alpha^2 + \beta^2 + \gamma^2 + 3(\alpha\beta + \beta\gamma + \gamma\alpha)\}$

$\text{RHS} = 2(b_1 b_2 + b_2 b_3 + b_3 b_1)$

$= 2\{(\alpha + \beta)(\alpha + \gamma) + (\alpha + \gamma)(\beta + \gamma) + (\beta + \gamma)(\alpha + \beta)\}$

$= 2\{(\alpha^2 + \alpha\gamma + \beta\alpha + \beta\gamma) + (\gamma^2 + \alpha\beta + \alpha\gamma + \gamma\beta) + (\beta^2 + \beta\alpha + \gamma\alpha + \gamma\beta)\}$

$= 2\{\alpha^2 + \beta^2 + \gamma^2 + 3(\alpha\beta + \beta\gamma + \gamma\alpha)\}$

$\therefore$ LHS = RHS.

73. If $x \in R$, prove that the maximum value of $2(a-x)\cdot(x+\sqrt{x^2+b^2})$ is a^2+b^2.

Let $y = 2(a-x)(x+\sqrt{x^2+b^2})$ ... (1)

Let $x \neq a$ for then the value of y is 0 which is not the maximum value.

$$\therefore \quad x+\sqrt{x^2+b^2} = \frac{y}{2(a-x)} \qquad \text{... (2)}$$

$$\therefore \quad (\sqrt{x^2+b^2}+x)(\sqrt{x^2+b^2}-x) = \frac{y}{2(a-x)}(\sqrt{x^2+b^2}-x)$$

$$\text{or} \quad (x^2+b^2)-x^2 = \frac{y}{2(a-x)}(\sqrt{x^2+b^2}-x)$$

$$\therefore \quad \sqrt{x^2+b^2}-x = \frac{2b^2(a-x)}{y} \qquad \text{... (3)}$$

$$\text{From (2) − (3), } 2x = \frac{y}{2(a-x)} - \frac{2b^2(a-x)}{y}$$

or $4xy(a-x) = y^2 - 4b^2(a-x)^2$

or $4axy - 4yx^2 = y^2 - 4b^2(a^2+x^2-2ax)$

or $4(y-b^2)x^2 + 4a(2b^2-y)x + y^2 - 4a^2b^2 = 0.$

But x is real; $\therefore$ $D \geq 0$

$\Rightarrow$ $16a^2(2b^2-y)^2 - 4\cdot 4(y-b^2)(y^2-4a^2b^2) \geq 0$

$\Rightarrow$ $a^2(4b^4-4b^2y+y^2)-(y-b^2)(y^2-4a^2b^2) \geq 0$

$\Rightarrow$ $y^3-(a^2+b^2)y^2 \leq 0$

or $y^2\{y-(a^2+b^2)\} \leq 0.$

But $y^2 \geq 0$; so $y-(a^2+b^2) \leq 0$

$\therefore$ $y \leq a^2+b^2$; so the maximum value of the given expression is a^2+b^2.

74. Show that the minimum value of $\dfrac{(x+a)(x+b)}{x+c}$, where $a>c$, $b>c$, is $(\sqrt{a-c}+\sqrt{b-c})^2$ for real values of $x > -c$.

$$\text{Let} \quad \frac{(x+a)(x+b)}{x+c} = y$$

or $x^2+(a+b)x+ab = yx+cy$

or $x^2+(a+b-y)x+ab-cy = 0.$

As x is real, $D \geq 0$

$\Rightarrow$ $(a+b-y)^2 - 4(ab-cy) \geq 0$

$\Rightarrow$ $y^2-2(a+b-2c)y+(a+b)^2-4ab \geq 0$

$\Rightarrow$ $y^2-2(a+b-2c)y+(a-b)^2 \geq 0.$

Roots of the corresponding equation are

$$y = \frac{2(a+b-2c) \pm \sqrt{4(a+b-2c)^2-4(a-b)^2}}{2}$$

$$= \frac{2(a+b-2c) \pm 2\sqrt{(a+b-2c+a-b)(a+b-2c-a+b)}}{2}$$

$= a+b-2c \pm \sqrt{2(a-c)2(b-c)}$

$= a+b-2c \pm 2\sqrt{(a-c)(b-c)}$

$= (\sqrt{a-c})^2 + (\sqrt{b-c})^2 \pm 2\sqrt{a-c}\,\sqrt{b-c}$

$= (\sqrt{a-c} \pm \sqrt{b-c})^2.$

$\therefore$ the sign-scheme of $y^2-2(a+b-2c)y+(a-b)^2$, $y \in R$ is as follows:

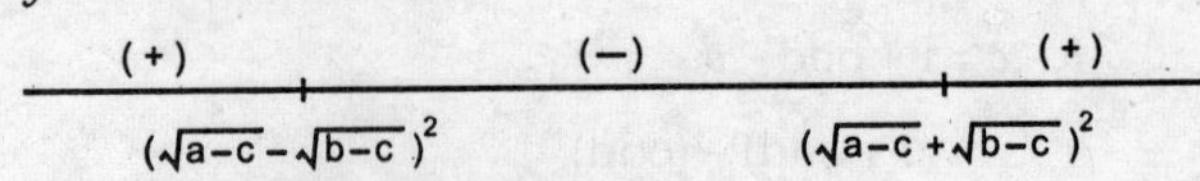

$\therefore$ $y \leq (\sqrt{a-c}-\sqrt{b-c})^2$ ... (1)

or $y \geq (\sqrt{a-c}+\sqrt{b-c})^2$ ... (2)

If $y \leq (\sqrt{a-c}-\sqrt{b-c})^2$ then

$$\frac{(x+a)(x+b)}{x+c} \leq (\sqrt{a-c}-\sqrt{b-c})^2$$

or $x^2+(a+b)x+ab$

$\leq \{a+b-2c-2\sqrt{(a-c)(b-c)}\}(x+c)$ $\{\because x+c>0\}$

or $x^2+\{a+b-a-b+2c+2\sqrt{(a-c)(b-c)}\}x$

$+ab-c(a+b-2c-2\sqrt{(a-c)(b-c)}) \leq 0$

or $x^2+2\{c+\sqrt{(a-c)(b-c)}\}x$

$+\{(a-c)(b-c)+c^2+2c\sqrt{(a-c)(b-c)}\} \leq 0$

or $x^2+2\{c+\sqrt{(a-c)(b-c)}\}x+\{c+\sqrt{(a-c)(b-c)}\}^2 \leq 0$

or $\{x+c+\sqrt{(a-c)(b-c)}\}^2 \leq 0.$

As $x+c>0$, $a>c$, $b>c$ the above is impossible.

Hence $y \nleq (\sqrt{a-c}-\sqrt{b-c})^2$. So from (2),

$y \geq (\sqrt{a-c}+\sqrt{b-c})^2$

$\therefore$ the minimum value of y is $(\sqrt{a-c}+\sqrt{b-c})^2$.

75. Show that for all real x, $|x| \neq |\cos\alpha|$, the expression $\dfrac{(x-1)\cos^2\alpha}{x^2-\cos^2 x}$ cannot lie between $\dfrac{1}{2}\left(\cos\dfrac{\alpha}{2}-\sin\dfrac{\alpha}{2}\right)^2$ and $\dfrac{1}{2}\left(\cos\dfrac{\alpha}{2}+\sin\dfrac{\alpha}{2}\right)^2$.

Let $\frac{(x-1)\cos^2\alpha}{x^2-\cos^2\alpha}=y$

or $(x-1)\cos^2\alpha = yx^2 - y\cos^2\alpha$

or $yx^2 - \cos^2\alpha \cdot x + \cos^2\alpha \cdot (1-y) = 0.$

As x is real, $D \ge 0$

$\Rightarrow \quad \cos^4\alpha - 4y(1-y)\cos^2\alpha \ge 0$

or $4y^2 - 4y + \cos^2\alpha \ge 0 \quad (\because \cos^2\alpha > 0).$

Roots of the corresponding equation are

$$y = \frac{4 \pm \sqrt{16-16\cos^2\alpha}}{8} = \frac{1 \pm \sqrt{1-\cos^2\alpha}}{2}$$

$$= \frac{1 \pm \sin\alpha}{2}$$

$\therefore$ by the sign-scheme we know that y must not lie between the roots, i.e.,

$$\frac{1}{2}(1-\sin\alpha) \text{ and } \frac{1}{2}(1+\sin\alpha), \text{ i.e.,}$$

$$\frac{1}{2}\left(\cos\frac{\alpha}{2} - \sin\frac{\alpha}{2}\right)^2 \text{ and } \frac{1}{2}\left(\cos\frac{\alpha}{2} + \sin\frac{\alpha}{2}\right)^2.$$

76. Find all the values of the parameter a for which $\frac{ax^2+3x-4}{a+3x-4x^2}$ takes all real values for $x \in R$.

Let $\frac{ax^2+3x-4}{a+3x-4x^2}=y$

or $ax^2+3x-4 = ay + 3yx - 4yx^2$

or $(a+4y)x^2 + 3(1-y)x - (4+ay) = 0.$

As x is real,

$D \ge 0 \quad \Rightarrow \quad 9(1-y)^2 + 4(a+4y)(4+ay) \ge 0$

or $9(1-2y+y^2) + 4(4a + a^2y + 16y + 4ay^2) \ge 0$

or $(9+16a)y^2 + \{4(a^2+16) - 18\}y + 9 + 16a \ge 0$

or $(9+16a)y^2 + 2(2a^2+23)y + 9 + 16a \ge 0 \quad \dots (1)$

Here $D = 4(2a^2+23)^2 - 4(9+16a)^2$

$= 4(2a^2+23+9+16a)(2a^2+23-9-16a)$

$= 4(2a^2+16a+32)(2a^2-16a+14)$

$= 16(a^2+8a+16)(a^2-8a+7)$

$= 16(a+4)^2(a-7)(a-1).$

The given expression takes all real values if y can take all real values satisfying (1).

For this, $D \le 0$ and coefficient of $y^2 = 9+16a > 0.$

$D \le 0 \quad \Rightarrow \quad (a-7)(a-1) \le 0$

$\Rightarrow \quad 1 \le a \le 7$

and $9+16a > 0 \quad \Rightarrow \quad a > \frac{-9}{16}.$

Hence $a \in [1, 7].$

77. Prove that $\left|\frac{12x}{4x^2+9}\right| \le 1$ for $x \in R$ and equality holds if $|x| = \frac{3}{2}$.

$\left|\frac{12x}{4x^2+9}\right| \le 1 \quad \Leftrightarrow \quad -1 \le \frac{12x}{4x^2+9} \le 1.$

Let $\frac{12x}{4x^2+9} = y$; we have to prove that $-1 \le y \le 1.$

Now, $12x = y(4x^2+9)$

or $4yx^2 - 12x + 9y = 0.$

As x is real, $D \ge 0 \quad \Rightarrow \quad 144 - 4 \cdot 4y \cdot 9y \ge 0$

or $1 - y^2 \ge 0$

or $y^2 \le 1; \quad \therefore \quad -1 \le y \le 1$

$\therefore \quad |y| \le 1;$ so, $\left|\frac{12x}{4x^2+9}\right| \le 1.$

Here the equality will hold if $\left|\frac{12x}{4x^2+9}\right| = 1$

or $\frac{12|x|}{4x^2+9} = 1 \quad \left\{\because \left|\frac{a}{b}\right| = \frac{|a|}{|b|}, |ab| = |a|\,|b|,\right.$

$\left.|4x^2+9| = 4x^2+9 \text{ because } 4x^2+9 \text{ is positive}\right\}$

or $12|x| = 4x^2 + 9$

or $4|x|^2 - 12|x| + 9 = 0$

or $(2|x| - 3)^2 = 0$

or $2|x| = 3; \quad \therefore \quad |x| = \frac{3}{2}.$

78. Prove that for all real x, y the value of $x^2+2xy+3y^2-6x-2y$ cannot be less than -11.

We have to prove that

$$x^2 + 2xy + 3y^2 - 6x - 2y \ge -11 \quad \dots (1)$$

i.e., $x^2 + 2xy + 3y^2 - 6x - 2y + 11 \ge 0$ for all $x, y \in R.$

or $x^2 + 2(y-3)x + 3y^2 - 2y + 11 \ge 0$ for all $x \in R.$

This is true if $D \le 0$ and coefficient of x^2 is positive.

Clearly coefficient of $x^2 = 1 > 0.$

Now $D \le 0$ if $4(y-3)^2 - 4(3y^2 - 2y + 11) \le 0$

or $(y-3)^2 - (3y^2 - 2y + 11) \le 0$

or $-2y^2 - 4y - 2 \le 0$

or $y^2 + 2y + 1 \ge 0$

or $\quad (y+1)^2 \geq 0,$ which is true for all real y.

Hence, (1) is true for all real x, y.

79. For what $m \in R$, the expression $2x^2 + mxy + 3y^2 - 5y - 2$ can be expressed as $(\lambda x + \mu y + \nu)(px + qy + r)$? Also for those values of m, factorize the expression.

$ax^2 + 2hxy + by^2 + 2gx + 2fy + c$ can be resolved into linear factors if $abc + 2fgh - af^2 - bg^2 - ch^2 = 0$

$\therefore$ the given expression can be factorized in linear factors

if $$2 \cdot 3 \cdot (-2) + 2 \cdot \left(-\frac{5}{2}\right) \cdot 0 \cdot \frac{m}{2} - 2 \cdot \left(-\frac{5}{2}\right)^2 - 3 \cdot 0^2 - (-2)\left(\frac{m}{2}\right)^2 = 0$$

or $\quad -12 - \frac{25}{2} + \frac{m^2}{2} = 0; \quad \therefore \quad m^2 = 49;$

$\therefore \quad m = 7, -7.$

Now, the corresponding equation is

$$2x^2 + mxy + 3y^2 - 5y - 2 = 0$$

or $$2x^2 + (my)x + 3y^2 - 5y - 2 = 0$$

$$\therefore \quad x = \frac{-my \pm \sqrt{m^2y^2 - 4.2(3y^2 - 5y - 2)}}{4}$$

When $m = 7$, $$x = \frac{-7y \pm \sqrt{49y^2 - 24y^2 + 40y + 16}}{4}$$

$$= \frac{-7y \pm \sqrt{25y^2 + 40y + 16}}{4}$$

$$= \frac{-7y \pm (5y + 4)}{4}$$

$$= \frac{-7y + 5y + 4}{4}, \frac{-7y - 5y - 4}{4}$$

$$= \frac{-y + 2}{2}, -3y - 1$$

$\therefore$ expression $= 2\left(x - \frac{-y+2}{2}\right)\{x - (-3y - 1)\}$

$= (2x + y - 2)(x + 3y + 1).$

When $m = -7$, $$x = \frac{7y \pm (5y + 4)}{4}$$

$$= \frac{7y + 5y + 4}{4}, \frac{7y - 5y - 4}{4}$$

$$= 3y + 1, \frac{y - 2}{2}.$$

$\therefore$ expression $= 2\{x - (3y + 1)\}\left(x - \frac{y - 2}{2}\right)$

$= (x - 3y - 1)(2x - y + 2).$

80. If α is a real root of the quadratic equation $ax^2 + bx + c = 0$ and β is a real root of $-ax^2 + bx + c = 0$ then show that there is a root γ of the equation $\frac{a}{2}x^2 + bx + c = 0$ which lies between α and β.

Let $f(x) = \frac{a}{2}x^2 + bx + c.$

α is a root of $ax^2 + bx + c = 0$

$\Rightarrow \quad a\alpha^2 + b\alpha + c = 0$

$\Rightarrow \quad \frac{a}{2}\alpha^2 + b\alpha + c = -\frac{a}{2}\alpha^2 \qquad \ldots (1)$

β is a root of $-ax^2 + bx + c = 0$

$\Rightarrow \quad -a\beta^2 + b\beta + c = 0$

$\Rightarrow \quad \frac{a}{2}\beta^2 + b\beta + c = \frac{3a}{2}\beta^2 \qquad \ldots (2)$

If $a > 0, \frac{a}{2}\alpha^2 + b\alpha + c < 0$

and $\quad \frac{a}{2}\beta^2 + b\beta + c > 0$ {using (1) and (2)}

$\therefore \quad f(\alpha) < 0$ and $f(\beta) > 0.$

If $a < 0, \frac{a}{2}\alpha^2 + b\alpha + c > 0$ and $\frac{a}{2}\beta^2 + b\beta + c < 0$

$\therefore \quad f(\alpha) > 0$ and $f(\beta) < 0$

$\therefore \quad f(\alpha)$ and $f(\beta)$ are of opposite signs.

$\therefore \quad f(\gamma) = 0$ where γ lies between α and β

or $\quad \gamma$ is a root of $\frac{a}{2}x^2 + bx + c = 0$

where γ lies between α and β.

81. If $a_1x^3 + b_1x^2 + c_1x + d_1 = 0$ and $a_2x^3 + b_2x^2 + c_2x + d_2 = 0$ have a pair of repeated roots common then prove that

$$\begin{vmatrix} 3a_1 & 2b_1 & c_1 \\ 3a_2 & 2b_2 & c_2 \\ a_2b_1 - a_1b_2 & c_1a_2 - c_2a_1 & d_1a_2 - d_2a_1 \end{vmatrix} = 0.$$

Let α, α be a pair of repeated roots common between the given cubic equations. Then α is a common root between the derived equations

$$3a_1x^2 + 2b_1x + c_1 = 0, 3a_2x^2 + 2b_2x + c_2 = 0$$

$\therefore \quad$ we get $a_1\alpha^3 + b_1\alpha^2 + c_1\alpha + d_1 = 0 \qquad \ldots (1)$

$a_2\alpha^3 + b_2\alpha^2 + c_2\alpha + d_2 = 0 \qquad \ldots (2)$

$3a_1\alpha^2 + 2b_1\alpha + c_1 = 0 \qquad \ldots (3)$

$3a_2\alpha^2 + 2b_2\alpha + c_2 = 0 \qquad \ldots (4)$

From (1) × a_2 − (2) × a_1, we get

$$(a_2b_1 - a_1b_2)\alpha^2 + (c_1a_2 - c_2a_1)\alpha$$
$$+ (d_1a_2 - d_2a_1) = 0 \quad \ldots (5)$$

Eliminating $\alpha^2, \alpha, \alpha^0$ from (3), (4) and (5) we get

$$\begin{vmatrix} 3a_1 & 2b_1 & c_1 \\ 3a_2 & 2b_2 & c_2 \\ a_2b_1 - a_1b_2 & c_1a_2 - c_2a_1 & d_1a_2 - d_2a_1 \end{vmatrix} = 0.$$

82. If $2a + 3b + 6c = 0$ then prove that the equation $ax^2 + bx + c = 0$ has both roots real of which one lies between 0 and 1 where a, b, c are real.

Consider the equation $2ax^3 + 3bx^2 + 6cx = 0 \quad \ldots (1)$

Let $f(x) = 2ax^3 + 3bx^2 + 6cx$.

Clearly $f(0) = 0, f(1) = 2a + 3b + 6c = 0$

(from the question)

$\therefore$ $x = 0, 1$ are two real roots of (1).

$\therefore$ the derived equation of (1) will have a real root lying between 0 and 1.

The derived equation of (1) is

$$6ax^2 + 6bx + 6c = 0$$

or $\quad ax^2 + bx + c = 0$.

$\therefore$ $ax^2 + bx + c = 0$ has a real root lying between 0 and 1.

As one root is real, the other root will also be real

($\because$ nonreal complex roots occur in conjugate pair).

Exercises

1. For what values of a the following will be an identity in x?
   $(x^2 - 1)a^2 - (2x + 3)a - (x^2 + 2x + 2) = 0$
2. Find the number of solutions of the triplet (a, b, c) for which $a\cos 2\theta + b\cos^2\theta + c = 0$ is satisfied for all values of θ.
3. Find the values of a for which the equation
   $(a^2 + 4a + 3)x^2 + (a^2 - a - 2)x + (a + 1)a = 0$
   in x has more than two roots.
4. Solve in R: $(x^2 + 2)^2 + 8x^2 = 6x(x^2 + 2)$.
5. Solve $(x + 4)(x + 7)(x + 8)(x + 11) + 20 = 0$.
6. Solve $\sqrt{5x^2 - 6x + 8} - \sqrt{5x^2 - 6x - 7} = 1$.
7. Solve $\sqrt{x} + \sqrt{x - \sqrt{1 - x}} = 1$.
8. Solve $(x + 3)^4 + (x + 2)^4 + (x + 1)^4 = 98$.
9. Solve for x: $2x^3 + 5x^2 - 2x + 3 = 0$.
10. Prove that the roots of the equation
    $x^3 - 12x^2 + 39x - 28 = 0$ are in AP.
11. Obtain the real solutions for x, y.
    $xy + 3y^2 - x + 4y - 7 = 0$ and
    $2xy + y^2 - 2x - 2y + 1 = 0$
12. Solve for x, y, z.
    $xy + x + y = 23$
    $xz + z + x = 41$
    $yz + y + z = 27$
13. Solve for x, y, z.
    $(b + c)(y + z) - ax = b - c$
    $(c + a)(z + x) - by = c - a$
    $(a + b)(x + y) - cz = a - b$
    where $a + b + c \neq 0$.
14. Solve for x, y.
    $x^2 + y(x + 1) = 17$ and
    $y^2 + x(y + 1) = 13$
15. Find x, y, z if
    $4x^2 + 25y^2 + 9z^2 - 10xy - 15yz - 6zx = 0$ and
    $x + y + z = 5$.
16. Solve $12^{5x-2} = (24\sqrt{3})^{3-x^2}$.
17. Solve $32^{\frac{x+5}{x-7}} = \frac{1}{4} \cdot 128^{\frac{x+17}{x-3}}$.
18. Solve $5^{x+1} + 5^{1-x} = 26$.
19. Solve $9^x + 78 = 3^{2x+3}$.
20. Solve $16^{\cos^2 x} + 16^{\sin^2 x} = 10$.
21. Solve $3^{\frac{x}{5}} + 3^{\frac{x-10}{10}} = 84$.
22. Solve in R: $2^x + (0.5)^{2x-3} - 6(0.5)^x = 1$.
23. Solve $6^{2x+4} = 3^{3x} \cdot 2^{x+8}$.
24. Solve $4^x - 3^{x-\frac{1}{2}} = 3^{x+\frac{1}{2}} - 2^{2x-1}$.
25. Solve $7 \cdot 3^{x+1} - 5^{x+3} = 3^{x+1} - 5^{x+2}$.
26. Solve $3^x \cdot 8^{\frac{x}{x+2}} = 6$.
27. Solve $(\sqrt{3} + \sqrt{2})^x + (\sqrt{3} - \sqrt{2})^x = 10$.
28. Solve $\{\sqrt{3 + 2\sqrt{2}}\}^x + \{\sqrt{3 - 2\sqrt{2}}\}^x = 6$.
29. Solve $(2 + \sqrt{3})^{x^2-2x+1} + (2 - \sqrt{3})^{x^2-2x-1} = \frac{2}{2 - \sqrt{3}}$.

30. Solve $5^{\frac{x}{2}} - 2^x = 1$.

31. Solve $(\sqrt{2}+1)^x + (\sqrt{2}-1)^x = 6^{x/2}$.

32. Solve $\log_{5-x}(x^2 - 2x + 65) = 2$.

33. Solve $\log_4 8 + \log_4 (x+3) - \log_4 (x-1) = 2$.

34. Solve $\log_7 \log_5 (\sqrt{x+5} + \sqrt{x}) = 0$.

35. Solve $\log_3 (3^x - 8) = 2 - x$.

36. Solve $\log_2 (25^{x+3} - 1) = 2 + \log_2 (5^{x+3} + 1)$.

37. Solve $\log_2 (4 \times 3^x - 6) - \log_2 (9^x - 6) = 1$.

38. Solve $\log_4 (2 \times 4^{x-2} - 1) + 4 = 2x$.

39. Solve $\log_5 (5^{\frac{1}{x}} + 125) = \log_5 6 + 1 + \frac{1}{2x}$.

40. Solve $x + \log_{10} (2^x + 1) = \log_{10} 6 + x\log_{10} 5$.

41. Solve in R: $1 + 2\log_{x+2} 5 = \log_5 (x+2)$.

42. Solve $\log_{1-x} (3-x) = \log_{3-x} (1-x)$.

43. Solve
$$\log_{1-2x} (6x^2 - 5x + 1) - \log_{1-3x} (4x^2 - 4x + 1) = 2.$$

44. Solve $\log_{x+1} (x^3 - 9x + 8) \cdot \log_{x-1} (x+1) = 3$.

45. Solve $6(\log_x 2 - \log_4 x) + 7 = 0$.

46. Solve $3\log_x 4 + 2\log_{4x} 4 + 3\log_{16x} 4 = 0$.

47. **Solve** $\log_{10} \sqrt{\log_{10} x} = \log_{x^2} x$.

48. Solve $7^{\log x} = 98 - x^{\log 7}$.

49. Solve $4^{\log x} = 32 - x^{\log 4}$.

50. Solve $4^{\log_2 2x} = 36$.

51. Solve $|3x - 1| = 2x, x \in R$.

52. Solve $|x + 4| = 3x - 2, x \in R$.

53. Solve $|x^2 + x| - 5 = 0$ for real x.

54. Solve $|x^2 - 2x - 8| = 2x, x \in R$.

55. Solve $|x^2 - x - 6| = x + 2, x \in R$.

56. Solve $|x^2 + x + 1| = 3, x \in R$.

57. Solve $2|x+1|^2 - |x+1| = 3, x \in R$.

58. Solve $2|x+1| + |x-3| = 4, x \in R$.

59. Solve $|x| + 2|x-1| - 3|x-2| = 4, x \in R$.

60. Solve $2|3-x| + 3|x-1| = 4, x \in R$.

61. Solve $|2x+1| - 2|x-2| = 5, x \in R$.

62. Solve $|x^2 - x| + |x+1| = 5, x \in R$.

63. Solve $|x^2 + x + 1| + |2x - 3| = 4x, x \in R$.

64. Solve $|2^x - 1| + |2^x + 1| = 2, x \in R$.

65. Solve $|2^{x+1} - 1| + |2^{x+1} + 1| = 2^{|x+1|}$.

66. Solve: (a) $|x|^{x^2 - x - 2} = 1$

(b) $2^{|y|} - |2^{y-1} - 1| = 2^{y-1} + 1$

67. Solve $(x^2 - 3x - 3)^{|x+1|} = 1$.

68. Solve $2^{|x+2|} - |2^{x+1} - 1| = 2^{x+1} + 1$.

69. Solve
$$\log_3 (\sqrt{x} + |\sqrt{x} - 1|) = \log_9 (4\sqrt{x} - 3 + 4|\sqrt{x} - 1|).$$

70. Solve $|x-2|^{\log_2 x^3 - 3\log_x 4} = (x-2)^3$.

71. If $((x))$ is defined as follows:
$$((x)) = \begin{cases} 2x + 1, & 0 \le x < \frac{1}{4} \\ 2x - 1, & \frac{1}{4} \le x < \frac{1}{3} \\ 2x, & \frac{1}{3} \le x \le \frac{1}{2} \end{cases}$$
then solve for x when $((x))^2 + x((x)) = x + ((x))$.

72. If x be a real number for which
(x) = the integer just greater than or equal to x,
$[x]$ = the integer just less than or equal to x
then solve $(x)^2 = [x]^2 + 2x$.

73. Solve in R: $[2x] - 2x = [x + 1]$
where $[x]$ = the greatest integer $\le x$.

74. Solve in R: $[x^2] + 2[x] = 3x, 0 \le x \le 2$.

75. If $y = 4 - [x]^2$ where x is real then find x for which $[y] + y = 6$ where $[x]$ = the greatest integer $\le x$.

76. Let $[x]$ = the greatest integer $\le x$, and $\{x\}$, (x) are defined as follows:
$$\{x\} = x - [x]; \quad (x) = \begin{cases} 2[x] - \{x\}, & x < 0 \\ [x] + 3\{x\}, & x \ge 0 \end{cases}$$
then solve the equation $(x) = x + \{x\}$.

77. Let x be a perfect square and a natural number. When x is divided by 5, the quotient is $[x]$ and the remainder is $\{x\}$. Then solve for x if $\sqrt{x} + \{x\} = [x]$.

78. If $xy = 2(x+y)$, $x \le y$ and $x, y \in N$ then find the number of solutions of the equation.

79. Find the integral solutions of the equation $|2x - y| + |x - y| = 3$.

80. Find the number of non-negative integral solutions of $x + y + z = n, n \in N$.

81. Find all positive integral solutions of $y^2 + 6xy - 16x = 0$ if such solutions exist.

82. Find the least possible two-digit number which is such that if the digits are reversed the number becomes $\frac{6}{5}$ times of the original number.

83. Solve $x^y = y^x$ and $x^2 = y^3$.

84. Solve $6^x + 6^y = 42$ and $x + y = 3$.

85. Solve $(3y^2 + 1)\log_3 x = 1$ and $x^{2y^2 + 10} = 27$.

86. Solve $3^x \cdot 2^y = 576$ and $\log_{\sqrt{2}} (y - x) = 4$.

87. Solve $\log_4 xy + 3\log_y x = 0$
and $\log_4 \frac{x}{y} - \log_4 x \cdot \log_4 y = 0$.

88. Solve for x, y, z: $z^x = x,\ z^y = y,\ y^y = x$.

89. Solve for x: $(15 + 4\sqrt{14})^t + (15 - 4\sqrt{14})^t = 30$
where $t = x^2 - 2|x|$.

90. Solve for x, y: $|y| = 1 - x^2$ and
$|x^2 - 2x| + y = 1$.

91. Solve

$$6^x \cdot \left(\frac{2}{3}\right)^y - 3 \cdot 2^{x+y} - 8 \cdot 3^{x-y} + 24 = 0,\ xy = 2.$$

92. Solve for x and y:

$$\log_{10} x + \log_{10} x^{1/2} + \log_{10} x^{1/4} + \ldots \text{ to } \infty = y$$

and $\dfrac{1 + 3 + 5 + \ldots + (2y - 1)}{4 + 7 + 10 + \ldots + (3y + 1)} = \dfrac{20}{7\log_{10} x}$.

93. Let $[a]$ denote the largest integer not exceeding the real number a. If x and y satisfy the equations $y = 2[x] + 3$ and $y = 3[x - 2] + 5$ simultaneously, find $[x + y]$.

94. Prove that the number of real solutions of the equation $2^x + x^2 = 1$ is two.

95. Find the number of real solutions of $e^x = |x - 1|$. Also find a solution if any exists.

96. Find the solution-set of the inequation
$x^2 + 6x - 27 > 0,\ x \in R$.

97. Solve in R: $2x^2 - 3x + 1 \le 0$.

98. Prove that $2x^2 - 6x + 9$ is positive for all real x.

99. If $6 + 10x - 2x^2 \le k$, a constant, for all $x \in R$ then find the minimum value of k.

100. Find all real values of x which satisfy
$x^2 - 3x + 2 > 0$ and $x^2 - 3x - 4 \le 0$.

101. Solve in the set of integers:
$5x - 1 < (x + 1)^2 < 7x - 3$.

102. Find the set of all real values of a such that $a^2 + 2a$, $2a + 3$ and $a^2 + 3a + 8$ are the sides of a triangle.

103. Solve $(x^2 - 4x + 3)(x - 2)^2 < 0,\ x \in R$.

104. Solve $(x + 3)^4 + (x - 1)^4 \ge 82,\ x \in R$.

105. Prove that $1 - x + 2x^2 - x^3 + x^4 > 0$ for all $x \in R$.

106. Solve in R: $\dfrac{4}{1 + x} + \dfrac{2}{1 - x} < 1$.

107. Solve in R: $\dfrac{(x - 1)(x - 2)}{x + 5} > 1$.

108. Solve $\dfrac{8x^2 + 16x - 51}{(2x - 3)(x + 4)} < 3,\ x \in R$.

109. Solve in R: $\dfrac{x^2 + 2x - 3}{x^2 + 1} < 0$.

110. Solve in R: $\dfrac{10x}{x^2 + 9} \le 1$.

111. Let $y = \sqrt{\dfrac{(x + 1)(x - 5)}{x - 3}}$.
Find all $x \in R$ for which y is a real function of x.

112. Solve $\dfrac{2x + 3}{x^2 + x - 12} \le \dfrac{1}{2}$.

113. Find the smallest integer x such that

$$\frac{x - 5}{x^2 + 5x - 14} > 0.$$

114. Solve $2^x + 2^{1-x} - 3 < 0$.

115. Solve $(0.5)^{2 - 3x} < (2^x)^x \cdot \left(\frac{1}{4}\right)^{x+4}$.

116. Solve $2|x + 1| > x + 4$.

117. Find the greatest and the least values of x for which $x^2 - 5|x| + 6 \le 0$ holds.

118. Solve in R: $\left|\dfrac{y - 3}{y + 1}\right| < 2$

or solve in R: $-2 < \dfrac{y - 3}{y + 1} < 2$.

119. Solve $\left|\dfrac{x^2 - 3x - 1}{x^2 + x + 1}\right| < 3$.

120. Solve $\left|\dfrac{x^2 - 1}{x^2 + x + 1}\right| < 1$.

121. Solve $\left|\dfrac{x^2 - 5x + 4}{x^2 - 4}\right| \le 1$.

122. Solve $\dfrac{|x - 1|}{x + 2} < 1$.

123. Solve $\dfrac{|x - 3|}{x^2 - 5x + 6} \ge 2$.

124. Solve $\dfrac{x^2 + 6x - 7}{|x + 4|} < 0$.

125. Solve in R: $\log_3 (2x^2 + 6x - 5) > 1$.

126. Solve $\log_{10} (x^2 - 2x - 2) \le 0$.

127. Solve $\log_{x-2} (3x^2 - x - 1) > 0$.

128. Solve for $x \in R$: $\log_x \left(2x - \frac{3}{4}\right) > 2$.

129. Solve $\log_{x^2 - 1} (x + 1) < 1$.

130. Solve $\log_{9x^2} (6 + 2x - x^2) \le 0.5$.

131. Solve $\log_x 2 \cdot \log_{2x} 2 \cdot \log_2 4x > 1$.

132. Solve $\log_2 \dfrac{x - 4}{2x + 5} < 1$.

133. Solve $\log_{0.2} \log_6 \left(\dfrac{x^2 - x}{x^2 + 1}\right) > 0$.

134. Solve $\dfrac{\log_2(4x^2 - x - 1)}{\log_2(x^2 + 1)} > 1$.

135. Solve $x^{(\log_{10} x)^2 - \log_{10} x^3 + 1} > 1000$.

136. Solve $2^x + 2^{|x|} \geq 2\sqrt{2}$.

137. Solve $||x - 1| + 2| \leq 4$.

138. Solve $||x - 1| - 1| \geq 3$.

139. Solve $\log_{|x+6|} 2 \cdot \log_2 (x^2 - x - 2) \geq 1$.

140. Solve $|\log_3 x| - \log_3 x - 3 < 0$.

141. Solve $|x|^{x^2 - x - 2} < 1$.

142. Solve $(x - 1)\sqrt{x^2 - x - 2} \geq 0$.

143. Solve in R: $\dfrac{x^2 + x + 1}{(x + 1)(x - 1)} \geq 0$ and $|x| \leq 2$.

144. Solve $||x - 1| - x| \leq 4$.

145. Solve in R: $[x] + |x - 2| \leq 0$ and $-1 \leq x \leq 3$ where $[x]$ is the greatest integer function.

146. There is a square whose sides are in full centimetre. Its area is more than that of a rectangle whose one side is 5 cm less than double the side of the square and the other side is 4 cm less than the side of the square. Also the area of the square is less than that of a rectangle for which one side is smaller by 2 cm and the other greater by 3 cm than the side of the square. Find the possible length of the side of the square.

147. In a two-digit number, the digit in the tens place is two more than the digit in the ones place. Also the square of the digit in the tens place exceeds 9 times the digit in the ones place. Find the number.

148. If the roots of $(1 + m^2)x^2 - 2(1 + 3m)x + (1 + 8m) = 0$ are equal then find the set of values of m.

149. If the roots of the equation

$$a(b - c)x^2 + b(c - a)x + c(a - b) = 0$$

are equal then prove that $\dfrac{2}{b} = \dfrac{1}{a} + \dfrac{1}{c}$, i.e., a, b, c are in HP.

150. If the roots of the equation

$$(a^2 + b^2)x^2 + 2(bc + ad)x + c^2 + d^2 = 0$$

are real then prove that they must be equal.

151. Prove that the roots of

$$abc^2x^2 + 3a^2cx + b^2cx - 6a^2 - ab + 2b^2 = 0$$

are rational where a, b, c are rational.

152. Show that if p, q, r and s are real numbers and $pr = 2(q + s)$ then at least one of the equations $x^2 + px + q = 0$, $x^2 + rx + s = 0$ has real roots.

153. If $x = 1$ is a root of the quadratic equation $ax^2 + bx + c = 0$ with real coefficients then prove that $4ax^2 + 3bx + 2c = 0$ must have real roots.

154. If $ax^2 + (b - c)x + a - b - c = 0$ has unequal real roots for all $c \in R$ then prove that either $b < a < 0$ or $b > a > 0$.

155. Prove that the equation $(bx + c)^2 - a^2x^4 = 0$ has at least two real roots for all $a, b, c \in R$.

156. Show that the equation $\dfrac{1}{x - a} + \dfrac{1}{x - b} + \dfrac{1}{x - c} = 0$ can have a pair of equal roots if $a = b = c$.

157. If the equations $ax + by = 1$, $cx^2 + dy^2 = 1$ have only one solution, prove that $\dfrac{a^2}{c} + \dfrac{b^2}{d} = 1$ and $x = \dfrac{a}{c}$, $y = \dfrac{b}{d}$.

158. If a, b, c are odd integers, $ax^2 + bx + c = 0$ cannot have rational roots. Prove it.

159. Show that the equation $e^{\sin x} - e^{-\sin x} - 4 = 0$ has no real solution.

160. In the equation $x^2 - px + q^2 = 0$ if x is real, prove that p cannot lie between $2q$ and $-2q$.

161. Determine the limits between which n must lie in order that the equation $2ax(ax + nc) + (n^2 - 2)c^2 = 0$ may have real roots.

162. Prove that for $-\dfrac{1}{4} < a < 0$, the equation $\sqrt{a + \sqrt{a + x}} = x$ has exactly two real solutions of which one is $\dfrac{1 + \sqrt{4a + 1}}{2}$.

163. Find the interval in which k must lie in order that the roots of

$$12(k + 2)x^2 - 12(2k - 1)x - 38k - 11 = 0$$

may be imaginary.

164. If the roots of $ax^2 + bx + c = 0$, where a, c are nonzero real numbers, be imaginary and the ratio of the roots be $r : 1$ where $r > 0$ then prove that a and c are of opposite signs.

165. If the ratio of the roots of $lx^2 + nx + n = 0$ is $p : q$ then prove that

$$\sqrt{\frac{p}{q}} + \sqrt{\frac{q}{p}} \pm \sqrt{\frac{n}{l}} = 0.$$

166. If one root of the equation $(l - m)x^2 + lx + 1 = 0$ is double the other and l is real, find the greatest value of m.

167. If the roots of $\dfrac{1}{x + p} + \dfrac{1}{x + q} = \dfrac{1}{r}$ are equal in magnitude but opposite in sign, show that $p + q = 2r$ and the product of roots $= -\dfrac{p^2 + q^2}{2}$.

168. If $ax^2+bx+c=0$ has one root equal to the nth power of the other root then prove that

$$(a^n c)^{\frac{1}{n+1}}+(ac^n)^{\frac{1}{n+1}}+b=0.$$

169. Prove that $\frac{b^2-4ac}{a^2}$ is invariant for the class of equations $ax^2+bx+c=0$ for which the difference of roots is a constant.

170. If the ratio of roots of $ax^2+bx+c=0$ is invariant then prove that $b^2 : ac$ is also invariant.

171. If a, b, c are in GP, find the ratio of the roots of $ax^2+bx+c=0$.

172. If a, b, c are in AP and the roots of $ax^2+bx+c=0$ are α, β then express β as a function of the other root if $4a+c \neq 2b$.

173. Let $a>0, b>0, c>0$. Prove that both roots of $ax^2+bx+c=0$ have negative real parts.

174. If one root of $4x^2+2x-1=0$ be α then prove that its other root is $4\alpha^3-3\alpha$.

175. If α, β be the roots of $x^2+px+q=0$ then prove that $\frac{\alpha}{\beta}$ is a root of $qx^2+(2q-p^2)x+q=0$.

176. If $\alpha, \beta, \gamma, \delta$ be in AP of which the first two terms are roots of $x^2-2px+q=0$ and the last two terms are roots of $x^2-2rx+s=0$ then prove that $s-q=r^2-p^2$.

177. Let a, b be the roots of $x^2-3x+A=0$ and c, d be the roots of $x^2-12x+B=0$. If a, b, c, d form an increasing GP, find A and B.

178. If α, β be the nonzero roots of $ax^2+bx+c=0$ and α^2, β^2 be the roots of $a^2x^2+b^2x+c^2=0$ then find the number of possible sets of values of α, β. Also prove that a, b, c are in GP.

179. Let α_1, α_2 be the roots of $ax^2+bx+c=0$ and β_1, β_2 be the roots of $px^2+qx+r=0$. If $\alpha_1 y+\alpha_2 z=0$ and $\beta_1 y+\beta_2 z=0$ have a nontrivial solution (nonzero solutions) then prove that $\frac{a}{p}, \frac{b}{q}$ and $\frac{c}{r}$ are in GP.

180. If α, β are the roots of $x^2+px+q=0$ then express $\alpha^5+\beta^5$ in terms of p and q.

181. If α, β are the roots of $ax^2+bx+c=0$ and $s_n=\alpha^n+\beta^n$, prove that $as_{n+1}+bs_n+cs_{n-1}=0$ and hence find s_4.

182. If α, β be the values of x in $m^2(x^2-x)+2mx+3=0$ and m_1, m_2 be two values of m for which α, β are connected by the relation $\frac{\alpha}{\beta}+\frac{\beta}{\alpha}=\frac{4}{3}$, find the value of $\frac{m_1^2}{m_2}+\frac{m_2^2}{m_1}$.

183. If α, β are the roots of the equation $x+1=\lambda x(1-\lambda x)$ and if λ_1, λ_2 are two values of λ determined from the relation $\frac{\alpha}{\beta}+\frac{\beta}{\alpha}=r-2$, show that $\frac{\lambda_1^2}{\lambda_2^2}+\frac{\lambda_2^2}{\lambda_1^2}+2=4\left(\frac{r+1}{r-1}\right)^2$.

184. If a, b are the roots of $x^2+px+1=0$ and c, d are the roots of $x^2+qx+1=0$ then show that

$$q^2-p^2=(a-c)(b-c)(a+d)(b+d).$$

185. If c, d are the roots of $(x-a)(x-b)=k$, show that a, b are roots of $(x-c)(x-d)+k=0$.

186. If α, β are the roots of $ax^2+bx+c=0$, find the quadratic equation whose roots are $(a\alpha+b)^{-1}$ and $(a\beta+b)^{-1}$.

187. If α, β are the roots of $x^2-px+q=0$ then find the quadratic equation whose roots are

$$(\alpha^2-\beta^2)(\alpha^3-\beta^3) \text{ and } \alpha^2\beta^3+\alpha^3\beta^2.$$

188. In the equation $x^2+px+q=0$ the coefficient of x was incorrectly written as 17 instead of 13 and the roots were found to be –2 and –15. Find the correct equation.

189. In copying an equation of the form $x^2+px+q=0$, the coefficient of x was written incorrectly and the roots were found to be 3 and 10; again the equation was written and this time the constant term was written incorrectly and the roots were found to be 4 and 7. Find the roots of the correct equation.

190. Find $p \in R$ for which $x^2-px+p+3=0$ has
(i) one positive and one negative root
(ii) both roots negative
(iii) one root > 2 and the other root < 2

191. If $x^2-(m+1)x+m^2+m-8=0$ has a root in the open interval $(1, +\infty)$ and the other in the open interval $(-\infty, 1)$ then find m.

192. For what values of $a \in R$ the quadratic equation $(a^2+1)x^2-(a+1)x+(a^2-a-2)=0$ have one root positive and the other root negative?

193. For what values of a does the equation $ax^2-(a+1)x+3=0$ have roots lying between 1 and 2?

194. Find the real values of the parameter a such that $(2a+1)x^2-a(x-1)=2$ has one root greater than 1 while the other less than 1.

195. Discuss the sign of $p^2+2p-3, p \in A$, where $A=\{p \mid x^2-px+1=0$ has one root <1 and the other root $>1\}$.

196. Find the integral values of a for which the equation
$$x^4 - (a^2 - 5a + 6)x^2 - (a^2 - 3a + 2) = 0$$
has real solutions only.

197. If the graph of the function
$$f(x) = x^2 + 2(k-1)x + k + 5,\ k \in R$$
cuts the x-axis at least at one point on the positive side, find the set of possible values of the constant k.

198. For what values of a does the equation
$$2(\log_3 x)^2 - |\log_3 x| + a = 0$$
possess four real solutions?

199. Find all real values of the parameter a for which the equation $x^4 + 2ax^3 + x^2 + 2ax + 1 = 0$ has
(i) exactly two distinct negative roots,
(ii) at least two distinct negative roots.

200. Find $a \in R$ if possible, so that
$$a(\log_{10} x)^2 - \log_{10} x^{a+2} + 2a = 0$$
may give two values of x whose integral parts are of two digits.

201. If $2(a\cos x - \cos 2x) = 1$ for $x \in R$ then find all possible real values of a.

202. Find $a \in R$ for which $a\left(\dfrac{1}{1+x^2}\right)^2 - 3\cdot\dfrac{a}{1+x^2} + 1 = 0$ will have all roots (i) real (ii) imaginary.

203. Find the integral values of a for which $(a+2)x^2 + 2(a+1)x + a = 0$ will have both roots integers.

204. If $a \in R$ and $0 < \theta < \dfrac{\pi}{2}$ or $\pi < \theta < \dfrac{3\pi}{2}$ always satisfy $a\tan\theta + (a-1)\cot\theta = 2a + 1$ then find all possible values of a.

205. Find all values of the parameter a for which the inequality $9^x - a\cdot 3^x - a + \dfrac{5}{4} \le 0$ is satisfied for at least one $x \in R$.

206. Find all values of the parameter a such that the roots α, β of the equation $2x^2 + 6x + a = 0$ satisfy the inequality $\dfrac{\alpha}{\beta} + \dfrac{\beta}{\alpha} < 2$.

207. Find all possible values of the common ratio if the lengths of the sides of a triangle are in GP.

208. If $\{a^3 + (1-\sqrt{2})a^2 - (3+\sqrt{2})a + 3\sqrt{2}\}x^2$
$$+ 2(a^2 - 2)x + a > -\sqrt{2}$$
is satisfied for all real $x > 0$ then obtain the possible values of the parameter.

209. Find all values of the parameter k for which the solution-set of the inequation
$$x^2 + 3k^2 - 1 \ge 2k(2x - 1)$$
is a subset of the solution-set of the inequation
$$x^2 - (2x-1)k + k^2 \ge 0.$$

210. (i) If the roots of $10x^3 - cx^2 - 54x - 27 = 0$ are in HP, find c.
(ii) If the roots of $x^3 - ax^2 + x - b = 0$ are in HP show that the mean root is $3b$.
(iii) Find the roots α, β, γ of $x^3 - 11x^2 + 36x - 36 = 0$ if $\dfrac{2}{\beta} = \dfrac{1}{\alpha} + \dfrac{1}{\gamma}$.

211. If the roots of $x^3 + px^2 + qx + r = 0$ are in GP, find the relation between p, q, r.

212. If $\left(m_r, \dfrac{1}{m_r}\right)$; $r = 1, 2, 3, 4$ be four pairs of values of x and y that satisfy the equation
$$x^2 + y^2 + 2gx + 2fy + c = 0$$
then prove that $m_1 \cdot m_2 \cdot m_3 \cdot m_4 = 1$.

213. (i) If two roots of the equation $x^3 + px^2 + qx + r = 0$ be equal in magnitude and of opposite signs, show that $pq = r$.
(ii) Solve $24x^3 - 14x^2 - 63x + 45 = 0$ if it is known that one root is double of another root.

214. Find the cubic equation with real coefficients whose two roots are $2 + 3i$ and 5.

215. If α, β, γ be the roots of $ax^3 + bx^2 + cx + d = 0$ then find the value of
(i) $\Sigma\alpha^2$ (ii) $\Sigma\dfrac{1}{\alpha}$ (iii) $\Sigma\,\alpha^2\beta$.

216. If $x^2 + px + q = 0$ and $x^2 + p'x + q' = 0$ have one root in common, prove that its value is
$$\frac{q - q'}{p' - p} \quad \text{or} \quad \frac{pq' - p'q}{q - q'}.$$

217. Find m so that the equations $3x^2 - 2mx - 4 = 0$ and $x^2 - 4mx + 2 = 0$ may have a common root. Can the equations have a common nonreal root?

218. Prove that the equations
$$(q - r)x^2 + (r - p)x + p - q = 0 \text{ and}$$
$$(r - p)x^2 + (p - q)x + q - r = 0$$
have a common root.

219. If the equations $x^2 + bx + ca = 0$ and $x^2 + cx + ab = 0$ have a nonzero common root, prove that $a + b + c = 0$ and their other roots satisfy $x^2 + ax + bc = 0$.

220. If the equations $x^2 - px + q = 0$ and $x^2 - ax + b = 0$ have a common root and roots of the second equation are equal then prove that $ap = 2(b + q)$.

221. If $ax^2 + 2bx + c = 0$ and $a_1x^2 + 2b_1x + c_1 = 0$ have a common root and $\dfrac{a}{a_1}, \dfrac{b}{b_1}, \dfrac{c}{c_1}$ are in AP, show that a_1, b_1, c_1 are in GP.

222. Prove that $ax^2 + bx + c = 0$ and $x^2 + 2x + 3 = 0$ cannot have a common root unless
$$a : b : c = 1 : 2 : 3.$$

223. If the equations $ax^2 + 2bx + c = 0$ and $a_1x^2 + 2b_1x + c_1 = 0$ with rational coefficients have one and only one root in common then prove that $b^2 - ac$ and $b_1^2 - a_1c_1$ will be both perfect squares.

224. If each pair of equations $x^2 = b_1x + c_1, x^2 = b_2x + c_2$ and $x^2 + b_3x = c_3$ have a common root, prove that
$$(b_1 + b_2 + b_3)^2 = 4(c_1 + c_2 + c_3 + b_1b_2).$$

225. If $a_rx^2 + b_rx + c_r = 0;\ r = 1, 2, 3$ have one root in common for each pair of equations then prove that
$$\frac{a_1a_2c_3}{a_3c_1c_2} = \left(\frac{a_1b_2 - a_2b_1}{c_1a_2 - c_2a_1}\right)^2 \text{ and}$$
$$\frac{c_1a_2 + c_2a_1}{c_1a_2 - c_2a_1} + \frac{a_1a_2b_3}{a_3(a_1b_2 - a_2b_1)} = 0.$$

226. Find the condition for the three equations $a_rx^2 + b_rx + c_r = 0; r = 1, 2, 3$ to have a common root.

227. If the equations $x^2 + ax + 12 = 0,\ x^2 + bx + 15 = 0$ and $x^2 + (a + b)x + 36 = 0$ have a common positive root, find a and b and the roots of the equations.

228. Find λ, μ so that the equations $4x^2 - 8x + 3 = 0$, $x^2 + \lambda x - 1 = 0$ and $2x^2 + x + \mu = 0$ may have a common root for each pair of equations.

229. Find the values of a for which $x^2 - ax + 1 - 2a^2 \geq 0$ for real values of x.

230. Find the minimum value of $3x^2 - 6x + 7, x \in R$ and also find the corresponding value of x.

231. Find the maximum value of $(2 + 3x)(1 - x), x \in R$ and also find the corresponding value of x.

232. Prove that the maximum value of xy takes place when $x + y =$ constant, if $x = y$.

233. Prove that $\frac{1}{7} \leq \frac{x^2 - 3x + 4}{x^2 + 3x + 4} \leq 7$ for all $x \in R$.

234. If $f(x) = \frac{x^2 - 2x + 4}{x^2 + 2x + 4}, x \in R$ then prove that the range of $f(x)$ is $\left[\frac{1}{3}, 3\right]$.

235. Find the greatest and the least values of $\frac{x^2 - x + 1}{x^2 + x + 1}$ for real values of x.

236. Show that the sum of a nonzero real number and its reciprocal cannot lie between –2 and 2.

237. If x is real, the expression $\frac{x^2 + 2x - 11}{x - 3}$ takes all values that do not lie between a and b, $(a < b)$. Find a and b.

238. If $a^2 + c^2 > ab$ and $b^2 > 4c^2$, for real x show that $\frac{x + a}{x^2 + bx + c^2}$ cannot lie between two limits.

239. Show that $\frac{(x + a)(1 + ax)}{(x - a)(1 - ax)}$ where $x \in R,\ x \neq a,\ \frac{1}{a}$ cannot lie between $\left(\frac{1 - a}{1 + a}\right)^2$ and $\left(\frac{1 + a}{1 - a}\right)^2$.

240. If $x \in R$, find m such that $\left|\frac{1 + mx + x^2}{1 + x + x^2}\right| < 3$.

241. Show that $\frac{4x^2 + 36x + 9}{12x^2 + 8x + 1}$ can be made to assume any real value by assigning suitable values of $x \in R$.

242. If $x \in R$ and a, b, c are in ascending or descending order of magnitude, show that $\frac{(x - a)(x - c)}{x - b}$, $(x \neq b)$ can assume any real value.

243. Prove that $f(x) = \frac{ax^2 + x - 2}{a + x - 2x^2}$ has the range R when $x \in R$ if $a \in [1, 3]$.

244. Let $f(x) = \frac{mx^2 + 3x + 4}{x^2 + 3x + 4}, m \in R$. If $f(x) < 5$ for all $x \in R$ then find the possible set of values of m.

245. Find the set of real values of the parameter c so that $\frac{x^2 + 2x + c}{x^2 + 4x + 3c}$ can take all real values for $x \in R$.

246. If y is a real function of the real variable x defined by $x^2 + y^2 - 6x - 8y = 0$ then find the range of the function y.

247. If (x, y) be a real solution of the equation $x^2 + 12xy + 4y^2 + 4x + 8y + 20 = 0$ then prove that $x \notin (-2, 1)$ and $y \notin \left(-1, \frac{1}{2}\right)$.

248. Show that $x^2 - xy + y^2 - 4x - 4y + 16 = 0$, an equation in x and y, has only one real solution. Find the solution.

249. If $p \neq 0, r \neq 0$ and $x^2 - \alpha^2$ divides the polynomial $(x^2 + px + q)(x^2 + rx + s)$ then prove that
$$(q - s)^2 + (ps + qr)(p + r) = 0.$$

250. If $x + \alpha$ is a common factor of $a_1x^2 + b_1x + c_1$ and $a_2x^2 + b_2x + c_1, (a_1 \neq a_2)$ then prove that $\alpha = \frac{b_1 - b_2}{a_1 - a_2}$.

251. If $y - mx$ is a factor of $ax^2 + bxy + cy^2$ and $my - x$ is a factor of $a_1x^2 + b_1xy + c_1y^2$, find the relation between a, b, c, a_1, b_1, c_1.

252. Prove that $lx^2 + mxy + ny^2$ and $l'x^2 + m'xy + n'y^2$ will have a common linear factor if
$$mn' - m'n,\ nl' - n'l,\ lm' - l'm \text{ are in GP.}$$

253. Prove that the expression
$$x^2 + 2(a + b + c)x + 3(bc + ca + ab)$$
cannot be a perfect square unless $a = b = c$.

254. If $(ax^2+bx+c)y+a'x^2+b'x+c'=0$, find the condition for x to be a rational function of y.

255. Show that $a(b-c)x^2+b(c-a)xy+c(a-b)y^2$ will be a perfect square if a, b, c are in HP.

256. For what values of k, $(x+y+z)^2+k(x^2+y^2+z^2)$ can be resolved into linear rational factors?

257. If $x^5-bx^3+cx^2+dx-e$ can be expressed as the product of a perfect square and a perfect cube then prove that $\frac{12b}{5}=\frac{9d}{b}=\frac{5e}{c}=\frac{d^2}{c^2}$.

258. If $(x-1)^2$ is a factor of ax^3+bx^2+c, find the roots of $cx^3+bx+a=0$.

259. Prove that the equation $x^3-8x^2+2x+8=0$ has a real root lying between 1 and 2.

260. Let a, b, c be real numbers and $a\neq 0$. If α is a real root of $a^2x^2+bx+c=0$ and β is a real root of $a^2x^2-bx-c=0$ then prove that the equation $a^2x^2+2bx+2c=0$ has a root lying between α and β.

261. If α is a nonzero real root of $a_1x^2+b_1x+c_1=0$ and α is a repeated root of the equation $a_2x^3+b_2x^2+c_2x+d_2=0$ then prove that

$$\begin{vmatrix} a_1 & b_1 & c_1 \\ 3a_2 & 2b_2 & c_2 \\ a_1b_2-a_2b_1 & a_1c_2-a_2c_1 & a_1d_2 \end{vmatrix}=0.$$

Objective Questions

Fill in the blanks.

262. If $|x|+|x-1|=3$ then $x=$ ______ .

263. The number of real solutions of $3^x+4^x=5^x$ is ______ .

264. The equation $3^x+4^x+5^x=6^x$ has the solution-set ______ .

265. The number of real solutions of the equation $2^{x+1}-3.2^{-x}+5=0$ is ______ .

266. The solution-set of the equation

$$\log_5\log_2(\sqrt{4x+2}+2\sqrt{x})=0 \text{ is ______ .}$$

267. If $x<0$, $y<0$, $x+y+\frac{x}{y}=\frac{1}{2}$ and $(x+y)\frac{x}{y}=\frac{-1}{2}$ then $x=$ ______ and $y=$ ______ .

268. The number of real solutions of the equation $|x|^2-3|x|+2=0$ is ______ .

269. If $x=2+2^{2/3}+2^{1/3}$ then the value of x^3-6x^2+6x is ______ .

270. The AM of the roots of $x^2-12x+3=0$ is ______ .

271. The GM of the roots of $2x^3-12x^2+3x-16=0$ is ______ .

272. The condition on p for the roots of $x^2+x+p=0$ to be both negative, is ______ .

273. The quadratic equation whose roots are twice the roots of $x^2-3x+3=0$ is ______ .

274. The quadratic equation whose roots are 2 more than the roots of $ax^2+bx+c=0$, is ______ .

275. If α, β are roots of $ax^2+2bx+c=0$ then

$$\sqrt{\frac{\alpha}{\beta}}+\sqrt{\frac{\beta}{\alpha}}= \text{______ .}$$

276. If $2+\sqrt{-3}$ is a root of $x^2+px+q=0$ where p, q are real then $p=$ ______ , $q=$ ______ .

277. The value of $\sqrt{2+\sqrt{2+\sqrt{2+\ldots \text{ to } \infty}}}$ is ______ .

278. If $[x+2]=5$, $x\in R$ where $[x]$ is the greatest integer function then the solution-set of the equation is ______ .

279. If the roots of $x(3-x^2)+2x^2=0$ are α, β, γ then the value of $\alpha+\beta+\gamma+1$ is ______ .

280. If the roots of $4-3x+5x^2+x^3=0$ are α, β, γ then the value of $\frac{1}{\alpha}+\frac{1}{\beta}+\frac{1}{\gamma}=$ ______ .

281. The equation $x^2+2(\lambda+1)x+9\lambda=5$ has negative roots only if λ belongs to the interval ______ .

282. If $2x^2-(a^3+8a-1)x+a(a-4)=0$ has one root positive and one root negative then a belongs to the interval ______ .

283. The solution-set of $x^4-1>0$, $x\in R$ is ______ .

284. The set of values of a for which $\frac{\alpha^2}{\beta}+\frac{\beta^2}{\alpha}>0$ where α, β are roots of $ax^2+3x+2=0$ is ______ .

285. If $x^2+ax+b=0$ and $x^2+bx+a=0$ $(a\neq b)$ have a common root then the numerical value of $a+b$ is ______ .

286. The root common to the equations

$$(a-b)x^2+(b-c)x+c-a=0 \text{ and}$$

$$(a-1)(b-c)x^2+(b-1)(c-a)x+(c-1)(a-b)=0$$

is $x=$ ______ .

287. The equations $x^2-3x+2=0$ and $x^2-ax+3=0$ have a common root. Then the value of a is either ______ or ______ .

288. (i) If $x\in R$, the minimum value of x^4-8x^2+18 is ______ .

(ii) If the sum of a real number x and the reciprocal of a number greater than 1 be 3 then x belongs to the interval ______ .

Choose the correct option(s).

289. The number of solutions for x from the equation $x^2-|x|-2=0$ is

(a) 0 (b) 4
(c) 2 (d) none of these

290. The solution of the equation $3^{x-1}+3^{1-x}=2$ is
(a) 1 (b) 2 (c) 0 (d) –1

291. The number of real solutions of the equation $\sin(e^x)=5^x+5^{-x}$ is
(a) 0 (b) 1
(c) 2 (d) none of these

292. The equation $x^{\frac{3}{4}(\log_2 x)^2+\log_2 x-\frac{5}{4}}=\sqrt{2}$ has
(a) only one real solution
(b) exactly one irrational solution
(c) exactly three real solutions
(d) nonreal complex roots only

293. The equation $x-\dfrac{2}{x-1}=1-\dfrac{2}{x-1}$ has
(a) no roots (b) one root
(c) two equal roots (d) infinite roots

294. The equation $2\cos^2\dfrac{x}{2}\sin^2 x=x^2+\dfrac{1}{x^2},\ 0\le x\le\dfrac{\pi}{2}$ has
(a) one real solution
(b) no real solution
(c) more than one real solution
(d) none of these

295. The equation $(c^2+a^2)x^2-2(a+b)cx+(b^2+c^2)=0$ has equal roots if
(a) a, b, c are in AP (b) a, b, c are in GP
(c) a, c, b are in AP (d) a, c, b are in GP

296. The roots of $x^2-px+q=0$ are reciprocal to each other if
(a) $p=-1$ (b) $q=1$ (c) $p=q$ (d) $p+q=0$

297. If l, m, n are real and $l\ne m$ then the roots of $(l-m)x^2-5(l+m)x-2(l-m)=0$ are
(a) real and equal (b) nonreal complex
(c) real and unequal (d) none of these

298. If the roots of $ax^2+bx+c=0$ are the reciprocals of those of $lx^2+mx+n=0$ where all the coefficients are nonzero real then
(a) $\dfrac{a}{l}=\dfrac{b}{m}=\dfrac{c}{n}$ (b) $\dfrac{a}{n}=\dfrac{b}{m}=\dfrac{c}{l}$
(c) $al=bm=cn$ (d) none of these

299. If $\dfrac{x^2-bx}{ax-c}=\dfrac{\lambda-1}{\lambda+1}$ has roots equal in magnitude but opposite in sign then the value of λ is
(a) $\dfrac{a-b}{a+b}$ (b) $\dfrac{a+b}{a-b}$ (c) c (d) $\dfrac{1}{c}$

300. If α, β are the roots of $x^2+px+q=0$ and α^4, β^4 are the roots of $x^2-rx+s=0$ then the equation $x^2-4qx+2q^2-r=0$ has always
(a) two real roots
(b) two negative roots
(c) two positive roots
(d) one positive and one negative roots

301. If $x^2+mx+1=0$ and $(a-b)x^2+(b-c)x+(c-a)=0$ have both roots common then which of the following is true?
(a) $m=-1$ (b) $m=2$
(c) $m=-2$ (d) b, a, c are in AP
(e) a, b, c are in AP

302. Given one root of $x^4-4x^2+8x+35=0$ is $2+\sqrt{3}i$, $i=\sqrt{-1}$ then
(a) all roots are imaginary
(b) two real and two imaginary roots
(c) one root real and three imaginary
(d) none of these

303. Let α, β be the roots of the equation $(x-a)(x-b)=c$, $c\ne 0$. Then the roots of the equation $(x-\alpha)(x-\beta)+c=0$ are
(a) a, c (b) b, c (c) a, b (d) $a+c, b+c$

304. The roots of the equation $x^3-12x^2+39x-28=0$ are in AP. The common difference of the AP is
(a) ± 1 (b) ± 2 (c) ± 3 (d) ± 4

305. If $P(x)=ax^2+bx+c$, $Q(x)=-ax^2+b'x+c$ where $ac\ne 0$ then $P(x)\cdot Q(x)=0$ has
(a) two real roots
(b) four real roots
(c) at most two real roots
(d) at least two real roots

306. The number of integral solutions (x, y) of the equations
$$x\sqrt{y}+y\sqrt{x}=20 \text{ and}$$
$$x\sqrt{x}+y\sqrt{y}=65 \text{ is}$$
(a) 0 (b) 1 (c) 2 (d) none of these

307. If $b^2\ge 4ac$ for the equation $ax^4+bx^2+c=0$ then all roots of the equation will be real if
(a) $b>0, a<0, c>0$ (b) $b<0, a>0, c>0$
(c) $b>0, a>0, c>0$ (d) $b>0, a<0, c<0$

308. The largest interval in which x lies satisfying $x^{12}-x^9+x^4-x+1>0$ is
(a) $[0, +\infty)$ (b) $(-\infty, 0]$
(c) $(-\infty, \infty)$ (d) none of these

309. If $a^2+b^2=2$ and a, b are real then $a+b$ lies in the interval

(a) $[-1, 1]$ (b) $[-2, 1]$
(c) $[1, 2]$ (d) $[-2, 2]$

310. If $a^2+b^2+c^2=1$ then the value of $ab+bc+ca$ lies in the interval

(a) $\left[\frac{1}{2}, 2\right]$ (b) $[-1, 2]$
(c) $\left[-\frac{1}{2}, 1\right]$ (d) $\left[-1, \frac{1}{2}\right]$

311. The set of common roots of $x^3+2x^2+2x+1=0$ and $1+x^{130}+x^{1988}=0$ is (w being nonreal cube root of unity)

(a) $\{w\}$ (b) $\{w^2\}$
(c) $\{w, w^2\}$ (d) none of these

312. If the difference between the roots of the equation $x^2+kx+1=0$ is less than $\sqrt{5}$ then the set of possible values of k is

(a) $(3, +\infty)$ (b) $(-\infty, -3)$
(c) $(-3, 3)$ (d) none of these

313. The real value of a for which $x^2+i(a-1)x+5=0$ will have a pair of conjugate complex roots, is

(a) values satisfying $a^2-2a+21>0$
(b) 1
(c) all values
(d) no value

314. If the equation $x^2-cx+d=0$ has roots equal to the fourth powers of the roots of $x^2+ax+b=0$, where $a^2>4b$ then the roots of $x^2-4bx+2b^2-c=0$ will be

(a) both real
(b) both negative
(c) both positive
(d) one positive and one negative

315. If the roots of $ax^2+bx+c=0$ are $\sin\phi$ and $\cos\phi$ for some ϕ then

(a) $a^2+b^2=2ac$ (b) $b^2-c^2=2ab$
(c) $b^2-a^2=2ac$ (d) none of these

316. If a, b, c are positive real numbers then the roots of $ax^2+bx+c=0$ are

(a) positive (b) negative
(c) with negative real parts (d) none of these

317. The expression a^2x^2+bx+1 will be positive for all $x \in R$ if

(a) $b^2>4a^2$ (b) $b^2<4a^2$
(c) $4b^2>a^2$ (d) $4b^2<a^2$

318. If x, y, z are real and unequal then the expression $x^2+4y^2+9z^2-6yz-3zx-2xy$ is always

(a) non-negative (b) non-positive
(c) zero (d) none of these

319. Let $g(x)=f(x)+f'(x)+f''(x)$ where
$$f(x)=ax^2+bx+c<0, \ (a \neq 0)$$
where a, b, c are real. Then

(a) $g(x)=0$ has real roots
(b) $g(x)=0$ has nonreal complex roots
(c) $g(x)>0$ for all $x \in R$
(d) $g(x)<0$ for all $x \in R$

320. The extreme value of $2x^2+x-3$ is

(a) $\frac{-25}{8}$ and it is the maximum value
(b) $\frac{-25}{8}$ and it is the minimum value
(c) either $\frac{-3}{2}$ or 1
(d) $-\frac{1}{4}$

321. The least value of ax^2+bx+c, $(a>0)$ is

(a) b^2-4ac (b) $\frac{-b}{2a}$
(c) $\frac{-b}{a}$ (d) $\frac{4ac-b^2}{4a}$

322. The maximum value of $5+4x-4x^2$ will be equal to

(a) 5 (b) 6 (c) 1 (d) 2

323. If x^2+px+1 is a factor of the expression ax^3+bx+c then

(a) $a^2+c^2=-ab$ (b) $a^2-c^2=-ab$
(c) $a^2-c^2=ab$ (d) none of these

324. If the equation $3ax^2+2bx+c=0$ has its coefficients such that $a+b+c=0$ where $a, b, c \in R$ then the equation has at least one real root in the interval

(a) $(-1, 1)$ (b) $(0, 2)$ (c) $\left(\frac{1}{2}, \frac{3}{2}\right)$ (d) $(0, 1)$

State whether the statements are true or false.

325. If the value of ax^2+bx+c is always an integer for integral values of x where a, b, c are rationals then $a+b$ must be an integer.

326. If $a>1$, both roots of the equation
$$(1-a)x^2+3ax-1=0 \text{ are negative.}$$

327. If $\sqrt{3}+\sqrt{2}$ is a root of the equation $x^2+px+q=0$ where p, q are rational then $\sqrt{3}-\sqrt{2}$ must be the other root of the equation.

328. ${}^pC_q x^2 - {}^{p+1}C_q x + {}^pC_{q-1}=0$, p and $q \in N$, $p>q$ has irrational roots for some p and q.

Answers

1. -1 **2.** infinite solutions **3.** -1

4. $2 \pm \sqrt{2}$ **5.** $-9, -6, \dfrac{-15 \pm \sqrt{41}}{2}$ **6.** $4, \dfrac{-14}{5}$

7. $\dfrac{16}{25}$ **8.** $0, -4, -2 \pm 2\sqrt{2}i$ **9.** $-3, \dfrac{1 \pm \sqrt{7}i}{4}$

11. $x = 2, y = -3$ or $x \in R, y = 1$

12. $(x, y, z) = (5, 3, 6)$ and $(-7, -5, -8)$

13. $x = \dfrac{c-b}{a+b+c}, y = \dfrac{a-c}{a+b+c}, z = \dfrac{b-a}{a+b+c}$

14. $x = 3, y = 2$ and $x = \dfrac{-23}{7}, y = \dfrac{-19}{7}$.

15. $x = 2\dfrac{13}{31}, y = \dfrac{30}{31}, z = 1\dfrac{19}{31}$

16. $1, \dfrac{-13}{3}$ **17.** 10 **18.** $1, -1$ **19.** $\dfrac{1}{2}$

20. $2n\pi \pm \dfrac{\pi}{3}, 2n\pi \pm \dfrac{2\pi}{3}, 2n\pi \pm \dfrac{\pi}{6}, 2n\pi \pm \dfrac{5\pi}{6}$

21. 20 **22.** $1, \log_2 \dfrac{\sqrt{17}-1}{2}$ **23.** 4

24. $\dfrac{3}{2}$ **25.** $\dfrac{\log \frac{50}{9}}{\log \frac{3}{5}}$ **26.** $1, -2 - \log_3 4$

27. $2, -2$ **28.** $2, -2$ **29.** $0, 2$ **30.** 2

31. 2 **32.** -5 **33.** 5 **34.** 4

35. 2 **36.** -2 **37.** 1 **38.** 2

39. $\dfrac{1}{2}, \dfrac{1}{4}$ **40.** 1 **41.** $23, -\dfrac{9}{5}$ **42.** $2 - \sqrt{2}$

43. $\dfrac{1}{4}$ **44.** 3 **45.** $8, 2^{-2/3}$ **46.** $\dfrac{1}{2}, \dfrac{1}{8}$

47. 10^{10} **48.** 100 **49.** 100 **50.** 3

51. $\dfrac{1}{5}, 1$ **52.** 3 **53.** $\dfrac{-1 \pm \sqrt{21}}{2}$

54. $2\sqrt{2}, 2(1 + \sqrt{3})$ **55.** $2, -2, 4$

56. $1, -2$ **57.** $\dfrac{1}{2}, \dfrac{-5}{2}$ **58.** -1 **59.** $[2, +\infty)$

60. 1 **61.** $[2, +\infty)$ **62.** $1 - \sqrt{7}, 2$ **63.** $2, 1$

64. $x \le 0$ **65.** -2

66. (a) $1, -1, 2$ (b) $y \ge 1, y = -1$ **67.** $4, -1$

68. $\{-3\} \cup [-1, +\infty)$ **69.** $\{4\} \cup [0, 1]$

70. $4, 3$ **71.** $0, \dfrac{1}{2}$ **72.** $0, n + \dfrac{1}{2}$ where $n \in Z$

73. $-1, -\dfrac{1}{2}$ **74.** $0, 1$

75. $1, -1, \pm 1 + k$ where k is any positive proper fraction

76. $x = 0, 1, 2, 3, \ldots$ **77.** $25, 36, 64$

78. two solutions

79. $(3, 6), (-3, -6), (1, 3), (-1, -3), (3, 5), (-3, -5), (1, 0), (-1, 0), (3, 4), (-3, -4), (3, 3), (-3, -3)$

80. $\dfrac{1}{2}(n+1)(n+2)$ **81.** $x = 1, y = 2$

82. 45 **83.** $x = 1, y = 1;\ x = \dfrac{27}{8}, y = \dfrac{9}{4}$

84. $(x, y) = (1, 2), (2, 1)$ **85.** $\left(3^{\frac{1}{4}}, \pm 1\right)$

86. $x = 2, y = 6$ **87.** $\left(\dfrac{1}{2}, \dfrac{1}{4}\right), \left(\dfrac{1}{8}, 64\right)$

88. $(1, 1, 1), (4, 2, \sqrt{2})$ **89.** $\pm 1, \pm(1 + \sqrt{2})$

90. $(0, 1), (1, 0), \left(\dfrac{1-\sqrt{5}}{2}, \dfrac{1-\sqrt{5}}{2}\right)$

91. $(2, 1), (1, 2), (-1, -2)$ **92.** $(100000, 10)$

93. 15 **95.** one solution, $x = 0$

96. $(-\infty, -9) \cup (3, +\infty)$ **97.** $\left[\dfrac{1}{2}, 1\right]$ **99.** $\dfrac{37}{2}$

100. $[-1, 1) \cup (2, 4]$ **101.** 3

102. $(5, +\infty)$ **103.** $(1, 2) \cup (2, 3)$

104. $(-\infty, -2] \cup [0, +\infty)$ **106.** $(-\infty, -1) \cup (1, +\infty)$

107. $(-5, 2 - \sqrt{7}) \cup (2 + \sqrt{7}, +\infty)$

108. $(-4, -3) \cup \left(\dfrac{3}{2}, \dfrac{5}{2}\right)$ **109.** $(-3, 1)$

110. $(-\infty, 1] \cup [9, +\infty)$ **111.** $[-1, 3) \cup [5, +\infty)$

112. $(-\infty, -4) \cup [-3, 3) \cup [6, +\infty)$ **113.** -6

114. $(0, 1)$ **115.** $(-\infty, -1) \cup (6, +\infty)$

116. $(-\infty, -2) \cup (2, +\infty)$ **117.** 3 and -3 respectively

118. $(-\infty, -5) \cup \left(\dfrac{1}{3}, +\infty\right)$ **119.** $(-\infty, -2) \cup (-1, +\infty)$

120. $\left(-2, -\dfrac{1}{2}\right) \cup (0, +\infty)$ **121.** $\left[0, \dfrac{8}{5}\right] \cup \left[\dfrac{5}{2}, +\infty\right)$

122. $(-\infty, -2) \cup \left(-\dfrac{1}{2}, +\infty\right)$ **123.** $\left[\dfrac{3}{2}, 2\right)$

124. $(-7, -4) \cup (-4, 1)$ **125.** $(-\infty, -4) \cup (1, +\infty)$

126. $[-1, 1 - \sqrt{3}) \cup (1 + \sqrt{3}, 3]$ **127.** $(3, +\infty)$

128. $\left(\dfrac{3}{8}, \dfrac{1}{2}\right) \cup \left(1, \dfrac{3}{2}\right)$ **129.** $(1, \sqrt{2}) \cup (2, +\infty)$

130. $(1 - \sqrt{7}, -1] \cup \left(-\dfrac{1}{3}, 0\right) \cup \left(0, \dfrac{1}{3}\right) \cup [2, 1 + \sqrt{7})$

131. $\left(\dfrac{1}{2^{\sqrt{2}}}, \dfrac{1}{2}\right) \cup (1, 2^{\sqrt{2}})$ **132.** $\left(-\infty, \dfrac{-14}{3}\right) \cup (4, +\infty)$

133. $(-\infty, -1)$ **134.** $\left(-\infty, \dfrac{-2}{3}\right) \cup (1, +\infty)$

135. $(1000, +\infty)$

136. $(-\infty, \log_2(\sqrt{2}-1)] \cup \left[\frac{1}{2}, +\infty\right)$

137. $[-1, 3]$ **138.** $(-\infty, -3] \cup [5, +\infty)$

139. $(-\infty, -7) \cup (-5, -2] \cup [4, +\infty)$

140. $(3^{-3/2}, +\infty)$ **141.** $(1, 2)$

142. $[2, +\infty)$ **143.** $[-2, -1) \cup (1, 2]$

144. $\left[-\frac{3}{2}, +\infty\right)$ **145.** no real solution

146. 7 or 8 or 9 or 10 or 11 cm

147. 20, 75, 86, 97 **148.** $\left\{0, \frac{1}{2}\right\}$

161. -2 and 2 **163.** $\left[-1, -\frac{1}{2}\right)$

166. $\frac{9}{8}$ **171.** w, w^2 where $\sqrt[3]{1} = w$

172. $\beta = \frac{-(2\alpha+1)}{\alpha+2}$ **177.** $A = 2, B = 32$

178. four **180.** $-p^5 + 5p^3q - 5pq^2$

181. $\frac{1}{a^4}(b^4 - 4ab^2c + 2c^2a^2)$ **182.** $\frac{-68}{3}$

186. $acx^2 - bx + 1 = 0$

187. $x^2 - p(p^4 - 5p^2q + 5q^2)x + p^2q^2(p^2 - 4q)(p^2 - q) = 0$

188. $x^2 + 13x + 30 = 0$ **189.** 5, 6

190. (i) $p < -3$ (ii) $-3 < p \le -2$ (iii) $p > 7$

191. $-2\sqrt{2} < m < 2\sqrt{2}$ **192.** $-1 < a < 2$

193. no real value **194.** $-\frac{1}{2} < a < \frac{1}{2}$

195. positive **196.** $a = 1, 2$ **197.** $(-\infty, -1]$

198. $0 < a < \frac{1}{8}$ **199.** (i) $a > \frac{3}{4}$ (ii) $a > \frac{3}{4}$

200. $1 < a < 2 \cdot \left(\frac{1+2\sqrt{2}}{7}\right)$ **201.** $-\frac{3}{2} \le a \le \frac{3}{2}$

202. (i) no value of a (ii) $\left(-\infty, \frac{1}{2}\right)$

203. 0, -1, -3, -4 **204.** $a > 1$

205. $a \ge 1$ **206.** $(-\infty, 0) \cup \left(\frac{9}{2}, \infty\right)$

207. $\frac{1}{2}(\sqrt{5}-1) < r < \frac{1}{2}(\sqrt{5}+1)$

208. $\left[-\sqrt{2}, \frac{-1+\sqrt{13}}{2}\right) \cup [\sqrt{2}, +\infty)$

209. $k \in [-1, \infty)$ **210.** (i) 9 (iii) 6, 3, 2

211. $rp^3 = q^3$ **213.** (ii) $\frac{3}{4}, \frac{3}{2}, \frac{-5}{3}$

214. $x^3 - 9x^2 + 33x - 65 = 0$

215. (i) $\frac{1}{a^2}(b^2 - 2ac)$ (ii) $\frac{-c}{d}$ (iii) $\frac{1}{a^2}(3ad - bc)$

217. $\pm\frac{1}{\sqrt{2}}$, no

226. $(c_1a_2 - c_2a_1)^2 = (b_1c_2 - b_2c_1)(a_1b_2 - a_2b_1)$ and

$$\begin{vmatrix} a_1 & b_1 & c_1 \\ a_2 & b_2 & c_2 \\ a_3 & b_3 & c_3 \end{vmatrix} = 0$$

227. $a = -7, b = -8$; roots are 3, 4; 3, 5; 3, 12 respectively

228. $\lambda = \frac{3}{2}, \mu = -6$ **229.** $\frac{-2}{3} \le a \le \frac{2}{3}$

230. 4, $x = 1$ **231.** $\frac{25}{12}, x = \frac{1}{6}$

235. 3, $\frac{1}{3}$ respectively **237.** 4, 12 respectively

240. $m \in (-1, 5)$ **244.** $m < \frac{11}{4}$

245. $[0, 1]$ **246.** $[-1, 9]$

248. $x = 4, y = 4$

251. $(aa_1 - cc_1)^2 = (bc_1 - ab_1)(cb_1 - ba_1)$

254. $(bb' - 2ac' - 2a'c)^2 = (b^2 - 4ac)(b'^2 - 4a'c')$

256. 0, -3 **258.** 1, 1, -2 **262.** -1, 2 **263.** one

264. $\{3\}$ **265.** one **266.** $\left\{\frac{1}{16}\right\}$ **267.** $\frac{-1}{4}, \frac{-1}{4}$

268. four **269.** 2 **270.** 6 **271.** 2

272. $0 < p \le \frac{1}{4}$ **273.** $x^2 - 6x + 12 = 0$

274. $ax^2 + (b - 4a)x + (4a - 2b + c) = 0$ **275.** $\frac{-2b}{\sqrt{ac}}$

276. -4, 7 respectively **277.** 2 **278.** $[3, 4)$

279. 3 **280.** $\frac{3}{4}$ **281.** $\left(\frac{5}{9}, 1\right] \cup [6, +\infty)$

282. $(0, 4)$ **283.** $(-\infty, -1) \cup (1, +\infty)$ **284.** $\left(\frac{3}{2}, +\infty\right)$

285. -1 **286.** 1 **287.** 4, $\frac{7}{2}$

288. (i) 2 (ii) (2, 3) **289.** (c)

290. (a) **291.** (a) **292.** (b), (c) **293.** (a)

294. (b) **295.** (d) **296.** (b) **297.** (c)

298. (b) **299.** (a) **300.** (a) **301.** (c), (d)

302. (a) **303.** (c) **304.** (c) **305.** (d)

306. (c) **307.** (b), (d) **308.** (c) **309.** (d)

310. (c) **311.** (c) **312.** (c) **313.** (b)

314. (a), (d) **315.** (c) **316.** (c) **317.** (b)

318. (a) **319.** (b), (d) **320.** (b) **321.** (d)

322. (b) **323.** (c) **324.** (a), (b), (d)

325. true **326.** false **327.** false **328.** false

□

Chapter Test

Time: 120 minutes

1. In the following, each statement is incomplete. Fill in the blank so that each resulting statement is correct.

(a) The minimum value of x for which $2x^3 - 5x^2 + x + 2 \geq 0$, is ______.

(b) If $f(x) = x + \frac{1}{x}$, $x \in R$ then the set of possible values of $f(x) =$ ______ .

(c) If $1 + i$ is a root of the equation $ix^2 - 3x + p = 0$ where $i = \sqrt{-1}$ then the other root is ______ .

2. In the following, each question has one or more than one correct answers. Indicate the correct answer(s):

(a) The equation $\cos 2x + 7 = a(2 - \sin x)$ can have a real solution for

A. all real a B. $a < 2$ C. $a > 6$ D. $2 \leq a \leq 6$

(b) If α, β be roots of $x^2 - x - 1 = 0$ and $A_n = \alpha^n + \beta^n$ then the AM of A_{n-1} and A_n is

A. $2A_{n+1}$ B. $\frac{1}{2}A_{n+1}$ C. $2A_{n-2}$ D. none of these

(c) For all $\lambda \in R$, the equation $ax^2 + (b - \lambda)x + (a - b - \lambda) = 0$, $a \neq 0$ has real roots. Then

A. $a = b$ B. $b < a < 0$ C. $b > a > 0$ D. $a > b > 0$

3. If $(x) = x - [x]$ where $[x] =$ the greatest integer $\leq x$ and $f(x) = (x)[x]$ then find the set $A = \{x \mid f(x) = 1, x \in R\}$.

4. If $f(x) = \dfrac{x^2 + 3x - 40}{x^2 + 5x - 14}$, $x \neq -7, x \neq 2$ then find the least and the greatest integral values of x for which $f(x) < 0$.

5. Consider the polynomial function $f(x) = (x + a_1)^2 + (x + a_2)^2 + \ldots + (x + a_n)^2$. What can you conclude about the value of $f(x)$ for $x \in R$? Hence prove that
$(a_1 + a_2 + \ldots + a_n)^2 \leq n(a_1^2 + a_2^2 + \ldots + a_n^2)$.

6. Prove that $(a^2 + b^2)x^2 - 2b(a + c)x + (b^2 + c^2) \geq 0$ for all $x \in R$. If equality holds, find the ratio of the roots of $ax^2 + bx + c = 0$.

7. Let $f(x) = x^4 - 3x^3 + 2x^2 - 3x + a$. If $f(\sqrt{-1}) = 0$ and $a \in R$ then prove that $f(x) < 0$ for $x \in (1, 2)$.

8. If the polynomials $a_1x^3 + b_1x^2 + c_1x + d_1$ and $a_2x^3 + b_2x^2 + c_2x + d_2$ have $x^2 - 2x + 1$ as a common factor then prove that $a_1 - a_2$, $d_1 - d_2$, $c_2 - c_1$ are in AP.

9. Find real values of a for which the inequations $x^2 - 4x - 6a \leq 0$ and $x^2 + 2x + a \leq 0$ have only one real solution common.

10. If a, b, c are nonzero real numbers and a pair of equations from $x^2 + ax + bc = 0$, $x^2 + bx + ca = 0$ and $x^2 + cx + ab = 0$ has a common root then prove that each pair of equations will have a common root.

11. Find $a \in R$ such that the least value of $4x^2 - 4ax + a^2 - 2a + 2$ on the interval $0 \leq x \leq 2$ is equal to 11.

12. Prove that the equation $x^3 + 2x^2 + x + 4 = 0$ has exactly one root in the open interval $(-3, -2)$.

13. Find $\lambda \in R$ for which the equation
$(x + 1)^4 - \lambda(x + 1)^2 \cdot (x^2 + 2x + 2) + (\lambda - 1) \cdot (x^2 + 2x + 2)^2 = 0$
has at least one real root.

14. Let α, β be real roots of $x^2 - 2px + pq = 0, p > q > 0$. Show that both roots are positive and

$$1 + 2 \cdot \frac{q}{p} + 3 \cdot \left(\frac{q}{p}\right)^2 + \ldots + n \cdot \left(\frac{q}{p}\right)^{n-1} < \left(\frac{\alpha + \beta}{\alpha - \beta}\right)^4.$$

15. The distances of three points A, B, C on a straight line from a fixed point on the line are roots of the equation

$$ax^3 + 3bx^2 + 3cx + d = 0.$$

Find the condition for one of the points A, B, C to be the middle point of the line segment joining the other two.

16. Prove that the largest number of real roots of the equation

$$x^5 + a_1x^4 + a_2x^3 + a_3x^2 + a_4x + a_5 = 0$$

whose coefficients are real, is three if $2a_1^2 - 5a_2 < 0$.

□

Answers

1. (a) $-\frac{1}{2}$ (b) $(-\infty, -2] \cup [2, +\infty)$ (c) $-1 - 4i$

2. (a) D (b) B (c) A, B, C

3. $A = \left\{n + \frac{1}{n} \,\middle|\, n \in N - \{1\}\right\}$

4. 3, 4 respectively

6. w, w^2

9. 0, 1

11. $a \in R - (1 - \sqrt{10},\ 5 + 3\sqrt{2})$

13. $\lambda \in (-\infty, 2)$

15. $a^2d + 2b^3 = 3abc$

□

4. Complex Numbers

Recap of Facts and Formulae

1. Algebra of complex numbers

- $z = x + iy$ is a complex number where $x \in R$, $y \in R$, $i = \sqrt{-1}$, i.e., $i^2 = -1$; real part of $z = \text{Re}(z) = x$, imaginary part of $z = \text{Im}(z) = y$.
- If $z_1 = x_1 + iy_1$, $z_2 = x_2 + iy_2$ then

$$z_1 \pm z_2 = (x_1 \pm x_2) + i(y_1 \pm y_2)$$

and $z_1 \cdot z_2 = (x_1x_2 - y_1y_2) + i(x_1y_2 + y_1x_2)$

$$\frac{1}{z_2} = \frac{1}{x_2 + iy_2} = \frac{x_2 - iy_2}{(x_2 + iy_2)(x_2 - iy_2)} = \frac{x_2 - iy_2}{x_2^2 - i^2y_2^2}$$

$$= \frac{x_2 - iy_2}{x_2^2 + y_2^2} = \frac{x_2}{x_2^2 + y_2^2} - i\frac{y_2}{x_2^2 + y_2^2}.$$

- $x_1 + iy_1 = x_2 + iy_2 \quad \Leftrightarrow \quad x_1 = x_2, y_1 = y_2$.

  Thus, one complex equation is equivalent to two real equations.
- $i^n = 1, i, -1, -i$

  according as $n = 4m, 4m + 1, 4m + 2, 4m + 3$.

Note The values of different integral powers of i are i or -1 or $-i$ or 1. The digit in the units place of the value of a positive integral power of a digit also follows a sequence of digits.

The digits in units places of $7^1, 7^2, 7^3, 7^4, 7^5$, etc., are 7, 9, 3, 1, 7, etc. Using this fact we can determine the digit in the units place of a power of a natural number.

For example What is the digit in the units place of $(193)^{50}$?

Consider the values of $3^1, 3^2, 3^3, 3^4, 3^5, 3^6$, etc. The digit in the units place will be in the sequence 3, 9, 7, 1, 3, 9, 7, 1, … . The 50th term in it is 9. So the digit in the units place of $(193)^{50}$ is 9.

2. Representation of complex numbers in Argand plane

- The complex number $z = x + iy$ is represented in a plane by the point (x, y). The plane in which complex numbers are represented by points is called the *Argand plane*.

If $x > 0, y > 0$ then the complex number will be represented by a point in the first quadrant. Similarly, for other possible signs of x and y the location of the point is as shown in the figure.

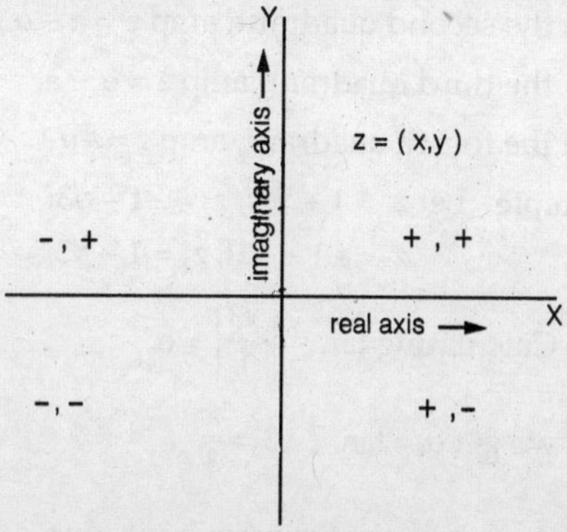

3. Modulus, amplitude (argument), conjugate of a complex number

If $z = x + iy$ then

- modulus of $z = |z| = +\sqrt{x^2 + y^2}$
- amplitude of z = amp z(or arg z) $= \tan^{-1}\frac{y}{x}$.

We know that $\tan^{-1}\frac{y}{x}$ has many values. The smallest numerical value falling in the quadrant of the complex number is called the *fundamental amplitude* (or simply amplitude). The general value is $2\pi r + \tan^{-1}\frac{y}{x}$ where $\tan^{-1}\frac{y}{x}$ is the fundamental amplitude, and this value is called the *general amplitude.*

The amp z lies between $-\pi$ and π, i.e., $0 \leq \text{amp } z \leq \pi$ or $-\pi < \text{amp } z < 0$.

If z belongs to the first quadrant then its amplitude is between 0 and $\frac{\pi}{2}$; if z belongs to the second quadrant then the amplitude is between $\frac{\pi}{2}$ and π; if z belongs to the third quadrant then the amplitude is between $-\pi$

and $\frac{-\pi}{2}$ and if z belongs to the fourth quadrant then the amplitude is between $\frac{-\pi}{2}$ and 0.

- *Method of calculating amp z is as follows.*

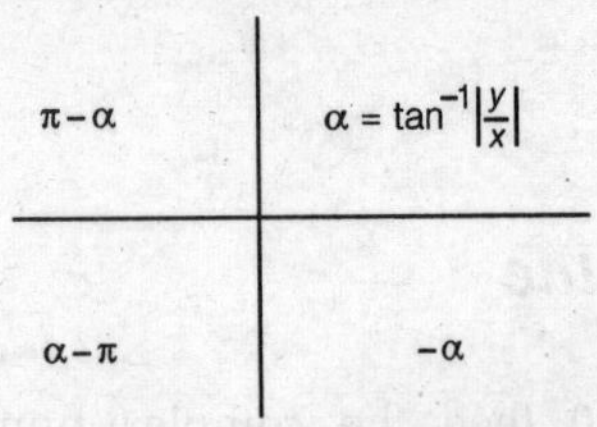

Calculate $\tan^{-1}\left|\frac{y}{x}\right|$ $(= \alpha)$ in the first quadrant.

If z is in the first quadrant, amp $z = \alpha$

If z is in the second quadrant, amp $z = \pi - \alpha$

If z is in the third quadrant, amp $z = \alpha - \pi$

If z is in the fourth quadrant, amp $z = -\alpha$.

For example Let $z_1 = 1 + \sqrt{3}i$, $z_2 = -1 - \sqrt{3}i$,
$z_3 = -1 + \sqrt{3}i$, $z_4 = 1 - \sqrt{3}i$

For z_1: Calculating $\tan^{-1}\left|\frac{\sqrt{3}}{1}\right| = \alpha$,

we get $\alpha = \tan^{-1}\sqrt{3} = \frac{\pi}{3}$.

As z_1 is in the first quadrant, amp $z_1 = \frac{\pi}{3}$.

For z_2: Calculating $\tan^{-1}\left|\frac{-\sqrt{3}}{-1}\right| = \alpha$,

we get $\alpha = \tan^{-1}\sqrt{3} = \frac{\pi}{3}$.

As z_2 is in the third quadrant,

amp $z_2 = \frac{\pi}{3} - \pi = \frac{-2\pi}{3}$.

For z_3: Calculating $\tan^{-1}\left|\frac{\sqrt{3}}{-1}\right| = \alpha$,

we get $\alpha = \tan^{-1}\sqrt{3} = \frac{\pi}{3}$.

As z_3 is in the second quadrant,

amp $z_3 = \pi - \frac{\pi}{3} = \frac{2\pi}{3}$.

For z_4: Calculating $\tan^{-1}\left|\frac{-\sqrt{3}}{1}\right| = \alpha$,

we get $\alpha = \tan^{-1}\sqrt{3} = \frac{\pi}{3}$.

As z_4 is in the fourth quadrant, amp $z_4 = -\frac{\pi}{3}$.

Note The complex number $z = 0$ has indeterminate amplitude.

- Conjugate of $z = \bar{z} = x - iy$.
- The *trigonometrical form* (or *polar form*) of a complex number is $z = |z| \{(\cos(\text{amp } z) + i\sin(\text{amp } z)\}$,

i.e., $z = r(\cos\theta + i\sin\theta)$ where $r = |z|$ and $\theta = \text{amp } z$.

- Unimodular complex number z is such that $|z| = 1$ and hence unimodular complex number $z = \cos\theta + i\sin\theta$ where $\theta = \text{amp } z$.

TIP While taking a complex number z in working out a problem or solving an equation we take $z = x + iy$ (in algebraic form) or $z = r(\cos\theta + i\sin\theta)$ (in trigonometrical form). If the modulus or amplitude of the complex number is known, it is always convenient to take z in the trigonometrical form.

Note (i) $\frac{z}{|z|}$ is always a unimodular complex number if $z \neq 0$.

(ii) If amp $z = \frac{\pi}{2}$ or $-\frac{\pi}{2}$, z is purely imaginary; if amp $z = 0$ or π, z is purely real.

4. To express real part and imaginary part in terms of the complex number z

- Let $z = x + iy$; then $\bar{z} = x - iy$

Adding these, $z + \bar{z} = 2x$; $\therefore$ $x = \frac{1}{2}(z + \bar{z})$

Subtracting these, $z - \bar{z} = 2iy$; $\therefore$ $y = \frac{1}{2i}(z - \bar{z})$.

Note (i) $z + \bar{z}$ is always real and $z - \bar{z}$ is always imaginary.

(ii) $z\bar{z}$ is always real.

5. Properties of conjugate, amplitude and modulus

- $\overline{z_1 \pm z_2} = \bar{z}_1 \pm \bar{z}_2$

$\overline{z_1 \cdot z_2} = \bar{z}_1 \cdot \bar{z}_2$

$\overline{\left(\frac{z_1}{z_2}\right)} = \frac{\bar{z}_1}{\bar{z}_2}$.

- $\text{amp}(z_1 \cdot z_2) = \text{amp } z_1 + \text{amp } z_2$

$\text{amp}\left(\frac{z_1}{z_2}\right) = \text{amp } z_1 - \text{amp } z_2$

$\text{amp } \frac{z}{\bar{z}} = 2\text{amp } z$

$\text{amp } z^2 = 2\text{amp } z$.

Note If $z_1 = r_1(\cos\theta_1 + i\sin\theta_1)$, $z_2 = r_2(\cos\theta_2 + i\sin\theta_2)$

then $z_1 \cdot z_2 = r_1 r_2(\cos\theta_1 + i\sin\theta_1)(\cos\theta_2 + i\sin\theta_2)$

$= r_1 r_2 \{\cos(\theta_1 + \theta_2) + i\sin(\theta_1 + \theta_2)\}$

and $\frac{z_1}{z_2} = \frac{r_1(\cos\theta_1 + i\sin\theta_1)}{r_2(\cos\theta_2 + i\sin\theta_2)}$

$$= \frac{r_1}{r_2}\{\cos(\theta_1 - \theta_2) + i\sin(\theta_1 - \theta_2)\}$$

$\therefore \quad \text{amp}(z_1 z_2) = \text{amp}\, z_1 + \text{amp}\, z_2$ and

$$\text{amp}\,\frac{z_1}{z_2} = \text{amp}\, z_1 - \text{amp}\, z_2$$

Clearly, these results on amplitude hold when we take fundamental amplitudes only.

- $|z_1 \cdot z_2| = |z_1| \cdot |z_2|$

  $\left|\frac{z_1}{z_2}\right| = \frac{|z_1|}{|z_2|}$

  $|z^n| = |z|^n$

- $|z_1 + z_2| \le |z_1| + |z_2|$,

  equality holding if $z = 0$, z_1, z_2 are collinear with $z = 0$ at one end.

  $|z_1 - z_2| \ge \big||z_1| - |z_2|\big|$.

- $z \cdot \bar{z} = |z|^2$

  $\left|\frac{z}{\bar{z}}\right| = 1.$

6. Power of a complex number (De Moivre's theorem)

- If $z_1 = \cos\theta_1 + i\sin\theta_1$, $z_2 = \cos\theta_2 + i\sin\theta_2$, etc., then

$$z_1 \cdot z_2 = (\cos\theta_1 + i\sin\theta_1)(\cos\theta_2 + i\sin\theta_2)$$
$$= \cos(\theta_1 + \theta_2) + i\sin(\theta_1 + \theta_2).$$
$$z_1 \cdot z_2 \cdot z_3 \ldots = \cos(\theta_1 + \theta_2 + \theta_3 + \ldots) + i\sin(\theta_1 + \theta_2 + \theta_3 + \ldots).$$

- $(\cos\theta + i\sin\theta)^n = \cos n\theta + i\sin n\theta$,

  where n is a positive integer.

  $(\cos\theta + i\sin\theta)^{-n} = \cos n\theta - i\sin n\theta.$

  $(\cos\theta - i\sin\theta)^n = \cos n\theta - i\sin n\theta.$

- $(\cos\theta + i\sin\theta)^{p/q} = \{\cos(2r\pi + \theta) + i\sin(2r\pi + \theta)\}^{p/q}$

$$= \cos\frac{(2r\pi + \theta)p}{q} + i\sin\frac{(2r\pi + \theta)p}{q},$$

  where $r = 0, 1, 2, \ldots, q - 1$.

- nth roots of unity $= 1^{1/n} = (\cos 0 + i\sin 0)^{1/n}$

$$= (\cos 2r\pi + i\sin 2r\pi)^{1/n}$$
$$= \cos\frac{2r\pi}{n} + i\sin\frac{2r\pi}{n},$$

  where $r = 0, 1, 2, \ldots, n - 1$.

If $\cos\frac{2\pi}{n} + i\sin\frac{2\pi}{n} = z_0$ then the nth roots of unity are

$1, z_0, z_0^2, z_0^3, \ldots, z_0^{n-1}$ which are in GP.

7. Cube roots of unity

- Cube roots of unity

$$= 1^{1/3} = (\cos 0 + i\sin 0)^{1/3} = (\cos 2r\pi + i\sin 2r\pi)^{1/3}$$
$$= \cos\frac{2r\pi}{3} + i\sin\frac{2r\pi}{3}, \text{ where } r = 0, 1, 2$$
$$= 1, -\frac{1}{2} + i\frac{\sqrt{3}}{2}, -\frac{1}{2} - i\frac{\sqrt{3}}{2}.$$

- If one of the nonreal complex roots be w then the other nonreal complex root will be w^2.
- $\sqrt[3]{1} = 1, w, w^2$ where $w^3 = 1$ and $1 + w + w^2 = 0$.
- The value of $1 + w^n + w^{2n}$

  $= 3$ if $n = 3m$, i.e., n is divisible by 3

  $= 0$ if $n \ne 3m$, i.e., n is not divisible by 3.

Note Any complex number for which

$$\left|\frac{\text{real part}}{\text{imaginary part}}\right| = 1 : \sqrt{3} \text{ or } \sqrt{3} : 1,$$

can be expressed in terms of w and i.

For example

$$1 + \sqrt{3}i = \frac{-1 - \sqrt{3}i}{2} \cdot (-2) = -2w \quad (\text{or } -2w^2)$$

$$i - \sqrt{3} = \frac{-1 - \sqrt{3}i}{i} = \frac{-1 - \sqrt{3}i}{2} \cdot \frac{2}{i} = \frac{2w}{i} \left(\text{or } \frac{2w^2}{i}\right).$$

8. Application of complex numbers in geometrical problems

The geometrical meaning of complex expressions, equations and inequations are as follows.

- $z = x + iy \Rightarrow z$ is a point whose coordinates are (x, y).

Note The complex number z is also represented by a vector. If P represents z in the Argand plane we say $\vec{OP}$ represents the complex number z, O being the origin. If P and Q represent complex numbers z and z_1 respectively then $\vec{QP} = z - z_1$.

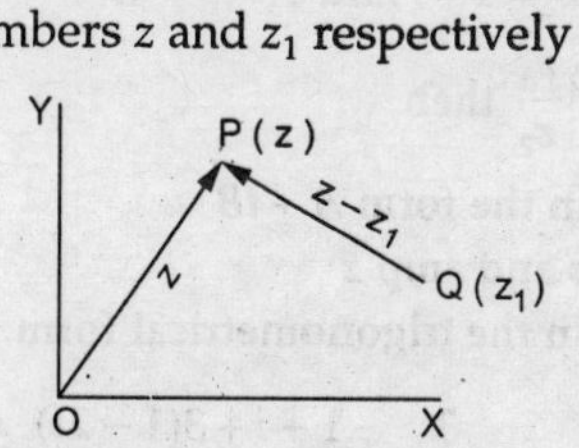

- $|z| =$ distance between the origin and the point z

  $|z - z_1| =$ distance between the points z and z_1.

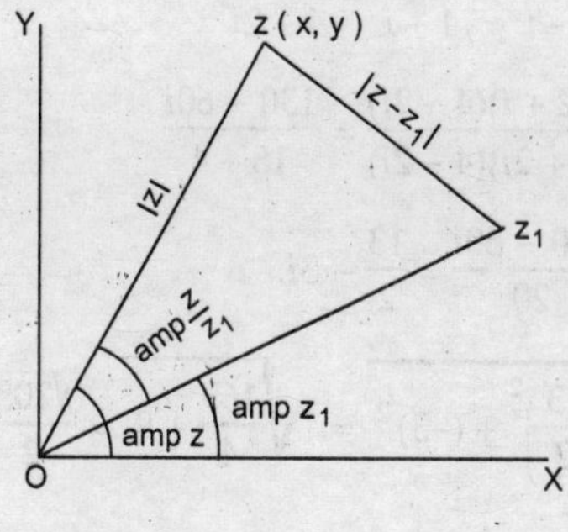

- amp $z = \angle ZOX$, where Z represents z

amp $\dfrac{z}{z_1} = \angle ZOZ_1$.

- If z_3, z_1, z_2 are three points taken in the anticlockwise sense then

$$\text{amp}\ \frac{z_3 - z_1}{z_2 - z_1} = \angle Z_3Z_1Z_2.$$

- The angle between two line segments joining the points z_1, z_2 and z_3, z_4 is

$$\text{amp}\ \frac{z_1 - z_2}{z_3 - z_4} \quad \text{or} \quad \pi - \text{amp}\ \frac{z_1 - z_2}{z_3 - z_4}.$$

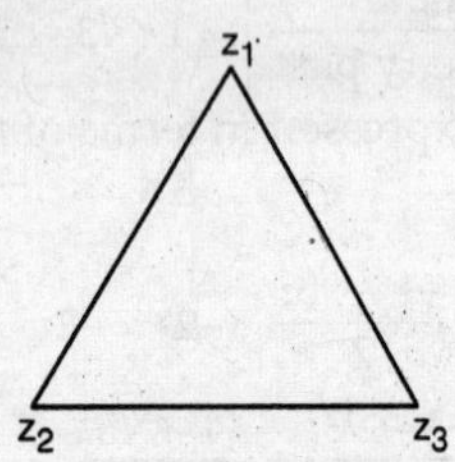

- Complex numbers z satisfying $|z - z_0| = \rho$ represent points on the circle whose centre is z_0 and radius = ρ.

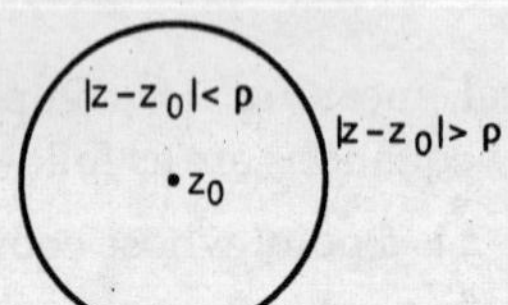

Complex numbers z satisfying $|z - z_0| < \rho$ represent points inside the circle whose centre is z_0 and radius = ρ.

Complex numbers z satisfying $|z - z_0| > \rho$ represent points outside the circle whose centre is z_0 and radius = ρ.

- If z is on the circle $|z| = \rho$ then iz is also on the circle, the radius vector being shifted by $\pi/2$ in the anticlockwise sense.

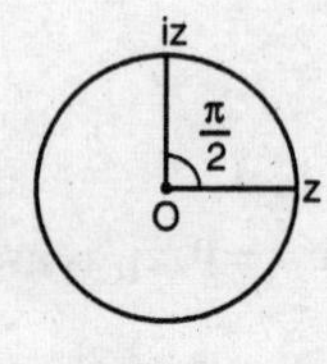

- The line segment joining the complex numbers z_1, z_2 is divided by the complex number z in the ratio $m : n$ if $z = \dfrac{mz_2 + nz_1}{m + n}$.

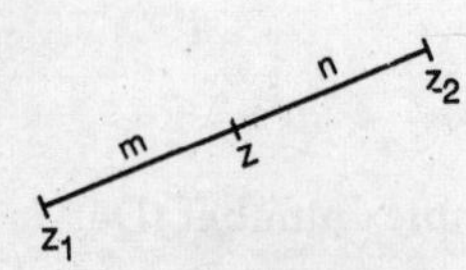

- General equation of a line is $\bar{\alpha} z + \alpha \bar{z} + \beta = 0$ where β is a real constant and α is a nonreal complex constant.
- General equation of a circle is $z\bar{z} + \alpha\bar{z} + \bar{\alpha}z + \beta = 0$ where β is a real constant and α is a nonreal complex constant.

Selected Solved Examples

1. Let $z_1 = 1 - 2i$, $z_2 = 1 + i$ and $z_3 = 3 + 4i$.

If $z = \left(\dfrac{1}{z_1} + \dfrac{3}{z_2}\right)\dfrac{z_3}{z_2}$ then

(i) express z in the form $A + iB$

(ii) find $|z|$, $\bar{z}$ and amp z

(iii) express z in the trigonometrical form.

(i) $\dfrac{1}{z_1} + \dfrac{3}{z_2} = \dfrac{1}{1 - 2i} + \dfrac{3}{1 + i} = \dfrac{1 + i + 3(1 - 2i)}{(1 - 2i)(1 + i)} = \dfrac{4 - 5i}{3 - i}$

$$\therefore \quad z = \frac{4 - 5i}{3 - i} \times \frac{3 + 4i}{1 + i} = \frac{32 + i}{4 + 2i}$$

$$= \frac{(32 + i)(4 - 2i)}{(4 + 2i)(4 - 2i)} = \frac{130 - 60i}{16 + 4}$$

$$= \frac{130 - 60i}{20} = \frac{13}{2} - 3i.$$

(ii) $|z| = \sqrt{\left(\dfrac{13}{2}\right)^2 + (-3)^2} = \sqrt{\dfrac{169}{4} + 9} = \dfrac{\sqrt{205}}{2}$

$$\bar{z} = \frac{13}{2} + 3i$$

$$\text{amp}\ z = \tan^{-1}\frac{-3}{\frac{13}{2}} = -\tan^{-1}\frac{6}{13}$$

{$\because$ z is in the fourth quadrant}.

(iii) $z = |z|\ \{\cos(\text{amp}\ z) + i\sin(\text{amp}\ z)\}$

$$= \frac{\sqrt{205}}{2}\left\{\cos\left(-\tan^{-1}\frac{6}{13}\right) + i\sin\left(-\tan^{-1}\frac{6}{13}\right)\right\}.$$

2. Find real $\theta \in (-\pi, \pi)$ so that $\dfrac{3 + 2i\sin\theta}{1 - 2i\sin\theta}$ is

(i) purely real (ii) purely imaginary.

Here $z = \dfrac{3 + 2i\sin\theta}{1 - 2i\sin\theta} = \dfrac{(3 + 2i\sin\theta)(1 + 2i\sin\theta)}{(1 + 2i\sin\theta)(1 - 2i\sin\theta)}$

$$= \frac{(3-4\sin^2\theta)+8i\sin\theta}{1^2+(2\sin\theta)^2}$$

$$= \frac{3-4\sin^2\theta}{1+4\sin^2\theta} + i\frac{8\sin\theta}{1+4\sin^2\theta}.$$

(i) z is purely real if $\text{Im}(z) = 0$,

i.e., $\frac{8\sin\theta}{1+4\sin^2\theta} = 0$; $\therefore$ $\sin\theta = 0$

$\therefore$ $\theta = 0, \pi$. But $\theta \neq \pi$ from the question.

So $\theta = 0$.

(ii) z is purely imaginary if $\text{Re}(z) = 0$,

i.e., $\frac{3-4\sin^2\theta}{1+4\sin^2\theta} = 0$ and $\sin\theta \neq 0$

or $3-4\sin^2\theta = 0$; $\therefore$ $\sin\theta = \pm\frac{\sqrt{3}}{2}$

$$\therefore \quad \theta = \frac{\pi}{3}, \pi - \frac{\pi}{3}, -\frac{\pi}{3}, \frac{\pi}{3} - \pi$$

$$= \frac{\pi}{3}, \frac{2\pi}{3}, \frac{-\pi}{3}, -\frac{2\pi}{3}.$$

3. Solve for $x, y \in R$:

$$(x^4 + 2xi) - (3x^2 + yi) = (1 + 2yi) + \frac{34}{3+5i}.$$

Here $(x^4 - 3x^2) + i(2x - y)$

$$= 1 + 2yi + \frac{34(3-5i)}{(3+5i)(3-5i)}$$

$$= 1 + 2yi + \frac{34(3-5i)}{3^2+5^2}$$

$$= 1 + 2yi + 3 - 5i$$

$\therefore$ $(x^4 - 3x^2) + i(2x - y) = 4 + i(2y - 5)$.

Equating real and imaginary parts,

$x^4 - 3x^2 = 4$ ... (1)

and $2x - y = 2y - 5$ ... (2)

From (1), $(x^2)^2 - 3x^2 - 4 = 0$

or $(x^2 - 4)(x^2 + 1) = 0$

$\therefore$ $x^2 - 4 = 0$ $(\because x^2 \neq -1)$; $\therefore$ $x = \pm 2$.

$\therefore$ from (2), $2(\pm 2) + 5 = 3y$

$\therefore$ $y = \frac{1}{3}\{\pm 4 + 5\} = 3, \frac{1}{3}$

$\therefore$ solutions are $x = 2, y = 3; x = -2, y = \frac{1}{3}$.

4. Prove that $\left(\frac{\sqrt{3}+i}{\sqrt{3}-i}\right)^{3n} + 1 = 0$ for all odd integral values of *n*.

$$\frac{\sqrt{3}+i}{\sqrt{3}-i} = \frac{(\sqrt{3}+i)(\sqrt{3}+i)}{(\sqrt{3}-i)(\sqrt{3}+i)} = \frac{3-1+2\sqrt{3}i}{(\sqrt{3})^2+1^2}$$

$$= \frac{2+2\sqrt{3}i}{4} = \frac{1+\sqrt{3}i}{2}$$

$$\therefore \quad \left(\frac{\sqrt{3}+i}{\sqrt{3}-i}\right)^{3n} + 1 = \left(\frac{1+\sqrt{3}i}{2}\right)^{3n} + 1$$

$$= \left\{\left(\frac{1+\sqrt{3}i}{2}\right)^3\right\}^n + 1$$

$$= \left\{\frac{1+3\cdot\sqrt{3}i+3\cdot(\sqrt{3}i)^2+(\sqrt{3}i)^3}{8}\right\}^n + 1$$

$$= \left\{\frac{1+3\sqrt{3}i-9-3\sqrt{3}i}{8}\right\}^n + 1$$

$$= \left(\frac{-8}{8}\right)^n + 1 = -1 + 1 \quad (\because n \text{ is odd})$$

$$= 0.$$

Second Method (using w, i.e., complex cube root of unity)

$$\frac{\sqrt{3}+i}{\sqrt{3}-i} = \frac{i(\sqrt{3}+i)}{i(\sqrt{3}-i)} = \frac{-1+\sqrt{3}i}{1+\sqrt{3}i}$$

$$= \frac{\frac{-1+\sqrt{3}i}{2}}{\frac{-1-\sqrt{3}i}{2}} \times (-1)$$

$$= \frac{w}{w^2} \times (-1) \quad \text{or} \quad \frac{w^2}{w} \times (-1)$$

$$= -\frac{1}{w} \quad \text{or} \quad -w$$

$$\therefore \quad \left(\frac{\sqrt{3}+i}{\sqrt{3}-i}\right)^{3n} + 1 = \left(-\frac{1}{w}\right)^{3n} + 1 \quad \text{or} \quad (-w)^{3n} + 1$$

$$= \frac{-1}{w^{3n}} + 1 \quad \text{or} \quad -w^{3n} + 1,$$

for n is odd

$$= -1 + 1 \quad \text{or} \quad -1 + 1$$

$$= 0.$$

5. Let $f(x) = x^4 + 9x^3 + 35x^2 - x + 4$. Find $f(-5 + 2\sqrt{-4})$.

Let $x = -5 + 2\sqrt{-4}$; $\therefore$ $x + 5 = 2\sqrt{-4}$

$\therefore$ $(x+5)^2 = 4(-4) = -16$

or $x^2 + 10x + 41 = 0$;

so, if $\phi(x) = x^2 + 10x + 41$ then $\phi(-5 + 2\sqrt{-4}) = 0$ ... (1)

Now, $x^4 + 9x^3 + 35x^2 - x + 4$

$$= x^2(x^2 + 10x + 41) - x^3 - 6x^2 - x + 4$$

$$= x^2(x^2 + 10x + 41) - x(x^2 + 10x + 41) + 4x^2 + 40x + 4$$

$$= (x^2 + 10x + 41)(x^2 - x) + 4(x^2 + 10x + 41) - 160$$

$$= (x^2 + 10x + 41)(x^2 - x + 4) - 160$$

$$\therefore \quad f(x) = \phi(x) \cdot (x^2 - x + 4) - 160$$

$$\therefore \quad f(-5 + 2\sqrt{-4})$$

$$= \phi(-5 + 2\sqrt{-4}) \cdot \{(-5 + 2\sqrt{-4})^2 - (-5 + 2\sqrt{-4}) + 4\} - 160$$

$= 0 - 160$, using (1)

$= -160.$

6. If $\sqrt{a + ib} = x + iy$, prove that $\sqrt{a - ib} = x - iy$.

Here $\sqrt{a + ib} = x + iy$

$\therefore \quad a + ib = (x + iy)^2 = x^2 - y^2 + 2ixy;$

equating real and imaginary parts,

$$a = x^2 - y^2 \quad \text{and} \quad b = 2xy$$

$$\therefore \quad a - ib = x^2 - y^2 - i \cdot 2xy$$

$$= x^2 + (iy)^2 - 2x(iy) = (x - iy)^2$$

$$\therefore \quad \sqrt{a - ib} = x - iy.$$

7. Express $(1 + a^2)(1 + b^2)(1 + c^2)$ as sum of two squares.

$1 + a^2 = 1^2 - (ia)^2 = (1 + ia)(1 - ia)$

$$\therefore \quad (1 + a^2)(1 + b^2)(1 + c^2)$$

$$= (1 + ia)(1 - ia)(1 + ib)(1 - ib)(1 + ic)(1 - ic)$$

$$= \{(1 + ia)(1 + ib)(1 + ic)\}\{(1 - ia)(1 - ib)(1 - ic)\}$$

$$= \{1 + i(a + b + c) + i^2(ab + bc + ca) + i^3abc\}\{1 - i(a + b + c) + i^2(ab + bc + ca) - i^3abc\}$$

$$= [\{1 - (ab + bc + ca)\} + i(a + b + c - abc)] \times [\{1 - (ab + bc + ca)\} - i(a + b + c - abc)]$$

$$= \{1 - (ab + bc + ca)\}^2 + (a + b + c - abc)^2.$$

8. If $z^2 = -4 - 3i$, find z. {Or Find the value of $\sqrt{-4 - 3i}$}.

TIP To find an unknown complex number, one of the methods is to find the real part and imaginary part of the complex number separately. For this we put $z = x + iy$ in the relation and separating real and imaginary parts we form two real equations, solving which we get x and y.

Let $z = x + iy$. Then $(x + iy)^2 = -4 - 3i$

or $\quad x^2 - y^2 + 2ixy = -4 - 3i;$

equating real and imaginary parts,

$$x^2 - y^2 = -4 \quad \ldots (1) \qquad 2xy = -3 \quad \ldots (2)$$

Now, $(x^2 + y^2)^2 = (x^2 - y^2)^2 + 4x^2y^2$

$$= (-4)^2 + (2xy)^2 = 16 + (-3)^2 = 25$$

$$\therefore \quad x^2 + y^2 = \sqrt{25} = 5 \quad \ldots (3)$$

$$(1) + (3) \Rightarrow 2x^2 = 1; \quad \text{or} \quad x = \pm\frac{1}{\sqrt{2}}$$

$$(3) - (1) \Rightarrow 2y^2 = 9; \quad \text{or} \quad y = \pm\frac{3}{\sqrt{2}}$$

From (2), x and y are of the opposite signs.

So $\quad x = \frac{1}{\sqrt{2}}, y = -\frac{3}{\sqrt{2}}$ and $x = \frac{-1}{\sqrt{2}}, y = \frac{3}{\sqrt{2}}.$

$$\therefore \quad z = x + iy = \frac{1}{\sqrt{2}} - \frac{3}{\sqrt{2}}i, -\frac{1}{\sqrt{2}} + \frac{3}{\sqrt{2}}i.$$

Second Method (converting into perfect square)

Here $z^2 = -4 - 3i = \frac{1}{2}(-8 - 6i),$

making the imaginary part even

$$= \frac{1}{2}\{-8 - 2 \cdot 3i\} = \frac{1}{2}\{1^2 + (3i)^2 - 2 \cdot 3i\}$$

$$= \frac{1}{2}(1 - 3i)^2$$

$$\therefore \quad z = \sqrt{-4 - 3i} = \pm\frac{1}{\sqrt{2}}(1 - 3i).$$

9. If $z = a + ib$, $|z| = 1$ and $b \neq 0$, show that z can be represented as $z = \dfrac{c + i}{c - i}$ where c is a real number.

As $|z| = 1$, we get $\sqrt{a^2 + b^2} = 1$.

Now, $\quad z = a + ib = \dfrac{a + ib}{1} = \dfrac{a + ib}{\sqrt{a^2 + b^2}}$

$$= \frac{a + ib}{\sqrt{a^2 - i^2b^2}} = \frac{a + ib}{\sqrt{(a + ib)(a - ib)}} = \frac{\sqrt{a + ib}}{\sqrt{a - ib}}.$$

Next, let $\sqrt{a + ib} = p + iq$, then $\sqrt{a - ib} = p - iq$

(see solved example 6).

As $\quad b \neq 0, q \neq 0.$

$$\therefore \quad z = \frac{p + iq}{p - iq} = \frac{\frac{p}{q} + i}{\frac{p}{q} - i} = \frac{c + i}{c - i}, \text{ putting } c \text{ for } \frac{p}{q}.$$

10. a, b, c are real numbers and z is a complex number such that $a^2 + b^2 + c^2 = 1$ and $b + ic = (1 + a)z$.

Prove that $\dfrac{a + ib}{1 + c} = \dfrac{1 + iz}{1 - iz}$.

Here $b + ic = (1 + a)z; \quad \therefore \quad z = \dfrac{b + ic}{1 + a}$

or $iz = \frac{-c + ib}{1 + a}$; by componendo and dividendo,

$$\frac{1 + iz}{1 - iz} = \frac{1 + a - c + ib}{1 + a + c - ib} = \frac{(a + ib) + (1 - c)}{(1 + c) + (a - ib)} \quad \dots (1)$$

Now, $a^2 + b^2 = 1 - c^2$

or $a^2 - i^2b^2 = 1 - c^2$

or $(a + ib)(a - ib) = (1 + c)((1 - c)$

or $\frac{a + ib}{1 + c} = \frac{1 - c}{a - ib} = \frac{a + ib + 1 - c}{1 + c + a - ib}$

(by property of ratio and proportion).

$\therefore$ from (1) we get, $\frac{1 + iz}{1 - iz} = \frac{a + ib}{1 + c}$.

11. Find all nonzero complex numbers z such that $\bar{z} = iz^2$.

Let $z = x + iy$. Putting in the equation we get,

$\overline{x + iy} = i(x + iy)^2$

or $x - iy = i(x^2 - y^2 + 2ixy)$.

Equating real and imaginary parts,

$x = -2xy \quad \dots (1)$

$-y = x^2 - y^2 \quad \dots (2)$

(1) $\Rightarrow$ $x + 2xy = 0$, i.e., $x(1 + 2y) = 0$

$\Rightarrow$ $x = 0$ or $y = -\frac{1}{2}$.

When $x = 0$, (2) gives $-y = -y^2$

or $y(1 - y) = 0$; $\therefore$ $y = 0, 1$

$\therefore$ $z = x + iy = 0 + i \cdot 0, 0 + i \cdot 1 = 0, i$.

When $y = -\frac{1}{2}$, (2) gives $\frac{1}{2} = x^2 - \frac{1}{4}$

or $x^2 = \frac{1}{2} + \frac{1}{4} = \frac{3}{4}$; $\therefore$ $x = \pm \frac{\sqrt{3}}{2}$

$$\therefore \quad z = x + iy = \frac{\sqrt{3}}{2} + i\left(-\frac{1}{2}\right), -\frac{\sqrt{3}}{2} + i\left(\frac{1}{2}\right)$$

$$= \frac{\sqrt{3} - i}{2}, \frac{-\sqrt{3} - i}{2} = \frac{-i^2\sqrt{3} - i}{2}, \frac{i^2\sqrt{3} - i}{2}$$

$$= i \cdot \frac{-1 - \sqrt{3}i}{2}, i \cdot \frac{-1 + \sqrt{3}i}{2} = iw, iw^2$$

$\therefore$ the nonzero solutions are $z = i, iw, iw^2$.

12. Solve for the complex number z such that $z^2 + z|z| + |z^2| = 0$.

Let $z = x + iy$. Putting in the equation we get,

$(x + iy)^2 + (x + iy)\,|x + iy| + |(x + iy)^2| = 0$

or $(x^2 - y^2 + 2ixy) + (x + iy)\sqrt{x^2 + y^2} + x^2 + y^2 = 0$

$\{\because |z^2| = |z|^2 = (\sqrt{x^2 + y^2})^2\}$

$\therefore$ $x^2 - y^2 + x\sqrt{x^2 + y^2} + x^2 + y^2 = 0$

or $x(2x + \sqrt{x^2 + y^2}) = 0 \quad \dots (1)$

and $2xy + y\sqrt{x^2 + y^2} = 0$

or $y(2x + \sqrt{x^2 + y^2}) = 0 \quad \dots (2)$

(2) $\Rightarrow$ $y = 0$ or $2x + \sqrt{x^2 + y^2} = 0$.

When $y = 0$, (1) gives $x(2x + |x|) = 0$

$\therefore$ $x = 0$ or $2x + |x| = 0$, i.e., $2x \pm x = 0$.

Thus $x = 0$.

So $z = x + iy = 0 + i.0 = 0$.

When $2x + \sqrt{x^2 + y^2} = 0$, (1) holds for all x.

$2x + \sqrt{x^2 + y^2} = 0$ gives $2x = -\sqrt{x^2 + y^2} \quad \dots (3)$

or $4x^2 = x^2 + y^2$

or $y^2 = 3x^2$; $\therefore$ $y = \pm\sqrt{3}x$

$\therefore$ $z = x + iy = x \pm i\sqrt{3}x = (1 \pm i\sqrt{3})x$

$$= -2 \cdot \frac{-1 \pm i\sqrt{3}}{2} \cdot x = -2wx, -2w^2x.$$

From (3), $x < 0$.

Hence, $z = 0, \lambda w, \lambda w^2$ where λ is a positive real number.

Second Method

Clearly $z = 0$ is a solution. When $z \neq 0$,

$z^2 + z|z| + |z \cdot z| = 0$

or $z^2 + z|z| + |z|^2 = 0$

or $\left(\frac{z}{|z|}\right)^2 + \frac{z}{|z|} + 1 = 0$

$$\therefore \quad \frac{z}{|z|} = \frac{-1 \pm \sqrt{1 - 4}}{2} = \frac{-1 \pm \sqrt{3}i}{2} = w, w^2$$

$\therefore$ $z = |z|w, |z|w^2$.

This holds for all $z = \lambda w, \lambda w^2$ where λ is any non-negative real number.

13. Find z such that $\left|\frac{z - 12}{z - 8i}\right| = \frac{5}{3}$ and $\left|\frac{z - 4}{z - 8}\right| = 1$.

Let $z = x + iy$; putting this in the equations,

$3|z - 12| = 5|z - 8i| \quad \dots (1)$

and $|z - 4| = |z - 8| \quad \dots (2)$

(1) $\Rightarrow$ $9|(x - 12) + iy|^2 = 25|x + i(y - 8)|^2$

or $9\{(x - 12)^2 + y^2\} = 25\{x^2 + (y - 8)^2\} \quad \dots (3)$

(2) $\Rightarrow$ $|(x - 4) + iy|^2 = |(x - 8)^2 + iy|^2$

or $(x-4)^2+y^2=(x-8)^2+y^2$ ... (4)

Solving (4),

$$x^2-8x+16=x^2-16x+64$$

or $8x=48;\quad \therefore\quad x=6.$

Putting in (3),

$$9(36+y^2)=25\{36+(y-8)^2\}$$

or $9\times 36+9y^2=900+25(y^2-16y+64)$

or $16y^2-400y+2176=0$

or $y^2-25y+136=0$

or $(y-8)(y-17)=0;\quad \therefore\quad y=8, 17.$

$\therefore\quad z=x+iy=6+i\cdot 8, 6+i\cdot 17.$

14. For every real $a\geq 0$, find all the complex numbers z that satisfy the equation $2|z|-4az+1+ia=0$.

Let $z=x+iy$. Putting it for z in the equation we get,

$$2\sqrt{x^2+y^2}-4a(x+iy)+1+ia=0$$

or $\{2\sqrt{x^2+y^2}-4ax+1\}+i(-4ay+a)=0+i\cdot 0.$

Equating real and imaginary parts on both sides,

$$2\sqrt{x^2+y^2}-4ax+1=0 \quad ...(1)$$

$$-4ay+a=0 \quad ...(2)$$

(2) $\Rightarrow$ $a(-4y+1)=0$, i.e., $a=0$ or $y=\frac{1}{4}\cdot$

But $a\neq 0$ for otherwise the given equation becomes $2|z|+1=0$ which is not possible for any z.

When $y=\frac{1}{4}$, (1) gives

$$2\sqrt{x^2+\frac{1}{16}}-4ax+1=0 \quad ...(3)$$

or $2\sqrt{x^2+\frac{1}{16}}=4ax-1>0$ because LHS >0

or $4\left(x^2+\frac{1}{16}\right)=(4ax-1)^2$

$\therefore\quad 4x^2+\frac{1}{4}=16a^2x^2-8ax+1$

or $4(1-4a^2)x^2+8ax-\frac{3}{4}=0$

$$\therefore\quad x=\frac{-8a\pm\sqrt{64a^2+4\cdot\frac{3}{4}\cdot 4(1-4a^2)}}{8(1-4a^2)},$$

provided $4(1-4a^2)\neq 0$, i.e., $a\neq\frac{1}{2}$

($\because$ $a\geq 0$ is given)

$$=\frac{-8a\pm\sqrt{16a^2+12}}{8(1-4a^2)}=\frac{4a\mp\sqrt{4a^2+3}}{4(4a^2-1)}\cdot$$

From (3), x must be positive as $a>0$, for otherwise sum of positive terms will be zero (absurd).

If $0<a<\frac{1}{2}$ then $4a^2-1<0$.

So in order that x is positive we must have

$$4a\mp\sqrt{4a^2+3}<0.$$

Clearly $4a+\sqrt{4a^2+3}\nless 0$ when $0<a<\frac{1}{2}\cdot$

So $4a-\sqrt{4a^2+3}<0$. It is true only if

$$0<4a<\sqrt{4a^2+3}$$

or $16a^2<4a^2+3$

or $12a^2<3;\quad \therefore\quad a^2<\frac{1}{4}$, i.e., $a<\frac{1}{2}\cdot$

Again $4ax-1>0$, i.e., $4a\cdot\frac{4a-\sqrt{4a^2+3}}{4(4a^2-1)}>1$ if $a>\frac{1}{2}\cdot$

So, for $0<a<\frac{1}{2}$ the condition $4ax-1>0$ does not hold.

If $a>\frac{1}{2}$ then $4a^2-1>0$. So in order that x is positive we must have $4a\mp\sqrt{4a^2+3}>0$.

Obviously, $4a+\sqrt{4a^2+3}>0$.

For $a>\frac{1}{2}$ we find $4a-\sqrt{4a^2+3}>0$.

But $4ax-1>0$ is not satisfied.

Thus we find that

if $0\leq a\leq\frac{1}{2}$, there is no solution

if $a>\frac{1}{2}$, solution is $z=x+iy=\frac{4a+\sqrt{4a^2+3}}{4(4a^2-1)}+i\cdot\frac{1}{4}\cdot$

15. Let $z_1=10+6i$ and $z_2=4+6i$. If z be a complex number such that $\arg\frac{z-z_1}{z-z_2}=\frac{\pi}{4}$ then prove that $|z-7-9i|=3\sqrt{2}$.

Let $\left|\frac{z-z_1}{z-z_2}\right|=r$; then

$$\frac{z-z_1}{z-z_2}=r\left(\cos\frac{\pi}{4}+i\sin\frac{\pi}{4}\right)=\frac{r(1+i)}{\sqrt{2}}\cdot$$

By componendo and dividendo,

$$\frac{2z-z_1-z_2}{(z-z_1)-(z-z_2)}=\frac{r(1+i)+\sqrt{2}}{r(1+i)-\sqrt{2}}$$

or $\quad \dfrac{2z-(10+6i+4+6i)}{z_2-z_1}=\dfrac{r(1+i)+\sqrt{2}}{r(1+i)-\sqrt{2}}$

or $\quad \dfrac{2\{z-7-6i\}}{4+6i-(10+6i)}=\dfrac{r(1+i)+\sqrt{2}}{r(1+i)-\sqrt{2}}$

or $\quad \dfrac{z-7-6i}{-3}=\dfrac{r(1+i)+\sqrt{2}}{r(1+i)-\sqrt{2}}$

or $\quad z-7-6i=\dfrac{-3r(1+i)-3\sqrt{2}}{r(1+i)-\sqrt{2}}$

$\therefore \quad z-7-9i=\dfrac{-3r(1+i)-3\sqrt{2}}{r(1+i)-\sqrt{2}}-3i$

$$=\frac{-3r(1+i)-3\sqrt{2}-3ir(1+i)+3\sqrt{2}i}{r(1+i)-\sqrt{2}}$$

$$=\frac{-3r-3ri-3\sqrt{2}-3ri+3r+3\sqrt{2}i}{r(1+i)-\sqrt{2}}$$

$$=\frac{-6ri-3\sqrt{2}+3\sqrt{2}i}{(r-\sqrt{2})+ir}=\frac{-3\sqrt{2}-3i(2r-\sqrt{2})}{(r-\sqrt{2})+ir}$$

$\therefore \quad |z-7-9i|=\left|\dfrac{-3\sqrt{2}-3i(2r-\sqrt{2})}{(r-\sqrt{2})+ir}\right|$

$$=\frac{|-3\sqrt{2}-3i(2r-\sqrt{2})|}{|(r-\sqrt{2})+ir|}$$

$$=\frac{\sqrt{18+9(2r-\sqrt{2})^2}}{\sqrt{(r-\sqrt{2})^2+r^2}}$$

$$=\frac{\sqrt{18+36r^2-36\sqrt{2}r+18}}{\sqrt{2r^2-2\sqrt{2}r+2}}$$

$$=\frac{\sqrt{18}\cdot\sqrt{2r^2-2\sqrt{2}r+2}}{\sqrt{2r^2-2\sqrt{2}r+2}}$$

$$=\sqrt{18}=3\sqrt{2}.$$

16. Let z_1, z_2 be two complex numbers such that $\dfrac{z_1-2z_2}{2-z_1\bar{z}_2}$ is unimodular. If z_2 is not unimodular then find $|z_1|$.

Here $\left|\dfrac{z_1-2z_2}{2-z_1\bar{z}_2}\right|=1$

or $\quad |z_1-2z_2|=|2-z_1\bar{z}_2| \qquad \left\{\because \left|\dfrac{z_1}{z_2}\right|=\dfrac{|z_1|}{|z_2|}\right\}$

or $\quad |z_1-2z_2|^2=|2-z_1\bar{z}_2|^2;$

or $\quad (z_1-2z_2)\overline{(z_1-2z_2)}=(2-z_1\bar{z}_2)\overline{(2-z_1\bar{z}_2)}$

$\{\because \ |z|^2=z\bar{z}\}$

or $\quad (z_1-2z_2)(\bar{z}_1-2\bar{z}_2)=(2-z_1\bar{z}_2)(2-\bar{z}_1z_2)$

$\{\because \ \overline{z_1\pm z_2}=\bar{z}_1\pm\bar{z}_2;\ \overline{z_1z_2}=\bar{z}_1\cdot\bar{z}_2 \text{ and } \bar{\bar{z}}=z\}$

or $\quad z_1\bar{z}_1-2z_1\bar{z}_2-2z_2\bar{z}_1+4z_2\bar{z}_2$

$$=4-2\bar{z}_1z_2-2z_1\bar{z}_2+z_1\bar{z}_2\bar{z}_1z_2$$

or $\quad |z_1|^2+4|z_2|^2=4+|z_1|^2|z_2|^2$

or $\quad |z_1|^2-|z_1|^2|z_2|^2+4|z_2|^2-4=0$

or $\quad |z_1|^2(1-|z_2|^2)-4(1-|z_2|^2)=0$

or $\quad (1-|z_2|^2)(|z_1|^2-4)=0.$

But $|z_2|\neq 1$ (given); so $|z_1|^2=4; \quad \therefore \quad |z_1|=2.$

17. If $u=\sqrt{z_1z_2}$, prove that

$$|z_1|+|z_2|=\left|\frac{z_1+z_2}{2}+u\right|+\left|\frac{z_1+z_2}{2}-u\right|.$$

$$\text{RHS}=\left|\frac{z_1+z_2}{2}+\sqrt{z_1z_2}\right|+\left|\frac{z_1+z_2}{2}-\sqrt{z_1z_2}\right|$$

$$=\frac{1}{2}|z_1+z_2+2\sqrt{z_1z_2}|+\frac{1}{2}|z_1+z_2-2\sqrt{z_1z_2}|$$

$$=\frac{1}{2}|(\sqrt{z_1}+\sqrt{z_2})^2|+\frac{1}{2}|(\sqrt{z_1}-\sqrt{z_2})^2|$$

$$=\frac{1}{2}|\sqrt{z_1}+\sqrt{z_2}|^2+\frac{1}{2}|\sqrt{z_1}-\sqrt{z_2}|^2$$

$\{\because \ |z^2|=|z\cdot z|=|z||z|=|z|^2\}.$

Now, for any two complex numbers α, β we get

$$|\alpha+\beta|^2+|\alpha-\beta|^2$$

$=(\alpha+\beta)\overline{(\alpha+\beta)}+(\alpha-\beta)\overline{(\alpha-\beta)} \quad \{\because \ |z|^2=z\bar{z}\}$

$=(\alpha+\beta)(\bar{\alpha}+\bar{\beta})+(\alpha-\beta)(\bar{\alpha}-\bar{\beta})$

$\{\because \ \overline{z_1\pm z_2}=\bar{z}_1\pm\bar{z}_2\}$

$$=\alpha\bar{\alpha}+\alpha\bar{\beta}+\beta\bar{\alpha}+\beta\bar{\beta}+\alpha\bar{\alpha}-\alpha\bar{\beta}-\beta\bar{\alpha}+\beta\bar{\beta}$$

$$=2|\alpha|^2+2|\beta|^2$$

$\therefore \quad \dfrac{1}{2}|\alpha+\beta|^2+\dfrac{1}{2}|\alpha-\beta|^2=|\alpha|^2+|\beta|^2.$

So, $\quad \text{RHS}=|\sqrt{z_1}|^2+|\sqrt{z_2}|^2=|(\sqrt{z_1})^2|+|(\sqrt{z_2})^2|$

$$=|z_1|+|z_2|=\text{LHS}.$$

18. If $|z-3i|<\sqrt{5}$ then prove that the complex number z also satisfies the inequality $|i(z+1)+1|<2\sqrt{5}$.

TIP: Here $i(z+1)+1$ has to be written as the sum of two complex numbers containing $z-3i$ because we have to use $|z-3i|<\sqrt{5}$.

$|i(z+1)+1|$

$=|i(z-3i)+i-2|$

$\leq|i(z-3i)|+|-2+i|$ (by triangle inequality)

$=|i||z-3i|+\sqrt{5}$

$=1\cdot|z-3i|+\sqrt{5}<\sqrt{5}+\sqrt{5}=2\sqrt{5}.$

19. Prove that $|z-1| \le ||z|-1| + |z|\,|\arg z|$.

Using triangle inequality,

$$|z-1| = |(z-|z|) + (|z|-1)|$$

$$\le |z-|z|| + ||z|-1| \quad \dots (1)$$

If we can prove that $|z-|z|| \le |z|\,|\arg z|$, our problem will be solved.

Let $z = r(\cos\theta + i\sin\theta)$ where $|z| = r$, $\arg z = \theta$.

$$\therefore \quad |z-|z||^2 = |r(\cos\theta + i\sin\theta) - r|^2$$

$$= r^2\{(\cos\theta - 1)^2 + \sin^2\theta\}$$

$$= |z|^2(2 - 2\cos\theta)$$

$$= |z|^2 \cdot 4\sin^2\frac{\theta}{2} \le |z|^2 \cdot 4 \cdot \left(\frac{\theta}{2}\right)^2$$

$$\{\because |\sin\phi| \le |\phi|\}$$

$$\therefore \quad |z-|z||^2 \le |z|^2\theta^2$$

$$\therefore \quad |z-|z|| \le |z|\,|\theta| = |z|\,|\arg z|$$

$\therefore$ (1) gives $|z-1| \le ||z|-1| + |z|\,|\arg z|$.

20. Prove that if z_1, z_2 are two complex numbers and $c > 0$ then $|z_1+z_2|^2 \le (1+c)|z_1|^2 + (1+c^{-1})|z_2|^2$.

Here $(1+c)|z_1|^2 + \left(1+\frac{1}{c}\right)|z_2|^2 - |z_1+z_2|^2$

$$= (1+c)z_1\bar{z}_1 + \left(1+\frac{1}{c}\right)z_2\bar{z}_2 - (z_1+z_2)\overline{(z_1+z_2)},$$

$$\{\because |z|^2 = z\bar{z}\}$$

$$= (1+c)z_1\bar{z}_1 + \left(1+\frac{1}{c}\right)z_2\bar{z}_2 - (z_1+z_2)(\bar{z}_1+\bar{z}_2),$$

$$\{\because \overline{z_1+z_2} = \bar{z}_1 + \bar{z}_2\}$$

$$= cz_1\bar{z}_1 + \frac{1}{c}z_2\bar{z}_2 - (z_1\bar{z}_2 + z_2\bar{z}_1)$$

$$= \frac{1}{c}\{c^2z_1\bar{z}_1 + z_2\bar{z}_2 - cz_1\bar{z}_2 - cz_2\bar{z}_1\}$$

$$= \frac{1}{c}\{cz_1(c\bar{z}_1 - \bar{z}_2) - z_2(c\bar{z}_1 - \bar{z}_2)\}$$

$$= \frac{1}{c}(c\bar{z}_1 - \bar{z}_2)(cz_1 - z_2) = \frac{1}{c}\overline{(cz_1 - z_2)}(cz_1 - z_2)$$

$$= \frac{1}{c}|cz_1 - z_2|^2 > 0 \text{ as } c > 0$$

$$\therefore \quad (1+c)|z_1|^2 + \left(1+\frac{1}{c}\right)|z_2|^2 - |z_1+z_2|^2 > 0$$

or $\quad |z_1+z_2|^2 < (1+c)|z_1|^2 + (1+c^{-1})|z_2|^2$.

21. Prove that $|z_1+z_2| \ge \frac{1}{2}(|z_1|+|z_2|)\left|\frac{z_1}{|z_1|} + \frac{z_2}{|z_2|}\right|$.

Let $z_1 = r_1(\cos\theta_1 + i\sin\theta_1)$, $z_2 = r_2(\cos\theta_2 + i\sin\theta_2)$

where $|z_1| = r_1$, amp $z_1 = \theta_1$, etc.

Then $z_1 + z_2 = (r_1\cos\theta_1 + r_2\cos\theta_2) + i(r_1\sin\theta_1 + r_2\sin\theta_2)$

$$\therefore \quad |z_1+z_2|^2 = (r_1\cos\theta_1 + r_2\cos\theta_2)^2 + (r_1\sin\theta_1 + r_2\sin\theta_2)^2$$

$$= r_1^2 + r_2^2 + 2r_1r_2\cos(\theta_1 - \theta_2)$$

Now, $\frac{1}{2}(|z_1|+|z_2|)\left|\frac{z_1}{|z_1|} + \frac{z_2}{|z_2|}\right|$

$$= \frac{1}{2}(r_1+r_2)\,|\cos\theta_1 + i\sin\theta_1 + \cos\theta_2 + i\sin\theta_2|$$

$$\therefore \quad \frac{1}{4}(|z_1|+|z_2|)^2\left|\frac{z_1}{|z_1|} + \frac{z_2}{|z_2|}\right|^2$$

$$= \frac{1}{4}(r_1+r_2)^2\{(\cos\theta_1 + \cos\theta_2)^2 + (\sin\theta_1 + \sin\theta_2)^2\}$$

$$= \frac{1}{4}(r_1+r_2)^2\{2 + 2\cos(\theta_1 - \theta_2)\}$$

$$\therefore \quad |z_1+z_2|^2 - \frac{1}{4}(|z_1|+|z_2|)^2\left|\frac{z_1}{|z_1|} + \frac{z_2}{|z_2|}\right|^2$$

$$= r_1^2 + r_2^2 + 2r_1r_2\cos(\theta_1 - \theta_2) - \frac{1}{4}(r_1+r_2)^2\{2 + 2\cos(\theta_1 - \theta_2)\}$$

$$= r_1^2 + r_2^2 - \frac{1}{2}(r_1+r_2)^2 + \cos(\theta_1 - \theta_2)\{2r_1r_2 - \frac{1}{2}(r_1+r_2)^2\}$$

$$= \frac{1}{2}(r_1-r_2)^2 - \cos(\theta_1 - \theta_2)\frac{1}{2}(r_1-r_2)^2$$

$$= \frac{1}{2}(r_1-r_2)^2 \cdot \{1 - \cos(\theta_1 - \theta_2)\}$$

$$= \frac{1}{2}(r_1-r_2)^2 \cdot 2\sin^2\frac{\theta_1 - \theta_2}{2} \ge 0$$

$$\therefore \quad |z_1+z_2|^2 \ge \frac{1}{4}(|z_1|+|z_2|)^2\left|\frac{z_1}{|z_1|} + \frac{z_2}{|z_2|}\right|^2$$

$$\therefore \quad |z_1+z_2| \ge \frac{1}{2}(|z_1|+|z_2|)\left|\frac{z_1}{|z_1|} + \frac{z_2}{|z_2|}\right|.$$

22. If $|z| \le 1$, $|w| \le 1$ then show that $|z-w|^2 \le (|z| - |w|)^2 + (\arg z - \arg w)^2$.

Let $z = r_1(\cos\theta_1 + i\sin\theta_1)$

where $r_1 = |z| < 1$, $\theta_1 = \arg z$

$w = r_2(\cos\theta_2 + i\sin\theta_2)$

where $r_2 = |w| \le 1$, $\theta_2 = \arg w$.

Now, $|z-w|^2-(|z|-|w|)^2-(\arg z-\arg w)^2$

$$= |r_1(\cos\theta_1+i\sin\theta_1)-r_2(\cos\theta_2+i\sin\theta_2)|^2 - (r_1-r_2)^2-(\theta_1-\theta_2)^2$$

$$= (r_1\cos\theta_1-r_2\cos\theta_2)^2+(r_1\sin\theta_1-r_2\sin\theta_2)^2 - (r_1-r_2)^2-(\theta_1-\theta_2)^2$$

$$= r_1^2+r_2^2-2r_1r_2\cos(\theta_1-\theta_2)-(r_1-r_2)^2-(\theta_1-\theta_2)^2$$

$$= 2r_1r_2-2r_1r_2\cos(\theta_1-\theta_2)-(\theta_1-\theta_2)^2$$

$$= 2r_1r_2\{1-\cos(\theta_1-\theta_2)\}-(\theta_1-\theta_2)^2$$

$$= 2r_1r_2\cdot 2\sin^2\frac{\theta_1-\theta_2}{2}-4\cdot\left(\frac{\theta_1-\theta_2}{2}\right)^2$$

$$\le 4\left\{\sin^2\frac{\theta_1-\theta_2}{2}-\left(\frac{\theta_1-\theta_2}{2}\right)^2\right\}$$

because $|r_1|\le 1, |r_2|\le 1$.

But $|\sin\theta|\le|\theta|$; $\therefore\ \sin^2\theta\le\theta^2$

$\therefore\quad \sin^2\frac{\theta_1-\theta_2}{2}\le\left(\frac{\theta_1-\theta_2}{2}\right)^2$

$\therefore\quad |z-w|^2-(|z|-|w|)^2-(\arg z-\arg w)^2\le 0$

$\therefore\quad |z-w|^2\le(|z|-|w|)^2+(\arg z-\arg w)^2.$

23. Show that all the roots of the equation

$$z^n\cos\theta_0+z^{n-1}\cos\theta_1+z^{n-2}\cos\theta_2+\ldots + z\cos\theta_{n-1}+\cos\theta_n=2$$

lie outside the circle $|z|=\frac{1}{2}$ where θ_0, θ_1, etc., are real.

Here $|z^n\cos\theta_0+z^{n-1}\cos\theta_1+z^{n-2}\cos\theta_2+\ldots + z\cos\theta_{n-1}+\cos\theta_n|=2 \quad \ldots (1)$

By triangle inequality,

$$2=|z^n\cos\theta_0+z^{n-1}\cos\theta_1+z^{n-2}\cos\theta_2+\ldots+z\cos\theta_{n-1}+\cos\theta_n|$$

$$\le|z^n\cos\theta_0|+|z^{n-1}\cos\theta_1|+|z^{n-2}\cos\theta_2|+\ldots+|z\cos\theta_{n-1}|+|\cos\theta_n|$$

$$=|z^n||\cos\theta_0|+|z^{n-1}||\cos\theta_1|+\ldots+|z||\cos\theta_{n-1}|+|\cos\theta_n|$$

$$\le|z|^n+|z|^{n-1}+\ldots+|z|+1$$

$\{\because\ |\cos\theta_i|\le 1$ and $|z^m|=|z|^m\}$

$$=\frac{1-|z|^{n+1}}{1-|z|}<\frac{1}{1-|z|}$$

$\therefore\quad 2<\frac{1}{1-|z|}.$

So $1-|z|$ is positive and $1-|z|<\frac{1}{2}$

$\therefore\quad |z|>1-\frac{1}{2}=\frac{1}{2}$

$\therefore\quad$ all z satisfying (1) lie outside the circle $|z|=\frac{1}{2}$.

24. Show that the condition for one of the roots of the equation $z^2+(p+ip')z+q+iq'=0$ to be real is $q'^2-pp'q'+qp'^2=0$ where p, q, p', q' are real. Also find the condition for both roots to be equal.

Let c be a real root of the equation. Then

$$c^2+(p+ip')c+q+iq'=0.$$

Equating real and imaginary parts,

$$c^2+pc+q=0 \qquad \ldots (1)$$

$$p'c+q'=0 \qquad \ldots (2)$$

From (2), $c=-\frac{q'}{p'}$. Putting in (1),

$$\left(-\frac{q'}{p'}\right)^2+p\cdot\left(-\frac{q'}{p'}\right)+q=0$$

or $\quad \frac{q'^2}{p'^2}-\frac{pq'}{p'}+q=0$

or $\quad q'^2-pp'q'+qp'^2=0.$

Let the equation have equal roots, each being $\alpha+i\beta$.

$\therefore\quad z^2+(p+ip')z+q+iq'$

$$=\{z-(\alpha+i\beta)\}\{z-(\alpha+i\beta)\}$$

$$=z^2-(\alpha+i\beta+\alpha+i\beta)z+(\alpha+i\beta)^2$$

$\therefore\quad p+ip'=-2(\alpha+i\beta) \qquad \ldots (1)$

and $\quad q+iq'=(\alpha+i\beta)^2=\alpha^2-\beta^2+2i\alpha\beta \qquad \ldots (2)$

(1) $\Rightarrow\quad p=-2\alpha, p'=-2\beta$

and (2) $\Rightarrow\quad q=\alpha^2-\beta^2, q'=2\alpha\beta.$

Eliminating α, β we get

$$q=\left(-\frac{p}{2}\right)^2-\left(-\frac{p'}{2}\right)^2=\frac{1}{4}(p^2-p'^2)$$

or $\quad p^2-p'^2=4q \qquad \ldots (3)$

and $\quad q'=2\left(-\frac{p}{2}\right)\left(\frac{-p'}{2}\right)=\frac{1}{2}pp'$ or $pp'=2q' \qquad \ldots (4)$

(3) and (4) give the required conditions.

25. If $\log_{1/\sqrt{3}} \dfrac{|z|^2 - |z| + 1}{2 + |z|} > -2$, show that $|z| < 5$.

Here $\log_{1/\sqrt{3}} \dfrac{|z|^2 - |z| + 1}{2 + |z|} > \log_{1/\sqrt{3}} \left(\dfrac{1}{\sqrt{3}}\right)^{-2}$

$$\Rightarrow \quad \frac{|z|^2 - |z| + 1}{2 + |z|} < \left(\frac{1}{\sqrt{3}}\right)^{-2}$$

$$\{\because \ \log_a x > \log_a y \ \Rightarrow \ x < y \ \text{ if } \ 0 < a < 1\}$$

$$\Rightarrow \quad \frac{|z|^2 - |z| + 1}{2 + |z|} < 3$$

or $\quad |z|^2 - |z| + 1 < 3(2 + |z|) \quad \{\because \ 2 + |z| > 0\}$

$\Rightarrow \quad |z|^2 - 4|z| - 5 < 0$

or $\quad (|z| - 5)(|z| + 1) < 0.$

But $|z| + 1$ is $+ve$; so $|z| - 5 < 0$

$\therefore \quad |z| < 5.$

26. If $\left|z + \dfrac{1}{z}\right| = a$ where z is a complex number and a is a real number, find the least and the greatest values of $|z|$. Also find a for which the greatest and the least values of $|z|$ are equal.

Here $a \geq 0$ because $\left|z + \dfrac{1}{z}\right| \geq 0.$

Let $z = r(\cos\theta + i\sin\theta)$ where $|z| = r$ and $\arg z = \theta$.

$$\therefore \quad \frac{1}{z} = \frac{1}{r(\cos\theta + i\sin\theta)} = \frac{1}{r}(\cos\theta + i\sin\theta)^{-1}$$

$$= \frac{1}{r}(\cos\theta - i\sin\theta).$$

$$\therefore \quad z + \frac{1}{z} = \left(r + \frac{1}{r}\right)\cos\theta + i\left(r - \frac{1}{r}\right)\sin\theta$$

$$\therefore \quad \left|z + \frac{1}{z}\right|^2 = \left(r + \frac{1}{r}\right)^2 \cos^2\theta + \left(r - \frac{1}{r}\right)^2 \sin^2\theta$$

$$= r^2 + \frac{1}{r^2} + 2(\cos^2\theta - \sin^2\theta)$$

$$= r^2 + \frac{1}{r^2} + 2\cos 2\theta$$

or $\quad a^2 = r^2 + \dfrac{1}{r^2} + 2 - 4\sin^2\theta$

or $\quad \left(r + \dfrac{1}{r}\right)^2 = a^2 + 4\sin^2\theta \leq a^2 + 4$

or $\quad r + \dfrac{1}{r} \leq \sqrt{a^2 + 4}$

or $\quad r^2 - \sqrt{a^2 + 4} \cdot r + 1 \leq 0$

or $\quad r^2 - 2 \cdot \dfrac{\sqrt{a^2 + 4}}{2} \cdot r + \dfrac{a^2 + 4}{4} + 1 - \dfrac{a^2 + 4}{4} \leq 0$

or $\quad \left(r - \dfrac{\sqrt{a^2 + 4}}{2}\right)^2 - \left(\dfrac{a}{2}\right)^2 \leq 0$

or $\quad \left(r - \dfrac{\sqrt{a^2 + 4} - a}{2}\right)\left(r - \dfrac{\sqrt{a^2 + 4} + a}{2}\right) \leq 0$

$\therefore \quad r$ lies between $\dfrac{\sqrt{a^2 + 4} - a}{2}$ and $\dfrac{\sqrt{a^2 + 4} + a}{2}$,

i.e., $\dfrac{\sqrt{a^2 + 4} - a}{2} \leq r \leq \dfrac{\sqrt{a^2 + 4} + a}{2}$ for all real positive a

$\therefore \quad$ maximum value of $r = |z|_{max} = \dfrac{\sqrt{a^2 + 4} + a}{2}$

minimum value of $r = |z|_{min} = \dfrac{\sqrt{a^2 + 4} - a}{2}$.

This is true for all real $a \geq 0$.

$|z|_{max} = |z|_{min}$ if

$$\frac{\sqrt{a^2 + 4} + a}{2} = \frac{\sqrt{a^2 + 4} - a}{2}, \text{ i.e., } a = 0.$$

Then $|z|_{max} = |z|_{min} = \dfrac{\sqrt{4}}{2} = 1.$

27. Locate the region in the Argand plane for z satisfying $|z + i| = |z - 2|$.

Let $z = x + iy$. Then

$$|x + i(y + 1)| = |(x - 2) + iy|$$

or $\quad |x + i(y + 1)|^2 = |(x - 2) + iy|^2$

or $\quad x^2 + (y + 1)^2 = (x - 2)^2 + y^2$

or $\quad 2y + 1 = -4x + 4$

or $\quad 4x + 2y = 3.$

Hence, the points on the line $4x + 2y = 3$ in Argand plane constitute the required region.

28. Locate the complex number z such that

$$\log_{\cos\frac{\pi}{6}} \frac{|z - 2| + 5}{4|z - 2| - 4} < 2.$$

Here $\log_{\frac{\sqrt{3}}{2}} \dfrac{|z - 2| + 5}{4|z - 2| - 4} < \log_{\frac{\sqrt{3}}{2}} \left(\dfrac{\sqrt{3}}{2}\right)^2$

or $\quad \dfrac{|z - 2| + 5}{4|z - 2| - 4} > \dfrac{3}{4} \left(\because \ \text{base } \dfrac{\sqrt{3}}{2} < 1\right).$

Also for the logarithm to be defined,

$$4|z - 2| - 4 > 0, \ \text{ i.e., } \ |z - 2| > 1.$$

When $|z - 2| > 1$, $\dfrac{|z - 2| + 5}{4|z - 2| - 4} > \dfrac{3}{4}$ gives

$$4(|z - 2| + 5) > 3 \cdot 4(|z - 2| - 1)$$

or $\quad |z-2|+5>3|z-2|-3$

or $\quad 8>2|z-2|; \quad \therefore \quad |z-2|<4.$

$\therefore$ z is such that $1<|z-2|<4$.

Hence the region for z in the Argand plane is as shown below by the shaded area:

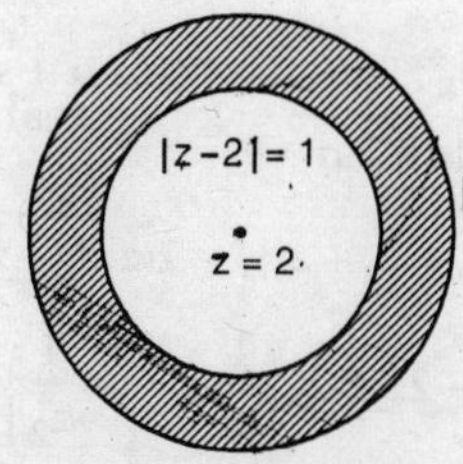

This is the area shaded between two circles with a common centre $z=2$ and radii 1 and 4.

29. Find the fourth roots of $-16i$.

Here $-16i = 16(0-i)$

$$\therefore \quad (-16i)^{1/4} = \{16(0-i)\}^{1/4}$$

$$= 16^{1/4}\left\{\cos\left(-\frac{\pi}{2}\right) + i\sin\left(-\frac{\pi}{2}\right)\right\}^{1/4}$$

$$= 2\left\{\cos\left(2r\pi - \frac{\pi}{2}\right) + i\sin\left(2r\pi - \frac{\pi}{2}\right)\right\}^{1/4}$$

$$= 2\left\{\cos\left(2r\pi - \frac{\pi}{2}\right)\frac{1}{4} + i\sin\left(2r\pi - \frac{\pi}{2}\right)\frac{1}{4}\right\}$$

where $r = 0, 1, 2, 3$

$$= 2\left\{\cos\frac{(4r-1)\pi}{8} + i\sin\frac{(4r-1)\pi}{8}\right\}$$

where $r = 0, 1, 2, 3.$

30. Find the roots common to the equations
$$x^5 - x^3 + x^2 - 1 = 0$$
$$x^4 = 1.$$

The first equation is $x^3(x^2-1) + 1(x^2-1) = 0$

or $\quad (x^3+1)(x^2-1) = 0$

$\therefore \quad x^3+1=0 \quad$ or $\quad x^2-1=0.$

If $x^3+1=0$ then

$$x = (-1)^{1/3} = (-1 + i\cdot 0)^{1/3} = (\cos\pi + i\sin\pi)^{1/3}$$

$$= \{\cos(2\pi r + \pi) + i\sin(2\pi r + \pi)\}^{1/3}$$

$$= \cos(2r+1)\frac{\pi}{3} + i\sin(2r+1)\frac{\pi}{3}, \text{ where } r = 0, 1, 2.$$

If $x^2-1=0$ then $x^2=1; \quad \therefore \quad x = \sqrt{1} = \pm 1.$

$\therefore \quad$ the roots of the first equation are

$$1, -1, \cos\frac{\pi}{3} + i\sin\frac{\pi}{3},$$

$$\cos\pi + i\sin\pi, \cos\frac{5\pi}{3} + i\sin\frac{5\pi}{3} \quad \ldots (1)$$

The second equation is $x^4 = 1$

or $\quad x = 1^{1/4}$

$$\therefore \quad x = (\cos 0 + i\sin 0)^{1/4} = (\cos 2\pi r + i\sin 2\pi r)^{1/4}$$

$$= \cos 2\pi r\cdot\frac{1}{4} + i\sin 2\pi r\cdot\frac{1}{4} \quad \text{where } r = 0, 1, 2, 3$$

$$= \cos\frac{\pi r}{2} + i\sin\frac{\pi r}{2} \quad \text{where } r = 0, 1, 2, 3$$

$$= 1, i, -1, -i \quad \ldots (2)$$

The roots common to (1) and (2) are $1, -1$.

TIP We could have solved one of the equations, say the second one, and tried the roots one by one in the other equation. Thus we could detect which one satisfies both the equations.

31. Find the integral solutions for x: $(1-i)^x = 2^x$.

Here $\left(\dfrac{1-i}{2}\right)^x = 1; \quad \therefore \quad \left(\dfrac{1}{2} - \dfrac{1}{2}i\right)^x = 1$

or $\quad \left\{\dfrac{1}{\sqrt{2}}\left(\cos\dfrac{-\pi}{4} + i\sin\dfrac{-\pi}{4}\right)\right\}^x = 1.$

As x is an integer, by De Moivre's theorem,

$$\left(\frac{1}{2}\right)^{\frac{x}{2}}\left(\cos\frac{-\pi x}{4} + i\sin\frac{-\pi x}{4}\right) = 1 + i0.$$

Equating real and imaginary parts,

$$\left(\frac{1}{2}\right)^{\frac{x}{2}}\cdot\cos\frac{-\pi x}{4} = 1 \quad \ldots (1)$$

and $\quad \left(\dfrac{1}{2}\right)^{\frac{x}{2}}\cdot\sin\dfrac{-\pi x}{4} = 0 \quad \ldots (2)$

From (2), $\sin\dfrac{\pi x}{4} = 0 \quad \Rightarrow \quad \dfrac{\pi x}{4} = n\pi; \quad \therefore \quad x = 4n.$

Putting in (1), $\left(\dfrac{1}{2}\right)^{2n}\cos n\pi = 1$

or $\quad \cos n\pi = 2^{2n}$; but $\cos n\pi = 1, -1$ and $2^{2n} \neq -1$

$\therefore \quad 1 = 2^{2n}; \quad \therefore \quad n = 0$

$\therefore \quad x = 4\times 0 = 0.$

So, 0 is the only integral solution.

Second Method

Taking modulus of both sides,

$$|(1-i)^x| = |2^x|.$$

$$|1-i|^x = |2|^x \quad \{\because |z|^n = |z^n|\}$$

or $(\sqrt{2})^x = 2^x$

or $2^{x/2}(2^{x/2}-1)=0$

$\Rightarrow$ $2^{x/2}-1=0 \quad (\because 2^{x/2}\neq 0)$

$\therefore$ $2^{x/2}=1; \quad \therefore \quad x=0.$

32. If $z+\dfrac{1}{z}=2\cos\theta$, prove that $\left|\dfrac{z^{2n}-1}{z^{2n}+1}\right| = |\tan n\theta|$.

Here $z+\dfrac{1}{z}=2\cos\theta$; $\quad\therefore\quad z^2-2\cos\theta\cdot z+1=0$

$$\therefore z=\frac{2\cos\theta\pm\sqrt{4\cos^2\theta-4}}{2}=\cos\theta\pm i\sin\theta.$$

Taking positive sign, $z=\cos\theta+i\sin\theta$

$$\therefore \quad \frac{1}{z}=(\cos\theta+i\sin\theta)^{-1}=\cos\theta-i\sin\theta$$

$$\therefore \quad \frac{z^{2n}-1}{z^{2n}+1}=\frac{z^n-\dfrac{1}{z^n}}{z^n+\dfrac{1}{z^n}}$$

$$=\frac{(\cos\theta+i\sin\theta)^n-(\cos\theta-i\sin\theta)^n}{(\cos\theta+i\sin\theta)^n+(\cos\theta-i\sin\theta)^n}$$

$$=\frac{\cos n\theta+i\sin n\theta-(\cos n\theta-i\sin n\theta)}{\cos n\theta+i\sin n\theta+(\cos n\theta-i\sin n\theta)}$$

$$=\frac{2i\sin n\theta}{2\cos n\theta}=i\tan n\theta.$$

Taking negative sign, similarly we get

$$\frac{z^{2n}-1}{z^{2n}+1}=\frac{-2i\sin n\theta}{2\cos n\theta}=-i\tan n\theta$$

$$\therefore \quad \left|\frac{z^{2n}-1}{z^{2n}+1}\right| = |\pm i\tan n\theta| = |\tan n\theta|, \text{ for } |\pm i|=1.$$

33. If $\cos\alpha+\cos\beta+\cos\gamma=0$, $\sin\alpha+\sin\beta+\sin\gamma=0$ then prove that

(i) $\cos 2\alpha+\cos 2\beta+\cos 2\gamma=0$

(ii) $\sin 3\alpha+\sin 3\beta+\sin 3\gamma=3\sin(\alpha+\beta+\gamma)$.

Let $z_1=\cos\alpha+i\sin\alpha$, $z_2=\cos\beta+i\sin\beta$,

$z_3=\cos\gamma+i\sin\gamma$.

$$\therefore \quad z_1+z_2+z_3=(\cos\alpha+\cos\beta+\cos\gamma) + i(\sin\alpha+\sin\beta+\sin\gamma)$$

$$=0+i\cdot 0=0 \quad \ldots (1)$$

(i) Also $\dfrac{1}{z_1}=(\cos\alpha+i\sin\alpha)^{-1}=\cos\alpha-i\sin\alpha$

$$\frac{1}{z_2}=\cos\beta-i\sin\beta, \quad \frac{1}{z_3}=\cos\gamma-i\sin\gamma$$

$$\therefore \quad \frac{1}{z_1}+\frac{1}{z_2}+\frac{1}{z_3}=(\cos\alpha+\cos\beta+\cos\gamma) - i(\sin\alpha+\sin\beta+\sin\gamma)$$

$$=0-i\cdot 0=0 \quad \ldots (2)$$

Now $z_1^2+z_2^2+z_3^2=(z_1+z_2+z_3)^2-2(z_1z_2+z_2z_3+z_3z_1)$

$$=0-2z_1z_2z_3\left(\frac{1}{z_3}+\frac{1}{z_1}+\frac{1}{z_2}\right)$$

$$=0-2z_1z_2z_3\cdot 0=0, \text{ using (1) and (2)}$$

or $(\cos\alpha+i\sin\alpha)^2+(\cos\beta+i\sin\beta)^2 + (\cos\gamma+i\sin\gamma)^2=0$

or $\cos 2\alpha+i\sin 2\alpha+\cos 2\beta+i\sin 2\beta + \cos 2\gamma+i\sin 2\gamma=0+i\cdot 0.$

Equating real parts on both sides,

$$\cos 2\alpha+\cos 2\beta+\cos 2\gamma=0.$$

(ii) $z_1^3+z_2^3+z_3^3=(z_1+z_2)^3-3z_1z_2(z_1+z_2)+z_3^3$

$$=(-z_3)^3-3z_1z_2(-z_3)+z_3^3, \text{ using (1)}$$

$$=3z_1z_2z_3$$

$\therefore$ $(\cos\alpha+i\sin\alpha)^3+(\cos\beta+i\sin\beta)^3 + (\cos\gamma+i\sin\gamma)^3$

$$=3(\cos\alpha+i\sin\alpha)(\cos\beta+i\sin\beta)(\cos\gamma+i\sin\gamma)$$

or $\cos 3\alpha+i\sin 3\alpha+\cos 3\beta+i\sin 3\beta+\cos 3\gamma+i\sin 3\gamma$

$$=3\{\cos(\alpha+\beta+\gamma)+i\sin(\alpha+\beta+\gamma)\}.$$

Equating imaginary parts on both sides,

$$\sin 3\alpha+\sin 3\beta+\sin 3\gamma=3\sin(\alpha+\beta+\gamma).$$

34. Solve $z^8+1=0$ and deduce that

(i) $\cos 4\theta=8\left(\cos\theta-\cos\dfrac{\pi}{8}\right)\left(\cos\theta-\cos\dfrac{3\pi}{8}\right) \times\left(\cos\theta-\cos\dfrac{5\pi}{8}\right)\left(\cos\theta-\cos\dfrac{7\pi}{8}\right)$

(ii) $\cos\dfrac{\pi}{8}\cdot\cos\dfrac{3\pi}{8}\cdot\cos\dfrac{5\pi}{8}\cdot\cos\dfrac{7\pi}{8}=\dfrac{1}{8}$

Here $z^8+1=0$; $\quad\therefore\quad z^8=-1$

or $z=(-1)^{\frac{1}{8}}$

$$\therefore \quad z=(-1+0\cdot i)^{\frac{1}{8}}=(\cos\pi+i\sin\pi)^{\frac{1}{8}}$$

$$=\{\cos(2r\pi+\pi)+i\sin(2r\pi+\pi)\}^{\frac{1}{8}}$$

$$= \cos(2r+1)\cdot\frac{\pi}{8} + i\sin(2r+1)\cdot\frac{\pi}{8}$$

where $r = 0, 1, 2, 3, \ldots; 7$

$$= \cos\frac{\pi}{8} \pm i\sin\frac{\pi}{8}, \cos\frac{3\pi}{8} \pm i\sin\frac{3\pi}{8},$$

$$\cos\frac{5\pi}{8} \pm i\sin\frac{5\pi}{8}, \cos\frac{7\pi}{8} \pm i\sin\frac{7\pi}{8}.$$

Now, using the relation between roots and factors,

$$z^8 + 1 = \left\{z - \left(\cos\frac{\pi}{8} + i\sin\frac{\pi}{8}\right)\right\}\left\{z - \left(\cos\frac{\pi}{8} - i\sin\frac{\pi}{8}\right)\right\}$$

$$\times\left\{z - \left(\cos\frac{3\pi}{8} + i\sin\frac{3\pi}{8}\right)\right\}\left\{z - \left(\cos\frac{3\pi}{8} - i\sin\frac{3\pi}{8}\right)\right\}$$

$$\times\left\{z - \left(\cos\frac{5\pi}{8} + i\sin\frac{5\pi}{8}\right)\right\}\left\{z - \left(\cos\frac{5\pi}{8} - i\sin\frac{5\pi}{8}\right)\right\}$$

$$\times\left\{z - \left(\cos\frac{7\pi}{8} + i\sin\frac{7\pi}{8}\right)\right\}\left\{z - \left(\cos\frac{7\pi}{8} - i\sin\frac{7\pi}{8}\right)\right\}$$

$$= \left\{\left(z - \cos\frac{\pi}{8}\right)^2 + \sin^2\frac{\pi}{8}\right\}\left\{\left(z - \cos\frac{3\pi}{8}\right)^2 + \sin^2\frac{3\pi}{8}\right\}$$

$$\times\left\{\left(z - \cos\frac{5\pi}{8}\right)^2 + \sin^2\frac{5\pi}{8}\right\}$$

$$\times\left\{\left(z - \cos\frac{7\pi}{8}\right)^2 + \sin^2\frac{7\pi}{8}\right\}$$

$$= \left(z^2 + 1 - 2z\cos\frac{\pi}{8}\right)\left(z^2 + 1 - 2z\cos\frac{3\pi}{8}\right)$$

$$\times\left(z^2 + 1 - 2z\cos\frac{5\pi}{8}\right)\left(z^2 + 1 - 2z\cos\frac{7\pi}{8}\right) \quad \ldots (1)$$

(i) Dividing both sides by z^4,

$$z^4 + \frac{1}{z^4} = \left(z + \frac{1}{z} - 2\cos\frac{\pi}{8}\right)\left(z + \frac{1}{z} - 2\cos\frac{3\pi}{8}\right)$$

$$\left(z + \frac{1}{z} - 2\cos\frac{5\pi}{8}\right)\left(z + \frac{1}{z} - 2\cos\frac{7\pi}{8}\right) \quad \ldots (2)$$

Let $z = \cos\theta + i\sin\theta$; then

$$\frac{1}{z} = (\cos\theta + i\sin\theta)^{-1} = \cos\theta - i\sin\theta$$

$$\therefore \quad z^4 = (\cos\theta + i\sin\theta)^4 = \cos 4\theta + i\sin 4\theta$$

$$\frac{1}{z^4} = (\cos\theta + i\sin\theta)^{-4} = \cos 4\theta - i\sin 4\theta$$

$$\therefore \quad z + \frac{1}{z} = 2\cos\theta \quad \text{and} \quad z^4 + \frac{1}{z^4} = 2\cos 4\theta.$$

$\therefore$ from (2),

$$2\cos 4\theta = \left(2\cos\theta - 2\cos\frac{\pi}{8}\right)\left(2\cos\theta - 2\cos\frac{3\pi}{8}\right)$$

$$\times\left(2\cos\theta - 2\cos\frac{5\pi}{8}\right)\left(2\cos\theta - 2\cos\frac{7\pi}{8}\right)$$

$$\therefore \quad \cos 4\theta = 8\left(\cos\theta - \cos\frac{\pi}{8}\right)\left(\cos\theta - \cos\frac{3\pi}{8}\right)$$

$$\times\left(\cos\theta - \cos\frac{5\pi}{8}\right)\left(\cos\theta - \cos\frac{7\pi}{8}\right).$$

(ii) Put $z = i$ in (1). Then

$$i^8 + 1 = \left(i^2 + 1 - 2i\cos\frac{\pi}{8}\right)\left(i^2 + 1 - 2i\cos\frac{3\pi}{8}\right)$$

$$\times\left(i^2 + 1 - 2i\cos\frac{5\pi}{8}\right)\left(i^2 + 1 - 2i\cos\frac{7\pi}{8}\right)$$

or $$1 + 1 = \left(-2i\cos\frac{\pi}{8}\right)\left(-2i\cos\frac{3\pi}{8}\right)$$

$$\times\left(-2i\cos\frac{5\pi}{8}\right)\left(-2i\cos\frac{7\pi}{8}\right)$$

or $$\frac{1}{8} = \cos\frac{\pi}{8}\cdot\cos\frac{3\pi}{8}\cos\frac{5\pi}{8}\cdot\cos\frac{7\pi}{8}.$$

35. If $(1+x)^n = a_0 + a_1x + a_2x^2 + \ldots + a_nx^n$, find the sum $a_0 + a_4 + a_8 + a_{12} + \ldots$.

Putting $x = \pm\sqrt{1}$, i.e., $x = 1, -1$ we get

$$(1+1)^n = a_0 + a_1 + a_2 + a_3 + a_4 + \ldots + a_n$$

$$(1-1)^n = a_0 - a_1 + a_2 - a_3 + a_4 - \ldots + (-1)^n a_n.$$

Adding these, $2^n = 2(a_0 + a_2 + a_4 + a_6 + \ldots)$

$$\therefore \quad a_0 + a_2 + a_4 + a_6 + \ldots = 2^{n-1}. \quad \ldots (1)$$

Putting $x = \pm\sqrt{-1}$, i.e., $x = i, -i$ we get

$$(1+i)^n = a_0 + a_1 i + a_2 i^2 + a_3 i^3 + a_4 i^4 + \ldots + a_n i^n$$

$$(1-i)^n = a_0 - a_1 i + a_2 i^2 - a_3 i^3 + a_4 i^4 - \ldots + a_n(-i)^n.$$

Adding these,

$$(1+i)^n + (1-i)^n = 2(a_0 + a_2 i^2 + a_4 i^4 + \ldots)$$

$$\therefore \quad 2(a_0 - a_2 + a_4 - a_6 + \ldots) = (1+i)^n + (1-i)^n$$

$$= \left\{\sqrt{2}\left(\cos\frac{\pi}{4} + i\sin\frac{\pi}{4}\right)\right\}^n + \left\{\sqrt{2}\left(\cos\frac{-\pi}{4} + i\sin\frac{-\pi}{4}\right)\right\}^n$$

(changing the complex numbers in polar forms)

$$= 2^{n/2}\left(\cos\frac{n\pi}{4} + i\sin\frac{n\pi}{4}\right) + 2^{n/2}\left(\cos\frac{n\pi}{4} - i\sin\frac{n\pi}{4}\right)$$

$$= 2\cdot 2^{n/2}\cos\frac{n\pi}{4}$$

$$\therefore \quad a_0 - a_2 + a_4 - a_6 + \ldots = 2^{n/2}\cdot\cos\frac{n\pi}{4} \quad \ldots (2)$$

Adding (1) and (2),

$$2(a_0 + a_4 + a_8 + a_{12} + \ldots) = 2^{n-1} + 2^{n/2}\cdot\cos\frac{n\pi}{4}$$

$\therefore \quad a_0 + a_4 + a_8 + a_{12} + \ldots = 2^{n-2} + 2^{n/2-1} \cdot \cos\frac{n\pi}{4}$.

36. If 1, $a_1, a_2, a_3, \ldots, a_{n-1}$ be the nth roots of unity and w be a nonreal complex cube root of unity then prove that $(w - a_1)(w - a_2) \ldots (w - a_{n-1})$

= 0, if n is of the form $3m$

1, if n is of the form $3m + 1$

$1 + w$, if n is of the form $3m + 2$.

Here $\sqrt[n]{1} = 1, a_1, a_2, \ldots, a_{n-1}$

$\therefore \quad x^n - 1 = 0$ has the roots $1, a_1, a_2, \ldots, a_{n-1}$

$\therefore \quad x^n - 1 = (x-1)(x-a_1)(x-a_2)\ldots(x-a_{n-1})$.

Putting $x = w$, we get

$$w^n - 1 = (w-1)(w-a_1)(w-a_2)\ldots(w-a_{n-1})$$

$$\therefore \quad (w-a_1)(w-a_2)(w-a_3)\ldots(w-a_{n-1}) = \frac{w^n - 1}{w-1} \quad \ldots (1)$$

If $\quad n = 3m,\ w^n - 1 = w^{3m} - 1 = 1 - 1 = 0$

$\therefore \quad \frac{w^n - 1}{w-1} = 0.$

If $\quad n = 3m+1,\ w^n - 1 = w^{3m+1} - 1 = w - 1$

$\therefore \quad \frac{w^n - 1}{w-1} = \frac{w-1}{w-1} = 1.$

If $\quad n = 3m+2,\ w^n - 1 = w^{3m+2} - 1 = w^2 - 1$

$\therefore \quad \frac{w^n - 1}{w-1} = \frac{w^2-1}{w-1} = w + 1.$

Hence, the problem follows from (1).

37. If n is an odd integer greater than 3 but not a multiple of 3, prove that

$xy(x^2 + xy + y^2)$ is a factor of $(x+y)^n - x^n - y^n$.

We have $\quad xy(x^2 + xy + y^2)$

$= xy\{x^2 - (w + w^2)xy + w^3y^2\} \quad (\because 1 + w + w^2 = 0)$

$= (x-0)(y-0)(x-wy)(x-w^2y) \quad \ldots (1)$

Putting $x = 0$,

the expression $= (0+y)^n - 0^n - y^n = y^n - y^n = 0$

$\therefore \quad x - 0$ is a factor.

Putting $y = 0$,

the expression $= (x+0)^n - x^n - 0 = x^n - x^n = 0$

$\therefore \quad y - 0$ is a factor.

Putting $x = wy$,

the expression $= (wy + y)^n - (wy)^n - y^n$

$= y^n\{(w+1)^n - w^n - 1\}$

$= y^n\{(-w^2)^n - w^n - 1\}$

$= -y^n\{w^{2n} + w^n + 1\} \quad (\because n$ is odd)

$= -y^n \cdot 0 = 0 \quad (\because n$ is not divisible by 3)

$\therefore \quad x - wy$ is a factor.

Putting $x = w^2y$,

the expression $= (w^2y + y)^n - (w^2y)^n - y^n$

$= y^n\{(w^2+1)^n - w^{2n} - 1\}$

$= y^n\{(-w)^n - w^{2n} - 1\}$

$= -y^n\{1 + w^n + w^{2n}\} \quad (\because n$ is odd)

$= -y^n \cdot 0 = 0 \quad (\because n$ is not divisible by 3)

$\therefore \quad x - w^2y$ is a factor.

Thus $(x-0)(y-0)(x-wy)(x-w^2y)$ is a factor of the expression.

$\therefore \quad$ by (1), $xy(x^2 + xy + y^2)$ is a factor of the expression.

38. Find the complex number z such that $|z - 2 + 2i| \le 1$ and z has the (i) least absolute value (ii) numerically least amplitude.

We know that $|z - 2(1-i)| \le 1$ is the set of complex numbers represented by the points inside or on the circle whose centre is $2(1-i)$ and radius is 1.

Let C be the centre and the line OC cuts the circle at the point B, nearest to O. Also let the tangents from O touch the circle at A and D such that $\angle AOX < \angle DOX$.

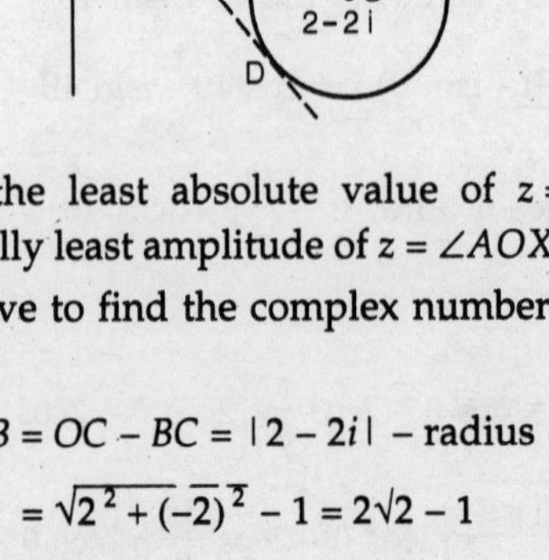

Clearly, the least absolute value of $z = OB$ and the numerically least amplitude of $z = \angle AOX$.

(i) We have to find the complex number z represented by B.

Now, $\quad OB = OC - BC = |2 - 2i| -$ radius

$= \sqrt{2^2 + (-2)^2} - 1 = 2\sqrt{2} - 1$

$\therefore \quad |z| = 2\sqrt{2} - 1.$

Also $\angle COX = -\frac{\pi}{4}$; $\therefore$ $\angle BOX = -\frac{\pi}{4}$

so amp $z = -\frac{\pi}{4}$

$\therefore$ $z = |z|\{\cos(\text{amp } z) + i\sin(\text{amp } z)\}$

$$= (2\sqrt{2}-1)\left\{\cos\left(-\frac{\pi}{4}\right) + i\sin\left(-\frac{\pi}{4}\right)\right\}$$

$$= \frac{(2\sqrt{2}-1)(1-i)}{\sqrt{2}} = \left(2 - \frac{1}{\sqrt{2}}\right)(1-i).$$

(ii) We have to find the complex number z represented by A.

Now, $OA = \sqrt{OC^2 - CA^2} = \sqrt{|2-2i|^2 - 1^2}$

$$= \sqrt{2^2 + (-2)^2 - 1} = \sqrt{7}$$

$\therefore$ $|z| = \sqrt{7}$

$$\angle AOX = -|\angle COX - \angle COA| = -\left|\frac{\pi}{4} - \sin^{-1}\frac{1}{OC}\right|$$

$$= -\left|\frac{\pi}{4} - \sin^{-1}\frac{1}{|2-2i|}\right| = -\left(\frac{\pi}{4} - \sin^{-1}\frac{1}{2\sqrt{2}}\right)$$

$\therefore$ amp $z = -\left(\frac{\pi}{4} - \sin^{-1}\frac{1}{2\sqrt{2}}\right)$

$\therefore$ $z = |z|\{\cos(\text{amp } z) + i\sin(\text{amp } z)\}$

$$= \sqrt{7}\left[\cos\left\{-\left(\frac{\pi}{4} - \sin^{-1}\frac{1}{2\sqrt{2}}\right)\right\} + i\sin\left\{-\left(\frac{\pi}{4} - \sin^{-1}\frac{1}{2\sqrt{2}}\right)\right\}\right]$$

$$= \sqrt{7}\left\{\cos\left(\frac{\pi}{4} - \sin^{-1}\frac{1}{2\sqrt{2}}\right) - i\sin\left(\frac{\pi}{4} - \sin^{-1}\frac{1}{2\sqrt{2}}\right)\right\}$$

39. If z_1, z_2, z_3 are three collinear points in the Argand plane then prove that $\begin{vmatrix} z_1 & \bar{z}_1 & 1 \\ z_2 & \bar{z}_2 & 1 \\ z_3 & \bar{z}_3 & 1 \end{vmatrix} = 0$.

First Method (by using section of line segment)

Let z_3 divides the line segment joining z_1, z_2 in the ratio $\lambda : 1$.

Then $z_3 = \frac{\lambda z_2 + z_1}{\lambda + 1}$

or $(\lambda+1)z_3 = \lambda z_2 + z_1$

or $\lambda(z_3 - z_2) = z_1 - z_3$

or $\frac{z_1 - z_3}{z_3 - z_2} = \lambda = \text{real}.$

$\therefore$ $\frac{z_1 - z_3}{z_3 - z_2} = \overline{\left(\frac{z_1 - z_3}{z_3 - z_2}\right)}$ $\{\because$ for real $z, z = \bar{z}\}$

or $\frac{z_1 - z_3}{z_3 - z_2} = \frac{\overline{z_1 - z_3}}{\overline{z_3 - z_2}} = \frac{\bar{z}_1 - \bar{z}_3}{\bar{z}_3 - \bar{z}_2}$

or $(z_1 - z_3)(\bar{z}_3 - \bar{z}_2) = (z_3 - z_2)(\bar{z}_1 - \bar{z}_3)$

or $z_1(\bar{z}_3 - \bar{z}_2) - z_3(\bar{z}_3 - \bar{z}_2) = z_3(\bar{z}_1 - \bar{z}_3) - z_2(\bar{z}_1 - \bar{z}_3)$

or $z_1(\bar{z}_2 - \bar{z}_3) - z_2(\bar{z}_1 - \bar{z}_3) + z_3(\bar{z}_1 - \bar{z}_2) = 0$

or $\begin{vmatrix} z_1 & \bar{z}_1 & 1 \\ z_2 & \bar{z}_2 & 1 \\ z_3 & \bar{z}_3 & 1 \end{vmatrix} = 0.$

Second Method (by using general equation of a straight line)

Let the equation of the line on which z_1, z_2, z_3 lie be

$$\bar{\alpha}z + \alpha\bar{z} + \beta = 0 \quad \dots (1)$$

where β is a real constant.

(1) is satisfied for $z = z_1, z_2, z_3$.

So $\bar{\alpha}z_1 + \alpha\bar{z}_1 + \beta = 0$ ... (2)

$\bar{\alpha}z_2 + \alpha\bar{z}_2 + \beta = 0$ ... (3)

$\bar{\alpha}z_3 + \alpha\bar{z}_3 + \beta = 0$ ... (4)

Eliminating $\bar{\alpha}, \alpha, \beta$ from (2), (3) and (4) we get

$$\begin{vmatrix} z_1 & \bar{z}_1 & 1 \\ z_2 & \bar{z}_2 & 1 \\ z_3 & \bar{z}_3 & 1 \end{vmatrix} = 0.$$

40. If $z_1 + z_2 + z_3 + \dots + z_n = 0$ then prove that all the points representing the complex numbers $z_1, z_2, \dots, z_n$ do not lie on the same side of any line passing through the origin.

As $z_1 + z_2 + \dots + z_n = 0,$

$\overline{z_1 + z_2 + \dots + z_n} = \bar{0} = 0,$

or $\bar{z}_1 + \bar{z}_2 + \dots + \bar{z}_n = 0.$

Any line passing through the origin has the complex equation $\alpha\bar{z} + \bar{\alpha}z = 0$.

If possible let all the points representing the complex numbers $z_1, z_2, \dots, z_n$ are on the same side of

$\alpha\bar{z} + \bar{\alpha}z = 0.$

Then $\alpha\bar{z}_i + \bar{\alpha}z_i$ is of the same sign for all $i = 1, 2, 3, \dots, n$.

Adding all these,

$$\alpha\Sigma\bar{z}_i + \bar{\alpha}\Sigma z_i \text{ is either} > 0 \text{ or } < 0 \quad \dots (1)$$

But $\Sigma z_i = 0, \Sigma\bar{z}_i = 0.$

So $\alpha\Sigma\bar{z}_i + \bar{\alpha}\Sigma z_i = 0$ ... (2)

(2) contradicts (1). Hence our assumption was wrong.

$\therefore$ all the points representing $z_1, z_2, \dots, z_n$ cannot lie on the same side of $\alpha\bar{z} + \bar{\alpha}z = 0$. Hence, the problem.

41. The complex slope of the line joining the points represented by z_1, z_2 in the Argand plane is defined as $\dfrac{z_1 - z_2}{\bar{z}_1 - \bar{z}_2}$. Find the condition for perpendicularity of two lines with complex slopes w_1 and w_2.

Let $w_1 = \dfrac{z_1 - z_2}{\bar{z}_1 - \bar{z}_2}$ and $w_2 = \dfrac{z_3 - z_4}{\bar{z}_3 - \bar{z}_4}$.

Now, $|w_1| = \left|\dfrac{z_1 - z_2}{\bar{z}_1 - \bar{z}_2}\right| = \left|\dfrac{z_1 - z_2}{\overline{z_1 - z_2}}\right| = 1$, for $\left|\dfrac{z}{\bar{z}}\right| = 1$.

Similarly, $|w_2| = 1$.

Now, $\text{amp } w_1 = \text{amp } \dfrac{z_1 - z_2}{\bar{z}_1 - \bar{z}_2} = \text{amp } \dfrac{z_1 - z_2}{\overline{z_1 - z_2}}$

$$= 2\text{ amp}(z_1 - z_2) \quad \ldots (1)$$

Similarly, $\text{amp } w_2 = 2\text{amp}(z_3 - z_4) \quad \ldots (2)$

Now $\text{amp}\dfrac{z_1 - z_2}{z_3 - z_4} = \pm\dfrac{\pi}{2}$ because the line segments joining z_1, z_2 and z_3, z_4 are at right angles.

From (1), (2) we get

$$\text{amp } w_1 - \text{amp } w_2 = 2\{\text{amp}(z_1 - z_2) - \text{amp}(z_3 - z_4)\}$$

$$= 2\text{amp } \frac{z_1 - z_2}{z_3 - z_4} = 2 \times \left(\pm\frac{\pi}{2}\right) = \pm\pi.$$

$\therefore \quad \text{amp } \dfrac{w_1}{w_2} = \pm\pi$

$\therefore \quad \dfrac{w_1}{w_2} = \left|\dfrac{w_1}{w_2}\right|\{\cos(\pm\pi) + i\sin(\pm\pi)\}$

$$= -\left|\frac{w_1}{w_2}\right| = -\frac{|w_1|}{|w_2|} = -\frac{1}{1} = -1$$

$\therefore \quad w_1 + w_2 = 0.$

42. If the two triangles whose vertices are z_1, z_2, z_3 and w_1, w_2, w_3 are similar then prove that $w_1(z_2 - z_3) + w_2(z_3 - z_1) + w_3(z_1 - z_2) = 0.$

As the triangles are similar, the angle at the z_1 and w_1 as shown in the figure must be equal

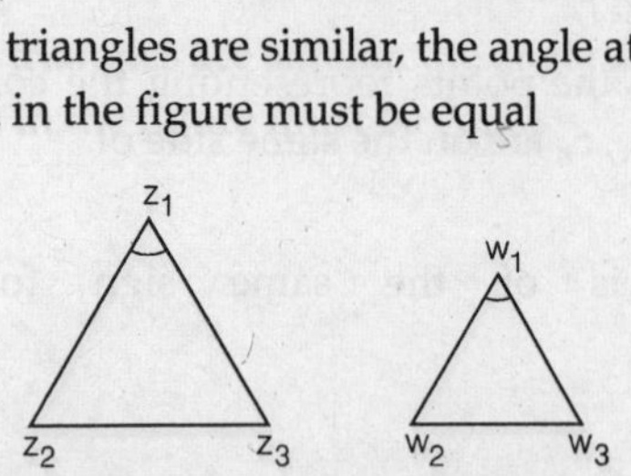

i.e., $\text{amp } \dfrac{z_2 - z_1}{z_3 - z_1} = \text{amp } \dfrac{w_2 - w_1}{w_3 - w_1} \quad \ldots (1)$

Also $\dfrac{|z_2 - z_1|}{|w_2 - w_1|} = \dfrac{|z_3 - z_1|}{|w_3 - w_1|}$

($\because$ for similar triangles corresponding sides are proportional)

or $\dfrac{|z_2 - z_1|}{|z_3 - z_1|} = \dfrac{|w_2 - w_1|}{|w_3 - w_1|}. \quad \ldots (2)$

From (1), (2) we get $\dfrac{z_2 - z_1}{z_3 - z_1} = \dfrac{w_2 - w_1}{w_3 - w_1}$

or $(z_2 - z_1)(w_3 - w_1) = (z_3 - z_1)(w_2 - w_1)$

or $w_3(z_2 - z_1) - w_1(z_2 - z_1) = w_2(z_3 - z_1) - w_1(z_3 - z_1)$

or $w_1(z_3 - z_1 - z_2 + z_1) - w_2(z_3 - z_1) + w_3(z_2 - z_1) = 0$

or $w_1(z_3 - z_2) - w_2(z_3 - z_1) - w_3(z_1 - z_2) = 0$

or $w_1(z_2 - z_3) + w_2(z_3 - z_1) + w_3(z_1 - z_2) = 0.$

43. Prove that a triangle, whose vertices represent the complex numbers z_1, z_2 and z_3 is equilateral if and only if $z_1^2 + z_2^2 + z_3^2 - z_1z_2 - z_2z_3 - z_3z_1 = 0.$

Let the vertices A, B, and C of the triangle ABC represent the complex numbers z_1, z_2 and z_3 respectively.

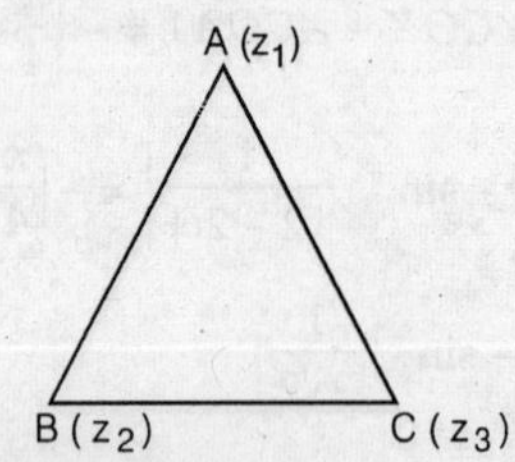

ΔABC is equilateral

$\Leftrightarrow \quad \angle BAC = \dfrac{\pi}{3}$ and $AB = AC$

$\Leftrightarrow \quad \text{amp } \dfrac{z_2 - z_1}{z_3 - z_1} = \dfrac{\pi}{3}$ and $|z_2 - z_1| = |z_3 - z_1|$

$\Leftrightarrow \quad \text{amp } \dfrac{z_2 - z_1}{z_3 - z_1} = \dfrac{\pi}{3}$ and $\left|\dfrac{z_2 - z_1}{z_3 - z_1}\right| = 1$

$\Leftrightarrow \quad \dfrac{z_2 - z_1}{z_3 - z_1}$

$$= \left|\frac{z_2 - z_1}{z_3 - z_1}\right|\left\{\cos\left(\text{amp } \frac{z_2 - z_1}{z_3 - z_1}\right) + i\sin\left(\text{amp } \frac{z_2 - z_1}{z_3 - z_1}\right)\right\}$$

$$= \cos\frac{\pi}{3} + i\sin\frac{\pi}{3} = \frac{1}{2} + i\frac{\sqrt{3}}{2}$$

$\Leftrightarrow \quad 2\dfrac{z_2 - z_1}{z_3 - z_1} - 1 = i\sqrt{3}$

$\Leftrightarrow \quad \left(\dfrac{2z_2 - z_3 - z_1}{z_3 - z_1}\right)^2 = (i\sqrt{3})^2$

$\Leftrightarrow \quad (2z_2 - z_3 - z_1)^2 = -3(z_3 - z_1)^2$

$\Leftrightarrow \quad 4z_2^2 + z_3^2 + z_1^2 - 4z_2z_3 - 4z_1z_2 + 2z_3z_1$

$$= -3(z_3^2 + z_1^2 - 2z_3z_1)$$

$\Leftrightarrow \quad 4(z_1^2 + z_2^2 + z_3^2 - z_2z_3 - z_3z_1 - z_1z_2) = 0$

$\Leftrightarrow \quad z_1^2 + z_2^2 + z_3^2 - z_1z_2 - z_2z_3 - z_3z_1 = 0.$

44. If the complex numbers z_1, z_2, z_3 are represented by the vertices of an equilateral triangle whose circumcentre represents the complex number z_0 then prove that $3z_0^2 = z_1^2 + z_2^2 + z_3^2$.

In an equilateral triangle the circumcentre and the centroid are the same point. So

$$z_0 = \frac{z_1 + z_2 + z_3}{3} \quad \ldots (1)$$

$\therefore \quad 3z_0 = z_1 + z_2 + z_3$

or $\quad (z_1 - z_0) + (z_2 - z_0) = -(z_3 - z_0)$

$\therefore$ squaring,

$$(z_1 - z_0)^2 + (z_2 - z_0)^2 + 2(z_1 - z_0)(z_2 - z_0) = (z_3 - z_0)^2 \quad \ldots (2)$$

Now, distances of the vertices from the circumcentre are equal and angles subtended by the sides at the circumcentre are each equal to $\frac{2\pi}{3}$.

$\therefore \quad |z_1 - z_0| = |z_2 - z_0| = |z_3 - z_0|$

i.e., $\left|\frac{z_3 - z_0}{z_1 - z_0}\right| = \left|\frac{z_2 - z_0}{z_3 - z_0}\right| = 1$

and amp $\frac{z_3 - z_0}{z_1 - z_0} = \text{amp}\, \frac{z_2 - z_0}{z_3 - z_0} = \frac{2\pi}{3}$.

$$\therefore \quad \frac{z_3 - z_0}{z_1 - z_0} = \frac{z_2 - z_0}{z_3 - z_0} = 1 \cdot \left(\cos \frac{2\pi}{3} + i \sin \frac{2\pi}{3}\right)$$

$\therefore \quad (z_1 - z_0)(z_2 - z_0) = (z_3 - z_0)^2$

$\therefore$ from (2),

$$(z_1 - z_0)^2 + (z_2 - z_0)^2 + (z_3 - z_0)^2 = 0$$

or $\quad z_1^2 + z_2^2 + z_3^2 + 3z_0^2 - 2z_0(z_1 + z_2 + z_3) = 0$

or $\quad z_1^2 + z_2^2 + z_3^2 + 3z_0^2 - 6z_0^2 = 0$, using (1)

$\therefore \quad 3z_0^2 = z_1^2 + z_2^2 + z_3^2$.

45. Prove that the orthocentre z of the triangle whose vertices represent the complex numbers z_1, z_2 and z_3 is given by

$$z\Sigma(z_1\bar{z}_2 - \bar{z}_1 z_2) = \Sigma z_1^2(\bar{z}_2 - \bar{z}_3) + \Sigma |z_1|^2 (z_2 - z_3).$$

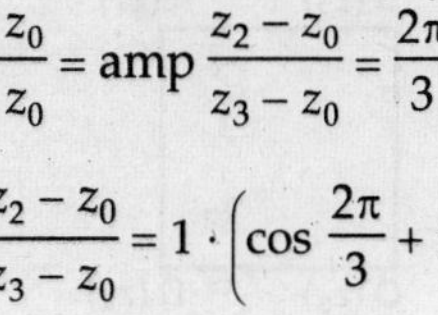

As $\quad \angle ADC = \frac{\pi}{2}$, amp $\frac{z_1 - z}{z_3 - z_2} = \frac{\pi}{2}$

$$\therefore \quad \frac{z_1 - z}{z_3 - z_2} = r\left(\cos \frac{\pi}{2} + i \sin \frac{\pi}{2}\right) = ri, \text{ where } r = \left|\frac{z_1 - z}{z_3 - z_2}\right|$$

$$\therefore \quad \overline{\left(\frac{z_1 - z}{z_3 - z_2}\right)} = \overline{ri} = r\,\bar{i} = -ri;$$

adding we get,

$$\frac{z_1 - z}{z_3 - z_2} + \overline{\left(\frac{z_1 - z}{z_3 - z_2}\right)} = 0$$

or $$\frac{z_1 - z}{z_3 - z_2} + \frac{\bar{z}_1 - \bar{z}}{\bar{z}_3 - \bar{z}_2} = 0$$

$$\therefore \quad \bar{z}_1 - \bar{z} = -\frac{z_1 - z}{z_3 - z_2}(\bar{z}_3 - \bar{z}_2) \quad \ldots (1)$$

Again $\angle BEA = \frac{\pi}{2}$; so replacing z_1, z_2, z_3 by z_2, z_3, z_1 respectively in (1) we find

$$\bar{z}_2 - \bar{z} = -\frac{z_2 - z}{z_1 - z_3}(\bar{z}_1 - \bar{z}_3) \quad \ldots (2)$$

From (1) – (2), we get

$$\bar{z}_1 - \bar{z}_2 = \frac{z_2 - z}{z_1 - z_3}(\bar{z}_1 - \bar{z}_3) - \frac{z_1 - z}{z_3 - z_2}(\bar{z}_3 - \bar{z}_2)$$

$$= \frac{z_2(\bar{z}_1 - \bar{z}_3)}{z_1 - z_3} - \frac{z_1(\bar{z}_3 - \bar{z}_2)}{z_3 - z_2} + z\left\{\frac{\bar{z}_3 - \bar{z}_2}{z_3 - z_2} - \frac{\bar{z}_1 - \bar{z}_3}{z_1 - z_3}\right\}$$

or $\quad (\bar{z}_1 - \bar{z}_2)(z_1 - z_3)(z_3 - z_2)$

$= (\bar{z}_1 - \bar{z}_3)z_2(z_3 - z_2) - (\bar{z}_3 - \bar{z}_2)z_1(z_1 - z_3)$

$\quad + z\{(\bar{z}_3 - \bar{z}_2)(z_1 - z_3) - (\bar{z}_1 - \bar{z}_3)(z_3 - z_2)\}$

$\therefore \quad z\{\bar{z}_3 z_1 - \bar{z}_3 z_3 - \bar{z}_2 z_1 + \bar{z}_2 z_3 - \bar{z}_1 z_3 + \bar{z}_1 z_2 + \bar{z}_3 z_3 - \bar{z}_3 z_2\}$

$= (\bar{z}_1 - \bar{z}_2)(z_1 - z_3)(z_3 - z_2) - (\bar{z}_1 - \bar{z}_3)z_2(z_3 - z_2)$

$\quad + (\bar{z}_3 - \bar{z}_2)z_1(z_1 - z_3)$

or $\quad z\{(\bar{z}_1 z_2 - z_1\bar{z}_2) + (\bar{z}_2 z_3 - z_2\bar{z}_3) + (\bar{z}_3 z_1 - \bar{z}_1 z_3)\}$

$= (\bar{z}_1 - \bar{z}_2)\{z_1 z_3 - z_1 z_2 - z_3^2 + z_2 z_3\}$

$\quad - (\bar{z}_1 - \bar{z}_3)(z_2 z_3 - z_2^2) + (\bar{z}_3 - \bar{z}_2)(z_1^2 - z_3 z_1)$

$= -\{z_1^2(\bar{z}_2 - \bar{z}_3) + z_2^2(\bar{z}_3 - \bar{z}_1) + z_3^2(\bar{z}_1 - \bar{z}_2)\}$

$\quad + (\bar{z}_1 - \bar{z}_2)(z_1 z_3 - z_1 z_2 + z_2 z_3)$

$\quad - (\bar{z}_1 - \bar{z}_3)z_2 z_3 - (\bar{z}_3 - \bar{z}_2)z_3 z_1$

$\therefore \quad z\Sigma(z_1\bar{z}_2 - \bar{z}_1 z_2)$

$= \Sigma z_1^2(\bar{z}_2 - \bar{z}_3) - (\bar{z}_1 - \bar{z}_2)(z_1 z_3 - z_1 z_2 + z_2 z_3)$

$\quad + (\bar{z}_1 - \bar{z}_3)z_2 z_3 + (\bar{z}_3 - \bar{z}_2)z_3 z_1$

$= \Sigma z_1^2(\bar{z}_2 - \bar{z}_3) - |z_1|^2(z_3 - z_2) - \bar{z}_1 z_2 z_3 + z_1\bar{z}_2 z_3$

$\quad - |z_2|^2(z_1 - z_3) + \bar{z}_1 z_2 z_3$

$\quad - |z_3|^2 z_2 + |z_3|^2 z_1 - z_1\bar{z}_2 z_3$.

$= \Sigma z_1^2(\bar{z}_2 - \bar{z}_3) + \Sigma |z_1|^2(z_2 - z_3)$.

46. The vertices of a triangle represent the complex numbers z_1, z_2 and z_3.

(i) Prove that the area of the triangle is

$$\Sigma \frac{(z_2 - z_3)|z_1|^2}{4iz_1}.$$

(ii) Find the altitude of the triangle through the vertex z_1.

(i) Let $z_1 = x_1 + iy_1$, $z_2 = x_2 + iy_2$, $z_3 = x_3 + iy_3$.

So the vertices of the triangle have the coordinates (x_1, y_1), (x_2, y_2) and (x_3, y_3)

$\therefore$ the area of the triangle

$$= \frac{1}{2}\{x_1(y_2 - y_3) + x_2(y_3 - y_1) + x_3(y_1 - y_2)\} \qquad \dots (1)$$

Now, $\bar{z}_1 = x_1 - iy_1$, $\bar{z}_2 = x_2 - iy_2$, $\bar{z}_3 = x_3 - iy_3$

$$\therefore \quad x_1 = \frac{z_1 + \bar{z}_1}{2}, \; y_1 = \frac{z_1 - \bar{z}_1}{2i}$$

$$x_2 = \frac{z_2 + \bar{z}_2}{2}, \; y_2 = \frac{z_2 - \bar{z}_2}{2i}$$

$$x_3 = \frac{z_3 + \bar{z}_3}{2}, \; y_3 = \frac{z_3 - \bar{z}_3}{2i}.$$

$\therefore$ from (1), the area of the triangle

$$= \frac{1}{2}\left\{\frac{z_1 + \bar{z}_1}{2}\left(\frac{z_2 - \bar{z}_2}{2i} - \frac{z_3 - \bar{z}_3}{2i}\right) + \frac{z_2 + \bar{z}_2}{2}\left(\frac{z_3 - \bar{z}_3}{2i} - \frac{z_1 - \bar{z}_1}{2i}\right) + \frac{z_3 + \bar{z}_3}{2}\left(\frac{z_1 - \bar{z}_1}{2i} - \frac{z_2 - \bar{z}_2}{2i}\right)\right\}$$

$$= \frac{1}{8i}[(z_1 + \bar{z}_1)\{(z_2 - z_3) - (\bar{z}_2 - \bar{z}_3)\} + (z_2 + \bar{z}_2)\{(z_3 - z_1) - (\bar{z}_3 - \bar{z}_1)\} + (z_3 + \bar{z}_3)\{(z_1 - z_2) - (\bar{z}_1 - \bar{z}_2)\}$$

$$= \frac{1}{8i}[\Sigma z_1(z_2 - z_3) - \Sigma \bar{z}_1(\bar{z}_2 - \bar{z}_3) - z_1(\bar{z}_2 - \bar{z}_3) + \bar{z}_1(z_2 - z_3) - z_2(\bar{z}_3 - \bar{z}_1) + \bar{z}_2(z_3 - z_1) - z_3(\bar{z}_1 - \bar{z}_2) + \bar{z}_3(z_1 - z_2)]$$

$$= \frac{1}{8i}[0 - 0 - z_1\bar{z}_2 + z_1\bar{z}_3 + \bar{z}_1(z_2 - z_3) - z_2\bar{z}_3 + z_2\bar{z}_1 + \bar{z}_2(z_3 - z_1) - z_3\bar{z}_1 + z_3\bar{z}_2 + \bar{z}_3(z_1 - z_2)]$$

$$= \frac{2}{8i}[\bar{z}_1(z_2 - z_3) + \bar{z}_2(z_3 - z_1) + \bar{z}_3(z_1 - z_2)]$$

$$= \frac{1}{4iz_1} z_1\bar{z}_1(z_2 - z_3) + \frac{1}{4iz_2} z_2\bar{z}_2(z_3 - z_1) + \frac{1}{4iz_3} z_3\bar{z}_3(z_1 - z_2)$$

$$= \frac{1}{4iz_1}|z_1|^2(z_2 - z_3) + \frac{1}{4iz_2}|z_2|^2(z_3 - z_1) + \frac{1}{4iz_3}|z_3|^2(z_1 - z_2)$$

$$= \Sigma \frac{(z_2 - z_3)|z_1|^2}{4iz_1}.$$

(ii) The area of the triangle

$$= \frac{1}{2}|z_2 - z_3| \cdot (\text{required altitude})$$

$$\therefore \quad \frac{1}{2}|z_2 - z_3| \cdot (\text{required altitude}) = \Sigma \frac{(z_2 - z_3)|z_1|^2}{4iz_1}$$

$$\therefore \quad \text{required altitude} = \frac{1}{|z_2 - z_3|} \Sigma \frac{(z_2 - z_3)|z_1|^2}{2iz_1}$$

47. If the vertices of a square are z_1, z_2, z_3, z_4 taken in order in the anticlockwise sense then prove that

(i) $z_3 = -iz_1 + (1 + i)z_2$

(ii) $z_4 = iz_2 + (1 - i)z_1$.

Let the vertices A, B, C and D of the square ABCD represent the complex numbers z_1, z_2, z_3 and z_4 respectively. Clearly in the square,

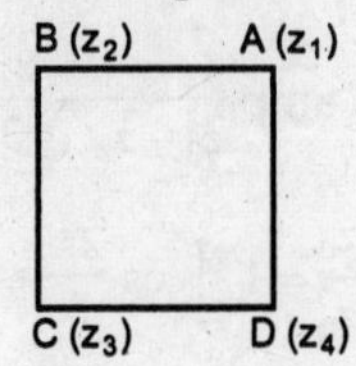

(i) $\angle ABC = \frac{\pi}{2} = \text{amp}\, \frac{z_1 - z_2}{z_3 - z_2}$

and $AB = |z_1 - z_2| = BC = |z_3 - z_2|$.

$$\therefore \quad \left|\frac{z_1 - z_2}{z_3 - z_2}\right| = 1$$

Hence $\frac{z_1 - z_2}{z_3 - z_2} = 1 \cdot \left(\cos\frac{\pi}{2} + i\sin\frac{\pi}{2}\right)$

or $\frac{z_1 - z_2}{z_3 - z_2} = i$

or $z_3 - z_2 = \frac{1}{i}(z_1 - z_2) = -i(z_1 - z_2)$

$$\therefore \quad z_3 = z_2 - i(z_1 - z_2) = -iz_1 + (1 + i)z_2 \qquad \dots (1)$$

(ii) As in (1), replacing z_1, z_2, z_3 by z_3, z_4, z_1 respectively we get

$$z_1 = -iz_3 + (1 + i)z_4.$$

Putting for z_3 from (1),

$$z_1 = -i\{-iz_1 + (1 + i)z_2\} + (1 + i)z_4$$

$$= -z_1 + (1 - i)z_2 + (1 + i)z_4$$

or $2z_1 = (1 - i)z_2 + (1 + i)z_4$

or $z_4 = \frac{i - 1}{i + 1} z_2 + \frac{2}{i + 1} z_1$

or $z_4 = \frac{(i - 1)^2}{i^2 - 1} z_2 + \frac{2(i - 1)}{i^2 - 1} z_1$

$= \frac{-2i}{-2} z_2 + \frac{2(i - 1)}{-2} z_1$

$\therefore$ $z_4 = iz_2 + (1 - i)z_1$.

48. If z_1, z_2 are two complex numbers representing consecutive vertices of a regular hexagon then find the complex number z_3 representing the vertex adjacent to z_2.

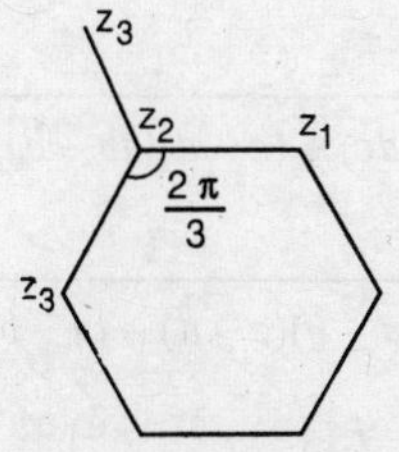

Clearly, $|z_1 - z_2| = |z_3 - z_2|$ ... (1)

and amp $\frac{z_1 - z_2}{z_3 - z_2} = \pm \frac{2\pi}{3} = \angle Z_1Z_2Z_3$,

where the point Z_1 represents z_1, etc.

$\therefore$ $\frac{z_1 - z_2}{z_3 - z_2} = \left|\frac{z_1 - z_2}{z_3 - z_2}\right| \left\{\cos \frac{2\pi}{3} \pm i\sin \frac{2\pi}{3}\right\}$

$= 1\left(-\frac{1}{2} \pm i \cdot \frac{\sqrt{3}}{2}\right)$, using (1)

$\therefore$ $z_1 - z_2 = \alpha(z_3 - z_2)$,

where $\alpha = \frac{-1 \pm i\sqrt{3}}{2} = w, w^2$ (and $\alpha^3 = 1$)

or $\alpha^2(z_1 - z_2) = \alpha^3(z_3 - z_2) = z_3 - z_2$

or $z_3 = z_2 + \alpha^2(z_1 - z_2)$

$\therefore$ $z_3 = \alpha^2 z_1 + (1 - \alpha^2)z_2$

$= \left(-\frac{1}{2} \mp i\frac{\sqrt{3}}{2}\right) z_1 + \left(1 - \frac{-1 \mp i\sqrt{3}}{2}\right) z_2$

$= -\frac{1}{2}(1 \pm i\sqrt{3})z_1 + \frac{1}{2}(3 \pm i\sqrt{3})z_2$.

Note If z_1, z_2, z_3 are given to be in anticlockwise sense then amp $\frac{z_1 - z_2}{z_3 - z_2} = \frac{2\pi}{3}$ will be taken.

49. Let z_1, z_2, z_3 be nonreal complex numbers and $z_1 \neq z_2$. If $\begin{vmatrix} |z_1| & |z_2| & |z_3| \\ |z_2| & |z_3| & |z_1| \\ |z_3| & |z_1| & |z_2| \end{vmatrix} = 0$ then prove that

(i) z_1, z_2, z_3 lie on a circle with the centre at the origin

(ii) amp $\left(\frac{z_3}{z_2}\right) =$ amp $\left(\frac{z_3 - z_1}{z_2 - z_1}\right)^2$.

Let $z_1 = r_1(\cos \alpha_1 + i\sin \alpha_1)$,

$z_2 = r_2(\cos \alpha_2 + i\sin \alpha_2)$, $z_3 = r_3(\cos \alpha_3 + i\sin \alpha_3)$.

From the question, $\begin{vmatrix} r_1 & r_2 & r_3 \\ r_2 & r_3 & r_1 \\ r_3 & r_1 & r_2 \end{vmatrix} = 0$

or $\begin{vmatrix} r_1 + r_2 + r_3 & r_2 & r_3 \\ r_1 + r_2 + r_3 & r_3 & r_1 \\ r_1 + r_2 + r_3 & r_1 & r_2 \end{vmatrix} = 0$

$\therefore$ $(r_1 + r_2 + r_3) \begin{vmatrix} 1 & r_2 & r_3 \\ 1 & r_3 & r_1 \\ 1 & r_1 & r_2 \end{vmatrix} = 0$

or $(r_1 + r_2 + r_3) \begin{vmatrix} 1 & r_2 & r_3 \\ 0 & r_3 - r_2 & r_1 - r_3 \\ 0 & r_1 - r_2 & r_2 - r_3 \end{vmatrix} = 0$,

using $R_2 \to R_2 - R_1$, $R_3 \to R_3 - R_1$

or $(r_1 + r_2 + r_3) \begin{vmatrix} r_3 - r_2 & r_1 - r_3 \\ r_1 - r_2 & r_2 - r_3 \end{vmatrix} = 0$

or $(r_1 + r_2 + r_3)\{(r_3 - r_2)(r_2 - r_3) - (r_1 - r_2)(r_1 - r_3)\} = 0$

or $(r_1 + r_2 + r_3)$

$\times \{r_2r_3 - r_3^2 - r_2^2 + r_2r_3 - r_1^2 + r_1r_3 + r_1r_2 - r_2r_3\} = 0$

or $(r_1 + r_2 + r_3)(-1)\{r_1^2 + r_2^2 + r_3^2 - r_2r_3 - r_3r_1 - r_1r_2\} = 0$

or $(-1)\frac{1}{2}(r_1 + r_2 + r_3)$

$\times \{(r_1 - r_2)^2 + (r_2 - r_3)^2 + (r_3 - r_1)^2\} = 0$.

Clearly $r_1 + r_2 + r_3 \neq 0$ as r_1, r_2, r_3 are each positive.

$\therefore$ $r_1 - r_2 = 0 = r_2 - r_3 = r_3 - r_1$

$\therefore$ $r_1 = r_2 = r_3$.

Therefore z_1, z_2, z_3 are three complex numbers whose distances from the origin are equal. So z_1, z_2, z_3 lie on a circle with the centre at the origin.

Let the points representing the complex numbers be Z_1, Z_2, Z_3 respectively.

Again we know amp $\frac{z_3}{z_2}$ is the angle that the line segment Z_3Z_2 makes at the origin and

$$\text{amp}\left(\frac{z_3 - z_1}{z_2 - z_1}\right)^2 = 2\text{amp}\frac{z_3 - z_1}{z_2 - z_1} = 2 \times \angle Z_3Z_1Z_2.$$

We know that if the chord Z_3Z_2 subtends an angle 2θ at the centre of a circle then the chord subtends an angle $\angle Z_3Z_1Z_2 = \theta$ at the circumference.

Hence, we get $\text{amp}\dfrac{z_3}{z_2} = 2\text{amp}\dfrac{z_3 - z_1}{z_2 - z_1} = \text{amp}\left(\dfrac{z_3 - z_1}{z_2 - z_1}\right)^2$.

50. If z_1, z_2, z_3 are nonzero complex numbers such that $\dfrac{2}{z_1} = \dfrac{1}{z_2} + \dfrac{1}{z_3}$ then prove that z_1, z_2, z_3 lie on a circle passing through the origin.

Here $\dfrac{2}{z_1} = \dfrac{1}{z_2} + \dfrac{1}{z_3}$

or $\dfrac{1}{z_1} - \dfrac{1}{z_2} = \dfrac{1}{z_3} - \dfrac{1}{z_1}$

or $\dfrac{z_2 - z_1}{z_1z_2} = \dfrac{z_1 - z_3}{z_1z_3}$

or $\dfrac{z_3 - z_1}{z_2 - z_1} = \dfrac{-z_3}{z_2}$

$\therefore \quad \text{amp}\dfrac{z_3 - z_1}{z_2 - z_1} = \text{amp}\dfrac{-z_3}{z_2} = \text{amp}\dfrac{-1}{z_2/z_3}$

$= \text{amp}(-1) - \text{amp}\dfrac{z_2}{z_3}$

$= \pi - \text{amp}\dfrac{z_2}{z_3}$. ... (1)

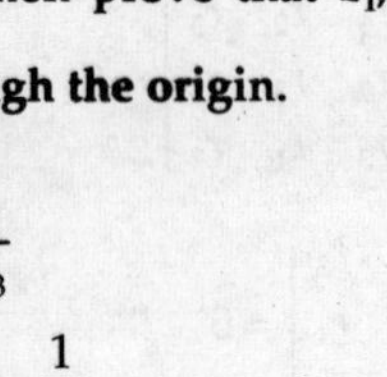

Let z_1, z_2, z_3 be represented by Z_1, Z_2, Z_3 respectively.

From (1), $\angle Z_3 Z_1 Z_2 = \pi - \angle Z_2OZ_3$

$\therefore \quad \angle Z_3Z_1Z_2 + \angle Z_2OZ_3 = \pi.$

Hence Z_1, Z_2, Z_3 and O are concyclic.

$\therefore \quad z_1, z_2, z_3$ lie on a circle passing through the origin.

51. If two lines cut the circle $|z| = r$ at the points representing the complex numbers a, b and c, d, and they meet at the point z, prove that

$$z(a^{-1}b^{-1} - c^{-1}d^{-1}) = a^{-1} + b^{-1} - c^{-1} - d^{-1}.$$

Clearly $|a| = |b| = |c| = |d| = r$.

$\therefore \quad a\bar{a} = b\bar{b} = c\bar{c} = d\bar{d} = r^2 \quad (\because \ |z|^2 = z\bar{z}).$

Now a, b and z are collinear. Let they lie on the line $\bar{\alpha}z + \alpha\bar{z} + \beta = 0, \beta \in R$.

$\therefore \quad \bar{\alpha}z + \alpha\bar{z} + \beta = 0$

$\bar{\alpha}a + \alpha\bar{a} + \beta = 0$

$\bar{\alpha}b + \alpha\bar{b} + \beta = 0.$

Eliminating $\bar{\alpha}, \alpha, \beta$ we get $\begin{vmatrix} z & \bar{z} & 1 \\ a & \bar{a} & 1 \\ b & \bar{b} & 1 \end{vmatrix} = 0.$

Similarly, $\begin{vmatrix} z & \bar{z} & 1 \\ c & \bar{c} & 1 \\ d & \bar{d} & 1 \end{vmatrix} = 0.$

On simplification,

$z(\bar{a} - \bar{b}) - \bar{z}(a - b) + (a\bar{b} - b\bar{a}) = 0$

and $z(\bar{c} - \bar{d}) - \bar{z}(c - d) + (c\bar{d} - d\bar{c}) = 0.$

By cross-multiplication,

$$\frac{z}{-(a-b)(c\bar{d} - d\bar{c}) + (c-d)(a\bar{b} - b\bar{a})}$$

$$= \frac{1}{-(\bar{a} - \bar{b})(c-d) + (a-b)(\bar{c} - \bar{d})}$$

$\therefore \quad z\left\{(a-b)\left(\dfrac{r^2}{c} - \dfrac{r^2}{d}\right) - \left(\dfrac{r^2}{a} - \dfrac{r^2}{b}\right)(c-d)\right\}$

$= (c-d)\left(a \cdot \dfrac{r^2}{b} - b\dfrac{r^2}{a}\right) - (a-b)\left(c\dfrac{r^2}{d} - d\dfrac{r^2}{c}\right)$

$\{\because \ a\bar{a} = r^2, \text{ etc.}\}$

or $\quad z\left\{(a-b) \cdot \dfrac{d-c}{cd} - \dfrac{b-a}{ab} \cdot (c-d)\right\}$

$= (c-d)\dfrac{a^2 - b^2}{ab} - (a-b)\dfrac{c^2 - d^2}{cd}$

or $\quad z(a-b)(c-d)\left(-\dfrac{1}{cd} + \dfrac{1}{ab}\right)$

$= (a-b)(c-d)\left\{\dfrac{a+b}{ab} - \dfrac{c+d}{cd}\right\}$

$\therefore \quad z(a^{-1}b^{-1} - c^{-1}d^{-1}) = a^{-1} + b^{-1} - c^{-1} - d^{-1}$

$\{\because \ a \neq b, c \neq d\}.$

52. If $b_1 + b_2 + b_3 + b_4 = 0$ where b_1, etc., are nonzero real numbers, sum of no two being zero, and $b_1z_1 + b_2z_2 + b_3z_3 + b_4z_4 = 0$ where no three of the points z_1, z_2, z_3, z_4 are collinear then prove that the four points will be concyclic if

$$b_1b_2|z_1 - z_2|^2 = b_3b_4|z_3 - z_4|^2.$$

Here $b_1 + b_2 = -(b_3 + b_4)$.

Also $b_1z_1 + b_2z_2 = -(b_3z_3 + b_4z_4)$.

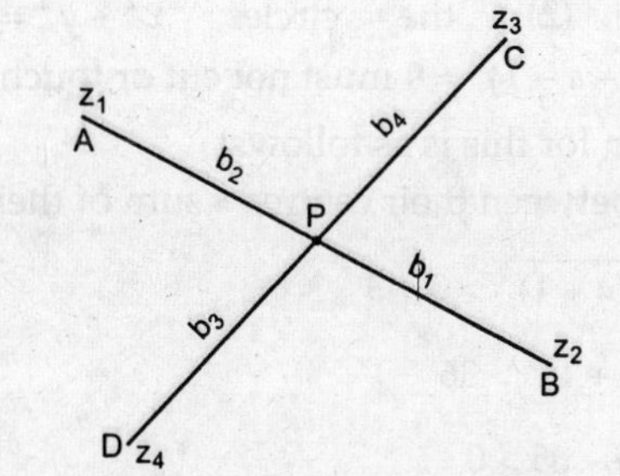

Dividing these, $\dfrac{b_1z_1 + b_2z_2}{b_1 + b_2} = \dfrac{b_3z_3 + b_4z_4}{b_3 + b_4}$.

The left side gives the point that divides the line segment joining $A(z_1)$, $B(z_2)$ in the ratio $b_2 : b_1$ and the right side gives the point that divides the line segment joining the points $C(z_3)$, $D(z_4)$ in the ratio $b_4 : b_3$. So the line segments intersect at P which represents $\dfrac{b_1z_1 + b_2z_2}{b_1 + b_2}$ as well as $\dfrac{b_3z_3 + b_4z_4}{b_3 + b_4}$.

Now, $AB = |z_1 - z_2|$

$\therefore \quad PA = \dfrac{b_2}{b_1 + b_2}|z_1 - z_2|$ and $PB = \dfrac{b_1}{b_1 + b_2}|z_1 - z_2|$

Also, $CD = |z_3 - z_4|$

$\therefore \quad PC = \dfrac{b_4}{b_3 + b_4}|z_3 - z_4|$ and $PD = \dfrac{b_3}{b_3 + b_4}|z_3 - z_4|$

The four points A, B, C, D will be concyclic if

$$PA \cdot PB = PC \cdot PD,$$

i.e., $\dfrac{b_1b_2}{(b_1 + b_2)^2}|z_1 - z_2|^2 = \dfrac{b_3b_4}{(b_3 + b_4)^2}|z_3 - z_4|^2$,

i.e., $b_1b_2|z_1 - z_2|^2 = b_3b_4|z_3 - z_4|^2$.

$$\{\because\ b_1 + b_2 = -(b_3 + b_4)\}.$$

53. If $z_1 + z_2 = 0$ and $z_1z_2 + z_3z_4 = 0$ then prove that the points representing the four complex numbers z_1, z_2, z_3 and z_4 in the Argand plane are concyclic.

[**Note** $\cos\theta + i\sin\theta$ may be written as $e^{i\theta}$ in Euler's notation. Thus $e^{i\theta} = \cos\theta + i\sin\theta$, $e^{-i\theta} = \cos\theta - i\sin\theta$; the complex number whose modulus is r and amplitude is θ may be written as $re^{i\theta}$. Also $e^{i\theta_1} \cdot e^{i\theta_2} = e^{i(\theta_1 + \theta_2)}$ because

$$(\cos\theta_1 + i\sin\theta_1)(\cos\theta_2 + i\sin\theta_2)$$
$$= \cos(\theta_1 + \theta_2) + i\sin(\theta_1 + \theta_2).]$$

Let $z_1 = ke^{i\theta}$; as $z_1 + z_2 = 0$ we get $z_2 = -ke^{i\theta}$.

Let $z_3 = re^{i\alpha}$ and $z_4 = Re^{i\beta}$.

$\because \quad z_1z_2 + z_3z_4 = 0,$

$$-k^2e^{i\theta} \cdot e^{i\theta} + rRe^{i\alpha} \cdot e^{i\beta} = 0$$

or $\quad rRe^{i(\alpha+\beta)} = k^2e^{2i\theta}$

or $\quad rR\{\cos(\alpha + \beta) + i\sin(\alpha + \beta)\} = k^2(\cos 2\theta + i\sin 2\theta)$

$\Rightarrow \quad rR\cos(\alpha + \beta) = k^2\cos 2\theta$ and

$\quad rR\sin(\alpha + \beta) = k^2\sin 2\theta$

Squaring and adding, $rR = k^2$ $\quad$ ... (1)

Dividing, $\tan(\alpha + \beta) = \tan 2\theta$,

i.e., $\alpha + \beta = 2\theta$ $\quad$ ... (2)

Let the equation of the circle passing through z_1, z_2, z_3 be

$$z\bar{z} + \bar{\lambda}z + \lambda\bar{z} + \mu = 0.$$

As $z_1 = ke^{i\theta}$, $z_2 = -ke^{i\theta}$ and $z_3 = re^{i\alpha}$ satisfy the equation

$$k^2 + k(\bar{\lambda}e^{i\theta} + \lambda e^{-i\theta}) + \mu = 0 \quad \text{... (3)}$$

$$k^2 - k(\bar{\lambda}e^{i\theta} + \lambda e^{-i\theta}) + \mu = 0 \quad \text{... (4)}$$

$$r^2 + r(\bar{\lambda}e^{i\alpha} + \lambda e^{-i\alpha}) + \mu = 0 \quad \text{... (5)}$$

Adding (3), (4) we get $2k^2 + 2\mu = 0$

i.e., $\mu = -k^2$ $\quad$ ... (6)

and $\bar{\lambda}e^{i\theta} + \lambda e^{-i\theta} = 0$

$\therefore \quad \bar{\lambda}e^{i\theta} + \dfrac{\lambda}{e^{i\theta}} = 0$

or $\quad \bar{\lambda}e^{2i\theta} + \lambda = 0$

or $\quad \bar{\lambda}e^{i(\alpha+\beta)} + \lambda = 0$, using (2)

or $\quad \bar{\lambda}e^{i\alpha} \cdot e^{i\beta} + \lambda = 0$

or $\quad \bar{\lambda}e^{i\alpha} + \lambda e^{-i\beta} = 0$ $\quad$ ... (7)

Taking conjugate, $\lambda e^{-i\alpha} + \bar{\lambda}e^{i\beta} = 0$ $\quad$ ... (8)

Adding (7), (8) we get $(\bar{\lambda}e^{i\alpha} + \lambda e^{-i\alpha}) + (\bar{\lambda}e^{i\beta} + \lambda e^{-i\beta}) = 0$

or $\quad \bar{\lambda}e^{i\alpha} + \lambda e^{-i\alpha} = -(\bar{\lambda}e^{i\beta} + \lambda e^{-i\beta})$ $\quad$ ... (9)

Putting in (5) from (1), (9) and (6),

$$\frac{k^4}{R^2} - \frac{k^2}{R}(\bar{\lambda}e^{i\beta} + \lambda e^{-i\beta}) - k^2 = 0$$

or $\quad k^2 - R(\bar{\lambda}e^{i\beta} + \lambda e^{-i\beta}) - R^2 = 0$

or $\quad R^2 + R(\bar{\lambda}e^{i\beta} + \lambda e^{-i\beta}) + \mu = 0$, using (6)

or $\quad z_4\bar{z}_4 + \bar{\lambda}z_4 + \lambda\bar{z}_4 + \mu = 0.$

Thus z_4 also satisfies the equation of the circle.

Hence the problem.

54. Find $a \in R$ if the complex number z is to satisfy
$|z| = 3$, $|z - \{a(1+i) - i\}| \le 3$ and
$|z + 2a - (a+1)i| > 3$
simultaneously for at least one z.

To satisfy all at a time z should be on the circle $|z| = 3$ but inside the circle $|z - \{a(1+i) - i\}| = 3$ and outside the circle $|z + 2a - (a+1)i| = 3$.

So $|z| = 3$ and $|z - \{a(1+i) - i\}| = 3$ should cut or touch ... (1)

and $|z| = 3$ and $|z + 2a - (a+1)i| = 3$ should not cut or touch ... (2)

Let $z = x + iy$. To satisfy (1), the circles $x^2 + y^2 = 9$ and $(x-a)^2 + (y-a+1)^2 = 9$ should cut or touch.

The condition for this is as follows:

distance between their centres ≤ sum of their radii

or $\sqrt{(a-0)^2 + (a-1-0)^2} \le 3 + 3$

or $\sqrt{a^2 + a^2 - 2a + 1} \le 6$

or $2a^2 - 2a + 1 \le 36$

or $2a^2 - 2a - 35 \le 0$ ... (3)

The corresponding equation is $2a^2 - 2a - 35 = 0$

or $a = \dfrac{2 \pm \sqrt{4 + 4 \cdot 2 \cdot 35}}{4} = \dfrac{2 \pm 2\sqrt{71}}{4} = \dfrac{1 \pm \sqrt{71}}{2}$

∴ sign-scheme for (3) is

(+) $\quad \dfrac{1-\sqrt{71}}{2}$ $\quad$ (−) $\quad \dfrac{1+\sqrt{71}}{2}$ $\quad$ (+)

∴ (3) is satisfied if $\dfrac{1-\sqrt{71}}{2} \le a \le \dfrac{1+\sqrt{71}}{2}$ ... (4)

To satisfy (2), the circles $x^2 + y^2 = 9$ and $(x+2a)^2 + (y-a-1)^2 = 9$ must not cut or touch.

The condition for this is as follows:

distance between their centres > sum of their radii

or $\sqrt{4a^2 + (a+1)^2} > 3 + 3$

or $4a^2 + (a+1)^2 > 36$

or $5a^2 + 2a - 35 > 0$ ... (5)

The corresponding equation is $5a^2 + 2a - 35 = 0$

or $a = \dfrac{-2 \pm \sqrt{4 + 4 \cdot 5 \cdot 35}}{10}$

$= \dfrac{-2 \pm 2\sqrt{176}}{10} = \dfrac{-1 \pm 4\sqrt{11}}{5}$

∴ the sign-scheme for (5) is

(+) $\quad \dfrac{-1-4\sqrt{11}}{5}$ $\quad$ (−) $\quad \dfrac{-1+4\sqrt{11}}{5}$ $\quad$ (+)

∴ (5) is satisfied if

$a < \dfrac{-1-4\sqrt{11}}{5}$ or $a > \dfrac{-1+4\sqrt{11}}{5}$ ... (6)

The common values of a satisfying (4) and (6) are

$\dfrac{-1+4\sqrt{11}}{5} < a \le \dfrac{1+\sqrt{71}}{2}$ or $\dfrac{1-\sqrt{71}}{2} \le a < \dfrac{-1-4\sqrt{11}}{5}$.

Exercises

1. Express $\dfrac{(2+i)^3}{2+3i}$ in the form $x + iy$.

2. If $z = \dfrac{3-4i}{(4-2i)(1+i)}$ then find Re(z).

3. If $\dfrac{\sin\frac{x}{2} + \cos\frac{x}{2} + i\tan x}{1 + 2i\sin\frac{x}{2}}$ is purely real, find the set of values of x.

4. Simplify $\dfrac{20}{\sqrt{3} - \sqrt{-2}} + \dfrac{30}{3\sqrt{-2} - 2\sqrt{3}} - \dfrac{14}{2\sqrt{3} - \sqrt{-2}}$ in the form $A + iB$.

5. Find all values of θ for which $\dfrac{1 - i\sin\theta}{1 + i\cos\theta}$ is purely real.

6. Find x, y if $(3x - 2iy)(2+i)^2 = 10(1+i)$.

7. Find z if $(x+iy)(2-3i) = 4 + i$ where $z = x + iy$.

8. Find all complex numbers z for which $\dfrac{z-2}{z+2}$ is purely imaginary.

9. Solve for x, y: $\dfrac{(1+i)x - 2i}{3+i} + \dfrac{(2-3i)y + i}{3-i} = i$.

10. If $z_1 = -3 + ixy$, $z_2 = x + y + 4i$ and $z_1 = \bar{z}_2$ then find x and y.

11. Evaluate $\left(i^{19} + \dfrac{1}{i^{25}}\right)^2$.

12. Evaluate $\left(i^{17} - \dfrac{1}{i^{34}}\right)^3$

13. If $\left(\dfrac{1+i}{1-i}\right)^n = 1$, find the least positive integral value of n.

14. Prove that

$(1+i)^{2n} + (1-i)^{2n} = 0$ if n is an odd integer

$\dfrac{2^{n+1}}{(-1)^{n/2}}$ if n is an even integer.

15. If $x = \sqrt{-2} - 1$, find the value of

$x^4 + 4x^3 + 6x^2 + 4x + 9$.

16. Find the value of $a^6 + a^4 + a^2 + 1$ when $a = \frac{1+i}{\sqrt{2}}$.

17. When $x = \frac{3 + 5\sqrt{-1}}{2}$, find the value of

$2x^3 + 2x^2 - 7x + 2$

and show that it will be unaltered if $x = \frac{3 - 5\sqrt{-1}}{2}$.

18. If $\sqrt{a - ib} = x - iy$, prove that $\sqrt{a + ib} = x + iy$.

19. If $\sqrt[3]{a + ib} = x + iy$, prove that

(i) $\sqrt[3]{a - ib} = x - iy$ (ii) $\frac{a}{x} + \frac{b}{y} = 4(x^2 - y^2)$.

20. Express $(1 + a^2)(1 + b^2)$ as sum of two squares.

21. If $(x^2 + a^2)(x^2 + b^2)(x^2 + c^2) = \{f(x)\}^2 + \{\phi(x)\}^2$, find $f(x)$ and $\phi(x)$ where $f(x)$ is a polynomial of the third degree.

22. Find $\sqrt{7 - 24i}$.

23. Find the square root of $8 - 15i$.

24. If $z^2 + 5 = 12\sqrt{-1}$, find the complex number z.

25. Find the square root of $\frac{x^2}{y^2} + \frac{y^2}{x^2} + \frac{1}{2i}\left(\frac{x}{y} + \frac{y}{x}\right) + \frac{31}{16}$.

26. Find the modulus, amplitude and conjugate of $\sqrt{2} - \sqrt{2}i$.

27. If $z = \frac{(1+i)(1+\sqrt{3}i)^2}{1-i}$, find $|z|$, arg z and $\bar{z}$.

28. Write $\frac{1+i}{2-i}$ in the trigonometrical form.

29. If $z_1 = 3i$ and $z_2 = -1 - i$, find the value of arg $\frac{z_1}{z_2}$.

30. Find the modulus and amplitude of

$$\frac{1}{(1-i)^2} - \frac{1}{(1+i)^2}$$

Also write the complex number in its polar form.

31. Solve for z: $z^2 + |z| = 0$.

32. Find the complex number z if $z^2 + \bar{z} = 0$.

33. Solve for z: $z^2 = (i\bar{z})^2$.

34. Find the solution of the equation

$|z + 1| = z + 2 + 2i$.

35. There are three nonzero complex numbers which satisfy the equation $i\bar{z} = z^2$. Determine these complex numbers.

36. Find all complex numbers satisfying

$2|z|^2 + z^2 - 5 + i\sqrt{3} = 0$.

37. Is there any z such that $\left|\frac{z+6}{z+4i}\right| = \frac{5}{3}$ and $\left|\frac{z+2}{z+4}\right| = 1$?

If so, find z.

38. For $c \geq 1$, find all complex numbers z satisfying the equation $z + c|z + 1| + i = 0$.

39. Find all complex numbers satisfying $\bar{z} = z^2$.

40. For every real $a > 0$, determine the complex numbers which will satisfy the equation

$|z|^2 - 2iz + 2a(1 + i) = 0$.

41. Solve for z:

$$\arg \frac{3z - 6 - 3i}{2z - 8 - 6i} = \frac{\pi}{4} \quad \text{and} \quad |z - 3 + i| = 3.$$

42. If $z = x + iy$ and $|z + 6| = |2z + 3|$, prove that $x^2 + y^2 = 9$.

43. If $z = x + iy$ and amp $\frac{z-1}{z+1} = \frac{\pi}{4}$ then prove that $x^2 + y^2 - 2y = 1$.

44. If z is a nonreal complex for which $|z| = 1$, prove that $\frac{z-1}{z+1}$ is a purely imaginary number.

45. If z is a unimodular complex number, prove that

$$\arg(z^2 + \bar{z}) = \frac{1}{2} \arg z.$$

46. If $|z| = 1$ and z is nonreal then prove that z can be expressed in the form $\frac{c+i}{c-i}$ where c is real. Also find c in terms of arg z.

47. If $\frac{1 - \bar{z}_1 z_2}{z_1 - z_2}$ is a unimodular complex number, prove that at least one of the two complex numbers z_1, z_2 must be unimodular.

48. Show that a real value of x will satisfy the equation

$$\frac{1 - ix}{1 + ix} = a - ib \text{ if } a^2 + b^2 = 1.$$

49. If $(a_1 + ib_1)(a_2 + ib_2) \ldots (a_n + ib_n) = x + iy$, prove that

(i) $(a_1^2 + b_1^2)(a_2^2 + b_2^2) \ldots (a_n^2 + b_n^2) = x^2 + y^2$

(ii) $\tan^{-1}\frac{b_1}{a_1} + \tan^{-1}\frac{b_2}{a_2} + \ldots + \tan^{-1}\frac{b_n}{a_n} = \tan^{-1}\frac{y}{x}$.

50. If $|z_1| = |z_2| = \ldots = |z_n| = 1$, prove that

$$|z_1 + z_2 + \ldots + z_n| = \left|\frac{1}{z_1} + \frac{1}{z_2} + \ldots + \frac{1}{z_n}\right|.$$

51. If $iz^3 + z^2 - z + i = 0$, show that $|z| = 1$.

52. If $f(z) = a_0 z^n + a_1 z^{n-1} + a_2 z^{n-2} + \ldots + a_n$ where z is a complex number and a_1, a_2, etc., are real, prove that $\overline{f(z)} = f(\bar{z})$.

53. If $|z_1 + z_2| = |z_1 - z_2|$, prove that

$$\text{amp } z_1 - \text{amp } z_2 = \frac{\pi}{2}.$$

54. Prove that $|z_1 + z_2|^2 = |z_1|^2 + |z_2|^2$ if and only if $\frac{z_1}{z_2}$ is purely imaginary.

55. Prove that

$$|z_1 + z_2|^2 + |z_1 - z_2|^2 = 2|z_1|^2 + 2|z_2|^2.$$

56. Prove that
$$|\alpha+\sqrt{\alpha^2-\beta^2}|+|\alpha-\sqrt{\alpha^2-\beta^2}| = |\alpha+\beta|+|\alpha-\beta|$$
where α, β are complex numbers.

57. Prove that
$$|1-\bar{z}_1 z_2|^2-|z_1-z_2|^2=(1-|z_1|^2)(1-|z_2|^2).$$

58. If $|z-1|<3$, prove that $|iz+3-5i|<8$.

59. If $|z|<\frac{1}{2}$, show that $|(1+i)z^3+iz|<\frac{3}{4}$.

60. Prove that for nonzero complex numbers
$$|z_1+z_2|\left|\frac{z_1}{|z_1|}+\frac{z_2}{|z_2|}\right|\le 2(|z_1|+|z_2|).$$

61. Prove that
$$|z_1-z_2|<|1-\bar{z}_1 z_2| \text{ if } |z_1|<1, |z_2|<1.$$

62. Prove that $\left|\frac{z}{|z|}-1\right|\le|\arg z|$.

63. Prove that
$$|z_1|^2+|z_2|^2-|z_1^2+z_2^2| \le \frac{2|az_1+bz_2|^2}{a^2+b^2}\le|z_1|^2+|z_2|^2+|z_1+z_2|^2,$$
where a, b are nonzero real numbers and z_1, z_2 are any two complex numbers.

64. If $|a_r|<2$; $r=1, 2, 3, \ldots, N$ then prove that there is no z inside the circle $|z|=\frac{1}{3}$ such that
$$a_N z^N+a_{N-1}z^{N-1}+\ldots+a_1 z+1=0.$$

65. Find the condition on the complex constants α, β if $z^2+\alpha z+\beta=0$ has a real root.

66. Find the range of values of $\alpha\in R$ for which the equation $z+\alpha|z-1|+2i=0$, $z=x+iy$ has a solution. Also find the solution.

67. If z is a complex number and $z^2+az+b=0$ $(a\ne 0)$ has roots both of which have unit modulus then prove that $|a|\le 2$, $|b|=1$, amp b = 2amp a.

68. If a, b are real, prove that the equation $z^2+az+b=0$ will not have any purely real solution if $a^2<4b$.

69. Find minimum $|z|$ satisfying $\left|z+\frac{1}{z}\right|=2$.

70. If $|z|\ge 3$, determine the least value of $\left|z+\frac{1}{z}\right|$.

71. If $\left|z+\frac{1}{z}\right|=1$, show that $\text{Re}(z)=0$ when $|z|$ is the maximum.

72. Locate the region in the Argand plane for the complex number z satisfying $|z-4|<|z-2|$.

73. Indicate the region represented by $\frac{\pi}{6}\le\arg z\le\frac{\pi}{4}$.

74. Indicate the region in the Argand plane represented by $|z+1|^2+|z-1|^2=4$.

75. Locate the complex number z in the Argand plane satisfying
$$\log_{\sec\frac{\pi}{4}}\frac{|z|^2+|z|+4}{2|z|-1}>2.$$

76. Prove that the product of any number of unimodular complex numbers is also a unimodular complex number whose amplitude is equal to the sum of their amplitudes.

77. Separate the real and imaginary parts of
$$\frac{(\cos\alpha+i\sin\alpha)(\cos\beta+i\sin\beta)}{(\cot\gamma+i)(1+i\tan\gamma)}.$$

78. Prove that $\left(\frac{1+\cos\theta+i\sin\theta}{1+\cos\theta-i\sin\theta}\right)^n=\cos n\theta+i\sin n\theta$.

79. If $\alpha=\frac{\pi}{12}$, prove that
$$\frac{(\cos\alpha+i\sin\alpha)(\cos 2\alpha+i\sin 2\alpha)}{\cos 3\alpha-i\sin 3\alpha}=i.$$

80. Prove that
$$\left(\frac{\cos x+i\sin x}{\sin x+i\cos x}\right)^5=\cos\left(10x-\frac{5\pi}{2}\right)+i\sin\left(10x-\frac{5\pi}{2}\right).$$

81. Find the values of

(i) $i^{1/4}$ (ii) $\sqrt[4]{1}$ (iii) $1^{1/5}$

82. Find all the values of

(i) $(1-i)^{1/3}$ (ii) $(1+\sqrt{3}i)^{2/5}$

(iii) $\sqrt[4]{-64}$ (iv) $(1+i)^n, n\in N$

83. Solve for x: (i) $x^5-1=0$ (ii) $x^6+1=0$.

84. Solve for x:

(i) $x^5+x^3-x^2-1=0$ (ii) $x^4+x^3+x^2+x+1=0$.

85. Solve for the common roots of
$$1+z^{100}+z^{1985}=0$$
$$1+2z+2z^2+z^3=0.$$

86. Solve: ${}^8C_0x^8+{}^8C_2x^6+{}^8C_4x^4+{}^8C_6x^2+1=0$.

87. Find the integral solutions of $(1+i)^x=(1-i)^x$.

88. Solve: $(1+i)^x=2^x$ where x is an integer $\ne 0$.

89. If $x^2-2x\cos\theta+1=0$, prove that
$$x^{2n}+1=2x^n\cos n\theta.$$

90. If $x+\frac{1}{x}=2\cos\alpha$, $y+\frac{1}{y}=2\cos\beta$ then prove that
$$\frac{x}{y}+\frac{y}{x}=2\cos(\alpha\pm\beta).$$

91. If $x_r=\cos\frac{\pi}{2^r}+i\sin\frac{\pi}{2^r}$, prove that
$$x_1\cdot x_2\cdot x_3\ldots \text{ to } \infty=-1.$$

92. If z_1, z_2 be the complex roots of $x^2 + 4 = 2x$, prove that $z_1^n + z_2^n = 2^{n+1} \cos \frac{n\pi}{3}$.

93. If $z = \cos\theta + i\sin\theta$ be a root of the equation
$$a_0 z^n + a_1 z^{n-1} + a_2 z^{n-2} + \dots + a_{n-1} z + a_n = 0$$
then prove that
(i) $a_0 + a_1\cos\theta + a_2\cos 2\theta + \dots + a_n\cos n\theta = 0$
(ii) $a_1\sin\theta + a_2\sin 2\theta + \dots + a_n\sin n\theta = 0$.

94. If $\cos\theta_1 + \cos\theta_2 + \cos\theta_3 = 0 = \sin\theta_1 + \sin\theta_2 + \sin\theta_3$ then prove that
(i) $\sin 2\theta_1 + \sin 2\theta_2 + \sin 2\theta_3 = 0$
(ii) $\cos 3\theta_1 + \cos 3\theta_2 + \cos 3\theta_3 = 3\cos(\theta_1 + \theta_2 + \theta_3)$.

95. Solve $z^7 + 1 = 0$ and deduce that
$$\cos\frac{\pi}{7} \cdot \cos\frac{3\pi}{7} \cdot \cos\frac{5\pi}{7} = -\frac{1}{8}.$$

96. Solve $z^9 - 1 = 0$ and deduce that
$$\cos\frac{2\pi}{9} \cdot \cos\frac{4\pi}{9} \cdot \cos\frac{6\pi}{9} \cdot \cos\frac{8\pi}{9} = \frac{1}{16}.$$

97. Prove that $(a + w + w^2)(a + w^2 + w^4)(a + w^4 + w^8) \dots$ to $2n$ factors $= (a-1)^{2n}$.

98. Prove that
$$(2 + w + w^2)^3 + (1 + w - w^2)^8 - (1 - 3w + w^2)^4 = 1.$$

99. Prove that $\alpha^4 + \beta^4 + \alpha^{-1} \cdot \beta^{-1} = 0$ where α, β are nonreal cube roots of unity.

100. If $x = a + b$, $y = aw + bw^2$, $z = aw^2 + bw$ then prove that
(i) $x^3 + y^3 + z^3 = 3(a^3 + b^3)$
(ii) $x^2 + y^2 + z^2 = 6ab$.

101. If $\frac{1}{a+w} + \frac{1}{b+w} + \frac{1}{c+w} + \frac{1}{d+w} = \frac{2}{w}$ where a, b, c are real and w is a nonreal cube root of unity then prove that
(i) $\frac{1}{a+w^2} + \frac{1}{b+w^2} + \frac{1}{c+w^2} + \frac{1}{d+w^2} = \frac{2}{w^2}$
(ii) $\Sigma abc = 2$
(iii) $a + b + c + d = 2abcd$
(iv) $\frac{1}{1+a} + \frac{1}{1+b} + \frac{1}{1+c} + \frac{1}{1+d} = 2$.

102. Prove that $\left(\frac{i+\sqrt{3}}{2}\right)^{100} + \left(\frac{i-\sqrt{3}}{2}\right)^{100} + 1 = 0$.

103. Prove that $\left(\frac{i+\sqrt{3}}{-i+\sqrt{3}}\right)^{200} + \left(\frac{i-\sqrt{3}}{i+\sqrt{3}}\right)^{200} = -1$.

104. Prove that $\sqrt{-1 - \sqrt{-1 - \sqrt{-1 - \dots \text{ to } \infty}}} = w$ or w^2.

105. Factorize into linear real or complex factors.
(i) $a^3 + b^3$ (ii) $a^3 + b^3 + c^3 - 3abc$

106. If $(1+x)^n = a_0 + a_1 x + a_2 x^2 + \dots + a_n x^n$ then find
(i) $a_0 - a_2 + a_4 - a_6 + \dots$
(ii) $a_1 - a_3 + a_5 - a_7 + \dots$
(iii) $(a_0 - a_2 + a_4 - \dots)^2 + (a_1 - a_3 + a_5 \dots)^2$
(iv) $a_0 + a_3 + a_6 + a_9 + \dots$

107. If $(1 + x + x^2)^n = a_0 + a_1 x + a_2 x^2 + \dots + a_{2n} x^{2n}$ then prove that $a_0 + a_3 + a_6 + \dots = 3^{n-1}$.

108. If $1, a_1, a_2, a_3, \dots, a_{n-1}$ are the nth roots of unity then prove that
(i) $(1 - a_1)(1 - a_2)(1 - a_3) \dots (1 - a_{n-1}) = n$
(ii) $1 + a_1 + a_2 + \dots + a_{n-1} = 0$
(iii) $\frac{1}{2-a_1} + \frac{1}{2-a_2} + \dots + \frac{1}{2-a_{n-1}} = \frac{(n-2)2^{n-1}+1}{2^n - 1}$.

109. (i) Compute the value of
$$1 + 2\alpha + 3\alpha^2 + \dots + n\alpha^{n-1}$$
in the form of a complex number where α is a nonreal complex nth root of unity.
(ii) Compute the value of
$$1 + 3\alpha + 5\alpha^2 + \dots \text{ to } n \text{ terms}$$
where α is a nonreal complex nth root of unity.

110. Evaluate: $\sum_{p=1}^{32} (3p+2) \left\{ \sum_{q=1}^{10} \left(\sin\frac{2q\pi}{11} - i\cos\frac{2q\pi}{11} \right) \right\}^p$.

111. Show that the polynomial
$$x^{4p} + x^{4q+1} + x^{4r+2} + x^{4s+3}$$
is divisible by $x^3 + x^2 + x + 1$ where $p, q, r, s \in N$.

112. If α is a root of $x^2 + x + 1 = 0$ then show that α is also a root of $x^{3p} + x^{3q+1} + x^{3r+2} = 0$ where p, q, r are integers.

113. If p, q, r are three consecutive integers ≥ 3, prove that $x^p + x^q + x^r$ is divisible by $x^3 + x^2 + x$.

114. If n is an odd integer greater than 3 but not a multiple of 3, prove that
$$(x+1)^n - x^n - 1 \text{ is divisible by } x^3 + x^2 + x.$$

115. If $z^2(\overline{w})^4 = 1$ where w is a nonreal complex cube root of 1 then find z.

116. Given z_1, z_2, z_3 are complex numbers such that
$$\alpha = z_1 + z_2 + z_3, \ \beta = z_1 + wz_2 + w^2 z_3,$$
$$\gamma = z_1 + w^2 z_2 + wz_3.$$
Prove that
$$|\alpha|^2 + |\beta|^2 + |\gamma|^2 = 3(|z_1|^2 + |z_2|^2 + |z_3|^2).$$

117. Find z satisfying $|z - 5i| \leq 1$ such that amp z is the minimum.

118. If $|z - 25i| \leq 15$, find the value of
$$|\text{maximum amp } z - \text{minimum amp } z|.$$

119. Find the complex number z, the greatest in absolute value which satisfies $|z-2+2i| = 1$.

120. If the complex numbers z_1, z_2, z_3 are in AP, prove that they will represent collinear points in the Argand plane.

121. If the distinct points z_1, z_2, z_3 are collinear, prove that

$$z_1|z_2-z_3| \pm z_2|z_3-z_1| \pm z_3|z_1-z_2| = 0.$$

122. If $b_1+b_2+b_3=0$, none of these real constants being zero, and $b_1z_1+b_2z_2+b_3z_3=0$ then prove that z_1, z_2, z_3 will represent collinear points.

123. If $\left|\dfrac{1-iz}{z-i}\right| = 1$, prove that the locus of the variable point z in the Argand plane is the real axis.

124. Prove that $b\bar{z}+\bar{b}z = c$ where c is a real constant, represents a straight line in the Argand plane.

125. Prove that the points represented by the roots of $z^n = (z+1)^n$ lie on a straight line. Find the complex equation of the line.

126. The complex slope of the line joining the points represented by z_1, z_2 in the Argand plane is defined as $\dfrac{z_1-z_2}{\bar{z}_1-\bar{z}_2}$. Find the condition for parallelism of two lines with complex slopes w_1, w_2.

127. Find the area of the triangle whose vertices in the Argand plane are $i, -i, 1+i$.

128. Show that the area of the triangle on the Argand diagram formed by the complex numbers z, iz and $z+iz$ is $\dfrac{1}{2}|z|^2$.

129. Find the area of the triangle whose vertices represent the three roots of the complex equation

$$z^4 = (z+1)^4.$$

130. If the complex numbers z, iz and $z+iz$ are three vertices of a triangle, find the altitude of the triangle through the vertex $z+iz$. Also prove that the triangle is isosceles.

131. Let A and B be two complex numbers such that $\dfrac{A}{B}+\dfrac{B}{A}=1$. Prove that the origin and the two points represented by A and B form vertices of an equilateral triangle.

132. If the complex numbers z_1, z_2 and origin are the vertices of a triangle such that $z_1^2+z_2^2+z_1z_2=0$ then the triangle is isosceles. Prove it.

133. Prove that the complex numbers z_1, z_2 and the origin form an equilateral triangle only if $z_1^2+z_2^2-z_1z_2=0$.

134. Complex numbers z_1, z_2 and z_3 are the vertices A, B and C respectively of an isosceles right-angled triangle with right angle at C. Show that

$$(z_1-z_2)^2 = 2(z_1-z_3)(z_3-z_2).$$

135. If z_1, z_2, z_3 be the vertices of an equilateral triangle occurring in the anticlockwise sense then prove that

(i) $z_1^2+z_2^2+z_3^2 = z_1z_2+z_2z_3+z_3z_1$

(ii) $\dfrac{1}{z_1-z_2}+\dfrac{1}{z_2-z_3}+\dfrac{1}{z_3-z_1}=0$

(iii) $z_1+wz_2+w^2z_3=0$ where w is a nonreal cube root of unity.

136. If z_1, z_2, z_3 are the vertices of an isosceles triangle, right angled at z_1 then prove that

$$2z_1^2+z_2^2+z_3^2 = 2z_1(z_2+z_3).$$

137. Prove that the triangle whose vertices are represented by cube roots of unity is an equilateral triangle.

138. Prove that z, w, w^2 are the vertices of an isosceles triangle in the Argand plane if $\text{Im}(z)=0$, w and w^2 being complex cube roots of 1. Find the condition for the triangle to be an equilateral triangle.

139. If $z_1+z_2+z_3=0$ and z_1, z_2, z_3 are nonzero complex numbers then prove that the difference of the amplitudes of at least two of the complex numbers is greater than or equal to $\dfrac{2\pi}{3}$.

140. If a, b, c and u, v, w are the complex numbers representing the vertices of two triangles such that $c=(1-r)a+rb$, $w=(1-r)u+rv$ where r is nonreal complex then prove that the triangles are similar.

141. If z_0 represents the orthocentre of the triangle whose vertices represent the complex numbers $0, z_1, z_2$ then prove that

$$z_0(z_1\bar{z}_2-\bar{z}_1z_2) = (z_1-z_2)(z_1\bar{z}_2+\bar{z}_1z_2).$$

142. Let z_1, z_2, z_3 represent vertices of a triangle whose circumcentre is represented by z_0. Prove that

$$z_0\Sigma\,\bar{z}_1(z_2-z_3) = \Sigma|z_1|^2\cdot(z_2-z_3).$$

143. Prove that the nth roots of unity are represented on a circle of unit radius and form the vertices of a regular polygon of n sides. If z_1, z_2 are two consecutive vertices and the centre of the circle is $z=0$ then find the value of $\left|\text{amp}\,\dfrac{z_1}{z_2}\right|$.

144. If the centre of a regular hexagon be $z=0$ and one vertex $z=z_1$ then find the complex numbers representing two vertices adjacent to z_1.

145. If two consecutive vertices of a regular hexagon be z and $\bar{z}$, find the other vertices if the centre is $z=0$.

146. If z_1, z_2 are two complex numbers representing two consecutive vertices of a square then what is the complex number represented by the centre of the square? Also find the other vertices.

147. If two vertices of an equilateral triangle are z_1, z_2 then find the complex number z_3 represented by the third vertex where z_1, z_2, z_3 are anticlockwise.

148. z_1, z_2 are two vertices of a parallelogram whose angle at z_2 is $\frac{\pi}{4}\cdot$ Find the complex number z_3 representing the vertex adjacent to z_2 if z_1, z_2, z_3 are in anticlockwise sense and the side joining z_3, z_2 is half the length of the side joining z_1, z_2.

149. Prove that the point represented by z satisfying $z\bar{z} + a\bar{z} + \bar{a}z + b = 0, b \in R$ lies on a circle. Also find the complex number representing its centre.

150. If $\sum_{i=1}^{4} b_i = 0$ and $\sum_{i=1}^{4} b_i z_i = 0$ where b_is are nonzero real and z_1, z_2, z_3, z_4 are concyclic then prove that $\sum_{i=1}^{4} b_i |z_i|^2 = 0$.

151. Prove that $|z - z_1|^2 + |z - z_2|^2 = k$ will represent a circle if $|z_1 - z_2|^2 \le 2k$.

152. Find the equations in complex variables of all the circles which are orthogonal to $|z| = 1$ and $|z - 1| = 4$.

153. Prove that the locus of the point z such that $\arg\frac{z-a}{z-b} = k$ = constant, is an arc of a circle and find the centre of the circle where a, b are given complex numbers.

154. If $z = z_1 + \frac{1}{z_1}$ and z_1 is any point on a fixed circle of radius $\ne 1$ with the centre at the origin then prove that z lies on an ellipse.

155. Find $t \in R$ such that there is at least one z satisfying $|z+3| \le t^2 - 2t + 3$ and $|z - 3\sqrt{3}i| < t^2$.

Objective Questions

Fill in the blanks.

156. If $z = 4i - 5$ then $|z| =$ _____ and $\bar{z} =$ _____ .

157. If the conjugate of $(x + iy)(1 + 2i)$ be $1 + i$ then $x =$ _____ , $y =$ _____ .

158. If $|z| = 2$ and $\arg z = \frac{\pi}{4}$ then $z =$ _____ .

159. In the Argand plane, the vector OP, where O is the origin and P represents the complex number $z = 4 - 3i$, is turned in the clockwise sense through 180° and stretched 3 times. The complex number represented by the new vector is _____ .

160. The value of $i^{103} + i^{-99} =$ _____ where $i = \sqrt{-1}$.

161. The greatest and the least absolute value of $z + 1$, where $|z + 4| \le 3$ are _____ and _____ .

162. If $a + ib = x - iy$ then in terms of x, y we have $a - ib =$ _____ .

163. If n is not a multiple of 3, $1 + \omega^{-n} + \omega^{-2n} =$ _____ where ω is a nonreal cube root of 1.

164. If α, β, γ be the cube roots of $p, p < 0$ then for any real x, y and z

$$\frac{x\alpha + y\beta + z\gamma}{x\beta + y\gamma + z\alpha} = \_\_\_\_\_ .$$

165. The value of $|\sqrt{\alpha}|$, where α is a nonreal cube root of unity, is _____ .

166. If ω is a nonreal cube root of unity then

$$\frac{a + b\omega + c\omega^2}{b + c\omega + a\omega^2} + \frac{a + b\omega + c\omega^2}{c + a\omega + b\omega^2} = \_\_\_\_\_ .$$

167. Let the point P represent the complex number z. If OP is turned round O by an angle $\frac{\pi}{2}$ in the clockwise sense where O is the origin then in the new position P represents the complex number _____ .

168. (i) $|\sqrt{z}|^4 =$ _____ where z is a unimodular complex number.

(ii) $\frac{z|\bar{z}|^2}{\bar{z}|z|^2} =$ _____ where z is a complex number of amplitude $\frac{\pi}{3}$.

169. (i) The diagonals AC and BD of the rhombus intersect at M and $BD : AC = 2 : 1$. If D and M represent the complex numbers $1 + i$ and $2 - i$ respectively then A represents the complex number _____ or _____ .

(ii) The angle that the vector representing the complex number $\frac{1}{(\sqrt{3}+i)^{100}}$ makes with the positive direction of real axis is _____ .

170. If $|z| = 2$ then the complex number $\frac{1}{z}$ is represented on a fixed _____ .

171. For any two complex numbers z_1, z_2 and any two real numbers a and b, $|az_1 - bz_2|^2 + |bz_1 + az_2|^2 =$ _____ .

172. If a and b are real numbers between 0 and 1 such that the points $z_1 = a + i$, $z_2 = 1 + bi$ and $z_3 = 0$ form an equilateral triangle then $a =$ _____ , $b =$ _____ .

173. Suppose z_1, z_2, z_3 are the vertices of an equilateral triangle inscribed in the circle $|z| = 2$. If $z_1 = 1 + i\sqrt{3}$ then $z_2 =$ _____ , $z_3 =$ _____ .

174. Any complex number z satisfying $|z|^2 + b\bar{z} + \bar{b}z + c = 0$ where $c \in R$, $b\bar{b} > c$ is represented by points in the Argand plane at a fixed distance from the point representing the complex number ______ .

175. If the real part of $\frac{z+4}{2z+i}$ is $\frac{1}{2}$ then the point z lies on ______ .

Choose the correct option(s).

176. If $a > 0$, $b > 0$ then $\sqrt{-a} \cdot \sqrt{b}$ is equal to

(a) $-\sqrt{ab}$ (b) $\sqrt{ab}\,i$

(c) $\sqrt{ab}$ (d) none of these

177. The value of $\sum_{n=1}^{200} i^n$ where $i = \sqrt{-1}$ is

(a) 0 (b) 50

(c) –50 (d) none of these

178. The smallest $n \in N$ for which $(1+i)^n \cdot (1-i)^{2-n}$ is a real number, is

(a) 3 (b) 2 (c) 1 (d) 5

179. The digit in the units place in the value of $(727)^{39}$ is

(a) 3 (b) 7 (c) 1 (d) 9

180. If n is an integer, not a multiple of 3, the sum of $w^n + w^{2n}$, w being a nonreal complex cube root of unity, is

(a) 2 (b) –1

(c) 2 or –1 (d) none of these

181. If $\alpha = \sqrt[3]{1}$; and α is not real then $\alpha^{3n+1} + \alpha^{3n+3} + \alpha^{3n+5}$ has the value

(a) –1 (b) 0 (c) 1 (d) 3

182. If $\sqrt[3]{1} = 1, w, w^2$ then the roots of $(x-1)^3 + 8 = 0$ are

(a) $-1, 1+2w, 1+2w^2$ (b) $-1, 1-2w, 1-2w^2$

(c) –1, –1, –1 (d) none of these

183. If z is a complex number then

(a) $z + \bar{z}$ and $z\bar{z}$ are both purely real

(b) $z + \bar{z}$ is purely real but $z\bar{z}$ is not

(c) $z\bar{z}$ is purely real but $z + \bar{z}$ is not

(d) neither $z + \bar{z}$ nor $z\bar{z}$ need be purely real

184. If $z = i\bar{z}$ then

(a) z is purely real (b) z is purely imaginary

(c) $z = x(1+i)$, $x \in R$ (d) $z = 0$

185. The value of $\sqrt{i}$ is

(a) $1-i$ (b) $1+i$ (c) $\pm(1+i)$

(d) $i-1$ (e) $\frac{\pm 1}{\sqrt{2}}(1+i)$

186. If $z = \left(\frac{\sqrt{3}}{2} + \frac{i}{2}\right)^5 + \left(\frac{\sqrt{3}}{2} - \frac{i}{2}\right)^5$ then

(a) Re(z) = 0 (b) Im(z) = 0

(c) Re(z) > 0, Im(z) > 0 (d) Re(z) > 0, Im(z) < 0

187. If $z_r = \sin\frac{2\pi r}{11} - i\cos\frac{2r\pi}{11}$ then

(i) the value of $\sum_{r=0}^{10} z_r$ is equal to

(a) –1 (b) 0 (c) $-i$ (d) i

(ii) the value of $\sum_{r=1}^{10} z_r$ is equal to

(a) –1 (b) 0 (c) $-i$ (d) i

188. The complex numbers $\sin x - i\cos 2x$ and $\cos x - i\sin 2x$ are conjugate to each other for

(a) $x = n\pi$ (b) $x = 0$

(c) $x = \left(n + \frac{1}{2}\right)\pi$ (d) no value of x

189. If $|z_1| = |z_2| = 1$ and amp z_1 + amp $z_2 = 0$ then

(a) $z_1 z_2 = 1$ (b) $z_1 + z_2 = 0$

(c) $z_1 = \bar{z}_2$ (d) none of these

190. If z_1 and z_2 are two nonzero complex numbers such that $|z_1 + z_2| = |z_1| + |z_2|$ then $\arg z_1 - \arg z_2$ is equal to

(a) $-\pi$ (b) $-\frac{\pi}{2}$ (c) 0 (d) $\frac{\pi}{2}$ (e) π

191. The inequality $|z+2| < |z-2|$ represents the region given by

(a) Re(z) > 0 (b) Re(z) < 0

(c) Re(z) > 2 (d) none of these

192. Let z_1 and z_2 be complex numbers such that $z_1 \neq z_2$ and $|z_1| = |z_2|$. If Re(z_1) > 0 and Im(z_2) < 0 then $\frac{z_1 + z_2}{z_1 - z_2}$ may be

(a) zero (b) real and positive

(c) real and negative (d) purely imaginary

(e) none of these

193. If $z_1 = a + ib$ and $z_2 = c + id$ are complex numbers such that $|z_1| = |z_2| = 1$ and Re($z_1\bar{z}_2$) = 0 then the pair of complex numbers $w_1 = a + ic$ and $w_2 = b + id$ satisfy

(a) $|w_1| = 1$ (b) $|w_2| = 1$

(c) Re($w_1\bar{w}_2$) = 0 (d) none of these

194. If $|z_1| = |z_2| \neq 0$ and amp $\frac{z_1}{z_2} = \pi$ then

(a) $z_1 = z_2$ (b) $z_1 + z_2 = 0$

(c) $z_1 z_2 = 1$ (d) none of these

195. If z_1 and z_2 are two nonzero complex numbers such that $|z_1 - z_2| = |z_1| - |z_2|$ then $\arg z_1 - \arg z_2$ is equal to

(a) 0 (b) π (c) $\frac{\pi}{2}$ (d) $-\frac{\pi}{2}$

196. If z_1, z_2 are nonreal complex and $\left|\frac{z_1 + z_2}{z_1 - z_2}\right| = 1$ then $\frac{z_1}{z_2}$ is

(a) real positive (b) purely imaginary
(c) negative real (d) none of these

197. $|z^2|^3$ is equal to

(a) $|z^3|^2$ (b) $|\bar{z}^3|^2$ (c) $|z|^6$ (d) $|z^6|$

198. The equation $z^5 + z^4 + z^3 + z^2 + z + 1 = 0$ is satisfied by

(a) $z = \pm 1$ (b) $z = -1$
(c) $\pm\frac{1}{2} + i\frac{\sqrt{3}}{2}$ (d) $\frac{1}{2} \pm i\frac{\sqrt{3}}{2}$

199. The points z_1, z_2, z_3, z_4 in the complex plane are the vertices of a parallelogram taken in order if

(a) $z_1 + z_4 = z_2 + z_3$ (b) $z_1 + z_3 = z_2 + z_4$
(c) $z_1 + z_2 = z_3 + z_4$ (d) none of these

200. If $z^4 = (z-1)^4$ then the roots are represented in the Argand plane by the points that are

(a) collinear (b) concyclic
(c) vertices of a parallelogram (d) none of these

201. If $|z_1| = |z_2| = |z_3| = 1$ and $z_1 + z_2 + z_3 = 0$ then the area of the triangle whose vertices are z_1, z_2, z_3 is

(a) $\frac{3\sqrt{3}}{4}$ (b) $\frac{\sqrt{3}}{4}$
(c) 1 (d) none of these

202. In the Argand plane, $\left|\frac{z-i}{z+i}\right| = 4$ represents a

(a) pair of distinct lines (b) circle
(c) a pair of coincident lines (d) none of these

203. If $|z_1| = |z_2| = |z_3| = |z_4|$ and $z_1 + z_2 + z_3 + z_4 = 0$ then the points z_1, z_2, z_3, z_4 in the Argand plane are the vertices of a

(a) trapezium (b) rectangle
(c) square (d) none of these

204. If $\arg\frac{z - z_1}{z_2 - z_1} = 0$ for three distinct complex numbers z, z_1, z_2 then the three points are

(a) concyclic
(b) vertices of an equilateral triangle
(c) collinear
(d) vertices of a right-angled triangle

205. The complex number $z = x + iy$ which satisfies the equation $\left|\frac{z - 5i}{z + 5i}\right| = 1$ lies on

(a) the x-axis
(b) the straight line $y = 5$
(c) a circle passing through the origin
(d) none of these

206. The locus of a point z in the Argand plane for which $\arg\frac{z-1}{z+1} = \frac{\pi}{3}$ is

(a) a circle (b) a straight line
(c) a pair of straight lines (d) none of these

207. If amp$\left(\frac{z-2}{z+2}\right) = \frac{\pi}{4}$ then z lies on

(a) a straight line (b) a pair of straight lines
(c) a circle (d) an ellipse

208. $z\bar{z} + a\bar{z} + \bar{a}z + b = 0$ where $b \in R$ represents a real circle of nonzero radius if

(a) $|\bar{a}|^2 > b$ (b) $|a|^2 < b$
(c) $|a|^2 \geq b$ (d) $|a|^2 \leq b$

209. The points representing the values of $\sqrt[3]{\sqrt{5} + i\sqrt{3}}$ lie

(a) on a circle whose centre is (0, 0) and radius is $2\sqrt{2}$
(b) on a straight line
(c) on a circle whose centre is (0, 0) and radius is $\sqrt{2}$
(d) none of these

State whether the statements are true or false.

210. If $i^p = i^q$ where $i^2 = -1$ then $p - q$ is divisible by 4.

211. If $x_1 < x_2$ and $y_1 < y_2$ and $z_1 = x_1 + iy_1, z_2 = x_2 + iy_2$ then $z_1 < z_2$.

212. The equation $(5 + 12i)^x = 13^x$ has a positive integral solution.

213. If z is a nonreal complex number and $|z| = 1$ then $z^2 + \frac{1}{z^2} = 2$.

214. $(a_1^2 + b_1^2)(a_2^2 + b_2^2)(a_3^2 + b_3^2) \ldots (a_n^2 + b_n^2)$ can be written as the sum of two squares.

215. If $z = \frac{\cos 2\theta + i\sin 2\theta}{\cos\theta - i\sin\theta}$ then z is unimodular.

216. If two nonzero complex numbers z_1, z_2 be such that $z_1 + z_2$ is real then they are necessarily conjugate to each other.

217. For complex numbers $z_1 = x_1 + iy_1$ and $z_2 = x_2 + iy_2$ we write $z_1 \cap z_2$ if $x_1 \leq x_2$ and $y_1 \leq y_2$. Then for all complex numbers z with $1 \cap z$ we have $\frac{1-z}{1+z} \cap 0$.

218. If z is a nonzero complex number then $(\overline{z^{-1}}) = (\bar{z})^{-1}$.

219. If the points P and Q represent the complex numbers z and iz then $\angle POQ$ is a right angle.

220. If $|z_1 + z_2| = \left|\frac{1}{z_1} + \frac{1}{z_2}\right|$ then at least one of z_1, z_2 is unimodular.

221. If the complex numbers z_1, z_2, z_3 represent the vertices of an equilateral triangle such that $|z_1| = |z_2| = |z_3|$ then $z_1 + z_2 + z_3 \neq 0$.

222. If in the Argand plane z_1, z_2, z_3 and z_4 are four points such that $|z_1| = |z_2| = |z_3| = |z_4|$ then the four points are vertices of a square.

223. If $z \neq 0$ then $\arg z + \arg \frac{1}{z} = 0$.

224. Let z_1 and z_2 be the roots of $z^2 + pz + q = 0$. Then the points represented by z_1, z_2 and the origin form an equilateral triangle if $p^2 = 3q$.

225. If z_1, z_2 are nonzero complex numbers then

$$\left|\frac{z_1}{|z_1|} + \frac{z_2}{|z_2|}\right| \leq 2.$$

226. The nth roots of -1 can be n terms of a GP.

Answers

1. $\frac{37}{13} + \frac{16}{13}i$ **2.** $\frac{1}{4}$

3. $2n\pi, n\pi + \frac{\pi}{4};\ n \in \mathbb{Z}$ **4.** $0 + i0$

5. $n\pi - \frac{\pi}{4}$ **6.** $x = \frac{14}{15}, y = \frac{1}{5}$

7. $\frac{5}{13} + \frac{14}{13}i$

8. $z = 2(\cos\theta + i\sin\theta)$ where $\theta \neq n\pi$

9. $x = 3, y = -1$ **10.** $x = -4, 1;\ y = 1, -4$

11. -4 **12.** $-2(1 - i)$

13. 4 **15.** 12

16. 0 **17.** -66 **20.** $(1 - ab)^2 + (a + b)^2$

21. $f(x) = x^3 - (ab + bc + ca)x,\ \phi(x) = x^2(a + b + c) - abc$

22. $\pm(4 - 3i)$ **23.** $\pm\frac{1}{\sqrt{2}}(5 - 3i)$

24. $\pm(2 + 3i)$ **25.** $\pm\left(\frac{x}{y} + \frac{y}{x} - \frac{i}{4}\right)$

26. $2, -\frac{\pi}{4}, \sqrt{2} + \sqrt{2}i$ respectively

27. $4, \frac{-5\pi}{6}, 2(-\sqrt{3} + i)$ respectively

28. $\sqrt{\frac{2}{5}}\{\cos(\tan^{-1}3) + i\sin(\tan^{-1}3)\}$ **29.** $\frac{-3\pi}{4}$

30. $1, \frac{\pi}{2}, \cos\frac{\pi}{2} + i\sin\frac{\pi}{2}$ respectively **31.** $z = 0, i, -i$

32. $0, -1, \frac{1 \pm \sqrt{3}i}{2}$ **33.** $x(1 \pm i)$ where $x \in R$

34. $\frac{1}{2} - 2i$ **35.** $-i, -iw, -iw^2$

36. $\pm\frac{1 - 3\sqrt{3}i}{\sqrt{6}}, \pm\frac{\sqrt{3} - i}{\sqrt{2}}$ **37.** $-3 - 4i,\ -3 - \frac{17}{2}i$

38. $z = \frac{-c^2 \pm c\sqrt{2 - c^2}}{c^2 - 1} - i$ where $1 < c \leq \sqrt{2}$ and $z = -1 - i$ when $c = 1$

39. $0, 1, w, w^2$

40. $a + i\{-1 \pm \sqrt{1 - a^2 - 2a}\}$ where $0 < a \leq \sqrt{2} - 1$ and for other values of a, z has no value

41. $\left(4 - \frac{4}{\sqrt{5}}\right) + i\left(1 + \frac{2}{\sqrt{5}}\right), \left(4 + \frac{4}{\sqrt{5}}\right) + i\left(1 - \frac{2}{\sqrt{5}}\right)$

46. $\cot\frac{1}{2}(\arg z)$

65. $(\alpha - \bar{\alpha})\cdot(\bar{\alpha}\beta - \alpha\bar{\beta}) = (\beta - \bar{\beta})^2$

66. range of values of $\alpha = \left[-\frac{\sqrt{5}}{2}, 1\right)$

$z = \frac{\alpha^2 \pm |\alpha|\sqrt{5 - 4\alpha^2}}{\alpha^2 - 1} - 2i$ when $\alpha \in \left[-\frac{\sqrt{5}}{2}, -1\right)$

$z = \frac{5}{2} - 2i$ when $\alpha = -1$

$z = \frac{\alpha^2 - |\alpha|\sqrt{5 - 4\alpha^2}}{\alpha^2 - 1} - 2i$ when $\alpha \in (-1, 0)$

$z = -2i$ when $\alpha = 0$

$z = \frac{\alpha^2 + |\alpha|\sqrt{5 - 4\alpha^2}}{\alpha^2 - 1} - 2i$ when $\alpha \in (0, 1)$

69. $(3 - 2\sqrt{2})^{\frac{1}{2}}$

70. $\frac{8}{3}$ **72.** half plane $x > 3$

73. region bounded by the rays $y = x$ and $y = \frac{x}{\sqrt{3}}$ except the origin

74. points on the circle whose radius is 1 and the centre is $(0, 0)$

75. region outside the circle whose centre is (0, 0) and the radius $= \frac{1}{2}$

77. $\frac{1}{2}\sin 2\gamma\,\{\cos(\alpha+\beta-2\gamma)+i\sin(\alpha+\beta-2\gamma)\}$

81. (i) $\pm\left(\cos\frac{\pi}{8}+i\sin\frac{\pi}{8}\right), \pm\left(\cos\frac{5\pi}{8}+i\sin\frac{5\pi}{8}\right)$

(ii) $1, -1, i, -i$

(iii) $\cos\frac{2r\pi}{5}+i\sin\frac{2r\pi}{5};\ r=0,1,2,3,4$

82. (i) $2^{\frac{1}{6}}\left(\cos\frac{\pi}{12}-i\sin\frac{\pi}{12}\right),\ 2^{\frac{1}{6}}\left(\cos\frac{7\pi}{12}+i\sin\frac{7\pi}{12}\right),$

$2^{\frac{1}{6}}\left(\cos\frac{5\pi}{4}+i\sin\frac{5\pi}{4}\right)$

(ii) $2^{\frac{2}{5}}\left\{\cos\frac{2}{15}(6r+1)\pi+i\sin\frac{2}{15}(6r+1)\pi\right\};$

$r=0,1,2,3,4$

(iii) $2^{\frac{3}{2}}\left\{\cos\frac{(2r+1)\pi}{4}+i\sin\frac{(2r+1)\pi}{4}\right\};\ r=0,1,2,3$

(iv) $2^{\frac{n}{2}}\left(\cos\frac{n\pi}{4}+i\sin\frac{n\pi}{4}\right)$

83. (i) $\cos\frac{2r\pi}{5}+i\sin\frac{2r\pi}{5};\ r=0,1,2,3,4$

(ii) $\cos\frac{(2r+1)\pi}{6}+i\sin\frac{(2r+1)\pi}{6};\ r=0,1,2,3,4,5$

84. (i) $i, -i, 1, w, w^2$ (ii) $\cos\frac{2r\pi}{5}+i\sin\frac{2r\pi}{5};\ r=1,2,3,4$

85. w, w^2

86. $\dfrac{\cot\frac{(2r+1)\pi}{16}+i}{-\tan\frac{(2r+1)\pi}{16}+i};\ r=0,1,2,\ldots,7$

87. $0, \pm 4, \pm 8, \pm 12, \ldots$ **88.** no solution

95. $\cos\frac{(2r+1)\pi}{7}+i\sin\frac{(2r+1)\pi}{7};\ r=0,1,2,\ldots,6$

96. $\cos\frac{2r\pi}{9}+i\sin\frac{2r\pi}{9};\ r=0,1,2,\ldots,8$

105. (i) $(a+b)(a+bw)(a+bw^2)$

(ii) $(a+b+c)(a+bw+cw^2)(a+bw^2+cw)$

106. (i) $2^{\frac{n}{2}}\cos\frac{n\pi}{4}$ (ii) $2^{\frac{n}{2}}\sin\frac{n\pi}{4}$

(iii) 2^n (iv) $\{2^n+(-1)^n\cdot(w^n+w^{2n})\}\cdot\frac{1}{3}$

109. (i) $-\frac{n}{2}-i\cdot\frac{n}{2}\cot\frac{r\pi}{n};\ r=1,2,\ldots,(n-1)$

(ii) $-n-i\cdot n\cot\frac{r\pi}{n};\ r=1,2,\ldots,(n-1)$

110. $48(1-i)$ **115.** $\pm\overline{w}$

117. $\frac{2\sqrt{6}}{5}(1+2\sqrt{6}i)$ **118.** $2\sin^{-1}\frac{3}{5}$

119. $\frac{2\sqrt{2}+1}{\sqrt{2}}-i\frac{2\sqrt{2}+1}{\sqrt{2}}$ **125.** $z+\overline{z}+1=0$

126. $w_1=w_2$ **127.** 1

129. 0 **130.** $\frac{1}{\sqrt{2}}|z|$

138. $z=1,-2$ **143.** $\frac{2\pi}{n}$

144. $\frac{1+\sqrt{3}i}{2}z_1, \frac{1-\sqrt{3}i}{2}z_1$

145. $i|z|, -\overline{z}, -z, -i|z|$

146. $\frac{1-i}{2}z_1+\frac{1+i}{2}z_2;\ -iz_1+(1+i)z_2,\ (1-i)z_1+iz_2$

147. $\frac{1-\sqrt{3}i}{2}z_1+\frac{1+\sqrt{3}i}{2}z_2$

148. $\frac{1-i}{2\sqrt{2}}z_1+\frac{(2\sqrt{2}-1)+i(1+\sqrt{2})}{2\sqrt{2}}z_2$

149. $-a$ **152.** $|z|^2+(7-i\lambda)z+(7+i\lambda)\overline{z}+1=0$

153. centre $=\frac{a+b}{2}+i\frac{a-b}{2}\cot k$

155. $t\in\left(-\infty,\frac{1-\sqrt{7}}{2}\right)\cup\left(\frac{1+\sqrt{7}}{2},+\infty\right)$

156. $\sqrt{41}, -4i-5$ respectively

157. $\frac{-1}{5}, \frac{-3}{5}$ respectively **158.** $\sqrt{2}(1+i)$

159. $3(-4+3i)$ **160.** 0

161. 6, 0 respectively **162.** $x+iy$

163. 0 **164.** w^2

165. 1 **166.** -1

167. $-iz$ **168.** (i) 1 (ii) $-\frac{1}{2}+i\frac{\sqrt{3}}{2}$

169. (i) $3-\frac{i}{2}, 1-\frac{3i}{2}$ respectively (ii) $\frac{-2\pi}{3}$

170. circle

171. $(a^2+b^2)(|z_1|^2+|z_1|^2)$

172. $2-\sqrt{3}, 2-\sqrt{3}$ **173.** $-2, 1-\sqrt{3}i$

174. $-b$ **175.** a straight line **176.** (b)

177. (a) **178.** (c) **179.** (a) **180.** (b)

181. (b) **182.** (b) **183.** (a) **184.** (c)

185. (e) **186.** (b) **187.** (i) (b) (ii) (d)

188. (d) **189.** (a), (c) **190.** (c) **191.** (b)

192. (a), (d) **193.** (a), (b), (c) **194.** (b)

195. (a) **196.** (b) **197.** (a), (b), (c), (d)

198. (b), (c), (d) **199.** (b) **200.** (a) **201.** (a)

202. (b) **203.** (b) **204.** (c) **205.** (a)

206. (a) 207. (c) 208. (a) 209. (c)
210. true 211. false 212. false 213. false
214. true 215. true 216. false 217. true
218. true 219. true 220. false 221. true
222. false 223. true 224. true 225. true
226. true

□

Chapter Test

Time: 120 minutes

1. In the following, each question has one or more than one correct answers. Indicate the correct answer(s).

(i) If amp $\dfrac{z_2 - z_1}{z_3 - z_1} = \dfrac{\pi}{2}$ for three distinct complex numbers z_1, z_2, z_3 then the three points are

A. concyclic B. collinear

C. vertices of an isosceles triangle D. vertices of a right-angled triangle

E. none of these

(ii) If a, b, c are multiples of 3 then $x^a + x^{b+1} + x^{c+2}$ is divisible by

A. $x + 1$ B. $x^2 + 1$ C. $x^2 + x + 1$ D. none of these

(iii) If $|z - 3i| = 3$ and amp $z \in \left(0, \dfrac{\pi}{2}\right)$ then cot(amp z) $- \dfrac{6}{z}$ is equal to

A. i B. 1 C. -1 D. $-i$ E. $\dfrac{-1}{i}$

2. In the following, each statement is incomplete. Fill in the blanks so that each resulting statement becomes correct.

(i) If $z \neq 0$ then amp z + amp $\bar{z}$ = ______ .

(ii) If $|z_1 - z_0| = |z_2 - z_0| = |z_3 - z_0|$ and z_1, z_2, z_3 are the vertices of an equilateral triangle then $z_1 + z_2 + z_3$ in terms of z_0 is ______ .

(iii) If $x + iy = \left(\dfrac{\sqrt{3} + i}{1 - \sqrt{3}i}\right)^{25}$ then x = ______, y = ______ .

3. If z is unimodular, prove that $\left(\dfrac{1+z}{1+\bar{z}}\right)^n + \left(\dfrac{1+\bar{z}}{1+z}\right)^n = 2\cos n(\text{amp } z)$.

4. If $1, \alpha_1, \alpha_2, \alpha_3, \alpha_4$ be the roots of $x^5 - 1 = 0$ then prove that

$$\frac{w - \alpha_1}{w^2 - \alpha_1} \cdot \frac{w - \alpha_2}{w^2 - \alpha_2} \cdot \frac{w - \alpha_3}{w^2 - \alpha_3} \cdot \frac{w - \alpha_4}{w^2 - \alpha_4} = w$$

where w is a nonreal complex cube root of unity.

5. Find the condition for the complex equation $z^2 + bz + c = 0$ to have exactly one nonreal complex root where b and c are complex constants. Also obtain the complex root.

6. If $|z^2 - 4| = 2|z|$ then find the greatest value of $|z|$.

7. If $z - 2 = t + i\sqrt{3 - t^2}$ where $t \in R$ and $-\sqrt{3} < t < \sqrt{3}$ then prove that $\left|\dfrac{z+1}{z-1}\right|$ is independent of t.

8. If $|z_1| = 1$, $|z_2| = 1$ then prove that $|z_1 + z_2|^2 + |z_1 - z_2|^2 = 4$.

9. z_1, z_2 are represented by two consecutive vertices of a rhombus, the angle at z_1 being $\dfrac{\pi}{4}$. Find the complex numbers z_3, z_4 represented by the other vertices, the vertices z_1, z_2, z_3, z_4 being in the anticlockwise sense.

10. Prove that $t^2 + 3t + 3$ is a factor of $(t+1)^{n+1} + (t+2)^{2n-1}$ for all integral values of n.

11. If $x^2 - x + 1 = 0$ then find the value of

$$\left(x + \frac{1}{x}\right)^2 + \left(x^2 + \frac{1}{x^2}\right)^2 + \left(x^3 + \frac{1}{x^3}\right)^2 + \ldots + \left(x^{24} + \frac{1}{x^{24}}\right)^2.$$

12. Let α, β be real and $|\alpha| \neq 1, |\beta| \neq 1$ and a be an arbitrary complex number $\neq 1$ such that

$$\{(\alpha+1)(\beta-1)+(\beta+1)(\alpha-1)\}\,a+(\alpha-1)(\beta-1)=0$$

and $a(\alpha+1)(\beta+1)-(\alpha-1)(\beta-1)=0.$

If $A \cap B \neq \phi$, where the sets $A=\left\{\frac{\alpha+1}{\alpha-1}, \frac{\beta+1}{\beta-1}\right\}$, $B=\left\{\frac{2\alpha}{\alpha+1}, \frac{2\beta}{\beta+1}\right\}$ then find the set of possible values of a.

13. Find the quadratic equation whose roots are nth powers of the roots of $x^2-\sqrt{2}\cdot x+1=0$.

14. If the point representing the complex number z_r is a point on or inside the circle whose centre is the complex number $r+i\cdot 0$ and the radius is r then prove that the greatest value of $|z_1+z_2+z_3+\ldots+z_n|$ is $n(n+1)$.

□

Answers

1. (i) A, D (ii) C (iii) A, E

2. (i) 0 (ii) $3z_0$ (iii) 0, 1, respectively

5. $(c-\bar{c})(b^2-\bar{b}^2-2c+2\bar{c})=(c+\bar{c})(b-\bar{b})^2, \frac{c(\bar{b}-b)}{c-\bar{c}}$

6. $1+\sqrt{5}$

9. $z_3=-\frac{1+i}{\sqrt{2}}z_1+\frac{\sqrt{2}+1+i}{\sqrt{2}}z_2,\quad z_4=\frac{\sqrt{2}-1-i}{\sqrt{2}}z_1+\frac{1+i}{\sqrt{2}}z_2$

11. 48

12. $\{-1, 1+i, 1-i\}$

13. $x^2-2x\cos\frac{n\pi}{4}+1=0$

□

5. Permutation and Combination

Recap of Facts and Formulae

1. Counting of number of ways to do some work

If a work W consists of two parts W_1, W_2 of which one part can be done in m ways and the other part in n ways then

- the work W can be done in $m + n$ ways, if by doing any of the parts the work W is done.
  (*Addition law of counting*)
- the work W can be done in mn ways, if both the parts are to be done one after the other to do the work W.
  (*Multiplication law of counting*)

Similar is the law for works that have 3 or more parts.

If a work is to be done under some restriction then

- the number of ways to do the work under the restriction = (the number of ways to do the work without restriction) – (the number of ways to do the work under opposite restriction).

2. Counting formulae for permutation

- The number of permutations (arrangements) of n *different things* taking r at a time

$$= {}^nP_r = \frac{n!}{(n-r)!} \quad \text{where } n! = 1.2.3. \dots n.$$

- The number of permutations of n things *taking all at a time* of which p things are identical, q things are identical of another type and the rest are different

$$= \frac{n!}{p!\,q!}.$$

- The number of arrangements of n *different things* round a closed curve
  $= (n-1)!$ if clockwise and anticlockwise arrangements are considered different,
  $\frac{1}{2}(n-1)!$ if clockwise and anticlockwise arrangements are considered identical.
  (*Circular permutation*)

3. Counting formulae for combination

- The number of combinations (selections) of n *different things* taking r at a time

$$= {}^nC_r = \frac{n!}{r!\,(n-r)!}.$$

- The total number of selections of one or more objects from n *different objects*

$$= 2^n - 1 \,(= {}^nC_1 + {}^nC_2 + {}^nC_3 + \dots + {}^nC_n).$$

- The total number of selections of any number of things from n identical things
  $= n + 1$ (when selection of 0 things is allowed)
  n (when at least one thing is to be selected).
- The total number of selections from p like things, q like things of another type and r distinct things
  $= (p+1)(q+1)2^r - 1$ (if at least one thing is to be selected)
  $(p+1)(q+1)2^r - 2$ (if none or all cannot be selected)
- The total number of selections of r things from n different things when each thing can be repeated unlimited number of times $= {}^{n+r-1}C_r$.

For example If 8 balls are to be selected from balls of 3 colours, balls of each colour being available unlimited number of times, then the number of selections

$$= {}^{3+8-1}C_8 \quad (\because \text{ here } r = 8, n = 3)$$

$$= {}^{10}C_8 = \frac{10!}{8!\,2!} = \frac{10 \times 9}{2} = 45.$$

4. Number of distributions

- The number of ways to distribute n *different things* between two persons, one receiving p things and the other q things, where $p + q = n$

$$= {}^nC_p \times {}^{n-p}C_q$$

$$= \frac{n!}{p!\,(n-p)!} \times \frac{(n-p)!}{q!\,(n-p-q)!}$$

$= \dfrac{n!}{p!\,q!}$ $\{\because\ n = p + q\}$

Similarly for 3 persons, the number of ways

$= \dfrac{n!}{p!\,q!\,r!}$, where $p + q + r = n$.

- The number of ways to distribute $m \times n$ *different things* among n **persons** equally $= \dfrac{(mn)!}{(m!)^n}$.
- The number of ways to divide n different things into three bundles of p, q and r things $= \dfrac{n!}{p!\,q!\,r!} \cdot \dfrac{1}{3!}$.
- The number of ways to divide $m \times n$ different things into n equal bundles $= \dfrac{(mn)!}{(m!)^n} \cdot \dfrac{1}{n!}$.
- The total number of ways to divide n *identical things* among r persons

  $= {}^{n+r-1}C_{r-1}$.

For example The number of different ways to divide 20 identical apples among 4 persons

$= {}^{20+4-1}C_{4-1}$ $(\because$ here $n = 20, r = 4)$

$= {}^{23}C_3 = \dfrac{23 \times 22 \times 21}{6}$

$= 23 \times 11 \times 7$

$= 1771$.

- The total number of ways to divide n *identical things* among r persons so that each gets at least one

  $= {}^{n-1}C_{r-1}$.

For example The number of different ways to distribute 20 identical balls in 4 different boxes so that no box remains empty

$= {}^{20-1}C_{4-1}$ $\{\because$ here $n = 20, r = 4\}$

$= {}^{19}C_3 = \dfrac{19 \cdot 18 \cdot 17}{6}$

$= 969$.

5. Use of solution of linear equations and coefficient of a power in expansions to find the number of ways of distribution

- The number of integral solutions of $x_1 + x_2 + x_3 + \ldots + x_r = n$ where $x_1 \ge 0, x_2 \ge 0, \ldots, x_r \ge 0$ is the same as the number of ways to distribute n identical things among r persons.

  This is also equal to the coefficient of x^n in the expansion of $(x^0 + x^1 + x^2 + x^3 + \ldots)^r$

  $= \text{coefficient of } x^n \text{ in } \left(\dfrac{1}{1-x}\right)^r$

  $= \text{coefficient of } x^n \text{ in } (1-x)^{-r}$

  $= \text{coefficient of } x^n$ in

  $$\left\{1 + rx + \frac{r(r+1)}{2!}x^2 + \ldots + \frac{r(r+1)(r+2)\ldots(r+n-1)}{n!}x^n + \ldots\right\}$$

  $$= \frac{r(r+1)(r+2)\ldots(r+n-1)}{n!}$$

  $$= \frac{(r+n-1)!}{n!\,(r-1)!} = {}^{n+r-1}C_{r-1}.$$

- The number of integral solutions of $x_1 + x_2 + x_3 + \ldots + x_r = n$ where $x_1 \ge 1, x_2 \ge 1, \ldots, x_r \ge 1$ is the same as the number of ways to distribute n identical things among r persons each getting at least 1.

  This is also equal to the coefficient of x^n in the expansion of $(x^1 + x^2 + x^3 + \ldots)^r$

  $= \text{coefficient of } x^n \text{ in } \left(\dfrac{x}{1-x}\right)^r$

  $= \text{coefficient of } x^n \text{ in } x^r(1-x)^{-r}$

  $= \text{coefficient of } x^n$ in

  $$x^r\left\{1 + rx + \frac{r(r+1)}{2!}x^2 + \ldots + \frac{r(r+1)(r+2)\ldots(r+n-1)}{n!}x^n + \ldots\right\}$$

  $= \text{coefficient of } x^{n-r}$ in

  $$\left\{1 + rx + \frac{r(r+1)}{2!}x^2 + \ldots + \frac{r(r+1)(r+2)\ldots(r+n-1)}{n!}x^n + \ldots\right\}$$

  $$= \frac{r(r+1)(r+2)\ldots(r+n-r-1)}{(n-r)!}$$

  $$= \frac{r(r+1)(r+2)\ldots(n-1)}{(n-r)!}$$

  $$= \frac{(n-1)!}{(n-r)!\,(r-1)!} = {}^{n-1}C_{r-1}.$$

Note The number of solutions of $x_1 + x_2 + x_3 + x_4 = 20$ where $x_1 \ge 0, x_2 \ge 1, x_3 \ge 3, x_4 \ge 5$ is equal to the coefficient of x^{20} in

$$(x^0 + x^1 + x^2 + \ldots)(x^1 + x^2 + x^3 + \ldots) \times (x^3 + x^4 + x^5 + \ldots)(x^5 + x^6 + x^7 + \ldots).$$

6. Some useful results

- ${}^nC_r = {}^nC_{n-r}$
- ${}^nC_r + {}^nC_{r-1} = {}^{n+1}C_r$

- $(1-x)^{-n} = 1 + nx + \dfrac{n(n+1)}{2\,!}x^2 + \dfrac{n(n+1)(n+2)}{3\,!}x^3 + \dots + \dfrac{n(n+1)(n+2)\dots(n+r-1)}{r\,!}x^r + \dots$

$$= {}^{n-1}C_0 + {}^nC_1x + {}^{n+1}C_2x^2 + {}^{n+2}C_3x^3 + \dots + {}^{n+r-1}C_rx^r + \dots,$$

where n is a positive integer.

Selected Solved Examples

1. Show that $(kn)\,!$ is divisible by $(n\,!)^k$.

Here $(kn)\,! = 1\cdot2\cdot3\cdot \dots (kn)$

$$= (1\cdot2\cdot3\cdot \dots n)\{(n+1)(n+2)\dots(n+n)\}$$
$$\{(2n+1)(2n+2)\dots(2n+n)\}\dots$$
$$\times[\{(k-1)n+1\}\{(k-1)n+2\}\dots\{(k-1)n+n\}] \quad \dots(1)$$

Now, $(rn+1)(rn+2)\dots(rn+n)$, where $r \in N$

$$= \frac{1\cdot2\cdot3\cdot \dots (rn+1)(rn+2)\dots(rn+n)}{1\cdot2\cdot3\cdot \dots (rn)} = \frac{(rn+n)\,!}{(rn)\,!}$$

$$\therefore \quad \frac{(rn+1)(rn+2)\dots(rn+n)}{n\,!} = \frac{(rn+n)\,!}{(rn)\,!\,n\,!}$$

$$= {}^{rn+n}C_n = \text{integer}$$

$\therefore$ $(rn+1)(rn+2)\dots(rn+n)$ is divisible by $n\,!$

$\therefore$ $(rn+1)(rn+2)\dots(rn+n) = n\,!\cdot M_r$,

where M_r is an integer

$\therefore$ by (1),

$$(kn)\,! = n\,!\cdot(n\,!\cdot M_1)\cdot(n\,!\cdot M_2)\dots(n\,!\cdot M_{k-1})$$
$$= (n\,!)^k\cdot M_1\cdot M_2\cdot\dots M_{k-1} = (n\,!)^k \times \text{integer}$$

$\therefore$ $(kn)\,!$ is divisible by $(n\,!)^k$.

Note Putting $k = n$, we get $(n^2)\,!$ is divisible by $(n\,!)^n$.

2. Evaluate ${}^nC_k + \sum_{j=0}^{m} {}^{n+j}C_{k-1}$.

$$\text{Value} = {}^nC_k + ({}^nC_{k-1} + {}^{n+1}C_{k-1} + {}^{n+2}C_{k-1} + \dots + {}^{n+m}C_{k-1})$$
$$= ({}^nC_k + {}^nC_{k-1}) + ({}^{n+1}C_{k-1} + {}^{n+2}C_{k-1} + \dots + {}^{n+m}C_{k-1})$$
$$= {}^{n+1}C_k + ({}^{n+1}C_{k-1} + {}^{n+2}C_{k-1} + \dots + {}^{n+m-1}C_{k-1} + {}^{n+m}C_{k-1})$$
$$= ({}^{n+1}C_k + {}^{n+1}C_{k-1}) + ({}^{n+2}C_{k-1} + \dots + {}^{n+m-1}C_{k-1} + {}^{n+m}C_{k-1})$$
$$= {}^{n+2}C_k + ({}^{n+2}C_{k-1} + \dots + {}^{n+m-1}C_{k-1} + {}^{n+m}C_{k-1})$$
$$= \dots\dots\dots\dots\dots\dots\dots\dots$$
$$= {}^{n+m-1}C_k + ({}^{n+m-1}C_k + {}^{n+m}C_{k-1})$$
$$= {}^{n+m}C_k + {}^{n+m}C_{k-1}$$
$$= {}^{n+m+1}C_k.$$

3. Show that two positive integers n, r cannot be found such that ${}^nC_r, {}^nC_{r+1}, {}^nC_{r+2}, {}^nC_{r+3}$ are in AP.

${}^nC_r, {}^nC_{r+1}, {}^nC_{r+2}, {}^nC_{r+3}$ are in AP

$$\Rightarrow \quad 2\cdot{}^nC_{r+1} = {}^nC_r + {}^nC_{r+2} \quad \dots(1)$$

$$\text{and} \quad 2\cdot{}^nC_{r+2} = {}^nC_{r+1} + {}^nC_{r+3} \quad \dots(2)$$

From (1), $2\cdot\dfrac{n\,!}{(r+1)\,!\,(n-r-1)\,!}$

$$= \frac{n\,!}{r\,!\,(n-r)\,!} + \frac{n\,!}{(r+2)\,!\,(n-r-2)\,!}$$

or $\quad \dfrac{2}{(r+1)(n-r-1)} = \dfrac{1}{(n-r-1)(n-r)} + \dfrac{1}{(r+2)(r+1)}$

or $\quad 2(r+2)(n-r) = (r+2)(r+1) + (n-r-1)(n-r)$

or $\quad (n-r)^2 - (2r+5)(n-r) + (r^2+3r+2) = 0 \quad \dots(3)$

Similarly, from (2) replacing r by $r+1$ in (3), we get

$$(n-r-1)^2 - (2r+7)(n-r-1) + \{(r+1)^2 + 3(r+1) + 2\} = 0$$

or $\quad (n-r)^2 - 2(n-r) + 1 - (2r+7)(n-r) + (2r+7+r^2+2r+1+3r+3+2) = 0$

or $\quad (n-r)^2 - (2r+9)(n-r) + (r^2+7r+14) = 0 \quad \dots(4)$

(3) and (4) must be identical quadratic in $(n-r)$,

$$\therefore \quad \frac{1}{1} = \frac{2r+5}{2r+9} = \frac{r^2+3r+2}{r^2+7r+14}$$

$\Rightarrow \quad 2r+9 = 2r+5 \quad$ and $\quad r^2+7r+14 = r^2+3r+2$

or $\quad 9 = 5 \quad$ or $\quad 4r = -12$, i.e., $r = -3$.

Both $9 = 5$ and $r = -3$ are absurd.

$\therefore$ (1) and (2) cannot hold at the same time.

$\therefore$ ${}^nC_r, {}^nC_{r+1}, {}^nC_{r+2}, {}^nC_{r+3}$ cannot be in AP.

4. How many numbers of n digits can be made with the nonzero digits in which no two consecutive digits are the same?

There are nine nonzero digits, namely 1,2,3, ... and 9.

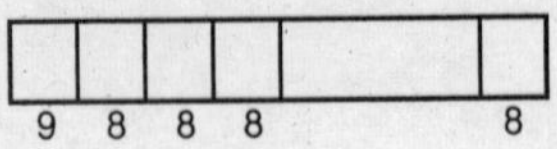

In order to make an n-digit number we have to fill n places by using the nine digits. As no two consecutive digits are to be the same, a digit used in a place cannot be used in the next place but it can be used again in the place coming after the next place.

So the first place can be filled in 9 ways;

the second place can be filled in 8 ways (rejecting the digit used in the first place)

the third place can be filled in 7 + 1, i.e., 8 ways (rejecting the digit used in the second place but including the digit used in the first place)

and so on.

$\therefore$ the required number of numbers

$= 9 \times 8 \times 8 \times 8 \times \ldots$ to n factors $= 9 \times 8^{n-1}$.

5. How many numbers of 4 digits can be made with the digits 0, 1, 2, 3, 4, 5 which are divisible by 3, digits being unrepeated in the same number? How many of these will be divisible by 6?

We know that a number is divisible by 3 if the sum of the digits in the number is divisible by 3.

Here $0 + 1 + 2 + 3 + 4 + 5 = 15$; so two digits are to be omitted whose sum is 3 or 6 or 9.

Hence, the number of four digits can be made by either

1, 2, 4, 5 or 0, 3, 4, 5 (omitting two digits whose sum is 3)

0, 1, 3, 5 or 0, 2, 3, 4 (omitting two digits whose sum is 6)

0, 1, 2, 3 (omitting two digits whose sum is 9)

The number of 4-digit numbers that can be made with 1, 2, 4, 5

$= {}^4P_4 = 4\,!$.

The number of 4-digit numbers that can be made by the digits in any one of remaining four groups (each containing 0)

$= 4\,! - 3\,!$ ($\because$ the number of numbers beginning with $0 = {}^3P_3 = 3\,!$)

$\therefore$ the required number of 4-digit numbers divisible by 3

$= 4\,! + 4(4\,! - 3\,!)$

$= 24 + 4(24 - 6) = 96$.

Now, a number is divisible by 6 if it is even as well as divisible by 3.

So, the number of 4-digit numbers divisible by 6 that can be made with 1, 2, 4, 5 $= 2 \times 3\,!$ ($\because$ the numbers should have an even digit in the units place).

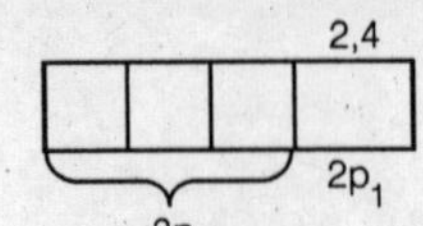

The number of numbers of 4 digits, divisible by 6, that can be made with 0, 3, 4, 5

$= (3\,! - 2\,!) + 3\,!$

($\because$ the numbers should have 4 or 0 in units place and 0 should not come in thousands place).

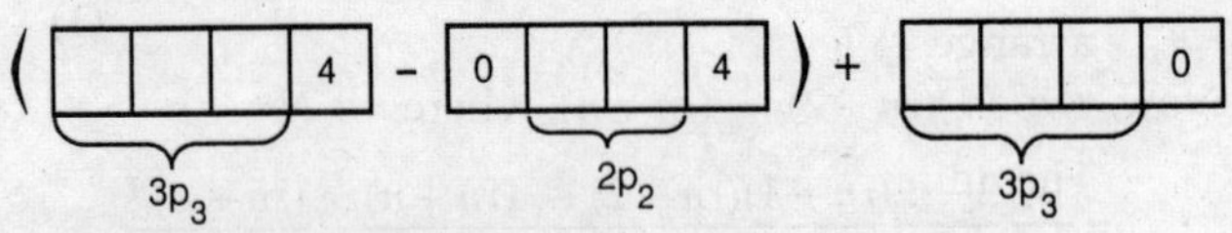

Similarly, the number of numbers of 4 digits, divisible by 6, that can be made with 0, 1, 2, 3

$= (3\,! - 2\,!) + 3\,!$

The number of 4-digit numbers divisible by 6 that can be made with the digits 0, 1, 3, 5 $= 3\,!$.

The number of numbers of 4 digits, divisible by 6, that can be made with 0, 2, 3, 4

$= (3\,! - 2\,!) + (3\,! - 2\,!) + 3\,!$

($\because$ the numbers should have 4 or 2 or 0 in units place and 0 should not come in thousands place).

$\therefore$ the required 4-digit numbers divisible by 6

$= 2 \times 3\,! + (3\,! - 2\,!) + 3\,! + (3\,! - 2\,!) + 3\,! + 3\,! + (3\,! - 2\,!) + (3\,! - 2\,!) + 3\,!$

$= 12 + 4 + 6 + 4 + 6 + 6 + 4 + 4 + 6 = 52$.

6. How many words can be made with letters of the word INTERMEDIATE if

(i) the words neither begin with I nor end with E

(ii) the vowels and consonants alternate in the words

(iii) the vowels are always consecutive

(iv) no vowel is between two consonants

(v) the relative order of vowels and consonants does not change

(vi) the order of vowels does not change?

(i) The required number of words

= (the number of words without restriction)

− (the number of words beginning with I)

− (the number of words ending with E)

+ (the number of words beginning with I and ending with E)

(because words beginning with I as well as words ending with E contain some words beginning with I and ending with E).

The number of words without restriction

$$= \frac{12!}{2!3!2!} \; (\because \text{ there are 12 letters in which there are two Is, three Es and two Ts}).$$

The number of words beginning with I $= \frac{11!}{3!2!}$

($\because$ with I in the extreme left place we are left to arrange 11 letters NTERMEDIATE in which there are two Ts and three Es).

The number of words ending with E $= \frac{11!}{2!2!2!}$

($\because$ with E in the extreme right place we are left to arrange 11 letters INTERMEDIAT in which there are two Is, two Es and two Ts).

The number of words beginning with I and ending with E $= \frac{10!}{2!2!}$

($\because$ with I in the extreme left and E in the extreme right places we are left to arrange 10 letters NTERMEDIAT in which there are two Ts and two Es).

$\therefore$ the required number of words

$$= \frac{12!}{2!3!2!} - \frac{11!}{3!2!} - \frac{11!}{2!2!2!} + \frac{10!}{2!2!}$$

$$= \frac{10!}{2!3!2!}(12 \cdot 11 - 11 \cdot 2 - 11 \cdot 3 + 6) = \frac{83 \times 10!}{24}.$$

(ii) There are 6 vowels and 6 consonants. So the number of words in which vowels and consonants alternate

= (the number of words in which vowels occupy odd places and consonants occupy even places)

+ (the number of words in which consonants occupy odd places and vowels occupy even places)

$$= \frac{6!}{2!3!} \times \frac{6!}{2!} + \frac{6!}{2!} \times \frac{6!}{2!3!}$$

$$= 2 \cdot \frac{6!}{2!3!} \cdot \frac{6!}{2!} = 43200.$$

(iii) Considering the 6 vowels IEEIAE as one thing, the number of arrangements of this with 6 consonants

$$= \frac{7!}{2!} \; (\because \text{ there are two Ts in the consonants}).$$

For each of these arrangements, the 6 consecutive vowels can be arranged among themselves in $\frac{6!}{2!3!}$ ways.

$\therefore$ the required number of words

$$= \frac{7!}{2!} \times \frac{6!}{2!3!} = 151200.$$

(iv) No vowel will be between two consonants if all the consonants become consecutive

$\therefore$ the required number of words

= the number of arrangements when all the consonants are consecutive

$$= \frac{7!}{2!3!} \times \frac{6!}{2!} \quad \text{(as above)}$$

$$= 151200.$$

(v) The relative order of vowels and consonants will not change if in the arrangement of letters the vowels occupy places of vowels, i.e., 1st, 4th, 7th, 9th, 10th, 12th places and consonants occupy their places, i.e., 2nd, 3rd, 5th, 6th, 8th, 11th places.

$\therefore$ the required number of words

$$= \frac{6!}{2!3!} \times \frac{6!}{2!} = 21600.$$

(vi) The order of vowels will not change if no two vowels interchange places, i.e., in the arrangement all the vowels are treated as identical.

(**For example** LATE, ATLE, TLAE, etc., have the same order of vowels A, E. But LETA, ETLA, TLEA, etc., have changed order of vowels A, E. So, LATE is counted but LETA is not.

If A, E, are taken as identical, say V then LVTV does not give a new arrangement by interchanging V, V.)

The required number of words

= the number of arrangements of 12 letters in which 6 vowels are treated as identical

$$= \frac{12!}{6!2!} \; (\because \text{ there are two Ts also})$$

7. In how many ways can 5 identical black balls, 7 identical red balls and 6 identical green balls be arranged in a row so that at least one ball is separated from balls of the same colour?

The required number of ways

= (the number of ways without restriction)

– (the number of ways when balls of each colours are consecutive)

$= \frac{18!}{5!7!6!} - 3!$ ($\because$ there are altogether 18 balls in which 5, 7 and 6 are identical; and considering balls of the same colour as one thing there are 3 things, there being no arrangement between balls of the same colour)

$$= \frac{18!}{5!6!7!} - 6.$$

8. Ten guests are to be seated in a row of which three are ladies. The ladies insist on sitting together while two of the gentlemen refuse to take consecutive seats. In how many ways can the guests be seated?

Considering 3 ladies as one object, the number of arrangements $= {}^8P_8$

($\because$ 7 gentlemen and 1 group of ladies, i.e., 8 things are to be arranged).

But the 3 ladies can be arranged among themselves in 3 ! ways.

$\therefore$ the number of arrangements in which 3 ladies are together $= {}^8P_8 \times 3\,!$.

Next, considering 3 ladies as one object and 2 gentlemen (who refused to sit together) as another object, the number of arrangements

$$= {}^7P_7 \times 3\,! \times 2\,!.$$

$\therefore$ the required number of arrangements

= (the number of arrangements in which 3 ladies are together)
− {the number of arrangements in which 3 ladies are together and 2 gentlemen (who refused to sit together) are together}

$$= {}^8P_8 \cdot 3\,! - {}^7P_7 \cdot 3\,! \cdot 2\,!$$

$$= 8\,!\,3\,! - 7\,! \cdot 3\,! \cdot 2\,!$$

$$= 7\,!(8 \times 3\,! - 3\,!\,2\,!) = 5040(48 - 12)$$

$$= 5040 \times 36 = 181440.$$

9. A dictionary is printed consisting of 7-lettered words that can be made with the letters of the word CRICKET. If the words are printed in the alphabetic order, as in an ordinary dictionary, find the position of the word CRICKET in that dictionary.

In alphabetic order the letters are C, C, E, I, K, R, T. Words beginning with C will be printed at first. Our word also begins with C. The number of words beginning with C and having C or E or I or K in the second place = 4 × 5 !.

```
C | C × × × × ×
  | E × × × × ×
  | I × × × × ×
  | K × × × × ×

CR | C × × × ×
   | E × × × ×

CRIC | E × ×
     | K E T
```

The number of words beginning with CR and having C or E in the third place = 2 × 4 !.

The number of words beginning with CRIC and having E in the fifth place = 2 !.

After printing all these words, the word CRICKET will be printed.

$\therefore$ the required place of the word CRICKET

$$= (4 \times 5\,! + 2 \times 4\,! + 2\,!) + 1 = 531.$$

$\therefore$ the word will be printed in the 531th place.

10. In a particular programming language, a valid variable name can consist of a sequence of one to six alphanumeric characters A, B, C, ..., Z, 0, 1, 2, ..., 9 beginning with a letter. Find the total number of valid variable names.

The number of valid variable names of one character

= 26 ($\because$ there are 26 letters and the variable name must begin with a letter).

The number of valid variable names of two characters

= 26 × 36 ($\because$ the first place should have a letter and the second place can have any one of the 26 + 10 alphanumeric characters, repetition being allowed).

The number of valid variable names of three characters = 26 × 36 × 36 and so on.

$\therefore$ the total number of valid variable names of one to six characters

$$= 26 + 26 \times 36 + 26 \times 36^2 + \ldots + 26 \times 36^5$$

$$= 26(1 + 36 + 36^2 + \ldots + 36^5)$$

$$= 26 \cdot \frac{36^6 - 1}{36 - 1} = \frac{26}{35}(36^6 - 1).$$

11. How many numbers of five digits can be made with at least one repeated digit?

Total number of numbers of five digits when digits can be repeated $= 9 \times 10^4$

($\because$ there are ten digits 0, 1, 2, ..., 9 and 0 cannot go in the ten thousands place)

~~0~~				
9	10	10	10	10

$\therefore$ the number of numbers which may or may not have repeated digits $= 9 \times 10^4$.

But the number of numbers which do not have repeated digits

$$= {}^{10}P_5 - {}^9P_4$$

($\because$ [□□□□□] − [0 □□□□] ; the first is $^{10}P_5$, the second $^{9}P_4$)

$\therefore$ the required number of numbers having at least one repeated digit

= (total number of numbers of five digits)
– (the number of numbers of five digits having no repeated digit)

$= 9 \times 10^4 - ({}^{10}P_5 - {}^{9}P_4)$

$= 9 \times 10^4 - 9.8.7.6(10 - 1)$

$= 90000 - 27216 = 62784.$

12. How many natural numbers are there lying between 20,000 and 60,000, the sum of digits being even?

The numbers will be of five digits beginning with 2, 3, 4 or 5.

So the ten thousands place can be filled in 4 ways.

Each of thousands, hundreds and tens places can be filled in 10 ways.

So the first four places can be filled in $4 \times 10 \times 10 \times 10$ ways.

After filling these the sum of the digits used is either even or odd.

∴ the last place can be filled in 5 ways (∵ if the sum of the digits is even, one of the digits 0, 2, 4, 6, 8 will be used and if the sum of the digits is odd, one of the digits 1, 3, 5, 7, 9 will be used).

∴ the required number of numbers (20,000 being excluded)

$= 4 \times 10 \times 10 \times 10 \times 5 - 1 = 19999.$

13. How many even numbers of four digits can be made with the digits 0, 3, 5, 4? Find the sum of the numbers.

The even numbers have 0 or 4 in the units place.

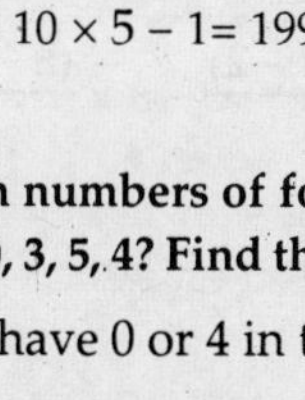

The number of even numbers of four digits with 0 in the units place $= {}^3P_3$ ... (1)

The number of even numbers of four digits with 4 in the units place $= {}^3P_3 - {}^2P_2$ ... (2)

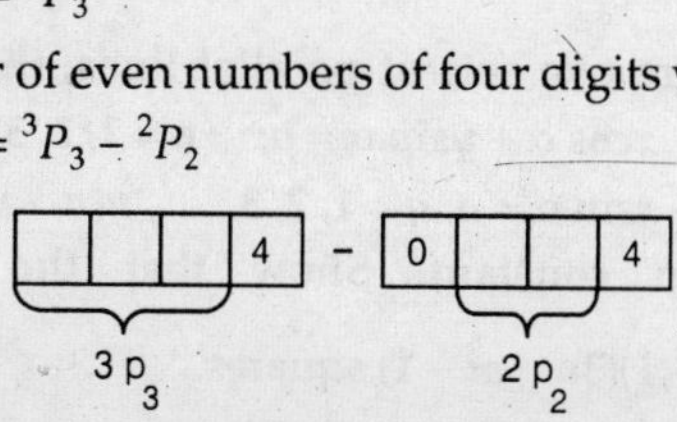

∴ total number of even numbers

$= {}^3P_3 + ({}^3P_3 - {}^2P_2) = 6 + 6 - 2 = 10.$

The even numbers have 0 or 4 in units place.

The number of numbers with 0 in units place

$= {}^3P_3 = 3\,! = 6$ [from (1)]

The number of numbers with 4 in units place

$= {}^3P_3 - {}^2P_2 = 4$ [from (2)]

∴ the sum of the digits in units place for all the numbers

$= 0 \times 6 + 4 \times 4 = 16.$

The number of numbers with 0 in tens place $= 2\,! = 2.$

The number of numbers with 3 in tens place

$= 2\,! + 1 = 3.$

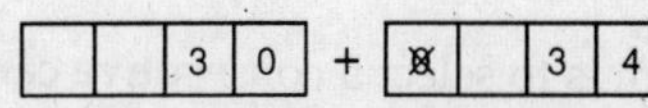

The number of numbers with 4 in tens place $= 2\,! = 2.$

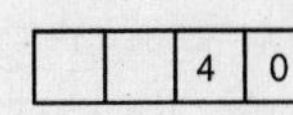

The number of numbers with 5 in tens place

$= 2\,! + 1 = 3$ (as in case of 3).

∴ the sum of the digits in tens place

$= 0 \times 2 + 3 \times 3 + 4 \times 2 + 5 \times 3 = 32.$

Similarly, sum of the digits in hundreds place = 32.

The number of numbers with 3 in thousands place

$= 2 \times {}^2P_2 = 4.$

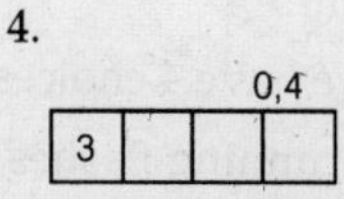

The number of numbers with 5 in thousands place

$= 2 \times {}^2P_2 = 4.$

The number of numbers with 4 in thousands place

$= 2\,! = 2.$

∴ the sum of the digits in thousands place

$= 3 \times 4 + 5 \times 4 + 4 \times 2 = 40.$

∴ the required sum of even numbers

$= 40 \times 1000 + 32 \times 100 + 32 \times 10 + 16 = 43536.$

14. There are unlimited number of identical balls of four different colours. How many arrangements of at most 8 balls in a row can be made by using them?

The number of arrangements of one ball = 4,

because there are only four different balls.

The number of arrangements of two balls

$= 4 \times 4 = 4^2$, etc.

∴ the required number of arrangements

$= 4 + 4^2 + 4^3 + \ldots + 4^8 = \dfrac{4(4^8 - 1)}{4 - 1}$

$= \dfrac{4}{3}(4^8 - 1) = \dfrac{4}{3} \times 65535 = 4 \times 21845 = 87380.$

15. From a well-shuffled pack of 52 cards, a player is given 5 cards. If the cards have consecutive values then it is said that the player has a *run* and if they are also from the same suit it is said that the player has a *running flash*. In how many different ways can a player hold a (i) run (ii) running flash?

There are 4 suits of cards in a pack of 52 cards, each suit having 13 cards, the values being

$$1(=A), 2, 3, \dots 10, 11(=J), 12(=Q), 13(=K).$$

In a game of cards A is also given the next value above K.

$\therefore$ our problem is to select 5 consecutive cards to have a running flash when the cards of a suit are arranged as follows:

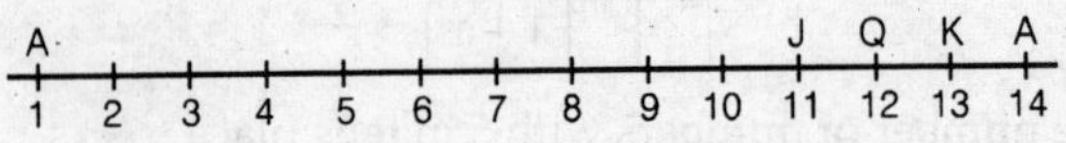

The number of selections of 5 consecutive cards of the same suit from the above arrangement = 10, because they are like (1, 2, 3, 4, 5), (2, 3, 4, 5, 6), … , (10, 11, 12, 13, 14).

The number of selections of 5 cards of consecutive values from the pack $= 10 \times 4^5$, because in each place we have 4 choices of suits.

$\therefore$ the total number of running flashes

$$= 10 + 10 + 10 + 10 = 40$$

and the total number of runs $= 10 \times 4^5 = 10240$.

16. There are m bags which are numbered by m consecutive positive integers starting with the number k. Each bag contains as many different flowers as the number levelled against the bag. A boy has to pick up k flowers from any one of the bags. In how many different ways can he do the work?

The total number of ways

$$= {}^kC_k + {}^{k+1}C_k + {}^{k+2}C_k + \dots + {}^{k+m-1}C_k$$

$$= {}^{k+1}C_{k+1} + ({}^{k+2}C_{k+1} - {}^{k+1}C_{k+1})$$
$$+ ({}^{k+3}C_{k+1} - {}^{k+2}C_{k+1}) + \dots$$
$$+ ({}^{k+m}C_{k+1} - {}^{k+m-1}C_{k+1})$$

because ${}^{n+1}C_{m+1} = {}^nC_m + {}^nC_{m+1}$,

i.e., ${}^nC_m = {}^{n+1}C_{m+1} - {}^nC_{m+1}$

$$= {}^{k+m}C_{k+1}.$$

17. There are n coplanar straight lines, no two being parallel and no three are concurrent. How many different new straight lines will be formed by joining the intersection points of the given lines?

The total number of points of intersection

= the number of selections of 2 lines from n lines

$= {}^nC_2$.

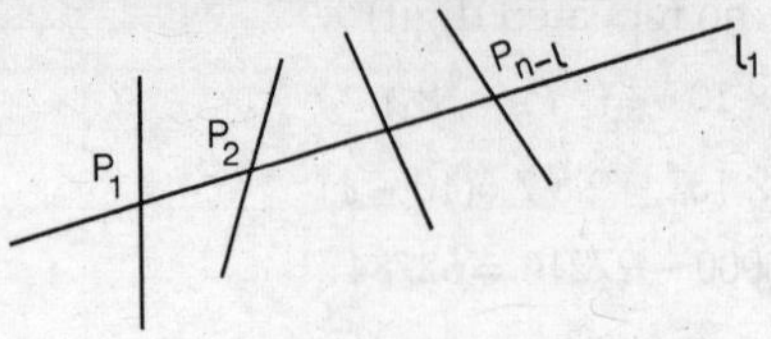

Clearly, each line will cut the remaining $n-1$ lines at $n-1$ points.

$\therefore$ out of ${}^nC_2$ points, $n-1$ points are on each line.

The number of lines obtained by joining any two of the ${}^nC_2$ points

$$= {}^{{}^nC_2}C_2 = \frac{\frac{n(n-1)}{2}\left\{\frac{n(n-1)}{2} - 1\right\}}{2}$$

$$= \frac{n(n-1)(n^2-n-2)}{8}.$$

But from each of the n sets of $(n-1)$ collinear points ${}^{n-1}C_2$ lines have been counted which are not new lines.

So $n \times {}^{n-1}C_2$ lines are not new lines.

$\therefore$ the required number of new lines

$$= \frac{n(n-1)(n^2-n-2)}{8} - n \times {}^{n-1}C_2$$

$$= \frac{n(n-1)(n^2-n-2)}{8} - n \times \frac{(n-1)(n-2)}{2}$$

$$= \frac{n(n-1)}{8}\{n^2 - n - 2 - 4(n-2)\}$$

$$= \frac{1}{8}n(n-1)(n^2 - 5n + 6)$$

$$= \frac{1}{8}n(n-1)(n-2)(n-3).$$

18. There are two sets of parallel lines, their equations being $x\cos\alpha + y\sin\alpha = p$; $p = 1, 2, 3, \dots, m$ and $y\cos\alpha - x\sin\alpha = q$; $q = 1, 2, 3, \dots, n (n > m)$ where α is a given constant. Show that the lines form $\frac{1}{6}m(m-1)(3n-m-1)$ squares.

Clearly $x\cos\alpha + y\sin\alpha = p$; $p = 1, 2, 3, \dots, m$ represent m parallel lines, the distance between consecutive lines being 1.

Also $y\cos\alpha - x\sin\alpha = q$; $q = 1, 2, 3, \dots, n$ represent n parallel lines, the distance between consecutive lines being 1.

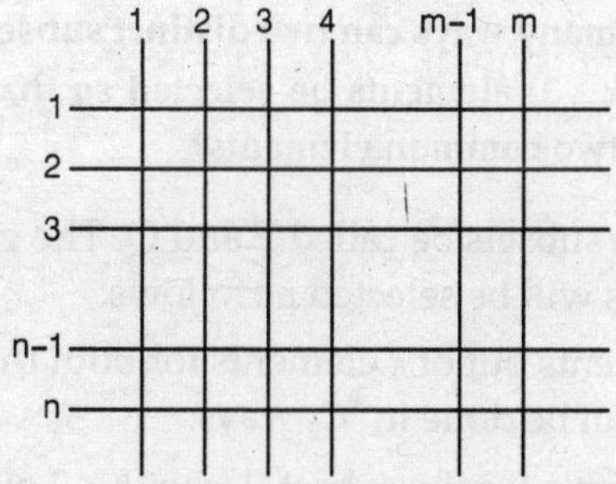

Again, the slope of each line of the first set = $-\cot\alpha$;
the slope of each line of the second set = $\tan\alpha$.

$\because$ $(-\cot\alpha)\cdot(\tan\alpha) = -1$, the two sets cut orthogonally, i.e., at 90°.

$\therefore$ 4 lines consisting of two lines of the first set and two lines of the second set will form a square, if they are taken at equal distances.

The number of squares of sides of length 1 unit

= the number of selections of 2 lines from m lines at unit distance and 2 lines from n lines at unit distance

$= (m-1)(n-1)$ {$\because$ selections will be (1, 2), (2, 3), … $(m-1, m)$ from the first set and (1, 2), (2, 3), … $(n-1, n)$ from the second set}.

The number of squares of sides of length 2 units

= the number of selections of 2 lines from m lines at a distance of 2 units and 2 lines from n lines at a distance of 2 units

$= (m-2)(n-2)$ {$\because$ selections will be (1, 3), (2, 4), …, $(m-2, m)$ and (1, 3), (2, 4), …, $(n-2, n)$}

and so on.

$\therefore$ total number of squares

$$= (m-1)(n-1) + (m-2)(n-2) + (m-3)(n-3) + \ldots + (m-\overline{m-1})\cdot(n-\overline{m-1})$$

$$= \{mn - 1\cdot(m+n) + 1^2\} + \{mn - 2(m+n) + 2^2\} + \{mn - 3(m+n) + 3^2\} + \ldots + \{mn - (m-1)(m+n) + (m-1)^2\}$$

$$= (m-1)mn - (m+n)(1+2+3+\ldots+\overline{m-1}) + \left(1^2+2^2+3^2+\ldots+(m-1)^2\right)$$

$$= (m-1)mn - (m+n)\frac{(m-1)m}{2} + \frac{(m-1)m(2m-1)}{6}$$

$$= \frac{m(m-1)}{6}\{6n - 3(m+n) + 2m - 1\}$$

$$= \frac{1}{6}m(m-1)(3n-m-1).$$

19. A man has 7 relatives, 4 of them are ladies and 3 gentlemen; his wife has 7 relatives, 3 of them are ladies and 4 gentlemen. In how many different ways can they invite a dinner party of 3 ladies and 3 gentlemen so that there are 3 of the man's relatives and 3 of the wife's relatives?

Let us denote the man's lady relatives by HL, man's gentleman relatives by HG, wife's gentleman relatives by WG and wife's lady relatives by WL.

Here $n(\text{HL}) = 4$, $n(\text{HG}) = 3$, $n(\text{WG}) = 4$, $n(\text{WL}) = 3$.

Possibilities	**Combinations**
3HL, 3WG	${}^4C_3 \times {}^4C_3 = 4 \times 4 = 16$
3HG, 3WL	${}^3C_3 \times {}^3C_3 = 1 \times 1 = 1$
2HG, 1HL, 1WG, 2WL	${}^3C_2 \times {}^4C_1 \times {}^4C_1 \times {}^3C_2 = 3 \times 4 \times 4 \times 3 = 144$
1HG, 2HL, 2WG, 1WL	${}^3C_1 \times {}^4C_2 \times {}^4C_2 \times {}^3C_1 = 3 \times 6 \times 6 \times 3 = 324$

Therefore, required number of ways

$$= 16 + 1 + 144 + 324 = 485.$$

20. In how many ways can a committee of 10 be selected with at least 4 women and 4 men from 9 women and 8 men if
(i) Ms X refuses to work with Mr Y
(ii) Ms X and Mr Y insist to work together?

The number of committees of 10 with at least 4 women and 4 men:

Possibilities	**Combinations**
4w, 6m	${}^9C_4 \times {}^8C_6 = \frac{9!}{4!5!} \times \frac{8!}{6!2!} = \frac{9\cdot8\cdot7\cdot6}{24} \times \frac{8\cdot7}{2} = 3528$
5w, 5m	${}^9C_5 \times {}^8C_5 = \frac{9\cdot8\cdot7\cdot6}{24} \times \frac{8\cdot7\cdot6}{6} = 7056$
6w, 4m	${}^9C_6 \times {}^8C_4 = \frac{9\cdot8\cdot7}{6} \times \frac{8\cdot7\cdot6\cdot5}{24} = 5880$

$\therefore$ total number of committees of 10 with at least 4 women and 4 men

$$= 3528 + 7056 + 5880 = 16464.$$

The number of such committees in which Ms X and Mr Y are present:

In this case we have to find the number of committees of 8 with at least 3 women and 3 men from 8 women and 7 men. As above, the number of committees in which Ms X and Mr Y work together

$$= {}^8C_3 \times {}^7C_5 + {}^8C_4 \times {}^7C_4 + {}^8C_5 \times {}^7C_3$$

$$= \frac{8 \cdot 7 \cdot 6}{6} \times \frac{7 \cdot 6}{2} + \frac{8 \cdot 7 \cdot 6 \cdot 5}{24} \times \frac{7 \cdot 6 \cdot 5}{6} + \frac{8 \cdot 7 \cdot 6}{6} \times \frac{7 \cdot 6 \cdot 5}{6}$$

$= 56 \times 21 + 70 \times 35 + 56 \times 35 = 5586$

[Answer to (ii)]

$\therefore$ the number of committees in which Ms X and Mr Y are not together $= 16464 - 5586 = 10878$. [Answer to (i)]

21. Find the number of subsets of a set of n elements if the number of elements in each subset is greater than or equal to half the elements of the given set.

The number of subsets of r elements

= the number of selections of r different things from n different things

$= {}^nC_r$.

$\therefore$ the total number of subsets having r or more elements

$= {}^nC_r + {}^nC_{r+1} + {}^nC_{r+2} + \ldots + {}^nC_n$.

If n is even, i.e., $n = 2m$ then $r = m$.

$\therefore$ the required number of subsets

$= {}^{2m}C_m + {}^{2m}C_{m+1} + {}^{2m}C_{m+2} + \ldots + {}^{2m}C_{2m}$

$= {}^{2m}C_0 + {}^{2m}C_1 + {}^{2m}C_2 + \ldots + {}^{2m}C_m$, writing in the reversed order and using ${}^nC_r = {}^nC_{n-r}$

$$= \left({}^{2m}C_0 + {}^{2m}C_1 + {}^{2m}C_2 + \ldots + {}^{2m}C_{m-1} + \frac{1}{2}\,{}^{2m}C_m\right) + \frac{1}{2}\,{}^{2m}C_m$$

= half of the series of binomial coefficients $+ \frac{1}{2} \cdot {}^{2m}C_m$

$$= 2^{2m-1} + \frac{1}{2} \cdot \frac{2m!}{m!\,m!} = 2^{n-1} + \frac{1}{2} \cdot \frac{n!}{\left(\frac{n}{2}!\right)^2}.$$

If n is odd, i.e., $n = 2m + 1$ then $r = m + 1$.

$\therefore$ the required number of subsets

$= {}^{2m+1}C_{m+1} + {}^{2m+1}C_{m+2} + \ldots + {}^{2m+1}C_{2m+1}$

$= {}^{2m+1}C_0 + {}^{2m+1}C_1 + {}^{2m+1}C_2 + \ldots + {}^{2m+1}C_m$

= half of the series of binomial coefficients

$= 2^{(2m+1)-1} = 2^{2m} = 2^{n-1}$.

22. In how many ways can two distinct subsets of the set A of $k(k \geq 3)$ elements be selected so that they have exactly two common elements?

Let the two subsets be called P and Q. The elements for the two sets will be selected as follows:

(i) 2 elements out of k elements for both the sets. This can be done in ${}^kC_2$ ways.

(ii) r elements for the subset P from $k - 2$ elements and any number of elements for Q from the remaining $k - 2 - r$ elements. Here r can vary from 0 to $k - 2$. For a fixed r, the number of selections

$= {}^{k-2}C_r \cdot 2^{k-2-r}$, because the number of selections of any number of things from n things is 2^n.

If r varies from 0 to $k - 2$, the total number of selections

$$= \sum_{r=0}^{k-2} {}^{k-2}C_r \cdot 2^{k-2-r} - 1,$$

excluding the case when both the subsets are equal having only the two common elements.

But every pair of P, Q is appearing twice like

$\{a_1, a_2, a_3\}, \{a_1, a_2, a_4, a_5\}$ and $\{a_1, a_2, a_4, a_5\}, \{a_1, a_2, a_3\}$.

Hence, the required number of ways

$$= {}^kC_2 \times \frac{1}{2}\left(\sum_{r=0}^{k-2} {}^{k-2}C_r \cdot 2^{k-2-r} - 1\right)$$

$$= \frac{1}{2} \cdot \frac{k(k-1)}{2} \cdot [({}^{k-2}C_0 \cdot 2^{k-2} + {}^{k-2}C_1 \cdot 2^{k-3} + {}^{k-2}C_2 \cdot 2^{k-4} + \ldots + {}^{k-2}C_{k-2}) - 1]$$

$$= \frac{k(k-1)}{4} \cdot \{(2+1)^{k-2} - 1\} = \frac{k(k-1)}{4} \cdot (3^{k-2} - 1).$$

23. A boat is to be manned by 9 crew with 4 on the stroke side, 4 on the row side and one to steer. There are 11 crew of which 2 can stroke only, 1 can row only while 3 can steer only. In how many ways the crew can be arranged for the boat?

The number of ways to select the man to steer $= {}^3C_1$. Of the remaining 8 persons, 2 (who can stroke only) are already selected for the stroke side and 1 is selected for the row side (who can row only). So, of the remaining 5 crew we have to select 2 for the stroke side and 3 for the row side.

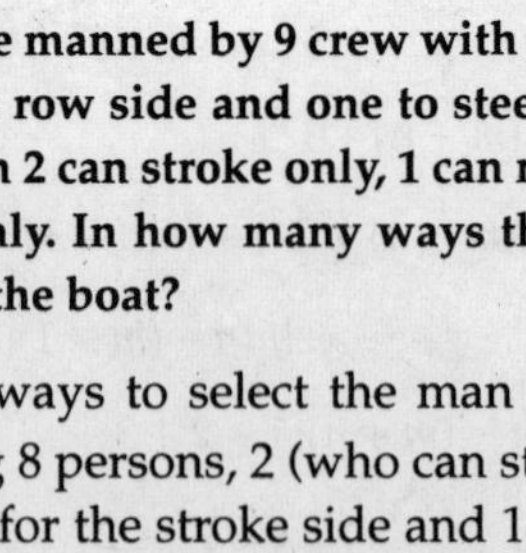

$\therefore$ the required number of selections $= {}^3C_1 \times {}^5C_2 \times {}^3C_3$.

But the four persons on the stroke side as well as row side can be arranged in 4 ! ways for every selection of crew.

$\therefore$ the required number of ways

$= {}^3C_1 \times {}^5C_2 \times {}^3C_3 \times 4! \times 4! = 17280.$

24. A is a set containing n elements. A subset P of A is chosen. The set A is reconstructed by replacing the elements of P. A subset Q of A is chosen. Find the number of ways of selecting P and Q so that P and Q are nonintersecting.

Each of the elements is available for selection in making P as well as Q. But P and Q must not have any common element. So each element can be taken in 3 ways (element $\in$ P, $\notin$ Q or $\notin$ P, $\in$ Q or $\notin$ P, $\notin$ Q)

$\therefore$ n elements can be taken in $3^n - 1$ ways which is equal to the number of ways to select nonintersecting subsets P and Q (excluding the case when P, Q are ϕ).

25. Prove that the number of ways to select n objects from $3n$ objects of which n are identical and the rest are different, is $2^{2n-1} + \frac{1}{2}\frac{2n!}{(n!)^2}$.

The number of selections of r objects from n identical objects and $n - r$ objects from $2n$ different objects

$= 1 \times {}^{2n}C_{n-r} = {}^{2n}C_{n-r}$.

Here r varies from 0 to n.

$\therefore$ the required number of selections

$$= \sum_{r=0}^{n} {}^{2n}C_{n-r} = {}^{2n}C_n + {}^{2n}C_{n-1} + {}^{2n}C_{n-2} + \ldots + {}^{2n}C_0$$

$$= {}^{2n}C_0 + {}^{2n}C_1 + {}^{2n}C_2 + \ldots + {}^{2n}C_n,$$

writing in reverse order

$$= \left({}^{2n}C_0 + {}^{2n}C_1 + {}^{2n}C_2 + \ldots + \frac{1}{2}\,{}^{2n}C_n\right) + \frac{1}{2}\cdot {}^{2n}C_n$$

= sum of the first half of the series of binomial coefficients $+ \frac{1}{2}\cdot {}^{2n}C_n$

$$= 2^{2n-1} + \frac{1}{2}\cdot\frac{2n!}{(n!)^2}.$$

26. Two different packs of cards are shuffled together. Cards are dealt equally among 4 players, each getting 13 cards. In how many ways can a player get his cards if no two cards are from the same suit with the same denomination?

Here, there are 52 distinct cards, each card being 2 in number. As no two cards are to be of the same suit with the same denomination (i.e., two cards are identical), 13 cards are to be selected from 52 cards where each card is two in number.

The number of selections of 13 cards from 52 distinct cards $= {}^{52}C_{13}$.

But each of the 13 cards can be selected in 2 ways (belonging to either of the two packs).

$\therefore$ the required number of ways

$$= {}^{52}C_{13} \times 2^{13} = \frac{52!}{13!\,39!}\cdot 2^{13}.$$

27. There are 5 mangoes and 4 apples. In how many different ways can a selection of fruits be made if
(i) fruits of the same kind are different
(ii) fruits of the same kind are identical?

(i) The number of ways to select any number of mangoes

$= {}^5C_0 + {}^5C_1 + {}^5C_2 + \ldots + {}^5C_5 = 2^5$.

The number of ways to select any number of apples

$= {}^4C_0 + {}^4C_1 + \ldots + {}^4C_4 = 2^4$.

$\therefore$ the required number of ways to select fruits

$= 2^5 \times 2^4 - 1$ {excluding the way in which 0 mangoes and 0 apples are selected}

$= 2^9 - 1$.

(ii) The required number of ways

$= (5 + 1)(4 + 1) - 1 = 6 \times 5 - 1 = 29$.

28. Find the number of factors of the number 37800. Also find the sum of the odd proper divisors of the number.

Here $37800 = 378 \times 100 = 3 \times 126 \times 2^2 \times 5^2$

$= 3 \times 3 \times 42 \times 2^2 \times 5^2$

$= 3 \times 3 \times 3 \times 7 \times 2^3 \times 5^2$

$= 2^3 \times 3^3 \times 5^2 \times 7$

$\therefore$ the required number of factors

= the total number of selections from

(2, 2, 2,), (3, 3, 3), (5, 5), (7)

$= (3 + 1)(3 + 1)(2 + 1)(1 + 1) - 2$

{excluding $2^0 \cdot 3^0 \cdot 5^0 \cdot 7^0$, i.e., 1 and $2^3 \cdot 3^3 \cdot 5^2 \cdot 7$, i.e., 37800 as factors}

$= 4 \times 4 \times 3 \times 2 - 2 = 96 - 2 = 94$.

The required sum

$= (3^0 + 3^1 + 3^2 + 3^3)(5^0 + 5^1 + 5^2)(7^0 + 7^1) - 1$

{2^n as a factor and 1 as a divisor are to be excluded}

$$= \frac{3^4 - 1}{3 - 1} \times \frac{5^3 - 1}{5 - 1} \times 8 - 1$$

$$= \frac{(3^4 - 1)(5^3 - 1)8}{2 \times 4} - 1$$

$$= (3^4 - 1)(5^3 - 1) - 1 = 80 \times 124 - 1 = 9919.$$

Note

(i) If we have to find the number of odd proper divisors then the required number

$$= (3 + 1)(2 + 1)(1 + 1) - 1$$

{$\because$ we have to exclude 2s and the number $3^0 \cdot 5^0 \cdot 7^0$}.

(ii) If we have to find the number of even proper divisors then the required number

$$= 3(3 + 1)(2 + 1)(1 + 1) - 1$$

{$\because$ we have to exclude $2^3 \cdot 3^3 \cdot 5^2 \cdot 7$ and at least one 2 is to be selected}.

(iii) If we have to find the number of proper divisors divisible by 10 then the required number

$$= 3 \times (3 + 1) \times 2 \times (1 + 1) - 1$$

{$\because$ at least one 2 and one 5 are to be selected}.

29. In how many ways can 3 girls and 9 boys be seated in two vans, each having numbered seats, 3 in the front and 4 at the back? How many sitting arrangements are possible if 3 girls should sit together in the back row on adjacent seats?

There are 7 seats in each of the vans. The total number of persons = 3 + 9 = 12. Now:

Possibilities		Combinations	Permutations
Van I	Van II		
7	5	${}^{12}C_7 \times {}^5C_5$	${}^{12}C_7 \times {}^5C_5 \times {}^7P_7 \times {}^7P_5$ $= \frac{12!}{7!5!} \cdot 7! \cdot \frac{7!}{2!}$
6	6	${}^{12}C_6 \times {}^6C_6$	${}^{12}C_6 \times {}^6C_6 \times {}^7P_6 \times {}^7P_6$ $= \frac{12!}{6!6!} \cdot 7! \cdot 7!$
5	7	${}^{12}C_5 \times {}^7C_7$	${}^{12}C_5 \times {}^7C_7 \times {}^7P_5 \times {}^7P_7$ $= \frac{12!}{5!7!} \cdot \frac{7!}{2!} \cdot 7!$

$\therefore$ total number of sitting arrangements

$$= \frac{12! \cdot 7 \cdot 6}{2} + 12! \cdot 7 \cdot 7 + \frac{12!}{2} \cdot 7 \cdot 6$$

$$= 21 \cdot (12!) + 49 \cdot (12!) + 21 \cdot (12!) = 91 \cdot (12!)$$

[**Note** We could have directly arranged 12 persons in 14 seats. The number of ways for this is ${}^{14}P_{12}$, i.e., $\frac{14!}{2!} = \frac{14 \cdot 13}{2} 12! = 91 \times 12!$. However, the first method is advised because it covers all situations.]

When 3 girls seat together in the back row of Van I:

Possibilities		Combinations	Permutations
Van I	Van II		
3g, 4b	5b	${}^9C_4 \times {}^5C_5$	${}^9C_4 \times {}^5C_5 \times (2 \times 3!)\, 4! \times {}^7P_5$
3g, 3b	6b	${}^9C_3 \times {}^6C_6$	${}^9C_3 \times {}^6C_6 \times (2 \times 3!)\, {}^4P_3 \times {}^7P_6$
3g, 2b	7b	${}^9C_2 \times {}^7C_7$	${}^9C_2 \times {}^7C_7 \times (2 \times 3!)\, {}^4P_2 \times {}^7P_7$

because the 3 girls in the back row can take place in 4 seats remaining together in 2 × 3 ! ways.

$\therefore$ the number of sitting arrangements with 3 girls in Van I

$$= {}^9C_4 \times 1 \times 2 \times 3!\, 4! \times \frac{7!}{2!} + {}^9C_3 \times 1 \times 2 \times 3! \times \frac{4!}{1!} \times \frac{7!}{1!}$$

$$+ {}^9C_2 \times 1 \times 2 \times 3! \times \frac{4!}{2!} \times 7!$$

$$= \frac{9!}{4!5!} \times 3!4!7! + \frac{9!}{3!6!} \times 2 \times 3!4!7!$$

$$+ \frac{9!}{2!7!} \times 2 \times 3 \times 4!7!$$

$$= 9!3! \times 7 \times 6 + 9! \times 14 \times 4! + 9! \times 3 \times 24$$

$$= 9!3!(42 + 56 + 12) = 660 \times 9!.$$

Similarly, the number of sitting arrangements with 3 girls in Van II = 660 × 9 !

$\therefore$ the required number of ways

$$= 2 \times 660 \times 9! = 1320 \times 9!.$$

30. In the given figure you have the road plan of a city. A man standing at X wants to reach the cinema hall at Y by the shortest path. What is the number of different paths that he can take?

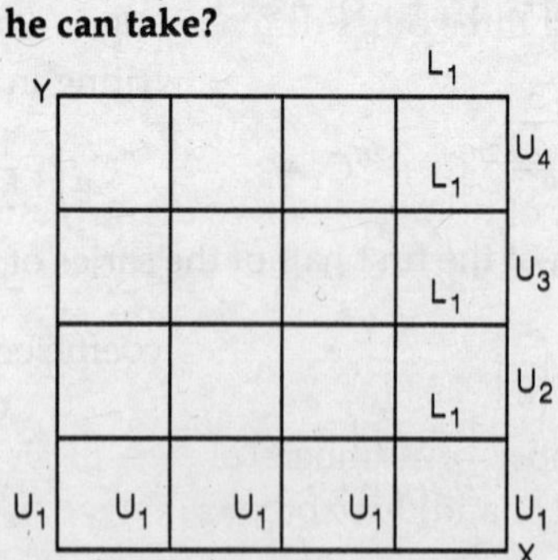

As the man wants to travel by one of the many possible shortest paths, he will never turn to the right or turn downwards. So a travel by one of the shortest paths is to take 4 horizontal pieces and 4 vertical pieces of roads. As he cannot take a right turn, he will use only one of the five horizonal pieces in the same vertical column. Similarly, he will use only one of the five vertical pieces in the same horizontal row.

$\therefore$ a shortest path is an arrangement of eight things $L_1, L_2, L_3, L_4, U_1, U_2, U_3, U_4$ so that the order of Ls and Us do not change ($\because$ clearly L_2 cannot be taken without taking L_1, U_2 cannot be taken without taking U_1, etc.)

Hence, the number of shortest paths

= the number of arrangements of $L_1, L_2, L_3, L_4, U_1, U_2, U_3, U_4$ where the order of Ls as well as the order of Us do not change

= the number of arrangements treating Ls as identical and Us as identical

$$= \frac{8!}{4!\,4!} = \frac{8 \cdot 7 \cdot 6 \cdot 5}{24} = 2 \cdot 7 \cdot 5 = 70.$$

31. How many numbers of five digits can be made having exactly two identical digits? How many of these will have the repeated digits in consecutive places?

If the identical digits be 0, 0 then three other digits can be selected from 1, 2, 3, ... , 9 in ${}^9C_3$ ways.

The number of numbers of five digits that can be made with 0, 0 and three other different digits

$$= \frac{5!}{2!} - 4!$$

($\because$ the number of arrangements without restriction is $\frac{5!}{2!}$ and the number of numbers beginning with 0 is 4 !)

$\therefore$ the number of numbers of five digits that can be made with 0, 0 and three other different digits

$$= {}^9C_3 \times \left(\frac{5!}{2!} - 4!\right).$$

The number of numbers of five digits that can be made with 1, 1 and three other different digits other than 0

$$= {}^8C_3 \times \frac{5!}{2!}.$$

The number of numbers of five digits that can be made with 1, 1, 0 and two other different digits

$$= {}^8C_2 \times \left(\frac{5!}{2!} - \frac{4!}{2!}\right).$$

$\therefore$ the number of numbers of five digits that can be made with 1, 1 and three other different digits

$$= {}^8C_3 \times \frac{5!}{2!} + {}^8C_2 \times \left(\frac{5!}{2!} - \frac{4!}{2!}\right).$$

Similarly for numbers having 2, 2 or 3, 3, etc.

$\therefore$ the required number of numbers having exactly two repeated digits

$$= {}^9C_3\left(\frac{5!}{2!} - 4!\right) + 9 \times \left\{{}^8C_3 \times \frac{5!}{2!} + {}^8C_2 \times \left(\frac{5!}{2!} - \frac{4!}{2!}\right)\right\}$$

$$= 84(60 - 24) + 9 \times \{56 \times 60 + 28 \times (60 - 12)\}$$

$$= 84 \times 36 + 9 \times (3360 + 1344)$$

$$= 3024 + 42336 = 45360.$$

Next, similarly as before, the number of numbers of five digits that can be made with 0, 0 and three other different digits when 0, 0 are consecutive

$$= {}^9C_3(4! - 3!) \qquad (\because 0, 0 \text{ are treated as one object}).$$

The number of numbers of five digits that can be made with 1, 1 (or 2, 2 or ... or 9, 9) and three other different digits other than 0, where 1, 1 are consecutive

$$= {}^8C_3 \times 4! \qquad (\because 1, 1 \text{ are treated as one object}).$$

The number of numbers of five digits that can be made with 1, 1, 0 and two other different digits where 1, 1 are consecutive

$$= {}^8C_2 \times (4! - 3!) \quad (\because 1, 1 \text{ are treated as one object}).$$

Hence, the required number of numbers with two identical digits consecutive

$$= {}^9C_3(4! - 3!) + 9\{{}^8C_3 \times 4! + {}^8C_2 \times (4! - 3!)\}$$

$$= 84 \times 18 + 9(56 \times 24 + 28 \times 18)$$

$$= 1512 + 16632 = 18144.$$

32. How many numbers of five digits can be made with the digits 1, 2, 3 each of which can be used at most thrice in a number?

We have the digits 1, 1, 1, 2, 2, 2, 3, 3, 3 to make numbers of five digits. The digits will be as follows (constructionwise):

(i) three identical, one pair (for example: 1, 1, 1, 2, 2, etc.)

(ii) three identical, two different (for example: 3, 3, 3, 1, 2, etc.)

(iii) two pairs, one different (for example: 2, 2, 1, 1, 3, etc.)

TIP: In order to cover all the possibilities start with all identical digits and go on reducing the number of identical digists, ultimately reaching all different digits.

(i) The number of selections of three identical digits, one pair = ${}^3C_1 \times {}^2C_1$.

Corresponding to each selection, the number of numbers that can be made = $\frac{5!}{3!\,2!}$

$\therefore$ the total number of numbers of three identical digits and one pair

$$= {}^3C_1 \times {}^2C_1 \times \frac{5!}{3!\,2!} = 3 \times 2 \times \frac{5 \times 4}{2} = 60 \qquad \dots (1)$$

(ii) The number of selections of three identical digits and two different digits = ${}^3C_1 \times {}^2C_2$

$\therefore$ the total number of numbers of three identical digits and two different digits

$$= {}^3C_1 \times {}^2C_2 \times \frac{5\,!}{3\,!} = 3 \times \frac{120}{6} = 60 \quad \ldots (2)$$

(iii) The number of selections of two pairs, one different digit $= {}^3C_2 \times {}^1C_1$

$\therefore$ the total number of numbers of two pairs and one different digit $= {}^3C_2 \times {}^1C_1 \times \frac{5\,!}{2\,!\,2\,!} = 3 \times 30 = 90.$

$\therefore$ the required number of numbers

$$= 60 + 60 + 90 = 210.$$

33. Find the total number of selections of 5 letters from five As, four Bs, three Cs and two Ds. Also find the number of 5-letter words that can be made.

There is one letter A to take 5 identical letters, two letters A, B to take 4 identical letters, three letters A, B, C to take 3 identical letters and four letters A, B, C, D to take pairs. There are only four different letters.

Possible structures, selections and arrangements are given in a tabular form as below:

Possibilities	Combinations	Permutations
5 identical	${}^1C_1$	${}^1C_1 \times \frac{5\,!}{5\,!}$
4 identical, 1 different	${}^2C_1 \times {}^3C_1$	${}^2C_1 \times {}^3C_1 \times \frac{5\,!}{4\,!}$
3 identical, 1 pair	${}^3C_1 \times {}^3C_1$	${}^3C_1 \times {}^3C_1 \times \frac{5\,!}{3\,!\,2\,!}$
3 identical, 2 different	${}^3C_1 \times {}^3C_2$	${}^3C_1 \times {}^3C_2 \times \frac{5\,!}{3\,!}$
2 pairs, 1 different	${}^4C_2 \times {}^2C_1$	${}^4C_2 \times {}^2C_1 \times \frac{5\,!}{2\,!\,2\,!}$
1 pair, 3 different	${}^4C_1 \times {}^3C_3$	${}^4C_1 \times {}^3C_3 \times \frac{5\,!}{2\,!}$

$\therefore$ the required number of selections

$$= {}^1C_1 + {}^2C_1 \times {}^3C_1 + {}^3C_1 \times {}^3C_1 + {}^3C_1 \times {}^3C_2 + {}^4C_2 \times {}^2C_1 + {}^4C_1 \times {}^3C_3$$

$$= 1 + 2 \times 3 + 3 \times 3 + 3 \times 3 + 6 \times 2 + 4 \times 1$$

$$= 1 + 6 + 9 + 9 + 12 + 4 = 41.$$

The required number of words (i.e., permutations)

$$= {}^1C_1 \times \frac{5\,!}{5\,!} + {}^2C_1 \times {}^3C_1 \times \frac{5\,!}{4\,!} + {}^3C_1 \times {}^3C_1 \times \frac{5\,!}{3\,!\,2\,!} + {}^3C_1 \times {}^3C_2 \times \frac{5\,!}{3\,!} + {}^4C_2 \times {}^2C_1 \times \frac{5\,!}{2\,!\,2\,!} + {}^4C_1 \times {}^3C_3 \times \frac{5\,!}{2\,!}$$

$$= 1 + 30 + 90 + 180 + 360 + 240 = 901.$$

Note The number of selections is also equal to the coefficient of x^5 in

$$(x^0 + x^1 + x^2 + \ldots + x^5)(x^0 + x^1 + x^2 + x^3 + x^4) \times (x^0 + x^1 + x^2 + x^3)(x^0 + x + x^2)$$

= coefficient of x^5 in

$$\frac{1 - x^6}{1 - x} \cdot \frac{1 - x^5}{1 - x} \cdot \frac{1 - x^4}{1 - x} \cdot \frac{1 - x^3}{1 - x}$$

= coefficient of x^5 in

$$(1 - x^6)(1 - x^5)(1 - x^4)(1 - x^3)(1 - x)^{-4}$$

= coefficient of x^5 in

$$(1 - x^5 - x^6 + x^{11})(1 - x^3 - x^4 + x^7) \times \left\{1 + 4x + \frac{4 \cdot 5}{2\,!}x^2 + \frac{4 \cdot 5 \cdot 6}{3\,!}x^3 + \frac{4 \cdot 5 \cdot 6 \cdot 7}{4\,!}x^4 + \frac{4 \cdot 5 \cdot 6 \cdot 7 \cdot 8}{5\,!}x^5 + \ldots\right\}$$

= coefficient of x^5 in

$$\{1 - x^3 - x^4 - x^5\} \times \{1 + 4x + 10x^2 + 20x^3 + 35x^4 + 56x^5\},$$

neglecting powers higher than x^5

$$= 56 - 10 - 4 - 1 = 41.$$

34. In how many different ways can a set A of $3n$ elements be partitioned into 3 subsets of equal number of elements? (The subsets P, Q, R form a partition if

$$\mathbf{P \cup Q \cup R = A, P \cap R = \phi, Q \cap R = \phi, R \cap P = \phi.)}$$

The number of ways to partition the set

= the number of ways in which $3n$ different things can be divided in 3 equal groups

= x (say).

Then $3n$ different things can be distributed among 3 persons equally in $x \times (3\,!)$ ways.

But the number of ways to distribute $3n$ different things equally among 3 persons

$$= {}^{3n}C_n \times {}^{2n}C_n \times {}^nC_n$$

$$= \frac{3n\,!}{n\,!\,2n\,!} \times \frac{2n\,!}{n\,!\,n\,!} \times \frac{n\,!}{n\,!\,0\,!} = \frac{3n\,!}{(n\,!)^3}.$$

$\therefore \quad x \times 3! = \frac{3n!}{(n!)^3}; \quad \therefore \quad x = \frac{3n!}{6(n!)^3}.$

35. Five balls are to be placed in three boxes. Each can hold all the five balls. In how many different ways can we place the balls so that no box remains empty, if
(i) balls and boxes are all different
(ii) balls are identical but boxes are different
(iii) balls are different but boxes are identical
(iv) balls as well as boxes are identical
(v) balls as well as boxes are identical but boxes are kept in a row?

As no box is to remain empty, boxes can have balls in the following numbers:

Possibilities 1, 1, 3 or 1, 2, 2.

(i) The number of ways to distribute the balls in groups of 1, 1, 3

$= {}^5C_1 \times {}^4C_1 \times {}^3C_3.$

But the boxes can interchange their content, no exchange giving a new way when boxes containing balls in equal numbers interchange.

$\therefore$ the total number of ways to distribute 1, 1, 3 balls to the boxes

$$= {}^5C_1 \times {}^4C_1 \times {}^3C_3 \times \frac{3!}{2!}$$

Similarly, the total number of ways to distribute 1, 2, 2 balls to the boxes

$$= {}^5C_1 \times {}^4C_2 \times {}^2C_2 \times \frac{3!}{2!}.$$

$\therefore$ the required number of ways

$$= {}^5C_1 \times {}^4C_1 \times {}^3C_3 \times \frac{3!}{2!} + {}^5C_1 \times {}^4C_2 \times {}^2C_2 \times \frac{3!}{2!}$$

$$= 5 \times 4 \times 3 + 5 \times 6 \times 3 = 60 + 90 = 150.$$

Note Writing the whole answer in tabular form,

possibilities	combinations	permutations
1, 1, 3	${}^5C_1 \times {}^4C_1 \times {}^3C_3$	${}^5C_1 \times {}^4C_1 \times {}^3C_3 \times \frac{3!}{2!}$
		$= 5 \times 4 \times 3 = 60$
1, 2, 2	${}^5C_1 \times {}^4C_2 \times {}^2C_2$	${}^5C_1 \times {}^4C_2 \times {}^2C_2 \times \frac{3!}{2!}$
		$= 5 \times 6 \times 3 = 90$

$\therefore$ the required number of ways = 60 + 90 = 150.

(ii) When balls are identical but boxes are different the number of combinations will be 1 in each case.

$\therefore$ the required number of ways

$$= 1 \times \frac{3!}{2!} + 1 \times \frac{3!}{2!} = 3 + 3 = 6.$$

(iii) When balls are different and boxes are identical, after giving 3 balls to a box, the remaining 2 are to be divided in two equal groups because the boxes are identical. Similarly, after giving 1 ball to a box, the remaining 4 are to be divided in two equal groups.

$\therefore$ the required number of ways

$$= {}^5C_3 \times \frac{{}^2C_1 \times {}^1C_1}{2!} + {}^5C_1 \times \frac{{}^4C_2 \times {}^2C_2}{2!} = 10 + 15 = 25.$$

(iv) When balls as well as boxes are identical, the number of combinations and arrangements will be 1 each in both cases.

$\therefore$ the required number of ways

$$= 1 \times 1 + 1 \times 1 = 2.$$

(v) When boxes are kept in a row, they will be treated as different. So, in this case the number of ways will be the same as in (ii).

36. Let n and k be positive integers such that $n \geq \frac{k(k+1)}{2}$. Find the number of solutions $(x_1, x_2, x_3, \ldots, x_k)$, $x_1 \geq 1$, $x_2 \geq 2, \ldots, x_k \geq k$, all integers satisfying the condition $x_1 + x_2 + x_3 + \ldots + x_k = n$.

Let $x_1 = y_1 + 1, x_2 = y_2 + 2, \ldots, x_k = y_k + k$.

Putting these in the equation,

$$(y_1 + 1) + (y_2 + 2) + (y_3 + 3) + \ldots + (y_k + k) = n$$

or $y_1 + y_2 + y_3 + \ldots + y_k$

$$= n - (1 + 2 + 3 + \ldots + k) = n - \frac{k(k+1)}{2}$$

$= m$ (say) where m is an integer ≥ 0.

Now, we have to find the number of non-negative integral solutions of the equation

$$y_1 + y_2 + y_3 + \ldots + y_k = m \qquad \ldots (1)$$

The number of solutions will be equal to the total number of ways in which m identical things (here the thing is the number 1) can be distributed among k persons (here k variables) and this is equal to ${}^{m+k-1}C_{k-1}$

$\therefore$ the number of non-negative integral solution of (1) and hence that of the given equation

$$= {}^{m+k-1}C_{k-1} = \frac{(m+k-1)!}{(k-1)!\,m!},$$

where $m = n - \frac{k(k+1)}{2}$.

Note

(1) The number of solutions of

$$y_1 + y_2 + y_3 + \ldots + y_k = m$$

where y_1, y_2, etc., and m are non-negative integers, can also be obtained by finding coefficient in a suitable expansion.

Consider the expansion of $(1 + x + x^2 + \ldots)^k$.

$$(1 + x + x^2 + x^3 + \dots)^k$$
$$= (x^0 + x + x^2 + \dots) \times (x^0 + x + x^2 + \dots) \times \dots$$
$$\dots \times (x^0 + x + x^2 + \dots)$$
{there being k factors}.

The coefficient of a power of x, say x^m in the RHS

= number of ways in which m can be split in k parts (non-negative integers), sum of the parts being m.

$\therefore$ the number of solutions of the equation

$= \text{coefficient of } x^m \text{ in } (1 + x + x^2 + x^3 + \dots)^k$

$= \text{coefficient of } x^m \text{ in } \left(\frac{1}{1-x}\right)^k$

$= \text{coefficient of } x^m \text{ in } (1-x)^{-k}$

$= \text{coefficient of } x^m$ in

$$\left[1 + kx + \frac{k(k+1)}{2!}x^2 + \dots + \frac{k(k+1)(k+2)\dots(k+m-1)}{m!}x^m + \dots\right]$$

(using binomial theorem for any index)

$$= \frac{k(k+1)(k+2)\dots(k+m-1)}{m!} = \frac{(m+k-1)!}{m!(k-1)!}.$$

(2) The number of non-negative integral solutions of $y_1 + y_2 + \dots + y_k \le m$ can be obtained by finding the number of non-negative integral solutions of $y_1 + y_2 + \dots + y_k + y_{k+1} = m$ (where y_{k+1} is also non-negative integer). Their number of non-negative integral solutions will be equal.

37. Find the number of non-negative integral solutions of $2x + y + z = 21$.

Clearly $x = 0, 1, 2, 3, \dots, 10$. Let $x = k$; then $0 \le k \le 10$.
When $x = k$, $y + z = 21 - 2k$.
The number of non-negative integral solutions

= the number of ways to distribute $(21 - 2k)$ identical things (each thing is the number 1) among 2 persons

$= {}^{21-2k+2-1}C_{2-1} = {}^{22-2k}C_1 = 22 - 2k.$

$\therefore$ the required number of solutions

$$= \sum_{k=0}^{10} (22 - 2k) = 22 + 20 + 18 + \dots + 2$$

$$= 2(1 + 2 + 3 + \dots + 11) = 2 \times \frac{11 \times 12}{2} = 132.$$

38. Show that the total number of non-negative integral solutions of 4-tuples (n_1, n_2, n_3, n_4) is ${}^{104}C_4$ such that $\sum_{i=1}^{4} n_i \le 100$.

Here $n_1 + n_2 + n_3 + n_4 \le 100$,

where $n_1 \ge 0, n_2 \ge 0, n_3 \ge 0, n_4 \ge 0$.

Then $n_1 + n_2 + n_3 + n_4 = 100 - k$,

where $0 \le k \le 100$ and k is integer.

The number of non-negative integral solutions of the equation for a particular k

$= \text{coefficient of } x^{100-k}$ in

$$(x^0 + x^1 + x^2 + \dots)(x^0 + {}^1 + x^2 + \dots) \times (x^0 + x^1 + x^2 + \dots)(x^0 + x^1 + x^2 + \dots)$$

$= \text{coefficient of } x^{100-k} \text{ in } (1 + x + x^2 + \dots)^4$

$= \text{coefficient of } x^{100-k} \text{ in } \left(\frac{1}{1-x}\right)^4$

$= \text{coefficient of } x^{100-k} \text{ in } (1-x)^{-4}$

$$= \frac{4 \cdot 5 \cdot 6 \dots \text{ to } (100-k) \text{ factors}}{(100-k)!}$$

$$= \frac{4 \cdot 5 \cdot 6 \dots (103-k)}{(100-k)!}$$

$$= \frac{(103-k)!}{(100-k)!\,3!} = {}^{103-k}C_3.$$

As $0 \le k \le 100$, summing up for all integers k, $0 \le k \le 100$ the required number of solutions

$$= \sum_{k=0}^{100} {}^{103-k}C_3$$

$$= {}^{103}C_3 + {}^{102}C_3 + \dots + {}^3C_3$$

$= \text{coefficient of } x^3$ in

$$\{(1+x)^{103} + (1+x)^{102} + \dots + (1+x)^3\}$$

$$= \text{coefficient of } x^3 \text{ in } (1+x)^3 \cdot \frac{(1+x)^{101} - 1}{(1+x) - 1}$$

$$= \text{coefficient of } x^4 \text{ in } (1+x)^3 \{(1+x)^{101} - 1\}$$

$$= \text{coefficient of } x^4 \text{ in } (1+x)^{104} = {}^{104}C_4.$$

39. How many triangles with sides of lengths in integral cm can we get if the length of the greatest side is given to be $2m$ cm? Also find how many of the triangles are isosceles and how many are equilateral.

If x and y be the other sides then we can have $x \le y \le 2m$.

Again, in a triangle, sum of any two sides > third side.

$\therefore$ the sum of two smaller sides > the longest side

$\therefore$ $x + y > 2m$.

As the greatest possible value of x, y is $2m$ each, we get

$$4m \ge x + y > 2m.$$

In order to find the number of triangles we have to find the number of positive integral solutions of the equations

$$x + y = 2m + 1 \qquad \dots (1)$$

$x + y = 2m + 2$ ... (2)

....

$x + y = 4m$ ... (2m)

so that $1 \le x \le 2m$, $1 \le y \le 2m$.

There are m solutions for (1) because the solutions are

$(1, 2m), (2, 2m - 1), \ldots, (m, m + 1)$ and $x \le y$.

Also (2) is $x + y = 2m + 2$.

There are m solutions because the solutions are

$(2, 2m), (3, 2m - 1), \ldots, (m + 1, m + 1)$.

Again, (3) is $x + y = 2m + 3$.

There are $m - 1$ solutions because the solutions are

$(3, 2m), (4, 2m - 1), \ldots, (m + 1, m + 2)$.

(4) is $x + y = 2m + 4$.

There are $m - 1$ solutions because the solutions are

$(4, 2m), (5, 2m - 1), \ldots, (m + 2, m + 2)$.

Thus, the total number of solutions

$= m + m + (m - 1) + (m - 1) + \ldots + 1 + 1$

$= 2\{m + (m - 1) + \ldots + 1\} = 2 \cdot \dfrac{m(m+1)}{2} = m(m + 1)$.

The triangle will be isosceles or equilateral if $x = y$ or any one of x, y is $2m$.

The number of solutions in which either $x = 2m$ or $y = 2m$

$= 1 + 1 + 1 + \ldots$ to $2m$ times

because each equation has one such solution

$= 2m$.

The number of solutions in which $x = y$

$= 1 + 1 + 1 + \ldots$ to $(m - 1)$ terms

$= m - 1$

because equations (2), (4), (6), ..., $(2m - 2)$ have each one such solution.

$\therefore$ total number of isosceles or equilateral triangles

$= 2m + m - 1 = 3m - 1$.

There is only one equilateral triangle (each side being $2m$ cm)

$\therefore$ the number of isosceles triangle (which are not equilateral) $= 3m - 2$.

Note We could have used combination to find the number of solutions of the equations (1), (2), etc.

40. A condolence meeting is being held in a big hall which has 7 doors by which the mourners enter the hall. One can use any one of the 7 doors to enter and can come at anytime during the meeting. At each door a register is kept in which a mourner has to put down his signature while entering the hall. If 200 people attend the meeting, how many different sets of 7 lists of signatures can arise?

200 people can be divided in seven groups consisting of any number of people in as many ways as the number of integral solutions of

$x_1 + x_2 + x_3 + x_4 + x_5 + x_6 + x_7 = 200$,

where each of x_1, x_2, etc. ≥ 0.

We know the number of integral solutions of the equation

= the number of ways in which 200 identical things can be distributed among 7 persons

$= {}^{200+7-1}C_{7-1}$

($\because$ here $n = 200$, $r = 7$ in the result ${}^{n+r-1}C_{r-1}$)

$= {}^{206}C_6$.

But for each distribution of 200 persons in the 200 places of signatures there are 200 ! arrangements.

$\therefore$ the required number of lists

$= {}^{206}C_6 \times 200\,!$

$= \dfrac{206\,!}{6\,!\,200\,!} \times 200\,! = \dfrac{206\,!}{6\,!} = {}^{206}P_{200}$.

41. In how many ways can 10 persons take seats in a row of 24 fixed seats so that no two persons take consecutive seats?

As no two persons take consecutive seats, there will be at least one seat vacant between any two persons with some or none before the first person and after the last person. Let the number of vacant seats before the first person $= x_0$, the number of vacant seats between the first and the second (persons) be x_1, etc., as shown in the figure.

Clearly, the total number of vacant seats $= 24 - 10 = 14$.

$\therefore \quad x_0 + x_1 + x_2 + \ldots + x_9 + x_{10} = 14$

where $x_0 \ge 0, x_1 \ge 1, x_2 \ge 1, x_3 \ge 1, \ldots, x_9 \ge 1, x_{10} \ge 0$.

Put $x_0 = y_0, x_1 = y_1 + 1, x_2 = y_2 + 1, \ldots,$

$x_9 = y_9 + 1, x_{10} = y_{10}$.

Then the equation becomes

$y_0 + (y_1 + 1) + (y_2 + 1) + \ldots + (y_9 + 1) + y_{10} = 14$

or $\quad y_0 + y_1 + y_2 + \ldots + y_9 + y_{10} = 14 - 9 = 5$

$\therefore$ the required number of ways

= the number of non-negative integral solutions of the equation $y_0 + y_1 + y_2 + \ldots + y_9 + y_{10} = 5$

$= {}^{5+11-1}C_{11-1} = {}^{15}C_{10}$.

But the 10 persons can arrange among themselves in ${}^{10}P_{10}$ ways.

$\therefore$ the required number of ways

$= {}^{15}C_{10} \times {}^{10}P_{10}$

$$= \frac{15\,!}{10\,!\,5\,!} \times 10\,! = \frac{15\,!}{5\,!}.$$

Note If 10 persons are to take seats such that each has exactly one neighbour then they have to be paired off and consider them as $\frac{10}{2}$, i.e., 5 persons and proceed as above.

42. An examination has four papers. Each paper has a maximum of *n* marks. Find the number of ways in which an examinee can secure 2*n* marks in the examination.

An examinee can secure marks 0, 1, 2, …, n in a paper and the sum of the marks secured in four papers should be $2n$.

$\therefore$ the required number of ways

= the number of ways to distribute $2n$ marks over four papers having maximum n marks each

$= \text{coefficient of } x^{2n} \text{ in } (x^0 + x + x^2 + \ldots + x^n)^4$

$= \text{coefficient of } x^{2n} \text{ in } \left(\frac{1 - x^{n+1}}{1 - x}\right)^4$

$= \text{coefficient of } x^{2n} \text{ in } (1 - x^{n+1})^4 (1 - x)^{-4}$

$= \text{coefficient of } x^{2n}$ in

$$({}^4C_0 - {}^4C_1\, x^{n+1} + {}^4C_2\, x^{2n+2} - {}^4C_3\, x^{3n+3} + {}^4C_4\, x^{4n+4})$$

$$\times \left\{1 + 4x + \frac{4 \cdot 5}{2\,!}x^2 + \frac{4 \cdot 5 \cdot 6}{3\,!}x^3 + \ldots + \frac{4 \cdot 5 \cdot 6 \ldots (n+3)}{n\,!}x^n + \ldots\right\}$$

$$= {}^4C_0 \cdot \frac{4 \cdot 5 \cdot 6 \ldots (2n+3)}{(2n)\,!} - {}^4C_1 \cdot \frac{4 \cdot 5 \cdot 6 \cdot \ldots (n+2)}{(n-1)\,!}$$

$$= \frac{4 \cdot 5 \cdot 6 \ldots (2n+3)}{(2n)\,!}\} - 4 \cdot \frac{4 \cdot 5 \cdot 6 \cdot \ldots (n+2)}{(n-1)\,!}$$

$$= \frac{(2n+1)(2n+2)(2n+3)}{1 \cdot 2 \cdot 3} - 4 \cdot \frac{n(n+1)(n+2)}{1 \cdot 2 \cdot 3}$$

$$= \frac{1}{3}(n+1)\,\{(2n+1)(2n+3) - 2n(n+2)\}$$

$$= \frac{1}{3}(n+1)(2n^2 + 4n + 3).$$

Note

(1) If at least one mark is to be obtained from each paper, the required number of ways

$= \text{coefficient of } x^{2n} \text{ in } (x + x^2 + x^3 + \ldots + x^n)^4.$

(2) If at least 1 mark from one paper, at least 2 marks from another, at least 3 marks from another and at least 4 marks from the fourth are to be obtained then the required number of ways

$= \text{coefficient of } x^{2n}$ in

$$(x + x^2 + x^3 + \ldots + x^n)(x^2 + x^3 + \ldots + x^n) \times (x^3 + x^4 + \ldots + x^n)(x^4 + x^5 + \ldots + x^n).$$

(3) If one paper has maximum marks $2n$ then the required number of ways

$= \text{coefficient of } x^{2n}$ in

$$(x^0 + x + x^2 + \ldots + x^n)^3 \times (x^0 + x + x^2 + \ldots + x^{2n}).$$

43. Find the number of ways in which five different letters can be put in their five addressed envelopes so that all the letters are in the wrong envelopes.

The number of ways to put 2 letters in 2 addressed envelopes so that all are in the wrong envelopes = 1.

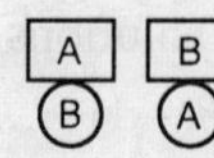

The number of ways to put 3 letters in 3 addressed envelopes so that all are in the wrong envelopes

= (the number of ways without restriction)

– (the number of ways in which all are in the correct envelopes)

– (the number of ways in which 1 letter is in the correct envelope)

$= 3\,! - 1 - {}^3C_1 \times 1 = 6 - 1 - 3 = 2.$

[**Note** two are in correct $\Rightarrow$ all are in correct]

The number of ways to put 4 letters in 4 addressed envelopes so that all are in the wrong envelopes

= (the number of ways without restriction)

– (the number of ways in which all are in the correct envelopes)

– (the number of ways in which 1 letter is in the correct envelope)

– (the number of ways in which 2 letters are in the correct envelopes)

$= 4\,! - 1 - {}^4C_1 \times 2 - {}^4C_2 \times 1$

$= 24 - 1 - 8 - 6 = 9.$

The number of ways to put 5 letters in 5 addressed envelopes so that all are in the wrong envelopes (as above)

$= 5\,! - 1 - {}^5C_1 \times 9 - {}^5C_2 \times 2 - {}^5C_3 \times 1$

$= 120 - 1 - 5 \times 9 - 10 \times 2 - 10 \times 1$

$= 120 - 76 = 44.$

44. An unlimited number of counters bearing the letters A, B and C are available. As a result any number of counters bearing A or B or C can be selected. Find the number of ways to select $m(\geq 3)$ counters so that they cannot be used to make the word CAB by using any three of the counters from the selected counters.

The word CAB cannot be spelled if
(i) only As or only Bs or only Cs are selected
(ii) m counters consist of two of the 3 letters A, B, C only.

The number of selections of m counters in which all are As or all are Bs or all are Cs

$$= {}^3C_1 \times {}^{1+m-1}C_m = {}^3C_1 \times {}^mC_m = 3.$$

($\because$ the total number of selections of r things from n things when each thing can be repeated any number of times $= {}^{n+r-1}C_r$. Here $n = 1$, $r = m$.)

The number of selections of m counters with two of the 3 letters A, B, C

$$= {}^3C_2 \times {}^{2+m-1}C_m = 3 \times {}^{m+1}C_m = 3(m+1).$$

This also includes selections from only one letter in three selections. So, the required number of selections

$$= 3(m+1) - 3 = 3m.$$

45. In how many ways can 10 gentlemen sit round a table so that in no two arrangements they have the same neighbours?

As the gentlemen are not to have the same two neighbours, clockwise and anticlockwise arrangements will be treated as identical.

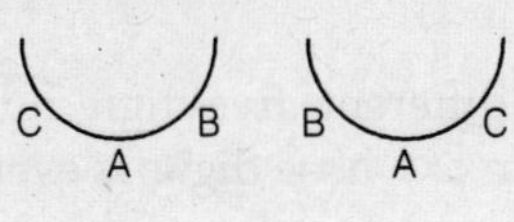

$\therefore$ the required number of arrangements

$$= \frac{1}{2}(10-1)! = \frac{1}{2}\,9!.$$

46. A regular polygon of 10 sides is constructed. In how many ways can 3 vertices be selected so that no two vertices are consecutive?

The required number of selections
= (the number of selections without restriction)
– (the number of selections when 3 vertices are consecutive)
– (the number of selections when 2 vertices are consecutive).

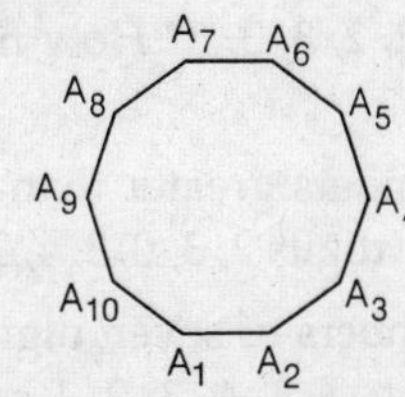

Now, the number of selections of 3 vertices without restriction $= {}^{10}C_3$.

The number of selections of 3 consecutive vertices = 10
(by observation $A_1A_2A_3$, $A_2A_3A_4$, … , $A_{10}A_1A_2$).

The number of selections when two vertices are consecutive $= 10 \times {}^6C_1$

($\because$ after selecting two consecutive vertices in 10 ways, the third can be selected from 6 vertices)

$\therefore$ the required number of selections

$$= {}^{10}C_3 - 10 - 10 \times {}^6C_1$$

$$= \frac{10 \times 9 \times 8}{6} - 10 - 60 = 120 - 70 = 50.$$

Exercises

1. Prove that
$${}^1P_1 + 2 \cdot {}^2P_2 + 3 \cdot {}^3P_3 + \ldots + n \cdot {}^nP_n = {}^{n+1}P_{n+1} - 1.$$

2. Evaluate ${}^{47}C_4 + \sum_{j=0}^{3} {}^{50-j}C_3$.

3. Prove that the product of r consecutive integers is divisible by $r!$.

4. Establish ${}^{n+1}C_{m+1} = {}^nC_m + {}^nC_{m+1}$.
Hence prove that
$${}^{n+1}C_{m+1} = {}^mC_m + {}^{m+1}C_m + {}^{m+2}C_m + \ldots + {}^nC_m,\ (n > m).$$

5. Prove that ${}^{n-1}C_3 + {}^{n-1}C_4 > {}^nC_3$ if $n > 7$.

6. Prove that $(n!)!$ is divisible by $(n!)^{(n-1)!}$.

7. Show that two positive integers n, r cannot be found such that ${}^nC_r$, ${}^nC_{r+1}$, ${}^nC_{r+2}$ are in GP.

8. Show that $1! + 2! + 3! + \ldots + n!$ cannot be a perfect square for any $n \in N$, $n \geq 4$.

9. A letter lock consists of three rings each marked with 10 different letters; find in how many ways it is possible to make an unsuccessful attempt to open the lock.

10. A telegraph has a number of arms and each arm is capable of taking 4 distinct positions, including the position of rest. If 1023 different signals can be sent in all then find the number of arms.

11. How many 10-digit numbers can be made with odd digits so that no two consecutive digits are the same?

12. How many odd numbers of five digits can be formed with the digits 1, 2, 3, 4, 5 if the digits cannot be repeated in the same number?

13. Let $n \in N$ and $300 < n < 3000$. If n is made of distinct digits by taking from 0, 1, 2, 3, 4, 5 then find the greatest possible number of values of n.

14. How many seven-digit numbers can be made from the digits 1, 2, 2, 2, 3, 3, 5? How many of these are odd?

15. How many numbers greater than a million can be formed with the digits 2, 3, 0, 3, 4, 2, 3?

16. How many numbers of seven digits can be formed with the digits 1, 2, 3, 4, 3, 2, 1 so that odd digits always occupy odd places?

17. How many words can be made with the letters of the word "TECHNOLOGY" which do not begin with T but end with Y?

18. Find the number of odd numbers greater than two million that can be made with the digits of the number 1315414.

19. How many words can be made with the letters of the word INSTITUTION such that vowels and consonants alternate?

20. In how many ways can 6 boys and 6 girls be arranged in a row so that no two boys and no two girls sit together?

21. How many words can be made with the letters of the word BHARATI so that all the vowels are consecutive?

22. How many numbers of six digits can be made by arranging the digits of the number 123425 so that all the even digits do not occupy consecutive places?

23. In how many ways can the letters of the word ALGEBRA be arranged without changing the relative order of vowels and consonants?

24. Find the number of ways in which 10 candidates in an examination can be ranked
    (i) if two particular candidates are always next to each other
    (ii) if one particular candidate is always above another particular candidate.

25. In how many ways can the letters of the word CINEMA be arranged so that the order of vowels do not change?

26. A dictionary is made of the words that can be made by arranging the letters of the word PARKAR. What is the position of the word "PARKAR" in that dictionary if words are printed in the same order as that of an ordinary dictionary?

27. In how many ways can 6 boys and 4 girls sit in a row so that no boy is between two girls?

28. There are three copies each of two books and two works each consisting of three volumes. In how many ways can a bookseller arrange the 12 books in a shelf of one row so that neither the copies of the same book nor the volumes of the same work are separated?

29. Show that the number of permutations of $2n$ letters which are either As or Bs is the greatest when the number of As is equal to the number of Bs.

30. How many numbers of four different digits can be made by the digits 1, 2, 3, 4, 5? How many of these are not multiples of 3?

31. How many five-digit numbers divisible by 4 can be formed by the digits 1, 2, 3, 4 and 5 if the
    (i) digits can be repeated in the same number
    (ii) digits cannot be repeated in the same number?

32. How many five-digit numbers divisible by 6 can be made with the digits 0, 1, 2, 3, 4 and 5 if the digits cannot be repeated in the same number?

33. How many numbers of four digits can be formed with the digits 1, 2, 3, 4, 5 if the digits can be repeated in the same number? How many of these have at least one repeated digit?

34. How many different four-digit numbers greater than 4000 can be made, the sum of whose digits is odd?

35. How many different five-digit numbers can be made, the sum of whose digits is even?

36. Find the sum of all the numbers greater than 10,000 that can be made with the digits 1, 3, 5, 7 and 9 if digits are not repeated in the same number.

37. Find the sum of all the odd numbers of five digits that can be made with the digits 0, 1, 4, 5, 4.

38. Find the sum of all natural numbers n such that $1000 \le n \le 4000$, that can be made with the digits 0, 1, 2, 3, 4 if repetition of digits in the same number is allowed.

39. Prove that the total number of permutations of $n(>1)$ different things taking at most r at a time when each thing can be repeated any number of times, is equal to $\dfrac{n(n^r-1)}{n-1}$.

40. In a plane, there are 10 points of which 4 are collinear. How many different
    (i) straight lines can be drawn joining them

(ii) triangles can be made with the points as vertices

(iii) quadrilaterals can be made with the points as vertices?

41. How many sides are there in a polygon which has 35 diagonals?

42. In how many interior points will the diagonals of a convex polygon of n sides intersect if no three diagonals pass through the same interior point?

43. In a plane there are two sets of parallel lines, one of m lines and the other of n lines. If the lines of one set cut the lines of the other set, how many different parallelograms will be formed?

44. How many different rectangles are there on a chessboard? How many of these have the area $= 3 \times$ area of a small square?

45. How many different squares are there on a chessboard?

46. There are 3, 4 and 5 points marked on the three sides of a triangle. How many triangles can be formed with these points as vertices if none of the marked points are at a vertex of the given triangle?

47. How many five-digit numbers of different digits can be made in which digits are in (i) ascending (ii) descending order?

48. Straight lines are drawn by joining m points on a straight line to n points on another line. Then excluding the given points, prove that the lines drawn will intersect at $\frac{1}{2}mn(m-1)(n-1)$ points if no two lines drawn are parallel and no three lines are concurrent.

49. How many different committees of 5 members can be formed from 6 men and 4 ladies if each committee is to contain at least one lady?

50. How many mixed committees of 5 members can be made from 5 ladies and 5 gentlemen? How many of these have ladies in majority? How many of these have ladies in minority?

51. From a panel of 5 lawyers, 5 chartered accountants and 1 lawyer who is also a chartered accountant, how many committees of four can be made if each committee is to contain at least one lawyer and one chartered accountant?

52. Rekha married Shivram and had 4 sons. Varsha married Ajoy and had 4 sons. Both the couples had divorce and after that Shivram married Varsha while Ajoy married Rekha. They too had 3 sons each from their wedlocks. How many selections of 8 children can be made from the 14 children so that each of them have equal number of sons in the selection?

53. To fill 12 vacancies, there are 25 candidates of which 5 are sportsmen. If 3 of the vacancies are reserved for sportsmen while the rest are open to all, find the number of ways in which the selection can be made.

54. To fill 8 vacancies there are 24 candidates of which 6 are from scheduled castes, 5 are from other backward castes. If 25% and $12\frac{1}{2}$% of the vacancies are reserved for the scheduled castes and other backward castes respectively while the rest are open to all, find the number ways to make the selection of candidates.

55. There are $n \geq 4$ points situated on a plane such that no two of the lines joining them are parallel and no three of the lines joining them are concurrent except at the given points. Find the number of points of intersection, other than the given points, of the lines obtained by joining them.

56. Find the number of triangles that can be formed with the vertices of a polygon of 10 sides as their vertices if

(i) the triangle cannot have more than one side common with the polygon.

(ii) the triangle cannot have any side common with the polygon.

57. There are 5 eligible Punjabi grooms of which 3 know Bengali and 5 eligible Bengali grooms of which 2 know Punjabi. There are 5 eligible Punjabi brides and 5 eligible Bengali brides. If an eligible groom is agreeable to marry a girl of his community or knowing her language and brides have no choice, in how many different ways 10 couples can be formed?

58. 5 Indian and 5 American couples meet at a party and shake hands. If no wife shakes hands with her husband and no Indian wife shakes hands with a male then find the number of handshakes that take place in the party.

59. There are 24 balls of 6 different sizes in a bag, there being 4 balls of each size in four different colours. In how many ways can 4 balls be selected so that they are of (i) different colours (ii) different sizes?

60. In a chess tournament, every participant played 2 games with each other. There are 2 lady participants in the tournament. The number of games that the men played among themselves exceeded the number of games that the men played with ladies by 66. How many participants were there in all. How many games were played all together?

61. In an examination a candidate has to pass in each of the papers. If the total number of different ways in which the candidate can fail is 63 then find the number of papers in the examination.

62. In an election of preferential votes the number of candidates is one more than the number of members to be elected. If the number of candidates be n, find the number of different ways in which a voter can cast his vote provided he has to vote for at least one candidate.

63. The total number of selections of at most n things from $2n+1$ different things is found to be 63. Find the value of n.

64. How many selections of at least one red ball can be made from 4 red balls and 3 green balls if balls of the same colour are different in size?

65. A man wishes to make up as many different parties as possible out of 20 friends, each party consisting of the same number of friends. How many should be invited at a time? In how many of these would the same friend be present?

66. A person has 32 teeth or less in the mouth. Prove that the largest sample of people will contain 16^8 people in which no two persons have the same setting of teeth.

67. There are 4 mangoes, 3 apples and 2 oranges in a bag, fruits of the same variety being identical. In how many different ways can a selection of fruits be made if

(i) there is no restriction

(ii) at least one mango is to be selected

(iii) at least one of each kind is to be selected?

68. A bag contains 4 one-rupee coins, 2 twenty-five-paisa coins and 5 ten-paisa coins. In how many ways can an amount not less than Re 1 be taken out from the bag? Consider coins of the same denomination to be identical.

69. If $N = a^p \cdot b^q \cdot c^r \cdot d^s$, where N, p, q, r, s are natural numbers and a, b, c, d are prime numbers then find the number of proper divisors of N. Also find the sum of these divisors.

70. Find the number of proper divisors of 2520. Also answer the following:

(i) How many of these are odd? Find their sum.

(ii) How many of these are even? Find their sum.

(iii) How many of these are divisible by 10? Find their sum.

71. In how many ways can a word of five letters beginning with a capital letter and consisting of two vowels and two consonants be formed with the capital letters C, A, C; vowels o, u, i, e and the consonants b, l, p, t, s?

72. A boat is to be manned by 8 men of which 3 can row only on one side and 2 only on the other. In how many ways can the men take their places on two sides of the boat?

73. Eighteen guests are to be seated, half on each side of a long table. Four particular guests desire to sit on one particular side and three others on the other side. Determine the number of different ways in which the sitting arrangement can be made.

74. Twenty guests are to be seated on two sides of a rectangular dinner table, half on each side. Five specific guests have to sit together on any side of the table and another set of four specific guests have to sit on the facing side. Determine the number of sitting arrangements possible.

75. There are 10 intermediary stations between two junctions where an express train stops. If 6 persons board the train at some intermediary station or other during the journey and each of the 6 passengers hold a different variety of ticket of the same class then find the number of ways in which they can hold their tickets.

76. Find the number of numbers of six digits that can be made with the digits 1, 2, 3, 4 if all the digits are to appear in the same number at least once.

77. A batsman scores exactly a century by hitting fours and sixes in twenty consecutive balls. In how many different ways can he do it if some balls may not yield runs and the order of boundaries and overboundaries are taken into account?

78. How many five digit numbers of different digits can be formed with two even and three odd digits?

79. In an exciting finish to an one-day cricket match between India and Pakisthan, the Indians require 10 runs in the last 3 balls to win. If any one of the scores 0, 1, 2, 3, 4, 6 can be made from a ball and no wides or no-balls are bowled then in how many different sequences can the batsmen make exactly 10 runs?

80. 16 persons are to be seated in a row so that in half of the seats from one side 4 particular persons sit consecutively and in the other half 3 particular persons sit together. Find the number of ways in which they can be seated.

81. How many different seven-digit numbers can be made of distinct digits if the sum of the digits is even?

82. Twelve different letters are to be put in twelve pockets in a row. If five of the pockets are too small for six of the letters then in how many different ways can the letters be put in the pockets?

83. A badminton club has 10 couples as members. They meet to organise a mixed double match. If each wife refuse to partner as well as oppose her husband in the match then in how many differnt ways can the match be arranged?

84. Six Xs are to be placed in the squares of the given figure, such that each row contains at least one X. In how many different ways can this be done?

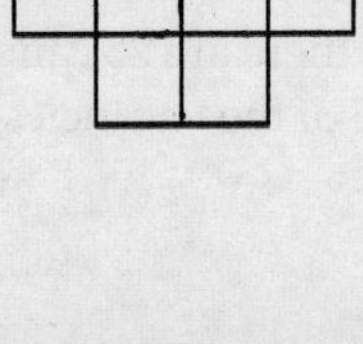

85. In how many different ways can the letters of the word KUMARI be placed in the 8 boxes of the given figure so that no row remains empty?

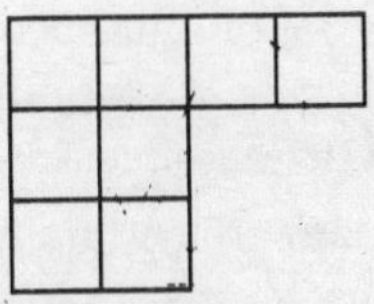

86. In how many different ways can the six digits 1, 2, 3, 4, 5, 6 be placed in the boxes of the given figure so that only even digits occupy top two boxes and the number read by the four bottom digits is even?

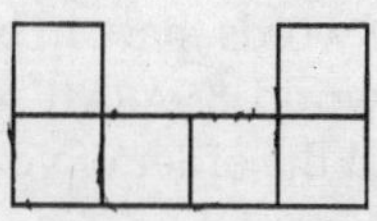

87. Find the total number of ways of selecting five letters from the letters of the word INDEPENDENT.

88. (a) Find the number of words of 4 letters that can be made with the letters of the word COMBINATION.

(b) Find the number of words of 5 letters that can be made with the letters of the word PROPOSITION.

89. Find the number of words of 6 letters that can be formed with the letters of the word ABRAKADABRA if the words begin and end with A.

90. Find the number of numbers of four digits that can be made with the digits 123432.

91. Find the number of numbers of five digits that can be made with the digits of the number 1203210.

92. Find the number of numbers of five significant digits (i.e., nonzero digits) which have at least three identical digits.

93. There are four bags containing identical balls. The first bag contains 4 red balls, the second contains 3 white balls, the third contains 2 green balls and the last contains 1 black ball. How many different selections of four balls from the bags are possible?

94. In a store there are eight varieties of articles and each article can be bought in any number. In how many ways can one buy 4 articles if

(i) there is no restriction

(ii) at least 2 of one variety are to be bought?

95. In how many ways can a pack of 52 cards be distributed among 4 players in a game of bridge? In how many of these does a particular player hold all the aces?

96. In how many ways can 12 different books be divided equally in 3 heaps of books?

97. In how many ways can 7 departments be distributed among 3 ministers if each minister gets at least one but not more than 3 departments?

98. In how many different ways can 8 different books be distributed among 3 students if each receives at least 2 books?

99. In how many ways can 16 identical mangoes be distributed among 4 persons if none gets less than 3 mangoes?

100. If x_r; $r = 1, 2, 3, \dots, n$ be non-negative integers, find the number of solutions of the equation

$$x_1 + x_2 + x_3 + \dots + x_n = m, \ (m > n).$$

101. Find the number of solutions of the equation

$$x_1 + x_2 + x_3 + \dots + x_n = m, \ (m > n)$$

if $x_1, x_2, \dots$ are natural numbers.

102. Find the number of integral solutions of the equation

$$x_1 + x_2 + x_3 + \dots + x_k = 4k^2 - 2k + 1$$

where $x_1 \geq 1, x_2 \geq 3, x_3 \geq 5, \dots, x_k \geq 2k - 1$.

103. If $n \in N$, find the number of non-negative integral solutions of the inequation

$$x_1 + x_2 + x_3 + \dots + x_k \leq n.$$

104. If $x_k \geq 0$; $k = 1, 2, 3, 4, 5$ and $x_1 + x_2 + x_3 + x_4 + x_5 = 20$ and $x_1 + x_2 + x_3 = 5$ then find the number of integral solutions of $(x_1, x_2, x_3, x_4, x_5)$.

105. Find the number of non-negative integral solutions of $x + y + 3z = 30$.

106. In how many ways can 30 identical apples be distributed among 4 persons? What is the number of ways when each gets at least one?

107. In how many different ways can 30 marks be allotted to 10 questions if each question is to carry at least two marks?

108. In how many ways can three persons throw a die each to have a total score of 11?

109. Prove that the number of triangles with sides in integral centimetre and the greatest side $= 2m + 1$ cm is $(m + 1)^2$ and the number of isosceles or equilateral triangles is $3m + 1$.

110. How many natural numbers below 1 million have the sum of digits equal to 18?

111. In how many ways can 10 boys and 6 girls be seated in a row so that no two girls may be consecutive?

112. In how many ways can 12 persons take seats in a row of 20 fixed seats so that each person has exactly one neighbour?

113. In how many ways can 4 persons holding a five-rupee note each and 4 persons holding a ten-rupee note each queue up for buying tickets, one each, if the booking clerk starts with an empty cash box and the price of a ticket is Rs 5?

114. In an entrance examination there are two papers each of 50 marks and one paper of 100 marks. A candidate will qualify to take admission if he secures at least 150 marks in aggregate, securing at least 50 marks in the last paper. In how many different ways can he secure (i) exactly 150 marks (ii) marks in three papers to qualify?

115. In how many different ways can five things be priced in full rupees so that their total price is Rs 50 and the minimum prices of the things are Re 1, Rs 2, Rs 3, Rs 4 and Rs 5 respectively?

116. In an examination, the maximum marks for each of 3 papers is n and for the fourth paper $2n$. Prove that the number of ways in which a candidate can get $3n$ marks is $\frac{1}{6}(n+1)(5n^2+10n+6)$.

117. Prove that the number of selections of n things from two sets of n identical things and n other distinct things is $(n+2)\cdot 2^{n-1}$.

118. In a certain examination of 6 papers each paper has 100 marks as maximum marks. Show that the number of ways in which a candidate can secure 40% marks in the whole examination is

$$\frac{1}{5!}\left\{\frac{245!}{240!}-6\cdot\frac{144!}{139!}+15\cdot\frac{43!}{38!}\right\}.$$

119. There are 4 pairs of hand gloves of 4 different colours. In how many ways can they be paired off so that a left-handed glove and a right-handed glove are not of the same colour?

120. Let A = {1, 2, 3, 4, 5} and B = $\{x_1, x_2, x_3, x_4, x_5\}$. How many one-one mapping $f: A \to B$ can be defined so that $f(i) \neq x_i$ for all i.

121. Find the number of selections of 10 balls from unlimited number of red, black, white and green balls. Also find how many of the selections contain balls of the four colours.

122. In how many ways can 6 gentlemen and 6 ladies take seats round a table if gentlemen and ladies alternate?

123. In how many ways can 4 gentlemen and 7 ladies form a round so that no two gentlemen are consecutive?

124. 10 different letters are printed round a circle. In how many ways can you select 3 letters so that no two are consecutive?

125. There are n persons sitting round a table. Prove that the number of different ways in which 3 persons can be selected so that no two are neighbours, is $\frac{1}{6}n(n-4)(n-5)$.

126. There are 8 line segments of lengths (in cm), 1, 2, 3, ..., 8. How many quadrilaterals can be made with these line segments if circles can be inscribed in the quadrilaterals made?

127. Prove that the total number of arrangements of things by taking any number of things from n different things is $[e \times n! - 1]$ where $[x]$ denotes the integer nearest to x.

Objective Questions

Fill in the blanks.

128. If. ${}^{k+5}P_{k+1} = \frac{11(k-1)}{2}\cdot {}^{k+3}P_k$ then the value of $k =$ ______ .

129. The number of words that can be made with all the letters of the world BHARATI so that the order of consonants and the order of vowels do not change is ______ .

130. If ${}^{15}C_{3r} = {}^{15}C_{r+3}$ then $r =$ ______ .

131. p positive and $q(\leq p)$ negative signs can be arranged in a row, so that no two negative signs occupy consecutive places, in ______ ways.

132. If ${}^{n+2}C_8 : {}^{n-2}P_4 = 57 : 16$ then $n =$ ______ .

133. The number of committees of five members that can be formed from 6 ladies and 4 gentlemen so that the committees contain at most two ladies, is equal to ______ .

134. The number of selections of 4 letters from the letters of the word MISSISSIPPI is ______ .

135. The number of positive integral solutions of $x+y+z=n$ is ______ .

136. The number of non-negative integral solutions of $2x+y=11$ is ______ .

137. The number of different words that can be made with the letters of the word PLATE so that there are exactly two letters between the two vowels, is ______ .

138. There are 16 points in a plane of which 8 are on one line and 8 are on another line. The number of triangles that can be formed with the points as vertices is ______ .

139. The number of ways in which 7 identical rings can be put on 5 fingers of a hand is ______ if each finger has at least one ring.

140. There are red, green and white identical balls, each being 10 in number. The number of selections of 10 balls in which the number of red balls is double the number of green balls is ______ .

141. If the number of selections of r things as well as $10+r$ things from 30 different things be equal then $r =$ ______ .

142. The total number of selections of at most 4 things from 9 different things is ______ .

143. The number of ways in which 6 identical things can be distributed among 3 persons so that each gets at least one, is ______ .

144. The number of distinct subsets of at least 9 elements of the set A where $n(A) = 18$ is ______ .

145. The number of numbers of six digits that can be made with the digits 4, 3, 3, 2, 5, 6 so that the odd and even digits alternate is ______ .

146. A gentleman has 9 children. He takes them to a circus 3 at a time, the same 3 not going together to the circus twice. The number of times the eldest child will go to the circus is ______ .

147. There are two sets $A = \{a_1, a_2, a_3, \dots, a_m\}$ and $B = \{b_1, b_2, b_3, \dots, b_n\}$ of real numbers. The minimum number of ordered pairs (x, y), $x \in A$, $y \in B$ that must be written, is ______ so that two pairs will be identical.

Choose the correct option(s).

148. The value of ${}^{18}C_2 + {}^{17}C_2 + {}^{16}C_2 + {}^{15}C_2 + {}^{14}C_2 + {}^{14}C_3$ is

(a) ${}^{17}C_3$ (b) ${}^{18}C_3$

(c) ${}^{19}C_{16}$ (d) none of these

149. On the occasion of Dipawali festival each student of a class sends greeting cards to others. If there are 20 students in the class, the number of cards sent by students is

(a) ${}^{20}C_2$ (b) ${}^{20}P_2$

(c) $2 \times {}^{20}C_2$ (d) $2 \times {}^{20}P_2$

150. If ${}^nC_{r-1} = 36$, ${}^nC_r = 84$, ${}^nC_{r+1} = 126$ then the value of r is

(a) 1 (b) 2

(c) 3 (d) none of these

151. In the first division football league, each team plays once against other teams. If 45 league matches are played in all, the number of teams is

(a) 9 (b) 10

(c) 5 (d) none of these

152. A five-digit number divisible by 3 is to be formed using the numerals 0, 1, 2, 3, 4 and 5 without repetition. The total number of ways in which this can be done is

(a) 216 (b) 600 (c) 240 (d) 3125

153. Ten different letters of an alphabet are given. Words with 5 letters are formed from these letters. The number of words which have at least one letter repeated is

(a) 69760 (b) 30240

(c) 99748 (d) none of these

154. The number of ways in which a set A where $n(A) = 12$ can be partitioned in three subsets P, Q, R each of 4 elements if $P \cup Q \cup R = A$, $P \cap Q = \phi$, $Q \cap R = \phi$, $R \cap P = \phi$, is

(a) $\dfrac{12!}{(4!)^3}$ (b) $\dfrac{12!}{(3!)^4}$

(c) $\dfrac{12!}{(4!)^3} \cdot \dfrac{1}{3!}$ (d) none of these

155. The number of diagonals in a polygon of 11 sides is

(a) 35 (b) 44

(c) 55 (d) none of these

156. The number of ways in which 6 men can be made to sit round a table in 6 numbered seats is

(a) 5 ! (b) 6 !

(c) $6 \times 6!$ (d) none of these

157. Five persons including one lady are to deliver lectures to an audience. The organiser can arrange the presentation of their lectures, so that the lady is always in the middle, in

(a) ${}^5P_5$ ways (b) $4 \cdot {}^4P_4$ ways

(c) 4 ! ways (d) ${}^5C_4$ ways

158. 150 students take admission. They are to be put in three sections A, B, C of equal size. The number of ways in which this can be done is

(a) $\dfrac{1}{3!} \dfrac{150!}{(50!)^3}$ (b) $\dfrac{150!}{(50!)^3}$

(c) $\dfrac{150!}{(50!)^3} \times 150!$ (d) none of these

159. From the sequence of the first 20 natural numbers, four are selected such that they are not all consecutive. The number of such selections is

(a) 284×17 (b) 285×17

(c) 284×16 (d) none of these

160. The number of ways in which 7 different things can be distributed among 4 persons, when each can receive as many as given, is

(a) 7^4 (b) ${}^7P_4$ (c) ${}^7C_4$ (d) 4^7

161. In a test of 10 multiple-choice questions of one correct answer, each having 4 alternative answers, the number of ways to put ticks at random for the answers to all the questions is

(a) 4^{10} (b) 10^4 (c) $4^{10} - 4$

(d) $10^4 - 10$ (e) 2^{40}

162. There are p points in space, no four of which are in the same plane with the exception of q points which are all in the same plane. The number of different planes determined by the points is

(a) $\frac{p(p-1)(p-2)}{6} - \frac{q(q-1)(q-2)}{6}$

(b) $\frac{p(p-1)(p-2)}{6} - \frac{q(q-1)(q-2)}{6} + 1$

(c) $\frac{p(p-1)(p-2)(p-3)}{24} - \frac{q(q-1)(q-2)(q-3)}{24} + 1$

(d) none of these

163. There are n different books and each book has p copies. The number of selection of books from these is

(a) $(p+1)^n - 1$ (b) $(p+1)^n - 2$

(c) $n(p+1) - 1$ (d) none of these

164. In the identity $\sum_{k=0}^{n} \frac{A_k}{x+k} = \frac{n!}{x(x+1)(x+2)\dots(x+n)}$

the value of A_k is

(a) ${}^nC_k$ (b) ${}^nC_{k+1}$

(c) $(-1)^k \cdot {}^nC_k$ (d) $(-1)^{k-1} \cdot {}^nC_{k-1}$

165. There are an even number of points in a plane of which half are lying on the same line and no other three are collinear. If the total number of triangles that can be made with vertices at these points is 110 then the number of given points is

(a) 6 (b) 10

(c) 20 (d) none of these

166. The number of different messages by signals with three dots and two dashes is

(a) 120 (b) 12 (c) 10 (d) 20

State whether the statements are true or false.

167. The number of diagonals of a regular polygon of interior angle 144° is 10.

168. $101 \times 102 \times 103 \times \dots \times 107$ is divisible by 5040.

169. The product of r consecutive positive even integers is divisible by $2^r \times r!$.

170. ${}^{n+1}C_k$ can be less than ${}^nC_k$ for some values of n and k.

171. The number of selections of r things from n different things is equal to the number of selections of all the things except those r things.

172. If $x = {}^nC_2$ then ${}^xC_2$ is equal to ${}^nC_4$.

173. If the numbers $a_1, a_2, \dots, a_{n+1}$ are different prime numbers then the number of different factors of the number $N = a_1^m \cdot a_2 \cdot a_3 \dots a_{n+1}$ is $(m+1)2^n - 1$.

174. If the number of ways in which m identical balls can be put in $2m$ boxes be p so that no box contains more than one ball then

$$p \in \left[\frac{4^m}{2\sqrt{m}}, \frac{4^m}{\sqrt{2m+1}}\right].$$

175. The number of ways to select 3 numbers in AP from the first n natural numbers is $\frac{1}{4}(n-1)^2$.

176. The number of ways to select three numbers in AP from the first $2n$ natural numbers is $n(n-1)$.

Answers

2. 249900 **9.** 999 **10.** 5 **11.** 5×4^9
12. 72 **13.** 180 **14.** 420, 240 **15.** 360
16. 18 **17.** 161280 **18.** 180 **19.** 1200
20. 1036800 **21.** 360 **22.** 288 **23.** 72
24. (i) 725760 (ii) 1814400 **25.** 120 **26.** 99th
27. 120960 **28.** 864 **30.** 120, 96
31. (i) 625 (ii) 24 **32.** 108 **33.** 625, 505
34. 3000 **35.** 45000 **36.** 6666600
37. 708894 **38.** 837250 **40.** (i) 40 (ii) 116 (iii) 185
41. 10 **42.** $\frac{n(n-1)(n-2)(n-3)}{24}$
43. $\frac{mn(m-1)(n-1)}{4}$ **44.** 1296, 96
45. 204 **46.** 205 **47.** (i) 126 (ii) 252
49. 246 **50.** 250, 125, 125
51. 320 **52.** 485 **53.** $10 \cdot \frac{22!}{9!\,13!}$
54. 1526175 **55.** $\frac{n}{8}(n-1)(n-2)(n-3)$
56. (i) 110 (ii) 50 **57.** 144000
58. 135 **59.** (i) 1296 (ii) 3840
60. 13, 156 **61.** 6 **62.** $2^n - 2$ **63.** 3
64. 120 **65.** $10, \frac{19!}{9!\,10!}$
67. (i) 59 (ii) 48 (iii) 24 **68.** 73
69. $(p+1)(q+1)(r+1)(s+1) - 2$,

$$\frac{a^{p+1}-1}{a-1} \cdot \frac{b^{q+1}-1}{b-1} \cdot \frac{c^{r+1}-1}{c-1} \cdot \frac{d^{s+1}-1}{d-1} - 1 - a^p b^q c^r d^s$$

70. 46, (i) 11, 623 (ii) 35, 6216 (iii) 17, 4760
71. 2880 **72.** 1728 **73.** ${}^{11}C_5 \cdot (9!)^2$
74. $2 \times 10! \times 11!$ **75.** 20872566000 **76.** 1560

77. $20!\left(\frac{1}{16!\,3!}+\frac{1}{14!\,4!\,2!}+\frac{1}{12!\,7!}+\frac{1}{10!\,10!}\right)$

78. 11040 **79.** 21 **80.** $12 \times 6!\,9!$

81. 273600 **82.** 3628800 **83.** 2520 **84.** 26

85. 18720 **86.** 36 **87.** 72

88. (a) 2454 (b) 6890 **89.** 370 **90.** 102

91. 258 **92.** 6129 **93.** 20

94. (i) 330 (ii) 260 **95.** $\frac{52!}{(13!)^4}, \frac{48!}{9!(13!)^3}$

96. $\frac{12!}{3!\,(4!)^3}$ **97.** 1050 **98.** 2940 **99.** 35

100. ${}^{m+n-1}C_{n-1}$ **101.** ${}^{m-1}C_{n-1}$

102. ${}^{k(3k-1)}C_{k-1}$ **103.** ${}^{n+k}C_k$

104. 336 **105.** 176 **106.** 5456, 3654

107. $\frac{19!}{9!\,10!}$ **108.** 27 **110.** 25927

111. $\frac{10!\,11!}{5!}$ **112.** $84 \times 12!$ **113.** $14\,(4!)^2$

114. (i) 1326 (ii) 23426 **115.** ${}^{39}C_4$ **119.** 9

120. 44 **121.** 286, 84 **122.** 86400 **123.** 604800

124. 50 **126.** 22 **128.** 6, 7 **129.** 35

130. 3 **131.** ${}^{p+1}C_q$ **132.** 19 **133.** 66

134. 21 **135.** $\frac{1}{2}(n-1)(n-2)$ **136.** 6

137. 24 **138.** 448 **139.** 15 **140.** 3

141. 10 **142.** 255 **143.** 10

144. $\frac{1}{2}\cdot\frac{18!}{9!\,9!}+2^{17}$ **145.** 36 **146.** 28

147. $mn+1$ **148.** (c) **149.** (c) **150.** (c)

151. (b) **152.** (a) **153.** (a) **154.** (c)

155. (b) **156.** (b) **157.** (c) **158.** (b)

159. (a) **160.** (d) **161.** (a) **162.** (b)

163. (a) **164.** (c) **165.** (b) **166.** (c)

167. false **168.** true **169.** true **170.** false

171. true **172.** false **173.** true **174.** true

175. false **176.** true

□

Chapter Test

Time: 120 minutes

1. In the following, each statement is incomplete. Fill in the blank so that each resulting statement becomes correct.

(i) The number of n-digit numbers in which no two consecutive digits are equal is ______ .

(ii) If the numbers of selections of r as well as $n + r$ things from $3n$ different things are equal then the value of r = ______ .

(iii) The number of different relations that can be defined from the set A to the set B is ______ where $n(A) = p$ and $n(B) = q$.

2. In the following, each question has one or more than one correct answers. Indicate the correct answer(s).

(i) There are 12 balls numbered from 1 to 12. The number of ways in which they can be used to fill 8 places in a row so that the balls are with numbers in ascending or descending order is equal to

A. ${}^{12}C_8$ B. ${}^{12}P_8$ C. $2 \times {}^{12}P_8$ D. $2 \times {}^{12}C_8$

(ii) The number of different ways to divide a set P of n elements into two nonempty disjoint subsets whose union is P, is

A. $2^{n-1} - 1$ B. $2^n - 1$ C. $\frac{1}{2}(2^n - 1)$ D. none of these

(iii) With 17 consonants and 5 vowels, the number of four-letter words that can be formed having 2 vowels in the middle and one consonant repeated or different consonants at each end is

A. 2890 B. 5440 C. 5780 D. none of these

3. A double-decker bus has 5 empty seats in the upstair and 5 empty seats in the downstair. 10 people board the bus of which 2 are old people and 3 are children. The children refuse to take seats downstair while old people insist to stay downstair. In how many different arrangements can the 10 people take their seats in the bus?

4. How many six-digit numbers of nonzero digits can be made if a digit can be repeated at most three times in the same number?

5. How many quadrilaterals can be made by joining the vertices of a convex polygon of $n(\geq 8)$ sides so that the quadrilaterals do not have any side common with the sides of the polygon?

6. In how many ways can 10 people take seats in 24 fixed seats so that out of every pair of seats equidistant from the beginning and end at least one seat is empty?

7. In how many different ways may 12 things, 4 each of three varieties, be distributed equally among two persons?

8. In how many ways can four tickets be selected from $n(n \geq 7)$ tickets numbered from 1 to n so that no two consecutive tickets are selected?

9. In how many different ways can a sum of Rs 20 be paid in one-rupee coins, 50-paisa coins and 25-paisa coins if each variety of coins is available in unlimited number?

10. In a Gaelic word no consonant can fall immediately between a strong vowel and a weak vowel, the strong vowels being a, o, u and the other vowels being weak. Show that the number of different Gaelic words of $n + 3$ different letters that can be made with n different consonants and the vowels a, e, o is $2 \cdot \frac{(n+3)!}{n+2}$.

11. In how many different ways can Rs 30 in one-rupee coins be distributed among four persons so that at least two of them receive the same amount and each gets at least one rupee?

12. Let a_i; $i = 1, 2, 3, \ldots, 100$ be hundred numbers in AP with a common difference $d > 0$. In how many different ways 20 numbers can be selected from them so that the least number is a_{40} or the greatest number is a_{70}?

13. There are 5 red balls, 6 black balls and 7 green balls, balls of the same colour being identical. In how many different ways can the balls be put equally in 3 different boxes so that each box contains balls of the three colours?

14. 100 marks are to be distributed over 10 questions in a paper. In how many ways this can be done if each question is to carry at least 8 marks and no question can be alloted more than 20 marks?

15. In how many different ways 24 apples can be given to 3 boys and 3 girls if the total number of apples received by boys is (a) equal to (b) double of that received by girls?

□

Answers

1. (i) 9^n (ii) n (iii) 2^{pq}

2. (i) D (ii) A (iii) C

3. 144000

4. 522360

5. $n\left\{\frac{(n-1)(n-2)(n-3)}{24} - \frac{(n-3)(n-5)}{2} - 1\right\}$

6. $512 \times (12\,!)$

7. 19

8. ${}^{n-3}C_4$

9. 441

11. 196

12. $\frac{1}{19\,!}\left\{\frac{60\,!}{41\,!} + \frac{69\,!}{50\,!} - \frac{19 \times (29\,!)}{11\,!}\right\}$

13. 42

14. $\frac{1}{9\,!}\left(\frac{29\,!}{20\,!} - \frac{10 \times (16\,!)}{7\,!}\right)$

15. (a) 8281 (b) 6885

□

6. Binomial Theorem for Positive Integral Index

Recap of Facts and Formulae

1. Binomial theorem for positive integral index

- $(a+x)^n = {}^nC_0\, a^n + {}^nC_1\, a^{n-1} x + {}^nC_2\, a^{n-2} x^2 + \ldots + {}^nC_n\, x^n$

  where ${}^nC_r = \dfrac{n!}{r!\,(n-r)!}$.
- $(1+x)^n = {}^nC_0 + {}^nC_1\, x + {}^nC_2\, x^2 + \ldots + {}^nC_n\, x^n$.
- $(1+x)^n = (x+1)^n$

  $= {}^nC_0\, x^n + {}^nC_1\, x^{n-1} + {}^nC_2\, x^{n-2} + \ldots + {}^nC_n$

  when expanded in descending powers of x.
- $(a-x)^n = {}^nC_0\, a^n - {}^nC_1\, a^{n-1} x + {}^nC_2\, a^{n-2} x^2 - \ldots + (-1)^n \cdot {}^nC_n\, x^n$.

2. Terms in expansion

In the expansion of $(a+x)^n$, $n \in N$
- the number of terms $= n+1$
- $(r+1)$th term $= t_{r+1} = {}^nC_r\, a^{n-r} \cdot x^r$
- kth term from the end $= \{(n+1)-(k-1)\}$th term $= t_{(n+1)-(k-1)}$
- middle term $= t_{\frac{(n+1)+1}{2}}$, i.e., $t_{\frac{n}{2}+1}$ when n is even

  middle term $= t_{\frac{n+1}{2}}$, $t_{\frac{n+3}{2}}$ when n is odd.

3. Properties of ${}^nC_r$ for simplification

- ${}^nC_0 = 1$, ${}^nC_n = 1$
- ${}^nC_r = {}^nC_{n-r}$
- ${}^nC_r + {}^nC_{r-1} = {}^{n+1}C_r$
- ${}^nC_r = {}^nC_k \Rightarrow r = k$ or $r + k = n$
- $r \cdot {}^nC_r = n \cdot {}^{n-1}C_{r-1}$
- $\dfrac{1}{r+1} \cdot {}^nC_r = \dfrac{1}{n+1} \cdot {}^{n+1}C_{r+1}$

- **The greatest among binomial coefficients**

  ${}^nC_0, {}^nC_1, {}^nC_2, \ldots, {}^nC_n$

  $= {}^nC_{\frac{n}{2}}$ when n is even

  ${}^nC_{\frac{n-1}{2}}$ or ${}^nC_{\frac{n+1}{2}}$ when n is odd.

4. Summation of series involving binomial coefficients

For $(1+x)^n = {}^nC_0 + {}^nC_1\, x + {}^nC_2\, x^2 + \ldots + {}^nC_n\, x^n$, the binomial coefficients are ${}^nC_0, {}^nC_1, {}^nC_2, \ldots, {}^nC_n$. A number of series may be formed with these coefficients figuring in the terms of a series. Standard series of the binomial coefficients are as follows:

- ${}^nC_0 + {}^nC_1 + {}^nC_2 + \ldots + {}^nC_n = 2^n$ ... (1)

  It is obtained by putting $x = 1$ in the binomial expansion for $(1+x)^n$.
- ${}^nC_0 - {}^nC_1 + {}^nC_2 - \ldots + (-1)^n \cdot {}^nC_n = 0$ ... (2)

  It is obtained by putting $x = -1$ in the binomial expansion for $(1+x)^n$.
- ${}^nC_0 + {}^nC_2 + {}^nC_4 + \ldots = 2^{n-1}$

  It is obtained by adding (1) and (2).
- ${}^nC_1 + {}^nC_3 + {}^nC_5 + \ldots = 2^{n-1}$

  It is obtained by subtracting (2) from (1).
- ${}^{2n}C_0 + {}^{2n}C_1 + {}^{2n}C_2 + \ldots + {}^{2n}C_{n-1} + \dfrac{1}{2} \cdot {}^{2n}C_n = 2^{2n-1}$

  We have

  $${}^{2n}C_0 + {}^{2n}C_1 + {}^{2n}C_2 + \ldots + {}^{2n}C_{2n-1} + {}^{2n}C_{2n} = 2^{2n}$$

  $\because$ ${}^{2n}C_0 = {}^{2n}C_{2n}$, ${}^{2n}C_1 = {}^{2n}C_{2n-1}$, etc., combining the terms equidistant from the beginning and end we get
- $2\left({}^{2n}C_0 + {}^{2n}C_1 + {}^{2n}C_2 + \ldots + {}^{2n}C_{n-1}\right) + {}^{2n}C_n = 2^{2n}$
- ${}^{2n+1}C_0 + {}^{2n+1}C_1 + {}^{2n+1}C_2 + \ldots + {}^{2n+1}C_n = 2^{2n}$ (as above)
- Sum of the first half of ${}^nC_0 + {}^nC_1 + {}^nC_2 + \ldots + {}^nC_n$

$= \text{sum of the last half of } {}^nC_0 + {}^nC_1 + {}^nC_2 + \dots + {}^nC_n$

$= 2^{n-1}$.

- **Bino-geometric series**

$${}^nC_0 + {}^nC_1 x + {}^nC_2 x^2 + \dots + {}^nC_n x^n = (1+x)^n.$$

- **Bino-arithmetic series**

$$a\,{}^nC_0 + (a+d)\,{}^nC_1 + (a+2d)\,{}^nC_2 + \dots + (a+nd)\,{}^nC_n$$

It is made by the sum of the products of corresponding terms of the sequences

$$a, a+d, a+2d, \dots, a+nd \quad \text{(AP)}$$

and ${}^nC_0, {}^nC_1, {}^nC_2, \dots, {}^nC_n$

(sequence of binomial coefficients).

Such series can be added in two ways:

(i) by elimination of r in the multiplier of binomial coefficient from the $(r+1)$th term of the series (using $r \cdot {}^nC_r = n\,{}^{n-1}C_{r-1}$)

(ii) by differentiating the expansion of $x^a(1+x^d)^n$.

For explanation see worked out example 20.

- **Bino-harmonic series**

$$\frac{{}^nC_0}{a} + \frac{{}^nC_1}{a+d} + \frac{{}^nC_2}{a+2d} + \dots + \frac{{}^nC_n}{a+nd}$$

It is made by the sum of the products of corresponding terms of the sequences

$$\frac{1}{a}, \frac{1}{a+d}, \frac{1}{a+2d}, \dots, \frac{1}{a+nd} \quad \text{(HP)}$$

and ${}^nC_0, {}^nC_1, {}^nC_2, \dots, {}^nC_n$

(sequence of binomial coefficients).

Such series are added in two ways:

(i) by elimination of r in the multiplier of binomial coefficient from the $(r+1)$th term of the series $\left(\text{using } \frac{1}{r+1}\,{}^nC_r = \frac{1}{n+1}\,{}^{n+1}C_{r+1}\right)$

(ii) by integrating suitable expansion.
For explanation see worked out example 24.

- **Bino-binomial series**

$${}^nC_0 \cdot {}^nC_r + {}^nC_1 \cdot {}^nC_{r+1} + {}^nC_2 \cdot {}^nC_{r+2} + \dots + {}^nC_{n-r} \cdot {}^nC_n$$

or $${}^mC_0 \cdot {}^nC_r + {}^mC_1 \cdot {}^nC_{r-1} + {}^mC_2 \cdot {}^nC_{r-2} + \dots + {}^mC_r \cdot {}^nC_0$$

Such series are added by multiplying two expansions, one involving the first factors as coefficients and the other involving the second factors as coefficients and finally equating coefficients of a suitable power of x on both sides.

For explanation see worked out examples 28 and 29.

Selected Solved Examples

1. Find the number of terms in the expansion of

$$(x + \sqrt{x^2-1})^6 + (x - \sqrt{x^2-1})^6.$$

Here, the expression

$$= \{{}^6C_0x^6 + {}^6C_1x^5\sqrt{x^2-1} + {}^6C_2x^4(\sqrt{x^2-1})^2 + \dots + {}^6C_6(\sqrt{x^2-1})^6\}$$

$$+ \{{}^6C_0x^6 - {}^6C_1x^5\sqrt{x^2-1} + {}^6C_2x^4(\sqrt{x^2-1})^2 - \dots + {}^6C_6(\sqrt{x^2-1})^6\}$$

$$= 2\{{}^6C_0x^6 + {}^6C_2x^4(\sqrt{x^2-1})^2 + {}^6C_4x^2(\sqrt{x^2-1})^4 + {}^6C_6(\sqrt{x^2-1})^6\}$$

$$= 2\{{}^6C_0x^6 + {}^6C_2x^4(x^2-1) + {}^6C_4x^2(x^2-1)^2 + {}^6C_6(x^2-1)^3\}$$

$$= 2[({}^6C_0 + {}^6C_2 + {}^6C_4 + {}^6C_6)x^6 + (-{}^6C_2 - 2\cdot{}^6C_4 - 3\cdot{}^6C_6)x^4 + ({}^6C_4 + {}^6C_6\cdot 3)x^2 - {}^6C_6]$$

$\therefore$ the number of terms = 4.

2. Prove that $13^{99} - 19^{93}$ is divisible by 162.

Here $13^{99} - 19^{93} = (1 + 3\cdot 4)^{99} - (1 + 9\cdot 2)^{93}$

$$= \{{}^{99}C_0 + {}^{99}C_1\cdot 3\cdot 4 + {}^{99}C_2\cdot(3\cdot 4)^2 + \dots + {}^{99}C_{99}(3\cdot 4)^{99}\}$$

$$- \{{}^{93}C_0 + {}^{93}C_1\cdot 9\cdot 2 + {}^{93}C_2\cdot(9\cdot 2)^2 + \dots + {}^{93}C_{93}(9\cdot 2)^{93}\}$$

$$= (99\times 12 - 93\times 18) + \{{}^{99}C_2 12^2 + {}^{99}C_3 12^3\} + 3^4\times\text{integer}$$

$$= 1188 - 1674 + 144\{{}^{99}C_2 + {}^{99}C_3\times 12\} + 81\times\text{integer}$$

$$= -486 + 144\{99\cdot 49 + 99\cdot 98\cdot 97\cdot 2\} + 81\times\text{integer}$$

$$= -81\cdot 6 + 16\cdot 81\{11\cdot 49 + 11\cdot 98\cdot 97\cdot 2\} + 81\times\text{integer}$$

$$= 81\{-6 + 16(11\cdot 49 + 11\cdot 98\cdot 97\cdot 2) + \text{integer}\}$$

$$= 81\times\text{integer}$$

$\therefore$ $13^{99} - 19^{93}$ is divisible by 81.

But $13^{99} - 19^{93}$

= odd number − odd number

= even number.

$\therefore$ $13^{99} - 19^{93}$ is divisible by 2.

But 2 and 81 are coprime.

$\therefore$ $13^{99} - 19^{93}$ is divisible by 81×2, i.e., 162.

3. Prove that $2 < f(n) < 3$ for all $n \in N, n \geq 2$ if $f(n) = \left(1 + \frac{1}{n}\right)^n$.

$$f(n) = \left(1 + \frac{1}{n}\right)^n = {}^nC_0 + {}^nC_1 \cdot \frac{1}{n} + {}^nC_2 \cdot \frac{1}{n^2} + \ldots + {}^nC_n \cdot \left(\frac{1}{n}\right)^n$$

$$= 1 + n \cdot \frac{1}{n} + \frac{n(n-1)}{2!} \cdot \frac{1}{n^2} + \frac{n(n-1)(n-2)}{3!} \cdot \frac{1}{n^3} + \ldots + \frac{n(n-1)(n-2)\ldots 1}{n!} \cdot \frac{1}{n^n}$$

$$= 1 + 1 + \frac{1}{2!}\left(1 - \frac{1}{n}\right) + \frac{1}{3!}\left(1 - \frac{1}{n}\right)\left(1 - \frac{2}{n}\right) + \ldots + \frac{1}{n!}\left(1 - \frac{1}{n}\right)\left(1 - \frac{2}{n}\right)\ldots\left(1 - \frac{n-1}{n}\right).$$

As each of $1 - \frac{1}{n}, 1 - \frac{2}{n}$, etc., is a positive proper fraction,

$f(n) = 2 +$ positive quantity, i.e., $f(n) > 2$

and $f(n) < 1 + 1 + \frac{1}{2!} + \frac{1}{3!} + \ldots + \frac{1}{n!}$

$$= 1 + 1 + \frac{1}{1 \cdot 2} + \frac{1}{1 \cdot 2 \cdot 3} + \ldots + \frac{1}{1 \cdot 2 \cdot 3 \cdot \ldots n}$$

$$< 1 + 1 + \frac{1}{2} + \frac{1}{2^2} + \frac{1}{2^3} + \ldots + \frac{1}{2^{n-1}}$$

$$\left\{\because \frac{1}{2 \cdot 3} < \frac{1}{2 \cdot 2}; \frac{1}{2 \cdot 3 \cdot 4} < \frac{1}{2 \cdot 2 \cdot 2}, \text{etc.}\right\}$$

$$= 1 + \frac{1 - \frac{1}{2^n}}{1 - \frac{1}{2}} = 1 + 2\left(1 - \frac{1}{2^n}\right)$$

$$= 3 - \frac{1}{2^{n-1}} < 3 \text{ when } n \geq 2$$

$\therefore$ $2 < f(n) < 3$ for all $n \in N, n \geq 2$.

4. In the binomial expansion of $\left(\sqrt[3]{2} + \frac{1}{\sqrt[3]{3}}\right)^n$, the ratio of the 7th term from the beginning to the 7th term from the end is 1 : 6; find n.

Here, 7th term from the beginning

$$= {}^nC_6(\sqrt[3]{2})^{n-6} \cdot \left(\frac{1}{\sqrt[3]{3}}\right)^6 = {}^nC_6 \cdot 2^{\frac{n-6}{3}} \cdot 3^{-2}.$$

Total number of terms in the expansion $= n + 1$.

$\therefore$ 7th term from the end

$= (n + 1) - (7 - 1)$, i.e., $(n - 5)$th term from the beginning

$$= {}^nC_{n-6}(\sqrt[3]{2})^{n-(n-6)} \cdot \left(\frac{1}{\sqrt[3]{3}}\right)^{n-6}$$

$$= {}^nC_{n-(n-6)} \cdot 2^2 \cdot 3^{-\frac{n-6}{3}}$$

$$= {}^nC_6 \cdot 2^2 \cdot 3^{-\frac{n-6}{3}}$$

From the question,

$$\frac{{}^nC_6 \cdot 2^{\frac{n-6}{3}} \cdot 3^{-2}}{{}^nC_6 \cdot 2^2 \cdot 3^{-\frac{n-6}{3}}} = \frac{1}{6}$$

or $\quad 2^{\frac{n-6}{3} - 2} \cdot 3^{\frac{n-6}{3} - 2} = 6^{-1}$

$\therefore$ $\quad 6^{\frac{n-12}{3}} = 6^{-1}$

$\therefore$ $\quad \frac{n-12}{3} = -1$

or $\quad n - 12 = -3; \quad \therefore \quad n = 9.$

5. Find the value of x for which the 6th term of $\left(\sqrt{2^{\log(10-3^x)}} + \sqrt[5]{2^{(x-2)\log 3}}\right)^m, m \in N$ is equal to 21 and binomial coefficients of the 2nd, 3rd and 4th terms are the first, third and fifth terms of an AP.

The binomial coefficients of the 2nd, 3rd and 4th terms are ${}^mC_1$, ${}^mC_2$ and ${}^mC_3$.

From the question, ${}^mC_1 = a$, ${}^mC_2 = a + 2d$, ${}^mC_3 = a + 4d$, the first term of AP $= a$, common difference $= d$.

$\therefore$ $\quad 2\ {}^mC_2 = 2a + 4d = {}^mC_1 + {}^mC_3$

or $\quad 2 \cdot \frac{m(m-1)}{2} = m + \frac{m(m-1)(m-2)}{6}$

or $\quad m(m-1) = m + \frac{m(m-1)(m-2)}{6}$

or $\quad m(m-2) = \frac{m(m-1)(m-2)}{6}$

or $\quad 1 = \frac{m-1}{6} \quad (\because m \neq 0, m \neq 2)$

$\therefore$ $\quad m = 7.$

Now, t_6 of $\left(\sqrt{2^{\log(10-3^x)}} + \sqrt[5]{2^{(x-2)\log 3}}\right)^7 = 21$

$\therefore$ $\quad {}^7C_5 \cdot \left\{\sqrt{2^{\log(10-3^x)}}\right\}^2 \cdot \left\{\sqrt[5]{2^{(x-2)\log 3}}\right\}^5 = 21$

or $\quad \frac{7 \cdot 6}{2} \cdot 2^{\log(10-3^x)} \cdot 2^{(x-2)\log 3} = 21$

or $\quad 2^{\log(10-3^x)+(x-2)\log 3} = 1 = 2^0$

or $\quad \log(10-3^x) + \log 3^{x-2} = 0$

or $\quad \log\{(10-3^x)\cdot 3^{x-2}\} = 0$

$\therefore \quad (10-3^x)\,3^{x-2} = 1$

or $\quad (10-3^x)\,3^x = 9$

or $\quad (3^x)^2 - 10\cdot 3^x + 9 = 0$

$\therefore \quad (3^x-1)(3^x-9) = 0$

$\therefore \quad 3^x = 1, 9 = 3^0, 3^2$

$\therefore \quad x = 0, 2.$

6. Find the coefficient of x^3 in the expansion of $(1+x+2x^2)\left(2x^2-\dfrac{1}{3x}\right)^9$.

The expression

$$= \left(2x^2-\frac{1}{3x}\right)^9 + x\left(2x^2-\frac{1}{3x}\right)^9 + 2x^2\left(2x^2-\frac{1}{3x}\right)^9.$$

In the expansion of $\left(2x^2-\dfrac{1}{3x}\right)^9$,

$$t_{r+1} = {}^9C_r\cdot(2x^2)^{9-r}\cdot\left(-\frac{1}{3x}\right)^r$$

$$= {}^9C_r\cdot 2^{9-r}\cdot\left(-\frac{1}{3}\right)^r\cdot x^{18-3r}$$

t_{r+1} is a term containing x^3

if $18-3r=3$, i.e., $r=5$

t_{r+1} is a term containing x^2

if $18-3r=2$, i.e., $r=\dfrac{16}{3}$ (absurd)

t_{r+1} is a term containing x

if $18-3r=1$, i.e., $r=\dfrac{17}{3}$ (absurd).

Now, coefficient of x^3 in

$$\left\{\left(2x^2-\frac{1}{3x}\right)^9 + x\left(2x^2-\frac{1}{3x}\right)^9 + 2x^2\left(2x^2-\frac{1}{3x}\right)^9\right\}$$

$= \text{coefficient of } x^3 \text{ in } \left(2x^2-\dfrac{1}{3x}\right)^9$

$\quad + \text{coefficient of } x^2 \text{ in } \left(2x^2-\dfrac{1}{3x}\right)^9$

$\quad + 2\times \text{coefficient of } x \text{ in } \left(2x^2-\dfrac{1}{3x}\right)^9$

$$= {}^9C_5\cdot 2^{9-5}\cdot\left(-\frac{1}{3}\right)^5 + 0 + 2\cdot 0$$

$$= {}^9C_5\cdot 2^4\cdot\frac{-1}{3^5}$$

$$= \frac{9\cdot 8\cdot 7\cdot 6}{4!}\cdot\frac{-16}{243} = -\frac{224}{27}.$$

7. Prove that the coefficient of x^r in the expansion of $(x+3)^{n-1}+(x+3)^{n-2}\cdot(x+2)+(x+3)^{n-3}\cdot(x+2)^2+\dots + (x+2)^{n-1}$ is $(3^{n-r}-2^{n-r})\cdot{}^nC_r$.

The expression

$$= (x+3)^{n-1}\cdot\frac{1-\left(\dfrac{x+2}{x+3}\right)^n}{1-\dfrac{x+2}{x+3}} \quad \text{\{by sum of GP\}}$$

$$= (x+3)^n\left\{1-\left(\frac{x+2}{x+3}\right)^n\right\}$$

$$= (x+3)^n-(x+2)^n = (3+x)^n-(2+x)^n$$

$\therefore \quad$ coefficient of $x^r = {}^nC_r\cdot 3^{n-r} - {}^nC_r\cdot 2^{n-r}$

$$= (3^{n-r}-2^{n-r})\cdot{}^nC_r.$$

8. Determine the term independent of a in the expansion of $\left(\dfrac{a+1}{a^{2/3}-a^{1/3}+1}-\dfrac{a-1}{a-a^{1/2}}\right)^{10}$.

Here, the expression

$$= \left\{\frac{(a^{1/3})^3+1}{(a^{1/3})^2-a^{1/3}+1}-\frac{(a^{1/2})^2-1}{a^{1/2}(a^{1/2}-1)}\right\}^{10}$$

$$= \left\{a^{1/3}+1-\frac{a^{1/2}+1}{a^{1/2}}\right\}^{10} = (a^{1/3}-a^{-1/2})^{10}.$$

Now, $t_{r+1} = {}^{10}C_r\,(a^{1/3})^{10-r}\cdot(-a^{-1/2})^r$

$$= {}^{10}C_r\cdot a^{\frac{10-r}{3}}\cdot(-1)^r\cdot a^{-\frac{r}{2}}$$

$$= (-1)^r\cdot{}^{10}C_r\cdot a^{\frac{10-r}{3}-\frac{r}{2}}.$$

It will be independent of a if $\dfrac{10-r}{3}-\dfrac{r}{2}=0$

or $\quad 2(10-r)-3r=0$

or $\quad 20-5r=0; \quad \therefore \quad r=4.$

$\therefore \quad$ the term independent of a

$$= t_5 = (-1)^4\cdot{}^{10}C_4\cdot a^0$$

$$= {}^{10}C_4 = \frac{10\cdot 9\cdot 8\cdot 7}{24} = 210.$$

9. Some three consecutive coefficients in the expansion of $(1+x)^n$ are in AP. Find two suitable positive integral values of n and the three coefficients.

Let the $(r+1)$th, $(r+2)$th and $(r+3)$th coefficients be in AP.

$\therefore \quad {}^nC_r, {}^nC_{r+1}, {}^nC_{r+2}$ are in AP

or $\quad 2 \cdot {}^nC_{r+1} = {}^nC_r + {}^nC_{r+2}$

or $\quad 2 \cdot \dfrac{n!}{(r+1)!\,(n-r-1)!}$

$$= \frac{n!}{r!\,(n-r)!} + \frac{n!}{(r+2)!\,(n-r-2)!}$$

or $\quad \dfrac{2}{(r+1)(n-r-1)} = \dfrac{1}{(n-r)(n-r-1)} + \dfrac{1}{(r+2)(r+1)}$

or $\quad 2(r+2)(n-r) = (r+1)(r+2) + (n-r)(n-r-1)$

or $\quad (n-r)^2 - (2r+5)(n-r) + (r^2+3r+2) = 0$

or $\quad n - r = \dfrac{2r+5 \pm \sqrt{(2r+5)^2 - 4(r^2+3r+2)}}{2}$

or $\quad n = r + \dfrac{2r+5 \pm \sqrt{8r+17}}{2}$.

Here n and r are positive integers. So $8r+17$ must be a perfect square for positive integral value of r.

When $r = 1$, $n = 1 + \dfrac{7 \pm \sqrt{25}}{2} = 1 + \dfrac{7 \pm 5}{2} = 7, 2$.

$r = 1, n = 2$ is not possible because if $n = 2$, there will be only three terms in the expansion and for $r = 1$, $r + 3 = 4$.

$r = 1, n = 7$ is possible. In that case the coefficients are ${}^7C_1, {}^7C_2, {}^7C_3$.

Again, when $r = 4$, $n = 4 + \dfrac{13 \pm \sqrt{49}}{2} = 4 + \dfrac{13 \pm 7}{2} = 14, 7$.

$n = 7$ has already been taken. So $r = 4, n = 14$ and the coefficients are ${}^{14}C_4, {}^{14}C_5, {}^{14}C_6$.

$\therefore$ two suitable values of n are 7 and 14; and the coefficients are ${}^7C_1, {}^7C_2, {}^7C_3$ and ${}^{14}C_4, {}^{14}C_5, {}^{14}C_6$ respectively.

10. If a, b, c, d be four consecutive coefficients in the binomial expansion of $(1+x)^n$ then prove that

$$\left(\frac{b}{b+c}\right)^2 > \frac{ac}{(a+b)(c+d)} \quad \textbf{if } x > 0.$$

Let the four coefficients be $(r+1)$th, $(r+2)$th, $(r+3)$th and $(r+4)$th coefficients respectively.

$\therefore \quad a = {}^nC_r, \; b = {}^nC_{r+1}, \; c = {}^nC_{r+2}, \; d = {}^nC_{r+3}$.

Now, $\dfrac{a}{a+b} = \dfrac{{}^nC_r}{{}^nC_r + {}^nC_{r+1}} = \dfrac{{}^nC_r}{{}^{n+1}C_{r+1}}$

$$= \frac{n!}{r!\,(n-r)!} \cdot \frac{(r+1)!\,(n-r)!}{(n+1)!} = \frac{r+1}{n+1}$$

$$\frac{b}{b+c} = \frac{{}^nC_{r+1}}{{}^nC_{r+1} + {}^nC_{r+2}} = \frac{(r+1)+1}{n+1}$$

(replacing r by $r+1$ in the above)

$$\frac{c}{c+d} = \frac{{}^nC_{r+2}}{{}^nC_{r+2} + {}^nC_{r+3}} = \frac{(r+2)+1}{n+1}$$

$$\therefore \quad \frac{a}{a+b} + \frac{c}{c+d} = \frac{r+1}{n+1} + \frac{r+3}{n+1}$$

$$= \frac{2(r+2)}{n+1} = 2 \cdot \frac{b}{b+c}$$

$\therefore \quad \dfrac{b}{b+c}$ is the AM of $\dfrac{a}{a+b}, \dfrac{c}{c+d}$.

As AM > GM,

$$\frac{b}{b+c} > \sqrt{\frac{a}{a+b} \cdot \frac{c}{c+d}}$$

$$\therefore \quad \left(\frac{b}{b+c}\right)^2 > \frac{ac}{(a+b)(c+d)}.$$

11. Let $R = (5\sqrt{5}+11)^{2n+1}$ and $f = R - [R]$ where [ ] denotes the greatest integral function. Prove that $Rf = 4^{2n+1}$.

$R = [R] + f = (5\sqrt{5}+11)^{2n+1}$

$= {}^{2n+1}C_0(5\sqrt{5})^{2n+1} + {}^{2n+1}C_1(5\sqrt{5})^{2n} \cdot 11 + \ldots$

$+ {}^{2n+1}C_{2n+1} \cdot 11^{2n+1}$ ... (1)

Let $\quad g = (5\sqrt{5}-11)^{2n+1}$

$= {}^{2n+1}C_0(5\sqrt{5})^{2n+1} - {}^{2n+1}C_1(5\sqrt{5})^{2n} \cdot 11 + \ldots$

$- {}^{2n+1}C_{2n+1} \cdot 11^{2n+1}$ ... (2)

Clearly, $\quad 5\sqrt{5} - 11 = \dfrac{(5\sqrt{5}-11)(5\sqrt{5}+11)}{5\sqrt{5}+11}$

$$= \frac{(5\sqrt{5})^2 - 11^2}{5\sqrt{5}+11} = \frac{4}{5\sqrt{5}+11}$$

$\therefore \quad 5\sqrt{5} - 11$ is a positive proper fraction;

$\therefore \quad g = (5\sqrt{5}-11)^{2n+1}$ is also a positive proper fraction.

Thus $0 < f < 1$ and $0 < g < 1$.

Now, $\quad [R] + f - g$

$= 2[{}^{2n+1}C_1(5\sqrt{5})^{2n} + {}^{2n+1}C_3(5\sqrt{5})^{2n-2} \cdot 11 + \ldots$

$+ {}^{2n+1}C_{2n+1} \cdot 11^{2n+1}]$

$= 2 \times$ integer = even integer

$\therefore \quad f - g$ must be an integer because $[R]$ is an integer.

But f and g are positive proper fractions.

So $f - g = 0$, i.e., $f = g$.

$\therefore \quad R \cdot f = \{[R] + f\} \cdot f = \{[R] + f\} \cdot g$

$= (5\sqrt{5}+11)^{2n+1} \cdot (5\sqrt{5}-11)^{2n+1}$

$= \{(5\sqrt{5} + 11)(5\sqrt{5} - 11)\}^{2n+1}$

$= \{(5\sqrt{5})^2 - 11^2\}^{2n+1} = 4^{2n+1}.$

12. Find the greatest term in the expansion of $(x+y)^{18}$ when $x = 2, y = 1$.

Here, $t_{r+1} = {}^{18}C_r\, x^{18-r} \cdot y^r$

$t_r = {}^{18}C_{r-1}\, x^{19-r} \cdot y^{r-1}$

$$\therefore \quad \frac{t_{r+1}}{t_r} = \frac{{}^{18}C_r\, x^{18-r} \cdot y^r}{{}^{18}C_{r-1}\, x^{19-r} \cdot y^{r-1}}$$

$$= \frac{18\,!}{r\,!\,(18-r)\,!} \times \frac{(r-1)\,!\,(19-r)\,!}{18\,!} \times \frac{y}{x}$$

$$= \frac{19-r}{r} \cdot \frac{1}{2}, \text{ putting } x = 2, y = 1.$$

$\therefore \quad t_{r+1} \ge t_r$ if $\dfrac{19-r}{2r} \ge 1$

or $\quad 19 - r \ge 2r$

or $\quad 19 \ge 3r$

$\therefore \quad \dfrac{19}{3} \ge r$

$\therefore \quad$ for $r = 1, 2, 3, \dots, 6$ we get $t_{r+1} > t_r$

for $r = 7, \dots, 17$ we get $t_{r+1} < t_r$

$\therefore \quad t_1 < t_2 < t_3 < \dots < t_7 > t_8 > t_9 > \dots$

$\therefore \quad t_7$ is the greatest term

$\therefore \quad$ the greatest term

$= t_7 = {}^{18}C_6\, x^{12} \cdot y^6 = {}^{18}C_6 \cdot 2^{12} \cdot 1^6$

$= {}^{18}C_6 \cdot 2^{12}.$

13. Given that the 4th term in the expansion of $\left(2 + \dfrac{3x}{8}\right)^{10}$ has the maximum numerical value, find the range of values of x for which this will be true.

According to the question, $|t_4| \ge |t_3|$, $|t_4| \ge |t_5|$.

Now $t_{r+1} = {}^{10}C_r \cdot 2^{10-r} \cdot \left(\dfrac{3x}{8}\right)^r$

$$\therefore \quad t_4 = {}^{10}C_3 \cdot 2^7 \cdot \left(\frac{3x}{8}\right)^3, \quad t_3 = {}^{10}C_2 \cdot 2^8 \cdot \left(\frac{3x}{8}\right)^2$$

$$t_5 = {}^{10}C_4 \cdot 2^6 \cdot \left(\frac{3x}{8}\right)^4$$

$|t_4| \ge |t_3|$

$$\Rightarrow \quad {}^{10}C_3 \cdot 2^7 \cdot \left(\frac{3}{8}\right)^3 \cdot |x|^3$$

$$\ge {}^{10}C_2 \cdot 2^8 \cdot \left(\frac{3}{8}\right)^2 \cdot |x|^2 \quad \dots (1)$$

$|t_4| \ge |t_5|$

$$\Rightarrow \quad {}^{10}C_3 \cdot 2^7 \cdot \left(\frac{3}{8}\right)^3 \cdot |x|^3$$

$$\ge {}^{10}C_4 \cdot 2^6 \cdot \left(\frac{3}{8}\right)^4 \cdot |x|^4 \quad \dots (2)$$

From (1),

$$\frac{10 \cdot 9 \cdot 8}{6} \cdot \frac{3}{8} \cdot |x|^3 \ge \frac{10 \cdot 9}{2} \cdot 2 \cdot |x|^2$$

or $\quad 45|x|^3 \ge 90|x|^2$

or $\quad |x|^3 - 2|x|^2 \ge 0$

or $\quad |x|^2(|x| - 2) \ge 0$

$\Rightarrow \quad |x| \ge 2 \quad \{\because\ |x| \ne 0$ for then $\left(2 + \dfrac{3x}{8}\right)^{10}$ is not the power of a binomial$\}$.

From (2),

$$\frac{10 \cdot 9 \cdot 8}{6} \cdot 2 \cdot |x|^3 \ge \frac{10 \cdot 9 \cdot 8 \cdot 7}{24} \cdot \frac{3}{8} \cdot |x|^4$$

or $\quad |x|^3 \ge \dfrac{7}{8} \cdot \dfrac{3}{8} |x|^4$

or $\quad |x|^3 \left\{1 - \dfrac{21}{64}|x|\right\} \ge 0$

$\Rightarrow \quad 1 - \dfrac{21}{64}|x| \ge 0$

$\therefore \quad |x| \le \dfrac{64}{21}.$

Thus, we get $|x| \ge 2$ and $|x| \le \dfrac{64}{21} = 3\dfrac{1}{21}$

$\therefore \quad -\dfrac{64}{21} \le x \le -2$ and $2 \le x \le \dfrac{64}{31}.$

So $\quad x \in \left[-\dfrac{64}{21}, -2\right] \cup \left[2, \dfrac{64}{21}\right].$

14. Find the coefficient of x^n in the expansion of

$$\left(1 + \frac{x}{1\,!} + \frac{x^2}{2\,!} + \frac{x^3}{3\,!} + \dots + \frac{x^n}{n\,!}\right)^2.$$

Here, $\left(1 + \dfrac{x}{1\,!} + \dfrac{x^2}{2\,!} + \dfrac{x^3}{3\,!} + \dots + \dfrac{x^n}{n\,!}\right)^2$

$$= \left(1 + \frac{x}{1\,!} + \frac{x^2}{2\,!} + \frac{x^3}{3\,!} + \dots + \frac{x^n}{n\,!}\right)$$

$$\times \left(1 + \frac{x}{1\,!} + \frac{x^2}{2\,!} + \frac{x^3}{3\,!} + \dots + \frac{x^n}{n\,!}\right)$$

$\therefore$ coefficient of x^n in $\left(1+\frac{x}{1!}+\frac{x^2}{2!}+\frac{x^3}{3!}+\ldots+\frac{x^n}{n!}\right)^2$

$$=1\cdot\frac{1}{n!}+\frac{1}{1!}\cdot\frac{1}{(n-1)!}+\frac{1}{2!}\cdot\frac{1}{(n-2)!}+\frac{1}{3!}\cdot\frac{1}{(n-3)!}+\ldots+\frac{1}{n!}\cdot 1$$

$$=\frac{1}{0!}\cdot\frac{1}{n!}+\frac{1}{1!}\cdot\frac{1}{(n-1)!}+\frac{1}{2!}\cdot\frac{1}{(n-2)!}+\ldots+\frac{1}{n!}\cdot\frac{1}{0!}$$

$$=\frac{1}{n!}\left(\frac{n!}{0!\,n!}+\frac{n!}{1!\,(n-1)!}+\frac{n!}{2!\,(n-2)!}+\ldots+\frac{n!}{n!\,0!}\right)$$

$$=\frac{1}{n!}({}^nC_0+{}^nC_1+{}^nC_2+\ldots+{}^nC_n)=\frac{1}{n!}\cdot 2^n.$$

15. If $(1+x+x^2)^n=a_0+a_1x+a_2x^2+a_3x^3+\ldots+a_{2n}x^{2n}$ then show that

$$a_0+a_3+a_6+\ldots=a_1+a_4+a_7+\ldots=a_2+a_5+a_8+\ldots=3^{n-1}.$$

Putting $x=1, w, w^2$, where w is a nonreal cube root of unity,

$$3^n=a_0+a_1+a_2+a_3+a_4+\ldots \quad \ldots (1)$$

$$0=a_0+a_1w+a_2w^2+a_3w^3+a_4w^4+\ldots \quad \ldots (2)$$

because $1+w+w^2=0$

$$0=a_0+a_1w^2+a_2w^4+a_3w^6+a_4w^8+\ldots, \quad \ldots (3)$$

Adding these,

$$3^n=3(a_0)+a_1(1+w+w^2)+a_2(1+w^2+w^4)+a_3(1+w^3+w^6)+\ldots$$

$$=3(a_0+a_3+a_6+\ldots)$$

$\therefore\quad a_0+a_3+a_6+\ldots=3^{n-1}.$

From (1) + (2) $\times w^2$ + (3) $\times w$, we get

$$3^n+0\times w^2+0\times w$$

$$=a_0(1+w^2+w)+a_1(1+w^3+w^3)+a_2(1+w^4+w^5)+a_3(1+w^5+w^7)+a_4(1+w^6+w^9)+\ldots$$

$\therefore\quad 3^n=3(a_1+a_4+a_7+\ldots),$

because coefficient of each is $1+w+w^2=0$, using $w^3=1$

$\therefore\quad a_1+a_4+a_7+\ldots=3^{n-1}.$

Again, from (1) + (2) $\times w$ + (3) $\times w^2$, we get

$$3^n=a_0(1+w+w^2)+a_1(1+w^2+w^4)+a_2(1+w^3+w^3)+\ldots$$

$$=3(a_2+a_5+a_8+\ldots)$$

$\therefore\quad a_2+a_5+a_8+\ldots=3^{n-1}.$

16. If $\sum_{r=0}^{2n} a_r(x-2)^r=\sum_{r=0}^{2n} b_r(x-3)^r$ and $a_k=1$ for all $k\ge n$ then show that $b_n={}^{2n+1}C_{n+1}$.

Here, $b_n=$ coefficient of $(x-3)^n$ in $\sum_{r=0}^{2n} b_r(x-3)^r$

$$=\text{coefficient of } (x-3)^n \text{ in } \sum_{r=0}^{2n} a_r(x-2)^r \quad \ldots (1)$$

Now $(x-2)^r=(1+\overline{x-3})^r$

$\therefore\quad \sum_{r=0}^{2n} a_r(x-2)^r=a_0(x-2)^0+a_1(x-2)^1+a_2(x-2)^2+\ldots+a_r(x-2)^r+\ldots+a_{2n}(x-2)^{2n}$

$$=a_0+a_1\{1+(x-3)\}^1+a_2\{1+(x-3)\}^2+\ldots+a_r\{1+(x-3)\}^r+\ldots+a_{2n}\{1+(x-3)\}^{2n}$$

$\therefore$ coefficient of $(x-3)^n$ in $\sum_{r=0}^{2n} a_r(x-2)^r$

$$=a_n\cdot{}^nC_n+a_{n+1}\cdot{}^{n+1}C_n+a_{n+2}\cdot{}^{n+2}C_n+\ldots+a_{2n}\cdot{}^{2n}C_n \quad \ldots (2)$$

As $a_k=1$ for all $k\ge n$, we get

$$a_n=a_{n+1}=a_{n+2}=\ldots=a_{2n}=1.$$

$\therefore$ from (1) and (2),

$$b_n={}^nC_n+{}^{n+1}C_n+{}^{n+2}C_n+\ldots+{}^{2n}C_n.$$

Clearly, ${}^{n+r}C_n$ is the coefficient of x^n in $(1+x)^{n+r}$

$\therefore\quad b_n=$ coefficient of x^n in

$$\{(1+x)^n+(1+x)^{n+1}+(1+x)^{n+2}+\ldots+(1+x)^{2n}\}$$

$$=\text{coefficient of } x^n \text{ in } \frac{(1+x)^n\{1-(1+x)^{n+1}\}}{1-(1+x)}$$

$$=\text{coefficient of } x^n \text{ in } \frac{(1+x)^{2n+1}-(1+x)^n}{x}$$

$$=\text{coefficient of } x^{n+1} \text{ in } \{(1+x)^{2n+1}-(1+x)^n\}$$

$\therefore\quad b_n={}^{2n+1}C_{n+1}.$

17. Sum the series:

$$^{1000}C_{50}+2\cdot{}^{999}C_{49}+3\cdot{}^{998}C_{48}+\ldots+51\cdot{}^{950}C_0.$$

$$\text{Sum} = {}^{1000}C_{950} + 2\cdot {}^{999}C_{950} + 3\cdot {}^{998}C_{950} + \ldots + 51\cdot {}^{950}C_{950}$$

$$(\because {}^nC_r = {}^nC_{n-r})$$

$$= \text{coefficient of } x^{950} \text{ in}$$

$$\{(1+x)^{1000} + 2(1+x)^{999} + 3(1+x)^{998} + \ldots + 51\cdot(1+x)^{950}\}.$$

Let $S = (1+x)^{1000} + 2(1+x)^{999} + 3(1+x)^{998} + \ldots + 51(1+x)^{950}$

$$\frac{S}{1+x} = (1+x)^{999} + 2(1+x)^{998} + \ldots + 50(1+x)^{950} + 51(1+x)^{949}$$

Subtracting,

$$S\left(1-\frac{1}{1+x}\right) = \{(1+x)^{1000} + (1+x)^{999} + (1+x)^{998} + \ldots + (1+x)^{950}\} - 51(1+x)^{949}$$

$$= (1+x)^{950}\frac{\{1-(1+x)^{51}\}}{1-(1+x)} - 51(1+x)^{949}$$

or $\quad \dfrac{x}{1+x}\cdot S = (1+x)^{950}\cdot\dfrac{(1+x)^{51}-1}{x} - 51(1+x)^{949}$

$$\therefore \quad S = \frac{(1+x)^{951}}{x^2}\{(1+x)^{51}-1\} - 51\cdot\frac{(1+x)^{950}}{x}$$

$\therefore$ required sum = coefficient of x^{950} in

$$\left[\frac{(1+x)^{951}}{x^2}\cdot\{(1+x)^{51}-1\} - 51\cdot\frac{(1+x)^{950}}{x}\right]$$

$$= \text{coefficient of } x^{952} \text{ in } (1+x)^{951}\cdot\{(1+x)^{51}-1\} - 51\times\text{coefficient of } x^{951} \text{ in } (1+x)^{950}$$

$$= \text{coefficient of } x^{952} \text{ in } \{(1+x)^{1002} - (1+x)^{951}\} - 51\times 0$$

$$= {}^{1002}C_{952} = {}^{1002}C_{50}.$$

18. Prove that

$$ {}^nC_1 \sin x\cdot\cos(n-1)x + {}^nC_2\sin 2x\cdot\cos(n-2)x + {}^nC_3\sin 3x\cdot\cos(n-3)x + \ldots + {}^nC_n\sin nx = 2^{n-1}\sin nx.$$

Let the sum of the series except the last term be S. Then

LHS $= S + {}^nC_n\sin nx \quad \ldots (1)$

Now $S = {}^nC_1\sin x\cdot\cos(n-1)x + {}^nC_2\sin 2x\cos(n-2)x + \ldots + {}^nC_{n-2}\sin(n-2)x\cdot\cos 2x + {}^nC_{n-1}\sin(n-1)x\cos x.$

Writing it in the reverse order,

$$S = {}^nC_{n-1}\sin(n-1)x\cos x + {}^nC_{n-2}\sin(n-2)x\cos 2x + \ldots + {}^nC_2\sin 2x\cos(n-2)x + {}^nC_1\sin x\cos(n-1)x.$$

Adding,

$$2S = {}^nC_1\{\sin x\cos(n-1)x + \sin(n-1)x\cos x\} + {}^nC_2\{\sin 2x\cos(n-2)x + \sin(n-2)x\cos 2x\} + \ldots + {}^nC_{n-2}\{\sin(n-2)x\cos 2x + \sin 2x\times\cos(n-2)x\} + {}^nC_{n-1}\{\sin(n-1)x\cos x + \sin x\cos(n-1)x\}$$

because ${}^nC_r = {}^nC_{n-r}$

$$= {}^nC_1\sin nx + {}^nC_2\sin nx + \ldots + {}^nC_{n-2}\sin nx + {}^nC_{n-1}\sin nx,$$

because $\sin A\cos B + \sin B\cos A = \sin(A+B)$.

$$= ({}^nC_1 + {}^nC_2 + \ldots + {}^nC_{n-2} + {}^nC_{n-1})\sin nx$$

$$= \{({}^nC_0 + {}^nC_1 + {}^nC_2 + \ldots + {}^nC_{n-1} + {}^nC_n) - {}^nC_0 - {}^nC_n\}\sin nx$$

$$= (2^n - 1 - 1)\sin nx = (2^n - 2)\sin nx$$

$\therefore \quad S = (2^{n-1} - 1)\sin nx;$

putting in (1),

$$\text{LHS} = (2^{n-1}-1)\sin nx + {}^nC_n\sin nx = 2^{n-1}\sin nx.$$

19. Evaluate:

$$\sum_{r=0}^{n}(-1)^r\cdot{}^nC_r\left\{\frac{1}{2^r} + \frac{3^r}{2^{2r}} + \frac{7^r}{2^{3r}} + \frac{15^r}{2^{4r}} + \ldots \text{ to } m \text{ terms}\right\}.$$

$$\text{Sum} = \sum_{r=0}^{n}(-1)^r\cdot{}^nC_r\frac{1}{2^r} + \sum_{r=0}^{n}(-1)^r\cdot{}^nC_r\frac{3^r}{2^{2r}} + \sum_{r=0}^{n}(-1)^r\cdot{}^nC_r\frac{7^r}{2^{3r}} + \ldots \text{ to } m \text{ sigmas}$$

$$= \sum_{r=0}^{n}(-1)^r\cdot{}^nC_r\cdot\frac{1}{2^r} + \sum_{r=0}^{n}(-1)^r\cdot{}^nC_r\left(\frac{3}{2^2}\right)^r + \sum_{r=0}^{n}(-1)^r\cdot{}^nC_r\left(\frac{7}{2^3}\right)^r + \ldots \text{ to } m \text{ sigmas}$$

$$= \left\{{}^nC_0 - {}^nC_1\cdot\frac{1}{2} + {}^nC_2\cdot\left(\frac{1}{2}\right)^2 - \ldots\right\} + \left\{{}^nC_0 - {}^nC_1\cdot\frac{3}{2^2} + {}^nC_2\cdot\left(\frac{3}{2^2}\right)^2 - \ldots\right\} + \left\{{}^nC_0 - {}^nC_1\cdot\frac{7}{2^3} + {}^nC_2\cdot\left(\frac{7}{2^3}\right)^2 - \ldots\right\} + \ldots \text{ to } m \text{ brackets}$$

$$= \left(1-\frac{1}{2}\right)^n + \left(1-\frac{3}{2^2}\right)^n + \left(1-\frac{7}{2^3}\right)^n + \ldots \text{ to } m \text{ terms}$$

$$= \left(\frac{1}{2}\right)^n + \left(\frac{1}{2^2}\right)^n + \left(\frac{1}{2^3}\right)^n + \ldots \text{ to } m \text{ terms}$$

$$= \left(\frac{1}{2}\right)^n + \left(\frac{1}{2}\right)^{2n} + \left(\frac{1}{2}\right)^{3n} + \ldots \text{ to } m \text{ terms}$$

$$= \frac{\left(\frac{1}{2}\right)^n \left\{1-\left(\frac{1}{2^n}\right)^m\right\}}{1-\frac{1}{2^n}} = \frac{1-\frac{1}{2^{mn}}}{2^n-1}$$

$$= \frac{2^{mn}-1}{2^{mn}(2^n-1)}.$$

20. If $s = a + (a+d) + (a+2d) + \ldots + (a+nd)$ **and**

$$S = a + (a+d)\cdot {}^nC_1 + (a+2d)\cdot {}^nC_2 + \ldots + (a+nd)\cdot {}^nC_n$$

then prove that $(n+1)S = 2^n \cdot s$.

Let the $(r+1)$th term of the second series be t_{r+1}. Then

$$t_{r+1} = (a+rd)\,{}^nC_r$$
$$= a\cdot {}^nC_r + d(r\cdot {}^nC_r)$$
$$= a\cdot {}^nC_r + d\cdot n\cdot {}^{n-1}C_{r-1}$$
$$(\because r\cdot {}^nC_r = n\cdot {}^{n-1}C_{r-1}).$$

Putting $r = 0, 1, 2, \ldots, n$, we get

$$t_1 = a\cdot {}^nC_0 + d\cdot n\cdot 0 \quad (\because {}^nC_r = 0 \text{ if } n<r)$$
$$t_2 = a\cdot {}^nC_1 + d\cdot n\cdot {}^{n-1}C_0$$
$$t_3 = a\cdot {}^nC_2 + d\cdot n\cdot {}^{n-1}C_1$$
$$\ldots\ldots\ldots\ldots\ldots\ldots$$
$$t_{n+1} = a\cdot {}^nC_n + d\cdot n\cdot {}^{n-1}C_{n-1}$$

Adding, $S = a({}^nC_0 + {}^nC_1 + {}^nC_2 + \ldots + {}^nC_n)$

$$+ d\cdot n\cdot({}^{n-1}C_0 + {}^{n-1}C_1 + \ldots + {}^{n-1}C_{n-1})$$
$$= a\cdot 2^n + d\cdot n\cdot 2^{n-1} = (2a+nd)2^{n-1}$$

$$\therefore \quad (n+1)S = \frac{n+1}{2}\{2a + (\overline{n+1}-1)d\}2^n = s\cdot 2^n.$$

Second Method (using calculus)

$$(1+x)^n = {}^nC_0 + {}^nC_1 x + {}^nC_2 x^2 + \ldots + {}^nC_n x^n$$

$$\therefore \quad (1+x^d)^n = {}^nC_0 + {}^nC_1\cdot x^d + {}^nC_2 x^{2d} + \ldots + {}^nC_n x^{nd}$$

$$\therefore \quad x^a(1+x^d)^n = {}^nC_0 x^a + {}^nC_1\cdot x^{a+d} + {}^nC_2\cdot x^{a+2d} + \ldots + {}^nC_n x^{a+nd}.$$

Differentiating w.r.t. x,

$$ax^{a-1}\cdot(1+x^d)^n + x^a\cdot n(1+x^d)^{n-1}\cdot dx^{d-1}$$
$$= a\,{}^nC_0 x^{a-1} + (a+d)\,{}^nC_1 x^{a+d-1} + (a+2d)\,{}^nC_2 x^{a+2d-1} + \ldots + (a+nd)\,{}^nC_n x^{a+nd-1}.$$

Putting $x = 1$,

$$a\cdot 2^n + n\cdot 2^{n-1}\cdot d = a\,{}^nC_0 + (a+d)\,{}^nC_1 + (a+2d)\,{}^nC_2 + \ldots + (a+nd)\,{}^nC_n$$

$$\therefore \quad (2a+nd)\,2^{n-1} = S$$

$$\text{or} \quad (n+1)S = \frac{n+1}{2}\{2a + (\overline{n+1}-1)d\}\,2^n$$
$$= s\cdot 2^n.$$

21. Prove that

$${}^nC_0 - 2^2\cdot {}^nC_1 + 3^2\cdot {}^nC_2 - \ldots + (-1)^n(n+1)^2\cdot {}^nC_n = 0, \quad n\ge 2.$$

Here $|t_{r+1}| = (r+1)^2\cdot {}^nC_r$

$$= (r^2+2r+1)\,{}^nC_r$$
$$= r(r\cdot {}^nC_r) + 2(r\cdot {}^nC_r) + {}^nC_r$$
$$= r\cdot n\cdot {}^{n-1}C_{r-1} + 2n\cdot {}^{n-1}C_{r-1} + {}^nC_r$$
$$(\because r\cdot {}^nC_r = n\cdot {}^{n-1}C_{r-1})$$
$$= n(\overline{r-1}+1)\cdot {}^{n-1}C_{r-1} + 2n\cdot {}^{n-1}C_{r-1} + {}^nC_r$$
$$= n(r-1)\cdot {}^{n-1}C_{r-1} + 3n\cdot {}^{n-1}C_{r-1} + {}^nC_r$$
$$= n(n-1)\cdot {}^{n-2}C_{r-2} + 3n\cdot {}^{n-1}C_{r-1} + {}^nC_r,$$

{$n\ge 2 \Rightarrow$ no index is negative}

Putting $r = 0, 1, 2, \ldots, n$ and taking note of the sign of the term, we get

$$t_1 = 0 + 0 + {}^nC_0$$
$$t_2 = -(0 + 3n\cdot {}^{n-1}C_0 + {}^nC_1)$$
$$t_3 = n(n-1)\cdot {}^{n-2}C_0 + 3n\cdot {}^{n-1}C_1 + {}^nC_2$$
$$t_4 = -\{n(n-1)\cdot {}^{n-2}C_1 + 3n\cdot {}^{n-1}C_2 + {}^nC_3\}$$
$$\ldots\ldots\ldots\ldots\ldots\ldots$$
$$t_{n+1} = (-1)^n\{n(n-1)\cdot {}^{n-2}C_{n-2} + 3n\cdot {}^{n-1}C_{n-1} + {}^nC_n\}$$

Adding,

$$\text{sum} = n(n-1)\{{}^{n-2}C_0 - {}^{n-2}C_1 + \ldots + (-1)^{n-2}\cdot {}^{n-2}C_{n-2}\}$$
$$- 3n\{{}^{n-1}C_0 - {}^{n-1}C_1 + \ldots + (-1)^{n-1}\cdot {}^{n-1}C_{n-1}\}$$
$$+ \{{}^nC_0 - {}^nC_1 + \ldots + (-1)^n\cdot {}^nC_n\}$$

$$= n(n-1) \times 0 - 3n \times 0 + 0 = 0,$$

because ${}^nC_0 - {}^nC_1 + {}^nC_2 - {}^nC_3 + \ldots = 0$ for all n.

Second Method (using calculus)

$$(1+x)^n = {}^nC_0 + {}^nC_1 x + {}^nC_2 x^2 + \ldots + {}^nC_n x^n$$

$$\therefore \quad x(1+x)^n = {}^nC_0 x + {}^nC_1 x^2 + {}^nC_2 x^3 + \ldots + {}^nC_n x^{n+1}.$$

Differentiating w.r.t. x,

$$(1+x)^n + x \cdot n(1+x)^{n-1}$$
$$= {}^nC_0 + 2 \cdot {}^nC_1 \cdot x + 3 \cdot {}^nC_2 x^2 + \ldots + (n+1) \cdot {}^nC_n x^n.$$

Multiplying by x,

$$x(1+x)^n + nx^2(1+x)^{n-1}$$
$$= {}^nC_0 x + 2 \cdot {}^nC_1 x^2 + 3 \cdot {}^nC_2 x^3 + \ldots + (n+1) \cdot {}^nC_n x^{n+1}.$$

Differentiating again w.r.t. x,

$$(1+x)^n + x \cdot n(1+x)^{n-1} + n\{2x(1+x)^{n-1} + x^2 \cdot (n-1)(1+x)^{n-2}\}$$
$$= {}^nC_0 + 2^2 \cdot {}^nC_1 x + 3^2 \cdot {}^nC_2 x^2 + \ldots + (n+1)^2 \cdot {}^nC_n x^n.$$

Putting $x = -1$ and remembering $n \geq 2$

$$0 + 0 + n\{0 + 0\} = {}^nC_0 - 2^2 \cdot {}^nC_1 + 3^2 \cdot {}^nC_2 - \ldots + (-1)^n (n+1)^2 \cdot {}^nC_n.$$

22. Sum the series $\sum\limits_{r=0}^{n} (r+1) \cdot {}^{2n}C_r$.

Here $t_{r+1} = (r+1) \cdot {}^{2n}C_r = r \cdot {}^{2n}C_r + {}^{2n}C_r$

$$= 2n \cdot {}^{2n-1}C_{r-1} + {}^{2n}C_r \qquad (\because r \cdot {}^nC_r = n \cdot {}^{n-1}C_{r-1})$$

$$\therefore \quad \sum_{r=0}^{n} t_{r+1} = \sum_{r=0}^{n} (2n \cdot {}^{2n-1}C_{r-1} + {}^{2n}C_r)$$

$$= 2n \sum_{r=0}^{n} {}^{2n-1}C_{r-1} + \sum_{r=0}^{n} {}^{2n}C_r$$

$$= 2n\{{}^{2n-1}C_0 + {}^{2n-1}C_1 + \ldots + {}^{2n-1}C_{n-1}\} + \{{}^{2n}C_0 + {}^{2n}C_1 + \ldots + {}^{2n}C_n\}$$

$$= n\{({}^{2n-1}C_0 + {}^{2n-1}C_0) + ({}^{2n-1}C_1 + {}^{2n-1}C_1) + \ldots + ({}^{2n-1}C_{n-1} + {}^{2n-1}C_{n-1})\}$$
$$+ \frac{1}{2}\{({}^{2n}C_0 + {}^{2n}C_0) + ({}^{2n}C_1 + {}^{2n}C_1) + \ldots + ({}^{2n}C_n + {}^{2n}C_n)\}$$

$$= n\{({}^{2n-1}C_0 + {}^{2n-1}C_{2n-1}) + ({}^{2n-1}C_1 + {}^{2n-1}C_{2n-2}) + \ldots + ({}^{2n-1}C_{n-1} + {}^{2n-1}C_n)\}$$
$$+ \frac{1}{2}\{({}^{2n}C_0 + {}^{2n}C_{2n}) + ({}^{2n}C_1 + {}^{2n}C_{2n-1}) + \ldots + ({}^{2n}C_{n-1} + {}^{2n}C_{n+1}) + {}^{2n}C_n + {}^{2n}C_n\}$$
$$(\because {}^nC_r = {}^nC_{n-r})$$

$$= n\{{}^{2n-1}C_0 + {}^{2n-1}C_1 + {}^{2n-1}C_2 + \ldots + {}^{2n-1}C_{2n-1}\} + \frac{1}{2}\{({}^{2n}C_0 + {}^{2n}C_1 + \ldots + {}^{2n}C_{2n}) + {}^{2n}C_n\}$$

$$= n \cdot 2^{2n-1} + \frac{1}{2}(2^{2n} + {}^{2n}C_n)$$

$$= (n+1)\, 2^{2n-1} + \frac{1}{2} \cdot {}^{2n}C_n$$

$$= (n+1)\, 2^{2n-1} + \frac{1}{2} \cdot \frac{(2n)!}{n!\, n!}.$$

23. Prove that
$${}^nC_0 - {}^nC_1 + {}^nC_2 - \ldots + (-1)^{m-1} \cdot {}^nC_{m-1} = (-1)^{m-1} \cdot {}^{n-1}C_{m-1}.$$

We have

$$(1-x)^n = {}^nC_0 - {}^nC_1 x + {}^nC_2 x^2 - \ldots + (-1)^{m-1} \cdot {}^nC_{m-1} \cdot x^{m-1} + \ldots + (-1)^n \cdot {}^nC_n x^n$$

$$\frac{1-x^m}{1-x} = 1 + x + x^2 + \ldots + x^{m-1} \quad \text{(from sum of GP).}$$

Multiplying these and equating coefficients of x^{m-1} on both sides, we get,

coefficient of x^{m-1} in $\dfrac{(1-x)^n \cdot (1-x^m)}{1-x}$

$$= {}^nC_0 - {}^nC_1 + {}^nC_2 - \ldots + (-1)^{m-1} \cdot {}^nC_{m-1}$$

$$\therefore \quad {}^nC_0 - {}^nC_1 + {}^nC_2 - \ldots + (-1)^{m-1} \cdot {}^nC_{m-1}$$

= coefficient of x^{m-1} in $(1-x)^{n-1} \cdot (1-x^m)$

= coefficient of x^{m-1} in $\{(1-x)^{n-1} - x^m(1-x)^{n-1}\}$

$$= {}^{n-1}C_{m-1}(-1)^{m-1} - 0$$

$$= (-1)^{m-1} \cdot {}^{n-1}C_{m-1}.$$

24. Evaluate $\sum\limits_{r=1}^{j} \dfrac{{}^nC_{2r-1}}{r}$

where $j = \left[\dfrac{n+1}{2}\right]$, i.e., greatest integer $\leq \dfrac{n+1}{2}$.

Here $t_r = \frac{{}^nC_{2r-1}}{r} = 2 \cdot \frac{{}^nC_{2r-1}}{2r} = 2 \cdot \frac{1}{n+1} \cdot {}^{n+1}C_{2r}$

$$\left\{\because \frac{1}{r+1}\,{}^nC_r = \frac{1}{n+1} \cdot {}^{n+1}C_{r+1}\right\}$$

$\therefore$ the required sum $= \frac{2}{n+1} \sum_{r=1}^{j} {}^{n+1}C_{2r}$

$$= \frac{2}{n+1}\{{}^{n+1}C_2 + {}^{n+1}C_4 + {}^{n+1}C_6 + \ldots + {}^{n+1}C_{2j}\}$$

$$= \frac{2}{n+1}\{({}^{n+1}C_0 + {}^{n+1}C_2 + {}^{n+1}C_4 + \ldots) - {}^{n+1}C_0\}$$

$$= \frac{2}{n+1}\{2^{(n+1)-1} - 1\} = \frac{2(2^n - 1)}{n+1}.$$

Second Method (using calculus)

$$(1+x)^n = {}^nC_0 + {}^nC_1 x + {}^nC_2 x^2 + {}^nC_3 x^3 + \ldots + {}^nC_n x^n$$

$$\therefore \quad \int_0^x (1+x)^n dx$$

$$= \int_0^x ({}^nC_0 + {}^nC_1 x + {}^nC_2 x^2 + {}^nC_3 x^3 + \ldots + {}^nC_n x^n) dx$$

or $$\left[\frac{(1+x)^{n+1}}{n+1}\right]_0^x = \left[{}^nC_0 x + \frac{{}^nC_1 x^2}{2} + \frac{{}^nC_2 x^3}{3} + \frac{{}^nC_3 x^4}{4} + \ldots + \frac{{}^nC_n x^{n+1}}{n+1}\right]_0^x$$

$$\therefore \quad \frac{(1+x)^{n+1}}{n+1} - \frac{1}{n+1}$$

$$= {}^nC_0 x + \frac{{}^nC_1 x^2}{2} + \frac{{}^nC_2 x^3}{3} + \frac{{}^nC_3 x^4}{4} + \ldots.$$

Putting $x = 1, -1$, we get

$$\frac{2^{n+1}}{n+1} - \frac{1}{n+1} = {}^nC_0 + \frac{{}^nC_1}{2} + \frac{{}^nC_2}{3} + \frac{{}^nC_3}{4} + \frac{{}^nC_4}{5} + \ldots$$

$$0 - \frac{1}{n+1} = -{}^nC_0 + \frac{{}^nC_1}{2} - \frac{{}^nC_2}{3} + \frac{{}^nC_3}{4} - \frac{{}^nC_4}{5} + \ldots.$$

Adding these,

$$\frac{2^{n+1}}{n+1} - \frac{2}{n+1} = 2\left(\frac{{}^nC_1}{2} + \frac{{}^nC_3}{4} + \frac{{}^nC_5}{6} + \ldots\right)$$

$$\therefore \quad \frac{{}^nC_1}{1} + \frac{{}^nC_3}{2} + \frac{{}^nC_5}{3} + \ldots = \frac{2(2^n - 1)}{n+1}.$$

25. Evaluate $\sum_{r=0}^{n} \frac{1}{(r+1)(r+3)} \cdot {}^nC_r x^r$.

Here $\frac{1}{(r+1)(r+3)}\,{}^nC_r x^r = \frac{1}{r+1}\,{}^nC_r \cdot \frac{1}{r+3} x^r$

$$= \frac{1}{n+1} \cdot {}^{n+1}C_{r+1} \cdot \frac{x^r}{r+3}$$

$$= \frac{1}{n+1} \cdot \frac{r+2}{(r+2)(r+3)}\,{}^{n+1}C_{r+1} x^r$$

$$= \frac{1}{n+1} \cdot \frac{r+2}{r+3} \cdot \frac{1}{n+2} \cdot {}^{n+2}C_{r+2} x^r$$

$$\left(\because \frac{1}{r+1} \cdot {}^nC_r = \frac{1}{n+1} \cdot {}^{n+1}C_{r+1}\right)$$

$$= \frac{1}{(n+1)(n+2)}(r+2)\frac{1}{n+3}\,{}^{n+3}C_{r+3} x^r$$

$$= \frac{1}{(n+1)(n+2)(n+3)}(\overline{r+3} - 1)\,{}^{n+3}C_{r+3} x^r$$

$$= \frac{1}{(n+1)(n+2)(n+3)} \times \{(n+3)\,{}^{n+2}C_{r+2} - {}^{n+3}C_{r+3}\} x^r$$

$$\therefore \text{ sum} = \sum_{r=0}^{n} \frac{1}{(n+1)(n+2)} \cdot {}^{n+2}C_{r+2} x^r - \sum_{r=0}^{n} \frac{1}{(n+1)(n+2)(n+3)} \cdot {}^{n+3}C_{r+3} x^r$$

$$= \frac{1}{(n+1)(n+2)} \times \{{}^{n+2}C_2 + {}^{n+2}C_3 x + {}^{n+2}C_4 x^2 + \ldots + {}^{n+2}C_{n+2} x^n\}$$
$$- \frac{1}{(n+1)(n+2)(n+3)} \times \{{}^{n+3}C_3 + {}^{n+3}C_4 x + {}^{n+3}C_5 x^2 + \ldots + {}^{n+3}C_{n+3} x^n\}$$

$$= \frac{1}{(n+1)(n+2)x^2} \times \{{}^{n+2}C_2 x^2 + {}^{n+2}C_3 x^3 + {}^{n+2}C_4 x^4 + \ldots + {}^{n+2}C_{n+2} x^{n+2}\}$$
$$- \frac{1}{(n+1)(n+2)(n+3)x^3} \times \{{}^{n+3}C_3 x^3 + {}^{n+3}C_4 x^4 + {}^{n+3}C_5 x^5 + \ldots + {}^{n+3}C_{n+3} x^{n+3}\}$$

$$= \frac{(1+x)^{n+2} - {}^{n+2}C_1\,x - {}^{n+2}C_0}{(n+1)(n+2)x^2}$$

$$- \frac{(1+x)^{n+3} - {}^{n+3}C_2\,x^2 - {}^{n+3}C_1\,x - {}^{n+3}C_0}{(n+1)(n+2)(n+3)x^3}.$$

26. Prove that $\boldsymbol{{}^nC_1 - \frac{1}{2}{}^nC_2 + \frac{1}{3}{}^nC_3 - \ldots + (-1)^{n-1}\frac{1}{n}\cdot{}^nC_n = 1 + \frac{1}{2} + \frac{1}{3} + \ldots + \frac{1}{n}}$.

Let $S_n = {}^nC_1 - \frac{1}{2}{}^nC_2 + \frac{1}{3}{}^nC_3 - \ldots + (-1)^{n-1}\cdot\frac{1}{n}{}^nC_n$

$$\therefore \quad S_{n-1} = {}^{n-1}C_1 - \frac{1}{2}{}^{n-1}C_2 + \frac{1}{3}{}^{n-1}C_3 - \ldots + (-1)^{n-2}\frac{1}{n-1}{}^{n-1}C_{n-1}$$

$$\therefore \quad S_n - S_{n-1} = ({}^nC_1 - {}^{n-1}C_1) - \frac{1}{2}({}^nC_2 - {}^{n-1}C_2) + \frac{1}{3}({}^nC_3 - {}^{n-1}C_3) - \ldots + (-1)^{n-2}\cdot\frac{1}{n-1}({}^nC_{n-1} - {}^{n-1}C_{n-1}) + (-1)^{n-1}\cdot\frac{1}{n}{}^nC_n$$

$$= {}^{n-1}C_0 - \frac{1}{2}{}^{n-1}C_1 + \frac{1}{3}{}^{n-1}C_2 - \ldots + (-1)^{n-2}\frac{1}{n-1}{}^{n-1}C_{n-2} + (-1)^{n-1}\cdot\frac{1}{n}{}^nC_n$$

$$= \frac{1}{n}{}^nC_1 - \frac{1}{n}{}^nC_2 + \frac{1}{n}{}^nC_3 - \ldots + (-1)^{n-2}\frac{1}{n}{}^nC_{n-1} + (-1)^{n-1}\cdot\frac{1}{n}{}^nC_n \quad \left(\because \frac{1}{r+1}{}^nC_r = \frac{1}{n+1}{}^{n+1}C_{r+1}\right)$$

$$= \frac{1}{n}[{}^nC_1 - {}^nC_2 + {}^nC_3 - \ldots + (-1)^{n-2}\,{}^nC_{n-1}] + (-1)^{n-1}\cdot\frac{1}{n}{}^nC_n$$

$$= -\frac{1}{n}[{}^nC_0 - {}^nC_1 + {}^nC_2 - \ldots + (-1)^{n-1}{}^nC_{n-1} + (-1)^n\cdot{}^nC_n] + \frac{1}{n}\cdot{}^nC_0$$

$$= -\frac{1}{n}\cdot(1-1)^n + \frac{1}{n} = \frac{1}{n}.$$

Thus, $S_n - S_{n-1} = \frac{1}{n}$; putting $n = 2, 3, \ldots, n$ we get

$$S_2 - S_1 = \frac{1}{2};\ S_3 - S_2 = \frac{1}{3};\ \ldots;\ S_n - S_{n-1} = \frac{1}{n}.$$

Adding these, $S_n - S_1 = \frac{1}{2} + \frac{1}{3} + \frac{1}{4} + \ldots + \frac{1}{n}$.

But $S_1 = {}^1C_1 = 1$

$$\therefore \quad S_n = 1 + \frac{1}{2} + \frac{1}{3} + \ldots + \frac{1}{n}.$$

Second Method (using calculus)

$$(1-x)^n = {}^nC_0 - {}^nC_1x + {}^nC_2x^2 - {}^nC_3x^3 + \ldots + (-1)^n\cdot{}^nC_nx^n$$

$$\therefore \quad {}^nC_1x - {}^nC_2x^2 + {}^nC_3x^3 - \ldots + (-1)^{n-1}\cdot{}^nC_nx^n = {}^nC_0 - (1-x)^n$$

or $\quad {}^nC_1 - {}^nC_2x + {}^nC_3x^2 - \ldots + (-1)^{n-1}\cdot{}^nC_n\,x^{n-1} = \dfrac{1-(1-x)^n}{x}$

$$\therefore \quad \int_0^1 \{{}^nC_1 - {}^nC_2x + {}^nC_3\,x^3 - \ldots + (-1)^{n-1}\cdot{}^nC_nx^{n-1}\}\,dx = \int_0^1 \frac{1-(1-x)^n}{x}\,dx$$

or $\quad \left[{}^nC_1x - {}^nC_2\dfrac{x^2}{2} + {}^nC_3\dfrac{x^3}{3} - \ldots + (-1)^{n-1}\,{}^nC_n\cdot\dfrac{x^n}{n}\right]_0^1$

$$= \int_0^1 \frac{1-(1-x)^n}{1-(1-x)}\,dx$$

$$= \int_0^1 \{1 + (1-x) + (1-x)^2 + \ldots + (1-x)^{n-1}\}\,dx$$

$$= \left[x + \frac{(1-x)^2}{-2} + \frac{(1-x)^3}{-3} + \ldots + \frac{(1-x)^n}{-n}\right]_0^1$$

$$= 1 + \frac{1}{2} + \frac{1}{3} + \ldots + \frac{1}{n}.$$

27. Prove that

$$\boldsymbol{\frac{{}^nC_0}{n} - \frac{{}^nC_1}{n+1} + \frac{{}^nC_2}{n+2} - \ldots + (-1)^n\frac{{}^nC_n}{2n} = \frac{1}{n\cdot{}^{2n}C_n}}.$$

We have,

$$(1-x)^n = {}^nC_0 - {}^nC_1x + {}^nC_2x^2 - \ldots + (-1)^n\cdot{}^nC_nx^n$$

$$\therefore \quad x^{n-1}\cdot(1-x)^n = {}^nC_0x^{n-1} - {}^nC_1x^n + {}^nC_2x^{n+1} - \ldots + (-1)^n\cdot{}^nC_nx^{2n-1}$$

$$\therefore \quad \int_0^1 x^{n-1}\cdot(1-x)^n\,dx$$

$$= \left[\frac{^nC_0}{n} \cdot x^n - \frac{^nC_1}{n+1} x^{n+1} + \frac{^nC_2}{n+2} x^{n+2} - \dots + (-1)^n \frac{^nC_n}{2n} x^{2n} \right]_0^1$$

$$= \frac{^nC_0}{n} - \frac{^nC_1}{n+1} + \frac{^nC_2}{n+2} - \dots + (-1)^n \frac{^nC_n}{2n} \quad \dots (1)$$

Now, $\int_0^1 x^{n-1} \cdot (1-x)^n \, dx$

$$= \left[(1-x)^n \cdot \frac{x^n}{n} \right]_0^1 - \int_0^1 \frac{x^n}{n} \cdot n(1-x)^{n-1} \cdot (-1) \, dx$$

(integrating by parts)

$$= \frac{n}{n} \int_0^1 x^n (1-x)^{n-1} \, dx$$

$$= \frac{n(n-1)}{n(n+1)} \int_0^1 x^{n+1} \cdot (1-x)^{n-2} \, dx$$

(integrating by parts again)

$$= \frac{n(n-1)(n-2)}{n(n+1)(n+2)} \int_0^1 x^{n+2} \cdot (1-x)^{n-3} \, dx,$$

(integrating by parts again)

....

$$= \frac{n(n-1)(n-2) \dots 1}{n(n+1)(n+2) \dots (n+n)} \int_0^1 x^{2n-1} \, dx$$

$$= \frac{n! \cdot (n-1)!}{(n+n)!} = \frac{1}{n} \cdot \frac{n!\,n!}{(2n)!} = \frac{1}{n \cdot {^{2n}C_n}}.$$

28. Prove that

$$^nC_0 \cdot {^nC_r} + {^nC_1} \cdot {^nC_{r+1}} + {^nC_2} \cdot {^nC_{r+2}} + \dots + {^nC_{n-r}}\,{^nC_n} = \frac{(2n)!}{(n-r)!\,(n+r)!}.$$

We have

$$(1+x)^n = {^nC_0} + {^nC_1}x + {^nC_2}x^2 + \dots + {^nC_r}x^r + \dots + {^nC_n}x^n$$

$$(x+1)^n = {^nC_0}x^n + {^nC_1}x^{n-1} + {^nC_2}x^{n-2} + \dots + {^nC_n}.$$

Multiplying these and equating coefficient of x^{n+r} on both sides,

coefficient of x^{n+r} in $(1+x)^n \cdot (x+1)^n$

$$= {^nC_0}\,{^nC_r} + {^nC_1}\,{^nC_{r+1}} + \dots + {^nC_{n-r}}\,{^nC_n}$$

$$\therefore \quad {^nC_0}\,{^nC_r} + {^nC_1}\,{^nC_{r+1}} + \dots + {^nC_{n-r}}\,{^nC_n}$$

$$= \text{coefficient of } x^{n+r} \text{ in } (1+x)^{2n}$$

$$= {^{2n}C_{n+r}} = \frac{(2n)!}{(n+r)!\,(n-r)!}.$$

Note

(i) Instead of taking the expansion of $(x+1)^n$ one can take the expansion

$$\left(1+\frac{1}{x}\right)^n = {^nC_0} + {^nC_1} \cdot \frac{1}{x} + {^nC_2} \cdot \frac{1}{x^2} + \dots + {^nC_n} \cdot \frac{1}{x^n}.$$

Then, series = coefficient of x^r in $(1+x)^n \cdot \left(1+\frac{1}{x}\right)^n$

= coefficient of x^{n+r} in $(1+x)^{2n}$.

(ii) Instead of taking the coefficient of x^{n+r} in the product one can equate the coefficients of x^{n-r} on both sides.

(iii) Taking $r = 0$ in the result we get

$${^nC_0^2} + {^nC_1^2} + {^nC_2^2} + \dots + {^nC_n^2} = \frac{(2n)!}{n!\,n!}.$$

(iv) $({^nC_0} + {^nC_1} + {^nC_2} + \dots + {^nC_n})^2$

$= {^nC_0^2} + {^nC_1^2} + \dots + {^nC_n^2} + 2$ {sum of the products of ${^nC_0}, {^nC_1}, \dots, {^nC_n}$ taking two at a time}

$\therefore \quad (2^n)^2 = \frac{(2n)!}{n!\,n!} + 2$ {sum of the products of binomial coefficients taking two at a time}

(v) Instead of taking the expansion of $(x+1)^n$ if we take the expansion of $(x-1)^n$ then equating the coefficients of x^{n+r} on both sides of the product we get

${^nC_0}\,{^nC_r} - {^nC_1}\,{^nC_{r+1}} + {^nC_2}\,{^nC_{r+2}} - \dots$

= coefficient of x^{n+r} in $(x^2-1)^n$

= coefficient of x^{n+r} in $(-1)^n \cdot (1-x^2)^n$.

If $n+r$ is odd, the coefficient = 0 because the expansion has powers of x^2 only.

(vi) For series like

${^nC_0} \cdot {^mC_r} + {^nC_1}\,{^mC_{r+1}} + {^nC_2}\,{^mC_{r+2}} + \dots,$

one can take expansions of $(1+x)^n$ and $(x+1)^m$ and then multiply.

(vii) In series like the above one, consisting of products of two binomial coefficients, the difference of suffixes is constant

{here $r - 0 = (r+1) - 1 = \dots = r$}.

29. Prove by binomial expansion that

$${^mC_r} \cdot {^nC_0} + {^mC_{r-1}} \cdot {^nC_1} + {^mC_{r-2}} \cdot {^nC_2} + \dots + {^mC_0} \cdot {^nC_r} = {^{m+n}C_r}, \text{ where } r < m \text{ or } n.$$

We have

$$(1+x)^m = {}^mC_0 + {}^mC_1x + {}^mC_2x^2 + \dots + {}^mC_rx^r + \dots + {}^mC_mx^m$$

$$(1+x)^n = {}^nC_0 + {}^nC_1x + {}^nC_2x^2 + \dots + {}^nC_rx^r + \dots + {}^nC_nx^n$$

Multiplying and equating coefficient of x^r on both sides, we get,

coefficient of x^r in $(1+x)^m \cdot (1+x)^n$

$$= {}^mC_r \cdot {}^nC_0 + {}^mC_{r-1}\,{}^nC_1 + \dots + {}^mC_0 \cdot {}^nC_r$$

$$\therefore \quad {}^mC_r\,{}^nC_0 + {}^mC_{r-1} \cdot {}^nC_1 + \dots + {}^mC_0 \cdot {}^nC_r$$

$= $ coefficient of x^r in $(1+x)^{m+n}$

$= {}^{m+n}C_r$.

Note In series like the above one, consisting of products of two binomial coefficients, the sum of suffixes is constant (here $r + 0 = (r-1) + 1 = \dots = r$).

30. Prove that

$${}^nC_0 \cdot {}^{2n}C_n - {}^nC_1 \cdot {}^{2n-2}C_n + {}^nC_2 \cdot {}^{2n-4}C_n - \dots = 2^n.$$

As ${}^{2n}C_n$, ${}^{2n-2}C_n$, ${}^{2n-4}C_n$, etc., are coefficients of the same power x^n in the expansions of $(1+x)^{2n}$, $(1+x)^{2n-2}$, etc., we consider the expression

$${}^nC_0(1+x)^{2n} - {}^nC_1 \cdot (1+x)^{2n-2} + {}^nC_2(1+x)^{2n-4} - \dots$$

Clearly LHS = coefficient of x^n in

$$[{}^nC_0(1+x)^{2n} - {}^nC_1(1+x)^{2n-2} + {}^nC_2(1+x)^{2n-4} - \dots]$$

$= $ coefficient of x^n in

$$[{}^nC_0\{(1+x)^2\}^n - {}^nC_1\{(1+x)^2\}^{n-1} + {}^nC_2\{(1+x)^2\}^{n-2} - \dots]$$

$= $ coefficient of x^n in $\{(1+x)^2 - 1\}^n$

$= $ coefficient of x^n in $(2x + x^2)^n$

$= $ coefficient of x^n in $\{x^n \cdot (2+x)^n\}$

$= $ constant term in $(2+x)^n = 2^n$.

31. Show that the HM of ${}^{2n+1}C_r$ and ${}^{2n+1}C_{r+1}$ is $\dfrac{2n+1}{n+1}$ times of ${}^{2n}C_r$. Also show that

$$\sum_{r=1}^{2n-1} (-1)^{r-1} \cdot \frac{r}{{}^{2n}C_r} = \frac{n}{n+1}.$$

$$\frac{1}{{}^{2n+1}C_r} + \frac{1}{{}^{2n+1}C_{r+1}}$$

$$= \frac{r!\,(2n-r+1)!}{(2n+1)!} + \frac{(r+1)!\,(2n-r)!}{(2n+1)!}$$

$$= \frac{r!\,(2n-r)!}{(2n+1)!}\{2n - r + 1 + r + 1\}$$

$$= \frac{r!\,(2n-r)!}{(2n)!} \cdot \frac{2n+2}{2n+1} = \frac{2(n+1)}{2n+1} \cdot \frac{1}{{}^{2n}C_r}$$

$$= \frac{n+1}{2n+1} \cdot \frac{2}{{}^{2n}C_r} = \frac{2}{\frac{2n+1}{n+1} \cdot {}^{2n}C_r} \quad \dots (1)$$

$\therefore \quad \dfrac{2n+1}{n+1} \cdot {}^{2n}C_r$ is the HM of ${}^{2n+1}C_r$ and ${}^{2n}C_r$.

Hence the result.

Now, $\displaystyle\sum_{r=1}^{2n-1} (-1)^{r-1} \cdot \frac{r}{{}^{2n}C_r}$

$$= \frac{1}{{}^{2n}C_1} - \frac{2}{{}^{2n}C_2} + \frac{3}{{}^{2n}C_3} - \dots + \frac{2n-1}{{}^{2n}C_{2n-1}}$$

$$= \left(\frac{1}{{}^{2n}C_1} + \frac{2n-1}{{}^{2n}C_{2n-1}}\right) - \left(\frac{2}{{}^{2n}C_2} + \frac{2n-2}{{}^{2n}C_{2n-2}}\right) + \left(\frac{3}{{}^{2n}C_3} + \frac{2n-3}{{}^{2n}C_{2n-3}}\right) - \dots + (-1)^{n-1}\frac{n}{{}^{2n}C_n}$$

(combining corresponding terms from the beginning and the end—the middle terms, i.e., nth term being alone)

$$= \left\{\frac{2n}{{}^{2n}C_1} - \frac{2n}{{}^{2n}C_2} + \frac{2n}{{}^{2n}C_3} - \dots \text{ to } (n-1) \text{ terms}\right\} + (-1)^{n-1} \cdot \frac{n}{{}^{2n}C_n} \quad (\because {}^{2n}C_r = {}^{2n}C_{2n-r})$$

$$= 2n\left\{\frac{1}{{}^{2n}C_1} - \frac{1}{{}^{2n}C_2} + \frac{1}{{}^{2n}C_3} - \dots + (-1)^n \cdot \frac{1}{{}^{2n}C_{n-1}}\right\} + (-1)^{n-1} \cdot \frac{n}{{}^{2n}C_n}$$

$$= 2n\left\{\frac{2n+1}{2(n+1)}\left(\frac{1}{{}^{2n+1}C_1} + \frac{1}{{}^{2n+1}C_2}\right) - \frac{2n+1}{2(n+1)}\left(\frac{1}{{}^{2n+1}C_2} + \frac{1}{{}^{2n+1}C_3}\right) + \frac{2n+1}{2(n+1)}\left(\frac{1}{{}^{2n+1}C_3} + \frac{1}{{}^{2n+1}C_4}\right) - \dots + (-1)^n \cdot \frac{2n+1}{2(n+1)}\left(\frac{1}{{}^{2n+1}C_{n-1}} + \frac{1}{{}^{2n+1}C_n}\right)\right\}$$

$$+ (-1)^{n-1} \cdot \frac{n}{{}^{2n}C_n}, \quad \text{using (1)}$$

$$= 2n \cdot \frac{2n+1}{2(n+1)} \left[\left(\frac{1}{^{2n+1}C_1} + \frac{1}{^{2n+1}C_2} \right) - \left(\frac{1}{^{2n+1}C_2} + \frac{1}{^{2n+1}C_3} \right) + \left(\frac{1}{^{2n+1}C_3} + \frac{1}{^{2n+1}C_4} \right) - \ldots + (-1)^n \cdot \left(\frac{1}{^{2n+1}C_{n-1}} + \frac{1}{^{2n+1}C_n} \right) \right] + (-1)^{n-1} \cdot \frac{n}{^{2n}C_n}$$

$$= \frac{n}{n+1} (2n+1) \left[\frac{1}{^{2n+1}C_1} + (-1)^n \cdot \frac{1}{^{2n+1}C_n} \right] + (-1)^{n-1} \cdot \frac{n}{^{2n}C_n}$$

$$= \frac{n}{n+1} (2n+1) \cdot \frac{1}{^{2n+1}C_1} + (-1)^n \cdot \frac{1}{^{2n+1}C_n} \left\{ \frac{n(2n+1)}{n+1} \right\} + (-1)^{n-1} \cdot \frac{n}{^{2n}C_n}$$

$$= \frac{n}{n+1} \cdot (2n+1) \cdot \frac{1}{2n+1} + (-1)^n \cdot \frac{n!\,(n+1)!}{(2n+1)!} \cdot \frac{n(2n+1)}{n+1} + (-1)^{n-1} \cdot n \cdot \frac{n!\,n!}{(2n)!}$$

$$= \frac{n}{n+1} + (-1)^n \cdot \frac{n!\,n!}{(2n)!} \cdot n + (-1)^{n-1} \cdot \frac{n!\,n!}{(2n)!} \cdot n$$

$$= \frac{n}{n+1} + \{(-1)^n + (-1)^{n-1}\} \frac{n!\,n!}{(2n)!} \cdot n$$

$$= \frac{n}{n+1}.$$

32. If $n \in N$ and $c_k = {}^nC_k$, find the value of $\sum_{k=1}^{n} k^3 \cdot \left(\frac{c_k}{c_{k-1}} \right)^2$.

$$\frac{c_k}{c_{k-1}} = \frac{^nC_k}{^nC_{k-1}} = \frac{n!}{k!\,(n-k)!} \cdot \frac{(k-1)!\,(n-k+1)!}{n!}$$

$$= \frac{n-k+1}{k}$$

$$\therefore \quad \sum_{k=1}^{n} k^3 \cdot \left(\frac{c_k}{c_{k-1}} \right)^2 = \sum_{k=1}^{n} k^3 \cdot \left(\frac{n-k+1}{k} \right)^2$$

$$= \sum_{k=1}^{n} k\,(n-k+1)^2$$

$$= \sum_{k=1}^{n} k\{(n+1)^2 - 2(n+1)k + k^2\}$$

$$= (n+1)^2 \sum_{k=1}^{n} k - 2(n+1) \sum_{k=1}^{n} k^2 + \sum_{k=1}^{n} k^3$$

$$= (n+1)^2 \cdot \frac{n(n+1)}{2} - 2(n+1) \cdot \frac{n(n+1)(2n+1)}{6} + \frac{n^2(n+1)^2}{4}$$

$$= \frac{n(n+1)^2}{2} \cdot \left\{ (n+1) - \frac{2}{3}(2n+1) + \frac{n}{2} \right\}$$

$$= \frac{n(n+1)^2}{2} \cdot \frac{6(n+1) - 4(2n+1) + 3n}{6}$$

$$= \frac{1}{12} n(n+1)^2 \cdot (n+2).$$

33. If $(1+x+x^2)^n = a_0 + a_1 x + a_2 x^2 + \ldots + a_{2n} x^{2n}$ then prove that

$$a_0^2 - a_1^2 + a_2^2 - a_3^2 + \ldots + (-1)^{n-1} a_{n-1} = \frac{1}{2} a_n \{1 - (-1)^n a_n\}.$$

We have

$$(1+x+x^2)^n = a_0 + a_1 x + a_2 x^2 + a_3 x^3 + \ldots + a_{2n} x^{2n} \quad \ldots (1)$$

$$\therefore \quad \left(1 - \frac{1}{x} + \frac{1}{x^2} \right)^n = a_0 - a_1 \cdot \frac{1}{x} + a_2 \cdot \frac{1}{x^2} - a_3 \cdot \frac{1}{x^3} + \ldots + a_{2n} \cdot \frac{1}{x^{2n}}.$$

Multiplying, and equating the coefficients of x^0 on both sides,

coefficient of x^0 in $(1+x+x^2)^n \cdot \left(1 - \frac{1}{x} + \frac{1}{x^2} \right)^n$

$$= a_0^2 - a_1^2 + a_2^2 - a_3^2 + \ldots + a_{2n}^2.$$

Now, $(1+x+x^2)^n \cdot \left(1 - \frac{1}{x} + \frac{1}{x^2} \right)^n$

$$= (1+x+x^2)^n \cdot \frac{(x^2-x+1)^n}{x^{2n}}$$

$$= \frac{\{(1+x^2)^2 - x^2\}^n}{x^{2n}} = \frac{(1+x^2+x^4)^n}{x^{2n}}.$$

$\therefore$ coefficient of x^0 in $\frac{(1+x^2+x^4)^n}{x^{2n}}$

$$= a_0^2 - a_1^2 + a_2^2 - \ldots + a_{2n}^2 \quad \ldots (2)$$

Putting x^2 in the given expansion,

$$(1+x^2+x^4)^n = a_0 + a_1 x^2 + a_2 x^4 + \ldots + a_n x^{2n} + \ldots + a_{2n} x^{4n}$$

or $\dfrac{(1+x^2+x^4)^2}{x^{2n}} = a_0x^{-2n} + a_1x^{2-2n} + \ldots$

$$+ a_n + \ldots + a_{2n}x^{2n}$$

$\therefore$ by (2), $a_0^2 - a_1^2 + a_2^2 - a_3^2 + \ldots + a_{2n}^2 = a_n$ ... (3)

Now, putting $\dfrac{1}{x}$ for x in (1),

$$\left(1+\frac{1}{x}+\frac{1}{x^2}\right)^n = a_0 + \frac{a_1}{x} + \frac{a_2}{x^2} + \frac{a_3}{x^3} + \ldots + \frac{a_{2n}}{x^{2n}}$$

$\therefore$ $x^{2n}\left(1+\dfrac{1}{x}+\dfrac{1}{x^2}\right)^n$

$$= a_0x^{2n} + a_1x^{2n-1} + a_2x^{2n-2} + \ldots + a_{2n}$$

or $(1+x+x^2)^n = a_{2n} + a_{2n-1}x + a_{2n-2}x^2$

$$+ \ldots + a_0x^{2n} \quad \ldots (4)$$

From (1) and (4),

$$a_{2n} = a_0, a_{2n-1} = a_1, a_{2n-2} = a_2, \ldots,$$

$$a_n = a_n, \ldots, a_0 = a_{2n}$$

$\therefore$ from (3),

$$(a_0^2 + a_{2n}^2) - (a_1^2 + a_{2n-1}^2) + (a_2^2 + a_{2n-2}^2) - \ldots$$

$$+ (-1)^{n-1}(a_{n-1}^2 + a_{n+1}^2) + (-1)^n a_n^2 = a_n$$

$\therefore$ $2a_0^2 - 2a_1^2 + 2a_2^2 - \ldots + (-1)^{n-1} \cdot 2a_{n-1}^2 + (-1)^n a_n^2$

$$= a_n$$

or $a_0^2 - a_1^2 + a_2^2 - \ldots + (-1)^{n-1}a_{n-1}^2 + \dfrac{1}{2}(-1)^n a_n^2 = \dfrac{1}{2}a_n$

$\therefore$ $a_0^2 - a_1^2 + a_2^2 - \ldots + (-1)^{n-1}a_{n-1}^2$

$$= \frac{1}{2}a_n\{1 - (-1)^n a_n\}.$$

34. If $(1+x+x^2)^n = a_0 + a_1x + a_2x^2 + \ldots + a_{2n}x^{2n}$, prove that

$$^nC_0 \cdot a_r - {}^nC_1 \cdot a_{r-1} + {}^nC_2 \cdot a_{r-2} - \ldots + (-1)^r \cdot {}^nC_r \cdot a_0$$

$= 0$ if r is not a multiple of 3

$(-1)^m \cdot {}^nC_m$ if $r = 3m$.

We have

$$(1+x+x^2)^n = a_0 + a_1x + a_2x^2 + \ldots$$

$$+ a_rx^r + \ldots + a_{2n}x^{2n}$$

$$(1-x)^n = {}^nC_0 - {}^nC_1x + {}^nC_2x^2 - \ldots + (-1)^r \cdot {}^nC_rx^r$$

$$+ \ldots + (-1)^n \cdot {}^nC_nx^n.$$

Multiplying these and equating the coefficients of x^r on both sides,

coefficient of x^r in $(1-x)^n \cdot (1+x+x^2)^n$

$$= {}^nC_0 \cdot a_r - {}^nC_1 a_{r-1} + {}^nC_2 a_{r-2} - \ldots + (-1)^r \cdot {}^nC_r a_0$$

$\therefore$ series = coefficient of x^r in $(1-x^3)^n$

= coefficient of x^r in

$$\{{}^nC_0 - {}^nC_1x^3 + {}^nC_2(x^3)^2 - \ldots$$

$$+ (-1)^n \cdot {}^nC_n(x^3)^n\}$$

$= 0$ when r is not a multiple of 3

because only powers of x^3 occur

$(-1)^m \cdot {}^nC_m$ where $r = 3m$.

35. Prove that the HCF of $^{2n}C_1, {}^{2n}C_3, {}^{2n}C_5, \ldots, {}^{2n}C_{2n-1}$ is 2^{m+1} where $n = 2^m \cdot q$, q being an odd integer.

We know that $^{2n}C_1 + {}^{2n}C_3 + {}^{2n}C_5 + \ldots + {}^{2n}C_{2n-1} = 2^{2n-1}$.

HCF of numbers divides each number and so it divides their sum also.

$\therefore$ HCF of $^{2n}C_1, {}^{2n}C_3, {}^{2n}C_5, \ldots, {}^{2n}C_{2n-1}$ also divides 2^{2n-1}.

So, the HCF $= 2^k$ where $0 \le k \le 2n-1$.

Now, every number n can be expressed as $2^m \cdot q$, q being an odd integer and m is an integer ≥ 0.

$\therefore$ $2n = {}^{2n}C_1 = 2 \cdot 2^m q = 2^{m+1} \cdot q$, where q is odd.

$\therefore$ HCF $= 2^k$ where $0 \le k \le m+1$.

Now $^{2n}C_p = \dfrac{p}{p} \cdot {}^{2n}C_p = \dfrac{1}{p} \cdot 2n \cdot {}^{2n-1}C_{p-1}$

$$(\because r \cdot {}^nC_r = n \cdot {}^{n-1}C_{r-1})$$

$$= 2n \cdot \left(\frac{1}{p}\,{}^{2n-1}C_{p-1}\right) = 2^{m+1}q \cdot \left(\frac{1}{p} \cdot {}^{2n-1}C_{p-1}\right)$$

$$= 2^{m+1} \cdot \frac{q \cdot {}^{2n-1}C_{p-1}}{p}.$$

For $p = 1, 3, 5, \ldots, 2n-1$; $^{2n}C_p$ is an integer. So $q \cdot {}^{2n-1}C_{p-1}$ is divisible by the integer p.

Let $\dfrac{q \cdot {}^{2n-1}C_{p-1}}{p} = Q_p =$ integer.

$\therefore$ $^{2n}C_p = 2^{m+1} \cdot 2^r \cdot M_p$, where $Q_p = 2^r M_p$ and M_p is an odd integer as well as $r \ge 0$.

$\therefore$ required HCF = HCF of $\{2^{m+1} \cdot q, 2^{m+1} \cdot 2^r \cdot M_p\}$

$$= 2^{m+1}, \text{ where } 2n = 2^{m+1} \cdot q, \text{ i.e., } n = 2^m \cdot q.$$

Exercises

1. If A be the sum of odd terms and B the sum of even terms in the expansion of $(x+a)^n$, prove that
(i) $A^2 - B^2 = (x^2 - a^2)^n$
(ii) $4AB = (x+a)^{2n} - (x-a)^{2n}$.

2. Evaluate $(\sqrt{2}+1)^6 + (\sqrt{2}-1)^6$.

3. Evaluate $(0.99)^5 + (1.01)^5$.

4. Find the remainder when $27^{10} + 7^{51}$ is divided by 10.

5. Prove that $32^{32^{32}}$ when divided by 7 leaves the remainder 4.

6. Prove that $(2n+1)^n \geq (2n-1)^n + (2n)^n$ for all $n \in N$.

7. Prove that $u_{n+1} > u_n$ where $u_n = \left(1 + \frac{1}{n}\right)^n$ for $n \in N$.

8. Find the 7th term of $\left(3x^2 - \frac{1}{3}\right)^{10}$.

9. Find the 7th term of $\left(\frac{4x}{5} - \frac{5}{2x}\right)^9$ when expanded in ascending powers of x.

10. Write the fourth term from the beginning and the fourth term from the end in the expansion of $\left(x + \frac{2}{x^2}\right)^9$.

11. Find the middle term or terms in the expansion of the following:

(i) $\left(\frac{2x}{3} - \frac{3}{2x}\right)^6$ (ii) $(1 - 2x + x^2)^n$

(iii) $\left(2x^2 - \frac{1}{x}\right)^7$ (iv) $\left(x + \frac{1}{x}\right)^{2n+1}$

(v) $\left(x - \frac{1}{x}\right)^n$, $n \in N$

12. Prove that the middle term in the expansion of $\left(x + \frac{1}{x}\right)^{2n}$ is $\frac{1 \cdot 3 \cdot 5 \cdot \ldots (2n-1)}{n!} \cdot 2^n$.

13. Find the second term after the middle term in the expansion of $\left(\frac{x}{2} + \frac{2}{x^2}\right)^{20}$.

14. Find the coefficient of (i) x^{-1} and (ii) x in the expansion of $\left(x + \frac{1}{x^2}\right)^8$.

15. Does the expansion of $\left(2x^2 - \frac{1}{x}\right)^{20}$ contain any term involving x^9? Give reasons.

16. Find the coefficient of x^{30} in $(x^3 + 3x^2 + 3x + 1)^{15}$.

17. Find the coefficient of $\frac{1}{x}$ in the expansion of $(1+x)^n \cdot \left(1 + \frac{1}{x}\right)^n$.

18. Find the coefficient of x^2 in the expansion of $(1 - x + 2x^2) \cdot \left(x + \frac{1}{x}\right)^{10}$.

19. Find the coefficient of $a^6 b^6$ in the expansion of $\left(a^2 - \frac{b}{a}\right)^{12}$.

20. Find the coefficient of $x^9 y^{-3}$ in the expansion of $\left(\frac{2x^2}{y} + \frac{y}{3x}\right)^{12}$.

21. Find the relation between a and b so that the coefficient of x^7 in $\left(ax^2 + \frac{1}{bx}\right)^{11}$ and that of x^{-7} in $\left(ax - \frac{1}{bx^2}\right)^{11}$ are equal.

22. If x^p occurs in the expansion of $\left(x^2 + \frac{1}{x}\right)^{2n}$, prove that its coefficient is $\dfrac{(2n)!}{\left\{\frac{1}{3}(4n-p)\right\}! \left\{\frac{1}{3}(2n+p)\right\}!}$.

23. If $\left(\sqrt{a} + \frac{1}{\sqrt{a+1}} + \frac{1}{\sqrt{a-1}}\right)^{2n}$
$= p_0 a^n + p_1 a^{n+1} + p_2 a^{n+2} + \ldots,$
where $a \geq 0$, find p_2.

24. Find the coefficient of x^r in the following:
$(x+2)^{n-1} + (x+2)^{n-2} \cdot (x+1)$
$+ (x+2)^{n-3} \cdot (x+1)^2 + \ldots + (x+1)^{n-1}$.

25. Find the expansion of $(1 + x + x^2)^4$ up to the term containing x^3. What is the coefficient of x^4?

26. If $n \in N$, prove that the coefficient of x^n in $(1 + x + x^2)^n$ is
$$1 + \frac{n(n-1)}{(1!)^2} + \frac{n(n-1)(n-2)(n-3)}{(2!)^2} + \ldots$$

27. Find the term independent of x in the expansion of the following:

(i) $\left(\sqrt{\frac{x}{3}} + \frac{\sqrt{3}}{2x^2}\right)^{10}$ (ii) $\left(2x^2 - \frac{3}{x^3}\right)^{25}$

(iii) $\left(x^2+\frac{1}{x^2}+2\right)^n$ (iv) $(1+x)^m \cdot \left(1+\frac{1}{x}\right)^n$

(v) $(2+3x-4x^2)\left(x+\frac{1}{x}\right)^{10}$

28. Determine the constant term in the expansion of $(1+x+x^{-2}+x^{-3})^{10}$.

29. Find the term independent of x in the expansion of $(x^{2/3}+4x^{1/3}+4)^5 \cdot \left(\frac{1}{x^{1/3}-1}+\frac{1}{x^{2/3}+x^{1/3}+1}\right)^{-9}$.

30. If the 4th term in the binomial expansion of $\left(px+\frac{1}{x}\right)^n$ is independent of x, find the value of n. Also calculate p if the fourth term is $\frac{5}{2}$.

31. Prove that there cannot be a constant term (independent of x) in $(x^a+x^{-b})^n$ unless na is an integral multiple of $a+b$.

32. Find the term which does not contain irrational expression in the expansion of $(\sqrt[5]{3}+\sqrt[7]{2})^{24}$.

33. If a, b, c and d be the 6th, the 7th, the 8th and the 9th terms respectively in a binomial expansion then prove that $\frac{b^2-ac}{c^2-bd}=\frac{4a}{3c}$.

34. If four consecutive terms in a binomial expansion with positive integral index be respectively 14, 84, 280 and 560 then find the expansion.

35. Determine the value of x in the expression $(x+x^{\log_{10} x})^5$ if the third term in the expansion is 10,00,000.

36. Prove that the coefficient of the middle term of $(1+x)^{2n}$ is equal to the sum of the coefficients of middle terms of $(1+x)^{2n-1}$.

37. In the expansion of $(1+x)^{43}$ the coefficient of the $(2r+1)$th and the $(r+2)$th terms are equal, find r.

38. Prove that in the expansion of $(1+x)^{2n}$, the coefficient of x^n is double the coefficient of x^n in the expansion of $\frac{(1+2x+x^2)^n}{1+x}$.

39. If in the expansion of $(1+x)^n$, the coefficients of the 5th, the 6th and the 7th terms are in AP, find n.

40. Prove that three consecutive binomial coefficients in the expansion of $(1+x)^n$, $n \in N$ cannot be in GP for any n. Is there any n such that they are in AP or HP?

41. Given positive integers $r>1, n>2$ and the coefficients of $3r$th term and $(r+2)$th term in the binomial expansion of $(1+x)^{2n}$ are equal, find r.

42. If in the expansion of $(1+x)^n$, the three consecutive coefficients are 165, 330 and 462 then find n.

43. If in the expansion of $(1+x)^n$, three consecutive coefficients be a, b, c then prove that $n=\frac{2ac+b(a+c)}{b^2-ac}$. Also show that a is the $\frac{a(b+c)}{b^2-ac}$ th coefficient in the expansion.

44. If a, b, c, d be four consecutive coefficients in the binomial expansion of $(1+x)^n$ then prove that

(i) $\frac{a+b}{a}, \frac{b+c}{b}, \frac{c+d}{c}$ are in HP

(ii) $\frac{a}{a+b}+\frac{c}{c+d}=\frac{2b}{b+c}$

45. If a, b, c, d be any four consecutive coefficients in a binomial expansion, show that

$$(bc+ad)(b-c)=2(ac^2-b^2d).$$

46. Let $(1+x^2)^2 \cdot (1+x)^n = \sum_{k=0}^{n+4} a_k x^k$. If a_1, a_2 and a_3 are in AP, find n.

47. If n be a positive integer then prove that the integral part P of $(5+2\sqrt{6})^n$ is an odd integer. If f be the fractional part of $(5+2\sqrt{6})^n$, prove that $P=\frac{1}{1-f}-f$.

48. If $(9+4\sqrt{5})^n=p+\beta$ where n and p are positive integers and β is a positive proper fraction, prove that $(1-\beta)(p+\beta)=1$.

49. If $((x))$ represents the least integer greater than x then prove that $((\{\sqrt{3}+1\}^{2n}))$, $n \in N$, is divisible by 2^{n+1}.

50. Find the greatest coefficient in the expansion of $(1+x)^{2n}$.

51. If $x=\frac{1}{3}$, find the greatest term in the expansion of $(1+4x)^8$.

52. (a) Find the numerically greatest term in the expansion of $(3-2x)^9$ when $x=1$.

(b) Find the value of the greatest term in the expansion of $\sqrt{3}\left(1+\frac{1}{\sqrt{3}}\right)^{20}$.

53. In the expansion of $(x+a)^{15}$, if the eleventh term is the geometric mean of the eighth and the twelfth terms, which term in the expansion is the greatest?

54. In the expansion of $\left(\frac{3}{2}+\frac{x}{3}\right)^n$ when $x=\frac{1}{2}$, it is known that the 6th term is the greatest term. Find the possible positive integral values of n.

55. Prove that the greatest coefficient in the expansion of $(1+x)^{2n}$ is double the greatest coefficient in the expansion of $(1+x)^{2n-1}$.

56. Find the sum:

$$^{2n+1}C_0 + {}^{2n+1}C_1 + {}^{2n+1}C_2 + \dots + {}^{2n+1}C_n.$$

57. Sum the series:

$$\frac{1}{1!(n-1)!} + \frac{1}{3!(n-3)!} + \frac{1}{5!(n-5)!} + \dots + \frac{1}{(n-1)!1!}$$

where n is an even positive integer.

58. Find the sum:

$$\frac{1}{2} \cdot {}^nC_0 + {}^nC_1 + 2 \cdot {}^nC_2 + 2^2 \cdot {}^nC_3 + \dots + 2^{n-1} \cdot {}^nC_n.$$

59. Prove that

$$(1 + {}^nC_1 + {}^nC_2 + {}^nC_3 + \dots + {}^nC_n)^2 = 1 + {}^{2n}C_1 + {}^{2n}C_2 + {}^{2n}C_3 + \dots + {}^{2n}C_{2n}.$$

60. Prove that ${}^nC_0 \cdot {}^nC_1 \cdot {}^nC_2 \cdot \dots {}^nC_n \le \left(\dfrac{2^n}{n+1}\right)^{n+1}$

61. If $t_0, t_1, t_2, \dots, t_n$ are the terms in the expansion of $(x+a)^n$ then prove that

$$(t_0 - t_2 + t_4 - \dots)^2 + (t_1 - t_3 + t_5 - \dots)^2 = (x^2 + a^2)^n.$$

62. If $\dfrac{(1+x-x^2)^{10}}{1+x^2} = a_0 + a_1x + a_2x^2 + \dots + a_nx^n + \dots$ then find

(i) $a_0 + a_1 + a_2 + \dots$ (ii) $a_0 - a_1 + a_2 - \dots$

(iii) $a_0 + a_2 + a_4 + \dots$ (iv) $a_1 + a_3 + a_5 + \dots$.

63. If $(1+x+x^2)^n = a_0 + a_1x + a_2x^2 + \dots + a_{2n}x^{2n}$ then find

(i) $a_0 + a_1 + a_2 + \dots + a_{2n}$

(ii) $a_0 - a_1 + a_2 - \dots + a_{2n}$

(iii) $a_0 + a_2 + a_4 + \dots + a_{2n}$

(iv) $a_1 + a_3 + a_5 + \dots + a_{2n-1}$

and prove that

(v) $a_0 + a_1 + a_2 + \dots + a_n = \dfrac{1}{2}(3^n + a_n)$.

64. The sum of the binomial coefficients in the expansion of $\left(x^2 + \dfrac{1}{x}\right)^n$ is 1024. Find the coefficient of x^{11} in the binomial expansion.

65. The exponent of a binomial exceeds that of another by 3. The sum of the binomial coefficients in expansions of both binomials taken together is 144. Find the smaller of the two exponents.

66. Find the coefficient of x^3 in the expansion of

$$1 + (1+x) + (1+x)^2 + (1+x)^3 + \dots + (1+x)^n.$$

67. Find the coefficient of x^{50} in the expression

$$(1+x)^{1000} + 2x(1+x)^{999} + 3x^2(1+x)^{998} + \dots + 1001\,x^{1000}$$

68. Sum the series: $\sum\limits_{r=m}^{n} {}^rC_m$.

69. Evaluate

$$({}^3C_3 + {}^4C_3 + {}^5C_3 + \dots + {}^nC_3) \times ({}^nC_3 + {}^nC_4 + {}^nC_5 + \dots + {}^nC_n).$$

70. Prove that

$$^nC_1 + {}^{n+1}C_2 + {}^{n+2}C_3 + \dots + {}^{n+m-1}C_m = {}^mC_1 + {}^{m+1}C_2 + {}^{m+2}C_3 + \dots + {}^{m+n-1}C_n.$$

71. Prove that ${}^{n+1}C_2 + 2 \cdot \sum\limits_{k=2}^{n} {}^kC_2 = \sum\limits_{k=1}^{n} k^2$.

72. If $(1+x)^n = C_0 + C_1x + C_2x^2 + \dots + C_nx^n$,

find the sum of the following series:

$$aC_1 + (a+d)C_2 + (a+2d)C_3 + \dots + (a+\overline{n-1}d)C_n.$$

73. Prove that

$$^nC_0 + 2 \cdot {}^nC_1 + 3 \cdot {}^nC_2 + \dots + (n+1)\,{}^nC_n = (n+2)2^{n-1}.$$

74. Prove that

$$^nC_0 + 3 \cdot {}^nC_1 + 5 \cdot {}^nC_2 + \dots + (2n+1)\,{}^nC_n = (n+1)2^n.$$

75. Prove that

$$^nC_0 - 2 \cdot {}^nC_1 + 3 \cdot {}^nC_2 - \dots + (-1)^n(n+1)\,{}^nC_n = 0.$$

76. Prove that

$$1 - (1+x)\,{}^nC_1 + (1+2x)\,{}^nC_2 - (1+3x)\,{}^nC_3 + \dots = 0.$$

77. If $s_n = {}^nC_1 + 2 \cdot {}^nC_2 + 3 \cdot {}^nC_3 + \dots + n \cdot {}^nC_n$

then find $\sum\limits_{n=1}^{n} s_n$. Also prove that

$$^nC_1 \cdot ({}^nC_2)^2 \cdot ({}^nC_3)^3 \dots ({}^nC_n)^n \le \left(\frac{2^n}{n+1}\right)^{{}^{n+1}C_2}$$

78. Find the sum:

$$1 \cdot {}^nC_0 + 2 \cdot {}^nC_2 + 3 \cdot {}^nC_4 + 4 \cdot {}^nC_6 + \dots .$$

79. Show that

$$^nC_0 \cdot m - {}^nC_1 \cdot (m-1) + {}^nC_2 \cdot (m-2) - \dots + (-1)^n \cdot {}^nC_n \cdot (m-n) = 0.$$

80. Sum the series:

$$3 \cdot {}^nC_0 - 8 \cdot {}^nC_1 + 13 \cdot {}^nC_2 - 18\,{}^nC_3 + \dots \text{ to } (n+1) \text{ terms.}$$

81. Evaluate $\sum_{r=1}^{n} \frac{p_r}{r} \cdot {}^nC_r$ where p_r denotes the sum of the first r natural numbers.

82. Prove by binomial expansion that

$$\sum_{k=1}^{n} k^2 \cdot {}^nC_k = n(n+1)2^{n-2}.$$

83. Prove that

$$ab - n(a-1)(b-1) + \frac{n(n-1)}{1 \cdot 2}(a-2)(b-2) - \dots + (-1)^n (a-n)(b-n) = 0, n \in N.$$

84. Prove that

$$1 + 3n + \frac{3 \cdot 4}{1 \cdot 2} \cdot \frac{n(n-1)}{2!} + \frac{4 \cdot 5}{1 \cdot 2} \frac{n(n-1)(n-2)}{3!} + \dots + \frac{(n+1)(n+2)}{1 \cdot 2} = (n^2 + 7n + 8) \cdot 2^{n-3},\ n \in N.$$

85. Evaluate $\sum_{r=0}^{n} (r+1)^2 \cdot {}^nC_r x^r$.

86. If $(1+x)^n = C_0 + C_1x + C_2x^2 + C_3x^3 + \dots + C_nx^n$ then prove that

(i) $C_0 + \frac{C_1}{2} + \frac{C_2}{3} + \frac{C_3}{4} + \dots + \frac{C_n}{n+1} = \frac{2^{n+1}-1}{n+1}$

(ii) $C_0 - \frac{1}{2}C_1 + \frac{1}{3}C_2 - \frac{1}{4}C_3 + \dots + (-1)^n \cdot \frac{C_n}{n+1} = \frac{1}{n+1}$.

87. (i) Prove that

$$3 \cdot {}^{10}C_0 + 3^2 \cdot \frac{{}^{10}C_1}{2} + 3^3 \cdot \frac{{}^{10}C_2}{3} + \dots + 3^{11} \cdot \frac{{}^{10}C_{10}}{11} = \frac{4^{11}-1}{11}.$$

(ii) Prove that

$$2 \cdot {}^nC_0 + 2^2 \cdot \frac{{}^nC_1}{2} + 2^3 \cdot \frac{{}^nC_2}{3} + \dots + 2^{n+1} \cdot \frac{{}^nC_n}{n+1} = \frac{3^{n+1}-1}{n+1}.$$

88. Find the sum:

(a) $\sum_{k=0}^{n} \frac{{}^nC_k}{(k+1)(k+2)}$ (b) $\sum_{r=0}^{n} (-1)^r \cdot \frac{{}^nC_r}{{}^{r+3}C_r}$

89. Find the sum:

(a) $\sum_{k=1}^{n} \frac{{}^nC_{2k-1}}{2k}$ (b) $\sum_{k=0}^{n} \frac{{}^nC_k}{k+2}$

90. If $(1+x)^n = \sum_{r=0}^{n} C_r x^r$, prove that

$$\sum_{r=0}^{n} \frac{C_r}{(r+1)2^{r+1}} = \frac{3^{n+1} - 2^{n+1}}{(n+1)2^{n+1}}.$$

91. Find $\sum_{r=0}^{n} \frac{r+1}{r+2} \cdot {}^nC_r x^r$.

92. Evaluate

$$\frac{{}^nC_0}{1} - \frac{{}^nC_1}{3} + \frac{{}^nC_2}{5} - \frac{{}^nC_3}{7} + \dots + (-1)^n \cdot \frac{{}^nC_n}{2n+1}.$$

93. Prove that ${}^nC_0^2 + {}^nC_1^2 + {}^nC_2^2 + \dots + {}^nC_n^2 = \frac{(2n)!}{n!\,n!}$.

94. Prove that $\sum_{k=0}^{n-2} {}^nC_k \cdot {}^nC_{k+2} = {}^{2n}C_{n+2}$.

95. Prove that ${}^nC_0^2 - {}^nC_1^2 + {}^nC_2^2 - \dots + (-1)^n \cdot {}^nC_n^2$

$= 0$ or $(-1)^{\frac{n}{2}} \cdot {}^nC_{\frac{n}{2}}$ according as n is odd or even.

96. Prove that

$${}^{2n+1}C_0^2 - {}^{2n+1}C_1^2 + {}^{2n+1}C_2^2 - \dots - {}^{2n+1}C_{2n+1}^2 = 0.$$

97. Prove that

$${}^{2n}C_0^2 - {}^{2n}C_1^2 + {}^{2n}C_2^2 - \dots + {}^{2n}C_{2n}^2 = (-1)^n \cdot \frac{(2n)!}{n!\,n!}.$$

98. Prove that

$${}^{2n}C_0^2 + {}^{2n}C_2^2 + {}^{2n}C_4^2 + \dots + {}^{2n}C_{2n}^2 = \frac{1}{2}\left[\frac{(4n)!}{\{(2n)!\}^2} + (-1)^n \cdot \frac{(2n)!}{(n!)^2}\right].$$

99. Find the sum of the products of binomial coefficients ${}^nC_0, {}^nC_1, {}^nC_2, \dots, {}^nC_n$ taken two at a time.

100. Prove that

$${}^{2n+1}C_0^2 + {}^{2n+1}C_2^2 + {}^{2n+1}C_4^2 + \dots + {}^{2n+1}C_{2n}^2 = {}^{2n+1}C_1^2 + {}^{2n+1}C_3^2 + {}^{2n+1}C_5^2 + \dots + {}^{2n+1}C_{2n+1}^2.$$

101. Find the sum

$${}^{20}C_{10} \cdot {}^{15}C_0 + {}^{20}C_9 \cdot {}^{15}C_1 + {}^{20}C_8 \cdot {}^{15}C_2 + \dots + {}^{20}C_0 \cdot {}^{15}C_{10}.$$

102. Prove that $\sum_{r=1}^{k} (-3)^{r-1} \cdot {}^{3n}C_{2r-1} = 0$

where $k = \frac{3n}{2}$ and n is an even integer.

103. Prove that

$$1 \cdot 2 \cdot {}^nC_r - 2 \cdot 3 \cdot {}^nC_{r-1} + 3 \cdot 4 \cdot {}^nC_{r-2} - \dots + (-1)^r (r+1)(r+2) = 2 \cdot {}^{n-3}C_r.$$

104. Show that

$${}^nC_1(1-x) - \frac{{}^nC_2}{2}(1-x)^2 + \frac{{}^nC_3}{3}(1-x)^3 - \dots + (-1)^{n-1} \cdot \frac{{}^nC_n}{n}(1-x)^n = \frac{1-x}{1} + \frac{1-x^2}{2} + \frac{1-x^3}{3} + \dots + \frac{1-x^n}{n}.$$

105. Prove that

$$\sum_{r=0}^{n} r^2 \cdot {}^nC_r\, p^r q^{n-r} = npq + n^2p^2 \text{ if } p+q=1.$$

106. Prove that

$${}^nC_1^2 + 2\cdot {}^nC_2^2 + 3\cdot {}^nC_3^2 + \dots + n\cdot {}^nC_n^2 = \frac{(2n-1)!}{\{(n-1)!\}^2}.$$

107. Prove that

$$\frac{^nC_1}{^nC_0} + 2\cdot\frac{^nC_2}{^nC_1} + 3\cdot\frac{^nC_3}{^nC_2} + \dots + n\cdot\frac{^nC_n}{^nC_{n-1}} = \frac{n(n+1)}{2}.$$

108. Prove that

$$({}^nC_0 + {}^nC_1)({}^nC_1 + {}^nC_2)({}^nC_2 + {}^nC_3)\dots({}^nC_{n-1} + {}^nC_n)$$
$$= \frac{(n+1)^n}{n!}\cdot {}^nC_1 \cdot {}^nC_2 \cdot {}^nC_3 \dots {}^nC_n.$$

109. Given

$$s_n = 1 + q + q^2 + \dots + q^n,$$
$$S_n = 1 + \frac{q+1}{2} + \left(\frac{q+1}{2}\right)^2 + \dots + \left(\frac{q+1}{2}\right)^n, q \neq -1.$$

Prove that

$${}^{n+1}C_1 + {}^{n+1}C_2\cdot s_1 + {}^{n+1}C_3\cdot s_2 + \dots + {}^{n+1}C_{n+1}\, s_n = 2^n \cdot S_n.$$

110. Prove that

$$\frac{^nC_0}{1^2} - \frac{^nC_1}{2^2} + \frac{^nC_2}{3^2} - \dots + (-1)^n\cdot\frac{^nC_n}{(n+1)^2} = \frac{1}{n+1}\left(1 + \frac{1}{2} + \frac{1}{3} + \dots + \frac{1}{n+1}\right).$$

111. Evaluate $\sum_{p=1}^{n}\left(\sum_{m=p}^{n} {}^nC_m \cdot {}^mC_p\right)$.

112. Evaluate ${}^{2n+1}C_0^2 + {}^{2n+1}C_1^2 + {}^{2n+1}C_2^2 + \dots + {}^{2n+1}C_n^2$.

113. Evaluate $\sum_{i\neq j} {}^{21}C_i \cdot {}^{21}C_j$ where $0 \le i \le 10, 0 \le j \le 10$.

114. If $(1 + x + x^2)^n = a_0 + a_1x + a_2x^2 + \dots + a_{2n}x^{2n}$

then prove that

(i) $a_0^2 - a_1^2 + a_2^2 - a_3^2 + \dots + a_{2n}^2 = a_n$

(ii) $a_0a_2 - a_1a_3 + a_2a_4 - \dots + a_{2n-2}\cdot a_{2n} = a_{n+1}$

(iii) $a_i = a_{2n-i}$.

115. If $(1 + x + x^2 + x^3)^n = a_0 + a_1x + a_2x^2 + \dots + a_{3n}x^{3n}$, prove that $a_r = a_{3n-r}$.

116. If $(1 + py + y^2)^n = a_0 + a_1y + a_2y^2 + \dots + a_{2n}y^{2n}$ then prove that $a_r = a_{2n-r}$.
Also prove that
$a_0 + 3a_1 + 5a_2 + \dots + (4n+1)a_{2n} = (2n+1)(2+p)^n$.

117. If $(1 + 2x + 2x^2)^n = a_0 + a_1x + a_2x^2 + \dots + a_{2n}x^{2n}$, $n \in N$ then prove that

(i) $a_r = {}^nC_0 \cdot {}^{2n}C_r + {}^nC_1 \cdot {}^{2n-2}C_{r-2} + {}^nC_2 \cdot {}^{2n-4}C_{r-4} + \dots$

(ii) $a_r = 2^{r-n}({}^{2n}C_r + {}^nC_1 \cdot {}^{2n-2}C_r + {}^nC_2 \cdot {}^{2n-4}C_r + \dots)$.

Objective Questions

Fill in the blanks.

118. The coefficient of a^4b^5 in the expansion of $(a+b)^9$ is ______ .

119. The coefficient in the third term of the expansion of $\left(x^2 - \frac{1}{4}\right)^n$ when expanded in descending powers of x is 31. Then n is equal to ______ .

120. The larger among $99^{50} + 100^{50}$ and 101^{50} is ______ .

121. (a) The sum of the coefficients in the polynomial expansion of $(1 + x - 3x^2)^{2163}$ is ______ .

(b) The sum of the numerical coefficients in the expansion of $(2x + 3y)^{10}$ is ______ .

122. If the 5th term in the expansion of $(a^{2/3} + a^{-1})^n$ does not contain a then the value of n is ______ .

123. The sum of the rational terms in the expansion of $(\sqrt{2} + 3^{\frac{1}{5}})^{10}$ is ______ .

124. The middle term in the expansion of $\left(x + \frac{1}{x}\right)^{10}$ is ______ .

125. Let n be a positive integer. If the coefficients of the 2nd, 3rd and 4th terms in the expansion of $(1+x)^n$ are in AP then the value of n is ______ .

126. If x^r occurs in the expansion of $\left(x + \frac{1}{x}\right)^n$ then its coefficient is ______ .

127. If the sum of the binomial coefficients in the expansion of $(a+b)^n$ is 4096 then the greatest binomial coefficient in the expansion is ______ .

128. The number of terms in the expansion of $(1 + x^{1/5})^{55}$ which are free from radicals is ______ .

129. If n is even then the coefficient of x in the expansion of $(1+x)^n \cdot \left(1 - \frac{1}{x}\right)^n$ is ______ .

130. The sum of ${}^{21}C_0 + {}^{21}C_1 + {}^{21}C_2 + \dots + {}^{21}C_{10}$ is equal to ______ .

131. The sum of ${}^{21}C_0 - {}^{21}C_1 + {}^{21}C_2 - \dots + {}^{21}C_{10}$ is equal to ______ .

132. Take the expansions

$(1+x)^n = {}^nC_0 + {}^nC_1x + {}^nC_2x^2 + \dots + {}^nC_nx^n$

$(y+1)^n = {}^nC_0y^n + {}^nC_1y^{n-1} + {}^nC_2y^{n-2} + \dots + {}^nC_n$

$$(x+y)^n = {}^nC_0x^n + {}^nC_1x^{n-1}y + {}^nC_2x^{n-2}y^2 + \ldots + {}^nC_ny^n.$$

The coefficient of x^ny^n in the expansion of $\{(1+x)(1+y)(x+y)\}^n$ is the series ______ .

Choose the correct option(s).

133. The number of terms in the expansion of $(1+2x+x^2)^n$ when expanded in descending powers of x, is

(a) $n+1$ (b) n (c) $2n$ (d) $2n+1$

134. The number of terms in the expansion of $(1+5\sqrt{2}x)^9 + (1-5\sqrt{2}x)^9$ is

(a) 5 (b) 7 (c) 9 (d) 10

135. In the expansion of $\left(x^3 - \frac{1}{x^2}\right)^{15}$, the constant term is

(a) ${}^{15}C_6$ (b) 0

(c) $-{}^{15}C_9$ (d) none of these

136. The largest coefficient in the expansion of $(1+x)^{24}$ is

(a) ${}^{24}C_{24}$ (b) ${}^{24}C_{13}$ (c) ${}^{24}C_{12}$ (d) ${}^{24}C_{11}$

137. 3^{51} when divided by 8 leaves the remainder

(a) 1 (b) 6 (c) 5 (d) 3

138. The sum of the series ${}^{20}C_0 + {}^{20}C_1 + {}^{20}C_2 + \ldots + {}^{20}C_{10}$ is

(a) 2^{20} (b) 2^{19}

(c) $2^{19} + \frac{1}{2}\cdot {}^{20}C_{10}$ (d) $2^{19} - \frac{1}{2}\cdot {}^{20}C_{10}$

139. The sum of the last eight coefficients in the expansion of $(1+x)^{16}$ is equal to

(a) 2^{15} (b) 2^{14}

(c) $2^{15} - \frac{1}{2}\cdot\frac{16!}{(8!)^2}$ (d) none of these

140. If $C_r = {}^nC_r$, the sum of the series

$$\frac{2\left[\left(\frac{n}{2}\right)!\right]^2}{n!}\{C_0^2 - 2C_1^2 + 3C_2^2 - \ldots + (-1)^n(n+1)C_n^2\}$$

where n is even integer, is

(a) 0 (b) $(-1)^{\frac{n}{2}}\cdot(n+1)$

(c) $(-1)^n\cdot(n+2)$ (d) $(-1)^n\cdot n$

(e) none of these

141. The coefficients of x^p and $x^q (p, q \in N)$ in the expansion of $(1+x)^{p+q}$ are

(a) equal

(b) equal but opposite in sign

(c) reciprocal to each other

(d) none of these

142. The number of dissimilar terms in the expansion of $(a+2b+3c)^8$ is

(a) 9 (b) 24 (c) 45 (d) 10

143. In the expansion of $(1+x)^{2m}\left(\frac{x}{1-x}\right)^{-2m}$, the term independent of x is

(a) ${}^{2m}C_m$ (b) ${}^{2m}C_0$

(c) $(-1)^m\cdot{}^{2m}C_m$ (d) none of these

State whether the statements are true or false.

144. The integral part of $(8+3\sqrt{7})^{20}$ is even.

145. In the expansion of $\left(\frac{x^2}{y}+\frac{y^2}{x}\right)^{15}$ there is a term independent of x and a term independent of y but no term independent of x and y both.

146. In the expansion of $(1+2x+x^2)^9$ there is exactly one term whose coefficient is not equal to coefficient of any other term.

147. In the expansion of $\left(x+\frac{1}{x}\right)^{13}$ every term is a function of x.

148. If $f(x) = \left(x+\frac{1}{x}\right)^{2n} + \left(x-\frac{1}{x}\right)^{2n}$ then $f(x)$ is a polynomial function which is an even function.

149. ${}^{16}C_0 - {}^{16}C_1 + {}^{16}C_2 - \ldots + {}^{16}C_8 = 0.$

Answers

2. 198 **3.** 2.0020001

4. 2 **8.** $\frac{70}{3}x^8$

9. $-\frac{84}{125}\cdot 2^9x^3$ **10.** 672, $\frac{5376}{x^9}$

11. (i) -20 (ii) $(-1)^n\cdot\frac{(2n)!}{n!\,n!}x^n$ (iii) $-560x^5, 280x^2$

(iv) $\frac{(2n+1)!}{n!\,(n+1)!}x$, $\frac{(2n+1)!}{(n+1)!\,n!}\cdot\frac{1}{x}$

(v) $(-1)^{\frac{n}{2}} \cdot \dfrac{n!}{\left(\frac{n}{2}\right)!\left(\frac{n}{2}\right)!}$ if n is even,

$(-1)^{\frac{n+1}{2}} \dfrac{n!}{\left\{\frac{1}{2}(n+1)\right\}!\left\{\frac{1}{2}(n-1)\right\}!} \cdot \dfrac{1}{x}$ and

$(-1)^{\frac{n-1}{2}} \dfrac{n!}{\left\{\frac{1}{2}(n-1)\right\}!\left\{\frac{1}{2}(n+1)\right\}!} x$ if n is odd

13. $\dfrac{20!}{12!\,8!} 2^4 \cdot \dfrac{1}{x^{16}}$ **14.** (i) 56 (ii) 0

15. no **16.** $\dfrac{45!}{30!\,15!}$

17. $\dfrac{(2n)!}{(n-1)!\,(n+1)!}$ **18.** 714

19. 924 **20.** 0

21. $ab = 1$ **23.** $8n^2$

24. $(2^{n-r} - 1)\, {}^nC_r$

25. $1 + 4x + 10x^2 + 16x^3 + \dots$; 19

27. (i) $\dfrac{5}{12}$ (ii) $\dfrac{25!}{10!\,15!} 2^{15} \cdot 3^{10}$

(iii) $\dfrac{(2n)!}{n!\,n!}$ (iv) $\dfrac{(m+n)!}{m!\,n!}$ (v) −336

28. 11851 **29.** 168

30. $n = 6, p = \dfrac{1}{2}$ **32.** $\dfrac{24!}{14!\,10!} 3^2 \cdot 2^2$

34. 1 + 14 + 84 + 280 + 560 + 672 + 448 + 128, i.e., $(1+2)^7$

35. $10, 10^{-\frac{5}{2}}$ **37.** 14 **39.** 7 or 14

40. possible to be in AP but not in HP

41. $\dfrac{1}{2} n$ provided n is even and $n > 2$ **42.** 11

46. 2, 3, 4 **50.** ${}^{2n}C_n$ **51.** 6th term, i.e., $\dfrac{57344}{243}$

52. (a) 4th and 5th, i.e., 489888 (b) $\dfrac{20!}{27 \times 7!\,13!}$

53. 8th term **54.** $n = 49, 50, 51, \dots, 59$

56. 2^{2n} **57.** $\dfrac{2^{n-1}}{n!}$ **58.** $\dfrac{1}{2} \cdot 3^n$

62. (i) $\dfrac{1}{2}$ (ii) $\dfrac{1}{2}$ (iii) $\dfrac{1}{2}$ (iv) 0

63. (i) 3^n (ii) 1 (iii) $\dfrac{1+3^n}{2}$ (iv) $\dfrac{3^n - 1}{2}$

64. 120 **65.** 4 **66.** $\dfrac{(n+1)!}{(n-3)!\,4!}$

67. $\dfrac{1002!}{50!\,952!}$ **68.** ${}^{n+1}C_{m+1}$

69. ${}^{n+1}C_4 \cdot [2^n - \frac{1}{2} n(n-1) - n - 1]$

72. $(2^n - 1)a + (n \cdot 2^{n-1} - 2^n + 1)d$

77. $(n-1)2^n + 1$ **78.** $(n+4)2^{n-3}$

80. 0 **81.** $(n+2)2^{n-2} - \dfrac{1}{2}$

85. $\{(n^2 + 2n + 1)x^2 + (3n+2)x + 1\}(1+x)^{n-2}$

88. (a) $\dfrac{2^{n+2} - n - 3}{(n+1)(n+2)}$ (b) $\dfrac{3}{n+3}$

89. (a) $\dfrac{2^n - 1}{n+1}$ (b) $\dfrac{n \cdot 2^{n+1} + 1}{(n+1)(n+2)}$

91. $\dfrac{(1+x)^{n+2} - (n+2)x - 1}{(n+1)(n+2)x^2} - \dfrac{(1+x)^{n+1} - 1}{(n+1)x} + (1+x)^n$

92. $\dfrac{2^{2n} \cdot (n!)^2}{(2n+1)!}$ **99.** $\dfrac{1}{2}\left\{2^{2n} - \dfrac{(2n)!}{n!\,n!}\right\}$

101. $\dfrac{35!}{10!\,25!}$ **111.** $3^n - 2^n$

112. $\dfrac{1}{2} \cdot \dfrac{(4n+2)!}{\{(2n+1)!\}^2}$ **113.** $\dfrac{1}{2}\left\{2^{40} - \dfrac{1}{2} \cdot \dfrac{42!}{(21!)^2}\right\}$

118. 126 **119.** 32 **120.** 101^{50}

121. (a) −1 (b) 5^{10} **122.** 10 **123.** 41

124. $\dfrac{10!}{(5!)^2}$ **125.** 7

126. $\dfrac{n!}{\left\{\frac{1}{2}(n-r)\right\}!\left\{\frac{1}{2}(n+r)\right\}!}$

127. 924 **128.** 12 **129.** 0 **130.** 2^{20}

131. $\dfrac{20!}{(10!)^2}$ **132.** ${}^nC_0^3 + {}^nC_1^3 + {}^nC_2^3 + \dots + {}^nC_n^3$

133. (d) **134.** (a) **135.** (c) **136.** (c)
137. (d) **138.** (c) **139.** (c) **140.** (e)
141. (a) **142.** (c) **143.** (c) **144.** false
145. true **146.** true **147.** true **148.** true
149. false

Chapter Test

Time: 75 minutes

1. In the following, each statement is incomplete. Fill in the blank so that each resulting statement becomes correct.

(i) If the sum of binomial coefficients in the expansion of $(x+y)^n$ is 1024 then the greatest binomial coefficient occurs in the ______ th term.

(ii) The coefficient of x^3 in the expansion of $(2-x+3x^2)^5$ is ______ .

(iii) If $n < (\sqrt{2}+1)^6 < n+1$, $n \in N$ then the value of n is ______ .

2. In the following, each question has one or more than one correct answers. Indicate the correct answer(s).

(i) The number of rational terms in the expansion of $(2^{1/5}+3^{1/10})^{45}$ is

A. 4 B. 10 C. 5 D. 11

(ii) If the coefficient of x^k, $0 \le k \le 15$ in the expansion of the expression $1+(1+x)+(1+x)^2+\ldots+(1+x)^{15}$ is the greatest then the value of k is

A. 9 B. 8 C. 7 D. none of these

(iii) If the coefficients of x^7 and x^4 are equal in magnitude but opposite in sign in the expansion of $\left(\frac{x^2}{a}-\frac{b}{x}\right)^{11}$ then

A. $ab=1$ B. $a=b$ C. $a+b=0$ D. none of these

(iv) The value of

$${}^nC_1\, x(1-x)^{n-1} + 2\cdot {}^nC_2\, x^2(1-x)^{n-2} + 3\cdot {}^nC_3\, x^3(1-x)^{n-3} + \ldots + n\cdot {}^nC_n\, x^n \text{ is}$$

A. n B. x C. nx D. none of these

3. Use binomial expansion to prove that the number of ways to select r things from two bags, one containing $m(m \ge r)$ things and the other $n(n \ge r)$ things, all the things being different, is equal to the number of ways to select r things from a bag when all the $m+n$ things are kept in a bag.

4. Evaluate $\sum_{r=1}^{n} \frac{s_r}{2r+1}\cdot {}^nC_r$, where s_r is the sum of the squares of the first r natural numbers.

5. Evaluate $\sum_{i=0}^{n}\sum_{j=1}^{n} {}^nC_j \cdot {}^jC_i$, $i \le j$.

6. If the coefficient of x^{2r} is greater than half of the coefficient of x^{2r+1} in the expansion of $(1+x)^{15}$ then find the possible values of r.

7. Prove that $\sum_{r=1}^{n} (-1)^{r-1}\cdot {}^nC_r(a-r)^2 = a^2$, $n \ge 3$.

8. If $(6-12x+12x^2)^n = \sum_{r=0}^{2n} a_r x^r$, prove that

$$a_r = (-1)^r\cdot 3^n\cdot 2^r\left\{{}^{2n}C_r + {}^nC_1\cdot {}^{2n-2}C_r + {}^nC_2\cdot {}^{2n-4}C_r + \ldots\right\}.$$

9. Sum the series: ${}^{6n}C_0 + {}^{6n}C_3 + {}^{6n}C_6 + \ldots + {}^{6n}C_{3n}$.

□

Answers

1. (i) 6 (ii) –520 (iii) 197

2. (i) C (ii) C (iii) A (iv) C

4. $n(n+3)2^{n-2}$

5. $3^n - 1$

6. 3, 4, 5, 6, 7

9. $\frac{1}{2} \cdot \frac{(6n)!}{\{(3n)!\}^2} + \frac{1}{3}(2^{6n-1} + 1)$

□

7. Principle of Mathematical Induction (PMI)

Recap of Facts and Formulae

1. Statement and predicate

- A sentence which is either definitely true or definitely false is called a statement. "Snow is white" is a statement but "Ram is a good boy" is not a statement.
- Some sentences depend on a variable for its truth value (i.e., true or false). "$1 + 2 + 3 + \ldots + n = 2n - 1$" is a mathematical sentence which is true for $n = 1, 2$ but false for $n = 3, 4$, etc. As the sentence is definitely true or definitely false for a particular positive integral value of n, the sentence is a statement and it depends on $n \in N$ for truth value. Such statements are called predicates and symbolized as $P(n)$.

2. Peano's axiom (Principle of mathematical induction)

- A statement $P(n)$ is true for all $n \in N$ if

  (i) $P(1)$ is true

  (ii) $P(m)$ is true $\Rightarrow$ $P(m + 1)$ is true.

3. Verification of truth of $P(n)$

PMI is a principle which can be used to verify whether a mathematical statement $P(n)$ is true for all $n \in N$ in any branch of mathematics. For this take the following steps:

- Put $n = 1$ on one side of the statement and then simplify it to take the shape of the other side where $n = 1$.
- Then assume $P(m)$ to be true. Use the mathematical result obtained by putting $n = m$ in the statement to establish $P(m + 1)$ is true.

4. Use of substitution in PMI

PMI may be used to prove whether a statement $P(n)$ is true for a particular infinite sequence of values of n.

- If the truth of $P(n)$ is to be established for *all positive even integral values* of n then take $\phi(n) = P(2n)$ and use PMI to prove that $\phi(n)$ is true for all $n \in N$.
- If the truth of $P(n)$ is to be established for *all positive odd integral values* of n then take $\phi(n) = P(2n - 1)$ and use PMI to prove that $\phi(n)$ is true for all $n \in N$.
- If the truth of $P(n)$ is to be established for *all* $n > k, n \in N, k \in N$ then take $\phi(n) = P(n + k)$ and use PMI to prove that $\phi(n)$ is true for all $n \in N$.
- Similarly use

  $\phi(n) = P(-n)$ for all negative integral values

  $\phi(n) = P\left(\frac{1}{n}\right)$ for all fractional values of the form $\frac{1}{r}$, $r \in N$

  $\phi(n) = P(3m)$ for all positive multiples of 3, etc.

5. Alternative forms of PMI

- $P(n)$ is true for all $n \in N$ if

  (i) $P(1)$ and $P(2)$ are true

  (ii) $P(m)$ and $P(m + 1)$ are true $\Rightarrow$ $P(m + 2)$ is true.
- $P(n)$ is true for all $n \in N$ if

  (i) $P(1)$, $P(2)$ and $P(3)$ are true

  (ii) $P(m)$, $P(m + 1)$ and $P(m + 2)$ are true $\Rightarrow$ $P(m + 3)$ is true.
- $P(n)$ is true for all $n \geq k$, $n \in N$ and k is a fixed positive integer, if

  (i) $P(k)$ is true

  (ii) $P(m)$ is true $\Rightarrow$ $P(m + 1)$ is true.
- $P(n)$ is true for all $n \in N$ if

  (i) $P(1), P(2), \ldots, P(k)$ are true

  (ii) $P(1), P(2), \ldots, P(m)$ are true $\Rightarrow$ $P(m + 1)$ is true.

6. Use of PMI in statements $P(m, n)$

Some mathematical statements are predicates of two arguments (i.e., truth value depending on two variables $m \in N, n \in N$). The method of establishing the truth of

$P(m, n)$ for all $m \in N$, $n \in N$ is as follows:

- Keep m fixed and treat the statement $P(m, n)$ as $\phi(n)$. Establish the truth of $\phi(n)$ for all $n \in N$ by using PMI.

Next keep n fixed and treat the statement $P(m, n)$ as $\psi(m)$.

Establish the truth of $\psi(n)$ for all $n \in N$ by using PMI.

Selected Solved Examples

1. Prove that

$$\frac{1}{1+x}+\frac{2}{1+x^2}+\frac{4}{1+x^4}+\ldots+\frac{2^n}{1+x^{2^n}}=\frac{1}{x-1}+\frac{2^{n+1}}{1-x^{2^{n+1}}}$$

where $x \neq \pm 1$ and $n \in N$.

Let $P(n)$: $\dfrac{1}{1+x}+\dfrac{2}{1+x^2}+\dfrac{4}{1+x^4}+\ldots+\dfrac{2^n}{1+x^{2^n}}$

$$=\frac{1}{x-1}+\frac{2^{n+1}}{1-x^{2^{n+1}}}.$$

When $n = 1$,

$$\frac{1}{1+x}+\frac{2}{1+x^2}=\frac{1}{x-1}+\left(\frac{1}{x+1}-\frac{1}{x-1}\right)+\frac{2}{1+x^2}$$

$$=\frac{1}{x-1}+\frac{x-1-x-1}{x^2-1}+\frac{2}{1+x^2}$$

$$=\frac{1}{x-1}+\frac{2}{1-x^2}+\frac{2}{1+x^2}$$

$$=\frac{1}{x-1}+\frac{2(1+x^2)+2(1-x^2)}{1-x^4}$$

$$=\frac{1}{x-1}+\frac{2^2}{1-x^{2^2}}$$

$\therefore$ $P(1)$ is true.

Let $P(m)$ be true, i.e.,

$$\frac{1}{1+x}+\frac{2}{1+x^2}+\frac{4}{1+x^4}+\ldots+\frac{2^m}{1+x^{2^m}}$$

$$=\frac{1}{x-1}+\frac{2^{m+1}}{1-x^{2^{m+1}}} \quad \ldots (1)$$

Now,

$$\frac{1}{1+x}+\frac{2}{1+x^2}+\frac{4}{1+x^4}+\ldots$$

$$+\frac{2^m}{1+x^{2^m}}+\frac{2^{m+1}}{1+x^{2^{m+1}}}$$

$$=\frac{1}{x-1}+\frac{2^{m+1}}{1-x^{2^{m+1}}}+\frac{2^{m+1}}{1+x^{2^{m+1}}}, \text{ using (1)}$$

$$=\frac{1}{x-1}+2^{m+1}\cdot\frac{1+x^{2^{m+1}}+1-x^{2^{m+1}}}{1-x^{2^{m+2}}}$$

$$=\frac{1}{x-1}+\frac{2^{m+2}}{1-x^{2^{m+2}}}$$

$\therefore$ $P(m+1)$ is true;

thus $P(m)$ is true $\Rightarrow$ $P(m+1)$ is true.

$\therefore$ by PMI, $P(n)$ is true for all $n \in N$.

Hence, the problem.

Note

(i) $x \neq \pm 1$ is only to ensure that no term of the series is infinite.

(ii) If the series on the left is asked to be summed (without the right-hand side), the PMI cannot be used.

2. Prove that, for all $n \in N$,

$$\left(1-\frac{1}{2^2}\right)\left(1-\frac{1}{3^2}\right)\left(1-\frac{1}{4^2}\right)\ldots\left\{1-\frac{1}{(n+1)^2}\right\}=\frac{n+2}{2n+2}.$$

Let $P(n)$: $\left(1-\dfrac{1}{2^2}\right)\left(1-\dfrac{1}{3^2}\right)\left(1-\dfrac{1}{4^2}\right)\ldots\left\{1-\dfrac{1}{(n+1)^2}\right\}$

$$=\frac{n+2}{2n+2}.$$

When $n=1$, $1-\dfrac{1}{2^2}=1-\dfrac{1}{4}=\dfrac{3}{4}=\dfrac{1+2}{2\cdot 1+2}$

$\therefore$ $P(1)$ is true.

Let $P(m)$ be true, i.e.,

$$\left(1-\frac{1}{2^2}\right)\left(1-\frac{1}{3^2}\right)\ldots\left\{1-\frac{1}{(m+1)^2}\right\}=\frac{m+2}{2m+2} \quad \ldots (1)$$

Now, $\left(1-\dfrac{1}{2^2}\right)\left(1-\dfrac{1}{3^2}\right)\ldots\left\{1-\dfrac{1}{(m+2)^2}\right\}$

$$=\left(1-\frac{1}{2^2}\right)\left(1-\frac{1}{3^2}\right)\ldots\left\{1-\frac{1}{(m+1)^2}\right\}\left\{1-\frac{1}{(m+2)^2}\right\}$$

$$=\frac{m+2}{2m+2}\cdot\left\{1-\frac{1}{(m+2)^2}\right\}, \text{ using (1)}$$

$$= \frac{m+2}{2m+2} \cdot \frac{(m+2)^2 - 1}{(m+2)^2} = \frac{(m+3)(m+1)}{2(m+1)(m+2)}$$

$$= \frac{m+3}{2m+4} = \frac{(m+1)+2}{2(m+1)+2}$$

$\therefore$ $P(m+1)$ is true.

$\therefore$ by PMI, $P(n)$ is true for all $n \in N$.

Hence, the problem.

3. If $S_n = 1 + \frac{1}{2} + \frac{1}{3} + \ldots + \frac{1}{n}$, $n \in N$ then prove that

$$S_1 + S_2 + S_3 + \ldots + S_n = (n+1)S_n - n.$$

Let $P(n)$: $S_1 + S_2 + S_3 + \ldots + S_n = (n+1)S_n - n$.

When $n = 1$, $S_1 = 1$ $\left\{\because S_n = 1 + \frac{1}{2} + \frac{1}{3} + \ldots + \frac{1}{n}\right\}$

$\therefore$ $S_1 = 1 = (1+1) \cdot 1 - 1 = (1+1)S_1 - 1$

$\therefore$ $P(1)$ is true.

Let $P(m)$ be true, i.e.,

$$S_1 + S_2 + S_3 + \ldots + S_m = (m+1)S_m - m \quad \ldots (1)$$

Now, $S_1 + S_2 + S_3 + \ldots + S_m + S_{m+1}$

$= (m+1)S_m - m + S_{m+1}$, using (1)

$$= (m+1)\left\{1 + \frac{1}{2} + \frac{1}{3} + \ldots + \frac{1}{m}\right\} - m + \left\{1 + \frac{1}{2} + \frac{1}{3} + \ldots + \frac{1}{m+1}\right\}$$

$$= (m+2)\left\{1 + \frac{1}{2} + \frac{1}{3} + \ldots + \frac{1}{m}\right\} - \left\{1 + \frac{1}{2} + \frac{1}{3} + \ldots + \frac{1}{m}\right\} - m + \left\{1 + \frac{1}{2} + \frac{1}{3} + \ldots + \frac{1}{m+1}\right\}$$

$$= (m+2)\left\{1 + \frac{1}{2} + \frac{1}{3} + \ldots + \frac{1}{m}\right\} + \frac{1}{m+1} - m$$

$$= (m+2)\left\{1 + \frac{1}{2} + \frac{1}{3} + \ldots + \frac{1}{m+1}\right\} - \frac{m+2}{m+1} + \frac{1}{m+1} - m$$

$$= (\overline{m+1} + 1)S_{m+1} + \frac{-(m+2)+1}{m+1} - m$$

$$= (\overline{m+1} + 1)S_{m+1} - (m+1)$$

$\therefore$ $P(m+1)$ is true;

thus $P(m)$ is true $\Rightarrow$ $P(m+1)$ is true.

$\therefore$ by PMI, $P(n)$ is true for all $n \in N$.

Hence, the problem.

4. Prove that $10^n + 3 \cdot 4^{n+2} + 5$ is divisible by 9 for all $n \in N$.

(or, Prove that $9 \mid (10^n + 3 \cdot 4^{n+2} + 5)$ for all $n \in N$.)

Let $P(n)$: $10^n + 3 \cdot 4^{n+2} + 5 = 9k$,

where k is an integer.

When $n = 1$,

$$10^1 + 3 \cdot 4^{1+2} + 5$$

$$= 10 + 3 \cdot 4^3 + 5 = 10 + 3 \cdot 64 + 5$$

$$= 10 + 192 + 5 = 207 = 9 \times 23$$

$\therefore$ $P(1)$ is true.

Let $P(m)$ be true, i.e., $10^m + 3 \cdot 4^{m+2} + 5 = 9k_1$ ... (1)

where k_1 is an integer

Now, $10^{m+1} + 3 \cdot 4^{m+1+2} + 5$

$$= 10 \cdot 10^m + 3 \cdot 4^{m+3} + 5$$

$$= 10(9k_1 - 3 \cdot 4^{m+2} - 5) + 3 \cdot 4^{m+3} + 5, \text{ using (i)}$$

$$= 90k_1 - 30 \cdot 4^{m+2} - 50 + 3 \cdot 4^{m+3} + 5$$

$$= 90k_1 - 30 \cdot 4^{m+2} + 12 \cdot 4^{m+2} - 45$$

$$= 90k_1 - 18 \cdot 4^{m+2} - 45$$

$$= 9\{10k_1 - 2 \cdot 4^{m+2} - 5\}$$

$$= 9 \times \text{integer}$$

$\therefore$ $P(m+1)$ is true;

thus $P(m)$ is true $\Rightarrow$ $P(m+1)$ is true.

$\therefore$ by PMI, $P(n)$ is true for all $n \in N$.

Hence, the problem.

5. Prove that $n^7 - 7n^5 + 14n^3 - 8n$ is divisible by 840 for all $n \in N$.

Here $n^7 - 7n^5 + 14n^3 - 8n$

$$= n(n^6 - 7n^4 + 14n^2 - 8)$$

$$= n(n-1)(n^5 + n^4 - 6n^3 - 6n^2 + 8n + 8)$$

$\{\because 1 - 7 + 14 - 8 = 0, n - 1 \text{ is a factor}\}$

$$= n(n-1)\{n^4(n+1) - 6n^2(n+1) + 8(n+1)\}$$

$$= n(n-1)(n+1)(n^4 - 6n^2 + 8)$$

$$= n(n-1)(n+1)(n^2 - 4)(n^2 - 2)$$

$$= n(n-1)(n+1)(n-2)(n+2)(n^2 - 2)$$

$$= (n-2)(n-1)\,n(n+1)(n+2)(n^2 - 2)$$

$$= \frac{(n+2)!}{(n-3)!\,5!} \cdot 5! \cdot (n^2 - 2)$$

$$= 5! \cdot (n^2 - 2) \cdot {}^{n+2}C_5 = 120 \times \text{integer} \times \text{integer}.$$

Now $840 = 7 \times 120$ and there is no factor common between 7 and 120.

$\therefore$ if we can prove that $n^7 - 7n^5 + 14n^3 - 8n$ is divisible by 7 our problem will be solved.

Let $P(n)$: $n^7 - 7n^5 + 14n^3 - 8n = 7k$,

where k is an integer.

When $n = 1$,

$$1^7 - 7 \cdot 1^5 + 14 \cdot 1^3 - 8 \cdot 1$$
$$= 1 - 7 + 14 - 8 = 0 = 7 \times 0$$

$\therefore$ $P(1)$ is true.

Let $P(m)$ be true, i.e., $m^7 - 7m^5 + 14m^3 - 8m = 7k$ ... (1)

where k_1 is an integer.

Now, $(m+1)^7 - 7(m+1)^5 + 14(m+1)^3 - 8(m+1)$

$$= (m^7 + {}^7C_1 m^6 + {}^7C_2 m^5 + \ldots + {}^7C_7)$$
$$- 7(m^5 + {}^5C_1 m^4 + {}^5C_2 m^3 + \ldots + {}^5C_5)$$
$$+ 14(m^3 + 3m^2 + 3m + 1) - 8m - 8$$
$$= (m^7 - 7m^5 + 14m^3 - 8m)$$
$$+ ({}^7C_1 m^6 + {}^7C_2 m^5 + \ldots + {}^7C_7)$$
$$- 7({}^5C_1 m^4 + {}^5C_2 m^3 + \ldots + {}^5C_5)$$
$$+ 14(3m^2 + 3m + 1) - 8$$
$$= 7k_1 + 7m^6 + 21m^5 + ({}^7C_3 - 7 \cdot {}^5C_1)m^4$$
$$+ ({}^7C_4 - 7 \cdot {}^5C_2)m^3 + ({}^7C_5 - 7 \cdot {}^5C_3 + 42)m^2$$
$$+ ({}^7C_6 - 7 \cdot {}^5C_4 + 42)m$$
$$+ ({}^7C_7 - 7 \cdot {}^5C_5 + 14 - 8), \text{ using (1)}$$
$$= 7k_1 + 7m^6 + 21m^5 - 35m^3 - 7m^2 + 14m$$
$$= 7\{k_1 + m^6 + 3m^5 - 5m^3 - m^2 + 2m\} = 7k_2$$

$\therefore$ $P(m+1)$ is true;

thus, $P(m)$ is true $\Rightarrow$ $P(m+1)$ is true.

$\therefore$ by PMI, $P(n)$ is true for all $n \in N$.

$\therefore$ $n^7 - 7n^5 + 14n^3 - 8n$ is divisible by 7 for all $n \in N$.

$\therefore$ $n^7 - 7n^5 + 14n^3 - 8n$ is divisible by 120 as well as 7, i.e., divisible by 840.

6. Prove by induction that $3^{2^n} - 1$ is divisible by 2^{n+2} but not by 2^{n+3} where $n \in N$.

Let $P(n)$: $3^{2^n} - 1 = 2^{n+2} \cdot k$, where k is an odd integer.

When $n = 1$, $3^{2^1} - 1 = 3^2 - 1 = 8 = 2^{1+2} \cdot 1$

$\therefore$ $P(1)$ is true.

Let $P(m)$ be true, i.e., $3^{2^m} - 1 = 2^{m+2} \cdot k_1$ ... (1)

where k_1 is an odd integer.

Now $3^{2^{m+1}} - 1 = 3^{2^m \cdot 2} - 1 = (3^{2^m})^2 - 1$

$$= (1 + k_1 \cdot 2^{m+2})^2 - 1$$
$$= 1 + k_1 \cdot 2 \cdot 2^{m+2} + (k_1 \cdot 2^{m+2})^2 - 1$$
$$= k_1 \cdot 2^{m+3} + k_1^2 \cdot 2^{2m+4}$$
$$= 2^{m+3}(k_1 + k_1^2 \cdot 2^{m+1})$$
$$= 2^{m+3} \cdot k_1(1 + k_1 \cdot 2^{m+1})$$
$$= k_2 \cdot 2^{(m+1)+2}, \text{ where } k_2 \text{ is an odd integer}$$

because $k_2 = k_1 \cdot (1 + k_1 \cdot 2^{m+1})$,

where k_1 is odd integer

= odd integer × (1 + even integer)

= odd integer × odd integer

= odd integer.

$\therefore$ $P(m+1)$ is true;

thus, $P(m)$ is true $\Rightarrow$ $P(m+1)$ is true.

$\therefore$ by PMI, $P(n)$ is true for all $n \in N$

$\therefore$ $3^{2^n} - 1 = 2^{n+2} \cdot k$, where k is an odd integer

$\therefore$ $3^{2^n} - 1$ is divisible by

2^{n+2} but not by $2^{n+2} \cdot 2$, i.e., 2^{n+3}.

7. Prove that $\dfrac{n^{11}}{11} + \dfrac{n^5}{5} + \dfrac{n^3}{3} + \dfrac{62n}{165}$ is a positive integer for all $n \in N$.

Let $P(n)$: $\dfrac{n^{11}}{11} + \dfrac{n^5}{5} + \dfrac{n^3}{3} + \dfrac{62}{165} n = k$,

where k is a positive integer.

When $n = 1$,

$$\frac{1}{11} + \frac{1}{5} + \frac{1}{3} + \frac{62}{165}$$
$$= \frac{15 + 33 + 55 + 62}{165} = \frac{165}{165} = 1.$$

$\therefore$ $P(1)$ is true.

Let $P(m)$ be true, i.e.,

$$\frac{m^{11}}{11} + \frac{m^5}{5} + \frac{m^3}{3} + \frac{62}{165} m = k_1 \quad \ldots (1)$$

where k_1 is a positive integer.

Now, $\dfrac{(m+1)^{11}}{11} + \dfrac{(m+1)^5}{5} + \dfrac{(m+1)^3}{3} + \dfrac{62}{165}(m+1)$

$$= \frac{1}{11}\{m^{11} + {}^{11}C_1 m^{10} + {}^{11}C_2 m^9 + \ldots + {}^{11}C_{11}\}$$

$$+ \frac{1}{5}\{m^5 + {}^5C_1 m^4 + {}^5C_2 m^3 + \ldots + {}^5C_5\}$$

$$+ \frac{1}{3}\{m^3 + 3m^2 + 3m + 1\} + \frac{62}{165}(m+1)$$

$$= \frac{m^{11}}{11} + \frac{m^5}{5} + \frac{m^3}{3} + \frac{62}{165}m$$

$$+ \frac{1}{11}\{{}^{11}C_1 m^{10} + {}^{11}C_2 m^9 + \ldots + {}^{11}C_{10} m + 1\}$$

$$+ \frac{1}{5}\{{}^5C_1 m^4 + {}^5C_2 m^3 + {}^5C_3 m^2 + {}^5C_4 m + 1\}$$

$$+ \frac{1}{3}(3m^2 + 3m + 1) + \frac{62}{165}$$

$$= k_1 + \frac{1}{11}\{11m^{10} + 11\cdot 5m^9 + 11\cdot 15m^8 + 11\cdot 30m^7$$

$$+ 11\cdot 42m^6 + 11\cdot 42m^5 + 11\cdot 30m^4$$

$$+ 11\cdot 15m^3 + 11\cdot 5m^2 + 11m + 1\}$$

$$+ \frac{1}{5}\{5m^4 + 5\cdot 2m^3 + 5\cdot 2m^2 + 5m + 1\}$$

$$+ \frac{1}{3}(3m^2 + 3m + 1) + \frac{62}{165}, \quad \text{using (1)}$$

$$= k_1 + (+\text{ve})\text{ integer} + \frac{1}{11} + (+\text{ve})\text{ integer}$$

$$+ \frac{1}{5} + (+\text{ve})\text{ integer} + \frac{1}{3} + \frac{62}{165}$$

$$= \text{positive integer} + \left(\frac{1}{11} + \frac{1}{5} + \frac{1}{3} + \frac{62}{165}\right)$$

= positive integer + 1 = positive integer.

$\therefore$ $P(m+1)$ is true;

thus, $P(m)$ is true $\Rightarrow$ $P(m+1)$ is true.

$\therefore$ by PMI, $P(n)$ is true for all $n \in N$.

Hence, the problem.

8. If r is a fixed positive integer, prove by induction that $(r+1)(r+2)(r+3)\ldots(r+n)$ is divisible by $n\,!$.

Let $P(n)$: $(r+1)(r+2)(r+3)\ldots(r+n) = n\,!\cdot k$, where k is an integer.

When $n = 1$, $r + 1 = 1\,!\cdot(r+1)$

$\therefore$ $P(1)$ is true.

Let $P(m)$ be true, i.e.,

$$(r+1)(r+2)(r+3)\ldots(r+m) = m\,!\cdot k \quad \ldots (1)$$

Now, $(r+1)(r+2)(r+3)\ldots(r+m)(r+\overline{m+1})$

$$= r(r+1)(r+2)(r+3)\ldots(r+m)$$

$$+ (m+1)(r+1)(r+2)(r+3)\ldots(r+m)$$

$$= \frac{(r+m)\,!}{(r-1)\,!} + (m+1)(m\,!)\,k, \text{ using (1)}$$

$$= (m+1)\,!\cdot\frac{(r+m)\,!}{(r-1)\,!\,(m+1)\,!} + (m+1)\,!\cdot k$$

$$= (m+1)\,!\cdot({}^{r+m}C_{r-1} + k)$$

$$= (m+1)\,!\,(\text{integer} + k)$$

$= (m+1)\,!\cdot k_1$, where k_1 is an integer.

$\therefore$ $P(m+1)$ is true;

thus, $P(m)$ is true $\Rightarrow$ $P(m+1)$ is true.

$\therefore$ $P(n)$ is true for all $n \in N$. Hence, the problem.

9. Prove by induction that $\sum_{r=0}^{n} r\cdot{}^nC_r = n\cdot 2^{n-1}$ for all $n \in N$.

Let $P(n)$: $1\cdot{}^nC_1 + 2\cdot{}^nC_2 + 3\cdot{}^nC_3 + \ldots + n\cdot{}^nC_n = n\cdot 2^{n-1}$.

When $n = 1$, $1\cdot{}^1C_1 = 1 = 1\cdot 2^0 = 1\cdot 2^{1-1}$

$\therefore$ $P(1)$ is true.

Let $P(m)$ be true, i.e.,

$$1\cdot{}^mC_1 + 2\cdot{}^mC_2 + 3\cdot{}^mC_3 + \ldots + m\cdot{}^mC_m = m\cdot 2^{m-1} \quad \ldots (1)$$

Now, $1\cdot{}^{m+1}C_1 + 2\cdot{}^{m+1}C_2 + 3\cdot{}^{m+1}C_3 + \ldots + m\cdot{}^{m+1}C_m + (m+1)\,{}^{m+1}C_{m+1}$

$$= 1\cdot({}^mC_1 + {}^mC_0) + 2\cdot({}^mC_2 + {}^mC_1) + 3\cdot({}^mC_3 + {}^mC_2)$$

$$+ \ldots + m({}^mC_m + {}^mC_{m-1}) + (m+1)\,{}^mC_m$$

$$= 1\cdot{}^mC_1 + 2\cdot{}^mC_2 + 3\cdot{}^mC_3 + \ldots + m\cdot{}^mC_m$$

$$+ (1\cdot{}^mC_0 + 2\cdot{}^mC_1 + 3\cdot{}^mC_2 + \ldots + m\cdot{}^mC_{m-1})$$

$$+ (m+1)\,{}^mC_m$$

$$= m\cdot 2^{m-1} + \{{}^mC_0 + (1+1)\,{}^mC_1 + (2+1)\,{}^mC_2 + \ldots$$

$$+ (\overline{m-1}+1)\,{}^mC_{m-1} + (m+1)\,{}^mC_m\}, \text{ using (1)}$$

$$= m\cdot 2^{m-1} + \{1\cdot{}^mC_1 + 2\cdot{}^mC_2 + \ldots$$

$$+ (m-1)\,{}^mC_{m-1} + m\,{}^mC_m\}$$

$$+ \{{}^mC_0 + {}^mC_1 + {}^mC_2 + \ldots + {}^mC_m\}$$

$$= m\cdot 2^{m-1} + m\cdot 2^{m-1} + 2^m,$$

using (1) and a property of ${}^nC_r$

$$= 2m\cdot 2^{m-1} + 2^m$$

$$= (m+1)2^m = (m+1)2^{(m+1)-1}.$$

$\therefore$ $P(m+1)$ is true;

thus, $P(m)$ is true $\Rightarrow$ $P(m+1)$ is true.

$\therefore$ by PMI, $P(n)$ is true for all $n \in N$.

Hence, the problem.

10. Using induction or otherwise prove that

$$\sum_{m=0}^{k} (n-m)\frac{(r+m)!}{m!} = \frac{(r+k+1)!}{k!}\left(\frac{n}{r+1} - \frac{k}{r+2}\right)$$

for all non-negative integral values of n, r and k.

We have to prove the truth of the statement

$$n \cdot \frac{r!}{0!} + (n-1)\frac{(r+1)!}{1!} + (n-2)\frac{(r+2)!}{2!} + \dots + (n-k)\frac{(r+k)!}{k!}$$

$$= \frac{(r+k+1)!}{k!}\left(\frac{n}{r+1} - \frac{k}{r+2}\right).$$

Let k be fixed. Then the number of terms do not change in LHS. Now,

$$\text{LHS} = r!\,\{n \cdot {}^rC_r + (n-1)\,{}^{r+1}C_r + (n-2)\,{}^{r+2}C_r + \dots + (n-k)\,{}^{r+k}C_r\}$$

$= r! \times$ coefficient of x^r in

$$\{n(1+x)^r + (n-1)(1+x)^{r+1} + (n-2)(1+x)^{r+2} + \dots + (n-k)(1+x)^{r+k}\} \quad \dots (1)$$

Let $S = n(1+x)^r + (n-1)(1+x)^{r+1} + (n-2)(1+x)^{r+2} + \dots + (n-k)(1+x)^{r+k}$

$\therefore$ $(1+x)S = n(1+x)^{r+1} + (n-1)(1+x)^{r+2} + \dots + (n-k)(1+x)^{r+k+1}$

Subtracting,

$$-xS = n(1+x)^r - (1+x)^{r+1} - (1+x)^{r+2} - \dots - (1+x)^{r+k} - (n-k)(1+x)^{r+k+1}$$

$$= n(1+x)^r - \{(1+x)^{r+1} + (1+x)^{r+2} + \dots + (1+x)^{r+k}\} - (n-k)(1+x)^{r+k+1}$$

$$= n(1+x)^r - (1+x)^{r+1} \cdot \frac{(1+x)^k - 1}{1+x-1} - (n-k)(1+x)^{r+k+1}$$

$$\therefore \quad S = -n\frac{(1+x)^r}{x} + \frac{(1+x)^k - 1}{x^2}(1+x)^{r+1} + (n-k)\frac{(1+x)^{r+k+1}}{x}$$

$\therefore$ from (1),

LHS $= r! \times$ coefficient of x^r in

$$\left\{-n \cdot \frac{(1+x)^r}{x} + \frac{(1+x)^k - 1}{x^2}(1+x)^{r+1} + (n-k)\frac{(1+x)^{r+k+1}}{x}\right\}$$

$= r! \times$ coefficient of x^r in

$$\left\{-n \cdot \frac{(1+x)^r}{x} + \frac{(1+x)^{k+r+1}}{x^2} - \frac{(1+x)^{r+1}}{x^2} + (n-k)\frac{(1+x)^{r+k+1}}{x}\right\}$$

$$= r! \times [-n \times 0 + {}^{r+k+1}C_{r+2} - 0 + (n-k)\,{}^{r+k+1}C_{r+1}]$$

$$= r! \times \left[\frac{(r+k+1)!}{(r+2)!\,(k-1)!} + (n-k)\frac{(r+k+1)!}{(r+1)!\,k!}\right]$$

$$= \frac{(r+k+1)!}{k!}\left\{\frac{k}{(r+2)(r+1)} + \frac{n-k}{r+1}\right\}$$

$$= \frac{(r+k+1)!}{k!}\left[\frac{n}{r+1} + k\left\{\frac{1}{(r+2)(r+1)} - \frac{1}{r+1}\right\}\right]$$

$$= \frac{(r+k+1)!}{k!}\left[\frac{n}{r+1} - \frac{k}{r+2}\right] = \text{RHS}.$$

By the principle of substitution this is true for all n, r (non-negative integers) if k is fixed.

Now let n, r have fixed non-negative integral values and k vary. Let the statement be denoted by $P(k)$.

When $k = 0$, LHS $= n \cdot (r!) =$ RHS; so $P(0)$ is true.

When $k = 1$,

$$\text{LHS} = \sum_{m=0}^{1} (n-m)\frac{(r+m)!}{m!} = n \cdot (r!) + (n-1) \cdot \{(r+1)!\}$$

$$= r!\,\{n(r+2)) - (r+1)\}$$

$$\text{RHS} = (r+2)!\left(\frac{n}{r+1} - \frac{1}{r+2}\right) = r!\,\{n(r+2) - (r+1)\}$$

$\therefore$ $P(1)$ is true.

Let $P(p)$ be true, i.e.,

$$\sum_{m=0}^{p} (n-m)\frac{(r+m)!}{m!} = \frac{(r+p+1)!}{p!}\left(\frac{n}{r+1} - \frac{p}{r+2}\right)$$

Now, $\displaystyle\sum_{m=0}^{p+1} (n-m)\frac{(r+m)!}{m!}$

$$= \frac{(r+p+1)!}{p!}\left(\frac{n}{r+1} - \frac{p}{r+2}\right) + (n-p-1)\frac{(r+p+1)!}{(p+1)!}$$

$$= \frac{(r+p+1)!}{(p+1)!}\left\{(p+1)\left(\frac{n}{r+1} - \frac{p}{r+2}\right) + (n-p-1)\right\}$$

$$= \frac{(r+p+1)!}{(p+1)!}\left\{n\left(\frac{p+1}{r+1}+1\right) - (p+1)\left(\frac{p}{r+2}+1\right)\right\}$$

$$= \frac{(r+p+2)!}{(p+1)!}\left\{\frac{n}{r+1} - \frac{p+1}{r+2}\right\};$$

so $P(m+1)$ is true.

Hence, by induction $P(k)$ is true for $k = 0$ and all $k \in N$. This establishes the result.

11. Let $a_1 = \frac{a+b}{2}, b_1 = \frac{a_1+b}{2}, a_2 = \frac{a_1+b_1}{2}, b_2 = \frac{a_2+b_1}{2}$, etc.

Prove, by induction, that

$$a_n = a + \frac{2}{3}(b-a)\cdot\left(1 - \frac{1}{4^n}\right) \text{ and}$$

$$b_n = a + \frac{2}{3}(b-a)\cdot\left(1 + \frac{1}{2\cdot 4^n}\right) \text{ for all } n \in N.$$

Let $P(n)$: $a_n = a + \frac{2}{3}(b-a)\left(1 - \frac{1}{4^n}\right)$ and

$$b_n = a + \frac{2}{3}(b-a)\left(1 + \frac{1}{2\cdot 4^n}\right).$$

When $n = 1$,

$$a + \frac{2}{3}(b-a)\left(1 - \frac{1}{4}\right) = a + \frac{2}{3}(b-a)\cdot\frac{3}{4}$$

$$= a + \frac{b-a}{2} = \frac{a+b}{2} = a_1$$

$$a + \frac{2}{3}(b-a)\left(1 + \frac{1}{2\cdot 4}\right) = a + \frac{2}{3}(b-a)\cdot\frac{9}{8}$$

$$= a + \frac{(b-a)3}{4} = \frac{a+3b}{4} = \frac{a+b}{4} + \frac{b}{2} = \frac{a_1+b}{2} = b_1$$

$\therefore$ $P(1)$ is true.

Let $P(m)$ be true, i.e.,

$$a_m = a + \frac{2}{3}(b-a)\cdot\left(1 - \frac{1}{4^m}\right),$$

$$b_m = a + \frac{2}{3}(b-a)\cdot\left(1 + \frac{1}{2\cdot 4^m}\right) \quad \ldots (1)$$

Now, $a + \frac{2}{3}(b-a)\cdot\left(1 - \frac{1}{4^{m+1}}\right)$

$$= \frac{1}{2}\left\{2a + \frac{4}{3}(b-a)\left(1 - \frac{1}{4^{m+1}}\right)\right\}$$

$$= \frac{1}{2}\left\{a + \frac{2}{3}(b-a)\left(1 - \frac{1}{4^{m+1}}\right)\right.$$

$$\left. + a + \frac{2}{3}(b-a)\left(1 - \frac{1}{4^{m+1}}\right)\right\}$$

$$= \frac{1}{2}\left\{a + \frac{2}{3}(b-a)\left(1 - \frac{1}{4^m} + \frac{1}{4^m} - \frac{1}{4^{m+1}}\right)\right.$$

$$\left. + a + \frac{2}{3}(b-a)\left(1 + \frac{1}{2\cdot 4^m} - \frac{1}{2\cdot 4^m} - \frac{1}{4^{m+1}}\right)\right\}$$

$$= \frac{1}{2}\left\{a_m + \frac{2}{3}(b-a)\left(\frac{1}{4^m} - \frac{1}{4^{m+1}}\right)\right.$$

$$\left. + b_m - \frac{2}{3}(b-a)\left(\frac{1}{2\cdot 4^m} + \frac{1}{4^{m+1}}\right)\right\}, \text{ using (1)}$$

$$= \frac{1}{2}(a_m + b_m)$$

$$+ \frac{1}{2}\left\{\frac{2}{3}(b-a)\frac{3}{4^{m+1}} - \frac{2}{3}(b-a)\cdot\frac{3}{4^{m+1}}\right\}$$

$$= \frac{a_m + b_m}{2} = a_{m+1} \text{ (from question)} \quad \ldots (2)$$

Again, $a + \frac{2}{3}(b-a)\left(1 + \frac{1}{2\cdot 4^{m+1}}\right)$

$$= \frac{1}{2}\left\{2a + \frac{4}{3}(b-a)\left(1 + \frac{1}{2\cdot 4^{m+1}}\right)\right\}$$

$$= \frac{1}{2}\left\{a + \frac{2}{3}(b-a)\left(1 + \frac{1}{2\cdot 4^{m+1}}\right)\right.$$

$$\left. + a + \frac{2}{3}(b-a)\left(1 + \frac{1}{2\cdot 4^{m+1}}\right)\right\}$$

$$= \frac{1}{2}\left\{a + \frac{2}{3}(b-a)\left(1 - \frac{1}{4^{m+1}} + \frac{1}{4^{m+1}} + \frac{1}{2\cdot 4^{m+1}}\right)\right.$$

$$\left. + a + \frac{2}{3}(b-a)\left(1 + \frac{1}{2\cdot 4^m} - \frac{1}{2\cdot 4^m} + \frac{1}{2\cdot 4^{m+1}}\right)\right\}$$

$$= \frac{1}{2}\left\{a_{m+1} + \frac{2}{3}(b-a)\left(\frac{1}{4^{m+1}} + \frac{1}{2\cdot 4^{m+1}}\right)\right.$$

$$\left. + b_m - \frac{2}{3}(b-a)\left(\frac{1}{2\cdot 4^m} - \frac{1}{2\cdot 4^{m+1}}\right)\right\}$$

$$= \frac{a_{m+1} + b_m}{2}$$

$$+ \frac{1}{2}\left\{\frac{2}{3}(b-a)\frac{3}{2\cdot 4^{m+1}} - \frac{2}{3}(b-a)\frac{3}{2\cdot 4^{m+1}}\right\}$$

$$= \frac{a_{m+1} + b_m}{2} = b_{m+1}, \text{ (from question)} \quad \ldots (3)$$

$\therefore$ by (2) and (3), $P(m+1)$ is true.

Hence, by PMI, $P(n)$ is true for all $n \in N$.

Hence, the problem.

12. If $x^3 = x + 1$, prove by induction that

$$x^{3n} = a_n x + b_n + \frac{c_n}{x}$$

where $a_n = a_{n-1} + b_{n-1}$,

$b_n = a_{n-1} + b_{n-1} + c_{n-1}$

and $c_n = a_{n-1} + c_{n-1}$

for $n \in N$ and also $a_0 = 0, b_0 = 1, c_0 = 0$.

Let $P(n)$: $x^{3n} = a_n x + b_n + \frac{c_n}{x}$.

When $n = 1$,

$$a_1 x + b_1 + \frac{c_1}{x} = (a_0 + b_0)x + (a_0 + b_0 + c_0) + \frac{a_0 + c_0}{x}$$

(from given relations)

$$= (0 + 1)x + (0 + 1 + 0) + \frac{0 + 0}{x}$$

$$= x + 1 = x^3 \quad (\because \text{ given that } x^3 = x + 1)$$

$\therefore$ $P(1)$ is true.

Let $P(m)$ be true, i.e., $x^{3m} = a_m x + b_m + \frac{c_m}{x}$ ... (1)

Now, $a_{m+1}x + b_{m+1} + \frac{c_{m+1}}{x}$

$$= (a_m + b_m)x + a_m + b_m + c_m + \frac{a_m + c_m}{x},$$

(from given relations $a_n = a_{n-1} + b_{n-1}$, etc.)

$$= \left(a_m x + b_m + \frac{c_m}{x}\right) + b_m x + a_m + c_m + \frac{a_m}{x}$$

$$= x^{3m} + a_m\left(1 + \frac{1}{x}\right) + b_m x + c_m, \text{ using (1)}$$

$$= x^{3m} + a_m x^2 + b_m x + c_m$$

$$\left(\because x^3 = x + 1 \Rightarrow x^2 = 1 + \frac{1}{x}\right)$$

$$= x^{3m} + x\left(a_m x + b_m + \frac{c_m}{x}\right)$$

$$= x^{3m} + x \cdot x^{3m}, \text{ using (1)}$$

$$= x^{3m}(1 + x) = x^{3m} \cdot x^3 \quad (\because 1 + x = x^3, \text{ given})$$

$$= x^{3(m+1)}.$$

$\therefore$ $P(m + 1)$ is true;

thus, $P(m)$ is true $\Rightarrow$ $P(m + 1)$ is true.

$\therefore$ by PMI, $P(n)$ is true for all $n \in N$.

Hence, the problem.

13. Let $a \geq 2$. Prove using induction that

$$\frac{1 + a + a^2 + \ldots + a^{n+1}}{a + a^2 + \ldots + a^n} > \frac{n + 2}{n} \quad \text{for all } n \in N.$$

Let $P(n)$: $\frac{1 + a + a^2 + \ldots + a^{n+1}}{a + a^2 + \ldots + a^n} > \frac{n + 2}{n}$.

When $n = 1$,

$$\frac{1 + a + a^2}{a} = \frac{(a^2 - 2a + 1) + 3a}{a}$$

$$= \frac{(a - 1)^2}{a} + 3 > 3 = \frac{1 + 2}{1}, \text{ because } a \geq 2$$

$\therefore$ $P(1)$ is true.

Next, let $P(m)$ be true, i.e.,

$$\frac{1 + a + a^2 + \ldots + a^{m+1}}{a + a^2 + \ldots + a^m} > \frac{m + 2}{m}$$

or $1 + a + a^2 + \ldots + a^{m+1} > \frac{m + 2}{m}(a + a^2 + \ldots + a^m)$ ... (1)

Now, we want to prove that $P(m + 1)$ is true. For this we have to prove that

$$\frac{1 + a + a^2 + \ldots + a^{m+2}}{a + a^2 + \ldots + a^{m+1}} > \frac{m + 3}{m + 1}$$

i.e., $1 + a + a^2 + \ldots + a^{m+2} > \frac{m + 3}{m + 1}(a + a^2 + \ldots + a^{m+1})$ ... (2)

From (1),

$$1 + a + a^2 + \ldots + a^{m+1} + a^{m+2}$$

$$> \frac{m + 2}{2}(a + a^2 + \ldots + a^m) + a^{m+2} \quad \ldots (3)$$

If (2) is not true then

$$\frac{m + 3}{m + 1}(a + a^2 + \ldots + a^{m+1}) \geq 1 + a + a^2 + \ldots + a^{m+2}.$$

Also from (3),

$$\frac{m + 3}{m + 1}(a + a^2 + \ldots + a^{m+1})$$

$$> \frac{m + 2}{m}(a + a^2 + \ldots + a^m) + a^{m+2}$$

i.e., $\left(\frac{m + 3}{m + 1} - \frac{m + 2}{m}\right)(a + a^2 + \ldots + a^m)$

$$+ \frac{m + 3}{m + 1}a^{m+1} - a^{m+2} > 0$$

or $\frac{-2}{m(m + 1)}(a + a^2 + \ldots + a^m)$

$$+ \frac{m + 3}{m + 1}a^{m+1} - a^{m+2} > 0$$

or $-2(a + a^2 + \ldots + a^m) + m(m + 3)a^{m+1}$

$$- m(m + 1)a^{m+2} > 0$$

or $\quad 2(a + a^2 + \dots + a^m) + ma^{m+1}\{(m+1)a - (m+3)\} < 0$

or $\quad 2(a + a^2 + \dots + a^m) + ma^{m+1}\{(m+1)(a-1) - 2\} < 0$

which is not possible for $a \geq 2$ and $m + 1 \geq 3$

($\because$ m is at least 2)

$\therefore$ (2) is true. Hence, $P(m+1)$ is true.

$\therefore$ by induction, $P(n)$ is true for all $n \in N$.

Hence, the problem.

14. Prove that $\dfrac{1}{n+1} + \dfrac{1}{n+2} + \dfrac{1}{n+3} + \dots + \dfrac{1}{3n+1} > 1$ **for all** $n \in N$.

Let $P(n)$: $\dfrac{1}{n+1} + \dfrac{1}{n+2} + \dfrac{1}{n+3} + \dots + \dfrac{1}{3n+1} > 1$.

When $n = 1$,

$$\frac{1}{1+1} + \frac{1}{1+2} + \frac{1}{1+3} = \frac{1}{2} + \frac{1}{3} + \frac{1}{4} = \frac{6+4+3}{12} = \frac{13}{12} > 1$$

[**Note** There are $2n+1$ terms in the series. So when $n = 1$, there will be 3 terms.]

Let $P(m)$ be true, i.e.,

$$\frac{1}{m+1} + \frac{1}{m+2} + \frac{1}{m+3} + \dots + \frac{1}{3m+1} > 1 \qquad \dots (1)$$

Now, $\dfrac{1}{(m+1)+1} + \dfrac{1}{(m+1)+2} + \dfrac{1}{(m+1)+3} + \dots + \dfrac{1}{3(m+1)+1}$

$$= \left(\frac{1}{m+2} + \frac{1}{m+3} + \frac{1}{m+4} + \dots + \frac{1}{3m+1}\right) + \frac{1}{3m+2} + \frac{1}{3m+3} + \frac{1}{3m+4}$$

$$> 1 - \frac{1}{m+1} + \frac{1}{3m+2} + \frac{1}{3m+3} + \frac{1}{3m+4}, \text{ using (1)}$$

$$= 1 + \frac{1}{3m+2} + \frac{1}{3m+4} - \frac{2}{3} \cdot \frac{1}{m+1}$$

$$= 1 + \frac{\{(3m+3)(3m+4) + (3m+2)(3m+3) - 2(3m+2)(3m+4)\}}{(3m+2)(3m+3)(3m+4)}$$

$$= 1 + \frac{18m^2 + 36m + 18 - (18m^2 + 36m + 16)}{(3m+2)(3m+3)(3m+4)}$$

$$= 1 + \frac{2}{(3m+2)(3m+3)(3m+4)} > 1$$

$\therefore$ $P(m+1)$ is true;

thus, $P(m)$ is true $\Rightarrow$ $P(m+1)$ is true.

$\therefore$ by PMI, $P(n)$ is true for all $n \in N$.

Hence, the problem.

15. Prove, by induction, that

$$2 + \sqrt{2 + \sqrt{2 + \sqrt{2 + \dots \text{ to } n \text{ times}}}} < 4 \text{ for all } n \geq 1.$$

Let $P(n)$: $2 + \sqrt{2 + \sqrt{2 + \sqrt{2 + \dots \text{ to } n \text{ times}}}} < 4$

When $n = 1$, $2 + \sqrt{2} = 2 + 1.41 = 3.41 < 4$

$\therefore$ $P(1)$ is true.

Let $P(m)$ be true, i.e.,

$$2 + \sqrt{2 + \sqrt{2 + \sqrt{2 + \dots \text{ to } m \text{ times}}}} < 4 \qquad \dots (1)$$

Now, $2 + \sqrt{2 + \sqrt{2 + \sqrt{2 + \dots \text{ to } (m+1) \text{ times}}}}$

$$= 2 + \sqrt{2 + \{\sqrt{2 + \sqrt{2 + \dots \text{ to } m \text{ times}}}\}}$$

$$< 2 + \sqrt{4} = 4, \text{ using (1)}$$

$\therefore$ $P(m+1)$ is true;

thus, $P(m)$ is true $\Rightarrow$ $P(m+1)$ is true.

$\therefore$ by PMI, $P(n)$ is true for all $n \in N$.

Hence, the problem.

16. Prove that $(n\,!)^2 \leq \left(\dfrac{1^2 + 2^2 + 3^2 + \dots + n^2}{n}\right)^n$.

Let $P(n)$: $(n\,!)^2 \leq \left(\dfrac{1^2 + 2^2 + 3^2 + \dots + n^2}{n}\right)^n$.

When $n = 1$, $\quad \left(\dfrac{1^2}{1}\right)^1 = 1 = (1\,!)^2 = 1$.

$\therefore$ $P(1)$ is true.

Let $P(m)$ be true, i.e.,

$$(m\,!)^2 \leq \left(\frac{1^2 + 2^2 + 3^2 + \dots + m^2}{m}\right)^m$$

or $\quad (m\,!)^2 \leq \left\{\dfrac{(m+1)(2m+1)}{6}\right\}^m$

$\therefore$ $\quad (m\,!)^2 = \left\{\dfrac{(m+1)(2m+1)}{6}\right\}^m - k, \quad (k \geq 0)$.

Now, $\left\{\dfrac{1^2 + 2^2 + 3^2 + \dots + (m+1)^2}{m+1}\right\}^{m+1} - \{(m+1)\,!\}^2$

$$= \left\{\frac{(m+1)(m+2)(2m+3)}{6(m+1)}\right\}^{m+1} - \{(m+1)\,!\}^2$$

$$= \left\{\frac{(m+2)(2m+3)}{6}\right\}^{m+1} - \{(m+1)\,!\}^2$$

$$= \left\{\frac{(m+2)(2m+3)}{6}\right\}^{m+1} - (m+1)^2 \cdot (m\,!)^2$$

$$= \left\{\frac{(m+2)(2m+3)}{6}\right\}^{m+1} - (m+1)^2\left[\left\{\frac{(m+1)(2m+1)}{6}\right\}^m - k\right]$$

$$= k(m+1)^2 + \left\{\frac{(m+2)(2m+3)}{6}\right\}^{m+1} - (m+1)^2\left\{\frac{(m+1)(2m+1)}{6}\right\}^m$$

$$= k(m+1)^2 + \left\{\frac{(m+1)(2m+1)}{6} + \frac{4m+5}{6}\right\}^{m+1} - (m+1)^2\left\{\frac{(m+1)(2m+1)}{6}\right\}^m$$

$$= k(m+1)^2 + \left\{\frac{(m+1)(2m+1)}{6}\right\}^{m+1} + {}^{m+1}C_1 \cdot \left\{\frac{(m+1)(2m+1)}{6}\right\}^m \cdot \frac{4m+5}{6} + \dots + \left(\frac{4m+5}{6}\right)^{m+1} - (m+1)^2\left\{\frac{(m+1)(2m+1)}{6}\right\}^m$$

$$= k(m+1)^2 + \left\{\frac{(m+1)(2m+1)}{6} + \frac{(m+1)(4m+5)}{6} - (m+1)^2\right\} \times \left\{\frac{(m+1)(2m+1)}{6}\right\}^m + \text{positive terms}$$

$$= k(m+1)^2 + 0 \cdot \left\{\frac{(m+1)(2m+1)}{6}\right\}^m + \text{positive terms} > 0, \text{ for } k \geq 0$$

$$\therefore \quad \left\{\frac{1^2+2^2+3^2+\dots+(m+1)^2}{m+1}\right\}^{m+1} > \{(m+1)!\}^2$$

$\therefore$ $P(m+1)$ is true;

thus, $P(m)$ is true $\Rightarrow$ $P(m+1)$ is true.

$\therefore$ $P(n)$ is true for all $n \in N$.

Hence, the problem.

17. If a_1, a_2, etc., are positive then prove that

$$\frac{a_1 + a_2 + a_3 + \dots + a_{2^n}}{2^n} \geq (a_1 \cdot a_2 \cdot a_3 \dots a_{2^n})^{\frac{1}{2^n}}$$

for all $n \in N$.

Let $P(n)$: $\dfrac{a_1 + a_2 + a_3 + \dots + a_{2^n}}{2^n} \geq (a_1 \cdot a_2 \cdot a_3 \dots a_{2^n})^{\frac{1}{2^n}}$.

When $n = 1$, we have to prove that

$$\frac{a_1 + a_2}{2} \geq (a_1 a_2)^{½}.$$

We have, $(a_1 + a_2)^2 - 4a_1a_2 = (a_1 - a_2)^2 \geq 0$

or $\quad (a_1 + a_2)^2 \geq 4a_1a_2$

or $\quad \left(\dfrac{a_1 + a_2}{2}\right)^2 \geq a_1a_2$

or $\quad \dfrac{a_1 + a_2}{2^1} \geq (a_1a_2)^{\frac{1}{2^1}}$

$\therefore$ $P(1)$ is true, i.e., $\dfrac{a_1 + a_2}{2} \geq (a_1a_2)^{1/2}$ ... (1)

Let $P(m)$ be true, i.e.,

$$\frac{a_1 + a_2 + a_3 + \dots + a_{2^m}}{2^m} \geq (a_1 \cdot a_2 \cdot a_3 \dots a_{2^m})^{\frac{1}{2^m}} \quad \dots (2)$$

Replacing, a_1, a_2, etc., by a_{2^m+1}, a_{2^m+2}, etc., in it,

$$\frac{a_{2^m+1} + a_{2^m+2} + \dots + a_{2^{m+1}}}{2^m} \geq (a_{2^m+1} \cdot a_{2^m+2} \cdot \dots a_{2^{m+1}})^{\frac{1}{2^m}} \quad \dots (3)$$

Now, replacing a_1, a_2 by

$$\frac{a_1 + a_2 + \dots + a_{2^m}}{2^m}, \ \frac{a_{2^m+1} + a_{2^m+2} + \dots + a_{2^{m+1}}}{2^m}$$

respectively in (1) we get

$$\frac{1}{2}\left(\frac{a_1 + a_2 + \dots + a_{2^m}}{2^m} + \frac{a_{2^m+1} + a_{2^m+2} + \dots + a_{2^{m+1}}}{2^m}\right)$$

$$\geq \left\{\frac{a_1 + a_2 + \dots + a_{2^m}}{2^m} \times \frac{a_{2^m+1} + a_{2^m+2} + \dots + a_{2^{m+1}}}{2^m}\right\}^{1/2}$$

$$\geq \left\{(a_1 \cdot a_2 \cdot \dots a_{2^m})^{\frac{1}{2^m}} \cdot (a_{2^m+1} \cdot a_{2^m+2} \cdot \dots a_{2^{m+1}})^{\frac{1}{2^m}}\right\}^{1/2},$$

using (2) and (3)

$$\therefore \quad \frac{a_1 + a_2 + a_3 + \dots + a_{2^{m+1}}}{2^{m+1}} \geq (a_1 \cdot a_2 \cdot a_3 \cdot \dots a_{2^{m+1}})^{\frac{1}{2^{m+1}}}$$

$\therefore$ $P(m+1)$ is true;

thus, $P(m)$ is true $\Rightarrow$ $P(m+1)$ is true.

$\therefore$ $P(n)$ is true for all $n \in N$.

Hence, the problem.

18. Use induction to show that

$$\sqrt{(x_1+x_2+\ldots+x_n)^2+(y_1+y_2+\ldots+y_n)^2} \le \sqrt{x_1^2+y_1^2}+\sqrt{x_2^2+y_2^2}+\ldots+\sqrt{x_n^2+y_n^2}.$$

Let the statement be $P(n)$.

When $n=1$, $\sqrt{x_1^2+y_1^2}=\sqrt{x_1^2+y_1^2}$. So $P(1)$ is true.

Let $P(m)$ be true, i.e.,

$$\sqrt{(x_1+x_2+\ldots+x_m)^2+(y_1+y_2+\ldots+y_m)^2} \le \sqrt{x_1^2+y_1^2}+\sqrt{x_2^2+y_2^2}+\ldots+\sqrt{x_m^2+y_m^2}$$

or $\quad (x_1+x_2+\ldots+x_m)^2+(y_1+y_2+\ldots+y_m)^2$

$$\le (\sqrt{x_1^2+y_1^2}+\sqrt{x_2^2+y_2^2}+\ldots+\sqrt{x_m^2+y_m^2})^2 \quad \ldots (1)$$

Now, $(x_1+x_2+\ldots+x_{m+1})^2+(y_1+y_2+\ldots+y_{m+1})^2$

$$-(\sqrt{x_1^2+y_1^2}+\sqrt{x_2^2+y_2^2}+\ldots+\sqrt{x_{m+1}^2+y_{m+1}^2})^2$$

$$=(x_1+\ldots+x_m)^2+x_{m+1}^2+2x_{m+1}(x_1+\ldots+x_m)$$

$$+(y_1+\ldots+y_m)^2+y_{m+1}^2+2y_{m+1}(y_1+\ldots+y_m)$$

$$-(\sqrt{x_1^2+y_1^2}+\ldots+\sqrt{x_m^2+y_m^2})^2-(x_{m+1}^2+y_{m+1}^2)$$

$$-2(\sqrt{x_1^2+y_1^2}+\ldots+\sqrt{x_m^2+y_m^2})\sqrt{x_{m+1}^2+y_{m+1}^2}$$

$$=\{(x_1+\ldots+x_m)^2+(y_1+\ldots+y_m)^2$$

$$-(\sqrt{x_1^2+y_1^2}+\ldots+\sqrt{x_m^2+y_m^2})^2\}$$

$$+2\{x_1x_{m+1}+y_1y_{m+1}-\sqrt{x_1^2+y_1^2}\sqrt{x_{m+1}^2+y_{m+1}^2}\}$$

$$+\ldots+2\{x_mx_{m+1}+y_my_{m+1}$$

$$-\sqrt{x_m^2+y_m^2}\cdot\sqrt{x_{m+1}^2+y_{m+1}^2}\} \quad \ldots (2)$$

But $(x_1x_{m+1}+y_1y_{m+1})^2-(x_1^2+y_1^2)(x_{m+1}^2+y_{m+1}^2)$

$$=2x_1y_1x_{m+1}y_{m+1}-(x_1^2y_{m+1}^2+y_1^2x_{m+1}^2)$$

$$=-(x_1y_{m+1}-y_1x_{m+1})^2\le 0$$

$\therefore \quad x_1x_{m+1}+y_1y_{m+1}\le\sqrt{x_1^2+y_1^2}\cdot\sqrt{x_{m+1}^2+y_{m+1}^2}$, etc. $\quad \ldots (3)$

Using (1) and inequalities like (3) we get from (2),

$$(x_1+x_2+\ldots+x_{m+1})^2+(y_1+y_2+\ldots+y_{m+1})^2$$

$$\le\sqrt{x_1^2+y_1^2}+\sqrt{x_2^2+y_2^2}+\ldots+\sqrt{x_{m+1}^2+y_{m+1}^2}$$

$\therefore \quad P(m+1)$ is true.

Hence, by induction, $P(n)$ is true for all $n\in N$.

19. Prove that $\dfrac{2n\,!}{2^{2n}(n\,!)^2}\le\dfrac{1}{\sqrt{3n+1}}$, **using induction.**

(Or Prove that $\dfrac{1}{2}\cdot\dfrac{3}{4}\cdot\dfrac{5}{6}\cdot\ldots\dfrac{2n-1}{2n}\le\dfrac{1}{\sqrt{3n+1}}$, **using induction.)**

Let $P(n)$: $\dfrac{(2n)\,!}{2^{2n}(n\,!)^2}\le\dfrac{1}{\sqrt{3n+1}}$.

When $n=1$, $\dfrac{(2\cdot1)\,!}{2^{2\cdot1}(1\,!)^2}=\dfrac{2}{4\times1}=\dfrac{1}{2}=\dfrac{1}{\sqrt{3\cdot1+1}}$

$\therefore \quad P(1)$ is true.

Let $P(m)$ be true, i.e., $\dfrac{(2m)\,!}{2^{2m}(m\,!)^2}\le\dfrac{1}{\sqrt{3m+1}} \quad \ldots (1)$

Now, $\dfrac{\{2(m+1)\}\,!}{2^{2(m+1)}\{(m+1)\,!\}^2}-\dfrac{1}{\sqrt{3(m+1)+1}}$

$$=\frac{2(m+1)\cdot(2m+1)\cdot(2m)\,!}{2^{2m}\cdot4(m+1)^2(m\,!)^2}-\frac{1}{\sqrt{3m+4}}$$

$$=\frac{2m+1}{2(m+1)}\cdot\frac{(2m)\,!}{2^{2m}(m\,!)^2}-\frac{1}{\sqrt{3m+4}}$$

$$<\frac{2m+1}{2(m+1)}\cdot\frac{1}{\sqrt{3m+1}}-\frac{1}{\sqrt{3m+4}}, \text{ using in (1)}$$

$$=\frac{(2m+1)\sqrt{3m+4}-2(m+1)\sqrt{3m+1}}{2(m+1)\sqrt{3m+1}\cdot\sqrt{3m+4}}$$

$$=\frac{(2m+1)^2(3m+4)-4(m+1)^2(3m+1)}{\left[2(m+1)\sqrt{3m+1}\sqrt{3m+4}\{(2m+1)\sqrt{3m+4}+2(m+1)\sqrt{3m+1}\}\right]}$$

$$=\frac{\{(4m^2+4m+1)(3m+4)-(4m^2+8m+4)(3m+1)\}}{\left[2(m+1)\sqrt{3m+1}\sqrt{3m+4}\{(2m+1)\sqrt{3m+4}+2(m+1)\sqrt{3m+1}\}\right]}$$

$$=\frac{-m}{\left[2(m+1)\sqrt{3m+1}\sqrt{3m+4}\{(2m+1)\sqrt{3m+4}+2(m+1)\sqrt{3m+1}\}\right]}$$

<0

$\therefore \quad \dfrac{\{2(m+1)\}!}{2^{2(m+1)}\{(m+1)!\}^2} < \dfrac{1}{\sqrt{3(m+1)+1}}$

$\therefore$ $P(m+1)$ is true;

thus, $P(m)$ is true $\Rightarrow$ $P(m+1)$ is true.

$\therefore$ by PMI, $P(n)$ is true for all $n \in N$.

Hence, the problem.

Note $\dfrac{1}{2}\cdot\dfrac{3}{4}\cdot\dfrac{5}{6}\dots\dfrac{2n-1}{2n} = \dfrac{1\cdot2\cdot3\cdot4\cdot\dots 2n}{(2\cdot4\cdot6\dots 2n)^2} = \dfrac{2n!}{2^{2n}(n!)^2}$.

20. Prove, by induction, that the value of $2^{2^n}+1$ has 7 in the units place for all $n \in N$ and $n \geq 2$.

Let $P(n)$: $2^{2^n}+1 = k$,

where k is a number with 7 in the units place.

When $n = 1$, $2^{2^1}+1 = 2^2+1 = 5$

$\neq$ a number with 7 in the units place.

When $n = 2$, $2^{2^2}+1 = 2^4+1 = 16+1$

$= 17$ which has 7 in the units place.

$\therefore$ $P(2)$ is true but $P(1)$ is not true.

Let $P(m)$ be true, i.e., $2^{2^m}+1 = k_1$,

where k_1 is a number with 7 in the units place.

Now, $2^{2^{m+1}}+1 = 2^{2^m\cdot 2}+1 = (2^{2^m})^2+1$

$= (k_1-1)^2+1$

$=$ (a number with 6 in the units place)$^2 + 1$

$=$ a number with 6 in the units place $+ 1$

$\{\because\ x6^2 = (10x+6)^2 = 100x^2+60x+6$

$=$a number with 6 in the units place$\}$

$=$ a number with 7 in the units place

$\therefore$ $P(m+1)$ is true;

thus, $P(m)$ is true $\Rightarrow$ $P(m+1)$ is true.

$\therefore$ by alternative PMI, $P(n)$ is true for all $n \in N$, $n \geq 2$.

21. If $a_n = \cos n\theta$, for $n \geq 1$ then show that

$a_n = 2a_{n-1}\cos\theta - a_{n-2}$

for all $n \geq 3$. Use induction.

Let $P(n)$: $a_n = 2a_{n-1}\cos\theta - a_{n-2}$.

Putting $n+2$ for n,

$\phi(n) = P(n+2)$: $a_{n+2} = 2a_{n+1}\cdot\cos\theta - a_n$.

When $n = 1$, $a_{n+2} = a_3 = \cos 3\theta$

$2a_{n+1}\cdot\cos\theta - a_n = 2a_2\cdot\cos\theta - a_1$

$= 2\cos 2\theta\cdot\cos\theta - \cos\theta$

$= \cos\theta + \cos 3\theta - \cos\theta = \cos 3\theta$

$\therefore$ $a_3 = 2a_2\cdot\cos\theta - a_1$

$\therefore$ $\phi(1)$ is true.

Let $\phi(m)$ be true, i.e., $a_{m+2} = 2a_{m+1}\cdot\cos\theta - a_m$.

Now, $2a_{m+2}\cos\theta - a_{m+1}$

$= 2(2a_{m+1}\cdot\cos\theta - a_m)\cos\theta - a_{m+1}$

$= 4a_{m+1}\cdot\cos^2\theta - 2a_m\cdot\cos\theta - a_{m+1}$

$= 4\cos(m+1)\theta\cdot\cos^2\theta$

$- 2\cos m\theta\cdot\cos\theta - \cos(m+1)\theta$

$= 2\cos(m+1)\theta\cdot(1+\cos 2\theta)$

$- \{\cos(m-1)\theta + \cos(m+1)\theta\} - \cos(m+1)\theta$

$= 2\cos(m+1)\theta\cdot\cos 2\theta - \cos(m-1)\theta$

$= \cos(m-1)\theta + \cos(m+3)\theta - \cos(m-1)\theta$

$= \cos(m+3)\theta = a_{m+3}$

$\therefore$ $a_{m+3} = 2a_{m+2}\cdot\cos\theta - a_{m+1}$;

so $\phi(m+1)$ is true.

$\therefore$ by PMI, $\phi(n)$ is true for all $n \in N$, i.e.,

$\phi(1), \phi(2), \phi(3), \dots$ are true;

$\therefore$ $P(3), P(4), P(5), \dots$ are true $\{\because\ \phi(n) = P(n+2)\}$

$\therefore$ $P(n)$ is true for all $n \geq 3$.

Hence, the problem.

22. Prove by induction that $x^n - y^n$ is divisible by $x+y$ when n is a positive even integer.

Let $P(n)$: $x^n - y^n = (x+y)\cdot\phi$,

where ϕ is a polynomial in x and y.

Substituting $2n$ for n,

$\phi(n) = P(2n)$: $x^{2n} - y^{2n} = (x+y)\phi_1$,

where ϕ_1 is a polynomial in x and y.

When $n = 1$, $x^{2\cdot1} - y^{2\cdot1} = x^2 - y^2 = (x+y)(x-y)$

$\therefore$ $\phi(1)$ is true.

Let $\phi(m)$ be true, i.e., $x^{2m} - y^{2m} = (x+y)\phi_2$ ... (1)

where ϕ_2 is a polynomial in x and y.

Now, $x^{2(m+1)} - y^{2(m+1)} = x^{2m}\cdot x^2 - y^{2m+2}$

$= \{y^{2m} + (x+y)\phi_2\}x^2 - y^{2m+2}$

$= y^{2m}(x^2-y^2) + (x+y)x^2\cdot\phi_2$.

$= (x + y)\{y^{2m}(x - y) + x^2 \cdot \phi_2\}$

$= (x + y)\phi_3,$

where ϕ_3 is a polynomial in x and y.

$\therefore$ $\phi(m + 1)$ is true;

thus, $\phi(m)$ is true $\Rightarrow$ $\phi(m + 1)$ is true.

$\therefore$ by PMI, $\phi(n)$ is true for all $n \in N$.

$\therefore$ $P(2n)$ is true for all $n \in N$, i.e.,

$P(2), P(4), P(6), \ldots$ are true.

$\therefore$ $P(n)$ is true for all positive even integral values of n.

Hence, the problem.

23. Prove that $(\sqrt{2} - 1)^{2n}$ can be expressed as the difference of the square roots of two consecutive integers, for all positive integral values of n. Apply induction.

Let $P(n)$: $(\sqrt{2} - 1)^{2n} = \sqrt{N} - \sqrt{N - 1}$,

where N is a positive integer.

When $n = 1$,

$$(\sqrt{2} - 1)^{2 \cdot 1} = (\sqrt{2})^2 + 1^2 - 2 \cdot \sqrt{2} \cdot 1$$

$$= 3 - 2\sqrt{2} = \sqrt{9} - \sqrt{8} = \sqrt{9} - \sqrt{9 - 1}.$$

$\therefore$ $P(1)$ is true.

Let $P(m)$ be true, i.e., $(\sqrt{2} - 1)^{2m} = \sqrt{p} - \sqrt{p - 1}$,

where p is a positive integer.

Now, $(\sqrt{2} - 1)^{2m} = (\sqrt{2})^{2m} - {}^{2m}C_1(\sqrt{2})^{2m - 1}$

$$+ {}^{2m}C_2(\sqrt{2})^{2m - 2} - \ldots + {}^{2m}C_{2m}$$

$$= (2^m + {}^{2m}C_2 \cdot 2^{m - 1} + \ldots)$$

$$- \sqrt{2}({}^{2m}C_1 \cdot 2^{m - 1} + {}^{2m}C_3 \cdot 2^{m - 2} + \ldots)$$

$$\equiv \sqrt{p} - \sqrt{p - 1} = \sqrt{p} - \sqrt{2}\sqrt{\frac{p - 1}{2}}$$

$\therefore$ $\sqrt{p} = 2^m + {}^{2m}C_2 \cdot 2^{m - 1} + \ldots = \text{integer}$ ... (1)

$$\sqrt{\frac{p - 1}{2}} = {}^{2m}C_1 \cdot 2^{m - 1} + {}^{2m}C_3 \cdot 2^{m - 2} + \ldots$$

$\therefore$ $\sqrt{\frac{p - 1}{2}} = \text{integer.}$ ... (2)

Next, $(\sqrt{2} - 1)^{2(m + 1)} = (\sqrt{2} - 1)^2 \cdot (\sqrt{2} - 1)^{2m}$

$= (3 - 2\sqrt{2})(\sqrt{p} - \sqrt{p - 1})$, using truth of $P(m)$

$$= 3\sqrt{p} + 2\sqrt{2}\sqrt{p - 1} - 2\sqrt{2}\sqrt{p} - 3\sqrt{p - 1}$$

$$= \left(3\sqrt{p} + 4\sqrt{\frac{p - 1}{2}}\right) - \sqrt{2}\left(2\sqrt{p} + 3\sqrt{\frac{p - 1}{2}}\right)$$

$$= \sqrt{\left(3\sqrt{p} + 4 \cdot \sqrt{\frac{p - 1}{2}}\right)^2} - \sqrt{2\left(2\sqrt{p} + 3 \cdot \sqrt{\frac{p - 1}{2}}\right)^2}$$

By (1) and (2), both

$$\left(3\sqrt{p} + 4 \cdot \sqrt{\frac{p - 1}{2}}\right)^2 \text{ and } 2\left(2\sqrt{p} + 3 \cdot \sqrt{\frac{p - 1}{2}}\right)^2$$

are integers.

But $\left(3\sqrt{p} + 4 \cdot \sqrt{\frac{p - 1}{2}}\right)^2 - 2\left(2\sqrt{p} + 3 \cdot \sqrt{\frac{p - 1}{2}}\right)^2$

$$= 9p + 16\,\frac{p - 1}{2} + 24\sqrt{p} \cdot \sqrt{\frac{p - 1}{2}}$$

$$- 2\left\{4p + 9 \cdot \frac{p - 1}{2} + 12\sqrt{p} \cdot \sqrt{\frac{p - 1}{2}}\right\}$$

$$= p - 2 \cdot \frac{p - 1}{2} = 1.$$

So, if we take $\left(3\sqrt{p} + 4 \cdot \sqrt{\frac{p - 1}{2}}\right)^2 = q$ then

$$2\left(2\sqrt{p} + 3 \cdot \sqrt{\frac{p - 1}{2}}\right)^2 = q - 1.$$

Thus $(\sqrt{2} - 1)^{2(m + 1)} = \sqrt{q} - \sqrt{q - 1}$, where q is an integer.

$\therefore$ $P(m + 1)$ is true.

Thus, $P(m)$ is true $\Rightarrow$ $P(m + 1)$ is true.

Hence, the problem.

24. Let $a_0 = 1$ and $a_n = na_{n - 1} + (-1)^n$ for all $n \geq 1$.

Prove that $a_n = n!\left(1 - \frac{1}{1!} + \frac{1}{2!} - \frac{1}{3!} + \ldots + (-1)^n \frac{1}{n!}\right)$ for all $n \geq 0$.

Let $P(n)$: $a_n = n!\left(1 - \frac{1}{1!} + \frac{1}{2!} - \frac{1}{3!} + \ldots + (-1)^n \frac{1}{n!}\right)$.

Putting $n - 1$ for n,

$$\phi(n) = P(n - 1): \ a_{n - 1} = (n - 1)!\left(1 - \frac{1}{1!} + \frac{1}{2!} - \frac{1}{3!} + \ldots + (-1)^{n - 1} \frac{1}{(n - 1)!}\right).$$

When $n = 1$, $(1-1)!\,(1) = 0! = 1 = a_0$.

$\therefore$ $\phi(1)$ is true.

Let $\phi(m)$ be true, i.e.,

$$a_{m-1} = (m-1)!\left(1 - \frac{1}{1!} + \frac{1}{2!} - \frac{1}{3!} + \dots + (-1)^{m-1}\frac{1}{(m-1)!}\right) \quad \dots (1)$$

Now, $a_m = ma_{m-1} + (-1)^m$

$\{\because a_n = na_{n-1} + (-1)^n, n \ge 1 \text{ (given)}\}$

$$= m\,(m-1)!\left(1 - \frac{1}{1!} + \frac{1}{2!} - \frac{1}{3!} + \dots + (-1)^{m-1}\frac{1}{(m-1)!}\right) + (-1)^m, \text{ using (1)}$$

$$= m!\left(1 - \frac{1}{1!} + \frac{1}{2!} - \frac{1}{3!} + \dots + (-1)^{m-1}\cdot\frac{1}{(m-1)!}\right) + (-1)^m\cdot\frac{1}{m!}\cdot m!$$

$$= m!\left(1 - \frac{1}{1!} + \frac{1}{2!} - \frac{1}{3!} + \dots + (-1)^{m-1}\cdot\frac{1}{(m-1)!} + (-1)^m\cdot\frac{1}{m!}\right)$$

$\therefore$ $\phi(m+1)$ is true;

thus, $\phi(m)$ is true $\Rightarrow$ $\phi(m+1)$ is true.

$\therefore$ by PMI, $\phi(n)$ is true for all $n \in N$

$\therefore$ $\phi(1), \phi(2), \phi(3), \phi(4), \dots$ are true

$\therefore$ $P(0), P(1), P(2), P(3), \dots$ are true

$\{\because \phi(n) = P(n-1)\}$

$\therefore$ $P(n)$ is true for all integers $n \ge 0$.

25. Apply induction to prove that for all $n \ge 2$, $n \in N$,

$$1 + \frac{1}{2} + \frac{1}{3} + \dots + \frac{1}{n} = \frac{k}{h}$$

where k is an odd integer and h is an even integer.

Let $P(n)$: $1 + \frac{1}{2} + \frac{1}{3} + \dots + \frac{1}{n} = \frac{k}{h}$

where k is odd and h is an even integer.

When $n = 1$, $1 = \frac{1}{1} = \frac{\text{odd}}{\text{odd}}$; $\therefore$ $P(1)$ is not true.

When $n = 2$, $1 + \frac{1}{2} = \frac{3}{2} = \frac{\text{odd}}{\text{even}}$; $\therefore$ $P(2)$ is true.

Let $P(m)$ be true, i.e., $1 + \frac{1}{2} + \frac{1}{3} + \dots + \frac{1}{m} = \frac{m_o}{m_e}$

where m_o is an odd integer and m_e is an even integer.

Now, $1 + \frac{1}{2} + \frac{1}{3} + \dots + \frac{1}{m} + \frac{1}{m+1}$

$$= \frac{m_o}{m_e} + \frac{1}{m+1} = \frac{m_o\cdot(m+1) + m_e}{m_e\cdot(m+1)}.$$

If $m+1$ is odd, $\dfrac{m_o(m+1) + m_e}{m_e\cdot(m+1)} = \dfrac{\text{odd} + \text{even}}{\text{even}} = \dfrac{\text{odd}}{\text{even}}$.

If $m+1$ is even, let $m+1 = p_e\cdot k_o$

where p_e = even and k_o = odd. Then

$$\frac{m_o\cdot(m+1) + m_e}{m_e\cdot(m+1)} = \frac{m_o\cdot p_e\cdot k_o + m_e}{m_e\cdot p_e\cdot k_o}.$$

After cancelling the factor of the form 2^t from the numerator and denominator one of p_e and m_e will become odd. So, after cancelling the common factor in the numerator and denominator, one of the terms in the numerator will be odd and the other even. So, the numerator will be odd. As both m_e and p_e are factors in the denominator, after cancelling the common factor, one of the factors in the denominator is even. Hence, the denominator will be even.

$\therefore$ $P(m+1)$ is true.

So, by PMI, $P(n)$ is true for all $n \in N$.

Hence, the problem.

26. If $n \in N$, prove that $\frac{2^n}{n} < \frac{n^n}{n!} < 3^{n-1}$ for $n \ge 3$.

Let $P(n)$: $\dfrac{2^n}{n} < \dfrac{n^n}{n!} < 3^{n-1}$.

Putting $n+2$ for n,

$$\phi(n) = P(n+2): \frac{2^{n+2}}{n+2} < \frac{(n+2)^{n+2}}{(n+2)!} < 3^{n+1}.$$

When $n = 1$, $\dfrac{2^{n+2}}{n+2} = \dfrac{2^3}{3} = \dfrac{8}{3}$

$$\frac{(n+2)^{n+2}}{(n+2)!} = \frac{3^3}{3!} = \frac{27}{6} = \frac{9}{2}$$

$$3^{n+1} = 3^2 = 9.$$

As $\frac{8}{3} < \frac{9}{2} < 9$; so $\phi(1)$ is true.

Let $\phi(m)$ be true, i.e.,

$$\frac{2^{m+2}}{m+2} < \frac{(m+2)^{m+2}}{(m+2)!} < 3^{m+1} \quad \dots (1)$$

Now, $\dfrac{(m+3)^{m+3}}{(m+3)!} - \dfrac{2^{m+3}}{m+3}$

$$= \frac{(m+3)^{m+3}}{(m+3)!} - \frac{2^{m+2}}{m+2} \cdot \frac{m+2}{m+3} \cdot 2$$

$$> \frac{(m+3)^{m+3}}{(m+3)!} - \frac{(m+2)^{m+2}}{(m+2)!} \cdot \frac{m+2}{m+3} \cdot 2,$$

using left part of (1)

$$= \frac{(m+3)^{m+3} - (m+2)^{m+3} \cdot 2}{(m+3)!}$$

$$= \frac{(m+2)^{m+3}}{(m+3)!} \left\{ \left(\frac{m+3}{m+2} \right)^{m+3} - 2 \right\}$$

$$= \frac{(m+2)^{m+3}}{(m+3)!} \left\{ \left(1 + \frac{1}{m+2} \right)^{m+3} - 2 \right\}$$

$$= \frac{(m+2)^{m+3}}{(m+3)!} \times \left\{ 1 + (m+3) \cdot \frac{1}{m+2} + \text{positive terms} - 2 \right\}$$

$$= \frac{(m+2)^{m+3}}{(m+3)!} \left\{ \frac{1}{m+2} + \text{positive terms} \right\} > 0$$

$$\therefore \quad \frac{2^{m+3}}{m+3} < \frac{(m+3)^{m+3}}{(m+3)!} \qquad \dots (2)$$

Now,

$$3^{m+2} - \frac{(m+3)^{m+3}}{(m+3)!}$$

$$> 3 \cdot \frac{(m+2)^{m+2}}{(m+2)!} - \frac{(m+3)^{m+2}}{(m+2)!}, \text{ by (1)}$$

$$= \frac{(m+2)^{m+2}}{(m+2)!} \cdot \left\{ 3 - \left(1 + \frac{1}{m+2} \right)^{m+2} \right\} \qquad \dots (3)$$

Also, $3 - \left(1 + \frac{1}{m+2} \right)^{m+2}$

$$= 3 - \left\{ 1 + 1 + \frac{1 - \frac{1}{m+2}}{2!} + \frac{\left(1 - \frac{1}{m+2}\right)\left(1 - \frac{2}{m+2}\right)}{3!} + \dots \text{ to } \overline{m+3} \text{ terms} \right\}$$

$$> 1 - \left(\frac{1}{2} + \frac{1}{2^2} + \frac{1}{2^3} + \dots \right) = 1 - \left(1 - \frac{1}{2^{m+1}} \right) = \frac{1}{2^{m+1}} > 0$$

$$\therefore \quad \text{from (3), } 3^{m+2} > \frac{(m+3)^{m+3}}{(m+3)!} \qquad \dots (4)$$

From (2) and (4), $\frac{2^{m+3}}{m+3} < \frac{(m+3)^{m+3}}{(m+3)!} < 3^{m+2}$

$\therefore$ $\phi(m+1)$ is true;

$\therefore$ by induction $\phi(n)$ is true for all $n \in N$.

$\therefore$ $\phi(1), \phi(2), \phi(3), \phi(4), \dots$ are true.

$\therefore$ $P(3), P(4), P(5), P(6), \dots$ are true

$\{\because \phi(n) = P(n+2)\}$

$\therefore$ $P(n)$ is true for all $n \geq 3$.

27. If $a_1 = 2$, $a_2 = 5$ and $a_n = 5a_{n-1} - 6a_{n-2}$, $n \geq 3$ then prove that $a_n = 2^{n-1} + 3^{n-1}$ for all $n \geq 1$.

Let $P(n)$: $a_n = 2^{n-1} + 3^{n-1}$.

When $n = 1$, $2^0 + 3^0 = 1 + 1 = 2 = a_1$.

When $n = 2$, $2^1 + 3^1 = 2 + 3 = 5 = a_2$.

$\therefore$ $P(1)$ and $P(2)$ are true.

Let $P(m)$ and $P(m+1)$ be true, i.e.,

$$a_m = 2^{m-1} + 3^{m-1} \quad \text{and} \quad a_{m+1} = 2^m + 3^m.$$

Now, $a_{m+2} = 5a_{m+1} - 6a_m$

$\{\because a_n = 5a_{n-1} - 6a_{n-2}, n \geq 3 \text{ (given)}\}$

$$= 5\{2^m + 3^m\} - 6\{2^{m-1} + 3^{m-1}\}$$

$$= 5 \cdot 2^m + 5 \cdot 3^m - 3 \cdot 2^m - 2 \cdot 3^m$$

$$= 2 \cdot 2^m + 3 \cdot 3^m = 2^{m+1} + 3^{m+1}$$

$\therefore$ $P(m+2)$ is true;

thus, $P(m)$ and $P(m+1)$ are true

$\Rightarrow$ $P(m+2)$ is true.

$\therefore$ by PMI (alternative form), $P(n)$ is true for all $n \in N$, i.e., for $n \geq 1$.

Hence, the problem.

28. Let $s_1 = 2$, $s_2 = 0$, $s_3 = -14$ and $s_{n+1} = 9s_n - 23s_{n-1} + 15s_{n-2}$, $n \geq 3$. Show, by induction, that $s_n = 3^{n-1} - 5^{n-1} + 2$, $n \geq 1$.

Let $P(n)$: $s_n = 3^{n-1} - 5^{n-1} + 2$.

When $n = 1$, $3^{1-1} - 5^{1-1} + 2 = 3^0 - 5^0 + 2 = 2 = s_1$

$n = 2$, $3^{2-1} - 5^{2-1} + 2 = 3 - 5 + 2 = 0 = s_2$

$n = 3$, $3^{3-1} - 5^{3-1} + 2 = 3^2 - 5^2 + 2$

$= 9 - 25 + 2 = -14 = s_3$

$\therefore$ $P(1), P(2), P(3)$ are true.

Let $P(m), P(m+1), P(m+2)$ be true, i.e.,

$$s_m = 3^{m-1} - 5^{m-1} + 2$$

$$s_{m+1} = 3^m - 5^m + 2$$

$$s_{m+2} = 3^{m+1} - 5^{m+1} + 2.$$

Now, $s_{m+3} = 9s_{m+2} - 23s_{m+1} + 15s_m$

$\{\because s_{n+1} = 9s_n - 23s_{n-1} + 15s_{n-2} \text{ (given)}\}$

$= 9(3^{m+1} - 5^{m+1} + 2) - 23(3^m - 5^m + 2)$

$+ 15(3^{m-1} - 5^{m-1} + 2)$

$= 27 \cdot 3^m - 45 \cdot 5^m + 18 - 23 \cdot 3^m + 23 \cdot 5^m$

$- 46 + 5 \cdot 3^m - 3 \cdot 5^m + 30$

$= 9 \cdot 3^m - 25 \cdot 5^m + 2$

$= 3^{m+2} - 5^{m+2} + 2.$

$\therefore$ $P(m+3)$ is true;

thus, $P(m)$, $P(m+1)$, $P(m+2)$ are true

$\Rightarrow$ $P(m+3)$ is true.

$\therefore$ by alternative PMI, $P(n)$ is true for all $n \in N$.

Hence, the problem.

29. Let $p \geq 3$ be an integer and α, β be the roots of $x^2 - (p+1)x + 1 = 0$. Using mathematical induction show that $\alpha^n + \beta^n$ is an integer which is not divisible by p.

Here $\alpha + \beta = p + 1$, $\alpha\beta = 1$.

Let $P(n)$: $\alpha^n + \beta^n = q$

where q is an integer not divisible by p.

When $n = 1$, $\alpha^1 + \beta^1 = p + 1$

= an integer not divisible by p

$(\because p < p + 1 < 2p)$.

When $n = 2$,

$\alpha^2 + \beta^2 = (\alpha + \beta)^2 - 2\alpha\beta = (p+1)^2 - 2$

$= p^2 + 2p - 1 = p(p+2) - 1$

= an integer not divisible by p

$\{\because p(p+1) < p(p+2) - 1 < p(p+2)\}$

$\therefore$ $P(1)$ and $P(2)$ are true.

Let $P(m)$, $m = 1, 2, 3, \ldots, m$ be true, i.e.,

$\alpha^m + \beta^m = r_m$ ... (1)

where r_m is an integer not divisible by p; $m = 1, 2, \ldots, m$.

Now, $\alpha^{m+1} + \beta^{m+1} = (\alpha + \beta)(\alpha^m + \beta^m) - \alpha\beta^m - \beta\alpha^m$

$= (\alpha + \beta)(\alpha^m + \beta^m) - \alpha\beta(\alpha^{m-1} + \beta^{m-1})$

$= (p+1)r_m - r_{m-1}$ = integer

$= pr_m + (r_m - r_{m-1})$ ... (2)

If possible, let $r_m - r_{m-1} = tp$ (i.e., divisible by p)

$\therefore$ $\alpha^m + \beta^m - (\alpha^{m-1} + \beta^{m-1}) = tp$

or $(\alpha + \beta)(\alpha^{m-1} + \beta^{m-1}) - \alpha\beta^{m-1} - \beta\alpha^{m-1}$

$- (\alpha^{m-1} + \beta^{m-1}) = tp$

or $p(\alpha^{m-1} + \beta^{m-1}) - \alpha\beta(\alpha^{m-2} + \beta^{m-2}) = tp$

$\{\because \alpha + \beta = p + 1\}$

or $p(\alpha^{m-1} + \beta^{m-1}) - (\alpha^{m-2} + \beta^{m-2}) = tp$, $\{\because \alpha\beta = 1\}$

$\therefore$ $\alpha^{m-2} + \beta^{m-2}$

is divisible by p which contradicts (1)

$\therefore$ $r_m - r_{m-1}$ is not divisible by p.

So, from (2), $\alpha^{m+1} + \beta^{m+1}$ is not divisible by p.

$\therefore$ $P(m+1)$ is true.

$\therefore$ by alternative form of induction, $P(n)$ is true for all positive integral values of n.

Hence, the problem.

30. Prove, by induction, that

$$\sin\alpha + \cos(\alpha + \beta) - \sin(\alpha + 2\beta) - \cos(\alpha + 3\beta) + \sin(\alpha + 4\beta) + \ldots \text{ to } n \text{ terms}$$

$$= \frac{\sin\frac{1}{4}n(2\beta + \pi)}{\sin\frac{1}{4}(2\beta + \pi)} \sin\left\{\alpha + \frac{1}{4}(n-1)(2\beta + \pi)\right\}$$

where $\sin\frac{1}{4}(2\beta + \pi) \neq 0$.

The series on the left side is

$$\sin\alpha + \sin\left(\alpha + \beta + \frac{\pi}{2}\right) + \sin(\alpha + 2\beta + \pi) + \sin\left(\alpha + 3\beta + \frac{3\pi}{2}\right) + \sin(\alpha + 4\beta + 2\pi) + \ldots \text{ to } n \text{ terms} \quad \ldots (1)$$

Let $P(n)$: $\sin\alpha + \sin\left(\alpha + \beta + \frac{\pi}{2}\right) + \sin(\alpha + 2\beta + \pi) + \sin\left(\alpha + 3\beta + \frac{3\pi}{2}\right) + \ldots$ to n terms

$$= \frac{\sin\frac{1}{4}n(2\beta + \pi)}{\sin\frac{1}{4}(2\beta + \pi)} \sin\left\{\alpha + \frac{1}{4}(n-1)(2\beta + \pi)\right\}.$$

When $n = 1$, $\sin\alpha = \dfrac{\sin\frac{1}{4}(2\beta + \pi)}{\sin\frac{1}{4}(2\beta + \pi)} \cdot \sin\alpha$

$$= \frac{\sin\frac{1}{4}(2\beta + \pi)}{\sin\frac{1}{4}(2\beta + \pi)} \cdot \sin\left\{\alpha + \frac{1}{4}(1-1)(2\beta + \pi)\right\}$$

$\therefore$ $P(1)$ is true.

Let $P(m)$ be true, i.e.,

$$\sin\alpha + \sin\left(\alpha+\beta+\frac{\pi}{2}\right) + \sin(\alpha+2\beta+\pi) + \sin\left(\alpha+3\beta+\frac{3\pi}{2}\right) + \dots \text{ to } m \text{ terms}$$

$$= \frac{\sin\frac{1}{4}m(2\beta+\pi)}{\sin\frac{1}{4}(2\beta+\pi)}\sin\left\{\alpha+\frac{1}{4}(m-1)(2\beta+\pi)\right\} \quad \dots (2)$$

Now,

$$\sin\alpha + \sin\left(\alpha+\beta+\frac{\pi}{2}\right) + \sin(\alpha+2\beta+\pi) + \sin\left(\alpha+3\beta+\frac{3\pi}{2}\right) + \dots \text{ to } (m+1) \text{ terms}$$

$$= \{\sin\alpha + \sin\left(\alpha+\beta+\frac{\pi}{2}\right) + \sin(\alpha+2\beta+\pi) + \sin\left(\alpha+3\beta+\frac{3\pi}{2}\right) + \dots \text{ to } m \text{ terms}\}$$
$$+ \sin\left(\alpha+m\beta+m\cdot\frac{\pi}{2}\right)$$

$$= \frac{\sin\frac{1}{4}m(2\beta+\pi)}{\sin\frac{1}{4}(2\beta+\pi)}\cdot\sin\left\{\alpha+\frac{1}{4}(m-1)(2\beta+\pi)\right\} + \sin\left(\alpha+m\beta+\frac{m\pi}{2}\right),$$

[using (2)]

$$= \frac{1}{\sin\frac{1}{4}(2\beta+\pi)} \times \left\{\sin\frac{1}{4}m(2\beta+\pi)\sin\left\{\alpha+\frac{1}{4}(m-1)(2\beta+\pi)\right\} + \sin\frac{1}{4}(2\beta+\pi)\sin\left(\alpha+m\beta+\frac{m\pi}{2}\right)\right\}$$

$$= \frac{1}{2\sin\frac{1}{4}(2\beta+\pi)} \times \Big[\cos\left\{\alpha+\frac{1}{4}(m-1)(2\beta+\pi)-\frac{1}{4}m(2\beta+\pi)\right\}$$
$$- \cos\left\{\alpha+\frac{1}{4}(m-1)(2\beta+\pi)+\frac{1}{4}m(2\beta+\pi)\right\}$$
$$+ \cos\left\{\alpha+m\beta+\frac{m\pi}{2}-\frac{1}{4}(2\beta+\pi)\right\}$$
$$- \cos\left\{\alpha+m\beta+\frac{m\pi}{2}+\frac{1}{4}(2\beta+\pi)\right\}\Big]$$

$$= \frac{1}{2\sin\frac{1}{4}(2\beta+\pi)}\Big[\cos\left\{\alpha-\frac{1}{4}(2\beta+\pi)\right\}$$
$$- \cos\left\{\alpha+\frac{1}{2}m(2\beta+\pi)-\frac{1}{4}(2\beta+\pi)\right\}$$
$$+ \cos\left\{\alpha+\frac{1}{2}m(2\beta+\pi)-\frac{1}{4}(2\beta+\pi)\right\}$$
$$- \cos\left\{\alpha+\frac{1}{2}m(2\beta+\pi)+\frac{1}{4}(2\beta+\pi)\right\}\Big]$$

$$= \frac{1}{2\sin\frac{1}{4}(2\beta+\pi)}\Big[\cos\left\{\alpha-\frac{1}{4}(2\beta+\pi)\right\} - \cos\left\{\alpha+\frac{1}{2}m(2\beta+\pi)+\frac{1}{4}(2\beta+\pi)\right\}\Big]$$

$$= \frac{1}{2\sin\frac{1}{4}(2\beta+\pi)} \times 2\sin\frac{\alpha-\frac{1}{4}(2\beta+\pi)+\alpha+\frac{1}{2}m(2\beta+\pi)+\frac{1}{4}(2\beta+\pi)}{2}$$
$$\times \sin\frac{\alpha+\frac{1}{2}m(2\beta+\pi)+\frac{1}{4}(2\beta+\pi)-\alpha+\frac{1}{4}(2\beta+\pi)}{2}$$

$$= \frac{\sin\frac{m+1}{4}(2\beta+\pi)}{\sin\frac{1}{4}(2\beta+\pi)}\cdot\sin\left\{\alpha+\frac{1}{4}(\overline{m+1}-1)(2\beta+\pi)\right\}$$

$\therefore$ $P(m+1)$ is true;

thus, $P(m)$ is true $\Rightarrow$ $P(m+1)$ is true.

$\therefore$ $P(n)$ is true for all $n \in N$.

Hence, the problem.

31. Prove, by induction, that

$$\sum_{r=1}^{n} \tan^{-1}\frac{2r}{2+r^2+r^4} = \tan^{-1}\{1+n(n+1)\} - \frac{\pi}{4},$$

for all $n \in N$.

Let $P(n)$: $\sum_{r=1}^{n} \tan^{-1}\dfrac{2r}{2+r^2+r^4} = \tan^{-1}\{1+n(n+1)\} - \dfrac{\pi}{4}$.

When $n = 1$,

$$\sum_{r=1}^{1} \tan^{-1}\frac{2r}{2+r^2+r^4} = \tan^{-1}\frac{2\cdot 1}{2+1^2+1^4}$$

$$= \tan^{-1}\frac{2}{4} = \tan^{-1}\frac{1}{2}$$

$$= \tan^{-1}\frac{2}{4} = \tan^{-1}\frac{3-1}{1+3\cdot 1} = \tan^{-1}3 - \tan^{-1}1$$

$$= \tan^{-1}\{1 + 1\cdot(1+1)\} - \frac{\pi}{4}.$$

$\therefore$ $P(1)$ is true.

Let $P(m)$ be true, i.e.,

$$\sum_{r=1}^{m} \tan^{-1}\frac{2r}{2+r^2+r^4} = \tan^{-1}\{1+m(m+1)\} - \frac{\pi}{4} \quad \ldots (1)$$

Now, $$\sum_{r=1}^{m+1} \tan^{-1}\frac{2r}{2+r^2+r^4} = \sum_{r=1}^{m} \tan^{-1}\frac{2r}{2+r^2+r^4} + \tan^{-1}\frac{2(m+1)}{2+(m+1)^2+(m+1)^4}$$

$$= \tan^{-1}\{1+m(m+1)\} - \frac{\pi}{4} + \tan^{-1}\frac{2(m+1)}{2+(m+1)^2+(m+1)^4} \quad \ldots (2)$$

Also, $\tan^{-1}\{1+(m+1)(m+2)\} - \tan^{-1}\{1+m(m+1)\}$

$$= \tan^{-1}\frac{1+(m+1)(m+2) - \{1+m(m+1)\}}{1+\{1+(m+1)(m+2)\}\{1+m(m+1)\}}$$

$$= \tan^{-1}\frac{2(m+1)}{1+1+m(m+1)+(m+1)(m+2)+m(m+1)^2(m+2)}$$

$$= \tan^{-1}\frac{2(m+1)}{2+(m+1)(2m+2)+(m+1)^2(m^2+2m)}$$

$$= \tan^{-1}\frac{2(m+1)}{2+(m+1)^2+(m+1)^2+(m+1)^2(m^2+2m)}$$

$$= \tan^{-1}\frac{2(m+1)}{2+(m+1)^2+(m+1)^2(1+m^2+2m)}$$

$$= \tan^{-1}\frac{2(m+1)}{2+(m+1)^2+(m+1)^4}$$

$\therefore$ from (2),

$$\sum_{r=1}^{m+1} \tan^{-1}\frac{2r}{2+r^2+r^4} = \tan^{-1}\{1+(m+1)(m+2)\} - \frac{\pi}{4}.$$

$\therefore$ $P(m+1)$ is true;

thus, $P(m)$ is true $\Rightarrow$ $P(m+1)$ is true.

$\therefore$ by PMI, $P(n)$ is true for all $n \in N$.

Hence, the problem.

32. If $y = \tan^{-1}x$, prove that

$$\frac{d^ny}{dx^n} = (n-1)!\cdot\cos\left\{ny + (n-1)\frac{\pi}{2}\right\}\cos^n y$$

for all $n \in N$.

Let $P(n)$: $\dfrac{d^ny}{dx^n} = (n-1)!\cos\left\{ny+(n-1)\dfrac{\pi}{2}\right\}\cos^n y.$

When $n = 1$, $\dfrac{dy}{dx} = \dfrac{d}{dx}(\tan^{-1}x) = \dfrac{1}{1+x^2} = \dfrac{1}{1+\tan^2 y}$ $\{\because y = \tan^{-1}x\}$

$$= \frac{1}{\sec^2 y} = \cos^2 y$$

$$= (1-1)!\cdot\cos\left\{1\cdot y + (1-1)\frac{\pi}{2}\right\}\cos y$$

$\therefore$ $P(1)$ is true.

Let $P(m)$ be true, i.e.,

$$\frac{d^my}{dx^m} = (m-1)!\cos\left\{my+(m-1)\frac{\pi}{2}\right\}\cos^m y \quad \ldots (1)$$

Now, $$\frac{d^{m+1}y}{dx^{m+1}} = \frac{d}{dx}\left(\frac{d^my}{dx^m}\right)$$

$$= \frac{d}{dx}\left[(m-1)!\cos\left\{my+(m-1)\frac{\pi}{2}\right\}\cos^m y\right]$$

$$= (m-1)!\left[-\sin\left\{my+(m-1)\frac{\pi}{2}\right\} m\frac{dy}{dx}\cdot\cos^m y + \cos\left\{my+(m-1)\frac{\pi}{2}\right\} m\cos^{m-1}y\cdot(-\sin y)\frac{dy}{dx}\right]$$

$$= m!\left[-\sin\left\{my+(m-1)\frac{\pi}{2}\right\}\cos^m y - \cos\left\{my+(m-1)\frac{\pi}{2}\right\}\cos^{m-1}y\sin y\right]\frac{dy}{dx}$$

$$= m!\left[-\sin\left\{my+(m-1)\frac{\pi}{2}\right\}\cos y - \cos\left\{my+(m-1)\frac{\pi}{2}\right\}\sin y\right]\cos^{m-1}y\cdot\cos^2 y,$$

$$\left(\because \frac{dy}{dx} = \cos^2 y\right)$$

$$= m!\left[-\sin\left\{my+(m-1)\frac{\pi}{2}+y\right\}\right]\cos^{m+1}y$$

$$= m!\left[\cos\left\{(m+1)y+(m-1)\frac{\pi}{2}+\frac{\pi}{2}\right\}\right]\cos^{m+1}y$$

$$= m!\cos\left\{(m+1)y+(\overline{m+1}-1)\frac{\pi}{2}\right\}\cos^{m+1}y$$

$\therefore$ $P(m+1)$ is true;

thus, $P(m)$ is true $\Rightarrow$ $P(m+1)$ is true.

$\therefore$ $P(n)$ is true for all $n \in N$.

Hence, the problem.

33. Prove, by induction, that $\int_0^{\pi/2} \cos^n x \cdot \cos(nx)dx = \frac{\pi}{2^{n+1}}$ for all $n \in N$.

Let $P(n)$: $\int_0^{\pi/2} \cos^n x \cdot \cos(nx)dx = \frac{\pi}{2^{n+1}}$.

If $n = 1$, $\int_0^{\pi/2} \cos x \cdot \cos x dx = \int_0^{\pi/2} \frac{1+\cos 2x}{2} dx$

$= \frac{1}{2}\int_0^{\pi/2} dx + \frac{1}{2}\int_0^{\pi/2} \cos 2x\, dx$

$= \frac{1}{2}[x]_0^{\pi/2} + \frac{1}{2}\left[\frac{\sin 2x}{2}\right]_0^{\pi/2}$

$= \frac{1}{2}\cdot\frac{\pi}{2} + \frac{1}{4}[\sin\pi - \sin 0] = \frac{\pi}{4} = \frac{\pi}{2^{1+1}}$.

$\therefore$ $P(1)$ is true.

Let $P(m)$ be true, i.e., $\int_0^{\pi/2} \cos^m x \cdot \cos(mx)dx = \frac{\pi}{2^{m+1}}$.

Now, $\int_0^{\pi/2} \cos^{m+1} x \cdot \cos(m+1)x\, dx$

$= \frac{1}{2}\int_0^{\pi/2} \cos^m x \cdot 2\cos x \cdot \cos(m+1)x\, dx$

$= \frac{1}{2}\int_0^{\pi/2} \cos^m x \cdot \{\cos mx + \cos(m+2)x\}dx$

$= \int_0^{\pi/2} \cos^m x \cos mx dx$

$\quad - \frac{1}{2}\int_0^{\pi/2} \cos^m x \{\cos mx - \cos(m+2)x\}dx$

$= \int_0^{\pi/2} \cos^m x \cos mx dx$

$\quad - \frac{1}{2}\int_0^{\pi/2} \cos^m x \cdot 2\sin(m+1)x \cdot \sin x\, dx$

$= \frac{\pi}{2^{m+1}} + \int_0^{\pi/2} \sin(m+1)x \cdot \cos^m x \cdot \frac{d}{dx}(\cos x)\, dx$

$= \frac{\pi}{2^{m+1}} + \left\{\left[\sin(m+1)x \cdot \frac{\cos^{m+1} x}{m+1}\right]_0^{\pi/2}\right.$

$\quad \left. - \int_0^{\pi/2} \frac{\cos^{m+1} x}{m+1} \cdot \cos(m+1)x \cdot (m+1)dx\right\}$

$= \frac{\pi}{2^{m+1}} - \int_0^{\pi/2} \cos^{m+1} x \cdot \cos(m+1)x\, dx$

$\therefore$ $2\int_0^{\pi/2} \cos^{m+1} x \cdot \cos(m+1)x\, dx = \frac{\pi}{2^{m+1}}$

$\therefore$ $\int_0^{\pi/2} \cos^{m+1} x \cdot \cos(m+1)x\, dx = \frac{\pi}{2^{m+2}}$;

$\therefore$ $P(m+1)$ is true;

thus, $P(m)$ is true $\Rightarrow$ $P(m+1)$ is true.

$\therefore$ by PMI, $P(n)$ is true for all $n \in N$.

Hence, the problem.

34. Prove that $\int_0^{\pi/2} \frac{\sin^2 nx}{\sin x} dx = 1 + \frac{1}{3} + \frac{1}{5} + \ldots + \frac{1}{2n-1}$ for all $n \in N$.

Let $P(n)$: $\int_0^{\pi/2} \frac{\sin^2 nx}{\sin x} dx = 1 + \frac{1}{3} + \frac{1}{5} + \ldots + \frac{1}{2n-1}$.

When $n = 1$,

$\int_0^{\pi/2} \frac{\sin^2 x}{\sin x} dx = \int_0^{\pi/2} \sin x\, dx = [-\cos x]_0^{\pi/2} = 0 + 1 = 1.$

$\therefore$ $P(1)$ is true.

Let $P(m)$ be true, i.e.,

$$\int_0^{\pi/2} \frac{\sin^2 mx}{\sin x} dx = 1 + \frac{1}{3} + \frac{1}{5} + \ldots + \frac{1}{2m-1} \quad \ldots (1)$$

Now, $\int_0^{\pi/2} \frac{\sin^2(m+1)x}{\sin x} dx - \int_0^{\pi/2} \frac{\sin^2 mx}{\sin x} dx$

$= \int_0^{\pi/2} \frac{\sin^2(m+1)x - \sin^2 mx}{\sin x} dx$

$= \int_0^{\pi/2} \frac{\sin(2m+1)x \cdot \sin x}{\sin x} dx$

$= \int_0^{\pi/2} \sin(2m+1)x\, dx$

$= \left[-\frac{\cos(2m+1)x}{2m+1}\right]_0^{\pi/2} = \frac{1}{2m+1}$

$\therefore$ $\int_0^{\pi/2} \frac{\sin^2(m+1)x}{\sin x} dx$

$= \int_0^{\pi/2} \frac{\sin^2 mx}{\sin x} dx + \frac{1}{2m+1}$

$= 1 + \frac{1}{3} + \frac{1}{5} + \ldots + \frac{1}{2m-1} + \frac{1}{2(m+1)-1}$

$\therefore$ $P(m+1)$ is true;

by PMI, $P(n)$ is true for all $n \in N$.

Hence, the problem.

35. Prove that $\int_0^1 x^{m-1}(\log x)^n dx = \frac{(-1)^n n!}{m^{n+1}}$ for all $n \in N$ where m is a fixed number.

Taking any fixed number m, let

$P(n)$: $\int_0^1 x^{m-1}(\log x)^n dx = \frac{(-1)^n n!}{m^{n+1}}$.

When $n = 1$,

$$\int_0^1 x^{m-1}\log x\,dx = \left[\log x \cdot \frac{x^m}{m}\right]_0^1 - \int_0^1 \frac{x^m}{m}\cdot\frac{1}{x}\,dx$$

$$= -\int_0^1 \frac{1}{m}x^{m-1}\,dx = -\left[\frac{x^m}{m^2}\right]_0^1 = -\frac{1}{m^2} = \frac{(-1)^1}{m^{1+1}},$$

because $\log 1 = 0$ and

$$\lim_{x\to 0}(\log x \cdot x^m)$$

$$= \lim_{x\to 0}\left(\frac{\log x}{x^{-m}}\right) = \lim_{x\to 0}\frac{\frac{1}{x}}{-mx^{-m-1}}$$

(using L' Hospital's rule)

$$= \lim_{x\to 0}\frac{x^m}{-m} = 0.$$

$\therefore$ $P(1)$ is true.

Let $P(k)$ be true, i.e.,

$$\int_0^1 x^{m-1}(\log x)^k\,dx = \frac{(-1)^k k!}{m^{k+1}} \quad \ldots (1)$$

Now, $\int_0^1 x^{m-1}(\log x)^{k+1}\,dx = \left[(\log x)^{k+1}\cdot\frac{x^m}{m}\right]_0^1$

$$-\int_0^1 \frac{x^m}{m}\cdot(k+1)(\log x)^k\cdot\frac{1}{x}\,dx$$

$$= 0 - \int_0^1 \frac{k+1}{m}\cdot x^{m-1}(\log x)^k\,dx$$

$$= -\frac{k+1}{m}\cdot\frac{(-1)^k k!}{m^{k+1}}, \text{ using (1)}$$

$$= (-1)^{k+1}\cdot\frac{(k+1)!}{m^{k+2}}$$

$\therefore$ $P(k+1)$ is true.

$\therefore$ by PMI, $P(n)$ is true for all $n \in N$.

Hence, the problem.

36. Let f be a function such that $f(x)$ is a positive real number when x is a positive real number. Let $f(xy) = f(x) + f(y)$ for all real $x > 0, y > 0$. Prove, by induction, that $f(a^n) = kn$ where a, k are positive constants and n is an integer ≥ 0.

Let $P(n)$: $f(a^n) = kn$.

Putting $n - 1$ for n, $\phi(n) = P(n-1)$: $f(a^{n-1}) = k(n-1)$.

When $n = 1$, $f(a^{1-1}) = f(a^0) = f(1) = 0 = k(1-1)$

{$\because$ putting $x = 1, y = 1$ in $f(xy) = f(x) + f(y)$ we get $f(1) = 0$}

$\therefore$ $\phi(1)$ is true.

Let $\phi(m)$ be true, i.e.,

$f(a^{m-1}) = k_1(m-1)$, k_1 is a positive constant.

But $f(a^{m-1}) = f(a^{m-2}\cdot a) = f(a^{m-2}) + f(a)$

$\{\because f(xy) = f(x) + f(y)\}$

$= f(a^{m-3}\cdot a) + f(a) = f(a^{m-3}) + f(a) + f(a)$

....

$= f(a^0) + (m-1)f(a) = (m-1)f(a)$, $\{\because f(1) = 0\}$

$\therefore$ $k_1(m-1) = (m-1)f(a)$;

$\therefore$ $f(a) = k_1$, $(\because m > 1)$

Now, $f(a^m) = f(a^{m-1}\cdot a) = f(a^{m-1}) + f(a)$,

$= k_1(m-1) + f(a) = k_1(m-1) + k_1$

$= k_1 m = k_1\overline{(m+1-1)}$

$\therefore$ $\phi(m+1)$ is true;

thus, $\phi(m)$ is true $\Rightarrow$ $\phi(m+1)$ is true.

$\therefore$ $\phi(n)$ is true for all $n \in N$.

$\therefore$ $P(n)$ is true for $n \in N, n \geq 0$.

Hence, the problem.

37. Use the principle of induction to prove that n straight lines in a plane divide the plane into $\frac{n^2+n+2}{2}$ parts if no two lines are parallel and no three lines are concurrent.

Let the number of parts made by n lines of the given variety be L_n. So the statement is

$$P(n):\ L_n = \frac{n^2+n+2}{2}.$$

When $n = 1$,

L_1 = the number of parts made by 1 line

$$= 2 = \frac{1^2+1+2}{2}$$

$\therefore$ $P(1)$ is true.

Let $P(m)$ be true, i.e., $L_m = \frac{m^2+m+2}{2}$.

Take the $(m+1)$th line. It is cut by the previous m lines at m points making $(m+1)$ subdivisions of the $(m+1)$th line. Each of these $(m+1)$ sections divide one of L_m regions into two parts. So the total number of regions will be increased by $(m+1)$. So

$$L_{m+1} = \frac{m^2+m+2}{2} + (m+1)$$

$$= \frac{m^2+3m+4}{2}$$

$$= \frac{(m+1)^2+(m+1)+2}{2}$$

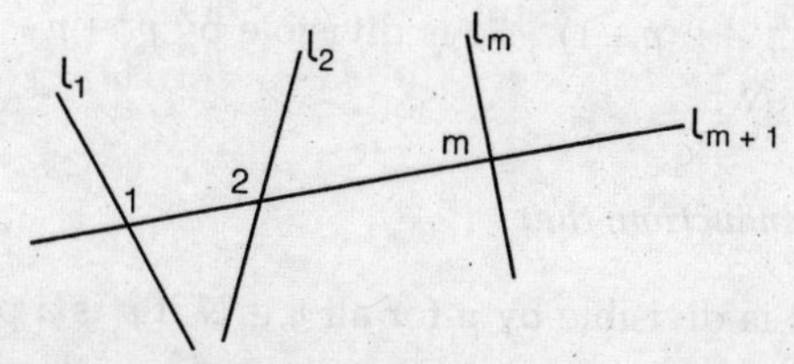

$\therefore$ $P(m+1)$ is true.

Hence, by induction, $P(n)$ is true for all $n \in N$.

Hence, the problem.

38. Prove, by induction, that any amount in full rupees greater than or equal to Rs 8 can be paid by stamps of denominations Rs 3 and Rs 5.

The problem is solved if we can prove that the equation $3x + 5y = n, n \geq 8$ and $n \in N$ has non-negative integral solutions.

Let $P(n)$: $3x + 5y = n + 7$.

When $n = 1, n + 7 = 1 + 7 = 8 = 3 \cdot 1 + 5 \cdot 1$;

so $x = 1, y = 1$.

$\therefore$ $P(1)$ is true.

Let $P(m)$ be true, i.e., x and y have non-negative integral solutions when $3x + 5y = m + 7$ ... (1)

Now, $(m+1) + 7 = (m+7) + 1 = 3x + 5y + 1$.

If in (1), $y = 0$ then $x \geq 3$ and $m + 7 = 3x$

$\therefore$ $m + 8 = 3x + 1 = 3(x-3) + 10 = 3(x-3) + 5 \cdot 2$.

If in (1), $x = 0$ then $y \geq 2$ and $m + 7 = 5y$.

$\therefore$ $m + 8 = 5y + 1 = 5(y-1) + 6 = 3 \cdot 2 + 5(y-1)$.

If in (1), $x \neq 0, y \neq 0$ then $x \geq 2, y \geq 1$ or $x \geq 1, y \geq 2$.

When $x \geq 2, y \geq 1, m + 7 = 3(x-2) + 5(y-1)$

$\therefore$ $m + 8 = 3(x-2) + 5(y-1) + 1$

$= 3(x-2) + 5(y-2) + 6 = 3x + 5(y-2)$

When $x \geq 1, y \geq 2$, $m + 7 = 3(x-1) + 5(y-2)$

$\therefore$ $m + 8 = 3(x-1) + 5(y-2) + 1$

$= 3(x+1) + 5(y-3)$.

Thus $3x + 5y = (m+1) + 7$ has non-negative integral solution.

$\therefore$ $P(m+1)$ is true.

So, by PMI, $P(n)$ is true for all $n \in N$ i.e., $3x + 5y = n + 7$ has non-negative integral solutions for all $n \in N$.

39. k is any positive integer. Prove by using induction that ${}^mC_0\,{}^nC_k + {}^mC_1\,{}^nC_{k-1} + \ldots + {}^mC_k\,{}^nC_0 = {}^{m+n}C_k$ for all $m, n \in N$ where ${}^pC_q = 0$ if $p < q$.

Taking n to be a fixed integer, the statement is

$P(m)$: ${}^mC_0\,{}^nC_k + {}^mC_1\,{}^nC_{k-1} + \ldots + {}^mC_k\,{}^nC_0 = {}^{m+n}C_k$

When $m = 1$, ${}^1C_0\,{}^nC_k + {}^1C_1\,{}^nC_{k-1} = {}^nC_k + {}^nC_{k-1} = {}^{1+n}C_k$

$\therefore$ $P(1)$ is true.

Let $P(r)$ be true, i.e.,

$${}^rC_0\,{}^nC_k + {}^rC_1\,{}^nC_{k-1} + \ldots + {}^rC_k\,{}^nC_0 = {}^{r+n}C_k \quad \ldots (1)$$

Now, ${}^{r+1}C_0\,{}^nC_k + {}^{r+1}C_1 \cdot {}^nC_{k-1} + \ldots + {}^{r+1}C_k \cdot {}^nC_0$

$= {}^rC_0\,{}^nC_k + ({}^rC_0 + {}^rC_1)\,{}^nC_{k-1} + ({}^rC_1 + {}^rC_2)\,{}^nC_{k-2}$

$+ \ldots + ({}^rC_{k-1} + {}^rC_k)\,{}^nC_0$

$\{\because {}^nC_r + {}^nC_{r-1} = {}^{n+1}C_r, {}^{r+1}C_0 = {}^rC_0 = 1\}$

$= ({}^rC_0\,{}^nC_k + {}^rC_1\,{}^nC_{k-1} + {}^rC_2\,{}^nC_{k-2} + \ldots + {}^rC_k\,{}^nC_0)$

$+ ({}^rC_0\,{}^nC_{k-1} + {}^rC_1 \cdot {}^nC_{k-2} + \ldots + {}^rC_{k-1}\,{}^nC_0)$

$= {}^{r+n}C_k + {}^{r+n}C_{k-1}$, using (1) for k and $k-1$

$= {}^{(r+1)+n}C_k$

$\therefore$ $P(r+1)$ is true.

Hence, $P(m)$ is true for all $n \in N$, by induction.

Taking m to be a fixed integer, the statement is

$Q(n)$: ${}^mC_0\,{}^nC_k + {}^mC_1\,{}^nC_{k-1} + \ldots + {}^mC_k\,{}^nC_0 = {}^{m+n}C_k$

When $n = 1$,

${}^mC_0\,{}^1C_k + {}^mC_1\,{}^1C_{k-1} + {}^mC_2\,{}^1C_{k-2}$

$+ \ldots + {}^mC_{k-1}\,{}^1C_1 + {}^mC_k\,{}^1C_0$

$= {}^mC_{k-1} + {}^mC_k = {}^{m+1}C_k$ $(\because {}^pC_q = 0 \text{ if } p < q)$

$\therefore$ $Q(1)$ is true.

Let $Q(t)$ be true, i.e.,

$${}^mC_0\,{}^tC_k + {}^mC_1\,{}^tC_{k-1} + \ldots + {}^mC_k\,{}^tC_0 = {}^{m+t}C_k \quad \ldots (2)$$

Now, ${}^mC_0\,{}^{t+1}C_k + {}^mC_1\,{}^{t+1}C_{k-1} + \ldots + {}^mC_k\,{}^{t+1}C_0$

$= {}^mC_0({}^tC_k + {}^tC_{k-1}) + {}^mC_1({}^tC_{k-1} + {}^tC_{k-2})$

$+ \ldots + {}^mC_{k-1}({}^tC_1 + {}^tC_0) + {}^mC_k\,{}^{t+1}C_0$

$= ({}^mC_0 \cdot {}^tC_k + {}^mC_1 \cdot {}^tC_{k-1} + \ldots + {}^mC_{k-1}\,{}^tC_1 + {}^mC_k\,{}^tC_0)$

$+ ({}^mC_0\,{}^tC_{k-1} + {}^mC_1 \cdot {}^tC_{k-2} + \ldots + {}^mC_{k-1}\,{}^tC_0)$

$= {}^{m+t}C_k + {}^{m+t}C_{k-1}$, using (2) for k and $k-1$

$= {}^{m+t+1}C_k$

$\therefore$ $Q(t+1)$ is true.

Hence, by induction $Q(n)$ is true for all $n \in N$.

Thus the statement is true for all $m, n \in N$.

Exercises

Prove, by mathematical induction, that

1. $1+2+3+\ldots+n=\dfrac{n(n+1)}{2}$.
2. $1+3+5+\ldots$ to n terms $=n^2$.
3. $1^2+2^2+3^2+\ldots+n^2=\dfrac{n(n+1)(2n+1)}{6}$.
4. $a+ar+ar^2+\ldots+ar^{n-1}=\dfrac{a(1-r^n)}{1-r}$.
5. $\dfrac{1}{1\cdot 2}+\dfrac{1}{2\cdot 3}+\dfrac{1}{3\cdot 4}+\ldots+\dfrac{1}{n(n+1)}=\dfrac{n}{n+1}$.
6. $2+222+22222+\ldots$ to n terms $=\dfrac{20}{891}(10^{2n}-1)-\dfrac{2n}{9}$ for all $n\in N$.
7. $\left(1-\dfrac{1}{2}\right)\left(1-\dfrac{1}{3}\right)\left(1-\dfrac{1}{4}\right)\ldots\left(1-\dfrac{1}{n+1}\right)=\dfrac{1}{n+1}$ for all $n\in N$.
8. $7^{2n}+2^{3n-3}\cdot 3^{n-1}$ is divisible by 25 for $n\in N$.
9. $10^{2n-1}+1$ is divisible by 11, where $n\in N$.
10. $2^{5n+5}-31n-32$ is divisible by 961 for all positive integral values of n.
11. Use induction to show that $11^{n+2}+12^{2n+1}$ is divisible by 133 for $n\in N$.
12. Prove that $3^{2n+1}+2^{n+2}$ is a multiple of 7 where $n\in N$.
13. Prove that $3^{4n+1}+16n-3$ is divisible by 256 for all $n\in N$.
14. Prove that n^7-n is divisible by 42 for $n\in N$.
15. Prove that n^3+11n is divisible by 6 for all $n\in N$.
16. For every natural number n, prove by induction that $4^n+15n-1$ is divisible by 9.
17. If $n\,|\,k$ denotes k is divisible by n then prove by induction, that

    $512\,|\,(3^{2n+5}+160n^2-56n-243)$ for all $n\in N$.
18. Prove, by induction, that $8\cdot 7^n+4^{n+2}$ is divisible by 24 but not by 48 for all $n\in N$.
19. Show that $5^{n+1}+4\cdot 6^n$, when divided by 20 leaves the remainder 9 for all $n\in N$. Use induction.
20. Prove that $f(n)=n^5-5n^3+60n^2-56n$ is a multiple of 120 for all $n\in N$.
21. Show, by induction, that $3^{2n}+1$ is even but not divisible by 4 for all $n\in N$.
22. Prove, by induction, that x^n-y^n is divisible by $x-y$ for all $n\in N$.
23. Prove that $(1+x)^n-nx-1$ is divisible by x^2 for all $n\in N$. Use induction.
24. If p is a fixed positive integer, prove by induction that $p^{n+1}+(p+1)^{2n-1}$ is divisible by p^2+p+1 for all $n\in N$.

Prove, by induction, that

25. n^p-n is divisible by p for all $n\in N$ if p is a prime number.
26. $\dfrac{n^5}{5}+\dfrac{n^3}{3}+\dfrac{7n}{15}\in N$ for all $n\in N$, N being the set of natural numbers.
27. $\dfrac{n^7}{7}+\dfrac{n^5}{5}+\dfrac{2n^3}{3}-\dfrac{n}{105}$ is an integer for every $n\in N$.
28. $\dfrac{n^7}{7}+\dfrac{n^5}{5}+\dfrac{n^3}{3}+\dfrac{n^2}{2}-\dfrac{37n}{210}$ is a positive integer for all $n\in N$.
29. $(r+1)(r+2)(r+3)(r+4)(r+5)$ is divisible by 120 for all $r\in N$.
30. Prove that $\sum\limits_{r=0}^{n}{}^nC_r=2^n$. Use induction.
31. Prove, by induction, that

    $$\sum_{k=1}^{n}k^2\cdot{}^nC_k=n(n+1)\cdot 2^{n-2}.$$
32. Prove, by induction, that

    $$\frac{{}^nC_0}{x}-\frac{{}^nC_1}{x+1}+\frac{{}^nC_2}{x+2}-\ldots+(-1)^n\cdot\frac{{}^nC_n}{x+n}=\frac{n!}{x(x+1)(x+2)\ldots(x+n)},\ x\in R, x\geq 0$$

    for all $n\in N$.
33. If $u_1=1$ and $u_{n+1}=\dfrac{1}{n+1}u_n$, $n\geq 1$ then show that

    $$u_{n+1}=\frac{1}{(n+1)!}.$$
34. If $a_0>0$ and $a_n=1-e^{-a_{n-1}}$, $n\geq 1$ then show by induction that $0<a_n<1$.
35. If $a_1=\dfrac{a^2+A}{2a}$ and $a_{n+1}=\dfrac{a_n^2+A}{2a_n}$ for $n\geq 1$ where a, A are positive, prove by induction that $\dfrac{a_n-\sqrt{A}}{a_n+\sqrt{A}}=\left(\dfrac{a-\sqrt{A}}{a+\sqrt{A}}\right)^{2^n}$ for all $n\in N$.

36. a and b are given numbers. Let

$$a_1 = \frac{2a+b}{3},\ b_1 = \frac{a+2b}{3},\ a_2 = \frac{2a_1+b_1}{3},\ b_2 = \frac{a_1+2b_1}{3}$$

and $a_n = \dfrac{2a_{n-1}+b_{n-1}}{3}$, $b_n = \dfrac{a_{n-1}+2b_{n-1}}{3}$ for $n \geq 3$.

Prove, by induction, that $a_n = a - \dfrac{a-b}{2}\left(1 - \dfrac{1}{3^n}\right)$,

and $b_n = b + \dfrac{a-b}{2}\left(1 - \dfrac{1}{3^n}\right)$ for all $n \in N$.

37. Prove, by induction, that $3^n > n$ for all $n \in N$.

38. Prove, by induction, that $n! \leq n^n$ for all $n \in N$.

39. Prove, by induction, that $\left(\dfrac{n+1}{2}\right)^n \geq n!$.

40. Prove that $(n+1)^{n-1} \cdot (n+2)^n \geq 3^n \cdot (n!)^2$.

41. Prove, by induction, that $\dfrac{1}{2} + \dfrac{1}{4} + \dfrac{1}{8} + \ldots + \dfrac{1}{2^n} < 1$ for all $n \in N$.

Prove by induction or otherwise that

42. $\dfrac{1}{3^2} + \dfrac{1}{5^2} + \dfrac{1}{7^2} + \ldots + \dfrac{1}{(2n+1)^2} < \dfrac{1}{4}$ for all $n \in N$.

43. $\dfrac{a_1}{a_2} + \dfrac{a_2}{a_3} + \dfrac{a_3}{a_4} + \ldots + \dfrac{a_{n-1}}{a_n} + \dfrac{a_n}{a_1} \geq n$ where a_1, a_2, etc., are positive integers.

44. $\sqrt{a + \sqrt{a + \sqrt{a + \ldots \text{ to } n \text{ times}}}} < \dfrac{1 + \sqrt{4a+1}}{2}$ where $a > 0$.

45. Prove, by induction, that

$$\frac{1}{2n} \leq \frac{1}{2} \cdot \frac{3}{4} \cdot \frac{5}{6} \cdots \frac{2n-1}{2n} < \frac{1}{\sqrt{2n+1}} \text{ for all } n \in N.$$

46. Prove that $\dfrac{1}{2} < \dfrac{1}{3n+1} + \dfrac{1}{3n+2} + \ldots + \dfrac{1}{5n} + \dfrac{1}{5n+1} < \dfrac{2}{3}$ for all $n \in N$.

47. Prove, by induction, that $4^n > {}^{2n}C_n \geq \dfrac{4^n}{n+1}$ for all $n \in N$.

48. Prove, by mathematical induction, that $x^n + y^n$ is divisible by $x + y$ for any positive odd integer n.

49. If n is any odd positive integer then prove that $n(n^2 - 1)$ is divisible by 24.

50. Prove, by induction, that

$$x(x^{n-1} - na^{n-1}) + (n-1)a^n$$

is divisible by $(x-a)^2$ for all $n \in N, n > 1$.

51. Prove that $(\sqrt{2} - 1)^n$ can be expressed as $\sqrt{t} - \sqrt{t-1}$, where t is a positive integer for all odd positive integral values of n.

52. Prove that $10^{n-2} > 81n$ for all integral values of $n \geq 5$.

53. Prove that $3^n < n!$ for $n \geq 7$.

54. Prove that $(n!)^2 > n^n$ for $n \geq 3$.

55. Prove that $2^n \times (n-1)! < n^n$ for $n > 2$.

56. Prove, by induction, that $(1+x)^n > 1 + nx$ for $n \geq 2, n \in N$ and $x > -1, x \neq 0$.

57. Prove by induction that for $n > 1$,

$$\frac{1}{n+1} + \frac{1}{n+2} + \frac{1}{n+3} + \ldots + \frac{1}{2n} > \frac{13}{24}.$$

58. Prove that $\dfrac{1}{\sqrt{1}} + \dfrac{1}{\sqrt{2}} + \dfrac{1}{\sqrt{3}} + \ldots + \dfrac{1}{\sqrt{n}} > \sqrt{n}$ for all $n > 1$.

59. If $x_1 \cdot x_2 \cdot x_3 \cdot \ldots x_n = 1$ where each $x_i > 0$ and at least one $x_i \neq 1$ then prove, by induction, that

$$x_1 + x_2 + x_3 + \ldots + x_n > n \text{ for all } n \geq 2.$$

60. Prove that

$$(a_1b_1 + a_2b_2 + \ldots + a_nb_n)^2 < (a_1^2 + a_2^2 + \ldots + a_n^2)(b_1^2 + b_2^2 + \ldots + b_n^2)$$

for $n > 1$.

61. If $a = \dfrac{1+\sqrt{r}}{2}$, $b = \dfrac{1-\sqrt{r}}{2}$ and $a_n = \dfrac{a^n - b^n}{\sqrt{r}}$ where r is not a perfect square and ab is an integer then prove that a_n is an integer. Use induction.

62. If $a + b = c + d$ and $a^2 + b^2 = c^2 + d^2$ for the real numbers a, b, c, d then prove by, induction, that $a^n + b^n = c^n + d^n$ where $n \in N$.

63. If α, β be the roots of $x^2 - 5x + 1 = 0$, prove by induction that $\alpha^n + \beta^n$ is an integer not divisible by 4 where $n \in N$.

64. Prove that $10^n - (5 + \sqrt{17})^n - (5 - \sqrt{17})^n$ is divisible by 2^{n+1} for all $n \in N$.

65. If $x + \dfrac{1}{x}$ is an integer, prove by induction, that $x^n + \dfrac{1}{x^n}$ is an integer for all $n \in N$.

66. If $u_1 = 1, u_2 = 1$ and $u_{n+2} = u_{n+1} + u_n$ for $n \geq 1$ then prove by induction,

$$u_n = \frac{1}{\sqrt{5}}\left\{\left(\frac{\sqrt{5}+1}{2}\right)^n - \left(\frac{1-\sqrt{5}}{2}\right)^n\right\} \text{ for all } n \geq 1.$$

67. Let $u_1 = u_2 = 5$ and $u_{n+1} = u_n + 6u_{n-1}, n \geq 2$. Show, by induction, that $u_n = 3^n - (-2)^n$ if $n \geq 1$.

68. If $3u_n - 7u_{n-1} + 5u_{n-2} - u_{n-3} = 0$, $n \geq 3$ and $u_0 = 1$, $u_1 = 8$, $u_2 = 17$ then prove that $2u_n = 20n - 7 + 3^{2-n}$ for $n \geq 0$.

69. If $a_0 = 2, a_1 = 3$ and $a_n = 3a_{n-1} - 2a_{n-2}$ for $n \geq 2$ then prove by induction that $a_n = 2^n + 1$ for all $n \geq 0$.

70. Use induction to prove that

$$a_n \leq \left(\frac{\sqrt{5}+1}{2}\right)^n \text{ for all } n \in N$$

if $a_n = a_{n-1} + a_{n-2}, n \geq 2$ and $a_0 = 1, a_1 = 1$.

71. Prove, by the method of induction, that for $n \geq 1$,

$$\frac{d^n}{dx^n}\{\sin(ax+b)\} = a^n \sin\left(ax + b + n \cdot \frac{\pi}{2}\right).$$

72. Prove, by induction, that for $n \geq 1$,

$$\frac{d^n}{dx^n}(uv) = \sum_{r=0}^{n} {}^nC_r \cdot \frac{d^{n-r}}{dx^{n-r}}(u) \cdot \frac{d^r}{dx^r}(v).$$

73. Let $\quad I_m = \int_0^{\pi} \frac{1-\cos mx}{1-\cos x} dx.$

Use mathematical induction to prove that

$I_m = m\pi, m = 0, 1, 2, \ldots$

74. Prove, by mathematical induction, that

$$\frac{d^n}{dx^n}(\log x) = (-1)^{n-1} \cdot \frac{(n-1)!}{x^n}.$$

75. Let $u_n = \int \sin^n x \, dx$. Prove, by induction, that

$$u_n = -\frac{1}{n}\cos x \cdot \sin^{n-1} x + \frac{n-1}{n} u_{n-2}$$

for all positive integral values of $n \geq 2$.

Prove, by induction, that

76. $\int_0^{\pi/2} \frac{\sin(2n-1)x}{\sin x} dx = \frac{\pi}{2}$ for $n \in N$

77. $\frac{d^n}{dx^n}\{e^x(\sin x + \cos x)\} = 2^{\frac{n+1}{2}} \cdot e^x \sin\left\{x + (n+1)\frac{\pi}{4}\right\}$

78. $\int_0^{\pi/2} \cos^n x \cdot \sin nx \, dx$

$$= \frac{1}{2^{n+1}} \cdot \left(2 + \frac{2^2}{2} + \frac{2^3}{3} + \ldots + \frac{2^n}{n}\right)$$

79. $\int_0^{\pi} \frac{\sin nx}{\sin x} dx = 0$ for all even positive integral values of n.

80. $\frac{d^n}{dx^n}(x^n \log x) = n!\left(\log x + \sum_{k=1}^{n} \frac{1}{k}\right)$ for all $n \in N$.

81. $|\sin nx| \leq n|\sin x|, n \in N$.

82. If x is not an integral multiple of 2π, use mathematical induction to prove that

$\cos x + \cos 2x + \cos 3x + \ldots + \cos nx$

$$= \cos\frac{n+1}{2}x \cdot \sin\frac{nx}{2} \cdot \operatorname{cosec}\frac{x}{2}.$$

83. Using mathematical induction, prove that

$$\tan^{-1}\frac{1}{3} + \tan^{-1}\frac{1}{7} + \ldots + \tan^{-1}\frac{1}{n^2+n+1} = \tan^{-1}\frac{n}{n+2}.$$

84. If $n \in N$, show by induction that

(a) $\frac{2\cos 2^n\theta + 1}{2\cos\theta + 1} = (2\cos\theta - 1)(2\cos 2\theta - 1)$

$$\times (2\cos 2^2\theta - 1) \ldots (2\cos 2^{n-1}\theta - 1)$$

(b) $\sin^2 x + \sin^2 2x + \sin^2 3x + \ldots + \sin^2 nx$

$$= \frac{1}{2}\left[n - \frac{\sin nx \cdot \cos(n+1)x}{\sin x}\right]$$

85. Let $0 < A_i < \pi$ for $i = 1, 2, 3, \ldots, n$. Use induction to prove that $\sum_{r=1}^{n} \sin A_r \leq n \sin\left(\frac{\sum_{r=1}^{n} A_r}{n}\right)$ where $n \in N$.

[Given the fact

$p \sin x + (1-p)\sin y \leq \sin\{px + (1-p)y\}$

where $0 \leq p \leq 1$ and $0 \leq x, y \leq \pi$.]

86. Prove, by induction, that

$(\cos\theta + i\sin\theta)^n = \cos n\theta + i\sin\theta$

for all positive as well as negative integral values of n.

87. Use induction to prove that

$\sin\theta - \sin 3\theta + \sin 5\theta - \ldots$

$$+ (-1)^{n-1}\sin(2n-1)\theta = (-1)^{n-1} \cdot \frac{\sin 2n\theta}{2\cos\theta}.$$

88. Use induction to prove that

$$\sqrt{2 + \sqrt{2 + \sqrt{2 + \ldots \text{ to } n \text{ times}}}} = 2\cos\left(\frac{\pi}{2^{n+1}}\right).$$

89. Using induction, prove that

$$\cos\theta \cdot \cos 2\theta \cdot \cos 2^2\theta \cdot \ldots \cos 2^{n-1}\theta = \frac{\sin 2^n\theta}{2^n \sin\theta}.$$

90. Prove, by induction, that

$$(1+i)^n = 2^{n/2} \cdot \left(\cos\frac{n\pi}{4} + i\sin\frac{n\pi}{4}\right)$$

where n is any integer.

91. Prove, by induction, that

$$\Sigma \cot^{-1}\{1 + n(n+1)\} = \tan^{-1}(n+1) - \frac{\pi}{4}, n \in N.$$

92. Prove, by induction, that

$$\sum_{r=1}^{n} \frac{1}{2^r}\tan\left(\frac{\theta}{2^r}\right) = \frac{1}{2^n}\cot\left(\frac{\theta}{2^n}\right) - \cot\theta$$

for all $n \in N, 0 < \theta < \frac{\pi}{2}$.

93. Prove, by induction, that the product of thre consecutive positive integers is divisible by 6.

94. Prove that the even power of every odd integer when divided by 8 leaves the remainder 1.

95. Prove, by induction, that the integer just greater than $(3+\sqrt{5})^n$ is divisible by 2^n.

96. Prove, by induction, that $(\sqrt{2}-1)^n$ can be represented as the difference of the square roots of two consecutive integers for all $n \in N$.

97. Prove, by induction, that the sum of cubes of three successive natural numbers is divisible by 9.

98. Show that the sum of the squares of three consecutive odd numbers increased by 1 is divisible by 12 but not by 24.

99. Prove, by induction, that n diameters of a circle divide it into $2n$ sectors.

100. Prove, by induction, that n distinct concurrent lines in a plane divide the plane into $2n$ parts.

101. Prove, by induction, that the equation $5x + 3y = n$, $n \in N$ and $n \geq 8$ has non- negative integral solution.

102. $f : R \to R$ is a function such that

$$f(x+y) = f(x) + f(y).$$

Use induction to prove that $f(n) = kn$ for all $n \in N$ where $k = f(1)$. Hence, prove that $f(x) = kx$ for all rational values of x.

103. Let $f : R \to R$ be a function such that $f(x+y) = f(x) \cdot f(y)$. Prove, by induction, that $f(n) = k^n$ where k is a fixed real number and $n \in N$, $n \geq 2$.

□

Chapter Test

Time: 120 minutes

1. Let $a_n = 3 + \cfrac{1}{3 + \cfrac{1}{3 + \cfrac{1}{3 + \dots \text{ to } n \text{ times.}}}}$

Prove that $a_n \cdot a_{n-1} = 3a_{n-1} + 1, \quad n \geq 2$. Also prove, by induction, that $a_n < 3.31$ for all $n \in N, n \geq 2$.

2. If $a_n = x^{\frac{1}{2^n}} + y^{\frac{1}{2^n}}$ and $b_n = x^{\frac{1}{2^n}} - y^{\frac{1}{2^n}}$ for $n \geq 1$, prove, by induction, that

$$a_1 \cdot a_2 \cdot a_3 \cdot \dots \cdot a_n = (x - y)\frac{1}{b_n}.$$

3. If $u_n = (3 + \sqrt{5})^n + (3 - \sqrt{5})^n$, prove, by induction, that

(i) $u_{n+1} = 6u_n - 4u_{n-1}$ for all $n \geq 2$

(ii) u_n is an integer for all $n \in N$.

4. Prove, by induction, that

$$(x+1)(x+2)(x+3)\dots(x+n) = x^n + P_1 x^{n-1} + P_2 x^{n-2} + \dots + P_n, \; n \in N$$

where P_r = sum of the products of 1, 2, 3, ... , n taking r at a time and $x \in R$.

5. Prove, by induction, that $n^6 + 3n^4 + 7n^2 - 11$ is divisible by 128 for all odd positive integral values of n.

6. Prove that $\int_0^{\pi} \frac{\sin nx}{\sin x}\,dx = \pi$ for all odd positive integral values of n.

7. Prove that $1! \cdot 3! \cdot 5! \cdot \dots (2n-1)! > (n!)^n$ for all $n \in N, n \geq 2$.

8. Prove that n and n^5 have the same digit in the units place of their values for all $n \in N$.

9. If $2f(x) = f(xy) + f\left(\frac{x}{y}\right)$ for all $x > 0, y > 0$ and $f(1) = 0$ then prove that $f(x^n) = nf(x)$ for all nonnegative integral values of n. Also show that it holds for all $n \in Z$, the set of integers.

10. If $n \in N$ then prove that

$$\frac{1}{n} + \frac{1}{n+1} + \frac{1}{n+2} + \dots + \frac{1}{2n-1} = 1 - \frac{1}{2} + \frac{1}{3} - \frac{1}{4} + \dots + \frac{1}{2n-1}$$

11. Let a, b, c be positive real numbers such that $b^2 - 4ac > 0$ and let $\alpha_1 = c$ and

$$\alpha_{n+1} = \frac{a\alpha_n^2}{b^2 - 2a(\alpha_1 + \alpha_2 + \dots + \alpha_n)}, \; n \geq 1.$$ Prove, by induction, that

(i) $b^2 - 2a(\alpha_1 + \alpha_2 + \dots + \alpha_n) \neq 0$ for all $n \in N$ (ii) $a_{n+1} < \frac{a_n}{2}$ for all $n \in N$.

12. If $\frac{x}{a} + \frac{y}{b} = 1$ and $\frac{x^2}{a} + \frac{y^2}{b} = \frac{ab}{a+b}$ then prove that

$$\frac{x^n}{a} + \frac{y^n}{b} = \left(\frac{ab}{a+b}\right)^{n-1}$$ for all $n \in N$.

□

8. Infinite Series

(Binomial Series, Exponential Series, Logarithmic Series)

Recap of Facts and Formulae

1. Binomial theorem for any index

- $(1+x)^n = 1 + nx + \frac{n(n-1)}{2!}x^2 + \frac{n(n-1)(n-2)}{3!}x^3 + \ldots$

  to $\propto$, provided $|x| < 1$.

  General term $\{(r+1)\text{th term}\}$ in the expansion of $(1+x)^n$ is

  $$t_{r+1} = \frac{n(n-1)(n-2)\ldots(n-r+1)}{r!}x^r.$$

2. Some useful binomial expansions for summation of series

- $(1-x)^{-n} = 1 + {}^nC_1x + {}^{n+1}C_2x^2 + {}^{n+2}C_3x^3 + \ldots$

  $+ {}^{n+r-1}C_rx^r + \ldots$ to $\propto$,

  where n is a positive integer.

  $(1+x)^{-n} = 1 - {}^nC_1x + {}^{n+1}C_2x^2 - {}^{n+2}C_3x^3 + \ldots$ to $\propto$

  where $n \in N$.

- $(1-x)^{-1} = 1 + x + x^2 + x^3 + x^4 + \ldots$ to $\propto$

  $(1+x)^{-1} = 1 - x + x^2 - x^3 + x^4 - \ldots$ to $\propto$

- $(1-x)^{-2} = 1 + 2x + 3x^2 + 4x^3 + \ldots$ to $\propto$

  $(1+x)^{-2} = 1 - 2x + 3x^2 - 4x^3 + \ldots$ to $\propto$

- $(1-x)^{-\frac{p}{q}} = 1 + \frac{p}{1!}\left(\frac{x}{q}\right) + \frac{p(p+q)}{2!}\left(\frac{x}{q}\right)^2 + \frac{p(p+q)(p+2q)}{3!}\left(\frac{x}{q}\right)^3 + \ldots$ to $\propto$

  $(1+x)^{-\frac{p}{q}} = 1 - \frac{p}{1!}\left(\frac{x}{q}\right) + \frac{p(p+q)}{2!}\left(\frac{x}{q}\right)^2 - \frac{p(p+q)(p+2q)}{3!}\left(\frac{x}{q}\right)^3 + \ldots$ to $\propto$

- $(1+x)^{\frac{p}{q}} = 1 + \frac{p}{1!}\left(\frac{x}{q}\right) + \frac{p(p-q)}{2!}\left(\frac{x}{q}\right)^2 + \frac{p(p-q)(p-2q)}{3!}\left(\frac{x}{q}\right)^3 + \ldots$ to $\propto$

  $(1-x)^{\frac{p}{q}} = 1 - \frac{p}{1!}\left(\frac{x}{q}\right) + \frac{p(p-q)}{2!}\left(\frac{x}{q}\right)^2 - \frac{p(p-q)(p-2q)}{3!}\left(\frac{x}{q}\right)^3 + \ldots$ to $\propto$

3. Exponential series

- $e^x = 1 + \frac{x}{1!} + \frac{x^2}{2!} + \frac{x^3}{3!} + \ldots + \frac{x^n}{n!} + \ldots$ to $\propto = \sum_{n=0}^{\infty} \frac{x^n}{n!}$

  $e^{-x} = 1 - \frac{x}{1!} + \frac{x^2}{2!} - \frac{x^3}{3!} + \ldots + (-1)^n\frac{x^n}{n!} + \ldots$ to $\propto$

  $= \sum_{n=0}^{\infty} (-1)^n \cdot \frac{x^n}{n!}$

- $\frac{1}{2}(e^x + e^{-x}) = 1 + \frac{x^2}{2!} + \frac{x^4}{4!} + \ldots + \frac{x^{2n}}{(2n)!} + \ldots$ to $\propto$

  $\frac{1}{2}(e^x - e^{-x}) = \frac{x}{1!} + \frac{x^3}{3!} + \frac{x^5}{5!} + \ldots + \frac{x^{2n-1}}{(2n-1)!} + \ldots$ to $\propto$

- $a^x = e^{\log_e a^x} = e^{x\log_e a}$

  $= 1 + \frac{x\log_e a}{1!} + \frac{(x\log_e a)^2}{2!} + \frac{(x\log_e a)^3}{3!} + \ldots$ to $\propto$

  $= \sum_{n=0}^{\infty} \frac{x^n}{n!}(\log_e a)^n$

- $e = 1 + \frac{1}{1!} + \frac{1}{2!} + \frac{1}{3!} + \ldots$ to $\propto$

  $e^{-1} = 1 - \frac{1}{1!} + \frac{1}{2!} - \frac{1}{3!} + \ldots$ to $\propto$

- $\frac{1}{2}(e+e^{-1}) = 1 + \frac{1}{2!} + \frac{1}{4!} + \frac{1}{6!} + \dots$ to ∞

$\frac{1}{2}(e-e^{-1}) = \frac{1}{1!} + \frac{1}{3!} + \frac{1}{5!} + \dots$ to ∞.

4. Logarithmic series

- $\log_e(1+x) = x - \frac{1}{2}x^2 + \frac{1}{3}x^3 - \dots + (-1)^{n-1}\cdot\frac{1}{n}x^n + \dots$ to ∞, provided $-1 < x \le 1$

$\log_e(1-x) = -x - \frac{1}{2}x^2 - \frac{1}{3}x^3 - \dots - \frac{1}{n}x^n - \dots$ to ∞

- $\log(1+x) + \log(1-x)$

$$= 2\left[-\frac{1}{2}x^2 - \frac{1}{4}x^4 - \frac{1}{6}x^6 - \dots \text{ to } \infty\right]$$

$\log(1+x) - \log(1-x)$

$$= 2\left[x + \frac{1}{3}x^3 + \frac{1}{5}x^5 + \dots \text{ to } \infty\right]$$

5. Recognising the type of the infinite series

- Each of the binomial, exponential and logarithmic series has infinite terms.
- Series involving n ! s in the denominators of terms are generally binomial or exponential series. But in a binomial series the number of factors in the numerators (other than the power of a fixed number) of terms goes on increasing.
- Logarithmic series do not contain n ! s in the denominators of terms. The terms contain $\frac{1}{n}$ s.

Selected Solved Examples

1. Find the coefficients of x^4 and x^{-4} in the expansion of $\frac{1}{2+x-x^2}$, $1 < x < 2$.

$$\frac{1}{2+x-x^2} = \frac{1}{(2-x)(1+x)} = \frac{1}{3}\cdot\frac{(2-x)+(1+x)}{(2-x)(1+x)}$$

$$= \frac{1}{3}\left[\frac{1}{1+x} + \frac{1}{2-x}\right]$$

$$= \frac{1}{3}\cdot\frac{1}{x\left(1+\frac{1}{x}\right)} + \frac{1}{3}\cdot\frac{1}{2\left(1-\frac{x}{2}\right)}$$

$$= \frac{1}{3x}\left(1+\frac{1}{x}\right)^{-1} + \frac{1}{6}\cdot\left(1-\frac{x}{2}\right)^{-1}, \qquad \left\{\frac{1}{x} < 1, \frac{x}{2} < 1\right\}$$

$$= \frac{1}{3x}\left(1 - \frac{1}{x} + \frac{1}{x^2} - \dots \text{ to } \infty\right)$$

$$+ \frac{1}{6}\left\{1 + \left(\frac{x}{2}\right) + \left(\frac{x}{2}\right)^2 + \dots \text{ to } \infty\right\}$$

$\therefore$ coefficient of $x^4 = 0 + \frac{1}{6}\cdot\left(\frac{1}{2}\right)^4 = \frac{1}{96}$

coefficient of $x^{-4} = \frac{1}{3}\cdot(-1) + 0 = -\frac{1}{3}$.

2. Sum the series: $1 + \frac{1}{3} + \frac{1\cdot3}{3\cdot6} + \frac{1\cdot3\cdot5}{3\cdot6\cdot9} + \dots$ to ∞.

$$\text{Series} = 1 + \frac{1}{1!}\cdot\frac{1}{3} + \frac{1(1+2)}{2!}\cdot\left(\frac{1}{3}\right)^2 + \frac{1(1+2)(1+2\cdot2)}{3!}\left(\frac{1}{3}\right)^3 + \dots \text{ to } \infty.$$

We know,

$$(1-x)^{-\frac{p}{q}} = 1 + \frac{p}{1!}\cdot\left(\frac{x}{q}\right) + \frac{p(p+q)}{2!}\left(\frac{x}{q}\right)^2 + \frac{p(p+q)(p+2q)}{3!}\left(\frac{x}{q}\right)^3 + \dots \text{ to } \infty.$$

Comparing the given series with it,

$$p = 1, \frac{x}{q} = \frac{1}{3}, \; q = 2; \qquad \therefore \qquad x = \frac{q}{3} = \frac{2}{3}.$$

$$\therefore \quad \text{required sum} = \left(1 - \frac{2}{3}\right)^{-\frac{1}{2}} = \left(\frac{1}{3}\right)^{-\frac{1}{2}} = 3^{\frac{1}{2}} = \sqrt{3}.$$

3. If $x = 1 + 3a + 6a^2 + 10a^3 + \dots$ to ∞, $|a| < 1$,
$y = 1 + 4b + 10b^2 + 20b^3 + \dots$ to ∞, $|b| < 1$
and $S = 1 + 3(ab) + 5(ab)^2 + \dots$ to ∞
then find S in terms of x and y.

Here $x = 1 + {}^3C_1a + {}^4C_2a^2 + {}^5C_3a^3 + \dots$ to ∞

$$= (1-a)^{-3}$$

$$\therefore \quad x = \frac{1}{(1-a)^3}; \qquad \therefore \quad \frac{1}{1-a} = x^{\frac{1}{3}}$$

$$\therefore \quad 1 - a = x^{-\frac{1}{3}}; \qquad \therefore \quad a = 1 - x^{-\frac{1}{3}}.$$

Again, $y = 1 + {}^4C_1b + {}^5C_2b^2 + {}^6C_3b^3 + \dots$ to ∞

$$= (1-b)^{-4}$$

$$\therefore \quad y = \frac{1}{(1-b)^4}; \qquad \therefore \quad \frac{1}{1-b} = y^{\frac{1}{4}}$$

$\therefore \quad 1 - b = y^{-\frac{1}{4}}; \quad \therefore \quad b = 1 - y^{-\frac{1}{4}}.$

Now, $S = 1 + 3(ab) + 5(ab)^2 + 7(ab)^3 + \ldots$ to ∞

$ab \cdot S = \quad ab + 3(ab)^2 + 5(ab)^3 + \ldots$ to ∞

Adding,

$$(1 - ab)S = 1 + 2ab + 2(ab)^2 + 2(ab)^3 + \ldots \text{ to } \infty$$

$$= 1 + 2ab\{1 + ab + (ab)^2 + \ldots \text{ to } \infty\}$$

$$= 1 + 2ab \cdot \frac{1}{1 - ab}$$

$$\{\because |ab| = |a|\,|b| < 1 \cdot 1 = 1\}$$

$$\therefore \quad S = \frac{1}{1 - ab} + \frac{2ab}{(1 - ab)^2} = \frac{1 + ab}{(1 - ab)^2}.$$

But $ab = (1 - x^{-1/3})(1 - y^{-1/4})$

$$= 1 - x^{-1/3} - y^{-1/4} + x^{-1/3} \cdot y^{-1/4}$$

$$\therefore \quad S = \frac{2 - x^{-\frac{1}{3}} - y^{-\frac{1}{4}} + x^{-\frac{1}{3}} \cdot y^{-\frac{1}{4}}}{(x^{-\frac{1}{3}} + y^{-\frac{1}{4}} - x^{-\frac{1}{3}} \cdot y^{-\frac{1}{4}})^2}.$$

4. If $k_n = \dfrac{1 \cdot 3 \cdot 5 \cdot \ldots (2n-1)}{2 \cdot 4 \cdot 6 \ldots 2n}$**, prove that**

$$k_{2n+1} + k_{2n} \cdot k_1 + k_{2n-1} \cdot k_2 + \ldots + k_{n+1} \cdot k_n = \frac{1}{2}.$$

Here $k_n = \dfrac{1 \cdot 3 \cdot 5 \cdot \ldots (2n-1)}{2 \cdot 4 \cdot 6 \cdot \ldots 2n}$

$$= \frac{1(1+2)(1+2 \cdot 2) \ldots (1 + \overline{n-1} \cdot 2)}{n!} \cdot \left(\frac{1}{2}\right)^n.$$

Now, $(1 - x)^{-1/2} = 1 + \dfrac{1}{1!} \cdot \left(\dfrac{x}{2}\right) + \dfrac{1 \cdot (1+2)}{2!}\left(\dfrac{x}{2}\right)^2$

$$+ \frac{1 \cdot (1+2) \cdot (1 + 2 \cdot 2)}{3!}\left(\frac{x}{2}\right)^3 + \ldots \text{ to } \infty$$

$$\therefore \quad (1 - x)^{-1/2} = 1 + k_1 x + k_2 x^2 + k_3 x^3$$

$$+ \ldots + k_{2n+1} x^{2n+1} + \ldots \text{ to } \infty.$$

Again, $(1 - x)^{-1/2} = 1 + k_1 x + k_2 x^2 + k_3 x^3 + \ldots$ to ∞.

Multiplying these and equating coefficients of x^{2n+1} on both sides,

coefficient of x^{2n+1} in $\{(1 - x)^{-1/2}\}^2$

$$= k_{2n+1} + k_{2n} \cdot k_1 + k_{2n-1} \cdot k_2 + \ldots + k_{n+1} \cdot k_n$$

$$+ k_n \cdot k_{n+1} + \ldots + k_1 \cdot k_{2n} + k_{2n+1}$$

$$= 2 \times (k_{2n+1} + k_{2n} \cdot k_1 + k_{2n-1} \cdot k_2 + \ldots + k_{n+1} \cdot k_n)$$

$\therefore \quad$ LHS $= \dfrac{1}{2} \times$ coefficient of x^{2n+1} in $(1 - x)^{-1}$

$= \dfrac{1}{2} \times$ coefficient of x^{2n+1} in $\{1 + x + x^2 + x^3 + \ldots\}$

$= \dfrac{1}{2} \times 1 = \dfrac{1}{2} =$ RHS.

5. Find the sum: $1^2 + \dfrac{3^2}{1!} + \dfrac{5^2}{3!} + \dfrac{7^2}{5!} + \ldots$ **to** ∞.

Consider the series $\dfrac{3^2}{1!} + \dfrac{5^2}{3!} + \dfrac{7^2}{5!} + \ldots$ to ∞.

Here $t_n = \dfrac{(2n+1)^2}{(2n-1)!} = \dfrac{4n^2 + 4n + 1}{(2n-1)!}$

$$= \frac{(2n-1)(2n-2) + 10n - 1}{(2n-1)!}$$

$$= \frac{1}{(2n-3)!} + \frac{10n - 1}{(2n-1)!}$$

$$= \frac{1}{(2n-3)!} + \frac{5(2n-1) + 4}{(2n-1)!}$$

$$= \frac{1}{(2n-3)!} + 5 \cdot \frac{1}{(2n-2)!} + 4 \cdot \frac{1}{(2n-1)!}.$$

If $\quad S = \dfrac{3^2}{1!} + \dfrac{5^2}{3!} + \dfrac{7^2}{5!} + \ldots$ to ∞

then $S = \sum_{n=1}^{\infty} t_n = \sum_{n=1}^{\infty} \dfrac{1}{(2n-3)!} + 5 \sum_{n=1}^{\infty} \dfrac{1}{(2n-2)!}$

$$+ 4 \sum_{n=1}^{\infty} \frac{1}{(2n-1)!}$$

$$= \left(\frac{1}{1!} + \frac{1}{3!} + \frac{1}{5!} + \ldots\right) + 5\left(1 + \frac{1}{2!} + \frac{1}{4!} + \ldots\right)$$

$$+ 4\left(\frac{1}{1!} + \frac{1}{3!} + \frac{1}{5!} + \ldots\right)$$

$$= \frac{1}{2}(e - e^{-1}) + 5 \cdot \frac{1}{2}(e + e^{-1}) + 4 \cdot \frac{1}{2}(e - e^{-1})$$

$$= \frac{5}{2}(e - e^{-1} + e + e^{-1}) = \frac{5}{2} \cdot 2e = 5e.$$

$\therefore$ the required sum $= 1 + S = 1 + 5e$.

6. Sum:

$$1 + \frac{1 + \frac{1}{1!}}{2} + \frac{1 + \frac{1}{1!} + \frac{1}{2!}}{2^2} + \frac{1 + \frac{1}{1!} + \frac{1}{2!} + \frac{1}{3!}}{2^3} + \ldots \textbf{ to } \infty.$$

Here, $t_{n+1} = \dfrac{1 + \frac{1}{1!} + \frac{1}{2!} + \frac{1}{3!} + \ldots + \frac{1}{n!}}{2^n}$

and $t_n = \dfrac{1 + \frac{1}{1!} + \frac{1}{2!} + \frac{1}{3!} + \ldots + \frac{1}{(n-1)!}}{2^{n-1}}$

$$\therefore \quad t_{n+1} = \frac{1}{2} \cdot \frac{1 + \frac{1}{1!} + \frac{1}{2!} + \frac{1}{3!} + \ldots + \frac{1}{(n-1)!}}{2^{n-1}} + \frac{1}{n!} \cdot \frac{1}{2^n}$$

$$= \frac{1}{2} t_n + \frac{1}{n!} \cdot \frac{1}{2^n}$$

$$\therefore \quad \sum_{n=0}^{\infty} t_{n+1} = \frac{1}{2} \sum_{n=1}^{\infty} t_n + \sum_{n=0}^{\infty} \frac{1}{n!} \cdot \left(\frac{1}{2}\right)^n$$

$$= \frac{1}{2} \sum_{n=1}^{\infty} t_n$$

$$+ \left[1 + \frac{1}{1!} \cdot \frac{1}{2} + \frac{1}{2!} \cdot \left(\frac{1}{2}\right)^2 + \frac{1}{3!} \cdot \left(\frac{1}{2}\right)^3 + \ldots\right]$$

$$\therefore \quad S = \frac{1}{2} S + e^{1/2}$$

where S is the required sum of the given series.

$$\therefore \quad \frac{1}{2} S = \sqrt{e}; \quad \therefore \quad S = 2\sqrt{e}.$$

7. If $\log_e \dfrac{1}{1 + x + x^2 + x^3}$ be expanded in ascending powers of x, show that the coefficient of x^n is

$-\dfrac{1}{n}$ if n is odd or of the form $4m + 2$ and

$\dfrac{3}{n}$ if n is of the form $4m$.

$$\log_e \frac{1}{1 + x + x^2 + x^3} = \log_e \frac{1}{(1 + x)(1 + x^2)}$$

$$= -\log_e\{(1 + x)(1 + x^2)\}$$

$$= -\log(1 + x) - \log(1 + x^2)$$

$$= -\left\{x - \frac{1}{2}x^2 + \frac{1}{3}x^3 - \ldots\right\} - \left\{x^2 - \frac{1}{2}x^4 + \frac{1}{3}x^6 - \ldots\right\}.$$

Let n be odd. Then there is no term containing x^n in the second series.

$$\therefore \quad \text{coefficient of } x^n = -\frac{1}{n}.$$

Let $n = 4m + 2$. Then

coefficient of x^n = coefficient of x^{4m+2}

$$= \frac{1}{4m + 2} - \frac{1}{2m + 1} = -\frac{1}{4m + 2} = -\frac{1}{n}.$$

Let $n = 4m$. Then

coefficient of x^n = coefficient of x^{4m}

$$= \frac{1}{4m} + \frac{1}{2m} = 3 \cdot \frac{1}{4m} = \frac{3}{n}.$$

Hence, the problem.

8. If $x^2 y = 2x - y$ and $|x| < 1$ then prove that

$$y + \frac{y^3}{3} + \frac{y^5}{5} + \ldots \text{ to } \infty = 2\left(x + \frac{x^3}{3} + \frac{x^5}{5} + \ldots \text{ to } \infty\right).$$

Here, $(x^2 + 1)y = 2x$; $\quad \therefore \quad y = \dfrac{2x}{1 + x^2}$.

$$\text{LHS} = \frac{1}{2}[\log(1 + y) - \log(1 - y)] = \frac{1}{2} \log \frac{1 + y}{1 - y}$$

$$= \frac{1}{2} \log \frac{1 + \frac{2x}{1 + x^2}}{1 - \frac{2x}{1 + x^2}} = \frac{1}{2} \log \frac{(1 + x)^2}{(1 - x)^2}$$

$$= \log \frac{1 + x}{1 - x} = \log(1 + x) - \log(1 - x)$$

$$= \left\{x - \frac{1}{2}x^2 + \frac{1}{3}x^3 - \ldots\right\} - \left\{-x - \frac{1}{2}x^2 - \frac{1}{3}x^3 - \ldots\right\}$$

$$= 2\left(x + \frac{x^3}{3} + \frac{x^5}{5} + \ldots \text{ to } \infty\right).$$

9. If $a + b + c = 0$ then using the expansion of the type

$$\log(1 - \alpha x) = -\alpha x - \frac{\alpha^2 x^2}{2} - \frac{\alpha^3 x^3}{3} - \ldots \text{ to } \infty,$$

prove that $\dfrac{a^5 + b^5 + c^5}{5} = \dfrac{a^2 + b^2 + c^2}{2} \cdot \dfrac{a^3 + b^3 + c^3}{3}$.

$$\log(1 - ax) = -ax - \frac{a^2 x^2}{2} - \frac{a^3 x^3}{3} - \ldots \text{ to } \infty$$

$$\log(1 - bx) = -bx - \frac{b^2 x^2}{2} - \frac{b^3 x^3}{3} - \ldots \text{ to } \infty$$

$$\log(1 - cx) = -cx - \frac{c^2 x^2}{2} - \frac{c^3 x^3}{3} - \ldots \text{ to } \infty.$$

Adding these, we get

$$\log\{(1 - ax)(1 - bx)(1 - cx)\}$$

$$= -(a + b + c)x - \frac{a^2 + b^2 + c^2}{2} x^2$$

$$- \frac{a^3 + b^3 + c^3}{3} x^3 - \ldots \text{ to } \infty$$

or $\quad \log\{1 - (a + b + c)x + (ab + bc + ca)x^2 - abc\, x^3\}$

$$= -(a + b + c)x - \frac{a^2 + b^2 + c^2}{2} x^2$$

$$- \frac{a^3 + b^3 + c^3}{3} x^3 - \ldots \text{ to } \infty.$$

But $a + b + c = 0$. So we get

$$\log\{1 + (ab + bc + ca)x^2 - abc\, x^3\}$$

$$= -\frac{a^2 + b^2 + c^2}{2} x^2 - \frac{a^3 + b^3 + c^3}{3} x^3 - \ldots \text{ to } \infty$$

or $\quad \log[1 - \{abc\, x^3 - (ab + bc + ca)x^2\}]$

$$= -\frac{a^2+b^2+c^2}{2}x^2 - \frac{a^3+b^3+c^3}{3}x^3 - \dots \text{ to } \infty$$

or $\quad -\{abc\,x^3 - (ab+bc+ca)x^2\}$

$$-\frac{1}{2}\{abc\,x^3 - (ab+bc+ca)x^2\}^2 - \dots \text{ to } \infty$$

$$= -\frac{a^2+b^2+c^2}{2}x^2 - \frac{a^3+b^3+c^3}{3}x^3 - \dots \text{ to } \infty$$

Equating coefficients of x^2, x^3 and x^5 on both sides we get

$$ab+bc+ca = -\frac{a^2+b^2+c^2}{2} \qquad \dots (1)$$

$$-abc = -\frac{a^3+b^3+c^3}{3} \qquad \dots (2)$$

$$\frac{1}{2}\cdot 2\cdot abc(ab+bc+ca) = -\frac{a^5+b^5+c^5}{5}$$

i.e., $\quad abc(ab+bc+ca) = -\dfrac{a^5+b^5+c^5}{5} \qquad \dots (3)$

using (1), (2) in (3),

$$\frac{a^3+b^3+c^3}{3}\cdot\left(-\frac{a^2+b^2+c^2}{2}\right) = -\frac{a^5+b^5+c^5}{5}$$

$$\therefore \quad \frac{a^2+b^2+c^2}{2}\cdot\frac{a^3+b^3+c^3}{3} = \frac{a^5+b^5+c^5}{5}.$$

Exercises

1. Find the fourth term in the expansion of $(1-2x)^{-3/4}$ if $|x| < \frac{1}{2}$.

2. Find the third term in the expansion of $(3+2x)^{3/5}$ if $|x| > \frac{3}{2}$.

3. Find the $(r+1)$th term in the expansion of $\sqrt{1-x}$ when $|x| < 1$.

4. Find the coefficient of x^3 in the expansion of $(1+2x)^{-3}$ when $|x| < \frac{1}{2}$.

5. Find the coefficient of x^n in the expansion of $(1+3x+6x^2+10x^3+\dots \text{ to } \infty)^{-n}, n \in N$.

6. Prove that the coefficient of x^n in the expansion of $(1-4x)^{-(n+1/2)}$ is equal to the middle term of $\left(x+\frac{1}{x}\right)^{4n}$, $n \in N$.

7. Find the coefficient of x^r in the expansion of $\dfrac{(1+x)^2}{(1-x)^3}$ when $|x| < 1$.

8. Find the coefficient of x^n in the expansion of
$$(1+x+3x^2+6x^3+10x^4+\dots \text{ to } \infty)^2.$$

9. Prove that the coefficient of x^n in the expansion of $(1+x+x^2)^{-1}$ is 1, 0 or -1 according as n is of the form $3m$, $3m-1$ or $3m+1$ when $|x| < 1$.

10. Prove that the coefficient of x^n in the expansion of $\dfrac{1}{1+x+x^2+x^3}$ is 1 or -1 according as n is of the form $4m$ or $4m+1$ and in all other cases it is zero when $|x| < 1$.

11. Prove that
$$(1+x+x^2+\dots \text{ to } \infty)\times(1+2x+3x^2+\dots \text{ to } \infty)$$
$$= \frac{1}{2}(1\cdot 2+2\cdot 3x+3\cdot 4x^2+\dots \text{ to } \infty).$$

12. Evaluate $\sqrt[3]{2}$ to four decimal places by using binomial expansion.

13. Sum the series:
$$1+\frac{3}{2^3}+\frac{1\cdot 3}{1\cdot 2}\cdot\frac{3^2}{2^6}+\frac{1\cdot 3\cdot 5}{1\cdot 2\cdot 3}\cdot\frac{3^3}{2^9}+\dots \text{ to } \infty.$$

14. Sum the series:
$$\frac{7}{5}\left(1+\frac{1}{10^2}+\frac{1\cdot 3}{1\cdot 2}\cdot\frac{1}{10^4}+\frac{1\cdot 3\cdot 5}{1\cdot 2\cdot 3}\cdot\frac{1}{10^6}+\dots \text{ to } \infty\right).$$

15. Sum the infinite series:
$$1+\frac{1}{4}+\frac{1\cdot 3}{4\cdot 8}+\frac{1\cdot 3\cdot 5}{4\cdot 8\cdot 12}+\dots$$

16. Sum the series to infinite terms:
$$1+\frac{2}{6}+\frac{2\cdot 5}{6\cdot 12}+\frac{2\cdot 5\cdot 8}{6\cdot 12\cdot 18}+\dots$$

17. Prove that
$$1-\frac{1}{2}\left(\frac{1}{2}\right)+\frac{1\cdot 3}{2\cdot 4}\left(\frac{1}{2}\right)^2-\frac{1\cdot 3\cdot 5}{2\cdot 4\cdot 6}\left(\frac{1}{2}\right)^3+\dots \text{ to } \infty = \sqrt{\frac{2}{3}}.$$

18. If $y = \dfrac{1}{4}+\dfrac{1\cdot 3}{4\cdot 8}+\dfrac{1\cdot 3\cdot 5}{4\cdot 8\cdot 12}+\dots$ to ∞, prove that $y^2+2y=1$.

19. Prove that
$$1+\frac{2n}{3}+\frac{2n(2n+2)}{3\cdot 6}+\frac{2n(2n+2)(2n+4)}{3\cdot 6\cdot 9}+\dots \text{ to } \infty$$
$$= 2^n\left\{1+\frac{n}{3}+\frac{n(n+1)}{3\cdot 6}+\frac{n(n+1)(n+2)}{3\cdot 6\cdot 9}+\dots \text{ to } \infty\right\}.$$

20. Find the sum: ${}^4C_1+{}^5C_2\cdot\frac{1}{2}+{}^6C_3\left(\frac{1}{2}\right)^2+\dots$ to ∞.

21. Find the sum:

$$1 - {}^3C_1 \cdot \frac{1}{3} + {}^4C_2 \cdot \left(\frac{1}{3}\right)^2 - {}^5C_3 \cdot \left(\frac{1}{3}\right)^3 + \ldots \text{ to } \infty.$$

22. If $p_n = \dfrac{1 \cdot 4 \cdot 7 \cdot \ldots (3n-2)}{3 \cdot 6 \cdot 9 \cdot \ldots 3n}$, prove that

$$p_{2n} - p_{2n-1} \cdot p_1 + p_{2n-2}\, p_2 - \ldots + (-1)^{n-1}\, p_{n+1}\, p_{n-1} = \frac{1}{2}\{p_n + (-1)^{n-1} p_n^2\}$$

23. Find the coefficient of x^n in the expression

$$\frac{a+bx}{1!} + \frac{(a+bx)^2}{2!} + \frac{(a+bx)^3}{3!} + \ldots \text{ to } \infty.$$

24. Find the coefficient of x^r in the expression

$$\frac{1 + \frac{1+x}{1!} + \frac{(1+x)^2}{2!} + \frac{(1+x)^3}{3!} + \ldots \text{ to } \infty}{1 + 2 \cdot \frac{x}{1!} + 4 \cdot \frac{x^2}{2!} + 8 \cdot \frac{x^3}{3!} + \ldots \text{ to } \infty}.$$

25. Prove that $\dfrac{1 + \frac{1}{2!} + \frac{1}{4!} + \frac{1}{6!} + \ldots \text{to } \infty}{\frac{1}{1!} + \frac{1}{3!} + \frac{1}{5!} + \frac{1}{7!} + \ldots \text{to } \infty} = \dfrac{e^2+1}{e^2-1}.$

26. Prove that

$$\left(1 + \frac{x^2}{2!} + \frac{x^4}{4!} + \frac{x^6}{6!} + \ldots \text{ to } \infty\right)^2 - \left(x + \frac{x^3}{3!} + \frac{x^5}{5!} + \frac{x^7}{7!} + \ldots \text{ to } \infty\right)^2 = 1.$$

27. Prove that

$$2\left[1 + \frac{(\log_e n)^2}{2!} + \frac{(\log_e n)^4}{4!} + \frac{(\log_e n)^6}{6!} + \ldots \text{ to } \infty\right] = n + 1/n.$$

28. Prove that $\dfrac{\frac{1}{2!} + \frac{1}{4!} + \frac{1}{6!} + \ldots \text{ to } \infty}{1 + \frac{1}{3!} + \frac{1}{5!} + \frac{1}{7!} + \ldots \text{ to } \infty} = \dfrac{e-1}{e+1}.$

29. Sum: $1 + \dfrac{3}{1!} + \dfrac{5}{2!} + \dfrac{7}{3!} + \ldots \text{ to } \infty.$

30. Sum the series:

$$(x^2 - y^2) + \frac{1}{2!}(x^4 - y^4) + \frac{1}{3!}(x^6 - y^6) + \ldots \text{ to } \infty.$$

31. Sum to infinite terms: $\dfrac{3}{2!} + \dfrac{5}{4!} + \dfrac{7}{6!} + \ldots \text{ to } \infty.$

32. Prove that $\dfrac{1}{3!} + \dfrac{2}{5!} + \dfrac{3}{7!} + \ldots \text{ to } \infty = \dfrac{1}{2e}.$

33. Show that, for $x \neq 0$,

$$1 + \frac{x}{2!} + \frac{2x^2}{3!} + \frac{3x^3}{4!} + \ldots \text{ to } \infty = \frac{1}{x}\{e^x(x-1) + x + 1\}.$$

34. Sum: $\displaystyle\sum_{n=1}^{\infty} \frac{n^2}{n!}.$

35. Sum: $\dfrac{1^2}{2!} + \dfrac{2^2}{3!} + \dfrac{3^2}{4!} + \ldots \text{ to } \infty.$

36. Prove that $1 + \dfrac{2^3}{2!} + \dfrac{3^3}{3!} + \dfrac{4^3}{4!} + \ldots \text{ to } \infty = 5e.$

37. Find the sum: $1 + \dfrac{2^4}{2!} + \dfrac{3^4}{3!} + \dfrac{4^4}{4!} + \ldots \text{ to } \infty.$

38. Find the sum: $\dfrac{1^2 \cdot 2^2}{1!} + \dfrac{2^2 \cdot 3^2}{2!} + \dfrac{3^2 \cdot 4^2}{3!} + \ldots \text{ to } \infty.$

39. Sum to infinite terms: $1 + \dfrac{1+2}{2!} + \dfrac{1+2+3}{3!} + \ldots$

40. Sum: $\dfrac{1}{1!} + \dfrac{1+3}{2!} + \dfrac{1+3+5}{3!} + \ldots \text{ to } \infty.$

41. Find the sum of $1 + \dfrac{1+2}{2!} + \dfrac{1+2+2^2}{3!} + \ldots \text{ to } \infty.$

42. Sum: $\dfrac{1}{2!} + \dfrac{1^2+2^2}{3!} + \dfrac{1^2+2^2+3^2}{4!} + \ldots \text{ to } \infty.$

43. Sum:

$$\frac{1+3}{1!}\log_e 3 + \frac{1+3^2}{2!}(\log_e 3)^2 + \frac{1+3^3}{3!}(\log_e 3)^3 + \ldots \text{ to } \infty.$$

44. Sum: $9 + \dfrac{16}{2!} + \dfrac{27}{3!} + \dfrac{42}{4!} + \ldots \text{ to } \infty.$

45. Sum: $\dfrac{4}{1!} + \dfrac{11}{2!} + \dfrac{22}{3!} + \dfrac{37}{4!} + \ldots \text{ to } \infty.$

46. Sum:

$$\sum_{n=0}^{\infty} \frac{1}{n!}\{n + (n-1)x^2\}^2 \cdot \frac{x^{2n}}{(1+x^2)^{n+2}} \cdot \exp\left(\frac{-x^2}{1+x^2}\right).$$

47. Find $\displaystyle\sum_{r=1}^{\infty} a_r$ where $a_n = (\log_e 3)^n \cdot \displaystyle\sum_{k=1}^{n} \frac{2k+1}{k!\,(n-k)!}.$

48. Prove that the coefficient of x^n in the expansion of $\log_e(1 + x + x^2)$ is $\dfrac{-2}{n}$ or $\dfrac{1}{n}$ according as n is a multiple of 3 or not.

49. Show that the coefficient of x^n in the expansion of $\log_e(1 + x + x^2 + x^3)$ in powers of x is $\dfrac{-3}{n}$ or $\dfrac{1}{n}$ according as n is a multiple of 4 or is not.

Prove that

50. (i) $\log_e 2 = \dfrac{1}{1 \cdot 2} + \dfrac{1}{3 \cdot 4} + \dfrac{1}{5 \cdot 6} + \ldots \text{ to } \infty.$

(ii) $\frac{1}{2 \cdot 3} + \frac{1}{4 \cdot 5} + \frac{1}{6 \cdot 7} + \ldots$ to $\infty = 1 - \log_e 2$.

51. $1 + \frac{1}{3 \cdot 2^2} + \frac{1}{5 \cdot 2^4} + \frac{1}{7 \cdot 2^6} + \ldots$ to $\infty = \log_e 3$.

52. $\frac{1}{4} + \frac{1}{2 \cdot 4^2} + \frac{1}{3 \cdot 4^3} + \ldots$ to ∞

$= \frac{1}{3} - \frac{1}{2 \cdot 3^2} + \frac{1}{3 \cdot 3^3} - \frac{1}{4 \cdot 3^4} + \ldots$ to ∞.

53. $\frac{1}{2}x^2 + \frac{2}{3}x^3 + \frac{3}{4}x^4 + \ldots$ to $\infty = \log_e(1 - x) + \frac{x}{1 - x}$

where $|x| < 1$.

54. $\log_e \frac{1}{1 - x} = x + \frac{x^2}{2} + \frac{x^3}{3} + \ldots$ to ∞.

55. $\log_e \frac{1 + x}{1 - x} = 2\left(x + \frac{x^2}{3} + \frac{x^5}{5} + \ldots \text{ to } \infty\right)$.

56. $\log_e n - \log_e(n - 1) = \frac{1}{n} + \frac{1}{2n^2} + \frac{1}{3n^3} + \ldots$ to ∞.

57. $\log_e(n + 1) - \log_e n$

$= 2\left[\frac{1}{2n + 1} + \frac{1}{3(2n + 1)^3} + \frac{1}{5(2n + 1)^5} + \ldots \text{ to } \infty\right]$.

58. $\log_e(n + 1) - \log_e(n - 1)$

$= 2\left\{\frac{1}{n} + \frac{1}{3n^3} + \frac{1}{5n^5} + \ldots \text{ to } \infty\right\}$.

59. $2\log_e x - \log_e(x + 1) - \log_e(x - 1)$

$= \frac{1}{x^2} + \frac{1}{2x^4} + \frac{1}{3x^6} + \ldots$ to ∞.

60. $\log_e\left(1 + \frac{1}{n}\right)^n = 1 - \frac{1}{2(n + 1)} - \frac{1}{2 \cdot 3(n + 1)^2}$

$- \frac{1}{3 \cdot 4 \cdot (n + 1)^3} - \ldots$ to ∞.

61. $\log_e\{(1 + x)^{1 + x} \cdot (1 - x)^{1 - x}\}$

$= 2\left(\frac{x^2}{1 \cdot 2} + \frac{x^4}{3 \cdot 4} + \frac{x^6}{5 \cdot 6} + \ldots \text{ to } \infty\right)$.

62. $\log(1 + 3x + 2x^2) = 3x - \frac{5x^2}{2} + \frac{9x^3}{3} - \frac{17x^4}{4} + \ldots$ to ∞.

63. If α and β be the roots of the equation $ax^2 + bx + c = 0$ then prove that

$$\log_e(a - bx + cx^2) = \log_e a + (\alpha + \beta)x - \frac{\alpha^2 + \beta^2}{2}x^2 + \frac{\alpha^3 + \beta^3}{3}x^3 - \ldots \text{ to } \infty.$$

64. If α, β be the roots of $ax^2 + bx + c = 0$ then prove that

$$\log_e(ax^2 + bx + c) = \log_e a + 2\log_e x - \frac{\alpha + \beta}{x} - \frac{\alpha^2 + \beta^2}{2x^2} - \frac{\alpha^3 + \beta^3}{3x^3} - \ldots \text{ to } \infty.$$

65. If $y = 1 - \frac{x}{1!} + \frac{x^2}{2!} - \frac{x^3}{3!} + \ldots$ to ∞ and

$z = -y - \frac{y^2}{2} - \frac{y^3}{3} - \ldots$ to ∞

then prove that $x + \log_e(1 - e^z) = 0$.

66. If $y = x - \frac{x^2}{2} + \frac{x^3}{3} - \frac{x^4}{4} + \ldots$ to ∞

then prove that $x = y + \frac{y^2}{2!} + \frac{y^3}{3!} + \ldots$ to ∞.

67. Find the sum of the series:

$1 + \frac{\sqrt{2} - 1}{2\sqrt{2}} + \frac{3 - 2\sqrt{2}}{12} + \frac{5\sqrt{2} - 7}{24\sqrt{2}} + \frac{17 - 12\sqrt{2}}{80} + \ldots$ to ∞.

Answers

1. $\frac{77}{16}x^3$ **2.** $\frac{-27}{50} \cdot 2^{-\frac{2}{5}} \cdot x^{-\frac{7}{5}}$

3. $-\frac{1 \cdot 3 \cdot 5 \cdot \ldots (2r - 3)}{r!}\left(\frac{x}{2}\right)^r$

4. -80 **5.** $(-1)^n \frac{(3n)!}{n!(2n)!}$

7. $2r^2 + 2r + 1$

8. $n(n + 1)\left\{1 + \frac{(n + 3)(n + 2)(n - 1)}{5!}\right\}$

12. 1.2599 **13.** 2 **14.** $\sqrt{2}$

15. $\sqrt{2}$ **16.** $\sqrt[3]{4}$ **20.** 30

21. $\frac{27}{64}$ **23.** $\frac{b^n e^a}{n!}$ **24.** $(-1)^r \cdot \frac{e}{r!}$

29. $3e$ **30.** $e^{x^2} - e^{y^2}$ **31.** $e - 1$

34. $2e$ **35.** $e - 1$ **37.** $15e$

38. $27e$ **39.** $\frac{3}{2}e$ **40.** $2e$

41. $e(e-1)$ **42.** $\frac{5e}{6}$ **43.** 28

44. $11e-6$ **45.** $6e-1$ **46.** $\frac{x^2}{1+x^2}$

47. $6+18\log_e 3$

67. $2-\frac{1}{2(\sqrt{2}-1)}\log 2$

□

Chapter Test

Time: 60 minutes

1. In the following, fill in the blanks so that the resulting statement becomes correct.

(i) $1 + {}^nC_1x + {}^{n+1}C_2x^2 + {}^{n+2}C_3x^3 + \ldots$ to ∞ = ______, where $n \in N$.

(ii) If $\dfrac{1+x}{1-2x+x^2} = a_0 + a_1x + a_2x^2 + \ldots$ to ∞, $|x| < 1$ then

$a_0 + \dfrac{a_1}{2} + \dfrac{a_2}{2^2} + \ldots$ to ∞ = ______ .

(iii) The coefficient of x^3 in $(1 + 2x + 3x^2 + 4x^3 + \ldots$ to $\infty)^3$ is ______ where $|x| < 1$.

(iv) The sum of $1 - \dfrac{1}{2} + \dfrac{1}{3} - \dfrac{1}{4} + \ldots$ to ∞ = ______ .

(v) In the expansion of $(e^x)^x$, the coefficient of x^{2n} is ______ and the coefficient of x^{2n+1} is ______ .

2. In the following, a number of options are given of which one or more than one are correct. Choose the correct option(s).

(i) $1 + {}^2C_1x + {}^3C_2x^2 + {}^4C_3x^3 + \ldots$ to ∞ can be summed up if

A. $x < 1$ B. $x > -1$ C. $-1 < x < 1$ D. none of these

(ii) The sum of the series $1 + \dfrac{2}{1!} + \dfrac{3}{2!} + \dfrac{4}{3!} + \ldots$ to ∞ is

A. $2e$ B. $2e - 1$ C. $2e + 1$ D. none of these

(iii) The sum of the series $1 - \dfrac{1}{2\cdot 2} + \dfrac{1}{3\cdot 2^2} - \dfrac{1}{4\cdot 2^3} + \ldots$ to ∞ is

A. $\log_e \dfrac{3}{2}$ B. $\log_e \dfrac{9}{4}$ C. $e^{\sqrt{2}}$ D. none of these

(iv) $\dfrac{a-b}{a} + \dfrac{1}{2}\left(\dfrac{a-b}{a}\right)^2 + \dfrac{1}{3}\left(\dfrac{a-b}{a}\right)^3 + \ldots$ to ∞ is equal to

A. $\log_e \dfrac{a}{b}$ B. $\log_e \dfrac{b}{a}$ C. $\log_e \dfrac{a-b}{a}$ D. none of these

3. If a_r = coefficient of x^r in the expansion of $\dfrac{(1+x)^2}{(1-x)^3}$ where $|x| < 1$ then find the value of $\sum\limits_{r=1}^{n} a_r$.

4. If $q_n = \dfrac{1\cdot 3\cdot 5\cdot 7\cdot \ldots (2n-1)}{2\cdot 4\cdot 6\cdot 8\cdot \ldots 2n}$, prove that

$$2\{q_{2n} - q_1 q_{2n-1} + q_2 q_{2n-2} - \ldots + (-1)^{n-1} q_{n-1} q_{n+1}\} = q_n + (-1)^{n-1}\cdot q_n^2.$$

5. Find the coefficient of $\dfrac{1}{n^4}$ in the expression $\dfrac{1}{n+1} + \dfrac{1}{2(n+1)^2} + \dfrac{1}{3(n+1)^3} + \ldots$, $n \in N$.

6. Find the sum: $1 - n^2 + \dfrac{n^2(n^2-1^2)}{(2!)^2} - \dfrac{n^2(n^2-1^2)(n^2-2^2)}{(3!)^2} + \ldots$, where $n \in N$.

□

Answers

1. (i) $(1-x)^{-n}$ (ii) 6 (iii) 56 (iv) $\log_e 2$ (v) $\frac{1}{n!}$, 0 respectively

2. (i) C (ii) A (iii) B (iv) A

3. $n\left\{1+\frac{2}{3}(n+1)(n+2)\right\}$

5. $-\frac{1}{4}$

6. 0

□

9. Matrices

Recap of Facts and Formulae

1. Matrix

- An array of mn numbers written in m rows and n columns is called a matrix of the order $m \times n$. A matrix has no numerical value. The numbers are the elements of the matrix.

For example $\begin{bmatrix} 3 & -2 & 1 \\ 4 & 1 & 0 \end{bmatrix}$ is a 2×3 matrix. It has two rows and three columns. Its elements are 3, −2, 1, 4, 1, 0.

$\begin{bmatrix} 2 & 5 \\ -1 & 2 \\ 7 & -4 \end{bmatrix}$ is a 3×2 matrix. It has three rows and two columns.

A matrix is denoted by a capital letter like A, B, X, Y, etc.

2. Rectangular matrix and square matrix

- If in an $m \times n$ matrix $m \neq n$ then it is a rectangular matrix. But if $m = n$, it is a square matrix.

For example $\begin{bmatrix} 1 & -3 \\ 0 & 5 \\ -6 & 9 \end{bmatrix}$ is a rectangular matrix, while $\begin{bmatrix} 1 & -3 & 4 \\ 0 & 5 & 8 \\ -6 & 9 & 7 \end{bmatrix}$ is a square matrix.

3. Row matrix and column matrix

- $1 \times n$ matrices are called row matrices, while $m \times 1$ matrices are called column matrices.

For example [1 2 3], [0 0] are row matrices, while $\begin{bmatrix} 1 \\ 2 \\ 3 \end{bmatrix}, \begin{bmatrix} 0 \\ 0 \end{bmatrix}$ are column matrices.

4. Zero matrix, unit (identity) matrix and diagonal matrix

- A matrix whose every element is zero is a zero matrix.

For example [0 0 0], $\begin{bmatrix} 0 & 0 \\ 0 & 0 \end{bmatrix}, \begin{bmatrix} 0 & 0 & 0 \\ 0 & 0 & 0 \end{bmatrix}$ are all zero matrices.

A square matrix whose elements in the leading diagonal are each equal to 1 and the rest are zeros, is a unit matrix.

For example $\begin{bmatrix} 1 & 0 \\ 0 & 1 \end{bmatrix}, \begin{bmatrix} 1 & 0 & 0 \\ 0 & 1 & 0 \\ 0 & 0 & 1 \end{bmatrix}$ are all unit matrices. A unit matrix of order $n \times n$ is denoted by I_n.

A square matrix is called a diagonal matrix if all its elements other than the elements in the leading diagonal are zeros.

5. Determinant of a square matrix

- The determinant of the square matrix $A = \begin{bmatrix} 7 & -9 \\ 4 & 2 \end{bmatrix}$ is $|A| = \begin{vmatrix} 7 & -9 \\ 4 & 2 \end{vmatrix} = 50$.
- The determinant of the square matrix

$$B = \begin{bmatrix} 1 & 0 & 0 \\ 0 & 1 & 0 \\ 0 & 0 & 1 \end{bmatrix} \text{ is } |B| = \begin{vmatrix} 1 & 0 & 0 \\ 0 & 1 & 0 \\ 0 & 0 & 1 \end{vmatrix} = 1.$$

6. Equality of matrices

- Two matrices A and B of the same order are equal when their corresponding elements are equal.

For example $\begin{bmatrix} -5 & 7 \\ 4 & 3 \end{bmatrix} = \begin{bmatrix} a & b \\ c & d \end{bmatrix}$ if $a = -5, b = 7, c = 4, d = 3$.

$\begin{bmatrix} -5 & 7 \\ 4 & 3 \end{bmatrix} \neq \begin{bmatrix} -5 & 7 & 1 \\ 4 & 3 & 2 \end{bmatrix}$ because they are not of the same order.

7. Singular matrix and nonsingular matrix

- A square matrix A is a singular matrix if $|A| = 0$. But if $|A| \neq 0$, the matrix is nonsingular.

8. Scalar multiplication of a matrix by a number

- If A is a matrix and k is a scalar, $k \cdot A$ is a matrix whose every element is k times the corresponding element of A.

For example If $A = \begin{bmatrix} a & b \\ c & d \\ e & f \end{bmatrix}$ then $kA = \begin{bmatrix} ka & kb \\ kc & kd \\ ke & kf \end{bmatrix}$

9. Negative of a matrix

- The negative of the matrix A is the matrix $(-1)A$, written as $-A$.

For example If $A = \begin{bmatrix} a & b & c \\ d & e & f \end{bmatrix}$ then $-A = \begin{bmatrix} -a & -b & -c \\ -d & -e & -f \end{bmatrix}$.

10. Addition of two matrices

- The addition (or sum) of two matrices A and B of the same order is a matrix, written as $A + B$, which is of the same order as that of A as well as B, and the elements of $A + B$ are obtained by adding the corresponding elements of A and B.

For example Let

$$A = \begin{bmatrix} 3 & -5 \\ 1 & 4 \\ 0 & 2 \end{bmatrix}, B = \begin{bmatrix} 8 & 11 \\ 10 & 6 \\ -1 & -3 \end{bmatrix}, C = \begin{bmatrix} 11 & 13 \\ -7 & 15 \end{bmatrix}.$$

A and B are matrices of order 3×2. So $A + B$ can be found.

$$A + B = \begin{bmatrix} 3 & -5 \\ 1 & 4 \\ 0 & 2 \end{bmatrix} + \begin{bmatrix} 8 & 11 \\ 10 & 6 \\ -1 & -3 \end{bmatrix} = \begin{bmatrix} 3+8 & -5+11 \\ 1+10 & 4+6 \\ 0-1 & 2-3 \end{bmatrix}$$

$$= \begin{bmatrix} 11 & 6 \\ 11 & 10 \\ -1 & -1 \end{bmatrix}.$$

A and C are not matrices of the same order. A is of the order 3×2, while C is of the order 2×2. So, $A + C$ cannot be found. Similarly, $B + C$ cannot be found.

11. Subtraction of two matrices

- If A and B are two matrices of the same order then the sum $A + (-B)$, written as $A - B$, is the matrix obtained by subtracting B from A.

12. Multiplication of two matrices

- For two matrices A and B, the product matrix AB can be obtained if the number of columns in A = the number of rows in B.

  If A is a 2×3 matrix and B is a 3×4 matrix then the product matrix AB can be obtained (number of columns in A = number of rows in $B = 3$). But the product matrix BA cannot be obtained (number of columns in $B = 4 \neq 2$ = number of rows in A).

Rule of multiplication

Let $A = \begin{bmatrix} a & b \\ c & d \\ e & f \end{bmatrix}, B = \begin{bmatrix} x & y & z \\ l & m & n \end{bmatrix}$.

A is of the order 3×2 and B is of the order 2×3.

So, AB as well as BA can be found. The rule for finding the product is as shown below.

$$AB = \begin{bmatrix} a & b \\ c & d \\ e & f \end{bmatrix} \times \begin{bmatrix} x & y & z \\ l & m & n \end{bmatrix} = \begin{bmatrix} ax+bl & ay+bm & az+bn \\ cx+dl & cy+dm & cz+dn \\ ex+fl & ey+fm & ez+fn \end{bmatrix}.$$

$$BA = \begin{bmatrix} x & y & z \\ l & m & n \end{bmatrix}\begin{bmatrix} a & b \\ c & d \\ e & f \end{bmatrix} = \begin{bmatrix} xa+yc+ze & xb+yd+zf \\ la+mc+ne & lb+md+nf \end{bmatrix}.$$

Observe that AB is a 3×3 matrix, while BA is a 2×2 matrix. So, $AB \neq BA$.

13. Transpose matrix

- The transpose matrix, written as A^T (or A'), of the matrix A is the matrix obtained by interchanging the rows with the corresponding columns of A.

For example If $A = \begin{bmatrix} a & b & c \\ d & e & f \end{bmatrix}$ then $A^T = \begin{bmatrix} a & d \\ b & e \\ c & f \end{bmatrix}$.

If $B = \begin{bmatrix} l & m & n \\ p & q & r \\ u & v & w \end{bmatrix}$ then $B^T = \begin{bmatrix} l & p & u \\ m & q & v \\ n & r & w \end{bmatrix}$.

14. Adjoint matrix

- Let A be a square matrix. The transpose matrix of the matrix of the corresponding cofactors of the elements of $|A|$, is called the adjoint matrix of A, written as *adj* A.

For example Let $A = \begin{bmatrix} a_1 & b_1 & c_1 \\ a_2 & b_2 & c_2 \\ a_3 & b_3 & c_3 \end{bmatrix}$. If A_1, B_1, etc.,

represent the cofactors of a_1, b_1, etc., respectively in $|A|$ then

$$adj\ A = \begin{bmatrix} A_1 & A_2 & A_3 \\ B_1 & B_2 & B_3 \\ C_1 & C_2 & C_3 \end{bmatrix}.$$

15. Inverse matrix (or reciprocal matrix)

- Let A be a square matrix which is nonsingular, i.e., $|A| \neq 0$.

  The matrix $\frac{1}{|A|} \cdot adj\ A$ is called the inverse matrix of A, written as A^{-1}.

16. Orthogonal matrix

- A square matrix A is called an orthogonal matrix if $A^TA = AA^T = I$.

17. Symmetric and skew-symmetric matrix

- A square matrix A is called symmetric if $A^T = A$. It is skew-symmetric if $A^T = -A$.

18. Properties of addition of matrices

- If A, B, C are three matrices of the same order then

(1) $A + B = B + A$ (commutative law of addition)

(2) $A + (B + C) = (A + B) + C$ (associative law of addition)

(3) $A + 0 = 0 + A = A$

(4) $A + B = A + C \Rightarrow B = C$

(5) $A + (-A) = (-A) + A = 0$.

19. Properties of multiplication of matrices

- If A, B, C are three matrices for which all the products are possible then

(1) $(AB)C = A(BC)$ (associative law of multiplication)

(2) $A(B + C) = AB + AC$

$(B + C)A = BA + CA$ (distributive laws)

(3) $AB \neq BA$ in general.

20. Other properties of matrices

(1) $AA^{-1} = A^{-1}A = I$

(2) $(AB)^{-1} = B^{-1}A^{-1}$

(3) $(A^T)^T = A$

(4) $(A + B)^T = A^T + B^T$ (if additions are possible)

(5) $(AB)^T = B^TA^T$ (if multiplications are possible)

(6) $|A^2| = |A|^2$

21. Rank of a matrix

- The rank of a matrix A is the order of the highest-order nonzero minors of A.

$\therefore$ rank of a matrix is r if

(1) every minor of order $(r + 1)$ is zero

(2) there is at least one nonzero minor of order r.

Selected Solved Examples

1. Let $A = \begin{bmatrix} 2 & 0 & 1 \\ 2 & 1 & 3 \\ -1 & -1 & 0 \end{bmatrix}$ and $f(x) = x^2 - 5x + 6I_3$. Find $f(A)$.

Here, $f(A) = A^2 - 5A + 6I_3$

$$= \begin{bmatrix} 2 & 0 & 1 \\ 2 & 1 & 3 \\ -1 & -1 & 0 \end{bmatrix}^2 - 5\begin{bmatrix} 2 & 0 & 1 \\ 2 & 1 & 3 \\ -1 & -1 & 0 \end{bmatrix} + 6\begin{bmatrix} 1 & 0 & 0 \\ 0 & 1 & 0 \\ 0 & 0 & 1 \end{bmatrix}$$

$$= \begin{bmatrix} 2 & 0 & 1 \\ 2 & 1 & 3 \\ -1 & -1 & 0 \end{bmatrix} \times \begin{bmatrix} 2 & 0 & 1 \\ 2 & 1 & 3 \\ -1 & -1 & 0 \end{bmatrix} - \begin{bmatrix} 10 & 0 & 5 \\ 10 & 5 & 15 \\ -5 & -5 & 0 \end{bmatrix} + \begin{bmatrix} 6 & 0 & 0 \\ 0 & 6 & 0 \\ 0 & 0 & 6 \end{bmatrix}$$

$$= \begin{bmatrix} 2\times2+0\times2+1(-1) & 2\times0+0\times1+1(-1) & 2\times1+0\times3+1\times0 \\ 2\times2+1\times2+3(-1) & 2\times0+1\times1+3(-1) & 2\times1+1\times3+3\times0 \\ (-1)2+(-1)2+0(-1) & (-1)0+(-1)1+0(-1) & (-1)1+(-1)3+0\times0 \end{bmatrix}$$

$$+ \begin{bmatrix} 6-10 & 0-0 & 0-5 \\ 0-10 & 6-5 & 0-15 \\ 0-(-5) & 0-(-5) & 6-0 \end{bmatrix}$$

$$= \begin{bmatrix} 3 & -1 & 2 \\ 3 & -2 & 5 \\ -4 & -1 & -4 \end{bmatrix} + \begin{bmatrix} -4 & 0 & -5 \\ -10 & 1 & -15 \\ 5 & 5 & 6 \end{bmatrix}$$

$$= \begin{bmatrix} 3-4 & -1+0 & 2-5 \\ 3-10 & -2+1 & 5-15 \\ -4+5 & -1+5 & -4+6 \end{bmatrix} = \begin{bmatrix} -1 & -1 & -3 \\ -7 & -1 & -10 \\ 1 & 4 & 2 \end{bmatrix}.$$

2. If $A = \begin{bmatrix} 2 & 0 & 3 \\ -3 & -1 & 2 \\ 1 & 1 & -1 \end{bmatrix}$, $B = \begin{bmatrix} 3 & 1 \\ 2 & 0 \\ 4 & -1 \end{bmatrix}$ and $C = \begin{bmatrix} 2 & 1 & -4 & 3 \\ 0 & 2 & 1 & -2 \end{bmatrix}$ then prove that $(AB)C = A(BC)$.

Here, $AB = \begin{bmatrix} 2 & 0 & 3 \\ -3 & -1 & 2 \\ 1 & 1 & -1 \end{bmatrix} \times \begin{bmatrix} 3 & 1 \\ 2 & 0 \\ 4 & -1 \end{bmatrix}$

$$= \begin{bmatrix} 2\times3+0\times2+3\times4 & 2\times1+0\times0+3(-1) \\ -3(3)+(-1)2+2\times4 & -3(1)+(-1)0+2(-1) \\ 1\times3+1\times2+(-1)4 & 1\times1+1\times0+(-1)(-1) \end{bmatrix}$$

$$= \begin{bmatrix} 18 & -1 \\ -3 & -5 \\ 1 & 2 \end{bmatrix}.$$

$$\therefore \quad (AB)C = \begin{bmatrix} 18 & -1 \\ -3 & -5 \\ 1 & 2 \end{bmatrix} \begin{bmatrix} 2 & 1 & -4 & 3 \\ 0 & 2 & 1 & -2 \end{bmatrix}$$

$$= \begin{bmatrix} 36 & 16 & -73 & 56 \\ -6 & -13 & 7 & 1 \\ 2 & 5 & -2 & -1 \end{bmatrix}.$$

Now, $BC = \begin{bmatrix} 3 & 1 \\ 2 & 0 \\ 4 & -1 \end{bmatrix} \begin{bmatrix} 2 & 1 & -4 & 3 \\ 0 & 2 & 1 & -2 \end{bmatrix} = \begin{bmatrix} 6 & 5 & -11 & 7 \\ 4 & 2 & -8 & 6 \\ 8 & 2 & -17 & 14 \end{bmatrix}.$

$$\therefore \quad A(BC)=\begin{bmatrix}2&0&3\\-3&-1&2\\1&1&-1\end{bmatrix}\begin{bmatrix}6&5&-11&7\\4&2&-8&6\\8&2&-17&14\end{bmatrix}$$

$$=\begin{bmatrix}36&16&-73&56\\-6&-13&7&1\\2&5&-2&-1\end{bmatrix}.$$

$\therefore \quad (AB)C = A(BC).$

3. If $f(x)=\begin{bmatrix}\cos x&-\sin x&0\\\sin x&\cos x&0\\0&0&1\end{bmatrix}$, prove that

$$f(\theta+\phi)=f(\theta)\cdot f(\phi).$$

Also, establish $|f(\theta+\phi)|=1$ for all values of θ and ϕ.

$$f(\theta)\cdot f(\phi)=\begin{bmatrix}\cos\theta&-\sin\theta&0\\\sin\theta&\cos\theta&0\\0&0&1\end{bmatrix}\begin{bmatrix}\cos\phi&-\sin\phi&0\\\sin\phi&\cos\phi&0\\0&0&1\end{bmatrix}$$

$$=\begin{bmatrix}\cos\theta\cdot\cos\phi-\sin\theta\cdot\sin\phi+0 & -\cos\theta\cdot\sin\phi-\sin\theta\cdot\cos\phi+0 & 0+0+0\\ \sin\theta\cdot\cos\phi+\cos\theta\cdot\sin\phi+0 & -\sin\theta\cdot\sin\phi+\cos\theta\cdot\cos\phi+0 & 0+0+0\\ 0+0+0 & 0+0+0 & 0+0+1\end{bmatrix}$$

$$=\begin{bmatrix}\cos(\theta+\phi)&-\sin(\theta+\phi)&0\\\sin(\theta+\phi)&\cos(\theta+\phi)&0\\0&0&1\end{bmatrix}=f(\theta+\phi).$$

$$\text{Now, } |f(\theta+\phi)|=\begin{vmatrix}\cos(\theta+\phi)&-\sin(\theta+\phi)&0\\\sin(\theta+\phi)&\cos(\theta+\phi)&0\\0&0&1\end{vmatrix}$$

$$=\begin{vmatrix}\cos(\theta+\phi)&-\sin(\theta+\phi)\\\sin(\theta+\phi)&\cos(\theta+\phi)\end{vmatrix}$$

$=\cos^2(\theta+\phi)+\sin^2(\theta+\phi)=1$ for all θ and ϕ.

4. Let $A=\begin{bmatrix}x+y&y\\2x&x-y\end{bmatrix}, B=\begin{bmatrix}2\\-1\end{bmatrix}$ and $C=\begin{bmatrix}3\\2\end{bmatrix}$. If $AB=C$ then find the matrix A^2.

$$\text{Here, }\begin{bmatrix}x+y&y\\2x&x-y\end{bmatrix}\begin{bmatrix}2\\-1\end{bmatrix}=\begin{bmatrix}3\\2\end{bmatrix}$$

$$\Rightarrow \quad \begin{bmatrix}2(x+y)-y\\2x\cdot2-(x-y)\end{bmatrix}=\begin{bmatrix}3\\2\end{bmatrix}$$

$\Rightarrow \quad 2(x+y)-y=3$ and $4x-(x-y)=2$

$\Rightarrow \quad 2x+y=3$ and $3x+y=2.$

Subtracting the two equations, we get, $x=-1$. So, $y=5$.

$$\therefore \quad A=\begin{bmatrix}-1+5&5\\2(-1)&-1-5\end{bmatrix}=\begin{bmatrix}4&5\\-2&-6\end{bmatrix}.$$

$$\therefore \quad A^2=\begin{bmatrix}4&5\\-2&-6\end{bmatrix}\begin{bmatrix}4&5\\-2&-6\end{bmatrix}$$

$$=\begin{bmatrix}4\times4+5(-2)&4\times5+5(-6)\\-2\times4+(-6)(-2)&-2\times5+(-6)(-6)\end{bmatrix}$$

$$=\begin{bmatrix}6&-10\\4&26\end{bmatrix}.$$

5. If $A=\begin{bmatrix}1&-2&3\\-4&2&5\end{bmatrix}$ and $B=\begin{bmatrix}1&3\\-1&0\\2&4\end{bmatrix}$ then prove that $(AB)^T=B^TA^T$.

$$\text{Here, } AB=\begin{bmatrix}1&-2&3\\-4&2&5\end{bmatrix}\begin{bmatrix}1&3\\-1&0\\2&4\end{bmatrix}$$

$$=\begin{bmatrix}1(1)-2(-1)+3(2)&1(3)-2(0)+3(4)\\-4(1)+2(-1)+5(2)&-4(3)+2(0)+5(4)\end{bmatrix}$$

$$=\begin{bmatrix}9&15\\4&8\end{bmatrix}. \qquad \therefore \quad (AB)^T=\begin{bmatrix}9&4\\15&8\end{bmatrix}.$$

$$B^TA^T=\begin{bmatrix}1&-1&2\\3&0&4\end{bmatrix}\begin{bmatrix}1&-4\\-2&2\\3&5\end{bmatrix}$$

$$=\begin{bmatrix}1(1)-1(-2)+2(3)&1(-4)-1(2)+2(5)\\3(1)+0(-2)+4(3)&3(-4)+0(2)+4(5)\end{bmatrix}$$

$$=\begin{bmatrix}9&4\\15&8\end{bmatrix}=(AB)^T.$$

6. If the matrix $A=\begin{bmatrix}3-x&2&2\\2&4-x&1\\-2&-4&-1-x\end{bmatrix}$ is a singular matrix then find x. Verify whether $AA^T=I$ for that value of x.

Here, A is a singular matrix if $|A|=0$, i.e.,

$$\begin{vmatrix}3-x&2&2\\2&4-x&1\\-2&-4&-1-x\end{vmatrix}=0$$

$$\text{or} \quad \begin{vmatrix}3-x&2&2\\2&4-x&1\\0&-x&-x\end{vmatrix}=0, \text{ using } R_3\to R_3+R_2$$

$$\text{or} \quad \begin{vmatrix}3-x&0&2\\2&3-x&1\\0&0&-x\end{vmatrix}=0, \text{ using } C_2\to C_2-C_3$$

or $\quad -x(3-x)^2=0. \qquad \therefore \quad x=0, 3.$

$$\text{When } x=0, A=\begin{bmatrix}3&2&2\\2&4&1\\-2&-4&-1\end{bmatrix}.$$

$$\therefore \quad AA^T=\begin{bmatrix}3&2&2\\2&4&1\\-2&-4&-1\end{bmatrix}\begin{bmatrix}3&2&-2\\2&4&-4\\2&1&-1\end{bmatrix}$$

$$=\begin{bmatrix}17&16&-18\\16&21&-21\\-16&-21&21\end{bmatrix}\neq I.$$

$$\text{When } x=3, A=\begin{bmatrix}0&2&2\\2&1&1\\-2&-4&-4\end{bmatrix}.$$

$$\therefore \quad AA^T = \begin{bmatrix} 0 & 2 & 2 \\ 2 & 1 & 1 \\ -2 & -4 & -4 \end{bmatrix} \begin{bmatrix} 0 & 2 & -2 \\ 2 & 1 & -4 \\ 2 & 1 & -4 \end{bmatrix}$$

$$= \begin{bmatrix} 8 & 4 & -16 \\ 4 & 6 & -10 \\ -16 & -12 & 36 \end{bmatrix} \neq I.$$

7. Let $A = \begin{bmatrix} 2 & 1 & -1 \\ 0 & 1 & 0 \\ 1 & 3 & -1 \end{bmatrix}$ and $B = \begin{bmatrix} 1 & 2 & 5 \\ 2 & 3 & 1 \\ -1 & 1 & 1 \end{bmatrix}$. Prove that $(AB)^{-1} = B^{-1}A^{-1}$.

Here, $AB = \begin{bmatrix} 2 & 1 & -1 \\ 0 & 1 & 0 \\ 1 & 3 & -1 \end{bmatrix} \begin{bmatrix} 1 & 2 & 5 \\ 2 & 3 & 1 \\ -1 & 1 & 1 \end{bmatrix}$

$$= \begin{bmatrix} 2+2+1 & 4+3-1 & 10+1-1 \\ 0+2+0 & 0+3+0 & 0+1+0 \\ 1+6+1 & 2+9-1 & 5+3-1 \end{bmatrix} = \begin{bmatrix} 5 & 6 & 10 \\ 2 & 3 & 1 \\ 8 & 10 & 7 \end{bmatrix}.$$

Now, $|AB| = \begin{vmatrix} 5 & 6 & 10 \\ 2 & 3 & 1 \\ 8 & 10 & 7 \end{vmatrix}$

$$= 5(21-10) - 6(14-8) + 10(20-24)$$

$$= 55 - 36 - 40 = -21.$$

The matrix of cofactors of $|AB|$ is

$$\begin{bmatrix} 3(7)-1(10) & -\{2(7)-8(1)\} & 2(10)-3(8) \\ -\{6(7)-10(10)\} & 5(7)-8(10) & -\{5(10-6(8)\} \\ 6(1)-10(3) & -\{5(1)-2(10)\} & 5(3)-6(2) \end{bmatrix}$$

$$= \begin{bmatrix} 11 & -6 & -4 \\ 58 & -45 & -2 \\ -24 & 15 & 3 \end{bmatrix}.$$

$$\therefore \quad adj\ AB = \begin{bmatrix} 11 & 58 & -24 \\ -6 & -45 & 15 \\ -4 & -2 & 3 \end{bmatrix}.$$

$$\therefore \quad (AB)^{-1} = \frac{adj\ AB}{|AB|} = \frac{-1}{21} \begin{bmatrix} 11 & 58 & -24 \\ -6 & -45 & 15 \\ -4 & -2 & 3 \end{bmatrix}.$$

Next, $|B| = \begin{vmatrix} 1 & 2 & 5 \\ 2 & 3 & 1 \\ -1 & 1 & 1 \end{vmatrix}$

$$= 1(3-1) - 2(2+1) + 5(2+3) = 21.$$

$$\therefore \quad B^{-1} = \frac{adj\ B}{|B|} = \frac{1}{21} \begin{bmatrix} 2 & 3 & -13 \\ -3 & 6 & 9 \\ 5 & -3 & -1 \end{bmatrix}.$$

$$|A| = \begin{vmatrix} 2 & 1 & -1 \\ 0 & 1 & 0 \\ 1 & 3 & -1 \end{vmatrix} = 1(-2+1) = -1.$$

$$\therefore \quad A^{-1} = \frac{adj\ A}{|A|} = \frac{1}{-1} \begin{bmatrix} -1 & -2 & 1 \\ 0 & -1 & 0 \\ -1 & -5 & 2 \end{bmatrix}.$$

$$\therefore \quad B^{-1}A^{-1} = -\frac{1}{21} \begin{bmatrix} 2 & 3 & -13 \\ -3 & 6 & 9 \\ 5 & -3 & -1 \end{bmatrix} \begin{bmatrix} -1 & -2 & 1 \\ 0 & -1 & 0 \\ -1 & -5 & 2 \end{bmatrix}$$

$$= -\frac{1}{21} \begin{bmatrix} 11 & 58 & -24 \\ -6 & -45 & 15 \\ -4 & -2 & 3 \end{bmatrix}.$$

Thus $(AB)^{-1} = B^{-1}A^{-1}$.

8. If $A = \begin{bmatrix} \sin\alpha & -\cos\alpha & 0 \\ \cos\alpha & \sin\alpha & 0 \\ 0 & 0 & 1 \end{bmatrix}$ then find $|A^T|$ and $|A^{-1}|$.

$$A^T = \text{transpose of } A = \begin{bmatrix} \sin\alpha & \cos\alpha & 0 \\ -\cos\alpha & \sin\alpha & 0 \\ 0 & 0 & 1 \end{bmatrix}.$$

$$\therefore \quad |A^T| = \begin{vmatrix} \sin\alpha & \cos\alpha & 0 \\ -\cos\alpha & \sin\alpha & 0 \\ 0 & 0 & 1 \end{vmatrix} = \begin{vmatrix} \sin\alpha & \cos\alpha \\ -\cos\alpha & \sin\alpha \end{vmatrix} = 1.$$

Also, $|A^T| = |A|$. So, $|A| = 1$.

$$A^{-1} = \frac{adj\ A}{|A|} = adj\ A \quad (\because\ |A| = 1).$$

In $|A|$, cofactor of $\sin\alpha$(in R_1, C_1) $= \begin{vmatrix} \sin\alpha & 0 \\ 0 & 1 \end{vmatrix} = \sin\alpha$

cofactor of $-\cos\alpha$(in R_1, C_2) $= -\begin{vmatrix} \cos\alpha & 0 \\ 0 & 1 \end{vmatrix} = -\cos\alpha$

cofactor of 0(in R_1, C_3) $= \begin{vmatrix} \cos\alpha & \sin\alpha \\ 0 & 0 \end{vmatrix} = 0$

cofactor of $\cos\alpha$(in R_2, C_1) $= -\begin{vmatrix} -\cos\alpha & 0 \\ 0 & 1 \end{vmatrix} = \cos\alpha$

cofactor of $\sin\alpha$(in R_2, C_2) $= \begin{vmatrix} \sin\alpha & 0 \\ 0 & 1 \end{vmatrix} = \sin\alpha$

cofactor of 0(in R_2, C_3) $= -\begin{vmatrix} \sin\alpha & -\cos\alpha \\ 0 & 0 \end{vmatrix} = 0$

cofactor of 0(in R_3, C_1) $= \begin{vmatrix} -\cos\alpha & 0 \\ \sin\alpha & 0 \end{vmatrix} = 0$

cofactor of 0(in R_3, C_2) $= -\begin{vmatrix} \sin\alpha & 0 \\ \cos\alpha & 0 \end{vmatrix} = 0$

cofactor of 1(in R_3, C_3) $= \begin{vmatrix} \sin\alpha & -\cos\alpha \\ \cos\alpha & \sin\alpha \end{vmatrix} = 1.$

$$\therefore \quad A^{-1} = adj\ A = \begin{bmatrix} \sin\alpha & \cos\alpha & 0 \\ -\cos\alpha & \sin\alpha & 0 \\ 0 & 0 & 1 \end{bmatrix}.$$

$$\therefore \quad |A^{-1}| = \begin{vmatrix} \sin\alpha & \cos\alpha & 0 \\ -\cos\alpha & \sin\alpha & 0 \\ 0 & 0 & 1 \end{vmatrix} = 1.$$

9. If the matrix $A = \begin{bmatrix} a & b & c \\ b & c & a \\ c & a & b \end{bmatrix}$ where a, b, c are positive real numbers such that $abc = 1$ and $A^TA = I$ then find the value of $a^3 + b^3 + c^3$.

Here, $A = \begin{bmatrix} a & b & c \\ b & c & a \\ c & a & b \end{bmatrix}$. So, $A^T = \begin{bmatrix} a & b & c \\ b & c & a \\ c & a & b \end{bmatrix}$, interchanging rows and columns.

$\therefore \quad A^T A = \begin{bmatrix} a & b & c \\ b & c & a \\ c & a & b \end{bmatrix}^2 = A^2.$

$\therefore \quad |A^T A| = |A^2|$; but $A^T A = I$ (given).

$\therefore \quad |I| = |A|^2 \Rightarrow 1 = |A|^2.$

Now, $|A| = \begin{vmatrix} a & b & c \\ b & c & a \\ c & a & b \end{vmatrix} = (a+b+c)\begin{vmatrix} 1 & 1 & 1 \\ b & c & a \\ c & a & b \end{vmatrix}$, $R_1 \to R_1 + R_2 + R_3$

$= (a+b+c)\begin{vmatrix} 1 & 0 & 0 \\ b & c-b & a-b \\ c & a-c & b-c \end{vmatrix}$, $C_2 \to C_2 - C_1$, $C_3 \to C_3 - C_1$

$= (a+b+c)\{(c-b)(b-c) - (a-b)(a-c)\}$

$= (a+b+c)(-b^2 - c^2 + 2bc - a^2 + ac + ab - bc)$

$= -(a+b+c)(a^2 + b^2 + c^2 - bc - ca - ab)$

$= -(a^3 + b^3 + c^3 - 3abc)$

$= -(a^3 + b^3 + c^3 - 3)$ ($\because$ $abc = 1$).

$\therefore \quad |A|^2 = 1 \Rightarrow (a^3 + b^3 + c^3 - 3)^2 = 1.$ ... (1)

As a, b, c are positive, $\dfrac{a^3 + b^3 + c^3}{3} > \sqrt[3]{a^3 b^3 c^3} = abc = 1.$

$\therefore \quad a^3 + b^3 + c^3 > 3.$

$\therefore \quad (1) \Rightarrow a^3 + b^3 + c^3 - 3 = 1.$

$\therefore \quad a^3 + b^3 + c^3 = 4.$

10. For what values of x the matrix $\begin{bmatrix} 3+x & 5 & 2 \\ 1 & 7+x & 6 \\ 2 & 5 & 3+x \end{bmatrix}$ has the rank 2?

There is only one 3rd-order minor. In order that the rank may be 2, we must have

$$\begin{vmatrix} 3+x & 5 & 2 \\ 1 & 7+x & 6 \\ 2 & 5 & 3+x \end{vmatrix} = 0. \quad \text{... (1)}$$

Now, $\begin{vmatrix} 3+x & 5 & 2 \\ 1 & 7+x & 6 \\ 2 & 5 & 3+x \end{vmatrix} = \begin{vmatrix} 1+x & 0 & -1-x \\ 1 & 7+x & 6 \\ 2 & 5 & 3+x \end{vmatrix}$, using $R_1 \to R_1 - R_3$

$= \begin{vmatrix} 1+x & 0 & 0 \\ 1 & 7+x & 7 \\ 2 & 5 & 5+x \end{vmatrix}$, using $C_3 \to C_3 + C_1$

$= (1+x)\begin{vmatrix} 7+x & 7 \\ 5 & 5+x \end{vmatrix} = (1+x)\{(7+x)(5+x) - 35\}$

$= (1+x)(x^2 + 12x) = x(1+x)(x+12).$

$\therefore \quad$ (1) holds for $x = 0, -1, -12.$

When $x = 0$, the matrix $= \begin{bmatrix} 3 & 5 & 2 \\ 1 & 7 & 6 \\ 2 & 5 & 3 \end{bmatrix}$.

Clearly, a minor $\begin{vmatrix} 3 & 5 \\ 1 & 7 \end{vmatrix} \neq 0$. So, the rank = 2.

When $x = -1$, the matrix $= \begin{bmatrix} 2 & 5 & 2 \\ 1 & 6 & 6 \\ 2 & 5 & 2 \end{bmatrix}$.

Clearly, a minor $\begin{vmatrix} 2 & 5 \\ 1 & 6 \end{vmatrix} \neq 0$. So, the rank = 2.

When $x = -12$, the matrix $= \begin{bmatrix} -9 & 5 & 2 \\ 1 & -5 & 6 \\ 2 & 5 & -9 \end{bmatrix}$.

Clearly, a minor $\begin{vmatrix} -9 & 5 \\ 1 & -5 \end{vmatrix} \neq 0$. So, the rank = 2.

$\therefore \quad$ the matrix has the rank 2 if $x = 0, -1, -12$.

11. Solve the following equations by matrix inversion.

$$2x + y + 2z = 0$$
$$2x - y + z = 10$$
$$x + 3y - z = 5.$$

In the matrix form, the equations can be written as

$$\begin{bmatrix} 2 & 1 & 2 \\ 2 & -1 & 1 \\ 1 & 3 & -1 \end{bmatrix}\begin{bmatrix} x \\ y \\ z \end{bmatrix} = \begin{bmatrix} 0 \\ 10 \\ 5 \end{bmatrix}.$$

$\therefore \quad AX = D$ where $A = \begin{bmatrix} 2 & 1 & 2 \\ 2 & -1 & 1 \\ 1 & 3 & -1 \end{bmatrix}$, $X = \begin{bmatrix} x \\ y \\ z \end{bmatrix}$, $D = \begin{bmatrix} 0 \\ 10 \\ 5 \end{bmatrix}$

$\Rightarrow \quad A^{-1}(AX) = A^{-1}D \Rightarrow (A^{-1}A)X = A^{-1}D$

$\Rightarrow \quad IX = A^{-1}D \Rightarrow X = A^{-1}D.$ ... (1)

Now $A^{-1} = \dfrac{adj\,A}{|A|}$.

$|A| = \begin{vmatrix} 2 & 1 & 2 \\ 2 & -1 & 1 \\ 1 & 3 & -1 \end{vmatrix}$

$= 2(1-3) - 1(-2-1) + 2(6+1) = 13.$

The matrix of cofactors of $|A|$ is

$\begin{bmatrix} -2 & 3 & 7 \\ 7 & -4 & -5 \\ 3 & 2 & -4 \end{bmatrix}$. So, $adj\,A = \begin{bmatrix} -2 & 7 & 3 \\ 3 & -4 & 2 \\ 7 & -5 & -4 \end{bmatrix}$.

$\therefore \quad A^{-1} = \dfrac{1}{13}\begin{bmatrix} -2 & 7 & 3 \\ 3 & -4 & 2 \\ 7 & -5 & -4 \end{bmatrix}$.

$\therefore \quad$ from (1), $X = \dfrac{1}{13}\begin{bmatrix} -2 & 7 & 3 \\ 3 & -4 & 2 \\ 7 & -5 & -4 \end{bmatrix}\begin{bmatrix} 0 \\ 10 \\ 5 \end{bmatrix}$

$= \dfrac{1}{13}\begin{bmatrix} 0 + 70 + 15 \\ 0 - 40 + 10 \\ 0 - 50 - 20 \end{bmatrix} = \begin{bmatrix} 85/13 \\ -30/13 \\ -70/13 \end{bmatrix}$.

$\therefore \quad \begin{bmatrix} x \\ y \\ z \end{bmatrix} = \begin{bmatrix} 85/13 \\ -30/13 \\ -70/13 \end{bmatrix} \Rightarrow x = \dfrac{85}{13}, y = \dfrac{-30}{13}, z = \dfrac{-70}{13}.$

Exercises

1. If $A = \begin{bmatrix} 1 & 0 & 2 \\ 0 & 1 & 1 \\ 2 & 1 & 0 \end{bmatrix}$ and $B = \begin{bmatrix} 1 & 0 & 2 \\ 2 & 0 & 1 \\ 2 & 3 & 1 \end{bmatrix}$ and

$2A + 3B = \begin{bmatrix} 5 & 0 & x+y \\ 6 & 2 & 5 \\ 10 & 2x+3y & 3 \end{bmatrix}$ then find x and y.

2. If A, B be two matrices such that

$A + B = \begin{bmatrix} 1 & 0 & 2 \\ 2 & 2 & 1 \end{bmatrix}$ and $A - B = \begin{bmatrix} 1 & 4 & 4 \\ 4 & 2 & 0 \end{bmatrix}$ then find the matrix $3A - 4B$.

3. If $A = \begin{bmatrix} 1 & 2 & 2 \\ 2 & 1 & 2 \\ 2 & 2 & 1 \end{bmatrix}$, prove that $A^2 - 4A = 5I_3$.

4. If $A = \begin{bmatrix} 2 & 3 & 4 \\ 1 & 2 & 3 \\ 1 & -1 & 2 \end{bmatrix}$ and $B = \begin{bmatrix} 1 & 3 & 2 \\ -1 & 0 & 1 \\ 0 & 0 & 3 \end{bmatrix}$, verify whether $AB = BA$.

5. If $A = \begin{bmatrix} 2 & -2 & -4 \\ -1 & 3 & 4 \\ 1 & -2 & -3 \end{bmatrix}$, prove that $A^2 = A$.

6. If $f(\alpha) = \begin{bmatrix} \cos\alpha & \sin\alpha \\ -\sin\alpha & \cos\alpha \end{bmatrix}$, prove that

$\{f(\alpha)\}^n = \begin{bmatrix} \cos n\alpha & \sin n\alpha \\ -\sin n\alpha & \cos n\alpha \end{bmatrix}$.

7. Let $A = \begin{bmatrix} 3-x & 2 & 2 \\ 2 & 4-x & 1 \\ -2 & -4 & -1-x \end{bmatrix}$. For what value of x, A^{-1} cannot be found?

8. If $A = \begin{bmatrix} 0 & -\tan \alpha/2 \\ \tan \alpha/2 & 0 \end{bmatrix}$ then prove that

$A + I = (I - A)\begin{bmatrix} \cos\alpha & -\sin\alpha \\ \sin\alpha & \cos\alpha \end{bmatrix}$ where I is a unit matrix of order 2×2.

9. For what values of x and y

$\begin{bmatrix} 2 & -3 \\ 7 & 4 \\ 9 & 0 \end{bmatrix} + \begin{bmatrix} 3 & x \\ 7 & -4 \\ y & 3 \end{bmatrix} = \begin{bmatrix} 5 & 2 \\ 14 & 0 \\ 11 & 3 \end{bmatrix}$?

10. If $A = \begin{bmatrix} 1 & 0 & 3 \\ 5 & 3 & -2 \end{bmatrix}$, $B = \begin{bmatrix} 3 & -1 & -3 \\ 1 & -3 & 2 \\ 4 & 5 & 7 \end{bmatrix}$ and

$C = \begin{bmatrix} 0 & 2 & -1 \\ 5 & -3 & 1 \\ -4 & 2 & -2 \end{bmatrix}$, verify the distributive law

$$A(B + C) = AB + AC.$$

11. For the matrices $A = \begin{bmatrix} 1 & 1 & 1 \\ 1 & 2 & 3 \\ 1 & 4 & 9 \end{bmatrix}$ and $B = \begin{bmatrix} 2 & 5 & 3 \\ 3 & 1 & 2 \\ 1 & 2 & 1 \end{bmatrix}$

show that

(i) $(AB)^T = B^T A^T$ (ii) $(AB)^{-1} = B^{-1}A^{-1}$

12. If $A = \begin{bmatrix} 0 & 2/\sqrt{6} & 1/\sqrt{3} \\ 1/\sqrt{2} & 1/\sqrt{6} & -1/\sqrt{3} \\ 1/\sqrt{2} & -1/\sqrt{6} & 1/\sqrt{3} \end{bmatrix}$, prove that $A^TA = I$.

13. Let $A = \begin{bmatrix} 0 & 2y & z \\ x & y & -z \\ x & -y & z \end{bmatrix}$ such that $A^TA = I$. Find the value of $x^2 + y^2 + z^2$.

14. Prove that the rank of the matrix $\begin{bmatrix} 15-2x & 11 & 10 \\ 11-3x & 17 & 16 \\ 7-x & 14 & 13 \end{bmatrix}$ is 3 for all real $x \in R - \{4\}$. If $x = 4$, what is the rank?

15. Solve the following system of equations by matrix inversion.

$$x + y + z = 7$$
$$5x + 4y - 3z = 1$$
$$6x - 3y + 2z = 8$$

16. Use matrix method to solve the following system of equations.

$$x^2 + y^2 + z^2 = 9$$
$$2x^2 + 5y^2 + 7z^2 = 52$$
$$2x^2 + y^2 - z^2 = 0$$

Objective Questions

17. If $A = \begin{bmatrix} a & b \\ b & a \end{bmatrix}$ and $A^2 = \begin{bmatrix} \alpha & \beta \\ \beta & \alpha \end{bmatrix}$ then

(a) $\alpha = a^2 + b^2, \beta = ab$ (b) $\beta = a^2 + b^2, \alpha = 2ab$
(c) $\alpha = a^2 + b^2, \beta = 2ab$ (d) none of these

18. If $A = \begin{bmatrix} \alpha & 0 \\ 1 & 1 \end{bmatrix}$ and $B = \begin{bmatrix} 1 & 0 \\ 5 & 1 \end{bmatrix}$ such that $A^2 = B$ ther α is

(a) 1 (b) −1
(c) 4 (d) none of these

19. If $A \times \begin{bmatrix} 1 & -2 \\ 1 & 4 \end{bmatrix} = 6I$ where I is a unit matrix of orde 2×2 then A is

(a) $\begin{bmatrix} 2 & 4 \\ 1 & -1 \end{bmatrix}$ (b) $\begin{bmatrix} -1 & 1 \\ 4 & 2 \end{bmatrix}$
(c) $\begin{bmatrix} 4 & 2 \\ -1 & 1 \end{bmatrix}$ (d) none of these

20. The rank of the matrix $\begin{bmatrix} 1 & 2 & 3 \\ \lambda & 2 & 4 \\ 2 & -3 & 1 \end{bmatrix}$ is 3 if

(a) $\lambda \neq \frac{18}{11}$ (b) $\lambda = \frac{18}{11}$
(c) $\lambda = \frac{-18}{11}$ (d) none of these

21. If $[1 \;\; x \;\; 1]\begin{bmatrix} 1 & 3 & 2 \\ 2 & 5 & 1 \\ 15 & 3 & 2 \end{bmatrix}\begin{bmatrix} 1 \\ 2 \\ x \end{bmatrix} = 0$ then x is

(a) 2 (b) –2 (c) 14 (d) 1

22. If A, B, C are nonsingular square matrices of order 3×3 then which of the following is not necessarily true?

(a) $A^TA = I \Rightarrow AA^T = I$

(b) $A^{-1}B^{-1} = (BA)^{-1}$

(c) $I^{-1} = I$

(d) $(A - B)C = AC - BC$

Answers

1. $x = 19, y = -9$ **2.** $\begin{bmatrix} 3 & 14 & 13 \\ 13 & 6 & -\frac{1}{2} \end{bmatrix}$

4. No **7.** 0, 3 **9.** $x = 5, y = 2$

13. 1 **14.** 2 **15.** $x = 1, y = 2, z = 4$

16. $x = \pm 1, y = \pm\sqrt{3}, z = \pm\sqrt{5}$ **17.** c **18.** d

19. c **20.** a **21.** b **22.** a

□

Chapter Test

Time: 45 minutes

1. In the following, a number of solutions are given of which one is correct. Choose the correct option.

(i) If $A = \begin{bmatrix} -1 & x \\ 4 & 2 \end{bmatrix}$, $B = \begin{bmatrix} x & 1 \\ -1 & 0 \end{bmatrix}$ and $AB = \begin{bmatrix} -6 & -1 \\ 10 & 4 \end{bmatrix}$ then the value of x is

A. 0 B. 2 C. 3 D. indeterminate

(ii) Let A and B be matrices of order $3 \times m$ and $n \times 3$ respectively. If $A + B$ as well as AB exist then

A. $m = 3$ B. $m = n = 3$ C. $m = n$ D. $n = 3$

(iii) If $A^2 - A + I = 0$ where A is a square matrix and I is the unit matrix of the same order then A^{-1} is

A. $A + I$ B. $A - I$ C. $I - A$ D. A

(iv) If $A = \begin{bmatrix} a & b & c \\ c & a & b \\ b & c & a \end{bmatrix}$ and a, b, c are roots of the equation $x^3 + x^2 - 4 = 0$ then AA^T is equal to

A. I B. $I + A$ C. A^2 D. $A - I$

2. If $A = \begin{bmatrix} 2 & -1 & 1 \\ -2 & 3 & -2 \\ -4 & 4 & -3 \end{bmatrix}$ then prove that $A^5 = A$.

3. If $A = \begin{bmatrix} \cos\alpha & -\sin\alpha & 0 \\ \sin\alpha & \cos\alpha & 0 \\ 0 & 0 & 1 \end{bmatrix}$ then prove that $A^{-1} = A^T$.

4. If $B = A + A^T$ and $C = A - A^T$ where A is a 3×3 matrix and A^T is the transpose of A then prove that B is a symmetric matrix and C is the skew-symmetric matrix. Also, prove that every square matrix can be written as the sum of a symmetric matrix and a skew-symmetric matrix.

5. If $P = \begin{bmatrix} \sqrt{3}/2 & 1/2 \\ -1/2 & \sqrt{3}/2 \end{bmatrix}$ and $A = \begin{bmatrix} 1 & 1 \\ 0 & 1 \end{bmatrix}$ then prove that P is an orthogonal matrix. Also, if $Q = PAP^T$ and $X = P^TQ^{2005}P$ then prove that $X = A^{2005} = \begin{bmatrix} 1 & 2005 \\ 0 & 1 \end{bmatrix}$.

6. If $A = \begin{bmatrix} 1 & 0 & 0 \\ 0 & 1 & 1 \\ 0 & -2 & 4 \end{bmatrix}$ and $A^{-1} = \frac{1}{6}(A^2 + cA + dI)$ where I is the unit matrix of the third order then find the values of c and d.

Answers

1. (i) C (ii) B (iii) C (iv) A **6.** $c = -6, d = 11$

□

Trigonometry

1. Circular Functions, Identities

Recap of Facts and Formulae

1. Identities

$\sin x$, $\cos x$ and $\tan x$ are (trigonometrical) circular functions. Their reciprocals $\operatorname{cosec} x$, $\sec x$ and $\cot x$ are also circular functions. These functions are related by the following identities.

- $\sin^2 x + \cos^2 x = 1$
- $\sin^2 x = 1 - \cos^2 x$
- $\cos^2 x = 1 - \sin^2 x$
- $\sec^2 x - \tan^2 x = 1$
- $1 + \tan^2 x = \sec^2 x$
- $\sec^2 x - 1 = \tan^2 x$
- $\operatorname{cosec}^2 x - \cot^2 x = 1$
- $1 + \cot^2 x = \operatorname{cosec}^2 x$
- $\operatorname{cosec}^2 x - 1 = \cot^2 x$
- $\sec x + \tan x = \dfrac{1}{\sec x - \tan x}$
- $\operatorname{cosec} x + \cot x = \dfrac{1}{\operatorname{cosec} x - \cot x}$

2. Domain and range of circular function

Function	*Domain*	*Range*
$\sin x$	R	$[-1, 1]$
$\cos x$	R	$[-1, 1]$
$\tan x$	$R - \left\{\frac{\pi}{2}, \frac{3\pi}{2}, \frac{5\pi}{2}, \dots\right\}$	$(-\infty, \infty)$
$\operatorname{cosec} x$	$R - \{0, \pi, 2\pi, 3\pi, \dots\}$	$(-\infty, -1] \cup [1, +\infty)$
$\sec x$	$R - \left\{\frac{\pi}{2}, \frac{3\pi}{2}, \frac{5\pi}{2}, \dots\right\}$	$(-\infty, -1] \cup [1, +\infty)$
$\cot x$	$R - \{0, \pi, 2\pi, 3\pi, \dots\}$	$(-\infty, \infty)$

Note For values of circular functions, the angle $x = -\dfrac{\pi}{2}$ and the angle $x = \dfrac{3\pi}{2}$ are the same angle.

3. Sign of value

$\sin x \ge 0$ if $0 \le x \le \pi$

$\sin x < 0$ if $\pi < x < 2\pi$

$\cos x \ge 0$ if $-\dfrac{\pi}{2} \le x \le \dfrac{\pi}{2}$

$\cos x < 0$ if $\dfrac{\pi}{2} < x < \dfrac{3\pi}{2}$

$\tan x \ge 0$ if $0 \le x < \dfrac{\pi}{2}$ or $\pi \le x < \dfrac{3\pi}{2}$

$\tan x < 0$ if $\dfrac{\pi}{2} < x < \pi$ or $\dfrac{3\pi}{2} < x < 2\pi$

4. Trend of values of circular functions in $[0, 2\pi]$ or $[-\pi, \pi]$

Function	*Domain of gradual increase*	*Domain of gradual decrease*
$\sin x$	$\left[-\frac{\pi}{2}, \frac{\pi}{2}\right]$ i.e., $\left[0, \frac{\pi}{2}\right] \cup \left[\frac{3\pi}{2}, 2\pi\right]$	$\left[\frac{\pi}{2}, \frac{3\pi}{2}\right]$
$\cos x$	$[-\pi, 0]$ i.e., $[\pi, 2\pi]$	$[0, \pi]$
$\tan x$	$\left[0, \frac{\pi}{2}\right), \left(\frac{\pi}{2}, \pi\right]$, $\left[\pi, \frac{3\pi}{2}\right), \left(\frac{3\pi}{2}, 2\pi\right]$	No gradual decrease; sudden fall of value at $\frac{\pi}{2}$ and $\frac{3\pi}{2}$

Pictorically the results can be remembered as follows:

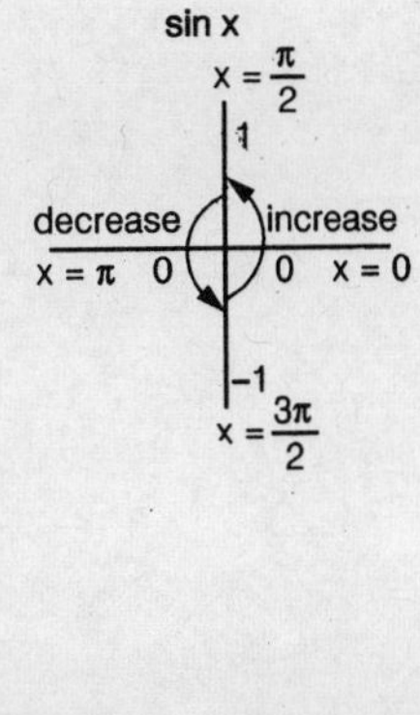

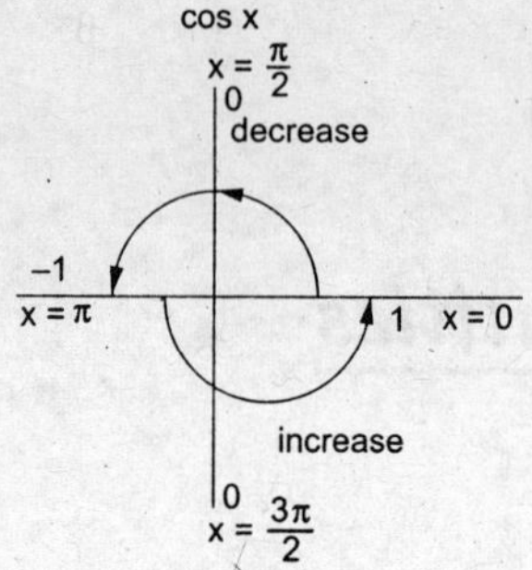

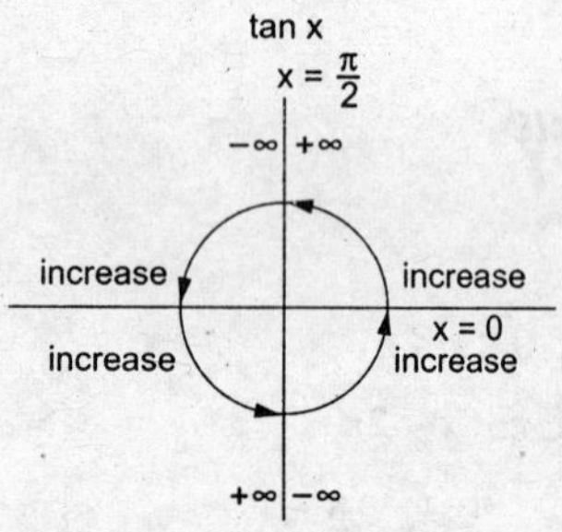

5. Conversion of circular functions of complementary angles, supplementary angles, etc.

- If $f(x)$ is a circular function then $f(2n\pi + x) = f(x)$ where n is an integer. That is why all circular functions are periodic functions.
- $\sin(-\theta) = -\sin\theta$
- $\cos(-\theta) = \cos\theta$
- $\tan(-\theta) = -\tan\theta$
- $\sin\left(\frac{\pi}{2} - \theta\right) = \cos\theta$
- $\cos\left(\frac{\pi}{2} - \theta\right) = \sin\theta$
- $\tan\left(\frac{\pi}{2} - \theta\right) = \cot\theta$
- $\sin\left(\frac{\pi}{2} + \theta\right) = \cos\theta$
- $\cos\left(\frac{\pi}{2} + \theta\right) = -\sin\theta$
- $\tan\left(\frac{\pi}{2} + \theta\right) = -\cot\theta$
- $\sin(\pi - \theta) = \sin\theta$
- $\cos(\pi - \theta) = -\cos\theta$
- $\tan(\pi - \theta) = -\tan\theta$
- $\sin(\pi + \theta) = -\sin\theta$
- $\cos(\pi + \theta) = -\cos\theta$
- $\tan(\pi + \theta) = \tan\theta$

6. Conversion of circular functions of compound, multiple and submultiple angles

Circular functions of the algebraic sum of two angles can be expressed as circular functions of separate angles.

- $\sin(A \pm B) = \sin A \cos B \pm \cos A \sin B$
- $\cos(A \pm B) = \cos A \cos B \mp \sin A \sin B$
- $\tan(A \pm B) = \dfrac{\tan A \pm \tan B}{1 \mp \tan A \cdot \tan B}$
- $\cot(A \pm B) = \dfrac{\cot A \cot B \mp 1}{\cot B \pm \cot A}$

Circular functions of the multiple of an angle can be expressed as circular functions of the angle.

- $\sin 2A = 2\sin A \cos A = \dfrac{2\tan A}{1 + \tan^2 A}$
- $\cos 2A = \cos^2 A - \sin^2 A = \dfrac{1 - \tan^2 A}{1 + \tan^2 A}$

  $= 2\cos^2 A - 1 = 1 - 2\sin^2 A$
- $\tan 2A = \dfrac{2\tan A}{1 - \tan^2 A}$
- $\sin 3A = 3\sin A - 4\sin^3 A$
- $\cos 3A = 4\cos^3 A - 3\cos A$
- $\tan 3A = \dfrac{3\tan A - \tan^3 A}{1 - 3\tan^2 A}$

Circular functions of half of an angle can be expressed as circular functions of the angle.

- $\sin\dfrac{A}{2} = \sqrt{\dfrac{1 - \cos A}{2}}$
- $\cos\dfrac{A}{2} = \sqrt{\dfrac{1 + \cos A}{2}}$
- $\tan\dfrac{A}{2} = \sqrt{\dfrac{1 - \cos A}{1 + \cos A}} = \dfrac{1 - \cos A}{\sin A} = \dfrac{\sin A}{1 + \cos A}$

7. Conversions of sums of sines and cosines into products of sines and cosines, and conversely

- $\sin C + \sin D = 2\sin\dfrac{C + D}{2} \cdot \cos\dfrac{C - D}{2}$

  $\sin C - \sin D = 2\cos\dfrac{C + D}{2} \cdot \sin\dfrac{C - D}{2}$
- $\cos C + \cos D = 2\cos\dfrac{C + D}{2} \cdot \cos\dfrac{C - D}{2}$

  $\cos C - \cos D = 2\sin\dfrac{C + D}{2} \cdot \sin\dfrac{D - C}{2}$

Note $\sin C + \cos D = \sin C + \sin\left(\frac{\pi}{2} - D\right)$

$$= 2\sin\frac{C + \pi/2 - D}{2}\cdot\cos\frac{C - \pi/2 + D}{2}$$

$$\tan C + \tan D = \frac{\sin C}{\cos C} + \frac{\sin D}{\cos D} = \frac{\sin(C + D)}{\cos C\cdot\cos D}$$

- $\sin A\cdot\cos B = \frac{1}{2}\{\sin(A + B) + \sin(A - B)\}$

  $\sin A\cdot\sin B = \frac{1}{2}\{\cos(A - B) - \cos(A + B)\}$

  $\cos A\cdot\cos B = \frac{1}{2}\{\cos(A - B) + \cos(A + B)\}$

- $\sin(A + B)\cdot\sin(A - B) = \sin^2 A - \sin^2 B$

  $\cos(A + B)\cdot\cos(A - B) = \cos^2 A - \sin^2 B$

8. Conversion of powers of a circular function in terms of circular functions of multiple angles

- $\sin^2 A = \frac{1}{2}(1 - \cos 2A)$
- $\cos^2 A = \frac{1}{2}(1 + \cos 2A)$
- $\tan^2 A = \frac{1 - \cos 2A}{1 + \cos 2A}$
- $\sin^3 A = \frac{3\sin A - \sin 3A}{4}$
- $\cos^3 A = \frac{3\cos A + \cos 3A}{4}$

9. Periodic form of $a\cos x + b\sin x$

- $a\cos x + b\sin x = \sqrt{a^2 + b^2}\cos(x - \alpha)$

  where $\cos\alpha = \frac{a}{\sqrt{a^2 + b^2}}$

  or $a\cos x + b\sin x = \sqrt{a^2 + b^2}\sin(x + \beta)$

  where $\sin\beta = \frac{a}{\sqrt{a^2 + b^2}}$

- $-\sqrt{a^2 + b^2} \le a\cos x + b\sin x \le \sqrt{a^2 + b^2}$

10. Values of circular functions when the angles are 18°, 36°

- $\sin 18° = \frac{\sqrt{5} - 1}{4}$
- $\cos 36° = \frac{\sqrt{5} + 1}{4}$

11. Some useful identities

- $\tan(A + B + C) = \frac{\Sigma\tan A - \tan A\cdot\tan B\cdot\tan C}{1 - \Sigma\tan A\cdot\tan B}$
- $\tan(A + B) - \tan A - \tan B = \tan A\cdot\tan B\cdot\tan(A + B)$
- $\tan\theta = \cot\theta - 2\cot 2\theta$
- $\sin\alpha + \sin(\alpha + \beta) + \sin(\alpha + 2\beta) + \ldots + \sin(\alpha + \overline{n - 1}\,\beta)$

  $$= \frac{\sin\frac{n\beta}{2}}{\sin\frac{\beta}{2}}\sin\frac{2\alpha + (n - 1)\beta}{2}$$

- $\cos\alpha + \cos(\alpha + \beta) + \cos(\alpha + 2\beta) + \ldots + \cos(\alpha + \overline{n - 1}\,\beta)$

  $$= \frac{\sin\frac{n\beta}{2}}{\sin\frac{\beta}{2}}\cos\frac{2\alpha + (n - 1)\beta}{2}$$

Selected Solved Examples

1. Eliminate θ and ϕ if

$$a\sin^2\theta + b\cos^2\theta = m$$
$$b\sin^2\phi + a\cos^2\phi = n$$

and $a\tan\theta = b\tan\phi$.

Dividing $a\sin^2\theta + b\cos^2\theta = m$ by $\cos^2\theta$, we get

$$a\tan^2\theta + b = m\sec^2\theta$$

or $(a - m)\tan^2\theta = m - b$ ...(1)

Dividing $b\sin^2\phi + a\cos^2\phi = n$ by $\cos^2\phi$, we get

$$b\tan^2\phi + a = n\sec^2\phi$$

or $(b - n)\tan^2\phi = n - a$ ...(2)

$(1) \div (2) \Rightarrow \frac{a - m}{b - n}\cdot\left(\frac{\tan\theta}{\tan\phi}\right)^2 = \frac{m - b}{n - a}$

$\therefore \frac{a - m}{b - n}\cdot\frac{b^2}{a^2} = \frac{m - b}{n - a}$ $\{\because a\tan\theta = b\tan\phi\}$

or $b^2(a - m)(n - a) = a^2(b - n)(m - b)$

or $b^2\{(m + n)a - a^2 - mn\} = a^2\{(m + n)b - b^2 - mn\}$

or $(m + n)(ab^2 - a^2 b) + mn(a^2 - b^2) = 0$

or $ab(m + n)(b - a) + mn(a - b)(a + b) = 0$

or $ab(m + n) = mn(a + b)$

$\{\because a - b \ne 0$ from the question$\}$;

or $\dfrac{1}{n}+\dfrac{1}{m}=\dfrac{1}{b}+\dfrac{1}{a}$ {dividing by $abmn$}

2. Show that for all real θ, the expression $a\sin^2\theta + b\sin\theta\cos\theta + c\cos^2\theta$ lies between $\frac{1}{2}(a+c)-\frac{1}{2}\sqrt{b^2+(a-c)^2}$ and $\frac{1}{2}(a+c)+\frac{1}{2}\sqrt{b^2+(a-c)^2}$.

Here, the expression

$$=\frac{a}{2}(1-\cos 2\theta)+\frac{b}{2}\sin 2\theta+\frac{c}{2}(1+\cos 2\theta)$$

$$=\frac{a+c}{2}+\left(\frac{b}{2}\sin 2\theta-\frac{a-c}{2}\cos 2\theta\right)$$

$$=\frac{a+c}{2}+\sqrt{\left(\frac{b}{2}\right)^2+\left(\frac{a-c}{2}\right)^2}\times\left\{\frac{\frac{b}{2}}{\sqrt{\left(\frac{b}{2}\right)^2+\left(\frac{a-c}{2}\right)^2}}\sin 2\theta-\frac{\frac{a-c}{2}}{\sqrt{\left(\frac{b}{2}\right)^2+\left(\frac{a-c}{2}\right)^2}}\cos 2\theta\right\}$$

$$=\frac{a+c}{2}+\frac{1}{2}\sqrt{b^2+(a-c)^2}\times\{\cos\alpha\cdot\sin 2\theta-\sin\alpha\cdot\cos 2\theta\}$$

where $\cos\alpha=\dfrac{\frac{b}{2}}{\sqrt{\left(\frac{b}{2}\right)^2+\left(\frac{a-c}{2}\right)^2}}$.

$$=\frac{a+c}{2}+\frac{1}{2}\sqrt{b^2+(a-c)^2}\cdot\sin(2\theta-\alpha)$$

But $-1\le\sin(2\theta-\alpha)\le 1$

∴ maximum value of the expression

$$=\frac{a+c}{2}+\frac{1}{2}\sqrt{b^2+(a-c)^2}$$

and minimum value of the expression

$$=\frac{a+c}{2}-\frac{1}{2}\sqrt{b^2+(a-c)^2}$$

Hence, the problem.

3. Prove that $2^{\sin x}+2^{\cos x}\ge 2^{1-1/\sqrt{2}}$ for all real x.

Clearly, $2^{\sin x}$ and $2^{\cos x}$ are positive; so their AM ≥ GM

$$\therefore\quad \frac{2^{\sin x}+2^{\cos x}}{2}\ge\sqrt{2^{\sin x}\cdot 2^{\cos x}}=\sqrt{2^{\sin x+\cos x}}\qquad\ldots(1)$$

Now, $\sin x+\cos x=\sqrt{2}\left(\frac{1}{\sqrt{2}}\sin x+\frac{1}{\sqrt{2}}\cos x\right)$

$$=\sqrt{2}\sin\left(x+\frac{\pi}{4}\right)$$

∴ minimum value of $\sin x+\cos x=-\sqrt{2}$

∴ from (1),

$$2^{\sin x}+2^{\cos x}\ge 2\cdot\sqrt{2^{\sin x+\cos x}}$$

$$\ge 2\sqrt{2^{-\sqrt{2}}}=2\cdot 2^{\frac{-\sqrt{2}}{2}}=2^{1-\frac{1}{\sqrt{2}}}.$$

4. If $\operatorname{cosec}\theta=\dfrac{x^2-y^2}{x^2+y^2}$ where x, y are two unequal nonzero real numbers then prove that θ has no real value.

Here $x\ne y$ and they are real. So, $x^2+y^2>x^2-y^2$

∴ $\operatorname{cosec}\theta<1$ if $x>y$ and

$\operatorname{cosec}\theta>-1$ if $x<y$

∴ $-1<\operatorname{cosec}\theta<1$

which is not possible because we know that

$\operatorname{cosec}\theta\ge 1$ or ≤ -1.

Hence, there is no real value of θ.

5. Show that the equation $\sec\theta+\operatorname{cosec}\theta=c$ has two roots between 0 and 2π if $2<c^2<8$.

Here $\dfrac{1}{\cos\theta}+\dfrac{1}{\sin\theta}=c$

or $\sin\theta+\cos\theta=c\sin\theta\cos\theta$

∴ $(\sin\theta+\cos\theta)^2=c^2\sin^2\theta\cos^2\theta$

or $1+\sin 2\theta=\dfrac{c^2}{4}\sin^2 2\theta$

or $c^2\sin^2 2\theta-4\sin 2\theta-4=0$

∴ $\sin 2\theta=\dfrac{4\pm\sqrt{16+16c^2}}{2c^2}=\dfrac{2\pm 2\sqrt{1+c^2}}{c^2}$

As $1+c^2>0$, $\sin 2\theta$ has two real solutions provided $-1\le\sin 2\theta\le 1$

i.e., $-1\le\dfrac{2\pm 2\sqrt{1+c^2}}{c^2}\le 1$ … (1)

Also, corresponding to each value of sin 2θ, we get two roots for θ between 0 and 2π.

Clearly, $-1<\dfrac{2+2\sqrt{1+c^2}}{c^2}$ and $\dfrac{2-2\sqrt{1+c^2}}{c^2}<1$

$\therefore$ (1) is satisfied if

$$-1 \le \frac{2-2\sqrt{1+c^2}}{c^2} \text{ and } \frac{2+2\sqrt{1+c^2}}{c^2} < 1$$

Now, $-1 \le \dfrac{2-2\sqrt{1+c^2}}{c^2}$

$\Rightarrow \quad -c^2 \le 2 - 2\sqrt{1+c^2}$

$\Rightarrow \quad 2\sqrt{1+c^2} \le 2 + c^2$

$\Rightarrow \quad 4(1+c^2) \le 4 + 4c^2 + c^4$

$\Rightarrow \quad 0 \le c^4$ which is true

and $\dfrac{2+2\sqrt{1+c^2}}{c^2} < 1$

$\Rightarrow \quad 2 + 2\sqrt{1+c^2} < c^2$

$\Rightarrow \quad 2\sqrt{1+c^2} < c^2 - 2 \qquad \{c^2 - 2$ is positive$\}$

$\Rightarrow \quad 4(1+c^2) < c^4 - 4c^2 + 4$

$\Rightarrow \quad c^4 - 8c^2 > 0$

$\Rightarrow \quad c^2 - 8 > 0 \quad \Rightarrow \quad c^2 > 8$

But $c^2 < 8$ (given). Hence only admissible value of $\sin 2\theta$ is

$$\frac{2-2\sqrt{1+c^2}}{c^2} \text{ when } 2 < c^2 < 8.$$

Hence, θ has two roots between 0 and 2π.

6. If $\dfrac{\sin^4\alpha}{a} + \dfrac{\cos^4\alpha}{b} = \dfrac{1}{a+b}$, show that

$$\frac{\sin^8\alpha}{a^3} + \frac{\cos^8\alpha}{b^3} = \frac{1}{(a+b)^3}.$$

Here $\dfrac{a+b}{a}\sin^4\alpha + \dfrac{a+b}{b}\cos^4\alpha = 1$

or $\quad \sin^4\alpha + \cos^4\alpha + \dfrac{b}{a}\sin^4\alpha + \dfrac{a}{b}\cos^4\alpha = 1$

or $\quad (\sin^2\alpha + \cos^2\alpha)^2 - 2\sin^2\alpha \cdot \cos^2\alpha + \dfrac{b}{a}\sin^4\alpha + \dfrac{a}{b}\cos^4\alpha = 1$

or $\quad \left(\sqrt{\dfrac{b}{a}}\sin^2\alpha\right)^2 - 2\cdot\sqrt{\dfrac{b}{a}}\sin^2\alpha \cdot \sqrt{\dfrac{a}{b}}\cos^2\alpha + \left(\sqrt{\dfrac{a}{b}}\cos^2\alpha\right)^2 = 0$

or $\quad \left(\sqrt{\dfrac{b}{a}}\sin^2\alpha - \sqrt{\dfrac{a}{b}}\cos^2\alpha\right)^2 = 0$

$\therefore \quad \sqrt{\dfrac{b}{a}}\sin^2\alpha = \sqrt{\dfrac{a}{b}}\cos^2\alpha$

or $\quad \sin^2\alpha = \dfrac{a}{b}\cos^2\alpha$

$\therefore \quad \dfrac{\sin^2\alpha}{a} = \dfrac{\cos^2\alpha}{b} = \dfrac{\sin^2\alpha + \cos^2\alpha}{a+b}$

$\therefore \quad \sin^2\alpha = \dfrac{a}{a+b}, \ \cos^2\alpha = \dfrac{b}{a+b}$

$$\therefore \quad \frac{\sin^8\alpha}{a^3} + \frac{\cos^8\alpha}{b^3} = \frac{1}{a^3}\cdot\frac{a^4}{(a+b)^4} + \frac{1}{b^3}\cdot\frac{b^4}{(a+b)^4}$$

$$= \frac{a}{(a+b)^4} + \frac{b}{(a+b)^4}$$

$$= \frac{a+b}{(a+b)^4} = \frac{1}{(a+b)^3}.$$

7. If $\dfrac{\tan(\alpha+\beta-\gamma)}{\tan(\alpha-\beta+\gamma)} = \dfrac{\tan\gamma}{\tan\beta}$, show that

$\sin(\beta-\gamma) = 0$ or $\sin 2\alpha + \sin 2\beta + \sin 2\gamma = 0$.

By componendo and dividendo,

$$\frac{\tan(\alpha+\beta-\gamma) + \tan(\alpha-\beta+\gamma)}{\tan(\alpha+\beta-\gamma) - \tan(\alpha-\beta+\gamma)} = \frac{\tan\gamma + \tan\beta}{\tan\gamma - \tan\beta}$$

or $$\frac{\dfrac{\sin(\alpha+\beta-\gamma+\alpha-\beta+\gamma)}{\cos(\alpha+\beta-\gamma)\cdot\cos(\alpha-\beta+\gamma)}}{\dfrac{\sin(\alpha+\beta-\gamma-\alpha-\beta+\gamma)}{\cos(\alpha+\beta-\gamma)\cdot\cos(\alpha-\beta+\gamma)}} = \frac{\dfrac{\sin(\gamma+\beta)}{\cos\gamma\cdot\cos\beta}}{\dfrac{\sin(\gamma-\beta)}{\cos\gamma\cdot\cos\beta}}$$

or $\quad \dfrac{\sin 2\alpha}{\sin 2(\beta-\gamma)} = \dfrac{\sin(\gamma+\beta)}{\sin(\gamma-\beta)}$

or $\quad \sin 2(\beta-\gamma)\cdot\sin(\beta+\gamma) + \sin 2\alpha\cdot\sin(\beta-\gamma) = 0$

or $\quad \sin(\beta-\gamma)\,\{2\cos(\beta-\gamma)\sin(\beta+\gamma) + \sin 2\alpha\} = 0$

or $\quad \sin(\beta-\gamma)\,\{\sin 2\beta + \sin 2\gamma + \sin 2\alpha\} = 0$

$\therefore \quad$ either $\sin(\beta-\gamma) = 0$

or $\quad \sin 2\alpha + \sin 2\beta + \sin 2\gamma = 0.$

8. If $\dfrac{x}{\tan(\theta+\alpha)} = \dfrac{y}{\tan(\theta+\beta)} = \dfrac{z}{\tan(\theta+\gamma)}$,

show that $\sum \dfrac{x+y}{x-y}\sin^2(\alpha-\beta) = 0$.

Here $\dfrac{x}{y} = \dfrac{\tan(\theta+\alpha)}{\tan(\theta+\beta)}$. By componendo and dividendo,

$$\frac{x+y}{x-y} = \frac{\tan(\theta+\alpha) + \tan(\theta+\beta)}{\tan(\theta+\alpha) - \tan(\theta+\beta)}$$

$$= \frac{\dfrac{\sin(\theta+\alpha)}{\cos(\theta+\alpha)} + \dfrac{\sin(\theta+\beta)}{\cos(\theta+\beta)}}{\dfrac{\sin(\theta+\alpha)}{\cos(\theta+\alpha)} - \dfrac{\sin(\theta+\beta)}{\cos(\theta+\beta)}} = \frac{\sin(\theta+\alpha+\theta+\beta)}{\sin(\overline{\theta+\alpha}-\overline{\theta+\beta})}$$

$$= \frac{\sin(2\theta+\alpha+\beta)}{\sin(\alpha-\beta)}$$

$$\therefore \quad \frac{x+y}{x-y}\sin^2(\alpha-\beta) = \sin(2\theta+\alpha+\beta)\cdot\sin(\alpha-\beta)$$

$$= \frac{1}{2}[\cos 2(\theta+\beta) - \cos 2(\theta+\alpha)] \qquad \dots (1)$$

Similarly, $\dfrac{y+z}{y-z}\sin^2(\beta-\gamma)$

$$= \frac{1}{2}[\cos 2(\theta+\gamma) - \cos 2(\theta+\beta)] \qquad \dots (2)$$

$$\frac{z+x}{z-x}\sin^2(\gamma-\alpha) = \frac{1}{2}[\cos 2(\theta+\alpha) - \cos 2(\theta+\gamma)] \qquad \dots (3)$$

Adding (1), (2) and (3) we get: LHS $= 0 =$ RHS.

9. If $\sin x + \sin y = a$ and $\cos x + \cos y = b$, show that

$$\sin(x+y) = \frac{2ab}{a^2+b^2} \text{ and } \tan\frac{x-y}{2} = \sqrt{\frac{4-a^2-b^2}{a^2+b^2}}.$$

$\sin x + \sin y = a$

$$\Rightarrow \quad 2\sin\frac{x+y}{2}\cdot\cos\frac{x-y}{2} = a \qquad \dots (1)$$

$\cos x + \cos y = b$

$$\Rightarrow \quad 2\cos\frac{x+y}{2}\cdot\cos\frac{x-y}{2} = b \qquad \dots (2)$$

$$(1) \div (2) \quad \Rightarrow \quad \tan\frac{x+y}{2} = \frac{a}{b}$$

$$\therefore \quad \sin(x+y) = \frac{2\tan\frac{x+y}{2}}{1+\tan^2\frac{x+y}{2}} = \frac{2\cdot\frac{a}{b}}{1+\frac{a^2}{b^2}} = \frac{2ab}{a^2+b^2}.$$

Squaring (1), (2) and adding,

$$4\cos^2\frac{x-y}{2}\left\{\sin^2\frac{x+y}{2} + \cos^2\frac{x+y}{2}\right\} = a^2+b^2$$

$$\text{or} \quad 4\cos^2\frac{x-y}{2} = a^2+b^2$$

$$\therefore \quad \cos^2\frac{x-y}{2} = \frac{a^2+b^2}{4}$$

$$\therefore \quad \sin^2\frac{x-y}{2} = 1 - \frac{a^2+b^2}{4} = \frac{4-a^2-b^2}{4}$$

$$\therefore \quad \tan^2\frac{x-y}{2} = \frac{\sin^2\frac{x-y}{2}}{\cos^2\frac{x-y}{2}}$$

$$= \frac{4-a^2-b^2}{4}\times\frac{4}{a^2+b^2}$$

$$= \frac{4-a^2-b^2}{a^2+b^2}$$

$$\therefore \quad \tan\frac{x-y}{2} = \sqrt{\frac{4-a^2-b^2}{a^2+b^2}}.$$

10. If α and β be two different roots of the equation $a\cos\theta + b\sin\theta = c$ then prove that

$$\cos(\alpha+\beta) = \frac{a^2-b^2}{a^2+b^2}.$$

$$\text{Here } a\cdot\frac{1-\tan^2\frac{\theta}{2}}{1+\tan^2\frac{\theta}{2}} + b\cdot\frac{2\tan\frac{\theta}{2}}{1+\tan^2\frac{\theta}{2}} = c$$

$$\text{or} \quad a\left(1-\tan^2\frac{\theta}{2}\right) + 2b\tan\frac{\theta}{2} = c\left(1+\tan^2\frac{\theta}{2}\right)$$

$$\text{or} \quad (a+c)\tan^2\frac{\theta}{2} - 2b\tan\frac{\theta}{2} + (c-a) = 0$$

Roots of the equation are $\tan\dfrac{\alpha}{2}$ and $\tan\dfrac{\beta}{2}$.

$$\therefore \quad \tan\frac{\alpha}{2} + \tan\frac{\beta}{2} = \frac{2b}{a+c}$$

$$\text{and} \quad \tan\frac{\alpha}{2}\cdot\tan\frac{\beta}{2} = \frac{c-a}{c+a}$$

$$\therefore \quad \tan\frac{\alpha+\beta}{2} = \frac{\tan\frac{\alpha}{2}+\tan\frac{\beta}{2}}{1-\tan\frac{\alpha}{2}\cdot\tan\frac{\beta}{2}} = \frac{\frac{2b}{a+c}}{1-\frac{c-a}{c+a}} = \frac{b}{a}$$

$$\therefore \quad \cos(\alpha+\beta) = \frac{1-\tan^2\frac{\alpha+\beta}{2}}{1+\tan^2\frac{\alpha+\beta}{2}} = \frac{1-\frac{b^2}{a^2}}{1+\frac{b^2}{a^2}} = \frac{a^2-b^2}{a^2+b^2}.$$

11. If $\cos\alpha + \cos\beta + \cos\gamma = 0 = \sin\alpha + \sin\beta + \sin\gamma$, show that

$$\sin^2\alpha + \sin^2\beta + \sin^2\gamma = \cos^2\alpha + \cos^2\beta + \cos^2\gamma = \frac{3}{2}.$$

Here $\cos\alpha + \cos\beta = -\cos\gamma$ ... (1)

$\sin\alpha + \sin\beta = -\sin\gamma$ ... (2)

Squaring these and adding,

$$2 + 2\cos(\alpha-\beta) = 1$$

or $\cos(\alpha - \beta) = -\frac{1}{2}$

Similarly, $\cos(\beta - \gamma) = -\frac{1}{2}$ and $\cos(\gamma - \alpha) = -\frac{1}{2}$

Again, squaring (1) and (2), and subtracting, we get

$$\cos^2\alpha - \sin^2\alpha + \cos^2\beta - \sin^2\beta + 2\cos(\alpha+\beta) = \cos^2\gamma - \sin^2\gamma$$

or $\cos 2\alpha + \cos 2\beta + 2\cos(\alpha+\beta) = \cos 2\gamma$

or $2\cos(\alpha+\beta)\cdot\cos(\alpha-\beta) + 2\cos(\alpha+\beta) = \cos 2\gamma$

or $-\cos(\alpha+\beta) + 2\cos(\alpha+\beta) = \cos 2\gamma$

$$\left\{\because \cos(\alpha-\beta) = -\frac{1}{2}\right\}$$

or $\cos(\alpha+\beta) = \cos 2\gamma$, etc.

Now, $(\cos\alpha + \cos\beta + \cos\gamma)^2 - (\sin\alpha + \sin\beta + \sin\gamma)^2 = 0$

or $\cos 2\alpha + \cos 2\beta + \cos 2\gamma + 2[\cos(\alpha+\beta) + \cos(\beta+\gamma) + \cos(\gamma+\alpha)] = 0$

or $\cos 2\alpha + \cos 2\beta + \cos 2\gamma + 2[\cos 2\gamma + \cos 2\alpha + \cos 2\beta] = 0$

$$\{\because \cos(\alpha+\beta) = \cos 2\gamma, \text{ etc.}\}$$

$\therefore$ $\cos 2\alpha + \cos 2\beta + \cos 2\gamma = 0$

or $2\cos^2\alpha - 1 + 2\cos^2\beta - 1 + 2\cos^2\gamma - 1 = 0$

or $2(\cos^2\alpha + \cos^2\beta + \cos^2\gamma) - 3 = 0$

$\therefore$ $\cos^2\alpha + \cos^2\beta + \cos^2\gamma = \frac{3}{2}$

$\therefore$ $1 - \sin^2\alpha + 1 - \sin^2\beta + 1 - \sin^2\gamma = \frac{3}{2}$

or $\sin^2\alpha + \sin^2\beta + \sin^2\gamma = \frac{3}{2}$.

12. If $\cos(\beta-\gamma) + \cos(\gamma-\alpha) + \cos(\alpha-\beta) = \frac{-3}{2}$, prove that $\cos\alpha + \cos\beta + \cos\gamma = 0$, $\sin\alpha + \sin\beta + \sin\gamma = 0$.

Here, $2\{\cos(\beta-\gamma) + \cos(\gamma-\alpha) + \cos(\alpha-\beta)\} + 3 = 0$

or $2\{\cos\beta\cos\gamma + \sin\beta\sin\gamma + \cos\gamma\cdot\cos\alpha + \sin\gamma\cdot\sin\alpha + \cos\alpha\cdot\cos\beta + \sin\alpha\cdot\sin\beta\} + (\sin^2\alpha + \cos^2\alpha) + (\sin^2\beta + \cos^2\beta) + (\sin^2\gamma + \cos^2\gamma) = 0$

or $\cos^2\alpha + \cos^2\beta + \cos^2\gamma + 2\cos\beta\cos\gamma + 2\cos\gamma\cos\alpha + 2\cos\alpha\cos\beta + \sin^2\alpha + \sin^2\beta + \sin^2\gamma + 2\sin\beta\sin\gamma + 2\sin\gamma\cdot\sin\alpha + 2\sin\alpha\cdot\sin\beta = 0$

or $(\cos\alpha + \cos\beta + \cos\gamma)^2 + (\sin\alpha + \sin\beta + \sin\gamma)^2 = 0$

$\Rightarrow$ $\cos\alpha + \cos\beta + \cos\gamma = 0$
and $\sin\alpha + \sin\beta + \sin\gamma = 0$

13. Show that $\tan\frac{\pi}{16} = \sqrt{4+2\sqrt{2}} - (\sqrt{2}+1)$.

We know, $\tan\theta = \dfrac{1-\cos 2\theta}{\sin 2\theta}$ where $\theta = \frac{\pi}{16}$.

Also, $\sin 2\theta = \sqrt{\dfrac{1-\cos 4\theta}{2}} = \sqrt{\dfrac{1-\cos\frac{\pi}{4}}{2}}$

$$= \sqrt{\frac{1-\frac{1}{\sqrt{2}}}{2}} = \sqrt{\frac{\sqrt{2}-1}{2\sqrt{2}}}$$

and $\cos 2\theta = \sqrt{\dfrac{1+\cos 4\theta}{2}} = \sqrt{\dfrac{1+\cos\frac{\pi}{4}}{2}}$

$$= \sqrt{\frac{1+\frac{1}{\sqrt{2}}}{2}} = \sqrt{\frac{\sqrt{2}+1}{2\sqrt{2}}}$$

$$\therefore \quad \tan\frac{\pi}{16} = \frac{1-\sqrt{\frac{\sqrt{2}+1}{2\sqrt{2}}}}{\sqrt{\frac{\sqrt{2}-1}{2\sqrt{2}}}} = \frac{\sqrt{2\sqrt{2}} - \sqrt{\sqrt{2}+1}}{\sqrt{\sqrt{2}-1}}$$

$$= \frac{\sqrt{\sqrt{2}+1}\,(\sqrt{2\sqrt{2}} - \sqrt{\sqrt{2}+1})}{\sqrt{\sqrt{2}+1}\cdot\sqrt{\sqrt{2}-1}}$$

$$= \frac{\sqrt{2\sqrt{2}\,(\sqrt{2}+1)} - \sqrt{(\sqrt{2}+1)^2}}{\sqrt{(\sqrt{2}+1)(\sqrt{2}-1)}}$$

$$= \frac{\sqrt{4+2\sqrt{2}} - (\sqrt{2}+1)}{\sqrt{2-1}} = \sqrt{4+2\sqrt{2}} - (\sqrt{2}+1).$$

14. Prove that $\frac{3}{4} \le \sin^2\theta + \cos^4\theta \le 1$ for all real θ.

$\sin^2\theta + \cos^4\theta = \cos^4\theta - \cos^2\theta + 1$

$$= \left\{(\cos^2\theta)^2 - 2\cdot\frac{1}{2}\cos^2\theta + \frac{1}{4}\right\} + \frac{3}{4}$$

$$= \left(\cos^2\theta - \frac{1}{2}\right)^2 + \frac{3}{4} \ge \frac{3}{4},$$

because $\left(\cos^2\theta - \frac{1}{2}\right)^2 \ge 0$

Also, $\sin^2\theta + \cos^4\theta = \cos^4\theta - \cos^2\theta + 1$

$$= \cos^2\theta(\cos^2\theta - 1) + 1$$

$$= 1 - \sin^2\theta\cos^2\theta \le 1,$$

because $\sin^2\theta\cos^2\theta \ge 0$

$\therefore \quad \frac{3}{4} \le \sin^2\theta + \cos^4\theta \le 1.$

Second Method

Let $\sin^2\theta + \cos^4\theta = y;$

then $\cos^4\theta - \cos^2\theta + 1 - y = 0$

$\therefore \quad \cos^2\theta = \frac{1 \pm \sqrt{1-4(1-y)}}{2}$

But $\cos^2\theta$ is real and non-negative.

$\therefore \quad 1 - 4(1-y) \ge 0 \quad$ and $\quad 1 \ge \sqrt{1-4(1-y)}$

$\Rightarrow \quad 4y - 3 \ge 0 \quad$ and $\quad 1 \ge 1 - 4(1-y)$

$\Rightarrow \quad y \ge \frac{3}{4} \quad$ and $\quad 4(1-y) \ge 0,$ i.e., $y \le 1$

$\therefore \quad \frac{3}{4} \le y \le 1,$ i.e., $\frac{3}{4} \le \sin^2\theta + \cos^4\theta \le 1.$

15. Let A, B, C be three angles such that $A = \frac{\pi}{4}$ and $\tan B \cdot \tan C = p$. Find all possible values of p such that A, B, C are three angles of a triangle.

$\because \quad A + B + C = \pi, \quad B + C = \pi - A = \pi - \frac{\pi}{4} = \frac{3\pi}{4}.$

Now $\tan B \cdot \tan C = p$

or $\quad \tan B \cdot \tan\left(\frac{3\pi}{4} - B\right) = p$

or $\quad \tan B \cdot \frac{\tan\frac{3\pi}{4} - \tan B}{1 + \tan\frac{3\pi}{4} \cdot \tan B} = p$

or $\quad \frac{\tan B \cdot (-1 - \tan B)}{1 - \tan B} = p$

or $\quad p(1 - \tan B) + \tan B(1 + \tan B) = 0$

or $\quad \tan^2 B + (1-p)\tan B + p = 0$

$\because \quad \tan B$ is real, $D \ge 0$

or $\quad (1-p)^2 - 4p \ge 0$

or $\quad p^2 - 6p + 1 \ge 0$

The corresponding equation is $p^2 - 6p + 1 = 0$

or $\quad p = \frac{6 \pm \sqrt{36-4}}{2} = \frac{6 \pm 4\sqrt{2}}{2} = 3 \pm 2\sqrt{2}$

$\therefore$ by sign-scheme, $p \in (-\infty, 3 - 2\sqrt{2}] \cup [3 + 2\sqrt{2}, +\infty)$

16. If $\frac{\tan 3A}{\tan A} = k$, show that $\frac{\sin 3A}{\sin A} = \frac{2k}{k-1}$ and k cannot lie between $\frac{1}{3}$ and 3.

Here $\frac{\tan 3A}{\tan A} = k$

or $\quad \frac{1}{\tan A} \cdot \frac{3\tan A - \tan^3 A}{1 - 3\tan^2 A} = k$

or $\quad \frac{3 - \tan^2 A}{1 - 3\tan^2 A} = k$

or $\quad 3 - \tan^2 A = k - 3k\tan^2 A$

or $\quad (3k-1)\tan^2 A = k - 3 \qquad \ldots (1)$

Now, $\frac{\sin 3A}{\sin A} = \frac{1}{\sin A}(3\sin A - 4\sin^3 A) = 3 - 4\sin^2 A$

$= 3 - \frac{4}{1 + \cot^2 A} = 3 - \frac{4}{1 + \left(\frac{3k-1}{k-3}\right)},$ using (1)

$= 3 - \frac{4(k-3)}{(k-3) + (3k-1)} = 3 - \frac{4(k-3)}{4(k-1)}$

$= 3 - \frac{k-3}{k-1} = \frac{3k - 3 - k + 3}{k-1} = \frac{2k}{k-1}.$

Also from (1), $0^2 - 4(3k-1)(3-k) \ge 0$

{$\because$ roots of $\tan A$ are real}

$\therefore \quad (3k-1)(3-k) \le 0$

$\therefore \quad$ by sign-scheme we get

$k \le \frac{1}{3} \quad$ or $\quad k \ge 3.$

But $k \ne \frac{1}{3}$ for then by (1), $\tan A = \infty$ (undefined)

and $k \ne 3$ for then by (1), $\tan A = 0$ which makes $\frac{\tan 3A}{\tan A}$ indeterminate.

Thus $k < \frac{1}{3}$ or $k > 3.$

Hence, k cannot lie between $\frac{1}{3}$ and 3.

17. If x be real, prove that $\frac{x^2 - 2x\cos\alpha + 1}{x^2 - 2x\cos\beta + 1}$ lies between $\sin^2\frac{\alpha}{2} \cdot \operatorname{cosec}^2\frac{\beta}{2}$ and $\cos^2\frac{\alpha}{2} \cdot \sec^2\frac{\beta}{2}$.

Let $\quad \frac{x^2 - 2x\cos\alpha + 1}{x^2 - 2x\cos\beta + 1} = y$

or $\quad (1-y)x^2 - 2(\cos\alpha - y\cos\beta)x + (1-y) = 0.$

As x is real, $D \ge 0$

$\therefore \quad 4(\cos\alpha - y\cos\beta)^2 - 4(1-y)(1-y) \ge 0$

or $\quad \cos^2\alpha + y^2\cos^2\beta - 2\cos\alpha\cdot\cos\beta\cdot y - (1-2y+y^2) \ge 0$

or $\quad (\cos^2\beta - 1)y^2 + 2(1-\cos\alpha\cdot\cos\beta)y + \cos^2\alpha - 1 \ge 0$

or $\quad \sin^2\beta\cdot y^2 + 2(\cos\alpha\cdot\cos\beta - 1)\,y + \sin^2\alpha \le 0 \quad \ldots(1)$

The corresponding equation is

$$\sin^2\beta\cdot y^2 + 2(\cos\alpha\cdot\cos\beta-1)\,y + \sin^2\alpha = 0$$

or $\quad y = \dfrac{\left\{2(1-\cos\alpha\cdot\cos\beta) \pm \sqrt{4(\cos\alpha\cdot\cos\beta-1)^2 - 4\sin^2\alpha\cdot\sin^2\beta}\right\}}{2\sin^2\beta}$

$$= \frac{\left\{1-\cos\alpha\cdot\cos\beta \pm \sqrt{(\cos\alpha\cdot\cos\beta-1)^2 - (\sin\alpha\cdot\sin\beta)^2}\right\}}{\sin^2\beta}$$

$$= \frac{\left\{1-\cos\alpha\cdot\cos\beta \pm \sqrt{\{\cos(\alpha-\beta)-1\{\cos(\alpha+\beta)-1}\right\}}{\sin^2\beta}$$

$$= \frac{1-\cos\alpha\cdot\cos\beta \pm 2\sin\dfrac{\alpha-\beta}{2}\cdot\sin\dfrac{\alpha+\beta}{2}}{\sin^2\beta}$$

$$= \frac{1-\cos\alpha\cdot\cos\beta \pm (\cos\beta-\cos\alpha)}{\sin^2\beta}$$

$$= \frac{1-\cos\alpha\cdot\cos\beta+\cos\beta-\cos\alpha}{\sin^2\beta},\quad \frac{1-\cos\alpha\cdot\cos\beta-\cos\beta+\cos\alpha}{\sin^2\beta}$$

$$= \frac{(1+\cos\beta)(1-\cos\alpha)}{\sin^2\beta},\quad \frac{(1-\cos\beta)(1+\cos\alpha)}{\sin^2\beta}$$

$$= \frac{2\cos^2\dfrac{\beta}{2}\cdot 2\sin^2\dfrac{\alpha}{2}}{4\sin^2\dfrac{\beta}{2}\cdot\cos^2\dfrac{\beta}{2}},\quad \frac{2\sin^2\dfrac{\beta}{2}\cdot 2\cos^2\dfrac{\alpha}{2}}{4\sin^2\dfrac{\beta}{2}\cdot\cos^2\dfrac{\beta}{2}}$$

$$= \sin^2\frac{\alpha}{2}\cdot\operatorname{cosec}^2\frac{\beta}{2},\quad \cos^2\frac{\alpha}{2}\cdot\sec^2\frac{\beta}{2}$$

$\therefore$ by sign-scheme, (1) holds for y lying between the two roots of the corresponding equation, i.e., the expression lies between

$$\sin^2\frac{\alpha}{2}\cdot\operatorname{cosec}^2\frac{\beta}{2} \text{ and } \cos^2\frac{\alpha}{2}\cdot\sec^2\frac{\beta}{2}.$$

18. Prove that $(1+\sec 2\theta)(1+\sec 2^2\theta)(1+\sec 2^3\theta)\ldots \times (1+\sec 2^n\theta) = \tan 2^n\theta\cdot\cot\theta$

$$\text{LHS} = \left(1+\frac{1}{\cos 2\theta}\right)\left(1+\frac{1}{\cos 2^2\theta}\right)\left(1+\frac{1}{\cos 2^3\theta}\right)\cdots \times\left(1+\frac{1}{\cos 2^n\theta}\right)$$

$$= \frac{1+\cos 2\theta}{\cos 2\theta}\cdot\frac{1+\cos 2^2\theta}{\cos 2^2\theta}\cdot\frac{1+\cos 2^3\theta}{\cos 2^3\theta}\cdots\times\frac{1+\cos 2^n\theta}{\cos 2^n\theta}$$

$$= \frac{2\cos^2\theta\cdot 2\cos^2 2\theta\cdot 2\cos^2 2^2\theta\cdot\ldots 2\cos^2 2^{n-1}\theta}{\cos 2\theta\cdot\cos 2^2\theta\cdot\ldots\cos 2^{n-1}\theta\cdot\cos 2^n\theta}$$

$$= \cos\theta\cdot(2\cos\theta)(2\cos 2\theta)(2\cos 2^2\theta)\ldots\times(2\cos 2^{n-1}\theta)\cdot\frac{1}{\cos 2^n\theta}$$

$$= \cos\theta\cdot\frac{1}{\sin\theta}(2\sin\theta\cos\theta)(2\cos 2\theta)(2\cos 2^2\theta)\ldots\times(2\cos 2^{n-1}\theta)\cdot\frac{1}{\cos 2^n\theta}$$

$$= \frac{\cos\theta}{\sin\theta}\cdot(2\sin 2\theta\cdot\cos 2\theta)(2\cos 2^2\theta)\ldots\times(2\cos 2^{n-1}\theta)\cdot\frac{1}{\cos 2^n\theta}$$

$$= \ldots$$

$$= \frac{\cos\theta}{\sin\theta}\cdot(2\sin 2^{n-1}\theta\cdot\cos 2^{n-1}\theta)\cdot\frac{1}{\cos 2^n\theta}$$

$$= \frac{\cos\theta}{\sin\theta}\cdot\sin 2^n\theta\cdot\frac{1}{\cos 2^n\theta}$$

$$= \tan 2^n\theta\cdot\cot\theta.$$

19. If $a_{r+1} = \sqrt{\frac{1}{2}(1+a_r)}$,

prove that $\cos\left(\dfrac{\sqrt{1-a_0^2}}{a_1\cdot a_2\cdot a_3\cdot\ldots\text{ to }\infty}\right) = a_0$.

Let $a_0 = \cos\theta$. Then $a_{r+1} = \sqrt{\frac{1}{2}(1+a_r)}$ gives

$$a_1 = \sqrt{\frac{1}{2}(1+a_0)} = \sqrt{\frac{1}{2}(1+\cos\theta)} = \cos\frac{\theta}{2}$$

$$a_2 = \sqrt{\frac{1}{2}(1 + a_1)} = \sqrt{\frac{1}{2}\left(1 + \cos\frac{\theta}{2}\right)} = \cos\frac{\theta}{2^2}$$

$$a_3 = \sqrt{\frac{1}{2}(1 + a_2)} = \sqrt{\frac{1}{2}\left(1 + \cos\frac{\theta}{2^2}\right)} = \cos\frac{\theta}{2^3}, \text{ etc.}$$

$$\therefore \quad a_1 \cdot a_2 \cdot a_3 \dots a_n = \cos\frac{\theta}{2} \cdot \cos\frac{\theta}{2^2} \cdot \cos\frac{\theta}{2^3} \cdot \dots \cos\frac{\theta}{2^n}$$

$$= \frac{\cos\frac{\theta}{2} \cdot \cos\frac{\theta}{2^2} \cdot \cos\frac{\theta}{2^3} \cdot \dots \cdot \cos\frac{\theta}{2^n} \cdot 2\sin\frac{\theta}{2^n}}{2\sin\frac{\theta}{2^n}}$$

$$= \frac{\cos\frac{\theta}{2} \cdot \cos\frac{\theta}{2^2} \cdot \cos\frac{\theta}{2^3} \cdot \dots \cdot \cos\frac{\theta}{2^{n-1}} \cdot \sin\frac{\theta}{2^{n-1}}}{2\sin\frac{\theta}{2^n}}$$

$$\{\because 2\sin\alpha \cdot \cos\alpha = \sin 2\alpha\}$$

$$= \frac{\cos\frac{\theta}{2} \cdot \cos\frac{\theta}{2^2} \cdot \dots \cdot \cos\frac{\theta}{2^{n-2}} \cdot \sin\frac{\theta}{2^{n-2}}}{2^2\sin\frac{\theta}{2^n}}$$

$$= \frac{\cos\frac{\theta}{2} \cdot \cos\frac{\theta}{2^2} \cdot \dots \cdot \cos\frac{\theta}{2^{n-3}} \cdot \sin\frac{\theta}{2^{n-3}}}{2^3\sin\frac{\theta}{2^n}}$$

$$= \dots$$

$$= \frac{\cos\frac{\theta}{2^{n-(n-1)}} \cdot \sin\frac{\theta}{2^{n-(n-1)}}}{2^{n-1}\sin\frac{\theta}{2^n}} = \frac{\cos\frac{\theta}{2} \cdot \sin\frac{\theta}{2}}{2^{n-1}\sin\frac{\theta}{2^n}}$$

$$= \frac{\sin\theta}{2^n \cdot \sin\frac{\theta}{2^n}}$$

$$a_1 \cdot a_2 \cdot a_3 \cdot \dots \text{ to } \infty = \lim_{n \to \infty} \frac{\sin\theta}{2^n \sin\frac{\theta}{2^n}}$$

$$= \lim_{n \to \infty} \frac{\sin\theta}{\theta \cdot \left(\frac{\sin\theta/2^n}{\theta/2^n}\right)} = \frac{\sin\theta}{\theta}$$

$$\frac{\sqrt{1 - a_0^2}}{a_1 a_2 a_3 \dots \text{ to } \infty} = \frac{\sqrt{1 - \cos^2\theta}}{\frac{\sin\theta}{\theta}} = \theta$$

$$\cos\left(\frac{\sqrt{1 - a_0^2}}{a_1 a_2 a_3 \dots \text{ to } \infty}\right) = \cos\theta = a_0.$$

20. Prove that

$$\frac{\sin\theta}{\cos 3\theta} + \frac{\sin 3\theta}{\cos 9\theta} + \frac{\sin 9\theta}{\cos 27\theta} = \frac{1}{2}(\tan 27\theta - \tan\theta).$$

Here, $\dfrac{\sin\theta}{\cos 3\theta} = \dfrac{1}{2}\dfrac{2\sin\theta \cdot \cos\theta}{\cos 3\theta \cdot \cos\theta} = \dfrac{1}{2}\dfrac{\sin 2\theta}{\cos 3\theta \cdot \cos\theta}$

$$= \frac{1}{2}\frac{\sin(3\theta - \theta)}{\cos 3\theta \cdot \cos\theta} \qquad \dots (1)$$

$$= \frac{1}{2} \cdot (\tan 3\theta - \tan\theta)$$

Similarly, $\dfrac{\sin 3\theta}{\cos 9\theta} = \dfrac{1}{2}(\tan 9\theta - \tan 3\theta)$,

replacing θ by 3θ in (1)

$$\frac{\sin 9\theta}{\cos 27\theta} = \frac{1}{2}(\tan 27\theta - \tan 9\theta),$$

replacing θ by 9θ in (1)

$$\therefore \quad \text{LHS} = \frac{1}{2}[(\tan 3\theta - \tan\theta) + (\tan 9\theta - \tan 3\theta) + (\tan 27\theta - \tan 9\theta)]$$

$$= \frac{1}{2}(\tan 27\theta - \tan\theta).$$

21. Prove that $\tan\alpha + 2\tan 2\alpha + 2^2\tan 2^2\alpha + \dots + 2^{n-1}\tan 2^{n-1}\alpha + 2^n\cot 2^n\alpha = \cot\alpha$.

We know that $\tan\theta = \cot\theta - 2\cot 2\theta \qquad \dots (1)$

because $\cot\theta - \tan\theta = \dfrac{\cos\theta}{\sin\theta} - \dfrac{\sin\theta}{\cos\theta} = \dfrac{\cos^2\theta - \sin^2\theta}{\sin\theta \cdot \cos\theta}$

$$= \frac{\cos 2\theta}{\sin\theta\cos\theta} = \frac{2\cos 2\theta}{\sin 2\theta} = 2\cot 2\theta.$$

Putting $\theta = \alpha,\ 2\alpha,\ 2^2\alpha, \dots$ in (1), we get

$$\tan\alpha = \cot\alpha - 2\cot 2\alpha$$

$$2\tan 2\alpha = 2(\cot 2\alpha - 2\cot 2^2\alpha)$$

$$2^2\tan 2^2\alpha = 2^2(\cot 2^2\alpha - 2\cot 2^3\alpha)$$

....

$$2^{n-1}\tan 2^{n-1}\alpha = 2^{n-1}(\cot 2^{n-1}\alpha - 2\cot 2^n\alpha)$$

Adding,

$$\tan\alpha + 2\tan 2\alpha + 2^2\tan 2^2\alpha + \dots + 2^{n-1}\tan 2^{n-1}\alpha$$

$$= \cot\alpha - 2^n\cot 2^n\alpha$$

$$\therefore \quad \tan\alpha + 2\tan 2\alpha + 2^2\tan 2^2\alpha + \dots + 2^{n-1}\tan 2^{n-1}\alpha + 2^n\cot 2^n\alpha = \cot\alpha.$$

22. Prove that

$$\sum_{r=1}^{n} \tan r\alpha \cdot \tan(r+1)\alpha = \cot \alpha \cdot \tan(n+1)\alpha - n - 1.$$

$\tan r\alpha \cdot \tan(r+1)\alpha + 1$

$$= \frac{\sin r\alpha \cdot \sin(r+1)\alpha + \cos r\alpha \cdot \cos(r+1)\alpha}{\cos r\alpha \cdot \cos(r+1)\alpha}$$

$$= \frac{\cos\{(r+1)\alpha - r\alpha\}}{\cos r\alpha \cdot \cos(r+1)\alpha}$$

$$= \frac{\cos \alpha}{\cos r\alpha \cdot \cos(r+1)\alpha}$$

$$= \cot \alpha \cdot \frac{\sin \alpha}{\cos r\alpha \cdot \cos(r+1)\alpha}$$

$$= \cot \alpha \,. \frac{\sin\{(r+1)\alpha - r\alpha\}}{\cos r\alpha \cdot \cos(r+1)\alpha}$$

$$= \cot \alpha \cdot \frac{\sin(r+1)\alpha \cdot \cos r\alpha - \cos(r+1)\alpha \cdot \sin r\alpha}{\cos r\alpha \cdot \cos(r+1)\alpha}$$

$$= \cot \alpha \cdot \{\tan(r+1)\alpha - \tan r\alpha\}$$

$$\therefore \quad \sum_{r=1}^{n} \tan r\alpha \cdot \tan(r+1)\alpha + \sum_{r=1}^{n} 1$$

$$= \cot \alpha \sum_{r=1}^{n} \{\tan(r+1)\alpha - \tan r\alpha\}$$

or $\displaystyle\sum_{r=1}^{n} \tan r\alpha \cdot \tan(r+1)\alpha + n$

$$= \cot \alpha[(\tan 2\alpha - \tan \alpha) + (\tan 3\alpha - \tan 2\alpha) + (\tan 4\alpha - \tan 3\alpha) + \ldots + \{\tan(n+1)\alpha - \tan n\alpha\}]$$

$\therefore$ LHS $= -n + \cot \alpha\{\tan(n+1)\alpha - \tan \alpha\}$

$$= -n + \cot \alpha \cdot \tan(n+1)\alpha - \cot \alpha \tan \alpha$$

$$= \cot \alpha \cdot \tan(n+1)\alpha - n - 1 = \text{RHS.}$$

23. Evaluate $\displaystyle\sum_{r=2}^{n} \sin r\alpha$, where $(n+2)\alpha = 2\pi$.

Let $S = \displaystyle\sum_{r=2}^{n} \sin r\alpha = \sin 2\alpha + \sin 3\alpha + \sin 4\alpha + \ldots + \sin n\alpha$

$$\therefore \quad 2\sin\frac{\alpha}{2} \cdot S = 2\sin\frac{\alpha}{2}\sin 2\alpha + 2\sin\frac{\alpha}{2}\sin 3\alpha + 2\sin\frac{\alpha}{2}\sin 4\alpha + \ldots + 2\sin\frac{\alpha}{2} \cdot \sin n\alpha$$

$$= \left\{\cos\frac{3\alpha}{2} - \cos\frac{5\alpha}{2}\right\} + \left\{\cos\frac{5\alpha}{2} - \cos\frac{7\alpha}{2}\right\} + \left\{\cos\frac{7\alpha}{2} - \cos\frac{9\alpha}{2}\right\} + \ldots + \left\{\cos\left(n-\frac{1}{2}\right)\alpha - \cos\left(n+\frac{1}{2}\right)\alpha\right\}$$

$$= \cos\frac{3\alpha}{2} - \cos\left(n+\frac{1}{2}\right)\alpha$$

$$= 2\sin\frac{\frac{3\alpha}{2} + \left(n+\frac{1}{2}\right)\alpha}{2} \cdot \sin\frac{\left(n+\frac{1}{2}\right)\alpha - \frac{3\alpha}{2}}{2}$$

$$= 2\sin\frac{(n+2)\alpha}{2} \cdot \sin\frac{(n-1)\alpha}{2}$$

$$\therefore \quad S = \frac{\sin\frac{(n-1)\alpha}{2}}{\sin\frac{\alpha}{2}} \cdot \sin\frac{(n+2)\alpha}{2}$$

$$= \frac{\sin\frac{(n-1)\alpha}{2}}{\sin\frac{\alpha}{2}} \cdot \sin\frac{2\pi}{2} = 0.$$

Note As we have a series of sines in which angles are in AP we can use the formula

$$S = \frac{\sin\frac{n\beta}{2}}{\sin\frac{\beta}{2}} \sin\frac{2\alpha + (n-1)\beta}{2}$$

where α = first angle,

β = common difference of angles.

24. Sum the series:

$$\sqrt{1 + \cos \alpha} + \sqrt{1 + \cos 2\alpha} + \sqrt{1 + \cos 3\alpha} + \ldots \text{ to } n \text{ terms.}$$

Series $= \sqrt{2\cos^2\frac{\alpha}{2}} + \sqrt{2\cos^2\alpha} + \sqrt{2\cos^2\frac{3\alpha}{2}} + \ldots$ to n terms

$$= \sqrt{2}\left\{\cos\frac{\alpha}{2} + \cos \alpha + \cos\frac{3\alpha}{2} + \ldots + \text{to } n \text{ terms}\right\}$$

$$= \sqrt{2}\,\frac{\sin\frac{n\alpha}{4}}{\sin\frac{\alpha}{4}} \cdot \cos\frac{\frac{\alpha}{2} + \frac{\alpha}{2} + (n-1)\frac{\alpha}{2}}{2}$$

(using formula)

$$= \sqrt{2}\,\frac{\sin\frac{n\alpha}{4}}{\sin\frac{\alpha}{4}} \cdot \cos(n+1)\frac{\alpha}{4}.$$

25. If $A + B + C = \pi$, show that

$$\cot A + \cot B + \cot C - \operatorname{cosec} A \cdot \operatorname{cosec} B \cdot \operatorname{cosec} C = \cot A \cdot \cot B \cdot \cot C.$$

$$\text{LHS} = \frac{\cos A}{\sin A} + \frac{\cos B}{\sin B} + \frac{\cos C}{\sin C} - \frac{1}{\sin A \cdot \sin B \cdot \sin C}$$

$$= \frac{\left\{\begin{matrix}\cos A \sin B \sin C + \cos B \sin C \sin A \\ + \cos C \sin A \sin B - 1\end{matrix}\right\}}{\sin A \sin B \sin C}$$

$$= \frac{\left\{\begin{matrix}\sin C(\cos A \sin B + \cos B \sin A) \\ + \cos C \sin A \sin B - 1\end{matrix}\right\}}{\sin A \sin B \sin C}$$

$$= \frac{\sin C \cdot \sin(A+B) + \cos C \sin A \sin B - 1}{\sin A \sin B \sin C}$$

$$= \frac{\sin^2 C + \cos C \sin A \sin B - 1}{\sin A \sin B \sin C}$$

$$\{\because \sin(A+B) = \sin(\pi - C) = \sin C\}$$

$$= \frac{\cos C \cdot \sin A \sin B - \cos^2 C}{\sin A \sin B \sin C}$$

$$= \frac{\cos C\{\sin A \sin B - \cos(\pi - \overline{A+B})\}}{\sin A \sin B \sin C}$$

$$= \frac{\cos C \cdot \{\sin A \sin B + \cos(A+B)\}}{\sin A \sin B \sin C}$$

$$= \frac{\cos C \cdot \cos A \cdot \cos B}{\sin A \sin B \sin C}$$

$$= \cot A \cdot \cot B \cdot \cot C.$$

26. If $A + B + C = 2\pi$, show that

$$\sin^3 A + \sin^3 B + \sin^3 C$$
$$= 3\sin\frac{A}{2} \cdot \sin\frac{B}{2} \cdot \sin\frac{C}{2} - \sin\frac{3A}{2} \cdot \sin\frac{3B}{2} \cdot \sin\frac{3C}{2}.$$

$$\text{LHS} = \frac{3\sin A - \sin 3A}{4} + \frac{3\sin B - \sin 3B}{4} + \frac{3\sin C - \sin 3C}{4}$$

$$= \frac{3}{4}(\sin A + \sin B + \sin C) - \frac{1}{4}(\sin 3A + \sin 3B + \sin 3C) \quad \dots (1)$$

Now, $\sin A + \sin B + \sin C$

$$= 2\sin\frac{A+B}{2} \cdot \cos\frac{A-B}{2} + \sin C$$

$$= 2\sin\frac{C}{2} \cdot \cos\frac{A-B}{2} + 2\sin\frac{C}{2} \cdot \cos\frac{C}{2}$$

$$\left\{\because \frac{A+B}{2} = \frac{2\pi - C}{2} = \pi - \frac{C}{2}\right\}$$

$$= 2\sin\frac{C}{2}\left\{\cos\frac{A-B}{2} + \cos\left(\pi - \frac{A+B}{2}\right)\right\}$$

$$= 2\sin\frac{C}{2}\left\{\cos\frac{A-B}{2} - \cos\frac{A+B}{2}\right\}$$

$$= 4\sin\frac{C}{2} \cdot \sin\frac{A}{2} \cdot \sin\frac{B}{2}.$$

Again, $\sin 3A + \sin 3B + \sin 3C$

$$= 2\sin\frac{3(A+B)}{2} \cdot \cos\frac{3(A-B)}{2} + \sin 3C$$

$$= 2\sin 3\left(\pi - \frac{C}{2}\right) \cdot \cos\frac{3(A-B)}{2} + 2\sin\frac{3C}{2} \cdot \cos\frac{3C}{2}$$

$$= 2\sin\left(3\pi - \frac{3C}{2}\right) \cdot \cos\frac{3(A-B)}{2} + 2\sin\frac{3C}{2} \cdot \cos\frac{3C}{2}$$

$$= 2\sin\frac{3C}{2}\left[\cos\frac{3(A-B)}{2} + \cos\frac{3C}{2}\right]$$

$$= 2\sin\frac{3C}{2} \cdot 2\cos\frac{3(A-B) - 3C}{4} \cdot \cos\frac{3(A-B) + 3C}{4}$$

$$= 4\sin\frac{3C}{2} \cdot \cos\frac{3A - 3(B+C)}{4} \cdot \cos\frac{3(A+C) - 3B}{4}$$

$$= 4\sin\frac{3C}{2} \cdot \cos\frac{3A - 3(2\pi - A)}{4} \cos\frac{3(2\pi - B) - 3B}{4}$$

$$= 4\sin\frac{3C}{2} \cdot \cos\left(\frac{3A}{2} - \frac{3\pi}{2}\right) \cos\left(\frac{3\pi}{2} - \frac{3B}{2}\right)$$

$$= 4\sin\frac{3C}{2} \cdot \cos\left(\frac{\pi}{2} + \frac{3A}{2}\right) \cos\left(-\frac{\pi}{2} - \frac{3B}{2}\right)$$

$$= 4\sin\frac{3C}{2} \cdot \left\{-\sin\frac{3A}{2}\right\}\left\{-\sin\frac{3B}{2}\right\}$$

$$= 4\sin\frac{3A}{2} \sin\frac{3B}{2} \sin\frac{3C}{2}$$

$\therefore$ from (1),

$$\text{LHS} = \frac{3}{4} \cdot 4\sin\frac{A}{2}\sin\frac{B}{2}\sin\frac{C}{2} - \frac{1}{4} \cdot 4\sin\frac{3A}{2} \cdot \sin\frac{3B}{2} \cdot \sin\frac{3C}{2}$$

$$= 3\sin\frac{A}{2}\sin\frac{B}{2}\sin\frac{C}{2} - \sin\frac{3A}{2} \cdot \sin\frac{3B}{2} \cdot \sin\frac{3C}{2}$$

$$= \text{RHS}.$$

27. If $4\cos(y-z)\cos(z-x)\cos(x-y) = 1$, prove that
$1 + 12\cos 2(y-z) \cdot \cos 2(z-x) \cdot \cos 2(x-y)$
$= 4\cos 3(y-z)\cos 3(z-x)\cos 3(x-y)$.

Let $y - z = A,\ z - x = B,\ x - y = C$. Then $A + B + C = 0$.

$\therefore$ the given relation is

$4\cos A \cos B \cos C = 1$ where $A + B + C = 0$

$\Rightarrow \quad 2\cos A \cdot \{\cos(B-C) + \cos(B+C)\} = 1$

$\Rightarrow \quad 2\cos A \cdot \cos(B-C) + 2\cos^2 A - 1 = 0$

$\Rightarrow \quad 2\cos(B+C) \cdot \cos(B-C) + 2\cos^2 A - 1 = 0$

$\Rightarrow \quad \cos 2B + \cos 2C + \cos 2A = 0$ ... (1)

$$\text{RHS} = 4\cos 3(y-z)\cdot\cos 3(z-x)\cdot\cos 3(x-y)$$
$$= 4\cos 3A\cdot\cos 3B\cdot\cos 3C$$
$$= 2\cos 3A\{\cos 3(B-C)+\cos 3(B+C)\}$$
$$= 2\cos 3A\cdot\{\cos 3(B-C)+\cos 3A\}$$
$$= 2\cos 3(B+C)\cdot\cos 3(B-C)+2\cos^2 3A$$
$$= 2(\cos^2 3B-\sin^2 3C)+2\cos^2 3A$$
$$= 2(\cos^2 3A+\cos^2 3B+\cos^2 2C)-2$$
$$= 1+\cos 6A+\cos 6B+\cos 6C$$
$$= 1+4(\cos^3 2A+\cos^3 2B+\cos^3 2C)$$
$$-3(\cos 2A+\cos 2B+\cos 2C)$$
$$= 1+4(\cos^3 2A+\cos^3 2B+\cos^3 2C),$$

because $\cos 2A+\cos 2B+\cos 2C=0$ from (1)

$$= 1+4\{\cos^3 2A+\cos^3 2B+\cos^3 2C$$
$$-3\cos 2A\cos 2B\cos 2C\}$$
$$+12\cos 2A\cos 2B\cos 2C$$
$$= 1+4\{(\cos 2A+\cos 2B+\cos 2C)(\cos^2 2A+\cos^2 2B$$
$$+\cos^2 2C-\cos 2A\cos 2B-\cos 2B\cos 2C$$
$$-\cos 2C\cdot\cos 2A)\}+12\cos 2A\cos 2B\cos 2C$$

$\{\because\ a^3+b^3+c^3-3abc = (a+b+c)(a^2+b^2+c^2-bc-ca-ab)\}$

$$= 1+12\cos 2A\cos 2B\cos 2C$$
$$= 1+12\cos 2(y-z)\cdot\cos 2(z-x)\cdot\cos 2(x-y)$$
$$= \text{LHS}$$

28. If $xy+yz+zx=1$, show that
$$\frac{x}{1-x^2}+\frac{y}{1-y^2}+\frac{z}{1-z^2}=\frac{4xyz}{(1-x^2)(1-y^2)(1-z^2)}.$$

Let $x=\tan A$, $y=\tan B$ and $z=\tan C$.

$\therefore$ from the question,

$\tan A\tan B+\tan B\tan C+\tan C\cdot\tan A=1$

or $\tan B(\tan A+\tan C)-(1-\tan C\cdot\tan A)=0$

or $(1-\tan C\tan A)\left\{\tan B\cdot\dfrac{\tan A+\tan C}{1-\tan C\cdot\tan A}-1\right\}=0$

$\therefore$ $\tan B\cdot\tan(A+C)-1=0$

$\{\because\ \tan C\cdot\tan A\neq 1$ as $zx\neq 1\}$

$\therefore$ $\tan(A+C)=\cot B=\tan\left(\dfrac{\pi}{2}-B\right)$

$\therefore$ $A+C=n\pi+\dfrac{\pi}{2}-B$; $\therefore$ $2A+2C=2n\pi+\pi-2B$

or $\tan(2A+2C)=\tan(2n\pi+\pi-2B)=-\tan 2B$

or $\dfrac{\tan 2A+\tan 2C}{1-\tan 2A\cdot\tan 2C}=-\tan 2B$

or $\tan 2A+\tan 2C=-\tan 2B+\tan 2A\cdot\tan 2B\cdot\tan 2C$

or $\tan 2A+\tan 2B+\tan 2C=\tan 2A\cdot\tan 2B\cdot\tan 2C$

or $\dfrac{2\tan A}{1-\tan^2 A}+\dfrac{2\tan B}{1-\tan^2 B}+\dfrac{2\tan C}{1-\tan^2 C}$
$$=\frac{2\tan A}{1-\tan^2 A}\cdot\frac{2\tan B}{1-\tan^2 B}\cdot\frac{2\tan C}{1-\tan^2 C}$$

or $\dfrac{2x}{1-x^2}+\dfrac{2y}{1-y^2}+\dfrac{2z}{1-z^2}=\dfrac{2x}{1-x^2}\cdot\dfrac{2y}{1-y^2}\cdot\dfrac{2z}{1-z^2}$

or $\dfrac{x}{1-x^2}+\dfrac{y}{1-y^2}+\dfrac{z}{1-z^2}=\dfrac{4xyz}{(1-x^2)(1-y^2)(1-z^2)}$

29. If $A+B+C=\pi$, $\sin A\sin B\sin C=p$ and $\cos A\cos B\cos C=q$ then obtain the cubic equation whose roots are $\tan A$, $\tan B$ and $\tan C$.

The required cubic equation is

$$(x-\tan A)(x-\tan B)(x-\tan C)=0$$

or $x^3-(\tan A+\tan B+\tan C)x^2$
$$+(\tan A\tan B+\tan B\tan C+\tan C\cdot\tan A)x$$
$$-\tan A\tan B\tan C=0 \quad \text{... (1)}$$

Now, $\tan A\cdot\tan B\cdot\tan C$
$$=\frac{\sin A}{\cos A}\cdot\frac{\sin B}{\cos B}\cdot\frac{\sin C}{\cos C}=\frac{p}{q} \quad \text{... (2)}$$

Also, $\tan(A+B)=\tan(\pi-C)=-\tan C$

or $\dfrac{\tan A+\tan B}{1-\tan A\tan B}=-\tan C$

or $\tan A+\tan B+\tan C$
$$=\tan A\tan B\tan C=\frac{p}{q} \quad \text{... (3)}$$

using (2).

Now, $\sin^2 A+\sin^2 B+\sin^2 C$
$$=\frac{1}{2}[3-(\cos 2A+\cos 2B+\cos 2C)]$$
$$=\frac{1}{2}[3-\{2\cos(A+B)\cdot\cos(A-B)+2\cos^2 C-1\}]$$
$$=\frac{1}{2}[4-\{-2\cos C\cdot\cos(A-B)+2\cos^2 C\}]$$
$$=\frac{1}{2}[4+2\cos C\{\cos(A-B)+\cos(A+B)\}]$$
$$=2(1+\cos A\cos B\cos C)=2(1+q)$$

$\therefore$ $\sin^2 A\sin^2 B\sin^2 C(\operatorname{cosec}^2 B\operatorname{cosec}^2 C$
$$+\operatorname{cosec}^2 C\cdot\operatorname{cosec}^2 A+\operatorname{cosec}^2 A\operatorname{cosec}^2 B)=2(1+q)$$

or $p^2\{(1+\cot^2 B)(1+\cot^2 C)+(1+\cot^2 C)(1+\cot^2 A)$

$+(1+\cot^2 A)(1+\cot^2 B)\}=2(1+q)$

or $p^2\{3+2(\cot^2 A+\cot^2 B+\cot^2 C)+(\cot^2 B\cot^2 C$

$+\cot^2 C\cot^2 A+\cot^2 A\cot^2 B)\}=2(1+q)$

or $p^2[3+2\{(\cot A+\cot B+\cot C)^2$

$-2(\cot A\cot B+\cot B\cot C+\cot C\cot A)\}$

$+\{(\cot B\cot C+\cot C\cot A+\cot A\cot B)^2$

$-2\cot A\cot B\cot C(\cot A+\cot B+\cot C)\}]$

$=2(1+q)$

or $p^2[3+2\{(\cot A+\cot B+\cot C)^2-2\}$

$+\{1-2\cdot\frac{q}{p}(\cot A+\cot B+\cot C)\}]=2(1+q)$ ... (4)

$\{\because\ \Sigma\cot A\cdot\cot B=1 \text{ if } A+B+C=\pi\}$

Let $y=\tan A\tan B+\tan B\tan C+\tan C\tan A$

$=\tan A\tan B\tan C(\cot C+\cot A+\cot B)$

$=\frac{p}{q}(\cot A+\cot B+\cot C)$

$\therefore$ $\cot A+\cot B+\cot C=\frac{yq}{p}$ ... (5)

$\therefore$ (4) gives

$$p^2\left[3+2\left\{\frac{y^2q^2}{p^2}-2\right\}+1-\frac{2q}{p}\cdot\frac{yq}{p}\right]=2(1+q)$$

or $p^2\left(\frac{2q^2y^2}{p^2}-\frac{2q^2y}{p^2}\right)-2(1+q)=0$

or $q^2y^2-q^2y-(1+q)=0$

$\therefore$ $y=\frac{q^2\pm\sqrt{q^4+4q^2(1+q)}}{2q^2}=\frac{q^2\pm q\sqrt{q^2+4q+4}}{2q^2}$

$=\frac{q\pm(q+2)}{2q}=\frac{q+1}{q},\ -\frac{1}{q}$ ... (6)

From (1), (2), (3), (4), (5) and (6),

$x^3-\frac{p}{q}x^2+\frac{q+1}{q}x-\frac{p}{q}=0$ when $y=\frac{q+1}{q}$

or $x^3-\frac{p}{q}x^2-\frac{1}{q}x-\frac{p}{q}=0$ when $y=-\frac{1}{q}$

$\therefore$ the required equation is

$qx^3-px^2+(q+1)x-p=0$

or $qx^3-px^2-x-p=0.$

30. Show that $\sin\frac{\pi}{7}\cdot\sin\frac{2\pi}{7}\cdot\sin\frac{3\pi}{7}=\frac{\sqrt7}{8}$.

Let $\theta=\frac{\pi}{7}$ or $\frac{3\pi}{7}$ or $\frac{5\pi}{7}$. (Note $\sin\frac{2\pi}{7}=\sin\frac{5\pi}{7}$)

$\therefore$ $7\theta=\pi,\ 3\pi,\ 5\pi$

$\therefore$ $4\theta=\pi-3\theta,\ 3\pi-3\theta,\ 5\pi-3\theta$

$\therefore$ $\sin 4\theta=\sin(\pi-3\theta),\ \sin(3\pi-3\theta),\ \sin(5\pi-3\theta)$

or $\sin 4\theta=\sin 3\theta$

or $2\sin 2\theta\cdot\cos 2\theta=3\sin\theta-4\sin^3\theta$

or $4\cos\theta\cdot\cos 2\theta=3-4\sin^2\theta$ ($\because\ \sin\theta\neq 0$)

or $4\cos\theta(2\cos^2\theta-1)=3-4(1-\cos^2\theta)$

or $8\cos^3\theta-4\cos^2\theta-4\cos\theta+1=0$ ... (1)

This is a third degree equation in $\cos\theta$ and each of $\cos\frac{\pi}{7}$, $\cos\frac{3\pi}{7}$, $\cos\frac{5\pi}{7}$ satisfies it.

$\therefore$ roots of (1) are $\cos\frac{\pi}{7}$, $\cos\frac{3\pi}{7}$ and $\cos\frac{5\pi}{7}$.

$\therefore$ $\cos\frac{\pi}{7}+\cos\frac{3\pi}{7}+\cos\frac{5\pi}{7}$

$=$ sum of the roots $=-\frac{-4}{8}=\frac{1}{2}$ ... (2)

$\cos\frac{\pi}{7}\cdot\cos\frac{3\pi}{7}+\cos\frac{3\pi}{7}\cdot\cos\frac{5\pi}{7}+\cos\frac{5\pi}{7}\cdot\cos\frac{\pi}{7}$

$=\frac{-4}{8}=\frac{-1}{2}$ ... (3)

$\cos\frac{\pi}{7}\cdot\cos\frac{3\pi}{7}\cdot\cos\frac{5\pi}{7}=-\frac{1}{8}$ ... (4)

Now, $\left(\sin\frac{\pi}{7}\cdot\sin\frac{2\pi}{7}\cdot\sin\frac{3\pi}{7}\right)^2=\left(\sin\frac{\pi}{7}\cdot\sin\frac{5\pi}{7}\cdot\sin\frac{3\pi}{7}\right)^2$

$=\left(1-\cos^2\frac{\pi}{7}\right)\left(1-\cos^2\frac{3\pi}{7}\right)\left(1-\cos^2\frac{5\pi}{7}\right)$

$=1-\left(\cos^2\frac{\pi}{7}+\cos^2\frac{3\pi}{7}+\cos^2\frac{5\pi}{7}\right)$

$+\left(\cos^2\frac{\pi}{7}\cdot\cos^2\frac{3\pi}{7}+\cos^2\frac{3\pi}{7}\cdot\cos^2\frac{5\pi}{7}\right.$

$\left.+\cos^2\frac{5\pi}{7}\cdot\cos^2\frac{\pi}{7}\right)-\cos^2\frac{\pi}{7}\cdot\cos^2\frac{3\pi}{7}\cdot\cos^2\frac{5\pi}{7}$

$=1-\left\{\left(\cos\frac{\pi}{7}+\cos\frac{3\pi}{7}+\cos\frac{5\pi}{7}\right)^2\right.$

$-2\left(\cos\frac{\pi}{7}\cdot\cos\frac{3\pi}{7}+\cos\frac{3\pi}{7}\cdot\cos\frac{5\pi}{7}\right.$

$\left.\left.+\cos\frac{5\pi}{7}\cdot\cos\frac{\pi}{7}\right)\right\}$

$+\left\{\left(\cos\frac{\pi}{7}\cdot\cos\frac{3\pi}{7}+\cos\frac{3\pi}{7}\cdot\cos\frac{5\pi}{7}+\cos\frac{5\pi}{7}\cdot\cos\frac{\pi}{7}\right)^2\right.$

$\left.-2\cos\frac{\pi}{7}\cdot\cos\frac{3\pi}{7}\cdot\cos\frac{5\pi}{7}\left(\cos\frac{\pi}{7}+\cos\frac{3\pi}{7}+\cos\frac{5\pi}{7}\right)\right\}$

$-\left(\cos\frac{\pi}{7}\cdot\cos\frac{3\pi}{7}\cdot\cos\frac{5\pi}{7}\right)^2$

$$= 1 - \left\{\left(\frac{1}{2}\right)^2 - 2\left(-\frac{1}{2}\right)\right\} + \left\{\left(-\frac{1}{2}\right)^2 - 2 \cdot \frac{-1}{8} \cdot \frac{1}{2}\right\} - \left(-\frac{1}{8}\right)^2$$

{using (2), (3), (4)}

$$= 1 - \frac{5}{4} + \frac{3}{8} - \frac{1}{64}$$

$$= \frac{64 - 80 + 24 - 1}{64} = \frac{7}{64} = \left(\frac{\sqrt{7}}{8}\right)^2$$

$$\therefore \quad \sin\frac{\pi}{7} \cdot \sin\frac{2\pi}{7} \cdot \sin\frac{3\pi}{7} = \frac{\sqrt{7}}{8}$$

{$\because$ all the factors are positive}.

Note (1) provides the answer to the following question:

Prove that $\cos\frac{\pi}{7} + \cos\frac{3\pi}{7} + \cos\frac{5\pi}{7} = \frac{1}{2}$.

Similarly, (2) and (3) also provide answers to two questions.

Exercises

1. Eliminate θ between $\operatorname{cosec}\theta - \sin\theta = m$ and $\sec\theta - \cos\theta = n$.

2. If $x \sin^2\theta + y \cos^2\theta = \frac{1}{2}\sin 2\theta$ and $x \sin\theta - y\cos\theta = 0$, $\theta \neq n\pi - \pi/4$ prove that the point (x, y) lies on a line in the x–y plane.

3. If $\cos\alpha + \cos 2\alpha = p$, $\sin\alpha + \sin 2\alpha = q$ then eliminate α.

4. If $\cos\alpha = \cos\beta \cdot \cos\phi = \cos\gamma \cdot \cos\theta$ and $\sin\alpha = 2\sin\frac{\phi}{2} \cdot \sin\frac{\theta}{2}$ then prove that $\tan\frac{\alpha}{2} = \pm\tan\frac{\beta}{2} \cdot \tan\frac{\gamma}{2}$.

5. Eliminate θ between $a = \cos(\theta - \alpha)$, $b = \sin(\theta - \beta)$ and prove that $a^2 - 2ab\sin(\alpha - \beta) + b^2 = \cos^2(\alpha - \beta)$.

6. Eliminating α from

$$x\cos\alpha = y\cos\left(\alpha + \frac{2\pi}{3}\right) = z\cos\left(\alpha + \frac{4\pi}{3}\right),$$

establish that $xy + yz + zx = 0$.

7. If $\cos^2\theta = \frac{m^2 - 1}{3}$ and $\tan\alpha = \tan^3\frac{\theta}{2}$ then prove that $(\sin\alpha)^{2/3} + (\cos\alpha)^{2/3} = \left(\frac{4}{m^2}\right)^{1/3}$.

8. If $\cos rx + \cos ry = a_r$; $r = 1, 2, 3$ then prove that $2a_1^3 + a_3 = 3a_1(a_2 + 1)$.

9. If $\frac{\cos(\alpha - 3\theta)}{\cos^3\theta} = \frac{\sin(\alpha - 3\theta)}{\sin^3\theta} = m$ then show that $\cos\alpha = \frac{2 - m^2}{m}$.

10. Eliminate θ, ϕ from $\sin\theta + \sin\phi = a$, $\cos\theta + \cos\phi = b$ and $\tan\frac{\theta}{2} \cdot \tan\frac{\phi}{2} = c$.

11. Eliminate θ from $\lambda\cos 2\theta = \cos(\theta + \alpha)$ and $\lambda\sin 2\theta = 2\sin(\theta + \alpha)$.

12. Eliminate θ from $a\tan\theta + b\cot 2\theta = c$, $a\cot\theta - b\tan 2\theta = c$.

13. If $f(\theta) = 1 + \sin\left(\frac{\pi}{4} + \theta\right) + \cos\left(\frac{\pi}{4} - \theta\right)$, find the range of values of $f(\theta)$.

14. Prove that the value of $5\cos\theta + 3\cos\left(\theta + \frac{\pi}{3}\right) + 3$ lies between -4 and 10.

15. If $f(\theta) = 11\cos^2\theta + 15\sin\theta\cos\theta - 9\sin^2\theta$, express $f(\theta)$ in the form $a\cos(2\theta + \alpha) + b$. Hence or otherwise find the range of $f(\theta)$.

16. Find the maximum and minimum values of $\sin^6 x + \cos^6 x$.

17. In $\triangle ABC$, $B = \frac{\pi}{3}$ and $\sin A \cdot \sin C = x$ then find the set of all possible values of x.

18. Prove that the relation $\sin^2\theta = \frac{(x+y)^2}{4xy}$ is not possible for any real θ where $x \in R$, $y \in R$ such that $|x| \neq |y|$.

19. Prove that $\sin\theta = x + \frac{p}{x}$ is possible for real x if $p \leq \frac{1}{4}$.

20. Show that there is no real x for which $e^{\sin x} - e^{-\sin x} = 4$.

21. If $f(x) = 3(\sin x - \cos x)^4 + 6(\sin x + \cos x)^2 + 4(\sin^6 x + \cos^6 x)$ and $f(\pi/7) = 13$ then find $f(2\pi/7)$.

22. Prove that $1 + \cos 2x + \cos 4x + \cos 6x = 4\cos x \cdot \cos 2x \cdot \cos 3x$.

23. If α, β are acute angles and $\cos 2\alpha = \frac{3\cos 2\beta - 1}{3 - \cos 2\beta}$, prove that $\tan\alpha : \tan\beta = \sqrt{2} : 1$.

24. If $\tan\theta\cdot\tan\phi=\sqrt{\dfrac{x-y}{x+y}}$,

prove that $(x-y\cos 2\theta)(x-y\cos 2\phi)=x^2-y^2$.

25. If $\dfrac{\sin(\theta-\alpha)}{\sin(\theta-\beta)}=\dfrac{x}{y}$ and $\dfrac{\cos(\theta-\alpha)}{\cos(\theta-\beta)}=\dfrac{u}{v}$ such that $uy+vx\neq 0$ then prove that $\cos(\alpha-\beta)=\dfrac{ux+vy}{uy+vx}$.

26. If $x\in(\pi,2\pi)$,

prove that $\dfrac{\sqrt{1+\cos x}+\sqrt{1-\cos x}}{\sqrt{1+\cos x}-\sqrt{1-\cos x}}=\cot\left(\dfrac{\pi}{4}+\dfrac{x}{2}\right)$.

27. Show that $\cot\theta\cdot\cot 2\theta+\cot 2\theta\cdot\cot 3\theta+2$

$=\cot\theta(\cot\theta-\cot 3\theta)$.

28. If $\tan^3\left(\dfrac{\alpha}{2}+\dfrac{\pi}{4}\right)=\tan\left(\dfrac{\beta}{2}+\dfrac{\pi}{4}\right)$ then

prove that $\sin\beta=\dfrac{(3+\sin^2\alpha)\sin\alpha}{1+3\sin^2\alpha}$.

29. Prove that

$\sin x\cdot\sin y\cdot\sin(x-y)+\sin y\cdot\sin z\cdot\sin(y-z)$

$+\sin z\cdot\sin x\sin(z-x)$

$+\sin(x-y)\cdot\sin(y-z)\cdot\sin(z-x)=0$.

30. Prove that $(3+\cos 4\theta)\cos 2\theta=4(\cos^8\theta-\sin^8\theta)$.

31. Prove that

$\cos^2\theta+\cos^2(\alpha+\theta)-2\cos\alpha\cdot\cos\theta\cdot\cos(\alpha+\theta)$

is independent of θ.

32. If $\sin\beta=\dfrac{1}{5}\sin(2\alpha+\beta)$,

show that $\tan(\alpha+\beta)=\dfrac{3}{2}\tan\alpha$.

33. If $T_n=\sin^n\theta+\cos^n\theta$

then prove that $(T_5-T_3):(T_7-T_5)=T_1:T_3$.

34. If $f(\alpha,\beta)=\dfrac{\cos^4\alpha}{\cos^2\beta}+\dfrac{\sin^4\alpha}{\sin^2\beta}$

then prove that $f(\alpha,\beta)=1 \Leftrightarrow f(\beta,\alpha)=1$.

35. Write $\tan 3A$ as the product of tangents of three angles in AP.

36. If $\cot A$, $\cot B$, $\cot C$ are in AP, prove that $\cot(B-A)$, $\cot B$, $\cot(B-C)$ are in AP.

37. If $\cos A=\tan B$, $\cos B=\tan C$ and $\cos C=\tan A$ then show that

$\sin A=\sin B=\sin C=2\sin 18°$.

38. Prove that $\cos\theta=\dfrac{\cos\alpha-\cos\beta}{1-\cos\alpha\cdot\cos\beta}$

$\Leftrightarrow \quad \tan\dfrac{\theta}{2}=\pm\tan\dfrac{\alpha}{2}\cdot\cot\dfrac{\beta}{2}$.

39. If $\sqrt{a+b}\tan\dfrac{\theta}{2}=\sqrt{a-b}\tan\dfrac{\phi}{2}$ then prove that

$a+b\cos\phi=(a\cos\phi+b)\sec\theta$.

40. If $\sin\theta-\sin\phi=a$ and $\cos\theta+\cos\phi=b$ then prove that

$\cos(\theta+\phi)=\dfrac{a^2+b^2-2}{2}$ and $\tan(\theta-\phi)=\dfrac{2ab}{b^2-a^2}$.

41. If $\sin\theta+\cos\phi=\dfrac{1}{\sqrt{2}}$ and $\cos\theta+\sin\phi=\sqrt{\dfrac{2}{3}}$.

then find the values of $\sin(\theta+\phi)$ and $\tan\dfrac{\theta-\phi}{2}$.

42. If $\sin x+\sin y+\sin z=0$ and

$\cos x+\cos y+\cos z=0$ then prove that

$\cot\dfrac{x+y}{2}=\cot z$ and $\cos\dfrac{x-y}{2}=\pm\dfrac{1}{2}$.

43. If $\sin\alpha+\sin\beta=p$, $\cos\alpha+\cos\beta=q$ then find the value of $\cos 2\alpha+\cos 2\beta$ in terms of p and q.

44. If $\cos(x-y)=-1$, show that

$\cos x+\cos y=0$ and $\sin x+\sin y=0$.

45. If α,β be the solutions of θ for the equation $a\tan\theta+b\sec\theta=c$ then prove that

$\tan(\alpha+\beta)=\dfrac{2ac}{a^2-c^2}$.

46. If $\alpha+\beta$ and $\alpha-\beta$ are solutions of θ for the equation $\tan^2\theta-4\tan\theta=-1$ where $0 \cdot \alpha<\dfrac{\pi}{2}$ and $0<\beta<\dfrac{\pi}{2}$

then find α and β.

47. Prove that $\tan 7\dfrac{1°}{2}=\sqrt{6}-\sqrt{3}+\sqrt{2}-2$.

48. Prove that $\cot 142.5°=\sqrt{2}+\sqrt{3}-\sqrt{4}-\sqrt{6}$.

49. Prove that $\sin\dfrac{23\pi}{24}=\sqrt{\dfrac{2\sqrt{2}-\sqrt{3}-1}{4\sqrt{2}}}$.

50. Prove that $\sin 27°=\dfrac{1}{4}\{\sqrt{5+\sqrt{5}}-\sqrt{3-\sqrt{5}}\}$.

51. Prove that $\sin 9°=\dfrac{1}{4}\{\sqrt{3+\sqrt{5}}-\sqrt{5-\sqrt{5}}\}$.

52. Prove that $\tan 20°\cdot\tan 80°=\sqrt{3}\tan 50°$.

53. Prove that

$\left(1+\cos\dfrac{\pi}{9}\right)\left(1+\cos\dfrac{3\pi}{9}\right)\left(1+\cos\dfrac{5\pi}{9}\right)\left(1+\cos\dfrac{7\pi}{9}\right)=\dfrac{9}{16}$.

54. Prove that

$\sin 55°-\sin 19°+\sin 53°-\sin 17°=\cos 1°$.

55. Prove that

$\cos^4\dfrac{\pi}{8}+\cos^4\dfrac{3\pi}{8}+\cos^4\dfrac{5\pi}{8}+\cos^4\dfrac{7\pi}{8}=\dfrac{3}{2}$.

56. Prove that $\tan 9°-\tan 27°-\tan 63°+\tan 81°=4$.

57. Prove that $\sin^4\dfrac{\pi}{8}+\sin^4\dfrac{3\pi}{8}+\sin^4\dfrac{5\pi}{8}+\sin^4\dfrac{7\pi}{8}=\dfrac{3}{2}$.

58. (a) Prove that

$\cos\theta\cdot\cos 2\theta\cdot\cos 2^2\theta\cdot\cos 2^3\theta$

$\times\ldots\cos 2^{n-1}\theta=\dfrac{1}{2^n}$, when $\theta=\dfrac{\pi}{2^n+1}$.

(b) Prove that

$$\cos\frac{\pi}{15}\cdot\cos\frac{2\pi}{15}\cdot\cos\frac{3\pi}{15}\cdot\ldots\text{ to 7 factors}=\frac{1}{2^7}.$$

59. Prove that

$$\sin\frac{\pi}{14}\cdot\sin\frac{3\pi}{14}\cdot\sin\frac{5\pi}{14}\cdot\ldots\text{ to 7 factors}=\frac{1}{2^6}.$$

60. If $A > 0$, $B > 0$ and $A + B = \frac{\pi}{3}$ then find the maximum value of $\tan A \cdot \tan B$.

61. Find the minimum value of $9\tan^2\theta + 4\cot^2\theta$.

62. Show that the maximum and minimum values of

$$\sqrt{a^2\cos^2\theta + b^2\sin^2\theta} + \sqrt{a^2\sin^2\theta + b^2\cos^2\theta}$$

are $\sqrt{2(a^2+b^2)}$ and $a + b$ respectively.

63. Find the set of possible values of $\dfrac{\tan\left(x+\frac{\pi}{6}\right)}{\tan x}$ where x is any real angle.

64. Prove that for real x, $\dfrac{\tan(x+\alpha)}{\tan(x-\alpha)}$ where $\alpha \neq \frac{\pi}{4}, \frac{3\pi}{4}$ cannot lie between

$$\left(\frac{1-\sin 2\alpha}{1+\sin 2\alpha}\right)\text{ and }\left(\frac{1+\sin 2\alpha}{1-\sin 2\alpha}\right).$$

65. Show that $\dfrac{\sec^2\theta - \tan\theta}{\sec^2\theta + \tan\theta}$ lies between $\frac{1}{3}$ and 3 for all real θ.

66. Prove that

$$\tan\alpha + 2\tan 2\alpha + 4\tan 4\alpha + 8\cot 8\alpha = \cot\alpha.$$

67. Find the sum:

$$\tan x + \frac{1}{2}\tan\frac{x}{2} + \frac{1}{2^2}\tan\frac{x}{2^2} + \ldots + \frac{1}{2^{n-1}}\tan\frac{x}{2^{n-1}}.$$

68. Prove that $\operatorname{cosec} 2\alpha = \cot\alpha - \cot 2\alpha$.

Hence or otherwise prove that

$$\operatorname{cosec}\theta + \operatorname{cosec}2\theta + \operatorname{cosec}4\theta + \operatorname{cosec}8\theta = \cot\frac{\theta}{2} - \cot 8\theta.$$

69. Prove that $\displaystyle\sum_{r=1}^{9}\sin^2\frac{r\pi}{18} = 5.$

70. Sum the series:

$\sin\alpha - \sin 2\alpha + \sin 3\alpha - \sin 4\alpha + \ldots$ to n terms.

71. Find $\displaystyle\sum_{r=1}^{10}\cos^3\frac{r\pi}{3}.$

72. Sum: $\sqrt{1+\sin 2\alpha} + \sqrt{1+\sin 4\alpha} + \sqrt{1+\sin 6\alpha} + \ldots$ to 20 terms.

73. Prove that $\sin^2 12° + \sin^2 21° + \sin^2 39° + \sin^2 48° = 1 + \sin^2 9° + \sin^2 18°.$

74. If $A + B + C = \pi$, prove that

$$\cos A\operatorname{cosec}B\operatorname{cosec}C + \cos B\operatorname{cosec}C\operatorname{cosec}A + \cos C\operatorname{cosec}A\operatorname{cosec}B = 2.$$

75. Prove that in a ΔABC,

$$\sin^2 A - \sin^2 B + \sin^2 C = 2\sin A\cdot\cos B\cdot\sin C.$$

76. Prove that in a ΔABC, $\displaystyle\sum\frac{\cot B + \cot C}{\tan B + \tan C} = 1.$

77. If $A + B + C = 180°$, prove that

$$\cos\frac{A}{2} + \cos\frac{B}{2} + \cos\frac{C}{2} = 4\cos\frac{\pi - A}{4}\cdot\cos\frac{\pi - B}{4}\cos\frac{\pi - C}{4}.$$

78. Prove that $\sin(B-C) + \sin(C-A) + \sin(A-B)$

$$+ 4\sin\frac{B-C}{2}\cdot\sin\frac{C-A}{2}\sin\frac{A-B}{2} = 0.$$

79. If $\alpha + \beta - \gamma = \pi$, prove that

$$\sin^2\alpha + \sin^2\beta - \sin^2\gamma = 2\sin\alpha\sin\beta\cos\gamma.$$

80. If $\alpha + \beta + \gamma = 2\theta$, prove that

$$\cos^2\theta + \cos^2(\theta - \alpha) + \cos^2(\theta - \beta) + \cos^2(\theta - \gamma) = 2(1 + \cos\alpha\cos\beta\cos\gamma)$$

81. In a quadrilateral ABCD, show that

$$\cos A - \cos B + \cos C - \cos D = 4\sin\frac{A+B}{2}\cdot\sin\frac{A+D}{2}\cdot\cos\frac{A+C}{2}.$$

82. In a quadrilateral ABCD, prove that

$$\cos A + \cos B + \cos C + \cos D + 4\cos\frac{A+B}{2}\cdot\cos\frac{A+C}{2}\cdot\cos\frac{A+D}{2} = 0.$$

83. If $A + B + C = \pi$, prove that $\begin{vmatrix}\sin^2 A & \cot A & 1\\ \sin^2 B & \cot B & 1\\ \sin^2 C & \cot C & 1\end{vmatrix} = 0.$

84. If $A + B + C = \pi$, show that

$$\cot A + \frac{\sin A}{\sin B\sin C} = \cot B + \frac{\sin B}{\sin C\sin A} = \cot C + \frac{\sin C}{\sin A\sin B}.$$

85. If $A + B + C = 0$,

prove that $\Sigma\sin 2A = 2\Sigma\sin A\cdot(1 + \Sigma\cos A)$.

86. Prove that $\tan(x-y) + \tan(y-z) + \tan(z-x) = \tan(x-y)\cdot\tan(y-z)\cdot\tan(z-x).$

87. If $\alpha + \beta + \gamma + \delta = 2\pi$ then prove that

$$\Sigma\tan\alpha = \tan\alpha\cdot\tan\beta\cdot\tan\gamma\cdot\tan\delta\cdot\Sigma\cot\alpha.$$

88. If $\cos(\beta-\gamma)\cdot\cos(\gamma-\alpha)\cdot\cos(\alpha-\beta) + 1 = 0$ then prove that at least one of the angles $\beta - \gamma$, $\gamma - \alpha$ and $\alpha - \beta$ is an odd multiple of π.

89. If $x + y + z = xyz$, prove that

$$x(1-y^2)(1-z^2)+y(1-z^2)(1-x^2) + z(1-x^2)(1-y^2) = 4xyz.$$

90. If $x+y+z=xyz$, prove that

$$\frac{3x-x^3}{1-3x^2}+\frac{3y-y^3}{1-3y^2}+\frac{3z-z^3}{1-3z^2} = \frac{3x-x^3}{1-3x^2}\cdot\frac{3y-y^3}{1-3y^2}\cdot\frac{3z-z^3}{1-3z^2}.$$

91. If $A+B+C=\pi$ then show that

$$\cot\theta = \cot A + \cot B + \cot C$$
$$\Leftrightarrow \quad \sin(A-\theta)\cdot\sin(B-\theta)\cdot\sin(C-\theta)=\sin^3\theta.$$

92. If $\theta=\frac{\pi}{7}$, prove that $\cos\theta\cdot\cos 2\theta\cdot\cos 3\theta=\frac{1}{8}$.

93. If $\alpha=\frac{2\pi}{7}$, prove that

$$\tan\alpha\cdot\tan 2\alpha+\tan 2\alpha\cdot\tan 4\alpha+\tan 4\alpha\cdot\tan\alpha=-7.$$

94. If $\alpha=\frac{2\pi}{7}$, prove that $\sec\alpha+\sec 2\alpha+\sec 4\alpha=-4$.

95. Prove that $\sin\frac{\pi}{14}\cdot\sin\frac{3\pi}{14}\cdot\sin\frac{5\pi}{14}=\frac{1}{8}$.

96. Show that $\sin\frac{2\pi}{7}+\sin\frac{4\pi}{7}+\sin\frac{8\pi}{7}=\frac{\sqrt{7}}{2}$.

Objective Questions

Fill in the blanks.

97. If $u=\sec\theta+\tan\theta$ then $\tan\theta$ in terms of u is ______ .

98. If $\sin x+\cos x=m(\neq\pm 1)$ and $\sec x+\operatorname{cosec} x=n$ then n in terms of m is ______ .

99. If $\alpha>0$ such that $\tan(x+\alpha)=\tan x$ for all $x\in R$ then the least value of α is ______ .

100. If $\theta=\frac{5\pi}{9}$, the sign of the expression $\cos\theta+\sin\theta$ is ______ .

101. If $5\cos 2t+12\sin 2t=\lambda\sin(2t-\tan^{-1}\mu)$ where λ,μ are constants then $\lambda=$ ______ and $\mu=$ ______ .

102. If $\cos A=\frac{3}{4}$ then the value of $\sin\frac{A}{2}\cdot\sin\frac{5A}{2}$ is ______ .

103. If $\sin(\theta+3\alpha)=3\sin(\alpha-\theta)$ then $\tan\theta$ in terms of $\tan\alpha$ is ______ .

104. If $0<\beta<\alpha<\frac{\pi}{4}$ and

$$\cos(\alpha+\beta)=\frac{4}{5},\ \sin(\alpha-\beta)=\frac{5}{13}$$

then $\tan 2\alpha=$ ______ .

105. If $4n\alpha=\pi$ then the numerical value of $\tan\alpha\cdot\tan 2\alpha\cdot\tan 3\alpha.\ldots\tan(2n-1)\alpha=$ ______ .

106. Let $\sin^3 x\cdot\sin 3x=\sum_{m=0}^{n} c_m\cos mx$ be an identity in x where c_1, c_2, etc., are constants, $c_n\neq 0$. Then $n=$ ______ and $c_5=$ ______ .

107. If $\frac{\cos(\alpha_1-\alpha_2)}{\cos(\alpha_1+\alpha_2)}+\frac{\cos(\alpha_3+\alpha_4)}{\cos(\alpha_3-\alpha_4)}=0$ then $\tan\alpha_1\cdot\tan\alpha_2\cdot\tan\alpha_3\cdot\tan\alpha_4=$ ______ .

108. If $\sin x+\sin y=a$, $\cos x-\cos y=b$ then the value of $\tan\frac{y-x}{2}$ is ______ in terms of a and b.

109. The numerical value of

$$\sin^2 5°+\sin^2 10°+\sin^2 15°+\ldots+\sin^2 90°= \text{______}.$$

110. The numerical value of

$$\cos\frac{2\pi}{7}+\cos\frac{4\pi}{7}+\cos\frac{6\pi}{7} \text{ is ______}.$$

111. The numerical value of $\sin\frac{\pi}{18}\cdot\sin\frac{5\pi}{18}\cdot\sin\frac{7\pi}{18}$ is ______ .

112. The numerical value of

$\tan 20°.\ \tan 40°.\ \tan 80°=$ ______ .

113. The numerical value of

$$\cos\frac{\pi}{15}\cdot\cos\frac{2\pi}{15}\cdot\cos\frac{4\pi}{15}\cos\frac{7\pi}{15}= \text{______}.$$

114. The numerical value of

$\sin 6°\cdot\sin 42°\cdot\sin 66°\cdot\sin 78°=$ ______ .

115. The numerical value of

$$\cos\frac{\pi}{11}+\cos\frac{3\pi}{11}+\cos\frac{5\pi}{11}+\cos\frac{7\pi}{11}+\cos\frac{9\pi}{11}= \text{______}.$$

116. The numerical value of (where $i=\sqrt{-1}$)

$$\left(\cos\frac{\pi}{10}+i\sin\frac{\pi}{10}\right)\left(\cos\frac{2\pi}{10}+i\sin\frac{2\pi}{10}\right)\times\left(\cos\frac{3\pi}{10}+i\sin\frac{3\pi}{10}\right)\left(\cos\frac{4\pi}{10}+i\sin\frac{4\pi}{10}\right)= \text{______}.$$

117. The numerical value of $\cot^2 36°\cdot\cot^2 72°=$ ______ .

118. The numerical value of

$\sin 12°\cdot\sin 48°\cdot\sin 54°=$ ______ .

119. If $\sin 54°=\frac{1}{2}+\sin\theta,\ 0<\theta<\frac{\pi}{2}$ then $\theta=$ ______ .

120. The numerical value of

$\tan 225°\cdot\cot 405°+\tan 765°\cdot\cot 675°=$ ______ .

121. The numerical value of

$$\cos\frac{\pi}{7}+\cos\frac{3\pi}{7}+\cos\frac{5\pi}{7}= \text{______}$$

122. If $\sin 3A$ is written as the product of sines of three angles then $\sin 3A=$ ______ .

123. If $\frac{1+\sin 2x}{1-\sin 2x}=\tan^2(\alpha+x)$ for all x then the numerically smallest value of α = ______ .

124. If $\cos(x-y)$, $\cos x$ and $\cos(x+y)$ are in HP then $\cos x\cdot\sec\left(\frac{y}{2}\right)$ = ______ .

125. If A, B, C are in AP and $B=\frac{\pi}{4}$ then $\tan A\cdot\tan B\cdot\tan C$ = ______ .

126. If $p\in(0,\pi)$ then the set of values of p for which $\sin p\cdot\cos^3 p>\sin^3 p\cdot\cos p$ holds, is ______ .

Choose the correct option(s).

127. $\sec^2\theta=\frac{4xy}{(x+y)^2}$ is true if and only if

(a) $x+y\neq 0$ (b) $x=y,\ x\neq 0$
(c) $x=y$ (d) $x\neq 0,\ y\neq 0$

128. If $\cos\theta+\sec\theta=2$ then $\cos^n\theta+\sec^n\theta$ is equal to

(a) 2^n (b) 2
(c) $(-2)^n$ (d) none of these

129. If $\left|5\sin\theta+3\sin\left(\theta-\frac{\pi}{3}\right)\right|\leq a$ for all θ then the least value of a is

(a) $\sqrt{34}$ (b) 5 (c) 7 (d) $2\sqrt{2}$ (e) 3

130. If $\alpha=\theta_1+\theta_2$ and $x=\theta_1-\theta_2$ and $\tan\theta_1=\lambda\tan\theta_2$ then $\sin x:\sin\alpha$ is equal to

(a) $\frac{\lambda+1}{\lambda-1}$ (b) $\frac{\lambda-1}{\lambda+1}$
(c) $\frac{1}{\lambda}$ (d) $\frac{\lambda}{\lambda+1}$

131. If ABCD is a cyclic quadrilateral then $\cos A+\cos B+\cos C+\cos D$ is equal to

(a) 1 (b) –1
(c) 0 (d) none of these

132. $\tan 40°+2\tan 10°$ is equal to

(a) $\tan 60°$ (b) $\tan 50°$
(c) $\cot 40°$ (d) none of these

133. If $\frac{\cos(\alpha+\gamma)}{\cos(\alpha-\gamma)}=\cos 2\beta$ then $\tan\alpha$, $\tan\beta$ and $\tan\gamma$ are in

(a) AP (b) GP
(c) HP (d) none of these

134. The value of $\sqrt{3}$ cosec $20°-\sec 20°$ is equal to

(a) 2 (b) 4
(c) $2\cdot\frac{\sin 20°}{\sin 40°}$ (d) $4\cdot\frac{\sin 20°}{\sin 40°}$

135. The number of real values of x for which $\sin(e^x)=5^x+5^{-x}$ is

(a) 0 (b) 1 (c) 2 (d) infinite

136. If $\cot y=\frac{\sin x-\sin z}{\cos z-\cos x}$ then which of the following is possible?

(a) x, y, z are in AP (b) $-x, y, z$ are in AP
(c) x, y, z are in HP (d) none of these

137. If $A+B=\frac{\pi}{4}$, $(\cot A-1)(\cot B-1)$ is equal to

(a) 0 (b) –2
(c) 2 (d) none of these

138. Which of the following is equal to

$$\sum_{k=1}^{6}\left(\sin\frac{2\pi k}{7}-i\cos\frac{2\pi k}{7}\right),\ \text{where } i=\sqrt{-1}?$$

(a) i (b) $-i$
(c) 0 (d) none of these

139. If α,β are two values of θ obtained from the equation $a\cos\theta+b\sin\theta=c$ then the value of $\tan\frac{\alpha+\beta}{2}$ is

(a) $\frac{a}{b}$ (b) $\frac{b}{a}$ (c) $\frac{c}{a}$ (d) $\frac{c}{b}$

140. The value of $\cos^2 x+\cos^2\left(\frac{\pi}{3}+x\right)-\cos x\cdot\cos\left(\frac{\pi}{3}+x\right)$ is

(a) $\cos 2x$ (b) $\sin^2 x$
(c) $\frac{3}{4}$ (d) none of these

141. The equation $(\cos p-1)x^2+\cos p\cdot x+\sin p=0$ where x is a variable, has real roots. Then the interval of possible values of p is

(a) $(0, 2\pi)$ (b) $(-\pi, 0)$
(c) $\left[-\frac{\pi}{2},\frac{\pi}{2}\right]$ (d) $(0,\pi)$

142. If $\tan\theta+\tan\left(\frac{\pi}{3}+\theta\right)+\tan\left(\frac{2\pi}{3}+\theta\right)=k\tan 3\theta$ then k is equal to

(a) 1 (b) $\frac{1}{3}$
(c) 3 (d) none of these

143. The value of $\sin\sqrt{x^2-\frac{\pi^2}{36}}$ lies in the interval

(a) $[-1, 1]$ (b) $\left[-\frac{1}{2},\frac{1}{2}\right]$
(c) $[-1, 0]$ (d) $[0, 1]$

144. If α, β are two values lying between 0 and 2π for which $\tan\theta = k$ then the value of $\tan\frac{\alpha}{2}\cdot\tan\frac{\beta}{2}$ is

(a) -1 (b) 1 (c) $-\frac{2}{k}$ (d) none of these

145. For all $x \in R_1$,

$$\frac{24\sin^3 x\cdot\cos x - 6\sin 2x + \sin 4x(1+4\cos^2 2x)}{2\cos 2x - 2\sin^2 2x + 4\sin^2 x - 1}$$

is equal to

(a) $\cos 4x$ (b) $2 - \cos 2x$

(c) $4\sin 2x$ (d) $2\sin 4x$

State whether the statements are true or false.

146. $\sin\theta = x + \frac{1}{x}$ is possible for some real values of x.

147. $(\sec x + \tan x + 1)(\sec x - \tan x - 1) - 2\tan x = 0$ is an identity.

148. If $a_1 + a_2\cos 2x + a_3\sin^2 x = 0$ be true for all x then a_1, a_2 and a_3 can be uniquely determined.

149. The maximum value of $5\cos\theta + 12\sin\theta$ is 17.

150. If $\cos x + \cos y - \cos(x+y) = \frac{3}{2}$ then $x = y$ is possible.

151. If $x = r\sin\theta\cdot\cos\phi$, $y = r\sin\theta\cdot\sin\phi$ and $z = r\cos\theta$ then $x^2 + y^2 + z^2 = r^2$.

152. The minimum value of $3\cos x + 4\sin x + 5$ is 0.

153. $\tan 22\frac{1°}{2}$ is a root of the equation $\frac{1+x^2}{1-x^2} = \sqrt{2}$.

154. If $\tan\theta = 2$ is satisfied by α, β then the value of $\tan\frac{\alpha}{2} + \tan\frac{\beta}{2} = 1$.

155. $\sin 1° < \sin 1$.

156. If $\sin\alpha = \frac{-3}{5}$, $\pi < \alpha < \frac{3\pi}{2}$ then $\cos\frac{\alpha}{2} = \frac{1}{\sqrt{10}}$.

Answers

1. $(m^2 n)^{2/3} + (mn^2)^{2/3} = 1 = m^2 n^2(m^2 + n^2 + 3)$

3. $\frac{4}{(p^2+q^2-2)^2} - \frac{q^2}{(1+p)^2} = 1$

10. $(a^2+b^2)(1-c) = 2b(1+c)$

11. $(\lambda^2 - 1)^3 = 27\lambda^2\cos^2\alpha\cdot\sin^2\alpha$

12. $(a^2+b^2-c^2)\sqrt{2a-b} = 2c\sqrt{b}(a-b)$

13. $[-1, 3]$ **15.** $\left[-\frac{23}{2}, \frac{27}{2}\right]$

16. $1, \frac{1}{4}$ respectively **17.** $\left[0, \frac{3}{4}\right]$ **21.** 13

35. $-\tan\left(A - \frac{\pi}{3}\right)\tan A\tan\left(\frac{\pi}{3} + A\right)$

41. $-\frac{5}{12}$ and $(\pm 4\sqrt{3} - 7)$ **43.** $\frac{(p^2+q^2-2)(q^2-p^2)}{p^2+q^2}$

46. $\frac{\pi}{4}, \frac{\pi}{6}$ respectively **60.** $\frac{1}{3}$ **61.** 12

63. all real values except those between $\frac{1}{3}$ and 3

67. $\frac{1}{2^{n-1}}\cot\frac{x}{2^{n-1}} - 2\cot 2x$

70. $\sec\frac{\alpha}{2}\cdot\sin\frac{n(\pi+\alpha)}{2}\cdot\sin\frac{(n+1)\alpha + (n-1)\pi}{2}$

71. $-\frac{9}{8}$ **72.** $\operatorname{cosec}\frac{\alpha}{2}\cdot\sin 10\alpha\cdot\sqrt{1+\sin 21\alpha}$

97. $\frac{u^2-1}{2u}$ **98.** $\frac{2m}{m^2-1}$ **99.** π **100.** positive

101. $\lambda = 13$, $\mu = \frac{-5}{12}$ **102.** $\frac{11}{32}$ **103.** $\tan^3\alpha$

104. $\frac{56}{33}$ **105.** 1 **106.** $n = 6$, $c_5 = 0$

107. -1 **108.** $\frac{b}{a}$ **109.** $\frac{19}{2}$ **110.** $\frac{-1}{2}$

111. $\frac{1}{8}$ **112.** $\sqrt{3}$ **113.** $\frac{1}{16}$ **114.** $\frac{1}{16}$

115. $\frac{1}{2}$ **116.** -1 **117.** $\frac{1}{5}$ **118.** $\frac{1}{8}$

119. $\frac{\pi}{10}$ **120.** 0 **121.** $\frac{1}{2}$

122. $4\sin A\cdot\sin\left(\frac{\pi}{3} + A\right)\sin\left(\frac{\pi}{3} - A\right)$ **123.** $\frac{\pi}{4}$

124. $\pm\sqrt{2}$ **125.** 1 **126.** $\left(0, \frac{\pi}{4}\right)\cup\left(\frac{\pi}{2}, \frac{3\pi}{4}\right)$

127. (b)	**128.** (b)	**129.** (c)	**130.** (b)
131. (c)	**132.** (b), (c)	**133.** (b)	**134.** (b)
135. (a)	**136.** (a)	**137.** (c)	**138.** (a)
139. (b)	**140.** (c)	**141.** (d)	**142.** (c)
143. (a)	**144.** (a)	**145.** (d)	**146.** false
147. false	**148.** false	**149.** false	**150.** true
151. true	**152.** true	**153.** true	**154.** false
155. true	**156.** false		

Chapter Test

Time: 120 minutes

1. In the following incomplete sentences fill in the blanks such that the resulting sentences become true.

(a) If $\sqrt{2}\cos x = \cos y + \cos^3 y$ and $\sqrt{2}\sin x = \sin y - \sin^3 y$ then $\sin(x - y) =$ ______ .

(b) If $\dfrac{\sin x}{\sin y} = \dfrac{1}{2}$ and $\dfrac{\cos x}{\cos y} = \dfrac{3}{2}$ then $\tan x =$ ______ and $\tan y =$ ______ where x and y are acute angles.

(c) If $\sin 17° \cdot \sin 43° \cdot \sin 77° = \lambda \sin 51°$ then λ is equal to ______ .

(d) The minimum value of $27^{\cos x} + 81^{\sin x}$ is ______ .

(e) Let $\cos 3x = 4\cos x \cdot \cos(x + \beta) \cdot \cos(x - \beta)$, $0 < \beta < \dfrac{\pi}{2}$. Then the value of $\beta =$ ______ .

(f) If $\tan \alpha$ is an integral solution of the inequation $4x^2 - 16x + 15 < 0$ and $\cos \beta$ is the slope of the bisector of the angle in the first quadrant between the x and y axes then the value of $\sin(\alpha + \beta) \cdot \sin(\alpha - \beta) =$ ______ .

2. In the following each question has multiple answers. Choose the correct answer(s).

(a) If $u_n = 2\cos n\theta$ then $u_1 u_n - u_{n-1}$ is equal to

A. u_{n+2} B. u_{n+1} C. $u_2 u_{n+1}$ D. none of these

(b) If $\dfrac{x}{\cos\theta} = \dfrac{y}{\cos\left(\theta - \dfrac{2\pi}{3}\right)} = \dfrac{z}{\cos\left(\theta + \dfrac{2\pi}{3}\right)}$ then $x + y + z$ is equal to

A. $3\cos\theta$ B. $\cos 3\theta$ C. 0 D. 1

(c) If $\cos\alpha + \cos\beta + \cos\gamma = 0$ then $\cos 3\alpha + \cos 3\beta + \cos 3\gamma = \lambda \cos\alpha \cdot \cos\beta \cdot \cos\gamma$ where λ is equal to

A. 12 B. 4 C. 1 D. none of these

(d) The numerical value of $4\sin 50° - \sqrt{3}\tan 50°$ is equal to

A. –1 B. 1 C. 0 D. none of these

(e) A factor of $\cos 4\theta - \cos 4\phi$ is

A. $\cos\theta + \cos\phi$ B. $\cos\theta + \sin\phi$ C. $\cos\theta - \cos\phi$ D. $\cos\theta - \sin\phi$

3. The chord AB of a circle subtends an angle of $\dfrac{2\pi}{3}$ at a point C on the circumference. Prove that the value of $\cos^2 A + \cos^2 B - \cos A \cdot \cos B$ is invariant for all such triangles ABC.

4. If $\{\cos A \cdot \sin B - \sin(A + B) \cdot \cos 2B\}^2 = 4\cos A \sin B \cdot \sin(A + B)$ then prove that the equation $x^2 - 2(2 + \cos 2B)\,x + \cos^2 2B = 0$ has a root $\dfrac{\tan B}{\tan A + \tan B}$.

5. If $\sin\alpha + \cos\alpha = \lambda$ then find $|\sin^6\alpha - \cos^6\alpha|$ in terms of λ.

6. Let $f(\theta) = \dfrac{\cos\theta + \cot\theta}{\sin\theta + \tan\theta}$ and the set $D = R - \left\{\theta \mid \theta \neq \dfrac{n\pi}{2}, n \in \mathbb{Z}\right\}$.

Show that $f(\theta) > 0$ for all $\theta \in D$.

7. Prove that $\sin\dfrac{\pi}{14}$ is a root of the equation $8x^3 - 4x^2 - 4x + 1 = 0$.

8. Find the sum of the products of sines of the angles $\alpha, 2\alpha, 3\alpha, \ldots, n\alpha$ taking two at a time.

9. If $A = \frac{\pi}{2^{n+1}}$, $B = \frac{\pi}{2^{n+2}}$ then prove that

$$(\cos A + \cos B)(\cos 2A + \cos 2B)(\cos 2^2 A + \cos 2^2 B) \ldots (\cos 2^n A + \cos 2^n B)$$

$$= \frac{1}{2^{n+1}\left(\cos \frac{\pi}{2^{n+2}} - \cos \frac{\pi}{2^{n+1}}\right)}.$$

10. In a convex polygon the interior angles are in AP, the smallest angle being 120° and the common difference 2°. Find the sum of the sines of the interior angles of the polygon if $\sin 1° = \lambda$..

11. AB and CD are two parallel chords of a circle of radius 6 cm lying on the same side of the centre. If the chords subtend angles of 72° and 144° at the centre, find the distance between the chords.

□

Answers

1. (a) $\pm \frac{1}{3}$ (b) $\frac{1}{3} \cdot \sqrt{\frac{5}{3}}, \sqrt{\frac{5}{3}}$ (c) $\frac{1}{4}$ (d) $\frac{2}{9\sqrt{3}}$ (e) $\frac{\pi}{3}$ (f) $\frac{4}{5}$

2. (a) B (b) C (c) A (d) B (e) A, B, C, D

5. $\frac{1}{4} \cdot (1 + \lambda^2)(3 - \lambda^2)\sqrt{2 - \lambda^2} \cdot |\lambda|$

8. $\frac{1}{4} \cdot \frac{\sin \frac{n\alpha}{2}}{\sin^2 \frac{\alpha}{2}} \left(\sin \frac{n\alpha}{2} - \frac{\sin \frac{n-1}{2}\alpha}{\cos \frac{\alpha}{2}} \cos(n+1)\alpha \right) - \frac{n}{4}$

10. $\frac{\sqrt{3}\lambda + \sqrt{1 - \lambda^2}}{4\lambda}$ **11.** 3 cm

□

2. Solution of Equations

(Involving Circular Functions)

Recap of Facts and Formulae

1. General values of unknown angles

- If $\sin x = k = \sin \alpha$, $(|k| \le 1)$ then $x = n\pi + (-1)^n \alpha$ where $n \in \mathbb{Z}$, $\mathbb{Z}$ being the set of integers.
- If $\cos x = k = \cos \alpha$, $(|k| \le 1)$ then $x = 2n\pi \pm \alpha$ where $n \in \mathbb{Z}$.
- If $\tan x = k = \tan \alpha$, $(k \in R)$ then $x = n\pi + \alpha$ where $n \in \mathbb{Z}$.

2. Condition for existence of real solutions

- The equation $\sin \theta = k$ can have real solutions for θ if $-1 \le k \le 1$.
- The equation $\cos \theta = k$ can have real solutions for θ if $-1 \le k \le 1$.
- The equation $\tan \theta = k$ can have real solutions for all real k.
- The equation $\cot \theta = k$ can have real solutions for all real k.
- The equation $\sec \theta = k$ can have real solutions for θ if $k \le -1$ or $k \ge 1$.
- The equation $\operatorname{cosec} \theta = k$ can have real solutions for θ if $k \le -1$ or $k \ge 1$.

Selected Solved Examples

1. Solve: $2\sin^3 x = \cos x$.

Here $2\sin^2 x = \dfrac{\cos x}{\sin x}$ $\{\because \sin x \ne 0$ because $\sin x = 0$ does not satisfy the given equation$\}$

or $\dfrac{2}{\operatorname{cosec}^2 x} = \cot x$

or $2 = \cot x \cdot (1 + \cot^2 x)$

or $\cot^3 x + \cot x - 2 = 0$

By trial, $\cot x = 1$ satisfies the equation.

$\therefore \quad (\cot x - 1)(\cot^2 x + \cot x + 2) = 0$

$\Rightarrow \quad \cot x = 1$ or $\cot^2 x + \cot x + 2 = 0$

When $\cot x = 1$, $\tan x = 1$

$\therefore \quad x = n\pi + \dfrac{\pi}{4}$, $n \in \mathbb{Z}$.

When $\cot^2 x + \cot x + 2 = 0$,

$\cot x$ has no real value because $D = 1 - 8 < 0$.

$\therefore$ solutions are $x = n\pi + \dfrac{\pi}{4}$, $n \in \mathbb{Z}$.

2. Solve: $\tan\left(\dfrac{\pi}{4} + x\right) + \tan\left(\dfrac{\pi}{4} - x\right) = 2$.

Here $\dfrac{\tan \frac{\pi}{4} + \tan x}{1 - \tan \frac{\pi}{4} \cdot \tan x} + \dfrac{\tan \frac{\pi}{4} - \tan x}{1 + \tan \frac{\pi}{4} \cdot \tan x} = 2$

or $\dfrac{1 + \tan x}{1 - \tan x} + \dfrac{1 - \tan x}{1 + \tan x} = 2$

or $\dfrac{(1 + \tan x)^2 + (1 - \tan x)^2}{1 - \tan^2 x} = 2$

or $2(1 + \tan^2 x) = 2(1 - \tan^2 x)$

or $1 + \tan^2 x = 1 - \tan^2 x$

$\therefore \quad \tan x = 0; \quad \therefore \quad x = n\pi, n \in \mathbb{Z}$.

3. Find the condition for $a\cos x + b\sin x = c$ to have a solution. Is the condition satisfied for $\sin x + \sqrt{3}\cos x = \sqrt{2}$? If so, solve the equation.

We know, $a\cos x + b\sin x$

$$= \sqrt{a^2+b^2}\left(\frac{a}{\sqrt{a^2+b^2}}\cos x + \frac{b}{\sqrt{a^2+b^2}}\sin x\right)$$

$$= \sqrt{a^2+b^2}\cos(x-\alpha), \text{ where } \cos\alpha = \frac{a}{\sqrt{a^2+b^2}}.$$

As $-1 \le \cos(x-\alpha) \le 1$ for all x,

$$-\sqrt{a^2+b^2} \le a\cos x + b\sin x \le \sqrt{a^2+b^2}$$

$\therefore$ $a\cos x + b\sin x = c$ will have solution if

$$-\sqrt{a^2+b^2} \le c \le \sqrt{a^2+b^2}, \text{ i.e., } |c| \le \sqrt{a^2+b^2}.$$

For the equation $\sin x + \sqrt{3}\cos x = \sqrt{2}$,

$$\sqrt{a^2+b^2} = \sqrt{1^2+(\sqrt{3})^2} = 2 \text{ and } |c| = |\sqrt{2}| < 2.$$

So, the equation has solutions.

Now, $\sin x + \sqrt{3}\cos x = \sqrt{2}$

or $\sqrt{1^2+(\sqrt{3})^2}\left\{\frac{1}{2}\sin x + \frac{\sqrt{3}}{2}\cos x\right\} = \sqrt{2}$

or $\cos\left(x - \frac{\pi}{6}\right) = \frac{\sqrt{2}}{2} = \frac{1}{\sqrt{2}} = \cos\frac{\pi}{4}$

$\therefore$ $x - \frac{\pi}{6} = 2n\pi \pm \frac{\pi}{4}$

$\therefore$ $x = 2n\pi + \frac{\pi}{6} \pm \frac{\pi}{4} = 2n\pi + \frac{5\pi}{12},\ 2n\pi - \frac{\pi}{12},\ n \in \mathbb{Z}.$

4. Solve: $\sin x + \cos x - 2\sqrt{2}\sin x \cdot \cos x = 0$.

Here $\sin x + \cos x = 2\sqrt{2}\sin x \cdot \cos x = \sqrt{2}\sin 2x$ ... (1)

or $\sqrt{2}\left(\frac{1}{\sqrt{2}}\sin x + \frac{1}{\sqrt{2}}\cos x\right) = \sqrt{2}\sin 2x$

or $\sin\left(x + \frac{\pi}{4}\right) = \sin 2x$

$\Rightarrow$ $2x = n\pi + (-1)^n\left(x + \frac{\pi}{4}\right)$

Taking n even, $n = 2m$, $m \in \mathbb{Z}$. Then $2x = 2m\pi + x + \frac{\pi}{4}$

$\therefore$ $x = 2m\pi + \frac{\pi}{4}$ where $m \in \mathbb{Z}$.

Taking n odd, $n = 2m+1$, $m \in \mathbb{Z}$.

Then, $2x = (2m+1)\pi - \left(x + \frac{\pi}{4}\right)$

$\therefore$ $3x = (2m+1)\pi - \frac{\pi}{4}$; $\therefore$ $x = \frac{2m+1}{3}\pi - \frac{\pi}{12}.$

Thus, $x = \left(2m + \frac{1}{4}\right)\pi,\ \frac{1}{3}\left(2m + \frac{3}{4}\right)\pi$ where $m \in \mathbb{Z}$.

TIP Squaring of both sides of an equation should be avoided in solving an equation because in that case the solutions found will correspond to the given equation as well as another equation. In the above equation, squaring (1), we get $1 + \sin 2x = 2\sin^2 2x$. The solutions of it will include the solutions of the two equations $\sin x + \cos x \pm 2\sqrt{2}\sin x \cdot \cos x = 0$. Hence, the solution corresponding to the equation $\sin x + \cos x + 2\sqrt{2}\sin x \cdot \cos x = 0$ must be excluded. It is not always easy to detect such solutions.

5. Solve: $\sec\theta - \operatorname{cosec}\theta = \frac{4}{3}$.

Here, $\sin\theta - \cos\theta = \frac{4}{3}\sin\theta\cos\theta$

or $\frac{3}{2}(\sin\theta - \cos\theta) = \sin 2\theta$

$$= 1 - (1 - \sin 2\theta) = 1 - (\sin\theta - \cos\theta)^2$$

or $2(\sin\theta - \cos\theta)^2 + 3(\sin\theta - \cos\theta) - 2 = 0.$

On factorization,

$(\sin\theta - \cos\theta + 2)\{2(\sin\theta - \cos\theta) - 1\} = 0$

$\therefore$ either $\sin\theta - \cos\theta + 2 = 0$ ... (1)

or $2(\sin\theta - \cos\theta) - 1 = 0$ ... (2)

(1) $\Rightarrow$ $\sin\theta - \cos\theta + 2 = 0$

or $\cos\theta - \sin\theta = 2.$

This is impossible because the maximum value of $\cos\theta - \sin\theta$ is $\sqrt{2}$.

(2) $\Rightarrow$ $2(\sin\theta - \cos\theta) - 1 = 0$

or $\sin\theta - \cos\theta = \frac{1}{2}$

or $\sqrt{2}\left(\frac{1}{\sqrt{2}}\sin\theta - \frac{1}{\sqrt{2}}\cos\theta\right) = \frac{1}{2}$

or $\sin\left(\theta - \frac{\pi}{4}\right) = \frac{1}{2\sqrt{2}}$

$\therefore$ $\theta - \frac{\pi}{4} = n\pi + (-1)^n \cdot \sin^{-1}\frac{1}{2\sqrt{2}}$

$\therefore$ $\theta = n\pi + \frac{\pi}{4} + (-1)^n \cdot \sin^{-1}\frac{1}{2\sqrt{2}},\ n \in \mathbb{Z}.$

6. Solve: $\cos\theta \cdot \cos 2\theta \cdot \cos 3\theta = \frac{1}{4}$, $0 \le \theta \le \pi$.

Here, $(2\cos\theta \cdot \cos 3\theta)\cos 2\theta = \frac{1}{2}$

or $(\cos 4\theta + \cos 2\theta)\cos 2\theta = \frac{1}{2}$

or $(2\cos^2 2\theta - 1 + \cos 2\theta)\cos 2\theta = \frac{1}{2}$

or $4\cos^3 2\theta + 2\cos^2 2\theta - 2\cos 2\theta - 1 = 0$

or $2\cos 2\theta(2\cos^2 2\theta - 1) + (2\cos^2 2\theta - 1) = 0$

or $(2\cos^2 2\theta - 1)(2\cos 2\theta + 1) = 0$

or $\cos 4\theta \cdot (2\cos 2\theta + 1) = 0$

$\therefore$ $\cos 4\theta = 0$ ... (1)

or $2\cos 2\theta + 1 = 0$ ... (2)

(1) $\Rightarrow$ $4\theta = 2n\pi \pm \frac{\pi}{2}$; $\therefore$ $\theta = \frac{n\pi}{2} \pm \frac{\pi}{8}$

$\therefore$ $\theta = \pm\frac{\pi}{8}, \frac{\pi}{2} \pm \frac{\pi}{8}, \pi \pm \frac{\pi}{8}$, etc.

(2) $\Rightarrow$ $\cos 2\theta = -\frac{1}{2}$

$\therefore$ $2\theta = 2n\pi \pm \frac{2\pi}{3}$

$\therefore$ $\theta = n\pi \pm \frac{\pi}{3}$

$\therefore$ $\theta = \pm\frac{\pi}{3}, \pi \pm \frac{\pi}{3}$, etc.

But $0 \le \theta \le \pi$; so, $\theta = \frac{\pi}{8}, \frac{\pi}{2} \pm \frac{\pi}{8}, \pi - \frac{\pi}{8}, \frac{\pi}{3}, \pi - \frac{\pi}{3}$

$\therefore$ $\theta = \frac{\pi}{8}, \frac{\pi}{3}, \frac{3\pi}{8}, \frac{2\pi}{3}, \frac{5\pi}{8}, \frac{7\pi}{8}$.

7. Find all values of α for which the equation $\sin^4 x + \cos^4 x + \sin 2x + \alpha = 0$ is valid. Also find the general solution of the equation.

Here, $(\sin^2 x + \cos^2 x)^2 - 2\sin^2 x \cdot \cos^2 x + \sin 2x + \alpha = 0$

or $1 - \frac{1}{2}\sin^2 2x + \sin 2x + \alpha = 0$

or $\sin^2 2x - 2\sin 2x - 2(1 + \alpha) = 0$

$\therefore$ $\sin 2x = \frac{2 \pm \sqrt{4 + 8(1+\alpha)}}{2} = 1 \pm \sqrt{2\alpha + 3}$ ... (1)

But $\sin 2x$ is real; so $2\alpha + 3 \ge 0$, i.e., $\alpha \ge \frac{-3}{2}$.

Also, $-1 \le \sin 2x \le 1$

$\therefore$ $-1 \le 1 \pm \sqrt{2\alpha + 3} \le 1$.

$1 - \sqrt{2\alpha + 3} \le 1$ is true for all $\alpha \ge \frac{-3}{2}$

$1 + \sqrt{2\alpha + 3} \ge -1$ is true for all $\alpha \ge \frac{-3}{2}$.

Hence, we get $-1 \le 1 - \sqrt{2\alpha + 3}$ ... (2)

and $1 + \sqrt{2\alpha + 3} \le 1$ ... (3)

(2) $\Rightarrow$ $\sqrt{2\alpha + 3} \le 2$

$\Rightarrow$ $2\alpha + 3 \le 4$

$\Rightarrow$ $\alpha \le \frac{1}{2}$.

(3) $\Rightarrow$ $\sqrt{2\alpha + 3} \le 0$

$\Rightarrow$ $2\alpha + 3 = 0$, because $\sqrt{2\alpha + 3} \ge 0$

$\Rightarrow$ $\alpha = -\frac{3}{2}$.

Thus, $-\frac{3}{2} \le \alpha \le \frac{1}{2}$; so $\alpha \in \left[-\frac{3}{2}, \frac{1}{2}\right]$

$\therefore$ from (1), $\sin 2x = 1 - \sqrt{2\alpha + 3}$ where $\alpha \in \left[-\frac{3}{2}, \frac{1}{2}\right]$

$\therefore$ $2x = n\pi + (-1)^n\theta, n \in \mathbb{Z}$ and $\sin\theta = 1 - \sqrt{2\alpha + 3}$

$\therefore$ $x = \frac{n\pi}{2} + \frac{1}{2}(-1)^n \cdot \sin^{-1}\{1 - \sqrt{2\alpha + 3}\}$

where $n \in \mathbb{Z}$ and $\alpha \in \left[-\frac{3}{2}, \frac{1}{2}\right]$.

8. Determine the smallest positive value of x (in degrees) for which

$$\tan(x + 100°) = \tan(x + 50°) \cdot \tan x \cdot \tan(x - 50°)$$

Here, $\frac{\tan(x + 100°)}{\tan x} = \tan(x + 50°) \cdot \tan(x - 50°)$

{$\because$ $\tan x \ne 0$ for $x = 0$ does not satisfy the equation}

or $\frac{\sin(x + 100°) \cdot \cos x}{\cos(x + 100°)\sin x} = \frac{\sin(x + 50°)\sin(x - 50°)}{\cos(x + 50°)\cos(x - 50°)}$.

By componendo and dividendo,

$$\frac{\sin(x + 100°)\cos x + \cos(x + 100°)\sin x}{\sin(x + 100°)\cos x - \cos(x + 100°)\sin x}$$

$$= \frac{\sin(x + 50°) \cdot \sin(x - 50°) + \cos(x + 50°)\cos(x - 50°)}{\sin(x + 50°)\sin(x - 50°) - \cos(x + 50°)\cos(x - 50°)}$$

or $\frac{\sin(2x + 100°)}{\sin(x + 100° - x)} = \frac{\cos(x + 50° - x + 50°)}{-\cos(x + 50° + x - 50°)}$

or $\frac{\sin(2x + 100°)}{\sin 100°} = -\frac{\cos 100°}{\cos 2x}$

or $\sin(2x + 100°) \cdot \cos 2x + \sin 100° \cdot \cos 100° = 0$

or $2\sin(2x + 100°) \cdot \cos 2x + \sin 200° = 0$

or $\sin(4x + 100°) + \sin 100° + \sin 200° = 0$

or $\sin(4x + 100°) + 2 \cdot \sin 150° \cdot \cos 50° = 0$

or $\sin(4x + 100°) + \cos 50° = 0$

or $\sin(4x + 100°) = -\cos 50°$

$= \sin(50° - 90°) = \sin(-40°)$

$\therefore\quad 4x + 100° = n\pi + (-1)^n(-40°)$

$= n\pi + (-1)^{n+1} \cdot 40°$

$= 2m\pi - 40°, (2m+1)\pi + 40°$

$\therefore\quad x = \frac{m\pi}{2} - 35°, \frac{(2m+1)\pi}{4} - 15°$, where $m \in \mathbb{Z}$.

Putting integral values of m, the positive values of x are

55°, 145°, ... and 30°, 120°, ...

$\therefore$ required smallest positive value of θ is 30°.

9. If $32\tan^8\theta = 2\cos^2\alpha - 3\cos\alpha$ and $3\cos 2\theta = 1$ then find the general value of α.

$3\cos 2\theta = 1 \quad \Rightarrow \quad 3(2\cos^2\theta - 1) = 1$

$\Rightarrow\quad 6\cos^2\theta = 4; \quad \therefore\ \cos^2\theta = \frac{2}{3}$

$\therefore\quad \tan^2\theta = \frac{\sin^2\theta}{\cos^2\theta} = \frac{1-\cos^2\theta}{\frac{2}{3}} = \frac{1-\frac{2}{3}}{\frac{2}{3}} = \frac{1}{2}$

$\therefore\quad 32\tan^8\theta = 2\cos^2\alpha - 3\cos\alpha$

$\Rightarrow\quad 32 \cdot \left(\frac{1}{2}\right)^4 = 2\cos^2\alpha - 3\cos\alpha$

$\therefore\quad 2\cos^2\alpha - 3\cos\alpha - 2 = 0$

or $\quad (2\cos\alpha + 1)(\cos\alpha - 2) = 0$

$\therefore\quad 2\cos\alpha + 1 = 0$ or $\cos\alpha - 2 = 0$

$\therefore\quad \cos\alpha = -\frac{1}{2}$. or $\cos\alpha = 2$ (impossible)

$\therefore\quad \alpha = 2n\pi \pm \left(\pi - \frac{\pi}{3}\right) = 2n\pi \pm \frac{2\pi}{3},\ n \in \mathbb{Z}$.

10. Find the smallest positive number p for which the equation $\cos(p\sin x) = \sin(p\cos x)$ has a solution $x \in [0, 2\pi]$.

Here $\cos(p\sin x) = \cos\left(\frac{\pi}{2} - p\cos x\right)$

$\therefore\quad p\sin x = 2n\pi \pm \left(\frac{\pi}{2} - p\cos x\right)$

$\therefore\quad p(\sin x \pm \cos x) = 2n\pi \pm \frac{\pi}{2} = \frac{(4n \pm 1)\pi}{2}$

or $\quad \sin x \pm \cos x = \frac{(4n \pm 1)\pi}{2p}$

$\therefore\quad \sqrt{2}\sin\left(x \pm \frac{\pi}{4}\right) = \frac{(4n \pm 1)\pi}{2p}$

or $\quad \sin\left(x \pm \frac{\pi}{4}\right) = \frac{(4n \pm 1)\pi}{2\sqrt{2}p}$ ... (1)

As $|\sin\theta| \le 1$, (1) will have solution if

$$\left|\frac{4n \pm 1}{p}\right| \frac{\pi}{2\sqrt{2}} \le 1$$

or $\quad |p| \ge |4n \pm 1| \cdot \frac{\pi}{2\sqrt{2}}$ where $n \in \mathbb{Z}$.

$\therefore$ the smallest positive value of p is $\frac{\pi}{2\sqrt{2}}$, i.e., $\frac{\pi}{4}\sqrt{2}$

and the smallest positive integral value of p is 2.

11. Solve: $|\cos x| = \cos x - 2\sin x$.

If $\cos x \ge 0$ then the equation becomes

$\cos x = \cos x - 2\sin x$

or $\quad \sin x = 0; \quad \therefore \quad x = n\pi = 2m\pi, (2m+1)\pi;\ m \in \mathbb{Z}$.

But $\cos x \ge 0$; so $x = 2m\pi,\ m \in \mathbb{Z}$.

If $\cos x < 0$ then the equation becomes

$-\cos x = \cos x - 2\sin x$

or $\quad \sin x = \cos x$

or $\quad \tan x = 1;$

$\therefore\quad x = n\pi + \frac{\pi}{4} = 2m\pi + \frac{\pi}{4},\ (2m+1)\pi + \frac{\pi}{4};\ m \in \mathbb{Z}$.

But $\cos x < 0$; so $x = (2m+1)\pi + \frac{\pi}{4},\ m \in \mathbb{Z}$.

Thus $x = 2m\pi, (2m+1)\pi + \frac{\pi}{4}$ where $m \in \mathbb{Z}$.

12. Find k if the equation $2\cos x + \cos 2kx = 3$ has only one solution.

$|2\cos x + \cos 2kx| \le |2\cos x| + |\cos 2kx| \le 2 + 1 = 3$

$\therefore\quad 2\cos x + \cos 2kx = 3$ is possible only when both $2\cos x$ and $\cos 2kx$ have greatest values,

i.e., $\cos x = 1$ and $\cos 2kx = 1$.

Now, $\cos x = 1 \quad \Rightarrow \quad x = 2n\pi,\ n \in \mathbb{Z}$

$\therefore\quad x = 0,\ \pm 2\pi,\ \pm 4\pi, \ldots$ ... (1)

In order to have only one solution $\cos 2kx = 1$ is to be satisfied by only one value of (1).

Clearly, if k is irrational then $x = 0$ is the only solution because if k is rational, say $\frac{p}{q}$, then $x = 2mq\pi$ will be a solution for every $m \in \mathbb{Z}$.

Hence, the equation will have only one solution $x = 0$ if k is irrational. So, the set of values of k = set of irrational numnbers.

13. Obtain the general solution for x and y if

$$\sin x(3 - 4\sin^2 x)\cos 2y - 2\cos x \sin y \cos y(4\cos^2 x - 3) = \frac{1}{\sqrt{2}}$$

and $$(2\cos^2 x - 1)\cos y(4\cos^2 y - 3) - 2\sin x \sin y \cos x(3 - 4\sin^2 y) = 1.$$

Here $\sin 3x \cos 2y - \sin 2y \cos 3x = \dfrac{1}{\sqrt{2}}$

or $\sin(3x - 2y) = \dfrac{1}{\sqrt{2}}$ ... (1)

and $\cos 2x \cos 3y - \sin 2x \sin 3y = 1$

or $\cos(2x + 3y) = 1$ ... (2)

(1) $\Rightarrow$ $3x - 2y = n\pi + (-1)^n \cdot \dfrac{\pi}{4}$ ... (3)

(2) $\Rightarrow$ $2x + 3y = 2m\pi$ ... (4)

From $3 \times (3) + 2 \times (4)$, we get

$$13x = 3n\pi + (-1)^n \frac{3\pi}{4} + 4m\pi$$

$\therefore$ $x = (3n + 4m)\dfrac{\pi}{13} + (-1)^n \cdot \dfrac{3\pi}{52}$ where $m, n \in N$.

From $3 \times (4) - 2 \times (3)$, we get

$$13y = 6m\pi - 2n\pi - (-1)^n \frac{\pi}{2}$$

$\therefore$ $y = (3m - n)\dfrac{2\pi}{13} - (-1)^n \cdot \dfrac{\pi}{26}$ where $m, n \in N$.

14. Solve for θ and ϕ: $\tan\theta + \tan\phi = 4$
$\tan 3\theta + \tan 3\phi = 2$.

Here, $\tan\theta + \tan\phi = 4$

or $(\tan\theta - 2) + (\tan\phi - 2) = 0$.

Putting $\tan\theta = x + 2$

and $\tan\phi = y + 2$, we get $x + y = 0$ ... (1)

Again, $\tan 3\theta + \tan 3\phi = 2$

or $\dfrac{3\tan\theta - \tan^3\theta}{1 - 3\tan^2\theta} + \dfrac{3\tan\phi - \tan^3\phi}{1 - 3\tan^2\phi} = 2$

or $3\tan\theta - \tan^3\theta - 9\tan\theta \cdot \tan^2\phi + 3\tan^3\theta \cdot \tan^2\phi$
$+ 3\tan\phi - \tan^3\phi - 9\tan^2\theta \cdot \tan\phi + 3\tan^2\theta \cdot \tan^3\phi$
$= 2(1 - 3\tan^2\phi - 3\tan^2\theta + 9\tan^2\theta \cdot \tan^2\phi)$

or $3(\tan\theta + \tan\phi) - (\tan^3\theta + \tan^3\phi)$
$- 9\tan\theta \tan\phi(\tan\theta + \tan\phi)$
$+ 3\tan^2\theta \tan^2\phi(\tan\theta + \tan\phi)$
$= 2 - 6(\tan^2\theta + \tan^2\phi) + 18\tan^2\theta \cdot \tan^2\phi$

or $3 \cdot 4 - \{(\tan\theta + \tan\phi)^3 - 3\tan\theta \cdot \tan\phi(\tan\theta + \tan\phi)\}$
$-9.4\tan\theta \tan\phi + 3.4\tan^2\theta \tan^2\phi$
$= 2 - 6\{(\tan\theta + \tan\phi)^2 - 2\tan\theta \cdot \tan\phi\}$
$+ 18\tan^2\theta \cdot \tan^2\phi$

or $12 - 64 + 12\tan\theta \tan\phi - 36\tan\theta \cdot \tan\phi$
$+ 12\tan^2\theta \cdot \tan^2\phi$
$= 2 - 96 + 12\tan\theta \cdot \tan\phi + 18\tan^2\theta \cdot \tan^2\phi$

or $6\tan^2\theta \cdot \tan^2\phi + 36\tan\theta \tan\phi - 42 = 0$

or $(\tan\theta + \tan\phi)^2 + 6\tan\theta \cdot \tan\phi - 7 = 0$

or $(\tan\theta \cdot \tan\phi + 7)(\tan\theta \cdot \tan\phi - 1) = 0$.

$\therefore$ $\tan\theta \cdot \tan\phi = -7, 1$

or $(x + 2)(-x + 2) = -7, 1$ {using (1)}

or $4 - x^2 = -7, 1$; $\therefore$ $x^2 = 11, 3$.

Now, $x^2 = 11$ $\Rightarrow$ $x = \pm\sqrt{11}$

$\therefore$ $\tan\theta = 2 \pm \sqrt{11}$

$\therefore$ $\theta = n\pi + \tan^{-1}(2 \pm \sqrt{11})$ where $n \in \mathbb{Z}$

$x^2 = 3$ $\Rightarrow$ $x = \pm\sqrt{3}$

$\therefore$ $\tan\theta = 2 \pm \sqrt{3}$

$\therefore$ $\theta = n\pi + \tan^{-1}(2 \pm \sqrt{3})$ where $n \in \mathbb{Z}$.

Also, $\tan\phi = 4 - \tan\theta = 2 \mp \sqrt{11},\ 2 \mp \sqrt{3}$

$\therefore$ $\phi = m\pi + \tan^{-1}(2 \mp \sqrt{11})$,
$m\pi + \tan^{-1}(2 \mp \sqrt{3})$ where $m \in \mathbb{Z}$.

15. Solve for x and y: $4^{\sin x} + 3^{\frac{1}{\cos y}} = 11$
$5.16^{\sin x} - 2.3^{\frac{1}{\cos y}} = 2$.

Let $4^{\sin x} = \lambda,\ 3^{\frac{1}{\cos y}} = \mu$.

Then the equations become

$\lambda + \mu = 11$ ... (1)

$5\lambda^2 - 2\mu = 2$ ... (2)

Now $2 \times (1) + (2)$ $\Rightarrow$ $2\lambda + 5\lambda^2 = 24$

or $5\lambda^2 + 2\lambda - 24 = 0$

or $5\lambda^2 + 12\lambda - 10\lambda - 24 = 0$

or $\lambda(5\lambda + 12) - 2(5\lambda + 12) = 0$

or $(5\lambda + 12)(\lambda - 2) = 0$.

So $\lambda = 2, -\dfrac{12}{5}$.

If $\lambda = 2$, $4^{\sin x} = 2$; $\therefore$ $2^{2\sin x} = 2$

$\therefore \quad 2\sin x = 1; \quad \therefore \quad \sin x = \frac{1}{2}.$

If $\lambda = \frac{-12}{5}$ then $4^{\sin x} = -\frac{12}{5}$,

which is impossible for $4^{\sin x} > 0$

When $\lambda = 2$, we get $\mu = 11 - 2 = 9$

$\therefore \quad 3^{\frac{1}{\cos y}} = 9 = 3^2$

$\therefore \quad \frac{1}{\cos y} = 2; \quad \therefore \quad \cos y = \frac{1}{2}$

When $\lambda = \frac{-12}{5}$, x has no solution. So y has no solution.

Thus we have $\sin x = \frac{1}{2}$, $\cos y = \frac{1}{2}$

$\therefore \quad x = n\pi + (-1)^n \frac{\pi}{6}$ and $y = 2m\pi \pm \frac{\pi}{3}$ where $m, n \in Z$.

16. Solve for θ: $\cos\left(\frac{\sqrt{2}+1}{2}\right)\theta \cdot \cos\left(\frac{\sqrt{2}-1}{2}\right)\theta = 1$.

Here, $\cos\left(\frac{\theta}{\sqrt{2}} + \frac{\theta}{2}\right) \cdot \cos\left(\frac{\theta}{\sqrt{2}} - \frac{\theta}{2}\right) = 1$

or $\quad \cos^2\frac{\theta}{\sqrt{2}} - \sin^2\frac{\theta}{2} = 1$

$\{\because \cos(A+B)\cdot\cos(A-B) = \cos^2 A - \sin^2 B\}$

or $\quad \cos^2\frac{\theta}{\sqrt{2}} = 1 + \sin^2\frac{\theta}{2} \geq 1$. But $\cos^2\frac{\theta}{\sqrt{2}} \leq 1$.

$\therefore$ only possibility is $\cos^2\frac{\theta}{\sqrt{2}} = 1$ and then $\sin^2\frac{\theta}{2} = 0$.

Now, $\cos^2\frac{\theta}{\sqrt{2}} = 1 \Rightarrow \cos\frac{\theta}{\sqrt{2}} = \pm 1$

$\therefore \quad \frac{\theta}{\sqrt{2}} = 2n\pi \pm 0, \; 2n\pi \pm \pi$

$\therefore \quad \theta = 2\sqrt{2}n\pi, \; \sqrt{2}(2n+1)\pi$ where $n \in Z$ ... (1)

Also, $\sin^2\frac{\theta}{2} = 0 \Rightarrow \theta = 2m\pi, m \in Z$ ... (2)

$\therefore$ only value common between (1) and (2) is $\theta = 0$.

Alternative Method As $-1 \leq \cos x \leq 1$, the equation holds only in either of the following situations.

(i) $\cos\left(\frac{\sqrt{2}+1}{2}\right)\theta = 1, \; \cos\left(\frac{\sqrt{2}-1}{2}\right)\theta = 1$

(ii) $\cos\left(\frac{\sqrt{2}+1}{2}\right)\theta = -1, \; \cos\left(\frac{\sqrt{2}-1}{2}\right)\theta = -1$

Now solve as given in the sum below.

17. Solve for x and y: $\cos x \cdot \cos y = 1$.

We know that $-1 \leq \cos x \leq 1, \; -1 \leq \cos y \leq 1$.

As the product of two fractions is a fraction,

$\cos x \cdot \cos y = 1$

$\Rightarrow \quad \cos x = 1, \; \cos y = 1$ ... (1)

or $\quad \cos x = -1, \; \cos y = -1$ ... (2)

If (1) is true then

$\cos x = 1 \quad \Rightarrow \quad x = 2n\pi, \; n \in Z$

$\cos y = 1 \quad \Rightarrow \quad y = 2m\pi, \; m \in Z$.

If (2) is true then

$\cos x = -1 \quad \Rightarrow \quad x = (2n \pm 1)\pi = (2n-1)\pi, n \in Z$

$\cos y = -1 \quad \Rightarrow \quad y = (2m \pm 1)\pi = (2m-1)\pi, m \in Z$.

$\therefore$ we get $x = n\pi, y = m\pi$ where n, m are both even integers or both odd integers.

18. Solve for x and y: $12\sin x + 5\cos x = 2y^2 - 8y + 21$.

Here $12\sin x + 5\cos x = 2y^2 - 8y + 21$

$\therefore \quad \sqrt{12^2 + 5^2}\left(\frac{12}{13}\sin x + \frac{5}{13}\cos x\right) = 2(y^2 - 4y + 4) + 13$

or $\quad 13\cos(x - \alpha) = 2(y-2)^2 + 13$, where $\cos\alpha = \frac{5}{13}$.

Clearly, LHS ≤ 13 because the greatest value of $\cos(x-\alpha)$ is 1 when $x = \alpha$.

Also RHS ≥ 13 because the least value of RHS is 13 when $y = 2$

$\therefore$ the equation can hold if the value of each side = 13.

Thus $\cos(x - \alpha) = 1$ and $y = 2$

$\therefore \quad x - \alpha = 2n\pi$ and $y = 2$

$\therefore \quad x = 2n\pi + \alpha = 2n\pi + \cos^{-1}\frac{5}{13}$ where $n \in Z$

and $y = 2$.

Exercises

1. Solve: $\tan^2 x = 3\,\text{cosec}^2 x - 1$.

2. Solve: $\tan x - \cot x = \text{cosec}\, x$.

3. Solve: $2\sin^2 x + \sin^2 2x = 2$.

4. Solve: $\sec\left(\frac{\pi}{4} + \theta\right) + \sec\left(\frac{\pi}{4} - \theta\right) = 2\sqrt{2}$.

5. Solve: $\sec x \cdot \cos 3x + 1 = 0$.

6. Solve: $\tan\theta + \tan\left(\theta + \frac{\pi}{3}\right) + \tan\left(\theta + \frac{2\pi}{3}\right) = 3.$

7. Prove that the general solution of

$$\cot x + \tan\left(\frac{\pi}{3} - x\right)\cdot\tan\left(\frac{\pi}{3} + x\right) = 0$$

is $x = \frac{n\pi}{3} + \frac{\pi}{4}$ where n is an integer.

8. Solve: $\sin x \cdot \sin 3x = \frac{1}{2}.$

9. Solve: $2\sin x + \cos x = 1.$

10. Solve: $5\cos x + 2\sin x = 2$ where $2\cot 21°\,48' = 5.$

11. Solve: $\operatorname{cosec} x + \cot x = \sqrt{3}.$

12. Solve: $\sin 3\theta = 4\sin\theta\cdot\sin 2\theta\cdot\sin 4\theta.$

13. Solve: $\cos\left(\frac{\pi}{3} - x\right)\cdot\cos x\cdot\cos\left(\frac{\pi}{3} + x\right) = \frac{1}{4}.$

14. Solve: $\cos 2\theta + \cos 4\theta + \cos 6\theta + 1 = 0$

15. Solve: $3\tan\theta + \cot\theta = 5\operatorname{cosec}\theta.$

16. Find the general solution of
$2(\sin x - \cos 2x) - \sin 2x(1 + 2\sin x) + 2\cos x = 0.$

17. Solve: $\sec x - 1 = \frac{1}{\sqrt{2} + 1}\tan x.$

18. Solve: $\cos 2x + \cos 4x = 2\cos x.$

19. Solve: $\cot\theta - \tan\theta - \cos\theta + \sin\theta = 0.$

20. Solve: $\cot 2x = \cos x + \sin x.$

21. Solve: $\sec 4\theta - \sec 2\theta - 2 = 0.$

22. Solve: $\sin^4 x + \cos^4 x = \sin x\cos x.$

23. Solve: $\cos 2x + \sin 3x = \cos 3x.$

24. Solve: $\cos x\left(1 + \sin\frac{x}{4} - 2\cos x\right) + \sin x\left(\cos\frac{x}{4} - 2\sin x\right) = 0.$

25. Solve: $2\cos^2 x + \cos 2x = 6\tan^2 x.$

26. Solve: $\tan\theta + \tan 2\theta + \tan 3\theta = 0$

27. Solve: $\sin 2x - \sin x + \cos x = 1.$

28. Solve: $(1 + \cos x)(2\sin x - \cos x) = \sin^2 x.$

29. Solve: $5(\sin 2x - \cos 2x) = \tan x + 5.$

30. Solve: $\sin\theta\cdot\cos\theta\cdot\cos 2\theta\cdot\cos 4\theta = \frac{1}{4}\sin 8\theta.$

31. Find the solutions of the following equation
$\sin 2x - 12(\sin x - \cos x) + 12 = 0.$

32. Solve: $\cos 2\theta - \sin 2\theta = \cos\theta - \sin\theta - 1,$
$90° \le \theta \le 540°.$

33. Solve: $\sin\theta + \sin 4\theta + \sin 7\theta = 0,\ \theta \in [0°, 90°].$

34. Find $x \in (-\pi, \pi)$ if
$8^{1 + |\cos x| + \cos^2 x + |\cos^3 x| + \dots \text{ to } \infty} = 64.$

35. Find the sum of all the solutions of
$7\cos 2x - \cos 4x = 6$ in the interval $[0, 99\pi]$.

36. Solve: $\cos 3x + \cos 2x = \sin\frac{3x}{2} + \sin\frac{x}{2},\ x \in [0, \pi].$

37. Solve: $(1 + \cot\theta)^2 = 2\cot\frac{\theta}{2},\ 0 \le \theta \le 2\pi.$

38. Solve: $\begin{vmatrix} 1 + \sin^2\theta & \cos^2\theta & 4\sin 4\theta \\ \sin^2\theta & 1 + \cos^2\theta & 4\sin 4\theta \\ \sin^2\theta & \cos^2\theta & 1 + 4\sin 4\theta \end{vmatrix} = 0,$

where $\theta \in \left[0, \frac{\pi}{2}\right].$

39. Find the general value of x if

$$\begin{vmatrix} 1 & a & a^2 \\ \cos(n-1)x & \cos nx & \cos(n+1)x \\ \sin(n-1)x & \sin nx & \sin(n+1)x \end{vmatrix} = 0 \text{ and } a \in R.$$

Also, prove that the values of x are independent of a.

40. Solve the equation $\cot\theta + \tan\theta = 2\operatorname{cosec}\theta.$
Does $\theta = n\pi$ satisfy the equation? Give reasons for your answer.

41. Determine all values of x for which
$\frac{1}{6}\sin x, \cos x, \tan x$ are consecutive terms of a GP.

42. If $\dfrac{\sin\frac{x}{2} + \cos\frac{x}{2} + i\tan x}{1 + 2i\sin\frac{x}{2}}$ is real, find the set of all possible values of x.

43. Find the general solution for x:

$$16^{\sin^2 x} + 16^{\cos^2 x} = 10$$

44. Solve: $(\cos x)^{\sin^2 x - \frac{3}{2}\sin x + \frac{1}{2}} = 1.$

45. Solve: $\sin^4 x + \cos^{15} x = 1.$

46. If $\sqrt{2}\cos x + \sqrt{6}\sin x = \max\limits_{\theta \in R}\{\sin\theta + \cos\theta\},$
find the general value of x.

47. Solve for x: $|\tan x| = \sin 2x.$

48. Find the general value of x satisfying both
$2\sin^2 x + \sin^2 2x = 2$ and $\sin 2x + \cos 2x = \tan x.$

49. If $\tan(\pi\cos\phi) = \cot(\pi\sin\phi)$, prove that

$$\cos\left(\frac{\pi}{4} - \phi\right) = \pm\frac{1}{2\sqrt{2}}.$$

50. If $\tan(\cot x) = \cot(\tan x)$, prove that

$$\sin 2x = \frac{4}{(2n + 1)\pi}.$$

51. If $\sin(\pi \cos \theta) = \cos(\pi \sin \theta)$, prove that
$$\cos\left(\theta \pm \frac{\pi}{4}\right) = \frac{1}{2\sqrt{2}}.$$

52. Prove that $\sin(\cos x) = \cos(\sin x)$ does not possess any real solution for x.

53. Prove that the roots of $x^3 - 2x + 1 = 0$ are
$$\sqrt{2}\sin\frac{\pi}{4}, 2\sin\frac{\pi}{10} \text{ and } 2\sin\frac{13\pi}{10}.$$

54. If $\sin(\pi \cot \theta) = \cos(\pi \tan \theta)$ then prove that $\cot 2\theta$ or $\operatorname{cosec} 2\theta$ is equal to $n + \frac{1}{4}$, $n \in Z$.

55. Solve for x, y: $x \cos^3 y + 3x \cos y \cdot \sin^2 y = 14$
$x \sin^3 y + 3x \cos^2 y \cdot \sin y = 13.$

56. Solve for x, y: $\sin x \cdot \cos y = -\frac{1}{2}$
$\cos x \cdot \sin y = \frac{1}{2}$

57. Solve for x, y: $\sin x \cdot \sin y = \frac{3}{4}$
$\tan x \cdot \tan y = 3$

58. Determine r, θ satisfying
$r \sin \theta = 3$
$r = 4(1 + \sin \theta)$ where $r > 0$, $\theta \in [0, 2\pi]$.

59. Solve for x, y: $\sin(x - y) = 3\sin x \cdot \cos y - 1$
$\sin(x + y) = -2\cos x \cdot \sin y.$

60. Solve for x, y: $2\sin x \cdot \sec y = 1$, $\sec y + \sin x = 2$.

61. Solve for x, y: $|\sin x| \sin y = -\frac{1}{4}$
$\cos(x - y) + \cos(x + y) = \frac{3}{2}$
if $x \in (0, 2\pi)$, $y \in (\pi, 2\pi)$.

62. Solve for rational x, y:
$\sin x + \sin y = \sin(x + y)$ and $|x| + |y| = 1.$

63. Solve for x, y: $\sin x \cdot \sin y = 1$.

64. Solve for x, y: $\sin x + \sin y = 2$.

65. Solve for θ: $\sin\left(\frac{a^2+1}{a}\right)\theta \cdot \sin\left(\frac{a^2-1}{a}\right)\theta = 1$
where a is a rational number.

66. Solve for x: $\sin x + \sin 2x + \sin 3x = 3$.

67. Solve for θ, ϕ: $\cos^2\theta + \cos\theta \cdot \cos\phi + \cos^2\phi = 0$.

68. Show that the equation $\sin x(\sin x + \cos x) = a$ will give real values of x if a lies between $\frac{1-\sqrt{2}}{2}$ and $\frac{1+\sqrt{2}}{2}$. Also find the solution of the equation for positive integral values of a.

69. Find $a \in R$ such that
$\cos^4 x - (a + 2)\cos^2 x - (a + 3) = 0$ has real solutions. Also find the solutions of the equation.

70. Find $a \in R$ for which $\cos x + \cos\left(\frac{\pi}{2} + x\right) = a$ has real solutions in x. Also find all the solutions in $[0, 2\pi]$ when a is a positive integer.

71. If the equation $1 + \sin^2 px = \cos x$ has exactly one solution then find p.

72. Prove that the equation $(\sin x + \sqrt{3}\cos x)\sin 4x = a$, where a is a constant, has no solution for $|a| \geq 2$.

Objective Questions

Fill in the blanks.

73. If $3\cot^2 x - 1 = 0$ then the general value of x is ______.

74. If $\tan 5\theta = \cot 2\theta$ then the general value of θ is ______.

75. If $\tan 3x \cdot \tan 7x + 1 = 0$ then the general value of θ is ______.

76. If $\sin 3\theta = \cos 2\theta$, $\theta \in \left[0, \frac{\pi}{2}\right]$ then $\theta =$ ______.

77. If $\sin x + \cos x = \frac{1}{\sqrt{2}}$ then the general value of x is ______.

78. If $\tan x + \tan 2x + \tan x \cdot \tan 2x = 1$ then the general value of x is ______.

79. If $\sec x \cdot \tan x = \sqrt{2}$, $x \in [0, 2\pi]$ then $x =$ ______.

80. If $\sin^2\theta - \cos\theta = \frac{1}{4}$, $0 \leq \theta \leq 2\pi$ then $\theta =$ ______.

81. If $\tan(A - B) = 1$ and $\sec(A + B) = \frac{2}{\sqrt{3}}$ then the smallest positive $A =$ ______ and the smallest positive $B =$ ______.

82. The solution set of the equations $x + y = \frac{2\pi}{3}$ and $\cos x + \cos y = \frac{3}{2}$ where x, y are real is ______.

Choose the correct option(s).

83. If $\sin a\theta + \cos b\theta = 0$ then the possible values of θ form
(a) an AP (b) two APs
(c) one GP (d) two GPs

84. If $\cos 2\theta = \sin \alpha$ then the most general relation between θ and α is (where $n \in Z$)
(a) $2\theta + \alpha = \frac{\pi}{2}$ (b) $\theta = n\pi \pm \left(\frac{\pi}{4} - \frac{\alpha}{2}\right)$
(c) $\theta = 2n\pi \pm \left(\frac{\pi}{2} - \alpha\right)$ (d) $\theta = \frac{n\pi + (-1)^n\alpha}{2}$

85. If $\cos\theta + \sqrt{3}\sin\theta = 2$ and $\theta \in [0, 2\pi]$ then θ is

(a) $\frac{\pi}{3}$ (b) $\frac{5\pi}{3}$ (c) $\frac{2\pi}{3}$ (d) $\frac{4\pi}{3}$

86. If $\tan\theta + \tan\left(\frac{\pi}{2} + \theta\right) = 0$ then the most general value of θ is (where $n \in Z$)

(a) $n\pi \pm \frac{\pi}{4}$ (b) $2n\pi + \frac{\pi}{4}$

(c) $2n\pi \pm \frac{\pi}{4}$ (d) $\frac{n\pi}{2} + (-1)^n \cdot \frac{\pi}{4}$

87. The number of solutions of the equation $8\tan^2\theta + 9 = 6\sec\theta$ in the interval $(-\pi/2, \pi/2)$ is

(a) two (b) four

(c) zero (d) none of these

88. The number of solutions of the equation $\tan x + \sec x = 2\cos x$ lying in the interval $[0, 2\pi]$ is

(a) 0 (b) 1 (c) 2 (d) 3

89. If $\sin(\pi\cos\theta) = \cos(\pi\sin\theta)$ then $\sin 2\theta$ is equal to

(a) $\frac{3}{4}$ (b) $\frac{1}{4}$

(c) $-\frac{3}{4}$ (d) none of these

90. If $\cos A = \cos B$ and $\sin A = \sin B$ then

(a) $A + B = 0$ (b) $A = B$

(c) $A + B = 2n\pi$ (d) $A = B + 2n\pi$

91. The general solution of $\sin x - 3\sin 2x + \sin 3x = \cos x - 3\cos 2x + \cos 3x$ is

(a) $n\pi + \frac{\pi}{8}$ (b) $\frac{n\pi}{2} + \frac{\pi}{8}$

(c) $(-1)^n \cdot \frac{n\pi}{2} + \frac{\pi}{8}$ (d) $2n\pi + \cos^{-1}\frac{3}{2}$

92. The general solution of $\sin x + \cos x = 1$ is given by

(a) $2n\pi$ (b) $2n\pi + \frac{\pi}{2}$

(c) $n\pi + (-1)^n \cdot \frac{\pi}{4} - \frac{\pi}{4}$ (d) none of these

93. If $\cot\theta = \sin 2\theta$ and $\theta \neq n\pi$, $n \in Z$ then θ is equal to

(a) 45°, 60° (b) 45°, 90°

(c) 45° only (d) 90° only

94. If $2\sin\theta + 1 = 0$ and $\sqrt{3}\tan\theta = 1$ then the most general value of θ is

(a) $n\pi \pm \frac{\pi}{6}$ (b) $n\pi + (-1)^n \cdot \frac{7\pi}{6}$

(c) $2n\pi + \frac{7\pi}{6}$ (d) $2n\pi + \frac{11\pi}{6}$

State whether the statements are true or false.

95. If $\sin x + 2\cos x = 0$ then $x = n\pi - \cos^{-1}\frac{1}{\sqrt{5}}$, $n \in Z$ is the general solution of the equation.

96. One value of θ which satisfies the equation $\sin^4\theta - 2\sin^2\theta - 1 = 0$ lies between 0 and 2π.

97. The smallest positive angle x in degrees, satisfying the equation

$$\sqrt{1 + \sin 2x} - \sqrt{2}\cos 3x = 0 \text{ is } 11.25°.$$

98. The equation $4\cos 2\theta = 7 - 4x + 4x^2$ has no solution for θ if $x \in R$.

Answers

1. $n\pi \pm \frac{\pi}{3}$; $n \in Z$

2. $(2n + 1)\pi$, $\left(2n \pm \frac{1}{3}\right)\pi$; $n \in Z$

3. $\left(n + \frac{1}{2}\right)\pi$, $n\pi \pm \frac{\pi}{4}$; $n \in Z$ **4.** $\frac{2n\pi}{3}$, $2n\pi$; $n \in Z$

5. $\frac{1}{4}(2n + 1)\pi$, $\frac{1}{2}(2n + 1)\pi$; $n \in Z$

6. $\frac{1}{3}n\pi + \frac{1}{12}\pi$; $n \in Z$

8. $(2n + 1)\frac{\pi}{4}$, $n\pi \pm \frac{\pi}{6}$; $n \in Z$

9. $2n\pi$, $2n\pi + 2\cos^{-1}\frac{1}{\sqrt{5}}$; $n \in Z$

10. $n \times 180° - 68°\,12' + (-1)^n\, 21°\,48'$; $n \in Z$

11. $\left(2n + \frac{1}{3}\right)\pi$; $n \in Z$

12. $n\pi$, $\frac{n\pi}{3} \pm \frac{\pi}{9}$; $n \in Z$ **13.** $\frac{2n\pi}{3}$; $n \in Z$

14. $(2n + 1)\frac{\pi}{2}$, $(2n + 1)\frac{\pi}{4}$, $(2n + 1)\frac{\pi}{6}$; $n \in Z$

15. $2n\pi \pm \frac{\pi}{3}$; $n \in Z$

16. $2n\pi$, $\left(2n + \frac{1}{2}\right)\frac{\pi}{3}$, $\left(2n - \frac{1}{2}\right)\pi$; $n \in Z$

17. $2n\pi$, $2n\pi + \frac{\pi}{4}$; $n \in Z$ **18.** $(2n + 1)\frac{\pi}{2}$, $\frac{2n\pi}{3}$; $n \in Z$

19. $n\pi + \frac{\pi}{4},\ 2n\pi + \frac{\pi}{4} \pm \cos^{-1}\frac{1-\sqrt{2}}{\sqrt{2}},\ n \in Z$

20. $n\pi - \frac{\pi}{4}, 2n\pi - \frac{\pi}{4} \pm \cos^{-1}\frac{\sqrt{5}-1}{2\sqrt{2}}; n \in Z$

21. $\frac{(2n+1)\pi}{10}, \frac{(2n+1)\pi}{2}; n \in Z$

22. $\frac{4n+1}{4}\pi; n \in Z$

23. $\frac{(4n-1)\pi}{4}, 2n\pi, \frac{(4n-1)\pi}{2},$

$\frac{(4n+1)\pi}{4} + (-1)^n \sin^{-1}\left(\frac{1}{2\sqrt{2}}\right); n \in Z$

24. $(4n+1)2\pi; n \in Z$

25. $\frac{(6n \pm 1)\pi}{6}; n \in Z$

26. $\frac{n\pi}{3}, n\pi \pm \tan^{-1}\frac{1}{\sqrt{2}}; n \in Z$

27. $\frac{(4n+1)\pi}{4}, 2n\pi, \frac{(4n-1)\pi}{2}; n \in Z$

28. $(2n+1)\pi, n\pi + (-1)^n \cdot \frac{\pi}{6}; n \in Z$

29. $n\pi + \tan^{-1}2, n\pi + \tan^{-1}(\sqrt{6}-1),$

$n\pi - \tan^{-1}(\sqrt{6}+1); n \in Z$

30. $\frac{n\pi}{8}; n \in Z$

31. $\frac{(4n+1)\pi}{2}, (2n+1)\pi; n \in Z$

32. 225°, 300°, 405°, 420°

33. 0°, 40°, 45°, 80°, 90°

34. $\pm\frac{\pi}{3}, \pm\frac{2\pi}{3}$

35. 4950π

36. $\pi, \frac{\pi}{7}, \frac{5\pi}{7}$

37. $\frac{\pi}{6}, \frac{5\pi}{6}$

38. $\frac{7\pi}{24}, \frac{11\pi}{24}$

39. $n\pi;\ n \in Z$

40. $2n\pi \pm \frac{\pi}{3}, n \in Z$; no, as $\theta = n\pi$ makes $\cot\theta$ and $\operatorname{cosec}\theta$ undefined

41. $2n\pi \pm \frac{\pi}{3}; n \in Z$

42. $n\pi + \frac{\pi}{4}, 2n\pi; n \in Z$

43. $n\pi \pm \frac{\pi}{6}, n\pi \pm \frac{\pi}{3}; n \in Z$

44. $n\pi + (-1)^n \frac{\pi}{6}; n \in Z$

45. $2n\pi, 2n\pi \pm \frac{\pi}{2}; n \in Z$

46. $2n\pi, 2n\pi + \frac{2\pi}{3}; n \in Z$

47. $n\pi, 2n\pi + \frac{\pi}{4}, 2n\pi - \frac{3\pi}{4}; n \in Z$

48. $n\pi \pm \frac{\pi}{4}; n \in Z$

55. $(x, y) = \left(5\sqrt{5}, 2n\pi + \tan^{-1}\frac{1}{2}\right),$

$\left(-5\sqrt{5}, (2n+1)\pi + \tan^{-1}\frac{1}{2}\right); n \in Z$

56. $x = \left(\frac{n}{2} - m - \frac{1}{4}\right)\pi, y = \left(\frac{n}{2} + m + \frac{1}{4}\right)\pi; m, n \in Z$

57. $x = \left(m + n \pm \frac{1}{3}\right)\pi, y = \left(m - n \pm \frac{1}{3}\right)\pi; m, n \in Z$

58. $\left(6, \frac{\pi}{6}\right), \left(6, \frac{5\pi}{6}\right)$

59. $x = \frac{1}{2}\left\{(n+m)\pi + (-1)^n \cdot \sin^{-1}\frac{4}{5} + (-1)^m \cdot \sin^{-1}\frac{2}{5}\right\}$

$y = \frac{1}{2}\left\{(m-n)\pi + (-1)^m \cdot \sin^{-1}\frac{2}{5} - (-1)^n \cdot \sin^{-1}\frac{4}{5}\right\}$

60. $x = n\pi + (-1)^n \sin^{-1}\left(1 - \frac{1}{\sqrt{2}}\right)$

$y = 2m\pi \pm \cos^{-1}(2-\sqrt{2}); m, n \in Z$

61. $(x, y) = \left(\frac{\pi}{6}, \frac{11\pi}{6}\right), \left(\frac{5\pi}{6}, \frac{7\pi}{6}\right), \left(\frac{7\pi}{6}, \frac{7\pi}{6}\right), \left(\frac{11\pi}{6}, \frac{11\pi}{6}\right)$

62. $(x, y) = \left(\frac{1}{2}, \frac{-1}{2}\right), \left(-\frac{1}{2}, \frac{1}{2}\right), (1, 0), (-1, 0), (0, 1), (0, -1)$

63. $(x, y) = \left((4n \pm 1)\frac{\pi}{2}, (4m \pm 1)\frac{\pi}{2}\right); n, m \in Z$

64. $(x, y) = \left(n\pi + (-1)^n \frac{\pi}{2}, m\pi + (-1)^m \frac{\pi}{2}\right); n, m \in Z$

65. $\frac{1}{a}\left(n\pi \pm \frac{\pi}{2}\right); n \in Z$

66. no solution

67. $(\theta, \phi) = \left(n\pi + \frac{\pi}{2}, m\pi + \frac{\pi}{2}\right); m, n \in Z$

68. $n\pi + \frac{\pi}{4}, (2n+1)\frac{\pi}{2}; n \in Z$

69. $a \in [-3, -2],\ 2n\pi \pm \cos^{-1}\sqrt{a+3}; n \in Z$

70. $|a| \le \sqrt{2}, x = 0, 2\pi, \frac{3\pi}{2}$

71. p is any irrational

73. $n\pi \pm \frac{\pi}{3}; n \in Z$

74. $(2n+1)\frac{\pi}{14}; n \in Z$

75. $\frac{2n+1}{8}\pi; n \in Z$

76. $\frac{\pi}{2}, \frac{\pi}{10}$

77. $2n\pi + \frac{\pi}{4} \pm \frac{\pi}{3}; n \in Z$

78. $\frac{(4n+1)\pi}{12}; n \in Z$

79. $\frac{\pi}{4}, \frac{3\pi}{4}$

80. $\frac{\pi}{3}, \frac{5\pi}{3}$

81. $\frac{25\pi}{24}, \frac{19\pi}{24}$ respectively

82. ϕ

83. (b) 84. (b) 85. (a) 86. (a)

87. (c) 88. (c) 89. (a), (c) 90. (b)

91. (b) 92. (c) 93. (b) 94. (c)

95. true 96. false 97. true 98. true

□

Chapter Test

Time: 90 minutes

1. In the following incomplete sentences fill in the blanks so that the resulting sentences become true.

(a) The values of α for which $\cos\alpha \cdot (1 + x^2) + (1 - \sin\alpha) \cdot x = 0$ is an identity in x are _____ .

(b) The number of real solutions for x from $2\cos\frac{x}{3} = 2^x + 2^{-x}$ is _____ .

(c) The set of values of θ for which $\frac{1 - i\sin\theta}{1 + i\cos\theta}$ is purely real is _____ .

(d) The coordinates of the point of intersection of the curves $y = \cos x$ and $y = \sin 3x$ where $x \in \left[-\frac{\pi}{2}, \frac{\pi}{2}\right]$ are _____ .

2. In the following, each question has multiple answers. Choose the correct answer(s).

(a) If $\cos\theta - \sin\theta = \cos\alpha - \sin\alpha$ then the minimum value of $|\theta + \alpha|$ is

A. 0 B. $\frac{\pi}{2}$ C. $\frac{3\pi}{2}$ D. none of these

(b) $2\cos x - 3\sin x = a$ has real solutions for x if

A. $a = -1$ B. $|a| \le \sqrt{13}$ C. $|a| \le \sqrt{5}$ D. $|a| = 5$

(c) The number of possible solutions of x such that $\sin^8 x + \cos^8 x = 1$ is

A. zero B. 8 C. 2 D. infinite

(d) The general solution-set of the equation $\cos x + \cos 5x = 2$ is

A. $\left\{x \mid x = \frac{2m\pi}{5}, m \in \mathbb{Z}\right\}$ B. $\{x \mid x = 2n\pi, n \in \mathbb{Z}\}$

C. $\left\{x \mid x = \frac{2m\pi}{5}, m \text{ is a multiple of } 5\right\}$ D. ϕ

3. Determine all real x for which

$$\tan\left(x + \frac{\pi}{6}\right) \cdot \tan\left(x + \frac{\pi}{4}\right) + \tan\left(x + \frac{\pi}{4}\right)\tan\left(x + \frac{\pi}{3}\right) + \tan\left(x + \frac{\pi}{3}\right)\tan\left(x + \frac{\pi}{6}\right) = 1.$$

4. Find the least distance between the positive values of x for which

$2\cos x$, $\sqrt{\frac{1}{2}(1 + \cos 2x)}$ and $1 - 3\cos^2 x$ are in GP.

5. Solve for A and B: $\sqrt{3}\sin 2A = \sin 2B$

$$\sqrt{3}\sin^2 A + \sin^2 B = \frac{\sqrt{3} - 1}{2}.$$

6. There are three angles in AP. If the tangent of the middle angle be 2 and the ratio of the tangents of the other angles is $11 : 1$ then find the possible values of the common difference of the angles, given that the common difference lies between 0 and $\frac{\pi}{4}$.

7. Solve for x, y: $4\cos x - 3\sin x = -y^2 + 6y - 14$.

8. Find the general solution of α such that the following system of equation in x, y, z has nonzero solutions.

$$\begin{aligned} \sin 3\alpha \cdot x + y - z &= 0 \\ \cos 2\alpha \cdot x + 3y + 4z &= 0 \\ 2x + 7y + 7z &= 0 \end{aligned}$$

9. Solve for θ : $|4\cos\theta - 3\sin\theta| \geq 5, -\pi \leq \theta \leq \pi$.

□

Answers

1. (a) $2n\pi + \frac{\pi}{2}, n \in Z$ (b) one (c) $\left\{\theta \mid \theta = n\pi - \frac{\pi}{4}, n \in Z\right\}$ (d) $\left(\frac{\pi}{4}, \frac{1}{\sqrt{2}}\right), \left(\frac{\pi}{8}, \cos\frac{\pi}{8}\right)$

2. (a) B (b) B (c) D (d) B

3. $\frac{n\pi}{3} - \frac{\pi}{12}, n \in Z$

4. $\frac{\pi}{6}$

5. $A = \frac{m\pi}{2} \pm \frac{\pi}{12}, B = \left(n - \frac{m}{2}\right)\pi \pm \frac{\pi}{6}$

6. $\tan^{-1}\frac{3 - \sqrt{5}}{2}$

7. $x = n\pi + (-1)^n \frac{\pi}{2} + \cos^{-1}\frac{3}{5}, y = 3$

8. $n\pi, n\pi + (-1)^n \frac{\pi}{6}, n \in Z$

9. $n\pi - \cos^{-1}\frac{4}{5}, n = 0, 1.$

□

3. Inverse Circular Functions

Recap of Facts and Formulae

1. Inverse circular functions and their domains

$\sin^{-1}x$ is an angle whose sine is equal to x, i.e., $\sin^{-1}x = \alpha$ if $\sin\alpha = x$. $\sin^{-1}x$ is an inverse circular function. Similarly, other inverse circular functions are $\cos^{-1}x$, $\tan^{-1}x$, $\sec^{-1}x$, $\operatorname{cosec}^{-1}x$ and $\cot^{-1}x$.

- $\sin^{-1}x$ and $\cos^{-1}x$ are defined if $|x| \le 1$, i.e., $-1 \le x \le 1$.
- $\tan^{-1}x$ and $\cot^{-1}x$ are defined for all $x \in R$.
- $\sec^{-1}x$ and $\operatorname{cosec}^{-1}x$ are defined for $|x| \ge 1$, i.e., $x \le -1$ or $x \ge 1$.

2. Principal value

We know that $\sin\theta = \sin\alpha \;\Rightarrow\; \theta = n\pi + (-1)^n\alpha,\ n \in \mathbb{Z}$.

Therefore, $\sin^{-1}x$ will have infinite number of values for a fixed value of x.

For example

$\sin^{-1}\left(\frac{1}{2}\right)$ has the values $n\pi + (-1)^n\frac{\pi}{6},\ n \in \mathbb{Z}$.

But $f(x) = \sin^{-1}x$ is a function if we have only one value $\sin^{-1}\frac{1}{2}$ for $x = \frac{1}{2}$. For this we take the principal value of $\sin^{-1}x$ for a fixed x. The numerically least value is taken as the principal value.

So the principal value of $\sin^{-1}\left(\frac{1}{2}\right)$ is $\frac{\pi}{6}$.

If there are two different angles of numerically least measure, one being positive and the other negative then the positive angle is taken as the principal value.

For example

$\cos^{-1}\left(\frac{1}{2}\right)$ has two values $\frac{\pi}{3}$ and $-\frac{\pi}{3}$ which are numerically least among the possible values.

So, $\frac{\pi}{3}$ is the principal value of $\cos^{-1}\left(\frac{1}{2}\right)$.

3. Intervals for principal values

The principal values of different inverse circular functions may belong to the intervals as shown below:

range of pr. value of $\sin^{-1}x$

$= \left[0, \frac{\pi}{2}\right],\ x \ge 0$

$\left[-\frac{\pi}{2}, 0\right),\ x < 0$

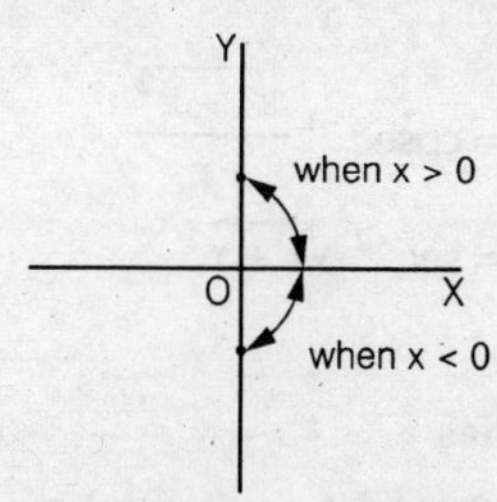

range of pr. value of $\cos^{-1}x$

$= \left[0, \frac{\pi}{2}\right],\ x \ge 0$

$\left(\frac{\pi}{2}, \pi\right],\ x < 0$

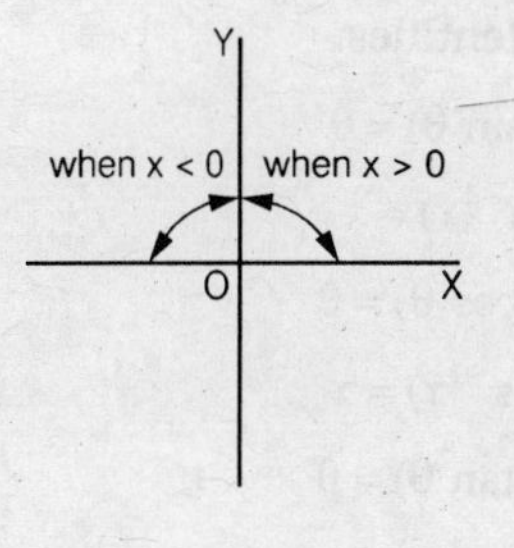

range of pr. value of $\tan^{-1}x$

$= \left[0, \frac{\pi}{2}\right),\ x \ge 0$

$\left(-\frac{\pi}{2}, 0\right),\ x < 0$

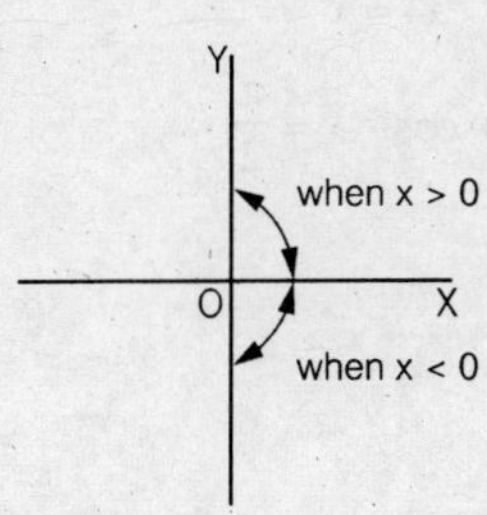

4. Conversions of inverse circular functions

- $\sin^{-1}x = \cos^{-1}\sqrt{1-x^2}$

$= \tan^{-1}\frac{x}{\sqrt{1-x^2}}$

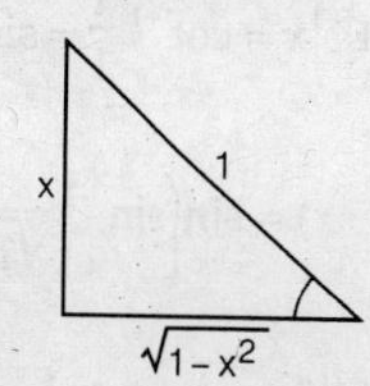

$= \cot^{-1}\frac{\sqrt{1-x^2}}{x}$

$= \operatorname{cosec}^{-1}\frac{1}{x} = \sec^{-1}\frac{1}{\sqrt{1-x^2}}$

- $\cos^{-1}x = \sin^{-1}\sqrt{1-x^2} = \tan^{-1}\dfrac{\sqrt{1-x^2}}{x}$

$= \cot^{-1}\dfrac{x}{\sqrt{1-x^2}}$

$= \operatorname{cosec}^{-1}\dfrac{1}{\sqrt{1-x^2}}$

$= \sec^{-1}\dfrac{1}{x}$

- $\tan^{-1}x = \sin^{-1}\dfrac{x}{\sqrt{1+x^2}}$

$= \cos^{-1}\dfrac{1}{\sqrt{1+x^2}}$

$= \cot^{-1}\dfrac{1}{x}$

$= \operatorname{cosec}^{-1}\dfrac{\sqrt{1+x^2}}{x}$

$= \sec^{-1}\sqrt{1+x^2}$

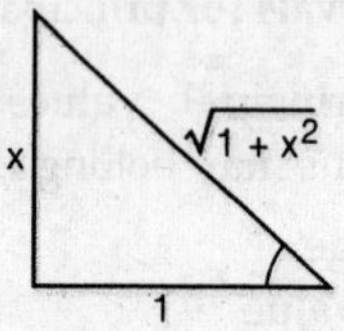

5. Identities

Taking principal value only, we have the following useful identities.

- $\sin^{-1}(\sin\theta) = \theta$
- $\sin(\sin^{-1}x) = x$
- $\cos^{-1}(\cos\theta) = \theta$
- $\cos(\cos^{-1}x) = x$
- $\tan^{-1}(\tan\theta) = \theta$
- $\tan(\tan^{-1}x) = x$
- $\sin^{-1}x + \cos^{-1}x = \dfrac{\pi}{2}$
- $\sec^{-1}x + \operatorname{cosec}^{-1}x = \dfrac{\pi}{2}$
- $\tan^{-1}x + \cot^{-1}x = \dfrac{\pi}{2}$ if $x > 0$ and

  $\dfrac{-\pi}{2}$ if $x < 0$

- $\tan^{-1}x + \tan^{-1}y = \tan^{-1}\dfrac{x+y}{1-xy}$ if $x > 0, y > 0, xy < 1$

  $\pi + \tan^{-1}\dfrac{x+y}{1-xy}$ if $x > 0, y > 0, xy > 1$

  $\tan^{-1}x - \tan^{-1}y = \tan^{-1}\dfrac{x-y}{1+xy}$

- $\sin^{-1}x + \sin^{-1}y = \sin^{-1}(x\sqrt{1-y^2} + y\sqrt{1-x^2})$

  if $x \ge 0,\ y \ge 0,\ x^2 + y^2 \le 1$

  $\pi - \sin^{-1}(x\sqrt{1-y^2} + y\sqrt{1-x^2})$

  if $x \ge 0,\ y \ge 0,\ x^2 + y^2 > 1$

  $\sin^{-1}x - \sin^{-1}y = \sin^{-1}(x\sqrt{1-y^2} - y\sqrt{1-x^2})$

- $\cos^{-1}x + \cos^{-1}y = \cos^{-1}(xy - \sqrt{1-x^2}\,\sqrt{1-y^2})$

  $\cos^{-1}x - \cos^{-1}y = \cos^{-1}(xy + \sqrt{1-x^2}\,\sqrt{1-y^2})$

- If $|x| \le 1$ then

  $2\tan^{-1}x = \sin^{-1}\dfrac{2x}{1+x^2} = \cos^{-1}\dfrac{1-x^2}{1+x^2} = \tan^{-1}\dfrac{2x}{1-x^2}.$

  If $|x| > 1$, change x into $\dfrac{1}{x}$ in the above.

Note In case of identities in inverse circular functions, principal values are to be taken. As such signs of x, y, etc., will determine the quadrant in which the angles will fall. In order to bring the angles of both sides in the same quadrant, adjustment by π is to be made.

Selected Solved Examples

1. Prove that $\cos[\tan^{-1}\{\sin(\cot^{-1}x)\}] = \left(\dfrac{x^2+1}{x^2+2}\right)^{1/2}$

We know, $\cot^{-1}x = \cot^{-1}\dfrac{x}{1} = \sin^{-1}\dfrac{1}{\sqrt{1+x^2}}$

$\therefore\quad \sin(\cot^{-1}x) = \sin\left(\sin^{-1}\dfrac{1}{\sqrt{1+x^2}}\right) = \dfrac{1}{\sqrt{1+x^2}}$

Now, $\tan^{-1}\dfrac{1}{\sqrt{1+x^2}} = \cos^{-1}\dfrac{\sqrt{1+x^2}}{\sqrt{1+(\sqrt{1+x^2})^2}}$

$= \cos^{-1}\dfrac{\sqrt{1+x^2}}{\sqrt{2+x^2}}$

$\therefore\quad \text{LHS} = \cos\left[\tan^{-1}\left\{\sin\left(\sin^{-1}\dfrac{1}{\sqrt{1+x^2}}\right)\right\}\right]$

$= \cos\left[\tan^{-1}\dfrac{1}{\sqrt{1+x^2}}\right]$

$= \cos\left[\cos^{-1}\dfrac{\sqrt{1+x^2}}{\sqrt{2+x^2}}\right]$

$= \dfrac{\sqrt{1+x^2}}{\sqrt{2+x^2}}$

$= \left(\dfrac{x^2+1}{x^2+2}\right)^{1/2}$

= RHS.

2. Prove that

$$\tan^{-1}\sqrt{\frac{a(a+b+c)}{bc}}+\tan^{-1}\sqrt{\frac{b(a+b+c)}{ca}}+\tan^{-1}\sqrt{\frac{c(a+b+c)}{ab}}=\pi$$

where a, b, c are positive.

Let $\tan^{-1}\sqrt{\frac{a(a+b+c)}{bc}}=\alpha,\ \tan^{-1}\sqrt{\frac{b(a+b+c)}{ca}}=\beta,$

$\tan^{-1}\sqrt{\frac{c(a+b+c)}{ab}}=\gamma$

$\therefore\quad \tan\alpha=\sqrt{\frac{a(a+b+c)}{bc}},\ \tan\beta=\sqrt{\frac{b(a+b+c)}{ca}},$

$\tan\gamma=\sqrt{\frac{c(a+b+c)}{ab}}$

$\therefore\quad \tan\alpha\cdot\tan\beta+\tan\beta\cdot\tan\gamma+\tan\gamma\cdot\tan\alpha$

$$=(a+b+c)\left(\sqrt{\frac{ab}{abc^2}}+\sqrt{\frac{bc}{a^2bc}}+\sqrt{\frac{ca}{ab^2c}}\right)$$

$$=(a+b+c)\left(\frac{1}{c}+\frac{1}{a}+\frac{1}{b}\right)\geq 2\sqrt[3]{abc}\cdot 2\sqrt[3]{\frac{1}{abc}}=4$$

{using AM ≥ GM}

Also, $\tan\alpha+\tan\beta+\tan\gamma$

$$=\sqrt{a+b+c}\left(\sqrt{\frac{a}{bc}}+\sqrt{\frac{b}{ca}}+\sqrt{\frac{c}{ab}}\right)$$

$$=\sqrt{a+b+c}\cdot\frac{a+b+c}{\sqrt{abc}}=\frac{(a+b+c)^{3/2}}{\sqrt{abc}} \quad \ldots (1)$$

and $\tan\alpha\cdot\tan\beta\cdot\tan\gamma$

$$=(a+b+c)^{3/2}\cdot\sqrt{\frac{a}{bc}}\cdot\sqrt{\frac{b}{ca}}\cdot\sqrt{\frac{c}{ab}}$$

$$=\frac{(a+b+c)^{3/2}}{\sqrt{abc}} \quad \ldots (2)$$

Now, $\tan(\alpha+\beta+\gamma)$

$$=\frac{\tan\alpha+\tan\beta+\tan\gamma-\tan\alpha\tan\beta\tan\gamma}{1-\tan\alpha\tan\beta-\tan\beta\tan\gamma-\tan\gamma\cdot\tan\alpha}$$

$=0$ {using (1) and (2)}

$\therefore\quad \alpha+\beta+\gamma=\pi$ {$\because$ sum of three positive angles $\neq 0$}

Hence, the problem.

3. Evaluate $\cos^{-1}x+\cos^{-1}\left\{\frac{x}{2}+\frac{1}{2}\sqrt{3-3x^2}\right\},\ x\in\left[\frac{1}{2},1\right]$.

Expression $=\cos^{-1}x+\cos^{-1}\left\{\frac{1}{2}\cdot x+\frac{\sqrt{3}}{2}\cdot\sqrt{1-x^2}\right\}$

$=\cos^{-1}x+\cos^{-1}\left\{\cos\frac{\pi}{3}\cdot\cos\alpha+\sin\frac{\pi}{3}\cdot\sin\alpha\right\},$

where $\cos\alpha=x$

$=\cos^{-1}x+\cos^{-1}\cos\left(\frac{\pi}{3}-\alpha\right)$

$\left\{\because \text{ from the question, } \frac{1}{2}\leq\cos\alpha\leq 1,\ \alpha<\frac{\pi}{3}\right\}$

$=\cos^{-1}x+\frac{\pi}{3}-\alpha=\cos^{-1}x+\frac{\pi}{3}-\cos^{-1}x=\frac{\pi}{3}.$

4. If $\sin^{-1}\frac{x}{a}+\sin^{-1}\frac{y}{b}=\sin^{-1}\frac{c^2}{ab}$, show that

$b^2x^2+2xy\sqrt{a^2b^2-c^4}+a^2y^2=c^4$.

We have $\sin^{-1}\frac{x}{a}=\cos^{-1}\frac{\sqrt{a^2-x^2}}{a}$

$\sin^{-1}\frac{y}{b}=\cos^{-1}\frac{\sqrt{b^2-y^2}}{b},$

$\sin^{-1}\frac{c^2}{ab}=\cos^{-1}\frac{\sqrt{a^2b^2-c^4}}{ab}.$

Thus, we get

$$\cos^{-1}\frac{\sqrt{a^2-x^2}}{a}+\cos^{-1}\frac{\sqrt{b^2-y^2}}{b}=\cos^{-1}\frac{\sqrt{a^2b^2-c^4}}{ab}$$

or $\cos^{-1}\left\{\frac{\sqrt{a^2-x^2}}{a}\cdot\frac{\sqrt{b^2-y^2}}{b}-\sqrt{1-\left(\frac{\sqrt{a^2-x^2}}{a}\right)^2}\cdot\sqrt{1-\left(\frac{\sqrt{b^2-y^2}}{b}\right)^2}\right\}$

$$=\cos^{-1}\frac{\sqrt{a^2b^2-c^4}}{ab}$$

[using $\cos^{-1}x+\cos^{-1}y=\cos^{-1}\{xy-\sqrt{1-x^2}\ \sqrt{1-y^2}\}$]

or $\frac{\sqrt{a^2-x^2}}{a}\cdot\frac{\sqrt{b^2-y^2}}{b}-\frac{x}{a}\cdot\frac{y}{b}=\frac{\sqrt{a^2b^2-c^4}}{ab}$

or $\sqrt{a^2-x^2}\cdot\sqrt{b^2-y^2}-xy=\sqrt{a^2b^2-c^4}$

or $(a^2-x^2)(b^2-y^2)=(xy+\sqrt{a^2b^2-c^4})^2$

or $a^2b^2-a^2y^2-b^2x^2+x^2y^2=x^2y^2+a^2b^2-c^4+2xy\sqrt{a^2b^2-c^4}$

or $b^2x^2+2xy\sqrt{a^2b^2-c^4}+a^2y^2=c^4.$

5. Prove that $f(x) = 2\tan^{-1}x + \sin^{-1}\frac{2x}{1+x^2}$ is a constant for all $x \geq 1$. Find that constant.

If $0 \leq x \leq 1$ then $\sin^{-1}\frac{2x}{1+x^2} = 2\tan^{-1}x$.

If $x \geq 1$, i.e., $0 \leq \frac{1}{x} \leq 1$ then $\sin^{-1}\frac{\frac{2}{x}}{1+\left(\frac{1}{x}\right)^2} = 2\tan^{-1}\frac{1}{x}$.

Here $x \geq 1$. So $f(x) = 2\tan^{-1}x + \sin^{-1}\frac{\frac{2}{x}}{1+\frac{1}{x^2}}$

$= 2\tan^{-1}x + 2\tan^{-1}\frac{1}{x} = 2(\tan^{-1}x + \cot^{-1}x)$

$= 2\cdot\frac{\pi}{2} = \pi =$ constant.

6. If $\cos^{-1}x + \cos^{-1}y + \cos^{-1}z = \pi$ then prove that $x^2 + y^2 + z^2 + 2xyz = 1$.

Let $\cos^{-1}x = A$, $\cos^{-1}y = B$, $\cos^{-1}z = C$.

Then, from the question, $A + B + C = \pi$.

Now, $x^2 + y^2 + z^2 + 2xyz$

$= \cos^2 A + \cos^2 B + \cos^2 C + 2\cos A\cos B\cos C$

$= \frac{1+\cos 2A}{2} + \frac{1+\cos 2B}{2} + \frac{1+\cos 2C}{2} + 2\cos A\cos B\cos C$

$= \frac{3}{2} + \frac{1}{2}(\cos 2A + \cos 2B) + \frac{1}{2}\cos 2C + 2\cos A\cos B\cos C$

$= \frac{3}{2} + \frac{1}{2}\cdot 2\cos(A+B)\cdot\cos(A-B) + \frac{1}{2}(2\cos^2 C - 1) + 2\cos A\cos B\cos C$

$= \frac{3}{2} - \cos C\cdot\cos(A-B) + \cos^2 C - \frac{1}{2} + 2\cos A\cos B\cos C \quad \{\because A + B = \pi - C\}$

$= 1 - \cos C\{\cos(A-B) - \cos C\} + 2\cos A\cos B\cos C$

$= 1 - \cos C\cdot\{\cos(A-B) + \cos(A+B)\} + 2\cos A\cos B\cos C$

$= 1 - 2\cos A\cos B\cos C + 2\cos A\cos B\cos C$

$= 1 =$ RHS.

7. Solve $\sin[2\cos^{-1}\{\cot(2\tan^{-1}x)\}] = 0$; express x in the form $a \pm \sqrt{b}$.

$2\tan^{-1}x = \tan^{-1}\frac{2x}{1-x^2}$

$\therefore \quad \cot(2\tan^{-1}x) = \cot\left(\tan^{-1}\frac{2x}{1-x^2}\right)$

or $\quad \cot(2\tan^{-1}x) = \cot\left(\cot^{-1}\frac{1-x^2}{2x}\right) = \frac{1-x^2}{2x}$

$\therefore \quad \cos^{-1}\{\cot(2\tan^{-1}x)\} = \cos^{-1}\left(\frac{1-x^2}{2x}\right)$

$\therefore \quad$ the given equation is $\sin\left[2\cos^{-1}\left(\frac{1-x^2}{2x}\right)\right] = 0$

$\therefore \quad 2\cos^{-1}\frac{1-x^2}{2x} = n\pi$

$\therefore \quad \frac{1-x^2}{2x} = \cos\frac{n\pi}{2} \quad$ {= 0 if n is odd; = ± 1 if n is even}

$\therefore \quad \frac{1-x^2}{2x} = 0, 1, -1$

or $\quad 1 - x^2 = 0, 2x, -2x$

Now, $1 - x^2 = 0 \Rightarrow x^2 = 1$, i.e., $x = \pm 1$

$1 - x^2 = 2x \Rightarrow x^2 + 2x - 1 = 0$,

i.e., $x = \frac{-2 \pm \sqrt{4+4}}{2} = -1 \pm \sqrt{2}$

$1 - x^2 = -2x \Rightarrow x^2 - 2x - 1 = 0$,

i.e., $x = \frac{2 \pm \sqrt{4+4}}{2} = 1 \pm \sqrt{2}$

Thus, $x = 1, -1, -1 + \sqrt{2}, -1 - \sqrt{2}, 1 + \sqrt{2}, 1 - \sqrt{2}$

$= 1 \pm \sqrt{0}, -1 \pm \sqrt{0}, -1 \pm \sqrt{2}, 1 \pm \sqrt{2}$.

8. If $\tan^{-1}y : \tan^{-1}x = 4 : 1$, express y as an algebraic function of x. Hence or otherwise prove that $\tan 22\frac{1°}{2}$ is a root of the equation $x^4 + 1 = 6x^2$.

Here $\tan^{-1}y = 4\tan^{-1}x = 2(2\tan^{-1}x) = 2\tan^{-1}\frac{2x}{1-x^2}$

$= \tan^{-1}\frac{2\cdot\frac{2x}{1-x^2}}{1-\left(\frac{2x}{1-x^2}\right)^2} = \tan^{-1}\frac{4x(1-x^2)}{(1-x^2)^2 - 4x^2}$

or $\quad y = \frac{4x(1-x^2)}{1+x^4-6x^2}$.

Let $\tan^{-1}x = 22\frac{1°}{2}$; then $4\tan^{-1}x = 90°$

$\therefore\quad \tan^{-1}y = 90° \quad \{\because\ \tan^{-1}y = 4\tan^{-1}x\}$

$\therefore\quad y = \tan 90° = \dfrac{4x(1-x^2)}{1+x^4-6x^2}$

$\therefore\quad 1+x^4-6x^2 = 0$

$\therefore\quad 1+x^4 = 6x^2$

$\therefore\quad x = \tan 22\frac{1°}{2}$ is a root of the equation $1+x^4 = 6x^2$.

9. Solve $\theta = \tan^{-1}(2\tan^2\theta) - \frac{1}{2}\sin^{-1}\left(\dfrac{3\sin 2\theta}{5+4\cos 2\theta}\right)$.

$$\frac{3\sin 2\theta}{5+4\cos 2\theta} = \frac{3\cdot\dfrac{2\tan\theta}{1+\tan^2\theta}}{5+4\cdot\dfrac{1-\tan^2\theta}{1+\tan^2\theta}} = \frac{6\tan\theta}{9+\tan^2\theta}$$

$$= \frac{2\left(\frac{1}{3}\tan\theta\right)}{1+\left(\frac{1}{3}\tan\theta\right)^2}$$

$\therefore$ the equation is

$$\tan^{-1}(2\tan^2\theta) - \frac{1}{2}\sin^{-1}\left\{\frac{2\left(\frac{1}{3}\tan\theta\right)}{1+\left(\frac{1}{3}\tan\theta\right)^2}\right\} = \theta$$

or $\quad \tan^{-1}(2\tan^2\theta) - \frac{1}{2}\cdot 2\tan^{-1}\left(\frac{1}{3}\tan\theta\right) = \theta$

$$\left\{\because\ 2\tan^{-1}x = \sin^{-1}\frac{2x}{1+x^2}\right\}$$

or $\quad \tan^{-1}(2\tan^2\theta) - \tan^{-1}\left(\frac{1}{3}\tan\theta\right) = \theta$

or $\quad \tan^{-1}\dfrac{2\tan^2\theta - \frac{1}{3}\tan\theta}{1+2\tan^2\theta\cdot\frac{1}{3}\tan\theta} = \theta$

or $\quad \dfrac{6\tan^2\theta - \tan\theta}{3+2\tan^3\theta} = \tan\theta$

or $\quad \tan\theta\left[\dfrac{6\tan\theta - 1}{3+2\tan^3\theta} - 1\right] = 0$

or $\quad \tan\theta(2\tan^3\theta - 6\tan\theta + 4) = 0$

or $\quad \tan\theta(\tan\theta - 1)(2\tan^2\theta + 2\tan\theta - 4) = 0$

$\Rightarrow\quad \tan\theta = 0 \quad \ldots (1)$

or $\quad \tan\theta = 1 \quad \ldots (2)$

or $\quad \tan^2\theta + \tan\theta - 2 = 0 \quad \ldots (3)$

(1) $\Rightarrow\quad \theta = n\pi$

(2) $\Rightarrow\quad \theta = n\pi + \dfrac{\pi}{4}$

(3) $\Rightarrow\quad (\tan\theta - 1)(\tan\theta + 2) = 0$

$\therefore\quad \tan\theta = -2$

$\therefore\quad \theta = n\pi + \tan^{-1}(-2)$

$\therefore\quad$ we get $\theta = n\pi,\ n\pi + \dfrac{\pi}{4},\ n\pi + \tan^{-1}(-2)$ where $n \in \mathbb{Z}$.

10. Find the positive integral solutions of the equation

$$\tan^{-1}x + \cos^{-1}\frac{y}{\sqrt{1+y^2}} = \sin^{-1}\frac{3}{\sqrt{10}}.$$

We have

$$\cos^{-1}\frac{y}{\sqrt{1+y^2}} = \tan^{-1}\frac{\sqrt{(\sqrt{1+y^2})^2 - y^2}}{y} = \tan^{-1}\frac{1}{y}$$

$$\sin^{-1}\frac{3}{\sqrt{10}} = \tan^{-1}\frac{3}{\sqrt{(\sqrt{10})^2 - 3^2}} = \tan^{-1}3$$

$\therefore\quad$ from the given equation

$$\tan^{-1}x + \tan^{-1}\frac{1}{y} = \tan^{-1}3$$

or $\quad \tan^{-1}\dfrac{1}{y} = \tan^{-1}3 - \tan^{-1}x$

or $\quad \tan^{-1}\dfrac{1}{y} = \tan^{-1}\dfrac{3-x}{1+3x};\quad \therefore\quad y = \dfrac{1+3x}{3-x}$

But x, y are positive integers; so $x = 1, 2$

When $x = 1,\ y = \dfrac{1+3}{3-1} = 2$

$x = 2,\ y = \dfrac{1+6}{3-2} = 7$

$\therefore\quad$ positive integral solutions of $(x, y) = (1, 2), (2, 7)$.

11. Solve for x: $\tan^{-1}x + \tan^{-1}(1-x) = \cot^{-1}\dfrac{7}{9},\ x \in (0, 1)$.

Here $\tan^{-1}x + \tan^{-1}(1-x) = \tan^{-1}\dfrac{9}{7} \quad \ldots (1)$

We know that $\tan^{-1}x + \tan^{-1}(1-x)$

$$= \tan^{-1}\frac{x+1-x}{1-x(1-x)} \quad \text{if } 0 < x < 1 \text{ and } x(1-x) < 1$$

$$\pi + \tan^{-1}\frac{x+1-x}{1-x(1-x)} \quad \text{if } 0 < x < 1 \text{ and } x(1-x) > 1.$$

Now, $x(1-x) < 1 \quad \Rightarrow \quad x^2 - x + 1 > 0$

$\Rightarrow\quad \left(x - \dfrac{1}{2}\right)^2 + \dfrac{3}{4} > 0$ which is true for all $x \in R$.

So, $x(1-x) > 1$ is not true for any $x \in R$.

$\therefore$ for all $0 < x < 1$, (1) becomes

$$\tan^{-1}\frac{x+1-x}{1-x(1-x)} = \tan^{-1}\frac{9}{7}$$

or $$\tan^{-1}\frac{1}{1-x+x^2} = \tan^{-1}\frac{9}{7}$$

or $$\frac{1}{1-x+x^2} = \frac{9}{7}, \text{ i.e., } 7 = 9 - 9x + 9x^2$$

$\therefore$ $9x^2 - 9x + 2 = 0$, i.e., $(3x-1)(3x-2) = 0$

$\therefore$ $x = \frac{1}{3}, \frac{2}{3}$.

Exercises

1. Prove that $\tan^{-1}\frac{1}{3} + \tan^{-1}\frac{1}{5} + \tan^{-1}\frac{1}{7} + \tan^{-1}\frac{1}{8} = \frac{\pi}{4}$.

2. Prove that $2\cot^{-1}5 + \cot^{-1}7 + 2\cot^{-1}8 = \frac{\pi}{4}$.

3. Prove that

$$\tan^{-1}\left(\frac{1}{2}\tan 2A\right) + \tan^{-1}(\cot A) + \tan^{-1}(\cot^3 A) = 0.$$

4. Prove that

$$\tan^{-1}\sqrt{\frac{xr}{yz}} + \tan^{-1}\sqrt{\frac{yr}{zx}} + \tan^{-1}\sqrt{\frac{zr}{xy}} = \pi$$

where $r = x + y + z$.

5. Prove that $\tan^{-1}\frac{yz}{xr} + \tan^{-1}\frac{zx}{yr} + \tan^{-1}\frac{xy}{zr} = \frac{\pi}{2}$

where $r^2 = x^2 + y^2 + z^2$.

6. Prove that $\sin^{-1}\frac{4}{5} + \sin^{-1}\frac{5}{13} + \sin^{-1}\frac{16}{65} = \frac{\pi}{2}$.

7. Prove that $\cos^{-1}\frac{3}{5} + \cos^{-1}\frac{12}{13} + \cos^{-1}\frac{63}{65} = \frac{\pi}{2}$.

8. Show that

$$\sin^{-1}\sqrt{\frac{u}{1+u}} - \sin^{-1}\frac{u-1}{u+1} = \sin^{-1}\frac{1}{\sqrt{1+u}}, \quad u > 1$$

is an identity.

9. If $\cos^{-1}\frac{x}{a} + \cos^{-1}\frac{y}{b} = \alpha$, prove that

$$b^2x^2 - 2xy\,ab\cos\alpha + a^2y^2 = a^2b^2\sin^2\alpha.$$

10. If $u = \cot^{-1}\sqrt{\cos 2\theta} - \tan^{-1}\sqrt{\cos 2\theta}$,

prove that $\sin u = \tan^2\theta$.

11. Prove that

$$\tan\left(\frac{\pi}{4} + \frac{1}{2}\cos^{-1}\frac{a}{b}\right) + \tan\left(\frac{\pi}{4} - \frac{1}{2}\cos^{-1}\frac{a}{b}\right) = \frac{2b}{a}.$$

12. Prove that

$$\tan^{-1}\left(\tan\frac{\theta}{2}\cdot\tan\frac{\phi}{2}\right) = \frac{1}{2}\cos^{-1}\left(\frac{\cos\theta + \cos\phi}{1 + \cos\theta\cos\phi}\right).$$

13. Prove that

$$2\tan^{-1}\left[\tan\frac{\alpha}{2}\cdot\tan\left(\frac{\pi}{4} - \frac{\beta}{2}\right)\right] = \tan^{-1}\frac{\sin\alpha\cdot\cos\beta}{\cos\alpha + \sin\beta}.$$

14. Express $\cot^{-1}\frac{y}{\sqrt{1-x^2-y^2}}$

$$= 2\tan^{-1}\sqrt{\frac{3}{4x^2} - 1} - \tan^{-1}\sqrt{\frac{3}{x^2} - 4}$$

as a rational integral equation between x and y.

15. If $\sin^{-1}x + \sin^{-1}y = \frac{\pi}{2}$, prove that

$$2(x^2 - x^2y^2 + y^2) = 1 + x^4 + y^4.$$

16. If $\tan^{-1}x + \tan^{-1}y + \tan^{-1}z = \pi$, prove that

$$x + y + z = xyz.$$

17. If $\cot^{-1}x + \cot^{-1}y + \cot^{-1}z = \pi$, prove that

$$xy + yz + zx = 1.$$

18. If $\sin^{-1}x + \sin^{-1}y + \sin^{-1}z = \pi$, prove that

$$x\sqrt{1-x^2} + y\sqrt{1-y^2} + z\sqrt{1-z^2} = 2xyz.$$

19. If in ΔABC, $A = 90°$ then prove that

$$\tan^{-1}\frac{b}{c+a} + \tan^{-1}\frac{c}{a+b} = \frac{\pi}{4}.$$

20. If in a triangle, two angles are $\tan^{-1}2$ and $\tan^{-1}3$ then prove that the third angle is $\frac{\pi}{4}$.

21. Solve: $\tan^{-1}2x + \tan^{-1}3x = \frac{\pi}{4}$.

22. Solve: $\sin^{-1}x + \sin^{-1}2x = \frac{\pi}{3}$.

23. Solve: $\cos^{-1}\frac{x^2-1}{x^2+1} + \tan^{-1}\frac{2x}{x^2-1} = \frac{2}{3}\pi$.

24. Solve: $\sin^{-1}6x + \sin^{-1}6\sqrt{3}\,x = -\frac{\pi}{2}$.

25. Solve: $\tan^{-1}(x+1) + \tan^{-1}(x-1) = \tan^{-1}\frac{8}{31}$.

Objective Questions

Fill in the blanks.

26. The principal value of $\cos^{-1}\left(\cos\frac{7\pi}{4}\right) =$ ______ .

27. The principal value of $\tan^{-1}(-\sqrt{3}) =$ ______ .

28. The value of $\sin\left(\cos^{-1}\frac{1}{2} + \sin^{-1}\frac{\sqrt{3}}{2}\right) =$ ______ .

29. The numerical value of $2\tan^{-1}\frac{1}{3} + \tan^{-1}\frac{1}{7} =$ ______ .

30. $\sin^{-1}\left(\sqrt{\frac{x}{x+a}}\right) = \tan^{-1}($______$)$.

31. The numerical value of

$4\left(\cot^{-1}3 + \sin^{-1}\frac{1}{\sqrt{5}}\right) =$ ______ .

32. The numerical value of

$\cos\{\tan^{-1}(-3) + \cot^{-1}(-3)\} =$ ______ .

33. The numerical value of

$\sec^2(\tan^{-1}2) + \text{cosec}^2(\cot^{-1}3) =$ ______ .

34. $2\sin^{-1}\sqrt{\frac{1-x}{2}} = \cos^{-1}($______$)$.

35. $\sin^{-1}(3x - 4x^3) = \lambda \sin^{-1}x$ then $\lambda =$ ______ .

36. The numerical value of

$\tan^{-1}1 + \tan^{-1}2 + \tan^{-1}3 =$ ______ .

37. The numerical value of

$\tan^{-1}1 + \tan^{-1}\frac{1}{2} + \tan^{-1}\frac{1}{3} =$ ______ .

38. If $x > 1$ then $2\tan^{-1}x - \tan^{-1}\frac{2x}{1-x^2} =$ ______ .

39. $3\tan^{-1}x = \tan^{-1}\frac{3x - x^3}{1 - 3x^2}$, $(x > 0)$ if $x <$ ______ .

40. $\cos^{-1}\frac{63}{65} + 2\tan^{-1}\frac{1}{5} = \sin^{-1}($______$)$.

41. $\cot^{-1}\frac{ab+1}{a-b} + \cot^{-1}\frac{bc+1}{b-c} + \cot^{-1}\frac{ca+1}{c-a} =$ ______ .

42. $\tan^{-1}\left(\frac{a_1x - y}{a_1y + x}\right) + \tan^{-1}\frac{a_2 - a_1}{a_1a_2 + 1} + \tan^{-1}\frac{a_3 - a_2}{a_2a_3 + 1} + \ldots$

$+ \tan^{-1}\frac{a_n - a_{n-1}}{a_{n-1}a_n + 1} + \tan^{-1}\frac{1}{a_n} =$ ______ .

Choose the correct option(s).

43. The principal value of $\cos^{-1}\left(\sin\frac{10\pi}{7}\right)$ is

(a) $\frac{3\pi}{7}$ (b) $\frac{\pi}{14}$

(c) $\frac{13\pi}{14}$ (d) none of these

44. The value of $\sin\left\{\tan^{-1}\left(\tan\frac{7\pi}{6}\right) + \cos^{-1}\left(\cos\frac{7\pi}{3}\right)\right\}$ is

(a) 0 (b) 1

(c) –1 (d) none of these

45. $\sin[\cot^{-1}\{\tan(\cos^{-1}x)\}]$ is equal to

(a) $-x$ (b) x^2

(c) x (d) none of these

46. $2\tan^{-1}x = \tan^{-1}\frac{2x}{1-x^2}$ is true if

(a) $x > 0,\ x < 1$ (b) $x > 1$

(c) $x < 0$ (d) $x \in R$

47. $\tan^{-1}x + \tan^{-1}\frac{2x}{1-x^2} = \pi + \tan^{-1}\frac{3x - x^3}{1 - 3x^2}$, $(x > 0)$ is true if

(a) $x < \frac{1}{\sqrt{3}}$ (b) $\frac{1}{\sqrt{3}} < x < 1$

(c) $x > \frac{1}{\sqrt{3}}$ (d) $\frac{1}{\sqrt{3}} < x < \sqrt{3}$

48. If $\cot^{-1}x + \cot^{-1}y + \cot^{-1}z = \frac{\pi}{2}$ then $x + y + z$ equals

(a) xyz (b) $xy + yz + zx$

(c) $2xyz$ (d) none of these

Answers

14. $27y^2 = x^2(9 - 8x^2)^2$ **21.** $\frac{1}{6}$ **22.** $\frac{1}{2}\sqrt{\frac{3}{7}}$

23. $\sqrt{3}, 2 - \sqrt{3}$ **24.** $\frac{-1}{12}$ **25.** $\frac{1}{4}$ **26.** $\frac{\pi}{4}$

27. $-\frac{\pi}{3}$ **28.** $\frac{\sqrt{3}}{2}$ **29.** $\frac{\pi}{4}$ **30.** $\sqrt{\frac{x}{a}}$

31. π **32.** 0 **33.** 15 **34.** x

35. 3 **36.** π **37.** $\frac{\pi}{2}$ **38.** π

39. $\frac{1}{\sqrt{3}}$ **40.** $\frac{3}{5}$ **41.** 0

42. $\tan^{-1}\frac{x}{y}$ **43.** (c) **44.** (b) **45.** (c)

46. (a) **47.** (b) **48.** (a)

□

Chapter Test

Time: 60 minutes

1. In the following incomplete sentences fill in the blanks such that the resulting sentences become true.

(a) $\cot^{-1}\dfrac{\sqrt{1-\sin x}+\sqrt{1+\sin x}}{\sqrt{1-\sin x}-\sqrt{1+\sin x}} =$ ______ if $0° < x < 90°$.

(b) $\sin^{-1}\dfrac{3}{5} - \cos^{-1}\dfrac{12}{13} = \sin^{-1}($ ______ $)$.

(c) If $x > 1$ then $\dfrac{1}{2}\cos^{-1}\dfrac{x^2-1}{x^2+1} = \tan^{-1}($ ______ $)$.

(d) If $\tan^{-1}\dfrac{1}{2} + \tan^{-1}a = \dfrac{\pi}{4}$ then $a =$ ______ .

(e) The value of $\sin^{-1}[\cos\{\cos^{-1}(\cos x) + \sin^{-1}(\sin x)\}] = \ldots$, where $\dfrac{\pi}{2} < x < \pi$.

2. In the following questions each has multiple answers. Choose the correct answer(s).

(a) If $\tan^{-1}\dfrac{\sqrt{1+x^2}-1}{x} = \lambda\tan^{-1}x$ then the value of λ is

A. 2 B. $\dfrac{1}{2}$ C. 1 D. none of these

(b) If $\phi = \tan^{-1}\dfrac{x\sqrt{3}}{2k-x}$ and $\psi = \tan^{-1}\dfrac{2x-k}{k\sqrt{3}}$ then one value of $\phi - \psi$ is

A. $\dfrac{\pi}{6}$ B. $\dfrac{\pi}{3}$ C. $\dfrac{7\pi}{6}$ D. $\dfrac{4\pi}{3}$

(c) The principal value of $\cos^{-1}\left(\cos\dfrac{2\pi}{3}\right) + \sin^{-1}\left(\sin\dfrac{2\pi}{3}\right)$ is

A. π B. $\dfrac{\pi}{2}$ C. $\dfrac{\pi}{3}$ D. $\dfrac{4\pi}{3}$

(d) If λ is a root of $x^2 + 3x + 1 = 0$ then $\tan^{-1}\lambda + \tan^{-1}\dfrac{1}{\lambda}$ is equal to

A. $\dfrac{\pi}{2}$ B. $-\dfrac{\pi}{2}$ C. $\dfrac{\pi}{3}$ D. none of these

(e) The equation $\sin^{-1}x - 3\sin^{-1}a = 0$ has real solutions for x if

A. $a \in R$ B. $a \in [-1, 1]$ C. $a \in \left[0, \dfrac{1}{2}\right]$ D. $a \in \left[-\dfrac{1}{2}, \dfrac{1}{2}\right]$

3. Solve: $\tan^{-1}3x - \tan^{-1}x = \cot^{-1}\dfrac{1}{x+1} + \cot^{-1}\dfrac{1}{x-1}$.

4. Find the integral values of n for which $n\pi - \tan^{-1}3$ is a solution of the equation

$1 + \sqrt{10}\sec x + 12\tan 2x = 0$.

5. Find the number of real solutions for x, y if

$|y| = \sin x,\ y = \cos^{-1}(\cos x),\ |x| \le 2\pi$. Also find the solutions.

6. Solve: $\tan^{-1}\dfrac{x-1}{x-2} + \tan^{-1}\dfrac{x+1}{x-2} = \dfrac{\pi}{4}$.

□

Answers

1. (a) $\frac{-x}{2}$ (b) $\frac{16}{65}$ (c) $\frac{1}{x}$ (d) $\frac{1}{3}$ (e) $-\frac{\pi}{2}$

2. (a) B (b) A (c) A (d) B (e) D

3. $0, \pm\frac{1}{2}$

4. $2k+1$ where $k \in Z$

5. three; $(0, 0)$, $(\pm 2\pi, 0)$

6. $-\sqrt{\frac{5}{2}}$

□

4. Trigonometrical Inequalities and Inequations

Recap of Facts and Formulae

1. Basic inequalities

- $-1 \le \sin x \le 1$
- $-1 \le \cos x \le 1$
- $\sec x \ge 1$ or $\sec x \le -1$
- $\operatorname{cosec} x \ge 1$ or $\operatorname{cosec} x \le -1$
- For positive quantities AM ≥ GM, equality holding if all the quantities are equal.

2. Solutions of trigonometrical inequations—*Value-scheme*

- If $\sin x > k$, where $|k| \le 1$, and $k = \sin \alpha$ then the solutions of the inequation are given by the following value-scheme.

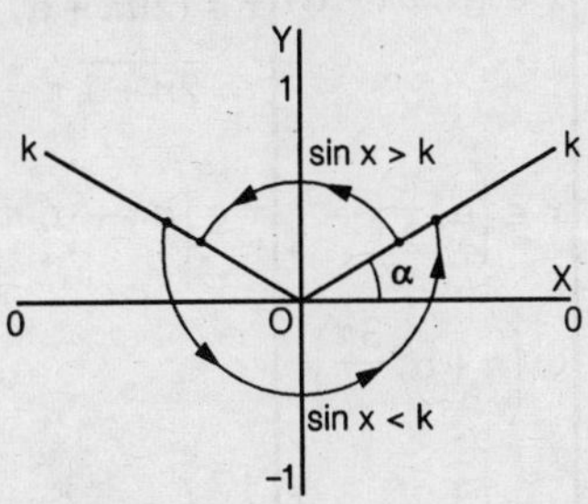

(i) The solutions of $\sin x > k = \sin \alpha$ are

$\alpha < x < \pi - \alpha$, i.e., $x \in (\alpha, \pi - \alpha)$.

General solutions will be

$2n\pi + \alpha < x < (2n + 1)\pi - \alpha$,

i.e., $x \in (2n\pi + \alpha, 2n\pi + \pi - \alpha)$, $n \in \mathbb{Z}$.

(ii) The solutions of $\sin x < k = \sin \alpha$ are

$0 < x < \alpha$, $\pi - \alpha < x < 2\pi$.

General solutions will be

$2n\pi < x < 2n\pi + \alpha$,

$(2n + 1)\pi - \alpha < x < (2n + 1)\pi + \pi$, $n \in \mathbb{Z}$

For example Solve $\sin x > -\dfrac{1}{2}$.

We know $-\dfrac{1}{2} = \sin\left(-\dfrac{\pi}{6}\right)$, $\sin\left(\pi + \dfrac{\pi}{6}\right)$

So, the value-scheme will be as follows.

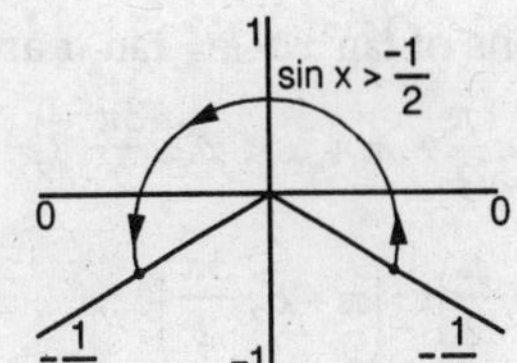

Solution-set of $\sin x > \dfrac{-1}{2}$ is $\left(-\dfrac{\pi}{6}, \pi + \dfrac{\pi}{6}\right)$ and the general solution-set is

$$\left(2n\pi - \frac{\pi}{6}, (2n + 1)\pi + \frac{\pi}{6}\right), n \in \mathbb{Z}.$$

Clearly, the general solution-set of the inequation $\sin x \ge -\dfrac{1}{2}$ will be

$$\left[2n\pi - \frac{\pi}{6}, (2n + 1)\pi + \frac{\pi}{6}\right], n \in \mathbb{Z}.$$

- If $\cos x > k$, where $|k| \le 1$ and $k = \cos \alpha$ then the solutions of the inequation are given by the following value-scheme.

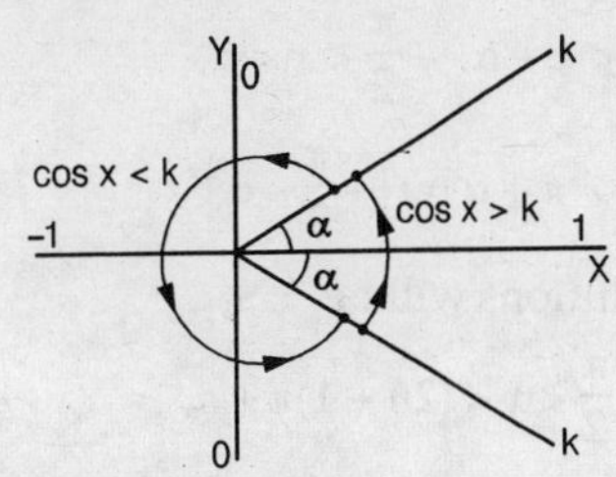

(i) The solutions of $\cos x > k = \cos \alpha$ are

$-\alpha < x < \alpha$, i.e., $x \in (-\alpha, \alpha)$.

General solutions will be

$2n\pi - \alpha < x < 2n\pi + \alpha$,

i.e., $x \in (2n\pi - \alpha, 2n\pi + \alpha)$, $n \in \mathbb{Z}$.

(ii) The solutions of $\cos x < k = \cos \alpha$ are

$\alpha < x < 2\pi - \alpha$, i.e., $x \in (\alpha, 2\pi - \alpha)$

General solutions will be

$2n\pi + \alpha < x < 2(n + 1)\pi - \alpha$,

i.e., $x \in (2n\pi + \alpha, \overline{2n + 1}\pi - \alpha)$, $n \in \mathbb{Z}$.

- If $\tan x > k$, where k is any real number and $k = \tan \alpha$ then the solutions of the inequation are given by the following value-scheme.

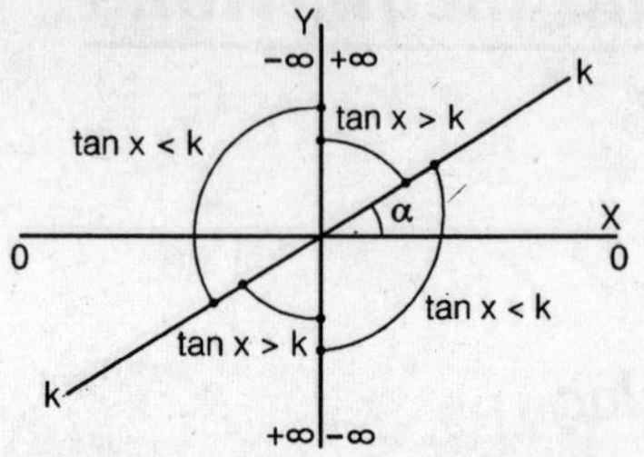

(i) The solutions of $\tan x > k = \tan \alpha$ are

$$\alpha < x < \frac{\pi}{2},\quad \pi + \alpha < x < \frac{3\pi}{2}$$

i.e., $x \in \left(\alpha, \frac{\pi}{2}\right) \cup \left(\pi + \alpha, \frac{3\pi}{2}\right)$.

General solutions will be

$$2n\pi + \alpha < x < 2n\pi + \frac{\pi}{2},$$

$$(2n+1)\pi + \alpha < x < (2n+1)\pi + \frac{\pi}{2}$$

i.e., $x \in \left(2n\pi + \alpha,\ 2n\pi + \frac{\pi}{2}\right)$

$$\cup \left((2n+1)\pi + \alpha, (2n+1)\pi + \frac{\pi}{2}\right).$$

Equivalently, $x \in \left(n\pi + \alpha, n\pi + \frac{\pi}{2}\right),\ n \in \mathbb{Z}$

(ii) The solutions of $\tan x < k = \tan \alpha$ are

$$\frac{\pi}{2} < x < \pi + \alpha,\quad -\frac{\pi}{2} < x < \alpha,$$

i.e., $x \in \left(\frac{\pi}{2}, \pi + \alpha\right) \cup \left(-\frac{\pi}{2}, \alpha\right)$

General solutions will be

$$2n\pi + \frac{\pi}{2} < x < (2n+1)\pi + \alpha,$$

$$2n\pi - \frac{\pi}{2} < x < 2n\pi + \alpha,$$

i.e., $x \in \left(2n\pi + \frac{\pi}{2},\ (2n+1)\pi + \alpha\right)$

$$\cup \left(2n\pi - \frac{\pi}{2},\ 2n\pi + \alpha\right).$$

Equivalently, $x \in \left(n\pi - \frac{\pi}{2},\ n\pi + \alpha\right),\ n \in \mathbb{Z}$.

Summarizing the above three in tabular form, we get:

Inequation	*Solution in* $[0, 2\pi]$ *or* $[-\pi, \pi]$	*General solution* $(n \in \mathbb{Z})$
$\sin x > k (= \sin \alpha)$	$x \in (\alpha, \pi - \alpha)$	$x \in (2n\pi + \alpha, \overline{2n+1}\,\pi - \alpha)$
$\sin x < k (= \sin \alpha)$	$x \in [0, \alpha) \cup (\pi - \alpha, 2\pi)$	$x \in [2n\pi, 2n\pi + \alpha) \cup (\overline{2n+1}\,\pi - \alpha, \overline{2n+1}\,\pi + \pi)$
$\cos x > k (= \cos \alpha)$	$x \in (-\alpha, \alpha)$	$x \in (2n\pi - \alpha, 2n\pi + \alpha)$
$\cos x < k (= \cos \alpha)$	$x \in (\alpha, 2\pi - \alpha)$	$x \in (2n\pi + \alpha, \overline{2n+1}\,\pi - \alpha)$
$\tan x > k (= \tan \alpha)$	$x \in \left(\alpha, \frac{\pi}{2}\right) \cup \left(\pi + \alpha, \frac{3\pi}{2}\right)$	$x \in \left(n\pi + \alpha, n\pi + \frac{\pi}{2}\right)$
$\tan x < k (= \tan \alpha)$	$x \in \left(\frac{\pi}{2}, \pi + \alpha\right) \cup \left(-\frac{\pi}{2}, \alpha\right)$	$x \in \left(n\pi - \frac{\pi}{2}, n\pi + \alpha\right)$

Selected Solved Examples

1. Prove that $\cos A + \cos B + \cos C \le \frac{3}{2}$ where $A + B + C = \pi$. Also prove that if in a $\triangle ABC$, $\cos A + \cos B + \cos C = \frac{3}{2}$ then the triangle is equilateral.

$\cos A + \cos B + \cos C$

$$= 2\cos\frac{A+B}{2} \cdot \cos\frac{A-B}{2} + 1 - 2\sin^2\frac{C}{2}$$

$$= 2\sin\frac{C}{2} \cdot \left(\cos\frac{A-B}{2} - \sin\frac{C}{2}\right) + 1 \quad (\because\ A + B = \pi - C)$$

$$\le 2\sin\frac{C}{2}\left(1 - \sin\frac{C}{2}\right) + 1$$

$$\left(\because \text{ the greatest value of } \cos\frac{A-B}{2} \text{ is } 1\right)$$

and equality holds when $\cos\frac{A-B}{2} = 1$ ...(1)

$$\therefore \quad \cos A + \cos B + \cos C \le 1 - 2\left(\sin^2\frac{C}{2} - \sin\frac{C}{2}\right)$$

$$= 1 - 2\left(\sin^2\frac{C}{2} - \sin\frac{C}{2} + \frac{1}{4}\right) + 2\cdot\frac{1}{4}$$

$$= \frac{3}{2} - 2\left(\sin\frac{C}{2} - \frac{1}{2}\right)^2 \le \frac{3}{2},$$

equality holding when $\sin\frac{C}{2} = \frac{1}{2}$ ... (2)

Thus, $\cos A + \cos B + \cos C \le \frac{3}{2}$ ... (3)

equality holding when (1) and (2) hold.

Now, in $\triangle ABC$, $A + B + C = \pi$; so (3) is true.

If $\cos A + \cos B + \cos C = \frac{3}{2}$ then in (3) equality holds.

$\therefore$ (1) and (2) are true,

i.e., $\cos\frac{A-B}{2} = 1$ and $\sin\frac{C}{2} - \frac{1}{2} = 0$.

$$\cos\frac{A-B}{2} = 1 \quad\Rightarrow\quad A - B = 0, \text{ i.e., } A = B$$

$$\sin\frac{C}{2} - \frac{1}{2} = 0 \quad\Rightarrow\quad \frac{C}{2} = 30°, \text{ i.e., } C = 60°.$$

Clearly, $A + B + C = 180°$, $A = B$ and $C = 60°$

$\Rightarrow$ $A = B = C = 60°$; so the $\triangle ABC$ is equilateral.

2. Prove that in an acute-angled $\triangle ABC$,

$$\tan^2\frac{A}{2} + \tan^2\frac{B}{2} + \tan^2\frac{C}{2} \ge 1.$$

Here $A + B + C = \pi$; $\therefore$ $\frac{B+C}{2} = \frac{\pi}{2} - \frac{A}{2}$

$$\therefore \quad \tan\frac{B+C}{2} = \tan\left(\frac{\pi}{2} - \frac{A}{2}\right) = \cot\frac{A}{2}$$

or $$\frac{\tan\frac{B}{2} + \tan\frac{C}{2}}{1 - \tan\frac{B}{2}\cdot\tan\frac{C}{2}} = \frac{1}{\tan\frac{A}{2}}$$

or $$\tan\frac{A}{2}\left(\tan\frac{B}{2} + \tan\frac{C}{2}\right) = 1 - \tan\frac{B}{2}\cdot\tan\frac{C}{2}$$

or $$\tan\frac{A}{2}\cdot\tan\frac{B}{2} + \tan\frac{B}{2}\cdot\tan\frac{C}{2} + \tan\frac{C}{2}\cdot\tan\frac{A}{2} = 1 \quad \text{... (1)}$$

Now, $\tan^2\frac{A}{2} + \tan^2\frac{B}{2} + \tan^2\frac{C}{2} - 1$

$$= \tan^2\frac{A}{2} + \tan^2\frac{B}{2} + \tan^2\frac{C}{2} - \left(\tan\frac{A}{2}\cdot\tan\frac{B}{2} + \tan\frac{B}{2}\cdot\tan\frac{C}{2} + \tan\frac{C}{2}\cdot\tan\frac{A}{2}\right), \text{ using (1)}$$

$$= \frac{1}{2}\left\{\left(\tan\frac{A}{2} - \tan\frac{B}{2}\right)^2 + \left(\tan\frac{B}{2} - \tan\frac{C}{2}\right)^2 + \left(\tan\frac{C}{2} - \tan\frac{A}{2}\right)^2\right\}$$

because $a^2 + b^2 + c^2 - bc - ca - ab$

$$= \frac{1}{2}\{(a-b)^2 + (b-c)^2 + (c-a)^2\} \ge 0,$$

and sum of squares of real numbers is positive. Here equality holds when

$$\tan\frac{A}{2} = \tan\frac{B}{2} = \tan\frac{C}{2}, \text{ i.e., } A = B = C.$$

$$\therefore \quad \tan^2\frac{A}{2} + \tan^2\frac{B}{2} + \tan^2\frac{C}{2} - 1 \ge 0,$$

i.e., $\tan^2\frac{A}{2} + \tan^2\frac{B}{2} + \tan^2\frac{C}{2} \ge 1$,

equality holding when $A = B = C$, i.e., the triangle is equilateral.

3. Prove that in an acute angled $\triangle ABC$,
$\tan A\cdot\tan B\cdot\tan C \ge 3\sqrt{3}$.
Also prove that the triangle is equilateral if the equality holds.

As the triangle is acute, $\tan A$, $\tan B$ and $\tan C$ are positive.

$\therefore$ AM $\ge$ GM

$$\Rightarrow \quad \frac{\tan A + \tan B + \tan C}{3} \ge \sqrt[3]{\tan A\tan B\cdot\tan C} \quad \text{... (1)}$$

equality holding if the numbers are equal, i.e.,

$$\tan A = \tan B = \tan C \quad \text{... (2)}$$

Now, $A + B + C = \pi$; $\therefore$ $B + C = \pi - A$

$\therefore$ $\tan(B + C) = \tan(\pi - A)$

or $$\frac{\tan B + \tan C}{1 - \tan B\cdot\tan C} = -\tan A$$

or $\tan A + \tan B + \tan C = \tan A\cdot\tan B\cdot\tan C = x$(say)

$\therefore$ from (1), $\frac{x}{3} \ge \sqrt[3]{x}$; or $\frac{x^3}{27} \ge x$.

As $x > 0$, $x^2 \ge 27$; $\therefore$ $x \ge \sqrt{27} = 3\sqrt{3}$

$\therefore$ $\tan A\cdot\tan B\cdot\tan C \ge 3\sqrt{3}$,

equality holding when (2) is true, i.e.,

$\tan A = \tan B = \tan C$

$\therefore$ $A = B = C$, i.e., the triangle is equilateral.

4. Prove that in a $\triangle ABC$, $\sin A\cdot\sin B\cdot\sin C \le \frac{3\sqrt{3}}{8}$.

$\sin^2 A + \sin^2 B + \sin^2 C$

$= \frac{1}{2}(1 - \cos 2A) + \frac{1}{2}(1 - \cos 2B) + \sin^2 C$

$= 1 - \frac{1}{2}(\cos 2A + \cos 2B) + \sin^2 C$

$= 1 - \frac{1}{2}\{2\cos(A + B) \cdot \cos(A - B)\} + \sin^2 C$

$= 1 + \cos C \cdot \cos(A - B) + 1 - \cos^2 C$

$= 2 - \cos^2 C + \cos C \cdot \cos(A - B)$

$\therefore \quad \sin^2 A + \sin^2 B + \sin^2 C \le 2 - \cos^2 C + \cos C$

$\{\because$ the greatest value of $\cos(A - B) = 1\}$

$= 2 - (\cos^2 C - \cos C)$

$= 2 - \left(\cos^2 C - \cos C + \frac{1}{4}\right) + \frac{1}{4}$

$= \frac{9}{4} - \left(\cos C - \frac{1}{2}\right)^2$

$\therefore \quad \sin^2 A + \sin^2 B + \sin^2 C \le \frac{9}{4}$ ... (1)

Now, for positive quantities, AM ≥ GM

$\therefore \quad \frac{\sin^2 A + \sin^2 B + \sin^2 C}{3} \ge \sqrt[3]{\sin^2 A \sin^2 B \sin^2 C}$

$\therefore \quad$ from (1), $\frac{\frac{9}{4}}{3} \ge \frac{\sin^2 A + \sin^2 B + \sin^2 C}{3}$

$\ge (\sin A \sin B \sin C)^{\frac{2}{3}}$

$\therefore \quad \frac{3}{4} \ge (\sin A \sin B \sin C)^{\frac{2}{3}}$

or $\quad \sin A \sin B \sin C \le \left(\frac{3}{4}\right)^{\frac{3}{2}}$, i.e., $\frac{3\sqrt{3}}{8}$.

5. For all $\theta \in \left[0, \frac{\pi}{2}\right]$, show that $\cos(\sin\theta) > \sin(\cos\theta)$.

$\cos(\sin\theta) - \sin(\cos\theta) = \cos(\sin\theta) - \cos\left(\frac{\pi}{2} - \cos\theta\right)$

$= 2\sin\frac{1}{2}\left(\frac{\pi}{2} + \sin\theta - \cos\theta\right) \cdot \sin\frac{1}{2}\left(\frac{\pi}{2} - \cos\theta - \sin\theta\right)$

$= 2\sin\left\{\frac{\pi}{4} + \frac{\sin\theta - \cos\theta}{2}\right\} \cdot \sin\left\{\frac{\pi}{4} - \frac{\sin\theta + \cos\theta}{2}\right\}$

$= 2\sin\left\{\frac{\pi}{4} + \frac{\sin\left(\theta - \frac{\pi}{4}\right)}{\sqrt{2}}\right\} \cdot \sin\left\{\frac{\pi}{4} - \frac{\sin\left(\theta + \frac{\pi}{4}\right)}{\sqrt{2}}\right\}$ ... (1)

Now, minimum value of

$\frac{\sin\left(\theta - \frac{\pi}{4}\right)}{\sqrt{2}} = \frac{\frac{-1}{\sqrt{2}}}{\sqrt{2}} = \frac{-1}{2}$, for $0 \le \theta \le \frac{\pi}{2}$

maximum value of $\frac{\sin\left(\theta - \frac{\pi}{4}\right)}{\sqrt{2}} = \frac{\frac{1}{\sqrt{2}}}{\sqrt{2}} = \frac{1}{2}$, for $0 \le \theta \le \frac{\pi}{2}$

$\therefore \quad \frac{\pi}{4} - \frac{1}{2} \le \frac{\pi}{4} + \frac{\sin\left(\theta - \frac{\pi}{4}\right)}{\sqrt{2}} \le \frac{\pi}{4} + \frac{1}{2}$.

Also, minimum value of

$\frac{\sin\left(\theta + \frac{\pi}{4}\right)}{\sqrt{2}} = \frac{\frac{1}{\sqrt{2}}}{\sqrt{2}} = \frac{1}{2}$, for $0 \le \theta \le \frac{\pi}{2}$

maximum value of $\frac{\sin\left(\theta + \frac{\pi}{4}\right)}{\sqrt{2}} = \frac{1}{\sqrt{2}}$, for $0 \le \theta \le \frac{\pi}{2}$

$\therefore \quad \frac{\pi}{4} - \frac{1}{\sqrt{2}} \le \frac{\pi}{4} - \frac{\sin\left(\theta + \frac{\pi}{4}\right)}{\sqrt{2}} \le \frac{\pi}{4} - \frac{1}{2}$.

Clearly, $\left(\frac{\pi}{4} - \frac{1}{2}\right)$, $\left(\frac{\pi}{4} + \frac{1}{2}\right)$, $\left(\frac{\pi}{4} - \frac{1}{\sqrt{2}}\right)$ and $\left(\frac{\pi}{4} - \frac{1}{2}\right)$ radians are all angles between 0 and π. Hence both the factors in (1) are positive.

Hence, $\cos(\sin\theta) - \sin(\cos\theta) > 0$

$\therefore \quad \cos(\sin\theta) > \sin(\cos\theta)$.

6. Prove that $-4 \le \cos 2x + 3\sin x \le \frac{17}{8}$ for all real x.

Here $\cos 2x + 3\sin x = 1 - 2\sin^2 x + 3\sin x$

$= 1 - 2\left(\sin^2 x - \frac{3}{2}\sin x\right)$

$= 1 - 2\left(\sin^2 x - 2 \cdot \frac{3}{4}\sin x + \frac{9}{16}\right) + 2 \cdot \frac{9}{16}$

$= 1 - 2\left(\sin x - \frac{3}{4}\right)^2 + \frac{9}{8} = \frac{17}{8} - 2\left(\sin x - \frac{3}{4}\right)^2$ ... (1)

$\therefore \quad \cos 2x + 3\sin x \le \frac{17}{8}$,

equality takes place when $\sin x = \frac{3}{4}$

and from (1), $\cos 2x + 3\sin x \ge \frac{17}{8} - 2\left(-1 - \frac{3}{4}\right)^2$,

for when $\sin x = -1$ (minimum value) is replaced, $2\left(\sin x - \frac{3}{4}\right)^2$ becomes the greatest.

So, $\quad \cos 2x + 3\sin x \ge \frac{17}{8} - 2 \cdot \left(\frac{-7}{4}\right)^2$

$= \frac{17}{8} - 2 \cdot \frac{49}{16} = \frac{17 - 49}{8} = -4$.

Thus, $-4 \le \cos 2x + 3\sin x \le \frac{17}{8}$.

7. Prove that $3(\cot^2\theta + \tan^2\theta) - 8(\cot\theta + \tan\theta) + 10 \ge 0$ for all real $\theta \ne n\pi,\ n\pi + \frac{\pi}{2}$ where $n \in \mathbb{Z}$.

Let $\cot\theta + \tan\theta = x$.

Then $\cot^2\theta + \tan^2\theta = (\cot\theta + \tan\theta)^2 - 2\cot\theta \cdot \tan\theta$
$= x^2 - 2$

$\therefore$ LHS $= 3(x^2 - 2) - 8x + 10$
$= 3x^2 - 8x + 4$
$= 3\left(x^2 - \frac{8}{3}x\right) + 4$
$= 3\left(x^2 - 2 \cdot \frac{4}{3}x + \frac{16}{9}\right) + 4 - 3 \cdot \frac{16}{9}$
$= 3\left(x - \frac{4}{3}\right)^2 + 4 - \frac{16}{3} = 3\left(x - \frac{4}{3}\right)^2 - \frac{4}{3}$ ... (1)

If $x < 0$, clearly LHS $> 3 \cdot \left(\frac{4}{3}\right)^2 - \frac{4}{3} > 0$ ... (2)

We know that, if θ is an angle in the second or the fourth quadrant then $\tan\theta$ and $\cot\theta$ are both negative and so $\cot\theta + \tan\theta$, i.e., x is negative. Thus, (2) shows that LHS > 0 when θ is an angle in the second or the fourth quadrant.

If θ is an angle in the first or the third quadrant, both $\cot\theta$ and $\tan\theta$ will be positive; so

AM $\ge$ GM

$\Rightarrow \frac{\cot\theta + \tan\theta}{2} \ge \sqrt{\cot\theta \cdot \tan\theta}$, i.e., $x \ge 2$.

When $x \ge 2$, from (1), LHS $\ge 3\left(2 - \frac{4}{3}\right)^2 - \frac{4}{3} = 3 \cdot \frac{4}{9} - \frac{4}{3} = 0$.

Thus, for all θ except quadrant angles, LHS ≥ 0.

8. Solve: $\sin x + \sin 2x > 0,\ x \in [-\pi, \pi]$.

Here $\sin x + 2\sin x \cdot \cos x > 0$

$\therefore$ $\sin x(1 + 2\cos x) > 0$

$\therefore$ either $\sin x > 0,\ 1 + 2\cos x > 0$, i.e., $\cos x > \frac{-1}{2}$ ... (1)

or $\sin x < 0,\ 1 + 2\cos x < 0$, i.e., $\cos x < -\frac{1}{2}$ ... (2)

When (1) is true:

$\sin x > 0 \Rightarrow 0 < x < \pi$

$\cos x > -\frac{1}{2} \Rightarrow -\frac{2\pi}{3} < x < \frac{2\pi}{3}$

$\therefore$ (1) gives $0 < x < \frac{2\pi}{3}$, taking common values of x.

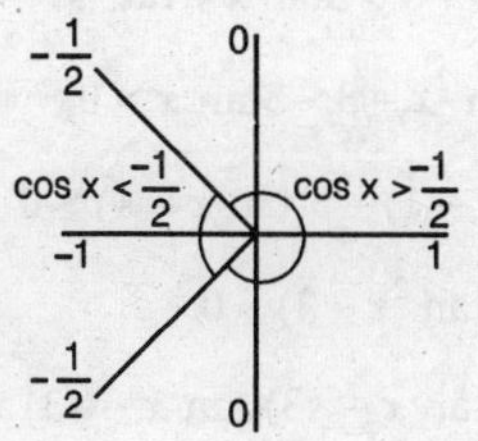

When (2) is true:

$\sin x < 0 \Rightarrow -\pi < x < 0$

$\cos x < -\frac{1}{2} \Rightarrow -\pi \le x < \frac{-2\pi}{3}$ or $\frac{2\pi}{3} < x \le \pi$

$\therefore$ (2) gives $-\pi < x < \frac{-2\pi}{3}$.

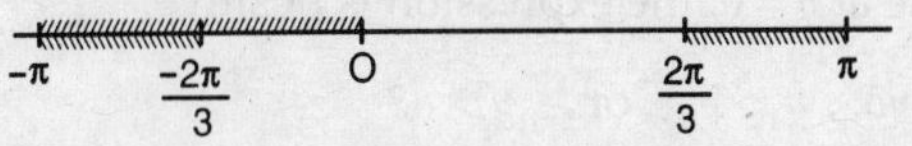

$\therefore$ the required solution-set is (taking union of solution-sets in the two cases)

$$\left(-\pi, -\frac{2\pi}{3}\right) \cup \left(0, \frac{2\pi}{3}\right).$$

9. Find $\theta \in \left[\frac{\pi}{2}, \frac{3\pi}{2}\right]$ satisfying $2\cos^2\theta + \sin\theta \le 2$.

Here $2(1 - \sin^2\theta) + \sin\theta \le 2$, i.e., $2\sin^2\theta - \sin\theta \ge 0$.

Roots of the corresponding equation are $\sin\theta = 0, \frac{1}{2}$.

$\therefore$ by sign-scheme:

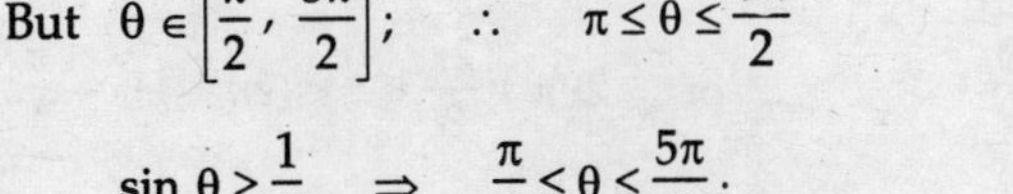

$\therefore$ $\sin\theta \le 0$ or $\sin\theta \ge \frac{1}{2}$.

Now, $\sin\theta \le 0 \Rightarrow \pi \le \theta \le 2\pi$.

But $\theta \in \left[\frac{\pi}{2}, \frac{3\pi}{2}\right]$; $\therefore$ $\pi \le \theta \le \frac{3\pi}{2}$

$\sin\theta \ge \frac{1}{2} \Rightarrow \frac{\pi}{6} \le \theta \le \frac{5\pi}{6}$.

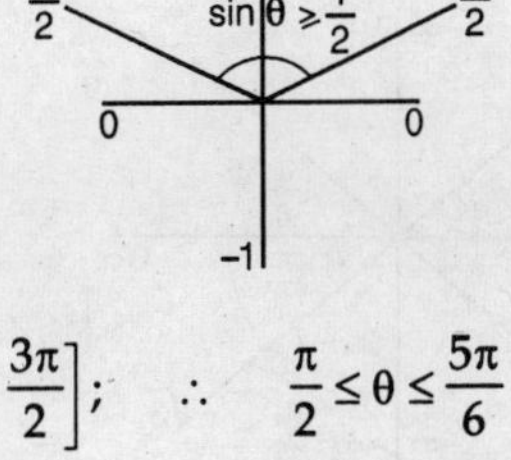

But $\theta \in \left[\frac{\pi}{2}, \frac{3\pi}{2}\right]$; $\therefore$ $\frac{\pi}{2} \le \theta \le \frac{5\pi}{6}$

$\therefore$ the solution-set $= \left[\frac{\pi}{2}, \frac{5\pi}{6}\right] \cup \left[\pi, \frac{3\pi}{2}\right]$.

10. Solve: $\tan^3 x + 3 > 3\tan x + \tan^2 x$.

Here, $\tan^3 x - \tan^2 x + 3 - 3\tan x > 0$

or $\tan^2 x(\tan x - 1) - 3(\tan x - 1) > 0$

or $(\tan x - 1)(\tan^2 x - 3) > 0$

or $(\tan x - 1)(\tan x + \sqrt{3})(\tan x - \sqrt{3}) > 0.$

Let $\tan x = y$. Then $(y - 1)(y + \sqrt{3})(y - \sqrt{3}) > 0$ ... (1)

Roots of the corresponding equation are $y = 1, -\sqrt{3}, \sqrt{3}$.

$\therefore$ the sign-scheme for (1) is as follows:

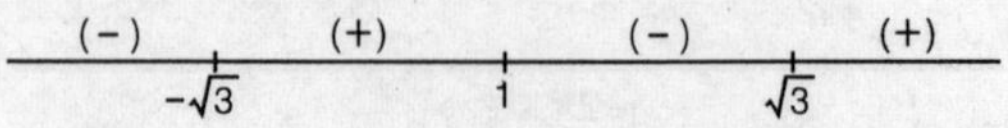

because at $y = 0$, the expression is positive.

$\therefore$ $-\sqrt{3} < y < 1$ or $y > \sqrt{3}$

$\Rightarrow$ $-\sqrt{3} < \tan x < 1$ or $\tan x > \sqrt{3}$.

Now, $\tan x > \sqrt{3}$

$\Rightarrow$ $2n\pi + \frac{\pi}{3} < x < 2n\pi + \frac{\pi}{2}$

or $(2n + 1)\pi + \frac{\pi}{3} < x < (2n + 1)\pi + \frac{\pi}{2}$,

i.e., $n\pi + \frac{\pi}{3} < x < n\pi + \frac{\pi}{2}$

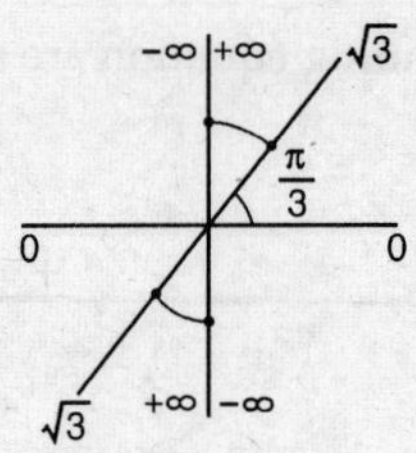

Also, $-\sqrt{3} < \tan x < 1$

$\Rightarrow$ $2n\pi - \frac{\pi}{3} < x < 2n\pi + \frac{\pi}{4}$

or $2n\pi + \frac{2\pi}{3} < x < (2n + 1)\pi + \frac{\pi}{4}$,

i.e., $n\pi - \frac{\pi}{3} < x < n\pi + \frac{\pi}{4}$

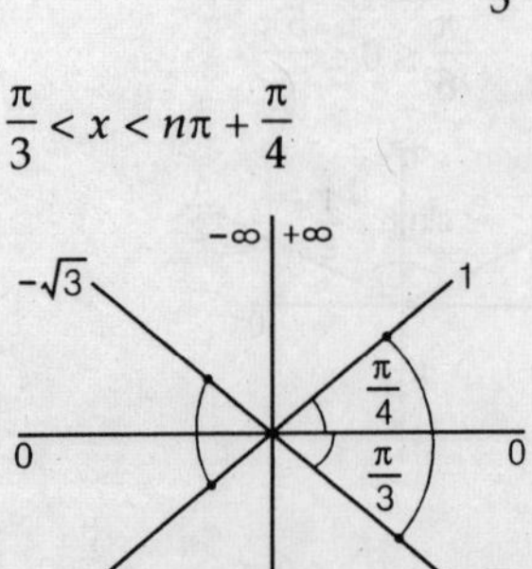

$\therefore$ the solution-set is

$$\left(2n\pi + \frac{\pi}{3},\ 2n\pi + \frac{\pi}{2}\right) \cup \left(2n\pi + \pi + \frac{\pi}{3},\ 2n\pi + \pi + \frac{\pi}{2}\right)$$

$$\cup \left(2n\pi - \frac{\pi}{3},\ 2n\pi + \frac{\pi}{4}\right)$$

$$\cup \left(2n\pi + \frac{2\pi}{3},\ 2n\pi + \pi + \frac{\pi}{4}\right) \text{ where } n \in \mathbb{Z},$$

equivalently,

$$\left(n\pi + \frac{\pi}{3},\ n\pi + \frac{\pi}{2}\right) \cup \left(n\pi - \frac{\pi}{3},\ n\pi + \frac{\pi}{4}\right) \text{ where } n \in \mathbb{Z}.$$

11. Solve: $\sin^{-1} x > \cos^{-1} x$.

Here $\sin^{-1} x > \cos^{-1} x = \frac{\pi}{2} - \sin^{-1} x$

$$\left(\because \sin^{-1} x + \cos^{-1} x = \frac{\pi}{2}\right)$$

$\therefore$ $2\sin^{-1} x > \frac{\pi}{2}$

or $\sin^{-1} x > \frac{\pi}{4}$

$\therefore$ $x > \sin\frac{\pi}{4}$, because larger the value of x, larger is the angle $\sin^{-1} x$ in $\left[-\frac{\pi}{2}, \frac{\pi}{2}\right]$. Remember that the principal value of $\sin^{-1} x$ lies in $\left[-\frac{\pi}{2}, \frac{\pi}{2}\right]$

$\therefore$ $x > \frac{1}{\sqrt{2}}$.

But $\sin^{-1} x$ is defined for $-1 \le x \le 1$

$\therefore$ $\frac{1}{\sqrt{2}} < x \le 1$. Hence $x \in \left(\frac{1}{\sqrt{2}}, 1\right]$.

12. Solve: $\cos 2x > |\sin x|$, $x \in [-\pi, \pi]$.

If $x \in (0, \pi)$ then $\sin x > 0$. So the inequation becomes

$\cos 2x > \sin x$

or $1 - 2\sin^2 x > \sin x$

or $2\sin^2 x + \sin x - 1 < 0$

or $(2\sin x - 1)(\sin x + 1) < 0$.

Roots of the corresponding equation are $\sin x = \frac{1}{2}, -1$.

So, the sign-scheme is

$\therefore \quad -1 < \sin x < \frac{1}{2}$. But $\sin x \geq -1$ for all x.

Thus, we get $\sin x < \frac{1}{2} = \sin \frac{\pi}{6}$ and $\sin x \neq -1$.

The value-scheme for $\sin x < \frac{1}{2}$ is as shown in the figure.

But $0 < x < \pi$.

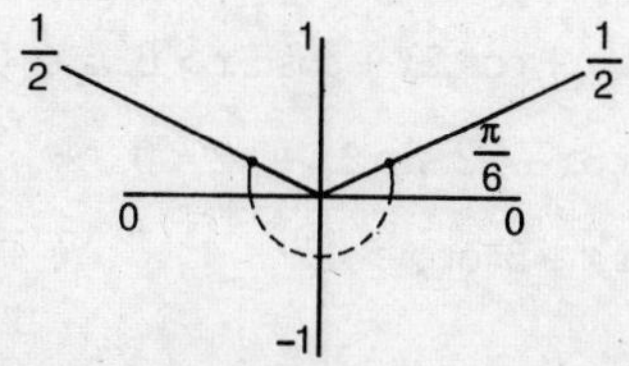

$$\therefore \quad 0 < x < \frac{\pi}{6},\ \pi - \frac{\pi}{6} < x < \pi \qquad \dots (1)$$

If $x \in (-\pi, 0)$ then $\sin x < 0$. So the inequation becomes

$$\cos 2x > -\sin x$$

or $\quad 1 - 2\sin^2 x > -\sin x$

or $\quad 2\sin^2 x - \sin x - 1 < 0$

or $\quad (2\sin x + 1)(\sin x - 1) < 0$.

Roots of the corresponding equation are $\sin x = -\frac{1}{2}, 1$.

So the sign-scheme is

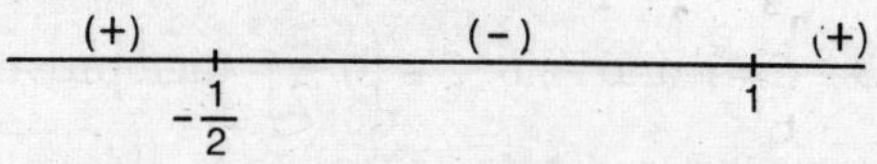

$\therefore \quad -\frac{1}{2} < \sin x < 1$. But $\sin x \leq 1$ for all x.

So we get $\sin x > \frac{-1}{2} = \sin\left(-\frac{\pi}{6}\right)$ and $\sin x \neq 1$.

The value-scheme for $\sin x > \frac{-1}{2}$ is as shown in the figure. But $-\pi < x < 0$

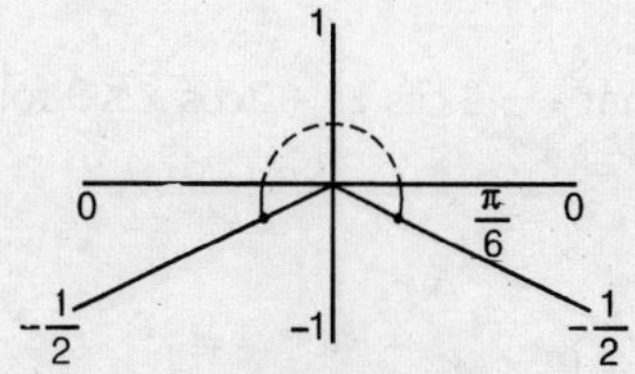

$$\therefore \quad -\frac{\pi}{6} < x < 0,\ -\pi < x < -\frac{5\pi}{6} \qquad \dots (2)$$

Also $x = -\pi,\ 0,\ \pi$ satisfies $\cos 2x > |\sin x|$.

Hence, collecting all the solutions in (1), (2) and the three values $-\pi, 0, \pi$ we get

$$0 \leq x < \frac{\pi}{6},\ \pi - \frac{\pi}{6} < x \leq \pi,\ -\frac{\pi}{6} < x \leq 0,\ -\pi \leq x < -\frac{5\pi}{6}$$

i.e., $x \in \left[-\pi, \frac{-5\pi}{6}\right) \cup \left(-\frac{\pi}{6}, \frac{\pi}{6}\right) \cup \left(\frac{5\pi}{6}, \pi\right]$.

Exercises

1. Prove that in a ΔABC, $\sin \frac{A}{2} \cdot \sin \frac{B}{2} \cdot \sin \frac{C}{2} \leq \frac{1}{8}$.
   Also prove that equality holds if the triangle is equilateral.
2. Prove that, in a triangle ABC,
   $\cos A \cdot \cos B \cdot \cos C \leq \frac{1}{8}$. Also prove that if
   $\cos A \cdot \cos B \cdot \cos C = \frac{1}{8}$, the triangle will be equilateral.
3. Prove that in ΔABC, $\cos^2 A + \cos^2 B + \cos^2 C \geq \frac{3}{4}$.
4. If ΔABC is acute angled then prove that
   $$\cos^2 A + \cos^2 B + \cos^2 C < \frac{3}{2}.$$
5. If $\alpha + \beta + \gamma = \pi$ where $\alpha, \beta, \gamma > 0$ then prove that
   $$(1 - \cos \alpha)(1 - \cos \beta)(1 - \cos \gamma) \leq \frac{1}{8}.$$
6. Prove that in a ΔABC, $\cot^2 A + \cot^2 B + \cot^2 C \geq 1$.
   Also prove that, if equality holds then the triangle is equilateral.
7. Prove that in a ΔABC, $\cot A + \cot B + \cot C \geq \sqrt{3}$.
   Also prove that if $\cot A + \cot B + \cot C = \sqrt{3}$, the triangle will be equilateral.
8. Prove that in an acute-angled triangle ABC,
   $$\tan A + \tan B + \tan C \geq 3\sqrt{3}.$$
   If equality holds, the triangle will be equilateral.
9. If $A + B + C = \pi$, prove that
   $$\cos \frac{A}{2} \cdot \cos \frac{B}{2} \cdot \cos \frac{C}{2} \leq \frac{3\sqrt{3}}{8}.$$
10. Show that the greatest value of
    $\sin 2A + \sin 2B + \sin 2C$ in a triangle ABC is $\frac{3\sqrt{3}}{2}$.
11. Prove that in an acute-angled ΔABC,
    $$\tan^2 A + \tan^2 B + \tan^2 C \geq 9.$$
12. Prove that in a ΔABC, $\cot \frac{A}{2} \cdot \cot \frac{B}{2} \cdot \cot \frac{C}{2} \geq 3\sqrt{3}$
    and the triangle is equilateral if equality holds.
13. Prove that in ΔABC,
    $$\operatorname{cosec} \frac{A}{2} + \operatorname{cosec} \frac{B}{2} + \operatorname{cosec} \frac{C}{2} \geq 6.$$

14. If $A + B + C = \pi$ and $A, B, C \in \left(0, \frac{\pi}{2}\right)$ then prove that $\sec A + \sec B + \sec C \geq 6$.

15. Prove that in a ΔABC, $\sin^2\frac{A}{2} + \sin^2\frac{B}{2} + \sin^2\frac{C}{2} \geq \frac{3}{4}$.

16. Prove that $\sin^{2n}\theta + \cos^{2n}\theta \leq 1$ for all θ and $n \geq 1$.

17. If $A, B \in [0, \pi]$ then prove that

$$\frac{1}{2}(\sin A + \sin B) \leq \sin\frac{A + B}{2}.$$

18. Prove that $-\frac{3}{2} \leq \cos 2x + 2\cos x \leq 3$ for all real x.

19. Prove that $\sin\alpha \cdot \sin 2\alpha \cdot \sin 3\alpha < \frac{9}{16}$ for all $\alpha \in R$.

20. Find the set of all $x \in [0, \pi]$ for which

$2\sin^2 x - 3\sin x + 1 \geq 0$.

21. Solve: $\sin x < \cos x,\ 0 \leq x \leq \pi$.

22. Solve: $\sin 2x > \cos x$.

23. Solve: $\cos x + \cos 2x + \cos 3x > 0,\ x \in [0, \pi]$.

24. Solve: $\cos 3x + \sqrt{3}\sin 3x + \sqrt{2} < 0$.

25. Solve: $\tan x + 3\cot x > 4$.

Answers

20. $\left[0, \frac{\pi}{6}\right] \cup \left\{\frac{\pi}{2}\right\} \cup \left[\frac{5\pi}{6}, \pi\right]$

21. $\left[0, \frac{\pi}{4}\right)$

22. $\left(2n\pi + \frac{\pi}{6}, 2n\pi + \frac{\pi}{2}\right) \cup \left(2n\pi + \frac{5\pi}{6}, (2n+1)\pi + \frac{\pi}{2}\right), n \in Z$

23. $\left[0, \frac{\pi}{4}\right) \cup \left(\frac{2\pi}{3}, \frac{3\pi}{4}\right), n \in Z$

24. $\left(\frac{2n\pi}{3} + \frac{13\pi}{36}, \frac{2n\pi}{3} + \frac{19\pi}{36}\right), n \in Z$

25. $\left(n\pi, n\pi + \frac{\pi}{4}\right) \cup \left(n\pi + \tan^{-1}3, n\pi + \frac{\pi}{2}\right),\ n \in Z$

□

Chapter Test

Time: 60 minutes

1. In the following incomplete sentences, fill in the blanks so that the resulting sentences may become true.

(a) For all real x, the value of $\sin^2 x \cdot \cos^2 x \le k$, k being the least possible. Then $k =$ ______ .

(b) If $\sin 3x \ge \cos 3x$ and $0 \le x \le 2\pi$ then the solution-set of x is ______ .

(c) If $\sin \theta < \cos \theta + \sqrt{2}$ and $-\pi \le \theta \le \pi$ then the solution-set of θ is ______ .

(d) If $|\tan x| < 1$ and $0 \le x \le \pi$, the solution-set of x is ______ .

2. In the following, each question has multiple answers. Choose the currect answer(s).

(a) If in a ΔABC, the angle C is obtuse then $\tan A \cdot \tan B$ is

A. greater than 1 B. less than 1

C. equal to 1 D. none of these

(b) The sum of cosines of three angles of a triangle is always

A. greater than 1 B. less than 1

C. equal to 1 D. none of these

3. Prove that in a ΔABC: $(\sin A + \sin B)(\sin B + \sin C)(\sin C + \sin A) > \sin A \sin B \sin C$.

4. In a ΔABC, prove that $a^2 + b^2 + c^2 \ge 4\sqrt{3} \times$ (area of ΔABC) and if equality holds then the ΔABC is equilateral.

5. Prove that in a triangle $r_1 \cdot r_2 \cdot r_3 \ge 27r^3$ where r_1, r_2, r_3 are exradii and r is the inradius.

6. Prove that the values of θ that satisfy $\cos 3\theta + \cos \theta \ge 2$ are in AP.

7. In a ΔABC whose area is Δ, prove that $\Delta \le \frac{1}{4}\sqrt{(a+b+c)abc}$. Also, prove that equality will take place if and only if $a = b = c$.

□

Answers

1. (a) $\frac{1}{4}$ (b) $\left[\frac{\pi}{12}, \frac{5\pi}{12}\right] \cup \left[\frac{3\pi}{4}, \frac{13\pi}{12}\right] \cup \left[\frac{17\pi}{12}, \frac{7\pi}{4}\right]$ (c) $[-\pi, \pi] - \left\{\frac{3\pi}{4}\right\}$ (d) $\left[0, \frac{\pi}{4}\right) \cup \left(\frac{3\pi}{4}, \pi\right]$

2. (a) B (b) A

□

5. Logarithm

Recap of Facts and Formulae

1. Logarithm and its properties

- $\log_a b = x \quad \Leftrightarrow \quad a^x = b$

  (a is known as the base of the logarithm)
- $\log_a b$ is real if $b > 0,\ a > 0,\ a \neq 1$
- $\log_a 1 = 0$
- $\log_a a = 1$
- $\log_a(m \times n) = \log_a m + \log_a n$
- $\log_a \frac{m}{n} = \log_a m - \log_a n$
- $\log_a m^n = n \log_a m$
- $\log_{a^n} m = \frac{1}{n} \log_a m$
- $\log_a b = \frac{\log_x b}{\log_x a}$
- $\log_a b \times \log_b c = \log_a c$
- $\log_a b = \frac{1}{\log_b a}$
- $a^{\log_a x} = x$

2. Equality and inequality in logarithm

- $\log_a x = \log_a y \quad \Leftrightarrow \quad x = y$
- $\log_a x > \log_a y \quad \Rightarrow \quad x > y$ when $a > 1$

  $x < y$ when $0 < a < 1$

3. Characteristic and mantissa

- If $\log_{10} x = a \cdot bcd \ldots$ where $0 \cdot bcd \ldots$ is positive, a being an integer and b, c, d, etc., being digits then a is the characteristic and $0 \cdot bcd \ldots$ is the mantissa of $\log_{10} x$.

  If $\log_{10} x = 3.15642$ then 3 is the characteristic and 0.15642 is the mantissa.

  If $\log_{10} x = -4.2354$ then

  $\log_{10} x = -4 - 0.2354 = -5 + (1 - 0.2354) = -5 + 0.7646$.

  We write $\log_{10} x = \bar{5}.7646$ in which -5 is the characteristic and 0.7646 is the mantissa.
- If the characteristic a of $\log x$ is positive then x is a number whose integral part is of $a + 1$ digits.

  If the characteristic a of $\log_{10} x$ is negative, say $-b$, then x is a proper fraction, in the decimal form of which there will be $b - 1$ zeros immediately after the decimal point before a significant digit appears.
- For each of $x, 10x, 100x$, etc., mantissa of logarithm is the same. But their characteristics go on increasing by 1.

Selected Solved Examples

1. Prove that $\log_{34} 5$ lies in the open interval $\left(\frac{1}{3}, \frac{1}{2}\right)$.

We have $25 < 34 < 125$,

i.e., $5^2 < 34 < 5^3$

$\therefore \quad \log_{34} 5^2 < \log_{34} 34 < \log_{34} 5^3 \quad (\because$ the base $34 > 1)$

or $\quad 2\log_{34} 5 < 1 < 3\log_{34} 5$

$\therefore \quad 2\log_{34} 5 < 1$ and $3\log_{34} 5 > 1$

$\therefore \quad \log_{34} 5 < \frac{1}{2}$ and $\log_{34} 5 > \frac{1}{3}$

$\therefore \quad \frac{1}{3} < \log_{34} 5 < \frac{1}{2}$,

i.e., $\log_{34} 5$ lies in the interval $\left(\frac{1}{3}, \frac{1}{2}\right)$.

2. Show that $\sum_{r=2}^{43} \frac{1}{\log_r n} = \log_n (43)!$.

$$\text{LHS} = \sum_{r=2}^{43} \log_n r \quad \left(\because \log_a b = \frac{1}{\log_b a} \right)$$

$$= \log_n 2 + \log_n 3 + \log_n 4 + \ldots + \log_n 43$$

$$= \log_n (2 \times 3 \times 4 \times \ldots \times 43) = \log_n (43)! = \text{RHS}.$$

3. Given $n^4 < 10^n$ for a fixed positive integer $n \geq 2$, prove that $(n+1)^4 < 10^{n+1}$.

$$n^4 < 10^n \quad \Rightarrow \quad \log_{10} n^4 < \log_{10} 10^n$$

$$\therefore \quad 4\log_{10} n < n \quad \ldots (1)$$

$$\text{Now, } \log_{10} \frac{(n+1)^4}{10^{n+1}} = \log_{10}(n+1)^4 - \log_{10} 10^{n+1}$$

$$= 4\log_{10}(n+1) - (n+1)$$

$$= 4\left\{\log_{10} n + \log_{10}\left(1 + \frac{1}{n}\right)\right\} - (n+1)$$

$$< n + 4\log_{10}\left(1 + \frac{1}{n}\right) - n - 1, \text{ using (1)}$$

$$= 4\log_{10}\left(1 + \frac{1}{n}\right) - 1$$

$$< 4\log_{10} \frac{3}{2} - 1, \quad \text{for } n \geq 2.$$

$$\text{But } 4\log_{10} \frac{3}{2} - 1 = \log_{10}\left(\frac{3}{2}\right)^4 - \log_{10} 10 = \log_{10} \frac{81}{160} < 0$$

$$\therefore \quad \log_{10}\left\{\frac{(n+1)^4}{10^{n+1}}\right\} < 0 \Rightarrow \frac{(n+1)^4}{10^{n+1}} < 1$$

$$\therefore \quad (n+1)^4 < 10^{n+1}.$$

4. If $\frac{\log x}{y - z} = \frac{\log y}{z - x} = \frac{\log z}{x - y}$, **prove that**

$$x^{y+z} \cdot y^{z+x} \cdot z^{x+y} = 1.$$

Also prove that $x^{y+z} + y^{z+x} + z^{x+y} \geq 3$.

For $\log x$, $\log y$ and $\log z$ to be real we must have

$$x > 0, \; y > 0, \; z > 0.$$

$$\text{Now } \frac{\log x}{y - z} = \frac{\log y}{z - x} = \frac{\log z}{x - y}$$

$$\Rightarrow \frac{(y+z)\log x}{y^2 - z^2} = \frac{(z+x)\log y}{z^2 - x^2} = \frac{(x+y)\log z}{x^2 - y^2}$$

$$= \frac{(y+z)\log x + (z+x)\log y + (x+y)\log z}{y^2 - z^2 + z^2 - x^2 + x^2 - y^2},$$

using ratio and proportion

$$= \frac{\log x^{y+z} + \log y^{z+x} + \log z^{x+y}}{0}$$

$$\therefore \quad \log x^{y+z} + \log y^{z+x} + \log z^{x+y} = 0$$

$$\text{or} \quad \log(x^{y+z} \cdot y^{z+x} \cdot z^{z+y}) = 0,$$

$$\text{i.e.,} \quad x^{y+z} \cdot y^{z+x} \cdot z^{x+y} = 1 \quad \ldots (1)$$

Again, take the positive numbers x^{y+z}, y^{z+x} and z^{x+y}.

AM $\geq$ GM

$$\Rightarrow \quad \frac{x^{y+z} + y^{z+x} + z^{x+y}}{3} \geq \sqrt[3]{x^{y+z} \cdot y^{z+x} \cdot z^{z+y}} = 1,$$

from (1)

$$\therefore \quad x^{y+z} + y^{z+x} + z^{x+y} \geq 3.$$

5. Prove that $(yz)^{\log \frac{y}{z}} \cdot (zx)^{\log \frac{z}{x}} \cdot (xy)^{\log \frac{x}{y}} = 1$.

$$\text{Let} \quad m = (yz)^{\log \frac{y}{z}} \cdot (zx)^{\log \frac{z}{x}} \cdot (xy)^{\log \frac{x}{y}}$$

$$\text{Then } \log m = \log(yz)^{\log \frac{y}{z}} + \log(zx)^{\log \frac{z}{x}} + \log(xy)^{\log \frac{x}{y}}$$

$$= \log \frac{y}{z} \cdot \log(yz) + \log \frac{z}{x} \cdot \log(zx) + \log \frac{x}{y} \cdot \log(xy)$$

$$= (\log y - \log z)(\log y + \log z) + (\log z - \log x)(\log z + \log x) + (\log x - \log y)(\log x + \log y)$$

$$= (\log y)^2 - (\log z)^2 + (\log z)^2 - (\log x)^2 + (\log x)^2 - (\log y)^2 = 0$$

$$\therefore \quad \log m = 0. \text{ So } m = 1, \text{ i.e., LHS} = 1.$$

6. Solve:

$$\frac{6}{5} a^{\log_a x \cdot \log_{10} a \cdot \log_a 5} - 3^{\log_{10} \frac{x}{10}} = 9^{\log_{100} x + \log_4 2}.$$

$$\text{Here, } \frac{6}{5} (a^{\log_a 5})^{\log_{10} a \cdot \log_a x} - 3^{\log_{10} x - \log_{10} 10}$$

$$= 9^{\log_{100} x} \cdot 9^{\log_4 2}$$

$$\text{or} \quad \frac{6}{5} \cdot 5^{\log_{10} x} - 3^{\log_{10} x - 1} = 9^{\frac{1}{2}\log_{10} x} \cdot 9^{\frac{1}{2}}$$

$$\left\{\because a^{\log_a x} = x, \log_b a \cdot \log_a c = \log_b c, \log_4 2 = \frac{1}{2}\right\}.$$

Taking $\log_{10} x = y$, we get

$$\frac{6}{5} \cdot 5^y - 3^{y-1} = 3^y \times 3$$

$$\text{or} \quad 2 \times 5^{y-1} - 3^{y-2} = 3^y$$

$$\text{or} \quad 2 \times 5^{y-1} = 3^y + 3^{y-2} = 3^{y-2}(3^2 + 1) = 3^{y-2} \times 10$$

or $\quad 5^{y-1} = 3^{y-2} \times 5$, i.e., $5^{y-2} = 3^{y-2}$

This is possible if $y - 2 = 0$, i.e., $y = 2$.

$\therefore \quad \log_{10} x = y = 2$; so $x = 10^2 = 100$.

7. Solve for x: $\log_{0.1} \sin 2x + \log_{10} \cos x = \log_{10} 7$.

Here $\log_{0.1} 10 \times \log_{10} \sin 2x + \log_{10} \cos x = \log_{10} 7$

or $\quad -\log_{10} \sin 2x + \log_{10} \cos x = \log_{10} 7$

$$\{\because \log_{0.1} 10 = \log_{0.1} (0.1)^{-1} = -1\}$$

or $\quad \log_{10} \dfrac{\cos x}{\sin 2x} = \log_{10} 7$, i.e., $\dfrac{\cos x}{\sin 2x} = 7$

or $\quad \dfrac{1}{2\sin x} = 7; \quad \therefore \sin x = \dfrac{1}{14}$

$$\therefore \quad x = n\pi + (-1)^n \sin^{-1}\left(\frac{1}{14}\right);\ n \in \mathbb{Z} \qquad \ldots (1)$$

But here the logarithms are defined only if

$\cos x > 0$ and $\sin 2x > 0$.

So we must have $\cos x > 0$ as well as $\sin x > 0$.

Here x will be an angle in the first quadrant.

$\therefore \quad$ in (1), n must be an even integer.

$$\therefore \quad x = 2m\pi + \sin^{-1}\left(\frac{1}{14}\right);\ m \in \mathbb{Z}.$$

8. Solve: $\log_{\cos x} \sin x + \log_{\sin x} \cos x = 2$.

Here $\dfrac{\log \sin x}{\log \cos x} + \dfrac{\log \cos x}{\log \sin x} = 2$

or $\quad (\log \sin x)^2 + (\log \cos x)^2$

$$-2(\log \cos x)(\log \sin x) = 0$$

or $\quad (\log \sin x - \log \cos x)^2 = 0$,

i.e., $\quad \log \sin x = \log \cos x$

$\therefore \quad \sin x = \cos x \quad \Rightarrow \quad \tan x = 1$

$$\Rightarrow \quad x = n\pi + \frac{\pi}{4} \qquad \ldots (1)$$

But here the logarithms are real if $\sin x > 0$, $\cos x > 0$, $\sin x \neq 1$, $\cos x \neq 1$

$\therefore \quad x$ belongs to the first quadrant and

$$x \neq n\pi + (-1)^n \frac{\pi}{2}, 2n\pi$$

$\therefore \quad$ from (1), $x = 2m\pi + \dfrac{\pi}{4}$, $m \in \mathbb{Z}$.

9. If $1 + \log_2 (1 + \cos 2x) > \log_4 2$ then find x.

Here, $1 + \log_2 (2\cos^2 x) > \dfrac{1}{2} \log_2 2 = \dfrac{1}{2}$

or $\quad 2 + \log_2 \cos^2 x > \dfrac{1}{2}$

or $\quad 2\log_2 \cos x > \dfrac{-3}{2}$

$\therefore \quad \log_2 \cos x > \dfrac{-3}{4} = \log_2 2^{\frac{-3}{4}}$

$\Rightarrow \quad \cos x > 2^{\frac{-3}{4}} = \dfrac{1}{2^{3/4}}$

Let $\cos \alpha = \dfrac{1}{2^{3/4}}$, i.e., $\alpha = \cos^{-1} \dfrac{1}{2^{3/4}}$.

So, from the value-scheme we get,

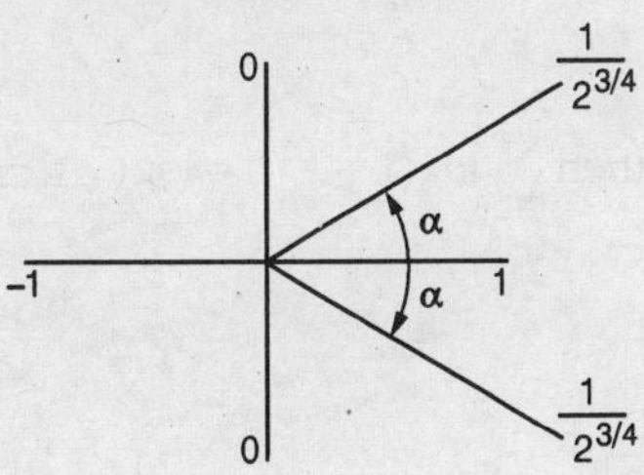

$$2n\pi - \alpha < x < 2n\pi + \alpha$$

$\therefore \quad 2n\pi - \cos^{-1} \dfrac{1}{2^{3/4}} < x < 2n\pi + \cos^{-1} \dfrac{1}{2^{3/4}}$ where $n \in \mathbb{Z}$.

Note $\log_2(1 + \cos 2x)$ is defined for all x obtained.

Exercises

1. Show that $\log_{300} 4$ lies in the interval $\left(\dfrac{1}{5}, \dfrac{1}{4}\right)$.

2. If $x = \log_{2a} a$, $y = \log_{3a} 2a$ and $z = \log_{4a} 3a$ then prove that $xyz + 1 = 2yz$.

3. If $x = \log_a (bc)$, $y = \log_b (ca)$ and $z = \log_c (ab)$, prove that

$$\frac{1}{x+1} + \frac{1}{y+1} + \frac{1}{z+1} = 1.$$

4. Prove that $\log_a(bc) \cdot \log_b(ca) \cdot \log_c(ab)$

$$= 2 + \log_a(bc) + \log_b(ca) + \log_c(ab).$$

5. If $\dfrac{x(y+z-x)}{\log x}=\dfrac{y(z+x-y)}{\log y}=\dfrac{z(x+y-z)}{\log z}$, prove that $y^z \cdot z^y = z^x \cdot x^z = x^y \cdot y^x$.

6. If $\dfrac{\log x}{y-z}=\dfrac{\log y}{z-x}=\dfrac{\log z}{x-y}$, prove that $x^x + y^y + z^z \geq 3$.

7. Prove that $x^{\log y - \log z} \cdot y^{\log z - \log x} \cdot z^{\log x - \log y} = 1$.

8. Solve: $\log_{10} \sin x + \log_{100} (8\cos x) = \dfrac{1}{4} \log_{\sqrt{10}} 3$.

9. Solve: $\log_2 \tan x + \log_2 \tan 2x = 0,\ 0 \leq x \leq 2\pi$

10. Solve: $1 + \log_4 \sin x + 2\log_{16} \cos x > 0$.

Objective Questions

Fill in the blanks.

11. $\log_3 \log_2 \log_{\sqrt{3}} 81 =$ ______ .

12. If $\log x + \log y = \log(x+y)$ then y as a function of x is given by $y =$ ______ .

13. The sum of the series $\sum\limits_{r=0}^{\infty} \left\{\log_3(e^{2^r})\right\}^{-1} =$ ______ .

14. If $a^2 + b^2 = 23ab$ then $\log \dfrac{a+b}{5}$ is the AM of ______ and ______ .

15. If $0 < x < 1$ then $\sum\limits_{r=0} \log(1 + x^{2^r}) = \log($ ______ $)$.

16. If $\log 2 = 0.30103$, $\log 3 = 0.4771213$ then the number of digits in the number equal to $3^{12} \times 2^8$ is ______ .

17. If $\log 3 = 0.4771213$ then the number of zeros between the decimal point and the first significant digit in the value of 9^{-16} is ______ .

18. If $\log_2 \left(\sin \dfrac{x}{2}\right) < -1$ then the set of general values of x is ______ .

Choose the correct option(s).

19. If $2\log a + 2\log a^2 + 2\log a^3 + \ldots + 2\log a^n = 55 \cdot \log a$ then the value of n is

(a) 10 (b) 11
(c) 20 (d) 5
(e) none of these

20. If $\dfrac{\log x}{b-c}=\dfrac{\log y}{c-a}=\dfrac{\log z}{a-b}$ then the value of xyz is

(a) 1 (b) 3
(c) 2 (d) none of these

21. $a^{\log_a b \times \log_b c \times \log_c d}$ is equal to

(a) b (b) c
(c) d (d) none of these

22. If a, b, c are positive numbers ($\neq 1$) in GP and $x > 0$ but $\neq 1$, the value of $\log_b x(\log_x a + \log_x c)$ is equal to

(a) 1 (b) 2
(c) 3 (d) none of these

Answers

8. $2n\pi + \dfrac{\pi}{3},\ 2n\pi + \cos^{-1}\dfrac{\sqrt{13}-1}{4};\ n \in Z$

9. $\dfrac{\pi}{6}, \dfrac{7\pi}{6}$

10. $\left(2n\pi + \dfrac{\pi}{12}, 2n\pi + \dfrac{5\pi}{12}\right), n \in Z$

11. 1

12. $\dfrac{x}{x-1}, x > 1$

13. $\log_e 9$

14. $\log a, \log b$

15. $\dfrac{1}{1-x}$

16. 9

17. 15

18. $\left(4n\pi, \dfrac{12n+1}{3}\pi\right) \cup \left(\dfrac{12n+5}{3}\pi, 4n\pi + 2\pi\right), n \in Z$

19. (e)

20. (a)

21. (c)

22. (b)

□

Chapter Test

Time: 30 minutes

1. In each of the following, fill in the blank so that the resulting sentence becomes true.

(a) If for a $\triangle ABC$, $\log_{0.1} \cos A > 1$ then $A \in$ ______ .

(b) $\log \cos \dfrac{x}{2} + \log \cos \dfrac{x}{2^2} + \log \cos \dfrac{x}{2^3} + \ldots + \log \cos \dfrac{x}{2^n} + \log \sin \dfrac{x}{2^n} = \log($ ______ $)$.

2. Prove that in an acute-angled $\triangle ABC$,

$$\log(\tan A + \tan B + \tan C) = \log \tan A + \log \tan B + \log \tan C.$$

Can this hold for any triangle ABC?

3. Prove that $4 + \log_2 \sin \dfrac{\pi}{5} + \log_2 \sin \dfrac{2\pi}{5} + \log_2 \sin \dfrac{3\pi}{5} + \log_2 \sin \dfrac{4\pi}{5} = \log_2 5$.

4. Solve: $1 + \log_2 \sin x + \log_2 \sin 3x > 0$, $0 \le x \le 2\pi$.

□

Answers

1. (a) $\left(\cos^{-1}\frac{1}{10}, \frac{\pi}{2}\right)$ (b) $\frac{1}{2^n}\sin x$ **2.** no

4. $\left(\frac{\pi}{6}, \frac{\pi}{4}\right)$

□

6. Properties of Triangle

Recap of Facts and Formulae

1. Angles of a triangle

In a triangle ABC, three angles are A, B and C.

- $A + B + C = \pi$
- $\sin(B + C) = \sin(\pi - A) = \sin A$
- $\cos(C + A) = \cos(\pi - B) = -\cos B$
- $\sin \dfrac{A+B}{2} = \sin\left(\dfrac{\pi}{2} - \dfrac{C}{2}\right) = \cos \dfrac{C}{2}$
- $\cos \dfrac{B+C}{2} = \cos\left(\dfrac{\pi}{2} - \dfrac{A}{2}\right) = \sin \dfrac{A}{2}$

2. Trigonometrical relations between sides and angles

For any $\triangle$ABC we have

- $\dfrac{a}{\sin A} = \dfrac{b}{\sin B} = \dfrac{c}{\sin C} = 2R$

  where R = circumradius (*Sine-rule*)
- $\cos A = \dfrac{b^2 + c^2 - a^2}{2bc}$, etc. (*Cosine formulae*)
- $a\cos B + b\cos A = c$, etc.
- $\sin \dfrac{A}{2} = \sqrt{\dfrac{(s-b)(s-c)}{bc}}$, etc., where $2s = a + b + c$
- $\cos \dfrac{A}{2} = \sqrt{\dfrac{s(s-a)}{bc}}$, etc.
- $\tan \dfrac{A}{2} = \sqrt{\dfrac{(s-b)(s-c)}{s(s-a)}} = \dfrac{(s-b)(s-c)}{\Delta} = \dfrac{\Delta}{s(s-a)}$, etc.,

  where Δ = area of $\triangle$ABC
- $\cot \dfrac{A}{2} = \dfrac{s(s-a)}{\Delta}$, etc.
- $\tan \dfrac{B-C}{2} = \dfrac{b-c}{b+c} \cot \dfrac{A}{2}$, etc.

3. Area of a triangle

If the area of a $\triangle$ABC is denoted by Δ then

- $\Delta = \dfrac{1}{2} bc \sin A$, etc.
- $\sin A = \dfrac{2\Delta}{bc}$, etc.
- $\Delta = \sqrt{s(s-a)(s-b)(s-c)}$

4. Ratio formula

In the $\triangle$ABC, AD divides BC in the ratio $m : n$ at D and $\angle BAC$ in two parts $\angle BAD = \alpha$, $\angle CAD = \beta$.

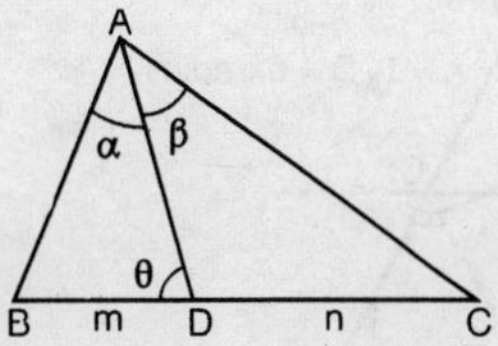

If $\angle ADB = \theta$ then

- $(m + n)\cot \theta = n \cot \beta - m \cot \alpha$
- $(m + n)\cot \theta = m \cot C - n \cot B$

5. Some important identities for angles of a triangle

In a $\triangle$ABC, we have

- $\sin 2A + \sin 2B + \sin 2C = 4\sin A \sin B \sin C$
- $1 + \cos 2A + \cos 2B + \cos 2C = -4\cos A \cos B \cos C$
- $\sin A + \sin B + \sin C = 4\cos \dfrac{A}{2} \cos \dfrac{B}{2} \cos \dfrac{C}{2}$
- $\cos A + \cos B + \cos C = 1 + 4\sin \dfrac{A}{2} \cdot \sin \dfrac{B}{2} \cdot \sin \dfrac{C}{2}$
- $\cos^2 A + \cos^2 B + \cos^2 C + 2\cos A \cos B \cos C = 1$
- $\tan A + \tan B + \tan C = \tan A \cdot \tan B \cdot \tan C$,

  i.e., $\cot B \cdot \cot C + \cot C \cdot \cot A + \cot A \cdot \cot B = 1$
- $\Sigma \tan \dfrac{A}{2} \cdot \tan \dfrac{B}{2} = 1$,

  i.e., $\cot \dfrac{A}{2} + \cot \dfrac{B}{2} + \cot \dfrac{C}{2} = \cot \dfrac{A}{2} \cot \dfrac{B}{2} \cot \dfrac{C}{2}$

6. Circumradius, inradius and exradii

In the $\triangle ABC$, let the circumradius = R, inradius = r and the three exradii corresponding to the vertices A, B and C be r_1, r_2 and r_3 respectively. Then

- $R = \dfrac{a}{2\sin A} = \dfrac{abc}{4\Delta}$
- $r = \dfrac{\Delta}{s} = (s-a)\tan\dfrac{A}{2} = (s-b)\tan\dfrac{B}{2} = (s-c)\tan\dfrac{C}{2}$

$= 4R\sin\dfrac{A}{2}\cdot\sin\dfrac{B}{2}\cdot\sin\dfrac{C}{2}$

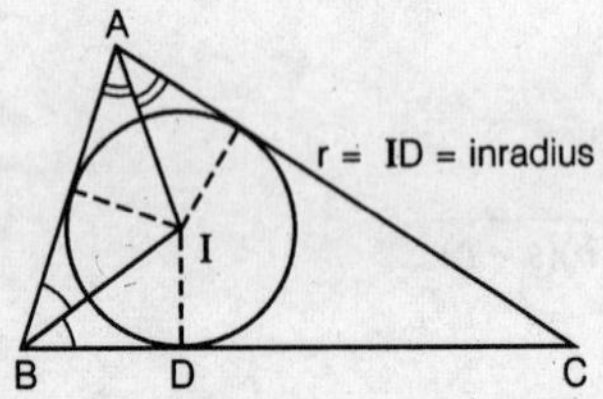

- $r_1 = \dfrac{\Delta}{s-a} = s\tan\dfrac{A}{2} = 4R\sin\dfrac{A}{2}\cos\dfrac{B}{2}\cos\dfrac{C}{2}$

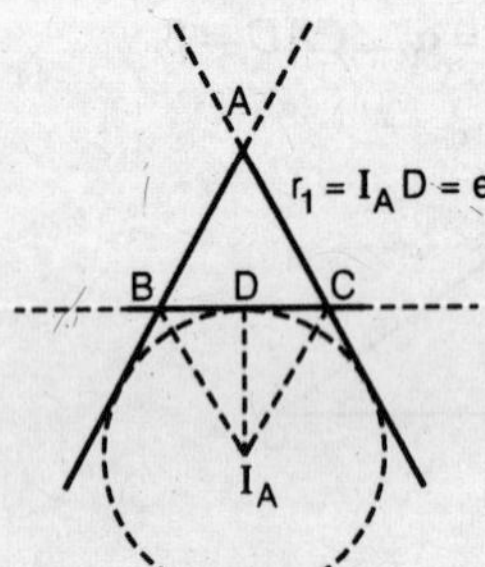

$r_2 = \dfrac{\Delta}{s-b} = s\tan\dfrac{B}{2} = 4R\sin\dfrac{B}{2}\cos\dfrac{C}{2}\cos\dfrac{A}{2}$

$r_3 = \dfrac{\Delta}{s-c} = s\tan\dfrac{C}{2} = 4R\sin\dfrac{C}{2}\cos\dfrac{A}{2}\cos\dfrac{B}{2}$

7. Regular polygon

- A regular polygon of n sides will have its vertices on a circle. If O be the centre and r be the radius of the circle, and a be the length of each side then clearly, in the $\triangle OA_1A_2$, $\angle A_1OA_2 = \dfrac{2\pi}{n}$ because $\triangle OA_1A_2$ will be one of the n equal triangles with a vertex at O.

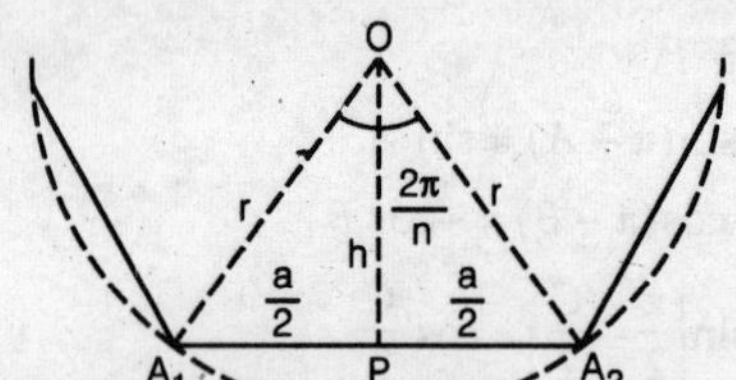

- Also, $\dfrac{\frac{a}{2}}{h} = \tan\dfrac{1}{2}\left(\dfrac{2\pi}{n}\right)$, $\dfrac{\frac{a}{2}}{r} = \sin\dfrac{1}{2}\left(\dfrac{2\pi}{n}\right)$, etc.
- Each interior angle of the regular polygon

$= \dfrac{2(n-2)}{n}\times 90°$.

Each exterior angle of the regular polygon $= \dfrac{2\pi}{n}$.

Selected Solved Examples

1. In a $\triangle ABC$, prove that
$\sum \sin^3 A\cdot\cos(B-C) = 3\sin A\sin B\sin C$.

$\sin^3 A\cdot\cos(B-C)$

$= \sin^2 A\cdot\sin A\cdot\cos(B-C)$

$= \sin^2 A\cdot\sin(B+C)\cos(B-C)$

$= \dfrac{1}{2}\sin^2 A\cdot 2\sin(B+C)\cos(B-C)$

$= \dfrac{1}{2}\sin^2 A\cdot(\sin 2B + \sin 2C)$.

Similarly for others.

$\therefore$ LHS $= \dfrac{1}{2}\sin^2 A(\sin 2B + \sin 2C)$

$+ \dfrac{1}{2}\sin^2 B(\sin 2C + \sin 2A)$

$+ \dfrac{1}{2}\sin^2 C(\sin 2A + \sin 2B)$

$= \dfrac{1}{2}\cdot\dfrac{a^2}{4R^2}\left(2\cdot\dfrac{b}{2R}\cos B + 2\cdot\dfrac{c}{2R}\cos C\right)$

$+ \dfrac{1}{2}\dfrac{b^2}{4R^2}\left(2\cdot\dfrac{c}{2R}\cos C + 2\cdot\dfrac{a}{2R}\cos A\right)$

$+ \dfrac{1}{2}\dfrac{c^2}{4R^2}\left(2\cdot\dfrac{a}{2R}\cos A + 2\cdot\dfrac{b}{2R}\cos B\right)$

$\left\{\because \dfrac{a}{\sin A} = \dfrac{b}{\sin B} = \dfrac{c}{\sin C} = 2R\right\}$

$= \dfrac{1}{8R^3}\{a^2(b\cos B + c\cos C) + b^2(c\cos C + a\cos A)$

$+ c^2(a\cos A + b\cos B)\}$

$= \dfrac{1}{8R^3}\{ab(a\cos B + b\cos A) + ac(a\cos C + c\cos A)$

$+ bc(b\cos C + c\cos B)\}$

$= \dfrac{1}{8R^3}\cdot 3abc$ $\{\because a\cos B + b\cos A = c$, etc.$\}$

$$= 3 \cdot \frac{a}{2R} \cdot \frac{b}{2R} \cdot \frac{c}{2R} = 3 \sin A \sin B \sin C.$$

2. Prove that in a $\triangle ABC$,

$$\left(\cot \frac{A}{2} + \cot \frac{B}{2}\right)\left(a \sin^2 \frac{B}{2} + b \sin^2 \frac{A}{2}\right) = c \cot \frac{C}{2}.$$

$$\text{LHS} = \left(\frac{s(s-a)}{\Delta} + \frac{s(s-b)}{\Delta}\right)\left\{\frac{a}{2}(1 - \cos B) + \frac{b}{2}(1 - \cos A)\right\}$$

$$= \frac{s}{\Delta}(s - a + s - b)\frac{1}{2}\{a + b - (a\cos B + b\cos A)\}$$

$$= \frac{s}{2\Delta}(2s - a - b)(a + b - c)$$

$$= \frac{s}{2\Delta}(a + b + c - a - b)(a + b + c - 2c)$$

$$= \frac{s}{2\Delta}(c)(2s - 2c) = \frac{cs(s-c)}{\Delta} = c\cot\frac{C}{2}.$$

3. In the $\triangle ABC$, prove that

$$\cot A + \cot B + \cot C = \frac{a^2 + b^2 + c^2}{4\Delta}.$$

$$\text{LHS} = \frac{\cos A}{\sin A} + \frac{\cos B}{\sin B} + \frac{\cos C}{\sin C}$$

$$= \frac{\frac{1}{2bc}(b^2 + c^2 - a^2)}{\frac{a}{2R}} + \frac{\frac{1}{2ca}(c^2 + a^2 - b^2)}{\frac{b}{2R}} + \frac{\frac{1}{2ab}(a^2 + b^2 - c^2)}{\frac{c}{2R}}$$

$$= \frac{R}{abc}\{b^2 + c^2 - a^2 + c^2 + a^2 - b^2 + a^2 + b^2 - c^2\}$$

$$= \frac{R}{abc}(a^2 + b^2 + c^2) = \frac{a^2 + b^2 + c^2}{\frac{a}{R} \cdot bc}$$

$$= \frac{a^2 + b^2 + c^2}{4 \cdot \frac{a}{2R} \cdot \frac{1}{2}bc} = \frac{a^2 + b^2 + c^2}{4 \cdot \frac{1}{2}bc \sin A}$$

$$= \frac{a^2 + b^2 + c^2}{4\Delta} = \text{RHS}.$$

4. If x, y and z are proportional to cosines of the angles A, B and C of a $\triangle ABC$ respectively, prove that

$$\Sigma (y + z) \tan \frac{B - C}{2} \cdot \cot \frac{A}{2} = 0.$$

Here $\frac{x}{\cos A} = \frac{y}{\cos B} = \frac{z}{\cos C} = k$ (given)

$$\therefore \quad (y + z)\tan\frac{B - C}{2} \cdot \cot\frac{A}{2}$$

$$= k(\cos B + \cos C) \cdot \tan\frac{B - C}{2} \cdot \cot\frac{A}{2}$$

$$= k \cdot 2\cos\frac{B + C}{2} \cdot \cos\frac{B - C}{2} \cdot \frac{\sin\frac{B - C}{2}}{\cos\frac{B - C}{2}} \cdot \cot\frac{A}{2}$$

$$= 2k\sin\frac{A}{2} \cdot \sin\frac{B - C}{2} \cdot \cot\frac{A}{2} \quad \left\{\because \frac{B + C}{2} = \frac{\pi}{2} - \frac{A}{2}\right\}$$

$$= 2k\cos\frac{A}{2} \cdot \sin\frac{B - C}{2}$$

$$= 2k\sin\frac{B + C}{2} \cdot \sin\frac{B - C}{2}$$

$$= k(\cos C - \cos B).$$

Similarly for others.

$$\therefore \quad \Sigma (y + z)\tan\frac{B - C}{2} \cdot \cot\frac{A}{2}$$

$$= \Sigma k(\cos C - \cos B)$$

$$= k\{(\cos C - \cos B) + (\cos A - \cos C) + (\cos B - \cos A)\}$$

$$= k \times 0 = 0.$$

5. In a $\triangle ABC$, the angles A, B, C are in AP. Show that

$$2\cos\frac{A - C}{2} = \frac{a + c}{\sqrt{a^2 - ac + c^2}}.$$

Here A, B, C are in AP; so $B = A - \alpha$, $C = A - 2\alpha$.

Also $B = \frac{A + C}{2} = \frac{180° - B}{2} = 90° - \frac{B}{2}$

$$\therefore \quad \frac{3B}{2} = 90°; \quad \therefore \; B = 60°$$

$$\therefore \quad A - \alpha = 60°, \text{ i.e., } A = 60° + \alpha$$

and $C = A - 2\alpha = 60° + \alpha - 2\alpha = 60° - \alpha$

$$\therefore \quad A - C = 2\alpha.$$

$$\therefore \quad \frac{a}{\sin(60° + \alpha)} = \frac{b}{\sin 60°} = \frac{c}{\sin(60° - \alpha)}$$

$$= \frac{a + c}{\sin(60° + \alpha) + \sin(60° - \alpha)} = \frac{a + c}{\sqrt{3}\cos\alpha}$$

$$\therefore \quad \frac{b}{\frac{\sqrt{3}}{2}} = \frac{a + c}{\sqrt{3}\cos\alpha}; \quad \therefore \quad 2\cos\alpha = \frac{a + c}{b}.$$

Now, $\dfrac{a+c}{\sqrt{a^2-ac+c^2}}=\dfrac{a+c}{b\sqrt{\left(\frac{a}{b}\right)^2-\frac{a}{b}\cdot\frac{c}{b}+\left(\frac{c}{b}\right)^2}}$

$$=2\cos\alpha\cdot\frac{1}{\left[\left\{\frac{\sin(60^\circ+\alpha)}{\sin 60^\circ}\right\}^2-\frac{\sin(60^\circ+\alpha)}{\sin 60^\circ}\times\frac{\sin(60^\circ-\alpha)}{\sin 60^\circ}+\left\{\frac{\sin(60^\circ-\alpha)}{\sin 60^\circ}\right\}^2\right]^{\frac{1}{2}}}$$

$$=\frac{2\cos\alpha}{\left[\left(\cos\alpha+\frac{1}{\sqrt3}\sin\alpha\right)^2-\left(\cos\alpha+\frac{1}{\sqrt3}\sin\alpha\right)\times\left(\cos\alpha-\frac{1}{\sqrt3}\sin\alpha\right)+\left(\cos\alpha-\frac{1}{\sqrt3}\sin\alpha\right)^2\right]^{\frac{1}{2}}}$$

$$=\frac{2\cos\alpha}{\sqrt{2\left(\cos^2\alpha+\frac{1}{3}\sin^2\alpha\right)-\left(\cos^2\alpha-\frac{1}{3}\sin^2\alpha\right)}}$$

$$=\frac{2\cos\alpha}{\sqrt{\cos^2\alpha+\sin^2\alpha}}=2\cos\alpha=2\cos\frac{A-C}{2}.$$

6. If in a $\triangle ABC$,

$$\cos a=\cos b\cdot\cos c+\sin b\cdot\sin c\cdot\cos A$$
$$\cos b=\cos c\cdot\cos a+\sin c\cdot\sin a\cdot\cos B$$
$$\cos c=\cos a\cdot\cos b+\sin a\cdot\sin b\cdot\cos C$$

then prove that $\left|\dfrac{a}{\sin a}\right|=\left|\dfrac{b}{\sin b}\right|=\left|\dfrac{c}{\sin c}\right|$.

Here, $\cos A=\dfrac{\cos a-\cos b\cdot\cos c}{\sin b\cdot\sin c}$

$$\therefore\quad \cos^2A=\frac{\cos^2a+\cos^2b\cdot\cos^2c-2\cos a\cos b\cos c}{\sin^2b\cdot\sin^2c}$$

$$\therefore\quad \sin^2A=1-\cos^2A$$

$$=1-\frac{\cos^2a+\cos^2b\cos^2c-2\cos a\cos b\cos c}{\sin^2b\cdot\sin^2c}$$

$$=\frac{\{\sin^2b\cdot\sin^2c-\cos^2a-\cos^2b\cos^2c+2\cos a\cos b\cos c\}}{\sin^2b\cdot\sin^2c}$$

$$=\frac{\{(1-\cos^2b)(1-\cos^2c)-\cos^2a-\cos^2b\cos^2c+2\cos a\cos b\cos c\}}{\sin^2b\cdot\sin^2c}$$

$$=\frac{1-\cos^2a-\cos^2b-\cos^2c+2\cos a\cos b\cos c}{\sin^2b\sin^2c}$$

$$\therefore\quad \frac{\sin^2A}{\sin^2a}=\frac{\{1-\cos^2a-\cos^2b-\cos^2c+2\cos a\cos b\cos c\}}{\sin^2a\cdot\sin^2b\sin^2c}$$

Similarly, $\dfrac{\sin^2B}{\sin^2b}=\dfrac{\sin^2C}{\sin^2c}$

$$=\frac{1-\cos^2a-\cos^2b-\cos^2c+2\cos a\cos b\cos c}{\sin^2a\sin^2b\sin^2c}$$

Thus, $\dfrac{\sin^2A}{\sin^2a}=\dfrac{\sin^2B}{\sin^2b}=\dfrac{\sin^2C}{\sin^2c}$ ... (1)

But in $\triangle ABC$, $\dfrac{a}{\sin A}=\dfrac{b}{\sin B}=\dfrac{c}{\sin C}$

$\therefore$ from (1), $\dfrac{a^2}{\sin^2a}=\dfrac{b^2}{\sin^2b}=\dfrac{c^2}{\sin^2c}$

$$\therefore\quad \left|\frac{a}{\sin a}\right|=\left|\frac{b}{\sin b}\right|=\left|\frac{c}{\sin c}\right|.$$

7. Consider the following statements concerning a $\triangle ABC$.

(i) The sides a, b, c and the area $\triangle$ are rational

(ii) a, $\tan\frac{B}{2}$, $\tan\frac{C}{2}$ are rational

(iii) a, $\sin A$, $\sin B$, $\sin C$ are rational.

Prove that (i) $\Rightarrow$ (ii) $\Rightarrow$ (iii) $\Rightarrow$ (i).

Let (i) be true, i.e., a, b, c and $\triangle$ be rational numbers.

Now, $\tan\dfrac{B}{2}=\dfrac{(s-c)(s-a)}{\triangle}$, $\tan\dfrac{C}{2}=\dfrac{(s-a)(s-b)}{\triangle}$

and $s=\dfrac{a+b+c}{2}$.

Now, (i) $\Rightarrow$ $a, b, c, \triangle, s$ are rational.

So $\tan\frac{B}{2}$ and $\tan\frac{C}{2}$ are rational because sum, difference, product and quotient of nonzero rational numbers are rational.

Thus (i) $\Rightarrow$ (ii).

Let (ii) be true, i.e., a, $\tan\frac{B}{2}$, $\tan\frac{C}{2}$ be rational.

Now, $\sin B=\dfrac{2\tan\frac{B}{2}}{1+\tan^2\frac{B}{2}}=$ rational,

because $\tan\frac{B}{2}$ is rational

$$\sin C = \frac{2\tan\frac{C}{2}}{1+\tan^2\frac{C}{2}} = \text{rational},$$

because $\tan\frac{C}{2}$ is rational.

Now, $\tan\frac{B}{2}\cdot\tan\frac{C}{2} = \frac{(s-c)(s-a)}{\Delta}\cdot\frac{(s-a)(s-b)}{\Delta}$

$$= \frac{(s-a)^2(s-b)(s-c)}{s(s-a)(s-b)(s-c)} = \frac{s-a}{s} = 1-\frac{a}{s}$$

$\therefore$ (ii) $\Rightarrow$ s is rational

$\Rightarrow$ $b+c$ is rational, because a is rational.

But $\frac{a}{\sin A} = \frac{b}{\sin B} = \frac{c}{\sin C}$

$\therefore$ $\frac{a}{\sin A} = \frac{b+c}{\sin B+\sin C} = \frac{\text{rational}}{\text{rational}}$

$\therefore$ $\frac{a}{\sin A}$ is rational. But a is rational. So $\sin A$ is rational.

Thus (ii) $\Rightarrow$ (iii).

Let (iii) be true, i.e., a, $\sin A$, $\sin B$, $\sin C$ be rational.

$\because$ $\frac{a}{\sin A} = \frac{b}{\sin B} = \frac{c}{\sin C}$

$\therefore$ $b = \frac{a\sin B}{\sin A} = \text{rational}$,

and $c = \frac{a\sin C}{\sin A} = \text{rational}$

$\therefore$ $\Delta = \frac{1}{2}bc\sin A = \text{rational}$.

Thus (iii) $\Rightarrow$ (i).

8. If in a triangle ABC, two sides are $a=6$, $b=3$ and $\cos(A-B)=\frac{4}{5}$, find the area of the triangle.

Here, $\cos(A-B) = \frac{4}{5}$; $\therefore$ $\frac{1-\tan^2\frac{A-B}{2}}{1+\tan^2\frac{A-B}{2}} = \frac{4}{5}$

By componendo and dividendo,

$$\frac{2\tan^2\frac{A-B}{2}}{2} = \frac{5-4}{5+4}$$

or $\tan^2\frac{A-B}{2} = \frac{1}{9}$

or $\tan\frac{A-B}{2} = \frac{1}{3}$ $\{\because A>B\}$.

But $\tan\frac{A-B}{2} = \frac{a-b}{a+b}\cot\frac{C}{2}$

$\therefore$ $\frac{1}{3} = \frac{6-3}{6+3}\cot\frac{C}{2}$

or $\cot\frac{C}{2} = 1$; $\therefore$ $C = \frac{\pi}{2}$.

$\therefore$ the area of the triangle

$$= \frac{1}{2}ab\sin C = \frac{1}{2}\cdot 6\cdot 3\cdot\sin\frac{\pi}{2} = 9.$$

9. If α, β and γ are the altitudes of the ΔABC from the vertices A, B and C respectively then show that

$$\frac{1}{\alpha^2}+\frac{1}{\beta^2}+\frac{1}{\gamma^2} = \frac{1}{\Delta}(\cot A+\cot B+\cot C).$$

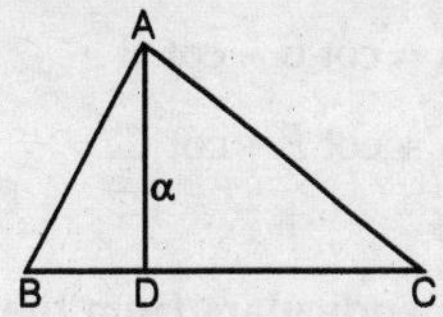

Clearly, $\frac{1}{2}\alpha\cdot a = \Delta$; $\therefore$ $\frac{1}{\alpha} = \frac{a}{2\Delta}$

Similarly, $\frac{1}{\beta} = \frac{b}{2\Delta}$ and $\frac{1}{\gamma} = \frac{c}{2\Delta}$

$$\therefore \quad \frac{1}{\alpha^2}+\frac{1}{\beta^2}+\frac{1}{\gamma^2} = \frac{a^2+b^2+c^2}{4\Delta^2} = \frac{1}{\Delta}\cdot\frac{a^2+b^2+c^2}{4\Delta}$$

$$= \frac{1}{\Delta}(\cot A+\cot B+\cot C)$$

{establish as in solved example 3}.

10. Let O be a point in the ΔABC such that $\angle OAC = \angle OCB = \angle OBA = \alpha$. Prove that $\cot\alpha = \cot A+\cot B+\cot C$.

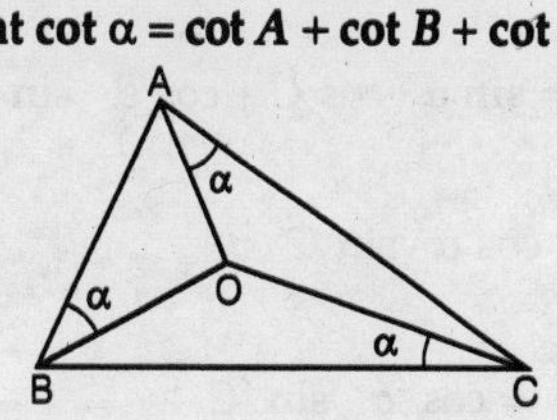

From ΔAOC, $\frac{AC}{\sin AOC} = \frac{OC}{\sin\alpha}$.

But $\angle AOC = \pi - (\angle OCA + \angle OAC)$

$= \pi - (C-\alpha+\alpha) = \pi - C$

$\therefore$ $\frac{AC}{\sin(\pi-C)} = \frac{OC}{\sin\alpha}$

or $\frac{b}{\sin C} = \frac{OC}{\sin\alpha}$ ... (1)

Similarly, from ΔOBC we get

$\frac{a}{\sin B} = \frac{OC}{\sin(B-\alpha)}$ ...(2)

(1) ÷ (2) $\Rightarrow$ $\frac{b\sin B}{a\sin C} = \frac{\sin(B-\alpha)}{\sin\alpha}$

or $\dfrac{\sin^2 B}{\sin A \sin C} = \dfrac{\sin B \cdot \cos\alpha - \cos B \cdot \sin\alpha}{\sin\alpha}$

$$\left\{\because \frac{b}{a} = \frac{\sin B}{\sin A}\right\}$$

$$= \cot\alpha \cdot \sin B - \cos B$$

or $\dfrac{\sin B}{\sin A \sin C} = \cot\alpha - \cot B$

or $\dfrac{\sin(A+C)}{\sin A \sin C} = \cot\alpha - \cot B \quad \{\because \pi - (A+C) = B\}$

or $\dfrac{\sin A \cdot \cos C + \cos A \cdot \sin C}{\sin A \cdot \sin C} = \cot\alpha - \cot B$

or $\cot C + \cot A = \cot\alpha - \cot B$

$\therefore \quad \cot\alpha = \cot A + \cot B + \cot C.$

11. If p, q are perpendiculars from the angular points A and B of the $\triangle$ABC drawn to any line through the vertex C then prove that

$$a^2b^2 \sin^2 C = a^2p^2 + b^2q^2 - 2abpq\cos C.$$

Let $\angle ACE = \alpha$. Clearly, from the figure, we get

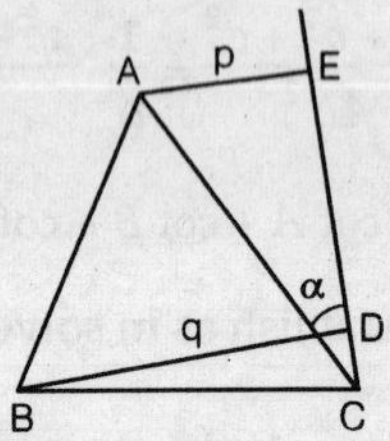

$$\frac{p}{AC} = \sin\alpha, \quad \frac{q}{BC} = \sin(\alpha + C)$$

$\Rightarrow \quad \dfrac{p}{b} = \sin\alpha, \quad \dfrac{q}{a} = \sin\alpha \cdot \cos C + \cos\alpha \cdot \sin C$

$\therefore \quad \dfrac{q}{a} = \dfrac{p}{b}\cos C + \cos\alpha \cdot \sin C$

or $\left(\dfrac{q}{a} - \dfrac{p}{b}\cos C\right)^2 = \cos^2\alpha \cdot \sin^2 C$

$$= \left(1 - \frac{p^2}{b^2}\right)(1 - \cos^2 C)$$

or $\dfrac{q^2}{a^2} + \dfrac{p^2}{b^2}\cos^2 C - \dfrac{2pq}{ab}\cos C = 1 - \dfrac{p^2}{b^2} - \left(1 - \dfrac{p^2}{b^2}\right)\cos^2 C$

or $\dfrac{q^2}{a^2} + \dfrac{p^2}{b^2} - \dfrac{2pq}{ab}\cos C = \sin^2 C$

or $a^2p^2 + b^2q^2 - 2abpq\cos C = a^2b^2\sin^2 C.$

12. If the distances of the sides of a $\triangle$ABC from its circumcentre be x, y, z then prove that

$$\frac{a}{x} + \frac{b}{y} + \frac{c}{z} = \frac{abc}{4xyz}.$$

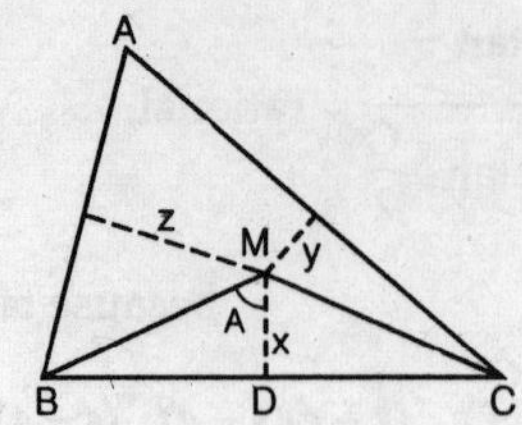

Let M be the circumcentre. MD $\perp$ BC. So $BD = DC = \dfrac{a}{2}$.

In $\triangle$BDM, $\dfrac{BD}{MD} = \tan A$

$$\left(\because \text{ from geometry, } \angle BMD = \frac{1}{2}\angle BMC = \frac{1}{2}\cdot 2A\right)$$

or $\dfrac{\frac{a}{2}}{x} = \tan A$, i.e., $\dfrac{a}{2x} = \tan A$.

Similarly, $\dfrac{b}{2y} = \tan B, \quad \dfrac{c}{2z} = \tan C$

$\therefore \quad \tan A + \tan B + \tan C = \dfrac{a}{2x} + \dfrac{b}{2y} + \dfrac{c}{2z}$

and $\tan A \cdot \tan B \cdot \tan C = \dfrac{a}{2x}\cdot\dfrac{b}{2y}\cdot\dfrac{c}{2z}$.

But in a triangle ABC,

$$\tan A + \tan B + \tan C = \tan A \cdot \tan B \cdot \tan C$$

$\therefore \quad \dfrac{a}{2x} + \dfrac{b}{2y} + \dfrac{c}{2z} = \dfrac{a}{2x}\cdot\dfrac{b}{2y}\cdot\dfrac{c}{2z}$

$\therefore \quad \dfrac{a}{x} + \dfrac{b}{y} + \dfrac{c}{z} = \dfrac{abc}{4xyz}$.

13. If x, y, z are the distances of the vertices of the $\triangle$ABC respectively from the orthocentre then prove that

$$\frac{a}{x} + \frac{b}{y} + \frac{c}{z} = \frac{abc}{xyz}.$$

Let H be the orthocentre. Then

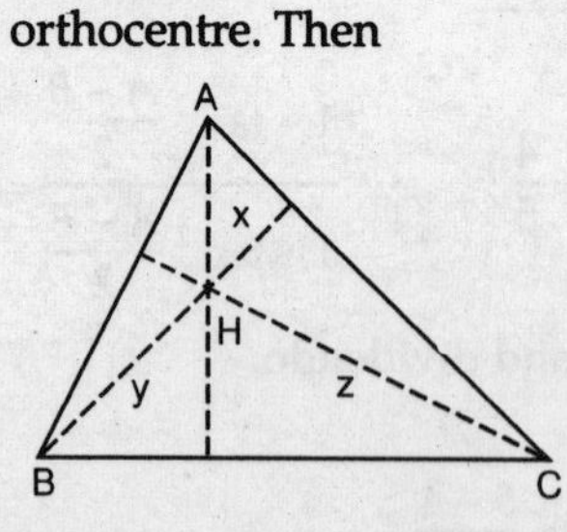

$$\angle BHC = 180° - \angle HBC - \angle HCB$$

$$= 180° - (90° - C) - (90° - B)$$

$$= B + C = \pi - A.$$

$\therefore \quad \text{ar}(\triangle BHC) = \dfrac{1}{2}BH \cdot CH \sin\angle BHC$

$$= \frac{1}{2}yz\sin(\pi - A) = \frac{1}{2}yz\sin A.$$

Similarly, $\text{ar}(\Delta\text{CHA}) = \frac{1}{2} zx \sin B$

$$\text{ar}(\Delta\text{AHB}) = \frac{1}{2} xy \sin C$$

$$\therefore \quad \text{ar}(\Delta\text{ABC}) = \frac{1}{2} yz \sin A + \frac{1}{2} zx \sin B + \frac{1}{2} xy \sin C$$

$$= \frac{1}{2} xyz\left(\frac{\sin A}{x} + \frac{\sin B}{y} + \frac{\sin C}{z}\right)$$

$$= \frac{1}{2} xyz \cdot \frac{1}{2R}\left(\frac{a}{x} + \frac{b}{y} + \frac{c}{z}\right) \quad \ldots (1)$$

Also, we know that $R = \frac{abc}{4\Delta}$, i.e., $\Delta = \frac{abc}{4R}$

$\therefore$ (1) gives, $\frac{abc}{4R} = \frac{1}{4R} xyz\left(\frac{a}{x} + \frac{b}{y} + \frac{c}{z}\right)$

$$\therefore \quad \frac{a}{x} + \frac{b}{y} + \frac{c}{z} = \frac{abc}{xyz}.$$

14. In a ΔABC, prove that $\cos A \cdot \cos C = \frac{2(c^2 - a^2)}{3ca}$ where AD is the median through A and AD $\perp$ AC.

From the ΔABC, $\cos A = \frac{b^2 + c^2 - a^2}{2bc}$ ... (1)

From the ΔCAD, $\cos C = \frac{AC}{CD} = \frac{b}{a/2} = \frac{2b}{a}$ ... (2)

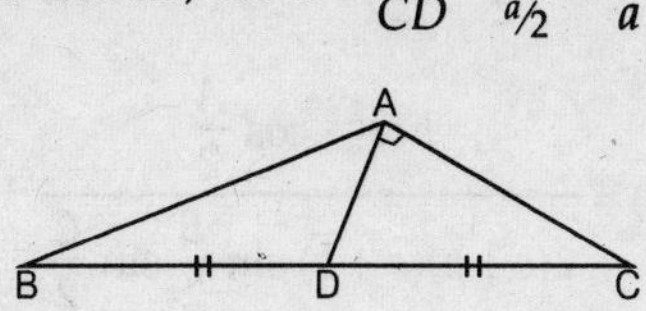

From the ΔABD, $\frac{BD}{\sin(A - 90°)} = \frac{AB}{\sin ADB}$

or $\frac{a/2}{-\cos A} = \frac{c}{\sin(90° + C)}$

or $\frac{a}{-2\cos A} = \frac{c}{\cos C}$

$\therefore \quad \cos A = \frac{a\cos C}{-2c} = \frac{a}{-2c} \cdot \frac{2b}{a} = -\frac{b}{c}$, from (2)

$\therefore$ from (1), $\frac{b^2 + c^2 - a^2}{2bc} = \frac{-b}{c}$

or $b^2 + c^2 - a^2 = -2b^2$

or $c^2 - a^2 = -3b^2$ ... (3)

Now, $\cos A \cdot \cos C = \frac{b^2 + c^2 - a^2}{2bc} \cdot \frac{2b}{a} = \frac{b^2 + c^2 - a^2}{ca}$

$$= \frac{3b^2 + 3(c^2 - a^2)}{3ca} = \frac{a^2 - c^2 + 3(c^2 - a^2)}{3ca}, \text{ from (3)}$$

$$= \frac{2(c^2 - a^2)}{3ca}.$$

15. The medians of a triangle ABC make angles α, β, γ with each other. Prove that

$\cot \alpha + \cot \beta + \cot \gamma + \cot A + \cot B + \cot C = 0$.

Here, G is the centroid of the ΔABC and $\angle BGC = \alpha$, etc.

We know from geometry,

$$AB^2 + AC^2 = 2(AD^2 + BD^2), \text{ etc.,}$$

and $AG = \frac{2}{3} AD,\ BG = \frac{2}{3} BE,\ CG = \frac{2}{3} CF.$

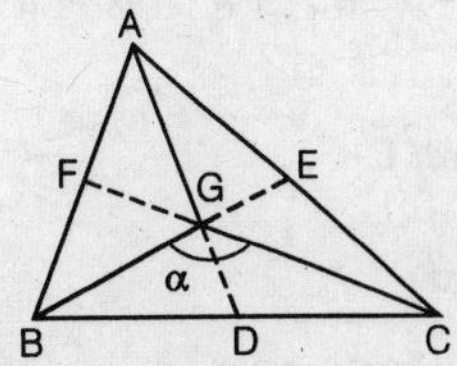

Now, $BC^2 + BA^2 = 2(BE^2 + CE^2)$

or $a^2 + c^2 = 2\left\{\left(\frac{3}{2} BG\right)^2 + \left(\frac{b}{2}\right)^2\right\}$

$$= \frac{1}{2}(9BG^2 + b^2)$$

$\therefore \quad 9BG^2 = 2(a^2 + c^2) - b^2$

or $BG^2 = \frac{1}{9}(2a^2 + 2c^2 - b^2).$

Similarly, $CG^2 = \frac{1}{9}(2a^2 + 2b^2 - c^2).$

Now, $\cos \alpha = \frac{BG^2 + CG^2 - BC^2}{2BG \cdot CG}$

$$= \frac{\frac{1}{9}(2a^2 + 2c^2 - b^2) + \frac{1}{9}(2a^2 + 2b^2 - c^2) - a^2}{2BG \cdot CG}$$

$$= \frac{1}{9} \cdot \frac{b^2 + c^2 - 5a^2}{2 \cdot \frac{1}{2} BG \cdot CG \cdot \sin \alpha} \cdot \frac{1}{2} \sin \alpha$$

$$= \frac{1}{9} \cdot \frac{1}{4} \cdot \frac{(b^2 + c^2 - 5a^2)\sin \alpha}{\text{ar}(\Delta\text{BGC})}$$

$$= \frac{1}{36} \cdot \frac{(b^2 + c^2 - 5a^2)\sin \alpha}{\frac{1}{3}\text{ar}(\Delta\text{ABC})}$$

$\{\because \text{ar}(\Delta\text{ABC}) = 3 \cdot \text{ar}(\Delta\text{BGC})\}$

$$\therefore \quad \frac{\cos \alpha}{\sin \alpha} = \frac{b^2 + c^2 - 5a^2}{12\Delta}$$

$$\therefore \quad \cot \alpha = \frac{b^2 + c^2 - 5a^2}{12\Delta}$$

Similarly, $\cot \beta = \frac{c^2 + a^2 - 5b^2}{12\Delta}$, $\cot \gamma = \frac{a^2 + b^2 - 5c^2}{12\Delta}$

$$\therefore \quad \cot\alpha + \cot\beta + \cot\gamma = -\frac{a^2+b^2+c^2}{4\Delta}.$$

Again $\cot A + \cot B + \cot C = \dfrac{\cos A}{\sin A} + \dfrac{\cos B}{\sin B} + \dfrac{\cos C}{\sin C}$

$$= \frac{b^2+c^2-a^2}{2bc}\cdot\frac{2R}{a} + \frac{c^2+a^2-b^2}{2ca}\cdot\frac{2R}{b} + \frac{a^2+b^2-c^2}{2ab}\cdot\frac{2R}{c}$$

$$= \frac{2R}{2abc}(b^2+c^2-a^2+c^2+a^2-b^2+a^2+b^2-c^2)$$

$$= \frac{R}{abc}(a^2+b^2+c^2)$$

$$= \frac{1}{4\Delta}(a^2+b^2+c^2) \qquad \left[\because R = \frac{abc}{4\Delta}\right]$$

$$\therefore \quad \cot\alpha + \cot\beta + \cot\gamma + \cot A + \cot B + \cot C$$

$$= -\frac{a^2+b^2+c^2}{4\Delta} + \frac{a^2+b^2+c^2}{4\Delta} = 0.$$

16. The internal bisectors of the angles of the ΔABC meet the sides BC, CA and AB at P, Q and R respectively. Show that the area of the ΔPQR is equal to

$$\frac{2abc\cdot\Delta}{(a+b)(b+c)(c+a)}.$$

Let the bisectors of the angles meet at I. Then I is the incentre. Let ID ⊥ BC. Then ID = inradius = r.

Clearly, $\angle BIP = \angle IBA + \angle IAB = \dfrac{B}{2} + \dfrac{A}{2}$.

Let $\angle DIP = \theta$, $PI = x$, $QI = y$ and $RI = z$. Then

$$\theta = \angle BIP - \angle BID = \frac{A}{2} + \frac{B}{2} - \left(90^\circ - \frac{B}{2}\right) = \frac{A}{2} + B - 90^\circ.$$

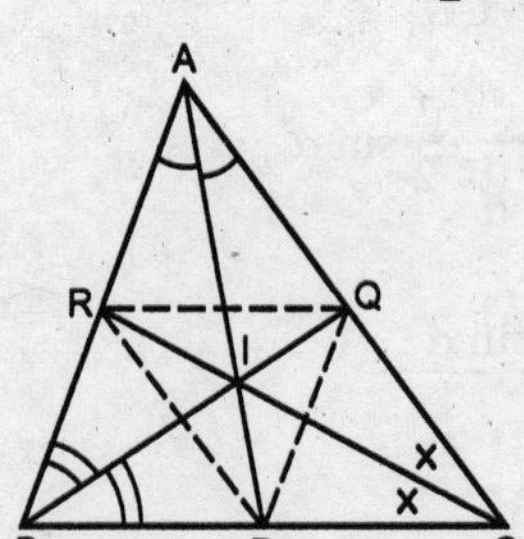

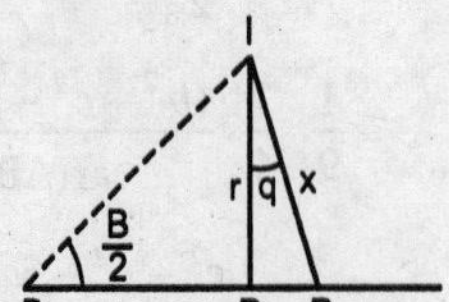

Now, from ΔDIP, $\dfrac{r}{x} = \cos\theta$

$$\therefore \quad x = \frac{r}{\cos\theta} = \frac{r}{\cos\left(\frac{A}{2}+B-90^\circ\right)} = \frac{r}{\sin\left(\frac{A}{2}+B\right)}$$

$$= \frac{r}{\frac{b+c}{a}\sin\frac{A}{2}}$$

$$\left[\because a\sin\left(\frac{A}{2}+B\right) = (b+c)\sin\frac{A}{2}, \text{ vide SN 4}\right]$$

$$\therefore \quad x = \frac{ar}{(b+c)\sin\frac{A}{2}}.$$

Similarly, $y = \dfrac{br}{(c+a)\sin\frac{B}{2}}$

and $z = \dfrac{cr}{(a+b)\sin\frac{C}{2}}$

Now, $\text{ar}(\Delta PIQ) = \dfrac{1}{2}PI\cdot IQ\sin\angle PIQ$

$$= \frac{1}{2}xy\sin(\angle PIC + \angle QIC)$$

$$= \frac{1}{2}\cdot xy\cdot\sin\left\{\left(\frac{A}{2}+\frac{C}{2}\right)+\left(\frac{C}{2}+\frac{B}{2}\right)\right\}$$

$$= \frac{1}{2}\frac{ar}{(b+c)\sin\frac{A}{2}}\cdot\frac{br}{(c+a)\sin\frac{B}{2}}\cdot\sin\left(\frac{A+B+C}{2}+\frac{C}{2}\right)$$

$$= \frac{abr^2}{2(b+c)(c+a)\sin\frac{A}{2}\sin\frac{B}{2}}\sin\left(90^\circ+\frac{C}{2}\right)$$

$$= \frac{abr^2\cdot\cos\frac{C}{2}}{2(b+c)(c+a)\sin\frac{A}{2}\sin\frac{B}{2}}.$$

Similarly, $\text{ar}(\Delta QIR) = \dfrac{bcr^2\cdot\cos\frac{A}{2}}{2(c+a)(a+b)\sin\frac{B}{2}\sin\frac{C}{2}}$

$$\text{ar}(\Delta RIP) = \frac{car^2\cos\frac{B}{2}}{2(a+b)(b+c)\sin\frac{C}{2}\sin\frac{A}{2}}$$

Now, $\text{ar}(\Delta PQR) = \text{ar}(\Delta PIQ) + \text{ar}(\Delta QIR) + \text{ar}(\Delta RIP)$

$$= \frac{abcr^2}{2(a+b)(b+c)(c+a)\sin\frac{A}{2}\sin\frac{B}{2}\cdot\sin\frac{C}{2}} \times\left\{\frac{a+b}{c}\sin\frac{C}{2}\cos\frac{C}{2}+\frac{b+c}{a}\sin\frac{A}{2}\cos\frac{A}{2}+\frac{c+a}{b}\sin\frac{B}{2}\cos\frac{B}{2}\right\}$$

$$= \frac{abcr^2}{2(a+b)(b+c)(c+a)\cdot\frac{r}{4R}} \times\frac{1}{2}\left\{\frac{a+b}{c}\sin C+\frac{b+c}{a}\sin A+\frac{c+a}{b}\sin B\right\}$$

$$\left[\because r = 4R\sin\frac{A}{2}\cdot\sin\frac{B}{2}\cdot\sin\frac{C}{2}\right]$$

$$= \frac{abcRr}{(a+b)(b+c)(c+a)} \times \left\{(a+b)\frac{1}{2R} + (b+c)\frac{1}{2R} + (c+a)\frac{1}{2R}\right\}$$

$$= \frac{abc \cdot r}{(a+b)(b+c)(c+a)}(a+b+c)$$

$$= \frac{abcr}{(a+b)(b+c)(c+a)} \cdot 2s$$

$$= \frac{2abc \cdot \Delta}{(a+b)(b+c)(c+a)} \qquad \left\{\because \ r = \frac{\Delta}{s}\right\}.$$

17. If a, b and c are sides of a $\triangle ABC$ and $3a = b + c$, show that $\cot \frac{B}{2} \cdot \cot \frac{C}{2} = 2$.

$$\text{LHS} = \frac{s(s-b)}{\Delta} \cdot \frac{s(s-c)}{\Delta} = \frac{s^2(s-b)(s-c)}{\Delta^2}$$

$$= \frac{s^2(s-b)(s-c)}{s(s-a)(s-b)(s-c)} = \frac{s}{s-a}$$

$$= \frac{2s}{2s-2a} = \frac{a+b+c}{a+b+c-2a} = \frac{a+3a}{b+c-a}$$

$$= \frac{4a}{3a-a} = \frac{4a}{2a} = 2.$$

18. If the sides of a triangle are in AP and its greatest angle exceeds the least by α, show that the sides are in the ratio $(1-x):1:(1+x)$ where $x = \sqrt{\frac{1-\cos\alpha}{7-\cos\alpha}}$.

Clearly $1-x, 1, 1+x$ are in AP and from the question the sides a, b, c of the triangle are in AP. Hence

$$\frac{a}{1-x} = \frac{b}{1} = \frac{c}{1+x} = k \text{ (suppose).}$$

We have to prove that $x = \sqrt{\frac{1-\cos\alpha}{7-\cos\alpha}}$.

Now, $a = k(1-x)$, $b = k$, $c = k(1+x)$

$$\therefore \quad s = \frac{a+b+c}{2} = \frac{k(1-x)+k+k(1+x)}{2} = \frac{3k}{2}.$$

Taking x positive, c is the greatest side. So C is the greatest angle and A is the least angle. Hence $C = A + \alpha$.

$$\therefore \quad \alpha = C - A$$

$$\text{or} \quad \frac{\alpha}{2} = \frac{C}{2} - \frac{A}{2}$$

$$\therefore \quad \cos\frac{\alpha}{2} = \cos\left(\frac{C}{2} - \frac{A}{2}\right) = \cos\frac{C}{2}\cos\frac{A}{2} + \sin\frac{C}{2} \cdot \sin\frac{A}{2}$$

$$= \sqrt{\frac{s(s-c)}{ab}} \cdot \sqrt{\frac{s(s-a)}{bc}} + \sqrt{\frac{(s-b)(s-a)}{ab}} \cdot \sqrt{\frac{(s-b)(s-c)}{bc}}$$

$$= \frac{1}{b}\sqrt{\frac{(s-c)(s-a)}{ac}} \cdot \{s + s - b\}$$

$$= \frac{1}{k}\sqrt{\frac{\left\{\frac{3k}{2} - k(1+x)\right\}\left\{\frac{3k}{2} - k(1-x)\right\}}{k(1-x) \cdot k(1+x)}} \cdot (3k-k)$$

$$= 2 \cdot \sqrt{\frac{\left(\frac{3}{2} - 1 - x\right)\left(\frac{3}{2} - 1 + x\right)}{1-x^2}}$$

$$= \sqrt{\frac{(1-2x)(1+2x)}{1-x^2}} = \sqrt{\frac{1-4x^2}{1-x^2}}.$$

$$\therefore \quad \cos\alpha = 2\cos^2\frac{\alpha}{2} - 1$$

$$= 2 \cdot \frac{1-4x^2}{1-x^2} - 1 = \frac{1-7x^2}{1-x^2}$$

$$\therefore \quad (1-x^2)\cos\alpha = 1 - 7x^2$$

$$\text{or} \quad (7 - \cos\alpha)x^2 = 1 - \cos\alpha$$

$$\therefore \quad x^2 = \frac{1-\cos\alpha}{7-\cos\alpha}; \quad \therefore \quad x = \sqrt{\frac{1-\cos\alpha}{7-\cos\alpha}}.$$

19. Prove that, for a scalene $\triangle ABC$,

$$\cos A + \cos B = 4\sin^2\frac{C}{2} \quad \Leftrightarrow \quad a, c, b \text{ are in AP.}$$

$$\cos A + \cos B = 4\sin^2\frac{C}{2}$$

$$\Leftrightarrow \quad 2\cos\frac{A+B}{2} \cdot \cos\frac{A-B}{2} = 4\sin^2\frac{C}{2}$$

$$\Leftrightarrow \quad 2\sin\frac{C}{2}\left(\cos\frac{A-B}{2} - 2\sin\frac{C}{2}\right) = 0 \quad \{\because \ A + B = \pi - C\}$$

$$\Leftrightarrow \quad \cos\frac{A-B}{2} = 2\sin\frac{C}{2}$$

$$\Leftrightarrow \quad 2\sin\frac{A+B}{2} \cdot \cos\frac{A-B}{2} = 4\sin\frac{A+B}{2}\sin\frac{C}{2}$$

$$\Leftrightarrow \quad \sin A + \sin B = 4\cos\frac{C}{2}\sin\frac{C}{2} = 2\sin C$$

$$\Leftrightarrow \quad \frac{a}{2R} + \frac{b}{2R} = \frac{2c}{2R} \quad \left\{\because \ \frac{a}{\sin A} = \frac{b}{\sin B} = \frac{c}{\sin C} = 2R\right\}$$

$$\Leftrightarrow \quad a + b = 2c \quad \Leftrightarrow \quad a, c, b \text{ are in AP.}$$

20. In a $\triangle ABC$, if $a\tan A + b\tan B = (a+b)\tan\dfrac{A+B}{2}$ then prove that the triangle is isosceles.

Here, $a\left\{\tan A - \tan\dfrac{A+B}{2}\right\} = b\left\{\tan\dfrac{A+B}{2} - \tan B\right\}$

or $a\left\{\dfrac{\sin A}{\cos A} - \dfrac{\sin\dfrac{A+B}{2}}{\cos\dfrac{A+B}{2}}\right\} = b\left\{\dfrac{\sin\dfrac{A+B}{2}}{\cos\dfrac{A+B}{2}} - \dfrac{\sin B}{\cos B}\right\}$

or $\dfrac{a\sin\left(A - \dfrac{A+B}{2}\right)}{\cos A\cos\dfrac{A+B}{2}} = \dfrac{b\sin\left(\dfrac{A+B}{2} - B\right)}{\cos\dfrac{A+B}{2}\cdot\cos B}$

or $\sin\dfrac{A-B}{2}\cdot\cos\dfrac{A+B}{2}(a\cos B - b\cos A) = 0.$

But $\cos\dfrac{A+B}{2} \neq 0.$

$\therefore$ $\sin\dfrac{A-B}{2}\{a\cos B - b\cos A\} = 0$

or $\sin\dfrac{A-B}{2}\{\sin A\cos B - \sin B\cos A\} = 0$

or $\sin\dfrac{A-B}{2}\cdot\sin(A-B) = 0$

$\Rightarrow$ $A - B = 0$, i.e., $A = B$.

Therefore, the triangle is isosceles.

21. If the tangents of the angles of a triangle are in AP, prove that the squares of the sides are in the ratio $x^2(x^2+9) : (x^2+3)^2 : 9(1+x^2)$ where x is the least or the greatest tangent.

Here $\tan A$, $\tan B$, $\tan C$ are in AP;

$\therefore$ $2\tan B = \tan A + \tan C = \dfrac{\sin(A+C)}{\cos A\cos C}$

$= \dfrac{\sin B}{\cos A\cos C}$ $\{\because A + C = \pi - B\}$

or $2\cos A\cos C = \sin B\cdot\cot B$

$\therefore$ $2\cos A\cos C = \cos B = -\cos(A+C)$ ... (1)

or $2\cos A\cos C = -\cos A\cdot\cos C + \sin A\sin C$

or $3\cos A\cos C = \sin A\sin C$

$\therefore$ $\tan A\tan C = 3$ ... (2)

Let $\tan A$ be the greatest or the least tangent.

Then $\tan A = x$.

Now, $\dfrac{a}{\sin A} = \dfrac{b}{\sin B} = \dfrac{c}{\sin C}$

$\therefore$ $\dfrac{a^2}{\sin^2 A} = \dfrac{b^2}{\sin^2 B} = \dfrac{c^2}{\sin^2 C}$ ... (3)

$\because$ $\tan A = x$,

$\sin^2 A = \dfrac{1}{1+\cot^2 A} = \dfrac{1}{1+\dfrac{1}{x^2}} = \dfrac{x^2}{1+x^2}$ ... (4)

$\sin^2 B = 1 - \cos^2 B = 1 - (2\cos A\cdot\cos C)^2$ {from (1)}

$= 1 - 4\cos^2 A\cos^2 C$

$= 1 - 4\cdot\dfrac{1}{1+\tan^2 A}\cdot\dfrac{1}{1+\tan^2 C}$

$= 1 - 4\cdot\dfrac{1}{1+x^2}\cdot\dfrac{1}{1+\left(\dfrac{3}{x}\right)^2}$ {from (2)}

$= 1 - \dfrac{4x^2}{(1+x^2)(9+x^2)} = \dfrac{(1+x^2)(9+x^2) - 4x^2}{(1+x^2)(9+x^2)}$

$= \dfrac{x^4+6x^2+9}{(1+x^2)(9+x^2)} = \dfrac{(x^2+3)^2}{(1+x^2)(9+x^2)}$ ... (5)

$\sin^2 C = \dfrac{1}{1+\cot^2 C} = \dfrac{1}{1+\left(\dfrac{x}{3}\right)^2}$ {from (2)}

$= \dfrac{9}{9+x^2}$ ... (6)

$\therefore$ from (3), (4), (5) and (6) we get

$$\frac{a^2}{\dfrac{x^2}{1+x^2}} = \frac{b^2}{\dfrac{(x^2+3)^2}{(1+x^2)(9+x^2)}} = \frac{c^2}{\dfrac{9}{9+x^2}}$$

or $\dfrac{a^2}{x^2(9+x^2)} = \dfrac{b^2}{(x^2+3)^2} = \dfrac{c^2}{9(1+x^2)}$.

Hence, the problem.

22. If a^2, b^2, c^2 are in AP then show that $\cot A$, $\cot B$, $\cot C$ are in AP.

a^2, b^2, c^2 are in AP

$\Rightarrow$ $-2a^2, -2b^2, -2c^2$ are in AP {multiplying by -2}

$\Rightarrow$ $b^2+c^2-a^2,\ c^2+a^2-b^2,\ a^2+b^2-c^2$ are in AP {adding $a^2+b^2+c^2$}

$\Rightarrow$ $2bc\cos A$, $2ca\cos B$, $2ab\cos C$ are in AP

$\Rightarrow$ $\dfrac{\cos A}{a}, \dfrac{\cos B}{b}, \dfrac{\cos C}{c}$ are in AP {dividing by $2abc$}

$\Rightarrow$ $\dfrac{\cos A}{2R\sin A}, \dfrac{\cos B}{2R\sin B}, \dfrac{\cos C}{2R\sin C}$ are in AP

$\Rightarrow$ $\cot A$, $\cot B$, $\cot C$ are in AP {multiplying by $2R$}

23. In a $\triangle ABC$, if $\tan\frac{A}{2}$, $\tan\frac{B}{2}$, $\tan\frac{C}{2}$ are in AP, prove that $\cos A$, $\cos B$, $\cos C$ are also in AP.

Here, $2\tan\frac{B}{2}=\tan\frac{A}{2}+\tan\frac{C}{2}$

or $\dfrac{2\sin\frac{B}{2}}{\cos\frac{B}{2}}=\dfrac{\sin\frac{A}{2}}{\cos\frac{A}{2}}+\dfrac{\sin\frac{C}{2}}{\cos\frac{C}{2}}=\dfrac{\sin\frac{A+C}{2}}{\cos\frac{A}{2}\cos\frac{C}{2}}$

or $2\cos\frac{A}{2}\cos\frac{C}{2}\sin\frac{B}{2}=\cos^2\frac{B}{2}\quad\left\{\because\ \frac{A+C}{2}=90^\circ-\frac{B}{2}\right\}$

or $\sin\frac{B}{2}\left\{\cos\frac{A+C}{2}+\cos\frac{A-C}{2}\right\}=\cos^2\frac{B}{2}$

or $\sin^2\frac{B}{2}+\sin\frac{B}{2}\cdot\cos\frac{A-C}{2}=\cos^2\frac{B}{2}$

or $\sin\frac{B}{2}\cdot\cos\frac{A-C}{2}=\cos B$

or $2\cos\frac{A+C}{2}\cdot\cos\frac{A-C}{2}=2\cos B$

or $\cos A+\cos C=2\cos B$

$\therefore$ $\cos A$, $\cos B$, $\cos C$ are in AP.

24. The lengths of sides of a triangle are three consecutive natural numbers and its largest angle is twice the smallest one. Determine the sides of the triangle.

Let the lengths of the sides be $n, n+1, n+2$ where $n\in N$.

From the question, the largest angle opposite to the side $n+2$ is 2θ while the smallest angle opposite to the side n is θ.

Now, $\cos\theta=\dfrac{(n+1)^2+(n+2)^2-n^2}{2(n+1)(n+2)}=\dfrac{n^2+6n+5}{2(n+1)(n+2)}$

$=\dfrac{(n+1)(n+5)}{2(n+1)(n+2)}=\dfrac{n+5}{2(n+2)}$

and $\cos 2\theta=\dfrac{n^2+(n+1)^2-(n+2)^2}{2n(n+1)}$

$=\dfrac{n^2-2n-3}{2n(n+1)}$

$=\dfrac{(n+1)(n-3)}{2n(n+1)}=\dfrac{n-3}{2n}$

But $\cos 2\theta=2\cos^2\theta-1$; so $\dfrac{n-3}{2n}=2\left\{\dfrac{n+5}{2(n+2)}\right\}^2-1$

or $\dfrac{n-3}{2n}=\dfrac{(n+5)^2}{2(n+2)^2}-1$

or $(n-3)(n+2)^2=n\{(n+5)^2-2(n+2)^2\}$

or $(n-3)(n^2+4n+4)=n(-n^2+2n+17)$

or $n^3+n^2-8n-12=-n^3+2n^2+17n$

or $2n^3-n^2-25n-12=0$

or $(n-4)(2n^2+7n+3)=0$

$\therefore$ $n=4$ or $2n^2+7n+3=0$.

Roots of $2n^2+7n+3=0$ are $\dfrac{-7\pm\sqrt{49-24}}{4}$,

i.e., $-\frac{1}{2}$ and -3 which are not natural numbers.

$\therefore$ $n=4$; hence sides are 4, 5, 6.

25. In a triangle of base a, the ratio of the two sides is $r(r<1)$. Show that the altitude of the triangle is less than or equal to $\dfrac{ar}{1-r^2}$.

Let the other sides be b, br. Let the altitude be h.

Now, (area of the Δ)$^2=\left(\frac{1}{2}ah\right)^2=\frac{1}{4}a^2h^2$ ... (1)

Also, $s=\dfrac{a+b+br}{2}$ and

(area of the Δ)$^2=\dfrac{a+b+br}{2}\left(\dfrac{a+b+br}{2}-a\right)\left(\dfrac{a+b+br}{2}-b\right)\times\left(\dfrac{a+b+br}{2}-br\right)$

$=\dfrac{a+b+br}{2}\cdot\dfrac{b+br-a}{2}\cdot\dfrac{a+br-b}{2}\cdot\dfrac{a+b-br}{2}$

$=\dfrac{(a+b)^2-b^2r^2}{4}\cdot\dfrac{b^2r^2-(a-b)^2}{4}$ ... (2)

From (1) and (2),

$\frac{1}{4}a^2h^2=\frac{1}{16}\{2ab+(a^2+b^2-b^2r^2)\}\times\{2ab-(a^2+b^2-b^2r^2)\}$

or $4a^2h^2=4a^2b^2-(a^2+b^2-b^2r^2)^2$

$=4a^2b^2-\{a^2+(1-r^2)b^2\}^2$

or $(1-r^2)^2b^4-\{4a^2-2(1-r^2)a^2\}b^2+a^4+4a^2h^2=0$

or $(1-r^2)^2(b^2)^2-2(1+r^2)a^2b^2+a^2(a^2+4h^2)=0.$

Now, the values of b^2 are real, so

$4(1+r^2)^2a^4-4(1-r^2)^2\cdot a^2(a^2+4h^2)\geq 0$

or $(1+r^2)^2a^2-(1-r^2)^2\cdot(a^2+4h^2)\geq 0$

or $\{(1+r^2)^2-(1-r^2)^2\}a^2-(1-r^2)^2.\,4h^2\geq 0$

or $r^2a^2-(1-r^2)^2h^2\geq 0$

or $\dfrac{r^2a^2}{(1-r^2)^2}\geq h^2$

$\therefore \quad \dfrac{ra}{1-r^2} \geq h.$

Hence, the altitude is less than or equal to $\dfrac{ar}{1-r^2}$.

26. If in a triangle ABC,
$\cos A \cdot \cos B + \sin A \cdot \sin B \cdot \sin^n C = 1,\ n \in N$
then prove that the sides are in the ratio 1 : 1 : √2.

We know that in $\triangle ABC$, $1 \geq \sin C > 0$

$\therefore \quad 0 < \sin^n C \leq 1$

$\therefore \quad \cos A \cdot \cos B + \sin A \cdot \sin B \cdot \sin^n C$
$\leq \cos A \cos B + \sin A \sin B = \cos(A - B),$

equality holding when $C = \dfrac{\pi}{2}$.

$\therefore$ from the question, $1 \leq \cos(A - B)$.

But $\cos(A - B) \leq 1$.

So, we get $\cos(A - B) = 1$ and then $C = \dfrac{\pi}{2}$.

Now, $\cos(A - B) = 1 \quad \Rightarrow \quad A - B = 0$

$\therefore \quad A = B$ and $C = \dfrac{\pi}{2}$

$\therefore \quad A = \dfrac{\pi}{4},\ B = \dfrac{\pi}{4},\ C = \dfrac{\pi}{2}.$

$$\therefore \quad \frac{a}{\sin\frac{\pi}{4}} = \frac{b}{\sin\frac{\pi}{4}} = \frac{c}{\sin\frac{\pi}{2}}$$

$$\text{or} \quad \frac{a}{1/\sqrt{2}} = \frac{b}{1/\sqrt{2}} = \frac{c}{1}$$

$$\therefore \quad \frac{a}{1} = \frac{b}{1} = \frac{c}{\sqrt{2}}$$

$\therefore$ the ratio of the sides = 1 : 1 : √2.

27. In the $\triangle ABC$, a similar $\triangle A'B'C'$ is inscribed so that $B'C' = \lambda \cdot BC$. If B'C' is inclined at an angle θ with BC, prove that $\lambda \cos\theta = \dfrac{1}{2}$.

$\triangle ABC$ and $\triangle A'B'C'$ are similar where

$\angle B'A'C' = \angle BAC = A,\ \angle A'B'C' = \angle ABC = B,$

$\angle B'C'A' = \angle BCA = C.$

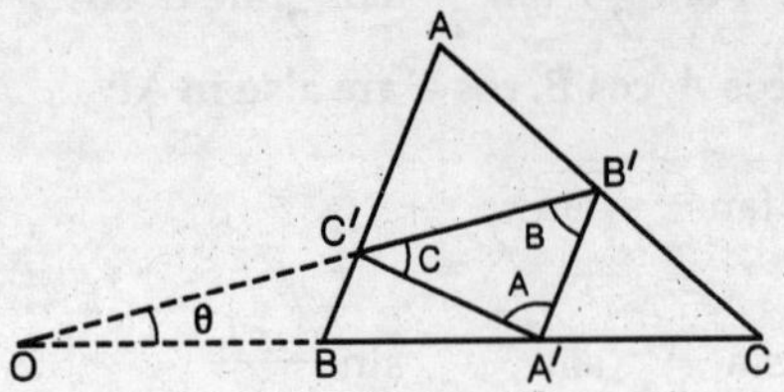

In $\triangle AC'B'$, $\angle AB'C' = \angle B'OC + \angle ACB = \theta + C$

$$\therefore \quad \frac{AC'}{\sin(\theta + C)} = \frac{B'C'}{\sin A}$$

$$\text{or} \quad \frac{AC'}{\sin(\theta + C)} = \frac{\lambda BC}{\sin A} = \lambda \cdot \frac{a}{\sin A} = 2\lambda R.$$

In $\triangle BA'C'$, $\angle BA'C' = \angle A'C'B' - \angle A'OC' = C - \theta$

$$\therefore \quad \frac{BC'}{\sin(C - \theta)} = \frac{A'C'}{\sin B}$$

$$\therefore \quad \frac{BC'}{\sin(C - \theta)} = \frac{\lambda \cdot AC}{\sin B} = \frac{\lambda b}{\sin B} = 2\lambda R$$

$$\left\{\because \frac{B'C'}{BC} = \frac{A'C'}{AC} = \lambda \text{ from similar triangles}\right\}$$

Thus, we get

$AC' = 2\lambda R \sin(\theta + C)$ and $BC' = 2\lambda R \sin(C - \theta)$

$\therefore \quad c = AB = AC' + BC' = 2\lambda R\{\sin(C + \theta) + \sin(C - \theta)\}$

$= 2\lambda R \cdot 2\sin C \cdot \cos\theta$

$$\therefore \quad \cos\theta = \frac{c}{4\lambda R \sin C} = \frac{1}{4\lambda R} \cdot \frac{c}{\sin C}$$

$$= \frac{1}{4\lambda R} \cdot 2R = \frac{1}{2\lambda}$$

$$\therefore \quad \lambda\cos\theta = \frac{1}{2}.$$

28. The two adjacent sides of a cyclic quadrilateral are 2 and 5 and the angle between them is 60°. If the area of the quadrilateral is 4√3, find the remaining sides.

Let the cyclic quadrilateral be ABCD in which ar(ABCD) = $4\sqrt{3}$, $AB = 2$, $BC = 5$ and $\angle B = 60°$.

Then $\angle D = 120°$.

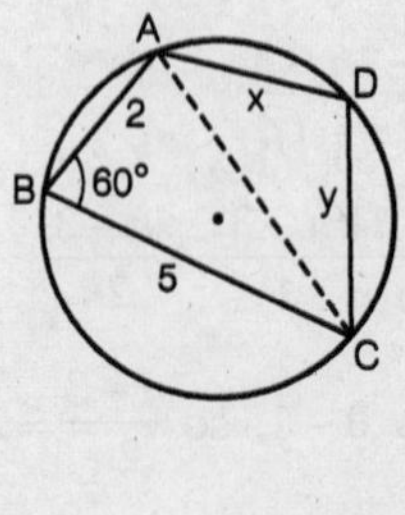

Let $AD = x$, $CD = y$.

Now, $\text{ar}(\triangle ABC) = \dfrac{1}{2} \cdot 2 \cdot 5 \cdot \sin 60° = \dfrac{5\sqrt{3}}{2}$

$$\text{ar}(\Delta ADC) = \frac{1}{2} \cdot x \cdot y \cdot \sin 120° = \frac{\sqrt{3}}{4} xy$$

$\therefore \quad \text{ar}(ABCD) = 4\sqrt{3} = \frac{5\sqrt{3}}{2} + \frac{\sqrt{3}}{4} xy$

or $\quad \frac{3\sqrt{3}}{2} = \frac{\sqrt{3}}{4} xy; \quad \therefore \quad xy = 6 \qquad \ldots (1)$

Also, $AC^2 = AB^2 + BC^2 - 2AB \cdot BC \cdot \cos 60°$

$$= 4 + 25 - 2 \cdot 2 \cdot 5 \cdot \frac{1}{2} = 19$$

$$AC^2 = AD^2 + DC^2 - 2AD \cdot DC \cdot \cos 120°$$

$$= x^2 + y^2 - 2xy\left(-\frac{1}{2}\right) = x^2 + y^2 + xy$$

$\therefore \quad x^2 + y^2 + xy = 19 \qquad \ldots (2)$

From (1) and (2), $x^2 + y^2 = 13$

or $\quad (x+y)^2 - 2xy = 13$

or $\quad (x+y)^2 = 13 + 2xy = 13 + 2 \cdot 6 = 25$

$\therefore \quad x + y = 5 \qquad \ldots (3)$

$(\because x > 0,\ y > 0)$

Also $(x-y)^2 = (x+y)^2 - 4xy = 5^2 - 4 \cdot 6 = 1$

$\therefore \quad x - y = \pm 1$

$\therefore \quad$ using (3), we get

$x = 3,\ y = 2$ or $x = 2,\ y = 3$.

29. Let $A_1, A_2, A_3, \ldots, A_n$ be the vertices of an n-sided regular polygon such that $\frac{1}{A_1A_2} = \frac{1}{A_1A_3} + \frac{1}{A_1A_4}$.

Find the value of n.

If O is the centre of the n-sided regular polygon, then

$$\angle A_1OA_2 = \angle A_2OA_3 = \ldots = \frac{2\pi}{n}.$$

Let $\quad OA_1 = OA_2 = \ldots = OA_n = r.$

Clearly we get $A_1A_2 = 2A_1M = 2r \sin \frac{\pi}{n}$.

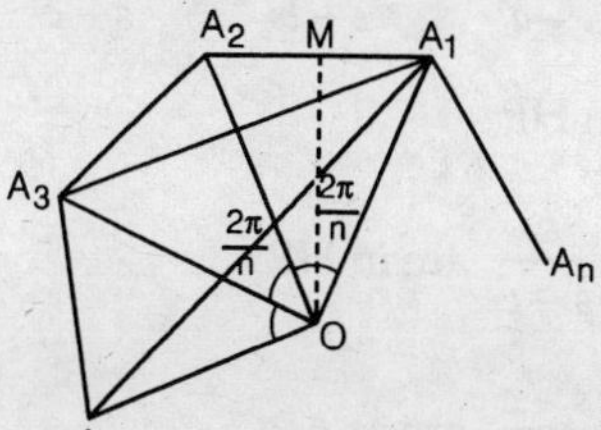

Similarly, $A_1A_3 = 2r \sin \frac{2\pi}{n}$

and $A_1A_4 = 2r \sin \frac{3\pi}{n}$.

$\therefore$ we get (from the question),

$$\frac{1}{2r\sin\frac{\pi}{n}} = \frac{1}{2r\sin\frac{2\pi}{n}} + \frac{1}{2r\sin\frac{3\pi}{n}}$$

or $\quad \frac{1}{\sin\frac{\pi}{n}} = \frac{1}{\sin\frac{2\pi}{n}} + \frac{1}{\sin\frac{3\pi}{n}}$

or $\quad \frac{1}{\sin\frac{\pi}{n}} - \frac{1}{\sin\frac{3\pi}{n}} = \frac{1}{\sin\frac{2\pi}{n}}$

or $\quad \sin\frac{2\pi}{n}\left(\sin\frac{3\pi}{n} - \sin\frac{\pi}{n}\right) = \sin\frac{\pi}{n} \cdot \sin\frac{3\pi}{n}$

or $\quad \sin\frac{2\pi}{n} \cdot 2\cos\frac{2\pi}{n} \cdot \sin\frac{\pi}{n} = \sin\frac{\pi}{n} \cdot \sin\frac{3\pi}{n}$

or $\quad \sin\frac{4\pi}{n} = \sin\frac{3\pi}{n} \quad \left\{\because \sin\frac{\pi}{n} \neq 0\right\}$

$\therefore \quad \frac{4\pi}{n} = \pi - \frac{3\pi}{n} \quad \{\because 4 \neq 3\}$

or $\quad \frac{7\pi}{n} = \pi; \quad \therefore \quad n = 7.$

30. If a regular polygon of n sides has the circumradius R and inradius r then prove that each side of the polygon is equal to $2(R+r)\tan\frac{\pi}{2n}$.

Let A_1A_2 be a side and O be the centre. Let $OB \perp A_1A_2$.

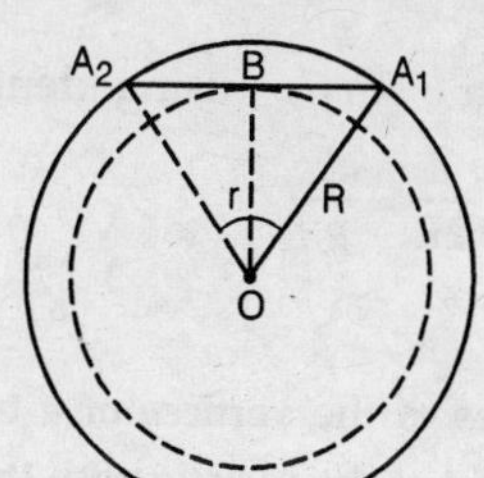

Clearly, $OA_1 = R$ and $OB = r$.

Also, $\angle A_1OA_2 = \frac{2\pi}{n}$ and $\angle A_1OB = \frac{\pi}{n}$.

In ΔA_1OB, $\frac{r}{R} = \cos \angle A_1OB = \cos\frac{\pi}{n} \qquad \ldots (1)$

and $\quad \frac{A_1B}{OB} = \tan\frac{\pi}{n}$, i.e., $A_1B = r\tan\frac{\pi}{n}$

$\therefore \quad A_1A_2 = 2A_1B = 2r\tan\frac{\pi}{n} \qquad \ldots (2)$

Now, $2(R+r)\tan\frac{\pi}{2n} = 2\left(\frac{r}{\cos\frac{\pi}{n}} + r\right)\tan\frac{\pi}{2n}$, from (1)

$$= 2r\left(\frac{1+\cos\frac{\pi}{n}}{\cos\frac{\pi}{n}}\right)\cdot\frac{1-\cos\frac{\pi}{n}}{\sin\frac{\pi}{n}} \quad \left\{\because \ \tan\frac{\theta}{2}=\frac{1-\cos\theta}{\sin\theta}\right\}$$

$$= 2r\cdot\frac{1-\cos^2\frac{\pi}{n}}{\cos\frac{\pi}{n}\cdot\sin\frac{\pi}{n}} = 2r\cdot\frac{\sin^2\frac{\pi}{n}}{\cos\frac{\pi}{n}\cdot\sin\frac{\pi}{n}}$$

$$= 2r\tan\frac{\pi}{n} = A_1A_2, \text{ from (2).}$$

31. If in the ΔABC, O is the circumcentre and R is the circumradius and R_1, R_2, R_3 are the circumradii of the triangles OBC, OCA and OAB respectively then prove that $\dfrac{a}{R_1}+\dfrac{b}{R_2}+\dfrac{c}{R_3}=\dfrac{abc}{R^3}$.

Clearly, in the ΔOBC, $\angle BOC = 2A$, $OB = OC = R$, $BC = a$.

$\therefore \quad 2R_1 = \dfrac{a}{\sin 2A}$ {using sine rule in ΔBOC}

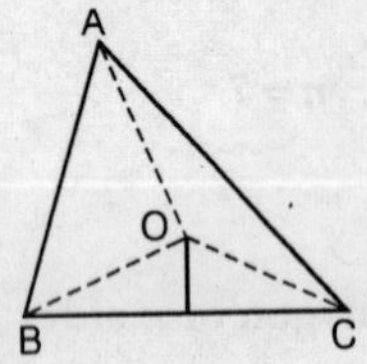

Similarly, $2R_2 = \dfrac{b}{\sin 2B}$ and $2R_3 = \dfrac{c}{\sin 2C}$

$\therefore \quad \dfrac{a}{R_1}+\dfrac{b}{R_2}+\dfrac{c}{R_3} = 2(\sin 2A + \sin 2B + \sin 2C)$

$= 2.4\sin A \sin B \sin C$, (using identity)

$= 8\dfrac{a}{2R}\cdot\dfrac{b}{2R}\cdot\dfrac{c}{2R}=\dfrac{abc}{R^3}$.

32. If the distances of the vertices of a triangle from the points of contact of the incircle with the sides be α, β, γ then prove that $r^2=\dfrac{\alpha\beta\gamma}{\alpha+\beta+\gamma}$ where r = inradius.

Let the incircle touches the side AB at P where $AP = \alpha$. Let I be the incentre. Then AI bisects the $\angle$BAC.

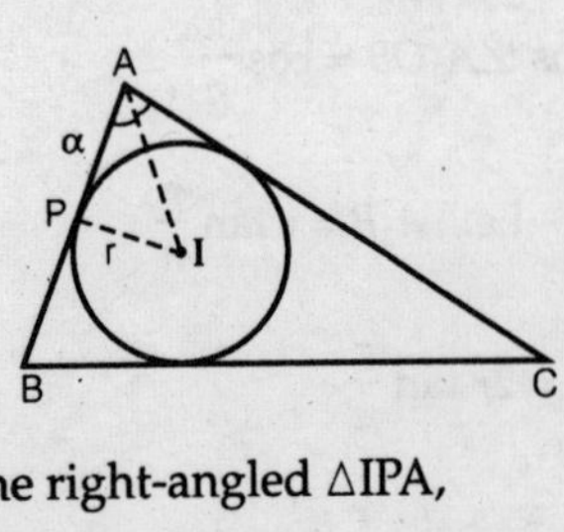

$\therefore$ from the right-angled ΔIPA,

$\dfrac{r}{\alpha} = \tan\dfrac{A}{2}$; $\quad \therefore \quad \alpha = r\cot\dfrac{A}{2}$.

Similarly, $\beta = r\cot\dfrac{B}{2}$ and $\gamma = r\cot\dfrac{C}{2}$.

In a ΔABC, we have the identity

$$\cot\frac{A}{2}+\cot\frac{B}{2}+\cot\frac{C}{2}=\cot\frac{A}{2}\cot\frac{B}{2}\cot\frac{C}{2}$$

$$\therefore \quad \frac{\alpha}{r}+\frac{\beta}{r}+\frac{\gamma}{r}=\frac{\alpha}{r}\cdot\frac{\beta}{r}\cdot\frac{\gamma}{r}$$

$$\text{or} \quad \frac{1}{r}(\alpha+\beta+\gamma)=\frac{1}{r^3}\alpha\beta\gamma$$

$$\therefore \quad r^2=\frac{\alpha\beta\gamma}{\alpha+\beta+\gamma}.$$

33. If r, r_1, r_2, r_3 are the inradius and the exradii respectively of a ΔABC then prove that $\dfrac{1}{r^2}+\dfrac{1}{r_1^2}+\dfrac{1}{r_2^2}+\dfrac{1}{r_3^2}=\dfrac{1}{\Delta^2}(a^2+b^2+c^2)$.

We know that $r=\dfrac{\Delta}{s}$, $r_1=\dfrac{\Delta}{s-a}$, $r_2=\dfrac{\Delta}{s-b}$, $r_3=\dfrac{\Delta}{s-c}$.

$$\therefore \quad \text{LHS}=\frac{s^2}{\Delta^2}+\frac{(s-a)^2}{\Delta^2}+\frac{(s-b)^2}{\Delta^2}+\frac{(s-c)^2}{\Delta^2}$$

$$=\frac{1}{\Delta^2}\{s^2+(s-a)^2+(s-b)^2+(s-c)^2\}$$

$$=\frac{1}{\Delta^2}\{4s^2-2s(a+b+c)+a^2+b^2+c^2\}$$

$$=\frac{1}{\Delta^2}\{4s^2-2s\cdot 2s+a^2+b^2+c^2\}$$

$$=\frac{1}{\Delta^2}(a^2+b^2+c^2)=\text{RHS}.$$

34. If the exradii r_1, r_2, r_3 of a ΔABC are in HP, show that its sides a, b, c are in AP.

We know that $r_1=\dfrac{\Delta}{s-a}$, $r_2=\dfrac{\Delta}{s-b}$, $r_3=\dfrac{\Delta}{s-c}$.

$\therefore \quad r_1, r_2, r_3$ are in HP

$\Rightarrow \quad \dfrac{\Delta}{s-a}, \dfrac{\Delta}{s-b}, \dfrac{\Delta}{s-c}$ are in HP

$\Rightarrow \quad \dfrac{s-a}{\Delta}, \dfrac{s-b}{\Delta}, \dfrac{s-c}{\Delta}$ are in AP

$\Rightarrow \quad s-a,\ s-b,\ s-c$ are in AP

$\Rightarrow \quad -a, -b, -c$ are in AP

$\Rightarrow \quad a, b, c$ are in AP.

35. Prove that the harmonic mean of the exradii of a triangle is three times the inradius.

Let the exradii be r_1, r_2, r_3 and inradius $= r$.

$$\text{HM of } r_1, r_2, r_3 = \frac{1}{\dfrac{\frac{1}{r_1}+\frac{1}{r_2}+\frac{1}{r_3}}{3}} = \frac{3}{\frac{1}{r_1}+\frac{1}{r_2}+\frac{1}{r_3}}$$

Thus, we have to prove that

$$\frac{3}{\frac{1}{r_1}+\frac{1}{r_2}+\frac{1}{r_3}} = 3r$$

$$\text{Now, } \frac{1}{r_1}+\frac{1}{r_2}+\frac{1}{r_3} = \frac{s-a}{\Delta}+\frac{s-b}{\Delta}+\frac{s-c}{\Delta}$$

$$= \frac{s-a+s-b+s-c}{\Delta} = \frac{3s-(a+b+c)}{\Delta}$$

$$= \frac{3s-2s}{\Delta} = \frac{s}{\Delta}.$$

$$\therefore \quad \frac{3}{\frac{1}{r_1}+\frac{1}{r_2}+\frac{1}{r_3}} = \frac{3}{\frac{s}{\Delta}} = \frac{3\Delta}{s} = 3r.$$

Hence, the problem.

36. Prove that in a triangle, the sum of exradii exceeds the inradius by twice the diameter of the circumcircle. (or, Prove that $r_1 + r_2 + r_3 = r + 4R$.)

Let the exradii be r_1, r_2, r_3 and inradius $= r$,

circumradius $= R$.

Then we have to prove that $r_1 + r_2 + r_3 = r + 4R$.

$$\text{Now, } r_1 + r_2 + r_3 - r = \frac{\Delta}{s-a}+\frac{\Delta}{s-b}+\frac{\Delta}{s-c}-\frac{\Delta}{s}$$

$$= \Delta\left\{\left(\frac{1}{s-a}-\frac{1}{s}\right)+\left(\frac{1}{s-b}+\frac{1}{s-c}\right)\right\}$$

$$= \Delta\left\{\frac{a}{s(s-a)}+\frac{2s-b-c}{(s-b)(s-c)}\right\}$$

$$= \Delta\left\{\frac{a}{s(s-a)}+\frac{a}{(s-b)(s-c)}\right\}, \text{ for } 2s = a+b+c$$

$$= \Delta\, a \cdot \frac{s^2 - s(b+c) + bc + s^2 - as}{s(s-a)(s-b)(s-c)}$$

$$= \Delta\, a \cdot \frac{2s^2 - s(a+b+c) + bc}{\Delta^2}$$

$$= \frac{a}{\Delta}\cdot(2s^2 - 2s^2 + bc) = \frac{abc}{\Delta} = 4R \quad \left\{\because R = \frac{abc}{4\Delta}\right\}$$

$$\therefore \quad r_1 + r_2 + r_3 = r + 4R.$$

37. Show that the $\triangle ABC$ is equilateral if its circumradius is double of the inradius.

Here $R = 2r$ (given).

$$\text{We know that } r = \frac{\Delta}{s} = \frac{\Delta^2}{s\Delta} = \frac{s(s-a)(s-b)(s-c)}{s\Delta}$$

$$= \frac{1}{\Delta}(s-a)(s-b)(s-c)$$

$$= \frac{1}{\Delta}\sqrt{(s-b(s-c)}\cdot\sqrt{(s-c)(s-a)}\cdot\sqrt{(s-a)(s-b)}$$

$$= \frac{abc}{\Delta}\sqrt{\frac{(s-b)(s-c)}{bc}}\cdot\sqrt{\frac{(s-c)(s-a)}{ca}} \times \sqrt{\frac{(s-a)(s-b)}{ab}}$$

$$= 4R\sin\frac{A}{2}\cdot\sin\frac{B}{2}\cdot\sin\frac{C}{2} \quad \left\{\because R = \frac{abc}{4\Delta}\right\}$$

$$\therefore \quad r = 8r\sin\frac{A}{2}\sin\frac{B}{2}\sin\frac{C}{2}$$

$$\text{or} \quad \sin\frac{A}{2}\cdot\sin\frac{B}{2}\cdot\sin\frac{C}{2} = \frac{1}{8} \qquad \ldots (1)$$

$$\text{Now, } \sin\frac{A}{2}\cdot\sin\frac{B}{2}\cdot\sin\frac{C}{2}$$

$$= \frac{1}{2}\sin\frac{A}{2}\left\{\cos\frac{B-C}{2} - \cos\frac{B+C}{2}\right\}$$

$$= \frac{1}{2}\sin\frac{A}{2}\left\{\cos\frac{B-C}{2} - \sin\frac{A}{2}\right\}$$

$$\le \frac{1}{2}\sin\frac{A}{2}\left(1 - \sin\frac{A}{2}\right) \quad \left\{\because \max\cos\frac{B-C}{2} = 1\right\}$$

equality holding when $B = C$.

$$\text{But } \frac{1}{2}\sin\frac{A}{2}\cdot\left(1-\sin\frac{A}{2}\right) = \frac{1}{2}\left[\sin\frac{A}{2} - \sin^2\frac{A}{2}\right]$$

$$= \frac{1}{2}\left[\frac{1}{4} - \left(\sin^2\frac{A}{2} - \sin\frac{A}{2} + \frac{1}{4}\right)\right]$$

$$= \frac{1}{2}\left[\frac{1}{4} - \left(\sin\frac{A}{2} - \frac{1}{2}\right)^2\right]$$

$$\le \frac{1}{2}\cdot\frac{1}{4} \quad \left\{\because \left(\sin\frac{A}{2} - \frac{1}{2}\right)^2 \ge 0\right\}$$

equality holding when $\sin\frac{A}{2} = \frac{1}{2}$.

$$\therefore \quad \sin\frac{A}{2}\sin\frac{B}{2}\sin\frac{C}{2} \le \frac{1}{8},$$

equality holding when $B = C$ and $\sin\frac{A}{2} = \frac{1}{2}$

From (1), equality must hold.

So $B = C$ and $\sin\frac{A}{2} = \frac{1}{2}$

i.e., $\frac{A}{2} = 30°$; $\therefore$ $A = 60°$.

$\therefore$ $B + C = 180° - A = 120°$;

but $B = C$; so $B = C = 60°$.

Hence $A = B = C = 60°$.

So the triangle is equilateral.

38. In a triangle ABC, the sides a, c and the angle A are given. If the side b has two possible values of which one is double the other then prove that

$$a = \frac{c}{3}\sqrt{1 + 8\sin^2 A}.$$

We have, $\cos A = \frac{b^2 + c^2 - a^2}{2bc}$

or $b^2 - (2c\cos A)b + (c^2 - a^2) = 0$.

This is a quadratic equation in b.

So, b will have two values if

$$D = 4c^2\cos^2 A - 4(c^2 - a^2) > 0$$

or $c^2\cos^2 A - c^2 + a^2 > 0$

or $a^2 - c^2\sin^2 A > 0$

or $a > c\sin A$.

From the question $a > c\sin A$ because

$$b^2 - (2c\cos A)b + (c^2 - a^2) = 0$$

has two real values.

According to the question, the values of b are b_1, $2b_1$

$\therefore$ $b_1 + 2b_1 = 2c\cos A$ and $b_1 \cdot 2b_1 = c^2 - a^2$

$\therefore$ $b_1 = \frac{2c\cos A}{3}$ and $2b_1^2 = c^2 - a^2$

$\therefore$ $2\cdot\left(\frac{2c\cos A}{3}\right)^2 = c^2 - a^2$

or $a^2 = c^2\left\{1 - \frac{8}{9}\cos^2 A\right\}$

or $a^2 = c^2\left\{1 - \frac{8}{9}(1 - \sin^2 A)\right) = \frac{c^2}{9}\{1 + 8\sin^2 A\}$

$\therefore$ $a = \frac{c}{3}\sqrt{1 + 8\sin^2 A}$.

Exercises

1. In a $\triangle ABC$, prove that $\Sigma\frac{a^2\sin(B - C)}{\sin B + \sin C} = 0$.

2. In a $\triangle ABC$, show that $\Sigma\frac{b^2 - c^2}{a^2}\sin 2A = 0$.

3. In a $\triangle ABC$, prove that

$$\frac{\cos A}{a} + \frac{a}{bc} = \frac{\cos B}{b} + \frac{b}{ca} = \frac{\cos C}{c} + \frac{c}{ab}.$$

4. In a $\triangle ABC$, prove that $a\sin\left(\frac{A}{2} + B\right) = (b + c)\sin\frac{A}{2}$.

5. In a $\triangle ABC$, prove that $\Sigma(b^2 - c^2)\cot A = 0$.

6. In a $\triangle ABC$, prove that

$$(a + b)\sin\frac{C}{2} = c\cos\frac{A - B}{2} = c\sin\left(B + \frac{C}{2}\right).$$

7. In a $\triangle ABC$, prove that

$$(b + c - a)\left(\cot\frac{B}{2} + \cot\frac{C}{2}\right) = 2a\cot\frac{A}{2}.$$

8. In a $\triangle ABC$, prove that

$$b\cos\alpha = c\cos(A - \alpha) + a\cos(C + \alpha),$$

where α is any angle.

9. Find real values of x for which $x^2 + x + 1$, $2x + 1$ and $x^2 - 1$ represent three sides of a triangle. Prove that, for each of the admissible values of x, the greatest angle of the triangle is $\frac{2\pi}{3}$.

10. If the angles of a right-angled triangle are in AP, show that the ratio of its sides will be $1 : \sqrt{3} : 2$.

11. If in the $\triangle ABC$, $\frac{A}{3} = \frac{B}{5} = \frac{C}{4}$ then prove that $a + c\sqrt{2} = 2b$.

12. If in a $\triangle ABC$, $\frac{b + c}{11} = \frac{c + a}{12} = \frac{a + b}{13}$ then prove that $\frac{\cos A}{7} = \frac{\cos B}{19} = \frac{\cos C}{25}$.

13. In a $\triangle ABC$, if $\tan\frac{A}{2} = \frac{5}{6}$ and $\tan\frac{B}{2} = \frac{20}{37}$ then prove that the sides a, b, c are in AP.

14. In a $\triangle ABC$, the angles A, B and C are in AP and they satisfy the conditions

$$\sin(2A + B) = \sin(C - A) = -\sin(B + 2C) = \frac{1}{2}.$$

Determine the measures of the angles A, B and C.

15. Can a triangle be constructed such that the cosines of its angles are $\frac{3}{5}$, $\frac{5}{13}$ and $\frac{7}{13}$? Give reason.

If α, β and γ be three angles given by $\cos\alpha = \frac{a}{b+c}$,

$\cos\beta = \frac{b}{c+a}$ and $\cos\gamma = \frac{c}{a+b}$ where a, b, c are the sides of a ΔABC then prove that

$$\tan\frac{\alpha}{2}\cdot\tan\frac{\beta}{2}\cdot\tan\frac{\gamma}{2} = \tan\frac{A}{2}\cdot\tan\frac{B}{2}\cdot\tan\frac{C}{2}.$$

17. The sides of a triangle are in AP and its area is $\frac{3}{5}$th of an equilateral triangle of the same perimeter. Prove that the sides are in the ratio 3 : 5 : 7.

18. In a ΔABC, if $\frac{1}{b+c} + \frac{1}{c+a} = \frac{3}{a+b+c}$ then prove that $C = \frac{\pi}{3}$.

19. In a ΔABC, if $c^4 - 2(a^2+b^2)c^2 + a^4 + a^2b^2 + b^4 = 0$ then prove that $C = \frac{\pi}{3}$ or $\frac{2\pi}{3}$.

20. Prove that ΔABC is isosceles if any one of the following holds:

(a) $2\cos A = \frac{b}{c}$

(b) cosines of two angles are proportional to the opposite sides.

21. Prove that ΔABC is isosceles or right angled if any one of the following holds:

(a) $(a^2+b^2)\sin(A-B) = (a^2-b^2)\sin(A+B)$

(b) cosines of two angles are inversely proportional to the opposite sides

(c) $\frac{\cos A + 2\cos C}{\cos A + 2\cos B} = \frac{\sin B}{\sin C}$.

22. If in a ΔABC, $\cot A + \cot B + \cot C = \sqrt{3}$ then prove that the triangle is equilateral.

23. If in a ΔABC, $\frac{a}{b+c} = \tan\frac{A}{2}$ then show that the triangle is right angled.

24. In a ΔABC, prove that

$\cot A, \cot B, \cot C$ are in AP $\Leftrightarrow a^2, b^2, c^2$ are in AP.

25. In a ΔABC, if the sides a, b, c are in HP, prove that

$\operatorname{cosec}^2\frac{A}{2}, \operatorname{cosec}^2\frac{B}{2}, \operatorname{cosec}^2\frac{C}{2}$ are in AP.

26. In a ΔABC, if $\sin A, \sin B, \sin C$ are in AP, show that

$$\tan\frac{A}{2}\cdot\tan\frac{C}{2} = \frac{1}{3}.$$

27. In a ΔABC, if $\cos A + 2\cos B + \cos C = 2$, prove that a, b, c are in AP.

28. In a ΔABC, if $a\sin(B-C) = c\sin(A-B)$, prove that a^2, b^2, c^2 are in AP.

29. If in a triangle ABC, the quantities

$$\tan\frac{A}{2}, \tan\frac{B}{2}, \tan\frac{C}{2}$$

are in HP then prove that the sides a, b, c are in AP.

30. If α, β, γ are the altitudes of a ΔABC from the vertices A, B and C respectively, prove that (where R is the circumradius)

(a) $\frac{1}{\alpha} + \frac{1}{\beta} + \frac{1}{\gamma} = \frac{s}{\Delta}$

(b) $\frac{\cos A}{\alpha} + \frac{\cos B}{\beta} + \frac{\cos C}{\gamma} = \frac{1}{R}$

(c) $\alpha\cos A + \beta\cos B + \gamma\cos C = \frac{a^2+b^2+c^2}{4R}$.

31. If p_1, p_2, p_3 are the altitudes of a triangle and R, the radius of its circumcircle, prove that

$$\Delta = \sqrt{\frac{1}{2}Rp_1p_2p_3},$$

where Δ denotes the area of the triangle.

32. Let O be a point inside a ΔABC such that $\angle OAC = \angle OCB = \angle OBA = \alpha$.

Prove that $\operatorname{cosec}^2\alpha = \operatorname{cosec}^2 A + \operatorname{cosec}^2 B + \operatorname{cosec}^2 C$.

33. In a ΔPQR, $QR = 12$, $RP = 16$ and $PQ = 20$. M is a point on PQ such that $MP : MQ = \lambda : 1$ $(\lambda > 1)$ and the line RM divides the $\angle PRQ$ in the ratio 2 : 1. Find the length RM and the value of λ.

34. In a ΔABC, AD is an altitude and H is the orthocentre. Prove that

$$AH : DH = (\tan B + \tan C) : \tan A.$$

35. In a ΔABC, the median $AD = \frac{1}{\sqrt{11 - 6\sqrt{3}}}$ and it divides the angle A into angles 30° and 45°. Find the length of the side BC.

36. In a ΔABC, $\angle C = 60°$, $\angle A = 75°$. If D is a point on AC such that the area of ΔBAD is $\sqrt{3}$ times the area of ΔBCD, find $\angle ABD$.

37. AD is a median of ΔABC and $\angle BAD = \theta$, $\angle CAD = \phi$ then prove that

$$\cot\theta + \cot C = \cot\phi + \cot B.$$

38. The bisector of the angle A of ΔABC meets BC at D. If $AD = m$ then prove that

(i) $\frac{a}{b+c} = \sqrt{1 - \frac{m^2}{bc}}$ (ii) $m = \frac{2bc}{b+c}\cos\frac{A}{2}$.

39. If the median AD of ΔABC be perpendicular to AB then prove that $\tan A + 2\tan B = 0$.

40. In a ΔABC, the lengths of the bisectors of the angles A, B and C are x, y and z respectively. Show that

$$\frac{1}{x}\cos\frac{A}{2} + \frac{1}{y}\cos\frac{B}{2} + \frac{1}{z}\cos\frac{C}{2} = \frac{1}{a} + \frac{1}{b} + \frac{1}{c}.$$

41. In a ΔABC, the median AD and the altitude AM divides the angle A in three equal parts. Show that

$$\Delta \cdot \cos\frac{A}{3} \cdot \sin^2\frac{A}{3} = \frac{3a^3}{128R}$$

where Δ = area of the triangle.

42. A cyclic quadrilateral ABCD of area $\frac{3\sqrt{3}}{4}$ is inscribed in a unit circle. If one of its sides $AB = 1$ and the diagonal $BD = \sqrt{3}$, find the lengths of other sides.

43. Let A_1, A_2, A_3, A_4 and A_5 be the vertices of a regular pentagon inscribed in a unit circle taken in order. Show that $A_1A_2 \cdot A_1A_3 = \sqrt{5}$.

44. Let Δ_1 and Δ_2 denote the areas of two regular polygons of 10 and 20 sides respectively, sides of each being of equal length. Prove that

$$\frac{4(3-\sqrt{5})}{\sqrt{10+2\sqrt{5}}}\Delta_1 = (4-\sqrt{10+2\sqrt{5}})\Delta_2.$$

45. In quadrilateral ABCD, if $AB = 10$, $\angle DAB = 75°$, $\angle CBA = 90°$, $\angle CAB = 45°$ and $\angle DBA = 30°$ then find CD.

46. Prove that the sum of the radii of the circles, inscribed in and circumscribed about a regular polygon of n sides, is $\frac{a}{2}\cot\frac{\pi}{2n}$ where a is the length of a side of the polygon.

47. The ratio of the area of the regular polygon of n sides circumscribed about a circle to the area of the regular polygon of equal number of sides inscribed in the circle is 4 : 3. Find the value of n.

48. The perimeter of a circle and a regular polygon of n sides are equal.

Prove that $\dfrac{\text{area of the circle}}{\text{area of the polygon}} = \tan\frac{\pi}{n} : \frac{\pi}{n}$.

49. Three points A′, B′ and C′ are taken on the plane of a ΔABC such that $\Delta A'BC$, $\Delta AB'C$ and $\Delta ABC'$ are equilateral with their circumradii R_a, R_b, R_c; inradii r_a, r_b, r_c and exradii r_a', r_b', r_c' respectively. Prove that $(r_a\, r_b\, r_c) : (R_a\, R_b\, R_c) : (r_a'\, r_b'\, r_c') = 1 : 8 : 27$.

50. If r_1, r_2, r_3 are exradii and r is the inradius of a triangle then prove that

(i) $r_1^{-1} + r_2^{-1} + r_3^{-1} = r^{-1}$ (ii) $r_2r_3 + r_3r_1 + r_1r_2 = \frac{\Delta^2}{r^2}$.

51. If A_1, A_2, A_3 are the areas of the excircles and A is the area of the incircle of a triangle then prove that

$$\frac{1}{\sqrt{A_1}} + \frac{1}{\sqrt{A_2}} + \frac{1}{\sqrt{A_3}} = \frac{1}{\sqrt{A}}.$$

52. If r_1, r_2, r_3 are exradii and r = inradius, R = circumradius then prove that

(i) $ar_1 + br_2 + cr_3 = abc\left(\frac{1}{r} - \frac{1}{2R}\right)$

(ii) $r_1^2 + r_2^2 + r_3^2 + r^2 = 16R^2 - (a^2 + b^2 + c^2)$.

53. Prove that the three exradii of a triangle are the roots of the equation

$$x^3 - x^2(4R + r) + (x - r)s^2 = 0$$

where R = circumradius, r = inradius and s = semiperimeter of the triangle.

54. If p_1, p_2, p_3 are the altitudes of a ΔABC from the vertices A, B, C respectively and the corresponding exradii are r_1, r_2, r_3 then prove that

$$\frac{1}{r_1} + \frac{1}{r_2} + \frac{1}{r_3} = \frac{1}{p_1} + \frac{1}{p_2} + \frac{1}{p_3}.$$

55. If I be the incentre of a ΔABC, prove that

$$IA \cdot IB \cdot IC = abc\tan\frac{A}{2} \cdot \tan\frac{B}{2} \cdot \tan\frac{C}{2}.$$

56. The internal bisectors of the angles of a ΔABC meet its circumcircle at D, E and F. Prove that ar(ΔABC) : ar(ΔDEF) = $2r : R$ where r is the inradius and R is the circumradius of the ΔABC.

57. In a triangle ABC, the distance of the circumcentre from the side BC is equal to the inradius then prove that $\cos B + \cos C = 1$.

58. If $r_1 = r + r_2 + r_3$ where r_1, r_2, r_3 are exradii and r is the inradius then show that the triangle is right angled.

59. In a ΔABC, $8R^2 = a^2 + b^2 + c^2$ where R is the circumradius. Show that the triangle is right angled.

60. If in a ΔABC, $\left(1 - \frac{r_1}{r_2}\right)\left(1 - \frac{r_1}{r_3}\right) = 2$ where r_1, r_2, r_3 are exradii then prove that the triangle is right angled.

61. If in a ΔABC, the exradii r_1, r_2, r_3 are such that $r_1 = 2r_2 = 3r_3$ then prove that the triangle is right angled. Also find $a : b : c$.

62. In a ΔABC, the sides a, b and the angle A are given. If the side c has two possible values c_1 and c_2 then prove that

(i) $(c_1 - c_2)^2 + (c_1 + c_2)^2\tan^2 A = 4a^2$

(ii) $|c_1 - c_2| = 2 \cdot \sqrt{a^2 - b^2\sin^2 A}$.

63. In a ΔABC, $a < b$ and $a : b = 2 : m$.

If $\cos A = \frac{1}{2}\sqrt{\frac{m^2 - 4}{m - 1}}$, prove that there are two values of c, one of which is $(m - 1)$ times the other.

Objective Questions

Fill in the blanks.

64. If in ΔABC, $a : b : c = 8 : 10 : 12$ and the least angle is α then the greatest angle in terms of α is ______.

65. If in ΔABC, $b = \sqrt{3}+1$, $c = \sqrt{3}-1$ and $\angle A = \frac{\pi}{3}$ then $\angle C =$ ______ .

66. If two angles of a triangle are $\frac{\pi}{4}$ and $\frac{\pi}{3}$ then the ratio between the largest side and the smallest side is ______ .

67. If two angles of a triangle are 30° and 45° and the included side is $(\sqrt{3}+1)$ cm then the area of the triangle = ______ .

68. In a ΔABC, if $b = \sqrt{3}$, $c = 1$, $A = \frac{\pi}{6}$ then $B =$ ______ .

69. In a ΔABC, the angles A, B and C are in AP and $b : c = \sqrt{3} : \sqrt{2}$. Then $A =$ ______ .

70. If in a ΔABC, $a = 5$, $b = 4$, $\cos(A - B) = \frac{31}{32}$ then $c =$ ______ .

71. The smallest angle of the triangle, whose sides are $7, 4\sqrt{3}$ and $\sqrt{13}$, is ______ .

72. If in a ΔABC, $2\tan\frac{B-C}{2} = \cot\frac{A}{2}$ then $b : c =$ ______ .

73. In any ΔABC,

$$\frac{a^2\sin(B-C)}{\sin A} + \frac{b^2\sin(C-A)}{\sin B} + \frac{c^2\sin(A-B)}{\sin C} = \_\_\_\_\_\_ .$$

74. If in a ΔABC, $\begin{vmatrix} a & b & c \\ b & c & a \\ c & a & b \end{vmatrix} = 0$ then the value of $\sin^3 A + \sin^3 B + \sin^3 C$ in terms of products of trigonometrical ratios of the angles, is ______ .

75. If in a ΔABC, $c = 2a$ then $\tan\frac{B}{2} \cdot \tan\frac{C-A}{2} =$ ______ .

76. If in the ΔABC, $\cos A = \frac{17}{22}$ and $\cos C = \frac{1}{14}$ then $a : b : c =$ ______ .

77. If in the ΔABC, $\frac{A}{1} = \frac{B}{2} = \frac{C}{7}$ then $c : a =$ ______ .

78. In a ΔABC, if $(a+b+c)(b+c-a) = 3bc$ then $A =$ ______ .

79. In a ΔABC, if $\frac{\tan A - \tan B}{\tan A + \tan B} = \frac{c-b}{c}$ then $A =$ ______ .

80. If in a ΔABC, $\frac{2\cos A}{a} + \frac{\cos B}{b} + \frac{2\cos C}{c} = \frac{a}{bc} + \frac{b}{ca}$ then $A =$ ______ .

81. In a ΔABC, AD is the altitude from A. Given $b > c$, $\angle C = 23°$ and $AD = \frac{abc}{b^2 - c^2}$ then $\angle B =$ ______ .

82. A circle is inscribed in an equilateral triangle of side a. The area of any square inscribed in the circle is ______ .

83. In a ΔABC, $a : b : c = 4 : 5 : 6$. The ratio of the radius of the circumcircle to that of the incircle is ______ .

Choose the correct option(s).

84. If in a ΔABC, $c = 2b$ and $C = B + \frac{\pi}{3}$ then the measure of $\angle A$ is

(a) $\frac{\pi}{2}$ (b) $\frac{\pi}{3}$ (c) $\frac{\pi}{6}$ (d) $\frac{\pi}{4}$

85. In a ΔABC, $(b-c)^2 + 4bc\sin^2\frac{A}{2}$ is equal to

(a) a^2 (b) $2a$

(c) $\left(\frac{b+c}{a}\right)^2$ (d) none of these

86. In a ΔABC, $C = 90°$. Then $\tan A$ and $\tan B$ are the roots of the equation

(a) $abx^2 - c^2x + ab = 0$ (b) $abx^2 - (a^2+b^2)x + ab = 0$

(c) $c^2x^2 - abx + c^2 = 0$ (d) $ax^2 - bx + a = 0$

87. In a ΔABC, $B = \frac{\pi}{8}$ and $C = \frac{5\pi}{8}$ and the altitude $AD = h$. Then $h : a$ is equal to

(a) 1 : 2 (b) 2 : 1

(c) 1 : 4 (d) none of these

88. If in a ΔABC, $a\cos^2\frac{C}{2} + c\cos^2\frac{A}{2} = \frac{3}{2}b$ then the sides a, b, c are in

(a) AP (b) GP

(c) HP (d) none of these

89. If in a ΔABC, $c(a+b)\cos\frac{B}{2} = b(c+a)\cos\frac{C}{2}$ then the triangle is

(a) isosceles

(b) right angled

(c) isosceles or right angled

(d) none of these

90. In a ΔABC, if $a^4 + b^4 + c^4 = 2c^2(a^2+b^2)$ then the measure of C is

(a) $\frac{\pi}{3}$ (b) $\frac{\pi}{4}$ (c) $\frac{3\pi}{4}$ (d) $\frac{2\pi}{3}$

91. Two adjacent sides of a cyclic quadrilateral are 2 and 5, and the angle between them is 60°. If a third side is 3, the fourth side is

(a) 2 (b) 3 (c) 4 (d) 5

92. If in a ΔABC, R is the circumradius then the distance of the orthocentre of the triangle from the vertex A is

(a) $a\cot A$ (b) $2R\sin A$

(c) $2R\cos A$ (d) $\frac{abc}{2\Delta}\cos A$

93. If in a ΔABC, $c = 150$, $b = 50\sqrt{3}$ and $B = 30°$ then C has the measure

(a) 60° (b) 120°

(c) 90° (d) none of these

94. The diameter of the circumcircle of a triangle with sides 5, 6 and 7 is

(a) $\frac{3\sqrt{6}}{2}$ (b) $2\sqrt{6}$

(c) $\frac{35}{48}$ (d) $\frac{9}{\sqrt{2}}$

(e) none of these

95. In the ΔABC, $AB = 5$ cm, $AC = 12$ cm and $BC = 13$ cm then the distance of A from the side BC is (in cm)

(a) $\frac{25}{13}$ (b) $\frac{60}{13}$ (c) $\frac{65}{12}$ (d) $\frac{144}{13}$

96. In the ΔABC, $A > B$. If the measures of A and B satisfy the equation $3\sin x - 4\sin^3 x - k = 0$, $0 < k < 1$ then the measure of C is

(a) $\frac{\pi}{3}$ (b) $\frac{\pi}{2}$ (c) $\frac{2\pi}{3}$ (d) $\frac{5\pi}{6}$

97. In a ΔABC, AD is the bisector of the angle A meeting BC at D. If I be the incentre of the triangle then $AI : DI$ is equal to

(a) $(\sin B + \sin C) : \sin A$

(b) $(\cos B + \cos C) : \cos A$

(c) $\cos\frac{B-C}{2} : \cos\frac{B+C}{2}$

(d) $\sin\frac{B-C}{2} : \sin\frac{B+C}{2}$

98. If in a ΔABC, $b : c = 2 : 1$ and $\sin(B - C) = \frac{3}{5}$ then the ΔABC is

(a) right angled (b) obtuse angled

(c) isosceles (d) none of these

99. ABC is a right-angled isosceles triangle in which $\angle A = 90°$. The midpoint D of AB is joined to C. The ratio of $\cot\angle DCA$ and $\cot\angle DCB$ is

(a) 3 : 2 (b) 1 : 2

(c) 2 : 3 (d) none of these

100. In a right-angled triangle the hypotenuse is four times as long as the distance of the hypotenuse from the opposite vertex. Its acute angles are

(a) 30°, 60° (b) 15°, 75°

(c) 45°, 45° (d) none of these

101. Two straight roads intersect at 30°. From the junction, two persons A and B start walking at the same time, one on each road. A walks at the rate of 5 km/h. At the end of 3 hours they are 9 km apart. If B walks at uniform rate then the speed of B is

(a) 5.99 km/h (nearly) (b) 2.67 km/h (nearly)

(c) 3 km/h (d) none of these

Answers

9. $x \in (1, +\infty)$ **14.** 45°, 60°, 75° **15.** no

33. $\frac{32}{13}(4\sqrt{3} - 3), \frac{4}{3}\sqrt{3}$ respectively **35.** 2

36. 30° **42.** $AD = 1$, others are 1, 2

45. 10 **47.** 6 **61.** 5 : 4 : 3 **64.** 2α

65. 15° **66.** $\frac{\sqrt{3}+1}{2}$ **67.** $\frac{\sqrt{3}+1}{2}$ cm^2

68. $\frac{2\pi}{3}$ **69.** 75° **70.** 6 **71.** 30°

72. 3 : 1 **73.** 0 **74.** $3\sin A\sin B\sin C$

75. $\frac{1}{3}$ **76.** 7 : 9 : 11

77. $(\sqrt{5}+1) : (\sqrt{5}-1)$ **78.** $\frac{\pi}{3}$ **79.** $\frac{\pi}{3}$

80. $\frac{\pi}{2}$ **81.** 113° **82.** $\frac{a^2}{6}$ **83.** 16 : 7

84. (b) **85.** (a) **86.** (a), (b) **87.** (a)

88. (a) **89.** (a) **90.** (b), (c) **91.** (a)

92. (a), (c), (d) **93.** (a), (b) **94.** (e) **95.** (b)

96. (c) **97.** (a), (c) **98.** (a) **99.** (c)

100. (b) **101.** (a), (b)

□

Chapter Test

Time: 120 minutes

1. In the following incomplete sentences, fill in the blanks such that the resulting sentences become true.

(a) In any ΔABC, $\Sigma\, a(b+c)\tan\dfrac{B-C}{2}\tan\dfrac{A}{2} =$ ______ .

(b) If in a ΔABC, $\sin A = p$ and $\sin B = q$ then $\cos C =$ ______ .

(c) In a ΔABC, $\tan A = 2$ and $\tan B = \dfrac{3}{2}$. If $c = 2\cdot\sqrt{65}$ then the circumradius of the triangle = ______ .

(d) If in a ΔABC, $\cos 3A + \cos 3B + \cos 3C = 1$ then the measure of one of the angles must be ______ .

(e) If in a ΔABC, $a : b = 3 : 4$ and $\angle C = \dfrac{\pi}{2}$ then $\cos A : \cos B =$ ______ .

2. In the following questions multiple answers are given. Choose the correct answer(s).

(a) If in the ΔABC, $\cos A \cos B + \sin A \sin B \sin C = 1$ then the triangle is

A. equilateral B. isosceles C. right angled D. none of these

(b) If the tangents of the angles of an acute angled triangle ABC are in AP then the geometric mean of the tangents of the greatest angle A and the least angle C is

A. $\sqrt{3}$ B. $\dfrac{1}{\sqrt{3}}$ C. $\tan\dfrac{B}{2}$ D. none of these

(c) If in a ΔABC, the sides a, b, c are in AP then $\cot\dfrac{A}{2} + \cot\dfrac{C}{2}$ is equal to

A. $\cot\dfrac{B}{2}$ B. $2\cot\dfrac{B}{2}$ C. $\tan\dfrac{B}{2}$ D. none of these

(d) In a ΔABC, if $A = \dfrac{4\pi}{9}$ and $B = \dfrac{2\pi}{9}$ then $a^2 - b^2$ is equal to

A. c^2 B. bc C. ab D. ca

(e) If ΔABC is right angled at B then the diameter of the inscribed circle of the ΔABC is

A. $a+b-c$ B. $b+c-a$ C. $c+a-b$ D. none of these

3. The sides of a ΔABC are in AP and, the greatest and the least angles are A and C respectively. Prove that $4(1-\cos A)(1-\cos C) = \cos A + \cos C$.

4. If the sides a, b, c of a ΔABC are in AP and $B = \dfrac{\pi}{4}$ then find $\tan\dfrac{A}{2} + \tan\dfrac{C}{2}$.

5. If in a ΔABC, $1 = 2\cos A \cos B \cos C + \Sigma \cos A \cdot \cos B$ then prove that the triangle will be equilateral.

6. Let Δ, Δ' denote the areas of a triangle and that of the incircle. Prove that

$$\Delta : \Delta' = \left(\cot\frac{A}{2}\cdot\cot\frac{B}{2}\cdot\cot\frac{C}{2}\right) : \pi.$$

7. If in a ΔABC, $A = 3B$ then prove that $\sin B = \dfrac{1}{2}\cdot\sqrt{\dfrac{3b-a}{b}}$.

8. If in a ΔABC, the perimeter is $4a$ then prove that the equation $rx^2 - ax + 2r = 0$ has the roots $\cot\dfrac{B}{2}$ and $\cot\dfrac{C}{2}$ where $r =$ inradius of the triangle.

9. If the internal bisectors of the angles of the ΔABC make angles α, β, γ with sides a, b, c respectively then show that

$$a \sin 2\alpha + b \sin 2\beta + c \sin 2\gamma = 0.$$

10. If the circumcentre of the ΔABC is on the incircle of the triangle then prove that

$$\cos A + \cos B + \cos C = \sqrt{2}.$$

11. In a right-angled triangle if a, b are the lengths of the sides forming the right-angle and R, r are respectively the circumradius and inradius then prove that $R = \dfrac{ab - (a + b)r}{2r}$.

12. Let D, E and F be the feet of the perpendiculars from the incentre I to the sides BC, CA and AB respectively. If r, r_1, r_2 and r_3 are the inradius of the triangle ABC and radii of the circles inscribed in the quadrilaterals AFIE, BDIF and CEID respectively, prove that

$$\frac{r_1}{r - r_1} + \frac{r_2}{r - r_2} + \frac{r_3}{r - r_3} = \frac{r_1 r_2 r_3}{(r - r_1)(r - r_2)(r - r_3)}.$$

13. Show that in a ΔABC, $\sin \dfrac{A}{2} \le \dfrac{a}{b + c}$. Hence or otherwise prove that $\operatorname{cosec}^n \dfrac{A}{2} + \operatorname{cosec}^n \dfrac{B}{2} + \operatorname{cosec}^n \dfrac{C}{2}$ has the minimum value 3.2^n for all $n \ge 1$.

□

Answers

1. (a) 0 (b) $pq - \sqrt{(1 - p^2)(1 - q^2)}$ (c) $\dfrac{65}{7}$ (d) $\dfrac{2\pi}{3}$ (e) 4 : 3

2. (a) B, C (b) A (c) B (d) B (e) C

4. $\dfrac{2}{3}(\sqrt{2} + 1)$

□

7. Heights and Distances

Recap of Facts and Formulae

1. Angle of elevation and angle of depression

- A and B are two positions, A being higher than B. AP and BQ are horizontal lines through A and B respectively. Then the angle of elevation of A as seen from B = $\angle ABQ$ and the angle of depression of B as seen from A = $\angle PAB$ (= $\angle ABQ$).

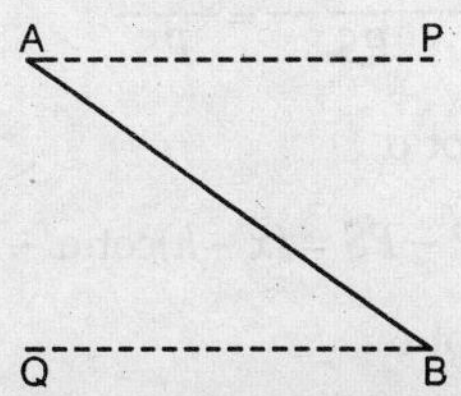

2. Vertical line and horizontal plane

- If AB is vertical and BC is a line on the horizontal plane then $\angle ABC = 90°$.

  If l is any other line in the horizontal plane then $AB \perp l$.

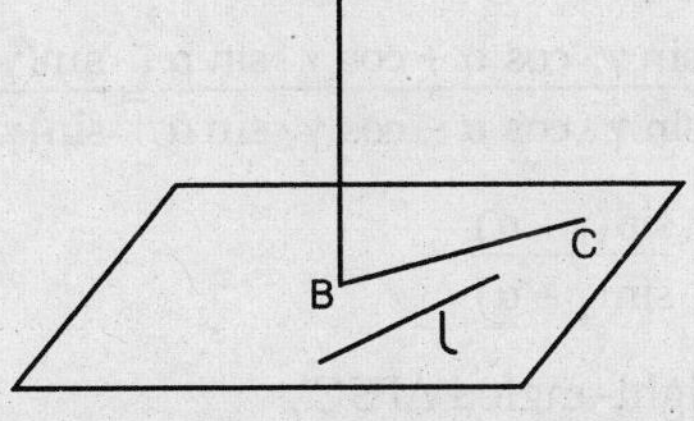

If AB is perpendicular to the lines BC and BD then AB is perpendicular to any line in the plane of BC and BD. So, $AB \perp CD$ also.

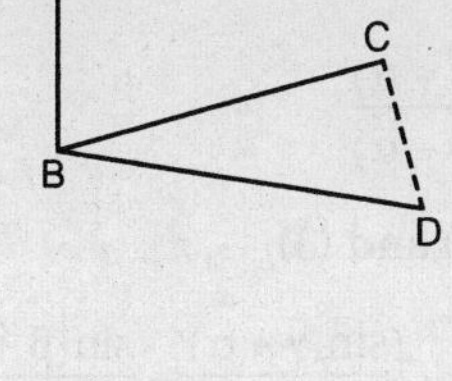

3. Bearing of a point

- At every point O on the horizontal there are four perpendicular directions ON, OE, OS and OW called north, east, south and west of O respectively.

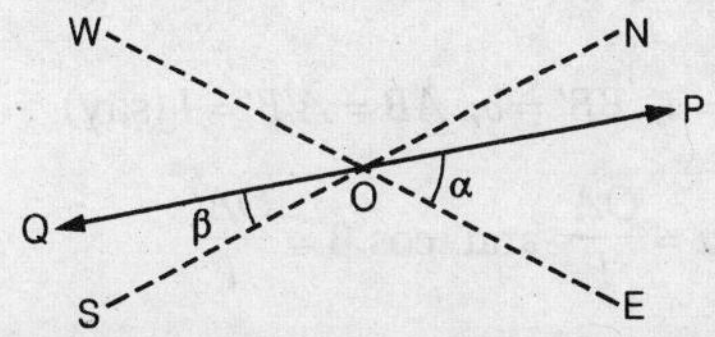

If OP falls in the angle NOE such that $\angle POE = \alpha$ then the point P is α north of east (or $90° - \alpha$ east of north). If $\alpha = 45°$ then P is north-east of O.

If OQ falls in the angle SOW such that $\angle SOQ = \beta$ then the point Q is β west of south (or $90° - \beta$ south of west). If $\beta = 45°$ then Q is south-west of O.

Some more descriptions of standard bearings are as follows.

NE means 45° north of east

NNE means $22\frac{1°}{2}$ north of NE

ENE means $22\frac{1°}{2}$ east of NE, etc.

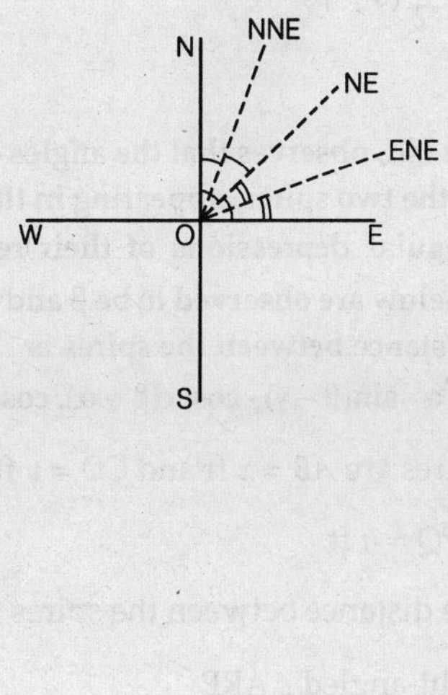

Selected Solved Examples

1. A ladder rests against a wall at an angle α to the horizontal. Its foot is pulled away from the wall through a distance *a*, so that it slides a distance *b* down the wall making an angle β with the horizontal.

Show that $a = b\tan\frac{1}{2}(\alpha+\beta)$.

Let AB be the original position of the ladder. After sliding, it takes the position A'B'.

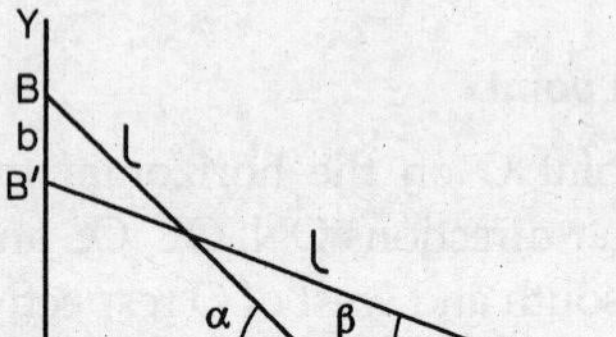

$\therefore \quad AA' = a,\ BB' = b,\ AB = A'B' = l$ (say)

Now, $\cos\alpha = \dfrac{OA}{l}$ and $\cos\beta = \dfrac{OA'}{l}$

$$\therefore \quad \cos\beta - \cos\alpha = \frac{OA' - OA}{l} = \frac{a}{l} \quad \ldots (1)$$

Also $\sin\alpha = \dfrac{OB}{l}$ and $\sin\beta = \dfrac{OB'}{l}$

$$\therefore \quad \sin\alpha - \sin\beta = \frac{OB - OB'}{l} = \frac{b}{l} \quad \ldots (2)$$

From (1) and (2), $\dfrac{\cos\beta - \cos\alpha}{\sin\alpha - \sin\beta} = \dfrac{\frac{a}{l}}{\frac{b}{l}} = \dfrac{a}{b}$

$$\therefore \quad a = b\cdot\frac{\cos\beta - \cos\alpha}{\sin\alpha - \sin\beta} = b\cdot\frac{2\sin\frac{\alpha+\beta}{2}\cdot\sin\frac{\alpha-\beta}{2}}{2\cos\frac{\alpha+\beta}{2}\cdot\sin\frac{\alpha-\beta}{2}}$$

or $\quad a = b\tan\frac{1}{2}(\alpha+\beta)$.

2. A man *h* ft tall, observes that the angles of elevation of the top of the two spires, appearing in the same line, is α. The angular depressions of their reflections in a lake just below are observed to be β and γ (γ < β). Show that the distance between the spires is

$2h\cos^2\alpha\cdot\sin(\beta-\gamma)\cdot\operatorname{cosec}(\beta-\alpha).\ \operatorname{cosec}(\gamma-\alpha)$ ft.

Here, the spires are $AB = x$ ft and $CD = y$ ft.

The man is $PQ = h$ ft.

Let $BD =$ the distance between the spires $= d$ ft.

From the right-angled ΔARP,

$$\tan\alpha = \frac{AR}{RP} = \frac{AB - RB}{RP} = \frac{x-h}{RP}$$

$\therefore \quad RP = (x-h)\cot\alpha$

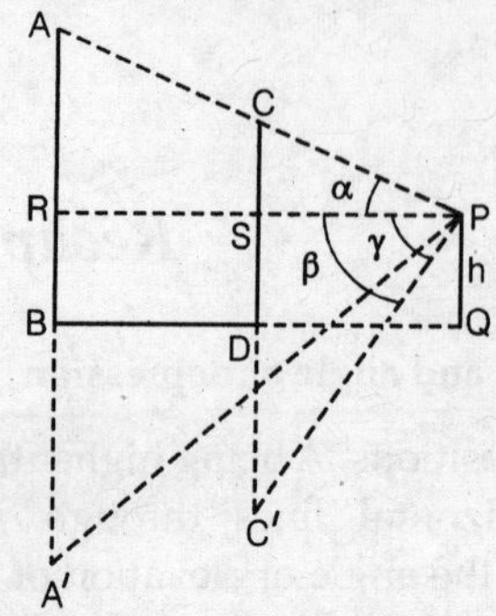

From the right-angled ΔCSP,

$$\tan\alpha = \frac{CS}{PS} = \frac{CD - SD}{PS} = \frac{y-h}{PS}$$

$\therefore \quad PS = (y-h)\cot\alpha$

$\therefore \quad BD = RS = RP - PS = (x-h)\cot\alpha - (y-h)\cot\alpha$

$$\therefore \quad BD = (x-y)\cot\alpha \quad \ldots (1)$$

From the right-angled ΔPRA′,

$$\tan\gamma = \frac{RA'}{RP} = \frac{RB + BA'}{RP} = \frac{h+x}{(x-h)\cot\alpha}$$

{∵ $AB = A'B$, A′ being image of A}

or $\quad \dfrac{x+h}{x-h} = \tan\gamma\cdot\cot\alpha = \dfrac{\sin\gamma\cdot\cos\alpha}{\cos\gamma\cdot\sin\alpha}$

By componendo and dividendo,

$$\frac{2x}{2h} = \frac{\sin\gamma\cdot\cos\alpha + \cos\gamma\cdot\sin\alpha}{\sin\gamma\cdot\cos\alpha - \cos\gamma\cdot\sin\alpha} = \frac{\sin(\gamma+\alpha)}{\sin(\gamma-\alpha)}$$

$$\therefore \quad x = h\cdot\frac{\sin(\gamma+\alpha)}{\sin(\gamma-\alpha)} \quad \ldots (2)$$

From the right-angled ΔPSC′,

$$\tan\beta = \frac{h+y}{(y-h)\cot\alpha}$$

or $\quad \dfrac{y+h}{y-h} = \tan\beta\cdot\cot\alpha = \dfrac{\sin\beta\cdot\cos\alpha}{\cos\beta\cdot\sin\alpha}$

∴ by componendo and dividendo as above,

$$y = h\cdot\frac{\sin(\beta+\alpha)}{\sin(\beta-\alpha)} \quad \ldots (3)$$

∴ from (1), (2) and (3),

$$BD = h\cot\alpha\cdot\left\{\frac{\sin(\gamma+\alpha)}{\sin(\gamma-\alpha)} - \frac{\sin(\beta+\alpha)}{\sin(\beta-\alpha)}\right\}$$

$$= h\cot\alpha\cdot\frac{\{\sin(\gamma+\alpha)\sin(\beta-\alpha)-\sin(\beta+\alpha)\sin(\gamma-\alpha)\}}{\sin(\gamma-\alpha)\cdot\sin(\beta-\alpha)}$$

$$=\frac{1}{2}h\cot\alpha\cdot\frac{\{\cos(\gamma-\beta+2\alpha)-\cos(\gamma+\beta)-\cos(\beta-\gamma+2\alpha)+\cos(\beta+\gamma)\}}{\sin(\gamma-\alpha)\cdot\sin(\beta-\alpha)}$$

$$=\frac{1}{2}h\cot\alpha\cdot\frac{2\sin 2\alpha\cdot\sin(\beta-\gamma)}{\sin(\gamma-\alpha)\sin(\beta-\alpha)}$$

$$= 2h\cos^2\alpha\sin(\beta-\gamma)\,\text{cosec}\,(\beta-\alpha)\cdot\text{cosec}\,(\gamma-\alpha)$$

$\therefore$ the distance between the spires

$$= 2h\cos^2\alpha\sin(\beta-\gamma)\,\text{cosec}\,(\beta-\alpha)\,\text{cosec}\,(\gamma-\alpha)\text{ ft.}$$

3. An object 2 m long is fired vertically upwards from the middle point of two locations A and B, 8 m apart. The speed of the object after t seconds, is given by $\frac{ds}{dt}=(2t+1)$ m/s. Let α and β be the respective angles subtended by the object at A and B after 1 and 2 seconds. Find the value of $\cos(\alpha-\beta)$.

Let the height of the lower point of the object after t seconds be s metres. Then

$$\frac{ds}{dt}=(2t+1) \quad\Rightarrow\quad ds=(2t+1)\,dt \qquad \text{... (1)}$$

$$\Rightarrow \int_{t=0}^{1} ds=\int_0^1(2t+1)dt=\left[t^2+t\right]_0^1=2$$

$\therefore\quad OQ = 2$ m

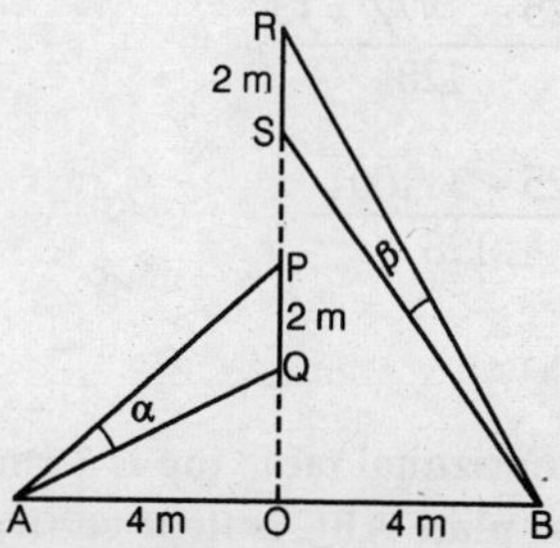

Again from (1), $\int_{t=0}^{2} dx=\int_0^2(2t+1)dt=\left[t^2+t\right]_0^2=6$

$\therefore\quad OS = 6$ m.

From the right-angled ΔQOA,

$$\tan\angle QAO=\frac{OQ}{AO}=\frac{2}{4}=\frac{1}{2}.$$

From the right-angled ΔPOA,

$$\tan\angle PAO=\frac{OP}{AO}=\frac{OQ+PQ}{4}=\frac{2+2}{4}=1.$$

$$\therefore\quad \tan\alpha=\tan(\angle PAO-\angle QAO)$$

$$=\frac{\tan\angle PAO-\tan\angle QAO}{1+\tan\angle PAO\cdot\tan\angle QAO}$$

$$=\frac{1-\frac{1}{2}}{1+1\cdot\frac{1}{2}}=\frac{\frac{1}{2}}{\frac{3}{2}}=\frac{1}{3}.$$

$$\therefore\quad \cos\alpha=\frac{3}{\sqrt{10}},\ \sin\alpha=\frac{1}{\sqrt{10}}.$$

From the right-angled ΔSOB, $\tan\angle SBO=\frac{OS}{BO}=\frac{6}{4}=\frac{3}{2}.$

From the right-angled ΔROB,

$$\tan\angle RBO=\frac{OR}{BO}=\frac{RS+SO}{4}=\frac{2+6}{4}=2$$

$$\therefore\quad \tan\beta=\tan(\angle RBO-\angle SBO)$$

$$=\frac{\tan\angle RBO-\tan\angle SBO}{1+\tan\angle RBO\cdot\tan\angle SBO}$$

$$=\frac{2-\frac{3}{2}}{1+2\cdot\frac{3}{2}}=\frac{\frac{1}{2}}{4}=\frac{1}{8}$$

$$\therefore\quad \cos\beta=\frac{8}{\sqrt{65}},\ \sin\beta=\frac{1}{\sqrt{65}}$$

$$\therefore\quad \cos(\alpha-\beta)=\cos\alpha\cdot\cos\beta+\sin\alpha\cdot\sin\beta$$

$$=\frac{3}{\sqrt{10}}\cdot\frac{8}{\sqrt{65}}+\frac{1}{\sqrt{10}}\cdot\frac{1}{\sqrt{65}}=\frac{25}{\sqrt{650}}$$

$$=\frac{25}{\sqrt{25\times 26}}=\frac{5}{\sqrt{26}}.$$

4. A 25-m-high vertical tower and a vertical pole stand on an inclined ground. The foot of the tower and the midpoint of the pole are in the same horizontal level. The angles of depression of the top and bottom of pole as seen from the top of the tower are 15° and 45° respectively. Find the length of the pole.

Let the tower be $AB = 25$ m and the pole be $CD = h$ m, whose middle point be E. So $CE = DE = \frac{h}{2}$ m.

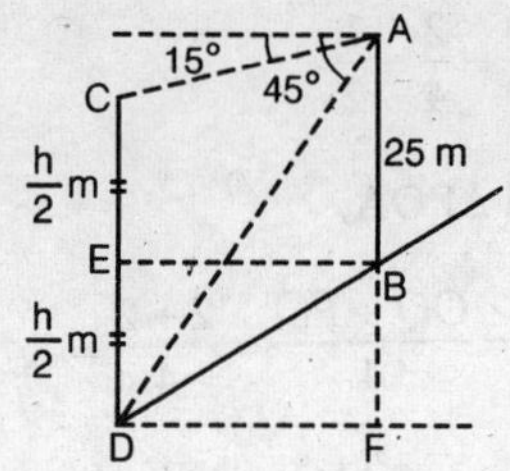

Clearly, from the figure, in the right-angled ΔAFD,

$$\angle ADF = 45°, \ AF = \left(25 + \frac{h}{2}\right) \text{ m}$$

$$\therefore \quad \sin 45° = \frac{AF}{AD} = \frac{25 + \frac{h}{2}}{AD}$$

$$\therefore \quad AD = \sqrt{2}\left(25 + \frac{h}{2}\right) \text{ m}$$

Now, in the ΔACD,

$$\angle DAC = 30°, \angle ACD = 15° + 90° = 105°$$

$$\therefore \quad \frac{AD}{\sin 105°} = \frac{CD}{\sin 30°}$$

or $$\frac{\sqrt{2}\left(25 + \frac{h}{2}\right)}{\sin(60° + 45°)} = \frac{h}{\frac{1}{2}}$$

or $$\sqrt{2}\left(25 + \frac{h}{2}\right) = 2h\left\{\frac{\sqrt{3}}{2} \cdot \frac{1}{\sqrt{2}} + \frac{1}{2} \cdot \frac{1}{\sqrt{2}}\right\}$$

or $$25 + \frac{h}{2} = \sqrt{2}h \cdot \frac{\sqrt{3}+1}{2\sqrt{2}}$$

or $$50 + h = h(\sqrt{3} + 1)$$

$$\therefore \quad h = \frac{50}{\sqrt{3}} = \frac{50\sqrt{3}}{3}$$

$\therefore$ the length of the pole $= \frac{50\sqrt{3}}{3}$ m.

5. OA is a crank which can rotate about the fixed point O. A rod AB is connecting A to a ring B which can slide freely on a straight rod passing through O, the lengths of the crank OA and the rod AB being 2 m and 5 m respectively. The ring B starts to slide from an extreme position. If α and β are the inclinations of the crank OA with OB when B has described $\frac{1}{4}$ and $\frac{3}{4}$ respectively of its total possible travel then find the value of cos(α – β).

Let E_1 and E_2 be two extreme positions of the ring B. Clearly,

$$OE_1 = AE_1 - AO = AB - AO = (5 - 2) \text{ m} = 3 \text{ m}$$

$$OE_2 = 2 \text{ m} + 5 \text{ m} = 7 \text{ m}$$

$$\therefore \quad E_1E_2 = 7 \text{ m} - 3 \text{ m} = 4 \text{ m}$$

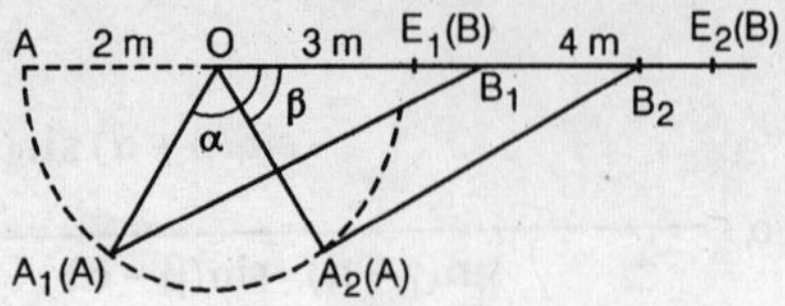

When B has described $\frac{1}{4}$ of E_1E_2 starting from E_1,

$$OB_1 = OE_1 + E_1B_1 = 3 \text{ m} + \frac{1}{4} \cdot 4 \text{ m} = 4 \text{ m}.$$

When B has described $\frac{3}{4}$ of E_1E_2 starting from E_1,

$$OB_2 = OE_1 + E_1B_2 = 3 \text{ m} + \frac{3}{4} \cdot 4 \text{ m} = 6 \text{ m}.$$

Now, in ΔOA_1B_1, $OA_1 = 2$ m, $A_1B_1 = 5$ m, $OB_1 = 4$ m and $\angle A_1OB_1 = \alpha$.

$$\therefore \quad \cos\alpha = \frac{2^2 + 4^2 - 5^2}{2 \cdot 2 \cdot 4} = \frac{-5}{16}$$

$$\therefore \quad \sin\alpha = \frac{\sqrt{16^2 - 5^2}}{16} = \frac{\sqrt{21 \times 11}}{16}.$$

Again, in ΔOA_2B_2, $OA_2 = 2$ m, $A_2B_2 = 5$ m, $OB_2 = 6$ m and $\angle A_2OB_2 = \beta$.

$$\therefore \quad \cos\beta = \frac{2^2 + 6^2 - 5^2}{2 \cdot 2 \cdot 6} = \frac{15}{24} = \frac{5}{8}$$

$$\therefore \quad \sin\beta = \frac{\sqrt{8^2 - 5^2}}{8} = \frac{\sqrt{39}}{8}.$$

$$\therefore \quad \cos(\alpha - \beta) = \cos\alpha \cdot \cos\beta + \sin\alpha \cdot \sin\beta$$

$$= \frac{-5}{16} \cdot \frac{5}{8} + \frac{\sqrt{21 \times 11}}{16} \cdot \frac{\sqrt{39}}{8}$$

$$= \frac{-25 + 3\sqrt{77 \times 13}}{128}$$

$$= \frac{-25 + 3\sqrt{1001}}{128}.$$

6. A peg fixed on a horizonal table top is 3 cm above the table. A triangular plate ABC is kept vertical with the end B of the side AB on the peg and C on the table at a distance 4 cm from the peg. If $AB = 4$ cm and $AC = 7$ cm, find the angles of inclination of the sides AB and AC with the vertical.

Here, the peg $BD = 3$ cm and ABC is the plate in which $AB = 4$ cm, $AC = 7$ cm.

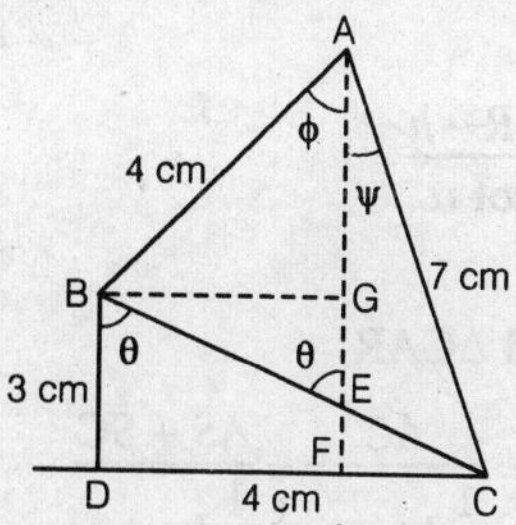

Let AE be the vertical through A and

$$\angle BAE = \phi, \angle CAE = \psi \text{ and } \angle DBC = \theta.$$

Then $\angle AEB = \theta \quad (\because AE \parallel BD)$

Now, in $\triangle AEB$, $\dfrac{BE}{\sin\phi} = \dfrac{4}{\sin\theta}$ ... (1)

In the right-angled $\triangle BDC$,

$$BC = \sqrt{3^2 + 4^2} \text{ cm} = 5 \text{ cm}.$$

In the $\triangle AEC$, $\dfrac{CE}{\sin\psi} = \dfrac{7}{\sin(\pi - \theta)}$, i.e., $\dfrac{7}{\sin\theta}$ ... (2)

Also, $\sin\theta = \dfrac{4}{5}\cdot$ So from (1), $BE = 5\sin\phi$.

From (2), $CE = \dfrac{35}{4}\sin\psi$

$$\therefore \quad BC = BE + CE = 5\sin\phi + \frac{35}{4}\sin\psi$$

$$\therefore \quad 5 = 5\sin\phi + \frac{35}{4}\sin\psi, \text{ i.e., } 4 = 4\sin\phi + 7\sin\psi$$

$$\therefore \quad 7\sin\psi = 4 - 4\sin\phi \qquad \text{... (3)}$$

Also, in the right-angled $\triangle BGA$,

$$GA = AB\cos\phi = 4\cos\phi.$$

In the right-angled $\triangle AFC$,

$$AF = AC\cos\psi = 7\cos\psi$$

$$\therefore \quad GF = AF - GA = 7\cos\psi - 4\cos\phi;$$

but $GF = BD = 3$

$$\therefore \quad 3 = 7\cos\psi - 4\cos\phi, \text{ i.e., } 7\cos\psi = 3 + 4\cos\phi \qquad \text{... (4)}$$

Squaring and adding (3) and (4), we get

$$49 = 16 + 9 + 16 - 32\sin\phi + 24\cos\phi$$

or $\quad 3\cos\phi - 4\sin\phi = 1$

or $\quad \dfrac{3}{5}\cos\phi - \dfrac{4}{5}\sin\phi = \dfrac{1}{5}$

or $\quad \cos(\phi + \alpha) = \dfrac{1}{5}$ where $\cos\alpha = \dfrac{3}{5}\cdot$

$$\therefore \quad \phi + \alpha = \cos^{-1}\frac{1}{5}$$

or $\quad \phi = \cos^{-1}\dfrac{1}{5} - \cos^{-1}\dfrac{3}{5}$

$$\therefore \quad \phi = \cos^{-1}\left\{\frac{1}{5}\cdot\frac{3}{5} + \sqrt{1-\left(\frac{1}{5}\right)^2}\cdot\sqrt{1-\left(\frac{3}{5}\right)^2}\right\}$$

$$= \cos^{-1}\frac{3 + 8\sqrt{6}}{25}$$

Again, from (3) and (4),

$$(4\sin\phi)^2 + (4\cos\phi)^2 = (4 - 7\sin\psi)^2 + (7\cos\psi - 3)^2$$

$$= 16 + 9 + 49 - 56\sin\psi - 42\cos\psi$$

or $\quad 16 = 74 - 56\sin\psi - 42\cos\psi$

or $\quad 42\cos\psi + 56\sin\psi = 58$

$$\therefore \quad 21\cos\psi + 28\sin\psi = 29$$

or $\quad \dfrac{21}{\sqrt{21^2 + 28^2}}\cos\psi + \dfrac{28}{\sqrt{21^2 + 28^2}}\sin\psi = \dfrac{29}{\sqrt{21^2 + 28^2}}$

or $\quad \dfrac{3}{5}\cos\psi + \dfrac{4}{5}\sin\psi = \dfrac{29}{35}$

$$\therefore \quad \cos(\psi - \alpha) = \frac{29}{35} \text{ where } \alpha = \cos^{-1}\frac{3}{5}$$

$$\therefore \quad \psi - \alpha = \cos^{-1}\frac{29}{35}$$

$$\therefore \quad \psi = \cos^{-1}\frac{3}{5} + \cos^{-1}\frac{29}{35} = \cos^{-1}\frac{87 - 32\sqrt{6}}{175}.$$

7. A train is moving at a constant speed at an angle θ east of north. Observations of the train are made from a fixed point. It is due north at some instant. Ten minutes earlier its bearing was α west of north, whereas ten minutes afterwards its bearing is β east of north. Find the value of tan θ.

The train is moving on the line AB. O is the point of observation. At the time of observation the train is at P. Ten minutes earlier it was at A and ten minutes afterwards it is at B. So $AP = PB$.

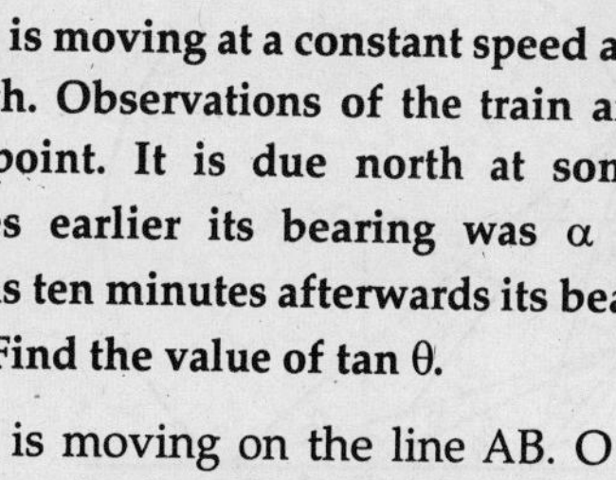

Also $\angle POA = \alpha$, $\angle POB = \beta$ and $\angle NPB = \theta$.

In the $\triangle OPB$, $\dfrac{PB}{\sin\beta} = \dfrac{OP}{\sin OBP}$

or $\quad \dfrac{PB}{\sin\beta} = \dfrac{OP}{\sin(\theta - \beta)}$ ... (1)

In the ΔOPA, $\dfrac{PA}{\sin\alpha}=\dfrac{OP}{\sin OAP}$

or $\quad \dfrac{PA}{\sin\alpha}=\dfrac{OP}{\sin(180°-\overline{\alpha+\theta})}$ ... (2)

From (1) and (2),

$$\frac{\sin(\theta-\beta)}{\sin\beta}=\frac{\sin(\alpha+\theta)}{\sin\alpha}\quad (\because PA=PB)$$

or $\quad \sin\theta\cdot\cot\beta-\cos\theta=\cos\theta+\cot\alpha\cdot\sin\theta$

or $\quad 2\cos\theta=\sin\theta(\cot\beta-\cot\alpha)$

$\therefore \quad \tan\theta=\dfrac{2}{\cot\beta-\cot\alpha}$.

8. A flagstaff stands on the top of a vertical tower. It is found that the flagstaff subtends the same angle α at two points, 2*d* apart, on a straight line through the base of the tower on the horizontal plane. Also the flagstaff subtends an angle β at the midpoint of these two points. Prove that the height of the flagstaff is

$$d\sin\alpha\cdot\sqrt{\frac{2\sin\beta}{\cos\alpha\cdot\sin(\beta-\alpha)}}.$$

Here, the flagstaff $BC=h$ stands on the tower $AB=x$.

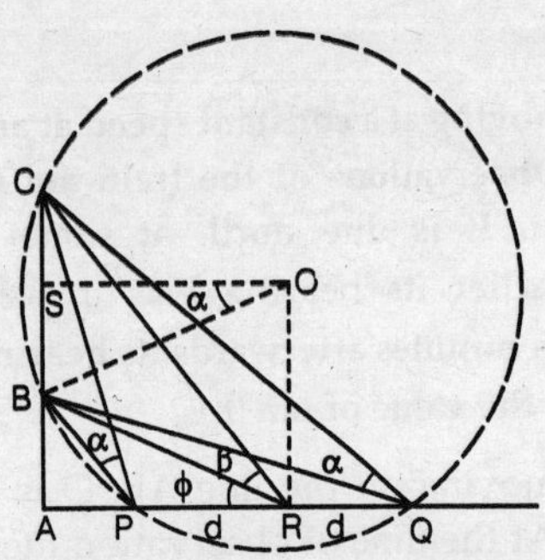

$\because\ \angle BPC=\angle BQC=\alpha$, we find that C, B, P and Q are concyclic. Let O be the centre of the circle.

From geometry,

$$\angle BOS=\frac{1}{2}\angle BOC=\frac{1}{2}\cdot 2\angle BPC=\alpha.$$

$\therefore \quad$ in the right-angled ΔOSB,

$$\tan\alpha=\frac{BS}{OS}=\frac{\frac{1}{2}h}{AR};\quad \therefore \quad AR=\frac{1}{2}h\cot\alpha.$$

Let $\angle ARB=\phi$.

From the right-angled ΔBAR,

$$\tan\phi=\frac{AB}{AR}=\frac{AB}{\frac{1}{2}h\cot\alpha}=\frac{AS-SB}{\frac{1}{2}h\cot\alpha}$$

$$=\frac{OR-\frac{1}{2}h}{\frac{1}{2}h\cot\alpha}=\frac{2OR-h}{h\cot\alpha}.$$

From the right-angled ΔCAR,

$$\tan(\beta+\phi)=\frac{AC}{AR}=\frac{AC}{\frac{1}{2}h\cot\alpha}=\frac{AS+SC}{\frac{1}{2}h\cot\alpha}$$

$$=\frac{OR+\frac{1}{2}h}{\frac{1}{2}h\cot\alpha}=\frac{2OR+h}{h\cot\alpha}.$$

$\therefore \quad \tan\beta=\tan(\beta+\phi-\phi)=\dfrac{\tan(\beta+\phi)-\tan\phi}{1+\tan(\beta+\phi)\cdot\tan\phi}$

$$=\frac{\dfrac{2OR+h}{h\cot\alpha}-\dfrac{2OR-h}{h\cot\alpha}}{1+\dfrac{2OR+h}{h\cot\alpha}\cdot\dfrac{2OR-h}{h\cot\alpha}}$$

$$=\frac{2h\cdot h\cot\alpha}{h^2\cot^2\alpha+4OR^2-h^2} \quad \text{... (1)}$$

Now, $OR^2=OP^2-PR^2=OB^2-d^2$ ... (2)

and $\dfrac{BS}{OB}=\sin\alpha$

or $\quad OB=BS\operatorname{cosec}\alpha=\dfrac{1}{2}h\operatorname{cosec}\alpha$

$\therefore \quad$ **(2) gives,** $OR^2=\left(\dfrac{1}{2}h\operatorname{cosec}\alpha\right)^2-d^2$

or $\quad 4OR^2=h^2\operatorname{cosec}^2\alpha-4d^2$

$\therefore \quad$ **(1) gives** $\tan\beta=\dfrac{2h^2\cot\alpha}{h^2\cot^2\alpha+h^2\operatorname{cosec}^2\alpha-4d^2-h^2}$

or $\quad \tan\beta\cdot\{h^2(\cot^2\alpha+\operatorname{cosec}^2\alpha-1)-4d^2\}=2h^2\cot\alpha$

or $\quad \tan\beta(2h^2\cot^2\alpha-4d^2)=2h^2\cot\alpha$

or $\quad 2h^2(\cot^2\alpha\cdot\tan\beta-\cot\alpha)=4d^2\tan\beta$

or $\quad h^2=\dfrac{2d^2\tan\beta}{\cot^2\alpha\cdot\tan\beta-\cot\alpha}$

$$=\frac{2d^2\cdot\sin\beta}{\cos\beta\left\{\dfrac{\cos^2\alpha}{\sin^2\alpha}\cdot\dfrac{\sin\beta}{\cos\beta}-\dfrac{\cos\alpha}{\sin\alpha}\right\}}$$

$$=\frac{2d^2\sin^2\alpha\cdot\sin\beta}{\cos^2\alpha\cdot\sin\beta-\cos\alpha\cdot\cos\beta\cdot\sin\alpha}$$

$$=d^2\sin^2\alpha\cdot\frac{2\sin\beta}{\cos\alpha\cdot\sin(\beta-\alpha)}$$

$\therefore \quad$ height of the flagstaff

$$=h=d\sin\alpha\cdot\sqrt{\frac{2\sin\beta}{\cos\alpha\cdot\sin(\beta-\alpha)}}.$$

9. A flag, on the top of a vertical tower whose bottom is at B, subtends equal angles α at two points C and D on the horizontal line through B. A is another point 100 m away from B on this horizontal line. There is a valley below. A rocket is fired vertically upwards from the base of the valley and at the same time a man starts running from A towards B. The distances covered by the rocket and the man at time *t* are given by

$$s_1 = (t^2 + t - 6) - 5(1+t)\log(1+t) \quad \text{and} \quad s_m = 1 + t$$

respectively. The man notes that the ratio between the distances covered by the rocket and himself is the maximum when he reaches the point C and the minimum when he reaches D. Find the height of the tower. (Note: Here s_r, s_m will represent numerically positive values when $t > 0$.)

Here, let the vertical tower $QB = h$ m and the flag $PQ = x$ m.

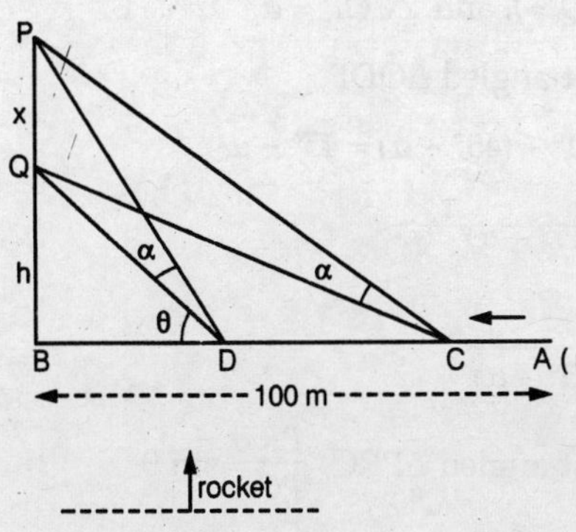

Also $\angle PCQ = \angle PDQ = \alpha$.

Let $y = \dfrac{s_r}{s_m}$. Then

$$y = \frac{(t^2 + t - 6) - 5(1+t)\log(1+t)}{1+t}$$

$$\therefore \quad \frac{dy}{dt} = \frac{\{2t + 1 - 5 - 5\log(1+t)\}(1+t) - \{t^2 + t - 6 - 5(1+t)\log(1+t)\}}{(1+t)^2}$$

For maximum or minimum y we have, $\dfrac{dy}{dt} = 0$

or $\quad (2t-4)(1+t) - (t^2 + t - 6) = 0$

or $\quad 2t^2 - 2t - 4 - t^2 - t + 6 = 0$

or $\quad t^2 - 3t + 2 = 0$

$\therefore \quad (t-1)(t-2) = 0; \quad \therefore \ t = 1, 2.$

Now, $\dfrac{dy}{dt} = \dfrac{t^2 - 3t + 2}{(1+t)^2}$

$$= \frac{(t^2 + 2t + 1) - 5t + 1}{(1+t)^2}$$

$$= 1 - \frac{5(t+1)}{(1+t)^2} + \frac{6}{(1+t)^2}$$

$$= 1 - \frac{5}{1+t} + \frac{6}{(1+t)^2}$$

$$\therefore \quad \frac{d^2y}{dt^2} = \frac{5}{(1+t)^2} - \frac{12}{(1+t)^3}$$

$$\therefore \quad \left(\frac{d^2y}{dt^2}\right)_{t=1} = \frac{5}{4} - \frac{12}{8} = \frac{-2}{8} = -\frac{1}{4} < 0$$

$\therefore \quad y$ is maximum when $t = 1$.

$$\left(\frac{d^2y}{dt^2}\right)_{t=2} = \frac{5}{9} - \frac{12}{27} = \frac{3}{27} = \frac{1}{9} > 0$$

$\therefore \quad y$ is minimum when $t = 2$

$\therefore \quad AC = s_m$, when $t = 1$

and $\quad AD = s_m$, when $t = 2$

$\therefore \quad AC = 2$ m and $AD = 3$ m.

So $\quad CD = (3 - 2)$ m $= 1$ m.

Thus we get $BD = 97$ m, $DC = 1$ m.

Let $\angle BDQ = \theta$; then $\angle PCB = \angle DQB = 90° - \theta$

{∵ PQDC is concyclic because $\angle PCQ = \angle PDQ = \alpha$}

Now, from ΔQBD, $\tan\theta = \dfrac{QB}{BD} = \dfrac{h}{97}$ ... (1)

From ΔQBC, $\tan\angle QCB = \dfrac{QB}{BC}$

or $\quad \tan\{90° - (\theta + \alpha)\} = \dfrac{h}{98}$, i.e., $\cot(\theta + \alpha) = \dfrac{h}{98}$

$$\therefore \quad \tan(\theta + \alpha) = \frac{98}{h}$$

or $\quad \dfrac{\tan\theta + \tan\alpha}{1 - \tan\theta \cdot \tan\alpha} = \dfrac{98}{h}$

or $\quad \dfrac{\dfrac{h}{97} + \tan\alpha}{1 - \dfrac{h}{97}\cdot\tan\alpha} = \dfrac{98}{h}$ {from (1)}

or $\quad \dfrac{h + 97\tan\alpha}{97 - h\tan\alpha} = \dfrac{98}{h}$

$$\therefore \quad h^2 + 195\tan\alpha \cdot h - 98 \times 97 = 0$$

$$\therefore \quad h = \frac{-195\tan\alpha \pm \sqrt{(195\tan\alpha)^2 + 4 \times 98 \times 97}}{2}$$

$$= \frac{1}{2}\{\sqrt{(195)^2\tan^2\alpha + 4 \times 98 \times 97} - 195\tan\alpha\}$$

10. A wheel with diameter AB touches the horizontal ground at A. The rod BC is fixed at B, ABC being vertical. A man from a point P on the ground at a distance d from A, finds that the angle of elevation of C is α. The wheel turns about the fixed centre O of the wheel such that C turns away from the man and its angle of elevation is β when it is about to disappear. Find PC when C is about to disappear.

When C is about to disappear, let it be at C′ and then PC′ touches the wheel at Q.

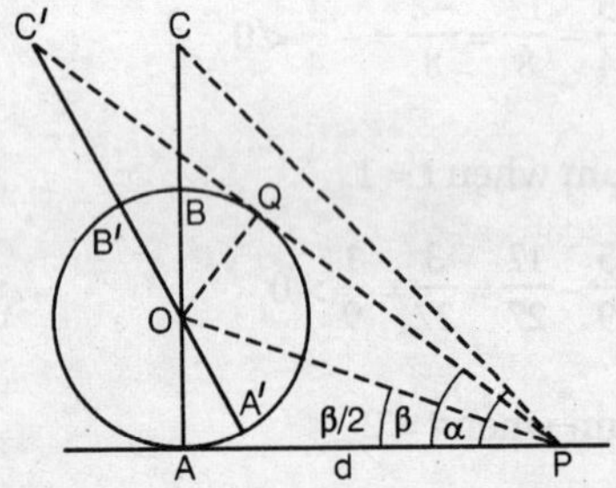

From the question, we have to find PC'.

Clearly, $PQ = PA = d$ and $OC' = OC = AC - r$,

r being the radius of the wheel.

In the ΔCAP, $\tan\alpha = \dfrac{AC}{d}$; $\quad\therefore\quad AC = d\tan\alpha$

$\therefore\quad OC + r = d\tan\alpha;\quad \therefore\quad OC' = d\tan\alpha - r$

$\therefore\quad C'Q = \sqrt{OC'^2 - OQ^2} = \sqrt{(d\tan\alpha - r)^2 - r^2}$

$\therefore\quad PC' = PQ + C'Q = d + \sqrt{(d\tan\alpha - r)^2 - r^2} \qquad \ldots (1)$

Also, from the ΔOAP,

$$\tan\frac{\beta}{2} = \frac{r}{d};\quad \therefore\quad r = d\tan\frac{\beta}{2}.$$

$$\therefore\quad PC' = d + \sqrt{d^2\tan^2\alpha - 2d\tan\alpha \cdot d\tan\frac{\beta}{2}} \text{ \{from (1)\}}$$

$$= d\left\{1 + \sqrt{\tan^2\alpha - 2\tan\alpha\cdot\tan\frac{\beta}{2}}\right\}$$

$$= d\left\{1 + \tan\alpha\cdot\sqrt{1 - 2\cot\alpha\cdot\tan\frac{\beta}{2}}\right\}.$$

11. A semicircular arch AB of length $2l$ and a vertical tower PQ are situated in the same vertical plane. The feet A and B of the arch and the base Q of the tower are on the same horizontal level, with B between A and Q. A man at A finds the tower hidden from his view due to the arch. He starts crawling up the arch and just sees the topmost point P of the tower after covering a distance $\dfrac{l}{2}$ along the arch. He crawls further to the topmost point of the arch and notes the angle of elevation of P to be θ. Compute the height of the tower in terms of l and θ.

Let arc $AD = \frac{l}{2}$ and C is the topmost point of the semicircular arch.

$\therefore\quad \angle AOD = \angle DOC = 45°$

and $OC = \text{radius } r = \dfrac{2l}{\pi}$ $\quad$ {∵ here $\pi r = 2l$}.

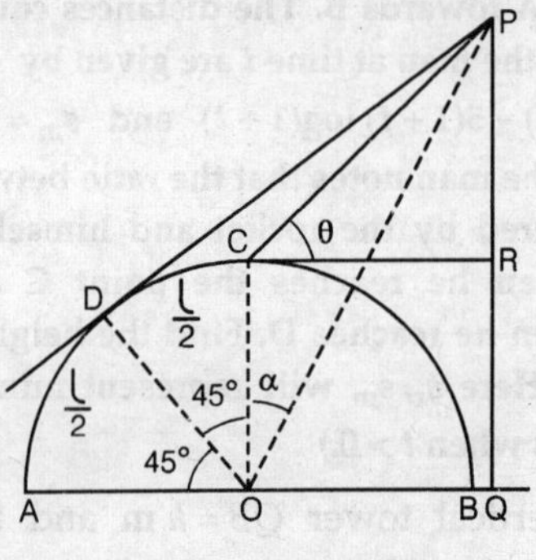

Let the tower $PQ = h$ and $\angle COP = \alpha$.

Then in the right-angled ΔODP,

$$\angle OPD = 90° - (45° + \alpha) = 45° - \alpha$$

and $\dfrac{OD}{OP} = \sin(45° - \alpha)$

$$\therefore\quad OP = \frac{r}{\sin(45° - \alpha)} \qquad \ldots (1)$$

Also, in the right-angled ΔPRC, $\dfrac{PR}{PC} = \sin\theta$

$$\therefore\quad PC = \frac{PR}{\sin\theta} = \frac{PQ - CO}{\sin\theta} = \frac{h - r}{\sin\theta} \qquad \ldots (2)$$

In ΔPCO, $\dfrac{PC}{\sin\alpha} = \dfrac{PO}{\sin(90° + \theta)} = \dfrac{OC}{\sin\{180° - \alpha - (90° + \theta)\}}$

$$\therefore\quad \frac{h - r}{\sin\alpha\cdot\sin\theta} = \frac{r}{\sin(45° - \alpha)\cos\theta}$$

$$= \frac{r}{\sin(90° - \theta + \alpha)} \quad \text{\{using (1), (2)\}}$$

or $\quad \dfrac{h - r}{\sin\alpha\sin\theta} = \dfrac{r}{\frac{1}{\sqrt{2}}(\cos\alpha - \sin\alpha)\cos\theta} = \dfrac{r}{\cos(\theta + \alpha)}$

$$\therefore\quad \frac{1}{\sqrt{2}}(\cos\alpha - \sin\alpha)\cos\theta = \cos\theta\cdot\cos\alpha - \sin\theta\cdot\sin\alpha$$

or $\quad \left(\dfrac{1}{\sqrt{2}}\cos\theta - \cos\theta\right)\cos\alpha = \left(\dfrac{1}{\sqrt{2}}\cos\theta - \sin\theta\right)\sin\alpha$

$$\therefore\quad \frac{\cos\alpha}{\cos\theta - \sqrt{2}\sin\theta} = \frac{\sin\alpha}{(1 - \sqrt{2})\cos\theta}$$

$$\therefore\quad \cot\alpha = \frac{\cos\theta - \sqrt{2}\sin\theta}{(1 - \sqrt{2})\cos\theta}.$$

Also $\dfrac{h - r}{\sin\alpha\cdot\sin\theta} = \dfrac{r}{\frac{1}{\sqrt{2}}(\cos\alpha - \sin\alpha)\cos\theta}$ gives

$$\frac{h-r}{r}=\frac{\sqrt{2}\sin\alpha\cdot\sin\theta}{(\cos\alpha-\sin\alpha)\cos\theta}$$

$$=\sqrt{2}\cdot\tan\theta\cdot\frac{1}{\cot\alpha-1}$$

$$=\sqrt{2}\tan\theta\cdot\frac{1}{\dfrac{\cos\theta-\sqrt{2}\sin\theta}{(1-\sqrt{2})\cos\theta}-1}$$

$$=\sqrt{2}\tan\theta\cdot\frac{(1-\sqrt{2})\cos\theta}{\sqrt{2}(\cos\theta-\sin\theta)}$$

$$\therefore\quad \frac{h}{r}=1+\frac{(\sqrt{2}-1)\sin\theta}{\sin\theta-\cos\theta}$$

$$=\frac{\sqrt{2}\sin\theta-\cos\theta}{\sin\theta-\cos\theta}$$

$$\therefore\quad h=\frac{2l}{\pi}\cdot\frac{\sqrt{2}\sin\theta-\cos\theta}{\sin\theta-\cos\theta}$$

$$=\frac{2l}{\pi}\cdot\frac{\sqrt{2}\tan\theta-1}{\tan\theta-1}.$$

12. A man moves along a bank of a canal and observes a tower on the other bank. He finds that the angle of elevation of the top of the tower to be α from each of two points A and B, 6*d* apart. From a third point C, between A and B, at a distance 2*d* from A, the angle of elevation is found to be β. Find the height of the tower and the width of the canal.

Here, the tower $PQ = h$; $AB = 6d$, $AC = 2d$, $BC = 4d$.

Let QD ⊥ AB. Then the width of the canal = QD.

From the right-angled triangles PQA, PQB and PQC we get, $QA = h\cot\alpha$, $QB = h\cot\alpha$, $QC = h\cot\beta$.

∵ $QA = QB$, ΔQAB is an isosceles triangle.

∵ QD ⊥ AB, D is the middle point of AB.

∴ $AD = 3d$; but $AC = 2d$, ∴ $CD = d$.

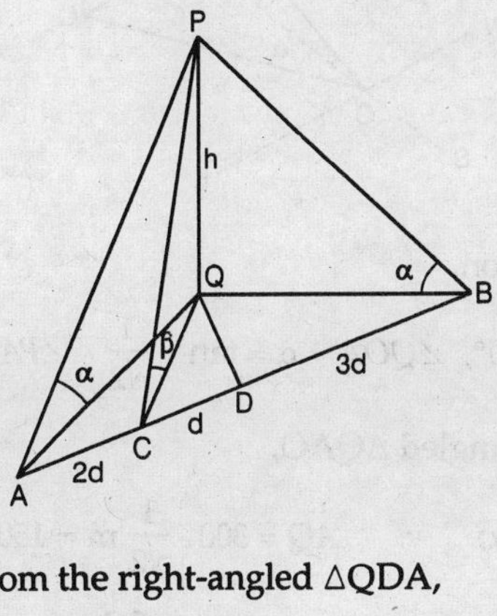

Now, from the right-angled ΔQDA,

$$QD=\sqrt{QA^2-AD^2};\quad\therefore\quad QD=\sqrt{h^2\cot^2\alpha-9d^2}.$$

Also, from the right-angled ΔQDC,

$$QD=\sqrt{QC^2-CD^2};\quad\therefore\quad QD=\sqrt{h^2\cot^2\beta-d^2}.$$

Thus, $\sqrt{h^2\cot^2\alpha-9d^2}=\sqrt{h^2\cot^2\beta-d^2}$

or $\quad h^2(\cot^2\alpha-\cot^2\beta)=8d^2$

$$\therefore\quad h=\frac{2\sqrt{2}d}{\sqrt{\cot^2\alpha-\cot^2\beta}}=\text{height of the tower}$$

$$\therefore\quad QD=\{h^2\cot^2\beta-d^2\}^{1/2}=\left\{\frac{8d^2\cot^2\beta}{\cot^2\alpha-\cot^2\beta}-d^2\right\}^{1/2}$$

$$=d\cdot\sqrt{\frac{8\cot^2\beta-\cot^2\alpha+\cot^2\beta}{\cot^2\alpha-\cot^2\beta}}$$

$$=d\cdot\sqrt{\frac{9\cot^2\beta-\cot^2\alpha}{\cot^2\alpha-\cot^2\beta}}$$

= width of the canal.

13. A person standing by the side of a road observes a row of equidistant telephone poles of equal height. Neglecting the height of the person's eye, the 10th and 17th poles subtend the same angle that they would have subtended if they were in the position of the first pole and were respectively $\frac{1}{2}$ and $\frac{1}{3}$ of their common height. Find the secant of the angle between the base line of the poles and the line drawn from the person's eye to the base of the first pole.

Let the distance between the consecutive poles = k and the height of each pole = h

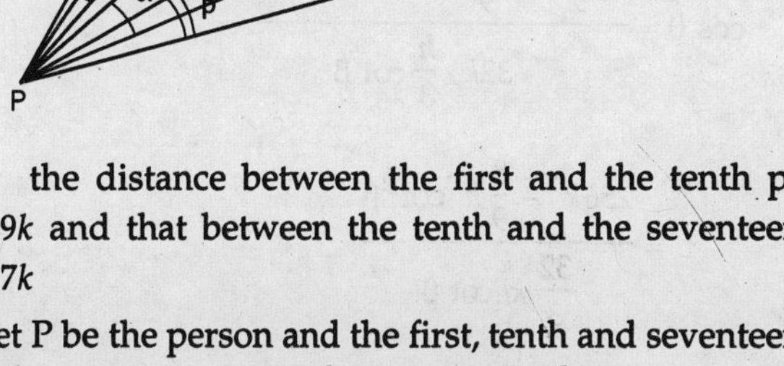

∴ the distance between the first and the tenth pole $= 9k$ and that between the tenth and the seventeenth $= 7k$

Let P be the person and the first, tenth and seventeenth poles are at A_1, A_{10} and A_{17} respectively.

Let $\angle PA_1A_{17} = \theta$. We have to find sec θ.

Clearly, from the right-angled triangles $B_{10}A_{10}P$ and $B_{17}A_{17}P$ we get

$$PA_{10} = h\cot\alpha,\ PA_{17} = h\cot\beta.$$

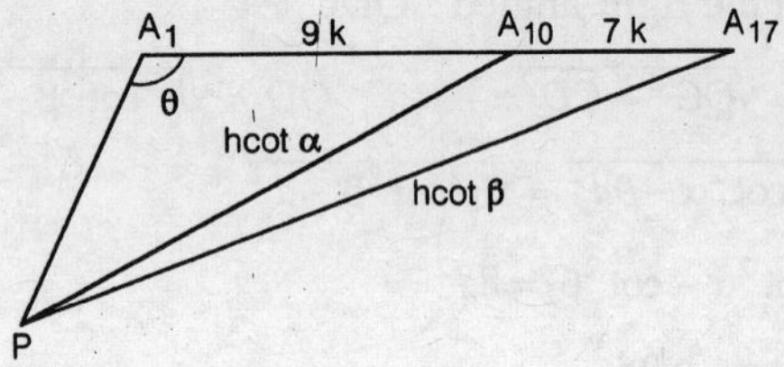

Also, from the question, $\dfrac{\frac{h}{2}}{PA_1} = \tan\alpha$ and $\dfrac{\frac{h}{3}}{PA_1} = \tan\beta$

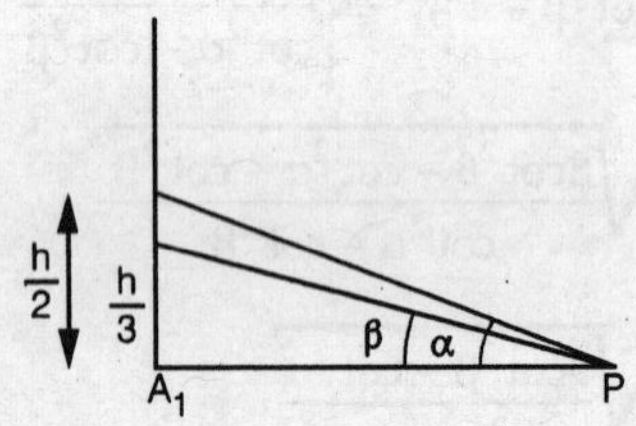

$$\therefore \quad PA_1 = \frac{h}{2}\cot\alpha = \frac{h}{3}\cot\beta$$

$$\therefore \quad 3\cot\alpha = 2\cot\beta \qquad \ldots (1)$$

Now, from the ΔPA_1A_{10},

$$\cos\theta = \frac{(9k)^2 + PA_1^2 - h^2\cot^2\alpha}{2\cdot 9k\cdot PA_1}$$

$$= \frac{81k^2 + \frac{h^2}{4}\cot^2\alpha - h^2\cot^2\alpha}{18k\cdot\frac{h}{2}\cot\alpha}$$

$$\therefore \quad \cos\theta = \frac{81k^2 - \frac{3}{4}h^2\cot^2\alpha}{9kh\cot\alpha} \qquad \ldots (2)$$

Also, from the ΔPA_1A_{17},

$$\cos\theta = \frac{(16k)^2 + PA_1^2 - h^2\cot^2\beta}{2\cdot 16k\cdot PA_1}$$

$$\therefore \quad \cos\theta = \frac{256k^2 + \frac{h^2}{9}\cot^2\beta - h^2\cot^2\beta}{32k\cdot\frac{h}{3}\cot\beta}$$

$$= \frac{256k^2 - \frac{8}{9}h^2\cot^2\beta}{\frac{32}{3}kh\cot\beta} \qquad \ldots (3)$$

(2) and (3) $\Rightarrow$

$$\frac{81k^2 - \frac{3}{4}h^2\cot^2\alpha}{9kh\cot\alpha} = \frac{256k^2 - \frac{8}{9}h^2\cot^2\beta}{\frac{32}{3}kh\cot\beta}$$

or $\quad \dfrac{81k^2 - \frac{3}{4}h^2\cot^2\alpha}{9\cot\alpha} = \dfrac{256k^2 - \frac{8}{9}h^2\cdot\frac{9}{4}\cot^2\alpha}{\frac{32}{3}\cdot\frac{3\cot\alpha}{2}}$, using (1)

or $\quad 16\left(81k^2 - \frac{3}{4}h^2\cot^2\alpha\right) = 9(256k^2 - 2h^2\cot^2\alpha)$

or $\quad 1296k^2 - 12h^2\cot^2\alpha = 2304k^2 - 18h^2\cot^2\alpha$

or $\quad 1008k^2 = 6h^2\cot^2\alpha$

$\therefore \quad h^2\cot^2\alpha = 168k^2$

$$\therefore \quad \text{from (2), } \cos\theta = \frac{81k^2 - \frac{3}{4}\cdot 168k^2}{9k\cdot\sqrt{168}k}$$

$$= \frac{81 - 126}{9\sqrt{168}}$$

$$= \frac{-45}{18\sqrt{42}} = \frac{-5}{2\sqrt{42}}$$

$$\therefore \quad \sec\theta = -\frac{2\sqrt{42}}{5}.$$

14. An observer at O notices that the angle of elevation of the top of a vertical tower is 30°. At O the bearing of the tower is $\tan^{-1}\frac{1}{\sqrt{2}}$ east of north. The observer travels a distance of 300 m towards north to a point A and finds the tower to his east. The angle of elevation of the top of the tower at A is ϕ. Find ϕ and the height of the tower.

Let the tower $PQ = h$ m.

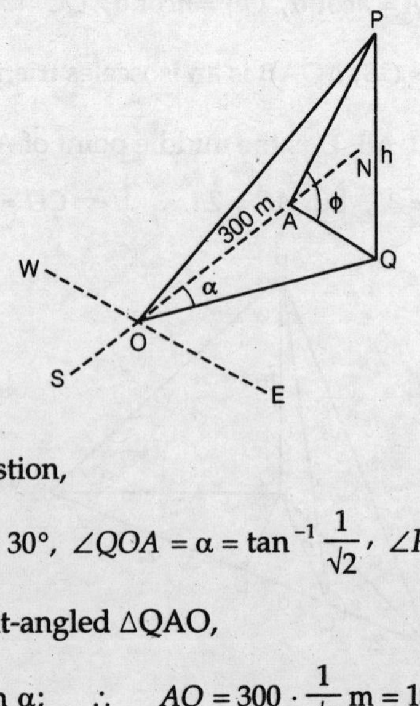

From the question,

$$\angle POQ = 30°,\ \angle QOA = \alpha = \tan^{-1}\frac{1}{\sqrt{2}},\ \angle PAQ = \phi.$$

From the right-angled ΔQAO,

$$\frac{AQ}{OA} = \tan\alpha; \quad \therefore \quad AQ = 300\cdot\frac{1}{\sqrt{2}}\text{ m} = 150\sqrt{2}\text{ m}.$$

From the right-angled ΔPQA, $\tan\phi = \dfrac{PQ}{AQ}$

$\therefore \quad \tan\phi = \dfrac{h}{150\sqrt{2}}; \quad \therefore \quad h = 150\sqrt{2} \cdot \tan\phi \qquad \ldots (1)$

From the right-angled ΔPQO, $\tan \angle POQ = \dfrac{PQ}{OQ}$

$$\therefore \quad \tan 30° = \frac{h}{\sqrt{OA^2 + AQ^2}}$$

$$\therefore \quad \frac{1}{\sqrt{3}} = \frac{h}{\sqrt{300^2 + (150\sqrt{2})^2}}$$

$$\therefore \quad h = \frac{\sqrt{150^2 \cdot 4 + 150^2 \cdot 2}}{\sqrt{3}} = \frac{150 \cdot \sqrt{6}}{\sqrt{3}} = 150\sqrt{2}$$

and from (1), $150\sqrt{2} = 150\sqrt{2} \cdot \tan\phi$

$\therefore \quad \tan\phi = 1; \quad \therefore \quad \phi = 45°$

Thus, $\phi = 45°$ and the height of the tower $= 150\sqrt{2}$ m.

15. An observer at an anti-aircraft post A identifies an enemy aircraft due east of his post at an angle of elevation 60°. At the same instant a detection post D situated 4 km due south of A, reports that the aircraft is at an angle of elevation 30°. Calculate the altitude at which the plane is flying.

Here, PQ = altitude of the aircraft P at which it is flying at the moment of observation

$= h$ km

$AD = 4$ km.

From the right-angled ΔPQA, $\dfrac{h}{AQ} = \tan 60°$

$\therefore \quad AQ = h \cot 60° = \dfrac{h}{\sqrt{3}}$.

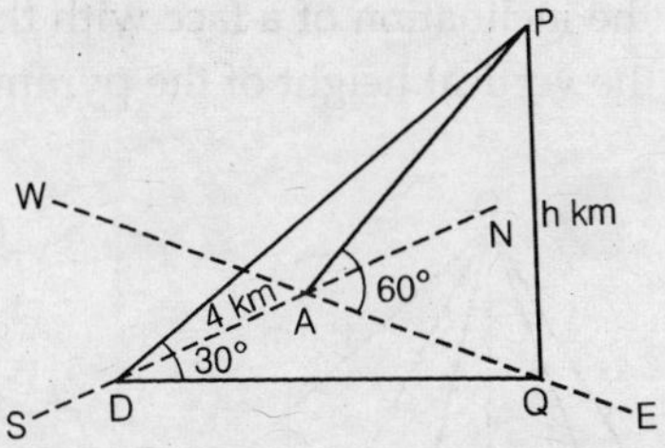

From the right-angled ΔPQD, $\dfrac{h}{DQ} = \tan 30°$

$\therefore \quad DQ = h\cot 30° = h\sqrt{3}$.

From the right-angled ΔDAQ,

$$DQ^2 = DA^2 + AQ^2$$

$$\therefore \quad 3h^2 = 4^2 + \frac{h^2}{3}$$

or $\quad \dfrac{8h^2}{3} = 4^2; \quad \therefore \quad h^2 = \dfrac{48}{8} = 6$

$\therefore \quad h = \sqrt{6}$

$\therefore \quad$ the required altitude $= \sqrt{6}$ km.

16. PQ is a vertical tower and A, B, C are three points in the horizontal plane through the bottom P of the tower. The angle of elevation of Q at each of the points A, B, C is θ. Show that the height of the tower is $\dfrac{abc}{4\Delta} \cdot \tan\theta$, where Δ is the area and *a, b, c* are sides of the triangle ABC.

Let the height of the tower $PQ = h$.

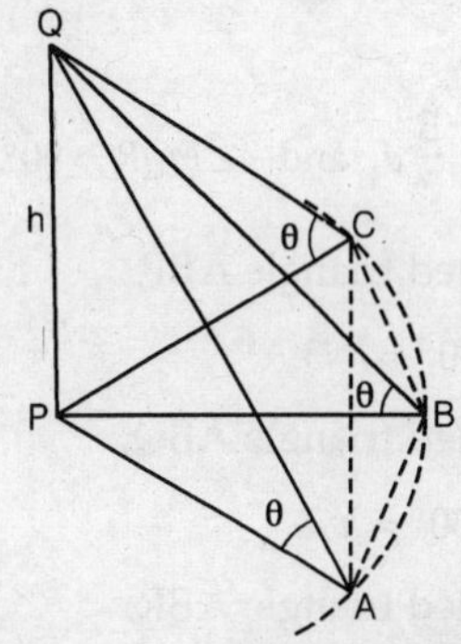

From the right-angled triangles QPA, QPB and QPC we get, $PA = h\cot\theta$, $PB = h\cot\theta$, $PC = h\cot\theta$.

$\therefore \quad PA = PB = PC = h\cot\theta$.

$\therefore$ A, B and C lie on a circle of radius $h\cot\theta$ whose centre is P.

$\therefore \quad$ circumradius of the $\Delta ABC = R = h\cot\theta$.

Now, $\Delta = \dfrac{1}{2} bc \sin A = \dfrac{1}{2} bc \cdot \dfrac{a}{2R}$

$$\left(\because \text{ in } \Delta ABC, \frac{a}{\sin A} = \frac{b}{\sin B} = \frac{c}{\sin C} = 2R\right)$$

$$= \frac{abc}{4R} = \frac{abc}{4h\cot\theta} = \frac{abc}{4h}\tan\theta$$

$$\therefore \quad h = \frac{abc}{4\Delta} \cdot \tan\theta.$$

17. A tower stands vertically in an inaccessible place. In order to ascertain its height an investigator stations himself on the ground and finds the angle of elevation of the top of the tower to be 30°. Then he moves a distance *d* in a certain direction and from there he finds the angle of elevation of the top to be the same as before. On moving a distance $\frac{5}{3}d$ at right angle to his former direction, he finds the elevation of the top to be 60°. Prove that the height of the tower is either $\sqrt{\frac{5}{6}} \cdot d$ or $\sqrt{\frac{85}{48}} \cdot d$.

Let the height of the tower $AB = h$.

The initial position of the investigator is P and the two subsequent positions of observation are Q and R respectively.

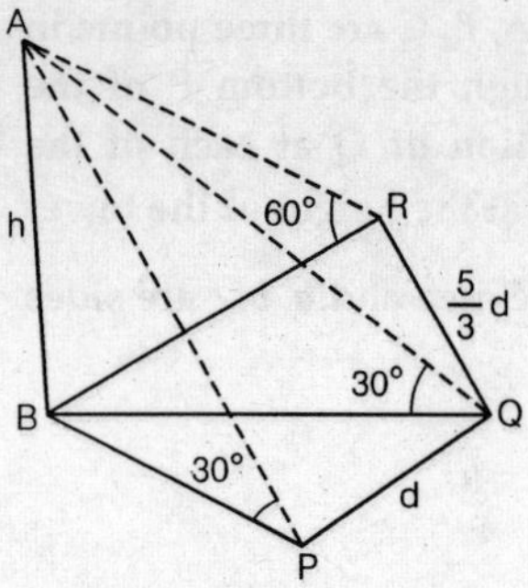

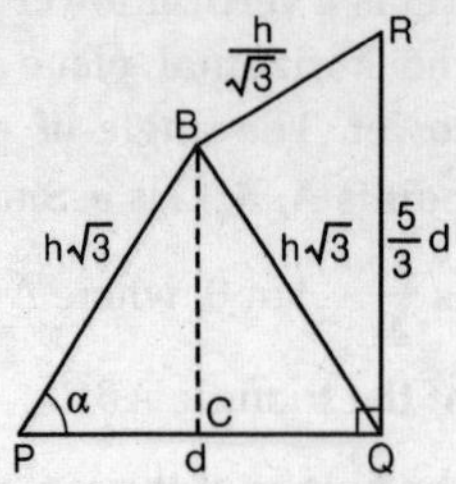

So $PQ = d$, $QR = \frac{5}{3}d$ and $\angle PQR = 90°$.

In the right-angled triangle ABP,

$$BP = h\cot 30° = h\sqrt{3}.$$

In the right-angled triangle ABQ,

$$BQ = h\cot 30° = h\sqrt{3}.$$

In the right-angled triangle ABR,

$$BR = h\cot 60° = \frac{h}{\sqrt{3}}.$$

The figure on the ground is as shown. Let BC ⊥ PQ.

Clearly, $\cos\alpha = \frac{PC}{BP} = \frac{d/2}{h\sqrt{3}} = \frac{d}{2\sqrt{3}\,h}$ ... (1)

In the $\triangle BQR$, $\angle BQR = 90° - \angle BQP = 90° - \alpha$

$$\therefore \quad \cos(90° - \alpha) = \frac{(h\sqrt{3})^2 + \left(\frac{5}{3}d\right)^2 - \left(\frac{h}{\sqrt{3}}\right)^2}{2h\sqrt{3}\cdot\frac{5}{3}d}$$

or $$\sin\alpha = \frac{3h^2 + \frac{25}{9}d^2 - \frac{h^2}{3}}{\frac{10}{\sqrt{3}}hd} = \frac{\frac{8h^2}{3} + \frac{25}{9}d^2}{\frac{10}{\sqrt{3}}hd}$$

$$= \frac{4}{5\sqrt{3}}\frac{h}{d} + \frac{5}{6\sqrt{3}}\frac{d}{h}$$

$$\therefore \quad \sin^2\alpha = \frac{16}{75}\frac{h^2}{d^2} + \frac{25}{108}\frac{d^2}{h^2} + \frac{4}{9}$$

or $$1 - \frac{d^2}{12h^2} = \frac{16}{75}\cdot\frac{h^2}{d^2} + \frac{25}{108}\cdot\frac{d^2}{h^2} + \frac{4}{9}, \quad \text{\{using (1)\}}$$

or $$\frac{16}{75}\frac{h^2}{d^2} + \frac{17}{54}\frac{d^2}{h^2} - \frac{5}{9} = 0;$$

if $\frac{h}{d} = x$ then

$$\frac{16}{75}x^2 + \frac{17}{54}\cdot\frac{1}{x^2} - \frac{5}{9} = 0$$

or $$\frac{16}{25}x^2 + \frac{17}{18}\cdot\frac{1}{x^2} - \frac{5}{3} = 0$$

or $$\frac{16}{25}x^4 - \frac{5}{3}x^2 + \frac{17}{18} = 0$$

$$\therefore \quad x^2 = \frac{\frac{5}{3} \pm \sqrt{\frac{25}{9} - 4\cdot\frac{16}{25}\cdot\frac{17}{18}}}{\frac{32}{25}}$$

$$\therefore \quad x^2 = \frac{\frac{5}{3} \pm \frac{3}{5}}{\frac{32}{25}} = \frac{25}{32}\left(\frac{5}{3} \pm \frac{3}{5}\right) = \frac{125}{96} \pm \frac{15}{32} = \frac{125 \pm 45}{96}$$

$$= \frac{170}{96}, \ \frac{80}{96} = \frac{85}{48}, \frac{5}{6}$$

$$\therefore \quad x = \frac{h}{d} = \sqrt{\frac{85}{48}}, \ \sqrt{\frac{5}{6}}$$

$$\therefore \quad h = \sqrt{\frac{5}{6}}\,d, \ \sqrt{\frac{85}{48}}\,d.$$

Hence, the problem.

18. A regular pyramid on a square base has an edge 150 m long and the length of the side of its base is 200 m. Find the inclination of a face to the base and the vertical height of the pyramid.

Here, edge = $OA = OB = OC = OD = 150$ m

and side of square base = $AB = BC = CD = DA = 200$ m.

Let OM ⊥ base. Then M is the centre of the square. MN ⊥ AB.

Clearly, $AN = NB = 100$ m.

Also $MN = \frac{1}{2}AD = 100$ m.

We have to find the inclination of a face with the base, i.e., $\angle ONM$ and the vertical height of the pyramid, i.e., OM.

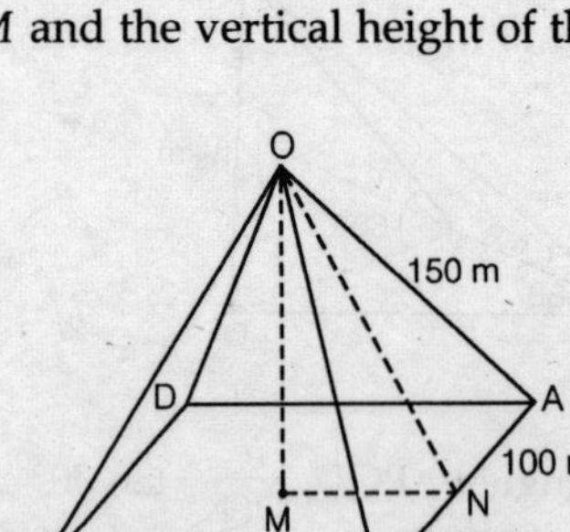

As $\triangle OAB$ is isosceles, ON ⊥ NA.

In $\triangle OAN$, $OA = 150$ m and $NA = 100$ m.

$$\therefore \quad ON = \sqrt{150^2 - 100^2} \text{ m} = 50\sqrt{5} \text{ m}.$$

In $\triangle OMN$, $OM = \sqrt{ON^2 - MN^2}$

$$= \sqrt{12500 - 10000} \text{ m}$$

$$= \sqrt{2500} \text{ m} = 50 \text{ m}$$

and $\tan \angle ONM = \frac{OM}{MN} = \frac{50 \text{ m}}{100 \text{ m}} = \frac{1}{2}$

$\therefore \quad \angle ONM = \tan^{-1}\frac{1}{2}$.

19. ABC is a triangular park with $AB = AC = 100$ m. A TV tower stands vertically at the middle point of BC. The angles of elevation of the top of the tower at A, B, C are 45°, 60°, 60° respectively. Find the height of the tower.

Here, the tower $PQ = h$ m.

From the right-angled ΔPQA, $QA = h\cot 45° = h$.

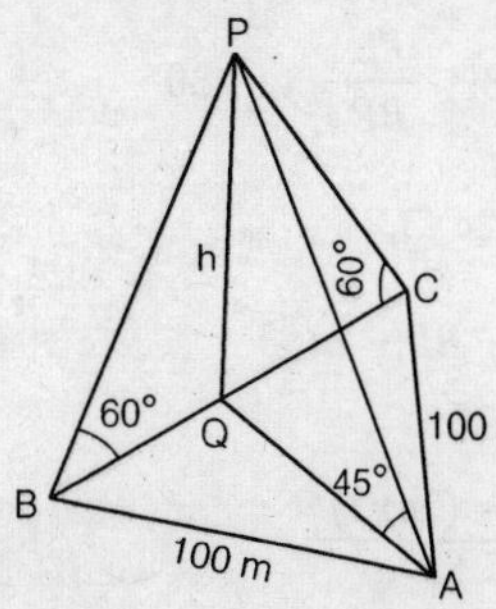

From the right-angled ΔPQB, $BQ = h\cot 60° = \frac{h}{\sqrt{3}}$.

From the right-angled ΔPQC, $CQ = h\cot 60° = \frac{h}{\sqrt{3}}$.

$\therefore \quad BQ = CQ = \frac{h}{\sqrt{3}}$;

so Q is the middle point of BC.

Also $AB = AC = 100$ m; so, ΔABC is isosceles.

$\therefore$ from geometry, $AQ \perp BC$.

Now, from the right-angled ΔAQB,

$$BQ^2 + AQ^2 = AB^2$$

$$\therefore \quad \frac{h^2}{3} + h^2 = 100^2$$

or $\quad \frac{4h^2}{3} = 100^2; \quad \therefore \ h = 50\sqrt{3}.$

$\therefore$ the height of the TV tower $= 50\sqrt{3}$ m.

20. From a certain station A the angular elevation of a mountain peak P, to the north of A, is α. From the top B of a hill of height h above the level of A, the angular elevation of P is found to be β, the bearing of A is γ west of south and the bearing of P is δ north of A. Show that the height of P above the level of A is

$$\frac{h \tan \alpha \cdot \sin \delta}{\tan \alpha \cdot \sin \delta - \tan \beta \sin \gamma}.$$

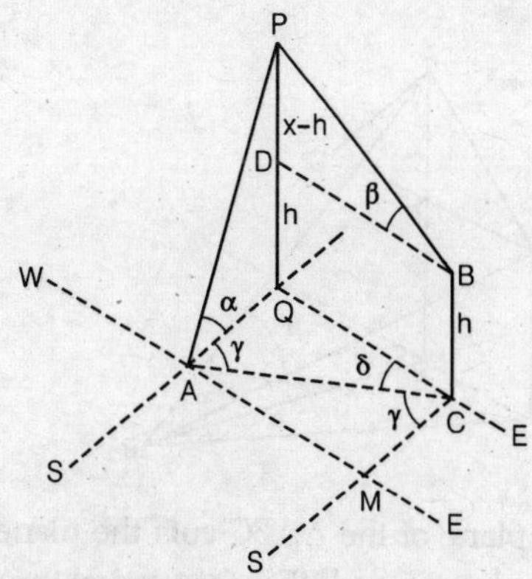

Let the height of the mountain $= PQ = x$;

the height of the hill $= BC = h$.

From the question,

$$\angle PAQ = \alpha, \ \angle ACQ = \delta \text{ and } \angle MCA = \angle CAQ = \gamma.$$

Let $BD \perp PD$. Then, from the question, $\angle PBD = \beta$.

Also $DQ = BC = h$.

Now, from the right-angled triangles PQA and PDB we get

$$AQ = PQ\cot \alpha = x\cot \alpha,$$

$$BD = PD\cot \beta = (PQ - DQ)\cot \beta = (x - h)\cot \beta$$

In $\quad \Delta AQC, \frac{QC}{\sin \gamma} = \frac{AQ}{\sin \delta}$

$$\therefore \quad \frac{BD}{\sin \gamma} = \frac{x\cot \alpha}{\sin \delta}$$

$$\therefore \quad \frac{(x-h)\cot \beta}{\sin \gamma} = \frac{x\cot \alpha}{\sin \delta}$$

or $\quad x\left\{\frac{\cot \beta}{\sin \gamma} - \frac{\cot \alpha}{\sin \delta}\right\} = \frac{h\cot \beta}{\sin \gamma}$

or $\quad x \cdot \left(\frac{1}{\tan \beta \cdot \sin \gamma} - \frac{1}{\tan \alpha \cdot \sin \delta}\right) = \frac{h}{\tan \beta \cdot \sin \gamma}$

or $\quad x \cdot \frac{\tan \alpha \cdot \sin \delta - \tan \beta \cdot \sin \gamma}{\tan \alpha \cdot \tan \beta \cdot \sin \gamma \cdot \sin \delta} = \frac{h}{\tan \beta \cdot \sin \gamma}$

$$\therefore \quad x = \frac{h\tan \alpha \cdot \sin \delta}{\tan \alpha \cdot \sin \delta - \tan \beta \cdot \sin \gamma}$$

21. A sign-post in the form of an isosceles triangle ABC is mounted on a pole of height h fixed to the ground. The base BC of the triangle is parallel to the ground. A man standing on the ground at a distance d from the sign-post finds that the top vertex A of the triangle subtends an angle β and either of the other two vertices subtends the same angle α at his feet. Find the area of the ΔABC.

Here, the sign-post is ΔABC and the pole $DE = h$

The man is at P such that $EP = d$.

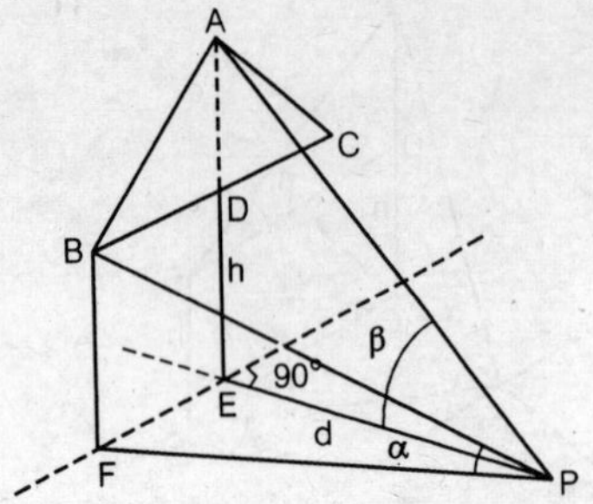

Let the vertical plane of the $\triangle ABC$ cuts the plane of the ground along the line FE and BF is the vertical through B.

Clearly, $BF = DE = h$.

$\therefore$ in the right-angled $\triangle BFP$, $\dfrac{h}{PF} = \tan\alpha$

$\therefore \quad PF = h\cot\alpha$

$\therefore \quad FE = \sqrt{PF^2 - PE^2} = \sqrt{h^2\cot^2\alpha - d^2}$

$\therefore \quad BD = FE = \sqrt{h^2\cot^2\alpha - d^2}$

$\therefore \quad BC = 2BD = 2\sqrt{h^2\cot^2\alpha - d^2}$

Also, in $\triangle AEP$, $\dfrac{AE}{PE} = \tan\beta$;

$\therefore \quad AE = d\tan\beta$

$\therefore \quad AD = AE - DE = d\tan\beta - h$

$$\therefore \quad \text{ar}(\triangle ABC) = \frac{1}{2} BC \cdot AD$$

$$= \frac{1}{2} \cdot 2\sqrt{h^2\cot^2\alpha - d^2} \cdot (d\tan\beta - h)$$

$$= (d\tan\beta - h)\sqrt{h^2\cot^2\alpha - d^2}.$$

22. The base of a vertical tower is a square. A man on a diagonal (produced) of the square is at a distance $2a$ from the tower. He observes that the angle of elevation of each of the two outer corners of the top of the tower is 30° while that of the nearest corner is 45°. Prove that the breadth of the tower is $a\sqrt{2}(\sqrt{5}-1)$.

P is on the diagonal AC (produced) of the square base ABCD such that $PC = 2a$.

The corner of the top of the tower nearest to P is E and F is one of the outer corners.

Let the breadth of the tower $= BC = x$.

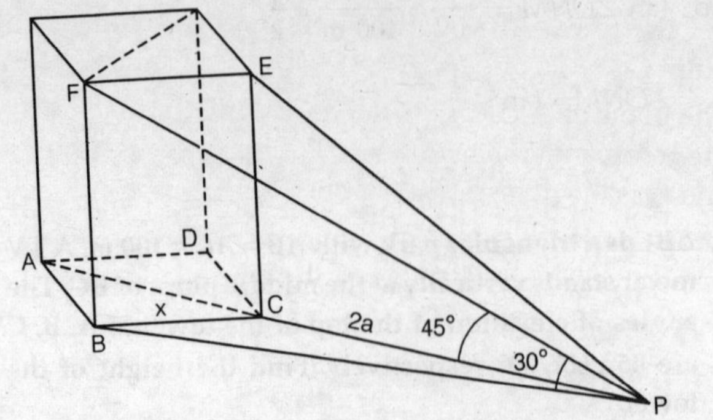

Now, from the right-angled $\triangle ECP$, $\dfrac{EC}{PC} = \tan 45°$

$\therefore \quad EC = PC = 2a = FB.$

From the right-angled $\triangle FBP$, $\dfrac{FB}{BP} = \tan 30°$

$\therefore \quad \dfrac{2a}{BP} = \dfrac{1}{\sqrt{3}}; \quad \therefore BP = 2a\sqrt{3}.$

In $\triangle BCP$, $BC = x$, $CP = 2a$, $BP = 2a\sqrt{3}$
and $\angle BCP = 180° - 45° = 135°$

$$\therefore \quad \cos 135° = \frac{x^2 + (2a)^2 - (2a\sqrt{3})^2}{2 \cdot x \cdot 2a}$$

$$\text{or} \quad -\frac{1}{\sqrt{2}} = \frac{x^2 - 8a^2}{4ax}$$

$$\text{or} \quad x^2 + 2\sqrt{2}\,ax - 8a^2 = 0$$

$$\therefore \quad x = \frac{-2\sqrt{2}a \pm \sqrt{8a^2 - 4\cdot(-8a^2)}}{2}$$

$$\text{or} \quad x = \frac{-2\sqrt{2}a + \sqrt{40a^2}}{2} \quad (\because\ x \text{ is positive})$$

$$= \frac{-2\sqrt{2}a + 2\sqrt{2}.\ \sqrt{5}a}{2}$$

$$= -\sqrt{2}a + \sqrt{2}.\ \sqrt{5}a = a\sqrt{2}(\sqrt{5}-1).$$

23. A man standing at a distance d from a tower of height h finds that an aeroplane is describing a horizontal circle of radius r with the top of the tower at its centre. The angle of elevation of the aeroplane when nearest to the man, is α and when farthest it is β. If $d > r$, prove that

$$\frac{r}{\sin(\alpha-\beta)} = \frac{d}{\sin(\alpha+\beta)} = \frac{h}{2\sin\alpha\cdot\sin\beta}.$$

(Neglect the height of the man)

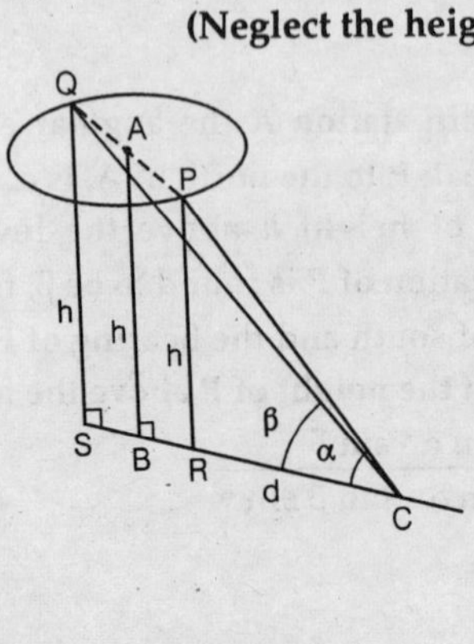

Let the tower $AB = h$ and the man observes from the point C.

The plane of A, B, C cuts the horizontal circular path of the aeroplane along PQ where P is the point nearest to C and Q is the point farthest from C.

Drop perpendiculars PR and QS from P and Q respectively on the line BC on the ground. Clearly,

ΔPRC and ΔQSC are right-angled triangles and $PR = AB = QS = h$. Given $BC = d$.

Also $BR = AP = r,\ SB = QA = r$.

Now, in ΔPRC we have, $\tan\alpha = \dfrac{PR}{RC} = \dfrac{h}{d-r}$ ... (1)

and in ΔQSC we have, $\tan\beta = \dfrac{QS}{SC} = \dfrac{h}{d+r}$ ... (2)

(1) $\Rightarrow$ $(d-r)\sin\alpha = h\cos\alpha$

or $r\sin\alpha - d\sin\alpha + h\cos\alpha = 0$ ... (3)

(2) $\Rightarrow$ $(d+r)\sin\beta = h\cos\beta$

or $r\sin\beta + d\sin\beta - h\cos\beta = 0$ ... (4)

By cross-multiplication we get from (3) and (4),

$$\frac{r}{\sin\alpha\cdot\cos\beta - \cos\alpha\cdot\sin\beta} = \frac{d}{\cos\alpha\cdot\sin\beta + \sin\alpha\cdot\cos\beta} = \frac{h}{\sin\alpha\cdot\sin\beta + \sin\alpha\sin\beta}$$

$$\therefore\quad \frac{r}{\sin(\alpha-\beta)} = \frac{d}{\sin(\alpha+\beta)} = \frac{h}{2\sin\alpha\cdot\sin\beta}.$$

24. A pole stands at the bank of a circular pond. A man walking along the bank finds that the angle of elevation of the top of the pole from the two points A and B is 30° each and from a third point C, it is 45°. If the distances from A to B and from B to C measured along the bank are 40 m and 20 m respectively, find the radius of the pond and the height of the pole.

Let the pole $PQ = h$ m. From the question,

$PA = h\cot 30°,\ PB = h\cot 30°,\ PC = h\cot 45°$

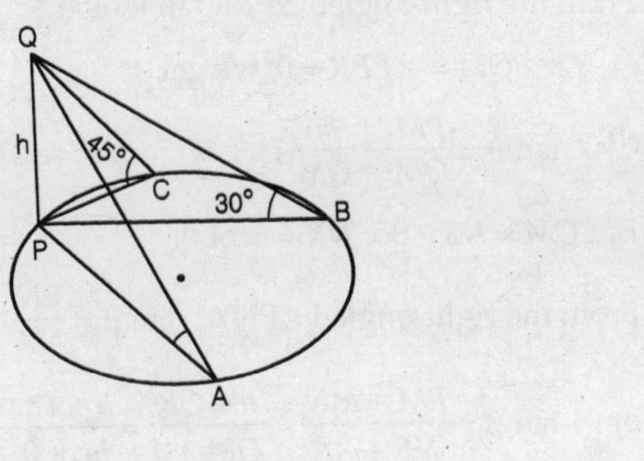

$\therefore\quad PA = PB = h\sqrt{3},\ PC = h.$

Let O be the centre and r m be the radius of the pond.

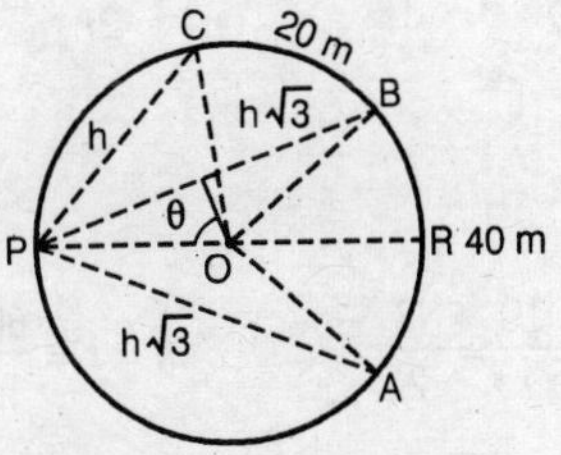

Let $\angle POB = 2\theta$ radian. Then, clearly

$$\frac{\frac{h\sqrt{3}}{2}}{r} = \sin\theta;\quad \therefore\quad \sin\theta = \frac{h\sqrt{3}}{2r} \quad\text{... (1)}$$

$\therefore\quad \angle AOB = 2\angle BOR = 2(180° - 2\theta) = 2\pi - 4\theta.$

From $\theta = \dfrac{l}{r}$ (formula) we get $2\pi - 4\theta = \dfrac{40}{r}$ ... (2)

or $\dfrac{\pi}{2} - \theta = \dfrac{10}{r};\quad \therefore\quad \cos\left(\dfrac{\pi}{2} - \theta\right) = \cos\dfrac{10}{r}$

or $\sin\theta = \cos\dfrac{10}{r}$ ...(3)

Again, from ΔPOC, if $\angle POC = 2\phi$ radian,

$$\frac{\frac{h}{2}}{r} = \sin\phi,\ \text{i.e.,}\ \frac{h}{2r} = \sin\phi \quad\text{... (4)}$$

$\therefore\quad \angle BOC = 2\theta - 2\phi = \pi - \dfrac{20}{r} - 2\phi,$ from (2)

$\therefore$ from $\theta = \dfrac{l}{r}$ we get $\pi - \dfrac{20}{r} - 2\phi = \dfrac{20}{r}$

or $\pi - 2\phi = \dfrac{40}{r}$; then $\dfrac{\pi}{2} - \phi = \dfrac{20}{r}$

$\therefore\quad \cos\left(\dfrac{\pi}{2} - \phi\right) = \cos\dfrac{20}{r}$

or $\sin\phi = \cos\dfrac{20}{r}.$

$\therefore$ (4) $\Rightarrow$ $\dfrac{h}{2r} = \cos\dfrac{20}{r}$

(1) and (3) $\Rightarrow$ $\dfrac{h\sqrt{3}}{2r} = \cos\dfrac{10}{r}$

$\therefore\quad \sqrt{3}\cos\dfrac{20}{r} = \cos\dfrac{10}{r}.$

Let $\cos\dfrac{10}{r} = t$; then $\sqrt{3}(2t^2 - 1) = t$

or $2\sqrt{3}\,t^2 - t - \sqrt{3} = 0$

$$\therefore\quad t = \frac{1 \pm \sqrt{1 + 4\sqrt{3}.\,2\sqrt{3}}}{4\sqrt{3}} = \frac{1 \pm 5}{4\sqrt{3}}$$

$$\therefore\quad \cos\frac{10}{r} = \frac{1+5}{4\sqrt{3}},\ \frac{1-5}{4\sqrt{3}} = \frac{\sqrt{3}}{2},\ \frac{-1}{\sqrt{3}}$$

$\therefore \quad \cos\frac{10}{r} = \frac{\sqrt{3}}{2} \qquad \left\{\because \cos\frac{10}{r} = \frac{h\sqrt{3}}{2r} \text{ is positive}\right\}$

$\therefore \quad \frac{10}{r} = \frac{\pi}{6}; \quad \therefore \quad r = \frac{60}{\pi}.$

Also, $\frac{h\sqrt{3}}{2r} = \cos\frac{10}{r} = \frac{\sqrt{3}}{2}; \quad \therefore \quad h = r = \frac{60}{\pi}.$

$\therefore$ radius of the pond = height of the pole $= \frac{60}{\pi}$ m.

25. A circular plate touches a vertical wall. The plate is fixed horizontally at a certain height above the ground. A lighted candle of a length equal to half of the height of the plate above the ground, stands at the centre of the plate. Prove that the breadth of the shadow thrown on the wall where it meets the ground is $4\sqrt{2}$ times the radius of the plate.

Here, the candle $AC = c$ (say) and

the height of the plate $= CO = 2c$.

Let the radius of the plate $= a$.

Clearly from the figure, the required breadth of the shadow is the length of the chord PQ of the circle whose centre is O and radius $= OD$.

Now, from similar triangles ACB and AOD, we get

$$\frac{AC}{AO} = \frac{BC}{OD}; \quad \therefore \quad \frac{c}{3c} = \frac{a}{OD}; \quad \therefore \quad OD = 3a.$$

Clearly, the distance of the chord PQ from the centre O is OR = radius of the plate $= a$.

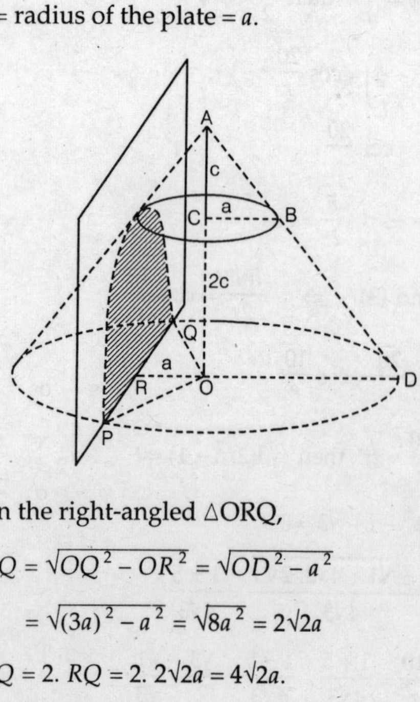

Now, in the right-angled ΔORQ,

$$RQ = \sqrt{OQ^2 - OR^2} = \sqrt{OD^2 - a^2}$$

$$= \sqrt{(3a)^2 - a^2} = \sqrt{8a^2} = 2\sqrt{2}a$$

$\therefore \quad PQ = 2.\ RQ = 2.\ 2\sqrt{2}a = 4\sqrt{2}a.$

Hence, the problem.

26. A flagstaff is mounted in the middle of a square-based vertical tower. A man standing on the ground midway in front of one side of the tower, finds the elevation of the top of the tower to be α and that of the flagstaff to be β from a distance d. After getting closer to the tower by a distance a, he finds that the top of the flagstaff just disappears. Prove that the height of the flagstaff is $\frac{d\sin\theta\cdot\sin(\beta-\alpha)}{\cos\alpha\cdot\sin(\theta-\beta)}$, where $(d-a)\tan\theta = d\tan\alpha$.

Here, QR is the height of the tower and the flagstaff $PM = h$ (say). Clearly, from the figure, $MN = QR$.

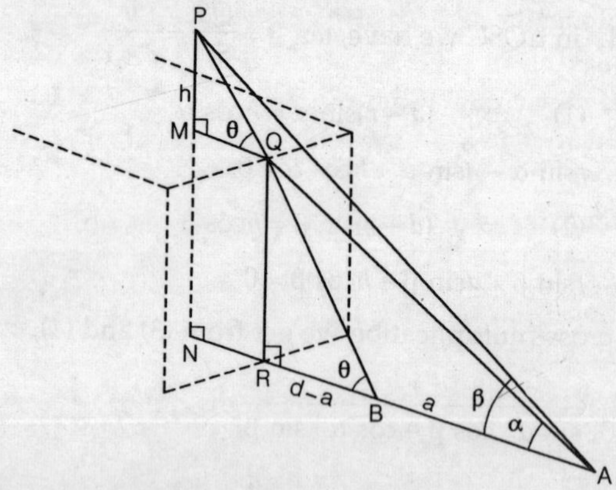

The man is at A initially and at B when the top P disappears.

So $AR = d$, $AB = a$.

Let $\angle QBR = \theta$.

From the right-angled ΔQRB,

$$\tan\theta = \frac{QR}{RB} = \frac{QR}{d-a} \qquad \dots (1)$$

From the right-angled ΔQRA,

$$\tan\alpha = \frac{QR}{RA} = \frac{QR}{d}; \quad \therefore \quad QR = d\tan\alpha \qquad \dots (2)$$

From (1) and (2), $\tan\theta = \frac{d\tan\alpha}{d-a}$

or $\quad (d-a)\tan\theta = d\tan\alpha.$

From the right-angled ΔPMQ in which

$\angle PQM = \angle PBR = \theta$, we get

$$\tan\theta = \frac{PM}{QM} = \frac{h}{QM}; \quad \therefore \quad QM = h\cot\theta.$$

But $QM = NR$. So $NR = h\cot\theta$.

From the right-angled ΔPNA, $\tan\beta = \frac{PN}{NA}$

or $\quad \tan\beta = \frac{PM + MN}{NR + RA} = \frac{h + QR}{QM + d} = \frac{h + d\tan\alpha}{h\cot\theta + d}$

$\therefore \quad \tan\beta(h\cot\theta + d) = h + d\tan\alpha$

or $\quad d(\tan\beta - \tan\alpha) = h(1 - \tan\beta \cdot \cot\theta)$

or $\quad d \cdot \dfrac{\sin(\beta-\alpha)}{\cos\beta\cos\alpha} = h\dfrac{\sin(\theta-\beta)}{\cos\beta \cdot \sin\theta}$

$\therefore \quad h = \dfrac{d\sin\theta \cdot \sin(\beta-\alpha)}{\cos\alpha \cdot \sin(\theta-\beta)}$

$\therefore$ the height of the flagstaff $= \dfrac{d\sin\theta \cdot \sin(\beta-\alpha)}{\cos\alpha \cdot \sin(\theta-\beta)}$

where $(d-a)\tan\theta = d\tan\alpha$.

27. Two men start climbing up the plane side of a hill starting from the same base position. One climbs along the line of the greatest slope and the other making an angle β with the line of the greatest slope. If they climb at the same speed and after some time the first man is *h* above the ground level then find the height of the other man above the ground at that time, the inclination of the hillside being α with the ground.

Starting from O, after some time the first man is at A whose height $AC = h$.

At that time the second man is at B where $OB = OA$ and $\angle BOA = \beta$ (from the question).

Let the height of B $= BD = x$.

From the right-angled $\triangle$ACO, $OA = h\text{cosec}\,\alpha$ ... (1)

Let BE $\perp$ OA. So BE $\parallel$ PQ because the line OA of the greatest slope is perpendicular to PQ.

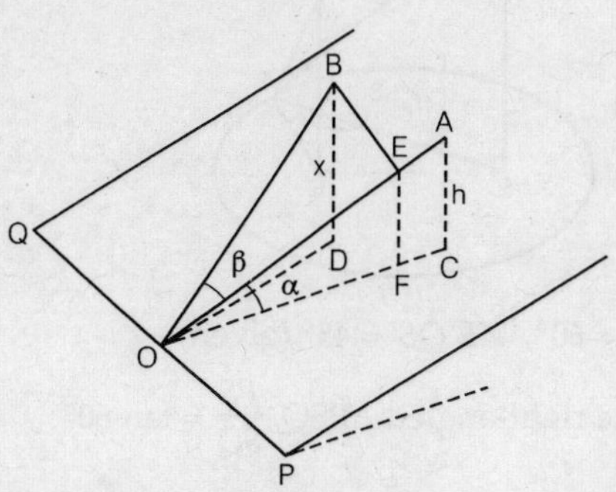

Clearly, the height of B = the height of E $= EF$.

From the right-angled $\triangle$BEO,

$OE = OB\cos\beta$

$\therefore$ from the right-angled $\triangle$EFO,

$EF = OE \cdot \sin\alpha$

$\therefore \quad EF = OB\cos\beta \cdot \sin\alpha$

$= OA\cos\beta \cdot \sin\alpha \quad (\because OA = OB)$

$= h\text{cosec}\,\alpha \cdot \cos\beta \cdot \sin\alpha$, from (1)

$= h\cos\beta$

$\therefore$ the height of other man $= h\cos\beta$.

28. Two vertical walls of equal heights *a* are inclined to each other at an angle α. At noon the breadths of their shadows are *b* and *c*. If θ be the altitude of the sun, prove that $a^2\sin^2\alpha \cdot \cot^2\theta = b^2 + c^2 - 2bc\cos\alpha$.

At noon the breadths of shadows of the two walls are $BC = b$, $QR = c$. When the altitude of the sun is θ, the breadths are BD and QS respectively along the cross-section of the shadows.

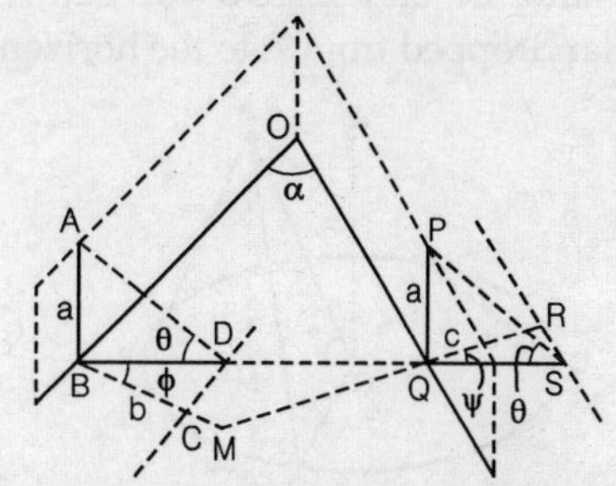

Then $BD = a\cot\theta$, {from the right-angled $\triangle$ABD}

$QS = a\cot\theta$, {from the right-angled $\triangle$PQS}

Let $\angle DBC = \phi$ and $\angle SQR = \psi$.

In the right-angled $\triangle$BCD, $\dfrac{BC}{BD} = \cos\phi$.

$\therefore \quad \dfrac{b}{a\cot\theta} = \cos\phi$ ... (1)

In the right-angled $\triangle$QRS, $\dfrac{QR}{QS} = \cos\psi$

$\therefore \quad \dfrac{c}{a\cot\theta} = \cos\psi$ ... (2)

Now, $\cos\alpha = \cos(180° - \angle BMQ)$

$= \cos\{180° - (180° - \phi - \psi)\}$

$= \cos(\phi + \psi) = \cos\phi \cdot \cos\psi - \sin\phi \cdot \sin\psi$

$\therefore \quad \sin\phi \cdot \sin\psi = \cos\phi\cos\psi - \cos\alpha;$

squaring, $(1-\cos^2\phi)(1-\cos^2\psi) = \cos^2\phi\cos^2\psi + \cos^2\alpha$
$- 2\cos\phi\cos\psi\cos\alpha$

or $\quad 1 - \cos^2\phi - \cos^2\psi = \cos^2\alpha - 2\cos\phi\cos\psi\cos\alpha$

or $\quad \sin^2\alpha = \cos^2\phi + \cos^2\psi - 2\cos\phi\cos\psi\cos\alpha$

$= \dfrac{b^2}{a^2\cot^2\theta} + \dfrac{c^2}{a^2\cot^2\theta}$

$- 2 \cdot \dfrac{b}{a\cot\phi} \cdot \dfrac{c}{a\cot\theta} \cdot \cos\alpha$ {from (1), (2)}

or $\quad a^2\sin^2\alpha \cdot \cot^2\theta = b^2 + c^2 - 2bc\cos\alpha$.

29. A man walks in a horizontal circle round the foot of a pole which is inclined to the vertical. The foot of the pole is at the cente of the circle. The greatest and the least angles of elevation of the top of the pole as observed by the man are α and β respectively. When he is midway between the two positions, the angle subtended at his eye by the pole is θ. Prove that

$$\tan\theta = \frac{1}{\sin(\alpha+\beta)}\sqrt{\sin^2(\alpha-\beta)+4\sin^2\alpha\cdot\sin^2\beta}.$$

Let $OA = l$ be the pole and the radius of the circle $= a$. Let the vertical plane through the inclined pole cut the circle at B and C, and $\angle AOB = \phi$. Let AM be the perpendicular dropped from A to the horizontal.

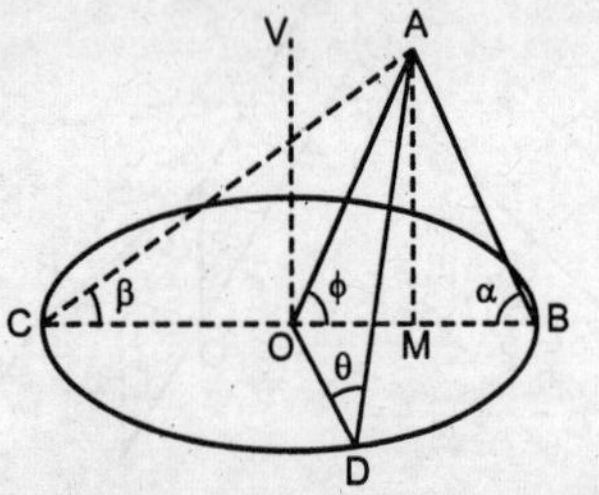

As B and C are the nearest and the farthest points from M, on the circle we get the angles of elevation at B and C are respectively α and β (from the question).

D is midway between B and C on the circle. Clearly OD ⊥ BC and OD ⊥ OV (the vertical line).

So, OD is perpendicular to the plane of VO and OA. So, OD ⊥ OA.

∵ OA subtends θ at D, $\tan\theta = \dfrac{OA}{OD} = \dfrac{l}{a}$ ... (1)

Now, $OM = OA\cos\phi$

∴ $BM = a - l\cos\phi$ and $CM = a + l\cos\phi$.

Also $AM = OA\sin\phi$; ∴ $AM = l\sin\phi$

∴ $\tan\alpha = \dfrac{AM}{BM} = \dfrac{l\sin\phi}{a - l\cos\phi}$

or $\cot\alpha = \dfrac{a}{l}\cdot\dfrac{1}{\sin\phi} - \cot\phi$ ... (2)

and $\tan\beta = \dfrac{AM}{CM} = \dfrac{l\sin\phi}{a + l\cos\phi}$

or $\cot\beta = \dfrac{a}{l}\cdot\dfrac{1}{\sin\phi} + \cot\phi$ ... (3)

From (3) – (2), $\cot\beta - \cot\alpha = 2\cot\phi$ ... (4)

From (3) + (2),

$$\cot\beta + \cot\alpha = \frac{2a}{l}\cdot\frac{1}{\sin\phi} = \frac{2}{\tan\theta}\cdot\operatorname{cosec}\phi, \text{ using (1)}$$

$$\therefore\ \tan\theta = \frac{2\operatorname{cosec}\phi}{\cot\beta+\cot\alpha} = \frac{2\sqrt{1+\cot^2\phi}}{\dfrac{\sin(\alpha+\beta)}{\sin\alpha\cdot\sin\beta}}$$

$$= \frac{2\sin\alpha\sin\beta\cdot\sqrt{1+\left(\dfrac{\cot\beta-\cot\alpha}{2}\right)^2}}{\sin(\alpha+\beta)}$$

$$= \frac{2\sin\alpha\sin\beta\sqrt{1+\dfrac{\sin^2(\alpha-\beta)}{4\sin^2\alpha\cdot\sin^2\beta}}}{\sin(\alpha+\beta)}, \text{ using (4)}$$

$$= \frac{\sqrt{\sin^2(\alpha-\beta)+4\sin^2\alpha\sin^2\beta}}{\sin(\alpha+\beta)}.$$

30. Two poles, one double of the other in length, stand vertically on a field. The sum of the distances of any point on the boundary from the two poles is always 50 m. A man standing at a place on the boundary where the feet of the poles subtend an angle of 30°, finds that the elevation of the top of the taller pole is 60° and that of the shorter is 45°. Find the distance between the poles and their heights. Also find the distance of the man from the line joining the feet of the poles if he moves to a point on the boundary from which the two poles are equidistant.

Here, the taller pole $PS = 2h$ m and the shorter pole $P'S' = h$ m. As the sum of QS and QS' is always 50 m, the boundary of the field is elliptical with S and S′ at foci of the ellipse.

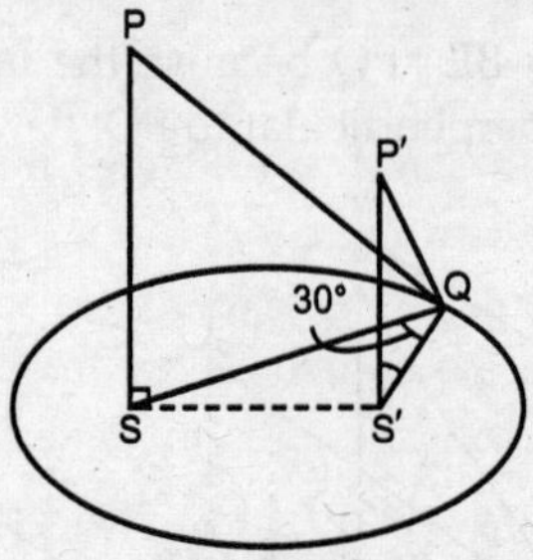

Here, $\angle PQS = 60°$, $\angle P'QS' = 45°$ (given)

∴ from the right-angled ΔPSQ, $\dfrac{PS}{SQ} = \tan 60°$

∴ $SQ = \dfrac{2h}{\sqrt3}$.

From the right-angled ΔP′S′Q, $\dfrac{P'S'}{S'Q} = \tan 45°$

∴ $S'Q = h$

∵ $SQ + S'Q = 50$, ∴ $\dfrac{2h}{\sqrt3} + h = 50$

or $h\cdot\dfrac{2+\sqrt3}{\sqrt3} = 50$

or $h = \dfrac{50\sqrt3}{2+\sqrt3} = \dfrac{50\sqrt3(2-\sqrt3)}{4-3} = 50(2\sqrt3-3)$.

∴ the height of the shorter pole $= 50(2\sqrt3 - 3)$ m

and that of the longer pole $= 100(2\sqrt{3}-3)$ m.

From $\Delta SQS'$,

$$SS'^2 = QS^2 + QS'^2 - 2QS \cdot QS' \cos 30°$$

$$\therefore \quad SS'^2 = \left(\frac{2h}{\sqrt{3}}\right)^2 + h^2 - 2 \cdot \frac{2h}{\sqrt{3}} \cdot h \cdot \frac{\sqrt{3}}{2}$$

$$= h^2\left\{\frac{4}{3} + 1 - 2\right\} = h^2 \cdot \frac{1}{3}$$

$$\therefore \quad SS' = \frac{1}{\sqrt{3}}h = \frac{1}{\sqrt{3}} \cdot 50(2\sqrt{3}-3) = 50(2-\sqrt{3})$$

$\therefore$ the distance between the poles $= 50(2-\sqrt{3})$ m.

Now, from coordinate geometry, the equation of the boundary in the standard form is $\frac{x^2}{a^2} + \frac{y^2}{b^2} = 1$

where $SS' = 2ae$, $SQ + S'Q = 2a$ and $b^2 = a^2(1-e^2)$, b being the distance from the line SS' of that point on the boundary which is equidistant from S and S'.

Here $2a = 50$; $\therefore a = 25$

$SS' = 2ae = 50e = 50(2-\sqrt{3})$; $\therefore e = 2-\sqrt{3}$

$\therefore b^2 = a^2(1-e^2)$

$= 25^2 \cdot \{1-(2-\sqrt{3})^2\} = 25^2 . \sqrt{3}(\sqrt{3}-1)^2$

$\therefore b = 25(\sqrt{3}-1).\ 3^{1/4}$.

Hence, the required distance $= 25(\sqrt{3}-1).\ 3^{1/4}$ m.

Note The reader must have the knowledge of ellipse from coordinate geometry to understand the solution of the above problem.

31. A vertical pole standing on horizontal ground is observed to bear an angle α east of north from a point P on the ground. The pole subtends an angle θ at P. Due to a storm the pole leans towards the east by an angle β. If the pole subtends an angle φ at P after the storm then prove that

$$\cos\phi\sqrt{1+\sin\alpha\sin\beta\cdot\sin 2\theta} = \cos\theta + \sin\alpha\sin\beta\sin\theta.$$

Let the vertical pole be AO. After the storm it takes the position OB.

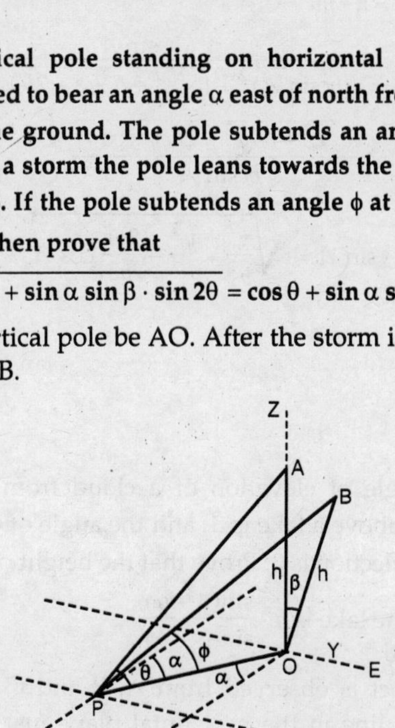

Let $OA = OB = h$. So, from the right-angled ΔAOP, we get

$OP = h\cot\theta$.

Taking X and Y axes along the south and east directions at O and the vertical through O as Z axis, we get

$$B = (0, h\sin\beta, h\cos\beta)$$

and $P = (OP\cos\alpha, -OP\sin\alpha, 0)$

$$\therefore \quad \overrightarrow{OB} = h\sin\beta\,\vec{j} + h\cos\beta\,\vec{k}$$

$$\overrightarrow{OP} = h\cot\theta\cos\alpha\,\vec{i} - h\cot\theta\sin\alpha\,\vec{j}$$

$$\{\because OP = h\cot\theta\}$$

$$\therefore \quad \overrightarrow{BP} = \overrightarrow{OP} - \overrightarrow{OB}$$

$$= h\cot\theta\cdot\cos\alpha\,\vec{i} - h(\cot\theta\cdot\sin\alpha + \sin\beta)\,\vec{j} - h\cos\beta\,\vec{k}$$

$$\therefore \quad \cos\phi = \frac{\overrightarrow{OP}\cdot\overrightarrow{BP}}{|\overrightarrow{OP}|\,|\overrightarrow{BP}|}$$

$$= \frac{\left\{h^2\cot^2\theta\cdot\cos^2\alpha + h^2\cot\theta\sin\alpha \times (\cot\theta\cdot\sin\alpha + \sin\beta)\right\}}{\left[\{h^2\cot^2\theta\cos^2\alpha + h^2\cot^2\theta\sin^2\alpha\}^{1/2}\{h^2\cot^2\theta\cos^2\alpha + h^2(\cot\theta\sin\alpha + \sin\beta)^2 + h^2\cos^2\beta\}^{1/2}\right]}$$

$$= \frac{h^2\cot^2\theta + h^2\cot\theta\sin\alpha\cdot\sin\beta}{\left[h\cot\theta\cdot\{h^2\cot^2\theta + h^2\sin^2\beta + 2h^2\cot\theta\sin\alpha\cdot\sin\beta + h^2\cos^2\beta\}^{1/2}\right]}$$

$$= \frac{\cot\theta + \sin\alpha\cdot\sin\beta}{\sqrt{\cot^2\theta + \sin^2\beta + 2\cot\theta\cdot\sin\alpha\sin\beta + \cos^2\beta}}$$

$$= \frac{\cos\theta + \sin\alpha\sin\beta\sin\theta}{\left\{\cos^2\theta + \sin^2\beta\cdot\sin^2\theta + 2\cos\theta\sin\theta\sin\alpha\sin\beta + \cos^2\beta\cdot\sin^2\theta\right\}^{\frac{1}{2}}}$$

$$= \frac{\cos\theta + \sin\alpha\cdot\sin\beta\cdot\sin\theta}{\sqrt{1+\sin\alpha\cdot\sin\beta\cdot\sin 2\theta}}$$

$$\therefore \quad \cos\phi\cdot\sqrt{1+\sin\alpha\sin\beta\sin 2\theta}$$

$$= \cos\theta + \sin\alpha\sin\beta\sin\theta.$$

Note In case of triangles not in any vertical plane, we have to use vectors to estimate angles and heights conveniently. The reader must know algebra of vectors.

32. In order to find the dip of a stratum of coal below the surface of the ground, vertical borings are made from the angular points A, B, C of a triangle ABC which is in a horizontal plane. The depths of the stratum at these points are found to be $x, x+y$ and $x+z$ respectively. Show that the dip θ of the stratum, which is assumed to be a plane, is given by

$$\tan\theta\cdot\sin A=\sqrt{\frac{y^2}{c^2}+\frac{z^2}{b^2}-\frac{2yz}{bc}\cos A}$$

where b and c are the lengths of the sides CA and AB respectively and A is the angle between CA and AB.

Here the stratum is the plane A′PQ where the points A′, P and Q are vertically below A, B and C respectively. From the question,

$$AA'=x,\ BP=x+y,\ CQ=x+z.$$

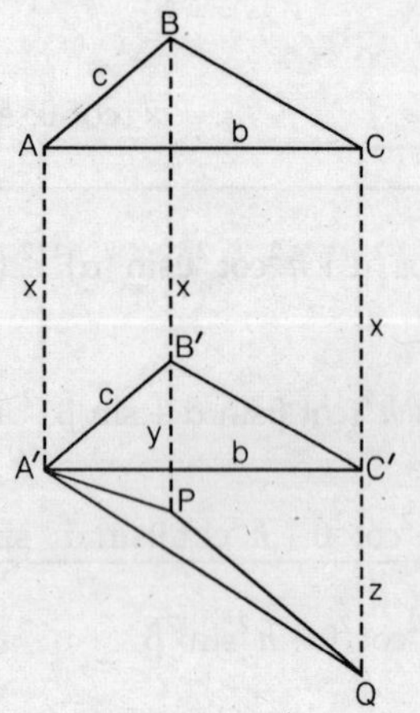

Let A′, B′, C′ be the pojections of A, B, C respectively on the horizontal plane through A′. Then

$$AA'=BB'=CC'=x,\ B'P=y,\ C'Q=z.$$

The dip θ of the stratum A′PQ is the angle between the planes A′PQ and A′B′C′. Also, the angle between the two planes is the angle between the normals to the two planes. For finding the angle we use vector method.

Now, take the plane A′B′C′ as the x-y plane with A′C′ as the x-axis, A′ as the origin and z-axis along A′A. Then

$$C'=(b,0,0),\ B'=(c\cos A, c\sin A, 0)$$

$$Q=(b,0,-z),\ P=(c\cos A, c\sin A, -y)$$

$$\therefore\quad \vec{A'C'}\times\vec{A'B'}=b\,\vec{i}\times(c\cos A\,\vec{i}+c\sin A\,\vec{j})$$

$$=bc\sin A\,\vec{k}$$

$$\vec{A'Q}\times\vec{A'P}=(b\,\vec{i}-z\,\vec{k})\times(c\cos A\,\vec{i}+c\sin A\,\vec{j}-y\,\vec{k})$$

$$=cz\sin A\,\vec{i}+(by-cz\cos A)\,\vec{j}+bc\sin A\,\vec{k}$$

$\therefore$ the unit vector perpendicular to the plane A′B′C′,

$$\hat{n}=\frac{\vec{A'C'}\times\vec{A'B'}}{|\vec{A'C'}\times\vec{A'B'}|}=\frac{bc\sin A\,\vec{k}}{bc\sin A}=\vec{k}.$$

The unit vector perpendicular to the stratum A′PQ,

$$\hat{m}=\frac{\vec{A'Q}\times\vec{A'P}}{|\vec{A'Q}\times\vec{A'P}|}$$

$$=\frac{cz\sin A\,\vec{i}+(by-cz\cos A)\,\vec{j}+bc\sin A\,\vec{k}}{\sqrt{c^2z^2\sin^2A+(by-cz\cos A)^2+b^2c^2\sin^2A}}$$

$$\therefore\quad \cos\theta=\hat{n}\cdot\hat{m}$$

$$=\frac{bc\sin A}{\sqrt{(bc\sin A)^2+\{c^2z^2+b^2y^2-2bcyz\cos A\}}}$$

$$\therefore\quad \tan\theta=\frac{\sqrt{c^2z^2+b^2y^2-2bcyz\cos A}}{bc\sin A}$$

$$\text{or}\quad \tan\theta\cdot\sin A=\sqrt{\frac{z^2}{b^2}+\frac{y^2}{c^2}-\frac{2yz}{bc}\cos A}.$$

Exercises

1. A vertical pole is divided in the ratio 1 : 9 by a mark on it, the lower part being shorter than the upper part. If the two parts subtend equal angles at a distance of 20 m from the base of the pole, find the height of the pole.

2. The angle of elevation of the top of a mountain from any point of a horizontal plane through the base of the mountain is 45°. After ascending 1500 m towards the mountain up a slope inclined at an angle of 15°, the elevation changes to 75°. Find the height of the mountain.

3. The angle of elevation of a cloud from a point h metres above a lake is α, and the angle of depression of its reflection is β. Prove that the height of the cloud above the lake is $h\cdot\dfrac{\sin(\beta+\alpha)}{\sin(\beta-\alpha)}$.

4. An object is observed from A, B and C lying in a straight line in the horizontal plane just below the object. The angular elevation at B is twice that at A, and at C three times that at A. If $AB=a,\ BC=b$ then show that the height of the object is $\dfrac{a}{2b}\{(a+b)(3b-a)\}^{1/2}$.

5. The angles of elevation of a bird flying in a horizontal straight line, from a fixed point at four successive observations are α, β, γ and δ, the observations being taken at equal intervals of time. If the speed of the bird is uniform, and the observer and the line of motion of the bird are in the same vertical plane, show that $\cot^2\alpha - \cot^2\delta = 3(\cot^2\beta - \cot^2\gamma)$.

6. AB is a vertical pole. The end A is on the level ground, C is the middle point of AB. P is a point on the level ground. The portion CB subtends an angle β at P. If $AP = nAB$, show that $n = (2n^2 + 1)\tan\beta$.

7. A spherical balloon whose radius is r, subtends at an observer's eye an angle α, where the angular elevation of its centre is β. Determine the height of the balloon.

8. A person standing on the bank of a river observes that the angle subtended by a tree on the opposite bank is 60°. When he retires 40 m from the bank perpendicular to it, he finds the angle to be 30°. Find the height of the tree and the breadth of the river.

9. The angle of elevation of the top of a tower at a point P due east of the tower is α while at a point Q due west of the tower it is $\beta(\alpha > \beta)$. If the angle of elevation of the tower at a point midway between P and Q be θ then prove that

$$\cot(\alpha - \beta) = \tan\theta + \cos(\alpha + \beta).\operatorname{cosec}(\alpha - \beta).$$

10. Two stations due south of a tower, on the same level with its base, which leans towards the north, are at distances a and $b(b > a)$ from the base of the tower. If α and β be the elevations of the top of the tower from these stations, show that the cotangent of the inclination of the tower with the horizontal is $\dfrac{b\cot\alpha - a\cot\beta}{b-a}$.

11. A man notices two objects in a straight line due west of him. After walking a distance c due north he observes that the objects subtend an angle α at his eye and after walking a further distance c due north, an angle β. Show that the distance between the objects is $\dfrac{3c}{2\cot\beta - \cot\alpha}$.

12. From a fixed station of observation, at an instant the bearing of a ship moving in a line is α west of north. Ten minutes later the ship is due north and after a further interval of ten minutes the bearing of the ship is β east of north. Show that the course of the ship is θ east of north where

$$\tan\theta = \frac{2\sin\alpha\sin\beta}{\sin(\alpha - \beta)}.$$

13. A man observes that when he has walked c metres up an inclination, the angular depression of an object in a horizontal plane through the foot of the slope is α, and also observes that when he has walked a further distance of c metres up the slope, the depression is β. Prove that the inclination of the slope to the horizontal is $\cot^{-1}(2\cot\beta - \cot\alpha)$.

14. A vertical tree stands on a hillside that makes an angle α with the horizontal. From a point directly up the hill from the tree, the angle of elevation of the tree-top is β. From a point m cm further up the hill the angle of depression of the tree-top is γ. If the tree is h cm tall, express h in terms of α, β, γ.

15. A man walking towards a building on which a flagstaff is fixed, observes the angle subtended by the flagstaff to be the greatest when he is at a distance d from the building. If θ be the greatest angle observed, find the length of the flagstaff and the height of the building.

16. A man walks along a straight road and observes that the greatest angle subtended by two objects is α. From the point where this angle is subtended, he walks a distance c along the road and finds that the two objects are now in a straight line which makes an angle β with the road. Prove that the distance between the objects is $\dfrac{2c\sin\alpha \cdot \sin\beta}{\cos\alpha + \cos\beta}$.

17. A flagstaff standing on the top of a tower of height b subtends the same angle at all points on the horizontal plane through the foot of the tower which are at the distance a or b from the tower. Prove that the height of the flagstaff is $|a - b|$.

18. A man standing on the straight shore of a sea observes two stationary boats in the same direction, making an angle θ with the shore. He then walks a distance x along the shore and finds the boats subtend an angle θ at his eye. Again on walking a further distance y he finds they again subtend an angle θ at his eye. Show that the distance between the boats is $\left(x + \dfrac{y}{2}\right)\sec\theta - \dfrac{2x(x+y)}{2x+y}\cos\theta$.

(Neglect the height of the man's eye above the shore.)

19. There are two inaccessible points P, Q and two stations S and T of observations, at each of which PQ is found to subtend the same angle α. PT subtends an angle β at S and QS subtends an angle γ at T. Prove that $PQ : ST = \sin\alpha : \sin(\beta + \gamma - \alpha)$.

20. A flagstaff on a tower stands vertically. A man on the horizontal through the foot of the tower observes the angles subtended at a point on the horizontal by the flagstaff and the tower to be α and β respectively. He then walks a distance a towards the tower and finds that the flagstaff subtends the same angle as before. Prove that length of the flagstaff is

$$a\sin\alpha \cdot \sec(\alpha + 2\beta).$$

21. AB and CD are two houses on two sides of a street. The angle subtended by CD at the foot B of AB is an angle whose tangent is 3. From two windows P, Q of the house AB, angles subtended by CD are found to be equal and their tangents are each equal to −3. If $CD = 30$ m, find the heights PB and QB of the windows.

22. The angles of elevation of the top of a distant hill in the forest as seen from three consecutive kilometre-stones on a straight horizontal road are 30°, 45° and 60°. Find the height of the hill.

23. A, B, C and D are four consecutive metre-stones on a straight road. A pole 4 m high, away from the road subtends equal angles of 30° at A and D. Find the angles that the pole subtends at B and C.

24. A, B, C are three consecutive kilometre-stones on a straight road from each of which a distant spire is visible. The spire is observed to bear north-east at A, east at B and 60° east of south at C. Show that the shortest distance of the spire from the road is $\frac{7+5\sqrt{3}}{13}$ km.

25. A man due south of a vertical tower is at A and finds the altitude of the tower to be 60°. When he walks to B due west of A the altitude is found to be 45°, and again at C due west of B he observes it to be 30°. Show that B is midway between A and C.

26. A man observes a balloon in the east at an elevation of 60° moving towards the north-west direction. On walking 400 metres towards the north he finds the balloon just above him. If the balloon always remains in the same horizontal plane, find the height of the balloon above him.

27. The angle of elevation of the top of a vertical tower AB from a station P due south of it and on the same level with the base A of the tower, is θ. From another station Q due west of the former the elevation is ϕ. Prove that the height of the tower is

$$\frac{PQ \sin\theta \sin\phi}{\sqrt{\sin(\theta+\phi)\cdot\sin(\theta-\phi)}}.$$

28. The angular elevation of a tower at a place A due south of it is 30° and at a place due west of A and at a distance a from A the elevation is 18°. Prove that the height of the tower is $\frac{a}{\sqrt{2(1+\sqrt{5})}}$.

29. A person on the bank of a canal observes that the angle subtended by a tree just in front on the opposite bank is θ. If he moves a distance a from the bank at an angle of 45° with the bank, he finds the angle to be 45° again. If the breadth of the river be b, prove that θ is $\tan^{-1}(a^2+b^2+\sqrt{2}ab)^{1/2}/b$.

30. A triangular field has its sides $30\sqrt{6}$, 60 and $30(\sqrt{3}+1)$ metres long. A vertical pole stands at the orthocentre of the triangle. The angle of elevation of the top of the pole at the vertex opposite to the largest side is 75°. Find the height of the pole.

31. A man standing on the bank of a river finds the angle of elevation of a vertical pillar in the same bank to be α. After walking perpendicularly to the bank through a distance a metres he finds the angle of elevation of the pillar to be β. Find the height of the pillar.

32. A man at the top of a hill observes three vertical towers of heights a, b and c on the horizontal plane below the hill which subtend equal angles at his eye. The angles of depression of their feet are α, β and γ respectively. Prove that

$$\frac{\sin(\beta-\gamma)}{a\sin\alpha}+\frac{\sin(\gamma-\alpha)}{b\sin\beta}+\frac{\sin(\alpha-\beta)}{c\sin\gamma}=0.$$

33. A vertical pole stands at a point O on the horizontal ground. A and B are points on the ground, d metres apart. The pole subtends angles α and β at A and B respectively. AB subtends an angle γ at O. Find the height of the pole.

34. A vertical tower is due north of A and north-west of B. B is due east of A at a distance 10 m. The angles of elevation of the tower as found from A and B are complementary to each other. Find the height of the tower.

35. A 10-m-high pole stands vertically at the centre of a horizontal equilateral triangle, each side of which subtends an angle 60° at the top of the pole. Find the side of the triangle.

36. Two vertical poles of heights a and b subtend the same angle α at a point on the line joining their feet. If they subtend angles β and γ at any point in the horizontal plane at which the line joining their feet subtends a right angle, prove that

$$\cot^2\alpha = \frac{a^2\cot^2\beta + b^2\cot^2\gamma}{(a+b)^2}.$$

37. A pole stands vertically at the vertex of the right angle of a triangular field in the shape of a right-angled triangle. If the length of the pole is equal to the longest side of the field then prove that the product of the tangents of the angles of elevation of the top of the pole as seen from the two ends of the longest side, cannot be less than 2.

38. A tower stands in a field whose shape is that of an equilateral triangle, each side being 80 metres. It subtends angles at the three corners whose tangents are respectively $\sqrt{3}+1$, $\sqrt{2}$ and $\sqrt{2}$. Find the height of the tower.

39. A tower stands upright at a point inside a rectangular field. The angles of elevation of the top of the tower from the four corners of the field taken in order are α, β, γ, δ. Prove that

$$\cot^2\alpha + \cot^2\gamma = \cot^2\beta + \cot^2\delta.$$

40. A tripod has three legs each 5 m long and the feet are planted in the ground at points 3 m apart. Find the height of the top of the tripod above the ground.

41. A tower PQ leans towards the east making an angle β with the vertical. At the point O on the ground, the base point P of the tower is found to bear north and at a distance a. If the angle of elevation of Q at O is found to be α then prove that the length of the tower is

$$\frac{a\sin\alpha}{\sqrt{\cos^2\alpha - \sin^2\beta}}.$$

42. A vertical wall touches a circular garden at A; there is a source of light at the centre of the garden. A man starts walking from A along the boundary of the garden. When he reaches B such that arc $AB < \frac{1}{8}$th of the length of the boundary, he finds that the length of his shadow on the wall is $\frac{4}{3}$ times his height. How many times of his height will the shadow on the wall become when he reaches C on the boundary such that $AB = BC$?

43. A lamp-post stands by the side of a straight road, the nearest point A on the road from the lamp-post being at a distance a. A man of the height b is walking along the road. When he is at B the length of his shadow is $\frac{3}{2}$ times the length of the shadow when he is at C, the middle point of AB. Find the height of the lamp-post if the length of the shadow when he is at C is a.

44. In a building there is a big window in the shape of a rectangle surmounted by a semicircle. A man on the ground d away, midway in front of the window, finds that the top in the middle of the rectangular portion has the angle of elevation 45° while the topmost point of the window has the elevation of 60°. If the angle of elevation of the lowest corner of the window farthest from the man is 30° then find the area of the window.

45. A tower stands on the edge of a circular lake. The angles of elevation of its top from three points A, B and C on the edge of the pond are found to be α, β and γ respectively. If $\angle BAC = \angle ACB = \theta$ then show that $\cos\theta \cdot \cot\beta$ is the arithmetic mean of $\cot\alpha$ and $\cot\gamma$.

46. A square-based tower stands on a horizontal plane. From a point in this plane, from where three of its top corners are visible, their angular elevations are respectively 45°, 60° and 45°. Show that the ratio of the height of the tower to its breadth is $\sqrt{3}(\sqrt{5}+1) : 2\sqrt{2}$.

47. A right circular cylindrical tower of height h and radius r stands on a horizontal plane. Let A be a point in the horizontal plane and PQR be the semicircular edge of the top of the tower on the side of A such that Q is the point nearest to A. The angles of elevation of the point P and Q from A are 45° and 60° respectively. Show that $\frac{h}{r} = \frac{\sqrt{3}}{2}(\sqrt{5}+1)$.

48. A flagstaff stands in the middle of a square-based tower. A man on the ground, opposite the middle of a face of the tower and at a distance of 100 m from it, can just see the top of the flagstaff; on receding another 100 m, the tangents of angular elevations of the tops of the tower and the flagstaff are found to be $\frac{1}{2}$ and $\frac{5}{9}$ respectively. Find the dimensions of the tower and the height of the flagstaff, the ground being horizontal.

49. A wooden lamina in the shape of an isosceles triangle is kept vertically facing the sun with its vertex upwards. If the base of the triangle is $2a$, the height h and the altitude of the sun 30° then prove that the angle at the apex of the shadow at noon is

$$\tan^{-1}\frac{2ah\cdot\sqrt{3}}{3h^2 - a^2}.$$

50. A rectangular vertical wall runs east to west. The sun is 30° east of south and its altitude is 60°. Find the ratio of the area of the wall to the area of its shadow.

51. A pillar is east of south-east (ESE) of an observer. At noon the end of the shadow is north-east of the observer. The shadow is 80 m long and the elevation of the column at the observer's station is 45°. Find the height of the pillar.

52. The extremity of the shadow of a flagstaff of 6 m standing on the top of a regular pyramid on a square base, just reaches a side of the base. The extremity of the shadow is at distances 56 m and 8 m from the two ends of the side. The height of the pyramid is 34 m. Find the altitude of the sun.

53. A curve in the shape of a quadrant of a circle, has n posts at equal distances along the curve with one post at each end. A man stationed at a point on one of the extreme radii produced sees the pth and the qth posts from the end nearest to him in a straight line. Show that the radius of the curve is

$$\frac{b}{2}\cos(p+q)\phi \cdot \operatorname{cosec} p\phi \cdot \operatorname{cosec} q\phi,$$

where $\phi = \frac{\pi}{4(n-1)}$ and b is the distance of the man from the nearest end of the curve.

54. A field is in the shape of a parabola bounded by a straight side of length 40 m. C is the vertex of the parabola and it is equidistant from the two ends A and B of the straight side. A vertical pillar PQ stands at the middle point Q of the straight side. If the angles of elevation of the top of the pillar from A and C be respectively α and β then find the angle of elevation of the pillar from the point D on the curved boundary such that $\angle DQA = 45°$. Also, find the distance of the pillar from the vertex.

55. A dome standing on the ground is in the shape of a segment of a sphere. The dome meets the horizontal ground everywhere at an angle 60°. From a point a metres away from the nearest end of the dome, the elevation of the highest visible point is 45°. Find the height of the dome.

56. A hill has a circular base on the horizontal ground and it is in the shape of a part of a sphere. At two points on the horizontal ground, at distances $(3\sqrt{3} - 2\sqrt{6})$ m and $(5\sqrt{2} - 2\sqrt{6} - 1)$ m from the base of the hill, the angular elevations of the highest visible points on the hill are 60° and 45° respectively. Prove that the height of the hill is 4 m.

57. The side of a hill is a plane inclined at α to the horizon. There is a straight road on the hillside starting from the base and making an angle β with the line of the greatest slope of the hillside. Show that the inclination of the road with the horizontal is $\sin^{-1}(\sin\alpha \cdot \cos\beta)$.

Objective Questions

Fill in the blanks.

58. At a distance a from the foot of a tower AB of height b, a flagstaff BC and the tower AB subtend equal angles. Then the height of the flagstaff is ______ .

59 From the top of a house 10 m high by the side of a road, the angle of depression of the base and the angle of elevation of the top of another house 20 m high just on the opposite side of the road are found to be complementary. Then the breadth of the road is ______ .

60. Two posts are 120 m apart, and the height of one is double that of the other. From the middle point of the line joining their feet, an observer finds the angular elevations of their tops to be complementary. The height of the shorter post is ______ .

61. An electric pole stands at the vertex A of the equilateral triangular field and subtends an angle of 60° at either of the other two vertices. If the altitude of the triangular field be 50 m then the height of the pole is ______ .

62. Two nonparallel straight roads are boundaries of a field on two sides. Two persons are walking along the roads, one on each road. They find the angles of elevation of a vertical tower on the field to be 30° and 45° respectively when they are nearest to the tower, and they are at distances a and b respectively from the junction of the two roads. Then the height of the tower is ______ .

63. The angle of elevation of an aeroplane and the angle of depression of its reflection in a lake just below are found to be 45° and 75° respectively from the top of a cliff 200 m above the lake. The height of the aeroplane above the lake is ______ .

64. A and B are two points, a metres apart on the ground. A balloon falling down in a straight line inclined to the horizontal passes over the two points A and B. When above A, its altitude as seen from B is 60° and when above B, its altitude as seen from A is 30°. The distance of the point C, where the balloon touches the ground, from the point A is ______ .

Answers

1. $80\sqrt{5}$ m **2.** $750\sqrt{6}$ m

7. $r\operatorname{cosec}\dfrac{\alpha}{2} \cdot \sin\beta$ **8.** $20\sqrt{3}$ m, 20 m

14. $\dfrac{m\sin(\alpha+\beta)\sin(\alpha-\gamma)}{\cos\gamma\sin(\alpha+\beta) - \cos\beta\sin(\alpha-\gamma)}$

15. $2d\tan\theta,\ d\tan\left(\dfrac{\pi}{4} - \dfrac{\theta}{2}\right)$ **21.** 10 m, 20 m

22. $\sqrt{\dfrac{3}{2}}$ km **23.** $\cot^{-1}\sqrt{\dfrac{23}{8}}$ each

26. $400\sqrt{3}$ m **30.** $30(\sqrt{3}+1)$ m

31. $\dfrac{a\sin\alpha \cdot \sin\beta}{\sqrt{\sin^2\alpha - \sin^2\beta}}$

33. $\dfrac{d}{\sqrt{\cot^2\alpha + \cot^2\beta - 2\cot\alpha\cot\beta\cos\gamma}}$ metres

34. $10\sqrt[4]{2}$ m **35.** $5\sqrt{6}$ m

38. 80 m **40.** $\sqrt{22}$ m

42. 8 times **43.** $\left(1 + \dfrac{2\sqrt{3}}{\sqrt{7}}\right)b$

44. $\left\{\pi(2-\sqrt{3}) + 2(\sqrt{3}-1)\left(1 - \sqrt{\dfrac{5-2\sqrt{3}}{3}}\right)\right\}d^2$

48. height of the tower = 100 m,
side of top of the tower = 50 m,
height of the flagstaff = 25 m

50. 2 : 1 **51.** 61.22 m **52.** 45°

54. $\cot^{-1}\left\{\frac{\sqrt{4\tan^2\alpha+\tan^2\beta}-\tan\beta}{\sqrt{2}\tan^2\alpha}\right\}$, $20\tan\alpha\cdot\cot\beta$

55. $\frac{a}{2}\{4\sqrt{2}+2\sqrt{6}+3\sqrt{3}+5\}$

58. $\frac{b(a^2+b^2)}{a^2-b^2}$ **59.** 10 m

60. $30\sqrt{2}$ m **61.** 100 m

62. $\sqrt{\frac{b^2-a^2}{2}}$ **63.** $200\sqrt{3}$ m

64. $\frac{3a}{2}$ m

□

Chapter Test

Time: 120 minutes

1. In the following incomplete sentences, fill in the blanks such that the resulting sentences become true.

(a) A pole on the ground leans at 60° with the vertical. At a point a metres away from the base of the pole on the ground, two halves of the pole subtend the same angle. If the pole and the point are in the same vertical plane then the length of the pole is ______ .

(b) A man on a boat moving on the river, perpendicularly to a horizonal bridge across the river, finds himself just below a man moving along the bridge, the height of the bridge above the man being c metres. If both the men have the same speed c m/min then the two men will be ______ m apart after 2 minutes.

(c) A vertical tower standing at the corner of a rectangular field subtends angles of 60° and 45° at the two nearer corners. The angle that the tower subtends at the farthest corner is ______ .

(d) A circular ring of radius 5 cm is suspended horizontally from a point, 12 cm vertically above the centre, by 6 strings attached to its circumference at equal intervals. Then the cosine of the angle between two consecutive strings is ______ .

2. A railway flyover in the shape of a parabola, runs from north to south. A man 2 m tall walking vertically on the flyover from north to south, is being watched from a point A on the ground on the north of the flyover. It is found that the man disappears from sight when the elevation of the top of his head is 30°. If the span of the flyover is 24 m and its height is 6 m, the highest point being just in the middle, find the distance of the point A from the flyover and the height of the man above the ground when he is about to disappear.

3. A vertical tower on the bank of a river is being observed by a man 2 m tall, standing on the opposite bank. He finds that the angle of elevation of the top of the tower is 45°. After walking a distance of 10 m along the bank he finds that the angle of elevation of the tower is 30°. If the angle of elevation of the top of the tower from a point on the bank midway between the two positions of the man be 45° then find the height of the tower.

4. AP and AQ are the lines of the greatest slopes of two hills on opposite sides of A whose inclinations with the horizontal through A are 30° and 60° respectively. A man is at A and an elephant is at B on the incline AQ whose height above the horizontal level through A is a m. The man starts climbing up the incline AP at a uniform speed (in metres/minute) and the elephant starts climbing down the incline AQ with uniform speed. After 1 minute the man finds that the angle of elevation of the elephant is 30° while after 2 minutes he finds it to be θ where $\tan\theta = \dfrac{6\sqrt{3}-3}{18+19\sqrt{3}}$. Find the speed of the man and that of the elephant, and the distance between the man and the elephant after 2 minutes.

5. A flagstaff stands vertically at the centre of a square-shaped roof of a building of height 10 m. The sides of the roof are 8 m long. A is a point on the ground midway in front of a side of the house at a distance 4 m from the house. A man at A can see $\frac{1}{3}$ rd of the flagstaff. Had he been at a distance a metres more he could have seen $\frac{2}{3}$ rd of the flagstaff. Find a and the length of the flagstaff.

6. A stick of length l is fixed at the bottom on the horizontal ground and it can turn about the fixed point so that the shadow changes in length on the same line in the ground. Find the length of the longest shadow when the altitude of the sun is α. Also find the altitude of the sun if the length of the longest shadow is $\frac{7}{2}l$.
7. An inclined pole stands at the centre O of a circular field, making an angle of 15° with the vertical. The pole leans towards the south and it subtends an angle of 60° at a point A on the boundary bearing south of O. B is a point on the boundary bearing south-west of O. Find the angle that the pole subtends at B.
8. Two persons of the same height h m start running from O along two roads meeting at O at an angle of 45°. After some time it is found that they subtend complementary angles at O. Prove that this cannot happen unless the distance between the two persons is greater or equal to $h \cdot \sqrt{2-\sqrt{2}}$.
9. A semielliptical arch bounded by an axis stands vertically on the level ground on the axis. There is a vertical pole by the side of the arch on one end. After climbing the pole to a certain height, a vertical flagstaff on the top of the arch is just visible to a boy. On further climbing the same height, the boy finds that the flagstaff is just totally visible. If the height of the arch is 4 metres then find the height of the flagstaff.

□

Answers

1. (a) $a\sqrt{3}$ m (b) $3c$ m (c) $\cot^{-1}\frac{2}{\sqrt{3}}$ (d) $\frac{313}{338}$
2. $4(2\sqrt{3}-3)$ m, 6 m respectively
3. $(4+\sqrt{33})$ m
4. $\frac{a(5+\sqrt{3})}{11}$ m/min, $\frac{a(7\sqrt{3}-9)}{33}$ m/min; $\frac{2a}{11\sqrt{3}}(127+54\sqrt{3})^{1/2}$ m
5. $a = 4$, 15 m
6. $\frac{l}{\sin\alpha}$, $\sin^{-1}\frac{2}{7}$
7. $\cos^{-1}\left\{\frac{1}{2^{7/4}}\cdot\frac{\sqrt{3}+4\sqrt{2}-3}{\sqrt{\sqrt{3}+5\sqrt{2}-3}}\right\}$
9. 1 m

□

Coordinate Geometry

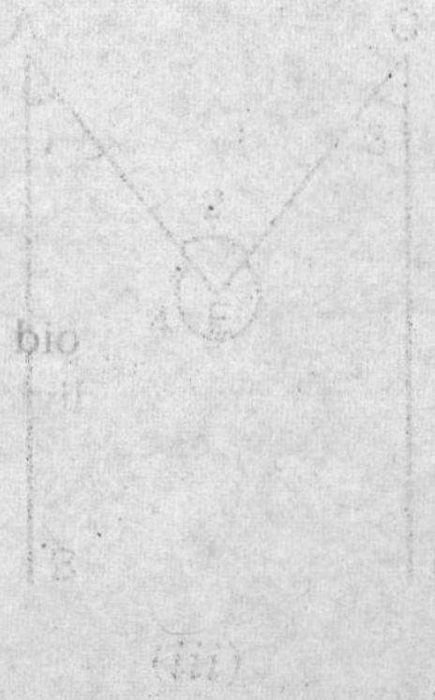

(iii)

(iv)

आकृति (i) में रेखाखंड हैं : AB, BC, CD, DL, DI DE, LK, LE, KE, EF, FG, GA, AH, HC, MN, NP, PO, OM, KI, IJ, JI.

इन कोणों को 1, 2, 3, 4, 5, 6, 7, 8, 9, 10, 11, 12, 13, 14, 15, 16, 17, 18, 19, 20, 21, 22, 23 से दर्शाया गया है।

1. Coordinates and Straight Lines

Recap of Facts and Formulae

1. Location of a point

(a) The point P = (x, y) in a plane is in

- the first quadrant $\Leftrightarrow$ $x > 0, y > 0$
- the second quadrant $\Leftrightarrow$ $x < 0, y > 0$
- the third quadrant $\Leftrightarrow$ $x < 0, y < 0$
- the fourth quadrant $\Leftrightarrow$ $x > 0, y < 0$

(b) The point P = (x, y) is on

- the x-axis $\Leftrightarrow$ $y = 0$
- the y-axis $\Leftrightarrow$ $x = 0$

(c) If the cartesian coordinates of a point are (x, y) and by taking the origin as pole and the x-axis as the initial line, the polar coordinates of the same point are (r, θ) then

- $x = r\cos\theta$, $y = r\sin\theta$
- $r = \sqrt{x^2 + y^2}$, $\tan\theta = \dfrac{y}{x}$

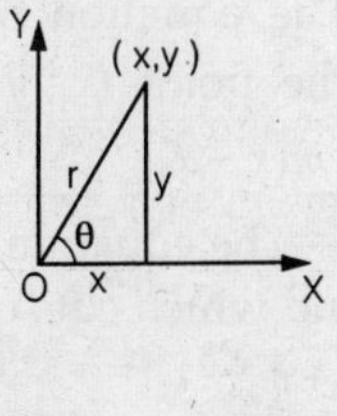

2. Distance formula

- If $P = (x_1, y_1)$ and $Q = (x_2, y_2)$ then the distance between P and Q = $PQ = \sqrt{(x_1 - x_2)^2 + (y_1 - y_2)^2}$.

3. Section formula

- If $A = (x_1, y_1)$, $B = (x_2, y_2)$ and $P = (x, y)$ divides the line segment AB in the ratio $\lambda : 1$ then

$$x = \frac{\lambda x_2 + x_1}{\lambda + 1}, \quad y = \frac{\lambda y_2 + y_1}{\lambda + 1}$$

where $\lambda > 0$ if P divides AB internally in the ratio $\lambda : 1$ and $\lambda < 0$ if P divides AB externally in the ratio $\lambda : 1$.

- The middle point M of the line segment joining $A = (x_1, y_1)$ and $B = (x_2, y_2)$ is $M = \left(\dfrac{x_1 + x_2}{2}, \dfrac{y_1 + y_2}{2}\right)$.

4. Centroid and incentre of a triangle

If the vertices of a ΔABC are

$A = (x_1, y_1)$, $B = (x_2, y_2)$ and $C = (x_3, y_3)$ then

- the centroid $G = \left(\dfrac{x_1 + x_2 + x_3}{3}, \dfrac{y_1 + y_2 + y_3}{3}\right)$

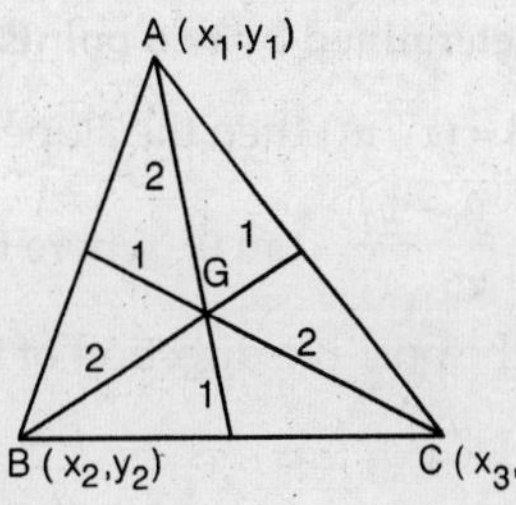

- the incentre

$$I = \left(\frac{ax_1 + bx_2 + cx_3}{a + b + c}, \frac{ay_1 + by_2 + cy_3}{a + b + c}\right)$$

where $BC = a$, $CA = b$, $AB = c$.

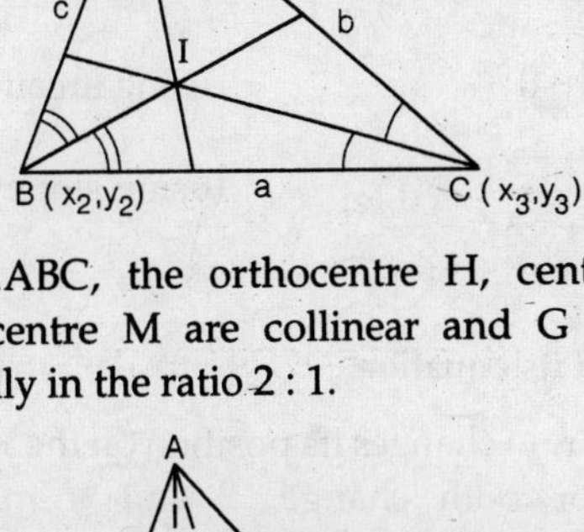

- In a ΔABC, the orthocentre H, centroid G and circumcentre M are collinear and G divides HM internally in the ratio 2 : 1.

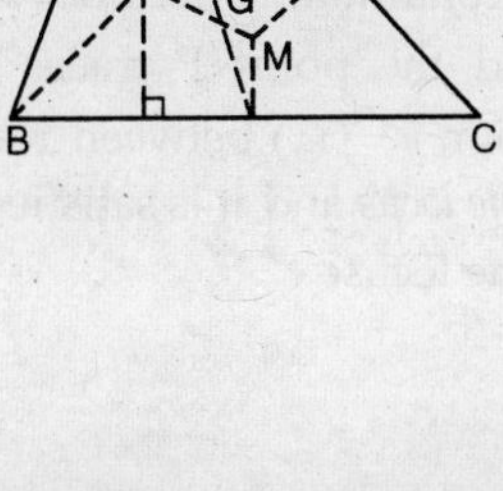

5. Areas of triangle and polygon

- If A = (x_1, y_1), B = (x_2, y_2) and C = (x_3, y_3) are the vertices of a triangle ABC then

$$\text{ar}(\Delta ABC) = \left| \frac{1}{2} \begin{vmatrix} x_1 & y_1 & 1 \\ x_2 & y_2 & 1 \\ x_3 & y_3 & 1 \end{vmatrix} \right|$$

$$= \left| \frac{1}{2} \left\{ \begin{vmatrix} x_1 & y_1 \\ x_2 & y_2 \end{vmatrix} + \begin{vmatrix} x_2 & y_2 \\ x_3 & y_3 \end{vmatrix} + \begin{vmatrix} x_3 & y_3 \\ x_1 & y_1 \end{vmatrix} \right\} \right|$$

$$= \left| \frac{1}{2} \{x_1(y_2 - y_3) + x_2(y_3 - y_1) + x_3(y_1 - y_2)\} \right|.$$

- If $A_r(x_r, y_r)$; $r = 1, 2, 3, \ldots, n$ be the vertices of a polygon, taken in order then the area of the polygon $A_1A_2A_3 \ldots A_n$

$$= \left| \frac{1}{2} \left\{ \begin{vmatrix} x_1 & y_1 \\ x_2 & y_2 \end{vmatrix} + \begin{vmatrix} x_2 & y_2 \\ x_3 & y_3 \end{vmatrix} + \begin{vmatrix} x_3 & y_3 \\ x_4 & y_4 \end{vmatrix} + \cdots + \begin{vmatrix} x_{n-1} & y_{n-1} \\ x_n & y_n \end{vmatrix} + \begin{vmatrix} x_n & y_n \\ x_1 & y_1 \end{vmatrix} \right\} \right|.$$

6. Slope of a line determined by two points

If A = (x_1, y_1) and B = (x_2, y_2) then the slope m of the line AB is given by $m = \dfrac{y_1 - y_2}{x_1 - x_2} = \tan\theta$, where θ is the angle made by AB with the positive direction of the x-axis.

7. Condition for collinearity

If A = (x_1, y_1), B = (x_2, y_2) and C = (x_3, y_3) are three points then they will be collinear if

- $AB \pm AC = BC$ (using distance formula)
- a real λ, positive or negative, can be obtained such that

$$x_2 = \frac{\lambda x_3 + x_1}{\lambda + 1}, \quad y_2 = \frac{\lambda y_3 + y_1}{\lambda + 1} \quad \text{(using section formula)}$$

- $\text{ar}(\Delta ABC) = 0$ (using area formula)
- $\dfrac{y_1 - y_2}{x_1 - x_2} = \dfrac{y_1 - y_3}{x_1 - x_3}$ (using slope formula).

8. Locus and its equation

- A point $P(x, y)$ changes its position on the x–y plane as x or y or both change. x and y may change independently or one depending on the other. When $P(x, y)$ changes its position under some rule (imposed by geometrical condition), y becomes a function of x, say $y = f(x)$ and the point P traces a locus. This functional relation $y = f(x)$ between x and y is called the *equation of the locus* and it is satisfied by all points (x, y) lying on the locus.

- In order to find the equation of a locus of a point P, take P = (α, β) in any position. Then obtain a relation between α, β using the condition under which the point P moves. Finally replace α, β, by x, y respectively in the relation to get the cartesian equation of the locus.

9. Parametric equations of a locus

- The rule by which a point $P(x, y)$ moves to trace the locus may be expressed by writing x as well as y as functions of a third variable (or arbitrary constant) θ.

For example:

$$\left.\begin{aligned} x &= \phi(\theta) \\ y &= \psi(\theta) \end{aligned}\right\}$$

Such equations of a locus give the parametric equations of the locus and any point on the locus can be taken as $(\phi(\theta), \psi(\theta))$.

- To obtain the cartesian equation of a locus whose parametric equations are $x = \phi(\theta)$, $y = \psi(\theta)$, eliminate θ from the relations $x = \phi(\theta)$, $y = \psi(\theta)$.

10. Equation of a straight line

- *Two-point form*—The equation of the straight line passing through the points (x_1, y_1) and (x_2, y_2) is

$$y - y_1 = \frac{y_1 - y_2}{x_1 - x_2}(x - x_1) \quad \text{or} \quad \begin{vmatrix} x & y & 1 \\ x_1 & y_1 & 1 \\ x_2 & y_2 & 1 \end{vmatrix} = 0.$$

- *Point-slope form*—The equation of the straight line passing through the point (x_1, y_1) and having the slope m is $y - y_1 = m(x - x_1)$.
- *Slope-intercept form*—The equation of the straight line whose slope is m and which cuts off an intercept c on the y-axis is $y = mx + c$.
- *Intercept form*—The equation of the straight line which cuts off intercepts a, b on the x and y axes respectively is $\dfrac{x}{a} + \dfrac{y}{b} = 1$.
- *Normal form*—The equation of the straight line for which the length of the normal from the origin to the line is p and the normal is inclined with the positive direction of the x-axis at an angle θ, is

$$x\cos\theta + y\sin\theta = p.$$

- *Parametric form (or distance form)*—The equation of the line passing through (x_1, y_1) and inclined at an angle θ with the positive direction of x-axis is

$$\frac{x - x_1}{\cos\theta} = \frac{y - y_1}{\sin\theta} (= r)$$

and any point on the line at a distance r from (x_1, y_1) is

$$(x_1 \pm r\cos\theta, y_1 \pm r\sin\theta).$$

- *General equation*—The general equation of a straight line is $ax + by + c = 0$ (a first degree equation in x, y), for which: the slope $m = \frac{-a}{b}$

  intercept on x-axis $= \frac{-c}{a}$ (obtained by putting $y = 0$ in $ax + by + c = 0$)

  intercept on y-axis $= \frac{-c}{b}$ (obtained by putting $x = 0$ in $ax + by + c = 0$).
- The equation of the x-axis is $y = 0$ and that of the y-axis is $x = 0$.

11. Angle between two lines

- If the acute angle between the straight lines $y = m_1x + c_1$ and $y = m_2x + c_2$ be ϕ then

$$\tan \phi = \left|\frac{m_1 - m_2}{1 + m_1m_2}\right|$$

  provided none of the lines is parallel to the y-axis.
- If the acute angle between the lines $y = mx + c$ and $x = k$ be ϕ then

$$\phi = \left|\frac{\pi}{2} - \theta\right| \quad \text{where } m = \tan \theta.$$

12. Conditions for perpendicularity and parallelism

Two straight lines $y = m_1x + c_1$ and $y = m_2x + c_2$ are

- parallel if $m_1 = m_2$
- perpendicular if $m_1m_2 = -1$, i.e., $m_2 = \frac{-1}{m_1}$.

13. Line parallel or perpendicular to a given line

Let a given line be $ax + by + c = 0$.

- Any line parallel to it has the equation $ax + by = k$ where k is an arbitrary constant whose value can be obtained by using an additional geometrical condition.
- Any line perpendicular to it has the equation $bx - ay = \lambda$ where λ is an arbitrary constant whose value can be obtained by using an additional geometrical condition.

14. Intersection of lines

- The point of intersection of the lines $a_1x + b_1y + c_1 = 0$ and $a_2x + b_2y + c_2 = 0$ is obtained by solving the equations simultaneously.
- The equation of any line passing through the intersection of the lines

  $L_1 \equiv a_1x + b_1y + c_1 = 0$ and $L_2 \equiv a_2x + b_2y + c_2 = 0$ is

  $L_1 + \lambda L_2 = 0$,

  i.e., $a_1x + b_1y + c_1 + \lambda(a_2x + b_2y + c_2) = 0$

  where λ is an arbitrary constant whose value can be obtained by using an additional geometrical condition.

  The equation $a_1x + b_1y + c_1 + \lambda(a_2x + b_2y + c_2) = 0$ where λ is a parameter, represents a family of straight lines passing through a fixed point which is the point of intersection of the fixed lines

$$a_1x + b_1y + c_1 = 0 \text{ and } a_2x + b_2y + c_2 = 0.$$

15. Concurrency of lines

The lines $L_1 \equiv a_1x + b_1y + c_1 = 0$

$L_2 \equiv a_2x + b_2y + c_2 = 0$

$L_3 \equiv a_3x + b_3y + c_3 = 0$

are concurrent lines if any one of the following holds:

- the point of intersection of $L_1 = 0$, $L_2 = 0$ satisfies the equation $L_3 = 0$.
- the three equations are consistent, i.e.,

$$\begin{vmatrix} a_1 & b_1 & c_1 \\ a_2 & b_2 & c_2 \\ a_3 & b_3 & c_3 \end{vmatrix} = 0.$$

- $L_1 + \lambda L_2 = 0$ is identical with $L_3 = 0$ for some real λ, i.e., $(a_1 + \lambda a_2)x + (b_1 + \lambda b_2)y + c_1 + \lambda c_2 = 0$ is identical with $a_3x + b_3y + c_3 = 0$, i.e.,

$$\frac{a_1 + \lambda a_2}{a_3} = \frac{b_1 + \lambda b_2}{b_3} = \frac{c_1 + \lambda c_2}{c_3}$$

  holds for some λ.
- if three nonzero constants λ, μ, ν can be found such that

  $\lambda L_1 + \mu L_2 + \nu L_3 \equiv 0$,

  i.e., $\lambda a_1 + \mu a_2 + \nu a_3 = 0$,

  $\lambda b_1 + \mu b_2 + \nu b_3 = 0$, $\lambda c_1 + \mu c_2 + \nu c_3 = 0$.

16. Location of points with respect to a given line

Let a given line be $L \equiv ax + by + c = 0$.

- The point $P(x_1, y_1)$ is on the above line if $L(x_1, y_1) = 0$, i.e., $ax_1 + by_1 + c = 0$.
- The points $P(x_1, y_1)$ and $Q(x_2, y_2)$ are on the same side of the above line if

  $L(x_1, y_1)$, i.e., $ax_1 + by_1 + c$ and

  $L(x_2, y_2)$, i.e., $ax_2 + by_2 + c$ are of the same sign.
- The points $P(x_1, y_1)$ and $Q(x_2, y_2)$ are on the opposite sides of the above line if

  $L(x_1, y_1)$ and $L(x_2, y_2)$ are of opposite signs.

- If $c > 0$ then $P(x_1, y_1)$ and the origin are on the same side of the line when $L(x_1, y_1) > 0$.
- If $c > 0$ then $P(x_1, y_1)$ and the origin are on the opposite sides of the line when $L(x_1, y_1) < 0$.

17. Length of the perpendicular from a point to a line

- The length of the perpendicular from the point $P(x_1, y_1)$ to the line $ax + by + c = 0$ is $\left|\dfrac{ax_1 + by_1 + c}{\sqrt{a^2 + b^2}}\right|$.

18. Equations of bisectors of angles between two lines

If $a_1x + b_1y + c_1 = 0$ and $a_2x + b_2y + c_2 = 0$ be two straight lines then

- the equation of bisectors of the angles between the two lines are

$$\frac{a_1x + b_1y + c_1}{\sqrt{a_1^2 + b_1^2}} = \pm \frac{a_2x + b_2y + c_2}{\sqrt{a_2^2 + b_2^2}}.$$

- if $c_1 > 0$, $c_2 > 0$ then the equation of the bisector of the angle containing the origin is

$$\frac{a_1x + b_1y + c_1}{\sqrt{a_1^2 + b_1^2}} = \frac{a_2x + b_2y + c_2}{\sqrt{a_2^2 + b_2^2}}$$

and the equation of the bisector of the angle not containing the origin is

$$\frac{a_1x + b_1y + c_1}{\sqrt{a_1^2 + b_1^2}} = -\frac{a_2x + b_2y + c_2}{\sqrt{a_2^2 + b_2^2}}.$$

- The bisector of the angle containing the origin is also the bisector of the acute angle between the lines $a_1x + b_1y + c_1 = 0$ and $a_2x + b_2y + c_2 = 0$ if (i) c_1, c_2 are of the same sign and (ii) $a_1a_2 + b_1b_2 < 0$.

Selected Solved Examples

1. If the coordinates of the middle points of the sides of a triangle are (1, 1), (2, –3) and (3, 4), find the vertices and the centroid of the triangle.

Let the vertices be $A(x_1, y_1)$, $B(x_2, y_2)$ and $C(x_3, y_3)$.

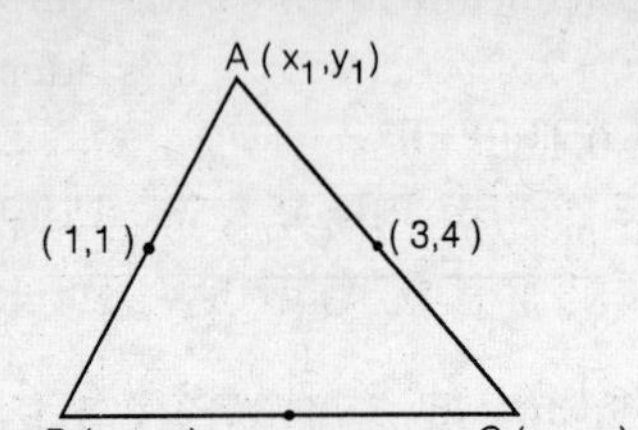

Then, $\dfrac{x_1 + x_2}{2} = 1$, $\dfrac{y_1 + y_2}{2} = 1$

$\dfrac{x_2 + x_3}{2} = 2$, $\dfrac{y_2 + y_3}{2} = -3$

$\dfrac{x_3 + x_1}{2} = 3$, $\dfrac{y_3 + y_1}{2} = 4$.

Adding, $x_1 + x_2 + x_3 = 6$ and $y_1 + y_2 + y_3 = 2$

$\therefore \quad x_1 + 4 = 6, x_2 + 6 = 6, x_3 + 2 = 6$

and $y_1 + (-6) = 2, y_2 + 8 = 2, y_3 + 2 = 2$

$\therefore \quad x_1 = 2, x_2 = 0, x_3 = 4$

$y_1 = 8, \; y_2 = -6, \; y_3 = 0$

$\therefore$ vertices are (2, 8), (0, – 6), (4, 0)

$\therefore$ centroid $= \left(\dfrac{2 + 0 + 4}{3}, \dfrac{8 - 6 + 0}{3}\right) = \left(2, \dfrac{2}{3}\right)$.

2. If $A = (at^2, 2at)$, $B = \left(\dfrac{a}{t^2}, -\dfrac{2a}{t}\right)$ and $S = (a, 0)$ then prove that $2a$ is the harmonic mean of SA and SB.

Here $SA^2 = (a - at^2)^2 + (0 - 2at)^2$

$= a^2\{(1 - t^2)^2 + 4t^2\}$

$\therefore \quad SA^2 = a^2(1 + t^2)^2$

$\therefore \quad SA = a(1 + t^2)$

and $SB^2 = \left(a - \dfrac{a}{t^2}\right)^2 + \left(0 + \dfrac{2a}{t}\right)^2$

$= a^2\left\{\left(1 - \dfrac{1}{t^2}\right)^2 + \dfrac{4}{t^2}\right\}$

$\therefore \quad SB^2 = a^2\left(1 + \dfrac{1}{t^2}\right)^2; \quad \therefore \quad SB = a\left(1 + \dfrac{1}{t^2}\right)$

$\therefore \quad \dfrac{1}{SA} + \dfrac{1}{SB} = \dfrac{1}{a(1 + t^2)} + \dfrac{1}{a\left(1 + \dfrac{1}{t^2}\right)}$

$= \dfrac{1}{a}\left\{\dfrac{1}{1 + t^2} + \dfrac{t^2}{t^2 + 1}\right\}$

$= \dfrac{1}{a} \cdot \dfrac{1 + t^2}{1 + t^2} = \dfrac{1}{a} = \dfrac{2}{2a}$

$\therefore$ $2a$ is the harmonic mean of SA and SB.

3. Find t if the area of the pentagon ABCDE be $\dfrac{45}{2}$ where $A = (1, 3)$, $B = (-2, 5)$, $C = (-3, -1)$, $D = (0, -2)$ and $E = (2, t)$.

Here the points are given in order.

The area of the pentagon ABCDE

$$= \left|\frac{1}{2}\left\{\begin{vmatrix}1 & 3\\ -2 & 5\end{vmatrix} + \begin{vmatrix}-2 & 5\\ -3 & -1\end{vmatrix} + \begin{vmatrix}-3 & -1\\ 0 & -2\end{vmatrix} + \begin{vmatrix}0 & -2\\ 2 & t\end{vmatrix} + \begin{vmatrix}2 & t\\ 1 & 3\end{vmatrix}\right\}\right|$$

$$\therefore \quad \frac{45}{2} = \left|\frac{1}{2}\{(5+6) + (2+15) + (6-0) + (0+4) + (6-t)\}\right|$$

$$= \left|\frac{1}{2}(11+17+6+4+6-t)\right| = \frac{1}{2}|44-t|$$

$$\therefore \quad 44 - t = \pm 45; \quad \therefore \quad t = 44 \mp 45 = -1, 99.$$

4. Show that the equation of the locus of a point, which moves so that the sum of its distances from two given points $(ae, 0)$ and $(-ae, 0)$ is equal to $2a$, is $\frac{x^2}{a^2} + \frac{y^2}{b^2} = 1$ where $b^2 = a^2(1 - e^2)$.

Let $P(\alpha, \beta)$ be any point on the locus and $S = (ae, 0)$, $S' = (-ae, 0)$.

The point P moves such that $PS + PS' = 2a$

$$\therefore \quad \sqrt{(\alpha - ae)^2 + (\beta - 0)^2} + \sqrt{(\alpha + ae)^2 + (\beta - 0)^2} = 2a \qquad \ldots (1)$$

Multiplying both sides by

$\sqrt{(\alpha - ae)^2 + \beta^2} - \sqrt{(\alpha + ae)^2 + \beta^2}$, we get

$$\{(\alpha - ae)^2 + \beta^2\} - \{(\alpha + ae)^2 + \beta^2\}$$

$$= 2a\{\sqrt{(\alpha - ae)^2 + \beta^2} - \sqrt{(\alpha + ae)^2 + \beta^2}\}$$

$$\therefore \quad -4ae\alpha = 2a\{\sqrt{(\alpha - ae)^2 + \beta^2} - \sqrt{(\alpha + ae)^2 + \beta^2}\}$$

$$\therefore \quad \sqrt{(\alpha - ae)^2 + \beta^2} - \sqrt{(\alpha + ae)^2 + \beta^2} = -2e\alpha \qquad \ldots (2)$$

$$(1) + (2) \quad \Rightarrow \quad 2\sqrt{(\alpha - ae)^2 + \beta^2} = 2a - 2e\alpha$$

$$\text{or} \quad (\alpha - ae)^2 + \beta^2 = (a - e\alpha)^2$$

$$\text{or} \quad \alpha^2 - 2ae\alpha + a^2e^2 + \beta^2 = a^2 + e^2\alpha^2 - 2ae\alpha$$

$$\text{or} \quad (1 - e^2)\alpha^2 + \beta^2 = a^2(1 - e^2)$$

$$\text{or} \quad \frac{\alpha^2}{a^2} + \frac{\beta^2}{a^2(1 - e^2)} = 1$$

$\therefore$ the equation of the locus is $\frac{x^2}{a^2} + \frac{y^2}{a^2(1 - e^2)} = 1$

$$\text{or} \quad \frac{x^2}{a^2} + \frac{y^2}{b^2} = 1.$$

5. ABC is a variable triangle with the fixed vertex C(1, 2) and A, B having the coordinates $(\cos t, \sin t)$, $(\sin t, -\cos t)$ respectively where t is a parameter. Find the locus of the centroid of the ΔABC.

Let G be the centroid in any position.

Then $(\alpha, \beta) = \left(\frac{1 + \cos t + \sin t}{3}, \frac{2 + \sin t - \cos t}{3}\right)$

$$\therefore \quad \alpha = \frac{1 + \cos t + \sin t}{3}, \ \beta = \frac{2 + \sin t - \cos t}{3}$$

$$\text{or} \quad 3\alpha - 1 = \cos t + \sin t, \ 3\beta - 2 = \sin t - \cos t$$

Squaring and adding,

$$(3\alpha - 1)^2 + (3\beta - 2)^2$$

$$= (\cos t + \sin t)^2 + (\sin t - \cos t)^2$$

$$= 2(\cos^2 t + \sin^2 t) = 2$$

$\therefore$ the equation of the locus of the centroid is

$$(3x - 1)^2 + (3y - 2)^2 = 2$$

$$\text{or} \quad 9(x^2 + y^2) - 6x - 12y + 3 = 0$$

$$\therefore \quad 3(x^2 + y^2) - 2x - 4y + 1 = 0.$$

6. A vertex of an equilateral triangle is (2, 3) and the equation of the opposite side is $x + y = 2$. Find the equations of the other two sides and the length of each side of the triangle.

Other sides pass through the vertex (2, 3).

$\therefore$ the equations of other sides will be of the form

$$y - 3 = m(x - 2) \qquad \ldots (1)$$

The equation of the third side is

$$x + y = 2 \qquad \ldots (2)$$

'm' of (2) is $-\frac{1}{1}$, i.e., -1

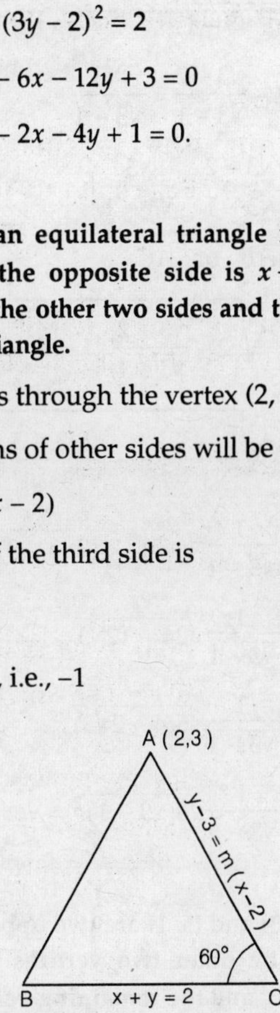

Being equilateral triangle, the angle between (1) and (2) will be 60°.

$$\therefore \quad \tan 60° = \left|\frac{m - (-1)}{1 + m(-1)}\right| \qquad \text{(by angle formula)}$$

$$\text{or} \quad \sqrt{3} = \left|\frac{m + 1}{1 - m}\right|; \quad \therefore \quad \frac{m + 1}{1 - m} = \pm\sqrt{3}$$

$$\therefore \quad m + 1 = \sqrt{3}(1 - m), -\sqrt{3}(1 - m)$$

$$\text{or} \quad (1 + \sqrt{3})m = \sqrt{3} - 1 \ \text{ and } \ (1 - \sqrt{3})m = -\sqrt{3} - 1$$

$$\text{or} \quad m = \frac{\sqrt{3} - 1}{\sqrt{3} + 1}, \frac{\sqrt{3} + 1}{\sqrt{3} - 1}$$

$$= \frac{(\sqrt{3}-1)^2}{2}, \frac{(\sqrt{3}+1)^2}{2}$$

$$= 2-\sqrt{3},\ 2+\sqrt{3}$$

∴ the equations of the other two sides are

$$y-3=(2-\sqrt{3})(x-2) \text{ and } y-3=(2+\sqrt{3})(x-2)$$

or $y=(2-\sqrt{3})x-1+2\sqrt{3}$ and $y=(2+\sqrt{3})x-1-2\sqrt{3}$.

Solving one of these, say,

$y=(2-\sqrt{3})x-1+2\sqrt{3}$ with $x+y=2$ we get

$$2-x=(2-\sqrt{3})x-1+2\sqrt{3}$$

or $3-2\sqrt{3}=(3-\sqrt{3})x$

$$\therefore \quad x=\frac{3-2\sqrt{3}}{3-\sqrt{3}}=\frac{\sqrt{3}-2}{\sqrt{3}-1}$$

$$\therefore \quad y=2-x=2-\frac{\sqrt{3}-2}{\sqrt{3}-1}=\frac{\sqrt{3}}{\sqrt{3}-1}$$

$$\therefore \quad \text{B or C}=\left(\frac{\sqrt{3}-2}{\sqrt{3}-1}, \frac{\sqrt{3}}{\sqrt{3}-1}\right)$$

∴ each side = AB or AC

$$=\sqrt{\left(2-\frac{\sqrt{3}-2}{\sqrt{3}-1}\right)^2+\left(3-\frac{\sqrt{3}}{\sqrt{3}-1}\right)^2}$$

$$=\sqrt{\left(\frac{\sqrt{3}}{\sqrt{3}-1}\right)^2+\left(\frac{2\sqrt{3}-3}{\sqrt{3}-1}\right)^2}$$

$$=\frac{1}{\sqrt{3}-1}\sqrt{(\sqrt{3})^2+(2\sqrt{3}-3)^2}$$

$$=\frac{1}{\sqrt{3}-1}\sqrt{24-12\sqrt{3}}$$

$$=\frac{1}{\sqrt{3}-1}\sqrt{6(4-2\sqrt{3})}$$

$$=\frac{1}{\sqrt{3}-1}\sqrt{6(\sqrt{3}-1)^2}=\sqrt{6}.$$

7. The points (1, 3) and (5, 1) are two opposite vertices of a rectangle. The other two vertices lie on the line $y = 2x + c$. Find c and the remaining vertices.

Let A = (1, 3) and C = (5, 1). The middle point P of AC is $\left(\frac{1+5}{2}, \frac{3+1}{2}\right)$, i.e., (3, 2). It lies on the other diagonal BD whose equation is $y = 2x + c$.

∴ $2 = 2 \times 3 + c$, i.e., $c = -4$.

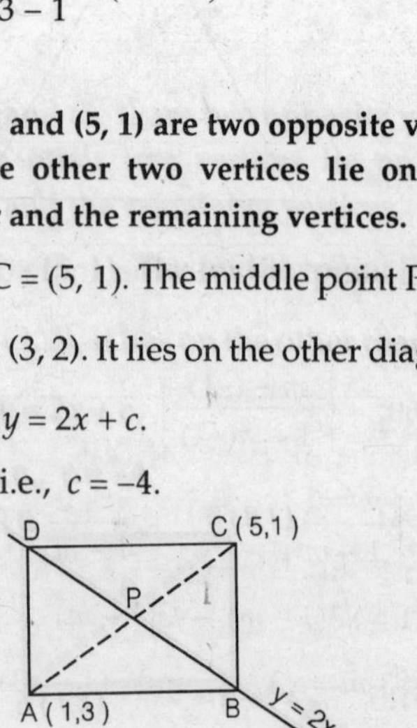

∴ the equation of BD is $y = 2x - 4$.

As ABCD is a rectangle,

$$PB = PD = \frac{1}{2}AC = \frac{1}{2}\sqrt{(5-1)^2+(1-3)^2}$$

$$= \frac{1}{2}\cdot\sqrt{20} = \sqrt{5}.$$

Now, the slope of DB is $\tan\theta = 2$.

Points on DB at distances $\sqrt{5}$ from P(3, 2) have the coordinates

$$(3 \pm \sqrt{5}\cos\theta,\ 2 \pm \sqrt{5}\sin\theta),$$

i.e., $\left(3 \pm \sqrt{5}\cdot\frac{1}{\sqrt{5}},\ 2 \pm \sqrt{5}\cdot\frac{2}{\sqrt{5}}\right)$

∴ the coordinates of the other vertices are $(3 \pm 1, 2 \pm 2)$, i.e., (4, 4) and (2, 0).

8. The equations of two sides of a square are $3x + 4y - 5 = 0$ and $3x + 4y - 15 = 0$. The line along the third side has a point (6, 5) on it. Find the equations of this and the remaining side of the square.

Clearly $3x + 4y - 5 = 0$ and $3x + 4y - 15 = 0$ are two parallel lines, their 'm' being equal to $\frac{-3}{4}$. Let these sides be AB and DC in the square ABCD and the point P(6, 5) be on the side AD.

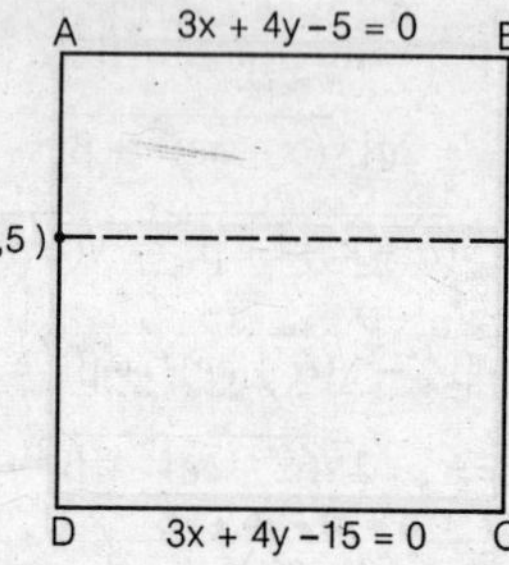

∴ the side AD is on a line through P(6, 5) perpendicular to $3x + 4y - 15 = 0$.

Any line perpendicular to $3x + 4y - 15 = 0$ has the equation $4x - 3y = \lambda$.

It passes through (6, 5). So $4 \times 6 - 3 \times 5 = \lambda$; ∴ $\lambda = 9$.

∴ the equation of AD is $4x - 3y = 9$.

As BC ∥ AD, the equation of BC is $4x - 3y = \lambda$

Let PQ ⊥ BC.

Then $PQ = \left|\frac{4.6 - 3.5 - \lambda}{\sqrt{4^2+(-3)^2}}\right| = \left|\frac{9-\lambda}{5}\right|$

= distance between the parallel lines AB and DC.

Now, any point on AB is $\left(0, \frac{5}{4}\right)$.

The distance of this point from DC

$$= \left| \frac{3\times 0 + 4\times \frac{5}{4} - 15}{\sqrt{3^2+4^2}} \right| = 2.$$

Hence, $\left| \frac{9-\lambda}{5} \right| = 2;$ or $|9-\lambda| = 10$

$\therefore$ $9-\lambda = \pm 10$

$\therefore$ $\lambda = 9 \mp 10 = -1, 19$

$\therefore$ the equation of the remaining side BC is

$4x-3y=-1$ or $4x-3y=19.$

Note As it has not been verified whether P lies between A and D or not, use of $PQ = AD = PA + PD$ may lead to wrong result.

9. Two sides of a rhombus, lying in the first quadrant, are given by $3x-4y=0$ and $12x-5y=0$. If the length of the longer diagonal is 12, find the equations of the other two sides of rhombus.

As the sides are in the first quadrant and their inclinations are $\tan^{-1}\frac{3}{4}$ and $\tan^{-1}\frac{12}{5}$, the angle at the vertex O(0, 0) is acute. So the longer diagonal passes through O(0, 0).

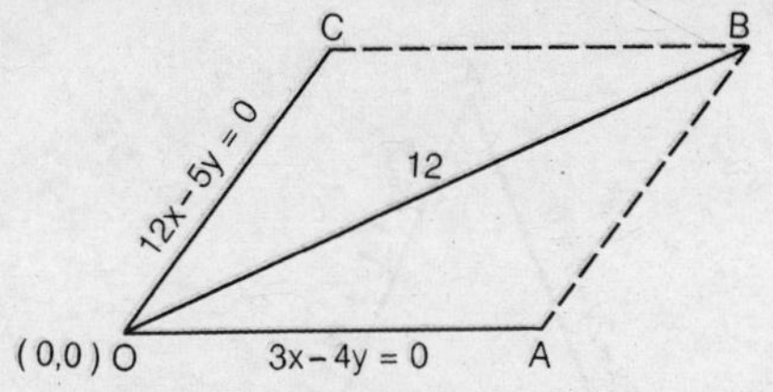

Let the length of each side be a.

The slope of OA is $\tan\theta = \frac{3}{4}$ and the slope of OC is $\tan\phi = \frac{12}{5}$.

$\therefore$ $A = (0 + a\cos\theta, 0 + a\sin\theta) = \left(\frac{4a}{5}, \frac{3a}{5}\right).$

Similarly, $C = (0 + a\cos\phi,\ 0 + a\sin\phi) = \left(\frac{5a}{13}, \frac{12a}{13}\right).$

Now, AB is a line parallel to OC and passing through $A\left(\frac{4a}{5}, \frac{3a}{5}\right).$

$\therefore$ the equation of AB is $12x - 5y = k$ where

$12\cdot\frac{4a}{5} - 5\cdot\frac{3a}{5} = k$ or $k = \frac{33}{5}a$

$\therefore$ the equation of AB is $12x - 5y = \frac{33}{5}a$ ... (1)

The equation of CB is $3x - 4y = \lambda$ where

$3\cdot\frac{5a}{13} - 4\cdot\frac{12a}{13} = \lambda$ or $\lambda = \frac{-33}{13}a$

$\therefore$ the equation of CB is $3x - 4y = \frac{-33}{13}a$ ... (2)

Solving (1) and (2), $y = \frac{99a}{65}$ and $x = \frac{77a}{65}$

$\therefore$ $B = \left(\frac{77a}{65}, \frac{99a}{65}\right).$

But $OB = 12;$

$\therefore$ $\left(\frac{77a}{65} - 0\right)^2 + \left(\frac{99a}{65} - 0\right)^2 = 12^2$

or $\left(\frac{11a}{65}\right)^2 \cdot (7^2 + 9^2) = 12^2$

$\therefore$ $\frac{11a}{65}\cdot\sqrt{130} = 12$ $\therefore$ $a = \frac{6\sqrt{130}}{11}$

$\therefore$ the equation of AB, from (1), is $12x - 5y = \frac{18\sqrt{130}}{5}$

the equation of CB, from (2), is $3x - 4y = -\frac{18}{13}\sqrt{130}.$

10. Find the condition for the quadrilateral to be concyclic whose consecutive sides are $a_r x + b_r y + c_r = 0;\ r = 1, 2, 3, 4.$

Let $L_r \equiv a_r x + b_r y + c_r = 0$ be the sides and the equations of AB, BC, CD, DA be respectively $L_1 = 0$, $L_2 = 0$, $L_3 = 0$, $L_4 = 0$.

Let $\angle ABC = \theta$; then $\angle ADC = \pi - \theta$.

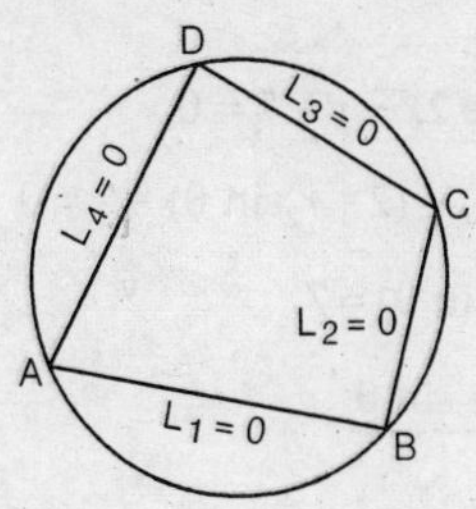

But $\tan\theta = \left|\frac{m_1 - m_2}{1 + m_1 m_2}\right|$

and $\tan(\pi - \theta) = -\left|\frac{m_3 - m_4}{1 + m_3 m_4}\right|$

($\because$ when θ is acute, $\pi - \theta$ is obtuse)

$\therefore$ $\left|\frac{m_1 - m_2}{1 + m_1 m_2}\right| = \left|\frac{m_3 - m_4}{1 + m_3 m_4}\right|$

or $\left|\dfrac{-\frac{a_1}{b_1}-\left(-\frac{a_2}{b_2}\right)}{1+\left(-\frac{a_1}{b_1}\right)\left(-\frac{a_2}{b_2}\right)}\right| = \left|\dfrac{-\frac{a_3}{b_3}-\left(-\frac{a_4}{b_4}\right)}{1+\left(-\frac{a_3}{b_3}\right)\left(-\frac{a_4}{b_4}\right)}\right|$

or $\left|\dfrac{b_1a_2-a_1b_2}{a_1a_2+b_1b_2}\right| = \left|\dfrac{b_3a_4-a_3b_4}{a_3a_4+b_3b_4}\right|$

or $\dfrac{a_1b_2-a_2b_1}{a_1a_2+b_1b_2} = \pm\dfrac{a_3b_4-a_4b_3}{a_3a_4+b_3b_4}$

or $(a_1b_2-a_2b_1)(a_3a_4+b_3b_4) \pm (a_3b_4-a_4b_3)(a_1a_2+b_1b_2) = 0.$

11. Find the equation of the line passing through the point P(1, 2) cutting the lines $x+y-5=0$ and $2x-y=7$ at A and B respectively such that the harmonic mean of *PA* and *PB* is 10.

Let the equation of the line passing through P(1, 2) be

$$y-2 = m(x-1) \quad \ldots (1)$$

Let $PA = r_1,\ PB = r_2$. Then

$$A = (1 + r_1\cos\theta,\ 2 + r_1\sin\theta)$$

$$B = (1 + r_2\cos\theta,\ 2 + r_2\sin\theta), \text{ where } \tan\theta = m.$$

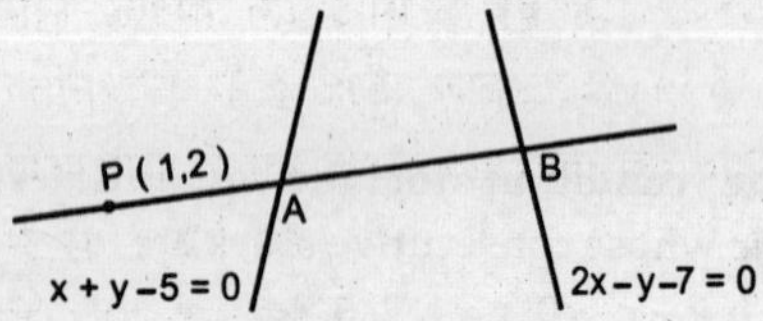

As A is on the line $x + y - 5 = 0$,

$$1 + r_1\cos\theta + 2 + r_1\sin\theta - 5 = 0$$

or $r_1(\cos\theta + \sin\theta) = 2$

$\therefore \quad \dfrac{1}{r_1} = \dfrac{\cos\theta+\sin\theta}{2} \quad \ldots (2)$

As B is on the line $2x - y - 7 = 0$,

$$2(1 + r_2\cos\theta) - (2 + r_2\sin\theta) - 7 = 0$$

or $r_2(2\cos\theta - \sin\theta) = 7$

$\therefore \quad \dfrac{1}{r_2} = \dfrac{2\cos\theta-\sin\theta}{7} \quad \ldots (3)$

Now, HM of *PA*, *PB* is 10

$\therefore \quad 10 = \dfrac{2PA\cdot PB}{PA+PB}; \quad \therefore \quad 10 = \dfrac{2r_1r_2}{r_1+r_2}$

or $\dfrac{r_1+r_2}{r_1r_2} = \dfrac{1}{5}; \quad \therefore \quad \dfrac{1}{r_2}+\dfrac{1}{r_1} = \dfrac{1}{5}$

or $\dfrac{2\cos\theta-\sin\theta}{7} + \dfrac{(\cos\theta+\sin\theta)}{2} = \dfrac{1}{5}$ [from (2), (3)]

or $10(2\cos\theta - \sin\theta) + 35(\cos\theta + \sin\theta) = 14$

or $55\cos\theta + 25\sin\theta = 14$

or $\sqrt{55^2+25^2}\cos(\theta-\alpha) = 14$

where $\cos\alpha = \dfrac{55}{\sqrt{55^2+25^2}}$

or $\cos(\theta-\alpha) = \dfrac{14}{5\sqrt{146}}$

$\therefore \quad \theta = \alpha + \cos^{-1}\dfrac{14}{5\sqrt{146}} = \cos^{-1}\dfrac{11}{\sqrt{146}} + \cos^{-1}\dfrac{14}{5\sqrt{146}}$

$\therefore$ from (1), the equation of the line is

$$y-2 = \tan\left(\cos^{-1}\frac{11}{\sqrt{146}} + \cos^{-1}\frac{14}{5\sqrt{146}}\right)\cdot(x-1).$$

12. Find the coordinates of the centroid, circumcentre and orthocentre of the triangle formed by the lines $3x-2y=6$, $3x+4y+12=0$ and $3x-8y+12=0$.

Let the sides AB, BC and CA have the equations

$$3x-2y-6=0 \quad \ldots (1)$$

$$3x-8y+12=0 \quad \ldots (2)$$

$$3x+4y+12=0 \quad \ldots (3)$$

respectively.

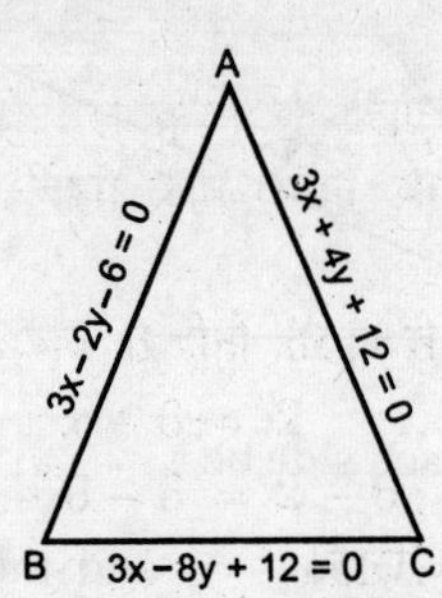

Solving (2), (3) we get $y = 0,\ x = -4$; $\quad \therefore \quad C = (-4, 0)$

Solving (1), (2) we get $y = 3,\ x = 4$; $\quad \therefore \quad B = (4, 3)$

Solving (1), (3) we get $y = -3,\ x = 0$; $\quad \therefore \quad A = (0, -3)$

To find the centroid

We know that centroid G

$$= \left(\frac{x_1+x_2+x_3}{3},\ \frac{y_1+y_2+y_3}{3}\right)$$

where vertices are (x_1, y_1), etc.

$$= \left(\frac{0+4-4}{3},\ \frac{-3+3+0}{3}\right) = (0, 0).$$

To find the circumcentre

Let M(α, β) be the circumcentre. Then $MA = MB = MC$.

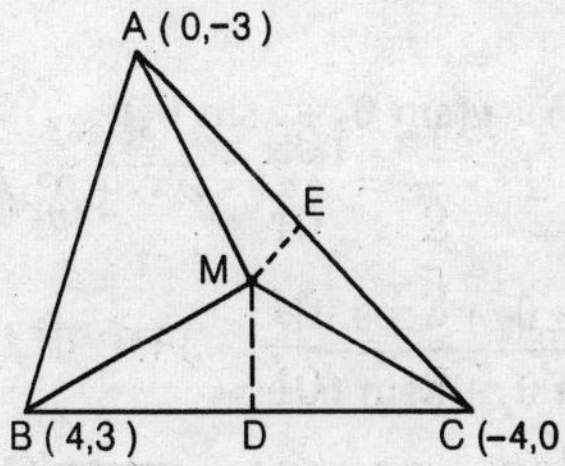

$\therefore \quad (\alpha - 0)^2 + (\beta + 3)^2$

$$= (\alpha - 4)^2 + (\beta - 3)^2$$

$$= (\alpha + 4)^2 + (\beta - 0)^2$$

or $\quad \alpha^2 + \beta^2 + 6\beta + 9$

$$= \alpha^2 + \beta^2 - 8\alpha - 6\beta + 25$$

$$= \alpha^2 + \beta^2 + 8\alpha + 16$$

or $\quad 6\beta + 9 = -8\alpha - 6\beta + 25 = 8\alpha + 16$

$\therefore \quad 6\beta + 9 = -8\alpha - 6\beta + 25 \quad$ and $\quad 6\beta + 9 = 8\alpha + 16$

or $\quad 8\alpha + 12\beta - 16 = 0$, i.e., $2\alpha + 3\beta - 4 = 0$ ... (i)

and $\quad 8\alpha - 6\beta + 7 = 0$ ... (ii)

Solving (i) and (ii) we get $\alpha = \frac{1}{12}, \beta = \frac{23}{18}$

$\therefore$ circumcentre $= \left(\frac{1}{12}, \frac{23}{18}\right)$.

Alternatively Take the perpendicular bisectors of the sides BC, CA which are respectively MD and ME.

Now, $\quad$ D = midpoint of BC $= \left(\frac{4-4}{2}, \frac{3+0}{2}\right) = \left(0, \frac{3}{2}\right)$

The slope of BC is $\frac{3-0}{4-(-4)}$, i.e., $\frac{3}{8}$.

So, the slope of MD is $\frac{-8}{3}$

$\therefore$ the equation of the perpendicular bisector MD of the side BC is

$$y - \frac{3}{2} = \frac{-8}{3}(x - 0); \quad \text{or} \quad 6y - 9 = -16x$$

or $\quad 16x + 6y - 9 = 0$ ... (iii)

Again, $\quad$ E = midpoint of AC

$$= \left(\frac{0-4}{2}, \frac{-3+0}{2}\right) = \left(-2, -\frac{3}{2}\right).$$

The slope of AC is $\frac{-3-0}{0-(-4)}$, i.e., $\frac{-3}{4}$.

So, the slope of ME is $\frac{4}{3}$.

$\therefore$ the equation of the perpendicular bisector ME of the side AC is

$$y + \frac{3}{2} = \frac{4}{3}(x + 2), \text{ i.e., } 6y + 9 = 8(x + 2)$$

$\therefore \quad 8x - 6y + 7 = 0$ ... (iv)

Solving (iii) and (iv) we get $x = \frac{1}{12}, y = \frac{23}{18}$.

$\therefore \quad$ the circumcentre $= \left(\frac{1}{12}, \frac{23}{18}\right)$.

To find the orthocentre

Let H(α, β) be the orthocentre. Then BH $\perp$ AC, AH $\perp$ BC.

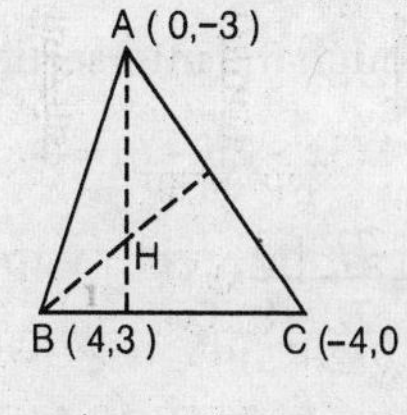

Now, $\quad$ 'm' of BH $= \frac{\beta - 3}{\alpha - 4}$, $\quad$ 'm' of AC $= \frac{-3-0}{0-(-4)} = \frac{-3}{4}$

'm' of AH $= \frac{\beta + 3}{\alpha - 0}$, $\quad$ 'm' of BC $= \frac{3-0}{4-(-4)} = \frac{3}{8}$

$\therefore \quad$ we get, BH $\perp$ AC $\quad \Rightarrow \quad \frac{\beta - 3}{\alpha - 4} \cdot \frac{-3}{4} = -1$

or $\quad 4(\alpha - 4) = 3(\beta - 3); \quad$ or $\quad 4\alpha - 3\beta = 7$ ... (I)

Again AH $\perp$ BC $\quad \Rightarrow \quad \frac{\beta + 3}{\alpha - 0} \cdot \frac{3}{8} = -1$

or $\quad 8\alpha + 3(\beta + 3) = 0$, i.e., $8\alpha + 3\beta = -9$ ... (II)

Solving (I) and (II) we get $\alpha = \frac{-1}{6}, \beta = \frac{-23}{9}$

$\therefore \quad$ the orthocentre is $\left(-\frac{1}{6}, \frac{-23}{9}\right)$.

Alternatively The equations of AB, AC are respectively

$$3x - 2y - 6 = 0 \text{ and } 3x + 4y + 12 = 0.$$

Any line through A is a line passing through the intersection of the above lines. So the equation of the altitude through A is

$$3x - 2y - 6 + \lambda(3x + 4y + 12) = 0 \quad \text{... (a)}$$

Being altitude, it is perpendicular to BC whose slope

$$= \frac{-3}{-8} = \frac{3}{8}.$$

The slope of (a) is $-\frac{3 + 3\lambda}{4\lambda - 2}$.

$\therefore \quad$ by condition of perpendicularity,

$$\frac{3}{8} \cdot \left(-\frac{3 + 3\lambda}{4\lambda - 2}\right) = -1$$

or $\quad 3(3 + 3\lambda) = 8(4\lambda - 2)$

or $\quad 9 + 9\lambda = 32\lambda - 16$

or $23\lambda = 25$; $\therefore$ $\lambda = \frac{25}{23}$.

$\therefore$ the equation of the altitude through A is

$$3x - 2y - 6 + \frac{25}{23}(3x + 4y + 12) = 0$$

or $69x - 46y - 138 + 75x + 100y + 300 = 0$

or $144x + 54y + 162 = 0$

or $8x + 3y + 9 = 0$ ... (b)

Again, any line through the intersection of BC and BA is

$$3x - 8y + 12 + \lambda(3x - 2y - 6) = 0$$

Its $'m' = -\frac{3\lambda + 3}{-8 - 2\lambda} = \frac{3\lambda + 3}{2\lambda + 8}$.

This line is perpendicular to AC whose $'m' = -\frac{3}{4}$.

$\therefore$ $\frac{3\lambda + 3}{2\lambda + 8} \cdot \left(-\frac{3}{4}\right) = -1$

or $3(3\lambda + 3) = 4(2\lambda + 8)$

or $9\lambda + 9 = 8\lambda + 32$; $\therefore$ $\lambda = 23$

$\therefore$ the equation of the altitude through B is

$$3x - 8y + 12 + 23(3x - 2y - 6) = 0$$

or $72x - 54y - 126 = 0$

or $4x - 3y - 7 = 0$ ... (c)

Solving (b) and (c) we get the coordinates of the orthocentre. We get $x = \frac{-1}{6}$, $y = \frac{-23}{9}$.

$\therefore$ the orthocentre $= \left(-\frac{1}{6}, \frac{-23}{9}\right)$.

Note By finding any two of the three centres namely, centroid, circumcentre and orthocentre we can find the third by using $HG : GM = 2 : 1$.

13. The three sides of a triangle are

$L_r \equiv x\cos\theta_r + y\sin\theta_r - p_r = 0; r = 1, 2, 3.$

Show that the orthocentre of the triangle is given by $L_1\cos(\theta_2 - \theta_3) = L_2\cos(\theta_3 - \theta_1) = L_3\cos(\theta_1 - \theta_2)$.

Let the equations of BC, CA, AB be respectively

$$L_1 = 0, L_2 = 0, L_3 = 0.$$

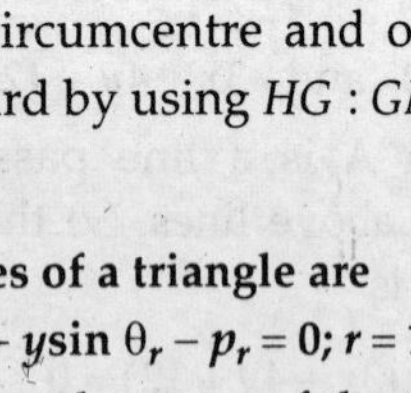

Any line passing through A, i.e., the intersection of the lines $L_2 = 0$, $L_3 = 0$ is

$$L_2 + \lambda L_3 = 0 \quad \text{... (1)}$$

or $x(\cos\theta_2 + \lambda\cos\theta_3) + y(\sin\theta_2 + \lambda\sin\theta_3) - (p_2 + \lambda p_3) = 0$

$'m'$ of this line $= -\frac{\cos\theta_2 + \lambda\cos\theta_3}{\sin\theta_2 + \lambda\sin\theta_3}$.

It is $\perp^r$ to BC whose $'m' = -\frac{\cos\theta_1}{\sin\theta_1}$

$\therefore$ $-\frac{\cos\theta_2 + \lambda\cos\theta_3}{\sin\theta_2 + \lambda\sin\theta_3} \cdot \left(-\frac{\cos\theta_1}{\sin\theta_1}\right) = -1$

or $\cos\theta_1(\cos\theta_2 + \lambda\cos\theta_3) + \sin\theta_1(\sin\theta_2 + \lambda\sin\theta_3) = 0$

or $\cos(\theta_1 - \theta_2) + \lambda\cos(\theta_3 - \theta_1) = 0$

$\therefore$ $\lambda = -\frac{\cos(\theta_1 - \theta_2)}{\cos(\theta_3 - \theta_1)}$.

Putting in (1), the equation of the altitude through A becomes

$$L_2 - \frac{\cos(\theta_1 - \theta_2)}{\cos(\theta_3 - \theta_1)}L_3 = 0$$

or $L_2\cos(\theta_3 - \theta_1) = L_3\cos(\theta_1 - \theta_2)$ ... (2)

Similarly, the equation of the altitudes through B and C are respectively

$L_3\cos(\theta_1 - \theta_2) = L_1\cos(\theta_2 - \theta_3)$ ... (3)

and $L_1\cos(\theta_2 - \theta_3) = L_2\cos(\theta_3 - \theta_1)$ ... (4)

The orthocentre being the intersection of altitudes, satisfies (2), (3) and (4). So the orthocentre is given by

$$L_1\cos(\theta_2 - \theta_3) = L_2\cos(\theta_3 - \theta_1) = L_3\cos(\theta_1 - \theta_2).$$

14. Vertices of a triangle are $A(x_1, x_1\tan\alpha_1)$, $B(x_2, x_2\tan\alpha_2)$ and $C(x_3, x_3\tan\alpha_3)$. If the circumcentre coincides with the origin and the orthocentre $H = (\bar{x}, \bar{y})$, prove that $\bar{y}(\cos\alpha_1 + \cos\alpha_2 + \cos\alpha_3) = \bar{x}(\sin\alpha_1 + \sin\alpha_2 + \sin\alpha_3)$.

Here, the circumcentre O = (0, 0). So $OA = OB = OC$.

$\therefore$ $x_1^2 + x_1^2\tan^2\alpha_1 = x_2^2 + x_2^2\tan^2\alpha_2 = x_3^2 + x_3^2\tan^2\alpha_3$

or $x_1^2\sec^2\alpha_1 = x_2^2\sec^2\alpha_2 = x_3^2\sec^2\alpha_3$

or $\frac{x_1}{\cos\alpha_1} = \frac{x_2}{\cos\alpha_2} = \frac{x_3}{\cos\alpha_3} = k$ (suppose)

$\therefore$ the vertices of the triangle become

$A = (k\cos\alpha_1, k\sin\alpha_1)$,

$B = (k\cos\alpha_2, k\sin\alpha_2)$,

$C = (k\cos\alpha_3, k\sin\alpha_3)$

$\therefore$ the centroid G

$$= \left(k\frac{\cos\alpha_1+\cos\alpha_2+\cos\alpha_3}{3},\ k\frac{\sin\alpha_1+\sin\alpha_2+\sin\alpha_3}{3}\right).$$

We know that the orthocentre H, centroid G and circumcentre O are collinear.

So, 'm' of HO = 'm' of GO

$\therefore$ 'm' of HO $= \dfrac{\bar{y}-0}{\bar{x}-0} = \dfrac{\bar{y}}{\bar{x}}$

and 'm' of GO $= \dfrac{k\cdot\dfrac{\sin\alpha_1+\sin\alpha_2+\sin\alpha_3}{3}-0}{k\cdot\dfrac{\cos\alpha_1+\cos\alpha_2+\cos\alpha_3}{3}-0}$

$$= \frac{\sin\alpha_1+\sin\alpha_2+\sin\alpha_3}{\cos\alpha_1+\cos\alpha_2+\cos\alpha_3}$$

$\therefore$ $\dfrac{\bar{y}}{\bar{x}} = \dfrac{\sin\alpha_1+\sin\alpha_2+\sin\alpha_3}{\cos\alpha_1+\cos\alpha_2+\cos\alpha_3}$

or $\bar{y}(\cos\alpha_1+\cos\alpha_2+\cos\alpha_3)$

$$= \bar{x}(\sin\alpha_1+\sin\alpha_2+\sin\alpha_3).$$

15. Show that the points (1, 3), (3, –2), (–1, –4) and (–4, 1) lie in four different compartments of *x*-*y* plane made by the lines $x-2y=2$ and $3x+2y+6=0$.

Let the lines be $L_1 \equiv x-2y-2=0$ and

$L_2 \equiv 3x+2y+6=0.$

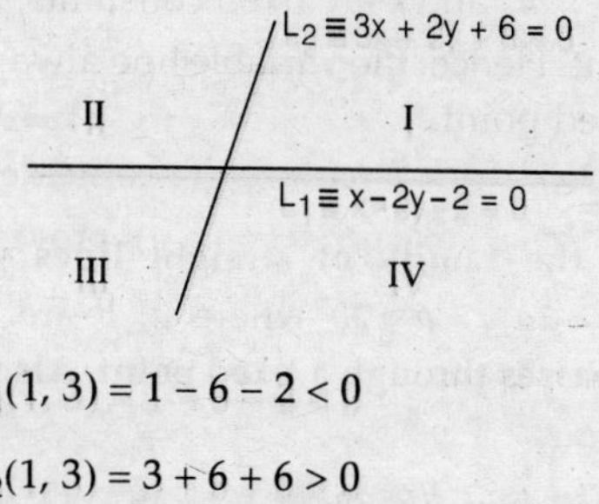

Now, $L_1(1, 3) = 1-6-2<0$

$L_2(1, 3) = 3+6+6>0$

$L_1(3, -2) = 3+4-2>0$

$L_2(3, -2) = 9-4+6>0$

$L_1(-1, -4) = -1+8-2>0$

$L_2(-1, -4) = -3-8+6<0$

$L_1(-4, 1) = -4-2-2<0$

$L_2(-4, 1) = -12+2+6<0.$

Also $L_1(0, 0) = 0-0-2<0$

$L_2(0, 0) = 0+0+6>0.$

Let us name the four compartments as I (the compartment containing origin), II, III, IV as shown.

$L_1(0, 0)<0,\ L_1(1, 3)<0$

$\Rightarrow$ (0, 0) and (1, 3) are on the same side of $L_1=0$

$L_2(0, 0)>0,\ L_2(1, 3)>0$

$\Rightarrow$ (0, 0) and (1, 3) are on the same side of $L_2=0$

$\therefore$ (1, 3) belongs to the compartment I.

$L_1(0, 0)<0,\ L_1(3, -2)>0$

$\Rightarrow$ (0, 0) and (3, –2) are on the opposite sides of $L_1=0$

$L_2(0, 0)>0,\ L_2(3, -2)>0$

$\Rightarrow$ (0, 0) and (3, –2) are on the same sides of $L_2=0$

$\therefore$ (3, –2) belongs to the compartment IV.

$L_1(0, 0)<0, L_1(-1, -4)>0$

$\Rightarrow$ (0, 0) and (–1, – 4) are on the opposite sides of $L_1=0$, $L_2(0, 0)>0$, $L_2(-1, -4)<0$

$\Rightarrow$ (0, 0) and (–1, – 4) are on the opposite sides of $L_2=0$

$\therefore$ (–1, – 4) belongs to the compartment III.

$L_1(0, 0)<0,\ L_1(-4, 1)<0$

$\Rightarrow$ (0, 0) and (– 4, 1) are on the same sides of $L_1=0$.

$L_2(0, 0)>0,\ L_2(-4, 1)<0$

$\Rightarrow$ (0, 0) and (– 4, 1) are on the opposite sides of $L_2=0$

$\therefore$ (– 4, –1) belongs to the compartment II.

Hence, the four points are in four different compartments.

16. Determine all values of α for which the point (α, α^2) lies inside the triangle formed by the lines $2x+3y-1=0$, $x+2y-3=0$, $5x-6y-1=0$.

Let the equations of BC, CA and AB be respectively

$L_1 \equiv 2x+3y-1=0$ ... (1)

$L_2 \equiv x+2y-3=0$ ... (2)

$L_3 \equiv 5x-6y-1=0$ ... (3)

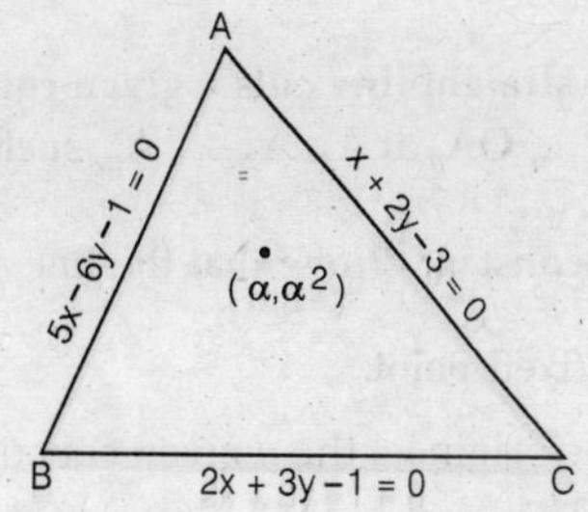

Solving (2), (3) we get $x = \frac{5}{4}, y = \frac{7}{8}$; $\therefore$ $A = \left(\frac{5}{4}, \frac{7}{8}\right)$

Solving (3), (1) we get $x = \frac{1}{3}, y = \frac{1}{9}$; $\therefore$ $B = \left(\frac{1}{3}, \frac{1}{9}\right)$

Solving (1), (2) we get $y = 5, x = -7$; $\therefore$ $C = (-7, 5)$

Now, $L_1(A) = L_1\left(\frac{5}{4}, \frac{7}{8}\right) = 2 \times \frac{5}{4} + 3 \times \frac{7}{8} - 1 = \frac{5}{2} + \frac{21}{8} - 1 > 0$

$L_2(B) = L_2\left(\frac{1}{3}, \frac{1}{9}\right) = \frac{1}{3} + 2 \cdot \frac{1}{9} - 3 = \frac{1}{3} + \frac{2}{9} - 3 < 0$

$L_3(C) = L_3(-7, 5) = 5 \times (-7) - 6 \times 5 - 1$

$= -35 - 30 - 1 < 0.$

As A and (α, α^2) are on the same side of $L_1 = 0$,

$$2\alpha + 3\alpha^2 - 1 > 0 \quad \dots (4)$$

Similarly, using for other vertices

$$\alpha + 2\alpha^2 - 3 < 0 \quad \dots (5)$$

$$5\alpha - 6\alpha^2 - 1 < 0 \quad \dots (6)$$

From (4), $3\alpha^2 + 2\alpha - 1 > 0$

or $(3\alpha - 1)(\alpha + 1) > 0$

$\therefore$ $\alpha < -1$ or $\alpha > \frac{1}{3}$ (using sign-scheme)

From (5), $2\alpha^2 + \alpha - 3 < 0$

or $(2\alpha + 3)(\alpha - 1) < 0$

$\therefore$ $-\frac{3}{2} < \alpha < 1$ (using sign-scheme)

From (6), $6\alpha^2 - 5\alpha + 1 > 0$ or $(3\alpha - 1)(2\alpha - 1) > 0$

$\therefore$ $\alpha < \frac{1}{3}$ or $\alpha > \frac{1}{2}$ (using sign-scheme)

$\therefore$ the common values of α satisfying (4), (5) and (6) are given by

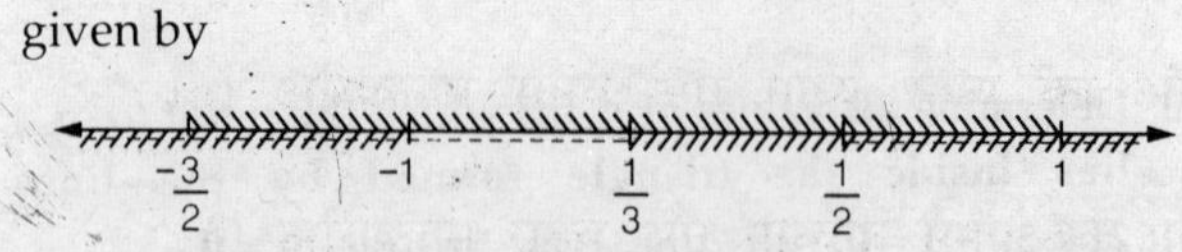

$\therefore$ $-\frac{3}{2} < \alpha < -1$ or $\frac{1}{2} < \alpha < 1$.

$\therefore$ the required values of $\alpha \in \left(-\frac{3}{2}, -1\right) \cup \left(\frac{1}{2}, 1\right)$.

17. A variable straight line cuts n given concurrent lines $OA_1, OA_2, \dots, OA_n$ at $A_1, A_2, \dots, A_n$ such that $\sum_{r=1}^{n} \frac{1}{OA_r} =$ constant. Prove that the line always passes through a fixed point.

Taking O as the origin let the n given concurrent lines be $y = m_r x$; $r = 1, 2, 3, \dots, n$.

Let the variable line be

$$y = mx + c \quad \dots (1)$$

Solving $y = m_r x$ and $y = mx + c$, we get

$m_r x = mx + c$, i.e., $(m_r - m)x = c$

$\therefore$ $x = \frac{c}{m_r - m}$

$\therefore$ $y = m_r \cdot \frac{c}{m_r - m}$; $\therefore$ $A_r = \left(\frac{c}{m_r - m}, \frac{cm_r}{m_r - m}\right)$

$\therefore$ $OA_r^2 = \frac{c^2}{(m_r - m)^2} + \frac{c^2 m_r^2}{(m_r - m)^2}$

$= \frac{c^2}{(m_r - m)^2}(1 + m_r^2)$.

$\therefore$ $\frac{1}{OA_r} = \frac{m_r - m}{c\sqrt{1 + m_r^2}}$

$\therefore$ from the given condition,

$$\sum_{r=1}^{n} \frac{m_r - m}{c\sqrt{1 + m_r^2}} = \text{constant} = k$$

or $\frac{1}{c} \sum_{r=1}^{n} \frac{m_r}{\sqrt{1 + m_r^2}} - \frac{m}{c} \sum_{r=1}^{n} \frac{1}{\sqrt{1 + m_r^2}} = k$

or $\frac{1}{k} \sum_{r=1}^{n} \frac{m_r}{\sqrt{1 + m_r^2}} = \frac{m}{k} \sum_{r=1}^{n} \frac{1}{\sqrt{1 + m_r^2}} + c$

$\therefore$ (1) is satisfied by the point

$$P\left(\frac{1}{k} \sum_{r=1}^{n} \frac{1}{\sqrt{1 + m_r^2}}, \frac{1}{k} \sum_{r=1}^{n} \frac{m_r}{\sqrt{1 + m_r^2}}\right)$$

As $m_1, m_2, m_3, \dots, m_n$ and k are fixed constants, the point P is a fixed point. Hence, the variable line always passes through the fixed point.

18. Prove that the family of straight lines given by $(a + b)x + (b - 2a)y = a + 2b$ where a, b are arbitrary constants, passes through a fixed point. Also find the point.

Here lines are $(a + b)x + (b - 2a)y = a + 2b$

or $a(x - 2y - 1) + b(x + y - 2) = 0 \quad \dots (1)$

Clearly, values of x, y for which $x - 2y - 1 = 0$ and $x + y - 2 = 0$, satisfy (1)

$\therefore$ (1) is a family of straight lines (being of first degree in x, y) passing through the intersection of the fixed lines

$$x - 2y - 1 = 0 \quad \dots (2)$$

and $x + y - 2 = 0 \quad \dots (3)$

Solving (2) and (3), $y = \frac{1}{3}$ and $x = \frac{5}{3}$.

$\therefore$ the lines pass through the fixed point $\left(\frac{5}{3}, \frac{1}{3}\right)$.

19. For what values of λ the following three lines are concurrent?

$$x + y = 1, \quad \lambda x + 2y = 3, \quad \lambda^2 x + 4y + 9 = 0$$

The three lines will be concurrent if

$$\begin{vmatrix} 1 & 1 & -1 \\ \lambda & 2 & -3 \\ \lambda^2 & 4 & 9 \end{vmatrix} = 0$$

or $$\begin{vmatrix} 0 & 0 & -1 \\ \lambda - 3 & -1 & -3 \\ \lambda^2 + 9 & 13 & 9 \end{vmatrix} = 0, \; C_1 \to C_1 + C_3, C_2 \to C_2 + C_3$$

or $13(\lambda - 3) + \lambda^2 + 9 = 0$

or $\lambda^2 + 13\lambda - 30 = 0$

$$\therefore \quad \lambda = \frac{-13 \pm \sqrt{13^2 + 4.30}}{2} = \frac{-13 \pm \sqrt{289}}{2}$$

$$= \frac{-13 \pm 17}{2} = 2, -15.$$

20. Let a given line L_1 intersect the x and y axes at P and Q respectively. Let another line L_2 perpendicular to L_1 cut the axes at R and S respectively. Find the locus of the point of intersection of the lines PS and QR.

Let the equation of the given line be

$$L_1 \equiv ax + by + c = 0$$

$$\therefore \quad P = \left(-\frac{c}{a}, 0\right), \quad Q = \left(0, -\frac{c}{b}\right).$$

Any line L_2 perpendicular to the line L_1 is

$$L_2 \equiv bx - ay + \lambda = 0$$

$$\therefore \quad R = \left(-\frac{\lambda}{b}, 0\right), \quad S = \left(0, \frac{\lambda}{a}\right)$$

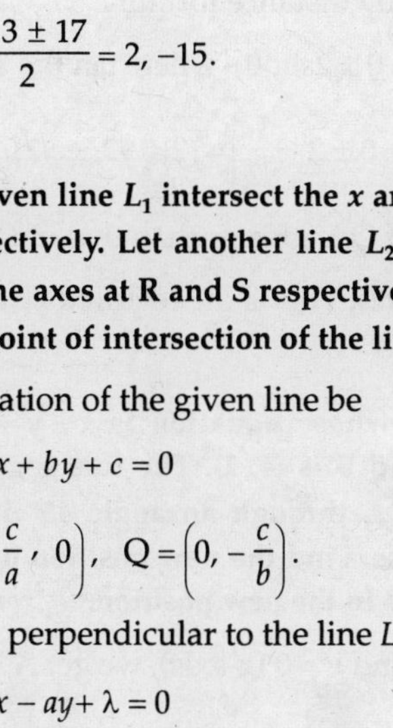

The equation of the line PS is

$$y - 0 = \frac{0 - \frac{\lambda}{a}}{-\frac{c}{a} - 0}\left(x + \frac{c}{a}\right),$$

i.e., $$y = \frac{\lambda}{c}\left(x + \frac{c}{a}\right) \qquad \ldots (1)$$

The equation of the line QR is

$$y - 0 = \frac{0 - \left(-\frac{c}{b}\right)}{-\frac{\lambda}{b} - 0}\left(x + \frac{\lambda}{b}\right)$$

or $$y = -\frac{c}{\lambda}\left(x + \frac{\lambda}{b}\right),$$

i.e., $$y = \frac{-c}{\lambda}x - \frac{c}{b} \qquad \ldots (2)$$

The locus of the point of intersection of (1) and (2) is obtained by eliminating λ from (1) and (2).

From (2), $y + \frac{c}{b} = -\frac{c}{\lambda}x$

Multiplying (1) and this, we get

$$y\left(y + \frac{c}{b}\right) = \frac{\lambda}{c}\left(x + \frac{c}{a}\right) \cdot \frac{-cx}{\lambda}$$

or $$y^2 + \frac{c}{b}y = -x\left(x + \frac{c}{a}\right)$$

or $$x^2 + y^2 + \frac{c}{a}x + \frac{c}{b}y = 0$$

Note The locus is a circle passing through the origin.

21. The ends A, B of a straight line segment of constant length c slide on the fixed rectangular axes OX, OY respectively. If the rectangle OAPB be completed, show that the locus of the foot of the perpendicular drawn from P upon AB is $x^{2/3} + y^{2/3} = c^{2/3}$.

Let $\angle BAO = \theta$. Then $OA = c\cos\theta$, $AP = OB = c\sin\theta$ because $AB = c$. Let M be the foot of the perpendicular from P on AB. Let MN ⊥ OX.

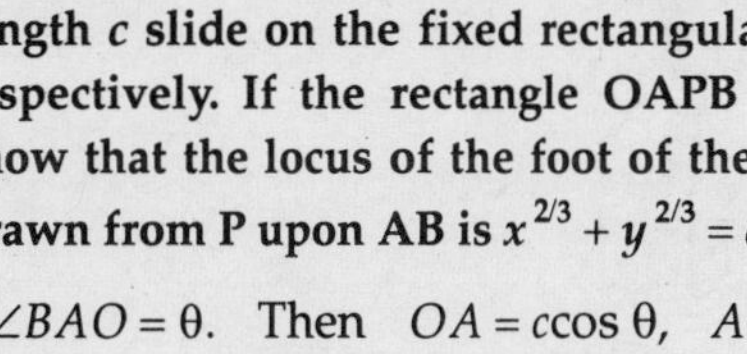

Let M = (α, β).

Now, $\alpha = ON = OA - NA$

$$= c\cos\theta - MA \cdot \cos\theta$$

$$= c\cos\theta - PA \cdot \cos(90° - \theta) \cdot \cos\theta$$

$$= c\cos\theta - c\sin\theta \cdot \sin\theta \cdot \cos\theta$$

$$\therefore \quad \alpha = c\cos\theta(1 - \sin^2\theta) = c\cos^3\theta \qquad \ldots (1)$$

Next, $\beta = MN = MA \cdot \sin\theta = PA \cdot \cos(90° - \theta)\sin\theta$

$= c\sin\theta \cdot \sin\theta \cdot \sin\theta$

$\therefore \quad \beta = c\sin^3\theta \quad \ldots (2)$

$\therefore$ from (1) and (2),

$$\left(\frac{\alpha}{c}\right)^{2/3} = \cos^2\theta \text{ and } \left(\frac{\beta}{c}\right)^{2/3} = \sin^2\theta$$

Adding these, $\left(\frac{\alpha}{c}\right)^{2/3} + \left(\frac{\beta}{c}\right)^{2/3} = \cos^2\theta + \sin^2\theta = 1.$

$\therefore$ the equation of the locus of the foot is

$$\left(\frac{x}{c}\right)^{2/3} + \left(\frac{y}{c}\right)^{2/3} = 1, \text{ i.e., } x^{2/3} + y^{2/3} = c^{2/3}.$$

22. If the image of the point (x_1, y_1) with respect to the mirror $ax + by + c = 0$ be (x_2, y_2), show that

$$\frac{x_2 - x_1}{a} = \frac{y_2 - y_1}{b} = \frac{-2(ax_1 + by_1 + c)}{a^2 + b^2}.$$

Clearly, the line PQ joining the points $P(x_1, y_1)$ and $Q(x_2, y_2)$ is perpendicular to the line $ax + by + c = 0$; also the midpoint M of PQ is on the line $ax + by + c = 0$.

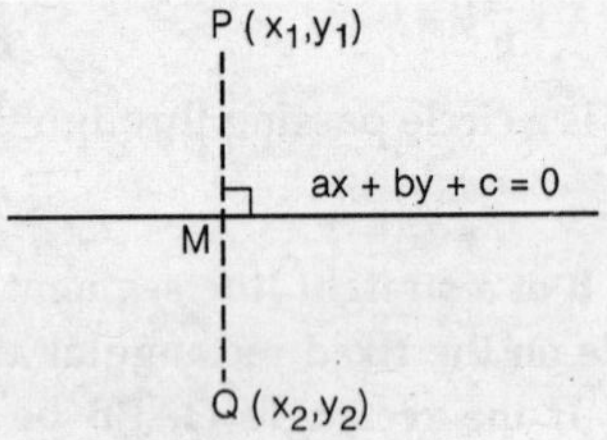

$$\therefore \quad \frac{y_1 - y_2}{x_1 - x_2} \cdot \left(-\frac{a}{b}\right) = -1 \quad \ldots (1)$$

$$\text{and } a \cdot \frac{x_1 + x_2}{2} + b \cdot \frac{y_1 + y_2}{2} + c = 0 \quad \ldots (2)$$

From (2), $a(x_1 + x_2) + b(y_1 + y_2) + 2c = 0$

or $\quad (ax_1 + by_1 + c) + (ax_2 + by_2 + c) = 0 \quad \ldots (3)$

From (1), $\dfrac{x_1 - x_2}{a} = \dfrac{y_1 - y_2}{b}$

$$= \frac{a(x_1 - x_2) + b(y_1 - y_2)}{a^2 + b^2},$$

by ratio and proportion

$$= \frac{(ax_1 + by_1 + c) - (ax_2 + by_2 + c)}{a^2 + b^2}$$

$$= \frac{2(ax_1 + by_1 + c)}{a^2 + b^2}, \text{ using (3)}$$

$$\therefore \quad \frac{x_2 - x_1}{a} = \frac{y_2 - y_1}{b} = \frac{-2(ax_1 + by_1 + c)}{a^2 + b^2}.$$

23. The middle point of the line segment joining (3, –1) and (1, 1) is shifted by two units (in the sense of increasing y) perpendicular to the line segment. Find the coordinates of the point in the new position.

Let P be the middle point of the line segment joining A(3, –1) and B(1, 1).

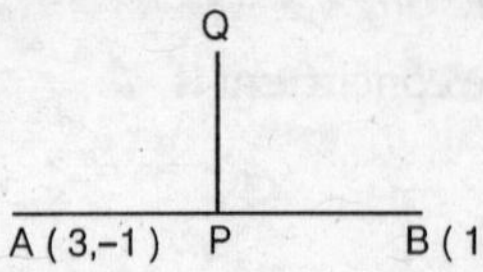

Then $P = \left(\frac{3+1}{2}, \frac{-1+1}{2}\right) = (2, 0).$

Let P be shifted to Q where PQ = 2 and y-coordinate of Q is greater than that of P (from question).

Now, 'm' of $AB = \dfrac{1-(-1)}{1-3} = \dfrac{2}{-2} = -1.$

$\therefore \quad$ 'm' of $PQ = -\dfrac{1}{-1} = 1.$

Coordinates of Q by distance formula

$= (2 \pm 2\cos\theta, 0 \pm 2\sin\theta)$ where $\tan\theta = 1$

$= \left(2 \pm 2 \cdot \frac{1}{\sqrt{2}}, 0 \pm 2 \cdot \frac{1}{\sqrt{2}}\right) = (2 \pm \sqrt{2}, \pm\sqrt{2}).$

As y-coordinate of Q is greater than that of P,

$Q = (2 + \sqrt{2}, \sqrt{2})$. This is the required point.

24. The line AB whose equation is $x - y = 2$ cuts the x-axis at A, and B is (4, 2). The line segment AB is rotated about A through an angle 45° in the anti-clockwise sense. Find the new position of B and the equation of AB in the new position.

Solving $x - y = 2$ and $y = 0$ (x-axis), we get A = (2, 0)

$$\angle BAX = \tan^{-1}\left(-\frac{1}{-1}\right) = \tan^{-1}1 = 45°.$$

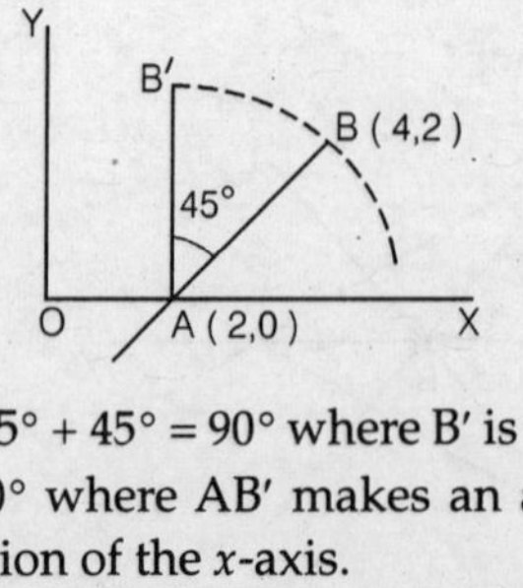

$\therefore \quad \angle B'AX = 45° + 45° = 90°$ where B′ is the new position of B. So $\theta = 90°$ where AB′ makes an angle θ with the positive direction of the x-axis.

Now $AB' = AB = \sqrt{(4-2)^2 + (2-0)^2} = \sqrt{4+4} = 2\sqrt{2}.$

$\therefore \quad B' = (2 + 2\sqrt{2}\cos\theta, 0 + 2\sqrt{2}\sin\theta)$

$= (2 + 2\sqrt{2}\cos 90°, 0 + 2\sqrt{2}\sin 90°)$

$= (2, 2\sqrt{2}).$

Also the equation of AB′, i.e., new position of AB, is

$y - 0 = \tan 90° \cdot (x - 2)$, i.e., $x - 2 = 0$.

$\therefore$ the new position of B = $(2, 2\sqrt{2})$ and the equation of AB in the new position is $x - 2 = 0$.

25. A ray of light is sent along the line $x - 2y - 3 = 0$. On reaching the line $3x - 2y - 5 = 0$ the ray is reflected from it. Find the equation of the line containing the reflected ray.

Let A be the point of incidence.

$\therefore$ A is the point of intersection of

$x - 2y - 3 = 0$ and $3x - 2y - 5 = 0$.

Solving these, we get $x = 1$, $y = -1$. So A = (1, –1).

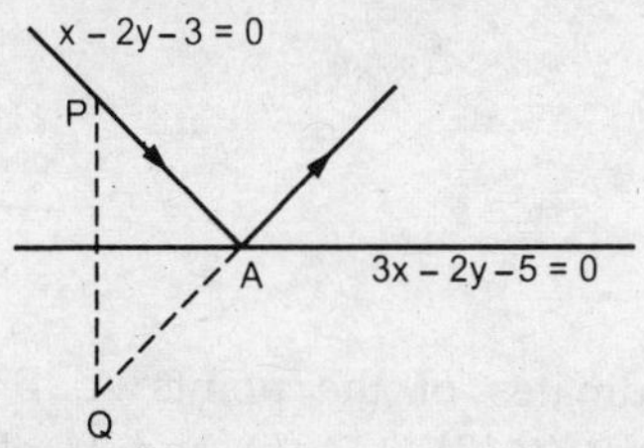

Let P be any point on the line of incidence $x - 2y - 3 = 0$.

So, we take P = (3, 0). Let Q be the image of P by the line $3x - 2y - 5 = 0$.

Let Q = (α, β). Clearly PQ ⊥ the line $3x - 2y - 5 = 0$ and the middle point of PQ is on $3x - 2y - 5 = 0$.

$$\therefore \quad \frac{\beta - 0}{\alpha - 3} \cdot \frac{3}{2} = -1 \qquad \text{... (1)}$$

$$\text{and } 3 \cdot \frac{\alpha + 3}{2} - 2 \cdot \frac{\beta + 0}{2} - 5 = 0 \qquad \text{... (2)}$$

(1) $\Rightarrow$ $3\beta + 2(\alpha - 3) = 0$

or $2\alpha + 3\beta - 6 = 0$ ... (3)

(2) $\Rightarrow$ $3(\alpha + 3) - 2\beta - 10 = 0$

or $3\alpha - 2\beta - 1 = 0$ ... (4)

Solving (3) and (4), $\dfrac{\alpha}{-3 - 12} = \dfrac{\beta}{-18 + 2} = \dfrac{1}{-4 - 9}$

$\therefore$ $\dfrac{\alpha}{-15} = \dfrac{\beta}{-16} = \dfrac{1}{-13}$; $\therefore$ $\alpha = \dfrac{15}{13}$, $\beta = \dfrac{16}{13}$.

$\therefore$ $Q = \left(\dfrac{15}{13}, \dfrac{16}{13}\right)$.

The line containing the reflected ray is the line joining the points A(1, –1) and $Q\left(\dfrac{15}{13}, \dfrac{16}{13}\right)$.

$$\therefore \text{ the required equation is } y + 1 = \frac{\frac{16}{13} + 1}{\frac{15}{13} - 1}(x - 1)$$

or $y + 1 = \dfrac{29}{2}(x - 1)$

or $2y + 2 = 29x - 29$; $\therefore$ $29x - 2y - 31 = 0$.

26. A ray of light is sent along the line $x - 2y = 8$. After refracting across the line $x + y = 1$ it enters the opposite side after turning by 15° away from the line $x + y = 1$. Find the equation of the line along which the refracted ray travels.

The point of incidence A is the intersection of

$x - 2y = 8$ ... (1)

$x + y = 1$ ... (2)

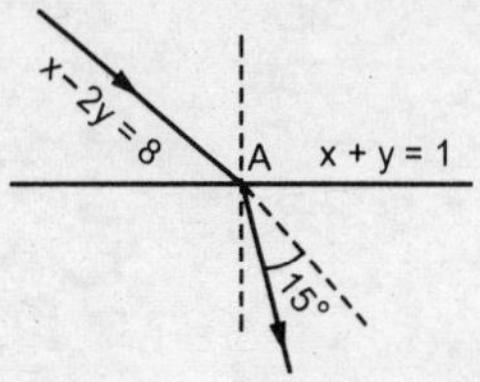

Solving (1), (2) we get

$y = -7/3$, $x = 10/3$; $\therefore$ A = (10/3, –7/3).

Let the refracted ray have the slope = m.

$$\text{Then } \tan 15° = \left|\frac{m - \frac{1}{2}}{1 + m \cdot \frac{1}{2}}\right| = \left|\frac{2m - 1}{m + 2}\right|$$

$$\text{or} \quad \tan(45° - 30°) = \frac{1 - \frac{1}{\sqrt{3}}}{1 + \frac{1}{\sqrt{3}}} = \left|\frac{2m - 1}{m + 2}\right|$$

$$\text{or} \quad \left|\frac{2m - 1}{m + 2}\right| = \frac{\sqrt{3} - 1}{\sqrt{3} + 1} = 2 - \sqrt{3}$$

$$\therefore \quad \frac{2m - 1}{m + 2} = 2 - \sqrt{3}, -2 + \sqrt{3}$$

$\therefore$ $2m - 1 = \pm (2 - \sqrt{3}) \cdot (m + 2)$

$\therefore$ $\sqrt{3}m = 5 - 2\sqrt{3}$, $(4 - \sqrt{3})m = 2\sqrt{3} - 3$

$\therefore$ $m = \dfrac{5\sqrt{3} - 6}{3}, \dfrac{5\sqrt{3} - 6}{13}$.

Let the angle between $x + y = 1$ and the line through A(10/3, –7/3) with the slope $\dfrac{5\sqrt{3} - 6}{3}$ be φ. Then

$$\tan \phi = \left|\frac{\frac{5\sqrt{3} - 6}{3} - (-1)}{1 + \frac{5\sqrt{3} - 6}{3}(-1)}\right| = \left|\frac{\frac{5\sqrt{3} - 6}{3} + 1}{1 - \frac{5\sqrt{3} - 6}{3}}\right|$$

$$= \left|\frac{5\sqrt{3} - 3}{9 - 5\sqrt{3}}\right| = \frac{(5\sqrt{3} - 3)(9 + 5\sqrt{3})}{81 - 75}$$

$$= \frac{30\sqrt{3}+48}{6} = 5\sqrt{3}+8.$$

If the angle between $x+y=1$ and the line through $A(10/3, -7/3)$ with the slope $\frac{5\sqrt{3}-6}{13}$ be ψ then

$$\tan\psi = \left|\frac{\frac{5\sqrt{3}-6}{13}-(-1)}{1+\frac{5\sqrt{3}-6}{13}(-1)}\right| = \left|\frac{\frac{5\sqrt{3}-6}{13}+1}{1-\frac{5\sqrt{3}-6}{13}}\right|$$

$$= \left|\frac{7+5\sqrt{3}}{19-5\sqrt{3}}\right| = \frac{7+5\sqrt{3}}{19-5\sqrt{3}}$$

$$= \frac{(7+5\sqrt{3})(19+5\sqrt{3})}{19^2-(5\sqrt{3})^2} = \frac{208+130\sqrt{3}}{286} = \frac{8+5\sqrt{3}}{11}.$$

As $5\sqrt{3}+8-\frac{8+5\sqrt{3}}{11} = \frac{55\sqrt{3}+88-8-5\sqrt{3}}{11}$

$$= \frac{80+50\sqrt{3}}{11} > 0$$

$\therefore$ $\tan\phi > \tan\psi$; $\therefore$ $\phi > \psi$

$\therefore$ the slope of the refracted ray $= \frac{5\sqrt{3}-6}{3}$

$\therefore$ the equation of the refracted ray is

$$y + \frac{7}{3} = \frac{5\sqrt{3}-6}{3}\left(x - \frac{10}{3}\right)$$

or $3(5-2\sqrt{3})x - 3\sqrt{3}y + 13\sqrt{3} - 50 = 0.$

Exercises

1. If a vertex of a triangle be $(1, 1)$ and the middle points of two sides through it be $(-2, 3)$ and $(5, 2)$ then find the centroid and the incentre of the triangle.
2. If two vertices of a parallelogram are $(3, 2)$, $(-1, 1)$ and the diagonals cut at $(2, -5)$, find the other vertices of the parallelogram.
3. If a triangle has its orthocentre at $(1, 1)$ and circumcentre at $\left(\frac{3}{2}, \frac{3}{4}\right)$ then find the centroid of the triangle.
4. Prove that the line $2x - 3y + 6 = 0$ does not cut the line segment joining the points $(-2, -3)$ and $(1, -2)$.
5. ABC is a triangle and the internal bisector of the angle A cuts the side BC at D. If $A = (3, -5)$, $B = (-3, 3)$, $C = (-1, -2)$ then find the length of AD.
6. Find the point P if $Q = (-2, 4)$ is the point on the line segment OP such that $OQ = \frac{1}{3}OP$, O being the origin.
7. One end of a thin straight elastic string is fixed at $A(4, -1)$ and the other end B is at $(1, 2)$ in the unstretched condition. If the string is stretched to triple its length, then find the new position of B.
8. The coordinates of the extremities of a rod are $A(1, 2)$ and $B(3, 4)$. $S(0, 0)$ is a point source of light. The rod AB is parallel to the wall and is midway between the point source and the wall. CD is the shadow of AB on the wall. S, AB and CD are in the same horizontal plane. Find the ends C, D of the shadow.
9. Find the circumcentre of the triangle whose vertices are $(2, 1)$, $(5, 2)$ and $(3, 4)$.
10. Find the centroid and the incentre of the triangle whose vertices are $(2, 4)$, $(6, 4)$ and $(2, 0)$.
11. The coordinates of the points A, B and C are respectively $(6, 3)$, $(-3, 5)$ and $(4, -2)$ and the coordinates of P are (x, y), prove that
$$ar(\Delta PBC) : ar(\Delta ABC) = |x + y - 2| : 7.$$
12. If the vertices of a triangle have integral coordinates, prove that the triangle cannot be equilateral.
13. Prove that the area of the triangle whose vertices are $(a\cos\theta_r, b\sin\theta_r)$; $r = 1, 2, 3$ is equal to
$$2ab\sin\frac{1}{2}(\theta_1-\theta_2)\cdot\sin\frac{1}{2}(\theta_2-\theta_3)\cdot\sin\frac{1}{2}(\theta_3-\theta_1).$$
Hence or otherwise prove that no three distinct points of the above type can be collinear.
14. Show that the equation of the locus of a point, which moves so that the difference of its distances from two given points $(ae, 0)$ and $(-ae, 0)$ is equal to $2a$, is
$$\frac{x^2}{a^2} - \frac{y^2}{a^2(e^2-1)} = 1.$$
15. Find the locus of the point at which the line segment joining the given points (α, β) and (γ, δ) subtends a right angle.
16. A rod of length l cm slides on two mutually perpendicular lines. Find the locus of the middle point.
17. The position of a moving point in the x-y plane at time t is given by $(u\cos\alpha\cdot t, u\sin\alpha\cdot t - pt^2)$ where u, α, p are constants. Find the equation of the locus of the moving point.
18. The straight line l is perpendicular to the line $5x - y = 1$. The area of the Δ formed by the line l and the coordinate axes is 5. Find the equation of the line.

19. Find the equations of straight lines passing through (2, 3) and having an intercept of length 2 units between the straight lines $y + 2x = 3$, $y + 2x = 5$.

20. Find the equation of the straight line through the point of intersection of lines $x - 3y + 1 = 0$ and $2x + 5y - 9 = 0$, and whose distance from the origin is $\sqrt{5}$.

21. Find the points on the line $x + y = 4$ that lie at a unit distance from the line $4x + 3y = 10$.

22. A line is such that its segment between the lines $5x - y + 4 = 0$ and $3x + 4y - 4 = 0$ is bisected at the point (1, 5). Obtain its equation.

23. Lines $L_1 \equiv ax + by + c = 0$ and $L_2 \equiv lx + my + n = 0$ intersect at the point P and make an angle θ with each other. Find the equation of a line different from L_2 which passes through P and makes the same angle θ with L_1.

24. Straight lines $3x + 4y = 5$ and $4x - 3y = 15$ intersect at A. Points B and C are chosen on these lines such that $AB = AC$. Determine the possible equations of the line BC passing through the point (1, 2).

25. The equations of perpendicular bisectors of the sides AB and AC of a ΔABC are $x - y + 5 = 0$ and $x + 2y = 0$ respectively. If the vertex A is (1, –2), find the equation of the line BC.

26. The equations of the sides AB, BC and CA of a ΔABC are respectively $x + y = 1$, $x - 2y = 3$ and $2x - y = 0$. D is the foot of the perpendicular from A to the side BC, and DE is drawn parallel to BA cutting AC at E. Find the point E.

27. The coordinates of the foot of the perpendicular drawn from the point (x_1, y_1) on the line $lx + my + n = 0$ are (h, k). Prove that

$$\frac{x_1 - h}{l} = \frac{y_1 - k}{m} = \frac{lx_1 + my_1 + n}{l^2 + m^2}.$$

28. Equations of two straight lines are

$x\cos\alpha + y\sin\alpha = p$ and $x\cos\beta + y\sin\beta = p'$.

Show that the area of the quadrilateral formed by the two lines and the perpendiculars drawn from the origin to the lines is

$$\frac{1}{2\sin(\beta - \alpha)}\{2pp' - (p^2 + p'^2)\cos(\alpha - \beta)\}.$$

29. The sides AB, BC, CD and DA of a quadrilateral have the equations $x + 2y = 3$, $x = 1$, $x - 3y = 4$ and $5x + y + 12 = 0$ respectively. Find the angle between the diagonals AC and BD.

30. Let P, Q and R be three collinear points on BC, CA and AB of ΔABC. Show that

$BP \cdot CQ \cdot AR + PC \cdot QA \cdot RB = 0$,

proper sign for external section being taken into account.

31. Let ABC be a triangle with $AB = AC$. If D is the midpoint of BC, E the foot of the perpendicular drawn from D to AC and F the midpoint of DE, prove that AF $\perp$ BE.

32. Let the line l_1 have intercepts a and b on the coordinate axes. Another line l_2 passes through the middle point of the intercept on l_1 between the axes. If l_2 cuts intercepts p, q on the axes respectively then prove that $aq + bp = 2pq$.

33. Show that the lines $4x + y - 9 = 0$, $5x - y = 6$ and $x - 2y + 3 = 0$ make equal intercepts on any line of gradient 2.

34. The equations of three lines are $7x + y = 16$, $5x - y - 8 = 0$ and $x - 5y + 8 = 0$. Any line through the point $A = (c + 1, 2c)$ meets the lines at P, Q, R respectively. Prove that AP is the harmonic mean between AQ and AR.

35. Prove that the lines $y = m_r x$; $r = 1, 2, 3$ cut off equal intercepts on the transversal $x + y = 1$ if $1 + m_1$, $1 + m_2$, $1 + m_3$ are in HP.

36. A straight line through A(–2, –3) cuts the lines $x + 3y = 9$ and $x + y + 1 = 0$ at B and C respectively. Find the equation of the line if $AB \cdot AC = 20$.

37. A line through the point A(–5, –4) meets the lines $x + 3y + 2 = 0$, $2x + y + 4 = 0$ and $x - y - 5 = 0$ at the points B, C and D respectively. If

$$\left(\frac{15}{AB}\right)^2 + \left(\frac{10}{AC}\right)^2 = \left(\frac{6}{AD}\right)^2,$$

find the equation of the line.

38. In the ΔABC, the equations of AB and AC are $2x + 3y = 29$ and $x + 2y = 16$. If the median through A cuts BC at (5, 6), find the equation of the side BC.

39. Find points equidistant from the lines $4x + 3y = 0$ and $3x - 4y = 3$, the points being at a distance 2 from the point of intersection of the given lines.

40. Find the equations of the lines which pass through (4, 5) and make equal angles with the lines $5y = 12x + 6$ and $3x = 4y + 7$.

41. Find the equations of the bisectors of the angles between the lines $x + y - 3 = 0$ and $7x - y + 5 = 0$, and indicate which of them bisects the acute angle between the lines.

42. The bisectors of the angle between the lines $y = \sqrt{3}x + 3$ and $\sqrt{3}y = x + 3\sqrt{3}$ meet the x-axis at P and Q. Find the length PQ.

43. Prove that the bisector of the angle containing the origin is also the bisector of the acute angle between the lines $a_1x + b_1y + c_1 = 0$ and $a_2x + b_2y + c_2 = 0$ provided c_1, c_2 are of the same sign and

$a_1a_2 + b_1b_2 < 0$.

44. Determine whether the origin lies inside or outside the triangle whose sides have the equations $7x - 5y - 11 = 0$, $8x + 3y + 31 = 0$ and $x + 8y - 19 = 0$.

45. Find the equation of the bisector of the angle between the lines $x + 2y - 11 = 0$, $3x - 6y - 5 = 0$ which contains the point $(1, -3)$.

46. Find the incentre of the triangle whose sides are $7x - y + 11 = 0$, $x + y - 15 = 0$ and $7x + 17y + 65 = 0$. Obtain the inradius of the triangle.

47. Derive the conditions to be imposed on β so that $(0, \beta)$ should lie on or inside the triangle having sides

$$y + 3x + 2 = 0, 3y - 2x - 5 = 0 \text{ and } 4y + x - 14 = 0.$$

48. Given vertices $A(1, 1)$, $B(4, -2)$ and $C(5, 5)$ of a triangle, find the equation of the perpendicular dropped from C to the interior bisector of the angle A.

49. The line $2x + y + 1 = 0$ cuts the sides OA and OB of the ΔOAB internally where O is the origin, $A = (a, 3)$ and $B = (2, 5a - 1)$. If a is a parameter find the set of possible values of a.

50. Find the centre of the excircle of the triangle whose sides are $x - y + 3 = 0$, $y = 7x + 3$ and $x + y + 1 = 0$, corresponding to the side $y = 7x + 3$.

51. Find the equations of the sides of a ΔABC with $A(1, 3)$ as a vertex and $x - 2y + 1 = 0$, $y - 1 = 0$ as the equations of two of its medians.

52. Find the equations of the sides of a triangle having $B(-4, -5)$ as a vertex and

$$5x + 3y - 4 = 0, 3x + 8y + 13 = 0$$

as the equations of two of its altitudes.

53. Find the equations of the legs of a right isosceles triangle if the equation of its hypotenuse is $x - 2y - 3 = 0$ and the vertex at the right angle is $(1, 6)$.

54. ABC is an equilateral triangle. AD is the altitude through A. If $A = (1, 2)$ and $D = (-2, 6)$ then find the equations of the sides of the triangle.

55. In a right-angled triangle the vertex at the right angle is $(1, 1)$. One of the sides of the triangle is $2x - y = 1$ and the midpoint of the hypotenuse is $(5, -2)$. Find the equations of the other sides of the triangle.

56. The altitudes of a ΔABC are respectively AD, BE and CF. If the points A, D, E and F have the coordinates $(-4, 5)$, $\left(\frac{16}{5}, \frac{-23}{5}\right)$, $(4, 1)$ and $(-1, -4)$ respectively, find the other vertices.

57. The extremities of the base of an isosceles triangle are the points $(2a, 0)$ and $(0, a)$. The equation of one of the sides is $x = 2a$. Find the equations of the other two sides and the area of the triangle.

58. Two equal sides of an isosceles triangle have the equations $7x - y + 3 = 0$ and $x + y - 3 = 0$ and its third side passes through the point $(1, -10)$. Determine the equation of the third side.

59. An equilateral triangle has its centroid at the origin and one side is on the line $x + y = 1$. Find the equations of the other sides.

60. Find the equations of the lines passing through $(5, 1)$ and forming isosceles triangles with the lines $12x - 5y + 3 = 0$ and $3x - 4y = 5$ if the vertical angle be the angle between the given lines.

61. ABC is an isosceles triangle, the equation of the base BC being $x + y = 2$ and vertex $A = (3, 1)$. If the lengths of equal sides AB and AC be each equal to 2, find the the vertical angle A and the area of the triangle.

62. The extremities of a diagonal of a square are $(1, 1)$ and $(-2, -1)$. Find the other vertices and the equation of the other diagonal.

63. Find the equations of the two straight lines through $(1, 2)$ forming the two sides of a square of which $4x + 7y = 12$ is a diagonal.

64. A square has its centre at the origin and a vertex at $(1, 2)$. Find the equations of its sides.

65. The point $(1, -1)$ is the centre of a square, one of whose sides lie on the line $x - 2y + 12 = 0$. Find the equations of the straight lines which contain the remaining sides of the square.

66. Two sides of a square lie on the lines

$$5x - 12y - 65 = 0 \text{ and } 5x - 12y + 26 = 0.$$

Find the area of the square.

67. The equations of two sides of a square whose area is 25 square units are $3x - 4y = 0$ and $4x + 3y = 0$. Find the possible equations of the other two sides.

68. Three sides of a square has the equations $2x - y - 4 = 0$, $x + 2y + 3 = 0$ and $x + 2y - 7 = 0$. Find the equation of the fourth side.

69. One side of a rectangle lies along the line $4x + 7y + 5 = 0$. Two of its vertices are $(-3, 1)$ and $(1, 1)$. Find the equations of the other three sides.

70. The equation of one side of a rectangle is $3x - 4y - 10 = 0$ and the coordinates of two of its vertices are $(-2, 1)$ and $(2, 4)$. Find the area of the rectangle and the equation of that diagonal of the rectangle which passes through the point $(2, 4)$.

71. The diagonals of a trapezium cut each other at $(3, 4)$ in the ratio $2 : 1$. The equation of the larger of the two parallel sides is $x + y = 3$ and that of one of the oblique sides is $2x - y = 3$. Find the equations of the other two sides of the trapezium.

72. Two consecutive sides of a parallelogram are $4x + 5y = 0$ and $7x + 2y = 0$. If the equation to one diagonal is $11x + 7y = 9$, find the equation of the other diagonal.

73. The equations of two sides of a parallelogram are $3x - 2y + 12 = 0$ and $x - 3y + 11 = 0$ and the point of intersection of its diagonals is (2, 2). Find the equations of other two sides and its diagonals.

74. Two opposite sides of a rhombus are $x + y = 1$ and $x + y = 5$. If one vertex is (2, –1) and the angle at that vertex is 45°, find the vertex opposite to the given vertex.

75. A rhombus has two of its vertices at (2, 3) and (–2, 6) and two of its sides are parallel to the line $2x + y = 1$. Find the other vertices of the rhombus if (0, 3) is an interior point of the rhombus.

76. Show that the diagonal of a parallelogram whose sides are $u = p$, $u = q$, $v = r$ and $v = s$ where $u \equiv ax + by + c$ and $v \equiv a'x + b'y + c'$, passing through the point of intersection of $u = p$ and $v = r$ is given by $\begin{vmatrix} u & v & 1 \\ p & r & 1 \\ q & s & 1 \end{vmatrix} = 0$.

77. A rhombus has two of its sides parallel to the lines $y = 2x + 3$ and $y = 7x + 2$. If the diagonals cut at (1, 2) and one vertex is on the y-axis, find the possible values of the ordinate of that vertex.

78. The area of a parallelogram is 12 square units. Two of its vertices are the points A(–1, 3) and B(–2, 4). Find the other two vertices of the parallelogram if the point of intersection of its diagonals lies on the x-axis.

79. Find the orthocentre of the triangle whose vertices are (6, 1), (2, 3) and the origin.

80. Find the centroid, orthocentre and circumcentre of the triangle whose sides have the equations $x + y = 1$, $2x + 3y = 6$ and $4x - y + 4 = 0$.

81. If the equations of the sides of a triangle are $a_r x + b_r y = 1$; $r = 1, 2, 3$ and the orthocentre is the origin then prove that

$$a_1a_2 + b_1b_2 = a_2a_3 + b_2b_3 = a_3a_1 + b_3b_1.$$

82. Two vertices of a triangle are (4, –3) and (–2, 5). If the orthocentre of the triangle is (1, 2), prove that the third vertex is (33, 26).

83. The equations of two sides of a triangle are $3x - 2y + 6 = 0$ and $4x + 5y = 20$ and the orthocentre is (1, 1). Find the equation of the third side.

84. A and B are two fixed points (3, 2) and (5, 1) respectively. ΔPAB is equilateral and situated on the side of AB which is opposite to the origin. Find the orthocentre of the ΔPAB.

85. Prove analytically that the ΔDEF has the same centroid as the ΔABC where D, E and F are the middle points of the sides of the latter triangle. Also prove that the orthocentre of the ΔDEF coincides with the circumcentre of the ΔABC.

86. ABC is a variable triangle whose one side is along the x-axis and the other sides are along the lines $(1 + t)x - ty + t(1 + t) = 0$ for any two values of $t \in R$. Show that the orthocentre of the triangle lies on a fixed line. Also find the equation of the line.

87. If P, Q and R are any three points on the curve whose equation is $xy = c^2$ then prove that the orthocentre of the ΔPQR also lies on that curve.

88. If two vertices of a triangle are (0, 1) and $\left(\frac{1}{4}, \frac{2 - \sqrt{7}}{4}\right)$ and its circumcentre is $\left(\frac{1}{2}, \frac{1}{2}\right)$, find the third vertex if it lies on the line $y = 2x$.

89. In a ΔABC, the equation of the perpendicular bisector of AC is $3x - 2y + 8 = 0$. If A = (1, –1) and B = (3, 1), find the equation of BC and the coordinates of the circumcentre.

90. Prove that the orthocentre of the triangle whose angular points are $(a\cos\alpha_r, a\sin\alpha_r)$; $r = 1, 2, 3$ is the point $(a\,\Sigma\cos\alpha_r, a\,\Sigma\sin\alpha_r)$.

91. A straight line moves so that the sum of the reciprocals of its intercepts on the axes of coordinates is constant. Show that it will always pass through a fixed point.

92. The vertices of a triangle lie on three concurrent lines $y = rx$; $r = 1, 2, 3$. Two of the sides of the triangle pass through the fixed points (1, 0), (0, –1) respectively. Prove that the third side will also pass through a fixed point.

93. If $al + bm + cn = 0$ then prove that the family of lines $lx + my + n = 0$ always passes through a fixed point if a, b, c are constants. Also find the fixed point.

94. A straight line is such that the algebraic sum of the lengths of the perpendiculars drawn on it from any number of fixed points is zero. Show that the line always passes through a fixed point.

95. A and B are two fixed points on the x-axis where $OA = a$, $OB = 2a$ and O is the origin. P is any point in the x-y plane. AP and BP meet the y-axis at C and D respectively. AD cuts OP at Q. Prove that CQ passes through a fixed point for all positions of P.

96. ABC is a given triangle in which $AB = AC$. The sides AB and AC are produced to P and Q respectively such that $BP \cdot CQ = AB^2$. Prove that the line PQ always passes through a fixed point.

97. Prove that $(\lambda - 1)x + \lambda y = 1 - 3\lambda$, where λ is an arbitrary constant, represents a family of straight lines passing through a fixed point. Also find the point.

98. If a, b, c are nonzero real numbers then prove that the four lines $x + y + 1 = 0$, $ax + by + c = 0$, $a^2x + b^2y + c^2 = 0$ and $a^3x + b^3y + c^3 = 0$ will pass through the same point if any two of a, b, c are equal.

99. The lines $ax + by + c = 0$, $bx + cy + a = 0$ and $cx + ay + b = 0$ are concurrent where a, b, c are the sides of the ΔABC in usual notations. Prove that $\sin^3 A + \sin^3 B + \sin^3 C = 3\sin A \cdot \sin B \cdot \sin C$.

100. The variable line $2x + 3y = c$ cuts the x and y axes at A and B respectively. A line PQ perpendicular to this line cuts the line segment AB at P and the y-axis at Q. If ar(ΔOAB) : ar(ΔPQA) = 13 : 10 where O is the origin then prove that the centroid of the ΔPQA lies on a fixed line.

101. ABC is a variable triangle with a fixed centroid (5, 5). The side $BC = 13$ and B and C move on the x and y axes respectively. Find the equation of the locus of the vertex A.

102. A straight line moves in such a way that the length of the perpendicular on it from the origin is always p. Find the locus of the centroid of the triangle which is formed by the line and the axes.

103. Find the locus of the mean point of the segment of the line $x\cos\alpha + y\sin\alpha = p$ between the axes where p is a constant and α is a parameter.

104. Find the locus of middle points of the intercepts made by the axes on the lines drawn through the point (α, β).

105. A line segment of length l moves with its ends on two mutually perpendicular lines. Find the locus of the point which divides the line segment in the ratio 1 : 2.

106. A line AB of length $2l$ moves with the end A always on the x-axis and the end B on the line $y = 6x$. Find the equation of the locus of the middle point of AB.

107. The variable line $x\cos\theta + y\sin\theta = 2$ cuts the x and y axes at A and B respectively. Find the locus of the vertex P of the rectangle OAPB, O being the origin.

108. A and B are two fixed points on the line $y = mx$ where $OA = a$, $OB = b$ and O is the origin. P is a variable point in the x-y plane. If the lines PA and PB make constant intercept c on the x-axis then find the locus of P.

109. A line cuts the x-axis at A(7, 0) and the y-axis at B(0, –5). A variable line PQ is drawn perpendicular to AB cutting the x-axis at P and the y-axis at Q. If AQ and BP intersect at R, find the locus of R.

110. A variable straight line passes through the point of intersection of the lines $x + 2y = 1$ and $2x - y = 1$, and meets the coordinate axes at A and B. Find the locus of the centroid of the ΔAOB where O is origin.

111. A variable line drawn through the point of intersection of the straight lines $\frac{x}{\alpha} + \frac{y}{\beta} = 1$ and $\frac{x}{\beta} + \frac{y}{\alpha} = 1$ meets the coordinate axes at A and B. If M is the midpoint of AB, show that the locus of M is the curve $2(\alpha + \beta)xy = \alpha\beta(x + y)$.

112. A variable line through the point P(–1, 2) cuts the axes at A and B respectively. Q is a point on AB such that PQ is the harmonic mean of PA and PB. Show that the locus of Q is the line $y = 2x$.

113. A rectangle PQRS has its side PQ parallel to the line $y = mx$ and vertices P, Q and S on the lines $y = a$, $x = b$ and $x = -b$ respectively. Find the locus of the vertex R.

114. A point P moves along the y-axis. Another point Q moves such that the fixed straight line $x\cos\alpha + y\sin\alpha = p$ is the perpendicular bisector of the line segment PQ. Find the locus of Q.

115. If a line moves in such a way that the sum of the squares of the reciprocals of the intercepts made by the line on the axes is a constant then prove that the locus of the foot of the perpendicular from the origin to the line is a circle.

116. Two fixed points A and B are taken on the x and y axes respectively such that $OA = a$, $OB = b$. Two variable points C, D are taken on the x and y axes respectively. Find the locus of the point of intersection of AD and CB when

$$\frac{1}{OC} - \frac{1}{OD} = \frac{1}{OA} - \frac{1}{OB}.$$

117. Given n straight lines and a fixed point O. Through O, a straight line is drawn meeting these lines in the points $R_1, R_2, \ldots, R_n$ and a point R is taken on it such that $\frac{n}{OR} = \sum_{r=1}^{n} \frac{1}{OR_r}$.

Show that the locus of R is a straight line.

118. Find the locus of the middle point of the intercept on the line $y = x + c$ made by the lines $2x + 3y = 5$ and $2x + 3y = 8$, c being a parameter.

119. If O is the origin, A = (4, 4) and B is any point on the y-axis, find the locus of the point of intersection of the perpendicular bisectors of OB and AB.

120. Find the equation of the locus of R if P, Q are two given points and R is a variable point on one side of the line PQ such that

$\angle RPQ - \angle RQP = 2\alpha$ (a positive constant). [Take $PQ = a$, $P = (0, 0)$ and PQ as the x-axis.]

121. Let $L_1 = 0$ and $L_2 = 0$ be two fixed lines. A variable line is drawn through the origin to cut the two lines at A and B. P is a point on the line AB such that

$$\frac{m+n}{OP} = \frac{m}{OA} + \frac{n}{OB}.$$

Show that the locus of P is a straight line passing through the point of intersection of the given lines.

122. A variable straight line passes through a fixed point (h, k). Find the locus of the foot of the perpendicular on it drawn from the origin.

123. Through a fixed point any straight line is drawn meeting two given parallel lines at A and B. Through A and B straight lines are drawn in given directions, meeting at P. Prove that the locus of P will be a straight line.

124. The equation of two sides of a triangle are $y = 3x$ and $y = 0$. The third side is on a variable line passing through $(1, 2)$. Find the locus of the (i) centroid (ii) orthocentre of the triangle.

125. A variable straight line OPQ passes through the fixed point O, meeting two fixed lines in points P, Q. In the straight line OPQ, a point R is taken such that OP, OR, OQ are in harmonic progression. Show that the locus of R is a straight line.

126. Let $P(2c-1, c+2)$ be a variable point. PM and PN are perpendiculars from P on two fixed lines which meet at O, the origin; MQ, NQ are drawn parallel to the fixed straight lines to meet at Q. Prove that the locus of Q is a straight line.

127. The product of perpendiculars drawn from a point P on the lines $x\cos\theta_r + y\sin\theta_r = a$; $r = 1, 2$ is equal to the square of the perpendicular drawn from the same point on the line

$$x\cos\frac{1}{2}(\theta_1+\theta_2) + y\sin\frac{1}{2}(\theta_1+\theta_2) = a\cos\frac{1}{2}(\theta_1-\theta_2).$$

Show that the locus of the point is a circle.

128. $A = (a, 0)$ and $B = (-a, 0)$ are two fixed points. P is a variable point such that lines PA and PB intercept a constant length k on a given line $y = x$. Find the equation of the locus of P.

129. The equation of a family of lines is given by $y + 2at = t(x - at^2)$ where t is a parameter. Find the locus of the intersection of perpendicular pairs of lines of the family.

130. O is a fixed point. The point P moves along a fixed straight line not passing through O. If Q is a point on OP such that $OP \cdot OQ = k^2$ then show that the locus of Q is a circle.

131. The point $(1, 1)$ is translated parallel to $2x = y$ in the first quadrant through a unit distance. Find the new position of the point.

132. Two particles start from the same point $(2, -1)$, one moving 2 units along the line $x + y = 1$ and the other 5 units along $x - 2y = 4$. If the particles move towards increasing y, find their new positions and the distance between them.

133. Is there a real value of λ for which the image of the point $(\lambda, \lambda - 1)$ by the line mirror $3x + y = 6\lambda$ is the point $(\lambda^2 + 1, \lambda)$? If so find λ.

134. The line $2x - y = 5$ turns about the point on it where the ordinate is equal to the abscissa through an angle of 45° in the anticlockwise direction. Find the equation of the line in the new position.

135. The line $x + 2y = 4$ is translated parallel to itself by 3 units in the sense of increasing x and then rotated by 30° in the clockwise direction about the point where the shifted line cuts the x-axis. Find the equation of the line in the new position.

136. A ray of light coming from the point $(1, 2)$ is reflected at a point A on the x-axis and then passes through the point $(5, 3)$. Find the point A.

137. A man starts from the point $P(-3, 4)$ and reaches the point $Q(0, 1)$ after touching the line $2x + y = 7$ at R. Find R on the line so that he travels along the shortest path.

138. A ray of light is sent along the straight line $y = \frac{2}{3}x - 4$. On reaching the x-axis it is reflected. Find the point of incidence and the equation of the reflected ray.

139. From the point $(2, 5)$ rays of light are sent at 45° with the line $2x + y = 1$. Find the equations of lines of reflected rays if the rays reflect from $2x + y = 1$.

140. A beam of light is sent along the line $x - y = 1$ which after refracting from the x-axis enters the opposite side by turning through 30° towards the normal at the point of incidence on the x-axis. Find the equation of the refracted ray.

Objective Questions

Fill in the blanks.

141. If two vertices of a triangle are $(-3, 1)$, $(0, -2)$ and the centroid is the origin then the third vertex is ______ .

142. The points $(-2, 3)$, $(3, 8)$ and $(4, 1)$ are the vertices of ______ triangle.

143. The points $(1, -2)$, $(3, 0)$, $(1, 2)$ and $(-1, 0)$ are vertices of a ______ .

144. If the area of a triangle with vertices (4, 0), (1, 1) and (3, a) be 2 then a = ______ .

145. A parallelogram has three consecutive vertices (– 3, 4), (0, – 4) and (5, 2). The fourth vertex = ______ .

146. If $mn = 1$ then the lines $mx + y = 1$ and $y - nx = 2$ will be ______ .

147. The equation of the line passing through (1, 2) and perpendicular to the line $x + y + 4 = 0$ is ______ .

148. The distance of (1, 2) from the line $3x - 4y + 15 = 0$ measured parallel to the line $4x + 3y = 0$ is ______ .

149. The set of lines $ax + by + c = 0$, where $3a + 2b + 4c = 0$, is concurrent at the point ______ .

150. The area of the quadrilateral whose sides are given by $|x| + |y| = 1$ is ______ .

151. If a and b are real numbers between 0 and 1 such that the points $(a, 1)$, $(1, b)$ and (0, 0) form an equilateral triangle then a = ______ and b = ______ .

152. The orthocentre of the triangle formed by the lines $x + y = 1$, $2x + 3y = 6$ and $4x - y + 9 = 0$ lies in the ______ quadrant.

153. If A, B, C are in AP then the straight line $Ax + By + C = 0$ will always pass through the fixed point ______ .

154. Let the algebraic sum of the perpendicular distances of the points (2, 0), (0, 2) and (1, 1) from a variable straight line be zero. Then the line passes through the fixed point ______ .

155. If the image of the point (–2, 1) by a line mirror be (2, –1) then the equation of the line mirror is ______ .

156. A line segment has its end points A(1, 1) and B($\sqrt{3}$, 3). The ray OP through the origin O, turns round O. The range of values of the inclination of the line OP with the x-axis so that OP cuts the line segment AB is ______ .

157. If the point $(2a - 3, a^2 - 1)$ is on the same side of the line $x + y - 4 = 0$ as that of the origin then the set of values of a is ______ .

158. If the vertices (x_r, y_r); $r = 1, 2, 3$ of a triangle lie on the curve $xy = 1$ and x_r satisfies the equation $x^3 - 2x^2 + 7x - 1 = 0$ for $r = 1, 2, 3$ then the centroid of the triangle is ______ .

159. If the point $(2, a)$ lies between the lines $x + y = 1$ and $2(x + y) = 5$ then a lies between ______ and ______ .

Choose the correct option(s).

160. The points $(-a, -b)$, (0, 0), (a, b) and (a^2, ab) are

(a) collinear

(b) vertices of a parallelogram

(c) vertices of a rectangle

(d) none of these

161. The points $\left(0, \frac{8}{3}\right)$, (1, 3) and (82, 30) are vertices of

(a) an obtuse-angled Δ (b) an acute-angled Δ

(c) a right-angled Δ (d) an isosceles Δ

(e) none of these

162. If the lines $3x - 4y + 7 = 0$ and $ax + 6y + 1 = 0$ are perpendicular then a is

(a) 4 (b) 5 (c) 10 (d) 8

163. The angle between the lines $\sqrt{3}x - y = 2$ and $3x = 2$ is

(a) 0° (b) 30°

(c) 60° (d) none of these

164. The equation of the line passing through (4, 0) and parallel to the line $2x - 2y - 3 = 0$ is

(a) $2x - 2y = 10$ (b) $2x - 2y - 8 = 0$

(c) $2x - 2y + 2 = 0$ (d) $x + y = 4$

165. The point on the line $3x - 4y = 25$ which is nearest to the origin is

(a) (– 4, 5) (b) (3, – 4)

(c) (3, 4) (d) (3, 5)

166. The distance between the lines $3x + 4y = 6$ and $6x + 8y = 15$ is

(a) $\frac{3}{2}$ (b) $\frac{3}{10}$

(c) 6 (d) none of these

167. The straight lines $x + y = 0$, $3x + y = 4$ and $x + 3y - 4 = 0$ form a triangle which is

(a) isosceles (b) equilateral

(c) right angled (d) none of these

168. The lines $px + qy + r = 0$, $qx + ry + p = 0$ and $rx + py + q = 0$ are concurrent if

(a) $p + q + r = 0$

(b) $p^2 + q^2 + r^2 = pq + qr + rp$

(c) $p^3 + q^3 + r^3 = 3pqr$

(d) none of these

169. If a, b, c are in AP then $ax + by + c = 0$ represents

(a) a single line

(b) a family of concurrent lines

(c) a family of parallel lin[illegible]

(d) none of these

170. Given four lines whose equations are $x + 2y - 3 = 0$, $2x + 3y - 4 = 0$, $3x + 4y - 7 = 0$ and $4x + 5y - 6 = 0$ then

(a) they are all concurrent

(b) they are sides of a quadrilateral

(c) they are sides of a trapezium

(d) none of these

171. PQRS is a rectangle whose sides are parallel to fixed directions. P lies on the x-axis while Q and S lie on the lines $x = a$ and $x = -a$ respectively. The locus of R is a

(a) circle

(b) straight line

(c) pair of straight lines

(d) none of these

172. If P = (1, 0), Q = (–1, 0) and R = (2, 0) are three given points, the locus of a point S satisfying the relation $SQ^2 + SR^2 = 2SP^2$ is

(a) a straight line parallel to the x-axis

(b) a circle passing through the origin

(c) a circle with the centre at the origin

(d) a straight line parallel to the y-axis

173. If the sum of the distances of a point from two perpendicular lines in a plane is 1 then its locus is

(a) a square (b) a circle

(c) a straight line (d) two intersecting lines

174. The image of the point (–1, 3) by the line $x - y = 0$ is

(a) (3, –1) (b) (1, –3) (c) (–1, –1) (d) (3, 3)

175. The point (4, 1) undergoes the following three transformations successively.

(i) reflection about the line $y = x$

(ii) translation through a distance of 2 units along the positive direction of the x-axis

(iii) rotation through an angle $\frac{\pi}{4}$ about the origin in the anticlockwise sense.

The final position of the point is given by

(a) $\left(\frac{1}{\sqrt{2}}, \frac{7}{\sqrt{2}}\right)$ (b) $(-\sqrt{2}, 7\sqrt{2})$

(c) $\left(-\frac{1}{\sqrt{2}}, \frac{7}{\sqrt{2}}\right)$ (d) $(\sqrt{2}, 7\sqrt{2})$

176. All points lying on or inside the triangle formed by the points (1, 3), (5, 0) and (–1, 2) satisfy

(a) $3x + 2y \geq 0$ (b) $2x + y - 13 \geq 0$

(c) $2x - 3y - 12 \leq 0$ (d) $-2x + y \geq 0$

177. The two line segments joining (–2, 7), (–5, –3) and (–8, –13), (1, 17) cut each other at

(a) only one point

(b) no point

(c) infinite number of points

(d) none of these

178. The points (1, 1), (2, 2), (3, –1) and (5, 1) are vertices of a

(a) rhombus (b) trapezium

(c) rectangle (d) none of these

179. If the line $y - x + 2 = 0$ is shifted parallel to itself towards the positive direction of the x-axis by a distance $3\sqrt{2}$ units then the equation of the line in the new position is

(a) $y = x - 8$ (b) $y = x + 4$

(c) $y = x - (2 + 3\sqrt{2})$ (d) none of these

180. The point $(a^2, a + 1)$ is a point in the angle between the lines $3x - y + 1 = 0$ and $x + 2y - 5 = 0$ containing the origin if

(a) $a \geq 1$ or $a \leq -3$ (b) $a \in (0, 1)$

(c) $a \in (-3, 0) \cup \left(\frac{1}{3}, 1\right)$ (d) none of these

181. Let $L_1 \equiv ax + by + a\sqrt[3]{b} = 0$ and

$L_2 \equiv bx - ay + b\sqrt[3]{a} = 0$

be two straight lines. The equations of the bisectors of the angle formed by the loci whose equations are $\lambda_1 L_1 - \lambda_2 L_2 = 0$ and $\lambda_1 L_1 + \lambda_2 L_2 = 0$, λ_1 and λ_2 being nonzero real numbers, are given by

(a) $L_1 = 0$ (b) $L_2 = 0$

(c) $\lambda_2 L_1 + \lambda_1 L_2 = 0$ (d) $\lambda_2 L_1 - \lambda_1 L_2 = 0$

182. The locus of a point which moves so that the sum of the perpendiculars let fall from it on two given straight lines is constant, is

(a) a circle (b) a straight line

(c) a pair of lines (d) none of these

State whether the statements are true or false.

183. If $\begin{vmatrix} x_1 & y_1 & 1 \\ x_2 & y_2 & 1 \\ x_3 & y_3 & 1 \end{vmatrix} = \begin{vmatrix} a_1 & b_1 & 1 \\ a_2 & b_2 & 1 \\ a_3 & b_3 & 1 \end{vmatrix}$ then the two triangles with vertices (x_1, y_1), (x_2, y_2), (x_3, y_3) and (a_1, b_1), (a_2, b_2), (a_3, b_3) must be congruent.

184. The straight line $5x + 4y = 0$ passes through the point of intersection of the straight lines $x + 2y - 10 = 0$ and $2x + y + 5 = 0$.

185. The perpendicular bisector of the line segment joining the points (1, 1) and (3, 5) passes through the point (0, 4).

186. The lines $y + 2x = 3$ and $y + 2x = 5$ cuts the axes at A, B and C, D respectively forming the quadrilateral ABDC. The point (2, 3) is inside the quadrilateral.

187. $(a, a - 2)$, $(b + 2, b)$ and (4, 2) are three points forming a triangle whose area depends on a or b.

Answers

1. centroid $= \left(\frac{5}{3}, 3\right)$,

incentre $= \left(\frac{5\sqrt{2} - 5\sqrt{17} + 9\sqrt{13}}{5\sqrt{2} + \sqrt{17} + \sqrt{13}}, \frac{5\sqrt{2} + 5\sqrt{17} + 3\sqrt{13}}{5\sqrt{2} + \sqrt{17} + \sqrt{13}}\right)$

2. (1, –12), (5, –11) **3.** $\left(\frac{4}{3}, \frac{5}{6}\right)$

5. $\frac{14\sqrt{2}}{3}$ **6.** (– 6, 12)

7. (–5, 8) **8.** C = (2, 4), D = (6, 8)

9. $\left(\frac{13}{4}, \frac{9}{4}\right)$

10. centroid $= \left(\frac{10}{3}, \frac{8}{3}\right)$, incentre $= (6 - 2\sqrt{2}, 2\sqrt{2})$

15. $x^2 + y^2 - (\alpha + \gamma)x - (\beta + \delta)y + \alpha\gamma + \beta\delta = 0$

16. $x^2 + y^2 = \left(\frac{l}{2}\right)^2$ where the given lines are axes

17. $y = x\tan\alpha - \frac{px^2}{u^2\cos^2\alpha}$ **18.** $x + 5y \pm 5\sqrt{2} = 0$

19. $3x + 4y = 18, x = 2$

20. $2x + y = 5$ **21.** (3, 1), (–7, 11)

22. $107x - 3y - 92 = 0$

23. $2(al + bm)(ax + by + c) - (a^2 + b^2)(lx + my + n) = 0$

24. $x - 7y + 13 = 0, 7x + y - 9 = 0$

25. $14x + 23y = 40$

26. $\left(\frac{1}{15}, \frac{2}{15}\right)$ **29.** $\pi/2$

36. $x - y = 1, 3x - y + 3 = 0$

37. $2x + 3y + 22 = 0$ **38.** $x + y = 11$

39. $\left(\frac{9 \pm 35\sqrt{2}}{25}, \frac{-12 \mp 5\sqrt{2}}{25}\right), \left(\frac{9 \pm 5\sqrt{2}}{25}, \frac{-12 \pm 35\sqrt{2}}{25}\right)$

40. $9x - 7y = 1, 7x + 9y = 73$

41. $x - 3y + 10 = 0, 6x + 2y - 5 = 0$
(bisector of acute angle)

42. 6 units **44.** inside **45.** $3x = 19$

46. incentre = (5, 1); $\frac{9}{\sqrt{2}}$ **47.** $\beta \in \left[\frac{5}{3}, \frac{7}{2}\right]$

48. $x = 5$ **49.** $(-\infty, -2)$ **50.** (4, 1)

51. $x + 2y - 7 = 0, x - 4y - 1 = 0, x - y + 2 = 0$

52. $3x - 5y - 13 = 0, 8x - 3y + 17 = 0, 5x + 2y - 1 = 0$

53. $3x - y + 3 = 0, x + 3y - 19 = 0$

54. $3x - 4y + 30 = 0, y - 2 = \frac{3 \mp 4\sqrt{3}}{4 \pm 3\sqrt{3}}(x - 1)$

55. $x + 2y = 3, 7x + 24y + 13 = 0$

56. B = (0, –7), C = (8, –1)

57. $x + 2y = 2a, 3x - 4y + 4a = 0;\ \frac{5}{2}a^2$

58. $3x + y + 7 = 0$ or $x - 3y - 31 = 0$

59. $(2 + \sqrt{3})x - y + \sqrt{3} + 1 = 0, (2 - \sqrt{3})x - y - \sqrt{3} + 1 = 0$

60. $9x - 7y = 38, 7x + 9y = 44$

61. $\pi/2$; 2 square units

62. $\left(\frac{1}{2}, \frac{-3}{2}\right), \left(\frac{-3}{2}, \frac{3}{2}\right)$; $6x + 4y + 3 = 0$

63. $3x - 11y + 19 = 0, 11x + 3y - 17 = 0$

64. $3x + y \pm 5 = 0, x - 3y \pm 5 = 0$

65. $x - 2y - 18 = 0, 2x + y + 14 = 0, 2x + y - 16 = 0$

66. 49 square units

67. $3x - 4y \pm 25 = 0, 4x + 3y \pm 25 = 0$

68. $2x - y - 14 = 0, 2x - y + 6 = 0$

69. $4x + 7y - 11 = 0, 7x - 4y - 3 = 0, 7x - 4y + 25 = 0$

70. 20 square units, $31x - 8y - 30 = 0$

71. $x + y = 9, 7x - 5y + 3 = 0$

72. $y - x = 0$

73. $x - 3y = 3, 3x - 2y = 16$ and
$x + 4y = 10; 5x - 8y + 6 = 0$

74. $(6 + 2\sqrt{2}, -1 - 2\sqrt{2})$ or $(2 - 2\sqrt{2}, 3 + 2\sqrt{2})$

75. $(-2 + \sqrt{5}, 6 - 2\sqrt{5}), (2 + \sqrt{5}, 3 - 2\sqrt{5})$

77. $\frac{5}{9}(1 \pm \sqrt{10})$

78. (–7, –3), (–6 , –4) or (17, –3), (18, –4)

79. $\left(\frac{15}{8}, \frac{15}{4}\right)$

80. $\left(-\frac{47}{35}, \frac{92}{35}\right), \left(\frac{3}{7}, \frac{22}{7}\right), \left(\frac{-78}{35}, \frac{83}{35}\right)$ respectively

83. $26x - 122y = 1675$

84. $\left(4 + \frac{1}{6}\sqrt{3}, \frac{3}{2} + \frac{1}{3}\sqrt{3}\right)$ **86.** $x + y = 0$

88. (0, 0) or $\left(\frac{3}{5}, \frac{6}{5}\right)$ **89.** $x + 4y = 7, \left(\frac{-4}{5}, \frac{14}{5}\right)$

93. $\left(\frac{a}{c}, \frac{b}{c}\right)$ **97.** (– 1, – 2)

101. $x^2 + y^2 - 30x - 30y + 281 = 0$

102. $\frac{1}{x^2} + \frac{1}{y^2} = \frac{9}{p^2}$ **103.** $\frac{1}{x^2} + \frac{1}{y^2} = \frac{4}{p^2}$

104. $\frac{\alpha}{x} + \frac{\beta}{y} = 2$

105. $9x^2 + 36y^2 = 4l^2$ with perpendicular lines as axes

106. $9x^2 + 10y^2 - 6xy - 9l^2 = 0$

107. $\frac{1}{x^2} + \frac{1}{y^2} = \frac{1}{4}$

108. $\frac{a-b}{\sqrt{1+m^2}} \cdot y(y-mx)$

$$\pm c\left(y-\frac{ma}{\sqrt{1+m^2}}\right)\left(y-\frac{mb}{\sqrt{1+m^2}}\right)=0$$

109. $x^2+y^2-7x+5y=0$ **110.** $x+3y=15xy$

113. $(1-m^2)x+my=am+b(1+m^2)$

114. $x\cos 2\alpha + y\sin 2\alpha = 2p\cos\alpha$

116. $x-y=0$ **118.** $4x+6y=13$

119. $x+y=4$

120. Part of the curve $(x^2-y^2)\tan 2\alpha - 2xy + a(y - x\tan 2\alpha) = 0$ where $x<\frac{a}{2}, y>0$ or $y<0$

122. $x^2+y^2-hx-ky=0$

124. (i) $3y^2-9xy+6x-y=0$ (ii) $x^2+3xy-x-2y=0$

128. $(x-y)^2-a^2=\pm\frac{2\sqrt{2}a}{k}\cdot y(x-y)$

129. $y^2=a(x-3a)$ **131.** $\left(1\pm\frac{1}{\sqrt{5}}, 1\pm\frac{2}{\sqrt{5}}\right)$

132. $(2-\sqrt{2}, \sqrt{2}-1), (2\sqrt{5}+2, \sqrt{5}-1)$ respectively; $\sqrt{29+2\sqrt{10}}$

133. 2 **134.** $3x+y=20$

135. $(2+\sqrt{3})x+(2\sqrt{3}-1)y=(4+3\sqrt{5})(\sqrt{3}+2)$

136. $\left(\frac{13}{5}, 0\right)$ **137.** $\left(\frac{42}{25}, \frac{91}{25}\right)$

138. $(6, 0); 3y+2x=12$

139. $x+3y-1=0, 3x-y+15=0$

140. $(2+\sqrt{3})x-y=2+\sqrt{3}$

141. (3, 1) **142.** an isosceles

143. square **144.** $-1, \frac{5}{3}$

145. (2, 10) **146.** perpendicular

147. $y-x-1=0$ **148.** 2

149. $\left(\frac{3}{4}, \frac{1}{2}\right)$ **150.** 2 sq units

151. $2-\sqrt{3}, 2-\sqrt{3}$ respectively

152. second **153.** (1, –2)

154. (1, 1) **155.** $y=2x$

156. $[\pi/4, \pi/3]$ **157.** (– 4, 2)

158. $\left(\frac{2}{3}, \frac{7}{3}\right)$ **159.** -1 and $\frac{1}{2}$

160. (a)	**161.** (e)	**162.** (d)	**163.** (b)
164. (b)	**165.** (b)	**166.** (b)	**167.** (a)
168. (a), (b), (c)	**169.** (b)	**170.** (d)	**171.** (b)
172. (d)	**173.** (a)	**174.** (a)	**175.** (c)
176. (a), (c)	**177.** (c)	**178.** (b)	**179.** (a)
180. (c)	**181.** (a), (b)	**182.** (b)	**183.** false
184. true	**185.** true	**186.** false	**187.** false

□

Chapter Test

Time: 120 minutes

1. In the following, each question has four options of which one or more options are correct. Indicate the correct option(s):

(a) $P(m, n)$ where m, n are natural numbers, is any point in the interior of the quadrilateral formed by the pair of lines $xy = 0$ and the two lines $2x + y - 2 = 0$ and $4x + 5y = 20$. The possible number of positions of the point P is

A. six B. four C. five D. none of these

(b) If (α, α^2) falls inside the angle made by the lines $2y = x, x > 0$ and $y = 3x, x > 0$ then the set of values of α is

A. $(-\alpha, 3)$ B. $\left(\frac{1}{2}, 3\right)$ C. $(0, 3)$ D. $(-\infty, 0) \cup \left(\frac{1}{2}, \infty\right)$

(c) The image of the point A(1, 2) by the line mirror $y = x$ is the point B and the image of B by the line mirror $y = 0$ is the point (α, β). Then

A. $\alpha = 1, \beta = -2$ B. $\alpha = 0, \beta = 0$ C. $\alpha = 2, \beta = -1$ D. none of these

(d) $A = (\sqrt{1 - t^2} + t, 0)$ and $B = (\sqrt{1 - t^2} - t, 2t)$ are two variable points where t is a parameter. The locus of the middle point of AB is

A. a straight line B. a pair of lines C. a circle D. none of these

2. Fill in the following incomplete statements so that the statements become true.

(a) The line through (0, 4) which together with the lines $y = (2 \pm \sqrt{3})x$ form an equilateral triangle, has the equation ______ .

(b) The set of real values of a for which the line segment AB where $A = \left(a^2 + 1, a + \frac{3}{7}\right)$ and $B = (5 - a, a^2 + 2)$ is divided into two segments by the line $2x + 7y = 9$ is ______ .

(c) A square is made with the segment on the line $\frac{x}{a} + \frac{y}{b} = 1$ cut off by the axes as a diagonal. The vertices of the square, not on this diagonal, are ______ and ______ .

(d) The intercepts made on the line $x + y = 5\sqrt{2}$ by the lines $y = x\tan\theta; \theta = 0, \frac{\pi}{4}, \alpha, \frac{\pi}{2}$ are in AP. Then α is ______ .

3. P is a point on the level ground. C is a point south-east of P at a distance $3\sqrt{2}$ m from P. Find a point A bearing north of P and a point B bearing west of P such that P is located at the centroid of the $\triangle ABC$.

4. ABC is a triangle and D, E are the feet of the perpendiculars from A and B to the opposite sides respectively. If D = (20, 25), E = (8, 16) and C = (10, 15) then find the orthocentre of the triangle ABC.

5. Q is any point on the line $x = a$. If $A = (a, 0)$ and the bisector QR of the angle OQA, meets the x-axis at R, find the locus of the foot of the perpendicular from R to OQ, O being the origin.

6. Two fixed lines have the equations $y = 1$ and $y = 4$. The line segment OB whose equation is $y = x$, is rotated about a point on it so that B reaches the lines after rotating through an angle 45°, once anticlockwise and the other time clockwise. If B is $\left(\frac{3+\sqrt{2}}{\sqrt{2}}, \frac{3+\sqrt{2}}{\sqrt{2}}\right)$, find the point about which OB must be rotated.

7. AB is a line joining the points (–1, 3) and (4, 1), and CD is a line whose slope is $\frac{-3}{4}$ and its distance from the origin is 1. Find λ so that the point P(0, λ) when translated parallel to the line $x - y = 1$ passes through the point of intersection Q of the lines AB and CD. Also find the distance through which P must be translated to reach Q.

8. Two mutually perpendicular lines are drawn through the point of intersection A of the lines $y - x = 0$ and $x + y = 1$ to cut the line $2x - y = 0$ at two points B and C. Find the equation of the locus of the centroid of the triangle ABC.

9. ΔOAB has the vertices O = (0, 0), A = $\left(\sqrt{\frac{3}{2}}, 0\right)$ and B = $\left(0, \sqrt{\frac{3}{2}}\right)$.

Find θ such that $0 \le \theta \le 2\pi$ and P = (sin θ, cos θ) is an interior point of the ΔOAB.

10. For the ΔABC, the vertex A = (4, 3), its orthocentre H = (3, 3) and the centroid G = (1, 1). Find the equations of the sides of the ΔABC.

11. ABC is an isosceles triangle in which AB = AC. The images of the vertices A, B, C by the lines, BC, CA, AB are respectively A′, B′ and C′. If the triangle A′B′C′ is equilateral then find the angle A.

12. Let a new distance d(P, Q) between the points P = (x_1, y_1) and Q = (x_2, y_2) be defined as $d(\text{P, Q}) = |x_1 - x_2| + |y_1 - y_2|$. Let O = (0, 0) and A = (3, 2) be two fixed points. Let R = (x, y), $x \ge 0$, $y \ge 0$ such that R is equidistant from the points O and A in the sense of the new distance. Prove that the locus of R consists of a line segment of finite length and an infinite ray. Also, sketch a labelled diagram to show the locus.

□

Answers

1. (a) A (b) B (c) C (d) C

2. (a) $\pm x + y = 4$ (b) $\left(-4, \frac{1}{2}\right)$ (c) $\left(\frac{a+b}{2}, \frac{a+b}{2}\right), \left(\frac{a-b}{2}, \frac{b-a}{2}\right)$ (d) $\tan^{-1} 5$

3. each at 3 m from P

4. (15, 30)

5. $(x^2 + y^2 - ax)^2 = y^2(x^2 + y^2)$

6. (1, 1)

7. $\lambda = 8, 18$; $\frac{27\sqrt{2}}{7}$, $11\sqrt{2}$

8. $12x - 6y = 1$

9. $0° < \theta < 15°$ or $75° < \theta < 90°$

10. $2x + 1 = 0$, $(\sqrt{11} \mp 2)x \pm 3y = 4\sqrt{11} \pm 1$

11. $\frac{\pi}{6}$

12.

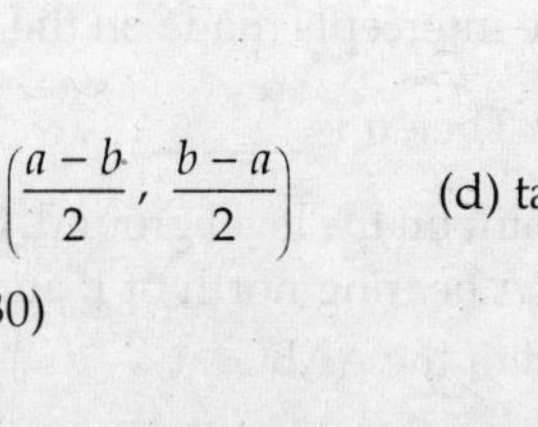

□

2. Pair of Straight Lines and Transformation of Axes

Recap of Facts and Formulae

1. Joint equation of two lines (Pair of lines)

- If $L_1 \equiv a_1x + b_1y + c_1 = 0$ and $L_2 \equiv a_2x + b_2y + c_2 = 0$ be the equations of two straight lines then their joint equation is

$$L_1 \cdot L_2 \equiv (a_1x + b_1y + c_1)(a_2x + b_2y + c_2) = 0,$$

which is an equation of the second degree in x and y.

- If $S \equiv ax^2 + 2hxy + by^2 + 2gx + 2fy + c = 0$ be the equation of a pair of lines then S can be factorized into two linear factors and each factor equated to zero gives the separate equation of a line belonging to the pair.
- The second degree homogeneous equation

$$ax^2 + 2hxy + by^2 = 0$$

represents a pair of straight lines passing through the origin.

- If the joint equation of a pair of lines be

$$ax^2 + 2hxy + by^2 = 0$$

then their separate equations are

$$y = m_1x \text{ and } y = m_2x$$

where $m_1 + m_2 = \dfrac{-2h}{b}$ and $m_1m_2 = \dfrac{a}{b}$.

- If the joint equation of a pair of lines be

$$ax^2 + 2hxy + by^2 + 2gx + 2fy + c = 0$$

then their separate equations are

$$l_1x + m_1y + n_1 = 0 \text{ and } l_2x + m_2y + n_2 = 0$$

where $l_1l_2 = a,\ m_1m_2 = b,\ n_1n_2 = c,$

$$l_1m_2 + l_2m_1 = 2h,\ m_1n_2 + m_2n_1 = 2f,$$
$$l_1n_2 + l_2n_1 = 2g.$$

2. Condition for the general equation of the second degree to represent a pair of lines

- The general equation of the second degree

$$ax^2 + 2hxy + by^2 + 2gx + 2fy + c = 0$$

will represent a pair of straight lines if

$$\Delta \equiv abc + 2fgh - af^2 - bg^2 - ch^2 = 0.$$

3. Formulae for the pair $ax^2 + 2hxy + by^2 = 0$

- For the pair of straight lines whose joint equation is $ax^2 + 2hxy + by^2 = 0$,
  (i) the angle between the lines $= \tan^{-1}\left|\dfrac{2\sqrt{h^2 - ab}}{a + b}\right|$
  (ii) the lines are perpendicular if $a + b = 0$
  (iii) the lines are coincident if $h^2 = ab$
  (iv) the joint equation of the bisectors of the angles between the lines is

$$\frac{x^2 - y^2}{a - b} = \frac{xy}{h}.$$

- The angle between the pair of lines

$$ax^2 + 2hxy + by^2 = 0$$

and that between the pair of lines

$$ax^2 + 2hxy + by^2 + 2gx + 2fy + c = 0$$

are equal.

4. To fined the point of intersection of the pair of lines $ax^2 + 2hxy + by^2 + 2gx + 2fy + c = 0$

- Denoting the equation by $S = 0$, the point of intersection is obtained by solving

$$\frac{\partial S}{\partial x} = 0 \text{ and } \frac{\partial S}{\partial y} = 0.$$

5. Pair of lines through the origin and intersection of a line with a second degree curve

- The equation of the pair of lines, passing through the origin and the points of intersection of the line $L = 0$ and the curve $S = 0$, is obtained by making $S = 0$ second degree homogeneous with the help of $L = 0$.

6. Transformation of equation due to a change of axes

- If $P = (x, y)$ and the axes are shifted parallel to the original axes so that the new origin is (α, β) and the new coordinates of P are (x', y') then the equations of transformation (i.e., relations between the old and the new coordinates) will be

$$x = x' + \alpha, \ y = y' + \beta.$$

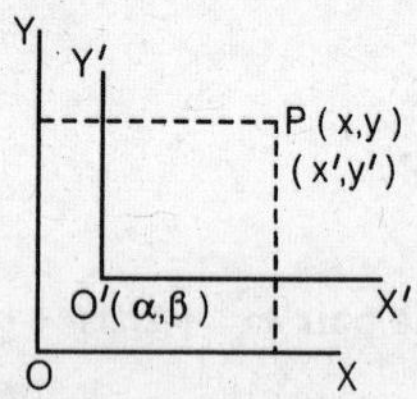

- If $P = (x, y)$ and the axes are rotated about the origin by an angle ϕ in the anticlockwise sense such that the new coordinates of P become (x', y') then the equations of transformation will be

$$x = x'\cos\phi - y'\sin\phi$$
$$y = x'\sin\phi + y'\cos\phi.$$

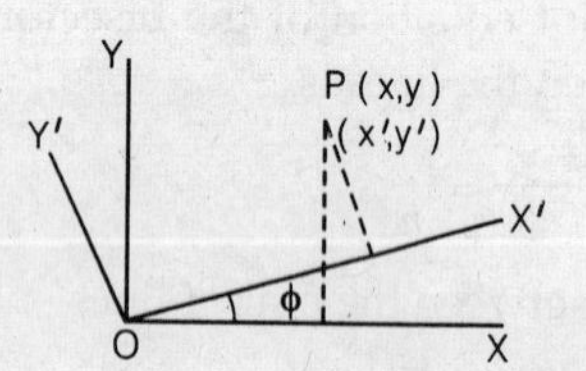

- If $P = (x, y)$ and the axes are shifted parallel to the original axes so that the new origin is (α, β) and then the axes are rotated about the new origin (α, β) by an angle ϕ in the anticlockwise sense such that the new coordinates of P become (x', y') then the equations of transformation will be

$$x = \alpha + x'\cos\phi - y'\sin\phi$$
$$y = \beta + x'\sin\phi + y'\cos\phi.$$

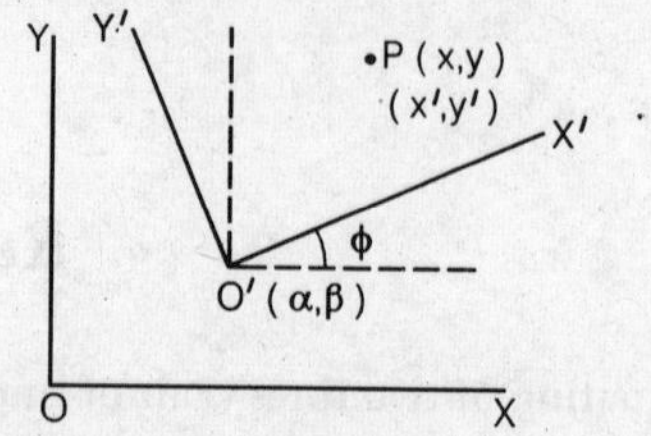

- If $P = (x, y)$ and two mutually perpendicular lines $ax + by + c = 0$ and $bx - ay + d = 0$ are taken as new axes such that new coordinates of P are (x', y') then the equations of transformation will be

$$x' = \frac{bx - ay + d}{\sqrt{b^2 + a^2}}, \ y' = \frac{ax + by + c}{\sqrt{a^2 + b^2}}.$$

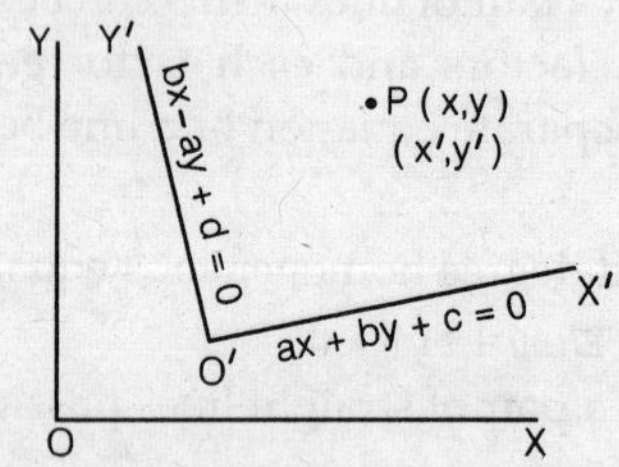

Selected Solved Examples

1. Show that the four lines given by
$12x^2 + 7xy - 12y^2 = 0$ and
$12x^2 + 7xy - 12y^2 - x + 7y - 1 = 0$
lie along sides of a square.

$$12x^2 + 7xy - 12y^2 = 12x^2 + 16xy - 9xy - 12y^2$$
$$= 4x(3x + 4y) - 3y(3x + 4y)$$
$$= (3x + 4y)(4x - 3y)$$

$\therefore$ the pair of lines given by the first equation has their separate equations $3x + 4y = 0$ and $4x - 3y = 0$ which are at right angle.

Again $12x^2 + 7xy - 12y^2 - x + 7y - 1$
$= (3x + 4y)(4x - 3y) - x + 7y - 1$

(factorized as above)

$= (3x + 4y - 1)(4x - 3y + 1)$, by trial

$\therefore$ the pair of lines given by the second equation has their separate equations $3x + 4y - 1 = 0$ and $4x - 3y + 1 = 0$ which are at right angle.

Thus the four lines are $3x + 4y = 0$, $4x - 3y = 0$, $3x + 4y - 1 = 0$ and $4x - 3y + 1 = 0$ of which the first and the third are parallel while the second and the fourth are also parallel. Moreover the first and the second are at right angle. Hence, the lines form a rectangle.

The distance between $3x + 4y = 0$ and $3x + 4y - 1 = 0$ is given by

d = distance of (0, 0) from $3x + 4y - 1 = 0$

{(0, 0) being a point on $3x + 4y = 0$}

$$= \left|\frac{-1}{\sqrt{3^2 + 4^2}}\right| = \frac{1}{5}.$$

The distance between $4x - 3y = 0$ and $4x - 3y + 1 = 0$ is given by

$$d' = \left|\frac{1}{\sqrt{3^2 + 4^2}}\right| = \frac{1}{5}; \quad \therefore \quad d = d'.$$

Hence, the lines form a square.

2. If the pairs of lines represented by $ax^2+2hxy+by^2=0$ and $a_1x^2+2h_1xy+b_1y^2=0$ have one line in common and the other lines represented by them be perpendicular, prove that

$$\frac{ha_1b_1}{b_1-a_1}=\frac{h_1ab}{b-a}=\frac{1}{2}\sqrt{-aba_1b_1}.$$

Let the lines of the first pair be $y-m_1x=0$, $y-m_2x=0$

where $m_1+m_2=\dfrac{-2h}{b}$ ... (1)

and $m_1m_2=\dfrac{a}{b}$ ... (2)

Let the common line be $y-m_1x=0$. Then the other line of the second pair is $y+\dfrac{x}{m_2}=0$, it being perpendicular to the other line of the first pair.

$$\therefore \quad a_1x^2+2h_1xy+b_1y^2\equiv b_1(y-m_1x)\left(y+\frac{x}{m_2}\right)$$

$$\equiv b_1\left\{y^2+\left(\frac{1}{m_2}-m_1\right)xy-\frac{m_1}{m_2}x^2\right\}$$

$$\therefore \quad a_1=-b_1\frac{m_1}{m_2} \qquad \text{... (3)}$$

$$2h_1=b_1\left(\frac{1}{m_2}-m_1\right) \qquad \text{... (4)}$$

From (4) and (2), $2h_1=b_1\cdot\dfrac{1-m_1m_2}{m_2}=b_1\cdot\dfrac{1-\dfrac{a}{b}}{m_2}$

or $\quad 2h_1=\dfrac{b_1(b-a)}{bm_2} \qquad \therefore \quad 2h_1ab=\dfrac{ab_1(b-a)}{m_2}$

$$\therefore \quad \frac{2h_1ab}{b-a}=\frac{ab_1}{m_2} \qquad \text{... (5)}$$

From (1) and (3), $-\dfrac{a_1m_2}{b_1}+m_2=\dfrac{-2h}{b}$

or $\quad \dfrac{(b_1-a_1)m_2}{b_1}=\dfrac{-2h}{b} \quad$ or $\quad m_2=\dfrac{-2hb_1}{b(b_1-a_1)}$

or $\quad \dfrac{2ha_1b_1}{a_1b(b_1-a_1)}=-m_2 \quad$ or $\quad \dfrac{2ha_1b_1}{b_1-a_1}=-a_1bm_2$ ... (6)

Again from (2) and (3),

$\dfrac{a_1m_2}{-b_1}\cdot m_2=\dfrac{a}{b}$; $\quad\therefore\quad m_2^2=\dfrac{-ab_1}{a_1b}$

(clearly ab_1 and a_1b are of opposite signs)

$\therefore$ from (5),

$$\frac{h_1ab}{b-a}=\frac{ab_1}{2m_2}=\frac{1}{2}\sqrt{\frac{a^2b_1^2}{m_2^2}}$$

$$=\frac{1}{2}\sqrt{a^2b_1^2\cdot\frac{-a_1b}{ab_1}}=\frac{1}{2}\sqrt{-aa_1bb_1}.$$

From (6), $\dfrac{ha_1b_1}{b_1-a_1}=\dfrac{1}{2}(-a_1b)m_2=\dfrac{1}{2}\sqrt{(-a_1b)^2m_2^2}$

$$=\frac{1}{2}\sqrt{a_1^2b^2\cdot\frac{-ab_1}{a_1b}}=\frac{1}{2}\sqrt{-aa_1bb_1}.$$

3. Show that the straight lines $(a^2-3b^2)x^2+8abxy+(b^2-3a^2)y^2=0$ form with the line $ax+by+c=0$, an equilateral triangle of area $\dfrac{c^2}{\sqrt{3}\cdot(a^2+b^2)}$.

Here $a^2x^2+2abxy+b^2y^2-3(b^2x^2-2abxy+a^2y^2)=0$

or $\quad (ax+by)^2-3(bx-ay)^2=0$

or $\quad \{ax+by+\sqrt{3}(bx-ay)\}\,\{(ax+by)-\sqrt{3}(bx-ay)\}=0$

or $\quad \{(a+b\sqrt{3})x+(b-a\sqrt{3})y\}\,\{(a-b\sqrt{3})x+(b+a\sqrt{3})y\}=0$

$\therefore$ the three lines are

$$(a+b\sqrt{3})x+(b-a\sqrt{3})y=0 \qquad \text{... (1)}$$

$$(a-b\sqrt{3})x+(b+a\sqrt{3})y=0 \qquad \text{... (2)}$$

$$ax+by+c=0 \qquad \text{... (3)}$$

'm' of (1) $=m_1=\dfrac{a+b\sqrt{3}}{a\sqrt{3}-b}$

'm' of (2) $=m_2=\dfrac{b\sqrt{3}-a}{b+a\sqrt{3}}$

'm' of (3) $=m_3=\dfrac{-a}{b}$.

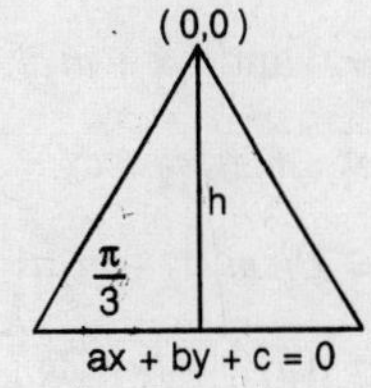

If the angle between (1) and (2) be θ and that between (2) and (3) be ϕ then

$$\tan\theta=\left|\frac{\dfrac{a+b\sqrt{3}}{a\sqrt{3}-b}-\dfrac{b\sqrt{3}-a}{b+a\sqrt{3}}}{1+\dfrac{a+b\sqrt{3}}{a\sqrt{3}-b}\cdot\dfrac{b\sqrt{3}-a}{b+a\sqrt{3}}}\right|$$

$$=\left|\frac{(a+b\sqrt{3})(b+a\sqrt{3})-(b\sqrt{3}-a)(a\sqrt{3}-b)}{(a\sqrt{3}-b)(b+a\sqrt{3})+(a+b\sqrt{3})(b\sqrt{3}-a)}\right|$$

$$=\left|\frac{ab+3ab+\sqrt{3}(a^2+b^2)-4ab+\sqrt{3}(a^2+b^2)}{3a^2-b^2-a^2+3b^2}\right|$$

$$=\left|\frac{2\sqrt{3}(a^2+b^2)}{2(a^2+b^2)}\right|=\sqrt{3};\quad\therefore\quad \theta=\frac{\pi}{3}$$

and $\tan\phi = \left|\dfrac{\dfrac{b\sqrt{3}-a}{b+a\sqrt{3}} - \dfrac{-a}{b}}{1+\dfrac{b\sqrt{3}-a}{b+a\sqrt{3}}\cdot\dfrac{-a}{b}}\right| = \left|\dfrac{b(b\sqrt{3}-a)+a(b+a\sqrt{3})}{b(b+a\sqrt{3})-a(b\sqrt{3}-a)}\right|$

$$= \left|\frac{\sqrt{3}(a^2+b^2)}{a^2+b^2}\right| = \sqrt{3}; \quad \phi = \frac{\pi}{3}$$

$\therefore$ the triangle formed is equilateral, each angle being equal to $\frac{\pi}{3}$. Now, the sides (1) and (2) meet at (0, 0).

The altitude through (0, 0) is

$$h = \frac{c}{\sqrt{a^2+b^2}}$$

$\therefore$ the area of the equilateral triangle

$$= \frac{1}{2}\times h\times\left(2h\cot\frac{\pi}{3}\right) \qquad \left\{\because \frac{h}{\frac{1}{2}\text{base}} = \tan\frac{\pi}{3}\right\}$$

$$= h^2\cdot\frac{1}{\sqrt{3}} = \frac{c^2}{\sqrt{3}\cdot(a^2+b^2)}.$$

4. If the equation $ax^2+2hxy+by^2+2gx+2fy+c=0$ represents two straight lines, prove that the product of the perpendiculars drawn from the origin to the lines is $\dfrac{c}{\sqrt{(a-b)^2+4h^2}}$.

Let $ax^2+2hxy+by^2+2gx+2fy+c$

$$\equiv (l_1x+m_1y+n_1)(l_2x+m_2y+n_2).$$

Then the separate lines are

$$l_1x+m_1y+n_1=0 \text{ and } l_2x+m_2y+n_2=0$$

where $l_1l_2=a,\ m_1m_2=b,\ n_1n_2=c,$

$$l_1m_2+l_2m_1=2h,\ m_1n_2+m_2n_1=2f,$$

$$n_1l_2+n_2l_1=2g.$$

Lengths of perpendiculars from (0, 0) to the lines are respectively

$$\frac{n_1}{\sqrt{l_1^2+m_1^2}} \text{ and } \frac{n_2}{\sqrt{l_2^2+m_2^2}}$$

$\therefore$ the required product of perpendiculars drawn from the origin

$$= \frac{n_1}{\sqrt{l_1^2+m_1^2}}\cdot\frac{n_2}{\sqrt{l_2^2+m_2^2}}$$

$$= \frac{n_1n_2}{\sqrt{l_1^2l_2^2+m_1^2m_2^2+l_1^2m_2^2+l_2^2m_1^2}}$$

$$= \frac{c}{\sqrt{a^2+b^2+(l_1m_2+l_2m_1)^2-2l_1l_2\cdot m_1m_2}}$$

$$= \frac{c}{\sqrt{a^2+b^2+(2h)^2-2ab}}$$

$$= \frac{c}{\sqrt{(a-b)^2+4h^2}}.$$

5. Show that the straight lines joining the origin to the points of intersection of the two curves

$$ax^2+2hxy+by^2+2gx=0$$

and $a'x^2+2h'xy+b'y^2+2g'x=0$

will be at right angles to one another if

$$g'(a+b)=g(a'+b').$$

Any curve passing through the intersection of the given curves is

$$ax^2+2hxy+by^2+2gx$$
$$+\lambda(a'x^2+2h'xy+b'y^2+2g'x)=0 \qquad \dots (1)$$

This will be a pair of straight lines through the origin if it is second degree homogeneous in x and y. For this the condition on (1) is

$$\text{coefficient of } x = 2g+\lambda\cdot 2g'=0; \quad \therefore \quad \lambda=\frac{-g}{g'}.$$

Putting in (1),

$$g'(ax^2+2hxy+by^2)-g(a'x^2+2h'xy+b'y^2)=0.$$

The pair of lines through the origin will be at right angles if coefficient of x^2 + coefficient of $y^2=0$.

or $(g'a-ga')+(g'b-gb')=0$

or $g'a+g'b=ga'+gb'$

or $g'(a+b)=g(a'+b')$.

6. Show that the lines joining the origin to the points of intersection of the line $fx-gy=\lambda$ and the curve $x^2+hxy-y^2+gx+fy+c=0$ are at right angles for all $\lambda\in R$ if $c=0$.

The equation of the given line is $fx-gy=\lambda$

or $\dfrac{fx-gy}{\lambda}=1.$

Making the equation $x^2+hxy-y^2+gx+fy+c=0$ second degree homogeneous with the help of the equation of the line, we get

$$x^2+hxy-y^2+(gx+fy)\cdot\frac{fx-gy}{\lambda}+c\cdot\left(\frac{fx-gy}{\lambda}\right)^2=0$$

or $\lambda^2(x^2+hxy-y^2)+\lambda(gx+fy)(fx-gy)$
$$+c(fx-gy)^2=0.$$

The pair of lines will be at right angles if

coefficient of x^2 + coefficient of $y^2=0$

or $(\lambda^2+\lambda gf+cf^2)+(-\lambda^2-\lambda fg+cg^2)=0$

or $\quad c(f^2+g^2)=0; \quad \therefore \quad c=0.$

7. The base BC of a triangle passes through a fixed point (f, g) and its sides AB and AC are bisected at right angles by the pair of straight lines $y^2-8xy-9x^2=0$. Determine the locus of its vertex A.

Let $A=(\alpha,\beta)$. Here $y^2-8xy-9x^2=0$

or $\quad y^2-9xy+xy-9x^2=0$

or $\quad y(y-9x)+x(y-9x)=0$

or $\quad (y-9x)(y+x)=0$

$\therefore$ the separate lines are $y-9x=0$, $y+x=0$

Let them be perpendicular bisectors of AC and AB respectively.

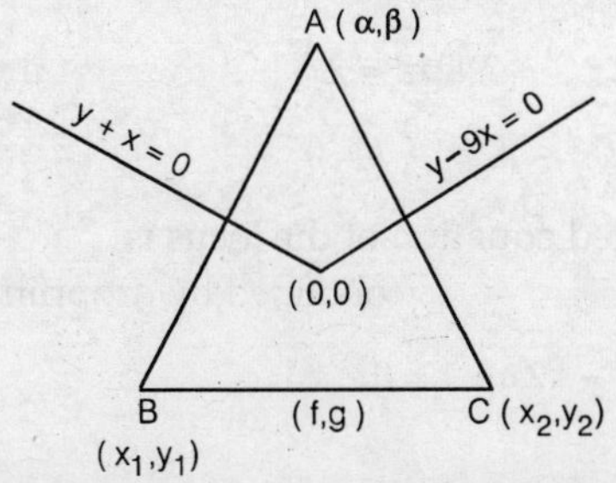

Let $B=(x_1, y_1)$, $C=(x_2, y_2)$. Clearly B is the image of A for the line $y+x=0$

$$\therefore \quad \frac{\beta-y_1}{\alpha-x_1}\cdot(-1)=-1$$

or $\quad \beta-y_1=\alpha-x_1 \qquad \dots (1)$

and $\quad \dfrac{\beta+y_1}{2}+\dfrac{\alpha+x_1}{2}=0$

or $\quad \beta+y_1+\alpha+x_1=0 \qquad \dots (2)$

Again, C is the image of A for the line $y-9x=0$. So

$$\frac{\beta-y_2}{\alpha-x_2}\cdot(9)=-1$$

or $\quad \beta-y_2=\dfrac{x_2-\alpha}{9} \qquad \dots (3)$

and $\quad \dfrac{\beta+y_2}{2}-9\cdot\dfrac{\alpha+x_2}{2}=0$

or $\quad \beta+y_2=9(\alpha+x_2) \qquad \dots (4)$

From (1) and (2), $x_1-y_1=\alpha-\beta$ and $x_1+y_1=-\alpha-\beta$.

Adding and subtracting, $x_1=-\beta$ and $y_1=-\alpha$;

$\therefore \quad B=(-\beta, -\alpha)$.

From (3) and (4),

$$2\beta=\frac{x_2-\alpha}{9}+9(\alpha+x_2)$$

$$=\frac{x_2-\alpha+81(\alpha+x_2)}{9}=\frac{80\alpha+82x_2}{9}$$

or $\quad 9\beta=40\alpha+41x_2; \quad \therefore \quad x_2=\dfrac{9\beta-40\alpha}{41}$

$\therefore$ from (3), $y_2=\beta-\dfrac{x_2-\alpha}{9}$

$$=\beta-\frac{1}{9}\left\{\frac{9\beta-40\alpha}{41}-\alpha\right\}$$

$$=\beta-\frac{1}{9}\cdot\frac{9\beta-81\alpha}{41}$$

$$=\beta-\frac{\beta-9\alpha}{41}=\frac{40\beta+9\alpha}{41}$$

$$\therefore \quad C=\left(\frac{9\beta-40\alpha}{41}, \frac{40\beta+9\alpha}{41}\right).$$

As the points B, (f, g) and C are collinear

$$\begin{vmatrix} -\beta & -\alpha & 1 \\ f & g & 1 \\ \dfrac{9\beta-40\alpha}{41} & \dfrac{40\beta+9\alpha}{41} & 1 \end{vmatrix}=0$$

or $$\begin{vmatrix} -\beta & -\alpha & 1 \\ f+\beta & g+\alpha & 0 \\ \dfrac{50\beta-40\alpha}{41} & \dfrac{40\beta+50\alpha}{41} & 0 \end{vmatrix}=0$$

or $\quad (f+\beta)(40\beta+50\alpha)-(g+\alpha)(50\beta-40\alpha)=0.$

So the locus of $A(\alpha, \beta)$ has the equation

$$(f+y)(4y+5x)-(g+x)(5y-4x)=0$$

or $\quad 4(x^2+y^2)+(4g+5f)x+(4f-5g)y=0,$

which is a circle.

8. Prove that two of the four lines represented by the joint equation

$$ax^4+bx^3y+cx^2y^2+dxy^3+ay^4=0$$

will bisect the angles between the other two if $c+6a=0$ and $b+d=0$.

Using $c+6a=0$, $b+d=0$, we get the joint equation as follows.

$$ax^4+bx^3y-6ax^2y^2-bxy^3+ay^4=0$$

or $\quad a(x^4-6x^2y^2+y^4)+bxy(x^2-y^2)=0$

or $\quad a(x^4-2x^2y^2+y^4)+bxy(x^2-y^2)-4ax^2y^2=0$

or $\quad a(x^2-y^2)^2+bxy(x^2-y^2)-4ax^2y^2=0$

or $\quad a\left(\dfrac{x^2-y^2}{xy}\right)^2+b\left(\dfrac{x^2-y^2}{xy}\right)-4a=0$

or $\left(\frac{x^2-y^2}{xy}\right)^2+\frac{b}{a}\left(\frac{x^2-y^2}{xy}\right)-4=0$

or $\left(\frac{x^2-y^2}{xy}-m_1\right)\left(\frac{x^2-y^2}{xy}-m_2\right)=0$

where $m_1m_2=-4,\ m_1+m_2=\frac{-b}{a}$

$\therefore$ the two pairs of lines will have the equations

$$x^2-y^2-m_1xy=0 \qquad \ldots (1)$$

and $x^2-y^2-m_2xy=0 \qquad \ldots (2)$

The equation of bisectors of the pair (1) is

$$\frac{x^2-y^2}{1-(-1)}=\frac{xy}{\frac{-m_1}{2}}$$

or $x^2-y^2=\frac{-4xy}{m_1}$

or $x^2-y^2=m_2xy\ \{\because\ m_1m_2=-4\}$

or $x^2-y^2-m_2xy=0$ which is the second pair.

9. Find the changed equation of the locus $x^2+6xy+y^2=1$ when the lines $x+y=0$ and $x-y+1=0$ are taken as the new x and y axes respectively.

Here the lines $x+y=0$ and $x-y+1=0$ are perpendicular to each other.

So, take $x'=\frac{x-y+1}{\sqrt{1^2+(-1)^2}} \qquad \ldots (1)$

and $y'=\frac{x+y}{\sqrt{1^2+1^2}} \qquad \ldots (2)$

From (1), $x-y+1=\sqrt{2}x' \qquad \ldots (3)$

From (2), $x+y=\sqrt{2}y' \qquad \ldots (4)$

(3) + (4) $\Rightarrow$ $2x+1=\sqrt{2}(x'+y')$

$\therefore$ $x=\frac{x'+y'}{\sqrt{2}}-\frac{1}{2}$

(4) – (3) $\Rightarrow$ $2y-1=\sqrt{2}(y'-x')$

$\therefore$ $y=\frac{y'-x'}{\sqrt{2}}+\frac{1}{2}$

$\therefore$ putting these in the equation of the locus, we get

$$\left(\frac{x'+y'}{\sqrt{2}}-\frac{1}{2}\right)^2+6\left(\frac{x'+y'}{\sqrt{2}}-\frac{1}{2}\right)\left(\frac{y'-x'}{\sqrt{2}}+\frac{1}{2}\right)+\left(\frac{y'-x'}{\sqrt{2}}+\frac{1}{2}\right)^2=1$$

or $\frac{(x'+y')^2}{2}+\frac{1}{4}-\frac{x'+y'}{\sqrt{2}}$

$$+6\left\{\frac{y'^2-x'^2}{2}-\frac{1}{4}+\frac{1}{2\sqrt{2}}(x'+y'-y'+x')\right\}$$

$$+\frac{(y'-x')^2}{2}+\frac{1}{4}+\frac{y'-x'}{\sqrt{2}}=1$$

or $\frac{1}{2}\{(x'+y')^2+(y'-x')^2\}$

$$+\frac{1}{2}+\frac{1}{\sqrt{2}}(y'-x'-x'-y')$$

$$+3(y'^2-x'^2)-\frac{3}{2}+\frac{3}{\sqrt{2}}\cdot 2x'=1$$

or $x'^2+y'^2-\sqrt{2}x'+3y'^2-3x'^2+3\sqrt{2}x'=2$

or $4y'^2-2x'^2+2\sqrt{2}x'=2$

or $x'^2-2y'^2-\sqrt{2}x'+1=0$

$\therefore$ the changed equation of the locus is (obtained by dropping dashes)

$$x^2-2y^2-\sqrt{2}x+1=0.$$

10. The equation of a pair of straight lines is $ax^2+2hxy+by^2=0$. By what angle must the axes be rotated so that the term containing xy in the equation may be removed?

Here the origin remains fixed. Let the axes be turned about the fixed origin through an angle ϕ in the anticlockwise sense and new coordinates of the point (x, y) become (x', y').

Then the equations of transformation will be

$$x=x'\cos\phi-y'\sin\phi$$

$$y=x'\sin\phi+y'\cos\phi$$

$\therefore$ the changed equation will be

$$a(x'\cos\phi-y'\sin\phi)^2$$
$$+2h(x'\cos\phi-y'\sin\phi)(x'\sin\phi+y'\cos\phi)$$
$$+b(x'\sin\phi+y'\cos\phi)^2=0$$

or $(a\cos^2\phi+2h\sin\phi\cos\phi+b\sin^2\phi)x'^2$

$$+(-a\sin 2\phi+2h\cos 2\phi+b\sin 2\phi)x'y'$$
$$+(a\sin^2\phi-2h\sin\phi\cos\phi+b\cos^2\phi)y'^2=0$$

$\therefore$ the transformed equation of the pair of lines for the new axes will be

$$(a\cos^2\phi+h\sin 2\phi+b\sin^2\phi)x^2$$
$$+\{(b-a)\sin 2\phi+2h\cos 2\phi\}xy$$
$$+(a\sin^2\phi-h\sin 2\phi+b\cos^2\phi)y^2=0.$$

This equation will not contain xy if

$$(b-a)\sin 2\phi + 2h\cos 2\phi = 0$$

or $\tan 2\phi = \dfrac{2h}{a-b}$

$\therefore \quad \phi = \dfrac{1}{2}\tan^{-1}\dfrac{2h}{a-b}$.

This is the required angle.

11. Transform the equation
$$\mathbf{14x^2 - 4xy + 11y^2 - 36x + 48y + 41 = 0}$$
in the form $ax^2 + by^2 = 1$ by suitable change of axes.

Let us take the following most general transformations of axes:

(i) the new origin is (α, β)

(ii) axes are turned through an angle ϕ in the anticlockwise sense.

Then equations of transformation are

$$x = \alpha + x'\cos\phi - y'\sin\phi$$
$$y = \beta + x'\sin\phi + y'\cos\phi$$

The changed equation in x', y' coordinates is

$$14(\alpha + x'\cos\phi - y'\sin\phi)^2$$
$$-4(\alpha + x'\cos\phi - y'\sin\phi)(\beta + x'\sin\phi + y'\cos\phi)$$
$$+11(\beta + x'\sin\phi + y'\cos\phi)^2$$
$$-36(\alpha + x'\cos\phi - y'\sin\phi)$$
$$+48(\beta + x'\sin\phi + y'\cos\phi) + 41 = 0 \quad \ldots (1)$$

The changed equation in x', y' has to be of the form $ax'^2 + by'^2 = 1$.

For this we must have in (1)

coefficient of $x'y' = 0$,

coefficient of $x' = 0$,

coefficient of $y' = 0$.

So (α, β) and ϕ should be such that

coefficient of $x'y'$

$$= -14\sin 2\phi - 4\cos 2\phi + 11\sin 2\phi = 0$$

or $3\sin 2\phi = -4\cos 2\phi$

$\therefore \quad \tan 2\phi = \dfrac{-4}{3}$

or $\dfrac{2\tan\phi}{1-\tan^2\phi} = \dfrac{-4}{3}$

or $2\tan^2\phi - 3\tan\phi - 2 = 0$

$\therefore \quad \tan\phi = -\dfrac{1}{2}, 2$

$\therefore$ smaller angle of rotation $= \tan^{-1}\left(-\dfrac{1}{2}\right)$

coefficient of x'

$$= 28\alpha\cos\phi - 4(\alpha\sin\phi + \beta\cos\phi)$$
$$+ 22\beta\sin\phi - 36\cos\phi + 48\sin\phi = 0$$

$\therefore \quad \cos\phi \cdot (28\alpha - 4\beta - 36) - \sin\phi \cdot (4\alpha - 22\beta - 48) = 0$

$\therefore \quad \tan\phi = \dfrac{28\alpha - 4\beta - 36}{4\alpha - 22\beta - 48} = -\dfrac{1}{2}$

or $28\alpha - 4\beta - 36 = -2\alpha + 11\beta + 24$

or $30\alpha - 15\beta - 60 = 0$

$\therefore \quad 2\alpha - \beta - 4 = 0 \quad \ldots (2)$

coefficient of y'

$$= -28\alpha\sin\phi - 4(\alpha\cos\phi - \beta\sin\phi) + 22\beta\cos\phi$$
$$+ 36\sin\phi + 48\cos\phi = 0$$

$\therefore \quad \cos\phi(-4\alpha + 22\beta + 48) - \sin\phi(28\alpha - 4\beta - 36) = 0$

$\therefore \quad \tan\phi = \dfrac{-4\alpha + 22\beta + 48}{28\alpha - 4\beta - 36} = -\dfrac{1}{2}$

or $-4\alpha + 22\beta + 48 = -14\alpha + 2\beta + 18$

or $10\alpha + 20\beta + 30 = 0$

$\therefore \quad \alpha + 2\beta + 3 = 0 \quad \ldots (3)$

Solving (2) and (3), $5\alpha - 5 = 0$; $\therefore \quad \alpha = 1$

$\therefore$ from (2), $\beta = 2\alpha - 4 = 2 \times 1 - 4 = -2$

$\therefore \quad (\alpha, \beta) = (1, -2)$.

As $\tan\phi = -\dfrac{1}{2}$, rotation is clockwise. So

$$\sin\phi = -\frac{1}{\sqrt{5}}$$

$$\cos\phi = \frac{2}{\sqrt{5}}$$

$\therefore$ from (1), the new equation is

$$(14\cos^2\phi - 4\cos\phi \cdot \sin\phi + 11\sin^2\phi)x'^2$$
$$+ (14\sin^2\phi + 4\sin\phi \cdot \cos\phi + 11\cos^2\phi)y'^2$$
$$+ 14\alpha^2 - 4\alpha\beta + 11\beta^2 - 36\alpha + 48\beta + 41 = 0$$

or $\left(14 \cdot \dfrac{4}{5} + 4 \cdot \dfrac{2}{5} + 11 \cdot \dfrac{1}{5}\right)x'^2$

$$+\left(14 \cdot \frac{1}{5} - 4 \cdot \frac{2}{5} + 11 \cdot \frac{4}{5}\right)y'^2$$

$$+ 14 + 8 + 11 \cdot 2^2 - 36 - 48 \cdot 2 + 41 = 0$$

or $15x'^2 + 10y'^2 - 25 = 0$

or $3x'^2 + 2y'^2 = 5$

$\therefore$ the new equation is $\dfrac{3}{5}x^2 + \dfrac{2}{5}y^2 = 1$.

Exercises

1. If L_1, L_2 are bisectors of the angles between the lines represented by $ax^2 + 2hxy + by^2 = 0$ then find the joint equation of the bisectors of the angle between the lines L_1 and L_2.

2. Prove that the pairs of lines $ax^2 + acxy + cy^2 = 0$ and $\left(3 + \frac{1}{c}\right)x^2 + xy + \left(3 + \frac{1}{a}\right)y^2 = 0$ are equally inclined for all real $a \neq 0$, $c \neq 0$.

3. Find the condition that the pairs of lines $ax^2 + 2hxy + by^2 = 0$ and $a'x^2 + 2h'xy + b'y^2 = 0$ should have one line in common.

4. Find the condition that one of the lines given by $ax^2 + 2hxy + by^2 = 0$ may be perpendicular to one of the lines given by $a'x^2 + 2h'xy + b'y^2 = 0$.

5. Prove that the equation
$$(a + 2h + b)x^2 - 2(a - b)xy + (a - 2h + b)y^2 = 0$$
denotes a pair of lines each inclined at 45° to one or other of the lines given by $ax^2 + 2hxy + by^2 = 0$.

6. Prove that the sum of the squares of the perpendiculars drawn from the point (x', y') on the lines given by $ax^2 + 2hxy + by^2 = 0$ is
$$\frac{2}{(a-b)^2 + 4h^2}\{2h^2(x'^2 + y'^2) + 2h(a+b)x'y' + (a-b)(ax'^2 - by'^2)\}.$$

7. Prove that the length of the straight line joining the feet of the perpendiculars let fall from the point (p, q) on the pair of lines $ax^2 + 2hxy + by^2 = 0$ is
$$\frac{2 \cdot \sqrt{(p^2 + q^2)(h^2 - ab)}}{\sqrt{(a-b)^2 + 4h^2}}.$$

8. Find the area of the triangle formed by the pair of lines $ax^2 + 2hxy + by^2 = 0$ and the line $lx + my = 1$.

9. Prove that the lines $x^2 + 4xy + y^2 = 0$ and $x - y = 4$ are sides of an equilateral triangle. Find its area.

10. A pair of mutually perpendicular lines is drawn through the origin forming an isosceles triangle with the line $2x + 3y = 6$, right angled at the origin. Find the pair of lines and the area of the triangle.

11. If (α, β) be the centroid of the triangle whose sides are the lines $ax^2 + 2hxy + by^2 = 0$ and $lx + my = 1$ then show that
$$\frac{\alpha}{bl - hm} = \frac{\beta}{am - hl} = \frac{2}{3} \cdot \frac{1}{bl^2 - 2hlm + am^2}.$$

12. A triangle has the lines $ax^2 + 2hxy + by^2 = 0$ for two of its sides and the point (l, m) for its orthocentre. Prove that the third side has the equation $(a + b)(lx + my) = am^2 - 2hlm + bl^2$.

13. Prove that the pairs of straight lines
$$(y - mx)^2 = a^2(1 + m^2) \text{ and } (y - nx)^2 = a^2(1 + n^2)$$
form a rhombus.

14. The sides of a triangle are given by $x^2 - 4xy + y^2 = 0$ and $x + y + 4\sqrt{6} = 0$. Find the internal angles of the triangle.

15. If $12x^2 + mxy + 2y^2 + 11x - 5y + 2 = 0$ represents a pair of lines then find the angle between the lines.

16. If $4x^2 + 2(3 + m)xy + 3my^2 - 4x + (m - 9)y - 3 = 0$ represents a pair of parallel lines then find the distance between the lines.

17. If the straight lines represented by
$$ax^2 + 2hxy + by^2 + 2gx + 2fy + c = 0$$
are equidistant from the origin then prove that
$$f^4 - g^4 = c(bf^2 - ag^2).$$

18. Find the pair of straight lines passing through (1, 1) and parallel to the lines represented by
$$x^2 - 5xy + 4y^2 + x + 2y - 2 = 0.$$

19. Find the equation of the pair of lines both of which pass through the point (1, 2) and are parallel to the bisectors of the angles between the lines given by
$$x^2 + xy - 2y^2 + 4x - y + 3 = 0.$$

20. If the equation $ax^2 + 2hxy + by^2 + 2gx + 2fy + c = 0$ represents a pair of parallel lines then prove that

(i) $ab = h^2$, $af^2 = bg^2$ and $\frac{a}{h} = \frac{h}{b} = \frac{g}{f}$

(ii) the distance between the lines $= 2\sqrt{\frac{g^2 - ac}{a(a+b)}}$.

21. The line $3x + 4y - 5 = 0$ cuts the curve $2x^2 + 3y^2 = 5$ at A and B. Find $\angle AOB$ where O is the origin.

22. The line $lx + my + n = 0$ cuts the parabola $y^2 = 4ax$ at P and Q. Find the condition for OP ⊥ OQ where O is the origin.

23. Find the equation of the pair of straight lines through the origin which passes through the intersection of the curves whose equations are
$$x^2 + y^2 - 2x - 2y - 2 = 0$$
and $x^2 + y^2 - 6x - 6y + 14 = 0$.

24. The circle $x^2 + y^2 = a^2$ cuts off an intercept on the straight line $lx + my = 1$, which subtends an angle of $\pi/4$ at the origin. Show that
$$4\{a^2(l^2 + m^2) - 1\} = \{a^2(l^2 + m^2) - 2\}^2.$$

25. The pair of lines $ax^2 + 2hxy + by^2 = 0$ is rotated about the origin through 90°. Find the equation of the pair in the new position.

26. The lines $x^2 - 3xy + 2y^2 = 0$ are shifted parallel to themselves so that their point of intersection comes to (1, 1). Find the combined equation of the lines in the new position.

27. The equation of the pair of incident and reflected rays is $2x^2 - xy - y^2 + x - y = 0$. Find the joint equation of the possible lines from which the incident ray has reflected.

28. If $|\lambda| \neq 1$ then prove that one of the lines represented by the equation $x^3 + \lambda x^2 y + \lambda xy^2 + y^3 = 0$ will bisect the angle between other two.

29. Reduce the equation of the pair of lines

$$12x^2 + 2y^2 - 10xy + 11x - 5y + 2 = 0$$

into a homogeneous equation of the second degree by changing the origin but keeping axes parallel to the original axes. Also find the new origin.

30. Find the angle through which the axes must be turned about the origin so that the equation of the curve $x^2 + 2\sqrt{3}xy - y^2 = 2$ may change into the form $ax^2 + by^2 = 1$.

31. Find the new equation of the curve

$$4(x - 2y + 1)^2 + 9(2x + y + 2)^2 = 25$$

if the lines $2x + y + 2 = 0$ and $x - 2y + 1 = 0$ are taken as the new x and y axes respectively.

Answers

1. $(a - b)(x^2 - y^2) + 4hxy = 0$

3. $(a'b - ab')^2 = 4(hb' - h'b)(ah' - a'h)$

4. $(bb' - aa')^2 + 4(h'a + b'h)(ha' + bh') = 0$

8. $\dfrac{\sqrt{h^2 - ab}}{bl^2 - 2hlm + am^2}$ square units

9. $\dfrac{8\sqrt{3}}{3}$ square units

10. $5x^2 - 24xy - 5y^2 = 0, \dfrac{36}{13}$ square units

14. 60° each

15. $\tan^{-1}\dfrac{1}{7}$ or $\tan^{-1}\dfrac{29}{28}$

16. $\dfrac{4}{\sqrt{13}}$

18. $x^2 - 5xy + 4y^2 + 3x - 3y = 0$

19. $x^2 - 6xy - y^2 + 10x + 10y - 15 = 0$

21. $\pi/2$

22. $4al + n = 0$

23. $3x^2 - 10xy + 3y^2 = 0$

25. $bx^2 - 2hxy + ay^2 = 0$

26. $x^2 - 3xy + 2y^2 + x - y = 0$

27. $3x^2 + 18xy - 3y^2 + 8x + 4y + 2 = 0$

29. $6x^2 - 5xy + y^2 = 0, \left(-\dfrac{3}{2}, -\dfrac{5}{2}\right)$

30. 30° anticlockwise

31. $4x^2 + 9y^2 = 5$

□

Chapter Test

Time: 60 minutes

1. Fill in the blanks so that the resulting sentences become true.

(a) The angle between the lines given by $x^2 - 2xy \sec\theta + y^2 = 0$ is ______ .

(b) The difference of the slopes of the lines $x^2(\sec^2\theta - \sin^2\theta) - 2xy\tan\theta + y^2\sin^2\theta = 0$ is ______ .

(c) If pairs of lines $x^2 - 2pxy - y^2 = 0$ and $x^2 - 2qxy - y^2 = 0$ are such that each pair bisects the angles between the other pair then the value of $p \cdot q$ is ______ .

(d) If the equation $12x^2 + 7xy - py^2 - 18x + qy + 6 = 0$ represents two perpendicular lines then $p =$ ______ , $q =$ ______ .

2. Choose the correct answer(s) from the given multiple options.

(a) The number of values of λ for which bisectors of the angle between the lines $ax^2 + 2hxy + by^2 + \lambda(x^2 + y^2) = 0$ are the same is

A. two B. one

C. zero D. infinite

(b) If one of the lines of $ax^2 + 2hxy + by^2 = 0$ be the bisector of the angle between the coordinate axes then

A. $(a - b)^2 = 4h^2$ B. $(a + b)^2 + 4h^2 = 0$

C. $(a + b)^2 = 4h^2$ D. $(a - b)^2 + 4h^2 = 0$

(c) The equation $(x + y + 1)^2 + k(x^2 + y^2 + 1) = 0$ will represent two straight lines only if

A. $k = 0$ B. $k = 3$

C. $k = 0$ or -3 D. none of these

(d) The equation $4x^2 - 24xy + 11y^2 = 0$ represents

A. two parallel lines B. a circle

C. two perpendicular lines D. two lines through the origin

(e) The value of c for which the lines joining the origin to the points of intersection of the line $y = \sqrt{3}x + c$ and the circle $x^2 + y^2 = 2$ are perpendicular to each other, is

A. 1 B. 0

C. –2 D. 2

3. The equations of pairs of opposite sides of a rectangle are $x^2 - 7x + 6 = 0$ and $y^2 - 14y + 40 = 0$. Find the equations to its diagonals.

4. Show that the orthocentre of the triangle formed by the pair of lines $ax^2 + 2hxy + by^2 = 0$ and the line $lx + my = 1$ is given by $\frac{x}{l} = \frac{y}{m} = \frac{a+b}{am^2 - 2hlm + bl^2}$.

5. If the angle between the lines joining the origin to the points of intersection of the circle $x^2 + y^2 = a^2$ and the line $x\cos\alpha + y\sin\alpha = 1$ be 45°, find the possible values of a where α is a constant.

6. Find the equation of the pair of lines joining the point (1, 1) to the points of intersection of the line $x + y = 0$ with the circle $x^2 + y^2 + 2x - 4y + 1 = 0$. Use change of the origin.

7. Prove that $9x^2 + 24xy + 16y^2 - 45x - 60y + 50 = 0$ represents two parallel lines L_1 and L_2, and the point (2, 0) lies between the two lines. A parallelogram is constructed with two sides on the lines L_1 and L_2, and a diagonal passing through (2, 0) such that the other two sides are perpendicular to this diagonal and the area of the parallelogram is the least possible. Prove that one of the angles of the parallelogram will be $\pi/4$. Also, find the joint equation of the two possible diagonals through the point (2, 0).

□

Answers

1. (a) θ or $\pi - \theta$ (b) 2 (c) –1 (d) 12; $\frac{-23}{2}$ or 1
2. (a) D (b) C (c) C (d) D (e) C, D
3. $6x - 5y + 14 = 0$, $6x + 5y - 56 = 0$
5. $\sqrt{4 \pm 2\sqrt{2}}$
6. $3x^2 + 2xy - 9y^2 - 8x + 16y - 4 = 0$
7. $7x^2 - 48xy - 7y^2 - 28x + 96y + 28 = 0$

□

3. Circles

Recap of Facts and Formulae

1. Equation of a circle

- The equation of a circle whose centre = (h, k) and radius = a, is

  $(x-h)^2+(y-k)^2=a^2$.
- The equation of a circle whose centre is the origin and radius is a, is

  $x^2+y^2=a^2$.
- The general equation of a circle is

  $x^2+y^2+2gx+2fy+c=0$

  whose centre = $(-g, -f)$ and radius = $\sqrt{g^2+f^2-c}$.
- The equation of a circle whose one diameter is the line segment joining the points (x_1, y_1) and (x_2, y_2), is

  $(x-x_1)(x-x_2)+(y-y_1)(y-y_2)=0$.

2. Condition for a circle

(i) The equation

$S \equiv x^2+y^2+2gx+2fy+c=0$ represents

- a real circle if $g^2+f^2 \geq c$
- a point circle (i.e., a circle of zero radius) if $g^2+f^2=c$
- no real circle if $g^2+f^2<c$

(ii) The general equation of the second degree in x and y, i.e., $ax^2+2hxy+by^2+2gx+2fy+c=0$ represents a circle if $a=b, h=0$

3. Location of a point in relation to a circle

(i) If $P=(x_1, y_1)$ and the circle is $S \equiv x^2+y^2-a^2=0$ then (here O is the origin)

- P is inside the circle if $PO<a$,

  i.e., $x_1^2+y_1^2-a^2<0$, i.e., $S(x_1, y_1)<0$
- P is on the circle if $PO=a$,

  i.e., $x_1^2+y_1^2-a^2=0$, i.e., $S(x_1, y_1)=0$
- P is outside the circle if $PO>a$,

  i.e., $x_1^2+y_1^2-a^2>0$, i.e., $S(x_1, y_1)>0$

(ii) If $P=(x_1, y_1)$ and the circle is

$S \equiv x^2+y^2+2gx+2fy+c=0$ then

- P is inside the circle if $S(x_1, y_1)<0$
- P is on the circle if $S(x_1, y_1)=0$
- P is outside the circle if $S(x_1, y_1)>0$

4. Location of a circle in relation to another circle

Let $S_1 \equiv x^2+y^2+2g_1x+2f_1y+c_1=0$ and

$S_2 \equiv x^2+y^2+2g_2x+2f_2y+c_2=0$

be two circles. Let D be the discriminant for the quadratic equation in x (or y) obtained by eliminating y (or x) from the two equations of the circle. Then

- they are two intersecting circles if $D>0$
- they are nonintersecting (no common points) if $D<0$
- they touch each other if $D=0$
- If $D<0$, i.e., the circles are nonintersecting then

  (a) $S_1=0$ is outside $S_2=0$ if

  $S_2(-g_1, -f_1)>0$ or $S_1(-g_2, -f_2)>0$;

  equivalently, $AB>r_1+r_2$ where A, B are centres and r_1, r_2 are radii respectively.

  (b) $S_1=0$ is inside $S_2=0$ if $S_2(-g_1, -f_1)<0$;

  equivalently, $AB<r_2-r_1$
- If $D=0$, i.e., the circles touch then

  (a) they touch externally if $AB=r_1+r_2$

  (b) they touch internally if $AB=|r_1-r_2|$

5. A line in relation to a circle

Let $L \equiv ax+by+c=0$ be a line and

$S \equiv x^2+y^2+2gx+2fy+c=0$ be a circle.

- The points of intersection is obtained by solving $L=0, S=0$.
- The line cuts the circle at two distinct real points if $D>0$ where D is the discriminant of the quadratic

equation in x (or y) obtained by eliminating y (or x) from $L=0, S=0$;

equivalently, the length of the perpendicular from the centre to the line is less than the radius.

- The line touches the circle at a point if $D=0$; equivalently, the length of the perpendicular from the centre to the line is equal to the radius.
- The line does not cut or touch the circle if $D<0$; equivalently, the length of the perpendicular from the centre to the line is greater than the radius.

6. Tangents and tangency

- The equation of the tangent to a curve $y=f(x)$ at the point (x_1, y_1) is

$$y-y_1=\left(\frac{dy}{dx}\right)_{x_1, y_1}\cdot(x-x_1).$$

- The equation of the tangent to the circle $x^2+y^2=a^2$ at the point (x_1, y_1) is

$$xx_1+yy_1=a^2.$$

- The equation of the tangent to the circle $x^2+y^2+2gx+2fy+c=0$ at the point (x_1, y_1) is

$$xx_1+yy_1+g(x+x_1)+f(y+y_1)+c=0.$$

Note In writing the equation of the tangent at (x_1, y_1) we replace
$x\cdot x$ (i.e., x^2) by xx_1, $y\cdot y$ (i.e., y^2) by yy_1,
$x+x$ (i.e., $2x$) by $x+x_1$, $y+y$ (i.e., $2y$) by $y+y_1$ in the equation of the circle. The same is true for other second degree curves together with the replacement of $2xy$ by x_1y+xy_1.

- The line $y=mx+c$ touches the circle $x^2+y^2=a^2$ if

$$c=\pm a\sqrt{1+m^2}.$$

- A line $L=0$ touches the circle if $p=r$ where p = length of the perpendicular from the centre of the circle to the line and r = radius of the circle.
- The equation of the pair of tangents to the circle $S=0$ from the point (x_1, y_1) is $S\cdot S_1=T^2$
where $S_1\equiv$ value of S at (x_1, y_1) and
$T=0$ is the equation of tangent at (x_1, y_1) when the point is on the circle.

7. Normal

- The normal to a circle $S=0$ at a point (x_1, y_1) is the line joining the point and the centre of the circle.
- The equation of the normal to a curve $y=f(x)$ at the point (x_1, y_1) is

$$y-y_1=\frac{-1}{\left(\frac{dy}{dx}\right)_{x_1, y_1}}\cdot(x-x_1).$$

- The equation of the normal to the circle $x^2+y^2=a^2$ at the point (x_1, y_1) is

$$\frac{x}{x_1}=\frac{y}{y_1}.$$

8. Any point on a circle, any tangent to a circle

- Any point on the circle $x^2+y^2=a^2$ can be taken as (x_1, y_1) where $x_1^2+y_1^2=a^2$, or $(a\cos\theta, a\sin\theta)$.
- Any tangent to the circle $x^2+y^2=a^2$ can be taken as

$$y=mx\pm a\sqrt{1+m^2}.$$

Here $\pm$ indicates that we can take two tangents to the circle in any direction.

- Any tangent to the circle $(x-h)^2+(y-k)^2=a^2$ is given by $y-k=m(x-h)\pm a\sqrt{1+m^2}$.

9. Length of tangent, chord of contact, polar and pole

- The length of tangent from (x_1, y_1) to the circle

$S\equiv x^2+y^2+2gx+2fy+c=0$ is

$$\sqrt{S_1}\equiv\sqrt{x_1^2+y_1^2+2gx_1+2fy_1+c}.$$

- The chord of contact of the tangents from the point (x_1, y_1) to the circle
  (i) $x^2+y^2=a^2$ is $xx_1+yy_1=a^2$
  (ii) $x^2+y^2+2gx+2fy+c=0$ is
  $xx_1+yy_1+g(x+x_1)+f(y+y_1)+c=0.$
- The polar of the point (x_1, y_1) with respect to the circle
  (i) $x^2+y^2=a^2$ is $xx_1+yy_1=a^2$
  (ii) $x^2+y^2+2gx+2fy+c=0$ is
  $xx_1+yy_1+g(x+x_1)+f(y+y_1)+c=0.$
- The pole of the line $L=0$ with respect to the circle $S=0$ is the point (α, β) whose polar with respect to the circle is $L=0$.

10. Chord with given middle point

- The chord of the circle $S=0$ whose middle point is (x_1, y_1) has the equation $S_1=T$.

11. Orthogonal circles

Let $S_1\equiv x^2+y^2+2g_1x+2f_1y+c_1=0$ and
$S_2\equiv x^2+y^2+2g_2x+2f_2y+c_2=0$
be two circles whose centres are A, B and radii r_1, r_2 respectively.

- The two circles cut orthogonally if $r_1^2+r_2^2=AB^2$; equivalently, $2g_1g_2+2f_1f_2=c_1+c_2$.

12. Equation of line/circles through intersection of a line/circle and a circle

- The equation of circles passing through the intersection of the line $L = 0$ and the circle $S = 0$ is
  $S + \lambda L = 0$

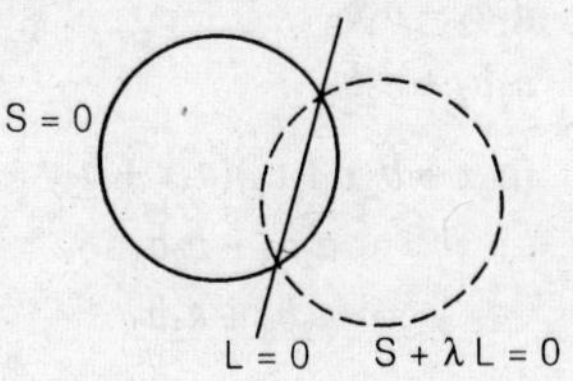

where λ is an arbitrary constant whose value is obtained using a given additional geometrical condition.

- The equation of circles passing through the intersection of the two circles
  $S_1 = 0$ and $S_2 = 0$ is
  $S_1 + \lambda S_2 = 0, \ (\lambda \neq -1)$

where λ is an arbitrary constant whose value is obtained using a given additional geometrical condition.

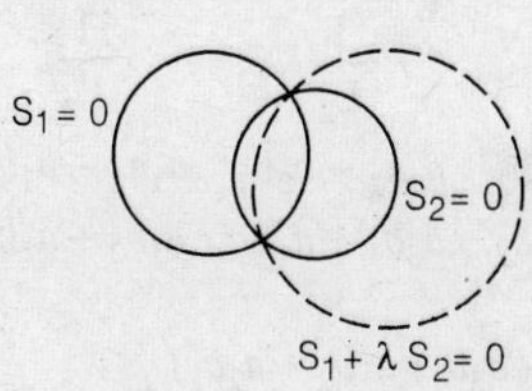

- The equation of the common chord of the circles $S_1 = 0$ and $S_2 = 0$ is given by $S_1 - S_2 = 0$ when the circles are intersecting.

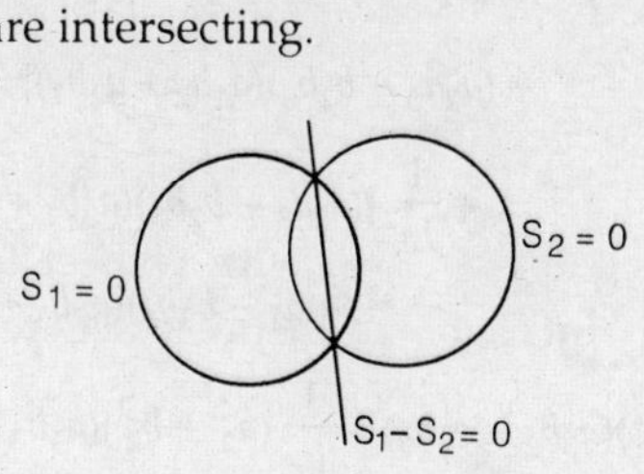

13. Radical axis, radical centre and coaxal circles

- The locus of the point, from which the lengths of tangents to the circles $S_1 = 0$ and $S_2 = 0$ are equal, is a straight line. It is called the radical axis of the two circles and its equation is $S_1 - S_2 = 0$.

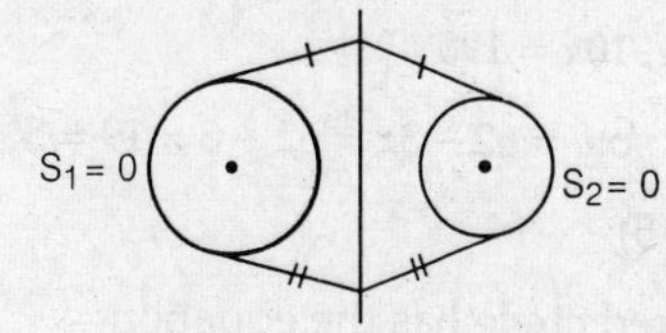

Note The equation of the common chord of two circles and their radical axis is the same when the circles are intersecting.

- The radical centre of three circles $S_1 = 0, S_2 = 0$ and $S_3 = 0$ is the point of intersection of the radical axes of the circles, taken in pairs.
  So, the radical centre is the point of intersection of the radical axes
  $$S_1 - S_2 = 0 \text{ and } S_2 - S_3 = 0.$$
- Three or more circles are called coaxal (or coaxial) circles if each pair has the same radical axis.
- The form of equation of a family of coaxal circles is
  $$x^2 + y^2 + 2\lambda x + c = 0,$$
  where λ is a parameter (c being fixed)
  and $x = 0$ (i.e. y-axis) is the common radical axis for each pair and the line joining centres of any two circles of the system is taken as the x-axis.
- The limiting point of a coaxal system is the circle belonging to the family whose radius is 0.
  So the limiting points of the coaxal system
  $x^2 + y^2 + 2\lambda x + c = 0$ are $(\pm \sqrt{c}, 0)$.

Selected Solved Examples

1. Find the equation of the circle circumscribing the quadrilateral whose sides are
$x - 3y = 4$, $3x + y = 22$, $x - 3y = 14$ and $3x + y = 62$.
Also find the centre and radius of the circumscribing circle.

Clearly, lines $x - 3y = 4, x - 3y = 14$ are parallel and $3x + y = 22$, $3x + y = 62$ are parallel. Also the line $x - 3y = 4$ is perpendicular to the line $3x + y = 22$. So the four lines form a rectangle.

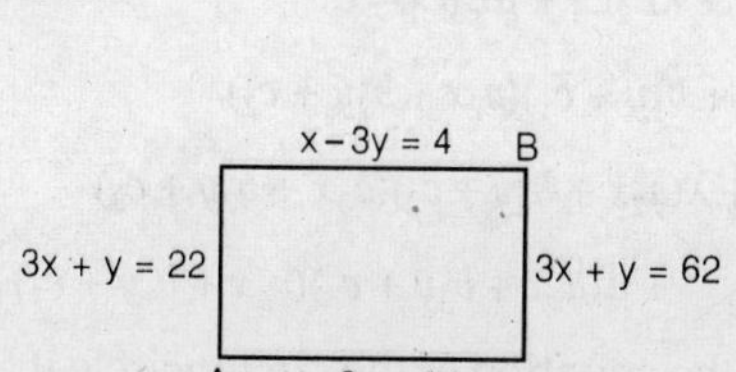

Let two ends of a diagonal be A, B as shown in the figure. Clearly, the circle circumscribing the rectangle is the circle whose one diameter has the ends A and B.

The point A is the intersection of
$$x - 3y = 14 \text{ and } 3x + y = 22.$$

Solving these, $10x = 80$

$\therefore \quad x = 8;\ \text{ so } y = 22 - 3x = 22 - 3 \times 8 = -2.$

$\therefore \quad A = (8, -2).$

The point B is the intersection of

$x - 3y = 4$ and $3x + y = 62.$

Solving these, $10x = 190$

$\therefore \quad x = 19;\ \text{ so } y = 62 - 3x = 62 - 3 \times 19 = 5$

$\therefore \quad B = (19, 5)$

$\therefore$ the required circle has the equation

$$(x-8)(x-19) + (y+2)(y-5) = 0$$

or $\quad x^2 - 27x + 152 + y^2 - 3y - 10 = 0$

or $\quad x^2 + y^2 - 27x - 3y + 142 = 0.$

Here $g = \frac{-27}{2}, f = \frac{-3}{2}, c = 142.$

$\therefore$ centre $= (-g, -f) = \left(\frac{27}{2}, \frac{3}{2}\right)$ and

radius $= \sqrt{g^2 + f^2 - c}$

$$= \sqrt{\left(-\frac{27}{2}\right)^2 + \left(-\frac{3}{2}\right)^2 - 142}$$

$$= \sqrt{\frac{729}{4} + \frac{9}{4} - 142}$$

$$= \sqrt{\frac{170}{4}} = \sqrt{\frac{85}{2}}.$$

2. Prove that the equation of the circumcircle of the triangle formed by the lines $L_1 \equiv a_1x + b_1y + c_1 = 0$, $L_2 \equiv a_2x + b_2y + c_2 = 0$ and $L_3 \equiv a_3x + b_3y + c_3 = 0$ is

$$\begin{vmatrix} \frac{1}{L_1}(a_1^2 + b_1^2) & \frac{1}{L_2}(a_2^2 + b_2^2) & \frac{1}{L_3}(a_3^2 + b_3^2) \\ a_1 & a_2 & a_3 \\ b_1 & b_2 & b_3 \end{vmatrix} = 0.$$

Any second degree curve passing through the points of intersection of $L_1 = 0, L_2 = 0, L_3 = 0$, two by two, is

$$L_1L_2 + \lambda L_2L_3 + \mu L_3L_1 = 0$$

or $\quad (a_1x + b_1y + c_1)(a_2x + b_2y + c_2)$

$$+ \lambda(a_2x + b_2y + c_2)(a_3x + b_3y + c_3)$$

$$+ \mu(a_3x + b_3y + c_3)(a_1x + b_1y + c_1) = 0 \quad \dots (1)$$

It will be the equation of the circumcircle if (1) is a circle.

For this the conditions are the following.

coefficient of x^2 = coefficient of y^2

and coefficient of $xy = 0$

$\Rightarrow \quad a_1a_2 + \lambda a_2a_3 + \mu a_3a_1 = b_1b_2 + \lambda b_2b_3 + \mu b_3b_1,$

i.e., $\quad a_1a_2 - b_1b_2 + \lambda(a_2a_3 - b_2b_3) + \mu(a_3a_1 - b_3b_1) = 0 \quad \dots (2)$

and $\quad a_1b_2 + a_2b_1 + \lambda(a_2b_3 + a_3b_2) + \mu(a_3b_1 + a_1b_3) = 0 \quad \dots (3)$

Eliminating λ, μ from (1), (2) and (3) we get

$$\begin{vmatrix} (a_1x + b_1y + c_1)(a_2x + b_2y + c_2) & (a_2x + b_2x + c_2)(a_3x + b_3y + c_3) & (a_3x + b_3y + c_3)(a_1x + b_1y + c_1) \\ a_1a_2 - b_1b_2 & a_2a_3 - b_2b_3 & a_3a_1 - b_3b_1 \\ a_1b_2 + a_2b_1 & a_2b_3 + a_3b_2 & a_3b_1 + a_1b_3 \end{vmatrix} = 0$$

or $$\begin{vmatrix} L_1L_2 & L_2L_3 & L_3L_1 \\ a_1a_2 - b_1b_2 & a_2a_3 - b_2b_3 & a_3a_1 - b_3b_1 \\ a_1b_2 + a_2b_1 & a_2b_3 + a_3b_2 & a_3b_1 + a_1b_3 \end{vmatrix} = 0$$

or $$\begin{vmatrix} \frac{1}{L_3} & \frac{1}{L_1} & \frac{1}{L_2} \\ a_1a_2 - b_1b_2 & a_2a_3 - b_2b_3 & a_3a_1 - b_3b_1 \\ a_1b_2 + a_2b_1 & a_2b_3 + a_3b_2 & a_3b_1 + a_1b_3 \end{vmatrix} = 0$$

or $$\begin{vmatrix} \frac{1}{L_1} & \frac{1}{L_2} & \frac{1}{L_3} \\ a_2a_3 - b_2b_3 & a_3a_1 - b_3b_1 & a_1a_2 - b_1b_2 \\ a_2b_3 + a_3b_2 & a_3b_1 + a_1b_3 & a_1b_2 + a_2b_1 \end{vmatrix} = 0$$

or $\quad \frac{1}{L_1}\{(a_3a_1 - b_3b_1)(a_1b_2 + a_2b_1)$

$$- (a_1a_2 - b_1b_2)(a_3b_1 + a_1b_3)\}$$

$$+ \frac{1}{L_2}\{(a_1a_2 - b_1b_2)(a_2b_3 + a_3b_2)$$

$$- (a_2a_3 - b_2b_3)(a_1b_2 + a_2b_1)\}$$

$$+ \frac{1}{L_3}\{(a_2a_3 - b_2b_3)(a_3b_1 + a_1b_3)$$

$$- (a_3a_1 - b_3b_1)(a_2b_3 + a_3b_2)\} = 0$$

or $\quad \frac{1}{L_1}(a_1^2 + b_1^2)(a_3b_2 - a_2b_3) + \frac{1}{L_2}(a_2^2 + b_2^2)(a_1b_3 - a_3b_1)$

$$+ \frac{1}{L_3}(a_3^2 + b_3^2)(a_2b_1 - a_1b_2) = 0$$

or $$\begin{vmatrix} \frac{a_1^2 + b_1^2}{L_1} & \frac{a_2^2 + b_2^2}{L_2} & \frac{a_3^2 + b_3^2}{L_3} \\ a_1 & a_2 & a_3 \\ b_1 & b_2 & b_3 \end{vmatrix} = 0$$

which is the required equation.

3. Find the equation of a circle which passes through the point (2, 0) and whose centre is the limit of the point of intersection of the lines

$3x + 5y = 1$ and $(2 + c)x + 5c^2y = 1$ as $c \to 1$.

The point of intersection of

$$3x + 5y = 1 \qquad \ldots (1)$$

$$(2 + c)x + 5c^2y = 1 \qquad \ldots (2)$$

is obtained by solving (1) and (2).

From (1), $y = \frac{1 - 3x}{5}$.

Putting in (2), $(2 + c)x + 5c^2 \cdot \frac{1 - 3x}{5} = 1$

or $(2 + c)x + c^2(1 - 3x) = 1$

or $(2 + c - 3c^2)x = 1 - c^2$

$\therefore \quad x = \frac{1 - c^2}{2 + c - 3c^2}$.

Now, $\lim_{c \to 1} \frac{1 - c^2}{2 + c - 3c^2} = \lim_{c \to 1} \frac{(1 + c)(1 - c)}{(2 + 3c)(1 - c)}$

$= \lim_{c \to 1} \frac{1 + c}{2 + 3c} = \frac{2}{5}$

$\therefore$ for the centre, $x = \frac{2}{5}$. So from (1),

$$y = \frac{1 - 3x}{5} = \frac{1 - 3 \cdot \frac{2}{5}}{5} = \frac{5 - 6}{25} = \frac{-1}{25}$$

$\therefore$ centre $= \left(\frac{2}{5}, \frac{-1}{25}\right)$.

$\therefore$ radius = distance between $\left(\frac{2}{5}, \frac{-1}{25}\right)$ and (2, 0)

$$= \sqrt{\left(2 - \frac{2}{5}\right)^2 + \left(\frac{1}{25}\right)^2} = \sqrt{\frac{64}{25} + \frac{1}{25^2}}$$

$$= \frac{1}{25}\sqrt{64 \times 25 + 1} = \frac{1}{25} \cdot \sqrt{1601}$$

$\therefore$ the required equation of the circle is

$$\left(x - \frac{2}{5}\right)^2 + \left(y + \frac{1}{25}\right)^2 = \left(\frac{1}{25}\sqrt{1601}\right)^2$$

or $\frac{(5x - 2)^2}{25} + \frac{(25y + 1)^2}{25^2} = \frac{1601}{25^2}$

or $25(5x - 2)^2 + (25y + 1)^2 = 1601$

or $625x^2 - 500x + 100 + 625y^2 + 50y + 1 = 1601$

or $625(x^2 + y^2) - 500x + 50y = 1500$

$\therefore \quad 25(x^2 + y^2) - 20x + 2y = 60.$

4. Find the equation of a circle passing through the points (1, 2) and (3, 4) and touching the line $3x + y - 3 = 0$.

Let the circle be $x^2 + y^2 + 2gx + 2fy + c = 0 \qquad \ldots (1)$

Here, centre $= (-g, -f)$ and radius $= \sqrt{g^2 + f^2 - c}$.

(1) passes through (1, 2) and (3, 4).

$\therefore \quad 1^2 + 2^2 + 2g + 4f + c = 0,$

i.e., $2g + 4f + c + 5 = 0 \qquad \ldots (2)$

and $3^2 + 4^2 + 6g + 8f + c = 0,$

i.e., $6g + 8f + c + 25 = 0 \qquad \ldots (3)$

As the line $3x + y - 3 = 0$ touches the circle (1),

radius = distance of the centre $(-g, -f)$ from the line

$$3x + y - 3 = 0$$

or $\sqrt{g^2 + f^2 - c} = \left|\frac{-3g - f - 3}{\sqrt{3^2 + 1^2}}\right|$

or $g^2 + f^2 - c = \frac{(3g + f + 3)^2}{10}$

or $10g^2 + 10f^2 - 10c = 9g^2 + f^2 + 9 + 6gf + 18g + 6f$

or $g^2 + 9f^2 - 6gf - 18g - 6f - 10c = 9 \qquad \ldots (4)$

From (3) – (2), $4g + 4f + 20 = 0$

or $g + f + 5 = 0$

$\therefore \quad g = -(f + 5) \qquad \ldots (5)$

Putting in (4),

$$(f + 5)^2 + 9f^2 + 6f(f + 5) + 18(f + 5) - 6f - 10c = 9 \qquad \ldots (6)$$

From (2) and (5), $-2(f + 5) + 4f + c + 5 = 0$

or $2f + c - 5 = 0 \quad \therefore \quad c = 5 - 2f \qquad \ldots (7)$

$\therefore$ from (6), (7) we get

$$(f + 5)^2 + 9f^2 + 6f(f + 5) + 18(f + 5) - 6f - 10(5 - 2f) = 9$$

or $16f^2 + 72f + 56 = 0$

or $2f^2 + 9f + 7 = 0$

or $(2f + 7)(f + 1) = 0 \quad \therefore \quad f = -1, -\frac{7}{2}$

$\therefore$ from (5), $g = -(-1 + 5), -\left(-\frac{7}{2} + 5\right) = -4, -\frac{3}{2}$

from (7), $c = 5 - 2(-1), 5 - 2\left(-\frac{7}{2}\right)$

$= 7, 12$

$\therefore$ we get $f = -1, g = -4, c = 7$

or $f = -\frac{7}{2}, g = \frac{-3}{2}, c = 12.$

$\therefore$ the required circle (1) is

$x^2 + y^2 - 8x - 2y + 7 = 0$

or $x^2 + y^2 - 3x - 7y + 12 = 0.$

5. If two lines $a_1x + b_1y + \lambda = 0$ and $a_2x + b_2y + \lambda = 0$, $(\lambda \neq 0)$ cut the coordinate axes in concyclic points then prove that $a_1a_2 = b_1b_2$. Also, find the equation of the circle.

Let the lines be AB and CD. Clearly,

$$A = \left(-\frac{\lambda}{a_1}, 0\right),\ B = \left(0, -\frac{\lambda}{b_1}\right)$$

$$C = \left(-\frac{\lambda}{a_2}, 0\right),\ D = \left(0, -\frac{\lambda}{b_2}\right)$$

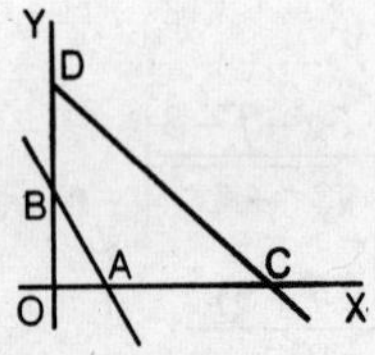

From the question, the points A, B, C and D are concyclic. (**Note** To prove $a_1a_2 = b_1b_2$ one can use the property that $OA \cdot OC = OB \cdot OD$.)

Let the circle through A, B, C and D be

$$x^2 + y^2 + 2gx + 2fy + c = 0 \qquad \dots (1)$$

Then $\frac{\lambda^2}{a_1^2} + 2g \cdot \left(-\frac{\lambda}{a_1}\right) + c = 0$,

i.e., $\lambda^2 - 2g\lambda a_1 + ca_1^2 = 0 \qquad \dots (2)$

and $\frac{\lambda^2}{b_1^2} + 2f\left(-\frac{\lambda}{b_1}\right) + c = 0$,

i.e., $\lambda^2 - 2f\lambda b_1 + cb_1^2 = 0 \qquad \dots (3)$

and $\frac{\lambda^2}{a_2^2} + 2g\left(-\frac{\lambda}{a_2}\right) + c = 0$,

i.e., $\lambda^2 - 2g\lambda a_2 + ca_2^2 = 0 \qquad \dots (4)$

and $\frac{\lambda^2}{b_2^2} + 2f\left(-\frac{\lambda}{b_2}\right) + c = 0$,

i.e., $\lambda^2 - 2f\lambda b_2 + cb_2^2 = 0 \qquad \dots (5)$

$(2) - (4) \quad \Rightarrow \quad 2g\lambda(a_2 - a_1) + c(a_1^2 - a_2^2) = 0$

or $\quad 2g\lambda - c(a_1 + a_2) = 0 \qquad \dots (6)$

$(3) - (5) \quad \Rightarrow \quad 2f\lambda(b_2 - b_1) + c(b_1^2 - b_2^2) = 0$

or $\quad 2f\lambda - c(b_1 + b_2) = 0 \qquad \dots (7)$

Also, $a_2 \times (2) - a_1 \times (4)$

$\Rightarrow \quad \lambda^2(a_2 - a_1) + c(a_1^2a_2 - a_2^2a_1) = 0$

or $\quad \lambda^2 - ca_1a_2 = 0 \qquad \dots (8)$

Again, $b_2 \times (3) - b_1 \times (5)$

$\Rightarrow \quad \lambda^2(b_2 - b_1) + c(b_1^2b_2 - b_2^2b_1) = 0$

or $\quad \lambda^2 - cb_1b_2 = 0 \qquad \dots (9)$

From (8) and (9), $a_1a_2 = b_1b_2$.

Also, $c = \frac{\lambda^2}{a_1a_2}$.

$\therefore$ from (6), $2g = \frac{c(a_1 + a_2)}{\lambda} = \frac{\lambda^2}{a_1a_2} \cdot \frac{a_1 + a_2}{\lambda} = \lambda \frac{a_1 + a_2}{a_1a_2}$

from (7), $2f = \frac{c(b_1 + b_2)}{\lambda} = \frac{\lambda^2}{a_1a_2} \cdot \frac{b_1 + b_2}{\lambda} = \lambda \frac{b_1 + b_2}{a_1a_2}$

$\therefore$ the equation of the circle, from (1) is

$$x^2 + y^2 + \frac{\lambda(a_1 + a_2)}{a_1a_2}x + \frac{\lambda(b_1 + b_2)}{a_1a_2}y + \frac{\lambda^2}{a_1a_2} = 0$$

or $\quad a_1a_2(x^2 + y^2) + \lambda(a_1 + a_2)x + \lambda(b_1 + b_2)y + \lambda^2 = 0.$

6. The point (1, 4) is inside the circle S whose equation is of the form $x^2 + y^2 - 6x - 10y + k = 0$, k being an arbitrary constant. Find the possible values of k if the circle S neither touches the axes nor cuts them.

As (1, 4) is interior point of the circle S,

$1^2 + 4^2 - 6 \cdot 1 - 10 \cdot 4 + k < 0; \quad \therefore \quad k < 29.$

Solving $y = 0$ and $x^2 + y^2 - 6x - 10y + k = 0$, we get

$$x^2 - 6x + k = 0$$

The circle S will neither cut the x-axis nor touch it if disc. < 0, i.e., $36 - 4k < 0$; $\therefore \quad k > 9$.

Solving $x = 0$ and $x^2 + y^2 - 6x - 10y + k = 0$, we get

$$y^2 - 10y + k = 0$$

The circle S will neither cut the y-axis nor touch it if disc. < 0, i.e., $10^2 - 4k < 0$; $\therefore \quad k > 25$

$\therefore \quad k$ should satisfy $k < 29$, $k > 9$, $k > 25$

$\therefore \quad 25 < k < 29$, i.e., $k \in (25, 29)$.

7. Find the values of α for which the point $(2\alpha, \alpha + 1)$ is an interior point of the larger segment of the circle $x^2 + y^2 - 2x - 2y - 8 = 0$ made by the chord whose equation is $x - y + 1 = 0$.

The point $(2\alpha, \alpha + 1)$ must be an interior point of the circle $x^2 + y^2 - 2x - 2y - 8 = 0$

$\therefore \quad (2\alpha)^2 + (\alpha + 1)^2 - 2 \cdot 2\alpha - 2(\alpha + 1) - 8 < 0$

or $\quad 5\alpha^2 - 4\alpha - 9 < 0$

or $\quad (5\alpha - 9)(\alpha + 1) < 0$

$\therefore \quad$ by sign-scheme we have $-1 < \alpha < 9/5 \qquad \dots (1)$

Also, the centre (1, 1) of the circle and the point $(2\alpha, \alpha + 1)$ must be on the same side of the line $x - y + 1 = 0$.

Clearly, $1 - 1 + 1 > 0$;

so, $2\alpha - (\alpha + 1) + 1 > 0$; $\therefore$ $\alpha > 0$ ... (2)

$\therefore$ the set of values of α satisfying (1) and (2) is $(0, 9/5)$.

8. Find the interval of values of a for which the line $y + x = 0$ bisects two chords drawn from a point $\left(\frac{1+\sqrt{2}a}{2}, \frac{1-\sqrt{2}a}{2}\right)$ to the circle whose equation is $2x^2 + 2y^2 - (1+\sqrt{2}a)x - (1-\sqrt{2}a)\,y = 0$.

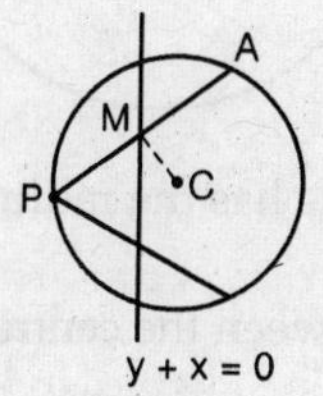

The point $P\left(\frac{1+\sqrt{2}a}{2}, \frac{1-\sqrt{2}a}{2}\right)$ lies on the circle because

$$2\cdot\left(\frac{1+\sqrt{2}a}{2}\right)^2 + 2\cdot\left(\frac{1-\sqrt{2}a}{2}\right)^2 - (1+\sqrt{2}a)\cdot\frac{1+\sqrt{2}a}{2} - (1-\sqrt{2}a)\cdot\frac{1-\sqrt{2}a}{2} = 0$$

Any chord PA through $P\left(\frac{1+\sqrt{2}a}{2}, \frac{1-\sqrt{2}a}{2}\right)$ is

$$y - \frac{1-\sqrt{2}a}{2} = m\left(x - \frac{1+\sqrt{2}a}{2}\right) \quad \text{... (1)}$$

Let M be the point of intersection of (1) and the line

$$y + x = 0 \quad \text{... (2)}$$

Solving (1) and (2), $-x - \frac{1-\sqrt{2}a}{2} = m\left(x - \frac{1+\sqrt{2}a}{2}\right)$

or $m\cdot\frac{1+\sqrt{2}a}{2} - \frac{1-\sqrt{2}a}{2} = (1+m)x$

$\therefore$ $x = \frac{(m-1)+\sqrt{2}a(m+1)}{2(m+1)}$

$\therefore$ $M = (\alpha, -\alpha)$ where $\alpha = \frac{m-1}{2(m+1)} + \frac{a}{\sqrt{2}}$.

Now, $C = \text{centre} = \left(\frac{1+\sqrt{2}a}{4}, \frac{1-\sqrt{2}a}{4}\right)$.

As M is the middle point of PA, $CM \perp PA$

$$\therefore \quad \frac{\frac{1-\sqrt{2}a}{4} + \alpha}{\frac{1+\sqrt{2}a}{4} - \alpha}\cdot m = -1$$

or $\frac{1-\sqrt{2}a}{4}\cdot m + \alpha m + \frac{1+\sqrt{2}a}{4} - \alpha = 0$

or $\frac{1-\sqrt{2}a}{4}\cdot m + \frac{1+\sqrt{2}a}{4} + (m-1)\left\{\frac{m-1}{2(m+1)} + \frac{a}{\sqrt{2}}\right\} = 0$, putting for α.

or $\frac{(m-1)^2}{2(m+1)} + \frac{a}{\sqrt{2}}(m-1) + \frac{1-\sqrt{2}a}{4}\cdot m + \frac{1+\sqrt{2}a}{4} = 0$

or $(m-1)^2 + a\sqrt{2}(m^2-1) + \frac{1-\sqrt{2}a}{2}m(m+1) + \frac{1+\sqrt{2}a}{2}(m+1) = 0$

or $\left\{1 + a\sqrt{2} + \frac{1-\sqrt{2}a}{2}\right\}m^2 + \left\{-2 + \frac{1-\sqrt{2}a}{2} + \frac{1+\sqrt{2}a}{2}\right\}m + \left\{1 - a\sqrt{2} + \frac{1+\sqrt{2}a}{2}\right\} = 0$

or $\left(\frac{3}{2} + \frac{a}{\sqrt{2}}\right)m^2 - m + \left(\frac{3}{2} - \frac{a}{\sqrt{2}}\right) = 0.$

For two chords like PA, the equation must have two real and unequal roots. So $D > 0$

$\therefore$ $1 - 4\left(\frac{3}{2} + \frac{a}{\sqrt{2}}\right)\left(\frac{3}{2} - \frac{a}{\sqrt{2}}\right) > 0$

or $1 - 4\left(\frac{9}{4} - \frac{a^2}{2}\right) > 0$

or $-8 + 2a^2 > 0$

or $a^2 > 4$

$\therefore$ $(a+2)(a-2) > 0$

$\therefore$ by sign-scheme for quadratic expression, we get

$a < -2$ or $a > 2$.

$\therefore$ $a \in (-\infty, -2) \cup (2, +\infty)$.

9. Let a circle be given by $2x(x-a) + y(2y-b) = 0$, $a \neq 0$, $b \neq 0$. Find the condition on a and b if two chords, each bisected by the x-axis, can be drawn to the circle from the point $\left(a, \frac{b}{2}\right)$.

Here, the equation of the circle is

$$2(x^2 + y^2) - 2ax - by = 0$$

or $x^2 + y^2 - ax - \frac{b}{2}y = 0$ ... (1)

Its centre $= \left(\frac{a}{2}, \frac{b}{4}\right)$ and

radius $= \sqrt{\left(\frac{a}{2}\right)^2 + \left(\frac{b}{4}\right)^2} = \frac{1}{4}\sqrt{4a^2 + b^2}$.

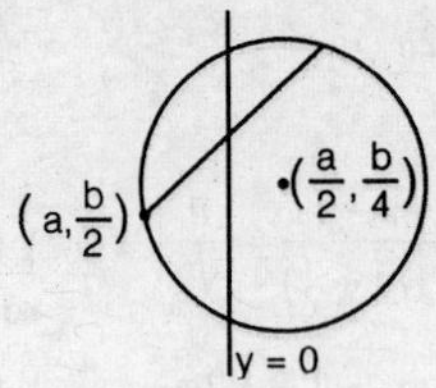

Any line through $\left(a, \frac{b}{2}\right)$ has the equation

$$y - \frac{b}{2} = m(x - a) \quad \ldots (2)$$

(2) cuts the x-axis, i.e., $y = 0$ at $\left(a - \frac{b}{2m}, 0\right)$.

As $\left(a, \frac{b}{2}\right)$ satisfies (1), $\left(a, \frac{b}{2}\right)$ lies on the circle.

The centre is $\left(\frac{a}{2}, \frac{b}{4}\right)$.

As the chord along (2) is bisected at $\left(a - \frac{b}{2m}, 0\right)$, the line joining $\left(a - \frac{b}{2m}, 0\right)$ and the centre $\left(\frac{a}{2}, \frac{b}{4}\right)$ will be perpendicular to the line (2).

$$\therefore \quad m \cdot \frac{\frac{b}{4} - 0}{\frac{a}{2} - \left(a - \frac{b}{2m}\right)} = -1$$

or $\quad \frac{mb}{4} + \frac{a}{2} - \left(a - \frac{b}{2m}\right) = 0$

or $\quad bm^2 + 2am - (4am - 2b) = 0$

or $\quad bm^2 - 2am + 2b = 0 \quad \ldots (3)$

Two chords are possible if (3) has two real and unequal roots, i.e., $4a^2 - 4 \cdot b \cdot 2b > 0$

or $\quad a^2 - 2b^2 > 0$

or $\quad a^2 > 2b^2$

$\therefore \quad a > \sqrt{2}b$ or $a < -\sqrt{2}b$.

10. Find the equation of the circle having the pair of lines $x^2 + 2xy + 3x + 6y = 0$ as its normals and having the size just sufficient to contain the circle $x(x - 4) + y(y - 3) = 0$.

Here the pair is $x(x + 3) + 2y(x + 3) = 0$

or $\quad (x + 3)(x + 2y) = 0$

$\therefore$ two normals of the circle are $x + 3 = 0$, $x + 2y = 0$.

$\therefore$ their intersection $\left(-3, \frac{3}{2}\right)$ is the centre of the circle.

The given circle is $x^2 + y^2 - 4x - 3y = 0$.

$\therefore \quad$ centre $= \left(2, \frac{3}{2}\right)$ and

radius $r = \sqrt{(-2)^2 + \left(\frac{-3}{2}\right)^2 - 0} = \frac{5}{2}$.

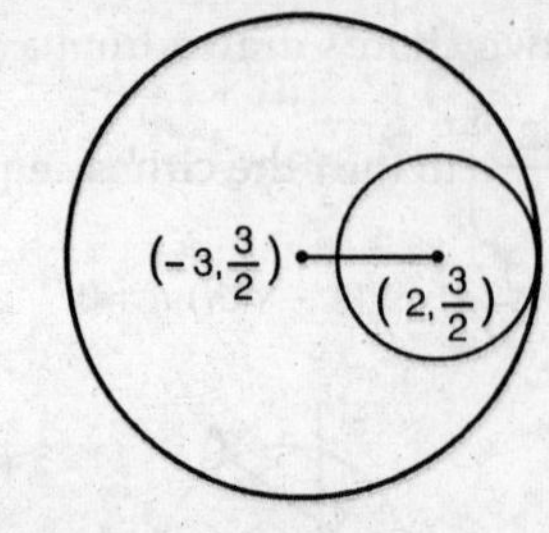

If the required circle has the radius a then

$a - r =$ distance between the centres $\left(-3, \frac{3}{2}\right)$ and $\left(2, \frac{3}{2}\right)$

$$\therefore \quad a - \frac{5}{2} = \sqrt{(-3 - 2)^2 + \left(\frac{3}{2} - \frac{3}{2}\right)^2} = 5$$

$$\therefore \quad a = 5 + \frac{5}{2} = \frac{15}{2}.$$

$\therefore \quad$ the required circle is

$$(x + 3)^2 + \left(y - \frac{3}{2}\right)^2 = \left(\frac{15}{2}\right)^2$$

or $\quad x^2 + 6x + 9 + y^2 - 3y + \frac{9}{4} = \frac{225}{4}$

or $\quad x^2 + y^2 + 6x - 3y = 45$.

11. If the curves $ax^2 + 2hxy + by^2 + 2gx + 2fy + c = 0$ and $Ax^2 + 2Hxy + By^2 + 2Gx + 2Fy + c = 0$ intersect at four concyclic points then prove that $\frac{a - b}{h} = \frac{A - B}{H}$.

Any second degree curve passing through the intersections of the given curves is

$$ax^2 + 2hxy + by^2 + 2gx + 2fy + c + \lambda(Ax^2 + 2Hxy + By^2 + 2Gx + 2Fy + C) = 0 \quad \ldots (1)$$

As the points of intersection of the two curves are concyclic, (1) must be a circle for some λ.

$\therefore \quad$ coefficient of x^2 = coefficient of y^2 and coefficient of $xy = 0$.

$\therefore \quad a + \lambda A = b + \lambda B$ and $2h + \lambda \cdot 2H = 0$

or $\quad a - b = \lambda(B - A)$ and $h = -\lambda H$

$$\therefore \quad \frac{a - b}{h} = \frac{\lambda(B - A)}{-\lambda H}; \quad \therefore \quad \frac{a - b}{h} = \frac{A - B}{H}.$$

12. Two circles have the equations

$$x^2+y^2+\lambda x+c=0 \text{ and } x^2+y^2+\mu x+c=0.$$

Prove that one of the circles will be within the other if $\lambda\mu>0$ and $c>0$.

Here the circles are $x^2+y^2+\lambda x+c=0 \quad \ldots (1)$

$x^2+y^2+\mu x+c=0 \quad \ldots (2)$

The points of intersection of the circles are obtained by solving (1) and (2).

$(1)-(2) \quad \Rightarrow \quad (\lambda-\mu)x=0; \quad \therefore \quad x=0 \quad (\because \lambda\neq\mu)$

Putting $x=0$ in (1),

$y^2+c=0; \quad \therefore \quad y=\pm\sqrt{-c}$

$\therefore$ if $c>0$, the points of intersection will be imaginary and hence the circles will be nonintersecting.

Hence, one circle will be within the other or fully outside the other. If (1) is within (2) then any chord through the centre $\left(-\frac{\lambda}{2}, 0\right)$ of the circle (1) will cut the circle (2) at real points.

Now, any chord through $\left(-\frac{\lambda}{2}, 0\right)$ is

$y=m\left(x+\frac{\lambda}{2}\right) \quad \ldots (3)$

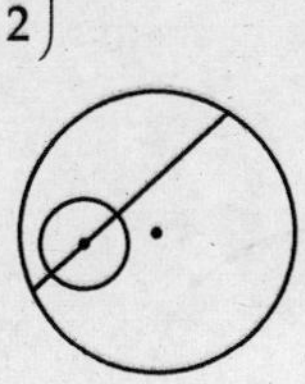

Solving (2) and (3),

$$x^2+m^2\left(x+\frac{\lambda}{2}\right)^2+\mu x+c=0$$

or $\quad (1+m^2)x^2+(\lambda m^2+\mu)x+\frac{m^2\lambda^2}{4}+c=0.$

Both roots must be real and unequal. So $D>0$

$\therefore \quad (\lambda m^2+\mu)^2-4(1+m^2)\cdot\left(\frac{m^2\lambda^2}{4}+c\right)>0$

or $\quad \lambda^2m^4+\mu^2+2m^2\lambda\mu-m^2\lambda^2(1+m^2)$
$\qquad -4c(1+m^2)>0$

or $\quad m^2(2\lambda\mu-\lambda^2-4c)+\mu^2-4c>0 \quad \ldots (4)$

This is true for all real m.

Also, the circle (2) is real if

radius $=\sqrt{(-\mu/2)^2-c}$ is real, i.e., $\mu^2-4c\geq 0$.

Hence, (4) is true for all real m if $2\lambda\mu-\lambda^2-4c>0$

or $\quad 2\lambda\mu>\lambda^2+4c;$ but $c>0 \quad \Rightarrow \quad \lambda^2+4c>0$

$\therefore \quad \lambda\mu>0.$

Hence, one circle is within the other if

$c>0, \lambda\mu>0.$

13. Find the equation of the circle whose radius is 5 and which touches the circle $x^2+y^2-2x-4y-20=0$ at the point (5, 5).

For the given circle, centre = (1, 2) and

radius $=\sqrt{1^2+2^2-(-20)}=5.$

So, the required circle of radius 5 must touch the given circle externally.

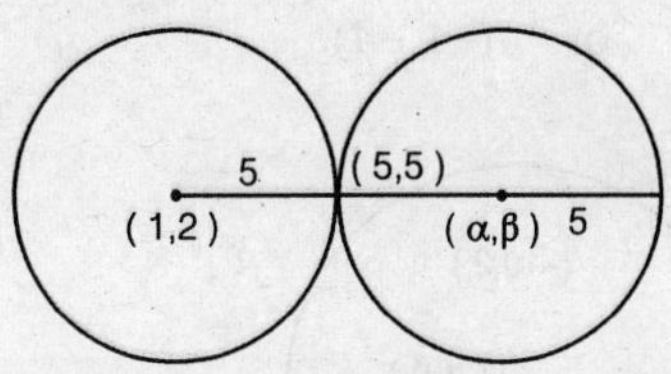

Let the centre of the required circle be (α, β).

$\therefore \quad \frac{\alpha+1}{2}=5$ and $\frac{\beta+2}{2}=5$

$\therefore \quad \alpha=9, \beta=8$

$\therefore$ for the circle, centre = (9, 8) and radius = 5.

$\therefore$ the equation of the required circle is

$(x-9)^2+(y-8)^2=5^2$

or $\quad x^2+y^2-18x-16y+120=0.$

14. Find the equation of the circle whose centre is (3, –1) and which cuts off a chord of length 6 on the line $2x-5y+18=0$.

The length of the perpendicular from the centre (3, –1) to the chord

$$=\left|\frac{2\times 3-5(-1)+18}{\sqrt{2^2+(-5)^2}}\right|=\frac{29}{\sqrt{29}}=\sqrt{29}$$

$\therefore \quad$ radius $=\sqrt{3^2+(\sqrt{29})^2}=\sqrt{38}.$

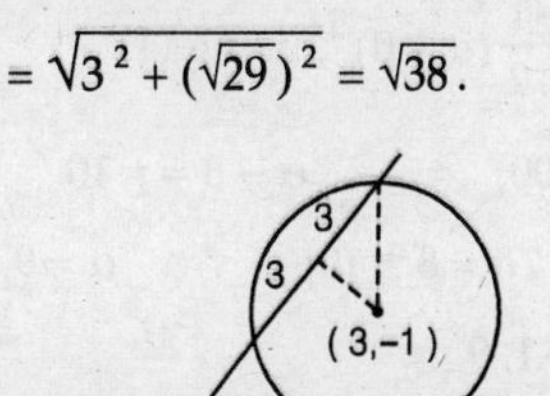

$\therefore$ the required equation of the circle is

$(x-3)^2+(y+1)^2=(\sqrt{38})^2$

or $\quad x^2-6x+9+y^2+2y+1=38$

or $\quad x^2+y^2-6x+2y-28=0.$

15. A circle touches the line $y = x$ at a point P such that OP $= 4\sqrt{2}$ where O is the origin. The circle contains the point $(-10, 2)$ in its interior and the length of its chord on the line $x + y = 0$ is $6\sqrt{2}$. Find the equation of the circle.

Clearly $y = x$ and $x + y = 0$ are two perpendicular lines passing through the origin O.

Let P $= (k, k)$ as it is on the line $y = x$.

$\therefore \quad OP = \sqrt{k^2 + k^2} = 4\sqrt{2}$

$\therefore \quad \pm\sqrt{2}k = 4\sqrt{2}$

$\therefore \quad k = \pm 4$

$\therefore \quad P = (4, 4) \quad$ or $\quad (-4, -4)$.

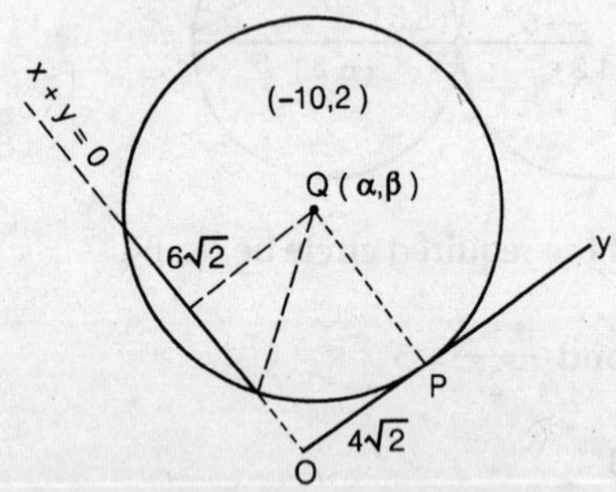

Let the centre Q of the circle $= (\alpha, \beta)$.

Taking P $= (4, 4)$,

$$QP \perp OP \quad \Rightarrow \quad \frac{\beta - 4}{\alpha - 4} \cdot 1 = -1$$

or $\quad \alpha + \beta = 8 \quad$ ... (1)

The length of perpendicular from Q (α, β) to the chord $x + y = 0$ is $\left|\frac{\alpha + \beta}{\sqrt{2}}\right|$

$$\therefore \quad (3\sqrt{2})^2 + \left(\frac{\alpha + \beta}{\sqrt{2}}\right)^2 = (\text{radius})^2 = QP^2 = \left|\frac{\alpha - \beta}{\sqrt{2}}\right|^2$$

or $\quad 18 + \frac{1}{2}(\alpha + \beta)^2 = \frac{1}{2}(\alpha - \beta)^2$

$\therefore \quad 18 + \frac{1}{2} \cdot 8^2 = \frac{1}{2}(\alpha - \beta)^2$, using (1)

or $\quad (\alpha - \beta)^2 = 100; \quad \therefore \quad \alpha - \beta = \pm 10 \quad$ ... (2)

From (1) and (2), $2\alpha = 8 \pm 10; \quad \therefore \quad \alpha = 9, -1$

$\therefore \quad \beta = 8 - \alpha = -1, 9.$

Hence, $(\alpha, \beta) = (9, -1)$ or $(-1, 9)$,

$$\therefore \quad \text{radius} = \left|\frac{\alpha - \beta}{\sqrt{2}}\right| = \frac{9 + 1}{\sqrt{2}} = 5\sqrt{2}.$$

If the centre $= (9, -1)$ then the distance between the points $(-10, 2)$ and $(9, -1)$ is $\sqrt{19^2 + 3^2} > 5\sqrt{2}$.

So $(-10, 2)$ will not be an interior point.

If the centre $= (-1, 9)$ then the distance between the points $(-10, 2)$ and $(-1, 9)$ is $\sqrt{9^2 + 7^2} > 5\sqrt{2}$.

So $(-10, 2)$ will not be an interior point.

Taking P $= (-4, -4)$ as before, we get

$$\alpha + \beta + 8 = 0 \quad \text{and} \quad \alpha - \beta = \pm 10$$

$\therefore \quad 2\alpha = -8 \pm 10$

$\therefore \quad \alpha = 1, -9$ and $\beta = -9, 1$

$\therefore \quad (\alpha, \beta) = (1, -9) \quad$ or $\quad (-9, 1)$.

If the centre $= (1, -9)$ then the distance between the points $(-10, 2)$ and $(1, -9)$ is $\sqrt{11^2 + 11^2} > 5\sqrt{2}$.

So $(-10, 2)$ will not be an interior point.

If the centre $= (-9, 1)$ then the distance between the points $(-10, 2)$ and $(-9, 1)$ is $\sqrt{1^2 + 1^2} < 5\sqrt{2}$.

So $(-10, 2)$ will be an interior point.

$\therefore$ the required circle has centre $= (-9, 1)$, radius $= 5\sqrt{2}$.

$\therefore$ the equation of the required circle is

$$(x + 9)^2 + (y - 1)^2 = (5\sqrt{2})^2$$

or $\quad x^2 + y^2 + 18x - 2y + 32 = 0.$

Second Method Here $x + y = 0$ and $y = x$ are two perpendicular lines.

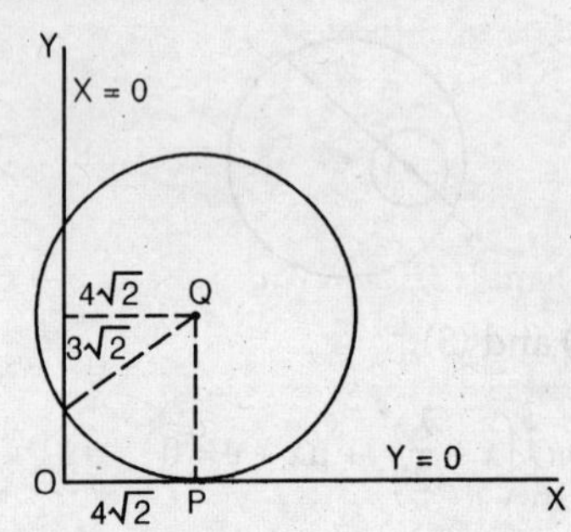

So, take the following equations of transformation:

$$X = \frac{x + y}{\sqrt{1^2 + 1^2}} \quad \text{and} \quad Y = \frac{x - y}{\sqrt{1^2 + (-1)^2}}.$$

Then we have to find a circle touching X-axis and cutting off $6\sqrt{2}$ from Y-axis.

Clearly, radius $= \sqrt{(4\sqrt{2})^2 + (3\sqrt{2})^2} = 5\sqrt{2}$.

Also, when $x = -10$, $y = 2$ we get

$$X = \frac{-10 + 2}{\sqrt{2}} = -4\sqrt{2}, \; Y = \frac{-10 - 2}{\sqrt{2}} = -6\sqrt{2}$$

$\therefore$ in the new coordinates the point $(-4\sqrt{2}, -6\sqrt{2})$ must be an interior point

$\therefore$ the required circle will be in the new third quadrant.

$\therefore$ the centre $= (-4\sqrt{2}, -5\sqrt{2})$ and radius $= 5\sqrt{2}$.

$\therefore$ the equation of the circle, in the new coordinates, is

$$(X + 4\sqrt{2})^2 + (Y + 5\sqrt{2})^2 = (5\sqrt{2})^2$$

$\therefore$ the equation of the circle, in the original coordinates, is

$$\left(\frac{x+y}{\sqrt{2}}+4\sqrt{2}\right)^2+\left(\frac{x-y}{\sqrt{2}}+5\sqrt{2}\right)^2=(5\sqrt{2})^2$$

or $(x+y+8)^2+(x-y+10)^2=100$

or $2x^2+2y^2+36x-4y+64=0$

or $x^2+y^2+18x-2y+32=0.$

16. Let A be the centre of the circle whose equation is $x^2+y^2-2x-4y-20=0$. Suppose that the tangents at the points B(1, 7) and D(4, –2) on the circle meet at the point C. Find the area of the quadrilateral ABCD.

The equation of the tangents at B, D are respectively

$$x\cdot 1+y\cdot 7-(x+1)-2(y+7)-20=0,$$

i.e., $5y-35=0 \quad \therefore \quad y-7=0$ ... (1)

and $x\cdot 4+y(-2)-(x+4)-2(y-2)-20=0,$

i.e., $3x-4y-20=0$ ... (2)

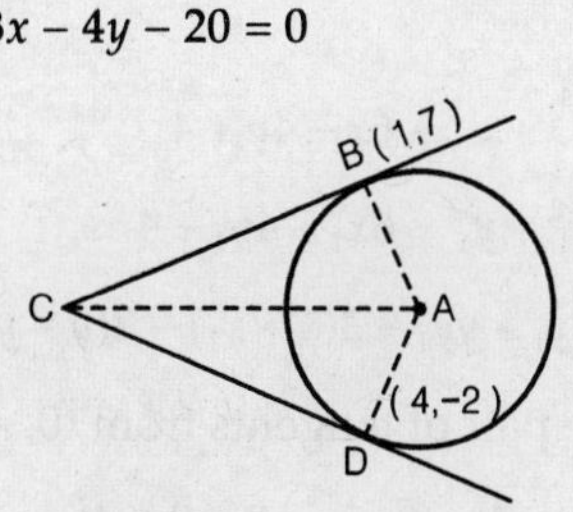

Solving (1) and (2), $y=7$ and

$$x=\frac{4y+20}{3}=\frac{4\times 7+20}{3}=16$$

$\therefore$ C = (16, 7).

Now, BC = length of the tangent from C(16, 7) to the circle

$$=\sqrt{S(16,7)}=\sqrt{16^2+7^2-2\cdot 16-4\cdot 7-20}$$

$$=\sqrt{225}=15.$$

So, ar(ABCD) = 2 ar($\triangle$ACB) = $2\cdot\frac{1}{2}BC\cdot AB$

= 15 × radius

$$=15\times\sqrt{1^2+2^2-(-20)}=75.$$

17. AP and BQ are fixed parallel tangents to a circle, and a tangent at any point C cuts them at P and Q respectively. Show that $CP\cdot CQ$ is independent of the position of C on the circle and $\angle POQ$ is a right angle.

Let the parallel tangents touch the circle at M and N respectively. Then MN is a diameter through the centre O.

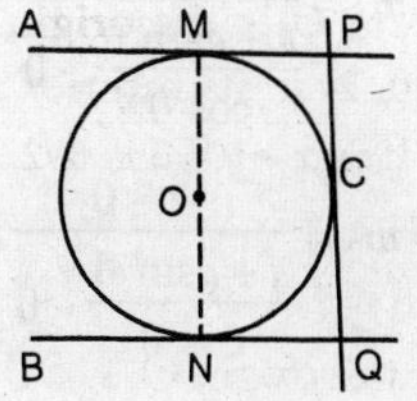

Taking the centre O as the origin and MN as the y-axis we can take the equation of the circle as

$$x^2+y^2=a^2 \quad \text{... (1)}$$

and the parallel tangents as $y=\pm a$ ... (2)

Any point C on the circle is ($a\cos\theta$, $a\sin\theta$).

The tangent to the circle at C is

$$x\cdot a\cos\theta+y\cdot a\sin\theta=a^2$$

or $x\cos\theta+y\sin\theta=a$ ... (3)

Solving (2) and (3),

$$y=a \text{ and } x=\frac{a-a\sin\theta}{\cos\theta}$$

$$y=-a \text{ and } x=\frac{a+a\sin\theta}{\cos\theta}$$

$\therefore$ $P=\left(\frac{a-a\sin\theta}{\cos\theta}, a\right)$, $Q=\left(\frac{a+a\sin\theta}{\cos\theta}, -a\right)$ and

C = ($a\cos\theta$, $a\sin\theta$)

$$\therefore \quad CP^2\cdot CQ^2=\left\{\left(a\cos\theta-\frac{a-a\sin\theta}{\cos\theta}\right)^2+(a\sin\theta-a)^2\right\}$$

$$\times\left\{\left(a\cos\theta-\frac{a+a\sin\theta}{\cos\theta}\right)^2+(a\sin\theta+a)^2\right\}$$

$$=\left\{a^2+a^2-2a(a-a\sin\theta)-2a^2\sin\theta+a^2\left(\frac{1-\sin\theta}{\cos\theta}\right)^2\right\}\times\left\{a^2+a^2-2a(a+a\sin\theta)+2a^2\sin\theta+a^2\left(\frac{1+\sin\theta}{\cos\theta}\right)^2\right\}$$

$$=a^2\left\{2-2(1-\sin\theta)-2\sin\theta+\left(\frac{1-\sin\theta}{\cos\theta}\right)^2\right\}\times a^2\left\{2-2(1+\sin\theta)+2\sin\theta+\left(\frac{1+\sin\theta}{\cos\theta}\right)^2\right\}$$

$$=a^4\left(\frac{1-\sin\theta}{\cos\theta}\right)^2\cdot\left(\frac{1+\sin\theta}{\cos\theta}\right)^2$$

$$=a^4\cdot\frac{(1-\sin^2\theta)^2}{\cos^4\theta}=a^4.$$

$\therefore \quad CP\cdot CQ=a^2=$ constant.

Also 'm' of PO $= m_1 = \dfrac{a-0}{\dfrac{a - a\sin\theta}{\cos\theta} - 0} = \dfrac{\cos\theta}{1-\sin\theta}$

'm' of QO $= m_2 = \dfrac{-a-0}{\dfrac{a + a\sin\theta}{\cos\theta} - 0} = -\dfrac{\cos\theta}{1+\sin\theta}$.

Clearly, $m_1 \cdot m_2 = \dfrac{\cos\theta}{1-\sin\theta} \cdot \left(-\dfrac{\cos\theta}{1+\sin\theta}\right) = -1;$

hence $\angle POQ = \dfrac{\pi}{2}$.

18. From a point on the line $4x - 3y = 6$ tangents are drawn to the circle $x^2 + y^2 - 6x - 4y + 4 = 0$ which make an angle of $\tan^{-1}\dfrac{24}{7}$ between them. Find the coordinates of all such points and the equations of tangents.

Let P(x_1, y_1) be a point on the line $4x - 3y = 6$ having the given property.

Let the angle between the tangents be θ.

Then $\tan\theta = \dfrac{24}{7}$.

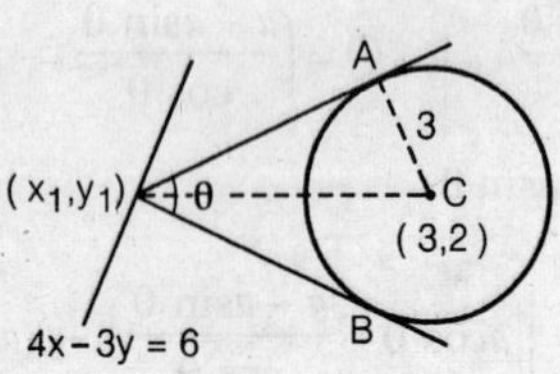

Now, centre C = (3, 2) and

radius $= CA = \sqrt{3^2 + 2^2 - 4} = 3$

and PA = length of tangent $= \sqrt{x_1^2 + y_1^2 - 6x_1 - 4y_1 + 4}$

$\therefore \quad \tan\dfrac{\theta}{2} = \dfrac{AC}{PA} = \dfrac{3}{\sqrt{x_1^2 + y_1^2 - 6x_1 - 4y_1 + 4}}$

$\therefore \quad \tan^2\dfrac{\theta}{2} = \dfrac{9}{x_1^2 + y_1^2 - 6x_1 - 4y_1 + 4}$

or $\quad \dfrac{1 - \tan^2\dfrac{\theta}{2}}{1 + \tan^2\dfrac{\theta}{2}} = \dfrac{x_1^2 + y_1^2 - 6x_1 - 4y_1 + 4 - 9}{x_1^2 + y_1^2 - 6x_1 - 4y_1 + 4 + 9}$

or $\quad \cos\theta = \dfrac{x_1^2 + y_1^2 - 6x_1 - 4y_1 - 5}{x_1^2 + y_1^2 - 6x_1 - 4y_1 + 13}$

$\therefore \quad \dfrac{7}{25} = \dfrac{x_1^2 + y_1^2 - 6x_1 - 4y_1 - 5}{x_1^2 + y_1^2 - 6x_1 - 4y_1 + 13} \quad \left(\because \tan\theta = \dfrac{24}{7}\right)$

or $\quad 7x_1^2 + 7y_1^2 - 42x_1 - 28y_1 + 91$

$\qquad = 25x_1^2 + 25y_1^2 - 150x_1 - 100y_1 - 125$

or $\quad 18x_1^2 + 18y_1^2 - 108x_1 - 72y_1 - 216 = 0$

or $\quad x_1^2 + y_1^2 - 6x_1 - 4y_1 - 12 = 0 \qquad \ldots (1)$

As (x_1, y_1) is on the line $4x - 3y = 6$, we get

$4x_1 - 3y_1 = 6 \qquad \ldots (2)$

Solving (1) and (2),

$x_1^2 + \left(\dfrac{4x_1 - 6}{3}\right)^2 - 6x_1 - 4 \cdot \dfrac{4x_1 - 6}{3} - 12 = 0$

or $\quad 9x_1^2 + (4x_1 - 6)^2 - 54x_1 - 12(4x_1 - 6) - 108 = 0$

or $\quad 25x_1^2 - 150x_1 = 0$

or $\quad x_1(x_1 - 6) = 0$

$\therefore \quad x_1 = 0, 6$

$\therefore \quad y_1 = \dfrac{4x_1 - 6}{3} = -\dfrac{6}{3}, \dfrac{18}{3} = -2, 6$

$\therefore \quad (x_1, y_1) = (0, -2), (6, 6).$

The equation of the pair of tangents from (x_1, y_1) is

$S \cdot S_1 = T^2$

where $S \equiv x^2 + y^2 - 6x - 4y + 4$

$S_1 = x_1^2 + y_1^2 - 6x_1 - 4y_1 + 4$

$T_1 = xx_1 + yy_1 - 3(x + x_1) - 2(y + y_1) + 4$

$\therefore$ the equation of the pair of tangents from $(0, -2)$ is

$(x^2 + y^2 - 6x - 4y + 4) \cdot (0 + 4 - 0 + 8 + 4)$

$\qquad = \{x \cdot 0 + y(-2) - 3(x + 0) - 2(y - 2) + 4\}^2$

or $\quad 16(x^2 + y^2 - 6x - 4y + 4)$

$\qquad = (-3x - 4y + 8)^2$

$\qquad = 9x^2 + 16y^2 + 64 + 24xy - 48x - 64y$

or $\quad 7x^2 - 24xy - 48x = 0$

or $\quad x(7x - 24y - 48) = 0$

$\therefore \quad$ tangents from $(0, -2)$ are $x = 0$, $7x - 24y - 48 = 0$.

Similarly, the equation of the pair of tangents from $(6, 6)$ is

$(x^2 + y^2 - 6x - 4y + 4) \cdot (36 + 36 - 6 \cdot 6 - 4 \cdot 6 + 4)$

$\qquad = \{x \cdot 6 + y \cdot 6 - 3(x + 6) - 2(y + 6) + 4\}^2$

or $\quad 16(x^2 + y^2 - 6x - 4y + 4)$

$\qquad = (3x + 4y - 26)^2$

$\qquad = 9x^2 + 16y^2 + 676 + 24xy - 156x - 208y$

or $\quad 7x^2 - 24xy + 60x + 144y - 612 = 0$

or $\quad (7x - 24y + 102)(x - 6) = 0$

$\therefore \quad$ the tangents are $x - 6 = 0$, $7x - 24y + 102 = 0$.

19. Show that the circles $x^2+y^2+2x-8y+8=0$ and $x^2+y^2+10x-2y+22=0$ touch each other. Do they touch internally? Also obtain the equations of circles of radius 1 which cut both the circles orthogonally.

For the first circle,

centre = (–1, 4) and radius $r_1=\sqrt{1^2+4^2-8}=3$.

For the second circle,

centre = (–5, 1) and radius $r_2=\sqrt{5^2+1^2-22}=2$.

The distance between the centres

$$d=\sqrt{(-1+5)^2+(4-1)^2}=\sqrt{16+9}=5$$

$\therefore\quad d=r_1+r_2$

$\therefore$ the circles touch externally, not internally.

Let the required circle of radius 1 be

$$(x-\alpha)^2+(y-\beta)^2=1^2$$

or $\quad x^2+y^2-2\alpha x-2\beta y+\alpha^2+\beta^2-1=0 \quad \ldots (1)$

(1) cuts the first circle orthogonally if

$$2\cdot(-\alpha)\cdot 1+2(-\beta)\cdot(-4)=(\alpha^2+\beta^2-1)+8$$

$\{$using $2g_1g_2+2f_1f_2=c_1+c_2\}$

or $\quad \alpha^2+\beta^2+2\alpha-8\beta+7=0 \quad \ldots (2)$

(1) cuts the second circle orthogonally if

$$2\cdot(-\alpha)\cdot 5+2(-\beta)(-1)=\alpha^2+\beta^2-1+22$$

or $\quad \alpha^2+\beta^2+10\alpha-2\beta+21=0 \quad \ldots (3)$

(3) – (2) $\Rightarrow\quad 8\alpha+6\beta+14=0$

$\therefore\quad 4\alpha+3\beta+7=0 \quad \ldots (4)$

Putting in (2) from (4),

$$\alpha^2+\left(-\frac{4\alpha+7}{3}\right)^2+2\alpha-8\left(-\frac{4\alpha+7}{3}\right)+7=0$$

or $\quad 9\alpha^2+(4\alpha+7)^2+18\alpha+24(4\alpha+7)+63=0$

or $\quad 25\alpha^2+170\alpha+280=0$

or $\quad 5\alpha^2+34\alpha+56=0$

or $\quad 5\alpha^2+20\alpha+14\alpha+56=0$

$\therefore\quad (5\alpha+14)(\alpha+4)=0$

$\therefore\quad \alpha=-4,\ -14/5$

$\therefore\quad$ (4) gives $3\beta=-(4\alpha+7)$

$$=-(-16+7),\ -\left(4\cdot\frac{-14}{5}+7\right)$$

$$=9,\ \frac{21}{5}$$

$\therefore\quad \beta=3,\ 7/5.$

$\therefore\quad$ from (1), the equations of required circles are

$$x^2+y^2+8x-6y+24=0$$

and $\quad x^2+y^2+\frac{28}{5}x-\frac{14}{5}y+\frac{44}{5}=0,$

i.e., $\quad 5(x^2+y^2)+28x-14y+44=0.$

20. Find the equation of the circle of minimum radius which contains the three circles

$x^2+y^2-4y-5=0,$

$x^2+y^2+12x+4y+31=0$ and

$x^2+y^2+6x+12y+36=0.$

For the first circle,

centre = (0, 2) and radius $=\sqrt{(-2)^2+5}=3.$

For the second circle,

centre = (–6, –2) and radius $=\sqrt{6^2+2^2-31}=3.$

For the third circle,

centre = (–3, –6) and radius $=\sqrt{3^2+6^2-36}=3.$

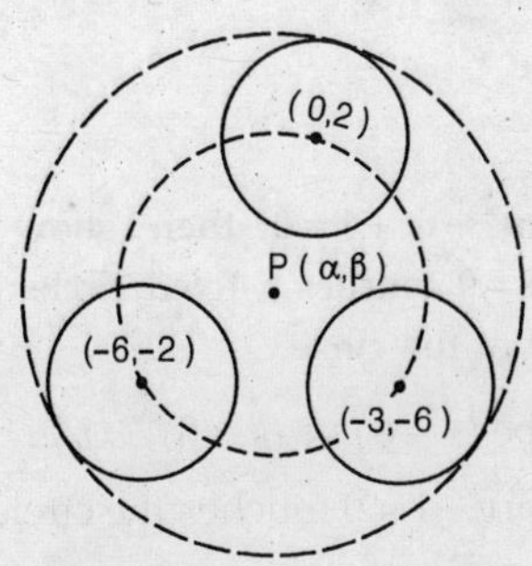

Let P(α, β) be the centre of the circle passing through the centres of the three given circles. Then

$$\alpha^2+(\beta-2)^2=(\alpha+3)^2+(\beta+6)^2$$

$$=(\alpha+6)^2+(\beta+2)^2,\ \{=r^2\}$$

or $\quad 4-4\beta=6\alpha+9+12\beta+36=12\alpha+36+4\beta+4$

$\therefore\quad 6\alpha+16\beta+41=0 \quad \ldots (1)$

and $\quad 6\alpha-8\beta-5=0 \quad \ldots (2)$

(1) – (2) $\Rightarrow\quad 24\beta+46=0$

$$\therefore\quad \beta=-\frac{23}{12}$$

$\therefore\quad$ (2) $\Rightarrow\quad 6\alpha-8\cdot\frac{-23}{12}-5=0$

or $\quad 6\alpha+\frac{46}{3}-5=0;\quad \therefore\quad \alpha=\frac{-31}{18}$

$\therefore$ the centre of the circle passing through centres of given circles is

$\left(\frac{-31}{18},\ \frac{-23}{12}\right)$ and

radius $r=\sqrt{\alpha^2+(\beta-2)^2}$

$$= \sqrt{\left(\frac{-31}{18}\right)^2 + \left(\frac{-23}{12} - 2\right)^2}$$

$$= \sqrt{\frac{31^2}{18^2} + \frac{47^2}{12^2}} = \sqrt{\frac{4 \times 31^2 + 9 \times 47^2}{36^2}}$$

$$= \frac{1}{36}\sqrt{23725} = \frac{5}{36}\sqrt{949}.$$

$\therefore$ the radius of the required circle $= 3 + \frac{5}{36}\sqrt{949}$ and

$$\text{centre} = \left(\frac{-31}{18}, \frac{-23}{12}\right)$$

$\therefore$ the equation of the required circle is

$$\left(x + \frac{31}{18}\right)^2 + \left(y + \frac{23}{12}\right)^2 = \left(3 + \frac{5}{36}\sqrt{949}\right)^2.$$

TIP In similar situations if the radii of the circles are different then add the largest radius to the radius of the circle passing through the centres.

21. If $4l^2 - 5m^2 + 6l + 1 = 0$ then show that the line $lx + my + 1 = 0$ touches a fixed circle. Find the radius and centre of the circle.

Let the circle be $(x - \alpha)^2 + (y - \beta)^2 = a^2$.

The line $lx + my + 1 = 0$ touches the circle if

$$a = \left|\frac{l\alpha + m\beta + 1}{\sqrt{l^2 + m^2}}\right|$$

or $\quad a^2(l^2 + m^2) = (l\alpha + m\beta + 1)^2$

or $\quad (a^2 - \alpha^2)l^2 + (a^2 - \beta^2)m^2$

$$- 2\alpha\beta lm - 2l\alpha - 2m\beta - 1 = 0 \quad \ldots (1)$$

But $4l^2 - 5m^2 + 6l + 1 = 0 \quad \ldots (2)$

It is possible to find α, β, a if (1) and (2) are identical.

The condition is

$$\frac{a^2 - \alpha^2}{4} = \frac{a^2 - \beta^2}{-5} = \frac{-2\alpha\beta}{0} = \frac{-2\alpha}{6} = \frac{-2\beta}{0} = \frac{-1}{1} \quad \ldots (3)$$

$\therefore \quad \beta = 0, \alpha = 3$

and $\quad \frac{a^2 - \alpha^2}{4} = -1$ which implies

$$a^2 - 3^2 = -4, \text{ i.e., } a = \sqrt{5}.$$

Also $\alpha = 3, \beta = 0, a = \sqrt{5}$ satisfies (3)

$\therefore$ the line touches the fixed circle,

$$(x - 3)^2 + (y - 0)^2 = (\sqrt{5})^2$$

or $\quad x^2 + y^2 - 6x + 4 = 0$

whose centre $= (\alpha, \beta) = (3, 0)$ and radius $= a = \sqrt{5}$.

22. Show that the common tangents to the circles $x^2 + y^2 - 6x = 0$ and $x^2 + y^2 + 2x = 0$ form an equilateral triangle.

Here the circle $x^2 + y^2 - 6x = 0$ has the

centre $= (3, 0)$ and radius $= 3$.

The circle $x^2 + y^2 + 2x = 0$ has the

centre $= (-1, 0)$ and radius $= 1$.

$\because \quad$ the distance between $(3, 0)$ and $(-1, 0)$

$$= \sqrt{(3 + 1)^2 + 0^2} = 4$$

and sum of radii $= 3 + 1 = 4$

$\therefore \quad$ the circles touch externally.

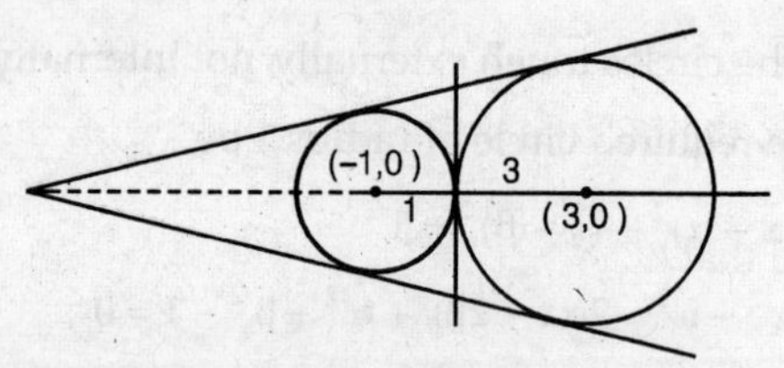

Obviously, the point of contact $= (0, 0)$.

Let $y = mx + c$ be a common tangent to the circles except at the point of contact.

Then, the length of the perpendicular from $(-1, 0)$ to $y = mx + c$ is equal to the radius of the second circle

$$\therefore \quad \left|\frac{-m + c}{\sqrt{1 + m^2}}\right| = 1$$

or $\quad (-m + c)^2 = 1 + m^2 \quad \ldots (1)$

and the length of the perpendicular from $(3, 0)$ to $y = mx + c$ is equal to the radius of the first circle

$$\therefore \quad \left|\frac{3m + c}{\sqrt{1 + m^2}}\right| = 3$$

or $\quad (3m + c)^2 = 9(1 + m^2) \quad \ldots (2)$

(1) $\Rightarrow \quad c^2 - 2mc = 1 \quad \ldots (3)$

(2) $\Rightarrow \quad 6mc + c^2 = 9 \quad \ldots (4)$

$\therefore \quad (6mc + c^2) - (c^2 - 2mc) = 9 - 1 = 8$

or $\quad 8mc = 8; \quad \therefore \quad mc = 1.$

$\therefore \quad$ (3) gives $c^2 = 3; \quad \therefore \quad c = \pm\sqrt{3}.$

When $c = \sqrt{3}$, from (3) we get

$$3 - 2m\sqrt{3} = 1; \quad \therefore \quad m = 1/\sqrt{3}.$$

When $c = -\sqrt{3}$, from (3) we get

$$3 + 2m\sqrt{3} = 1; \quad \therefore \quad m = -1/\sqrt{3}$$

$\therefore \quad$ two common tangents are

$$y = \frac{1}{\sqrt{3}}x + \sqrt{3} \text{ and } y = \frac{-1}{\sqrt{3}}x - \sqrt{3}.$$

The tangent at the point of contact (0, 0) is

$$x \cdot 0 + y \cdot 0 + (x + 0) = 0; \quad \therefore \quad x = 0.$$

Now, we have to prove that the lines

$$x = 0,\ y = \frac{-x}{\sqrt{3}} - \sqrt{3} \text{ and } y = \frac{x}{\sqrt{3}} + \sqrt{3}$$

form an equilateral triangle.

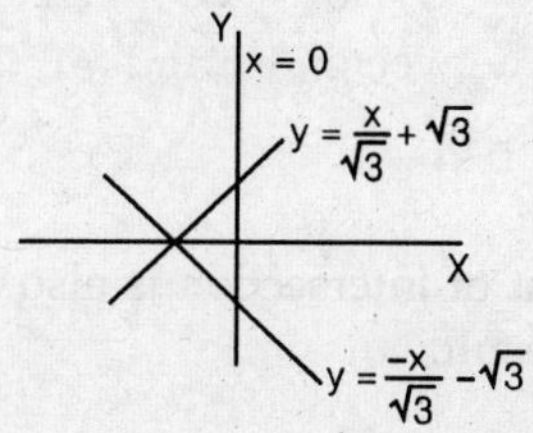

Clearly, the inclination of $y = \frac{-x}{\sqrt{3}} - \sqrt{3}$ with the x-axis is $\frac{5\pi}{6}$ and that of $y = \frac{x}{\sqrt{3}} + \sqrt{3}$ is $\frac{\pi}{6}$ as shown in the figure.

Also, their point of intersection is (– 3,0).

$\therefore$ we get from the figure that the triangle is equilateral.

23. Tangents PQ, PR are drawn to the circle $x^2 + y^2 = a^2$ from the point $P(x_1, y_1)$ touching the circle at Q, R respectively. Find the equation of the circumcircle of the ΔPQR.

QR is the chord of contact of the tangents to the circle

$$x^2 + y^2 - a^2 = 0 \quad \text{... (1)}$$

from the point $P(x_1, y_1)$.

$\therefore$ the equation of QR is $xx_1 + yy_1 - a^2 = 0$ ... (2)

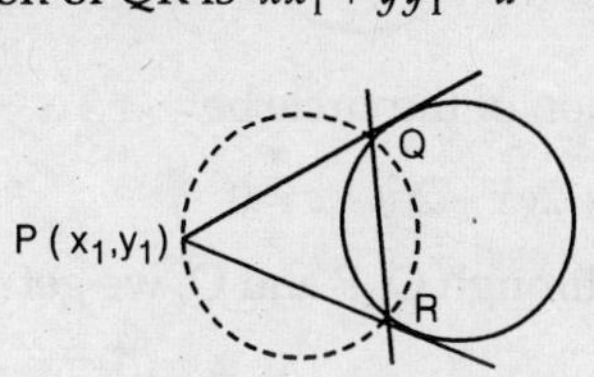

The circumcircle of ΔPQR is a circle passing through the intersection of the circle (1) and the line (2), and the point $P(x_1, y_1)$.

Any circle through the intersection of (1) and (2), is

$$x^2 + y^2 - a^2 + \lambda(xx_1 + yy_1 - a^2) = 0 \quad \text{... (3)}$$

It will pass through (x_1, y_1) if

$$x_1^2 + y_1^2 - a^2 + \lambda(x_1^2 + y_1^2 - a^2) = 0$$

or $\quad (x_1^2 + y_1^2 - a^2)(1 + \lambda) = 0.$

But $x_1^2 + y_1^2 \neq a^2$, for (x_1, y_1) is not on the circle.

$\therefore \quad \lambda = -1.$

$\therefore$ from (3), the circumcircle of ΔPQR is

$$x^2 + y^2 - a^2 - 1(xx_1 + yy_1 - a^2) = 0$$

or $\quad x^2 + y^2 - xx_1 - yy_1 = 0.$

24. Prove that, for all $c \in R$, the pole of the line $\frac{x}{a} + \frac{y}{b} = 1$ with respect to the circle $x^2 + y^2 = c^2$ lies on a fixed line.

Let the pole be (x_1, y_1). Then the polar of (x_1, y_1) with respect to the circle $x^2 + y^2 = c^2$ is

$$xx_1 + yy_1 = c^2 \quad \text{... (1)}$$

Now, the line (1) and $\frac{x}{a} + \frac{y}{b} = 1$ must be the same line.

$\therefore$ comparing coefficients, $\frac{x_1}{1/a} = \frac{y_1}{1/b} = \frac{c^2}{1}$

or $\quad ax_1 = by_1 = c^2; \quad \therefore \quad ax_1 = by_1$

$\therefore$ (x_1, y_1) always lies on the line $ax = by$ which is a fixed line.

25. Find the equation of the circle described on the common chord of the circles $x^2 + y^2 - 4x - 5 = 0$ and $x^2 + y^2 + 8y + 7 = 0$ as a diameter.

Here the circles are $S_1 \equiv x^2 + y^2 - 4x - 5 = 0$ ... (1)

and $\quad S_2 \equiv x^2 + y^2 + 8y + 7 = 0$ ... (2)

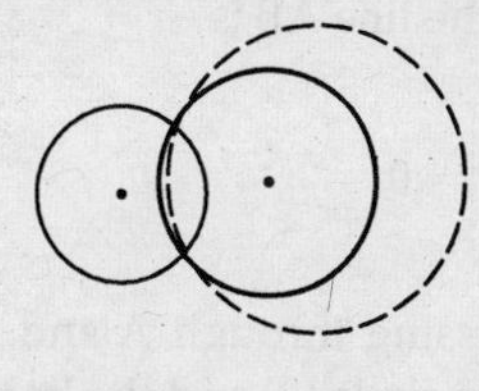

Any circle through the intersection of the circles (1) and (2) is given by $S_1 + \lambda S_2 = 0$ $(\lambda \neq -1)$

or $\quad x^2 + y^2 - 4x - 5 + \lambda(x^2 + y^2 + 8y + 7) = 0$ ... (3)

or $\quad (1 + \lambda)(x^2 + y^2) - 4x + 8\lambda y + 7\lambda - 5 = 0$

or $\quad x^2 + y^2 - \frac{4}{1 + \lambda}x + \frac{8\lambda}{1 + \lambda}y + \frac{7\lambda - 5}{1 + \lambda} = 0.$

Its centre $= \left(\frac{2}{1 + \lambda}, \frac{-4\lambda}{1 + \lambda}\right)$ ... (4)

The equation of the common chord of (1) and (2) is

$$S_1 - S_2 = 0$$

i.e., $\quad (x^2 + y^2 - 4x - 5) - (x^2 + y^2 + 8y + 7) = 0$

or $\quad 4x + 8y + 12 = 0$

or $\quad x + 2y + 3 = 0$ ... (5)

The centre will lie on the common chord, i.e., (3) will be a circle described on the common chord as a diameter if

$$\frac{2}{1+\lambda}+2\cdot\frac{-4\lambda}{1+\lambda}+3=0 \quad \text{\{from (4) and (5)\}}$$

or $\quad 2-8\lambda+3(1+\lambda)=0; \quad \therefore \quad \lambda=1$

$\therefore$ putting this value in (3), the equation of the required circle is

$$x^2+y^2-4x-5+(x^2+y^2+8y+7)=0$$

or $\quad 2(x^2+y^2)-4x+8y+2=0$

or $\quad x^2+y^2-2x+4y+1=0.$

26. A fixed circle is cut by circles passing through two fixed points A(x_1, y_1) and B(x_2, y_2). Show that the chord of intersection of the fixed circle with any one of the circles, passes through a fixed point.

Let $S \equiv x^2+y^2+2gx+2fy+c=0$ ... (1)

be the fixed circle, and $A=(x_1, y_1)$, $B=(x_2, y_2)$.

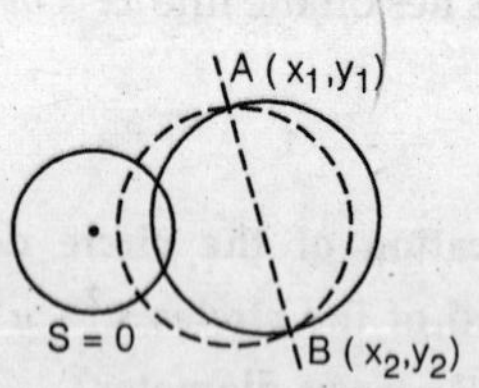

The equation of the circle with AB as a diameter, is

$$(x-x_1)(x-x_2)+(y-y_1)(y-y_2)=0 \qquad \text{... (2)}$$

The equation of the line AB is

$$\begin{vmatrix} x & y & 1 \\ x_1 & y_1 & 1 \\ x_2 & y_2 & 1 \end{vmatrix}=0 \qquad \text{... (3)}$$

$\therefore$ any circle passing through A and B (the points of intersection of the circle (2) and the line (3) is given by

$$(x-x_1)(x-x_2)+(y-y_1)(y-y_2)+\lambda\begin{vmatrix} x & y & 1 \\ x_1 & y_1 & 1 \\ x_2 & y_2 & 1 \end{vmatrix}=0 \qquad \text{... (4)}$$

The chord of intersection of the fixed circle $S=0$ and any one of the circles (4) is given by

$$x^2+y^2+2gx+2fy+c-\left\{(x-x_1)(x-x_2)+(y-y_1)(y-y_2)+\lambda\begin{vmatrix} x & y & 1 \\ x_1 & y_1 & 1 \\ x_2 & y_2 & 1 \end{vmatrix}\right\}=0$$

or $\quad (x_1+x_2)x+(y_1+y_2)y-x_1x_2-y_1y_2$

$$+2gx+2fy+c-\lambda\begin{vmatrix} x & y & 1 \\ x_1 & y_1 & 1 \\ x_2 & y_2 & 1 \end{vmatrix}=0$$

or $\quad (x_1+x_2+2g)x+(y_1+y_2+2f)y$

$$-x_1x_2-y_1y_2+c-\lambda\begin{vmatrix} x & y & 1 \\ x_1 & y_1 & 1 \\ x_2 & y_2 & 1 \end{vmatrix}=0.$$

This represents a family of lines passing through the intersection of the two fixed straight lines

$$(x_1+x_2+2g)x+(y_1+y_2+2f)y-x_1x_2-y_1y_2+c=0$$

and $\begin{vmatrix} x & y & 1 \\ x_1 & y_1 & 1 \\ x_2 & y_2 & 1 \end{vmatrix}=0.$

Hence, their point of intersection is also a fixed point. This proves the result.

27. A variable circle passes through the point of intersection O of two given straight lines and cuts off from them portions OP and OQ such that $m\cdot OP+n\cdot OQ=1$; prove that the circle always passes through a fixed point other than O.

Let O be selected as the origin and the lines be $y=0$ and $y=\alpha x$ where P is on $y=0$ and Q is on $y=\alpha x$.

Now, $\quad P=(OP, 0)$ and

$$Q=(OQ\cos\theta, OQ\sin\theta)$$

$$=\left(OQ\cdot\frac{1}{\sqrt{1+\alpha^2}}, OQ\cdot\frac{\alpha}{\sqrt{1+\alpha^2}}\right) \quad (\because \tan\theta=\alpha)$$

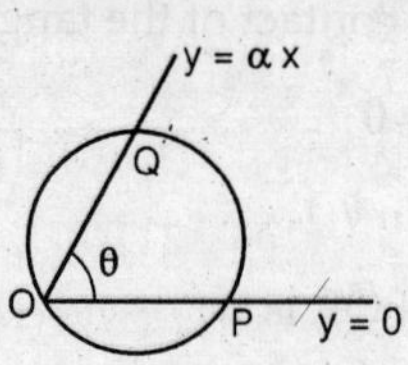

Let the equation of the circle be

$$x^2+y^2+2gx+2fy+c=0 \qquad \text{... (1)}$$

As it passes through O, P and Q, we get

$$c=0 \qquad \text{... (2)}$$

$$OP^2+2g\cdot OP=0; \quad \therefore \quad OP=-2g \qquad \text{... (3)}$$

$$OQ^2\cdot\frac{1}{1+\alpha^2}+OQ^2\cdot\frac{\alpha^2}{1+\alpha^2}+2g\cdot OQ\cdot\frac{1}{\sqrt{1+\alpha^2}}+2f\cdot OQ\cdot\frac{\alpha}{\sqrt{1+\alpha^2}}=0$$

or $\quad OQ^2+OQ\cdot\dfrac{2g+2f\alpha}{\sqrt{1+\alpha^2}}=0$

$\therefore \quad OQ=-\dfrac{2(g+f\alpha)}{\sqrt{1+\alpha^2}}$

But $\quad m\cdot OP+n\cdot OQ=1$

$\therefore \quad m(-2g)+n\cdot\left\{-\dfrac{2(g+f\alpha)}{\sqrt{1+\alpha^2}}\right\}=1$

or $\left(-2m - \frac{2n}{\sqrt{1+\alpha^2}}\right)g - 2n \cdot \frac{\alpha}{\sqrt{1+\alpha^2}} f = 1$

$\therefore \quad -2n \cdot \frac{\alpha}{\sqrt{1+\alpha^2}} f = 1 + 2\left(m + \frac{n}{\sqrt{1+\alpha^2}}\right)g$

$\therefore \quad 2f = -\frac{\sqrt{1+\alpha^2}}{n\alpha}\left\{1 + 2\left(m + \frac{n}{\sqrt{1+\alpha^2}}\right)g\right\}$

$= kg + p$ where k and p are constants.

$\therefore$ from (1) and (2), the circle has the equation

$x^2 + y^2 + 2gx + (kg + p)y = 0$

or $\quad x^2 + y^2 + py + g(2x + ky) = 0$

which is of the form $S + \lambda L = 0$.

$\therefore$ this circle passes through the intersection of the circle $x^2 + y^2 + py = 0$ and the line $2x + ky = 0$, both being fixed.

But the line $2x + ky = 0$ and the circle $x^2 + y^2 + py = 0$ cut at two fixed points, one of them being the origin O.

Hence, the circles pass through a fixed point other than O.

28. If two circles cut a third circle orthogonally, prove that their common chord will pass through the centre of the third circle.

Let us take the equations of the two circles as

$x^2 + y^2 + 2\lambda_1 x + a = 0 \quad \ldots (1)$

and $\quad x^2 + y^2 + 2\lambda_2 x + a = 0 \quad \ldots (2)$

TIP For a general statement in geometry concerning two circles we can select axes suitably (the line of centres as x-axis and the point midway between the centres as origin) to get the above form of equations.

Let the third circle be $x^2 + y^2 + 2gx + 2fy + c = 0 \quad \ldots (3)$

The circles (1) and (3) cut orthogonally

$\Rightarrow \quad 2\lambda_1 g = a + c \quad \ldots (4)$

The circles (2) and (3) cut orthogonally

$\Rightarrow \quad 2\lambda_2 g = a + c \quad \ldots (5)$

$\therefore$ from (4) and (5), $2g(\lambda_1 - \lambda_2) = 0$. But $\lambda_1 \neq \lambda_2$

$\therefore \quad g = 0$.

Hence, the centre of the third circle $= (0, -f)$.

The common chord of (1) and (2) has the equation

$S_1 - S_2 = 0$

i.e., $\quad (x^2 + y^2 + 2\lambda_1 x + a) - (x^2 + y^2 + 2\lambda_2 x + a) = 0$

or $\quad 2(\lambda_1 - \lambda_2)x = 0 \quad \therefore \quad x = 0$

$\therefore$ $(0, -f)$ satisfies the equation $x = 0$.

Hence, the problem.

29. Find the equation of the circle which cuts each of the circles $x^2 + y^2 = 4$, $x^2 + y^2 - 6x - 8y + 10 = 0$ and $x^2 + y^2 + 2x - 4y - 2 = 0$ at the extremities of a diameter.

Here the circles are

$S_1 \equiv x^2 + y^2 - 4 = 0 \quad \ldots (1)$

$S_2 \equiv x^2 + y^2 - 6x - 8y + 10 = 0 \quad \ldots (2)$

$S_3 \equiv x^2 + y^2 + 2x - 4y - 2 = 0 \quad \ldots (3)$

Let the required circle be

$S \equiv x^2 + y^2 + 2gx + 2fy + c = 0 \quad \ldots (4)$

The circle (4) will cut the circle (1) at extremities of a diameter if the common chord of (4) and (1) passes through the centre of (1).

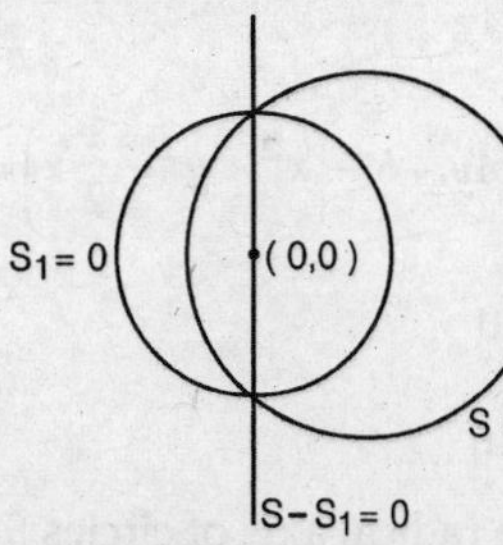

The equation of the common chord of (1) and (4) is

$S - S_1 = 0,$

i.e., $\quad 2gx + 2fy + c + 4 = 0$

It has to pass through the centre (0, 0) of the circle (1).

$\therefore \quad c + 4 = 0 \quad \ldots (5)$

The equation of the common chord of (2) and (4) is

$S - S_2 = 0,$

i.e., $\quad (2g + 6)x + (2f + 8)y + c - 10 = 0.$

It has to pass through the centre (3, 4) of the circle (2)

$\therefore \quad (2g + 6)3 + (2f + 8)4 + c - 10 = 0$

or $\quad 6g + 8f + c + 40 = 0 \quad \ldots (6)$

The equation of the common chord of (3) and (4) is

$S - S_3 = 0,$

i.e., $\quad (2g - 2)x + (2f + 4)y + c + 2 = 0.$

It has to pass through the centre (−1, 2) of the circle (3)

$\therefore \quad (2g - 2)(-1) + (2f + 4)2 + c + 2 = 0$

or $\quad -2g + 4f + c + 12 = 0 \quad \ldots (7)$

Now, (5) $\Rightarrow$ $c = -4$; so, (6) and (7) give

$6g + 8f + 36 = 0$, i.e., $3g + 4f + 18 = 0 \quad \ldots (8)$

and $\quad -2g + 4f + 8 = 0 \quad \ldots (9)$

$\therefore \quad (8) - (9) \quad \Rightarrow \quad 5g + 10 = 0$

$\therefore \quad g = -2$ and so $f = -3$.

$\therefore$ the required circle is $x^2 + y^2 - 4x - 6y - 4 = 0$.

30. For what values of l and m the circle $5(x^2+y^2)+ly-m=0$ belongs to the coaxal system determined by the circles $x^2+y^2+2x+4y-6=0$ and $2(x^2+y^2)-x=0$?

If the radical axis for each pair of the three given circles is the same then the result is established.

Let the circles be

$$S_1 \equiv x^2+y^2+2x+4y-6=0$$

$$S_2 \equiv x^2+y^2-\frac{1}{2}x=0$$

$$S_3 \equiv x^2+y^2+\frac{l}{5}y-\frac{m}{5}=0.$$

The equation of the radical axis of circles $S_1=0, S_2=0$ is

$$S_1-S_2=0,$$

i.e., $x^2+y^2+2x+4y-6-\left(x^2+y^2-\frac{1}{2}x\right)=0$

or $\frac{5}{2}x+4y-6=0$

or $5x+8y-12=0$ ... (1)

The equation of the radical axis of circles $S_2=0, S_3=0$ is

$$S_2-S_3=0,$$

i.e., $x^2+y^2-\frac{1}{2}x-\left(x^2+y^2+\frac{l}{5}y-\frac{m}{5}\right)=0$

or $-\frac{1}{2}x-\frac{l}{5}y+\frac{m}{5}=0$

or $5x+2ly-2m=0$ ... (2)

(1) and (2) must be identical. So, comparing them,

$$\frac{5}{5}=\frac{8}{2l}=\frac{-12}{-2m}$$

or $1=\frac{4}{l}=\frac{6}{m}$

$\therefore$ $l=4, m=6.$

Note The system of circles coaxal with the circles $S_1=0$ and $S_2=0$ has the equation of the form $S_1+\lambda S_2=0, (\lambda \neq -1)$.

If we can find a $\lambda \in R$ such that $S_1+\lambda S_2=0$ and $S_3=0$ are identical equations then we can say $S_3=0$, is a circle belonging to the coaxal system determined by $S_1=0, S_3=0$.

31. Find the locus of the middle points of the chords of the circle $x^2+y^2=a^2$ which pass through a given point (x_1, y_1).

Let $P(\alpha, \beta)$ be the middle point of any chord AB through the given point (x_1, y_1).

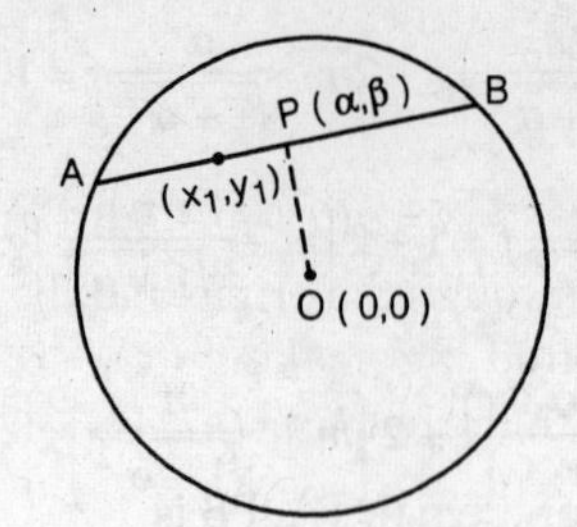

The centre of the circle is O(0, 0). Clearly PO is perpendicular to AB.

Now, $'m'$ of AB $=\frac{\beta-y_1}{\alpha-x_1}$

$'m'$ of OP $=\frac{\beta-0}{\alpha-0}=\frac{\beta}{\alpha}$

$\therefore$ $\frac{\beta-y_1}{\alpha-x_1}\cdot\frac{\beta}{\alpha}=-1$

or $\alpha(\alpha-x_1)+\beta(\beta-y_1)=0$

$\therefore$ the equation of the locus of $P(\alpha, \beta)$ is

$$x(x-x_1)+y(y-y_1)=0.$$

32. Find the locus of the middle points of chords of a given circle $x^2+y^2=a^2$ which subtend a right angle at the fixed point (p, q).

Change the origin to the point $P(p, q)$.

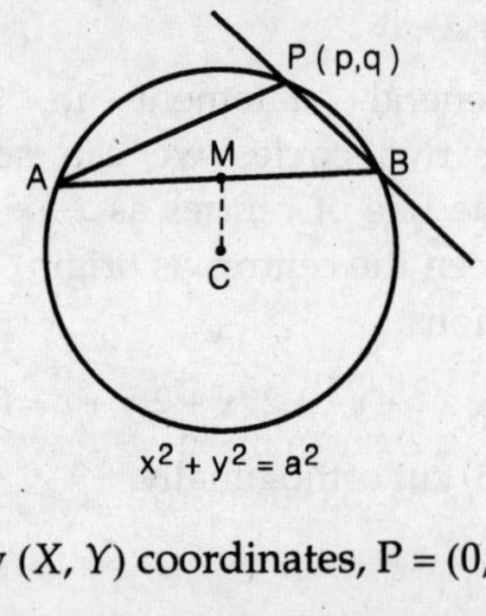

$\therefore$ in new (X, Y) coordinates, $P=(0, 0)$ and equations of transformation are

$$x=p+X, y=q+Y. \quad ... (0)$$

$\therefore$ the equation of the circle becomes

$$(p+X)^2+(q+Y)^2=a^2 \quad ... (1)$$

whose centre is $C=(-p, -q)$.

Let the middle point M of the chord AB be (α, β) in the new coordinates.

Then $'m'$ of CM $=\frac{\beta+q}{\alpha+p}$.

$\because$ AB is perpendicular to MC, the equation of MC will be

$$Y - \beta = -\frac{\alpha + p}{\beta + q}(X - \alpha)$$

or $\quad (\beta + q)Y + (\alpha + p)X = \alpha(\alpha + p) + \beta(\beta + q) \qquad \dots (2)$

The pair of lines through the origin and the intersection of (1), (2) is found by making (1), second degree homogeneous with the help of (2).

$\therefore$ the equation of the pair PA, PB is

$$X^2 + Y^2 + 2(pX + qY) \cdot \frac{(\beta + q)Y + (\alpha + p)X}{\alpha(\alpha + p) + \beta(\beta + q)}$$

$$= (a^2 - p^2 - q^2)\left\{\frac{(\beta + q)Y + (\alpha + p)X}{\alpha(\alpha + p) + \beta(\beta + q)}\right\}^2.$$

The pair has a right angle between the lines if

coefficient of X^2 + coefficient of $Y^2 = 0$

or $$\left[1 + \frac{2p(\alpha + p)}{\alpha(\alpha + p) + \beta(\beta + q)} - (a^2 - p^2 - q^2)\frac{(\alpha + p)^2}{\{\alpha(\alpha + p) + \beta(\beta + q)\}^2}\right]$$

$$+\left[1 + \frac{2q(\beta + q)}{\alpha(\alpha + p) + \beta(\beta + q)} - (a^2 - p^2 - q^2)\frac{(\beta + q)^2}{\{\alpha(\alpha + p) + \beta(\beta + q)\}^2}\right] = 0$$

or $$2\{\alpha(\alpha + p) + \beta(\beta + q)\}^2 + 2\{p(\alpha + p) + q(\beta + q)\}\{\alpha(\alpha + p) + \beta(\beta + q)\} + (p^2 + q^2 - a^2)\{(\alpha + p)^2 + (\beta + q)^2\} = 0$$

$\therefore$ the equation of the locus in X, Y coordinates is

$$2\{X(X + p) + Y(Y + q)\}^2 + 2\{p(X + p) + q(Y + q)\}\{X(X + p) + Y(Y + q)\} + (p^2 + q^2 - a^2)\{(X + p)^2 + (Y + q)^2\} = 0$$

Using equations of transformation (0), we get the equation of the locus of M in old coordinates x, y as

$$2\{(x - p)x + (y - q)y\}^2 + 2\{px + qy\}\{(x - p)x + (y - q)y\} + (p^2 + q^2 - a^2)(x^2 + y^2) = 0$$

or $$2\{(x^2 + y^2) - (px + qy)\}^2 + 2(px + qy)\{(x^2 + y^2) - (px + qy)\} + (p^2 + q^2 - a^2)(x^2 + y^2) = 0$$

or $$2(x^2 + y^2)^2 - 2(px + qy)(x^2 + y^2) + (p^2 + q^2 - a^2)(x^2 + y^2) = 0$$

or $$2(x^2 + y^2) - 2(px + qy) + (p^2 + q^2 - a^2) = 0 \qquad (\because x^2 + y^2 \neq 0)$$

or $$x^2 + y^2 - px - qy + \frac{1}{2}(p^2 + q^2 - a^2) = 0.$$

33. Find the locus of the feet of the perpendiculars from the point (h, k) to the tangents to the circle $x^2 + y^2 = 2ax$.

Let (x_1, y_1) be any point on the circle. The equation of the tangent to the circle at (x_1, y_1) is

$$xx_1 + yy_1 = a(x + x_1)$$

i.e., $\quad (x_1 - a)x + y_1 y - ax_1 = 0 \qquad \dots (1)$

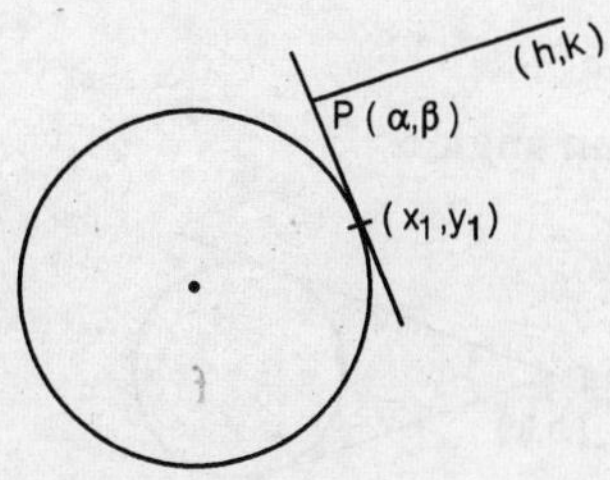

Let P(α, β) be the foot of the perpendicular to (1) from (h, k). Then (α, β) satisfies (1),

i.e., $\quad (x_1 - a)\alpha + y_1\beta - ax_1 = 0 \qquad \dots (2)$

and $\quad \dfrac{y_1 - \beta}{x_1 - \alpha} \cdot \dfrac{k - \beta}{h - \alpha} = -1$ (using perpendicularity)

or $\quad (y_1 - \beta)(k - \beta) + (x_1 - \alpha)(h - \alpha) = 0 \qquad \dots (3)$

From (2), $(\alpha - a)x_1 + \beta y_1 - a\alpha = 0$.

From (3), $(h - \alpha)x_1 + (k - \beta)y_1 - \alpha(h - \alpha) - \beta(k - \beta) = 0$

$$\therefore \quad \frac{x_1}{a\alpha(k - \beta) - \beta\{\alpha(h - \alpha) + \beta(k - \beta)\}} = \frac{y_1}{-a\alpha(h - \alpha) + (\alpha - a)\{\alpha(h - \alpha) + \beta(k - \beta)\}} = \frac{1}{(\alpha - a)(k - \beta) - \beta(h - \alpha)}.$$

But $x_1^2 + y_1^2 = 2ax_1$

$$\therefore \quad \left[\frac{a\alpha(k - \beta) - \beta\{\alpha(h - \alpha) + \beta(k - \beta)\}}{(\alpha - a)(k - \beta) - \beta(h - \alpha)}\right]^2 + \left[\frac{-a\alpha(h - \alpha) + (\alpha - a)\{\alpha(h - \alpha) + \beta(k - \beta)\}}{(\alpha - a)(k - \beta) - \beta(h - \alpha)}\right]^2$$

$$= 2a \cdot \frac{a\alpha(k - \beta) - \beta\{\alpha(h - \alpha) + \beta(k - \beta)\}}{(\alpha - a)(k - \beta) - \beta(h - \alpha)}$$

or $$\{(a\alpha - \beta^2)(k - \beta) - \alpha\beta(h - \alpha)\}^2 + \{(-a\alpha + \alpha^2 - a\alpha)(h - \alpha) + \beta(\alpha - a)(k - \beta)\}^2$$

$$= 2a\{(a\alpha - \beta^2)(k - \beta) - \alpha\beta(h - \alpha)\} \times \{k(\alpha - a) - \beta(h - a)\}$$

$\therefore$ the equation of the locus of the feet is

$$\{(ax - y^2)(k - y) - xy(h - x)\}^2 + \{x(x - 2a)(h - x) + y(x - a)(k - y)\}^2$$

$= 2a\{(ax - y^2)(k - y) - xy(h - x)\}\{k(x - a) - y(h - a)\}.$

34. Show that the locus of the point, the tangents from which to the circle $x^2 + y^2 = a^2$ include a constant angle α is

$$(x^2 + y^2 - 2a^2)^2 \tan^2\alpha = 4a^2(x^2 + y^2 - a^2).$$

Let P(h, k) be a point from which the tangents drawn to the circle

$$x^2 + y^2 = a^2 \quad \ldots (1)$$

be inclined at an angle α.

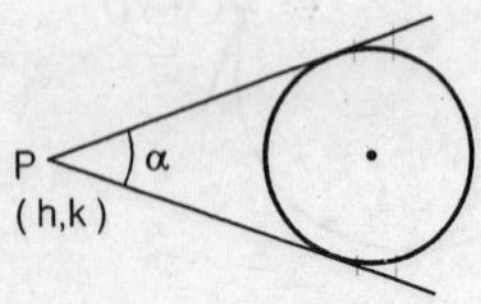

Any line through P(h, k) is $y - k = m(x - h)$ ... (2)

(2) will touch (1) if

radius of (1) = length of the perpendicular from (0, 0) to the line (2)

or $a = \left|\dfrac{k - mh}{\sqrt{1 + m^2}}\right|$

or $a^2(1 + m^2) = (k - mh)^2$

or $(a^2 - h^2)m^2 + 2hkm + a^2 - k^2 = 0$ ... (3)

This is quadratic equation in m whose roots m_1, m_2 are the slopes of the two tangents. As the angle between the two tangents is α,

$$\tan\alpha = \left|\frac{m_1 - m_2}{1 + m_1m_2}\right|$$

or $\tan^2\alpha = \dfrac{(m_1 - m_2)^2}{(1 + m_1m_2)^2}$

or $(1 + m_1m_2)^2 \cdot \tan^2\alpha = (m_1 + m_2)^2 - 4m_1m_2$

or $\left(1 + \dfrac{a^2 - k^2}{a^2 - h^2}\right)^2 \cdot \tan^2\alpha = \left(-\dfrac{2hk}{a^2 - h^2}\right)^2 - 4 \cdot \dfrac{a^2 - k^2}{a^2 - h^2}$ {from (3)}

or $(2a^2 - h^2 - k^2)^2 \tan^2\alpha = 4h^2k^2 - 4(a^2 - h^2)(a^2 - k^2)$

$= 4a^2(h^2 + k^2 - a^2)$

$\therefore$ the equation of the locus is

$$(x^2 + y^2 - 2a^2)^2 \cdot \tan^2\alpha = 4a^2(x^2 + y^2 - a^2).$$

Second Method

Any tangent to $x^2 + y^2 = a^2$ is $y = mx + a\sqrt{1 + m^2}$.

It passes through P(h, k) if $k = mh + a\sqrt{1 + m^2}$

or $(k - mh)^2 = a^2(1 + m^2)$ which is the same as (3).

Then proceed as above.

35. Find the equation of the system of coaxal circles that are tangent at $(\sqrt{2}, 4)$ to the locus of point of intersection of mutually perpendicular tangents to the circle $x^2 + y^2 = 9$.

At first we find the locus of point of intersection of mutually perpendicular tangents.

Let (x_1, y_1) be the point of intersection of perpendicular tangents.

$\therefore$ the equation of the pair of tangents is $S \cdot S_1 = T^2$

i.e., $(x^2 + y^2 - 9)(x_1^2 + y_1^2 - 9) = (xx_1 + yy_1 - 9)^2$

or $(y_1^2 - 9)x^2 + (x_1^2 - 9)y^2 - 9(x_1^2 + y_1^2)$

$= 2x_1y_1xy - 18xx_1 - 18yy_1.$

As the pair is at right angle, $a + b = 0$,

i.e., sum of coefficients of x^2 and y^2 is zero

$\therefore \quad y_1^2 - 9 + x_1^2 - 9 = 0$

or $\quad x_1^2 + y_1^2 = 18$

$\therefore$ the equation of the locus of point of intersection of perpendicular tangents is

$$x^2 + y^2 = 18 \quad \ldots (1)$$

Clearly, $(\sqrt{2}, 4)$ is a point on $x^2 + y^2 = 18$.

$\therefore$ the tangent at $(\sqrt{2}, 4)$ to the circle $x^2 + y^2 = 18$ is

$$x \cdot \sqrt{2} + y \cdot 4 = 18 \quad \ldots (2)$$

The equation of circles passing through the intersection of (1) and (2) is

$$x^2 + y^2 - 18 + \lambda(\sqrt{2}x + 4y - 18) = 0$$

or $x^2 + y^2 + \sqrt{2}\lambda x + 4\lambda y - 18(\lambda + 1) = 0$ ... (3)

As (2) touches (1),

(3) is the family of circles touching (1) at $(\sqrt{2}, 4)$.

Also any two circles of (3) have the same radical axis

$$\sqrt{2}x + 4y - 18 = 0$$

$\therefore$ the required equation of coaxal circles is (3).

36. A ball moving around the circle

$$x^2 + y^2 - 2x - 4y - 20 = 0$$

in anticlockwise direction leaves it tangentially at the point (–2, –2). After getting reflected from a straight line it passes through the centre of the circle. Find the equation of the straight line if its perpendicular distance from P is $\frac{5}{2}$.

Here, centre C = (1, 2) and the point where the ball leaves the circle is P = (–2, –2).

Let AB be the required line and N be the point of incidence. PN is the line of incidence and NC is the line of reflection.

Let NM be the normal to the line AB at N.

So $\angle PNM = \angle CNM = \theta$ (say).

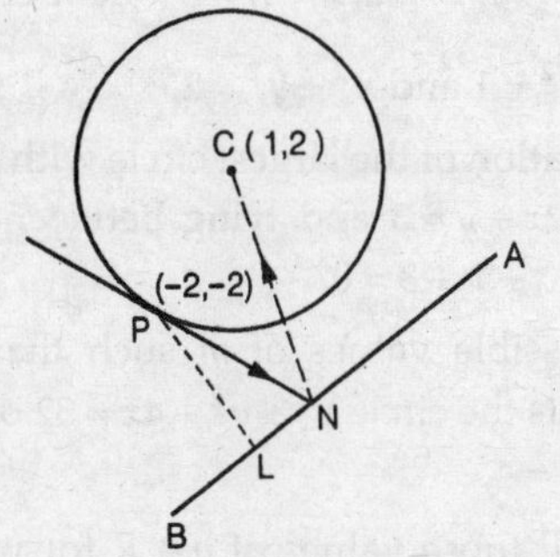

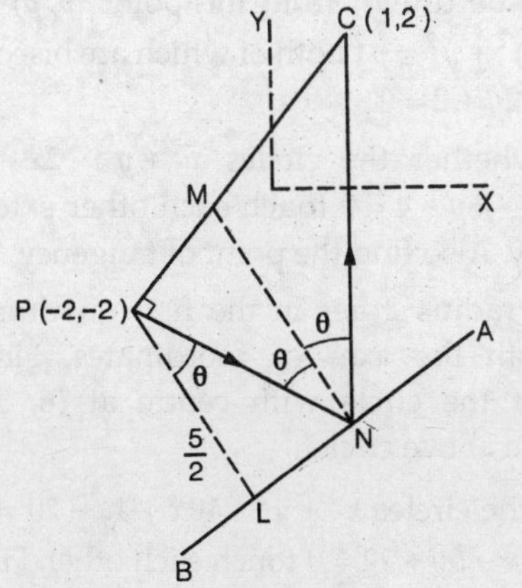

Now, $PN \perp PC$ and $PL \perp AB$.

So $\angle NPL = \angle PNM = \theta$.

From the right-angled ΔPLN,

$$PN = PL \sec\theta = \frac{5}{2}\sec\theta.$$

From the right-angled ΔCPN,

$$\frac{PC}{PN} = \tan 2\theta$$

$$\therefore \quad \frac{\sqrt{(1+2)^2 + (2+2)^2}}{\frac{5}{2}\cdot\sec\theta} = \tan 2\theta$$

$$\text{or} \quad 2\cos\theta = \frac{\sin 2\theta}{\cos 2\theta}$$

$$\text{or} \quad 2\cos\theta\cdot\cos 2\theta = 2\sin\theta\cdot\cos\theta$$

$$\therefore \quad \cos 2\theta = \sin\theta \quad \left(\because \theta \neq \frac{\pi}{2}\right)$$

$$= \cos(90° - \theta)$$

$$\therefore \quad 2\theta = 90° - \theta; \qquad \therefore \quad \theta = 30°.$$

Now, 'm' of PC $= \dfrac{2-(-2)}{1-(-2)} = \dfrac{4}{3}$.

Let 'm' of MN $= m$

$$\therefore \quad \tan(180° - \overline{90° + 30°}) = \left|\frac{\frac{4}{3} - m}{1 + \frac{4}{3}\cdot m}\right| = \left|\frac{4-3m}{3+4m}\right|$$

$$\therefore \quad \tan 60° = \left|\frac{4-3m}{3+4m}\right|$$

$$\therefore \quad \frac{4-3m}{3+4m} = \pm\sqrt{3}$$

$$\text{or} \quad 4 - 3m = \pm\sqrt{3}(3+4m)$$

$$\text{or} \quad 4 \mp 3\sqrt{3} = 3m \pm 4\sqrt{3}m$$

$$\therefore \quad m = \frac{4 \mp 3\sqrt{3}}{3 \pm 4\sqrt{3}}$$

$$= \frac{4-3\sqrt{3}}{3+4\sqrt{3}}, \frac{4+3\sqrt{3}}{3-4\sqrt{3}}.$$

$$\text{'}m\text{' of AB} = -\frac{1}{m} = \frac{3+4\sqrt{3}}{3\sqrt{3}-4}, \frac{4\sqrt{3}-3}{4+3\sqrt{3}}.$$

Let the equation of AB be $y = mx + c$ where

$$m = \frac{4\sqrt{3}-3}{4+3\sqrt{3}} (> 0)$$

$$\therefore \quad \frac{2 - 2m + c}{\sqrt{1+m^2}} = \pm\frac{5}{2}$$

$$\text{or} \quad \frac{2 - 2\cdot\frac{4\sqrt{3}-3}{4+3\sqrt{3}} + c}{\sqrt{1 + \left(\frac{4\sqrt{3}-3}{4+3\sqrt{3}}\right)^2}} = \pm\frac{5}{2}$$

$$\text{or} \quad 8 + 6\sqrt{3} - 8\sqrt{3} + 6 + c(4+3\sqrt{3})$$

$$= \pm\frac{5}{2}\sqrt{16 + 27 + 24\sqrt{3} + 48 + 9 - 24\sqrt{3}}$$

$$\text{or} \quad 14 - 2\sqrt{3} + c(4+3\sqrt{3}) = \pm\frac{5}{2}\cdot 10 = \pm 25$$

$$\therefore \quad c = \frac{2\sqrt{3} - 14 \pm 25}{4+3\sqrt{3}}.$$

As the ball is projected from a point in the third quadrant while moving in the anticlockwise sense, clearly the slope of the AB will be positive and it should cut negative intercept on the y-axis.

$$\therefore \quad c = \frac{2\sqrt{3}-39}{4+3\sqrt{3}}$$

$\therefore$ the equation of the line is $y = \dfrac{4\sqrt{3}-3}{4+3\sqrt{3}}x + \dfrac{2\sqrt{3}-39}{4+3\sqrt{3}}$.

Similarly, we can obtain the equation of the line when 'm' of AB is $\dfrac{3+4\sqrt{3}}{3\sqrt{3}-4}$ which is also positive.

Exercises

1. Find the equation of a circle whose diameter has the length 20 and the equations of two of its diameters are $2x + y = 6$ and $3x + 2y = 4$.
2. Find the equations of circles touching the y-axis at $(0, 3)$ and making an intercept of 8 units on the x-axis.
3. Find the equation of the circle passing through the points $(1, -2)$ and $(4, -3)$, and whose centre lies on the line $3x + 4y = 7$.
4. Find the equation of the circle passing through the intersection of the lines $3x + y = 4$ and $x - 3y + 2 = 0$ and concentric with the circle
$$2(x^2 + y^2) - 3x + 8y - 1 = 0.$$
5. The abscissa of two points A and B are the roots of the equation $x^2 + 2ax - b^2 = 0$ and their ordinates are the roots of the equation $x^2 + 2px - q^2 = 0$. Find the equation and radius of the circle with AB as diameter.
6. Let $A = (0, 1)$ and $B = \left(-\frac{p}{2}, \frac{q+1}{2}\right)$ be two fixed points in a plane. Let C denote a circle with centre B and passing through A. Prove that the real roots of the equation $x^2 + px + q = 0$ are given by the abscissae of the points of intersection of C with the x-axis.
7. Find the equation of the circumcircle of the triangle formed by the lines $y = x$, $y = 2x$ and $y = 3x + 2$.
8. Prove that the equation of the circle passing through the noncollinear points (x_r, y_r); $r = 1, 2, 3$ is
$$\begin{vmatrix} x^2 + y^2 & x_1^2 + y_1^2 & x_2^2 + y_2^2 & x_3^2 + y_3^2 \\ x & x_1 & x_2 & x_3 \\ y & y_1 & y_2 & y_3 \\ 1 & 1 & 1 & 1 \end{vmatrix} = 0.$$
9. Of the two concentric circles the smaller one has the equation $x^2 + y^2 = 4$. If each of the two intercepts on the line $x + y = 2$ made between the two circles is 1, find the equation of the larger circle.
10. If the four distinct points $\left(m_r, \frac{1}{m_r}\right)$; $r = 1, 2, 3, 4$ are concyclic then prove that the product of their ordinates is equal to 1.
11. Find the angle that the chord of the circle $x^2 + y^2 - 4y = 0$ along the line $x + y = 1$ subtends at the circumference of the larger segment.
12. Prove that the circles $x^2 + y^2 - 9 = 4r^2$; $r = 1, 2, 3$ cut off equal intercepts between the circles on the line $3x + 4y + 15 = 0$.
13. Can the four lines $4x - 3y = 5$, $x - 2y = 10$, $7x + y = 40$ and $x + 3y + 10 = 0$ form the sides of a cyclic quadrilateral? Justify your answer.
14. Find the range of values of r for which the points on the line $\frac{2}{\sqrt{3}}(x - 2) = 2(y - 1) = r$ lie between the circles $x^2 + y^2 = 1$ and $x^2 + y^2 = 2$.
15. Find the equation of the largest circle with the centre on the line $2x + y = 3$ and lying between the lines $x - y = 1$ and $x - y + 3 = 0$.
16. Find the possible values of m such that the line $y = mx + 4$ cuts the circle $x^2 + y^2 - 4x = 32$ but not the circle $x^2 + y^2 = 4$.
17. Determine all those values of $p \in R$ for which two chords can be drawn from the point (p, p) to the circle $(x - p)^2 + y^2 = p^2$ both of which are bisected by the line $x - 2y + 2 = 0$.
18. Examine whether the circles $x^2 + y^2 - 2x - 4y = 0$ and $x^2 + y^2 - 8y - 4 = 0$ touch each other externally or internally. Also find the point of tangency.
19. A circle of radius 2 lies in the first quadrant and touches both the axes of coordinates. Find the equation of the circle with centre at $(6, 5)$ and touching the above circle.
20. Show that the circles $x^2 + y^2 - 10x + 4y - 20 = 0$ and $x^2 + y^2 + 14x - 6y + 22 = 0$ touch each other. Find the equation of the common tangent at the point of contact.
21. Two circles, each of radius 5, touch each other at $(1, 2)$. If the equation of the common tangent is $4x + 3y = 10$, find the equation of the circles.
22. Prove that $x^2 + y^2 = a^2$ and $(x - 2a)^2 + y^2 = a^2$ are two equal circles touching each other. Find the equation of the circle(s) of equal radius touching both the circles.
23. Find the equation of the circle which touches the circle $x^2 + y^2 - 6x + 6y + 17 = 0$ externally and to which the lines $x^2 - 3xy - 3x + 9y = 0$ are normals.
24. Find the area of an equilateral triangle inscribed in the circle
$$x^2 + y^2 + 2gx + 2fy + c = 0.$$
25. Find the area of a square circumscribing the circle
$$3(x^2 + y^2) - 6x + 8y = 0.$$
26. Show that the area of the triangle formed by tangents from the point $(4, 3)$ to the circle $x^2 + y^2 = 9$ and the line segment joining their points of contact is $7\frac{17}{25}$ square units.
27. Find the equations of the circles passing through $(-4, 3)$ and touching the lines $x + y = 2$ and $x - y = 2$.

28. Find the equation of the family of circles which touch the pair of lines $x^2 - y^2 + 2y - 1 = 0$.

29. Lines $5x + 12y - 10 = 0$ and $5x - 12y - 40 = 0$ touch the circle C_1 of diameter 6. If the centre of C_1 lies in the first quadrant, find the equation of the circle C_2 which is concentric with C_1 and cuts intercepts of length 8 on these lines.

30. Find the equation of the circle passing through the points A(4, 3) and B(2, 5), and touching the y-axis. Also find the point P on the y-axis such that $\angle APB$ has the greatest magnitude.

31. A circle touches both the x-axis and the line $4x - 3y + 4 = 0$. Its centre is in the third quadrant and lies on the line $x - y - 1 = 0$. Find the equation of the circle.

32. Obtain the equations of the straight lines passing through the point A(2, 0) and making an angle 45° with the tangent at A to the circle $(x + 2)^2 + (y - 3)^2 = 25$. Find the equations of the circles, each of radius 3 whose centres are on these straight lines at a distance $5\sqrt{2}$ from A.

33. The extremities of a diagonal of a rectangle are (– 4, 4) and (6, –1). A circle circumscribes the rectangle and cuts an intercept AB on the y-axis. Find the area of the triangle formed by AB and the tangents to the circle at A and B.

34. Tangents are drawn to the circle $x^2 + y^2 = 12$ at the points where it is met by the circle whose equation is $x^2 + y^2 - 5x + 3y - 2 = 0$; find the point of intersection of these tangents.

35. Find the equation of the tangent to the circle $x^2 + y^2 - 80x - 60y + 2100 = 0$ at the point nearest to the origin.

36. Prove that the line $x + y = 2\sqrt{2}$ touches the circle $x^2 + y^2 = 4$. Also, if the point of contact be P then find the equations of the tangents to this circle which are at a distance $2 - \sqrt{2}$ from the point P.

37. Prove that the tangent to the circle $x^2 + y^2 = 5$ at the point (1, –2) also touches the circle

$$x^2 + y^2 - 8x + 6y + 20 = 0$$

and find its point of contact.

38. AB is a diameter of a circle. CD is a chord parallel to AB and $2CD = AB$. The tangent at B meets AC produced at E. Prove that $AE = 2AB$.

39. From the point P(1, 0), a tangent PA is drawn to the circle $x^2 + y^2 - 8x + 12 = 0$, A being the point of contact. Find the equations of the tangents to the circle from the middle point of PA if A is in the first quadrant.

40. A tangent drawn from the point (4, 0) to the circle $x^2 + y^2 = 8$ touches it at a point A in the first quadrant. Find the coordinates of another point B on the circle such that $AB = 4$.

41. A tangent is drawn to the circle $(x - a)^2 + y^2 = b^2$ and a perpendicular tangent to the circle $(x + a)^2 + y^2 = c^2$; find the locus of their point of intersection, and prove that the bisectors of the angles between them always touch one or other of two other fixed circles.

42. Find the coordinates of the point at which the circles $x^2 + y^2 - 4x - 2y = 4$ and $x^2 + y^2 - 12x - 8y + 48 = 0$ touch each other. Also find equations of common tangents touching the circles in distinct points.

43. A circle C_1 of radius 5 has its centre at the origin A. Circles C_2 and C_3 with centres at B and C, and of radii 3 and 4 metres respectively touch the circle C_1 and also touch the x-axis to the right of A, C_2 being in the first quadrant and C_3 in the fourth quadrant. Find the equations of the common tangents to the circles C_2 and C_3 except the x-axis.

44. A line segment AB is divided at C so that $AC = 3CB$. Circles are described on AC and CB as diameters and a common tangent meets AB produced at D. Show that BD is equal to the radius of the smaller circle.

45. Find the equations of common tangents to the circles

$$x^2 + y^2 = 1 \text{ and } (x - 1)^2 + (y - 3)^2 = 4.$$

46. From any point on the circle

$$x^2 + y^2 + 2gx + 2fy + c = 0$$

tangents are drawn to the circle

$$x^2 + y^2 + 2gx + 2fy + c\sin^2\alpha + (g^2 + f^2)\cos^2\alpha = 0.$$

Prove that the angle between the tangents is 2α.

47. The chord of contact of tangents from a point on the circle $x^2 + y^2 = a^2$ to the circle $x^2 + y^2 = b^2$ touches the circle $x^2 + y^2 = c^2$. Show that a, b, c are in GP.

48. Show that the polar of the origin with respect to the circle $x^2 + y^2 + 2gx + 2fy + c = 0$ touches the circle $x^2 + y^2 = a^2$ if $c^2 = a^2(f^2 + g^2)$.

49. If the pole of a straight line with respect to the circle $x^2 + y^2 = c^2$ lies on the circle $x^2 + y^2 = 9c^2$ then prove that the line is a tangent to the circle $9x^2 + 9y^2 = c^2$.

50. Tangents are drawn to the circle $x^2 + y^2 = a^2$ from the point (h, k). Prove that the area of the triangle formed by them and their chord of contact is

$$\frac{a(h^2 + k^2 - a^2)^{3/2}}{h^2 + k^2}.$$

51. Let C be the centre of a circle. Lines L_1 and L_2 are the polars of the points A and B with respect to the circle respectively. Perpendiculars AM and BN are dropped from A to the line L_2 and from B to the line L_1 respectively. Prove that $CA : CB = AM : BN$.

52. Find the pole of the chord of the circle $x^2 + y^2 = 81$, the chord being bisected at the point $(-2, 3)$.

53. Find the equation of the circle which has for its diameter the chord cut off on the line $px + qy - 1 = 0$ by the circle $x^2 + y^2 - a^2 = 0$.

54. If $y = mx$ be the equation of a chord of the circle $x^2 + y^2 - 2ax = 0$, prove that the circle of which this chord is a diameter has the equation

$$(1 + m^2)(x^2 + y^2) - 2a(x + my) = 0.$$

55. Prove that the equation

$$x^2 + y^2 + 2(3 + p)x + 2(3 - p)y + 4 = 0$$

represents a circle for all values of p, passing through two fixed points. Find the fixed points.

56. Prove that the circle $x^2 + y^2 - 6x - 4y + 9 = 0$ bisects the circumference of the circle

$$x^2 + y^2 - 8x - 6y + 23 = 0.$$

57. The polar lines of a point P with respect to two given intersecting circles meet in Q. Show that the common chord of the circles bisects PQ.

58. Find the equation to the circle whose diameter is the common chord of the circles $(x - a)^2 + y^2 = a^2$ and $x^2 + (y - b)^2 = b^2$. Also find the length of the common chord.

59. Prove that the length of the common chord of the circles

$$x^2 + y^2 + 2hx + a^2 = 0$$

and $x^2 + y^2 - 2ky - a^2 = 0$ is

$$2 \cdot \left\{\frac{(h^2 - a^2)(k^2 + a^2)}{h^2 + k^2}\right\}^{1/2}.$$

60. Find the equation of the circle passing through the points of intersection of the circles $x^2 + y^2 = 4a^2$ and $x^2 + y^2 - 2x - 4y + 4 = 0$ and touching the line $x + 2y = 0$.

61. Find the equation of the circle which passes through the points of intersection of the circles

$$x^2 + y^2 - 6x + 2y + 4 = 0$$

and $x^2 + y^2 + 2x - 4y - 6 = 0$

and whose centre lies on the line $y = x$.

62. Find the equation of the circle whose diameter is the common chord of the circles

$x^2 + y^2 + 2x + 3y + 1 = 0$ and $x^2 + y^2 + 4x + 3y + 2 = 0$.

63. Do the circles $x^2 + y^2 - 6x - 4y + 9 = 0$ and $x^2 + y^2 - 8x - 6y + 23 = 0$ intersect each other? If they intersect, find the length of the common chord. If they touch, find the common tangent at the point of contact. If they do not intersect or touch, find the radius of the smallest circle touching both the circles.

64. The line $Ax + By + C = 0$ cuts the circle

$$x^2 + y^2 + ax + by + c = 0$$

in P and Q. Another line $A'x + B'y + C' = 0$ cuts the circle $x^2 + y^2 + a'x + b'y + c' = 0$ in R and S. If P, Q, R and S are concyclic then show that

$$\begin{vmatrix} a - a' & b - b' & c - c' \\ A & B & C \\ A' & B' & C' \end{vmatrix} = 0.$$

65. Prove that $x^2 + y^2 + \lambda(y - x - 2\sqrt{2}) = 4$ represents circles touching each other at a common point for all real λ. Also find the common point.

66. Prove that the equation $x^2 + y^2 - 2x - 2\lambda y - 8 = 0$ represents a circle passing through two fixed points A, B for all real λ. If the tangents at A, B to the circle meet on the line $x + 2y + 5 = 0$ then find the equation of that particular circle.

67. Find the equations of circles passing through the intersection of the circles $x^2 + y^2 = 4$ and

$$x^2 + y^2 - 2x - 2y + 1 = 0$$

whose centre is at a distance $2\sqrt{2}$ from the origin.

68. Prove that the common chord of the circle $x^2 + y^2 + 6x + 8y + 7 = 0$ and the circle which always touches the line $y = x$ and passes through $(0, 0)$, will always pass through a fixed point.

69. Consider a family of circles passing through two fixed points A(3, 7) and B(6, 5). Show that the chords in which the circle $x^2 + y^2 - 4x - 6y - 3 = 0$ cuts the members of the family are concurrent. Find the coordinates of the point of concurrency.

70. If the circle $x^2 + y^2 = 16$ intersects another circle C of radius 5 in such a manner that the common chord is of the maximum length and has a slope $\frac{3}{4}$, show that the coordinates of the centre of the circle C are either $\left(\frac{9}{5}, \frac{-12}{5}\right)$ or $\left(-\frac{9}{5}, \frac{12}{5}\right)$.

71. Find the equation of the circle passing through the origin and cutting the circles

$x^2 + y^2 - 4x + 6y + 10 = 0$ and $x^2 + y^2 + 12y + 6 = 0$

at right angles.

72. Find the equation of the circle which passes through the origin, has its centre on the line $x + y + 4 = 0$ and cuts the circle $x^2 + y^2 - 4x + 2y + 4 = 0$ orthogonally.

73. Prove that the two circles which pass through the points $(0, a)$ and $(0, -a)$ and touch the line $y = mx + c$, will cut each other orthogonally if $c^2 = a^2(2 + m^2)$.

74. Circles $S = 0$, $S' = 0$ have radii r and r' respectively. Prove that the circles $\frac{S}{r} \pm \frac{S'}{r'} = 0$ will cut each other orthogonally.

75. Prove that the family of circles whose centres lie on the line $2x - 2y + 9 = 0$ and each of which cuts the circle $x^2 + y^2 = 4$ orthogonally, passes through two fixed points. Find the fixed points.

76. Prove that the general equation of circles cutting the two circles $x^2 + y^2 + 2g_r x + 2f_r y + c_r = 0$; $r = 1, 2$ orthogonally is

$$\begin{vmatrix} x^2+y^2 & -x & -y \\ c_1 & g_1 & f_1 \\ c_2 & g_2 & f_2 \end{vmatrix} + \lambda \begin{vmatrix} -x & -y & 1 \\ g_1 & f_1 & 1 \\ g_2 & f_2 & 1 \end{vmatrix} = 0.$$

77. Prove that a common tangent to two circles is bisected by the radical axis of the circles.

78. Find the equation of the circle coaxal with the circles

$$2x^2 + 2y^2 - 2x + 6y - 3 = 0$$

and $x^2 + y^2 + 4x + 2y + 1 = 0$,
and whose centre is on the radical axis of the circles.

79. Three circles are given:

$$x^2 + y^2 = 1, x^2 + y^2 - 8x + 15 = 0$$

and $x^2 + y^2 + 10y + 24 = 0$.
Determine the radical centre (i.e., the point such that the tangents drawn from it to three circles are equal in length).

80. From a point P, tangents drawn to the circles

$$x^2 + y^2 + x - 3 = 0,\ 3x^2 + 3y^2 - 5x + 3y = 0$$

and $4x^2 + 4y^2 + 8x + 7y + 9 = 0$ are equal. Find the equation of the circle through P which touches the line $x + y = 5$ at the point $(6, -1)$.

81. Prove that a common tangent to two circles of a coaxal system subtends a right angle at either limiting point of the system.

82. Let A = (–2, 0) and B = (1, 0) and P is a variable point such that $\angle APB = 60°$. Prove analytically that the locus of P is a circle. Find its radius and centre.

83. A circle passes through the origin O and cuts two lines $x + y = 0$ and $x - y = 0$ in P and Q respectively. If the straight line PQ always passes through a fixed point, find the locus of the centre of the circle.

84. A point moves such that the sum of the squares of its distances from the sides of a square of side unity is equal to 9. Show that the locus of the point is a circle. Find its centre and radius.

85. Two rods of lengths a and b slide along the coordinate axes in a manner that their ends are always concyclic. Find the locus of the centre of the circle passing through these ends.

86. The point A(1, 5) is joined to any point of the circle $x^2 + y^2 = 4$. Find the locus of the middle point of AP as P moves on the circle.

87. A variable circle passes through the point P(1, 2) and touches the x-axis. Show that the locus of the other end of the diameter through P is $(x - 1)^2 = 8y$.

88. From the point A(0, 3) on the circle

$$x^2 + 4x + (y - 3)^2 = 0,$$

a chord AB is drawn and extended to a point M such that $AM = 2AB$. Find the equation of the locus of M.

89. From the origin, chords are drawn to the circle $(x - 1)^2 + y^2 = 1$. Find the locus of the middle points of the chords.

90. The base of a triangle passes through a fixed point (a, b) and its sides are respectively bisected at right angles by the lines $y^2 - 8xy - 9x^2 = 0$. Prove that the locus of the vertex is a circle. Find its equation.

91. A circle of radius r passes through the origin O and cuts the axes at A and B. Find the locus of the centroid of $\triangle$OAB.

92. A circle of radius r passes through the origin O and cuts the axes at A and B. Let P be the foot of the perpendicular from the origin to the line AB. Find the locus of P.

93. Find the locus of the middle points of the chords of the circle $x^2 + y^2 = a^2$ subtending a right angle at the point $(\alpha, 0)$.

94. Through a fixed point P, a line is drawn perpendicular to any diameter l of a given circle to meet the diameter m at the point Q where m makes 45° with l in anticlockwise sense. Prove that the locus of Q is a circle passing through the centre of the given circle.

95. The point A is one of the points of intersection of two given intersecting circles. Any line is drawn through A to cut the circles again at P and Q. Prove that the locus of the middle point of PQ is a circle.

96. Prove that the locus of a point which moves such that the sum of the squares of its distances from three vertices of a triangle is constant, is a circle whose centre is at the centroid of the triangle.

97. The tangent at any point P on the circle $x^2 + y^2 = 2$ cuts the axes in L and M. Find the locus of the middle point of LM.

98. A triangle has two of its sides along the axes; its third side touches the circle $x^2 + y^2 - 2ax - 2ay + a^2 = 0$. Find the equation of the locus of the circumcentre of the triangle.

99. Find the locus of the foot of the perpendicular drawn from a fixed point on the x-axis to any tangent to the circle $x^2 + y^2 = a^2$.

100. Let $S \equiv x^2 + y^2 + 2gx + 2fy + c = 0$ be a given circle. Find the locus of the foot of the perpendicular drawn from the origin upon any chord which subtends a right angle at the origin.

101. Show that the locus of points from which the tangents drawn to a circle are orthogonal, is a concentric circle.

102. Find the locus of the point of intersection of tangents to the circle $x = a\cos\theta$, $y = a\sin\theta$ at two points whose parametric angles θ differ by $\frac{\pi}{3}$.

103. The circle $x^2 + y^2 - 4x - 4y + 4 = 0$ is inscribed in a triangle which has two of its sides along the axes of coordinates. If the locus of the circumcentre of the triangle is $x + y - xy + \lambda\sqrt{x^2 + y^2} = 0$, find λ.

104. A system of circles is drawn through two fixed points. Tangents are drawn to these circles parallel to a given line. Find the equation of the locus of the points of contact.

105. Find the locus of centres of the circles which touch the two circles $x^2 + y^2 = a^2$ and $x^2 + y^2 = 4ax$ externally.

106. A line through the origin O cuts the circle $x^2 + y^2 = 2(x + y)$ at A. Prove that the locus of the centre of the circle drawn on OA as diameter is also a circle which touches the given circle internally.

107. Find the locus of the point from which the chord of contact of tangents drawn to the circle $x^2 + y^2 = 16$ subtends a right angle at the centre of the circle.

108. Find the locus of a point whose polar with respect to the circle $x^2 + y^2 = a^2$ touches the circle
$(x - \alpha)^2 + (y - \beta)^2 = b^2$.

109. Two equal circles touch one another. Find the locus of a point which moves so that the sum of the tangents from it to the two circles is the constant k.

110. Find the locus of the centres of circles which cut the two circles

$$x^2 + y^2 + 4x - 6y + 9 = 0$$

and $x^2 + y^2 - 4x + 6y + 4 = 0$ orthogonally.

111. Find the locus of the point of contact of parallel tangents which are drawn to each of a series of coaxal circles.

112. Prove that the circle $(x - a)^2 + (y - a)^2 = a^2$ touches the x-axis. If the circle is rolled on the x-axis in the positive direction through a complete revolution, find the equation of the circle in the new position.

113. The circle $x^2 + y^2 - 4x - 8y + 16 = 0$ rolls up the tangent to it at $(2 + \sqrt{3}, 3)$ by 2 units in the direction of increasing x. Find the equation of the circle in the new position.

114. The centre of a circle is (1, 1) and its radius is 5 units. If the centre is shifted along the line $y - x = 0$ through a distance $\sqrt{2}$ units, find the equation of the circle in the new position.

Objective Questions

Fill in the blanks.

115. The point on a circle nearest to the exterior point P(1, 2), is 4 units away from P while the point farthest is (–3, –2). Then the equation of the circle is ______ .

116. The equation of the circle which touches the circle $x^2 + y^2 + 2x + 2y - 7 = 0$ at (2, –1) and passes through (1, 0), is ______ .

117. The point of intersection of the lines $ax + by - a = 0$ and $bx - ay + b = 0$ is P. A circle with the centre at (1, 0) passes through P. The tangent to the circle at P meets the x-axis at $(k, 0)$. Then $k =$ ______ .

118. The equation of the circle which is concentric with the circle $x^2 + y^2 - 5x - 4y - 1 = 0$ and touches the straight line $4x - 3y - 6 = 0$ is ______ .

119. A circle touches the x-axis at (2, 0) and has an intercept of 4 units on y-axis. Then the equation of the circle is ______ .

120. The line L passes through the points of intersection of the circles $x^2 + y^2 = 25$ and $x^2 + y^2 - 8x + 7 = 0$. The length of perpendicular from the centre of the second circle on the line L is ______ .

121. The equation of the locus of midpoints of the chords of the circle $4x^2 + 4y^2 - 12x + 4y + 1 = 0$ that subtend an angle of $\frac{2\pi}{3}$ at its centre is ______ .

122. The sides of a square are $x = 4, x = 7, y = 1$ and $y = 4$. The equation of the circumcircle of the square is ______ .

123. If (4, 1) be an extremity of a diameter of the circle $x^2 + y^2 - 2x + 6y - 15 = 0$ then the other extremity of the diameter is ______ .

124. The length of the chord $4x - 3y = 5$ of the circle $x^2 + y^2 + 3x = y + 10$ is ______ .

125. The line $3x + 4y = k$ touches the circle $x^2 + y^2 = 10x$. Then $k =$ ______ and the point of contact = ______ .

126. The line $7y - x = 5$ is a tangent to the circle $x^2 + y^2 - 5x + 5y = 0$. The other parallel tangent is ______ .

127. The length of the tangent to the circle
$x^2 + y^2 + 6x - 4y - 3 = 0$

drawn from the point (5, 1) is ______ .

128. The equations of tangents to the circle $x^2 + y^2 = 4$ which are inclined at 60° with the positive direction of the x-axis are ______ and ______ .

129. The pole of the line $3x + 5y + 17 = 0$ with respect to the circle $x^2 + y^2 + 4x + 6y + 9 = 0$ is ______ .

130. The circles

$(x - a)^2 + (y - b)^2 = c^2$ and $(x - b)^2 + (y - a)^2 = c^2$ will touch each other if ______ .

131. If a circle passes through the points of intersection of the coordinates axes with the lines $\lambda x - y + 1 = 0$ and $x - 2y + 3 = 0$ then the value of λ is ______ .

132. The area of the triangle formed by the positive x-axis, the normal and the tangent to the circle $x^2 + y^2 = 4$ at $(1, \sqrt{3})$ is ______ .

133. The equation of the circle passing through (2, 0) and (0, 4), and having the minimum radius is ______ .

134. The equation of the diameter of the circle $(x - 2)^2 + (y + 1)^2 = 16$ which bisects the chord cut off by the circle on the line $x - 2y - 3 = 0$, is ______ .

135. The equation of the line passing through the points of intersection of the circles

$$3x^2 + 3y^2 - 2x + 12y - 9 = 0$$

and $x^2 + y^2 + 6x + 2y - 15 = 0$ is ______ .

136. The area bounded by the circles $x^2 + y^2 = 1$ and $x^2 + y^2 = 4$ and the rays given by $2x^2 - 3xy - 2y^2 = 0, y > 0$ is ______ .

137. The set of values of λ for which the circles $x^2 + y^2 = 2$ and $x^2 + y^2 - 4x - 4y + \lambda = 0$ have exactly three real common tangents is ______ .

138. If the distances between the origin and centres of the three circles $x^2 + y^2 - 2\lambda_r x - a^2 = 0;\ r = 1, 2, 3$ are in GP then the lengths of tangents to the circles from any point on the circle $x^2 + y^2 = a^2$ are in ______ .

139. If lines $4x - 3y + 2 = 0$ and $8x = 6y + 5$ are two tangents of the same circle then the radius of the circle will be ______ .

140. The limiting points of the coaxal circles $x^2 + y^2 + 4\lambda x + 9 = 0$ are ______ .

141. (a) The circle $x^2 + y^2 - 2ax - 2ay + a^2 = 0$ touches the axes of coordinates at ______ and ______ .

(b) The line $(x - 1)\cos\theta + (y - 1)\sin\theta = 1$ touches the circle whose equation is ______, for all values of θ.

142. The equation of the circle passing through the points of intersection of the circles $x^2 + y^2 = 6$ and $x^2 + y^2 - 6x + 8 = 0$, and the point (1, 1) is ______ .

143. The point on the circle

$$16(x^2 + y^2) + 48x - 8y - 43 = 0$$

nearest to the line $8x - 4y + 73 = 0$ is ______ .

144. The chord of contact of the pair of tangents drawn from each point on the line $2x + y = 4$ to the circle $x^2 + y^2 = 1$ passes through the point ______ .

145. The chord whose equation is $y - x = 3$, subtends an angle 30° in the major segment of the circle $x^2 + y^2 = k^2$ if $k =$ ______ .

Choose the correct option(s).

146. The radius of the circle $4x^2 + 4y^2 - 10x + 5y + 5 = 0$ is

(a) $\frac{3\sqrt{5}}{8}$ (b) $\frac{3\sqrt{5}}{7}$ (c) $3\sqrt{5}$ (d) 5

147. The line $3x - 4y = k$ will cut the circle $x^2 + y^2 - 4x - 8y - 5 = 0$ at distinct points if

(a) $k < -35$ or $k > 15$ (b) $-35 < k < 15$

(c) $k > 15$ (d) none of these

148. The line $x\sin\alpha - y\cos\alpha = a$ touches the circle $x^2 + y^2 = a^2$. Then

(a) $\alpha \in \left[-\frac{\pi}{2}, \frac{\pi}{2}\right]$ (b) $\alpha \in [0, \pi]$

(c) $\alpha \in [-\pi, \pi]$ (d) α is any angle

149. The equation of a tangent to the circle

$$x^2 + y^2 - 6x + 4y - 3 = 0$$

which is perpendicular to the line $y = 2x - 1$, is

(a) $x + 2y + 1 = 0$ (b) $x + 2y + 1 \pm 4\sqrt{5} = 0$

(c) $x + 2y \pm 4\sqrt{5} = 0$ (d) none of these

150. The equation of a tangent drawn to the circle $x^2 + y^2 - 2x + 4y = 0$ from the point (0, 1) is

(a) $2x - y + 1 = 0$ (b) $2x - y - 2 = 0$

(c) $x + 2y - 2 = 0$ (d) $x + 2y + 1 = 0$

151. The centre of the circle passing through the origin and cutting off intercepts a and b on the x and y axes, is

(a) $\left(\frac{a}{2}, \frac{b}{2}\right)$ (b) $\left(-\frac{a}{2}, -\frac{b}{2}\right)$

(c) $\left(\frac{a}{2}, \frac{-b}{2}\right)$ (d) (a, b) (e) $\left(-\frac{a}{2}, \frac{b}{2}\right)$

152. A square is inscribed in the circle

$$x^2 + y^2 - 2x + 4y + 3 = 0$$

and its sides are parallel to coordinate axes. Then one vertex of the square is

(a) $(1 + \sqrt{2}, -2)$ (b) $(1 - \sqrt{2}, -2)$

(c) $(1, -2 + \sqrt{2})$ (d) none of these

153. Two circles $(x - 1)^2 + (y - 3)^2 = r^2$ and

$$x^2 + y^2 - 8x + 2y + 8 = 0$$

intersect in two distinct points if

(a) $2 < r < 8$ (b) $r < 2$

(c) $r = 2$ (d) $r > 2$

154. The centre of a circle passing through the points (0, 0), (1, 0) and touching the circle $x^2 + y^2 = 9$ is

(a) $\left(\frac{3}{2}, \frac{1}{2}\right)$ (b) $\left(\frac{1}{2}, \frac{3}{2}\right)$

(c) $\left(\frac{1}{2}, \frac{1}{2}\right)$ (d) $\left(\frac{1}{2}, -\sqrt{2}\right)$

155. If exactly two real common tangents can be drawn to the circles $x^2 + y^2 - 2x - 2y = 0$ and $x^2 + y^2 - 8x - 8y + \lambda = 0$ then

(a) $\lambda < 32$ (b) $24 < \lambda < 32$

(c) $\lambda < 24$ (d) $0 < \lambda < 24$

156. If a circle passes through the point (a, b) and cuts the circle $x^2 + y^2 = k^2$ orthogonally, the equation of the locus of its centre is

(a) $2ax + 2by - (a^2 + b^2 + k^2) = 0$

(b) $2ax + 2by - (a^2 - b^2 + k^2) = 0$

(c) $x^2 + y^2 - 3ax - 4by + (a^2 + b^2 - k^2) = 0$

(d) $x^2 + y^2 - 2ax - 3by + (a^2 - b^2 - k^2) = 0$

157. For any $\lambda \in R$, the locus

$$x^2 + y^2 - 2\lambda x - 2\lambda y + \lambda^2 = 0$$

touches the line

(a) $x = 0$ (b) $x = y$ (c) $x + y = 0$ (d) $y = 0$

158. The equation of a circle which touches the line $2x - y = 1$ at (1, 1) and also touches the line $2x + y = 4$ is

(a) $x^2 + y^2 - 3y + 1 = 0$

(b) $x^2 + y^2 + 3y + 1 = 0$

(c) $4(x^2 + y^2) + 10x + 7y - 9 = 0$

(d) $4(x^2 + y^2) - 10x - 7y + 9 = 0$

159. The circle for which the line joining the points $(am^2, 2am)$ and $\left(\frac{a}{m^2}, -\frac{2a}{m}\right)$ is a diameter, is touched, for all values of m, by the line

(a) $x = a$ (b) $x + a = 0$ (c) $x = 2a$ (d) $x + 2a = 0$

160. The radius of the circle passing through the point (6, 2) and having $x + y = 6$ as a normal and $x + 2y = 4$ as a diameter is

(a) 10 (b) $2\sqrt{5}$ (c) $5\sqrt{2}$ (d) $4\sqrt{5}$

161. A circle passing through the point $(3, \sqrt{7/2})$ touches the pair of lines $x^2 - y^2 - 2x + 1 = 0$. The centre of the circle is

(a) (4, 0) (b) (6, 0) (c) (0, 4) (d) (5, 0)

162. AB is the diameter of the circle along the x-axis whose centre is the origin and $A = (a, 0)$. $P(a\cos\alpha, a\sin\alpha)$ and $Q(a\cos\beta, a\sin\beta)$ are any two points on the circle such that $\alpha = \beta + 2\gamma$, $(\gamma > 0)$ Then the locus of the point of intersection of AP and BQ is

(a) $x^2 - y^2 - 2ay\tan\gamma = a^2$

(b) $x^2 - y^2 - 2ay\cot\gamma = a^2$

(c) $x^2 + y^2 + 2ay\tan\gamma = a^2$

(d) $x^2 + y^2 + 2ay\cot\gamma = a^2$

163. The angle between a pair of tangents drawn from a point P to the circle

$$x^2 + y^2 + 4x - 6y + 9\sin^2\alpha + 13\cos^2\alpha = 0$$

is 2α. The equation of the locus of the point P is

(a) $x^2 + y^2 + 4x - 6y + 4 = 0$

(b) $x^2 + y^2 + 4x - 6y - 9 = 0$

(c) $x^2 + y^2 + 4x - 6y - 4 = 0$

(d) $x^2 + y^2 + 4x - 6y + 9 = 0$

164. Let AB be a line segment of length 4 with A on the line $y = 2x$ and B on the line $y = x$. The locus of the middle point of the line segment is

(a) a line (b) a pair of lines

(c) a circle (d) none of these

165. The locus of the centre of a circle which touches the circle $x^2 + y^2 - 6x - 6y + 14 = 0$ externally and also touches the y-axis, is given by the equation

(a) $x^2 - 6x - 10y + 14 = 0$

(b) $x^2 - 10x - 6y + 14 = 0$

(c) $y^2 - 6x - 10y + 14 = 0$

(d) $y^2 - 10x - 6y + 14 = 0$

166. The circles $x^2 + y^2 + 2ux + 2vy = 0$ and $x^2 + y^2 + 2u_1x + 2v_1y = 0$ touch each other if

(a) $uu_1 = vv_1$ (b) $uv_1 = u_1v$

(c) $uv = u_1v_1$ (d) none of these

167. The centres of the circles $x^2 + y^2 = 1$, $x^2 + y^2 + 6x - 2y - 1 = 0$ and $x^2 + y^2 - 12x + 4y = 1$ are

(a) vertices of an equilateral Δ

(b) vertices of a right-angled triangle

(c) collinear

(d) none of these

168. The line $y = mx + c$ touches the circle $(x - \alpha)^2 + (y - \beta)^2 = a^2$ if

(a) $\beta = m\alpha + c + a\sqrt{1 + m^2}$

(b) $m\beta = \alpha + c + a\sqrt{1 + m^2}$

(c) $\beta + a\sqrt{1 + m^2} = m\alpha + c$

(d) $c = \pm a\sqrt{1 + m^2}$

169. A circle passing through the origin and cutting off equal chords of length $\sqrt{2}$ from the straight lines

$y = |x|$, is

(a) $x^2 + y^2 + 2y = 0$ (b) $x^2 + y^2 - 2x = 0$

(c) $x^2 + y^2 - 2y = 0$ (d) $x^2 + y^2 + 2x = 0$

(e) $x^2 + y^2 - 2x - 2y = 0$

State whether the statements are true or false.

170. The line $x + 3y = 0$ is a diameter of the circle $x^2 + y^2 - 6x + 2y = 0$.

171. There are two real tangents from the point (2, 1) to the circle $x^2 + y^2 = 8$.

172. The length of the tangent from the point (3, 4) to the circle $2(x^2 + y^2) + x + 2y - 1 = 0$ is $2\sqrt{15}$.

173. The straight line $y = mx + c$ cuts the circle $x^2 + y^2 - a^2 = 0$ in real points if $a\sqrt{1 + m^2} > |c|$.

174. The lines $2x + 3y + 19 = 0$ and $9x + 6y - 17 = 0$ cut the coordinate axes in concyclic points.

Answers

1. $x^2 + y^2 - 16x + 20y + 64 = 0$

2. $(x \pm 5)^2 + (y - 3)^2 = 25$

3. $15(x^2 + y^2) - 94x + 18y + 55 = 0$

4. $2(x^2 + y^2) - 3x + 8y - 9 = 0$

5. $x^2 + y^2 + 2(ax + py) = b^2 + q^2, \sqrt{a^2 + b^2 + p^2 + q^2}$

7. $x^2 + y^2 - 6x + 8y = 0$ **9.** $x^2 + y^2 = 5 + 2\sqrt{2}$

11. $\cos^{-1}\left(\frac{1}{2\sqrt{2}}\right)$ **13.** yes

14. $\left(-\frac{1 + 2\sqrt{3} + \sqrt{4\sqrt{3} + 1}}{2}, -\frac{1 + 2\sqrt{3} + \sqrt{4\sqrt{3} - 3}}{2}\right)$

$\cup\left(-\frac{1 + 2\sqrt{3} - \sqrt{4\sqrt{3} - 3}}{2}, -\frac{1 + 2\sqrt{3} - \sqrt{4\sqrt{3} + 1}}{2}\right)$

15. $9(x^2 + y^2) - 12x - 30y + 11 = 0$

16. $-\sqrt{3} < m < \sqrt{3}$

17. $\left(-\infty, -\frac{4}{\sqrt{5}}\right) \cup \left(\frac{4}{\sqrt{5}}, +\infty\right)$

18. internally, (2, 0)

19. $x^2 + y^2 - 12x - 10y + (52 \text{ or } 12) = 0$

20. $12x - 5y + 21 = 0$

21. $x^2 + y^2 - 10x - 10y + 25 = 0$,
$x^2 + y^2 + 6x + 2y - 15 = 0$

22. $x^2 + y^2 - 2ax \pm 2\sqrt{3}ay + 3a^2 = 0$

23. $x^2 + y^2 - 6x - 2y + 1 = 0$

24. $\frac{3\sqrt{3}}{4}(g^2 + f^2 - c)$ unit^2

25. $11\frac{1}{9}$ unit^2

27. $x^2 + y^2 - 2(-10 \pm 3\sqrt{6})x + (55 \mp 24\sqrt{6}) = 0$

28. $x^2 + y^2 \pm 2\sqrt{2}rx - 2y + r^2 + 1 = 0$ and
$x^2 + y^2 - 2(1 \pm \sqrt{2}r)y + r^2 + 1 \pm 2\sqrt{2}r = 0$
where r is a parameter

29. $x^2 + y^2 - 10x - 4y + 4 = 0$

30. $x^2 + y^2 - 20x - 22y + 121 = 0$,
$x^2 + y^2 - 4x - 6y + 9 = 0$; (0, 11), (0, 3)

31. $9(x^2 + y^2) + 6x + 24y + 1 = 0$

32. $7x + y = 14$, $x - 7y = 2$, $x^2 + y^2 - 2x - 14y + 41 = 0$,
$x^2 + y^2 - 6x + 14y + 49 = 0$,
$x^2 + y^2 - 18x - 2y + 73 = 0$,
$x^2 + y^2 + 10x + 2y + 17 = 0$

33. 166.375 unit^2

34. $\left(6, \frac{-18}{5}\right)$ **35.** $4x + 3y - 150 = 0$

36. $y = 2, x = 2$ **37.** (3, –1)

39. $2x - \sqrt{5}y = 2$, $62x + 5\sqrt{5}y = 122$

40. (2, –2), (–2, 2)

41. $(x^2 + y^2 - a^2)^2 = (b^2 + c^2)(x^2 + y^2 + a^2)$
$+ 2a(b^2 - c^2)x - 4abcy$

42. $\left(\frac{22}{5}, \frac{14}{5}\right)$; $y - 1 = m(x - 2) + 3\sqrt{1 + m^2}$
where $m = \frac{2\sqrt{6}(\sqrt{6} \pm 1)}{15}$

43. $y = m(x - \sqrt{55}) + 3(1 + \sqrt{1 + m^2})$ where
$m = \frac{14(\sqrt{65} - \sqrt{55})}{10\sqrt{143} - 71}, \frac{-7(\sqrt{65} - \sqrt{55}) \pm \sqrt{168 - 10\sqrt{143}}}{119 - 10\sqrt{143}}$

45. $3y = 4x - 5$, $3x + 4y = 5$, $y = 1$, $x = -1$

52. $\left(-\frac{162}{13}, \frac{243}{13}\right)$

53. $(p^2 + q^2)(x^2 + y^2 - a^2) - 2(px + qy - 1) = 0$

55. $(-3 + \sqrt{7}, -3 + \sqrt{7})$, $(-3 - \sqrt{7}, -3 - \sqrt{7})$

58. $(a^2 + b^2)(x^2 + y^2) - 2ab(bx + ay) = 0, \frac{2ab}{\sqrt{a^2 + b^2}}$

60. $(1 + a^2)(x^2 + y^2) - 2a^2x - 4a^2y = 0$

61. $7(x^2 + y^2) - 10x - 10y - 12 = 0$

62. $2(x^2 + y^2) + 2x + 6y + 1 = 0$

63. intersect, $2\sqrt{2}$

65. $(-\sqrt{2}, \sqrt{2})$

66. $x^2 + y^2 - 2x - 6y - 8 = 0$

67. $x^2 + y^2 - 4x - 4y + 6 = 0$, $x^2 + y^2 + 4x + 4y - 14 = 0$

69. $\left(2, \frac{23}{3}\right)$

71. $2(x^2+y^2)-7x+2y=0$

72. $3(x^2+y^2)+4x+20y=0$

75. $\left(-\frac{1}{2}, \frac{1}{2}\right), (-4, 4)$

78. $4(x^2+y^2)+6x+10y-1=0$

79. $\left(2, -\frac{5}{2}\right)$ **80.** $x^2+y^2-7x+7y+12=0$

82. radius $= \sqrt{3}$, centre $= \left(-\frac{1}{2}, \pm\frac{\sqrt{3}}{2}\right)$

83. $x^2-y^2-\alpha x+y\beta=0$ where (α, β) is the fixed point

84. centre is the centre of the square, radius = 2

85. $4(x^2-y^2)=a^2-b^2$

86. $2(x^2+y^2)-2x-10y+11=0$

88. $x^2+y^2+8x-6y+9=0$

89. the circle $x^2+y^2=x$

90. $4(x^2+y^2)+(5a+4b)x+(4a-5b)y=0$

91. $9(x^2+y^2)=4r^2$

92. $(x^2+y^2)^2 \cdot \left(\frac{1}{x^2}+\frac{1}{y^2}\right)=4r^2$

93. $2(x^2+y^2)-2\alpha x+\alpha^2-a^2=0$

97. $\frac{1}{x^2}+\frac{1}{y^2}=2$ **98.** $2(x+y)-a=\frac{2xy}{a}$

99. $(x^2+y^2-bx)^2=a^2\{y^2+(b-x)^2\}$
where the fixed point $= (b, 0)$

100. $x^2+y^2+gx+fy+\frac{1}{2}c=0$

102. $x^2+y^2=\frac{4}{3}a^2$ **103.** 1

104. $x^2-y^2-\frac{2}{m}xy-ax+\frac{a}{m}y=0$ where fixed points are $(0, 0), (a, 0)$ and slope of the given line is m

105. $12x^2-4y^2-24ax+9a^2=0$

107. $x^2+y^2=32$

108. $b^2(x^2+y^2)=(\alpha x+\beta y-a^2)^2$

109. $(k^2-4a^2)x^2+k^2y^2-\frac{1}{4}k^4=0$
where radius $= a$ and the line joining the centres is the x-axis and the point of contact is the origin

110. $8x-12y+5=0$

111. $x^2-y^2+2mxy-c=0$ where slope of tangents is m

112. $\{x-a(1+2\pi)\}^2+(y-a)^2=a^2$

113. $x^2+y^2-6x-2(4+\sqrt{3})y+8\sqrt{3}(\sqrt{3}+1)=0$

114. $x^2+y^2=25, x^2+y^2-4x-4y-17=0$

115. $(x+3)(x-1+2\sqrt{2})+(y+2)(y-2+2\sqrt{2})=0$

116. $x^2+y^2-2x+2y+1=0$

117. -1 **118.** $x^2+y^2-5x-4y+10.09=0$

119. $x^2+y^2-4x-4\sqrt{2}y+4=0$

120. 0 **121.** $x^2+y^2-3x+y+\frac{31}{16}=0$

122. $x^2+y^2-11x-5y+32=0$

123. $(-2, -7)$ **124.** 5

125. 40, −10; (8, 4), (2, − 4) **126.** $7y-x+45=0$

127. 7 **128.** $\sqrt{3}x-y+4=0, \sqrt{3}x-y-4=0$

129. (1, 2) **130.** $a=b\pm\sqrt{2c}$

131. $2, \frac{1}{3}$ **132.** $2\sqrt{3}$ unit^2

133. $x^2+y^2-2x-4y=0$

134. $2x+y-3=0$ **135.** $10x=3y+18$

136. $\frac{3\pi}{4}$ unit^2 **137.** 6

138. GP **139.** $\frac{9}{20}$ **140.** $(\pm 3, 0)$

141. (a) $(a, 0), (0, a)$ respectively
(b) $x^2+y^2-2x-2y+1=0$

142. $x^2+y^2-3x+1=0$ **143.** $\left(-\frac{7}{2}, \frac{5}{4}\right)$

144. $\left(\frac{1}{2}, \frac{1}{4}\right)$ **145.** $\pm\sqrt{6}$ **146.** (a) **147.** (b)

148. (d) **149.** (b) **150.** (a), (c)

151. (a), (b), (c), (e) **152.** (d)

153. (a) **154.** (d) **155.** (d) **156.** (a)

157. (a), (d) **158.** (a), (d) **159.** (b) **160.** (b)

161. (a), (b) **162.** (c) **163.** (d) **164.** (d)

165. (d) **166.** (b) **167.** (c) **168.** (a), (c)

169. (c) **170.** true **171.** false

172. false **173.** true **174.** true

□

Chapter Test

Time: 120 minutes

1. In each of the following, fill in the blanks so that the resulting statement becomes correct.

(a) The coordinates of the points on the circle $x^2 + y^2 = 4$ which are at a distance 5 from the line $4x - 3y = 25$ are ______ and ______ .

(b) The equation to the circle on the other side of the line $x + y = 2$ similarly situated as the circle $x^2 + y^2 - 2x = 0$ is ______ .

(c) The sum of the squares of the lengths of the chords intercepted by the lines $y = x + n, n \in N$ on the circle $x^2 + y^2 = 100$ is equal to ______ .

(d) The equation of the circle of radius 2 containing the point (1, –2) and touching the lines $|x| = |y|$ is ______ .

2. In each of the following one or more options are correct. Choose the correct option(s).

(a) The straight line $mx - y = 1 + 2m$ cuts the circle $x^2 + y^2 = 1$ at one point at least. Then the set of values of m is

A. $\left[-\frac{4}{3}, 0\right]$ B. $\left[-\frac{4}{3}, \frac{4}{3}\right]$ C. $\left[0, \frac{4}{3}\right]$ D. none of these

(b) If (2, 5) is an interior point of the circle $x^2 + y^2 - 8x - 12y + p = 0$ and the circle neither cuts nor touches any one of the axes of coordinates then

A. $p \in (36, 47)$ B. $p \in (16, 47)$ C. (16, 36) D. none of these

(c) The range of values of a such that the angle θ between the pair of tangents drawn from $(a, 0)$ to the circle $x^2 + y^2 = 1$ satisfies $\frac{\pi}{2} < \theta < \pi$, is

A. (1, 2) B. $(1, \sqrt{2})$ C. $(-\sqrt{2}, -1)$ D. $(-\sqrt{2}, -1) \cup (1, \sqrt{2})$

(d) The circle of radius 1, touching the pair of lines $12x^2 - 25xy + 12y^2 = 0, x > 0$ has the equation

A. $x^2 + y^2 + 10x + 10y + 49 = 0$ B. $x^2 + y^2 - 10x - 10y + 49 = 0$

C. $x^2 + y^2 + 10x - 10y + 49 = 0$ D. none of these

(e) The number of points $(a + 1, a)$, where a is an integer, lying inside the region bounded by the circles $x^2 + y^2 - 2x - 1 = 0$ and $x^2 + y^2 - 2x - 17 = 0$, is

A. 2 B. 4 C. 1 D. none of these

3. Find the equations of the sides of a square inscribed in the circle $3(x^2 + y^2) = 4$, one of whose sides is parallel to the line $x - y + 7 = 0$.

4. Let A = (–1, 0), B = (3, 0) and PQ be any line passing through (4, 1). Find the range of values of the slope of PQ for which there are two points on PQ at which AB subtends a right angle.

5. Three concentric circles of which the biggest is $x^2 + y^2 = 1$, have their radii in AP. If the line $y = x + 1$ cuts all the circles in real and distinct points then find the interval in which the common difference of the AP will lie.

6. The equation of the common chord of two circles is $x + y = 1$. One of the circles has the ends of a diameter at the points (1, –3) and (4, 1), and the other passes through the point (1, 2). Find the equations of the two circles.

7. Prove that one-third of the circumference of the circle $x^2 + y^2 - 2x = 0$ falls inside the circle $\sqrt{3}(x^2 + y^2) - 2\sqrt{3}x + 2y + 1 = 0$.

8. Find the locus of the centres of the circles $x^2 + y^2 - 2\lambda x - 2\mu y + 2 = 0$ if the tangents from the origin to each of the circles are orthogonal.

9. Let $ax^2 + 2hxy + by^2 = 1$ be the equation of a conic section and P be a point not on the curve. Any line through P cuts the conic section at Q and R such that $PQ \cdot PR$ is a constant. Show that the conic section is a circle.
10. Find the range of values of λ for which the variable line $y = 2x + \lambda$ lies between the circles $x^2 + y^2 - 2x - 2y + 1 = 0$ and $x^2 + y^2 - 16x - 2y + 61 = 0$ without intercepting a chord on either circle.
11. Find the equation of the circle touching the pair of lines $7x^2 - 18xy + 7y^2 = 0$ and the circle $x^2 + y^2 - 8x - 8y = 0$, and contained in the given circle.
12. If a vertex of an equilateral triangle is $(2\sqrt{2}, -1)$ and its centroid is $(0, 0)$ then find the equation of its incircle.
13. P(2, 2) is a point on a circle $x^2 + y^2 - 2x - 2y = 0$. If P travels on the circle and reaches Q such that arc $PQ = \frac{1}{6} \times$ circumference of the circle then find the coordinates of Q.
14. A regular hexagon is inscribed in the circle $x^2 + y^2 - 2(x + y + 1) = 0$ with one vertex at (3, 1). Find the coordinates of the two vertices of the hexagon consecutive to the vertex (3, 1).
15. Let C_2 be a circle lying inside the circle C_1. The circle C is inside C_1 but outside C_2 such that C touches C_1 internally but C_2 externally. Identify the locus of the centre of C.

□

Answers

1. (a) $\left(\frac{6}{5}, \frac{8}{5}\right), \left(-\frac{6}{5}, -\frac{8}{5}\right)$ (b) $x^2 + y^2 - 4x - 2y + 4 = 0$ (c) 3570 (d) $x^2 + y^2 + 4\sqrt{2}x + 4 = 0$

2. (a) A (b) A (c) D (d) B (e) A

3. $x - y = \pm\frac{2}{\sqrt{3}}, x + y = \pm\frac{2}{\sqrt{3}}$

4. $\left(\frac{3 - 2\sqrt{6}}{5}, \frac{3 + 2\sqrt{6}}{5}\right)$

5. $\left(0, \frac{2 - \sqrt{2}}{4}\right)$

6. $x^2 + y^2 - 5x + 2y + 1 = 0,\ 2(x^2 + y^2) - 15x - y + 7 = 0$

8. $x^2 + y^2 = 4$

10. $(-15 + 2\sqrt{5}, -1 - \sqrt{5})$

11. $x^2 + y^2 - 12x - 12y + 64 = 0$

12. $4(x^2 + y^2) = 9$

13. $\left(\frac{3}{2} \mp \frac{\sqrt{3}}{2}, \frac{3}{2} \pm \frac{\sqrt{3}}{2}\right)$

14. $(2, 1 \pm \sqrt{3})$

15. circle if concentric; ellipse if not concentric

□

4. Parabola

Recap of Facts and Formulae

1. Intersection of a right circular cone and a plane

- The curve/lines of intersection of a right circular cone and a plane is a *conic section*. Conic sections are of different varieties for different orientation of the plane.
- The intersection of a right circular cone and a plane passing through the axis of the cone is a *pair of lines*.
- The intersection of a right circular cone and a plane perpendicular to the axis of the cone is a *circle*.
- The intersection of a right circular cone and a plane parallel to a generator of the cone is a *parabola*.
- The intersection of a right circular cone and a plane cutting the axis at an angle $\alpha\left(0 < \alpha < \dfrac{\pi}{2}\right)$ is an *ellipse*.
- The intersection of a right circular cone and a plane parallel to the axis of the cone is a *hyperbola*.

2. Conic section as a locus of a point

- If a point moves in a plane such that its distances from a fixed point and a fixed line always bear a constant ratio *e* then the locus of the point is a conic section of the *eccentricity e* (*focus–directrix property*). The fixed point is the *focus* and the fixed line is the *directrix*.

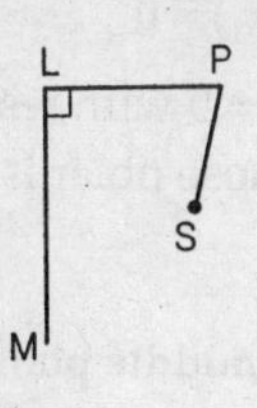

- If $e = 1$, it is a parabola.
- If $e < 1$, it is an ellipse.
- If $e > 1$, it is a hyperbola.

3. Equation of conic section by focus–directrix property

- If the focus is (α, β) and the directrix is $ax + by + c = 0$ then the equation of the conic section whose eccentricity $= e$, is

$$(x-\alpha)^2 + (y-\beta)^2 = e^2 \cdot \frac{(ax+by+c)^2}{a^2+b^2}.$$

- If the focus is (α, β) and the directrix is $ax + by + c = 0$ then the equation of the parabola is

$$(x-\alpha)^2 + (y-\beta)^2 = \frac{(ax+by+c)^2}{a^2+b^2}.$$

4. Standard equation of a parabola and its parts

The standard equation of a parabola is $y^2 = 4ax$ for which

- vertex $V = (0, 0)$, focus $S = (a, 0)$.

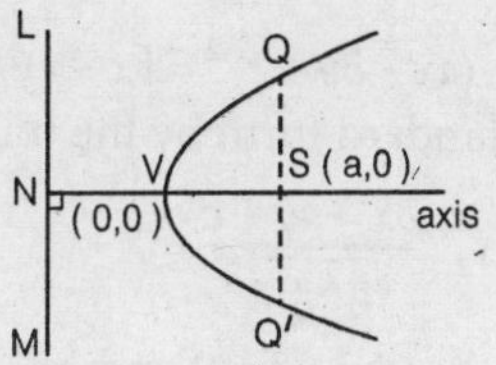

- the equation of the directrix LM is $x + a = 0$ and that of the axis (or axis of symmetry) VS of the parabola is $y = 0$.
- latus rectum $QSQ' = 4a (= 4VS)$.
- $VS = VN = a$.

5. Location of a point in relation to a parabola

If $S \equiv y^2 - 4ax = 0$ be a parabola and $P(x_1, y_1)$ be a point then

- P is in the interior of the parabola if
  $S_1 < 0$, i.e., $y_1^2 - 4ax_1 < 0$.
- P is on the parabola if
  $S_1 = 0$, i.e., $y_1^2 - 4ax_1 = 0$.

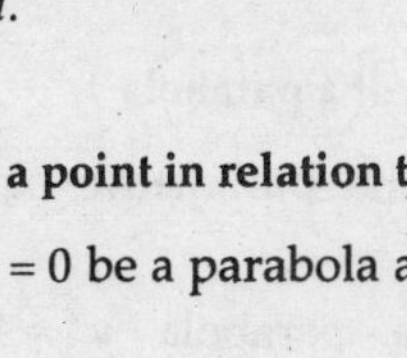

- P is in the exterior of the parabola if

$S_1 > 0$, i.e., $y_1^2 - 4ax_1 > 0$.

6. Condition for general equation of the second degree to represent a parabola

- The equation

$$ax^2 + 2hxy + by^2 + 2gx + 2fy + c = 0$$

represents a parabola if $h^2 = ab$,

i.e., the second degree terms form a perfect square provided

$$\Delta \equiv abc + 2fgh - af^2 - bg^2 - ch^2 \neq 0.$$

7. Reduction of equation of a parabola into standard form

- The equation $(y - \beta)^2 = 4a(x - \alpha)$ can be reduced to the standard form by the transformations

$$x - \alpha = X, \quad y - \beta = Y.$$

The equation becomes $Y^2 = 4aX$. which is the standard form in X, Y coordinates.

$(y - \beta)^2 = 4a(x - \alpha)$ is the form of equation of a parabola whose axis of symmetry is parallel to the x-axis.

- The equation $(ax + by + c)^2 = bx - ay + c'$ can be reduced to the standard form by the transformations

$$\frac{ax + by + c}{\sqrt{a^2 + b^2}} = Y, \quad \frac{bx - ay + c'}{\sqrt{b^2 + a^2}} = X$$

[**Note** The lines $ax + by + c = 0$ and $bx - ay + c' = 0$ are perpendicular to each other.]

The equation becomes $Y^2 = \frac{1}{\sqrt{a^2 + b^2}} X$ which is the standard form in X, Y coordinates.

8. Parametric equations of a parabola

- $x = at^2$, $y = 2at$ are the parametric equations of a parabola.
- Any point on the parabola $y^2 = 4ax$ has the coordinates

$$(at^2, 2at) \text{ or } \left(\frac{a}{m^2}, \frac{-2a}{m}\right) \text{ or } \left(\frac{a}{m^2}, \frac{2a}{m}\right)$$

9. Tangents and normals

Let the equation of a parabola be $y^2 = 4ax$.

- The equation of the tangent at (x_1, y_1) to the parabola is

$$yy_1 = 2a(x + x_1)$$

- The equation of the tangent at $(at^2, 2at)$ is

$$ty = x + at^2$$

- The equation of the tangent at $\left(\frac{a}{m^2}, \frac{2a}{m}\right)$ is

$$y = mx + \frac{a}{m}$$

- The line $y = mx + c$ touches the parabola if $c = \frac{a}{m}$ and so any tangent to the parabola can be taken as

$$y = mx + \frac{a}{m}$$

- The equation of the normal at (x_1, y_1) is

$$\frac{x - x_1}{-2a} = \frac{y - y_1}{y_1}$$

- The equation of the normal at $(at^2, 2at)$ is

$$y + tx = 2at + at^3$$

Let $S \equiv ax^2 + 2hxy + by^2 + 2gx + 2fy + c = 0$ be a parabola.

- The equation of the tangent at (x_1, y_1) is

$$y - y_1 = \left(\frac{dy}{dx}\right)_{x_1, y_1} \cdot (x - x_1), \text{ i.e.,}$$

$$axx_1 + h(xy_1 + x_1y) + byy_1 + g(x + x_1) + f(y + y_1) + c = 0$$

- The equation of the normal at (x_1, y_1) is

$$y - y_1 = \frac{-1}{\left(\frac{dy}{dx}\right)_{x_1, y_1}} \cdot (x - x_1)$$

10. Chord of contact, polar line, pole

Let the equation of a parabola be $y^2 - 4ax = 0$.

- The chord of contact of tangents from the exterior point $P(x_1, y_1)$ to the parabola is

$$T \equiv yy_1 - 2a(x + x_1) = 0$$

- The polar line of the point $P(x_1, y_1)$ with respect to the parabola is

$$T \equiv yy_1 - 2a(x + x_1) = 0$$

- The pole of a line $L = 0$ with respect to the parabola is the point (x_1, y_1) whose polar is the line $L = 0$.

11. Chord with given middle point

The equation of a chord of a second degree curve $S = 0$ whose middle point is (x_1, y_1) is $S_1 = T$.

So, for the parabola $y^2 - 4ax = 0$ it is

$$y_1^2 - 4ax_1 = yy_1 - 2a(x + x_1).$$

12. Diameter of a parabola

The locus of the middle points of parallel chords of a parabola is a line which is called a *diameter* of the parabola.

Selected Solved Examples

1. Find the equation of the parabola whose focus is (1, 1) and the tangent at the vertex is $x + y = 1$. Also find its latus rectum.

The directrix is parallel to the tangent at the vertex V.

$\therefore$ the directrix will be of the form $x + y = \lambda$ ... (1)

Now, V is the foot of the perpendicular from S(1, 1) to the line $x + y = 1$.

Let $V = (\alpha, \beta)$. Then $\alpha + \beta = 1$ ... (2)

and $\dfrac{\beta - 1}{\alpha - 1} \cdot (-1) = -1$, i.e., $\alpha = \beta$ ... (3)

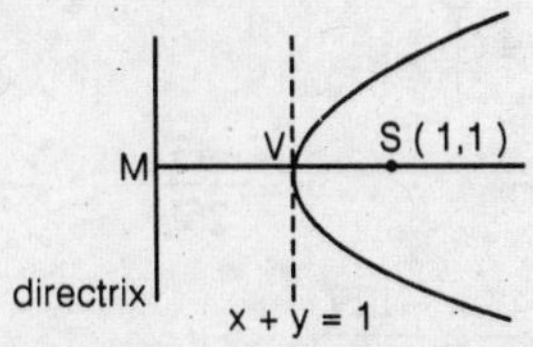

Solving (2), (3) we get $\alpha = \beta = \frac{1}{2}$. So $V = \left(\frac{1}{2}, \frac{1}{2}\right)$.

Let $M = (x_1, y_1)$. As $MV = VS$, V is the middle point of MS.

$$\therefore \quad \frac{x_1 + 1}{2} = \frac{1}{2}, \quad \frac{y_1 + 1}{2} = \frac{1}{2}$$

$\therefore \quad x_1 = 0, y_1 = 0$. So $M = (0, 0)$.

As M is on the directrix, (0, 0) satisfies (1).

Hence, $\lambda = 0$.

$\therefore$ the equation of the directrix is $x + y = 0$.

Using focus-directrix properly, the equation of the parabola is

$$(x-1)^2 + (y-1)^2 = \left(\frac{x+y}{\sqrt{1^2 + 1^2}}\right)^2$$

or $\quad 2\{(x-1)^2 + (y-1)^2\} = (x+y)^2$

or $\quad 2(x^2 + y^2 - 2x - 2y + 2) = x^2 + y^2 + 2xy$

or $\quad x^2 + y^2 - 2xy - 4x - 4y + 4 = 0$

or $\quad (x-y)^2 = 4(x + y - 1)$.

The latus rectum

$$= 4 \times VS = 4\sqrt{\left(\frac{1}{2} - 1\right)^2 + \left(\frac{1}{2} - 1\right)^2} = 4 \cdot \frac{1}{\sqrt{2}} = 2\sqrt{2}.$$

2. Find the equation of the parabola whose axis is parallel to the y-axis and which passes through the points (0, 4), (1, 9) and (–2, 6). Also find its latus rectum.

As the axis is parallel to the y-axis, it will be $x - \alpha = 0$ for some α and the tangent to the vertex (which is perpendicular to the axis) will be $y - \beta = 0$ for some β.

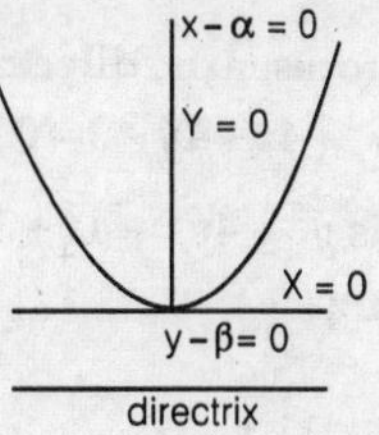

Hence the equation of the parabola will be of the form

$$(x - \alpha)^2 = 4a(y - \beta) \quad \text{... (1)}$$

when α, β, a are unknown constants, $4a$ being latus rectum.

(1) passes through (0, 4), (1, 9) and (–2, 6). So

$(0 - \alpha)^2 = 4a(4 - \beta)$,

i.e., $\alpha^2 = 4a(4 - \beta)$ ... (2)

and $(1 - \alpha)^2 = 4a(9 - \beta)$,

i.e., $1 - 2\alpha + \alpha^2 = 4a(9 - \beta)$ ... (3)

and $(-2 - \alpha)^2 = 4a(6 - \beta)$,

i.e., $4 + 4\alpha + \alpha^2 = 4a(6 - \beta)$ ... (4)

(3) – (2) $\Rightarrow$ $1 - 2\alpha = 20a$ ... (5)

(4) – (3) $\Rightarrow$ $3 + 6\alpha = -12a$

$$\therefore \quad \frac{1 - 2\alpha}{3 + 6\alpha} = \frac{20a}{-12a}$$

or $\quad -3(1 - 2\alpha) = 5(3 + 6\alpha)$

or $\quad -3 + 6\alpha = 15 + 30\alpha$

or $\quad 24\alpha = -18; \quad \therefore \quad \alpha = -\frac{3}{4}$

$\therefore \quad (5) \Rightarrow 1 + \frac{3}{2} = 20a; \quad \therefore \quad a = \frac{5}{40} = \frac{1}{8}$

$\therefore \quad (2) \Rightarrow \left(-\frac{3}{4}\right)^2 = 4 \cdot \frac{1}{8}(4 - \beta)$

or $\quad \frac{9}{16} = \frac{1}{2}(4 - \beta)$, i.e., $4 - \beta = \frac{9}{8}$

or $\quad \beta = 4 - \frac{9}{8} = \frac{23}{8}$.

$\therefore$ from (1), the equation of the parabola is

$$\left(x + \frac{3}{4}\right)^2 = 4 \cdot \frac{1}{8} \cdot \left(y - \frac{23}{8}\right)$$

or $\quad x^2 + \frac{3}{2}x + \frac{9}{16} = \frac{1}{2}y - \frac{23}{16}$

or $\quad x^2 + \frac{3}{2}x - \frac{1}{2}y + 2 = 0$

$\therefore \quad 2x^2 + 3x - y + 4 = 0$

and its latus rectum $= 4a = 4 \cdot \frac{1}{8} = \frac{1}{2}$.

3. Find the vertex, focus, axis, directrix and latus rectum of the parabola $y^2 + 4x + 4y - 3 = 0$.

Here the equation is $y^2 + 4y = -4x + 3$

or $\quad y^2 + 4y + 4 = -4x + 7$

or $\quad (y+2)^2 = 4\left(\frac{7}{4} - x\right)$.

Put $\frac{y+2}{\sqrt{1^2}} = Y, \frac{\frac{7}{4} - x}{\sqrt{(-1)^2}} = X$ {here $y + 2 = 0, \frac{7}{4} - x = 0$ are perpendicular lines}

i.e., $y + 2 = Y, \frac{7}{4} - x = X$.

Then the equation of the parabola in the X, Y coordinates becomes $Y^2 = 4X$.

Comparing with the standard equation, here $a = 1$.

$\therefore \quad$ vertex $= (0, 0)_{X, Y}$, i.e., in X, Y coordinates the vertex is (0, 0)

$= \left(\frac{7}{4}, -2\right)$ {$\because$ when $X = 0, \frac{7}{4} - x = X$ gives $x = \frac{7}{4}$; when $Y = 0, y + 2 = Y$ gives $y = -2$}

focus $= (a, 0)_{X, Y} = (1, 0)_{X, Y}$

$= \left(\frac{3}{4}, -2\right)$ {$\because$ when $X = 1, \frac{7}{4} - x = X$ gives $x = \frac{3}{4}$ when $Y = 0, y + 2 = Y$ gives $y = -2$}

The equation of the axis is $Y = 0$, i.e., $y + 2 = 0$.

The equation of the directrix is $X + a = 0$

or $\quad \frac{7}{4} - x + 1 = 0$, i.e., $x = \frac{11}{4}$.

The latus rectum $= 4a = 4 \times 1 = 4$.

4. Prove that

$$9x^2 - 24xy + 16y^2 - 20x - 15y - 60 = 0$$

represents a parabola. Also find its focus and directrix.

Here $h^2 - ab = (-12)^2 - 9 \cdot 16 = 144 - 144 = 0$. Also, $\Delta \neq 0$

$\therefore$ the equation represents a parabola.

Now, the equation is $(3x - 4y)^2 = 5(4x + 3y + 12)$.

Clearly, the lines $3x - 4y = 0$ and $4x + 3y + 12 = 0$ are perpendicular to each other. So, let

$$\frac{3x - 4y}{\sqrt{3^2 + (-4)^2}} = Y, \quad \frac{4x + 3y + 12}{\sqrt{4^2 + 3^2}} = X \quad \ldots (1)$$

The equation of the parabola becomes

$$Y^2 = X = 4 \cdot \frac{1}{4} X$$

$\therefore$ here $a = \frac{1}{4}$ in the standard equation.

$\therefore$ the focus $= (a, 0)_{X, Y} = \left(\frac{1}{4}, 0\right)_{X, Y}$

Now, if $X = \frac{1}{4}, Y = 0$ then we have from the equations of transformation in (1),

$$\frac{3x - 4y}{5} = 0, \frac{4x + 3y + 12}{5} = \frac{1}{4}$$

$\Rightarrow \quad 3x - 4y = 0, 4x + 3y = \frac{-43}{4}$

$\Rightarrow \quad y = \frac{3x}{4}, y = \frac{1}{3}\left(\frac{-43}{4} - 4x\right)$

$\therefore \quad \frac{3x}{4} = \frac{1}{3}\left(-\frac{43}{4} - 4x\right)$

or $\quad 9x = -43 - 16x; \quad \therefore \quad x = \frac{-43}{25}$

and $\quad y = \frac{3}{4} x = \frac{3}{4} \cdot \frac{-43}{25} = \frac{-129}{100}$

$\therefore \quad$ focus $= \left(\frac{-43}{25}, \frac{-129}{100}\right)$.

The equation of the directrix is

$$X + a = 0, \text{ i.e., } X + \frac{1}{4} = 0$$

or $\quad \frac{4x + 3y + 12}{5} + \frac{1}{4} = 0$

or $\quad 4x + 3y = -\frac{5}{4} - 12 = -\frac{53}{4}$

$\therefore \quad$ the directrix is $\quad 4x + 3y + \frac{53}{4} = 0$.

5. Prove that for a suitable point K on the axis of the parabola the chord PQ, passing through K, can be drawn such that $\frac{1}{PK^2} + \frac{1}{QK^2}$ is the same for all positions of the chord.

The chord PQ passes through $K(k, 0)$ where k is to be suitably taken. Any point on PQ is $(k + r\cos\theta, r\sin\theta)$ where $\tan\theta$ is the slope of PQ. The equation of the parabola is $y^2 = 4ax$.

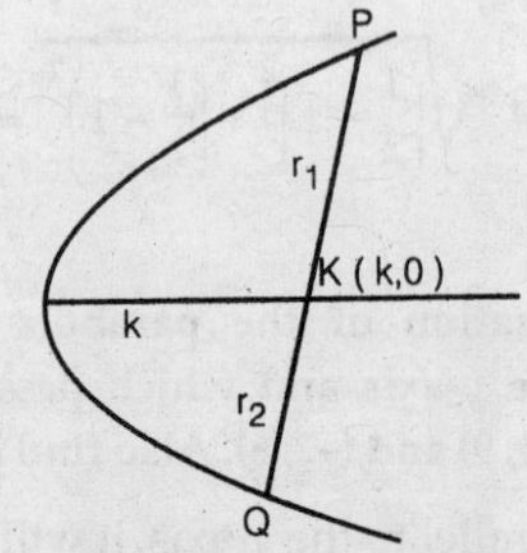

The point $(k + r\cos\theta, r\sin\theta)$ is on $y^2 = 4ax$ if

$$r^2\sin^2\theta = 4a(k + r\cos\theta)$$

or $$r^2\sin^2\theta - 4ar\cos\theta - 4ak = 0 \quad \dots (1)$$

If $KP = r_1$, $KQ = r_2$ then r_1, r_2 are roots of (1).

$$\therefore \quad r_1 + r_2 = \frac{4a\cos\theta}{\sin^2\theta},\ r_1r_2 = \frac{-4ak}{\sin^2\theta}.$$

Now, $$\frac{1}{PK^2} + \frac{1}{QK^2} = \frac{1}{r_1^2} + \frac{1}{r_2^2}$$

$$= \frac{r_1^2 + r_2^2}{r_1^2r_2^2} = \frac{(r_1 + r_2)^2 - 2r_1r_2}{(r_1r_2)^2}$$

$$= \frac{\cos^2\theta}{k^2} - 2\cdot\frac{\sin^2\theta}{-4ak} = \frac{\cos^2\theta}{k^2} + \frac{\sin^2\theta}{2ak}$$

$$= \frac{\cos^2\theta}{4a^2} + \frac{\sin^2\theta}{4a^2} \text{ if } k = 2a$$

$$= \frac{1}{4a^2}, \text{ independent of } \theta.$$

$\therefore \quad \dfrac{1}{PK^2} + \dfrac{1}{QK^2}$ is the same for all positions of the chord if $K = (2a, 0)$.

6. For what real values of a, the point $(-2a, a + 1)$ will be an interior point of the smaller region bounded by the circle $x^2 + y^2 = 4$ and the parabola $y^2 = 4x$?

The point $P(-2a, a + 1)$ will be an interior point of both the circle $x^2 + y^2 - 4 = 0$ and the parabola $y^2 - 4x = 0$.

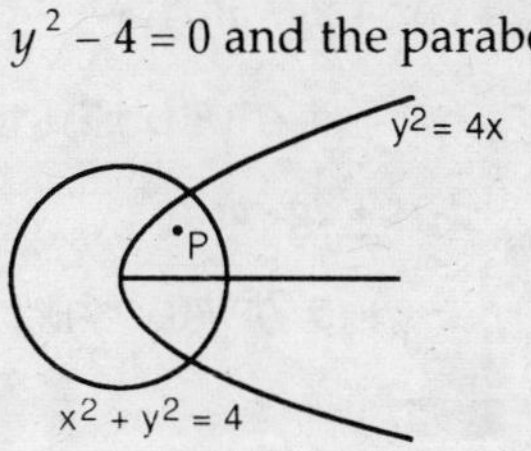

$$\therefore \quad (-2a)^2 + (a + 1)^2 - 4 < 0,$$

i.e., $5a^2 + 2a - 3 < 0 \quad \dots (1)$

and $(a + 1)^2 - 4(-2a) < 0,$

i.e., $a^2 + 10a + 1 < 0 \quad \dots (2)$

The required values of a will satisfy both (1) and (2).

From (1), $(5a - 3)(a + 1) < 0$

$\therefore$ by sign-scheme we get, $-1 < a < \dfrac{3}{5}$. $\quad \dots (3)$

Solving (2), the corresponding equation is

$$a^2 + 10a + 1 = 0$$

or $$a = \frac{-10 \pm \sqrt{100 - 4}}{2} = -5 \pm 2\sqrt{6}.$$

$\therefore$ by sign-scheme for (2),

$$-5 - 2\sqrt{6} < a < -5 + 2\sqrt{6} \quad \dots (4)$$

The set of values of a satisfying (3) and (4) is

$$-1 < a < -5 + 2\sqrt{6}.$$

7. The tangents to the parabola $y^2 = 4ax$ at $P(t_1)$ and $Q(t_2)$ intersect at R. Prove that the area of the triangle PQR is $\frac{1}{2}a^2(t_2 - t_1)^3$.

The tangent to the parabola $y^2 = 4ax$ at $P(at_1^2, 2at_1)$ is

$$yt_1 = x + at_1^2 \quad \dots (1)$$

and that at $Q(at_2^2, 2at_2)$ is $yt_2 = x + at_2^2 \quad \dots (2)$

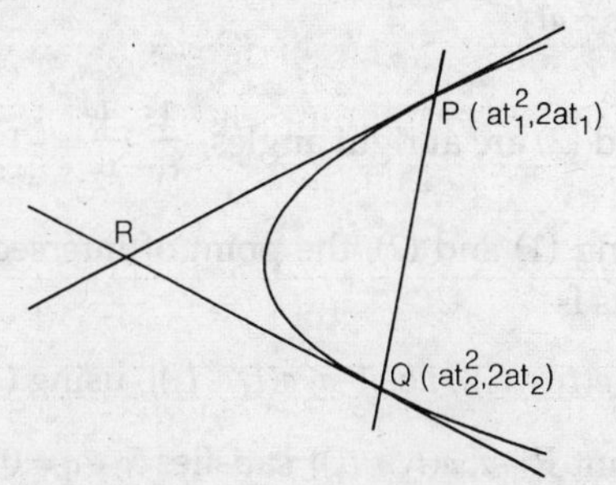

Solving (1) and (2) we get the coordinates of R.

$(1) - (2) \quad \Rightarrow \quad y(t_1 - t_2) = a(t_1^2 - t_2^2)$

or $\quad y = a(t_1 + t_2) \quad \{\because t_1 \neq t_2\}.$

From $t_2 \times (1) - t_1 \times (2)$,

$$0 = x(t_2 - t_1) + a(t_1^2t_2 - t_2^2t_1)$$

or $$0 = x(t_2 - t_1) + at_1t_2(t_1 - t_2)$$

$\therefore \quad x = at_1t_2 \quad \{\because t_1 \neq t_2\}$

$\therefore \quad R = \{at_1t_2, a(t_1 + t_2)\}$

$\therefore \quad$ the area of ΔPQR

$$= \frac{1}{2}\begin{vmatrix} at_1^2 & 2at_1 & 1 \\ at_2^2 & 2at_2 & 1 \\ at_1t_2 & a(t_1 + t_2) & 1 \end{vmatrix}$$

$$= \frac{1}{2}a^2\begin{vmatrix} t_1^2 & 2t_1 & 1 \\ t_2^2 & 2t_2 & 1 \\ t_1t_2 & t_1 + t_2 & 1 \end{vmatrix}$$

$$= \frac{1}{2}a^2\begin{vmatrix} t_1^2 & 2t_1 & 1 \\ (t_2 + t_1)(t_2 - t_1) & 2(t_2 - t_1) & 0 \\ t_1(t_2 - t_1) & t_2 - t_1 & 0 \end{vmatrix},$$

$$R_2 \to R_2 - R_1, R_3 \to R_3 - R_1$$

$$= \frac{1}{2}a^2(t_2 - t_1)^2\begin{vmatrix} t_2 + t_1 & 2 \\ t_1 & 1 \end{vmatrix} = \frac{1}{2}a^2(t_2 - t_1)^3.$$

8. If the tangents at two points of a parabola are at right angles then they intersect at a point on the directrix and the chord joining the two points passes through the focus. Prove this.

Let the parabola be $y^2 = 4ax$ whose focus is $(a, 0)$ and the equation of directrix is $x + a = 0$.

Let $P(at_1^2, 2at_1)$ and $Q(at_2^2, 2at_2)$ be two points where the tangents are at right angles.

The tangent to the parabola $y^2 = 4ax$ at $(at_1^2, 2at_1)$ is

$$yt_1 = x + at_1^2 \qquad \ldots (1)$$

and the tangent to the parabola $y^2 = 4ax$ at $(at_2^2, 2at_2)$ is

$$yt_2 = x + at_2^2 \qquad \ldots (2)$$

$\because$ (1) and (2) are at right angles, $\dfrac{1}{t_1} \cdot \dfrac{1}{t_2} = -1 \qquad \ldots (3)$

Also, solving (1) and (2), the point of intersection R of the tangents is

$\{at_1t_2, a(t_1 + t_2)\}$, i.e., $\{-a, a(t_1 + t_2)\}$, using (3).

But the point $R\{-a, a(t_1 + t_2)\}$ satisfies $x + a = 0$ which is the directrix.

$\therefore$ the tangents at right angles meet on the directrix.

Also, the equation of the chord PQ is

$$y - 2at_1 = \frac{2at_1 - 2at_2}{at_1^2 - at_2^2} \cdot (x - at_1^2)$$

or $\quad y - 2at_1 = \dfrac{2}{t_1 + t_2} \cdot (x - at_1^2)$.

This equation is satisfied by $(a, 0)$ because by putting $x = a, y = 0$ we have

$$-2at_1 = \frac{2}{t_1 + t_2} \cdot (a - at_1^2)$$

or $\quad -t_1(t_1 + t_2) = 1 - t_1^2$

or $\quad t_1t_2 = -1$, which is true from (3).

Hence, the chord joining the points passes through the focus.

9. Prove that the circle circumscribing the triangle formed by any three tangents to a parabola passes through the focus.

Let $P(at_1^2, 2at_1)$, $Q(at_2^2, 2at_2)$ and $R(at_3^2, 2at_3)$ be the three points on the parabola $y^2 = 4ax$.

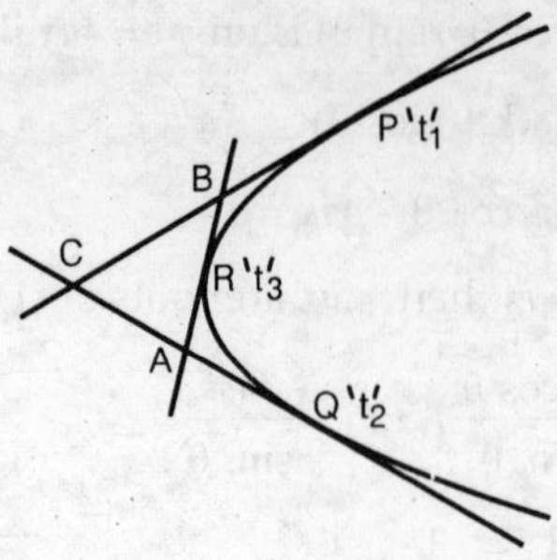

The equations of tangents at P 't_1' and Q 't_2' are

$$yt_1 = x + at_1^2$$

$$yt_2 = x + at_2^2.$$

Solving these, $x = at_1t_2$ and $y = a(t_1 + t_2)$.

$\therefore$ the point of intersection C of the tangents at P and Q is $\{at_1t_2, a(t_1 + t_2)\}$.

Similarly, other points of intersection of tangents are

$$B = \{(at_3t_1, a(t_3 + t_1)\},\ A = \{(at_2t_3, a(t_2 + t_3)\}$$

Let the equation of the circumcircle of the $\triangle ABC$ be

$$x^2 + y^2 + 2gx + 2fy + c = 0 \qquad \ldots (1)$$

(1) will pass through the focus $(a, 0)$ if

$$a^2 + 2ga + c = 0 \qquad \ldots (2)$$

A, B, C are points on (1). So

$$a^2t_1^2t_2^2 + a^2(t_1 + t_2)^2 + 2g \cdot at_1t_2 + 2f \cdot a(t_1 + t_2) + c = 0 \qquad \ldots (3)$$

$$a^2t_2^2t_3^2 + a^2(t_2 + t_3)^2 + 2g \cdot at_2t_3 + 2f \cdot a(t_2 + t_3) + c = 0 \qquad \ldots (4)$$

$$a^2t_3^2t_1^2 + a^2(t_3 + t_1)^2 + 2g \cdot at_3t_1 + 2f \cdot a(t_3 + t_1) + c = 0 \qquad \ldots (5)$$

(3) – (4) $\Rightarrow$

$$a^2t_2^2(t_1^2 - t_3^2) + a^2(t_1 - t_3)(t_1 + 2t_2 + t_3) + 2gat_2(t_1 - t_3) + 2fa \cdot (t_1 - t_3) = 0$$

or $\quad a^2t_2^2(t_1 + t_3) + a^2(t_1 + 2t_2 + t_3) + 2gat_2 + 2fa = 0 \quad \ldots (6)$

Similarly, (4) – (5) $\Rightarrow$

$$a^2t_3^2(t_2 + t_1) + a^2(t_1 + t_2 + 2t_3) + 2gat_3 + 2fa = 0 \quad \ldots (7)$$

(6) – (7) $\Rightarrow$

$$a^2[t_2^2(t_1 + t_3) - t_3^2(t_2 + t_1) + t_2 - t_3] + 2ga(t_2 - t_3) = 0$$

or $\quad a[t_2t_3(t_2 - t_3) + t_1(t_2^2 - t_3^2) + (t_2 - t_3)] + 2g(t_2 - t_3) = 0$

or $\quad a[t_2t_3 + t_1t_2 + t_1t_3 + 1] + 2g = 0$

$\therefore \quad 2g = -a(1 + t_1t_2 + t_2t_3 + t_3t_1)$.

From $t_3 \times (6) - t_2 \times (7)$, we get

$$2f = -a(t_1 + t_2 + t_3 - t_1t_2t_3).$$

Putting values of $2g$, $2f$ in (3), we get

$$c = a^2(t_1t_2 + t_2t_3 + t_3t_1)$$

$\therefore$ the equation of the circumcircle is

$$x^2 + y^2 - a(1 + \Sigma t_1t_2)x$$
$$- a(\Sigma t_1 - t_1t_2t_3)y + a^2\Sigma t_1t_2 = 0.$$

It passes through $(a, 0)$ because

$$a^2 - a^2(1 + \Sigma t_1t_2) + a^2 \Sigma t_1t_2 = 0.$$

Hence, the problem.

10. Prove that three normals can be drawn from the point $(c, 0)$ to the parabola $y^2 = x$ if $c > \frac{1}{2}$ and then one of the normals is always the axis of the parabola. Also find c for which the other two normals will be perpendicular to each other.

Let (t^2, t) be a foot of one of the normals to the parabola $y^2 = x$ from the point $(c, 0)$.

Now, the equation of the normal to $y^2 = x$ at (t^2, t) is

$$y - t = \frac{-1}{\left(\frac{dy}{dx}\right)_{t^2, t}} \cdot (x - t^2)$$

or $\quad y - t = \dfrac{-1}{\frac{1}{2t}}(x - t^2)$

$$\left\{\because\ y^2 = x \quad \Rightarrow \quad 2y\frac{dy}{dx} = 1; \quad \therefore \quad \frac{dy}{dx} = \frac{1}{2y}\right\}$$

or $\quad y - t = -2t(x - t^2)$

or $\quad y + 2tx = t + 2t^3 \qquad \ldots (1)$

It passes through $(c, 0)$ if $0 + 2tc = t + 2t^3$

or $\quad 2t^3 + t(1 - 2c) = 0$

or $\quad t\{2t^2 - (2c - 1)\} = 0$

$\therefore \quad t = 0, \pm\sqrt{\dfrac{2c-1}{2}}.$

Three normals can be drawn if t has three real distinct values.

So, $2c - 1 > 0$, i.e., $c > \frac{1}{2}$.

The foot of one of the normals is (t^2, t) where $t = 0$, i.e., the foot is $(0, 0)$.

By (1), the corresponding normal is $y = 0$, i.e., the x-axis which is the axis of the parabola.

For the other two normals $t = \pm\sqrt{\dfrac{2c-1}{2}}$.

By (1), 'm' of a normal $= -2t$

$\therefore$ 'm' of the other two normals are

$$-2 \cdot \sqrt{\frac{2c-1}{2}},\ 2 \cdot \sqrt{\frac{2c-1}{2}}.$$

These are perpendicular if

$$-2 \cdot \sqrt{\frac{2c-1}{2}} \times 2 \cdot \sqrt{\frac{2c-1}{2}} = -1$$

or $\quad -2(2c - 1) = -1$

or $\quad -4c = -3; \quad \therefore \quad c = \dfrac{3}{4}.$

11. From the point, where any normal to the parabola $y^2 = 4ax$ meets the axis, a line perpendicular to the normal is drawn. Prove that this line always touches the parabola $y^2 + 4a(x - 2a) = 0$.

The normal at any point $(at^2, 2at)$ of the parabola $y^2 = 4ax$ is

$$y + tx = 2at + at^3.$$

It cuts the axis $y = 0$ of the parabola at $(2a + at^2, 0)$.

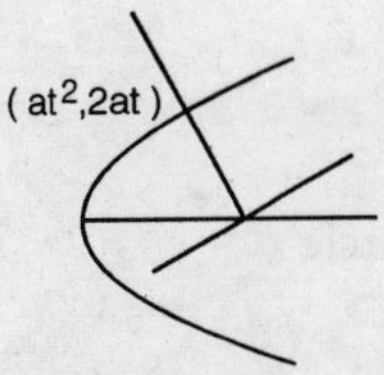

$\therefore$ the equation of the line through this point drawn perpendicular to the normal is

$$y - 0 = \frac{1}{t}(x - 2a + at^2) \quad \{\because\ \text{'}m\text{' of normal} = -t\}$$

or $\quad ty = x - 2a - at^2 \qquad \ldots (1)$

We have to prove that (1) touches the parabola

$$y^2 + 4a(x - 2a) = 0 \qquad \ldots (2)$$

Solving (1) and (2), $y^2 + 4a(ty + at^2) = 0$

or $\quad y^2 + 4aty + (2at)^2 = 0$

or $\quad (y + 2at)^2 = 0$

$\therefore \quad y = -2at, -2at.$

$\therefore \quad$ (1) cuts (2) at coincident points, i.e., (1) touches (2).

12. Let (x_r, y_r); $r = 1, 2, 3, 4$ be the points of intersection of the parabola $y^2 = 4ax$ and the circle $x^2 + y^2 + 2gx + 2fy + c = 0$. Prove that $y_1 + y_2 + y_3 + y_4 = 0$.

Let $\quad x^2 + y^2 + 2gx + 2fy + c = 0 \qquad \ldots (1)$

$$y^2 = 4ax \qquad \ldots (2)$$

Solving (1) and (2) we get the coordinates of points of intersection.

From (2), $x = \dfrac{y^2}{4a}$. Putting in (1),

$$\left(\frac{y^2}{4a}\right)^2 + y^2 + 2g \cdot \frac{y^2}{4a} + 2fy + c = 0$$

or $$\frac{1}{(4a)^2} y^4 + \left(1 + \frac{g}{2a}\right) y^2 + 2fy + c = 0.$$

It has four roots.

Its roots are y_1, y_2, y_3 and y_4.

Now, sum of roots $= -\dfrac{\text{coefficient of } y^3}{\text{coefficient of } y^4}$

$$\therefore \quad y_1 + y_2 + y_3 + y_4 = -\frac{0}{1/(4a)^2} = 0.$$

13. Find the equations to the common tangents to the circle $x^2 + y^2 = 2a^2$ and the parabola $y^2 = 8ax$.

The equation of any tangent to the parabola $y^2 = 8ax$, i.e., $y^2 = 4(2a)x$ is

$$y = mx + \frac{2a}{m} \qquad \ldots (1)$$

(1) will touch the circle $x^2 + y^2 = 2a^2$ if

radius $= \sqrt{2}a$

= length of the perpendicular from the centre (0, 0) to the line (1)

or $$\sqrt{2}a = \left|\frac{\frac{2a}{m}}{\sqrt{1 + m^2}}\right|$$

or $$2a^2(1 + m^2) = \frac{4a^2}{m^2}$$

or $$m^2(1 + m^2) = 2$$

or $$m^4 + m^2 - 2 = 0$$

or $$(m^2 + 2)(m^2 - 1) = 0$$

But $m^2 + 2 = 0$ gives nonreal values of m.

$\therefore \quad m^2 - 1 = 0; \quad \therefore \quad m = \pm 1.$

$\therefore$ putting $m = \pm 1$ in (1) we get the equations of common tangents.

$\therefore$ the equations of common tangents are

$$y = \pm x \pm 2a$$

or $y = x + 2a$ and $y = -(x + 2a)$.

14. A parabola is drawn to pass through A and B, the ends of a diameter of a given circle of radius a, and to have as directrix a tangent to a concentric circle of radius b; the axes of reference being AB and a perpendicular diameter, prove that the locus of the focus of the parabola is $\dfrac{x^2}{b^2} + \dfrac{y^2}{b^2 - a^2} = 1$.

Let A = $(-a, 0)$ and B = $(a, 0)$;

the centre of the circle = (0, 0).

The equation of the concentric circle will be

$$x^2 + y^2 = b^2.$$

Any tangent to $x^2 + y^2 = b^2$ is $y = mx + b\sqrt{1 + m^2}$ which is the directrix of the parabola.

Let (α, β) be the focus.

Then by focus–directrix property, the equation of the parabola will be

$$(x - \alpha)^2 + (y - \beta)^2 = \left(\frac{y - mx - b\sqrt{1 + m^2}}{\sqrt{1 + m^2}}\right)^2.$$

It passes through A$(-a, 0)$, B$(a, 0)$; so

$$(a + \alpha)^2 + \beta^2 = \left(\frac{ma - b\sqrt{1 + m^2}}{\sqrt{1 + m^2}}\right)^2$$

$$= \frac{m^2a^2 + b^2(1 + m^2) - 2abm\sqrt{1 + m^2}}{1 + m^2} \qquad \ldots (1)$$

$$(a - \alpha)^2 + \beta^2 = \left(\frac{-ma - b\sqrt{1 + m^2}}{\sqrt{1 + m^2}}\right)^2$$

$$= \frac{m^2a^2 + b^2(1 + m^2) + 2abm\sqrt{1 + m^2}}{1 + m^2} \qquad \ldots (2)$$

$$(1) + (2) \quad \Rightarrow \quad 2(a^2 + \alpha^2 + \beta^2) = 2 \cdot \frac{m^2a^2 + b^2(1 + m^2)}{1 + m^2}$$

or $$a^2 + \alpha^2 + \beta^2 = \frac{m^2}{1 + m^2} a^2 + b^2 \qquad \ldots (3)$$

$$(2) - (1) \quad \Rightarrow \quad -4a\alpha = \frac{4abm\sqrt{1 + m^2}}{1 + m^2}$$

or $$\alpha^2 = \frac{b^2m^2}{1 + m^2}$$

$$\therefore \quad \frac{m^2}{1 + m^2} = \frac{\alpha^2}{b^2} \qquad \ldots (4)$$

Putting in (3) from (4),

$$a^2 + \alpha^2 + \beta^2 = \frac{a^2\alpha^2}{b^2} + b^2$$

$$\therefore \quad \left(1 - \frac{a^2}{b^2}\right)\alpha^2 + \beta^2 = b^2 - a^2$$

or $\quad \dfrac{b^2-a^2}{b^2}\alpha^2+\beta^2=b^2-a^2$

$\therefore \quad \dfrac{\alpha^2}{b^2}+\dfrac{\beta^2}{b^2-a^2}=1.$

$\therefore$ the equation of the locus of the focus (α, β) is

$$\frac{x^2}{b^2}+\frac{y^2}{b^2-a^2}=1.$$

15. Find the locus of the middle points of the normal chords of the parabola $y^2 = 4ax$.

Let PQ be a normal chord to the parabola $y^2 = 4ax$, which is normal at $P(at_1^2, 2at_1)$.

Let the chord cut the parabola again at $Q(at_2^2, 2at_2)$.

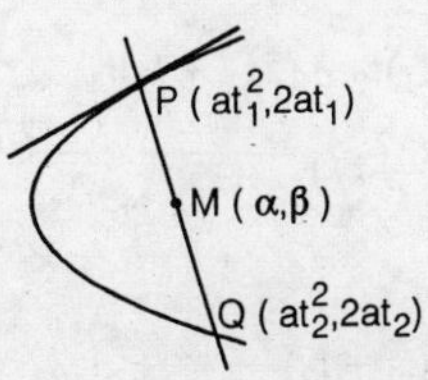

The equation of the normal at $P(at_1^2, 2at_1)$ is

$$y + t_1x = 2at_1 + at_1^3 \qquad \text{... (1)}$$

$Q(at_2^2, 2at_2)$ satisfies it. So

$$2at_2 + t_1 \cdot at_2^2 = 2at_1 + at_1^3$$

or $\quad 2a(t_2 - t_1) + at_1(t_2^2 - t_1^2) = 0$. As $t_1 \neq t_2$, we get

$$2 + t_1(t_2 + t_1) = 0. \qquad \text{... (2)}$$

Let $M(\alpha, \beta)$ be the middle point of the normal chord. Then

$$\alpha = \frac{at_1^2 + at_2^2}{2} = \frac{a}{2}(t_1^2 + t_2^2) \qquad \text{... (3)}$$

$$\beta = \frac{2at_1 + 2at_2}{2} = a(t_1 + t_2) \qquad \text{... (4)}$$

(4) $\Rightarrow \quad t_1 + t_2 = \dfrac{\beta}{a}$

$\therefore$ (2) $\Rightarrow \quad 2 + t_1 \cdot \dfrac{\beta}{a} = 0$

$\therefore \quad t_1 = \dfrac{-2a}{\beta}$

$\therefore$ (2) $\Rightarrow \quad 2 + \dfrac{-2a}{\beta}\left(t_2 - \dfrac{2a}{\beta}\right) = 0$

or $\quad \dfrac{\beta}{a} - \left(t_2 - \dfrac{2a}{\beta}\right) = 0$

$\therefore \quad t_2 = \dfrac{\beta}{a} + \dfrac{2a}{\beta}.$

Putting for t_1, t_2 in (3) we get

$$\alpha = \frac{a}{2}\left[\left(\frac{-2a}{\beta}\right)^2 + \left(\frac{\beta}{a} + \frac{2a}{\beta}\right)^2\right]$$

$$= \frac{a}{2}\left[\frac{4a^2}{\beta^2} + \frac{\beta^2}{a^2} + \frac{4a^2}{\beta^2} + 4\right]$$

$\therefore$ the equation of the required locus is

$$x = \frac{a}{2}\left[\frac{8a^2}{y^2} + \frac{y^2}{a^2} + 4\right]$$

or $\quad x = \dfrac{4a^3}{y^2} + \dfrac{y^2}{2a} + 2a$

$\therefore \quad 2ay^2(x - 2a) = 8a^4 + y^4.$

16. Show that the locus of a point that divides a chord of slope 2 of the parabola $y^2 = 4x$ internally in the ratio 1 : 2 is a parabola. Also find the vertex of this parabola.

Let PQ be a chord of the parabola $y^2 = 4x$ having the slope 2.

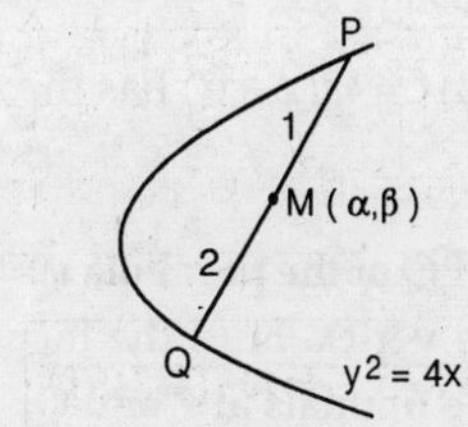

Let the ends of the chord be $P(at_1^2, 2at_1)$ and $Q(at_2^2, 2at_2)$.

Let M divide PQ in the ratio 1 : 2 and $M = (\alpha, \beta)$. Then

$$\alpha = \frac{at_2^2 + 2at_1^2}{3} = \frac{a}{3}(2t_1^2 + t_2^2) \qquad \text{... (1)}$$

$$\beta = \frac{2at_2 + 2 \cdot 2at_1}{3} = \frac{2a}{3}(2t_1 + t_2) \qquad \text{... (2)}$$

Again, slope of $PQ = \dfrac{2at_1 - 2at_2}{at_1^2 - at_2^2} = \dfrac{2}{t_1 + t_2}$

$\therefore$ from the question, $\dfrac{2}{t_1 + t_2} = 2$

$$\therefore \quad t_1 + t_2 = 1 \qquad \text{... (3)}$$

From (2) and (3), $\beta = \dfrac{2a}{3}(t_1 + 1)$; $\quad \therefore \quad t_1 = \dfrac{3\beta}{2a} - 1$

$\therefore \quad t_2 = 1 - t_1 = 1 - \left(\dfrac{3\beta}{2a} - 1\right) = 2 - \dfrac{3\beta}{2a}.$

Putting these in (1),

$$\alpha = \frac{a}{3}\left[2\left(\frac{3\beta}{2a} - 1\right)^2 + \left(2 - \frac{3\beta}{2a}\right)^2\right]$$

or $\quad \alpha = \frac{a}{3}\left[2\left(\frac{9\beta^2}{4a^2} - \frac{3\beta}{a} + 1\right) + 4 - \frac{6\beta}{a} + \frac{9\beta^2}{4a^2}\right]$

$= \frac{a}{3}\left[\frac{27\beta^2}{4a^2} - \frac{12\beta}{a} + 6\right]$

$\therefore$ the equation of the required locus is

$$\frac{3x}{a} = \frac{27}{4a^2}y^2 - \frac{12}{a}y + 6$$

or $\quad 12ax = 27y^2 - 48ay + 24a^2.$

Now, $27y^2 - 48ay = 12ax - 24a^2$

or $\quad 9y^2 - 16ay = 4ax - 8a^2$

or $\quad 9y^2 - 16ay + \left(\frac{8}{3}a\right)^2 = 4ax - 8a^2 + \frac{64a^2}{9}$

or $\quad \left(3y - \frac{8a}{3}\right)^2 = 4ax - \frac{8}{9}a^2 = 4a\left(x - \frac{2a}{9}\right)$

or $\quad \left(y - \frac{8a}{9}\right)^2 = \frac{4a}{9}\left(x - \frac{2a}{9}\right)$

which is a parabola with the vertex $\left(\frac{2a}{9}, \frac{8a}{9}\right)$

{because $(y-\beta)^2 = 4a(x-\alpha)$ has the vertex (α, β)}.

17. A variable chord PQ of the parabola $y = x^2$ subtends a right angle at the vertex. Find the locus of points of intersection of the normals at P and Q.

The vertex V of the parabola is (0, 0) and any point on $y = x^2$ has the coordinates (t, t^2).
So let us take $P = (t_1, t_1^2)$, $Q = (t_2, t_2^2)$ and $\angle PVQ = 90°$.

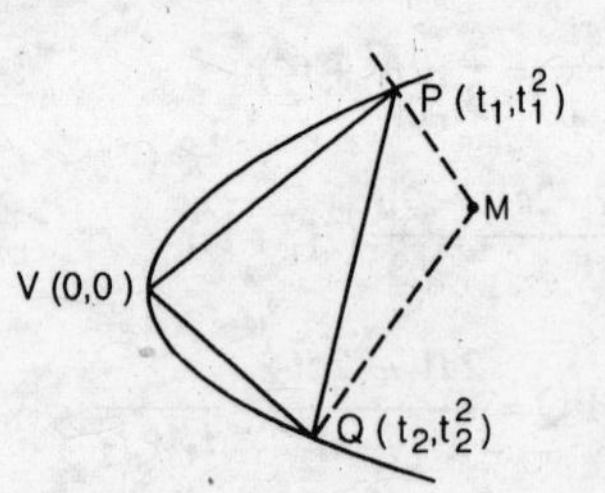

As 'm' of VP $= \frac{t_1^2 - 0}{t_1 - 0} = t_1$

and 'm' of VQ $= \frac{t_2^2 - 0}{t_2 - 0} = t_2$,

VP $\perp$ VQ $\quad \Rightarrow \quad t_1 \cdot t_2 = -1 \quad \ldots (1)$

The equation of the normal to a curve at (x_1, y_1) is

$$y - y_1 = \frac{-1}{\left(\frac{dy}{dx}\right)_{x_1, y_1}} \cdot (x - x_1).$$

Here, $y = x^2 \quad \therefore \quad \frac{dy}{dx} = 2x.$

$\therefore$ the equation of the normal at $P(t_1, t_1^2)$ is

$$y - t_1^2 = \frac{-1}{2t_1}(x - t_1)$$

or $\quad 2t_1y + x = 2t_1^3 + t_1 \quad \ldots (2)$

Similarly, the equation of the normal at $Q(t_2, t_2^2)$ is

$2t_2y + x = 2t_2^3 + t_2 \quad \ldots (3)$

Eliminating t_1, t_2 from (1), (2) and (3) we get the locus of M.

(2) – (3) $\quad \Rightarrow \quad 2y(t_1 - t_2) = 2(t_1^3 - t_2^3) + (t_1 - t_2)$

or $\quad 2y = 2(t_1^2 + t_1t_2 + t_2^2) + 1 \quad \ldots (4)$

Also, $t_2 \times (2) - t_1 \times (3)$

$\Rightarrow \quad (t_2 - t_1)x = (2t_1^3 + t_1)t_2 - (2t_2^3 + t_2)t_1$

$= 2t_1t_2(t_1^2 - t_2^2)$

or $\quad x = -2t_1t_2(t_1 + t_2) \quad \ldots (5)$

From (1) and (5), $x = 2(t_1 + t_2)$

or $\quad t_1 + t_2 = \frac{1}{2}x \quad \ldots (6)$

From (4), $2y = 2\{(t_1 + t_2)^2 - t_1t_2\} + 1$

$= 2\left\{\left(\frac{1}{2}x\right)^2 + 1\right\} + 1$, using (1) and (6)

$\therefore$ the equation of the required locus is

$$2y = \frac{x^2}{2} + 3$$

or $\quad x^2 = 2(2y - 3)$, which is a parabola.

18. Find the locus of the foot of the perpendicular drawn from a fixed point to any tangent to a parabola.

Let the parabola be $y^2 = 4ax$ and the fixed point be (h, k).

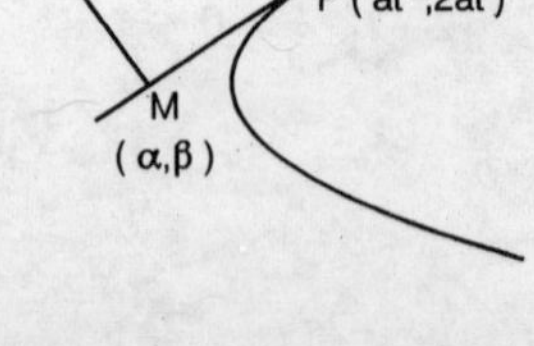

The tangent at any point $P(at^2, 2at)$ is $ty = x + at^2 \quad \ldots (1)$

Let $M(\alpha, \beta)$ be the foot of the perpendicular to (1) from the point (h, k).

Using perpendicularity, $\frac{\beta - k}{\alpha - h} \cdot \frac{1}{t} = -1 \quad \ldots (2)$

As $M(\alpha, \beta)$ is on (1), $t\beta = \alpha + at^2$ ... (3)

We have to eliminate t from (2) and (3).

From (2), $t = -\dfrac{\beta - k}{\alpha - h}$. Putting in (3),

$$\beta\left(-\frac{\beta - k}{\alpha - h}\right) = \alpha + a \cdot \left(\frac{\beta - k}{\alpha - h}\right)^2$$

or $\quad -\beta(\beta - k)(\alpha - h) = \alpha(\alpha - h)^2 + a(\beta - k)^2$

$\therefore$ the equation of the locus of the foot M is

$$x(x - h)^2 + y(x - h)(y - k) + a(y - k)^2 = 0.$$

19. Show that the locus of the poles of the tangents to the parabola $y^2 = 4ax$ with respect to the parabola $y^2 = 4bx$ is the parabola $ay^2 = 4b^2x$.

Any tangent to the parabola

$y^2 = 4ax$ is $ty = x + at^2$ ... (1)

Let (α, β) be the pole of (1) with respect to the parabola

$y^2 = 4bx$.

Then (1) is the polar of (α, β) with respect to $y^2 = 4bx$

$\therefore$ (1) and $y\beta = 2b(x + \alpha)$ are identical.

So, comparing these,

$$\frac{t}{\beta} = \frac{1}{2b} = \frac{at^2}{2b\alpha}; \quad \therefore \quad t = \frac{\beta}{2b},\ t^2 = \frac{\alpha}{a}$$

$$\therefore \quad \left(\frac{\beta}{2b}\right)^2 = \frac{\alpha}{a}$$

or $\quad a\beta^2 = 4b^2\alpha$

$\therefore$ the equation of the required locus of the poles is

$$ay^2 = 4b^2x.$$

20. The general equation to a system of parallel chords of the parabola $y^2 = \frac{25}{7}x$ is $4x - y + k = 0$. What is the equation to the corresponding diameter?

Let PQ be a chord of the system whose equation is

$4x - y + k = 0$ ... (1)

where k is a parameter.

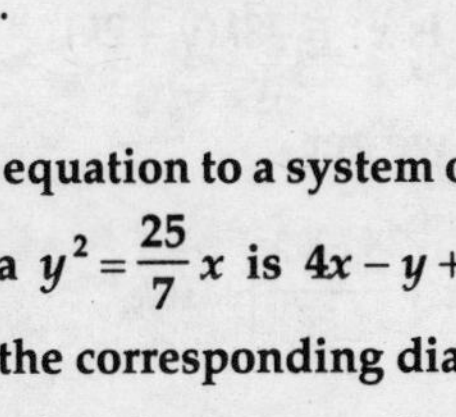

Let $M(\alpha, \beta)$ be the middle point of PQ. The locus of M is the required diameter.

The equation of the parabola is $y^2 = \dfrac{25}{7}x$ ... (2)

Solving (1) and (2), $4 \times \dfrac{7y^2}{25} - y + k = 0$

or $\quad 28y^2 - 25y + 25k = 0.$

Let its roots be y_1, y_2.

Then $\beta = \dfrac{y_1 + y_2}{2} = \dfrac{25}{2 \cdot 28} = \dfrac{25}{56}$

$\therefore$ the equation of the locus of $M(\alpha, \beta)$ is $y = \dfrac{25}{56}$.

21. A ray of light is coming along the line $y = b$, $(b > 0)$ from the positive direction of the x-axis and strikes a concave mirror whose intersection with the x-y plane is the parabola $y^2 = 4ax$, $(a > 0)$. Find the equation of the reflected ray and show that it passes through the focus of the parabola.

Let P be the point of incidence.

Then P is the intersection of the line $y = b$ and the parabola $y^2 = 4ax$.

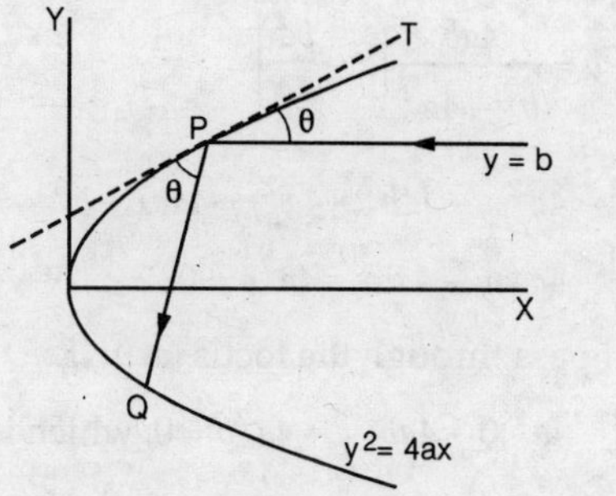

$$\therefore \quad P = \left(\frac{b^2}{4a},\ b\right)$$

$\therefore$ the equation of the tangent PT at P is

$$y \cdot b = 2a\left(x + \frac{b^2}{4a}\right)$$

or $\quad by = 2ax + \dfrac{b^2}{2}$ ... (1)

'm' of (1) is $\dfrac{2a}{b}$. So, $\tan\theta = \dfrac{2a}{b}$.

Let the slope of the reflected ray PQ be m.

$$\therefore \quad \tan\theta = \left|\frac{m - \dfrac{2a}{b}}{1 + m \cdot \dfrac{2a}{b}}\right|$$

or $\quad \dfrac{2a}{b} = \left|\dfrac{m - \dfrac{2a}{b}}{1 + m \cdot \dfrac{2a}{b}}\right|$

$$\therefore \quad \frac{m - \frac{2a}{b}}{1 + m \cdot \frac{2a}{b}} = \pm \frac{2a}{b}$$

or $$m - \frac{2a}{b} = \pm \left(\frac{2a}{b} + m \cdot \frac{4a^2}{b^2} \right)$$

$$\therefore \quad m - \frac{2a}{b} = \frac{2a}{b} + m \cdot \frac{4a^2}{b^2}$$

and $$m - \frac{2a}{b} = -\frac{2a}{b} - m \cdot \frac{4a^2}{b^2}$$

$$\therefore \quad m\left(1 - \frac{4a^2}{b^2}\right) = \frac{4a}{b} \text{ and } m\left(1 + \frac{4a^2}{b^2}\right) = 0.$$

But $m \neq 0$; $\therefore \quad m\left(1 - \frac{4a^2}{b^2}\right) = \frac{4a}{b}$

$$\therefore \quad m = \frac{4ab}{b^2 - 4a^2}$$

$\therefore$ the equation of the reflected ray PQ is

$$y - b = \frac{4ab}{b^2 - 4a^2}\left(x - \frac{b^2}{4a}\right)$$

or $$(b^2 - 4a^2)y - b(b^2 - 4a^2) = 4abx - b^3$$

or $$(b^2 - 4a^2)y - 4abx + 4a^2b = 0.$$

This will pass through the focus $(a, 0)$ if

$(b^2 - 4a^2)0 - 4ab \cdot a + 4a^2b = 0$, which is true.

$\therefore$ the reflected ray passes through the focus.

22. A parabola is drawn touching the axis of x at the origin and having its vertex at a given distance k from the x-axis. Prove that the axis of the parabola is a tangent to the parabola $x^2 + 8k(y - 2k) = 0$.

Let the equation of the parabola be $Y^2 = 4aX$.

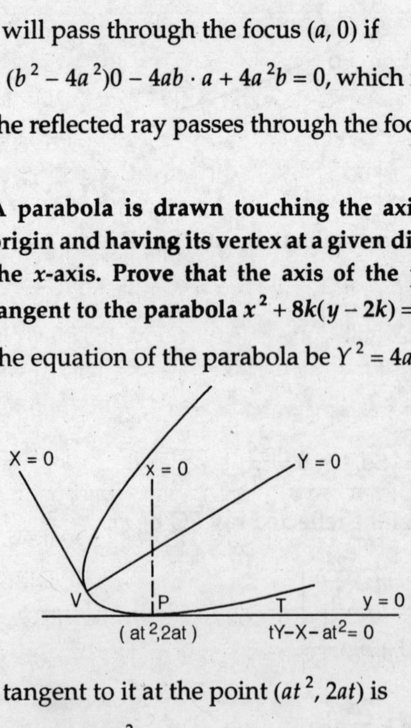

Any tangent to it at the point $(at^2, 2at)$ is

$$Yt = X + at^2 \quad \dots (1)$$

The normal at the point $(at^2, 2at)$ is

$$Y + t\,X = 2at + at^3 \quad \dots (2)$$

Take the equations of transformation

$$\frac{tY - X - at^2}{\sqrt{1 + t^2}} = y \quad \dots (3)$$

and $$\frac{Y + tX - 2at - at^3}{\sqrt{1 + t^2}} = x \quad \dots (4)$$

$\therefore$ in x, y coordinates P = (0, 0) and PT is the x-axis which is tangent to the parabola at the origin.

Now, (3) $\Rightarrow$ $tY - X - at^2 = y\sqrt{1 + t^2}$ ... (5)

(4) $\Rightarrow$ $Y + tX - 2at - at^3 = x\sqrt{1 + t^2}$ ... (6)

$\therefore$ (5) $\times t$ + (6) $\Rightarrow$ $(t^2 + 1)Y - 2at^3 - 2at$
$= yt\sqrt{1 + t^2} + x\sqrt{1 + t^2}$

$\therefore$ the axis of the parabola ($Y = 0$) becomes

$$-2at^3 - 2at = (yt + x)\sqrt{1 + t^2}$$

or $$yt + x = \frac{-2at(1 + t^2)}{\sqrt{1 + t^2}} = -2at\sqrt{1 + t^2} \quad \dots (7)$$

The distance of the vertex V(0, 0) in the X, Y coordinates from PT

$$= \frac{-at^2}{\sqrt{1 + t^2}} = k \quad \text{(from the question)}$$

$\therefore$ from (7), the equation of the axis of the parabola in x, y coordinates becomes

$$yt + x = -2at\left(\frac{-at^2}{k}\right)$$

or $$yt + x - \frac{2a^2t^3}{k} = 0 \quad \dots (8)$$

The given parabola is $x^2 = -8k(y - 2k)$ ... (9)

Solving (8) and (9), we get

$$x^2 = -8k \cdot \frac{1}{t}\left(-x + \frac{2a^2t^3}{k}\right) + 16k^2$$

or $$x^2 = 8\frac{k}{t}x - 16a^2t^2 + 16k^2$$

or $$tx^2 - 8kx + 16t(a^2t^2 - k^2) = 0.$$

Here $D = 64k^2 - 64t^2(a^2t^2 - k^2)$

$$= 64[k^2 - a^2t^4 + t^2k^2]$$

$$= 64\left[(1 + t^2) \cdot \frac{a^2t^4}{1 + t^2} - a^2t^4\right] = 0.$$

$\therefore$ the axis, given by (8) touches the given parabola.

Note If we take $k = \frac{at^2}{\sqrt{1 + t^2}}$, the points of intersection of the axis and the given parabola will be imaginary.

Exercises

1. Find the equation of the parabola whose focus is at (–6, –6) and the vertex is at (–2, 2). Also find the latus rectum of the parabola.

2. Prove that the equation $y^2 + 2ax + 2by + c = 0$ represents a parabola whose axis is parallel to x-axis. Find its vertex and the equation of the double ordinate through the focus.

3. If the latus rectum = 4, the vertex is (–2, 0) and the equation of the axis is $3x + 4y + 6 = 0$ then find the equation of a parabola satisfying these.

4. Prove that the equation $y^2 - 2y + 8x - 23 = 0$ represents a parabola and find its focus and directrix.

5. Prove that $x^2 + 2xy + x = 3 + y - y^2$ represents a parabola and find its latus rectum, vertex and equation of the axis.

6. Prove that $P(t^2, 2t)$ is a point on the parabola $y^2 = 4x$ for all $t \in R$. Hence find the range of values of t for which the point P is an interior point of the circle $x^2 + y^2 = 5$.

7. If $(a^2, a - 2)$ be a point interior to the region of the parabola $y^2 = 2x$ bounded by the chord joining the points (2, 2) and (8, –4) then find the set of all possible real values of a.

8. If the line $2x + 3y = 1$ touches the parabola $y^2 = 4ax$, find the point of contact and the latus rectum of the parabola.

9. Find the equations of the tangents to the parabola $y^2 + 4 = 4x$ which are equally inclined with the axes of reference and also find the tangent at the vertex of the parabola.

10. Prove that the tangents at the extremities of any focal chord of a parabola intersect at right angles on the directrix.

11. Prove that the orthocentre of the triangle formed by three tangents to a parabola lie on the directrix.

12. Show that if tangents are drawn from any point on the line $x + 4a = 0$ to the parabola $y^2 = 4ax$ then their chord of contact will subtend a right angle at the vertex.

13. Show that the circle described on a focal chord of a parabola as diameter touches its directrix.

14. Prove that the area of the triangle formed by the tangents to the parabola $y^2 = 4ax$ from the point (x_1, y_1) and the chord of contact is

$$\frac{1}{2a}(y_1^2 - 4ax_1)^{3/2} \text{ unit}^2.$$

15. Points A, B and C lie on the parabola $y^2 = 4ax$. The tangents to the parabola at A, B and C taken in pairs, intersect at P, Q and R. Determine the ratio of the areas of the triangles ABC and PQR.

16. P and Q are two variable points on the parabola $y^2 = 4ax$ whose sum of ordinates is k, a constant. Prove that the centroid of the triangle formed by the tangents at P, Q and the chord PQ always lies on a straight line parallel to the axis of the parabola.

17. P, Q and R are three points on a parabola whose ordinates are in GP. Show that the tangents at P and R will meet on the ordinate of Q.

18. Three tangents are drawn to a parabola such that the tangents of their inclinations to the axis of the parabola are three consecutive terms of a fixed HP. Prove that the tangents will form a triangle of constant area.

19. Of the two perpendicular lines one touches the parabola $y^2 = 4a(x + a)$ and the other touches the parabola $y^2 = 4b(x + b)$. Prove that the lines will intersect on the line $x + a + b = 0$.

20. Tangents at the three points of a parabola $y^2 = 4ax$ form an equilateral triangle. Prove that vertices of the triangle lie on the curve $(3x + a)(3a + x) = y^2$.

21. Find the equation of the chord to the parabola $y^2 = 4ax$ whose middle point is (h, k).

22. Prove that a normal chord to a parabola at the point whose ordinate is equal to the abscissa, subtends a right angle at the focus.

23. Prove that the area of the triangle formed by the normals to the parabola $y^2 = 4ax$ at points 't_1', 't_2', 't_3' is $\frac{1}{2} \cdot a^2(t_1 - t_2)(t_2 - t_3)(t_3 - t_1)(t_1 + t_2 + t_3)^2$.

24. Prove that in general three normals can be drawn to a parabola from a given point and the centroid of the triangle formed by the feet of the three normals lies on the axis of the parabola.

25. Prove that from any point $P(at^2, 2at)$ on the parabola $y^2 = 4ax$, two normals can be drawn to the parabola and their feet Q and R have the parameters satisfying the equation $\lambda^2 + \lambda t + 2 = 0$.

26. Prove that the normal to the parabola $y^2 = 4ax$ at $(am^2, -2am)$ intersects the parabola again at an angle $\tan^{-1}(m/2)$.

27. The tangents from the point T to the parabola $y^2 = 4ax$ touch at P and Q. If the chord of contact PQ is a normal to the parabola at P, prove that TP is bisected by the directrix of the parabola.

28. For what values of a will the tangents drawn to the parabola $y^2 = 4ax$ from a point, not on the y-axis, will be normals to the parabola $x^2 = 4y$?

29. Find the equations of the two circles that touch the parabola $75y^2 = 64(5x - 3)$ at $(6/5, 8/5)$ and the x-axis.

30. Circles are described on any two focal chords of a parabola as diameters. Prove that their common chord passes through the vertex of the parabola.

31. Find the equation of common tangents to the circle $x^2 + y^2 = 1$ and the parabola $y^2 = 4x$, if any such tangent exists.

32. Prove that the length of the intercept on the normal to the parabola $y^2 = 4ax$ at the point $(at^2, 2at)$, made by the circle which is described on the focal distance of the given point as diameter, is $a \cdot \sqrt{1 + t^2}$.

33. Find the angle subtended by the common chord of the circle $x^2 + y^2 - 2x - 3 = 0$ and the parabola $y^2 = 3(x - 1)$ at the focus of the parabola.

34. Prove that the locus of the middle points of chords of the parabola $y^2 = 4ax$ through the vertex is also a parabola. Find focus and latus rectum of the locus.

35. Prove that the locus of a point which moves so that its distance from a given line is equal to the length of the tangent from the point to a given circle, is a parabola. Find the position of its focus and directrix.

36. Let two parabolas be $y^2 = 4ax$ and $y^2 + 4bx = 0$, $a \neq b$. Any line parallel to their common axis cuts the parabolas at P and P′ respectively. Prove that the locus of the middle point of PP′ is another parabola.

37. Find the locus of the point of intersection of two tangents to a parabola such that their chord of contact subtends a right angle at the vertex.

38. Find the locus of the intersection of normals to the parabola $y^2 = 4ax$ at the extremities of a focal chord.

39. Find the locus of the point of intersection of the normals to the parabola $x^2 = 8y$ which are at right angles to each other.

40. Prove that the locus of the foot of the perpendicular from the focus of a parabola on the tangent at any point is the tangent at the vertex.

41. Find the locus of the point of intersection of two tangents to the parabola $y^2 = 4ax$ which are at right angles.

42. Find the locus of the point from which two of the normals to the parabola $y^2 = 4ax$ are at right angles.

43. Show that the locus of points such that two of the three normals drawn from them to the parabola $y^2 = 4ax$ coincide is $27ay^2 = 4(x - 2a)^3$.

44. Prove that the locus of the intersection of the normals at the ends of a system of parallel chords of a parabola is a straight line which is normal to the parabola.

45. Find the locus of middle points of chords of the parabola $y^2 = 4ax$ which subtend a right angle at the vertex.

46. Find the locus of middle points of chords of the parabola $y^2 = 4ax$ that pass through the fixed point (α, β).

47. If two tangents to the parabola $y^2 = 4ax$ from a point P make angles θ_1 and θ_2 with the axis such that

$$\tan^2\theta_1 + \tan^2\theta_2 = \lambda \text{ (a constant)}$$

then find the locus of P.

48. Find the locus of the feet of the perpendiculars drawn from the vertex upon chords of the parabola $y^2 = 4ax$, which subtend an angle of $\pi/4$ at the vertex.

49. Prove that the locus of the poles of focal chords of a parabola is the directrix.

50. Prove that the locus of the poles of normal chords to the parabola $y^2 = 4ax$ is the curve

$$y^2(x + 2a) + 4a^3 = 0.$$

51. Prove that the locus of the poles of the tangents to the parabola $y^2 = 4ax$ with respect to the circle $x^2 + y^2 = 2ax$ is the circle $x^2 + y^2 = ax$.

Objective Questions

Fill in the blanks.

52. If the vertex = (1, −2) and the focus = (0, 0) then the equation of the tangent to the parabola at the vertex is ______ .

53. If the focus = (2, 3) and directrix is $x + y = 1$ then the equation of the parabola is ______ .

54. If the focus = (1, 1) and the axis meets the directrix at (−2, 3) then the equation of the parabola is ______ .

55. The line $x + y + 1 = 0$ touches the parabola $y^2 = kx$ if k = ______

56. The normals at the ends of the latus rectum of the parabola $y^2 = 4ax$ meet the parabola again at Q and Q′. Then QQ′ = ______ .

57. The point of contact of the tangent to $y^2 = x$, inclined at 45° to the axis is ______ .

58. The equation of the normal to the parabola $y^2 = 4x$ which is parallel to the line $y = 2x - 5$ is ______ .

59. The equation of the tangent to the parabola $y^2 = 16x$ which is perpendicular to the line $2x - y + 5 = 0$ is ______ .

60. The length of the chord of the parabola $y^2 = 4x$ which passes through the vertex and whose slope is 0.5, is ______ .

61. The line $x\cos\alpha + y\sin\alpha = p$ touches the parabola $y^2 = 4a(x+a)$ if ______ .

62. The point of intersection of the tangents at the ends of the latus rectum of the parabola $y^2 = 4x$ is ______ .

63. The angle between the tangents drawn from (1, 3) to the parabola $y^2 = 4x$ is ______ .

64. The equation of the circle described on the line segment joining the foci of the parabolas $x^2 = 4ay$ and $y^2 = 4a(x-a)$ as diameter, is ______ .

Choose the correct option(s).

65. A double ordinate PQ of the parabola $y^2 = 4px$ has the length $8p$. If V be the vertex of the parabola then $\angle PVQ$ has the measure

(a) $\pi/4$ (b) $\pi/2$
(c) $\pi/3$ (d) none of these

66. The equation to the parabola whose vertex and focus are on the axis of x at distances a and a' from the origin respectively, is

(a) $y^2 = 4(a'-a)(x-a)$ (b) $y^2 = 4(a'-a)(x-a')$
(c) $x^2 = 4(a'-a)(y-a)$ (d) $y^2 = 4(a-a')(x-a)$

67. If the normal at $(at_1^2, 2at_1)$ to the parabola $y^2 = 4ax$ meets the parabola again at $(at_2^2, 2at_2)$ then

(a) $t_1 = -t_2 - \dfrac{2}{t_2}$ (b) $\dfrac{t_1+t_2}{2} = \dfrac{1}{t_1}$
(c) $t_2 = -t_1 - \dfrac{2}{t_1}$ (d) $\dfrac{t_1+t_2}{2} = \dfrac{1}{t_2}$

68. If the tangents to the parabola $y^2 = 4ax$ at the points (x_1, y_1) and (x_2, y_2) meet at the point (x_3, y_3) then

(a) $y_3 = \sqrt{y_1 y_2}$ (b) $2y_3 = y_1 + y_2$
(c) $\dfrac{2}{y_3} = \dfrac{1}{y_1} + \dfrac{1}{y_2}$ (d) none of these

69. The line $x\cos\alpha + y\sin\alpha = p$ touches the parabola $y^2 = 4ax$ if

(a) $p\cos\alpha + a\sin^2\alpha = 0$ (b) $p\sin\alpha + a\cos^2\alpha = 0$
(c) $a\cos\alpha + p\sin^2\alpha = 0$ (d) $a\sin\alpha + p\cos^2\alpha = 0$

70. The angle between the tangents drawn from the origin to the parabola $y^2 = 4a(x-a)$ is

(a) 90° (b) 30°
(c) $\tan^{-1}(1/2)$ (d) 45°

71. The angle subtended at the focus by the normal chord at the point (λ, λ), $\lambda \neq 0$ on the parabola $y^2 = 4ax$ is

(a) $\pi/4$ (b) $\tan^{-1}(\lambda/a)$
(c) $\tan^{-1}(1/4)$ (d) $\pi/2$

72. The equation of a tangent to the parabola $y^2 = 8x$ is $y = x + 2$. The point on this line from which the other tangent to the parabola is perpendicular to the given tangent, is

(a) (2, 4) (b) (–2, 0)
(c) (–1, 1) (d) none of these

73. If 't_1' and 't_2' be the ends of a focal chord of the parabola $y^2 = 4ax$ then $t_1 t_2$ is equal to

(a) 1 (b) –1
(c) 2 (d) none of these

Answers

1. $4x^2 + y^2 - 4xy + 104x + 148y - 124 = 0, 16\sqrt{5}$

2. $\left(\dfrac{b^2-c}{2a}, -b\right)$; $2ax = b^2 - a^2 - c$

3. $9x^2 + 24xy + 16y^2 - 44x + 108y - 124 = 0$,
$9x^2 + 24xy + 16y^2 + 36x + 48y + 108 = 0$

4. $(1, 1), x - 5 = 0$

5. $\dfrac{1}{\sqrt{2}}\left(\dfrac{3}{2}, \dfrac{-3}{2}\right)$, $x + y = 0$ 6. (–1, 1)

7. $(-2+2\sqrt{2}, 2)$ 8. $\left(-\dfrac{1}{2}, \dfrac{2}{3}\right), \dfrac{8}{9}$

9. $y = x, x + y = 0; x = 1$ 15. 2 : 1

21. $k(y-k) = 2a(x-h)$ 28. $a \le -2\sqrt{2}$ or $a \ge 2\sqrt{2}$

29. $x^2 + y^2 - 4x - 2y + 4 = 0, x^2 + y^2 + 4x - 8y + 4 = 0$

31. $x \pm \sqrt{\dfrac{1}{2}(1+\sqrt{5})} \cdot y + \dfrac{1}{2}(1+\sqrt{5}) = 0$

33. $\pi - \tan^{-1}\left(\dfrac{8\sqrt{3}}{47}\right)$ 34. $\left(\dfrac{a}{2}, 0\right), 2a$

35. $\left(\dfrac{-a^2}{2b}, 0\right)$, $x + \dfrac{a^2+2b^2}{2b} = 0$

if the circle is $x^2 + y^2 = a^2$, the line is $x + b = 0$

37. $x + 4a = 0$ when the parabola is $y^2 = 4ax$

38. $y^2 = a(x-3a)$ 39. $x^2 = 2(y-6)$

41. $x + a = 0$ 42. $y^2 = a(x-3a)$

45. $y^2 = 2a(x-4a)$ 46. $\left(y - \dfrac{\beta}{2}\right)^2 = 2a\left(x - \alpha + \dfrac{\beta^2}{8a}\right)$

47. $y^2 = 2ax + \lambda x^2$

48. $(x^2+y^2)^2 - 24ax(x^2+y^2) + 16a^2(x^2-y^2) = 0$

52. $x - 2y - 5 = 0$

53. $x^2 + y^2 - 2xy - 6x - 10y + 25 = 0$

54. $4x^2 + 12xy + 9y^2 - 98x + 22y - 118 = 0$

55. 4 **56.** $12a$

57. $\left(\frac{1}{4}, \pm\frac{1}{2}\right)$ **58.** $y = 2(x-6)$

59. $x + 2y + 16 = 0$ **60.** $8\sqrt{5}$

61. $p\cos\alpha + a = 0$ **62.** $(-1, 0)$

63. $\tan^{-1}\left(\frac{\sqrt{5}}{2}\right)$ **64.** $x^2 + y^2 - 2ax - ay = 0$

65. (b) **66.** (a) **67.** (c) **68.** (b)

69. (a) **70.** (a) **71.** (d) **72.** (b)

73. (b)

□

Chapter Test

Time: 120 minutes

1. Fill in the blanks of the following incomplete statements so that the statements become true.

(a) The focus of a parabola is (0, 2) and the ends of its double ordinate through the focus are (–1, 2) and (1, 2). Then the equation of the parabola is ______ .

(b) The general equation to a system of parallel chords of the parabola $y^2 = 4x$ is $y = 2x + k$. The equation of the corresponding diameter is ______ .

2. In the following, each question has four options of which one or more options are correct. Indicate the correct option(s):

(a) If the normals to the parabola $y^2 = 8ax$ at the points (x_1, y_1), (x_2, y_2) and (x_3, y_3) are concurrent then

A. $x_1 + x_2 + x_3 = 0$ B. $y_1 + y_2 + y_3 = 0$

C. $x_1y_1 + x_2y_2 + x_3y_3 = 0$ D. none of these

(b) S is the focus of a parabola. PQ is any focal chord of the parabola. Then the semilatus rectum of the parabola is

A. AM of PS and QS B. GM of PS and QS

C. HM of PS and QS D. none of these

(c) P is the point 't' on the parabola $y^2 = 4ax$ and PQ is a focal chord. PT is the tangent at P and QN is the normal at Q. If the angle between PT and QN be α and the distance between PT and QN be d then

A. $0° < \alpha < 90°$ B. $\alpha = 0°$

C. $d = 0$ D. $d = \dfrac{a(1+t^2)^{3/2}}{t^2}$

3. Find the equation of the normal to the parabola $y^2 = 4x$ at the point (4, 4). Also find the point on this normal from which the other two normals drawn to the parabola will be at right angles.

4. Find the equation of the parabola whose axis is along x-axis and which touches the pair of lines $x^2 - y^2 - 2x + 1 = 0$, focus being at (4, 0).

5. What is the radius of the circle whose centre is (–4, 0) and which cuts the parabola $y^2 = 8x$ at A and B such that the common chord AB subtends a right angle at the vertex of the parabola?

6. Find the number of points with integral coordinates $(2a, a - 1)$ that fall in the interior of the larger segment of the circle $x^2 + y^2 = 25$ cut off by the parabola $x^2 + 4y = 0$. Also find their coordinates.

7. Prove that the point (α, β) which satisfies the equation $\frac{1}{2}\sqrt{x} + \frac{1}{3}\sqrt{y} = 1$ lies on that part of a parabola touching the positive x and y axes at distances 4 and 9 respectively from the origin which falls in the rectangle bounded by the lines $x = 0$, $x = 4$, $y = 0$, $y = 9$.

8. The equation of a family of straight lines is $y + 2at = t(x - at^2)$ where t is a parameter. Prove that the locus of the points of intersection of pairs of perpendicular lines of the family is a parabola. Find the vertex, focus and the latus rectum of the parabola.

9. Prove that the foot of any perpendicular from the point $(0, -c)$, $c > 0$, to any normal to the parabola $x^2 = 4ay$ lies on the curve whose equation is $x^4 = (y + c)\{x^2(2a - y) + a(y + c)^2\}$.

10. S is the focus and V is the vertex of the parabola $y^2 = 4ax$. P is any point on the parabola. Prove that the circle drawn on SP as diameter touches the tangent at V to the parabola.

11. Find the equation of the common tangent to the parabolas $y^2 = 32x$ and $x^2 = 108y$.

12. P, Q, R are three points on the parabola $y^2 = 4ax$ such that the ordinates of the points are in AP. If ar(ΔPQR) $= 2a^2$, prove that the centroid of the ΔPQR always lies on a parabola having the same latus rectum.

13. Let P be a fixed point. Any line through P cuts a parabola at Q and R. If S is a point on the line PQR such that PS is the HM between PQ and PR then prove that the locus of S is a straight line whose slope is independent of the abscissa of the point P.

14. Consider the two parabolas $C_1 : x^2 + 1 = y$ and $C_2 : y^2 + 1 = x$. Let P and Q be any two points on the parabolas C_1 and C_2 respectively. Let P′ and Q′ be the reflections of P and Q respectively with respect to the line $y = x$. Prove that P′ lies on C_2, Q′ lies on C_1 and $PQ \geq \min\{PP', QQ'\}$. Also determine the points P. Q such that PQ is the minimum.

15. P, Q and R be three feet of normals to the parabola $y^2 = x + 4y - 5$ drawn from any point. Prove that the circumcircle of the Δ PQR always passes through a fixed point. Also prove that the fixed point is the vertex of the parabola.

Answers

1. (a) $x^2 = 2y - 3$, $-2y + 5$ (b) $y = 1$

2. (a) B (b) C (c) B, D

3. $y + 2x = 12$, $(7, -2)$

4. $y^2 = 6x - 15$

5. $4\sqrt{13}$

6. two; $(2, 0)$, $(4, 1)$

8. $(3a, 0)$, $\left(\frac{13a}{4}, 0\right)$ and a respectively

11. $2x + 3y + 36 = 0$

14. $Q = \left(\frac{5}{4}, \frac{1}{2}\right)$, $P = \left(\frac{1}{2}, \frac{5}{4}\right)$

□

5. Ellipse and Hyperbola

Recap of Facts and Formulae

1. Standard equation of an ellipse

The standard equation of an ellipse is $\frac{x^2}{a^2}+\frac{y^2}{b^2}=1$ for which

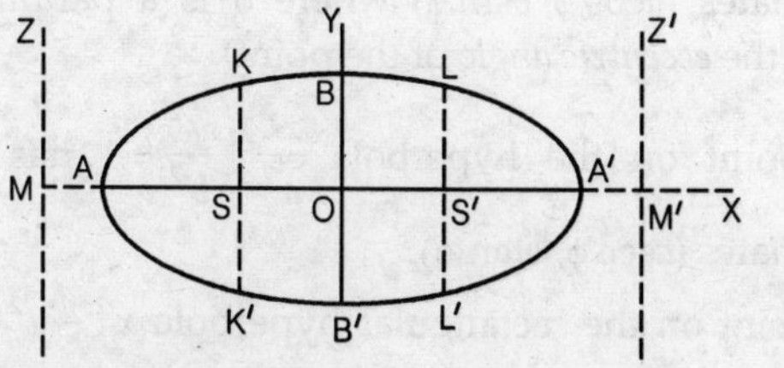

- centre O = (0, 0)
- foci S, S′ = $(\mp ae, 0)$
- vertices A, A′ = $(\mp a, 0)$
- the equation of the major axis AA′ is $y = 0$
  the length of the major axis = $AA' = 2a$
- the equation of the minor axis BB′ is $x = 0$
  the length of the minor axis = $BB' = 2b$
- the relation between semimajor axis a, semiminor axis b and eccentricity e is
  $b^2 = a^2(1-e^2), e < 1$
- the equations of directrices are $x \pm \frac{a}{e} = 0$
- latus rectum LL' (or KK') = $\frac{2b^2}{a}$.

2. Standard equation of a hyperbola

The standard equation of a hyperbola is $\frac{x^2}{a^2}-\frac{y^2}{b^2}=1$

for which

- centre O = (0, 0)
- foci S, S′ = $(\pm ae, 0)$
- vertices A, A′ = $(\pm a, 0)$
- the equation of the transverse axis A′A is $y = 0$
  the length of the transverse axis = $A'A = 2a$
- the equation of the conjugate axis B′B is $x = 0$

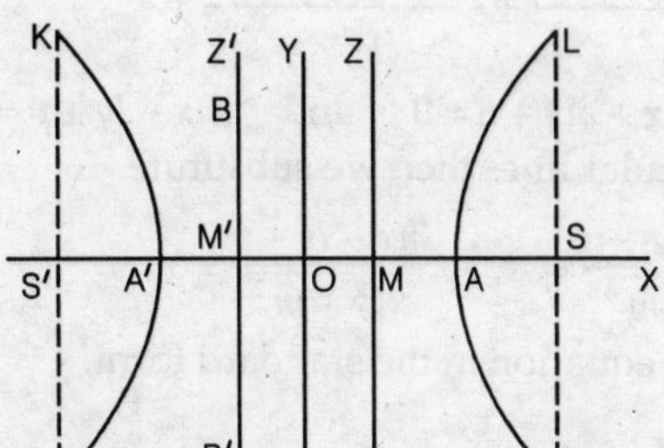

- the length of the conjugate axis = $BB' = 2b$
- the relation between semitransverse axis a, semiconjugate axis b and eccentricity e is
  $b^2 = a^2(e^2-1), e > 1$
- the equation of directrices are $x \pm \frac{a}{e} = 0$
- latus rectum LL' (or KK') = $\frac{2b^2}{a}$.

Note Comparing the results for ellipses and hyperbolas we find that for coordinates, lengths or equations results are the same, only difference being in the relation between a, b, e ($-b^2$ takes place of b^2).

3. Equation of ellipse/hyperbola when one directrix and the corresponding focus are given

If a directrix has the equation $ax+by+c=0$ and the corresponding focus is (α, β) then the equation of the ellipse/hyperbola is

$$(x-\alpha)^2+(y-\beta)^2 = e^2\cdot\left(\frac{ax+by+c}{\sqrt{a^2+b^2}}\right)^2,$$

e being given eccentricity of the ellipse/hyperbola.

4. Transformation into standard form

- If the equation of the curve is $\frac{(x-\alpha)^2}{a^2} \pm \frac{(y-\beta)^2}{b^2} = 1$
  then by taking the equations of transformation
  $x = \alpha + X, y = \beta + Y$

the equation changes in the standard form

$$\frac{X^2}{a^2} \pm \frac{Y^2}{b^2} = 1$$

- In case of the ellipse $\frac{x^2}{a^2} + \frac{y^2}{b^2} = 1,\ a > b.$

  If the equation of the ellipse is

  $\frac{(x-\alpha)^2}{b^2} + \frac{(y-\beta)^2}{a^2} = 1$ then we substitute

  $y = \beta + X,\ x = \alpha + Y$

- If the equation of the curve is

  $$\frac{(lx + my + n)^2}{a^2} \pm \frac{(mx - ly + p)^2}{b^2} = 1$$

  where $lx + my + n = 0$ and $mx - ly + p = 0$ are perpendicular lines then we substitute

  $$\frac{lx + my + n}{\sqrt{l^2 + m^2}} = X, \quad \frac{mx - ly + p}{\sqrt{l^2 + m^2}} = Y$$

  to put the equation in the standard form.

5. Location of a point

(a) If the ellipse is $S \equiv \frac{x^2}{a} + \frac{y^2}{b^2} - 1 = 0$ and $P = (x_1, y_1)$ then

- P is in the interior of the ellipse if

  $S(x_1, y_1)$, i.e., $\frac{x_1^2}{a^2} + \frac{y_1^2}{b^2} - 1 < 0$
- P is on the ellipse if $S(x_1, y_1) = 0$
- P is in the exterior of the ellipse if $S(x_1, y_1) > 0$.

(b) If the hyperbola is $S \equiv \frac{x^2}{a^2} - \frac{y^2}{b^2} - 1 = 0$ and $P = (x_1, y_1)$ then

- P is in the interior of the hyperbola if $S(x_1, y_1) < 0$
- P is on the hyperbola if $S(x_1, y_1) = 0$
- P is in the exterior of the hyperbola if $S(x_1, y_1) > 0$.

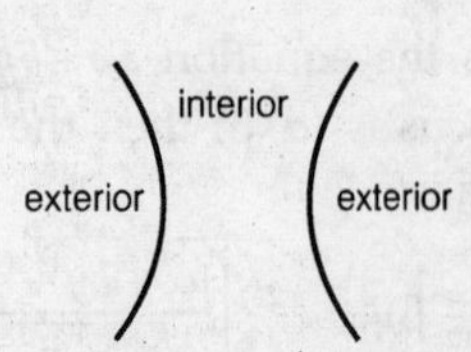

6. Condition for the general equation of the second degree in x, y to represent an ellipse/hyperbola

The equation $ax^2 + 2hxy + by^2 + 2gx + 2fy + c = 0$ represents

- an ellipse if $h^2 - ab < 0,\ \Delta \neq 0$ and it is not a circle
- a hyperbola if $h^2 - ab > 0,\ \Delta \neq 0$
- a *rectangular* (or *equilateral*) hyperbola {in a rectangular hyperbola, transverse axis = conjugate axis} if $h^2 - ab > 0$ and $a + b = 0,\ \Delta \neq 0$.

7. Standard equation of a rectangular hyperbola

The standard equation of a rectangular hyperbola is

- $x^2 - y^2 = a^2$ whose eccentricity $e = \sqrt{2}$, the transverse axis and conjugate axis being the x and y axes respectively.
- $xy = c^2$ when the bisectors of the angles between the transverse and conjugate axes are taken as x and y axes.

8. Coordinates of any point on an ellipse/hyperbola

- Any point on the ellipse $\frac{x^2}{a^2} + \frac{y^2}{b^2} = 1$ has the coordinates $(a\cos\phi, b\sin\phi)$ where ϕ is a parameter (called the *eccentric angle* of the point).
- Any point on the hyperbola $\frac{x^2}{a^2} - \frac{y^2}{b^2} = 1$ has the coordinates $(a\sec\phi, b\tan\phi)$.
- Any point on the rectangular hyperbola $x^2 - y^2 = a^2$ has the coordinates $(a\sec\phi, a\tan\phi)$.
- Any point on the rectangular hyperbola $xy = c^2$ has the coordinates $\left(ct, \frac{c}{t}\right)$

9. Equation of the chord joining two points of an ellipse

- The equation of the chord of the ellipse $\frac{x^2}{a^2} + \frac{y^2}{b^2} = 1$ joining the points 'ϕ_1' and 'ϕ_2' is

  $$\frac{x}{a}\cos\frac{\phi_1 + \phi_2}{2} + \frac{y}{b}\sin\frac{\phi_1 + \phi_2}{2} = \cos\frac{\phi_1 - \phi_2}{2}.$$

10. Equation of tangent and condition of tangency

- The equation of the tangent at (x_1, y_1) to the curve

  $\frac{x^2}{a^2} \pm \frac{y^2}{b^2} = 1$ is $\frac{xx_1}{a^2} \pm \frac{yy_1}{b^2} = 1.$
- The equation of the tangent at $(a\cos\phi, b\sin\phi)$ to the ellipse $\frac{x^2}{a^2} + \frac{y^2}{b^2} = 1$ is

  $$\frac{x\cos\phi}{a} + \frac{y\sin\phi}{b} = 1$$
- The equation of the tangent at $(a\sec\phi, b\tan\phi)$ to the hyperbola $\frac{x^2}{a^2} - \frac{y^2}{b^2} = 1$ is

  $$\frac{x\sec\phi}{a} - \frac{y\tan\phi}{b} = 1$$

- The line $y = mx + c$ touches the ellipse $\frac{x^2}{a^2} + \frac{y^2}{b^2} = 1$ if $c^2 = a^2m^2 + b^2$.

  So, the line $y = mx \pm \sqrt{a^2m^2 + b^2}$ touches the ellipse for all real m.

- The line $y = mx + c$ touches the hyperbola $\frac{x^2}{a^2} - \frac{y^2}{b^2} = 1$ if $c^2 = a^2m^2 - b^2$.

  So, the line $y = mx \pm \sqrt{a^2m^2 - b^2}$ touches the hyperbola for all real m.

11. Equation of normal

The equation of the normal at (x_1, y_1) to a curve is

$$y - y_1 = \frac{-1}{\left(\frac{dy}{dx}\right)_{x_1, y_1}} \cdot (x - x_1).$$

- For the ellipse $\frac{x^2}{a^2} + \frac{y^2}{b^2} = 1$,

  the equation of the normal at (x_1, y_1) is

  $$\frac{x - x_1}{x_1/a^2} = \frac{y - y_1}{y_1/b^2}$$

  and that at $(a\cos\phi, b\sin\phi)$ is

  $$ax\sec\phi - by\,\mathrm{cosec}\,\phi = a^2 - b^2.$$

- For the hyperbola $\frac{x^2}{a^2} - \frac{y^2}{b^2} = 1$,

  the equation of the normal at (x_1, y_1) is

  $$\frac{x - x_1}{x_1/a^2} = \frac{y - y_1}{-y_1/b^2}$$

  and that at $(a\sec\phi, b\tan\phi)$ is

  $$ax\cos\phi + by\cot\phi = a^2 + b^2.$$

12. Chord of contact, polar line, pole

- The chord of contact of tangents from (x_1, y_1) to the curve

  $\frac{x^2}{a^2} \pm \frac{y^2}{b^2} = 1$ is $\frac{xx_1}{a^2} \pm \frac{yy_1}{b^2} = 1.$

- The polar of the point (x_1, y_1) with respect to the curve

  $\frac{x^2}{a^2} \pm \frac{y^2}{b^2} = 1$ is $\frac{xx_1}{a^2} \pm \frac{yy_1}{b^2} = 1.$

- The pole of the line $L = 0$ with respect to the ellipse or hyperbola $S = 0$ is the point (x_1, y_1) whose polar is the line $L = 0$.

13. Equation of chords and the pair of tangents from a point

- The equation of the chord of $S \equiv \frac{x^2}{a^2} \pm \frac{y^2}{b^2} - 1 = 0$ whose middle point is (x_1, y_1), is $T = S_1$ where

  $$T = \frac{xx_1}{a^2} \pm \frac{yy_1}{b^2} - 1, \quad S_1 \equiv \frac{x_1^2}{a^2} \pm \frac{y_1^2}{b^2} - 1$$

- The equation of the pair of tangents from (x_1, y_1) to the curve $\frac{x^2}{a^2} \pm \frac{y^2}{b^2} = 1$ is

  $$S \cdot S_1 = T^2 \text{ where } S \equiv \frac{x^2}{a^2} \pm \frac{y^2}{b^2} - 1,$$

  $$S_1 \equiv \frac{x_1^2}{a^2} \pm \frac{y_1^2}{b^2} - 1, \quad T = \frac{xx_1}{a^2} \pm \frac{yy_1}{b^2} - 1.$$

14. Some properties of ellipse

- If S, S′ are foci, major axis = $2a$ then for any point P on the ellipse, $SP + S'P = 2a$.
- A chord PP′ passing through the centre O is a *diameter* of the ellipse. Two diameters PP′ and QQ′ are *conjugate diameters* if chords parallel to PP′ are bisected by QQ′ and chords parallel to QQ′ are bisected by PP′.

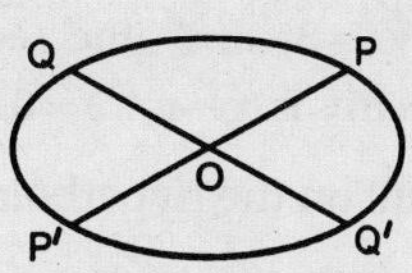

- If the eccentric angle of P is ϕ then the other end P′ of the diameter PP′ will have the eccentric angle $\pi + \phi$ and the ends of the conjugate diameter have the eccentric angles $\phi \pm \frac{\pi}{2}$.
- Two diameters $y = m_1x, y = m_2x$ are conjugate diameters of $\frac{x^2}{a^2} + \frac{y^2}{b^2} = 1$ if $m_1m_2 = -\frac{b^2}{a^2}$.

15. Some properties of hyperbola

- If S, S′ are foci, transverse axis = $2a$ then for any point P on the hyperbola, $|SP - S'P| = 2a$.
- The hyperbola $\frac{x^2}{a^2} - \frac{y^2}{b^2} = -1$ is the conjugate hyperbola of the hyperbola $\frac{x^2}{a^2} - \frac{y^2}{b^2} = 1$.
- The asymptotes of the hyperbola $\frac{x^2}{a^2} - \frac{y^2}{b^2} = 1$ are the lines $\frac{x^2}{a^2} - \frac{y^2}{b^2} = 0$.

Selected Solved Examples

1. Find the equation of the ellipse whose foci are (4, 0) and (–4, 0) and whose eccentricity is 1/3.

Clearly, the foci are on the x-axis and the centre is (0, 0), being midway between the foci. So the equation will be in the standard form.

Let it be $\frac{x^2}{a^2}+\frac{y^2}{b^2}=1$.

Foci are $(\pm ae, 0)$. Here they are $(\pm 4, 0)$.

$\therefore \quad ae = 4.$

But $e=\frac{1}{3}$

$\therefore \quad a\cdot\frac{1}{3}=4$, i.e., $a = 12$.

Again, $b^2 = a^2(1-e^2)$

$\Rightarrow \quad b^2 = 12^2\cdot\left(1-\frac{1}{3^2}\right)=12^2\cdot\frac{8}{3^2}=128$

$\therefore$ the equation of the ellipse is $\frac{x^2}{144}+\frac{y^2}{128}=1$.

2. Find the equation of the hyperbola whose one focus is (–1, 1), eccentricity = 3 and the equation of the corresponding directrix is $x-y+3=0$.

Let P(x, y) be any point on the hyperbola.

Then by focus–directrix property,

$$\frac{\text{distance of P from the focus}}{\text{distance of P from the directrix}}=e=3$$

$$\therefore \quad \frac{\sqrt{(x+1)^2+(y-1)^2}}{\frac{x-y+3}{\sqrt{1^2+(-1)^2}}}=3$$

or $\quad (x+1)^2+(y-1)^2=9\cdot\left(\frac{x-y+3}{\sqrt{2}}\right)^2$

or $\quad 2[x^2+2x+1+y^2-2y+1]=9(x-y+3)^2$

or $\quad 2(x^2+y^2+2x-2y+2)$
$\qquad = 9(x^2+y^2+9-2xy+6x-6y)$

or $\quad 7x^2-18xy+7y^2+50x-50y+77=0.$

3. Find the latus rectum, eccentricity, foci, centre and the lengths of axes of the ellipse $9x^2+5y^2-30y=0$.

Here, $\quad 9x^2+5(y^2-6y+9)=45$

or $\quad 9x^2+5(y-3)^2=45$

$\therefore \quad \frac{x^2}{5}+\frac{(y-3)^2}{9}=1.$

Putting $x = Y$ and $y-3=X$, the equation becomes

$$\frac{X^2}{9}+\frac{Y^2}{5}=1$$

$\therefore \quad a^2=9, \quad b^2=5$

$\because \quad b^2=a^2(1-e^2); \quad \therefore \quad 5=9(1-e^2)$

or $\quad 1-e^2=\frac{5}{9}$,

or $\quad e^2=1-\frac{5}{9}=\frac{4}{9}; \quad \therefore \quad e=\frac{2}{3}$;

latus rectum $=\frac{2b^2}{a}=\frac{2\cdot 5}{3}=\frac{10}{3}$

the length of major axis $= 2a = 2\times 3 = 6$

the length of minor axis $= 2b = 2\times\sqrt{5}=2\sqrt{5}$

centre $=(0, 0)_{X,Y}$

$=(0, 3)$ {$\because$ when $X = 0$, $y-3 = X$ gives $y = 3$ and when $Y = 0$, $x = Y$ gives $x = 0$}

foci $=(\pm ae, 0)_{X,Y}=\left(\pm 3\cdot\frac{2}{3}, 0\right)_{X,Y}=(\pm 2, 0)_{X,Y}$

$=(0, 3\pm 2)=(0, 5), (0, 1)$

{$\because$ when $X=\pm 2$, $y-3=X$ gives $y=3\pm 2$ and when $Y = 0$, $x = Y$ gives $x = 0$}

4. Find the centre, eccentricity, foci and directrices of the hyperbola

$$\mathbf{16x^2-9y^2+32x+36y-164=0.}$$

Here, $16x^2+32x+16-(9y^2-36y+36)-144=0$

or $\quad 16(x+1)^2-9(y-2)^2=144$

$\therefore \quad \frac{(x+1)^2}{9}-\frac{(y-2)^2}{16}=1.$

Putting $x+1=X$ and $y-2=Y$, the equation becomes

$$\frac{X^2}{9}-\frac{Y^2}{16}=1$$

which is in the standard form.

Here, $a^2=9$ and $b^2=16$;

$\because \quad b^2=a^2(e^2-1)$, we get $16=9(e^2-1)$

$\therefore \quad e^2-1=\frac{16}{9}$

$\therefore \quad e^2=\frac{25}{9}$, i.e., $e=\frac{5}{3}$

Now, centre $=(0, 0)_{X,Y}=(-1, 2)$

{$\because$ when $X = 0$, $x+1 = X$ gives $x = -1$ and when $Y = 0$, $y-2 = Y$ gives $y = 2$}

foci $= (\pm ae, 0)_{X,Y} = \left(\pm 3 \cdot \frac{5}{3}, 0\right)_{X,Y} = (\pm 5, 0)_{X,Y}$

$= (-1 \pm 5, 2) = (4, 2), (-6, 2).$

Directrices in X, Y coordinates have the equations

$$X \pm \frac{a}{e} = 0$$

or $\quad x + 1 \pm \frac{3}{5/3} = 0$

i.e., $\quad x + 1 \pm \frac{9}{5} = 0$

$\therefore \quad x = -\frac{14}{5}$ and $x = \frac{4}{5}$.

5. Show that the following equation represents an ellipse. Also find its centre and eccentricity.

$$\mathbf{4(x - 2y + 1)^2 + 9(2x + y + 2)^2 = 25.}$$

On simplification the equation becomes

$$4(x^2 + 4y^2 + 1 - 4xy + 2x - 4y)$$
$$+ 9(4x^2 + y^2 + 4 + 4xy + 8x + 4y) = 25$$

or $\quad 40x^2 + 20xy + 25y^2 + 80x + 20y + 15 = 0$

or $\quad 8x^2 + 4xy + 5y^2 + 16x + 4y + 3 = 0.$

Comparing with the general equation of the second degree

$$h^2 - ab = 2^2 - 8.5 = 4 - 40 < 0$$

$$\Delta = 8 \cdot 5 \cdot 3 + 2 \cdot 2 \cdot 8 \cdot 2 - 8 \cdot 2^2 - 5 \cdot 8^2 - 3 \cdot 2^2$$
$$= 120 + 64 - 32 - 320 - 12 \neq 0, \ h \neq 0$$

$\therefore$ the curve is an ellipse.

For the equation $4(x - 2y + 1)^2 + 9(2x + y + 2)^2 = 25$, we find that $x - 2y + 1 = 0$ and $2x + y + 2 = 0$ are mutually perpendicular lines.

$\therefore$ substituting $\frac{x - 2y + 1}{\sqrt{1^2 + (-2)^2}} = X$ ... (1)

and $\frac{2x + y + 2}{\sqrt{2^2 + 1^2}} = Y$ ... (2)

the equation changes to $4(\sqrt{5}X)^2 + 9(\sqrt{5}Y)^2 = 25$

or $\quad 4X^2 + 9Y^2 = 5$

or $\quad \frac{X^2}{5/4} + \frac{Y^2}{5/9} = 1$

which is in the standard form of equation of an ellipse.

$\therefore \quad a^2 = \frac{5}{4}$ and $b^2 = \frac{5}{9}$;

but $b^2 = a^2(1 - e^2)$

$\therefore \quad \frac{5}{9} = \frac{5}{4}(1 - e^2)$

or $\quad \frac{4}{9} = 1 - e^2$

$\therefore \quad e^2 = 1 - \frac{4}{9} = \frac{5}{9}$;

$\therefore \quad e = \frac{\sqrt{5}}{3}$.

Now, centre $= (0, 0)_{X,Y}$;

when $X = 0, Y = 0$ we have from (1), (2)

$$x - 2y + 1 = 0 \text{ and } 2x + y + 2 = 0.$$

Solving these, $5x + 5 = 0$

$\therefore \quad x = -1$; so $y = 0$

$\therefore \quad$ centre $= (-1, 0)$.

6. Prove that the sum of the squares of the reciprocals of two perpendicular diameters of an ellipse is constant.

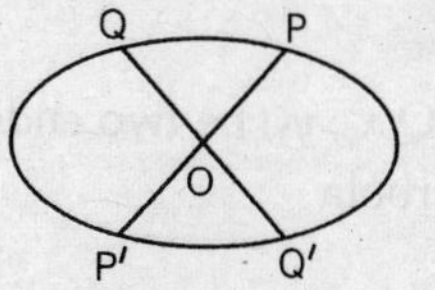

Let the ellipse be $\frac{x^2}{a^2} + \frac{y^2}{b^2} = 1$ ... (1)

and PP′, QQ′ be two perpendicular diameters.

Let the equation of PP′ be $y = mx$ ... (2)

Solving (1) and (2),

$$x^2\left(\frac{1}{a^2} + \frac{m^2}{b^2}\right) = 1$$

$\therefore \quad x = \pm \frac{ab}{\sqrt{a^2m^2 + b^2}}$

$\therefore \quad y = \frac{\pm mab}{\sqrt{a^2m^2 + b^2}}$

$\therefore \quad PP'^2 = (2 \times OP)^2 = 4\left\{\frac{a^2b^2}{a^2m^2 + b^2} + \frac{m^2a^2b^2}{a^2m^2 + b^2}\right\}$

$$= \frac{4a^2b^2}{a^2m^2 + b^2}(1 + m^2).$$

As QQ′ ⊥ PP′ and QQ′ passes through the centre O(0, 0), the equation of QQ′ will be

$$y = \frac{-1}{m}x \quad \text{... (3)}$$

Solving (1), (3) and proceeding as above,

$$QQ'^2 = \frac{4a^2b^2}{a^2\left(-\frac{1}{m}\right)^2 + b^2}\left\{1 + \left(-\frac{1}{m}\right)^2\right\}$$

$$= \frac{4a^2b^2m^2}{a^2+b^2m^2} \cdot \frac{1+m^2}{m^2}$$

$$= \frac{4a^2b^2}{a^2+b^2m^2}(1+m^2)$$

$$\therefore \quad \frac{1}{PP'^2} + \frac{1}{QQ'^2} = \frac{a^2m^2+b^2}{4a^2b^2(1+m^2)} + \frac{a^2+b^2m^2}{4a^2b^2(1+m^2)}$$

$$= \frac{1}{4a^2b^2(1+m^2)}\{a^2m^2+b^2+a^2+b^2m^2\}$$

$$= \frac{(a^2+b^2)(1+m^2)}{4a^2b^2(1+m^2)}$$

$$= \frac{a^2+b^2}{4a^2b^2} = \text{constant.}$$

7. Prove that the angle subtended by any chord of a rectangular hyperbola at the centre is the supplement of the angle between the tangents at the ends of the chord.

Let $P(x_1, y_1)$ and $Q(x_2, y_2)$ be two ends of a chord of the rectangular hyperbola

$$x^2 - y^2 = 1 \qquad \dots (1)$$

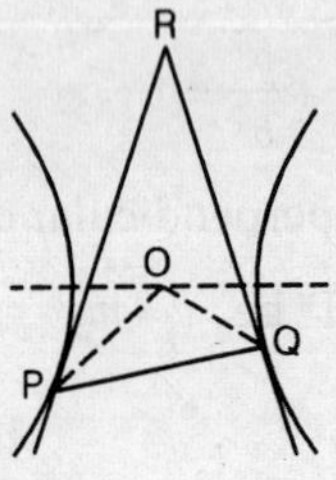

Now, $'m'$ of $OP = \dfrac{y_1}{x_1}$

$'m'$ of $OQ = \dfrac{y_2}{x_2}$

$$\therefore \quad \tan\theta = \frac{\dfrac{y_1}{x_1} - \dfrac{y_2}{x_2}}{1 + \dfrac{y_1}{x_1}\cdot\dfrac{y_2}{x_2}} = \frac{x_2y_1 - x_1y_2}{x_1x_2 + y_1y_2}, \text{ where } \angle POQ = \theta.$$

The equations of tangents at P and Q are

$$xx_1 - yy_1 = 1 \text{ and } xx_2 - yy_2 = 1.$$

Their slopes are $\dfrac{x_1}{y_1}$ and $\dfrac{x_2}{y_2}$.

$$\therefore \quad \tan\phi = \frac{\dfrac{x_1}{y_1} - \dfrac{x_2}{y_2}}{1 + \dfrac{x_1}{y_1}\cdot\dfrac{x_2}{y_2}} = \frac{x_1y_2 - x_2y_1}{y_1y_2 + x_1x_2}$$

$\therefore$ $\tan\theta$ and $\tan\phi$ are equal in magnitude but opposite in sign

$$\therefore \quad \tan\theta = -\tan\phi = \tan(\pi - \phi)$$

$\therefore$ $\theta + \phi = \pi$. Hence, the problem.

8. Prove that $(0, 3+\sqrt{5})$ is a point on the ellipse whose foci are $(2, 3)$, $(-2, 3)$ and semimajor axis is 3. Also find the equation of the tangent at the point $(0, 3+\sqrt{5})$ to the ellipse.

Let $P = (0, 3+\sqrt{5})$ and foci $S = (2, 3)$, $S' = (-2, 3)$.

Here major axis $= 2 \times 3 = 6$.

P is a point on the ellipse if $PS + PS' = 6 (= 2a)$

Now, $PS = \sqrt{(-2)^2 + (\sqrt{5})^2} = \sqrt{9} = 3$, $PS' = \sqrt{2^2 + (\sqrt{5})^2} = 3$

$\therefore$ $PS + PS' = 6$.

Hence $(0, 3+\sqrt{5})$ is a point on the ellipse.

At first we have to find the equation of the ellipse to find the tangent. Let any point Q on the ellipse be (x, y).

$\because$ $QS + QS' = 2a$, we get

$$\sqrt{(x-2)^2 + (y-3)^2} + \sqrt{(x+2)^2 + (y-3)^2} = 6 \quad \dots (1)$$

Multiplying both sides by the conjugate of the left-hand side,

$$\{(x-2)^2 + (y-3)^2\} - \{(x+2)^2 + (y-3)^2\}$$
$$= 6\{\sqrt{(x-2)^2 + (y-3)^2} - \sqrt{(x+2)^2 + (y-3)^2}\}$$

$$\therefore \quad \sqrt{(x-2)^2 + (y-3)^2} - \sqrt{(x+2)^2 + (y-3)^2} = \frac{-8x}{6} \quad \dots (2)$$

$$\therefore \quad (1) + (2) \Rightarrow 2\sqrt{(x-2)^2 + (y-3)^2} = 6 - \frac{8x}{6}$$

or $\quad (x-2)^2 + (y-3)^2 = \left(3 - \dfrac{2x}{3}\right)^2$

or $\quad x^2 - 4x + 4 + y^2 - 6y + 9 = 9 - 4x + \dfrac{4x^2}{9}$

or $\quad \dfrac{5x^2}{9} + y^2 - 6y + 4 = 0$

or $\quad 5x^2 + 9y^2 - 54y + 36 = 0 \qquad \dots (1)$

The equation of the tangent at (x_1, y_1) to the curve

$$ax^2 + 2hxy + by^2 + 2gx + 2fy + c = 0$$

is $\quad axx_1 + h(xy_1 + x_1y) + byy_1$

$$+ g(x + x_1) + f(y + y_1) + c = 0$$

$\therefore$ the equation of the tangent to (1) at $(0, 3+\sqrt{5})$ is

$$5x \cdot 0 + 9y \cdot (3+\sqrt{5}) - 27(y + 3 + \sqrt{5}) + 36 = 0$$

or $\quad 9\sqrt{5}y - 27(3+\sqrt{5}) + 36 = 0$

or $\quad \sqrt{5}y - 3(3+\sqrt{5}) + 4 = 0$

$\therefore$ $\sqrt{5}y = 5 + 3\sqrt{5}$; $\quad\therefore\quad y = 3 + \sqrt{5}$.

9. Let d be the perpendicular distance from the centre of the ellipse $\frac{x^2}{a^2}+\frac{y^2}{b^2}=1$ to the tangent drawn at a point P on the ellipse. If F_1 and F_2 are the two foci of the ellipse then show that

$$(PF_1 - PF_2)^2 = 4a^2\left(1-\frac{b^2}{d^2}\right).$$

Let $P = (a\cos\phi, b\sin\phi)$.

The equation of the tangent at P is

$$\frac{x\cos\phi}{a}+\frac{y\sin\phi}{b}=1 \qquad \dots(1)$$

$$\therefore \quad d=\frac{1}{\sqrt{\dfrac{\cos^2\phi}{a^2}+\dfrac{\sin^2\phi}{b^2}}} \qquad \dots(2)$$

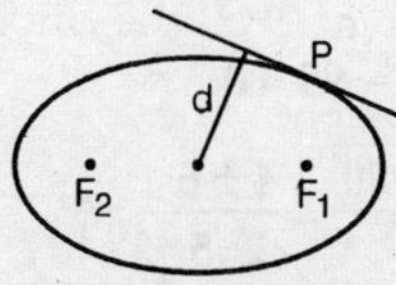

Here $F_1 = (ae, 0)$ and $F_2 = (-ae, 0)$

$$\therefore \quad PF_1 - PF_2 = \sqrt{(a\cos\phi - ae)^2 + b^2\sin^2\phi} - \sqrt{(a\cos\phi + ae)^2 + b^2\sin^2\phi}$$

$$= \sqrt{a^2\cos^2\phi - 2a^2e\cos\phi + a^2e^2 + b^2\sin^2\phi} - \sqrt{a^2\cos^2\phi + 2a^2e\cos\phi + a^2e^2 + b^2\sin^2\phi}$$

$$= \sqrt{a^2\cos^2\phi - 2a^2e\cos\phi + a^2 - b^2 + b^2\sin^2\phi} - \sqrt{a^2\cos^2\phi + 2a^2e\cos\phi + a^2 - b^2 + b^2\sin^2\phi}$$

$$\{\because\ b^2 = a^2(1-e^2)\}$$

$$= \sqrt{(a^2-b^2)\cos^2\phi - 2a^2e\cos\phi + a^2} - \sqrt{(a^2-b^2)\cos^2\phi + 2a^2e\cos\phi + a^2}$$

$$= \sqrt{a^2e^2\cos^2\phi - 2a^2e\cos\phi + a^2} - \sqrt{a^2e^2\cos^2\phi + 2a^2e\cos\phi + a^2}$$

$$= a|e\cos\phi - 1| - a|e\cos\phi + 1|$$

$$= a(1 - e\cos\phi) - a(1 + e\cos\phi) \quad \{\because\ e\cos\phi \le 1\}$$

$$= -2ae\cos\phi$$

$$\therefore \quad (PF_1 - PF_2)^2 = 4a^2e^2\cos^2\phi \qquad \dots(3)$$

Also, $1-\frac{b^2}{d^2} = 1 - b^2\cdot\left(\frac{\cos^2\phi}{a^2}+\frac{\sin^2\phi}{b^2}\right)$, from (2)

$$= 1 - \frac{b^2}{a^2}\cos^2\phi - \sin^2\phi$$

$$= \cos^2\phi - \frac{b^2}{a^2}\cos^2\phi$$

$$= \frac{a^2-b^2}{a^2}\cos^2\phi = \frac{a^2e^2}{a^2}\cos^2\phi$$

$$= e^2\cos^2\phi \qquad \dots(4)$$

$\therefore$ (3) and (4) $\Rightarrow$

$$(PF_1 - PF_2)^2 = 4a^2\left(1-\frac{b^2}{d^2}\right).$$

10. An ellipse has the eccentricity $\frac{1}{2}$ and one focus is $F\left(\frac{1}{2}, 1\right)$. The corresponding directrix is the common tangent, nearer to P, to the circle $x^2+y^2=1$ and the hyperbola $x^2-y^2=1$. Find the equation of the ellipse and its centre and axes.

Any point on the hyperbola $x^2 - y^2 = 1$ is $(\sec\theta, \tan\theta)$. The tangent at $(\sec\theta, \tan\theta)$ to $x^2 - y^2 = 1$ is

$$x\sec\theta - y\tan\theta = 1 \qquad \dots(1)$$

This will also be a tangent to $x^2 + y^2 = 1$ if

radius = length of perpendicular from (0, 0) to (1)

or $\quad 1 = \dfrac{1}{\sqrt{\sec^2\theta + \tan^2\theta}}$

or $\quad \sec^2\theta + \tan^2\theta = 1$

or $\quad 2\tan^2\theta = 0$

or $\quad \tan\theta = 0; \quad \therefore \quad \theta = 0, \pi$

$\therefore$ putting for θ in (1), the common tangents are

$\pm x = 1$, i.e., $x = 1$ and $x + 1 = 0$.

$x = 1$ is nearer to $F\left(\frac{1}{2}, 1\right)$ than $x + 1 = 0$.

$\therefore$ the ellipse has the following:

focus $= \left(\frac{1}{2}, 1\right)$,

corresponding directrix is $x - 1 = 0$

and $e = \frac{1}{2}$.

$\therefore$ by focus–directrix property, the equation of the ellipse is

$$\left(x-\frac{1}{2}\right)^2 + (y-1)^2 = \left(\frac{1}{2}\right)^2\cdot\left(\frac{x-1}{\sqrt{1}}\right)^2$$

or $\quad x^2 - x + \frac{1}{4} + y^2 - 2y + 1 = \frac{1}{4}(x^2 - 2x + 1)$

or $\quad 4x^2 + 4y^2 - 4x - 8y + 5 = x^2 - 2x + 1$

or $\quad 3x^2 + 4y^2 - 2x - 8y + 4 = 0$

or $\quad 3\left(x^2 - 2\cdot\frac{1}{3}x + \frac{1}{9}\right) + 4(y^2 - 2y + 1) = \frac{1}{3}$

or $\quad 3\left(x - \frac{1}{3}\right)^2 + 4(y-1)^2 = \frac{1}{3}$

or $\quad \dfrac{\left(x-\frac{1}{3}\right)^2}{\frac{1}{9}} + \dfrac{(y-1)^2}{\frac{1}{12}} = 1$

$\therefore\quad$ centre $= \left(\frac{1}{3}, 1\right)$, semimajor axis $= \frac{1}{3}$,

semiminor axis $= \dfrac{1}{2\sqrt{3}}$.

11. If the tangent drawn at a point $(t^2, 2t)$ on the parabola $y^2 = 4x$ is the same as the normal drawn at a point $(\sqrt{5}\cos\phi, 2\sin\phi)$ on the ellipse $4x^2 + 5y^2 = 20$, find the values of t and ϕ.

The equation of the tangent at $P(t^2, 2t)$ on the parabola $y^2 = 4x$ is

$$y\cdot 2t = 2(x + t^2)$$

or $\quad yt = x + t^2 \quad \ldots(1)$

The equation of the normal at $Q(\sqrt{5}\cos\phi, 2\sin\phi)$ on the ellipse $4x^2 + 5y^2 = 20$ is

$$y - 2\sin\phi = \frac{-1}{\left(\frac{dy}{dx}\right)_Q}\cdot(x - \sqrt{5}\cos\phi) \quad \ldots(2)$$

Differentiating $4x^2 + 5y^2 = 20$,

$$8x + 10y\frac{dy}{dx} = 0, \text{ i.e., } \frac{dy}{dx} = \frac{-4x}{5y}$$

$$\therefore\quad \left(\frac{dy}{dx}\right)_Q = \frac{-4\cdot\sqrt{5}\cos\phi}{5\cdot 2\sin\phi} = -\frac{2\cos\phi}{\sqrt{5}\sin\phi}$$

$\therefore\quad$ (2) gives

$$y - 2\sin\phi = \frac{\sqrt{5}\sin\phi}{2\cos\phi}(x - \sqrt{5}\cos\phi)$$

or $\quad 2y\cos\phi = \sqrt{5}x\sin\phi - \sin\phi\cdot\cos\phi \quad \ldots(3)$

From the question, (1) and (3) represent the same line. So,

$$\frac{t}{2\cos\phi} = \frac{1}{\sqrt{5}\sin\phi} = \frac{t^2}{-\sin\phi\cdot\cos\phi}$$

$$\Rightarrow\quad t = \frac{2}{\sqrt{5}}\frac{\cos\phi}{\sin\phi} = \frac{2}{\sqrt{5}}\cot\phi$$

and $\quad t^2 = -\dfrac{\cos\phi}{\sqrt{5}} \quad$ {clearly $\phi \neq 0, \pi$}

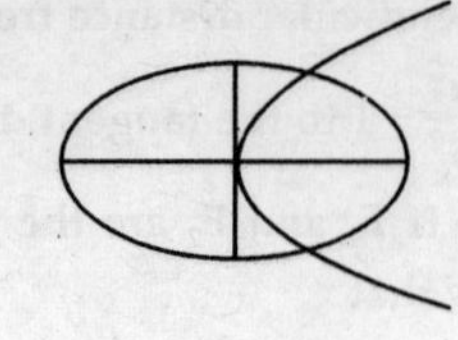

$$\therefore\quad \frac{4}{5}\cot^2\phi = -\frac{\cos\phi}{\sqrt{5}}$$

or $\quad 4\cot^2\phi + \sqrt{5}\cos\phi = 0$

or $\quad \cos\phi\left(\dfrac{4\cos\phi}{\sin^2\phi} + \sqrt{5}\right) = 0$

$\therefore\quad \cos\phi = 0$ or $4\cos\phi + \sqrt{5}\sin^2\phi = 0$.

If $\cos\phi = 0$ then $\phi = \frac{\pi}{2}, \frac{3\pi}{2}$.

If $4\cos\phi + \sqrt{5}\sin^2\phi = 0$

then $\sqrt{5}\cos^2\phi - 4\cos\phi - \sqrt{5} = 0$

or $\quad \cos\phi = \dfrac{4 \pm \sqrt{16+20}}{2\sqrt{5}} = \dfrac{4 \pm 6}{2\sqrt{5}}$

$$= \sqrt{5}, -\frac{1}{\sqrt{5}}$$

$$\therefore\quad \cos\phi = -\frac{1}{\sqrt{5}} \quad (\because \sqrt{5} > 1)$$

$$\therefore\quad \phi = \frac{\pi}{2}, \frac{3\pi}{2}, \pi - \cos^{-1}\frac{1}{\sqrt{5}}, \pi + \cos^{-1}\frac{1}{\sqrt{5}}$$

and $\quad t = \dfrac{2}{\sqrt{5}}\dfrac{\cos\phi}{\sin\phi}$

$$= \frac{2}{\sqrt{5}}\cdot\frac{0}{1}, \frac{2}{\sqrt{5}}\cdot\frac{0}{-1}, \frac{2}{\sqrt{5}}\cdot\frac{-\frac{1}{\sqrt{5}}}{\sqrt{1-\left(\frac{1}{\sqrt{5}}\right)^2}}$$

$$= 0, 0, \frac{2}{\sqrt{5}}\cdot\frac{-1}{\sqrt{5}}\cdot\frac{\pm\sqrt{5}}{2} = 0, \frac{1}{\sqrt{5}}, -\frac{1}{\sqrt{5}}$$

12. Prove that the greatest value of $\tan\theta$ is $\dfrac{a^2-b^2}{2ab}$ where θ is the angle between the diameter through any point on the ellipse $\dfrac{x^2}{a^2} + \dfrac{y^2}{b^2} = 1$ and the normal at that point.

Let the diameter pass through the point $P(a\cos\phi, b\sin\phi)$ on the ellipse.

$\therefore\quad$ the equation of the diameter is

$$y - 0 = \frac{b\sin\phi - 0}{a\cos\phi - 0}(x - 0)$$

or $\quad y = \left(\dfrac{b}{a}\tan\phi\right)x \quad \ldots(1)$

The normal to the ellipse at P($a\cos\phi$, $b\sin\phi$) is

$$\frac{x-a\cos\phi}{\frac{a\cos\phi}{a^2}}=\frac{y-b\sin\phi}{\frac{b\sin\phi}{b^2}}$$

or $\quad y-b\sin\phi=\frac{a}{b}\tan\phi(x-a\cos\phi)$ ... (2)

'm' of (1) $=\frac{b}{a}\tan\phi$

'm' of (2) $=\frac{a}{b}\tan\phi$

$$\therefore\quad \tan\theta=\frac{\frac{a}{b}\tan\phi-\frac{b}{a}\tan\phi}{1+\frac{a}{b}\tan\phi\cdot\frac{b}{a}\tan\phi}$$

or $\quad \tan\theta=\frac{\left(\frac{a}{b}-\frac{b}{a}\right)\tan\phi}{1+\tan^2\phi}$

or $\quad \tan\theta=\frac{a^2-b^2}{ab}\frac{\tan\phi}{\sec^2\phi}$

$$=\frac{a^2-b^2}{ab}\sin\phi\cos\phi$$

$$=\frac{a^2-b^2}{2ab}\sin 2\phi$$

whose greatest value $=\frac{a^2-b^2}{2ab}\cdot 1=\frac{a^2-b^2}{2ab}$

Hence, the problem.

13. Prove that the straight line $lx+my=n$ is a normal to the ellipse

$$\frac{x^2}{a^2}+\frac{y^2}{b^2}=1 \text{ if } \frac{a^2}{l^2}+\frac{b^2}{m^2}=\frac{(a^2-b^2)^2}{n^2}.$$

Let the line be a normal at (x_1, y_1).

The equation of the normal at (x_1, y_1) to the ellipse is

$$\frac{x-x_1}{\frac{x_1}{a^2}}=\frac{y-y_1}{\frac{y_1}{b^2}}$$

or $\quad \frac{y_1}{b^2}(x-x_1)=\frac{x_1}{a^2}(y-y_1)$

or $\quad \frac{y_1}{b^2}x-\frac{x_1}{a^2}y=\left(\frac{1}{b^2}-\frac{1}{a^2}\right)x_1y_1.$

This line and the line $lx+my=n$ are the same; so

$$\frac{l}{\frac{y_1}{b^2}}=\frac{m}{\frac{-x_1}{a^2}}=\frac{n}{\left(\frac{1}{b^2}-\frac{1}{a^2}\right)x_1y_1}$$

$$\therefore\quad \frac{n}{l}=\frac{\left(\frac{1}{b^2}-\frac{1}{a^2}\right)x_1y_1}{y_1/b^2}=\left(1-\frac{b^2}{a^2}\right)x_1$$

or $\quad \frac{a}{l}=\frac{a^2-b^2}{n}\cdot\frac{x_1}{a}.$

Similarly, $\quad \frac{b}{m}=-\frac{a^2-b^2}{n}\cdot\frac{y_1}{b}.$

$$\therefore\quad \frac{a^2}{l^2}+\frac{b^2}{m^2}=\frac{(a^2-b^2)^2}{n^2}\left(\frac{x_1^2}{a^2}+\frac{y_1^2}{b^2}\right)$$

$$=\frac{(a^2-b^2)^2}{n^2}$$

{$\because$ (x_1, y_1) satisfies the equation of the ellipse}

14. If the normals at (x_r, y_r); $r=1, 2, 3, 4$ on the rectangular hyperbola $xy=c^2$ meet at the point Q(h, k), prove that the sum of the ordinates of the four points is k. Also prove that the product of the ordinates is $-c^4$.

Any point on the curve $xy=c^2$ is $\left(ct, \frac{c}{t}\right)$.

The equation of the normal to the hyperbola at the point $\left(ct, \frac{c}{t}\right)$ is

$$y-\frac{c}{t}=\frac{-1}{\left(\frac{dy}{dx}\right)_{ct,\frac{c}{t}}}\cdot(x-ct).$$

Here, $xy=c^2$; or $y=\frac{c^2}{x}$; $\quad\therefore\quad \frac{dy}{dx}=\frac{-c^2}{x^2}$

$$\therefore\quad \left(\frac{dy}{dx}\right)_{ct,\frac{c}{t}}=-\frac{c^2}{c^2t^2}=-\frac{1}{t^2}$$

$\therefore$ the equation of the normal at $\left(ct, \frac{c}{t}\right)$ is

$$y-\frac{c}{t}=t^2(x-ct)$$

or $\quad ty-c=t^3(x-ct)$

or $\quad ct^4-t^3x+ty-c=0$

The normal passes through (h, k). So

$$ct^4-t^3h+tk-c=0 \quad \text{...(1)}$$

Let the roots of (1) be t_1, t_2, t_3, t_4. Then $x_r=ct_r$, $y_r=\frac{c}{t_r}$.

$\therefore$ sum of ordinates

$$=y_1+y_2+y_3+y_4$$

$$=\frac{c}{t_1}+\frac{c}{t_2}+\frac{c}{t_3}+\frac{c}{t_4}$$

$$= c\,\frac{t_2t_3t_4 + t_3t_4t_1 + t_4t_1t_2 + t_1t_2t_3}{t_1t_2t_3t_4}$$

$$= c \cdot \frac{-k/c}{-c/c} = k, \quad \text{\{from roots of the equation (1)\}}$$

and, product of the ordinates

$$= y_1y_2y_3y_4$$

$$= \frac{c}{t_1} \cdot \frac{c}{t_2} \cdot \frac{c}{t_3} \cdot \frac{c}{t_4}$$

$$= \frac{c^4}{t_1t_2t_3t_4} = \frac{c^4}{-c/c} = -c^4.$$

15. Find the equation of that diameter which bisects the chord $7x + y - 20 = 0$ of the hyperbola $\frac{x^2}{3} - \frac{y^2}{7} = 1$.

The centre of the hyperbola is (0, 0). Let the diameter be

$$y = mx \quad \dots (1)$$

The ends of the chord are found by solving

$$7x + y - 20 = 0 \quad \dots (2)$$

$$\text{and } \frac{x^2}{3} - \frac{y^2}{7} = 1 \quad \dots (3)$$

Solving (2), (3) we get $\frac{x^2}{3} - \frac{1}{7}(20 - 7x)^2 = 1$

or $\quad 7x^2 - 3(400 - 280x + 49x^2) = 21$

or $\quad 140x^2 - 840x + 1221 = 0.$

Let the roots be x_1, x_2.

Then $x_1 + x_2 = \frac{840}{140} = 6 \quad \dots (4)$

If $(x_1, y_1), (x_2, y_2)$ be ends then

$$7x_1 + y_1 - 20 = 0, \quad 7x_2 + y_2 - 20 = 0.$$

Adding, $7(x_1 + x_2) + (y_1 + y_2) - 40 = 0$

or $\quad 42 + y_1 + y_2 - 40 = 0,$ using (4)

$\therefore \quad y_1 + y_2 = -2$

$\therefore \quad$ the middle point of the chord

$$= \left(\frac{x_1 + x_2}{2}, \frac{y_1 + y_2}{2}\right)$$

$$= \left(\frac{6}{2}, \frac{-2}{2}\right) = (3, -1).$$

This lies on (1). So $-1 = 3m$; $\quad \therefore \quad m = -\frac{1}{3}$

$\therefore$ the equation of the diameter is $y = -\frac{1}{3}x$.

16. Find the locus of the foot of the perpendicular drawn from the centre on any tangent to the ellipse $\frac{x^2}{a^2} + \frac{y^2}{b^2} = 1$.

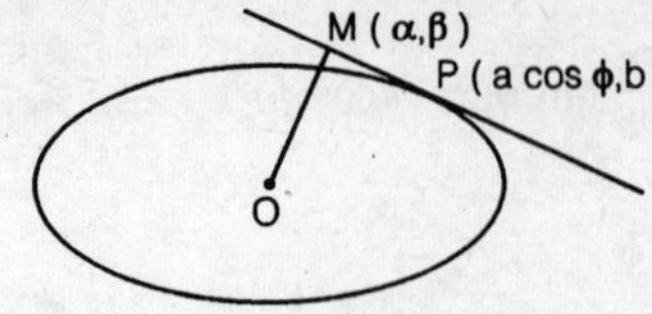

The equation of the tangent at any point $(a\cos\phi, b\sin\phi)$ is

$$\therefore \quad \frac{x\cos\phi}{a} + \frac{y\sin\phi}{b} = 1 \quad \dots (1)$$

Let $M(\alpha, \beta)$ be the foot of the perpendicular drawn from the centre (0, 0) to the tangent (1).

$\because \quad$ M is on the tangent, $\frac{\alpha\cos\phi}{a} + \frac{\beta\sin\phi}{b} = 1 \quad \dots (2)$

$\because \quad OM \perp PM, \; \frac{\beta}{\alpha} \cdot \left(-\frac{b\cos\phi}{a\sin\phi}\right) = -1$

or $\quad b\beta\cos\phi = a\alpha\sin\phi$

$$\therefore \quad \frac{\cos\phi}{a\alpha} = \frac{\sin\phi}{b\beta} = \frac{1}{\sqrt{a^2\alpha^2 + b^2\beta^2}}.$$

Putting in (2), $\frac{\alpha}{a} \cdot \frac{a\alpha}{\sqrt{a^2\alpha^2 + b^2\beta^2}} + \frac{\beta}{b} \cdot \frac{b\beta}{\sqrt{a^2\alpha^2 + b^2\beta^2}} = 1$

or $\quad \alpha^2 + \beta^2 = \sqrt{a^2\alpha^2 + b^2\beta^2}$

$\therefore \quad (\alpha^2 + \beta^2)^2 = a^2\alpha^2 + b^2\beta^2.$

$\therefore \quad$ the equation of the required locus is

$$(x^2 + y^2)^2 = a^2x^2 + b^2y^2.$$

17. The perpendicular from the centre on the normal at any point of the hyperbola $\frac{x^2}{a^2} - \frac{y^2}{b^2} = 1$ meet at R. Find the locus of R.

Let (x_1, y_1) be any point on the hyperbola.

So, $\frac{x_1^2}{a^2} - \frac{y_1^2}{b^2} = 1 \quad \dots (1)$

The equation of the normal at (x_1, y_1) is

$$\frac{x - x_1}{\frac{x_1}{a^2}} = \frac{y - y_1}{-\frac{y_1}{b^2}}$$

or $\quad \frac{x_1}{a^2}(y - y_1) + \frac{y_1}{b^2}(x - x_1) = 0 \quad \dots (2)$

'm' of the normal $= -\frac{a^2y_1}{b^2x_1}$

$\therefore$ the equation of the perpendicular from the centre (0, 0) on (2) is

$$y = \frac{b^2 x_1}{a^2 y_1} \cdot x \qquad \dots (3)$$

The intersection of (2) and (3) is R and the required locus is obtained by eliminating x_1, y_1 from (1), (2) and (3).

From (3), $\frac{x_1}{a^2 y} = \frac{y_1}{b^2 x} = t$ (say)

Putting in (2), $yt(y - b^2 xt) + xt(x - a^2 yt) = 0$

or $(x^2 + y^2)t - (a^2 + b^2)xyt^2 = 0.$

But $t \neq 0$ for then $(x_1, y_1) = (0, 0)$ which is not true.

$$\therefore \quad t = \frac{x^2 + y^2}{xy(a^2 + b^2)}$$

$$\therefore \quad x_1 = \frac{x^2 + y^2}{xy(a^2 + b^2)} a^2 y = \frac{a^2(x^2 + y^2)}{x(a^2 + b^2)}$$

and $$y_1 = \frac{x^2 + y^2}{xy(a^2 + b^2)} b^2 x = \frac{b^2(x^2 + y^2)}{y(a^2 + b^2)}$$

$\therefore$ from (1),

$$\frac{1}{a^2} \cdot \frac{a^4(x^2 + y^2)^2}{x^2(a^2 + b^2)^2} - \frac{1}{b^2} \cdot \frac{b^4(x^2 + y^2)^2}{y^2(a^2 + b^2)^2} = 1$$

or $$(x^2 + y^2)^2 \cdot \left(\frac{a^2}{x^2} - \frac{b^2}{y^2}\right) = (a^2 + b^2)^2.$$

18. A point P moves such that the tangents PT_1 and PT_2 from it to the hyperbola $4x^2 - 9y^2 = 36$ are mutually perpendicular. Find the equation of the locus of P.

Let $P = (\alpha, \beta)$. Any line through P has the equation

$$y - \beta = m(x - \alpha) \qquad \dots (1)$$

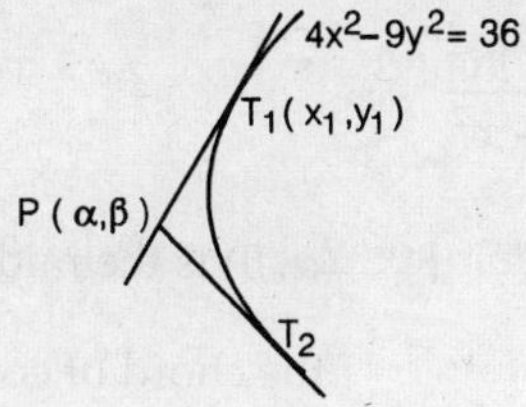

This will touch the hyperbola

$$4x^2 - 9y^2 = 36 \qquad \dots (2)$$

if (1) cuts (2) at coincident points.

Solving (1), (2) we get

$$4x^2 - 9\{\beta + m(x - \alpha)\}^2 = 36$$

or $4x^2 - 9\{m^2 x^2 + 2m(\beta - \alpha m)x + (\beta - \alpha m)^2\} - 36 = 0$

or $(4 - 9m^2)x^2 - 18m(\beta - \alpha m)x - 9(\beta - \alpha m)^2 - 36 = 0$

(1) will touch (2) if this equation has equal roots.

For this,

$$\{18m(\beta - \alpha m)\}^2 + 4(4 - 9m^2)\{9(\beta - \alpha m)^2 + 36\} = 0$$

or $9m^2(\beta - \alpha m)^2 + (4 - 9m^2)\{(\beta - \alpha m)^2 + 4\} = 0$

or $(\beta - \alpha m)^2(9m^2 + 4 - 9m^2) + 4(4 - 9m^2) = 0$

or $(\beta - \alpha m)^2 + 4 - 9m^2 = 0$

or $(\alpha^2 - 9)m^2 - 2\alpha\beta m + \beta^2 + 4 = 0 \qquad \dots (3)$

The two tangents PT_1, PT_2 will be perpendicular if the product of roots of (3) is -1,

i.e., $\frac{\beta^2 + 4}{\alpha^2 - 9} = -1; \quad \therefore \quad \alpha^2 + \beta^2 = 5$

$\therefore$ the equation of the locus of $P(\alpha, \beta)$ is

$$x^2 + y^2 = 5.$$

(This is the director circle of the hyperbola.)

19. Find the locus of the intersection of two normals to the ellipse $\frac{x^2}{a^2} + \frac{y^2}{b^2} = 1$ which are perpendicular to each other.

Let the two perpendicular normals cut at (α, β). If the foot of one of the normals be 'ϕ' then the normal is

$$ax\sec\phi - by\operatorname{cosec}\phi = a^2 - b^2 \qquad \dots (1)$$

and it passes through (α, β).

$\therefore \quad a\alpha\sec\phi - b\beta\operatorname{cosec}\phi = a^2 - b^2$

or $a\alpha\sin\phi - b\beta\cos\phi = (a^2 - b^2)\sin\phi \cdot \cos\phi \qquad \dots (2)$

'm' of (1) is given by $m = \frac{a\sec\phi}{b\operatorname{cosec}\phi} = \frac{a}{b}\tan\phi$

$\therefore \quad \cot\phi = \frac{a}{bm}$.

From (2), $a\alpha\tan\phi - b\beta = (a^2 - b^2)\sin\phi$

or $$\left(a\alpha \cdot \frac{bm}{a} - b\beta\right)^2 = (a^2 - b^2)^2\sin^2\phi$$

$$= \frac{(a^2 - b^2)^2}{1 + \frac{a^2}{b^2 m^2}}$$

or $b^2(m\alpha - \beta)^2 = (a^2 - b^2)^2 \cdot \frac{b^2 m^2}{a^2 + b^2 m^2}$

or $(m\alpha - \beta)^2 \cdot (a^2 + b^2 m^2) = (a^2 - b^2)^2 \cdot m^2$

or $(m^2\alpha^2 - 2\alpha\beta m + \beta^2)(a^2 + b^2 m^2) - (a^2 - b^2)^2 m^2 = 0$

or $b^2\alpha^2 m^4 - 2b^2\alpha\beta m^3 + \{a^2\alpha^2 + b^2\beta^2 - (a^2 - b^2)^2\}m^2$
$- 2a^2\alpha\beta m + a^2\beta^2 = 0$

Let m_1, m_2, m_3, m_4 be four roots of the equation giving the slopes of the four normals that will pass through (α, β). As two of them are perpendicular to each other, let

$$m_1m_2 = -1. \quad \dots (3)$$

Now, $m_1 + m_2 + m_3 + m_4 = \dfrac{2b^2\alpha\beta}{b^2\alpha^2} = \dfrac{2\beta}{\alpha}$ ... (4)

$$m_1m_2 + (m_1+m_2)(m_3+m_4) + m_3m_4$$

$$= \frac{a^2\alpha^2 + b^2\beta^2 - (a^2-b^2)^2}{b^2\alpha^2} \quad \dots (5)$$

$$m_1m_2(m_3+m_4) + m_3m_4(m_1+m_2)$$

$$= \frac{2a^2\alpha\beta}{b^2\alpha^2} = \frac{2a^2\beta}{b^2\alpha} \quad \dots (6)$$

$$m_1m_2m_3m_4 = \frac{a^2\beta^2}{b^2\alpha^2} \quad \dots (7)$$

From (3) and (7),

$$m_3m_4 = -\frac{a^2\beta^2}{b^2\alpha^2}.$$

Putting $m_1 + m_2 = \lambda$, $m_3 + m_4 = \mu$ we get the following:

$$\lambda + \mu = \frac{2\beta}{\alpha} \quad \dots (8)$$

$$-1 + \lambda\mu - \frac{a^2\beta^2}{b^2\alpha^2} = \frac{a^2\alpha^2 + b^2\beta^2 - (a^2-b^2)^2}{b^2\alpha^2} \quad \dots (9)$$

$$-\mu - \lambda \cdot \frac{a^2\beta^2}{b^2\alpha^2} = \frac{2a^2\beta}{b^2\alpha} \quad \dots (10)$$

From (8) + (10), we get

$$\left(1 - \frac{a^2\beta^2}{b^2\alpha^2}\right)\lambda = \frac{2\beta}{\alpha} + \frac{2a^2\beta}{b^2\alpha} = \frac{2\beta}{\alpha}\left(1 + \frac{a^2}{b^2}\right)$$

or $\lambda = \dfrac{2\beta}{\alpha} \cdot \dfrac{a^2+b^2}{b^2} \cdot \dfrac{b^2\alpha^2}{b^2\alpha^2 - a^2\beta^2} = \dfrac{2\alpha\beta(a^2+b^2)}{b^2\alpha^2 - a^2\beta^2}$

$\therefore$ by (8),

$$\mu = \frac{2\beta}{\alpha} - \frac{2\alpha\beta(a^2+b^2)}{b^2\alpha^2 - a^2\beta^2}$$

$$= \frac{b^2\alpha^2 - a^2\beta^2 - \alpha^2(a^2+b^2)}{\alpha(b^2\alpha^2 - a^2\beta^2)} \cdot 2\beta$$

$$= -\frac{2\beta}{\alpha} \frac{a^2(\alpha^2+\beta^2)}{b^2\alpha^2 - a^2\beta^2}$$

$\therefore$ from (9),

$$\frac{2\alpha\beta(a^2+b^2)}{b^2\alpha^2 - a^2\beta^2} \cdot \frac{-2\beta a^2(\alpha^2+\beta^2)}{\alpha(b^2\alpha^2 - a^2\beta^2)}$$

$$= 1 + \frac{a^2\beta^2}{b^2\alpha^2} + \frac{a^2\alpha^2 + b^2\beta^2 - (a^2-b^2)^2}{b^2\alpha^2}$$

or $\dfrac{-4a^2\beta^2(a^2+b^2)(\alpha^2+\beta^2)}{(b^2\alpha^2 - a^2\beta^2)^2}$

$$= \frac{b^2\alpha^2 + a^2\beta^2 + a^2\alpha^2 + b^2\beta^2 - (a^2-b^2)^2}{b^2\alpha^2}$$

or $-4a^2b^2\alpha^2\beta^2(a^2+b^2)(\alpha^2+\beta^2)$

$$= (b^2\alpha^2 - a^2\beta^2)^2\{(a^2+b^2)(\alpha^2+\beta^2) - (a^2-b^2)^2\}$$

or $(a^2-b^2)^2(b^2\alpha^2 - a^2\beta^2)^2$

$$= (a^2+b^2)(\alpha^2+\beta^2)(b^2\alpha^2 + a^2\beta^2)^2$$

$\therefore$ the equation of the locus is

$$(a^2-b^2)^2\left(\frac{x^2}{a^2} - \frac{y^2}{b^2}\right)^2 = (a^2+b^2)(x^2+y^2)\left(\frac{x^2}{a^2} + \frac{y^2}{b^2}\right)^2.$$

20. From points on the circle $x^2 + y^2 = a^2$ tangents are drawn to the hyperbola $x^2 - y^2 = a^2$. Prove that the locus of the middle points of the chord of contact is the curve $(x^2-y^2)^2 = a^2(x^2+y^2)$.

Let (x_1, y_1) be any point on the circle $x^2 + y^2 = a^2$.

Then $x_1^2 + y_1^2 = a^2$ ... (1)

The equation of the chord of contact of tangents from (x_1, y_1) to the hyperbola is

$$xx_1 - yy_1 = a^2 \quad \dots (2)$$

Let (α, β) be the middle point of the chord of contact.

The points of intersection of (2) and the hyperbola

$$x^2 - y^2 = a^2 \quad \dots (3)$$

are obtained by solving (2) and (3). Solving we get,

$$\left(\frac{a^2 + yy_1}{x_1}\right)^2 - y^2 = a^2$$

or $(a^2 + yy_1)^2 - x_1^2y^2 = a^2x_1^2$

or $(y_1^2 - x_1^2)y^2 + 2a^2y_1y + a^2(a^2 - x_1^2) = 0$

The roots of this equation give the ordinates of the points of contact.

$$\therefore \quad y_2 + y_3 = -\frac{2a^2y_1}{y_1^2 - x_1^2}$$

$$\therefore \quad 2\beta = -\frac{2a^2y_1}{y_1^2 - x_1^2}$$ {$\because$ (α, β) is the middle point of the chord of contact}

$$\therefore \quad \beta = -\frac{a^2y_1}{y_1^2 - x_1^2} \quad \dots (4)$$

Also, (α, β) satisfies (2).

So $\alpha x_1 - \beta y_1 = a^2$ ... (5)

$$\therefore \quad \alpha x_1 - \left(-\frac{a^2y_1}{y_1^2 - x_1^2}\right)y_1 = a^2$$

or $\alpha x_1 + \dfrac{a^2y_1^2}{y_1^2 - x_1^2} = a^2$

$$\therefore \quad \alpha x_1 = a^2 - \frac{a^2 y_1^2}{y_1^2 - x_1^2} = \frac{-a^2 x_1^2}{y_1^2 - x_1^2}$$

$$\therefore \quad \alpha = \frac{-a^2 x_1}{y_1^2 - x_1^2} \quad \ldots (6)$$

Thus we get from (4) and (6), $\dfrac{\alpha}{-a^2 x_1} = \dfrac{\beta}{-a^2 y_1}$

$$\text{or} \quad \frac{\alpha}{x_1} = \frac{\beta}{y_1} = \frac{\alpha^2 - \beta^2}{\alpha x_1 - \beta y_1} = \frac{\alpha^2 - \beta^2}{a^2} \quad \{\text{from (5)}\}$$

$$\therefore \quad x_1 = \frac{\alpha a^2}{\alpha^2 - \beta^2}, \quad y_1 = \frac{\beta a^2}{\alpha^2 - \beta^2}.$$

Putting in (1), we get

$$\frac{\alpha^2 a^4}{(\alpha^2 - \beta^2)^2} + \frac{\beta^2 a^4}{(\alpha^2 - \beta^2)^2} = a^2$$

$$\text{or} \quad \frac{a^2(\alpha^2 + \beta^2)}{(\alpha^2 - \beta^2)^2} = 1$$

$$\therefore \quad (\alpha^2 - \beta^2)^2 = a^2(\alpha^2 + \beta^2)$$

$\therefore$ the equation of the locus of the middle point (α, β) is

$$(x^2 - y^2)^2 = a^2(x^2 + y^2).$$

21. If a chord of the ellipse $\dfrac{x^2}{a^2} + \dfrac{y^2}{b^2} = 1$ touches the hyperbola $\dfrac{x^2}{a^2} - \dfrac{y^2}{b^2} = 1$, show that the locus of its middle point is $\left(\dfrac{x^2}{a^2} + \dfrac{y^2}{b^2}\right)^2 = \dfrac{x^2}{a^2} - \dfrac{y^2}{b^2}$.

Let $M(\alpha, \beta)$ be the middle point of the chord PQ of the ellipse

$$\frac{x^2}{a^2} + \frac{y^2}{b^2} = 1 \quad \ldots (1)$$

The equation of the chord is $y - \beta = m(x - \alpha)$ ... (2)

The equation of the hyperbola is $\dfrac{x^2}{a^2} - \dfrac{y^2}{b^2} = 1$ ... (3)

(2) touches (3). Solving these

$$\frac{x^2}{a^2} - \frac{1}{b^2}\{\beta + m(x - \alpha)\}^2 = 1$$

$$\text{or} \quad \frac{x^2}{a^2} - \frac{1}{b^2}\{m^2 x^2 + (\beta - m\alpha)^2 + 2mx(\beta - m\alpha)\} - 1 = 0$$

$$\text{or} \quad \left(\frac{1}{a^2} - \frac{m^2}{b^2}\right)x^2 - \frac{2m}{b^2}(\beta - m\alpha)x - \left\{\frac{(\beta - m\alpha)^2}{b^2} + 1\right\} = 0 \quad \ldots (4)$$

It will have equal roots. So

$$\frac{4m^2}{b^4}(\beta - m\alpha)^2 + 4\left(\frac{1}{a^2} - \frac{m^2}{b^2}\right)\left\{\frac{(\beta - m\alpha)^2}{b^2} + 1\right\} = 0 \quad \ldots (5)$$

Again solving (1), (2) we get as in (4):

$$\left(\frac{1}{a^2} + \frac{m^2}{b^2}\right)x^2 + \frac{2m}{b^2}(\beta - m\alpha)x - \left\{1 - \frac{(\beta - m\alpha)^2}{b^2}\right\} = 0 \quad \ldots (6)$$

If $P = (x_1, y_1)$, $Q = (x_2, y_2)$ then x_1, x_2 will be roots of (6)

$$\therefore \quad x_1 + x_2 = \frac{-\frac{2m}{b^2}(\beta - m\alpha)}{\frac{1}{a^2} + \frac{m^2}{b^2}}$$

$$\therefore \quad 2\alpha = -\frac{\frac{2m}{b^2}(\beta - m\alpha)}{\frac{1}{a^2} + \frac{m^2}{b^2}}$$

$\{\because \; (\alpha, \beta)$ is middle point of PQ$\}$

$$\text{or} \quad \alpha\left(\frac{1}{a^2} + \frac{m^2}{b^2}\right) + \frac{m}{b^2}(\beta - m\alpha) = 0$$

$$\text{or} \quad \frac{\alpha}{a^2} + \frac{m\beta}{b^2} = 0$$

$$\therefore \quad m = -\frac{b^2\alpha}{a^2\beta}.$$

Putting this value of m in (5), we get

$$\frac{4}{b^4}\cdot\left(-\frac{b^2\alpha}{a^2\beta}\right)^2\left\{\beta + \frac{b^2\alpha^2}{a^2\beta}\right\}^2 + 4\left\{\frac{1}{a^2} - \frac{1}{b^2}\cdot\frac{b^4\alpha^2}{a^4\beta^2}\right\}\left\{\frac{1}{b^2}\left(\beta + \frac{b^2\alpha^2}{a^2\beta}\right)^2 + 1\right\} = 0$$

$$\text{or} \quad \frac{\alpha^2}{a^4\beta^2}\cdot\frac{(a^2\beta^2 + b^2\alpha^2)^2}{a^4\beta^2} + \frac{(a^2\beta^2 - b^2\alpha^2)}{a^4\beta^2}\left\{\frac{(a^2\beta^2 + b^2\alpha^2)^2}{a^4b^2\beta^2} + 1\right\} = 0$$

$$\text{or} \quad \alpha^2(a^2\beta^2 + b^2\alpha^2)^2 + (a^2\beta^2 - b^2\alpha^2) \times \frac{1}{b^2}\{(a^2\beta^2 + b^2\alpha^2)^2 + a^4b^2\beta^2\} = 0$$

$$\text{or} \quad \alpha^2(a^2\beta^2 + b^2\alpha^2)^2 + \frac{a^2\beta^2}{b^2}(a^2\beta^2 + b^2\alpha^2)^2 - \alpha^2(a^2\beta^2 + b^2\alpha^2)^2 + (a^2\beta^2 - b^2\alpha^2)a^4\beta^2 = 0$$

$$\text{or} \quad \frac{a^2\beta^2}{b^2}(a^2\beta^2 + b^2\alpha^2)^2 = (b^2\alpha^2 - a^2\beta^2)a^4\beta^2$$

$$\text{or} \quad (a^2\beta^2 + b^2\alpha^2)^2 = a^2b^2(b^2\alpha^2 - a^2\beta^2)$$

or $\left(\frac{\alpha^2}{a^2}+\frac{\beta^2}{b^2}\right)^2=\frac{\alpha^2}{a^2}-\frac{\beta^2}{b^2}$

$\therefore$ the equation of the required locus of the middle point (α, β) is

$$\left(\frac{x^2}{a^2}+\frac{y^2}{b^2}\right)^2=\frac{x^2}{a^2}-\frac{y^2}{b^2}.$$

22. Find the locus of the point of intersection of normals at two points on an ellipse which are extremities of conjugate diameters.

Let PP′ and QQ′ be two conjugate diameters of the ellipse $\frac{x^2}{a^2}+\frac{y^2}{b^2}=1$.

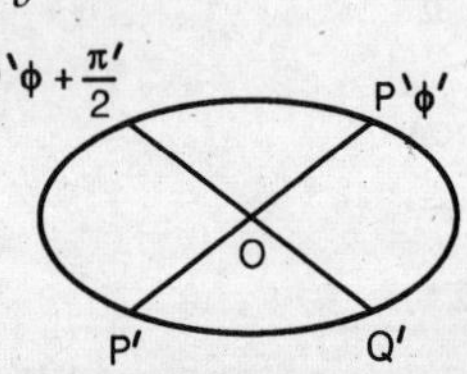

Let the eccentric angle of the point P be 'ϕ'.

Then the eccentric angle of Q is '$\phi+\pi/2$'.

$\therefore$ $P=(a\cos\phi, b\sin\phi)$

$$Q=\left\{a\cos\left(\phi+\frac{\pi}{2}\right), b\sin\left(\phi+\frac{\pi}{2}\right)\right\}$$

The equation of the normal at $P=(a\cos\phi, b\sin\phi)$ is

$$\frac{x-a\cos\phi}{\frac{a\cos\phi}{a^2}}=\frac{y-b\sin\phi}{\frac{b\sin\phi}{b^2}}$$

or $\quad ax\sec\phi-by\,\text{cosec}\,\phi=a^2-b^2 \quad \ldots (1)$

Similarly, the equation of the normal at Q is

$$ax\sec\left(\phi+\frac{\pi}{2}\right)-by\,\text{cosec}\left(\phi+\frac{\pi}{2}\right)=a^2-b^2$$

or $\quad -ax\,\text{cosec}\,\phi-by\sec\phi=a^2-b^2 \quad \ldots (2)$

The locus of the point of intersection of (1) and (2) is obtained by eliminating ϕ from them.

Now, we have

$$ax\sec\phi-by\,\text{cosec}\,\phi-(a^2-b^2)=0$$

$$by\sec\phi+ax\,\text{cosec}\,\phi+(a^2-b^2)=0$$

By cross-multiplication,

$$\frac{\sec\phi}{-by+ax}=\frac{\text{cosec}\,\phi}{-by-ax}=\frac{a^2-b^2}{a^2x^2+b^2y^2}$$

$$\therefore \quad \cos\phi=\frac{a^2x^2+b^2y^2}{ax-by}\cdot\frac{1}{a^2-b^2}$$

$$\sin\phi=\frac{a^2x^2+b^2y^2}{-(ax+by)}\cdot\frac{1}{a^2-b^2}$$

Squaring and adding,

$$1=\frac{(a^2x^2+b^2y^2)^2}{(a^2-b^2)^2}\left\{\frac{1}{(ax-by)^2}+\frac{1}{(ax+by)^2}\right\}$$

$$=\left(\frac{a^2x^2+b^2y^2}{a^2-b^2}\right)^2\cdot\frac{2(a^2x^2+b^2y^2)}{(a^2x^2-b^2y^2)^2}$$

or $\quad 2(a^2x^2+b^2y^2)^3=(a^2-b^2)^2\cdot(a^2x^2-b^2y^2)^2.$

23. If $\phi_1, \phi_2, \phi_3, \phi_4$ are the eccentric angles of four concyclic points on the ellipse $\frac{x^2}{a^2}+\frac{y^2}{b^2}=1$ then prove that $\phi_1+\phi_2+\phi_3+\phi_4=2n\pi, n\in N$.

The equation of the chord joining the points with eccentric angles ϕ_1, ϕ_4 is

$$\frac{x}{a}\cos\frac{\phi_4+\phi_1}{2}+\frac{y}{b}\sin\frac{\phi_4+\phi_1}{2}=\cos\frac{\phi_4-\phi_1}{2} \quad \ldots (1)$$

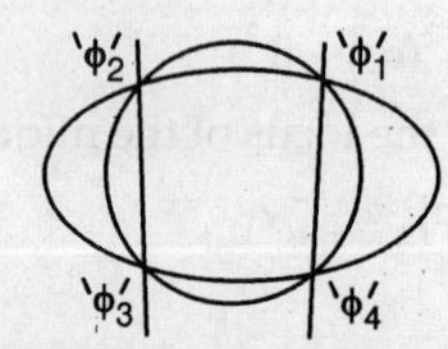

The equation of the chord joining the points with eccentric angles ϕ_2, ϕ_3 is

$$\frac{x}{a}\cos\frac{\phi_2+\phi_3}{2}+\frac{y}{b}\sin\frac{\phi_2+\phi_3}{2}=\cos\frac{\phi_2-\phi_3}{2} \quad \ldots (2)$$

The equation of any second degree conic passing through the intersection of the ellipse $\frac{x^2}{a^2}+\frac{y^2}{b^2}-1=0$ and the chord (1), (2) is

$$\frac{x^2}{a^2}+\frac{y^2}{b^2}-1$$

$$+\lambda\left(\frac{x}{a}\cos\frac{\phi_4+\phi_1}{2}+\frac{y}{b}\sin\frac{\phi_4+\phi_1}{2}-\cos\frac{\phi_4-\phi_1}{2}\right)$$

$$\times\left(\frac{x}{a}\cos\frac{\phi_2+\phi_3}{2}+\frac{y}{b}\sin\frac{\phi_2+\phi_3}{2}-\cos\frac{\phi_2-\phi_3}{2}\right)=0$$

If this is a circle then the four points 'ϕ_1', 'ϕ_2', 'ϕ_3', 'ϕ_4' will be concyclic. The conditions for that, are

coefficient of x^2 = coefficient of y^2,

i.e., $\frac{1}{a^2}+\frac{\lambda}{a^2}\cos\frac{\phi_4+\phi_1}{2}\cos\frac{\phi_2+\phi_3}{2}$

$$=\frac{1}{b^2}+\frac{\lambda}{b^2}\sin\frac{\phi_4+\phi_1}{2}\sin\frac{\phi_2+\phi_3}{2} \quad \ldots (3)$$

and, coefficient of $xy=0$,

i.e., $\frac{\lambda}{ab}\left(\cos\frac{\phi_4+\phi_1}{2}\cdot\sin\frac{\phi_2+\phi_3}{2} + \sin\frac{\phi_4+\phi_1}{2}\cos\frac{\phi_2+\phi_3}{2}\right)=0 \quad \ldots (4)$

$\therefore \quad \sin\left(\frac{\phi_4+\phi_1}{2}+\frac{\phi_2+\phi_3}{2}\right)=0$

or $\quad \sin\left(\frac{\phi_1+\phi_2+\phi_3+\phi_4}{2}\right)=0$

$\therefore \quad \frac{\phi_1+\phi_2+\phi_3+\phi_4}{2}=n\pi$

$\therefore \quad \phi_1+\phi_2+\phi_3+\phi_4=2n\pi.$

Exercises

1. Find the equation to the ellipse whose one vertex is (3, 1), the nearer focus is (1, 1) and the eccentricity is 2/3.

2. Find the equation to the ellipse whose centre is (–2, 3) and whose semiaxes are 3 and 2 and the major axis is parallel to the y-axis.

3. Find the equation of the hyperbola whose foci are (6, 4) and (–4, 4), and the eccentricity is 2.

4. If e and e' are the eccentricities of the hyperbolas $\frac{x^2}{a^2}-\frac{y^2}{b^2}=1$ and $\frac{y^2}{b^2}-\frac{x^2}{a^2}=1$ then prove that $\frac{1}{e^2}+\frac{1}{e'^2}=1.$

5. Find the latus rectum, eccentricity and foci of the curve $4x^2+9y^2-8x-36y+4=0.$

6. Find the centre, eccentricity and the lengths of the axes for the curve $x^2-3y^2-2x=8.$

7. Find the inclination to the major axis of the diameter of the ellipse, the square of whose length is the harmonic mean between the squares of the major and minor axes.

8. Show that the length of the focal chord of the ellipse $\frac{x^2}{a^2}+\frac{y^2}{b^2}=1$ which makes an angle θ with the major axis is $\frac{2ab^2}{a^2\sin^2\theta+b^2\cos^2\theta}.$

9. If the eccentric angles of the ends of a focal chord of the ellipse $\frac{x^2}{a^2}+\frac{y^2}{b^2}=1$ be θ and ϕ, show that

$$\tan\frac{\theta}{2}\cdot\tan\frac{\phi}{2}=\frac{e-1}{e+1}.$$

10. The hyperbola $\frac{x^2}{a^2}-\frac{y^2}{b^2}=1$ passes through the point of intersection of the lines $7x+13y-87=0$ and $5x-8y+7=0$ and its latus rectum is $\frac{32\sqrt{2}}{5}$. Find a and b.

11. PN is the ordinate of any point P on the hyperbola $\frac{x^2}{a^2}-\frac{y^2}{b^2}=1$. If Q divides AP in the ratio $a^2:b^2$, show that NQ is perpendicular to A′P where AA′ is the transverse axis of the hyperbola.

12. If a rectangular hyperbola circumscribes a triangle, prove that it also passes through the orthocentre of the triangle.

13. Show that the product of the perpendiculars drawn from the foci on any tangent to an ellipse is constant.

14. Prove the line joining two points of an ellipse, the difference of whose eccentric angles is constant, touches a fixed ellipse.

15. (a) Any tangent to an ellipse is cut by the tangents at the ends of the major axis in the points T, T′. Prove that the circle whose diameter is TT′ passes through the foci of the ellipse.

(b) A tangent to the ellipse $x^2+4y^2=4$ meets the ellipse $x^2+2y^2=6$ at P and Q. Prove that the tangents at P and Q to the second ellipse are at right angles.

16. Find the equations of the tangents from the point (2, 2) to the ellipse $4x^2+9y^2=36$. Also find the angle between the tangents.

17. If the normals be drawn at the extremities of a focal chord of an ellipse, a line through their point of intersection parallel to the major axis will bisect the chord. Prove this.

18. The normals at four points on the ellipse $\frac{x^2}{a^2}+\frac{y^2}{b^2}=1$ meet at (α, β). Find the mean position of the four points.

19. If the normal at an end of a latus rectum of an ellipse passes through an extremity of the minor axis, show that the eccentricity of the curve is given by the equation $e^4+e^2-1=0.$

20. If the normal at P to the hyperbola $\frac{x^2}{a^2} - \frac{y^2}{b^2} = 1$ meets the transverse axis in G and the conjugate axis in G′ and CF be the perpendicular to the normal from the centre C then show that
$PF \cdot PG = b^2$ and $PF \cdot PG' = a^2$.

21. Tangents from the point T to the ellipse $\frac{x^2}{a^2} + \frac{y^2}{b^2} = 1$, $(a > b)$ cut off a length equal to the minor axis from the tangent at $(a, 0)$. Prove that T lies on a parabola.

22. P, Q are two points on the ellipse $x^2 + 4y^2 = 4$ whose eccentric angles α, β satisfy the relation $\sec\alpha + \sec\beta = 2$. Prove that PQ touches the circle $x^2 + y^2 - 2x = 0$.

23. The lines $|x| = a$ make an intercept on a variable line such that the intercept subtends a right angle at $(2a, 0)$. Prove that the line always touches the fixed hyperbola $3x^2 - y^2 = 3a^2$.

24. Find the equations of common tangents to the ellipse $4x^2 + 9y^2 = 36$ and the parabola $y^2 = 4x$.

25. A tangent to the ellipse $\frac{x^2}{a^2} + \frac{y^2}{b^2} = 1$ cuts the coordinate axes in A and B. Show that the equation of the locus of the middle point of AB is
$$\frac{a^2}{x^2} + \frac{b^2}{y^2} = 4.$$

26. P is a variable point such that the two tangents drawn to the ellipse $\frac{x^2}{a^2} + \frac{y^2}{b^2} = 1$ from the point P are perpendicular to each other. Find the equation of the locus of P.

27. Show that the locus of the point of intersection of tangents to an ellipse at two points whose eccentric angles differ by a constant, is an ellipse.

28. If the normal at any point P of an ellipse cuts the major axis at G, prove that the locus of the middle point of PG will be an ellipse.

29. Find the equation of the locus of a point from which two tangents can be drawn to the ellipse
$$\frac{x^2}{a^2} + \frac{y^2}{b^2} = 1$$
making angles θ_1, θ_2 with the major axis such that $\tan\theta_1 + \tan\theta_2 = c$ (a constant).

30. Find the locus of the middle points of all chords of the ellipse $\frac{x^2}{a^2} + \frac{y^2}{b^2} = 1$ which are of constant length c.

31. P is any point on an ellipse. Prove that the locus of the centroid G of the points P and the two foci of the ellipse is a concentric ellipse of the same eccentricity.

32. (a) A variable straight line of slope 4, intersects the rectangular hyperbola $xy = 1$ at two points. Find the equation of the locus of the point which divides the line segment between these two points in the ratio 1 : 2.

(b) Show that the locus of the middle points of the chords of the hyperbola $\frac{x^2}{a^2} - \frac{y^2}{b^2} = 1$ which passes through a fixed point (α, β), is a hyperbola whose centre is $(\alpha/2, \beta/2)$.

33. Find the locus of the middle points of the portions of the tangents to the hyperbola $\frac{x^2}{a^2} - \frac{y^2}{b^2} = 1$ included between the axes.

34. A normal to the hyperbola $x^2 - 4y^2 = 4$ meets the x and y axes at A and B respectively. Prove that the locus of the point of intersection of the straight lines drawn through A and B perpendicular to x and y axes respectively is a hyperbola. Find its equation and eccentricity.

35. A circle cuts two fixed perpendicular lines so that each intercept is of a given length which are unequal. Prove that the locus of the centre of the circle is a rectangular hyperbola.

36. A circle with centre $(3\alpha, 3\beta)$ and of variable radius cuts the rectangular hyperbola $x^2 - y^2 = 9a^2$ at the points P, Q, R and S. Prove that the locus of centroid of the triangle PQR is $(x - 2\alpha)^2 - (y - 2\beta)^2 = a^2$.

37. (a) Show that the locus of points of intersection of tangents at the extremities of normal chords of the hyperbola $\frac{x^2}{a^2} - \frac{y^2}{b^2} = 1$ is
$$\frac{a^6}{x^2} - \frac{b^6}{y^2} = (a^2 + b^2)^2.$$

(b) Find the locus of the middle points of the chords of the circle $x^2 + y^2 = 16$ which are tangents to the hyperbola $9x^2 - 16y^2 = 144$.

38. If α, β be the angles subtended by the major axis of an ellipse at the extremities of a pair of conjugate diameters then show that $\cot^2\alpha + \cot^2\beta$ is a constant.

39. If P and Q be the two ends of a pair of conjugate diameters of the ellipse $\frac{x^2}{a^2} + \frac{y^2}{b^2} = 1$, find the locus of the intersection of the tangents at P and Q.

Objective Questions

Fill in the blanks.

40. The equation $x^2 - 3xy + y^2 + 10x - 10y + 21 = 0$ represents a ______ having its centre at ______ .

41. The eccentricity of a rectangular hyperbola is ______ .

42. If a focus = (6, 7), the corresponding directrix is $x + y + 2 = 0$ and eccentricity $= \frac{1}{\sqrt{3}}$ then the equation of the ellipse is ______ .

43. If the minor axis = the distance between foci, and latus rectum = 10 then the equation of the ellipse in the standard form is ______ .

44. The ellipse $\frac{x^2}{a^2} + \frac{y^2}{b^2} = 1$ passes through the point (–3, 1) and has the eccentricity $\sqrt{2/5}$. Then the major axis = ______ , minor axis = ______ .

45. If eccentricity $= \sqrt{3}$, the distance between foci = 9 then the equation of the hyperbola in the standard form is ______ .

46. The foci of the hyperbola $25x^2 - 36y^2 = 225$ are ______ and ______.

47. The equation $x^2 - 2y^2 - 2x + 8y - 1 = 0$ represents a(an) ______ and its centre is ______ .

48. The equation of the tangent at the point 'ϕ' on the hyperbola $4x^2 - 3y^2 = 12$ is ______ where $\phi = \pi/4$.

49. The equation of the normal to the ellipse $x^2 + 4y^2 = 16$ at the end of the latus rectum in the first quadrant is ______ .

50. The inclination to the major axis of the diameter of the ellipse $\frac{x^2}{a^2} + \frac{y^2}{b^2} = 1$ whose length is the GM between the major and minor axes, is ______ .

51. Let P be a variable point on the ellipse $\frac{x^2}{a^2} + \frac{y^2}{b^2} = 2$ with foci F_1 and F_2. If A is the area of the ΔPF_1F_2 then the maximum value of A is ______ .

52. The number of normals that can be drawn to an ellipse $\frac{x^2}{a^2} + \frac{y^2}{b^2} = 1$ from an exterior point is ______ in general.

53. If a tangent to the ellipse $\frac{x^2}{a^2} + \frac{y^2}{b^2} = 1$, whose centre is C, meets the major and minor axes produced at P and Q respectively then $\frac{a^2}{CP^2} + \frac{b^2}{CQ^2}$ is equal to ______ .

54. If the middle point of a chord of the ellipse $\frac{x^2}{25} + \frac{y^2}{16} = 1$ be $\left(\frac{1}{2}, \frac{2}{5}\right)$ then the length of the chord will be ______ .

55. The line $lx + my + n = 0$ touches the hyperbola $b^2x^2 - a^2y^2 = a^2b^2$ if ______ .

56. If $y = x$ and $3y + 2x = 0$ are the equations of a pair of conjugate diameters of an ellipse $\frac{x^2}{a^2} + \frac{y^2}{b^2} = 1$ then the eccentricity = ______ .

57. The equation $3x^2 - 3y^2 - 18x + 12y + 2 = 0$ represents one ______ whose eccentricity is ______ .

58. The coordinates of the point P on the ellipse $4x^2 + 9y^2 = 36$ are ______ if the area of the $\Delta PSS' = \sqrt{10}$, S and S' being the foci.

59. If the tangent and normal to a rectangular hyperbola at a point cut off intercepts a_1, a_2 on one axis and b_1, b_2 on the other axis then $a_1a_2 + b_1b_2 =$ ______ .

Choose the correct option(s).

60. The equation $14x^2 - 4xy + 11y^2 - 44x - 58y + 71 = 0$ represents

(i) a parabola (ii) an ellipse
(iii) a hyperbola (iv) a rectangular hyperbola

whose centre is at

(a) (2, 3) (b) (2, –3) (c) (–2, 3) (d) (–2, –3)

61. The latus rectum of the curve $\frac{x^2}{a^2} + \frac{y^2}{b^2} = 1$ whose eccentricity = e, is

(a) $2a^2/b$ (b) $2b/a^2$ (c) $2a(1 - e^2)$ (d) $2b(1 - e^2)$

62. The eccentric angle of a point on the ellipse $x^2 + 3y^2 = 6$ at a distance 2 units from the centre of the ellipse is

(a) $\pi/4$ (b) $5\pi/4$ (c) $3\pi/4$ (d) $7\pi/4$

63. Which of the following is an exterior point of the ellipse $16x^2 + 9y^2 - 16x - 32 = 0$?

(a) (1/2, 2) (b) (1/4, 1)
(c) (3, –2) (d) none of these

64. If the inclination of the diameter PP′ of the ellipse $\frac{x^2}{a^2} + \frac{y^2}{b^2} = 1$ to the major axis is θ and PP'^2 is the AM of squares of major and minor axes then $\tan\theta$ is equal to

(a) b/a (b) a/b
(c) $\pi/4$ (d) none of these

65. The line $3x + 5y = k$ touches the ellipse $16x^2 + 25y^2 = 400$ if k is

(a) $\pm\sqrt{5}$ (b) $\pm\sqrt{15}$
(c) ± 25 (d) none of these

66. The equation of a tangent to the hyperbola $x^2 - 2y^2 = 18$ which is perpendicular to the line $x - y = 0$ is

(a) $x + y = 3$ (b) $x + y + 3 = 0$
(c) $x + y = 3\sqrt{2}$ (d) $x + y + 3\sqrt{2} = 0$

67. A point on the ellipse $x^2 + 3y^2 = 37$ where the normal is parallel to the line $6x - 5y = 2$, is
(a) $(5, -2)$ (b) $(5, 2)$
(c) $(-5, 2)$ (d) $(-5, -2)$

68. If the ordinate of the point of contact be 2 then the equation of the tangent to $x^2 + 4y^2 = 25$ is
(a) $3x + 8y = 25$ (b) $8x + 3y = 25$
(c) $8y - 3x = 25$ (d) $3x - 8y = 25$

69. A tangent to the ellipse $16x^2 + 9y^2 = 144$ making equal intercepts on both the axes is
(a) $y = x + 5$ (b) $y = x - 5$
(c) $y = -x + 5$ (d) $y = -x - 5$

70. If the tangent to the ellipse $x^2 + 4y^2 = 16$ at the point P'ϕ' is a normal to the circle $x^2 + y^2 - 8x - 4y = 0$ then ϕ is equal to
(a) $\pi/2$ (b) $\pi/4$ (c) 0 (d) $-\pi/4$

71. An ellipse having foci (3, 1) and (1, 1) passes through the point (1, 3) has the eccentricity
(a) $\sqrt{2} - 1$ (b) $\sqrt{3} - 1$ (c) $\frac{\sqrt{2}-1}{2}$ (d) $\frac{\sqrt{3}-1}{2}$

Answers

1. $5x^2 + 9y^2 + 30x - 18y - 126 = 0$

2. $9x^2 + 4y^2 + 36x - 24y + 36 = 0$

3. $12x^2 - 4y^2 - 24x + 32y - 127 = 0$

5. $\frac{8}{3}, \frac{\sqrt{5}}{3}, (\pm\sqrt{5} + 1, 2)$

6. $(1, 0), \frac{2\sqrt{3}}{3}; 6, 2\sqrt{3}$

7. $\frac{\pi}{4}$ **10.** $a = \frac{5}{2}\sqrt{2},\ b = 4$

16. $y - 2 = 0,\ 8x + 5y - 26 = 0;\ \tan^{-1}\frac{8}{5}$

18. $\left(\frac{a^2\alpha}{2(a^2 - b^2)}, \frac{b^2\beta}{2(b^2 - a^2)}\right)$

24. $x \pm \sqrt{2 + \sqrt{13}} \cdot y + 2 + \sqrt{13} = 0$

26. $x^2 + y^2 = a^2 + b^2$

29. $c(x^2 - a^2) - 2xy = 0$

30. $\left(\frac{x^2}{a^2} + \frac{y^2}{b^2} - 1\right)(b^4x^2 + a^4y^2) + \frac{c^2}{4}(b^2x^2 + a^2y^2) = 0$

32. (a) $16x^2 + 10xy + y^2 = 2$

33. $\frac{a^2}{x^2} - \frac{b^2}{y^2} = 4$ **34.** $4x^2 - y^2 = 25, \sqrt{5}$

37. (b) $(x^2 + y^2)^2 = 16x^2 - 9y^2$

39. $\frac{x^2}{a^2} + \frac{y^2}{b^2} = 2$ **40.** hyperbola, $(-2, 2)$

41. $\sqrt{2}$

42. $5x^2 - 2xy + 5y^2 - 76x - 88y + 506 = 0$

43. $\frac{x^2}{10^2} + \frac{y^2}{(5\sqrt{2})^2} = 1$ **44.** $\frac{8\sqrt{2}}{\sqrt{3}}, \frac{8\sqrt{2}}{\sqrt{5}}$ respectively

45. $\frac{x^2}{\left(\frac{3\sqrt{3}}{2}\right)^2} - \frac{y^2}{\left(\frac{3\sqrt{3}}{\sqrt{2}}\right)^2} = 1$ **46.** $\left(\frac{\sqrt{61}}{2}, 0\right), \left(\frac{-\sqrt{61}}{2}, 0\right)$

47. hyperbola, (1, 2) **48.** $2\sqrt{2}x - \sqrt{3}y = 2\sqrt{3}$

49. $2x - \sqrt{3}y - 3\sqrt{3} = 0$ **50.** $\tan^{-1}\sqrt{b/a}$

51. $2b\sqrt{a^2 - b^2}$ unit^2 **52.** 4

53. 1 **54.** $\frac{7}{5}\sqrt{41}$

55. $a^2l^2 - b^2m^2 = n^2$ **56.** $\frac{1}{\sqrt{3}}$

57. rectangular hyperbola, $\sqrt{2}$

58. $\left(\pm\frac{3}{\sqrt{2}}, \pm\sqrt{2}\right)$ **59.** 0

60. (ii), (a) **61.** (c)

62. (a), (b), (c), (d) **63.** (c)

64. (a) **65.** (c)

66. (a), (b) **67.** (b), (d)

68. (a), (c) **69.** (a), (b), (c), (d)

70. (a), (c) **71.** (a)

□

Chapter Test

Time: 120 minutes

1. Fill in the following incomplete sentences so that the sentences become true.

(a) If P and Q are the ends of a pair of conjugate diameters and C is the centre of the ellipse $4x^2 + 9y^2 = 36$ then the area of the ΔCPQ is ______ .

(b) The angle of intersection of the ellipse $x^2 + 4y^2 = 4$ and the parabola $x^2 + 1 = y$ is ______ .

(c) If any point on a hyperbola is $(3\tan\phi,\ 2\sec\phi)$ then the eccentricity of the hyperbola is ______ .

2. In the following, each question has four options of which one or more options are correct. Indicate the correct option(s).

(a) The normal to the rectangular hyperbola $xy = c^2$ at the point 't_1' meets the curve again at the point 't_2'. Then the value of $t_1^3 t_2$ is

A. 1 B. c C. $-c$ D. -1

(b) If $\cos\alpha = \dfrac{2}{3}$ then the range of values of ϕ for which the point 'ϕ' on the ellipse $x^2 + 4y^2 = 4$ falls inside the circle $x^2 + y^2 - 4x + 3 = 0$ is

A. $(-\alpha, \alpha)$ B. $(0, \alpha)$ C. (α, π) D. none of these

(c) The set of real values of λ for which the equation $(1 + \lambda)x^2 - 2\lambda xy + (\lambda - 2)y^2 - 4x + 3 = 0$ represents a hyperbola is

A. $(-2, +\infty)$ B. $(-\infty, -2)$ C. $\left(-2, \dfrac{2}{7}\right) \cup \left(\dfrac{2}{7}, +\infty\right)$ D. none of these

3. Tangents at right angles are drawn to the ellipse $x^2 + 4y^2 = 16$. Find the equation of the locus of the middle point of the chord of contact of the tangents.

4. From the point P(1, 3) tangents PA and PB are drawn to the ellipse $4x^2 + 9y^2 = 36$. Find the area of the quadrilateral PAOB where O is the centre of the ellipse.

5. If P and Q are two ends of conjugate diameters and R is the foot of the perpendicular to the chord PQ from the centre of the ellipse $9x^2 + 25y^2 = 225$ then find the equation of the locus of R.

6. For what real values of a, the point $(a, a - 2)$ falls in the interior of the ellipse $4x^2 + 9y^2 = 36$ as well as the parabola $y^2 = x$?

7. The sides of a triangle touch the circle $x^2 + y^2 = a^2$ and two of the vertices are on the lines $|y| = b$. Prove that the locus of the third vertex is $x^2 + y^2 = a^2 + \dfrac{4a^2b^2x^2}{(a^2 + b^2)^2}$.

8. Prove that the locus of the middle points of normal chords of the rectangular hyperbola $x^2 - y^2 = a^2$ is $(y^2 - x^2)^3 = 4a^2x^2y^2$.

9. Prove that $y = mx - \dfrac{(a^2 - b^2)m}{\sqrt{a^2 + b^2m^2}}$ is a normal to the ellipse $\dfrac{x^2}{a^2} + \dfrac{y^2}{b^2} = 1$ for all $m \in R$. Hence or otherwise prove that the sum of the angles that the four normals drawn from the point (α, β) to the ellipse $x^2 + 2y^2 = 4$ make with the major axis is equal to the sum of the angles that the two tangents from the same point make with the major axis.

10. A rectangular hyperbola whose centre is C is cut by a circle of radius r in four points P, Q, R, S. Prove that $CP^2 + CQ^2 + CR^2 + CS^2 = 4r^2$.

11. Let P, Q be two points on the ellipse $\frac{x^2}{a^2}+\frac{y^2}{b^2}=1$ whose eccentric angles differ by a right angle. The tangents at P and Q meet at R. Prove that the chord PQ bisects the line segment CR where C is the centre of the ellipse.

12. Any normal to the hyperbola $\frac{x^2}{a^2}-\frac{y^2}{b^2}=1$ meets its axes in M and N. The rectangle OMPN is completed where O is the centre of the hyperbola. Prove that the locus of P is $a^2x^2-b^2y^2=(a^2+b^2)^2$.

13. P, Q, R are points on a rectangular hyperbola, and PQ ⊥ PR. Prove that the tangent at P is perpendicular to QR.

14. Let P be a point such that the sum of the slopes of normals drawn from the point P to the rectangular hyperbola $xy=c^2$ is equal to the sum of ordinates of the feet of normals. Prove that the locus of P is a parabola. Find its focus and latus rectum.

15. Let A, B and C be vertices of an equilateral triangle inscribed in the circle of radius a, the origin being at the centre. Perpendiculars are drawn from A, B and C to the major axis of the ellipse $\frac{x^2}{a^2}+\frac{y^2}{b^2}=1$, $(a>b)$ and they meet the ellipse at P, Q and R respectively. If P, Q and R lie on the same side of the major axis as A, B and C respectively then prove that the normals to the ellipse drawn at the point P, Q and R are concurrent.
[**Note** If A = $(a\cos\theta, a\sin\theta)$ then P = $(a\cos\theta, b\sin\theta)$.]

□

Answers

1. (a) 3 unit^2 (b) 0° (c) $\frac{\sqrt{13}}{2}$

2. (a) D (b) A (c) C

3. $x^2+y^2=20\left(\frac{x^2}{16}+\frac{y^2}{4}\right)^2$

4. $\frac{1793}{340}$unit^2

5. $25x^2+9y^2=2(x^2+y^2)^2$

6. $\left(1,\frac{36}{13}\right)$

14. focus = $\left(0,\frac{c^2}{4}\right)$, latus rectum = c^2

□

Calculus

1. Function

Recap of Facts and Formulae

1. Function and its value

If $f: A \to B$ where $A \subseteq R$, $B \subseteq R$ is a real function then

- f is a rule according to which, corresponding to each $x \in A$ there is a unique real $f(x) \in B$.
- $f(a)$ is the value of the function at $x = a$, i.e., $x = a \in A$ corresponds to $f(a) \in B$.
- $y = f(x)$, i.e., $x \in A$ corresponds to $y \in B$ then x is the independent variable and y is the dependent variable.

2. Domain and range of a real-valued function

If $y = f(x)$ be a real function then

- domain of f = the set of real x for which $f(x)$ is real
- range of f = the set of real values of $f(x)$ for $x \in$ domain f
  $= [\min f(x), \max f(x)]$

Note The above result on the range of f is true for continuous functions.

If the domain of $\phi(x)$ be D_1 and the domain of $\psi(x)$ be D_2 then

- the domain of $f(x) = \phi(x) \pm \psi(x)$ is $D_1 \cap D_2$
- the domain of $f(x) = \phi(x) \times \psi(x)$ is $D_1 \cap D_2$
- the domain of $f(x) = \dfrac{\phi(x)}{\psi(x)}$ is $D_1 \cap D_2 - E$

where E = the set of zeros of $\psi(x)$.

3. Equality of functions

- Two functions $f(x)$ and $\phi(x)$ are equal if
  (i) domain of f = domain of ϕ and
  (ii) $f(x) = \phi(x)$ for all $x \in$ the common domain

For example

$f(x) = \log x^2$, $x > 0$ and $\phi(x) = 2\log x$, $x > 0$ are equal functions because they have the same domain $(0, +\infty)$ and for each $x \in (0, +\infty)$ we have $f(x) = \phi(x)$ as $\log x^2 = 2\log x$.

But $F(x) = \log x^2$, $x \in R$ and $\phi(x) = 2\log x$, $x > 0$ are not equal because they do not have the same domain.

4. Different ways of defining a function

A function may be defined in any one of the following ways:

(i) uniform definition
(ii) piecewise definition
(iii) general definition given by a property of the function.

For example Let (a) $f(x) = x^2 + 1$

(b) $g(x) = 2x - 1$, $x < 0$
$x + 3$, $x \geq 0$

(c) $h(x + y) = h(x) \cdot h(y)$ for all $x, y \in R$.

In (a), the definition is uniform. For every $x \in R, f(x) = x^2 + 1$.

In (b), the definition is piecewise. For negative values of x, $g(x) = 2x - 1$ is to be used while for non-negative values of x, $g(x) = x + 3$ is to be used.

In (c), the definition is general. The function is described by no rules but by a property of $h(x)$. Clearly $h(x) = e^x$ obeys the property, but there may be other functions satisfying the same property.

5. Some special piecewise functions

- *Modulus function* $f(x) = |x|$,
  i.e., $f(x) = x$, $x > 0$
  0, $x = 0$
  $-x$, $x < 0$
- *Sign function* $f(x) = 1$, $x > 0$
  0, $x = 0$
  -1, $x < 0$

- *Step function* or *greatest integer function*
  $f(x) = [x]$, where x = greatest integer less than or equal to x, i.e, $f(x) = n, n \le x < n+1$ where n is any integer
- *Dirichlet function* $f(x) = 1$, x is rational
  0, x is irrational

6. Even, odd and periodic functions

- A function $f(x)$ is even if
  $f(-x) = f(x)$ for all $x \in$ domain
- A function $f(x)$ is odd if
  $f(-x) = -f(x)$ for all $x \in$ domain
- A function $f(x)$ is periodic if
  $f(x+k) = f(x)$ for all $x \in$ domain and k is a positive constant.
  If k is the least possible positive constant then k is the *period* of the function.

For example $\cos x$, $|x|$, $x^2 - 1$ are even functions
$\sin x$, x^3 are odd functions
$x + 2$, e^x are neither even nor odd functions
$\sin x$, $\tan x$, $x - [x]$ are periodic functions

- Some standard periodic functions with their periods:

Function	Period	Function	Period
$\sin x$, $\cos x$	2π	$\sin ax$, $\cos ax$	$\frac{2\pi}{a}$
$\sec x$, $\operatorname{cosec} x$	2π	$\sec ax$, $\operatorname{cosec} ax$	$\frac{2\pi}{a}$
$\tan x$, $\cot x$	π	$\tan ax$, $\cot ax$	$\frac{\pi}{a}$
$x - [x]$	1		
constant	indeterminate		

If $f(x)$ and $g(x)$ are two trigonometrical functions of periods λ and μ then $af(x) + bg(x)$ is periodic and has the period = l.c.m. of $\{\lambda, \mu\}$

7. One-one (injective) and many-one functions

- A function $f(x)$ is one-one (injective) if $f(x_1) = f(x_2) \Rightarrow x_1 = x_2$; otherwise it is many-one.

For example $f(x) = 3x - 1$ is one-one but
$g(x) = x^2 + 1$ is many-one
because $f(x_1) = f(x_2) \Rightarrow 3x_1 - 1 = 3x_2 - 1$
$\Rightarrow x_1 = x_2$
and $g(x_1) = g(x_2) \Rightarrow x_1^2 + 1 = x_2^2 + 1$
$\Rightarrow x_1^2 = x_2^2$
$\Rightarrow x_1 = x_2, -x_2$

8. Into and onto (surjective) functions

- A function $f : A \to B$ is an *into* function if for at least one $\beta \in B$ there is no $\alpha \in A$ such that $f(\alpha) = \beta$.
  If there is no such $\beta \in B$ then f is an *onto* (surjective) function.

For example $y = f(x) = 2x - 1$ is an onto function from R to R but $y = g(x) = x^2 + 1$ is not onto, i.e., into function from R to R because,
$x = \frac{f(x)+1}{2} = \frac{y+1}{2}$ and so for every $y \in R$ we have the real $x \left(= \frac{y+1}{2}\right)$, and $x = \sqrt{g(x) - 1} = \sqrt{y-1}$ and so for $y < 1$ we have no real $x(= \sqrt{y-1})$.

- In order to check whether $y = f(x)$ from the set A to B is onto or not, write x in terms of y and see if for every $y \in B$, $x \in A$. If so, it is onto. Otherwise, it is into.
- A function f which is one-one and onto (i.e., injective as well as surjective) is called *bijective.*

9. Inverse function

- Let $y = f(x)$ be a function whose domain is A and whose range is B. If for every $y \in B$ there exists a single value of x such that $f(x) = y$ then an inverse function of $f(x)$ from B to A is defined, given by $x = f^{-1}(y)$.

For example If $y = f(x) = 4x - 7$ then the inverse of f is $g(= f^{-1})$ such that $x = \frac{7+y}{4}$.

- The necessary condition for existence of the inverse function of f is that f is one-one and onto, i.e., bijective.
- A sufficient condition for the function f to have an inverse is that f is strictly monotonic increasing (or monotonic decreasing) function over the whole domain.
  (See the chapter *Monotonic Function*)

10. Composite function

If $f(x)$ be a function whose domain is A and range is B, and $g(y)$ be a function whose domain is B and the range is C then $(g \circ f)(x)$ is a composite function whose domain is A and the range is C such that $(g \circ f)(x) = z \in C$ where $f(x) = y$ and $g(y) = z$.

For example Let $y = f(x) = 2x + 3$ and $g(x) = \sin x$
then $(g \circ f)(x) = g\{f(x)\} = g(y)$
$= g(2x+3) = \sin(2x+3)$.

Selected Solved Examples

1. ABCD is a square of side a. A line parallel to the diagonal BD at a distance x from the vertex A cuts two adjacent sides. Express the area of the segment of the square, with A at a vertex, as a function of x.

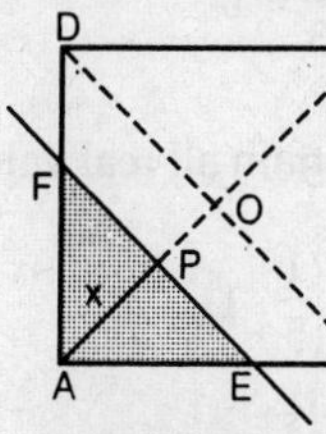

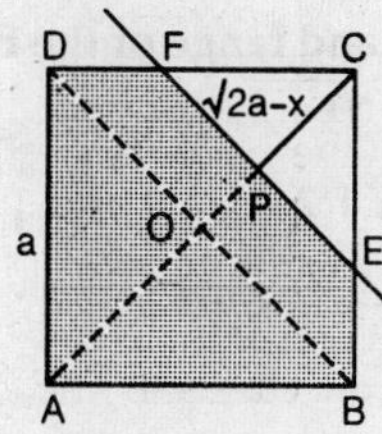

There are two different situations:

(i) when $x = AP \le OA$, i.e., $x \le \dfrac{a}{\sqrt{2}}$

(ii) when $x = AP > OA$, i.e., $x > \dfrac{a}{\sqrt{2}}$ but $x \le \sqrt{2}a$.

Case (i) $\text{ar}(\Delta AEF) = x^2 = \dfrac{1}{2} x \cdot 2x$

$\{\because PE = PF = AP = x\}$

Case (ii) $\text{ar}(ABEFDA) = \text{ar}(ABCD) - \text{ar}(\Delta CFE)$

$$= a^2 - \frac{1}{2}\cdot(\sqrt{2}a - x)\cdot 2(\sqrt{2}a - x)$$

$$= a^2 - (\sqrt{2}a - x)^2$$

$$= 2\sqrt{2}ax - x^2 - a^2$$

$\therefore$ the required function $f(x)$ is as follows:

$$f(x) = x^2,\ 0 \le x \le \frac{a}{\sqrt{2}}$$

$$2\sqrt{2}ax - x^2 - a^2,\ \frac{a}{\sqrt{2}} < x \le \sqrt{2}a.$$

2. If $f(x)$ be a function satisfying

$f(x)\cdot f\left(\dfrac{1}{x}\right) = f(x) + f\left(\dfrac{1}{x}\right)$ and $f(4) = 65$ then find $f(6)$.

Here, $f(x)\cdot f\left(\dfrac{1}{x}\right) - f(x) = f\left(\dfrac{1}{x}\right)$

$\therefore \quad f(x)\left\{f\left(\dfrac{1}{x}\right) - 1\right\} = f\left(\dfrac{1}{x}\right)$

$\therefore \quad f(x)\left\{f\left(\dfrac{1}{x}\right) - 1\right\} = f\left(\dfrac{1}{x}\right) - 1 + 1$

$\therefore \quad f(x)\left\{f\left(\dfrac{1}{x}\right) - 1\right\} - \left\{f\left(\dfrac{1}{x}\right) - 1\right\} = 1$

$\therefore \quad \left\{f\left(\dfrac{1}{x}\right) - 1\right\}\{f(x) - 1\} = 1 \qquad \text{... (1)}$

So, $f(x) - 1$ and $f\left(\dfrac{1}{x}\right) - 1$ are reciprocal to each other.

Also $x, \dfrac{1}{x}$ are reciprocal to each other.

Thus, (1) can hold only when

$f(x) - 1 = \pm x^n$ where $n \in R$.

$\therefore \quad f(x) = \pm x^n + 1;$

but $f(4) = 65$

$\therefore \quad 65 = \pm 4^n + 1$

$\therefore \quad 4^n = 64 \quad (\because 4^n > 0)$

$\therefore \quad 4^n = 4^3$

$\therefore \quad n = 3.$

So, $f(x) = x^3 + 1$

$\therefore \quad f(6) = 6^3 + 1$

$= 217.$

3. Find the domain of the function

$$f(x) = \frac{1}{\sqrt{[x]^2 - [x] - 6}}.$$

$\text{dom} f = \{x \in R \mid [x]^2 - [x] - 6 > 0\}$.

Now, we have to solve the inequation

$[x]^2 - [x] - 6 > 0$

or $\{[x] - 3\}\{[x] + 2\} > 0.$

Using sign-scheme,

$\therefore \quad [x] < -2 \quad$ or $\quad [x] > 3.$

But $[x] < -2 \quad \Rightarrow \quad [x] = -3, -4, -5, \ldots$

$\therefore \quad x < -2$

$[x] > 3 \quad \Rightarrow \quad [x] = 4, 5, 6, \ldots$

$\therefore \quad x \ge 4$

$\therefore \quad \text{dom} f = (-\infty, -2) \cup [4, +\infty).$

4. Find the domain of the function
$$f(x) = \sqrt{3 - 2^x - 2^{1-x}} + \sqrt{\sin^{-1} x}.$$

Let $\phi(x) = \sqrt{3 - 2^x - 2^{1-x}}$

and $\psi(x) = \sqrt{\sin^{-1} x}$.

So $f(x) = \phi(x) + \psi(x)$.

Let dom $\phi = D_1$, dom $\psi = D_2$. Then dom $f = D_1 \cap D_2$.

So, $D_1 = \{x \in R \mid 3 - 2^x - 2^{1-x} \geq 0\}$ and

$D_2 = \{x \in R \mid \sin^{-1} x \geq 0\}$.

Now, $3 - 2^x - 2^{1-x} \geq 0$

or $3 - 2^x - \dfrac{2}{2^x} \geq 0$

or $3 \cdot 2^x - (2^x)^2 - 2 \geq 0$

or $(2^x)^2 - 3 \cdot 2^x + 2 \leq 0$

$\therefore$ $(2^x - 2)(2^x - 1) \leq 0$

By sign-scheme,

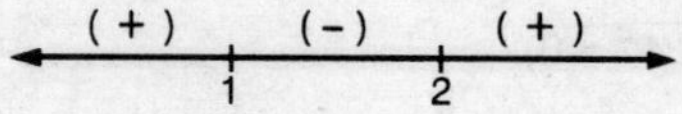

$\therefore$ $2^x \in [1, 2]$, i.e., $2^0 \leq 2^x \leq 2$; $\therefore$ $0 \leq x \leq 1$

$\therefore$ $D_1 = [0, 1]$.

Now, $\sin^{-1} x \geq 0 \Rightarrow 0 \leq \sin^{-1} x \leq \dfrac{\pi}{2}$

{$\because$ the principal value of $\sin^{-1} x$ is positive in the first quadrant}

$\Rightarrow 0 \leq x \leq 1$

$\therefore$ $D_2 = [0, 1]$

$\therefore$ dom $f = D_1 \cap D_2 = [0, 1] \cap [0, 1] = [0, 1]$.

5. Find the domain and range of $f(x) = \dfrac{x^2 - 3x + 2}{x^2 + x - 6}$.

Here, $f(x) = \dfrac{x^2 - 3x + 2}{x^2 + x - 6}$

$f(x)$ is not defined for those values of x for which

$x^2 + x - 6 = 0$

or $(x + 3)(x - 2) = 0$, i.e., $x = 2, -3$

$\therefore$ domain of $f = R - \{2, -3\}$.

Now $y = f(x) = \dfrac{(x-2)(x-1)}{(x+3)(x-2)} = \dfrac{x-1}{x+3}$

$\therefore$ $f(-3) = \infty$ and

$\lim\limits_{x \to 2} f(x) = \left(\text{value of } \dfrac{x-1}{x+3} \text{ at } x = 2\right)$

$= \dfrac{1}{5}$. (See *Limit*)

Also, $y = \dfrac{x-1}{x+3} \Rightarrow x = -\dfrac{3y+1}{y-1}$

$\therefore$ for all real $y (\neq 1)$, x is real, and

$f(\infty) = \lim\limits_{x \to \infty} \dfrac{x-1}{x+3} = 1$ (See *Limit*)

$\therefore$ for real x, $f(x)$ can attain all real values except $\dfrac{1}{5}$ and 1.

$\therefore$ range of $f(x) = R - \left\{\dfrac{1}{5}, 1\right\}$.

6. Find the domain and range of the function
$$f(x) = \sqrt{2 - x} + \sqrt{1 + x}.$$

Domain of $f(x) = \{x \mid 2 - x \geq 0 \text{ and } 1 + x \geq 0\}$.

Now, $2 - x \geq 0 \Rightarrow x \leq 2$

$1 + x \geq 0 \Rightarrow x \geq -1$

$\therefore$ domain of $f(x) = [-1, 2]$.

Again, $\{f(x)\}^2 = (\sqrt{2 - x} + \sqrt{1 + x})^2$

$= 3 + 2\sqrt{(2 - x)(1 + x)}$

$= 3 + 2\sqrt{2 + x - x^2}$

$= 3 + 2\sqrt{\dfrac{9}{4} - \left(x - \dfrac{1}{2}\right)^2}$

$\therefore$ the greatest value of $\{f(x)\}^2 = 3 + 2 \cdot \sqrt{\dfrac{9}{4}} = 6$,

when $x = \dfrac{1}{2}$

the least value of $\{f(x)\}^2 = 3 + 0 = 3$,

when $x - \dfrac{1}{2} = \dfrac{3}{2}$, i.e., $x = 2$

$\therefore$ the greatest value of $f(x) = \sqrt{6}$ and

the least value of $f(x) = \sqrt{3}$

$\therefore$ range of $f(x) = [\sqrt{3}, \sqrt{6}]$.

7. A function $f: R \to R$ is defined by $f(x) = \dfrac{\alpha x^2 + 6x - 8}{\alpha + 6x - 8x^2}$. Find the interval of values of α for which f is onto. Is the function one-to-one for $\alpha = 3$? Justify your answer.

As $f: R \to R$, the function will be onto if

$\dfrac{\alpha x^2 + 6x - 8}{\alpha + 6x - 8x^2}$

can assume any real value for real values of x.

Let $\dfrac{\alpha x^2 + 6x - 8}{\alpha + 6x - 8x^2} = y$

$\therefore \quad (\alpha + 8y)x^2 + 6(1-y)x - (8 + \alpha y) = 0.$

As x is real, $D \geq 0$

$\therefore \quad 36(1-y)^2 + 4(\alpha + 8y)(8 + \alpha y) \geq 0$

or $\quad 9(1 - 2y + y^2) + 8\alpha + (\alpha^2 + 64)y + 8\alpha y^2 \geq 0$

or $\quad (9 + 8\alpha)y^2 + (\alpha^2 + 46)y + (9 + 8\alpha) \geq 0.$

This has to be true for all real y.

Hence its $D \leq 0$ and $9 + 8\alpha > 0$

i.e., $(\alpha^2 + 46)^2 - 4(9 + 8\alpha)^2 \leq 0$ ... (1)

and $9 + 8\alpha > 0$, i.e., $\alpha > \frac{-9}{8}$ ... (2)

$\therefore$ from (1),

$(\alpha^2 + 46 + 18 + 16\alpha)(\alpha^2 + 46 - 18 - 16\alpha) \leq 0$

or $\quad (\alpha^2 + 16\alpha + 64)(\alpha^2 - 16\alpha + 28) \leq 0$

or $\quad (\alpha + 8)^2(\alpha - 14)(\alpha - 2) \leq 0$

$\therefore \quad (\alpha - 14)(\alpha - 2) \leq 0$

$$\left\{\because \ \alpha \neq -8 \text{ because by (2), } \alpha > \frac{-9}{8}\right\}$$

By sign-scheme, $2 \leq \alpha \leq 14$.

When $\alpha = 3$, $f(x) = \dfrac{3x^2 + 6x - 8}{3 + 6x - 8x^2}$.

Let $f(x_1) = f(x_2)$. Then

$$\frac{3x_1^2 + 6x_1 - 8}{3 + 6x_1 - 8x_1^2} = \frac{3x_2^2 + 6x_2 - 8}{3 + 6x_2 - 8x_2^2}$$

$$= \frac{3(x_1^2 - x_2^2) + 6(x_1 - x_2)}{6(x_1 - x_2) - 8(x_1^2 - x_2^2)},$$

(using ratio and proportion)

$\therefore \quad \{6(x_1 - x_2) - 8(x_1^2 - x_2^2)\}\{3x_1^2 + 6x_1 - 8\}$

$= \{3(x_1^2 - x_2^2) + 6(x_1 - x_2)\}(3 + 6x_1 - 8x_1^2)$

or $\quad (x_1 - x_2)[(6 - 8x_1 - 8x_2)(3x_1^2 + 6x_1 - 8)$

$- (3x_1 + 3x_2 + 6)(3 + 6x_1 - 8x_1^2)] = 0$

or $\quad (x_1 - x_2)[(6 - 8x_1)(3x_1^2 + 6x_1 - 8)$

$- (3x_1 + 6)(3 + 6x_1 - 8x_1^2) - 8x_2(3x_1^2 + 6x_1 - 8)$

$- 3x_2(3 + 6x_1 - 8x_1^2)] = 0$

or $\quad (x_1 - x_2)[-24x_1^3 - 30x_1^2 + 100x_1 - 48$

$- (-24x_1^3 - 30x_1^2 + 45x_1 + 18)$

$-x_2(-66x_1 + 55)] = 0$

or $\quad (x_1 - x_2)[55x_1 - 66 - x_2(-66x_1 + 55)] = 0$

$\therefore \quad x_1 = x_2$ or $55x_1 - 66 = x_2(55 - 66x_1)$

i.e., $x_2 = x_1$ or $\dfrac{5x_1 - 6}{5 - 6x_1}$ $\quad \left(x_1 \neq \dfrac{5}{6}\right)$

$\therefore$ the function is not one-to-one for $\alpha = 3$.

8. If $f(x)$ satisfies the relation $f(x + y) = f(x) + f(y)$ for all $x, y \in R$ and $f(1) = 5$ then find $\sum_{n=1}^{m} f(n)$.

Also prove that $f(x)$ is odd.

Here $f(r) = f(\overline{r - 1 + 1})$

$= f(r - 1) + f(1)$ {using the given property}

$\therefore \quad f(r) = f(r-1) + 5$ ... (1)

$= \{f(r-2) + 5\} + 5$, using (1)

$= f(r-2) + 2 \cdot 5 = \{f(r-3) + 5\} + 2 \cdot 5$

$= f(r-3) + 3 \cdot 5$

$= \ldots$

$= f(r - \overline{r-1}) + (r-1) \cdot 5$

$= f(1) + 5(r-1) = 5 + 5(r-1) = 5r$

$$\therefore \quad \sum_{n=1}^{m} f(n) = \sum_{n=1}^{m} 5n = 5(1 + 2 + 3 + \ldots + m)$$

$$= \frac{5m(m+1)}{2}.$$

Now, putting $x = 0$ and $y = 0$ in the given relation,

$f(0 + 0) = f(0) + f(0); \quad \therefore \quad f(0) = 0.$

Also, putting $-x$ for y in the given relation

$f(x - x) = f(x) + f(-x)$

$\therefore \quad f(0) = f(x) + f(-x)$

$\therefore \quad 0 = f(x) + f(-x)$, i.e., $f(-x) = -f(x)$.

So, the function is odd.

9. If $a, b \in R$ be fixed positive numbers such that

$$f(a + x) = b + [b^3 + 1 - 3b^2 \cdot f(x) + 3b\{f(x)\}^2 - \{f(x)\}^3]^{1/3}$$

for all $x \in R$ then prove that $f(x)$ is a periodic function.

Here, $\{f(a + x) - b\}^3$

$= b^3 + 1 - 3b^2 f(x) + 3b\{f(x)\}^2 - \{f(x)\}^3$

$= 1 - [\{f(x)\}^3 - 3b \cdot \{f(x)\}^2 + 3b^2 \cdot f(x) - b^3]$

$= 1 - \{f(x) - b\}^3$

$\therefore \quad \{f(a + x) - b\}^3 + \{f(x) - b\}^3 = 1$ ... (1)

This is true for all x.

Putting $a + x$ for x in (1), we get,

$$\{f(2a+x)-b\}^3+\{f(a+x)-b\}^3=1 \qquad \dots (2)$$

Subtracting (1) from (2),

$$\{f(2a+x)-b\}^3-\{f(x)-b\}^3=0$$

or $\quad \{f(2a+x)-b\}^3=\{f(x)-b\}^3$

or $\quad f(2a+x)-b=f(x)-b$

or $\quad f(2a+x)=f(x)$

$\therefore \quad f(x)$ is a periodic function.

10. If f, g, h are functions from R to R such that

$f(x)=x^2-1,\ g(x)=\sqrt{x^2+1},$

$h(x)=0$, if $\quad 0$

x, if $\quad \geq 0$

then find the composite function $h\circ(f\circ g)$ and determine whether the function $f\circ g$ is invertible and the function h is the identity function.

Here $f(x)=x^2-1$ for all x

and $g(x)=\sqrt{x^2+1}$ for all x

$\therefore \quad f\{g(x)\}=\{g(x)\}^2-1$

$=x^2+1-1=x^2$ for all x

$\therefore \quad h\{f(g(x))\}=h(x^2)=x^2$

because $x^2\geq 0$ [from definition of $h(x)$].

Now, $f\{g(x)\}=x^2$ for all x

$\Rightarrow \quad (f\circ g)(x)=x^2$ for all x.

As $x^2\geq 0$, $(f\circ g)(x)$ cannot be negative.

So $f\circ g$ is not an onto function. Hence $f\circ g$ is not invertible.

Again, $h(x)=x$ for $x\geq 0$.

But, by definition $h(x)\neq x$ for $x<0$.

Hence h is not the identity function.

Note $f(x)$ is an identity function if $f(x)=x$ for all $x\in$ domain.

Exercises

1. Let $f(x)=\dfrac{ax+b}{cx+d},\ x\neq-\dfrac{d}{c}$.

   If $d=-a$, show that $f\{f(x)\}=x$ is an identity.

2. If $f(x)=(a-x^n)^{\frac{1}{n}}, a>0, n\in N$ then show that $f\{f(x)\}=x$.

3. If $f(x)=\log_e(x+\sqrt{1+x^2})$,

   $g(x)=\log_e(1+\sqrt{1+x^2})$ and

   $h(x)=f(x)-g(x)$ then prove that $h\left(\dfrac{1}{x}\right)=-h(x)$.

4. In the figure, $BC=b$ and $AH=h$ where AH ⊥ BC. If $EF=x$ where EF ⊥ BC then express the area and perimeter of the rectangle DEFG as functions.

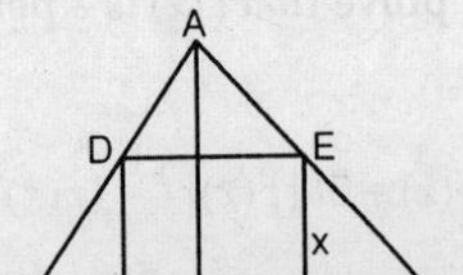

5. O is a fixed point on a circle of unit radius whose centre is P. A chord perpendicular to OP at a distance x from O cuts the circle at A and B. Express the area of the segment AOB as a function of x.

6. Find the natural number a for which

   $$\sum_{k=1}^{n} f(a+k)=16(2^n-1)$$

   where the function f satisfies the relation $f(x+y)=f(x)\cdot f(y)$ for all natural numbers x, y and also $f(1)=2$.

Find the domain of the following functions:

7. $f(x)=\sqrt{x^2-4x+3}$

8. $f(x)=\dfrac{1}{\sqrt{x^2-x-2}}$

9. $f(x)=\dfrac{1}{\sqrt{|x|-x}}$

10. $f(x)=\sqrt{\dfrac{(x+1)(x-3)}{x-2}}$

11. $f(x)=\sqrt{\log_{10}\dfrac{3-x}{x}}$

12. $f(x)=\sin^{-1}\left\{\log_2\left(\dfrac{1}{2}x^2\right)\right\}$

13. $y=\cos^{-1}\dfrac{2}{2+\sin x}$

14. $f(x)=\sqrt{5|x|-x^2-6}$

15. $y = \cos^{-1}\{x - [x]\}$

16. $f(x) = \log(x - [x])$ where $[x]$ is the usual step function

17. $y = \dfrac{\log_e x}{\sin^{-1}(x-3)}$

18. $f(x) = \log_{2x-5}(x^2 - 3x - 10)$

Find the domain of the real-valued functions:

19. $f(x) = \sqrt{4-x} + \sqrt{x-5}$

20. $f(x) = \sqrt{x^2 - 1} + \dfrac{1}{\sqrt{x^2 - 3x + 2}}$

21. $f(x) = \sqrt{3-x} + \cos^{-1}\dfrac{3-2x}{5}$

22. $f(x) = \cos^{-1}\dfrac{1+x^2}{2x} + \sqrt{\sin(\cos x)}$

23. $y = \sin^{-1}x + \sqrt{2x - x^2} - \dfrac{1}{\sqrt{8x - 4x^2 - 3}}$

24. $f(x) = \cos^{-1}x \cdot \sqrt{1 - \log_3(2x^2 + 6x - 5)}$

25. If $\log_x y \ \log_{xy} y \cdot \log_{x^2 y} y = \dfrac{1}{6}$ then express y as a real function of x and indicate the domain of definition.

26. $f(x)$ is defined over $[0, 1]$. Find the domain of the functions
    (i) $f(2x+3)$ (ii) $f(\sin x)$ (iii) $f(\cos x)$.

Find the domain and range of the following:

27. $f(x) = \log_{10}(3x^2 - 4x + 5)$

28. $f(x) = \dfrac{x}{1+x^2}$

29. $f(x) = \dfrac{x}{|x|}$

30. $y = |x-4| + |x-3| + |x-2| + |x-1|$

31. $f(x) = \dfrac{1}{3 - \cos 2x}$

32. If $f(x) = \dfrac{x^2 - 3x + 4}{(x+4)(x-1)}$ and domain $= R - \{1, -4\}$ then find the range of $f(x)$.

33. If $0 < c \le 1$ then prove that the range of
$$f(x) = \frac{x^2 + 2x + c}{x^2 + 4x + 3c} \text{ is } R.$$

34. Find the equations of the lines parallel to the x-axis between which the graph of $y = \left|\dfrac{12x}{4x^2 + 9}\right|$ lies.

35. Let $f(x) = \dfrac{x-1}{m - x^2 + 1}$, $x \in R$. If the range of $f(x)$ does not contain the open interval $\left(-1, \dfrac{-1}{3}\right)$ then prove that $m \le \dfrac{-1}{4}$.

36. Let $f: \{x, y, z\} \to \{a, b, c\}$ be an one-one function. It is known that only one of the following statements is true:
    (i) $f(x) \ne b$ (ii) $f(y) = b$ (iii) $f(z) \ne a$.
    Find the function f.

37. Verify if $f(x) = \dfrac{x^2 - 8x + 18}{x^2 + 4x + 30}$ is an one-one function.

38. Check whether $f(x) = \dfrac{x}{1 + |x|}$, $x \in R$ is one-one or onto or both or none.

39. Prove that $f: (-1, 1) \to R$ defined by
$$f(x) = \frac{x}{1+x}, \ -1 < x \le 0$$
$$\frac{x}{1-x}, \ 0 < x < 1$$
is a bijective function.

40. Prove that, if $f(x)$ is any function, $f(x) + f(-x)$ is an even function while $f(x) - f(-x)$ is an odd function. Hence, prove that every function can be expressed as the sum of an even function and an odd function.

41. Prove that the product of two odd functions is an even function while their sum is an odd function.

42. (a) If $f(t) = \dfrac{t}{e^t - 1} + \dfrac{t}{2} + 1$, show that $f(t)$ is an even function.
    (b) Prove that $f(x) = \sqrt[3]{(1-x)^2} + \sqrt[3]{(1+x)^2}$ is an even function.

43. If the function f satisfies the relation
$$f(x+y) + f(x-y) = 2f(x) \cdot f(y)$$
for all $x, y \in R$ and $f(0) \ne 0$ then prove that $f(x)$ is an even function.

44. Define a function $f(x)$ suitably in the interval $[0, \propto)$ so that $f(x)$ may be an
    (i) even real function
    (ii) odd real function
    whose definition is as follows:
$$f(x) = 1, \ x < -1$$
$$-x, \ -1 \le x \le 0$$

45. Extend $f(x) = x^2 + x$ defined in $[0, 3]$ onto the interval $[-3, 3]$ so that $f(x)$ is (i) even (ii) odd.

Mention whether the following functions are periodic. If yes, mention their periods:

46. $f(x) = 2x - [2x]$
where $[x]$ denotes the greatest integer function

47. $g(x) = 1 + \dfrac{3}{2 - \sin^2 x}$

48. $|\sin x| + |\cos x|$

49. $\log(2 + \cos 3x)$

50. $\sin^4 x + \cos^4 x$

51. Prove that the function $f(x) = 1$, x is rational
0, x is irrational
is a periodic function with indeterminate period.

52. Prove that (i) $f(x) = x + \sin x$
(ii) $f(x) = \cos x^2$ are nonperiodic.

53. If $f(x) \in [1, 2]$ when $x \in R$ and for a fixed positive real number p,
$$f(x + p) = 1 + \sqrt{2f(x) - \{f(x)\}^2} \text{ for all } x \in R$$
then prove that $f(x)$ is a periodic function.

54. Let $f(x) = (1 + \tan x)\left\{1 + \tan\left(\dfrac{\pi}{4} - x\right)\right\}$ and let $g(x)$ be defined for every real x. Prove that the function $(g \circ f)(x)$ is constant for all those x for which $f(x)$ is defined.

55. Let f be increasing and g be decreasing functions from $[0, +\infty)$ to $[0, +\infty)$. Let $h(x) = (f \circ g)(x)$ and $h(0) = 0$. Prove that $h(x) = 0$ for all $x \in [0, +\infty)$.

56. Find the inverse of the function
$$y = \log_e(x + \sqrt{x^2 + 1}).$$

57. If $f(x) = 1, x > 0$
$0, x = 0$
$-1, x < 0$
and $g(x) = (1 + x^2) f(x)$
then find the inverse function of $g(x)$.

Objective Questions

Fill in the blanks.

58. If $f(x) = \dfrac{1 - x}{1 + x}$, $x \neq 0$ then $f\left\{f\left(\dfrac{1}{x}\right)\right\} =$ ______ .

59. If $f(x) = \cos(\log x)$ then
$$f(x) \cdot f(y) - \frac{1}{2}\left[f\left(\frac{x}{y}\right) + f(xy)\right] = \text{______} .$$

60. If $f(x) = \dfrac{36}{x^2} + x^2$ then $f(\alpha) =$ ______
where α is a root of $\dfrac{6}{x} + x = 2$.

61. If $f(x) = x + \dfrac{1}{x}$ and $\phi(x) = x^4 + \dfrac{1}{x^4}$ then $\phi(a) =$ ______
where $f(a) = 5$.

62. If $f(x)$ is a polynomial function of the second degree such that $f(0) = 5$, $f(-1) = 10$ and $f(1) = 6$ then $f(x) =$ ______ .

63. If $f(x + 1) = x^2 - 3x + 2$ then $f(x) =$ ______ .

64. The functions $f(x) = \sqrt{x} \cdot \sqrt{x - 1}$ and $\phi(x) = \sqrt{x(x - 1)}$ are identical in the interval ______ .

65. P is a point on a circle of radius a and centre O. The distance of a chord AB from P, perpendicular to OP, is x. $\angle AOB (\leq \pi)$ as a function of x is ______ .

66. If $f(x) + g(x) = e^{-x}$ where $f(x)$ is an even function and $g(x)$ is an odd function then $f(x) =$ ______ and $g(x) =$ ______ .

67. If f is an even function on the interval $(-5, 5)$ then four real values of x satisfying the equation $f(x) = f\left(\dfrac{x + 1}{x + 2}\right)$ are
______ , ______ , ______ and ______ .

68. If $f(x)$ is a periodic function of the period k then $f(ax + b)$ is a periodic function of the period ______ .

69. The period of the function $f(x) = 2\sin 2x + 3\cos 3x$ is ______ .

70. The domain of $f(x) = \sqrt{\log_{10}(\sin \pi x)}$ is ______ .

71. If $f(x) = 3\sin\sqrt{\dfrac{\pi^2}{16} - x^2}$ then its domain = ______ and range = ______ .

72. If $f(x) = \sin x - \cos x$ then its domain = ______ and range = ______ .

73. The domain of the function $f(x) = \log_2\log_3\log_4 x$ is the interval ______ .

74. If $f : R \to R$ be a function given by
$$f(x) = \sqrt{1 - \sqrt{1 - x^2}}$$
then the domain of the function is the interval ______ .

75. If the function $f : [2, +\infty) \to X$ be bijective where $f(x) = 5 - 4x + x^2$ then $X =$ ______ .

76. If $y = 5^{\log_e x}$ then the inverse function is $x =$ ______.

77. If $f(x) = \log_{100x}\left(\dfrac{2\log_{10} x + 2}{-x}\right)$ and $g(x) = \{x\} =$ fractional part of x and $(f \circ g)(x)$ exists then the range of $g(x) =$ ______ .

78. If $f(x) = \sin^2 x + \sin^2\left(x + \frac{\pi}{3}\right) + \cos x \cdot \cos\left(x + \frac{\pi}{3}\right)$ and $g\left(\frac{5}{4}\right) = 1$ then $(g \circ f)(x) =$ ______ .

Choose the correct option(s).

79. If $y = f(x) = \frac{x+2}{x-1}$ then $f(y)$ is equal to

(a) x (b) $\frac{y+2}{y-1}$ (c) $\frac{x-1}{x+2}$ (d) $\frac{y-1}{y+2}$

80. If $f(x) = \cos[\pi^2]x + \cos[-\pi^2]x$ where $[x]$ is the greatest integer function then which one is correct?

(a) $f(\pi/2) = -1$ (b) $f(\pi) = 1$
(c) $f(-\pi) = 0$ (d) $f(\pi/4) = 2$

81. If $f(x) = \log(x + \sqrt{1+x^2})$ then $f(x)$ is
(a) periodic (b) even function
(c) odd function (d) none of these

82. If $f(x) = \log \frac{1+x}{1-x}$ then
(a) $f(x_1) \cdot f(x_2) = f(x_1 + x_2)$
(b) $f(x)$ is odd
(c) $f(x_1) + f(x_2) = f\left(\frac{x_1 + x_2}{1 + x_1 x_2}\right)$
(d) $f(x)$ is even

83. The period of $f(x) = 2\cos \frac{x - \pi}{3}$ is
(a) $2\pi/3$ (b) $4\pi/3$ (c) 2π (d) 6π

84. Let $g(x)$ be a function defined on $[-1, 1]$. If the area of the equilateral triangle with two of its vertices at $(0, 0)$ and $\{x, g(x)\}$ is $\frac{\sqrt{3}}{4}$ then the function $g(x)$ is
(a) $\pm\sqrt{1-x^2}$ (b) $\sqrt{1-x^2}$
(c) $-\sqrt{1-x^2}$ (d) $\sqrt{1+x^2}$

85. The domain of the real-valued function $f(x) = \frac{x-3}{(x-1)\sqrt{x^2-4}}$ is
(a) $(1, 2)$ (b) $R - [-2, 2]$
(c) $(-\infty, -2) \cup (1, +\infty)$ (d) $R - \{1, 2, -2\}$

86. The real function $f(x) = \cos^{-1}\sqrt{x^2 + 3x + 1} + \cos^{-1}\sqrt{x^2 + 3x}$ is defined on the set
(a) $\{0, 3\}$ (b) interval $(0, 3)$
(c) $\{0, -3\}$ (d) interval $[-3, 0]$

87. The domain of $f(x) = \sqrt{\log \frac{1}{|\sin x|}}$ is
(a) R (b) $R - \{-\pi, \pi\}$
(c) $R - \{x \mid x = n\pi, n \in \mathbb{Z}\}$ (d) none of these

88. The functions $f(x) = \cos^{-1}\sqrt{1-x^2}$ and $g(x) = -\sin^{-1} x$ are identical for x belonging to
(a) $[-1, 1]$ (b) $[0, 1]$
(c) $[-1, 0]$ (d) none of these

89. The domain of the real valued function $f(x)$ for which $4^{f(x)} + 4^{1-f(x)} = 4^x$ is
(a) $(-1, 1]$ (b) $[1, +\infty)$
(c) $(-\infty, 1]$ (d) $(-\infty, -1]$

State whether the statements are true or false.

90. If $f(x) = \sin x + \cos ax$ is periodic then a is rational.

91. The product of an even function and odd function is even.

92. The function $f(x) = x[x]$ is periodic.

93. If $f(x) = \frac{|x|}{x}$, $x \neq 0$ then $|f(\alpha) - f(-\alpha)| = 2$ where $\alpha \neq 0$.

94. The domain of the real-valued function
$$f(x) = \log_{10}\sqrt{\frac{(3-x)(x+2)}{(x+1)(x-4)}}$$
does not contain the intervals $(-\infty, -5)$ and $(5, 6)$.

Answers

4. $A = \frac{bx(h-x)}{h}$, $P = 2\left\{x + \frac{b(h-x)}{h}\right\}$

5. $\cos^{-1}(1-x) - (1-x)\sqrt{x(2-x)}$, $0 \le x \le 1$ and $\pi - \cos^{-1}(x-1) + (x-1)\sqrt{(2-x)x}$, $1 < x \le 2$

6. 3 7. $(-\infty, 1] \cup [3, +\infty)$
8. $(-\infty, -1) \cup (2, +\infty)$ 9. $(-\infty, 0)$
10. $[-1, 2) \cup [3, +\infty)$ 11. $(0, 3/2]$
12. $[-2, -1] \cup [1, 2]$ 13. $\bigcup_{n \in \mathbb{Z}} [2n\pi, (2n+1)\pi]$
14. $[-3, -2] \cup [2, 3]$ 15. R
16. $R - \mathbb{Z}$ 17. $[2, 3) \cup (3, 4]$
18. $(5, +\infty)$ 19. not defined
20. $(-\infty, -1] \cup (2, +\infty)$ 21. $[-1, 3]$
22. $\{1, -1\}$ 23. $(1/2, 1]$

24. $\left(\frac{\sqrt{19}-3}{2}, 1\right]$

25. $y = x$, domain $= (0, 1) \cup (1, +\infty)$

26. (i) $[-3/2, -1]$ (ii) $\bigcup_{n \in Z} [2n\pi, (2n+1)\pi]$

(iii) $\bigcup_{n \in Z} \left[\left(2n - \frac{1}{2}\right)\pi, \left(2n + \frac{1}{2}\right)\pi\right]$

27. domain $= R$, range $= [\log_{10} 11/3, +\infty)$

28. domain $= R$, range $= \left[-\frac{1}{2}, \frac{1}{2}\right]$

29. domain $= R - \{0\}$, range $= \{1, -1\}$

30. domain $= R$, range $= [4, +\infty)$

31. domain $= R$, range $= [1/4, 1/2]$

32. $(-\infty, -1] \cup [7/25, +\infty)$

34. $y = 0$ and $y = 1$

36. f is such that $f(x) = b, f(y) = a, f(z) = c$

37. no

38. one-one, not onto

44. (i) $f(x) = x, 0 \le x \le 1$; $1, x > 1$ (ii) $f(x) = -x, 0 \le x \le 1$; $-1, x > 1$

45. (i) $f(x) = x^2 - x, -3 \le x < 0$

(ii) $f(x) = -x^2 + x, -3 \le x < 0$

46. period $= \frac{1}{2}$

47. period $= \pi$

48. period $= \frac{\pi}{2}$

49. period $= \frac{2\pi}{3}$

50. period $= \frac{\pi}{2}$

56. $x = \frac{1}{2}(e^y - e^{-y})$

57. $g^{-1}(y) = -\sqrt{-y-1}, y < -1$; $0, y = 0$; $\sqrt{y-1}, y > 1$

58. $\frac{1}{x}$

59. 0

60. -8

61. 527

62. $3x^2 - 2x + 5$

63. $x^2 - 5x + 6$

64. $[1, +\infty)$

65. $2\cos^{-1}\left(\frac{1}{a}|x - a|\right), 0 \le x \le 2a$

66. $\frac{e^{-x} + e^x}{2}, \frac{e^{-x} - e^x}{2}$ respectively

67. $\frac{-1 \pm \sqrt{5}}{2}, \frac{-3 \pm \sqrt{5}}{2}$

68. $\frac{k}{a}$

69. 2π

70. $\left\{x \mid x = 2n + \frac{1}{2}, n \in Z\right\}$

71. $\left[-\frac{\pi}{4}, \frac{\pi}{4}\right]$, $[0, 3/\sqrt{2}]$ respectively

72. R, $[-\sqrt{2}, \sqrt{2}]$ respectively

73. $(4, +\infty)$

74. $[-1, 1]$

75. $[1, +\infty)$

76. $y^{\log_5 e}, y > 0$

77. $\left(0, \frac{1}{100}\right) \cup \left(\frac{1}{100}, \frac{1}{10}\right)$

78. 1

79. (a), (b)

80. (a), (c)

81. (c)

82. (b), (c)

83. (d)

84. (b), (c)

85. (b)

86. (c)

87. (c)

88. (c)

89. (b)

90. true

91. false

92. false

93. true

94. true

□

Chapter Test

Time: 90 minutes

1. In each of the following, fill in the blanks so that the resulting sentence becomes true.

(a) If $f(x) = \frac{1-x}{1+x}$, $x > 0$ then the least value of $f\{f(x)\} + f\left\{f\left(\frac{1}{x}\right)\right\}$ is ______ .

(b) If $f(x) = \sqrt{\log \frac{5x - x^2}{4}}$ then domain of f = ______ .

(c) If $f(x) = \frac{1}{3\cos x - 4\sin x + 5}$ then domain = ______ , range = ______ .

2. In each of following there are multiple options. Choose the correct option(s).

(a) $f(x) = x - [x] + n,\ n \in N$ is a periodic function whose period is

A. indeterminate B. 1 C. n D. none of these

(b) If $f(x) = 1, x > 0$

$0, x = 0$

$-1, x < 0$ then which of the following are true?

A. $xf(x) = |x|$ B. $x = |x|\ f(x)$ C. $f\{f(x)\} = f(x)$ D. none of these

(c) If $f(x) = \frac{1}{1 + x + x^2}$ then the function $f : R \to R$ is

A. one-one B. many-one C. into D. onto

3. Let f be a function satisfying $2f(xy) = \{f(x)\}^y + \{f(y)\}^x$ and $f(1) = k \neq 1$. Prove that $(k-1) \sum_{n=1}^{n} f(n) = k^{n+1} - k$.

4. If $f(a - x) = f(a + x)$ and $f(b - x) = f(b + x)$ for all real x where $a, b (a > b)$ are constants then prove that $f(x)$ is a periodic function.

5. If $f(x + y) = f(x) \cdot f(y)$ for all real x, y and $f(0) \neq 0$ then prove that the function $F(x) = \frac{f(x)}{1 + \{f(x)\}^2}$ is an even function.

6. Find the domain and the range of the function $f(x) = \sqrt{a - x} + \sqrt{x - b}$ where $a > b > 0$.

7. Let $f(x) = \frac{x^3 - 3x + 1}{x(x-1)}$, $x \neq 0, 1$. Prove that $f(x) = f\left(\frac{1}{1-x}\right) = f\left(1 - \frac{1}{x}\right)$. Also find the set of values of x for which $f(x) = \frac{a^3 - 3a^2 + 1}{a(1-a)}$, $a \neq 0, 1$ and a is a constant.

8. Let y be a function of x given by $10x^2 - 8xy - 2y^2 - 4x - 2y = 5$. Prove that the function is two-one and among the two real values of x that correspond to a real value of y, one is $\leq -\frac{1}{2}$ and the other is $\geq \frac{1}{2}$.

9. If $r^2 - 4x + 3 = 0$ where $r = \sqrt{x^2 + y^2}, x \in R, y \in R$ then find the greatest and the least values of r.

10. Let $f(x) = 2x - x^2, x \leq 1$. Prove that the function $f^{-1}(x)$ exists and find the solution-set of the equation $f^{-1}(x) = f(x)$.

11. Let f be a function defined by $2f(x-1) - f\left(\dfrac{1-x}{x}\right) = x, x \neq 0$. Find the domain and range of the function $f(x)$.

□

Answers

1. (a) 2 (b) [1, 4] (c) $R - \left\{x \mid x = n\pi + (-1)^n \dfrac{\pi}{2} + \sin^{-1}\dfrac{3}{5}, n \in Z\right\}, \left[\dfrac{1}{10}, +\infty\right)$ respectively

2. (a) B (b) A, B, C (c) B, C

6. domain = $[b, a]$, range = $[\sqrt{a-b}, \sqrt{2(a-b)}]$ **7.** $\left\{\dfrac{1}{a}, \dfrac{a}{a-1}, 1-a\right\}$

9. greatest value = 3, least value = 1. **10.** $\{0, 1\}$

11. $R - \{-1\}$, $R - \left(\dfrac{-2\sqrt{2}}{3}, \dfrac{2\sqrt{2}}{3}\right)$ respectively

□

2. Differentiation

Recap of Facts and Formulae

1. Definition of differential coefficient

If $y = f(x)$ be a function then

- differential coefficient of y w.r.t. x is

$$\frac{dy}{dx} \text{ or } f'(x) = \lim_{h \to 0} \frac{f(x+h) - f(x)}{h}$$

- differential coefficient of y w.r.t. x at $x = a$ is

$$\left(\frac{dy}{dx}\right)_{x=a} \text{ or } f'(a) = \lim_{h \to 0} \frac{f(a+h) - f(a)}{h}.$$

2. Differential coefficients of standard functions

- $\frac{d(x^n)}{dx} = nx^{n-1}$
- $\frac{d(c)}{dx} = 0$
- $\frac{d(\log_e x)}{dx} = \frac{1}{x}$ where $x > 0$
- $\frac{d(e^x)}{dx} = e^x$
- $\frac{d(a^x)}{dx} = a^x \cdot \log_e a$ where $a > 0$
- $\frac{d(\sin x)}{dx} = \cos x$
- $\frac{d(\cos x)}{dx} = -\sin x$
- $\frac{d(\tan x)}{dx} = \sec^2 x$
- $\frac{d(\sec x)}{dx} = \sec x \cdot \tan x$
- $\frac{d(\cot x)}{dx} = -\operatorname{cosec}^2 x$
- $\frac{d(\operatorname{cosec} x)}{dx} = -\operatorname{cosec} x \cdot \cot x$
- $\frac{d(\sin^{-1} x)}{dx} = \frac{1}{\sqrt{1-x^2}}$
- $\frac{d(\cos^{-1} x)}{dx} = \frac{-1}{\sqrt{1-x^2}}$
- $\frac{d(\tan^{-1} x)}{dx} = \frac{1}{1+x^2}$
- $\frac{d(\cot^{-1} x)}{dx} = \frac{-1}{1+x^2}$
- $\frac{d(\sec^{-1} x)}{dx} = \frac{1}{|x|\sqrt{x^2-1}}$
- $\frac{d(\operatorname{cosec}^{-1} x)}{dx} = \frac{-1}{|x|\sqrt{x^2-1}}$

Note $\frac{d(\sec^{-1} x)}{dx} = \frac{d\cos^{-1}(1/x)}{dx} = \frac{-1}{\sqrt{1-(1/x)^2}} \cdot \left(-\frac{1}{x^2}\right)$, using chain rule

$$= \frac{1}{x^2\sqrt{1-(1/x^2)}} = \frac{|x|}{x^2\sqrt{x^2-1}}$$

$$= \frac{1}{|x|\sqrt{x^2-1}}.$$

3. Rules for differentiation

- If $y = u(x) \pm v(x) \pm w(x) \pm \ldots$ then

$$\frac{dy}{dx} = \frac{du(x)}{dx} \pm \frac{dv(x)}{dx} \pm \frac{dw(x)}{dx} \pm \ldots$$

(term-by-term differentiation)

Differentiation of a term:

- $\frac{d}{dx}\{kf(x)\} = k\frac{df(x)}{dx}$
- $\frac{d}{dx}\{u(x) \cdot v(x)\} = \frac{du(x)}{dx} \cdot v(x) + u(x) \cdot \frac{dv(x)}{dx}$ *(product rule)*
- $\frac{d}{dx}\left\{\frac{u(x)}{v(x)}\right\} = \frac{\frac{du(x)}{dx} \cdot v(x) - u(x) \cdot \frac{dv(x)}{dx}}{\{v(x)\}^2}$ *(quotient rule)*
- $\frac{d}{dx}[u\{v(x)\}] = \frac{du\{v(x)\}}{d\{v(x)\}} \times \frac{dv(x)}{dx}$ *(chain rule)*

$$\frac{d}{dx}u[v\{w(x)\}] = \frac{du[v\{w(x)\}]}{dv\{w(x)\}} \times \frac{dv\{w(x)\}}{dw(x)} \times \frac{dw(x)}{dx}$$

- $\frac{d}{dx}\{f(x)\}^{\phi(x)}$ = d.c. of $\{f(x)\}^{\phi(x)}$ w.r.t. x taking $\phi(x)$ as a constant + d.c. of $\{f(x)\}^{\phi(x)}$ w.r.t. x taking $f(x)$ as a constant

$$= \phi(x)\cdot\{f(x)\}^{\phi(x)-1}\cdot\frac{df(x)}{dx} + \{f(x)\}^{\phi(x)}\cdot\log f(x)\cdot\frac{d\phi(x)}{dx}$$

(rule of logarithmic differentiation).

4. Differentiation of function represented parametrically

- If y is a function of x such that $x = \phi(t), y = \psi(t)$ where t is the parameter then

$$\frac{dy}{dx} = \frac{\frac{dy}{dt}}{\frac{dx}{dt}}, \text{ i.e., } \frac{\psi'(t)}{\phi'(t)}.$$

5. Differentiation of one function w.r.t. another function

- If $y = \phi(x)$ and $z = \psi(x)$ then

$$\frac{dy}{dz} = \frac{\frac{dy}{dx}}{\frac{dz}{dx}}, \text{ i.e., } \frac{\phi'(x)}{\psi'(x)}.$$

6. Relation between $\frac{dy}{dx}$ and $\frac{dx}{dy}$

- $\frac{dy}{dx}\times\frac{dx}{dy} = 1.$

7. Higher derivatives of a function

If $y = f(x)$ then the derivative of $\frac{dy}{dx}$ w.r.t. x is called the second derivative of y w.r.t. x and it is denoted by $\frac{d^2y}{dx^2}$.

- $\frac{d^2y}{dx^2} = \frac{d}{dx}\left(\frac{dy}{dx}\right);\ \frac{d^3y}{dx^3} = \frac{d}{dx}\left(\frac{d^2y}{dx^2}\right)$
- If y is a function of x given parametrically by $y = \phi(t), x = \psi(t)$ then

$$\frac{d^2y}{dx^2} = \frac{d}{dx}\left(\frac{dy}{dx}\right) = \frac{d}{dx}\left(\frac{\phi'(t)}{\psi'(t)}\right) = \frac{\frac{d}{dt}\left(\frac{\phi'(t)}{\psi'(t)}\right)}{\frac{dx}{dt}}.$$

8. Differentiation of a function given in the form of a determinant

If $y = \begin{vmatrix} u(x) & v(x) & w(x) \\ p(x) & q(x) & r(x) \\ \lambda(x) & \mu(x) & \nu(x) \end{vmatrix}$ then

$$\frac{dy}{dx} = \begin{vmatrix} u'(x) & v'(x) & w'(x) \\ p(x) & q(x) & r(x) \\ \lambda(x) & \mu(x) & \nu(x) \end{vmatrix} + \begin{vmatrix} u(x) & v(x) & w(x) \\ p'(x) & q'(x) & r'(x) \\ \lambda(x) & \mu(x) & \nu(x) \end{vmatrix} + \begin{vmatrix} u(x) & v(x) & w(x) \\ p(x) & q(x) & r(x) \\ \lambda'(x) & \mu'(x) & \nu'(x) \end{vmatrix}.$$

The differentiation can also be done columnwise.

Selected Solved Examples

1. If $f(x) = \log_{\sec 2x}|\cos 4x| + |\sin x|$ then find $\frac{dy}{dx}$ at $x = \frac{-\pi}{6}$ from the first principle.

Clearly, the definition of the function $f(x)$ in

$$\left[-\frac{\pi}{6}-h, -\frac{\pi}{6}+h\right],$$

where h is a very small positive quantity, remains the same. So, $x = \frac{-\pi}{6}$ is not a turning point of definition.

$$\text{Now, } \left(\frac{dy}{dx}\right)_{x=\frac{-\pi}{6}} = \lim_{h\to 0}\frac{f\left(-\frac{\pi}{6}+h\right)-f\left(-\frac{\pi}{6}\right)}{h}$$

$$= \lim_{h\to 0}\frac{\left\{\log_{\sec\left(-\frac{\pi}{3}+2h\right)}\left|\cos 4\left(-\frac{\pi}{6}+h\right)\right| + \left|\sin\left(-\frac{\pi}{6}+h\right)\right| - \log_{\sec\frac{-\pi}{3}}\left|\cos\frac{-2\pi}{3}\right| - \left|\sin\frac{-\pi}{6}\right|\right\}}{h}$$

$$= \lim_{h\to 0}\frac{1}{h}\left\{\log_{\sec\left(-\frac{\pi}{3}+2h\right)}\left|\cos\left(\frac{-2\pi}{3}+4h\right)\right| - \log_{\sec\frac{-\pi}{3}}\left|\cos\frac{-2\pi}{3}\right|\right\}$$

$$+ \lim_{h\to 0}\frac{1}{h}\left\{\left|\sin\left(-\frac{\pi}{6}+h\right)\right| - \left|\sin\frac{-\pi}{6}\right|\right\}$$

$$= \lim_{h \to 0} \frac{1}{h}\left[\frac{\log\left\{-\cos\left(-\frac{2\pi}{3}+4h\right)\right\}}{\log \sec\left(-\frac{\pi}{3}+2h\right)} - \log_2\left(\frac{1}{2}\right)\right]$$

$$+ \lim_{h \to 0} \frac{1}{h}\left\{-\sin\left(-\frac{\pi}{6}+h\right)-\frac{1}{2}\right\},$$

(using definition of mod)

$$= \lim_{h \to 0} \frac{1}{h}\cdot\left[\frac{\log \cos\left(\frac{\pi}{3}+4h\right)}{-\log \cos\left(-\frac{\pi}{3}+2h\right)}+1\right]$$

$$+ \lim_{h \to 0} \frac{1}{h}\left\{\frac{1}{2}\cos h - \frac{\sqrt{3}}{2}\sin h - \frac{1}{2}\right\}.$$

Now, $\lim_{h \to 0} \frac{1}{h}\left\{\frac{1}{2}\cos h - \frac{\sqrt{3}}{2}\sin h - \frac{1}{2}\right\}$

$$= \lim_{h \to 0} \frac{1}{h}\left\{-\frac{\sqrt{3}}{2}\sin h - \frac{1}{2}\cdot 2\sin^2\frac{h}{2}\right\}$$

$$= \lim_{h \to 0}\left[-\frac{\sqrt{3}}{2}\cdot\frac{\sin h}{h} - \frac{1}{2}\cdot\left(\frac{\sin\frac{h}{2}}{\frac{h}{2}}\right)\cdot\sin\frac{h}{2}\right]$$

$$= -\frac{\sqrt{3}}{2}.$$

$$\lim_{h \to 0} \frac{1}{h}\left[\frac{\log \cos\left(\frac{\pi}{3}+4h\right)}{-\log \cos\left(-\frac{\pi}{3}+2h\right)}+1\right]$$

$$= \lim_{h \to 0} \frac{1}{h}\,\frac{\left\{-\log\left(\frac{1}{2}\cos 4h - \frac{\sqrt{3}}{2}\sin 4h\right) + \log\left(\frac{1}{2}\cos 2h + \frac{\sqrt{3}}{2}\sin 2h\right)\right\}}{\log\left(\frac{1}{2}\cos 2h + \frac{\sqrt{3}}{2}\sin 2h\right)}$$

$$= \lim_{h \to 0} \frac{\left\{-\frac{-2\sin 4h - 2\sqrt{3}\cos 4h}{\frac{1}{2}\cos 4h - \frac{\sqrt{3}}{2}\sin 4h} + \frac{-\sin 2h + \sqrt{3}\cos 2h}{\frac{1}{2}\cos 2h + \frac{\sqrt{3}}{2}\sin 2h}\right\}}{\left\{\log\left(\frac{1}{2}\cos 2h + \frac{\sqrt{3}}{2}\sin 2h\right) + h\times\frac{-\sin 2h + \sqrt{3}\cos 2h}{\frac{1}{2}\cos 2h + \frac{\sqrt{3}}{2}\sin 2h}\right\}}$$

(using L' Hospital's rule)

$$= \frac{4\sqrt{3}+2\sqrt{3}}{\log\frac{1}{2}} = \frac{6\sqrt{3}}{-\log 2}$$

$$\therefore \quad \left(\frac{dy}{dx}\right)_{x=\frac{-\pi}{6}} = \frac{-6\sqrt{3}}{\log 2} - \frac{\sqrt{3}}{2}.$$

2. If $\sqrt{1-x^{2n}} + \sqrt{1-y^{2n}} = a^n(x^n - y^n)$, prove that

$$y^{n-1}\cdot\sqrt{1-x^{2n}}\;dy = x^{n-1}\sqrt{1-y^{2n}}\;dx.$$

Differentiating both sides w.r.t. x,

$$\frac{1}{2}\cdot\frac{1}{\sqrt{1-x^{2n}}}\cdot(-2nx^{2n-1}) + \frac{1}{2}\cdot\frac{1}{\sqrt{1-y^{2n}}}(-2ny^{2n-1})\frac{dy}{dx}$$

$$= a^n\left\{nx^{n-1} - ny^{n-1}\frac{dy}{dx}\right\}$$

or $\quad \dfrac{x^{2n-1}}{\sqrt{1-x^{2n}}} + \dfrac{y^{2n-1}}{\sqrt{1-y^{2n}}}\dfrac{dy}{dx} = a^n\left(y^{n-1}\dfrac{dy}{dx} - x^{n-1}\right)$

or $\quad \dfrac{x^{2n-1}}{\sqrt{1-x^{2n}}} + a^n x^{n-1} = \left(a^n y^{n-1} - \dfrac{y^{2n-1}}{\sqrt{1-y^{2n}}}\right)\dfrac{dy}{dx}$

or $\quad x^{n-1}\cdot\dfrac{x^n + a^n\cdot\sqrt{1-x^{2n}}}{\sqrt{1-x^{2n}}}$

$$= y^{n-1}\cdot\frac{a^n\cdot\sqrt{1-y^{2n}} - y^n}{\sqrt{1-y^{2n}}}\cdot\frac{dy}{dx}$$

or $\quad \dfrac{dy}{dx} = \dfrac{x^{n-1}}{y^{n-1}}\cdot\sqrt{\dfrac{1-y^{2n}}{1-x^{2n}}}\cdot\dfrac{x^n + a^n\cdot\sqrt{1-x^{2n}}}{a^n\cdot\sqrt{1-y^{2n}} - y^n} \quad \ldots (1)$

Now, we have

$$\sqrt{1-x^{2n}} + \sqrt{1-y^{2n}} = a^n(x^n - y^n)$$

or $\quad (1-x^{2n}) - (1-y^{2n})$

$$= a^n(x^n - y^n)\{\sqrt{1-x^{2n}} - \sqrt{1-y^{2n}}\}$$

or $\quad y^{2n} - x^{2n} = a^n(x^n - y^n)\{\sqrt{1-x^{2n}} - \sqrt{1-y^{2n}}\}$

$\therefore \quad \sqrt{1-x^{2n}} - \sqrt{1-y^{2n}} = \dfrac{-1}{a^n}(y^n + x^n)$

or $\quad a^n\sqrt{1-x^{2n}} + x^n = a^n\sqrt{1-y^{2n}} - y^n \quad \ldots (2)$

Using (2) in (1),

$$\frac{dy}{dx} = \frac{x^{n-1}}{y^{n-1}}\cdot\sqrt{\frac{1-y^{2n}}{1-x^{2n}}}.$$

Writing in the differential form,

$$y^{n-1}\cdot\sqrt{1-x^{2n}}\;dy = x^{n-1}\sqrt{1-y^{2n}}\;dx.$$

3. If $u=f(x^2)$, $v=g(x^3)$, $f'(x)=\sin x$ and $g'(x)=\cos x$ then find $\frac{du}{dv}$.

Differentiating $u=f(x^2)$ and $v=g(x^3)$ w.r.t. x we get

$$\frac{du}{dx}=f'(x^2)\cdot 2x=\sin(x^2)\cdot 2x$$

$$\{\because f'(x)=\sin x \quad \Rightarrow \quad f'(x^2)=\sin(x^2)\}$$

$$\frac{dv}{dx}=g'(x^3)\cdot 3x^2=\cos(x^3)\cdot 3x^2$$

$$\{\because g'(x)=\cos x \quad \Rightarrow \quad g'(x^3)=\cos(x^3)\}$$

$$\therefore \quad \frac{du}{dv}=\frac{\frac{du}{dx}}{\frac{dv}{dx}}=\frac{\sin(x^2)\cdot 2x}{\cos(x^3)\cdot 3x^2}=\frac{2}{3x}\cdot\frac{\sin x^2}{\cos x^3}.$$

4. Find $\frac{dy}{dx}$ at $x=-1$ when

$$(\sin y)^{\sin\frac{\pi x}{2}}+\frac{\sqrt{3}}{2}\sec^{-1}(2x)+2^x\tan\{\log_e(x+2)\}=0.$$

We know,

$\frac{d(u^v)}{dx}$ = diff. coeff. of u^v w.r.t. x when v is a constant
+ diff. coeff. of u^v w.r.t. x when u is a constant.

$\therefore$ differentiating the given relation w.r.t. x,

$$\sin\frac{\pi x}{2}\cdot(\sin y)^{\sin\frac{\pi x}{2}-1}\cdot\cos y\,\frac{dy}{dx}$$

$$+(\sin y)^{\sin\frac{\pi x}{2}}\cdot\log\sin y\cdot\frac{\pi}{2}\cos\frac{\pi x}{2}$$

$$+\frac{\sqrt{3}}{2}\cdot\frac{1}{|2x|\sqrt{4x^2-1}}\cdot 2$$

$$+2^x\log 2\cdot\tan\{\log_e(x+2)\}$$

$$+2^x\cdot\sec^2\{\log_e(x+2)\}\cdot\frac{1}{x+2}=0.$$

Putting $x=-1$, we get (assuming $y=y_0$ when $x=-1$)

$$-(\sin y_0)^{-2}\cdot\cos y_0\cdot\left[\frac{dy}{dx}\right]_{x=-1}$$

$$+(\sin y_0)^{-1}\cdot\log\sin y_0\cdot\frac{\pi}{2}\cdot 0$$

$$+\frac{\sqrt{3}}{2}\cdot\frac{1}{|-2|\sqrt{3}}\cdot 2+0+\frac{1}{2}\cdot 1\cdot\frac{1}{1}=0$$

or $\quad -\cos y_0(\sin y_0)^{-2}\cdot\left[\frac{dy}{dx}\right]_{x=-1}+\frac{1}{2}+\frac{1}{2}=0$

$$\therefore \quad \left[\frac{dy}{dx}\right]_{x=-1}=\frac{\sin^2 y_0}{\cos y_0}.$$

Now, putting $x=-1$ in the original relation,

$$(\sin y_0)^{-1}+\frac{\sqrt{3}}{2}\sec^{-1}(-2)+\frac{1}{2}\tan\log_e 1=0$$

or $\quad (\sin y_0)^{-1}+\frac{\sqrt{3}}{2}\left(\pi-\frac{\pi}{3}\right)=0$

$$\therefore \quad \frac{1}{\sin y_0}+\frac{\pi}{\sqrt{3}}=0$$

$$\therefore \quad \sin y_0=-\frac{\sqrt{3}}{\pi}.$$

Hence,
$$\left[\frac{dy}{dx}\right]_{x=-1}=\frac{\left(-\frac{\sqrt{3}}{\pi}\right)^2}{\sqrt{1-\left(\frac{\sqrt{3}}{\pi}\right)^2}}$$

$$=\frac{3}{\pi\sqrt{\pi^2-3}}.$$

5. If f, g, h are differentiable functions of x and

$$\Delta(x)=\begin{vmatrix} f & g & h \\ (xf)' & (xg)' & (xh)' \\ (x^2f)'' & (x^2g)'' & (x^2h)'' \end{vmatrix}, \text{ prove that}$$

$$\frac{d\Delta(x)}{dx}=\begin{vmatrix} f & g & h \\ f' & g' & h' \\ (x^3f'')' & (x^3g'')' & (x^3h'')' \end{vmatrix}.$$

$$\Delta(x)=\begin{vmatrix} f & g & h \\ f+xf' & g+xg' & h+xh' \\ 2f+4xf'+x^2f'' & 2g+4xg'+x^2g'' & 2h+4xh'+x^2h'' \end{vmatrix}$$

$$\{\because (x^2f)'=2xf+x^2f'$$
$$(x^2f)''=2(f+xf')+2xf'+x^2f''\}$$

$$=\begin{vmatrix} f & g & h \\ xf' & xg' & xh' \\ 2xf'+x^2f'' & 2xg'+x^2g'' & 2xh'+x^2h'' \end{vmatrix},$$

$R_3\to R_3-2R_2$, $R_2\to R_2-R_1$

$$=\begin{vmatrix} f & g & h \\ xf' & xg' & xh' \\ x^2f'' & x^2g'' & x^2h'' \end{vmatrix}, \quad R_3\to R_3-2R_2$$

$$=\begin{vmatrix} f & g & h \\ f' & g' & h' \\ x^3f'' & x^3g'' & x^3h'' \end{vmatrix},$$

taking the common factor x from R_2 to R_3

$$\therefore \quad \frac{d\Delta(x)}{dx}=\begin{vmatrix} f' & g' & h' \\ f' & g' & h' \\ x^3f'' & x^3g'' & x^3h'' \end{vmatrix}$$

$$+\begin{vmatrix} f & g & h \\ f'' & g'' & h'' \\ x^3f'' & x^3g'' & x^3h'' \end{vmatrix}$$

$$+\begin{vmatrix} f & g & h \\ f' & g' & h' \\ (x^3f'')' & (x^3g'')' & (x^3h'')' \end{vmatrix}$$

$$=0+x^3\begin{vmatrix} f & g & h \\ f'' & g'' & h'' \\ f'' & g'' & h'' \end{vmatrix}$$

$$+\begin{vmatrix} f & g & h \\ f' & g' & h' \\ (x^3f'')' & (x^3g'')' & (x^3h'')' \end{vmatrix}$$

$$=0+x^3\times 0+\begin{vmatrix} f & g & h \\ f' & g' & h' \\ (x^3f'')' & (x^3g'')' & (x^3h'')' \end{vmatrix}$$

$$=\begin{vmatrix} f & g & h \\ f' & g' & h' \\ (x^3f'')' & (x^3g'')' & (x^3h'')' \end{vmatrix}.$$

6. If $2x=y^{\frac{1}{5}}+y^{-\frac{1}{5}}$ then express y as an explicit function of x and prove that

$$(x^2-1)\frac{d^2y}{dx^2}+x\frac{dy}{dx}=25y.$$

$$\left(y^{\frac{1}{5}}-y^{-\frac{1}{5}}\right)^2=\left(y^{\frac{1}{5}}+y^{-\frac{1}{5}}\right)^2-4$$

$$=4x^2-4 \qquad \left\{\because\ y^{\frac{1}{5}}+y^{-\frac{1}{5}}=2x\right\}$$

$$\therefore \quad y^{\frac{1}{5}}-y^{-\frac{1}{5}}=2\sqrt{x^2-1}$$

$$\therefore \quad 2y^{\frac{1}{5}}=\left(y^{\frac{1}{5}}+y^{-\frac{1}{5}}\right)+\left(y^{\frac{1}{5}}-y^{-\frac{1}{5}}\right)$$

$$=2x+2\sqrt{x^2-1}$$

$$\therefore \quad y=(x+\sqrt{x^2-1})^5. \qquad \dots (1)$$

Differentiating w.r.t. x,

$$\frac{dy}{dx}=5(x+\sqrt{x^2-1})^4\cdot\left\{1+\frac{1}{2}\cdot\frac{2x}{\sqrt{x^2-1}}\right\}$$

$$=5\,\frac{(x+\sqrt{x^2-1})^5}{\sqrt{x^2-1}}=\frac{5y}{\sqrt{x^2-1}} \quad \{\text{using (1)}\}$$

$$\therefore \quad (x^2-1)\left(\frac{dy}{dx}\right)^2=25y^2;$$

differentiating again w.r.t. x,

$$2x\left(\frac{dy}{dx}\right)^2+(x^2-1)\cdot 2\frac{dy}{dx}\cdot\frac{d^2y}{dx^2}=50y\frac{dy}{dx}$$

or $\quad x\dfrac{dy}{dx}+(x^2-1)\dfrac{d^2y}{dx^2}=25y$

$$\left\{\because\ \frac{dy}{dx}\neq 0 \text{ for then } y \text{ is a constant}\right\}.$$

7. Let f be twice differentiable such that $f''(x)=-f(x)$ and $f'(x)=g(x)$. If $h(x)=\{f(x)\}^2+\{g(x)\}^2$ where $h(5)=11$, find $h(10)$.

Differentiating $h(x)=\{f(x)\}^2+\{g(x)\}^2$ w.r.t. x,

$$h'(x)=2f(x)\cdot f'(x)+2g(x)\cdot g'(x) \qquad \dots (1)$$

Now, $f'(x)=g(x) \Rightarrow f''(x)=g'(x)$

$$\Rightarrow -f(x)=g'(x) \quad \{\because\ f''(x)=-f(x)\}$$

$\therefore \quad (1) \Rightarrow h'(x)=2f(x)\cdot g(x)+2g(x)\cdot\{-f(x)\}=0$

$\therefore \quad h(x)=c$; but $h(5)=11$. So $11=c$.

$\therefore \quad h(x)=11$

$\therefore \quad h(10)=11.$

8. A function $f: R\to R$ satisfies the equation given by $f(x+y)=f(x)\cdot f(y)$, for all x, y in R and $f(x)\neq 0$ for any x in R. Let $f(x)$ be differentiable and $f'(0)=2$. Show that $f'(x)=2f(x)$ for all $x\in R$. Hence find $f(x)$.

Differentiating $f(x+y)=f(x)\cdot f(y)$ w.r.t. x,

$$f'(x+y)\cdot\left(1+\frac{dy}{dx}\right)=f'(x)f(y)+f(x)\cdot f'(y)\frac{dy}{dx}\cdot$$

But y is independent of x. So $\dfrac{dy}{dx}=0$.

Thus we get $f'(x+y)=f'(x)\cdot f(y)$.

Putting $x=0$ and x for y, $f'(x)=f'(0)\cdot f(x)$.

$\therefore \quad f'(x)=2f(x) \quad \{\because\ f'(0)=2\}$

or $\quad \dfrac{f'(x)}{f(x)}=2$

$\therefore \quad$ integrating, $\log f(x)=2x+c \qquad \dots (1)$

Again, putting $x=0$, $y=0$ in the given relation,

$$f(0+0)=f(0)\cdot f(0)$$

$\therefore \quad f(0)\{f(0)-1\}=0.$

But $f(0)\neq 0$ (given); so $f(0)=1$.

Putting $x=0$ in (1), $\log 1=0+c$; $\quad \therefore \quad c=0.$

$\therefore \quad \log f(x)=2x$

$\therefore \quad f(x)=e^{2x}.$

Note If the differentiability of $f(x)$ is not given, the correct method will be as given ahead.

$$f'(x) = \lim_{h \to 0} \frac{f(x+h) - f(x)}{h}$$

$$= \lim_{h \to 0} \frac{f(x) \cdot f(h) - f(x)}{h}$$

$$= \lim_{h \to 0} \frac{f(x)\{f(h) - 1\}}{h}$$

$$= f(x) \cdot \lim_{h \to 0} \frac{f(h) - 1}{h}$$

$$= f(x) \cdot \lim_{h \to 0} \frac{f(h) - f(0)}{h}$$

$\{\because f(0) = 1$ as obtained above$\}$

$$= f(x) \cdot f'(0) = 2f(x).$$

9. Let $f(x)$ be a real function not identically zero such that $f(x + y^{2n+1}) = f(x) + \{f(y)\}^{2n+1}$, $n \in N$ and x, y are any real numbers and $f'(0) \geq 0$. Find the values of $f(5)$ and $f'(10)$.

Here, $f(x + y^{2n+1}) = f(x) + \{f(y)\}^{2n+1}$ ... (1)

Putting $x = 0, y = 0$ we get $f(0) = f(0) + \{f(0)\}^{2n+1}$

$\therefore \quad f(0) = 0.$

$$f'(0) \geq 0 \quad \Rightarrow \quad \lim_{h \to 0} \frac{f(x) - f(0)}{x} \geq 0$$

$$\Rightarrow \quad \lim_{x \to 0} \frac{f(x)}{x} \geq 0$$

$\therefore \quad$ if $x > 0, f(x) \geq 0$ ... (2)

Putting $x = 0, y = 1$ in (1),

$$f(1) = f(0) + \{f(1)\}^{2n+1}$$

or $\quad f(1)\,[1 - \{f(1)\}^{2n}] = 0$

$\therefore \quad f(1) = 0$ or 1, {from (2)}.

Putting $y = 1$ in (1), for all real x,

$f(x + 1) = f(x) + \{f(1)\}^{2n+1}$ ... (3)

$\therefore \quad f(1) = 0 \quad \Rightarrow \quad f(x + 1) = f(x)$

$\Rightarrow \quad f(1) = f(2) = f(3) = \ldots = 0,$

i.e., $f(x)$ is identically zero.

$\therefore \quad f(1) \neq 0.$ Hence $f(1) = 1.$

So from (3), $f(x + 1) = f(x) + 1$ ... (4)

$\therefore \quad f(5) = f(4) + 1 = \{f(3) + 1\} + 1$

$= \{f(2) + 1)\} + 2$

$= \{f(1) + 1\} + 3$

$= f(1) + 4 = 1 + 4 = 5$

Also (4) $\Rightarrow$ $f(x)$ is a function whose value increases by 1 when variable x is increased by 1.

$\therefore \quad f(x) = x; \quad \therefore \quad f'(x) = 1$

$\therefore \quad f'(10) = 1.$

10. Let $f\left(\frac{x_1 + x_2 + x_3 + \ldots + x_n}{n}\right) = \frac{f(x_1) + f(x_2) + f(x_3) + \ldots + f(x_n)}{n}$

where x_i are any real numbers and $n \in N$.

If $f(n)$ is differentiable and $f'(0) = a$ and $f(0) = b$, find $\frac{d^k f(x)}{dx^k}$.

From the question,

$$f\left(\frac{x+y}{2}\right) = \frac{f(x) + f(y)}{2} \quad \{\text{taking } x_1 = x, x_2 = y, n = 2\}$$

This holds for any real x, y. So y is independent of x, i.e., $\frac{dy}{dx} = 0.$

Differentiating w.r.t. x,

$$f'\left(\frac{x+y}{2}\right) \cdot \frac{1}{2}\left(1 + \frac{dy}{dx}\right) = \frac{1}{2}\left\{f'(x) + f'(y)\frac{dy}{dx}\right\}$$

$$\therefore \quad \frac{1}{2} f'\left(\frac{x+y}{2}\right) = \frac{1}{2} f'(x) \quad \left\{\because \frac{dy}{dx} = 0\right\}.$$

Taking $x = 0$ and x in place of y we get

$$\frac{1}{2} f'\left(\frac{0+x}{2}\right) = \frac{1}{2} f'(0)$$

$$\therefore \quad f'\left(\frac{x}{2}\right) = f'(0) = a \quad \text{(given)}.$$

Integrating w.r.t. $\frac{x}{2}$, $f\left(\frac{x}{2}\right) = a \cdot \frac{x}{2} + c$

When $x = 0, f(0) = 0 + c$, i.e., $b = c$ $\quad [\because f(0) = b(\text{given})]$

Thus, $f\left(\frac{x}{2}\right) = \frac{ax}{2} + b$ ... (1)

Again, putting $y = 0$ in $f\left(\frac{x+y}{2}\right) = \frac{f(x) + f(y)}{2}$ we get

$$f\left(\frac{x}{2}\right) = \frac{f(x) + f(0)}{2} = \frac{f(x) + b}{2}$$

$\therefore \quad$ from (1),

$$\frac{f(x) + b}{2} = \frac{ax}{2} + b$$

$\therefore \quad f(x) = ax + b.$

$$\therefore \quad f'(x) = a,\ f''(x) = 0,\ f'''(x) = 0,\ \ldots,\ \frac{d^k f(x)}{dx^n} = 0.$$

11. Find the sum of
$\sin x + 3\sin 3x + 5\sin 5x + \dots + (2k-1)\sin(2k-1)x$.

Let $S = \cos x + \cos 3x + \cos 5x + \dots + \cos(2k-1)x$.

Here the angles are in AP whose first term $= x$,

common diff. $= 2x$.

$$\therefore \quad S = \frac{\sin \frac{k \cdot 2x}{2}}{\sin \frac{2x}{2}} \cos\left\{\frac{x + (2k-1)x}{2}\right\}$$

$$= \frac{\sin kx}{\sin x} \cos kx = \frac{\sin 2kx}{2\sin x}$$

$$\therefore \quad \cos x + \cos 3x + \cos 5x + \dots + \cos(2k-1)x = \frac{\sin 2kx}{2\sin x}$$

Differentiating w.r.t. x,

$$-\{\sin x + 3\sin 3x + 5\sin 5x + \dots + (2k-1)\sin(2k-1)x\}$$

$$= \frac{1}{2} \cdot \frac{2k\cos 2kx \cdot \sin x - \sin 2kx \cdot \cos x}{\sin^2 x}$$

$$\therefore \quad \sin x + 3\sin 3x + 5\sin 5x + \dots + (2k-1)\sin(2k-1)x$$

$$= \frac{-1}{2\sin^2 x}\left[k\{\sin(2k+1)x - \sin(2k-1)x\} - \frac{1}{2}\{\sin(2k+1)x + \sin(2k-1)x\}\right]$$

$$= \frac{1}{4\sin^2 x}[(2k+1)\sin(2k-1)x - (2k-1)\sin(2k+1)x].$$

Exercises

Differentiate the following functions w.r.t. x from the first principle:

1. $x^2 \cos x$
2. $\cos^{-1} x^2$
3. $\cos^{-1}(e^{\sqrt{\tan x}})$
4. $\sqrt[3]{\sin x}$
5. e^{x^2}
6. $\dfrac{\log x}{x}$

Find the derivatives w.r.t. x from the first principle at the indicated points:

7. $x\tan^{-1} x$ at $x = 1$
8. $\log |x|$ at $x = -2$
9. $|x-1| + |x-3|$ at $x = 2$
10. $|\cos x|$ at $x = \dfrac{3\pi}{4}$
11. $x[x] + \sqrt{x}$ at $x = \dfrac{3}{2}$
12. $\sin(\log x)$ at $x = e$
13. If $f(x) = x + \log |x|, x \neq 0$ find $f'(x)$.
14. If $y = \dfrac{2}{\sqrt{a^2 - b^2}} \tan^{-1}\left\{\sqrt{\dfrac{a-b}{a+b}} \tan \dfrac{x}{2}\right\}$, show that $\dfrac{dy}{dx} = \dfrac{1}{a + b\cos x}$ and $\dfrac{d^2y}{dx^2} = \dfrac{b \sin x}{(a + b\cos x)^2}$.
15. Show that $y = A\sin x + B\cos x + x\sin x$ is a solution of the differential equation $y + \dfrac{d^2y}{dx^2} = 2\cos x$, where A, B are constants.
16. Show that $v = \dfrac{A}{r} + B$ is a solution of the differential equation $\dfrac{d^2v}{dr^2} + \dfrac{2}{r} \cdot \dfrac{dv}{dr} = 0$, where A, B are constants.
17. If $y = \dfrac{x^2}{2} + \dfrac{x\sqrt{x^2+1}}{2} + \log \sqrt{x + \sqrt{x^2+1}}$, prove that $2y = x\dfrac{dy}{dx} + \log\left(\dfrac{dy}{dx}\right)$.
18. If $y = 1 + \dfrac{a}{x-a} + \dfrac{bx}{(x-a)(x-b)} + \dfrac{cx^2}{(x-a)(x-b)(x-c)}$, show that $\dfrac{dy}{dx} = \dfrac{y}{x}\left\{\dfrac{a}{a-x} + \dfrac{b}{b-x} + \dfrac{c}{c-x}\right\}$.
19. If $y = \sqrt{a^2 - x^2} + \dfrac{a}{2} \log \dfrac{a - \sqrt{a^2 - x^2}}{a + \sqrt{a^2 - x^2}}$, show that $\dfrac{dx}{dy} = \dfrac{x}{\sqrt{a^2 - x^2}}$.

Find $\dfrac{dy}{dx}$ if y is equal to:

20. $\sin^{-1}(x^2 \cdot \sqrt{1-x} - \sqrt{x} \cdot \sqrt{1-x^4})$
21. $\cos^{-1}\left\{\dfrac{5x + 12\sqrt{1-x^2}}{13}\right\}$, where $0 < x < \dfrac{1}{4}$
22. $(x\log x)^{\log \log x}$

23. $x^{x^x} + (\cot 4x)^x$

24. $(\tan x)^{\sin x} + (\log x)^{ex}$

25. $x^{x^n} + x^{n^x}$

26. $\{(\tan x)^{\tan x}\}^{\tan x}$ at $x = \frac{\pi}{4}$

27. $(\log_{\cos x}\sin x)(\log_{\sin x}\cos x)^{-1} + \sin^{-1}\frac{2x}{1+x^2}$ at $x = \frac{\pi}{4}$

28. $2^{\log_2 x^{2x}} + \left(\tan\frac{\pi x}{4}\right)^{\frac{4}{\pi x}}$ at $x = 1$

Find $\frac{dy}{dx}$ *if*

29. $x^{\sin y} + y^{\cos x} = 1$

30. $x^y + y^x = (x+y)^{x+y}$

31. $y = \frac{x}{a+}\,\frac{x}{b+}\,\frac{x}{a+}\,\frac{x}{b+}\cdots$

32. $y = (\sqrt{x})^{x^{x^{x^{\cdots \text{ to } \infty}}}}$

33. If $\sqrt{1-x^4} + \sqrt{1-y^4} = k(x^2 - y^2)$, prove that

$$\frac{dy}{dx} = \frac{x\sqrt{1-y^4}}{y\sqrt{1-x^4}}.$$

34. Find the derivative of

$f(x) = \log_x \sin x^2 + (\sin x^2)^{\log_e x}$

w.r.t. $\phi(x) = \log_e x$.

35. Find the differential coefficient of $\sec^{-1}\frac{1}{2x^2-1}$ w.r.t. $\sqrt{1-x^2}$ at $x = \frac{1}{2}$.

36. Find the differential coefficient of $\log_{(1-\sqrt{x})}\sin^{-1}(1-\sqrt{x})$ with respect to $2^{2(1-\sqrt{x})}$ and also find its value at $x = 0.25$.

37. If $y = \sqrt{\frac{t-\alpha}{\beta - t}}$ and $x = \sqrt{(t-\alpha)(\beta - t)}$ then find $\frac{dy}{dx}$.

38. If $x = f(t)$ and $y = \phi(t)$, prove that

$$\frac{d^2y}{dx^2} = \frac{f_1\phi_2 - f_2\phi_1}{f_1^3}$$

where suffixes denote differentiation w.r.t. t.

39. If $x = \sec\theta - \cos\theta$ and $y = \sec^n\theta - \cos^n\theta$, show that

$$(4+x^2)\left(\frac{dy}{dx}\right)^2 = n^2(4+y^2).$$

40. If $x = \cos\theta$, $y = \sin^3\theta$ then show that

$$y\frac{d^2y}{dx^2} + \left(\frac{dy}{dx}\right)^2 = 3\sin^2\theta\cdot(5\cos^2\theta - 1).$$

41. If $y = 2\cos t - \cos 2t$ and $x = 2\sin t - \sin 2t$ then prove that $\left(\frac{d^2y}{dx^2}\right)_{t=\pi/2} = \frac{-3}{2}$.

42. Find $\frac{dy}{dx}$ when $\sqrt{1-y^2} + \sqrt{1-t^2} = \alpha(y - t)$ and $x = \sin^{-1}(t\cdot\sqrt{1-t} + \sqrt{t}\cdot\sqrt{1-t^2})$.

Express your result as a function of y and t independent of α.

43. If $x^2 - y^2 = t - \frac{1}{t}$ and $x^4 + y^4 = t^2 + \frac{1}{t^2}$

then prove that $x^3y\frac{dy}{dx} + 1 = 0$.

44. Prove that $\frac{d}{dx}\begin{vmatrix} u_1 & v_1 & w_1 \\ u_2 & v_2 & w_2 \\ u_3 & v_3 & w_3 \end{vmatrix} = \begin{vmatrix} u_1 & v_1 & w_1 \\ u_2 & v_2 & w_2 \\ u_4 & v_4 & w_4 \end{vmatrix}$

where u, v, w are functions of x and

$\frac{du}{dx} = u_1, \frac{d^2u}{dx^2} = u_2$, etc.

45. If $f_r(x), g_r(x), h_r(x); r = 1, 2, 3$ are polynomials in x such that $f_r(a) = g_r(a) = h_r(a); r = 1, 2, 3$ and

$$F(x) = \begin{vmatrix} f_1(x) & f_2(x) & f_3(x) \\ g_1(x) & g_2(x) & g_3(x) \\ h_1(x) & h_2(x) & h_3(x) \end{vmatrix}$$ then find $F'(a)$.

46. If $y = \cos ax$, show that $\begin{vmatrix} y & y_1 & y_2 \\ y_3 & y_4 & y_5 \\ y_6 & y_7 & y_8 \end{vmatrix} = 0$

where $y_r = \frac{d^r}{dx^r}(\cos ax)$.

47. If $\phi(x) = \lambda(x)\cdot f(x)$ and $\psi(x) = \mu(x)\cdot f(x)$ then prove that

$$\begin{vmatrix} f(x) & \phi(x) & \psi(x) \\ f'(x) & \phi'(x) & \psi'(x) \\ f''(x) & \phi''(x) & \psi''(x) \end{vmatrix} = \{f(x)\}^3\cdot\begin{vmatrix} \lambda'(x) & \mu'(x) \\ \lambda''(x) & \mu''(x) \end{vmatrix}.$$

48. Let $\Delta(x) = \begin{vmatrix} x^2-1 & x+1 & x-2 \\ 2x^2-1 & 3x & 3x-3 \\ x^2+4 & 2x-1 & 2x-1 \end{vmatrix}$.

Prove, by using calculus, that $\Delta(x)$ is a first degree polynomial.

49. If $y = \sin(2\sin^{-1}x)$, show that

$$(1-x^2)\frac{d^2y}{dx^2} = x\frac{dy}{dx} - 4y.$$

50. If $y = a\cos(\log x) + b\sin(\log x)$ where a, b are arbitrary constants then obtain the relation between $x, y, \frac{dy}{dx}$ and $\frac{d^2y}{dx^2}$ which does not contain a and b.

51. (a) If $y = x^{n-1}\log x$, prove that
$$x^2y_2 + (3-2n)xy_1 + (n-1)^2 \cdot y = 0$$
where $y_1 = \frac{dy}{dx}$ and $y_2 = \frac{d^2y}{dx^2}$.

(b) Let $y = \tan^{-1}\sqrt{x^2-1}$. Prove that
$$(2x^2-1)\frac{dy}{dx} + x(x^2-1)\frac{d^2y}{dx^2} = 0.$$

52. If $y = \frac{ax+b}{x^2+c}$, prove that $(2xy_1 + y)y_3 = 3(xy_2 + y_1)y_2$.

53. Let $y = f(x)\cdot\phi(x)$ and $z = f'(x)\cdot\phi'(x)$.

Prove that
$$\frac{1}{y}\cdot\frac{d^2y}{dx^2} = \frac{1}{f}\cdot\frac{d^2f}{dx^2} + \frac{1}{\phi}\cdot\frac{d^2\phi}{dx^2} + \frac{2z}{f\phi}.$$

54. Let $f(x+y) = f(x)\cdot f(y)$ for all x and y and $f(5) = -2$, $f'(0) = 3$. Find $f'(5)$.

55. Let $f'(x)$ exists for all $x \neq 0$ and $f(xy) = f(x) + f(y)$ for all real x, y. Prove that $f(x) = k\log x$ where k is a constant.

56. A function f is defined such that for all real x, y

(a) $f(x+y) = f(x)\cdot f(y)$

(b) $f(x) = 1 + xg(x)$ where $\lim_{x\to 0} g(x) = 1$.

Prove that $f'(x) = f(x)$ and $f(x) = e^x$.

57. Let $f(x+y) = f(x) + f(y)$ for all real x, y and $f'(0)$ exists. Prove that $f'(x) = f'(0)$ for all $x \in R$ and $2f(x) = xf(2)$.

58. Let $f\left(\frac{x+y}{2}\right) = \frac{1}{2}\{f(x) + f(y)\}$ for all real x and y. If $f'(0)$ exists and it is equal to -1, and $f(0) = 1$ then find $f(2)$.

59. A function $f(x)$ is so defined that for all real x, $\{f(x)\}^n = f(nx)$. Prove that $f(x)\cdot f'(nx) = f'(x)\cdot f(nx)$.

60. Prove that the derivative of

(i) an odd function is an even function;

(ii) a periodic function of period T is a periodic function of period T.

61. Let $f(x)$ be a function satisfying $f(-x) = f(x)$ for all $x \in R$. If $f'(0)$ exists find its value.

62. A function $f: R \to R$ satisfies the relation
$$f\left(\frac{x+y}{3}\right) = \frac{1}{3}\{f(x) + f(y) + f(0)\}$$
for all $x, y \in R$. If $f'(0)$ exists, prove that $f'(x)$ exists for all $x \in R$.

63. Find the sum of $\sum_{r=1}^{n} rx^{r-1}$, using calculus.

64. Find the sum of $\sum_{n=1}^{\infty} n^2x^{n-1}$, $|x| < 1$, using calculus.

65. Find the sum:
$$\cos\theta + 2^2\cos 2\theta + 3^2\cos 3\theta + \ldots + n^2\cos n\theta.$$

Objective Questions

Fill in the blanks.

66. Differential coefficient of $\log_2(\log_2 x)$ w.r.t. x is ______ .

67. If $f(x) = \log_x(\log_e x)$ then $f'(e) =$ ______ .

68. If $x = a\left(\cos t + \log\tan\frac{t}{2}\right)$ and $y = a\sin t$ then $\frac{dy}{dx} =$ ______ .

69. If $f(x) = \sin(\log x)$ and $y = f\left(\frac{2x+3}{3-2x}\right)$ then $\frac{dy}{dx} =$ ______ at $x = 1$.

70. If $f'(x) = \sin x^2$ and $y = f(x^2+1)$ then $\frac{dy}{dx} =$ ______ .

71. If $x = \cos t$, $y = \log_e t$ then at $t = \frac{\pi}{2}$,
$$\frac{d^2y}{dx^2} + \left(\frac{dy}{dx}\right)^2 = \text{______} .$$

72. If $\tan y = \frac{2t}{1-t^2}$ and $\sin x = \frac{2t}{1+t^2}$ then $\frac{dy}{dx} =$ ______ .

73. Let $F(x) = f(x)g(x)h(x)$. At some point x_0, it is given that $F'(x_0) = 21F(x_0)$, $f'(x_0) = 4f(x_0)$, $g'(x_0) = -7g(x_0)$ and $h'(x_0) = k\cdot h(x_0)$. Then $k =$ ______ .

74. If $xe^{xy} = y + \sin^2 x$ then at $x = 0$, $\frac{dy}{dx} =$ ______ .

75. If $x^2y + y^3 = 2$, the value of $\frac{d^2y}{dx^2}$ at the point (1, 1) is ______ .

76. Let $f(x) = \left\{\tan\frac{\pi}{4} + \tan x\right\}\left\{\tan\frac{\pi}{4} + \tan\left(\frac{\pi}{4} - x\right)\right\}$ and $g(x) = x^2 + 1$. Then $g\{f(x)\} + f'(x) =$ ______ .

77. If $y = \frac{1}{1 + x + x^2 + x^3}$ then
$$\left(\frac{dy}{dx}\right)_{x=0} = \text{______}, \quad \left(\frac{d^2y}{dx^2}\right)_{x=0} = \text{______}.$$

78. If $f'(x) = \sin x + \sin 4x \cdot \cos x$ then
$$f'\left(2x^2 + \frac{\pi}{2}\right) \text{ at } x = \sqrt{\frac{\pi}{2}} \text{ is ______}$$
where ($'$) denotes differentiation w.r.t. x.

79. If $y = \log|x+5|$, $x \neq -5$, then $\frac{dy}{dx} =$ ______ .

80. If $x^y = e^{x-y}$ then $\frac{dy}{dx} =$ ______ in terms of x.

81. If $x = e^t \cos t, y = e^t \sin t$ then $\frac{d^2y}{dx^2} =$ ______ .

82. If $\phi(x) = \begin{vmatrix} 1 & 2x & 3x^2 \\ x & x^2 & x^3 \\ 0 & 2 & 6x \end{vmatrix}$ then $\phi'(1) =$ ______ .

Choose the correct option(s).

83. The value of $\left(\frac{dy}{dx}\right)_{(1,1)}$ for the curve $2y = 3 - x^2$ is

(a) 1 (b) –1 (c) 0 (d) 2

84. Let $y = \log_e \sin(x^2), 0 < x \le \frac{\pi}{2}$. The value of $\frac{dy}{dx}$ at $x = \frac{\sqrt{\pi}}{2}$ is

(a) 0 (b) 1 (c) $\pi/4$ (d) $\sqrt{\pi}$

85. If $f(x) = \log_e x$ then the differential coefficient of $f(\log_e x)$ with respect to x is

(a) $\frac{x}{\log_e x}$ (b) $\frac{\log_e x}{x}$

(c) $\frac{1}{x \log_e x}$ (d) none of these

86. If $y^2 = P(x)$ which is a polynomial of degree 3 then $2\frac{d}{dx}\left(y^3 \frac{d^2y}{dx^2}\right)$ equals

(a) $P'''(x) + P'(x)$ (b) $P''(x) \cdot P'''(x)$

(c) $P(x) \cdot P'''(x)$ (d) a constant

87. If $f(x) = \tan^{-1}\left\{\frac{1}{x}(\sqrt{1+x^2} - 1)\right\}$ then $f'(0)$ is

(a) $\frac{1}{2}$ (b) 0

(c) 1 (d) does not exist

88. If $t = e^x$ and $y = t^2 - 1$ then $\left(\frac{dx}{dy}\right)_{t=1}$ is

(a) $\frac{1}{2e^2}$ (b) $\frac{1}{2}$ (c) 2 (d) $2e^2$

89. Let $x^3 - 2x^2y^2 + 5x + y - 5 = 0$ and at $x = 1, y = 1$. Then $\frac{d^2y}{dx^2}$ at $y = 1$ is

(a) $-8\frac{22}{27}$ (b) $-7\frac{21}{28}$ (c) 8 (d) $\frac{22}{7}$

90. Let $f(x)$ be a polynomial of the second degree which is positive definite. If $g(x) = f(x) + f'(x) + f''(x)$ then for any real x

(a) $g(x) < 0$ (b) $g(x) > 0$ (c) $g(x) = 0$ (d) $g(x) \ge 0$

91. If $f(0) = 1, f'(0) = -1, f(x) > 0$ for all x then there exists a function $f(x)$ such that

(a) $f'(x) < 0$ for all x

(b) $-1 < f''(x) < 0$ for all x

(c) $-2 \le f''(x) \le -1$ for all x

(d) $f''(x) \le -2$ for all x

92. Let $f(x) = \begin{vmatrix} x^3 & \sin x & \cos x \\ 6 & -1 & 0 \\ p & p^2 & p^3 \end{vmatrix}$ where p is a constant. Then $\frac{d^3}{dx^3}\{f(x)\}$ at $x = 0$ is

(a) p (b) $p + p^2$

(c) $p + p^3$ (d) independent of p

State whether the statements are true or false.

93. If $y = e^{\sin^{-1}x}$ and $z = e^{-\cos^{-1}x}$ then $\frac{d^2y}{dz^2} = 0$.

94. The derivative of an even function is an odd function.

95. If $f(x)$ and $g(x)$ have no derivative at $x = a$ then $f(x) + g(x)$ may have a derivative at $x = a$.

96. If $x^2 + y^2 = r^2$ and $z = \frac{1}{r}$ then $z = \frac{\left|\frac{d^2y}{dx^2}\right|}{\sqrt{1 + \left(\frac{dy}{dx}\right)^2}}$.

Answers

1. $x(2\cos x - x\sin x)$

2. $\frac{-2x}{\sqrt{1-x^4}}$

3. $\frac{-\sec^2 x \cdot e^{\sqrt{\tan x}}}{2\sqrt{\tan x} \cdot \sqrt{1 - e^{2\sqrt{\tan x}}}}$

4. $\frac{\cos x}{3 \cdot \sqrt[3]{\sin^2 x}}$

5. $2xe^{x^2}$

6. $\frac{1 - \log x}{x^2}$

7. $\frac{\pi}{4} + \frac{1}{2}$

8. $-\frac{1}{2}$

9. 0

10. $\frac{1}{\sqrt{2}}$

11. $1 + \frac{1}{\sqrt{6}}$

12. $\frac{1}{e}\cos 1$

13. $1 + \frac{1}{x}$

20. $\dfrac{2x}{\sqrt{1-x^4}}-\dfrac{1}{2\sqrt{x}\sqrt{1-x}}$ **21.** $\dfrac{-1}{\sqrt{1-x^2}}$

22. $(x\log x)^{\log\log x}\cdot\dfrac{1}{x}\left\{1+\log(\log x)\left(1+\dfrac{2}{\log x}\right)\right\}$

23. $x^{x^x}\cdot x^x\log x\left(1+\log x+\dfrac{1}{x\log x}\right)$

$+(\cot 4x)^x\{\log\cot 4x-8x\operatorname{cosec}8x\}$

24. $(\tan x)^{\sin x-1}\cdot\{\sec x\cdot\tan x+\sin x\cdot\log\tan x\}$

$+(\log x)^{ex}\cdot e\left\{\dfrac{1}{\log x}+\log\log x\right\}$

25. $x^{x^n+n-1}(n\log x+1)+x^{n^x}\cdot n^x\left(\log n\cdot\log x+\dfrac{1}{x}\right)$

26. 2 **27.** $8\left(\dfrac{4}{16+\pi^2}-\dfrac{1}{\log 2}\right)$ **28.** 4

29. $\dfrac{y^{\cos x}\cdot\sin x\cdot\log y-x^{\sin y-1}\cdot\sin y}{x^{\sin y}\cdot\cos y\cdot\log x+y^{\cos x-1}\cdot\cos x}$

30. $\dfrac{(x+y)^{x+y}\{1+\log(x+y)\}-y(x^{y-1}+y^{x-1}\cdot\log y)}{x\{x^{y-1}\log x+y^{x-1}\}-(x+y)^{x+y}\{1+\log(x+y)\}}$

31. $\dfrac{b}{a(b+2y)}$ **32.** $\dfrac{y}{x}\cdot\dfrac{\log y}{\log x}\cdot\dfrac{1}{1-2\log y}$

34. $\dfrac{2x^2}{\log x}\cot x^2-\dfrac{1}{(\log x)^2}\log\sin x^2$

$+2x^2\log x\cdot(\sin x^2)^{\log x-1}\cdot\cos x^2$

$+\log\sin x^2\cdot(\sin x^2)^{\log x}$

35. 4

36. $\dfrac{\left\{\dfrac{\log(1-\sqrt{x})}{\sqrt{2\sqrt{x}-x}}\cdot\dfrac{1}{\sin^{-1}(1-\sqrt{x})}-\dfrac{\log\sin^{-1}(1-\sqrt{x})}{(1-\sqrt{x})}\right\}}{\left[\log 4\cdot(-4^{1-\sqrt{x}})\cdot\{\log(1-\sqrt{x})\}^2\right]}$,

$\dfrac{-\dfrac{2\sqrt{3}}{\pi}\log 2-\log\dfrac{\pi}{6}}{2(\log 2)^3}$

37. $\dfrac{\alpha-\beta}{(\beta-t)(2t-\alpha-\beta)}$ **42.** $\dfrac{2\sqrt{t}\cdot\sqrt{1-y^2}}{2\sqrt{t}+\sqrt{1+t}}$

45. 0 **50.** $x^2\dfrac{d^2y}{dx^2}+x\dfrac{dy}{dx}+y=0$

54. −6 **58.** −1 **61.** 0

63. $\dfrac{1}{(1-x)^2}\{1-(n+1)x^n+nx^{n+1}\}$

64. $\dfrac{1+x}{(1-x)^3}$

65. $\dfrac{n^2}{2}\sin\left(n+\dfrac{1}{2}\right)\theta\cdot\operatorname{cosec}\dfrac{\theta}{2}+\dfrac{n}{2}\cos n\theta\cdot\operatorname{cosec}^2\dfrac{\theta}{2}$

$-\dfrac{1}{4}\sin n\theta\cdot\operatorname{cosec}^2\dfrac{\theta}{2}\cdot\cot\dfrac{\theta}{2}$

66. $\dfrac{\log_x e}{x\log_e 2}$ **67.** $\dfrac{1}{e}$ **68.** $\tan t$

69. $\dfrac{12}{5}\cos(\log 5)$ **70.** $2x\sin(x^2+1)^2$

71. 0 **72.** 1 **73.** 24 **74.** 1

75. $\dfrac{-3}{8}$ **76.** 5 **77.** −1, 0 respectively

78. $-2\sqrt{2\pi}$ **79.** $\dfrac{1}{x+5}$ **80.** $\dfrac{\log x}{(1+\log x)^2}$

81. $\dfrac{1}{\sqrt{2e^t}}\sec^3\left(\dfrac{\pi}{4}+t\right)$ **82.** −6 **83.** (b)

84. (d) **85.** (c) **86.** (c) **87.** (a)

88. (b) **89.** (a) **90.** (b) **91.** (a)

92. (d) **93.** true **94.** true **95.** true

96. false

□

Chapter Test

Time: 60 minutes

1. In each of the following, fill in the blanks so that the resulting sentence becomes true.

(a) If $y = f\left(\dfrac{2x-1}{x^2+1}\right)$ and $f'(x) = \sin x^2$ then $\dfrac{dy}{dx} =$ ______ .

(b) If $f(x) = (ax+b)\cos x + (cx+d)\sin x$ and $f'(x) = x\cos x$ is an identity in x then $a =$ ______ , $b =$ ______ , $c =$ ______ , $d =$ ______ .

(c) If $f(x) = |x-1|$ and $g(x) = f\{f(x)\}$ then $g'(4) =$ ______ .

(d) $\dfrac{d(x^{\log x})}{d(\log x)} =$ ______ .

2. Find $\dfrac{dy}{dx}$ in terms of x if $t = \dfrac{x}{1+x^2}$, $y = x^2 + t^2$.

3. If g is the inverse function of f and $f'(x) = \dfrac{1}{1+x^5}$, prove that $g'(x) = 1 + \{g(x)\}^5$.

4. If $y = (1+x)(1+x^{\frac{1}{2}})(1+x^{\frac{1}{4}}) \ldots (1+x^{\frac{1}{2^{n-1}}})$, find $\dfrac{dy}{dx}$ in the simplest form.

5. If $y = f(x)$, $p = \dfrac{dy}{dx}$ and $q = \dfrac{d^2y}{dx^2}$ then express $\dfrac{d^2x}{dy^2}$ as a function of p and q.

6. Let $f(x+y) = f(x)\cdot f(y)$ for all x, y and $f(0) \neq 0$ then prove that $f'(x)$ can be obtained at all $x \in R$ if $f'(0)$ exists.

7. Transform the differential equation $(1-x^2)\dfrac{d^2y}{dx^2} + y = x\dfrac{dy}{dx}$ by substituting $x = \cos t$.

8. Let $y = \dfrac{f(x)}{\phi(x)}$ and $z = \dfrac{f'(x)}{\phi'(x)}$. Prove that $\dfrac{1}{y}\dfrac{d^2y}{dx^2} = \dfrac{1}{f}\dfrac{d^2f}{dx^2} - \dfrac{1}{\phi}\cdot\dfrac{d^2\phi}{dx^2} + 2\dfrac{y-z}{f\phi}\cdot\left(\dfrac{d\phi}{dx}\right)^2$.

□

Answers

1. (a) $\dfrac{2(1+x-x^2)}{(1+x^2)^2}\cdot\sin\left(\dfrac{2x-1}{x^2+1}\right)^2$ (b) 0, 1, 1, 0 respectively

(c) 1 (d) $2\log x\cdot x^{\log x}$

2. $2x\left\{1+\dfrac{1-x^2}{(1+x^2)^3}\right\}$

4. $\dfrac{2x}{x^{1/2^{n-1}}-1}+\dfrac{1-x^2}{(x^{1/2^{n-1}}-1)^2}\cdot\dfrac{1}{2^{n-1}}\cdot x^{\frac{1}{2^{n-1}}-1}$

5. $\dfrac{-q}{p^3}$

7. $\dfrac{d^2y}{dt^2}+y=0$

□

3. Limit, Indeterminate Form

Recap of Facts and Formulae

1. Indeterminante forms of value

- The indeterminate forms of values of a function at a point are

$\frac{0}{0}, \frac{\infty}{\infty}, 0 \times \infty, \infty - \infty; 1^{\infty}, 0^{0}, \infty^{0}.$

2. Standard limits

- $\lim\limits_{x \to \infty} \frac{1}{x} = 0$
- $\lim\limits_{x \to 0} \frac{1}{|x|} = \infty$
- $\lim\limits_{x \to 0} |x|^{n} = 0,$ where $n > 0$
- $\lim\limits_{n \to \infty} x^{n} = 0$ if $|x| < 1$
- $\lim\limits_{n \to \infty} |x|^{n} = \infty$ if $|x| > 1$
- $\lim\limits_{x \to a} \frac{x^{n} - a^{n}}{x - a} = na^{n-1}$
- $\lim\limits_{x \to 0} \frac{\sin x}{x} = 1,$ where x is in radian measure
- $\lim\limits_{x \to 0} \cos x = 1$
- $\lim\limits_{x \to 0} \frac{\tan x}{x} = 1,$ where x is in radian measure
- $\lim\limits_{x \to 0} \frac{e^{x} - 1}{x} = 1$
- $\lim\limits_{x \to 0} \frac{a^{x} - 1}{x} = \log_{e} a$
- $\lim\limits_{x \to \infty} \left(1 + \frac{a}{x}\right)^{x} = e^{a}$
- $\lim\limits_{x \to 0} (1 + ax)^{1/x} = e^{a}$

3. Properties of limits

- $\lim\limits_{x \to a} \{f(x) \pm \phi(x)\} = \lim\limits_{x \to a} f(x) \pm \lim\limits_{x \to a} \phi(x)$
- $\lim\limits_{x \to a} \{f(x) \times \phi(x)\} = \lim\limits_{x \to a} f(x) \times \lim\limits_{x \to a} \phi(x)$
- $\lim\limits_{x \to a} \frac{f(x)}{\phi(x)} = \frac{\lim\limits_{x \to a} f(x)}{\lim\limits_{x \to a} \phi(x)}$
- $\lim\limits_{x \to a} f(x) = e^{\lim\limits_{x \to a} \log_{e} f(x)}$
- $\lim\limits_{x \to a} \{f(x)\}^{\phi(x)} = \left\{\lim\limits_{x \to a} f(x)\right\}^{\lim\limits_{x \to a} \phi(x)}$
- $\lim\limits_{x \to a} f\{\phi(x)\} = f\left(\lim\limits_{x \to a} \phi(x)\right)$ if $f(x)$ is continuous

4. L' Hospital's rule

- If $f(a) = 0$ and $\phi(a) = 0$ then

$$\lim_{x \to a} \frac{f(x)}{\phi(x)} = \lim_{x \to a} \frac{f'(x)}{\phi'(x)}$$

- If $\lim\limits_{x \to a} f(x) = \infty$ and $\lim\limits_{x \to a} \phi(x) = \infty$ then

$$\lim_{x \to a} \frac{f(x)}{\phi(x)} = \lim_{x \to a} \frac{f'(x)}{\phi'(x)}$$

Selected Solved Examples

1. Evaluate $\lim\limits_{x \to 0} \dfrac{\sqrt[3]{8+x} - \sqrt[3]{8+x^2-x^3}}{\sqrt[3]{8+x} - \sqrt[3]{8+x^2+x^3}}$.

We know, $(a-b)(a^2+ab+b^2) = a^3 - b^3$.

$\therefore$ limit

$$= \lim_{x \to 0} \frac{(8+x) - (8+x^2-x^3)}{\left\{(\sqrt[3]{8+x})^2 + \sqrt[3]{8+x}\ \sqrt[3]{8+x^2-x^3} + (\sqrt[3]{8+x^2-x^3})^2\right\}}$$

$$\times \frac{\left\{(\sqrt[3]{8+x})^2 + \sqrt[3]{8+x}\ \sqrt[3]{8+x^2+x^3} + (\sqrt[3]{8+x^2+x^3})^2\right\}}{(8+x) - (8+x^2+x^3)}$$

$$= \lim_{x \to 0} \frac{x - x^2 + x^3}{x - x^2 - x^3} \times \frac{(\sqrt[3]{8+x})^2 + \sqrt[3]{8+x} \cdot \sqrt[3]{8+x^2+x^3} + (\sqrt[3]{8+x^2+x^3})^2}{(\sqrt[3]{8+x})^2 + \sqrt[3]{8+x} \cdot \sqrt[3]{8+x^2-x^3} + (\sqrt[3]{8+x^2-x^3})^2}$$

$$= \lim_{x \to 0} \frac{1 - x + x^2}{1 - x - x^2} \times \frac{(\sqrt[3]{8+x})^2 + \sqrt[3]{8+x} \cdot \sqrt[3]{8+x^2+x^3} + (\sqrt[3]{8+x^2+x^3})^2}{(\sqrt[3]{8+x})^2 + \sqrt[3]{8+x} \cdot \sqrt[3]{8+x^2-x^3} + (\sqrt[3]{8+x^2-x^3})^2}$$

$$= \frac{1}{1} \cdot \frac{2^2 + 2 \cdot 2 + 2^2}{2^2 + 2 \cdot 2 + 2^2} = 1.$$

2. Evaluate $\lim\limits_{x \to \infty} \left(\sqrt{x + \sqrt{x + \sqrt{x}}} - \sqrt{x}\right)$.

$$\text{Limit} = \lim_{x \to \infty} \frac{\left(\sqrt{x+\sqrt{x+\sqrt{x}}} - \sqrt{x}\right)\left(\sqrt{x+\sqrt{x+\sqrt{x}}} + \sqrt{x}\right)}{\sqrt{x+\sqrt{x+\sqrt{x}}} + \sqrt{x}}$$

$$= \lim_{x \to \infty} \frac{x + \sqrt{x+\sqrt{x}} - x}{\sqrt{x+\sqrt{x+\sqrt{x}}} + \sqrt{x}}$$

$$= \lim_{x \to \infty} \frac{\sqrt{x+\sqrt{x}}}{\sqrt{x+\sqrt{x+\sqrt{x}}} + \sqrt{x}}$$

$$= \lim_{x \to \infty} \frac{x^{1/2} \cdot \sqrt{1 + \dfrac{1}{\sqrt{x}}}}{x^{1/2} \cdot \left\{\sqrt{1 + \sqrt{\dfrac{x+\sqrt{x}}{x^2}}} + 1\right\}}$$

$$= \lim_{x \to \infty} \frac{\sqrt{1 + \dfrac{1}{\sqrt{x}}}}{\sqrt{1 + \sqrt{\dfrac{1}{x} + \dfrac{1}{x^{3/2}}}} + 1}$$

$$= \frac{1}{1+1} = \frac{1}{2}.$$

3. Evaluate

$$\lim_{n \to \infty} \frac{\mathbf{1} \cdot \sum\limits_{1}^{n} r + \mathbf{2} \cdot \sum\limits_{1}^{n-1} r + \mathbf{3} \cdot \sum\limits_{1}^{n-2} r + \ldots + n \cdot \mathbf{1}}{n^4}.$$

Consider

$$m \cdot \sum_{1}^{n-m+1} r$$

$$= m\{1 + 2 + 3 + \ldots + \text{to } (n-m+1) \text{ terms}\}$$

$$= m \cdot \frac{(n-m+1)(n-m+2)}{2}$$

$$= \frac{m}{2}\{n^2 - (2m-3)n + (m-1)(m-2)\}$$

$$= \frac{n^2}{2} \cdot m - \frac{n}{2} m(2m-3) + \frac{m}{2}(m^2 - 3m + 2)$$

$$= \frac{n^2}{2} \cdot m - n \cdot m^2 + \frac{3n}{2} \cdot m + \frac{m^3}{2} - \frac{3}{2} m^2 + m$$

$$= \left(\frac{n^2}{2} + \frac{3n}{2} + 1\right) m - \left(n + \frac{3}{2}\right) m^2 + \frac{1}{2} m^3$$

$$\therefore \quad \sum_{1}^{n} \left\{ m \sum_{1}^{n-m+1} r \right\}$$

$$= \left(\frac{n^2}{2} + \frac{3n}{2} + 1\right) \sum_{1}^{n} m - \left(n + \frac{3}{2}\right) \sum_{1}^{n} m^2 + \frac{1}{2} \sum_{1}^{n} m^3$$

$$= \frac{n^2 + 3n + 2}{2} \cdot \frac{n(n+1)}{2} - \frac{2n+3}{2} \cdot \frac{n(n+1)(2n+1)}{6} + \frac{1}{2} \cdot \frac{n^2(n+1)^2}{4}$$

$$= \frac{n(n+1)^2(n+2)}{4} - \frac{n(n+1)(2n+1)(2n+3)}{12} + \frac{n^2(n+1)^2}{8}$$

$\therefore$ limit

$$= \lim_{n \to \infty} \frac{1}{n^4} \cdot n^4 \left\{ \frac{\left(\frac{1}{n}+1\right)^2 \left(\frac{2}{n}+1\right)}{4} - \frac{\left(1+\frac{1}{n}\right)\left(2+\frac{1}{n}\right)\left(2+\frac{3}{n}\right)}{12} + \frac{\left(1+\frac{1}{n}\right)^2}{8} \right\}$$

$$= \frac{1}{4} - \frac{4}{12} + \frac{1}{8}$$

$$= \frac{1}{4} - \frac{1}{3} + \frac{1}{8} = \frac{1}{24}.$$

4. Evaluate $\lim\limits_{x \to -\infty} \dfrac{x^4 \sin \frac{1}{x} + x^2}{1 + |x|^3}$.

Let $x = -y$. Then $y \to \infty$ when $x \to -\infty$.

$$\therefore \text{ limit} = \lim_{y \to \infty} \frac{y^4 \sin \frac{-1}{y} + y^2}{1 + |-y|^3} = \lim_{y \to \infty} \frac{-y^4 \sin \frac{1}{y} + y^2}{1 + y^3}$$

$\{\because\ y$ is positive$\}$

$$= \lim_{y \to \infty} \left\{ \frac{\sin \frac{1}{y}}{\frac{1}{y}} \cdot \frac{-y^3}{1+y^3} + \frac{y^2}{1+y^3} \right\}$$

$$= \lim_{y \to \infty} \left\{ \frac{\sin \frac{1}{y}}{\frac{1}{y}} \cdot \frac{-1}{\frac{1}{y^3}+1} + \frac{1}{y\left(\frac{1}{y^3}+1\right)} \right\}$$

$$= 1 \cdot \frac{-1}{1} + 0 = -1.$$

5. Prove that

$$\lim_{m \to \infty} \lim_{n \to \infty} \{1 + \cos^{2m}(n\,!\,\pi x)\} = 2, \text{ if } x \text{ is rational}$$

$$= 1, \text{ if } x \text{ is irrational.}$$

We know that $|\cos \theta| \le 1$ for all θ.

If $|\cos (n\,!\,\pi x)| = p < 1$,

$$\text{limit} = \lim_{m \to \infty} (1 + p^{2m}) = 1 \qquad \dots (1)$$

as $p^{2m} \to 0$ when $m \to \infty$.

If $|\cos (n\,!\,\pi x)| = 1$,

$$\text{limit} = \lim_{m \to \infty} (1 + 1^{2m}) = \lim_{m \to \infty} 2 = 2. \qquad \dots (2)$$

Now, $|\cos (n\,!\,\pi x)| = 1$

$\Rightarrow$ $n\,!\,\pi x = k\pi$, where k is an integer

$\therefore$ $x = \dfrac{k}{n\,!}$ which is a rational number.

$\therefore$ from (2), limit = 2 if x is rational.

Again, $|\cos (n\,!\,\pi x)| < 1$

$\Rightarrow$ $|\cos (n\,!\,\pi x)| \neq 1$ for all $n \in N$

$\Rightarrow$ $n\,!\,\pi x \neq k\pi$ for all $n \in N, k \in Z$

$\Rightarrow$ $x \neq \dfrac{k}{n\,!}$ for all $n \in N, k \in Z$

$\Rightarrow$ x is not rational, i.e., x is irrational.

$\therefore$ from (1), limit = 1 if x is irrational.

6. Without expansion or using L' Hospital's rule, prove that $\lim\limits_{\theta \to 0} \dfrac{\theta - \sin \theta}{\theta^3} = \dfrac{1}{6}$.

Let the limit be p. Then

$$p = \lim_{\theta \to 0} \frac{\theta - \sin \theta}{\theta^3}$$

$$= \lim_{\theta \to 0} \frac{3\theta - \sin 3\theta}{(3\theta)^3} \quad \text{(replacing } \theta \text{ by } 3\theta\text{)}$$

$$= \lim_{\theta \to 0} \frac{3\theta - 3\sin \theta + 4\sin^3 \theta}{(3\theta)^3}$$

$$= \lim_{\theta \to 0} \left\{ \frac{\theta - \sin \theta}{9\theta^3} + \frac{4}{27} \left(\frac{\sin \theta}{\theta} \right)^3 \right\}$$

$$= \frac{1}{9} \lim_{\theta \to 0} \frac{\theta - \sin \theta}{\theta^3} + \frac{4}{27} \lim_{\theta \to 0} \left(\frac{\sin \theta}{\theta} \right)^3$$

$$= \frac{1}{9} p + \frac{4}{27} \cdot 1^3$$

$$\therefore \quad p - \frac{1}{9} p = \frac{4}{27}$$

$$\therefore \quad \frac{8p}{9} = \frac{4}{27}; \qquad \therefore \quad p = \frac{1}{6}.$$

7. Evaluate: $\lim\limits_{x \to \pi/2} \tan x \cdot \log_e \sin x$.

$$\text{Limit} = \lim_{x \to \pi/2} \frac{\log_e \sin x}{\cot x} \qquad \left\{\text{form: } \frac{0}{0}\right\}$$

$$= \lim_{x \to \pi/2} \frac{\frac{1}{\sin x} \cdot \cos x}{-\operatorname{cosec}^2 x} \quad \text{\{using L' Hospital's rule\}}$$

$$= \lim_{x \to \pi/2} (-\sin x \cdot \cos x) = 0.$$

8. Evaluate: $\lim_{x\to\infty}\left(\dfrac{3x^2+1}{4x^2-1}\right)^{\frac{x^3}{1+x}}$.

$$\text{Limit}=\lim_{x\to\infty}\left(\frac{3}{4}\cdot\frac{x^2+\frac{1}{3}}{x^2-\frac{1}{4}}\right)^{\frac{x^3}{1+x}}$$

$$=\lim_{x\to\infty}\left(\frac{3}{4}\right)^{\frac{x^3}{1+x}}\left(\frac{x^2+\frac{1}{3}}{x^2-\frac{1}{4}}\right)^{\frac{x^3}{1+x}}$$

$$=\lim_{x\to\infty}\left(\frac{3}{4}\right)^{\frac{x^3}{1+x}}\cdot\lim_{x\to\infty}\left(1+\frac{\frac{7}{12}}{x^2-\frac{1}{4}}\right)^{\frac{x^3}{1+x}}$$

$$=\left(\frac{3}{4}\right)^{\lim_{x\to\infty}\frac{x^3}{1+x}}\cdot\lim_{x\to\infty}\left(1+\frac{\frac{7}{12}}{x^2-\frac{1}{4}}\right)^{\frac{x^3}{1+x}}$$

Now, $\lim_{x\to\infty}\dfrac{x^3}{1+x}=\lim_{x\to\infty}\dfrac{x^2}{\frac{1}{x}+1}=\infty.$

$$\therefore\quad\left(\frac{3}{4}\right)^{\lim_{x\to\infty}\frac{x^3}{1+x}}=0\quad\left(\because\ 0<\frac{3}{4}<1\right)$$

and, $\lim_{x\to\infty}\left(1+\dfrac{\frac{7}{12}}{x^2-\frac{1}{4}}\right)^{\frac{x^3}{1+x}}$

$$=\lim_{x\to\infty}\left\{\left(1+\frac{\frac{7}{12}}{x^2-\frac{1}{4}}\right)^{x^2-\frac{1}{4}}\right\}^{\frac{x^3}{\left(x^2-\frac{1}{4}\right)(1+x)}}$$

$$=\left\{e^{7/12}\right\}^{\lim_{x\to\infty}\frac{1}{\left(1-\frac{1}{4x^2}\right)\left(\frac{1}{x}+1\right)}}=\left(e^{\frac{7}{12}}\right)^{1}=e^{\frac{7}{12}}$$

because $\lim_{x\to\infty}\left(1+\dfrac{a}{f(x)}\right)^{f(x)}=e^a,$

when $f(x)\to\infty$ as $x\to\infty$.

Hence, the required limit $=0\cdot e^{7/12}=0$.

9. Evaluate: $\lim_{x\to0}\left(\dfrac{\tan x}{x}\right)^{\frac{1}{x}}$.

Here, the form is 1^{∞}.

$$\text{Limit}=e^{\lim_{x\to0}\log\left(\frac{\tan x}{x}\right)^{\frac{1}{x}}}$$

$$=e^{\lim_{x\to0}\frac{1}{x}\log\left(\frac{\tan x}{x}\right)}\quad\dots(1)$$

Now, $\lim_{x\to0}\dfrac{1}{x}\log\dfrac{\tan x}{x}=\lim_{x\to0}\dfrac{\log\frac{\tan x}{x}}{x}$ $\left(\text{form: }\dfrac{0}{0}\right)$

$$=\lim_{x\to0}\frac{\frac{x}{\tan x}\cdot\frac{x\sec^2x-\tan x}{x^2}}{1}$$

(using L' Hospital's rule)

$$=\lim_{x\to0}\frac{x\sec^2x-\tan x}{x\tan x}\quad\left(\text{form: }\frac{0}{0}\right)$$

$$=\lim_{x\to0}\frac{\sec^2x+x\cdot2\sec^2x\tan x-\sec^2x}{\tan x+x\sec^2x}$$

(using L' Hospital's rule)

$$=\lim_{x\to0}\frac{2x\sec^2x\cdot\tan x}{\tan x+x\sec^2x}\quad\left(\text{form: }\frac{0}{0}\right)$$

$$=\lim_{x\to0}\frac{2\sec^2x\cdot\tan x+2x(\sec^4x+2\sec^2x\cdot\tan^2x)}{2\sec^2x+x\cdot2\sec^2x\tan x}$$

$$=\frac{0}{2}=0$$

$\therefore$ from (1), limit $=e^0=1$.

10. Evaluate: $\lim_{x\to0}\left(\dfrac{\sin x}{x}\right)^{\frac{\sin x}{x-\sin x}}$.

$$\text{Limit}=\lim_{x\to0}\left(\frac{\sin x}{x}\right)^{\frac{\frac{\sin x}{x}}{1-\frac{\sin x}{x}}}$$

Put $\dfrac{\sin x}{x}=t$; then $t\to1$ when $x\to0$.

$\therefore$ limit $=\lim_{t\to1}t^{\frac{t}{1-t}}$, which is of the form 1^{∞}

$$=e^{\lim_{t\to1}\log t^{\frac{t}{1-t}}}=e^{\lim_{t\to1}\frac{t}{1-t}\log t}$$

$$= e^{\lim\limits_{t \to 1} \frac{\log t + 1}{-1}} \quad \text{(using L' Hospital's rule)}$$

$$= e^{-1}.$$

Second Method:

$$\text{Limit} = \lim_{x \to 0} \left(1 - \frac{x - \sin x}{x}\right)^{\frac{\sin x}{x - \sin x}}$$

$$= \lim_{x \to 0} \left\{\left(1 + \frac{-1}{\frac{x}{x - \sin x}}\right)^{\frac{x}{x - \sin x}}\right\}^{\frac{\sin x}{x}}$$

$$= \left\{\lim_{y \to \infty} \left(1 + \frac{-1}{y}\right)^{y}\right\}^{\lim\limits_{x \to 0} \frac{\sin x}{x}}$$

$$\left\{\text{putting } \frac{x}{x - \sin x} = y \to \infty \text{ as } x \to 0\right\}$$

$$= \left(e^{-1}\right)^{1} = e^{-1}.$$

11. Find $\lim\limits_{x \to \infty} \left(\dfrac{1^{1/x} + 2^{1/x} + 3^{1/x} + \ldots + n^{1/x}}{n}\right)^{nx}$.

Here the form is 1^{∞}.

$$\text{Limit} = e^{\lim\limits_{x \to \infty} \log\left(\frac{1^{1/x} + 2^{1/x} + 3^{1/x} + \ldots + n^{1/x}}{n}\right)^{nx}} \quad \ldots (1)$$

Now, $\lim\limits_{x \to \infty} \log\left(\dfrac{1^{1/x} + 2^{1/x} + 3^{1/x} + \ldots + n^{1/x}}{n}\right)^{nx}$

$$= \lim_{x \to \infty} nx \cdot \log \frac{1^{1/x} + 2^{1/x} + 3^{1/x} + \ldots + n^{1/x}}{n}$$

$$= \lim_{y \to 0} n \frac{\log(1^{y} + 2^{y} + \ldots + n^{y}) - \log n}{y}$$

$$\left\{\text{putting } y = \frac{1}{x}\right\}$$

$$= n \cdot \lim_{y \to 0} \frac{\left[\frac{1}{1^{y} + 2^{y} + \ldots + n^{y}} \cdot \{2^{y}\log 2 + 3^{y}\log 3 + \ldots + n^{y}\log n\} - 0\right]}{1}$$

{using L' Hospital's rule}

$$= n \cdot \frac{1}{1^{0} + 2^{0} + \ldots + n^{0}} \{\log 2 + \log 3 + \ldots + \log n\}$$

$$= \log n\,!$$

$\therefore$ by (1), limit $= e^{\log n\,!} = n\,!$.

12. If $f(1) = g(1) = 2$ and $f'(1)$, $g'(1)$ exist then evaluate

$$\lim_{x \to 1} \frac{f(1)g(x) - f(1) - g(1)f(x) + g(1)}{g(x) - f(x)}.$$

Here, the form is $\frac{0}{0}$.

$$\therefore \text{ limit} = \lim_{x \to 1} \frac{f(1)g'(x) - g(1)f'(x)}{g'(x) - f'(x)},$$

{using L' Hospital's rule}

$$= \lim_{x \to 1} \frac{2\{g'(x) - f'(x)\}}{g'(x) - f'(x)} = \lim_{x \to 1} 2 = 2.$$

13. Find a, b and c such that

$$\lim_{x \to 0} \frac{axe^{x} - b\log(1 + x) + cxe^{-x}}{x^{2}\sin x} = 2.$$

$$\text{Limit} = \lim_{x \to 0} \frac{a(e^{x} + xe^{x}) - b \cdot \frac{1}{1 + x} + c(e^{-x} - xe^{-x})}{2x \sin x + x^{2}\cos x}$$

{using L' Hospital's rule}

$$= \frac{a - b + c}{0} = \infty \text{ unless } a - b + c = 0.$$

But, the limit is given to be 2. So, the indeterminate form $\frac{0}{0}$ should continue.

So, $a - b + c = 0$. ... (1)

Then, limit

$$= \lim_{x \to 0} \frac{\left\{a(1 + x)e^{x} + ae^{x} + \frac{b}{(1 + x)^{2}} + c(-1)e^{-x} + c(1 - x)e^{-x} \cdot (-1)\right\}}{2\sin x + 2x \cos x + 2x \cos x - x^{2}\sin x}$$

{using L' Hospital's rule}

$$= \lim_{x \to 0} \frac{a(2 + x)e^{x} + \frac{b}{(1 + x)^{2}} - c(2 - x)e^{-x}}{2\sin x + 4x \cos x - x^{2}\sin x}$$

$$= \frac{2a + b - 2c}{0} = \infty \text{ unless } 2a + b - 2c = 0.$$

But, the limit is given to be 2.

So, $2a + b - 2c = 0$ ... (2)

Then, limit $= \lim\limits_{x \to 0} \dfrac{\left\{a(2+x)e^x + ae^x - \dfrac{2b}{(1+x)^3} + c(2-x)e^{-x} + ce^{-x}\right\}}{6\cos x - 4x \sin x - 2x \sin x - x^2 \cos x}$

$= \dfrac{2a + a - 2b + 2c + c}{6} = 2$ (given)

$\therefore \quad 3a - 2b + 3c = 12 \qquad \ldots (3)$

(1) $\Rightarrow \quad b = a + c$

(2) $\Rightarrow \quad 2a + a + c - 2c = 0$, i.e., $c = 3a$

$\therefore \quad b = a + c$

$= a + 3a = 4a$

$\therefore$ (3) $\Rightarrow \quad 3a - 2 \times 4a + 3 \times 3a = 12; \quad \therefore \quad a = 3$

$\therefore \quad b = 4a = 4 \times 3 = 12$

$c = 3a = 3 \times 3 = 9.$

Thus, $a = 3$, $b = 12$, $c = 9$.

14. Find $\lim\limits_{\substack{x \to 1 \\ y \to 0}} \dfrac{y^3}{x^3 - y^2 - 1}$ as $(x, y) \to (1, 0)$ along the line $y = x - 1$.

When $y = x - 1$, we get $x = y + 1$. As (x, y) tends to $(1, 0)$ along the line, $x = y + 1$ holds throughout.

$$\therefore \quad \frac{y^3}{x^3 - y^2 - 1} = \frac{y^3}{(y+1)^3 - y^2 - 1}$$

$$= \frac{y^3}{y^3 + 2y^2 + 3y} = \frac{y^2}{y^2 + 2y + 3}$$

$$\therefore \quad \text{limit} = \lim_{y \to 0} \frac{y^2}{y^2 + 2y + 3} = \frac{0}{3} = 0.$$

Exercises

Evaluate the following:

1. $\lim\limits_{x \to 1} \dfrac{\sqrt{x^2+8} - \sqrt{10 - x^2}}{\sqrt{x^2+3} - \sqrt{5 - x^2}}$

2. $\lim\limits_{x \to 0} \dfrac{\sqrt[3]{1 + x^2} - 1}{x^2}$

3. $\lim\limits_{x \to 0} \dfrac{(1+x)^{1/3} - (1-x)^{1/3}}{x}$

4. $\lim\limits_{x \to 0} \dfrac{\sqrt[3]{1 + \sin x} - \sqrt[3]{1 - \sin x}}{x}$

5. $\lim\limits_{x \to 0} \dfrac{\sqrt{\cos x} - \sqrt[3]{\cos x}}{\sin^2 x}$

6. $\lim\limits_{x \to \frac{\pi}{4}} \dfrac{\sqrt{1 - \sqrt{\sin 2x}}}{\pi - 4x}$ where $x < \dfrac{\pi}{4}$.

7. $\lim\limits_{x \to \infty} x^3\{\sqrt{x^2 + \sqrt{1 + x^4}} - x\sqrt{2}\}$

8. $\lim\limits_{x \to \infty} \dfrac{2\sqrt{x} + 3\sqrt[3]{x} + 5\sqrt[5]{x}}{\sqrt{3x - 2} + \sqrt[3]{2x - 3}}$

9. $\lim\limits_{x \to -1} \dfrac{x + 1}{\sqrt[4]{x + 17} - 2}$

10. $\lim\limits_{x \to \frac{\pi}{2}} \dfrac{\cos x}{(1 - \sin x)^{2/3}}$

Evaluate:

11. $\lim\limits_{x \to 0} \dfrac{\tan^{-1} x - \sin^{-1} x}{x^3}$

12. $\lim\limits_{x \to 2} \dfrac{2^x + 2^{3-x} - 6}{2^{-x/2} - 2^{1-x}}$

13. $\lim\limits_{x \to \frac{\pi}{4}} \dfrac{2\sqrt{2} - (\cos x + \sin x)^3}{1 - \sin 2x}$

14. $\lim\limits_{x \to 0} \dfrac{27^x - 9^x - 3^x + 1}{\sqrt{2} - \sqrt{1 + \cos x}}$

Find the limits:

15. $\lim\limits_{x \to \infty} \left(\dfrac{x^2 - 2x + 2}{x^2 - 4x + 1}\right)^x$

16. $\lim\limits_{x \to 0} \left\{\tan\left(\dfrac{\pi}{4} + x\right)\right\}^{\frac{1}{x}}$

17. $\lim\limits_{x \to \frac{\pi}{2}} (\sin x)^{\tan x}$

18. $\lim\limits_{x \to 0} \left(\dfrac{\sin x}{x}\right)^{\frac{1}{x}}$

19. $\lim\limits_{x \to 0} (\sin x + \cos x)^{1/x}$

20. $\lim\limits_{x \to 0} (\sec \sqrt{x})^{10/x}$

21. $\lim_{x\to 1} (2-x)^{\tan\frac{\pi x}{2}}$

22. $\lim_{x\to 0} \left(\frac{5}{2+\sqrt{9+x}}\right)^{\frac{1}{\sin x}}$

23. $\lim_{x\to +0} (\log \cot x)^{\tan x}$

24. $\lim_{x\to 1} \left(\frac{1+x}{2+x}\right)^{\frac{1-\sqrt{x}}{1-x}}$

25. If $f(x)$ is differentiable at $x=a$, show that

$$\lim_{x\to a} \frac{xf(a)-af(x)}{x-a} + af'(a) = f(a).$$

26. Let $\Delta(h) = \begin{vmatrix} \sin\alpha & \sin(\alpha+h) & \sin(\alpha+2h) \\ \sin(\alpha+2h) & \sin\alpha & \sin(\alpha+h) \\ \sin(\alpha+h) & \sin(\alpha+2h) & \sin\alpha \end{vmatrix}$.

Find $\lim_{h\to 0} \frac{\Delta(h)}{h^2}$.

27. If $t_r = \frac{1^2+2^2+3^2+\ldots+r^2}{1^3+2^3+3^3+\ldots+r^3}$ and $S_n = \sum_{r=1}^{n} (-1)^r \cdot t_r$

then find $\lim_{n\to\infty} S_n$.

28. If α, β are the roots of the equation $x^2+bx+c=0$ then find

$$\lim_{x\to\beta} \frac{1-\cos(x^2+bx+c)}{(\beta-x)^2}.$$

29. Let

$$P_n = \frac{(1+x)(1-x^2)(1+x^3)(1-x^4)\ldots(1-x^{4n})}{\{(1+x)(1-x^2)(1+x^3)(1-x^4)\ldots(1-x^{2n})\}^2}.$$

Prove that the term independent of x in the expansion of $\left(x^2+\frac{1}{x^2}+2\right)^{2n}$ is equal to $\lim_{x\to -1} P_n$.

30. Let $P_n = \cos\frac{x}{2}\cdot\cos\frac{x}{2^2}\cdot\cos\frac{x}{2^3}\ldots\cos\frac{x}{2^n}$.

Show that $P_n = \frac{1}{2^n}\sin x \cdot \operatorname{cosec}\frac{x}{2^n}$.

Hence, show that $\lim_{n\to\infty} \sum_{r=1}^{n} \frac{1}{2^r}\tan\left(\frac{x}{2^r}\right) = \frac{1}{x} - \cot x$.

31. If $\lim_{x\to 0} \frac{\sin 2x + a\sin x}{x^3} = b$ where $a, b \in R$ then find a and b.

32. If $\lim_{n\to\infty} \left(\frac{n^2+1}{n+1} - an\right)$ be finite, find $a \in R$ and the limit.

33. If $\lim_{x\to -\infty} (\sqrt{x^2-x+1} + ax - b) = 0$ then find the constants a and b.

Objective Questions

Fill in the blanks.

34. $\lim_{x\to 1} \frac{x^n+x^{n-1}+x^{n-2}+\ldots+x-n}{x-1} = $ ______ .

35. $\lim_{n\to\infty} \frac{\sum_{r=1}^{100} (n+r)^{10}}{n^{10}+10^{10}} = $ ______ .

36. $\lim_{n\to\infty} \frac{\sum_{r=0}^{n} 2^{-r}}{\sum_{r=0}^{n} 3^{-r}} = $ ______ .

37. $\lim_{x\to 1} (1-x)\tan\frac{\pi x}{2} = $ ______ .

38. $\lim_{h\to 0} \frac{(a+h)^2\cdot\sin(a+h) - a^2\sin a}{h} = $ ______ .

39. $\lim_{x\to 0} \frac{2^x-1}{\sqrt{1+x}-1} = $ ______ .

40. $\lim_{n\to\infty} \frac{n!}{(n+1)!-n!} = $ ______ .

41. $\lim_{\theta\to\pi/2} \frac{\sec\theta-\tan\theta}{\pi-2\theta} = $ ______ .

42. $\lim_{n\to\infty} \frac{4^n+3^n}{4^n-3^n} = $ ______ .

43. $\lim_{n\to\infty} \frac{3^{1/n}-1}{2^{1/n}-1} = $ ______ .

44. $\lim_{x\to\infty} (x-\sqrt{x^2+x}) = $ ______ .

45. $\lim_{x\to\infty} \frac{\sqrt{x}}{\sqrt{x+\sqrt{x+\sqrt{x}}}} = $ ______ .

46. $\lim_{x\to 0} \left(\frac{1}{\sin x} - \frac{1}{x}\right) = $ ______ .

47. $\lim_{h\to 0} \frac{\log_e(1+2h) - 2\log_e(1+h)}{h^2} = $ ______ .

48. $\lim_{n\to\infty} \frac{n^k \cos(n!)}{n+1}$, $0<k<1$ is equal to ______ .

49. $\lim_{x \to 0} (1+\sin x)^{\operatorname{cosec} x} =$ ______ .

50. $\lim_{x \to \infty} \left(\frac{x+6}{x+1}\right)^{x+4} =$ ______ .

51. $\lim_{x \to \infty} \left(\frac{3x-4}{3x+2}\right)^{\frac{x+1}{3}} =$ ______ .

52. $\lim_{x \to 0} \left(\frac{1+5x^2}{1+3x^2}\right)^{\frac{1}{x^2}} =$ ______ .

53. $\lim_{x \to \infty} \left(\frac{x^2+1}{x^2-1}\right)^{x^2} =$ ______ .

54. If $\lim_{x \to 0} \frac{x(1+m\cos x)-n\sin x}{x^3} = 1$ then

$m =$ ______ , $n =$ ______ .

55. If $f(2)=2, f'(2)=1$ then $\lim_{x \to 2} \frac{2x^2-4f(x)}{x-2} =$ ______ .

56. If $\lim_{x \to 0} \frac{a+bx\sin x + c\cos x}{x^4} = 2$ then

$a =$ ______ , $b =$ ______ , $c =$ ______ .

57. If $f(9)=9, f'(9)=4$ then $\lim_{x \to 9} \frac{\sqrt{f(x)}-3}{\sqrt{x}-3} =$ ______ .

58. If $G(x) = -\sqrt{25-x^2}$ then $\lim_{x \to 1} \frac{G(x)-G(1)}{x-1} =$ ______ .

59. If $f(a)=2, f'(a)=1, g(a)=-1, g'(a)=2$ then

$\lim_{x \to a} \frac{g(x)f(a)-g(a)f(x)}{x-a} =$ ______ .

Choose the correct option(s).

60. If $\lim_{x \to \infty} \frac{px+q}{qx+p} = l$ and $\lim_{x \to 0} \frac{px+q}{qx+p} = m$ where $p, q \neq 0$ then lm is

(a) 1 (b) $\frac{p^2}{q^2}$

(c) $\frac{q^2}{p^2}$ (d) none of these

61. If $f(x) = \sqrt{\frac{x-\sin x}{x+\cos^2 x}}$ then $\lim_{x \to \infty} f(x)$ is

(a) 0 (b) ∞

(c) 1 (d) none of these

62. $\lim_{x \to 0} \frac{xe^x - \log(1+x)}{x^2}$ is equal to

(a) $\frac{3}{2}$ (b) $\frac{2}{3}$

(c) 0 (d) 1

(e) none of these

63. If $q > p > 0$ then $\lim_{x \to \infty} \frac{ax^p + bx^{p-2}+c}{dx^q + ex^{q-2}+b}$ is

(a) ∞ (b) 0

(c) $\frac{a}{d}$ (d) none of these

64. If $\lim_{x \to 0} \frac{x(1+a\cos x)+b\sin x}{x} = 1$ then

(a) $a = b$ (b) $a+b=0$

(c) $2a = b$ (d) none of these

65. If $f(x)$ is twice differentiable and $f''(0) = p$ then $\lim_{x \to 0} \frac{2f(x)-3f(2x)+f(4x)}{x^2}$ is

(a) $2p$ (b) $3p$ (c) p (d) $-3p$

66. $\lim_{x \to \pi/2} \frac{2x-\pi}{\cos x}$ is equal to

(a) 5 (b) -2

(c) $\frac{1}{2}$ (d) none of these

67. $\lim_{\theta \to \pi/4} \frac{\sqrt{2}-\cos\theta-\sin\theta}{(4\theta-\pi)^2}$ is equal to

(a) $\frac{\sqrt{2}}{4}$ (b) $\frac{\sqrt{2}}{8}$ (c) $\frac{\sqrt{2}}{16}$ (d) $\frac{\sqrt{2}}{32}$

68. If $f(x) = \cot^{-1}\frac{3x-x^3}{1-3x^2}$ and $g(x) = \cos^{-1}\frac{1-x^2}{1+x^2}$ then $\lim_{x \to a} \frac{f(x)-f(a)}{g(x)-g(a)}, \left(0 < a < \frac{1}{2}\right)$ is

(a) $\frac{3}{2(1+x^2)}$ (b) $\frac{3}{2}$

(c) $\frac{3}{2(1+a^2)}$ (d) $\frac{-3}{2}$

(e) none of these

69. $\lim_{h \to 0} \left(\frac{1}{h \cdot \sqrt[3]{8+h}} - \frac{1}{2h}\right)$ is equal to

(a) $\frac{1}{12}$ (b) $-\frac{4}{3}$ (c) $\frac{-16}{3}$ (d) $-\frac{1}{48}$

70. $\lim_{h \to 0} \frac{2\left\{\sqrt{3}\sin\left(\frac{\pi}{6}+h\right) - \cos\left(\frac{\pi}{6}+h\right)\right\}}{\sqrt{3}h(\sqrt{3}\cos h - \sin h)}$ is equal to

(a) $\frac{2}{3}$ (b) $\frac{4}{3}$ (c) $-2\sqrt{3}$ (d) $-\frac{4}{3}$

71. $\lim_{x \to 0} \frac{\sqrt{\frac{1}{2}(1-\cos 2x)}}{x}$ is equal to

(a) 1 (b) −1 (c) 0 (d) none of these

State whether the statements are true or false.

72. $\lim_{x \to 0} \frac{\log(1+x)}{3^x - 1} = \frac{1}{\log_e 3}$

73. $\lim_{x \to a} \frac{\cos x - \cos a}{\cot x - \cot a} = \sin^3 a$

74. $\lim_{x \to \infty} \frac{\sin x}{x} = 1$

75. $\lim_{\substack{x \to 0 \\ y \to 0}} \frac{x^2}{y} = 0$

when $(x, y) \to (0, 0)$ along the curve $y^2 = x^3$.

Answers

1. $\frac{2}{3}$ 2. $\frac{1}{3}$ 3. $\frac{2}{3}$ 4. $\frac{2}{3}$

5. $-\frac{1}{12}$ 6. $\frac{1}{4}$ 7. $\frac{1}{4\sqrt{2}}$ 8. $\frac{2}{\sqrt{3}}$

9. 32 10. ∞ 11. $-\frac{1}{2}$ 12. 8

13. $\frac{3}{\sqrt{2}}$ 14. $8\sqrt{2}(\log 3)^2$ 15. e^2

16. e^2 17. 1 18. 1 19. e

20. e^5 21. $e^{2/\pi}$ 22. $e^{-1/30}$ 23. 1

24. $\sqrt{\frac{2}{3}}$ 26. $9\sin\alpha \cdot \cos^2\alpha$ 27. $-\frac{2}{3}$

28. $\frac{(\beta-\alpha)^2}{2}$ 31. $a = -2, b = -1$

32. $a = 1$, limit $= -1$ 33. $a = 1, b = \frac{1}{2}$

34. $\frac{n(n+1)}{2}$ 35. 100

36. $\frac{4}{3}$ 37. $\frac{2}{\pi}$

38. $2a\sin a + a^2\cos a$ 39. $2\log 2$

40. 0 41. $\frac{1}{4}$ 42. 1 43. $\log_2 3$

44. $-\frac{1}{2}$ 45. $\frac{1}{2}$ 46. 0 47. −1

48. 0 49. e 50. e^5 51. $e^{-2/3}$

52. e^2 53. e^2

54. $-\frac{5}{2}, \frac{-3}{2}$ respectively 55. 4

56. 48, −24, −48 respectively

57. 4 58. $\frac{1}{\sqrt{24}}$ 59. 5 60. (a)

61. (c) 62. (a) 63. (b) 64. (b)

65. (b) 66. (b) 67. (d) 68. (d)

69. (d) 70. (b) 71. (d) 72. true

73. true 74. false 75. true

□

Chapter Test

Time: 60 minutes

1. In the following, fill in the blanks so that the resulting sentences become true.

(a) $\lim\limits_{x \to \infty} (\sin\sqrt{x+1} - \sin\sqrt{x}) = \_\_\_\_\_\_$.

(b) $\lim\limits_{x \to 0} x^x = \_\_\_\_\_\_$.

2. In each of the following, multiple options are given. Select the correct option(s).

(a) If $\lim\limits_{x \to a} \dfrac{a^x - x^a}{x^x - a^a} = -1$ then a is

A. e B. 0 C. 1 D. none of these

(b) $\lim\limits_{n \to \infty} \log_{(n-1)} n \cdot \log_n (n+1) \cdot \log_{(n+1)}(n+2) \ldots \log_{(n^5 - 1)} n^5$ equals

A. $\dfrac{1}{5}$ B. 5 C. 1 D. none of these

3. If $|x| < 1$, prove that $\lim\limits_{n \to \infty} (1+x)(1+x^2)(1+x^4) \ldots (1+x^{2^n}) = \dfrac{1}{1-x}$.

4. Find $\lim\limits_{x \to 0.4} \dfrac{f(x) - f(0.4)}{g(x) - g(0.4)}$

where $f(x) = \tan^{-1} \dfrac{3x - x^3}{1 - 3x^2}$ and $g(x) = \sin^{-1} \dfrac{1 - x^2}{1 + x^2}$.

5. Evaluate $\lim\limits_{n \to \infty} ({}^nC_r\, p^r q^{n-r})$ if $np = m$, $p + q = 1$ and m, r are constants.

6. Evaluate $\lim\limits_{x \to \infty} \log_{\frac{x^2+1}{x+1}} (\sqrt{x^2+1} - x)$.

7. Evaluate $\lim\limits_{x \to 1} (1 + \sin \pi x)^{\cot \pi x}$.

8. Find the constants a and b if $\lim\limits_{x \to \infty} \left(\dfrac{x^2+1}{x+1} - ax - b \right) = 0$.

9. Find $\lim\limits_{\substack{x \to 0 \\ y \to 0}} \dfrac{x^2 + \sin y}{y^2 + \sin x^2}$ when $(x, y) \to (0, 0)$ along the curve $y = x^2$.

□

Answers

1. (a) 0 (b) 1
2. (a) C (b) B
4. $\frac{-3}{2}$
5. $\frac{e^{-m} \cdot m^r}{r!}$
6. -1
7. e^{-1}
8. $a = 1, b = -1$
9. 2

□

4. Continuity, Differentiability and Graph of Function

Recap of Facts and Formulae

1. Existence of limit

- $\lim_{h \to 0} f(a+h)$ is the *right-hand limit* of $f(x)$ at $x = a$ and it is denoted by $f(a+0)$ or $f(a+)$ or $\lim_{x \to a+0} f(x)$.

  $\lim_{h \to 0} f(a-h)$ is the *left-hand limit* of $f(x)$ at $x = a$ and it is denoted by $f(a-0)$ or $f(a-)$ or $\lim_{x \to a-0} f(x)$.

- $\lim_{x \to a} f(x)$ exists if $f(a+0) = f(a-0)$ and the value of $\lim_{x \to a} f(x)$ is equal to the common value of $f(a+0)$ and $f(a-0)$.

2. Continuity of a function at a point

- A function $f(x)$ is continuous at $x = a$ iff

  $$f(a+0) = f(a-0) = f(a),$$

  i.e., $\lim_{h \to 0} f(a+h) = \lim_{h \to 0} f(a-h) = f(a)$.

  If any two of the above three are unequal then $f(x)$ is *discontinuous* at $x = a$.

Note

(i) If $\lim_{x \to a} f(x)$ does not exist then $f(x)$ cannot be continuous at $x = a$.

(ii) If $x = a$ is the left-end point of the domain of definition then x cannot tend to a from the left and so the question of getting $f(a-0)$ does not arise. In this case $f(x)$ will be continuous at $x = a$ if $f(a+0) = f(a)$. Similarly, if $x = a$ is the right-end point of the domain of definition then $f(x)$ will be continuous at $x = a$ provided $f(a-0) = f(a)$.

3. Differentiability of a function at a point

- *Right-hand derivative* of $f(x)$ at $x = a$, denoted by $f'(a+0)$ or $f'(a+)$, is the $\lim_{h \to 0} \frac{f(a+h) - f(a)}{h}$.

- *Left-hand derivative* of $f(x)$ at $x = a$, denoted by $f'(a-0)$ or $f'(a-)$, is the $\lim_{h \to 0} \frac{f(a-h) - f(a)}{-h}$.

- A function $f(x)$ is said to be differentiable (finitely) at $x = a$ if

  $$f'(a+0) = f'(a-0) = \text{finite},$$

  i.e., $\lim_{h \to 0} \frac{f(a+h) - f(a)}{h} = \lim_{h \to 0} \frac{f(a-h) - f(a)}{-h} = \text{finite}$

  and the common limit is called the derivative of $f(x)$ at $x = a$, denoted by $f'(a)$.

  Clearly, $f'(a) = \lim_{x \to a} \frac{f(x) - f(a)}{x - a}$,

  {$x \to a$ from the left as well as from the right}.

4. Relation between continuity and differentiability

- $f(x)$ is differentiable (finitely) at $x = a$

  $\Rightarrow$ $f(x)$ is continuous at $x = a$

- $f(x)$ is not continuous at $x = a$

  $\Rightarrow$ $f(x)$ is not differentiable (finitely) at $x = a$.

TIP

While examining the continuity and differentiability of a function $f(x)$ at a point $x = a$, if you start with the differentiability and find that $f(x)$ is differentiable then you can conclude that the function is also continuous. But if you find $f(x)$ is not differentiable at $x = a$, you will also have to check the continuity separately.

Instead, if you start with continuity and find that the function is not continuous then you can conclude that the function is also nondifferentiable. But if you find $f(x)$ is continuous, you will also have to check the differentiability separately.

5. Continuity and differentiability of some standard functions

- Polynomial functions (i.e., $a_0x^n + a_1x^{n-1} + \ldots + a_n$), $\sin x$, $\cos x$ and e^x are continuous and differentiable at all points of the set R of real numbers.

- $\log_e x$ is continuous and differentiable at all points of $(0, +\infty)$.
- $\tan x$ and $\sec x$ are continuous and differentiable at all points of $R - \left\{\frac{\pi}{2}, \frac{3\pi}{2}, \frac{5\pi}{2}, \ldots\right\}$.
- $\cot x$ and $\text{cosec}\, x$ are continuous and differentiable at all points of $R - \{0, \pi, 2\pi, 3\pi, \ldots\}$.
- $|x - a|$ is continuous everywhere and differentiable at all points except at $x = a$.
- $[x]$ is continuous and differentiable everywhere except at $x =$ an integer.

6. Algebraic property of continuity and differentiability

- If $f(x)$ and $\phi(x)$ are both continuous (or differentiable) at $x = a$ then $f(x) \pm \phi(x)$ and $f(x) \times \phi(x)$ are continuous (or differentiable) at $x = a$.
- If $f(x)$ is continuous (or differentiable) and $\phi(x)$ is discontinuous (or nondifferentiable) then $f(x) \pm \phi(x)$ and $f(x) \times \phi(x)$ are discontinuous (or nondifferentiable) at $x = a$.

7. Continuity and differentiability in an interval

- $f(x)$ is continuous in an interval if it is continuous at each point of the interval.
- $f(x)$ is differentiable in an interval if it is differentiable at each point of the interval.

Method of examining continuity and differentiability in an interval:

(i) Detect all the pieces of the intervals of definition of the function.

(ii) Apply continuity and differentiability of standard functions in each of the intervals.

(iii) Examine the continuity and differentiability at each turning point of definition of the function.

8. Some properties of a function continuous over a closed interval

If $f(x)$ is continuous in the closed interval $[a, b]$ then

- the value of $f(x)$ at all points in $[a, b]$ will lie between two fixed real numbers.
- in the interval $[a, b]$, $f(x)$ has a greatest value and a least value
- $f(x)$ will obtain all the values lying between $\min f(x)$ and $\max f(x)$ for points in $[a, b]$
- the range of $f(x) = [\min f(x), \max f(x)]$ when the domain of $f(x) = [a, b]$.

9. Rough sketch of a function in an interval

- If a function $f(x)$ is continuous at $x = a$, the graph of $f(x)$ at the corresponding point $\{a, f(a)\}$ will not be broken. But if $f(x)$ is discontinuous at $x = a$, the graph will be broken at the corresponding point.

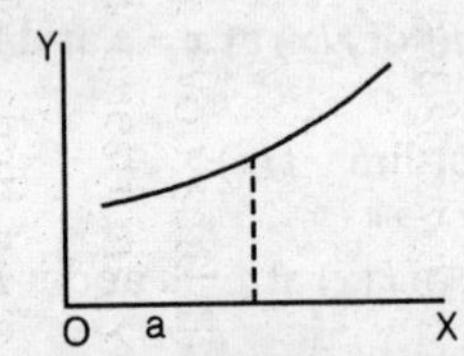

(When continuous at x = a)

Fig. 1

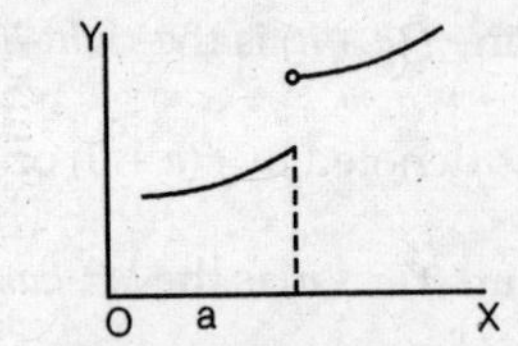

(When discontinuous at x = a)

Fig. 2

- If a function $f(x)$ is differentiable at $x = a$, the graph of $f(x)$ will be such that there is only one tangent to the graph at the corresponding point. But if $f(x)$ is nondifferentiable at $x = a$, there will not be unique tangent at the corresponding point of the graph.

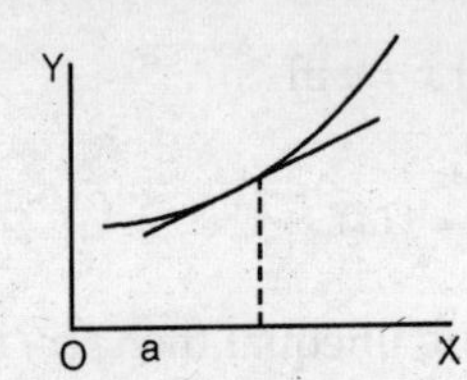

(When differentiable at x = a)

Fig. 3

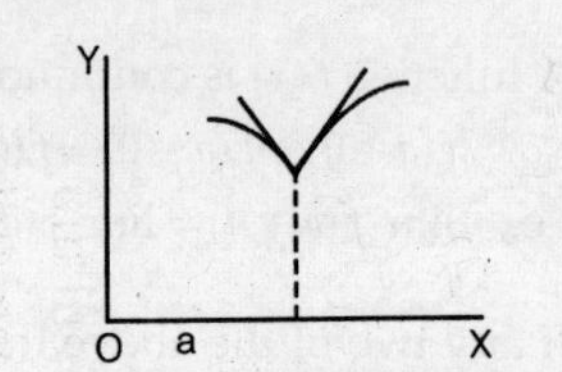

(When nondifferentiable at x = a)

Fig. 4

Method of drawing a rough sketch of a function in an interval:

(i) Take the different pieces of intervals of definition of the function and draw graph in each of the intervals.

(ii) Examine the continuity and differentiability at each of the turning points of definition.

Let $x = a$ be a turning point. Let $f(x)$ be defined in $[b, a]$ and $(a, c]$. If $f(x)$ is continuous at a then the graphs for $f(x)$ in $[b, a]$ and $(a, c]$ will be joined at $x = a$ as in Figure 1. If $f(x)$ is discontinuous at $x = a$, draw the graph of $f(x)$ in $[b, a]$ and the graph of $f(x)$ in $(a, c]$ with o at the beginning of the graph in $(a, c]$ as in Figure 2. If $f(x)$ is continuous at $x = a$ but not differentiable at $x = a$ then the graph of $f(x)$ at $x = a$ will be as shown in Figure 4.

Selected Solved Examples

1. If $f(x) = x^4, x^2 < 1$
$x, x^2 \geq 1$
discuss the existence of limit at $x = 1, -1$.

We have $x^2 < 1 \Rightarrow (x+1)(x-1) < 0$.

From the sign-scheme, $-1 < x < 1$.

Also $x^2 \geq 1 \Rightarrow (x+1)(x-1) \geq 0$.

From the sign-scheme, $x \leq -1$ or $x \geq 1$.

Thus, the function is $f(x) = x, x \leq -1$

$x^4, -1 < x < 1$

$x, x \geq 1$

Now, $\lim_{x \to 1+0} f(x) = \lim_{h \to 0} f(1+h) = \lim_{h \to 0} \{1+h\} = 1$

$\lim_{x \to 1-0} f(x) = \lim_{h \to 0} f(1-h) = \lim_{h \to 0} (1-h)^4 = 1$

$\therefore \lim_{x \to 1} f(x)$ exists and it is equal to 1.

Next, $\lim_{x \to -1+0} f(x) = \lim_{h \to 0} f(-1+h) = \lim_{h \to 0} (-1+h)^4 = 1$

$\lim_{x \to -1-0} f(x) = \lim_{h \to 0} f(-1-h) = \lim_{h \to 0} \{-1-h\} = -1$

$\therefore \lim_{x \to -1} f(x)$ does not exist.

2. The function $f(x)$ is defined as follows:

$f(x) = x - 1, x < 0$

$\frac{1}{4}, x = 0$

$x^2, x > 0.$

Find $\lim_{x \to 0+0} f(x)$ and $\lim_{x \to 0-0} f(x)$.

Examine the continuity of $f(x)$ at $x = 0$. Also, construct a rough graph of the function.

$\lim_{x \to 0+0} f(x) = \lim_{h \to 0} f(0+h) = \lim_{h \to 0} (0+h)^2 = 0$

$\lim_{x \to 0-0} f(x) = \lim_{h \to 0} f(0-h) = \lim_{h \to 0} \{(0-h) - 1\} = -1.$

As $\lim_{x \to 0+0} f(x) \neq \lim_{x \to 0-0} f(x)$, $f(x)$ is not continuous at $x = 0$.

We know that a polynomial function is continuous everywhere.

So $f(x) = x - 1, x < 0$ is continuous

and $f(x) = x^2, x > 0$ is continuous.

So, the graph of $f(x)$ will be as follows:

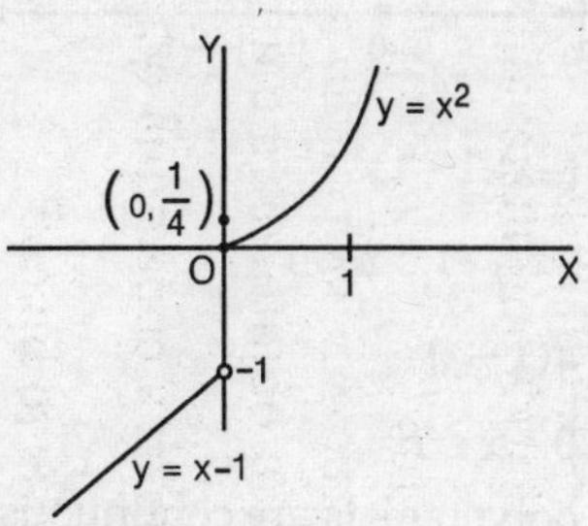

3. Let $f(x) = \frac{1 - \cos 4x}{x^2}, x < 0$

$a, x = 0$

$\frac{\sqrt{x}}{\sqrt{16 + \sqrt{x}} - 4}, x > 0.$

If possible, find the value of a so that the function may be continuous at $x = 0$.

$f(x)$ will be continuous at $x = 0$ if

$$\lim_{h \to 0} f(0+h) = \lim_{h \to 0} f(0-h) = f(0) \quad \ldots (1)$$

Now, $\lim_{h \to 0} f(0+h) = \lim_{h \to 0} \frac{\sqrt{0+h}}{\sqrt{16 + \sqrt{0+h}} - 4}$

$= \lim_{h \to 0} \frac{\sqrt{h}}{\sqrt{16 + \sqrt{h}} - 4}$

$= \lim_{h \to 0} \frac{\sqrt{h}\{\sqrt{16 + \sqrt{h}} + 4\}}{16 + \sqrt{h} - 16}$

$= \lim_{h \to 0} \{\sqrt{16 + \sqrt{h}} + 4\} = 8$

$\lim_{h \to 0} f(0-h) = \lim_{h \to 0} \frac{1 - \cos 4(0-h)}{(0-h)^2}$

$= \lim_{h \to 0} \frac{1 - \cos 4h}{h^2}$

$= \lim_{h \to 0} \frac{2\sin^2 2h}{h^2}$

$= \lim_{h \to 0} 2 \cdot \left(\frac{\sin 2h}{2h}\right)^2 \times 4 = 8$

$f(0) = a$

$\therefore$ by (1), $8 = 8 = a$; $\therefore a = 8$.

4. Let $f(x) = x - |x - x^2|, -1 \leq x \leq 1$.
Discuss the continuity of $f(x)$ in the closed interval $[-1, 1]$. Draw the graph of $f(x)$ in the interval.

Here, $f(x) = x - |x(1-x)|$

$= x - |x|\ |x-1|, -1 \le x \le 1.$

Now, using definition of modulus function we have:

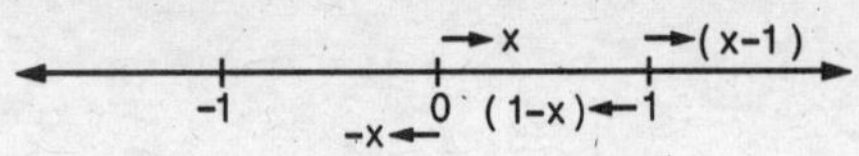

$\therefore \quad f(x) = x - (-x)(1-x)$

$= 2x - x^2, -1 \le x < 0$

$f(x) = x - x(1-x)$

$= x^2, 0 \le x \le 1.$

We know that polynomials are continuous everywhere. So, only doubtful point is the turning point $x = 0$ of definition.

$$\lim_{h \to 0} f(0+h) = \lim_{h \to 0} (0+h)^2 = 0$$

$$\lim_{h \to 0} f(0-h) = \lim_{h \to 0} 2(0-h) - (0-h)^2 = 0$$

$f(0) = 0^2 = 0;$

so, $f(x)$ is continuous at $x = 0$.

Hence, $f(x)$ is continuous in $[-1, 1]$.

Hence, the graph of $f(x)$ is continuous in $[-1, 1]$ and

$f(x) = 2x - x^2$ in $-1 \le x < 0$

x^2 in $0 \le x \le 1$.

$\therefore$ the graph of $f(x)$ is as follows:

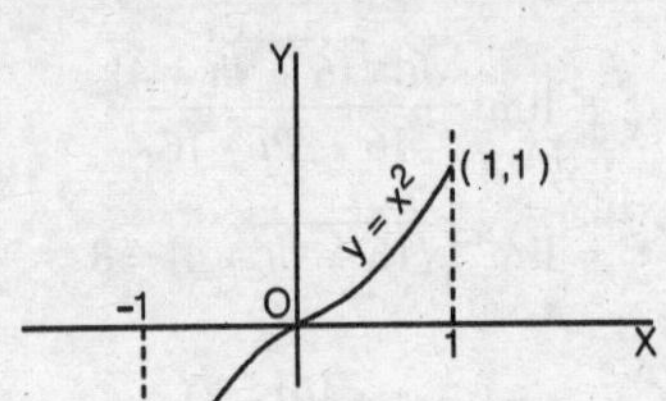

5. If $f(x) = \lim_{n \to \infty} \dfrac{\log(x+2) - x^{2n} \sin x}{x^{2n} + 1}$,

examine the continuity of $f(x)$ at $x = 1$.

TIP: In order to examine the continuity at $x = 1$, we are required to derive the definition of $f(x)$ in the intervals $x < 1, x > 1$ and at $x = 1$, i.e., on and around $x = 1$.

Now, if $0 < x < 1$,

$$f(x) = \lim_{n \to \infty} \frac{\log(x+2) - x^{2n} \sin x}{x^{2n} + 1}$$

$$= \frac{\log(x+2) - 0 \cdot \sin x}{0 + 1} = \log(x+2)$$

if $\quad x = 1, f(x) = \lim_{n \to \infty} \dfrac{\log(x+2) - 1 \cdot \sin x}{1 + 1}$

$$= \frac{\log(x+2) - \sin x}{2}$$

if $\quad x > 1, f(x) = \lim_{n \to \infty} \dfrac{\log(x+2) - x^{2n} \sin x}{x^{2n} + 1}$

$$= \lim_{n \to \infty} \frac{\frac{1}{x^{2n}} \log(x+2) - \sin x}{1 + \frac{1}{x^{2n}}} = -\sin x.$$

Thus, we have $f(x) = \log(x+2), 0 < x < 1$

$\frac{1}{2}\{\log(x+2) - \sin x\}, x = 1$

$-\sin x, x > 1.$

$\therefore \quad f(1+0) = \lim_{h \to 0} f(1+h)$

$= \lim_{h \to 0} \{-\sin(1+h)\} = -\sin 1$

$f(1-0) = \lim_{h \to 0} f(1-h)$

$= \lim_{h \to 0} \log(1-h+2) = \log 3$

$\therefore \quad f(1+0) \ne f(1-0).$

So $f(x)$ is not continuous at $x = 1$.

6. Examine the continuity of

$$f(x) = \lim_{n \to \infty} \frac{x}{(2\sin x)^{2n} + 1} \text{ for } x \in R.$$

If $|2\sin x| < 1, f(x) = \lim_{n \to \infty} \dfrac{x}{(2\sin x)^{2n} + 1} = \dfrac{x}{0+1} = x.$

If $|2\sin x| = 1, f(x) = \lim_{n \to \infty} \dfrac{x}{(2\sin x)^{2n} + 1} = \dfrac{x}{1+1} = \dfrac{1}{2}x.$

If $|2\sin x| > 1, f(x) = \lim_{n \to \infty} \dfrac{x}{(2\sin x)^{2n} + 1} = \dfrac{x}{\infty + 1} = 0.$

But $\quad |2\sin x| < 1 \quad \Rightarrow \quad |\sin x| < \dfrac{1}{2}$

$\Rightarrow \quad -\dfrac{1}{2} < \sin x < \dfrac{1}{2}$

$\Rightarrow \quad n\pi - \dfrac{\pi}{6} < x < n\pi + \dfrac{\pi}{6}$

$|2\sin x| = 1 \quad \Rightarrow \quad |\sin x| = \dfrac{1}{2}$

$\Rightarrow \quad \sin x = \pm \dfrac{1}{2}$

$$\Rightarrow \quad x = n\pi + (-1)^n \cdot \left(\pm\frac{\pi}{6}\right) = n\pi \pm \frac{\pi}{6}$$

$$|2\sin x| > 1 \quad \Rightarrow \quad |\sin x| > \frac{1}{2}$$

$$\Rightarrow \quad \sin x > \frac{1}{2} \quad \text{or} \quad \sin x < -\frac{1}{2}$$

$$\Rightarrow \quad n\pi + \frac{\pi}{6} < x < n\pi + \frac{5\pi}{6}.$$

Thus we have,

$$f(x) = x, \; n\pi - \frac{\pi}{6} < x < n\pi + \frac{\pi}{6}$$

$$\frac{1}{2}x, \; x = n\pi \pm \frac{\pi}{6}$$

$$0, \; n\pi + \frac{\pi}{6} < x < n\pi + \frac{5\pi}{6}.$$

As polynomial functions are continuous everywhere, only doubtful points are $x = n\pi \pm \frac{\pi}{6}$. Clearly,

$$f\left(n\pi - \frac{\pi}{6} + 0\right) \neq f\left(n\pi - \frac{\pi}{6} - 0\right) \text{ because } n\pi - \frac{\pi}{6} \neq 0$$

$$f\left(n\pi + \frac{\pi}{6} + 0\right) \neq f\left(n\pi + \frac{\pi}{6} - 0\right) \text{ because } 0 \neq n\pi + \frac{\pi}{6}$$

$\therefore$ $f(x)$ is not continuous at $x = n\pi \pm \frac{\pi}{6}$.

$\therefore$ the function $f(x)$ is continuous everywhere in R except the set of points $\left\{x \mid x = n\pi \pm \frac{\pi}{6}, n \in Z\right\}$.

7. If $f(x+y) = f(x) \cdot f(y)$ for all $x, y \in R$ and $f(x) = 1 + g(x) \cdot G(x)$ where $\lim_{x \to 0} g(x) = 0$ and $\lim_{x \to 0} G(x)$ exists, prove that $f(x)$ is continuous at all $x \in R$.

$$\lim_{x \to 0} g(x) = 0 \quad \Rightarrow \quad \lim_{h \to 0} g(0+h) = \lim_{h \to 0} g(0-h) = 0$$

$$\Rightarrow \quad \lim_{h \to 0} g(h) = \lim_{h \to 0} g(-h) = 0 \qquad \ldots (1)$$

$$\lim_{x \to 0} G(x) \text{ exists} \quad \Rightarrow \quad \lim_{h \to 0} G(0+h) = \lim_{h \to 0} G(0-h)$$

$$\Rightarrow \lim_{h \to 0} G(h) = \lim_{h \to 0} G(-h) = \text{finite} \qquad \ldots (2)$$

Now, $\lim_{h \to 0} f(x+h) = \lim_{h \to 0} f(x) \cdot f(h) = f(x) \lim_{h \to 0} f(h)$

$$\{\because f(x+y) = f(x) \cdot f(y)\}$$

$$= f(x) \cdot \lim_{h \to 0} \{1 + g(h)\, G(h)\},$$

using given relation

$$= f(x) \cdot \left\{1 + \lim_{h \to 0} g(h) \cdot \lim_{h \to 0} G(h)\right\}$$

$= f(x) \cdot \{1 + 0 \cdot \text{finite}\}$, using (1) and (2)

$$= f(x)$$

Also, $\lim_{h \to 0} f(x-h) = \lim_{h \to 0} f(x) \cdot f(-h) = f(x) \cdot \lim_{h \to 0} f(-h)$

$= f(x) \cdot \lim_{h \to 0} \{1 + g(-h) \cdot G(-h)\}$, using given relation

$$= f(x) \cdot \left\{1 + \lim_{h \to 0} g(-h) \cdot \lim_{h \to 0} G(-h)\right\}$$

$= f(x) \cdot \{1 + 0 \cdot \text{finite}\}$, using (1) and (2)

$$= f(x)$$

$$\therefore \quad \lim_{h \to 0} f(x+h) = \lim_{h \to 0} f(x-h) = f(x).$$

$\therefore$ $f(x)$ is continuous everywhere.

8. Is $f(x)$ differentiable at $x = 0$ if $f(x)$ is defined as follows:

$$f(x) = \frac{x}{1 + e^{1/x}}, \; x \neq 0$$
$$0, \; x = 0\,?$$

$$f'(0+0) = \lim_{h \to 0} \frac{f(0+h) - f(0)}{h} = \lim_{h \to 0} \frac{\dfrac{0+h}{1 + e^{\frac{1}{0+h}}} - 0}{h}$$

$$= \lim_{h \to 0} \frac{1}{1 + e^{1/h}} = \frac{1}{1 + e^{\infty}} = 0$$

$$f'(0-0) = \lim_{h \to 0} \frac{f(0-h) - f(0)}{-h} = \lim_{h \to 0} \frac{\dfrac{0-h}{1 + e^{\frac{1}{0-h}}} - 0}{-h}$$

$$= \lim_{h \to 0} \frac{1}{1 + e^{-1/h}} = \frac{1}{1 + e^{-\infty}} = \frac{1}{1+0} = 1$$

$$\therefore \quad f'(0+0) \neq f'(0-0).$$

Hence, $f(x)$ is not differentiable at $x = 0$.

9. Find $\frac{dy}{dx}$ at $x = \frac{\pi}{4}$ if possible where $y = |\cos x - \sin x|$.

Here $x = \frac{\pi}{4}$ is a turning point of definition of the function. We have

$$y = f(x) = \cos x - \sin x, \; 0 \leq x \leq \frac{\pi}{4}$$

$$\sin x - \cos x, \; \frac{\pi}{4} < x \leq \frac{\pi}{2}$$

$$\therefore \quad f'\left(\frac{\pi}{4}-0\right) = \lim_{h \to 0} \frac{f\left(\frac{\pi}{4}-h\right)-f\left(\frac{\pi}{4}\right)}{-h}$$

$$= \lim_{h \to 0} \frac{\cos\left(\frac{\pi}{4}-h\right)-\sin\left(\frac{\pi}{4}-h\right)-0}{-h}$$

$$= \lim_{h \to 0} \frac{\frac{1}{\sqrt{2}}\cos h + \frac{1}{\sqrt{2}}\sin h - \left(\frac{1}{\sqrt{2}}\cos h - \frac{1}{\sqrt{2}}\sin h\right)}{-h}$$

$$= \lim_{h \to 0} \frac{\sqrt{2}\sin h}{-h} = -\sqrt{2}$$

$$f'\left(\frac{\pi}{4}+0\right) = \lim_{h \to 0} \frac{f\left(\frac{\pi}{4}+h\right)-f\left(\frac{\pi}{4}\right)}{h}$$

$$= \lim_{h \to 0} \frac{\sin\left(\frac{\pi}{4}+h\right)-\cos\left(\frac{\pi}{4}+h\right)-0}{h}$$

$$= \lim_{h \to 0} \frac{\frac{1}{\sqrt{2}}\cos h + \frac{1}{\sqrt{2}}\sin h - \left(\frac{1}{\sqrt{2}}\cos h - \frac{1}{\sqrt{2}}\sin h\right)}{h}$$

$$= \lim_{h \to 0} \frac{\sqrt{2}\sin h}{h} = \sqrt{2}$$

$\therefore \quad f'\left(\frac{\pi}{4}-0\right) \neq f'\left(\frac{\pi}{4}+0\right)$. Hence, f' [illegible] s not exist.

TIP If the derivative [illegible] $f(x)$ is to be obtained at a r[illegible] ıich is a turning point of de[illegible] find $f'(a)$ use the following:

$$f'(a) = f'(a-0) = f'(a+0).$$

10. Let R be the set of real numbers and $f: R \to R$ such that for all $x, y \in R$ we have $|f(x)-f(y)| \le |x-y|^3$. Prove that $f(x)$ is a constant.

Here, $|f(x)-f(y)| \le |x-y|^3$ ... (1)

We know that $f'(x) = \lim_{y \to x} \frac{f(y)-f(x)}{y-x}$

$$\therefore \quad |f'(x)| = \left|\lim_{y \to x} \frac{f(y)-f(x)}{y-x}\right| = \lim_{y \to x} \left|\frac{f(y)-f(x)}{y-x}\right|$$

$$\le \lim_{y \to x} |x-y|^2 = 0, \text{ using (1)}$$

$\therefore \quad |f'(x)| \le 0$. So $f'(x) = 0$.

Hence, $f(x)$ is a constant.

11. The function $f(x)$ is defined as follows:

$$f(x) = [x] + |1-x|, -1 \le x \le 3$$

where $[x]$ is the greatest integer function. Determine the points, if any, where $f(x)$ is not differentiable. Draw a graph of the function.

TIP Here $f(x)$ is the sum of two functions which are themselves piecewisely defined. In such cases, at first break the domain of definition in subintervals keeping an eye on the turning points of definition of the constituent functions.

Here, the domain is the interval $[-1, 3]$ which is equal to

$$[-1, 0) \cup [0, 1) \cup [1, 2) \cup [2, 3) \cup \{3\}.$$

If $-1 \le x < 0$,

$[x] = -1$ and $|1-x| = |x-1| = -(x-1)$.

If $0 \le x < 1$,

$[x] = 0$ and $|1-x| = |x-1| = -(x-1)$.

If $1 \le x < 2$,

$[x] = 1$ and $|1-x| = |x-1| = x-1$.

If $2 \le x < 3$,

$[x] = 2$ and $|1-x| = |x-1| = x-1$.

If $x = 3$, $[x] = 3$ and $|1-x| = 2$.

Thus, $f(x) = -1-(x-1) = -x, -1 \le x < 0$

$0-(x-1) = 1-x, 0 \le x < 1$

$1+(x-1) = x, 1 \le x < 2$

$2+x-1 = x+1, 2 \le x < 3$

$3+2 = 5, x = 3.$

We know that every polynomial function is continuous and differentiable everywhere. Hence $f(x)$ is differentiable except perhaps at the turning points of definition, namely, $x = 0, 1, 2, 3$.

Now, $f(0-0) = 0$ and $f(0+0) = 1$

$f(1-0) = 1-1 = 0$ and $f(1+0) = 1$

$f(2-0) = 2$ and $f(2+0) = 2+1 = 3$

$f(3-0) = 3+1 = 4$ and $f(3) = 5$

($\because x = 3$ is the right-end point of definition)

Hence $f(x)$ is not continuous at $x = 0, 1, 2, 3$.

So $f(x)$ is not differentiable at $x = 0, 1, 2, 3$.

In order to draw the graph of the function $y = f(x)$ in $[-1, 3]$ we have to draw the graphs of

$y = -x$ in $[-1, 0)$,

$y = 1-x$ in $[0, 1)$,

$y = x$ in $[1, 2)$,

$y = x+1$ in $[2, 3)$,

$y = 5$ at $x = 3$.

Hence, the required graph is as below:

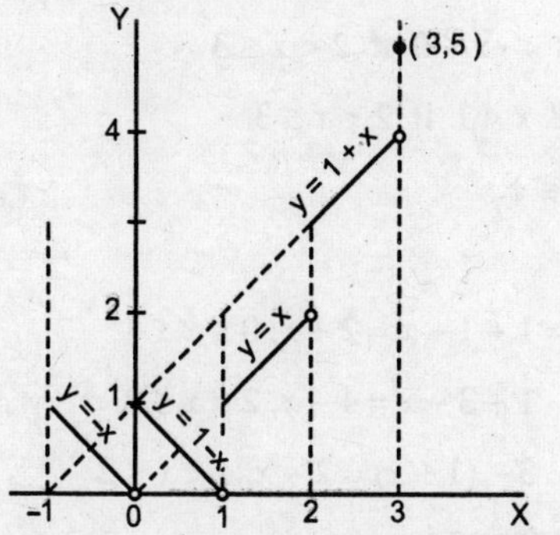

12. The function $y = f(x)$ is defined as follows:

$x = 2t - |t|, y = t^2 + t|t|, t \in R.$

Discuss the continuity and differentiability of the function at $x = 0$. Also, draw the graph of the function in the interval $[-1, 1]$.

When $t < 0$, $x = 2t - (-t) = 3t < 0$

$y = t^2 + t(-t) = 0$

$\therefore \quad y = 0$ when $x < 0$.

When $t \geq 0$, $x = 2t - t = t \geq 0$

$y = t^2 + t(t) = 2t^2 = 2x^2$

$\therefore \quad y = 2x^2$ when $x \geq 0$.

Thus the function is as follows:

$$f(x) = 0, x < 0$$
$$2x^2, x \geq 0$$

$$\text{Now, } f'(0+0) = \lim_{h \to 0} \frac{f(0+h) - f(0)}{h}$$

$$= \lim_{h \to 0} \frac{2h^2 - 0}{h} = \lim_{h \to 0} 2h = 0$$

$$f'(0-0) = \lim_{h \to 0} \frac{f(0-h) - f(0)}{-h}$$

$$= \lim_{h \to 0} \frac{0-0}{-h}$$

$$= \lim_{h \to 0} 0 = 0$$

$\therefore \quad f'(0+0) = f'(0-0)$

$\therefore \quad f(x)$ is differentiable (finitely) at $x = 0$.

$\therefore \quad f(x)$ is also continuous at $x = 0$.

Being a polynomial function, $f(x)$ is continuous at other points of the interval $[-1, 1]$.

Now, in the interval $[-1, 1]$ we have

$y = 0$ in $[-1, 0)$, $y = 2x^2$ in $[0, 1]$.

So, the graph is as below:

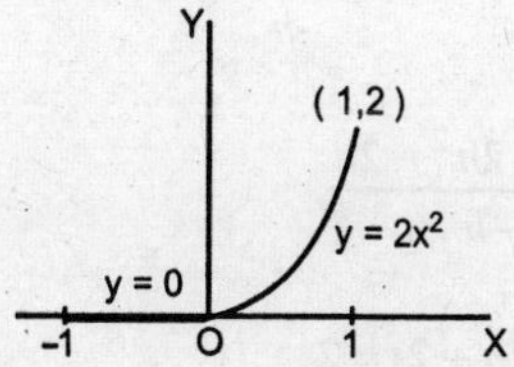

13. Let $f(x) = x^3 - x^2 + x + 1$ and

$g(x) = \max\{f(t), 0 \leq t \leq x\}, 0 \leq x \leq 1,$

$3 - x, 1 < x \leq 2.$

Discuss the continuity and differentiability of $g(x)$ in the interval $[0, 2]$.

Here, $f(x) = x^3 - x^2 + x + 1$

$\therefore \quad f'(x) = 3x^2 - 2x + 1$

$$\text{or} \quad f'(x) = 3\left\{x^2 - \frac{2}{3}x\right\} + 1$$

$$= 3\left\{x^2 - \frac{2}{3}x + \frac{1}{9}\right\} + 1 - \frac{1}{3}$$

$$= 3\left(x - \frac{1}{3}\right)^2 + \frac{2}{3} > 0 \text{ for all } x.$$

$\therefore \quad f(x)$ is an increasing function of x

(See *monotone function*).

$\therefore \quad$ in $0 \leq x \leq 1$, $\max\{f(t), 0 \leq t \leq x\} = f(x)$

$\therefore \quad g(x) = f(x) = x^3 - x^2 + x + 1, 0 \leq x \leq 1$

$3 - x, 1 < x \leq 2$

As polynomial functions are continuous and differentiable everywhere, $g(x)$ is also continuous and differentiable everywhere except perhaps at the turning point of definition $x = 1$.

$$\text{Now, } g(1+0) = \lim_{h \to 0} g(1+h) = \lim_{h \to 0} \{3 - (1+h)\} = 2$$

$$g(1-0) = \lim_{h \to 0} g(1-h)$$

$$= \lim_{h \to 0} \{(1-h)^3 - (1-h)^2 + (1+h) + 1\} = 2$$

$$g(1) = 1^3 - 1^2 + 1 + 1 = 2$$

$\therefore \quad g(1+0) = g(1-0) = g(1)$

$\therefore \quad g(x)$ is continuous at $x = 1$.

$$\text{Next, } g'(1+0) = \lim_{h \to 0} \frac{g(1+h) - g(1)}{h}$$

$$= \lim_{h \to 0} \frac{\{3 - (1+h)\} - 2}{h} = -1$$

$$g'(1-0) = \lim_{h \to 0} \frac{g(1-h) - g(1)}{-h}$$

$$= \lim_{h \to 0} \frac{(1-h)^3 - (1-h)^2 + (1-h) + 1 - 2}{-h}$$

$$= \lim_{h \to 0} \frac{-h^3 + 2h^2 - 2h}{-h}$$

$$= \lim_{h \to 0} (h^2 - 2h + 2) = 2$$

$\therefore \quad g'(1+0) \neq g'(1-0);$

so, $g(x)$ is not differentiable at $x = 1$.

$\therefore$ $g(x)$ is continuous in [0, 2], but differentiable in [0, 2] except at the point $x = 1$.

14. Let $f(x) = 1 + x, 0 \le x \le 2$
$3 - x, 2 < x \le 3$.

Determine $f\{f(x)\}$ and hence find the points of discontinuity and nondifferentiability. Also, draw the graph of $f\{f(x)\}$ in [0, 3].

Clearly, $f\{f(x)\} = 1 + f(x), 0 \le f(x) \le 2$

$3 - f(x), 2 < f(x) \le 3$

When $0 \le f(x) \le 2$,

$0 \le 1 + x \le 2$ if $0 \le x \le 2$

(using the first piece of definition)

$\Rightarrow \quad -1 \le x \le 1$ if $0 \le x \le 2$

$\Rightarrow \quad 0 \le x \le 1$

(taking the interval of common points)

$\therefore \quad 0 \le f(x) \le 2$ when $0 \le x \le 1$ and $f(x) = 1 + x$ ... (1)

When $0 \le f(x) \le 2$,

$0 \le 3 - x \le 2$ if $2 < x \le 3$

(using the second piece of definition)

$\Rightarrow \quad -3 \le -x \le -1$ if $2 < x \le 3$

$\Rightarrow \quad 1 \le x \le 3$ if $2 < x \le 3$

$\Rightarrow \quad 2 < x \le 3$

(taking the interval of common points)

$\therefore \quad 0 \le f(x) \le 2$ when $2 < x \le 3$ and $f(x) = 3 - x$ ... (2)

When $2 < f(x) \le 3$,

$2 < 1 + x \le 3$ if $0 \le x \le 2$

(using the first piece of definition)

$\Rightarrow \quad 1 < x \le 2$ if $0 \le x \le 2$

$\Rightarrow \quad 1 < x \le 2$

$\therefore \quad 2 < f(x) \le 3$ when $1 < x \le 2$ and $f(x) = 1 + x$ ... (3)

When $2 < f(x) \le 3$,

$2 < 3 - x \le 3$ if $2 < x \le 3$

(using the second piece of definition)

$\Rightarrow \quad -1 < -x \le 0$ if $2 < x \le 3$

$\Rightarrow \quad 0 \le x < 1$ if $2 < x \le 3$

$\Rightarrow \quad x \in \phi$.

Thus, we get

$f\{f(x)\} = 1 + 1 + x = 2 + x, 0 \le x \le 1$ {from (1)}

$1 + 3 - x = 4 - x, 2 < x \le 3$ {from (2)}

$3 - (1 + x) = 2 - x, 1 < x \le 2$ {from (3)}

Hence, the function is

$f\{f(x)\} = 2 + x, 0 \le x \le 1$

$2 - x, 1 < x \le 2$

$4 - x, 2 < x \le 3$

As polynomial functions are continuous and differentiable everywhere, $f\{f(x)\}$ is continuous and differentiable in [0, 3] except perhaps at the turning points of definition, namely, $x = 1, 2$. Denote $f\{f(x)\}$ by $g(x)$.

Now, $g(1+0) = 2 - 1 = 1$

$g(1-0) = 2 + 1 = 3$.

So $g(x)$ is not continuous at $x = 1$.

$g(2+0) = 4 - 2 = 2$

$g(2-0) = 2 - 2 = 0$.

So $g(x)$ is not continuous at $x = 2$.

Hence, $g(x)$ is also not differentiable at $x = 1, 2$.

$\therefore$ the points of discontinuity and nondifferentiability are $x = 1, 2$.

Now, we have to draw the graph of $y = f\{f(x)\}$ where

$y = 2 + x$ in [0, 1],

$y = 2 - x$ in (1, 2] and

$y = 4 - x$ in (2, 3],

the graph being discontinuous at $x = 1, 2$.

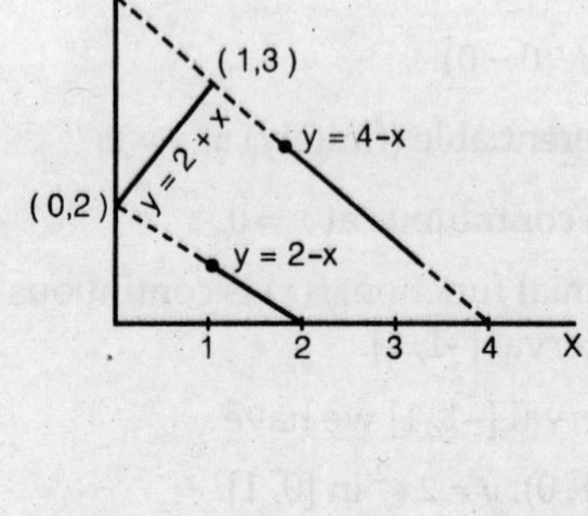

Exercises

1. Evaluate $\lim_{h \to 0} \frac{|-1+h|-|-1|}{h}$ if the limit exists.

2. $f(x)$ is defined as follows:

$$f(x) = \sqrt{1-x^2}, \ 0 \le x < 1$$
$$1, \ 1 \le x < 2$$
$$2, \ x = 2.$$

Answer the following:

(a) What is the domain and range of the function?

(b) At what point c does $\lim_{x \to c} f(x)$ exist?

(c) At what point does only the left-hand limit exist?

(d) At what point does only the right-hand limit exist?

3. If $f(x) = \frac{x^2}{a}, \ 0 \le x \le 1$

$ax, \ 1 < x < 2$

then assign suitable values to a so that $\lim_{x \to 1} f(x)$ exists.

4. If $f(x) = \frac{e^{1/x}}{1+e^{1/x}}, \ x \ne 0$

$0, \ x = 0$

then show that $f(x)$ is discontinuous at $x = 0$.

5. Examine the continuity of $f(x) = |x-1| + |x+1|$ at $x = 1, 0$.

6. By defining $f\left(\frac{\pi}{4}\right)$ suitably, obtain a continuous extension of the function

$$f(x) = \frac{\sqrt{2}\cos x - 1}{\cot x - 1}, \ x \ne \frac{\pi}{4}.$$

7. Find a, b and c such that the function

$$f(x) = \frac{\sin(a+1)x + \sin x}{x}, \ x < 0$$
$$c, \ x = 0$$
$$\frac{\sqrt{x+bx^2} - \sqrt{x}}{bx^{3/2}}, \ x > 0$$

is continuous at $x = 0$.

8. Discuss the continuity of $f(x)$ at $x = 0$ where

$$f(x) = x\left\{\frac{1}{x(1+x)} + \frac{1}{(1+x)(1+2x)} + \frac{1}{(1+2x)(1+3x)} + \dots \text{ to } \infty\right\}.$$

9. Evaluate $\lim_{x \to 0} \frac{1}{x}\log_e \frac{1+ax}{1-bx}$.

Use this limit to show that the function

$$f(x) = \frac{1}{x}\log_e \frac{1+ax}{1-bx}, \ x \ne 0$$
$$a+b, \ x = 0$$

is continuous at $x = 0$.

10. $f(x)$ is defined as follows:

$$f(x) = (\sin x + \cos x)^{\operatorname{cosec} x}, \ -\frac{\pi}{2} < x < 0$$
$$a, \ x = 0$$
$$\frac{e^{\frac{1}{x}} + e^{\frac{2}{x}} + e^{\frac{3}{|x|}}}{ae^{\frac{2}{x}} + be^{\frac{3}{|x|}}}, \ 0 < x < \frac{\pi}{2}$$

If $f(x)$ is continuous at $x = 0$, find a and b.

11. If $g(x) = \frac{1 - a^x + xa^x \log a}{x^2 \cdot a^x}, \ x < 0$

$$\frac{(2a)^x - x\log(2a) - 1}{x^2}, \ x > 0 \quad (\text{where } a > 0)$$

then find a and $g(0)$ so that $g(x)$ is continuous at $x = 0$.

12. Show that the function $f(x)$ defined as follows:

$$f(x) = 2x + 3, \ x \in (-3, -2)$$
$$x + 1, \ x \in [-2, 0]$$
$$x + 2, \ x \in (0, 1)$$

is discontinuous at $x = 0$ and continuous at every other point of the open interval $(-3, 1)$. Draw a rough sketch of the function.

13. Let $f(x) = \frac{x^2}{2}, \ 0 \le x \le 1$

$$2x^2 - 3x + \frac{3}{2}, \ 1 < x \le 2.$$

Discuss the continuity of $f(x)$ in $[0, 2]$. Give a rough sketch of the function.

14. Discuss the continuity of $f(x) = \lim_{n \to \infty} \frac{x^{2n} - 1}{x^{2n} + 1}$ for real x.

15. If $f(x)$ and $g(x)$ are two functions continuous everywhere and

$$F(x) = \lim_{n \to \infty} \frac{f(x) + x^{2n} g(x)}{1 + x^{2n}}$$

then prove that $F(x)$ is continuous everywhere except at $x = 1, -1$. Find the condition on $f(x)$, $g(x)$ which makes $F(x)$ continuous everywhere.

16. Consider the function f defined by $f(x) = x - [x]$, where $[x]$ denotes the greatest integral function. Show that the function is discontinuous for integral values of x and continuous for all other values.

17. Find the values of a and b so that the function

$$f(x) = x + a\sqrt{2} \sin x, 0 \le x < \frac{\pi}{4}$$
$$2x\cot x + b, \frac{\pi}{4} \le x \le \frac{\pi}{2}$$
$$a\cos 2x - b\sin x, \frac{\pi}{2} < x \le \pi$$

is continuous for $0 \le x \le \pi$.

18. Determine the set of all points where the function $f(x) = \dfrac{x^3}{4 + |x|}$ is continuous.

19. Let $f(x)$ be a continuous function and $g(x)$ be a discontinuous function. Prove that $f(x) + g(x)$ is a discontinuous function.

20. Let $f(x + y) = f(x) + f(y)$ for all x, y. If $f(x)$ is continuous at $x = 0$ then show that $f(x)$ is continuous at all x.

21. Let f be a function such that $f(xy) = f(x)f(y)$ for all x and y. If $f(x)$ is continuous at $x = 1$, show that $f(x)$ is continuous at all $x \ne 0$.

22. Let $f: [0, 1] \to [0, 1]$ be continuous. Show that there exists a point $x \in [0, 1]$ such that $f(x) = x$.

23. If $\phi(x) = |\sin x| - |\cos x|$, find $\phi'\left(\frac{\pi}{2}\right)$ if it exists.

24. If $f(x) = e^{|x|}$, find $f'(0)$.

25. If $f(x) = x^2 \sin \frac{1}{x}, x \ne 0$
$x, x = 0$
then find $f'(0)$ if it exists.

26. Do $f'(1)$ and $f'(2)$ exist for the function
$$f(x) = x[x], 0 \le x < 2$$
$$x^2 - x, 2 \le x \le 3?$$

27. Given $f(x) = |x|$ and $\phi(x) = |x^3|$. Do the derivatives $f'(0)$ and $\phi'(0)$ exist? Explain.

28. If $f(x) = x^2 + 4, 0 \le x < 1$
$2x + 3, 1 \le x \le 2$ and,
$\phi(x) = 3x - 2, 0 \le x < 1$
$x^3, 1 \le x \le 2$
then find $\dfrac{df}{d\phi}$ at $x = 1$ if it exists.

29. Determine the set of all points where the function $f(x) = \dfrac{x}{1 + |x|}$ is differentiable.

30. Find the set of points where $f(x) = x|x|$ is twice differentiable.

31. Let $f(x)$ be defined in the interval $[-2, 2]$ such that
$$f(x) = -1, -2 \le x \le 0$$
$$x - 1, 0 < x \le 2$$
and $g(x) = f(|x|) + |f(x)|$.
Test the differentiability of $g(x)$ in $[-2, 2]$.

32. A function is defined as follows:
$$f(x) = x^3, x^2 < 1$$
$$x, x^2 \ge 1.$$
Discuss limit, continuity and differentiability of $f(x)$ at $x = 1$. Also draw the rough sketch of the function.

33. Determine the values of x for which the following function fails to be continuous or differentiable:
$$f(x) = 1 - x, x < 1$$
$$(1 - x)(2 - x), 1 \le x \le 2$$
$$3 - x, x > 2.$$

34. Examine the continuity and differentiability of $f(x)$ at $x = 0$ if
$$f(x) = xe^{-\left(\frac{1}{|x|} + \frac{1}{x}\right)}, x \ne 0$$
$$0, x = 0.$$

35. If $f(x)$ be defined as follows:
$$f(x) = \frac{\sin x}{x}, x > 0$$
$$1 - x\cos x, x \le 0$$
then discuss the continuity and differentiability of $f(x)$ at $x = 0$.

36. Let $g(x)$ be a polynomial of degree one and $f(x)$ is defined by
$$f(x) = g(x), x \le 0$$
$$= \left(\frac{1 + x}{2 + x}\right)^{\frac{1}{x}}, x > 0.$$
Find $g(x)$ such that $f(x)$ is continuous and $f'(1) = f(-1)$.

37. Discuss the continuity and differentiability of $f(x) = |x - 2|$ in the interval $[1, 3]$. Also draw a rough sketch of the function in the interval.

38. Examine the continuity and differentiability of $f(x)$ at $x = 3$ if
$$f(x) = x[x], 0 \le x < 3$$
$$(x - 1)[x], 3 \le x \le 4$$
where $[x]$ is the greatest integer function. Give the qualitative sketch of the function in $[0, 4]$.

39. Let $f(x) = \dfrac{x^2}{2}, 0 \le x < 1$
$$2x^2 - 3x + \frac{3}{2}, 1 \le x \le 2.$$
Discuss the continuity of f, f', f'' in $[0, 2]$.

40. Let $f(x) = 2x - 2, x < -1$
$Ax + B, -1 \le x \le 1$
$5x + 7, x > 1.$
Determine A and B so that $f(x)$ is continuous for all x. Is f, so obtained, differentiable for all x?

41. Find the value of $f(0)$ so that the function
$$f(x) = \frac{1}{x} - \frac{2}{e^{2x} - 1}, x \ne 0$$
is continuous at $x = 0$. Also examine the differentiability of $f(x)$ at $x = 0$ for that value of $f(0)$.

42. Discuss the continuity and differentiability at $x = 1$ of the function
$f(x) = 3^x, -1 \le x \le 1$
$4 - x, 1 < x < 4.$
Also, draw a graph of the function.

43. Prove that $f(x) = \sin \pi |x|$ is continuous at $x = 0$ but is not differentiable there. Draw a rough sketch of the function in the closed interval $[-1, 1]$.

44. If $f(x) = \sin x, x \ne 0, \pm \pi, \pm 2\pi, \pm 3\pi, \ldots$
$= 2$ otherwise, and
$g(x) = x^2 + 1, x \ne 0, 2$
$= 4, x = 0$
$= 5, x = 2$ then find $\lim_{x \to 0} g\{f(x)\}$.

45. Let $f(x) = \frac{1}{1 - x}$. Find the points of discontinuity of the functions $f(x), f\{f(x)\}$ and $f\{f(f(x)\}$.

46. If $f(x) = \sqrt{|x - 1|}$ and $g(x) = \sin x$ then calculate $(f \circ g)(x)$ and $(g \circ f)(x)$ and discuss the differentiability of $(g \circ f)(x)$ at $x = 1$.

47. If $f(x) = -1 + |x - 1|, -1 \le x \le 3$
$g(x) = 2 - |x + 1|, -2 \le x \le 2$
then calculate $(f \circ g)(x)$ and $(g \circ f)(x)$. Discuss the continuity of $(f \circ g)(x)$ at $x = -1$ and differentiability of $(g \circ f)(x)$ at $x = 1$. Also draw the graphs of $(f \circ g)(x)$ and $(g \circ f)(x)$ in $[-1, 2]$.

48. If $f(x) = 1 + x, x \ge 0$
$1 - x, x < 0$
then discuss the continuity and differentiability of $f\{f(x)\}$. Also give a rough sketch of the curve $y = f\{f(x)\}$.

49. If $\phi(x) = |x|, x \le 1$
$2 - x, x > 1$
then discuss the continuity and differentiability of the function $\phi\{\phi(x)\}$. Also give a rough sketch of the function.

50. If $f(x) = 1 + x, x \in [0, 1)$
$g(x) = 2 - x, x \in [1, 2]$
then draw a rough sketch of the function $g\{f(x)\}$.

Objective Questions

Fill in the blanks.

51. If $\lim_{x \to 0} f(x)$ exists where
$$f(x) = \frac{\sin(3p - 1)x}{3x}, x < 0$$
$$\frac{\tan(3p + 1)x}{2x}, x > 0$$ then $p =$ ______ .

52. If $f(x) = ax^2 - b, 0 \le x < 1$
$2, x = 1$
$x + 1, 1 < x \le 2$
is continuous at $x = 1$ then $a - b =$ ______ .

53. If $f(x) = \frac{1 - \sin^3 x}{3\cos^2 x}, x < \frac{\pi}{2}$
$a, x = \frac{\pi}{2}$
$\frac{b(1 - \sin x)}{(\pi - 2x)^2}, x > \frac{\pi}{2}$
is continuous at $x = \frac{\pi}{2}$ then $a =$ ______ , $b =$ ______ .

54. If $f(x) = x - [x]$ then
$\lim_{x \to 2+0} f(x) =$ ______ , $\lim_{x \to 2-0} f(x) =$ ______ .

55. The points at which $f(x) = \cos^{-1} \frac{x + 1}{2}$ has no finite derivative, are ______ .

56. Let $f(x) = (x - 1)^2 \sin \frac{1}{x - 1} - |x|, x \ne 1$
$= -1, x = 1$
be a real-valued function. The set of points where $f(x)$ is not differentiable is ______ .

57. Let $f(x) = \frac{x^2 + 3x - 10}{x - 2}, x \ne 2.$
The value $f(2) =$ ______ will make the function $f(x)$ continuous at $x = 2$.

58. If $f(x)$ is continuous in $[0, 1]$ and $f\left(\frac{1}{2}\right) = 1$ then
$$\lim_{n \to \infty} f\left(\frac{\sqrt{n}}{2\sqrt{n} + 1}\right) = \_\_\_\_\_\_ .$$

59. Let $f(x) = \frac{\sin(e^{x-2} - 1)}{\log(x - 1)}, x \ne 2.$
$f(x)$ will be continuous at $x = 2$ if $f(2) =$ ______ .

60. If $f(x) = \dfrac{\sqrt{1+px} - \sqrt{1-px}}{x}, -1 \le x < 0$

$\dfrac{2x+1}{x-2}, 0 \le x \le 1$

is continuous in $[-1, 1]$ then $p =$ _____ .

61. Let $f(x) = \sin x, x < \pi$

$mx + b, x \ge \pi.$

(a) If $f(x)$ is continuous for all x then $m : b =$ _____ .

(b) If $f(x)$ is differentiable for all x then $m =$ _____ , $b =$ _____ .

62. If $f(x) = x, x \le 1$

$ax, 1 < x < 2$

$x^2 + bx + 4, x \ge 2$

and $f(x)$ is differentiable everywhere then $a =$ _____ , $b =$ _____ .

Choose the correct option(s).

63. If $f(x) = \left[\tan^2\left(x - \dfrac{\pi}{4}\right)\right]$ where $[x]$ is the greatest integer function then

(a) $f(x)$ is continuous at $x = 0$

(b) $f(x)$ is differentiable at $x = \dfrac{\pi}{2}$

(c) $f(x)$ is continuous in $\left(0, \dfrac{\pi}{2}\right)$

(d) $f(x)$ is differentiable in $\left(0, \dfrac{\pi}{2}\right)$

64. If $f(x) = \dfrac{\sin [x]}{[x]}, [x] \ne 0$

$= 0, [x] = 0$

where $[x]$ is the greatest integer function then $\lim\limits_{x \to 0} f(x)$ equals

(a) 1 (b) 0

(c) –1 (d) none of these

65. Let $f(x) = \dfrac{\log_e(1+ax) - \log_e(1-bx)}{x}, x \ne 0.$

The value to be assigned to $f(x)$ at $x = 0$ so that $f(x)$ is continuous at $x = 0$, is

(a) $a - b$ (b) $a + b$

(c) $\log_e(ab)$ (d) none of these

66. The function $f(x) = |x - 3|, x \ge 1$

$\dfrac{x^2}{4} - \dfrac{3x}{2} + \dfrac{13}{4}, x < 1$

(a) is continuous at $x = 1$ (b) is continuous at $x = 3$

(c) is differentiable at $x = 1$ (d) $f'(3)$ exists

67. The function $f(x) = \dfrac{\tan(\pi[x - \pi])}{1 + [x]^2}$, where $[x]$ is the greatest integer function,

(a) is discontinuous at some x

(b) $f'(x)$ exists for all x

(c) $f'(x)$ exists for all x but $f''(x)$ does not exist

(d) is continuous for all x but $f'(x)$ does not exist for some x

68. $f(x) = 1 + |\sin x|$

(a) is continuous nowhere

(b) is continuous everywhere

(c) is differentiable nowhere

(d) $f'(0)$ does not exist

(e) nondifferentiable at infinite number of points

69. If $x + |y| = 2y$ then y as a function of x is

(a) defined for all x

(b) continuous at $x = 0$

(c) differentiable for all x

(d) such that $\dfrac{dy}{dx} = \dfrac{1}{3}$ for $x < 0$

70. The graph of the function $f(x) = \cos x \cdot \cos(x+2) - \cos^2(x+1)$ is

(a) a straight line passing through $(0, -\sin^2 1)$ with slope 2

(b) a straight line passing through (0, 0)

(c) a parabola with vertex $(1, -\sin^2 1)$

(d) a straight line passing through the point $\left(\dfrac{\pi}{2}, -\sin^2 1\right)$ and parallel to the x-axis

71. If $f(x) = x(\sqrt{x} - \sqrt{x+1})$ then

(a) $f(x)$ is continuous but not differentiable at $x = 0$

(b) $f'(0)$ exists

(c) $f(x)$ is nondifferentiable at $x = 0$

(d) none of these

72. The set of points where $f(x) = \dfrac{x}{4 + |x|}$ is differentiable, is

(a) $(-\infty, +\infty)$ (b) $(0, +\infty)$

(c) $(-\infty, 0) \cup (0, +\infty)$ (d) none of these

State whether the statements are true or false.

73. $\lim\limits_{x \to 1} f(x)$ exists if $f(x)$ is defined as follows:

$f(x) = x^2, x < 1$

$\dfrac{5}{2}, x = 1$

$x^2 + 2x, x > 1.$

74. $\lim\limits_{x \to 2} [x]$ exists where $[x]$ denotes the integral part of x.

75. If $f(x) = \lim_{n \to \infty} \frac{x^n - 1}{x^n + 1}$ then $f(x)$ is discontinuous at $x = 1$.

76. If $f(x)$ and $g(x)$ are not differentiable finitely at a point then $f(x) \cdot g(x)$ will also be nondifferentiable finitely at that point.

77. The inverse function of a continuous function is continuous.

78. If $f(x) = |x|$ then $f'(x) = \frac{|x|}{x}, x \neq 0$.

79. $f(x) = 2x^2 + 3, x \leq 1$

$3x + 2, x > 1$

is neither differentiable nor continuous at $x = 1$.

80. If $\lim_{x \to a} \{f(x) \cdot g(x)\}$ exists then both $\lim_{x \to a} f(x)$ and $\lim_{x \to a} g(x)$ exist.

81. If $\lim_{x \to a} f(x)$ and $\lim_{x \to a} g(x)$ exist then $\lim_{x \to a} \frac{f(x)}{g(x)}$ exists.

82. $\lim_{x \to 1} 2^{\frac{1}{x-1}}$ exists.

83. $\lim_{x \to 0+0} \frac{(1 - \cos 2x)\sin 5x}{x^2 \sin 3x} = \frac{10}{3}$.

84. If $f(x)$ be a continuous function in $[1, 3]$ and $f(x)$ takes rational values for all x in the interval, and $f(2) = 10$ then $f(x) = 10$ for all $x \in [1, 3]$.

85. The function $f(x) = 1$, if x is rational

0, if x is irrational

is discontinuous at all points x.

Answers

1. -1
2. (a) domain $= [0, 2]$, range $= \{1, 2\} \cup (0, 1)$
   (b) $[0, 1) \cup (1, 2]$ (c) $x = 2$ (d) $x = 0$
3. ± 1
5. continuous at $x = 1, 0$
6. $\frac{1}{2}$
7. $a = \frac{-3}{2}$, $b =$ any constant, $c = \frac{1}{2}$
8. discontinuous
9. $a + b$
10. $a = e, b = \frac{1}{e}$
11. $a = \frac{1}{\sqrt{2}}, g(0) = \frac{1}{8}(\log 2)^2$
13. continuous in $[0, 2]$
14. continuous everywhere except at $x = \pm 1$
15. $g(1) = f(1), g(-1) = f(-1)$
17. $a = \frac{\pi}{6}, b = \frac{-\pi}{12}$
18. R
23. $\phi'\left(\frac{\pi}{2}\right)$ does not exist
24. $f'(0)$ does not exist
25. 0
26. do not exist
27. $f'(0)$ does not exist
28. $\frac{2}{3}$
29. R
30. $R - \{0\}$
31. differentiable in $R - \{0, 1\}$
32. $\lim_{x \to 1} f(x) = 1$, continuous but nondifferentiable at $x = 1$
33. $x = 2$ at which $f(x)$ is neither continuous nor differentiable
34. continuous but not differentiable
35. continuous but not differentiable
36. $g(x) = \frac{1}{9}\left(6\log\frac{2}{3} - 1\right)x, x \leq 0$
37. continuous everywhere but not differentiable at $x = 2$
38. continuous but nondifferentiable at $x = 3$
39. f, f' are continuous in $[0, 2]$ and f'' is continuous in $[0, 2]$ except at $x = 1$
40. $A = 8, B = 4$, not differentiable at $x = \pm 1$
41. $f(0) = 1$, differentiable at $x = 0$
42. continuous but not differentiable
44. 1
45. for $f(x) : x = 1$; for $f\{f(x)\} : x = 1, 0$
    for $f\{f(f(x))\} : x = 1, 0$
46. $(f \circ g)(x) = \sqrt{1 - \sin x}$,
    $(g \circ f)(x) = \sin \sqrt{|x - 1|}$,
    not differentiable at $x = 1$
47. $(f \circ g)(x) = 1 + x, -2 \leq x < -1$
    $-1 - x, -1 \leq x < 0$
    $-1 + x, 0 \leq x \leq 2$
    $(g \circ f)(x) = 1 + x, -1 \leq x < 1$
    $3 - x, 1 \leq x \leq 3$
    $(f \circ g)(x)$ is not continuous at $x = -1$,
    $(g \circ f)(x)$ is not differentiable at $x = 1$
48. continuous in R, differentiable in $R - \{0\}$

49. continuous in R, differentiable in $R - \{-1, 0, 1, 2\}$

51. $-\frac{5}{3}$ **52.** 2

53. $\frac{1}{2}$, 4 respectively **54.** 0, 1 respectively

55. $x = -3, 1$ **56.** $\{0\}$

57. 7 **58.** 1

59. 1 **60.** $-\frac{1}{2}$

61. (a) $-1 : \pi$ (b) $-1, \pi$ respectively

62. 1, −3 respectively **63.** (c), (d)

64. (d) **65.** (b)

66. (a), (b), (c) **67.** (b)

68. (b), (d), (e) **69.** (a), (b), (d)

70. (d) **71.** (b) **72.** (a) **73.** false

74. false **75.** true **76.** false **77.** true

78. true **79.** false **80.** false **81.** false

82. false **83.** true **84.** true **85.** true

□

Chapter Test

Time: 90 minutes

1. In the following, fill in the blanks so that the resulting sentences become true.

(a) If $f(x) = x^2 + bx, 0 \le x \le 1$
$3 - ax^2, 1 < x \le 2$
such that $\lim_{x \to 1} f(x) = 4$ then $a =$ ______ , $b =$ ______ .

(b) If $f(x) = \tan x$ then $f'\left(\frac{\pi}{2} - 0\right) =$ ______ , $f'\left(\frac{\pi}{2} + 0\right) =$ ______ .

(c) If $f(x)$ be a continuous function in $[2, 3]$ which takes only irrational values for all $x \in [2, 3]$ and $f(2.5) = \sqrt{5}$ then $f(2.8) =$ ______ .

2. In each of the following, multiple options are given. Choose the correct option(s).

(a) If $f(x) = x^p \cos \frac{1}{x}, x \ne 0$ and $f(0) = 0$ then $f(x)$ will be continuous and differentiable at $x = 0$ if

A. $p > 0$ B. $p > 1$ C. $0 < p < 1$ D. $p > \frac{1}{2}$

(b) $\lim_{x \to 1} \{[1 - x] + [x - 1] - x + 1\}$ is equal to

A. 0 B. 1 C. -1 D. none of these

(c) If $f(x) = (x) - x$ where $(x) =$ the least integer greater than or equal to x then

A. $f(x)$ is continuous at all x B. $f(x)$ is continuous at all nonintegral x

C. $f'(0)$ exists D. $\lim_{x \to 0-0} f(x) = 0$

3. A function $f(x)$ is defined as follows:

$f(x) = \alpha x - \beta, x \le 1$
$3x, 1 < x < 2$
$\beta x^2 - \alpha, x \ge 2.$

Prove that if $f(x)$ is continuous at $x = 1$ but discontinuous at $x = 2$ then the locus of the point (α, β) is a straight line excluding the point where it cuts the line $y = 3$.

4. Let $\phi(x)$ be a function defined as below:

$\phi(x) = \frac{1}{2} - x, 0 < x < \frac{1}{2}$
$-\frac{1}{2} + x, \frac{1}{2} < x < 1$

and $\phi(0) = 0$, $\phi\left(\frac{1}{2}\right) = \frac{1}{2}$, $\phi(1) = 1$.

Verify the following statements:

(a) $\phi(x)$ assumes every value between 0 and 1 as x increases from 0 to 1.

(b) $\phi(x)$ is continuous at $x = \frac{1}{2}$.

(c) $\phi(x)$ is differentiable in $[0, 1]$.

5. $f(x)$ is a function such that

$$f'(x) = \begin{cases} x + 1, & 0 \le x < 1 \\ |x - 3|, & 1 \le x \le 4. \end{cases}$$

Examine the continuity of $f(x)$ and $f\{f'(x)\}$ at $x = 1, 2$.

6. If $|p(x)| \le |e^{x-1} - 1|$ for all $x > 0$ where $p(x) = \sum_{r=0}^{n} a_r x^r$ then prove that $\left|\sum_{r=1}^{n} r a_r\right| \le 1$.

7. Show that the function $f(x) = |x - 1|\ \phi(x)$ where $\phi(x)$ is continuous and $\phi(1) \ne 0$, has no derivative at $x = 1$. Find $f'(1 + 0)$ and $f'(1 - 0)$.

8. Discuss the continuity and differentiability of $f(x)$ in $[-2, 4]$ where

$$f(x) = \begin{cases} 1 - x^2, & -2 \le x < 1 \\ |x - 2|, & 1 \le x < 3 \\ [x + 1], & 3 \le x \le 4. \end{cases}$$

Also, give a rough sketch of the function in $[-2, 4]$.

9. Let $\vec{A}(t) = f_1(t)\vec{i} + f_2(t)\vec{j}$ and $\vec{B}(t) = g_1(t)\vec{i} + g_2(t)\vec{j}$, $t \in [0, 1]$ where f_1, f_2, g_1, g_2 are all continuous functions. If $\vec{A}(t)$ and $\vec{B}(t)$ are nonzero vectors for all t and $\vec{A}(0) = 2\vec{i} + 3\vec{j}$, $\vec{A}(1) = 6\vec{i} + 2\vec{j}$, $\vec{B}(0) = 3\vec{i} + 2\vec{j}$, and $\vec{B}(1) = 2\vec{i} + 6\vec{j}$, then prove that $\vec{A}(t) \parallel \vec{B}(t)$ for at least one $t \in [0, 1]$.

10. Let $f(x) = \begin{cases} x + a, & x < 0 \\ |x - 1|, & x \ge 0 \end{cases}$ and $g(x) = \begin{cases} x + 1, & x < 0 \\ (x - 1)^2 + b, & x \ge 0 \end{cases}$

where, $a \ge 0$, $b \ge 0$. Determine the composite function $g \circ f$. If $(g \circ f)(x)$ is continuous for all real x, determine the constants a and b. For these values of a, b is the function $g \circ f$ differentiable at $x = 0$? Give reasons.

□

Answers

1. (a) –1, 3 respectively (b) ∞, ∞ (c) √5
2. (a) B (b) C (c) B, D
4. (a) wrong (b) wrong (c) wrong
5. both are continuous at $x = 1, 2$
7. $\phi(1), -\phi(1)$ respectively
8. discontinuous only at $x = 1, 3, 4$; not differentiable only at $x = 1, 2, 3, 4$
9. $(g \circ f)(x) = \begin{cases} x + a + 1, & x < -a \\ (x + a - 1)^2 + b, & -a \le x < 0 \\ (|x - 1| - 1)^2 + b, & x \ge 0 \end{cases}$ $a = 1,\ b = 0$; differentiable at $x = 0$

□

5. Application of dy/dx

Recap of Facts and Formulae

1. Geometrical meaning of $\frac{dy}{dx}$ and its use

- $\left(\frac{dy}{dx}\right)_{\alpha, \beta}$ = slope of the tangent to the curve $y = f(x)$ at the point (α, β)
- The equation of the tangent to the curve $y = f(x)$ at the point (α, β) is

$$y - \beta = \left(\frac{dy}{dx}\right)_{\alpha, \beta} \cdot (x - \alpha)$$

- The equation of the normal to the curve $y = f(x)$ at the point (α, β) is

$$y - \beta = \frac{-1}{\left(\frac{dy}{dx}\right)_{\alpha, \beta}} \cdot (x - \alpha)$$

2. $\frac{dy}{dx}$ as rate measurer

- If $x = x(t)$ where t is the time then

$$\frac{dx}{dt} = \text{rate of change of } x.$$

- If $y = f(x)$ then $\frac{dy}{dx} = \frac{\text{rate of change of } y}{\text{rate of change of } x}$.

- If x is the displacement then

$$\frac{dx}{dt} = \text{rate of change of displacement at time } t$$

$$= \text{velocity at time } t$$

$$\frac{d}{dt}\left(\frac{dx}{dt}\right), \text{ i.e., } \frac{d^2x}{dt^2} = \text{rate of change of velocity at time } t$$

$$= \text{acceleration at time } t$$

- $\frac{dx}{dt}$ = rate of increase iff $\frac{dx}{dt} > 0$

  $\frac{dx}{dt}$ = rate of decrease iff $\frac{dx}{dt} < 0$

3. Approximate estimation

- If $y = f(x)$ and a very small change δx takes place in the value of x when it is a then the corresponding change δy in the value of y is

$$\delta y = \left(\frac{dy}{dx}\right)_{x=a} \cdot \delta x \quad \text{(approximately)}$$

- If x is a variable then

  the per cent error in calculation of $x = \frac{dx}{x} \times 100$.

Selected Solved Examples

1. The curve $y = ax^3 + bx^2 + cx + 5$ touches the x-axis at P(−2, 0) and cuts the y-axis at a point Q where its gradient is 3. Find the equation of the curve completely.

As the x-axis touches the curve at P(−2, 0), we have

$$\left(\frac{dy}{dx}\right)_{-2, 0} = \tan 0° = 0 \qquad \text{... (1)}$$

and $(y)_{x=-2} = 0$ ... (2)

As the curve cuts the y-axis at Q where the gradient of the curve is 3,

$$\left(\frac{dy}{dx}\right)_{x=0} = 3 \qquad \text{... (3)}$$

Differentiating $y = ax^3 + bx^2 + cx + 5$ w.r.t x,

$$\frac{dy}{dx} = 3ax^2 + 2bx + c$$

$\therefore$ (1) $\Rightarrow$ $0 = 3a(-2)^2 + 2b(-2) + c$

i.e., $12a - 4b + c = 0$ ... (4)

$(2) \Rightarrow \quad 0 = a(-2)^3 + b(-2)^2 + c(-2) + 5,$

i.e., $-8a + 4b - 2c + 5 = 0 \quad \ldots (5)$

$(3) \Rightarrow \quad 3 = 3a \cdot 0^2 + 2b \cdot 0 + c$, i.e., $c = 3$

$\therefore$ (4) and (5) give, $12a - 4b + 3 = 0$

and $-8a + 4b - 1 = 0$.

Solving these, $a = \dfrac{-1}{2}$ and $b = \dfrac{-3}{4}$.

$\therefore$ the curve is $y = -\dfrac{1}{2}x^3 - \dfrac{3}{4}x^2 + 3x + 5$.

2. Show that the curves $\dfrac{x^2}{a} + \dfrac{y^2}{b} = 1$ and $\dfrac{x^2}{a_1} + \dfrac{y^2}{b_1} = 1$ will cut orthogonally if $a - b = a_1 - b_1$.

Let the curves intersect at (α, β). Then

$$\frac{\alpha^2}{a} + \frac{\beta^2}{b} = 1 \quad \ldots (1)$$

and
$$\frac{\alpha^2}{a_1} + \frac{\beta^2}{b_1} = 1 \quad \ldots (2)$$

Differentiating $\dfrac{x^2}{a} + \dfrac{y^2}{b} = 1$ w.r.t. x,

$$\frac{2x}{a} + \frac{2y}{b}\frac{dy}{dx} = 0$$

$$\therefore \quad \frac{dy}{dx} = -\frac{bx}{ay}$$

$$\therefore \quad \left(\frac{dy}{dx}\right)_{\alpha, \beta} = \frac{-b\alpha}{a\beta} \ (= m_1).$$

Similarly, from $\dfrac{x^2}{a_1} + \dfrac{y^2}{b_1} = 1$,

$$\left(\frac{dy}{dx}\right)_{\alpha, \beta} = \frac{-b_1\alpha}{a_1\beta} \ (= m_2).$$

Two curves will cut orthogonally if $m_1 \cdot m_2 = -1$,

i.e., $\dfrac{-b\alpha}{a\beta} \cdot \dfrac{-b_1\alpha}{a_1\beta} = -1$

$$\therefore \quad bb_1\alpha^2 + aa_1\beta^2 = 0 \quad \ldots (3)$$

$$(1) - (2) \Rightarrow \quad \alpha^2\left(\frac{1}{a} - \frac{1}{a_1}\right) + \beta^2\left(\frac{1}{b} - \frac{1}{b_1}\right) = 0 \quad \ldots (4)$$

$$(3) \text{ and } (4) \Rightarrow \quad \frac{bb_1}{\frac{1}{a} - \frac{1}{a_1}} = \frac{aa_1}{\frac{1}{b} - \frac{1}{b_1}}$$

or $\quad bb_1\left(\dfrac{1}{b} - \dfrac{1}{b_1}\right) = aa_1\left(\dfrac{1}{a} - \dfrac{1}{a_1}\right)$,

i.e., $b_1 - b = a_1 - a$

$\therefore \quad a - b = a_1 - b_1$.

3. Tangents are drawn from the origin to the curve $y = \sin x$. Prove that their points of contact lie on the curve $x^2y^2 = x^2 - y^2$.

Let (x_1, y_1) be a point of contact of tangents from the origin (0, 0).

Here, $y = \sin x$

$$\therefore \quad \frac{dy}{dx} = \cos x; \quad \therefore \quad \left(\frac{dy}{dx}\right)_{x_1, y_1} = \cos x_1$$

$\therefore$ the equation of the tangent at (x_1, y_1) is

$$y - y_1 = \left(\frac{dy}{dx}\right)_{x_1, y_1} \cdot (x - x_1)$$

or $\quad y - y_1 = \cos x_1 \cdot (x - x_1)$.

It passes through (0, 0).

So, $-y_1 = -x_1 \cos x_1 \quad \ldots (1)$

Also, (x_1, y_1) is on the curve.

So, $y_1 = \sin x_1 \quad \ldots (2)$

$$(1) \text{ and } (2) \Rightarrow \quad \left(\frac{y_1}{x_1}\right)^2 + y_1^2 = \cos^2 x_1 + \sin^2 x_1 = 1$$

or $\quad y_1^2 + x_1^2y_1^2 = x_1^2$, i.e., $x_1^2y_1^2 = x_1^2 - y_1^2$

$\therefore$ the point of contact (x_1, y_1) lies on the curve

$$x^2y^2 = x^2 - y^2.$$

4. In the curve $x^a y^b = k^{a+b}$, $(a, b > 0)$ prove that the portion of the tangent intercepted between the coordinate axes is divided at its point of contact into segments which are in constant ratio.

Let $P(x_1, y_1)$ be the point of contact of the tangent.

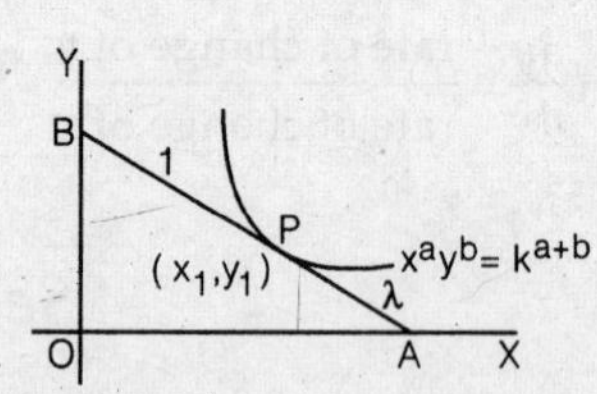

Here, $x^a y^b = k^{a+b}$

$\therefore \quad a\log x + b\log y = (a + b)\log k$.

Differentiating, $\dfrac{a}{x} + \dfrac{b}{y}\dfrac{dy}{dx} = 0$

or $\quad \dfrac{dy}{dx} = -\dfrac{ay}{bx}; \quad \therefore \quad \left(\dfrac{dy}{dx}\right)_{x_1, y_1} = \dfrac{-ay_1}{bx_1}$

$\therefore$ the equation of the tangent at $P(x_1, y_1)$ is

$$y - y_1 = \left(\frac{dy}{dx}\right)_{x_1, y_1} \cdot (x - x_1)$$

or $y - y_1 = \dfrac{-ay_1}{bx_1}(x - x_1)$ ... (1)

Solving with $y = 0$, $-y_1 = \dfrac{-ay_1}{bx_1}(x - x_1)$

or $bx_1 = a(x - x_1)$; $\therefore$ $x = \dfrac{(a+b)x_1}{a}$

Solving (1) with $x = 0$, $y - y_1 = \dfrac{-ay_1}{bx_1}(-x_1)$

or $y = y_1 + \dfrac{ay_1}{b} = \dfrac{(a+b)y_1}{b}$

$\therefore$ $A = \left(\dfrac{a+b}{a}x_1, 0\right)$ and $B = \left(0, \dfrac{a+b}{b}y_1\right)$.

Let P divide AB in the ratio $\lambda : 1$.

Then $P = \left(\dfrac{\lambda \cdot 0 + 1 \cdot \frac{a+b}{a}x_1}{\lambda + 1}, \dfrac{\lambda \cdot \frac{a+b}{b}y_1 + 1 \cdot 0}{\lambda + 1}\right)$

$= \left(\dfrac{a+b}{a(\lambda+1)}x_1, \dfrac{\lambda(a+b)}{b(\lambda+1)}y_1\right)$

$\therefore$ $x_1 = \dfrac{a+b}{a(\lambda+1)}x_1$ and $y_1 = \dfrac{\lambda(a+b)}{b(\lambda+1)}y_1$

$\Rightarrow$ $a(\lambda+1) = a + b$ and $b(\lambda+1) = \lambda(a+b)$

$\Rightarrow$ $\lambda = \dfrac{b}{a}$ and $b = \lambda a$, i.e., $\lambda = \dfrac{b}{a}$

$\therefore$ P divides AB in the constant ratio $b : a$.

5. Prove that the condition that $x\cos\alpha + y\sin\alpha = p$ should touch the curve $x^m y^n = a^{m+n}$ is $p^{m+n} \cdot m^m \cdot n^n = (m+n)^{m+n} \cdot a^{m+n} \cdot \sin^n\alpha \cdot \cos^m\alpha$.

Let the line $x\cos\alpha + y\sin\alpha = p$ ... (1)

touch the curve at (x_1, y_1).

$x^m y^n = a^{m+n}$ $\Rightarrow$ $m\log x + n\log y = (m+n)\log a$.

Differentiating w.r.t. x,

$\dfrac{m}{x} + \dfrac{n}{y}\dfrac{dy}{dx} = 0$; $\therefore$ $\dfrac{dy}{dx} = -\dfrac{my}{nx}$

$\therefore$ the equation of the tangent at (x_1, y_1) is

$y - y_1 = \left(\dfrac{dy}{dx}\right)_{x_1, y_1} \cdot (x - x_1)$

or $y - y_1 = \dfrac{-my_1}{nx_1}(x - x_1)$

$\therefore$ $nx_1y - nx_1y_1 = -my_1x + mx_1y_1$

or $my_1x + nx_1y = (m+n)x_1y_1$ ... (2)

(1) and (2) must be the same equation. So, comparing these

$\dfrac{\cos\alpha}{my_1} = \dfrac{\sin\alpha}{nx_1} = \dfrac{p}{(m+n)x_1y_1}$

$\therefore$ $\dfrac{\cos\alpha}{my_1} = \dfrac{p}{(m+n)x_1y_1}$, i.e., $x_1 = \dfrac{mp}{(m+n)\cos\alpha}$

and $\dfrac{\sin\alpha}{nx_1} = \dfrac{p}{(m+n)x_1y_1}$, i.e., $y_1 = \dfrac{np}{(m+n)\sin\alpha}$.

As (x_1, y_1) is on the curve, $x_1^m y_1^n = a^{m+n}$

or $\left\{\dfrac{mp}{(m+n)\cos\alpha}\right\}^m \cdot \left\{\dfrac{np}{(m+n)\sin\alpha}\right\}^n = a^{m+n}$

or $\dfrac{p^{m+n} \cdot m^m \cdot n^n}{(m+n)^{m+n} \cdot \cos^m\alpha \cdot \sin^n\alpha} = a^{m+n}$

$\therefore$ $p^{m+n}m^m n^n = (m+n)^{m+n}a^{m+n}\cos^m\alpha\sin^n\alpha$.

6. Show that the normal to the curve

$$5x^5 - 10x^3 + x + 2y + 6 = 0$$

at P(0, –3) meets the curve again at two points. Find the equations of the tangents to the curve at these points.

Here, $5x^5 - 10x^3 + x + 2y + 6 = 0$ ... (1)

Differentiating w.r.t. x,

$25x^4 - 30x^2 + 1 + 2\dfrac{dy}{dx} = 0$

$\therefore$ $\dfrac{dy}{dx} = -\dfrac{1}{2}(25x^4 - 30x^2 + 1)$; $\therefore$ $\left(\dfrac{dy}{dx}\right)_{0,-3} = -\dfrac{1}{2}$.

$\therefore$ the equation of the normal at P(0, –3) is

$y + 3 = \dfrac{-1}{\left(\frac{dy}{dx}\right)_{0,-3}} \cdot (x - 0)$, i.e., $y + 3 = 2x$ ... (2)

Solving (1) and (2),

$5x^5 - 10x^3 + x + 2(2x - 3) + 6 = 0$

or $5x^5 - 10x^3 + 5x = 0$; $\therefore$ $x(x^4 - 2x^2 + 1) = 0$

or $x(x^2 - 1)^2 = 0$; $\therefore$ $x = 0, 1, -1$

and so corresponding $y = -3, -1, -5$.

$\therefore$ the normal cuts the curve at (0, –3), (1, –1) and (–1, –5).

$\therefore$ the normal at (0, –3) cuts the curve again at the two points (1, –1) and (–1, –5).

The equation of the tangent at (1, –1) is

$y - (-1) = \left(\dfrac{dy}{dx}\right)_{1,-1} \cdot (x - 1)$

or $y + 1 = -\dfrac{1}{2}(25 - 30 + 1) \cdot (x - 1)$

i.e., $y + 1 = 2(x - 1)$; $\therefore$ $y + 3 = 2x$.

The equation of the tangent at (–1, –5) is

$$y-(-5)=\left(\frac{dy}{dx}\right)_{-1,\,-5}\cdot(x+1)$$

or $\quad y+5=-\frac{1}{2}(25-30+1)\cdot(x+1),$

i.e., $\quad y+5=2(x+1);\quad \therefore\quad y+3=2x.$

7. Find the equation of the straight line which is a tangent at one point and normal at another point to the curve $y=8t^3-1,\ x=4t^2+3$.

Let the tangent to the curve at P't_1', i.e., $(4t_1^2+3,\ 8t_1^3-1)$ be normal to the curve at Q't_2', i.e., $(4t_2^2+3,\ 8t_2^3-1)$.

The equation of the tangent at 't_1' is

$$y-(8t_1^3-1)=\left(\frac{dy}{dx}\right)_{t_1}\cdot\{x-(4t_1^2+3)\}$$

or $\quad y-(8t_1^3-1)=\left(\frac{dy/dt}{dx/dt}\right)_{t_1}\cdot\{x-(4t_1^2+3)\}$

or $\quad y-(8t_1^3-1)=\frac{24t_1^2}{8t_1}\cdot\{x-(4t_1^2+3)\}$

or $\quad y-(8t_1^3-1)=3t_1\{x-(4t_1^2+3)\}\qquad\ldots(1)$

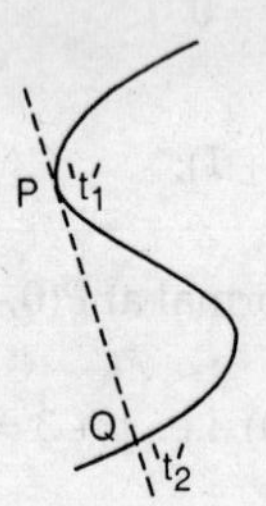

Clearly, slope of the tangent at 't_1'

= slope of the normal at 't_2'

and (1) passes through the point 't_2'.

$$\therefore\quad \left(\frac{dy}{dx}\right)_{t_1}=\frac{-1}{\left(\frac{dy}{dx}\right)_{t_2}},\ \text{i.e., } 3t_1=\frac{-1}{3t_2}\qquad\ldots(2)$$

and $\quad (8t_2^3-1)-(8t_1^3-1)=3t_1\{(4t_2^2+3)-(4t_1^2+3)\}$

$\Rightarrow\quad 8(t_2^3-t_1^3)=3t_1\cdot4(t_2^2-t_1^2)$

$\Rightarrow\quad 2(t_2^2+t_2t_1+t_1^2)=3t_1(t_2+t_1)$

$\Rightarrow\quad 2t_2^2=t_1t_2+t_1^2$, i.e., $2\cdot\left(\frac{-1}{9t_1}\right)^2=\frac{-1}{9}+t_1^2$ {using (2)}

$\Rightarrow\quad 2=-9t_1^2+81t_1^4$

$\therefore\quad 81t_1^4-9t_1^2-2=0$

or $\quad (9t_1^2+1)(9t_1^2-2)=0$

$\therefore\quad 9t_1^2-2=0;\quad \therefore\quad t_1=\pm\frac{\sqrt2}{3}$

∴ putting in (1), the equations of the required lines are

$$y-\left(8\cdot\frac{\pm2\sqrt2}{27}-1\right)=3\left(\pm\frac{\sqrt2}{3}\right)\left\{x-\left(4\cdot\frac{2}{9}+3\right)\right\}$$

or $\quad y\mp\frac{16\sqrt2}{27}+1=\pm\sqrt2\left(x-\frac{35}{9}\right)$

or $\quad 27(y+1)\mp16\sqrt2=\pm\sqrt2(27x-105).$

8. A ladder 15 m long, leans against a wall 7 m high, and a portion of the ladder protrudes over the wall such that its projection along the vertical is 3 m. How fast does the bottom start to slip away from the wall if the ladder slides down along the top edge of the wall at 2 m/s?

AB is the ladder of which the portion BC is protruded over the wall whose projection along the vertical is $CD=3$ m.

Let $OA=x$ m, $BC=y$ m and $\angle OCA=\theta$.

As $AB=15$ m, we get $AC=(15-y)$ m.

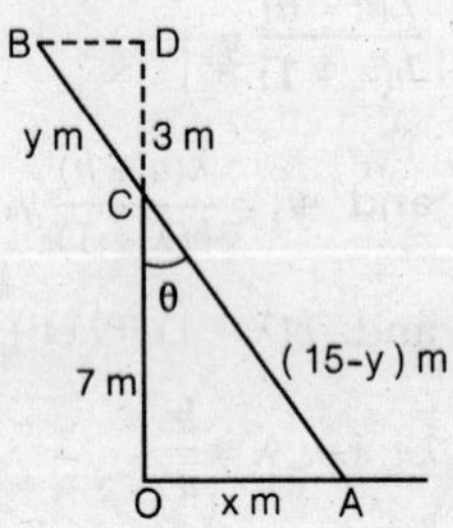

Also, from the question, $\frac{dy}{dt}=-2$ m/s

(as y decreases with time t).

We have to find $\frac{dx}{dt}$.

Now, from ΔOCA, $x=7\tan\theta\qquad\ldots(1)$

From ΔBCD, $y\cos\theta=3\qquad\ldots(2)$

Also, $7\sec\theta=AC=15-y$

$\therefore\quad y=15-7\sec\theta\qquad\ldots(3)$

Differentiating (3), $\frac{dy}{dt}=-7\sec\theta\tan\theta\,\frac{d\theta}{dt}$

or $\quad -2=-7\sec\theta\tan\theta\,\frac{d\theta}{dt}$

$\therefore\quad \frac{d\theta}{dt}=\frac{2}{7\sec\theta\tan\theta}.$

Differentiating (1), $\frac{dx}{dt}=7\sec^2\theta\,\frac{d\theta}{dt}$

$\therefore\quad \frac{dx}{dt}=7\sec^2\theta\cdot\frac{2}{7\sec\theta\cdot\tan\theta}=2\operatorname{cosec}\theta\qquad\ldots(4)$

From (2) and (3), $3\sec\theta=15-7\sec\theta$

$\therefore \quad \sec\theta = \frac{3}{2}; \quad \therefore \quad \cos\theta = \frac{2}{3}$

$\therefore \quad$ from (4), $\frac{dx}{dt} = 2 \cdot \frac{3}{\sqrt{3^2 - 2^2}} = \frac{6}{\sqrt{5}}$

$\therefore$ the bottom starts to slip away from the wall at the rate of $\frac{6}{\sqrt{5}}$ m/s.

9. A lamp-post, 4.5 m high, is 4.2 m away from a straight footpath. A boy, 1.5 m tall, walks along the path at the rate of 1.8 m/s. Find the rate at which his shadow is shortening when he has 14.4 m more to cover in order to arrive at the point on the path nearest to the lamp-post.

The height of the lamp-post = AB = 4.5 m;

the height of the man = PQ = 1.5 m;

the distance of the lamp-post from the path

$= BC = 4.2$ m.

Let $BQ = x$ m, $BD = y$ m and $QC = z$ m

$\therefore$ the length of the shadow = $QD = (y - x)$ m.

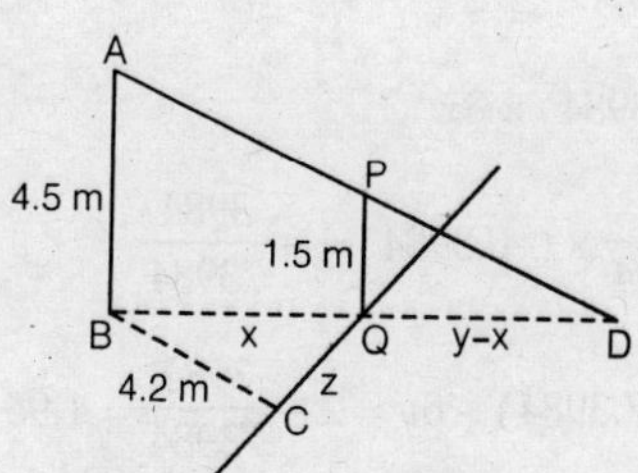

We have to find $\frac{d(y-x)}{dt}$ when $z = 14.4$ m.

From the question, $\frac{dz}{dt} = -1.8$ m/s

Now, from $\triangle$BCQ, $x^2 = z^2 + (4.2)^2$ ... (1)

As $\triangle ABD \sim \triangle PQD$, $\frac{4.5}{1.5} = \frac{BD}{QD}$, i.e., $3 = \frac{y}{y-x}$

or $\quad 3(y-x) = y = (y-x) + x; \quad \therefore \quad 2(y-x) = x$

$\therefore \quad 2\frac{d(y-x)}{dt} = \frac{dx}{dt}$ ... (2)

Differentiating (1), $2x\frac{dx}{dt} = 2z\frac{dz}{dt} = 2z(-1.8)$

$\therefore \quad \frac{dx}{dt} = \frac{-1.8z}{x} = -\frac{1.8z}{\sqrt{z^2 + (4.2)^2}}.$

When $z = 14.4$ m, $\frac{dx}{dt} = -\frac{1.8 \times 14.4}{\sqrt{(14.4)^2 + (4.2)^2}}$ m/s

$\therefore \quad$ from (2), $\frac{d(y-x)}{dx} = \frac{1}{2} \cdot \frac{-1.8 \times 14.4}{\sqrt{(14.4)^2 + (4.2)^2}}$ m/s

$= \frac{1}{2} \cdot \frac{-1.8 \times 14.4}{15}$ m/s

$= -0.06 \times 14.4$ m/s

$= -0.864$ m/s.

$\therefore$ his shadow is shortening at the rate of 0.864 m/s.

10. A man is standing on a straight bridge over a river and another man on a boat is on the river just below the man on the bridge. If the first man starts walking at the uniform speed of 4 m/min and the boat moves perpendicularly to the bridge at the speed of 5 m/min then at what rate are they separating after 4 minutes if the height of the bridge above the boat is 3 m?

In the beginning, the man is at A on the bridge and the boat is at B on the river. After t minutes, the man is at C and the boat is at D.

From the question, AB $\perp$ AC and BD $\perp$ AC.

$\therefore$ AC is perpendicular to the plane of AB and BD.

$\therefore$ AC is perpendicular to each line in the plane ABD.

$\therefore$ AC $\perp$ AD.

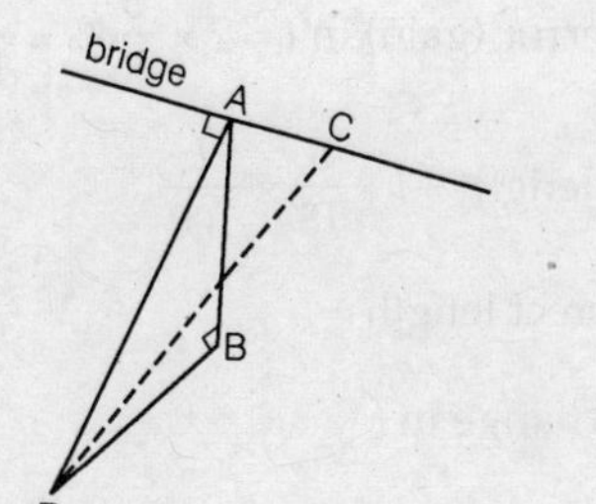

Also, $AC = 4t$ m and $BD = 5t$ m and $AB = 3$ m

$\therefore$ from the right-angled $\triangle$ABD,

$AD = \sqrt{AB^2 + BD^2} = \sqrt{3^2 + (5t)^2}$ m

and from the right-angled $\triangle$DAC,

$DC = \sqrt{AD^2 + AC^2}.$

If $DC = z$ m, $z = \sqrt{3^2 + (5t)^2 + (4t)^2} = \sqrt{9 + 41t^2}$

$\therefore \quad \frac{dz}{dt} = \frac{1}{2\sqrt{9 + 41t^2}} \times 41 \times 2t$

$\therefore$ at $t = 4$ min.,

$\frac{dz}{dt} = \frac{41 \times 2 \times 4}{2\sqrt{9 + 41 \times 4^2}}$ m/min $= \frac{164}{\sqrt{665}}$ m/min.

11. The time of a complete oscillation of a simple pendulum of length l is given by the relation $T = 2\pi \cdot \sqrt{\frac{l}{g}}$, where g is a constant. By what per cent should the length be changed in order to correct a loss of 2 minutes per day?

Here, $T = 2\pi\sqrt{\dfrac{l}{g}}$; $\therefore \quad \dfrac{dT}{dl} = \dfrac{2\pi}{\sqrt{g}} \cdot \dfrac{1}{2\sqrt{l}}$

$$\therefore \quad dT = \left(\frac{dT}{dl}\right) dl = \frac{\pi}{\sqrt{g}} \cdot \frac{1}{\sqrt{l}}\, dl$$

$$\therefore \quad \frac{dT}{T} = \frac{1}{2\pi\sqrt{l/g}} \cdot \frac{\pi}{\sqrt{g}} \cdot \frac{1}{\sqrt{l}}\, dl = \frac{1}{2l}\, dl$$

$$\therefore \quad \frac{dT}{T} \times 100 = \frac{1}{2} \cdot \frac{dl}{l} \times 100$$

or per cent error in $T = \dfrac{1}{2} \times$ per cent error in l ... (1)

Now, in per cent, 2 minutes in a day, i.e., 24 × 60 minutes

$$= \frac{2}{24 \times 60} \times 100\% = \frac{5}{36}\%$$

$\therefore$ loss of 2 minutes per day $= \dfrac{5}{36}\%$ error (gain) in T

$\therefore$ by (1), $\dfrac{5}{36}\% = \dfrac{1}{2} \times$ per cent error (gain) in l

$\therefore$ per cent error (gain) in $l = 2 \times \dfrac{5}{36}\% = \dfrac{5}{18}\%$

$\therefore$ changed length $= l + \dfrac{5}{18} \times \dfrac{l}{100}$

while the correct length $= l$

$\therefore$ per cent change in l

$$= \frac{\left(l + \dfrac{5}{18} \times \dfrac{l}{100}\right) - l}{l + \dfrac{5}{18} \times \dfrac{l}{100}} \times 100$$

$$= \frac{\dfrac{l}{360}}{\dfrac{361l}{360}} \times 100 = \frac{100}{361}$$

$\therefore$ the per cent change required in the length $= \dfrac{100}{361}\%$.

12. Without use of tables and any standard value of logarithm, find the value of $\log_e 7$.

Let $f(x) = \log_e x$.

We know $e = 2.72$ (nearly) and so $e^2 = 7.3984$

$\therefore$ when $x = 7.3984$, $f(e^2$, i.e., $7.3984) = \log_e e^2 = 2$

and change in x, i.e., $\delta x = -0.3984$

Now $f'(x) = \dfrac{1}{x}$

$$\therefore \quad \delta y = f'(7.3984) \times \delta x$$

$$\therefore \quad \delta y = \frac{1}{7.3984} \times (-0.3984) = -\frac{3984}{73984}$$

$$\therefore \quad \log_e 7 = f(7.3984) + \delta y = 2 - \frac{3984}{73984} = 1.95 \text{ (appr.)}$$

Exercises

1. Determine the quadratic curve $y = f(x)$ if it touches the line $y = x$ at the point $x = 1$ and passes through the point $(-1, 0)$.
2. Find the constants a, b and c so that the two curves $y = x^2 + ax + b$ and $y = cx - x^2$ will touch each other at the point $(1, 0)$.
3. Prove that the curves $2y^2 = x^3$ and $y^2 = 32x$ cut each other orthogonally at the origin.
4. Prove that the curves $xy = 4$ and $x^2 + y^2 = 8$ touch each other.
5. Find the tangents to the curve $y = (x^3 - 1)(x - 2)$ at the points where the curve cuts the x-axis.
6. Show that the line $\dfrac{x}{a} + \dfrac{y}{b} = 1$ touches the curve $y = be^{-x/a}$ at the point where the curve crosses the y-axis.
7. Find all tangents to the curve $y = \cos(x + y)$, $-2\pi \le x \le 2\pi$, that are parallel to the line $x + 2y = 0$.
8. Find the equations of tangents drawn to the curve $y^2 - 2x^3 - 4y + 8 = 0$ from the point $(1, 2)$.
9. If the tangent at (x_1, y_1) to the curve $x^3 + y^3 = a^3$ meets the curve again at (x_2, y_2) then prove that
$$\frac{x_2}{x_1} + \frac{y_2}{y_1} + 1 = 0.$$
10. Prove that the portion of the tangent to the curve $x^{2/3} + y^{2/3} = a^{2/3}$ which is intercepted between the axes, is of constant length.
11. Show that the sum of the intercepts made on the axes of coordinates by any tangent to the curve $\sqrt{x} + \sqrt{y} = \sqrt{a}$ is equal to a.

12. Prove that the tangent at any point on the rectangular hyperbola $xy = c^2$, makes a triangle of constant area with coordinate axes.

13. Prove that the segment of the tangent to the curve

$$y = \frac{a}{2}\log\frac{a+\sqrt{a^2-x^2}}{a-\sqrt{a^2-x^2}} - \sqrt{a^2-x^2}$$

contained between the point of contact and the y-axis has a constant length.

14. If α, β are the intercepts made on the axes by the tangent at any point of the curve $x = a\cos^3\theta$ and $y = b\sin^3\theta$, prove that $\frac{\alpha^2}{a^2} + \frac{\beta^2}{b^2} = 1$.

15. Show that the segment of the tangent to the hyperbola $y = \frac{a^2}{x}$ intercepted between the axes of coordinates is bisected at the point of contact.

16. Show that the curve $\left(\frac{x}{a}\right)^n + \left(\frac{y}{b}\right)^n = 2$ touches the straight line $\frac{x}{a} + \frac{y}{b} = 2$ at the point (a, b) whatever the values of n may be.

17. Find the condition that the line $x\cos\alpha + y\sin\alpha = p$ may touch the curve

$$\left(\frac{x-a}{a}\right)^{\frac{n}{n-1}} + \left(\frac{y-b}{b}\right)^{\frac{n}{n-1}} = 1.$$

18. Find the equation of the normal to the curve $y = (1+x)^y + \sin^{-1}(\sin^2 x)$ at the point $x = 0$.

19. Find the equation to the normal to the curve $x^{2/3} + y^{2/3} = a^{2/3}$ at the point where θ is the angle that the normal makes with the x-axis.

20. Find the equation to the normal to the parabola $x^2 = 4y$ which passes through the point $(1, 2)$.

21. Find points on the curve $9y^2 = x^3$ where the normal to the curve makes equal intercepts with the axes.

22. Show that the normal at any point of the curve

$$x = a(\cos t + t\sin t),\ y = a(\sin t - t\cos t)$$

is at a constant distance from the origin.

23. Show that for any point of the curve $x^2 - y^2 = a^2$ the segment of the normal from the point to the point of intersection of the normal with the x-axis is equal to the distance of the point from the origin.

24. Use calculus to find the condition that the line $x\cos\theta + y\sin\theta = p$ is a normal to the curve $b^2x^2 + a^2y^2 = a^2b^2$.

25. The volume of the cylindrical trunk of a tree is proportional to the cube of its diameter and the diameter increases uniformly from year to year, prove that the rate of increase in the volume when the diameter is equal to 90 cm is 25 times the rate when it is 18 cm.

26. A stick of length 100 cm rests against a vertical wall and the horizontal floor. If the foot of the stick slides with the constant velocity of 10 cm/s then find the magnitude of the velocity of the middle point of the stick when it is inclined at $\frac{\pi}{6}$ with the floor.

27. A man running along a circular track has the speed of 10 km per hour. A source of light is at the centre of the circular track. A wall is along the tangent to the circular track at the point from which he starts. What is the speed of the shadow of the man on the wall when he covers $\frac{1}{8}$th of the track? (Assume the track to be a line.)

28. A point is moving along the curve $y^3 = 27x$. Find interval of values of x in which the ordinate changes faster than the abscissa.

29. A particle is moving along the parabola $y^2 = 12x$ at the uniform rate of 10 cm/s. Find the components of velocity parallel to each of the axes when the particle is at the point $(3, 6)$.

30. A spherical balloon is being inflated so that its volume increases uniformly at the rate of 40 cm^3/min. How fast is its surface area increasing when the radius is 8 cm? Find how much approximately the radius will increase during the next $\frac{1}{2}$ minute.

31. The time T of a complete oscillation of a simple pendulum of length l is given by $T = 2\pi\sqrt{\frac{l}{g}}$ where g is a constant.

(a) Find the approximate error in the calculated value of T corresponding to an error of 2% in the value of l.

(b) Find the approximate percentage error in the calculated value of l corresponding to an error of 0.25% in the value of T.

32. A spherical iceball is melting, such that the radius is decreasing at a constant rate of 0.1 cm/s. Find the amount of water formed in one second when the radius of the sphere is 7 cm. (Given $\pi = \frac{22}{7}$, sp. gr. of ice = 0.9.)

33. Find the approximate value of $f(x) = \left(\frac{2-x}{2+x}\right)^{\frac{1}{5}}$ at $x = 0.15$.

34. Use differentials to obtain a reasonable approximation to $(8.01)^{4/3} + (8.01)^2$.

Objective Questions

Fill in the blanks.

35. Let C be the curve $y^3 - 3xy + 2 = 0$. If H is the set of points on the curve C where the tangent is parallel to the x-axis and V is the set of points on C where the tangent is parallel to the y-axis then $H =$ ______ , $V =$ ______ .

36. A spherical raindrop evaporates at a rate proportional to its surface area at any instant. The differential equation giving the rate of change of the radius of the raindrop is ______ .

37. The law of rectilinear motion of a body of mass 10 kg is given by $s = 2t^2 + 3t + 4$. Then the kinetic energy $\frac{1}{2}mv^2$ of the body after 5 s from the start is ______ .

38. The slope of the tangent to the curve $r^2 = a^2\cos 2\theta$, where $x = r\cos\theta$ and $y = r\sin\theta$, at the point $\theta = \frac{\pi}{6}$ is ______ .

39. A point moves along the curve $xy = 10$ such that its abscissa grows at the speed of 1 unit/s. The speed of the change of ordinate at (5, 2) is ______ .

40. A balloon is pumped at the rate of 10 cu cm/min. The rate of increase of its radius will be ______ when its radius is 15 cm.

Choose the correct option(s).

41. The line $ax + by + c = 0$ is a normal to the curve $xy = 1$. Then

(a) $a > 0, b > 0$ (b) $a > 0, b < 0$

(c) $a < 0, b > 0$ (d) $a < 0, b < 0$

42. The angle between the tangents to the curves $y = \sin x$ and $y = \cos x$ at a point of intersection is

(a) $\frac{\pi}{4}$ (b) $\tan^{-1}(2\sqrt{2})$

(c) $\tan^{-1}\left(\frac{1}{2\sqrt{2}}\right)$ (d) none of these

43. If the law of linear motion of a particle is given by $s = \frac{1}{3}t^3 - 16t$ then the acceleration at the time when the velocity vanishes, is

(a) 0 (b) 4

(c) 8 (d) none of these

Answers

1. $4y = (x + 1)^2$ **2.** $a = -3, b = 2, c = 1$

5. $3x + y = 3, 7x - y = 14$ **7.** $2y + x = \frac{\pi}{2}, 2y + x + \frac{3\pi}{2} = 0$

8. $y + 2\sqrt{3}x = 2(1 + \sqrt{3}),\ y - 2\sqrt{3}x = 2(1 - \sqrt{3})$

17. $a^n\cos^n\alpha + b^n\sin^n\alpha = (p - a\cos\alpha - b\sin\alpha)^n$

18. $x + y = 1$

19. $x\sin\theta - y\cos\theta + a\cos 2\theta = 0$

20. $x + y = 3$ **21.** $(0, 0), \left(4, \frac{8}{3}\right)$

24. $p^2(a^2\sin^2\theta + b^2\cos^2\theta) = (a^2 - b^2)^2\sin^2\theta\cos^2\theta$

26. 10 cm/s **27.** 20 km/h

28. (–1, 1) **29.** $5\sqrt{2}$ cm/s each

30. 10 cm^2/min, $\frac{5}{64\pi}$ cm **31.** (a) 1% (b) 0.5%

32. 55.44 cu cm **33.** 0.97

34. 80.1867

35. $\{(x, y) \mid y = 0, x \in R\}, \{(1, 1)\}$ respectively

36. $\frac{dr}{dt} + k = 0$ **37.** 2645

38. 0 **39.** reduces at $\frac{2}{5}$ unit/s

40. $\frac{1}{90\pi}$ cm/min **41.** (b), (c)

42. (b) **43.** (c)

Chapter Test

Time: 45 minutes

1. In the following, fill in the blanks so that the resulting sentences become true.

(a) The equation of the normal to the parabola $y = x^2 + x + 1$ which is perpendicular to the line joining the origin and the vertex of the parabola, is ______ .

(b) Points on the ellipse $\frac{x^2}{25} + \frac{9y^2}{400} = 1$ at which the abscissa increases at the same rate as the ordinate decreases, are ______ and ______ .

2. In each of the following, multiple options are given. Choose the correct option(s).

(a) A point is in motion on the curve $12y = x^3$. The ordinate is changing at a slower rate than the abscissa in the interval

A. $(-2, 2)$ B. $(-\infty, -2) \cup (2, +\infty)$ C. $(2, +\infty)$ D. none of these

(b) If the slope of the tangent to the curve $y = x^3$ at a point is equal to the ordinate of the point then the point is

A. $(0, 0)$ B. $(27, 3)$ C. $(3, 27)$ D. none of these

(c) Let $y = f(x)$ be an even function. If $f'(2) = -\sqrt{3}$ then the inclination of the tangent to the curve $y = f(x)$ at $x = -2$ with the positive direction of the x-axis is

A. $\frac{\pi}{6}$ B. $\frac{\pi}{3}$ C. $\frac{2\pi}{3}$ D. $\frac{5\pi}{6}$

3. Find the equation of the normal at any point to the curve

$$x = 3\cos\theta - \cos^3\theta,\ y = 3\sin\theta - \sin^3\theta.$$

Also show that at the point P where $\theta = \frac{\pi}{4}$, the normal passes through the origin.

4. Let $f(x)$ be a differentiable function and $f(x) > 0$ for all x. Prove that the curves $y = f(x)$ and $y = \sin kx \cdot f(x)$ touch each other at the points of intersection of the two curves.

5. Prove that the curve $y = e^{|x|}$ cannot have a unique tangent line at the point $x = 0$. Find the angle between the one-sided tangents to the curve at the point $x = 0$.

□

Answers

1. (a) $32x - 48y + 103 = 0$ (b) $\left(3, \frac{16}{3}\right)$, $\left(-3, \frac{-16}{3}\right)$

2. (a) A (b) A, C (c) B

3. $y - (3\sin\theta - \sin^3\theta) = \tan^3\theta \cdot \{x - (3\cos\theta - \cos^3\theta)\}$

5. $\frac{\pi}{2}$

□

6. Maxima and Minima

Recap of Facts and Formulae

1. Relative (local) maximum and minimum of functions of one variable

- A function $f(x)$ is said to have a local maximum (minimum) at $x = a$ if for $a - h < x < a + h$, $f(a)$ is the greatest (least) where h is an arbitrary positive number (which may be taken very small).

$\therefore \quad f(a) > f(x),\ x \in (a - h, a + h)$

$\Rightarrow \quad f(x)$ has a local maximum at $x = a$

and $f(a) < f(x),\ x \in (a - h, a + h)$

$\Rightarrow \quad f(x)$ has a local minimum at $x = a$.

Note

(i) A function may have local maximum (minimum) at more than one point.

(ii) The value of the function at a local maximum may be less than the value of the function at a local minimum.

2. Global maximum and minimum

- A function $f(x)$ is said to have a global maximum (minimum) at a point $x = a$ if $f(a)$ is the greatest (least) among the values of $f(x)$ at all the points of the interval of values of x.

3. Determination of points of local maximum or minimum

If the function $f(x)$ is differentiable in R then at the points of extremum, i.e., local maximum or minimum, the following will hold:

- $f'(x)$, i.e., $\dfrac{dy}{dx} = 0$
- If at $x = \alpha, f'(\alpha) = 0$ and

(i) $f''(\alpha) < 0$ then at $x = \alpha$, the function $f(x)$ has a local maximum and the maximum value is $f(\alpha)$.

(ii) $f''(\alpha) > 0$ then at $x = \alpha, f(x)$ has a local minimum and the minimum value is $f(\alpha)$.

Note If $f''(\alpha) = 0$, nothing can be concluded. In this case we have to take help of other methods.

- If at $x = \alpha, f'(\alpha) = 0$ and

(i) $f'(\alpha - \epsilon) > 0, f'(\alpha + \epsilon) < 0$ then $x = \alpha$ is a point of local maximum.

(ii) $f'(\alpha - \epsilon) < 0, f'(\alpha + \epsilon) > 0$ then $x = \alpha$ is a point of local minimum.

(iii) $f'(\alpha - \epsilon)$ and $f'(\alpha + \epsilon)$ both have the same sign then $x = \alpha$ is a point of neither maximum nor minimum.

TIP The above knowledge points give the method of finding maximum or minimum value of a function of one variable. If a student is required to find the maximum or minimum value of a function of two or more variables, where the variables are connected by one or more relations, then change the function into a function of one variable by using the relation(s), and then work with the resulting function of one variable.

4. Critical points, point of inflection

- The function $y = f(x)$ has a critical point at $x = \alpha$ if $f'(\alpha) = 0$ or $f'(\alpha)$ does not exist.
- A function can have maximum or minimum at critical points.
- The function $y = f(x)$ has a point of inflection at $x = \alpha$ if

(i) $f''(\alpha) = 0$ or does not exist and

(ii) $f'(\alpha)$ exists and $f''(\alpha - \epsilon), f''(\alpha + \epsilon)$ have opposite signs.

5. Maximum or minimum value of a function in a closed interval

The maximum or minimum value of a function in a closed interval is generally used in the global sense of the term.

For the function $y = f(x),\ x \in [a, b]$

- $\max f(x) =$ greatest among $f(a), f(b), f(\alpha)$

  $\min f(x) =$ least among $f(a), f(b), f(\alpha)$

  where α is a critical point of $f(x)$ over $[a, b]$.

6. Convexity and concavity of a curve

- If $f''(x) < 0,\ x \in (a, b)$ then the curve $y = f(x)$ is convex in (a, b).
- If $f''(x) > 0,\ x \in (a, b)$ then the curve $y = f(x)$ is concave in (a, b).

Selected Solved Examples

1. Determine the points of maxima and minima of the function $f(x) = \frac{1}{8}\log_e x - bx + x^2, x > 0$ where $b \geq 0$ is a constant.

Here, $f'(x) = \frac{1}{8x} - b + 2x$ ... (1)

For maximum or minimum, $f'(x) = 0$

or $\frac{1}{8x} - b + 2x = 0$

or $16x^2 - 8bx + 1 = 0$ ... (2)

$\therefore\ x = \frac{8b \pm \sqrt{64b^2 - 64}}{32} = \frac{b \pm \sqrt{b^2 - 1}}{4}$

If $b^2 - 1 < 0$, i.e., $0 \leq b < 1$ (given $b \geq 0$), the roots are imaginary. So, $f(x)$ will have no point of maxima or minima.

If $b^2 - 1 > 0$, i.e., $b > 1$ (given $b \geq 0$), the roots are real and unequal, and they are

$$\frac{b + \sqrt{b^2 - 1}}{4},\ \frac{b - \sqrt{b^2 - 1}}{4}$$

both of which are positive.

Differentiating (1), $f''(x) = -\frac{1}{8x^2} + 2$ ... (3)

At $x = \frac{b + \sqrt{b^2 - 1}}{4}$,

$$f''(x) = 2 - \frac{1}{8} \cdot \frac{16}{(b + \sqrt{b^2 - 1})^2}$$

$$\therefore\ f''(x) = 2 - 2(b - \sqrt{b^2 - 1})^2$$

$$= 2[1 - (b^2 + b^2 - 1 - 2b\sqrt{b^2 - 1})]$$

$$= 4[1 - b^2 + b\sqrt{b^2 - 1}]$$

$$= 4\sqrt{b^2 - 1}(b - \sqrt{b^2 - 1}) > 0$$

$\therefore\ f(x)$ has a minimum at $x = \frac{b + \sqrt{b^2 - 1}}{4}, b > 1$.

At $x = \frac{b - \sqrt{b^2 - 1}}{4}$,

$$f''(x) = 2 - \frac{1}{8} \cdot \frac{16}{(b - \sqrt{b^2 - 1})^2}$$

$$\therefore\ f''(x) = 2 - 2(b + \sqrt{b^2 - 1})^2$$

$$= 4[1 - b^2 - b\sqrt{b^2 - 1}]$$

$$= -4\sqrt{b^2 - 1}(b + \sqrt{b^2 - 1}) < 0$$

$\therefore\ f(x)$ has a maximum at $x = \frac{b - \sqrt{b^2 - 1}}{4}, b > 1$.

If $b^2 - 1 = 0$, i.e., $b = 1$ (given $b \geq 0$) then the roots of (2) are real and equal and it is $x = \frac{1}{4}$ (> 0).

$\therefore$ from (3), $f''\left(\frac{1}{4}\right) = -\frac{1}{8} \cdot 4^2 + 2 = 0$.

Now, we use $\in$-test for $x = \frac{1}{4}$.

$$f'\left(\frac{1}{4} + \in\right) = \frac{1}{8\left(\frac{1}{4} + \in\right)} - 1 + 2\left(\frac{1}{4} + \in\right)$$

{using (1) and $b = 1$}

$$= \frac{1}{2(1 + 4\in)} - 1 + \frac{1 + 4\in}{2}$$

$$= \frac{1 - 2(1 + 4\in) + (1 + 4\in)^2}{2(1 + 4\in)}$$

$$= \frac{16\in^2}{2(1 + 4\in)} > 0$$

$$f'\left(\frac{1}{4} - \in\right) = \frac{16(-\in)^2}{2(1 - 4\in)}$$

$$= \frac{16\in^2}{2 - 8\in} > 0$$

{$\because\ \in$ is a very small positive quantity}

$\therefore\ f(x)$ has neither a point of maximum nor a point of minimum at $x = \frac{1}{4}$.

Thus, if $b > 1, f(x)$ has

a point of maximum at $x = \frac{b - \sqrt{b^2 - 1}}{4}$ and a point of minimum at $x = \dfrac{b + \sqrt{b^2 - 1}}{4}$.

2. Let $f(x) = \sin^3 x + \lambda \sin^2 x$, $\dfrac{-\pi}{2} < x < \dfrac{\pi}{2}$.

Find the interval in which λ should lie so that $f(x)$ has exactly one minimum and exactly one maximum.

Here, $f'(x) = 3\sin^2 x \cdot \cos x + \lambda \cdot 2\sin x \cdot \cos x$

$$= \frac{3}{2}\sin x \cdot \sin 2x + \lambda \sin 2x \quad \dots (1)$$

For maximum or minimum, $f'(x) = 0$

or $\quad \sin x \cdot \cos x(3\sin x + 2\lambda) = 0$

$\therefore \quad x = 0$ or $\sin x_0 = \dfrac{-2\lambda}{3} \quad \left(\because x_0 \neq \pm\dfrac{\pi}{2}\right)$

Differentiating (1),

$$f''(x) = \cos 2x(3\sin x + 2\lambda) + \frac{1}{2}\sin 2x(3\cos x) \quad \dots (2)$$

When $x = 0$, $f''(0) = 2\lambda$.

When $\sin x_0 = -2\lambda/3$, x_0 will have real values provided

$-1 \le -2\lambda/3 \le 1$, i.e., $3/2 \ge \lambda \ge -3/2$.

If $\lambda = 0$, $\sin x_0 = 0$ which gives $x_0 = 0$. So, for $x = 0$ there will be only one point of extremum.

$\therefore \quad$ we have $-3/2 \le \lambda < 0$ or $0 < \lambda \le 3/2$

Using (2), $\sin x_0 = -2\lambda/3$ gives

$$f''(x_0) = (1 - 2\sin^2 x_0)(3\sin x_0 + 2\lambda) + 3\sin x_0(1 - \sin^2 x_0)$$

$$= \left(1 - 2\cdot\frac{4\lambda^2}{9}\right)\left(3\cdot\frac{-2\lambda}{3} + 2\lambda\right) + 3\,\frac{-2\lambda}{3}\left(1 - \frac{4\lambda^2}{9}\right)$$

$$= 0 - 2\lambda\cdot\frac{9 - 4\lambda^2}{9} = -\frac{2\lambda}{9}(9 - 4\lambda^2)$$

$$= -\frac{2\lambda}{9}(3 + 2\lambda)(3 - 2\lambda)$$

$\therefore \quad$ when $-3/2 \le \lambda < 0$,

$$f''(x_0) = (-\text{ve})(-\text{ve})(+\text{ve})(+\text{ve}) > 0 \quad \dots (3)$$

When $0 < \lambda \le 3/2$,

$$f''(x_0) = (-\text{ve})(+\text{ve})(+\text{ve})(+\text{ve}) < 0 \quad \dots (4)$$

Also, when $-3/2 \le \lambda < 0$, $f''(0) = 2\lambda < 0 \quad \dots (5)$

When $0 < \lambda \le 3/2$, $f''(0) = 2\lambda > 0 \quad \dots (6)$

$\therefore$ from (3) and (5),

$f(x)$ has exactly one minimum and one maximum

if $\quad -3/2 \le \lambda < 0$.

From (4) and (6),

$f(x)$ has exactly one maximum and one minimum

if $\quad 0 < \lambda \le 3/2$.

But, when $\lambda = \pm\dfrac{3}{2}$, $x_0 = \mp\dfrac{\pi}{2}$.

Hence, the required interval of values of λ is

$$(-3/2, 0) \cup (0, 3/2).$$

3. Find the polynomial function $f(x)$ of degree 6 which satisfies $\lim\limits_{x\to 0}\left\{1 + \dfrac{f(x)}{x^3}\right\}^{\frac{1}{x}} = e^2$ and has local maximum at $x = 1$ and local minima at $x = 0$ and 2.

Let $f(x) = a_0x^6 + a_1x^5 + a_2x^4 + a_3x^3 + a_4x^2 + a_5x + a_6$.

$$\therefore \quad \frac{f(x)}{x^3} = a_0x^3 + a_1x^2 + a_2x + a_3 + \frac{a_4}{x} + \frac{a_5}{x^2} + \frac{a_6}{x^3}$$

Now, $\lim\limits_{x\to 0}\left\{1 + \dfrac{f(x)}{x^3}\right\}^{\frac{1}{x}} = e^2$

$$\Rightarrow e^{\lim\limits_{x\to 0}\frac{1}{x}\log\left\{1 + \frac{f(x)}{x^3}\right\}} = e^2$$

$$\Rightarrow \lim_{x\to 0}\frac{\log\left\{1 + \frac{f(x)}{x^3}\right\}}{x} = 2$$

$$\Rightarrow \lim_{x\to 0}\frac{\log\left\{1 + a_0x^3 + a_1x^2 + a_2x + a_3 + \frac{a_4}{x} + \frac{a_5}{x^2} + \frac{a_6}{x^3}\right\}}{x} = 2.$$

The limit is ∞ unless $a_3 = 0, a_4 = 0, a_5 = 0, a_6 = 0$,

i.e., $f(x) = a_0x^6 + a_1x^5 + a_2x^4$.

In that case, $\lim\limits_{x\to 0}\dfrac{\log\{1 + (a_0x^3 + a_1x^2 + a_2x)\}}{x} = 2$

$$\text{or} \quad \lim_{x\to 0}\frac{\left\{(a_0x^3 + a_1x^2 + a_2x) - \frac{1}{2}(a_0x^3 + a_1x^2 + a_2x)^2 + \dots\right\}}{x} = 2$$

or $\quad a_2 = 2$

$\therefore \quad f(x) = a_0x^6 + a_1x^5 + 2x^4$

$$\therefore \quad f'(x) = 6a_0x^5 + 5a_1x^4 + 8x^3 \quad \dots (1)$$

From the question,

$f'(0) = 0$, $f'(2) = 0$ and $f'(1) = 0$.

Using (1), we get

$f'(0) = 0 \Rightarrow a_0$ and a_1 can have any real value

$f'(2) = 0 \Rightarrow 6 \times 32a_0 + 5 \times 16a_1 + 8 \times 8 = 0$

$\therefore \quad 12a_0 + 5a_1 + 4 = 0 \qquad \ldots (2)$

$f'(1) = 0 \quad \Rightarrow \quad 6a_0 + 5a_1 + 8 = 0 \qquad \ldots (3)$

$(2) - (3) \quad \Rightarrow \quad 6a_0 - 4 = 0$

$\therefore \quad a_0 = \frac{2}{3}$. So, $a_1 = \frac{-12}{5}$

$\therefore \quad f(x) = \frac{2}{3}x^6 - \frac{12}{5}x^5 + 2x^4.$

Now, verify that $f(x)$ has maximum at $x = 1$ and minima at $x = 0, 2$.

4. Find the point on the curve

$$4x^2 + a^2y^2 = 4a^2, 4 < a^2 < 8$$

that is farthest from the point (0, –2).

Here, the curve is $\frac{x^2}{a^2} + \frac{y^2}{4} = 1$.

Any point on it is ($a\cos\phi$, $2\sin\phi$).

Let the distance between (0, –2) and the point be z. Then

$$z^2 = (a\cos\phi - 0)^2 + (2\sin\phi + 2)^2$$

$$= a^2\cos^2\phi + 4\sin^2\phi + 4 + 8\sin\phi$$

$\therefore \quad \frac{d(z^2)}{d\phi} = -2a^2\cos\phi\sin\phi + 8\sin\phi\cos\phi + 8\cos\phi \quad \ldots (1)$

z^2 is maximum or minimum $\Rightarrow \quad \frac{d(z^2)}{d\phi} = 0$

or $\quad (8 - 2a^2)\cos\phi\sin\phi + 8\cos\phi = 0$

or $\quad \cos\phi\{(8 - 2a^2)\sin\phi + 8\} = 0$

$\Rightarrow \quad \cos\phi = 0$ or $\sin\phi = \frac{4}{a^2 - 4}$.

But $4 < a^2 < 8$.

So, $\sin\phi = \frac{4}{a^2 - 4} > 1$ which is impossible.

$\therefore \quad \frac{d(z^2)}{d\phi} = 0$ when $\cos\phi = 0$, i.e., $\phi = \frac{\pi}{2}$ or $\frac{3\pi}{2}$.

Differentiating (1),

$$\frac{d^2(z^2)}{d\phi^2} = (8 - 2a^2)(\cos^2\phi - \sin^2\phi) - 8\sin\phi$$

$\therefore \quad$ at $\phi = \frac{\pi}{2}$, $\frac{d^2(z^2)}{d\phi^2} = 2a^2 - 8 - 8$

$$= 2(a^2 - 8) < 0 \text{ for } 4 < a^2 < 8$$

$\therefore \quad z^2$ is the maximum when $\phi = \frac{\pi}{2}$.

Clearly, at $\phi = \frac{3\pi}{2}$, z^2 is the minimum.

$\therefore \quad z$ is the maximum when $\phi = \frac{\pi}{2}$,

i.e., the point is $\left(a\cos\frac{\pi}{2}, 2\sin\frac{\pi}{2}\right)$

$\therefore \quad$ the required point is (0, 2).

5. Let A(p^2, –p), B(q^2, q) and C(r^2, –r) be the vertices of a triangle ABC. A parallelogram AFDE is drawn with D, E and F on the line segments BC, CA and AB respectively. Using calculus, show that the maximum area of such a parallelogram is $\frac{1}{4}(p+q)(q+r)(p-r)$.

Clearly, DE ∥ = AF and ΔCDE ~ ΔCBA.

$\therefore \quad \frac{CE}{CA} = \frac{DE}{AB}$

$\therefore \quad \frac{b - y}{b} = \frac{x}{c} \qquad \ldots (1)$

where $DE = x$, $AE = y$.

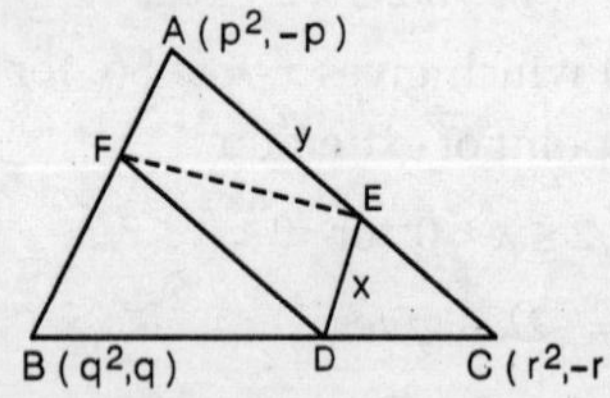

Now, ar(AFDE) = 2 × ar(ΔFAE)

$$= 2 \times \frac{1}{2}xy\sin A.$$

Denoting the area of the parallelogram by Δ,

$$\Delta = xy\sin A = \frac{c}{b}(b - y)y\sin A \qquad \ldots (1)$$

$\therefore \quad \frac{d\Delta}{dy} = \frac{c\sin A}{b}\{b - 2y\}$

$\therefore \quad \frac{d\Delta}{dy} = 0$ if $y = \frac{b}{2}$.

The question of minimum area does not arise (being 0).

$\therefore \quad \Delta$ is the maximum when $y = \frac{b}{2}$.

$\therefore \quad \Delta_{\max} = \frac{c}{b}\left(b - \frac{b}{2}\right)\frac{b}{2}\sin A$, from (1)

$$= \frac{1}{2}\cdot\frac{1}{2}bc\sin A = \frac{1}{2}\text{ar}(\Delta ABC)$$

$$= \frac{1}{2}\cdot\frac{1}{2}\begin{vmatrix} p^2 & -p & 1 \\ q^2 & q & 1 \\ r^2 & -r & 1 \end{vmatrix}$$

$$= \frac{1}{4}\begin{vmatrix} p^2 & -p & 1 \\ q^2-p^2 & q+p & 0 \\ r^2-q^2 & -r-q & 0 \end{vmatrix}, \begin{matrix} R_3 \to R_3 - R_2 \\ R_2 \to R_2 - R_1 \end{matrix}$$

$$= \frac{1}{4}\begin{vmatrix} (q-p)(q+p) & q+p \\ (r-q)(r+q) & -(r+q) \end{vmatrix}$$

$$= \frac{1}{4}(p+q)(q+r)\begin{vmatrix} q-p & 1 \\ r-q & -1 \end{vmatrix}$$

$$= \frac{1}{4}(p+q)(q+r)(-q+p-r+q)$$

$$= \frac{1}{4}(p+q)(q+r)(p-r).$$

6. What normal to the curve $y = x^2$ forms the shortest chord?

Any point on the curve is (t, t^2). The normal at (t, t^2) has the equation

$$y - t^2 = \frac{-1}{\left(\frac{dy}{dx}\right)_{t,t^2}}(x-t) \quad \dots (1)$$

But $y = x^2 \quad \Rightarrow \quad \frac{dy}{dx} = 2x.$

So, (1) becomes

$$y - t^2 = \frac{-1}{2t}(x-t)$$

or $\quad x + 2ty = t + 2t^3 \quad \dots (2)$

Solving (2) and $y = x^2$, we get

$$x + 2tx^2 = t + 2t^3$$

or $\quad 2tx^2 + x - t(1+2t^2) = 0.$

$$\therefore \quad x = \frac{-1 \pm \sqrt{1 + 8t^2(1+2t^2)}}{4t}$$

$$= \frac{-1 \pm (1+4t^2)}{4t}$$

$$= t, -\frac{1+2t^2}{2t}$$

$$\therefore \quad y = t^2, \left(-\frac{1+2t^2}{2t}\right)^2$$

$\therefore$ the normal chord is the line segment joining the points

$$(t, t^2) \text{ and } \left\{-\frac{1+2t^2}{2t}, \left(\frac{1+2t^2}{2t}\right)^2\right\}.$$

So, if z be the length of the chord,

$$z^2 = \left(t + \frac{1+2t^2}{2t}\right)^2 + \left\{t^2 - \left(\frac{1+2t^2}{2t}\right)^2\right\}^2$$

$$= \left(2t + \frac{1}{2t}\right)^2 + \left\{t^2 - \left(\frac{1}{2t} + t\right)^2\right\}^2$$

$$= 4t^2 + \frac{1}{4t^2} + 2 + \left(-\frac{1}{4t^2} - 1\right)^2$$

$$= 4t^2 + 3 + \frac{3}{4t^2} + \frac{1}{16t^4}.$$

$$\therefore \quad \frac{d(z^2)}{dt} = 8t - \frac{3}{2t^3} - \frac{1}{4t^5} \quad \dots (3)$$

For the maximum or minimum value of z^2,

$$\frac{d(z^2)}{dt} = 0$$

$$\Rightarrow \quad 8t - \frac{3}{2t^3} - \frac{1}{4t^5} = 0$$

$\therefore \quad 32t^6 - 6t^2 - 1 = 0$

or $\quad 4(2t^2)^3 - 3(2t^2) - 1 = 0$

$\therefore \quad (2t^2 - 1)(16t^4 + 8t^2 + 1) = 0$

or $\quad (2t^2 - 1)(4t^2 + 1)^2 = 0$

$\therefore \quad 2t^2 = 1 \quad \dots (4)$

From (3),

$$\frac{d^2(z^2)}{dt^2} = 8 + \frac{9}{2t^4} + \frac{5}{4t^6} > 0 \text{ when (4) holds.}$$

$\therefore \quad z^2$ is the minimum when

$$2t^2 = 1, \text{ i.e., } t = \pm\frac{1}{\sqrt{2}}.$$

$\therefore$ the normals at $\left(\pm\frac{1}{\sqrt{2}}, \frac{1}{2}\right)$ form the shortest chords.

7. A conical vessel is to be prepared out of a circular sheet of gold of unit radius. How much sectorial area is to be removed from the sheet so that the vessel has maximum volume?

Let the angle AOB of the sectorial area removed be α.

The arc AB $= \alpha \cdot 1 \quad \left(\text{using } \theta = \frac{l}{r}\right)$

$\therefore$ the length of the remaining arc ACB $= 2\pi \cdot 1 - \alpha$

$\therefore$ the circumference of the base of the cone

$$= \text{arc ACB} = 2\pi - \alpha$$

If the radius of the base of the cone be λ, then

$$2\pi\lambda = 2\pi - \alpha$$

$$\therefore \quad \lambda = \frac{2\pi - \alpha}{2\pi} = 1 - \frac{\alpha}{2\pi}.$$

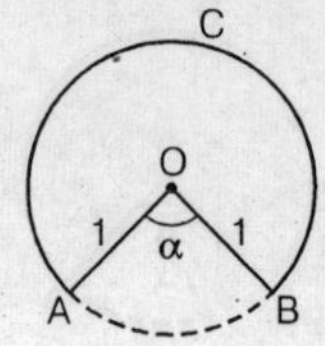

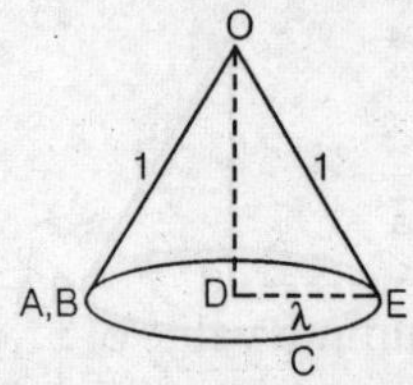

Let V be the volume of the vessel. Then

$$V = \frac{1}{3}\pi \cdot \lambda^2 \cdot OD$$

$$= \frac{1}{3}\pi \cdot \lambda^2 \cdot \sqrt{1-\lambda^2}$$

$$= \frac{\pi}{3}\cdot\left(1-\frac{\alpha}{2\pi}\right)^2 \cdot \sqrt{1-\left(1-\frac{\alpha}{2\pi}\right)^2}$$

$$= \frac{\pi}{3}\cdot\left(1-\frac{\alpha}{2\pi}\right)^2 \cdot \sqrt{\frac{\alpha}{\pi}-\frac{\alpha^2}{4\pi^2}}$$

$$\therefore \quad V = \frac{\pi}{3}\cdot\frac{1}{4\pi^2}\cdot\frac{1}{2\pi}(2\pi-\alpha)^2\sqrt{4\pi\alpha-\alpha^2}.$$

For maximum or minimum value of V, $\frac{dV}{dx} = 0$.

$$\frac{dV}{dx} = 0 \quad \Rightarrow$$

$$-2(2\pi-\alpha)\cdot\sqrt{4\pi\alpha-\alpha^2} + (2\pi-\alpha)^2\cdot\frac{1}{2\sqrt{4\pi\alpha-\alpha^2}}\cdot(4\pi-2\alpha) = 0$$

or $$(2\pi-\alpha)\frac{-2(4\pi\alpha-\alpha^2)+(2\pi-\alpha)^2}{\sqrt{4\pi\alpha-\alpha^2}} = 0$$

or $$(2\pi-\alpha)\{3\alpha^2-12\pi\alpha+4\pi^2\} = 0$$

$$\therefore \quad \alpha = 2\pi, \frac{12\pi \pm \sqrt{144\pi^2-48\pi^2}}{6}$$

$$= 2\pi, 2\pi \pm \frac{2\sqrt{2}}{\sqrt{3}}\pi.$$

Clearly, $\alpha = 2\pi + \frac{2\sqrt{2}}{\sqrt{3}}\pi$ is impossible because $0 \le \alpha \le 2\pi$.

Also, when $\alpha = 2\pi$, nothing will be left for the cone, i.e., the cone will have 0 volume (minimum) when $\alpha = 2\pi$.

$\therefore$ the volume will be maximum when

$$\alpha = 2\pi - \frac{2\sqrt{2}}{\sqrt{3}}\pi$$

$\therefore$ the required sectorial area $= \frac{\pi \cdot 1^2}{2\pi}\cdot\alpha$

$$= \frac{1}{2}\alpha = \frac{1}{2}\left(2\pi - \frac{2\sqrt{2}}{\sqrt{3}}\pi\right)$$

$$= \left(1-\frac{\sqrt{2}}{\sqrt{3}}\right)\pi = \frac{\sqrt{3}-\sqrt{2}}{\sqrt{3}}\pi.$$

8. Find the vertical angle of a right circular cone of minimum curved surface area that circumscribes a given sphere.

Let the radius of the given sphere be a.

Let the radius of the circumscribing cone be r and its semivertical angle be θ.

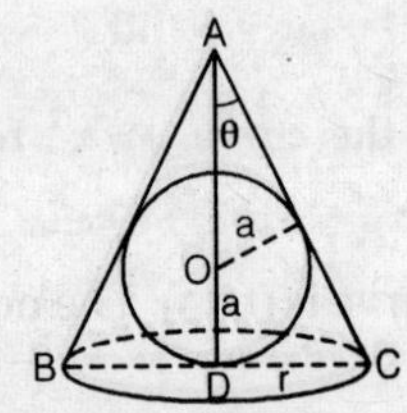

Then, $AD = r\cot\theta$ and $OA = a\,\text{cosec}\,\theta$

$\therefore \quad OD = a = AD - OA = r\cot\theta - a\,\text{cosec}\,\theta$

$\therefore \quad a(1+\text{cosec}\,\theta) = r\cot\theta \qquad \ldots (1)$

Let the curved surface area of the cone be S. Then

$$S = \pi r l = \pi r \cdot r\,\text{cosec}\,\theta$$

$$= \pi\,\text{cosec}\,\theta \cdot \left\{\frac{a(1+\text{cosec}\,\theta)}{\cot\theta}\right\}^2, \text{ using (1)}$$

$$= \pi a^2 \cdot \text{cosec}\,\theta \cdot \frac{(\text{cosec}\,\theta+1)^2}{\text{cosec}^2\theta - 1}$$

$$= \pi a^2 \cdot \text{cosec}\,\theta \cdot \frac{\text{cosec}\,\theta+1}{\text{cosec}\,\theta-1}$$

$$\therefore \quad S = \pi a^2 \frac{1+\sin\theta}{\sin\theta(1-\sin\theta)}$$

$$\therefore \quad \frac{dS}{d\theta} = \pi a^2 \cdot \frac{\left[\begin{array}{l}\cos\theta\sin\theta(1-\sin\theta) \\ \qquad -(1+\sin\theta)(\cos\theta - 2\sin\theta\cos\theta)\end{array}\right]}{\sin\theta(1-\sin\theta)}.$$

For maximum or minimum S, $\frac{dS}{d\theta} = 0$

or $$\cos\theta\sin\theta(1-\sin\theta) - (1+\sin\theta)\cos\theta(1-2\sin\theta) = 0$$

or $$\cos\theta\,[\sin\theta - \sin^2\theta - (1 - \sin\theta - 2\sin^2\theta)] = 0$$

or $$\cos\theta[\sin^2\theta + 2\sin\theta - 1] = 0$$

$$\therefore \quad \cos\theta = 0 \quad \text{or} \quad \sin\theta = \frac{-2\pm\sqrt{4+4}}{2} = -1\pm\sqrt{2}$$

$\therefore \quad \theta = \frac{\pi}{2}$ or $\sin^{-1}(\sqrt{2} - 1)$

$\{\because \sin\theta = -1 - \sqrt{2} < -1$ is absurd$\}$

Now, $\theta = \frac{\pi}{2} \Rightarrow$ A is on the sphere and in that case the radius of the cone is infinite indicating the maximum surface area $\propto$.

$\therefore$ S is minimum when $\theta = \sin^{-1}(\sqrt{2} - 1)$

$\therefore$ the required vertical angle $= 2\sin^{-1}(\sqrt{2} - 1)$.

9. Towns A and B are situated on the same side of a straight road at distances *a* and *b* respectively from it. Perpendiculars drawn from A and B meet the road at the points C and D respectively. The distance between C and D is *c*. A hospital is to be built at a point P on the road between C and D such that the distance APB is minimum. Find the position of P.

Let $CP = x$; then $PD = c - x$.

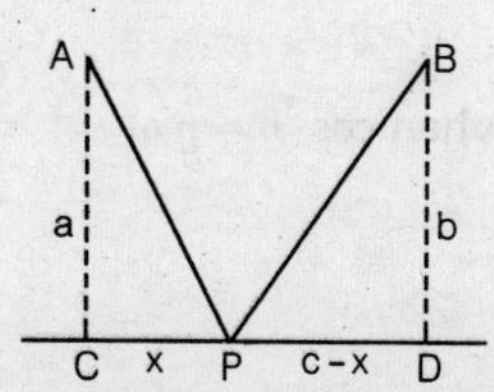

Let the distance $APB = z$. Then

$$z = AP + PB = \sqrt{a^2 + x^2} + \sqrt{b^2 + (c - x)^2}$$

$$\therefore \quad \frac{dz}{dx} = \frac{2x}{2\sqrt{a^2 + x^2}} - \frac{2(c - x)}{2\sqrt{b^2 + (c - x)^2}}$$

z is maximum or minimum $\Rightarrow \frac{dz}{dx} = 0$

$$\therefore \quad \frac{x}{\sqrt{a^2 + x^2}} = \frac{c - x}{\sqrt{b^2 + (c - x)^2}}$$

or $\quad x^2\{b^2 + (c - x)^2\} = (c - x)^2(a^2 + x^2)$

or $\quad b^2x^2 + x^2(c - x)^2 = a^2(c - x)^2 + x^2(c - x)^2$

or $\quad b^2x^2 - a^2(c - x)^2 = 0$

or $\quad \{bx + a(c - x)\}\{bx - a(c - x)\} = 0$

or $\quad \{(b - a)x + ac\}\{(a + b)x - ac\} = 0$

$\therefore \quad x = \frac{ac}{a - b}, \frac{ac}{a + b}$

If $\quad x = \frac{ac}{a - b}$,

$$c - x = c - \frac{ac}{a - b} = \frac{-bc}{a - b} = \frac{bc}{b - a}.$$

$\therefore$ either CP or PD is negative, i.e., P will be outside CD.

Hence z is minimum when $x = \frac{ac}{a + b}$, because z is greater for $x = 0$ (or c). This can be verified as follows.

When $x = \frac{ac}{a + b}$,

$$z = \sqrt{a^2 + \frac{a^2c^2}{(a + b)^2}} + \sqrt{b^2 + \left(c - \frac{ac}{a + b}\right)^2}$$

$$= \sqrt{a^2 + b^2 + c^2 + 2ab} \quad \text{(on simplification)}$$

When $x = 0$ (or c),

$$z = a + \sqrt{b^2 + c^2}$$

$$= \sqrt{a^2 + b^2 + c^2 + 2a\sqrt{b^2 + c^2}} > \sqrt{a^2 + b^2 + c^2 + 2ab}$$

Thus, z is minimum when

$$x = CP = \frac{ac}{a + b}, \quad c - x = PD = \frac{bc}{a + b}.$$

10. A window of fixed perimeter (including the base of the arch) is in the form of a rectangle surmounted by a semicircle. The semicircular portion is fitted with coloured glass while the rectangular part is fitted with clear glass. The clear glass transmits three times as much light per square metre as the coloured glass does. What is the ratio of the sides of the rectangle so that the window transmits maximum light?

Let the length of the rectangular portion = x m,

the breadth of the rectangular portion = kx m.

$\therefore$ the area of the rectangle $= kx^2$ sq m

the area of the semicircle $= \frac{1}{2}\pi \cdot \left(\frac{x}{2}\right)^2$ sq m.

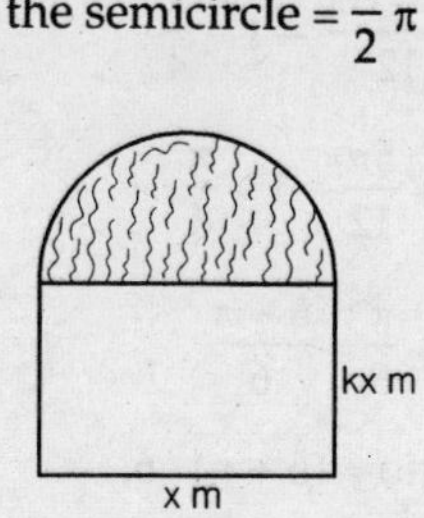

Let the light transmitted by coloured glass = a / sq m

$\therefore$ the amount of light transmitted = L

$$= \frac{1}{2}\pi \cdot \left(\frac{x}{2}\right)^2 a + kx^2 \cdot 3a.$$

Thus, $\quad L = \frac{\pi a}{8}x^2 + 3kax^2 \quad \dots (1)$

From the question,

$$2x + 2kx + \pi \cdot \frac{x}{2} = p = \text{constant}$$

$$\therefore \quad 2kx = p - 2x - \frac{\pi}{2}x$$

$$\therefore \quad kx = \frac{p}{2} - x - \frac{\pi}{4}x \qquad \ldots (2)$$

$$\therefore \quad \text{by (1)}, \; L = \frac{\pi a}{8}x^2 + 3a\left(\frac{p}{2}x - x^2 - \frac{\pi}{4}x^2\right)$$

$$= \left(\frac{\pi a}{8} - 3a - \frac{3a\pi}{4}\right)x^2 + \frac{3ap}{2}x$$

$$\therefore \quad \frac{dL}{dx} = 2\left(\frac{\pi a}{8} - 3a - \frac{3a\pi}{4}\right)x + \frac{3ap}{2}$$

$\therefore$ L is maximum or minimum if $\frac{dL}{dx} = 0$

$$\Rightarrow \quad 2\left(\frac{-5a\pi}{8} - 3a\right)x + \frac{3ap}{2} = 0$$

$$\text{or} \quad x = \frac{\frac{3ap}{2}}{2\left(3a + \frac{5a\pi}{8}\right)} = \frac{3p}{4} \cdot \frac{1}{3 + \frac{5\pi}{8}}$$

$$= \frac{3p}{12 + \frac{5\pi}{2}} = \frac{6p}{24 + 5\pi}.$$

The question of minimum light transmitted does not arise ($\because$ it is 0 when $x = 0$, $kx = \infty$).

$\therefore$ L is maximum when $x = \frac{6p}{24 + 5\pi}$.

$$\therefore \quad \text{from (2)}, \; k = \frac{p}{2x} - 1 - \frac{\pi}{4}$$

$$= \frac{p}{2} \cdot \frac{24 + 5\pi}{6p} - 1 - \frac{\pi}{4}$$

$$= \frac{24 + 5\pi}{12} - 1 - \frac{\pi}{4}$$

$$= 2 + \frac{5\pi}{12} - 1 - \frac{\pi}{4}$$

$$= 1 + \frac{\pi}{6} = \frac{6 + \pi}{6}$$

$\therefore$ the required ratio $= (6 + \pi) : 6$.

11. A lane runs at right angles out of a road a metres wide. Find how many metres wide the lane should be if it is just possible to carry a pole b metres long $(b > a)$ from the road into the lane keeping it horizontal.

Let AB be the pole and $AB = b$ m.

Let the inclination of the pole with the lane be θ when the pole is about to clear and the width of the lane $= w$ m.

Then $AB = BC + AC$

$$= w \operatorname{cosec} \theta + a \sec \theta$$

$$\therefore \quad b = w \operatorname{cosec} \theta + a \sec \theta$$

$$\therefore \quad w = b \sin \theta - a \tan \theta \qquad \ldots (1)$$

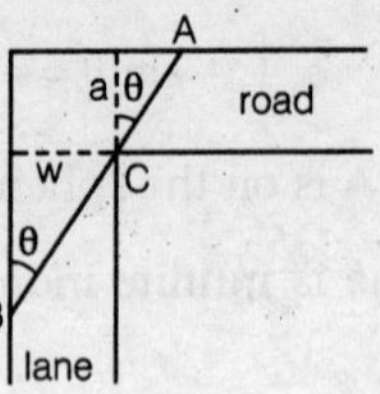

Here we have to find the largest w for $\theta \in \left(0, \frac{\pi}{2}\right)$ with ends of the pole on the lines as shown in the figure.

$$\text{Now,} \quad \frac{dw}{d\theta} = b \cos \theta - a \sec^2 \theta$$

$$\therefore \quad \frac{dw}{d\theta} = 0 \text{ if } b \cos \theta - a \sec^2 \theta = 0, \text{ i.e., } \cos^3 \theta = \frac{a}{b}$$

$$\text{and} \quad \frac{d^2w}{d\theta^2} = -b \sin \theta - 2a \sec^2 \theta \tan \theta.$$

$$\text{As} \quad 0 < \theta < \frac{\pi}{2}, \; \frac{d^2w}{d\theta^2} < 0$$

$\therefore$ w is maximum when $\cos^3 \theta = \frac{a}{b}$,

$$\text{i.e.,} \quad \cos \theta = \left(\frac{a}{b}\right)^{1/3}$$

$\therefore$ from (1), the maximum value of w

$$= b \cdot \sqrt{1 - \cos^2 \theta} - a \cdot \sqrt{\sec^2 \theta - 1}$$

$$= b\sqrt{1 - \left(\frac{a}{b}\right)^{2/3}} - a \cdot \sqrt{\left(\frac{b}{a}\right)^{2/3} - 1}$$

$$= b^{2/3}\sqrt{b^{2/3} - a^{2/3}} - a^{2/3}\sqrt{b^{2/3} - a^{2/3}}$$

$$= (b^{2/3} - a^{2/3})^{3/2}.$$

Note Here $b > a$. If $b \le a$ then whatever might be the width of the lane, the pole can be cleared. Also $\cos^3 \theta = \frac{a}{b}$ will give no θ other than 90° in which case the pole is cleared irrespective of the width of the lane.

12. A tyre company is able to manufacture x (hundred) grade A tyres and y (hundred) grade B tyres per day, where $y(5 - x) = 10(4 - x)$, $0 \le x \le 4$. If the profit on each grade A tyre is twice the profit on grade B tyre, what is the most profitable number of grade A tyres per day to manufacture?

Let the profit function be P per day. From the question,

$$P = 2xp + yp$$

where p is the profit/hundred of grade B

$$\therefore \quad P = 2xp + \frac{10(4 - x)}{5 - x}p = 2p\left\{x + \frac{20 - 5x}{5 - x}\right\}$$

$$\therefore \quad P = 2p \cdot \frac{20 - x^2}{5 - x}.$$

Now, $\frac{dP}{dx} = 0$

$$\Rightarrow \quad 2p \cdot \frac{-2x(5-x) - (20 - x^2)(-1)}{(5-x)^2} = 0$$

$$\Rightarrow \quad -10x + 2x^2 + 20 - x^2 = 0$$

$$\Rightarrow \quad x^2 - 10x + 20 = 0$$

$$\therefore \quad x = \frac{10 \pm \sqrt{100 - 80}}{2} = 5 \pm \sqrt{5}.$$

But $x \le 4$. So, $x = 5 - \sqrt{5}$.

The question of minimum profit does not arise at $x = 5 - \sqrt{5}$.

$\therefore$ P is maximum when $x = 5 - \sqrt{5} = 2.76$

$\therefore$ the most profitable number of grade A tyres $= 2.76$ hundreds $= 276$.

13. A man walks in a horizontal circle round the foot of a pole which is inclined to the vertical. The foot of the pole is at the centre of the circle. The greatest and the least angles which the pole subtends at his eyes are $\tan^{-1}\frac{9}{5}$ and $\tan^{-1}\frac{6}{5}$ respectively, and when he is midway between the corresponding positions, the angle is θ. If the man's height be neglected, find the length of the pole and θ if the radius of the circle is a, and the length of the pole > a.

Let O be the centre of the circle and the pole OA be inclined to the vertical OZ at an angle α. Let $OA = l$.

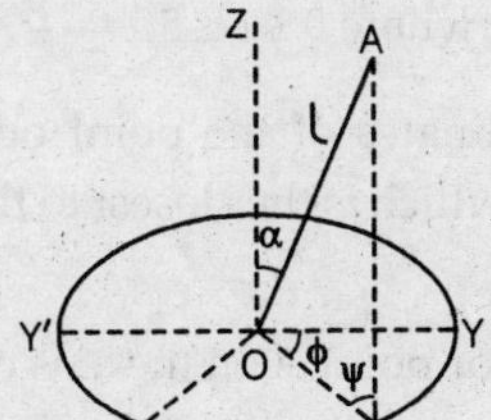

Let the plane of OZ and OA cut the circle at Y and Y′, Y being closer to A. OX ⊥ OY in the plane of the circle. Take OX, OY and OZ as the x, y and z axes.

P be any point on the circle. Let $\angle POY = \phi$ and $\angle OPA = \psi$.

Now, $A = (0, l\sin\alpha, l\cos\alpha)$ and $P = (a\sin\phi, a\cos\phi, 0)$ where the radius $= a$

$$\therefore \quad \vec{OP} = a\sin\phi\,\vec{i} + a\cos\phi\,\vec{j}$$

$$\vec{OA} = l\sin\alpha\,\vec{j} + l\cos\alpha\,\vec{k}$$

$$\vec{AP} = \vec{OP} - \vec{OA}$$

$$= a\sin\phi\,\vec{i} + (a\cos\phi - l\sin\alpha)\,\vec{j} - l\cos\alpha\,\vec{k}$$

$$\therefore \quad AP = |\vec{AP}|$$

$$= \sqrt{a^2\sin^2\phi + (a\cos\phi - l\sin\alpha)^2 + l^2\cos^2\alpha}$$

$$= \sqrt{a^2 + l^2 - 2al\cos\phi\cdot\sin\alpha}$$

$$\therefore \quad \cos\psi = \frac{OP^2 + AP^2 - OA^2}{2OP\cdot AP}$$

$$= \frac{a^2 + a^2 + l^2 - 2al\cos\phi\cdot\sin\alpha - l^2}{2a\sqrt{a^2 + l^2 - 2al\cos\phi\cdot\sin\alpha}}$$

$$= \frac{a - l\cos\phi\cdot\sin\alpha}{\sqrt{a^2 + l^2 - 2al\cos\phi\cdot\sin\alpha}}$$

$$\therefore \quad \tan\psi = \frac{\left\{a^2 + l^2 - 2al\cos\phi\sin\alpha - (a - l\cos\phi\cdot\sin\alpha)^2\right\}^{1/2}}{a - l\cos\phi\cdot\sin\alpha}$$

$$= \frac{\sqrt{l^2 - l^2\cos^2\phi\cdot\sin^2\alpha}}{a - l\cos\phi\cdot\sin\alpha} \qquad \ldots (1)$$

ψ and so $\tan\psi$ is maximum or minimum when

$$\frac{d(\tan\psi)}{d\phi} = 0$$

or $$\frac{1\cdot(2l^2\cos\phi\cdot\sin\phi\cdot\sin^2\alpha)}{2\sqrt{l^2 - l^2\cos^2\phi\cdot\sin^2\alpha}}\cdot(a - l\cos\phi\cdot\sin\alpha) - \sqrt{l^2 - l^2\cos^2\phi\cdot\sin^2\alpha}\,(l\sin\phi\sin\alpha) = 0$$

or $$l^2\cos\phi\cdot\sin\phi\sin^2\alpha\,(a - l\cos\phi\sin\alpha) - (l^2 - l^2\cos^2\phi\cdot\sin^2\alpha)l\sin\phi\sin\alpha = 0$$

or $$\cos\phi\cdot\sin\phi\sin\alpha\,(a - l\cos\phi\sin\alpha) - (1 - \cos^2\phi\cdot\sin^2\alpha)l\sin\phi = 0$$

or $$\sin\phi[a\cos\phi\sin\alpha - l\cos^2\phi\sin^2\alpha - l + l\cos^2\phi\cdot\sin^2\alpha] = 0$$

or $$\sin\phi[a\cos\phi\sin\alpha - l] = 0$$

$\therefore$ $\sin\phi = 0$, i.e., $\phi = 0$ or π

or $a\cos\phi\sin\alpha = l$, i.e., $\cos\phi = \frac{l}{a\sin\alpha}$ (not possible)

$\therefore$ ψ is maximum when $\phi = 0$ and it is minimum when $\phi = \pi$.

When $\phi = 0$, from (1),

$$\tan\psi = \frac{9}{5} = sise11\frac{\sqrt{l^2 - l^2\sin^2\alpha}}{a - l\sin\alpha} = \frac{l\cos\alpha}{a - l\sin\alpha}$$

$\therefore \quad 9a - 9l\sin\alpha = 5l\cos\alpha$

or $\quad 9a = l(5\cos\alpha + 9\sin\alpha) \qquad \dots (2)$

When $\phi = \pi$, from (1),

$$\tan\psi = \frac{6}{5} = \frac{\sqrt{l^2 - l^2\sin^2\alpha}}{a + l\sin\alpha} = \frac{l\cos\alpha}{a + l\sin\alpha}$$

$\therefore \quad 6a + 6l\sin\alpha = 5l\cos\alpha$

or $\quad 6a = l(5\cos\alpha - 6\sin\alpha) \qquad \dots (3)$

When the man is midway, $\phi = \frac{\pi}{2}$. So, from (1),

$$\tan\theta = \frac{l}{a} \qquad \dots (4)$$

From (2) and (3),

$$\frac{3}{2} = \frac{5\cos\alpha + 9\sin\alpha}{5\cos\alpha - 6\sin\alpha}$$

or $\quad 15\cos\alpha - 18\sin\alpha = 10\cos\alpha + 18\sin\alpha$

or $\quad 5\cos\alpha = 36\sin\alpha \quad \therefore \quad \tan\alpha = \frac{5}{36}.$

$$\therefore \quad \cos\alpha = \frac{36}{\sqrt{1321}},\ \sin\alpha = \frac{5}{\sqrt{1321}}$$

$$\therefore \quad (3) \Rightarrow \quad l = \frac{6a}{5\cos\alpha - 6\sin\alpha}$$

$$= \frac{6a}{\frac{5\times 36 - 6\times 5}{\sqrt{1321}}} = \frac{\sqrt{1321}}{25}a.$$

$$\therefore \quad \text{by (4), } \tan\theta = \frac{\sqrt{1321}}{25} \quad \therefore \quad \theta = \tan^{-1}\frac{\sqrt{1321}}{25}.$$

Exercises

Find the maximum or minimum value if there be any:

1. $\dfrac{40}{3x^4 + 8x^3 - 18x^2 + 60}$
2. $\dfrac{x}{1 + x\tan x}$
3. $2x - \tan^{-1}x - \log(x + \sqrt{1 + x^2})$
4. (a) $x(\log x)^2$ (b) $(2x^3 + 3x^2 - 36x)^{1/3}$
5. Show that $f(x) = \sqrt{e^{x^2} - 1}$ has its extremum at $x = 0$. Also mention whether it is a maximum or minimum.

Find the maximum and the minimum values of the following:

6. $x^2\log x,\ x \in [1, e]$
7. $\{(1 - x^2)(2x^2 + 1)\}^{1/2},\ x \in [-1, 1]$
8. If $xy(y - x) = 2a^3$, $(a > 0)$ then show that y has a minimum when $x = a$. Determine the minimum value. Also, show that y has a second value at $x = a$ which is less than the minimum. How do you explain this paradox?
9. Using examples, prove that the condition $\frac{dy}{dx} = 0$ is neither a necessary nor a sufficient condition for a function $y = f(x)$ to have a maximum or a minimum at a point.
10. Let $f(x) = x^3 - 12x$. Prove that the equation $x^3 - 12x - k = 0$ has a root -2 if $k = \max f(x)$ while it will have a root 2 if $k = \min f(x)$.
11. Find the value of a and b if $y = a\log x + bx^2 + x$ has its extreme values at $x = -1$ and $x = 2$.
12. If the function $y = \dfrac{px + q}{(x - 4)(x - 1)}$ has an extremum at P(2, −1), find p and q. Also, show that P is a point of maximum.
13. For what value of k does the function
$$y = x^3 - 3(7 - k)x^2 - 3(9 - k^2)x + 2,\ x > 0$$
have a point of maximum?
14. Find the shortest distance of the point $(0, c)$ from the parabola $y = x^2$ where $0 \le c \le 5$.
15. Find the coordinates of the point on the parabola $y = x^2 + 7x + 2$ which is the closest to the straight line $y = 3x - 3$.
16. Given an interior point A on the axis of the parabola $y^2 = 2px$, $(p > 0)$ at a distance a from its vertex. Find the abscissa of the point on the curve nearest to the given point.
17. Find the coordinates of the point on the curve $y = \dfrac{x}{1 + x^2}$ where the tangent to the curve has the greatest slope.
18. (a) The parametric equations of a curve are
$$x = 2t(t^2 + 3) - 3t^2,\ y = 2t(t^2 + 3) + 3t^2.$$
Find the maximum slope of the curve and the corresponding point on it.

(b) A function is defined parametrically as follows:

$$\left.\begin{aligned} x &= t^5 - 5t^3 - 20t + 7 \\ y &= 4t^3 - 3t^2 - 18t + 3 \end{aligned}\right\}, \ |t| < 2.$$

Find the maximum and the minimum of the function and mention the values of t where they are attained.

19. Let (h, k) be a fixed point, where $h > 0, k > 0$. A straight line passing through this point cuts the positive direction of the coordinate axes at points P and Q. Find the minimum area of the triangle OPQ, O being the origin.

20. A 12-cm-long wire is bent to form a triangle with one of the angles as 60°. Find the sides of the triangle if its area is the largest.

21. Find the largest area of the rectangle whose base is on the x-axis and two of its vertices lie on the curve $y = e^{-x^2}$.

22. Three sides of a trapezium are each equal to k cm. Find the greatest possible area of the trapezium.

23. A 4-m-high flagstaff stands erect on a vertical pillar, 5.6 m high. A man, 1.6 m tall, stands facing the pillar. How far from the pillar should the man stand so that he can see the flagstaff at the greatest angle?

24. One corner of a rectangular sheet of paper of width a cm is folded over so as to reach the opposite edge of the sheet. Find the minimum length of the crease.

25. In the printed page of a book, the combined width of the margin at the top and bottom is 9 cm and that on the sides is 6 cm. Find the dimensions of the page so that the area of the printed matter may be maximum if the total area of the page is 1536 sq cm.

26. A wire of length l is to be cut into two pieces, one of which is bent to form a circle and the other to form a square. How should the wire be cut if the sum of the areas enclosed by the two pieces is to be minimum? Can it be cut to make the above area a maximum?

27. A point P is given on the circumference of a circle of radius r. The chord QR is parallel to the tangent at P. Find the maximum area of ΔPQR.

28. A chord of length $2l$ divides a circular area of radius r into two segments. Find the sides of the rectangle with the largest area that can be inscribed in the smaller segment.

29. The circle $x^2 + y^2 = 1$ cuts the x-axis at P and Q. Another circle with centre at Q and radius variable intersects the first circle at R above the x-axis and the line segment PQ at S. Find the maximum area of ΔQSR.

30. Find the point (α, β) on the ellipse $4x^2 + 3y^2 = 12$, in the first quadrant, so that the area of the trapezium enclosed by the lines $y = x$, $y = \beta$, $x = \alpha$ and the x-axis is maximum.

31. Find the maximum area of an isosceles triangle inscribed in the ellipse $\frac{x^2}{a^2} + \frac{y^2}{b^2} = 1$ with its vertex at one end of the major axis.

32. Prove that the least intercept made on the tangents to the ellipse $\frac{x^2}{a^2} + \frac{y^2}{b^2} = 1$ by the axes is $a + b$. Also find the point of contact of the corresponding tangent.

33. Let PQ and P′Q′ be two double ordinates of the parabola $y^2 = 36ax$ such that PQ < P′Q′ and P′Q′ passes through the focus. Show that the area of trapezium made by the two double ordinates is maximum when the distance of PQ from the vertex is a.

34. Show that the semivertical angle of a right circular cone of given total surface area and maximum volume, is $\sin^{-1}\frac{1}{3}$.

35. Prove that a conical tent of given capacity will require the least amount of canvas when the height is $\sqrt{2}$ times the radius of the base.

36. It is desired to construct a cylindrical vessel of capacity 500 cu m, open at the top. What should be the dimensions of the vessel so that the material used is minimum, given that the thickness of the material used is 2 cm?

37. A manufacturer plans to construct a cylindrical can to hold one cu m of liquid. If the cost of constructing the top and bottom of the can is twice the cost of constructing the sides, what are the dimensions of the most economical can?

38. A square-based tank of capacity 250 cu m has to be dug out. The cost of land is Rs 50 per sq m. The cost of digging increases with the depth and for the whole tank the cost is Rs 400 × (depth)^2. Find the dimensions of the tank for the least total cost.

39. A box of maximum volume with top open is to be made out of a square tin sheet of sides 2 m in length by cutting out small squares from the four corners of the sheet. Find the height of the box.

40. A gutter is to be made out of a long sheet of metal, a cm wide, by turning up strips of width $\frac{a}{3}$ cm on each side so that they make equal angles θ with the vertical. For what value of θ will the containing capacity of the gutter be the greatest?

41. Find the altitude of the cylinder of maximum volume that can be inscribed in a given sphere.

42. Show that the cone of the greatest volume which can be inscribed in a given sphere has an altitude equal to 2/3 of the diameter of the sphere.

43. Prove that the cone, circumscribing a sphere of radius r, has the minimum volume if its altitude is $4r$ and its semivertical angle is $\sin^{-1}(1/3)$.

44. Show that the radius of the right circular cylinder of the greatest curved surface which can be inscribed in a given cone is half that of the cone.

45. A tree trunk, l long, is in the shape of a frustum of a cone, the radii of its ends being a and $b (a > b)$. It is required to cut from it a beam of uniform square section. Show that the beam of the greatest volume that can be cut is $\frac{al}{3(a-b)}$ long.

46. A metal block of volume V is to be cast into a half cylinder bounded by a rectangular base and two semicircular ends. Show that the half-cylinder so made will have the least total surface area if the ratio of the height of the cylinder to the diameter of the semicircular ends is $\pi : (\pi + 2)$.

47. The stiffness of a rectangular beam is proportional to the product of its breadth and the cube of its depth. Find the stiffest beam that can be cut from a cylindrical log of given diameter.

48. A swimmer S is in the sea at a distance d km from the closet point A on a straight shore. The house of the swimmer is on the shore at a distance L km from A. He can swim at a speed of U km/h and walk at a speed of V km/h $(V > U)$. At what point on the shore should he land so that he reaches his house in the shortest possible time?

49. A cable will be laid from a small island I, which is 3 km from the nearest point P on a straight shore, to a town T on the shore which is 12 km from P. If the cable costs Rs 4000 per km for laying under water and Rs 2000 per km for laying on the land, at what point R should the cable leave the water in order to give the minimum cost?

50. A firm has a branch store in each of the three cities A, B and C. A and B are 320 km apart and C is 200 km from each of them. A godown is to be built equidistant from A and B. In order to minimize the time of transportation, it should be so located that the sum of the distances from the godown to each of the cities is the minimum. Where should the godown be built?

51. Two townships are to get their water supply from a river. The townships are on the same side of the river and their reservoirs are 6 km and 18 km away from the nearest river bank which is straight. Two points on the river bank are 10 km apart which are nearest to the two reservoirs of the townships. Find the location of the single pumping station by the river bank supplying water to the townships so that the amount of pipe required will be the minimum.

52. A private telephone company serving a small community makes a profit of Rs 12 per subscriber, if it has 725 subscribers. It decides to reduce the rate by a fixed amount for each subscriber over 725, thereby reducing the profit by 1 paisa per subscriber. Thus there will be a profit of Rs 11.99 on each of the 726 subscribers, Rs 11.98 on each of the 727 subscribers, etc. What is the number of subscribers which will give the company the maximum profit?

53. A factory D is to be connected by a highway with a straight railway on which a town A is situated. The distance DB from the factory to the railway is equal to a and the segment AB of the railway is l long. Freight charges on the highway are $m(> 1)$ times higher than that on the railway. Locate the point P between A and B such that the highway DP will ensure the least freight charges from factory to the town.

Objective Questions

Fill in the blanks.

54. $f(x) = x^2 + ax^{-1}$ will have a minimum at $x = 2$ if $a =$ ______.

55. The maximum height of the curve $y = 6\cos x - 8\sin x$ above the x-axis is ______ .

56. The area of the rectangle of the greatest area that can be inscribed in the semicircle $y = \sqrt{a^2 - x^2}, y \geq 0$ is ______ .

57. Let x, y be such that $x > 0$ and $xy = 1$. Then the minimum value of $x + y$ is ______ .

58. If $x + y = 12$ then the maximum value of xy is ______ .

59. The sum of the squares of the roots of the equation $x^2 + (2 - a)x + 1 = a$ has the least value when $a =$ ______ .

60. 20 m of wire is available for fencing off a flower-bed in the shape of a sector of a circle. The bed will have the greatest area when the radius is ______ .

61. The total cost of making n pocket radios per day is Rs $\left\{\frac{n^2}{4} + 35n + 25\right\}$ and the rate at which they may be sold to a distributor is Rs $\left\{\frac{1}{2}(100 - n)\right\}$. For a maximum total profit, the daily output should be ______ .

62. A telephone company of a small town has 500 subscribers, from each of which the company charges Rs 300 per annum. The company decides upon a surcharge on annual subscription, knowing fully well that every increase of rupee 1 will force one subscriber to drop out. The company will have the maximum revenue if the surcharge is ______ .

Choose the correct option(s).

63. If $P(x) = a_0 + a_1x^2 + a_2x^4 + \ldots + a_nx^{2n}$ be a polynomial in $x \in R$ with $0 < a_1 < a_2 < \ldots < a_n$ then $P(x)$ has

(a) exactly one maximum

(b) exactly one minimum

(c) one maximum and one minimum

(d) none of these

64. The minimum radius vector of the curve $\frac{a^2}{x^2} + \frac{b^2}{y^2} = 1$ is of the length

(a) $|a - b|$ (b) ab

(c) $a + b$ (d) none of these

65. A right cylindrical vessel of a given capacity is formed using least possible material. Then the ratio of the height to the radius of the base is

(a) $\pi : 1$ (b) $\sqrt{\pi} : 1$ (c) $2 : 1$ (d) $1 : 1$

66. Let $f(x) = (x-1)^p \cdot (x-2)^q$ where $p > 1, q > 1$. Each critical point of $f(x)$ is a point of extremum when

(a) $p = 3, q = 4$ (b) $p = 4, q = 2$

(c) $p = 2, q = 3$ (d) $p = 2, q = 4$

67. The point $(0, 3)$ is closest to the curve $x^2 = 2y$ at

(a) $(2\sqrt{2}, 0)$ (b) $(0, 0)$

(c) $(2, 2)$ (d) none of these

68. For the function $f(x) = xe^x$, the point

(a) $x = -1$ is a minimum (b) $x = 0$ is a minimum

(c) $x = -1$ is a maximum (d) $x = 0$ is a maximum

69. If $f(x)$ be such that $f'(a) \neq 0$ then

(a) $f(x)$ cannot have maximum at $x = a$

(b) $f(x)$ cannot be minimum at $x = a$

(c) $f(x)$ must have neither a maximum nor a minimum at $x = a$

(d) none of these

70. N characters of information are held on a magnetic tape in batches of x characters each, the batch processing time being $\alpha + \beta x^2$ seconds where α, β are constants. The optimal value of x for fast processing is

(a) $\frac{\alpha}{\beta}$ (b) $\frac{\beta}{\alpha}$ (c) $\sqrt{\frac{\alpha}{\beta}}$ (d) $\sqrt{\frac{\beta}{\alpha}}$

State whether the statements are true or false.

71. The function $f(x) = x^2 + \frac{a}{x}$ cannot have a maximum at $x = -3$ for any real value of a.

72. A function $f(x)$ cannot have a minimum at a point where $f'(x) \neq 0$.

73. If $f(x)$ is a differentiable function and two of its points of maxima are $x = a, x = b, (a < b)$ then $f(x)$ has a point of minimum at $x = c$ where $a < c < b$.

Answers

1. max. value $= \frac{40}{53}, \frac{-8}{15}$; min. value $= \frac{2}{3}$

2. max. value $= \frac{\alpha}{1 + \sin \alpha}$ where $\cos \alpha = \alpha$;

min. value $= \frac{\beta}{1 - \sin \beta}$ where $\cos \beta = -\beta$

3. no maximum or minimum point

4. (a) max. value $= 4e^{-2}$, min. value $= 0$

(b) maximum value $= 3 \cdot \sqrt[3]{3}$,

minimum value $= -\sqrt[3]{44}$

5. minimum

6. max. value $= e^2$, min. value $= 0$

7. max. value $= \frac{3}{\sqrt{8}}$, min. value $= 0$

8. 2a **11.** $a = 2, b = -\frac{1}{2}$

12. $p = 1, q = 0$

13. $k < -3$ or $3 < k \le \frac{29}{7}$

14. $\sqrt{c - \frac{1}{4}}$ where $\frac{1}{2} \le c \le 5$

15. $(-2, -8)$

16. $a - p$ if $a > p$ and 0 if $a \le p$

17. $(0, 0)$

18. (a) 3, (5, 11) respectively

(b) maximum value $= 14$ at $t = -1$,

minimum value $= -17.25$ at $t = \frac{3}{2}$

19. $2hk$ **20.** 4 cm each

21. $\sqrt{\frac{2}{e}}$ **22.** $\frac{3\sqrt{3}}{4} k^2$ sq cm

23. $4\sqrt{2}$ m **24.** $\frac{3\sqrt{3}}{4} a$ cm

25. 48 cm × 32 cm

26. $\dfrac{\pi l}{4+\pi}, \dfrac{4l}{4+\pi}$; no

27. $\dfrac{3\sqrt{3}}{4} r^2$

28. $\dfrac{1}{4}\{\sqrt{9r^2 - l^2} - 3\sqrt{r^2 - l^2}\}$,

$\dfrac{1}{\sqrt{2}}\sqrt{3r^2 + l^2 - \sqrt{(r^2 - l^2)(9r^2 - l^2)}}$

29. $\dfrac{4}{3\sqrt{3}}$

30. $\left(\dfrac{3}{2}, 1\right)$

31. $\dfrac{3\sqrt{3}}{4} ab$

32. $\left(\dfrac{\pm a^{3/2}}{\sqrt{a+b}}, \dfrac{\pm b^{3/2}}{\sqrt{a+b}}\right)$

36. height = radius = $\sqrt[3]{\dfrac{500}{\pi}}$ m

37. radius = $\dfrac{1}{2}\left(\dfrac{2}{\pi}\right)^{1/3}$ m, height = $2\left(\dfrac{2}{\pi}\right)^{1/3}$ m

38. depth = 2.5 m, side = 10 m

39. $\dfrac{1}{3}$ m

40. $\dfrac{\pi}{6}$

41. $\dfrac{2}{\sqrt{3}} \times$ (radius)

47. depth = $\dfrac{\sqrt{3}}{2}$ (diameter), breadth = $\dfrac{\text{diameter}}{2}$

48. $\dfrac{Ud}{\sqrt{V^2 - U^2}}$ km from the point A towards his house

49. $\sqrt{3}$ km from P towards T

50. at a distance $\dfrac{320}{\sqrt{3}}$ km from A(or B) on the perpendicular bisector of AB

51. between the two nearest points, being $\dfrac{5}{2}$ km away from the point nearest to the first township

52. 962 or 963

53. $AP = l - \dfrac{a}{\sqrt{m^2 - 1}}$ if $\dfrac{a}{l} \le \sqrt{m^2 - 1}$

0 if $\dfrac{a}{l} > \sqrt{m^2 - 1}$,

because then freight charge is increasing in $[0, l]$

54. 16	**55.** 10	**56.** a^2	**57.** 2
58. 36	**59.** 1	**60.** 5 m	**61.** 10
62. Rs 100	**63.** (b)	**64.** (c)	**65.** (d)
66. (b), (d)	**67.** (c)	**68.** (a)	**69.** (d)
70. (c)	**71.** true	**72.** false	**73.** true

□

Chapter Test

Time: 60 minutes

1. In each of the following, fill in the blanks so that the resulting sentence becomes true.

(a) The curve $y = -x^3 + 3x^2 + 2x - 27$ has the maximum slope at the point ______ .

(b) If the area of ΔABC be maximum where A(1, 4) and B(3, 0) are fixed points on the ellipse $2x^2 + y^2 = 18$ and C is a variable point on the ellipse then C = ______ .

2. In each of the following, multiple options are given. Choose the correct option(s).

(a) $f(x) = \sum_{n=1}^{m} (x-n)^2$ has a minimum at x_0. Then x_0 is

(A) $\sqrt[m]{m!}$ (B) $\dfrac{1}{\dfrac{1 + \frac{1}{2} + \frac{1}{3} + \ldots + \frac{1}{m}}{m}}$

(C) $\dfrac{m+1}{2}$ (D) none of these

(b) If $y = a\log|x| + bx^2 + x$ has its extreme values at $x = -2$ and $x = -1$ then

(A) $a = -\frac{1}{2}, b = \frac{1}{4}$ (B) $a = -2, b = \frac{1}{4}$

(C) $a = 2, b = -\frac{1}{2}$ (D) $a = -2, b = -\frac{1}{2}$

3. Let $f(x)$ be a polynomial function with negative coefficients and $f(x) = f(-x)$ for all x. Prove that $f(x)$ has only one point of maximum.

4. P is a variable point on a continuous curve in a plane and A is a given point in the plane outside the curve. If the distance of P from the point A be a maximum or minimum then prove that the line AP will be perpendicular to the tangent to the curve at P.

5. A transport company runs long-distance buses. The total fixed cost on salaries, maintenance, etc. on each bus is Rs 300 per hour. The fuel charges are proportional to the square of the speed generated in km/h and it is Rs 48 per hour at a speed of 16 km/h. Find the most economical speed.

6. Two cars, moving with the same uniform speed of v km/h, are travelling along two straight roads crossing at 120°. At a certain moment one car is at the junction of the roads while the other car on the other road, travelling towards the junction, is at a distance a km from the junction. When will the distance between the cars be the least and what is the least distance?

□

Answers

1. (a) $(1, -23)$ (b) $(-\sqrt{6}, -\sqrt{6})$

2. (a) C (b) A

5. 40 km/h

6. $\frac{a}{2v}$ h, $\frac{\sqrt{3}a}{2}$ km

□

7. Monotonic Function and Lagrange's Theorem

Recap of Facts and Formulae

1. Increasing and decreasing functions at a point

- A function $f(x)$ is an *increasing function* at $x = a$ if $f'(a) > 0$.
  A function $f(x)$ is a *decreasing function* at $x = a$ if $f'(a) < 0$.

2. Monotonic (or monotone) function in an interval

- If a function $f(x)$, x belonging to an interval, be such that $f(x)$ goes on increasing as x increases in the interval then $f(x)$ is said to be monotone increasing or monotonically increasing or simply *monotonic increasing* (m.i.) in the interval.
  If $f'(x) > 0$ for $a < x < b$ then $f(x)$ is m.i. in (a, b).
  If $f'(x) > 0$ for $a \le x \le b$ then $f(x)$ is m.i. in $[a, b]$.
- If a function $f(x)$, x belonging to an interval, be such that $f(x)$ goes on decreasing as x increases in the interval then $f(x)$ is said to be *monotonic decreasing* (m.d.) in the interval.
  If $f'(x) < 0$ for $a < x < b$ then $f(x)$ is m.d. in (a, b).
  If $f'(x) < 0$ for $a \le x \le b$ then $f(x)$ is m.d. in $[a, b]$.
- $f(x)$ is m.i. in (a, b)
  $\Rightarrow \quad f(x_1) > f(x_2)$ if $x_1 > x_2$ in (a, b).
  $f(x)$ is m.d. in (a, b)
  $\Rightarrow \quad f(x_1) < f(x_2)$ if $x_1 > x_2$ in (a, b).

Note

(i) In the above, monotonicity is in the strict sense.
If $f'(x) \ge 0$ then also we say $f(x)$ is m.i.

(ii) If $f''(x) > 0$ in (a, b) then $f'(x)$ is m.i. in (a, b)
If $f''(x) < 0$ in (a, b) then $f'(x)$ is m.d. in (a, b)

(iii) The sketch of monotonic functions are as follows:

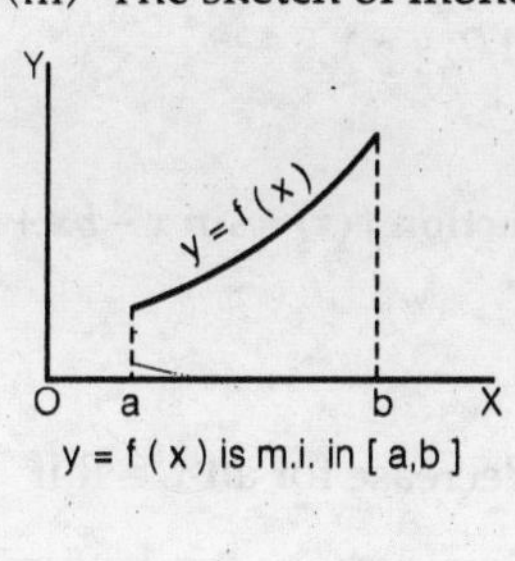

y = f (x) is m.i. in [a,b]

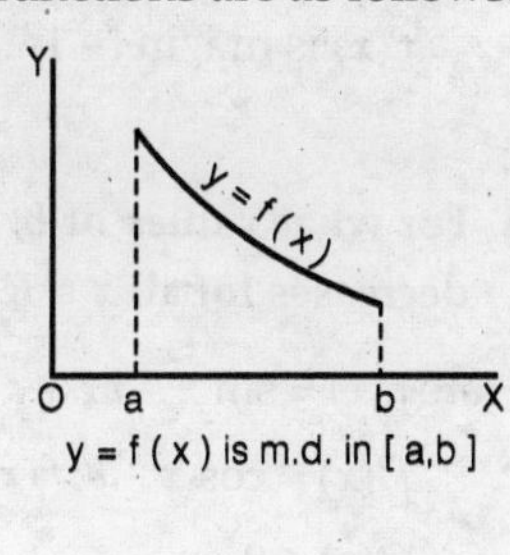

y = f (x) is m.d. in [a,b]

3. Greatest and least values of a monotonic function in a closed interval

- If $f(x)$ is m.i. in $[a, b]$ then in the interval,
  the greatest value of $f(x) = f(b)$
  the least value of $f(x) = f(a)$.
- If $f(x)$ is m.d. in $[a, b]$ then in the interval,
  the greatest value of $f(x) = f(a)$
  the least value of $f(x) = f(b)$.
- If $f(x)$ is m.i. in $[a, b]$ and m.d. in $[b, c]$ then in $[a, c]$,
  the greatest value of $f(x) = f(b)$
  the least value of $f(x) = \min \{f(a), f(c)\}$.
- If $f(x)$ is m.d. in $[a, b]$ and m.i. in $[b, c]$ then in $[a, c]$,
  the greatest value of $f(x) = \max \{f(a), f(c)\}$
  the least value of $f(x) = f(b)$.

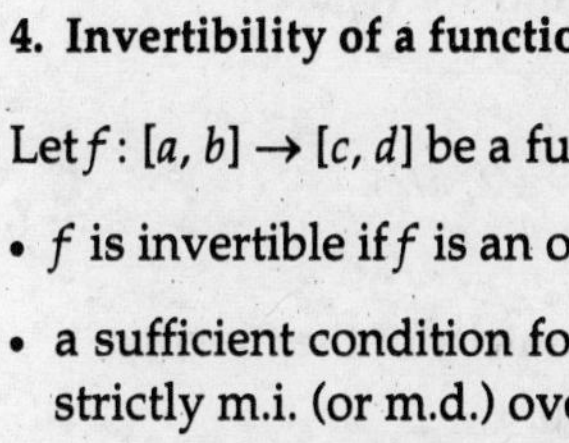

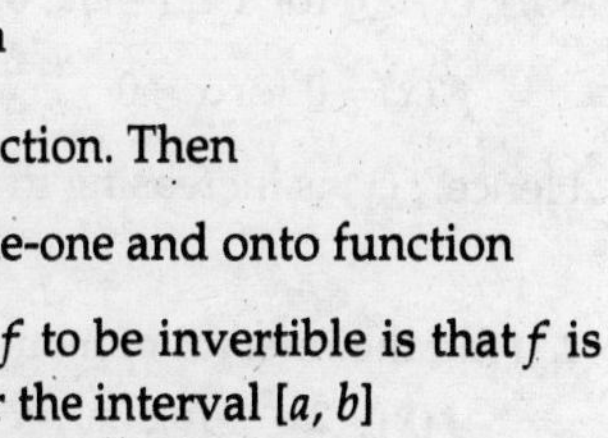

4. Invertibility of a function

Let $f : [a, b] \to [c, d]$ be a function. Then

- f is invertible if f is an one-one and onto function
- a sufficient condition for f to be invertible is that f is strictly m.i. (or m.d.) over the interval $[a, b]$

Note A function may be invertible though it is nonmonotonic.

For example $y = x$, x is rational
$\qquad\qquad -x$, x is irrational.

The inverse function is $x = y$, y is rational
$\qquad\qquad -y$, y is irrational.

5. Rolle's and Lagrange's theorems

If $f(x)$ be a continuous function in $[a, b]$ and $f(x)$ is differentiable in (a, b) then

- $f(a) = f(b) \quad \Rightarrow \quad f'(c) = 0$ where $a < c < b$ (*Rolle's Theorem*)
- $\dfrac{f(b) - f(a)}{b - a} = f'(c)$ where $a < c < b$ (*Lagrange's Theorem*)

6. Use of Rolle's and Lagrange's theorems in equations

- If the equation $f(x) = 0$ has two unequal real roots α and β $(\alpha < \beta)$ then $f'(x) = 0$ will have a real root γ such that $\alpha < \gamma < \beta$.
- If α, β are real such that $f(\alpha)$ and $f(\beta)$ are of opposite signs then $f(x) = 0$ has a real root lying between α and β.
- If $f(x) = 0$ has two real roots α, α (repeated root) then $f'(x) = 0$ will have a root α.

Selected Solved Examples

1. Prove that the function $f(x) = 2\sin x + \tan x - 3x$ is increasing in the interval $(-\pi/2, \pi/2)$.

Here, $f(x) = 2\sin x + \tan x - 3x$.

Clearly, the function $f(x)$ is defined at all points in $(-\pi/2, \pi/2)$.

Now, $f(x) = 2\cos x + \sec^2 x - 3$

$$= 2\cos x + \frac{1}{\cos^2 x} - 3$$

$$= \frac{2\cos^3 x - 3\cos^2 x + 1}{\cos^2 x}$$

$$= \frac{(\cos x - 1)(2\cos^2 x - \cos x - 1)}{\cos^2 x}$$

$$= \frac{(\cos x - 1)(2\cos x + 1)(\cos x - 1)}{\cos^2 x}$$

$$= \frac{(\cos x - 1)^2 \cdot (2\cos x + 1)}{\cos^2 x}.$$

Clearly, $2\cos x + 1 > 0$ for all $x \in (-\pi/2, \pi/2)$.

So, $f'(x) > 0$ for $x \in (-\pi/2, 0)$ and $x \in (0, \pi/2)$

$f'(x) = 0$ for $x = 0$

Hence, $f(x)$ is increasing in $(-\pi/2, \pi/2)$.

2. Find the interval of monotonicity of the function $f(x) = 2x^2 - \log|x|, x \neq 0$.

If $x < 0$, $f(x) = 2x^2 - \log(-x)$

$\therefore \quad f'(x) = 4x - \dfrac{1}{-x}(-1) = 4x - \dfrac{1}{x}$

$\therefore \quad f'(x) > 0 \quad \Rightarrow \quad 4x - \dfrac{1}{x} > 0$

or $\quad 4x^2 - 1 < 0 \quad (\because x < 0)$

or $\quad (2x + 1)(2x - 1) < 0$

$\therefore$ from sign-scheme

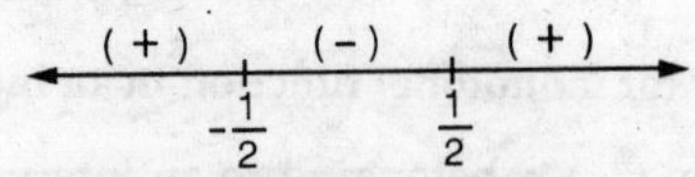

$\therefore \quad f'(x) > 0$ when $-1/2 < x < 1/2$.

But $x < 0$. So $-1/2 < x < 0$.

Similarly, $f'(x) < 0$ when $x < -1/2$ or $x > 1/2$.

But $x < 0$. So $x < -1/2$.

$\therefore$ $f(x)$ is monotonic increasing in $(-1/2, 0)$ and monotonic decreasing in $(-\infty, -1/2)$.

If $x > 0$,

$f'(x) > 0 \quad \Rightarrow \quad 4x^2 - 1 > 0$ (as above)

or $\quad (2x + 1)(2x - 1) > 0$

$\therefore$ from sign-scheme

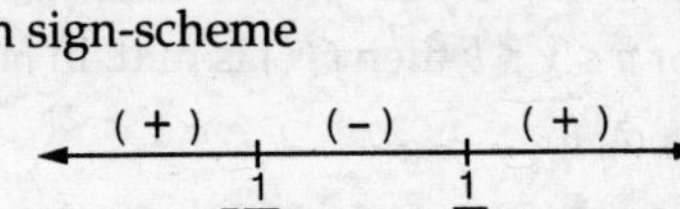

$\therefore \quad f'(x) > 0$ when $x < -1/2$ or $x > 1/2$.

But $x > 0$. So $x > 1/2$.

Similarly, $f'(x) < 0$ when $-1/2 < x < 1/2$.

But $x > 0$. So $0 < x < 1/2$.

$\therefore \quad f(x)$ is m.i. in $(1/2, +\infty)$ and m.d. in $(0, 1/2)$.

Thus, $f(x)$ is m.d. in $(-\infty, -1/2) \cup (0, 1/2)$

$f(x)$ is m.i. in $(-1/2, 0) \cup (1/2, +\infty)$.

3. For what values of b, the function $f(x) = \sin x - bx + c$ decreases for all $x \in R$?

Here $f(x) = \sin x - bx + c$

$\therefore \quad f'(x) = \cos x - b$; $f(x)$ will decrease for all $x \in R$ if $f'(x) \leq 0$

or $\cos x - b \le 0$, i.e., $\cos x \le b$ for all $x \in R$.

$\therefore$ $b \ge$ the greatest value of $\cos x = 1$

$\therefore$ $b \in [1, +\infty)$.

4. Let $f(x) = xe^{ax}$, $x \le 0$

$x + ax^2 - x^3$, $x > 0$

where a is a positive constant. Find the interval in which $f'(x)$ is increasing.

As x^n (where n is a non-negative integer) and e^{ax} are continuous and differentiable everywhere,

$f(x)$ is differentiable everywhere except perhaps at $x = 0$.

But $f'(0-0) = [e^{ax} + xae^{ax}]_{x=0} = 1$

$f'(0+0) = [1 + 2ax - 3x^2]_{x=0} = 1$

$\therefore$ we get $f'(0-0) = f'(0+0)$.

So, $f(x)$ is differentiable at $x = 0$ also.

$\therefore$ $f'(x) = e^{ax} + axe^{ax}$, $x \le 0$

$1 + 2ax - 3x^2$, $x > 0$.

We know that e^{ax} and polynomial functions are differentiable everywhere and the product and the sum of differentiable functions are differentiable.

So, $f'(x)$ is differentiable everywhere except perhaps at $x = 0$.

But $f''(0-0) = [ae^{ax} + ae^{ax} + a^2xe^{ax}]_{x=0} = 2a$

$f''(0+0) = [2a - 6x]_{x=0} = 2a$

$\therefore$ we get $f''(0-0) = f''(0+0)$.

So, $f'(x)$ is differentiable at $x = 0$ also.

$\therefore$ $f''(x) = 2ae^{ax} + a^2xe^{ax}$, $x \le 0$

$2a - 6x$, $x > 0$.

Now, $f'(x)$ is increasing if $f''(x) > 0$.

So, when $x \le 0$,

$2ae^{ax} + a^2xe^{ax} > 0$, i.e., $2 + ax > 0$ ($\because$ $a > 0$, $e^{ax} > 0$)

$\therefore$ $x \le 0$ as well as $x > -2/a$

$\therefore$ $-2/a < x \le 0$.

When $x > 0$,

$2a - 6x > 0$, i.e., $x < a/3$

$\therefore$ $x > 0$ as well as $x < a/3$

$\therefore$ $0 < x < a/3$

$\therefore$ $f'(x)$ increases in $(-2/a, 0] \cup (0, a/3)$, i.e., $(-2/a, a/3)$.

5. Prove that $x > \sin x > x - \dfrac{x^3}{6}$ for all $x > 0$.

Let $f(x) = x - \sin x$

$\therefore$ $f'(x) = 1 - \cos x > 0$ for all $x > 0$, $x \ne 2n\pi$, $n \in N$

$\therefore$ $f(x)$ is m.i.

$\therefore$ $x > 0 \Rightarrow f(x) > f(0) = 0$

for all $x > 0$, $x \ne 2n\pi$, $n \in N$.

But, for $x = 2n\pi$, $f(x) = 2n\pi - \sin 2n\pi > 0$

$\therefore$ $f(x) > 0$ for all $x > 0$

$\therefore$ $x - \sin x > 0$

or $x > \sin x$ for all $x > 0$ ... (1)

Next, let $\phi(x) = \sin x - x + \dfrac{x^3}{6}$ ... (2)

$\therefore$ $\phi'(x) = \cos x - 1 + \dfrac{1}{2}x^2$ ... (3)

$\therefore$ $\phi''(x) = -\sin x + x > 0$ for all $x > 0$, using (1)

$\therefore$ $\phi'(x)$ is m.i.

$\therefore$ $x > 0 \Rightarrow \phi'(x) > \phi'(0) = 0$, using (3)

$\therefore$ $\phi(x)$ is m.i. for all $x > 0$

$\therefore$ $x > 0 \Rightarrow \phi(x) > \phi(0) = 0$, using (2)

$\therefore$ $\sin x - x + \dfrac{x^3}{6} > 0$ for all $x > 0$

or $\sin x > x - \dfrac{x^3}{6}$ for all $x > 0$.

Hence, combining with (1),

$x > \sin x > x - \dfrac{x^3}{6}$ for all $x > 0$.

6. Prove that $\tan x > x + \dfrac{x^3}{3}$ for all $x \in \left(0, \dfrac{\pi}{2}\right)$.

Let $f(x) = \tan x - x - \dfrac{x^3}{3}$ ... (1)

Clearly, $f(x)$ is defined at all $x \in (0, \pi/2)$.

Now, $f'(x) = \sec^2 x - 1 - x^2$ ... (2)

$f''(x) = 2\sec^2 x \cdot \tan x - 2x$ ... (3)

$f'''(x) = 2\sec^4 x + 4\sec^2 x \cdot \tan^2 x - 2$

$= 2(1 + \tan^2 x)^2 + 4\sec^2 x \cdot \tan^2 x - 2$

$= 2\tan^4 x + 4\tan^2 x + 4\sec^2 x \cdot \tan^2 x > 0$

for all $x \in (0, \pi/2)$.

$\therefore$ $f'''(x) > 0$ in the interval $(0, \pi/2)$

$\therefore$ $f''(x)$ is monotonic increasing in $(0, \pi/2)$

$\therefore \quad f''(x) > f''(0)$ when $x \in (0, \pi/2)$.

But from (3), $f''(0) = 0$. Thus, $f''(x) > 0$ for all $x \in (0, \pi/2)$.

$\therefore \quad f'(x)$ is monotonic increasing in $(0, \pi/2)$

$\therefore \quad f'(x) > f'(0)$ when $x \in (0, \pi/2)$.

But from (2), $f'(0) = 1 - 1 - 0 = 0$.

Thus, $f'(x) > 0$ for all $x \in (0, \pi/2)$

$\therefore \quad f(x)$ is monotonic increasing in $(0, \pi/2)$

$\therefore \quad f(x) > f(0)$ when $x \in (0, \pi/2)$.

But from (1), $f(0) = 0$.

Thus, $f(x) > 0$ for all $x \in (0, \pi/2)$

$\therefore \quad \tan x - x - \frac{x^3}{3} > 0$ for all $x \in (0, \pi/2)$

or $\quad \tan x > x + \frac{x^3}{3}$ for all $x \in (0, \pi/2)$.

7. Prove that $\frac{\tan \beta}{\tan \alpha} > \frac{\alpha}{\beta}$ for $0 < \alpha < \beta < \frac{\pi}{2}$.

As α and β are positive, we are required to prove that $\beta \tan \beta > \alpha \tan \alpha$. Clearly, two sides are values of the $x \tan x$ at the points $x = \alpha, \beta$.

Consider the function, $f(x) = x \tan x$.

$\therefore \quad f'(x) = \tan x + x \sec^2 x > 0$ for all x in $(0, \pi/2)$

$\therefore \quad f(x)$ is m.i. in the interval $(0, \pi/2)$

$\therefore \quad \alpha < \beta \quad \Rightarrow \quad f(\alpha) < f(\beta)$

$\Rightarrow \quad \alpha \tan \alpha < \beta \tan \beta$

$\Rightarrow \quad \frac{\alpha}{\beta} < \frac{\tan \beta}{\tan \alpha}$ for $\alpha, \beta \in \left(0, \frac{\pi}{2}\right)$

Hence, $\frac{\tan \beta}{\tan \alpha} > \frac{\alpha}{\beta}$ for $0 < \alpha < \beta < \frac{\pi}{2}$.

8. Prove that the function $f(x) = -2x^3 + 21x^2 - 60x + 41$ is strictly positive in the interval $(-\infty, 1)$.

$f(x) = -2x^3 + 21x^2 - 60x + 41$

$\therefore \quad f'(x) = -6x^2 + 42x - 60$

$= -6(x^2 - 7x + 10) = -6(x - 5)(x - 2)$.

The sign-scheme for $x^2 - 7x + 10$, $x \in R$ is as follows:

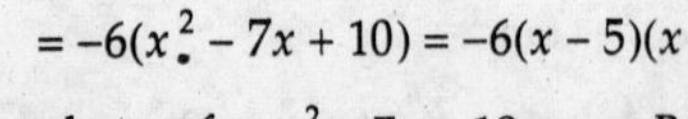

$\therefore \quad x \in (2, 5) \quad \Rightarrow \quad f'(x) > 0$, i.e., $f(x)$ is m.i.

and $x \notin (2, 5) \quad \Rightarrow \quad f'(x) < 0$, i.e., $f(x)$ is m.d.

$\therefore \quad x \in (-\infty, 1) \quad \Rightarrow \quad f(x)$ is m.d.

When $x \in (-\infty, 1)$, $x < 1$; so, $f(x) > f(1)$.

But $f(1) = -2 + 21 - 60 + 41 = 0$.

$\therefore \quad x \in (-\infty, 1) \quad \Rightarrow \quad f(x) > f(1) = 0$

$\therefore \quad f(x)$ is strictly positive in the interval $(-\infty, 1)$.

9. Find the intervals of monotonicity of

$$f(x) = \text{arc} \tan x - \frac{1}{2} \ln x, x \neq 0.$$

Also, find the greatest and the least values of $f(x)$ in the interval $\left[\frac{1}{\sqrt{3}}, \sqrt{3}\right]$.

Here, $f(x) = \tan^{-1} x - \frac{1}{2} \log_e x, x \neq 0$

$\therefore \quad f'(x) = \frac{1}{1 + x^2} - \frac{1}{2x}$

$= \frac{2x - (1 + x^2)}{2x(1 + x^2)} = -\frac{(1 - x)^2}{2x(1 + x^2)}$

$\therefore \quad x > 0 \quad \Rightarrow \quad f'(x) < 0$, i.e., $f(x)$ is m.d.

$x < 0 \quad \Rightarrow \quad f'(x) > 0$, i.e., $f(x)$ is m.i.

$\therefore \quad f(x)$ is m.d. in $(0, +\infty)$ and m.i. in $(-\infty, 0)$.

Clearly, $\left[\frac{1}{\sqrt{3}}, \sqrt{3}\right] \subset (0, +\infty)$

$\therefore \quad f(x)$ is m.d. in the closed interval $\left[\frac{1}{\sqrt{3}}, \sqrt{3}\right]$.

$\therefore \quad$ the greatest value of $f(x)$ in the interval $= f\left(\frac{1}{\sqrt{3}}\right)$

and the least value of $f(x)$ in the interval $= f(\sqrt{3})$

$\therefore \quad$ the greatest value of $f(x)$

$= f\left(\frac{1}{\sqrt{3}}\right) = \tan^{-1} \frac{1}{\sqrt{3}} - \frac{1}{2} \log \frac{1}{\sqrt{3}}$

$= \frac{\pi}{6} + \frac{1}{4} \log 3$

and, the least value of $f(x)$

$= f(\sqrt{3}) = \tan^{-1} \sqrt{3} - \frac{1}{2} \log \sqrt{3}$

$= \frac{\pi}{3} - \frac{1}{4} \log 3$.

10. Find the greatest and the least values of the function

$$f(x) = \left| \frac{x}{2} + 1 \right| - \left| \frac{x}{2} - 1 \right| + 1.$$

Hence, find the range of $f(x)$.

$f(x) = \frac{1}{2} \,|x + 2|\, - \frac{1}{2} \,|x - 2|\, + 1$.

When $x < -2$,

$$f(x) = \frac{1}{2}\{-(x+2)\} - \frac{1}{2}\{-(x-2)\} + 1$$

$$= -\frac{x}{2} - 1 + \frac{x}{2} - 1 + 1 = -1.$$

When $-2 \le x \le 2$,

$$f(x) = \frac{1}{2}(x+2) - \frac{1}{2}\{-(x-2)\} + 1$$

$$= \frac{x}{2} + 1 + \frac{x}{2} - 1 + 1 = x + 1.$$

When $x > 2$,

$$f(x) = \frac{1}{2}(x+2) - \frac{1}{2}(x-2) + 1 = 3.$$

Thus, the function $f(x)$ is defined as follows:

$$f(x) = \begin{cases} -1, & x < -2 \\ x+1, & -2 \le x \le 2 \\ 3, & x > 2. \end{cases}$$

In $-2 \le x \le 2, f(x) = x + 1$;

so, $f'(x) = 1 > 0$

$\therefore$ $f(x)$ is m.i. in $[-2, 2]$.

Also $f(x) = -1$ in $(-\infty, -2)$ and $f(x) = 3$ in $(2, +\infty)$

$\therefore$ the greatest value = maximum $\{-1, f(2), 3\}$

$= \max \{-1, 3, 3\} = 3$

and the least value = minimum $\{-1, f(-2), 3\}$

$= \min \{-1, -1, 3\} = -1.$

As polynomial functions are continuous everywhere, here $f(x)$ is continuous in $(-\infty, -2), (-2, 2)$ and $(2, +\infty)$.

By the test of continuity, clearly $f(x)$ is continuous at $x = -2, 2$.

Hence, $f(x)$ is continuous in R.

$\therefore$ range of $f(x)$ = [least $f(x)$, greatest $f(x)$]

$= [-1, 3].$

11. Find the maximum and the minimum values of $f(x) = x(1-x)^2, 0 \le x \le 2$.

As $f(x)$ is a polynomial function and every polynomial function is continuous and differentiable everywhere, $f(x)$ is continuous and differentiable in $0 \le x \le 2$.

We know that the maximum (or minimum) value of $f(x)$ in a closed interval = the greatest (or least) among the values at the critical points and the values at end points of definition.

Now, $f'(x) = (1-x)^2 + x \cdot 2(1-x)(-1)$

$= (1-x)(1-3x).$

At critical points, $f'(x) = 0$

or $(1-x)(1-3x) = 0 \quad \Rightarrow \quad x = 1, 1/3.$

Now, $f(0) = 0$,

$$f\left(\frac{1}{3}\right) = \frac{1}{3}\left(1 - \frac{1}{3}\right)^2 = \frac{1}{3} \cdot \frac{4}{9} = \frac{4}{27}$$

$f(1) = 0$

$f(2) = 2 \cdot (1-2)^2 = 2.$

$\therefore$ the maximum value of $f(x) = 2$

and the minimum value of $f(x) = 0$.

Second Method

As above, $f'(x) = (1-x)(1-3x)$.

sign-scheme for $(1-x)(1-3x), x \in R$:

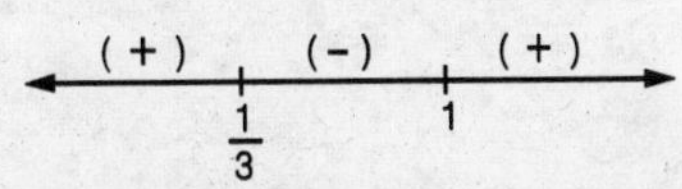

$\therefore$ in $(0, 1/3)$ and $(1, 2)$ the function $f(x)$ is m.i., and in $(1/3, 1)$ the function is m.d.

But the function is continuous throughout.

$\therefore$ the maximum value of $f(x)$

$$= \max\left\{f\left(\frac{1}{3}\right), f(2)\right\}$$

$$= \max\left\{\frac{4}{27}, 2\right\} = 2$$

and the minimum value of $f(x)$

$= \min \{f(0), f(1)\}$

$= \min \{0, 0\} = 0.$

12. Let $f(x) = -x^3 + \dfrac{b^3 - b^2 + b - 1}{b^2 + 3b + 2}, 0 \le x < 1,$

$2x - 3, 1 \le x \le 3.$

Find all possible values of b such that $f(x)$ has the smallest value at $x = 1$.

In $0 \le x < 1, f'(x) = -3x^2 \le 0$

$\therefore$ $f(x)$ is m.d. in $0 \le x < 1$.

In $1 \le x \le 3, f'(x) = 2 > 0$

$\therefore$ $f(x)$ is m.i. in $1 \le x \le 3$

$\therefore$ $f(1)$ will be the smallest value if $f(1-0) \ge f(1)$

or $$-(1-0)^3 + \frac{b^3 - b^2 + b - 1}{b^2 + 3b + 2} \ge 2 \cdot 1 - 3$$

or $$-1 + \frac{b^3 - b^2 + b - 1}{b^2 + 3b + 2} \ge -1$$

or $\frac{b^3 - b^2 + b - 1}{b^2 + 3b + 2} \geq 0$

or $(b^2 + 1)(b - 1)(b^2 + 3b + 2) \geq 0$

{multiplying by $(b^2 + 3b + 2)^2$}

or $(b^2 + 1)(b - 1)(b + 1)(b + 2) \geq 0$

$\therefore$ $(b - 1)(b + 1)(b + 2) \geq 0$, because $b^2 + 1 > 0$

$\therefore$ by sign-scheme:

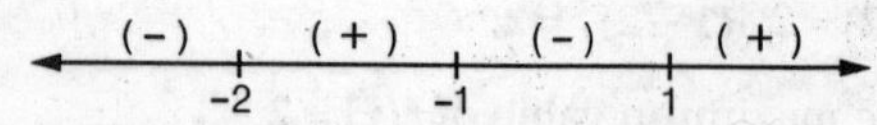

because at $b = 0$, $(b - 1)(b + 1)(b + 2)$ is negative.

$\therefore$ $-2 \leq b \leq -1$ or $b \geq 1$.

But $b^2 + 3b + 2 \neq 0$; so $b \neq -1, -2$

$\therefore$ $-2 < b < -1$ or $b \geq 1$.

So, $b \in (-2, -1) \cup [1, +\infty)$.

13. Find the range of values of a for which the function $f(x) = x^3 + (a + 2)x^2 + 3ax + 5$ is monotonic in R. Hence, find the set of values of a for which $f(x)$ is invertible.

Here, $f(x) = x^3 + (a + 2)x^2 + 3ax + 5$

$\therefore$ $f'(x) = 3x^2 + 2(a + 2)x + 3a$

For monotonic function in R,

$f'(x) > 0$ for all x or $f'(x) < 0$ for all x.

As $3x^2 + 2(a + 2)x + 3a$ is quadratic in x and coefficient of $x^2 = 3 > 0$, it cannot be always negative.

$\therefore$ $f'(x) > 0$ for all x.

For this, the conditions is $D < 0$,

i.e., $4(a + 2)^2 - 4 \cdot 3 \cdot 3a < 0$

or $a^2 - 5a + 4 < 0$

or $(a - 1)(a - 4) < 0$.

By sign-scheme:

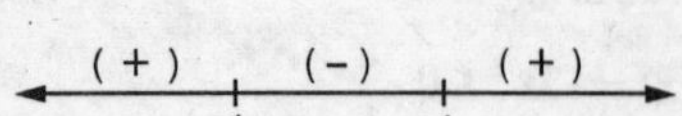

$\therefore$ $1 < a < 4$.

Thus $f(x)$ is m.i. in R for $a \in (1, 4)$.

We know that a monotonic increasing (or monotonic decreasing) function is necessarily invertible. So for $1 < a < 4$, the function is invertible.

When $a = 1$, $y = f(x) = x^3 + 3x^2 + 3x + 5 = (x + 1)^3 + 4$

$\therefore$ $(x + 1)^3 = y - 4$

$\therefore$ $x = -1 + \sqrt[3]{y - 4}$, which is defined for all $y \in R$.

When $a = 4$, $y = f(x) = x^3 + 6x^2 + 12x + 5$

$= (x + 2)^3 - 3$

$\therefore$ $(x + 2)^3 = y + 3$

$\therefore$ $x = -2 + \sqrt[3]{y + 3}$ **which is defined for all $y \in R$.**

Hence, $f(x)$ **is invertible for $a \in [1, 4]$.**

14. Prove that the equation $x + e^x = 0$ has only one real root.

Let $f(x) \equiv x + e^x = 0$.

Clearly $f(-\infty) = -\infty + e^{-\infty} = -\infty$

$f(+\infty) = \infty + e^{\infty} = \infty$

$\therefore$ $f(x) = 0$ has a real root (lying between $-\infty$ and $+\infty$).

Let the real root be α. Then $f(\alpha) = 0$.

Now, $f'(x) = 1 + e^x > 0$ for all real x;

so, $f(x)$ is m.i. in R.

$\therefore$ for any other real number β,

$f(\beta) > f(\alpha)$ or $f(\beta) < f(\alpha)$.

But $f(\alpha) = 0$; so, $f(\beta) \neq 0$

$\therefore$ $f(x) = 0$ has no other real root.

Hence, the equation has only one real root.

15. If $f(x)$ and $g(x)$ are continuous functions in $[a, b]$ and they are differentiable in (a, b) then prove that

$$\begin{vmatrix} f(a) & f(b) \\ g(a) & g(b) \end{vmatrix} = (b - a) \begin{vmatrix} f(a) & f'(c) \\ g(a) & g'(c) \end{vmatrix}$$

where $a < c < b$.

Consider the function $\phi(x) = f(a)\,g(x) - g(a)\,f(x)$ ... (1)

Clearly, from the question we get $\phi(x)$ is continuous in $[a, b]$ and differentiable in (a, b)

$\therefore$ by Lagrange's theorem,

$$\frac{\phi(b) - \phi(a)}{b - a} = \phi'(c) \text{ where } a < c < b. \quad \text{... (2)}$$

Now, $\phi(b) - \phi(a) = \{f(a)\,g(b) - g(a)\,f(b)\}$

$- \{f(a)\,g(a) - g(a)\,f(a)\}$

$$= \begin{vmatrix} f(a) & f(b) \\ g(a) & g(b) \end{vmatrix} \quad \text{... (3)}$$

$\phi'(x) = f(a)\,g'(x) - g(a)\,f'(x)$ {differentiating (1)}

$$\phi'(c) = f(a)\,g'(c) - g(a)\,f'(c) = \begin{vmatrix} f(a) & f'(c) \\ g(a) & g'(c) \end{vmatrix} \quad \text{... (4)}$$

Putting in (2) from (3) and (4), we get the result.

16. Use Rolle's theorem to find the condition for the polynomial equation $f(x) = 0$ to have a repeated real root. Hence or otherwise prove that

$$1 + \frac{x}{1!} + \frac{x^2}{2!} + \frac{x^3}{3!} + \ldots + \frac{x^n}{n!} = 0$$

cannot have a repeated root.

$f(x) = 0$ must have two real roots. Let them be α and $\beta (\alpha < \beta)$.

$\therefore\ f(\alpha) = 0, f(\beta) = 0$

$f(x)$, being a polynomial, is continuous and differentiable everywhere.

$\therefore$ by Rolle's theorem, there exists γ such that $\alpha < \gamma < \beta$ and $f'(\gamma) = 0$

$\therefore$ if $f(x) = 0$ has two real roots α, β then $f(x) = 0$ will have a real root γ lying between α, β.

$\therefore$ if $f(x) = 0$ has two real roots α, α then $f'(x) = 0$ will have a root α, i.e.,

$f(x) = 0$ will have a repeated root α if

$f(\alpha) = 0$ and $f'(\alpha) = 0$.

If possible let

$$\phi(x) \equiv 1 + \frac{x}{1!} + \frac{x^2}{2!} + \frac{x^3}{3!} + \ldots + \frac{x^n}{n!} = 0 \qquad \ldots (1)$$

have a repeated root α.

Then $\phi(\alpha) = 0$ and $\phi'(\alpha) = 0$.

$$\phi(\alpha) = 0 \Rightarrow 1 + \frac{\alpha}{1!} + \frac{\alpha^2}{2!} + \ldots + \frac{\alpha^n}{n!} = 0 \qquad \ldots (2)$$

$$\phi'(\alpha) = 0 \Rightarrow \frac{1}{1!} + \frac{2\alpha}{2!} + \frac{3\alpha^2}{3!} + \ldots + \frac{n\alpha^{n-1}}{n!} = 0$$

$$\text{or} \quad 1 + \frac{\alpha}{1!} + \frac{\alpha^2}{2!} + \ldots + \frac{\alpha^{n-1}}{(n-1)!} = 0 \qquad \ldots (3)$$

$$(2) \text{ and } (3) \Rightarrow \frac{\alpha^n}{n!} = 0; \quad \therefore \quad \alpha = 0$$

$\therefore$ the repeated root must be 0.

But 0 is not a root of (1) because it does not satisfy (1).

Hence, (1) cannot have a repeated root.

17. If $\frac{a_n}{n+1} + \frac{a_{n-1}}{n} + \ldots + a_0 = 0, (n \neq -1)$ then prove that the equation $a_n x^n + a_{n-1} x^{n-1} + \ldots + a_0 = 0$ has a real root lying between 0 and 1.

Consider the equation

$$\phi(x) \equiv \frac{1}{n+1} a_n x^{n+1} + \frac{1}{n} a_{n-1} x^n + \ldots + a_0 x = 0 \qquad \ldots (1)$$

Clearly, $x = 0$ satisfies it and due to the given relation $x = 1$ also satisfies it.

So $x = 0, 1$ are real roots of (1).

$\therefore\ \phi(0) = \phi(1) = 0$. Also $\phi(x)$ is continuous and differentiable, being a polynomial function.

$\therefore$ there exists $c \in (0, 1)$ such that $\phi'(c) = 0$ (by Rolle's theorem)

Differentiating (1),

$$\phi'(x) \equiv a_n x^n + a_{n-1} x^{n-1} + \ldots + a_0 = 0 \qquad \ldots (2)$$

$\therefore\quad x = c \in (0, 1)$ is a root of (2).

18. Prove that

$$(b - a)\sec^2 a < \tan b - \tan a < (b - a)\sec^2 b$$

where $0 < a < b < \frac{\pi}{2}$.

Consider the function $f(x) = \tan x, x \in (a, b)$.

Clearly, $f(x)$ is continuous and differentiable in $\left(0, \frac{\pi}{2}\right)$ and so in (a, b). Therefore, by Lagrange's theorem

$$\frac{f(b) - f(a)}{b - a} = f'(c) \text{ where } a < c < b$$

we get, $\dfrac{\tan b - \tan a}{b - a} = \sec^2 c$

or $\quad \tan b - \tan a = (b - a)\sec^2 c$, where $a < c < b$.

But $a < c < b \Rightarrow \sec a < \sec c < \sec b$

$\therefore\quad (b - a)\sec^2 a < (b - a)\sec^2 c < (b - a)\sec^2 b$

$\therefore\quad (b - a)\sec^2 a < \tan b - \tan a < (b - a)\sec^2 b$.

Exercises

Find the intervals of monotonicity of the functions:

1. $\sqrt{2x - x^2}$
2. $x^2 e^{-x}$
3. $2x^2 - \log x$
4. $\dfrac{1 - x + x^2}{1 + x + x^2}$

5. $x^3(x-2)^2$

6. $4x^3 - 21x^2 + 18x + 20$

Find the intervals of increase and decrease for the functions:

7. $\log(1-x^2)$

8. $\dfrac{2x}{1+x^2}$

9. $\sin x + \cos x, x \in [0, 2\pi]$

10. $3\cos^4 x + 10\cos^3 x + 6\cos^2 x - 3, 0 \le x \le \pi$

11. If $f(x) = (x-1)e^x + 1$, show that $f(x)$ is m.i. for $x > 0$.

12. Prove that the sum of two m.i. functions on the interval $[a, b]$ is a m.i. on $[a, b]$. Will the difference be a monotonic function?

13. Show that $\log_e(1+x) < x$ for all $x > 0$.

14. Show that $1 + x\log_e(x + \sqrt{x^2+1}) \ge \sqrt{1+x^2}$ for all $x \ge 0$.

15. Show that $2x\tan^{-1}x \ge \log_e(1+x^2)$ for all $x \in R$.

16. Show that $x - \dfrac{x^3}{3} \le \tan^{-1}x$ for all $x \ge 0$.

17. Prove that $2\sin x + \tan x \ge 3x$ where $0 \le x < \pi/2$.

18. Prove that $\sin x + \tan x > 2x$ for $0 < x < \pi/2$.

19. If $x \in (0, \pi/2)$, show that

(a) $\cos x > 1 - \dfrac{x^2}{2}$

(b) $\sin x > x - \dfrac{x^3}{6}$.

20. Prove that for positive values of x,
$$x^2 > (1+x)\{\log(1+x)\}^2.$$

21. If $0 < \alpha < \beta < \dfrac{\pi}{2}$, prove that $\tan\alpha - \tan\beta < \alpha - \beta$.

22. If $a, b \in [0, \pi/2]$, prove that

(i) $|\cos a - \cos b| \le |a - b|$

(ii) $|\sin a - \sin b| \le |a - b|$.

23. Show that $(1+x)^p \le 1 + x^p$ for $0 \le p \le 1, x > 0$. Also prove that $(a+b)^p \le a^p + b^p$ where $a, b > 0$.

24. Use the function $f(x) = x^{1/x}, x > 0$ to determine the bigger of the two numbers e^π and π^e.

25. At what value of x, the greatest value of
$$2x^3 - 3x^2 - 12x + 5 \text{ for } -2 \le x \le 4 \text{ occurs?}$$
Also find the greatest value.

26. Find the maximum and the minimum values of the function
$$f(x) = |x^2 - 5x + 6| \text{ on the closed interval } \left[0, \frac{9}{4}\right].$$

27. Find the range of the function
$$f(x) = |2x+1| - 2|x-1| - 3.$$

28. Find the maximum and the minimum values of
$$f(x) = 1 + 2\sin x + 3\cos^2 x, 0 \le x \le \frac{2\pi}{3}.$$

29. Find the greatest and the least values of the function $f(x)$ defined as below:
$$f(x) = \text{minimum of } \{3t^4 - 8t^3 - 6t^2 + 24t;\ 1 \le t \le x\},\quad 1 \le x < 2$$
$$\text{maximum of } \left\{3t + \frac{1}{4}\sin^2 \pi t + 2;\ 2 \le t \le x\right\},\quad 2 \le x \le 4.$$

30. (a) Prove that the equation $3x^5 + 15x - 8 = 0$ has only one real root in $(-1, 1)$.

(b) Let $-1 \le p \le 1$. Prove that the equation $4x^3 - 3x - p = 0$ has a unique root in the closed interval $\left[\frac{1}{2}, 1\right]$ and identify it.

31. Verify whether the function $y = x^3 + 3x$ is invertible.

32. $P(x_1, y_1)$ and $Q(x_2, y_2)$ are two points on the parabola $y = 4ax^2$ and $R(x_3, y_3)$ is a point on the arc PQ such that the tangent at R to the parabola is parallel to the chord PQ. Show that x_3 is the AM of x_1, x_2.

33. Prove that the equation $4x^3 - 9x^2 + 6x - 3 = 0$ has a real root lying between 1 and 2.

34. Prove that the equation
$$x^4 + 4x^3 + 12x^2 + 24x + 24 = 0$$
does not have a repeated real root.

35. If $a_0x^n + a_1x^{n-1} + \ldots + a_{n-1}x = 0$ has a real positive root α then prove that
$$na_0x^{n-1} + (n-1)a_1x^{n-2} + \ldots + a_{n-1} = 0$$
has a positive real root less than α.

36. Let $f(x)$ and $\phi(x)$ be continuous in $[a, b]$ and differentiable in (a, b). If $f(a) = \phi(a)$ and
$$f'(x) > \phi'(x), x \in (a, b)$$
then prove that $f(x) > \phi(x), x \in (a, b)$.

37. If $f(x)$ and $g(x)$ are differentiable functions for $0 \le x \le 1$ such that $f(0) = 2, g(0) = 0, f(1) = 6, g(1) = 2$ then show that there exists c satisfying $0 < c < 1$ and $f'(c) = 2g'(c)$.

38. If $f'(x) \ge 0$ for all $x \in [a, b]$ then prove that $f(x_1) \ge f(x_2)$ for all x_1, x_2 such that $a < x_2 < x_1 < b$.

39. Prove that $\dfrac{b-a}{b} < \log\dfrac{b}{a} < \dfrac{b-a}{a}$, where $0 < a < b$.

40. Prove that $\sqrt{n+1} - \sqrt{n} < \dfrac{1}{2N}$ if $n > N^2$, n being a natural number.

Objective Questions

Fill in the blanks.

41. $f(x) = x^2 - 2x$ is m.i. in the interval ______ .

42. $f(x) = \tan^{-1}x - x$ is m.d. everywhere except at the point $x =$ ______ .

43. For $f(x) = \frac{4}{3}x^3 - 4x, 0 \le x \le 2$, the greatest value of $f(x) =$ ______ and the least value = ______ .

44. The function $y = 1 - x^{4/5}$ becomes zero at the end points of the interval $[-1, 1]$. The Rolle's theorem is not applicable here because ______ . However, the maximum value occurs at $x =$ ______ for the given interval.

Choose the correct option(s).

45. If $f(x) = \frac{x}{\sin x}$ and $g(x) = \frac{x}{\tan x}$ where $0 < x \le 1$ then in the interval

(a) both $f(x)$ and $g(x)$ are increasing functions

(b) both $f(x)$ and $g(x)$ are decreasing functions

(c) $f(x)$ is an increasing function

(d) $g(x)$ is an increasing function

46. A function $f(x)$ which is invertible must be

(a) monotonic

(b) one-one and onto

(c) continuous

(d) none of these

47. $f(x) = \tan^{-1}(\sin x + \cos x)$ is an increasing function in

(a) $(0, \pi/4)$ (b) $(0, \pi/2)$

(c) $(-\pi/4, \pi/4)$ (d) none of these

Answers

1. m.i. in (0, 1), m.d. in (1, 2)

2. m.d. in $(-\infty, 0) \cup (2, +\infty)$, m.i. in (0, 2)

3. m.d. in $\left(0, \frac{1}{2}\right)$, m.i. in $\left(\frac{1}{2}, +\infty\right)$

4. m.d. in $[-1, 1)$, m.i. in $(-\infty, -1) \cup (1, +\infty)$

5. m.i. in $\left(-\infty, \frac{6}{5}\right) \cup (2, +\infty)$, m.d. in $\left(\frac{6}{5}, 2\right)$

6. m.i. in $\left(-\infty, \frac{1}{2}\right) \cup (3, +\infty)$, m.d. in $\left(\frac{1}{2}, 3\right)$

7. increasing in (–1, 0), decreasing in (0, 1)

8. decreasing in $(-\infty, -1) \cup (1, \infty)$, increasing in (–1, 1)

9. increasing in $\left(0, \frac{\pi}{4}\right) \cup \left(\frac{5\pi}{4}, 2\pi\right)$, decreasing in $\left(\frac{\pi}{4}, \frac{5\pi}{4}\right)$

10. increasing in $\left(\frac{\pi}{2}, \frac{2\pi}{3}\right)$, decreasing in $\left(0, \frac{\pi}{2}\right) \cup \left(\frac{2\pi}{3}, \pi\right)$

12. not necessary; take $f(x) = x, g(x) = x^2, (x > 0)$ and verify

24. e^{π} **25.** $x = 4, 37$

26. maximum = 6, minimum = 0

27. [– 6, 0]

28. maximum = $\frac{13}{3}$, minimum = 3

29. greatest = 14, least = 8

30. (b) $\cos\left(\frac{1}{3}\cos^{-1}p\right)$ where $-1 \le p \le 1$.

31. yes **41.** $(1, +\infty)$

42. 0 **43.** $\frac{8}{3}, -\frac{8}{3}$ respectively

44. y is nondifferentiable at $x = 0$, 0 respectively

45. (c) **46.** (b)

47. (a), (c)

□

Chapter Test

Time: 120 minutes

1. Fill in the blank in each of the following so that the resulting sentence becomes true.

(a) The set of values of a for which $f(x) = x^3 - ax$ is m.i. in R is ______ .

(b) If $f(x) = 2x^2 + \frac{2}{x^2}, 0 < |x| \le 2$

$1, x = 0$

then the least value of $f(x)$ = ______ .

2. Choose the correct option(s) from the given alternative answers.

(a) Let $f(x) = x - \sin x$ and $g(x) = x - \tan x$ where $0 \le x < \frac{\pi}{2}$. Then in this interval,

A. $f(x) \ge 0, g(x) \ge 0$ B. $f(x) \le 0, g(x) \le 0$

C. $f(x) \ge 0, g(x) \le 0$ D. $f(x) \le 0, g(x) \ge 0$

(b) The number of real roots of the equation $x^3 - 3x + 1 = 0$ in $[1, 2]$ is

A. one B. two C. three D. none of these

(c) If $I = \{y \in R \mid y = 5x^3 - 15x, -1 \le x \le 3\}$ then the length of the interval I is

A. 80 B. 20 C. 100 D. none of these

3. Let $F(x) = f\{g(x)\}$ where $f(x)$ is an increasing function of x. Let $g(x)$ attain its greatest value at x_0. Prove that $F(x)$ attains its greatest value at x_0.

4. Prove that $e^x \ge 1 + x$ for all $x \in R$.

5. Prove that $x - \frac{x^3}{3} < \tan^{-1} x < x - \frac{x^3}{6}$, for $0 < x \le 1$.

6. If $0 < x < 2\pi$, prove that $\lim_{y \to x} \frac{\cos y - \cos x}{x - y} < x$.

7. Prove that $\sin x < x - \frac{x^3}{6} + \frac{x^5}{120}$ for all $x > 0$.

8. Prove that for $a \in \left(0, \frac{20}{9}\right)$ the real function $f(x) = x^5 + a(x^3 + x) + 1$ is invertible.

9. Let the coefficients in the cubic equation $ax^3 + bx^2 + cx + d = 0$ be related by $-a + b - c + d = 3$ and $8a + 4b + 2c + d = 6$. Show that the quadratic equation $3ax^2 + 2bx + c - 1 = 0$ has a root in $(-1, 2)$.

10. If the graph of $f(x) = 2x^3 + ax^2 + bx$, $a \in N$, $b \in N$, cuts the x-axis at three distinct points then find the minimum possible value of $a + b$. Also prove that there is only one such function for that value of $a + b$ and find it.

11. Let $f(x) = x^3 + 2x^2 + x + 6$. Prove that the equation $f(x) = 0$ has exactly one real root α such that $(\alpha) = -2$ where (α) is the least integer greater than or equal to α.

12. Let $f(x) = \frac{x^4 - x^2 - 2(x-3)}{x^4 - x^2 - 2(x-2)}$, $x \in R$. Find the range of $f(x)$.

13. Find $p \in R, p \ne 1$ for which the function $f(x) = \left(\frac{\sqrt{p+11}}{1-p} - 1\right) x^5 - 3x + \log 2$ is a monotonic decreasing function of x.

□

Answers

1. (a) $(-\infty, 0]$ (b) $f(0)$, i.e., 1

2. (a) C (b) A (c) C

10. 4, $f(x) = 2x^3 + 3x^2 + x$

12. $(1, 2]$

13. $p \in [-11, -2] \cup (1, +\infty)$

8. Indefinite Integration of Elementary Functions

Recap of Facts and Formulae

1. Primitive functions, antiderivatives, indefinite integration

- If $\frac{df(x)}{dx} = F(x)$ then the derivative of $f(x)$ is $F(x)$, i.e., $f'(x) = F(x)$. Equivalently, $f(x)$ is the *primitive function* or *antiderivative* of $F(x)$.
- The indefinite integration of $f'(x)$ is $f(x) + c$, symbolically, $\int f'(x)dx = f(x) + c$, when c is an arbitrary constant, called *constant of integration*.

$\int \phi(x)dx$ is an *integral* and $\phi(x)$ is the *integrand*.

2. Standard integrals of elementary functions

Standard integrals are as follows (without writing the constant of integration):

- $\int x^n dx = \frac{x^{n+1}}{n+1} \quad (n \neq -1)$
- $\int \frac{1}{x} dx = \log_e x$ (where x is positive)
- $\int e^x dx = e^x$
- $\int a^x dx = \frac{a^x}{\log_e a}$
- $\int \sin x \, dx = -\cos x$
- $\int \cos x \, dx = \sin x$
- $\int \tan x \, dx = \log \sec x$
- $\int \cot x \, dx = \log \sin x$
- $\int \sec x \, dx = \log(\sec x + \tan x)$ or $\log \tan\left(\frac{\pi}{4} + \frac{x}{2}\right)$
- $\int \operatorname{cosec} x \, dx = \log \tan \frac{x}{2}$ or $-\log(\operatorname{cosec} x + \cot x)$
- $\int \sec x \cdot \tan x \, dx = \sec x$
- $\int \operatorname{cosec} x \cdot \cot x \, dx = -\operatorname{cosec} x$
- $\int \sec^2 x \, dx = \tan x$
- $\int \operatorname{cosec}^2 x \, dx = -\cot x$
- $\int \frac{dx}{\sqrt{1 - x^2}} = \sin^{-1} x$
- $\int \frac{dx}{1 + x^2} = \tan^{-1} x$
- $\int \frac{dx}{x\sqrt{x^2 - 1}} = \sec^{-1} x$

3. Derivative of indefinite integral, integral of derivative

- $\frac{d}{dx} \int f(x)dx = f(x)$
- $\int \frac{d}{dx} f(x)dx = f(x) + c$
- $\int \{f(x) \cdot \phi'(x) + f'(x) \cdot \phi(x)\}dx = f(x) \cdot \phi(x) + c$
- $\int \frac{f'(x) \cdot \phi(x) - f(x) \cdot \phi'(x)}{\{\phi(x)\}^2} dx = \frac{f(x)}{\phi(x)} + c$

4. Rules of integration

- $\int \{f_1(x) \pm f_2(x) \pm \ldots \pm f_n(x)\}dx$
$= \int f_1(x)dx \pm \int f_2(x)dx \pm \ldots \pm \int f_n(x)dx$
(rule of *term-by-term integration*)
- $\int kf(x)dx = k \int f(x)dx$, where k is a constant
- $\int f'(ax + b)dx = \frac{f(ax + b)}{a}$

5. Methods of integration

Given the integrand our primary objective is to change the integrand into algebraic sum of standard integrands. This objective can be achieved in any of the following three methods:

- *Simplification or transformation*—We change the integrand into algebraic sum of standard integrands by simplification, using algebraic or trigonometrical simplifications.
- *Substitution*—We make suitable substitution for the given variable in terms of some other variable so that the integrand changes into a standard integrand or algebraic sum of standard integrands in the new variable.
- *By parts*—

$\int f(x)\cdot\phi(x)dx = f(x)\cdot\int\phi(x)dx - \int[\int\phi(x)dx]\cdot f'(x)dx$

This method is applicable when the integrand can be put as the product of two functions of which one can be integrated easily.

6. Forms of integrands suitable for specific substitution

There is no fixed rule for selecting a function as the new variable. While selecting it, we always remember that the resulting integrand after substitution must change the original integrand in algebraic sum of standard integrands in the new variable. However, there are certain forms of the integrand which indicate the appropriate substitution.

- Form

$\int f\{\phi(x)\}\cdot\phi'(x)dx$ substitute $\phi(x) = z$.

$\int \dfrac{\phi'(x)}{\phi(x)}dx$ substitute $\phi(x) = z$

- Form

$\int \sin^p x\cdot\cos^{2n-1}x\,dx, n\in N$ substitute $\sin x = z$

$\int \cos^p x\cdot\sin^{2n-1}x\,dx, n\in N$ substitute $\cos x = z$

$\int \sin^p x\cdot\cos^q x\,dx$, where $p+q =$ negative even integer substitute $\tan x = z$

- Form

$\int f(x, \sqrt{a^2-x^2})dx$ substitute $x = a\sin\theta$ or $a\cos\theta$

$\int f(x, \sqrt{x^2-a^2})dx$ substitute $x = a\sec\theta$ or $a\text{cosec}\,\theta$

$\int f(x, \sqrt{x^2+a^2})dx$ substitute $x = a\tan\theta$ or $a\cot\theta$

$\int f(x^2, \sqrt{a^2-x^2}, \sqrt{a^2+x^2})dx$ substitute $x^2 = a^2\cos 2\theta$

7. Choice of *u*, *v* in "by parts"

The method of "by parts",

$$\int uv dx = u\int v dx - \int\left[\int v dx\right]\frac{du}{dx}dx,$$

is used when the integrand is thought of as a product of two functions. While using this method, one has to take care of the following:

(i) the second part v must be a standard integrand or can be easily integrated by simplification or substitution.

(ii) the integral $\int\left[\int v dx\right]\frac{du}{dx}dx$ must not be more complicated than the original integrand.

TIP Though there is no hard and fast rule for selecting the two parts, we can observe the following:

Follow the order of letters in *LIATE*—*L* standing for logarithmic function, *I* for inverse function, *A* for algebraic function, *T* for trigonometrical function and *E* for exponential function. If the two parts in an integrand are *L* and *A*, select *L* (the first one among *L* and *A* in *LIATE*) as *u* and the other as *v*. Similarly for other pairs of functions.

8. An important result

$\int e^x\{f(x)+f'(x)\}\,dx = e^x\cdot f(x)$

Selected Solved Examples

1. Evaluate $\displaystyle\int\left\{\frac{1+x+x^2}{x^3(1+x)}+\frac{2x-1}{(x+1)^2}+\frac{4-5\sin x}{\cos^2 x}+\frac{1}{\sin^2 x\cdot\cos^2 x}+\sin 2x\cdot\cos 3x\right\}dx.$

Here, $I = \displaystyle\int\frac{1+x+x^2}{x^3(1+x)}dx + \int\frac{2x-1}{(x+1)^2}dx + \int\frac{4-5\sin x}{\cos^2 x}dx + \int\frac{dx}{\sin^2 x\cdot\cos^2 x} + \int\sin 2x\cdot\cos 3x\,dx$

$$= \int\frac{(1+x)+x^2}{x^3(1+x)}dx + \int\frac{2(x+1)-3}{(x+1)^2}dx + \int(4\sec^2 x - 5\tan x\cdot\sec x)\,dx + \int\frac{4}{\sin^2 2x}dx + \frac{1}{2}\int(\sin 5x - \sin x)\,dx$$

$$= \int \left\{ \frac{1}{x^3} + \frac{1}{x(1+x)} \right\} dx$$

$$+ \int \left\{ \frac{2}{x+1} - \frac{3}{(x+1)^2} \right\} dx$$

$$+ 4 \int \sec^2 x \, dx - 5 \int \tan x \sec x \, dx$$

$$+ 4 \int \operatorname{cosec}^2 2x \, dx + \frac{1}{2} \int \sin 5x \, dx - \frac{1}{2} \int \sin x \, dx$$

$$= \frac{x^{-2}}{-2} + \int \frac{(1+x) - x}{x(1+x)} dx + 2\log(x+1) + \frac{3}{x+1}$$

$$+ 4\tan x - 5\sec x + 4 \cdot \frac{-\cot 2x}{2}$$

$$+ \frac{1}{2} \cdot \frac{-\cos 5x}{5} + \frac{1}{2} \cos x + c_1.$$

But $\int \frac{(1+x) - x}{x(1+x)} dx$

$$= \int \left(\frac{1}{x} - \frac{1}{1+x} \right) dx$$

$$= \log x - \log(1+x) + c_2$$

$$\therefore \quad I = -\frac{1}{2x^2} + \log \frac{x}{1+x} + 2\log(x+1) + \frac{3}{x+1}$$

$$+ 4\tan x - 5\sec x - 2\cot 2x$$

$$- \frac{1}{10} \cos 5x + \frac{1}{2} \cos x + c$$

$$= -\frac{1}{2x^2} + \log x(x+1) + \frac{3}{x+1} + 5(\tan x - \sec x)$$

$$- \cot x - \frac{1}{10} \cos 5x + \frac{1}{2} \cos x + c$$

(because $2\cot 2x + \tan x = \cot x$ from trigonometry)

2. Evaluate $\int \frac{\sin x}{\sin (x - \alpha)} dx$.

$$I = \int \frac{\sin(\overline{x - \alpha + \alpha})}{\sin(x - \alpha)} dx$$

$$= \int \frac{\sin(x-\alpha) \cdot \cos \alpha + \cos(x - \alpha) \cdot \sin \alpha}{\sin(x - \alpha)} dx$$

$$= \int \{\cos \alpha + \sin \alpha \cdot \cot(x - \alpha)\} \, dx$$

$$= x\cos \alpha + \sin \alpha \cdot \log \sin(x - \alpha) + c.$$

3. Evaluate (i) $\int \frac{dx}{\sin(x-\alpha) \cdot \cos(x - \beta)}$

(ii) $\int \{1 + \tan x \cdot \tan(x + \alpha)\} \, dx.$

(i) $I = \int \frac{\cos(\overline{x - \alpha} - \overline{x - \beta})}{\sin(x-\alpha) \cdot \cos(x - \beta)} \cdot \frac{1}{\cos(\beta - \alpha)} dx$

$$= \frac{1}{\cos(\beta - \alpha)}$$

$$\times \int \frac{\cos(x-\alpha) \cdot \cos(x - \beta) + \sin(x - \alpha) \cdot \sin(x - \beta)}{\sin(x - \alpha) \cdot \cos(x - \beta)} dx$$

$$= \frac{1}{\cos(\beta - \alpha)} \int \{\cot(x - \alpha) + \tan(x - \beta)\} \, dx$$

$$= \frac{1}{\cos(\alpha - \beta)} [\log \sin(x - \alpha) + \log \sec(x - \beta)] + c$$

$$= \sec(\alpha - \beta) \cdot \log \frac{\sin(x - \alpha)}{\cos(x - \beta)} + c$$

(ii) $I = \int \left\{ 1 + \frac{\sin x \cdot \sin(x + \alpha)}{\cos x \cdot \cos(x + \alpha)} \right\} dx$

$$= \int \frac{\cos(\overline{x + \alpha - x})}{\cos x \cdot \cos(x + \alpha)} dx$$

$$= \cos \alpha \int \frac{1}{\cos x \cdot \cos(x + \alpha)} dx$$

$$= \frac{\cos \alpha}{\sin \alpha} \int \frac{\sin(\overline{x + \alpha - x})}{\cos x \cdot \cos(x + \alpha)} dx$$

$$= \cot \alpha \int \frac{\sin(x + \alpha) \cdot \cos x - \cos(x + \alpha) \cdot \sin x}{\cos x \cdot \cos(x + \alpha)} dx$$

$$= \cot \alpha \int \{\tan(x + \alpha) - \tan x\} dx$$

$$= \cot \alpha [\log \sec(x + \alpha) - \log \sec x] + c$$

$$= \cot \alpha \cdot \log \frac{\cos x}{\cos(x + \alpha)} + c.$$

4. Integrate $\int \frac{dx}{a\cos x + b\sin x}$.

$$I = \int \frac{dx}{\sqrt{a^2 + b^2} \left(\frac{a}{\sqrt{a^2 + b^2}} \cos x + \frac{b}{\sqrt{a^2 + b^2}} \sin x \right)}$$

$$= \int \frac{1}{\sqrt{a^2 + b^2}} \cdot \frac{dx}{\cos(x - \alpha)},$$

where $\alpha = \cos^{-1} \frac{a}{\sqrt{a^2 + b^2}}$

$$= \frac{1}{\sqrt{a^2 + b^2}} \int \sec(x - \alpha) dx$$

$$= \frac{1}{\sqrt{a^2 + b^2}} \log \tan \left(\frac{\pi}{4} + \frac{x - \alpha}{2} \right) + c.$$

5. Integrate $\int \tan(x - \alpha) \cdot \tan(x + \alpha) \cdot \tan 2x \, dx.$

$$\tan 2x = \tan(\overline{x + \alpha} + \overline{x - \alpha}) = \frac{\tan(x + \alpha) + \tan(x - \alpha)}{1 - \tan(x + \alpha) \cdot (x - \alpha)}$$

or $\quad \tan 2x - \tan 2x \cdot \tan(x + \alpha)\tan(x - \alpha)$

$$= \tan(x + \alpha) + \tan(x - \alpha)$$

$\therefore \quad \tan(x-\alpha)\tan(x+\alpha)\tan 2x$

$= \tan 2x - \tan(x+\alpha) - \tan(x-\alpha)$

$\therefore \quad I = \int \{\tan 2x - \tan(x+\alpha) - \tan(x-\alpha)\}\, dx$

$= \frac{1}{2}\log \sec 2x - \log \sec(x+\alpha) - \log \sec(x-\alpha) + c$

$= \log\{\sqrt{\sec 2x}\cdot \cos(x+\alpha)\cdot \cos(x-\alpha)\} + c.$

6. Evaluate (a) $\int \frac{dx}{x^2(1+x^4)^{3/4}}$

(b) $\int \frac{\sqrt{x}}{\sqrt{x}+\sqrt[3]{x}}\, dx.$

(a) $I = \int \frac{dx}{x^2\cdot x^3\left(\frac{1}{x^4}+1\right)^{3/4}}$

$= \int \frac{1}{\left(1+\frac{1}{x^4}\right)^{3/4}}\cdot \frac{dx}{x^5}.$

Putting $1+\frac{1}{x^4} = z,$

$\frac{-4}{x^5}dx = dz$, i.e., $\frac{dx}{x^5} = -\frac{1}{4}dz$

$\therefore \quad I = \int \frac{1}{z^{3/4}}\cdot\frac{-1}{4}dz = -\frac{1}{4}\int z^{-3/4}dz$

$= -\frac{1}{4}\cdot\frac{z^{\frac{1}{4}}}{\frac{1}{4}} + c$

$= -z^{\frac{1}{4}} + c = -\left(1+\frac{1}{x^4}\right)^{\frac{1}{4}} + c$

$= -\frac{(1+x^4)^{\frac{1}{4}}}{x} + c.$

(b) Here we have second and third roots of x.

The LCM of 2 and 3 = 6.

So, put $x = z^6$; then $dx = 6z^5dz$

$\therefore \quad I = \int \frac{z^3}{z^3+z^2}\cdot 6z^5dz$

$= 6\int \frac{z^6}{z+1}dz$

$= 6\int \frac{(z^6-1)+1}{z+1}dz$

$= 6\int \frac{(z^3+1)(z^3-1)}{z+1}dz + 6\int\frac{dz}{z+1}$

$= 6\int (z^2-z+1)(z^3-1)dz + 6\log(1+z)$

$= 6\int (z^5-z^4+z^3-z^2+z-1)dz + 6\log(1+z)$

$= 6\left[\frac{z^6}{6}-\frac{z^5}{5}+\frac{z^4}{4}-\frac{z^3}{3}+\frac{z^2}{2}-z\right] + 6\log(1+z)+c$

$= x - \frac{6}{5}x^{5/6} + \frac{3}{2}x^{2/3} - 2x^{1/2} + 3x^{1/3} - 6x^{1/6}$

$+ 6\log(1+x^{1/6}) + c.$

7. Evaluate $\int \cos\left(2\cot^{-1}\sqrt{\frac{1-x}{1+x}}\right)dx.$

Put $x = \cos 2\theta$; then $dx = -2\sin 2\theta\, d\theta$

$\therefore \quad I = \int \cos\left(2\cot^{-1}\sqrt{\frac{1-\cos 2\theta}{1+\cos 2\theta}}\right)(-2\sin 2\theta)d\theta$

$= \int \cos(2\cot^{-1}\tan\theta)(-2\sin 2\theta)d\theta$

$= \int -2\sin 2\theta\cdot\cos\left\{2\left(\frac{\pi}{2}-\tan^{-1}(\tan\theta)\right)\right\}d\theta$

$= -2\int \sin 2\theta\cdot\cos(\pi-2\theta)d\theta$

$= \int 2\sin 2\theta\cdot\cos 2\theta\, d\theta = \int \sin 4\theta\, d\theta$

$= \frac{-\cos 4\theta}{4} + c = -\frac{1}{4}(2\cos^2 2\theta - 1) + c$

$= -\frac{1}{2}x^2 + c_1.$

8. Evaluate $\int x^x(1+\log x)\,dx.$

Let $x^x = z$; then $x\log x = \log z$

$\Rightarrow \quad \left(\log x + x\cdot\frac{1}{x}\right)dx = \frac{1}{z}dz$

$\therefore \quad I = \int z\cdot\frac{1}{z}dz = \int dz = z + c$

$= x^x + c.$

9. Evaluate $\int \frac{\sin^{-1}\sqrt{x}-\cos^{-1}\sqrt{x}}{\sin^{-1}\sqrt{x}+\cos^{-1}\sqrt{x}}\,dx.$

We have $\sin^{-1}\sqrt{x}+\cos^{-1}\sqrt{x} = \frac{\pi}{2}$, from trigonometry.

$\therefore \quad I = \int \frac{\frac{\pi}{2} - 2\cos^{-1}\sqrt{x}}{\frac{\pi}{2}}dx$

$$= \int \left(1 - \frac{4}{\pi} \cos^{-1}\sqrt{x}\right) dx$$

$$= x - \frac{4}{\pi} \int x^0 \cdot \cos^{-1}\sqrt{x}\, dx$$

$$= x - \frac{4}{\pi}\left[\cos^{-1}\sqrt{x} \cdot x - \int x \cdot \frac{-1}{\sqrt{1-(\sqrt{x})^2}} \cdot \frac{1}{2\sqrt{x}} dx\right],$$

using by parts

$$\therefore \quad I = x - \frac{4x}{\pi} \cos^{-1}\sqrt{x} - \frac{2}{\pi} \int \frac{\sqrt{x}}{\sqrt{1-x}} dx \quad \ldots (1)$$

Put $x = \cos^2\theta$; then $dx = -2\cos\theta \cdot \sin\theta\, d\theta$

$$\therefore \quad \int \frac{\sqrt{x}}{\sqrt{1-x}} dx = \int \frac{\cos\theta}{\sqrt{1-\cos^2\theta}} (-2\cos\theta \cdot \sin\theta) d\theta$$

$$= -\int 2\cos^2\theta\, d\theta = -\int (1 + \cos 2\theta) d\theta$$

$$= -\left\{\theta + \frac{\sin 2\theta}{2}\right\} + c$$

$$= -\cos^{-1}\sqrt{x} - \sqrt{x} \cdot \sqrt{1-x} + c$$

$\therefore$ from (1),

$$I = x - \frac{4x}{\pi} \cos^{-1}\sqrt{x} - \frac{2}{\pi} \{-\cos^{-1}\sqrt{x} - \sqrt{x}\sqrt{1-x} + c\}$$

$$= x + \frac{2}{\pi}(1-2x)\cos^{-1}\sqrt{x} + \frac{2}{\pi}\sqrt{x(1-x)} + c_1.$$

10. Integrate (a) $\int \frac{\log x}{(1+x)^3} dx$

(b) $\int \log(x + \sqrt{x^2 + a^2})\, dx.$

(a) $I = \int \log x \cdot (1+x)^{-3} dx$; using by parts we get

$$I = \log x \cdot \int (1+x)^{-3} dx - \int \frac{(1+x)^{-2}}{-2} \cdot \frac{1}{x} dx$$

$$= \log x \cdot \frac{(1+x)^{-2}}{-2} + \frac{1}{2} \int \frac{1}{x(1+x)^2} dx \quad \ldots (1)$$

Now, $\int \frac{1}{x(1+x)^2} dx = \int \frac{(1+x) - x}{x(1+x)^2} dx$

$$= \int \left\{\frac{1}{x(1+x)} - \frac{1}{(1+x)^2}\right\} dx$$

$$= \int \left\{\frac{1}{x} - \frac{1}{1+x} - \frac{1}{(1+x)^2}\right\} dx$$

$$= \log x - \log(1+x) + \frac{1}{1+x} + c$$

$\therefore$ from (1),

$$I = -\frac{1}{2(1+x)^2} \cdot \log x + \frac{1}{2} \log \frac{x}{1+x} + \frac{1}{2} \cdot \frac{1}{1+x} + c_1.$$

(b) $I = \int x^0 \cdot \log(x + \sqrt{x^2 + a^2}) dx$; using by parts we get

$$I = \log(x + \sqrt{x^2 + a^2}) \cdot \int x^0 dx - \int x \cdot \frac{1}{x + \sqrt{x^2 + a^2}} \left(1 + \frac{2x}{2\sqrt{x^2 + a^2}}\right) dx$$

$$= x\log(x + \sqrt{x^2 + a^2}) - \int \frac{x}{\sqrt{x^2 + a^2}} dx$$

$$= x\log(x + \sqrt{x^2 + a^2}) - \frac{1}{2} \int (x^2 + a^2)^{-1/2} d(x^2 + a^2)$$

$$= x\log(x + \sqrt{x^2 + a^2}) - \frac{1}{2} \cdot \frac{(x^2 + a^2)^{1/2}}{\frac{1}{2}} + c$$

$$= x\log(x + \sqrt{x^2 + a^2}) - \sqrt{x^2 + a^2} + c.$$

11. Find $\int \left\{\frac{1}{\sqrt[3]{x} + \sqrt[4]{x}} + \frac{\log(1 + \sqrt[6]{x})}{\sqrt[3]{x} + \sqrt{x}} dx\right\} dx.$

Here, $I = \int \frac{1}{\sqrt[3]{x} + \sqrt[4]{x}} dx + \int \frac{\log(1 + \sqrt[6]{x})}{\sqrt[3]{x} + \sqrt{x}} dx = I_1 + I_2$

Now, $I_1 = \int \frac{dx}{\sqrt[3]{x} + \sqrt[4]{x}}$; putting $x = z^{12}$, we get

$$I_1 = \int \frac{12z^{11} dz}{z^4 + z^3} = \int \frac{12z^8 dz}{z+1}$$

$$= 12 \int \frac{z^8 - (-1)^8 + 1}{z - (-1)} dz$$

$$= 12 \int \frac{z^8 - (-1)^8}{z - (-1)} dz + 12 \int \frac{dz}{z+1}$$

$$= 12 \int \frac{\{z^4 + 1\}\{z^4 - (-1)^4\}}{z - (-1)} dz + 12\log(z+1) + c_1$$

$$= 12 \int (z^4 + 1) \frac{\{z^2 + (-1)^2\}\{z^2 - (-1)^2\}}{z - (-1)} dz + 12\log(z+1) + c_1$$

$$= 12 \int (z^4 + 1)(z^2 + 1)(z - 1) dz + 12\log(z+1) + c_1$$

$$= 12 \int (z^6 + z^4 + z^2 + 1)(z - 1) dz + 12\log(z+1) + c_1$$

$$= 12 \int (z^7 - z^6 + z^5 - z^4 + z^3 - z^2 + z - 1)\, dz + 12\log(z+1) + c_1$$

$$= 12\left[\frac{z^8}{8} - \frac{z^7}{7} + \frac{z^6}{6} - \frac{z^5}{5} + \frac{z^4}{4} - \frac{z^3}{3} + \frac{z^2}{2} - z\right] + 12\log(z+1) + c_2$$

$I_2 = \int \frac{\log(1 + \sqrt[6]{x})}{\sqrt[3]{x} + \sqrt{x}} dx$; putting $x = t^6$,

$I_2 = \int \frac{\log(1+t)}{t^2 + t^3} \cdot 6t^5 dt = 6 \int \frac{\log(1+t)}{1+t} t^3 dt;$

putting $1 + t = e^\theta$,

$$I_2 = 6\int \frac{\log e^{\theta}}{e^{\theta}}(e^{\theta}-1)^3 e^{\theta} d\theta$$

$$= 6\int \theta(e^{\theta}-1)^3 d\theta$$

$$= 6\int \theta(e^{3\theta} - 3e^{2\theta} + 3e^{\theta} - 1)d\theta$$

$$= 6\left[\left(\theta\cdot\frac{e^{3\theta}}{3} - \int\frac{e^{3\theta}}{3}d\theta\right) - 3\left(\theta\cdot\frac{e^{2\theta}}{2} - \int\frac{e^{2\theta}}{2}d\theta\right)\right.$$

$$\left. + 3\left(\theta\cdot e^{\theta} - \int e^{\theta}d\theta\right) - \frac{\theta^2}{2}\right] + c_2$$

$$= 6\left[\frac{1}{3}\theta e^{3\theta} - \frac{1}{9}e^{3\theta} - 3\left(\theta\cdot\frac{e^{2\theta}}{2} - \frac{e^{2\theta}}{4}\right)\right.$$

$$\left. + 3(\theta\cdot e^{\theta} - e^{\theta}) - \frac{\theta^2}{2}\right] + c_3$$

$$\therefore \quad I = 12\left[\frac{z^8}{8} - \frac{z^7}{7} + \frac{z^6}{6} - \frac{z^5}{5} + \frac{z^4}{4} - \frac{z^3}{3} + \frac{z^2}{2} - z\right]$$

$$+ 12\log(z+1) + 6\left[\frac{1}{3}\theta e^{3\theta} - \frac{1}{9}e^{3\theta} - \frac{3}{2}\theta e^{2\theta} + \frac{3}{4}e^{2\theta}\right.$$

$$\left. + 3\theta\cdot e^{\theta} - 3e^{\theta} - \frac{\theta^2}{2}\right] + c$$

where $z = x^{\frac{1}{12}}$, $\theta = \log(1+\sqrt[6]{x})$ and $e^{\theta} = 1 + t$.

12. Integrate

$$\int \frac{\sqrt{1+x^{2n}}\,\{\log(1+x^{2n}) - 2n\log x\}}{x^{3n+1}}\,dx,\ n \in R.$$

Here $I = \int \frac{\sqrt{1+x^{2n}}}{x^n}\cdot\frac{\log(1+x^{2n}) - \log x^{2n}}{x^{2n+1}}dx$

$$= \int\sqrt{\frac{1}{x^{2n}}+1}\cdot\frac{\log\left(\frac{1}{x^{2n}}+1\right)}{x^{2n+1}}dx.$$

Put $\frac{1}{x^{2n}} + 1 = z^2$; then $\frac{-2n}{x^{2n+1}}dx = 2z\,dz$

$$\therefore \quad I = \int z\cdot\log z^2\cdot\frac{z\,dz}{-n} = -\frac{2}{n}\int z^2\log z\,dz$$

$$= -\frac{2}{n}\left[\log z\cdot\frac{z^3}{3} - \int\frac{z^3}{3}\cdot\frac{1}{z}dz\right]$$

$$= \frac{-2}{n}\left[\frac{1}{3}z^3\log z - \frac{1}{3}\cdot\frac{z^3}{3}\right] + c$$

$$= -\frac{2z^3}{9n}\{3\log z - 1\} + c$$

$$= -\frac{2}{9n}\left(\frac{1}{x^{2n}}+1\right)^{3/2}\cdot\left\{3\log\left(\frac{1}{x^{2n}}+1\right)^{1/2} - 1\right\} + c$$

$$= -\frac{2}{9n}\frac{(1+x^{2n})^{3/2}}{x^{3n}}\left\{\frac{3}{2}\log\frac{1+x^{2n}}{x^{2n}} - 1\right\} + c$$

$$= -\frac{2}{9n}\cdot\frac{(1+x^{2n})^{3/2}}{x^{3n}}$$

$$\times\left\{\frac{3}{2}\log(1+x^{2n}) - 3n\log x - 1\right\} + c.$$

13. Evaluate $\int\sqrt{\frac{\cos^3 x}{\sin^{11} x}}\,dx.$

$I = \int \cos^{3/2}x\cdot\sin^{-11/2}x\,dx.$

Here $\frac{3}{2} + \frac{-11}{2} = -4$ = negative even integer.

So, put $\tan x = z$; then $\sec^2 x\,dx = dz$

$$\therefore \quad I = \int\cos^{3/2}x\cdot\sin^{-11/2}x\,\frac{dz}{\sec^2 x}$$

$$= \int\cos^{7/2}x\cdot\sin^{-11/2}x\,dz$$

$$= \int\frac{\cos^{7/2}x}{\sin^{7/2}x}\cdot\operatorname{cosec}^2 x\,dz$$

$$= \int\cot^{7/2}x\cdot\operatorname{cosec}^2 x\,dz$$

$$= \int z^{-\frac{7}{2}}\cdot\left(1+\frac{1}{z^2}\right)dz = \int\left(z^{-\frac{7}{2}} + z^{-\frac{11}{2}}\right)dz$$

$$= \frac{z^{-\frac{5}{2}}}{-\frac{5}{2}} + \frac{z^{-\frac{9}{2}}}{-\frac{9}{2}} + c$$

$$= \frac{-2}{5}\cot^{5/2}x + \frac{-2}{9}\cot^{9/2}x + c.$$

14. Evaluate $\int\frac{e^{m\tan^{-1}x}}{(1+x^2)^{3/2}}dx.$

Put $\tan^{-1}x = \theta$; then $x = \tan\theta$ and so, $dx = \sec^2\theta\,d\theta$

$$\therefore \quad I = \int\frac{e^{m\theta}}{(1+\tan^2\theta)^{3/2}}\cdot\sec^2\theta\,d\theta$$

$$= \int e^{m\theta}\cdot\frac{1}{\sec^3\theta}\cdot\sec^2\theta\,d\theta$$

$$= \int e^{m\theta}\cdot\cos\theta\,d\theta \qquad \ldots(1)$$

$$= e^{m\theta}\cdot(\sin\theta) - \int\sin\theta\cdot e^{m\theta}\cdot m\,d\theta$$

$$= e^{m\theta}\cdot\sin\theta - m\int e^{m\theta}\cdot\sin\theta\,d\theta$$

$$= e^{m\theta} \cdot \sin\theta - m\left\{e^{m\theta} \cdot (-\cos\theta) - \int -\cos\theta \cdot e^{m\theta} \cdot m\, d\theta\right\}$$

$\therefore \quad I = e^{m\theta}(\sin\theta + m\cos\theta) - m^2 \int e^{m\theta} \cdot \cos\theta\, d\theta.$

$= e^{m\theta}(\sin\theta + m\cos\theta) - m^2 I$, using (1)

or $\quad (1 + m^2)I = e^{m\theta}(\sin\theta + m\cos\theta)$

or $\quad I = \dfrac{e^{m\theta}(\sin\theta + m\cos\theta)}{1 + m^2}$

$\therefore$ the general integral is

$$I = \frac{e^{m\theta}(\sin\theta + m\cos\theta)}{1 + m^2} + c$$

or $\quad I = e^{m\tan^{-1}x} \cdot \dfrac{\dfrac{x}{\sqrt{1+x^2}} + m \cdot \dfrac{1}{\sqrt{1+x^2}}}{1 + m^2} + c$

$$= \frac{(x + m)e^{m\tan^{-1}x}}{(1 + m^2)\sqrt{1 + x^2}} + c.$$

15. Find (a) $\int e^x \left(\dfrac{1 - x}{1 + x^2}\right)^2 dx$

(b) $\int \left\{\log(1 + \cos x) - x\tan\dfrac{x}{2}\right\} dx.$

(a) $\quad I = \int e^x \cdot \left\{\dfrac{1 - 2x + x^2}{(1 + x^2)^2}\right\} dx$

$$= \int e^x \cdot \left\{\frac{1}{1 + x^2} - \frac{2x}{(1 + x^2)^2}\right\} dx$$

$$= \int e^x \left\{\frac{1}{1 + x^2} + \frac{d}{dx}\left(\frac{1}{1 + x^2}\right)\right\} dx$$

$$= e^x \cdot \frac{1}{1 + x^2} + c$$

$\left[\because \text{ we know that } \int e^x\{f(x) + f'(x)\}dx = e^x f(x)\right]$

(b) $\quad I = \int x^0 \log(1 + \cos x)dx - \int x\tan\dfrac{x}{2}\, dx$

$$= \log(1 + \cos x) \cdot x - \int x \cdot \frac{1}{1 + \cos x}(-\sin x)dx - \int x\tan\frac{x}{2}\, dx \text{ (using by parts)}$$

$$= x\log(1 + \cos x) + \int x \cdot \frac{2 \cdot \sin\frac{x}{2} \cdot \cos\frac{x}{2}}{2\cos^2\frac{x}{2}} dx - \int x\tan\frac{x}{2}\, dx$$

$$= x\log(1 + \cos x) + \int x\tan\frac{x}{2}\, dx - \int x\tan\frac{x}{2}\, dx$$

$$= x\log(1 + \cos x)$$

$\therefore$ the general integral $= x\log(1 + \cos x) + c$.

16. Integrate the function $f(x) = \log(\log x) + \dfrac{1}{(\log x)^2}$ **with respect to** x.

$$\int f(x)dx = \int \left\{\log(\log x) + \frac{1}{(\log x)^2}\right\} dx$$

Put $\log x = z$, i.e., $x = e^z$; then $dx = e^z dz$

$\therefore \quad \int f(x)dx = \int \left(\log z + \dfrac{1}{z^2}\right) e^z dz$

$$= \int \left\{\left(\log z + \frac{1}{z}\right) - \left(\frac{1}{z} - \frac{1}{z^2}\right)\right\} e^z dz$$

$$= \int \left(\log z + \frac{1}{z}\right) e^z dz - \int \left(\frac{1}{z} - \frac{1}{z^2}\right) e^z dz$$

$$= \int \left\{\log z + \frac{d}{dz}(\log z)\right\} e^z dz - \int \left\{\frac{1}{z} + \frac{d}{dz}\left(\frac{1}{z}\right)\right\} e^z dz$$

$$= \log z \cdot e^z - \frac{1}{z} e^z + c$$

$$= x\log(\log x) - \frac{x}{\log x} + c.$$

17. If $I_n = \int \dfrac{dx}{(x^2 + a^2)^n}$**, show that**

$$2(n - 1)a^2 \cdot I_n = (2n - 3)I_{n-1} + \frac{x}{(x^2 + a^2)^{n-1}}.$$

Here, $\quad I_n = \int \dfrac{dx}{(x^2 + a^2)^n} = \int \dfrac{1}{(x^2 + a^2)^n} \cdot x^0 dx$

$$= \frac{1}{(x^2 + a^2)^n} \cdot x - \int x \cdot \frac{d}{dx}\left\{\frac{1}{(x^2 + a^2)^n}\right\} dx$$

$$= \frac{x}{(x^2 + a^2)^n} - \int x \cdot \frac{-n \cdot 2x}{(x^2 + a^2)^{n+1}} dx$$

$$= \frac{x}{(x^2 + a^2)^n} + 2n\int \frac{x^2}{(x^2 + a^2)^{n+1}} dx$$

$$= \frac{x}{(x^2 + a^2)^n} + 2n\int \frac{(x^2 + a^2) - a^2}{(x^2 + a^2)^{n+1}} dx$$

$$= \frac{x}{(x^2 + a^2)^n} + 2n\int \frac{dx}{(x^2 + a^2)^n} - 2na^2 \int \frac{dx}{(x^2 + a^2)^{n+1}}$$

$$\therefore \quad I_n = \frac{x}{(x^2+a^2)^n} + 2n\, I_n - 2na^2\, I_{n+1}$$

$$\text{or} \quad 2na^2\, I_{n+1} = \frac{x}{(x^2+a^2)^n} + (2n-1)\, I_n.$$

This is true for all $n \in N$. So, taking $n-1$ in place of n, we get

$$2(n-1)a^2 \cdot I_n = \frac{x}{(x^2+a^2)^{n-1}} + (2n-3)\, I_{n-1}.$$

Exercises

Evaluate the following:

1. $\int \sqrt{1+\sin\frac{x}{n}}\, dx$ **2.** $\int \sqrt{1+\cos\frac{x^\circ}{n}}\, dx$

3. $\int \frac{1}{1+\sec x}\, dx$ **4.** $\int \frac{\sin x}{1-\sin x}\, dx$

5. $\int \frac{\cos x}{\sqrt{1+\cos x}}\, dx$ **6.** $\int \frac{1}{\operatorname{cosec} x - 1}\, dx$

Integrate the following:

7. $\int \left\{ \frac{5\cos^3 x + 2\sin^3 x}{2\sin^2 x \cdot \cos^2 x} + \sqrt{1+\sin 2x} + \frac{1+2\sin x}{\cos^2 x} + \frac{1-\cos 2x}{1+\cos 2x} \right\} dx$

Evaluate the following:

8. $\int \frac{\cos x\, dx}{\cos(x+\alpha)}$ **9.** $\int \frac{dx}{\sin x \cdot \sin(x+\alpha)}$

10. $\int \frac{dx}{\cos(x+\alpha)\cos x}$ **11.** $\int \frac{dx}{\sin\left(x-\frac{\pi}{3}\right)\cos x}$

12. $\int \frac{\sin 2x\, dx}{\sin^2 x - \sin^2 \alpha}$

13. $\int \{1+\cot(x-\alpha)\cdot\cot(x+\alpha)\}dx$

Integrate the following:

14. $\int \frac{dx}{3\cos x - 4\sin x}$ **15.** $\int \frac{dx}{\cos\frac{x}{2}+\sin\frac{x}{2}}$

16. $\int \tan x \cdot \tan 2x \cdot \tan 3x\, dx$

Integrate the following:

17. $\int \frac{(1+x)e^x}{\cos^2(xe^x)}\, dx$ **18.** $\int \frac{\sin 2x\, dx}{a^2\cos^2 x + b^2 \sin^2 x}$

19. $\int \frac{dx}{1+3\cos^2 x}$ **20.** $\int \frac{d\theta}{1-\sin^4\theta}$

21. $\int \sec^4 3x\, dx$ **22.** $\int \operatorname{cosec}^6 x\, dx$

23. $\int \frac{\cos^4 x}{\sin^3 x}\, dx$ **24.** $\int \frac{x\tan^{-1}(x^2)}{1+x^4}\, dx$

25. $\int \frac{2x-\sqrt{\sin^{-1} x}}{\sqrt{1-x^2}}\, dx$ **26.** $\int \frac{x^2 dx}{\sqrt{1-x}}$

27. $\int \frac{1+x-x^2}{(1-x^2)^{3/2}}\, dx$ **28.** $\int \frac{\sqrt{t^3-1}}{t}\, dt$

29. $\int \frac{dx}{x^4 \cdot \sqrt{x^2-3}}$ **30.** $\int \frac{dx}{x\log x \cdot \log(\log x)}$

31. $\int \frac{dx}{x\sin^2(1+\log x)}$ **32.** $\int \frac{dx}{x(x^n+1)}$

33. $\int \frac{dx}{(1+x^4)^{1/4}}$ **34.** $\int \frac{dx}{x^4 \cdot \sqrt{1+x^2}}$

Evaluate the following:

35. $\int \frac{dx}{x+\sqrt{x}}$ **36.** $\int \frac{dx}{\sqrt[3]{x}+\sqrt{x}}$

37. $\int \frac{dx}{\sqrt{x}+\sqrt[4]{x}}$ **38.** $\int x^{\frac{13}{2}} \cdot \left(1+x^{\frac{5}{2}}\right)^{\frac{1}{2}} dx$

39. $\int \frac{dx}{1+\sqrt[3]{1+x}}$ **40.** $\int \frac{x+\sqrt[3]{x^2}+2\cdot\sqrt[6]{x}}{x(1+\sqrt[3]{x})}\, dx$

41. $\int \sqrt[3]{x} \cdot \sqrt[7]{1+\sqrt[3]{x^4}}\, dx$

Integrate the following:

42. $\int \frac{1+x^2}{\sqrt{1-x^2}}\, dx$ **43.** $\int \frac{x^2-2}{x^3\sqrt{x^2-1}}\, dx$

44. $\int \frac{x^3 dx}{(x^2-a^2)^{3/2}}$ **45.** $\int \frac{\sqrt{1+x^2}}{x^4}\, dx$

46. $\int \frac{x^2-x+1}{(x^2+1)^{3/2}}\, dx$ **47.** $\int \sqrt{\frac{1-\sqrt{x}}{1+\sqrt{x}}}\, dx$

48. $\int \left(\frac{1-\sqrt{x}}{1+\sqrt{x}}\right)^{\frac{1}{2}} \cdot \frac{dx}{x}$

Integrate the following:

49. $\int \frac{dx}{\cos^3 x\sqrt{\sin 2x}}$ **50.** $\int \sec^5 x \cdot \operatorname{cosec}^3 x\, dx$

51. $\int \text{cosec } 2x \cdot \sqrt{\tan x}\, dx$ 52. $\int \frac{dx}{\sqrt[3]{\sin^{11} x \cdot \cos x}}$

Integrate the following:

53. $\int x \log(1+x) dx$ 54. $\int \sqrt{x}(\log x)^2 dx$

55. $\int \frac{1}{x^2} \log(a^2+x^2) dx$ 56. $\int \log(1+x^2) dx$

57. $\int \cos\theta \cdot \log \tan \frac{\theta}{2}\, d\theta$

58. $\int \sin\theta \cdot \log(\sec\theta + \tan\theta) d\theta$

59. $\int \frac{\log \cos x}{\cos^2 x} dx$ 60. $\int \sin x \cdot \log \tan x\, dx$

61. $\int \cos 2\theta \cdot \log\left(\frac{\cos\theta + \sin\theta}{\cos\theta - \sin\theta}\right) d\theta$

Evaluate the following:

62. $\int \frac{\log(x+\sqrt{1+x^2})}{\sqrt{1+x^2}} dx$

63. $\int \frac{\sqrt{x^2+1}\{\log(x^2+1) - 2\log x\}}{x^4} dx$

64. $\int \sec^{-1}\sqrt{x}\, dx$ 65. $\int (\sin^{-1} x)^2 dx$

66. $\int \sin^{-1} \sqrt{\frac{x}{a+x}}\, dx$ 67. $\int \tan^{-1} \sqrt{\frac{1-x}{1+x}}\, dx$

68. $\int \frac{\tan^{-1} x}{x^4} dx$ 69. $\int \frac{\tan^{-1}(e^x)}{e^x} dx$

70. $\int \sin^3(\sqrt{x})\, dx$

Integrate the following:

71. $\int \text{cosec } x \cdot \cot^2 x\, dx$ 72. $\int \sec(\tan^{-1} x) dx$

Evaluate the following:

73. $\int \frac{x \tan^{-1} x}{(1+x^2)^{3/2}} dx$ 74. $\int \frac{x^3 \sin^{-1} x}{\sqrt{1-x^2}} dx$

75. $\int e^{m \sin^{-1} x} dx$ 76. $\int \sin^2(\log x) dx$

77. $\int \cos\left(b \log \frac{x}{a}\right) dx$

Find the following:

78. $\int \frac{x-1}{(x+1)^3} \cdot e^x dx$ 79. $\int \left\{\frac{1}{\log x} - \frac{1}{(\log x)^2}\right\} dx$

80. $\int \left(\frac{\cos x}{x} - \log x^{\sin x}\right) dx$

81. $\int \frac{\sqrt{1-\sin x}}{1+\cos x} e^{-x/2} dx$

82. $\int \{\sin(\log x) + \cos(\log x)\}\, dx$

83. $\int \frac{x^2 e^x}{(x+2)^2} dx$ 84. $\int \frac{\log x}{(1+\log x)^2} dx$

85. $\int e^x(\tan x - \log \cos x) dx$

86. If $I_n = \int e^{mx} \cos^n x\, dx$ then prove that
$$(m^2+n^2) I_n = e^{mx} \cdot (m \cos x + n \sin x) \cos^{n-1} x + n(n-1) I_{n-2}.$$

87. If $I_n = \int \sin^n x\, dx$ then prove that
$$nI_n - (n-1)I_{n-2} + \cos x \cdot \sin^{n-1} x = 0.$$

88. If $I_{m,n} = \int \cos^m x \cdot \sin nx\, dx$ then prove that
$$(m+n) I_{m,n} = mI_{m-1,n-1} - \cos^m x \cdot \cos nx.$$

89. If $I_{n,m} = \int x^{n-1} \cdot (1-x)^m dx$ then prove that
$$I_{n,m} = \frac{m}{n} I_{n+1,m-1} + \frac{x^n(1-x)^m}{n}.$$

90. If $I_{m,n} = \int \cos^m\theta \cdot \cos n\theta\, d\theta$, prove that
$$(m+n) I_{m,n} - mI_{m-1,n-1} = \cos^m\theta \cdot \sin n\theta.$$

91. If $I_n = \int (\log x)^n dx$ then prove that
$$I_n + n I_{n-1} = x(\log x)^n.$$

92. If $I_n = \int \frac{x^n dx}{\sqrt{x^2+a}}$ then prove that
$$I_n + \frac{n-1}{n} a I_{n-2} = \frac{1}{n} x^{n-1} \cdot \sqrt{x^2+a}.$$

Objective Questions

Fill in the blanks.

93. If $\int (\sin 2x + \cos 2x) dx = \frac{1}{\sqrt{2}} \sin(2x - c) + a$ then $c =$ ______ .

94. If $\int \frac{dx}{1+\sin x} = \tan\left(\frac{x}{2} + a\right) + b$ then $a =$ ______ and $b =$ ______ .

95. $\int \frac{dx}{(\cos^{-1} x + \sin^{-1} x)^2} =$ ______ .

96. $\int \frac{3 \cdot 2^x - 2 \cdot 3^x}{4^x} dx =$ ______ .

97. $\int \frac{x^4 dx}{x^2+1} =$ ______ .

98. The primitive function of $\frac{(1+x)^2}{x^3+x}$ is ______ .

99. $\int \frac{dx}{a^2 \cos^2 x + b^2 \sin^2 x} =$ ______ .

100. $\int \frac{dx}{(a\cos x - b\sin x)^2} =$ ______ .

101. $\int \sqrt{2 + \sin 3t} \cdot \cos 3t\, dt =$ ______ .

102. $\int \frac{(1+x^2)\, e^{\frac{x^2-1}{x}}}{x^2} dx =$ ______ .

103. $\int \frac{x^5}{1+x^3} dx =$ ______ .

104. $\int \frac{2^x dx}{\sqrt{1-4^x}} =$ ______ .

105. $\int \frac{d(1+\log x)}{\cos^2(1+\log x)} =$ ______ .

106. If $y = \int \frac{dx}{(1+x^2)^{3/2}}$ and $y = 0$ when $x = 0$ then the value of y when $x = 1$ is ______.

107. $\int \log_{10} x\, dx =$ ______ .

108. $\int x^3 \cdot \sqrt{1+x^2}\, dx =$ ______ .

109. $\int \frac{dx}{x(1+x^4)} =$ ______ .

110. $\int \frac{dx}{\sqrt{x} + \sqrt{x-1}} =$ ______ .

111. $\int \frac{\sqrt{1+\sqrt[3]{x}}}{\sqrt[3]{x^2}} dx =$ ______ .

112. If $\int \frac{4e^x + 6e^{-x}}{9e^x - 4e^{-x}} dx = Ax + B\log_e(9e^{2x} - 4) + C$ then $A =$ ______, $B =$ ______, $C =$ ______.

113. The value of $\int x^2 \sec x^3 dx =$ ______ .

114. The value of $\int \frac{d(x^2-1)}{\sqrt{2-x^2}} =$ ______ .

115. $\int \tan^{-1} \frac{x}{1+\sqrt{1-x^2}} dx$ is equal to ______.

Choose the correct option(s).

116. If $\int x^{-3} \cdot 5^{\frac{1}{x^2}} dx = k \cdot 5^{\frac{1}{x^2}}$ then k is

(a) $\frac{-1}{2\log 5}$ (b) $-2\log 5$

(c) $\frac{2}{\log 5}$ (d) $\frac{-2}{\log 5}$

117. If $\phi(x) = f(x) + xf'(x)$ then $\int \phi(x)dx$ is equal to

(a) $(x+1)f(x) + k$ (b) $(x-1)f(x) + k$

(c) $xf(x) + k$ (d) none of these

Answers

1. $2\sqrt{2n} \sin\left(\frac{x}{2n} - \frac{\pi}{4}\right) + c$

2. $\frac{360\sqrt{2}}{\pi} n\sin \frac{\pi x}{360n} + c$

3. $x - \tan \frac{x}{2} + c$

4. $-x + \tan\left(\frac{\pi}{4} + \frac{x}{2}\right) + c$

5. $2\sqrt{2} \sin \frac{x}{2} - \sqrt{2} \log\left(\sec \frac{x}{2} + \tan \frac{x}{2}\right) + c$

6. $-x + \tan\left(\frac{\pi}{4} + \frac{x}{2}\right) + c$

7. $\frac{-5}{2} \operatorname{cosec} x + 3\sec x + (\sin x - \cos x) + 2\tan x - x + c$

8. $(x+\alpha)\cos\alpha + \sin\alpha \cdot \log \sec(x+\alpha) + c$

9. $\operatorname{cosec} \alpha\{\log \sin x - \log \sin(x+\alpha)\} + c$

10. $\operatorname{cosec} \alpha \cdot \{\log \sec(x+\alpha) - \log \sec x\} + c$

11. $2\left\{\log \sin\left(x - \frac{\pi}{3}\right) - \log \cos x\right\} + c$

12. $\log(\sin^2 x - \sin^2 \alpha) + c$

13. $c - \cot 2\alpha\{\log \operatorname{cosec}(x-\alpha) + \log \sin(x+\alpha)\}$

14. $\frac{1}{5} \log\{\sec(x+\alpha) + \tan(x+\alpha)\} + c$, where $\alpha = \cos^{-1} \frac{3}{5}$

15. $\sqrt{2} \log\left\{\sec \frac{2x-\pi}{4} + \tan \frac{2x-\pi}{4}\right\} + c$

16. $\frac{1}{3} \log \sec 3x - \frac{1}{2} \log \sec 2x - \log \sec x + c$

17. $\tan(xe^x) + c$

18. $\frac{1}{b^2 - a^2} \log(a^2\cos^2 x + b^2 \sin^2 x) + c$

19. $\frac{1}{2} \tan^{-1}\left(\frac{1}{2} \tan x\right) + c$

20. $\frac{1}{2} \tan\theta + \frac{1}{2\sqrt{2}} \tan^{-1}(\sqrt{2}\tan\theta) + c$

21. $\frac{1}{3}\left(\tan 3x + \frac{1}{3} \tan^3 3x\right) + c$

22. $-\cot x\left\{\frac{1}{5} \cot^4 x + \frac{2}{3} \cot^2 x + 1\right\} + c$

23. $-\frac{\cos^3 x}{2\sin^2 x} - \frac{3}{2} \cos x + \frac{3}{4} \log \frac{1+\cos x}{1-\cos x} + c$

24. $\frac{1}{4} (\tan^{-1} x^2)^2 + c$

25. $-2\sqrt{1-x^2} - \frac{2}{3} (\sin^{-1} x)^{3/2} + c$

26. $-\frac{2}{15}\sqrt{1-x}(8+4x+3x^2)+c$

27. $\sin^{-1}x+(1-x^2)^{-1/2}+c$

28. $\frac{2}{3}\{\sqrt{t^3-1}-\tan^{-1}\sqrt{t^3-1}\}+c$

29. $\frac{1}{12}\sin\theta+\frac{1}{108}\sin 3\theta+c$, where $\cos\theta=\frac{\sqrt{3}}{x}$

30. $\log\{\log(\log x)\}+c$ **31.** $-\cot(1+\log x)+c$

32. $\frac{1}{n}\log\frac{x^n}{1+x^n}+c$

33. $\frac{1}{4}\log\frac{\sqrt[4]{1+x^4}+x}{\sqrt[4]{1+x^4}-x}-\frac{1}{2}\tan^{-1}\left(\frac{1+x^4}{x^4}\right)^{1/4}+c$

34. $\frac{(2x^2-1)\sqrt{1+x^2}}{3x^3}+c$ **35.** $2\log(1+\sqrt{x})+c$

36. $2\sqrt{x}-3\sqrt[3]{x}+6\sqrt[6]{x}-6\log(1+\sqrt[6]{x})+c$

37. $2\{\sqrt{x}-2x^{1/4}+2\log(1+x^{1/4})\}+c$

38. $\frac{4}{35}(1+x^{5/2})^{7/2}-\frac{8}{25}(1+x^{5/2})^{5/2}+\frac{4}{15}(1+x^{5/2})^{3/2}+c$

39. $\frac{3}{2}(1+x)^{2/3}-3(1+x)^{1/3}+3\log(1+\sqrt[3]{1+x})+c$

40. $12\tan^{-1}\sqrt[6]{x}+\frac{3}{2}\sqrt[3]{x^2}+c$

41. $\frac{21}{32}\cdot\left\{\sqrt[7]{1+\sqrt[3]{x^4}}\right\}^8+c$

42. $\frac{3}{2}\sin^{-1}x-\frac{1}{2}x\sqrt{1-x^2}+c$

43. $-\frac{1}{x^2}\sqrt{x^2-1}+c$

44. $\frac{x^2-2a^2}{\sqrt{x^2-a^2}}+c$ **45.** $-\frac{1}{3}\left(\frac{\sqrt{1+x^2}}{x}\right)^3+c$

46. $\log(x+\sqrt{1+x^2})+\frac{1}{\sqrt{1+x^2}}+c$

47. $-2\sqrt{1-x}+\cos^{-1}\sqrt{x}+\sqrt{x}\cdot\sqrt{1-x}+c$

48. $2\cos^{-1}\sqrt{x}-2\log\frac{1+\sqrt{1-x}}{\sqrt{x}}+c$

49. $\sqrt{2}\left(1+\frac{1}{5}\tan^2 x\right)\sqrt{\tan x}+c$

50. $\frac{3}{2}\tan^2 x-\frac{1}{2}\cot^2 x+3\log\tan x+\frac{1}{4}\tan^4 x+c$

51. $\sqrt{\tan x}+c$ **52.** $-\frac{3(1+4\tan^2 x)}{8\tan^2 x\cdot\sqrt[3]{\tan^2 x}}+c$

53. $\frac{1}{2}(x^2-1)\log(1+x)-\frac{x^2}{4}+\frac{x}{2}+c$

54. $\frac{2}{3}x^{3/2}(\log x)^2-\frac{8}{9}x^{3/2}\cdot\log x+\frac{16}{27}x^{3/2}+c$

55. $-\frac{1}{x}\log(a^2+x^2)+\frac{2}{a}\tan^{-1}\frac{x}{a}+c$

56. $x\log(1+x^2)-2x+2\tan^{-1}x+c$

57. $\sin\theta\cdot\log\tan\frac{\theta}{2}-\theta+c$

58. $\theta-\cos\theta\cdot\log(\sec\theta+\tan\theta)+c$

59. $\tan x\cdot\log\cos x+\tan x-x+c$

60. $-\cos x\cdot\log\tan x+\log\tan\frac{x}{2}+c$

61. $\frac{1}{2}\sin 2\theta\log\left(\frac{\cos\theta+\sin\theta}{\cos\theta-\sin\theta}\right)+\frac{1}{2}\log\cos 2\theta+c$

62. $\frac{1}{2}\{\log(x+\sqrt{1+x^2})\}^2+c$

63. $\frac{(x^2+1)^{3/2}}{3x^3}\left\{\frac{2}{3}-\log(1+x^2)+2\log x\right\}+c$

64. $x\sec^{-1}\sqrt{x}-\sqrt{x-1}+c$

65. $x(\sin^{-1}x)^2+2\sin^{-1}x\cdot\sqrt{1-x^2}-2x+c$

66. $(x+a)\tan^{-1}\sqrt{\frac{x}{a}}-\sqrt{ax}+c$

67. $\frac{x}{2}\cos^{-1}x-\frac{1}{2}\sqrt{1-x^2}+c$

68. $-\frac{1}{3x^3}\tan^{-1}x-\frac{1}{6x^2}-\frac{1}{3}\log\frac{x}{\sqrt{1+x^2}}+c$

69. $-e^{-x}\cdot\tan^{-1}(e^x)+x-\frac{1}{2}\log(1+e^{2x})+c$

70. $-\frac{3}{2}\sqrt{x}\cdot\cos\sqrt{x}+\frac{3}{2}\sin\sqrt{x}+\frac{1}{6}\sqrt{x}\cdot\cos 3\sqrt{x}-\frac{1}{18}\sin 3\sqrt{x}+c$

71. $\frac{1}{2}\log\cot\frac{x}{2}-\frac{1}{2}\operatorname{cosec}x\cdot\cot x+c$

72. $\frac{1}{2}x\sqrt{1+x^2}+\frac{1}{2}\log(x+\sqrt{1+x^2})+c$

73. $\frac{x-\tan^{-1}x}{\sqrt{1+x^2}}+c$

74. $\frac{3}{4}\{x-\sqrt{1-x^2}\cdot\sin^{-1}x\}+\frac{1}{4}\left\{\frac{1}{3}\sin^{-1}x\cdot(1-4x^2)\sqrt{1-x^2}-\frac{1}{9}(3x-4x^3)\right\}+c$

75. $\frac{x+m\sqrt{1-x^2}}{1+m^2}\cdot e^{m\sin^{-1}x}+c$

76. $\frac{x}{10}\{5-\cos(2\log x)-2\sin(2\log x)\}+c$

77. $\frac{x}{b^2+1}\left\{\cos\left(b\log\frac{x}{a}\right)+b\sin\left(b\log\frac{x}{a}\right)\right\}+c$

78. $\frac{e^x}{(1+x)^2}+c$ **79.** $\frac{x}{\log x}+c$

80. $\cos x \cdot \log x + c$

81. $-\sec\frac{x}{2} \cdot e^{-x/2} + c$

82. $x\sin(\log x) + c$

83. $\frac{x-2}{x+2}e^x + c$

84. $\frac{x}{1+\log x} + c$

85. $-e^x \cdot \log(\cos x) + c$

93. $\frac{\pi}{4}$

94. $\frac{3\pi}{4}$, arbitrary constant

95. $\frac{4x}{\pi^2} + c$

96. $\frac{2\left(\frac{3}{4}\right)^x}{\log\frac{4}{3}} - \frac{3}{2^x \log 2} + c$

97. $\frac{1}{3}x^3 - x + \tan^{-1}x + c$

98. $\log x + 2\tan^{-1}x$

99. $\frac{1}{ab}\tan^{-1}\left(\frac{b}{a}\tan x\right) + c$

100. $\frac{1}{b(a - b\tan x)} + c$

101. $\frac{2}{9}(2 + \sin 3t)^{3/2} + c$

102. $e^{x - 1/x} + c$

103. $\frac{1}{3}\{1 + x^3 - \log(1 + x^3)\} + c$

104. $\log_2 e \cdot \sin^{-1}(2^x) + c$

105. $\tan(1 + \log x) + c$

106. $\frac{1}{\sqrt{2}}$

107. $x\log_{10}\frac{x}{e} + c$

108. $\frac{3x^2 - 2}{15}(1 + x^2)^{3/2} + c$

109. $\log x - \frac{1}{4}\log(1 + x^4) + c$

110. $\frac{2}{3}\{x^{3/2} - (x-1)^{3/2}\} + c$

111. $2(1 + x^{1/3})^{3/2} + c$

112. $-\frac{3}{2}, \frac{35}{36}$, arbitrary constant c respectively

113. $\frac{1}{3}\log\tan\left(\frac{\pi}{4} + \frac{x^3}{2}\right) + c$

114. $-2\sqrt{2 - x^2} + c$

115. $\frac{1}{2}\{x\sin^{-1}x + \sqrt{1 - x^2}\} + c$

116. (a)

117. (c)

□

Chapter Test

Time: 120 minutes

1. In each of the following, fill in the blank so that the resulting sentence becomes true.

(a) The value of $\int \frac{dx}{x^n(1+x^n)^{1/n}}$, $n \in N$ is equal to ______ .

(b) The value of $\int \cos^{3/2}x \cdot \sin^3 x\, dx =$ ______ .

(c) The value of $\int \frac{\sin(x+\alpha)}{\cos(x+\alpha)+\cos(x-\alpha)}\, dx =$ ______ .

(d) The primitive function of the function $f(x) = \frac{\sqrt{a^2-x^2}}{x^4}$ is ______ .

(e) $\int \frac{dx}{\sin^4 x + \cos^4 x} =$ ______ .

Evaluate the following:

2. $\int \frac{1}{x} \cdot \sqrt{\frac{1-x^2}{1+x^2}}\, dx$

3. $\int \frac{\sin^{-1}\sqrt{x}}{\sqrt{1-x}}\, dx$

4. $\int x^{\log_e x^2} \cdot \log_{e^x}(x)\, dx$

5. $\int \frac{dx}{\sqrt{1+x} - \sqrt[3]{1+x}}$

6. $\int \frac{dx}{x^4+x^6}$

7. $\int \frac{x \log x}{(x^2-1)^{3/2}}\, dx$

8. $\int \sec 2x \cdot \sqrt{\tan x}\, dx$

9. $\int \sqrt{4+\sqrt{1+\frac{x}{3}}}\, dx$

10. $\int \cos^{-1}\left(\frac{x+1}{\sqrt{x^2+2x+5}}\right) dx$

11. $\int \frac{\sec^2 x}{(\sec x + \tan x)^n}\, dx,\ (n>1)$

12. Prove that $\int \frac{dx}{(1+x^2)^n} = \frac{1}{2(n-1)}\left[\frac{x}{(1+x^2)^{n-1}} + (2n-3)\int \frac{dx}{(1+x^2)^{n-1}}\right]$, $n \in N$.

Hence, compute the value of $\int \cos^4 x\, dx$.

13. Prove that $\int e^{f(x)} \cdot \left[xf'(x) + \frac{f''(x)}{\{f'(x)\}^2}\right] dx = e^{f(x)} \cdot \left(x - \frac{1}{f'(x)}\right) + c.$

Hence or otherwise evaluate $\int e^{x\sin x + \cos x} \cdot \left(\frac{x^4\cos^3 x - x\sin x + \cos x}{x^2 \cdot \cos^2 x}\right) dx.$

14. Evaluate $\int (x^{3m} + x^{2m} + x^m)(2x^{2m} + 3x^m + 6)^{\frac{1}{m}}\, dx,\ x > 0,\ m \in N.$

□

Answers

1. (a) $\frac{1}{1-n}\left(1+\frac{1}{x^n}\right)^{1-1/n}+c$ (b) $\frac{2}{45}\cos^{5/2}x\cdot(5\cos^2 x-9)+c$

(c) $\frac{1}{2}\log\sec x+\frac{1}{2}x\tan\alpha+c$ (d) $-\frac{(a^2-x^2)^{3/2}}{3a^2x^3}$

(e) $\frac{1}{\sqrt{2}}\tan^{-1}\left(\frac{1}{\sqrt{2}}\tan 2x\right)+c$

2. $\frac{1}{2}\left\{\cos^{-1}x^2-\log\frac{1+\sqrt{1-x^4}}{x^2}\right\}+c$

3. $2(\sqrt{x}-\sqrt{1-x}\cdot\sin^{-1}\sqrt{x})+c$

4. $\frac{1}{4}x^{\log x^2}+c$

5. $2\sqrt{1+x}+3\cdot\sqrt[3]{1+x}+6\cdot\sqrt[6]{1+x}+6\log(\sqrt[6]{1+x}-1)+c$

6. $-\frac{1}{3x^3}+\frac{1}{x}+\tan^{-1}x+c$

7. $\sec^{-1}x-\frac{\log x}{\sqrt{x^2-1}}+c$

8. $\frac{1}{2}\log\frac{1+\sqrt{\tan x}}{1-\sqrt{\tan x}}-\tan^{-1}\sqrt{\tan x}+c$

9. $\frac{4}{5\cdot\sqrt[4]{27}}(4\sqrt{3}+\sqrt{3+x})^{3/2}\cdot(\sqrt{3x+9}-8)+c$

10. $(x+1)\cot^{-1}\left(\frac{x+1}{2}\right)+\log(x^2+2x+5)+c$

11. $-\frac{1}{2}\cdot\frac{1}{(\sec x+\tan x)^{n-1}}\left\{\frac{1}{n-1}+\frac{1}{n+1}\cdot\frac{1}{(\sec x+\tan x)^2}\right\}+c$

12. $\frac{1}{4}\sin x\cdot\cos^3 x+\frac{3}{8}(x+\sin x\cdot\cos x)+c$

13. $e^{x\sin x+\cos x}\cdot\left(x-\frac{1}{x\cos x}\right)+c$

14. $\frac{1}{6(m+1)}(2x^{3m}+3x^{2m}+6x^m)^{\frac{m+1}{m}}+c$

□

9. Indefinite Integration of Rational and Irrational Functions

Recap of Facts and Formulae

1. Integration of some standard rational and irrational functions (fractions)

Standard integrals are as follows (without writing the constant of integration):

- $\int \frac{dx}{x^2+a^2} = \frac{1}{a}\tan^{-1}\frac{x}{a}$

  $\int \frac{dx}{x^2-a^2} = \frac{1}{2a}\log\frac{x-a}{x+a}, (x > a)$

  $\int \frac{dx}{a^2-x^2} = \frac{1}{2a}\log\frac{a+x}{a-x}, (x < a)$

- $\int \frac{dx}{\sqrt{x^2-a^2}} = \log(x+\sqrt{x^2-a^2})$

  $\int \frac{dx}{\sqrt{a^2-x^2}} = \sin^{-1}\frac{x}{a}$

  $\int \frac{dx}{\sqrt{a^2+x^2}} = \log(x+\sqrt{a^2+x^2})$

- $\int \sqrt{a^2+x^2}\,dx = \frac{x}{2}\sqrt{x^2+a^2}+\frac{a^2}{2}\log(x+\sqrt{x^2+a^2})$

  $\int \sqrt{a^2-x^2}\,dx = \frac{a^2}{2}\sin^{-1}\frac{x}{a}+\frac{x\sqrt{a^2-x^2}}{2}$

  $\int \sqrt{x^2-a^2}\,dx = \frac{x}{2}\sqrt{x^2-a^2}-\frac{a^2}{2}\log(x+\sqrt{x^2-a^2})$

2. Integration of rational fractions

Let L_1, L_2, etc., denote polynomials of the first degree and Q_1, Q_2, etc., denote polynomials of the second degree, and N denote the numerator. Usual methods of integration of rational fractions are as follows:

- Form $\int \frac{dx}{L_1^n}$ Substitute $L_1 = z$.
- Form $\int \frac{dx}{Q_1}$ Put $Q_1 = \pm a^2 \pm (L_1)^2$ and apply the suitable result from Section 1.

3. Use of partial fractions in the integration of rational fractions

- Form

  $\int \frac{Ndx}{Q_1}$, where $Q_1 = L_1 \cdot L_2$ and degree of $N \le 1$

  $\int \frac{Ndx}{L_1 \cdot Q_1}$, where Q_1 cannot be factorized and degree of $N \le 2$

  $\int \frac{Ndx}{L_1^2 \cdot Q_1}$, where Q_1 cannot be factorized and degree of $N \le 3$

  $\int \frac{Ndx}{Q_1 \cdot Q_2}$, where Q_1, Q_2 cannot be factorized and degree of $N \le 3$

  $\int \frac{Ndx}{L_1 \cdot Q_1^2}$, where Q_1 cannot be factorized and degree of $N \le 4$

  In each of these we use partial fractions to change the integrand in algebraic sum of standard integrands.

- *Method of changing* $\frac{N}{Q_1}$ in partial fractions, where $Q_1 = L_1 \cdot L_2$ and the degree of $N \le 1$:

$$\frac{N}{Q_1} = \frac{N}{L_1 \cdot L_2} = \frac{A}{L_1}+\frac{B}{L_2} \text{ where } A, B \text{ are constants}$$

$$\Rightarrow \quad N = AL_2 + BL_1.$$

Equate the coefficients of similar powers of the variable on both sides and obtain two equations in A, B to obtain their values.

Or, put two suitable values for the variable on both sides successively and obtain two equations in A, B to find their values.

- *Method of changing* $\frac{N}{L_1 \cdot Q_1}$ in partial fractions, where Q_1 cannot be factorized and the degree of $N \le 2$:

$$\frac{N}{L_1 \cdot Q_1} \equiv \frac{A}{L_1} + \frac{Bx+C}{Q_1}$$

where A, B, C are constants and x is the variable $\Rightarrow N \equiv AQ_1 + (Bx+C)L_1$. Obtain three equations in A, B, C in any one of the two ways as given in the above point and solve.

- *Method of changing* $\dfrac{N}{L_1^2 \cdot Q_1}$ in partial fractions, where Q_1 cannot be factorized and the degree of $N \le 3$:

$$\frac{N}{L_1^2 \cdot Q_1} = \frac{A}{L_1} + \frac{B}{L_1^2} + \frac{Cx+D}{Q_1}$$

Find A, B, C, D as in the above points.

- *Method of changing* $\dfrac{N}{L_1 \cdot Q_1^2}$ in partial fractions, where Q_1 cannot be factorized and the degree of $N \le 4$:

$$\frac{N}{L_1 \cdot Q_1^2} = \frac{A}{L_1} + \frac{Bx+C}{Q_1} + \frac{Dx+E}{Q_1^2}$$

Find A, B, C, D, E as in the above points.

Note If the fraction is $\dfrac{N}{D}$ where the degree of $N \ge$ the degree of D then write $\dfrac{N}{D} = P + \dfrac{N_1}{D}$ where the degree of $N_1 <$ the degree of D. Then express $\dfrac{N_1}{D}$ in partial fractions.

- *Method of integrating* $\int \frac{N}{D}dx$, where $N = AD + B\dfrac{dD}{dx} + N_1$ and A, B are constants, and the degree of $N_1 <$ the degree of D:

$$\int \frac{N}{D}dx = \int \frac{AD + B\dfrac{dD}{dx} + N_1}{D}dx$$

$$= A\int dx + B\int \frac{dD}{D} + \int \frac{N_1}{D}dx$$

$$= Ax + B\log D + \int \frac{N_1}{D}dx$$

where $\int \frac{N_1}{D}dx$ can be integrated by using partial fractions.

4. Integration of irrational fractions

- Form $\int \dfrac{dx}{\sqrt{L_1}}$ Substitute $L_1 = z$.
- Form $\int \dfrac{dx}{\sqrt{Q_1}}$ Put $Q_1 = \pm a^2 \pm (L_1)^2$ and apply the suitable result from Section 1.
- Form $\int \dfrac{L_1 dx}{\sqrt{Q_1}}$ Express $L_1 = A\dfrac{dQ_1}{dx} + B$ where A, B are constants.
- Form $\int \dfrac{Q_2 dx}{\sqrt{Q_1}}$ Express $Q_2 = AQ_1 + B\dfrac{dQ_1}{dx} + C$ where A, B, C are constants.
- Form $\int \dfrac{dx}{L_1\sqrt{L_2}}$ Substitute $\sqrt{L_2} = t$.
- Form $\int \dfrac{dx}{L_1\sqrt{Q_1}}$ Substitute $L_1 = \dfrac{1}{t}$.
- Form $\int \dfrac{dx}{Q_1\sqrt{L_2}}$ Substitute $\sqrt{L_2} = t$.
- Form $\int \dfrac{dx}{Q_2\sqrt{Q_1}}$ Substitute $\sqrt{\dfrac{Q_1}{Q_2}} = t$.

Selected Solved Examples

1. Integrate $\int \dfrac{x^2+x+3}{(x-2)(x+1)}dx.$

$$\frac{x^2+x+3}{(x-2)(x+1)} = \frac{x^2+x+3}{x^2-x-2} = \frac{(x^2-x-2)+(2x+5)}{x^2-x-2}$$

$$= 1 + \frac{2x+5}{x^2-x-2} = 1 + \frac{2x+5}{(x-2)(x+1)}.$$

Now, $\dfrac{2x+5}{(x-2)(x+1)} \equiv \dfrac{A}{x-2} + \dfrac{B}{x+1}$

$\Rightarrow \quad 2x+5 \equiv A(x+1) + B(x-2).$

Putting $x = -1, 2$ successively we get

$3 = B(-3)$ and $9 = A(3)$; $\quad \therefore \quad A = 3, B = -1$

$$\therefore \quad I = \int\left(1 + \frac{A}{x-2} + \frac{B}{x+1}\right)dx = \int\left(1 + \frac{3}{x-2} - \frac{1}{x+1}\right)dx$$

$$= x + 3\log(x-2) - \log(x+1) + c.$$

2. Evaluate $\int \dfrac{dx}{(x-1)(1+x^2)}.$

Let $\dfrac{1}{(x-1)(1+x^2)} \equiv \dfrac{A}{x-1} + \dfrac{Bx+C}{1+x^2}$

$\Rightarrow \quad 1 \equiv A(1+x^2) + (Bx+C)(x-1)$ ... (1)

'ting $x = 1$, $1 = A(2)$; $\quad \therefore \quad A = \dfrac{1}{2}$.

Equating coefficients of x^2 on both sides, $0 = A + B$.

Equating coefficients of x^0 on both sides, $1 = A - C$

$$\therefore \quad B = -A = -\frac{1}{2}, C = A - 1 = \frac{1}{2} - 1 = -\frac{1}{2}$$

$$\therefore \quad I = \int \left(\frac{A}{x-1} + \frac{Bx+C}{1+x^2} \right) dx$$

$$= \int \left(\frac{1}{2} \cdot \frac{1}{x-1} - \frac{1}{2} \cdot \frac{x+1}{1+x^2} \right) dx$$

$$= \frac{1}{2} \int \frac{dx}{x-1} - \frac{1}{2} \int \frac{x\,dx}{1+x^2} - \frac{1}{2} \int \frac{dx}{1+x^2}$$

$$= \frac{1}{2} \log(x-1) - \frac{1}{4} \log(1+x^2) - \frac{1}{2} \tan^{-1} x + c.$$

3. Integrate $\int \frac{3x+1}{(x-1)^3 \cdot (x+1)} dx.$

Let $\frac{3x+1}{(x-1)^3 \cdot (x+1)} \equiv \frac{A}{x-1} + \frac{B}{(x-1)^2} + \frac{C}{(x-1)^3} + \frac{D}{x+1}$ (partial fractions)

$$\Rightarrow \quad 3x + 1 \equiv A(x-1)^2(x+1) + B(x-1)(x+1) + C(x+1) + D(x-1)^3.$$

Putting $x = -1$, $-2 = D(-2)^3$; $\therefore \quad D = \frac{1}{4}$.

Equating coefficients of x^3 on both sides, $0 = A + D$

Equating coefficients of x^2 on both sides,

$$0 = -A + B - 3D$$

Equating coefficients of x^0 on both sides,

$$1 = A - B + C - D$$

$$\therefore \quad A = -D = -\frac{1}{4}; \quad B = A + 3D = -\frac{1}{4} + 3 \cdot \frac{1}{4} = \frac{1}{2}$$

$$C = 1 - A + B + D = 1 + \frac{1}{4} + \frac{1}{2} + \frac{1}{4} = 2.$$

$$\therefore \quad I = \int \left\{ \frac{A}{x-1} + \frac{B}{(x-1)^2} + \frac{C}{(x-1)^3} + \frac{D}{x+1} \right\} dx$$

$$= \int -\frac{1}{4} \cdot \frac{dx}{x-1} + \int \frac{1}{2} \cdot \frac{dx}{(x-1)^2} + \int 2 \cdot \frac{dx}{(x-1)^3} + \int \frac{1}{4} \cdot \frac{dx}{x+1}$$

$$= -\frac{1}{4} \log(x-1) - \frac{1}{2} \cdot \frac{1}{x-1} - \frac{1}{(x-1)^2} + \frac{1}{4} \log(x+1) + c$$

$$= \frac{1}{4} \log \frac{x+1}{x-1} - \frac{1}{2(x-1)} - \frac{1}{(x-1)^2} + c.$$

Second Method

Put $x - 1 = \frac{1}{y}$; then $dx = \frac{-1}{y^2} dy$

$$\therefore \quad I = \int \frac{3\left(1 + \frac{1}{y}\right) + 1}{\frac{1}{y^3}\left(1 + \frac{1}{y} + 1\right)} \cdot \frac{-1}{y^2} dy = -\int \frac{\left(4 + \frac{3}{y}\right) y}{2 + \frac{1}{y}} dy$$

$$= -\int \frac{4y^2 + 3y}{2y+1} dy$$

$$= -\int \frac{2y(2y+1) + y}{2y+1} dy$$

$$= -\int \left(2y + \frac{y}{2y+1}\right) dy = -\int \left\{2y + \frac{1}{2} \cdot \frac{(2y+1) - 1}{2y+1}\right\} dy$$

$$= -2 \int y\,dy - \frac{1}{2} \int dy + \frac{1}{2} \int \frac{1}{2y+1} dy$$

$$= -y^2 - \frac{1}{2} y + \frac{1}{4} \log(2y+1) + c.$$

$$= -\frac{1}{(x-1)^2} - \frac{1}{2(x-1)} + \frac{1}{4} \log\left(\frac{2}{x-1} + 1\right) + c$$

$$= \frac{1}{4} \log \frac{x+1}{x-1} - \frac{1}{2(x-1)} - \frac{1}{(x-1)^2} + c.$$

4. Integrate $\frac{x^5}{1-x^9}$ with respect to x.

Here, $I = \int \frac{x^5}{1-x^9} dx.$

Putting $x^3 = z$, $3x^2 dx = dz$.

$$\therefore \quad I = \frac{1}{3} \int \frac{x^3}{1-x^9} dz = \frac{1}{3} \int \frac{z}{1-z^3} dz$$

$$= \frac{1}{3} \int \frac{z\,dz}{(1-z)(1+z+z^2)}.$$

Let $\frac{z}{(1-z)(1+z+z^2)} \equiv \frac{A}{1-z} + \frac{Bz+C}{1+z+z^2}$

$$\Rightarrow \quad z \equiv A(1+z+z^2) + (Bz+C)(1-z).$$

Putting $z = 1$, $1 = 3A$; $\therefore \quad A = \frac{1}{3}$.

Equating coefficients of z^2 on both sides,

$$0 = A - B; \quad \therefore \quad B = A = \frac{1}{3}$$

Equating coefficients of z^0 on both sides,

$$0 = A + C; \quad \therefore \quad C = -A = -\frac{1}{3}$$

$$\therefore \quad I = \frac{1}{3} \int \left(\frac{A}{1-z} + \frac{Bz+C}{1+z+z^2} \right) dz$$

$$= \frac{1}{3}\int \frac{1}{3}\cdot\frac{dz}{1-z} + \frac{1}{3}\int \frac{\frac{1}{3}(z-1)}{1+z+z^2}dz$$

$$= \frac{-1}{9}\log(z-1) + \frac{1}{9}\int \frac{z-1}{z^2+z+1}dz$$

$$= \frac{-1}{9}\log(z-1) + \frac{1}{18}\int \frac{(2z+1)-3}{z^2+z+1}dz$$

$$= \frac{-1}{9}\log(z-1) + \frac{1}{18}\int \frac{d(z^2+z+1)}{z^2+z+1} - \frac{1}{6}\int \frac{dz}{\left(z+\frac{1}{2}\right)^2+\left(\frac{\sqrt{3}}{2}\right)^2}$$

$$= -\frac{1}{9}\log(z-1) + \frac{1}{18}\log(z^2+z+1) - \frac{1}{6}\cdot\frac{1}{\frac{\sqrt{3}}{2}}\tan^{-1}\frac{z+\frac{1}{2}}{\frac{\sqrt{3}}{2}} + c$$

$$= -\frac{1}{9}\log(x^3-1) + \frac{1}{18}\log(x^6+x^3+1) - \frac{1}{3\sqrt{3}}\tan^{-1}\frac{2x^3+1}{\sqrt{3}} + c.$$

5. Integrate $\frac{1+x^{2/3}}{1+x}$ with respect to x.

Here, $I = \int \frac{1+x^{2/3}}{1+x}dx$; putting $x = z^3$, $dx = 3z^2dz$.

$$\therefore \quad I = \int \frac{1+z^2}{1+z^3}\cdot 3z^2dz$$

$$= 3\int \frac{z^4+z^2}{z^3+1}dz$$

$$= 3\int \left(z + \frac{z^2-z}{z^3+1}\right)dz$$

$$= 3\cdot\frac{z^2}{2} + 3\int \frac{z^2-z}{(z+1)(z^2-z+1)}dz.$$

Let $\frac{z^2-z}{(z+1)(z^2-z+1)} \equiv \frac{A}{z+1} + \frac{Bz+C}{z^2-z+1}$

$\therefore \quad z^2 - z \equiv A(z^2-z+1) + (Bz+C)(z+1).$

Putting $z = -1$, $2 = A(1+1+1)$; $\quad\therefore\quad A = \frac{2}{3}$.

Equating coefficients of z^2 on both sides,

$1 = A + B; \quad \therefore \quad B = 1 - A = 1 - \frac{2}{3} = \frac{1}{3}$.

Equating coefficients of z^0 on both sides,

$0 = A + C; \quad \therefore \quad C = \frac{-2}{3}$

$$\therefore \quad I = \frac{3}{2}z^2 + 3\int \frac{2}{3}\cdot\frac{dz}{z+1} + 3\int \frac{1}{3}\cdot\frac{z-2}{z^2-z+1}dz$$

$$= \frac{3}{2}z^2 + 2\log(z+1) + \frac{1}{2}\int \frac{(2z-1)-3}{z^2-z+1}dz$$

$$= \frac{3}{2}z^2 + 2\log(z+1) + \frac{1}{2}\int \frac{d(z^2-z+1)}{z^2-z+1} - \frac{3}{2}\int \frac{dz}{\left(z-\frac{1}{2}\right)^2+\left(\frac{\sqrt{3}}{2}\right)^2}$$

$$= \frac{3}{2}z^2 + 2\log(z+1) + \frac{1}{2}\log(z^2-z+1) - \frac{3}{2}\cdot\frac{1}{\frac{\sqrt{3}}{2}}\tan^{-1}\frac{z-\frac{1}{2}}{\frac{\sqrt{3}}{2}} + c$$

$$= \frac{3}{2}x^{2/3} + 2\log(x^{1/3}+1) + \frac{1}{2}\log(x^{2/3} - x^{1/3} + 1) - \sqrt{3}\tan^{-1}\frac{2x^{1/3}-1}{\sqrt{3}} + c.$$

6. Evaluate $\int \frac{dx}{\sin x\cdot(3+2\cos x)}$.

$$I = \int \frac{\sin x\, dx}{\sin^2 x(3+2\cos x)} = \int \frac{\sin x\, dx}{(1-\cos^2 x)(3+2\cos x)}.$$

Put $\cos x = z$; then $-\sin x\, dx = dz$

$$\therefore \quad I = \int \frac{-dz}{(1-z^2)(3+2z)} = \int \frac{dz}{(z+1)(z-1)(2z+3)}.$$

Let $\frac{1}{(z+1)(z-1)(2z+3)} \equiv \frac{A}{z+1} + \frac{B}{z-1} + \frac{C}{2z+3}$

or $\quad 1 \equiv A(z-1)(2z+3) + B(z+1)(2z+3) + C(z+1)(z-1).$

Putting $z = 1, -1, -\frac{3}{2}$ successively, we get

$$1 = B\cdot 2\cdot 5, \quad 1 = A(-2)\cdot 1, \quad 1 = C\cdot\left(-\frac{1}{2}\right)\left(-\frac{5}{2}\right)$$

$$\therefore \quad B = \frac{1}{10}, A = -\frac{1}{2}, C = \frac{4}{5}.$$

$$\therefore \quad I = \int\left(\frac{A}{z+1} + \frac{B}{z-1} + \frac{C}{2z+3}\right)dz$$

$$= -\frac{1}{2}\log(z+1) + \frac{1}{10}\log(z-1) + \frac{2}{5}\log(2z+3) + c$$

$$= -\frac{1}{2}\log(\cos x + 1) + \frac{1}{10}\log(\cos x - 1) + \frac{2}{5}\log(2\cos x + 3) + c.$$

7. Integrate $\int \frac{dx}{1+x^4}$.

Here $I = \frac{1}{2}\int \frac{(x^2+1)+(1-x^2)}{1+x^4}\,dx$

$$= \frac{1}{2}\int \frac{x^2+1}{x^4+1}\,dx + \frac{1}{2}\int \frac{1-x^2}{1+x^4}\,dx = I_1 + I_2 \qquad \dots (1)$$

Now, $I_1 = \frac{1}{2}\int \frac{x^2+1}{x^4+1}\,dx = \frac{1}{2}\int \frac{1+\frac{1}{x^2}}{x^2+\frac{1}{x^2}}\,dx.$

Put $x - \frac{1}{x} = z$; then $\left(1+\frac{1}{x^2}\right)dx = dz$

$$\therefore \quad I_1 = \frac{1}{2}\int \frac{dz}{\left(x-\frac{1}{x}\right)^2 + 2} = \frac{1}{2}\int \frac{dz}{z^2 + (\sqrt{2})^2}$$

$$= \frac{1}{2}\cdot\frac{1}{\sqrt{2}}\tan^{-1}\frac{z}{\sqrt{2}} + c_1$$

$$= \frac{1}{2\sqrt{2}}\tan^{-1}\frac{1}{\sqrt{2}}\left(x - \frac{1}{x}\right) + c_1 \qquad \dots (2)$$

Next, $I_2 = \frac{1}{2}\int \frac{1-x^2}{1+x^4}\,dx = \frac{1}{2}\int \frac{\frac{1}{x^2}-1}{\frac{1}{x^2}+x^2}\,dx.$

Put $x + \frac{1}{x} = t$; then $\left(1 - \frac{1}{x^2}\right)dx = dt$

$$\therefore \quad I_2 = \frac{1}{2}\int \frac{-dt}{\left(x+\frac{1}{x}\right)^2 - 2} = -\frac{1}{2}\int \frac{dt}{t^2 - (\sqrt{2})^2}$$

$$= -\frac{1}{2}\cdot\frac{1}{2\sqrt{2}}\log\frac{x+\frac{1}{x}-\sqrt{2}}{x+\frac{1}{x}+\sqrt{2}} + c_2$$

$$= -\frac{1}{4\sqrt{2}}\log\frac{x^2-\sqrt{2}x+1}{x^2+\sqrt{2}x+1} + c_2 \qquad \dots (3)$$

$$\therefore \quad I = \frac{1}{2\sqrt{2}}\tan^{-1}\frac{1}{\sqrt{2}}\left(x-\frac{1}{x}\right) - \frac{1}{4\sqrt{2}}\log\frac{x^2-\sqrt{2}x+1}{x^2+\sqrt{2}x+1} + c.$$

8. Evaluate $\int \frac{dx}{\cos\alpha + \cos x}$**, where α is a constant.**

Let $\cos\alpha = a(\neq \pm 1)$. Then

$$I = \int \frac{dx}{a + \frac{1-\tan^2\frac{x}{2}}{1+\tan^2\frac{x}{2}}} = \int \frac{\sec^2\frac{x}{2}\,dx}{a\left(1+\tan^2\frac{x}{2}\right) + 1 - \tan^2\frac{x}{2}}.$$

Put $\tan\frac{x}{2} = z$; then $\frac{1}{2}\sec^2\frac{x}{2}\,dx = dz$

$$\therefore \quad I = \int \frac{2dz}{a(1+z^2)+1-z^2} = \int \frac{2dz}{(a-1)z^2+(a+1)}$$

$$= 2\int \frac{dz}{(a+1)-(1-a)z^2}$$

$$= \frac{2}{1-a}\int \frac{dz}{\left(\sqrt{\frac{1+a}{1-a}}\right)^2 - z^2}$$

$$\left(\because \frac{1+a}{1-a} \text{ is positive for } -1 < \cos\alpha < 1\right)$$

$$= \frac{2}{1-a}\cdot\frac{1}{2\cdot\sqrt{\frac{1+a}{1-a}}}\log\frac{\sqrt{\frac{1+a}{1-a}}+z}{\sqrt{\frac{1+a}{1-a}}-z} + c$$

$$= \frac{1}{\sqrt{1-a^2}}\log\frac{\sqrt{1+a}+\sqrt{1-a}\cdot\tan\frac{x}{2}}{\sqrt{1+a}-\sqrt{1-a}\cdot\tan\frac{x}{2}} + c$$

$$= \frac{1}{\sin\alpha}\cdot\log\frac{\cos\frac{\alpha}{2}+\sin\frac{\alpha}{2}\cdot\tan\frac{x}{2}}{\cos\frac{\alpha}{2}-\sin\frac{\alpha}{2}\cdot\tan\frac{x}{2}}\, c \quad (\because a = \cos\alpha)$$

If $\cos\alpha = 1$, $I = \int \frac{dx}{1+\cos x} = \frac{1}{2}\int \sec^2\frac{x}{2}\,dx = \tan\frac{x}{2} + c_1$.

If $\cos\alpha = -1$, $I = \int \frac{dx}{-1+\cos x} = \frac{1}{2}\int -\frac{dx}{\sin^2\frac{x}{2}} = \cot\frac{x}{2} + c_2$.

Thus, we get $I = \frac{1}{\sin\alpha}\log\frac{\cos\frac{\alpha-x}{2}}{\cos\frac{\alpha+x}{2}} + c$ if $\alpha \neq n\pi$

$\tan\frac{x}{2} + c_1$ if $\alpha = 2n\pi$

$\cot\frac{x}{2} + c_2$ if $\alpha = (2n+1)\pi$.

9. Evaluate $\int \frac{dx}{1-\cot x}$ $\left(\textbf{or Find} \int \frac{\sin x}{\sin x - \cos x}\,dx\right)$.

$$I=\int \frac{dx}{1-\cot x}=\int \frac{dx}{1-\dfrac{\cos x}{\sin x}}=\int \frac{\sin x\, dx}{\sin x-\cos x}.$$

Let $\sin x \equiv A \dfrac{d}{dx}(\sin x-\cos x)+B(\sin x-\cos x)$

$\Rightarrow \quad \sin x \equiv A(\cos x+\sin x)+B(\sin x-\cos x).$

Equating coefficients of $\sin x$ and $\cos x$ successively on both sides,

$$1=A+B \text{ and } 0=A-B.$$

Solving these we get $A=B=\dfrac{1}{2}$.

$$\therefore \quad I=\int \frac{A\dfrac{d}{dx}(\sin x-\cos x)+B(\sin x-\cos x)}{\sin x-\cos x}dx$$

$$=\frac{1}{2}\int \frac{d(\sin x-\cos x)}{\sin x-\cos x}+\frac{1}{2}\int dx$$

$$=\frac{1}{2}\log(\sin x-\cos x)+\frac{1}{2}x+c.$$

10. Evaluate $\displaystyle\int \frac{3+2\cos x+\sin x}{6+3\cos x+4\sin x}dx.$

Let $3+2\cos x+\sin x \equiv A\dfrac{d}{dx}(6+3\cos x+4\sin x)$
$+B(6+3\cos x+4\sin x)+C$

$\Rightarrow \quad 3+2\cos x+\sin x \equiv A(-3\sin x+4\cos x)$
$+B(6+3\cos x+4\sin x)+C.$

Equating constant terms, coefficients of $\cos x$ and coefficients of $\sin x$ successively on both the sides, we get

$$3=6B+C,\ 2=4A+3B \text{ and } 1=-3A+4B.$$

From the last two we get, $B=\dfrac{2}{5}$ and $A=\dfrac{1}{5}$

$$\therefore \quad C=3-6B=3-6\times\frac{2}{5}=\frac{3}{5}.$$

$$\therefore \quad I=\int \frac{\left\{A\dfrac{d}{dx}(6+3\cos x+4\sin x)+B(6+3\cos x+4\sin x)+C\right\}}{6+3\cos x+4\sin x}dx$$

$$=\frac{1}{5}\int \frac{d(6+3\cos x+4\sin x)}{6+3\cos x+4\sin x}+\frac{2}{5}\int dx+\frac{3}{5}\int \frac{dx}{6+3\cos x+4\sin x}$$

$$=\frac{1}{5}\log(6+3\cos x+4\sin x)+\frac{2}{5}x+\frac{3}{5}\int \frac{dx}{6+3\cos x+4\sin x}.$$

Now, $\displaystyle\int \frac{dx}{6+3\cos x+4\sin x}$

$$=\int \frac{dx}{6+3\cdot\dfrac{1-\tan^2\frac{x}{2}}{1+\tan^2\frac{x}{2}}+4\cdot\dfrac{2\tan\frac{x}{2}}{1+\tan^2\frac{x}{2}}}$$

$$=\int \frac{\sec^2\frac{x}{2}\,dx}{6\left(1+\tan^2\frac{x}{2}\right)+3\left(1-\tan^2\frac{x}{2}\right)+8\tan\frac{x}{2}}$$

$$=\int \frac{2dz}{6(1+z^2)+3(1-z^2)+8z}$$

$\left\{\text{putting } \tan\dfrac{x}{2}=z, \text{ we get } \dfrac{1}{2}\sec^2\dfrac{x}{2}dx=dz\right\}$

$$=\int \frac{2dz}{3z^2+8z+9}=\frac{2}{3}\int \frac{dz}{z^2+\frac{8}{3}z+3}$$

$$=\frac{2}{3}\int \frac{dz}{\left(z+\frac{4}{3}\right)^2+3-\frac{16}{9}}=\frac{2}{3}\int \frac{dz}{\left(z+\frac{4}{3}\right)^2+\left(\frac{\sqrt{11}}{3}\right)^2}$$

$$=\frac{2}{3}\cdot\frac{1}{\frac{\sqrt{11}}{3}}\tan^{-1}\frac{z+\frac{4}{3}}{\frac{\sqrt{11}}{3}}+c$$

$$=\frac{2}{\sqrt{11}}\tan^{-1}\frac{3\tan\frac{x}{2}+4}{\sqrt{11}}+c$$

$$\therefore \quad I=\frac{1}{5}\log(6+3\cos x+4\sin x)+\frac{2}{5}x+\frac{6}{5\sqrt{11}}\tan^{-1}\frac{3\tan\frac{x}{2}+4}{\sqrt{11}}+c.$$

11. Evaluate $\displaystyle\int \frac{dx}{\cos x+\operatorname{cosec} x}.$

$$I=\int \frac{dx}{\cos x+\dfrac{1}{\sin x}}=\int \frac{\sin x}{\sin x\cdot\cos x+1}dx$$

$$=\int \frac{2\sin x}{2+\sin 2x}dx=\int \frac{(\sin x+\cos x)+(\sin x-\cos x)}{2+\sin 2x}dx$$

$$=\int \frac{\sin x+\cos x}{2+\sin 2x}dx+\int \frac{\sin x-\cos x}{2+\sin 2x}dx$$

$$=\int \frac{\frac{d}{dx}(\sin x-\cos x)}{3-(1-\sin 2x)}dx-\int \frac{\frac{d}{dx}(\sin x+\cos x)}{1+(1+\sin 2x)}dx$$

$$= \int \frac{\frac{d}{dx}(\sin x - \cos x)}{3 - (\sin x - \cos x)^2} dx - \int \frac{\frac{d}{dx}(\sin x + \cos x)}{1 + (\sin x + \cos x)^2} dx$$

$= I_1 - I_2.$

For I_1, put $\sin x - \cos x = z$. Then

$$I_1 = \int \frac{dz}{3 - z^2} = \int \frac{dz}{(\sqrt{3})^2 - z^2}, \quad (|\sin x - \cos x| \le \sqrt{2})$$

$$= \frac{1}{2\sqrt{3}} \log \frac{\sqrt{3} + z}{\sqrt{3} - z} + c_1$$

$$= \frac{1}{2\sqrt{3}} \log \frac{\sqrt{3} + \sin x - \cos x}{\sqrt{3} - \sin x + \cos x} + c_1.$$

For I_2, put $\sin x + \cos x = t$. Then

$$I_2 = \int \frac{dt}{1 + t^2} = \tan^{-1} t + c_2 = \tan^{-1}(\sin x + \cos x) + c_2$$

$\therefore \quad I = I_1 - I_2$

$$= \frac{1}{2\sqrt{3}} \log \frac{\sqrt{3} + \sin x - \cos x}{\sqrt{3} - \sin x + \cos x} - \tan^{-1}(\sin x + \cos x) + c.$$

12. Evaluate $\displaystyle\int \frac{x^2 dx}{(x\sin x + \cos x)^2}$.

$$\frac{d}{dx}\left(\frac{1}{x\sin x + \cos x}\right) = \frac{-\{\sin x + x\cos x - \sin x\}}{(x\sin x + \cos x)^2} = \frac{-x\cos x}{(x\sin x + \cos x)^2}$$

$$\therefore \quad \int \frac{-x\cos x}{(x\sin x + \cos x)^2} dx = \frac{1}{x\sin x + \cos x} \quad \ldots (1)$$

$$\therefore \quad I = \int \frac{-x}{\cos x} \cdot \frac{-x\cos x}{(x\sin x + \cos x)^2} dx$$

$$= \frac{-x}{\cos x} \cdot \frac{1}{x\sin x + \cos x} - \int \frac{1}{x\sin x + \cos x} \cdot \frac{d}{dx}\left(\frac{-x}{\cos x}\right) dx,$$

{using by parts and (1)}

$$= \frac{-x}{\cos x(x\sin x + \cos x)} - \int \frac{1}{x\sin x + \cos x} \cdot \frac{-\cos x - x\sin x}{\cos^2 x} dx$$

$$= \frac{-x}{\cos x(x\sin x + \cos x)} + \int \sec^2 x \, dx$$

$$= \frac{-x}{\cos x(x\sin x + \cos x)} + \tan x + c$$

$$= \frac{-x}{\cos x(x\sin x + \cos x)} + \frac{\sin x}{\cos x} + c$$

$$= \frac{-x + x\sin^2 x + \sin x \cos x}{\cos x(x\sin x + \cos x)} + c$$

$$= \frac{\sin x - x\cos x}{x\sin x + \cos x} + c.$$

13. Evaluate $\displaystyle\int \frac{dx}{\sqrt{(x-\alpha)(\beta - x)}}$ **where** $\beta > \alpha$.

Here, $\displaystyle I = \int \frac{dx}{\sqrt{-x^2 + (\alpha + \beta)x - \alpha\beta}}$

$$= \int \frac{dx}{\sqrt{\frac{(\alpha+\beta)^2}{4} - \left\{x^2 - (\alpha+\beta)x + \frac{(\alpha+\beta)^2}{4}\right\} - \alpha\beta}}$$

$$= \int \frac{dx}{\sqrt{\left(\frac{\beta - \alpha}{2}\right)^2 - \left\{x - \frac{\alpha + \beta}{2}\right\}^2}}$$

$$= \sin^{-1} \frac{x - \frac{\alpha+\beta}{2}}{\frac{\beta - \alpha}{2}} + c = \sin^{-1} \frac{2x - \alpha - \beta}{\beta - \alpha} + c.$$

Second Method

Put $x = \alpha\cos^2\theta + \beta\sin^2\theta$; then $dx = (\beta - \alpha)\sin 2\theta \, d\theta$

$$\therefore \quad I = \int \frac{(\beta - \alpha)\sin 2\theta d\theta}{\sqrt{(\alpha\cos^2\theta + \beta\sin^2\theta - \alpha)(\beta - \alpha\cos^2\theta - \beta\sin^2\theta)}}$$

$$= \int \frac{(\beta - \alpha)2\sin\theta\cos\theta \, d\theta}{\sqrt{(\beta - \alpha)\sin^2\theta \cdot (\beta - \alpha)\cos^2\theta}}$$

$$= \int 2d\theta = 2\theta + c.$$

But $x - \alpha = (\beta - \alpha)\sin^2\theta$; $\quad \therefore \quad \sin\theta = \sqrt{\frac{x - \alpha}{\beta - \alpha}}$

$$\therefore \quad I = 2\theta + c = 2\sin^{-1}\sqrt{\frac{x - \alpha}{\beta - \alpha}} + c.$$

14. Evaluate $\displaystyle\int \frac{2x + 3}{\sqrt{1 + x + x^2}} dx$.

$$I = \int \frac{(2x+1) + 2}{\sqrt{1 + x + x^2}} dx = \int \frac{2x+1}{\sqrt{1 + x + x^2}} dx + \int \frac{2}{\sqrt{1 + x + x^2}} dx$$

$$= \int \frac{d(1 + x + x^2)}{(1 + x + x^2)^{1/2}} + 2\int \frac{dx}{\sqrt{\left(x + \frac{1}{2}\right)^2 + \left(\frac{\sqrt{3}}{2}\right)^2}}$$

$$= \frac{(1+x+x^2)^{-1/2+1}}{-\frac{1}{2}+1}$$

$$+ 2\log\left(x+\frac{1}{2}+\sqrt{\left(x+\frac{1}{2}\right)^2+\left(\frac{\sqrt{3}}{2}\right)^2}\right)+c$$

$$= 2\sqrt{1+x+x^2} + 2\log\left(x+\frac{1}{2}+\sqrt{1+x+x^2}\right)+c.$$

15. Evaluate $\int (3x-5)\sqrt{2+3x-x^2}\,dx.$

$$I = \frac{3}{2}\int\left(2x-\frac{10}{3}\right)\sqrt{2+3x-x^2}\,dx$$

$$= \frac{3}{2}\int\left\{(2x-3)-\frac{1}{3}\right\}\sqrt{2+3x-x^2}\,dx$$

$$= \frac{3}{2}\int(2x-3)\sqrt{2+3x-x^2}\,dx - \frac{1}{2}\int\sqrt{2+3x-x^2}\,dx$$

$$= \frac{3}{2}\int -(2+3x-x^2)^{1/2}\,d(2+3x-x^2)$$

$$-\frac{1}{2}\int\sqrt{2+\frac{9}{4}-\left(x^2-3x+\frac{9}{4}\right)}\,dx$$

$$= -\frac{3}{2}\,\frac{(2+3x-x^2)^{3/2}}{\frac{3}{2}} - \frac{1}{2}\int\sqrt{\left(\frac{\sqrt{17}}{2}\right)^2-\left(x-\frac{3}{2}\right)^2}\,dx$$

$$= -(2+3x-x^2)^{3/2}$$

$$-\frac{1}{2}\left\{\frac{17}{8}\sin^{-1}\frac{x-\frac{3}{2}}{\frac{\sqrt{17}}{2}}+\frac{\left(x-\frac{3}{2}\right)\sqrt{\frac{17}{4}-\left(x-\frac{3}{2}\right)^2}}{2}\right\}+c$$

$$= -(2+3x-x^2)^{3/2}$$

$$-\frac{1}{16}\left\{17\sin^{-1}\frac{2x-3}{\sqrt{17}}+(2x-3)\sqrt{17-(2x-3)^2}\right\}+c.$$

16. Integrate $\int \frac{\sin x+\sin 2x}{\sqrt{\cos x+\cos 2x}}\,dx.$

$$I = \int\frac{\sin x(1+2\cos x)}{\sqrt{\cos x+2\cos^2 x-1}}\,dx.$$

Put $\cos x = t$; then $-\sin x\,dx = dt$

$$\therefore\quad I = -\int\frac{1+2t}{\sqrt{2t^2+t-1}}\,dt = -\frac{1}{2}\int\frac{(4t+1)+1}{\sqrt{2t^2+t-1}}\,dt$$

$$= -\frac{1}{2}\int\frac{d(2t^2+t-1)}{(2t^2+t-1)^{1/2}} - \frac{1}{2}\int\frac{dt}{\sqrt{2\left(t^2+\frac{t}{2}-\frac{1}{2}\right)}}$$

$$= -\frac{1}{2}\cdot\frac{(2t^2+t-1)^{-1/2+1}}{-\frac{1}{2}+1}$$

$$-\frac{1}{2\sqrt{2}}\int\frac{dt}{\sqrt{\left(t+\frac{1}{4}\right)^2-\frac{1}{16}-\frac{1}{2}}}$$

$$= -\sqrt{2t^2+t-1} - \frac{1}{2\sqrt{2}}\int\frac{dt}{\sqrt{\left(t+\frac{1}{4}\right)^2-\left(\frac{3}{4}\right)^2}}$$

$$= -\sqrt{2t^2+t-1}$$

$$-\frac{1}{2\sqrt{2}}\log\left(t+\frac{1}{4}+\sqrt{\left(t+\frac{1}{4}\right)^2-\left(\frac{3}{4}\right)^2}\right)+c$$

$$= -\sqrt{\cos x+\cos 2x}$$

$$-\frac{1}{2\sqrt{2}}\log\left(\cos x+\frac{1}{4}+\sqrt{\cos^2 x+\frac{\cos x}{2}-\frac{1}{2}}\right)+c.$$

17. Evaluate $\int\frac{dx}{(3+x^2)\sqrt{1+x}}.$

Put $\sqrt{1+x} = t$, i.e., $1+x = t^2$; then $dx = 2t\,dt$

$$\therefore\quad I = \int\frac{2t\,dt}{\{3+(t^2-1)^2\}\,t}$$

$$= \int\frac{2\,dt}{t^4-2t^2+4}$$

$$= \int\frac{2\,dt}{(t^2+2)^2-6t^2}$$

$$= \int\frac{2\,dt}{(t^2+\sqrt{6}t+2)(t^2-\sqrt{6}t+2)}.$$

Let $\frac{1}{(t^2+\sqrt{6}t+2)(t^2-\sqrt{6}t+2)}$

$$\equiv \frac{At+B}{t^2+\sqrt{6}t+2}+\frac{Ct+D}{t^2-\sqrt{6}t+2}$$

$$\Rightarrow\quad 1 \equiv (At+B)(t^2-\sqrt{6}t+2)+(Ct+D)(t^2+\sqrt{6}t+2).$$

Equating coefficients of t^3, t^2, t, t^0 on both sides,

$$0 = A+C,\ 0 = -\sqrt{6}A+B+\sqrt{6}C+D,$$

$$0 = 2A-\sqrt{6}B+2C+\sqrt{6}D,\ 1 = 2B+2D.$$

Solving these, $A = \frac{1}{4\sqrt{6}}$, $B = \frac{1}{4}$, $C = -\frac{1}{4\sqrt{6}}$, $D = \frac{1}{4}$

$$\therefore \quad I = 2\int \left(\frac{\frac{1}{4\sqrt{6}}t + \frac{1}{4}}{t^2+\sqrt{6}t+2} + \frac{-\frac{1}{4\sqrt{6}}t + \frac{1}{4}}{t^2-\sqrt{6}t+2} \right) dt$$

$$= \frac{1}{2\sqrt{6}}\int \frac{t+\sqrt{6}}{t^2+\sqrt{6}t+2}\,dt + \frac{1}{2\sqrt{6}}\int \frac{-t+\sqrt{6}}{t^2-\sqrt{6}t+2}\,dt$$

$$= \frac{1}{4\sqrt{6}}\int \frac{(2t+\sqrt{6})+\sqrt{6}}{t^2+\sqrt{6}t+2}\,dt - \frac{1}{4\sqrt{6}}\int \frac{(2t-\sqrt{6})-\sqrt{6}}{t^2-\sqrt{6}t+2}\,dt$$

$$= \frac{1}{4\sqrt{6}}\log(t^2+\sqrt{6}t+2)$$

$$+\frac{1}{4}\int \frac{dt}{\left(t^2+2\cdot\frac{\sqrt{6}}{2}t+\frac{6}{4}\right)+2-\frac{6}{4}}$$

$$-\frac{1}{4\sqrt{6}}\log(t^2-\sqrt{6}t+2)$$

$$+\frac{1}{4}\int \frac{dt}{\left(t^2-2\cdot\frac{\sqrt{6}}{2}t+\frac{6}{4}\right)+2-\frac{6}{4}}$$

$$= \frac{1}{4\sqrt{6}}\log\frac{t^2+\sqrt{6}t+2}{t^2-\sqrt{6}t+2} + \frac{1}{4}\int \frac{dt}{\left(t+\sqrt{\frac{3}{2}}\right)^2+\left(\frac{1}{\sqrt{2}}\right)^2}$$

$$+\frac{1}{4}\int \frac{dt}{\left(t-\sqrt{\frac{3}{2}}\right)^2+\left(\frac{1}{\sqrt{2}}\right)^2}$$

$$= \frac{1}{4\sqrt{6}}\log\frac{t^2+\sqrt{6}t+2}{t^2-\sqrt{6}t+2}$$

$$+\frac{1}{4}\left\{\frac{1}{1/\sqrt{2}}\tan^{-1}\frac{t+\sqrt{\frac{3}{2}}}{\frac{1}{\sqrt{2}}} + \frac{1}{\frac{1}{\sqrt{2}}}\tan^{-1}\frac{t-\sqrt{\frac{3}{2}}}{\frac{1}{\sqrt{2}}}\right\} + c$$

$$= \frac{1}{4\sqrt{6}}\log\frac{t^2+\sqrt{6}t+2}{t^2-\sqrt{6}t+2}$$

$$+\frac{1}{2\sqrt{2}}\{\tan^{-1}(\sqrt{2}t+\sqrt{3}) + \tan^{-1}(\sqrt{2}t-\sqrt{3})\} + c$$

$$= \frac{1}{4\sqrt{6}}\log\frac{3+x+\sqrt{6+6x}}{3+x-\sqrt{6+6x}}$$

$$+\frac{1}{2\sqrt{2}}\{\tan^{-1}(\sqrt{2+2x}+\sqrt{3})$$

$$+\tan^{-1}(\sqrt{2+2x}-\sqrt{3})\} + c.$$

18. Integrate $\int \frac{dx}{(1-x)\sqrt{1+x^2}}$.

Put $1-x=\frac{1}{t}$; then $dx=\frac{1}{t^2}dt$

$$\therefore \quad I = \int \frac{\frac{1}{t^2}dt}{\frac{1}{t}\sqrt{1+\left(1-\frac{1}{t}\right)^2}} = \int \frac{dt}{t\sqrt{2-\frac{2}{t}+\frac{1}{t^2}}}$$

$$= \int \frac{dt}{\sqrt{2t^2-2t+1}} = \frac{1}{\sqrt{2}}\int \frac{dt}{\sqrt{t^2-t+\frac{1}{2}}}$$

$$= \frac{1}{\sqrt{2}}\int \frac{dt}{\sqrt{\left(t-\frac{1}{2}\right)^2+\frac{1}{2}-\frac{1}{4}}} = \frac{1}{\sqrt{2}}\int \frac{dt}{\sqrt{\left(t-\frac{1}{2}\right)^2+\left(\frac{1}{2}\right)^2}}$$

$$= \frac{1}{\sqrt{2}}\cdot\log\left(t-\frac{1}{2}+\sqrt{\left(t-\frac{1}{2}\right)^2+\left(\frac{1}{2}\right)^2}\right)+c$$

$$= \frac{1}{\sqrt{2}}\log\left(t-\frac{1}{2}+\sqrt{t^2-t+\frac{1}{2}}\right)+c$$

$$= \frac{1}{\sqrt{2}}\log\left(\frac{1}{1-x}-\frac{1}{2}+\sqrt{\left(\frac{1}{1-x}\right)^2-\frac{1}{1-x}+\frac{1}{2}}\right)+c$$

$$= \frac{1}{\sqrt{2}}\log\left(\frac{1+x}{2(1-x)}+\frac{1}{\sqrt{2}}\frac{\sqrt{x^2+1}}{1-x}\right)+c.$$

19. Evaluate $\int \frac{e^x dx}{(1+e^{2x})\sqrt{1-e^{2x}}}$.

Put $e^x = z$; then $e^x dx = dz$

$$\therefore \quad I = \int \frac{dz}{(1+z^2)\sqrt{1-z^2}} \qquad \left(\text{Form: } \int \frac{dx}{Q_1\sqrt{Q_2}}\right)$$

Let $\frac{1-z^2}{1+z^2}=t^2$, i.e., $z^2=\frac{1-t^2}{1+t^2}$

$$\therefore \quad 2z\,dz = \frac{-2t(1+t^2)-(1-t^2)\cdot 2t}{(1+t^2)^2}dt$$

$$\text{or} \quad z\,dz = \frac{-2t\,dt}{(1+t^2)^2}$$

$$\therefore \quad I = \int \frac{1}{\left(1+\frac{1-t^2}{1+t^2}\right)\sqrt{1-\frac{1-t^2}{1+t^2}}}\cdot\frac{-2t\,dt}{z(1+t^2)^2}$$

$$= \int \frac{(1+t^2)^{3/2}}{2\sqrt{2t^2}}\cdot\frac{-2t\,dt}{(1+t^2)^2}\cdot\sqrt{\frac{1+t^2}{1-t^2}}\,dt$$

$$= -\frac{1}{\sqrt{2}}\int \frac{dt}{\sqrt{1-t^2}} = -\frac{1}{\sqrt{2}}\sin^{-1}t + c$$

$$= -\frac{1}{\sqrt{2}}\sin^{-1}\sqrt{\frac{1-z^2}{1+z^2}} + c$$

$$= \frac{-1}{\sqrt{2}}\sin^{-1}\sqrt{\frac{1-e^{2x}}{1+e^{2x}}} + c.$$

20. Evaluate $\int \frac{dx}{x+\sqrt{x^2-x+1}}$.

Put $x+\sqrt{x^2-x+1} = z$ ... (1)

$\therefore \quad x^2-(x^2-x+1) = z(x-\sqrt{x^2-x+1})$

or $\quad x-\sqrt{x^2-x+1} = \frac{x-1}{z}$ ... (2)

(1) + (2) $\Rightarrow \quad 2x = z + \frac{x-1}{z}$; so, $2xz = z^2+x-1$

or $\quad (2z-1)x = z^2-1$, i.e., $x = \frac{z^2-1}{2z-1}$

$$\therefore \quad dx = \frac{2z(2z-1)-(z^2-1)2}{(2z-1)^2}dz = \frac{2z^2-2z+2}{(2z-1)^2}dz$$

$$\therefore \quad I = \int \frac{1}{z}\cdot\frac{2z^2-2z+2}{(2z-1)^2}dz = 2\int \frac{z^2-z+1}{z(2z-1)^2}dz.$$

Let $\frac{z^2-z+1}{z(2z-1)^2} \equiv \frac{A}{z} + \frac{B}{2z-1} + \frac{C}{(2z-1)^2}$

$\Rightarrow \quad z^2-z+1 \equiv A(2z-1)^2 + Bz(2z-1) + Cz.$

Putting $z = 0, \frac{1}{2}, 1$ successively we get

$$1 = A, \ \frac{3}{4} = \frac{1}{2}C, \ 1 = A+B+C.$$

These give $A = 1, C = \frac{3}{2}, B = -\frac{3}{2}$

$$\therefore \quad I = 2\int\frac{dz}{z} + 2\cdot\frac{-3}{2}\int\frac{dz}{2z-1} + 2\cdot\frac{3}{2}\int\frac{dz}{(2z-1)^2}$$

$$= 2\log z - 3\cdot\frac{\log(2z-1)}{2} + 3\cdot\left\{-\frac{1}{2z-1}\right\}\cdot\frac{1}{2} + c$$

$$= 2\log(x+\sqrt{x^2-x+1})$$

$$-\frac{3}{2}\log(2x-1+2\sqrt{x^2-x+1})$$

$$-\frac{3}{2}\cdot\frac{1}{2x-1+2\sqrt{x^2-x+1}} + c.$$

21. Evaluate $\int \sqrt{\tan x}\, dx.$

$I = \int \sqrt{\tan x}\, dx$

$$= \frac{1}{2}\left[\int(\sqrt{\tan x}+\sqrt{\cot x})\,dx + \int(\sqrt{\tan x}-\sqrt{\cot x})\,dx\right].$$

Now, $I_1 = \int(\sqrt{\tan x}+\sqrt{\cot x})\,dx$

$$= \int\frac{\sin x+\cos x}{\sqrt{\sin x\cos x}}dx = \int\frac{\sqrt{2}\,d(\sin x-\cos x)}{\sqrt{\sin 2x}}$$

$$= \sqrt{2}\int\frac{d(\sin x-\cos x)}{\sqrt{1-(1-\sin 2x)}}$$

$$= \sqrt{2}\int\frac{d(\sin x-\cos x)}{\sqrt{1-(\sin x-\cos x)^2}}$$

$$= \sqrt{2}\int\frac{dz}{\sqrt{1-z^2}}, \text{ where } z = \sin x-\cos x$$

$$= \sqrt{2}\sin^{-1}z + c_1 = \sqrt{2}\sin^{-1}(\sin x-\cos x) + c_1.$$

$$I_2 = \int(\sqrt{\tan x}-\sqrt{\cot x})\,dx = \int\frac{\sin x-\cos x}{\sqrt{\sin x\cos x}}dx$$

$$= \sqrt{2}\int\frac{-d(\sin x+\cos x)}{\sqrt{\sin 2x}} = -\sqrt{2}\int\frac{d(\sin x+\cos x)}{\sqrt{(1+\sin 2x)-1}}$$

$$= -\sqrt{2}\int\frac{d(\sin x+\cos x)}{\sqrt{(\sin x+\cos x)^2-1}}$$

$$= -\sqrt{2}\int\frac{dz}{\sqrt{z^2-1}}, \text{ where } z = \sin x+\cos x$$

$$= -\sqrt{2}\log(z+\sqrt{z^2-1}) + c_2$$

$$= -\sqrt{2}\log(\sin x+\cos x+\sqrt{\sin 2x}) + c_2$$

$$\therefore \quad I = \frac{1}{2}I_1 + \frac{1}{2}I_2$$

$$= \frac{1}{\sqrt{2}}\sin^{-1}(\sin x-\cos x)$$

$$-\frac{1}{\sqrt{2}}\log(\sin x+\cos x+\sqrt{\sin 2x}) + c.$$

22. Evaluate $\int \sqrt[3]{\tan x}\, dx.$

Let $\tan x = z^{3/2}$; then $\sec^2 x\, dx = \frac{3}{2}z^{1/2}\,dz$

$$\therefore \quad I = \int\sqrt[3]{z^{3/2}}\cdot\frac{3}{2}z^{1/2}\cdot\frac{1}{\sec^2 x}dz = \frac{3}{2}\int\frac{z}{1+z^3}dz$$

$$= \frac{3}{2}\int\frac{z\,dz}{(1+z)(1-z+z^2)}.$$

Let $\frac{z}{(1+z)(1-z+z^2)} \equiv \frac{A}{1+z} + \frac{Bz+C}{z^2-z+1}$

$\Rightarrow \quad z \equiv A(z^2-z+1) + (Bz+C)(1+z).$

Put $z = -1$. Then $-1 = A(1+1+1)$; so, $A = \frac{-1}{3}$.

Equating coefficients of z^2 on both sides,

$0 = A + B; \quad \therefore \quad B = \frac{1}{3}$

Equating coefficients of z^0 on both sides,

$0 = A + C; \quad \therefore \quad C = \frac{1}{3}$

$$\therefore \quad I = \frac{3}{2}\int -\frac{1}{3}\cdot\frac{dz}{1+z} + \frac{3}{2}\int \frac{\frac{1}{3}z+\frac{1}{3}}{z^2-z+1}\,dz$$

$$= -\frac{1}{2}\log(1+z) + \frac{1}{2}\int\frac{z+1}{z^2-z+1}\,dz$$

$$= -\frac{1}{2}\log(1+z) + \frac{1}{4}\int\frac{(2z-1)+3}{z^2-z+1}\,dz$$

$$= -\frac{1}{2}\log(1+z) + \frac{1}{4}\int\frac{d(z^2-z+1)}{z^2-z+1} + \frac{3}{4}\int\frac{dz}{\left(z-\frac{1}{2}\right)^2+\left(\frac{\sqrt{3}}{2}\right)^2}$$

$$= -\frac{1}{2}\log(1+z) + \frac{1}{4}\log(z^2-z+1) + \frac{3}{4}\cdot\frac{1}{\frac{\sqrt{3}}{2}}\tan^{-1}\frac{z-\frac{1}{2}}{\frac{\sqrt{3}}{2}} + c$$

$$= -\frac{1}{2}\log(1+\tan^{2/3}x) + \frac{1}{4}\log(\tan^{4/3}x - \tan^{2/3}x + 1) + \frac{\sqrt{3}}{2}\tan^{-1}\frac{2\tan^{2/3}x - 1}{\sqrt{3}} + c.$$

23. Evaluate $\int\sqrt{1+\sec x}\,dx$.

Here, $I = \int\sqrt{1+\frac{1}{\cos x}}\,dx$

$$= \int\frac{\sqrt{1+\cos x}}{\sqrt{\cos x}}\,dx$$

$$= \int\frac{\sqrt{2}\cos\frac{x}{2}}{\sqrt{1-2\sin^2\frac{x}{2}}}\,dx.$$

Put $\sin\frac{x}{2} = z$; then $\frac{1}{2}\cos\frac{x}{2}\,dx = dz$

$$\therefore \quad I = \int\frac{\sqrt{2}\cdot 2dz}{\sqrt{1-2z^2}}$$

$$= \frac{2\sqrt{2}}{\sqrt{2}}\int\frac{dz}{\sqrt{\left(\frac{1}{\sqrt{2}}\right)^2 - z^2}}$$

$$= 2\sin^{-1}\frac{z}{\frac{1}{\sqrt{2}}} + c = 2\sin^{-1}(\sqrt{2}z) + c$$

$$= 2\sin^{-1}\left(\sqrt{2}\sin\frac{x}{2}\right) + c.$$

24. Evaluate $\int\frac{\sqrt{\cos 2x}}{\sin x}\,dx$.

Here, $I = \int\frac{\cos 2x}{\sin x\sqrt{\cos 2x}}\,dx$

$$= \int\frac{1-2\sin^2 x}{\sin x\sqrt{\cos 2x}}\,dx$$

$$= \int\frac{dx}{\sin x\sqrt{\cos^2 x - \sin^2 x}} - 2\int\frac{\sin x}{\sqrt{2\cos^2 x - 1}}\,dx$$

$$= I_1 - 2I_2 \quad \ldots (1)$$

Now, $I_1 = \int\frac{dx}{\sin^2 x\sqrt{\cot^2 x - 1}}$

$$= \int\frac{\operatorname{cosec}^2 x\,dx}{\sqrt{\cot^2 x - 1}}.$$

Put $\cot x = z$; then $-\operatorname{cosec}^2 x\,dx = dz$

$$\therefore \quad I_1 = \int\frac{-dz}{\sqrt{z^2-1}} = -\log(z+\sqrt{z^2-1}) + c_1$$

$$= -\log(\cot x + \sqrt{\cot^2 x - 1}) + c_1.$$

Next, $I_2 = \int\frac{\sin x}{\sqrt{2\cos^2 x - 1}}\,dx;$

putting $\cos x = z$, $-\sin x\,dx = dz$

$$\therefore \quad I_2 = \int\frac{-dz}{\sqrt{2z^2-1}}$$

$$= -\frac{1}{\sqrt{2}}\int\frac{dz}{\sqrt{z^2-\left(\frac{1}{\sqrt{2}}\right)^2}}$$

$$= -\frac{1}{\sqrt{2}}\log\left|z+\sqrt{z^2-\left(\frac{1}{\sqrt{2}}\right)^2}\right| + c_2$$

$$= -\frac{1}{\sqrt{2}}\log\left|\cos x + \sqrt{\cos^2 x - \frac{1}{2}}\right| + c_2$$

$\therefore$ from (1),

$$I = -\log(\cot x + \sqrt{\cot^2 x - 1}) + \sqrt{2}\log\left|\cos x + \sqrt{\cos^2 x - \frac{1}{2}}\right| + c.$$

Exercises

Integrate the following:

1. $\int \frac{x-1}{x^2-5x+6}\,dx$

2. $\int \frac{x\,dx}{(x^2-a^2)(x^2-b^2)}$

3. $\int \frac{dx}{x^3-2x^2-x+2}$

4. $\int \frac{dx}{(x-2)(x-1)^2}$

5. $\int \frac{x^4dx}{(x-1)(x^2+1)}$

6. $\int \frac{dx}{(x-a)^2\cdot(x-b)^3}$

7. $\int \frac{dx}{1-x^4}$

8. $\int \frac{dx}{x^3+1}$

9. $\int \frac{x^3-1}{x^3+x}\,dx$

10. $\int \frac{dx}{x(1+x+x^2+x^3)}$

11. Evaluate $\int \frac{f(x)}{x^3-1}\,dx$, where $f(x)$ is a polynomial of the second degree in x such that $f(0)=f(1)=3f(2)=-3$.

Evaluate the following:

12. $\int \frac{\cos x\,dx}{(1+\sin x)^2\cdot(2+\sin x)}$

13. $\int \frac{\sin 2x\,dx}{3+3\cos x-\sin^2x}$

14. $\int \frac{dx}{\cos x(1+\cos^2x)}$

15. $\int \frac{dx}{\sin 2x-\sin x}$

16. $\int \sec x\cdot\sec 2x\,dx$

17. $\int \sin x\cdot\operatorname{cosec} 4x\,dx$

18. $\int \frac{\cos 2x\cdot\sin 4x\,dx}{\cos^4x(1+\cos^22x)}$

19. $\int \frac{\cos^3x+\cos^5x}{\sin^2x+\sin^4x}\,dx$

20. $\int \frac{\cos^2x+\sin 2x}{(2\cos x-\sin x)^2}\,dx$ or $\int \frac{1+2\tan x}{(2-\tan x)^2}\,dx$

21. $\int \frac{dx}{x\{10+7\log x+(\log x)^2\}}$

22. $\int \frac{dt}{e^t+3+2e^{-t}}$

23. $\int \frac{(x+1)dx}{x(1+xe^x)^2}$

Evaluate the following:

24. $\int \frac{x^2+1}{x^4+x^2+1}\,dx$

25. $\int \frac{x^2-1}{x^4-x^2+1}\,dx$

26. $\int \frac{x^2-1}{x^4+1}\,dx$

27. $\int \frac{x^2}{x^4+1}\,dx$

Integrate the following:

28. $\int \frac{dx}{3+5\sin x}$

29. $\int \frac{dx}{1+\sin\alpha\cdot\cos x}$, $\alpha\in\left(0,\frac{\pi}{2}\right)$

30. $\int \frac{dx}{(2+\sin x)(1-\sin x)}$

31. $\int \frac{dx}{5-4\sin x+3\cos x}$

32. $\int \frac{d\theta}{3\cos\theta+\sin\theta+2}$

33. $\int \frac{dx}{1+\sin x+\cos x}$

Find the following:

34. $\int \frac{\cos x}{\sin x+\cos x}\,dx$ or $\int \frac{dx}{1+\tan x}$

35. $\int \frac{2\sin x+3\cos x}{3\sin x+4\cos x}\,dx$

36. $\int \frac{1+\tan x}{1-\tan x}\,dx$

37. $\int \frac{dx}{1+\tan x+4\cot x}$

Find the following:

38. $\int \frac{dx}{a\tan x+b\cot x}$

39. $\int \frac{dx}{\sec x+\operatorname{cosec} x}$

40. $\int \frac{dx}{\sin x+\sec x}$ or $\int \frac{\cos x\,dx}{1+\sin x\cdot\cos x}$

Evaluate the following:

41. $\int \frac{dx}{(1+e\cos x)^2}$, $e<1$

42. $\int \frac{dx}{(2+\sin x)^2}$

43. $\int \frac{x^2\cos x}{(1+\sin x)^2}\,dx$

44. $\int \frac{x^2dx}{(x\cos x-\sin x)^2}$

Integrate the following:

45. $\int \frac{dx}{\sqrt{2x-x^2}}$

46. $\int \frac{dx}{\sqrt{(x-1)(2-x)}}$

47. $\int \frac{x\,dx}{\sqrt{(2x+1)(2x-3)}}$

48. $\int \sqrt{1-2x-x^2}\,dx$

49. $\int (x-1)\sqrt{1+x+x^2}\,dx$

50. $\int \frac{(x^2+1)}{\sqrt{x^2+x+1}}\,dx$

51. $\int \frac{\sin 2x}{\sqrt{1+\cos x+\cos^2x}}\,dx$

52. $\int \sqrt{x+a\sqrt{ax-a^2}}\,dx$, $|a|<2$

53. $\int \sqrt{x+\sqrt{x^2+a^2}}\,dx$

Evaluate:

54. $\int \sqrt[3]{\frac{1-x}{1+x}}\cdot\frac{dx}{x}$

55. $\int \frac{dx}{(1-x^3)^{1/3}}$

56. $\int \frac{dx}{(1+x)\sqrt{1-x}}$, $|x|<1$

57. $\int \frac{dx}{(1+x)\sqrt{1-x^2}}$

58. $\int \frac{dx}{(1-x)^{3/2}\sqrt{1+x}}$

59. $\int \frac{dx}{(1+x^2)\sqrt{1-x^2}}$

60. $\int \frac{dx}{(1+x)\sqrt{1+3x+3x^2}}$

61. $\int \frac{dx}{\sqrt{1+e^x+e^{2x}}}$

62. $\int \frac{x^2+1}{x\sqrt{x^2-1}}\,dx$

63. $\int \frac{(x^2-1)dx}{x\sqrt{x^4+3x^2+1}}$

64. $\int \frac{(1-x)dx}{(1+x)\sqrt{x+x^2+x^3}}$

65. $\int \frac{2-x^2}{(1+x)\sqrt{1-x^2}}\,dx$

66. $\int \frac{x^3+1}{\sqrt{x^2+x}}\,dx$

Integrate the following:

67. $\int (\sqrt{\tan x}+\sqrt{\cot x})dx$

68. $\int \sqrt{\cot x}\,dx$

69. $\int \frac{\sin x+\cos x}{\sqrt{\sin x\cos x}}\,dx$

70. $\int \frac{\sin x-\cos x}{\sqrt{2+\sin 2x}}\,dx$

Evaluate the following:

71. $\int \sqrt{1+\operatorname{cosec} x}\,dx$

72. $\int \sqrt{\sec x-1}\,dx$

73. $\int \sec\theta\cdot\sqrt{\cos 2\theta}\,d\theta$

74. $\int \frac{dx}{\cos x\sqrt{\cos 2x}}$

75. $\int \sqrt{\frac{\sin(x-\alpha)}{\sin(x+\alpha)}}\,dx$

76. $\int \frac{dx}{\sqrt{13+5\cos x-12\sin x}}$

77. $\int \frac{dx}{\sqrt{\sin^3 x(\sin x+2\cos x)}}$

Answers

1. $\log\frac{(x-3)^2}{x-2}+c$

2. $\frac{1}{2(a^2-b^2)}\log\frac{x^2-a^2}{x^2-b^2}+c$

3. $\frac{1}{3}\log(x-2)-\frac{1}{2}\log(x-1)+\frac{1}{6}\log(x+1)+c$

4. $\log\frac{x-2}{x-1}+\frac{1}{x-1}+c$

5. $\frac{x^2}{2}+x+\frac{1}{2}\log(x-1)-\frac{1}{4}\log(x^2+1)-\frac{1}{2}\tan^{-1}x+c$

6. $\frac{1}{(a-b)^4}\left\{-\frac{a-b}{x-a}-3\log\frac{x-a}{x-b}+2\frac{b-a}{x-b}-\frac{1}{2}\left(\frac{b-a}{x-b}\right)^2\right\}+c$

7. $\frac{1}{4}\log\frac{1+x}{1-x}+\frac{1}{2}\tan^{-1}x+c$

8. $\frac{1}{3}\log(x+1)-\frac{1}{6}\log(x^2-x+1)+\frac{1}{\sqrt{3}}\tan^{-1}\frac{2x-1}{\sqrt{3}}+c$

9. $x-\log x+\frac{1}{2}\log(x^2+1)-\tan^{-1}x+c$

10. $\log x-\frac{1}{2}\log(1+x)-\frac{1}{2}\tan^{-1}x-\frac{1}{4}\log(1+x^2)+c$

11. $\log\frac{1+x+x^2}{x-1}+\frac{2}{\sqrt{3}}\tan^{-1}\frac{2x+1}{\sqrt{3}}+c$

12. $\frac{-1}{1+\sin x}-\log(1+\sin x)+\log(2+\sin x)+c$

13. $2\log(1+\cos x)-4\log(2+\cos x)+c$

14. $\frac{1}{2}\log\frac{1+\sin x}{1-\sin x}-\frac{1}{2\sqrt{2}}\log\frac{\sqrt{2}+\sin x}{\sqrt{2}-\sin x}+c$

15. $\frac{1}{6}\log(1+\cos x)+\frac{1}{2}\log(1-\cos x)$
$-\frac{2}{3}\log(2\cos x-1)+c$

16. $\frac{1}{\sqrt{2}}\log\frac{1+\sqrt{2}\sin x}{1-\sqrt{2}\sin x}-\frac{1}{2}\log\frac{1+\sin x}{1-\sin x}+c$

17. $\frac{1}{4\sqrt{2}}\log\frac{1+\sqrt{2}\sin x}{1-\sqrt{2}\sin x}-\frac{1}{8}\log\frac{1+\sin x}{1-\sin x}+c$

18. $4\log\cos x+\sec^2 x-\log(3+\cos 4x)+c$

19. $\sin x-2\operatorname{cosec} x-6\tan^{-1}(\sin x)+c$

20. $\frac{2}{5}\log\frac{\sec x}{\tan x-2}-\frac{1}{\tan x-2}-\frac{1}{5}x+c$

21. $\frac{1}{3}\log\frac{2+\log x}{5+\log x}+c$

22. $\log\frac{e^t+1}{e^t+2}+c$

23. $\log x+x-\log(1+xe^x)+\frac{1}{1+xe^x}+c$

24. $\frac{1}{\sqrt{3}}\tan^{-1}\left\{\frac{\sqrt{3}\cdot x}{1-x^2}\right\}+c$

25. $\frac{1}{2\sqrt{3}}\log\frac{x^2-\sqrt{3}x+1}{x^2+\sqrt{3}x+1}+c$

26. $\frac{1}{2\sqrt{2}}\log\frac{x^2-\sqrt{2}x+1}{x^2+\sqrt{2}x+1}+c$

27. $\frac{1}{2\sqrt{2}}\tan^{-1}\frac{x^2-1}{x\sqrt{2}}+\frac{1}{4\sqrt{2}}\log\frac{x^2-\sqrt{2}x+1}{x^2+\sqrt{2}x+1}+c$

28. $\frac{1}{4}\log\frac{3\tan\frac{x}{2}+1}{\tan\frac{x}{2}+3}+c$

29. $2\sec\alpha\cdot\tan^{-1}\left\{\tan\frac{x}{2}\cdot\tan\left(\frac{\pi}{4}-\frac{\alpha}{2}\right)\right\}+c$

30. $\frac{1}{3}\tan\left(\frac{\pi}{4}+\frac{x}{2}\right)+\frac{2}{3\sqrt{3}}\tan^{-1}\frac{2\tan\frac{x}{2}+1}{\sqrt{3}}+c$

31. $\frac{1}{5}\tan\left(\frac{x}{2}+\frac{1}{2}\cos^{-1}\frac{3}{5}\right)+c$

32. $\frac{1}{\sqrt{6}}\log\frac{\sqrt{6}-1+\tan\frac{\theta}{2}}{\sqrt{6}+1-\tan\frac{\theta}{2}}+c$

33. $\log\left(1+\tan\frac{x}{2}\right)+c$

34. $\frac{x}{2}+\frac{1}{2}\log(\sin x+\cos x)+c$

35. $\frac{18}{25}x+\frac{1}{25}\log(3\sin x+4\cos x)+c$

36. $-\log(\cos x-\sin x)+c$

37. $\frac{x}{10}-\frac{3}{20}\log(5+3\cos 2x+\sin 2x)$
$-\frac{1}{2\sqrt{15}}\tan^{-1}\left\{\frac{2\tan x+1}{\sqrt{15}}\right\}+c$

38. $\frac{1}{2(a-b)}\log\{(a+b)-(a-b)\cos 2x\}+c$

39. $\frac{1}{2}(\sin x-\cos x)$
$-\frac{1}{2\sqrt{2}}\log\left\{\sec\left(x-\frac{\pi}{4}\right)+\tan\left(x-\frac{\pi}{4}\right)\right\}+c$

40. $\frac{1}{2\sqrt{3}}\log\frac{\sqrt{3}+\sin x-\cos x}{\sqrt{3}-\sin x+\cos x}+\tan^{-1}(\sin x+\cos x)+c$

41. $\frac{-e\sin x}{(1-e^2)(1+e\cos x)}$
$+\frac{2}{(1-e^2)^{3/2}}\tan^{-1}\left\{\sqrt{\frac{1-e}{1+e}}\tan\frac{x}{2}\right\}+c$

42. $\frac{1}{3}\cdot\frac{\cos x}{2+\sin x}+\frac{4}{3\sqrt{3}}\tan^{-1}\left\{\frac{1}{\sqrt{3}}\left(2\tan\frac{x}{2}+1\right)\right\}+c$

43. $\frac{-x^2}{1+\sin x}-2x\tan\frac{\pi-2x}{4}-4\log\sec\frac{\pi-2x}{4}+c$

44. $\frac{x\sin x+\cos x}{x\cos x-\sin x}+c$ 45. $\sin^{-1}(x-1)+c$

46. $\sin^{-1}(2x-3)+c$

47. $\frac{1}{4}\cdot\sqrt{4x^2-4x-3}+\frac{1}{4}\log\{2x-1+\sqrt{4x^2-4x-3}\}+c$

48. $\sin^{-1}\left(\frac{x+1}{\sqrt{2}}\right)+\frac{x+1}{2}\sqrt{1-2x-x^2}+c$

49. $\frac{1}{3}(x^2+x+1)^{3/2}-\frac{3}{8}(2x+1)\sqrt{1+x+x^2}$
$-\frac{9}{16}\log(2x+1+2\sqrt{x^2+x+1})+c$

50. $\frac{1}{4}(2x-3)\sqrt{x^2+x+1}$
$+\frac{7}{8}\log(2x+1+2\sqrt{x^2+x+1})+c$

51. $-2\sqrt{1+\cos x+\cos^2 x}$
$+\log(2\cos x+1+2\sqrt{1+\cos x+\cos^2 x})+c$

52. $\frac{2}{3a^{3/2}}(t^2+a^2t+a^2)^{3/2}$
$-a^{1/2}\left[\frac{2t+a^2}{4}\sqrt{t^2+a^2t+a^2}\right.$
$\left.+\frac{a^2(4-a^2)}{8}\log\left\{2t+a^2+4\sqrt{t^2+a^2t+a^2}\right\}\right]+c,$
where $t=\sqrt{ax-a^2}$

53. $\frac{1}{3}(x+\sqrt{x^2+a^2})^{3/2}-\frac{a^2}{\sqrt{x+\sqrt{x^2+a^2}}}+c$

54. $\log(1-t^2)-\frac{1}{2}\log(t^4+t^2+1)-\sqrt{3}\tan^{-1}\left(\frac{\sqrt{3}}{1+2t^2}\right)+c$
where $t=\sqrt[3]{\frac{1-x}{1+x}}$

55. $-\frac{1}{\sqrt{3}}\tan^{-1}\left(\frac{2z-1}{\sqrt{3}}\right)+\frac{1}{3}\log(z+1)$
$-\frac{1}{6}\log(z^2-z+1)+c,$ where $z^3=\frac{1}{x^3}-1$

56. $-\frac{1}{\sqrt{2}}\log\frac{\sqrt{2}+\sqrt{1-x}}{\sqrt{2}-\sqrt{1-x}}+c$

57. $-\sqrt{\frac{1-x}{1+x}}+c$ 58. $\sqrt{\frac{1+x}{1-x}}+c$

59. $-\frac{1}{\sqrt{2}}\sin^{-1}\sqrt{\frac{1-x^2}{1+x^2}}+c$

60. $-\log\left(\frac{-3x-1+2\sqrt{3x^2+3x+1}}{1+x}\right)+c$

61. $-\log(2e^{-x}+1+2\sqrt{e^{-2x}+e^{-x}+1})+c$

62. $\sqrt{x^2-1}+\sec^{-1}x+c$

63. $\log\left(x+\frac{1}{x}+\sqrt{x^2+\frac{1}{x^2}+3}\right)+c$

64. $-2\tan^{-1}\sqrt{x+\frac{1}{x}+1}+c$

65. $\sqrt{1-x^2}-\cos^{-1}x-\sqrt{\frac{1-x}{1+x}}+c$

66. $\frac{1}{3}(x^2+x+3)\sqrt{x^2+x}$
$-\frac{3}{2}\left\{\frac{2x+1}{4}\cdot\sqrt{x^2+x}-\frac{11}{24}\log(2x+1+2\sqrt{x^2+x})\right\}+c$

67. $\sqrt{2}\sin^{-1}(\sin x-\cos x)+c$

68. $\frac{1}{\sqrt{2}}\sin^{-1}(\sin x - \cos x)$

$$+\frac{1}{\sqrt{2}}\log(\sin x + \cos x + \sqrt{\sin 2x}) + c$$

69. $\sqrt{2}\sin^{-1}(\sin x - \cos x) + c$

70. $-\log\{\sin x + \cos x + \sqrt{2 + \sin 2x}\} + c$

71. $-2\sin^{-1}\left\{\sqrt{2}\sin\left(\frac{\pi}{4} - \frac{x}{2}\right)\right\} + c$

72. $-2\log\left|\cos\frac{x}{2} + \sqrt{\frac{1}{2}\cos x}\right| + c$

73. $\sqrt{2}\sin^{-1}(\sqrt{2}\sin\theta) - \sin^{-1}(\tan\theta) + c$

74. $-\sin^{-1}\sqrt{\frac{\cos 2x}{\cos^2 x}} + c$ or $\sin^{-1}(\tan x) + c$

75. $-\cos\alpha \cdot \sin^{-1}(\cos x \cdot \sec\alpha)$

$$-\sin\alpha \cdot \log(\sin x + \sqrt{\sin^2 x - \sin^2\alpha}) + c$$

76. $\frac{2}{\sqrt{26}}\log\left\{\sec\frac{x+\alpha}{2} + \tan\frac{x+\alpha}{2}\right\} + c,$

where $\alpha = \cos^{-1}\frac{5}{13}$

77. $-\sqrt{1 - \tan\frac{x}{2} + \cot\frac{x}{2}} + c$

□

Chapter Test

Time: 75 minutes

1. In each of the following, fill in the blank so that the resulting sentence becomes true.

(a) $\int \frac{dx}{x - \sqrt{x^2 - 1}} =$ ______ .

(b) $\int \sqrt{4x^2 - 4x + 5}\, dx =$ ______ .

(c) $\int \frac{x^2 + x + 2}{(1 + x^2)(1 + x)}\, dx =$ ______ .

(d) $\int \frac{dx}{(1 + x^2)^{3/2}} =$ ______ .

Evaluate the following:

2. $\int \frac{dx}{(1 + \sqrt{x})\sqrt{x + x^2}}$

3. $\int \frac{dx}{(1 + e^x)^2}$

4. $\int \frac{dx}{1 + \tan x + \cot x}$

5. $\int \frac{\sin x + \sin 3x}{\sqrt{1 + \sin^2 x}}\, dx$

6. $\int \frac{x^2 - 1}{(x^2 + 1)\sqrt{1 + x^4}}\, dx$

7. $\int \left\{ \int \frac{\sqrt{\tan x}}{\sin 2x}\, dx \right\} dx$

8. $\int \frac{\sec^3\theta\, d\theta}{2 - \sec^2\theta}$

9. $\int \frac{dx}{x + \sqrt{x - 1}}$

10. $\int \frac{x^3 - x}{1 + x^6}\, dx$

□

Answers

1. (a) $\frac{x^2}{2} + \frac{x\sqrt{x^2-1}}{2} - \frac{1}{2}\log(x+\sqrt{x^2-1}) + c$

(b) $\frac{2x-1}{4}\sqrt{4x^2-4x+5} + \log(2x-1+\sqrt{4x^2-4x+5}) + c$

(c) $\log(1+x) + \tan^{-1}x + c$

(d) $\frac{x}{\sqrt{1+x^2}} + c$

2. $\sqrt{2}\log\tan\left\{\frac{1}{2}(\tan^{-1}\sqrt{x}) + \frac{\pi}{8}\right\} + c$

3. $x + \frac{1}{1+e^x} - \log(1+e^x) + c$

4. $x - \frac{2}{\sqrt{3}}\tan^{-1}\left\{\frac{1}{\sqrt{3}}(2\tan x + 1)\right\} + c$

5. $2\cos x\sqrt{2-\cos^2 x} - 4\sin^{-1}\left(\frac{1}{\sqrt{2}}\cos x\right) + c$

6. $\frac{1}{\sqrt{2}}\sec^{-1}\left\{\frac{1}{\sqrt{2}}\left(x+\frac{1}{x}\right)\right\} + c$

7. $\frac{1}{2}\sin^{-1}(\sin x - \cos x) - \frac{1}{\sqrt{2}}\log(\sin x + \cos x + \sqrt{\sin 2x}) + c_1 x + c_2$

8. $\frac{1}{\sqrt{2}}\log\frac{\sqrt{2}\sin\theta + 1}{1-\sqrt{2}\sin\theta} - \frac{1}{2}\log\frac{1+\sin\theta}{1-\sin\theta} + c$

9. $\log(x+\sqrt{x-1}) - \frac{2}{\sqrt{3}}\tan^{-1}\frac{2\sqrt{x-1}+1}{\sqrt{3}} + c$

10. $\frac{1}{6}\log\frac{x^4-x^2+1}{(1+x^2)^2} + c$

□

10. Definite Integration

Recap of Facts and Formulae

1. Definition of definite integral

- The definite integral of $f(x)$ over the interval $[a, b]$, denoted by $\int_a^b f(x)dx$, is defined as the limit of a sum as follows:

$$\int_a^b f(x)dx = \lim_{n \to \infty} \sum_{r=1}^{n} hf(a + rh)$$

$$\text{or} \quad \lim_{n \to \infty} \sum_{r=0}^{n-1} hf(a + rh),$$

where $nh = b - a$

$f(x)$ is said to be integrable over $[a, b]$ if the above two limits exist and are equal.

2. Fundamental theorem of definite integration

- $\lim_{n \to \infty} \sum_{r=1}^{n} hf(a + rh)$ or $\lim_{n \to \infty} \sum_{r=0}^{n-1} hf(a + rh)$

$$= F(b) - F(a)$$

where $nh = b - a$ and $F'(x) = f(x)$.

Consequently, $\int_a^b f(x)dx = F(b) - F(a)$, i.e., $[F(x)]_{x=a}^{x=b}$

where $F'(x) = f(x)$, i.e., $F(x)$ is the primitive function of $f(x)$.

TIP In computing a definite integral $\int_a^b f(x)dx$ where $f(a) = f(b)$, it is convenient to break the interval in two parts and use

$$\int_a^b f(x)dx = \int_a^{\frac{a+b}{2}} f(x)dx + \int_{\frac{a+b}{2}}^b f(x)dx.$$

3. Evaluation of integral from the first principle

- To find the value of $\int_a^b f(x)dx$ from the first principle i.e., definition, obtain

$$\lim_{n \to \infty} \sum_{r=1}^{n} hf(a + rh)$$

$$\text{or} \quad \lim_{n \to \infty} \sum_{r=0}^{n-1} hf(a + rh) \text{ where } nh = b - a.$$

- $\int_0^1 f(x)dx = \lim_{n \to \infty} \sum_{r=1}^{n} \frac{1}{n} f\left(\frac{r}{n}\right)$ or $\lim_{n \to \infty} \sum_{r=0}^{n-1} \frac{1}{n} f\left(\frac{r}{n}\right)$

4. Limit of a sum as definite integral

- $\lim_{n \to \infty} \sum_{r=1}^{n} \frac{b-a}{n} f\left(a + r \cdot \frac{b-a}{n}\right)$

$$\text{or} \quad \lim_{n \to \infty} \sum_{r=0}^{n-1} \frac{b-a}{n} f\left(a + r \cdot \frac{b-a}{n}\right) = \int_a^b f(x)dx$$

- $\lim_{n \to \infty} \sum_{r=1}^{n} \frac{1}{n} f\left(\frac{r}{n}\right)$

$$\text{or} \quad \lim_{n \to \infty} \sum_{r=0}^{n-1} \frac{1}{n} f\left(\frac{r}{n}\right) = \int_0^1 f(x)dx$$

- $\lim_{n \to \infty} \sum_{r=1}^{pn} \frac{1}{n} f\left(\frac{r}{n}\right) = \int_\alpha^\beta f(x)dx$

where $\alpha = \lim_{n \to \infty} \frac{r}{n}$ when $r = 1$,

$\beta = \lim_{n \to \infty} \frac{r}{n}$ when $r = pn$

Selected Solved Examples

1. Use the method of summation to evaluate $\int_0^1 e^{2x}dx$.

Or Integrate from the first principle: $\int_0^1 e^{2x}dx$.

We know that $\int_0^1 f(x)dx = \lim_{n\to\infty} \sum_{r=1}^{n} \frac{1}{n} f\left(\frac{r}{n}\right)$

$$\therefore \int_0^1 e^{2x}dx = \lim_{n\to\infty} \sum_{r=1}^{n} \frac{1}{n} e^{2\cdot\frac{r}{n}}$$

$$= \lim_{n\to\infty} \left[\frac{1}{n}\left(e^{\frac{2}{n}} + e^{\frac{4}{n}} + e^{\frac{6}{n}} + \ldots + e^{\frac{2n}{n}}\right)\right]$$

$$= \lim_{h\to 0} \left[h\left(e^{2h} + e^{4h} + e^{6h} + \ldots + e^{2nh}\right)\right],$$

where $h = \frac{1}{n}$

$$= \lim_{h\to 0} h \frac{e^{2h}(1-e^{2nh})}{1-e^{2h}} = \lim_{h\to 0} h \cdot \frac{e^{2h}(1-e^2)}{1-e^{2h}}$$

$$= (1-e^2) \lim_{h\to 0} \frac{h}{1-e^{2h}} \quad \{\because e^0 = 1\}$$

$$= (1-e^2) \lim_{h\to 0} \frac{h}{1-\left\{1+\frac{2h}{1!}+\frac{(2h)^2}{2!}+\ldots\right\}}$$

$$= (1-e^2) \lim_{h\to 0} \frac{h}{-\frac{2h}{1!}-\frac{(2h)^2}{2!}-\ldots}$$

$$= \frac{1-e^2}{-2} = \frac{e^2-1}{2}.$$

2. Express $\int_1^2 x\,dx$ as the limit of a sum and evaluate it.

We know that $\int_a^b f(x)dx = \lim_{n\to\infty} \sum_{r=1}^{n} \frac{b-a}{n} f\left(a + r\cdot\frac{b-a}{n}\right)$

$$\therefore \int_1^2 x\,dx = \lim_{n\to\infty} \sum_{r=1}^{n} \frac{2-1}{n} f\left(1 + r\cdot\frac{2-1}{n}\right)$$

$$= \lim_{n\to\infty} \sum_{r=1}^{n} \frac{1}{n} f\left(1+\frac{r}{n}\right)$$

$$= \lim_{n\to\infty} \sum_{r=1}^{n} \frac{1}{n}\cdot\left(1+\frac{r}{n}\right) \quad \{\because f(x) = x\}$$

$$= \lim_{n\to\infty} \left\{\sum_{r=1}^{n} \frac{1}{n} + \sum_{r=1}^{n} \frac{r}{n^2}\right\}$$

$$= \lim_{n\to\infty} \left\{\frac{1}{n}\sum_{r=1}^{n} 1 + \frac{1}{n^2}\sum_{r=1}^{n} r\right\}$$

$$= \lim_{n\to\infty} \left\{\frac{1}{n}\cdot n + \frac{1}{n^2}\cdot\frac{n(n+1)}{2}\right\}$$

$$= \lim_{n\to\infty} \left\{1 + \frac{1}{2}\cdot\left(1+\frac{1}{n}\right)\right\} = \frac{3}{2}.$$

3. Find $\lim_{n\to\infty} S_n$ if

$$S_n = \frac{1}{2n} + \frac{1}{\sqrt{4n^2-1}} + \frac{1}{\sqrt{4n^2-4}} + \ldots + \frac{1}{\sqrt{3n^2+2n-1}}.$$

Here, $S_n = \sum_{r=1}^{n} \frac{1}{\sqrt{(2n)^2-(r-1)^2}}$

$$= \sum_{r=0}^{n-1} \frac{1}{\sqrt{(2n)^2-r^2}}$$

$$= \sum_{r=0}^{n-1} \frac{1}{n}\cdot\frac{1}{\sqrt{4-\left(\frac{r}{n}\right)^2}}.$$

But we know that $\lim_{n\to\infty} \sum_{r=0}^{n-1} \frac{1}{n}\cdot f\left(\frac{r}{n}\right) = \int_0^1 f(x)dx$

$$\therefore \lim_{n\to\infty} S_n = \lim_{n\to\infty} \sum_{r=0}^{n-1} \frac{1}{n}\cdot\frac{1}{\sqrt{4-\left(\frac{r}{n}\right)^2}}$$

$$= \int_0^1 \frac{dx}{\sqrt{4-x^2}} = \frac{1}{2}\int_0^1 \frac{dx}{\sqrt{1-\left(\frac{x}{2}\right)^2}}$$

$$= \frac{1}{2}\left[\frac{\sin^{-1}\frac{x}{2}}{\frac{1}{2}}\right]_0^1 = \sin^{-1}\frac{1}{2}$$

$$= \frac{\pi}{6}.$$

4. Show that $\lim_{n\to\infty}\left(\frac{1}{n+1}+\frac{1}{n+2}+\frac{1}{n+3}+\ldots+\frac{1}{6n}\right) = \log 6.$

Here, limit $= \sum_{r=1}^{5n} \frac{1}{n+r} = \sum_{r=1}^{5n} \frac{1}{n}\cdot\frac{1}{1+\frac{r}{n}} = \int_\alpha^\beta \frac{1}{1+x}dx$

where $\alpha = \lim_{n\to\infty} \frac{r}{n}$ when $r = 1$

$$\beta = \lim_{n\to\infty} \frac{r}{n} \text{ when } r = 5n,$$

i.e., $\alpha = \lim_{n\to\infty} \frac{1}{n} = 0$ and $\beta = \lim_{n\to\infty} \frac{5n}{n} = 5$

$$\therefore \quad \text{limit} = \int_0^5 \frac{dx}{1+x}$$

$$= [\log(1+x)]_{x=0}^{5} = \log 6.$$

5. Evaluate $\lim_{n\to\infty} \left(\frac{n!}{n^n}\right)^{\frac{1}{n}}$.

We know that $\lim_{x\to a} f(x) = e^{\lim_{x\to a} \log f(x)}$ ... (1)

Here, $\lim_{n\to\infty} \log\left(\frac{n!}{n^n}\right)^{\frac{1}{n}}$

$$= \lim_{n\to\infty} \frac{1}{n} \log\left(\frac{1\cdot 2\cdot 3\cdot \ldots \cdot n}{n\cdot n\cdot n\cdot \ldots \cdot n}\right)$$

$$= \lim_{n\to\infty} \frac{1}{n} \sum_{r=1}^{n} \log\frac{r}{n}$$

$$= \int_0^1 \log x\, dx$$

$$= \left[\log x\cdot x - \int x\cdot \frac{1}{x}\, dx\right]_0^1 = [x\log x - x]_0^1$$

$$= -1 - \lim_{x\to 0} x\log x$$

$$= -1 - \lim_{x\to 0} \frac{\log x}{\frac{1}{x}}$$

$$= -1 - \lim_{x\to 0} \frac{\frac{1}{x}}{-\frac{1}{x^2}}, \text{ using L' Hospital's rule}$$

$$= -1 - \lim_{x\to 0}(-x) = -1$$

$$\therefore \quad \text{using (1)}, \lim_{n\to\infty}\left(\frac{n!}{n^n}\right)^{\frac{1}{n}} = e^{-1} = \frac{1}{e}.$$

6. Evaluate $\int_0^1 \log(\sqrt{1-x} + \sqrt{1+x})\, dx$.

Let $x = \cos 2\theta$; then $dx = -2\sin 2\theta\, d\theta$;

also, $x = 0 \Rightarrow \theta = \frac{\pi}{4}$; $x = 1 \Rightarrow \theta = 0$

$$\therefore \quad I = \int_{\pi/4}^{0} \log(\sqrt{1-\cos 2\theta} + \sqrt{1+\cos 2\theta})(-2\sin 2\theta)d\theta$$

$$= -2\int_{\pi/4}^{0} \log\{\sqrt{2}(\sin\theta + \cos\theta)\}\sin 2\theta\, d\theta$$

$$= -2\left[\log\{\sqrt{2}(\sin\theta+\cos\theta)\}\cdot\frac{-\cos 2\theta}{2} - \int \frac{-\cos 2\theta}{2}\cdot\frac{\sqrt{2}(\cos\theta - \sin\theta)}{\sqrt{2}(\sin\theta+\cos\theta)}\, d\theta\right]_{\pi/4}^{0}$$

$$= -2\left[-\frac{1}{2}\log\sqrt{2} + \frac{1}{2}\int_{\pi/4}^{0} \frac{(\cos^2\theta - \sin^2\theta)(\cos\theta - \sin\theta)}{\cos\theta + \sin\theta}\, d\theta\right]$$

$$= \log\sqrt{2} - \int_{\pi/4}^{0} (\cos\theta - \sin\theta)^2 d\theta$$

$$= \log\sqrt{2} - \int_{\pi/4}^{0} (1 - \sin 2\theta)d\theta$$

$$= \log\sqrt{2} - \left[\theta + \frac{\cos 2\theta}{2}\right]_{\pi/4}^{0}$$

$$= \log\sqrt{2} - \frac{1}{2} + \frac{\pi}{4}.$$

7. Prove that $\int_0^{\pi/2} \frac{x\, dx}{\sin x} = 2\int_0^1 \frac{\tan^{-1} z}{z}\, dz.$

Let $z = \tan\theta$; then $dz = \sec^2\theta\, d\theta$

Also $z = 0 \Rightarrow \theta = 0$; $z = 1 \Rightarrow \theta = \frac{\pi}{4}$

$$\therefore \quad \text{RHS} = 2\int_0^{\pi/4} \frac{\theta}{\tan\theta}\cdot\sec^2\theta\, d\theta$$

$$= 2\int_0^{\pi/4} \frac{\theta}{\sin\theta\cdot\cos\theta}\, d\theta$$

$$= 2\int_0^{\pi/4} \frac{2\theta}{\sin 2\theta}\, d\theta$$

$$= 2\int_0^{\pi/2} \frac{1}{2}\cdot\frac{x\, dx}{\sin x} \quad (\text{putting } 2\theta = x)$$

$$= \text{LHS}.$$

8. Prove that $\int_0^{\pi/2} \frac{dx}{(\sqrt{\cos x} + \sqrt{\sin x})^4} = \frac{1}{3}.$

$$I = \int_0^{\pi/2} \frac{dx}{\cos^2 x(1+\sqrt{\tan x})^4} = \int_0^{\pi/2} \frac{\sec^2 x\, dx}{(1+\sqrt{\tan x})^4}.$$

Put $\tan x = z$; then $\sec^2 x\, dx = dz$

and $x = 0 \Rightarrow z = 0$; $x = \frac{\pi}{2} \Rightarrow z = \infty$.

$\therefore \quad I = \int_0^{\infty} \frac{dz}{(1+\sqrt{z})^4}$.

Let $1 + \sqrt{z} = t$; then $\frac{1}{2\sqrt{z}} dz = dt$

and $z = 0 \Rightarrow t = 1$; $z = \infty \Rightarrow t = \infty$.

$$\therefore \quad I = \int_1^{\infty} \frac{2\sqrt{z}\,dt}{t^4} = \int_1^{\infty} \frac{2(t-1)}{t^4} dt$$

$$= 2\int_1^{\infty} \left(\frac{1}{t^3} - \frac{1}{t^4}\right) dt$$

$$= 2\left[\frac{1}{-2t^2} + \frac{1}{3t^3}\right]_1^{\infty} = -2\left(-\frac{1}{2} + \frac{1}{3}\right)$$

$$= 1 - \frac{2}{3} = \frac{1}{3}.$$

9. Show that $\int_0^{\pi/2} \sin x \cdot \log \sin x \, dx = \log_e 2 - 1$.

Let $I = \int_{\theta}^{\pi/2} \sin x \cdot \log \sin x \, dx$.

Clearly, the required integral $= \lim_{\theta \to 0} I$.

Now, $I = [\log \sin x \cdot (-\cos x)]_{\theta}^{\pi/2} - \int_{\theta}^{\pi/2} -\cos x \cdot \frac{\cos x}{\sin x} dx$, using by parts

$$= \cos\theta \cdot \log \sin\theta + \int_{\theta}^{\pi/2} \frac{1 - \sin^2 x}{\sin x} dx$$

$$= \cos\theta \cdot \log \sin\theta + \int_{\theta}^{\pi/2} (\operatorname{cosec} x - \sin x) dx$$

$$= \cos\theta \cdot \log \sin\theta + [-\log(\operatorname{cosec} x + \cot x) + \cos x]_{\theta}^{\pi/2}$$

$$= \cos\theta \cdot \log \sin\theta + \log(\operatorname{cosec}\theta + \cot\theta) - \cos\theta$$

$$= \cos\theta \cdot \log \sin\theta + \log \frac{1 + \cos\theta}{\sin\theta} - \cos\theta$$

$$= (\cos\theta - 1)\log \sin\theta + \log(1 + \cos\theta) - \cos\theta$$

$$\therefore \quad \text{LHS} = \lim_{\theta \to 0} I$$

$$= \lim_{\theta \to 0} \{(\cos\theta - 1)\log \sin\theta\} + \log_e 2 - 1$$

$$= \lim_{\theta \to 0} \frac{\log \sin\theta}{\frac{1}{\cos\theta - 1}} + \log_e 2 - 1$$

$$= \lim_{\theta \to 0} \frac{\frac{\cos\theta}{\sin\theta}}{\frac{-1}{(\cos\theta - 1)^2} \cdot (-\sin\theta)} + \log_e 2 - 1$$

$$= \lim_{\theta \to 0} \frac{\cos\theta(\cos\theta - 1)^2}{\sin^2\theta} + \log_e 2 - 1$$

$$= \lim_{\theta \to 0} \frac{\cos\theta(1 - \cos\theta)^2}{1 - \cos^2\theta} + \log_e 2 - 1$$

$$= \lim_{\theta \to 0} \frac{\cos\theta(1 - \cos\theta)}{1 + \cos\theta} + \log_e 2 - 1$$

$$= 0 + \log_e 2 - 1 = \log_e 2 - 1.$$

TIP In this problem we did not take 0 as the lower limit to find the value of the integral because at 0, $\log \sin x$ becomes $-\infty$. In such cases, $\int_0^{\pi/2} f(x) dx$ is found by obtaining $\lim_{\theta \to 0} \int_{\theta}^{\pi/2} f(x) dx$. Thus, if $f(a) \to \infty$ or $-\infty$ then $\int_a^b f(x) dx$ is obtained by finding $\lim_{\lambda \to a} \int_{\lambda}^{b} f(x) dx$.

10. Evaluate $\int_0^{\pi} \frac{dx}{1 - 2a\cos x + a^2}$, $a \neq 1$.

$$I = \int_0^{\pi} \frac{dx}{1 - 2a\left(1 - 2\sin^2 \frac{x}{2}\right) + a^2}$$

$$= \int_0^{\pi} \frac{dx}{(1-a)^2 + 4a\sin^2 \frac{x}{2}}$$

$$= \int_0^{\pi} \frac{\operatorname{cosec}^2 \frac{x}{2} dx}{(1-a)^2 \operatorname{cosec}^2 \frac{x}{2} + 4a}$$

$$= \int_0^{\pi} \frac{\operatorname{cosec}^2 \frac{x}{2} dx}{(1-a)^2 \cdot \left(1 + \cot^2 \frac{x}{2}\right) + 4a}$$

$$= \int_0^{\pi} \frac{\operatorname{cosec}^2 \frac{x}{2} dx}{(1+a)^2 + (1-a)^2 \cot^2 \frac{x}{2}}.$$

Put $\cot \frac{x}{2} = z$; then $-\frac{1}{2}\operatorname{cosec}^2 \frac{x}{2} dx = dz$;

also $x = 0 \Rightarrow z = \infty$; $x = \pi \Rightarrow z = 0$

$$\therefore \quad I = \int_{\infty}^{0} \frac{-2dz}{(1+a)^2 + (1-a)^2 z^2}$$

$$= \frac{-2}{(1-a)^2}\int_{\infty}^{0} \frac{dz}{z^2 + \left(\frac{1+a}{1-a}\right)^2}, \; a \neq 1$$

$$= \frac{-2}{(1-a)^2} \cdot \frac{1}{\frac{1+a}{1-a}}\left[\tan^{1} \frac{z}{\frac{1+a}{1-a}}\right]_{\infty}^{0}$$

$$= \frac{-2}{1-a^2} \cdot \left[\tan^{-1} \frac{1-a}{1+a} z\right]_{\infty}^{0}$$

$$= \frac{-2}{1-a^2}\{\tan^{-1}0 - \tan^{-1}\infty\}, \text{ when } a < 1$$

$$\frac{-2}{1-a^2}\{\tan^{-1}0 - \tan^{-1}(-\infty)\}, \text{ when } a > 1$$

$$= \frac{-2}{1-a^2}\left(-\frac{\pi}{2}\right), \text{ when } a < 1$$

$$\frac{-2}{1-a^2}\left(\frac{\pi}{2}\right), \text{ when } a > 1$$

$$= \frac{\pi}{1-a^2} \text{ when } a < 1 \text{ and}$$

$$\frac{\pi}{a^2-1} \text{ when } a > 1.$$

11. Find the value of $\int_0^{\pi/4} \frac{\sin x + \cos x}{9 + 16\sin 2x}\, dx.$

$$I = \int_0^{\pi/4} \frac{d(\sin x - \cos x)}{9 - 16(1 - \sin 2x) + 16}$$

$$= \int_0^{\pi/4} \frac{d(\sin x - \cos x)}{25 - 16(\sin x - \cos x)^2}$$

$$= \frac{1}{16}\int_0^{\pi/4} \frac{d(\sin x - \cos x)}{\left(\frac{5}{4}\right)^2 - (\sin x - \cos x)^2}.$$

Put $\sin x - \cos x = z$; then

$$x = 0 \quad \Rightarrow \quad z = -1; \; x = \frac{\pi}{4} \quad \Rightarrow \quad z = 0$$

$$\therefore \quad I = \frac{1}{16}\int_{-1}^{0} \frac{dz}{\left(\frac{5}{4}\right)^2 - z^2} = \frac{1}{16} \cdot \left[\frac{1}{2 \cdot \frac{5}{4}} \log \frac{\frac{5}{4} + z}{\frac{5}{4} - z}\right]_{-1}^{0}$$

$$= \frac{1}{40}\left[\log \frac{5+4z}{5-4z}\right]_{-1}^{0} = \frac{1}{40}\left\{\log 1 - \log \frac{1}{9}\right\}$$

$$= \frac{1}{40}\log 9 = \frac{1}{20}\log 3.$$

12. Evaluate $\int_0^{\infty} \frac{dx}{(x + \sqrt{1+x^2})^n}$**, where n is an integer > 1.**

Let $x + \sqrt{1+x^2} = z$; then $\left(1 + \frac{x}{\sqrt{1+x^2}}\right)dx = dz.$

Also, $x = 0 \quad \Rightarrow \quad z = 1; \; x = \infty \quad \Rightarrow \quad z = \infty$

$$\therefore \quad I = \int_1^{\infty} \frac{\sqrt{1+x^2}}{(x+\sqrt{1+x^2})} \cdot \frac{1}{(x+\sqrt{1+x^2})^n}\, dz$$

$$= \int_1^{\infty} \frac{\sqrt{1+x^2}}{z^{n+1}}\, dz \qquad \dots (1)$$

Again, $z = x + \sqrt{1+x^2} \quad \Rightarrow \quad \frac{1}{z} = \frac{1}{x+\sqrt{1+x^2}}$

$$\therefore \quad \frac{1}{z} = \frac{x - \sqrt{1+x^2}}{(x+\sqrt{1+x^2})(x-\sqrt{1+x^2})} = \sqrt{1+x^2} - x$$

$\therefore \quad z + \frac{1}{z} = 2\sqrt{1+x^2}$; so (1) gives

$$I = \int_1^{\infty} \frac{1}{2} \cdot \frac{z + \frac{1}{z}}{z^{n+1}}\, dz = \frac{1}{2}\int_1^{\infty}\left(\frac{1}{z^n} + \frac{1}{z^{n+2}}\right)dz$$

$$= \frac{1}{2}\left[\frac{z^{-n+1}}{-n+1} + \frac{z^{-n-2+1}}{-n-2+1}\right]_1^{\infty}$$

$$= \frac{1}{2}\left[\frac{-1}{n-1} \cdot \frac{1}{z^{n-1}} - \frac{1}{n+1} \cdot \frac{1}{z^{n+1}}\right]_1^{\infty}$$

$$= \frac{1}{2}\left\{0 + \frac{1}{n-1} + \frac{1}{n+1}\right\} = \frac{n}{n^2-1}.$$

13. Evaluate $\int_0^{\pi/4} \sqrt{2 + \tan^2\theta}\, d\theta.$

Put $2 + \tan^2\theta = z^2$; then $2\tan\theta \cdot \sec^2\theta\, d\theta = 2z dz.$

Also, $\theta = 0 \quad \Rightarrow \quad z = \sqrt{2}; \; \theta = \frac{\pi}{4} \quad \Rightarrow \quad z = \sqrt{3}$

$$\therefore \quad I = \int_{\sqrt{2}}^{\sqrt{3}} z \cdot \frac{z dz}{\tan\theta \cdot \sec^2\theta} = \int_{\sqrt{2}}^{\sqrt{3}} \frac{z^2 dz}{\sqrt{z^2-2} \cdot (z^2-1)}$$

$$= \int_{\sqrt{2}}^{\sqrt{3}} \frac{(z^2-1)+1}{\sqrt{z^2-2}(z^2-1)}\, dz$$

$$= \int_{\sqrt{2}}^{\sqrt{3}} \frac{dz}{\sqrt{z^2-2}} + \int_{\sqrt{2}}^{\sqrt{3}} \frac{dz}{\sqrt{z^2-2}(z^2-1)} = I_1 + I_2.$$

Now, $I_1 = \int_{\sqrt{2}}^{\sqrt{3}} \frac{dz}{\sqrt{z^2 - (\sqrt{2})^2}}$

$$= \left[\log\left(z + \sqrt{z^2-2}\right)\right]_{\sqrt{2}}^{\sqrt{3}}$$

$$= \log(\sqrt{3}+1) - \log\sqrt{2} = \log\frac{\sqrt{3}+1}{\sqrt{2}}.$$

$$I_2 = \int_{\sqrt{2}}^{\sqrt{3}} \frac{dz}{\sqrt{z^2-2}(z^2-1)}.$$

For this integral, put $\sqrt{\dfrac{z^2-2}{z^2-1}} = t.$

So, $\dfrac{z^2-2}{z^2-1} = t^2$

$\therefore \quad 1 - \dfrac{1}{z^2-1} = t^2$, i.e., $z^2 - 1 = \dfrac{1}{1-t^2}$

$\therefore \quad z^2 = 1 + \dfrac{1}{1-t^2} = \dfrac{2-t^2}{1-t^2}$

$\therefore \quad 2z\,dz = \dfrac{-2t(1-t^2) - (2-t^2)(-2t)}{(1-t^2)^2}\,dt$

or $\quad z\,dz = \dfrac{t\,dt}{(1-t^2)^2}.$

Also, $z = \sqrt{2} \Rightarrow t = 0;\ z = \sqrt{3} \Rightarrow t = \dfrac{1}{\sqrt{2}}$

$$\therefore \quad I_2 = \int_0^{1/\sqrt{2}} \frac{1}{\sqrt{\dfrac{2-t^2}{1-t^2}}} \cdot \frac{t}{(1-t^2)^2} \times \frac{1}{\sqrt{\dfrac{2-t^2}{1-t^2} - 2}\left(\dfrac{2-t^2}{1-t^2} - 1\right)}\,dt$$

$$= \int_0^{1/\sqrt{2}} \frac{\sqrt{1-t^2}}{\sqrt{2-t^2}} \cdot \frac{t}{(1-t^2)^2} \cdot \frac{\sqrt{1-t^2}}{\sqrt{t^2}} \cdot \frac{1-t^2}{1}\,dt$$

$$= \int_0^{1/\sqrt{2}} \frac{dt}{\sqrt{2-t^2}} = \int_0^{1/\sqrt{2}} \frac{dt}{\sqrt{(\sqrt{2})^2 - t^2}}$$

$$= \left[\sin^{-1}\frac{t}{\sqrt{2}}\right]_0^{1/\sqrt{2}}$$

$$= \sin^{-1}\frac{1}{2} - \sin^{-1}0 = \frac{\pi}{6}.$$

$$\therefore \quad I = I_1 + I_2 = \log\frac{\sqrt{3}+1}{\sqrt{2}} + \frac{\pi}{6}.$$

14. Evaluate $\displaystyle\int_0^{\pi/3} \frac{x}{1+\sec x}\,dx.$

$$I = \int_0^{\pi/3} \frac{x\cos x}{1+\cos x}\,dx = \int_0^{\pi/3} x \cdot \frac{2\cos^2\frac{x}{2} - 1}{2\cos^2\frac{x}{2}}\,dx$$

$$= \int_0^{\pi/3} x\left(1 - \frac{1}{2}\sec^2\frac{x}{2}\right)dx$$

$$= \int_0^{\pi/3} x\,dx - \frac{1}{2}\int_0^{\pi/3} x\sec^2\frac{x}{2}\,dx$$

$$= \left[\frac{x^2}{2}\right]_0^{\pi/3} - \frac{1}{2}\left\{\left[x \cdot \frac{\tan\frac{x}{2}}{\frac{1}{2}}\right]_0^{\pi/3} - \int_0^{\pi/3} 2\tan\frac{x}{2}\,dx\right\}$$

$$= \frac{\pi^2}{18} - \frac{1}{2}\left\{2 \cdot \frac{\pi}{3} \cdot \tan\frac{\pi}{6} - 0\right\} + \left[\frac{\log\sec\frac{x}{2}}{\frac{1}{2}}\right]_0^{\pi/3}$$

$$= \frac{\pi^2}{18} - \frac{\pi}{3\sqrt{3}} + 2\left\{\log\sec\frac{\pi}{6} - \log 1\right\}$$

$$= \frac{\pi^2}{18} - \frac{\pi}{3\sqrt{3}} + 2\log\frac{2}{\sqrt{3}}.$$

15. Evaluate $\displaystyle\int_0^1 \cot^{-1}(1 + x^2 - x)dx.$

$$I = \left[\cot^{-1}(1+x^2-x) \cdot x\right]_0^1 - \int_0^1 x \cdot \frac{-1}{1+(1+x^2-x)^2}(2x-1)dx$$

(using by parts)

$$= \cot^{-1}1 + \int_0^1 \frac{x(2x-1)}{x^4 - 2x^3 + 3x^2 - 2x + 2}\,dx$$

$$= \frac{\pi}{4} + \int_0^1 \frac{x(2x-1)dx}{x^4 + 3x^2 + 2 - 2x(x^2+1)}$$

$$= \frac{\pi}{4} + \int_0^1 \frac{x(2x-1)dx}{(x^2+1)(x^2+2) - 2x(x^2+1)}$$

$$= \frac{\pi}{4} + \int_0^1 \frac{x(2x-1)dx}{(x^2+1)(x^2-2x+2)}$$

$$= \frac{\pi}{4} + \int_0^1 \frac{x\{(x^2+1) - (x^2-2x+2)\}}{(x^2+1)(x^2-2x+2)}\,dx$$

$$= \frac{\pi}{4} + \int_0^1 \frac{x\,dx}{x^2-2x+2} - \int_0^1 \frac{x\,dx}{x^2+1}$$

$$= \frac{\pi}{4} + \frac{1}{2}\int_0^1 \frac{(2x-2)+2}{x^2-2x+2}\,dx - \frac{1}{2}\int_0^1 \frac{2x\,dx}{x^2+1}$$

$$= \frac{\pi}{4} + \frac{1}{2}\int_0^1 \frac{d(x^2-2x+2)}{x^2-2x+2} + \int_0^1 \frac{dx}{x^2-2x+2} - \frac{1}{2}\int_0^1 \frac{d(x^2+1)}{x^2+1}$$

$$= \frac{\pi}{4} + \frac{1}{2}\left[\log(x^2-2x+2)\right]_0^1 - \frac{1}{2}\left[\log(x^2+1)\right]_0^1 + \int_0^1 \frac{dx}{(x-1)^2+1}$$

$$= \frac{\pi}{4} + \frac{1}{2}\{\log 1 - \log 2\} - \frac{1}{2}\{\log 2 - \log 1\} + \left[\tan^{-1}(x-1)\right]_0^1$$

$$= \frac{\pi}{4} - \frac{1}{2}\log 2 - \frac{1}{2}\log 2 - \tan^{-1}(-1)$$

$$= \frac{\pi}{4} - \log 2 - \left(-\frac{\pi}{4}\right) = \frac{\pi}{2} - \log 2.$$

Exercises

Integrate by using summation (or Integrate from the first principle):

1. $\int_0^{\pi/2} \sin x\, dx$
2. $\int_0^{\pi/4} \cos^2 x\, dx$
3. $\int_0^1 x^2\, dx$
4. $\int_0^1 2^x\, dx$
5. $\int_a^b e^x\, dx$
6. $\int_1^2 x^3\, dx$

Evaluate the following:

7. $\lim_{n\to\infty}\left\{\frac{1}{1+n^3} + \frac{4}{8+n^3} + \ldots + \frac{r^2}{r^3+n^3} + \ldots + \frac{1}{2n}\right\}$

8. $\lim_{n\to\infty}\frac{1}{n}\left(\sin\frac{\pi}{2n} + \sin\frac{2\pi}{2n} + \sin\frac{3\pi}{2n} + \ldots + \sin\frac{n\pi}{2n}\right)$

9. $\lim_{n\to\infty}\sum_{r=1}^{n}\frac{1}{n}\tan\frac{r\pi}{4n}$

10. $\lim_{n\to\infty}\left(\frac{n}{n^2+1^2} + \frac{n}{n^2+2^2} + \frac{n}{n^2+3^2} + \ldots + \frac{n}{n^2+n^2}\right)$

11. $\lim_{n\to\infty}\frac{1}{n}(e^{1/n} + e^{2/n} + e^{3/n} + \ldots + e^{n/n})$

12. $\lim_{n\to\infty}\left\{\frac{1}{n} + \frac{1}{\sqrt{n^2-1^2}} + \frac{1}{\sqrt{n^2-2^2}} + \ldots + \frac{1}{\sqrt{n^2-(n-1)^2}}\right\}$

13. $\lim_{n\to\infty}\left(\frac{1}{n+1} + \frac{1}{n+2} + \frac{1}{n+3} + \ldots + \frac{1}{pn}\right)$, where $p \in N$ and $p \geq 2$.

14. $\lim_{n\to\infty}\frac{1}{n}\left(\frac{1}{n+1} + \frac{2}{n+2} + \frac{3}{n+3} + \ldots + \frac{3n}{4n}\right)$

Find the following limits:

15. $\lim_{n\to\infty}\left\{\left(1+\frac{1}{n}\right)\left(1+\frac{2}{n}\right)\left(1+\frac{3}{n}\right)\ldots\left(1+\frac{n}{n}\right)\right\}^{\frac{1}{n}}$

16. $\lim_{n\to\infty}\left\{\left(1+\frac{1^2}{n^2}\right)\left(1+\frac{2^2}{n^2}\right)\left(1+\frac{3^2}{n^2}\right)\ldots\left(1+\frac{n^2}{n^2}\right)\right\}^{\frac{1}{n}}$

Evaluate the following:

17. $\int_0^{\pi/2} \cos^5\frac{x}{2}\cdot\sin x\, dx$

18. $\int_0^{\pi/2} \frac{dx}{(a^2\cos^2 x + b^2\sin^2 x)^2}$

19. $\int_0^{\pi/4} \frac{\sin^2 x\cdot\cos^2 x}{(\sin^3 x + \cos^3 x)^2}\, dx$

20. $\int_0^1 x\log\sqrt{x+2}\, dx$

21. $\int_1^{16} \tan^{-1}\sqrt{\sqrt{x}-1}\, dx$

22. $\int_0^{\alpha} \frac{\sin x}{\cos(x+\alpha)}\, dx,\ 0 < \alpha < \frac{\pi}{4}$

23. $\int_0^1 \frac{x^5 dx}{1+x^4}$

24. $\int_0^1 \frac{x\, dx}{\sqrt{x+\lambda} + \sqrt{x+\mu}},\ \lambda \neq \mu$

25. $\int_0^{\pi/4} \cos 2x\cdot\log(1+\tan x)dx$

26. Show that $\int_0^{\pi} \cos 2x\cdot\log\sin x\, dx = -\frac{\pi}{2}$.

Evaluate the following:

27. $\int_{\pi/2}^{\pi} e^x\cdot\frac{1-\sin x}{1-\cos x}\, dx$

28. $\int_0^{2\pi} e^x\cdot\sin\left(\frac{\pi}{4}+\frac{x}{2}\right)dx$

29. $\int_0^{\pi/4} \frac{e^{\sin x}(x\cos^3 x - \sin x)}{\cos^2 x}\, dx$

30. $\int_0^{2a} \sqrt{2ax - x^2}\, dx$

31. $\int_0^a \frac{dx}{a + \sqrt{a^2 - x^2}}$

32. $\int_2^3 \frac{2x^5 + x^4 - 2x^3 + 2x^2 + 1}{(x^2+1)(x^4-1)}\, dx$

33. $\int_0^1 \frac{x\,dx}{1+x^3}$

34. $\int_0^{\pi/2} \frac{\cos x\,dx}{3-2\cos x}$

35. $\int_0^{\pi/2} \frac{\cos^2\phi\,d\phi}{\cos^2\phi+4\sin^2\phi}$

36. $\int_0^{\pi/2} \frac{dx}{1+\frac{1}{6}\sin^2 x}$

37. $\int_0^{\pi/4} \frac{\sec x\,dx}{1+2\sin^2 x}$

38. $\int_0^{\pi/4} \frac{\sin x\cdot\cos x}{\cos^2 x+\sin^4 x}\,dx$

39. $\int_{\pi/6}^{\pi/3} \frac{\sin x+\cos x}{\sqrt{\sin 2x}}\,dx$

40. $\int_{-\pi/4}^{0} \sin x\cdot\sec^2\left(x+\frac{\pi}{4}\right)dx$

41. $\int_0^1 \frac{dx}{(\sqrt{1+x^2}-x)^2}$

42. $\int_0^{\pi/2} \frac{\sec^2 x\,dx}{(\sec x+\tan x)^n},\ n>1$

43. If $f(x)=x^3$, show that

$$\int_a^b f(x)dx=\frac{b-a}{6}\left\{f(a)+4f\left(\frac{a+b}{2}\right)+f(b)\right\}.$$

44. Prove that

$$\int_0^{\pi/2} \frac{x\sin x\cdot\cos x}{(a^2\cos^2 x+b^2\sin^2 x)^2}\,dx=\frac{\pi}{4ab^2(a+b)}.$$

45. If $f(x)=\begin{vmatrix}\sec x & \cos x & \sec^2 x+\cot x\cdot\operatorname{cosec} x\\ \cos^2 x & \cos^2 x & \operatorname{cosec}^2 x\\ 1 & \cos^2 x & \cos^2 x\end{vmatrix}$,

find $\int_0^{\pi/2} f(x)dx$.

46. If $A=\int_0^{\pi}\frac{\cos x}{(x+2)^2}\,dx$, show that

$$\int_0^{\pi/2}\frac{\sin 2x}{x+1}\,dx=\frac{1}{2}+\frac{1}{\pi+2}-A.$$

Objective Questions

Fill in the blanks.

47. The value of $\int_0^{\pi}\frac{dx}{a^2\cos^2 x+b^2\sin^2 x}$ is ______ .

48. The value of $\int_0^{\infty}\frac{54a}{(3+a+4y)^4}\,dy=$ ______ .

49. If $\int_0^1 x\log\left(1+\frac{x}{2}\right)dx=a+b\cdot\log\frac{2}{3}$

then $a=$ ______ .

50. The value of $\int_1^{e^{37}}\frac{\pi\sin(\pi\log x)}{x}\,dx=$ ______ .

51. $\lim_{x\to\infty}\int_0^x\left\{\frac{1}{\sqrt{1+t^2}}-\frac{1}{1+t}\right\}dt=$ ______ .

52. $\int_0^{\pi/2}\left\{\int_{\pi/2}^{y}\cos x\,dx\right\}\sin y\,dy=$ ______ .

Choose the correct option(s).

53. The value of $\int_1^e\frac{1}{x}(1+\log x)dx$ is

(a) $\frac{3}{2}$ (b) $\frac{1}{2}$ (c) e (d) $\frac{1}{e}$

54. The value of $\int_0^1\frac{\sin^{-1}x}{\sqrt{1-x^2}}\,dx$ is

(a) 0 (b) $\frac{3}{2}$

(c) $\frac{\pi^2}{72}$ (d) none of these

55. $\int_0^2\sqrt{\frac{2+x}{2-x}}\,dx$ is equal to

(a) $\frac{\pi}{2}+1$ (b) $\pi+\frac{3}{2}$

(c) $\pi+1$ (d) none of these

56. The value of $\int_{-2}^{2}(px^3+qx+r)dx$, where p, q, r are constants, depends on the value of

(a) q (b) r (c) p (d) p and q

57. $\lim_{n\to\infty}\frac{1}{n}\sum_{r=1}^{2n}\frac{r}{\sqrt{n^2+r^2}}$ equals

(a) $1+\sqrt{5}$ (b) $-1+\sqrt{5}$

(c) $-1+\sqrt{2}$ (d) $1+\sqrt{2}$

58. $\lim_{n\to\infty}\frac{1}{n}\sum_{r=n+1}^{2n}\log\left(1+\frac{r}{n}\right)$ equals

(a) $\log\frac{27}{4e}$ (b) $\log\frac{27}{e^2}$

(c) $\log\frac{4}{e}$ (d) none of these

Answers

1. 1
2. $\frac{\pi + 2}{8}$
3. $\frac{1}{3}$
4. $\log_2 e$
5. $e^b - e^a$
6. $\frac{15}{4}$
7. $\frac{1}{3}\log 2$
8. $\frac{2}{\pi}$
9. $\frac{2}{\pi}\log 2$
10. $\frac{\pi}{4}$
11. $e - 1$
12. $\frac{\pi}{2}$
13. $\log_e p$
14. $3 - \log 4$
15. $\frac{4}{e}$
16. $2e^{\frac{\pi - 4}{2}}$
17. $\frac{4}{7}\left(1 - \frac{1}{8\sqrt{2}}\right)$
18. $\frac{\pi(a^2 + b^2)}{4a^3b^3}$
19. $\frac{1}{6}$
20. $\frac{1}{4}\log\frac{16}{27} + \frac{3}{8}$
21. $\frac{16\pi}{3} - 2\sqrt{3}$
22. $\cos\alpha \cdot \log\frac{\cos\alpha}{\cos 2\alpha} - \alpha\sin\alpha$
23. $\frac{4 - \pi}{8}$
24. $\frac{2}{15(\lambda - \mu)}\{(3 - 2\lambda)(1 + \lambda)^{3/2} + (2\mu - 3)(1 + \mu)^{3/2} + 2(\lambda^{5/2} - \mu^{5/2})\}$
25. $\frac{6\log 2 - \pi}{8}$
27. $e^{\pi/2}$
28. $-\frac{\sqrt{2}}{5}(1 + e^{2\pi})$
29. $1 + e^{\frac{1}{\sqrt{2}}}\left(\frac{\pi}{4} - \sqrt{2}\right)$
30. $\frac{1}{2}\pi a^2$
31. $\frac{\pi}{2} - 1$
32. $\frac{1}{2}\left(\log 6 - \frac{1}{5}\right)$
33. $\frac{\pi}{3\sqrt{3}} - \frac{1}{3}\log 2$
34. $-\frac{\pi}{4} + \frac{3}{\sqrt{5}}\tan^{-1}\sqrt{5}$
35. $\frac{\pi}{6}$
36. $\frac{\pi}{2}\cdot\sqrt{\frac{6}{7}}$
37. $\frac{\pi}{6\sqrt{2}} + \frac{1}{3}\log(1 + \sqrt{2})$
38. $\frac{\pi}{6\sqrt{3}}$
39. $2\sin^{-1}\frac{\sqrt{3} - 1}{2}$
40. $\frac{(\sqrt{2} - 1) - \log(\sqrt{2} + 1)}{\sqrt{2}}$
41. $1 + \frac{4}{3}\sqrt{2}$
42. $\frac{n}{n^2 - 1}$
45. $-\frac{32 + 15\pi}{60}$
47. $\frac{\pi}{ab}$
48. $\frac{9a}{2(3 + a)^3}$
49. $\frac{3}{4}$
50. 2
51. $\log_e 2$
52. $\frac{\pi - 4}{4}$
53. (a)
54. (d)
55. (d)
56. (b)
57. (b)
58. (a)

□

Chapter Test

Time: 75 minutes

1. In each of the following, fill in the blanks so that the resulting sentence becomes true.

(a) If $\int_a^b f(x)dx = \lambda \int_{ac}^{bc} f\left(\frac{x}{c}\right) dx$, $c \neq 0$ then $\lambda =$ _____ .

(b) $\int_2^3 \frac{\sqrt{x}}{\sqrt{5-x}+\sqrt{x}} dx =$ _____ .

(c) $\int_0^{4/\pi} \left(\sin\frac{1}{x} + 2x\cos\frac{1}{x}\right) dx =$ _____ .

(d) If $f(x) = A \cdot 2^x + B$ such that $f'(1) = 2$ and $\int_0^3 f(x)dx = 7$ then the constants

$A =$ _____ , $B =$ _____ .

(e) $\int_{-1}^1 \log\frac{2-x}{2+x} dx =$ _____ .

(f) If for $x \neq 0$, $\lambda f(x) + \mu f\left(\frac{1}{x}\right) = \frac{1}{x} - 5$ where $\lambda \neq \mu$ then $\int_1^2 x f(x) dx =$ _____ .

Evaluate the following:

2. $\int_1^8 \frac{1+x^{-2/3}}{1+x} dx$

3. $\int_0^{\pi/2} \sin 2x \cdot \tan^{-1}(\cos x) dx$

4. $\int_{\pi/4}^{\pi/2} \frac{\sin x - x\cos x}{x(x-\sin x)} dx$

5. $\int_0^{\pi} \frac{dx}{1+2\cos^2 x}$

6. $\int_0^{\pi/2} xe^x \sin x\, dx$

7. $\int_0^1 \log(1+x+x^2) dx$

8. If $S_n = \frac{1}{1+\sqrt{n}} + \frac{1}{2+\sqrt{2n}} + \ldots + \frac{1}{n+\sqrt{n^2}}$, find $\lim_{n\to\infty} S_n$.

□

Answers

1. (a) $\frac{1}{c}$ (b) $\frac{1}{2}$ (c) $\frac{8\sqrt{2}}{\pi^2}$ (d) $\log_2 e, \frac{7}{3}\{1-(\log_2 e)^2\}$ respectively

   (e) 0 (f) $\frac{31\mu - 39\lambda}{6(\lambda^2 - \mu^2)}$

2. $\frac{5}{2}\log 3 - 2\log 2 + \frac{\pi}{2\sqrt{3}}$

3. $\frac{\pi - 2}{2}$

4. $\log \frac{\pi - 2}{\pi - 2\sqrt{2}}$

5. $\frac{\pi}{\sqrt{3}}$

6. $\frac{\pi}{4} e^{\pi/2} - \frac{1}{2}$

7. $\frac{3}{2}\log 3 - 2 + \frac{\pi}{2\sqrt{3}}$

8. $2\log 2$

□

11. Properties and Application of Definite Integrals

Recap of Facts and Formulae

1. Properties of definite integral

- $\int_a^b f(x)dx = \int_a^b f(z)dz$
- $\int_a^b f(x)dx = -\int_b^a f(x)\,dx$
- $\int_a^b f(x)dx = \int_a^c f(x)dx + \int_c^b f(x)dx$, where $a < c < b$.

TIP: This property must be used for integrating functions of piecewise definition.

- $\int_0^a f(x)dx = \int_0^a f(a-x)dx$
- $\int_{-a}^a f(x)dx = 2\int_0^a f(x)dx$ if $f(x)$ is even, i.e., $f(-x) = f(x)$

  0 if $f(x)$ is odd, i.e., $f(-x) = -f(x)$
- $\int_0^{na} f(x)dx = n\int_0^a f(x)dx$

  if $f(x)$ is a periodic function of the period a, i.e., $f(a+x) = f(x)$.
- $\int_{ma}^{na} f(x)dx = (n-m)\int_0^a f(x)dx$

  if $f(x)$ is a periodic function of the period a.

2. Derivative of indefinite integral

- If $y = \int_a^x f(t)dt$ then $\frac{dy}{dx} = f(t)\Big]_{t=x} = f(x)$
- If $y = \int_a^{\phi(x)} f(t)dt$ then $\frac{dy}{d\phi} = f(t)\Big]_{t=\phi(x)} = f\{\phi(x)\}$

3. Inequalities in definite integrals

- If $f(x) \geq 0$ for all $x \in [a, b]$ then $\int_a^b f(x)\,dx \geq 0$,

  equality holding if $f(x) = 0$ at all points of $[a, b]$.
- $\left|\int_a^b f(x)dx\right| \leq \int_a^b |f(x)|\,dx$
- If $f(x) \geq g(x)$ for all $x \in [a, b]$ then $\int_a^b f(x)dx \geq \int_a^b g(x)dx$,

  equality holding if $f(x) = g(x)$ at all points of $[a, b]$.
- If minimum $f(x) = m$, maximum $f(x) = M$ in $[a, b]$ then

  $$m(b-a) \leq \int_a^b f(x)dx \leq M(b-a),$$

  equality holding for constant functions.

4. Average value of a function over an interval

- The *average value* of $f(x)$ over the interval $[a, b]$

  $$= \frac{1}{b-a}\int_a^b f(x)dx.$$

For example. The average value of $y = \sin x$ over the interval $[0, \pi] = \frac{1}{\pi - 0}\int_0^\pi \sin x\,dx$

$$= \frac{1}{\pi}[-\cos x]_0^\pi = \frac{1}{\pi}(1+1) = \frac{2}{\pi}.$$

Selected Solved Examples

1. Show that $\int_0^\pi xf(\sin x)\,dx = \frac{\pi}{2}\int_0^\pi f(\sin x)\,dx$.

$I = \int_0^\pi xf(\sin x)\,dx$

$= \int_0^\pi (\pi - x)f(\sin \overline{\pi - x})\,dx$

$$\left\{\because \int_0^a f(x)\,dx = \int_0^a f(a-x)\,dx\right\}$$

$$= \int_0^{\pi} (\pi - x) f(\sin x) dx$$

$$= \int_0^{\pi} \pi f(\sin x) dx - \int_0^{\pi} x f(\sin x) dx$$

$$\therefore \quad I = \pi \int_0^{\pi} f(\sin x) dx - I$$

$$\therefore \quad 2I = \pi \int_0^{\pi} f(\sin x) dx$$

$$\therefore \quad I = \frac{\pi}{2} \int_0^{\pi} f(\sin x) dx.$$

2. Prove that

$$\int_0^{\pi/2} f(\sin 2x) \sin x \, dx = \int_0^{\pi/2} f(\sin 2x) \cos x \, dx = \sqrt{2} \int_0^{\pi/4} f(\cos 2x) \cos x \, dx.$$

$$\int_0^{\pi/2} f(\sin 2x) \sin x dx = \int_0^{\pi/2} f\left\{\sin 2\left(\frac{\pi}{2} - x\right)\right\} \sin\left(\frac{\pi}{2} - x\right) dx$$

$$= \int_0^{\pi/2} f\{\sin(\pi - 2x)\} \cos x \, dx$$

$$= \int_0^{\pi/2} f(\sin 2x) \cos x \, dx.$$

Denoting each of the first two integrals by I,

$$2I = I + I = \int_0^{\pi/2} f(\sin 2x) \sin x \, dx + \int_0^{\pi/2} f(\sin 2x) \cos x \, dx$$

$$= \int_0^{\pi/2} f(\sin 2x) \cdot (\sin x + \cos x) dx$$

$$= \sqrt{2} \int_0^{\pi/2} f(\sin 2x) \cos\left(\frac{\pi}{4} - x\right) dx.$$

Put $\frac{\pi}{4} - x = z$; then $-dx = dz$.

Also $x = 0 \Rightarrow z = \frac{\pi}{4}$; $x = \frac{\pi}{2} \Rightarrow z = \frac{-\pi}{4}$.

$$\therefore \quad 2I = \sqrt{2} \int_{\pi/4}^{-\pi/4} f\left\{\sin 2\left(\frac{\pi}{4} - z\right)\right\} \cos z(-dz)$$

$$= \sqrt{2} \int_{-\pi/4}^{\pi/4} f\left\{\sin\left(\frac{\pi}{2} - 2z\right)\right\} \cos z dz$$

$$\left(\because \int_a^b f(x) dx = -\int_b^a f(x) dx\right)$$

$$= \sqrt{2} \int_{-\pi/4}^{\pi/4} f(\cos 2z) \cos z dz$$

$$= \sqrt{2} \times 2 \int_0^{\pi/4} f(\cos 2z) \cos z dz$$

$\{\because f(\cos 2z) \cos z$ is an even function$\}$

$$= 2\sqrt{2} \int_0^{\pi/4} f(\cos 2x) \cos x \, dx$$

$$\left\{\because \int_a^b f(x) dx = \int_a^b f(z) dz\right\}$$

$$\therefore \quad I = \sqrt{2} \int_0^{\pi/4} f(\cos 2x) \cos x \, dx.$$

3. $f(x)$ is a monotonic and differentiable real-valued function and a, b are two real numbers. Show that

$$\int_a^b \{f(x) + f(a)\}\{f(x) - f(a)\} dx = 2 \int_{f(a)}^{f(b)} x\{b - f^{-1}(x)\} dx.$$

As $f(x)$ is monotonic, $f^{-1}(x)$ exists.

Also, if $x = f(z)$ then $f^{-1}(x) = z$

and $x = f(a) \Rightarrow z = f^{-1}\{f(a)\} = a$

$x = f(b) \Rightarrow z = f^{-1}\{f(b)\} = b$. Then

$$\text{RHS} = 2 \int_a^b f(z)(b - z) f'(z) dz$$

$\{\because x = f(z) \Rightarrow dx = f'(z) dz\}$

$$= \int_a^b (b - z) \, d[\{f(z)\}^2]$$

$$= \left[(b - z)\{f(z)\}^2\right]_a^b - \int_a^b \{f(z)\}^2 \cdot (-1) dz,$$

using by parts

$$= -(b - a)\{f(a)\}^2 + \int_a^b \{f(z)\}^2 dz$$

$$= -\int_a^b \{f(a)\}^2 dz + \int_a^b \{f(z)\}^2 dz$$

$$= \int_a^b [\{f(z)\}^2 - \{f(a)\}^2] dz$$

$$= \int_a^b [\{f(x)\}^2 - \{f(a)\}^2] dx, \text{ using property}$$

$$= \int_a^b \{f(x) + f(a)\}\{f(x) - f(a)\} dx.$$

4. Show that $\int_0^1 \frac{\log x}{\sqrt{1 - x^2}} dx = -\frac{\pi}{2} \log 2.$

Put $x = \sin\theta$; then $dx = \cos\theta \, d\theta$

and $x = 0 \Rightarrow \theta = 0$; $x = 1 \Rightarrow \theta = \frac{\pi}{2}$

$$\therefore \quad I = \int_0^{\pi/2} \frac{\log \sin\theta}{\sqrt{1 - \sin^2\theta}} \cdot \cos\theta \, d\theta = \int_0^{\pi/2} \log \sin\theta \, d\theta$$

$$= \int_0^{\pi/2} \log \sin\left(\frac{\pi}{2} - \theta\right) d\theta,$$

using $\int_0^a f(x)\,dx = \int_0^a f(a-x)dx$

$$\therefore \quad I = \int_0^{\pi/2} \log \cos\theta\, d\theta$$

$$\therefore \quad 2I = \int_0^{\pi/2} \log \sin\theta\, d\theta + \int_0^{\pi/2} \log\cos\theta\, d\theta$$

$$= \int_0^{\pi/2} \log(\sin\theta \cdot \cos\theta) d\theta$$

$$= \int_0^{\pi/2} \log \frac{\sin 2\theta}{2} d\theta$$

$$= \int_0^{\pi/2} \log \sin 2\theta\, d\theta - \int_0^{\pi/2} \log 2 d\theta$$

$$= \frac{1}{2}\int_0^{\pi} \log \sin\phi\, d\phi - \log 2\, [\theta]_0^{\pi/2} \quad \{\text{putting } 2\theta = \phi\}$$

$$= \frac{1}{2}\int_0^{\pi/2} \log\sin\phi\, d\phi + \frac{1}{2}\int_{\pi/2}^{\pi} \log\sin\phi\, d\phi$$

$$- \log 2 \cdot \frac{\pi}{2}, \text{ using property.}$$

$$\text{Now, } \int_{\pi/2}^{\pi} \log\sin\phi\, d\phi = \int_0^{\pi/2} \log \sin\left(\frac{\pi}{2} + z\right) dz$$

$$\left\{\text{putting } \phi = \frac{\pi}{2} + z\right\}$$

$$= \int_0^{\pi/2} \log\cos z\, dz$$

$$= \int_0^{\pi/2} \log\cos\phi\, d\phi, \text{ (using property)}$$

$$\therefore \quad 2I = \frac{1}{2}\int_0^{\pi/2} \log\sin\phi\, d\phi + \frac{1}{2}\int_0^{\pi/2} \log\cos\phi\, d\phi - \frac{\pi}{2}\log 2$$

$$= \frac{1}{2}\int_0^{\pi/2} \log\sin\theta\, d\theta + \frac{1}{2}\int_0^{\pi/2} \log\cos\theta\, d\theta - \frac{\pi}{2}\log 2$$

(using property)

$$\therefore \quad 2I = \frac{1}{2}I + \frac{1}{2}I - \frac{\pi}{2}\log 2; \quad \therefore \quad I = -\frac{\pi}{2}\log 2.$$

5. Evaluate $\int_0^{\pi} \frac{x\sin 2x \sin\left(\frac{\pi}{2}\cos x\right)}{2x - \pi} dx.$

$$I = \frac{1}{2}\int_0^{\pi} \frac{2x\sin 2x \sin\left(\frac{\pi}{2}\cos x\right)}{2x-\pi} dx$$

$$= \frac{1}{2}\int_0^{\pi} \frac{\left\{(2x-\pi)\sin 2x \sin\left(\frac{\pi}{2}\cos x\right) + \pi\sin 2x \cdot \sin\left(\frac{\pi}{2}\cos x\right)\right\}}{2x-\pi} dx$$

$$= \frac{1}{2}\int_0^{\pi} \sin 2x \cdot \sin\left(\frac{\pi}{2}\cos x\right) dx$$

$$+ \frac{\pi}{2}\int_0^{\pi} \frac{\sin 2x \cdot \sin\left(\frac{\pi}{2}\cos x\right)}{2x-\pi} dx = I_1 + I_2.$$

$$\text{Now, } I_1 = \frac{1}{2}\int_0^{\pi} \sin 2x \cdot \sin\left(\frac{\pi}{2}\cos x\right) dx.$$

Put $\frac{\pi}{2}\cos x = t$; then $\frac{-\pi}{2}\sin x\, dx = dt$

and $x = 0 \Rightarrow t = \frac{\pi}{2}$; $x = \pi \Rightarrow t = -\frac{\pi}{2}$

$$\therefore \quad I_1 = \frac{1}{2}\int_{\pi/2}^{-\pi/2} 2\sin x \cdot \cos x \cdot \sin t \cdot \frac{dt}{-\frac{\pi}{2}\sin x}$$

$$= \frac{2}{\pi}\int_{-\pi/2}^{\pi/2} \frac{2t}{\pi} \cdot \sin t\, dt = \frac{4}{\pi^2}\int_{-\pi/2}^{\pi/2} t\sin t dt$$

$$= \frac{8}{\pi^2}\int_0^{\pi/2} t\sin t\, dt$$

$$\{\because f(t) = t\sin t = f(-t), \text{ i.e., } f(t) \text{ is even}\}$$

$$= \frac{8}{\pi^2}\left[t \cdot (-\cos t) - \int -\cos t\, dt\right]_0^{\pi/2}$$

$$= \frac{8}{\pi^2}\int_0^{\pi/2} \cos t\, dt$$

$$= \frac{8}{\pi^2}[\sin t]_0^{\pi/2} = \frac{8}{\pi^2}.$$

$$\text{Again, } I_2 = \frac{\pi}{2}\int_0^{\pi} \frac{\sin 2x \cdot \sin\left(\frac{\pi}{2}\cos x\right)}{2x-\pi} dx$$

$$= \frac{\pi}{2}\int_0^{\pi} \frac{\sin 2(\pi - x) \cdot \sin\left\{\frac{\pi}{2}\cos(\pi - x)\right\}}{2(\pi - x) - \pi} dx$$

(using property)

$$= \frac{\pi}{2}\int_0^{\pi} \frac{\sin 2x \cdot \sin\left(\frac{\pi}{2}\cos x\right)}{\pi - 2x} dx = -I_2$$

$$\therefore \quad 2I_2 = 0; \quad \therefore \quad I_2 = 0.$$

$$\therefore \quad I = I_1 + I_2 = \frac{8}{\pi^2} + 0 = \frac{8}{\pi^2}.$$

6. Evaluate $\displaystyle\int_{-\pi/4}^{\pi/4} \frac{x + \frac{\pi}{4}}{2 - \cos 2x}\, dx.$

$$I = \int_{-\pi/4}^{\pi/4} \frac{x}{2 - \cos 2x}\, dx + \frac{\pi}{4}\int_{-\pi/4}^{\pi/4} \frac{dx}{2 - \cos 2x}$$

$= I_1 + I_2.$

Let $f(x) = \dfrac{x}{2 - \cos 2x}$

$$\therefore \quad f(-x) = \frac{-x}{2 - \cos 2(-x)}$$

$$= -\frac{x}{2 - \cos 2x} = -f(x)$$

$\therefore$ $f(x)$ is an odd function. Hence $I_1 = 0$ (using property).

Let $\phi(x) = \dfrac{1}{2 - \cos 2x}$

$$\therefore \quad \phi(-x) = \frac{1}{2 - \cos 2(-x)}$$

$$= \frac{1}{2 - \cos 2x} = \phi(x)$$

$\therefore$ $\phi(x)$ is an even function. So, by property of definite integrals,

$$I_2 = \frac{\pi}{4}\int_{-\pi/4}^{\pi/4} \frac{dx}{2 - \cos 2x}$$

$$= \frac{\pi}{4} \times 2\int_0^{\pi/4} \frac{dx}{2 - \cos 2x}$$

$$= \frac{\pi}{2}\int_0^{\pi/4} \frac{dx}{2 - \frac{1 - \tan^2 x}{1 + \tan^2 x}}$$

$$= \frac{\pi}{2}\int_0^{\pi/4} \frac{\sec^2 x\, dx}{1 + 3\tan^2 x}$$

$$= \frac{\pi}{2}\int_0^1 \frac{dz}{1 + 3z^2} \quad \{\text{putting } \tan x = z\}$$

$$= \frac{\pi}{2}\int_0^1 \frac{dz}{1 + (\sqrt{3}z)^2} = \frac{\pi}{2}\left[\frac{\tan^{-1}\sqrt{3}z}{\sqrt{3}}\right]_0^1$$

$$= \frac{\pi}{2\sqrt{3}} \cdot \frac{\pi}{3} = \frac{\pi^2}{6\sqrt{3}}.$$

7. Evaluate $\displaystyle\int_{\frac{-1}{\sqrt{3}}}^{\frac{1}{\sqrt{3}}} \frac{x^4}{1 - x^4} \cdot \cos^{-1}\frac{2x}{1 + x^2}\, dx.$

$$I = \int_{-\frac{1}{\sqrt{3}}}^{0} \frac{x^4}{1 - x^4} \cos^{-1}\frac{2x}{1 + x^2}\, dx + \int_0^{\frac{1}{\sqrt{3}}} \frac{x^4}{1 - x^4} \cos^{-1}\frac{2x}{1 + x^2}\, dx$$

$= I_1 + I_2.$

For I_1, put $x = -y$; then $dx = -dy$

and $x = -\dfrac{1}{\sqrt{3}} \Rightarrow y = \dfrac{1}{\sqrt{3}}$; $x = 0 \Rightarrow y = 0.$

$$\therefore \quad I_1 = \int_{\frac{1}{\sqrt{3}}}^{0} \frac{(-y)^4}{1 - (-y)^4} \cos^{-1}\frac{2(-y)}{1 + (-y)^2}(-dy)$$

$$= \int_0^{\frac{1}{\sqrt{3}}} \cdot \frac{y^4}{1 - y^4} \cos^{-1}\frac{-2y}{1 + y^2}\, dy$$

$$= \int_0^{\frac{1}{\sqrt{3}}} \frac{x^4}{1 - x^4} \cos^{-1}\frac{-2x}{1 + x^2}\, dx$$

$$\therefore \quad I = \int_0^{\frac{1}{\sqrt{3}}} \frac{x^4}{1 - x^4}\left\{\cos^{-1}\frac{2x}{1 + x^2} + \cos^{-1}\frac{-2x}{1 + x^2}\right\} dx$$

$$= \int_0^{\frac{1}{\sqrt{3}}} \frac{x^4}{1 - x^4}\left\{\cos^{-1}\frac{2x}{1 + x^2} + \left(\pi - \cos^{-1}\frac{2x}{1 + x^2}\right)\right\} dx,$$

because for $x \in \left[0, \frac{1}{\sqrt{3}}\right]$, $\cos^{-1}\dfrac{2x}{1 + x^2} \in$ first quadrant

whereas $\cos^{-1}\dfrac{-2x}{1 + x^2} \in$ second quadrant,

considering principal values.

$$\therefore \quad I = \pi\int_0^{\frac{1}{\sqrt{3}}} \frac{x^4}{1 - x^4}\, dx = \pi\int_0^{\frac{1}{\sqrt{3}}} \frac{1 - (1 - x^4)}{1 - x^4}\, dx$$

$$= \pi\int_0^{\frac{1}{\sqrt{3}}} \frac{dx}{1 - x^4} - \pi\int_0^{\frac{1}{\sqrt{3}}} dx$$

$$= \pi\int_0^{\frac{1}{\sqrt{3}}} \frac{dx}{(1 + x^2)(1 - x^2)} - \pi\,[x]_0^{1/\sqrt{3}}$$

$$= \frac{\pi}{2}\int_0^{\frac{1}{\sqrt{3}}} \frac{(1 + x^2) + (1 - x^2)}{(1 + x^2)(1 - x^2)}\, dx - \frac{\pi}{\sqrt{3}}$$

$$= \frac{\pi}{2}\int_0^{\frac{1}{\sqrt{3}}} \frac{1}{1 - x^2}\, dx + \frac{\pi}{2}\int_0^{\frac{1}{\sqrt{3}}} \frac{dx}{1 + x^2} - \frac{\pi}{\sqrt{3}}$$

$$= \frac{\pi}{2} \cdot \frac{1}{2}\left[\log\frac{1 + x}{1 - x}\right]_0^{\frac{1}{\sqrt{3}}} + \frac{\pi}{2}\left[\tan^{-1}x\right]_0^{\frac{1}{\sqrt{3}}} - \frac{\pi}{\sqrt{3}}$$

$$= \frac{\pi}{4}\log\frac{\sqrt{3}+1}{\sqrt{3}-1} + \frac{\pi}{2}\tan^{-1}\frac{1}{\sqrt{3}} - \frac{\pi}{\sqrt{3}}$$

$$= \frac{\pi}{4}\log\frac{(\sqrt{3}+1)^2}{2} + \frac{\pi}{12} - \frac{\pi}{\sqrt{3}}$$

$$= \frac{\pi}{2}\log\frac{\sqrt{3}+1}{\sqrt{2}} + \pi\left(\frac{1}{12} - \frac{1}{\sqrt{3}}\right).$$

8. Prove that $\int_0^\infty \frac{dx}{1+x^4} = \int_0^\infty \frac{x^2dx}{1+x^4} = \frac{\pi}{2\sqrt{2}}$.

Put $x = \frac{1}{z}$ in the first integral $\int_0^\infty \frac{dx}{1+x^4}$.

Then $dx = \frac{-1}{z^2}dz$; also

$$x = 0 \Rightarrow z = \infty;\; x = \infty \Rightarrow z = 0$$

$$\therefore \quad I = \int_0^\infty \frac{dx}{1+x^4} = \int_\infty^0 -\frac{1}{z^2}\cdot\frac{1}{1+\frac{1}{z^4}}dz$$

$$= \int_0^\infty \frac{z^2dz}{1+z^4} = \int_0^\infty \frac{x^2dx}{1+x^4}, \text{ using property}$$

$$\therefore \quad 2I = \int_0^\infty \frac{dx}{1+x^4} + \int_0^\infty \frac{x^2dx}{1+x^4}$$

$$= \int_0^\infty \frac{1+x^2}{1+x^4}dx$$

$$= \int_0^\infty \frac{1+x^2}{x^2}\cdot\frac{1}{x^2+\frac{1}{x^2}}dx$$

$$= \int_0^\infty \frac{1+\frac{1}{x^2}}{\left(x-\frac{1}{x}\right)^2+2}dx.$$

Put $x - \frac{1}{x} = t$; then $\left(1+\frac{1}{x^2}\right)dx = dt$.

Also, $x = 0 \quad\Rightarrow\quad t = -\infty$ and $x = \infty \quad\Rightarrow\quad t = \infty$

$$\therefore \quad 2I = \int_{-\infty}^\infty \frac{dz}{z^2+2} = \left[\frac{1}{\sqrt{2}}\tan^{-1}\frac{z}{\sqrt{2}}\right]_{-\infty}^\infty$$

$$= \frac{1}{\sqrt{2}}[\tan^{-1}\infty - \tan(-\infty)]$$

$$= \frac{1}{\sqrt{2}}\left[\frac{\pi}{2} - \left(-\frac{\pi}{2}\right)\right] = \frac{\pi}{\sqrt{2}}.$$

$$\therefore \quad I = \frac{1}{2}\cdot\frac{\pi}{\sqrt{2}} = \frac{\pi}{2\sqrt{2}}.$$

9. Evaluate $\int_{-1}^{\frac{3}{2}} |x\sin\pi x|\,dx$.

TIP

The modulus of $f(x)$, i.e., $|f(x)|$ has two definitions, one for $f(x) \geq 0$ and the other for $f(x) < 0$. So, in case the integrand is a modulus function we have to divide the interval of the limits, here $\left[-1, \frac{3}{2}\right]$, in subintervals such that in each subinterval $f(x)$ has the same sign and then use the following property of definite integral:

$$\int_a^b f(x) = \int_a^c f(x)dx + \int_c^b f(x)dx,\; a < c < b.$$

Similar is the case for all integrands whose definitions are piecewise, like step functions, etc.

Here, $x\sin\pi x \geq 0$ if $x \in [-1, 1]$ and

$$x\sin\pi x < 0 \text{ if } x \in \left(1, \frac{3}{2}\right]$$

$$\therefore \quad I = \int_{-1}^1 |x\sin\pi x|\,dx + \int_1^{\frac{3}{2}} |x\sin\pi x|\,dx$$

$$= \int_{-1}^1 x\sin\pi x\,dx + \int_1^{\frac{3}{2}} -x\sin\pi x\,dx.$$

Now, $\int x\sin\pi x\,dx = x\cdot\frac{-\cos\pi x}{\pi} - \int\frac{-\cos\pi x}{\pi}dx$

$$= -\frac{1}{\pi}x\cos\pi x + \frac{1}{\pi}\cdot\frac{\sin\pi x}{\pi}$$

$$\therefore \quad I = \left[-\frac{1}{\pi}x\cos\pi x + \frac{\sin\pi x}{\pi^2}\right]_{-1}^1$$

$$-\left[-\frac{1}{\pi}x\cos\pi x + \frac{\sin\pi x}{\pi^2}\right]_1^{3/2}$$

$$= -\frac{1}{\pi}\cos\pi - \left\{\frac{1}{\pi}\cos(-\pi)\right\}$$

$$-\left\{\left(-\frac{3}{2\pi}\cos\frac{3\pi}{2} + \frac{1}{\pi^2}\sin\frac{3\pi}{2}\right) - \left(-\frac{1}{\pi}\cos\pi\right)\right\}$$

$$= \frac{1}{\pi} + \frac{1}{\pi} - \left\{\left(0 + \frac{1}{\pi^2}(-1)\right) - \frac{1}{\pi}\right\}$$

$$= \frac{1}{\pi^2} + \frac{3}{\pi} = \frac{1}{\pi}\left(\frac{1}{\pi} + 3\right).$$

10. Evaluate $\int_0^{1.5} [x^2]dx$**, where** $[x]$ **is the greatest integral function.**

Here, $I = \int_0^{1.5} [x^2]dx = \int_0^1 [x^2]dx + \int_1^{\sqrt{2}} [x^2]dx + \int_{\sqrt{2}}^{1.5} [x^2]dx$

because $[x^2]$ changes its definition in $[0, 1)$, $[1, \sqrt{2})$, $[\sqrt{2}, 1.5]$

$$\therefore \quad I = \int_0^1 0\,dx + \int_1^{\sqrt{2}} 1dx + \int_{\sqrt{2}}^{1.5} 2\,dx$$

because in $[0, 1)$, $[x^2] = 0$; in $[1, \sqrt{2})$, $[x^2] = 1$; in $[\sqrt{2}, 1.5]$, $[x^2] = 2$

$$\therefore \quad I = 0 + x]_1^{\sqrt{2}} + 2 \times x]_{\sqrt{2}}^{1.5} = \sqrt{2} - 1 + 2(1.5 - \sqrt{2})$$

$$= 2 - \sqrt{2}.$$

11. The function $f(x)$ is defined as follows:

$$f(x) = x^2, \; -1 \le x < 1$$
$$\sqrt{x}, \quad 1 \le x < 2$$
$$\sqrt{2}, \quad 2 \le x \le 4.$$

Evaluate $\int_{-1}^{4} f(x)dx$.

Here $I = \int_{-1}^{4} f(x)dx = \int_{-1}^{1} f(x)dx + \int_1^2 f(x)dx + \int_2^4 f(x)dx$

$$= \int_{-1}^{1} x^2 dx + \int_1^2 \sqrt{x}\,dx + \int_2^4 \sqrt{2}\,dx$$

{using the given definition of $f(x)$}

$$= \left[\frac{x^3}{3}\right]_{-1}^{1} + \left[\frac{x^{3/2}}{\frac{3}{2}}\right]_1^2 + \sqrt{2}\,[x]_2^4$$

$$= \frac{1}{3} - \left(-\frac{1}{3}\right) + \frac{2}{3}(2^{3/2} - 1^{3/2}) + \sqrt{2}(4 - 2)$$

$$= \frac{2}{3} + \frac{2}{3}(2\sqrt{2} - 1) + 2\sqrt{2} = \frac{10}{3}\sqrt{2}.$$

12. Evaluate (a) $\int_{-20\pi}^{20\pi} |\cos x|\,dx$

(b) $\int_{10}^{100} (x - [x])dx$.

(a) Clearly $|\cos x|$ is an even function.

$$\therefore \quad I = \int_{-20\pi}^{20\pi} |\cos x|\,dx$$

$$= 2\int_0^{20\pi} |\cos x|\,dx, \text{ using property.}$$

Now, $|\cos x|$ is a periodic function of the period π because $|\cos x| = |\cos(x + \pi)|$.

$$\therefore \quad I = 2 \times 20 \int_0^{\pi} |\cos x|\,dx, \text{ using property}$$

$$= 40\left\{\int_0^{\pi/2} |\cos x|\,dx + \int_{\pi/2}^{\pi} |\cos x|\,dx\right\},$$

using property

$$= 40\left\{\int_0^{\pi/2} \cos x dx + \int_{\pi/2}^{\pi} -\cos x dx\right\}$$

$\left(\because \cos x \text{ is negative if } \frac{\pi}{2} < x \le \pi\right)$

$$= 40[\sin x]_0^{\pi/2} - 40[\sin x]_{\pi/2}^{\pi}$$

$$= 40 + 40 = 80.$$

(b) Clearly, $f(x) = x - [x]$ is a periodic function of the period 1 because

$$f(x + 1) = x + 1 - [x + 1] = x + 1 - \{[x] + 1\}$$

$$= x - [x] = f(x).$$

$$\therefore \quad \int_{10}^{100} (x - [x])dx = (100 - 10)\int_0^1 (x - [x])dx$$

$$= 90\int_0^1 xdx - 90\int_0^1 [x]dx$$

$$= 90\left[\frac{x^2}{2}\right]_0^1 - 90\int_0^1 0dx,$$

because $[x] = 0$ in $[0, 1)$

$$= 90 \times \frac{1}{2} = 45.$$

13. Show that $\int_0^{n\pi+\theta} |\sin x|\,dx = 2n + 1 - \cos\theta$,

where $n \in N$ and $0 \le \theta < \pi$.

$$I = \int_0^{n\pi+\theta} |\sin x|\,dx$$

$$= \int_0^{n\pi} |\sin x|\,dx + \int_{n\pi}^{n\pi+\theta} |\sin x|\,dx = I_1 + I_2.$$

Now, $I_1 = \int_0^{n\pi} |\sin x|\,dx = n\int_0^{\pi} |\sin x|\,dx,$

($\because$ $|\sin x|$ is a periodic function of the period π)

$$= n\int_0^{\pi} \sin x\,dx$$

($\because$ $\sin x$ is positive in the interval)

$$= n[-\cos x]_0^{\pi} = n(1 + 1) = 2n.$$

$I_2 = \int_{n\pi}^{n\pi+\theta} |\sin x|\,dx$. Putting $x = n\pi + z$,

$$I_2 = \int_0^{\theta} |\sin(n\pi + z)|\,dz = \int_0^{\theta} |\sin z|\,dz$$

$$= \int_0^{\theta} |\sin x|\,dx$$

$$= \int_0^{\theta} \sin x\,dx \quad (\because \text{ in } 0 \le \theta < \pi, |\sin x| = \sin x)$$

$$= [-\cos x]_0^{\theta} = -\cos\theta + 1$$

$\therefore \quad I = I_1 + I_2 = 2n + 1 - \cos\theta.$

14. If $f(x) = -\int_0^x \log\cos\theta\,d\theta$, prove that

$$f\left(\frac{\pi}{4}+\frac{x}{2}\right) - f\left(\frac{\pi}{4}-\frac{x}{2}\right) = \frac{1}{2}\{f(x) + x\log 2\}.$$

Here $f\left(\frac{\pi}{4}+\frac{x}{2}\right) = -\int_0^{\frac{\pi}{4}+\frac{x}{2}} \log\cos\theta\,d\theta$

and $f\left(\frac{\pi}{4}-\frac{x}{2}\right) = -\int_0^{\frac{\pi}{4}-\frac{x}{2}} \log\cos\theta\,d\theta$

$\therefore \quad f\left(\frac{\pi}{4}+\frac{x}{2}\right) - f\left(\frac{\pi}{4}-\frac{x}{2}\right)$

$$= -\int_0^{\frac{\pi}{4}+\frac{x}{2}} \log\cos\theta\,d\theta + \int_0^{\frac{\pi}{4}-\frac{x}{2}} \log\cos\theta\,d\theta$$

$$= -\int_{\frac{\pi}{4}-\frac{x}{2}}^{\frac{\pi}{4}+\frac{x}{2}} \log\cos\theta\,d\theta$$

$$= -\int_{\frac{x}{2}}^{-\frac{x}{2}} \log\cos\left(\frac{\pi}{4}-z\right)(-dz) \quad \left(\text{putting } \theta = \frac{\pi}{4}-z\right)$$

$$= -\int_{-\frac{x}{2}}^{\frac{x}{2}} \log\left\{(\cos z + \sin z)\cdot\frac{1}{\sqrt{2}}\right\} dz$$

$$= -\int_{-\frac{x}{2}}^{\frac{x}{2}} \left\{\log(\cos z + \sin z) - \frac{1}{2}\log 2\right\} dz$$

$$= \frac{1}{2}\log 2\,[z]_{-x/2}^{x/2} - \int_{-\frac{x}{2}}^{\frac{x}{2}} \log(\cos z + \sin z)dz$$

$$= \frac{1}{2}\left[x\log 2 - 2\int_{-\frac{x}{2}}^{\frac{x}{2}} \log(\cos z + \sin z)dz\right]$$

$$= \frac{1}{2}\left[x\log 2 - \int_{-x}^{x} \log\left(\cos\frac{t}{2} + \sin\frac{t}{2}\right)dt\right] \quad \ldots (1)$$

$$\left(\text{putting } z = \frac{t}{2}\right)$$

Now, $\int_{-x}^{x} \log\left(\cos\frac{t}{2} + \sin\frac{t}{2}\right)dt$

$$= \int_{-x}^{0} \log\left(\cos\frac{t}{2} + \sin\frac{t}{2}\right)dt + \int_0^x \log\left(\cos\frac{t}{2} + \sin\frac{t}{2}\right)dt$$

$$= -\int_x^0 \log\left(\cos\frac{\phi}{2} - \sin\frac{\phi}{2}\right)d\phi$$

$$+ \int_0^x \log\left(\cos\frac{t}{2} + \sin\frac{t}{2}\right)dt \quad (\text{putting } t = -\phi)$$

$$= \int_0^x \log\left(\cos\frac{\theta}{2} - \sin\frac{\theta}{2}\right)d\theta$$

$$+ \int_6^x \log\left(\cos\frac{\theta}{2} + \sin\frac{\theta}{2}\right)d\theta \quad (\text{using property})$$

$$= \int_0^x \log\left(\cos\frac{\theta}{2} - \sin\frac{\theta}{2}\right)\left(\cos\frac{\theta}{2} + \sin\frac{\theta}{2}\right)d\theta$$

$$= \int_0^x \log\left(\cos^2\frac{\theta}{2} - \sin^2\frac{\theta}{2}\right)d\theta$$

$$= \int_0^x \log\cos\theta\,d\theta = -f(x), \text{ from the question.}$$

$\therefore$ from (1), we get

$$f\left(\frac{\pi}{4}+\frac{x}{2}\right) - f\left(\frac{\pi}{4}-\frac{x}{2}\right) = \frac{1}{2}[x\log 2 + f(x)].$$

15. If f and g are continuous functions in $[0, a]$ satisfying $f(x) = f(a-x)$ and $g(x) + g(a-x) = 2$ then prove that $\int_0^a f(x)\,g(x)dx = \int_0^a f(x)dx.$

$$\int_0^a f(x)\,g(x)dx = \int_0^a f(a-x)\cdot g(a-x)dx, \text{ using property}$$

$$= \int_0^a f(x)\cdot\{2 - g(x)\}dx \quad (\text{from given conditions})$$

$$= 2\int_0^a f(x)dx - \int_0^a f(x)\cdot g(x)dx$$

$$\therefore \quad 2\int_0^a f(x)\,g(x)dx = 2\int_0^a f(x)dx$$

$$\therefore \quad \int_0^a f(x)\,g(x)dx = \int_0^a f(x)dx.$$

16. If $f(t)$ is an odd function then prove that $\int_a^x f(t)dt$ will be an even function.

Let $\phi(x) = \int_a^x f(t)dt$ ... (1)

As $f(t)$ is an odd function, $f(-t) = -f(t)$.

Now, $\phi(-x) = \int_a^{-x} f(t)dt$, putting $-x$ for x in (1)

$$\therefore \quad \phi(x) - \phi(-x) = \int_a^x f(t)dt - \int_a^{-x} f(t)dt$$

$= \int_a^x f(t)dt + \int_{-x}^a f(t)dt$, using property

$= \int_{-x}^x f(t)dt = 0$, because $f(t)$ is an odd function

$\therefore \quad \phi(x) - \phi(-x) = 0$; so, $\phi(-x) = \phi(x)$.

Hence $\phi(x)$, i.e., $\int_a^x f(t)dt$ is an even function.

17. Given a function $f(x)$ such that
(i) it is integrable over every interval in R
(ii) $f(x+t) = f(x)$ for every x and a real t.
Show that the integral $\int_a^{a+t} f(x)dx$ is independent of a.

$$I = \int_a^{a+t} f(x)dx = \int_a^0 f(x)dx + \int_0^t f(x)dx + \int_t^{a+t} f(x)dx \quad \dots (1)$$

Now, $\int_t^{a+t} f(x)dx = \int_0^a f(t+z)dz$ (putting $x = t+z$)

$= \int_0^a f(z)dz$, using (ii)

$= \int_0^a f(x)dx$, using property

$\therefore \quad I = \int_a^0 f(x)dx + \int_0^t f(x)dx + \int_0^a f(x)dx$, from (1)

$= -\int_0^a f(x)dx + \int_0^t f(x)dx + \int_0^a f(x)dx$, using property

$= \int_0^t f(x)dx$, which is independent of a.

18. Prove that for $n \in N$,
$\dfrac{\sin 2nx}{\sin x} = 2\,[\cos x + \cos 3x + \cos 5x + \dots + \cos(2n-1)x]$.
Hence prove that $\int_0^{\pi/2} \sin 2nx \cdot \cot x\, dx = \dfrac{\pi}{2}$.

Let $S = \cos x + \cos 3x + \cos 5x + \dots + \cos(2n-1)x$

$\therefore \quad 2\sin x \cdot S = 2\sin x \cdot \cos x + 2\sin x \cdot \cos 3x$
$+ 2\sin x \cos 5x + \dots + 2\sin x \cdot \cos(2n-1)x$
$= \sin 2x + (\sin 4x - \sin 2x)$
$+ (\sin 6x - \sin 4x) + \dots$
$+ \{\sin 2nx - \sin(2n-2)x\}$
$= \sin 2nx$

$\therefore \quad \dfrac{\sin 2nx}{\sin x} = 2S$

$= 2[\cos x + \cos 3x + \cos 5x + \dots + \cos(2n-1)x]$

$\therefore \quad \int_0^{\frac{\pi}{2}} \sin 2nx \cdot \cot x\, dx = \int_0^{\frac{\pi}{2}} \dfrac{\sin 2nx}{\sin x} \cos x\, dx$

$= \int_0^{\frac{\pi}{2}} 2\cos x \cdot \{\cos x + \cos 3x + \cos 5x + \dots$
$+ \cos(2n-1)x\}dx$

$= \int_0^{\frac{\pi}{2}} [(1 + \cos 2x) + (\cos 2x + \cos 4x)$
$+ (\cos 4x + \cos 6x) + \dots$
$+ \{\cos(2n-2)x + \cos 2nx\}]dx$

$= \int_0^{\frac{\pi}{2}} [1 + 2\{\cos 2x + \cos 4x + \cos 6x + \dots$
$+ \cos(2n-2)x\} + \cos 2nx]dx$

$= [x]_0^{\pi/2} + 2\left[\dfrac{\sin 2x}{2} + \dfrac{\sin 4x}{4} + \dfrac{\sin 6x}{6} + \dots\right.$
$\left. + \dfrac{\sin(2n-2)x}{2n-2}\right]_0^{\pi/2} + \left[\dfrac{\sin 2nx}{2n}\right]_0^{\pi/2}$

$= \dfrac{\pi}{2}$.

19. $I_n = \int_0^{\frac{\pi}{2}} \sin^n x\,dx$, $n \in N$, show that $I_n = \dfrac{n-1}{n} I_{n-2}$.
Using this reduction formula, evaluate $\int_0^{\frac{\pi}{2}} \sin^9 x\, dx$.

Here $I_n = \int_0^{\frac{\pi}{2}} \sin^n x\,dx = \int_0^{\pi/2} \sin^{n-1} x \cdot \sin x\, dx$

$= \Big[\sin^{n-1} x \cdot (-\cos x)$
$- \int -\cos x \cdot (n-1)\sin^{n-2} x \cdot \cos x\, dx\Big]_0^{\pi/2}$

$= \int_0^{\pi/2} (n-1)\sin^{n-2} x \cdot \cos^2 x\, dx$

$= (n-1)\int_0^{\pi/2} \sin^{n-2} x \cdot (1 - \sin^2 x)dx$

$= (n-1)\int_0^{\frac{\pi}{2}} \sin^{n-2} x\,dx - (n-1)\int_0^{\frac{\pi}{2}} \sin^n x\, dx$

$= (n-1)I_{n-2} - (n-1)I_n$

or $\quad I_n + (n-1)I_n = (n-1)I_{n-2}$

$\therefore \quad I_n = \dfrac{n-1}{n} I_{n-2} \quad \dots (1)$

Now, $\int_0^{\frac{\pi}{2}} \sin^9 x\, dx = I_9 = \frac{9-1}{9} I_7$, using (1)

$$= \frac{9-1}{9} \cdot \frac{7-1}{7} I_5 = \frac{8 \times 6}{9 \times 7} \cdot \frac{5-1}{5} I_3$$

$$= \frac{8 \times 6 \times 4}{9 \times 7 \times 5} \cdot \frac{3-1}{3} I_1 = \frac{8 \times 6 \times 4 \times 2}{9 \times 7 \times 5 \times 3} \int_0^{\pi/2} \sin x\, dx$$

$$= \frac{128}{315} [-\cos x]_0^{\pi/2} = \frac{128}{315}.$$

Note All integrals of the form $\int_0^{\frac{\pi}{2}} \sin^n x\, dx, n \in N$ can be evaluated by the reduction formula

$$I_n = \frac{n-1}{n} I_{n-2}.$$

20. Evaluate $\int_0^1 (tx + 1 - x)^n dx, n \in N$ and t is a parameter independent of x. Also show that

$$\int_0^1 x^k (1-x)^{n-k} dx = \frac{1}{(n+1) \cdot {}^nC_k}.$$

$$\int_0^1 (tx + 1 - x)^n dx = \int_0^1 \{(t-1)x + 1\}^n dx$$

$$= \left[\frac{\{(t-1)x + 1\}^{n+1}}{(n+1)(t-1)} \right]_0^1 = \frac{t^{n+1} - 1}{(n+1)(t-1)}$$

$$= \frac{1}{n+1} \{t^n + t^{n-1} + \ldots + t + 1\} \quad \ldots (1)$$

using sum of GP

Now, $\{tx + (1-x)\}^n = {}^nC_0 (tx)^n + {}^nC_1 (tx)^{n-1} \cdot (1-x)$

$$+ {}^nC_2 (tx)^{n-2} \cdot (1-x)^2 + \ldots$$

$$+ {}^nC_k (tx)^{n-k} \cdot (1-x)^k + \ldots + {}^nC_n (1-x)^n$$

$$\therefore \quad \int_0^1 (tx + 1 - x)^n dx = \int_0^1 \sum_{k=0}^{n} {}^nC_k (tx)^{n-k} \cdot (1-x)^k dx$$

$$= \sum_{k=0}^{n} t^{n-k} \int_0^1 {}^nC_k\, x^{n-k} (1-x)^k dx$$

$\therefore$ using (1), we get

$$\frac{1}{n+1} \{t^n + t^{n-1} + \ldots + t + 1\}$$

$$= \sum_{k=0}^{n} t^{n-k} \cdot \int_0^1 {}^nC_k \cdot x^{n-k} (1-x)^k dx.$$

Equating coefficients of t^{n-k} on both sides

$$\frac{1}{n+1} = {}^nC_k \cdot \int_0^1 x^{n-k} (1-x)^k dx$$

$$= {}^nC_k \cdot \int_0^1 (1-x)^{n-k} \cdot (1 - \overline{1-x})^k dx,$$

using property

$$\therefore \quad \int_0^1 x^k \cdot (1-x)^{n-k} dx = \frac{1}{(n+1) \cdot {}^nC_k}.$$

21. If $y = \int_1^{x^3} \frac{dt}{1+t^4}$, find $\frac{dy}{dx}$.

If $y = \int_a^{\phi(x)} f(t) dt$ then we have $\frac{dy}{d\phi} = f(t)\Big]_{t = \phi(x)}$

$$\therefore \quad y = \int_0^{x^3} \frac{dt}{1+t^4}$$

$$\Rightarrow \quad \frac{dy}{d(x^3)} = \frac{1}{1+t^4}\Big]_{t = x^3}$$

$$\text{or} \quad \frac{dy}{d(x^3)} = \frac{1}{1+(x^3)^4} = \frac{1}{1+x^{12}}.$$

Now, $\frac{dy}{dx} = \frac{dy}{d(x^3)} \cdot \frac{d(x^3)}{dx} = \frac{1}{1+x^{12}} \cdot 3x^2 = \frac{3x^2}{1+x^{12}}.$

22. If $\int_0^y \cos t^2 dt = \int_0^{x^2} \frac{\sin t}{t} dt$, find $\frac{dy}{dx}$.

Differentiating both sides w.r.t. x,

$$\frac{d}{dy}\left(\int_0^y \cos t^2 dt\right) \cdot \frac{dy}{dx} = \frac{d}{d(x^2)}\left(\int_0^{x^2} \frac{\sin t}{t} dt\right) \cdot \frac{d(x^2)}{dx}$$

$$\text{or} \quad \cos y^2 \cdot \frac{dy}{dx} = \frac{\sin x^2}{x^2} \cdot 2x$$

$$\text{or} \quad \frac{dy}{dx} = \frac{2\sin x^2}{x \cos y^2}.$$

23. Find $\lim_{x \to \infty} \frac{\left(\int_0^x e^{x^2} dx\right)^2}{\int_0^x e^{2x^2} dx}$.

As $e^{x^2} > 0$ in $[0, x]$, $\int_0^x e^{x^2} dx \to \infty$ as $x \to \infty$.

Hence, the indeterminacy is of the form $\frac{\infty}{\infty}$.

So, L'Hospital's rule is applicable.

$$\therefore \quad \text{limit} = \lim_{x \to \infty} \frac{\frac{d}{dx}\left\{\left(\int_0^x e^{x^2} dx\right)^2\right\}}{\frac{d}{dx}\left\{\int_0^x e^{2x^2} dx\right\}}$$

$$= \lim_{x \to \infty} \frac{\left(2\int_0^x e^{x^2} dx\right) \cdot e^{x^2}}{e^{2x^2}}$$

$$= \lim_{x \to \infty} 2 \cdot \frac{\int_0^x e^{x^2} dx}{e^{x^2}}, \left(\text{form } \frac{\infty}{\infty}\right)$$

$$= \lim_{x \to \infty} 2 \cdot \frac{e^{x^2}}{2xe^{x^2}}, \text{ using L'Hospital's rule}$$

$$= \lim_{x \to \infty} \frac{1}{x} = 0.$$

24. Show that $\frac{\pi}{6} < \int_0^1 \frac{dx}{\sqrt{4 - x^2 - x^3}} < \frac{\pi}{4\sqrt{2}}$.

Clearly, $4 - x^2 - x^3 > 0$ for $x \in [0, 1]$.

Also, $4 - x^2 > 4 - x^2 - x^3$ because $0 < x \le 1$ and at $x = 0$ they are equal. Again, both sides are positive.

$$\therefore \quad \frac{1}{4 - x^2} < \frac{1}{4 - x^2 - x^3}$$

$$\therefore \quad \int_0^1 \frac{dx}{\sqrt{4 - x^2}} < \int_0^1 \frac{dx}{\sqrt{4 - x^2 - x^3}} \quad \dots (1)$$

Again, $4 - x^2 - x^3 > 4 - x^2 - x^2$ for $0 < x < 1$ and at $x = 0, 1$ they are equal. Also, both sides are positive.

$$\therefore \quad \frac{1}{4 - x^2 - x^3} < \frac{1}{4 - 2x^2}$$

$$\therefore \quad \int_0^1 \frac{dx}{\sqrt{4 - x^2 - x^3}} < \int_0^1 \frac{dx}{\sqrt{4 - 2x^2}} \quad \dots (2)$$

Now, $\int_0^1 \frac{dx}{\sqrt{4 - x^2}} = \left[\sin^{-1}\frac{x}{2}\right]_0^1 = \sin^{-1}\frac{1}{2} = \frac{\pi}{6}$ ... (3)

and $\int_0^1 \frac{dx}{\sqrt{4 - 2x^2}} = \frac{1}{\sqrt{2}}\int_0^1 \frac{dx}{\sqrt{2 - x^2}}$

$$= \frac{1}{\sqrt{2}}\left[\sin^{-1}\frac{x}{\sqrt{2}}\right]_0^1 = \frac{1}{\sqrt{2}} \cdot \frac{\pi}{4} \quad \dots (4)$$

$\therefore$ from (1), (2), (3) and (4) we get

$$\frac{\pi}{6} < \int_0^1 \frac{dx}{\sqrt{4 - x^2 - x^3}} < \frac{\pi}{4\sqrt{2}}.$$

25. Let $f: \left[0, \frac{\pi}{2}\right] \to R$ **such that** $f(0) = 3$ **and**

$f'(x) = \frac{1}{1 + \cos x}$. **Show that** $3 + \frac{\pi}{4} < f\left(\frac{\pi}{2}\right) < 3 + \frac{\pi}{2}$.

$$\int_0^{\frac{\pi}{2}} f'(x)dx = [f(x)]_0^{\pi/2} = f\left(\frac{\pi}{2}\right) - f(0) = f\left(\frac{\pi}{2}\right) - 3 \quad \dots (1)$$

Also, $\int_0^{\frac{\pi}{2}} f'(x)dx = \int_0^{\frac{\pi}{2}} \frac{1}{1 + \cos x} dx < \int_0^{\frac{\pi}{2}} \frac{dx}{1 + 0}$

for $\cos x > 0$ when $x \in \left[0, \frac{\pi}{2}\right)$

and $\int_0^{\frac{\pi}{2}} f'(x)dx = \int_0^{\frac{\pi}{2}} \frac{1}{1 + \cos x} dx > \int_0^{\frac{\pi}{2}} \frac{1}{1 + 1} dx$

for $\cos x < 1$ when $x \in \left(0, \frac{\pi}{2}\right]$.

From the above two we get,

$$\int_0^{\frac{\pi}{2}} \frac{dx}{2} < \int_0^{\pi/2} f'(x)dx < \int_0^{\frac{\pi}{2}} dx$$

$$\therefore \quad \frac{1}{2}[x]_0^{\pi/2} < f\left(\frac{\pi}{2}\right) - 3 < [x]_0^{\pi/2}, \text{ from (1)}$$

$$\therefore \quad \frac{\pi}{4} < f\left(\frac{\pi}{2}\right) - 3 < \frac{\pi}{2}.$$

So, $3 + \frac{\pi}{4} < f\left(\frac{\pi}{2}\right) < 3 + \frac{\pi}{2}$.

26. A cubic function $f(x)$ vanishes at $x = -2$ and has relative maximum/minimum at $x = -1$ and $x = \frac{1}{3}$.

If $\int_{-1}^1 f(x)dx = \frac{14}{3}$, find $f(x)$.

Let $f(x) = ax^3 + bx^2 + cx + \lambda$;

then $f'(x) = 3ax^2 + 2bx + c$.

From the question we have $f(-2) = 0$

$$\therefore \quad -8a + 4b - 2c + \lambda = 0 \quad \dots (1)$$

Also, $f'(x) = 0$ when $x = -1, \frac{1}{3}$ (using condition for relative maximum/minumum).

$$\therefore \quad f'(-1) = 0 \Rightarrow 3a - 2b + c = 0 \quad \dots (2)$$

and $f'\left(\frac{1}{3}\right) = 0 \Rightarrow \frac{a}{3} + \frac{2b}{3} + c = 0$, i. e.,

$$a + 2b + 3c = 0 \quad \dots (3)$$

$(2) + (3) \Rightarrow 4a + 4c = 0$, i.e., $c = -a$

$\therefore$ (3) $\Rightarrow a + 2b - 3a = 0$, i.e., $b = a$

$\therefore$ from (1), $-8a + 4a + 2a + \lambda = 0$, i.e., $\lambda = 2a$.

$$\therefore \quad f(x) = ax^3 + ax^2 - ax + 2a = a(x^3 + x^2 - x + 2) \quad \dots (4)$$

$\therefore$ $\int_{-1}^1 f(x)dx = \frac{14}{3}$ gives

$$\int_{-1}^{1} a(x^3 + x^2 - x + 2)dx = \frac{14}{3}$$

$$\therefore \quad a\left[\frac{x^4}{4} + \frac{x^3}{3} - \frac{x^2}{2} + 2x\right]_{-1}^{1} = \frac{14}{3}$$

$$\text{or} \quad a\left\{\left(\frac{1}{4} + \frac{1}{3} - \frac{1}{2} + 2\right) - \left(\frac{1}{4} - \frac{1}{3} - \frac{1}{2} - 2\right)\right\} = \frac{14}{3}$$

$$\therefore \quad a \cdot \frac{14}{3} = \frac{14}{3}, \text{ i.e., } a = 1.$$

$\therefore$ from (4), $f(x) = x^3 + x^2 - x + 2.$

27. Let $a + b = 4$ where $a < 2$ and let $g(x)$ be a differentiable function of x. If $\frac{dg}{dx} > 0$ for all x, prove that $\int_0^a g(x)dx + \int_0^b g(x)dx$ increases as $(b - a)$ increases.

Let $y = \int_0^a g(x)dx + \int_0^b g(x)dx$ ... (1)

$z = b - a = (4 - a) - a = 4 - 2a$ ... (2)

We have to prove that $\frac{dy}{dz} > 0.$

Differentiating (1) w.r.t. a, $\frac{dy}{da} = g(a) + g(b) \cdot \frac{db}{da}$

or $\frac{dy}{da} = g(a) + g(b)\frac{d}{da}(4 - a) = g(a) - g(b).$

Differentiating (2) w.r.t. a, $\frac{dz}{da} = -2.$

$$\therefore \quad \frac{dy}{dz} = \frac{\frac{dy}{da}}{\frac{dz}{da}} = \frac{g(a) - g(b)}{-2} = \frac{g(4 - a) - g(a)}{2} \quad \text{... (3)}$$

As $a < 2$, $4 - a > a$. Also $\frac{dg}{dx} > 0$ for all x, implies that g is an increasing function.

$\therefore$ $g(4 - a) > g(a).$

$\therefore$ (3) $\Rightarrow$ $\frac{dy}{dz} > 0$, i.e., y is an increasing function of z.

$\therefore$ $\int_0^a g(x)dx + \int_0^b g(x)dx$ increases as $b - a$ increases.

28. Let $f(x) = \int_{-2}^{x} |t + 1|\,dt$. Examine the continuity and differentiability of $f(x)$ in the interval $\left[-\frac{3}{2}, 1\right]$.

If $-2 \le x \le -1$ then

$$f(x) = \int_{-2}^{x} -(t + 1)dt = -\left[\frac{t^2}{2} + t\right]_{-2}^{x}$$

$$= -\frac{x^2}{2} - x + \left(\frac{4}{2} - 2\right) = -\frac{x^2}{2} - x.$$

If $-1 \le x \le 1$ then

$$f(x) = \int_{-2}^{x} |t + 1|\,dt$$

$$= \int_{-2}^{-1} |t + 1|\,dt + \int_{-1}^{x} |t + 1|\,dt$$

$$= \int_{-2}^{-1} -(t + 1)dt + \int_{-1}^{x} (t + 1)dt$$

$$= \left[-\frac{t^2}{2} - t\right]_{-2}^{-1} + \left[\frac{t^2}{2} + t\right]_{-1}^{x}$$

$$= \frac{1}{2} + \frac{x^2}{2} + x - \left(\frac{1}{2} - 1\right)$$

$$= \frac{x^2}{2} + x + 1.$$

Thus, the definition of $f(x)$ is as follows:

$$f(x) = -\frac{x^2}{2} - x, \qquad -2 \le x \le -1$$

$$\frac{x^2}{2} + x + 1, \quad -1 \le x \le 1$$

As both $\left(-\frac{x^2}{2} - x\right)$ and $\left(\frac{x^2}{2} + x + 1\right)$ are polynomial functions which are continuous and differentiable everywhere, $f(x)$ is continuous and differentiable in $\left[-\frac{3}{2}, 1\right]$ except perhaps at the turning point of definition, i.e., $x = -1$.

Now, $\lim_{h \to 0} f(-1 - h) = \lim_{h \to 0} \left\{-\frac{1}{2}(-1 - h)^2 - (-1 - h)\right\} = \frac{1}{2}$

$\lim_{h \to 0} f(-1 + h) = \lim_{h \to 0} \left\{\frac{1}{2}(-1 + h)^2 + (-1 + h) + 1\right\} = \frac{1}{2}$

and $f(-1) = \frac{1}{2}$; so, $f(x)$ is continuous at $x = -1$.

Again, $\lim_{h \to 0} \frac{f(-1 + h) - f(-1)}{h}$

$$= \lim_{h \to 0} \frac{\left\{\frac{(-1 + h)^2}{2} + (-1 + h) + 1\right\} - \frac{1}{2}}{h}$$

$$= \lim_{h \to 0} \frac{\frac{1}{2} - h + \frac{h^2}{2} - 1 + h + 1 - \frac{1}{2}}{h}$$

$$= \lim_{h \to 0} \frac{\frac{h^2}{2}}{h} = 0.$$

$$\lim_{h \to 0} \frac{f(-1-h)-f(-1)}{-h}$$

$$= \lim_{h \to 0} \frac{\left\{-\frac{1}{2}(-1-h)^2-(-1-h)\right\}-\frac{1}{2}}{-h}$$

$$= \lim_{h \to 0} \frac{-\frac{1}{2}-h-\frac{h^2}{2}+1+h-\frac{1}{2}}{-h}$$

$$= \lim_{h \to 0} \frac{h}{2} = 0$$

So, $f(x)$ is differentiable at $x = -1$.

$\therefore$ $f(x)$ is continuous and differentiable in $\left[-\frac{3}{2}, 1\right]$.

29. If $f(x)$ is a continuous function with $\int_0^x f(t)dt \to \infty$ as $|x| \to \infty$ then show that every line $y = mx$ intersects the curve $y^2 + \int_0^x f(t)dt = 2$.

Putting $y = mx$ in $y^2 + \int_0^x f(t)dt$ we get $m^2x^2 + \int_0^x f(t)dt$.

Consider the function $\phi(x) = m^2x^2 + \int_0^x f(t)dt$.

As $f(x)$ is continuous and $\frac{d}{dx}\int_0^x f(t)dt = f(x)$,

$\frac{d}{dx}\int_0^x f(t)dt$ is also continuous. The derivative of a discontinuous function cannot be continuous. So $\int_0^x f(t)dt$ is also continuous. Again, m^2x^2 is also continuous, being a polynomial function. As the sum of continuous functions is also continuous, $\phi(x)$ is also continuous.

Now, $\phi(0) = 0 + \int_0^0 f(t)dt = 0$

$\phi(x) \to \infty$ as $|x| \to \infty$, because $m^2x^2 \to \infty$

and $\int^x f(t)dt \to \infty$ when $|x| \to \infty$

$\therefore$ for all real values of x between $-\infty$ and ∞, $\phi(x)$ attains values between 0 and ∞.

[**Note** If $\phi(a) = p$, $\phi(b) = q$ and $\phi(x)$ is continuous then every value between p and q will be attained by $\phi(x)$ for x between a and b—this is clear from the graph of a continuous function]

$\therefore$ $\phi(x)$ will also attain the value 2 for some real x and all real m.

$\therefore$ $\phi(x) = 2$, i.e., $m^2x^2 + \int_0^x f(t)dt = 2$,

i.e., $y^2 + \int_0^x f(t)dt = 2$

will have real solutions. Hence, the problem.

Exercises

1. Prove that $\int_0^a x^m(a-x)^n dx = \int_0^a x^n(a-x)^m dx$.

2. Prove that $\int_0^{\frac{\pi}{2}} \frac{\cos\theta - \sin\theta}{1+\sin\theta \cdot \cos\theta} d\theta = 0$.

3. Evaluate $\int_0^{\frac{\pi}{2}} \frac{\sin^2 x}{\cos x + \sin x} dx$.

4. Prove that $\int_0^{\pi} \log(1+\cos x)dx = -\pi \log 2$.

5. Prove that $\int_0^{2t} \frac{f(x)}{f(x)+f(2t-x)} dx = t$.

6. Evaluate $\int_0^{\frac{\pi}{4}} \log(1+\tan x)dx$.

7. Prove that $\int_0^{\frac{\pi}{2}} \frac{x\sin 2x \, dx}{\cos^4 x + \sin^4 x} = \frac{\pi^2}{8}$.

Evaluate the following:

8. $\int_0^{\pi} \frac{x\sin x}{1+\cos^2 x} dx$

9. $\int_0^{\pi} \frac{xdx}{1+\cos\alpha \cdot \sin x}$, $\alpha \in \left(0, \frac{\pi}{2}\right)$

10. $\int_0^{\pi} \frac{xdx}{a^2\cos^2 x + b^2\sin^2 x}$

11. $\int_0^{\pi} \frac{xdx}{1+\cos^2 x}$

12. Prove that $\int_0^{\pi} \frac{x\tan x}{\tan x + \sec x} dx = \frac{1}{2}\pi(\pi - 2)$.

13. Prove that $\int_0^{\pi} x\log \sin x\, dx = -\frac{\pi^2}{2}\log 2$.

14. Prove that $\int_0^{\frac{\pi}{2}} x(\sqrt{\tan x} + \sqrt{\cot x})dx = \frac{\pi^2}{2\sqrt{2}}$.

15. Prove that $\int_0^{\frac{\pi}{2}} \log \sin x\, dx = \int_0^{\frac{\pi}{2}} \log \cos x\, dx = -\frac{\pi}{2}\log 2$.

16. Prove that $\int_0^{\infty} \log\left(x + \frac{1}{x}\right) \cdot \frac{dx}{1 + x^2} = \pi\log 2$.

17. Prove that $\int_0^{\pi/2} \log(\tan x + \cot x)dx = \pi\log 2$.

18. Prove that $\int_0^{\infty} \frac{x\log x}{(1 + x^2)^2} dx = 0$.

19. Prove that $\int_{\frac{\pi}{6}}^{\frac{\pi}{3}} \frac{dx}{1 + \sqrt{\tan x}} = \frac{\pi}{12}$.

20. Evaluate $\int_0^{\pi} e^{\cos^2 x} \cdot \cos^3(2n + 1)x dx$,

n being an integer.

21. Prove that $\int_0^{2\pi} \frac{\sin^2\theta}{a - b\cos\theta} d\theta = \frac{2\pi}{b^2}(a - \sqrt{a^2 - b^2})$.

22. Prove that $\int_{-1}^{1} \frac{x^3}{\sqrt{1 - x^2}} \log \frac{1 + x}{1 - x} dx = \frac{5\pi}{3}$.

23. Prove that $8\int_0^{1} \frac{\log(1 + x)}{1 + x^2} dx = \int_0^{\infty} \frac{\log(1 + x^2)}{1 + x^2} dx$.

24. Prove that $y = \int_{\frac{1}{8}}^{\sin^2 x} \sin^{-1}\sqrt{t}\, dt + \int_{\frac{1}{8}}^{\cos^2 x} \cos^{-1}\sqrt{t}\, dt$,

where $x \in \left[0, \frac{\pi}{2}\right]$, is the equation of a straight line parallel to the x-axis. Find its equation.

Evaluate the following:

25. $\int_0^{\pi} |\cos x - \sin x|\, dx$

26. $\int_{-2}^{2} (|x| + |x - 1|)dx$

27. If $f(x) = |x^2 - x| + |x - 2|$, evaluate $\int_0^{3} f(x)dx$.

28. If $[x]$ is the greatest integer function then evaluate $\int_0^{5} [\sqrt{x}]dx$.

29. Prove that $\int_a^{b} \{[x] + [-x]\}dx = a - b$,

where $[x]$ is the greatest integer $\le x$.

30. Prove that $\int_0^{n^2} [\sqrt{x}]dx = \frac{n(n - 1)(4n + 1)}{6}$,

where $n \in N$ and $[x]$ is the greatest integer $\le x$.

31. If $f(x) = \int_0^{x} [x]dx$ then define $f(x)$ as polynomials in the interval $[0, 3)$.

32. If $x > 0$, prove that

$$\int_0^{x} [x]dx = \frac{1}{2}[x][[x] - 1] + [x]\{x - [x]\}.$$

Evaluate the following:

33. $\int_0^{100\pi} \sqrt{1 - \cos 2x}\, dx$

34. $\int_{-10\pi}^{10\pi} \{|\sin x| + |\cos x|\}dx$

35. $\int_0^{50\pi} \sin^3 x dx$

36. $\int_{\pi}^{10\pi} |\cos x|\, dx$

37. Show that $\int_0^{a + b\pi} |\cos x|\, dx = 2b + \sin a$,

where $b \in N$ and $a \in \left(-\frac{\pi}{2}, \frac{\pi}{2}\right)$.

38. Prove that $\int_a^{b} f(x)dx = \int_{a + c}^{b + c} f(x - c)dx$.

Hence, show that $\int_0^{\pi} \sin^{100}x \cdot \cos^{99}x dx = 0$.

39. Show that $\int_0^{2a} f(x)dx = \int_0^{a} f(x)dx + \int_0^{a} f(2a - x)dx$.

40. If $f(t)$ is an even function, prove that $\int_0^{x} f(t)dt$ is odd.

Is $\int_a^{x} f(t)dt, a \ne 0$ odd? Justify your answer.

41. If $f(-x) + f(x) = 0$ in the interval $\left[-\frac{\lambda}{2}, \frac{\lambda}{2}\right]$ and $f(x)$ is periodic with period λ, prove that $\int_a^{x} f(t)dt$ is a periodic function of the period λ.

42. If $f(a + x) = f(x)$, prove that

$$\int_a^{na} f(x)dx = (n - 1)\int_0^{a} f(x)dx.$$

43. Prove that $f(x) = \frac{\sin 2x}{\sin^4 x + \cos^4 x}$ is a periodic function

and evaluate $\int_{\pi}^{\frac{5\pi}{4}} \frac{\sin 2x}{\sin^4 x + \cos^4 x} dx$.

44. If $f(x)$ is a periodic function with the period t then show that $\int_{a+nt}^{b+nt} f(x)dx = \int_a^b f(x)dx$ where $n \in N$.

45. If $I_n = \int_0^{\frac{\pi}{2}} \cos^n x\, dx,\ n \in Z$, show that $I_n = \frac{n-1}{n} I_{n-2}$.

Hence, evaluate $\int_0^{\frac{\pi}{2}} \cos^{11} x\, dx$.

46. If $n \in N$, prove that $2\int_0^{\frac{\pi}{2}} \sin^{2n} x\, dx = \pi \times t_m$, where t_m is the middle term in the expansion of $\left(x + \frac{1}{4x}\right)^{2n}$

47. If $I_n = \int_0^{\frac{\pi}{4}} \tan^n x\, dx,\ n \in N$, show that

$I_n + I_{n-2} = \frac{1}{n-1}$. Hence, evaluate $\int_0^{\pi/4} \tan^8 x\, dx$.

48. If $I_n = \int_0^1 e^x \cdot (x-1)^n dx,\ n \in N$, show that

$I_n + nI_{n-1} = (-1)^{n+1}$.

Hence, find $n \le 5$ such that

$\int_0^1 e^x (x-1)^n dx = 16 - 6e$.

49. If $I_n = \int_0^1 (1 - x^3)^n dx$, prove that $I_n = \frac{3n}{3n+1} I_{n-1}$.

Hence, evaluate $\frac{{}^nC_0}{1} - \frac{{}^nC_1}{4} + \frac{{}^nC_2}{7} - \ldots$.

50. If $y = \int_0^{x^2} \log(1+x)dx$, find $\frac{dy}{dx}$ and $\frac{d^2y}{dx^2}$.

51. If $\int_0^y e^{-x^2} dx + \int_0^{x^2} \sin y\, dy = 0$, find $\frac{dy}{dx}$.

52. Find $\frac{d^2}{dx^2}\left\{\int_4^{4x^2} t^4 e^{4t} dt\right\}$

53. Evaluate $\int_0^{\frac{\pi}{4}} \frac{d}{dx}\left\{\int_1^x \sec^4\theta\, d\theta\right\} dx$.

Evaluate the following limits:

54. $\lim_{\alpha \to 0} \frac{\int_0^\alpha x dx}{\alpha \sin \alpha}$

55. $\lim_{x \to 0} \frac{1}{x^3}\int_0^{x^2} \sin \sqrt{t}\, dt$

56. $\lim_{x \to a} \frac{x\int_a^x f(t)dt}{x - a}$

57. $\lim_{x \to 0} \frac{\int_0^{x^2} \cos t^2\, dt}{x \sin x}$

Establish the following inequalities:

58. $\left|\int_2^5 \frac{\sin x dx}{(1+x)^2}\right| < \frac{1}{6}$

59. $\left|\int_{10}^{19} \frac{\cos x}{1+x^8} dx\right| < \frac{1}{10^7}$

60. $1 < \int_0^{\frac{\pi}{2}} \frac{\sin x}{x} dx < \frac{\pi}{2}$

61. $\int_1^\infty e^{-x^2} dx < \frac{1}{2e}$

62. Show that

$\frac{\pi}{2} < \int_0^{\frac{\pi}{2}} \frac{dx}{\sqrt{1 - k^2 \sin^2 x}} < \frac{\pi}{2\sqrt{1-k^2}}$, $(0 < k < 1)$.

63. If $\int_0^1 e^{x^2}(x - \lambda)dx = 0$, prove that $0 < \lambda < 1$.

64. Find the maximum or minimum value of

$\int_0^x xe^{-x^2} dx$.

65. Investigate for maxima and minima of the function

$f(x) = \int_1^x \{2(t-1)(t-2)^3 + 3(t-1)^2 \cdot (t-2)^2\}dt$.

66. Find the intervals of increase of $f(x)$ defined by

$f(x) = \int_{-1}^x (t^2 + 2t)(t^2 - 1)dt$.

67. If $f(x) = \int_{-1}^x |t - 2|\, dt$, examine the continuity and differentiability of $f(x)$ at $x = 2$.

68. Find a function of the form $f(x) = \lambda e^{2x} + \mu e^x - \nu x$ where λ, μ, ν are independent of x and $f(0) = -1$, $f'(\log_e 2) = 30$ and $\int_0^{\log_e 4} \{f(x) + \nu x\}dx = 28.5$.

69. Find the equations of the tangent lines to the curve $y = \int_0^x 2|t|\, dt$ which are equally inclined to the x and y axes.

70. If $f(x) = \int_a^x \{f(x)\}^{-1} dx$ and $\int_a^1 \{f(x)\}^{-1}dx = \sqrt{2}$ then find $f(x)$.

Objective Questions

Fill in the blanks.

71. The value of $\int_0^{\pi} \sin^n x \cdot \cos^{2n-1} x\, dx, n \in N$ is ______ .

72. The value of $\int_0^{\frac{\pi}{2}} \frac{\sin 8x \cdot \log(\cot x)}{\cos 2x} dx$ is ______ .

73. $\int_0^{\pi} \exp(\cos^2 x) \cdot \cos^5 3x\, dx =$ ______ .

74. The value of $\int_{\frac{\pi}{4}}^{\frac{3\pi}{4}} \frac{x}{1+\sin x} dx =$ ______ .

75. If $f(x) = \int_{-\frac{\pi}{2}}^{x} \frac{2\sin x - \sin 2x}{x^3} dx, x \neq 0$ then $\lim_{x \to 0} f'(x) =$ ______ .

76. For $n \in N$, $\int_0^{2\pi} \frac{x\sin^{2n} x}{\sin^{2n} x + \cos^{2n} x} dx =$ ______ .

77. $\int_{-1}^{1} \{[x] + |x|\}\, dx$ has the value ______ .

78. If $\phi(x) = \frac{1}{2} a_0 + \sum_{k=1}^{n} a_k \cos kx + \sum_{k=1}^{n} b_k \sin kx$ then $\int_0^{2\pi} \phi(x) dx$ is ______ .

79. The value of $\int_1^4 \log[x] dx =$ ______ , where $[x]$ is the greatest integer function.

80. If $f(x) = x^2, 0 \le x < 1$ while $f(x) = \sqrt{x}, 1 \le x \le 2$ then $\int_0^2 f(x) dx =$ ______ .

81. If $\int_0^3 (3ax^2 + 2bx + c) dx = \int_1^3 (3ax^2 + 2bx + c) dx$, where a, b, c are constants then $a + b + c =$ ______ .

82. The value of $\int_{-10}^{10} x\left(|x| - e^{x^2}\right) dx =$ ______ .

83. $\lim_{x \to \infty} e^{-x^2} \cdot \int_0^x e^{t^2} dt =$ ______ .

Choose the correct option(s).

84. The value of $\int_{-2}^{2} |1 - x^2|\, dx$ is

(a) 4 (b) 2 (c) –2 (d) 0

85. The value of $\int_0^{\pi} \sec^2 x\, dx$ is

(a) 0 (b) 1
(c) 2π (d) none of these

86. If $\int_0^{\pi} f(\tan x) dx = \lambda$ then $\int_0^{2\pi} f(\tan x) dx$ is equal to

(a) $\frac{1}{2}\lambda$ (b) -2λ
(c) 2λ (d) none of these

87. $\int_0^{2a} f(x) dx = 2\int_0^a f(x) dx$ can hold if

(a) $f(x)$ is even (b) $f(x)$ is odd
(c) $f(x)$ is periodic (d) $f(2a - x) = -f(x)$
(e) $f(2a - x) = f(x)$

88. The value of $\int_0^{\frac{\pi}{2}} \frac{dx}{1 + \tan^3 x}$ is

(a) 0 (b) 1 (c) $\frac{\pi}{2}$ (d) $\frac{\pi}{4}$

89. Let f be a positive function. If

$$I_1 = \int_{1-k}^{k} xf\{x(1-x)\}dx, \; I_2 = \int_{1-k}^{k} f\{x(1-x)\}dx$$

where $2k - 1 > 0$, then $I_1 : I_2$ is equal to

(a) 2 (b) k (c) $\frac{1}{2}$ (d) 1

90. If $f(x)$ is an even function then the value of $\int_{-2}^{2} \{x^3 f(x) + xf''(x) + \lambda\} dx$, where λ is a constant, is

(a) 4λ (b) $f(2) + 4\lambda$
(c) $2f'(2) + 4\lambda$ (d) none of these

91. The value of $\int_0^1 (1 + e^{-x^2}) dx$ is

(a) –1 (b) 2
(c) $1 + e^{-1}$ (d) none of these

92. Let $f : R \to R$ be a differentiable function and $f(1) = 4$. Then the value of $\lim_{x \to 1} \int_4^{f(x)} \frac{2t\, dt}{x - 1}$ is

(a) $8f'(1)$ (b) $4f'(1)$ (c) $2f'(1)$ (d) $f'(1)$

93. Let $f : R \to R$ and $g : R \to R$ be continuous functions. Then the value of $\int_{-\frac{\pi}{2}}^{\frac{\pi}{2}} \{f(x) + f(-x)\} \cdot \{g(x) - g(-x)\} dx$ is

(a) π (b) 1 (c) –1 (d) 0

94. If $f(x) = \int_{-1}^{x} |t|\, dt$ then

(a) $f(x)$ and $f'(x)$ are continuous in $(-1, +\infty)$
(b) $f(x)$ is continuous but $f'(x)$ is not continuous in $(-1, +\infty)$
(c) $f(x), f'(x)$ are not continuous at $x = 0$
(d) $f(x)$ is continuous at $x = 0$ but $f'(x)$ is not so

95. If a, b, c be nonzero real numbers such that

$$\int_0^1 (1+\cos^8 x)(ax^2+bx+c)dx = \int_0^2 (1+\cos^8 x)(ax^2+bx+c)dx$$

then the quadratic equation $ax^2+bx+c=0$ has
(a) no root in $(0, 2)$
(b) at least one root in $(1, 2)$
(c) at least one root in $(0,1)$
(d) two imaginary roots

96. The value of $\int_{-\frac{\pi}{2}}^{\frac{\pi}{2}} (p\sin^3 x + q\sin^4 x + r\sin^5 x)dx$ depends on
(a) p (b) q (c) r (d) p and r

97. If $f(x)$ is continuous and $n \in N$ then the value of $\int_{-2}^{2} \{f(x)-f(-x)\}x^{2n}dx$ is
(a) $nf(2)$ (b) $2n$
(c) 0 (d) none of these

98. The point of extremum of $f(x) = \int_0^x (t-2)^2(t-1)dt$ is a
(a) maximum at $x=1$ (b) maximum at $x=2$
(c) minimum at $x=1$ (d) minimum at $x=2$
(e) none of these

99. The average value of the electromotive force E_m over the duration $t \in [0, T]$ where the electromotive force E is computed by $E = E_0 \sin\frac{2\pi t}{T}$, E_0 and T being constants, is equal to
(a) E_0 (b) 0
(c) $\frac{E_0}{2\pi}$ (d) none of these

100. If $g(x) = \int_0^x \cos^4 t\, dt$ then $g(x+\pi)$ equals
(a) $g(x)+g(\pi)$ (b) $g(x)-g(\pi)$
(c) $g(x)\,g(\pi)$ (d) $\frac{g(x)}{g(\pi)}$

101. The value of $\int_{-\pi}^{\pi} \frac{2x(1+\sin x)}{1+\cos^2 x}dx$ is
(a) $2\pi^2$ (b) $\frac{\pi^2}{2}$ (c) π^2 (d) 0

State whether the statements are true or false.

102. $f(x), g(x)$ are defined in $[a, b]$.
If $\int_a^b f(x)dx > \int_a^b g(x)dx$ then $f(x) > g(x)$ in the interval $[a, b]$.

103. If $y = \int_0^x f(t)dt$ then $\frac{dy}{dt} = f(t)$.

104. $\int_{-a}^{a} f(x)\cdot\cos x\, dx = \int_0^a \{f(x)+f(-x)\}\cos x\, dx$.

105. $\int_0^{\frac{\pi}{2}} \frac{dx}{\sqrt{\sin x}-\sqrt{\cos x}} = 0$.

Answers

3. $\frac{1}{\sqrt{2}}\log(\sqrt{2}+1)$ **6.** $\frac{\pi}{8}\log 2$ **8.** $\frac{\pi^2}{4}$

9. $\pi\alpha\operatorname{cosec}\alpha$ **10.** $\frac{\pi^2}{2ab}$ **11.** $\frac{\pi^2}{2\sqrt{2}}$ **20.** 0

24. $y = \frac{3\pi}{16}$ **25.** $2\sqrt{2}$ **26.** 9 **27.** $7\frac{1}{3}$

28. 5

31. $f(x) = 0, 0 \le x < 1;\ f(x) = x-1, 1 \le x < 2;$
$f(x) = 2x-3, 2 \le x < 3$

33. $200\sqrt{2}$ **34.** 80 **35.** 0 **36.** 18

40. no, because

$$\phi(x)+\phi(-x) = \int_a^x f(t)dt + \int_a^{-x} f(t)dt$$

and putting $t=-z$ in the second we get,

$$\phi(x)+\phi(-x) = -\int_{-a}^{a} f(t)dt$$

which is zero when $a=0$

43. $\frac{\pi}{4}$ **45.** $\frac{256}{693}$ **47.** $\frac{\pi}{4}-\frac{76}{105}$ **48.** 3

49. $\frac{3^n \cdot n!}{1\cdot 4\cdot 7\cdot \ldots \cdot(3n+1)}$

50. $2x\log(1+x^2)$, $\frac{2}{1+x^2}\{2x^2+(1+x^2)\log(1+x^2)\}$

51. $-2x\sin x^2 \cdot e^{y^2}$ **52.** $2048e^{16x^2}\cdot x^8(9+32x^2)$

53. $\frac{4}{3}$ **54.** $\frac{1}{2}$ **55.** $\frac{2}{3}$ **56.** $af(a)$

57. 1 **64.** minimum value = 0

65. maximum at $x = 1$, minimum at $x = \frac{7}{5}$, neither maximum nor minimum at $x = 2$

66. $(-\infty, -2) \cup (-1, 0) \cup (1, +\infty)$

67. continuous and differentiable

68. $7e^{2x} - 8e^x - 10x$ **69.** $y = x \pm \frac{1}{4}$

70. $\sqrt{2x}$ **71.** 0 **72.** 0 **73.** 0

74. $(\sqrt{2} - 1)\pi$ **75.** 1 **76.** π^2 **77.** 0

78. πa_0 **79.** log 6 **80.** $\frac{4\sqrt{2} - 1}{3}$ **81.** 0

82. 0 **83.** 0 **84.** (a) **85.** (d)

86. (c) **87.** (c), (e) **88.** (d) **89.** (c)

90. (a) **91.** (d) **92.** (a) **93.** (d)

94. (a) **95.** (b) **96.** (b) **97.** (c)

98. (c) **99.** (b) **100.** (a) **101.** (c)

102. false **103.** false **104.** true **105.** true

□

Chapter Test

Time: 120 minutues

1. In each of the following, fill in the blank so that the resulting sentence becomes true.

(a) If $\int_0^a \{f(x)+f(-x)\}dx = \int_{-a}^{a} \phi(x)dx$ then $\phi(x) =$ ______ .

(b) The value of $\int_0^{\pi} \frac{\cos(2n+1)x}{\sin x} dx =$ ______ , where $n \in Z$.

(c) The value $\int_{-\frac{\pi}{4}}^{\frac{\pi}{4}} x^2 \sin^{-1} x\, dx =$ ______ .

(d) The value of $\int_0^{\frac{\pi}{2}} \frac{dx}{1+e^{\sqrt{2}\sin(x-\pi/4)}} =$ ______ .

2. In each of the following, a number of alternative answers are given of which one or more are correct. Choose the correct answer(s).

(a) $\int_0^{\pi} f(\sin x)dx : \int_0^{\frac{\pi}{2}} f(\sin x)dx$ is equal to

A. $\frac{1}{2}$ B. 2 C. $\frac{\pi}{2}$ D. none of these

(b) If $\int_0^{n\pi} \frac{\cos^2 x}{1+\cos^4 x} dx = k \int_0^{\pi} \frac{\cos^2 x}{1+\cos^4 x} dx$ then the value of k is

A. $n-1$ B. $\frac{n}{2}$ C. n D. none of these

(c) The value of $\int_0^{\frac{\pi}{2}} \sin 2x \cdot \log \tan x dx$ is

A. 0 B. π C. $\frac{\pi}{2}$ D. none of these

(d) If $\int_0^1 \frac{e^t dt}{t+1} = a$ then $\int_{b-1}^{b} \frac{e^{-t} dt}{t-b-1}$ is equal to

A. ae^{-b} B. $-ae^{-b}$ C. $-be^{-a}$ D. ae^{b}

3. Prove that $\int_{-x}^{x} \log\left(\cos\frac{\theta}{2} + \sin\frac{\theta}{2}\right) d\theta = \int_0^x \log\cos\theta\, d\theta$.

4. Evaluate $\int_{-\frac{\pi}{2}}^{\frac{\pi}{2}} \frac{\sin 2x}{1+\cos^3 x} dx$.

5. Let $a < c < b$ and $f(x) \ge 0$ for $x \in [a, c]$ while $f(x) \le 0$ for $x \in [c, b]$ such that $\int_a^b \left|f(x)\right| dx = 0$.

Prove that there exists a function $F(x)$ such that the value of $F(x)$ at $x = c$ is the AM of its values at $x = a$ and $x = b$.

6. Let $I_n = \int_1^e (\log x)^n dx$ then prove that $I_n + nI_{n-1} = e$. Also find the value of $\int_1^e (\log x)^5 dx$.

7. If $f(x) = |2^x - 1| + |x - 1|$ then evaluate $\int_{-2}^{2} f(x)dx$.

8. Let $f(x) = \int_0^x |\cos t|\, dt$. Examine the continuity of $f(x)$ at $x = \frac{\pi}{2}$. Also find $f'\left(\frac{\pi}{2}\right)$ if it exists.

9. If $f(x)$ and $\phi(x)$ are continuous functions on the interval [0, 4] satisfying $f(x) = f(4-x)$, $\phi(x) + \phi(4-x) = 3$ and $\int_0^4 f(x)dx = 2$ then find the value of $\int_0^4 f(x)\,\phi(x)dx$.

10. $f(x)$ is an odd function in the interval $[-a, a]$. If $f(x)$ is also a periodic function of the period $2a$ then prove that $\int_a^x f(t)dt$ is also a periodic function of the period $2a$.

11. Find $x \in R$ such that $\int_0^x [x]dx = \int_0^{[x]} x\, dx$ where $[x]$ is the greatest integer less than or equal to x.

12. Find the set of possible real values of a such that the function

$$f(x) = \{b^2 + (a-1)b + 2\}x + \int_0^x (\sin^4 x + \cos^4 x)dx$$

is an increasing function of x for all $x \in R$ and any $b \in R$.

13. Evaluate: $\int_0^x [\sin x]dx$ where $2n\pi < x < (2n+1)\pi$.

14. Let $f(x) = \int_1^x \frac{\log t}{1+t} dt, x > 0$. Find the function $\phi(x) = f(x) + f\left(\frac{1}{x}\right)$. Show that $\phi(e) = \frac{1}{2}$.

15. Given $\int_0^1 \frac{\sin t}{1+t} dt = \alpha$, find $\int_{4\pi-2}^{4\pi} \frac{\sin \frac{t}{2}}{4\pi + 2 - t} dt$ in terms of α.

□

Answers

1. (a) $f(x)$ or $f(-x)$ (b) 0 (c) 0 (d) $\frac{\pi}{4}$

2. (a) B (b) C (c) A (d) B

4. 0

6. $4(30 - 11e)$

7. $5 + \frac{9}{4\log_e 2}$

8. continuous; 0

9. 3

11. $0 \le x < 1$, $x = n + \frac{1}{2}$ where $n \in Z - \{0\}$

12. $[1 - \sqrt{10}, 1 + \sqrt{10}]$

13. $-n\pi$

14. $\frac{1}{2}(\log x)^2$

15. $-\alpha$

□

12. Area

Recap of Facts and Formulae

1. Standard areas

- The area bounded by the curve $y = f(x)$, the x-axis (i.e., $y = 0$), the ordinates $x = a$, $x = b (b > a)$ is given by

  area $= \int_a^b y dx$

  where y is to be expressed in terms of x from the equation of the curve $y = f(x)$.

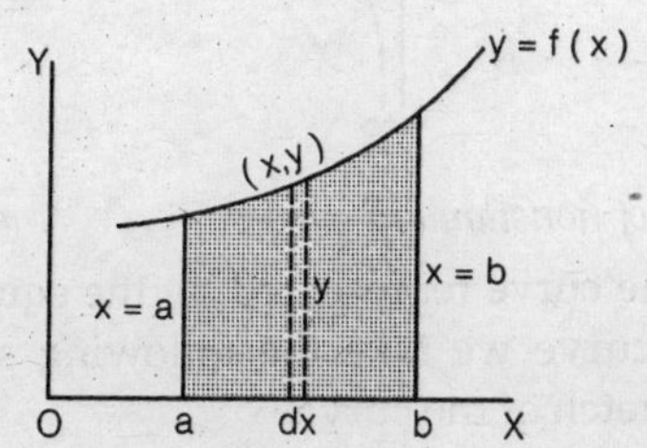

- The area bounded by the curve $x = \phi(y)$, the y-axis (i.e., $x = 0$), the lines (abscissae) $y = p$, $y = q$, $(q > p)$ is given by

  area $= \int_p^q x dy$

  where x is to be expressed in terms of y from the equation of the curve $x = \phi(y)$.

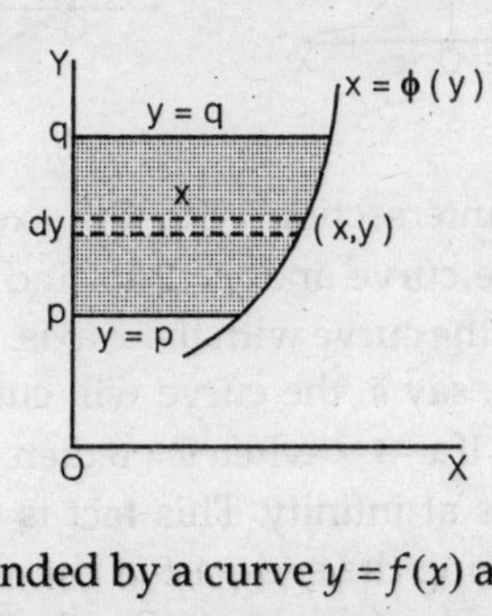

- The area bounded by a curve $y = f(x)$ and the x-axis is given by

  area $= \int_a^b y dx = \int_a^b (y)_{y = f(x)} dx$

  $= \int_a^b f(x) dx$

  where $y = f(x)$ cuts the x-axis at $(a, 0)$ and $(b, 0)$.

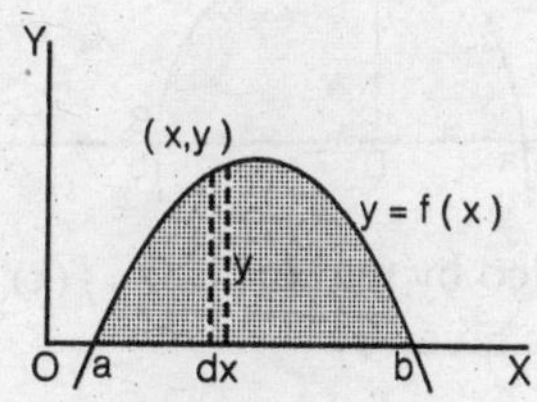

- The area bounded by a curve $x = \phi(y)$ and the y-axis is given by

  area $= \int_p^q x dy = \int_p^q (x)_{x = \phi(y)} dy = \int_p^q \phi(y) dy$

  where $x = \phi(y)$ cuts the y-axis at $(0, p)$ and $(0, q)$.

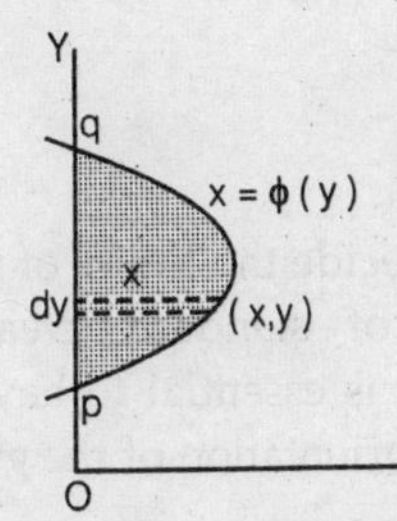

2. Nonstandard areas

An area may be bounded by two curves, or a curve and a line, or two curves and a line, etc. Such areas are not standard areas. In order to compute such areas, do the following:

- Divide the whole area into a number of standard areas whose algebraic sum (i.e., addition/subtraction/or both) gives the required area. Some nonstandard areas are given below. Observe their computation.

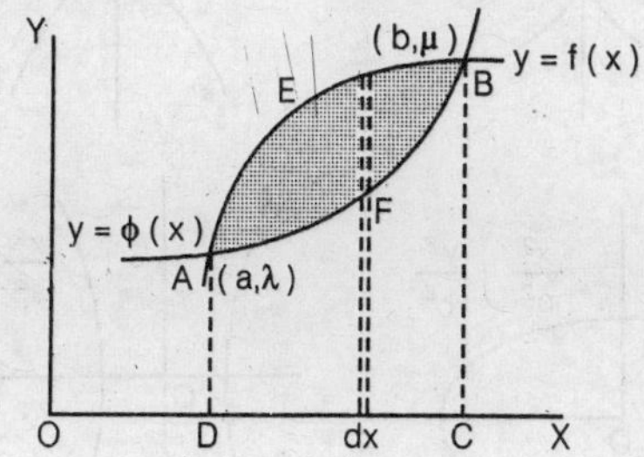

The area bounded by the curves $y = f(x)$ and $y = \phi(x)$ is given by

$$\text{area} = \int_a^b \{f(x) - \phi(x)\}\, dx$$

$$= \text{ar(AEBCDA)} - \text{ar(AFBCDA)},$$

where the x-coordinates of the points of intersection of the curves $y = f(x)$ and $y = \phi(x)$ are a and b.

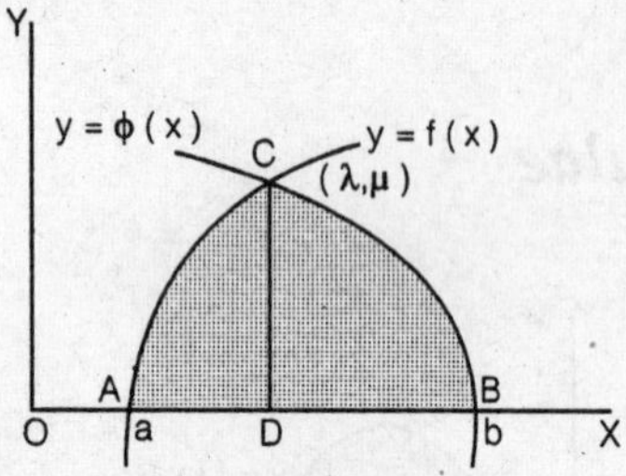

The area bounded by the curves $y = f(x)$ and $y = \phi(x)$, and the x-axis is given by

$$\text{area} = \int_a^{\lambda} f(x)dx + \int_{\lambda}^{b} \phi(x)dx$$

$$= \text{ar(ACDA)} + \text{ar(CDBC)},$$

where the points of intersection of the curves with the x-axis are $(a, 0)$ and $(b, 0)$ respectively and the two curves intersect at (λ, μ).

3. Sketches of curves

Clearly, in order to decide the limits of integration and sum or difference of standard areas to compute nonstandard areas, it is essential to have a rough idea about the shape and orientation of the given curves.

- *Sketches of standard curves*

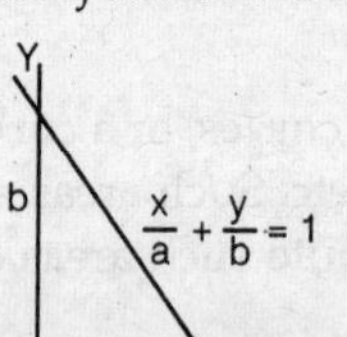

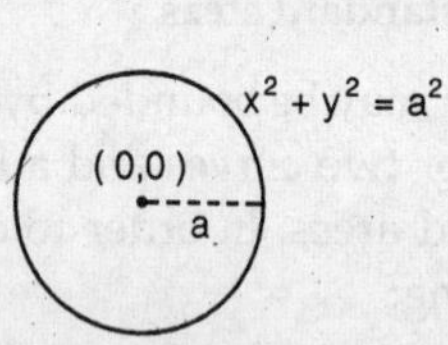

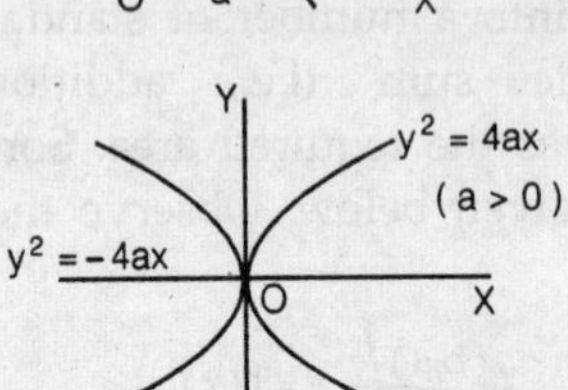

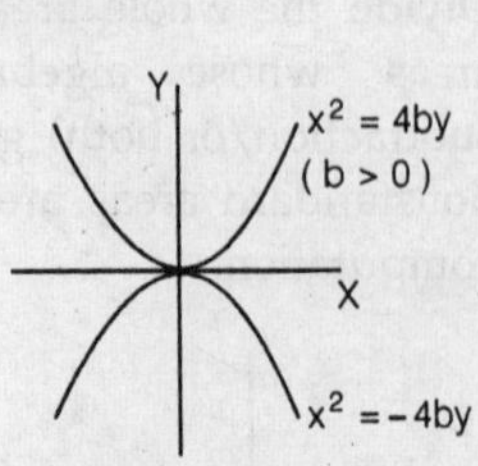

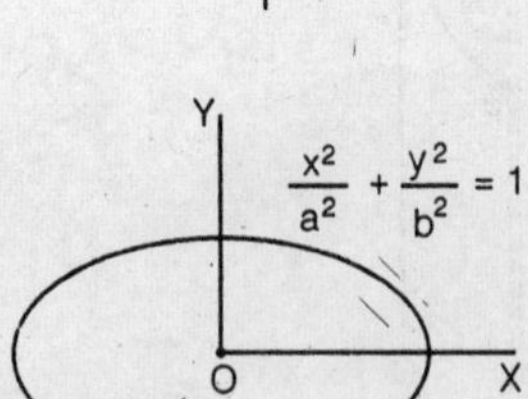

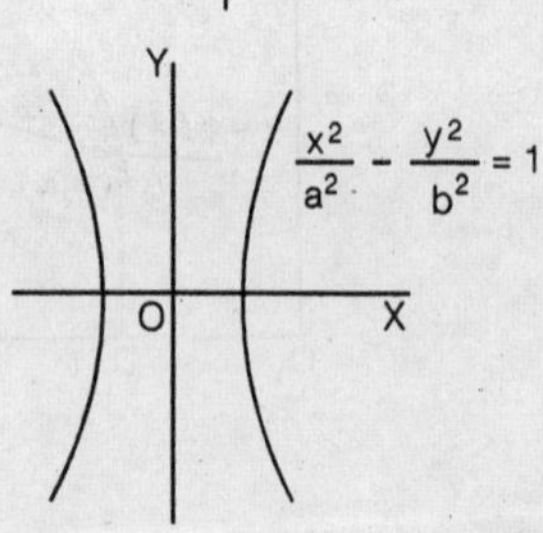

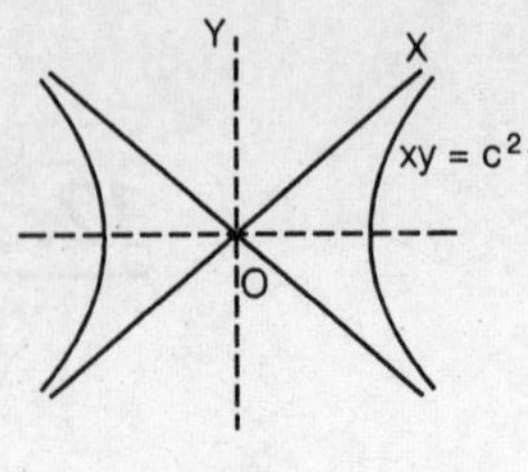

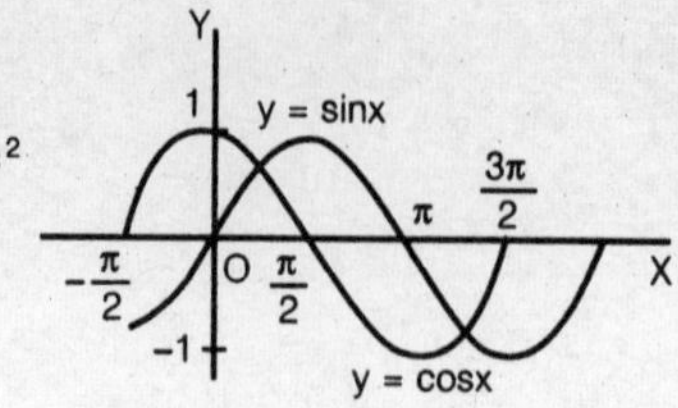

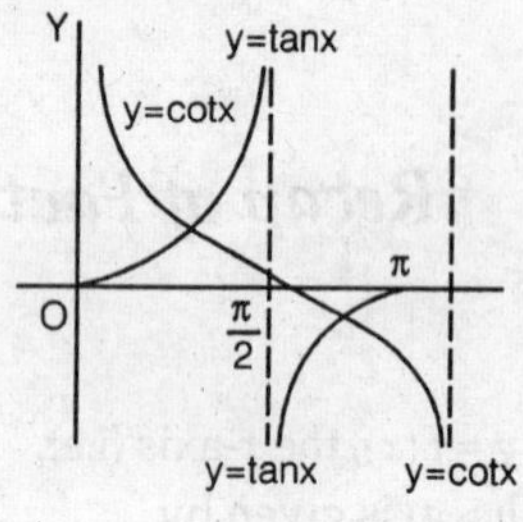

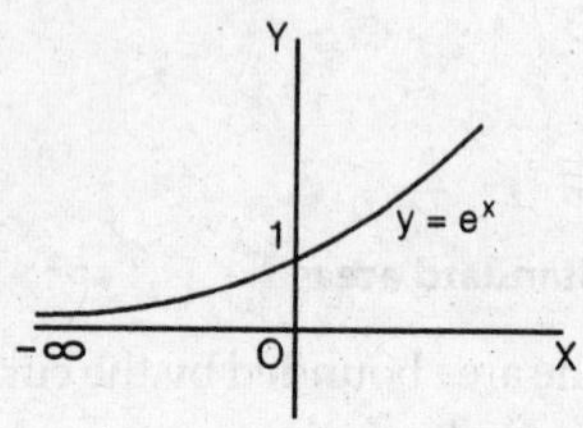

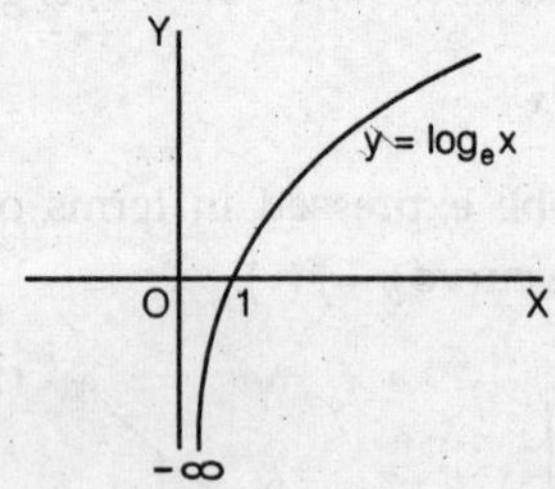

- *Sketches of nonstandard curves*

When the curve represented by the equation is not a known curve we take the following steps to get a rough sketch of the curve.

(a) Symmetry about the axes—If the equation of the curve contains only even powers of x then the curve is symmetrical about the y-axis. Again, if only even powers of y appear in the equation then the curve is symmetrical about the x-axis.

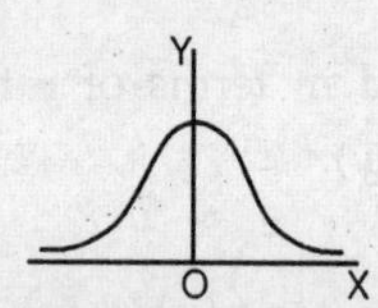

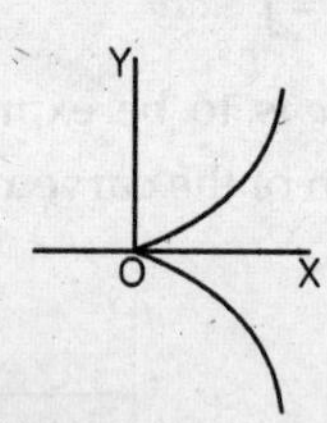

(b) Points of intersection with the axes—Solve the equation of the curve and $y = 0$ to find the points of intersection of the curve with the x-axis. If we get only one value of x, say a, the curve will cut the x-axis at one point only. If $x \to \infty$ when $y = 0$ then the curve will meet the x-axis at infinity. This fact is shown in the sketch by drawing the end part of the sketch parallel to the positive side of the x-axis. Similarly on the other side if $x \to -\infty$ when $y = 0$.

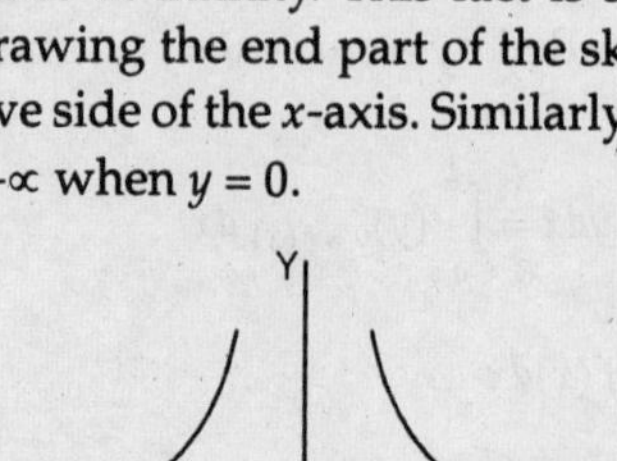

Solve the equation of the curve and $x = 0$ to find the points of intersection of the curve with the y-axis. If $y \to \infty$ (or $y \to -\infty$) when $x = 0$ then the curve will meet the y-axis at infinity.

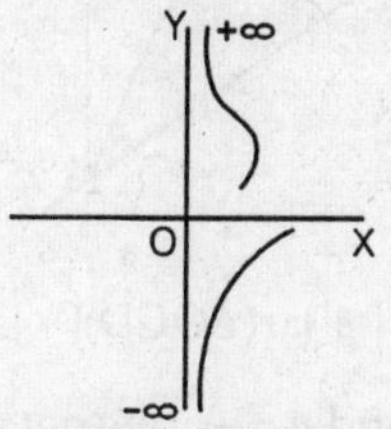

(c) Trend of values—Observe the effect on values of y when x changes. See the sign of y for values of x in different intervals. If $y > 0$ for $a < x < b$, the curve will be above the x-axis between the points $(a, f(a))$ and $(b, f(b))$. Instead, if $y < 0$ then the curve will be below the x-axis.

If y goes on increasing when $a < x < b$, then portion of the graph in $[a, b]$ will be rising lower to higher. But if y goes on decreasing in $a < x < b$, the graph will be falling higher to lower.

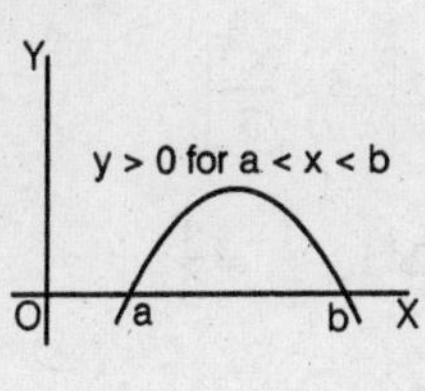

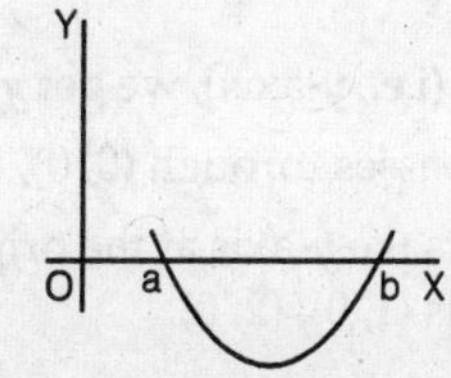

Observe the effect on values of x when y changes and make similar conclusions.

For example Rough sketch of the curve

$$y^2(a^2 + x^2) = x^2(a^2 - x^2).$$

Here only even powers of x as well as y appear. So the curve is *symmetrical about both the axes.*

Put $y = 0$ in the equation of the curve.

We get $0 = x^2(a^2 - x^2)$ i.e. $x = 0, a, -a$.

So, the curve *cuts the x-axis at* (0, 0), *(a, 0) and (–a, 0).*

Similarly, putting $x = 0$ we get $y = 0$; so, the curve *cuts the y-axis only at* (0, 0).

As $y^2 = \dfrac{x^2(a^2 - x^2)}{a^2 + x^2}$, we get, for $x > a$ or $x < -a$ the value of y^2 is negative and so y will not be real. Thus, we find the curve cannot go to the right of $x = a$ or to the left of $x = -a$.

Hence, the rough sketch of the curve is as given below.

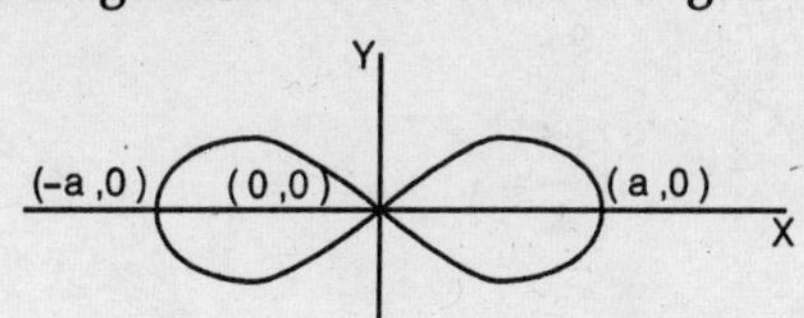

Selected Solved Examples

1. Find the area bounded by the curve $y = \sin^{-1}x$ and the lines $x = 0$, $|y| = \frac{\pi}{2}$.

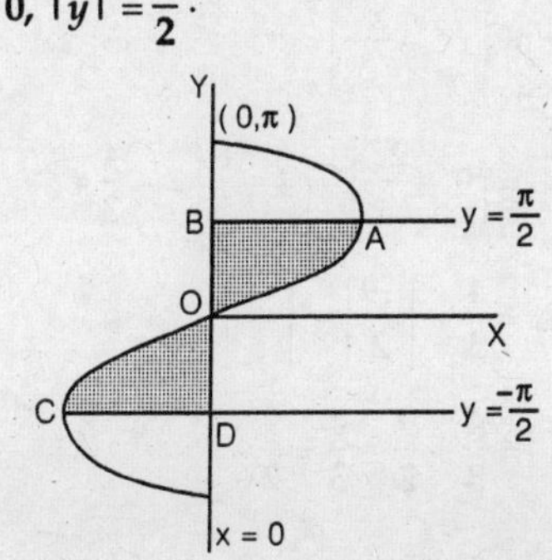

The curve is $y = \sin^{-1}x$, i.e., $x = \sin y$. This is a standard curve. Lines $x = 0$, $y = \frac{\pi}{2}$ and $y = -\frac{\pi}{2}$ are the y-axis and two lines parallel to the x-axis at a distance $\frac{\pi}{2}$, one above and the other below the x-axis respectively.

Hence, the shaded part is the required area Δ. By symmetry of the curve and the lines,

ar(OABO) = ar(OCDO)

$\therefore \quad \Delta = 2 \times$ ar(OABO)

$$= 2\int_0^{\frac{\pi}{2}} (x)_{\text{curve}}\, dy = 2\int_0^{\frac{\pi}{2}} \sin y\, dy,$$

($\because$ the equation of the curve is $x = \sin y$)

$$\therefore \quad \Delta = 2\,[-\cos y]_0^{\pi/2} = 2[0 + 1] = 2.$$

2. Find the area bounded by the curve $y = x(x-1)(x-2)$ and the x-axis.

Here the curve is not a standard curve. We make a rough sketch of the curve.

Solving with $y = 0$ (i.e., x-axis), we get $x = 0, 1, 2$.

Solving with $x = 0$ (i.e., y-axis), we get $y = 0$.

Hence, the curve passes through (0, 0), (1, 0), (2, 0).

Thus the curve cuts the y-axis at the origin only and the x-axis at two points (1, 0), (2, 0).

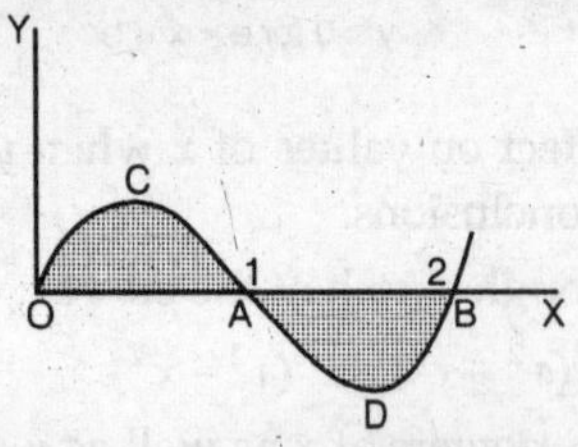

When $0 < x < 1$, $y = x(x-1)(x-2) = (+)(-)(-) = (+)$.

When $1 < x < 2$, $y = x(x-1)(x-2) = (+)(+)(-) = (-)$.

$\therefore$ in the interval (0, 1), the curve is above the x-axis and in (1, 2) the curve is below the x-axis. When $x > 2$, $y > 0$.

Thus, the shaded portion is the required area.

$\therefore$ the required area = | ar(OCAO) | + | ar(ADBA) | ... (1)

$$\text{ar(OCAO)} = \int_0^1 (y)_{\text{curve}}\, dx = \int_0^1 x(x-1)(x-2)dx$$

$$= \int_0^1 (x^3 - 3x^2 + 2x)dx$$

$$= \left[\frac{x^4}{4} - x^3 + x^2\right]_0^1 = \frac{1}{4}$$

$$\text{ar(ADBA)} = \int_1^2 (y)_{\text{curve}}\, dx = \left[\frac{x^4}{4} - x^3 + x^2\right]_1^2$$

$$= (4 - 8 + 4) - \frac{1}{4} = -\frac{1}{4}$$

$\therefore$ by (1), the required area

$$= \left|\frac{1}{4}\right| + \left|-\frac{1}{4}\right| = \frac{1}{4} + \frac{1}{4} = \frac{1}{2}$$

3. Find the area bounded by the curve $y = 2x - x^2$ and the straight line $y = -x$.

Here the curve is $y = 2x - x^2$.

$\therefore \quad y = 1 - (x^2 - 2x + 1) = 1 - (x-1)^2;$

so, $\quad (x-1)^2 = -(y-1)$.

This is the equation of a parabola whose vertex = (1, 1) and the axis is $x - 1 = 0$, i.e., $x = 1$. Also (0, 0) is a point on it. When $y = 0$, $2x - x^2 = 0$. So, $x = 0, 2$.

$\therefore \quad$ A = (0, 1) and B = (2, 0).

Thus, the required area Δ is the area of the portion shaded in the figure.

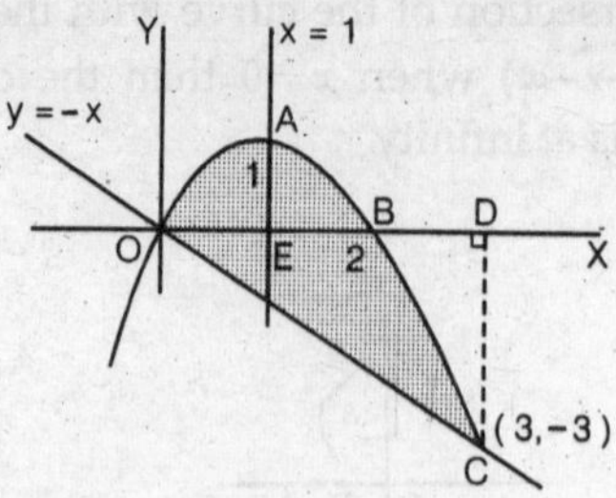

$\therefore \quad \Delta = |\text{ar(OABO)}| + |\text{ar}(\Delta\text{OCD})| - |\text{ar(BCDB)}| \quad ... (1)$

Solving $y = 2x - x^2$ and $y = -x$, we get

$$-x = 2x - x^2, \text{ i.e., } x^2 - 3x = 0$$

$\therefore \quad x = 0, 3$ and $y = 0, -3$

$\therefore \quad$ C = (3, −3).

Now, ar(OABO) = 2(ar OAEO), by symmetry

$$= 2\int_0^1 (y)_{\text{curve}}\, dx$$

$$= 2\int_0^1 (2x - x^2)dx$$

$$= 2\left[x^2 - \frac{x^3}{3}\right]_0^1$$

$$= 2\left(1 - \frac{1}{3}\right) = \frac{4}{3}.$$

$$\text{ar}(\Delta\text{OCD}) = \int_0^3 (y)_{\text{line}}\, dx = \int_0^3 -x dx$$

$$= \left[-\frac{x^2}{2}\right]_0^3 = -\frac{9}{2}.$$

$$\text{ar(BCDB)} = \int_2^3 (y)_{\text{curve}}\, dx = \int_2^3 (2x - x^2)dx$$

$$= \left[x^2 - \frac{x^3}{3}\right]_2^3$$

$$= \left(9 - \frac{27}{3}\right) - \left(4 - \frac{8}{3}\right) = -\frac{4}{3}.$$

$\therefore \quad$ from (1), $\Delta = \frac{4}{3} + \left|\frac{-9}{2}\right| - \left|\frac{-4}{3}\right|$

$$= \frac{4}{3} + \frac{9}{2} - \frac{4}{3} = \frac{9}{2}.$$

4. Prove that for any real t, $x = \frac{e^t + e^{-t}}{2}$ and $y = \frac{e^t - e^{-t}}{2}$ is a point on the hyperbola $x^2 - y^2 = 1$. Show that the area bounded by the hyperbola and the lines joining its centre to the points corresponding to t_1 and $-t_1$ is t_1.

Clearly, $x^2 - y^2 = \left(\frac{e^t + e^{-t}}{2}\right)^2 - \left(\frac{e^t - e^{-t}}{2}\right)^2$

$$= \frac{e^{2t} + e^{-2t} + 2}{4} - \frac{e^{2t} + e^{-2t} - 2}{4} = 1.$$

The hyperbola $x^2 - y^2 = 1$ is a standard curve whose centre is O(0, 0), x-axis (i.e., transverse axis) cuts the hyperbola at A(1, 0) and A′(–1, 0). Also, the curve is symmetrical about both x and y axes.

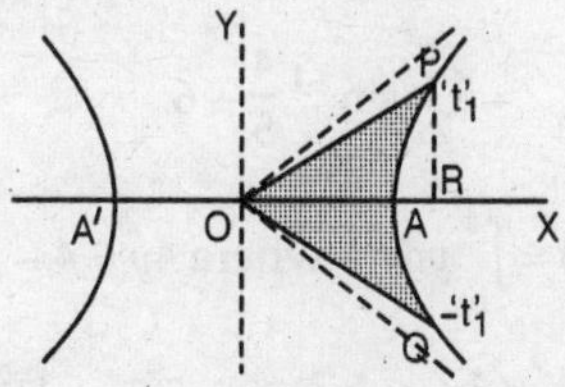

Let P ′t_1′ and Q ′$-t_1$′, i.e., $P = \left(\frac{e^{t_1} + e^{-t_1}}{2}, \frac{e^{t_1} - e^{-t_1}}{2}\right)$,

$$Q = \left(\frac{e^{-t_1} + e^{t_1}}{2}, \frac{e^{-t_1} - e^{t_1}}{2}\right)$$

$\therefore$ the area in question

= ar(POQAP) = 2ar(POAP)

= 2{ar(△POR) – ar(PARP)}

$$= 2 \cdot \frac{1}{2} \cdot OR \cdot PR - 2\int_1^{\frac{1}{2}(e^{t_1} + e^{-t_1})} y dx$$

$$= \frac{e^{t_1} + e^{-t_1}}{2} \cdot \frac{e^{t_1} - e^{-t_1}}{2} - 2\int_1^{\frac{1}{2}(e^{t_1} + e^{-t_1})} \sqrt{x^2 - 1}\, dx$$

$$= \frac{1}{4}(e^{2t_1} - e^{-2t_1})$$

$$- 2\left[\frac{x}{2}\sqrt{x^2 - 1} - \frac{1}{2}\log(x + \sqrt{x^2 - 1})\right]_1^{\frac{1}{2}(e^{t_1} + e^{-t_1})}$$

$$= \frac{1}{4}(e^{2t_1} - e^{-2t_1}) - \left[\frac{1}{2}(e^{t_1} + e^{-t_1})\sqrt{\left(\frac{e^{t_1} + e^{-t_1}}{2}\right)^2 - 1}\right.$$

$$\left. - \log\left\{\frac{e^{t_1} + e^{-t_1}}{2} + \sqrt{\left(\frac{e^{t_1} + e^{-t_1}}{2}\right)^2 - 1}\right\}\right]$$

$$= \frac{e^{2t_1} - e^{-2t_1}}{4} - \frac{(e^{t_1} + e^{-t_1})(e^{t_1} - e^{-t_1})}{4}$$

$$+ \log\left\{\frac{e^{t_1} + e^{-t_1}}{2} + \frac{e^{t_1} - e^{-t_1}}{2}\right\}$$

$= \log(e^{t_1}) = t_1.$

Note We could have found ar(PARP) as follows:

Any point on the hyperbola is (x, y) where

$$x = \frac{e^t + e^{-t}}{2},\ y = \frac{e^t - e^{-t}}{2}$$

$\therefore$ at A(1, 0), $t = 0$; at P, $t = t_1$.

$$\therefore \quad \text{ar(PARP)} = \int_0^{t_1} y dx = \int_0^{t_1} \frac{e^t - e^{-t}}{2}\, d\left(\frac{e^t + e^{-t}}{2}\right)$$

$$= \int_0^{t_1} \frac{e^t - e^{-t_1}}{2} \cdot \frac{e^t - e^{-t}}{2}\, dt$$

$$= \frac{1}{4}\int_0^{t_1} (e^{2t} + e^{-2t} - 2) dt.$$

5. Sketch the region bounded by the curves $y = \log_e x$ and $y = (\log_e x)^2$. Also find the area of the region.

We know $\log_e x$ is defined for $x > 0$. So both the curves will be on the positive side of the x-axis.

Solving with $y = 0$ (i.e., x-axis),

$y = \log_e x$ gives $\log_e x = 0$. So $x = 1$

$y = (\log_e x)^2$ gives $(\log_e x)^2 = 0$. So $x = 1$.

So, both the curves cut the x-axis at (1, 0).

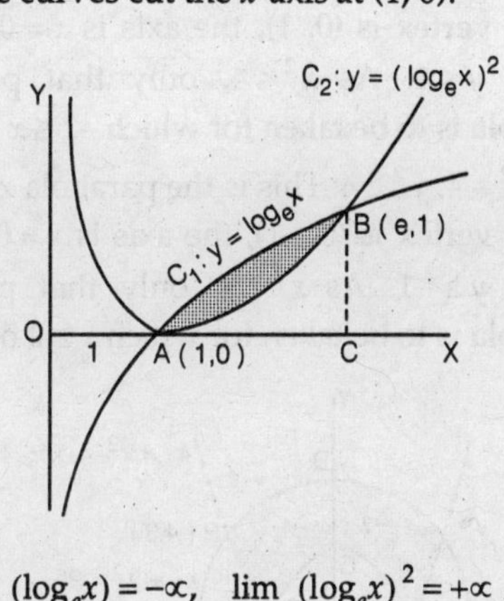

Now, $\lim_{x \to 0+0} (\log_e x) = -\infty$, $\lim_{x \to 0+0} (\log_e x)^2 = +\infty$

and $\log_e x < 0$, $(\log_e x)^2 > 0$ when $0 < x < 1$.

Points of intersection of the curves are given by

$\log_e x = (\log_e x)^2$

$\therefore$ $\log_e x = 0, 1$, i.e., $x = 1, e$.

$\therefore$ the points of intersection are A(1, 0) and B(e, 1).

Also, when $1 < x < e$ we get $(\log_e x)^2 < \log_e x$.

Hence, the required region is as shaded in the figure.

Now, the area of the region bounded by the curves

$$= \int_1^e (y)_{c_1} dx - \int_1^e (y)_{c_2} dx$$

$$= \int_1^e \log_e x dx - \int_1^e (\log_e x)^2 dx$$

$$= \left\{\left[(\log_e x)\cdot x\right]_1^e - \int_1^e x\cdot\frac{1}{x}\,dx\right\}$$

$$- \left\{\left[(\log_e x)^2 x\right]_1^e - \int_1^e x\cdot 2\log_e x\cdot\frac{1}{x}\,dx\right\}$$

$$= e - \int_1^e dx - (e-0) + 2\int_1^e \log_e x\,dx$$

$$= -(e-1) + 2\left\{\left[(\log_e x)x\right]_1^e - \int_1^e x\cdot\frac{1}{x}\,dx\right\}$$

$$= -e+1+2\{e-(e-1)\}$$

$$= 3-e.$$

6. Find the area bounded by the curves $x^2+y^2=25$, $4y=|4-x^2|$ and $x=0$, above the x-axis.

Here the curves are

(i) $x^2+y^2=25$ which is a circle of radius 5 and its centre is (0, 0),

(ii) $4y=4-x^2$, $x^2\le 4$. This is the parabola $x^2=4(1-y)$ whose vertex is (0, 1), the axis is $x=0$ and at all points $y\le 1$. As $x^2\le 4$, only that part of the parabola is to be taken for which $-2\le x\le 2$.

(iii) $4y=x^2-4$, $x^2\ge 4$. This is the parabola $x^2=4(1+y)$ whose vertex is (0, –1), the axis is $x=0$ and at all points $y\ge -1$. As $x^2\ge 4$, only that part of the parabola is to be taken for which $x\ge 2$ or $x\le -2$.

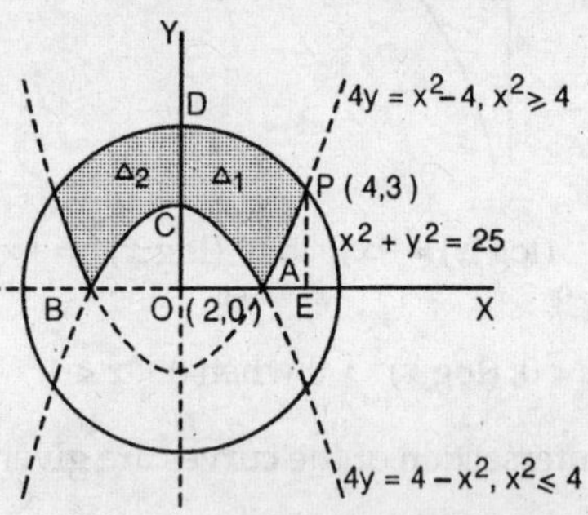

Clearly, there are two areas bounded by the curves and the y-axis, and by symmetry they are equal. So $\Delta_1=\Delta_2$.

Now, $\Delta_1 =$ ar(ODPEO) – ar(OCAO) – ar(APEA) ... (1)

Solving $x^2+y^2=25$ and $4y=x^2-4$, we get

$4+4y+y^2=25$

or $y^2+4y-21=0$.

So $(y+7)(y-3)=0$.

$\therefore$ $y=3,-7$. But y cannot be –7.

So $y=3$ and $x=4$.

$\therefore$ P = (4, 3).

Solving $4y=4-x^2$ and $y=0$, we get

$4-x^2=0$, i.e., $x=\pm 2$. So A = (2, 0) and B = (–2, 0).

Now, $\text{ar(ODPEO)} = \int_0^4 (y)_{\text{circle}}\,dx$

$$= \int_0^4 \sqrt{25-x^2}\,dx = \int_0^4 \sqrt{5^2-x^2}\,dx$$

$$= \left[\frac{25}{2}\sin^{-1}\frac{x}{5} + \frac{x\sqrt{25-x^2}}{2}\right]_0^4$$

$$= \frac{25}{2}\sin^{-1}\frac{4}{5}+6.$$

$\text{ar(OCAO)} = \int_0^2 y\,dx$, where $4y=4-x^2$

$$= \int_0^2\left(1-\frac{x^2}{4}\right)dx = \left[x-\frac{x^3}{12}\right]_0^2$$

$$= 2-\frac{8}{12}=\frac{4}{3}.$$

$\text{ar(APEA)} = \int_2^4 y\,dx$, where $4y=x^2-4$

$$= \int_2^4\left(\frac{x^2}{4}-1\right)dx = \left[\frac{x^3}{12}-x\right]_2^4.$$

$$= \left(\frac{64}{12}-4\right)-\left(\frac{8}{12}-2\right)=\frac{8}{3}.$$

$\therefore$ by (1), $\Delta_1 = \frac{25}{2}\sin^{-1}\frac{4}{5}+6-\frac{4}{3}-\frac{8}{3}$

$$= \frac{25}{2}\sin^{-1}\frac{4}{5}+2$$

Also, $\Delta_2=\Delta_1=\frac{25}{2}\sin^{-1}\frac{4}{5}+2.$

7. Sketch the region bounded by the curves $y=x^2$ and $y=\frac{2}{1+x^2}$. Find its area.

Here $y=x^2$ is a parabola whose vertex is (0, 0), axis is $x=0$ and every point has positive y-coordinate. Also, the parabola is symmetrical about y-axis.

The curve $y=\frac{2}{1+x^2}$ is not a standard curve. We make a rough sketch of the curve as follows.

When $x=0$, $y=\frac{2}{1+0}=2$.

For all real x, y is always positive and ≤ 2.

When $y=0$, $1+x^2=\infty$, i.e., $x\to+\infty$ or $-\infty$ when $y\to 0$.

Also $y=\frac{2}{1+x^2}=$ even function of x. So the curve is symmetrical about the y-axis.

Hence, the required region is as shaded in the figure.

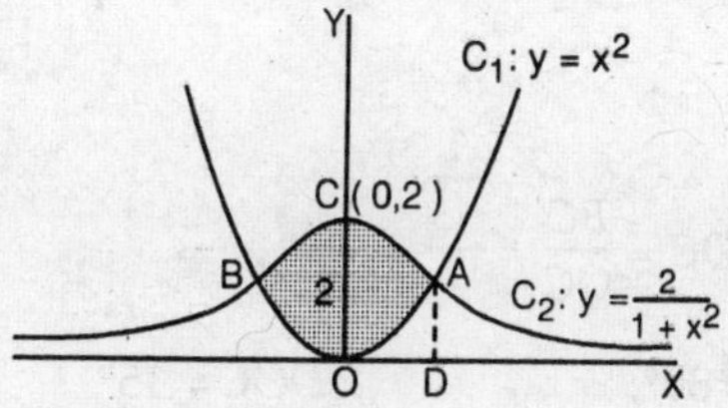

The required area= ar(OACBO)

$$= 2 \times \text{ar(OACO)} \quad \ldots (1)$$

(by symmetry)

Solving $y = x^2$ and $y = \frac{2}{1+x^2}$, we get

$$y = \frac{2}{1+y}; \text{ so, } y^2 + y - 2 = 0$$

or $(y+2)(y-1) = 0 \quad \Rightarrow \quad y = 1, -2.$

But y cannot be negative.

$\therefore$ $y = 1$; so, $x = \pm 1$. Hence, A = (1, 1) and B = (−1, 1).

$\therefore$ from (1), the required area

$= 2\{\text{ar(ODACO)} - \text{ar(ODAO)}\}$

$$= 2\left\{\int_0^1 (y)_{c_2}\, dx - \int_0^1 (y)_{c_1}\, dx\right\}$$

$$= 2\int_0^1 \frac{2}{1+x^2}\, dx - 2\int_0^1 x^2 dx$$

$$= 4\left[\tan^{-1}x\right]_0^1 - 2\cdot\left[\frac{x^3}{3}\right]_0^1$$

$$= 4\left(\frac{\pi}{4} - 0\right) - \frac{2}{3}(1-0) = \pi - \frac{2}{3}.$$

8. Find the area of the region bounded by the parabola $(y-2)^2 = x - 1$, the tangent to it at the point where the ordinate is 3, and the x-axis.

The vertex of the parabola is (1, 2) and the axis is $y - 2 = 0$.

When $y = 0$, $(0-2)^2 = x - 1$, i.e., $x = 5$.

$\therefore$ the parabola cuts the x-axis at (5, 0).

When $y = 3$, $(3-2)^2 = x - 1$, i.e., $x = 2$

$\therefore$ the point of contact of the tangent is (2, 3).

The equation of the tangent at (2, 3) is

$$y - 3 = \left(\frac{dy}{dx}\right)_{2,3} \cdot (x - 2) \quad \ldots (1)$$

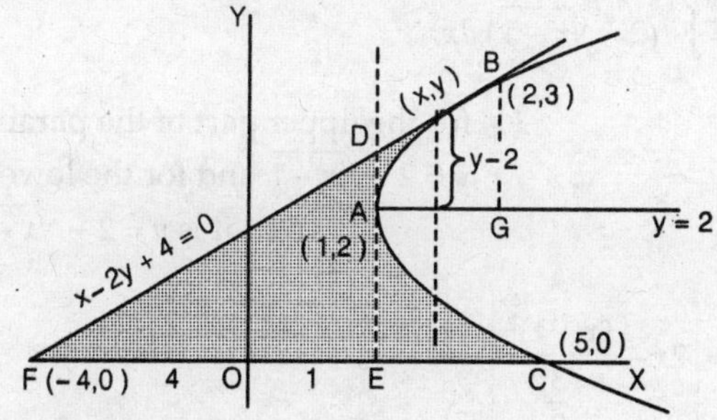

Differentiating $(y-2)^2 = x - 1$ w.r.t. x,

$$2(y-2)\frac{dy}{dx} = 1$$

$$\therefore \quad \frac{dy}{dx} = \frac{1}{2(y-2)}$$

$$\therefore \quad \left(\frac{dy}{dx}\right)_{2,3} = \frac{1}{2(3-2)} = \frac{1}{2}.$$

$\therefore$ from (1) the equation of the tangent at (2, 3) is

$$y - 3 = \frac{1}{2}(x-2)$$

or $x - 2y + 4 = 0 \quad \ldots (2)$

The required area

$$= \text{ar}(\triangle\text{DEF}) + \text{ar(ADBA)} + \text{ar(AECA)} \quad \ldots (3)$$

The tangent cuts the x-axis at F = (−4, 0), obtained by solving (2) with $y = 0$.

Also, the tangent cuts the line $x = 1$ (i.e., DE) at $D = \left(1, \frac{5}{2}\right)$. So, $DE = \frac{5}{2}$.

$$\therefore \quad \text{ar}(\triangle\text{DEF}) = \frac{1}{2}FE \cdot DE = \frac{1}{2}\cdot 5\cdot\frac{5}{2} = \frac{25}{4}.$$

$$\text{ar(ADBA)} = \text{ar(ADBGA)} - \text{ar(ABGA)}$$

$$= \int_1^2 (y-2)_{\text{line}}\, dx - \int_1^2 (y-2)_{\text{parabola}}\, dx$$

$$= \int_1^2 \left(\frac{x+4}{2} - 2\right)dx - \int_1^2 \sqrt{x-1}\, dx,$$

$\{\because$ for the tangent, $y = \frac{x+4}{2}$, by (1)

and for the parabola, $y - 2 = \sqrt{x-1}\}$

$$= \left[\frac{1}{2}\cdot\frac{x^2}{2}\right]_1^2 - \left[\frac{(x-1)^{\frac{3}{2}}}{\frac{3}{2}}\right]_1^2$$

$$= \left(1 - \frac{1}{4}\right) - \frac{2}{3}(1-0) = \frac{1}{12}.$$

$$\text{ar(AECA)} = \int_1^5 (y)_{\text{parabola}}\, dx$$

$$= \int_1^5 (2 - \sqrt{x-1})\, dx,$$

{∵ for the upper part of the parabola $y = 2 + \sqrt{x-1}$ and for the lower part it is $y = 2 - \sqrt{x-1}$}

$$= \left[2x - \frac{(x-1)^{\frac{3}{2}}}{\frac{3}{2}} \right]_1^5$$

$$= \left(10 - \frac{2}{3} \cdot 8\right) - (2 - 0) = \frac{8}{3}.$$

∴ by (3), the required area $= \frac{25}{4} + \frac{1}{12} + \frac{8}{3} = 9.$

9. Let O(0, 0), A(2, 0) and B$\left(1, \frac{1}{\sqrt{3}}\right)$ be the vertices of a triangle. Let R be the region consisting of all those points P inside △OAB which satisfy

$d(\text{P, OA}) \le \min \{d(\text{P, OB}), d(\text{P, AB})\}$

where d denotes the distance from the point to the corresponding line. Sketch the region R and find its area.

Clearly △OAB is isoceles because $AB = OB = \frac{2}{\sqrt{3}}$.

Let BC ⊥ OA.

BC will be the bisector of ∠OBA.

Each point on BC is equidistant from OB and BA. Again each point inside △BCA is nearer to AB than OB while each point inside △BOC is nearer to OB than AB.

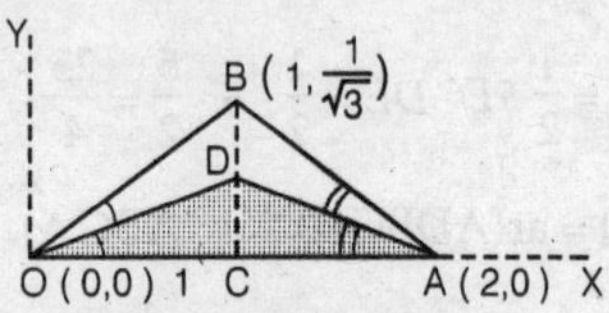

So, for the point P ∈ interior of △BCA,

$$d(\text{P, AB}) < d(\text{P, OB})$$

and for the point P ∈ interior of △BOC,

$$d(\text{P, OB}) < d(\text{P, AB}).$$

∴ from the given inequality,

if P ∈ △BCA, $d(\text{P, OA}) \le d(\text{P, AB})$ ... (1)

if P ∈ △BOC, $d(\text{P, OA}) \le d(\text{P, OB})$ ... (2)

Now, let AD and OD be the internal bisectors of the angles BAC and BOC respectively.

Any point P on AD is equidistant from AB and OA.

So, P satisfies (1) if P is a point of △DAC.

Any point P on OD is equidistant from OB and OA.

So, P satisfies (2) if P is a point of △DOC.

∴ the points of required region R are points of the △DOA.

Now, $\tan \angle BOC = \frac{BC}{OC} = \frac{\frac{1}{\sqrt{3}}}{1} = \frac{1}{\sqrt{3}}$

∴ $\angle BOC = 30°$; ∴ $\angle DOC = 15°$

∴ the equation of OD is $y = x \tan 15°$.

Similarly, the equation of DA is

$$y - 0 = \tan 165° \cdot (x - 2)$$

or $y = (2 - x)\tan 15°$.

Now, ar(R) = ar(△ODC) + ar(△DAC)

$$= \int_0^1 (y)_{OD}\, dx + \int_1^2 (y)_{DA}\, dx$$

$$= \int_0^1 x \tan 15°\, dx + \int_1^2 (2 - x)\tan 15°\, dx$$

$$= \tan 15° \left\{ \left[\frac{x^2}{2}\right]_0^1 + \left[2x - \frac{x^2}{2}\right]_1^2 \right\}$$

$$= \frac{\tan 60° - \tan 45°}{1 + \tan 60° \cdot \tan 45°} \left\{\frac{1}{2} + (4 - 2) - \left(2 - \frac{1}{2}\right)\right\}$$

$$= \frac{\sqrt{3} - 1}{\sqrt{3} + 1} \cdot 1$$

$$= \frac{(\sqrt{3} - 1)^2}{3 - 1}$$

$$= 2 - \sqrt{3}.$$

Note The areas of △DOA can be found by the formula,

$$\text{area} = \frac{1}{2} \cdot OA \cdot DC \text{ where}$$

$$DC = \text{inradius} = \frac{\text{ar}(\triangle\text{OBA})}{s}$$

10. Consider a square with vertices at (1, 1), (–1, 1), (–1, –1) and (1, –1). Let S be the region consisting of all points inside the square which are nearer to the origin than any side. Sketch the region S and find its area.

Let P(x, y) be any point in the region S. Then, according to the question,

$$\sqrt{x^2 + y^2} < |1 - x|, \quad \sqrt{x^2 + y^2} < |1 + x|$$

$$\sqrt{x^2 + y^2} < |1 - y|, \quad \sqrt{x^2 + y^2} < |1 + y|$$

$$\Rightarrow \quad y^2 < 1 - 2x, \quad y^2 < 1 + 2x$$

$$x^2 < 1 - 2y, \quad x^2 < 1 + 2y$$

$\therefore$ the point $P(x, y)$ is in the interior of each of the parabolas $y^2 = 1-2x$, $y^2 = 1+2x$, $x^2 = 1-2y$ and $x^2 = 1+2y$.

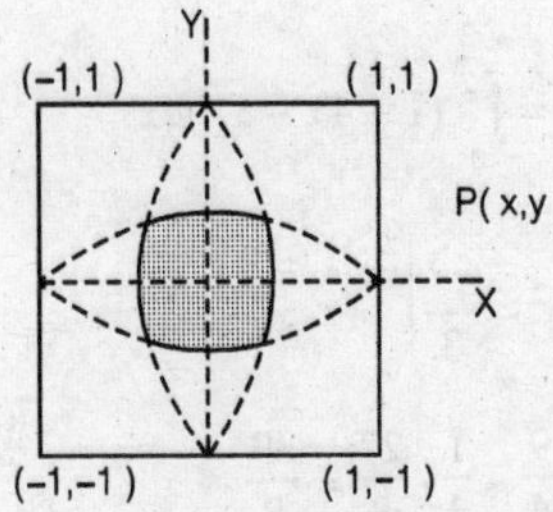

Now, $y^2 = 1-2x \Rightarrow y^2 = -2\left(x - \frac{1}{2}\right)$.

So, its vertex $A = \left(\frac{1}{2}, 0\right)$, axis is $y = 0$ and at all points $x \le \frac{1}{2}$.

$$y^2 = 1+2x \Rightarrow y^2 = 2\left(x + \frac{1}{2}\right)$$

So, its vertex $B = \left(-\frac{1}{2}, 0\right)$, axis is $y = 0$ and at all points $x \ge -\frac{1}{2}$.

$$x^2 = 1-2y \Rightarrow x^2 = -2\left(y - \frac{1}{2}\right)$$

So, its vertex $C = \left(0, \frac{1}{2}\right)$, axis is $x = 0$ and at all points $y \le \frac{1}{2}$.

$$x^2 = 1+2y \Rightarrow x^2 = 2\left(y + \frac{1}{2}\right)$$

So, its vertex $D = \left(0, -\frac{1}{2}\right)$, axis is $x = 0$ and at all points $y \ge -\frac{1}{2}$.

Hence, the region S as shown in the figure.

Clearly the region is symmetrical about both x and y axes.

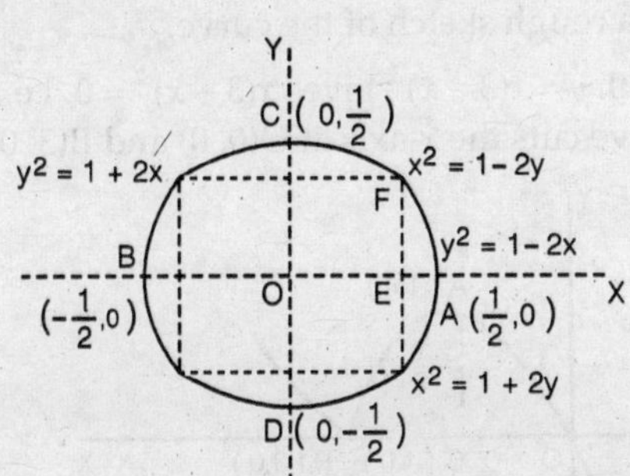

$\therefore$ area of $S = 4\{\text{ar}(CFEOC) + \text{ar}(FAEF)\}$ ... (1)

Now, solving $x^2 = 1-2y$ and $y^2 = 1-2x$,

on subtraction, we get $x^2 - y^2 = 2(x-y)$

or $x = y$ or $x + y = 2$.

When $x = y$, $x^2 = 1 - 2y \Rightarrow x^2 + 2x - 1 = 0$

$$\therefore \quad x = \frac{-2 \pm \sqrt{4+4}}{2} = -1 \pm \sqrt{2}$$

$x = -1-\sqrt{2}$ is not possible as the region is inside the square.

When $x + y = 2$,

$$x^2 = 1-2y \Rightarrow x^2 = 1 - 2(2-x) = 2x - 3$$

$\therefore$ $x^2 - 2x + 3 = 0$ which has imaginary roots.

So, $F = (\sqrt{2}-1, \sqrt{2}-1)$

$\therefore$ from (1),

$$\text{area } S = 4\left\{\int_0^{\sqrt{2}-1} \frac{1-x^2}{2}\,dx + \int_{\sqrt{2}-1}^{\frac{1}{2}} \sqrt{1-2x}\,dx\right\},$$

replacing y by the corresponding equations of the curves.

$$\therefore \text{ area } S = 4\left\{\frac{1}{2}\left[x - \frac{x^3}{3}\right]_0^{\sqrt{2}-1} + \left[\frac{(1-2x)^{3/2}}{\frac{3}{2}(-2)}\right]_{\sqrt{2}-1}^{\frac{1}{2}}\right\}$$

$$= 2\left\{\sqrt{2} - 1 - \frac{(\sqrt{2}-1)^3}{3}\right\} - \frac{4}{3}\{0 - (3-2\sqrt{2})^{3/2}\}$$

$$= \frac{2}{3}(\sqrt{2}-1)\{3 - (\sqrt{2}-1)^2\} + \frac{4}{3}(\sqrt{2}-1)^3$$

$$= \frac{1}{3}(16\sqrt{2} - 20).$$

11. Let A_n be the area bounded by the curve $y = (\tan x)^n$ and the lines $x = 0$, $y = 0$ and $x = \frac{\pi}{4}$.

Prove that for $n > 2$, $A_n + A_{n-2} = \frac{1}{n-1}$ and deduce that $\frac{1}{2n+2} < A_n < \frac{1}{2n-2}$.

Here $A_n = \int_0^{\frac{\pi}{4}} (\tan x)^n\,dx$

$$= \int_0^{\frac{\pi}{4}} \tan^{n-2} x \cdot (\sec^2 x - 1)\,dx$$

$$= \int_0^{\frac{\pi}{4}} \tan^{n-2} x \cdot d(\tan x) - \int_0^{\frac{\pi}{4}} \tan^{n-2} x\,dx$$

$$= \left[\frac{(\tan x)^{n-2+1}}{n-2+1}\right]_0^{\frac{\pi}{4}} - A_{n-2}$$

$$= \frac{1}{n-1} - A_{n-2}$$

$\therefore \quad A_n + A_{n-2} = \dfrac{1}{n-1}$ ... (1)

In $\left[0, \dfrac{\pi}{4}\right]$, $0 \le \tan x \le 1$

$\therefore \quad (\tan x)^{n-2} \ge (\tan x)^n \ge 0$

$\therefore \quad \int_0^{\frac{\pi}{4}} (\tan x)^{n-2}\, dx > \int_0^{\frac{\pi}{4}} (\tan x)^n\, dx$

$\therefore \quad A_{n-2} > A_n$

$\therefore \quad$ from (1), $A_n + A_n < \dfrac{1}{n-1}$

and $A_{n-2} + A_{n-2} > \dfrac{1}{n-1}$

or $\quad 2A_n < \dfrac{1}{n-1}$ and $2A_{n-2} > \dfrac{1}{n-1}$

$\therefore \quad A_n < \dfrac{1}{2n-2}$

and $2A_{(n+2)-2} > \dfrac{1}{(n+2)-1}$, i.e., $A_n > \dfrac{1}{2(n+1)}$.

Thus, $\dfrac{1}{2n+2} < A_n < \dfrac{1}{2n-2}$.

12. The line $y = mx$ bisects the area enclosed by the lines $x = 0, y = 0, x = \dfrac{3}{2}$ and the curve $y = 1 + 4x - x^2$. Find m.

Here, the curve is $x^2 - 4x = 1 - y$,

i.e., $(x-2)^2 = -(y-5)$.

It is a parabola whose vertex is (2, 5), axis is $x - 2 = 0$ and $y \le 5$ at every point of the curve.

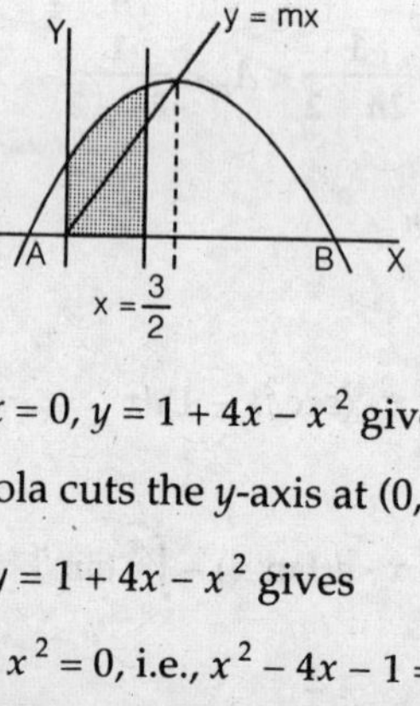

Also, when $x = 0$, $y = 1 + 4x - x^2$ gives $y = 1$.

So the parabola cuts the y-axis at (0, 1).

When $y = 0$, $y = 1 + 4x - x^2$ gives

$1 + 4x - x^2 = 0$, i.e., $x^2 - 4x - 1 = 0$

$\therefore \quad x = \dfrac{4 \pm \sqrt{20}}{2} = 2 \pm \sqrt{5}$

$\therefore \quad$ the parabola cuts the x-axis at

$A(2 - \sqrt{5}, 0)$ and $B(2 + \sqrt{5}, 0)$.

The area enclosed by the lines $x = 0, y = 0, x = \dfrac{3}{2}$ and the curve

$$= \int_0^{\frac{3}{2}} y\,dx = \int_0^{\frac{3}{2}} (1 + 4x - x^2)\,dx$$

$$= \left[x + 2x^2 - \frac{x^3}{3}\right]_0^{\frac{3}{2}}$$

$$= \frac{3}{2} + 2 \cdot \frac{9}{4} - \frac{1}{3} \cdot \frac{27}{8} = \frac{39}{8}.$$

The area bounded by the lines $\left(y = mx, y = 0 \text{ and } x = \dfrac{3}{2}\right)$

$$= \int_0^{\frac{3}{2}} y\,dx = \int_0^{\frac{3}{2}} mx\,dx$$

$$= m\left[\frac{x^2}{2}\right]_0^{\frac{3}{2}} = \frac{9}{8}m.$$

According to the question,

$\dfrac{9}{8}m = \dfrac{1}{2} \cdot \dfrac{39}{8}$; $\quad \therefore \quad m = \dfrac{13}{6}$.

13. Calculate the area bounded by the curve $y = x(3-x)^2$, the x-axis and the ordinates of the maximum and minimum points of the curve.

Here $y = x(3-x)^2$.

For maximum and minimum points, $\dfrac{dy}{dx} = 0$.

Now, $\dfrac{dy}{dx} = (3-x)^2 + x \cdot 2(3-x)(-1)$.

So, $\dfrac{dy}{dx} = 0 \quad \Rightarrow \quad (3-x)(3-x-2x) = 0$

$\therefore \quad x = 3, 1.$

$\therefore$ ordinates of maximum and minimum points are $x = 1, x = 3$.

We draw a rough sketch of the curve.

When $y = 0$, $y = x(3-x)^2$ gives $x(3-x)^2 = 0$, i.e., $x = 0, 3$. So the curve cuts the x-axis at O(0, 0) and B(3, 0).

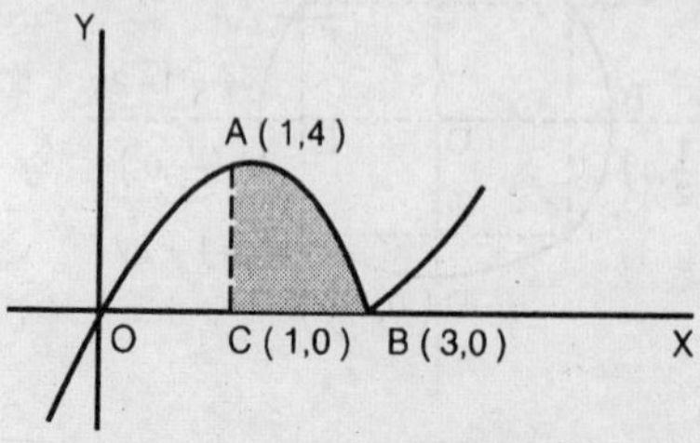

When $x = 0$, $y = x(3-x)^2$ gives $y = 0$.

So the curve cuts the y-axis at the origin O(0, 0) only.

When $0 < x < 3$,

$y = x(3-x)^2 \quad \Rightarrow \quad y > 0.$

$\therefore$ in $0 < x < 3$, the curve is always above the x-axis.

As the points of maximum and minimum are respectively at A(1, 4) and B(3, 0), the required area is as shown by shades in the figure.

The required area = ar(ACBA) $= \int_1^3 y\,dx$

$$= \int_1^3 x(3-x)^2\,dx$$

$$= \int_1^3 (9x - 6x^2 + x^3)dx$$

$$= \left[\frac{9x^2}{2} - 2x^3 + \frac{x^4}{4}\right]_1^3$$

$$= \left(\frac{81}{2} - 54 + \frac{81}{4}\right) - \left(\frac{9}{2} - 2 + \frac{1}{4}\right)$$

$$= \frac{27}{4} - \frac{11}{4} = \frac{16}{4} = 4.$$

14. Let $f(x) =$ maximum $\{x^2, (1-x)^2, 2x(1-x)\}$ where $x \in [0, 1]$. Determine the area of the region bounded by the curve $y = f(x)$ and the lines $y = 0$, $x = 0, x = 1$.

Consider the curves

$y = x^2, y = (1-x)^2$ and $y = 2x(1-x)$.

Clearly, $y = x^2$ is a parabola whose vertex = (0, 0), $x = 0$, i.e., y-axis is the axis and it cuts $x = 1$ at (1, 1).

$y = (1-x)^2$ is also a parabola whose vertex is (1, 0) and axis is $1 - x = 0$, i.e., $x = 1$ and it cuts $x = 0$ at (0, 1).

These parabolas cut at $\left(\frac{1}{2}, \frac{1}{4}\right)$.

Also $y = 2x(1-x)$ is a parabola. Changing the equation we get $x^2 - x = -\frac{1}{2}y$; or $\left(x - \frac{1}{2}\right)^2 = -\frac{1}{2}\left(y - \frac{1}{2}\right)$.

So, its vertex is $\left(\frac{1}{2}, \frac{1}{2}\right)$ and axis is $x = \frac{1}{2}$. It passes through (0, 0) and (1, 0).

Let this parabola cut $y = (1-x)^2$ and $y = x^2$ at D and E respectively.

Solving $y = 2x(1-x)$ and $y = (1-x)^2$, we get

$2x(1-x) = (1-x)^2$

or $\quad (1-x)(2x - 1 + x) = 0$

$\therefore \quad x = 1, \frac{1}{3}$ and so $y = 0, \frac{4}{9}$.

$\therefore \quad D = \left(\frac{1}{3}, \frac{4}{9}\right).$

Solving $y = 2x(1-x)$ and $y = x^2$, we get

$2x(1-x) = x^2$

or $\quad x(2 - 2x - x) = 0$

$\therefore \quad x = 0, \frac{2}{3}$ and so $y = 0, \frac{4}{9}$.

$\therefore \quad E = \left(\frac{2}{3}, \frac{4}{9}\right).$

$\therefore$ the equation of the curve on the boundary, is

$$y = f(x) = \begin{cases}(1-x)^2, & 0 \le x \le \frac{1}{3}\\ 2x(1-x), & \frac{1}{3} \le x \le \frac{2}{3}\\ x^2, & \frac{2}{3} \le x \le 1.\end{cases}$$

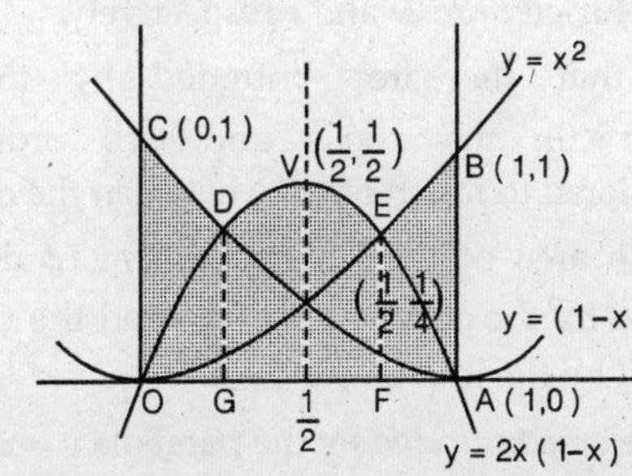

$\therefore$ the required area

$$= \int_0^{\frac{1}{3}} (1-x)^2\,dx + \int_{\frac{1}{3}}^{\frac{2}{3}} 2x(1-x)dx + \int_{\frac{2}{3}}^{1} x^2 dx$$

$$= \left[\frac{(1-x)^3}{-3}\right]_0^{\frac{1}{3}} + 2\left[\frac{x^2}{2} - \frac{x^3}{3}\right]_{\frac{1}{3}}^{\frac{2}{3}} + \left[\frac{x^3}{3}\right]_{\frac{2}{3}}^{1}$$

$$= -\frac{1}{3}\left(\frac{8}{27} - 1\right) + 2\left\{\frac{1}{2}\left(\frac{4}{9} - \frac{1}{9}\right) - \frac{1}{3}\left(\frac{8}{27} - \frac{1}{27}\right)\right\} + \frac{1}{3}\left(1 - \frac{8}{27}\right)$$

$$= \frac{19}{81} + 2\left(\frac{1}{6} - \frac{7}{81}\right) + \frac{19}{81}$$

$$= \frac{38}{81} + \frac{13}{81} = \frac{51}{81} = \frac{17}{27}.$$

15. The area bounded by the curve $y = f(x)$, x-axis and the ordinates $x = 1$ and $x = b$ is $(b-1)\sin(3b+4)$. Find $f(x)$.

We know that the area bounded by the curve $y = f(x)$, the x-axis and the ordinates $x = 1$, $x = b$ is $\int_1^b f(x)dx$.

$\therefore$ from the question,

$$\int_1^b f(x)dx = (b-1)\sin(3b+4).$$

Differentiating w.r.t. b,

$$f(x)]_{x=b} = \sin(3b+4) + (b-1)3\cos(3b+4)$$

or $f(b) = \sin(3b+4) + 3(b-1)\cos(3b+4)$

$\therefore$ $f(x) = \sin(3x+4) + 3(x-1)\cos(3x+4)$.

Exercises

1. Find the area bounded by the curves $y = |\sin x|$, the x-axis and the lines $|x| = \pi$.

2. Find the area bounded by the curve $x = 2 - y - y^2$ and y-axis.

3. Find the area of the region bounded by an ellipse and the two lines through the foci which are perpendicular to the major axis if the major axis and the eccentricity are $2a$ and e respectively.

4. Show that the area bounded by the curve $y = ce^x (c > 0)$, the x-axis and two ordinates is proportional to the difference between the ordinates.

5. Find the area of the larger segment of the region bounded by the curve $y = \sqrt{9 - x^2}$ and the x-axis that is cut off by the line $x - y = 1$.

6. Find the area bounded by the parabola $y = x^2$ and the line $y - 2x = 2$.

7. Find the area bounded by the curve $x = 2y - y^2$ and the line $y = 2 + x$.

8. Find the area bounded by the curve $y = \sqrt{5 - x^2}$ and the lines $y = |x - 1|$.

9. Find the area enclosed by the curve

$$9(x+2)^2 + 16(y+3)^2 = 144.$$

10. Find the area of the region bounded by the curve $xy = 1$ and the lines $y = x, y = 0, x = e$.

11. Find the area bounded by the curves

$$y = \tan x, -\frac{\pi}{3} \le x \le \frac{\pi}{3} \text{ and } y = \cot x, \frac{\pi}{6} \le x \le \frac{3\pi}{2},$$

and the x-axis.

12. Find the area of one of the curvilinear triangles formed by the curves $y = \sin x$, $x = \cos^{-1} y$ and the x-axis.

13. Determine the area of the portion of the circle $x^2 + y^2 = 64$ which is exterior to the parabola $y^2 = 12x$.

14. Find the area of the region bounded by the parabolas $x + 2y^2 = 0$ and $3y^2 + x = 1$.

15. Find the area of the region situated below the x-axis, bounded by the curves $x^2 + y^2 = 4$, $x^2 = -\sqrt{2}y$ and the line $y = x$.

16. Find the ratio of the areas into which the parabola $y^2 = 6x$ divides the circle $x^2 + y^2 = 16$.

17. Find the area bounded by the curves $y = \sqrt{9 - x^2}$ and $x^2 + y^2 = 6x$, and the line $x = 0$.

18. Find the area given by $x + y \le 6$, $x^2 + y^2 \le 6y$ and $y^2 \le 8x$.

19. In what ratio does the x-axis divide the area of the region bounded by the parabolas $y = 4x - x^2$ and $y = x^2 - x$?

20. Sketch the region bounded by the curve $x^{2/3} + y^{2/3} = a^{2/3}$, the x and y axes in the first quadrant. Find its area.
(Or Find the area enclosed by the curve $x = a\cos^3\theta$, $y = a\sin^3\theta$ in the first quadrant.)

21. Find the area bounded by the curve $y = x\sin x$ and x-axis in the interval $0 \le x \le 2\pi$.

22. Find the ratio of the areas of two regions bounded by the curve $y = x^2 \sin 2x$ and the x-axis in the interval $0 \le x \le \pi$.

23. Determine the area bounded by the curve $y = x(x-1)^2$, the y-axis and the line $y = 2$.

24. Sketch the region bounded by the curve $y = (x-1)(x-2)(x-3)$, $0 \le x \le 3$ and the line $4y = 3(x-1)$. Also find its area.

25. Determine the area of the region in the x-y plane enclosed by the curve $a^2y^2 = x^2(a^2 - x^2)$, using integration.

26. Find the area bounded by the curve $y = (x-1)(x-2)(x-3)$ and the x-axis in the interval $1 \le x \le 3$.

27. Find the area included between the parabola $x^2 = 4ay$, $a > 0$ and the witch of Agnesi $y = \dfrac{8a^3}{x^2 + 4a^2}$.

28. Sketch the region enclosed by the curve

$$y^2(a^2 + x^2) = x^2(a^2 - x^2) \text{ and find its area.}$$

29. Sketch the curves and identify the region bounded by $x = \frac{1}{2}$, $x = 2$, $y = \log_e x$ and $y = 2^x$.

Find the area of this region.

30. Find the area enclosed by the curves $y = \log_e(x + e)$, $x = \log_e \frac{1}{y}$ and the x-axis.

31. Compute the area of the region bounded by the curves $y = ex\log_e x$ and $y = \frac{1}{ex}\log_e x$.

32. Calculate the area of the region bounded by the curves $x = \log_2 y$, $y = 2x - x^2$ and the lines $x(x - 2) = 0$.

33. Find the area of the region bounded by the curves $y = \log_e x$, $y = \sin^4 \pi x$ and the line $x = 0$.

34. Find the area of the region bounded by the curve $y = \tan x$, the tangent to it at $x = \frac{\pi}{4}$ and the x-axis.

35. Find the area bounded by the normal at (1, 2) to the parabola $y^2 = 4x$ $(y \geq 0)$, the curve itself and the axis of the parabola.

36. Find the area bounded by the parabola $y = 6 + 4x - x^2$ and the chord joining the points $(-2, -6)$ and $(4, 6)$.

37. Find the area enclosed by the curves $x^2 + y^2 = 4$, $x^2 + y^2 - 6x + 8 = 0$ and a common tangent to the curves.

38. Find the area enclosed by the parabola $y^2 = 2x$ and two tangents to the parabola from the point $(-2, 0)$.

39. Find the area bounded by the curve $y = 2x^4 - x^2$, the x-axis and the two ordinates corresponding to the minima of the function.

40. If the ordinate $x = a$ divides the area bounded by x-axis, part of the curve $y = 1 + \frac{8}{x^2}$ and the ordinates $x = 2$, $x = 4$ into two equal parts then find a.

41. Find the continuous function $f(x)$ such that the area bounded by the x-axis, the curve $y = f(x)$ and the ordinates $x = 1$, $x = a$ is equal to $\sqrt{1 + a^2} - \sqrt{2}$ for all $a > 1$.

42. The area bounded by the continuous curve $y = f(x)$ and the lines $y = 2$, $x = 1$ and $x = a$, $a > 1$ is equal to $\frac{2}{3}\{(2a)^{3/2} - 3a + 3 - 2\sqrt{2}\}$. Find $f(x)$ if $f(x) > 2$ for all $x \in [1, a]$.

43. Find $b \in R_+$ so that the area bounded by the curves $y = x - bx^2$ and $y = \frac{1}{b}x^2$ is the maximum.

44. The vertices of a rectangle are (0, 0), (2, 0), (2, 1) and (0, 1). Find two parabolas having vertices at the origin, one having x-axis as the axis and the other y-axis as the axis, such that each of the parabola divides the rectangle into two equal parts.

45. Prove that the area bounded by $y = \frac{x^2 - x + 1}{x^2 + x + 1}$ and the lines $x = 0$, $y = 0$, $x = a$ is a monotonic increasing function of a. What is the increase in the area as a changes from 2 to 4?

46. Find the area of the region in the Argand plane in which the point z will be located if

$$\sqrt{5} \leq |z| \leq 2\sqrt{3} \text{ and } z^2 + \bar{z}^2 - 2z\bar{z} + 8z + 8\bar{z} > 0.$$

Objective Questions

Fill in the blanks.

47. The area bounded by the lines $|x| + |y| = 1$ is ______ .

48. The total area enclosed by the lines $y = |x|$, $|x| = 1$ and $y = 0$ is ______ .

49. The area in the region $x > 0$, $y > 0$ bounded by $y = 4x^2$, $x = 0$, $y = 1$ and $y = 4$ is ______ .

50. The area bounded by $y = \log_e x$, the ordinate $x = e$ and the x-axis is ______ .

51. The area bounded by $y = \frac{|x|}{x}$, $x \neq 0$ and the lines $yx(x - 1)(x - 3) = 0$ is ______ .

52. The area bounded by the curve $x = \cos^{-1} y$ and the lines $|x| = 1$ is ______ .

53. The area of the region bounded by $x^2 = y$, $y = x + 2$ and the x-axis is ______ .

Choose the correct option(s).

54. The area bounded by $y = \sqrt{4 - x^2}$, $y \geq 0$ and the x-axis is

(a) 4π (b) 2π
(c) π (d) none of these

55. The area bounded by the curve $y = \sin x$, $x \in [0, 2\pi]$ and the x-axis is

(a) 4 (b) 0
(c) 2 (d) none of these

56. The ratio in which the area bounded by the curves $y^2 = 12x$ and $x^2 = 12y$ is divided by the line $x = 3$ is

(a) 15 : 16 (b) 15 : 49
(c) 1 : 2 (d) none of these

Answers

1. 4 **2.** $\frac{9}{2}$

3. $2a^2\sqrt{1-e^2}(e\sqrt{1-e^2}+\sin^{-1}e)$

5. $\frac{9\pi}{4}+\frac{\sqrt{17}-1}{4}+\frac{9}{2}\sin^{-1}\frac{\sqrt{17}+1}{6}$

6. $4\sqrt{3}$ **7.** $\frac{9}{2}$ **8.** $\frac{5\pi}{4}-\frac{1}{2}$

9. 12π **10.** $\frac{3}{2}$ **11.** $\log_e 2$

12. $2-\sqrt{2}$ **13.** $\frac{16}{3}(8\pi-\sqrt{3})$

14. $\frac{4}{3}$ **15.** $\pi+\frac{1}{3}$

16. $\frac{4\pi+\sqrt{3}}{8\pi-\sqrt{3}}$ **17.** $\frac{3}{4}(3\sqrt{3}-\pi)$

18. $\frac{9\pi}{4}-\frac{1}{6}$ **19.** 121 : 4

20.

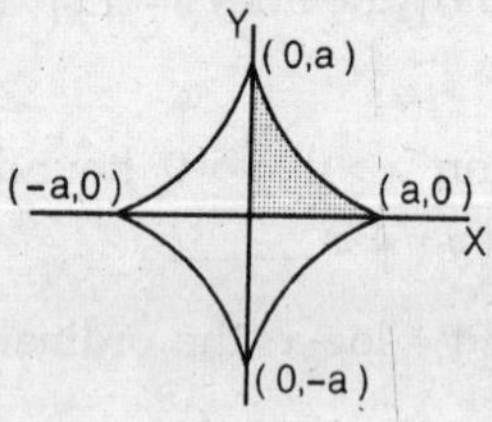

area $=\frac{3}{32}\pi a^2$

21. 4π **22.** $\frac{\pi^2-4}{5\pi^2-4}$

23. $\frac{10}{3}$

24.

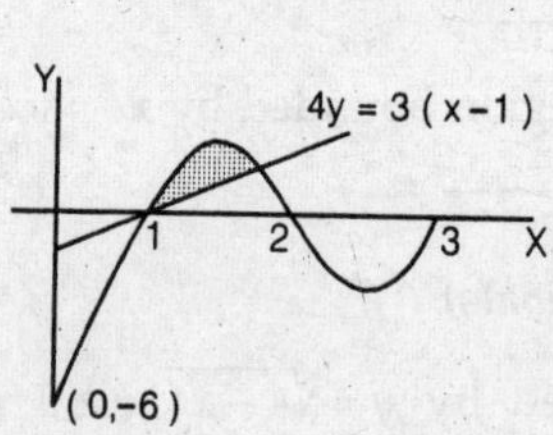

area $=\frac{3}{64}$

25. $\frac{4}{3}a^2$ **26.** $\frac{1}{2}$ **27.** $a^2\left(2\pi-\frac{4}{3}\right)$

28.

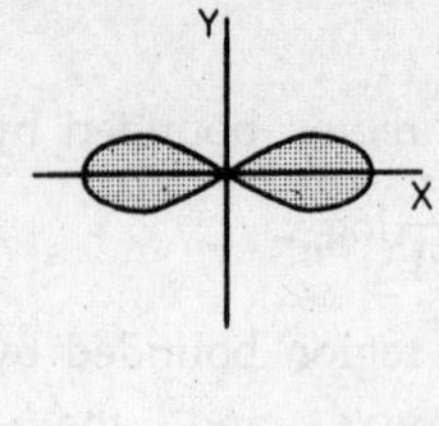

area $=a^2(\pi-2)$

29.

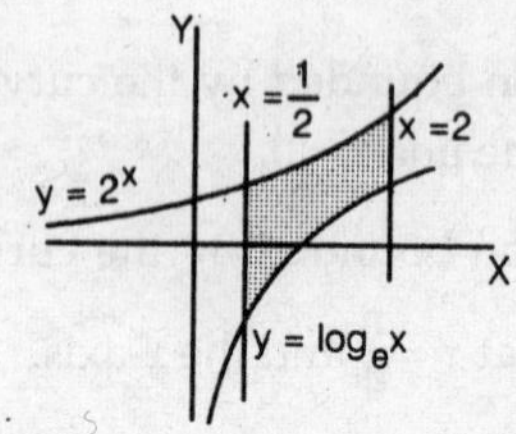

area $=\frac{4-\sqrt{2}}{\log 2}-\frac{5}{2}\log 2+\frac{3}{2}$

30. 2 **31.** $\frac{e^2-5}{4e}$

32. $\frac{3}{\log_e 2}-\frac{4}{3}$ **33.** $\frac{11}{8}$

34. $\frac{1}{2}\log 2-\frac{1}{4}$ **35.** $\frac{10}{3}$

36. 36 **37.** $3\sqrt{2}+\frac{3}{2}\sin^{-1}\frac{1}{3}-\frac{5\pi}{4}$

38. $\frac{8}{3}$ **39.** $\frac{7}{120}$

40. $2\sqrt{2}$ **41.** $\frac{x}{\sqrt{1+x^2}}$, $x>1$

42. $2\sqrt{2x}$, $x\ge 1$ **43.** ± 1

44. $32y^2=9x$, $4x^2=9y$

45. $2-\log 3+\frac{2}{\sqrt{3}}\tan^{-1}\frac{1}{4\sqrt{3}}$

46. $\frac{7\pi}{2}-12\sin^{-1}\frac{1}{\sqrt{3}}+5\sin^{-1}\frac{1}{\sqrt{5}}+\frac{4\sqrt{2}-2}{3}$

47. 2 **48.** 1 **49.** $\frac{7}{3}$ **50.** 1

51. 2 **52.** 2sin 1 **53.** $\frac{5}{6}$ **54.** (b)

55. (a) **56.** (b)

□

Chapter Test

Time: 120 minutes

1. In each of the following, fill in the blank so that the resulting sentence becomes true.
   (a) The area of the portion enclosed by the curve $\sqrt{x} + \sqrt{y} = \sqrt{a}$ and the axes of reference, is ______.
   (b) If $f(x)$ is monotonic increasing in $[a, b]$ then the area bounded by the curve $y = f(x)$ and the lines $y = f(a)$, $y = f(b)$ and the y-axis is given by the integral ______.
2. In each of the following, alternative answers are given of which one or more answers are correct. Choose the correct answer(s).
   (a) The latus rectum of the parabola $y^2 = 4ax$ for which the area bounded by the curve and the double ordinate $x = 6$ is three times the area bounded by the curve and the double ordinate through the focus, is
   A. 2 B. $2\sqrt[3]{3}$ C. $8\sqrt[3]{3}$ D. none of these
   (b) Let $f(x)$ be a continuous function such that the area bounded by the curve $y = f(x)$, the x-axis and the two ordinates $x = 0$ and $x = a$ is $\frac{a^2}{2} + \frac{a}{2}\sin a + \frac{\pi}{2}\cos a$. Then $f\left(\frac{\pi}{2}\right)$ is
   A. $\frac{1}{2}$ B. $\frac{\pi^2}{8} + \frac{\pi}{4}$ C. $\frac{\pi + 1}{2}$ D. none of these
3. Find the area of the region exterior to the ellipse $4x^2 + 9y^2 = 36$ bounded by the tangents to the curve from the point (3, 2) and the ellipse.
4. z is a variable point in the Argand plane such that $3 \le |z| \le 5$ and $\frac{\pi}{4} \le \arg z \le \frac{\pi}{3}$. Find the area of the region in the plane in which z can be located, using the method of integration.
5. The vertices of a square are (0, 0), (0, 4), (4, 4) and (4, 0). Show that the parabolas $y^2 = 4x$ and $x^2 = 4y$ divide the area of the square in three equal parts.
6. Prove that the function $f(x) = \sqrt{2x - x^2}$ is monotonic increasing in one-half of its domain of definition and monotonic decreasing in the other half. Find the ratio of the areas of the two regions bounded by the curve and either ordinate through the end points of the domain and its midpoint, and the x-axis.
7. Find the least value of the area bounded by the line $y = mx + 1$ and the parabola $y = x^2 + 2x - 3$, where m is a parameter.
8. Find the area of the portion of the region bounded by the parabolas $y^2 = 4x$ and $x^2 = 4y$ that falls inside the circle $x^2 + y^2 = 12$.
9. Consider the closed figure C made by the lines $|x| + |y| = \sqrt{2}$. Let S be the region inside the figure such that any point in it is nearer to the side $x + y = \sqrt{2}$ than the origin. Find the area of S.
10. Sketch the curve whose equation is $y^2 = x^4(a - x^2)$, $a > 0$ and calculate the whole area enclosed by the curve.
11. Let the sequence $a_1, a_2, a_3, \ldots$ be in GP. If the area bounded by the parabolas $y^2 = 4a_n x$ and $y^2 + 4a_n(x - a_n) = 0$ be Δ_n, prove that the sequence $\Delta_1, \Delta_2, \Delta_3, \ldots$ is also in GP.
12. Give a rough sketch of the region H consisting of points (x, y) satisfying $|x \pm y| \le 2$ and $x^2 + y^2 \ge 2$ and find the area of the region.
13. The ellipse $\frac{x^2}{16} + \frac{y^2}{7} = 1$ has its foci at S and S′. Find the area of the ellipse falling outside the circle whose diameter is SS′.

14. In an isosceles triangle whose one side is equal to the corresponding altitude, each being 2, find the maximum area of the ellipse inscribed in the triangle whose one axis is along the altitude. Also find the major and minor axis of the ellipse of maximum area.

15. Find the area bounded by the curves $x^2 = y$, $y + x^2 = 0$ and $y^2 + 3 = 4x$.

□

Answers

1. (a) $\frac{a^2}{6}$ (b) $\int_{f(a)}^{f(b)} f^{-1}(y)dy$

2. (a) C (b) A

3. $\frac{3}{2}(4 - \pi)$

4. $\frac{2\pi}{3}$

6. 1 : 1

7. $\frac{32}{3}$

8. $4\left(\frac{\sqrt{2}}{3} + \frac{3}{2}\sin^{-1}\frac{1}{3}\right)$

9. $\frac{4}{3}$

10. $\frac{a^2\pi}{4}$

12. $2(4 - \pi)$

13. $(4\sqrt{7} - 9)\pi - 8\sqrt{7}\sin^{-1}\frac{\sqrt{2}}{3} + 18\sin^{-1}\frac{4\sqrt{2}}{9}$

14. max. area $= \frac{2\pi}{3\sqrt{3}}$; $\frac{4}{3}$, $\frac{2}{\sqrt{3}}$ respectively

15. $\frac{1}{3}$

□

13. Differential Equation of the First Order

Recap of Facts and Formulae

1. Differential equation, its degree and order

- An equation involving derivatives (i.e., $\frac{dy}{dx}$, $\frac{d^2y}{dx^2}$, etc.) in x and y is a differential equation.
- The order of the highest order derivative in the equation is the order of the equation.
- The degree of the highest order derivative (when put in rational form) is the degree of the equation.

For example

(i) $y\frac{dy}{dx}+x^2=2$ is a differential equation of the first order and the first degree (linear).

(ii) $\frac{d^2y}{dx^2}+\left(\frac{dy}{dx}\right)^2=0$ is a differential equation of the second order and the first degree.

(iii) $\sqrt{1+\left(\frac{dy}{dx}\right)^2}=x^3$ is a differential equation of the first order and the second degree, it being $1+\left(\frac{dy}{dx}\right)^2=x^6$ in the rationalised form.

2. Formation of differential equation

We know $x^2+y^2=a^2$ is a circle whose centre is the origin and a is the radius. If a is a parameter, it will represent a family of concentric circles with the common centre (0, 0).

Differentiating $x^2+y^2=a^2$,

$$2x+2y\frac{dy}{dx}=0$$

i.e., $xdx+ydy=0$ (in differential form).

This is a differential equation for all the members of the family and it does not contain any parameter (arbitrary constant).

- The differential equation of a family of curves of one parameter is a differential equation of the first order, obtained by eliminating the parameter by differentiation.
- The differential equation of a family of curves (like $\frac{x^2}{a^2}+\frac{y^2}{b^2}=1$) of two parameters is a differential equation of the second order, obtained by eliminating the parameters by differentiating the algebraic equation twice. Similar procedure is used to find differential equation of a family of curves of three or more parameters.

3. Solution of differential equation

- The general solution of a differential equation is the relation in the variables x, y obtained by integrating (removing derivatives) where the relation contains as many arbitrary constants as the order of the equation.
- The general solution of a differential equation of the first order contains one arbitrary constant while that of the second order contains two arbitrary constants. So, the general solution of a differential equation of the first order gives one-parameter family of curves.
- In the general solution, if particular values of the arbitrary constants are put, we get a particular solution which will give one member of the family of curves.

4. To solve differential equation of the first order and the first degree

Simple standard forms of differential equation of the first order and the first degree are as follows.

- Variable separable. **Form:** $f(x)dx+\phi(y)dy=0$

  Method Integrate it, i.e., find $\int f(x)dx+\int\phi(y)dy=c$
- Reducible into variable separable.

  Method Make a suitable substitution so that the equation becomes variable separable in the new variables.

For example Take the equation $(x-y)^2\frac{dy}{dx}=1$.

This is not in the form of variable separable because we cannot write it in the form $f(x)dx+\phi(y)dy=0$.

Put $x - y = z$; then $1 - \frac{dy}{dx} = \frac{dz}{dx}$

$\therefore \quad (x-y)^2 \frac{dy}{dx} = 1 \quad \Rightarrow \quad z^2 \cdot \left(1 - \frac{dz}{dx}\right) = 1$

$\Rightarrow \quad z^2 - 1 = z^2 \frac{dz}{dx} \quad \Rightarrow \quad dx = \frac{z^2}{z^2 - 1} dz$

which is in the form of variable separable.

- Homogeneous equation. **Form:** $\frac{dy}{dx} = F\left(\frac{y}{x}\right)$

  Method Put $y = vx$ and solve the equation in the new variables x, v.

- Nonhomogeneous equations.

  Form $\frac{dy}{dx} = \frac{a_1x + b_1y + c_1}{a_2x + b_2y + c_2}$

  Method If $\frac{a_1}{a_2} \neq \frac{b_1}{b_2}$, put $x = X + h, y = Y + k$ such that $a_1h + b_1k + c_1 = 0,\ a_2h + b_2k + c_2 = 0$. In this way, the equation becomes homogeneous in X, Y. Then use the method for homogeneous equations.

  If $\frac{a_1}{a_2} = \frac{b_1}{b_2}$, put $a_1x + b_1y = v$ or $a_2x + b_2y = v$. The equation changes in the form of variable separable in x, v.

- Linear equations. **Form** $\frac{dy}{dx} + P(x) \cdot y = Q(x)$

  Method Multiply the equation by $e^{\int P(x)dx}$, called integrating factor. Then the equation becomes

  $\frac{d}{dx}\left\{y \cdot e^{\int P(x)dx}\right\} = Q(x) \cdot e^{\int P(x)dx}$

  $\therefore$ integrating, $y \cdot e^{\int P(x)dx} = \int Q(x) \cdot e^{\int P(x)dx}\, dx$

- Reducible into linear equation.

  Form $R(y) \cdot \frac{dy}{dx} + P(x) \cdot S(y) = Q(x)$

  such that $\frac{dS(y)}{dy} = R(y)$

  Method Put $S(y) = z$; then $\frac{dS(y)}{dx} = \frac{dS(y)}{dy} \cdot \frac{dy}{dx}$, i.e.,

  $\frac{dz}{dx} = R(y)\frac{dy}{dx}$. The equation becomes $\frac{dz}{dx} + P(x)z = Q(x)$ which is in the linear form.

- Exact differential equations. **Form** $\Sigma f\{\phi(x, y)\}d\phi = 0$.

  Method In order to put a differential equation in this form, one has to remember the following differentials.

$d(xy) = xdy + ydx \qquad d\left(\frac{y}{x}\right) = \frac{xdy - ydx}{x^2}$

$d\left(\frac{x}{y}\right) = \frac{ydx - xdy}{y^2} \qquad d\log\left(\frac{y}{x}\right) = \frac{xdy - ydx}{xy}$

$d\tan^{-1}\left(\frac{y}{x}\right) = \frac{xdy - ydx}{x^2 + y^2} \qquad d\tan^{-1}\left(\frac{x}{y}\right) = \frac{ydx - xdy}{x^2 + y^2}$

$d\sin^{-1}(xy) = \frac{xdy + ydx}{\sqrt{1 - (xy)^2}}$

5. To solve differential equation of the first order but of higher degree

Standard forms of differential equation of the first order and higher degree are as follows (here we denote $\frac{dy}{dx}$ by p).

- Solvable for p.

  Form $\{p - f_1(x, y)\}\{p - f_2(x, y)\} \dots \{p - f_n(x, y)\} = 0$

  Method Solve the first order and first degree equations $p - f_1(x, y) = 0$, etc. If $\phi_1(x, y, c) = 0$, etc., are solutions (taking the same arbitrary constant c for each) then $\phi_1(x, y, c) \cdot \phi_2(x, y, c) \cdot \dots \cdot \phi_n(x, y, c) = 0$ is the solution.

- Solvable for y. **Form** $y = f(x, p)$

  Method Differentiate $y = f(x, p)$ w.r.t. x which gives a first order and first degree equation in p and x. Solve it to get $\phi(x, p, c) = 0$.

  Eliminate p from $y = f(x, p)$ and $\phi(x, p, c) = 0$.

- Solvable for x. **Form** $x = f(y, p)$

  Method Differentiate $x = f(y, p)$ w.r.t y which gives a first order and first degree equation in p and y. Solve it to get $\psi(y, p, c) = 0$. Eliminate p from $x = f(y, p)$ and $\psi(y, p, c) = 0$.

- Clairaut's equation. **Form** $y = px + f(p)$

  Method Differentiating w.r.t. x, we get

  $\{x + f'(p)\}\frac{dp}{dx} = 0.$

  $\therefore \quad p = c$ or $f'(p) + x = 0$.

  When $p = c$, the general solution is $y = cx + f(c)$ which gives a family of straight lines.

  When $f'(p) + x = 0$, eliminating p from $y = px + f(p)$ and $f'(p) + x = 0$ we get a solution which is a curve (without any arbitrary constant) touching all the lines given by $y = cx + f(c)$.

  This solution is called the *singular solution.*

6. Some results on tangents and normals

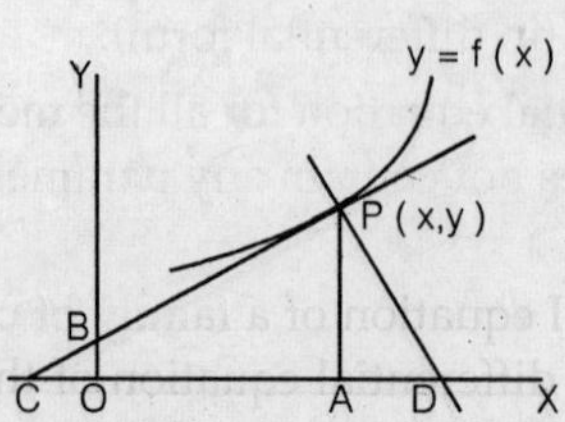

- The equation of the tangent at $P(x, y)$ to the curve $y = f(x)$ is $Y - y = \frac{dy}{dx}(X - x)$.

- The equation of the normal at $P(x, y)$ to the curve $y = f(x)$ is $Y - y = \frac{-1}{\frac{dy}{dx}}(X - x)$.

- The length of the tangent $= CP = y\sqrt{1 + \left(\frac{dx}{dy}\right)^2}$.

- The length of the normal $= PD = y\sqrt{1 + \left(\frac{dy}{dx}\right)^2}$.

- The length of the cartesian subtangent $= CA = y\frac{dx}{dy}$.

- The length of the cartesian subnormal $= AD = y\frac{dy}{dx}$.

- The initial ordinate of the tangent $= OB = y - x\frac{dy}{dx}$.

7. Orthogonal trajectory

The orthogonal trajectories of a family of curves form another family of curves such that each curve of one family cuts all the curves of the other family at right angles.

- The differential equation of the orthogonal trajectories of the curves $f\left(x, y, \frac{dy}{dx}\right) = 0$ is the family of curves whose differential equation is $f\left(x, y, -\frac{dx}{dy}\right) = 0$.

Method To find the orthogonal trajectories of a family of curves whose differential equation is known, put $-\frac{dx}{dy}$ in place of $\frac{dy}{dx}$ in the equation. The resulting differential equation is the equation of the orthogonal trajectories.

Selected Solved Examples

1. Find the differential equation of the family of curves $y = e^x(A\cos x + B\sin x)$ where A, B are arbitrary constants. Also write its order and degree.

Here, the equation has two arbitrary constants. So its differential equation will be of the second order.

We have $y = e^x(A\cos x + B\sin x)$ ... (1)

Differentiating (1) w.r.t. x,

$$\frac{dy}{dx} = e^x(A\cos x + B\sin x) + e^x(-A\sin x + B\cos x)$$

or $$\frac{dy}{dx} = y + e^x(-A\sin x + B\cos x) \quad \text{... (2)}$$

Differentiating (2) w.r.t. x,

$$\frac{d^2y}{dx^2} = \frac{dy}{dx} + e^x(-A\sin x + B\cos x) + e^x(-A\cos x - B\sin x)$$

or $$\frac{d^2y}{dx^2} = \frac{dy}{dx} + \left(\frac{dy}{dx} - y\right) - y, \text{ using (2) and (1)}$$

$$\therefore \quad \frac{d^2y}{dx^2} = 2\left(\frac{dy}{dx} - y\right).$$

This is the differential equation.

2. Solve $\log_e\left(\frac{dy}{dx}\right) = ax + by$.

Here $\frac{dy}{dx} = e^{ax + by}$; $\quad \therefore \quad e^{-by}dy = e^{ax}dx$

or $\int e^{-by}dy = \int e^{ax}dx$; $\quad \therefore \quad \frac{e^{-by}}{-b} = \frac{e^{ax}}{a} + c$

$$\therefore \quad \frac{1}{a}e^{ax} + \frac{1}{b}e^{-by} + c = 0.$$

3. Determine the equation of the curve passing through the origin, in the form $y = f(x)$, which satisfies the differential equation $\frac{dy}{dx} = \sin(10x + 6y)$.

Put $10x + 6y = z$; then $10 + 6\frac{dy}{dx} = \frac{dz}{dx}$

$$\therefore \quad \frac{dy}{dx} = \frac{1}{6}\left(\frac{dz}{dx} - 10\right)$$

$\therefore$ the equation becomes $\frac{1}{6}\left(\frac{dz}{dx} - 10\right) = \sin z$

or $\frac{dz}{dx} = 6\sin z + 10$; $\quad \therefore \quad \frac{dz}{6\sin z + 10} = dx$

or $$\frac{1}{2}\int \frac{dz}{3\sin z + 5} = \int dx$$

$$\therefore \quad \frac{1}{2}\int \frac{dz}{3 \cdot \frac{2\tan\frac{z}{2}}{1 + \tan^2\frac{z}{2}} + 5} = x + c$$

or $$\frac{1}{2}\int \frac{\sec^2\frac{z}{2}\,dz}{5\tan^2\frac{z}{2} + 6\tan\frac{z}{2} + 5} = x + c$$

or $\frac{1}{2}\int \frac{2dt}{5t^2+6t+5} = x + c$ $\left\{\text{putting } \tan\frac{z}{2} = t\right\}$

or $\frac{1}{5}\int \frac{dt}{t^2+\frac{6}{5}t+1} = x + c$

or $\frac{1}{5}\int \frac{dt}{\left(t+\frac{3}{5}\right)^2 + 1 - \frac{9}{25}} = x + c$

or $\frac{1}{5}\int \frac{dt}{\left(t+\frac{3}{5}\right)^2 + \left(\frac{4}{5}\right)^2} = x + c$

$\therefore$ $\frac{1}{5}\cdot\frac{1}{\frac{4}{5}}\tan^{-1}\frac{t+\frac{3}{5}}{\frac{4}{5}} = x + c$

or $\frac{1}{4}\tan^{-1}\frac{5t+3}{4} = x + c$

$\therefore$ $\frac{5t+3}{4} = \tan 4(x+c)$

or $5\tan\frac{z}{2} + 3 = 4\tan 4(x+c)$

$\therefore$ $5\tan(5x+3y) + 3 = 4\tan 4(x+c)$ ... (1)

It passes through (0, 0). So, $5\tan 0 + 3 = 4\tan 4c$

$\therefore$ $\frac{3}{4} = \tan 4c;$ $\therefore$ $\tan^{-1}\frac{3}{4} = 4c$

$\therefore$ $c = \frac{1}{4}\tan^{-1}\frac{3}{4}$.

Putting in (1),

$$5\tan(5x+3y) + 3 = 4\tan\left(4x + \tan^{-1}\frac{3}{4}\right)$$

or $\tan(5x+3y) = \frac{1}{5}\left\{4\tan\left(4x+\tan^{-1}\frac{3}{4}\right) - 3\right\}$

$$= \frac{4}{5}\,\frac{\tan 4x + \frac{3}{4}}{1 - \tan 4x\cdot\frac{3}{4}} - \frac{3}{5}$$

$$= \frac{4(3+4\tan 4x)}{5(4-3\tan 4x)} - \frac{3}{5} = \frac{25\tan 4x}{5(4-3\tan 4x)}$$

$\therefore$ $y = \frac{1}{3}\left\{\tan^{-1}\frac{5\tan 4x}{4-3\tan 4x} - 5x\right\}$.

4. Solve $\frac{dy}{dx} + yf'(x) = f(x)\cdot f'(x)$

where $f(x)$ is a given function of x.

Here $\frac{dy}{dx} = \{f(x) - y\}f'(x)$ ... (1)

Put $f(x) - y = z$; then $f'(x) - \frac{dy}{dx} = \frac{dz}{dx}$

$\therefore$ (1) becomes, $f'(x) - \frac{dz}{dx} = zf'(x)$

$\therefore$ $\frac{dz}{dx} = (1-z)f'(x)$

or $\frac{dz}{1-z} = f'(x)dx$

$\therefore$ $\int\frac{dz}{1-z} = \int f'(x)dx$

or $-\log(1-z) = f(x) - c$

$\therefore$ $f(x) + \log(1-z) = c$

$\therefore$ $f(x) + \log\{1 + y - f(x)\} = c$.

5. Solve $xdy - ydx = \sqrt{x^2+y^2}\,dx$.

It is a homogeneous equation. Put $y = vx$.

Then $\frac{dy}{dx} = v + x\frac{dv}{dx}$. ... (1)

We have the equation $x\frac{dy}{dx} - y = \sqrt{x^2+y^2}$

Putting in the equation from (1), we get

$$x\left(v + x\frac{dv}{dx}\right) - vx = \sqrt{x^2+v^2x^2}$$

or $x^2\frac{dv}{dx} = x\sqrt{1+v^2}$

$\therefore$ $\frac{dv}{\sqrt{1+v^2}} = \frac{dx}{x}$

or $\int\frac{dv}{\sqrt{1+v^2}} = \int\frac{dx}{x}$

$\therefore$ $\log\{v + \sqrt{v^2+1}\} = \log x + \log c$

or $v + \sqrt{v^2+1} = cx$

$\therefore$ $\frac{y}{x} + \sqrt{\frac{y^2}{x^2}+1} = cx$

$\therefore$ $y + \sqrt{x^2+y^2} = cx^2$.

6. Solve $\left(xe^{y/x} - y\sin\frac{y}{x}\right)dx + x\sin\frac{y}{x}\,dy = 0.$

This is a homogeneous equation. So, put $y = vx$.

Then $\frac{dy}{dx} = v + x\frac{dv}{dx}$. So, the equation becomes,

$$xe^{y/x} - y\sin\frac{y}{x} + x\sin\frac{y}{x}\cdot\frac{dy}{dx} = 0$$

or $xe^v - vx\sin v + x\sin v\cdot\left(v + x\frac{dv}{dx}\right) = 0$

or $\quad xe^v + x^2 \sin v \dfrac{dv}{dx} = 0$

$\therefore \quad \dfrac{dx}{x} + e^{-v} \sin v dv = 0$

or $\quad \int \dfrac{dx}{x} + \int e^{-v} \sin v \, dv = c \quad \ldots (1)$

Now, $I = \int e^{-v} \sin v \, dv = \sin v \cdot \dfrac{e^{-v}}{-1} - \int \dfrac{e^{-v}}{-1} \cdot \cos v \, dv$

$= -e^{-v} \sin v + \int e^{-v} \cos v \, dv$

$= -e^{-v} \sin v + \cos v \cdot \dfrac{e^{-v}}{-1} - \int \dfrac{e^{-v}}{-1} \cdot (-\sin v) dv$

$= -e^{-v} (\sin v + \cos v) - I$

$\therefore \quad 2I = -e^{-v} (\sin v + \cos v)$

$\therefore \quad I = \dfrac{-1}{2} e^{-v} (\sin v + \cos v)$

$\therefore$ from (1), $\log x + \dfrac{-1}{2} \cdot e^{-v} (\sin v + \cos v) = c$

or $\quad \log x = c + \dfrac{1}{2} e^{-y/x} \left\{ \sin \dfrac{y}{x} + \cos \dfrac{y}{x} \right\}$.

7. Show that the differential equation $y^3 dy + (x + y^2) dx = 0$ can be reduced to a homogeneous equation. Hence, solve it.

Put $x = X$, $y^2 = Y$. Then $dx = dX$, $2y dy = dY$.

$\therefore \quad 2y \dfrac{dy}{dx} = \dfrac{dY}{dX}$. Using this in the given equation,

we get

$$y^3 \frac{dy}{dx} + x + y^2 = 0; \quad \text{or} \quad \frac{1}{2} Y \frac{dY}{dX} + X + Y = 0$$

which is a homogenous equation.

So, in order to solve the equation,

put $y^2 = vx$; then $2y \dfrac{dy}{dx} = v + x \dfrac{dv}{dx}$.

$\therefore$ the equation becomes, $y^3 \dfrac{dy}{dx} + x + y^2 = 0$

or $\quad y^2 \cdot \dfrac{1}{2} \left(v + x \dfrac{dv}{dx} \right) + (x + y^2) = 0$

or $\quad vx \cdot \dfrac{1}{2} \left(v + x \dfrac{dv}{dx} \right) + x + vx = 0$

or $\quad \dfrac{v}{2} \left(v + x \dfrac{dv}{dx} \right) + 1 + v = 0$

$\therefore \quad 1 + v + \dfrac{v^2}{2} + \dfrac{vx}{2} \dfrac{dv}{dx} = 0$

or $\quad \dfrac{dx}{x} + \dfrac{vdv}{2\left(1 + v + \dfrac{v^2}{2}\right)} = 0$

$\therefore \quad \dfrac{dx}{x} + \dfrac{vdv}{v^2 + 2v + 2} = 0$

or $\quad \int \dfrac{dx}{x} + \dfrac{1}{2} \int \dfrac{(2v + 2) - 2}{v^2 + 2v + 2} dv = 0$

or $\quad \log x + \dfrac{1}{2} \int \dfrac{2v + 2}{v^2 + 2v + 2} dv - \int \dfrac{dv}{(v + 1)^2 + 1} = 0$

or $\quad \log x + \dfrac{1}{2} \log(v^2 + 2v + 2) - \tan^{-1}(v + 1) = c$

or $\quad \log x + \dfrac{1}{2} \log \left(\dfrac{y^4}{x^2} + 2 \dfrac{y^2}{x} + 2 \right) - \tan^{-1} \left(\dfrac{y^2}{x} + 1 \right) = c$

or $\quad \dfrac{1}{2} \log(y^4 + 2y^2 x + 2x^2) - \tan^{-1} \dfrac{y^2 + x}{x} = c.$

8. Solve $(y - 3x + 3) \dfrac{dy}{dx} = 2y - x - 4$.

Here, $\dfrac{dy}{dx} = \dfrac{2y - x - 4}{y - 3x + 3}$.

Put $x = X + h$, $y = Y + k$; then $dx = dX$, $dy = dY$

$\therefore$ the equation becomes

$$\frac{dY}{dX} = \frac{2(Y + k) - (X + h) - 4}{Y + k - 3(X + h) + 3}$$

or $\quad \dfrac{dY}{dX} = \dfrac{2Y - X + (2k - h - 4)}{Y - 3X + (k - 3h + 3)}$

Select h, k such that $2k - h - 4 = 0$, $k - 3h + 3 = 0$.

Solving these, $h = 2$, $k = 3$.

$\therefore$ on substitution of $x = X + 2$, $y = Y + 3$, the equation becomes $\dfrac{dY}{dX} = \dfrac{2Y - X}{Y - 3X}$ which is homogeneous.

Put $Y = VX$; then $\dfrac{dY}{dX} = V + X \dfrac{dV}{dX}$

$\therefore \quad V + X \dfrac{dV}{dX} = \dfrac{2VX - X}{VX - 3X} \quad$ or $\quad X \dfrac{dV}{dX} = \dfrac{2V - 1}{V - 3} - V$

or $\quad X \dfrac{dV}{dX} = \dfrac{-V^2 + 5V - 1}{V - 3} \quad \therefore \quad \dfrac{V - 3}{-V^2 + 5V - 1} dV = \dfrac{dX}{X}$.

Integrating, $\int \dfrac{V - 3}{-V^2 + 5V - 1} dV = \log X + \log c$

or $\quad \dfrac{1}{-2} \int \dfrac{(-2V + 5) + 1}{-V^2 + 5V - 1} dV = \log cX$

or $\quad -\dfrac{1}{2} \log(-V^2 + 5V - 1) + \dfrac{1}{2} \int \dfrac{dV}{(V - 5/2)^2 - (\sqrt{21}/2)^2}$

$= \log cX$

or $$-\frac{1}{2}\log(-V^2+5V-1)+\frac{1}{2}\cdot\frac{1}{2\cdot\frac{\sqrt{21}}{2}}\log\frac{V-\frac{5}{2}-\frac{\sqrt{21}}{2}}{V-\frac{5}{2}+\frac{\sqrt{21}}{2}}$$
$$=\log cX$$

or $$-\frac{1}{2}\log\left(-\frac{Y^2}{X^2}+5\frac{Y}{X}-1\right)+\frac{1}{2\sqrt{21}}\cdot\log\frac{\frac{Y}{X}-\frac{5+\sqrt{21}}{2}}{\frac{Y}{X}-\frac{5-\sqrt{21}}{2}}$$
$$=\log cX$$

or $$-\frac{1}{2}\log\frac{5XY-X^2-Y^2}{X^2}+\frac{1}{2\sqrt{21}}\log\frac{2Y-(5+\sqrt{21})X}{2Y-(5-\sqrt{21})X}$$
$$=\log cX$$

or $$\left\{\frac{5(x-2)(y-3)-(x-2)^2-(y-3)^2}{(x-2)^2}\right\}^{-1/2}$$
$$\times\left\{\frac{2(y-3)-(5+\sqrt{21})(x-2)}{2(y-3)-(5-\sqrt{21})(x-2)}\right\}^{\frac{1}{2\sqrt{21}}}=c(x-2)$$

9. Solve $(x+2y)(dx-dy)=dx+dy$.

$(x+2y-1)dx=(x+2y+1)dy$; $\therefore$ $\dfrac{dy}{dx}=\dfrac{x+2y-1}{x+2y+1}$.

Put $x+2y=z$; then $1+2\dfrac{dy}{dx}=\dfrac{dz}{dx}$

$\therefore$ the equation becomes $\dfrac{1}{2}\left(\dfrac{dz}{dx}-1\right)=\dfrac{z-1}{z+1}$

or $\dfrac{dz}{dx}=\dfrac{2z-2}{z+1}+1=\dfrac{3z-1}{z+1}$; $\therefore$ $\dfrac{z+1}{3z-1}dz=dx$

$\therefore$ integrating,

$$x+c=\int\frac{z+1}{3z-1}dz=\frac{1}{3}\int\frac{(3z-1)+4}{3z-1}dz$$
$$=\frac{1}{3}\int\left(1+\frac{4}{3z-1}\right)dz=\frac{1}{3}\left[z+\frac{4}{3}\log(3z-1)\right]$$

$\therefore$ $$x+c=\frac{1}{3}\left[x+2y+\frac{4}{3}\log(3x+6y-1)\right].$$

10. Solve $\dfrac{2x^2+3y^2-7}{y}dx=\dfrac{3x^2+2y^2-8}{x}dy$.

Here, $(2x^2+3y^2-7)xdx=(3x^2+2y^2-8)y\,dy$.

Put $x^2=X$, $y^2=Y$; then $2xdx=dX$, $2y\,dy=dY$

$\therefore$ the equation becomes

$$(2X+3Y-7)dX=(3X+2Y-8)dY$$

or $$\frac{dY}{dX}=\frac{2X+3Y-7}{3X+2Y-8} \qquad \text{... (1)}$$

Put $X=u+h$, $Y=v+k$; then $dX=du$, $dY=dv$

$\therefore$ (1) becomes,

$$\frac{dv}{du}=\frac{2(u+h)+3(v+k)-7}{3(u+h)+2(v+k)-8}$$
$$=\frac{2u+3v+(2h+3k-7)}{3u+2v+(3h+2k-8)}$$

Select h, k such that $2h+3k-7=0$ ... (2)

$3h+2k-8=0$ ... (3)

From $2\times(2)-3\times(3)$, $-5h-14+24=0$

$\therefore$ $h=2$ and so $k=1$.

$\therefore$ (1) becomes, $\dfrac{dv}{du}=\dfrac{2u+3v}{3u+2v}$ where $X=u+2$, $Y=v+1$

Put $v=uz$; then $\dfrac{dv}{du}=z+u\dfrac{dz}{du}$

$\therefore$ $$z+u\frac{dz}{du}=\frac{2u+3uz}{3u+2uz}=\frac{2+3z}{3+2z}$$

or $$u\frac{dz}{du}=\frac{2+3z}{3+2z}-z=\frac{2-2z^2}{3+2z}$$

or $\dfrac{3+2z}{2-2z^2}dz=\dfrac{du}{u}$ $\quad\therefore\quad$ $\displaystyle\int\frac{3+2z}{2(1-z^2)}dz=\int\frac{du}{u}$

or $$\frac{1}{2}\int\left(\frac{3}{1-z^2}+\frac{2z}{1-z^2}\right)dz=\log u+\log c$$

or $$\frac{1}{2}\left[3\cdot\frac{1}{2}\log\frac{z+1}{1-z}-\log(1-z^2)\right]=\log u+\log c$$

or $$\frac{3}{4}\log\frac{1+\frac{v}{u}}{1-\frac{v}{u}}-\frac{1}{2}\log\left(1-\frac{v^2}{u^2}\right)=\log u+\log c$$

or $$\frac{3}{4}\log\frac{u+v}{u-v}-\frac{1}{2}\log\frac{(u+v)(u-v)}{u^2}=\log u+\log c$$

or $$\frac{3}{4}\log\frac{X-2+Y-1}{X-2-(Y-1)}$$
$$-\frac{1}{2}\log\frac{(X-2+Y-1)(X-2-Y+1)}{(X-2)^2}$$
$$=\log(X-2)+\log c$$

or $$\frac{3}{4}\log\frac{x^2+y^2-3}{x^2-y^2-1}-\frac{1}{2}\cdot\log\frac{(x^2+y^2-3)(x^2-y^2-1)}{(x^2-2)^2}$$
$$=\log(x^2-2)+\log c$$

or $$\frac{3}{4}\log(x^2+y^2-3)-\frac{3}{4}\log(x^2-y^2-1)$$
$$-\frac{1}{2}\log(x^2+y^2-3)$$
$$-\frac{1}{2}\log(x^2-y^2-1)+\log(x^2-2)$$
$$=\log(x^2-2)+\log c$$

or $\frac{1}{4}\log(x^2+y^2-3)-\frac{5}{4}\log(x^2-y^2-1)=\log c$

or $\log\frac{(x^2+y^2-3)}{(x^2-y^2-1)^5}=4\log c=\log c^4=\log k$ (say)

$\therefore \quad x^2+y^2-3=k(x^2-y^2-1)^5.$

11. Solve $x\frac{dy}{dx}=y-\cos\left(\frac{1}{x}\right)$.

Here, $x\frac{dy}{dx}-y=-\cos\left(\frac{1}{x}\right)$

$\therefore \quad \frac{dy}{dx}-\frac{1}{x}\cdot y=-\frac{1}{x}\cos\left(\frac{1}{x}\right)$; this is in the linear form.

Integrating factor $=e^{\int -\frac{1}{x}dx}=e^{-\log x}=e^{\log x^{-1}}=x^{-1}.$

Multiplying by the integrating factor,

$$\frac{1}{x}\cdot\frac{dy}{dx}-\frac{1}{x^2}y=-\frac{1}{x^2}\cos\left(\frac{1}{x}\right)$$

or $\frac{d}{dx}\left\{\frac{1}{x}\cdot y\right\}=-\frac{1}{x^2}\cos\left(\frac{1}{x}\right)$

or $d\left(\frac{y}{x}\right)=-\frac{1}{x^2}\cos\left(\frac{1}{x}\right)dx$

or $\int d\left(\frac{y}{x}\right)=\int -\frac{1}{x^2}\cos\left(\frac{1}{x}\right)dx$

$\therefore \quad \frac{y}{x}=\int\cos\left(\frac{1}{x}\right)d\left(\frac{1}{x}\right)$

or $\frac{y}{x}=\sin\left(\frac{1}{x}\right)+c$

$\therefore \quad y-x\sin\left(\frac{1}{x}\right)=cx.$

12. Solve $(1+y^2)dx=(\tan^{-1}y-x)dy$.

Here, $(1+y^2)\frac{dx}{dy}=\tan^{-1}y-x$

or $\frac{dx}{dy}+\frac{1}{1+y^2}x=\frac{\tan^{-1}y}{1+y^2}.$

This is in the linear form. Clearly,

integrating factor $=e^{\int\frac{1}{1+y^2}dy}=e^{\tan^{-1}y}.$

Multiplying by it,

$$e^{\tan^{-1}y}\left(\frac{dx}{dy}+\frac{1}{1+y^2}x\right)=\frac{\tan^{-1}y}{1+y^2}\cdot e^{\tan^{-1}y}$$

or $\frac{d}{dy}\left(x\cdot e^{\tan^{-1}y}\right)=\frac{\tan^{-1}y}{1+y^2}\cdot e^{\tan^{-1}y}.$

Integrating, we get

$$xe^{\tan^{-1}y}+c=\int\frac{\tan^{-1}y}{1+y^2}\cdot e^{\tan^{-1}y}dy$$

$$=\int ze^z dz,\text{ where }\tan^{-1}y=z$$

$$=ze^z-\int e^z dz=(z-1)e^z$$

$\therefore \quad xe^{\tan^{-1}y}+c=(\tan^{-1}y-1)e^{\tan^{-1}y}.$

13. Solve $\frac{dy}{dx}=x^3y^3-xy$.

Here, $\frac{dy}{dx}+xy=x^3y^3$

$\therefore \quad \frac{1}{y^3}\frac{dy}{dx}+x\cdot\frac{1}{y^2}=x^3 \qquad \dots (1)$

Put $\frac{1}{y^2}=z$; then $\frac{-2}{y^3}\frac{dy}{dx}=\frac{dz}{dx}$

$\therefore$ (1) becomes $-\frac{1}{2}\frac{dz}{dx}+xz=x^3$

or $\frac{dz}{dx}-2xz=-2x^3.$

It is in the linear form. So,

integrating factor $=e^{\int -2x dx}=e^{-x^2}$

Multiplying by it,

$$e^{-x^2}\frac{dz}{dx}-2xe^{-x^2}\cdot z=-2x^3\cdot e^{-x^2}$$

or $\frac{d}{dx}\left(ze^{-x^2}\right)=-2x^3e^{-x^2}$

$\therefore \quad d\left(ze^{-x^2}\right)=-2x^3e^{-x^2}dx$

or $\int d\left(ze^{-x^2}\right)=\int -2x^3\cdot e^{-x^2}dx$

or $ze^{-x^2}=-\int te^{-t}dt$ (putting $x^2=t$)

$$=-\left[t\cdot\frac{e^{-t}}{-1}-\int\frac{e^{-t}}{-1}dt\right]=te^{-t}-\int e^{-t}dt$$

or $\frac{1}{y^2}e^{-x^2}=x^2\cdot e^{-x^2}+e^{-x^2}+c$

$\therefore \quad (x^2+1+ce^{x^2})y^2=1.$

14. Solve $\frac{dy}{dx}=\frac{2xy}{x^2-2y-1}$.

Here $x^2-2y-1=2xy\frac{dx}{dy}$

or $\quad 2xy\dfrac{dx}{dy}-x^2=-(1+2y)$

or $\quad 2x\dfrac{dx}{dy}-\dfrac{1}{y}\cdot x^2=-\left(2+\dfrac{1}{y}\right).$

Put $x^2=z$; then $2x\dfrac{dx}{dy}=\dfrac{dz}{dy}$

$\therefore$ the equation becomes

$$\frac{dz}{dy}-\frac{1}{y}\cdot z=-\left(2+\frac{1}{y}\right)\text{\{linear form\}}$$

$\therefore$ integrating factor $=e^{\int -\frac{1}{y}dy}=e^{-\log y}=e^{\log\frac{1}{y}}=\dfrac{1}{y}.$

Multiplying by $\dfrac{1}{y}$,

$$\frac{1}{y}\frac{dz}{dy}-\frac{1}{y^2}z=-\frac{1}{y}\left(2+\frac{1}{y}\right)$$

or $\quad \dfrac{d}{dy}\left(z\cdot\dfrac{1}{y}\right)=-\left(\dfrac{2}{y}+\dfrac{1}{y^2}\right)$

$\therefore \quad \dfrac{z}{y}=-\int\left(\dfrac{2}{y}+\dfrac{1}{y^2}\right)dy$

or $\quad \dfrac{z}{y}=-\left(2\log y-\dfrac{1}{y}\right)+c$

$\therefore \quad \dfrac{x^2}{y}+2\log y-\dfrac{1}{y}=c.$

15. Solve $xdx+ydy+\dfrac{xdy-ydx}{x^2+y^2}=0.$

We know, $d(x^2+y^2)=2(xdx+ydy)$

$$d\tan^{-1}\left(\frac{y}{x}\right)=\frac{1}{1+\left(\frac{y}{x}\right)^2}\cdot\frac{xdy-ydx}{x^2}=\frac{xdy-ydx}{x^2+y^2}$$

$\therefore$ the equation becomes

$$\frac{1}{2}d(x^2+y^2)+d\tan^{-1}\left(\frac{y}{x}\right)=0 \quad \text{(exact equation)}$$

Integrating, $\dfrac{1}{2}(x^2+y^2)+\tan^{-1}\dfrac{y}{x}=\dfrac{c}{2}$

or $\quad x^2+y^2+2\tan^{-1}\dfrac{y}{x}=c$

$\therefore \quad y=x\tan\dfrac{c-x^2-y^2}{2}.$

Second Method Put $x=r\cos\theta$ and $y=r\sin\theta$

$$\left(\text{i.e., } r^2=x^2+y^2,\ \tan\theta=\frac{y}{x}\right)$$

$\therefore \quad dx=dr\cos\theta+r(-\sin\theta)d\theta$ and

$$dy=dr\sin\theta+r\cos\theta d\theta$$

$\therefore \quad xdx+ydy=r\cos\theta(\cos\theta dr-r\sin\theta d\theta)$
$\quad +r\sin\theta(\sin\theta dr+r\cos\theta d\theta)$
$\quad =rdr$

$xdy-ydx=r\cos\theta(\sin\theta dr+r\cos\theta d\theta)$
$\quad -r\sin\theta(\cos\theta dr-r\sin\theta\, d\theta)$
$\quad =r^2d\theta$

$\therefore$ the equation becomes

$rdr+\dfrac{r^2d\theta}{r^2}=0; \quad \therefore \quad rdr+d\theta=0$

$\therefore \quad \int rdr+\int d\theta=c; \quad \therefore \quad \dfrac{r^2}{2}+\theta=c$

$\therefore \quad \dfrac{1}{2}(x^2+y^2)+\tan^{-1}\dfrac{y}{x}=\dfrac{c}{2}$

$\therefore \quad y=x\tan\dfrac{c-x^2-y^2}{2}.$

16. Solve $xdy-\{y+xy^3(1+\log x)\}\,dx=0.$

Here, $xdy-ydx=xy^3(1+\log x)dx$

or $\quad -\dfrac{ydx-xdy}{y^2}=xy(1+\log x)dx$

or $\quad -d\left(\dfrac{x}{y}\right)=xy(1+\log x)dx$

$\therefore \quad -\dfrac{x}{y}d\left(\dfrac{x}{y}\right)=x^2(1+\log x)dx$

or $\quad -\int\dfrac{x}{y}d\left(\dfrac{x}{y}\right)=\int x^2(1+\log x)dx$

or $\quad -\dfrac{\left(\frac{x}{y}\right)^2}{2}=(1+\log x)\dfrac{x^3}{3}-\int\dfrac{x^3}{3}\cdot\dfrac{1}{x}dx$

or $\quad -\dfrac{x^2}{2y^2}=\dfrac{x^3}{3}(1+\log x)-\dfrac{1}{3}\cdot\dfrac{x^3}{3}+\dfrac{c}{2}$

or $\quad -\dfrac{x^2}{y^2}=\dfrac{2x^3}{3}\left(\dfrac{2}{3}+\log x\right)+c.$

17. Find the equation of the curve passing through (1, 2) whose differential equation is
$y(x+y^3)dx=x(y^3-x)dy.$

Here, $(xy+y^4)dx=(xy^3-x^2)dy$

or $\quad y^3(ydx-xdy)+x(ydx+xdy)=0$

or $\quad -x^2y^3\cdot\dfrac{xdy-ydx}{x^2}+xd(xy)=0$

or $-\frac{y}{x}d\left(\frac{y}{x}\right)+\frac{d(xy)}{x^2y^2}=0$ (dividing by x^3y^2)

or $-\int\frac{y}{x}d\left(\frac{y}{x}\right)+\int\frac{d(xy)}{(xy)^2}=c$

$\therefore$ $-\frac{\left(\frac{y}{x}\right)^2}{2}+\frac{(xy)^{-1}}{-1}=c$

$\therefore$ $\frac{1}{2}\cdot\frac{y^2}{x^2}+\frac{1}{xy}+c=0$

$\therefore$ $y^3+2x+2cx^2y=0$

It passes through (1, 2). So

$$2^3+2+2c\cdot 2=0$$

$\therefore$ $c=\frac{-10}{4}=\frac{-5}{2}$.

$\therefore$ the curve is $y^3+2x-5x^2y=0$.

18. If y_1, y_2 are two solutions of the differential equation $\frac{dy}{dx}+P(x)\cdot y=Q(x)$ then prove that $y=y_1+c(y_1-y_2)$ is the general solution of the equation where c is any constant. For what relation between the constants α, β will the linear combination $\alpha y_1+\beta y_2$ also be a solution?

As y_1, y_2 are two solutions of the differential equation

$$\frac{dy}{dx}+P(x)\cdot y=Q(x) \quad \text{... (1)}$$

$\therefore$ $$\frac{dy_1}{dx}+P(x)\cdot y_1=Q(x) \quad \text{... (2)}$$

and $$\frac{dy_2}{dx}+P(x)\cdot y_2=Q(x) \quad \text{... (3)}$$

From (1) – (2),

$$\left(\frac{dy}{dx}-\frac{dy_1}{dx}\right)+P(x)\cdot(y-y_1)=0$$

$\therefore$ $$\frac{d}{dx}(y-y_1)+P(x)\cdot(y-y_1)=0 \quad \text{... (4)}$$

From (2) – (3),

$$\frac{d}{dx}(y_1-y_2)+P(x)\cdot(y_1-y_2)=0 \quad \text{... (5)}$$

From (4) and (5), $\dfrac{\frac{d}{dx}(y-y_1)}{\frac{d}{dx}(y_1-y_2)}=\dfrac{y-y_1}{y_1-y_2}$

$\therefore$ $$\frac{\frac{d}{dx}(y-y_1)}{y-y_1}=\frac{\frac{d}{dx}(y_1-y_2)}{y_1-y_2}$$

$\therefore$ integrating, $\log(y-y_1)=\log(y_1-y_2)+\log c$

or $\log(y-y_1)=\log c(y_1-y_2)$

$\therefore$ $y=y_1+c(y_1-y_2)$.

Now, $y=\alpha y_1+\beta y_2$ will be a solution if

$$\frac{d}{dx}(\alpha y_1+\beta y_2)+P(x)\cdot(\alpha y_1+\beta y_2)=Q(x)$$

or $\alpha\left\{\frac{dy_1}{dx}+P(x)y_1\right\}+\beta\left\{\frac{dy_2}{dx}+P(x)y_2\right\}=Q(x)$

or $\alpha\cdot Q(x)+\beta\cdot Q(x)=Q(x)$, using (2) and (3)

$\therefore$ $(\alpha+\beta)\,Q(x)=Q(x)$.

Hence $\alpha+\beta=1$.

19. Let $u(x)$ and $v(x)$ satisfy the differential equations $\frac{du}{dx}+p(x)u=f(x)$ and $\frac{dv}{dx}+p(x)v=g(x)$ respectively where $p(x)$, $f(x)$ and $g(x)$ are continuous functions. If $u(x_1)>v(x_1)$ for some x_1 and $f(x)>g(x)$ for all $x>x_1$, prove that any point (x, y), where $x>x_1$, does not satisfy the equations $y=u(x)$ and $y=v(x)$.

Here, $\frac{du(x)}{dx}+p(x)\,u(x)=f(x)$

and $\frac{dv(x)}{dx}+p(x)\,v(x)=g(x)$

$\therefore$ $$\frac{d\{u(x)-v(x)\}}{dx}+p(x)\{u(x)-v(x)\}=f(x)-g(x)$$

or $$\frac{d}{dx}\left[\{u(x)-v(x)\}\cdot e^{\int p(x)dx}\right]=\{f(x)-g(x)\}\,e^{\int p(x)dx}$$

($\because$ it was in linear form)

$\therefore$ $$\left[\{u(x)-v(x)\}\,e^{\int p(x)dx}\right]_{x_1}^{x}=\int_{x_1}^{x}\{f(x)-g(x)\}\,e^{\int p(x)dx}\,dx \quad \text{... (1)}$$

Let $\left(e^{\int p(x)dx}\right)_x=\lambda$, $\left(e^{\int p(x)dx}\right)_{x_1}=\mu$.

Clearly λ, μ are positive.

$\therefore$ $\{u(x)-v(x)\}\lambda-\{u(x_1)-v(x_1)\}\mu>0$

[$\because f(x)>g(x)$ for $x>x_1$ and $e^{\int p(x)dx}>0$]

$\therefore$ $\{u(x)-v(x)\}\lambda>\{u(x_1)-v(x_1)\}\mu>0$

because $u(x_1)>v(x_1)$

$\therefore$ $u(x)-v(x)>0$ for $x>x_1$

$\therefore$ $u(x)\neq v(x)$ when $x>x_1$. Hence the problem.

20. Solve $x\left(\frac{dy}{dx}\right)^2 + (y-x)\frac{dy}{dx} - y = 0.$

Here $xp^2 + (y-x)p - y = 0$ where $p = \frac{dy}{dx}$

or $\quad xp^2 - xp + yp - y = 0$

$\therefore \quad xp(p-1) + y(p-1) = 0$

or $\quad (p-1)(xp+y) = 0$

$\therefore \quad p - 1 = 0 \quad \text{or} \quad xp + y = 0.$

Now, $p - 1 = 0 \quad \Rightarrow \quad \frac{dy}{dx} = 1$

$\Rightarrow \quad y = x + c$

$xp + y = 0 \quad \Rightarrow \quad x\frac{dy}{dx} + y = 0$

$\Rightarrow \quad xdy + ydx = 0$

$\Rightarrow \quad d(xy) = 0$

$\Rightarrow \quad xy = c$

$\therefore$ the general solution is $(y - x - c)(xy - c) = 0$.

21. Solve $y = 2x\frac{dy}{dx} + \left(\frac{dy}{dx}\right)^2.$

Here, $y = 2xp + p^2$ where $p = \frac{dy}{dx}$.

It is in the form $y = f(x, p)$ {solvable for y}

Differentiating w.r.t. x,

$$p = 2p + 2x\frac{dp}{dx} + 2p\frac{dp}{dx}$$

$\therefore \quad 2(x+p)\frac{dp}{dx} + p = 0$

or $\quad p\frac{dx}{dp} + 2x + 2p = 0$

$\therefore \quad \frac{dx}{dp} + \frac{2}{p}x = -2 \qquad \ldots (1)$

The integrating factor

$$= e^{\int \frac{2}{p}dp} = e^{2\log p} = e^{\log p^2} = p^2.$$

Multiplying (1) by p^2,

$$p^2\frac{dx}{dp} + 2px = -2p^2$$

$\therefore \quad \frac{d}{dp}(p^2x) = -2p^2$

or $\quad d(p^2x) = -2p^2dp$

$\therefore \quad p^2x = \int -2p^2dp = -\frac{2p^3}{3} + \frac{c}{3}.$

Thus, we have $y = 2xp + p^2$

and $\ 3p^2x + 2p^3 = c$

$\therefore \quad 2py - c = 2p(2xp + p^2) - (3p^2x + 2p^3)$

$= 4xp^2 - 3p^2x$

$= xp^2 = x(y - 2xp)$

$\therefore \quad p(2y + 2x^2) = xy + c$

$\therefore \quad p = \frac{xy + c}{2y + 2x^2}.$

Putting this in $y = 2xp + p^2$

$$y = 2x \cdot \frac{xy+c}{2y+2x^2} + \left(\frac{xy+c}{2y+2x^2}\right)^2$$

$\therefore \quad 4y(y + x^2)^2 = 4x(xy + c)(y + x^2) + (xy + c)^2.$

22. Solve $1 + \left(\frac{dy}{dx}\right)^2 = x\frac{dy}{dx}.$

Here, $xp = 1 + p^2$ where $p = \frac{dy}{dx}$

$\therefore \quad x = \frac{1}{p} + p \qquad \ldots (1)$

which is of the form $x = f(y, p)$ {solvable for x}

Differentiating (1) w.r.t. y, $\frac{1}{p} = \left(\frac{-1}{p^2} + 1\right)\frac{dp}{dy}$

or $\quad dy = p\left(1 - \frac{1}{p^2}\right)dp$

$\therefore \quad y + c = \int\left(p - \frac{1}{p}\right)dp$

or $\quad y + c = \frac{p^2}{2} - \log p \qquad \ldots (2)$

$\therefore$ the p-eliminant, obtained by eliminating p from (1) and (2), is the general solution.

23. Solve $y = x\frac{dy}{dx} + \left(\frac{dy}{dx}\right)^2.$

Here, $y = xp + p^2$ where $p = \frac{dy}{dx}$;

this is of the form $y = xp + f(p)$.

Differentiating w.r.t. x,

$$p = p + x\frac{dp}{dx} + 2p\frac{dp}{dx}$$

$\therefore \quad (x + 2p)\frac{dp}{dx} = 0; \quad \therefore \quad x + 2p = 0 \ \text{ or } \ \frac{dp}{dx} = 0$

$\frac{dp}{dx} = 0 \quad \Rightarrow \quad p = c$

$\therefore \quad y = xp + p^2$ gives $y = cx + c^2$

$x + 2p = 0 \quad \Rightarrow \quad p = -\frac{x}{2}.$

But $y = xp + p^2$; $\therefore$ $y = x\left(-\frac{x}{2}\right) + \left(-\frac{x}{2}\right)^2$

$\therefore$ $4y = -2x^2 + x^2$; $\therefore$ $4y + x^2 = 0$.

$\therefore$ solutions are $y = cx + c^2$, a family of straight lines with a singular solution $4y + x^2 = 0$.

24. Use the methods of solving first order differential equation to find the general solution of

$$\frac{dy}{dx} \cdot \frac{d^3y}{dx^3} = 3\left(\frac{d^2y}{dx^2}\right)^2.$$

Put $\frac{dy}{dx} = p$. Then the equation is $p\frac{d^2p}{dx^2} = 3\left(\frac{dp}{dx}\right)^2$

$$\therefore \quad \frac{\frac{d^2p}{dx^2}}{\frac{dp}{dx}} = \frac{3}{p}\frac{dp}{dx}$$

or $\frac{1}{\frac{dp}{dx}} \cdot \frac{d}{dx}\left(\frac{dp}{dx}\right) = \frac{3}{p}\frac{dp}{dx}$; $\therefore$ $\frac{d}{dx}\left\{\log\frac{dp}{dx}\right\} = \frac{3}{p}\frac{dp}{dx}$

or $d\left\{\log\frac{dp}{dx}\right\} = \frac{3}{p}dp$; integrating, we get

$$\log\frac{dp}{dx} = \int\frac{3}{p}dp = 3\log p + \log a,$$

a being an arbitrary constant

$$\therefore \quad \frac{dp}{dx} = e^{3\log p + \log a} = e^{\log(ap^3)} = ap^3$$

or $\frac{dp}{p^3} = a\,dx$;

integrating, $-\frac{1}{2p^2} = ax + b$

or $\frac{-1}{2(ax+b)} = p^2$; $\therefore$ $p = \sqrt{-\frac{1}{2(ax+b)}}$

or $\frac{dy}{dx} = \sqrt{\frac{-1}{2(ax+b)}}$

$$\therefore \quad y = \int\frac{1}{\sqrt{-2ax-b}}dx$$

$$\therefore \quad y = \frac{\sqrt{-2ax-b}}{-a} + c.$$

25. The rate of cooling of a substance in moving air is proportional to the difference of temperatures of the substance and the air. A substance cools from 36°C to 34°C in 15 minutes. Find when the substance will have the temperature 32°C, it being known that the constant temperature of air is 30°C.

Let the temperature of the substance at time t minutes be T. The rate of cooling of the substance

$= k(T - 30)$°C/minute (from the question).

But the rate of cooling

$= $ rate of decrease of temperature $= -\frac{dT}{dt}$

$\therefore$ from the question, $-\frac{dT}{dt} = k(T - 30)$

or $-\frac{dT}{T-30} = k\,dt$

$$\therefore \quad \int_{36}^{34} -\frac{dT}{T-30} = \int_0^{15} k\,dt$$

or $[-\log(T-30)]_{36}^{34} = k[t]_0^{15}$

$\therefore$ $\log\frac{6}{4} = k \cdot 15$; $\therefore$ $k = \frac{1}{15}\log\frac{3}{2}$.

Again, $\int_{36}^{32} -\frac{dT}{T-30} = \int_0^t k\,dt$,

where t minutes is the required time.

$\therefore$ $[-\log(T-30)]_{36}^{32} = \frac{1}{15}\log\frac{3}{2} \cdot [t]_0^t$

$\therefore$ $\log 3 = \frac{1}{15}\log\frac{3}{2} \cdot t$

$\therefore$ $t = \frac{15\log 3}{\log\frac{3}{2}}$ minutes.

26. A and B are two separate reservoirs of water. The capacity of A is double that of B. Both the reservoirs are filled completely with water. Water is released simultaneously from both the reservoirs. For each of the reservoirs, the rate of flow out at any instant is proportional to the quantity of water left in the reservoir. After one hour, the quantity of water in A is 1.5 times the quantity of water in B. After how many hours from the time of release of water, do both A and B have the same quantity of water?

Let at time t hours, the volume of water in A and B be u and v respectively. From the question,

$$\frac{du}{dt} = -k_1u \quad \dots(1)$$

and $$\frac{dv}{dt} = -k_2v \quad \dots(2)$$

At $t = 0$, $u = 2V$ and $v = V$ (from the question).

Solving (1), $\frac{du}{u} = -k \cdot dt$; $\therefore$ $\log u = -k_1t + c_1$

$$\therefore \quad u = e^{-k_1t + c_1} \quad \dots(3)$$

When $t = 0$, $u = 2V$ $\Rightarrow$ $2V = e^{c_1}$

$\therefore$ (3) gives, $u = 2Ve^{-k_1 t}$ ... (4)

Similarly, from (2) we get $v = Ve^{-k_2 t}$ ... (5)

From the question, if $v = v_0$ when $t = 1$ then $u = \frac{3}{2} v_0$

$\therefore$ (4) $\Rightarrow$ $\frac{3}{2} v_0 = 2Ve^{-k_1}$

and (5) $\Rightarrow$ $v_0 = Ve^{-k_2}$

Dividing these, $\frac{3}{2} = \frac{2e^{-k_1}}{e^{-k_2}} = 2e^{k_2 - k_1}$

$\therefore$ $e^{k_2 - k_1} = \frac{3}{4}$ ... (6)

Let, after T hours, the volume of water in A and B be equal.

$\therefore$ (4) and (5) $\Rightarrow$ $2Ve^{-k_1 T} = Ve^{-k_2 T}$

$\therefore$ $2e^{-k_1 T} = e^{-k_2 T}$, i.e., $2e^{(k_2 - k_1)T} = 1$

Using (6), $2 \cdot \left(\frac{3}{4}\right)^T = 1$, i.e., $\left(\frac{4}{3}\right)^T = 2$

$\therefore$ $T \log \frac{4}{3} = \log 2$, i.e., $T = \dfrac{\log 2}{\log \frac{4}{3}} = \log_{4/3} 2$

$\therefore$ after $\log_{4/3} 2$ hours the reservoirs will have the same quantity of water.

27. A curve (or line) passes through (1, 1) such that the triangle formed by the coordinate axes and the tangent at any point of the curve is in the first quadrant and has its area equal to 2. Form the differential equation and find the equations of the possible curves.

Let the curve be $y = f(x)$.
The equation of the tangent at the point (x, y) of the curve is

$$Y - y = \frac{dy}{dx}(X - x) \qquad \text{... (1)}$$

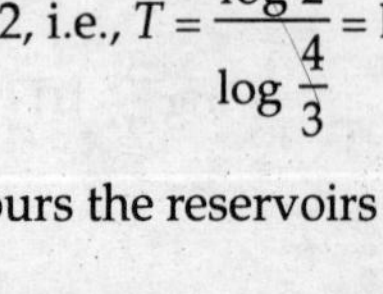

Solving (1) with $Y = 0$ and $X = 0$ successively, we get

$$-y = \frac{dy}{dx}(X - x), \text{ i.e., } X = \frac{x\frac{dy}{dx} - y}{\frac{dy}{dx}}, (= OA)$$

and $$Y - y = \frac{dy}{dx}(-x), \text{ i.e., } Y = y - x\frac{dy}{dx}, (= OB)$$

$\therefore$ the area of the triangle formed by the tangent and the axes in the first quadrant $= 2 = \frac{1}{2} OA \cdot OB$

or $$2 = \frac{1}{2} \cdot \frac{x\frac{dy}{dx} - y}{\frac{dy}{dx}} \cdot \left(y - x\frac{dy}{dx}\right)$$

or $$4\frac{dy}{dx} = \left(x\frac{dy}{dx} - y\right)\left(y - x\frac{dy}{dx}\right)$$

or $4p = (xp - y)(y - xp)$, taking $\frac{dy}{dx} = p$

$= -(y - xp)^2$; $\therefore$ $y - xp = \pm 2\sqrt{-p}$

or $y = xp \pm 2\sqrt{-p}$ ... (2)

Differentiating w.r.t. x,

$$p = p + x\frac{dp}{dx} \pm 2 \cdot \frac{1}{2\sqrt{-p}}(-1)\frac{dp}{dx}$$

or $$\frac{dp}{dx}\left\{x \mp \frac{1}{\sqrt{-p}}\right\} = 0$$

$\therefore$ $\frac{dp}{dx} = 0$ or $x = \pm \frac{1}{\sqrt{-p}}$.

Now, $\frac{dp}{dx} = 0$ $\Rightarrow$ $p = c$

$\therefore$ putting in (2), $y = cx \pm 2\sqrt{-c}$.

This gives a family of lines. If it passes through (1, 1) then $1 = c \pm 2\sqrt{-c}$, i.e., $(1 - c)^2 = -4c$

$\Rightarrow$ $(1 + c)^2 = 0$ $\Rightarrow$ $c = -1$

$\therefore$ the line is $y = -x \pm 2$, i.e., $x + y = \pm 2$.

But in the first quadrant x, y are positive. So $x + y = 2$.

Now, $x = \pm \frac{1}{\sqrt{-p}}$ $\Rightarrow$ $p = -\frac{1}{x^2}$. Putting this in (2),

$$y = x \cdot \left(-\frac{1}{x^2}\right) \pm 2\sqrt{\frac{1}{x^2}}$$

$\therefore$ $y = \frac{1}{x}, \frac{-3}{x}$, i.e., $xy = 1, -3$.

But in the first quadrant x, y are positive. So $xy = 1$. This is the singular solution of the equation (2) which is a curve and it passes through (1, 1).

$\therefore$ the required curves (or lines) are $x + y = 2$ and $xy = 1$.

28. A curve passing through (1, 2) has its slope at any point (x, y) equal to $\frac{2}{y - 2}$. Find the area of the region bounded by the curve and the line $2x - y - 4 = 0$.

Here, slope $= \frac{dy}{dx} = \frac{2}{y - 2}$

$\therefore \quad (y-2)dy = 2dx$

$\therefore \quad \int (y-2)dy = 2\int dx; \quad \therefore \quad \frac{y^2}{2} - 2y = 2x + c.$

It passes through (1, 2). So $\frac{4}{2} - 4 = 2 + c$, i.e., $c = -4$

$\therefore$ the equation of the curve is $\frac{y^2}{2} - 2y = 2x - 4$

or $y^2 - 4y = 4x - 8$, i.e., $y^2 = 4(x+y) - 8$.

To find the area, we have to draw a rough sketch of the curve.

When $y = 0$, $0 = 4(x+0) - 8$, i.e., $x = 2$

$\therefore$ the curve cuts the x-axis at A(2, 0) only.

When $x = 0$, $y^2 = 4(0+y) - 8$

or $y^2 - 4y + 8 = 0$; its roots are imaginary.

$\therefore$ the curve does not cut the y-axis.

Again, $y^2 = 4(x+y) - 8 \quad \Rightarrow \quad y^2 - 4y + 4 = 4x - 4$

or $(y-2)^2 = 4(x-1)$. So it is a parabola.

Its vertex C is (1, 2) and the axis is $y - 2 = 0$.

Now, the line is $2x - y - 4 = 0$, i.e., $\frac{x}{2} + \frac{y}{-4} = 1$.

Hence it passes through A(2, 0) and (0, −4).

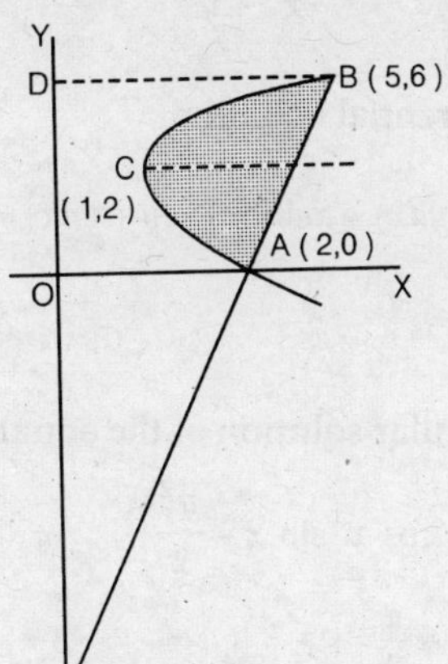

Solving $y^2 = 4(x+y) - 8$ and $2x - y - 4 = 0$, we get

$$y^2 = 2(y+4) + 4y - 8$$

$\therefore \quad y^2 - 6y = 0$, i.e., $y = 0, 6$.

When $y = 6$, $x = \frac{y+4}{2} = \frac{6+4}{2} = 5$. So B = (5, 6).

Clearly, the required area

$$= \text{ar(OABDO)} - \text{ar(BCAODB)} \qquad \dots (1)$$

Now, $\text{ar(OABDO)} = \int_0^6 (x)_{\text{line}}\, dy = \int_0^6 \frac{y+4}{2}\, dy$

$$= \frac{1}{2}\left[\frac{y^2}{2} + 4y\right]_0^6 = \frac{1}{2}[18 + 24] = 21$$

$$\text{ar(BCAODB)} = \int_0^6 (x)_{\text{curve}}\, dy = \int_0^6 \frac{y^2 - 4y + 8}{4}\, dy$$

$$= \left[\frac{y^3}{12} - \frac{y^2}{2} + 2y\right]_0^6 = 18 - 18 + 12 = 12$$

$\therefore$ from (1), the required area = 21 − 12 = 9.

29. Find the orthogonal trajectories of the circles $x^2 + y^2 - ay = 0$ where a is a parameter.

Here, $x^2 + y^2 - ay = 0$; differentiating w.r.t. x,

$$2x + 2y\frac{dy}{dx} - a\frac{dy}{dx} = 0$$

$$\therefore \quad 2x + 2y\frac{dy}{dx} - \frac{x^2+y^2}{y}\frac{dy}{dx} = 0$$

$$\left(\because \quad x^2 + y^2 - ay = 0 \quad \Rightarrow \quad a = \frac{x^2+y^2}{y}\right)$$

or $\quad 2x + \frac{y^2 - x^2}{y}\frac{dy}{dx} = 0.$

This is the differential equation of the circles.

The equation of orthogonal trajectories is

$$2x + \frac{y^2 - x^2}{y} \cdot \left(-\frac{dx}{dy}\right) = 0,$$

$$\left(\text{putting } \frac{-dx}{dy} \text{ in place of } \frac{dy}{dx}\right)$$

or $\quad 2xy\, dy + (x^2 - y^2)dx = 0.$

It is a homogeneous equation.

Put $y = vx$; then $\frac{dy}{dx} = v + x\frac{dv}{dx}$

$$\therefore \quad 2x \cdot vx \cdot \left(v + x\frac{dv}{dx}\right) + x^2 - v^2x^2 = 0$$

or $\quad 2v\left(v + x\frac{dv}{dx}\right) + 1 - v^2 = 0$

$$\therefore \quad 1 + v^2 + 2vx\frac{dv}{dx} = 0$$

or $\quad \frac{dx}{x} + \frac{2v}{1+v^2}dv = 0$; integrating, we get

$$\log x + \log(1 + v^2) = \log c$$

$$\therefore \quad x(1 + v^2) = c$$

or $\quad x\left(1 + \frac{y^2}{x^2}\right) = c$, i.e., $x^2 + y^2 = cx$.

Note If the orthogonal trajectories form the same family of curves as the given family of curves then the given system of curves is called *self-orthogonal*.

Exercises

1. Find the differential equation corresponding to the family of curves $y = c(x - c)^2$, where c is an arbitrary constant.
2. Obtain the differential equation of all conic sections $ax^2 + by^2 = 1$, where a, b are parameters.
3. Find the differential equation of the family of circles touching the x-axis at the origin.
4. Solve $x^2dy + y^2dx = xy(xdy - ydx)$.
5. Solve $\frac{dy}{dx}(1 + x^2)\tan^{-1}x + y = 0$.
6. Solve $\sec^2x \cdot \tan y dx + \sec^2y \cdot \tan x dy = 0$.
7. Solve $\frac{dy}{dx} + x\tan(x + y) + 1 = 0$.
8. Solve $(x^2 + y^2 + 2xy)\frac{dy}{dx} = a^2$.
9. Solve $x + y = \sin^{-1}\left(\frac{dy}{dx}\right)$.
10. Solve $\sqrt{x + y + 1} \cdot \frac{dy}{dx} = x + y - 1$.
11. Solve $ydx - xdy = \sqrt{x^2 - y^2}\, dx$.
12. Solve $y^2dx + x(x + y)dy = 0$.
13. Solve $\frac{x}{y} \cdot \frac{dy}{dx} = \frac{x - 2y}{x - 3y}$.
14. Solve $(1 + e^{x/y})\, dx + \left(1 - \frac{x}{y}\right)e^{x/y}\, dy = 0$.
15. Solve $\frac{dy}{dx} = \frac{y}{x} + \tan\frac{y}{x}$.
16. By making suitable substitutions, put the equation $\frac{dy}{dx} = \frac{2x^3 + 2xy^2 - 4x}{x^2y - 3y^3 + 2y}$ in the homogeneous form. Hence, solve it.
17. Solve $(4x - y + 3)dy + (2x - 3y - 1)dx = 0$.
18. Solve $\frac{dy}{dx} = \frac{3x - 4y + 2}{4x - 5y + 3}$.
19. Solve $(2x + 4y + 1)dy = (2y + x - 1)dx$.
20. Solve $x(1 + x^2)\frac{dy}{dx} = y(1 - x^2) + x^2\log x$.
21. Solve $(1 - x^2)^2\frac{dy}{dx} + y\sqrt{1 - x^2} = x + \sqrt{1 - x^2}$.
22. Solve $\frac{dy}{dx} = \frac{1 + y^2}{y(x + y^2 + 1)}$.
23. Solve $(1 + y^2) + (x - e^{-\tan^{-1}y})\frac{dy}{dx} = 0$.
24. If $y + \frac{d}{dx}(xy) = x(\sin x + \log x)$, find $y(x)$.
25. Solve $x(1 - x^2)dy + (2x^2y - y - 5x^3)dx = 0$.
26. Solve the equation $\cos^2x \cdot \frac{dy}{dx} - (\tan 2x)y = \cos^4x$, where $|x| < \frac{\pi}{4}$ and $y\left(\frac{\pi}{6}\right) = \frac{3\sqrt{3}}{8}$.
27. Solve $\frac{dy}{dx} + \frac{y}{\sqrt{x^2 + 4}} = 3x$ where $y(0) = 4$.
28. Solve $$\sin y\frac{dy}{dx} = \cos(x + y) + \cos(x - y) - \frac{1}{2}\sin x \cdot \sin 2x.$$
29. Solve $(1 + x^2)dy = y(x - y)dx$.
30. Solve $ydx + (x - 1)dy = x(x - 1)\sqrt[3]{y}\, dx$.
31. Solve $ydx + x(1 + xy)dy = 0$.
32. Solve $xdy = y(xy^3 + 1)dx$.
33. Solve $xdx + ydy + (x^2 + y^2)dy = 0$.
34. Solve $\dfrac{x + y\frac{dy}{dx}}{x\frac{dy}{dx} - y} = \sqrt{\dfrac{1 - x^2 - y^2}{x^2 + y^2}}$.
35. Solve the differential equation $$y\cos\frac{y}{x} \cdot (xdy - ydx) + x\sin\frac{y}{x}(xdy + ydx) = 0$$ when $y(1) = \frac{\pi}{2}$.
36. Find the particular solution of the equation $$y\frac{dy}{dx}\sin x = \cos x\left(\sin x - \frac{y^2}{2}\right)$$ if $y = 1$ when $x = \frac{\pi}{2}$.
37. Solve $\left(\frac{dy}{dx}\right)^2 - (e^x + e^{-x})\frac{dy}{dx} + 1 = 0$.
38. Solve $\left(\frac{dy}{dx}\right)^2 + 2x\frac{dy}{dx} = 3x^2$.
39. Solve $xy\left\{\left(\frac{dy}{dx}\right)^2 - 1\right\} = (x^2 - y^2)\frac{dy}{dx}$.
40. Solve $x\left\{\left(\frac{dy}{dx}\right)^2 + \frac{dy}{dx}\right\} = y$.
41. Solve $y = 2px + y^2p^3$ where $p = \frac{dy}{dx}$.

42. Prove that the general solution of the differential equation $y = (x+1)\frac{dy}{dx} - \left(\frac{dy}{dx}\right)^2$ form a family of straight lines and it has a solution which is a parabola.

43. Obtain all the solutions of $(y+1)\frac{dy}{dx} + 2 = x\left(\frac{dy}{dx}\right)^2$.

44. Using methods of solving first order equations, solve the differential equation $x\frac{d^2y}{dx^2} + 2\frac{dy}{dx} = 0$.

45. Reduce the differential equation $\frac{d^2y}{dx^2} - 2\frac{dy}{dx} + y = xe^x$, using the substitution $y = ve^x$ where v is a function of x. Hence, solve the equation when $y = 1$ and $\frac{dy}{dx} = 0$, for $x = 0$.

46. (a) A radioactive substance decays with time such that at any moment the rate of decay of volume is proportional to the volume at that time. The half-life of the substance is the time it takes for half the substance to disappear. Calculate the half-life of the substance if 20% of it disappears in 15 years.

(b) A country has a food deficit of 10%. Its population grows continuously at a rate of 3% per year. Its annual food production every year is 4% more than that of the previous year. Assuming that the average food requirement per person remains constant, prove that the country will become self-sufficient in food after n years where n is the smallest integer bigger than or equal to

$$\frac{\log 10 - \log 9}{\log (1.04) - 0.03}.$$

47. Show that the equation of the curve passing through (1, 0) and satisfying the differential equation $(1+y^2)dx - xy\,dy = 0$ will be $x^2 - y^2 = 1$.

48. Prove that the equation of a curve whose slope at (x, y) is $\frac{-(x+y)}{x}$ and which passes through the point (2, 1), is $x^2 + 2xy = 8$.

49. If the square of the intercept cut by any tangent on the y-axis is equal to the product of the coordinates of the point of contact then find the equation of such curves.

50. Find the curves for which the length of the normal is equal to the radius vector.

51. Find the cartesian equation of the curves for which the length of the tangent is of constant length.

52. Find the curves for which the portion of the tangent at any point intercepted between the axes of reference, is of constant length.

53. Find the equation of the curve for which the length of the normal is constant (a) and the curve passes through the point (1, 0).

54. Find the family of curves which are such that the area of the rectangle constructed on the abscissa of any point and the initial ordinate of the tangent at that point is a constant.

55. Find the curves for which the portion of the tangent included between the coordinate axes is bisected at the point of contact.

56. Find the curves for which the distances of the origin from the tangent and the normal at any point (x, y) are equal.

57. The normal at any point $P(x, y)$ of a curve meets the x-axis at G. If the distance of G from the origin be twice the abscissa of P, prove that the curve is a rectangular hyperbola.

58. Find the equation of curves for which the cartesian subtangent varies as the abscissa.

59. Find the equation of curves for which the cartesian subnormal is constant.

60. Find the orthogonal trajectory of the family of parabolas $y^2 = 4ax$ where a is a parameter.

61. Prove that the family of curves $\frac{x^2}{a^2+\lambda} + \frac{y^2}{b^2+\lambda} = 1$, where λ is a parameter, is self orthogonal.

Objective Questions

Fill in the blanks.

62. For the differential equation $\sqrt{1 + \frac{d^2y}{dx^2}} = x$, the degree = ______ and order = ______ .

63. If the general solution of a differential equation is $(y+c)^2 = cx$, where c is an arbitrary constant then the order of the differential equation is ______ .

64. The general solution of $\frac{dy}{dx} + \frac{y(x+y)}{x^2} = 0$ is ______ .

65. The particular solution of $(x + \log y)dy + y\,dx = 0$ where $y(0) = 1$, is ______ .

66. Let $\frac{dF(x)}{dx} = \frac{e^{\sin x}}{x}$, $x > 0$. If $\int_1^4 \frac{2e^{\sin x^2}}{x}\,dx = F(k) - F(1)$ then one of the possible values of k is ______ .

67. If $y(t)$ is a solution of $(1+t)\frac{dy}{dt} - ty = 1$ and $y(0) = -1$ then $y(1) =$ ______.

Answers

1. $\left(\frac{dy}{dx}\right)^3 = 4y\left(x\frac{dy}{dx} - 2y\right)$

2. $xy\frac{d^2y}{dx^2} + x\left(\frac{dy}{dx}\right)^2 = y\frac{dy}{dx}$

3. $(x^2 - y^2)\frac{dy}{dx} = 2xy$

4. $\log\frac{x}{y} = \frac{x+y}{xy} + c$

5. $x = \tan\frac{c}{y}$

6. $\tan x \cdot \tan y = c$

7. $\log\sin(x+y) + \frac{1}{2}x^2 = c$

8. $x + y = a\tan\frac{y+c}{a}$

9. $\tan(x+y) - \sec(x+y) = x + c$

10. $x + c = 2\sqrt{x+y+1} + \frac{2}{3}\log(\sqrt{x+y+1} - 1) - \frac{8}{3}\log(\sqrt{x+y+1} + 2)$

11. $\log x + \sin^{-1}\frac{y}{x} = c$

12. $kxy^2 = x + 2y$

13. $y^3 e^{x/y} = kx^2$

14. $x + ye^{x/y} = c$

15. $cx = \sin\frac{y}{x}$

16. Substitutions: $x^2 = X + 1, y^2 = Y + 1$;

$$\frac{3}{\sqrt{23}}\tan^{-1}\frac{6\sqrt{y^2-1} + \sqrt{x^2-1}}{\sqrt{23}\cdot\sqrt{x^2-1}} - \frac{1}{2}\log\{2x^2 + 3y^2 - 5 + \sqrt{(x^2-1)(y^2-1)}\} = c$$

17. $(x + y + 2)^5 = c(y - 2x - 1)^2$

18. $(y - x - 1)(5y - 3x - 1) = c$

19. $3\log(4x + 8y - 1) = 4(x - 2y + c)$

20. $y(1 + x^2) = x^2\log x - x^2 + cx$

21. $y = \frac{x}{\sqrt{1-x^2}} + ce^{-x/\sqrt{1+x^2}}$

22. $x = 1 + y^2 + c\sqrt{1+y^2}$

23. $xe^{\tan^{-1}y} = c + \tan^{-1}y$

24. $\left(\frac{2}{x^2} - 1\right)\cos x + \frac{2\sin x}{x} + \frac{1}{3}x\left(\log x - \frac{1}{3}\right) + \frac{c}{x^2}$

25. $y = 5x + cx\sqrt{1-x^2}$

26. $y\cos 2x = \sin x \cdot \cos^3 x$

27. $y(x + \sqrt{x^2+4}) = x^3 + (x^2+4)^{3/2}$

28. $\cos y = \frac{1}{2}\sin^2 x - \frac{1}{2}\sin x + \frac{1}{4} + ce^{-2\sin x}$

29. $\sqrt{1+x^2} = y\log(x + \sqrt{1+x^2}) + cy$

30. $y^{2/3} = \frac{1}{4}(x-1)^2 + \frac{2}{5}(x-1) + c(x-1)^{-2/3}$

31. $\log y = \frac{1}{xy} + c$

32. $3x^4 + \frac{4x^3}{y^3} = c$

33. $2y + \log(x^2 + y^2) = c$

34. $\sin^{-1}(2x^2 + 2y^2 - 1) = 2\tan^{-1}\frac{y}{x} + c$

35. $2xy\sin\frac{y}{x} = \pi$

36. $y^2 = \sin x$

37. $(y - e^x + c)(y + e^{-x} + c) = 0$

38. $(2y - x^2 - c)(2y + 3x^2 - c) = 0$

39. $(xy - c)(x^2 - y^2 - c) = 0$

40. p eliminant of $xp^2 = ce^{1/p}$ and $y = x(p + p^2)$

41. $y^2 = 2cx + c^3$

43. $(y+1)c + 2 = c^2x, (y+1)^2 + 8x = 0$

44. $xy + c = kx$

45. $y = e^x\left(\frac{x^3}{6} - x + 1\right)$

46. nearly 46.6 years

49. $x = ce^{\pm 2\sqrt{y/x}}$

50. $x^2 \pm y^2 = a^2$

51. $x = a\log\frac{a - \sqrt{a^2-y^2}}{y} + \sqrt{a^2-y^2} + c$

52. $x^{2/3} + y^{2/3} = a^{2/3}$

53. $y^2 + (x - 1 \pm a)^2 = a^2$

54. $y = cx \pm \frac{a^2}{2x}$

55. $xy = c^2$

56. $x^2 + y^2 = ce^{-2\tan^{-1}\frac{y}{x}}$

58. $cx = y^k$

59. $y^2 = 2ax + c$

60. $2x^2 + y^2 = c^2$

62. 1, 2 respectively

63. one

64. $yx^2 = c(y + 2x)$

65. $y(x - 1 + \log y) + 1 = 0$

66. 16

67. $-\frac{1}{2}$

□

Chapter Test

Time: 120 minutes

1. In each of the following, fill in the blank so that the resulting sentence becomes true.

   (a) The differential equation is _____ whose general solution is $y = Ae^x + Be^{-x}$ where A, B are arbitrary constants.

   (b) If $\phi(x) = \int_a^x \{\phi(x)\}^{-2} dx$ and $\phi(1) = 0$ then $\phi(x) =$ _____ .

2. In each of the following, alternative answers are given of which one or more are correct. Choose the correct answer(s).

   (a) If $x\dfrac{dy}{dx} + y = x \cdot \dfrac{\phi(xy)}{\phi'(xy)}$ then $\phi(xy)$ is equal to (where k is an arbitrary constant)

   A. $ke^{x^2/2}$ B. $ke^{y^2/2}$ C. $ke^{xy/2}$ D. none of these

   (b) The differential equation of the family of curves $cy^2 = 2x + c$ where c is an arbitrary constant is

   A. $y\dfrac{dy}{dx} = 1$ B. $\left(\dfrac{dy}{dx}\right)^2 + y\dfrac{d^2y}{dx^2} = 0$ C. $y^2 = 2xy\dfrac{dy}{dx} + 1$ D. none of these

3. Express the differential equation $\phi(x)dy = y\{\phi'(x) - y\}dx$ in the form $dF(x, y) = 0$. Hence or otherwise solve it.

4. Solve $\dfrac{dy}{dx} = 1 - x(y - x) - x^3(y - x)^2$.

5. Using solution of first order differential equation, find the general solution of the equation $\dfrac{d^2y}{dx^2} = \sqrt{1 + \left(\dfrac{dy}{dx}\right)^2}$.

6. There are two poles A and B on a straight road. A man starts running from A towards B with a variable velocity. At any instant his speed is found to be inversely proportional to his distance from the pole B. If he covers one-third of the distance between the poles in 10 minutes, after what time will he be halfway between the poles?

7. Find the area bounded by the curves passing through the point (1, 1) whose differential equations are $xdy - 2ydx = 0$ and $y = 2\left|x\dfrac{dy}{dx} - 1\right| + 3\sqrt{2}$.

8. x and y are two variables. At any instant the rate of increase of x is proportional to the square root of the value of x at that instant and that of decrease of y is proportional to the square of the value of y. If at $t = 0$, $x = 4$ and $y = 16$ while at $t = 4$, $x = 9$ and $y = 10$ then obtain the functional relation between x and y.

9. Find the curve passing through the point $(0, k)$ for which the perpendicular from the foot of the ordinate of the point of contact to the tangent is of the constant length k.

10. The equations of motion of a particle in a plane are given by $\dfrac{dx}{dt} = t(t + 1)$ and $(1 + t)\dfrac{dy}{dt} = 1$ where the particle is at the point (x, y) at time t. If the particle is at the origin at time $t = 0$ then obtain the cartesian equation of the path of the particle.

11. Solve $(1 - xy - x^5y^5)dx = x^2(x^4y^4 + 1)dy$.

12. If $f(x)$ is a function such that $x\int_0^x (1 - t)f(t)dt = \int_0^x t \cdot f(t)dt$ and $f(1) = 1$ then find $f(x)$.

13. Let $f(x)$ be a function of x for which $f(0) = 0$ and $\int_0^{f(x)} t \cdot \cos t \, dt = \int_0^{x^2} \sqrt{z} \cdot \sin \sqrt{z} \, dz$.

Prove that $f(x) \cdot \sin f(x) + \cos f(x) + 3 = -2x^2 \cos x + 4x \cos x + 4 \cos x$.

14. If $f(x)$ be a twice differentiable function from $R \to R$ such that the equation $\lambda^2 f(x) - 2\lambda f'(x) + f''(x) = 0$ provides two equal values of λ for all x and $f(0) = 1, f'(0) = 2$ then find $f(x)$.

15. Find the equation to the curve $y = f(x)$ such that $\frac{d^2y}{dx^2} = \sqrt{1 - \left(\frac{dy}{dx}\right)^2}$ and the tangent to the curve at the origin is inclined at $\frac{\pi}{4}$ with the positive direction of the x-axis.

□

Answers

1. (a) $\frac{d^2y}{dx^2} - y = 0$ (b) $\{3(x-1)\}^{1/3}$

2. (a) A (b) C

3. $y(x + c) = \phi(x)$

4. $1 + (x^2 + 2)(y - x) = c(y - x)e^{x^2/2}$

5. $2y = e^{x+c} + e^{-x-c} + k$

6. 13.5 minutes

7. $\frac{5}{3} + \sqrt{2}$

8. $(3\sqrt{x} - 1)y = 80$

9. $e^{2x/k} - \frac{2y}{k} e^{x/k} + 1 = 0$

10. $6x = (2e^y + 1)(e^y - 1)^2$

11. $x = ce^{xy + \frac{1}{5}x^5 y^5}$

12. $\frac{1}{x^3} \cdot e^{1 - \frac{1}{x}}$

14. e^{2x}

15. $y + \sin x = 0$

□

Vectors

1. Vectors and Addition of Vectors

Recap of Facts and Formulae

1. Vectors and their representation

- Vector quantities are specified by definite magnitude and definite direction.
- Vector quantities are represented by directed line segments.

  Vector AB, i.e. $\vec{AB}$ is a vector whose magnitude is represented by the length AB and its direction is represented by the direction from A to B along the line segment.

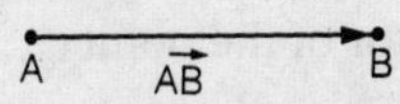

- The magnitude of the vector $\vec{AB}$ is written as $|\vec{AB}|$ or simply AB.
- Unit vector $\vec{a}$ is a vector whose magnitude $|\vec{a}| = 1$. Zero vector is a vector of 0 magnitude. The direction of $\vec{0}$ (zero vector) is indeterminate.
- If $AB = CD$ and AB || CD then $\vec{AB} = \vec{CD}$.
- $\vec{AB} = \vec{CD} \Rightarrow AB = CD$ and AB || CD.

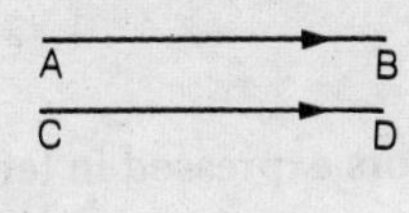

2. Addition of vectors

- If $\vec{AB} = \vec{a}$, $\vec{BC} = \vec{b}$ then $\vec{a} + \vec{b}$ is the vector represented by $\vec{AC}$.

  $\therefore \quad \vec{AB} + \vec{BC} = \vec{AC}$.

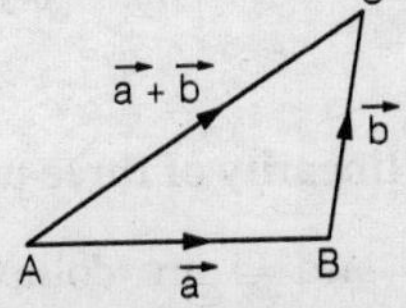

- If A, B, C, D and E are in order then

  $\vec{AB} + \vec{BC} + \vec{CD} + \vec{DE} = \vec{AE}$.
- $\vec{AB} + \vec{AC} = \vec{AB} + \vec{BD} = \vec{AD} = 2\vec{AE}$

  where ABDC is a parallelogram, E being the point of intersection of the diagonals.

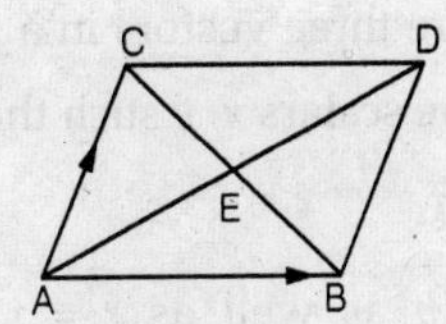

- $\vec{a} + \vec{0} = \vec{a}$.
- $-\vec{a}$ is the negative of the vector $\vec{a}$ such that $\vec{a} + (-\vec{a}) = 0$.

3. Scalar multiplication of a vector by a scalar

- If k is a scalar then $k\vec{a}$ is a vector having the same (or opposite) direction as that of $\vec{a}$ whose magnitude is $|k|$ times that of $\vec{a}$.

  $\therefore \quad k\vec{a} \parallel \vec{a} \quad$ and $\quad |k\vec{a}| = |k|\,|\vec{a}|$.
- $k(\vec{a} + \vec{b}) = k\vec{a} + k\vec{b}$.
- $p(q\vec{a}) = (pq)\vec{a} = q(p\vec{a})$.
- Unit vector along $\vec{a} = \hat{a} = \dfrac{\vec{a}}{|\vec{a}|}$.

4. Position vector of a point

- If O is the origin and P is any point then the position vector of P = $\vec{OP}$.

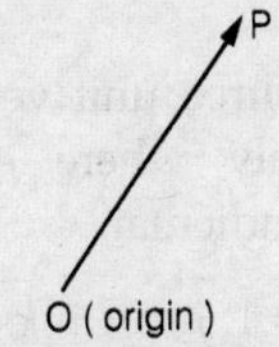

- If the position vectors of points A and B are $\vec{a}$ and $\vec{b}$ respectively then the position vector of the point P dividing AB in the ratio $m : n$ is $\dfrac{m\vec{b} + n\vec{a}}{m + n}$.

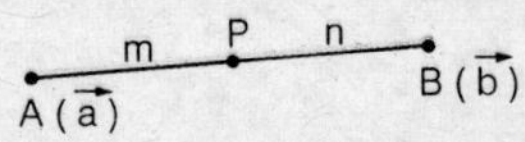

5. Relation between two collinear vectors

- If $\vec{a}$ and $\vec{b}$ are two collinear (or parallel) vectors then there exists a scalar λ such that $\vec{b} = \lambda \vec{a}$.

6. Relation between three coplanar vectors

- If $\vec{a}$, $\vec{b}$ and $\vec{r}$ are three vectors in a plane then there exists two unique scalars x, y such that $\vec{r} = x\vec{a} + y\vec{b}$.
- If $\vec{r} = x_1\vec{a} + y_1\vec{b}$ as well as $\vec{r} = x_2\vec{a} + y_2\vec{b}$ then $x_1 = x_2$, $y_1 = y_2$.

7. Relation between four vectors in space

- If $\vec{a}$, $\vec{b}$, $\vec{c}$ and $\vec{r}$ are four vectors in space (of which no three are coplanar) then there exists three unique scalars x, y, z such that $\vec{r} = x\vec{a} + y\vec{b} + z\vec{c}$.
- If $\vec{r} = x_1\vec{a} + y_1\vec{b} + z_1\vec{c}$ as well as $\vec{r} = x_2\vec{a} + y_2\vec{b} + z_2\vec{c}$ then $x_1 = x_2$, $y_1 = y_2$, $z_1 = z_2$.

8. Vectors $\vec{i}$, $\vec{j}$, $\vec{k}$

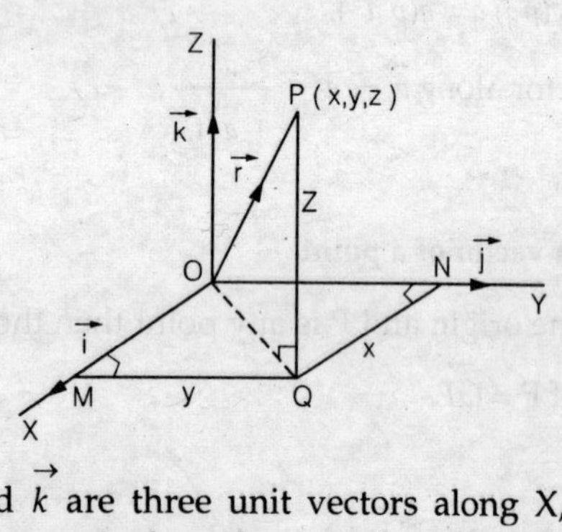

$\vec{i}$, $\vec{j}$ and $\vec{k}$ are three unit vectors along X, Y and Z axes respectively where each pair of axes are mutually perpendicular.

$$\therefore \quad |\vec{i}| = 1 = |\vec{j}| = |\vec{k}|.$$

Let P be a point whose coordinates are (x, y, z). The position vector of P is $\vec{r}$ where

$$\vec{r} = \vec{OQ} + \vec{QP} = (\vec{OM} + \vec{ON}) + \vec{QP}$$

$$= x\vec{i} + y\vec{j} + z\vec{k} = (x, y, z)$$

$$|\vec{r}| = |\vec{OP}| = \sqrt{OQ^2 + QP^2}$$

$$= \sqrt{(QN^2 + ON^2) + QP^2}$$

$$= \sqrt{x^2 + y^2 + z^2}$$

$$\cos \angle POZ = \cos(90° - \angle POQ)$$

$$= \sin \angle POQ = \frac{z}{OP} = \frac{z}{|\vec{r}|}.$$

Similarly, $\cos \angle POX = \dfrac{x}{|\vec{r}|}$, $\cos \angle POY = \dfrac{y}{|\vec{r}|}$.

The direction of $\vec{r}$ is given by the direction cosines

$$\frac{x}{|\vec{r}|}, \frac{y}{|\vec{r}|}, \frac{z}{|\vec{r}|}.$$

For example The magnitude of the vector $\vec{i} + 2\vec{j} - 3\vec{k}$, i.e.,

$$|\vec{i} + 2\vec{j} - 3\vec{k}| = \sqrt{1^2 + 2^2 + (-3)^2} = \sqrt{14}.$$

The direction of the vector $\vec{i} + 2\vec{j} - 3\vec{k}$ is given by the direction cosines

$$\frac{1}{\sqrt{14}}, \frac{2}{\sqrt{14}}, \frac{-3}{\sqrt{14}}.$$

- The position vector of the point (x, y, z) is $\vec{r} = x\vec{i} + y\vec{j} + z\vec{k}$.
- Any vector $\vec{a}$ in space can be written linearly in $\vec{i}$, $\vec{j}$, $\vec{k}$, i.e., $\vec{a} = \lambda\vec{i} + \mu\vec{j} + \nu\vec{k}$.
- If $\vec{r} = x\vec{i} + y\vec{j} + z\vec{k}$ is equally inclined with axes then

$$\frac{x}{|\vec{r}|} = \frac{y}{|\vec{r}|} = \frac{z}{|\vec{r}|}, \text{ i.e., } x = y = z.$$

9. Addition of vectors expressed in terms of $\vec{i}$, $\vec{j}$, $\vec{k}$

If $\vec{a} = a_1\vec{i} + a_2\vec{j} + a_3\vec{k}$ and $\vec{b} = b_1\vec{i} + b_2\vec{j} + b_3\vec{k}$ then

- $\vec{a} + \vec{b} = (a_1 + b_1)\vec{i} + (a_2 + b_2)\vec{j} + (a_3 + b_3)\vec{k}$

  $\vec{a} - \vec{b} = (a_1 - b_1)\vec{i} + (a_2 - b_2)\vec{j} + (a_3 - b_3)\vec{k}$.

10. Condition of collinearity of three points

- The points A, B and C are collinear if $\vec{AB} = \lambda \vec{AC}$ where λ is a suitable scalar.

11. Condition of coplanarity of four points or three vectors

- The points A, B, C and D are coplanar if the three vectors $\vec{AB}$, $\vec{AC}$ and $\vec{AD}$ are coplanar, i.e., if two scalars x, y can be found such that $\vec{AB} = x\vec{AC} + y\vec{AD}$.

Selected Worked Examples

1. G is the centroid of the triangle ABC and if G_1 is the centroid of another triangle $A_1B_1C_1$ then prove that $\vec{AA_1} + \vec{BB_1} + \vec{CC_1} = 3\vec{GG_1}$.

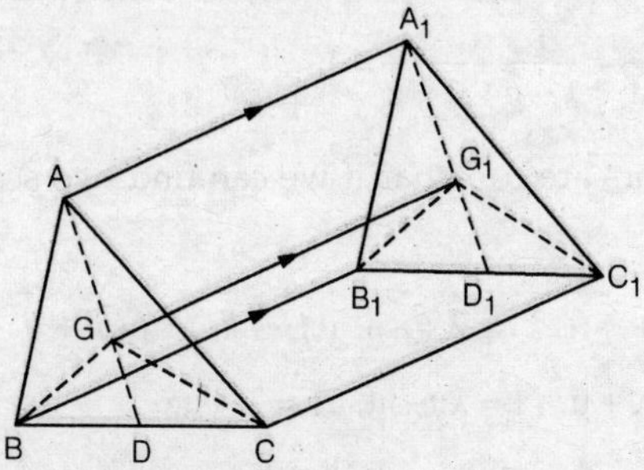

Clearly, $\vec{AA_1} = \vec{AG} + \vec{GG_1} + \vec{G_1A_1}$

$\vec{BB_1} = \vec{BG} + \vec{GG_1} + \vec{G_1B_1}$

$\vec{CC_1} = \vec{CG} + \vec{GG_1} + \vec{G_1C_1}$.

Adding these, $\vec{AA_1} + \vec{BB_1} + \vec{CC_1}$

$= 3\,\vec{GG_1} + (\vec{AG} + \vec{BG} + \vec{CG}) + (\vec{G_1A_1} + \vec{G_1B_1} + \vec{G_1C_1})$

$= 3\,\vec{GG_1} + (\vec{AG} + 2\,\vec{DG}) + (\vec{G_1A_1} + 2\,\vec{G_1D_1})$

(where AD and A_1D_1 are medians of the triangles ABC and $A_1B_1C_1$ respectively)

$= 3\,\vec{GG_1} + (\vec{AG} + \vec{GA}) + (\vec{G_1A_1} + \vec{A_1G_1})$

$= 3\,\vec{GG_1} + \vec{0} + \vec{0} = 3\,\vec{GG_1}$.

2. Prove that the vectors $3\vec{i} + 5\vec{j} + 2\vec{k}$, $2\vec{i} - 3\vec{j} - 5\vec{k}$ and $5\vec{i} + 2\vec{j} - 3\vec{k}$ form the sides of an equilateral triangle.

Let $\vec{a} = 3\vec{i} + 5\vec{j} + 2\vec{k}$, $\quad \vec{b} = 2\vec{i} - 3\vec{j} - 5\vec{k}$

and $\vec{c} = 5\vec{i} + 2\vec{j} - 3\vec{k}$.

Clearly, $\vec{a} + \vec{b} = (3+2)\vec{i} + (5-3)\vec{j} + (2-5)\vec{k}$

$= 5\vec{i} + 2\vec{j} - 3\vec{k} = \vec{c}$

$\therefore$ the vectors $\vec{a}, \vec{b}, \vec{c}$ are represented by three sides of a triangle.

Now $|\vec{a}| = |3\vec{i} + 5\vec{j} + 2\vec{k}|$

$= \sqrt{3^2 + 5^2 + 2^2} = \sqrt{38}$

$|\vec{b}| = |2\vec{i} - 3\vec{j} - 5\vec{k}|$

$= \sqrt{2^2 + (-3)^2 + (-5)^2} = \sqrt{38}$

$|\vec{c}| = |5\vec{i} + 2\vec{j} - 3\vec{k}|$

$= \sqrt{5^2 + 2^2 + (-3)^2} = \sqrt{38}$

$\therefore \quad |\vec{a}| = |\vec{b}| = |\vec{c}|$;

so, the vectors form the sides of an equilateral $\triangle$.

3. Show that the points A(–1, 6, 6), B(– 4, 9, 6) and C(0, 7, 10) form an isosceles right-angled triangle.

Let O be the origin. Then we have

$\vec{OA} = -\vec{i} + 6\vec{j} + 6\vec{k}$,

$\vec{OB} = -4\vec{i} + 9\vec{j} + 6\vec{k}$,

$\vec{OC} = 7\vec{j} + 10\vec{k}$

$\therefore \quad \vec{AB} = \vec{AO} + \vec{OB} = \vec{OB} - \vec{OA}$

$= (-4\vec{i} + 9\vec{j} + 6\vec{k}) - (-\vec{i} + 6\vec{j} + 6\vec{k})$

$= -3\vec{i} + 3\vec{j}$

$\vec{AC} = \vec{AO} + \vec{OC} = \vec{OC} - \vec{OA}$

$= (7\vec{j} + 10\vec{k}) - (-\vec{i} + 6\vec{j} + 6\vec{k})$

$= \vec{i} + \vec{j} + 4\vec{k}$

$\vec{BC} = \vec{BO} + \vec{OC} = \vec{OC} - \vec{OB}$

$= 7\vec{j} + 10\vec{k} - (-4\vec{i} + 9\vec{j} + 6\vec{k})$

$= 4\vec{i} - 2\vec{j} + 4\vec{k}$

$\therefore \quad AB = |\vec{AB}| = |-3\vec{i} + 3\vec{j}|$

$= \sqrt{(-3)^2 + 3^2} = 3\sqrt{2}$

$AC = |\vec{AC}| = |\vec{i} + \vec{j} + 4\vec{k}|$

$= \sqrt{1^2 + 1^2 + 4^2} = 3\sqrt{2}$

$BC = |\vec{BC}| = |4\vec{i} - 2\vec{j} + 4\vec{k}|$

$= \sqrt{4^2 + (-2)^2 + 4^2} = 6$

$\because \quad AB = AC = 3\sqrt{2}$, the triangle is isosceles.

$\because \quad AB^2 + AC^2 = (3\sqrt{2})^2 + (3\sqrt{2})^2 = 36 = BC^2$,

the triangle is right angled at A.

4. Given three points whose position vectors are $x\vec{i} + y\vec{j} + z\vec{k}$, $\vec{i} + z\vec{j}$ and $-\vec{i} - \vec{j}$. Find the condition for the points to be collinear.

Let the points be A, B and C respectively. Then

$\vec{OA} = x\vec{i} + y\vec{j} + z\vec{k}$, $\vec{OB} = \vec{i} + z\vec{j}$,

$\vec{OC} = -\vec{i} - \vec{j}$.

Now, $\vec{AB} = \vec{AO} + \vec{OB} = \vec{OB} - \vec{OA}$

$= (\vec{i} + z\vec{j}) - (x\vec{i} + y\vec{j} + z\vec{k})$

$= (1-x)\vec{i} + (z-y)\vec{j} - z\vec{k}$

$$\vec{AC} = \vec{AO} + \vec{OC} = \vec{OC} - \vec{OA}$$

$$= -\vec{i} - \vec{j} - (x\vec{i} + y\vec{j} + z\vec{k})$$

$$= -(1+x)\vec{i} - (1+y)\vec{j} - z\vec{k}$$

A, B, C are collinear if

$$\vec{AB} = \lambda\,\vec{AC}$$

or $(1-x)\vec{i} + (z-y)\vec{j} - z\vec{k}$

$$= \lambda\{-(1+x)\vec{i} - (1+y)\vec{j} - z\vec{k}\}$$

$\Rightarrow$ $1 - x = -\lambda(1+x)$ ... (1)

$z - y = -\lambda(1+y)$ ... (2)

$-z = -\lambda z$ ... (3)

Clearly $\lambda \neq 1$, for otherwise B and C are coincident.

$\therefore$ (3) $\Rightarrow$ $z = 0$

$\therefore$ (2) $\Rightarrow$ $-y = -\lambda(1+y)$

or $\lambda = \dfrac{y}{1+y}$

and (1) $\Rightarrow$ $\lambda = \dfrac{x-1}{x+1}$

$\therefore$ $\dfrac{y}{1+y} = \dfrac{x-1}{x+1}$

or $xy + y = x - 1 + xy - y$

or $2y = x - 1$.

Hence, the conditions are $z = 0$ and $x = 1 + 2y$.

5. If the vectors

$$x\vec{i} + \vec{j} + \vec{k},\ \vec{i} + y\vec{j} + \vec{k} \text{ and } \vec{i} + \vec{j} + z\vec{k}$$

are coplanar where $x \neq 1, y \neq 1, z \neq 1$ then prove that

$$\frac{1}{1-x} + \frac{1}{1-y} + \frac{1}{1-z} = 1.$$

The vectors are coplanar if we can find two scalars λ, μ such that

$$x\vec{i} + \vec{j} + \vec{k} = \lambda(\vec{i} + y\vec{j} + \vec{k}) + \mu(\vec{i} + \vec{j} + z\vec{k})$$

$\Rightarrow$ $x = \lambda + \mu,\ 1 = \lambda y + \mu,\ 1 = \lambda + \mu z$

$\therefore$ $1 - x = 1 - \lambda - \mu,\ y = \dfrac{1-\mu}{\lambda},\ z = \dfrac{1-\lambda}{\mu}$

$\therefore$ $1 - y = \dfrac{\lambda - 1 + \mu}{\lambda}$ and $1 - z = \dfrac{\mu - 1 + \lambda}{\mu}$

$\therefore$ $\dfrac{1}{1-x} + \dfrac{1}{1-y} + \dfrac{1}{1-z}$

$$= \frac{1}{1-\lambda-\mu} + \frac{\lambda}{\lambda+\mu-1} + \frac{\mu}{\lambda+\mu-1}$$

$$= \frac{1-\lambda-\mu}{1-\lambda-\mu} = 1.$$

Exercises

1. If O is the circumcentre and O′ the orthocentre of ΔABC then prove that
   (i) $\vec{OA} + \vec{OB} + \vec{OC} = \vec{OO'}$
   (ii) $\vec{O'A} + \vec{O'B} + \vec{O'C} = 2\,\vec{O'O}$
   (iii) $\vec{OA}\sin 2A + \vec{OB}\sin 2B + \vec{OC}\sin 2C = \vec{0}$.
2. If ABCDEF is a regular hexagon then prove that
   (i) $\vec{AC} + \vec{AD} + \vec{EA} + \vec{FA} = 3\,\vec{AB}$
   (ii) $\vec{AB} + \vec{AC} + \vec{AD} + \vec{AE} + \vec{AF} = 6\,\vec{AO}$, where O is the centre of the hexagon.
3. Prove that the points (1, 2, 3), (–1, –1, 8) and (–4, 4, 6) form the vertices of an equilateral triangle.
4. Prove that the vectors $4(\vec{i} + \vec{j} + \vec{k})$, $7\vec{i} + 6\vec{j} - \vec{k}$ and $3\vec{i} + 2\vec{j} - 5\vec{k}$ form a right-angled triangle.
5. The position vectors of A, B, C and D are respectively $2\vec{i} + 3\vec{j} + 5\vec{k}$, $\vec{i} + 2\vec{j} + 3\vec{k}$, $-5\vec{i} + 4\vec{j} - 2\vec{k}$ and $\vec{i} + 10\vec{j} + 10\vec{k}$. Show that AB and CD are parallel.
6. Find all $\lambda \in R$ such that $(x, y, z) \neq (0, 0, 0)$ and
   $$(\vec{i} + \vec{j} + 3\vec{k})x + (3\vec{i} - 3\vec{j} + \vec{k})y + (-4\vec{i} + 5\vec{j})z = \lambda(x\vec{i} + y\vec{j} + z\vec{k}).$$
7. What is the unit vector $\hat{a}$ parallel to the vector $\vec{a} = 3\vec{i} + 4\vec{j} - 2\vec{k}$? Find the vector $\vec{b}$ such that $\hat{a} + \vec{b}$ becomes the unit vector $\vec{i}$?
8. Vectors $\vec{a}, \vec{b}$ are noncollinear. Find x so that the vectors $(x-1)\vec{a} + \vec{b}$ and $(2+3x)\vec{a} - 2\vec{b}$ are collinear.
9. If $\vec{a}, \vec{b}, \vec{c}$ are three noncoplanar vectors then prove that the following points are collinear.
   $$\vec{a} - 2\vec{b} + 3\vec{c},\ 2\vec{a} + 3\vec{b} - 4\vec{c} \text{ and } -7\vec{b} + 10\vec{c}$$
10. The position vectors of the points A, B and C are $\vec{a}, \vec{b}, \vec{c}$ respectively. If $3\vec{a} + 2\vec{c} = 5\vec{b}$, are the points A, B and C collinear? If so find $AB : BC$.

11. Find the value of a such that the vectors $2\vec{i} - \vec{j} + \vec{k}$, $\vec{i} + 2\vec{j} - 3\vec{k}$ and $3\vec{i} + a\vec{j} + 5\vec{k}$ are coplanar.

12. If $\vec{a}, \vec{b}, \vec{c}$ are three noncoplanar vectors then show that the following points are coplanar.

$6\vec{a} - 4\vec{b} + 10\vec{c}$, $-5\vec{a} + 3\vec{b} - 10\vec{c}$,

$4\vec{a} - 6\vec{b} - 10\vec{c}$ and $2\vec{b} + 10\vec{c}$

13. If $\vec{a} = (1, -2, 3)$, $\vec{b} = (-2, 3, -4)$, $\vec{c} = (1, -1, \lambda)$ then for what value of $\lambda \in R$ do the vectors form a linearly dependent system?

14. Prove that the points

$(4, 5, 1), (0, -1, -1), (3, 9, 4)$ and $(-4, 4, 4)$

are coplanar.

15. If the position vectors of the points A and B be $2\vec{i} - \vec{j} + 2\vec{k}$ and $4\vec{i} - 3\vec{k}$ respectively then find the unit vectors along the bisectors of the angle AOB, O being the origin.

16. Vectors $\vec{a}, \vec{b}, \vec{c}$ are such that every pair is noncollinear, and the vector $\vec{a} + \vec{b}$ is collinear with $\vec{c}$ and the vector $\vec{b} + \vec{c}$ is collinear with $\vec{a}$.

Prove that $\vec{a} + \vec{b} + \vec{c} = \vec{0}$.

Answers

6. 0, −1

7. $\frac{1}{\sqrt{29}}(3\vec{i} + 4\vec{j} - 2\vec{k})$, $\vec{i} - \frac{1}{\sqrt{29}}(3\vec{i} + 4\vec{j} - 2\vec{k})$

8. 0

10. yes; 2 : 3

11. − 4

13. 1

15. $\frac{1}{\sqrt{510}}(22\vec{i} - 5\vec{j} + \vec{k})$, $\frac{1}{\sqrt{390}}(-2\vec{i} - 5\vec{j} + 19\vec{k})$

□

Chapter Test

Time: 35 minutes

1. In each of the following, fill in the blank so that the resulting statement becomes correct.

(a) If $\vec{a}$ and $\vec{b}$ are noncollinear then vectors $\vec{p} = (x-2)\vec{a} + \vec{b}$ and $\vec{q} = (2x+1)\vec{a} - \vec{b}$ are collinear when the value of x is ______ .

(b) A vector of magnitude 3 equally inclined to the axes is ______ .

(c) If AB || CD where A = (1, 1, 0), B = (–λ, 0, 1) and C = (λ, 1, 0), D = (11, λ, –2) then the set of possible values of λ is ______ .

(d) The points where position vectors are $60\vec{i} + 3\vec{j}$, $40\vec{i} - 8\vec{j}$ and $a\vec{i} - 52\vec{j}$, are collinear if a = ______ .

(e) A vector $\vec{a}$ has components $2p$ and 1 with respect to a rectangular cartesian system. The axes are rotated through a certain angle about the origin in the counterclockwise sense. If with respect to the new cartesian system, $\vec{a}$ has components $p + 1$ and 1 then p = ______ .

2. In each of the following, one or more options are correct. Choose the correct option(s).

(a) The position vectors of the points P and Q are $\vec{p}$ and $\vec{q}$ respectively. If O is the origin and R is a point in the interior of ∠POQ such that OR bisects the ∠POQ then what is the unit vector along OR?

A. $\dfrac{\vec{p}}{|\vec{p}|} + \dfrac{\vec{q}}{|\vec{q}|}$ B. $\dfrac{\vec{p}}{|\vec{p}|} - \dfrac{\vec{q}}{|\vec{q}|}$

C. $\left(\dfrac{\vec{p}}{|\vec{p}|} + \dfrac{\vec{q}}{|\vec{q}|}\right) \Big/ \left|\dfrac{\vec{p}}{|\vec{p}|} + \dfrac{\vec{q}}{|\vec{q}|}\right|$ D. none of these

(b) Points whose position vectors are $\vec{a} + \vec{b}$, $\vec{a} - \vec{b}$ and $\vec{a} + k\vec{b}$ are collinear for

A. only one integral value of k B. all integral values of k

C. all real values of k D. none of these

(c) If the points A, B, C and D have the position vectors $\vec{a}$, $\vec{b}$, $\vec{c}$ and $\vec{d}$ respectively such that $(\lambda + \mu + \nu)\vec{a} = \lambda\vec{b} + \mu\vec{c} + \nu\vec{d}$ then

A. lines AC and BD are parallel B. points A, B, C, D are coplanar

C. points A, B, C, D are collinear D. none of these

(d) The position vectors of three poins A, B and C are $3\vec{i} - \vec{j} + 2\vec{k}$, $\vec{i} + \lambda\vec{j} - 3\vec{k}$ and $4\vec{i} - 3\vec{j} - \lambda\vec{k}$ respectively. The value of $\lambda \in R$ for which ΔABC is equilateral, is

A. 3 B. –1 C. $\dfrac{21}{2}$ D. no value

3. If $\vec{a}$, $\vec{b}$ are noncollinear vectors, show that the points $l_r\vec{a} + m_r\vec{b}$; r = 1, 2, 3 are collinear if

$$\begin{vmatrix} 1 & 1 & 1 \\ l_1 & l_2 & l_3 \\ m_1 & m_2 & m_3 \end{vmatrix} = 0.$$

4. Let the points $P_1, P_2, \ldots, P_n$ have the position vectors $\vec{r_1}, \vec{r_2}, \ldots, \vec{r_n}$ with reference to the origin O. If these satisfy the vector equation $\sum_{\lambda=1}^{n} a_\lambda \vec{r_\lambda} = \vec{0}$ then prove that a similar equation $\sum_{\lambda=1}^{n} a_k \vec{r_\lambda}' = \vec{0}$ will hold with another origin if $\sum_{\lambda=1}^{n} a_\lambda = 0$.

5. If $\vec{a}$, $\vec{b}$ and $\vec{c}$ are position vectors of the vertices A, B and C respectively of ΔABC and the position vector of the incentre is $\vec{r}$ then prove that

$$(|\vec{b}-\vec{c}| + |\vec{c}-\vec{a}| + |\vec{a}-\vec{b}|)\vec{r} = |\vec{b}-\vec{c}|\,\vec{a} + |\vec{c}-\vec{a}|\,\vec{b} + |\vec{a}-\vec{b}|\,\vec{c}.$$

□

Answers

1. (a) $\frac{1}{3}$ (b) $\pm\sqrt{3}(\vec{i}+\vec{j}+\vec{k})$ (c) {3} (d) −40 (e) $1, -\frac{1}{3}$

2. (a) C (b) B, C (c) B (d) D

□

2. Product of two Vectors

Recap of Facts and Formulae

1. Scalar product of two vectors

- The scalar product (or dot product) $\vec{a} \cdot \vec{b}$ of two vectors $\vec{a}$ and $\vec{b}$ is a scalar quantity such that

  $$\vec{a} \cdot \vec{b} = |\vec{a}|\,|\vec{b}| \cos\theta$$

  where θ is the angle between $\vec{a}$ and $\vec{b}$.
- $\vec{a} \perp \vec{b} \iff \vec{a} \cdot \vec{b} = 0.$
- The angle θ between two unit vectors $\hat{a}, \hat{b}$ is given by $\cos\theta = \hat{a} \cdot \hat{b}$.
- The projection of $\hat{b}$ on the vector $\vec{a}$

  $$= \vec{b} \cdot \frac{\vec{a}}{|\vec{a}|}, \text{ i.e., } \vec{b} \cdot \hat{a}.$$
- $\vec{i} \cdot \vec{i} = 1 = \vec{j} \cdot \vec{j} = \vec{k} \cdot \vec{k}$ and $\vec{i} \cdot \vec{j} = 0 = \vec{j} \cdot \vec{k} = \vec{k} \cdot \vec{i}$.
- $\vec{a} \cdot \vec{b} = \vec{b} \cdot \vec{a}$.
- $\vec{a} \cdot (\vec{b} + \vec{c}) = \vec{a} \cdot \vec{b} + \vec{a} \cdot \vec{c}$.
- If $\vec{a} = a_1 \vec{i} + a_2 \vec{j} + a_3 \vec{k}$ and $\vec{b} = b_1 \vec{i} + b_2 \vec{j} + b_3 \vec{k}$ then

  $$\vec{a} \cdot \vec{b} = a_1 b_1 + a_2 b_2 + a_3 b_3.$$
- $(\vec{a} + \vec{b})^2 = \vec{a}^{\,2} + \vec{b}^{\,2} + 2\,\vec{a} \cdot \vec{b}$ where $\vec{a} \cdot \vec{a} = \vec{a}^{\,2}$, etc.

2. Vector product of two vectors

- The vector product (or cross product) $\vec{a} \times \vec{b}$ of two vectors $\vec{a}$ and $\vec{b}$ is a vector quantity such that

  $$\vec{a} \times \vec{b} = |\vec{a}|\,|\vec{b}| \sin\theta\, \hat{n}$$

  where θ = angle between $\vec{a}$ and $\vec{b}$ and $\hat{n}$ is a unit vector perpendicular to the plane of $\vec{a}$ and $\vec{b}$ such that $\vec{a}, \vec{b}, \hat{n}$ form a right-handed triad.

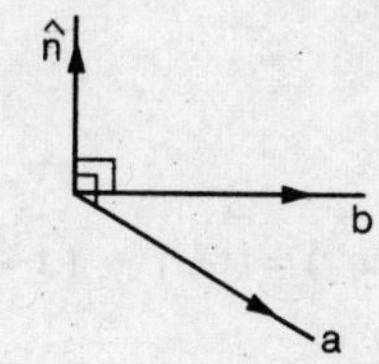

- $\vec{a} \parallel \vec{b} \iff \vec{a} \times \vec{b} = 0.$
- $\vec{a} \times \vec{b} \perp \vec{a}$ and $\vec{a} \times \vec{b} \perp \vec{b}$.
- Unit vector perpendicular to both $\vec{a}$ and $\vec{b}$ is

  $$\frac{\vec{a} \times \vec{b}}{|\vec{a} \times \vec{b}|}.$$
- The angle θ between $\vec{a}, \vec{b}$ is given by $\sin\theta = |\hat{a} \times \hat{b}|$.
- The vector area $\vec{\Delta}$ of the triangle whose two sides represent vectors $\vec{a}$ and $\vec{b} = \frac{1}{2}(\vec{a} \times \vec{b})$ and area (scalar) $= \frac{1}{2}|\vec{a} \times \vec{b}|$.

  The vector area of the parallelogram whose two adjacent sides represent vectors $\vec{a}$ and $\vec{b} = \vec{a} \times \vec{b}$ and area (scalar) $= |\vec{a} \times \vec{b}|$.
- $\vec{i} \times \vec{i} = 0 = \vec{j} \times \vec{j} = \vec{k} \times \vec{k}$

  $\vec{i} \times \vec{j} = \vec{k}, \quad \vec{j} \times \vec{k} = \vec{i}, \quad \vec{k} \times \vec{i} = \vec{j}.$
- $\vec{a} \times \vec{b} = -\vec{b} \times \vec{a}$.

  $\vec{i} \times \vec{k} = -\vec{k} \times \vec{i} = -\vec{j}, \quad \vec{k} \times \vec{j} = -\vec{j} \times \vec{k} = -\vec{i},$

  $\vec{j} \times \vec{i} = -\vec{i} \times \vec{j} = -\vec{k}.$
- $\vec{a} \times (\vec{b} + \vec{c}) = \vec{a} \times \vec{b} + \vec{a} \times \vec{c}$.
- If $\vec{a} = a_1 \vec{i} + a_2 \vec{j} + a_3 \vec{k}$ and $\vec{b} = b_1 \vec{i} + b_2 \vec{j} + b_3 \vec{k}$ then

  $$\vec{a} \times \vec{b} = \begin{vmatrix} \vec{i} & \vec{j} & \vec{k} \\ a_1 & a_2 & a_3 \\ b_1 & b_2 & b_3 \end{vmatrix}.$$

Selected Solved Examples

1. Find the cosine of the angle between the vectors $\vec{p}$ and $\vec{q}$ such that

$$2\vec{p} + \vec{q} = \vec{i} + \vec{j}$$
$$\vec{p} + 2\vec{q} = \vec{i} - \vec{j}.$$

Here, $(2\vec{p} + \vec{q}) + (\vec{p} + 2\vec{q}) = \vec{i} + \vec{j} + \vec{i} - \vec{j}$

$\therefore \quad 3(\vec{p} + \vec{q}) = 2\vec{i}$

$\therefore \quad \vec{p} + \vec{q} = \frac{2}{3}\vec{i}$ ...(1)

Also $(2\vec{p} + \vec{q}) - (\vec{p} + 2\vec{q}) = \vec{i} + \vec{j} - (\vec{i} - \vec{j})$

$\therefore \quad \vec{p} - \vec{q} = 2\vec{j}$ ... (2)

From (1), (2) we get

$2\vec{p} = \frac{2}{3}\vec{i} + 2\vec{j}$ ... (3)

$2\vec{q} = \frac{2}{3}\vec{i} - 2\vec{j}$ ... (4)

$\therefore \quad 4\vec{p} \cdot \vec{q} = \frac{2}{3} \cdot \frac{2}{3} + 2 \cdot (-2) = \frac{4}{9} - 4 = -\frac{32}{9}$

$\therefore \vec{p} \cdot \vec{q} = \frac{-8}{9}$

Also, from (3), $2|\vec{p}| = \sqrt{\left(\frac{2}{3}\right)^2 + (2)^2} = \frac{2}{3}\sqrt{10}$

$\therefore \quad |\vec{p}| = \frac{1}{3}\sqrt{10}$

and from (4), $2|\vec{q}| = \sqrt{\left(\frac{2}{3}\right)^2 + (-2)^2} = \frac{2}{3}\sqrt{10}$

$\therefore \quad |\vec{q}| = \frac{1}{3}\sqrt{10}$

$$\therefore \quad \cos\theta = \frac{\vec{p} \cdot \vec{q}}{|\vec{p}|\,|\vec{q}|} = \frac{-\frac{8}{9}}{\frac{1}{3}\sqrt{10} \cdot \frac{1}{3}\sqrt{10}} = -\frac{4}{5}$$

$\therefore \quad \theta = \cos^{-1}\left(-\frac{4}{5}\right).$

2. Determine the values of c so that for all real x, the vectors $cx\vec{i} - 6\vec{j} + 3\vec{k}$ and $x\vec{i} + 2\vec{j} + 2cx\vec{k}$ make an obtuse angle with each other.

If θ is the angle between the vectors then

$$\cos\theta = \frac{(x\vec{i} + 2\vec{j} + 2cx\vec{k}) \cdot (cx\vec{i} - 6\vec{j} + 3\vec{k})}{|x\vec{i} + 2\vec{j} + 2cx\vec{k}|\;|cx\vec{i} - 6\vec{j} + 3\vec{k}|}$$

$$= \frac{cx^2 - 12 + 6cx}{\sqrt{x^2 + 4 + 4c^2x^2}\;\sqrt{c^2x^2 + 36 + 9}}$$

$$\therefore \quad \cos\theta = \frac{cx^2 + 6cx - 12}{\sqrt{(1 + 4c^2)x^2 + 4}\;\sqrt{c^2x^2 + 45}}.$$

From the question, $\cos\theta < 0$

$$\therefore \quad \frac{cx^2 + 6cx - 12}{\sqrt{(1 + 4c^2)x^2 + 4} \cdot \sqrt{c^2x^2 + 45}} < 0$$

$\therefore \quad cx^2 + 6cx - 12 < 0$, for all x ...(1)

(1) is possible if $c < 0$ and $36c^2 - 4c \cdot (-12) < 0$

or $\quad c < 0$ and $3c^2 + 4c < 0$

or $\quad c < 0$ and $c(3c + 4) < 0$

$\therefore \quad c < 0$ and $3c + 4 > 0$, i.e., $c > -\frac{4}{3}$; so $-\frac{4}{3} < c < 0$.

Note $(a_1b_1 + a_2b_2 + a_3b_3)^2 \le (a_1^2 + a_2^2 + a_3^2)(b_1^2 + b_2^2 + b_3^2)$; so, $\cos\theta \ge -1$ need not be applied.

3. If $|\vec{a}| = 3$, $|\vec{b}| = 1$, $|\vec{c}| = 4$ and $\vec{a} + \vec{b} + \vec{c} = 0$, find the value of $\vec{a} \cdot \vec{b} + \vec{b} \cdot \vec{c} + \vec{c} \cdot \vec{a}$.

$(\vec{a} + \vec{b} + \vec{c})^2 = (\vec{a} + \vec{b} + \vec{c}) \cdot (\vec{a} + \vec{b} + \vec{c})$

$\therefore \quad 0 = \vec{a}^2 + \vec{b}^2 + \vec{c}^2 + 2\vec{a} \cdot \vec{b} + 2\vec{b} \cdot \vec{c} + 2\vec{c} \cdot \vec{a}$

or $\quad 0 = |\vec{a}|^2 + |\vec{b}|^2 + |\vec{c}|^2 + 2(\vec{a} \cdot \vec{b} + \vec{b} \cdot \vec{c} + \vec{c} \cdot \vec{a})$

$= 9 + 1 + 16 + 2(\vec{a} \cdot \vec{b} + \vec{b} \cdot \vec{c} + \vec{c} \cdot \vec{a})$

$\therefore \quad \vec{a} \cdot \vec{b} + \vec{b} \cdot \vec{c} + \vec{c} \cdot \vec{a} = -\frac{26}{2} = -13.$

4. If $\vec{b} = 2\vec{i} + 3\vec{j} - \vec{k}$ and $\vec{c} = \vec{i} + 4\vec{j} + 5\vec{k}$ then find a vector a such that $\vec{b} \cdot \vec{a} = 0$ and $\vec{c} \cdot \vec{a} = 0$. Also find the unit vector along $\vec{a}$.

Let $\vec{a} = x\vec{i} + y\vec{j} + z\vec{k}$. Then

$\vec{b} \cdot \vec{a} = 0 \Rightarrow 2x + 3y - z = 0$ ... (1)

$\vec{c} \cdot \vec{a} = 0 \Rightarrow x + 4y + 5z = 0$ ... (2)

From (1) and (2), $\frac{x}{15 + 4} = \frac{y}{-1 - 10} = \frac{z}{8 - 3}$

or $\quad \frac{x}{19} = \frac{y}{-11} = \frac{z}{5} (= \lambda)$

$\therefore \quad \vec{a} = \lambda(19\vec{i} - 11\vec{j} + 5\vec{k}).$

The unit vector along $\vec{a} = \dfrac{\lambda(19\vec{i} - 11\vec{j} + 5\vec{k})}{\sqrt{\lambda^2(19^2 + 11^2 + 5^2)}}$

$$= \frac{1}{\sqrt{507}}(19\vec{i} - 11\vec{j} + 5\vec{k}).$$

Second Method

As $\vec{b} \cdot \vec{a} = 0$, $\vec{a} \perp \vec{b}$; as $\vec{c} \cdot \vec{a} = 0$, $\vec{a} \perp \vec{c}$

$\therefore$ $\vec{a} \perp$ the plane of $\vec{b}$ and $\vec{c}$.

$\therefore$ $\vec{a} \parallel \vec{b} \times \vec{c}$.

Now, $\vec{b} \times \vec{c} = \begin{vmatrix} \vec{i} & \vec{j} & \vec{k} \\ 2 & 3 & -1 \\ 1 & 4 & 5 \end{vmatrix} = 19\vec{i} - 11\vec{j} + 5\vec{k}$

$\therefore$ $\vec{a} = \lambda(19\vec{i} - 11\vec{j} + 5\vec{k})$.

As above find the unit vector along $\vec{a}$.

5. Let $\vec{A}$, $\vec{B}$ and $\vec{C}$ be unit vectors. Suppose $\vec{A} \cdot \vec{B} = \vec{A} \cdot \vec{C} = 0$ and the angle between $\vec{B}$ and $\vec{C}$ is $\frac{\pi}{6}$. Prove that $\vec{A} = \pm 2(\vec{B} \times \vec{C})$.

$\vec{A} \cdot \vec{B} = 0 \Rightarrow \vec{A} \perp \vec{B}$

and $\vec{A} \cdot \vec{C} = 0 \Rightarrow \vec{A} \perp \vec{C}$

$\therefore$ $\vec{A} \parallel \vec{B} \times \vec{C}$, i.e., $\vec{A} = k\vec{B} \times \vec{C}$ ... (1)

$\therefore$ $|\vec{A}| = |k\vec{B} \times \vec{C}| = |k|\,|\vec{B} \times \vec{C}|$ ... (2)

But $|\vec{B} \times \vec{C}| = |\vec{B}|\,|\vec{C}| \sin\frac{\pi}{6}$

$$= 1 \cdot 1 \cdot \frac{1}{2} = \frac{1}{2};$$

also $|\vec{A}| = 1$

$\therefore$ from (2), $1 = |k|\frac{1}{2}$; $\therefore$ $|k| = 2$; $\therefore$ $k = \pm 2$.

Putting in (1), $\vec{A} = \pm 2(\vec{B} \times \vec{C})$.

6. If $A_1, A_2, \ldots, A_n$ are the vertices of a regular plane polygon of n sides whose centre is O then prove that

$$\sum_{i=1}^{n-1} (\vec{OA_i} \times \vec{OA_{i+1}}) = (1-n)(\vec{OA_2} \times \vec{OA_1}).$$

Clearly, $\angle A_i O A_{i+1} = \frac{2\pi}{n}$ for $i = 1, 2, \ldots$

Also $|\vec{OA_1}| = |\vec{OA_2}| = \ldots = |\vec{OA_n}| = r$.

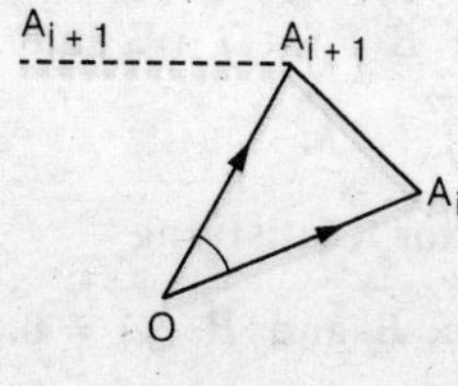

Now, $\vec{OA_i} \times \vec{OA_{i+1}} = |\vec{OA_i}|\,|\vec{OA_{i+1}}| \sin\frac{2\pi}{n}\hat{p}$

$$= r^2 \sin\frac{2\pi}{n}\hat{p}, \text{ for all } i$$

where $\hat{p}$ is the unit vector perpendicular to the plane of the polygon.

$\therefore$ $\sum_{i=1}^{n-1} (\vec{OA_i} \times \vec{OA_{i+1}})$

$$= \left(\sum_{i=1}^{n-1} r^2 \sin\frac{2\pi}{n}\right)\hat{p} = \left(r^2 \sin\frac{2\pi}{n}\right)(n-1)\hat{p}.$$

Also, $(1-n)(\vec{OA_2} \times \vec{OA_1})$

$$= (1-n)|\vec{OA_2}|\,|\vec{OA_1}| \sin\frac{2\pi}{n}(-\hat{p})$$

$$= (n-1)\left(r^2 \sin\frac{2\pi}{n}\right)\hat{p}$$

$\therefore$ $\sum_{i=1}^{n-1} (\vec{OA_i} \times \vec{OA_{i+1}}) = (1-n)(\vec{OA_2} \times \vec{OA_1})$.

7. If $\vec{a}, \vec{b}, \vec{c}$ are unit vectors such that $\vec{a}$ is perpendicular to the plane of $\vec{b}, \vec{c}$ and the angle between $\vec{b}, \vec{c}$ is $\frac{\pi}{3}$ then find $|\vec{a} + \vec{b} + \vec{c}|$.

Here $|\vec{a}| = 1$, $|\vec{b}| = 1$, $|\vec{c}| = 1$,

$\vec{a} \cdot \vec{b} = 0$ and $\vec{a} \cdot \vec{c} = 0$.

Also, $\vec{b} \cdot \vec{c} = |\vec{b}|\,|\vec{c}| \cos\frac{\pi}{3} = 1 \cdot 1 \cdot \frac{1}{2} = \frac{1}{2}$.

Now, $|\vec{a} + \vec{b} + \vec{c}|^2$

$$= (\vec{a} + \vec{b} + \vec{c})^2$$

$$= \vec{a}^2 + \vec{b}^2 + \vec{c}^2 + 2\vec{a} \cdot \vec{b} + 2\vec{a} \cdot \vec{c} + 2\vec{b} \cdot \vec{c}$$

$$= |\vec{a}|^2 + |\vec{b}|^2 + |\vec{c}|^2 + 2\left(0 + 0 + \frac{1}{2}\right)$$

$$= 1^2 + 1^2 + 1^2 + 2 \cdot \frac{1}{2} = 4.$$

$\therefore$ $|\vec{a} + \vec{b} + \vec{c}| = 2$.

8. Let $\vec{A} = 2\vec{i} + \vec{k},\ \vec{B} = \vec{i} + \vec{j} + \vec{k}$ and

$$\vec{C} = 4\vec{i} - 3\vec{j} + 7\vec{k}.$$

Determine a vector $\vec{R}$ satisfying

$$\vec{R} \times \vec{B} = \vec{C} \times \vec{B} \text{ and } \vec{R} \cdot \vec{A} = 0.$$

Let $\vec{R} = x\vec{i} + y\vec{j} + z\vec{k}$.

Now, $\vec{R} \times \vec{B} = \vec{C} \times \vec{B}$

$$\Rightarrow \begin{vmatrix} \vec{i} & \vec{j} & \vec{k} \\ x & y & z \\ 1 & 1 & 1 \end{vmatrix} = \begin{vmatrix} \vec{i} & \vec{j} & \vec{k} \\ 4 & -3 & 7 \\ 1 & 1 & 1 \end{vmatrix}$$

or $\quad (y - z)\vec{i} + (z - x)\vec{j} + (x - y)\vec{k} = -10\vec{i} + 3\vec{j} + 7\vec{k}$

$\Rightarrow \quad y - z = -10,\ z - x = 3,\ x - y = 7 \qquad \ldots (1)$

$\vec{R} \cdot \vec{A} = 0 \quad \Rightarrow \quad 2x + z = 0$

As any two equations of (1) imply the third, we have the following equations to solve for x, y, z:

$z - x = 3$

$x - y = 7$

$2x + z = 0$

$\therefore \quad z = -2x;$ so $-2x - x = 3$

$\therefore \quad x = -1$ and $z = 2$

$\therefore \quad -1 - y = 7; \qquad \therefore \quad y = -8$

$\therefore \quad \vec{R} = -\vec{i} - 8\vec{j} + 2\vec{k}.$

9. The vectors $\vec{AB} = 3\vec{i} - 2\vec{j} + 2\vec{k}$ and $\vec{BC} = -\vec{i} - 2\vec{k}$ are adjacent sides of a parallelogram. Find the area of the parallelogram ABCD. Also find the angle between its diagonals.

Here $\vec{AB} = 3\vec{i} - 2\vec{j} + 2\vec{k},\ \vec{AD} = \vec{BC} = -\vec{i} - 2\vec{k}.$

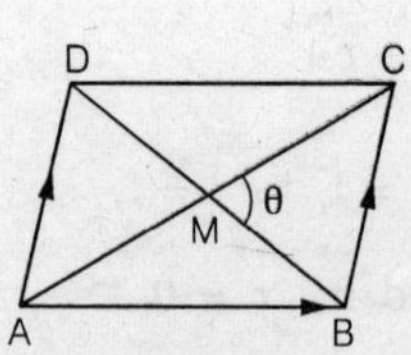

$\therefore$ vector area of the parallelogram

$$= \vec{AB} \times \vec{AD} = (3\vec{i} - 2\vec{j} + 2\vec{k}) \times (-\vec{i} - 2\vec{k})$$

$$= \begin{vmatrix} \vec{i} & \vec{j} & \vec{k} \\ 3 & -2 & 2 \\ -1 & 0 & -2 \end{vmatrix} = 4\vec{i} + 4\vec{j} - 2\vec{k}$$

$\therefore$ area of the parallelogram

$$= |4\vec{i} + 4\vec{j} - 2\vec{k}| = \sqrt{4^2 + 4^2 + (-2)^2} = 6.$$

Now, $\vec{AC} = \vec{AB} + \vec{BC} = 3\vec{i} - 2\vec{j} + 2\vec{k} + (-\vec{i} - 2\vec{k})$

$$= 2\vec{i} - 2\vec{j}$$

$$\vec{DB} = \vec{DA} + \vec{AB} = \vec{i} + 2\vec{k} + 3\vec{i} - 2\vec{j} + 2\vec{k}$$

$$= 4\vec{i} - 2\vec{j} + 4\vec{k}.$$

If the angle between the diagonals is θ then

$$\cos\theta = \frac{\vec{AC}}{|\vec{AC}|} \cdot \frac{\vec{DB}}{|\vec{DB}|}$$

$$= \frac{(2\vec{i} - 2\vec{j}) \cdot (4\vec{i} - 2\vec{j} + 4\vec{k})}{\sqrt{2^2 + (-2)^2}\ \sqrt{4^2 + (-2)^2 + 4^2}}$$

$$= \frac{8 + 4}{2\sqrt{2} \cdot 6} = \frac{1}{\sqrt{2}}; \qquad \therefore \quad \theta = \frac{\pi}{4}.$$

10. The position vectors of the points A and B are $\vec{i} + 2\vec{j} + 3\vec{k}$ and $2\vec{i} - \vec{j} - \vec{k}$ respectively. Find the projection of $\vec{AB}$ on the vector $\vec{i} + \vec{j} + \vec{k}$. Also find the resolved part of $\vec{AB}$ in that direction.

Here $\vec{AB} = \vec{AO} + \vec{OB} = \vec{OB} - \vec{OA}$

$$= 2\vec{i} - \vec{j}\ \vec{k} - (\vec{i} + 2\vec{j} + 3\vec{k})$$

$$= \vec{i} - 3\vec{j} - 4\vec{k}.$$

Unit vector along $\vec{i} + \vec{j} + \vec{k}$

$$= \frac{\vec{i} + \vec{j} + \vec{k}}{|\vec{i} + \vec{j} + \vec{k}|}$$

$$= \frac{\vec{i} + \vec{j} + \vec{k}}{\sqrt{3}}$$

$\therefore$ the projection of $\vec{AB}$ on the vector $\vec{i} + \vec{j} + \vec{k}$

$$= (\vec{i} - 3\vec{j} - 4\vec{k}) \cdot \frac{\vec{i} + \vec{j} + \vec{k}}{\sqrt{3}} \qquad \ldots (1)$$

$$= \frac{1 - 3 - 4}{\sqrt{3}} = \frac{-6}{\sqrt{3}} = -2\sqrt{3}.$$

$\therefore$ the required projection $= 2\sqrt{3}$.

The required resolved part of $\vec{AB}$ {using (1)}

$$= \left\{(\vec{i} - 3\vec{j} - 4\vec{k}) \cdot \frac{\vec{i} + \vec{j} + \vec{k}}{\sqrt{3}}\right\} \frac{\vec{i} + \vec{j} + \vec{k}}{\sqrt{3}}$$

$$= (-2\sqrt{3})\,\frac{\vec{i} + \vec{j} + \vec{k}}{\sqrt{3}} = -2(\vec{i} + \vec{j} + \vec{k}).$$

11. The vector $\vec{a}$ is expressed as sum of two vectors $\vec{a}'$ and $\vec{a}''$ respectively along and perpendicular to $\vec{b}$; show that $\vec{a}' = \dfrac{(\vec{a} \cdot \vec{b})\vec{b}}{\vec{b}^2}$ and $\vec{a}'' = \vec{a} - \dfrac{(\vec{a} \cdot \vec{b})\vec{b}}{\vec{b}^2}$.

Here $\vec{AD} = \vec{a} = \vec{AB} + \vec{AC} = \vec{a}\,' + \vec{a}\,''$.

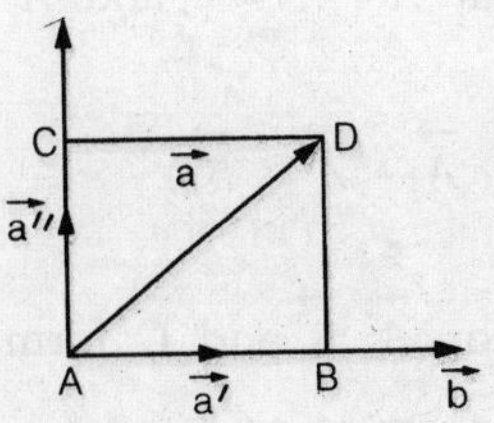

Now, $\vec{a}\,'$ = (projection of $\vec{AD}$ on $\vec{b}$) $\hat{b}$

$$= \left(\vec{a} \cdot \frac{\vec{b}}{|\vec{b}|}\right)\frac{\vec{b}}{|\vec{b}|} = \frac{(\vec{a} \cdot \vec{b})\vec{b}}{|\vec{b}|^2} = \frac{(\vec{a} \cdot \vec{b})\vec{b}}{\vec{b}^2}.$$

$$\vec{a}\,'' = \vec{AC} = \vec{AD} - \vec{AB} = \vec{a} - \vec{a}\,' = \vec{a} - \frac{(\vec{a} \cdot \vec{b})\vec{b}}{\vec{b}^2}.$$

12. Show that the altitude of ΔABC through the vertex A is equal to $\dfrac{|\vec{a} \times \vec{b} + \vec{b} \times \vec{c} + \vec{c} \times \vec{a}|}{|\vec{b} - \vec{c}|}$ where the position vectors of A, B and C are respectively $\vec{a}$, $\vec{b}$ and $\vec{c}$.

Here, $\vec{BC} = \vec{BO} + \vec{OC} = \vec{OC} - \vec{OB} = \vec{c} - \vec{b}$

$\vec{BA} = \vec{BO} + \vec{OA} = \vec{OA} - \vec{OB} = \vec{a} - \vec{b}$

$\therefore \quad \text{ar}(\Delta\text{ABC}) = \frac{1}{2} |\vec{BC} \times \vec{BA}|$

$$= \frac{1}{2} |(\vec{c} - \vec{b}) \times (\vec{a} - \vec{b})|$$

$$= \frac{1}{2} |\vec{c} \times \vec{a} - \vec{c} \times \vec{b} - \vec{b} \times \vec{a}|.$$

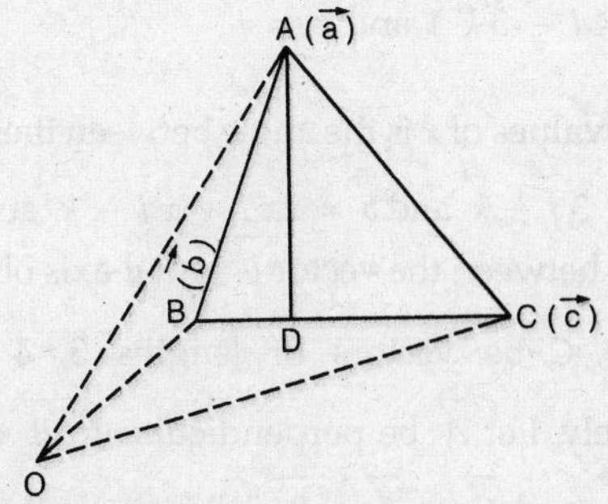

Now, $BC = |\vec{BC}| = |\vec{c} - \vec{b}|$ and

$$\text{ar}(\Delta\text{ABC}) = \frac{1}{2} BC \cdot AD$$

$$\therefore \quad AD = \frac{2\,\text{ar}(\Delta\text{ABC})}{BC}$$

$$= \frac{2 \cdot \frac{1}{2} |\vec{c} \times \vec{a} - \vec{c} \times \vec{b} - \vec{b} \times \vec{a}|}{|\vec{c} - \vec{b}|}$$

$$= \frac{|\vec{c} \times \vec{a} + \vec{b} \times \vec{c} + \vec{a} \times \vec{b}|}{|\vec{b} - \vec{c}|}.$$

Exercises

1. Prove that $\vec{r} = (\vec{r} \cdot \vec{i})\vec{i} + (\vec{r} \cdot \vec{j})\vec{j} + (\vec{r} \cdot \vec{k})\vec{k}$.

2. If $|\vec{a} + \vec{b}| = |\vec{a} - \vec{b}|$ where $\vec{a}, \vec{b}$ are not zero vectors then prove that $\vec{a}$ and $\vec{b}$ are perpendicular to each other.

3. If $\hat{a}$ and $\hat{b}$ be unit vectors and θ be the angle between them, prove that $\sin\frac{\theta}{2} = \frac{1}{2} |\hat{a} - \hat{b}|$.

4. Prove that

$$|\vec{a} + \vec{b}| = \sqrt{|\vec{a}|^2 + |\vec{b}|^2 + 2\vec{a} \cdot \vec{b}}.$$

5. Prove that $|\vec{a} - \vec{b}|^2 + 4\vec{a} \cdot \vec{b} = |\vec{a} + \vec{b}|^2$.

6. Prove that $(\vec{a} \times \vec{b})^2 + (\vec{a} \cdot \vec{b})^2 = a^2 b^2$.

7. If $\vec{a} \times \vec{b} = \vec{c} \times \vec{d}$ and $\vec{a} \times \vec{c} = \vec{b} \times \vec{d}$ then prove that $(\vec{a} - \vec{d}) \parallel (\vec{b} - \vec{c})$.

8. Find the angle between the vectors $\vec{a} = 2\vec{p} + 4\vec{q}$ and $\vec{b} = \vec{p} - \vec{q}$ where $\vec{p}$ and $\vec{q}$ are unit vectors forming an angle of 120°.

9. If a, b and c are unit vectors such that $\vec{a} + \vec{b} = \vec{c}$ then prove that $|\vec{a} - \vec{b}| = \sqrt{3}$. Also find the angle between $\vec{a}$ and $\vec{b}$.

10. If $|\vec{a}| = 1 = |\vec{b}|$ and $|\vec{a} + \vec{b}| = \sqrt{3}$ then evaluate $(2\vec{a} - 5\vec{b}) \cdot (3\vec{a} + \vec{b})$.

11. Find λ such that the scalar product of the vector $\vec{i} + \vec{j} + \vec{k}$ with the unit vector parallel to the sum of

the vectors $2\vec{i} + 4\vec{j} - 5\vec{k}$ and $\lambda\vec{i} + 2\vec{j} + 3\vec{k}$ is equal to 1.

12. Find a vector of magnitude $\sqrt{51}$ which makes equal angles with the vectors $\vec{a} = \frac{1}{3}(\vec{i} - 2\vec{j} + 2\vec{k})$, $\vec{b} = \frac{1}{5}(-4\vec{i} - 3\vec{k})$ and $\vec{c} = \vec{j}$.

13. For what values of x is the angle between the vectors $\vec{a} = x\vec{i} - 3\vec{j} - \vec{k}$ and $\vec{b} = 2x\vec{i} + x\vec{j} - \vec{k}$ acute and the angle between the vector $\vec{b}$ and y-axis obtuse?

14. Let $\vec{A}, \vec{B}, \vec{C}$ be vectors of lengths 3, 4 and 5 respectively. Let $\vec{A}$ be perpendicular to $\vec{B} + \vec{C}$, $\vec{B}$ to $\vec{C} + \vec{A}$ and $\vec{C}$ to $\vec{A} + \vec{B}$. What is the length of the vector $\vec{A} + \vec{B} + \vec{C}$?

15. If $3\vec{a} - 5\vec{b}$ and $2\vec{a} + \vec{b}$ are perpendicular to each other, and $\vec{a} + 4\vec{b}$ and $-\vec{a} + \vec{b}$ are also mutually perpendicular then prove that the cosine of the angle between $\vec{a}$ and $\vec{b}$ is $\frac{19}{5\sqrt{43}}$.

16. The pth, qth and rth terms of a GP are the positive numbers a, b and c respectively. Show that the vectors $\vec{i}\log_e a + \vec{j}\log_e b + \vec{k}\log_e c$ and

$\vec{i}(q - r) + \vec{j}(r - p) + \vec{k}(p - q)$

are mutually perpendicular.

17. If $\vec{a}, \vec{b}$ and $\vec{c}$ are mutually perpendicular vectors of equal magnitude, show that $\vec{a} + \vec{b} + \vec{c}$ is equally inclined to $\vec{a}, \vec{b}$ and $\vec{c}$.

18. A line makes angles α, β, γ and δ with the diagonals of a cube. Prove that

$$\cos^2\alpha + \cos^2\beta + \cos^2\gamma + \cos^2\delta = \frac{4}{3}.$$

19. If $\vec{A} = \vec{i} + \vec{j} + \vec{k}$, $\vec{C} = \vec{j} - \vec{k}$ then find $\vec{B}$ so that $\vec{A} \times \vec{B} = \vec{C}$ and $\vec{A} \cdot \vec{B} = 3$.

20. If $\vec{a}, \vec{b}$ and $\vec{c}$ are the position vectors of three noncollinear points, prove that

$$\vec{a} \times \vec{b} + \vec{b} \times \vec{c} + \vec{c} \times \vec{a}$$

will be perpendicular to the plane of the points.

21. If $\vec{A} = \vec{a} + \vec{b}$, $\vec{B} = \vec{a} - \vec{b}$ where $\vec{a}$ and $\vec{b}$ are two unit vectors, show that

$\vec{A} \cdot \vec{B} = 0$ and $|\vec{A} \times \vec{B}| = 2\sin\theta$

where θ is the angle between $\vec{a}$ and $\vec{b}$.

22. If $\vec{A}$ and $\vec{B}$ are nonzero vectors, $c \neq 0$ and the vector $\vec{X}$ is such that $\vec{A} \cdot \vec{X} = c$ and $\vec{A} \times \vec{X} = \vec{B}$ then prove that

$$\vec{X} = \frac{1}{\vec{A}^2}(c\vec{A} - \vec{A} \times \vec{B}).$$

23. Given the vectors $\vec{A}, \vec{B}$ and $\vec{C}$ form a triangle such that $\vec{A} = \vec{B} + \vec{C}$. Find a, b, c and d such that the area of the triangle is $5\sqrt{6}$ where

$\vec{A} = a\vec{i} + b\vec{j} + c\vec{k}$, $\vec{B} = d\vec{i} + 3\vec{j} + 4\vec{k}$ and $\vec{C} = 3\vec{i} + \vec{j} - 2\vec{k}$.

24. The position vector of the vertices of ΔABC are respectively

$\vec{i} - \vec{j} + 2\vec{k}$, $\vec{i} + 2\vec{j} - \vec{k}$ and $2\vec{i} - \vec{j} + 3\vec{k}$.

Find the length of the altitude AD.

25. If AC and BD are the diagonals of a quadrilateral ABCD, prove that its vectorial area is equal to $\frac{1}{2}\vec{AC} \times \vec{BD}$.

26. If A, B, C, D are four points in space, prove that

$$\text{ar}(\Delta ABC) = \frac{1}{4}|\vec{AB} \times \vec{CD} + \vec{BC} \times \vec{AD} + \vec{CA} \times \vec{BD}|.$$

Objective Questions

Fill in the blanks.

27. If the angle between the vectors $x\vec{i} + \vec{j} - \vec{k}$ and $\vec{i} + x\vec{j} + \vec{k}$ is equal to $\frac{\pi}{3}$ then $x =$ ______.

28. If the vector $\vec{a} = x\vec{i} + y\vec{j} + 2\vec{k}$ is perpendicular to the vector $\vec{b} = \vec{i} - \vec{j} + \vec{k}$ and the scalar product of $\vec{a}$ and $\vec{c} = \vec{i} + 2\vec{j}$ is equal to 4 then $x =$ ______ and $y =$ ______.

29. If $\vec{a}$ is a vector of magnitude 50 and parallel to $\vec{b} = 6\vec{i} - 8\vec{j} - \frac{15}{2}\vec{k}$ and makes an acute angle with the z-axis then $\vec{a} =$ ______.

30. If $\vec{a}, \vec{b}, \vec{c}$ are unit vectors such that $\vec{a} + \vec{b} + \vec{c} = 0$ then $\vec{a} \cdot \vec{b} + \vec{b} \cdot \vec{c} + \vec{c} \cdot \vec{a} =$ ______.

31. If $\vec{a}, \vec{b}$ are two vectors such that $|\vec{a}| = 11$, $|\vec{b}| = 23$, $|\vec{a} - \vec{b}| = 30$ then $|\vec{a} + \vec{b}| =$ ______ and the angle between $\vec{a}$ and $\vec{b}$ is ______.

32. $\vec{a} \times (\vec{b} + \vec{c}) + \vec{b} \times (\vec{c} + \vec{a}) + \vec{c} \times (\vec{a} + \vec{b})$ is equal to ______ .

33. If $\vec{a}$ and $\vec{b}$ are vectors, each of magnitude 3 then $|\vec{a} + \vec{b}|^2 + |\vec{a} - \vec{b}|^2 =$ ______ .

34. Vectors of magnitude 5 which are perpendicular to the vectors $2\vec{i} + \vec{j} - 3\vec{k}$ and $\vec{i} - 2\vec{j} + \vec{k}$ are ______ .

35. If $\vec{a} + \lambda\vec{b}$ and $\vec{a} - \lambda\vec{b}$ are perpendicular to each other and $|\vec{a}| = 3$, $|\vec{b}| = 4$ then $\lambda =$ ______ .

36. If $\vec{A} = (1, 1, 1)$, $\vec{C} = (0, 1, -1)$ are given then a vector $\vec{B}$ satisfying the equations $\vec{A} \times \vec{B} = \vec{C}$ and $\vec{A} \cdot \vec{B} = 3$ is ______ .

37. The unit vector perpendicular to the plane determined by the points

P(1, –1, 2), Q(2, 0, –1) and R(0, 2, 1) is ______ .

38. Let $\vec{b} = 4\vec{i} + 3\vec{j}$ and $\vec{c}$ be two vectors perpendicular to each other in *x-y* plane. All the vectors in the same plane having projections 1 and 2 along $\vec{b}$ and $\vec{c}$ are given by ______ and ______ respectively.

39. A unit vector coplanar with $\vec{i} + \vec{j} + 2\vec{k}$ and $\vec{i} + 2\vec{j} + \vec{k}$ and perpendicular to $\vec{i} + \vec{j} + \vec{k}$ is ______ .

40. If $|\vec{a} + \vec{b}| = |\vec{a} - \vec{b}|$ then $|\vec{a} \times \vec{b}| =$ ______ if $|\vec{a}| = 3$, $|\vec{b}| = 1$.

41. If $|\vec{a}| = 3$, $|\vec{b}| = 4$, $|\vec{c}| = 5$ and $\vec{a} + \vec{b} + \vec{c} = 0$ then the angle between the vectors $\vec{a}$ and $\vec{b}$ is ______ .

42. $\vec{a}$ is perpendicular to $\vec{b} = \vec{i} + 2\vec{j} + 3\vec{k}$ and $\vec{c} = -2\vec{i} + 4\vec{j} + \vec{k}$, and satisfies the condition $\vec{a} \cdot (\vec{i} - 2\vec{j} + \vec{k}) = -6$. Then $\vec{a} =$ ______ .

43. (a) A parallelogram has two adjacent sides which represent the vectors $5\vec{p} + 2\vec{q}$ and $\vec{p} - 3\vec{q}$ where $|\vec{p}| = 2\sqrt{2}$, $|\vec{q}| = 3$ and $\vec{p}, \vec{q}$ are inclined at an angle $\frac{\pi}{4}$. Then the length of the shorter diagonal is ______ .

(b) Let $\vec{OA} = \vec{a}$, $\vec{OB} = 10\vec{a} + 2\vec{b}$ and $\vec{OC} = \vec{b}$ where O, A and C are noncollinear points. Let p denote ar(OABC) and q denote the area of the parallelogram with OA and OC as adjacent sides. If $p = kq$ then $k =$ ______ .

44. $\vec{a}, \vec{b}, \vec{c}$ are three nonzero vectors such that $\vec{a} = \vec{b} \times \vec{c}$, $\vec{b} = \vec{c} \times \vec{a}$ and $\vec{c} = \vec{a} \times \vec{b}$. Then $|\vec{a}| =$ ______ , $|\vec{b}| =$ ______ , $|\vec{c}| =$ ______ .

45. A vector $\vec{OA} = \vec{i} + 2\vec{j} + 2\vec{k}$ turns through a right angle, passing through the positive *x*-axis on the way. The vector in the new position is ______ .

46. OX, OY, OZ form a right-handed cartesian system and along them $\vec{A}$ has the components $\vec{A_1}, \vec{A_2}, \vec{A_3}$ repectively. The coordinate system is rotated about the *z*-axis through an angle $\frac{\pi}{2}$. The components of $\vec{A}$ in new coordinate system are ______ , ______ and ______ .

Choose the correct option(s).

47. If $\vec{a} \cdot \vec{b} = \vec{a} \cdot \vec{c}$ then

(a) $\vec{a} = 0$ (b) $\vec{b} = \vec{c}$

(c) $\vec{a} \perp (\vec{b} - \vec{c})$ (d) any one of these

48. If $\vec{a} + \vec{b} + \vec{c} = 0$ then $\vec{a} \times \vec{b}$ is equal to

(a) $\vec{c} \times \vec{a}$ (b) $\vec{c}$ (c) $\vec{b} \times \vec{c}$ (d) $\vec{a} \times \vec{c}$

49. If $\vec{a} \times \vec{b} = \vec{b} \times \vec{c} \neq 0$ then which one is true where k is a suitable scalar?

(a) $\vec{a} + \vec{b} = k\vec{c}$ (b) $\vec{a} + \vec{c} = k\vec{b}$

(c) $\vec{b} + \vec{c} = k\vec{a}$ (d) none of these

50. A, B, C and D are four points in a plane with position vectors $\vec{a}, \vec{b}, \vec{c}$ and $\vec{d}$ respectively such that $(\vec{a} - \vec{d}) \cdot (\vec{b} - \vec{c}) = (\vec{b} - \vec{d}) \cdot (\vec{c} - \vec{a}) = 0$. Then for the $\triangle$ABC, D is its

(a) incentre (b) circumcentre

(c) orthocentre (d) centroid

51. If $|\vec{a}| = |\vec{b}| = 1$ and $|\vec{a} \times \vec{b}| = 1$ then

(a) $\vec{a} \parallel \vec{b}$ (b) $\vec{a} \perp \vec{b}$

(c) $\vec{a} \cdot \vec{b} = 1$ (d) none of these

52. The vector $\vec{B} = 3\vec{j} + 4\vec{k}$ is to be written as $\vec{B_1} + \vec{B_2}$ where $\vec{B_1}$ is parallel to $\vec{A} = \vec{i} + \vec{j}$ and $\vec{B_2}$ is perpendicular to $\vec{A}$. Then $\vec{B_1}$ is equal to

(a) $\frac{3}{2}(\vec{i} + \vec{j})$ (b) $\frac{2}{3}(\vec{i} + \vec{j})$

(c) $\frac{1}{2}(\vec{i} + \vec{j})$ (d) none of these

53. If $(\vec{R} - \vec{r}) \cdot \vec{t} = 0$, $(\vec{R} - \vec{r}) \cdot \vec{n} = 0$ and $(\vec{R} - \vec{r}) \cdot \vec{b} = 0$, where $\vec{t}, \vec{n}, \vec{b}$ form a right-handed triad of orthogonal vectors then

(a) $\vec{R} = \vec{0}$ (b) $\vec{R} = \vec{r}$

(c) $\vec{R} \cdot \vec{r} = 0$ (d) $\vec{R} \times \vec{r} = \vec{0}$

54. Let $\vec{a} = 2\vec{i} - \vec{j} + \vec{k}$, $\vec{b} = \vec{i} + 2\vec{j} - \vec{k}$ and $\vec{c} = \vec{i} + \vec{j} - 2\vec{k}$ be three vectors. A vector in the plane of $\vec{b}$ and $\vec{c}$ whose projection on $\vec{a}$ is of the magnitude $\sqrt{\frac{2}{3}}$, is

(a) $2\vec{i} + 3\vec{j} - 3\vec{k}$

(b) $2\vec{i} + 3\vec{j} + 3\vec{k}$

(c) $-2\vec{i} - \vec{j} + 5\vec{k}$

(d) $2\vec{i} + \vec{j} + 5\vec{k}$

55. If $|\vec{a}| = 3$, $|\vec{b}| = 5$, $|\vec{c}| = 7$ and $\vec{a} + \vec{b} + \vec{c} = \vec{0}$ then the angle between $\vec{a}$ and $\vec{b}$ is

(a) $\frac{\pi}{3}$ (b) $\frac{5\pi}{3}$

(c) $\frac{2\pi}{3}$ (d) $\frac{\pi}{6}$

56. If $\vec{a} \cdot \vec{b} = 0$ and $\vec{a} \times \vec{b} = \vec{0}$ then

(a) $\vec{a} \parallel \vec{b}$ (b) $\vec{a} \perp \vec{b}$

(c) $\vec{a} = \vec{0}$ or $\vec{b} = \vec{0}$ (d) none of these

57. If $\vec{a} = 4\vec{i} + 6\vec{j}$ and $\vec{b} = 3\vec{j} + 4\vec{k}$ then the resolved part of $\vec{a}$ along $\vec{b}$ is

(a) $\frac{9}{5\sqrt{3}}(3\vec{j} + 4\vec{k})$ (b) $\frac{18}{25}$

(c) $\frac{18}{25}(3\vec{j} + 4\vec{k})$ (d) $\frac{18}{\sqrt{13}}(3\vec{j} + 4\vec{k})$

Answers

8. 120° **9.** 120° **10.** $\frac{-11}{2}$ **11.** 1

12. $-5\vec{i} + \vec{j} + 5\vec{k}$ **13.** $x < 0$ **14.** $5\sqrt{2}$

19. $\frac{1}{3}(5\vec{i} + 2\vec{j} + 2\vec{k})$

23. $a = 8,\ b = 4,\ c = 2,\ d = 5$;
$a = -8,\ b = 4,\ c = 2,\ d = -11$

24. $\frac{3\sqrt{3}}{\sqrt{26}}$ **27.** 2 **28.** 0, 2 respectively

29. $-24\vec{i} + 32\vec{j} + 30\vec{k}$ **30.** $-\frac{3}{2}$

31. 20, $\cos^{-1}\left(-\frac{125}{253}\right)$ **32.** $\vec{0}$ **33.** 36

34. $\pm\frac{5\sqrt{3}}{3}(\vec{i} + \vec{j} + \vec{k})$ **35.** $\pm\frac{3}{4}$

36. $\frac{1}{3}(5\vec{i} + 2\vec{j} + 2\vec{k})$ **37.** $\frac{1}{\sqrt{6}}(2\vec{i} + \vec{j} + \vec{k})$

38. $2\vec{i} - \vec{j}$, $\frac{-2\vec{i} + 11\vec{j}}{5}$ **39.** $\pm\frac{1}{\sqrt{2}}(-\vec{j} + \vec{k})$

40. 3 **41.** $\frac{\pi}{2}$

42. $5\vec{i} + \frac{7}{2}\vec{j} - 4\vec{k}$ **43.** (a) 15 (b) 6

44. 1, 1, 1 **45.** $\frac{1}{\sqrt{2}}(4\vec{i} - \vec{j} - \vec{k})$

46. $A_2, -A_1, A_3$ respectively

47. (d) **48.** (a), (c) **49.** (b) **50.** (c)

51. (b) **52.** (a) **53.** (b), (d) **54.** (c)

55. (a) **56.** (c) **57.** (c)

□

Chapter Test

Time: 60 minutes

1. In each of the following, fill in the blank so that the resulting statement is correct.

(a) If $\vec{a}, \vec{b}, \vec{c}$ are three unit vectors such that $\vec{a} = \lambda\vec{b} + \mu\vec{c}$ and $\vec{a}\cdot\vec{b} = 0$, $\vec{a}\cdot\vec{c} = \frac{1}{2}$ then $|\vec{b}\cdot\vec{c}|$ = ______ .

(b) If $|\vec{a}\times\vec{b}| = |\vec{a}\cdot\vec{b}|$ then the angle between $\vec{a}$ and $\vec{b}$ is ______ .

(c) The vector $-\vec{i}+\vec{j}-\vec{k}$ bisects the angle between the vectors $\vec{c}$ and $3\vec{i}+4\vec{j}$. Then the unit vector along $\vec{c}$ = ______ .

(d) If $\vec{a}\perp\vec{b}$ and $|\vec{a}+\vec{b}| = 4$ then $|\vec{a}-\vec{b}|$ = ______ .

2. In each of the following, one or more options are correct. Choose the correct option(s).

(a) If θ is the angle between the unit vectors $\hat{a}$ and $\hat{b}$ then $\cos\frac{\theta}{2}$ is equal to

A. $|\hat{a}+\hat{b}|$ B. $\frac{1}{2}|\hat{a}-\hat{b}|$ C. $\frac{1}{2}\hat{a}\cdot\hat{b}$ D. $\frac{1}{2}|\hat{a}+\hat{b}|$

(b) If $\vec{a} = (1, 2, 3)$, $\vec{b} = (-1, 2, 1)$ and $\vec{c} = (3, 1, 0)$ then the acute angle between $\vec{c}$ and a vector perpendicular to both $\vec{a}$ and $\vec{b}$ is

A. $\cos^{-1}\sqrt{\frac{8}{15}}$ B. $\sin^{-1}\sqrt{\frac{8}{15}}$ C. $\pi - \cos^{-1}\sqrt{\frac{8}{15}}$ D. none of these

(c) If $|\vec{a}| = |\vec{b}|$ then $\vec{a}+\vec{b}$ is perpendicular to the vector

A. $(\vec{a}\cdot\vec{b})\vec{a}$ B. $\vec{a}-\vec{b}$ C. $\vec{a}\times\vec{b}$ D. $(\vec{a}\cdot\vec{b})\vec{b}$

(d) The vertices of a triangle have the position vectors $\vec{a}, \vec{b}, \vec{c}$ and $P(\vec{r})$ is a point in the plane of the triangle such that $\vec{a}\cdot\vec{b}+\vec{c}\cdot\vec{r} = \vec{a}\cdot\vec{c}+\vec{b}\cdot\vec{r} = \vec{b}\cdot\vec{c}+\vec{a}\cdot\vec{r}$ then for the triangle, P is the

A. circumcentre B. centroid C. orthocentre D. incentre

3. If $\vec{a}$, $\vec{b}$ and $\vec{c}$ are three vectors of equal magnitude which are pairwise equally inclined then find $\vec{c}$ if $\vec{a} = \vec{i}+\vec{j}$ and $\vec{b} = \vec{j}+\vec{k}$.

4. Find λ if the angle between $\vec{a} = -3\vec{i}+\lambda\vec{j}+\vec{k}$ and $\vec{b} = \lambda\vec{i}+2\lambda\vec{j}+\vec{k}$ is acute and the inclination of $\vec{b}$ with the x-axis lies between $\frac{\pi}{2}$ and π.

5. $\hat{a}, \hat{b}$ are two given unit vectors at right angle. Find a unit vector equally inclined with $\hat{a}, \hat{b}$ and $\hat{a}\times\hat{b}$.

6. Let $\vec{\lambda} = a\vec{i}+b\vec{j}+c\vec{k}$, $\vec{\mu} = x\vec{i}+y\vec{j}+z\vec{k}$, $|\vec{\lambda}| = 5$, $|\vec{\mu}| = 6$, $\vec{\lambda}\cdot\vec{\mu} = 30$.

Find the ratio of projections of $\vec{\lambda}$ and $\vec{\mu}$ on the vector making equal angles with the positive directions of axes.

□

Answers

1. (a) $\frac{\sqrt{3}}{2}$ (b) $\frac{\pi}{4}$ or $\frac{3\pi}{4}$ (c) $\frac{-1}{15}(11\vec{i} + 10\vec{j} + 2\vec{k})$ (d) 4

2. (a) D (b) A (c) B, C (d) C

3. $\vec{i} + \vec{k}$ or $-\frac{1}{3}\vec{i} + \frac{4}{3}\vec{j} - \frac{1}{3}\vec{k}$ **4.** $\lambda < 0$

5. $\frac{1}{\sqrt{3}}(\hat{a} + \hat{b} + \hat{a} \times \hat{b})$ **6.** $\frac{5}{6}$

□

3. Product of three or more Vectors

Recap of Facts and Formulae

1. Scalar triple product

- The dot product of the vector $\vec{a} \times \vec{b}$ with the vector $\vec{c}$ is a scalar triple product of the three vectors $\vec{a}, \vec{b}, \vec{c}$ and it is written as $(\vec{a} \times \vec{b}) \cdot \vec{c}$. It is a scalar quantity.

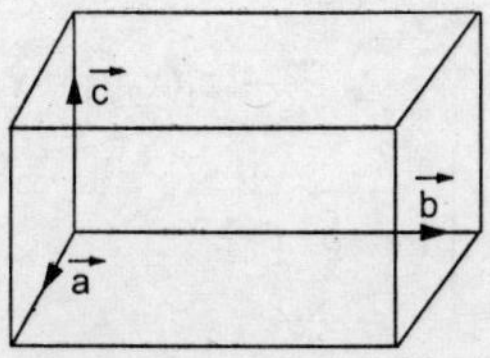

The magnitude of $(\vec{a} \times \vec{b}) \cdot \vec{c}$ is equal to the volume V of a parallelepiped whose three concurrent edges are represented by $\vec{a}, \vec{b}$ and $\vec{c}$ as shown in the figure.

$\therefore$ volume of the parallelepiped whose three concurrent edges are $\vec{a}, \vec{b}, \vec{c} = (\vec{a} \times \vec{b}) \cdot \vec{c}$.

- $(\vec{a} \times \vec{b}) \cdot \vec{c} = \vec{a} \cdot (\vec{b} \times \vec{c})$, i.e., dot and cross can be interchanged in a scalar triple product and each scalar triple product is written as $[\vec{a}\ \vec{b}\ \vec{c}\,]$.
- $[\vec{a}\ \vec{b}\ \vec{c}\,] = [\vec{b}\ \vec{c}\ \vec{a}\,] = [\vec{c}\ \vec{a}\ \vec{b}\,]$
- $[\vec{a}\ \vec{b}\ \vec{c}\,] = -[\vec{b}\ \vec{a}\ \vec{c}\,] = -[\vec{a}\ \vec{c}\ \vec{b}\,] = -[\vec{c}\ \vec{b}\ \vec{a}\,]$
- $[\vec{a}\ \vec{b}\ \vec{c}\,] = 0$ if $\vec{a} = \vec{b}$ or $\vec{b} = \vec{c}$ or $\vec{c} = \vec{a}$

  or at least one vector $= \vec{0}$

  or $\vec{a}, \vec{b}$ and $\vec{c}$ are coplanar
- $[\vec{i}\ \vec{j}\ \vec{k}\,] = 1$
- If $\vec{a} = a_1\vec{i} + a_2\vec{j} + a_3\vec{k}$, $\vec{b} = b_1\vec{i} + b_2\vec{j} + b_3\vec{k}$, $\vec{c} = c_1\vec{i} + c_2\vec{j} + c_3\vec{k}$ then

$$[\vec{a}\ \vec{b}\ \vec{c}\,] = \begin{vmatrix} a_1 & a_2 & a_3 \\ b_1 & b_2 & b_3 \\ c_1 & c_2 & c_3 \end{vmatrix}.$$

2. Condition for coplanarity of three vectors

- $\vec{a}, \vec{b}$ and $\vec{c}$ are coplanar $\Leftrightarrow$ $[\vec{a}\ \vec{b}\ \vec{c}\,] = 0$.

3. Vector triple product

- The vector product of $\vec{a} \times \vec{b}$ and $\vec{c}$ is a vector triple product of the three vectors $\vec{a}, \vec{b}$ and $\vec{c}$.

  $(\vec{a} \times \vec{b}) \times \vec{c}$, $\vec{a} \times (\vec{b} \times \vec{c})$ are vector triple products.
- $\vec{a} \times (\vec{b} \times \vec{c}) = (\vec{a} \cdot \vec{c})\vec{b} - (\vec{a} \cdot \vec{b})\vec{c}$

  This is a vector in the plane of $\vec{b}$ and $\vec{c}$, and perpendicular to $\vec{a}$.
- $(\vec{a} \times \vec{b}) \times \vec{c} = (\vec{a} \cdot \vec{c})\vec{b} - (\vec{b} \cdot \vec{c})\vec{a}$

  This is a vector in the plane of $\vec{a}$ and $\vec{b}$, and perpendicular to $\vec{c}$.

4. Scalar product of four vectors

- $(\vec{a} \times \vec{b}) \cdot (\vec{c} \times \vec{d})$ is a scalar product of four vectors.

  It is a dot product of the vectors $\vec{a} \times \vec{b}$ and $\vec{c} \times \vec{d}$.

  It is scalar triple product of the vectors $\vec{a}, \vec{b}$ and $\vec{c} \times \vec{d}$ as well as a scalar triple product of the vectors $\vec{a} \times \vec{b}$, $\vec{c}$ and $\vec{d}$.

$$(\vec{a} \times \vec{b}) \cdot (\vec{c} \times \vec{d}) = \begin{vmatrix} \vec{a} \cdot \vec{c} & \vec{a} \cdot \vec{d} \\ \vec{b} \cdot \vec{c} & \vec{b} \cdot \vec{d} \end{vmatrix}.$$

5. Vector product of four vectors

- $(\vec{a} \times \vec{b}) \times (\vec{c} \times \vec{d})$ is a vector product of four vectors.

  It is the cross product of the vectors $\vec{a} \times \vec{b}$ and $\vec{c} \times \vec{d}$.
- $\vec{a} \times \{\vec{b} \times (\vec{c} \times \vec{d})\}$, $\{(\vec{a} \times \vec{b}) \times \vec{c}\} \times \vec{d}$ are also different vector products of four vectors $\vec{a}, \vec{b}, \vec{c}$ and $\vec{d}$.

6. To express a vector in terms of 3 noncoplanar vectors

- If $\vec{a}, \vec{b}$ are two known noncollinear vectors then $\vec{a}, \vec{b}, \vec{a} \times \vec{b}$ are three noncoplanar vectors. So any vector $\vec{x}$ can be written as
$$\vec{x} = \lambda \vec{a} + \mu \vec{b} + \nu \vec{a} \times \vec{b}$$
where λ, μ, ν are unknown scalars.

- If $\vec{a}, \vec{b}$ and $\vec{c}$ are three known noncoplanar vectors then any vector $\vec{x}$ can be written as
$$\vec{x} = \lambda \vec{a} + \mu \vec{b} + \nu \vec{c}$$
where λ, μ, ν are unknown scalars.

Note The above two ways of expressing $\vec{x}$ is useful in solving vector equations.

Selected Solved Examples

1. The position vectors of three points A, B and C in space are respectively

$$2\vec{i} + 3\vec{j} - \vec{k},\ \vec{i} - 2\vec{j} + 3\vec{k} \text{ and } 4\vec{i} + \vec{j} + \vec{k}.$$

Find the volume of the parallelepiped whose three concurrent edges are OA, OB and OC where O is the origin.

The volume of the parallelepiped $= [\vec{OA}\ \vec{OB}\ \vec{OC}]$

$$= \begin{vmatrix} 2 & 3 & -1 \\ 1 & -2 & 3 \\ 4 & 1 & 1 \end{vmatrix} \begin{aligned} &= 2(-5) - 3(1-12) - 1(1+8) \\ &= -10 + 33 - 9 = 14. \end{aligned}$$

2. If $\vec{A} = x_1\vec{a} + y_1\vec{b} + z_1\vec{c}$, $\vec{B} = x_2\vec{a} + y_2\vec{b} + z_2\vec{c}$ and $\vec{C} = x_3\vec{a} + y_3\vec{b} + z_3\vec{c}$ then prove that

$$[\vec{A}\ \vec{B}\ \vec{C}] = \begin{vmatrix} x_1 & y_1 & z_1 \\ x_2 & y_2 & z_2 \\ x_3 & y_3 & z_3 \end{vmatrix} [\vec{a}\ \vec{b}\ \vec{c}].$$

We have $[\vec{A}\ \vec{B}\ \vec{C}] = (\vec{A} \times \vec{B}) \cdot \vec{C}$.

Now, $\vec{A} \times \vec{B}$

$$= (x_1\vec{a} + y_1\vec{b} + z_1\vec{c}) \times (x_2\vec{a} + y_2\vec{b} + z_2\vec{c})$$

$$= (x_1y_2 - x_2y_1)(\vec{a} \times \vec{b}) + (z_1x_2 - z_2x_1)(\vec{c} \times \vec{a}) + (y_1z_2 - y_2z_1)(\vec{b} \times \vec{c})$$

$\therefore \quad (\vec{A} \times \vec{B}) \cdot \vec{C}$

$$= \{(x_1y_2 - x_2y_1)(\vec{a} \times \vec{b}) + (z_1x_2 - z_2x_1)(\vec{c} \times \vec{a}) + (y_1z_2 - y_2z_1)(\vec{b} \times \vec{c})\} \cdot (x_3\vec{a} + y_3\vec{b} + z_3\vec{c})$$

$$= (x_1y_2 - x_2y_1)z_3[\vec{a}\ \vec{b}\ \vec{c}] + (z_1x_2 - z_2x_1)y_3[\vec{c}\ \vec{a}\ \vec{b}] + (y_1z_2 - y_2z_1)x_3[\vec{b}\ \vec{c}\ \vec{a}],$$

because $(\vec{a} \times \vec{b}) \cdot \vec{a} = [\vec{a}\ \vec{b}\ \vec{a}] = 0$, etc.

$$= \{z_3(x_1y_2 - x_2y_1) - y_3(x_1z_2 - x_2z_1) + x_3(y_1z_2 - y_2z_1)\}[\vec{a}\ \vec{b}\ \vec{c}],$$

because $[\vec{a}\ \vec{b}\ \vec{c}] = [\vec{b}\ \vec{c}\ \vec{a}] = [\vec{c}\ \vec{a}\ \vec{b}]$

$$= \{x_1(y_2z_3 - y_3z_2) + y_1(z_2x_3 - z_3x_2) + z_1(x_2y_3 - x_3y_2)\}[\vec{a}\ \vec{b}\ \vec{c}]$$

$$= \begin{vmatrix} x_1 & y_1 & z_1 \\ x_2 & y_2 & z_2 \\ x_3 & y_3 & z_3 \end{vmatrix} [\vec{a}\ \vec{b}\ \vec{c}].$$

3. Prove that $[\vec{a} + \vec{b}\quad \vec{b} + \vec{c}\quad \vec{c} + \vec{a}] = 2[\vec{a}\ \vec{b}\ \vec{c}]$.

$$\text{LHS} = \{(\vec{a} + \vec{b}) \times (\vec{b} + \vec{c})\} \cdot (\vec{c} + \vec{a})$$

$$= (\vec{a} \times \vec{b} + \vec{a} \times \vec{c} + \vec{b} \times \vec{c}) \cdot (\vec{c} + \vec{a}) \quad \{\because \vec{b} \times \vec{b} = \vec{0}\}$$

$$= [\vec{a}\ \vec{b}\ \vec{c}] + [\vec{a}\ \vec{b}\ \vec{a}] + [\vec{a}\ \vec{c}\ \vec{c}] + [\vec{a}\ \vec{c}\ \vec{a}] + [\vec{b}\ \vec{c}\ \vec{c}] + [\vec{b}\ \vec{c}\ \vec{a}]$$

$$= [\vec{a}\ \vec{b}\ \vec{c}] + [\vec{b}\ \vec{c}\ \vec{a}] \quad \{\because [\vec{a}\ \vec{b}\ \vec{a}] = 0, \text{ etc.}\}$$

$$= 2[\vec{a}\ \vec{b}\ \vec{c}].$$

4. If $\vec{a}, \vec{b}, \vec{c}$ are three vectors such that $\vec{a} \times \vec{b} = \vec{c}$, $\vec{b} \times \vec{c} = \vec{a}$, $\vec{c} \times \vec{a} = \vec{b}$ then prove that

$$|\vec{a}| = |\vec{b}| = |\vec{c}|.$$

$$\vec{a} \times \vec{b} = \vec{c} \quad \Rightarrow \quad (\vec{a} \times \vec{b}) \cdot \vec{c} = \vec{c} \cdot \vec{c}$$

$$\therefore \quad [\vec{a}\ \vec{b}\ \vec{c}] = |\vec{c}|^2$$

$$\vec{b} \times \vec{c} = \vec{a} \quad \Rightarrow \quad (\vec{b} \times \vec{c}) \cdot \vec{a} = \vec{a} \cdot \vec{a}$$

$$\therefore \quad [\vec{b}\ \vec{c}\ \vec{a}] = |\vec{a}|^2$$

$$\vec{c} \times \vec{a} = \vec{b} \quad \Rightarrow \quad (\vec{c} \times \vec{a}) \cdot \vec{b} = \vec{b} \cdot \vec{b}$$

$$\therefore \quad [\vec{c}\ \vec{a}\ \vec{b}] = |\vec{b}|^2$$

But $[\vec{a}\ \vec{b}\ \vec{c}] = [\vec{b}\ \vec{c}\ \vec{a}] = [\vec{c}\ \vec{a}\ \vec{b}]$

$\therefore\quad |\vec{c}|^2 = |\vec{a}|^2 = |\vec{b}|^2$

$\therefore\quad |\vec{a}| = |\vec{b}| = |\vec{c}|.$

5. If $\vec{a}, \vec{b}, \vec{c}, \vec{d}$ are four coplanar points then show that

$$[\vec{a}\ \vec{b}\ \vec{c}] = [\vec{b}\ \vec{c}\ \vec{d}] + [\vec{c}\ \vec{a}\ \vec{d}] + [\vec{a}\ \vec{b}\ \vec{d}].$$

As $\vec{a}, \vec{b}, \vec{c}, \vec{d}$ are coplanar points we have $\vec{b} - \vec{a}$, $\vec{c} - \vec{a}, \vec{d} - \vec{a}$ are coplanar.

$\therefore\quad [\vec{b} - \vec{a}\quad \vec{c} - \vec{a}\quad \vec{d} - \vec{a}] = 0$

or $\quad \{(\vec{b} - \vec{a}) \times (\vec{c} - \vec{a})\} \cdot (\vec{d} - \vec{a}) = 0$

or $\quad (\vec{b} \times \vec{c} - \vec{b} \times \vec{a} - \vec{a} \times \vec{c}) \cdot (\vec{d} - \vec{a}) = 0$

or $\quad (\vec{b} \times \vec{c}) \cdot \vec{d} - (\vec{b} \times \vec{c}) \cdot \vec{a} - (\vec{b} \times \vec{a}) \cdot \vec{d}$
$\quad + (\vec{b} \times \vec{a}) \cdot \vec{a} - (\vec{a} \times \vec{c}) \cdot \vec{d} + (\vec{a} \times \vec{c}) \cdot \vec{a} = 0$

or $\quad [\vec{b}\ \vec{c}\ \vec{d}] - [\vec{b}\ \vec{c}\ \vec{a}] - [\vec{b}\ \vec{a}\ \vec{d}] - [\vec{a}\ \vec{c}\ \vec{d}] = 0$

$\{\because\ [\vec{b}\ \vec{a}\ \vec{a}] = 0$, etc.$\}$

or $\quad [\vec{b}\ \vec{c}\ \vec{d}] - [\vec{a}\ \vec{b}\ \vec{c}] + [\vec{a}\ \vec{b}\ \vec{d}] + [\vec{c}\ \vec{a}\ \vec{d}] = 0$

{using property of scalar triple product}

$\therefore\quad [\vec{a}\ \vec{b}\ \vec{c}] = [\vec{b}\ \vec{c}\ \vec{d}] + [\vec{a}\ \vec{b}\ \vec{d}] + [\vec{c}\ \vec{a}\ \vec{d}].$

6. If $\vec{x} \times \vec{a} + (\vec{x} \cdot \vec{b})\vec{c} = \vec{d}$ then prove that

$$\vec{x} = \lambda \vec{a} + \vec{a} \times \frac{\vec{a} \times (\vec{d} \times \vec{c})}{(\vec{a} \cdot \vec{c})\vec{a}^2},$$ **where λ is a scalar.**

Here $\vec{x} \times \vec{a} + (\vec{x} \cdot \vec{b})\vec{c} = \vec{d}$

$\therefore\quad \{\vec{x} \times \vec{a} + (\vec{x} \cdot \vec{b})\vec{c}\} \times \vec{c} = \vec{d} \times \vec{c}$

or $\quad (\vec{x} \times \vec{a}) \times \vec{c} = \vec{d} \times \vec{c}$, for $\vec{c} \times \vec{c} = \vec{0}$

or $\quad (\vec{x} \cdot \vec{c})\vec{a} - (\vec{a} \cdot \vec{c})\vec{x} = \vec{d} \times \vec{c}$

$\therefore\quad \vec{a} \times \{(\vec{x} \cdot \vec{c})\vec{a} - (\vec{a} \cdot \vec{c})\vec{x}\} = \vec{a} \times (\vec{d} \times \vec{c})$

or $\quad -(\vec{a} \cdot \vec{c})(\vec{a} \times \vec{x}) = \vec{a} \times (\vec{d} \times \vec{c})$

$\{\because\ \vec{a} \times \vec{a} = \vec{0}\}$

or $\quad \vec{x} \times \vec{a} = \dfrac{\vec{a} \times (\vec{d} \times \vec{c})}{\vec{a} \cdot \vec{c}}$

or $\quad \vec{a} \times (\vec{x} \times \vec{a}) = \vec{a} \times \dfrac{\vec{a} \times (\vec{d} \times \vec{c})}{\vec{a} \cdot \vec{c}}$

or $\quad (\vec{a} \cdot \vec{a})\vec{x} - (\vec{a} \cdot \vec{x})\vec{a} = \vec{a} \times \dfrac{\vec{a} \times (\vec{d} \times \vec{c})}{\vec{a} \cdot \vec{c}}$

or $\quad \vec{a}^2\ \vec{x} = (\vec{a} \cdot \vec{x})\vec{a} + \vec{a} \times \dfrac{\vec{a} \times (\vec{d} \times \vec{c})}{\vec{a} \cdot \vec{c}}$

$\therefore\quad \vec{x} = \dfrac{\vec{a} \cdot \vec{x}}{\vec{a}^2}\ \vec{a} + \vec{a} \times \dfrac{\vec{a} \times (\vec{d} \times \vec{c})}{(\vec{a} \cdot \vec{c})\vec{a}^2}$

$\quad = \lambda \vec{a} + \vec{a} \times \dfrac{\vec{a} \times (\vec{d} \times \vec{c})}{(\vec{a} \cdot \vec{c})\vec{a}^2}$

where $\lambda = \dfrac{\vec{a} \cdot \vec{x}}{\vec{a}^2} =$ scalar.

7. If $\vec{x} \cdot \vec{a} = 0,\ \vec{x} \cdot \vec{b} = 1,\ [\vec{x}\ \vec{a}\ \vec{b}] = 1,\ \vec{a} \cdot \vec{b} \neq 0$ then find $\vec{x}$ in terms of $\vec{a}$ and $\vec{b}$.

As $\vec{a}, \vec{b}$ and $\vec{a} \times \vec{b}$ are noncoplanar vectors, let

$\quad \vec{x} = \lambda \vec{a} + \mu \vec{b} + \nu \vec{a} \times \vec{b}$ ... (1)

$\therefore\quad \vec{x} \cdot \vec{a} = \lambda \vec{a}^2 + \mu \vec{b} \cdot \vec{a} + \nu[\vec{a}\ \vec{b}\ \vec{a}]$

or $\quad 0 = \lambda \vec{a}^2 + \mu \vec{a} \cdot \vec{b}$ ... (2)

Again from (1),

$\quad \vec{x} \cdot \vec{b} = \lambda \vec{a} \cdot \vec{b} + \mu \vec{b}^2 + \nu[\vec{a}\ \vec{b}\ \vec{b}]$

$\therefore\quad 1 = \lambda \vec{a} \cdot \vec{b} + \mu \vec{b}^2$ ... (3)

Again from (1),

$\quad \vec{x} \cdot (\vec{a} \times \vec{b}) = \lambda \vec{a} \cdot (\vec{a} \times \vec{b})$
$\quad + \mu \vec{b} \cdot (\vec{a} \times \vec{b}) + \nu(\vec{a} \times \vec{b})^2$

$\therefore\quad [\vec{x}\ \vec{a}\ \vec{b}] = \lambda[\vec{a}\ \vec{a}\ \vec{b}]$
$\quad + \mu[\vec{b}\ \vec{a}\ \vec{b}] + \nu(\vec{a} \times \vec{b})^2$

$\therefore\quad 1 = \nu(\vec{a} \times \vec{b})^2$ ... (4)

From (2), $\mu \vec{a} \cdot \vec{b} = -\lambda \vec{a}^2$

$\therefore\quad \mu = -\lambda \dfrac{\vec{a}^2}{\vec{a} \cdot \vec{b}}$

$\therefore\quad (3) \Rightarrow\quad 1 = \lambda \vec{a} \cdot \vec{b} - \lambda \dfrac{\vec{a}^2\ \vec{b}^2}{\vec{a} \cdot \vec{b}}$

or $\quad 1 = \lambda \left\{ \dfrac{(\vec{a} \cdot \vec{b})^2 - \vec{a}^2\ \vec{b}^2}{\vec{a} \cdot \vec{b}} \right\}$

$\therefore \quad \lambda = \dfrac{\vec{a}\cdot\vec{b}}{(\vec{a}\cdot\vec{b})^2 - \vec{a}^2\,\vec{b}^2}$

$\therefore \quad \mu = -\dfrac{\vec{a}^2}{(\vec{a}\cdot\vec{b})^2 - \vec{a}^2\,\vec{b}^2}.$

Also, (4) $\Rightarrow \quad \nu = \dfrac{1}{(\vec{a}\times\vec{b})^2}.$

Putting values of $\lambda,\ \mu,\ \nu$ in (1), we get

$$\vec{x} = \frac{(\vec{a}\cdot\vec{b})\,\vec{a}}{(\vec{a}\cdot\vec{b})^2 - \vec{a}^2\,\vec{b}^2} - \frac{(\vec{a}^2)\,\vec{b}}{(\vec{a}\cdot\vec{b})^2 - \vec{a}^2\,\vec{b}^2} + \frac{\vec{a}\times\vec{b}}{(\vec{a}\times\vec{b})^2}$$

$$\therefore \quad \vec{x} = \frac{(\vec{a}\cdot\vec{b})\,\vec{a} - (\vec{a}\cdot\vec{a})\,\vec{b}}{(\vec{a}\cdot\vec{b})^2 - \vec{a}^2\,\vec{b}^2} + \frac{\vec{a}\times\vec{b}}{(\vec{a}\times\vec{b})^2}$$

$$= \frac{\vec{a}\times(\vec{a}\times\vec{b})}{(\vec{a}\cdot\vec{b})^2 - \vec{a}^2\,\vec{b}^2} + \frac{\vec{a}\times\vec{b}}{(\vec{a}\times\vec{b})^2}.$$

8. Given two noncollinear vectors $\vec{a},\ \vec{b}$ solve for $\vec{x},\ \vec{y}$ where $\vec{x}+\vec{y}=\vec{a},\quad \vec{x}\times\vec{y}=\vec{b},\quad \vec{x}\cdot\vec{a}=1.$

Here $(\vec{x}+\vec{y})\cdot(\vec{x}\times\vec{y}) = \vec{a}\cdot\vec{b}$

$\therefore \quad \vec{a}\cdot\vec{b} = [\vec{x}\ \vec{x}\ \vec{y}] + [\vec{y}\ \vec{x}\ \vec{y}] = 0 + 0 = 0 \qquad \dots (1)$

$\vec{x}\times\vec{y} = \vec{b}$

$\Rightarrow \quad \vec{x}\cdot\vec{b} = \vec{x}\cdot(\vec{x}\times\vec{y}) = [\vec{x}\ \vec{x}\ \vec{y}] = 0 \qquad \dots (2)$

Also $\vec{y}\cdot\vec{b} = \vec{y}\cdot(\vec{x}\times\vec{y}) = [\vec{y}\ \vec{x}\ \vec{y}] = 0 \qquad \dots (3)$

$\vec{x}+\vec{y} = \vec{a} \quad \Rightarrow \quad (\vec{x}+\vec{y})\cdot\vec{a} = \vec{a}\cdot\vec{a}$

or $\quad \vec{x}\cdot\vec{a} + \vec{y}\cdot\vec{a} = \vec{a}^2$

$\therefore \quad \vec{y}\cdot\vec{a} = \vec{a}^2 - 1 \qquad \dots (4)$

because $\vec{x}\cdot\vec{a} = 1$

Now, let $\vec{x} = \lambda\vec{a} + \mu\vec{b} + \nu\vec{a}\times\vec{b} \qquad \dots (5)$

$\{\because\ \vec{a},\ \vec{b},\ \vec{a}\times\vec{b}$ are noncoplanar$\}$

$\therefore \quad \vec{x}\cdot\vec{a} = \lambda\vec{a}^2 + \mu\vec{b}\cdot\vec{a} + \nu(\vec{a}\times\vec{b})\cdot\vec{a}$

$\therefore \quad 1 = \lambda\vec{a}^2$, using (1), and $[\vec{a}\ \vec{b}\ \vec{a}] = 0$

$\therefore \quad \lambda = \dfrac{1}{\vec{a}^2}.$

Again, $\vec{x}\cdot\vec{b} = \lambda\vec{a}\cdot\vec{b} + \mu\vec{b}^2 + \nu\vec{a}\times\vec{b}\cdot\vec{b}$ {from (5)}

$\therefore \quad$ from (2), $0 = \lambda\cdot 0 + \mu\vec{b}^2; \quad \therefore \quad \mu = 0.$

Thus, $\vec{x} = \dfrac{1}{\vec{a}^2}\vec{a} + \nu\vec{a}\times\vec{b} \qquad \dots (6)$

Let $\vec{y} = \lambda'\vec{a} + \mu'\vec{b} + \nu'\vec{a}\times\vec{b}$

$\therefore \quad$ as above, $\vec{y}\cdot\vec{a} = \lambda'\vec{a}^2$

$\therefore \quad$ from (4), $\vec{a}^2 - 1 = \lambda'\vec{a}^2; \quad \therefore \quad \lambda' = \dfrac{\vec{a}^2 - 1}{\vec{a}^2}.$

Also $\vec{y}\cdot\vec{b} = \lambda'\vec{a}\cdot\vec{b} + \mu'\vec{b}^2$

$\therefore \quad$ from (3), $0 = \mu'\vec{b}^2; \quad \therefore \quad \mu' = 0.$

Thus, $\vec{y} = \dfrac{\vec{a}^2 - 1}{\vec{a}^2}\vec{a} + \nu'\vec{a}\times\vec{b} \qquad \dots (7)$

(6) + (7) gives,

$$\vec{x}+\vec{y} = \left(\frac{1}{\vec{a}^2} + \frac{\vec{a}^2 - 1}{\vec{a}^2}\right)\vec{a} + (\nu+\nu')(\vec{a}\times\vec{b})$$

or $\quad \vec{a} = \vec{a} + (\nu+\nu')(\vec{a}\times\vec{b})$ (given $\vec{x}+\vec{y}=\vec{a}$)

$\therefore \quad \nu+\nu' = 0; \quad \therefore \quad \nu' = -\nu.$

So, $\vec{y} = \dfrac{\vec{a}^2 - 1}{\vec{a}^2}\vec{a} - \nu\vec{a}\times\vec{b} \qquad \dots (8)$

From (6) and (8),

$$\vec{x}\times\vec{y} = \left(\frac{1}{\vec{a}^2}\vec{a} + \nu\vec{a}\times\vec{b}\right)\times\left(\frac{\vec{a}^2-1}{\vec{a}^2}\vec{a} - \nu\vec{a}\times\vec{b}\right)$$

$$= -\frac{\nu}{\vec{a}^2}\vec{a}\times(\vec{a}\times\vec{b}) + \frac{\nu(\vec{a}^2-1)}{\vec{a}^2}(\vec{a}\times\vec{b})\times\vec{a}$$

$$= \left\{\frac{\nu}{\vec{a}^2} + \frac{\nu(\vec{a}^2-1)}{\vec{a}^2}\right\}(\vec{a}\times\vec{b})\times\vec{a}$$

$$= \nu(\vec{a}\times\vec{b})\times\vec{a} = \nu\{(\vec{a}\cdot\vec{a})\vec{b} - (\vec{b}\cdot\vec{a})\vec{a}\}$$

$\therefore \quad \vec{b} = (\nu\vec{a}^2)\vec{b}$

because $\vec{b}\cdot\vec{a} = 0$ by (1), and $\vec{x}\times\vec{y} = \vec{b}$

$\therefore \quad 1 = \nu\vec{a}^2; \quad \therefore \quad \nu = \dfrac{1}{\vec{a}^2}$

$\therefore \quad$ from (6) and (8),

$$\vec{x} = \frac{1}{\vec{a}^2}(\vec{a} + \vec{a}\times\vec{b})$$

and $\vec{y} = \frac{1}{\vec{a}^{\,2}}\{(\vec{a}^{\,2} - 1)\vec{a} - \vec{a} \times \vec{b}\}$.

9. If $\vec{b}$ and $\vec{c}$ are any two noncollinear unit vectors and $\vec{a}$ is any other vector then prove that

$$(\vec{a} \cdot \vec{b})\vec{b} + (\vec{a} \cdot \vec{c})\vec{c} + \frac{\vec{a} \cdot \vec{b} \times \vec{c}}{|\vec{b} \times \vec{c}|^2}(\vec{b} \times \vec{c})$$
$$= \vec{a} - (\vec{b} \cdot \vec{c})(\vec{b} + \vec{c}).$$

The resolved part of $\vec{a}$ along $\vec{b} \times \vec{c}$

$$= \left(\vec{a} \cdot \frac{\vec{b} \times \vec{c}}{|\vec{b} \times \vec{c}|}\right)\frac{\vec{b} \times \vec{c}}{|\vec{b} \times \vec{c}|}$$

$$\therefore \quad \vec{OQ} = \vec{r} = \vec{a} - \frac{\vec{a} \cdot (\vec{b} \times \vec{c})}{|\vec{b} \times \vec{c}|^2}(\vec{b} \times \vec{c}) \qquad \dots (1)$$

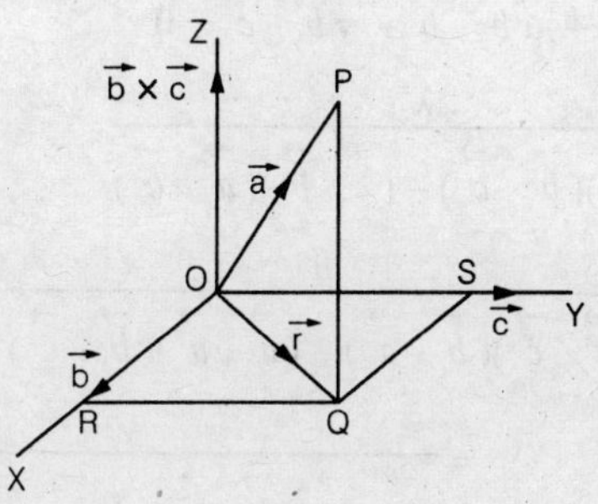

Now, $\vec{OQ} = \vec{OR} + \vec{OS}$

and $\vec{OR} = |\vec{OR}|\,\vec{b} = (\vec{r} \cdot \vec{b} + \vec{c} \cdot \vec{b})\vec{b}$

$\vec{OS} = |\vec{OS}|\,\vec{c} = (\vec{r} \cdot \vec{c} + \vec{b} \cdot \vec{c})\vec{c}$

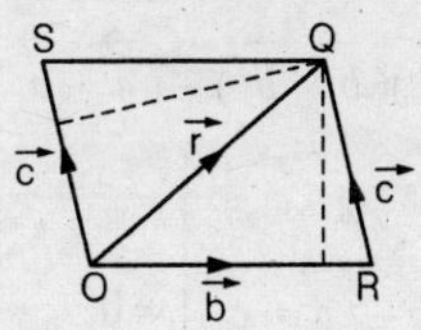

$$\therefore \quad \vec{OQ} = (\vec{r} \cdot \vec{b})\vec{b} + (\vec{r} \cdot \vec{c})\vec{c} + (\vec{b} \cdot \vec{c})(\vec{b} + \vec{c}) \qquad \dots (2)$$

From (1),

$$\vec{r} \cdot \vec{b} = \vec{a} \cdot \vec{b} - \frac{\vec{a} \cdot (\vec{b} \times \vec{c})}{|\vec{b} \times \vec{c}|^2}[\vec{b}\ \vec{c}\ \vec{b}]$$

$$= \vec{a} \cdot \vec{b} \qquad (\because [\vec{b}\ \vec{c}\ \vec{b}] = 0)$$

$$\vec{r} \cdot \vec{c} = \vec{a} \cdot \vec{c} - \frac{\vec{a} \cdot (\vec{b} \times \vec{c})}{|\vec{b} \times \vec{c}|^2}[\vec{b}\ \vec{c}\ \vec{c}]$$

$$= \vec{a} \cdot \vec{c}$$

$\therefore$ from (2),

$$\vec{OQ} = (\vec{a} \cdot \vec{b})\vec{b} + (\vec{a} \cdot \vec{c})\vec{c} + (\vec{b} \cdot \vec{c})(\vec{b} + \vec{c})$$

$\therefore$ from (1),

$$(\vec{a} \cdot \vec{b})\vec{b} + (\vec{a} \cdot \vec{c})\vec{c} + (\vec{b} \cdot \vec{c})(\vec{b} + \vec{c})$$
$$= \vec{a} - \frac{\vec{a} \cdot (\vec{b} \times \vec{c})}{|\vec{b} \times \vec{c}|^2}(\vec{b} \times \vec{c})$$

$$\therefore \quad (\vec{a} \cdot \vec{b})\vec{b} + (\vec{a} \cdot \vec{c})\vec{c} + \frac{\vec{a} \cdot (\vec{b} \times \vec{c})}{|\vec{b} \times \vec{c}|^2}(\vec{b} \times \vec{c})$$
$$= \vec{a} - (\vec{b} \cdot \vec{c})(\vec{b} + \vec{c}).$$

10. Let $\hat{x}$, $\hat{y}$ and $\hat{z}$ be unit vectors such that

$$\hat{x} + \hat{y} + \hat{z} = \vec{a}, \quad \hat{x} \times (\hat{y} \times \hat{z}) = \vec{b},$$
$$(\hat{x} \times \hat{y}) \times \hat{z} = \vec{c}, \quad \vec{a} \cdot \hat{x} = \frac{3}{2},$$
$$\vec{a} \cdot \hat{y} = \frac{7}{4} \text{ and } |\vec{a}| = 2.$$

Find $\hat{x}$, $\hat{y}$ and $\hat{z}$ in terms of $\vec{a}$, $\vec{b}$ and $\vec{c}$.

$$\hat{x} + \hat{y} + \hat{z} = \vec{a} \qquad \dots (1)$$

$$\therefore \quad \vec{a} \cdot \hat{x} + \vec{a} \cdot \hat{y} + \vec{a} \cdot \hat{z} = \vec{a} \cdot \vec{a} = |\vec{a}|^2 = 4$$

$$\therefore \quad \frac{3}{2} + \frac{7}{4} + \vec{a} \cdot \hat{z} = 4$$

or $\quad \vec{a} \cdot \hat{z} = \frac{3}{4}$

Thus we get, $\vec{a} \cdot \hat{x} = \frac{3}{2}$, $\vec{a} \cdot \hat{y} = \frac{7}{4}$, $\vec{a} \cdot \hat{z} = \frac{3}{4} \qquad \dots (2)$

From (1), $\hat{x} \cdot (\hat{x} + \hat{y} + \hat{z}) = \hat{x} \cdot \vec{a} = \vec{a} \cdot \hat{x} = \frac{3}{2}$

or $\quad \hat{x}^2 + \hat{x} \cdot \hat{y} + \hat{x} \cdot \hat{z} = \frac{3}{2}$

or $\quad 1 + \hat{x} \cdot \hat{y} + \hat{x} \cdot \hat{z} = \frac{3}{2}$

$$\therefore \quad \hat{x} \cdot \hat{y} + \hat{x} \cdot \hat{z} = \frac{1}{2} \qquad \dots (3)$$

From (1), $\hat{y} \cdot (\hat{x} + \hat{y} + \hat{z}) = \hat{y} \cdot \vec{a} = \vec{a} \cdot \hat{y} = \frac{7}{4}$

or $\quad \hat{y} \cdot \hat{x} + \hat{y}^2 + \hat{y} \cdot \hat{z} = \frac{7}{4}$

or $\quad \hat{x} \cdot \hat{y} + 1 + \hat{y} \cdot \hat{z} = \frac{7}{4}$

$\therefore \quad \hat{x}\cdot\hat{y}+\hat{y}\cdot\hat{z}=\frac{3}{4}$ ... (4)

From (1), $(\hat{x}+\hat{y}+\hat{z})^2=\vec{a}^2$

or $\hat{x}^2+\hat{y}^2+\hat{z}^2+2(\hat{x}\cdot\hat{y}+\hat{y}\cdot\hat{z}+\hat{x}\cdot\hat{z})=|\vec{a}|^2=4$

or $3+2(\hat{x}\cdot\hat{y}+\hat{y}\cdot\hat{z}+\hat{x}\cdot\hat{z})=4$

$\therefore \quad \hat{x}\cdot\hat{y}+\hat{y}\cdot\hat{z}+\hat{x}\cdot\hat{z}=\frac{1}{2}$ ... (5)

Fom (3), (4), (5) we get

$\hat{y}\cdot\hat{z}=0, \quad \hat{x}\cdot\hat{z}=\frac{-1}{4}, \quad \hat{x}\cdot\hat{y}=\frac{3}{4}.$

Now, $\hat{x}\times(\hat{y}\times\hat{z})=\vec{b}$

$\therefore \quad (\hat{x}\cdot\hat{z})\hat{y}-(\hat{x}\cdot\hat{y})\hat{z}=\vec{b}$

or $-\frac{1}{4}\hat{y}-\frac{3}{4}\hat{z}=\vec{b}$ ... (6)

Again, $(\hat{x}\times\hat{y})\times\hat{z}=\vec{c}$

$\therefore \quad (\hat{x}\cdot\hat{z})\hat{y}-(\hat{y}\cdot\hat{z})\hat{x}=\vec{c}$

$-\frac{1}{4}\hat{y}=\vec{c}; \quad \therefore \quad \hat{y}=-4\vec{c}.$

$\therefore$ from (6), $\vec{c}-\frac{3}{4}\hat{z}=\vec{b}$

or $\vec{c}-\vec{b}=\frac{3}{4}\hat{z}; \quad \therefore \quad \hat{z}=\frac{4}{3}(\vec{c}-\vec{b}).$

$\therefore$ from (1), $\hat{x}=\vec{a}-\hat{y}-\hat{z}=\vec{a}+4\vec{c}-\frac{4}{3}(\vec{c}-\vec{b})$

$=\vec{a}+\frac{4}{3}\vec{b}+\frac{8}{3}\vec{c}$

$=\frac{1}{3}(3\vec{a}+4\vec{b}+8\vec{c}).$

Thus, $\hat{x}=\frac{1}{3}(3\vec{a}+4\vec{b}+8\vec{c}),$

$\hat{y}=-4\vec{c}, \quad \hat{z}=\frac{4}{3}(\vec{c}-\vec{b}).$

11. Prove that $(\vec{a}\times\vec{b})^2=\begin{vmatrix}\vec{a}\cdot\vec{a} & \vec{a}\cdot\vec{b}\\ \vec{b}\cdot\vec{a} & \vec{b}\cdot\vec{b}\end{vmatrix}.$

$(\vec{a}\times\vec{b})^2=(\vec{a}\times\vec{b})\cdot(\vec{a}\times\vec{b})$

$=\vec{a}\cdot\{\vec{b}\times(\vec{a}\times\vec{b})\}$

($\because$ in a scalar triple product, dot and cross can be interchanged)

$=\vec{a}\cdot\{(\vec{b}\cdot\vec{b})\vec{a}-(\vec{b}\cdot\vec{a})\vec{b}\}$

$=(\vec{b}\cdot\vec{b})(\vec{a}\cdot\vec{a})-(\vec{b}\cdot\vec{a})(\vec{a}\cdot\vec{b})$

$=\begin{vmatrix}\vec{a}\cdot\vec{a} & \vec{a}\cdot\vec{b}\\ \vec{b}\cdot\vec{a} & \vec{b}\cdot\vec{b}\end{vmatrix}$

12. If $\vec{a}$, $\vec{b}$ and $\vec{c}$ are coplanar vectors, prove that

$$\begin{vmatrix}\vec{a} & \vec{b} & \vec{c}\\ \vec{a}\cdot\vec{a} & \vec{a}\cdot\vec{b} & \vec{a}\cdot\vec{c}\\ \vec{b}\cdot\vec{a} & \vec{b}\cdot\vec{b} & \vec{b}\cdot\vec{c}\end{vmatrix}=0.$$

As $\vec{a}$, $\vec{b}$ and $\vec{c}$ are coplanar, they are linearly dependent.

So, $\lambda\vec{a}+\mu\vec{b}+\nu\vec{c}=0$ ... (1)

for some scalars λ, μ, ν.

Taking dot product with $\vec{a}$ and $\vec{b}$ respectively,

$\lambda\vec{a}\cdot\vec{a}+\mu\vec{a}\cdot\vec{b}+\nu\vec{a}\cdot\vec{c}=0$

and $\lambda\vec{b}\cdot\vec{a}+\mu\vec{b}\cdot\vec{b}+\nu\vec{b}\cdot\vec{c}=0$

$$\therefore \quad \frac{\lambda}{(\vec{a}\cdot\vec{b})(\vec{b}\cdot\vec{c})-(\vec{b}\cdot\vec{b})(\vec{a}\cdot\vec{c})}=\frac{\mu}{(\vec{a}\cdot\vec{c})(\vec{b}\cdot\vec{a})-(\vec{a}\cdot\vec{a})(\vec{b}\cdot\vec{c})}=\frac{\nu}{(\vec{a}\cdot\vec{a})(\vec{b}\cdot\vec{b})-(\vec{a}\cdot\vec{b})(\vec{b}\cdot\vec{a})}$$

$\therefore$ from (1),

$\vec{a}\{(\vec{a}\cdot\vec{b})(\vec{b}\cdot\vec{c})-(\vec{b}\cdot\vec{b})(\vec{a}\cdot\vec{c})\}$

$+\vec{b}\{(\vec{a}\cdot\vec{c})(\vec{b}\cdot\vec{a})-(\vec{a}\cdot\vec{a})(\vec{b}\cdot\vec{c})\}$

$+\vec{c}\{(\vec{a}\cdot\vec{a})(\vec{b}\cdot\vec{b})-(\vec{a}\cdot\vec{b})(\vec{b}\cdot\vec{a})\}=0$

$$\therefore \quad \begin{vmatrix}\vec{a} & \vec{b} & \vec{c}\\ \vec{a}\cdot\vec{a} & \vec{a}\cdot\vec{b} & \vec{a}\cdot\vec{c}\\ \vec{b}\cdot\vec{a} & \vec{b}\cdot\vec{b} & \vec{b}\cdot\vec{c}\end{vmatrix}=0.$$

13. If $\vec{a}$, $\vec{b}$, $\vec{c}$ are noncoplanar vectors, prove that

$$\vec{p}\times\vec{q}=\frac{1}{[\vec{a}\ \vec{b}\ \vec{c}]}\begin{vmatrix}\vec{p}\cdot\vec{a} & \vec{q}\cdot\vec{a} & \vec{a}\\ \vec{p}\cdot\vec{b} & \vec{q}\cdot\vec{b} & \vec{b}\\ \vec{p}\cdot\vec{c} & \vec{q}\cdot\vec{c} & \vec{c}\end{vmatrix}.$$

As $\vec{a}$, $\vec{b}$, $\vec{c}$ are noncoplanar, we can write

$\vec{p}\times\vec{q}=\lambda\vec{a}+\mu\vec{b}+\nu\vec{c}$ ... (1)

$\therefore \quad (\vec{p} \times \vec{q}) \cdot (\vec{b} \times \vec{c})$

$= \lambda \vec{a} \cdot (\vec{b} \times \vec{c}) + \mu \vec{b} \cdot (\vec{b} \times \vec{c}) + \nu \vec{c} \cdot (\vec{b} \times \vec{c})$

or $\quad \vec{p} \cdot \{\vec{q} \times (\vec{b} \times \vec{c})\} = \lambda [\vec{a}\ \vec{b}\ \vec{c}]$

or $\quad \vec{p} \cdot \{(\vec{q} \cdot \vec{c})\vec{b} - (\vec{q} \cdot \vec{b})\vec{c}\} = \lambda [\vec{a}\ \vec{b}\ \vec{c}]$

or $\quad (\vec{q} \cdot \vec{c})(\vec{p} \cdot \vec{b}) - (\vec{q} \cdot \vec{b})(\vec{p} \cdot \vec{c}) = \lambda [\vec{a}\ \vec{b}\ \vec{c}]$

$$\therefore \quad \lambda [\vec{a}\ \vec{b}\ \vec{c}] = \begin{vmatrix} \vec{p} \cdot \vec{b} & \vec{q} \cdot \vec{b} \\ \vec{p} \cdot \vec{c} & \vec{q} \cdot \vec{c} \end{vmatrix}.$$

Similarly, taking dot product of (1) with $\vec{c} \times \vec{a}$ and $\vec{a} \times \vec{b}$ respectively we get

$$\mu [\vec{b}\ \vec{c}\ \vec{a}] = \begin{vmatrix} \vec{p} \cdot \vec{c} & \vec{q} \cdot \vec{c} \\ \vec{p} \cdot \vec{a} & \vec{q} \cdot \vec{a} \end{vmatrix}$$

$$\nu [\vec{c}\ \vec{a}\ \vec{b}] = \begin{vmatrix} \vec{p} \cdot \vec{a} & \vec{q} \cdot \vec{a} \\ \vec{p} \cdot \vec{b} & \vec{q} \cdot \vec{b} \end{vmatrix}$$

But $[\vec{a}\ \vec{b}\ \vec{c}] = [\vec{b}\ \vec{c}\ \vec{a}] = [\vec{c}\ \vec{a}\ \vec{b}]$.

$\therefore \quad$ putting values of λ, μ, ν in (1) we get

$$\vec{p} \times \vec{q} = \frac{1}{[\vec{a}\ \vec{b}\ \vec{c}]} \left\{ \begin{vmatrix} \vec{p} \cdot \vec{b} & \vec{q} \cdot \vec{b} \\ \vec{p} \cdot \vec{c} & \vec{q} \cdot \vec{c} \end{vmatrix} \vec{a} + \begin{vmatrix} \vec{p} \cdot \vec{c} & \vec{q} \cdot \vec{c} \\ \vec{p} \cdot \vec{a} & \vec{q} \cdot \vec{a} \end{vmatrix} \vec{b} + \begin{vmatrix} \vec{p} \cdot \vec{a} & \vec{q} \cdot \vec{a} \\ \vec{p} \cdot \vec{b} & \vec{q} \cdot \vec{b} \end{vmatrix} \vec{c} \right\}$$

$$= \frac{1}{[\vec{a}\ \vec{b}\ \vec{c}]} \begin{vmatrix} \vec{p} \cdot \vec{a} & \vec{q} \cdot \vec{a} & \vec{a} \\ \vec{p} \cdot \vec{b} & \vec{q} \cdot \vec{b} & \vec{b} \\ \vec{p} \cdot \vec{c} & \vec{q} \cdot \vec{c} & \vec{c} \end{vmatrix}.$$

14. If $\vec{a}' = \dfrac{\vec{b} \times \vec{c}}{[\vec{a}\ \vec{b}\ \vec{c}]}$, $\vec{b}' = \dfrac{\vec{c} \times \vec{a}}{[\vec{a}\ \vec{b}\ \vec{c}]}$, $\vec{c}' = \dfrac{\vec{a} \times \vec{b}}{[\vec{a}\ \vec{b}\ \vec{c}]}$

then show that $\vec{a} \times \vec{a}' + \vec{b} \times \vec{b}' + \vec{c} \times \vec{c}' = 0$

where $\vec{a}, \vec{b}, \vec{c}$ **are noncoplanar vectors.**

Here $\vec{a} \times \vec{a}' = \vec{a} \times \dfrac{\vec{b} \times \vec{c}}{[\vec{a}\ \vec{b}\ \vec{c}]}$, etc.

$\therefore$ LHS $= \dfrac{1}{[\vec{a}\ \vec{b}\ \vec{c}]} \{\vec{a} \times (\vec{b} \times \vec{c})$

$+ \vec{b} \times (\vec{c} \times \vec{a}) + \vec{c} \times (\vec{a} \times \vec{b})\}$.

Now, $\vec{a} \times (\vec{b} \times \vec{c}) + \vec{b} \times (\vec{c} \times \vec{a}) + \vec{c} \times (\vec{a} \times \vec{b})$

$= \{(\vec{a} \cdot \vec{c})\vec{b} - (\vec{a} \cdot \vec{b})\vec{c}\}$

$+ \{(\vec{b} \cdot \vec{a})\vec{c} - (\vec{b} \cdot \vec{c})\vec{a}\}$

$+ \{(\vec{c} \cdot \vec{b})\vec{a} - (\vec{c} \cdot \vec{a})\vec{b}\}$

$= 0 \quad (\because \vec{a} \cdot \vec{b} = \vec{b} \cdot \vec{a}$, etc.).

15. Prove that $[\vec{a} \times \vec{b}\ \ \vec{b} \times \vec{c}\ \ \vec{c} \times \vec{a}] = [\vec{a}\ \vec{b}\ \vec{c}]^2$**.**

LHS $= \{(\vec{a} \times \vec{b}) \times (\vec{b} \times \vec{c})\} \cdot (\vec{c} \times \vec{a})$

$= [\{(\vec{a} \times \vec{b}) \cdot \vec{c}\}\vec{b} - \{(\vec{a} \times \vec{b}) \cdot \vec{b}\}\vec{c}] \cdot (\vec{c} \times \vec{a})$

$= ([\vec{a}\ \vec{b}\ \vec{c}]\vec{b}) \cdot (\vec{c} \times \vec{a})$

$\{\because (\vec{a} \times \vec{b}) \cdot \vec{b} = [\vec{a}\ \vec{b}\ \vec{b}] = 0\}$

$= [\vec{a}\ \vec{b}\ \vec{c}][\vec{b}\ \vec{c}\ \vec{a}] = [\vec{a}\ \vec{b}\ \vec{c}]^2$.

16. If $(\vec{e_1}, \vec{e_2}, \vec{e_3})$ **and** $(\vec{e_1}', \vec{e_2}', \vec{e_3}')$ **are two sets of noncoplanar vectors such that for** $i = 1, 2, 3$ **we have**

$\vec{e_i} \cdot \vec{e_j}' = 1$ **if** $i = j$

0 **if** $i \neq j$

then show that $[\vec{e_1}\ \vec{e_2}\ \vec{e_3}][\vec{e_1}'\ \vec{e_2}'\ \vec{e_3}'] = 1$**.**

Here $\vec{e_1} \cdot \vec{e_2}' = 0$, $\vec{e_1} \cdot \vec{e_3}' = 0$;

$\therefore \quad \vec{e_1} \perp \vec{e_2}'$, $\vec{e_1} \perp \vec{e_3}'$; $\quad \therefore \quad \vec{e_1} \parallel \vec{e_2}' \times \vec{e_3}'$

$\therefore \quad \vec{e_1} = \lambda (\vec{e_2}' \times \vec{e_3}')$ ... (1)

$\therefore \quad \vec{e_1} \cdot \vec{e_1}' = \lambda (\vec{e_2}' \times \vec{e_3}') \cdot \vec{e_1}'$

$\therefore \quad 1 = \lambda [\vec{e_2}'\ \vec{e_3}'\ \vec{e_1}']$

$(\because \vec{e_1} \cdot \vec{e_1}' = 1$, from the question)

$\therefore \quad \lambda = \dfrac{1}{[\vec{e_1}',\ \vec{e_2}',\ \vec{e_3}']}$

Thus $\vec{e_1} = \dfrac{\vec{e_2}' \times \vec{e_3}'}{[\vec{e_1}',\ \vec{e_2}',\ \vec{e_3}']}$, from (1)

Similarly, $\vec{e_2} = \dfrac{\vec{e_3}' \times \vec{e_1}'}{[\vec{e_2}'\ \vec{e_3}'\ \vec{e_1}']} = \dfrac{\vec{e_3}' \times \vec{e_1}'}{[\vec{e_1}'\ \vec{e_2}'\ \vec{e_3}']}$

and $\vec{e_3} = \dfrac{\vec{e_1}' \times \vec{e_2}'}{[\vec{e_3}'\ \vec{e_1}'\ \vec{e_2}']} = \dfrac{\vec{e_1}' \times \vec{e_2}'}{[\vec{e_1}'\ \vec{e_2}'\ \vec{e_3}']}$

$\therefore \quad [\vec{e_1}\ \vec{e_2}\ \vec{e_3}] = \dfrac{[\vec{e_2}' \times \vec{e_3}'\ \ \vec{e_3}' \times \vec{e_1}'\ \ \vec{e_1}' \times \vec{e_2}']}{[\vec{e_1}'\ \vec{e_2}'\ \vec{e_3}']^3}$

or $\quad [\vec{e}_1\ \vec{e}_2\ \vec{e}_3]\,[\vec{e}_1{}'\ \vec{e}_2{}'\ \vec{e}_3{}']^3$

$= [\vec{e}_2{}'\times\vec{e}_3{}'\quad \vec{e}_3{}'\times\vec{e}_1{}'\quad \vec{e}_1{}'\times\vec{e}_2{}']$ ... (2)

Now, $[\vec{e}_2{}'\times\vec{e}_3{}'\quad \vec{e}_3{}'\times\vec{e}_1{}'\quad \vec{e}_1{}'\times\vec{e}_2{}']$

$= \{(\vec{e}_2{}'\times\vec{e}_3{}')\times(\vec{e}_3{}'\times\vec{e}_1{}')\}\cdot(\vec{e}_1{}'\times\vec{e}_2{}')$

$= \{(\vec{e}_2{}'\times\vec{e}_3{}'\cdot\vec{e}_1{}')\,\vec{e}_3{}' - (\vec{e}_2{}'\times\vec{e}_3{}'\cdot\vec{e}_3{}')\,\vec{e}_1{}'\}$

$\cdot(\vec{e}_1{}'\times\vec{e}_2{}')$

$[\because\ \vec{a}\times(\vec{b}\times\vec{c}) = (\vec{a}\cdot\vec{c})\,\vec{b} - (\vec{a}\cdot\vec{b})\,\vec{c}\,]$

$= [\vec{e}_2{}'\ \vec{e}_3{}'\ \vec{e}_1{}']\,\vec{e}_3{}'\cdot(\vec{e}_1{}'\times\vec{e}_2{}')$

$= [\vec{e}_2{}'\ \vec{e}_3{}'\ \vec{e}_1{}']\,[\vec{e}_3{}'\ \vec{e}_1{}'\ \vec{e}_2{}']$

$= [\vec{e}_1{}'\ \vec{e}_2{}'\ \vec{e}_3{}']^2$

$\therefore$ From (2), $[\vec{e}_1\ \vec{e}_2\ \vec{e}_3]\,[\vec{e}_1{}'\ \vec{e}_2{}'\ \vec{e}_3{}'] = 1$

$\{\because\ [\vec{e}_1{}'\ \vec{e}_2{}'\ \vec{e}_3{}'] \neq 0\}$.

Exercises

1. The vectors $\vec{a} = (12, 4, 3)$, $\vec{b} = (8, -12, -9)$ and $\vec{c} = (33, -4, -24)$ define a parallelepiped. Evaluate the lengths of its edges, areas of its faces and its volume.

2. The volume of a parallelepiped is 4 whose three concurrent edges are represented by the vectors $\vec{i} + x\vec{j} - x^2\vec{k}$, $\vec{i} + \vec{j} - \vec{k}$ and $\vec{i} - \vec{j} + \vec{k}$. Find the possible real values of x.

3. V_1 is the volume of a parallelepiped and V_2 is the volume of the parallelepiped formed with three concurrent diagonals of three faces of the original parallelepiped. Find the relation between V_1 and V_2.

4. The position vectors of the points A, B, C, D are respectively $2\vec{i} + \vec{j} - \vec{k}$, $\vec{i} + \vec{j} + \vec{k}$, $\vec{i} - 2\vec{j} + 3\vec{k}$ and $3\vec{i} - \vec{j} + 2\vec{k}$. Evaluate $[\,\vec{AB}\ \vec{AC}\ \vec{AD}\,]$.
What can you conclude about the four points?

5. Prove that

$$\{(\vec{a} - \vec{b})\times(\vec{a} - \vec{b} - \vec{c})\}\cdot(\vec{a} + 2\vec{b} - \vec{c}) = 3\,[\vec{a}\ \vec{b}\ \vec{c}\,].$$

6. If $\vec{a}, \vec{b}, \vec{c}$ are noncoplanar then prove that $\vec{a} + \vec{b}$, $\vec{b} + \vec{c}$ and $\vec{c} + \vec{a}$ are also noncoplanar.

7. Prove that $[\,\vec{a} - \vec{b}\quad \vec{b} - \vec{c}\quad \vec{c} - \vec{a}\,] = 0$.

8. If the vectors $\vec{A} = (1, a, a^2)$, $\vec{B} = (1, b, b^2)$ and $\vec{C} = (1, c, c^2)$ are noncoplanar, and $\vec{P} = (a, b, c)$, $\vec{Q} = (a^2, b^2, c^2)$ and $\vec{R} = (1 + a^3, 1 + b^3, 1 + c^3)$ are coplanar then find the value of abc.

9. Let $\vec{a} = a_1\vec{i} + a_2\vec{j} + a_3\vec{k}$, $\vec{b} = b_1\vec{i} + b_2\vec{j} + b_3\vec{k}$, $\vec{c} = c_1\vec{i} + c_2\vec{j} + c_3\vec{k}$ be three nonzero vectors such that $\vec{c}$ is a unit vector perpendicular to both the vectors $\vec{a}$ and $\vec{b}$. If the angle between $\vec{a}$ and $\vec{b}$ is $\frac{\pi}{6}$, show that

$$\begin{vmatrix} a_1 & b_1 & c_1 \\ a_2 & b_2 & c_2 \\ a_3 & b_3 & c_3 \end{vmatrix}^2 = \frac{1}{4}(a_1^2 + a_2^2 + a_3^2)(b_1^2 + b_2^2 + b_3^2).$$

10. Let $\vec{a}, \vec{b}, \vec{c}$ be three noncoplanar vectors and $\vec{p}, \vec{q}, \vec{r}$ are vectors defined by

$$\vec{p} = \frac{\vec{b}\times\vec{c}}{[\vec{a}\ \vec{b}\ \vec{c}\,]},\quad \vec{q} = \frac{\vec{c}\times\vec{a}}{[\vec{a}\ \vec{b}\ \vec{c}\,]},\quad \vec{r} = \frac{\vec{a}\times\vec{b}}{[\vec{a}\ \vec{b}\ \vec{c}\,]}$$

then find the value of

(i) $(\vec{a} + \vec{b})\cdot\vec{p} + (\vec{b} + \vec{c})\cdot\vec{q} + (\vec{c} + \vec{a})\cdot\vec{r}$

(ii) $(\vec{b} + \vec{c})\cdot\vec{p} + (\vec{c} + \vec{a})\cdot\vec{q} + (\vec{a} + \vec{b})\cdot\vec{r}$.

11. If $\vec{a}, \vec{b}, \vec{c}$ are three vectors such that $\vec{a}\times\vec{b} = \vec{c}$ and $\vec{b}\times\vec{c} = \vec{a}$, show that $\vec{a}, \vec{b}, \vec{c}$ are orthogonal in pairs and $|\vec{b}| = 1$, $|\vec{c}| = |\vec{a}|$.

12. If $\vec{r}$ is a vector satisfying $\lambda\vec{r} + \vec{r}\times\vec{a} = \vec{b}$ where λ is a scalar $\neq 0$ and $\vec{a}, \vec{b}$ are two given vectors then prove that

$$\vec{r} = \frac{1}{a^2 + \lambda^2}\left(\frac{\vec{a}\cdot\vec{b}}{\lambda}\,\vec{a} + \lambda\vec{b} + \vec{a}\times\vec{b}\right).$$

13. (i) Let $\vec{a}, \vec{b}, \vec{c}$ be given vectors in space with $\vec{a}\cdot\vec{c} \neq 0$ and let m be a given scalar. Find $\vec{d}$ which satisfies $\vec{d}\times\vec{a} = \vec{b}$ as well as $\vec{d}\cdot\vec{c} = m$.

(ii) If $\vec{r}\times\vec{b} = \vec{c}\times\vec{b}$ and $\vec{r}\cdot\vec{a} = 0$ then find $\vec{r}$ when $\vec{a}$ is not orthogonal to $\vec{b}$.

14. Prove that
$\vec{i} \times (\vec{a} \times \vec{i}) + \vec{j} \times (\vec{a} \times \vec{j}) + \vec{k} \times (\vec{a} \times \vec{k}) = 2\vec{a}$.

15. If $\vec{a}, \vec{b}, \vec{c}$ are nonzero vectors, of which no two are collinear or perpendicular, such that
$\vec{a} \times (\vec{b} \times \vec{c}) = (\vec{a} \times \vec{b}) \times \vec{c}$
then prove that
(i) $(\vec{c} \times \vec{a}) \times \vec{b} = 0$
(ii) $\vec{c} \times \vec{a}$ and $\vec{b}$ are collinear.

16. If $\vec{a}' = \dfrac{\vec{b} \times \vec{c}}{[\vec{a}\ \vec{b}\ \vec{c}]}$, $\vec{b}' = \dfrac{\vec{c} \times \vec{a}}{[\vec{a}\ \vec{b}\ \vec{c}]}$, $\vec{c}' = \dfrac{\vec{a} \times \vec{b}}{[\vec{a}\ \vec{b}\ \vec{c}]}$
where $\vec{a}, \vec{b}, \vec{c}$ are noncoplanar then prove that
(i) $[\vec{a}'\ \vec{b}'\ \vec{c}'] = \dfrac{1}{[\vec{a}\ \vec{b}\ \vec{c}]}$
(ii) $\vec{a}'\ \vec{b}'\ \vec{c}'$ are noncoplanar.

17. If $\vec{a}, \vec{b}, \vec{c}$ are noncoplanar, prove that $\vec{a} \times \vec{b}, \vec{b} \times \vec{c}$ and $\vec{c} \times \vec{a}$ are also noncoplanar.

18. If $\vec{b} = \vec{a} \times (\vec{i} \times \vec{a}) + \vec{a} \times (\vec{j} \times \vec{a}) + \vec{a} \times (\vec{k} \times \vec{a})$ and $\vec{a} \perp (\vec{i} + \vec{j} + \vec{k})$ then prove that
$|\vec{b}| = \sqrt{3}\ |\vec{a}|^2$.

19. Prove that
$[\{(\vec{a} + \vec{b}) \times (\vec{a} + \vec{c})\} \times (\vec{b} \times \vec{c})] \cdot (\vec{b} + \vec{c}) = 0$
if $|\vec{b}| = |\vec{c}|$.

20. If $\vec{a} \times (\vec{b} \times \vec{c}) + (\vec{a} \cdot \vec{b})\vec{b}$
$= (4 - 2\lambda - \sin\alpha)\vec{b} + (\lambda^2 - 1)\vec{c}$
and $(\vec{c} \cdot \vec{c})\vec{a} = \vec{c}$ where $\vec{b}, \vec{c}$ are noncollinear then find the values of the scalars λ, α.

21. If $\vec{AB} = \vec{a} \times (\vec{a} \times \vec{b})$ and $\vec{AC} = \vec{a} \times \vec{b}$ where $|\vec{a}| = \sqrt{3}$ then find the angles of the triangle ABC.

22. (i) If $\vec{a} \times (\vec{a} \times \vec{b}) = \vec{b} \times (\vec{b} \times \vec{c})$ then prove that $\vec{a}, \vec{b}, \vec{c}$ are coplanar.
(ii) Prove that $\vec{a} \times (\vec{b} \times \vec{c})$ is perpendicular to $(\vec{a} \times \vec{b}) \times \vec{c}$ if $\vec{a}$ and $\vec{c}$ are perpendicular to each other.

23. Let $\vec{a}, \vec{b}$ be noncollinear vectors of which $\vec{a}$ is a unit vector. Find the angles of the triangle whose two sides represent $\sqrt{3}(\vec{a} \times \vec{b})$ and $\vec{b} - (\vec{a} \cdot \vec{b})\vec{a}$.

24. Prove that
$|\vec{a} \times \vec{i}|^2 + |\vec{a} \times \vec{j}|^2 + |\vec{a} \times \vec{k}|^2 = 2|\vec{a}|^2$.

25. Prove that
$(\vec{a} \times \vec{b}) \times (\vec{c} \times \vec{d}) = [\vec{a}\ \vec{b}\ \vec{d}]\vec{c} - [\vec{a}\ \vec{b}\ \vec{c}]\vec{d}$
$= [\vec{a}\ \vec{c}\ \vec{d}]\vec{b} - [\vec{b}\ \vec{c}\ \vec{d}]\vec{a}$.

26. If the vectors $\vec{b}, \vec{c}, \vec{d}$ are noncoplanar then prove that the vector
$(\vec{a} \times \vec{b}) \times (\vec{c} \times \vec{d}) + (\vec{a} \times \vec{c}) \times (\vec{d} \times \vec{b})$
$+ (\vec{a} \times \vec{d}) \times (\vec{b} \times \vec{c})$ is parallel to $\vec{a}$.

27. If $l\vec{a} + m\vec{b} = \vec{c}$ then prove that

$$l = \frac{\begin{vmatrix} \vec{c} \cdot \vec{a} & \vec{b} \cdot \vec{a} \\ \vec{c} \cdot \vec{b} & \vec{b} \cdot \vec{b} \end{vmatrix}}{\begin{vmatrix} \vec{a} \cdot \vec{a} & \vec{a} \cdot \vec{b} \\ \vec{b} \cdot \vec{a} & \vec{b} \cdot \vec{b} \end{vmatrix}} \text{ and } m = \frac{\begin{vmatrix} \vec{c} \cdot \vec{b} & \vec{a} \cdot \vec{b} \\ \vec{c} \cdot \vec{a} & \vec{a} \cdot \vec{a} \end{vmatrix}}{\begin{vmatrix} \vec{a} \cdot \vec{a} & \vec{a} \cdot \vec{b} \\ \vec{b} \cdot \vec{a} & \vec{b} \cdot \vec{b} \end{vmatrix}}.$$

28. If $\vec{a}, \vec{b}$ and $\vec{c}$ are three conterminuous edges of a parallelepiped of the volume 6 then find the value of $[\vec{a} \times \vec{b}\ \ \vec{a} \times \vec{c}\ \ \vec{b} \times \vec{c}]$.

29. Let $\vec{a}, \vec{b}$ and $\vec{c}$ be noncoplanar unit vectors, equally inclined to one another at an angle $\theta \neq \dfrac{2\pi}{3}$. If $\vec{a} \times \vec{b} + \vec{b} \times \vec{c} = p\vec{a} + q\vec{b} + r\vec{c}$, find scalars p, q, r in terms of θ.

Objective Questions

Fill in the blanks.

30. If $\vec{a} = \vec{i} + \vec{j} + \vec{k}$, $\vec{b} = \vec{i} - \vec{j} + \vec{k}$ and $\vec{c} = -\vec{i} + \vec{j} + \vec{k}$ then $[\vec{a}\ \vec{b}\ \vec{c}] =$ ______ .

31. If the volume of a parallelepiped, whose three concurrent edges are $-12\vec{i} + \lambda\vec{k}$, $3\vec{j} - \vec{k}$ and $2\vec{i} + \vec{j} - 15\vec{k}$, is 546 then $\lambda =$ ______ .

32. The volume of the parallelepiped whose edges are represented by
$\vec{a} = 2\vec{i} - 3\vec{j} + 4\vec{k}$, $\vec{b} = \vec{i} + 2\vec{j} - \vec{k}$ and
$\vec{c} = 3\vec{i} - \vec{j} + 2\vec{k}$, is ______ .

33. The vectors $2\vec{i} - \vec{j} + \vec{k}$, $\vec{i} + 2\vec{j} - 3\vec{k}$ and $3\vec{i} - a\vec{j} + 2\vec{k}$ are coplanar if $a =$ ______ .

34. The points (3, –2, –1), (2, 3, –4), (–1, 1, 2) and (4, 5, λ) are coplanar if λ = ______ .

35. Let $\vec{a}$, $\vec{b}$ and $\vec{c}$ be three unit vectors and $\vec{a} \cdot \vec{b} = \vec{a} \cdot \vec{c} = 0$. If the angle between $\vec{b}$ and $\vec{c}$ is 30° then the absolute value of $[\vec{a}\ \vec{b}\ \vec{c}]$ is ______ .

36. If $\vec{A}, \vec{B}, \vec{C}$ are three noncoplanar vectors then

$$\frac{\vec{A} \cdot \vec{B} \times \vec{C}}{\vec{C} \times \vec{A} \cdot \vec{B}} + \frac{\vec{B} \cdot \vec{A} \times \vec{C}}{\vec{C} \cdot \vec{A} \times \vec{B}} = \_\_\_\_\_\_ .$$

37. $(\vec{a} \times \vec{i}) \cdot (\vec{j} \times \vec{k}) + (\vec{a} \times \vec{j}) \cdot (\vec{k} \times \vec{i}) + (\vec{a} \times \vec{k}) \cdot (\vec{i} \times \vec{j}) =$ ______ .

38. If $\vec{a} \cdot \vec{b} = \vec{b} \cdot \vec{c} = \vec{c} \cdot \vec{a} = 0$ and $|\vec{a}| = x$, $|\vec{b}| = y$, $|\vec{c}| = z$ then the volume of the parallelepiped whose three coterminous edges are $\vec{a}, \vec{b}, \vec{c}$, is ______ .

39. If $\vec{b}$ and $\vec{c}$ are two orthogonal vectors and $\vec{a}$ is any other vector then

$$\frac{(\vec{a} \cdot \vec{b})\vec{b}}{\vec{b}^2} + \frac{(\vec{a} \cdot \vec{c})\vec{c}}{\vec{c}^2} + \frac{\vec{a} \cdot \vec{b} \times \vec{c}}{|\vec{b} \times \vec{c}|^2}(\vec{b} \times \vec{c}) = \_\_\_\_\_\_ .$$

40. The value of

$$\vec{i} \times (\vec{i} \times \vec{j}) + \vec{j} \times (\vec{j} \times \vec{k}) + \vec{k} \times (\vec{k} \times \vec{i})$$

in the simplest form is ______ .

41. If $\vec{a}, \vec{b}, \vec{c}$ are three unit vectors of which $\vec{b}, \vec{c}$ are noncollinear and $\vec{a} \times (\vec{b} \times \vec{c}) = \frac{1}{2}\vec{b}$ then the angle between $\vec{a}$ and $\vec{b}$ is ______ and that between $\vec{a}$ and $\vec{c}$ is ______ .

42. If $\vec{x} \cdot \vec{a} = \vec{x} \cdot \vec{b} = \vec{x} \cdot \vec{c} = 0$ and $\vec{x} \neq \vec{0}$ then $[\vec{a}\ \vec{b}\ \vec{c}] =$ ______ .

43. $\vec{i} \cdot (\vec{j} \times \vec{k}) + \vec{j} \cdot (\vec{i} \times \vec{k}) + \vec{k} \cdot (\vec{i} \times \vec{j}) =$ ______ .

44. Let $\vec{a}$, $\vec{b}$ and $\vec{c}$ be three vectors having magnitudes 1, 1 and 2 respectively. If $\vec{a} \times (\vec{a} \times \vec{c}) + \vec{b} = \vec{0}$ then the acute angle between $\vec{a}$ and $\vec{c}$ is ______ .

Choose the correct option(s).

45. $\vec{A} \cdot (\vec{B} + \vec{C}) \times (\vec{A} + \vec{B} + \vec{C})$ is equal to

(a) 0

(b) $[\vec{A}\ \vec{B}\ \vec{C}] + [\vec{B}\ \vec{C}\ \vec{A}]$

(c) $[\vec{A}\ \vec{B}\ \vec{C}]$

(d) none of these

46. Let a, b, c be distinct nonzero numbers. If the vectors $a\vec{i} - a\vec{j} + b\vec{k}$, $\vec{i} + \vec{k}$ and $c\vec{i} + c\vec{j} + b\vec{k}$ lie in a plane then b is

(a) the AM of a and c (b) the GM of a and c

(c) the HM of a and c (d) equal to zero

47. For nonzero vectors $\vec{a}, \vec{b}, \vec{c}$ the equality $|(\vec{a} \times \vec{b}) \cdot \vec{c}| = |\vec{a}|\,|\vec{b}|\,|\vec{c}|$ holds if and only if

(a) $\vec{a} \cdot \vec{b} = 0$, $\vec{b} \cdot \vec{c} = 0$

(b) $\vec{b} \cdot \vec{c} = 0$, $\vec{c} \cdot \vec{a} = 0$

(c) $\vec{c} \cdot \vec{a} = 0$, $\vec{a} \cdot \vec{b} = 0$

(d) $\vec{a} \cdot \vec{b} = \vec{b} \cdot \vec{c} = \vec{c} \cdot \vec{a} = 0$

48. The vector $(\vec{a} \cdot \vec{b})\vec{c} - (\vec{a} \cdot \vec{c})\vec{b}$ is perpendicular to the vector

(a) $\vec{b}$ (b) $\vec{a}$ (c) $\vec{b} \times \vec{c}$ (d) $\vec{c}$

49. $\vec{a} \times (\vec{b} \times \vec{c}) + \vec{b} \times (\vec{c} \times \vec{a}) + \vec{c} \times (\vec{a} \times \vec{b})$ is a

(a) unit vector

(b) null vector

(c) vector of magnitude $3|\vec{a}|\,|\vec{b}|\,|\vec{c}|$

(d) none of these

50. If $|\vec{a}| = |\vec{b}| = |\vec{c}| = 1$ and $\vec{a} \cdot \vec{b} = 0$, $\vec{a} \cdot \vec{c} = \vec{b} \cdot \vec{c}$ where $\vec{c} = \lambda\vec{a} + \mu\vec{b} + \nu\vec{a} \times \vec{b}$ then

(a) $\nu^2 + 2\lambda^2 = 1$ (b) $\lambda = \mu$

(c) $\lambda = \mu = \nu$ (d) none of these

51. Let $\vec{p}$, $\vec{q}$ and $\vec{r}$ be three mutually perpendicular vectors of the same magnitude. If $\vec{x}$ satisfies the equation

$$\vec{p} \times \{(\vec{x} - \vec{q}) \times \vec{p}\} + \vec{q} \times \{(\vec{x} - \vec{r}) \times \vec{q}\} + \vec{r} \times \{(\vec{x} - \vec{p}) \times \vec{r}\} = \vec{0}$$ then $\vec{x}$ is

(a) $\frac{1}{2}(\vec{p} + \vec{q} - 2\vec{r})$ (b) $\frac{1}{2}(\vec{p} + \vec{q} + \vec{r})$

(c) $\frac{1}{3}(\vec{p} + \vec{q} + \vec{r})$ (d) $\frac{1}{3}(2\vec{p} + \vec{q} - \vec{r})$

Answers

1. 13, 17, 41; 220, 435, 455; 3696

2. −1, 2

3. $V_2 = 2V_1$

4. 11, noncoplanar points

8. −1

10. (i) 3 (ii) 0

13. (i) $\vec{d} = \dfrac{1}{\vec{a}\cdot\vec{c}}\left\{m + \dfrac{(\vec{b}\cdot\vec{c})^2}{[\vec{a}\ \vec{b}\ \vec{c}]} - \dfrac{\vec{b}^2\ \vec{c}^2}{[\vec{a}\ \vec{b}\ \vec{c}]}\right\}\vec{a}$

$$- \frac{\vec{b}\cdot\vec{c}}{[\vec{a}\ \vec{b}\ \vec{c}]}\vec{b} + \frac{\vec{b}^2}{[\vec{a}\ \vec{b}\ \vec{c}]}\vec{c}$$

(ii) $\vec{r} = \vec{c} - \dfrac{\vec{c}\cdot\vec{a}}{\vec{b}\cdot\vec{a}}\vec{b}$

20. $\alpha = \dfrac{\pi}{2},\ \lambda = 1$

21. $A = \dfrac{\pi}{2},\ B = \dfrac{\pi}{6},\ C = \dfrac{\pi}{3}$

23. $\dfrac{\pi}{2}, \dfrac{\pi}{3}, \dfrac{\pi}{6}$

28. 36

29. $p = r = 1/\sqrt{1 + 2\cos\theta},\ q = -2\cos\theta/\sqrt{1 + 2\cos\theta}$

30. −4 **31.** −3, 179 **32.** 7 **33.** $\dfrac{13}{7}$

34. $-\dfrac{146}{17}$ **35.** $\dfrac{1}{2}$ **36.** 0 **37.** 0

38. xyz **39.** $\vec{a}$ **40.** $-(\vec{i} + \vec{j} + \vec{k})$

41. $\dfrac{\pi}{2}, \dfrac{\pi}{3}$ respectively **42.** 0

43. 1 **44.** $\dfrac{\pi}{6}$ **45.** (a)

46. (c) **47.** (d) **48.** (b), (c)

49. (b) **50.** (a), (b) **51.** (b)

□

Chapter Test

Time: 90 minutes

1. In each of the following, fill in the blank so that the resulting statement is correct.

(a) If $\vec{a}, \vec{b}$ are orthogonal unit vectors then

$$[\vec{a} + \vec{a} \times \vec{b} \quad \vec{b} + \vec{a} \times \vec{b} \quad \vec{a} \times \vec{b}\,] = \underline{\qquad}.$$

(b) If α, β, γ are angles of a triangle then the value of

$$[\tan\alpha\,\vec{i} + \vec{j} + \vec{k} \quad \vec{i} + \tan\beta\,\vec{j} + \vec{k} \quad \vec{i} + \vec{j} + \tan\gamma\,\vec{k}\,]$$

in the simplest form is ______ .

(c) If $\vec{a} \times \vec{r} = \vec{b} + \lambda\,\vec{a}$ where $\vec{a}, \vec{b}$ are known vectors and λ is a scalar parameter, and $\vec{a} \cdot \vec{r} = 3$ then $\vec{b} \cdot \vec{r}$ in terms of $\vec{a}$ and $\vec{b}$ is equal to ______ .

2. In each of the following, one or more options are correct. Choose the correct option(s).

(a) $\vec{c} \times (\vec{a} \times \vec{b}\,)$ is a vector which is

A. perpendicular to $\vec{c}$ B. perpendicular to $\vec{a}$

C. coplanar with $\vec{a}$ and $\vec{b}$ D. perpendicular to $\vec{b}$

(b) If $\vec{a}, \vec{b}, \vec{c}$ represent three concurrent edges of a rectangular parallelepiped whose lengths are 4, 3, 2 respectively then the value of

$$(\vec{a} + \vec{b} + \vec{c}\,)\cdot(\vec{a} \times \vec{b} + \vec{b} \times \vec{c} + \vec{c} \times \vec{a}\,) \text{ is}$$

A. 0 B. 48 C. 72 D. none of these

(c) If $\vec{a} = \vec{p} + \vec{q},\ \vec{p} \times \vec{b} = \vec{0}$ and $\vec{q} \cdot \vec{b} = 0$ then $\dfrac{\vec{b} \times (\vec{a} \times \vec{b}\,)}{\vec{b} \cdot \vec{b}}$ is equal to

A. $\vec{q}$ B. $\vec{p}$ C. $\vec{p} \times \vec{q}$ D. none of these

(d) If $\vec{\lambda} = \begin{vmatrix} \vec{i} & \vec{j} & \vec{k} \\ a & b & c \\ x & y & z \end{vmatrix}$, $\vec{\mu} = (a+x)\,\vec{i} + (b+y)\,\vec{j} + (c+z)\,\vec{k}$

$\vec{\nu} = (a-x)\,\vec{i} + (b-y)\,\vec{j} + (c-z)\,\vec{k}$ then

A. $\vec{\lambda} \perp \vec{\nu},\ \vec{\lambda} \cdot \vec{\mu} \neq 0$ B. $\vec{\lambda} \cdot \vec{\nu} = 0,\ \vec{\lambda} \cdot \vec{\mu} = 0$

C. $\vec{\lambda} \times \vec{\nu} = 0,\ \vec{\mu} \times \vec{\lambda} = 0$ D. $\vec{\lambda} \parallel \vec{\nu},\ \vec{\lambda} \perp \vec{\mu}$

3. If $\vec{x} = \dfrac{\vec{a} \times \vec{b} - \vec{a} \times (\vec{a} \times \vec{b}\,)}{(\vec{a} \times \vec{b}\,)^2}$ then prove that

$\vec{x} \cdot \vec{a} = 0,\ \vec{x} \cdot \vec{b} = 1$ and $[\,\vec{x}\ \vec{a}\ \vec{b}\,] = 1$.

4. If $\vec{a}, \vec{b}, \vec{c}$ are coplanar and $\vec{a} \times \vec{b} \neq 0$, show that

$$\vec{c} = \frac{\begin{vmatrix} \vec{c}\cdot\vec{a} & \vec{a}\cdot\vec{b} \\ \vec{c}\cdot\vec{b} & \vec{b}\cdot\vec{b} \end{vmatrix} \vec{a} + \begin{vmatrix} \vec{a}\cdot\vec{a} & \vec{c}\cdot\vec{a} \\ \vec{a}\cdot\vec{b} & \vec{c}\cdot\vec{b} \end{vmatrix} \vec{b}}{(\vec{a} \times \vec{b})^2}.$$

5. Prove that $(\vec{a} \times \vec{i}) \cdot (\vec{b} \times \vec{i}) + (\vec{a} \times \vec{j}) \cdot (\vec{b} \times \vec{j}) + (\vec{a} \times \vec{k}) \cdot (\vec{b} \times \vec{k}) = 2\vec{a} \cdot \vec{b}$.

6. If $\hat{n}$ is a unit vector perpendicular to the plane containing the points whose position vectors are $\vec{a}, \vec{b}$ and $\vec{r}$ then prove that $[\vec{r} - \vec{a} \;\; \vec{b} - \vec{a} \;\; \hat{n}] = |\vec{r} \times (\vec{b} - \vec{a}) - \vec{a} \times \vec{b}|$.

7. Let $\hat{a}, \hat{b}, \hat{c}$ be three unit vectors such that $\hat{a} \times (\hat{b} \times \hat{c}) = \frac{1}{2}(\hat{b} + \hat{c})$. Find the angles between the vectors $\hat{a}, \hat{b}$ and $\hat{a}, \hat{c}$ if $\hat{b}$ and $\hat{c}$ are non-parallel.

8. Let $|\vec{a}| = |\vec{b}| = 1$ and $|\vec{a} + \vec{b}| = \sqrt{3}$. If $\vec{c}$ be a vector such that $\vec{c} - \vec{a} - 2\vec{b} = 3\,\vec{a} \times \vec{b}$ then find $|(\vec{a} \times \vec{b}) \times \vec{c}|$.

9. If $\vec{v_1}$ is a vector along the x-axis, $\vec{v_2}$ is a vector in the xy plane and $\vec{v_3}$ is a vector such that $[\vec{v_1}\,\vec{v_2}\,\vec{v_3}] \neq 0$ and $\vec{v_1} \cdot \vec{v_1} = 4, \vec{v_1} \cdot \vec{v_2} = -2, \vec{v_1} \cdot \vec{v_3} = 6, \vec{v_2} \cdot \vec{v_2} = 2, \vec{v_2} \cdot \vec{v_3} = -5, \vec{v_3} \cdot \vec{v_3} = 29$ then find $\vec{v_1}, \vec{v_2}$ and $\vec{v_3}$. If $\vec{v_1}, \vec{v_2}, \vec{v_3}$ are any three vecotrs in space satisfying the above six conditions on dot products then find the relation between $\vec{v_1}, \vec{v_2}$ and $\vec{v_3}$.

10. Let $|\vec{a}| = 2$, $|\vec{b}| = 3$ and $\vec{c}$ is a vector such that $[\vec{a}\,\vec{b}\,\vec{c}] = 2$ as well as $\vec{a} + \vec{c} \times \vec{b} + 3\vec{c} = 0$. Prove that $\vec{b} = \frac{9}{2}\vec{c} \times \vec{a}$ and $|\vec{a} \times \vec{c}| = \frac{2}{3}$, and $\vec{a}, \vec{b}$ are orthogonal vectors. Also, prove that the minimum value of $|\vec{c}|$ is $\frac{1}{3}$ and in that case $\vec{a}, \vec{b}, \vec{c}$ form an orthogonal triad.

11. Let a_r, b_r, c_r; $r = 1, 2, 3$ are nonnegative real numbers and $\sum_{r=1}^{3} (a_r + b_r + c_r) = 3\lambda$. If V be the volume of the parallelopiped formed by three coterminuous edges representing the vectors $a_1\vec{i} + a_2\vec{j} + a_3\vec{k}$, $b_1\vec{i} + b_2\vec{j} + b_3\vec{k}$ and $c_1\vec{i} + c_2\vec{j} + c_3\vec{k}$ then prove that the greatest value of V is λ^3.

□

Answers

1. (a) 1 (b) 2 (c) $3\dfrac{\vec{a}\cdot\vec{b}}{\vec{a}^2}$

2. (a) A, C (b) C (c) A (d) B

7. $\frac{2\pi}{3}, \frac{\pi}{3}$ respectively

8. $\frac{1}{2}\sqrt{21}$

9. $\vec{v_1} = 2\vec{i}$, $\vec{v_2} = -\vec{i} \pm \vec{j}$, $\vec{v_3} = 3\vec{i} \pm 2\vec{j} \pm 4\vec{k}$ and $\vec{v_3} = \frac{1}{2}\vec{v_1} - 2\vec{v_2} \pm 2(\vec{v_1} \times \vec{v_2})$

□

4. Application of Vectors

Recap of Facts and Formulae

1. Position vector of a point of section

- If A($\vec{a}$), B($\vec{b}$) be two points then the position vector of the middle point P of AB is $\frac{\vec{a}+\vec{b}}{2}$.
- If A($\vec{a}$), B($\vec{b}$) and C($\vec{c}$) be the vertices of a triangle ABC then the position vector of the centroid G is $\frac{\vec{a}+\vec{b}+\vec{c}}{3}$.

2. Vector conditions for geometrical results

- Line segments AB and CD are equal
  $\Leftrightarrow \quad |\vec{AB}| = |\vec{CD}|$.
- Lines AB and CD are parallel
  $\Leftrightarrow \quad \vec{AB} = k\,\vec{CD} \quad$ or $\quad \vec{AB} \times \vec{CD} = \vec{0}$.
- Lines AB and CD are perpendicular
  $\Leftrightarrow \quad \vec{AB} \cdot \vec{CD} = 0$.
- A, B, C will be collinear
  $\Leftrightarrow \vec{AB} = k\,\vec{AC}$.
- A, B, C, D will be coplanar
  $\Leftrightarrow \quad \vec{AB} = \lambda\,\vec{AC} + \mu\,\vec{AD} \quad$ or $\quad [\vec{AB}\ \vec{AC}\ \vec{AD}] = 0$.
- A vector perpendicular to the plane passing through the points A, B and C is $\vec{AB} \times \vec{AC}$.

3. Equation of straight lines

- The equation of the straight line passing through the point ($\vec{a}$) and parallel to the vector $\vec{b}$ is
  $\vec{r} = \vec{a} + t\,\vec{b}$
  where t is an arbitrary scalar and $\vec{r}$ is the position vector of any point on the line.
- The equation of the straight line passing through the points ($\vec{a}$) and ($\vec{b}$) is
  $\vec{r} = (1-t)\,\vec{a} + t\,\vec{b}$
  where t is an arbitrary scalar.
- Three points whose position vectors are $\vec{a}, \vec{b}, \vec{c}$ are collinear if
  $\lambda\,\vec{a} + \mu\,\vec{b} + \nu\,\vec{c} = 0$ where $\lambda + \mu + \nu = 0$.

4. Equations of bisectors of the angle between two lines

Let A($\vec{a}$) be the vertex of the angle BAC where $\vec{b}$ and $\vec{c}$ are vectors along AB and AC respectively. The equations of the bisectors of ∠BAC are

$$\vec{r} = \vec{a} + t\left(\frac{\vec{b}}{|\vec{b}|} \pm \frac{\vec{c}}{|\vec{c}|}\right)$$

where t is a scalar parameter.

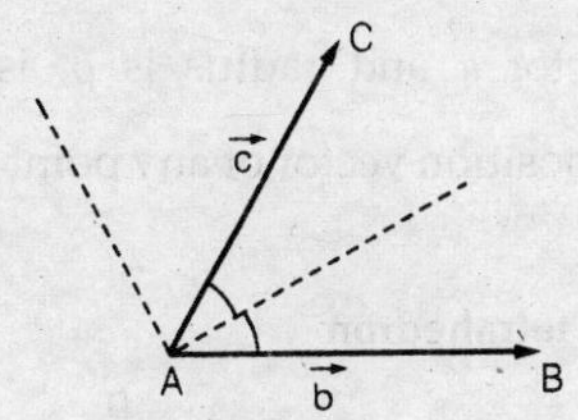

5. Shortest distance between two noncoplanar lines

- If two lines AB, CD do not intersect, there is always a line cutting both the lines perpendicularly.

  The intercept on this line made by AB and CD is called the *shortest distance* between the lines AB and CD. In the figure, the shortest distance = LM where $\angle ALM = \angle CML = 90°$.

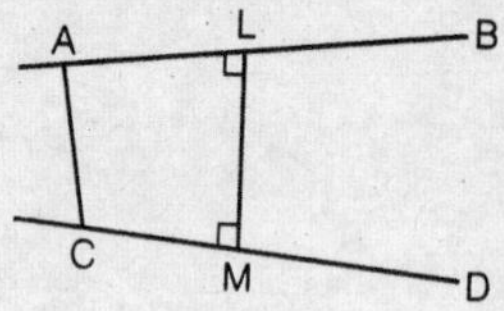

- In the above figure, shortest distance LM

$= |\text{ projection of } \vec{AC} \text{ along } \vec{ML} \,|$

$= \left| \vec{AC} \cdot \dfrac{\vec{ML}}{|\vec{ML}|} \right|.$

6. Equation of a plane

- The equation of the plane passing through the points $A(\vec{a})$, $B(\vec{b})$ and $C(\vec{c})$ is

(a) $\vec{r} = x\vec{a} + y\vec{b} + (1-x-y)\vec{c}$

where $\vec{r}$ is the position vector of any point on the plane and x, y are parameters

(b) $[\vec{r}-\vec{a} \quad \vec{r}-\vec{b} \quad \vec{r}-\vec{c}\,] = 0$

- Four points whose position vectors are $\vec{a}, \vec{b}, \vec{c}$ and $\vec{d}$ will be coplanar if

$x\vec{a} + y\vec{b} + z\vec{c} + w\vec{d} = 0$

where $x+y+z+w=0$.

- The equation of the plane passing through $A(\vec{a})$ and perpendicular to the vector $\vec{p}$ is

$(\vec{r} - \vec{a}) \cdot \vec{p} = 0.$

7. Vector equation of a sphere

- The vector equation of a sphere whose centre has the position vector $\vec{a}$ and radius is ρ, is $|\vec{r} - \vec{a}| = \rho$, $\vec{r}$ being the position vector of any point on the sphere.

8. Volume of a tetrahedron

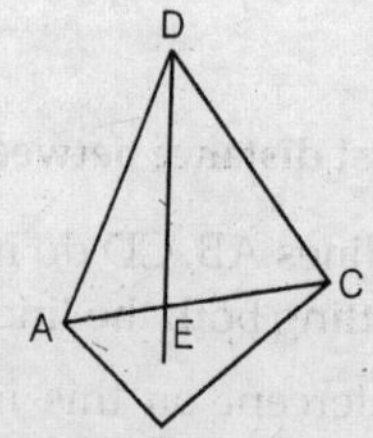

- The volume of a tetrahedron

$= \dfrac{1}{3}$ (area of the base) (corresponding altitude)

$= \dfrac{1}{3} \cdot \dfrac{1}{2} \, |\vec{AB} \times \vec{AC}| \, |\vec{ED}|$

$= \dfrac{1}{6} \, |\vec{AB} \times \vec{AC}| \, |\vec{ED}| \cos 0°$ for $\vec{AB} \times \vec{AC} \parallel \vec{ED}$

$= \dfrac{1}{6} (\vec{AB} \times \vec{AC}) \cdot \vec{ED} = \dfrac{1}{6} [\vec{AB} \;\; \vec{AC} \;\; \vec{ED}\,]$

$= \dfrac{1}{6} [\vec{AB} \;\; \vec{AC} \;\; \vec{EA} + \vec{AD}\,] = \dfrac{1}{6} [\vec{AB} \;\; \vec{AC} \;\; \vec{AD}\,]$

because $\vec{AB}, \vec{AC}, \vec{EA}$ are coplanar and so

$[\vec{AB} \;\; \vec{AC} \;\; \vec{EA}\,] = 0.$

9. Work done by a force vector

- The work done by the force $\vec{F}$ in shifting a particle from the point A to the point B

$= \vec{AB} \cdot \vec{F} = \vec{r} \cdot \vec{F}.$

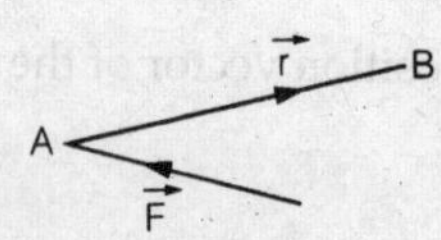

10. Vector moment (torque) of a force vector about a point

- If the force $\vec{F}$ acts at the point P then the vector moment of $\vec{F}$ about the point A

$= \vec{AP} \times \vec{F} = \vec{r} \times \vec{F}.$

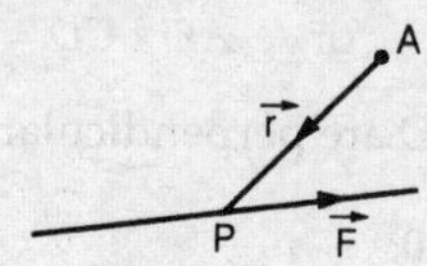

11. Rotation about an axis

- When a rigid body rotates about a fixed axis ON with an angular velocity ω then the velocity $\vec{v}$ of a particle P is given by

$\vec{v} = \vec{w} \times \vec{r}$

where $\vec{r} = \vec{OP}$ and $\vec{w} = w\,\hat{w}$ (unit vector along ON).

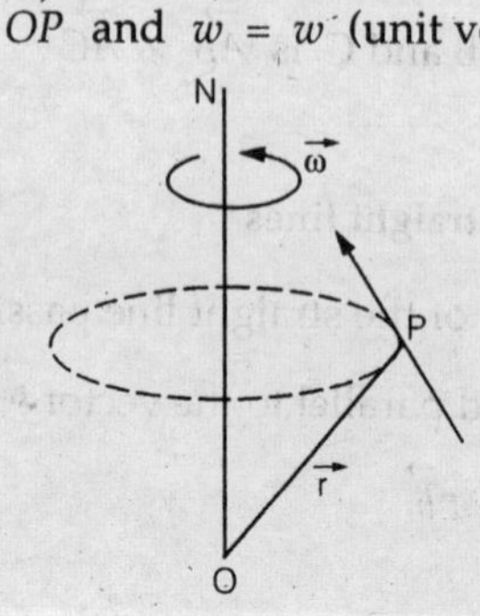

Selected Solved Examples

1. Prove, by vector method, that the line segments joining the midpoints of consecutive sides of a quadrilateral form a parallelogram.

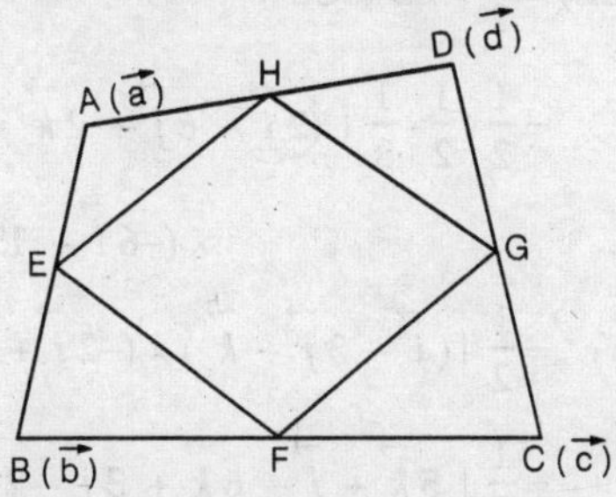

Let the vertices be A($\vec{a}$), B($\vec{b}$), C($\vec{c}$) and D($\vec{d}$). E, F, G, H are the midpoints of AB, BC, CD, DA respectively.

Clearly, the position vector of E = $\vec{OE} = \dfrac{\vec{a}+\vec{b}}{2}$,

the position vector of H = $\vec{OH} = \dfrac{\vec{a}+\vec{d}}{2}$.

$\therefore \quad \vec{EH} = \vec{EO} + \vec{OH} = \vec{OH} - \vec{OE} = \dfrac{\vec{d}-\vec{b}}{2}$,

O being the origin.

The position vector of F = $\vec{OF} = \dfrac{\vec{b}+\vec{c}}{2}$,

the position vector of G = $\vec{OG} = \dfrac{\vec{c}+\vec{d}}{2}$.

$\therefore \quad \vec{FG} = \vec{FO} + \vec{OG} = \vec{OG} - \vec{OF} = \dfrac{\vec{d}-\vec{b}}{2}$

$\therefore \quad \vec{EH} = \vec{FG} = \dfrac{\vec{d}-\vec{b}}{2}$

$\therefore \quad EH = FG$ and $EH \parallel FG$.

$\therefore$ EFGH is a parallelogram.

2. In the parallelogram ABCD the interior bisectors of the consecutive angles B and C intersect at P. Use vector method to find $\angle BPC$.

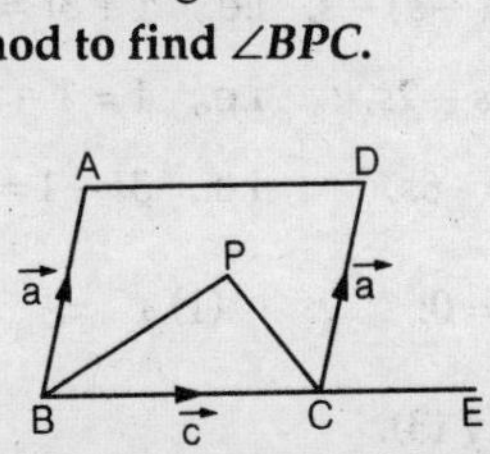

Let $\vec{BC} = \vec{c}, \quad \vec{BA} = \vec{a};$ then $\vec{CD} = \vec{BA} = \vec{a}$.

A vector along the interior bisector BP of the $\angle ABC$

$$= \frac{\vec{c}}{|\vec{c}|} + \frac{\vec{a}}{|\vec{a}|}$$

$$\therefore \quad \vec{BP} = t\left(\frac{\vec{c}}{|\vec{c}|} + \frac{\vec{a}}{|\vec{a}|}\right).$$

A vector along the interior bisector CP of $\angle BCD$

= a vector along the exterior bisector CP of $\angle DCE$

$$= \frac{\vec{c}}{|\vec{c}|} - \frac{\vec{a}}{|\vec{a}|}$$

$$\therefore \quad \vec{CP} = s\left(\frac{\vec{c}}{|\vec{c}|} - \frac{\vec{a}}{|\vec{a}|}\right).$$

$$\text{Now,}\quad \vec{BP}\cdot\vec{CP} = st\left(\frac{\vec{c}}{|\vec{c}|} + \frac{\vec{a}}{|\vec{a}|}\right)\cdot\left(\frac{\vec{c}}{|\vec{c}|} - \frac{\vec{a}}{|\vec{a}|}\right)$$

$$= st\left(\frac{\vec{c}^{\,2}}{|\vec{c}|^2} - \frac{\vec{a}^{\,2}}{|\vec{a}|^2}\right)$$

$$= st(1-1) = 0.$$

$\therefore \quad \vec{BP} \perp \vec{CP}; \quad \therefore \quad \angle BPC = \dfrac{\pi}{2}$.

3. Find the distance of the point B($\vec{i} + 2\vec{j} + 3\vec{k}$) from the line which is passing through A($4\vec{i} + 2\vec{j} + 2\vec{k}$) and which is parallel to the vector $\vec{c} = 2\vec{i} + 3\vec{j} + 6\vec{k}$.

Clearly, the distance $BD = AB \sin\theta$ ... (1)

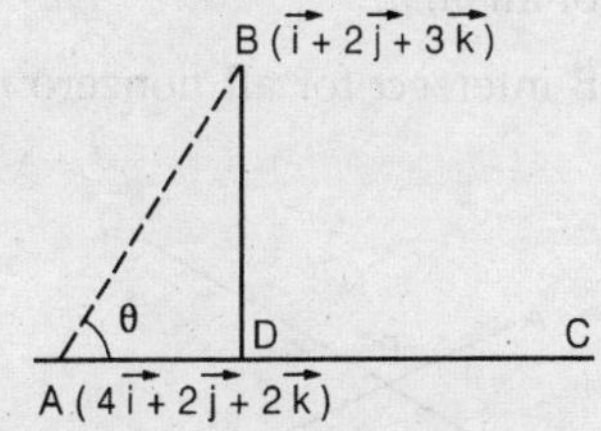

Now, $\vec{AB} = \vec{AO} + \vec{OB} = \vec{OB} - \vec{OA}$, where O is the origin.

$$= \vec{i} + 2\vec{j} + 3\vec{k} - (4\vec{i} + 2\vec{j} + 2\vec{k})$$

$$= -3\vec{i} + \vec{k}$$

$\therefore$ unit vector along $\vec{AB} = \dfrac{1}{\sqrt{10}}(-3\vec{i} + \vec{k})$;

unit vector along the line AC

= unit vector along $\vec{c}$

$$= \frac{2\vec{i} + 3\vec{j} + 6\vec{k}}{\sqrt{2^2 + 3^2 + 6^2}}$$

$$= \frac{2\vec{i} + 3\vec{j} + 6\vec{k}}{7}$$

$\therefore \quad \sin\theta = \left|\frac{1}{\sqrt{10}}(-3\vec{i} + \vec{k}) \times \frac{1}{7}(2\vec{i} + 3\vec{j} + 6\vec{k})\right|$

$= \frac{1}{7\sqrt{10}} \mid -9\vec{k} + 20\vec{j} - 3\vec{i} \mid$

$= \frac{1}{7\sqrt{10}} \sqrt{(-3)^2 + 20^2 + (-9)^2}$

$= \frac{\sqrt{490}}{7\sqrt{10}} = 1$

$\therefore$ from (1), we get

$BD = AB = \mid -3\vec{i} + \vec{k} \mid = \sqrt{10}.$

4. A, B, C and D are four points such that $\vec{AB} = m(2\vec{i} - 6\vec{j} + 2\vec{k})$, $\vec{BC} = \vec{i} - 2\vec{j}$ and $\vec{CD} = n(-6\vec{i} + 15\vec{j} - 3\vec{k})$.

Find the conditions on the scalars *m* and *n* so that CD intersects AB at some point E. Also find the area of the triangle BCE.

AB and CD will intersect if $\vec{AB}$, $\vec{CD}$ and $\vec{BC}$ are coplanar.

$\therefore \quad [\, m(2\vec{i} - 6\vec{j} + 2\vec{k}) \quad n(-6\vec{i} + 15\vec{j} - 3\vec{k}) \quad \vec{i} - 2\vec{j}\,] = 0$

or $\quad mn\begin{vmatrix} 2 & -6 & 2 \\ -6 & 15 & -3 \\ 1 & -2 & 0 \end{vmatrix} = 0$

or $\quad mn\{-12 + 18 - 6\} = 0$

which is true for all *m*, *n*.

$\therefore$ CD and AB intersect for all nonzero real values of *m*, *n*.

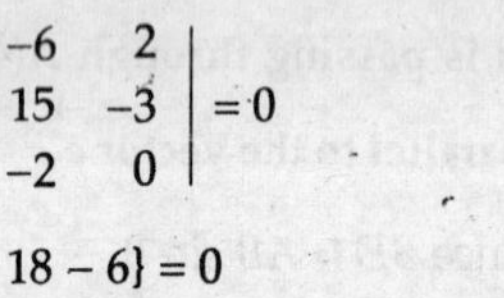

The point E is on both AB and CD. So

$\vec{EB} = \lambda\, \vec{AB} = \lambda(2\vec{i} - 6\vec{j} + 2\vec{k})$

and $\vec{CE} = \mu(-6\vec{i} + 15\vec{j} - 3\vec{k})$. But $\vec{EB} + \vec{CE} = \vec{CB}$

$\therefore \quad \lambda(2\vec{i} - 6\vec{j} + 2\vec{k}) + \mu(-6\vec{i} + 15\vec{j} - 3\vec{k}) = -\vec{i} + 2\vec{j}$

$\therefore \quad 2\lambda - 6\mu = -1$

$-6\lambda + 15\mu = 2$

$2\lambda - 3\mu = 0$

$\therefore \quad 3\mu - 6\mu = -1$, i.e., $\mu = \frac{1}{3}$; $\quad \therefore \quad \lambda = \frac{1}{2}$.

$\therefore \quad \vec{EB} = \frac{1}{2}(2\vec{i} - 6\vec{j} + 2\vec{k}),$

$\vec{CE} = \frac{1}{3}(-6\vec{i} + 15\vec{j} - 3\vec{k})$

$\therefore \quad \text{ar}(\Delta BCE) = \frac{1}{2} \mid \vec{EB} \times \vec{CE} \mid$

$= \frac{1}{2}\cdot\frac{1}{2}\cdot\frac{1}{3} \mid (2\vec{i} - 6\vec{j} + 2\vec{k}) \times (-6\vec{i} + 15\vec{j} - 3\vec{k}) \mid$

$= \frac{1}{2} \mid (\vec{i} - 3\vec{j} + \vec{k}) \times (-2\vec{i} + 5\vec{j} - \vec{k}) \mid$

$= \frac{1}{2} \mid 5\vec{k} + \vec{j} - 6\vec{k} + 3\vec{i} - 2\vec{j} - 5\vec{i} \mid$

$= \frac{1}{2} \mid -2\vec{i} - \vec{j} - \vec{k} \mid$

$= \frac{1}{2}\sqrt{2^2 + 1^2 + 1^2} = \frac{1}{2}\sqrt{6}.$

5. Find the point of intersection of the lines joining points (1, 1, –1), (4, 1, 2) and points (2, 1, 0), (–1, –2, –6).

The equation of the line passing through the points

$A(\vec{i} + \vec{j} - \vec{k})$ and $B(4\vec{i} + \vec{j} + 2\vec{k})$ is

$\vec{r} = (1 - t)(\vec{i} + \vec{j} - \vec{k}) + t(4\vec{i} + \vec{j} + 2\vec{k}).$

The equation of the line joining the points $C(2\vec{i} + \vec{j})$ and $D(-\vec{i} - 2\vec{j} - 6\vec{k})$ is

$\vec{r} = (1 - s)(2\vec{i} + \vec{j}) + s(-\vec{i} - 2\vec{j} - 6\vec{k}).$

At the point of intersection $P(\vec{r})$, we have

$\vec{r} = (1 - t)(\vec{i} + \vec{j} - \vec{k}) + t(4\vec{i} + \vec{j} + 2\vec{k})$

$= (1 - s)(2\vec{i} + \vec{j}) + s(-\vec{i} - 2\vec{j} - 6\vec{k})$

or $\quad \{(1 - t) + 4t\}\vec{i} + \{1 - t + t\}\vec{j} + \{-(1 - t) + 2t\}\vec{k}$

$= \{2(1 - s) - s\}\vec{i} + \{1 - s - 2s\}\vec{j} - 6s\vec{k}$

$\therefore \quad 1 - t + 4t = 2(1 - s) - s$, i.e., $1 + 3t = 2 - 3s$ $\quad$ ... (1)

$1 - t + t = 1 - s - 2s$, i.e., $1 = 1 - 3s$ $\quad$ ... (2)

$-(1 - t) + 2t = -6s$, i.e., $3t - 1 = -6s$ $\quad$ ... (3)

(2) $\Rightarrow$ $s = 0$; $\quad \therefore \quad$ (1) $\Rightarrow$ $t = \frac{1}{3}$.

These values satisfy (3).

$\therefore \quad \vec{r} = (1 - s)(2\vec{i} + \vec{j}) + s(-\vec{i} - 2\vec{j} - 6\vec{k})$

$= 2\vec{i} + \vec{j}$ because $s = 0$.

$\therefore$ the point of intersection has the position vector $2\vec{i} + \vec{j}$.

6. If A(1, 2, 3), B(3, 1, 2), C(0, 1, 1) and D(2, 0, 4) are four points then find the shortest distance between the lines AB and CD.

Here, $\vec{AB} = \vec{AO} + \vec{OB} = \vec{OB} - \vec{OA}$, O being the origin

$$= 3\vec{i} + \vec{j} + 2\vec{k} - (\vec{i} + 2\vec{j} + 3\vec{k})$$

$$= 2\vec{i} - \vec{j} - \vec{k}$$

$$\vec{CD} = \vec{CO} + \vec{OD} = \vec{OD} - \vec{OC}$$

$$= 2\vec{i} + 4\vec{k} - (\vec{j} + \vec{k}) = 2\vec{i} - \vec{j} + 3\vec{k}.$$

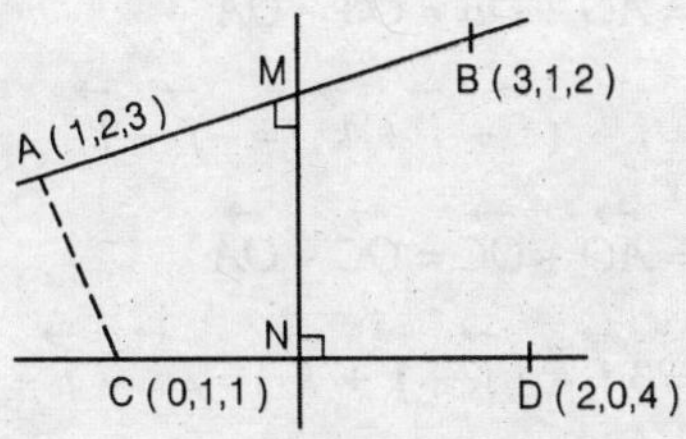

A unit vector along the shortest distance between AB and CD

= a unit vector perpendicular to both $\vec{AB}$ and $\vec{CD}$

$$= \frac{\vec{AB} \times \vec{CD}}{|\vec{AB} \times \vec{CD}|}$$

$$= \frac{(2\vec{i} - \vec{j} - \vec{k}) \times (2\vec{i} - \vec{j} + 3\vec{k})}{|(2\vec{i} - \vec{j} - \vec{k}) \times (2\vec{i} - \vec{j} + 3\vec{k})|}.$$

Now, $(2\vec{i} - \vec{j} - \vec{k}) \times (2\vec{i} - \vec{j} + 3\vec{k})$

$$= \begin{vmatrix} \vec{i} & \vec{j} & \vec{k} \\ 2 & -1 & -1 \\ 2 & -1 & 3 \end{vmatrix} = -4\vec{i} - 8\vec{j}.$$

$\therefore$ a unit vector along the shortest distance $= \dfrac{\vec{i} + 2\vec{j}}{\sqrt{5}}$.

Now, the shortest distance MN

$$= \left| \text{projection of } \vec{AC} \text{ on } \frac{\vec{i} + 2\vec{j}}{\sqrt{5}} \right|$$

$$= \left| \text{projection of } \{\vec{j} + \vec{k} - (\vec{i} + 2\vec{j} + 3\vec{k})\} \text{ on } \frac{\vec{i} + 2\vec{j}}{\sqrt{5}} \right|$$

$$= \left| \text{projection of } (-\vec{i} - \vec{j} - 2\vec{k}) \text{ on } \frac{\vec{i} + 2\vec{j}}{\sqrt{5}} \right|$$

$$= \left| (-\vec{i} - \vec{j} - 2\vec{k}) \cdot \frac{\vec{i} + 2\vec{j}}{\sqrt{5}} \right|$$

$$= \left| \frac{-1 - 2}{\sqrt{5}} \right| = \frac{3}{\sqrt{5}}.$$

7. Find the distance of the point $\vec{P}$ $(\vec{i} + \vec{j} + \vec{k})$ from the plane π which passes through the points $\vec{A}$ $(2\vec{i} + \vec{j} + \vec{k})$, $\vec{B}$ $(\vec{i} + 2\vec{j} + \vec{k})$ and $C(\vec{i} + \vec{j} + 2\vec{k})$.

Also find the position vector of the foot of the perpendicular from P on the plane π.

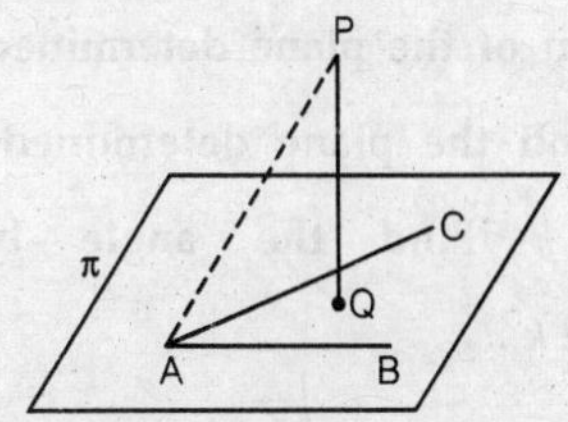

Here, $\vec{AB} = \vec{AO} + \vec{OB} = \vec{OB} - \vec{OA}$, where O is the origin

$$= \vec{i} + 2\vec{j} + \vec{k} - (2\vec{i} + \vec{j} + \vec{k}) = -\vec{i} + \vec{j}$$

$$\vec{AC} = \vec{AO} + \vec{OC} = \vec{OC} - \vec{OA}$$

$$= \vec{i} + \vec{j} + 2\vec{k} - (2\vec{i} + \vec{j} + \vec{k}) = -\vec{i} + \vec{k}$$

$\therefore$ $\vec{AB} \times \vec{AC} = (-\vec{i} + \vec{j}) \times (-\vec{i} + \vec{k})$

$= \vec{j} + \vec{k} + \vec{i}$ which is $\perp^r$ to the plane π

$\therefore$ unit vector perpendicular the plane π

$$= \pm \frac{\vec{i} + \vec{j} + \vec{k}}{\sqrt{3}}.$$

Let PQ = λ; then $\vec{QP} = \pm \dfrac{\lambda}{\sqrt{3}} (\vec{i} + \vec{j} + \vec{k})$ ... (1)

Clearly, $\vec{PA} = \vec{PO} + \vec{OA} = \vec{OA} - \vec{OP}$

$$= 2\vec{i} + \vec{j} + \vec{k} - (\vec{i} + \vec{j} + \vec{k}) = \vec{i}.$$

λ = PQ = | projection of $\vec{PA}$ in the direction of $\vec{PQ}$ |

= | $\vec{i}$ · (unit vector along $\vec{PQ}$) |

$$= \left| \vec{i} \cdot \left(\pm \frac{\vec{i} + \vec{j} + \vec{k}}{\sqrt{3}} \right) \right|$$

$$= \left| \pm \frac{1}{\sqrt{3}} \right| = \frac{1}{\sqrt{3}} \quad \text{... (2)}$$

Clearly, $\vec{OQ} = \vec{OP} + \vec{PQ}$

$= \vec{i} + \vec{j} + \vec{k} \pm \dfrac{\lambda}{\sqrt{3}} (\vec{i} + \vec{j} + \vec{k})$, from (1)

$= \vec{i} + \vec{j} + \vec{k} \pm \dfrac{1}{3} (\vec{i} + \vec{j} + \vec{k})$, using (2)

$$= \frac{2}{3} (\vec{i} + \vec{j} + \vec{k}) \text{ or } \frac{4}{3} (\vec{i} + \vec{j} + \vec{k}).$$

But A, B, C and Q$\left(\frac{2}{3}\vec{i} + \frac{2}{3}\vec{j} + \frac{2}{3}\vec{k}\right)$ are not coplanar

because $[\vec{AB} \ \ \vec{AC} \ \ \vec{AQ}] \neq 0$

$\therefore$ the position vector of the foot of the perpendicular

$= \frac{4}{3}(\vec{i} + \vec{j} + \vec{k})$.

8. A nonzero vector $\vec{a}$ is parallel to the line of intersection of the plane determined by the vectors $\vec{i}$, $\vec{i} + \vec{j}$ and the plane determined by the vectors $\vec{i} - \vec{j}$, $\vec{i} + \vec{k}$. Find the angle between $\vec{a}$ and $\vec{i} - 2\vec{j} + 2\vec{k}$.

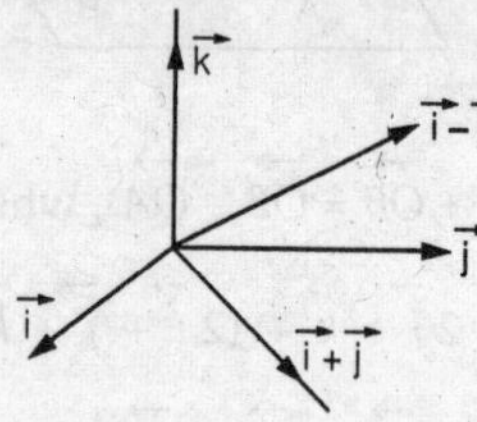

Let $\vec{a} = x\vec{i} + y\vec{j} + z\vec{k}$.

Now, $\vec{a}$, $\vec{i}$, $\vec{i} + \vec{j}$ are coplanar

and $\vec{a}$, $\vec{i} - \vec{j}$, $\vec{i} + \vec{k}$ are coplanar.

$\therefore \quad [\vec{a} \ \ \vec{i} \ \ \vec{i} + \vec{j}] = 0, \ [\vec{a} \ \ \vec{i} - \vec{j} \ \ \vec{i} + \vec{k}] = 0$

$\therefore \quad \begin{vmatrix} x & y & z \\ 1 & 0 & 0 \\ 1 & 1 & 0 \end{vmatrix} = 0$ and $\begin{vmatrix} x & y & z \\ 1 & -1 & 0 \\ 1 & 0 & 1 \end{vmatrix} = 0$

$\therefore \quad z = 0$ and $-x - y + z = 0$

$\Rightarrow \quad z = 0, \ x + y = 0$, i.e., $y = -x$

$\therefore \quad \vec{a} = x\vec{i} - x\vec{j}; \quad \therefore \quad \hat{a} = \frac{\vec{i} - \vec{j}}{\sqrt{2}}$.

Let the angle between $\vec{a}$ and $\vec{i} - 2\vec{j} + 2\vec{k}$ be θ.

Then $\cos\theta = \hat{a} \cdot \frac{\vec{i} - 2\vec{j} + 2\vec{k}}{|\vec{i} - 2\vec{j} + 2\vec{k}|}$

$= \frac{\vec{i} - \vec{j}}{\sqrt{2}} \cdot \frac{\vec{i} - 2\vec{j} + 2\vec{k}}{3} = \frac{1+2}{3\sqrt{2}} = \frac{1}{\sqrt{2}}$

$\therefore \quad \theta = \frac{\pi}{4}$.

9. The position vectors of the vertices A, B and C of a tetrahedron ABCD are $\vec{i} + \vec{j} + \vec{k}$, $\vec{i}$, $3\vec{i}$ respectively. The altitude from the vertex D to the opposite face ABC meets the median line AF through A of the △ABC at the point E. If AD = 4 and the volume of the tetrahedron is $\frac{2\sqrt{2}}{3}$, find the position vector of E for all its positions.

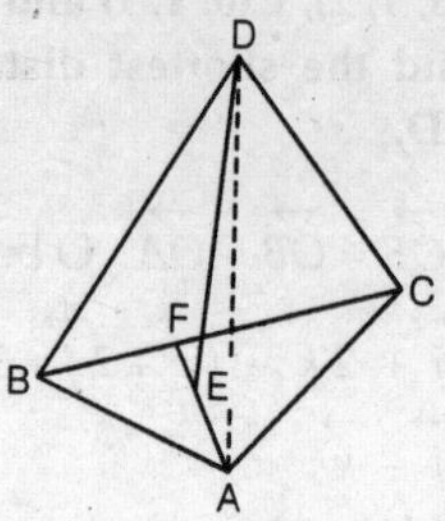

Here, $\vec{AF} = \frac{1}{2}(\vec{AB} + \vec{AC})$

$\vec{AB} = \vec{AO} + \vec{OB} = \vec{OB} - \vec{OA}$

$= \vec{i} - (\vec{i} + \vec{j} + \vec{k}) = -\vec{j} - \vec{k}$

$\vec{AC} = \vec{AO} + \vec{OC} = \vec{OC} - \vec{OA}$

$= 3\vec{i} - (\vec{i} + \vec{j} + \vec{k}) = 2\vec{i} - \vec{j} - \vec{k}$

where O is the origin

$\therefore \quad \vec{AF} = \frac{1}{2}(-\vec{j} - \vec{k} + 2\vec{i} - \vec{j} - \vec{k}) = \vec{i} - \vec{j} - \vec{k}$.

Let $FE : AE = \lambda : 1; \quad \therefore \quad AE = \frac{1}{\lambda + 1} AF$

$\therefore \quad \vec{AE} = \frac{1}{\lambda + 1}\vec{AF} = \frac{1}{\lambda + 1}(\vec{i} - \vec{j} - \vec{k})$

$\therefore \quad \vec{OE} = \vec{OA} + \vec{AE}$

$= \vec{i} + \vec{j} + \vec{k} + \frac{1}{\lambda + 1}(\vec{i} - \vec{j} - \vec{k}) \quad \ldots (1)$

Now, $\vec{AC} \times \vec{AB} = (2\vec{i} - \vec{j} - \vec{k}) \times (-\vec{j} - \vec{k})$

$= \begin{vmatrix} \vec{i} & \vec{j} & \vec{k} \\ 2 & -1 & -1 \\ 0 & -1 & -1 \end{vmatrix} = 2(\vec{j} - \vec{k}) \quad \ldots (2)$

which is perpendicular to the plane ABC and hence parallel to ED.

$\therefore$ unit vector along ED $= \frac{2(\vec{j} - \vec{k})}{|2(\vec{j} - \vec{k})|} = \frac{1}{\sqrt{2}}(\vec{j} - \vec{k})$.

Let $ED = \mu$. Then $\vec{ED} = \frac{\mu}{\sqrt{2}}(\vec{j} - \vec{k}) \quad \ldots (3)$

$\therefore \quad \vec{AD} = \vec{AE} + \vec{ED}$

$= \frac{1}{\lambda + 1}(\vec{i} - \vec{j} - \vec{k}) + \frac{\mu}{\sqrt{2}}(\vec{j} - \vec{k})$

$= \frac{1}{\lambda + 1}\vec{i} + \left(\frac{\mu}{\sqrt{2}} - \frac{1}{\lambda + 1}\right)\vec{j} - \left(\frac{\mu}{\sqrt{2}} + \frac{1}{\lambda + 1}\right)\vec{k}$

$\therefore \quad |\vec{AD}| = AD = 4$

$= \sqrt{\left(\frac{1}{\lambda + 1}\right)^2 + \left(\frac{\mu}{\sqrt{2}} - \frac{1}{\lambda + 1}\right)^2 + \left(\frac{\mu}{\sqrt{2}} + \frac{1}{\lambda + 1}\right)^2}$

or $$16 = \left(\frac{1}{\lambda+1}\right)^2 + 2\left\{\frac{\mu^2}{2} + \left(\frac{1}{\lambda+1}\right)^2\right\}$$

$$= 3 \cdot \frac{1}{(\lambda+1)^2} + \mu^2 \qquad \ldots (4)$$

Also, volume of the tetrahedron

$$= \frac{1}{3} \text{ (area of the base) altitude}$$

$$\therefore \quad \frac{2\sqrt{2}}{3} = \frac{1}{3} \cdot \frac{1}{2} \mid \vec{AC} \times \vec{AB} \mid ED$$

or $$4\sqrt{2} = \mid 2(\vec{j} - \vec{k}) \mid \left| \frac{\mu}{\sqrt{2}} (\vec{j} - \vec{k}) \right| \text{ \{using (2), (3)\}}$$

or $4 = \mid \mu \mid 2; \quad \therefore \quad \mid \mu \mid = 2.$

$$\therefore \quad (4) \quad \Rightarrow \quad 16 = \frac{3}{(\lambda+1)^2} + 4$$

or $$4 = \frac{1}{(\lambda+1)^2}$$

$$\therefore \quad (\lambda+1)^2 = \frac{1}{4}$$

$$\therefore \quad \lambda + 1 = \pm \frac{1}{2}$$

$\therefore$ from (1),

$$\vec{OE} = \vec{i} + \vec{j} + \vec{k} \pm 2(\vec{i} - \vec{j} - \vec{k})$$

$$= (1 \pm 2)\vec{i} + (1 \mp 2)\vec{j} + (1 \mp 2)\vec{k}$$

$$= 3\vec{i} - \vec{j} - \vec{k}, \quad -\vec{i} + 3\vec{j} + 3\vec{k}.$$

10. Constant forces $\vec{P_1} = \vec{i} - \vec{j} + \vec{k}$, $\vec{P_2} = -\vec{i} + 2\vec{j} - \vec{k}$ and $\vec{P_3} = \vec{j} - \vec{k}$ act on a particle at the point A. Determine the work done when the particle is displaced from the point A to B where $\vec{OA} = 4\vec{i} - 3\vec{j} - 2\vec{k}$ and $\vec{OB} = 6\vec{i} + \vec{j} - 3\vec{k}$ are the position vectors of A and B. Also find the moment of $\vec{P_1}$ about the point C(1, 0, 1).

The resultant force $\vec{P}$ acting at A $= \vec{P_1} + \vec{P_2} + \vec{P_3}$

$$= (\vec{i} - \vec{j} + \vec{k}) + (-\vec{i} + 2\vec{j} - \vec{k}) + (\vec{j} - \vec{k})$$

$$= 2\vec{j} - \vec{k}$$

$$\vec{AB} = \vec{AO} + \vec{OB} = \vec{OB} - \vec{OA}$$

$$= 6\vec{i} + \vec{j} - 3\vec{k} - (4\vec{i} - 3\vec{j} - 2\vec{k})$$

$$= 2\vec{i} + 4\vec{j} - \vec{k}$$

$\therefore$ the required work done $= \vec{P} \cdot \vec{AB}$

$$= (2\vec{j} - \vec{k}) \cdot (2\vec{i} + 4\vec{j} - \vec{k})$$

$$= 8 + 1 = 9.$$

The required moment vector

$$= \vec{CA} \times \vec{P_1}$$

$$= (\vec{OA} - \vec{OC}) \times (\vec{i} - \vec{j} + \vec{k})$$

$$= \{(4\vec{i} - 3\vec{j} - 2\vec{k}) - (\vec{i} + \vec{k})\} \times (\vec{i} - \vec{j} + \vec{k})$$

$$= 3(\vec{i} - \vec{j} - \vec{k}) \times (\vec{i} - \vec{j} + \vec{k})$$

$$= 3 \begin{vmatrix} \vec{i} & \vec{j} & \vec{k} \\ 1 & -1 & -1 \\ 1 & -1 & 1 \end{vmatrix}$$

$$= 3(-2\vec{i} - 2\vec{j}) = -6(\vec{i} + \vec{j})$$

and the moment $= \mid -6(\vec{i} + \vec{j}) \mid = 6\sqrt{2}.$

11. A rigid body is rotating about an axis through the point (3, –1, –2). If the particle at the point (4, 1, 0) has the velocity $4\vec{i} - 4\vec{j} + 2\vec{k}$ and that at the point (3, 2, 1) has the velocity $6\vec{i} - 4\vec{j} + 4\vec{k}$, find the magnitude and direction of the angular velocity of the body.

Let the angular velocity be $\vec{w}$.

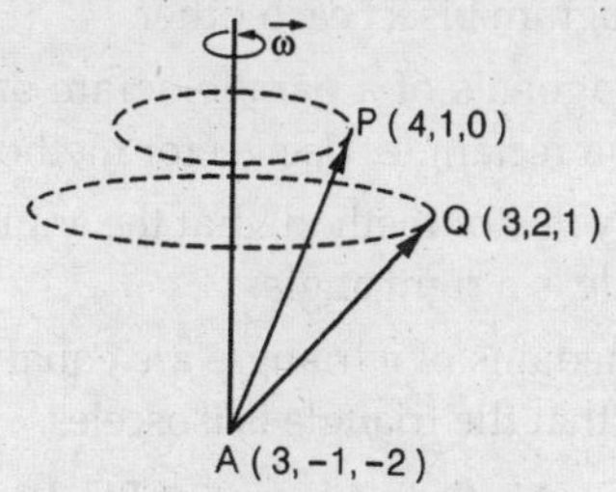

The velocity at a point is given by $\vec{v} = \vec{w} \times \vec{r}$

$\therefore$ from the question, $4\vec{i} - 4\vec{j} + 2\vec{k} = \vec{w} \times \vec{AP}$

and $6\vec{i} - 4\vec{j} + 4\vec{k} = \vec{w} \times \vec{AQ}$.

But, $\vec{AP} = \vec{AO} + \vec{OP} = \vec{OP} - \vec{OA}$

$$= (4\vec{i} + \vec{j}) - (3\vec{i} - \vec{j} - 2\vec{k})$$

$$= \vec{i} + 2\vec{j} + 2\vec{k}$$

$$\vec{AQ} = \vec{AO} + \vec{OQ} = \vec{OQ} - \vec{OA}$$

$$= (3\vec{i} + 2\vec{j} + \vec{k}) - (3\vec{i} - \vec{j} - 2\vec{k})$$

$$= 3\vec{j} + 3\vec{k}$$

$\therefore$ we get,

$$4\vec{i} - 4\vec{j} + 2\vec{k} = \vec{w} \times (\vec{i} + 2\vec{j} + 2\vec{k}) \qquad \ldots (1)$$

$$6\vec{i} - 4\vec{j} + 4\vec{k} = \vec{w} \times (3\vec{j} + 3\vec{k}) \quad \dots (2)$$

Let $\vec{w} = x\vec{i} + y\vec{j} + z\vec{k}$. Then from (1),

$$4\vec{i} - 4\vec{j} + 2\vec{k} = \begin{vmatrix} \vec{i} & \vec{j} & \vec{k} \\ x & y & z \\ 1 & 2 & 2 \end{vmatrix}$$

$$= (2y - 2z)\vec{i} - (2x - z)\vec{j} + (2x - y)\vec{k}$$

$\therefore \quad 2y - 2z = 4, \quad 2x - z = 4, \quad 2x - y = 2$

$\Rightarrow \quad y - z = 2, \quad 2x - y = 2$

($\because$ the second equation can be obtained from the other two)

From (2), $6\vec{i} - 4\vec{j} + 4\vec{k} = \begin{vmatrix} \vec{i} & \vec{j} & \vec{k} \\ x & y & z \\ 0 & 3 & 3 \end{vmatrix}$

$$= (3y - 3z)\vec{i} - 3x\vec{j} + 3x\vec{k}$$

$\therefore \quad 3y - 3z = 6, \quad -3x = -4, \quad 3x = 4$

$\Rightarrow \quad y - z = 2, \quad x = \frac{4}{3}$.

Thus, we have the following three different equations in x, y, z:

$$y - z = 2, \quad 2x - y = 2, \quad x = \frac{4}{3}$$

Solving these $x = \frac{4}{3}, \quad y = \frac{2}{3}, \quad z = \frac{-4}{3}$

$\therefore$ angular velocity $\vec{w} = \frac{4}{3}\vec{i} + \frac{2}{3}\vec{j} - \frac{4}{3}\vec{k}$

$\therefore \quad |\vec{w}| = \sqrt{\left(\frac{4}{3}\right)^2 + \left(\frac{2}{3}\right)^2 + \left(\frac{-4}{3}\right)^2}$

$$= \sqrt{4} = 2$$

and unit vector along $\vec{w} = \frac{\vec{w}}{|\vec{w}|}$

$$= \frac{2}{3}\vec{i} + \frac{1}{3}\vec{j} - \frac{2}{3}\vec{k}.$$

Exercises

1. Prove, by using vector, that the diagonals of a parallelogram bisect each other.
2. If the diagonals of a parallelogram are equal, prove that it is a rectangle. Use vector method.
3. Prove, by vector method, that the angle inscribed in a semicircle is a right angle.
4. If two medians of a triangle are equal then prove, by vectors, that the triangle is isosceles.
5. In ΔABC, D is the midpoint of BC. Prove, by vectors, that $AB^2 + AC^2 = 2(AD^2 + BD^2)$.
6. Prove, by vector method, that the line segment joining the midpoints of nonparallel sides of a trapezium is parallel to the parallel sides and is equal to AM of the lengths of the parallel sides.
7. If two pairs of opposite sides of tetrahedron are perpendicular, show by vectors that the third pair of opposite sides are also perpendicular.
8. Prove vectorially that $\sin 2A = 2 \sin A \cdot \cos A$.
9. ABCD is a parallelogram such that $\vec{AB} = \vec{a}$ and $\vec{AD} = \vec{c}$. E is the midpoint of AB and F is the point of intersection of AC and DE. Prove that F divides AC in the ratio 1 : 2. Use vector method.
10. The vertices of a triangle ABC are A($\vec{i} + \vec{j} + 2\vec{k}$), B($4\vec{i} + 3\vec{j} + \vec{k}$) and C($2\vec{i} + 3\vec{j} + 5\vec{k}$); AD is the bisector of the angle BAC and it cuts BC at D. Find the vector equation of AD and the length of AD.
11. In a triangle OAB, E is the middle point of OB and D is a point on AB such that $AD : DB = 2 : 1$. If OD and AE intersect at P, determine the ratio $OP : PD$, using vector method.
12. If collinear point P, Q and R divide the sides BC, CA and AB in the ratio $\lambda : 1$ externally, $\mu : 1$ internally and $\nu : 1$ internally respectively then prove that $\lambda \cdot \mu \cdot \nu + 1 = 0$.
13. The position vectors of three points A, B and C are $\vec{a}, \vec{b}$ and $\vec{c}$ respectively such that
$$3\vec{a} - 5\vec{b} + 2\vec{c} = \vec{0}.$$
Are the points A, B, C collinear? If so find $AB : BC$.
14. Show that the perpendicular distance of any point $\vec{a}$ from the line $\vec{r} = \vec{b} + t\vec{c}$ is $\frac{|(\vec{b} - \vec{a}) \times \vec{c}|}{|\vec{c}|}$.
15. Verify whether the line joining points given by $-\vec{a} - 2\vec{b} - 3\vec{c}$, $\vec{a} + 2\vec{b} - 5\vec{c}$ intersects the line joining the points $-4\vec{c}$, $6\vec{a} - 4\vec{b} + 4\vec{c}$. If they intersect find the point of intersection. If they do not intersect find the shortest distance between the lines.

16. Find the shortest distance between the lines joining the points (2, 0, 1), (2, 3, 0) and (1, 1, 6), (5, 5, 1).

17. Find the shortest distance between a line passing through A(1, –1, 2) and parallel to the vector $\vec{i} + 2\vec{j}$, and a line passing through B(2, 0, 3) and parallel to the vector $2\vec{i} - \vec{j}$.

18. (a) The position vector of two points A and C are $9\vec{i} - \vec{j} + 7\vec{k}$ and $7\vec{i} - 2\vec{j} + 7\vec{k}$ respectively. The point of intersection of vectors $\vec{AB} = 4\vec{i} - \vec{j} + 3\vec{k}$ and $\vec{CD} = 2\vec{i} - \vec{j} + 2\vec{k}$ is P. If $\vec{PQ}$ is perpendicular to $\vec{AB}$ and $\vec{CD}$ and $PQ = 15$, find the position vector of Q.

(b) The position vectors of points P and Q are $5\vec{i} + 7\vec{j} - 2\vec{k}$ and $-3\vec{i} + 3\vec{j} + 6\vec{k}$ respectively. Vectors $\vec{A} = 3\vec{i} - \vec{j} + \vec{k}$, $\vec{B} = -3\vec{i} + 2\vec{j} + 4\vec{k}$ pass through P and Q respectively. Another vector $\vec{C} = 2\vec{i} + 7\vec{j} - 5\vec{k}$ intersects $\vec{A}$ and $\vec{B}$ at M and N respectively. Find the position vectors of M and N.

19. Show that the plane passing through the points $\vec{r_1}, \vec{r_2}, \vec{r_3}$ has the equation

$$[\vec{r}\ \vec{r_2}\ \vec{r_3}] + [\vec{r}\ \vec{r_3}\ \vec{r_1}] + [\vec{r}\ \vec{r_1}\ \vec{r_2}] = [\vec{r_1}\ \vec{r_2}\ \vec{r_3}].$$

20. Find the equation of the plane passing through the point $2\vec{i} + 3\vec{j} - \vec{k}$ and perpendicular to the vector $3\vec{i} + 4\vec{j} + 7\vec{k}$.

21. The position vector of the points A, B, C, D are respectively $2\vec{i} + \vec{j} - \vec{k}$, $\vec{i} + \vec{j} + \vec{k}$, $\vec{i} - 2\vec{j} + 3\vec{k}$ and $3\vec{i} - \vec{j} + 2\vec{k}$. Find the length of the perpendicular let fall from A to the plane containing B, C and D.

22. ABC is a triangle whose vertices have the position vectors $\vec{i}$, $\vec{j}$ and $\vec{k}$ respectively. A line parallel to the vector $\vec{i} + \vec{j}$ and passing through the point $(\vec{i} - 2\vec{j} + 3\vec{k})$ cuts the plane of the triangle at P. Find the position vector of the point P.

23. The position vector of the points A, B and C are respectively $\vec{i} + \vec{j}$, $\vec{j} + \vec{k}$ and $\lambda\vec{k} + \vec{i}$. If the volume of the tetrahedron OABC is 6, find λ where O is the origin.

24. The forces $\vec{P}$, $\vec{Q}$ and $\vec{R}$ of 15 N each act along AB, BC and CA respectively. The position vectors of A, B and C are respectively $2\vec{i} - 4\vec{j} + 3\vec{k}$, $5\vec{i} + 3\vec{j} - 2\vec{k}$ and $-2\vec{i} + 2\vec{j} + 3\vec{k}$. Find the resultant force vector.

25. Constant forces

$$\vec{P} = 2\vec{i} - 5\vec{j} + 6\vec{k} \text{ and } \vec{Q} = -\vec{i} + 2\vec{j} - \vec{k}$$

act on a particle. Determine the work done when the particle is displaced from a point A($4\vec{i} - 3\vec{j} - 2\vec{k}$) to the point B($6\vec{i} + \vec{j} - 3\vec{k}$).

26. A force $\vec{F} = 3\vec{i} + 2\vec{j} - 4\vec{k}$ is acting at the point (1, –1, 2). Find the vector moment of $\vec{F}$ about the point (2, –1, 3).

27. The vertices of a ΔABC are A(–1, 0, 2), B(1, 2, 0) and C(2, 3, 4). Find the moment of a force of magnitude 10 N acting at A along AB, about C.

28. n forces $\vec{F_1}, \vec{F_2}, \vec{F_3}, \ldots, \vec{F_n}$ act at points $R_1, R_2, R_3, \ldots, R_n$ respectively. If $\sum_{i=1}^{n} \vec{F_i} = \vec{0}$ and the sum of their moments about a point O is $\vec{0}$ then show that the sum of their moments about another point O′ is also $\vec{0}$.

29. A force $\vec{F_1} = 3(\vec{i} + \vec{j} + \vec{k})$ acts at the point A(1, 3, 5). The position vector of the point P is $\vec{i} + \vec{j} + \vec{k}$. Another force $\vec{F_2}$ of magnitude 6 acts at the point A in the plane of the force $\vec{F_1}$ and the point P. Find the unit vector along $\vec{F_2}$ if the moment of $\vec{F_1}$ and $\vec{F_2}$ about the point P are equal.

30. A rigid body is rotating at 5 radians per second about an axis AB where A and B are the points $2\vec{i} + \vec{j} + \vec{k}$ and $8\vec{i} - 2\vec{j} + 3\vec{k}$ respectively. Find the velocity of the particle P of the body at the point $5\vec{i} + \vec{j} - \vec{k}$.

31. Find the vector equation of the sphere whose radius is $|\vec{i} \times \vec{j} + \vec{j} \times \vec{k} + \vec{k} \times \vec{i}|$ and the centre has the position vector $(\vec{a} \cdot \vec{i})\vec{i} + (\vec{a} \cdot \vec{j})\vec{j} + (\vec{a} \cdot \vec{k})\vec{k}$.

Answers

10. $\vec{r} = \vec{i} + \vec{j} + 2\vec{k} + t(2\vec{i} + 2\vec{j} + \vec{k})$; 3

11. 3 : 2 **13.** yes, 2 : 3

15. intersect at the point whose position vector is $-4\vec{c}$

16. $\frac{53}{\sqrt{281}}$ **17.** 1

18. (a) $6\vec{i} - 9\vec{j} - 9\vec{k}$ or $-4\vec{i} + 11\vec{j} + 11\vec{k}$

(b) $2\vec{i} + 8\vec{j} - 3\vec{k}$, $\vec{j} + 2\vec{k}$ respectively

20. $\vec{r} \cdot (3\vec{i} + 4\vec{j} + 7\vec{k}) = 11$

21. $\frac{11}{\sqrt{53}}$ **22.** $\frac{1}{2}(\vec{i} - 5\vec{j} + 6\vec{k})$

23. 35, −37

24. $15\left[\left(\frac{3}{\sqrt{83}} - \frac{7}{5\sqrt{3}} + \frac{2}{\sqrt{13}}\right)\vec{i} + \left(\frac{7}{\sqrt{83}} - \frac{1}{5\sqrt{3}} - \frac{3}{\sqrt{13}}\right)\vec{j} + 5\left(\frac{1}{5\sqrt{3}} - \frac{1}{\sqrt{83}}\right)\vec{k}\right]$

25. 15 **26.** $2\vec{i} - 7\vec{j} - 2\vec{k}$

27. $\frac{50}{3}\sqrt{6}$ units

29. $3\left(\vec{i} + \frac{2 \pm \sqrt{14}}{5}\vec{j} + \frac{-1 \pm \sqrt{14}}{5}\vec{k}\right)$

30. $\frac{15}{7}(2\vec{i} + 6\vec{j} + 3\vec{k})$ **31.** $|\vec{r} - \vec{a}| = \sqrt{3}$

□

Chapter Test

Time: 90 minutes

1. In each of the following, fill in the blank so that the resulting statement is correct.

(a) If three consecutive vertices of a parallelogram are $\vec{i} + \vec{j} + \vec{k}$, $2\vec{i} - \vec{j} + 7\vec{k}$ and $\vec{i} - 2\vec{j}$ then the position vector of the fourth vertex is ______ .

(b) In the tetrahedron OABC, three vertices are $A(\vec{i} + \vec{j} - \vec{k})$, $B(\vec{j} + \vec{k} - \vec{i})$, $C(\vec{k} + \vec{i} - \vec{j})$ and the fourth vertex O is the origin. Then the volume of the tetrahedron is ______ .

(c) Two vertices of a triangle and its centroid have the position vectors $2\vec{i} + \vec{j} + \vec{k}$, $\vec{i} - 3\vec{j} - \vec{k}$ and $\vec{i} + \vec{j}$ respectively. Then the position vector of the third vertex is ______ .

2. In each of the following, one or more options are correct. Choose the correct option(s).

(a) The position vectors of four points are $\vec{a}, \vec{b}, \vec{c}$ and $\vec{d}$ such that $5\vec{a} - 4\vec{b} + 2\vec{c} - 3\vec{d} = 0$. Then

A. the four points are coplanar B. the four points are collinear

C. the line segment joining the points $\vec{a}, \vec{c}$ is divided in the ratio 5 : 2 by the line segment joining the points $\vec{b}, \vec{d}$

D. none of these

(b) If $\vec{r} = \sin^2\theta\, \vec{a} + \cos^2\theta\, \vec{b}$ where θ is a parameter then the points whose position vectors are $\vec{a}, \vec{b}, \vec{r}$, are collinear for

A. all values of θ B. $0 \le \theta \le \pi/2$

C. $0 \le \theta \le \pi$ D. none of these

(c) Let (P, QRS) be a regular tetrahedran and A, B, C be respectively the middle points of the edges PQ, PR and PS respectively. The ratio in which the plane ABC divides the solid is

A. 1 : 3 B. 1 : 2 C. 1 : 7 D. 1 : 8

3. A force $\vec{F_1} = 3(-\vec{i} + \vec{j})$ acts through the point A(1, 1, 1). The moment of $\vec{F_1}$ about the point (2, –3, 4) is equal to that of a force $\vec{F_2}$ through the point A parallel to the vector $2\vec{i} - 3\vec{j} + \vec{k}$. Find the force $\vec{F_2}$.

4. A plane passes through the points (0, 0, 0), (1, 2, 3) and (–1, 1, 1). Another plane passing through the origin is perpendicular to the vector $\vec{i} + \vec{j} - 2\vec{k}$. Find the vector equation of the line of intersection of the two planes.

5. Prove that the lines $\vec{r} = \vec{a} + t\vec{b}$ and $\vec{r} = \vec{c} + s\vec{d}$, where t and s are parameters, are coplanar if $[\vec{a}\ \vec{b}\ \vec{d}] = [\vec{b}\ \vec{d}\ \vec{c}]$. If the lines are nonintersecting, find the shortest distance between them when $\vec{a}, \vec{b}, \vec{c}, \vec{d}$ are known.

6. Find the volume of a regular tetrahedron whose each edge is of length a. Use vector method.

7. The plane base of a solid in the x-y plane is in the shape of a triangle made of the points (x, y) such that $0 \le x \le 1$ and $0 \le y \le x$. The cross section of the solid, perpendicular to the x-axis, is always a square. Find the volume of the solid, using vectors.
8. If (O, PQR) be a regular tetrahedron of edge a with O as the origin and the position vectors of P, Q, R are $\vec{p}, \vec{q}, \vec{r}$ respectively then find the value of $|\vec{p} \times \vec{q} + \vec{q} \times \vec{r} + \vec{r} \times \vec{p}|$.
9. Let BC be a line segment on which D, E are two points such that $BE > BD$ and $BC \cdot DE = BD \cdot CE$. A is any other point not on the line BC then prove that BE subtends a right angle at A if AE bisects the angle DAC.
10. Let P be any point in the plane of the triangle ABC, whose orthocentre is H, such that $\angle BAC = \angle BPC$. If D is the middle point of BC and E is the middle point of PH then prove by vector method or otherwise that $DE \perp AP$.
11. Show, by vector methods, that the angular bisectors of a triangle are concurrent and find an expression for the position vector of the point of concurrency in terms of the position vectors of the vertices.

Answers

1. (a) $-6\vec{k}$ (b) $\frac{2}{3}$ (c) $5\vec{j}$

2. (a) A, C (b) A, B, C (c) C

3. $-\frac{9}{5}(2\vec{i} - 3\vec{j} + \vec{k})$

4. $\vec{r} = t(5\vec{i} + \vec{j} + 3\vec{k})$

5. $\dfrac{|[\vec{c}\ \vec{b}\ \vec{d}] - [\vec{a}\ \vec{b}\ \vec{d}]|}{|\vec{b} \times \vec{d}|}$

6. $\frac{1}{3}\left(\frac{a}{\sqrt{2}}\right)^3$

7. $\frac{1}{3}$

8. $\frac{\sqrt{3}}{2}a^2$

□

Probability

1. Elementary Probability

Recap of Facts and Formulae

1. Sample space and event

- The set S of all possible outcomes of a random experiment is called the *sample space* or *probability space*.
- Every subset of a sample space is an *event*.

For example In throwing a dice, the sample space $S = \{1, 2, 3, 4, 5, 6\}$ and $n(S) = 6$.

$E_1 = \{1, 3, 5\} \subset S$. So E_1 is an event and $n(E_1) = 3$.

The event $E_1 = \{1, 3, 5\}$ is also expressed as the event of getting an odd number in throwing a dice.

- $\phi \subset S$ and $S \subseteq S$. So ϕ and S are also events for the sample space S. ϕ is the *impossible event* (null event) and S is the *certain event*.

2. Probability of an event

- If the sample space S is discrete (i.e. $n(S)$ is finite) then the probability $P(E)$ of the event E is given by

$$P(E) = \frac{n(E)}{n(S)}$$

$$= \frac{\text{number of favourable outcomes}}{\text{total number of outcomes}}$$

TIP The problems of restricted permutation or combination are convertible into problems of probability. *Students are advised to work out such problems by using exercises in the corresponding chapter in Algebra.* For such problems,

$$P(E) = \frac{\text{number of ways under the restriction}}{\text{number of ways without restriction}} = \frac{n(E)}{n(S)}.$$

For example How many numbers of four distinct digits can be made with the digits 1, 2, 3 and 4 which begin with 4?

It is a problem on restricted permutation. A similar sum in probability will be as the following:

"A four-digit number of distinct digits is written down at random by using digits 1, 2, 3 and 4. What is the probability (chance) that the number begins with the digit 4?"

Here $n(S)$ = total number of numbers of four distinct digits that can be made with 1, 2, 3 and 4 (without restriction)

$= {}^4P_4 = 4\,!$.

$n(E)$ = the number of numbers of four distinct digits beginning with 4 that can be made with 1, 2, 3 and 4

$= 1 \times {}^3P_3 = 3\,!$.

$\therefore$ required probability $= \dfrac{n(E)}{n(S)} = \dfrac{3\,!}{4\,!} = \dfrac{1}{4}$.

- If $n(S)$ and $n(E)$ are both infinite, the probability can be estimated by geometrical method as below:

$$P(E) = \frac{\text{favourable length}}{\text{total length}}$$

$$\text{or} \quad \frac{\text{favourable area}}{\text{total area}}$$

$$\text{or} \quad \frac{\text{favourable volume}}{\text{total volume}}$$

according as the sample points (outcomes) are distributed over a length (one dimension) or an area (two dimensions) or a volume (three dimensions).

For example A football match is played from 4 PM to 6 PM. A boy arrives to see the match (not before the match starts). What is the probability that he will miss the only goal of the match which takes place at the 15th minute of the match?

To arrive at the match, the boy has to reach between 4 PM and 6 PM. So, the total possible length of time = 2 hours = 120 minutes. He will miss the goal if he arrives between 4.15 PM and 6 PM. So, the favourable length of time = 1 hour 45 minutes = 105 minutes

$\therefore$ the probability of his missing the goal

$$= \frac{105}{120} = \frac{7}{8}.$$

- $0 \le P(E) \le 1$.

3. Complementary event

- The *complementary event* E' (or $\overline{E}$ or E^c) of the event E is the event of E not happening and
$P(E') = 1 - P(E)$, i.e., $P(E) + P(E') = 1$.

4. Odds in favour, odds against

- The *odds in favour* of the event $E = \frac{P(E)}{P(E')}$.
- The *odds against* the event $E = \frac{P(E')}{P(E)}$
- If odds in favour of the event $E = a : b$ then
$P(E) = \frac{a}{a+b}$.
- If odds against the event $E = a : b$ then $P(E) = \frac{b}{a+b}$.

5. Union and intersection of events

- The *union* $E_1 \cup E_2$ of events E_1 and E_2 is the event of at least one of the events E_1, E_2 happening.
- The *intersection* $E_1 \cap E_2$ of events E_1 and E_2 is the event of both the events E_1, E_2 happening.
- $P(E_1 \cup E_2) = P(E_1) + P(E_2) - P(E_1 \cap E_2)$ *{Addition theorem}*

$$P(E_1 \cup E_2 \cup E_3) = P(E_1) + P(E_2) + P(E_3) - P(E_1 \cap E_2) - P(E_2 \cap E_3) - P(E_3 \cap E_1) + P(E_1 \cap E_2 \cap E_3)$$

$$P(E_1 \cup E_2 \cup E_3 \cup \ldots \cup E_n) = \sum_{i=1}^{n} P(E_i) - \sum_{1 \le i \ne j \le n} P(E_i \cap E_j) + \sum_{1 \le i \ne j \ne k \le n} P(E_i \cap E_j \cap E_k) - \ldots + (-1)^{n-1} P(E_1 \cap E_2 \cap \ldots \cap E_n)$$

6. Mutually exclusive and independent events

- Two events E_1 and E_2 are *mutually exclusive* if $P(E_1 \cap E_2) = 0$. So, for mutually exclusive events
$P(E_1 \cup E_2) = P(E_1) + P(E_2)$
$P(E_1 \cup E_2 \cup E_3) = P(E_1) + P(E_2) + P(E_3)$
- Two events E_1 and E_2 are *independent iff*
$P(E_1 \cap E_2) = P(E_1) \cdot P(E_2)$.
Thus, for independent events:
$P(E_1 \cap E_2) = P(E_1) \cdot P(E_2)$ *(Multiplication theorem)*
$P(E_1 \cap E_2 \cap E_3) = P(E_1) \cdot P(E_2) \cdot P(E_3)$.
- Two non-null events E_1, E_2 cannot be mutually exclusive and independent at the same time.
- If E_1, E_2 are independent events then E_1, E_2'; E_1', E_2 and E_1', E_2' are also independent.
- If E_1, E_2 are independent events then
$P(E_1 \cup E_2)' = \{1 - P(E_1)\}\{1 - P(E_2)\}$.
- For any two events A and B we have
$P(A) = P(A \cap B) + P(A \cap B')$.

7. Conditional probability

- If E_1, E_2 are two events which occur in conjunction and E_2 takes place after E_1 has taken place (i.e., E_2 is dependent on E_1) then the probability of E_2 taking place after E_1 has taken place, denoted by $P(E_2/E_1)$, is given by $P(E_2/E_1) = \frac{P(E_1 \cap E_2)}{P(E_1)}$.

8. Expectation

- The *expectation* of the random variable x to have the value x_1 is $p_1 x_1$ if the probability of the value x_1 is p_1.
- If the values $x_1, x_2, x_3, \ldots x_n$ of the random variable x have the probabilities $p_1, p_2, p_3, \ldots, p_n$ respectively then the expectation of x, denoted by $E(x)$, is given by
$$E(x) = \sum_{i=1}^{n} x_i p_i.$$

9. Repeated trial

If the probability of success in one trial is p and that of failure is q so that $p + q = 1$ then

- the probability of r successes in n trials $= {}^nC_r \cdot p^r q^{n-r}$, i.e., $(r+1)$th term in the binomial expansion of $(q+p)^n$.
- the probability of at least r successes in n trials
$= {}^nC_r p^r q^{n-r} + {}^nC_{r+1} p^{r+1} q^{n-(r+1)} + \ldots + {}^nC_n p^n$.
- the probability of at most r successes in n trials
$= {}^nC_0 q^n + {}^nC_1 \cdot pq^{n-1} + \ldots + {}^nC_r p^r q^{n-r}$.

Selected Solved Examples

1. If two numbers p and q are chosen at random from the set {1, 2, 3, 4, 5, 6, 7, 8, 9, 10} with replacement, determine the probability that the roots of the equation $x^2 + px + q = 0$ are real.

Roots of $x^2 + px + q = 0$ are real if $p^2 - 4q \ge 0$

or $p^2 \ge 4q$, i.e., $\left(\frac{p}{2}\right)^2 \ge q$

$\therefore$ if $p = 2$ then $q = 1$

if $p = 3$ then $q = 1, 2$

if $p = 4$ then $q = 1, 2, 3, 4$

if $p = 5$ then $q = 1, 2, 3, 4, 5, 6$

if $p = 6$ then $q = 1, 2, 3, \ldots, 9$

if $p = 7$ then $q = 1, 2, 3, \ldots, 10$

if $p = 8$ then $q = 1, 2, 3, \ldots, 10$

if $p = 9$ then $q = 1, 2, 3, \ldots, 10$

if $p = 10$ then $q = 1, 2, 3, \ldots, 10$.

Now, $n(S) = 10 \times 10 = 100$

and $E = \{(2, 1), (3, 1), (3, 2), (4, 1), (4, 2), (4, 3), (4,4),$
$(5, 1), (5, 2), \ldots, (5, 6), (6, 1), (6, 2), \ldots, (6, 9),$
$(7, 1), (7, 2), \ldots, (7, 10), (8, 1), (8, 2), \ldots, (8, 10),$
$(9, 1), (9, 2), \ldots, (9, 10),$
$(10, 1), (10, 2), \ldots, (10, 10)\}$

$\therefore$ $n(E) = 1 + 2 + 4 + 6 + 9 + 10 + 10 + 10 + 10 = 62.$

$\therefore$ the required probability $= \dfrac{n(E)}{n(S)} = \dfrac{62}{100} = \dfrac{31}{50}.$

2. A determinant of the second order is made with the elements 0 and 1. What is the probability that the determinant made is non-negative?

The number of determinants of the second order that can be made with 0, 1

$= 2 \times 2 \times 2 \times 2 = 16$

($\because$ each of the four places of elements can be filled in 2 ways—by 0 or 1).

$\therefore$ $n(S) = 16.$

Let E = the event of getting non-negative determinants. Then E' = the event of getting negative determinants. Clearly, negative determinants of the second order that can be made with 0, 1 are

$$\begin{vmatrix} 0 & 1 \\ 1 & 0 \end{vmatrix}, \begin{vmatrix} 0 & 1 \\ 1 & 1 \end{vmatrix}, \begin{vmatrix} 1 & 1 \\ 1 & 0 \end{vmatrix}.$$

$\therefore$ $P(E') = \dfrac{n(E')}{n(S)} = \dfrac{3}{16}$

$\therefore$ $P(E) = 1 - P(E') = 1 - \dfrac{3}{16} = \dfrac{13}{16}.$

Hence, the required probability $= \dfrac{13}{16}.$

3. Two squares are chosen at random from the small squares drawn on a chessboard. What is the chance that the two squares chosen have exactly one corner in common?

Total number of small squares on a chessboard = 64.

$\therefore$ the total number of ways to select 2 squares $= n(S) = {}^{64}C_2.$

Two squares selected can have a corner common if they are selected from two consecutive rows (or columns). The number of ways to select two consecutive rows (or columns) = 7, because there are 8 rows (or columns) of small squares.

For each pair of two consecutive rows (or columns), the number of pairs of squares having exactly one common corner $= 2 \times 7 = 14.$

$\therefore$ the number of favourable selections $= n(E) = 7 \times 14.$

$\therefore$ the required probability

$$= \frac{n(E)}{n(S)} = \frac{7 \times 14}{{}^{64}C_2} = \frac{7 \times 14 \times 2}{64 \times 63} = \frac{7}{144}.$$

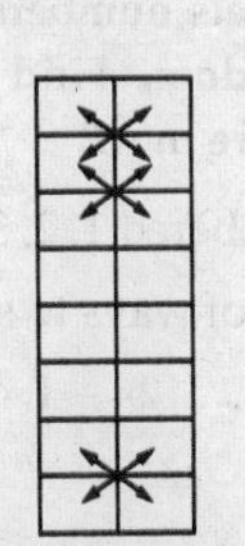

4. Five ordinary dice are rolled at random and the sum of the numbers shown on them is 16. What is the probability that the numbers shown on each is any one from 2, 3, 4 or 5?

If the numbers shown be integers x_1, x_2, x_3, x_4 and x_5 then

$$x_1 + x_2 + x_3 + x_4 + x_5 = 16$$

where $1 \le x_1 \le 6, 1 \le x_2 \le 6, \ldots, 1 \le x_5 \le 6.$

The number of solutions of this equation

$= \text{coefficient of } x^{16} \text{ in } (x + x^2 + x^3 + x^4 + x^5 + x^6)^5$

$= \text{coefficient of } x^{11} \text{ in } (1 + x + x^2 + x^3 + x^4 + x^5)^5$

$= \text{coefficient of } x^{11} \text{ in } \left(\dfrac{1 - x^6}{1 - x}\right)^5$

$= \text{coefficient of } x^{11} \text{ in } (1 - x^6)^5 \cdot (1 - x)^{-5}$

$= \text{coefficient of } x^{11} \text{ in } \{{}^5C_0 - {}^5C_1 x^6 + {}^5C_2 x^{12} - \ldots\}$
$\times \{{}^4C_0 + {}^5C_1 x + {}^6C_2 x^2 + \ldots + {}^{15}C_{11} x^{11} + \ldots \text{ to } \infty\}$

$= {}^5C_0 \cdot {}^{15}C_{11} - {}^5C_1 \cdot {}^9C_5$

$= \dfrac{15 \cdot 14 \cdot 13 \cdot 12}{24} - 5 \cdot \dfrac{9 \cdot 8 \cdot 7 \cdot 6}{24}$

$= 15 \cdot 7 \cdot 13 - 10 \cdot 9 \cdot 7 = 1365 - 630 = 735$

$\therefore$ $n(S) = 735.$

Now, $n(E)$ = the number of integral solutions of

$$x_1 + x_2 + \ldots + x_5 = 16 \text{ where}$$

$$2 \le x_1 \le 5, 2 \le x_2 \le 5, \ldots, 2 \le x_5 \le 5$$

$= \text{coefficient of } x^{16} \text{ in } (x^2 + x^3 + x^4 + x^5)^5$

$= \text{coefficient of } x^6 \text{ in } (1 + x + x^2 + x^3)^5$

$= \text{coefficient of } x^6 \text{ in } (1 + x)^5 \cdot (1 + x^2)^5$

$= \text{coefficient of } x^6$ in

$\{{}^5C_0 + {}^5C_1 x + {}^5C_2 x^2 + \ldots + {}^5C_5 x^5\}$

$\times \{{}^5C_0 + {}^5C_1 x^2 + {}^5C_2 x^4 + \ldots + {}^5C_5 x^{10}\}$

$= {}^5C_0 \cdot {}^5C_3 + {}^5C_2 \cdot {}^5C_2 + {}^5C_4 \cdot {}^5C_1$

$= 10 + 10 \times 10 + 5 \times 5 = 135.$

$\therefore$ the required probability $= \dfrac{n(E)}{n(S)} = \dfrac{135}{735} = \dfrac{9}{49}$.

5. Out of $(2n + 1)$ tickets numbered consecutively, three are drawn at random. Find the chance that the numbers on them are in AP.

Let the tickets be numbered $1, 2, 3, \ldots, 2n + 1$.

$\therefore$ $n(S)$ = the number of ways to select three numbers from $1, 2, 3, \ldots, 2n + 1$

$= {}^{2n+1}C_3$.

Now, possible APs with common difference 1 are

$(1, 2, 3), (2, 3, 4), \ldots, (2n - 1, 2n, 2n + 1)$,

i.e., $2n - 1$ possibilities.

Possible APs with common difference 2 are

$(1, 3, 5), (2, 4, 6), \ldots, (2n - 3, 2n - 1, 2n + 1)$,

i.e., $2n - 3$ possibilities.

....

Possible AP with common difference n is

$(1, n + 1, 2n + 1)$, i.e., 1 possibility.

$\therefore \quad n(E) = (2n - 1) + (2n - 3) + (2n - 5) + \ldots + 1$

$= 1 + 3 + 5 + \ldots + (2n - 1)$

$= \dfrac{n}{2}\{2 \times 1 + (n - 1)2\} = n^2.$

$\therefore$ the required probability

$$= \frac{n(E)}{n(S)} = \frac{n^2}{{}^{2n+1}C_3} = \frac{n^2 \cdot 3!}{(2n + 1)2n(2n - 1)}$$

$$= \frac{3n}{(2n + 1)(2n - 1)} = \frac{3n}{4n^2 - 1}.$$

6. A man parks his car among n cars standing in a row, his car not being parked at an end. On his return he finds that exactly m of the n cars are still there. What is the probability that both the cars parked on two sides of his car, have left?

Clearly, his car is at one of the crosses (×).

| × × × · · · × × × |

The number of the ways in which the remaining $m - 1$ cars can take their places (excluding the car of the man) $= {}^{n-1}C_{m-1}$ {$\because$ there are $n - 1$ places for the $m - 1$ cars}.

The number of ways in which the remaining $m - 1$ cars can take places keeping the two places on two sides of his car vacant $= {}^{n-3}C_{m-1}$

$\therefore$ the required probability

$$= \frac{n(E)}{n(S)} = \frac{{}^{n-3}C_{m-1}}{{}^{n-1}C_{m-1}}$$

$$= \frac{(n - 3)!}{(m - 1)!(n - m - 2)!} \times \frac{(m - 1)!(n - m)!}{(n - 1)!}$$

$$= \frac{(n - m)(n - m - 1)}{(n - 1)(n - 2)}.$$

7. Two non-negative integers are chosen at random from the set of non-negative integers with replacement. What is the probability that the sum of their squares is divisible by 10?

Let the two non-negative integers be x, y.

Now $x^2 + y^2$ will be divisible by 10 if the sum of digits in the units places of x and y is 0 or 10.

The units place of x can be filled in 10 ways

($\because$ any one of $0, 1, 2, 3, \ldots, 9$ can be used).

Also, the units place of y can be filled in 10 ways.

$\therefore$ the number of ways for the numbers x, y to have a digit each in their units places $= 10 \times 10 = 100$.

The digits in the units place of x^2 as well as y^2 can be from 0, 1, 4, 5, 6, 9 because the square of any number can have only one of these digits in the units place.

Sum of the digits in units places of x^2 and y^2 should be 0 or 10.

So, if x has 0 in units place, y must have 0 in units place

if x has 1 or 9 in units place, y must have 3 or 7 in units place

if x has 2 or 8 in units place, y must have 4 or 6 in units place

if x has 5 in units place, y must have 5 in units place

if x has 4 or 6 in units place, y must have 2 or 8 in units place

if x has 3 or 7 in units place, y must have 1 or 9 in units place

$\therefore$ the number of ways for the numbers x, y to have a digit in their units places whose sum is 0 or 10

$= 1 \times 1 + 2 \times 2 + 2 \times 2 + 1 \times 1 + 2 \times 2 + 2 \times 2 = 18$

$\therefore$ the required probability $= \dfrac{18}{100} = \dfrac{9}{50}$.

8. From an ordinary pack of 52 cards an even number of cards are drawn at random. Find the probability of getting equal number of black and red cards.

In an ordinary pack of 52 cards, 26 are black and 26 are red.

$n(S)$ = total number of ways of drawing even number of cards

$$= {}^{52}C_2 + {}^{52}C_4 + {}^{52}C_6 + \dots + {}^{52}C_{52} = 2^{52-1} - 1 = 2^{51} - 1$$

(using property of binomial coefficients)

$n(E)$ = number of ways of drawing $2n$ cards of which n are white and n are black; $n = 1, 2, \dots 26$

$$= \sum_{n=1}^{26} {}^{26}C_n \times {}^{26}C_n = \sum_{n=1}^{26} ({}^{26}C_n)^2$$

$$= {}^{26}C_1^2 + {}^{26}C_2^2 + {}^{26}C_3^2 + \dots + {}^{26}C_{26}^2$$

Now, $(1+x)^{26} = {}^{26}C_0 + {}^{26}C_1 \cdot x + {}^{26}C_2 \cdot x^2 + \dots + {}^{26}C_{26}x^{26}$

$$\left(1+\frac{1}{x}\right)^{26} = {}^{26}C_0 + {}^{26}C_1 \cdot \frac{1}{x} + {}^{26}C_2 \cdot \frac{1}{x^2} + \dots + {}^{26}C_{26} \frac{1}{x^{26}}.$$

Multiplying these and equating the constant terms on both sides,

constant term in $(1+x)^{26} \cdot \left(1+\dfrac{1}{x}\right)^{26}$

$$= {}^{26}C_0^2 + {}^{26}C_1^2 + {}^{26}C_2^2 + \dots + {}^{26}C_{26}^2.$$

But constant term in $(1+x)^{26} \cdot \left(1+\dfrac{1}{x}\right)^{26}$

$$= \text{coefficient of } x^0 \text{ in } \frac{(1+x)^{52}}{x^{26}}$$

$$= \text{coefficient of } x^{26} \text{ in } (1+x)^{52} = {}^{52}C_{26}$$

$$\therefore \quad {}^{52}C_{26} = {}^{26}C_0^2 + {}^{26}C_1^2 + {}^{26}C_2^2 + \dots + {}^{26}C_{26}^2$$

$$\therefore \quad n(E) = {}^{52}C_{26} - {}^{26}C_0^2 = \frac{52!}{(26!)^2} - 1$$

$$\therefore \quad \text{the required probability} = \frac{\dfrac{52!}{(26!)^2} - 1}{2^{51} - 1}.$$

9. The sum of the digits of a seven-digit number is 59. Find the probability that this number is divisible by 11.

As $7 \times 8 = 56$ and the sum of seven digits is 59, clearly at least three of the digits must be 9.

Obviously, the seven digits of the number will be as follows:

(a) 9, 9, 9, 8, 8, 8, 8 (b) 9, 9, 9, 9, 8, 8, 7

(c) 9, 9, 9, 9, 9, 7, 7 (d) 9, 9, 9, 9, 9, 8, 6

(e) 9, 9, 9, 9, 9, 9, 5

$\therefore$ the total number of ways to form a seven-digit number whose sum of digits is 59

$$= \frac{7!}{3!\,4!} + \frac{7!}{4!\,2!} + \frac{7!}{5!\,2!} + \frac{7!}{5!} + \frac{7!}{6!}$$

$$= \frac{7 \cdot 6 \cdot 5}{3 \cdot 2} + \frac{7 \cdot 6 \cdot 5}{2} + \frac{7 \cdot 6}{2} + 7 \cdot 6 + 7$$

$$= 210 \qquad \dots (1)$$

A number is divisible by 11 if the difference of the sum of the digits in odd places and that of the digits in even places is divisible by 11.

As the number is of seven digits we must have (for favourable cases), sum of four digits in odd places – sum of three digits in even places

$$= 0, 11, 22, 33, 44, 55.$$

If the two sums are denoted by x and y respectively then

$x + y = 59$, $x - y = 0$ (i)	or	$x + y = 59$, $x - y = 11$ (ii)	or	$x + y = 59$, $x - y = 22$ (iii)
or $x + y = 59$, $x - y = 33$ (iv)	or	$x + y = 59$, $x - y = 44$ (v)	or	$x + y = 59$, $x - y = 55$ (vi)

Clearly, (i), (iii) and (iv) do not give integral values of x and y.

(ii) $\Rightarrow$ $x = 35, y = 24$; (iv) $\Rightarrow$ $x = 46, y = 13$;

(vi) $\Rightarrow$ $x = 57, y = 2$.

Obviously, from (a), (b), (c), (d) and (e) we get, sum of three digits y cannot be 2 or 13.

Hence, only favourable case takes place when the sum of four digits in the odd places = 35 and the sum of the three digits in even places = 24.

$\therefore$ in the favourable numbers we will get,

9, 9, 9, 8 in odd places and 8, 8, 8 in even places

or 9, 9, 9, 8 in odd places and 9, 8, 7 in even places

or 9, 9, 9, 8 in odd places and 9, 9, 6 in even places

$\therefore$ the number of numbers divisible by 11

$$= \frac{4!}{3!} + \frac{4!}{3!} \times 3! + \frac{4!}{3!} \times \frac{3!}{2!}$$

$$= 4 + 24 + 12 = 40 \qquad \dots (2)$$

$\therefore$ from (1) and (2),

the required probability $= \dfrac{40}{210} = \dfrac{4}{21}$.

10. A, B, C are events such that $P(A) = 0.3$, $P(B) = 0.4$, $P(C) = 0.8$, $P(AB) = 0.08$, $P(AC) = 0.28$ and $P(ABC) = 0.09$. If $P(A \cup B \cup C) \geq 0.75$ then show that $P(BC)$ lies in the interval $0.23 \leq x \leq 0.48$.

We know, $P(A \cup B \cup C) \leq 1$.

Also, given that

$P(A \cup B \cup C) \geq 0.75$

$\therefore \quad 0.75 \leq P(A \cup B \cup C) \leq 1 \quad \dots (1)$

But, $P(A \cup B \cup C) = P(A) + P(B) + P(C) - P(A \cap B)$
$- P(B \cap C) - P(C \cap A) + P(A \cap B \cap C)$

$= 0.3 + 0.4 + 0.8 - 0.08 - P(BC) - 0.28 + 0.09$

$= 1.59 - 0.36 - P(BC)$

$= 1.23 - P(BC)$

$\therefore$ by (1), $0.75 \leq 1.23 - P(BC) \leq 1$

$\therefore \quad -0.48 \leq -P(BC) \leq -0.23$

$\therefore \quad 0.48 \geq P(BC) \geq 0.23$

$\therefore \quad 0.23 \leq P(BC) \leq 0.48.$

11. Two numbers x and y are chosen at random from the set $\{1, 2, 3, \dots, 3n\}$. Find the probability that $x^2 - y^2$ is divisible by 3.

$x^2 - y^2 = (x+y)(x-y)$ and 3 is a prime number.

$\therefore \quad x^2 - y^2$ is divisible by 3 if

$x + y$ or $x - y$ is divisible by 3.

Now, $\{1, 2, 3, \dots, 3n\}$

$= \{3, 6, 9, \dots, 3n\} \cup \{1, 4, 7, \dots, 3n-2\}$
$\cup \{2, 5, 8, \dots, 3n-1\}$

$= A \cup B \cup C$ (say).

Clearly, if x, y are selected from A or B or C then $x + y$ or $x - y$ are divisible by 3; and, if x, y are selected one from B and the other from C then $x + y$ is divisible by 3.

$\therefore$ the probability of $x^2 - y^2$ is divisible by 3

= probability of selecting both x, y from A or B or C
+ probability of selecting x, y one from B and the other from C

$$= \frac{{}^nC_2}{{}^{3n}C_2} \times 3 + \frac{{}^nC_1 \times {}^nC_1}{{}^{3n}C_2}$$

$$= \frac{3n(n-1)}{3n(3n-1)} + \frac{2n^2}{3n(3n-1)}$$

$$= \frac{3n-3}{3(3n-1)} + \frac{2n}{3(3n-1)}$$

$$= \frac{5n-3}{3(3n-1)}.$$

12. A coin is tossed $(m + n)$ times, $m > n$. Show that the probability of getting (at least) m consecutive heads is $(n + 2)/2^{m+1}$.

The sequences of outcomes in which there are m consecutive heads and their probabilities, are as follows (here × stands for any one of H or T).

Sequences of outcomes	*Probabilities*
$(HH \dots m \text{ times}) \times\times \dots \times$	$\left(\frac{1}{2}\right)^m \cdot 1$
$T(HH \dots m \text{ times}) \times\times \dots \times$	$\frac{1}{2} \cdot \left(\frac{1}{2}\right)^m \cdot 1$
$\times T(HH \dots m \text{ times}) \times\times \dots \times$	$1 \cdot \frac{1}{2} \cdot \left(\frac{1}{2}\right)^m \cdot 1$
$\times\times T(HH \dots m \text{ times}) \times\times \dots \times$	$1 \cdot \frac{1}{2} \cdot \left(\frac{1}{2}\right)^m \cdot 1$
....	
$\times\times \dots \times T(HH \dots m \text{ times})$	$1 \cdot \frac{1}{2} \cdot \left(\frac{1}{2}\right)^m$

$\therefore$ the required probability

$$= \left(\frac{1}{2}\right)^m + \left\{\left(\frac{1}{2}\right)^{m+1} + \left(\frac{1}{2}\right)^{m+1} + \dots + \text{to } n \text{ terms}\right\}$$

$$= \left(\frac{1}{2}\right)^m + n \cdot \left(\frac{1}{2}\right)^{m+1}$$

$$= \left(\frac{1}{2}\right)^{m+1} \cdot (2+n)$$

$$= \frac{n+2}{2^{m+1}}.$$

13. For the three independent events A, B and C, the probability of exactly one of the events A or B occurring = the probability of exactly one of the events B or C occuring = the probability of exactly one of the events C or A occurring $= p$. If the probability of all the events occurring simultaneously be p^2 where $0 < p < 0.5$ then find the probability of at least one of the events A, B and C occurring.

The probability of exactly one of the events A or B occurring

$= P(A \cap B') + P(A' \cap B)$

$= P(A) \cdot P(B') + P(A') \cdot P(B)$

$= P(A) \cdot \{1 - P(B)\} + \{1 - P(A)\}P(B)$

$\therefore \quad p = P(A) + P(B) - 2P(A) \cdot P(B) \quad \dots (1)$

Similarly,

$p = P(B) + P(C) - 2P(B) \cdot P(C) \quad \dots (2)$

$p = P(C) + P(A) - 2P(C) \cdot P(A) \quad \dots (3)$

$(1) - (2) \Rightarrow 0 = P(A) - P(C) - 2P(B) \cdot \{P(A) - P(C)\}$

$\therefore \quad \{P(A) - P(C)\}\{1 - 2P(B)\} = 0$

$\therefore \quad P(A) = P(C)$ or $P(B) = \frac{1}{2}$.

If $P(B) = \frac{1}{2}$,

(1) $\Rightarrow \quad p = \frac{1}{2}$.

But $0 < p < \frac{1}{2}$ (given)

$\therefore \quad P(A) = P(C)$;

similarly, $P(B) = P(C)$.

So $\quad P(A) = P(B) = P(C) = x$ (say).

Now, $P(A \cup B \cup C)$

$$= \Sigma P(A) - \Sigma\{P(A) \cdot P(B)\} + P(A \cap B \cap C)$$

$$= 3x - 3x^2 + p^2 \text{ (from question).}$$

Also, from (1), $p = x + x - 2x \cdot x$

$$= 2x - 2x^2$$

$\therefore \quad x - x^2 = \frac{p}{2}$

$\therefore \quad P(A \cup B \cup C) = 3(x - x^2) + p^2$

$$= \frac{3p}{2} + p^2 = \frac{1}{2}p(3 + 2p).$$

14. The decimal parts of the logarithms of two numbers taken at random are found to six places of decimal. What is the chance that the second can be subtracted from the first without "borrowing"?

For each column of the two numbers,

$n(S)$ = number of ways to fill the two places by the digits 0, 1, 2, ..., 9

$= 10 \times 10 = 100$.

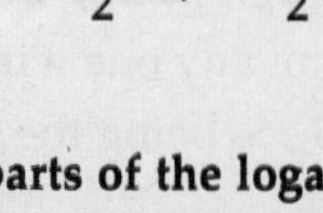

Let E be the event of subtracting in a column without borrowing. If the pair of digits be (x, y) in the column where x is in the first number and y is in the second number then

$E = \{(0, 0), (1, 0), (2, 0), \ldots, (9, 0),$
$(1, 1), (2, 1), \ldots, (9, 1),$
$(2, 2), (3, 2), \ldots, (9, 2),$
$(3, 3), (4, 3), \ldots, (9, 3),$
......
$(8, 8), (9, 8),$
$(9, 9)\}$

$\therefore \quad n(E) = 10 + 9 + 8 + \ldots + 2 + 1 = \frac{10 \cdot 11}{2} = 55.$

$\therefore$ the probability of subtracting without borrowing in each column $= \frac{55}{100}$.

$\therefore$ the required probability $= \left(\frac{55}{100}\right)^6 = \left(\frac{11}{20}\right)^6$.

15. In a multiple-choice question, there are four alternative answers of which one or more answers are correct. A candidate gets marks if he ticks all the correct answers. The candidate, being ignorant about the answers, decides to tick at random. How many attempts at least should he be allowed so that the probability of his getting marks in the question may exceed $\frac{1}{5}$?

Each answer may be marked or unmarked.

$\therefore$ the number of ways to tick answers $= 2^4 - 1$,

($\because$ the candidate cannot keep all answers unmarked)

As there is only one way of marking correctly, the probability of getting marks in one attempt

$$= \frac{1}{2^4 - 1} = \frac{1}{15}$$

$\therefore$ the probability of getting no marks in one attempt

$$= 1 - \frac{1}{15} = \frac{14}{15}.$$

$\therefore$ the probability of getting no marks in n attempts

$$= \left(\frac{14}{15}\right)^n.$$

$\therefore$ the probability of getting marks in n attempts

$$= 1 - \left(\frac{14}{15}\right)^n.$$

Now, $1 - \left(\frac{14}{15}\right)^n > \frac{1}{5}$

or $\quad \frac{4}{5} > \left(\frac{14}{15}\right)^n \quad \ldots (1)$

Clearly $\frac{4}{5} < \frac{14}{15}, \frac{4}{5} < \left(\frac{14}{15}\right)^2, \frac{4}{5} < \left(\frac{14}{15}\right)^3,$

$\frac{4}{5} > \left(\frac{14}{15}\right)^4, \frac{4}{5} > \left(\frac{14}{15}\right)^5, \ldots$

$\therefore$ from (1), $n = 4, 5, \ldots$;

so, the least value of $n = 4$.

$\therefore$ the candidate should be allowed at least 4 attempts.

16. A consignment of 15 radios contains 4 defectives. The radios are taken out one by one at random and examined. The ones examined are not put back. What is the probability that the ninth one examined is the last defective?

The ninth one will be the last defective if the first eight radios examined contain 3 defectives and 5 nondefectives, and the ninth radio examined is defective.

The probability of drawing 3 defectives and 5 non-defectives in the first 8 examinations

$$= P(E_3) = \frac{{}^4C_3 \times {}^{11}C_5}{{}^{15}C_8}$$

The probability of drawing 1 defective from the remaining 1 defective and 6 nondefectives in the ninth examination $= \frac{1}{{}^7C_1}$

∴ the required probability

$$= \frac{{}^4C_3 \times {}^{11}C_5}{{}^{15}C_8} \times \frac{1}{{}^7C_1} = \frac{4 \times \frac{11!}{5!6!}}{\frac{15!}{8!7!}} \times \frac{1}{7}$$

$$= \frac{4}{7} \times \frac{8!7!}{5!6!} \times \frac{11!}{15!}$$

$$= \frac{4}{7} \times 8 \times 7 \times 6 \times 7 \times \frac{1}{15 \times 14 \times 13 \times 12}$$

$$= \frac{8}{195}.$$

17. Three critics review a book. Odds in favour of the book are 5 : 2, 4 : 3 and 3 : 4 respectively for the three critics. Find the probability that majority are in favour of the book.

Let the critics be A, B and C. Let $P(A)$ denote the probability of the critic A to be in favour of the book, etc.

∵ the odds in favour of the book for the critic A = 5 : 2,

$$P(A) = \frac{5}{5+2} = \frac{5}{7}.$$

Similarly, $P(B) = \frac{4}{7}$ and $P(C) = \frac{3}{7}$.

Clearly, the event of majority being in favour

= the event of at least two critics being in favour.

∴ the required probability

$$= P(ABC) + P(AB\overline{C}) + P(A\overline{B}C) + P(\overline{A}BC)$$

$$= P(A) \cdot P(B) \cdot P(C) + P(A) \cdot P(B) \cdot P(\overline{C}) + P(A) \cdot P(\overline{B}) \cdot P(C) + P(\overline{A}) \cdot P(B) \cdot P(C)$$

$$= \frac{5}{7} \cdot \frac{4}{7} \cdot \frac{3}{7} + \frac{5}{7} \cdot \frac{4}{7} \cdot \left(1 - \frac{3}{7}\right) + \frac{5}{7} \cdot \left(1 - \frac{4}{7}\right) \cdot \frac{3}{7} + \left(1 - \frac{5}{7}\right) \cdot \frac{4}{7} \cdot \frac{3}{7}$$

$$= \frac{1}{7^3}\{60 + 80 + 45 + 24\} = \frac{209}{343}.$$

18. Sixteen players $S_1, S_2, S_3, \ldots, S_{16}$ play in a tournament. They are divided into eight pairs at random. From each pair a winner is decided on the basis of a game played between the two players of the pair. Assume that all the players are of equal strength.

(a) Find the probability that the player S_1 is among the eight winners.

(b) Find the probability that exactly one of the two players S_1 and S_2 is among the eight winners.

(a) Probability of S_1 to be among the eight winners

= (probability of S_1 being in a pair) × (probability of S_1 winning in the group)

$= 1 \times \frac{1}{2}$ (∵ S_1 is defnitely in a group, i.e., it is a certainity)

$= \frac{1}{2}$.

(b) If S_1, S_2 are in the same pair then exactly one wins. If S_1, S_2 are in two pairs separately then exactly one of S_1, S_2 will be among the eight winners if S_1 wins and S_2 loses or S_1 loses and S_2 wins.

Now the probability of S_1, S_2 being in the same pair and one wins

= (probability of S_1, S_2 being in the same pair) × (probability of any one winning in the pair)

The probability of S_1, S_2 being the same pair

$= \frac{n(E)}{n(S)}$ where $n(E)$ = the number of ways in which 16 persons can be divided in 8 pairs when S_1, S_2 are in the same pair and $n(S)$ = the number of ways in which 16 persons can be divided in 8 pairs

$$\therefore \ n(E) = \frac{14!}{(2!)^7 \cdot 7!} \ \text{and} \ n(S) = \frac{16!}{(2!)^8 \cdot 8!}$$

∴ the probability of S_1, S_2 being in the same pair

$$= \frac{\frac{14!}{(2!)^7 \cdot 7!}}{\frac{16!}{(2!)^8 \cdot 8!}} = \frac{2! \cdot 8}{16 \cdot 15} = \frac{1}{15}.$$

The probability of any one winning in the pair of S_1, S_2 = certainly = 1.

∴ the probability of S_1, S_2 being in the same pair and one wins $= \frac{1}{15} \times 1 = \frac{1}{15}$.

The probability of S_1, S_2 being in two pairs separately and one of S_1, S_2 wins

= the probability of S_1, S_2 being in two pairs separately and S_1 wins, S_2 loses

+ the probability of S_1, S_2 being two pairs separately and S_1 loses, S_2 wins

$$= \frac{\dfrac{14 \times 13 \times \dfrac{12\,!}{(2\,!)^6 \cdot 6\,!}}{\dfrac{16\,!}{(2\,!)^8 \cdot 8\,!}} \cdot \frac{1}{2} \cdot \frac{1}{2} + \frac{\dfrac{14 \times 13 \times \dfrac{12\,!}{(2\,!)^6 \cdot 6\,!}}{\dfrac{16\,!}{(2\,!)^8 \cdot 8\,!}} \cdot \frac{1}{2} \cdot \frac{1}{2}$$

$$= \frac{14 \cdot 13 \cdot (2\,!)^2 \cdot 8 \cdot 7}{16 \cdot 15 \cdot 14 \cdot 13} \cdot \left(\frac{1}{4} + \frac{1}{4}\right) = \frac{14}{15} \cdot \frac{1}{2} = \frac{7}{15}.$$

$\therefore$ the required probability $= \frac{1}{15} + \frac{7}{15} = \frac{8}{15}$

19. A pair of dice is rolled together till a sum of either 5 or 7 is obtained. Find the probability that 5 comes before 7.

Let E_1 = the event of getting 5 in a roll of two dice

$= \{(1, 4), (2, 3), (3, 2), (4, 1)\}$

$\therefore \quad P(E_1) = \frac{n(E_1)}{n(S)} = \frac{4}{6 \times 6} = \frac{1}{9}.$

Let E_2 = the event of getting either 5 or 7

$= \{(1, 4), (2, 3), (3, 2), (4, 1),$
$(1, 6), (2, 5), (3, 4), (4, 3), (5, 2), (6, 1)\}$

$\therefore \quad P(E_2) = \frac{n(E_2)}{n(S)} = \frac{10}{6 \times 6} = \frac{5}{18}.$

$\therefore$ the probability of getting neither 5 nor 7

$= P(\overline{E}_2) = 1 - P(E_2) = 1 - \frac{5}{18} = \frac{13}{18}.$

The event of getting 5 before 7

$= E_1 \cup (\overline{E}_2 E_1) \cup (\overline{E}_2 \overline{E}_2 E_1) \cup \ldots$ to ∞

$\therefore$ the probability of getting 5 before 7

$= P(E_1) + P(\overline{E}_2 E_1) + P(\overline{E}_2 \overline{E}_2 E_1) + \ldots$ to ∞

$= P(E_1) + P(\overline{E}_2)P(E_1) + P(\overline{E}_2)P(\overline{E}_2)P(E_1) + \ldots$ to ∞

$= \frac{1}{9} + \frac{13}{18} \cdot \frac{1}{9} + \frac{13}{18} \cdot \frac{13}{18} \cdot \frac{1}{9} + \ldots$ to ∞

$= \frac{1}{9}\left[1 + \frac{13}{18} + \left(\frac{13}{18}\right)^2 + \ldots \text{ to } \infty\right]$

$= \frac{1}{9} \cdot \frac{1}{1 - \frac{13}{18}} = \frac{1}{9} \cdot \frac{18}{5} = \frac{2}{5}.$

20. In a game A throws two ordinary dice. If he throws 7 or 11 he wins. If he throws 2, 3 or 12 he loses. If he throws any other number, he throws again and continues to throw until either the number he threw first or 7 turns up. In the first case he wins and in the second he loses. Show that the odds against his winning is 251 : 244.

Let A_r = the event of throwing r by A in a throw of two dice.

Clearly, $A_2 = \{(1, 1)\}$

$A_3 = \{(1, 2), (2, 1)\}$

$A_4 = \{(1, 3), (2, 2), (3, 1)\}$

$A_5 = \{(1, 4), (2, 3), (3, 2), (4, 1)\}$

$A_6 = \{(1, 5), (2, 4), (3, 3), (4, 2), (5, 1)\}$

$A_7 = \{(1, 6), (2, 5), (3, 4), (4, 3), (5, 2), (6, 1)\}$

$A_8 = \{(2, 6), (3, 5), (4, 4), (5, 3), (6, 2)\}$

$A_9 = \{(3, 6), (4, 5), (5, 4), (6, 3)\}$

$A_{10} = \{(4, 6), (5, 5), (6, 4)\}$

$A_{11} = \{(5, 6), (6, 5)\}$

$A_{12} = \{(6, 6)\}.$

Now, the probability of A winning

= (probability of throwing 7 or 11 in the first throw)
+ (probability of throwing 4 first and then throwing any number except 7 or 4 any number of times followed by 4)
+ (probability of throwing 5 first and then throwing any number except 7 or 5 any number of times followed by 5)
+ (probability of throwing 6 first and then throwing any number except 7 or 6 any number of times followed by 6)
+ (probability of throwing 8 first and then throwing any number except 7 or 8 any number of times followed by 8)
+ (probability of throwing 9 first and then throwing any number except 7 or 9 any number of times followed by 9)
+ (probability of throwing 10 first and then throwing any number except 7 or 10 any number of times followed by 10)

$$= \frac{8}{36} + \left\{\frac{3}{36} \times \frac{3}{36} + \frac{3}{36} \times \frac{27}{36} \times \frac{3}{36} + \frac{3}{36}\left(\frac{27}{36}\right)^2 \times \frac{3}{36} + \ldots \text{to } \infty\right\}$$

$$+ \left\{\frac{4}{36} \times \frac{4}{36} + \frac{4}{36} \times \frac{26}{36} \times \frac{4}{36} + \frac{4}{36}\left(\frac{26}{36}\right)^2 \times \frac{4}{36} + \ldots \text{to } \infty\right\}$$

$$+ \left\{\frac{5}{36} \times \frac{5}{36} + \frac{5}{36} \times \frac{25}{36} \times \frac{5}{36} + \frac{5}{36}\left(\frac{25}{36}\right)^2 \times \frac{5}{36} + \ldots \text{to } \infty\right\}$$

$$+ \left\{\frac{5}{36} \times \frac{5}{36} + \frac{5}{36} \times \frac{25}{36} \times \frac{5}{36} + \frac{5}{36}\left(\frac{25}{36}\right)^2 \times \frac{5}{36} + \ldots \text{to } \infty\right\}$$

$$+\left\{\frac{4}{36}\times\frac{4}{36}+\frac{4}{36}\times\frac{26}{36}\times\frac{4}{36}+\frac{4}{36}\left(\frac{26}{36}\right)^2\times\frac{4}{36}+\ldots\text{to }\infty\right\}$$

$$+\left\{\frac{3}{36}\times\frac{3}{36}+\frac{3}{36}\times\frac{27}{36}\times\frac{3}{36}+\frac{3}{36}\left(\frac{27}{36}\right)^2\times\frac{3}{36}+\ldots\text{to }\infty\right\}$$

$$=\frac{2}{9}+2\left\{\frac{9}{36^2}\cdot\frac{1}{1-\frac{27}{36}}+\frac{16}{36^2}\cdot\frac{1}{1-\frac{26}{36}}+\frac{25}{36^2}\cdot\frac{1}{1-\frac{25}{36}}\right\}$$

$$=\frac{2}{9}+2\left\{\frac{9}{36^2}\cdot\frac{36}{9}+\frac{16}{36^2}\cdot\frac{36}{10}+\frac{25}{36^2}\cdot\frac{36}{11}\right\}$$

$$=\frac{2}{9}+2\left\{\frac{1}{36}+\frac{2}{45}+\frac{25}{396}\right\}=\frac{2}{9}+\frac{1}{18}+\frac{4}{45}+\frac{25}{198}$$

$$=\frac{11}{30}+\frac{25}{198}=\frac{488}{990}=\frac{244}{495}$$

$\therefore$ odds against his winning

$$=\frac{P(E')}{P(E)}=\frac{1-\frac{244}{495}}{\frac{244}{495}}=\frac{251}{244}=251:244.$$

21. A and B play a game of tennis. The situation of the game is as follows:
If one scores two consecutive points after a deuce he wins; if loss of a point is followed or preceded by win of a point, it is deuce. The chance of a server to win a point is $\frac{2}{3}$. The game is at a deuce and A is serving. What is the chance that A will win the game?

Let E = the event of A winning a point.

Here $P(E)=\frac{2}{3}$; $\quad\therefore\quad P(\bar{E})=1-\frac{2}{3}=\frac{1}{3}$.

The event of deuce $=D=(E\cap\bar{E})\cup(\bar{E}\cap E)$

$\therefore$ the probability of a deuce

$=P(D)=P\{(E\cap\bar{E})\cup(\bar{E}\cap E)\}$

$=P(E)\cdot P(\bar{E})+P(\bar{E})\cdot P(E)$

$=\frac{2}{3}\cdot\frac{1}{3}+\frac{1}{3}\cdot\frac{2}{3}=\frac{4}{9}.$

The event of A winning the game = the event of any number of deuce followed by two consecutive points for A

$\therefore$ the required probability

$$=\left(\frac{2}{3}\right)^2+\frac{4}{9}\cdot\left(\frac{2}{3}\right)^2+\left(\frac{4}{9}\right)^2\cdot\left(\frac{2}{3}\right)^2+\left(\frac{4}{9}\right)^3\cdot\left(\frac{2}{3}\right)^2+\ldots\text{ to }\infty$$

$$=\left(\frac{2}{3}\right)^2\cdot\left\{1+\frac{4}{9}+\left(\frac{4}{9}\right)^2+\left(\frac{4}{9}\right)^3+\ldots\text{ to }\infty\right\}$$

$$=\frac{4}{9}\cdot\frac{1}{1-\frac{4}{9}}=\frac{4}{9}\cdot\frac{9}{5}=\frac{4}{5}.$$

22. Two students A and B attempt to solve the same question. Their chances of solving the question are $\frac{1}{8}$ and $\frac{1}{12}$ respectively. If the odds against making the same mistake by them be 1000 : 1, find the probability of their results being correct if they obtain the same result.

Here, the probability of A getting correct result $=\frac{1}{8}$

and the probability of B getting correct result $=\frac{1}{12}$

$\therefore$ the probability of both A and B getting correct result

$=\frac{1}{8}\times\frac{1}{12}$.

The probability of both A and B getting incorrect result

$=\left(1-\frac{1}{8}\right)\left(1-\frac{1}{12}\right)=\frac{7}{8}\cdot\frac{11}{12}$.

Odds against making the same mistake by A and B = 1000 : 1

$\therefore$ the probability of A and B making the same mistake (incorrect result) $=\frac{1}{1001}$.

$\therefore$ the probability of A and B being incorrect making the same mistake (incorrect result)$=\frac{1}{1001}\times\frac{7}{8}\times\frac{11}{12}$

$$\therefore\ \frac{\{\text{probability of being correct by obtaining the same result}\}}{\{\text{probability of being incorrect by obtaining the same result}\}}$$

$$=\frac{\frac{1}{8}\times\frac{1}{12}}{\frac{1}{1001}\times\frac{7}{8}\times\frac{11}{12}}$$

$$=\frac{1}{8}\times\frac{1}{12}\times1001\times\frac{8}{7}\times\frac{12}{11}=\frac{1001}{7\times11}=\frac{13}{1}.$$

But, by obtaining the same result both are correct or incorrect.

∴ odds in favour of being correct obtaining the same result = 13 : 1

∴ probability of being correct obtaining the same result $= \frac{13}{14}$.

23. In a series of five one-day cricket matches between India and Pakistan, the probability of India winning or drawing are respectively $\frac{1}{3}$ and $\frac{1}{6}$. If a win, loss or draw gives 2, 0 or 1 point respectively then find the probability that India will score 5 points in the series.

For scoring 5 points the result should be as follows:

(i) 2 wins, 2 losses, 1 draw
(ii) 1 win, 1 loss, 3 draws
(iii) 5 draws

∵ in a match either India win or lose or draw, the probability of a loss $= 1 - \left(\frac{1}{3} + \frac{1}{6}\right) = 1 - \frac{1}{2} = \frac{1}{2}$.

In the five matches, the probability of (i) happening

$$= {}^5C_2 \times {}^3C_2 \times {}^1C_1 \times \left(\frac{1}{3}\right)^2 \times \left(\frac{1}{2}\right)^2 \times \frac{1}{6}$$

(win can take place in any of the two matches, etc.)

$$= 30 \times \frac{1}{9} \times \frac{1}{4} \times \frac{1}{6} = \frac{5}{36}.$$

In the five matches, the probability of (ii) happening

$$= {}^5C_1 \times {}^4C_1 \times {}^3C_3 \times \frac{1}{3} \times \frac{1}{2} \times \left(\frac{1}{6}\right)^3$$

$$= 20 \times \frac{1}{6} \times \frac{1}{6^3} = \frac{5}{324}.$$

In the five matches, the probability of (iii) happening

$$= {}^5C_5 \cdot \left(\frac{1}{6}\right)^5 \cdot \left(\frac{1}{3}\right)^0 \cdot \left(\frac{1}{2}\right)^0 = \frac{1}{6^5} = \frac{1}{7776}$$

∴ the required probability

$$= \frac{5}{36} + \frac{5}{324} + \frac{1}{7776} = \frac{1201}{7776}.$$

24. A, B, C and D cut a pack of 52 cards successively in the order given. If the person who cuts a spade first receives Rs 350, what are their respective expectations?

Let E = the event of anyone cutting a spade in one cut.

$$\therefore \quad P(E) = p = \frac{{}^{13}C_1}{{}^{52}C_1} = \frac{13}{52} = \frac{1}{4}$$

and $P(\bar{E}) = q = 1 - P(E) = 1 - \frac{1}{4} = \frac{3}{4}$.

The probability of A winning (when A starts the game)

$$= p + qqqqp + (qqqq)^2 p + \dots \text{ to } \infty$$

(∵ A wins if he cuts spade in the first attempt or all fail and A cuts in the second attempt, etc.)

$$= \frac{1}{4} + \left(\frac{3}{4}\right)^4 \cdot \frac{1}{4} + \left(\frac{3}{4}\right)^8 \cdot \frac{1}{4} + \dots \text{ to } \infty$$

$$= \frac{\frac{1}{4}}{1 - \left(\frac{3}{4}\right)^4} = \frac{64}{175}.$$

∴ the expectation of A = Rs 350 × probability

$$= \text{Rs } 350 \times \frac{64}{175} = \text{Rs } 128.$$

The probability of B winning

$$= qp + qqqq(qp) + (qqqq)^2 qp + \dots \text{ to } \infty$$

$$= \frac{3}{4} \cdot \frac{1}{4} + \left(\frac{3}{4}\right)^4 \frac{3}{4} \cdot \frac{1}{4} + \left(\frac{3}{4}\right)^8 \cdot \frac{3}{4} \cdot \frac{1}{4} + \dots \text{ to } \infty$$

$$= \frac{3}{4} \cdot \frac{1}{4}\left\{1 + \left(\frac{3}{4}\right)^4 + \left(\frac{3}{4}\right)^8 + \dots \text{ to } \infty\right\}$$

$$= \frac{3}{4} \cdot \frac{1}{4} \cdot \frac{1}{1 - \left(\frac{3}{4}\right)^4}$$

$$= \frac{3}{4^2} \cdot \frac{4^4}{4^4 - 3^4} = \frac{3 \cdot 16}{175} = \frac{48}{175}.$$

∴ the expectation of B = Rs 350 × $\frac{48}{175}$ = Rs 96.

The probability of C winning

$$= qqp + qqqq(qqp) + (qqqq)^2 qqp + \dots \text{ to } \infty\}$$

$$= \left(\frac{3}{4}\right)^2 \cdot \frac{1}{4} + \left(\frac{3}{4}\right)^4 \cdot \left(\frac{3}{4}\right)^2 \cdot \frac{1}{4} + \left(\frac{3}{4}\right)^8 \cdot \left(\frac{3}{4}\right)^2 \cdot \frac{1}{4} + \dots \text{ to } \infty$$

$$= \left(\frac{3}{4}\right)^2 \cdot \frac{1}{4}\left\{1 + \left(\frac{3}{4}\right)^4 + \left(\frac{3}{4}\right)^8 + \dots \text{ to } \infty\right\}$$

$$= \frac{9}{4^3} \cdot \frac{1}{1 - \left(\frac{3}{4}\right)^4}$$

$$= \frac{9}{4^3} \cdot \frac{4^4}{4^4 - 3^4} = \frac{36}{175}.$$

∴ the expectation of C = Rs 350 × $\frac{36}{175}$ = Rs 72.

The expectation of D

= Rs 350 – (sum of the expectations of A, B, C)

= Rs 350 – (Rs 128 + Rs 96 + Rs 72) = Rs 54.

25. There are $n+3(n>1)$ seats numbered 1, 2, 3, ... , $n+3$. There are also $n+3$ persons who are holding tickets numbered 1, 2, 3, ... , $n+3$. They take seats at random. Find the probability that exactly three persons take seats having the same numbers as that in their tickets.

The number of ways in which 3 persons can take seats having the same numbers as that in their tickets $= {}^{n+3}C_3$. Of these, let in one of the ways, persons having the tickets numbered $n+1$, $n+2$ and $n+3$ take their corresponding seats.

The remaining n persons are to take n seats such that all of them are in the wrong seats.

Let E_i = event of the person having the ticket numbered i to take the seat numbered i

$\therefore$ the probability that the remaining n persons take wrong seats

$$= P(\overline{E}_1 \cap \overline{E}_2 \cap \ldots \cap \overline{E}_n) = P(\overline{E_1 \cup E_2 \cup \ldots \cup E_n})$$

$$= 1 - P(E_1 \cup E_2 \cup \ldots \cup E_n)$$

$$= 1 - \left\{ \sum_{i=1}^{n} P(E_i) - \sum_{1 \le i \ne j \le n} P(E_i \cap E_j) + \sum_{1 \le i \ne j \ne k \le n} P(E_i \cap E_j \cap E_k) - \ldots + (-1)^{n-1} P(E_1 \cap E_2 \cap \ldots \cap E_n) \right\} \quad \ldots (1)$$

Now, $P(E_i) = \frac{1}{n}$ ($\because$ ith person can take any one of n seats, only one being favourable)

$$\therefore \quad \sum_{i=1}^{n} P(E_i) = \sum_{i=1}^{n} \frac{1}{n} = \frac{1}{n} \sum_{i=1}^{n} 1$$

$$= \frac{1}{n}(1 + 1 + \ldots \text{ to } n \text{ times})$$

$$= \frac{1}{n} \times n = 1.$$

$$P(E_i \cap E_j) = \frac{{}^{n-2}P_{n-2}}{{}^{n}P_n}$$

($\because$ n persons can be arranged in n seats in ${}^{n}P_n$ ways without restriction and in ${}^{n-2}P_{n-2}$ ways when ith and jth persons are in their respective seats)

$$= \frac{(n-2)!}{n!} = \frac{1}{n(n-1)}$$

$$\therefore \quad \sum_{1 \le i \ne j \le n} P(E_i \cap E_j) = \frac{1}{n(n-1)} + \frac{1}{n(n-1)} + \ldots \text{ to } {}^{n}C_2 \text{ times}$$

$$= {}^{n}C_2 \times \frac{1}{n(n-1)} = \frac{1}{2!}$$

$$P(E_i \cap E_j \cap E_k) = \frac{{}^{n-3}P_{n-3}}{{}^{n}P_n} = \frac{(n-3)!}{n!}$$

$$= \frac{1}{n(n-1)(n-2)}$$

$$\therefore \quad \sum_{1 \le i \ne j \ne k \le n} P(E_i \cap E_j \cap E_k)$$

$$= \frac{1}{n(n-1)(n-2)} + \frac{1}{n(n-1)(n-2)} + \ldots \text{ to } {}^{n}C_3 \text{ times}$$

$$= {}^{n}C_3 \times \frac{1}{n(n-1)(n-2)}$$

$$= \frac{1}{3!} \text{ and so on.}$$

$\therefore$ from (1), the probability that the remaining n persons take wrong seats when any three take corresponding seats

$$= 1 - \left(1 - \frac{1}{2!} + \frac{1}{3!} - \ldots + (-1)^{n-1} \cdot \frac{1}{n!}\right)$$

$$= \frac{1}{2!} - \frac{1}{3!} + \ldots + (-1)^n \frac{1}{n!}.$$

The probability that 3 persons have taken correct seats

$$= \frac{1}{{}^{n+3}P_3} \times {}^{n+3}C_3 = \frac{1}{3!}.$$

$\therefore$ the required probability that exactly 3 persons take the seats corresponding to their ticket numbers

$$= \frac{1}{3!} \times \left(\frac{1}{2!} - \frac{1}{3!} - \ldots + (-1)^n \cdot \frac{1}{n!}\right)$$

$$= \frac{1}{6}\left(\frac{1}{2!} - \frac{1}{3!} - \ldots + (-1)^n \cdot \frac{1}{n!}\right).$$

26. A coin is tossed 10 times. Find the probability of getting (i) exactly six heads (ii) at least six heads (iii) at most six heads.

The probability of getting a head (success) in one trial

$$= p = \frac{1}{2}.$$

$\therefore$ the probability of not getting a head (failure) in one trial $= q = 1 - \frac{1}{2} = \frac{1}{2}$.

(i) The probability of getting exactly 6 heads in 10 trials

$$= {}^{10}C_6\, p^6 q^4 = \frac{10!}{6!\,4!} \cdot \left(\frac{1}{2}\right)^6 \cdot \left(\frac{1}{2}\right)^4$$

$$= \frac{105}{512}.$$

(ii) The probability of getting at least 6 heads

$$= {}^{10}C_6\, p^6 q^4 + {}^{10}C_7 p^7 q^3 + {}^{10}C_8\, p^8 q^2 + {}^{10}C_9\, p^9 q + {}^{10}C_{10}\, p^{10}$$

$$= ({}^{10}C_6 + {}^{10}C_7 + {}^{10}C_8 + {}^{10}C_9 + {}^{10}C_{10})\left(\frac{1}{2}\right)^{10}$$

$$= \left(\frac{10\cdot 9\cdot 8\cdot 7}{24} + \frac{10\cdot 9\cdot 8}{6} + \frac{10\cdot 9}{2} + 10 + 1\right)\left(\frac{1}{2}\right)^{10}$$

$$= (210 + 120 + 45 + 10 + 1)\frac{1}{2^{10}}$$

$$= \frac{386}{2^{10}} = \frac{193}{2^9} = \frac{193}{512}.$$

(iii) The probability of getting at most 6 heads

$$= {}^{10}C_0\, p^0 q^{10} + {}^{10}C_1\, p^1 q^9 + {}^{10}C_2\, p^2 q^8 + {}^{10}C_3\, p^3 q^7 + {}^{10}C_4\, p^4 q^6 + {}^{10}C_5\, p^5 q^5 + {}^{10}C_6\, p^6 q^4$$

$$= ({}^{10}C_0 + {}^{10}C_1 + {}^{10}C_2 + {}^{10}C_3 + {}^{10}C_4 + {}^{10}C_5 + {}^{10}C_6)\left(\frac{1}{2}\right)^{10}$$

$$= (1 + 10 + 45 + 120 + 210 + 252 + 210)\frac{1}{2^{10}}$$

$$= \frac{848}{2^{10}} = \frac{53}{64}.$$

27. A man takes a step forward with probability 0.4 and backward with probability 0.6. Find the probability that at the end of eleven steps he is one step away from the starting point.

As $0.4 + 0.6 = 1$, the man either takes a step forward or a step backward. Let a step forward be a success and a step backward be a failure.

Then, the probability of success in one step $= p = 0.4 = \frac{2}{5}$

the probability of failure in one step $= q = 0.6 = \frac{3}{5}$.

In 11 steps he will be one step away from the starting point if the numbers of successes and failures differ by 1.

So, the number of successes = 6,

the number of failures = 5

or the number of successes = 5,

the number of failures = 6

$\therefore$ the required probability

$$= {}^{11}C_6\, p^6 q^5 + {}^{11}C_5\, p^5 q^6$$

$$= {}^{11}C_6\left(\frac{2}{5}\right)^6\cdot\left(\frac{3}{5}\right)^5 + {}^{11}C_5\left(\frac{2}{5}\right)^5\cdot\left(\frac{3}{5}\right)^6$$

$$= \frac{11!}{6!\,5!}\cdot\left(\frac{2}{5}\right)^5\cdot\left(\frac{3}{5}\right)^5\cdot\left\{\frac{2}{5}+\frac{3}{5}\right\}$$

$$= \frac{11\cdot 10\cdot 9\cdot 8\cdot 7}{120}\cdot\frac{2^5\cdot 3^5}{5^{10}}$$

$$= 462\times\left(\frac{6}{25}\right)^5.$$

28. The probability of a man hitting a target in one fire is $\frac{1}{4}$. How many times at least must he fire at the target in order that his chance of hitting the target at least once will exceed $\frac{2}{3}$?

Here, the probability of hitting the target in one fire

$$= p = \frac{1}{4}$$

$\therefore$ the probability of not hitting the target in one fire

$$= q = 1 - \frac{1}{4} = \frac{3}{4}.$$

Let the man take n fires. Then

the probability of hitting the target at least once

= 1 – (probability of not hitting the target n times)

$$= 1 - {}^nC_0\, p^0 q^n = 1 - \left(\frac{3}{4}\right)^n.$$

Now, $1 - \left(\frac{3}{4}\right)^n > \frac{2}{3}$; or $\left(\frac{3}{4}\right)^n < \frac{1}{3}$.

But $\frac{3}{4} > \frac{1}{3}$, $\left(\frac{3}{4}\right)^2 > \frac{1}{3}$, $\left(\frac{3}{4}\right)^3 > \frac{1}{3}$,

$\left(\frac{3}{4}\right)^4 < \frac{1}{3}$, $\left(\frac{3}{4}\right)^5 < \frac{1}{3}$, etc.

$\therefore$ the least value of $n = 4$.

$\therefore$ he must take at least 4 shots to make his chance greater than $\frac{2}{3}$.

29. In a sequence of independent trials, the probability of success in one trial is $\frac{1}{4}$. Find the probability that the second success takes place on or after the fourth trial.

The probability of the second success taking place on the fourth trial

= probability of one success in the first three trials and success on the fourth trial

$$= \left\{{}^3C_1\times\frac{1}{4}\times\left(\frac{3}{4}\right)^2\right\}\times\frac{1}{4}.$$

The probability of the second success taking place on the fifth trial

= probability of one success in the first four trials and success on the fifth trial

$$= \left\{{}^4C_1\times\frac{1}{4}\times\left(\frac{3}{4}\right)^3\right\}\times\frac{1}{4} \text{ and so on.}$$

$\therefore$ the required probability

$$= {}^3C_1 \cdot \frac{1}{4} \cdot \left(\frac{3}{4}\right)^2 \cdot \frac{1}{4} + {}^4C_1 \cdot \frac{1}{4} \cdot \left(\frac{3}{4}\right)^3 \cdot \frac{1}{4}$$

$$+ {}^5C_1 \cdot \frac{1}{4} \left(\frac{3}{4}\right)^4 \cdot \frac{1}{4} + \dots$$

$$= \frac{9}{256}\left\{{}^3C_1 + {}^4C_1 \cdot \frac{3}{4} + {}^5C_1 \cdot \left(\frac{3}{4}\right)^2 + \dots \text{ to } \infty\right\}$$

$$= \frac{9}{256}\left\{3 + 4 \cdot \frac{3}{4} + 5 \cdot \left(\frac{3}{4}\right)^2 + \dots \text{ to } \infty\right\} \quad \dots (1)$$

Let $S = 3 + 4 \cdot \frac{3}{4} + 5 \cdot \left(\frac{3}{4}\right)^2 + \dots$ to ∞

$\therefore \quad \frac{3}{4} S = \quad 3 \cdot \frac{3}{4} + 4 \cdot \left(\frac{3}{4}\right)^2 + \dots$ to ∞

Subtracting,

$$\frac{1}{4} S = 3 + \frac{3}{4} + \left(\frac{3}{4}\right)^2 + \dots \text{ to } \infty$$

$$= 3 + \frac{\frac{3}{4}}{1 - \frac{3}{4}} = 6$$

$\therefore \quad S = 24.$

$\therefore$ from (1), the required probability $= \frac{9}{256} \times 24 = \frac{27}{32}$.

30. If p is chosen at random in the closed interval [0, 5], what is the probability of the equation $x^2 + px + \frac{1}{4}(p + 2) = 0$ to have real roots?

TIP: Here $n(S)$ is infinite because a real number p can be selected from the closed interval [0, 5] in infinite ways. So, we apply geometrical method.

Here $n(S)$ = length of the interval [0, 5] = 5

$n(E)$ = length of the interval $\subseteq$ [0, 5] in which p belongs such that the given equation has real roots.

Now, $x^2 + px + \frac{1}{4}(p + 2) = 0$ will have real roots

if $\quad D \geq 0.$

$D \geq 0 \Rightarrow p^2 - 4 \cdot 1 \cdot \frac{1}{4}(p + 2) \geq 0$

or $\quad p^2 - p - 2 \geq 0$

or $\quad (p - 2)(p + 1) \geq 0.$

So, by sign-scheme, $p \leq -1$ or $p \geq 2$.

But $p \in [0, 5]$. Hence $E = [2, 5]$.

$\therefore \quad n(E)$ = length of the interval [2, 5] = 3.

$\therefore$ the required probability $= \frac{n(E)}{n(S)} = \frac{3}{5}$.

31. Two numbers $x \in R$ and $y \in R$ are selected such that $x \in [0, 4]$ and $y \in [0, 4]$. Find the probability that the selected numbers satisfy the inequation $y^2 \leq x$.

Corresponding to a pair of real numbers x, y where $x \in [0, 4]$, $y \in [0, 4]$ we get a point (x, y) in the square bounded by the lines $x = 0, x = 4, y = 0, y = 4$.

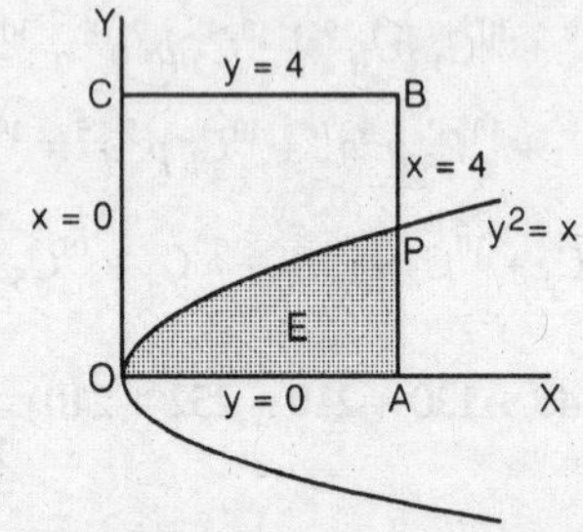

$\therefore \quad n(S)$ = area of the square = 4 × 4 = 16.

The selected numbers x, y will satisfy $y^2 \leq x$ if the point (x, y) is an interior point of the parabola $y^2 = x$.

$\therefore \quad n(E)$ = the area of that portion of the square which falls in the interior of the parabola

$$= \int_0^4 (y)_{\text{parabola}}\, dx = \int_0^4 \sqrt{x}\, dx$$

$$= \left[\frac{2}{3} x^{3/2}\right]_0^4 = \frac{2}{3} \cdot 4^{3/2} = \frac{16}{3}.$$

$\therefore$ the required probability $= \frac{n(E)}{n(S)} = \frac{\frac{16}{3}}{16} = \frac{1}{3}$.

32. There is 30% chance that it rains on any particular day. What is the probability that there is at least one rainy day within a period of 7 days? Given that there is at least one rainy day in a week, what is the probability that there are at least two rainy days in a week?

Let R be the event of having rain in a day.

From the question, $P(R) = \frac{3}{10}$. So $P(\overline{R}) = \frac{7}{10}$.

The probability of having at least one rainy day in 7 days is

$P(E_1) = 1 -$ (probability of not having rain in 7 days)

$$= 1 - \left(\frac{7}{10}\right)^7.$$

The second probability is a conditional probability. Clearly, we have to find the probability of having at least one rainy day in 6 remaining days when a rainy day has taken place.

$\therefore$ the required probability

$$= \frac{\left\{\text{probability of having atleast one rainy day in 6 remaining days}\right\}}{\text{probability of having one rainy day}}$$

$$= \frac{1-\left(\frac{7}{10}\right)^6}{{}^7C_1 \times \frac{3}{10}} = \frac{10}{21}\left\{1-\left(\frac{7}{10}\right)^6\right\}.$$

33. If for the events A and B, $P(A) = P(A/B) = \frac{1}{4}$ and $P(B/A) = \frac{1}{2}$ then verify which of the following statements are correct

(a) A and B are mutually exclusive

(b) A and B are independent

(c) $P(A'/B) = \frac{3}{4}$ (d) $P(B'/A') = \frac{1}{2}$.

We have $P(A/B) = \frac{P(A \cap B)}{P(B)}$

$\therefore \quad P(A) \cdot P(B) = P(A \cap B) \quad \{\because \quad P(A) = P(A/B)\}$

$\therefore \quad A$ and B are independent. So (b) is correct.

Two independent events cannot be mutually exclusive. So (a) is incorrect.

As A and B are independent, A' and B are independent.

$$\therefore \quad P(A'/B) = P(A') = 1 - P(A)$$

$$= 1 - \frac{1}{4} = \frac{3}{4}.$$

So (c) is correct.

As A and B are independent, A' and B' are independent

$$\therefore \quad P(B'/A') = P(B') = 1 - P(B)$$

$$= 1 - P(B/A) \quad \{\because \ A, B \text{ are independent}\}$$

$$= 1 - \frac{1}{2} = \frac{1}{2}.$$

So (d) is correct.

Exercises

1. A has 3 shares in a lottery containing 3 prizes and , blanks; B has two shares in a lottery containing 2 prizes and 6 blanks. Compare their chances of success.
2. Three dice named A, B and C are thrown and the numbers shown on them are put for a, b, c respectively in the quadratic equation $ax^2 + bx + c = 0$. What is the probability that the equation so formed will have real roots?
3. Out of 21 tickets marked with numbers 1 to 21, three are drawn at random. Find the probability that the three numbers on them are in AP.
4. A box contains 2 fifty-paisa coins, 5 twenty-five-paisa coins and a certain fixed number $N(\geq 2)$ of ten-paisa and five-paisa coins. Five coins are taken out of the box at random. Find the probability that the total value of these 5 coins is less than one rupee and fifty paise.
5. A ten-digit number is formed using the digits 0 to 9, every digit being used only once in the same number. Find the probability that the number is divisible by 4.
6. A bag contains 36 tickets, numbered from 0 to 35. Three of the tickets are drawn at random. Find the probability of the sum of the numbers in the three tickets drawn to be 36.
7. A four-letter word is written down by taking letters from the word KANYAKUMARI. What is the chance of the word to be of all different letters?
8. A card is drawn from a pack of 52 cards. The card is then replaced and the pack is reshuffled. It is done six times. Find the chance that the cards drawn are two hearts, two diamonds and two black cards.
9. A word of 5 letters is written down at random with the letters of the word COLLEGE. What is the probability that all the repeated letters were used?
10. 6 different balls are put in 3 different boxes, no box being empty. Find the probability of putting the balls in the boxes equally.
11. A six-digit number is written down at random. What is the chance that the sum of the digits of the number is odd? Also find the chance that the number is divisible by 3 but not by 10.
12. A bag contains 6 coupons numbered 1, 2, 3, 4, 5 and 6. Five coupons are drawn and the coupons are placed in a row at random. What is the probability that the number read will be divisible by 6?
13. Let $X = \{x \mid 1 \leq x \leq 50, x \in N\}$. A member of the set X is selected at random. Find the probability that the

selected number is a solution of the inequation $\frac{x^2 - 30x + 200}{x - 15} < 0$.

14. A bag contains 10 balls. Six balls are drawn in an attempt and replaced. Another draw of 5 balls is made from the bag. What is the chance that exactly 2 balls are common to both the draws? Also prove that the chance remains unaltered if the two draws are interchanged.

15. A question paper is split in two groups—A and B. The group A contains 4 questions, each question having an alternative. The group B also contains 4 questions. A student has to answer at least one question from each group and he can answer up to 8 questions. What is the probability that a student will answer 3 questions?

16. A dice is thrown six times and the sum of the digits shown is found to be 24. What is the probability that each time the digit shown is
(a) at least 3, (b) at most 5?

17. A and B are two candidates seeking admission in IIT. The probability that A is selected is 0.5 and the probability that both A and B are selected is at most 0.3. Is it possible that the probability of B getting selected is 0.9? Give reason.

18. A man and a woman appear in an interview for two vacancies for the same post. The probability that the man will be selected is $\frac{1}{4}$ and that of the woman to be selected is $\frac{1}{3}$. What is the probability that
(i) at least one of them will be selected
(ii) none of them will be selected.

19. An anti-aircraft gun can take a maximum of four shots at an enemy plane moving away from it. The probabilities of hitting the plane at the first, second, third and fourth shot are 0.4, 0.3, 0.2 and 0.1 respectively. What is the probability that the gun hits the plane?

20. An article manufactured by a company consists of two parts X and Y. In the process of manufacturing the part X, 9 out of 104 parts may be defective. Similarly, 5 out of 100 are likely to be defective in manufacturing the part Y. Calculate the probability that the assembled product will not be defective.

21. Each of two persons makes a single throw with a pair of unbiased dice. What is the probability that the throws are equal?

22. A coin is tossed 10 times. What is the probability of getting exactly 7 consecutive tails, no two other tails being consecutive?

23. A dice is thrown 10 times. What is the chance of getting exactly 6 (i.e., 6 but not more) consecutive sixes?

24. A bag contains 8 balls of two colours, red and green. 3 balls are taken out at random. The probability of getting more red balls than green is $\frac{15}{28}$ and that of getting more green balls than red is $\frac{15}{56}$. What is the number of green balls in the bag?

25. A hunter has a chance $\frac{1}{r^2}$ of shooting down a tiger from a distance r metres. He takes his first shot when the tiger is at a distance of 2 metres. If he misses then he takes the second shot when the tiger is at a distance of 3 metres. If he again misses, he takes another shot when the tiger is at a distance of 4 metres and so on. If he misses at his sixth shot also then the tiger escapes. Find the probability that the hunter will be successful to shoot down the tiger.

26. A and B play a game of throwing a pair of dice alternately. A wins if he obtains a sum of 6 before B obtains a sum of 7 in the pair of dice. If A begins the game, find the probability of
(i) A winning the game (ii) B winning the game.

27. A, B and C toss a coin in succession on the understanding that the first one to throw a head wins. What are their respective chances?

28. A and B throw with one dice for a stake of Rs 11 which is to be won by the player who first throws 6. If A has the first throw, what are their respective expectations?

29. In a set of five games in tennis between two players A and B, the probability of a player winning a game is $\frac{2}{3}$ who has won the earlier game. A wins the first game. What is the probability that A will win at least three of the next four games?

30. Three persons A, B and C in order cut a pack of well-shuffled cards, replacing them after each cut on condition that the first who cuts a spade will win a prize; find their respective chances.

31. A box contains 2 black, 4 white and 3 red balls. One ball is drawn at random from the box and kept aside. From the remaining balls in the box another ball is drawn at random and kept beside the first. This process is continued till all the balls are drawn from the box. Find the probability that balls drawn are in the sequence of 2 black, 4 white and 3 red.

32. A bag contains 9 white balls and 1 black ball. Another bag contains 10 white balls. Eight balls are drawn from the first bag at random and put in the second bag. Again, eight balls are drawn from the second bag and put in the first. What is the chance that the black ball is still in the first bag?

33. In a multiple-choice question there are four alternative answers, of which one or more are correct. A candidate will get marks in the question only if he ticks all the correct answers. The candidate decides to tick answers at random. If he is allowed up to 3 chances to answer the question, find the probability that he will get marks in the question.

34. Cards are drawn one-by-one at random from a well-shuffled pack of 52 cards.

(a) If N is the number of cards required to be drawn until two aces are obtained for the first time then show that the probability of N being equal to n is $\frac{(n-1)(52-n)(51-n)}{50 \times 49 \times 17 \times 13}$ where $2 \le n \le 50$.

(b) Find the probability that exactly n cards are to be drawn before the first king appears.

35. n different letters are put at random in n addressed envelopes. What is the probability that no letter is placed in the correct envelope?

36. A private tutor admits 10 students, boys or girls, in a batch. The probability that a student admitted in his batch is a girl, is 0.25. What is the probability that his batch has

(a) equal number of boys and girls

(b) more girls than boys?

37. If in 6 trials of an experiment the probability of 2 successes is 9 times the probability of 4 successes then find the probability of success in one trial.

38. Suppose the probability for A to win a game against B is 0.4. If A has an option of playing either a "best of 3 games" or a "best of 5 games" match against B, which option should A prefer so that the probability of his winning the match is higher? (No game ends in a draw.)

39. A hotel has 5 rooms for rental and they are always occupied. The owner has 3 TV sets available on the rental Rs 20 per day per set. Only one set will be supplied to one occupant if he wants it. The chance that an occupant will want to hire a set is $\frac{3}{5}$. Find the probability that the hotel owner will get Rs 60 per day by renting the sets.

40. Numbers are selected at random, one at a time, from the two-digit numbers, 00, 01, 02, ..., 99 with replacement. An event E occurs iff the product of the two digits of a selected number is 18. If four numbers are selected, find the probability that the event E occurs at least 3 times.

41. Find the minimum number of tosses of a pair of dice so that the probability of getting the sum of the digits on the dice equal to 7 in at least one toss is greater than 0.95.

(Given $\log_{10} 2 = 0.3010$, $\log_{10} 3 = 0.4771$.)

42. There are two circles in the x-y plane whose equations are $x^2 + y^2 - 2y = 0$ and $x^2 + y^2 - 2y - 3 = 0$. A point is chosen at random inside the larger circle. Find the probability that the point has been taken

(a) from the smaller circle

(b) from outside the smaller circle.

43. Two persons A and B agree to meet at a place between 11 noon to 12 noon. The first one to come waits for 20 minutes and then leaves. If the time of their arrival are independent and at random then what is the probability of a meeting between A and B?

44. A real number x is selected at random from the solution set of the inequation $y^2 - y - 6 \le 0$. What is the probability that $\frac{(x+1)(x-2)}{x-4} \ge 0$?

45. Three tangents are drawn at random to a given circle. Show that the odds are 3 : 1 against the circle being inscribed in the triangle formed by the tangents.

46. A floor is paved with rectangular marble blocks, each of length a and breadth b. A circular block of diameter $c(c < a, b)$ is thrown on the floor at random. Show that the chance that it falls entirely on one rectangular block is $\frac{(a-c)(b-c)}{ab}$.

47. A plane is ruled with parallel straight lines at equal distances of $2a$. A needle, $2l$ long ($l < a$), is thrown on the plane at random. Find the probability that the needle will hit any of the lines.

Objective Questions

Fill in the blanks.

48. A fair dice is rolled once. The probability that an odd number less than 5 turns up is ______ and the probability that a number greater than or equal to 3 turns up is ______ .

49. The probabilities of two independent events A and B are respectively ______ and ______ , $\{P(A) > P(B)\}$ if the probability that both occur is $\frac{1}{6}$ and the probability that none of them occur is $\frac{1}{3}$.

50. Two students A and B solve a problem. Their respective probabilities of solving the problem are $\frac{2}{3}$ and $\frac{1}{2}$. The probability of the problem being solved by at least one of them is ______ .

51. If A, B are two equiprobable events such that
$P(A \cap B) = \frac{1}{4}$ and $P(A' \cap B') = \frac{1}{5}$ then
$P(A) =$ ______ .

52. If $\frac{1+3p}{3}$, $\frac{1-p}{4}$ and $\frac{1-2p}{2}$ are the probabilities of three mutually exclusive events then the set of possible values of p is ______ .

53. Three faces of a fair dice are yellow, two faces red and one blue. The dice is tossed three times. The probability that yellow, red and blue faces appear in the first, second and third tosses respectively is ______ .

54. Items coming of a production line are marked defective (D) or nondefective (N). Items are observed and their conditions listed. This is continued until two consecutive defectives are produced or four items have been checked, whichever occurs first. Then
(a) the number of elements in the sample space is ______ .
(b) the probability of getting two consecutive defectives in the experiment is ______ .
(c) the probability of getting at least two defectives in the experiment is ______ .

55. Three letters are written to different persons and their addresses are also written on three envelopes. Letters are put at random in the envelopes. Then
(a) the probability that letters are put in correct envelopes, is ______ .
(b) the probability that only one letter is put in the correct envelope, is ______ .
(c) the probability that all the letters are put in wrong envelopes, is ______ .

56. The probability that in a group of $N(< 365)$ people at least two will have the same birthday, is ______ .

57. The probability of a leap year to have 53 Mondays is ______ .

58. Two small squares on a chessboard are selected at random. The probability that they have a common side is ______ .

59. 12 persons are asked a question in a random order of which exactly 2 persons know the answer. The probability that the first four do not know the answer is ______ and the probability of the sixth person being the first to know the answer is ______ .

60. Two numbers are selected at random from 1, 2, 3, ..., 100 and multiplied. The probability that the product thus obtained is divisible by 3 is ______ .

61. Three dice are thrown simultaneously. The probability of obtaining a sum of 6 is ______ .

62. Urn A contains 6 red and 4 black balls and urn B contains 4 red and 6 black balls. One ball is drawn at random from the urn A and placed in the urn B. Then one ball is drawn at random from urn B and placed in the urn A. If one ball is now drawn from the urn A, the probability that it is found to be red is ______ .

63. A and B are two events such that the probability that both A and B occur is $\frac{1}{6}$ and the probability that at least one of them occurs is $\frac{2}{3}$. The probability of the occurrence of $A =$ ______ if $P(A) = 2P(B)$.

64. The probabilities that on a certain day the two trains from Calcutta and Mumbai arrive on time at Delhi are 0.93 and 0.89 respectively. The probability that both trains arrive on time is 0.87. Then the probability that at least one train is on time is ______ .

65. If four whole numbers taken at random are multiplied together then the chance that the last digit in the product is 1, 3, 7 or 9, is ______ .

66. Three ordinary dice are thrown together. The probability of getting a sum $m (3 \le m \le 18)$ on them is ______ .

67. Four small squares on a chessboard are selected at random. The probability that they form a square of the size 2×2 is ______ .

68. A coin is tossed 3 times. The probability of getting head and tail alternately is ______ .

69. A room has 3 electric lamps. From a collection of 10 electric bulbs of which 6 are in order, 3 are selected at random and put in the lamps. The probability that the room is lighted, is ______ .

70. A bag contains 17 tickets numbered 1 to 17. A ticket is drawn at random and replaced. A second draw is carried out. The chance that the first number drawn is even and the second odd, is ______ .

71. 8 persons entered the lift cabin on the ground floor of an 8-floor house. If any person can leave the cabin at any one of the eight floors then the probability that all the persons leave the cabin at different floors, is ______ .

72. A card is drawn at random from a pack of 52 cards. The chance that it is either an ace or a heart is ______ .

73. $P(A \cup B) = P(A \cap B)$ if and only if the relation between $P(A)$, $P(B)$ and $P(A/B)$ is ______ .

74. India play two matches each with West Indies and Australia. In any match the probabilities of India getting points 0, 1 and 2 are 0.45, 0.05 and 0.50 respectively.

Assuming that the outcomes are independent, the probability of India getting at least 7 points is ______ .

75. The probabilities of different faces of a biased dice to appear are as follows:

face number	1	2	3	4	5	6
probability	0.1	0.32	0.21	0.15	0.05	0.17

The dice is thrown and it is known that either the face number 1 or 2 will appear. Then the probability of the face number 1 to appear is ______ .

76. If X denotes the number of sixes in four consecutive throws of a dice then $P(X = 4)$ is ______ .

77. Let an ordinary coin be tossed 15 times. If P_r denotes the probability of getting r tails and P_k is maximum then $k =$ ______ .

78. Two coins are tossed 4 times. The chance that there will be equal number of heads and tails, is ______ .

79. The probability that an event A happens in one trial of an experiment is 0.4. Three independent trials of the experiment are performed. The probability that the event A happens at least once is ______ .

80. The odds in favour of A winning a game against B are 4 : 3. The probability of B winning 3 games out of 7 played is ______ .

81. The probability of a missile hitting a target is $\frac{1}{5}$. Two missiles are enough to destroy a bridge. If six missiles are fired at the bridge, the chance of the bridge being destroyed is ______ .

82. A dice is thrown six times. The probability of getting three sixes, the third six appearing in the sixth throw, is ______ .

83. Let A = $\{1, 2, 3, \ldots, n\}$ and f be any mapping from the set A to itself. If one such f is selected at random then the probability of f being invertible is ______ .

84. Two integers x and y are chosen one by one with replacement, at random from the set $\{x \mid 0 \le x \le 10, x \text{ is an integer}\}$.
The probability that $|x - y| \le 5$ is ______ .

85. A trunk-call is booked. It is known that a trunk-call booked materialises any time within an hour. The chance that the call will take place within the first 15 minutes of the booking, is ______ .

86. If A and B are two events such that $P(A') = 0.3$, $P(B) = 0.4$ and $P(A \cap B') = 0.5$ then $P\{B/(A \cup B')\} =$ ______ .

Choose the correct option(s).

87. The probabilities of two events are 0.25 and 0.50. The probability of both happening together is 0.14. Which of the following is the probability of none of the events happening?

(a) 0.39 (b) 0.25
(c) 0.11 (d) none of these

88. The probability that at least one of the events A and B occurs is 0.60. If A and B occur simultaneously with probability 0.20 then $P(A') + P(B')$ is equal to

(a) 0.40 (b) 0.80 (c) 1.20 (d) 1.40

89. The letters of the word ASSASSIN are written down at random in a row. The probability that no two S occur together is

(a) $\frac{1}{35}$ (b) $\frac{1}{14}$
(c) $\frac{1}{15}$ (d) none of these

90. If $P(A) = \frac{2}{3}$, $P(B) = \frac{1}{2}$ and $P(A \cup B) = \frac{5}{6}$ then the events A and B are

(a) mutually exclusive
(b) independent
(c) independent as well as mutually exclusive
(d) none of these

91. Three identical dice are rolled. The probability that the same number will appear on each of them is

(a) $\frac{1}{6}$ (b) $\frac{1}{36}$ (c) $\frac{1}{18}$ (d) $\frac{3}{28}$

92. The probability of the event A occurring is 0.5 and of B occurring is 0.3. If A and B are mutually exclusive events then the probability of neither A nor B occurring is

(a) 0.6 (b) 0.5
(c) 0.7 (d) none of these

93. A and B are two independent events such that $P(A' \cap B) = \frac{2}{15}$ and $P(A \cap B') = \frac{1}{6}$ then $P(B)$ is equal to

(a) $\frac{4}{5}$ (b) $\frac{1}{6}$ (c) $\frac{1}{5}$ (d) $\frac{5}{6}$

94. A student appears for tests I, II and III. The student is successful if he passes in tests I, II or I, III. The probabilities of the student passing in tests I, II and III are respectively p, q and $\frac{1}{2}$. If the probability of the student to be successful is $\frac{1}{2}$ then

(a) $p(1 + q) = 1$ (b) $q(1 + p) = 1$
(c) $pq = 1$ (d) $\frac{1}{p} + \frac{1}{q} = 1$

95. The probability that the wife will be alive 10 years hence is $\frac{7}{15}$ and that of the husband is $\frac{7}{10}$. What is the probability that at least one of them will be alive 10 years hence?

(a) $\frac{129}{150}$ (b) $\frac{126}{150}$ (c) $\frac{101}{150}$ (d) $\frac{94}{150}$

96. A soldier is firing at a moving target. He fires four shots. The probability of hitting the target at the first, second, third and fourth shots are 0.6, 0.4, 0.2 and 0.1 respectively. What is the probability that he hits the target?

(a) $\frac{517}{625}$ (b) $\frac{3}{625}$

(c) $\frac{105}{625}$ (d) none of these

97. A and B are two independent events such that $P(A) = 0.3$ and $P(A \cup \overline{B}) = 0.8$. Then

(a) $P(B) = \frac{1}{2}$ (b) $P(B) = \frac{2}{7}$

(c) $P(A \cap B) = \frac{3}{35}$ (d) $P(A \cap B) = \frac{3}{20}$

98. A and B are two independent events. The probability that both A and B occur is $\frac{1}{6}$ and the probability that neither of them occur is $\frac{1}{3}$. Then $P(A)$ is equal to

(a) $\frac{1}{2}$ (b) $\frac{1}{3}$ (c) $\frac{5}{6}$ (d) $\frac{1}{2}$ or $\frac{1}{3}$

99. The probability of at least one of A and B occurring is $\frac{3}{5}$ and that of A and B occurring together is $\frac{3}{10}$ then $P(A') + P(B')$ is equal to

(a) $\frac{6}{5}$ (b) $\frac{11}{10}$ (c) $\frac{9}{10}$ (d) $\frac{23}{20}$

100. The probability that a rectangle picked up from a chessboard has the area 6 cm^2 where the distance between consecutive parallel lines on the board is 1 cm, is

(a) $\frac{3}{56}$ (b) $\frac{3}{28}$

(c) $\frac{9}{56}$ (d) none of these

101. Two persons A and B throw two dice each. If A throws a sum of 9 then the probability of Y throwing a sum greater than that of X is

(a) $\frac{1}{6}$ (b) $\frac{1}{9}$

(c) $\frac{1}{54}$ (d) none of these

102. On a chessboard small squares are either black or white, set alternately. Three pawn are placed at random on three squares of the chessboard. The probability that two are on the squares of the same colour, is

(a) $\frac{8}{21}$ (b) $\frac{5}{21}$

(c) $\frac{16}{21}$ (d) none of these

103. A natural number is selected at random from the set $X = \{x \mid 1 \le x \le 100\}$. The probability that the number satisfies the inequation $x^2 - 13x \le 30$, is

(a) $\frac{9}{50}$ (b) $\frac{3}{20}$

(c) $\frac{2}{11}$ (d) none of these

104. If A and B are two events then

(a) $P(A \cap B) \ge P(A) + P(B) - 1$

(b) $P(A \cap B) \ge P(A) + P(B)$

(c) $P(A \cap B) = P(A) + P(B) - P(A \cup B)$

(d) $P(A \cap B) = P(A) \cdot P(B)$

105. If E and F are independent events such that $0 < P(E) < 1$ and $0 < P(F) < 1$ then

(a) E, F are mutually exclusive

(b) E, $\overline{F}$ are independent

(c) $\overline{E}$, $\overline{F}$ are independent

(d) $P(E/F) + P(\overline{E}/F) = 1$

106. If $0 < P(A) \le 1$, $0 \le P(B) < 1$ then $P(\overline{A}/\overline{B})$ is equal to

(a) $1 - P(A/B)$ (b) $1 - P(A/\overline{B})$

(c) $\frac{1 - P(A \cup B)}{P(\overline{B})}$ (d) $\frac{P(\overline{A})}{P(\overline{B})}$

107. For any two events A and B,

(a) $P(A/B) \ge \frac{P(A) + P(B) - 1}{P(B)}$, $P(B) \ne 0$

(b) $P(A \cap \overline{B}) \ne P(A) - P(A \cap B)$

(c) $P(A \cup B) = 1 - P(\overline{A})P(\overline{B})$ if A, B are independent

(d) $P(A \cup B) = 1 - P(\overline{A})P(\overline{B})$ if A, B are mutually exclusive

108. The probability that exactly one of the two events A, B occurs is

(a) $P(A) + P(B) - 2 \cdot P(A \cap B)$

(b) $P(A) + P(B) - P(A \cap B)$

(c) $P(\overline{A}) + P(\overline{B}) - 2 \cdot P(\overline{A} \cap \overline{B})$

(d) $P(A \cap \overline{B}) + P(\overline{A} \cap B)$

109. Fifteen coupons are numbered from 1 to 15. Seven coupons are selected at random one at a time with replacement. The probability that the largest number appearing on a selected coupon is 9, is

(a) $\left(\frac{9}{16}\right)^6$ (b) $\left(\frac{8}{15}\right)^7$

(c) $\left(\frac{3}{5}\right)^7$ (d) none of these

110. In a box containing 100 bulbs, 10 are defective. What is the probability that out of a sample of 5 bulbs, none is defective?

(a) 10^{-5} (b) $\left(\frac{1}{2}\right)^5$ (c) $\left(\frac{9}{10}\right)^5$ (d) $\frac{9}{10}$

111. One hundred identical coins, each with probability p of showing up head, are tossed. If $0<p<1$ and the probability of heads showing on 50 coins is equal to that of heads showing on 51 coins then the value of p is

(a) $\frac{1}{2}$ (b) $\frac{49}{101}$ (c) $\frac{50}{101}$ (d) $\frac{51}{101}$

112. The probability that an event A happens in one trial of an experiment is 0.4. Three independent trials of the experiment are performed. The probability that the event A happens at least once is

(a) 0.936 (b) 0.784

(c) 0.904 (d) none of these

113. A fair coin is tossed n times. If the probabilities of getting 4, 5 and 6 heads be in AP then n is equal to

(a) 12 (b) 7 (c) 15 (d) 14

114. An unbiased ordinary dice is rolled four times. Out of the four face-values obtained, the probability that the minimum face-value is not less than 2 and the maximum face-value is not greater than 5, is

(a) $\frac{16}{81}$ (b) $\frac{1}{81}$ (c) $\frac{80}{81}$ (d) $\frac{65}{81}$

State whether the statements are true or false.

115. The probabilities of the events A, B and C are respectively $\frac{2}{3}, \frac{1}{4}$ and $\frac{1}{6}$. Then the events are mutually exclusive.

116. If the probability for A to fail in an examination is 0.2 and that for B is 0.3 then the probability that either A or B fails is 0.5.

117. The probability of getting exactly 2 heads in tossing a coin thrice is $\left(\frac{1}{2}\right)^3$.

118. For any three events A, B and C defined on the sample space R such that $B \subset C$ and $P(A) \neq 0$, $P(B/A) \leq P(C/A)$.

119. In a lottery all the tickets are blank except one, on which there is a prize. n persons draw a ticket each one after another without replacement. The probabilities of the 6th person to win the prize is $\frac{6}{n}$.

120. If A and B are two independent events then $P(A \cap \overline{B}) + P(A)P(B) = P(A)$.

Answers

1. 952 : 715 **2.** $\frac{43}{216}$ **3.** $\frac{10}{133}$ **4.** $1-\frac{10(N+2)}{^{N+7}C_5}$

5. $\frac{20}{81}$ **6.** $\frac{9}{595}$ **7.** $\frac{840}{1109}$ **8.** $\frac{45}{512}$

9. $\frac{3}{23}$ **10.** $\frac{1}{6}$ **11.** $\frac{1}{2}, \frac{3}{10}$ respectively

12. $\frac{1}{6}$ **13.** $\frac{13}{50}$ **14.** $\frac{5}{21}$ **15.** $\frac{2}{25}$

16. (a) $\frac{336}{3431}$ (b) $\frac{426}{3431}$ **17.** no

18. (i) $\frac{1}{2}$ (ii) $\frac{1}{2}$ **19.** $\frac{436}{625}$

20. $\frac{361}{416}$ **21.** $\frac{73}{648}$ **22.** $\frac{5}{2^9}$ **23.** $\frac{135}{6^8}$

24. 3 **25.** $\frac{3}{7}$ **26.** (i) $\frac{30}{61}$ (ii) $\frac{31}{61}$

27. $\frac{4}{7}, \frac{2}{7}, \frac{1}{7}$ respectively **28.** Rs 6, Rs 5 respectively

29. $\frac{4}{9}$ **30.** $\frac{16}{37}, \frac{12}{37}, \frac{9}{37}$ respectively

31. $\frac{1}{1260}$ **32.** $\frac{5}{9}$ **33.** $\frac{631}{3375}$

34. (b) $\frac{(49-n)(50-n)(51-n)}{13 \cdot 49 \cdot 50 \cdot 51}$

35. $\frac{1}{2!}-\frac{1}{3!}+\ldots+(-1)^n\frac{1}{n!}$

36. (a) $\frac{7 \times 3^7}{4^9}$ (b) $\frac{10343}{2 \times 4^9}$ **37.** $\frac{1}{4}$

38. best of 3 games **39.** $\frac{2133}{3125}$

40. $\frac{97}{390625}$ **41.** 17

42. (a) $\frac{1}{4}$ (b) $\frac{3}{4}$ **43.** $\frac{5}{9}$ **44.** $\frac{3}{5}$

47. $\frac{2l}{\pi a}$ **48.** $\frac{1}{3}, \frac{2}{3}$ respectively

49. $\frac{1}{2}, \frac{1}{3}$ respectively **50.** $\frac{5}{6}$ **51.** $\frac{21}{40}$

52. $\left[\frac{1}{3}, \frac{1}{2}\right]$ **53.** $\frac{1}{36}$ **54.** (a) 12 (b) $\frac{1}{3}$ (c) $\frac{7}{12}$

55. (a) $\frac{1}{6}$ (b) $\frac{1}{2}$ (c) $\frac{1}{3}$ **56.** $1 - \frac{365\,!}{365^N \cdot (365 - N)\,!}$

57. $\frac{2}{7}$ **58.** $\frac{1}{18}$ **59.** $\frac{14}{33}, \frac{1}{11}$ respectively

60. 0.55(nearly) **61.** $\frac{5}{108}$ **62.** $\frac{32}{55}$ **63.** $\frac{5}{9}$

64. 0.95 **65.** $\frac{16}{625}$ **66.** $\frac{(m-1)(m-2)}{432}$

67. $\frac{7}{90768}$ **68.** $\frac{1}{4}$ **69.** $\frac{29}{30}$ **70.** $\frac{72}{289}$

71. $\frac{7\,!}{8^7}$ **72.** $\frac{4}{13}$

73. $P(A) + P(B) = 2P(B) \cdot P(A/B)$

74. 0.0875 **75.** $\frac{5}{21}$ **76.** $\frac{1}{1296}$ **77.** 7 or 8

78. $\frac{35}{128}$ **79.** $\frac{98}{125}$ **80.** $\frac{34560}{117649}$ **81.** $\frac{1077}{3125}$

82. $\frac{625}{23328}$ **83.** $\frac{(n-1)\,!}{n^{n-1}}$ **84.** $\frac{91}{121}$ **85.** $\frac{1}{4}$

86. 0.25 **87.** (a) **88.** (c) **89.** (b)

90. (b) **91.** (b) **92.** (d) **93.** (a), (b)

94. (a) **95.** (b) **96.** (a) **97.** (b), (c)

98. (d) **99.** (b) **100.** (d) **101.** (a)

102. (c) **103.** (b) **104.** (a), (c)

105. (b), (c), (d) **106.** (c) **107.** (a), (c)

108. (a), (c), (d) **109.** (c) **110.** (c)

111. (d) **112.** (b) **113.** (b), (d)

114. (a) **115.** false **116.** false **117.** false

118. true **119.** false **120.** true

□

Chapter Test

Time: 120 minutes

1. In each of the following incomplete sentences, fill in the blank so that the resulting sentence becomes correct.

(a) Two numbers of five digits made by the digits 3, 4, 5, 6 are taken at random to add. The chance that the two numbers can be added without carrying at any place, is ______ .

(b) Out of twenty consecutive positive integers, two are chosen at random. The probability that the sum of the chosen numbers is odd, is ______ .

(c) The probability that a man fishing at a particular place will catch 1, 2, 3 and 4 fishes are 0.4, 0.3, 0.2 and 0.1 respectively. Then the expected number of fishes caught is ______ .

(d) A dice is rolled four times. The probability of having more sixes than any other number at any stage is ______ .

2. Each of the following has multiple answers. Choose the correct answer(s).

(a) The probability that a positive integral power of 137 will have the digit 7 in units place of the value of the power, is

A. $\frac{1}{3}$ B. $\frac{1}{2}$ C. $\frac{1}{4}$ D. $\frac{3}{4}$

(b) A coin is tossed 8 times. The probability of having 4 heads in the even places of tosses is

A. $\left(\frac{1}{2}\right)^3$ B. $\left(\frac{1}{2}\right)^8$ C. ${}^8C_4\left(\frac{1}{2}\right)^8$ D. none of these

(c) A five-digit number is written down at random. The probability that the number written is such that when the digits are put in the reversed order the new number is also a five-digit number equal to the original number, is upto two places of decimal

A. 0.01 B. 0.90 C. 0.81 D. none of these

(d) The adjoining figure gives the road plan of lanes connecting two parallel roads AB and FJ. A man walking on the road AB takes a turn at random to reach the road FJ. It is known that he reaches the road FJ from O by taking a straight lane. The chance that he moves on a straight lane from the road AB to the road FJ, is

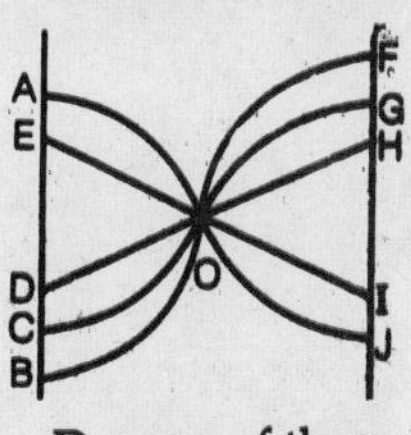

A. 0.25 B. 0.04 C. 0.2 D. none of these

3. Two numbers λ, μ are chosen one after another at random (with replacement) from the set $\{a \mid 1 \le a \le 10, a \in N\}$. Find the probability that $x^2 + \lambda x + \mu > 0$ for all $x \in R$.

4. Two teams A and B play an one-day cricket match. The match is in the last over and the situation of the match is as follows:
Four balls are to be bowled by the team B; no wides or no-balls are bowled; the team A has to score 16 runs to win. Runs that can be scored off a ball are 0, 1, 2, 3, 4, 5 and 6.
Find the probability that the team A will win by scoring in the last ball.

5. On a piece of paper certain odd number of horizontal lines at equal distances are drawn. An equal number of vertical lines at equal distances are also drawn, forming rectangles. The rectangles are coloured black or white, set alternately. Three coins are put at random in these rectangles. Let p be the probability of having two coins in rectangles of the same colour and the third in a rectangle of the other colour; and p' be the probability of having all the coins in the rectangles of the same colour. If $p : p' = 16 : 5$ then find the number of horizontal lines drawn on the paper.

6. A coin is tossed 4 times whose two faces are marked with the numbers 5 and 3. What is the odds in favour of getting a sum less than 15?

7. Let $(\alpha_1, \alpha_2, \ldots \alpha_6)$ be a positive integral solution of the equation $x_1 + x_2 + \ldots + x_6 = 18$. Find the probability of $(\alpha_1, \alpha_2, \ldots \alpha_5)$ to be a positive integral solution of the inequation
$$8 \le x_1 + x_2 + \ldots + x_5 \le 12.$$
8. A number of five different digits is written down at random. What is the probability that the number written has the largest digit in the middle?
9. Eight dice are thrown at random at a time. Find the probability of getting a sum 24.
10. If the sum of five natural numbers is 50, find the probability that the five numbers are even.
11. A and B are two persons standing in the frame as shown in the figure. A can take a step eastward or northward with equal probabilities while B can take a step westward or southward with equal probabilities. In each move they can step only to an adjoining square (having a common side). Both A and B take 3 steps at random. What is the probability that they will meet in a shaded square?

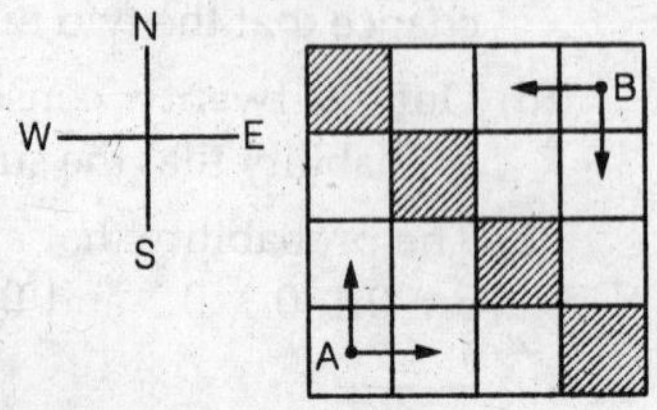

12. A wooden rod is 6 m long. It is cut at random at two places A and B on the two sides to obtain a smaller rod AB. What is the chance that the rod AB will be of at least 1 m in length?
13. A coin has probability p of showing head when tossed. It is tossed n times. Let p_n denote the probability that no two or more consecutive heads occur. Prove that $p_1 = 1$, $p_2 = 1 - p^2$ and $p_n = (1 - p)p_{n-1} + p(1 - p)p_{n-2}$, $n \ge 3$.

□

Answers

1. (a) $\frac{3125}{32768}$ (b) $\frac{10}{19}$ (c) 2 (d) $\frac{11}{1296}$
2. (a) C (b) B (c) A (d) C
3. $\frac{19}{50}$
4. $\frac{405}{7^4}$
5. 9
6. 5 : 11
7. $\frac{771}{6188}$
8. $\frac{7}{36}$
9. $\frac{1}{6^8 \cdot 7!}\left(\frac{23!}{16!} - \frac{8 \times (17!)}{10!} + \frac{7 \times (11!)}{6}\right)$
10. $\frac{33}{658}$
11. $\frac{5}{16}$
12. $\frac{25}{36}$

□

2. Total Conditional Probability and Bayes' Theorem

Recap of Facts and Formulae

1. Total conditional probability

- Events $E_1, E_2, \dots, E_n$ are mutually exclusive if $E_i \cap E_j = 0$ for $i \neq j = 1, 2, 3, \dots, n$.
- Events $E_1, E_2, \dots, E_n$ are exhaustive if
$$P(E_1) + P(E_2) + \dots + P(E_n) = 1.$$
- If $E_1, E_2, \dots, E_n$ are equiprobable and exhaustive then
$$P(E_1) = P(E_2) = \dots = P(E_n) = \frac{1}{n}.$$
- If $E_1, E_2, E_3, \dots, E_n$ are n mutually exclusive and exhaustive events and A is an event which can take place in conjunction with any one of E_i then
$$P(A) = P(E_1) \cdot P(A/E_1) + P(E_2) \cdot P \cdot (A/E_2) + \dots + P(E_n) \cdot P(A/E_n)$$

2. Bayes' theorem

- If $E_1, E_2, \dots, E_n$ are n mutually exclusive and exhaustive events and A is an event which takes place in conjunction with any one of E_i then the probability of the event E_i happening when the event A takes place, is given by
$$P(E_i/A) = \frac{P(E_i) \cdot P(A/E_i)}{\sum_{i=1}^{n} P(E_i) \cdot P(A/E_i)}.$$
In case of equiprobable events,
$$P(E_i/A) = \frac{P(A/E_i)}{\sum_{i=1}^{n} P(A/E_i)}.$$

Note $P(E_i/A)$ is calculated after the experiment has taken place. Such probabilities are called *posteriori* probabilities. $P(E_i)$, etc., which are known before the experiment takes place are called *priori* probabilities.

Selected Solved Examples

1. An urn contains 2 white and 2 black balls. A ball is drawn at random. If it is white, it is not replaced into the urn, otherwise it is replaced along with another ball of the same colour. The process is repeated. Find the probability that the third ball drawn is black.

Let E_1 = event of getting white, white in the first two draws

E_2 = event of getting white, black in the first two draws

E_3 = event of getting black, white in the first two draws

E_4 = event of getting black, black in the first two draws.

Clearly, $P(E_1) = P(W) \cdot P(W/W) = \frac{{}^2C_1}{{}^4C_1} \times \frac{{}^1C_1}{{}^3C_1} = \frac{1}{6}$,

because after drawing the first ball which is white, the ball is not replaced, leaving 1 white ball in 3 balls.

$$P(E_2) = P(W) \cdot P(B/W) = \frac{{}^2C_1}{{}^4C_1} \cdot \frac{{}^2C_1}{{}^3C_1} = \frac{1}{3},$$

because after drawing the first ball which is white, the ball is not replaced, leaving 2 black balls in 3 balls.

$$P(E_3) = P(B) \cdot P(W/B) = \frac{{}^2C_1}{{}^4C_1} \times \frac{{}^2C_1}{{}^5C_1} = \frac{1}{5},$$

because after drawing the first ball which is black, 2 black balls are replaced, leaving 3 black balls and 2 white balls.

$$P(E_4) = P(B) \cdot P(B/B) = \frac{{}^2C_1}{{}^4C_1} \times \frac{{}^3C_1}{{}^5C_1} = \frac{3}{10} \quad \text{(as above).}$$

Clearly, $P(E_1)+P(E_2)+P(E_3)+P(E_4)$

$$=\frac{1}{6}+\frac{1}{3}+\frac{1}{5}+\frac{3}{10}=1.$$

So E_1, E_2, E_3 and E_4 are exhaustive. Clearly, they are also mutually exclusive.

$\therefore$ the required probability

$$= P(E_1)\cdot P(B/E_1)+P(E_2)\cdot P(B/E_2)+P(E_3)\cdot P(B/E_3)$$
$$+P(E_4)\cdot P(B/E_4) \quad \ldots(1)$$

Now, $P(B/E_1)$ = probability of drawing a black ball when white balls are drawn in the first two draws

$$=\frac{^2C_1}{^2C_1}=1,$$

because only 2 black balls are left for the next draw.

$P(B/E_2)$ = probability of drawing a black ball when white, black balls are drawn in the first two draws

$$=\frac{^3C_1}{^4C_1}=\frac{3}{4},$$ because 3 black balls and 1 white ball are left.

$P(B/E_3)$ = probability of drawing a black ball when black, white balls are drawn in the first two draws

$$=\frac{^3C_1}{^4C_1}=\frac{3}{4} \text{ (as above).}$$

$$P(B/E_4)=\frac{^4C_1}{^6C_1}=\frac{2}{3},$$

because after drawing 2 black balls in the first draw there are 4 black balls and 2 white balls for the next draw.

$\therefore$ from (1), the required probability

$$=\frac{1}{6}\times 1+\frac{1}{3}\times\frac{3}{4}+\frac{1}{5}\times\frac{3}{4}+\frac{3}{10}\times\frac{2}{3}$$

$$=\frac{1}{6}+\frac{1}{4}+\frac{3}{20}+\frac{1}{5}=\frac{23}{30}.$$

2. Three factories A, B and C produce the same product. The factory A produces twice as many as B produces while the factories B and C produce in the same quantity. It is known that 2% of the products of A as well as C are defective while 4% of the products of B are defective. All the products of the three factories are stocked together. If a product is selected at random from the stock, what is the probability that the product is defective?

Let A, B and C produce $2n$, n and n products.

Let E_1 = event of a product being produced at A

E_2 = event of a product being produced at B

E_3 = event of a product being produced at C.

$$\therefore \quad P(E_1)=\frac{2n}{2n+n+n}=\frac{1}{2}$$

$$P(E_2)=\frac{n}{2n+n+n}=\frac{1}{4}$$

$$P(E_3)=\frac{1}{4}.$$

Also, $P(D/E_1)$ = probability of a product being defective when produced at A

$$=\frac{2}{100}=\frac{1}{50}.$$

Similarly, $P(D/E_2)=\frac{4}{100}=\frac{1}{25}$

$$P(D/E_3)=\frac{2}{100}=\frac{1}{50}.$$

As $P(E_1)+P(E_2)+P(E_3)=\frac{1}{2}+\frac{1}{4}+\frac{1}{4}=1,$

events E_1, E_2, E_3 are exhaustive.

Also, they are mutually exclusive.

The probability that the product is defective

$$=P(E_1)\cdot P(D/E_1)+P(E_2)\cdot P(D/E_2)+P(E_3)\cdot P(D/E_3)$$

$$=\frac{1}{2}\cdot\frac{1}{50}+\frac{1}{4}\cdot\frac{1}{25}+\frac{1}{4}\cdot\frac{1}{50}$$

$$=\frac{1}{100}+\frac{1}{100}+\frac{1}{200}=\frac{1}{40}.$$

3. A lot contains 20 articles. The probability that the lot contains exactly 2 defective articles is 0.4 and the probability that the lot contains exactly 3 defective articles is 0.6. Articles are drawn from the lot at random one by one without replacement and tested till all the defective articles are found. What is the probability that the testing procedure ends at the 12th testing?

Let E_2 = event of containing exactly 2 defective articles

E_3 = event of containing exactly 3 defective articles

E = event of getting the last defective at the 12th testing.

From the question, $P(E_2)=0.4$ and $P(E_3)=0.6$

$$\therefore \quad P(E_2)+P(E_3)=0.4+0.6=1.$$

So, E_2 and E_3 are exhaustive.

Also, E_2 and E_3 are mutually exclusive.

$\therefore$ the required probability

$$P(E) = P(E_2) \cdot P(E/E_2) + P(E_3) \cdot P(E/E_3) \quad \dots (1)$$

Now, $P(E/E_2)$ = probability of getting one defective and 10 nondefectives in the first 11 draws and the last defective in the 12th draw

$$= \frac{{}^2C_1 \times {}^{18}C_{10}}{{}^{20}C_{11}} \times \frac{1}{9}$$

$$= 2 \times \frac{18\,!}{10\,!\,8\,!} \times \frac{11\,!\,9\,!}{20\,!} \times \frac{1}{9}$$

$$= 2 \times \frac{11 \times 9}{20 \times 19} \times \frac{1}{9} = \frac{11}{190}.$$

$P(E/E_3)$ = probability of getting two defectives and 9 nondefectives in the first 11 draws and the last defective in the 12th draw

$$= \frac{{}^3C_2 \times {}^{17}C_9}{{}^{20}C_{11}} \times \frac{1}{9}$$

$$= 3 \times \frac{17\,!}{9\,!\,8\,!} \times \frac{11\,!\,9\,!}{20\,!} \times \frac{1}{9}$$

$$= \frac{3 \times 11 \times 10 \times 9}{20 \times 19 \times 18} \times \frac{1}{9} = \frac{11}{228}.$$

$\therefore$ by (1), the required probability

$$= \frac{2}{5} \times \frac{11}{190} + \frac{3}{5} \times \frac{11}{228}$$

$$= \frac{11}{5}\left(\frac{1}{95} + \frac{1}{76}\right) = \frac{11}{5} \times \frac{45}{1900} = \frac{99}{1900}.$$

4. A card from a pack of 52 cards is lost. From the remaining cards of the pack two cards are drawn and are found to be spades. Find the probability of the missing card to be a spade.

Let E_S = event of losing a card from spades

E_H = event of losing a card from hearts

E_D = event of losing a card from diamonds

E_C = event of losing a card from clubs

and E = event of drawing 2 spades from the remaining cards.

Clearly, $P(E_S) = \dfrac{{}^{13}C_1}{{}^{52}C_1} = \dfrac{13}{52} = \dfrac{1}{4} = P(E_H) = P(E_D) = P(E_C).$

$\therefore$ E_S, E_H, E_D and E_C are exhaustive and they are also mutually exclusive.

$P(E/E_S)$ = probability of drawing 2 spades when one spade is missing

$$= \frac{{}^{12}C_2}{{}^{51}C_2} = \frac{12 \cdot 11}{2} \cdot \frac{2}{51 \cdot 50} = \frac{66}{1275}$$

$P(E/E_H)$ = probability of drawing 2 spades when one heart is missing

$$= \frac{{}^{13}C_2}{{}^{51}C_2} = \frac{13 \cdot 12}{2} \cdot \frac{2}{51 \cdot 50} = \frac{78}{1275}.$$

Similarly, $P(E/E_D) = P(E/E_C) = \dfrac{78}{1275}.$

We have to find $P(E_S/E)$, i.e., the probability of the missing card being spade when the two cards drawn are spades.

By Bayes' theorem,

$$P(E_S/E) = \frac{P(E_S) \cdot P(E/E_S)}{\{P(E_S) \cdot P(E/E_S) + P(E_H) \cdot P(E/E_H) + P(E_D) \cdot P(E/E_D) + P(E_C) \cdot P(E/E_C)\}}$$

$$= \frac{\frac{1}{4} \cdot \frac{66}{1275}}{\frac{1}{4} \cdot \frac{66}{1275} + \frac{1}{4} \cdot \frac{78}{1275} + \frac{1}{4} \cdot \frac{78}{1275} + \frac{1}{4} \cdot \frac{78}{1275}}$$

$$= \frac{66}{66 + 78 + 78 + 78} = \frac{66}{300} = \frac{11}{50}.$$

5. In a test, an examinee either guesses or copies or knows the answer to a multiple-choice question with four choices, only one answer being correct. The probability that he makes a guess is $\frac{1}{3}$ and the probability that he copies the answer is $\frac{1}{6}$. The probability that his answer is correct, given that he copies it, is $\frac{1}{8}$. Find the probability that he knew the answer to the question, given that he correctly answers it.

Let G = the event of answering by guess work

C = the event of answering by copying

K = the event of answering by knowing

A = the event of answering correctly.

From the question, $P(G) = \dfrac{1}{3}$, $P(C) = \dfrac{1}{6}$.

As the question is answered by either guessing or copying or knowing,

$$P(K) = 1 - \frac{1}{3} - \frac{1}{6} = \frac{1}{2}.$$

Now, $P(A/G)$ = probability of answering correctly by guessing

$= \dfrac{1}{4}$ {$\because$ there are 4 choices and only one of them is correct}

$P(A/C)$ = probability of answering correctly by copying

$= \frac{1}{8}$ (given)

$P(A/K)$ = probability of answering correctly by knowing

$= 1$ ($\because$ if he knows he certainly answers correctly).

We have to find $P(K/A)$, i.e., the probability that he knew when he answered correctly.

By Bayes' theorem,

$$P(K/A) = \frac{P(K) \cdot P(A/K)}{P(G) \cdot P(A/G) + P(C) \cdot P(A/C) + P(K) \cdot P(A/K)}$$

$$= \frac{\frac{1}{2} \cdot 1}{\frac{1}{3} \cdot \frac{1}{4} + \frac{1}{6} \cdot \frac{1}{8} + \frac{1}{2} \cdot 1}$$

$$= \frac{\frac{1}{2}}{\frac{1}{12} + \frac{1}{48} + \frac{1}{2}} = \frac{\frac{1}{2}}{\frac{29}{48}}$$

$$= \frac{24}{29}.$$

6. A speaks truth 3 times out of 4 while B, 7 times out of 10. A ball is drawn at random from a bag containing one black ball and five other balls of different colours. Both A and B report that a black ball has been drawn from the bag. Find the probability of their assertion being true?

Let B = the event of a black ball drawn from the bag

T = the event of both A and B speaking truth

L = the event of both A and B speaking lies.

Clearly, $P(B) = \frac{1}{6}$ and $P(\overline{B}) = 1 - \frac{1}{6} = \frac{5}{6}$.

Obviously, B and $\overline{B}$ are exhaustive and mutually exclusive.

Now, $P(T/B)$ = probability of both telling truth when a black ball is drawn

$= \frac{3}{4} \times \frac{7}{10}$

$\{\because$ probability of A telling truth $= \frac{3}{4}$,

probability of B telling truth $= \frac{7}{10}\}$

$P(L/\overline{B})$ = probability of both telling lies when a black ball is not drawn

$$= \left(1 - \frac{3}{4}\right)\left(1 - \frac{7}{10}\right) = \frac{1}{4} \cdot \frac{3}{10}.$$

We have to find $P(B/T)$, i.e., the probability that a black ball was drawn when they spoke truth.

By Bayes' theorem,

$$P(B/T) = \frac{P(B) \cdot P(T/B)}{P(B) \cdot P(T/B) + P(\overline{B}) \cdot P(L/\overline{B})}$$

$$= \frac{\frac{1}{6} \cdot \frac{3}{4} \cdot \frac{7}{10}}{\frac{1}{6} \cdot \frac{3}{4} \cdot \frac{7}{10} + \frac{5}{6} \cdot \frac{1}{4} \cdot \frac{3}{10}}$$

$$= \frac{21}{21 + 15} = \frac{21}{36} = \frac{7}{12}.$$

7. In a bag there are six balls of unknown colours; three balls are drawn at random and found to be all black. Find the probability that no black ball is left in the bag (Or Find the probability that the bag contained exactly 3 black balls.)

Let E_3 = the event of the bag containing 3 black balls and 3 balls of other colours

E_4 = the event of the bag containing 4 black balls and 2 balls of other colours

E_5 = the event of the bag containing 5 black balls and 1 ball of another colour

E_6 = the event of the bag containing 6 black balls.

Clearly, they are equally probable and no other possibilities exist as 3 balls drawn at random are black means the bag has at least 3 black balls.

$\therefore$ $P(E_1) = P(E_2) = P(E_3) = P(E_4) = \frac{1}{4}$ ($\because$ their sum = 1).

Let B = the event of drawing 3 black balls.

Now, $P(B/E_3)$ = probability of drawing 3 black balls when E_3 happens

$$= \frac{{}^3C_3}{{}^6C_3} = \frac{1}{20}.$$

Similarly, $P(B/E_4) = \frac{{}^4C_3}{{}^6C_3} = \frac{1}{5}$

$P(B/E_5) = \frac{{}^5C_3}{{}^6C_3} = \frac{1}{2}$

$P(B/E_6) = \frac{{}^6C_3}{{}^6C_3} = 1.$

The bag will contain no black balls after drawing 3 black balls if B took place in conjunction with E_3.

So, we have to find $P(E_3/B)$.

By Bayes' theorem for equiprobable events,

$$P(E_3/B) = \frac{P(B/E_3)}{P(B/E_3) + P(B/E_4) + P(B/E_5) + P(B/E_6)}$$

$$= \frac{\frac{1}{20}}{\frac{1}{20} + \frac{1}{5} + \frac{1}{2} + 1} = \frac{\frac{1}{20}}{\frac{35}{20}} = \frac{1}{35}.$$

8. An urn contains 6 black balls and unknown number (≤ 6) of white balls. Three balls are drawn successively and not replaced and are all found to be white. Prove that the chance that a black ball will be drawn in the next draw is $\frac{677}{909}$.

Clearly, the urn contains at least 3 white balls.

Let E_3 = the event of the urn containing 6 black balls and 3 white balls

E_4 = the event of the urn containing 6 black balls and 4 white balls

E_5 = the event of the urn containing 6 black balls and 5 white balls

E_6 = the event of the urn containing 6 black balls and 6 white balls.

Each of E_3, E_4, E_5, E_6 are equiprobable and they are exhaustive.

Now, $P(W/E_3)$ = probability of 3 balls drawn being white in case of E_3

$$= \frac{{}^3C_3}{{}^9C_3} = \frac{1}{\frac{9 \cdot 8 \cdot 7}{6}} = \frac{1}{84}.$$

Similarly, $P(W/E_4) = \frac{{}^4C_3}{{}^{10}C_3} = \frac{1}{30}$

$$P(W/E_5) = \frac{{}^5C_3}{{}^{11}C_3} = \frac{2}{33}$$

$$P(W/E_6) = \frac{{}^6C_3}{{}^{12}C_3} = \frac{1}{11}.$$

$\therefore$ by Bayes' theorem for equiprobable events,

$$P(E_3/W) = \frac{P(W/E_3)}{P(W/E_3) + P(W/E_4) + P(W/E_5) + P(W/E_6)}$$

$$= \frac{\frac{1}{84}}{\frac{1}{84} + \frac{1}{30} + \frac{2}{33} + \frac{1}{11}}$$

$$= \frac{1}{84} \times \frac{66 \times 70}{909} = \frac{55}{909}.$$

Similarly, $P(E_4/W) = \frac{\frac{1}{30}}{\frac{1}{84} + \frac{1}{30} + \frac{2}{33} + \frac{1}{11}}$

$$= \frac{1}{30} \times \frac{66 \times 70}{909} = \frac{154}{909}$$

$$P(E_5/W) = \frac{\frac{2}{33}}{\frac{1}{84} + \frac{1}{30} + \frac{2}{33} + \frac{1}{11}}$$

$$= \frac{2}{33} \times \frac{66 \times 70}{909} = \frac{280}{909}$$

$$P(E_6/W) = \frac{\frac{1}{11}}{\frac{1}{84} + \frac{1}{30} + \frac{2}{33} + \frac{1}{11}}$$

$$= \frac{1}{11} \times \frac{66 \times 70}{909} = \frac{420}{909} = \frac{140}{303}.$$

Let B = the event of drawing a black ball in the next draw.

Now, the ball in the next draw may be black after the 3 white balls were drawn from the bag containing 6 black and 3 white balls. Its probability is

$$P(E_3/W) \cdot P\{B/(E_3/W)\}, \text{ etc.}$$

$\therefore$ by the theorem on total conditional probability,

$$P(B) = P(E_3/W) \cdot P\{B/(E_3/W)\} + P(E_4/W) \cdot P\{B/(E_4/W)\} + P(E_5/W) \cdot P\{B/(E_5/W)\} + P(E_6/W) \cdot P\{B/(E_6/W)\} \quad \ldots (1)$$

Now, $P\{B/(E_3/W)\}$ = the probability of drawing 1 black in case of E_3 when 3 whites are already drawn

$$= \frac{{}^6C_1}{{}^6C_1} = 1$$

$$P\{B/(E_4/W)\} = \frac{{}^6C_1}{{}^7C_1} = \frac{6}{7}$$

$$P\{B/(E_5/W)\} = \frac{{}^6C_1}{{}^8C_1} = \frac{3}{4}$$

$$P\{B/(E_6/W)\} = \frac{{}^6C_1}{{}^9C_1} = \frac{2}{3}.$$

$\therefore$ using (1),

$$P(B) = \frac{55}{909} \times 1 + \frac{154}{909} \times \frac{6}{7} + \frac{280}{909} \times \frac{3}{4} + \frac{140}{303} \times \frac{2}{3}$$

$$= \frac{1}{909}(55 + 132 + 210 + 280) = \frac{677}{909}.$$

9. A purse contains 5 coins, each either a fifty-paisa coin or one-rupee coin. Two coins are drawn at random and are found to be one-rupee coins. Find the expected (probable) total value of the remaining coins.

Clearly, at least two of the coins are one-rupee coins. Let r stand for one-rupee coin and f stand for fifty-paisa coin.

The coins in the purse can be any one of the following:

$2r, 3f;\ \ 3r, 2f;\ \ 4r, 1f;\ \ 5r.$

Let E_2 = the event of the purse containing $2r$ and $3f$.

E_3 = the event of the purse containing $3r$ and $2f$; etc.

Clearly, the events are E_2, E_3, E_4 and E_5, and they are equally probable.

$$\therefore \quad P(E_2) = P(E_3) = P(E_4) = P(E_5) = \frac{1}{4}.$$

Also E_2, E_3, E_4 and E_5 are exhaustive and mutually exclusive.

Now, $P(E/E_2)$ = probability of drawing two one-rupee coins when E_2 happens

$$= \frac{{}^2C_2}{{}^5C_2} = \frac{1}{10}.$$

$$\text{Similarly, } P(E/E_3) = \frac{{}^3C_2}{{}^5C_2} = \frac{3}{10}$$

$$P(E/E_4) = \frac{{}^4C_2}{{}^5C_2} = \frac{6}{10} = \frac{3}{5}$$

$$P(E/E_5) = \frac{{}^5C_2}{{}^5C_2} = 1.$$

Now, $P(E_2/E)$ = probability of the purse containing $2r$ and $3f$ when $2r$ are drawn, etc.

By Bayes' theorem for equiprobable events,

$$P(E_2/E) = \frac{P(E/E_2)}{P(E/E_2) + P(E/E_3) + P(E/E_4) + P(E/E_5)}$$

$$= \frac{\frac{1}{10}}{\frac{1}{10} + \frac{3}{10} + \frac{3}{5} + 1} = \frac{1}{10} \cdot \frac{10}{20} = \frac{1}{20}.$$

$$\text{Similarly, } P(E_3/E) = \frac{\frac{3}{10}}{\frac{1}{10} + \frac{3}{10} + \frac{3}{5} + 1} = \frac{3}{10} \cdot \frac{10}{20} = \frac{3}{20}$$

$$P(E_4/E) = \frac{\frac{3}{5}}{\frac{1}{10} + \frac{3}{10} + \frac{3}{5} + 1} = \frac{3}{5} \cdot \frac{10}{20} = \frac{3}{10}$$

$$P(E_5/E) = \frac{1}{\frac{1}{10} + \frac{3}{10} + \frac{3}{5} + 1} = \frac{10}{20} = \frac{1}{2}.$$

$\therefore$ the expected total value of the remaining coins

$= P(E_2/E) \cdot (\text{value of } 3f)$

$+ P(E_3/E) \cdot (\text{value of } 1r \text{ and } 2f)$

$+ P(E_4/E) \cdot (\text{value of } 2r \text{ and } 1f)$

$+ P(E_5/E) \cdot (\text{value of } 3r)$

$$= \frac{1}{20} \cdot (\text{Rs } 1.50) + \frac{3}{20} (\text{Rs } 2) + \frac{3}{10} (\text{Rs } 2.50) + \frac{1}{2} (\text{Rs } 3)$$

$$= \text{Rs}\left(\frac{1}{20} \times \frac{3}{2} + \frac{3}{20} \times 2 + \frac{3}{10} \times \frac{5}{2} + \frac{3}{2}\right)$$

$$= \text{Rs}\left(\frac{3}{40} + \frac{3}{10} + \frac{3}{4} + \frac{3}{2}\right) = \text{Rs } 2.625.$$

Exercises

1. There are two bags, one of which contains three black and four white balls while the other contains four black and three white balls. A dice is cast. If the face 1 or 3 turns up, a ball is taken out from the first bag. But if any other face turns up, a ball is taken from the second bag. Find the probability of getting a black ball.

2. Three urns contain 2 white and 3 black balls, 3 white and 2 black balls; and 4 white and 1 black ball respectively. A ball is drawn from an urn chosen at random. What is the probability that a white ball drawn if the choices of urns are equiprobable?

3. An urn contains two balls each of which is either white or black. A white ball is added to the urn. What is the probability of drawing a white ball from the urn now?

4. The probability that a certain electronic component fails when first used is 0.10. If it does not fail immediately, the probability that it lasts for one year is 0.99. What is the probability that a new component will last one year?

5. A factory A produces 10% defective valves and another factory B produces 20% defective valves. A bag contains 4 valves of factory A and 5 valves of factory B. If two valves are drawn at random from the

bag, find the probability that at least one valve is defective.

6. An unbiased coin is tossed. If the result is a head, a pair of unbiased dice is rolled and the number obtained by adding the numbers shown on them is noted. If the result is a tail, a card from a well-shuffled pack of eleven cards numbered 2, 3, 4, ..., 12 is picked and the number of the card is noted. What is the chance that the noted number is either 7 or 8?

7. A bolt factory has three machines A, B and C manufacturing respectively 25%, 35% and 40% of the total production. Of these the machines produce 5%, 4% and 2% defective bolts respectively. A bolt is selected at random and it is found to be defective. Find the probability that it was manufactured by the machine (a) A (b) B (c) C.

8. There are three bags each containing 5 white and 2 black balls. Also, there are 2 bags each containing 1 white and 4 black balls. A ball is drawn at random from a bag and it is found to be black. What is the chance that the black ball came from the first group?

9. A can hit a target 4 times in 5 shots; B can hit 3 times in 4 shots and C twice in 3 shots. They fire once each. If two of them hit, what is the chance that C has missed it?

10. A bag contains 5 balls of unknown colours. A ball is drawn and replaced twice. On each occasion it is found to be red. Again, two balls are drawn at a time. What is the probability of both the balls being red?

11. A letter is known to have come from either MAHARASTRA or MADRAS. On the postmark only consecutive letters RA can be read clearly. What is the chance that the letter came from MAHARASTRA?

12. A bag contains 10 coins of which at least 2 are one-rupee coins. Two coins are drawn and both are found to be not one-rupee coins. What is the probability of the bag to contain exactly 2 one-rupee coins?

13. A man has three coins A, B, C. The coin A is unbiased. The probability that a head will show when B is tossed is $\frac{2}{3}$ while it is $\frac{1}{3}$ in case of the coin C. A coin is chosen at random and tossed 3 times giving 2 heads and 1 tail. Find the probability that the coin A was chosen.

14. A person goes to office either by car, scooter, bus or train and the probabilities of these are $\frac{1}{7}, \frac{3}{7}, \frac{2}{7}$ and $\frac{1}{7}$ respectively. The probability that he reaches office late if he takes car, scooter, bus or train is $\frac{2}{9}, \frac{1}{9}, \frac{4}{9}$ or $\frac{1}{9}$. Given that he reached office in time, find the probability that he travelled by a car.

Objective Questions

Fill in the blanks.

15. The probability that in a year of the 22nd century, chosen at random there will be 53 Sundays, is ______ .

16. There are two bags X and Y; X contains 3 white balls and 2 black balls, and Y contains 2 white balls and 4 black balls. A bag and a ball out of that bag are picked at random. The probability that the ball is white, is ______ .

17. There are two coupons each with a word written on them. On one of them the word PATNA is written and on the other PLATE. A coupon is taken at random and 3 letters are selected at random from the letters of the word on the coupon. The probability that the selection contains two vowels, is ______ .

18. Two persons X and Y throw two dice each. The probability that Y throws a sum greater than X is ______ when it is known that X throws a sum of at least 8.

Answers

1. $\frac{11}{21}$ 2. $\frac{3}{5}$ 3. $\frac{2}{3}$ 4. 0.891

5. 0.29 6. $\frac{193}{792}$ 7. (a) $\frac{25}{69}$ (b) $\frac{28}{69}$ (c) $\frac{16}{69}$

8. $\frac{15}{43}$ 9. $\frac{6}{13}$ 10. $\frac{377}{550}$ 11. $\frac{10}{19}$

12. $\frac{1}{3}$ 13. $\frac{9}{25}$ 14. $\frac{1}{7}$

15. $\frac{5}{28}$ 16. $\frac{7}{15}$ 17. $\frac{51}{140}$

18. $\frac{17}{108}$

□

Chapter Test

Time: 75 minutes

1. A bag contains 3 white and 3 black balls. A person draws 3 balls at random from it. He then drops 3 red balls into the bag and again draws out 3 balls at random. What is the chance that the latter 3 balls will be of different colours?
2. A bag contains n coins of unknown values. A coin is drawn at random and it is found to be a rupee. What is the chance that it is the only rupee coin in the bag?
3. In a city filmland there are 3 male superstars and 3 female superstars. A director decides to make a film with 3 superstars. If the film has 2 male superstars and 1 female superstar, the chance of the film to be a hit is $\frac{1}{3}$ while the chance is $\frac{2}{5}$ if there are 1 male and 2 female superstars. If the 3 superstars are from the same sex, the chance of the film being a hit is $\frac{2}{15}$. At the time of making the film, only 3 superstars are available to sign for the film. If the director takes them all then what is the probability of his film being a hit?
4. Three groups A, B and C are competing for positions on the Board of Directors of a company. The probabilities of their winning are 0.5, 0.3 and 0.2 respectively. The probabilities of introducing a new product if a particular group wins are respectively 0.7, 0.6 and 0.5. Find the probability that a new product will be introduced.
5. A purse containing 16 notes, four each of denominations Rs 10, Rs 20, Rs 50 and Rs 100, is left on a table. A maid servant steals two of the notes. After that the man takes out two notes at random from the purse and finds that they amount to Rs 200. What is the probability that an amount of Rs 200 was stolen by the maid servant from his purse?
6. An urn containing 6 white balls and 4 black balls. A ball is drawn at random and is put back into the urn along with 5 balls of the same colour as that of the ball drawn. A ball is drawn again at random. What is the probability that the ball drawn now is white?
7. A bag contains p white and q red balls. $m(m < p, m < q)$ balls are drawn from the bag one by one without replacement. Prove that the probability of the mth ball being red is $\frac{q}{p+q}$.
8. A box contains N coins, m of which are fair and the rest are biased. The probability of getting a head when a biased coin is tossed is $\frac{2}{3}$. A coin is drawn at random from the box and tossed twice. The first time it shows head and the second time it shows tail. What is the probability that the coin drawn is fair?
9. For a student to qualify, he must pass at least two out of three examinations. The probability that he will pass the first examination is p. If he fails in one of the examinations then the probability of his passing the next examination is $\frac{p}{2}$, otherwise it remains the same. Find the probability that he will qualify.

□

Answers

1. $\dfrac{27}{100}$

2. $\dfrac{2}{n(n+1)}$

3. $\dfrac{103}{300}$

4. 0.63

5. $\dfrac{1}{91}$

6. $\dfrac{3}{5}$

8. $\dfrac{9m}{m+8N}$

9. $2p^2 - p^3$

□

Coordinate Geometry of Three Dimensions

1. Points, Direction Cosines and Direction Ratios

Recap of Facts and Formulae

1. Position of a point in 3 dimensions

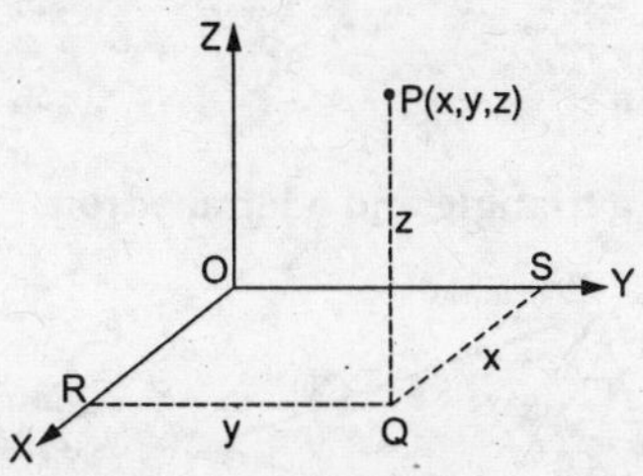

- The coordinates of a point P are (x, y, z) means that
  the distance of P from the y–z plane $= x$ $(= QS)$
  the distance of P from the z–x plane $= y$ $(= QR)$
  the distance of P from the x–y plane $= z$ $(= PQ)$

where three mutually perpendicular lines OX, OY and OZ are the x, y and z axes respectively, O being the origin.

2. Location of a point

(a) The three planes of axes divide the space in eight quadrants. The location of a point is decided by the signs as well as magnitudes of its coordinates.

(b) The point $P(x, y, z)$ is on the

x-axis $\Rightarrow y = 0,\ z = 0$

y-axis $\Rightarrow z = 0,\ x = 0$

z-axis $\Rightarrow x = 0,\ y = 0$

3. Relation between cartesian coordinates (x, y, z) and polar spherical coordinates (r, θ, ϕ)

- If $P = (x, y, z)$ be a point whose polar spherical coordinates are (r, θ, ϕ) then

$x = r\sin\theta \cdot \cos\phi$,

$y = r\sin\theta \cdot \sin\phi,\ z = r\cos\theta$

where the distance of the point P from the origin O is r, the inclination of OP with the positive direction of the z-axis is θ, and the inclination of OQ (Q being the foot of the perpendicular from P to the x–y plane) with the positive direction of the x-axis is ϕ.

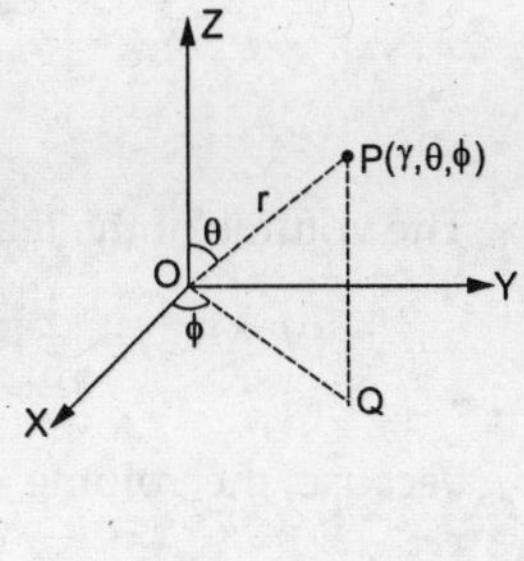

4. Relation between the position vector of a point and its cartesian coordinates

- If P be a point whose cartesian coordinates are (x, y, z) and the position vector is $\vec{r}$ then

$$\vec{r} = x\vec{i} + y\vec{j} + z\vec{k}$$

where $\vec{i}, \vec{j}, \vec{k}$ are unit vectors along x, y, z axes respectively.

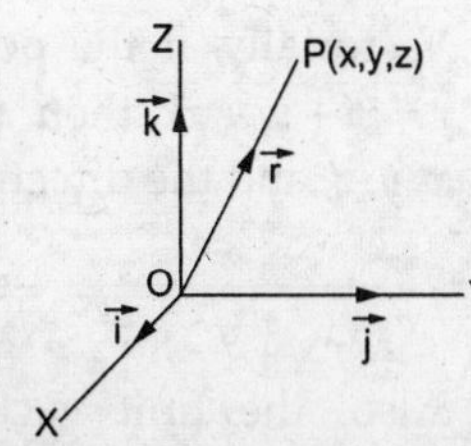

5. Distance between two points

- If $P = (x_1, y_1, z_1)$ and $Q = (x_2, y_2, z_2)$ be two points then their position vectors are $\vec{r_1} = x_1\vec{i} + y_1\vec{j} + z_1\vec{k}$ and $\vec{r_2} = x_2\vec{i} + y_2\vec{j} + z_2\vec{k}$ respectively, and

$$PQ = |\vec{r_1} - \vec{r_2}| = |(x_1 - x_2)\vec{i} + (y_1 - y_2)\vec{j} + (z_1 - z_2)\vec{k}|$$

$$= \sqrt{(x_1 - x_2)^2 + (y_1 - y_2)^2 + (z_1 - z_2)^2}.$$

6. Section formula

- If $P = (x_1, y_1, z_1)$ and $Q = (x_2, y_2, z_2)$ be two points and $R = (x, y, z)$ divides PQ in the ratio $\lambda : 1$ then

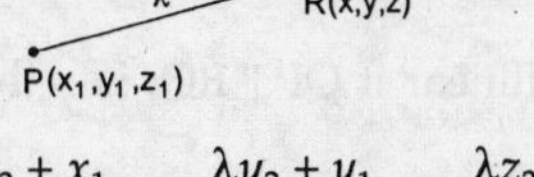

$$x = \frac{\lambda x_2 + x_1}{\lambda + 1},\ y = \frac{\lambda y_2 + y_1}{\lambda + 1},\ z = \frac{\lambda z_2 + z_1}{\lambda + 1}.$$

Vectorially, $\vec{r} = \dfrac{\lambda\vec{r_2} + \vec{r_1}}{\lambda + 1}$ where $\vec{r_1}, \vec{r_2}$ and $\vec{r}$ are the position vectors of P, Q and R respectively.

$\lambda > 0$ if R divides PQ internally, and $\lambda < 0$ if the division is external.

7. Direction cosines and direction ratios of a line

- The cosines of the angles that a line makes with the positive directions of x, y and z axes are called its direction cosines (or d.c.). So, the direction cosines of a line making angles of α, β and γ with the three axes respectively, are $\cos\alpha$, $\cos\beta$, $\cos\gamma$, and $\cos^2\alpha + \cos^2\beta + \cos^2\gamma = 1$.

  If λ, μ, ν are three numbers such that
  $$\frac{\lambda}{\cos\alpha} = \frac{\mu}{\cos\beta} = \frac{\nu}{\cos\gamma}$$
  then λ, μ, ν {also written as (λ, μ, ν)} are called the direction ratios (or d.r.) or direction numbers of the line.

  If λ, μ, ν are direction ratios of a line then its direction cosines are
  $$\frac{\lambda}{\sqrt{\lambda^2+\mu^2+\nu^2}}, \frac{\mu}{\sqrt{\lambda^2+\mu^2+\nu^2}}, \frac{\nu}{\sqrt{\lambda^2+\mu^2+\nu^2}}.$$
  Vectorially, if the position vector of the point P be $\vec{r} = x\vec{i} + y\vec{j} + z\vec{k}$ then the direction ratios of OP are x, y, z, and the direction cosines are
  $$\frac{x}{\sqrt{x^2+y^2+z^2}}, \frac{y}{\sqrt{x^2+y^2+z^2}}, \frac{z}{\sqrt{x^2+y^2+z^2}}.$$
  Also, the unit vector along OP whose direction cosines are l, m, n is $l\vec{i} + m\vec{j} + n\vec{k}$, and a vector along OP whose direction ratios are λ, μ, ν is $\lambda\vec{i} + \mu\vec{j} + \nu\vec{k}$.

8. Direction ratios and direction cosines of a line joining two points

- Let $P = (x_1, y_1, z_1)$ and $Q = (x_2, y_2, z_2)$. Then
  $$\vec{PQ} = (x_2 - x_1)\vec{i} + (y_2 - y_1)\vec{j} + (z_2 - z_1)\vec{k}.$$
  So, the direction ratios of the line PQ are $x_2 - x_1$, $y_2 - y_1$, $z_2 - z_1$. Hence, the direction cosines of PQ are
  $$\frac{x_2 - x_1}{\sqrt{\Sigma(x_2 - x_1)^2}}, \frac{y_2 - y_1}{\sqrt{\Sigma(x_2 - x_1)^2}}, \frac{z_2 - z_1}{\sqrt{\Sigma(x_2 - x_1)^2}}.$$

9. Condition of collinearity of three points

- Let $P(x_1, y_1, z_1)$, $Q(x_2, y_2, z_2)$ and $R(x_3, y_3, z_3)$ be three points.

  They will be collinear if QP || RQ, i.e., they have the same direction.

  Direction ratios of QP = $x_1 - x_2, y_1 - y_2, z_1 - z_2$.

  Direction ratios of RQ = $x_2 - x_3, y_2 - y_3, z_2 - z_3$.

  For the same direction, the direction numbers will be proportional.

  $\therefore$ the condition of collinearity is
  $$\frac{x_1 - x_2}{x_2 - x_3} = \frac{y_1 - y_2}{y_2 - y_3} = \frac{z_1 - z_2}{z_2 - z_3}.$$

10. Area of a triangle whose vertices are given

- Let $A(x_1, y_1, z_1)$, $B(x_2, y_2, z_2)$ and $C(x_3, y_3, z_3)$ be vertices of a ΔABC.
  $$\therefore\ \vec{AB} = (x_2 - x_1)\vec{i} + (y_2 - y_1)\vec{j} + (z_2 - z_1)\vec{k},$$
  $$\vec{AC} = (x_3 - x_1)\vec{i} + (y_3 - y_1)\vec{j} + (z_3 - z_1)\vec{k}.$$
  $$\text{ar}(\Delta ABC) = \left|\frac{1}{2}\vec{AB} \times \vec{AC}\right|$$
  $$= \frac{1}{2}\left|\begin{vmatrix} \vec{i} & \vec{j} & \vec{k} \\ x_2 - x_1 & y_2 - y_1 & z_2 - z_1 \\ x_3 - x_1 & y_3 - y_1 & z_3 - z_1 \end{vmatrix}\right|.$$
  The points A, B and C will be collinear if and only if $\text{ar}(\Delta ABC) = 0$.

11. Centroids of a triangle and a tetrahedron

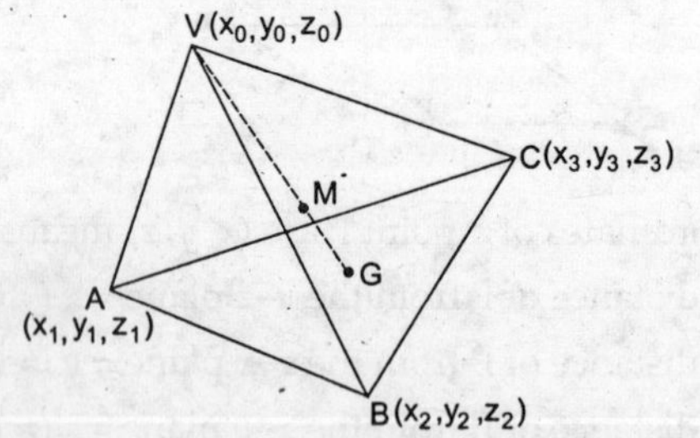

- Let $A = (x_1, y_1, z_1)$, $B = (x_2, y_2, z_2)$ and $C = (x_3, y_3, z_3)$ be the vertices of a triangle. The centroid of ΔABC is
  $$G = \left(\frac{x_1 + x_2 + x_3}{3}, \frac{y_1 + y_2 + y_3}{3}, \frac{z_1 + z_2 + z_3}{3}\right).$$
  If $V(x_0, y_0, z_0)$ be a point, not in the plane of the ΔABC, then we get a tetrahedron (V, ABC) whose centroid is
  $$M = \left(\frac{x_0 + x_1 + x_2 + x_3}{4}, \frac{y_0 + y_1 + y_2 + y_3}{4}, \frac{z_0 + z_1 + z_2 + z_3}{4}\right).$$
  Remember, $VM : MG = 3 : 1$.

12. Volume of a tetrahedron

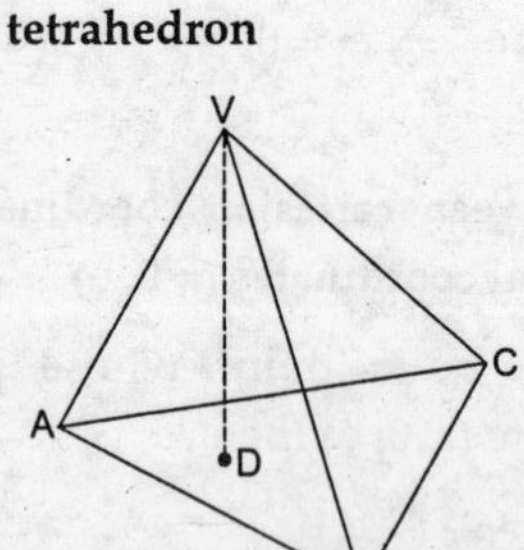

- The volume of the tetrahedron
  $$(V, ABC) = \left|\frac{1}{6}[\vec{VA}\ \vec{VB}\ \vec{VC}]\right|$$
  because, the volume $= \frac{1}{3}\text{ar}(\Delta ABC) \times$ height VD

$$= \frac{1}{3} \cdot \frac{1}{2} \, |\overrightarrow{AB} \times \overrightarrow{AC}| \cdot \text{ Projection of VA on VD}$$

$$= \frac{1}{6} \, |\overrightarrow{AB} \times \overrightarrow{AC}| \left(\overrightarrow{VA} \cdot \frac{\overrightarrow{AB} \times \overrightarrow{AC}}{|\overrightarrow{AB} \times \overrightarrow{AC}|} \right)$$

$$= \frac{1}{6} \overrightarrow{VA} \cdot (\overrightarrow{AB} \times \overrightarrow{AC}) = \frac{1}{6} \overrightarrow{VA} \cdot \{(\overrightarrow{VB} - \overrightarrow{VA}) \times (\overrightarrow{VC} - \overrightarrow{VA})\}$$

$$= \frac{1}{6} \overrightarrow{VA} \cdot \{\overrightarrow{VB} \times \overrightarrow{VC} - \overrightarrow{VB} \times \overrightarrow{VA} - \overrightarrow{VA} \times \overrightarrow{VC}\}$$

$$= \frac{1}{6} \overrightarrow{VA} \cdot (\overrightarrow{VB} \times \overrightarrow{VC}) = \frac{1}{6} [\overrightarrow{VA} \;\; \overrightarrow{VB} \;\; \overrightarrow{VC}].$$

$\therefore$ if the vertices are (x_r, y_r, z_r); $r = 1, 2, 3, 4$; then the volume of the tetrahedron

$$= \left| \frac{1}{6} \begin{vmatrix} x_2 - x_1 & y_2 - y_1 & z_2 - z_1 \\ x_3 - x_1 & y_3 - y_1 & z_3 - z_1 \\ x_4 - x_1 & y_4 - y_1 & z_4 - z_1 \end{vmatrix} \right|.$$

13. Condition of coplanarity of four points

- Four points (x_r, y_r, z_r); $r = 1, 2, 3, 4$; will be coplanar if the volume of the tetrahedron with the points as vertices is zero.

$\therefore$ the condition of coplanarity of the points (x_1, y_1, z_1), (x_2, y_2, z_2), (x_3, y_3, z_3) and (x_4, y_4, z_4) is

$$\begin{vmatrix} x_2 - x_1 & y_2 - y_1 & z_2 - z_1 \\ x_3 - x_1 & y_3 - y_1 & z_3 - z_1 \\ x_4 - x_1 & y_4 - y_1 & z_4 - z_1 \end{vmatrix} = 0.$$

14. Locus and its equation

- If a point P(x, y, z) changes its position in space under some geometrical condition(s) then the path traced out by the point is called a locus. The functional relation(s) between x, y, z imposed by the geometrical condition(s) constitute the equation of the locus of the point. A locus of a point in three dimensions is a surface or an intersection of surfaces.

15. Angle between two lines

(a) If two lines have d.c. l_1, m_1, n_1 and l_2, m_2, n_2 then unit vectors along the lines are $\hat{a}_1 = l_1\vec{i} + m_1\vec{j} + n_1\vec{k}$ and $\hat{a}_2 = l_2\vec{i} + m_2\vec{j} + n_2\vec{k}$. If θ be the angle between the lines then

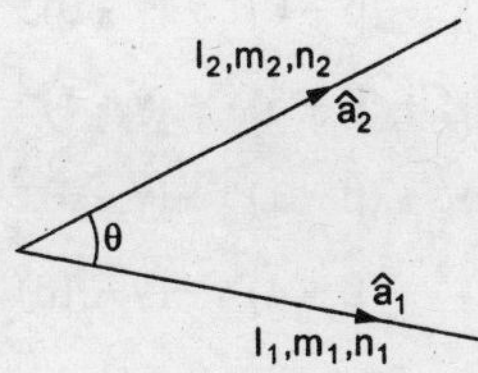

$$\cos\theta = \hat{a}_1 \cdot \hat{a}_2 = l_1 l_2 + m_1 m_2 + n_1 n_2,$$

$$\sin\theta = |\hat{a}_1 \times \hat{a}_2| = \left| \begin{vmatrix} \vec{i} & \vec{j} & \vec{k} \\ l_1 & m_1 & n_1 \\ l_2 & m_2 & n_2 \end{vmatrix} \right|$$

$$= |(m_1 n_2 - m_2 n_1)\vec{i} + (n_1 l_2 - n_2 l_1)\vec{j} + (l_1 m_2 - l_2 m_1)\vec{k}|$$

$$= \sqrt{(m_1 n_2 - m_2 n_1)^2 + (n_1 l_2 - n_2 l_1)^2 + (l_1 m_2 - l_2 m_1)^2}.$$

(b) *Condition of perpendicularity* Two lines with d.c. (or d.r.) l_1, m_1, n_1 and l_2, m_2, n_2 are perpendicular ($\theta = \pi/2$) if $l_1 l_2 + m_1 m_2 + n_1 n_2 = 0$.

Condition of parallelism Two lines with d.c. (or d.r.) l_1, m_1, n_1 and l_2, m_2, n_2 are parallel ($\theta = 0, \pi$) if $m_1 n_2 - m_2 n_1 = 0$, $n_1 l_2 - n_2 l_1 = 0$, $l_1 m_2 - l_2 m_1 = 0$

i.e., $\dfrac{l_1}{l_2} = \dfrac{m_1}{m_2} = \dfrac{n_1}{n_2}$.

16. Projection of the join of two points on a line

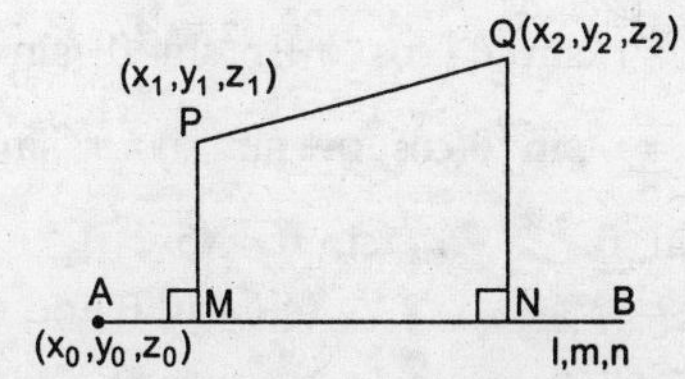

- Let P(x_1, y_1, z_1) and Q(x_2, y_2, z_2) be two points and AB be a given line passing through A(x_0, y_0, z_0) and having d.c. l, m, n. PM and QN are perpendiculars to AB. Then the projection of PQ on the line AB

$$= MN = |\overrightarrow{PQ} \cdot \hat{n}| \text{ where } \hat{n} = \text{unit vector along AB}$$

$$= |\{(x_2 - x_1)\vec{i} + (y_2 - y_1)\vec{j} + (z_2 - z_1)\vec{k}\} \cdot (l\vec{i} + m\vec{j} + n\vec{k})|$$

$$= |(x_2 - x_1) l + (y_2 - y_1) m + (z_2 - z_1) n|.$$

Note If the projection of the line segment PQ on the line AB is zero then PQ $\perp$ AB.

17. Condition of coplanarity of four points (using direction numbers)

- The points A(x_1, y_1, z_1), B(x_2, y_2, z_2), C(x_3, y_3, z_3) and D(x_4, y_4, z_4) are coplanar if AB, AC, AD are coplanar. It will be so if a line with direction ratios l, m, n can be found which is perpendicular to AB, AC and AD, i.e.,

$$l(x_2 - x_1) + m(y_2 - y_1) + n(z_2 - z_1) = 0$$

$$l(x_3 - x_1) + m(y_3 - y_1) + n(z_3 - z_1) = 0$$

$$l(x_4 - x_1) + m(y_4 - y_1) + n(z_4 - z_1) = 0$$

Eliminating l, m, n, the condition of coplanarity is

$$\begin{vmatrix} x_2 - x_1 & y_2 - y_1 & z_2 - z_1 \\ x_3 - x_1 & y_3 - y_1 & z_3 - z_1 \\ x_4 - x_1 & y_4 - y_1 & z_4 - z_1 \end{vmatrix} = 0.$$

Selected Solved Examples

1. Let PM be the perpendicular from the point P(1, 2, 3) to the *x-y* plane. If OP makes an angle θ with the positive direction of the *z*-axis and OM makes an angle φ with the positive direction of the *x*-axis, where O is the origin, then find θ and φ.

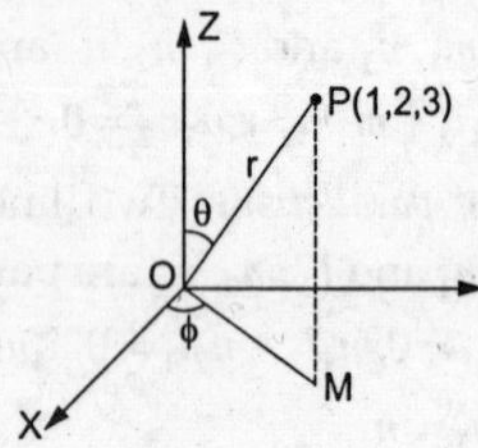

We know that if P = (*x*, *y*, *z*) then

$$x = r\sin\theta\cdot\cos\phi,\ y = r\sin\theta\cdot\sin\phi,\ z = r\cos\theta.$$

$\therefore\quad 1 = r\sin\theta\cdot\cos\phi,\ 2 = r\sin\theta\cdot\sin\phi,\ 3 = r\cos\theta.$

$\Rightarrow\quad 1^2 + 2^2 = r^2\sin^2\theta\cdot\cos^2\phi + r^2\sin^2\theta\cdot\sin^2\phi$

$= r^2\sin^2\theta(\cos^2\phi + \sin^2\phi) = r^2\sin^2\theta.$

$\therefore\quad 5 = r^2\sin^2\theta.\qquad \therefore\ r\sin\theta = \sqrt{5}$

(clearly, θ and φ are acute).

$\therefore\quad \dfrac{r\sin\theta}{r\cos\theta} = \dfrac{\sqrt{5}}{3}\quad$ or $\quad\tan\theta = \dfrac{\sqrt{5}}{3}.$

$\therefore\quad \theta = \tan^{-1}\left(\dfrac{\sqrt{5}}{3}\right).$

Also, $\dfrac{1}{2} = \dfrac{r\sin\theta\cdot\cos\phi}{r\sin\theta\cdot\sin\phi}.\quad \therefore\ \cot\phi = \dfrac{1}{2}.\quad \therefore\ \tan\phi = 2.$

$\therefore\quad \phi = \tan^{-1}2.$

2. Find the possible values of λ for which the triangle with vertices (6, 10, 10), (1, 0, –5) and (6, –10, λ) will be a right-angled triangle. Also, find the centroid of the right-angled triangle in each case.

Let the points be A, B and C respectively. Three points are always coplanar.

$AB^2 = (6-1)^2 + (10-0)^2 + (10+5)^2$

$= 25 + 100 + 225 = 350.$

$AC^2 = (6-6)^2 + (10+10)^2 + (10-\lambda)^2$

$= 400 + 100 - 20\lambda + \lambda^2 = 500 - 20\lambda + \lambda^2.$

$BC^2 = (1-6)^2 + (0+10)^2 + (-5-\lambda)^2$

$= 125 + 25 + 10\lambda + \lambda^2 = 150 + 10\lambda + \lambda^2.$

Now, $AB^2 + AC^2 = BC^2$

$\Rightarrow\quad 350 + 500 - 20\lambda + \lambda^2 = 150 + 10\lambda + \lambda^2$

$\Rightarrow\quad 30\lambda = 700.\qquad \therefore\ \lambda = \dfrac{70}{3}.$

Next, $AB^2 + BC^2 = AC^2$

$\Rightarrow\quad 350 + 150 + 10\lambda + \lambda^2 = 500 - 20\lambda + \lambda^2$

$\Rightarrow\quad 30\lambda = 0.\qquad \therefore\ \lambda = 0.$

Again, $BC^2 + AC^2 = AB^2$

$\Rightarrow\quad 150 + 10\lambda + \lambda^2 + 500 - 20\lambda + \lambda^2 = 350$

$\Rightarrow\quad \lambda^2 - 5\lambda + 150 = 0$, which gives imaginary values of λ.

$\therefore\quad$ the triangle is right angled for $\lambda = 0, \dfrac{70}{3}.$

Now, centroid $= \left(\dfrac{x_1+x_2+x_3}{3}, \dfrac{y_1+y_2+y_3}{3}, \dfrac{z_1+z_2+z_3}{3}\right)$

$= \left(\dfrac{6+1+6}{3}, \dfrac{10+0-10}{3}, \dfrac{10-5+\lambda}{3}\right) = \left(\dfrac{13}{3}, 0, \dfrac{5+\lambda}{3}\right)$

$= \left(\dfrac{13}{3}, 0, \dfrac{5+0}{3}\right)$ or $\left(\dfrac{13}{3}, 0, \dfrac{5+\frac{70}{3}}{3}\right)$

$= \left(\dfrac{13}{3}, 0, \dfrac{5}{3}\right)$ or $\left(\dfrac{13}{3}, 0, \dfrac{85}{9}\right).$

3. Three vertices of a tetrahedron are (0, 0, 0), (6, –5, –1) and (–4, 1, 3). If the centroid of the tetrahedron be (1, –2, 5) then find the fourth vertex. Also, find the radius of the sphere circumscribing the tetrahedron and the volume of the tetrahedron.

Let the fourth vertex be (*x*, *y*, *z*). Then the centroid of the tetrahedron is $\left(\dfrac{0+6-4+x}{4}, \dfrac{0-5+1+y}{4}, \dfrac{0-1+3+z}{4}\right)$

i.e., $\left(\dfrac{2+x}{4}, \dfrac{y-4}{4}, \dfrac{z+2}{4}\right)$

$\therefore\quad$ from the question, $\left(\dfrac{2+x}{4}, \dfrac{y-4}{4}, \dfrac{z+2}{4}\right) = (1, -2, 5)$

$\Rightarrow\quad \dfrac{2+x}{4} = 1,\ \dfrac{y-4}{4} = -2,\ \dfrac{z+2}{4} = 5$

$\Rightarrow\quad x = 2,\ y = -4,\ z = 18.$

$\therefore\quad$ the fourth vertex is (2, –4, 18).

Let *r* be the radius of the sphere and (α, β, γ) be the centre. Then

$r^2 = (\alpha-0)^2 + (\beta-0)^2 + (\gamma-0)^2$

$= (\alpha-6)^2 + (\beta+5)^2 + (\gamma+1)^2$

$= (\alpha+4)^2 + (\beta-1)^2 + (\gamma-3)^2$

$= (\alpha-2)^2 + (\beta+4)^2 + (\gamma-18)^2$

$\therefore\quad -12\alpha + 10\beta + 2\gamma + 62 = 0$

i.e., $-6\alpha + 5\beta + \gamma + 31 = 0$ ... (1)

and $8\alpha - 2\beta - 6\gamma + 26 = 0$ i.e., 4 13 = 0 ... (2)

and $-4\alpha + 8\beta - 36\gamma + 344 = 0$

i.e., $-\alpha + 2\beta - 9\gamma + 86 = 0$ ... (3)

$(1) - 6 \times (3) \Rightarrow -7\beta + 55\gamma - 485 = 0$ ... (4)

$(2) + 4 \times (3) \Rightarrow 7\beta - 39\gamma + 357 = 0$ ... (5)

$(4) + (5) \Rightarrow 16\gamma - 128 = 0 \Rightarrow \gamma = \frac{128}{16} = 8.$

Putting this value in (5), $7\beta - 39 \times 8 + 357 = 0$

$\Rightarrow 7\beta + 45 = 0. \quad \therefore \beta = \frac{-45}{7}.$

$\therefore$ (3) $\Rightarrow -\alpha + 2\left(-\frac{45}{7}\right) - 9 \times 8 + 86 = 0$

$\Rightarrow \alpha = -\frac{90}{7} - 72 + 86 = \frac{8}{7}.$

$$\therefore r^2 = \alpha^2 + \beta^2 + \gamma^2 = \left(\frac{8}{7}\right)^2 + \left(-\frac{45}{7}\right)^2 + 8^2$$

$$= \frac{8^2 + 45^2}{49} + 64 = \frac{64 + 2025}{49} + 64$$

$$= \frac{64 \times 50 + 2025}{49} = \frac{3200 + 2025}{49}$$

$$= \frac{5225}{49} = \frac{209 \times 25}{49}$$

$$\therefore r = \sqrt{\frac{209 \times 25}{49}} = \frac{5}{7} \cdot \sqrt{209}.$$

Now, the four vertices are (0, 0, 0), (6, –5, –1), (–4, 1, 3) and (2, –4, 18).

So, the volume of the tetrahedron

$$= \left| \frac{1}{6} \begin{vmatrix} 6-0 & -5-0 & -1-0 \\ -4-0 & 1-0 & 3-0 \\ 2-0 & -4-0 & 18-0 \end{vmatrix} \right|$$

$$= \left| \frac{1}{6} \begin{vmatrix} 6 & -5 & -1 \\ -4 & 1 & 3 \\ 2 & -4 & 18 \end{vmatrix} \right|$$

$$= \left| \frac{1}{6} \begin{vmatrix} 0 & 7 & -55 \\ 0 & -7 & 39 \\ 2 & -4 & 18 \end{vmatrix} \right|, \quad R_1 \to R_1 - 3 \times R_3, \; R_2 \to R_2 + 2 \times R_3$$

$$= \left| \frac{1}{6} \cdot 2 \cdot \begin{vmatrix} 7 & -55 \\ -7 & 39 \end{vmatrix} \right| = \left| \frac{1}{3} \cdot \begin{vmatrix} 7 & -55 \\ 0 & -16 \end{vmatrix} \right|$$

$$= \frac{1}{3} \cdot 7 \cdot 16 = \frac{112}{3} = 37\frac{1}{3}.$$

4. If the points (0, 1, –2), (3, λ, –1) and (μ, –3, –4) are collinear, verify whether the point (12, 9, 2) is also on the same line.

Let the points be A, B and C respectively.
Let D = (12, 9, 2).

Now, direction ratios of

$AB = 3 - 0, \lambda - 1, -1 - (-2) = 3, \lambda - 1, 1,$

direction ratios of AC

$= \mu - 0, -3 - 1, -4 - (-2) = \mu, -4, -2.$

A, B and C are collinear $\Rightarrow$ AB || AC

$$\Rightarrow \frac{3}{\mu} = \frac{\lambda - 1}{-4} = \frac{1}{-2}$$

$\therefore \mu = -6, \lambda = 3.$

$\therefore$ direction ratios of AB are 3, 3 – 1, 1, i.e., 3, 2, 1.

Now, direction ratios of AD are 12 – 0, 9 – 1, 2 – (–2), i.e., 12, 8, 4.

AB || AD if $\frac{3}{12} = \frac{2}{8} = \frac{1}{4}$, which is true.

So, AB and AD lie on the same line, and hence the point (12, 9, 2) is on the same line.

5. Show that the line segments joining the points (4, 7, 8), (–1, –2, 1), and (2, 3, 4), (1, 2, 5) intersect. Verify whether the four points are concyclic.

Let the points be A, B and C, D respectively. If possible let them intersect at the point P and $AP : PB = \lambda : 1$

and $CP : PD = \mu : 1$.

$$\text{Then } P = \left(\frac{\lambda(-1) + 1.4}{\lambda + 1}, \frac{\lambda(-2) + 1.7}{\lambda + 1}, \frac{\lambda(1) + 1.8}{\lambda + 1}\right)$$

$$= \left(\frac{\mu(1) + 1.2}{\mu + 1}, \frac{\mu(2) + 1.3}{\mu + 1}, \frac{\mu(5) + 1.4}{\mu + 1}\right)$$

$$\Rightarrow \frac{-\lambda + 4}{\lambda + 1} = \frac{\mu + 2}{\mu + 1}, \frac{-2\lambda + 7}{\lambda + 1} = \frac{2\mu + 3}{\mu + 1}, \frac{\lambda + 8}{\lambda + 1} = \frac{5\mu + 4}{\mu + 1}$$

$$\Rightarrow \frac{-\lambda - 1 + 5}{\lambda + 1} = \frac{\mu + 1 + 1}{\mu + 1}, \frac{-2\lambda - 2 + 9}{\lambda + 1} = \frac{2\mu + 2 + 1}{\mu + 1}, \frac{\lambda + 1 + 7}{\lambda + 1} = \frac{5\mu + 5 - 1}{\mu + 1}$$

$$\Rightarrow -1 + \frac{5}{\lambda + 1} = 1 + \frac{1}{\mu + 1}, -2 + \frac{9}{\lambda + 1} = 2 + \frac{1}{\mu + 1}, 1 + \frac{7}{\lambda + 1} = 5 - \frac{1}{\mu + 1}$$

$$\Rightarrow -1 + 5p = 1 + q, -2 + 9p = 2 + q, 1 + 7p = 5 - q$$

where $p = \frac{1}{\lambda + 1}, q = \frac{1}{\mu + 1}.$

Thus we have, $5p - q - 2 = 0$ ...(1)

$9p - q - 4 = 0$ ... (2)

$7p + q - 4 = 0$ ... (3)

$(1) - (2) \Rightarrow -4p + 2 = 0$ or $p = \frac{1}{2}$

$\Rightarrow \quad \dfrac{1}{\lambda+1}=\dfrac{1}{2} \quad \Rightarrow \lambda=1.$

$\therefore$ from (1), $5\times\dfrac{1}{2}-q-2=0 \quad \Rightarrow q=\dfrac{1}{2}$

$\Rightarrow \quad \dfrac{1}{\mu+1}=\dfrac{1}{2} \quad \Rightarrow \mu=1.$

Clearly, $p=\dfrac{1}{2}, q=\dfrac{1}{2}$ satisfies (3).

$\therefore$ the line segments AB and CD bisect each other and

$$P=\left(\frac{1(-1)+1.4}{1+1}, \frac{1(-2)+1.7}{1+1}, \frac{1(1)+1.8}{1+1}\right)$$

$\therefore \quad P=\left(\dfrac{3}{2}, \dfrac{5}{2}, \dfrac{9}{2}\right).$

Now, $AP=\sqrt{\left(4-\dfrac{3}{2}\right)^2+\left(7-\dfrac{5}{2}\right)^2+\left(8-\dfrac{9}{2}\right)^2}$

$$=\sqrt{\frac{25}{4}+\frac{81}{4}+\frac{49}{4}}=\sqrt{\frac{155}{4}}=\frac{1}{2}\sqrt{155}=BP$$

because P is the middle point of AB.

$\therefore \quad AP\cdot BP=\dfrac{1}{2}\sqrt{155}\cdot\dfrac{1}{2}\sqrt{155}=\dfrac{155}{4}.$

Next, $CP=\sqrt{\left(2-\dfrac{3}{2}\right)^2+\left(3-\dfrac{5}{2}\right)^2+\left(4-\dfrac{9}{2}\right)^2}$

$$=\sqrt{\frac{1}{4}+\frac{1}{4}+\frac{1}{4}}=\sqrt{\frac{3}{4}}=\frac{1}{2}\sqrt{3}=DP$$

because P is the middle point of CD.

$\therefore \quad CP\cdot DP=\dfrac{1}{2}\sqrt{3}\cdot\dfrac{1}{2}\sqrt{3}=\dfrac{3}{4}.$ So, $AP\cdot BP\neq CP\cdot DP$.

$\therefore$ the points A, B, C and D are not concyclic.

6. Let A = (1, 2, –3) and B = (4, 1, 5). P is a variable point such that PAB is an equilateral triangle. Find the equation of the locus of the point P.

Let P = (x, y, z). As PAB is an equilateral triangle, $PA^2=PB^2=AB^2$.

$$\Rightarrow (x-1)^2+(y-2)^2+(z+3)^2$$
$$=(x-4)^2+(y-1)^2+(z-5)^2$$
$$=(1-4)^2+(2-1)^2+(-3-5)^2.$$

Now, $(x-1)^2+(y-2)^2+(z+3)^2$

$$=(x-4)^2+(y-1)^2+(z-5)^2$$

$$\Rightarrow -2x+1-4y+4+6z+9$$
$$=-8x+16-2y+1-10z+25$$

$\Rightarrow \quad 6x-2y+16z=28 \quad \Rightarrow 3x-y+8z=14.$

Again, $(x-1)^2+(y-2)^2+(z+3)^2$

$$=(1-4)^2+(2-1)^2+(-3-5)^2$$

$\Rightarrow \quad x^2+y^2+z^2-2x-4y+6z+14=9+1+64$

$\Rightarrow \quad x^2+y^2+z^2-2x-4y+6z-60=0.$

$\therefore$ the equation of the locus of P is given by

$$x^2+y^2+z^2-2x-4y+6z-60=0,$$
$$3x-y+8z-14=0.$$

7. Prove that the two lines whose direction cosines l, m, n are given by $al+bm+cn=0$ and $mn+nl+lm=0$ are
(i) perpendicular if $ab+bc+ca=0$
(ii) parallel if $a^2+b^2+c^2=2(ab+bc+ca)$.

We have $al+bm+cn=0$ ... (1)

$mn+nl+lm=0$ ... (2)

(1) $\Rightarrow \quad al=-bm-cn \quad$ or $\quad l=-\dfrac{b}{a}m-\dfrac{c}{a}n.$

Putting in (2), $mn+(n+m)l=0$

$\Rightarrow \quad mn+(n+m)\cdot\left(-\dfrac{b}{a}m-\dfrac{c}{a}n\right)=0$

$\Rightarrow \quad mn-\dfrac{b}{a}(mn+m^2)-\dfrac{c}{a}(n^2+mn)=0$

or $\quad \left(1-\dfrac{b}{a}-\dfrac{c}{a}\right)mn-\dfrac{b}{a}m^2-\dfrac{c}{a}n^2+0$

or $\quad bm^2+(b+c-a)\,mn+cn^2=0$

or $\quad b\left(\dfrac{m}{n}\right)^2+(b+c-a)\dfrac{m}{n}+c=0.$ ... (1)

This is a quadratic equation in $\dfrac{m}{n}$ giving two values $\dfrac{m_1}{n_1}$ and $\dfrac{m_2}{n_2}$ where (l_1, m_1, n_1) and (l_2, m_2, n_2) are the direction cosines of the two lines.

$\therefore \quad \dfrac{m_1}{n_1}+\dfrac{m_2}{n_2}=-\dfrac{b+c-a}{b}$

and $\quad \dfrac{m_1}{n_1}\cdot\dfrac{m_2}{n_2}=\dfrac{c}{b} \quad \Rightarrow \quad \dfrac{m_1m_2}{\frac{1}{b}}=\dfrac{n_1n_2}{\frac{1}{c}}.$

By symmetry we get, $\dfrac{l_1l_2}{\frac{1}{a}}=\dfrac{m_1m_2}{\frac{1}{b}}=\dfrac{n_1n_2}{\frac{1}{c}}\ (=k).$

We know that two lines with direction cosines are perpendicular if

$$l_1l_2+m_1m_2+n_1n_2=0$$

$\Rightarrow \quad \dfrac{1}{a}\cdot k+\dfrac{1}{b}\cdot k+\dfrac{1}{c}\cdot k=0 \quad \Rightarrow \quad \dfrac{1}{a}+\dfrac{1}{b}+\dfrac{1}{c}=0$

$\Rightarrow \quad bc+ca+ab=0.$

The two lines are parallel if $\frac{l_1}{l_2}=\frac{m_1}{m_2}=\frac{n_1}{n_2} \quad \Rightarrow \quad \frac{m_1}{n_1}=\frac{m_2}{n_2}$.

So, the roots of (1) must be equal.

$\Rightarrow \quad D=0 \quad \Rightarrow \quad (b+c-a)^2-4\cdot b\cdot c=0$

$\Rightarrow \quad b^2+c^2+a^2+2bc-2ab-2ca-4bc=0$

$\Rightarrow \quad a^2+b^2+c^2=2(ab+bc+ca).$

8. Prove that the angle between the straight lines whose direction cosines l, m, n are given by $l+m+n=0$ and $fmn+gnl+hlm=0$ is $\frac{\pi}{3}$ if $\frac{1}{f}+\frac{1}{g}+\frac{1}{h}=0$. Also, find the condition for the two lines to be perpendicular.

Here, $l+m+n=0 \quad \Rightarrow \quad n=-(l+m).$

Putting this in $fmn+gnl+hlm=0$, we get

$(fm+gl)\{-(l+m)\}+hlm=0$

$\Rightarrow \quad -gl^2-fm^2-(f+g)\,lm+hlm=0$

or $\quad gl^2+(f+g-h)\,lm+fm^2=0$

$\Rightarrow \quad g\left(\frac{l}{m}\right)^2+(f+g-h)\left(\frac{l}{m}\right)+f=0.$

This is a quadratic equation whose roots are $\frac{l_1}{m_1}$ and $\frac{l_2}{m_2}$ where the direction cosines of the given lines are l_1, m_1, n_1 and l_2, m_2, n_2.

Then $\frac{l_1}{m_1}+\frac{l_2}{m_2}=-\frac{f+g-h}{g}$

$\Rightarrow \quad \frac{l_1m_2+l_2m_1}{m_1m_2}=-\frac{f+g-h}{g}$ ... (1)

and $\quad \frac{l_1}{m_1}\cdot\frac{l_2}{m_2}=\frac{f}{g} \quad \Rightarrow \quad \frac{l_1l_2}{f}=\frac{m_1m_2}{g}.$ ... (2)

(1) and (2) $\Rightarrow \quad \frac{l_1l_2}{f}=\frac{m_1m_2}{g}=\frac{l_1m_2+l_2m_1}{h-f-g}=t$ (say).

Similarly, by symmetry, $\frac{m_1m_2}{g}=\frac{n_1n_2}{h}=\frac{m_1n_2+m_2n_1}{f-g-h}=t$

and $\quad \frac{l_1l_2}{f}=\frac{n_1n_2}{h}=\frac{l_1n_2+l_2n_1}{g-f-h}=t.$

Now, $(l_1m_2-l_2m_1)^2=(l_1m_2+l_2m_1)^2-4l_1l_2m_1m_2$

$=t^2(h-f-g)^2-4\cdot ft\cdot gt=t^2\{(h-f-g)^2-4fg\}$

$=t^2\{f^2+g^2+h^2-2fg-2gh-2hf\}.$

By symmetry, $(m_1n_2-m_2n_1)^2=(n_1l_2-n_2l_1)^2$

$=t^2\{f^2+g^2+h^2-2fg-2gh-2hf\}.$

Now, $\cos\theta=l_1l_2+m_1m_2+n_1n_2$ where θ is the angle between the lines.

$\therefore \quad \tan\theta=\frac{\sqrt{1-(l_1l_2+m_1m_2+n_1n_2)^2}}{l_1l_2+m_1m_2+n_1n_2}$

$=\frac{\sqrt{(l_1^2+m_1^2+n_1^2)(l_2^2+m_2^2+n_2^2)-(l_1l_2+m_1m_2+n_1n_2)^2}}{l_1l_2+m_1m_2+n_1n_2},$

{being direction cosines, $l_1^2+m_1^2+n_1^2=1,\ l_2^2+m_2^2+n_2^2=1$}

$=\frac{\sqrt{(l_1m_2-l_2m_1)^2+(m_1n_2-m_2n_1)^2+(n_1l_2-n_2l_1)^2}}{l_1l_2+m_1m_2+n_1n_2}$

or $\quad \tan\frac{\pi}{3}=\frac{\sqrt{3t^2(f^2+g^2+h^2-2fg-2gh-2hf)}}{t(f+g+h)},$

using the above results

or $\quad \sqrt{3}=\sqrt{3}t\cdot\frac{\sqrt{f^2+g^2+h^2-2fg-2gh-2hf}}{t(f+g+h)}$

or $\quad (f+g+h)^2=f^2+g^2+h^2-2fg-2gh-2hf$

or $\quad 4(fg+gh+hf)=0 \quad \Rightarrow \quad \frac{1}{h}+\frac{1}{f}+\frac{1}{g}=0.$

Two lines are perpendicular if $l_1l_2+m_1m_2+n_1n_2=0$

$\Rightarrow \quad ft+gt+ht=0 \quad \Rightarrow \quad f+g+h=0.$

9. If l_r, m_r, n_r; $r=1,2,3$; be direction cosines of three mutually perpendicular lines then prove that

(a) $\frac{l_1}{m_2n_3-m_3n_2}=\frac{m_1}{n_2l_3-n_3l_2}=\frac{n_1}{l_2m_3-l_3m_2}=\pm1$

(b) the line whose direction ratios are $l_1+l_2+l_3$, $m_1+m_2+m_3$, $n_1+n_2+n_3$ makes equal angles with the given three lines.

(a) From the condition of perpendicularity of two lines we get,

$l_1l_2+m_1m_2+n_1n_2=0$

$l_1l_3+m_1m_3+n_1n_3=0$

$\Rightarrow \quad \frac{l_1}{m_2n_3-m_3n_2}=\frac{m_1}{n_2l_3-n_3l_2}=\frac{n_1}{l_2m_3-l_3m_2}$

$=\frac{\sqrt{l_1^2+m_1^2+n_1^2}}{\sqrt{(m_2n_3-m_3n_2)^2+(n_2l_3-n_3l_2)^2+(l_2m_3-l_3m_2)^2}},$

using ratio-proportion.

But the lines with direction cosines l_2, m_2, n_2 and l_3, m_3, n_3 are perpendicular.

$\therefore \quad \sin 90°=\sqrt{(m_2n_3-m_3n_2)^2+(n_2l_3-n_3l_2)^2+(l_2m_3-l_3m_2)^2}.$

Also, $l_1^2+m_1^2+n_1^2=1$ (by property of direction cosines).

$\therefore$ we get, $\frac{l_1}{m_2n_3-m_3n_2}=\frac{m_1}{n_2l_3-n_3l_2}$

$=\frac{n_1}{l_2m_3-l_3m_2}=\frac{\sqrt{1}}{\sqrt{\sin 90°}}=\pm1.$

(b) From the condition of perpendicularity of lines we get

$$l_1l_2 + m_1m_2 + n_1n_2 = 0 = l_2l_3 + m_2m_3 + n_2n_3$$
$$= l_3l_1 + m_3m_1 + n_3n_1.$$

The direction cosines of the lines are

$$k(l_1 + l_2 + l_3),\ k(m_1 + m_2 + m_3),\ k(n_1 + n_2 + n_3).$$

If θ_1 be the angle between this line and the line whose direction cosines are l_1, m_1, n_1 then

$$\cos\theta_1 = l_1 \cdot k(l_1 + l_2 + l_3) + m_1 \cdot k(m_1 + m_2 + m_3) + n_1 \cdot k(n_1 + n_2 + n_3)$$
$$= k\{l_1^2 + m_1^2 + n_1^2 + (l_1l_2 + m_1m_2 + n_1n_2) + (l_1l_3 + m_1m_3 + n_1n_3)\}$$
$$= k\{1 + 0 + 0\} = k.$$

Similarly, $\cos\theta_2 = k = \cos\theta_3$ where θ_2 and θ_3 are the angles between the line and the lines with direction cosines l_2, m_2, n_2 and l_3, m_3, n_3 respectively.

Hence, $\cos\theta_1 = \cos\theta_2 = \cos\theta_3 = k \quad \Rightarrow \quad \theta_1 = \theta_2 = \theta_3$.

Hence the problem.

10. Prove that three lines through the origin with direction cosines l_1, m_1, n_1; l_2, m_2, n_2; l_3, m_3, n_3 are coplanar if

$$\begin{vmatrix} l_1 & m_1 & n_1 \\ l_2 & m_2 & n_2 \\ l_3 & m_3 & n_3 \end{vmatrix} = 0.$$

Three given lines are coplanar if there is a line perpendicular to each of the three lines. Let the line have the direction cosines l, m, n.

Then using the condition of perpendicularity,

$$ll_1 + mm_1 + nn_1 = 0$$
$$ll_2 + mm_2 + nn_2 = 0$$
$$ll_3 + mm_3 + nn_3 = 0$$

Eliminating l, m, n from these, we have,

$$\begin{vmatrix} l_1 & m_1 & n_1 \\ l_2 & m_2 & n_2 \\ l_3 & m_3 & n_3 \end{vmatrix} = 0.$$

11. Show that the points A(1, 2, –1), B(2, 5, –2), C(4, 4, –3) and D(3, 1, –2) are the vertices of a rectangle ABCD and find its area.

ABCD is a rectangle if A, B, C and D are coplanar and $AB = CD$, $AD = BC$ and one angle $ABC = 90°$.

Now, direction ratios of AB are

(2 – 1, 5 – 2, –2 + 1), i.e., (1, 3, –1),

direction ratios of AC are

(4 – 1, 4 – 2, –3 + 1), i.e., (3, 2, –2),

direction ratios of AD are

(3 – 1, 1 – 2, –2 + 1), i.e., (2, –1, –1).

The four points are coplanar if AB, AC and AD are coplanar. It is so if there is a line with direction ratios l, m, n which is perpendicular to each of AB, AC and AD. For this the conditions are

$$l \cdot 1 + m \cdot 3 + n \cdot (-1) = 0$$
$$l \cdot 3 + m \cdot 2 + n(-2) = 0$$
$$l \cdot 2 + m(-1) + n(-1) = 0$$

$$\Rightarrow \begin{vmatrix} 1 & 3 & -1 \\ 3 & 2 & -2 \\ 2 & -1 & -1 \end{vmatrix} = 0 \text{ which is true because}$$

$$\Delta = 1(-2-2) - 3(-3+4) - 1(-3-4)$$
$$= -4 - 3 + 7 = 0.$$

$\therefore$ the four points are coplanar.

Now, $AB = \sqrt{(1-2)^2 + (2-5)^2 + (-1+2)^2}$
$= \sqrt{1+9+1} = \sqrt{11}$.

$CD = \sqrt{(4-3)^2 + (4-1)^2 + (-3+2)^2}$
$= \sqrt{1+9+1} = \sqrt{11}$.

$AD = \sqrt{(1-3)^2 + (2-1)^2 + (-1+2)^2}$
$= \sqrt{4+1+1} = \sqrt{6}$.

$BC = \sqrt{(2-4)^2 + (5-4)^2 + (-2+3)^2}$
$= \sqrt{4+1+1} = \sqrt{6}$.

So, $AB = CD, AD = BC$.

Now, $\angle ABC = 90°$ if AB ⊥ BC. The direction ratios of AB are (1, 3, –1) and those of BC are (4 – 2, 4 – 5, –3 + 2), i.e., (2, –1, –1).

As $1 \cdot 2 + 3 \cdot (-1) + (-1) \cdot (-1) = 2 - 3 + 1 = 0$, AB ⊥ BC.

$\therefore$ ABCD is a rectangle.

Its area $= AB \cdot AD = \sqrt{11} \cdot \sqrt{6} = \sqrt{66}$.

12. The projection of a line segment on the axes of reference are 3, 4 and 12 respectively. Find the length and the direction cosines of the line segment.

Let AB be the line segment where $A = (x_1, y_1, z_1)$ and $B = (x_2, y_2, z_2)$.

Then $\vec{AB} = \vec{OB} - \vec{OA} = (x_2\vec{i} + y_2\vec{j} + z_2\vec{k}) - (x_1\vec{i} + y_1\vec{j} + z_1\vec{k})$.
$= (x_2 - x_1)\vec{i} + (y_2 - y_1)\vec{j} + (z_2 - z_1)\vec{k}$.

Unit vectors along the axes are $\vec{i}, \vec{j}$ and $\vec{k}$ respectively.

$\therefore$ projection of AB on the x-axis $= \vec{AB} \cdot \vec{i} = x_2 - x_1 = 3$

projection of AB on the y-axis $= \vec{AB} \cdot \vec{j} = y_2 - y_1 = 4$

projection of AB on the z-axis $= \vec{AB} \cdot \vec{k} = z_2 - z_1 = 12$

The length of AB $= \sqrt{(x_2 - x_1)^2 + (y_2 - y_1)^2 + (z_2 - z_1)^2}$
$= \sqrt{3^2 + 4^2 + 12^2} = 13$.

The direction ratios of AB are $x_2 - x_1, y_2 - y_1, z_2 - z_1$, i.e., 3, 4, 12

$\therefore$ direction cosines are

$$\frac{3}{\sqrt{3^2+4^2+12^2}}, \frac{4}{\sqrt{3^2+4^2+12^2}}, \frac{12}{\sqrt{3^2+4^2+12^2}},$$

i.e., $\frac{3}{13}, \frac{4}{13}, \frac{12}{13}$.

13. Find the projection of the line segment joining the points A(2, 3, 4) and B(3, 4, 5) on the line *l* whose direction ratios are (6, 6, –7). Also, find the angle between the lines AB and *l*.

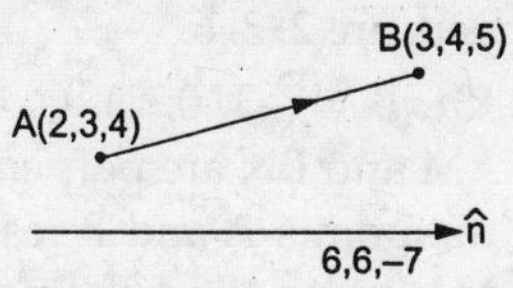

Here, $\vec{AB} = \vec{OB} - \vec{OA} = (3\vec{i} + 4\vec{j} + 5\vec{k}) - (2\vec{i} + 3\vec{j} + 4\vec{k})$

$= \vec{i} + \vec{j} + \vec{k}.$

A vector along the line $l = 6\vec{i} + 6\vec{j} - 7\vec{k}$.

$\therefore$ unit vector $\hat{n}$ along $l = \frac{6\vec{i} + 6\vec{j} - 7\vec{k}}{|6\vec{i} + 6\vec{j} - 7\vec{k}|}$

$= \frac{1}{11}(6\vec{i} + 6\vec{j} - 7\vec{k}).$

$\therefore$ required projection

$= \vec{AB} \cdot \hat{n} = (\vec{i} + \vec{j} + \vec{k}) \cdot \frac{1}{11}(6\vec{i} + 6\vec{j} - 7\vec{k})$

$= \frac{1}{11}(6 + 6 - 7) = \frac{5}{11}.$

If the angle be θ,

$$\cos\theta = \frac{\vec{AB}}{|\vec{AB}|} \cdot \hat{n} = \frac{\vec{i} + \vec{j} + \vec{k}}{|\vec{i} + \vec{j} + \vec{k}|} \cdot \frac{1}{11}(6\vec{i} + 6\vec{j} - 7\vec{k})$$

$$= \frac{1}{11\sqrt{3}}(6 + 6 - 7) = \frac{5}{11\sqrt{3}}.$$

$\therefore$ $\theta = \cos^{-1}\left(\frac{5}{11\sqrt{3}}\right).$

14. ABC is a triangle where A = (2, 3, 5), B = (–1, 3, 2) and C = (λ, 5, μ). If the median through A is equally inclined to the axes then find the values of λ and μ. Also, find the area of the ΔABC.

The middle point of BC is $\left(\frac{-1+\lambda}{2}, \frac{3+5}{2}, \frac{2+\mu}{2}\right)$,

i.e., $\left(\frac{\lambda-1}{2}, 4, \frac{\mu+2}{2}\right)$.

$\therefore$ the direction ratios of the median through A are

$\frac{\lambda-1}{2} - 2, 4 - 3, \frac{\mu+2}{2} - 5$, i.e., $\frac{\lambda-5}{2}, 1, \frac{\mu-8}{2}$.

As the median through A is equally inclined to the axes, its direction ratios will be *k*, *k*, *k* where *k* is a constant.

$$\therefore \quad \frac{\frac{\lambda-5}{2}}{k} = \frac{1}{k} = \frac{\frac{\mu-8}{2}}{k} \quad \Rightarrow \quad \frac{\lambda-5}{2} = 1 = \frac{\mu-8}{2}$$

$\Rightarrow$ λ = 7, μ = 10.

$\therefore$ C = (7, 5, 10).

Now, $\text{ar}(\Delta ABC) = \frac{1}{2}|\vec{AB} \times \vec{AC}|$

$= \frac{1}{2}|\{(-\vec{i} + 3\vec{j} + 2\vec{k}) - (2\vec{i} + 3\vec{j} + 5\vec{k})\}$

$\times \{(7\vec{i} + 5\vec{j} + 10\vec{k}) - (2\vec{i} + 3\vec{j} + 5\vec{k})\}|$

$= \frac{1}{2}|(-3\vec{i} - 3\vec{k}) \times (5\vec{i} + 2\vec{j} + 5\vec{k})|$

$= \frac{1}{2}|-6\vec{k} + 15\vec{j} - 15\vec{j} + 6\vec{i}| = \frac{1}{2}|6\vec{i} - 6\vec{k}|$

$= \frac{1}{2}\sqrt{6^2 + (-6)^2} = 3\sqrt{2}.$

Exercises

1. Let P be the point (3, 5, 1) and PN be the perpendicular from P to the *z*–*x* plane. If OP, where O is the origin, makes an angle θ with the positive direction of the *y*-axis and ON makes an angle ϕ with the positive direction of the *x*-axis then find θ and ϕ.
2. Show that the points (4, 2, 4), (10, 2, –2) and (2, 0, –4) are the vertices of an equilateral triangle. Also, find the length of its altitude and its area.
3. If the points A(0, 7, 10), B(–1, 6, 6) and C(–4, *a*, *b*) form an isosceles triangle right angled at *B* then find the values of *a* and *b*.
4. The *x*, *y* and *z* coordinates of each vertex of a triangle are in AP. The *x* and *y* coordinates of the centroid of the triangle are 1 and 3 respectively. Find the distance of the centroid from the origin.
5. A, B and C are three points on the axes of *x*, *y* and *z* respectively such that $OA = a$, $OB = b$ and $OC = c$, where O is the origin. Find the centroid of the tetrahedron (O, ABC). Also, find the distance of the centroid of ΔABC from the origin O. Again, obtain the point which is equidistant from A, B, C and O.

6. (O, ABC) is a tetrahedron whose vertex O is the origin, and the middle points of the edges OA, OB and OC are (6, –1, 2), (2, 3, 0) and (–1, 3, 5) respectively. Find the centroid of the tetrahedron. Also, find its volume.

7. Two vertices of a triangle are (3, 5, –2) and (5, –1, 4). The medians through these vertices intersect at (2, –1, 1). Find the third vertex.

8. Determine the vertices of the triangle whose sides have the middle points (3, 2, 3), (–1, 1, 5) and (0, 3, 4).

9. Show that the joins of the points A(0, 2, –1), B(1, 0, 0) and C(1, 2, –2), D(3, –1, –1) intersect. Also, find the point of intersection.

10. Find the equation of the locus of points which are equidistant from the points (1, 2, 3) and (3, 2, –1).

11. Find the locus of the point, the sum of whose distances from (4, 0, 0) and (–4, 0, 0) is equal to 10.

12. Find the acute angle between the lines whose direction cosines l, m, n satisfy the equations $l+m+n=0$ and $2l(m+n)=mn$.

13. Show that the pair of straight lines whose direction ratios are given by the relations $l+2m+3n=0$ and $3lm+mn-4nl=0$ are at right angles.

14. Show that the angle between the lines whose direction cosines satisfy the equations $l+m+n=0$ and $l^2+m^2=n^2$ is $\frac{\pi}{3}$.

15. Prove that the straight lines whose direction cosines are given by $al+bm+cn=0$ and $\frac{f}{l}+\frac{g}{m}+\frac{h}{n}=0$ are perpendicular to each other if $\frac{f}{a}+\frac{g}{b}+\frac{h}{c}=0$, and are parallel if $\sqrt{fa}+\sqrt{gb}+\sqrt{hc}=0$.

16. Find the direction cosines of a line that is perpendicular to each of the two lines whose direction numbers are (2, –1, 2) and (3, 0, 1)

17. Find the direction cosines of the line perpendicular to a pair of perpendicular lines whose direction cosines are l_1, m_1, n_1 and l_2, m_2, n_2.

18. Three straight lines passing through the origin have direction ratios (1, 2, 2), (2, 3, 6) and (3, 4, 12). Find the direction cosines of a straight line equally inclined to them.

19. Show that a line can be found perpendicular to three lines with direction ratios 2, 1, 5; 4, –2, 2 and –6, 4, –1.

20. If three lines joining the point (3, 4, 2) and the points (1, 1, 0), (2, 1, 1) and (λ, 1, 3) are coplanar then find the value of λ.

21. With the help of direction cosines or ratios, prove that the points (–1, 2, 3), (2, 1, 4), (3, 0, 3) and (2, 2, 6) are coplanar.

22. Show that the points A(1, 2, 3), B(–1, –2, –1), C(2, 3, 2) and D(4, 7, 6) are coplanar and ABCD is a parallelogram. Is it a rectangle?

23. Prove that there is exactly one sphere passing through the points (0, 0, 0), (0, 2, 0), (1, 0, 0) and (0, 0, 4). Find its centre.

24. Prove that the points A(6, 0, 2), B(8, –3, 8), C(2, –5, 11) and D(0, –2, 5) form a rhombus which is not a square.

25. Prove that the points (0, 1, 2,), (3, 0, 1), (4, 3, 6) and (2, 3, 2) are noncoplanar. Hence, find the volume of the tetrahedron whose vertices are the given points.

26. Find the projection of the line segment joining the points (1, 1, 2) and (3, 4, 1) on the line whose direction numbers are 2, 3, 6.

27. Let A = (1, 2, 3), B = (–3, 0, 1), C = (2, –4, 0) and D = (–1, 5, 1). AM and BN are perpendiculars to the line CD from the points A and B respectively. Find the length of MN. Also, find the projection of the line segment CD on the line AB and the angle between the lines AB and CD.

28. Let P, Q, R and S be (–2, 3, 4), (–4, 4, 6), (4, 3, 5) and (0, 1, 2) respectively. Use projection to prove that PQ is perpendicular to RS.

Objective Questions

29. If a line makes angles α, β, γ with the positive directions of the coordinate axes then the value of $\sin^2\alpha+\sin^2\beta+\sin^2\gamma$ is

(a) 1 (b) $\frac{3}{2}$

(c) 2 (d) $\frac{3}{4}$

30. If the line joining the origin and the point (–2, 1, 2) makes angles θ_1, θ_2 and θ_3 with the positive directions of the coordinate axes then the value of $\cos 2\theta_1+\cos 2\theta_2+\cos 2\theta_3$ is

(a) –1 (b) 1

(c) 2 (d) –2

31. Two straight lines whose direction ratios are 2, 3, 4 and 5, λ, 2 are perpendicular to each other. Then λ is

(a) 6 (b) –16 (c) –6 (d) 16

32. Which of the following triplets can be the direction cosines of a line?

(a) $\left(\frac{1}{7}, \frac{4}{7}, \frac{2}{7}\right)$ (b) $\left(\frac{-1}{5}, \frac{4}{5}, \frac{-2\sqrt{2}}{5}\right)$

(c) (–3, 2, 1) (d) $\left(\frac{2}{3}, 1, \frac{\sqrt{5}}{9}\right)$

33. For what value of λ, the three numbers $2\lambda-1, \frac{1}{4}, \lambda-\frac{1}{2}$ can be the direction cosines of a straight line?

(a) $\frac{1}{2} \pm \frac{\sqrt{3}}{4}$ (b) $\frac{3}{4}$ (c) $\pm \frac{3}{4}$ (d) $\frac{\sqrt{3}}{2} \pm \frac{1}{4}$

34. If the points (1, –1, –3), (2, 0, –1) and $(x, y, 3)$ lie on a straight line then the value of the pair (x, y) is

(a) (–4, 2) (b) (4, 2) (c) (2, 1) (d) (2, 4)

35. The direction ratios of two lines are 1, –3, 4 and –2, 0, 5. The direction cosines of a line perpendicular to both the lines are

(a) 15, 13, 6 (b) 6, –13, –15

(c) $\frac{6}{\sqrt{26 \times 29}}, \frac{-13}{\sqrt{26 \times 29}}, \frac{-15}{\sqrt{26 \times 29}}$

(d) $\frac{15}{\sqrt{430}}, \frac{13}{\sqrt{430}}, \frac{6}{\sqrt{430}}$

36. Let P = (1, 1, 1) and Q = (3, 2, –1). The direction cosines of the line PQ are

(a) 2, 1, –2 (b) $\frac{2}{3}, \frac{-1}{3}, \frac{2}{3}$

(c) –2, –1, 2 (d) $-\frac{2}{3}, \frac{-1}{3}, \frac{2}{3}$

37. If P = (1, 2, 3), Q = (2, –1, 4), R = (–4, 1, 5) and S = (2, λ, μ) are such that PQ || RS then the values of λ, μ are

(a) λ = –17, μ = –11 (b) λ = –17, μ = 11

(c) λ = 17, μ = –11 (d) λ = 11, μ = –17

38. If three lines drawn through the origin O with direction ratios 1, –1, 1; 2, –3, 0 and 1, 0, λ lie in a plane then λ is

(a) 3 (b) –3 (c) 0 (d) 2

39. The line joining the points (1, 2, 3) and (4, 5, 7), and the line joining the points (–4, 3, –6) and (2, 9, 2) are

(a) parallel (b) mutually perpendicular

(c) inclined at $\frac{\pi}{4}$ (d) none of these

40. If $l_1, m_1, n_1; l_2, m_2, n_2$ and l_3, m_3, n_3 be the direction cosines of three mutually perpendicular lines then the direction cosines of the line equally inclined to them are

(a) $l_1 + l_2 + l_3, m_1 + m_2 + m_3, n_1 + n_2 + n_3$

(b) $\frac{l_1 + l_2 + l_3}{\sqrt{3}}, \frac{m_1 + m_2 + m_3}{\sqrt{3}}, \frac{n_1 + n_2 + n_3}{\sqrt{3}}$

(c) $\frac{l_1 l_2 l_3}{\sqrt{3}}, \frac{m_1 m_2 m_3}{\sqrt{3}}, \frac{n_1 n_2 n_3}{\sqrt{3}}$

(d) $\frac{1}{3}(l_1 + l_2 + l_3), \frac{1}{3}(m_1 + m_2 + m_3), \frac{1}{3}(n_1 + n_2 + n_3)$

41. The angle between two lines that have the direction numbers 1, 1, 0 and 2, 1, 2 is

(a) $\frac{\pi}{3}$ (b) $\frac{\pi}{2}$ (c) $\frac{\pi}{4}$ (d) $\frac{\pi}{6}$

42. The projections of a line segment on the axes are 6, 2, 3 respectively. The length of the line segment is

(a) 7 (b) 11 (c) 5 (d) 1

43. If the line joining the points (4, –2, 3) and (–5, 4, 0) passes through $(a, b, 2)$ then

(a) $a = 0, b = 1$ (b) $a = 1, b = 1$

(c) $a = 1, b = 0$ (d) $a = 0, b = 0$

44. If A = $(p, 1, -1)$, B = $(2p, 0, 2)$ and C = $(2 + 2p, p, 1)$ are vertices of a triangle right angled at B then p is

(a) 3 (b) 1 (c) –3 (d) –2

45. If a line makes an angle of $\frac{\pi}{4}$ with the positive directions of each of x and y axes, the angle that the line makes with the positive direction of the z-axis is

(a) $\frac{\pi}{4}$ (b) $\frac{\pi}{2}$ (c) $\frac{\pi}{6}$ (d) $\frac{\pi}{3}$

46. P and Q are two points such that OP and OQ have the direction numbers –1, 2, –2 and 3, 1, 0. R(λ, 3 + λ, 1 – λ) is a point such that OP, OQ and OR are coplanar. Then the value of λ is

(a) 1 (b) –2 (c) $-\frac{22}{3}$ (d) $\frac{25}{3}$

Answers

1. $\theta = \tan^{-1}\sqrt{\frac{2}{5}}, \phi = \tan^{-1}\frac{1}{3}$ **2.** $3\sqrt{6}$ units, $18\sqrt{3}$ unit^2

3. $a = 9, b = 6; a = \frac{57}{17}, b = \frac{126}{17}$ **4.** $\sqrt{35}$

5. $\left(\frac{a}{4}, \frac{b}{4}, \frac{c}{4}\right); \frac{1}{3}\sqrt{a^2 + b^2 + c^2}, \left(\frac{a}{2}, \frac{b}{2}, \frac{c}{2}\right)$

6. $\left(\frac{7}{2}, \frac{5}{2}, \frac{7}{2}\right); 157\frac{1}{3}$ unit^3 **7.** (–2, –7, 1)

8. (4, 4, 2), (2, 0, 4), (–4, 2, 6) **9.** (–3, 8, –4)

10. $x = 2z$ **11.** $9x^2 + 25y^2 + 25z^2 = 225$

12. $\frac{\pi}{3}$ **16.** $\frac{-1}{\sqrt{26}}, \frac{4}{\sqrt{26}}, \frac{3}{\sqrt{26}}$

17. $m_1 n_2 - m_2 n_1, n_1 l_2 - n_2 l_1, l_1 m_2 - l_2 m_1$

18. $-\frac{1}{\sqrt{3}}, \frac{1}{\sqrt{3}}, \frac{1}{\sqrt{3}}$ **20.** 4

22. No **23.** $\left(\frac{1}{2}, 1, 2\right)$ **25.** 6 unit^3 **26.** 1

27. $\frac{8}{\sqrt{91}}$ units; $\frac{2}{3}\sqrt{6}$; $\cos^{-1}\left(\frac{2}{3}\cdot\sqrt{\frac{6}{91}}\right)$

29. (c) **30.** (a) **31.** (c) **32.** (b)

33. (a) **34.** (b) **35.** (d) **36.** (d)
37. (b) **38.** (a) **39.** (a) **40.** (b)
41. (c) **42.** (a) **43.** (c) **44.** (a)
45. (b) **46.** (d)

□

Chapter Test

Time: 45 minutes

1. (i) A straight line passing through (1, 1, 1) makes an angle of 60° with the positive direction of the z-axis, and the cosines of the angles made with the positive directions of the y-axis and x-axis are in the ratio $\sqrt{3} : 1$. Then the acute angle between the two possible positions of the line is

A. $\frac{\pi}{2}$ B. $\frac{\pi}{6}$ C. $\frac{\pi}{3}$ D. $\frac{\pi}{4}$

(ii) Any point on a line has the coordinates $(r+1, r-3, r\sqrt{2}+4)$ where r is any real number. The direction cosines of the line are

A. $\frac{1}{2}, \frac{1}{2}, \frac{1}{\sqrt{2}}$ B. $\frac{1}{2}, -\frac{1}{2}, \frac{1}{\sqrt{2}}$ C. $\frac{1}{2}, \frac{1}{2}, -\frac{1}{\sqrt{2}}$ D. $1, 1, \sqrt{2}$

(iii) If A(1, –1, 0), B(–2, 1, 8) and C(–1, 2, 7) are three consecutive vertices of a parallelogram then the fourth vertex is

A. (0, –2, 1) B. (2, 0, –1) C. (1, 0, –1) D. (1, –2, 0)

2. If a line makes angles $\alpha, \beta, \gamma, \delta$ with the four diagonals of a cube then prove that $\sin^2\alpha + \sin^2\beta + \sin^2\gamma + \sin^2\delta = \frac{8}{3}$.

3. If the straight lines whose direction cosines l, m, n are given by $l + m = \lambda n$ and $mn + nl + lm = 0$ be at right angles, find λ.

4. If the line joining the points (1, 4, 2) and (–2, 1, 2) is inclined at an angle of $\frac{\pi}{3}$ to the line joining the points (1, 2, 3) and (2, λ, 1) then find λ.

5. AP, BQ, CR and DS are perpendiculars drawn from the points A(4, 3, –2), B(6, 4, 1), C(5, 6, –3) and D(3, 5, 2) to the x–y plane. Prove that the feet P, Q, R and S of the perpendiculars form a square. Also, find the area of the square.

□

Answers

1. (i) C (ii) A (iii) B

3. $\frac{1}{2}$ **4.** $\pm\sqrt{7}$ **5.** 5 unit ^{2}

□

2. *Equations of a Straight Line in Space*

Recap of Facts and Formulae

1. Equations of a line passing through a given point having given direction cosines

- Let the given point be $A = (x_1, y_1, z_1)$ and the d.c. of the line AB be l, m, n. If any point on the line be $P(x, y, z)$ then d.r. of AP are $x - x_1, y - y_1, z - z_1$. But its d.c. are l, m, n.

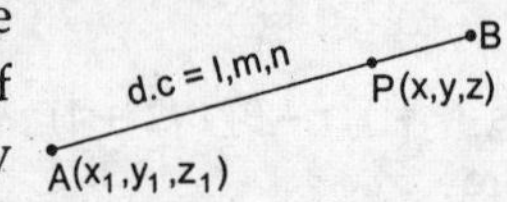

So, $\dfrac{x - x_1}{l} = \dfrac{y - y_1}{m} = \dfrac{z - z_1}{n}$. These are the equations of the line AB.

Vectorially, if the position vector of A be $\vec{r_1}$ and that of any point P be $\vec{r}$ then $\vec{AP} = \vec{r} - \vec{r_1}$, and the unit vector parallel to AB is $l\,\vec{i} + m\,\vec{j} + n\,\vec{k}$.

So, $\vec{AP} = t(l\,\vec{i} + m\,\vec{j} + n\,\vec{k})$, t being the length of AP.

$\therefore\ \vec{r} - \vec{r_1} = t(l\,\vec{i} + m\,\vec{i} + n\,\vec{k})$

$\Rightarrow\ x - x_1 = tl,\ y - y_1 = tm,\ z - z_1 = tn.$

$\therefore\ \dfrac{x - x_1}{l} = \dfrac{y - y_1}{m} = \dfrac{z - z_1}{n} (= t)$ (symmetric form).

Any point at a distance t from A on the line is

$$(x_1 + tl, y_1 + tm, z_1 + tn).$$

So, the parametric equations of the line passing through (x_1, y_1, z_1), having the direction ratios l, m, n, are $x = x_1 + \lambda l, y = y_1 + \lambda m, z = z_1 + \lambda n$ where λ is the parameter.

2. Equations of a line passing through two given points

- If $A = (x_1, y_1, z_1)$ and $B = (x_2, y_2, z_2)$, the d.r. of AB are $x_1 - x_2, y_1 - y_2, z_1 - z_2$.

$\therefore$ the equations of the line are

$$\frac{x - x_1}{x_1 - x_2} = \frac{y - y_1}{y_1 - y_2} = \frac{z - z_1}{z_1 - z_2}.$$

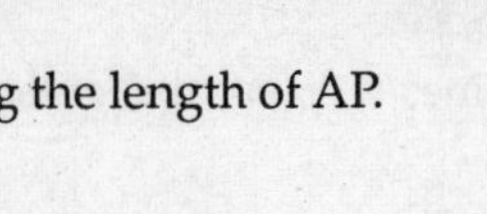

Vectorially, if the position vectors of A and B be $\vec{r_1}$ and $\vec{r_2}$, and P is any point on the line AB whose position vector is $\vec{r}$ then

$$\vec{AP} = \lambda\vec{AB} \quad \Rightarrow \quad \vec{r} - \vec{r_1} = \lambda(\vec{r_2} - \vec{r_1})$$

or $\vec{r} = (1 - \lambda)\,\vec{r_1} + \lambda\vec{r_2}$.

$$\therefore\ x\,\vec{i} + y\,\vec{j} + z\,\vec{k} = (1 - \lambda)(x_1\,\vec{i} + y_1\,\vec{j} + z_1\,\vec{k}) + \lambda(x_2\,\vec{i} + y_2\,\vec{j} + z_2\,\vec{k})$$

$$\Rightarrow\ x = (1 - \lambda)\,x_1 + \lambda x_2,\ y = (1 - \lambda)y_1 + \lambda y_2,\ z = (1 - \lambda)z_1 + \lambda z_2.$$

$$\therefore\ \frac{x - x_1}{x_2 - x_1} = \frac{y - y_1}{y_2 - y_1} = \frac{z - z_1}{z_2 - z_1} (= \lambda)$$

$$\Rightarrow\ \frac{x - x_1}{x_1 - x_2} = \frac{y - y_1}{y_1 - y_2} = \frac{z - z_1}{z_1 - z_2}.$$

3. Condition for two lines to be intersecting and their point of intersection

- Two lines $\dfrac{x - \alpha_1}{l_1} = \dfrac{y - \beta_1}{m_1} = \dfrac{z - \gamma_1}{n_1}$ and $\dfrac{x - \alpha_2}{l_2} = \dfrac{y - \beta_2}{m_2} = \dfrac{z - \gamma_2}{n_2}$

will intersect at $P(x_1, y_1, z_1)$ if

$$x_1 = \alpha_1 + l_1 r = \alpha_2 + l_2 r',\ y_1 = \beta_1 + m_1 r = \beta_2 + m_2 r'$$

and $z_1 = \gamma_1 + n_1 r = \gamma_2 + n_2 r'$

$\Rightarrow\ l_1 r - l_2 r' + \alpha_1 - \alpha_2 = 0$ ... (1)

$m_1 r - m_2 r' + \beta_1 - \beta_2 = 0$ ... (2)

$n_1 r - n_2 r' + \gamma_1 - \gamma_2 = 0$ ... (3)

Eliminating $r, -r'$ from (1), (2) and (3), we get

$$\begin{vmatrix} l_1 & l_2 & \alpha_1 - \alpha_2 \\ m_1 & m_2 & \beta_1 - \beta_2 \\ n_1 & n_2 & \gamma_1 - \gamma_2 \end{vmatrix} = 0.$$

This is the condition for the two lines to be intersecting. Solving any two of (1), (2) and (3), we get r (as well as r'). Putting this value of r in $x_1 = \alpha_1 + l_1 r$, $y_1 = \beta_1 + m_1 r$, $z_1 = \gamma_1 + n_1 r$, we get $P(x_1, y_1, z_1)$.

4. Skew lines, shortest distance between two skew lines and the equation of the line of shortest distance

- Two lines $\dfrac{x-x_1}{l_1}=\dfrac{y-y_1}{m_1}=\dfrac{z-z_1}{n_1}$

and $\dfrac{x-x_2}{l_2}=\dfrac{y-y_2}{m_2}=\dfrac{z-z_2}{n_2}$ are skew lines if they are neither parallel nor intersecting. As such, no plane containing the two lines exists.

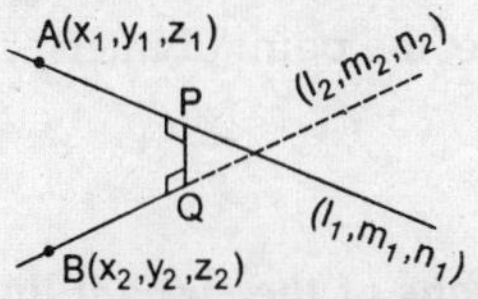

For every pair of skew lines there is a line which cuts both the skew lines perpendicularly. This line is called the line of shortest distance (s.d.). The intercept PQ made on the line of s.d. by the skew lines is called the shortest distance between the skew lines.

Vectorially, the shortest distance between two skew lines whose equations are

$\vec{r}=\vec{r_1}+\lambda\vec{u}$ and $\vec{r}=\vec{r_2}+\mu\vec{v}$ is given by

shortest distance $=\dfrac{(\vec{r_2}-\vec{r_1})\cdot(\vec{u}\times\vec{v})}{|\vec{u}\times\vec{v}|}$, i.e., $\dfrac{[\vec{r_2}-\vec{r_1}\ \ \vec{u}\ \ \vec{v}]}{|\vec{u}\times\vec{v}|}$

because the shortest distance PQ = projection of $\vec{AB}$ on the line PQ

$=$ projection of $(\vec{r_2}-\vec{r_1})$ on the vector $\vec{u}\times\vec{v}$ along PQ

$$=(\vec{r_2}-\vec{r_1})\cdot\frac{\vec{u}\times\vec{v}}{|\vec{u}\times\vec{v}|}=\frac{[\vec{r_2}-\vec{r_1}\ \ \vec{u}\ \ \vec{v}]}{|\vec{u}\times\vec{v}|}.$$

Note Being a distance, the above value will be taken with a positive sign.

$\therefore$ in the cartesian form, s.d.

$$=\frac{\left[(x_2-x_1)\vec{i}+(y_2-y_1)\vec{j}+(z_2-z_1)\vec{k}\quad l_1\vec{i}+m_1\vec{j}+n_1\vec{k}\quad l_2\vec{i}+m_2\vec{j}+n_2\vec{k}\right]}{|(l_1\vec{i}+m_1\vec{j}+n_1\vec{k})\times(l_2\vec{i}+m_2\vec{j}+n_2\vec{k})|}$$

$$=\frac{\begin{vmatrix} x_2-x_1 & y_2-y_1 & z_2-z_1\\ l_1 & m_1 & n_1\\ l_2 & m_2 & n_2\end{vmatrix}}{\{(m_1n_2-m_2n_1)^2+(n_1l_2-n_2l_1)^2+(l_1m_2-l_2m_1)^2\}^{1/2}}.$$

5. Condition for coplanarity of two lines (using shortest distance)

- The lines $\dfrac{x-x_1}{l_1}=\dfrac{y-y_1}{m_1}=\dfrac{z-z_1}{n_1}$

and $\dfrac{x-x_2}{l_2}=\dfrac{y-y_2}{m_2}=\dfrac{z-z_2}{n_2}$

are coplanar if the shortest distance between the lines is zero, i.e.,

$$\frac{\begin{vmatrix} x_2-x_1 & y_2-y_1 & z_2-z_1\\ l_1 & m_1 & n_1\\ l_2 & m_2 & n_2\end{vmatrix}}{\{(m_1n_2-m_2n_1)^2+(n_1l_2-n_2l_1)^2+(l_1m_2-l_2m_1)^2\}^{1/2}}=0$$

$$\Rightarrow\quad \begin{vmatrix} x_2-x_1 & y_2-y_1 & z_2-z_1\\ l_1 & m_1 & n_1\\ l_2 & m_2 & n_2\end{vmatrix}=0.$$

6. Distance of a point from a line

- The distance of the point (x_1, y_1, z_1) from the line $\dfrac{x-\alpha}{l}=\dfrac{y-\beta}{m}=\dfrac{z-\gamma}{n}$ is equal to

$$\left\{\begin{vmatrix} x_1-\alpha & y_1-\beta\\ l & m\end{vmatrix}^2+\begin{vmatrix} y_1-\beta & z_1-\gamma\\ m & n\end{vmatrix}^2+\begin{vmatrix} z_1-\gamma & x_1-\alpha\\ n & l\end{vmatrix}^2\right\}^{1/2}$$

where l, m, n are direction cosines of the line. This is also the length of the perpendicular from (x_1, y_1, z_1) to the line.

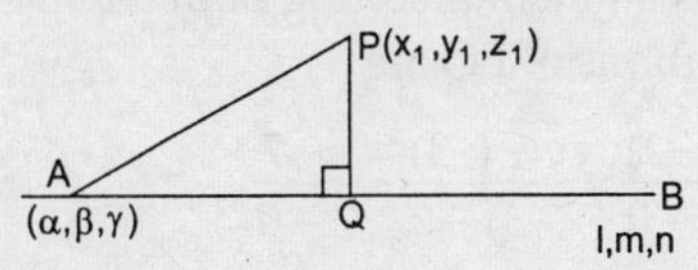

Clearly, $PQ^2=PA^2-AQ^2$

$=(x_1-\alpha)^2+(y_1-\beta)^2+(z_1-\gamma)^2$

$\qquad -\{\text{projection of } \vec{AP} \text{ on the line AB}\}^2$

$=(x_1-\alpha)^2+(y_1-\beta)^2+(z_1-\gamma)^2$

$\qquad -[\{(x_1-\alpha)\vec{i}+(y_1-\beta)\vec{j}+(z_1-\gamma)\vec{k}\}\cdot(l\vec{i}+m\vec{j}+n\vec{k})]^2$

$=(x_1-\alpha)^2+(y_1-\beta)^2+(z_1-\gamma)^2$

$\qquad -\{l(x_1-\alpha)+m(y_1-\beta)+n(z_1-\gamma)\}^2$

$=\{(x_1-\alpha)^2+(y_1-\beta)^2+(z_1-\gamma)^2\}(l^2+m^2+n^2)$

$\qquad -\{l(x_1-\alpha)+m(y_1-\beta)+n(z_1-\gamma)\}^2$

$=\{m(x_1-\alpha)-l(y_1-\beta)\}^2+\{n(y_1-\beta)-m(z_1-\gamma)\}^2$

$\qquad +\{l(z_1-\gamma)-n(x_1-\alpha)\}^2.$

Selected Solved Examples

1. Find the angle between the lines

$$\frac{x+2}{-1}=\frac{y+3}{1}=\frac{z-7}{0} \text{ and } \frac{x}{1}=\frac{y-2}{-2}=\frac{z}{1}.$$

Also, find the equations of a line cutting the given lines.

The direction cosines of the lines are

$$\frac{-1}{\sqrt{(-1)^2+1^2+0^2}}, \frac{1}{\sqrt{(-1)^2+1^2+0^2}}, \frac{0}{\sqrt{(-1)^2+1^2+0^2}}$$

and

$$\frac{1}{\sqrt{1^2+(-2)^2+1^2}}, \frac{-2}{\sqrt{1^2+(-2)^2+1^2}}, \frac{1}{\sqrt{1^2+(-2)^2+1^2}}$$

respectively, i.e., their direction cosines are $\frac{-1}{\sqrt{2}}, \frac{1}{\sqrt{2}}, 0$

and $\frac{1}{\sqrt{6}}, \frac{-2}{\sqrt{6}}, \frac{1}{\sqrt{6}}$ respectively.

If θ be the angle between the lines then

$$\cos\theta = l_1l_2+m_1m_2+n_1n_2 = \frac{-1}{\sqrt{2}}\cdot\frac{1}{\sqrt{6}}+\frac{1}{\sqrt{2}}\cdot\frac{-2}{\sqrt{6}}+0\cdot\frac{1}{\sqrt{6}}$$

$$=\frac{-1-2+0}{\sqrt{12}}=\frac{-3}{2\sqrt{3}}=-\frac{\sqrt{3}}{2}. \quad \text{So, } \theta=\frac{5\pi}{6}.$$

$\therefore$ the acute angle between the lines is $\pi-\frac{5\pi}{6}$, i.e., $\frac{\pi}{6}$.

Clearly, P(–2, –3, 7) is a point on the first line, and Q(0, 2, 0) is a point on the second line.

$\therefore$ PQ is a line intersecting both the lines and the equations of PQ are

$$\frac{x-(-2)}{(-2)-0}=\frac{y-(-3)}{(-3)-2}=\frac{z-7}{7-0}$$

or $$\frac{x+2}{-2}=\frac{y+3}{-5}=\frac{z-7}{7}.$$

2. Prove that the lines $\frac{x-\alpha}{\alpha'}=\frac{y-\beta}{\beta'}=\frac{z-\gamma}{\gamma'}$ **and** $\frac{x-\alpha'}{\alpha}=\frac{y-\beta'}{\beta}=\frac{z-\gamma'}{\gamma}$ **intersect (or lie in a plane) if they are not parallel and find their point of intersection.**

Any point on the first line is $(\alpha+\alpha'r, \beta+\beta'r, \gamma+\gamma'r)$, and any point on the second line is $(\alpha'+\alpha\lambda, \beta'+\beta\lambda, \gamma'+\gamma\lambda)$. Two lines will intersect if for some r and λ,

$$(\alpha+\alpha'r, \beta+\beta'r, \gamma+\gamma'r)=(\alpha'+\alpha\lambda, \beta'+\beta\lambda, \gamma'+\gamma\lambda)$$

$\Rightarrow \quad \alpha+\alpha'r=\alpha'+\alpha\lambda, \beta+\beta'r=\beta'+\beta\lambda, \gamma+\gamma'r=\gamma'+\gamma\lambda.$

From $\alpha+\alpha'r=\alpha'+\alpha\lambda$

$\beta+\beta'r=\beta'+\beta\lambda$

we get $\alpha\beta+\alpha'\beta r-\alpha\beta-\alpha\beta' r=\alpha'\beta+\alpha\beta\lambda-\alpha\beta'-\alpha\beta\lambda$

or $\quad (\alpha'\beta-\alpha\beta')r=\alpha'\beta-\alpha\beta' \quad \Rightarrow r=1$

($\because \alpha'\beta\neq\alpha\beta'$, because the lines are not parallel).

When $r=1$, $\alpha+\alpha'r=\alpha'+\alpha\lambda \quad \Rightarrow \alpha+\alpha'=\alpha'+\alpha\lambda$

$\Rightarrow \quad \lambda=1.$

$\therefore \quad \gamma+\gamma'r=\gamma'+r\lambda$ holds.

So, the lines intersect and they intersect at

$$(\alpha+\alpha'r, \beta+\beta'r, \gamma+\gamma'r)$$

where $r=1$. Hence, the point of intersection is

$$(\alpha+\alpha', \beta+\beta', \gamma+\gamma').$$

3. Find the equations of the straight line perpendicular to the lines $\frac{x+1}{-3}=\frac{y-3}{2}=\frac{z+2}{1}$ **and** $\frac{x}{1}=\frac{y-7}{-3}=\frac{z+7}{2}$, **and passing through their point of intersection.**

Any point on the first line is $(-1-3r, 3+2r, -2+r)$. It is the point of intersection of the lines if the point is on the second line.

Then $$\frac{-1-3r}{1}=\frac{3+2r-7}{-3}=\frac{-2+r+7}{2}.$$

Now, $$\frac{-1-3r}{1}=\frac{3+2r-7}{-3} \quad \Rightarrow 3+9r=2r-4$$

$\Rightarrow \quad 7r=-7 \quad \Rightarrow r=-1.$

This value of r satisfies $\frac{3+2r-7}{-3}=\frac{-2+r+7}{2}$

because $\frac{2(-1)-4}{-3}=\frac{-1+5}{2}$ or $\frac{-6}{-3}=\frac{4}{2}$, which is true.

$\therefore$ the point of intersection is

$(-1-3r, 3+2r, -2+r)$ where $r=-1$.

$\therefore$ the point of intersection is

$(-1+3, 3-2, -2-1)$, i.e., $(2, 1, -3)$.

$\therefore$ the required line is $\frac{x-2}{l}=\frac{y-1}{m}=\frac{z+3}{n}$ ... (1)

where $(-3)l+2m+1\cdot n=0$

and $1\cdot l+(-3)m+2\cdot n=0$

(using the condition of perpendicularity).

$$\therefore \quad \frac{l}{4-(-3)}=\frac{m}{1-(-3)2}=\frac{n}{(-3)(-3)-1.2}$$

$\Rightarrow \quad \frac{l}{7}=\frac{m}{7}=\frac{n}{7}=k$ (suppose).

$\therefore$ from (1), the required line is $\frac{x-2}{7k}=\frac{y-1}{7k}=\frac{z+3}{7k}$

or $$\frac{x-2}{1}=\frac{y-1}{1}=\frac{z+3}{1}.$$

4. Find the distance of the point (–2, 3, –5) from the line $\dfrac{x+2}{1}=\dfrac{y-3}{2}=\dfrac{z}{3}$.

Here, the direction cosines of the line are

$$\frac{1}{\sqrt{1^2+2^2+3^2}},\ \frac{2}{\sqrt{1^2+2^2+3^2}},\ \frac{3}{\sqrt{1^2+2^2+3^2}},$$

i.e., $\dfrac{1}{\sqrt{14}},\ \dfrac{2}{\sqrt{14}},\ \dfrac{3}{\sqrt{14}}$.

$\therefore$ the distance

$$=\left\{\begin{vmatrix}\dfrac{x_1-\alpha}{l} & \dfrac{y_1-\beta}{m}\end{vmatrix}^2+\begin{vmatrix}\dfrac{y_1-\beta}{m} & \dfrac{z_1-\gamma}{n}\end{vmatrix}^2+\begin{vmatrix}\dfrac{z_1-\gamma}{n} & \dfrac{x_1-\alpha}{l}\end{vmatrix}^2\right\}^{1/2}$$

$$=\left\{\begin{vmatrix}\dfrac{-2-(-2)}{1/\sqrt{14}} & \dfrac{3-3}{2/\sqrt{14}}\end{vmatrix}^2+\begin{vmatrix}\dfrac{3-3}{2/\sqrt{14}} & \dfrac{-5-0}{3/\sqrt{14}}\end{vmatrix}^2+\begin{vmatrix}\dfrac{-5-0}{3/\sqrt{14}} & \dfrac{-2-(-2)}{1/\sqrt{14}}\end{vmatrix}^2\right\}^{1/2}$$

$$=\left\{\begin{vmatrix}\dfrac{0}{1/\sqrt{14}} & \dfrac{0}{2/\sqrt{14}}\end{vmatrix}^2+\begin{vmatrix}\dfrac{0}{2/\sqrt{14}} & \dfrac{-5}{3/\sqrt{14}}\end{vmatrix}^2+\begin{vmatrix}\dfrac{-5}{3/\sqrt{14}} & \dfrac{0}{1/\sqrt{14}}\end{vmatrix}^2\right\}^{1/2}$$

$$=\left\{0^2+\left(\frac{10}{\sqrt{14}}\right)^2+\left(\frac{-5}{\sqrt{14}}\right)^2\right\}^{1/2}=\left(\frac{100+25}{14}\right)^{1/2}=\frac{5\sqrt{5}}{\sqrt{14}}.$$

5. Prove that the line passing through (1, –2, 0), having the direction ratios 2, 3, 6, and the line

$$\mathbf{3(x+3)=2(y-5)=z+1}$$

are parallel. Find the distance between the lines.

The lines are $\dfrac{x-1}{2}=\dfrac{y+2}{3}=\dfrac{z}{6}$ and

$$\frac{x+3}{1/3}=\frac{y-5}{1/2}=\frac{z+1}{1}.$$

Now, $\dfrac{2}{1/3}=\dfrac{3}{1/2}=\dfrac{6}{1}$. So, the lines are parallel.

The distance between the lines

= length of the perpendicular from (1, –2, 0)

to the line $\dfrac{x+3}{1/3}=\dfrac{y-5}{1/2}=\dfrac{z+1}{1}$.

The direction cosines of $\dfrac{x+3}{1/3}=\dfrac{y-5}{1/2}=\dfrac{z+1}{1}$ are

$$\frac{1/3}{\sqrt{\left(\frac{1}{3}\right)^2+\left(\frac{1}{2}\right)^2+1^2}},\ \frac{1/2}{\sqrt{\left(\frac{1}{3}\right)^2+\left(\frac{1}{2}\right)^2+1^2}},\ \frac{1}{\sqrt{\left(\frac{1}{3}\right)^2+\left(\frac{1}{2}\right)^2+1^2}},$$

i.e., $\dfrac{1/3}{7/6},\ \dfrac{1/2}{7/6},\ \dfrac{1}{7/6}$, i.e., $\dfrac{2}{7},\dfrac{3}{7},\dfrac{6}{7}$.

$\therefore$ the required length of the perpendicular

$$=\left\{\begin{vmatrix}\dfrac{x_1-\alpha}{l} & \dfrac{y_1-\beta}{m}\end{vmatrix}^2+\begin{vmatrix}\dfrac{y_1-\beta}{m} & \dfrac{z_1-\gamma}{n}\end{vmatrix}^2+\begin{vmatrix}\dfrac{z_1-\gamma}{n} & \dfrac{x_1-\alpha}{l}\end{vmatrix}^2\right\}^{1/2}$$

$$=\left\{\begin{vmatrix}\dfrac{1-(-3)}{2/7} & \dfrac{-2-5}{3/7}\end{vmatrix}^2+\begin{vmatrix}\dfrac{-2-5}{3/7} & \dfrac{0-(-1)}{6/7}\end{vmatrix}^2+\begin{vmatrix}\dfrac{0-(-1)}{6/7} & \dfrac{1-(-3)}{2/7}\end{vmatrix}^2\right\}^{1/2}$$

$$=\left\{\begin{vmatrix}\dfrac{4}{2/7} & \dfrac{-7}{3/7}\end{vmatrix}^2+\begin{vmatrix}\dfrac{-7}{3/7} & \dfrac{1}{6/7}\end{vmatrix}^2+\begin{vmatrix}\dfrac{1}{6/7} & \dfrac{4}{2/7}\end{vmatrix}^2\right\}^{1/2}$$

$$=\left\{\left(\frac{12}{7}+2\right)^2+\left(-6-\frac{3}{7}\right)^2+\left(\frac{2}{7}-\frac{24}{7}\right)^2\right\}^{1/2}$$

$$=\left\{\left(\frac{26}{7}\right)^2+\left(\frac{45}{7}\right)^2+\left(\frac{22}{7}\right)^2\right\}^{1/2}$$

$$=\frac{\sqrt{26^2+45^2+22^2}}{7}=\frac{\sqrt{3185}}{7}=\frac{7\sqrt{65}}{7}=\sqrt{65}.$$

$\therefore$ the distance $=\sqrt{65}$.

6. Find the foot of the perpendicular drawn from the point (6, –2, 5) to the straight line passing through the point (4, –1, 2), having direction numbers 1, 4, –3.

The line is $\dfrac{x-4}{1}=\dfrac{y+1}{4}=\dfrac{z-2}{-3}\ (=r)$

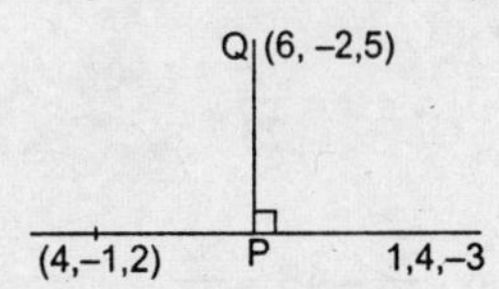

Any point on it is

$$P(4+r,\ -1+4r,\ 2-3r).$$

$\therefore$ direction ratios of PQ are

$$4+r-6,\ -1+4r-(-2),\ 2-3r-5,$$

i.e., $r-2,\ 4r+1,\ -3r-3$.

P is the foot of the perpendicular if PQ is perpendicular to the line. So,

$$1\cdot(r-2)+4(4r+1)+(-3)(-3r-3)=0$$

or $\quad r-2+16r+4+9r+9=0$

or $\quad 26r+11=0. \qquad \therefore\ r=\dfrac{-11}{26}.$

$$\therefore\quad P=\left(4-\frac{11}{26},\ -1-\frac{44}{26},\ 2+\frac{33}{26}\right)=\left(-\frac{93}{26},\ -\frac{35}{13},\ \frac{85}{26}\right)$$

Note By finding the foot P of the perpendicular, we can also find the length of the perpendicular.

7. Find the image of the point (0, 3, 2) with respect to the line $3(x-1)=\frac{y}{2}=\frac{z+1}{1}$.

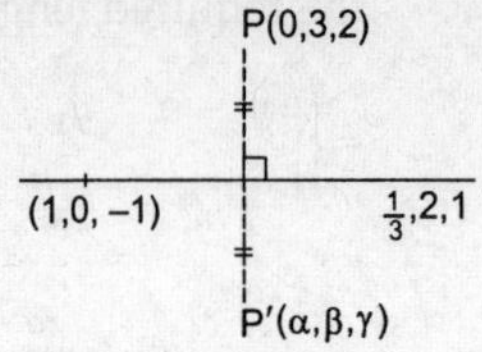

The equations of the line are

$$\frac{x-1}{1/3}=\frac{y}{2}=\frac{z+1}{1}.$$

Let the image of P(0, 3, 2) be P′(α, β, γ)

The line bisects PP′. So, $\left(\frac{0+\alpha}{2},\frac{3+\beta}{2},\frac{2+\gamma}{2}\right)$ is on the line.

$$\therefore \quad \frac{\frac{\alpha}{2}-1}{1/3}=\frac{\frac{3+\beta}{2}}{2}=\frac{\frac{2+\gamma}{2}+1}{1}$$

$$\Rightarrow \quad \frac{\alpha-2}{2/3}=\frac{3+\beta}{4}=\frac{\gamma+4}{2}. \qquad \ldots (1)$$

Also, PP′ is perpendicular to the line.

$$\therefore \quad (\alpha-0)\cdot\frac{1}{3}+(\beta-3)\cdot 2+(\gamma-2)\cdot 1=0$$

$$\text{or} \quad \frac{\alpha}{3}+2\beta+\gamma=8. \qquad \ldots (2)$$

$$(1)\Rightarrow \; 3+\beta=4\cdot\frac{3}{2}(\alpha-2) \quad \text{or} \quad \beta=6\alpha-15$$

$$\text{and} \quad \gamma+4=2\cdot\frac{3}{2}(\alpha-2) \quad \text{or} \quad \gamma=3\alpha-10.$$

$$\therefore \quad (2)\Rightarrow \; \frac{\alpha}{3}+2(6\alpha-15)+3\alpha-10=8$$

$$\text{or} \quad \frac{46}{3}\alpha=48. \qquad \therefore\; \alpha=\frac{72}{23}.$$

$$\therefore \quad \beta=6\alpha-15=6\cdot\frac{72}{23}-15=\frac{87}{23}$$

$$\gamma=3\alpha-10=3\cdot\frac{72}{23}-10=-\frac{14}{23}.$$

So, the image is $\left(\frac{72}{23},\frac{87}{23},-\frac{14}{23}\right)$.

8. Prove that the lines

$$\frac{x-1}{1}=\frac{y-1}{-2}=\frac{z-1}{1},\; \frac{x-2}{5}=\frac{y+1}{1}=\frac{z-2}{-6} \text{ and}$$

$$\frac{x+3}{8}=\frac{y+2}{-1}=\frac{z-8}{-7} \text{ are coplanar.}$$

Also, find the equations of a line perpendicular to the plane of the three lines, and passing through the intersection of the first two lines.

The lines $\frac{x-1}{1}=\frac{y-1}{-2}=\frac{z-1}{1}$ and $\frac{x-2}{5}=\frac{y+1}{1}=\frac{z-2}{-6}$

are coplanar if $\begin{vmatrix} x_1-x_2 & y_1-y_2 & z_1-z_2 \\ l_1 & m_1 & n_1 \\ l_2 & m_2 & n_2 \end{vmatrix}=0,$

$$\text{i.e.,} \quad \begin{vmatrix} 1-2 & 1-(-1) & 1-2 \\ 1 & -2 & 1 \\ 5 & 1 & 6 \end{vmatrix}=0$$

$$\text{or} \quad \begin{vmatrix} -1 & 2 & -1 \\ 1 & -2 & 1 \\ 5 & 1 & 6 \end{vmatrix}=0$$

or $\quad -1(-12-1)-2(6-5)-1(1+10)=0$, which is true.

$\therefore$ the first two lines are coplanar.

The lines $\frac{x-2}{5}=\frac{y+1}{1}=\frac{z-2}{-6}$

and $\frac{x+3}{8}=\frac{y+2}{-1}=\frac{z-8}{-7}$ are coplanar if

$$\begin{vmatrix} 2-(-3) & -1-(-2) & 2-8 \\ 5 & 1 & -6 \\ 8 & -1 & -7 \end{vmatrix}=0$$

or $\begin{vmatrix} 5 & 1 & -6 \\ 5 & 1 & -6 \\ 8 & -1 & -7 \end{vmatrix}=0$, which is true ($R_1\equiv R_2$).

$\therefore$ the last two lines are coplanar.

The lines $\frac{x-1}{1}=\frac{y-1}{-2}=\frac{z-1}{1}$

and $\frac{x+3}{8}=\frac{y+2}{-1}=\frac{z-8}{-7}$ are coplanar if

$$\begin{vmatrix} 1-(-3) & 1-(-2) & 1-8 \\ 1 & -2 & 1 \\ 8 & -1 & -7 \end{vmatrix}=0$$

$$\text{or} \quad \begin{vmatrix} 4 & 3 & -7 \\ 1 & -2 & 1 \\ 8 & -1 & -7 \end{vmatrix}=0$$

or $\quad 4(14+1)-3(-7-8)-7(-1+16)=0$, which is true.

$\therefore$ the first and the third lines are coplanar.

Hence, the three lines are coplanar.

Any point on the first line is $(1+r, 1-2r, 1+r)$, and any point on the second line is $(2+5r', -1+r', 2-6r')$. At the point of intersection of the first two lines,

$$1+r=2+5r',\; 1-2r=-1+r',\; 1+r=2-6r'$$

$$\Rightarrow \quad 1+r=2+5r'=2-6r' \text{ and } 1-2r=-1+r'$$

$$\Rightarrow \quad r'=0,\; r=1 \text{ which satisfy } 1-2r=-1+r'.$$

$\therefore$ the point of intersection of the first two lines is (2, −1, 2).

If l, m, n be the direction ratios of the perpendicular to the plane of the first two lines then

$$l(1)+m(-2)+n(1)=0,\; \text{i.e., } l-2m+n=0$$

$$l(5)+m(1)+n(-6)=0,\; \text{i.e., } 5l+m-6n=0$$

$\therefore \quad \dfrac{l}{12-1} = \dfrac{m}{5+6} = \dfrac{n}{1+10}$ or $\dfrac{l}{1} = \dfrac{m}{1} = \dfrac{n}{1}$.

$\therefore$ the required perpendicular is $\dfrac{x-2}{1} = \dfrac{y+1}{1} = \dfrac{z-2}{1}$.

9. Find the shortest distance between the lines $\dfrac{x-3}{2} = \dfrac{y-4}{1} = \dfrac{z+1}{-3}$ and $\dfrac{x-1}{-1} = \dfrac{y-3}{3} = \dfrac{z-1}{2}$.

Also, find the equations of the line of the shortest distance and the points of intersection of this line with the given lines.

Let the direction cosines of the line of shortest distance be l, m, n.

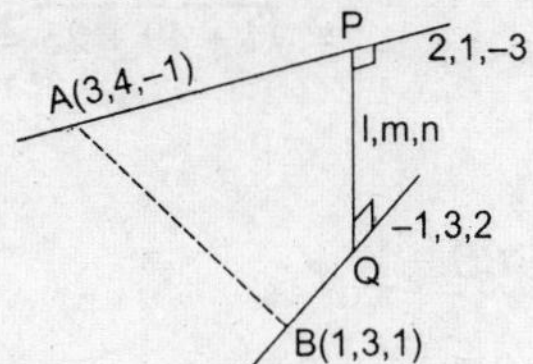

PQ ⊥ AP

$\Rightarrow \quad 2l + m - 3n = 0.$ ... (1)

PQ ⊥ BQ

$\Rightarrow \quad -l + 3m + 2n = 0.$ ... (2)

(1) and (2) $\Rightarrow \dfrac{1}{2+9} = \dfrac{m}{3-4} = \dfrac{n}{6+1}$

$\Rightarrow \dfrac{l}{11} = \dfrac{m}{-1} = \dfrac{n}{7}$.

$\therefore \quad l = \dfrac{11}{\sqrt{11^2\ (-1)^2 + 7^2}},\ m = \dfrac{-1}{\sqrt{11^2 + (-1)^2 + 7^2}},$

$n = \dfrac{7}{\sqrt{11^2 + (-1)^2 + 7^2}}$

or $\quad l = \dfrac{11}{3\sqrt{19}},\ m = \dfrac{-1}{3\sqrt{19}},\ n = \dfrac{7}{3\sqrt{19}}$.

$\therefore$ the shortest distance PQ

= projection of AB in the direction of PQ

$= |(x_2 - x_1)\, l + (y_2 - y_1)\, m + (z_2 - z_1)\, n|$

$= \left| (3-1)\dfrac{11}{3\sqrt{19}} + (4-3)\dfrac{-1}{3\sqrt{19}} + (-1-1)\dfrac{7}{3\sqrt{19}} \right|$

$= \dfrac{1}{3\sqrt{19}}\,|22 - 1 - 14| = \dfrac{7}{3\sqrt{19}}$.

Now, any point P on the line AP has coordinates $(3 + 2r, 4 + r, -1 - 3r)$, and any point Q on the line BQ has coordinates $(1 - r', 3 + 3r', 1 + 2r')$.

$\therefore$ direction ratios of PQ are

$3 + 2r - (1 - r'), 4 + r - (3 + 3r'), -1 - 3r - (1 + 2r'),$

i.e., $2r + r' + 2, r - 3r' + 1, -3r - 2r' - 2$.

But direction ratios of PQ are 11, –1, 7.

$\therefore \quad \dfrac{2r + r' + 2}{11} = \dfrac{r - 3r' + 1}{-1} = \dfrac{-3r - 2r' - 2}{7}$

$\Rightarrow \quad -2r - r' - 2 = 11r - 33r' + 11$

and $\quad 7r - 21r' + 7 = 3r + 2r' + 2$

$\Rightarrow \quad 13r - 32r' + 13 = 0$ and $4r - 23r' + 5 = 0$

$\Rightarrow \quad 13(r + 1) = 32r'$ and $4(r + 1) = 23r' - 1$

$\therefore \quad \dfrac{13}{4} = \dfrac{32r'}{23r' - 1}$ or $299r' - 13 = 128r'$

or $\quad 171r' = 13. \quad \therefore\ r' = \dfrac{13}{171}$.

$\therefore \quad r = \dfrac{128}{171 \times 4} - 1 = \dfrac{32}{171} - 1 = -\dfrac{139}{171}$.

$\therefore \quad P = \left(3 + 2 \cdot \dfrac{-139}{171}, 4 - \dfrac{139}{171}, -1 - 3 \cdot \dfrac{-139}{171}\right)$

$= \left(\dfrac{235}{171}, \dfrac{545}{171}, \dfrac{82}{57}\right)$

and $Q = \left(1 - \dfrac{13}{171}, 3 + 3 \cdot \dfrac{13}{171}, 1 + 2 \cdot \dfrac{13}{171}\right)$

$= \left(\dfrac{158}{171}, \dfrac{184}{57}, \dfrac{197}{171}\right)$.

[**Note:** We can also find s.d. = *PQ*.]

$\therefore$ the equations of the line of the shortest distance are

$$\frac{x - \frac{235}{171}}{11} = \frac{y - \frac{545}{171}}{-1} = \frac{z - \frac{82}{57}}{7}.$$

10. Are the following lines

$$\frac{x-5}{3} = \frac{y-7}{-1} = \frac{z+2}{1} \quad \text{and} \quad \frac{x+3}{-3} = \frac{y-3}{2} = \frac{z-6}{4}$$

skew? A line with direction ratios 2, 7, –5 is drawn to intersect them. Find the coordinates of the points of intersection and the length intercepted on it.

The straight lines $\dfrac{x-5}{3} = \dfrac{y-7}{-1} = \dfrac{z+2}{1}$

and $\dfrac{x+3}{-3} = \dfrac{y-3}{2} = \dfrac{z-6}{4}$ are coplanar if

$$\begin{vmatrix} x_2 - x_1 & y_2 - y_1 & z_2 - z_1 \\ l_1 & m_1 & n_1 \\ l_2 & m_2 & n_2 \end{vmatrix} = 0$$

i.e., $\begin{vmatrix} 5-(-3) & 7-3 & -2-6 \\ 3 & -1 & 1 \\ -3 & 2 & 4 \end{vmatrix} = 0$

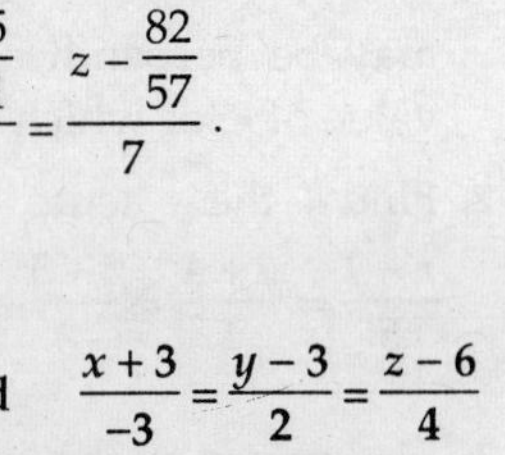

or $\begin{vmatrix} 8 & 4 & -8 \\ 3 & -1 & 1 \\ -3 & 2 & 4 \end{vmatrix} = 0$

or $\quad 8(-4 - 2) - 4(12 + 3) - 8(6 - 3) = 0$

or $\quad -48 - 60 - 24 = 0$, which is not true.

So, the lines are not coplanar. Hence, the lines are skew.

Let the line PQ with direction ratios 2, 7, –5 cut the lines at P and Q. Then P being a point on the first line has coordinates $(5 + 3r, 7 - r, -2 + r)$, and Q being a point on the second line has coordinates $(-3 - 3r', 3 + 2r', 6 + 4r')$.

$\therefore$ direction ratios of PQ are

$5 + 3r - (-3 - 3r'), 7 - r - (3 + 2r'), -2 + r - (6 + 4r')$

i.e., $3r + 3r' + 8, -r - 2r' + 4, r - 4r' - 8.$

As the direction ratios of PQ are given to be 2, 7, –5, we have

$$\frac{3r + 3r' + 8}{2} = \frac{-r - 2r' + 4}{7} = \frac{r - 4r' - 8}{-5}$$
$$= \frac{3r + 3r' + 8 + 3(-r - 2r' + 4)}{2 + 3 \times 7}$$
$$= \frac{-r - 2r' + 4 + r - 4r' - 8}{7 + (-5)}.$$

$\therefore \quad \frac{-3r' + 20}{23} = \frac{-6r' - 4}{2} \quad \Rightarrow -6r' + 40 = -138r' - 92$

$\Rightarrow \quad 132r' = -132 \quad \Rightarrow r' = -1.$

$\therefore \quad \frac{-r - 2r' + 4}{7} = \frac{r - 4r' - 8}{-5} \quad \Rightarrow \frac{-r + 6}{7} = \frac{r - 4}{-5}$

$\Rightarrow \quad 5r - 30 = 7r - 28 \quad \Rightarrow r = -1.$

$\therefore \quad P = (5 + 3(-1), 7 - (-1), -2 + (-1)) = (2, 8, -3)$

$Q = (-3 - 3(-1), 3 + 2(-1), 6 + 4(-1)) = (0, 1, 2)$

$\therefore \quad PQ = \sqrt{(2 - 0)^2 + (8 - 1)^2 + (-3 - 2)^2}$

$= \sqrt{4 + 49 + 25} = \sqrt{78}.$

Exercises

1. Find the value of k so that the lines $\frac{x - 1}{-1} = \frac{y + 3}{k} = \frac{z - 7}{2}$ and $\frac{x + 1}{k} = \frac{y}{2} = \frac{z + 6}{-3}$ may be perpendicular to each other. Is there any value of k for which the lines will be parallel?

2. Find the acute angle between the lines $\frac{x - 1}{5} = \frac{y + 4}{4} = \frac{z - 3}{-3}$ and $x = y = 2z$.

3. Find points on the line $\frac{x}{2} = \frac{y - 1}{3} = \frac{z + 2}{4}$ which are at a distance 5 from the point (4, 7, 6). Also, find a point on the line which is at a distance $2\sqrt{59}$ from the origin.

4. Show that the lines $\frac{x - 5}{4} = \frac{y - 7}{4} = \frac{z + 3}{-5}$ and $\frac{x - 8}{7} = \frac{y - 4}{1} = \frac{z - 5}{3}$ lie in the same plane. Also, find their point of intersection.

5. Find the equations of the straight line perpendicular to both the lines $\frac{x - 1}{1} = \frac{y - 1}{2} = \frac{z + 2}{3}$ and $\frac{x + 2}{2} = \frac{y - 5}{-1} = \frac{z + 3}{2}$ and passing through their point of intersection.

6. Find the equations of the perpendicular from the point (1, 6, 3) to the line $\frac{x}{1} = \frac{y - 1}{2} = \frac{z - 2}{3}$. Also, find the coordinates of the foot of the perpendicular.

7. Find the length of the perpendicular from the point (1, 1, 1) to the line joining the points (3, 2, 0) and (2, –1, 1). Also, find the foot of the perpendicular.

8. Prove that the lines $\frac{x - 1}{1} = \frac{y - 2}{2} = \frac{z - 3}{3}$ and $6(x + 3) = 3(y + 2) = 2(z + 1)$ are parallel. Find the distance between the lines. Also, find the distance between the lines in the direction parallel to the line $\frac{x + 2}{3} = \frac{y - 1}{2} = \frac{z}{1}$.

9. Find the equations of the line parallel to the line $\frac{x}{1} = \frac{y}{1} = \frac{z}{3}$, and which meets the lines $\frac{x - 1}{2} = \frac{y}{3} = \frac{z - 1}{4}$ and $\frac{x - 2}{1} = \frac{y - 1}{2} = \frac{z}{2}$. Also, find the intercept made on the line by the two lines.

10. Find the equations of the line intersecting the lines $\frac{x - a}{1} = \frac{y}{1} = \frac{z - a}{1}$ and $\frac{x + a}{1} = \frac{y}{1} = \frac{z + a}{2}$, and parallel to the line $\frac{x - a}{2} = \frac{y - a}{1} = \frac{z - 2a}{3}$.

11. Prove that the straight lines $\frac{x}{\alpha} = \frac{y}{\beta} = \frac{z}{\gamma}$, $\frac{x}{l} = \frac{y}{m} = \frac{z}{n}$ and $\frac{x}{a\alpha} = \frac{y}{b\beta} = \frac{z}{c\gamma}$ are coplanar if $\frac{l}{\alpha}(b - c) + \frac{m}{\beta}(c - a) + \frac{n}{\gamma}(a - b) = 0.$

12. If the lines $\frac{x}{\alpha/a} = \frac{y}{\beta/b} = \frac{z}{\gamma/c}$, $\frac{x}{\alpha} = \frac{y}{\beta} = \frac{z}{\gamma}$ and $\frac{x}{a\alpha} = \frac{y}{b\beta} = \frac{z}{c\gamma}$ are coplanar then prove that at least two of a, b, c must be equal.

13. Find the shortest distance between the lines $\frac{x - 1}{2} = \frac{y + 8}{-7} = \frac{z - 4}{5}$ and $\frac{x - 1}{2} = \frac{y - 2}{1} = \frac{z - 6}{-3}$.

14. Find the length and the equations of the line of shortest distance between the lines $\frac{x + 3}{-4} = \frac{y - 6}{3} = \frac{z}{2}$ and $\frac{x + 2}{-4} = \frac{y}{1} = \frac{z - 7}{1}$.

15. Find the equations of a line intersecting both the lines

$$\frac{x-3}{1}=\frac{y-5}{-2}=\frac{z-7}{1} \text{ and } \frac{x+1}{7}=\frac{y+1}{-6}=\frac{z+1}{1}$$

perpendicularly. Also, find the points of intersection of the required line with the given lines, and the length of the intercept on it by the given lines.

16. Two straight lines

$$\frac{x-\alpha}{l}=\frac{y-\beta}{m}=\frac{z-\gamma}{n} \text{ and } \frac{x-\alpha'}{l'}=\frac{y-\beta'}{m'}=\frac{z-\gamma'}{n'} \text{ are}$$

cut by a third line whose direction cosines are λ, μ, ν. Show that the length intercepted on the third line is given by

$$\begin{vmatrix} \alpha-\alpha' & \beta-\beta' & \gamma-\gamma' \\ l & m & n \\ l' & m' & n' \end{vmatrix} \div \begin{vmatrix} l & m & n \\ l' & m' & n' \\ \lambda & \mu & \nu \end{vmatrix}$$

17. Find the image Q of the point P(4, –2, 1) with respect to the line $3(x-1)=2(y+1)=z$. Also, find the point of intersection of the line PQ and the given line.

18. The image of the point P(0, –6, 5) is Q(–2, 4, –3) with respect to a line l whose direction numbers are $\lambda, \lambda-1, 2$. Find the equations of the line l independent of λ.

19. Prove that the lines $\frac{x+1}{1}=\frac{y+1}{1}=\frac{z}{2}$ and $\frac{x}{\sqrt{3}-1}=\frac{y}{-\sqrt{3}-1}=\frac{z-2}{4}$ intersect at an angle $\pi/3$. Find the equations of a line passing through (–1, –1, 0) which form an equilateral triangle with the given lines.

20. Prove that the lines $\frac{x-2}{1}=\frac{y-2}{2}=\frac{z+2}{3}$, $\frac{x-1}{1}=\frac{y}{-2}=\frac{z+5}{1}$ and $\frac{x-6}{0}=\frac{y+10}{2}=\frac{z}{1}$ are coplanar and they form a right-angled triangle. Also, find the area of the triangle.

21. Prove that the four lines

$$\frac{x}{1}=\frac{y-1}{7}=\frac{z-2}{0}, \frac{x}{1}=\frac{y-1}{-3}=\frac{z-2}{0},$$

$$\frac{x-1}{1}=\frac{y+2}{1}=\frac{z-2}{0} \text{ and } \frac{x}{1}=\frac{y+3}{3}=\frac{z-2}{0}$$

form the four sides, taken in order, of a concyclic quadrilateral.

22. Prove that the lines $\frac{x+1}{-3}=\frac{y-2}{1}=\frac{z-5}{1}$, $\frac{x-3}{-2}=\frac{y}{1}=\frac{z-3}{1}$ and $\frac{x-2}{3}=\frac{y-3}{1}=\frac{z-6}{1}$ are coplanar and concurrent. Also, find the equations of the line through the point of concurrency, perpendicular to each of the given lines.

23. Prove that the lines $\frac{x-4}{3}=\frac{y-3}{1}=\frac{z-1}{-2}$, $\frac{x+2}{-2}=\frac{y-1}{4}=\frac{z-5}{1}$ and $\frac{x-1}{1}=\frac{y-2}{5}=\frac{z-3}{-1}$ form a triangle. Find its centroid.

Objective Questions

24. The direction cosines of the line $\frac{x-2}{2}=\frac{y+1}{\sqrt{3}}=\frac{z}{3}$ are

(a) 2, –1, 0 (b) 2, $\sqrt{3}$, 3

(c) $\frac{1}{2}, \frac{\sqrt{3}}{4}, \frac{3}{4}$ (d) $-\frac{1}{2}, \frac{\sqrt{3}}{4}, 0$

25. A point on the line $\frac{x+1}{1}=\frac{y-3}{2}=\frac{z-1}{3}$ has the coordinates

(a) (–1, –3, 5) (b) (–1, 3, 3)

(c) (–3, 5, 1) (d) (–3, –1, –5)

26. A point on the line $\frac{x+2}{1}=\frac{y-3}{-4}=\frac{z-1}{2\sqrt{2}}$ at a distance 6 from the point (–2, 3, 1) is

(a) $(4, -21, 1+12\sqrt{2})$ (b) $\left(\frac{-4}{5}, \frac{-9}{5}, 1\right)$

(c) $\left(-\frac{16}{5}, \frac{39}{5}, \frac{5-12\sqrt{2}}{5}\right)$ (d) $\left(-\frac{16}{5}, -21, 1+12\sqrt{2}\right)$

27. If two lines with direction numbers 1, –2, 3 and $\lambda, 1-\lambda, 1$ are perpendicular to each other then λ is

(a) $-\frac{1}{3}$ (b) $\frac{1}{3}$ (c) –1 (d) $\frac{5}{3}$

28. If the lines $\frac{x-1}{-3}=\frac{y+1}{5\mu-3}=\frac{z}{6}$ and $\frac{x+2}{4\lambda+1}=\frac{y}{4}=\frac{z-2}{-18}$ are parallel to each other then the value of the pair (λ, μ) is

(a) $\left(2, -\frac{1}{3}\right)$ (b) $\left(2, \frac{1}{3}\right)$

(c) $\left(-2, \frac{1}{3}\right)$ (d) cannot be found

29. The angle between the lines $\frac{x}{3}=\frac{y}{4}=\frac{z}{-12}$ and $3x=-y=-4z$ is

(a) $\pi/2$ (b) $\pi/4$ (c) $\pi/3$ (d) $\pi/6$

30. The direction cosines of the line passing through the points (–1, 3, 4) and (4, 3, 9) are

(a) 5, 0, 5 (b) $\frac{1}{\sqrt{2}}, 0, \frac{1}{\sqrt{2}}$

(c) $\frac{5}{\sqrt{2}}, 0, \frac{5}{\sqrt{2}}$ (d) $\frac{-2}{13}, \frac{9}{26}, \frac{18}{13}$

31. The lines $\frac{x-1}{1}=\frac{y+3}{-3}=\frac{z-6}{6}$ and $\frac{x+1}{-1}=\frac{y-1}{1}=\frac{z-6}{6}$ intersect at

(a) (1, –3, 6) (b) (–1, 1, 6)
(c) (0, 0, 0) (d) no point

32. The foot of the perpendicular from the origin to the line $\frac{x-1}{0}=\frac{y-1}{2}=\frac{z-1}{-1}$ is

(a) $\left(1, \frac{3}{5}, \frac{6}{5}\right)$ (b) $\left(1, 0, \frac{6}{5}\right)$
(c) $\left(-1, 0, \frac{6}{5}\right)$ (d) $\left(-1, \frac{3}{5}, \frac{6}{5}\right)$

33. The lines $\frac{x-2}{1}=\frac{y-3}{1}=\frac{z-4}{-k}$ and $\frac{x-1}{k}=\frac{y-4}{2}=\frac{z-5}{1}$ are coplanar if

(a) $k = 1$ or -1 (b) $k = 0$ or -3
(c) $k = 3$ or -3 (d) $k = 0$ or -1

Answers

1. 6; no 2. $\pi/4$

3. $\left(4 \pm \frac{10}{\sqrt{29}}, 7 \pm \frac{15}{\sqrt{29}}, 6 \pm \frac{20}{\sqrt{29}}\right)$; (6, 10, 10) 4. (1, 3, 2)

5. $\frac{x-2}{7}=\frac{y-3}{4}=\frac{z-1}{-5}$

6. $\frac{x-1}{0}=\frac{y-6}{-3}=\frac{z-3}{2}$; (1, 3, 5)

7. $\sqrt{\frac{30}{11}}; \left(\frac{27}{11}, \frac{4}{11}, \frac{6}{11}\right)$ 8. $4\cdot\sqrt{\frac{3}{7}}$; $\sqrt{14}$

9. $\frac{x-9}{1}=\frac{y-12}{1}=\frac{z-17}{3}$; $3\cdot\sqrt{11}$

10. $\frac{x-a}{2}=\frac{y}{1}=\frac{z-a}{3}$ 13. $4\sqrt{3}$

14. 9; $\frac{x-1}{1}=\frac{y-3}{-4}=\frac{z+2}{8}$

15. $\frac{x-3}{2}=\frac{y-5}{3}=\frac{z-7}{4}$; (3, 5, 7), (–1, –1, –1); $2\sqrt{29}$

17. $\left(-\frac{62}{49}, \frac{54}{49}, \frac{59}{49}\right)$; $\left(\frac{67}{49}, \frac{-22}{49}, \frac{54}{49}\right)$

18. $\frac{x+1}{13}=\frac{y+1}{9}=\frac{z-1}{8}$

19. $\frac{x+1}{1}=\frac{y+1}{2+\sqrt{3}}=\frac{z}{0}$ 20. $25\sqrt{21}$

22. $\frac{x+1}{0}=\frac{y-2}{-1}=\frac{z-5}{1}$ 23. $\left(-\frac{1}{3}, 0, 4\right)$

24. (c) 25. (d) 26. (c) 27. (a)
28. (b) 29. (a) 30. (b) 31. (c)
32. (a) 33. (b)

Chapter Test

Time: 45 minutes

1. (i) The equations of a line are $\frac{x+4}{l} = \frac{y-2}{m} = \frac{z-1}{n}$ where $l = 2m = 3n$. Then the angle that the line makes with the positive direction of the y-axis is

A. $\cos^{-1}\frac{2}{7}$ B. 0° C. $\cos^{-1}\frac{3}{7}$ D. $\cos^{-1}\frac{6}{7}$

(ii) The projection of the line segment AB where A = (3, 7, –5) and B = (–2, 4, 8) on the x-axis is

A. 5 B. 1 C. 3 D. 13

(iii) The point P(0, 1, –1) is shifted parallel to the line $\frac{x+1}{3} = \frac{y}{4} = \frac{z-1}{0}$ by a distance 5. The coordinates of P in the new position are

A. (3, 5, 1) B. (–3, –3, –1) C. (–3, 5, –1) D. (3, –5, –1)

(iv) The number of possible values of λ for which the line joining the points $(\lambda + 1, -2, 1)$ and $(3, 1, \lambda)$ also passes through the point (–4, 2, 5) is

A. 1 B. 0 C. 2 D. infinite

2. The points (x_1, y_1, z_1) and (x_2, y_2, z_2) are at distances r_1 and r_2 respectively from the origin. If the two points are collinear with the origin then prove that $|x_1x_2 + y_1y_2 + z_1z_2| = r_1r_2$.

3. The line joining the points (1, 3, 5) and (2, –1, 4) is projected in the line L_1 in the plane $z = 2$ and in the line L_2 in the plane $x + y = 1$. Are the lines L_1 and L_2 intersecting? If not, find the shortest distance between lines L_1 and L_2.

4. A line $\frac{x+2}{1} = \frac{y-3}{2} = \frac{z-k}{3}$ cuts the y–z plane and the x–y plane at A and B respectively. If $\angle AOB = \pi/2$ then find k, where O is the origin.

5. Prove that the lines $\frac{x-1}{2} = \frac{y+1}{1} = \frac{z-k}{4}$ and $\frac{x}{-1} = \frac{y-1}{2} = \frac{z-(k+1)}{1}$ are intersecting for all values of k. Also, find the point of intersection of the lines for $k = 2$.

6. Find the angle between the lines $\frac{x-1}{l} = \frac{y+1}{m} = \frac{z}{n}$ and $\frac{x+1}{m} = \frac{y-3}{n} = \frac{z-1}{l}$ where $l > m > n$, and l, m, n are the roots of the cubic equation $x^3 + x^2 - 4x = 4$.

□

Answers

1. (i) C (ii) A (iii) B (iv) B

3. No; $\frac{30}{\sqrt{293}}$ **4.** $\frac{9}{2}$

5. (1, –1, 2) **6.** $\cos^{-1}\frac{4}{9}$

□

3. Planes

Recap of Facts and Formulae

1. Some properties of a plane

(i) Three noncollinear points determine a plane, i.e., given three noncollinear points, there is one and only one plane passing through them. Hence, through two intersecting lines there is one and only one plane.

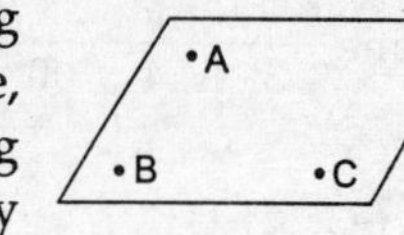

(ii) A line perpendicular to a plane is perpendicular to all lines in the plane.

(iii) If a line l is perpendicular to two lines in a plane then the line l is perpendicular to the plane, and hence, the line l is perpendicular to all lines in the plane.

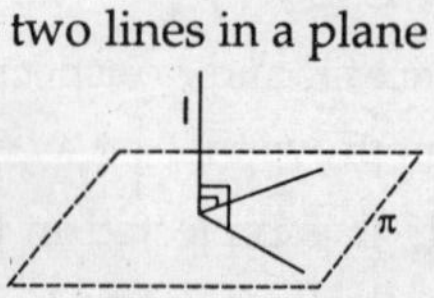

2. Equation of a plane passing through a point and perpendicular to a direction

- Let π be a plane passing through $A(x_1, y_1, z_1)$ and d.r. of its normal be a, b, c. Let $P(x, y, z)$ be any point on the plane.

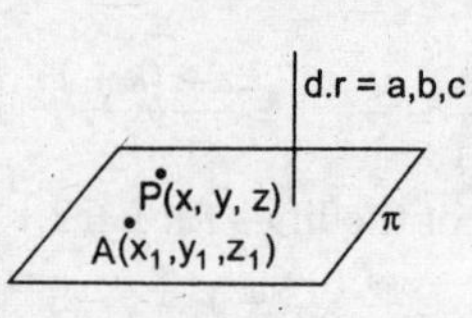

Direction ratios of AP are $x - x_1, y - y_1, z - z_1$.

Direction ratios of the normal are a, b, c.

AP $\perp$ normal $\Rightarrow$ $a(x - x_1) + b(y - y_1) + c(z - z_1) = 0$.

This is the equation of the plane.

Vectorially, if a plane passes through the point whose position vector is $\vec{r}_1 = x_1\vec{i} + y_1\vec{j} + z_1\vec{k}$ and $\vec{n}$ = a vector perpendicular to the plane, the equation of the plane is $(\vec{r} - \vec{r}_1) \cdot \vec{n} = 0$.

3. General equation of a plane

- The general equation of a plane is $ax + by + cz + d = 0$ where a, b, c are d.r. of the normal to the plane. This is a linear equation in three variables x, y, z.

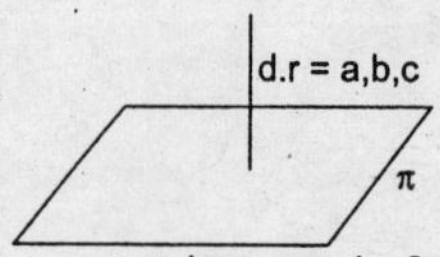

4. Equation of planes of reference

- The equation of the x–y plane is $z = 0$; that of the y–z plane is $x = 0$; that of the z–x plane is $y = 0$.

5. Equation of a plane through three given points

- Let $A(x_1, y_1, z_1)$, $B(x_2, y_2, z_2)$ and $C(x_3, y_3, z_3)$ be three given points.

  The equation of any plane through $A(x_1, y_1, z_1)$ is

  $$a(x - x_1) + b(y - y_1) + c(z - z_1) = 0. \quad \ldots (1)$$

  B is on (1) $\Rightarrow a(x_2 - x_1) + b(y_2 - y_1) + c(z_2 - z_1) = 0. \ldots(2)$

  C is on (1) $\Rightarrow a(x_3 - x_1) + b(y_3 - y_1) + c(z_3 - z_1) = 0. \ldots(3)$

  Eliminating a, b and c from (1), (2) and (3), the equation of the plane ABC is

  $$\begin{vmatrix} x - x_1 & y - y_1 & z - z_1 \\ x_2 - x_1 & y_2 - y_1 & z_2 - z_1 \\ x_3 - x_1 & y_3 - y_1 & z_3 - z_1 \end{vmatrix} = 0.$$

  Vectorially, if the position vectors of A, B and C are respectively $\vec{r}_1, \vec{r}_2$ and $\vec{r}_3$ then the equation of the plane ABC is $[\vec{r} - \vec{r}_1 \;\; \vec{r} - \vec{r}_2 \;\; \vec{r} - \vec{r}_3] = 0$,

  $\vec{r}$ being the position vector of any point P on the plane and applying the fact that $\vec{AP}, \vec{BP}$ and $\vec{CP}$ are coplanar.

6. Equation of a plane in the intercept form

- The equation of a plane cutting intercepts a, b, c on the x-axis, y-axis and z-axis respectively is $\dfrac{x}{a} + \dfrac{y}{b} + \dfrac{z}{c} = 1$.

7. Equation of a plane in the normal form

- The equation of a plane whose distance from the origin is p, and the direction cosines of the perpendicular from the origin to the plane are l, m, n, is $lx + my + nz = p$.

8. Coplanarity of four points

- The points $A(x_1, y_1, z_1)$, $B(x_2, y_2, z_2)$, $C(x_3, y_3, z_3)$ and $D(x_4, y_4, z_4)$ are coplanar if

$$\begin{vmatrix} x_2 - x_1 & y_2 - y_1 & z_2 - z_1 \\ x_3 - x_1 & y_3 - y_1 & z_3 - z_1 \\ x_4 - x_1 & y_4 - y_1 & z_4 - z_1 \end{vmatrix} = 0.$$

Vectorially, the points $A(\vec{r_1})$, $B(\vec{r_2})$, $C(\vec{r_3})$ and $D(\vec{r_4})$ are coplanar if $[\vec{r_4} - \vec{r_1} \;\; \vec{r_4} - \vec{r_2} \;\; \vec{r_4} - \vec{r_3}] = 0$ because $\vec{AD}$, $\vec{BD}$ and $\vec{CD}$ are coplanar.

9. Line of intersection of two planes

- Let the equation of two intersecting planes be $a_1x + b_1y + c_1z + d_1 = 0$ and $a_2x + b_2y + c_2z + d_2 = 0$.

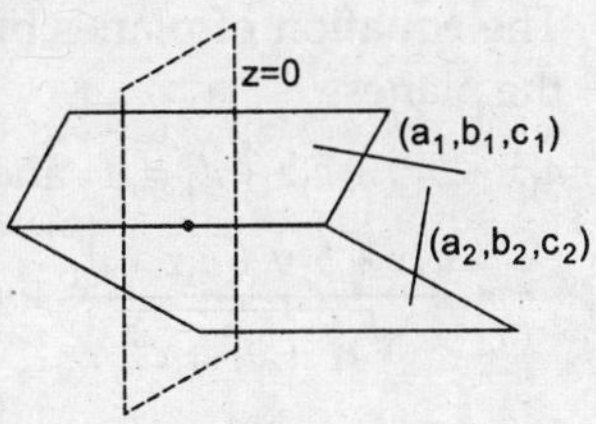

Then the equation of the line of intersection is given by

$$\begin{cases} a_1x + b_1y + c_1z + d_1 = 0 & \dots (1) \\ a_2x + b_2y + c_2z + d_2 = 0 & \dots (2) \end{cases}$$

A point on the line of intersection is obtained by solving (1) and (2) with $z = 0$ (i.e., the x–y plane). We get

$$\left.\begin{aligned} a_1x + b_1y + d_1 = 0 \\ a_2x + b_2y + d_2 = 0 \end{aligned}\right\} \Rightarrow \frac{x}{b_1d_2 - b_2d_1} = \frac{y}{d_1a_2 - d_2a_1} = \frac{1}{a_1b_2 - a_2b_1}$$

$\therefore$ a point on the line is $\left(\dfrac{b_1d_2 - b_2d_1}{a_1b_2 - a_2b_1}, \dfrac{d_1a_2 - d_2a_1}{a_1b_2 - a_2b_1}, 0\right)$.

If l, m, n are direction ratios of the line then by perpendicularity,

$$\left.\begin{aligned} a_1l + b_1m + c_1n = 0 \\ a_2l + b_2m + c_2n = 0 \end{aligned}\right\} \Rightarrow \frac{l}{b_1c_2 - b_2c_1} = \frac{m}{c_1a_2 - c_2a_1} = \frac{n}{a_1b_2 - a_2b_1}.$$

$\therefore$ the equation of the line in symmetric form is

$$\frac{x - \dfrac{b_1d_2 - b_2d_1}{a_1b_2 - a_2b_1}}{b_1c_2 - b_2c_1} = \frac{y - \dfrac{d_1a_2 - d_2a_1}{a_1b_2 - a_2b_1}}{c_1a_2 - c_2a_1} = \frac{z - 0}{a_1b_2 - a_2b_1}.$$

10. Condition for a line to lie in a plane

- The line $\dfrac{x - \alpha}{l} = \dfrac{y - \beta}{m} = \dfrac{z - \gamma}{n}$ lies in the plane $ax + by + cz + d = 0$ if
  (i) the point (α, β, γ) of the line lies in the plane, i.e.,
  $a\alpha + b\beta + c\gamma + d = 0$, and
  (ii) the normal to the plane is perpendicular to the line, i.e.,
  $al + bm + cn = 0$.

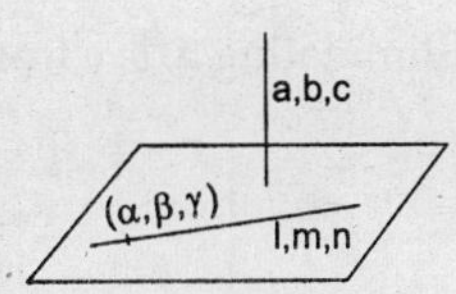

11. Point of intersection of a line and a plane

- The point of intersection of the line $\dfrac{x - \alpha}{l} = \dfrac{y - \beta}{m} = \dfrac{z - \gamma}{n}$ and the plane $ax + by + cz + d = 0$ is $P = (x_1, y_1, z_1)$ where
  $x_1 = \alpha + lr, y_1 = \beta + mr, z_1 = \gamma + nr$
  and $a(\alpha + lr) + b(\beta + mr) + c(\gamma + nr) + d = 0$.

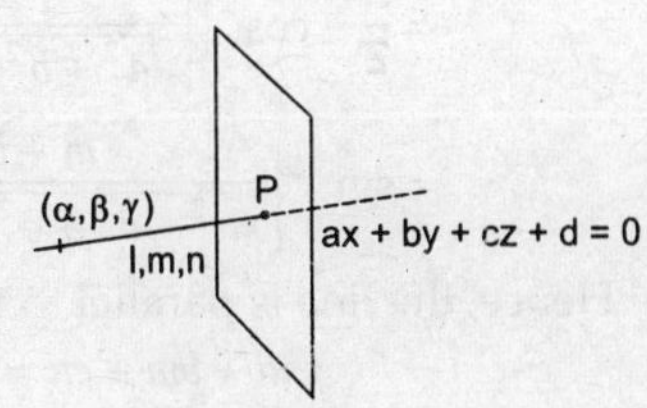

12. Length of perpendicular from a point to a plane

- The length of the perpendicular from the point $P(x_1, y_1, z_1)$ to the plane $ax + by + cz + d = 0$ is

$$p = \left|\frac{ax_1 + by_1 + cz_1 + d}{\sqrt{a^2 + b^2 + c^2}}\right|.$$

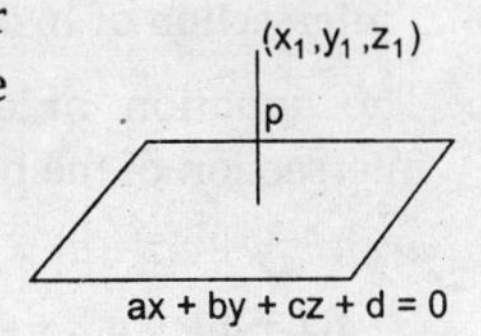

13. Angle between two planes

- The angle between two planes $a_1x + b_1y + c_1z + d_1 = 0$ and $a_2x + b_2y + c_2z + d_2 = 0$ is the angle between the normals to the two planes. So, it is the angle between two lines whose d.r. are a_1, b_1, c_1 and a_2, b_2, c_2.

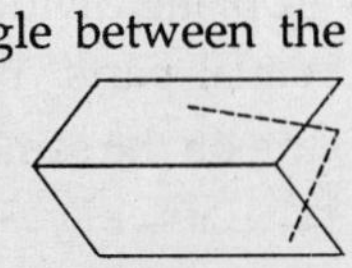

$\therefore$ the angle θ between the planes is given by

$$\cos\theta = \left|\frac{a_1a_2 + b_1b_2 + c_1c_2}{\sqrt{a_1^2 + b_1^2 + c_1^2}\,\sqrt{a_2^2 + b_2^2 + c_2^2}}\right|.$$

14. Condition for parallelism and perpendicularity of two planes

- The condition of parallelism and perpendicularity of two planes $a_1x + b_1y + c_1z + d_1 = 0$ and $a_2x + b_2y + c_2z + d_2 = 0$ are the same as those for two lines whose d.c. are a_1, b_1, c_1 and a_2, b_2, c_2.

$\therefore$ the condition for parallelism is $\dfrac{a_1}{a_2} = \dfrac{b_1}{b_2} = \dfrac{c_1}{c_2}$

and the condition for perpendicularity is

$$a_1a_2 + b_1b_2 + c_1c_2 = 0.$$

15. Angle between a line and a plane

- The angle between a line $\dfrac{x - \alpha}{l} = \dfrac{y - \beta}{m} = \dfrac{z - \gamma}{n}$ and the plane $ax + by + cz + d = 0$ is $\dfrac{\pi}{2} - \theta$ where θ is the angle between the line and the normal to the plane. Therefore,

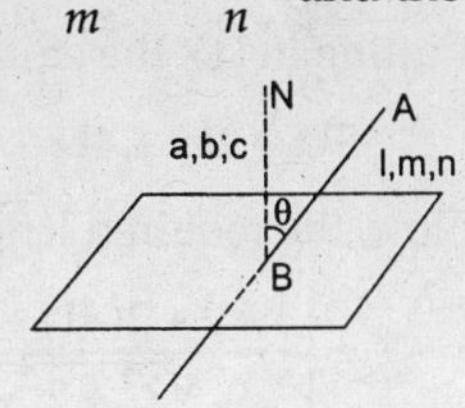

the angle between the line and the plane

$$= \frac{\pi}{2} - \cos^{-1}\left|\frac{al + bm + cn}{\sqrt{a^2+b^2+c^2}\ \sqrt{l^2+m^2+n^2}}\right|$$

$$= \sin^{-1}\left|\frac{al + bm + cn}{\sqrt{a^2+b^2+c^2}\ \sqrt{l^2+m^2+n^2}}\right|.$$

Hence, the line is parallel to the plane if

$$al + bm + cn = 0$$

and the line is perpendicular to the plane if

AB || BN which implies $\frac{a}{l} = \frac{b}{m} = \frac{c}{n}$.

16. Equation of any plane passing through the intersection of two planes

- The equation of any plane passing through the intersection of the planes

$$a_1x + b_1y + c_1z + d_1 = 0 \text{ and } a_2x + b_2y + c_2z + d_2 = 0 \text{ is}$$

$$a_1x + b_1y + c_1z + d_1 + \lambda(a_2x + b_2y + c_2z + d_2) = 0$$

where λ is an arbitrary constant.

17. Sides of a plane

- A plane divides the three-dimensional space in two equal parts. Two points $A(x_1, y_1, z_1)$ and $B(x_2, y_2, z_2)$ are on the same side of the plane $ax + by + cz + d = 0$ if

$$ax_1 + by_1 + cz_1 + d \text{ and } ax_2 + by_2 + cz_2 + d$$

are both positive or both negative.

Two points $A(x_1, y_1, z_1)$ and $C(x_3, y_3, z_3)$ are on opposite sides of the plane

$ax + by + cz + d = 0$ if

$$ax_1 + by_1 + cz_1 + d$$

and $ax_3 + by_3 + cz_3 + d$

are of opposite signs.

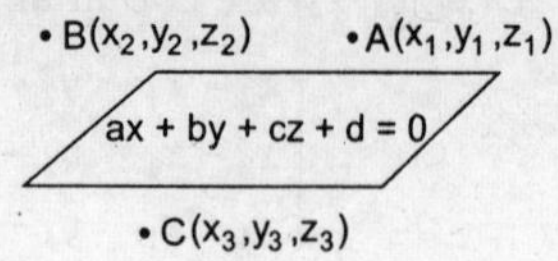

18. Equations of planes bisecting the angles between two planes

- The equation of planes bisecting the angles between the planes

$a_1x + b_1y + c_1z + d_1 = 0$ and $a_2x + b_2y + c_2z + d_2 = 0$ are

$$\frac{a_1x + b_1y + c_1z + d_1}{\sqrt{a_1^2 + b_1^2 + c_1^2}} = \pm \frac{a_2x + b_2y + c_2z + d_2}{\sqrt{a_2^2 + b_2^2 + c_2^2}}. \quad \dots (1)$$

If $d_1 > 0$, $d_2 > 0$ then the equation of the plane bisecting the angle between the given planes that contains the origin, is obtained by taking the positive sign in (1). Also, the origin is in the acute angle between the planes

$$a_1x + b_1y + c_1z + d_1 = 0, a_2x + b_2y + c_2z + d_2 = 0$$

where $d_1 > 0$, $d_2 > 0$ if $a_1a_2 + b_1b_2 + c_1c_2 < 0$.

Selected Solved Examples

1. Find the equation of the plane passing through the point (1, –2, 3) and perpendicular to the line $\frac{x-2}{3} = \frac{y+1}{2} = \frac{z-1}{4}$. Also, find the length of the perpendicular from (2, –1, 1) to the plane.

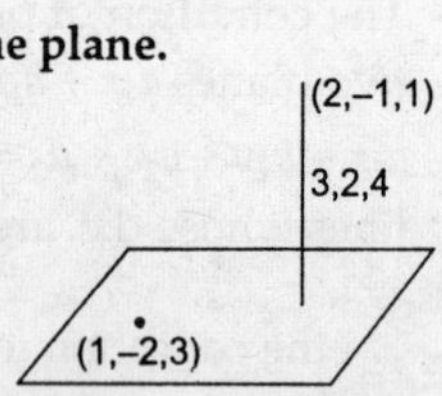

Let the equation of the plane be

$$ax + by + cz + d = 0. \quad \dots (1)$$

(1, –2, 3) is a point on it. So,

$$a - 2b + 3c + d = 0. \quad \dots (2)$$

The direction ratios of the perpendicular to the plane (1), which is parallel to the given line whose direction ratios are 3, 2, 4, are a, b, c.

$\therefore \quad \frac{a}{3} = \frac{b}{2} = \frac{c}{4} (= k)$. Then $a = 3k, b = 2k, c = 4k$.

$\therefore \quad$ from (2), $3k - 4k + 12k + d = 0$. $\quad \therefore\ d = -11k$.

Putting in (1), the equation of the plane is

$$3kx + 2ky + 4kz - 11k = 0 \quad \text{or} \quad 3x + 2y + 4z = 11.$$

Now, the required length of the perpendicular

$$= \left|\frac{3 \times 2 + 2(-1) + 4(1) - 11}{\sqrt{3^2 + 2^2 + 4^2}}\right| = \left|\frac{-3}{\sqrt{29}}\right| = \frac{3}{\sqrt{29}}.$$

2. Find the equation of the plane passing through the points (4, 7, 8), (2, 3, 4) and (–2, –4, 2). Also, prove that the point (2, 4, 10) is coplanar with the three given points.

Any plane passing through the point (4, 7, 8) has the equation

$$a(x - 4) + b(y - 7) + c(z - 8) = 0. \quad \dots (1)$$

It passes through the points (2, 3, 4) and (–2, –4, 2). So,

$$a(-2) + b(-4) + c(-4) = 0,$$

i.e., $a + 2b + 2c = 0$ $\quad \dots (2)$

and $a(-2 - 4) + b(-4 - 7) + c(2 - 8) = 0$

i.e., $6a + 11b + 6c = 0$. $\quad \dots (3)$

Eliminating a, b, c from (1), (2) and (3), we get

$$\begin{vmatrix} x-4 & y-7 & z-8 \\ 1 & 2 & 2 \\ 6 & 11 & 6 \end{vmatrix} = 0. \quad \dots (4)$$

This is the required equation of the plane.

The point (2, 4, 10) will be coplanar with the other three points if it satisfies (4).

$$\text{Now, } \begin{vmatrix} 2-4 & 4-7 & 10-8 \\ 1 & 2 & 2 \\ 6 & 11 & 6 \end{vmatrix} = \begin{vmatrix} -2 & -3 & 2 \\ 1 & 2 & 2 \\ 6 & 11 & 6 \end{vmatrix}$$

$$= -2(12-22) + 3(6-12) + 2(11-12) = 20 - 18 - 2 = 0$$

$\therefore$ (2, 4, 10) is a point on the plane of the other three points.

3. A plane passing through the line $\dfrac{x+3}{-1} = \dfrac{y-1}{2} = \dfrac{z+1}{k}$ and perpendicular to the line $\dfrac{x}{3} = \dfrac{y-1}{5} = \dfrac{z+1}{7}$ exists. Find the value of k and the equation of the plane.

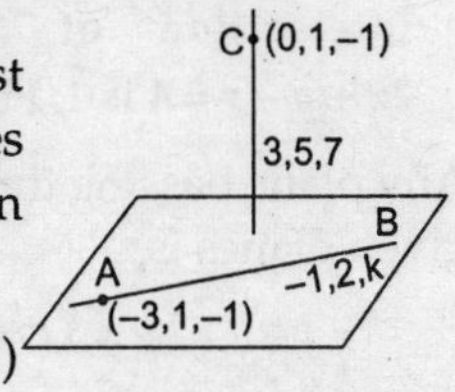

As (−3, 1, −1) is a point on the first line, the required plane passes through (−3, 1, −1). So, its equation will be of the form

$a(x+3) + b(y-1) + c(z+1) = 0.$ ...(1)

The direction ratios of the normal to (1) are a, b, c. This normal is perpendicular to AB and parallel to CD.

$\therefore \quad a(-1) + b \cdot 2 + c \cdot k = 0$ ... (2)

and $\dfrac{a}{3} = \dfrac{b}{5} = \dfrac{c}{7}.$ ... (3)

From (3) and (2), $3(-1) + 5 \cdot 2 + 7 \cdot k = 0 \quad \Rightarrow \quad 7k = -7$

$\therefore \quad k = -1.$

From (1) and (3), $3(x+3) + 5(y-1) + 7(z+1) = 0$

$\Rightarrow \quad 3x + 5y + 7z + 11 = 0.$

This is the required equation of the plane.

4. A plane meets the coordinate axes at A, B and C such that the centroid of the ΔABC is the point (a, b, c). Find the equation of the plane ABC.

Let the equation of the plane be $\dfrac{x}{\lambda} + \dfrac{y}{\mu} + \dfrac{z}{\nu} = 1$ (intercept form).

Then $OA = \lambda$, $OB = \mu$, $OC = \nu$

$\Rightarrow \quad A = (\lambda, 0, 0), B = (0, \mu, 0), C = (0, 0, \nu).$

$\therefore$ the centroid of the ΔABC

$$= \left(\frac{\lambda+0+0}{3}, \frac{0+\mu+0}{3}, \frac{0+0+\nu}{3}\right)$$

$= (a, b, c)$ (given).

$\therefore \quad \dfrac{\lambda}{3} = a, \dfrac{\mu}{3} = b, \dfrac{\nu}{3} = c.$ So, $\lambda = 3a$, $\mu = 3b$, $\nu = 3c$.

$\therefore$ the equation of the plane ABC is

$$\frac{x}{3a} + \frac{y}{3b} + \frac{z}{3c} = 1, \text{ i.e., } \frac{x}{a} + \frac{y}{b} + \frac{z}{c} = 3.$$

5. Find the equation of the plane containing the line $\dfrac{x-\alpha}{l} = \dfrac{y-\beta}{m} = \dfrac{z-\gamma}{n}$ and passing through the point (a, b, c).

Any plane passing through the point (a, b, c) has the equation

$A(x-a) + B(y-b) + C(z-c) = 0.$ ... (1)

As the line $\dfrac{x-\alpha}{l} = \dfrac{y-\beta}{m} = \dfrac{z-\gamma}{n}$ is in the plane, the point (α, β, γ) is on the plane.

So, $A(\alpha - a) + B(\beta - b) + C(\gamma - c) = 0.$ ... (2)

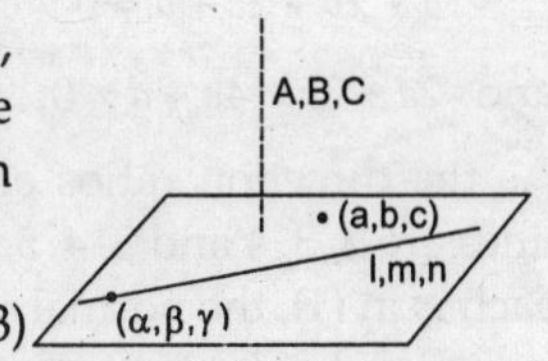

The normal to the plane (1), which is perpendicular to the given line, has the direction ratios A, B, C. So,

$Al + Bm + Cn = 0.$ ... (3)

Eliminating A, B, C from (1), (2), (3), the required equation of the plane is

$$\begin{vmatrix} x-a & y-b & z-c \\ \alpha-a & \beta-b & \gamma-c \\ l & m & n \end{vmatrix} = 0.$$

6. Find the equation of the plane containing the line $\dfrac{x-\alpha}{l} = \dfrac{y-\beta}{m} = \dfrac{z-\gamma}{n}$ and perpendicular to the plane $ax + by + cz + d = 0$.

Let the equation of the plane be

$Ax + By + Cz + D = 0.$ ... (1)

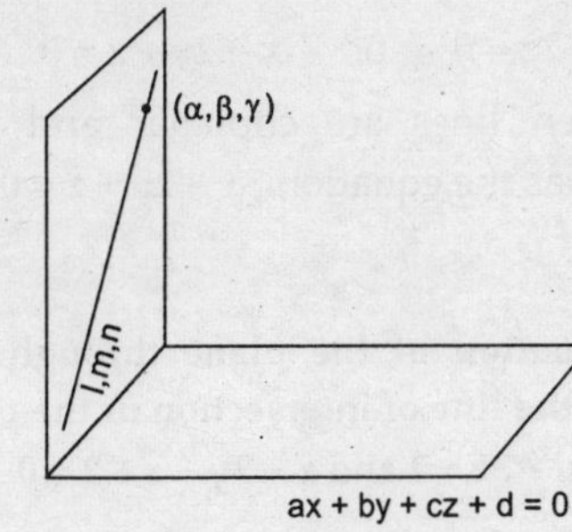

The line $\dfrac{x-\alpha}{l} = \dfrac{y-\beta}{m} = \dfrac{z-\gamma}{n}$ is in it.

So, $A\alpha + B\beta + C\gamma + D = 0$ ... (2)

and $Al + Bm + Cn = 0.$ ... (3)

Also, (1) is perpendicular to the plane $ax + by + cz + d = 0$.

$\therefore \quad Aa + Bb + Cc = 0.$ ... (4)

From (1) − (2), $A(x-\alpha) + B(y-\beta) + C(z-\gamma) = 0.$ ... (5)

Eliminating A, B, C from (5), (3), (4), we get

$$\begin{vmatrix} x-\alpha & y-\beta & z-\gamma \\ l & m & n \\ a & b & c \end{vmatrix} = 0.$$

This is the required equation of the plane

7. Prove that the lines

$$\frac{x-1}{2}=\frac{y-2}{3}=\frac{z-3}{4} \text{ and } \frac{x-2}{3}=\frac{y-3}{4}=\frac{z-4}{5}$$

are coplanar. Find the equation of the plane passing through the two lines.

If the lines are coplanar, they will lie in a plane.

Let the plane be $ax+by+cz+d=0$. ... (1)

It passes through (1, 2, 3) and (2, 3, 4). So,

$a+2b+3c+d=0$ ... (2)

and $2a+3b+4c+d=0$. ... (3)

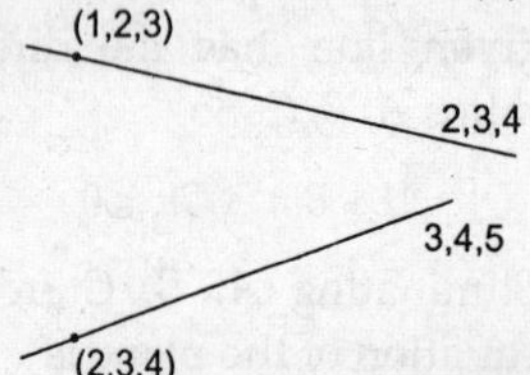

As the direction ratios of the lines are 2, 3, 4 and 3, 4, 5, and each is in (1), the normal to (1) is perpendicular to each of the lines.

$\therefore \quad 2a+3b+4c=0$ ... (4)

and $3a+4b+5c=0$. ... (5)

(3) and (4) $\Rightarrow$ $d=0$. So, the equation of the plane is

$ax+by+cz=0$

$a+2b+3c=0$

and $3a+4b+5c=0$.

Eliminating a, b, c, we get $\begin{vmatrix} x & y & z \\ 1 & 2 & 3 \\ 3 & 4 & 5 \end{vmatrix}=0$

or $-2x+4y-2z=0$ or $x-2y+z=0$.

Thus, the given lines are coplanar and the plane through them has the equation $x-2y+z=0$.

8. Find the equation of the plane through the point (1, –2, 3) and the line of intersection of the planes $4x+y-3z=1$ and $x-3y+z+2=0$.

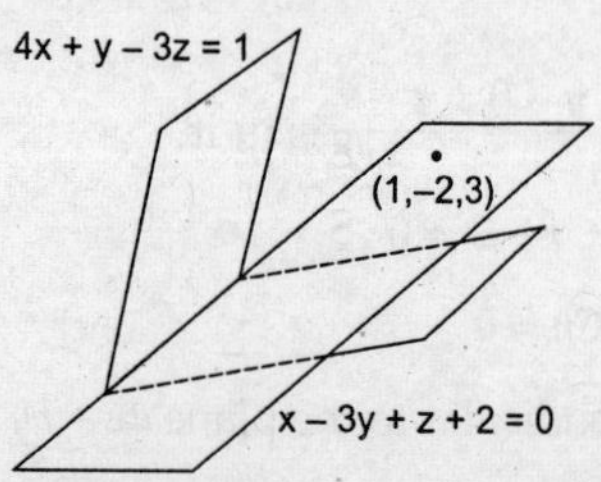

Any plane through the line of intersection of the given planes is

$4x+y-3z-1+\lambda(x-3y+z+2)=0$. ... (1)

It will pass through the point (1, –2, 3) if

$4.1+(-2)-3.3-1+\lambda\{1-3(-2)+3+2\}=0$

or $-8+\lambda\cdot 12=0$. $\therefore \lambda=\frac{2}{3}$.

Putting $\lambda=\frac{2}{3}$ in (1), the required equation of the plane is

$$4x+y-3z-1+\frac{2}{3}(x-3y+z+2)=0$$

or $12x+3y-9z-3+2x-6y+2z+4=0$

or $14x-3y-7z+1=0$.

9. The length of the perpendicular from the point (1, –1, 1) to a plane passing through the line of intersection of the planes $x+3y+5z=0$ and $2x+y-z=4$ is 1. Find the equation of the plane(s).

Any plane passing through the line of intersection of the given planes is

$x+3y+5z+\lambda(2x+y-z-4)=0$

or $(1+2\lambda)x+(3+\lambda)y+(5-\lambda)z-4\lambda=0$.

$\therefore$ the length of the perpendicular from the point (1, –1, 1) to the plane

$$=\left|\frac{(1+2\lambda)\cdot 1+(3+\lambda)\cdot(-1)+(5-\lambda)\cdot 1-4\lambda}{\sqrt{(1+2\lambda)^2+(3+\lambda)^2+(5-\lambda)^2}}\right|=2$$

(from the question).

$$\therefore \left|\frac{3-4\lambda}{\sqrt{6\lambda^2+35}}\right|=1 \quad \Rightarrow \quad (3-4\lambda)^2=6\lambda^2+35$$

or $9-24\lambda+16\lambda^2=6\lambda^2+35$ or $10\lambda^2-24\lambda-26=0$

or $5\lambda^2-12\lambda-13=0$

$$\therefore \lambda=\frac{12\pm\sqrt{144+260}}{10}=\frac{12\pm 2\sqrt{101}}{10}=\frac{6\pm\sqrt{101}}{5}.$$

$\therefore$ the equation of the plane is

$$x+3y+5z+\frac{6\pm\sqrt{101}}{5}(2x+y-z-4)=0$$

or $$\left(1+\frac{12\pm 2\sqrt{101}}{5}\right)x+\left(3+\frac{6\pm\sqrt{101}}{5}\right)y+\left(5-\frac{6\pm\sqrt{101}}{5}\right)z-\frac{24\pm 4\sqrt{101}}{5}=0$$

or $$(17\pm 2\sqrt{101})x+(21\pm\sqrt{101})y+(19\mp\sqrt{101})z=24\pm 4\sqrt{101}.$$

10. Show that the line of intersection of the planes $7x-4y+7z+16=0$, $4x+3y-2z+3=0$ is coplanar with the line of intersection of the planes $x-3y+4z+6=0$, $x-y+z+1=0$. Also, find the equation of the plane of the two lines of intersection.

Any plane through the line of intersection of the planes

$7x-4y+7z+16=0$, $4x+3y-2z+3=0$

is $\quad 7x - 4y + 7z + 16 + \lambda(4x + 3y - 2z + 3) = 0. \quad \dots (1)$

Any plane through the line of intersection of the planes

$$x - 3y + 4z + 6 = 0,\ x - y + z + 1 = 0$$

is $\quad x - 3y + 4z + 6 + \mu(x - y + z + 1) = 0. \quad \dots (2)$

If for some λ and μ, the planes (1) and (2) are the same then the two lines of intersection will be coplanar, and for that value of λ, (1) will be the plane of the two lines of intersection. Now, (1) and (2) are the same if the corresponding coefficients are proportional, i.e.,

$$\frac{7+4\lambda}{1+\mu} = \frac{-4+3\lambda}{-3-\mu} = \frac{7-2\lambda}{4+\mu} = \frac{16+3\lambda}{6+\mu}$$

$$\Rightarrow \frac{7+4\lambda}{1+\mu} = \frac{3+7\lambda}{-2} = \frac{-6\lambda}{3} = \frac{9-\lambda}{5} \text{ (by ratio-proportion).}$$

Now, $\dfrac{3+7\lambda}{-2} = \dfrac{-6\lambda}{3} \quad \Rightarrow 3 + 7\lambda = 4\lambda \quad \Rightarrow \lambda = -1$

$\dfrac{-6\lambda}{3} = \dfrac{9-\lambda}{5} \quad \Rightarrow -30\lambda = 2\lambda - 3\lambda \quad \Rightarrow \lambda = -1$

and $\dfrac{7+4\lambda}{1+\mu} + \dfrac{-6\lambda}{3} \quad \Rightarrow \dfrac{7-4}{1+\mu} = 2 \quad \Rightarrow 3 = 2 + 2\mu$

$\Rightarrow \quad \mu = \dfrac{1}{2}.$

$\therefore$ for $\lambda = -1$ and $\mu = \frac{1}{2}$, both (1) and (2) are the same.

$\therefore$ the two lines of intersection are coplanar and lie on the plane

$$7x - 4y + 7z + 16 + (-1)(4x + 3y - 2z + 3) = 0$$

or $\quad 3x - 7y + 9z + 13 = 0.$

11. Prove that the three planes $4x + 3y - 3z = 5$, $5x + 5y - 2z = 10$ and $2x - y - 5z + 5 = 0$ intersect along a common line.

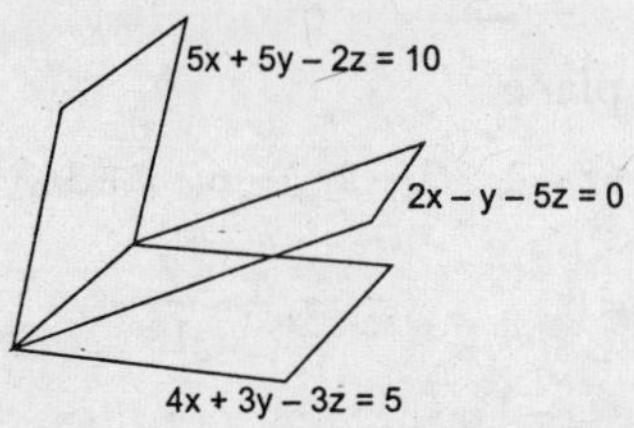

Any plane passing through the intersection of the first two planes has the equation

$$4x + 3y - 3z - 5 + \lambda(5x + 5y - 2z - 10) = 0$$

or $\quad (5\lambda + 4)x + (5\lambda + 3)y - (2\lambda + 3)z - 5(2\lambda + 1) = 0. \dots (1)$

If for some value of λ, (1) becomes

$2x - y - 5z + 5 = 0$ then the result is established.

In that case we must have

$$\frac{5\lambda+4}{2} = \frac{5\lambda+3}{-1} = \frac{-(2\lambda+3)}{-5} = \frac{-5(2\lambda+1)}{5}. \quad \dots (2)$$

Now, $\dfrac{5\lambda+4}{2} = \dfrac{5\lambda+3}{-1} \quad \Rightarrow -5\lambda - 4 = 10\lambda + 6$

$\Rightarrow \quad 15\lambda = -10. \quad \therefore \lambda = \dfrac{-2}{3}.$

For $\lambda = -\frac{2}{3}$, each of the ratios in (2) is $\frac{1}{3}$.

$\therefore \quad$ (2) holds for $\lambda = \dfrac{-2}{3}.$

Hence, the three planes have a common line of intersection.

12. Prove that the equations of the perpendicular drawn from the origin upon the line $\dfrac{x-\alpha}{l} = \dfrac{y-\beta}{m} = \dfrac{z-\gamma}{n}$ are given by $\begin{vmatrix} x & y & z \\ \alpha & \beta & \gamma \\ l & m & n \end{vmatrix} = 0 = lx + my + nz.$

The required perpendicular OP is the intersection of the following two planes:

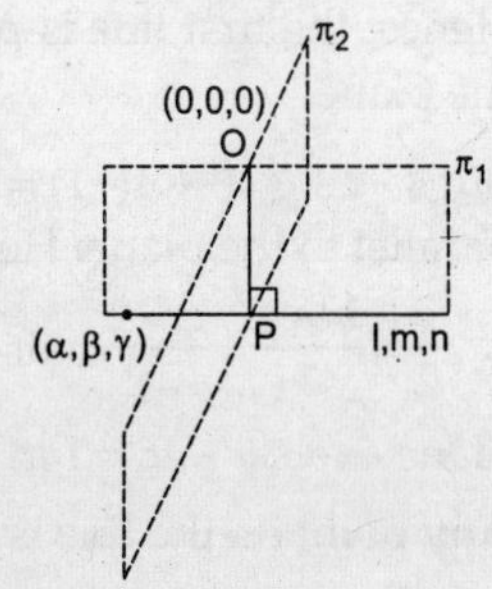

(i) the plane π_1 through the origin O and the given line

(ii) the plane π_2 through O, perpendicular to the given line

The equation of π_2—

Any plane through the origin (0, 0, 0) has the equation

$$ax + by + cz = 0. \quad \dots (1)$$

The line $\dfrac{x-\alpha}{l} = \dfrac{y-\beta}{m} = \dfrac{z-\gamma}{n}$

is perpendicular to the plane if

$$\frac{l}{a} = \frac{m}{b} = \frac{n}{c}$$

($\because$ normal to the plane is parallel to the line).

$\therefore \quad$ from (1), $\quad lx + my + nz = 0, \quad \dots (3)$

which is the equation of the plane π_2.

The equation of π_1—

Any plane through the origin is $Ax + By + Cz = 0. \quad \dots (4)$

It passes through (α, β, γ). So, $A\alpha + B\beta + C\gamma = 0. \quad \dots (5)$

Also, the line $\dfrac{x-\alpha}{l} = \dfrac{y-\beta}{m} = \dfrac{z-\gamma}{n}$ is on the plane.

So, $Al + Bm + Cn = 0. \quad \dots (6)$

Eliminating A, B, C from (4), (5), (6), we get

$$\begin{vmatrix} x & y & z \\ \alpha & \beta & \gamma \\ l & m & n \end{vmatrix} = 0,$$ which is the equation of the plane π_1.

$\therefore \quad$ the equations of the perpendicular OP are given by

$$\begin{vmatrix} x & y & z \\ \alpha & \beta & \gamma \\ l & m & n \end{vmatrix} = 0 = lx + my + nz.$$

13. Prove that among the two lines whose equations are

$$\frac{x-2}{1} = \frac{y}{-4} = \frac{z+1}{2} \text{ and } \frac{x}{2} = \frac{y-1}{3} = \frac{z-2}{-1},$$

one is parallel to the plane $4x + 3y + 4z = 1$, and the other intersects this plane. Find the point of intersection and the angle that the line makes with the plane.

The direction ratios of the normal to the plane

$4x + 3y + 4z = 1$ are 4, 3, 4, and

the direction ratios of the first line are 1, –4, 2 and those of the second line are 2, 3, –1.

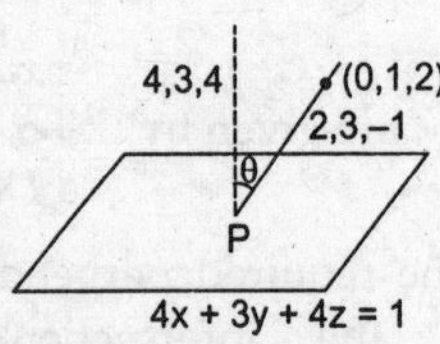

Clearly, $4 \cdot 1 + 3 \cdot (-4) + 4 \cdot 2 = 0$.

So, the first line is perpendicular to the normal to the plane. Hence, the first line is parallel to the plane.

But $4 \cdot 2 + 3 \cdot 3 + 4(-1) = 13 \neq 0$. So, the second line is not parallel to the plane. Hence, the line

$\frac{x}{2} = \frac{y-1}{3} = \frac{z-2}{-1}$ cuts the

plane $4x + 3y + 4z = 1$ at an angle.

Any point on the line is $(0 + 2r, 1 + 3r, 2 - r)$.

It is the point of intersection P if

$$4(0 + 2r) + 3(1 + 3r) + 4(2 - r) = 1$$

$$\Rightarrow \quad 13r + 11 = 0. \quad \therefore \ r = \frac{-11}{13}.$$

$$\therefore \quad P = \left(-\frac{22}{13}, 1 - \frac{33}{13}, 2 + \frac{11}{13}\right) = \left(-\frac{22}{13}, -\frac{20}{13}, \frac{37}{13}\right).$$

The angle between the line and the plane is $\frac{\pi}{2} - \theta$ where θ is the angle between the lines having direction ratios 2, 3, –1 and 4, 3, 4, i.e., the direction cosines are

$$\frac{2}{\sqrt{2^2 + 3^2 + (-1)^2}}, \frac{3}{\sqrt{2^2 + 3^2 + (-1)^2}}, \frac{-1}{\sqrt{2^2 + 3^2 + (-1)^2}}$$

$$\text{and } \frac{4}{\sqrt{4^2 + 3^2 + 4^2}}, \frac{3}{\sqrt{4^2 + 3^2 + 4^2}}, \frac{4}{\sqrt{4^2 + 3^2 + 4^2}}$$

$$\text{i.e., } \frac{2}{\sqrt{14}}, \frac{3}{\sqrt{14}}, \frac{-1}{\sqrt{14}} \text{ and } \frac{4}{\sqrt{41}}, \frac{3}{\sqrt{41}}, \frac{4}{\sqrt{41}}.$$

$$\therefore \quad \cos\theta = l_1 l_2 + m_1 n_2 + n_1 n_2$$

$$= \frac{2}{\sqrt{14}} \cdot \frac{4}{\sqrt{41}} + \frac{3}{\sqrt{14}} \cdot \frac{3}{\sqrt{41}} + \frac{-1}{\sqrt{14}} \cdot \frac{4}{\sqrt{41}}$$

$$= \frac{8 + 9 - 4}{\sqrt{14} \cdot \sqrt{41}} = \frac{13}{\sqrt{14} \cdot \sqrt{41}}.$$

$$\therefore \quad \theta = \cos^{-1}\left(\frac{13}{\sqrt{14 \times 41}}\right).$$

$$\therefore \text{ the required angle} = \frac{\pi}{2} - \cos^{-1}\left(\frac{13}{\sqrt{574}}\right) = \sin^{-1}\left(\frac{13}{\sqrt{574}}\right).$$

14. Prove that the two planes $x - 2y + 2z = 6$ and $3x - 6y + 6z = 2$ are parallel. Also,

(i) find the distance between the planes

(ii) find the intercept on the line $\frac{x-1}{2} = \frac{y+1}{3} = \frac{z}{-1}$ between the two planes.

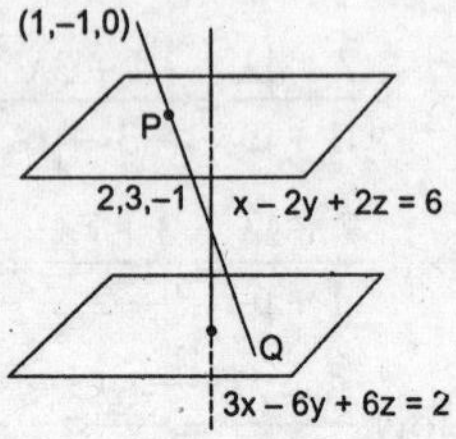

The direction ratios of normals to the planes are 1, –2, 2 and 3, –6, 6 respectively. Clearly, $\frac{1}{3} = \frac{-2}{-6} = \frac{2}{6}$.

$\therefore$ the two planes have the same direction ratios of normals.

$\therefore$ the two planes are parallel.

(i) A point on $x - 2y + 2z = 6$ is (0, 0, 3) (by trial).

$\therefore$ the required distance

= the length of the perpendicular from (0, 0, 3) to the plane $3x - 6y + 6z = 2$

$$= \left|\frac{3 \times 0 - 6 \times 0 + 6 \times 3 - 2}{\sqrt{3^2 + (-6)^2 + 6^2}}\right| = \left|\frac{16}{\sqrt{81}}\right| = \frac{16}{9}.$$

(ii) Let the line cut the planes at P and Q. Then using any point on the line,

$P = (1 + 2r_1, -1 + 3r_1, -r_1)$ and $Q = (1 + 2r_2, -1 + 3r_2, -r_2)$.

As P is on the plane

$$x - 2y + 2z = 6, \ 1 + 2r_1 - 2(-1 + 3r_1) + 2(-r_1) = 6$$

$$\text{or} \quad -6r_1 = 3 \quad \text{or} \quad r_1 = -\frac{1}{2}.$$

$$\therefore \quad P = \left(1 + 2 \cdot \frac{-1}{2}, -1 + 3 \cdot \frac{-1}{2}, -\frac{-1}{2}\right) = \left(0, -\frac{5}{2}, \frac{1}{2}\right).$$

As Q is on the plane

$$3x - 6y + 6z = 2, \ 3(1 + 2r_2) - 6(-1 + 3r_2) + 6(-r_2) = 2$$

$$\text{or} \quad -18r_2 + 7 = 0. \quad \therefore \ r_2 = -\frac{7}{18}.$$

$$\therefore \quad Q = \left(1 + 2 \cdot \frac{-7}{18}, -1 + 3 \cdot \frac{-7}{18}, -\frac{-7}{18}\right) = \left(\frac{2}{9}, -\frac{13}{6}, \frac{7}{18}\right).$$

$\therefore$ the length of the required intercept = PQ

$$= \sqrt{\left(0 - \frac{2}{9}\right)^2 + \left(-\frac{5}{2} + \frac{13}{6}\right)^2 + \left(\frac{1}{2} - \frac{7}{18}\right)^2}$$

$$= \sqrt{\frac{4}{81} + \frac{1}{9} + \frac{1}{81}} = \sqrt{\frac{14}{81}} = \frac{1}{9}\sqrt{14}.$$

15. The plane $ax + by = 0$ is rotated about its line of intersection with the plane $z = 0$ through an angle α.

Prove that the equation of the plane in its new position is $ax + by \pm \sqrt{a^2+b^2}\tan\alpha \cdot z = 0$.

Any plane through the line of intersection of the planes $ax + by = 0$ and $z = 0$ has the equation

$$ax + by + \lambda z = 0. \quad \ldots (1)$$

If the plane (1) makes an angle α with $ax + by = 0$ then (1) is the required equation.

The direction ratios of the normal to (1) are a, b, λ and that for $ax + by = 0$ are $a, b, 0$.

$$\therefore \quad \tan\alpha = \left|\frac{\sqrt{(l_1m_2 - l_2m_1)^2 + (m_1n_2 - m_2n_1)^2 + (n_1l_2 - n_2l_1)^2}}{l_1l_2 + m_1m_2 + n_1n_2}\right|$$

$$= \left|\frac{\sqrt{(ab - ab)^2 + (b\cdot 0 - b\lambda)^2 + (\lambda\cdot a - 0\cdot a)^2}}{a\cdot a + b\cdot b + \lambda\cdot 0}\right|$$

$$= \left|\frac{\sqrt{(a^2+b^2)\lambda^2}}{a^2+b^2}\right| = \frac{|\lambda|}{\sqrt{a^2+b^2}}.$$

$$\therefore \quad \lambda = \pm\sqrt{a^2+b^2}\cdot\tan\alpha.$$

$\therefore$ from (1), the required equation of the plane in the new position is $ax + by \pm \sqrt{a^2+b^2}\cdot\tan\alpha\cdot z = 0$.

16. Prove that the line $\frac{x}{1} = \frac{y}{1} = \frac{z-1}{-2}$ lies in the plane $x + y + z = 1$. Find the lines in the plane through the point (0, 0, 1), which are inclined at an angle $\cos^{-1}\left(\frac{1}{\sqrt{6}}\right)$ with the line.

Clearly, (0, 0, 1) is a point on the plane $x + y + z = 1$.

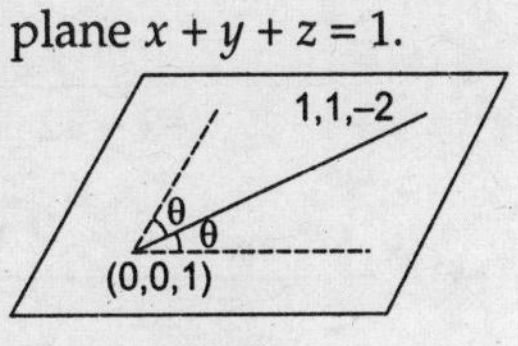

The direction ratios of the normal to the plane are 1, 1, 1 and those of the line are 1, 1, –2. As $1\cdot 1 + 1\cdot 1 + 1\cdot(-2) = 0$, the line $\frac{x}{1} = \frac{y}{1} = \frac{z-1}{-2}$ lies in the plane $x + y + z = 1$.

Let l, m, n be the direction cosines of a line in the plane making an angle θ with the line where $\cos\theta = \frac{1}{\sqrt{6}}$.

$$\text{Then } \cos\theta = l\cdot\frac{1}{\sqrt{1^2+1^2+(-2)^2}} + m\cdot\frac{1}{\sqrt{1^2+1^2+(-2)^2}} + n\cdot\frac{-2}{\sqrt{1^2+1^2+(-2)^2}}$$

$$\text{or} \quad \frac{1}{\sqrt{6}} = \frac{l}{\sqrt{6}} + \frac{m}{\sqrt{6}} - \frac{2n}{\sqrt{6}} \quad \Rightarrow \quad l + m - 2n = 1. \quad \ldots (1)$$

Also, the normal to the plane having direction ratios 1, 1, 1 is perpendicular to the line having direction cosines l, m, n.

$$\therefore \quad l\cdot 1 + m\cdot 1 + n\cdot 1 = 0. \quad \ldots (2)$$

$$\text{Again } l^2 + m^2 + n^2 = 1. \quad \ldots (3)$$

From (1) – (2), $-3n = 1$ or $n = -\frac{1}{3}$.

Then $l + m = \frac{1}{3} \Rightarrow m = \frac{1}{3} - l$.

$$\therefore \quad (3) \Rightarrow l^2 + \left(\frac{1}{3} - l\right)^2 + \left(-\frac{1}{3}\right)^2 = 1$$

$$\text{or} \quad l^2 + \frac{1}{9} + l^2 - \frac{2}{3}l + \frac{1}{9} - 1 = 0$$

$$\text{or} \quad 2l^2 - \frac{2}{3}l - \frac{7}{9} = 0 \quad \text{or} \quad 18l^2 - 6l - 7 = 0.$$

$$\therefore \quad l = \frac{6 \pm \sqrt{36 + 4\cdot 18\cdot 7}}{36} = \frac{6 \pm 6\sqrt{1+14}}{36} = \frac{1\pm\sqrt{15}}{6}.$$

$$\therefore \quad m = \frac{1}{3} - l = \frac{1}{3} - \frac{1\pm\sqrt{15}}{6} = \frac{1\mp\sqrt{15}}{6}.$$

$\therefore$ the direction cosines of the required lines are

$$\frac{1+\sqrt{15}}{6}, \frac{1-\sqrt{15}}{6}, -\frac{1}{3} \quad \text{and} \quad \frac{1-\sqrt{15}}{6}, \frac{1+\sqrt{15}}{6}, -\frac{1}{3}.$$

$$\therefore \quad \text{the lines are } \frac{x}{\frac{1+\sqrt{15}}{6}} = \frac{y}{\frac{1-\sqrt{15}}{6}} = \frac{z-1}{-\frac{1}{3}}$$

$$\text{and} \quad \frac{x}{\frac{1-\sqrt{15}}{6}} = \frac{y}{\frac{1+\sqrt{15}}{6}} = \frac{z-1}{-\frac{1}{3}}.$$

17. P(–3, 2, 5) is a point and $2x + 3y - z + 2 = 0$ is a plane. Find

(i) the length of the perpendicular from P to the plane

(ii) the foot of the perpendicular from P to the plane

(iii) the image of P with respect to the plane.

(i) The length of the perpendicular

$$= \left|\frac{2(-3) + 3(2) + (-1)\cdot 5 + 2}{\sqrt{2^2 + 3^2 + (-1)^2}}\right|$$

$$= \left|\frac{-3}{\sqrt{14}}\right| = \frac{3}{\sqrt{14}}.$$

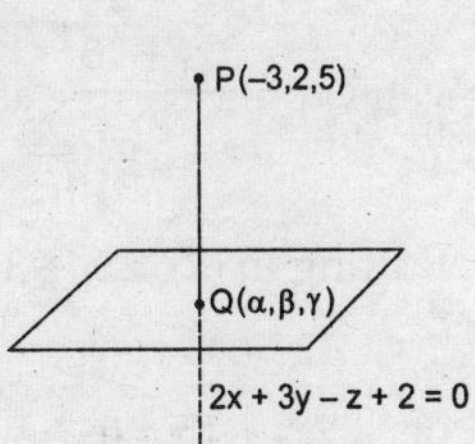

(ii) Let the foot of the perpendicular be Q(α, β, γ).

As Q is on the plane,

$$2\alpha + 3\beta - \gamma + 2 = 0. \quad \ldots (1)$$

Also, PQ ⊥ the plane. So, $(\alpha + 3), (\beta - 2), (\gamma - 5)$ are proportional to 2, 3, –1 (the normal to the plane being parallel to PQ).

$$\therefore \quad \frac{\alpha+3}{2} = \frac{\beta-2}{3} = \frac{\gamma-5}{-1}$$

$$\Rightarrow \quad 3(\alpha+3) = 2(\beta-2) \quad \text{and} \quad -1(\alpha+3) = 2(\gamma-5)$$

$$\Rightarrow \quad 2\beta = 3\alpha + 13 \quad \text{and} \quad 2\gamma = -\alpha + 7$$

$$\Rightarrow \quad \beta = \frac{3\alpha + 13}{2}, \gamma = \frac{-\alpha + 7}{2}.$$

Putting in (1), $2\alpha + 3 \cdot \frac{3\alpha + 13}{2} - \frac{-\alpha + 7}{2} + 2 = 0$

or $\left(2 + \frac{9}{2} + \frac{1}{2}\right)\alpha + \frac{39}{2} - \frac{7}{2} + 2 = 0$ or $7\alpha + 18 = 0.$

$$\therefore \quad \alpha = \frac{-18}{7}.$$

$$\therefore \quad \beta = \frac{3\alpha + 13}{2} = \frac{3 \times \frac{-18}{7} + 13}{2} = \frac{37}{14}$$

and $\gamma = \frac{-\alpha + 7}{2} = \frac{\frac{18}{7} + 7}{2} = \frac{67}{14}.$

$\therefore$ the foot Q is $\left(-\frac{18}{7}, \frac{37}{14}, \frac{67}{14}\right).$

(iii) Let the image P′ be (x_1, y_1, z_1). The middle point Q of PP′ is $\left(\frac{x_1 - 3}{2}, \frac{y_1 + 2}{2}, \frac{z_1 + 5}{2}\right).$

It is on the plane. So,

$$2 \cdot \frac{x_1 - 3}{2} + 3 \cdot \frac{y_1 + 2}{2} - \frac{z_1 + 5}{2} + 2 = 0$$

or $2(x_1 - 3) + 3(y_1 + 2) - (z_1 + 5) + 4 = 0$

or $2x_1 + 3y_1 - z_1 - 1 = 0.$ (2)

Also, PP′ ⊥ the plane. So, $\frac{x_1 + 3}{2} = \frac{y_1 - 2}{3} = \frac{z_1 - 5}{-1}$

$$\Rightarrow \quad 3(x_1 + 3) = 2(y_1 - 2) \quad \text{and} \quad -(x_1 + 3) = 2(z_1 - 5)$$

$$\Rightarrow \quad 2y_1 = 3x_1 + 13 \quad \text{and} \quad 2z_1 = -x_1 + 7.$$

$$\therefore \quad y_1 = \frac{3x_1 + 13}{2} \quad \text{and} \quad z_1 = \frac{-x_1 + 7}{2}.$$

Putting in (2), $2x_1 + 3 \cdot \frac{3x_1 + 13}{2} - \frac{-x_1 + 7}{2} - 1 = 0$

or $7x_1 + 15 = 0.$ $\therefore \; x_1 = \frac{-15}{7}.$

$$\therefore \quad y_1 = \frac{3x_1 + 13}{2} = \frac{3 \times \frac{-15}{7} + 13}{2} = \frac{23}{7}$$

and $z_1 = \frac{-x_1 + 7}{2} = \frac{\frac{15}{7} + 7}{2} = \frac{32}{7}.$

$\therefore$ the required image is $\left(-\frac{15}{7}, \frac{23}{7}, \frac{32}{7}\right).$

18. Find the shortest distance between the line $\frac{x-3}{3} = \frac{y-8}{-1} = \frac{z-3}{1}$ and the line of intersection of the planes $2x + 5y - z + 47 = 0$ and $2x + y + z + 7 = 0$. Also, find the equation of the plane parallel to both the lines and which lies midway between the lines.

The equations of the line of intersection of the planes

$$2x + 5y - z + 47 = 0 \quad \text{and} \quad 2x + y + z + 7 = 0$$

in the symmetric form are

$$\frac{x - \frac{b_1d_2 - b_2d_1}{a_1b_2 - a_2b_1}}{b_1c_2 - b_2c_1} = \frac{y - \frac{d_1a_2 - d_2a_1}{a_1b_2 - a_2b_1}}{c_1a_2 - c_2a_1} = \frac{z - 0}{a_1b_2 - a_2b_1}$$

or $$\frac{x - \frac{5 \times 7 - 1 \times 47}{2 \times 1 - 2 \times 5}}{5 \times 1 - 1 \times (-1)} = \frac{y - \frac{47 \times 2 - 7 \times 2}{2 \times 1 - 2 \times 5}}{(-1) \times 2 - 1 \times 2} = \frac{z}{2 \times 1 - 2 \times 5}$$

or $\frac{x - \frac{3}{2}}{6} = \frac{y + 10}{-4} = \frac{z}{-8}$ or $\frac{x - \frac{3}{2}}{-3} = \frac{y + 10}{2} = \frac{z}{4}.$ ... (1)

The other line is $\frac{x - 3}{3} = \frac{y - 8}{-1} = \frac{z - 3}{4}.$ ... (2)

Let the line of the shortest distance have the direction cosines l, m, n. Here, PQ is the shortest distance.

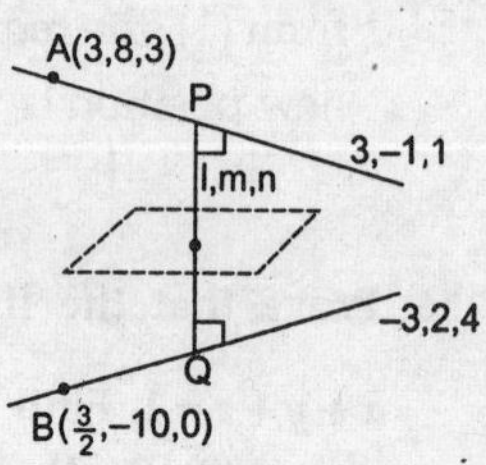

PQ ⊥ AP

$$\Rightarrow \quad 3l - m + n = 0.$$

PQ ⊥ BQ

$$\Rightarrow \quad -3l + 2m + 4n = 0.$$

$\therefore \quad \frac{l}{-4-2} = \frac{m}{-3-12} = \frac{n}{6-3}$ or $\frac{l}{-2} = \frac{m}{-5} = \frac{n}{1}$ ... (3)

or $\frac{l}{-2} = \frac{m}{-5} = \frac{n}{1} = \frac{\sqrt{l^2 + m^2 + n^2}}{\sqrt{(-2)^2 + (-5)^2 + 1^2}} = \frac{1}{\sqrt{30}}.$

$$\therefore \quad l = \frac{-2}{\sqrt{30}}, m = \frac{-5}{\sqrt{30}}, n = \frac{1}{\sqrt{30}}.$$

$\therefore$ the shortest distance PQ

= projection of AB on the direction PQ

$= |(x_2 - x_1)l + (y_2 - y_1)m + (z_2 - z_1)n|$

$= \left|\left(\frac{3}{2} - 3\right)\frac{-2}{\sqrt{30}} + (-10 - 8)\frac{-5}{\sqrt{30}} + (0 - 3)\frac{1}{\sqrt{30}}\right|$

$= \left|\frac{3}{\sqrt{30}} + \frac{90}{\sqrt{30}} - \frac{3}{\sqrt{30}}\right| = 3\sqrt{30}.$

Now, P is a point on AP. So, $P = (3 + 3r, 8 - r, 3 + r)$.

Q is a point on BQ. So, $Q = \left(\frac{3}{2} - 3r', -10 + 2r', 0 + 4r'\right).$

$\therefore$ direction ratios of PQ are

$3 + 3r - \frac{3}{2} + 3r', 8 - r + 10 - 2r', 3 + r - 4r'.$

$\therefore$ from (3), $\frac{\frac{3}{2} + 3r + 3r'}{-2} = \frac{18 - r - 2r'}{-5} = \frac{3 + r - 4r'}{1}$

$\Rightarrow$ $17r + 19r' = \frac{57}{2}$ and $4r - 22r' + 33 = 0$

$\Rightarrow$ $\frac{34}{19}r + 2r' - 3 = 0$ and $\frac{4}{11}r - 2r' + 3 = 0$

$\Rightarrow$ $r = 0, r' = \frac{3}{2}$.

$\therefore$ $P = (3 + 3 \times 0, 8 - 0, 3 + 0) = (3, 8, 3)$

and $Q = \left(\frac{3}{2} - 3 \times \frac{3}{2}, -10 + 2 \times \frac{3}{2}, 0 + 4 \times \frac{3}{2}\right) = (-3, -7, 6).$

$\therefore$ the middle point of PQ is $\left(\frac{3-3}{2}, \frac{8-7}{2}, \frac{3+6}{2}\right)$,

i.e., $\left(0, \frac{1}{2}, \frac{9}{2}\right)$

$\Rightarrow$ the equation of the required plane is

$(x - 0)l + \left(y - \frac{1}{2}\right)m + \left(z - \frac{9}{2}\right)n = 0$

or $x \cdot \left(-\frac{2}{\sqrt{30}}\right) + \left(y - \frac{1}{2}\right)\left(\frac{-5}{\sqrt{30}}\right) + \left(z - \frac{9}{2}\right)\frac{1}{\sqrt{30}} = 0$

or $2x + 5y - z + 2 = 0.$

19. Show that the shortest distance between any two opposite edges of the tetrahedron formed by the planes

$$y + z = 0, z + x = 0, x + y = 0 \text{ and } x + y + z = a$$

is $\frac{2a}{\sqrt{6}}$.

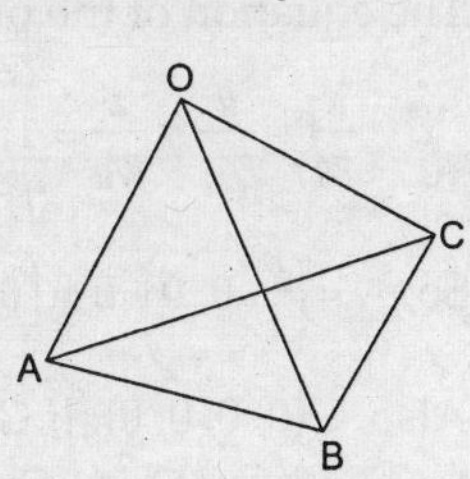

Clearly, the planes $y + z = 0$, $z + x = 0$ and $x + y = 0$ pass through the origin O(0, 0, 0).

The line of intersection OA of the planes OAC and OAB,

i.e., $y + z = 0$ and $z + x = 0$,

is given by $x = y = -z$, i.e.,

$$\frac{x-0}{1} = \frac{y-0}{1} = \frac{z-0}{-1}. \quad \dots (1)$$

The line of intersection OC of the planes OAC and OBC,

i.e., $y + z = 0$ and $x + y = 0$, is given by $x = -y = z$,

i.e., $$\frac{x-0}{1} = \frac{y-0}{-1} = \frac{z-0}{1}. \quad \dots (2)$$

The line of intersection OB of the planes OBC and OAB,

i.e., $x + y = 0$ and $z + x = 0$, is given by $-x = y = z$,

i.e., $$\frac{x-0}{-1} = \frac{y-0}{1} = \frac{z-0}{1}. \quad \dots (3)$$

The line BC is the intersection of the planes

$x + y = 0$ and $x + y + z = a$

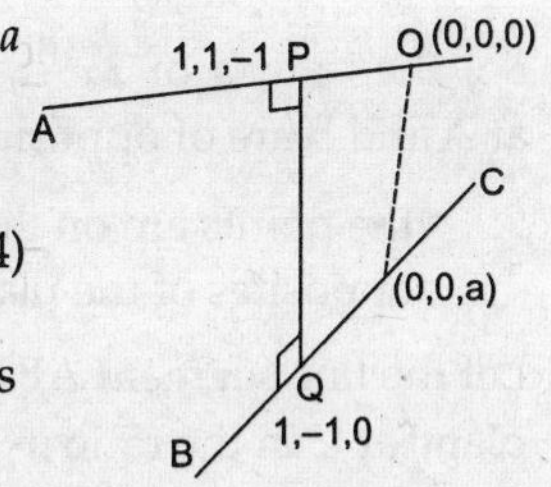

$\Rightarrow$ $y = -x, z = a$

$\Rightarrow$ $$\frac{x}{1} = \frac{y}{-1} = \frac{z-a}{0}. \quad \dots (4)$$

Similarly, CA has the equations

$y + z = 0, x + y + z = a$

$\Rightarrow$ $$\frac{x-a}{0} = \frac{y}{1} = \frac{z}{-1}. \quad \dots (5)$$

AB has the equations $z + x = 0, x + y + z = a$

$\Rightarrow$ $$\frac{x}{-1} = \frac{y-a}{0} = \frac{z}{1}. \quad \dots (6)$$

Let PQ be the shortest distance between OA and BC, and its line has the direction cosines l, m, n.

Then using perpendicularity, $l \cdot 1 + m \cdot 1 + n \cdot (-1) = 0$

and $l \cdot 1 + m(-1) + n \cdot 0 = 0$

$\Rightarrow$ $l + m - n = 0$ and $l - m = 0$

$\Rightarrow$ $\frac{l}{1} = \frac{m}{1} = \frac{n}{2} = \frac{\sqrt{l^2 + m^2 + n^2}}{\sqrt{1^2 + 1^2 + 2^2}} = \frac{1}{\sqrt{6}}$.

$\therefore$ $l = \frac{1}{\sqrt{6}}, m = \frac{1}{\sqrt{6}}, n = \frac{2}{\sqrt{6}}$.

$\therefore$ the shortest distance PQ

= projection of OC on the line PQ

$= |(x_2 - x_1)l + (y_2 - y_1)m + (z_2 - z_1)n|$

$= \left|(0 - 0)\frac{1}{\sqrt{6}} + (0 - 0)\frac{1}{\sqrt{6}} + (a - 0)\frac{2}{\sqrt{6}}\right| = \frac{2a}{\sqrt{6}}$.

Similarly, other shortest distances between opposite edges are

$\left|(0 - 0)\frac{1}{\sqrt{6}} + (a - 0)\frac{2}{\sqrt{6}} + (0 - 0)\frac{1}{\sqrt{6}}\right|$, i.e., $\frac{2a}{\sqrt{6}}$,

and $\left|(a - 0)\frac{2}{\sqrt{6}} + (0 - 0)\frac{1}{\sqrt{6}} + (0 - 0)\frac{1}{\sqrt{6}}\right|$, i.e., $\frac{2a}{\sqrt{6}}$.

Hence, the problem.

20. Prove that the points A(1, –4, 5) and B(0, 6, 1) are on the opposite sides of the plane $3x - y + 2z = 7$. Find the ratio in which the line segment AB is divided by the plane. Also, if $P(\lambda^2 + 1, \lambda, \lambda - 1)$ is a point on the same side of the plane as the point A then find the set of values of λ.

The equation of the plane is $3x - y + 2z - 7 = 0$.

At A(1, –4, 5),

$$3x - y + 2z - 7 = 3 \times 1 - (-4) + 2(5) - 7 = 3 + 4 + 10 - 7$$
$$= 10 > 0. \qquad \dots (1)$$

At B(0, 6, 1),

$$3x - y + 2z - 7 = 3 \times 0 - 6 + 2 \times 1 - 7 = -11 < 0.$$

So, the values of $3x - y + 2z - 7$ at A and B are of opposite signs.

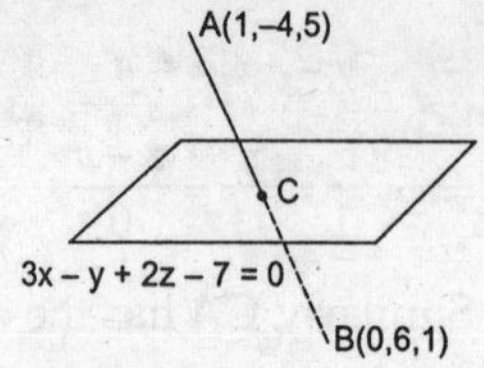

∴ the points are on the opposites of the plane.

Let the line segment AB cut the plane at C in the ratio $\mu : 1$,

i.e., $AC : CB = \mu : 1$.

$$\text{Then } C = \left(\frac{\mu \cdot 0 + 1 \cdot 1}{\mu + 1}, \frac{6\mu - 4 \cdot 1}{\mu + 1}, \frac{1 \cdot \mu + 5 \cdot 1}{\mu + 1}\right)$$
$$= \left(\frac{1}{\mu + 1}, \frac{6\mu - 4}{\mu + 1}, \frac{\mu + 5}{\mu + 1}\right)$$

is on the plane $3x - y + 2z = 7$.

$$\therefore \quad 3 \cdot \frac{1}{\mu + 1} - \frac{6\mu - 4}{\mu + 1} + 2 \cdot \frac{\mu + 5}{\mu + 1} = 7$$

or $3 - (6\mu - 4) + 2(\mu + 5) = 7(\mu + 1)$

or $3 - 6\mu + 4 + 2\mu + 10 = 7\mu + 7$ or $11\mu = 10$.

$$\therefore \quad \mu = \frac{10}{11}.$$

∴ the ratio is $\mu : 1$, i.e., $\frac{10}{11} : 1$ or $10 : 11$.

As A and P are on the same side of the plane, the value of $3x - y + 2z - 7$ has the same sign at A and P.

At A, it is positive (from (1)). So,

$$3(\lambda^2 + 1) - \lambda + 2(\lambda - 1) - 7 > 0 \text{ or } 3\lambda^2 + \lambda - 6 > 0.$$

Roots of the corresponding equation $3\lambda^2 + \lambda - 6 = 0$ are

$$\lambda = \frac{-1 \pm \sqrt{1 + 72}}{6}.$$

∴ the sign scheme is

$$\underset{}{(+)} \quad \frac{-1 - \sqrt{73}}{6} \quad (-) \quad \frac{-1 + \sqrt{73}}{6} \quad (+)$$

$$\therefore \quad \lambda < \frac{-1 - \sqrt{73}}{6} \quad \text{or} \quad \lambda > \frac{-1 + \sqrt{73}}{6}.$$

$$\text{So, } \lambda \in \left(-\infty, \frac{-1 - \sqrt{73}}{6}\right) \cup \left(\frac{-1 + \sqrt{73}}{6}, +\infty\right)$$

21. Show that the origin lies in the acute angle between the planes $x + 2y + 2z = 9$, $4x - 3y + 12z + 13 = 0$. Also, find the plane bisecting the acute angle between the planes.

The equations of the planes are

$$-x - 2y - 2z + 9 = 0 \text{ and } 4x - 3y + 12z + 13 = 0.$$

Here $d_1 = 9 > 0$, $d_2 = 13 > 0$.

Also, $a_1 = -1$, $b_1 = -2$, $c_1 = -2$, $a_2 = 4$, $b_2 = -3$, $c_2 = 12$.

∴ the origin lies in the acute angle if

$$a_1a_2 + b_1b_2 + c_1c_2 < 0$$

or $(-1)4 + (-2)(-3) + (-2)(12) < 0$

⇒ $-4 + 6 - 24 < 0$ which is true.

∴ the origin lies in the acute angle.

∴ the plane bisecting the acute angle between the given planes is the plane bisecting the angle containing the origin.

∴ the required plane bisecting the acute angle is

$$\frac{-x - 2y - 2z + 9}{\sqrt{(-1)^2 + (-2)^2 + (-2)^2}} = + \frac{4x - 3y + 12z + 13}{\sqrt{4^2 + (-3)^2 + 12^2}}$$

or $\frac{1}{3}(-x - 2y - 2z + 9) = \frac{1}{13}(4x - 3y + 12z + 13)$

or $-13x - 26y - 26z + 117 = 12x - 9y + 36z + 39$

or $25x + 17y + 62z - 78 = 0.$

22. A variable plane is at a constant distance p from the origin and meets the coordinate axes at A, B and C. Show that the locus of the centroid of the tetrahedron (O, ABC), where O is the origin, is

$$x^{-2} + y^{-2} + z^{-2} = 16p^{-2}.$$

Let the equation of the plane be $lx + my + nz = p$ (normal form) where l, m, n are direction cosines of the perpendicular from the origin to the plane and p is the length of this perpendicular. Here, p is a constant and l, m, n are variables.

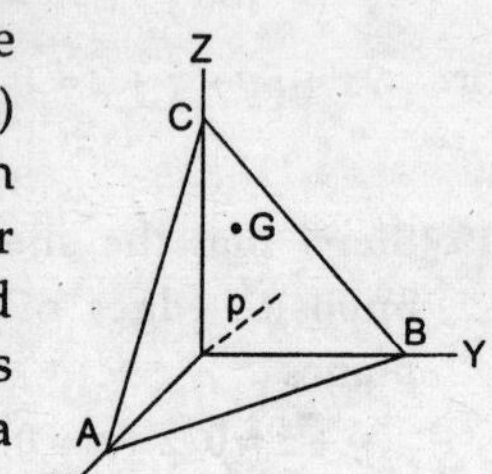

The equation of the plane can be written as

$$\frac{x}{p/l} + \frac{y}{p/m} + \frac{z}{p/n} = 1.$$

$$\text{So, } A = \left(\frac{p}{l}, 0, 0\right), B = \left(0, \frac{p}{m}, 0\right), C = \left(0, 0, \frac{p}{n}\right).$$

Also, O(0, 0, 0, 0). If $G(x_1, y_1, z_1)$ be the centroid of the tetrahedron (O, ABC) then

$$x_1 = \frac{\frac{p}{l} + 0 + 0 + 0}{4}, \; y_1 = \frac{0 + \frac{p}{m} + 0 + 0}{4},$$

$$z_1 = \frac{0 + 0 + \frac{p}{n} + 0}{4}.$$

$$\therefore \quad x_1 = \frac{p}{4l}, \; y_1 = \frac{p}{4m}, \; z_1 = \frac{p}{4n}$$

or $l = \frac{p}{4x_1}, \; m = \frac{p}{4y_1}, \; n = \frac{p}{4z_1}.$

But l, m, n are direction cosines of a line.

So, $l^2 + m^2 + n^2 = 1$.

$\therefore \quad \left(\frac{p}{4x_1}\right)^2 + \left(\frac{p}{4y_1}\right)^2 + \left(\frac{p}{4z_1}\right)^2 = 1$

or $\quad x_1^{-2} + y_1^{-2} + z_1^{-2} = 16p^{-2}$.

$\therefore$ the equation of the locus of G is

$$x^{-2} + y^{-2} + z^{-2} = 16p^{-2}.$$

23. Prove that the locus of a variable line which intersects the three lines $y = mx, z = c; y = -mx, z = -c$ and $y = z, mx = -c$ is the surface $y^2 - m^2x^2 = z^2 - c^2$.

The given lines are in the form of intersection of two planes.

The lines are given by

$$y - mx = 0, z - c = 0 \quad \text{... (1)}$$

$$y + mx = 0, z + c = 0 \quad \text{... (2)}$$

$$y - z = 0, mx + c = 0 \quad \text{... (3)}$$

Any plane through the intersection of the planes in (1) has the equation

$$y - mx + \lambda(z - c) = 0. \quad \text{... (4)}$$

Any plane through the intersection of the planes in (2) has the equation

$$y + mx + \mu(z + c) = 0. \quad \text{... (5)}$$

The line of intersection of the planes (4) and (5) will cut the lines (1) as well as (2) because two lines in the same plane intersect, being nonskew.

$\therefore$ the equation of a line intersecting the lines (1) and (2) has the equations

$$y - mx + \lambda(z - c) = 0, y + mx + \mu(z + c) = 0. \quad \text{... (6)}$$

This line will intersect the third line, if $y - z = 0$, $mx + c = 0$ also hold.

$\therefore$ the locus of the line (6) under the conditions $y - z = 0, mx + c = 0$ is the required locus.

Hence, we have to eliminate λ, μ and obtain a relation between x, y and z from

$$y - mx + \lambda(z - c) = 0 \quad \text{... (A)}$$

$$y + mx + \mu(z + c) = 0 \quad \text{... (B)}$$

$$y - z = 0 \quad \text{... (C)}$$

$$mx + c = 0 \quad \text{... (D)}$$

From (A), (C) and (D), $y + c + \lambda(y - c) = 0$

$\Rightarrow \quad y = \frac{\lambda c - c}{\lambda + 1}$.

From (B), (C) and (D), $y - c + \mu(y + c) = 0$

$\Rightarrow \quad y = \frac{c - \mu c}{1 + \mu}$.

$\therefore \quad \frac{\lambda c - c}{\lambda + 1} = \frac{c - \mu c}{1 + \mu} \quad \Rightarrow \quad \frac{\lambda - 1}{\lambda + 1} = \frac{1 - \mu}{1 + \mu}$

$\Rightarrow \quad \frac{2\lambda}{2} = \frac{2}{2\mu}$ (by componendo-dividendo)

$\Rightarrow \quad \lambda\mu = 1$.

$\therefore$ from (A), $\lambda\mu = \left(-\frac{y - mx}{z - c}\right) \cdot \left(-\frac{y + mx}{z + c}\right)$.

$\therefore \quad 1 = \frac{(y - mx)(y + mx)}{(z - c)(z + c)}$ or $z^2 - c^2 = y^2 - m^2x^2$.

$\therefore$ the equation of the locus is $y^2 - m^2x^2 = z^2 - c^2$.

24. Find the locus of lines which meet the lines $y = 2x, z = 1; \; y + 2x = 0, z + 1 = 0$ and the circle $x^2 + y^2 = 4, z = 0$.

Any plane through the line $y = 2x, z = 1$ has the equation

$$y - 2x + \lambda(z - 1) = 0. \quad \text{... (1)}$$

Any plane through the line $y + 2x = 0, z + 1 = 0$ has the equation

$$y + 2x + \mu(z + 1) = 0. \quad \text{... (2)}$$

The line of intersection of the planes (1) and (2) is a line intersecting both the given lines.

So, the equations of a line cutting the given lines are

$$y - 2x + \lambda(z - 1) = 0, y + 2x + \mu(z + 1) = 0.$$

It meets the circle $x^2 + y^2 = 4, z = 0$.

$\therefore$ the required locus is obtained by establishing the relation between x, y, z by eliminating λ, μ from

$$y - 2x + \lambda(z - 1) = 0 \quad \text{... (A)}$$

$$y + 2x + \mu(z + 1) = 0 \quad \text{... (B)}$$

$$x^2 + y^2 = 4 \quad \text{... (C)}$$

$$z = 0 \quad \text{... (D)}$$

From (A), (B) and (D), $y - 2x - \lambda = 0$
$\qquad y + 2x + \mu = 0$

$\therefore \quad 2y - \lambda + \mu = 0$ and $4x + \mu + \lambda = 0$

$\Rightarrow \quad y = \frac{\lambda - \mu}{2}$ and $x = -\frac{\lambda + \mu}{4}$.

$\therefore \quad$ (C) $\Rightarrow \left(-\frac{\lambda + \mu}{4}\right)^2 + \left(\frac{\lambda - \mu}{2}\right)^2 = 4$

or $\quad (\lambda + \mu)^2 + 4(\lambda - \mu)^2 = 64$

or $\quad (\lambda + \mu)^2 + 4\{(\lambda + \mu)^2 - 4\lambda\mu\} = 64$

or $\quad 5(\lambda + \mu)^2 - 16\lambda\mu = 64$

or $\quad 5\left\{\frac{y - 2x}{1 - z} - \frac{y + 2x}{1 + z}\right\}^2 + 16 \cdot \frac{y - 2x}{1 - z} \cdot \frac{y + 2x}{1 + z} = 64$

or $\quad 5\{(1 + z)(y - 2x) - (1 - z)(y + 2x)\}^2 + 16(y^2 - 4x^2)(1 - z^2) = 64(1 - z^2)^2$

or $5\{y-2x-y-2x+z(y-2x+y+2x)\}^2$

$+16(1-z^2)\{y^2-4x^2-4(1-z^2)\}=0$

or $5(-4x+2yz)^2+16(1-z^2)(y^2-4x^2+4z^2-4)=0$

or $5(yz-2x)^2+4(1-z^2)(y^2-4x^2+4z^2-4)=0.$

Exercises

1. Find the equation of the plane perpendicular to the line joining the points (2, –1, 2) and (3, 2, –1) and passing through the point (4, –3, 1).
2. Find the equation of the plane bisecting the line segment joining the points (–3, –2, 1) and (1, 6, –5) perpendicularly.
3. Find the equation of the plane passing through the points P(1, 1, 1), Q(3, –1, 2) and R(–3, 5, –4). Also, find the equation of the normal at the centroid of the ΔPQR to the plane of the points.
4. If the points (0, –1, –1), (4, 5, 1), (3, 9, 4) and (–4, 4, a) are coplanar, find the value of a.
5. Prove that the point P(–1, 4, –3) lies on the plane passing through the points A(3, 2, –5), B(–3, 8, –5) and C(–3, 2, 1). Also, show that P is the circumcentre of the ΔABC.
6. Find the equation of the plane through the points (0, 4, –3) and (6, –4, 3), other than the plane through the origin, and which cuts-off intercepts on the axes whose sum is zero.
7. A triangle is placed with the middle points of its sides on the three axes of reference. The lengths of its sides are a, b, c and the middle points are on the x, y, z axes respectively. Prove that the equation of the plane of the triangle is $\frac{x}{\alpha}+\frac{y}{\beta}+\frac{z}{\gamma}=1$ where $8\alpha^2=b^2+c^2-a^2$,
$8\beta^2=c^2+a^2-b^2, 8\gamma^2=a^2+b^2-c^2.$
8. Through a point P(α, β, γ), a plane is drawn at right angles to $\vec{OP}$, where O is the origin, to meet the axes at A, B and C respectively. Prove that the area of the ΔABC is $\frac{r^5}{2\alpha\beta\gamma}$ where $OP=r$.
9. Two systems of rectangular axes have the same origin. If a plane cuts them at distances a, b, c and a', b', c' respectively from the origin, prove that
$$a^{-2}+b^{-2}+c^{-2}=a'^{-2}+b'^{-2}+c'^{-2}.$$
10. If from the point P(1, 2, 1) perpendiculars PL and PM are drawn to the y–z and z–x planes, find the equation of the plane OLM, O being the origin.
11. Find the equation of the plane through the point (1, 2, 3) which is parallel to the plane $3x+4y-5z=0$. Also, find the equation of the plane perpendicular to the given plane which passes through the line joining the given point and the point (–2, 1, 0).
12. Find the value of k so that the plane $(k+1)x-y+(2-k)z=5$ is perpendicular to the plane $2x+6y-z+3=0$. Can the planes be parallel for any value of k?
13. Find the equation of the plane through the points (2, 2, 1), (9, 3, 6) and perpendicular to the plane
$$2x+6y+6z=9.$$
14. Find the equation of the plane which cuts off intercepts –4 and 6 on the x and z axes respectively and is perpendicular to the plane $2x+3y-5z=6$.
15. Find the sum of the intercepts made on the axes by the plane passing through the point (2, 5, –6) and perpendicular to each of the planes
$$2x-3y+4z+1=0 \text{ and } 4x+y-2z+6=0.$$
16. Find the direction cosines of the normal to the plane passing through the point (–1, 3, 2) and perpendicular to each of the planes $x+2y+2z=5$ and $3x+3y+2z=8$.
17. Find the equation of the plane through the line of intersection of the planes $2x+5y-5z-6=0$ and $2x+7y-8z-7=0$ and passing through the point (–1, 4, 3).
18. Find the equation of the plane through the line $x+4y=0=3y-z$ and perpendicular to the plane $5x+3y-4z=8$.
19. Prove that the plane
$$2(1+\lambda)x+3(1-\lambda)y+(10-7\lambda)z=2(\lambda+4)$$
always passes through a fixed straight line. Find the equation of the fixed straight line in the symmetric form.
20. Prove that the planes
$2x-3y-7z=0$, $3x-14y-13z=0$ and
$8x-31y-33z=0$ pass through the same line.
21. Show that the planes $bx-ay=n$, $cy-bz=l$ and $az-cx=m$ will intersect in a line if $al+bm+cn=0$.
22. Find the point of intersection of the planes
$x+2y-z=6$, $2x-y+3z+13=0$ and
$3x-2y+3z+16=0$.
23. Find the equation of the plane which passes through the line of intersection of the planes $2x+3y-4z+5=0$ and $x-5y+z+6=0$ and is parallel to the line $x+y-z+8=0=x-y+z+2$.

24. Prove that the equation of the plane through the line

$u_1 \equiv a_1x + b_1y + c_1z + d_1 = 0,$

$u_2 \equiv a_2x + b_2y + c_2z + d_2 = 0$

and parallel to the line $\frac{x}{l} = \frac{y}{m} = \frac{z}{n}$ is

$$\frac{u_1}{a_1l + b_1m + c_1n} = \frac{u_2}{a_2l + b_2m + c_2n}.$$

25. Find the equations of the line in symmetric form which is drawn parallel to the line $\frac{x-1}{2} = \frac{y-2}{3} = \frac{z+4}{4}$ such that it cuts the lines

$9x + y + z + 4 = 0 = 5x + y + 3z$

and $x + 2y - 3z - 3 = 0 = 2x - 5y + 3z + 3$.

26. Find the length and direction cosines of the normal from the origin to the plane passing through the points (1, 2, 3), (–2, 1, –4) and (3, 4, –2).

27. Find the length and the foot of the perpendicular drawn from the point (1, 2, 3) to the plane

$x + y + z - 3 = 0.$

28. Find the equations of the planes through the intersection of the planes $x + 3y + 6 = 0$ and $3x - y - 4z = 0$ whose perpendicular distance from the origin is unity.

29. Find the distance between the planes

$ax + by + cz + d = 0$ and $ax + by + cz + d' = 0$.

30. Find the equations of the planes parallel to the plane $4x - 4y + 7z - 3 = 0$ and at a distance 4 units from the point (4, 1, –2).

31. Find the equations of the planes perpendicular to the plane $x + y + z = 3$, passing through the point (–1, 2, 0) and at a distance 1 from the origin.

32. Which of the pairs of points from A(1, –1, 1), B(3, 1, 4) and C(–2, 3, 0) are on the opposite sides of the plane $4x - 3y + z = 5$? Find the point where the plane cuts the line joining two of the above points on the same side of the plane.

33. Prove that for all values of $\lambda \in R$, the point $(\lambda, 3, 2\lambda - 1)$ is on that side of the plane $2x - y - z + 6 = 0$ which is opposite to the side of the point (–3, 1, 1).

34. If the acute angle between the planes $x - y + 2z = 4$ and $\lambda x + y - z = 1$, $(\lambda > 0)$ is 60°, find the equations of planes bisecting the angles between the two planes.

35. Find the equation of the plane bisecting the angle between the planes $x + 2y + 2z - 3 = 0$ and $3x + 4y + 12z + 1 = 0$ that
(i) contains the origin (ii) contains the point (1, 1, 1).

36. Find the equation of the plane bisecting the (i) acute (ii) obtuse angle between the planes $2x - y - 2z = 3$ and $3x + 12y - 4z + 5 = 0$.

37. Show that the lines

$x + y + z - 3 = 0,\ 2x + 3y + 4z = 5$

and $4x - y + 5z - 7 = 0 = 2x - 5y - z - 3$

are coplanar. Find the equation of the plane in which they lie.

38. Prove that the lines $\frac{x-2}{4} = \frac{y+1}{3} = \frac{z-3}{5}$ and

$x + 2y + 3z - 9 = 0 = 2x - y + 2z - 11$

intersect. Find the equations of the line passing through the point of intersection and perpendicular to the plane of the given lines.

39. Prove that the lines $x = ay + b = cz + d$ and $x = \alpha y + \beta = \gamma z + \delta$ are coplanar if

$(\alpha - a)(c\delta - d\gamma) = (\gamma - c)(a\beta - b\alpha).$

40. Find the equations of the normal to the plane $4x - 3y + 2z + 6 = 0$ at the point where the line $\frac{x-1}{2} = \frac{y-1}{3} = \frac{z-1}{6}$ cuts the plane.

41. Prove that the plane through the point (α, β, γ) and the line $x = py + q = rz + s$ is given by

$$\begin{vmatrix} x & py+q & rz+s \\ \alpha & p\beta+q & r\gamma+s \\ 1 & 1 & 1 \end{vmatrix} = 0.$$

42. Find the distance of the point (4, 1, 1) from the line of intersection of the planes $x + y + z = 4$ and $x - 2y - z = 4$.

43. Find the equation of the perpendicular from the origin to the line $x + 4y + 4z - 27 = 0,\ 2x + 2y + 5z - 21 = 0$. Find also the coordinates of the foot of the perpendicular.

44. Find the equation of the plane containing the line $\frac{x+3}{3} = \frac{y-1}{4} = \frac{z-2}{-2}$ and passing through the point (0, 2, 4).

45. Prove that the line $\frac{x-1}{-2} = \frac{y+1}{3} = \frac{z}{1}$ does not lie in the plane $2x - 3y + z = 5$. Find the angle between the line and the plane.

46. Find the image of the point (1, 2, 3) by the plane $x + y + z + 3 = 0$.

47. The image of the point (6, –1, 3) by a plane is the point (–2, 3, 1). Find the equation of the plane.

48. AB is a line segment where A = (1, 1, 2), AB = 5 and the direction ratios of the line AB are 3, $\sqrt{6}$, 1. If the line segment PQ is the image of the line segment AB by the plane $x + y + z = 1$, A and B being on the same side of the plane, then find the equation of the line PQ.

49. Find the length and the equations of the shortest distance between the lines

$3x - 9y + 5z = 0 = x + y - z$ and

$6x + 8y + 3z - 13 = 0 = x + 2y + z - 3.$

50. Show that the plane containing the line $\frac{y}{b}+\frac{z}{c}=1$, $x=0$ and parallel to the line $\frac{x}{a}-\frac{z}{c}=1, y=0$ is $\frac{x}{a}-\frac{y}{b}-\frac{z}{c}+1=0$. If $2d$ be the shortest distance between the lines then prove that

$$\frac{1}{a^2}+\frac{1}{b^2}+\frac{1}{c^2}=\frac{1}{d^2}.$$

51. Prove that the line $\frac{x-1}{2}=\frac{y-2}{-3}=\frac{z+3}{1}$ lies in the plane $3x+4y+6z+7=0$. If the plane is rotated about the line till the plane passes through the origin then find the equation of the plane in the new position.

52. The line $\frac{x+6}{-1}=\frac{y}{\lambda}=\frac{z-1}{4}$ lies in the plane $2x-5y+3z+9=0$. Through what acute angle the plane should be turned about the line in order to pass through the point (1, 1, 1)?

53. Find the equation of the plane π through the point P(2, 4, –1) which is parallel to the plane $x-y+z=3$. If P is translated through a distance 3 in the plane π in the direction whose direction ratios are 3, 2, –1, find the coordinates of P in the new position.

54. A variable plane passes through a fixed point (a, b, c) and meets the coordinate axes in A, B, C. Show that the locus of the point common to the planes through A, B and C, parallel to the coordinate planes, is $\frac{a}{x}+\frac{b}{y}+\frac{c}{z}=1$.

55. Find the locus of the point whose distance from the point (1, 2, 3) is half its distance from the plane $2x+y-2z=7$.

56. Obtain the locus of a point, the sum of the squares of whose distances from the planes $x+y+z=0$, $x-z=0$, $x-2y+z=0$ is 9.

57. A variable plane is at a constant distance p from the origin and meets the axes at A, B and C. Find the locus of the centroid of the ΔABC.

58. Find the locus of the lines which intersect the three lines given by $y-z=1$, $x=0$; $z-x=1$, $y=0$ and $x-y=1, z=0$.

59. Find the locus of the lines which intersect the three lines given by

$$y=b, z+c=0; z=c, x+a=0 \text{ and } x=a, y+b=0.$$

60. Prove that the locus of a line which meets the lines $y=mx, z=c$; $y+mx=0, z+c=0$ and the circle $x^2+y^2=a^2, z=0$ is

$$c^2[m^2(cy-mzx)^2+(yz-cmx)^2]=a^2m^2(z^2-c^2)^2.$$

Objective Questions

61. The direction cosines of the normal to the plane $\frac{x}{3}+\frac{y}{2}+z=1$ are

(a) $\frac{1}{3}, \frac{1}{2}, 1$ (b) $\frac{2}{7}, \frac{3}{7}, \frac{6}{7}$

(c) $\frac{3}{\sqrt{14}}, \frac{2}{\sqrt{14}}, \frac{1}{\sqrt{14}}$ (d) $\frac{6}{7}, \frac{3}{7}, \frac{2}{7}$

62. The sum of the intercepts made by the plane $3x-4y+2z=1$ on the three axes is

(a) 1 (b) –1 (c) $\frac{7}{12}$ (d) $\frac{13}{12}$

63. The equation of the plane passing through the point (1, 2, 3) and perpendicular to the line whose direction numbers are (3, 2, 1), is

(a) $3x+2y+z=10$ (b) $x+2y+3z=10$

(c) $2x+3y+z=7$ (d) $2x+3y+z=11$

64. The length of the perpendicular from the origin to the plane $x-2y+3z=4$ is

(a) 4 (b) $2\sqrt{2/7}$

(c) 2 (d) $\frac{4}{3}$

65. The equations of the perpendicular from the point (1, 2, –3) to the plane $x+3y+z=4$ are

(a) $\frac{x+1}{1}=\frac{y+2}{3}=\frac{z-3}{1}$ (b) $x-1=3(y-2)=z+3$

(c) $\frac{x}{1}=\frac{y+1}{3}=\frac{z+4}{1}$ (d) none of these

66. The angle between the line $x=y=z$ and the plane $x+y+z=4$ is

(a) 0° (b) 60° (c) 45° (d) 90°

67. The equations of the line of intersection of the planes $4x-4y-z+11=0$ and $x+2y-z=1$ are

(a) $\frac{x-3}{4}=\frac{y-2}{1}=\frac{z}{2}$ (b) $\frac{x}{2}=\frac{y-2}{1}=\frac{z-3}{4}$

(c) $\frac{x-1}{2}=\frac{y-1}{1}=\frac{z-2}{4}$ (d) $\frac{x+2}{2}=\frac{y}{1}=\frac{z-3}{4}$

68. The line $\frac{x-1}{\lambda}=\frac{y+2}{-3}=\frac{z-2}{1}$ lies in the plane $2x-3y+z=10$ if λ is

(a) –5 (b) 0

(c) 5 (d) any real number

69. The two lines $x=ay+b, z=cy+d$ and $x=a'y+b', z=c'y+d'$ will be at right angles if

(a) $aa'+bb'+cc'=0$ (b) $a+b+c=a'+b'+d'$

(c) $aa'+bb'+cc'=1$ (d) $aa'+cc'+1=0$

70. The angle at which the line $\frac{x}{2} = \frac{y}{3} = \frac{z-1}{1}$ cuts the plane $x + y + z = 1$ is

(a) $\cos^{-1}\sqrt{\frac{6}{7}}$ (b) $\sin^{-1}\sqrt{\frac{6}{7}}$

(c) $0°$ (d) $90°$

71. A tetrahedron has vertices at O(0, 0, 0), A(1, 2, 1), B(2, 1, 3) and C(–1, 1, 2). Then the angle between the faces OAB and ABC will be

(a) $\cos^{-1}\left(\frac{17}{31}\right)$ (b) $\frac{\pi}{6}$

(c) $\frac{\pi}{2}$ (d) $\cos^{-1}\left(\frac{19}{35}\right)$

72. If $\frac{x-4}{1} = \frac{y-2}{1} = \frac{z-k}{2}$ is a line lying in the plane $2x - 4y + z = 7$ then k is

(a) 7 (b) –7

(c) 4 (d) no real value

Answers

1. $x + 3y - 3z + 8 = 0$ **2.** $2x + 4y - 3z = 12$

3. $x + y = 2; \frac{x - 1/3}{1} = \frac{y - 5/3}{1} = \frac{z + 1/3}{0}$ **4.** 4

6. $2x - 3y - 6z = 6$ or $6x + 3y - 2z = 18$

10. $2x + y - 2z = 0$

11. $3x + 4y - 5z + 4 = 0; 17x - 24y - 9z + 58 = 0$

12. 2; no value **13.** $3x + 4y - 5z = 9$

14. $9x - 16y - 6z + 36 = 0$

15. $12\frac{3}{7}$ **16.** $\frac{2}{\sqrt{29}}, \frac{-4}{\sqrt{29}}, \frac{3}{\sqrt{29}}$

17. $4x + 4y - z = 9$ **18.** $13x + y + 17z = 0$

19. $\frac{x - 5/2}{9} = \frac{y-1}{34} = \frac{z}{-12}$ **22.** $(-1, 2, -3)$

23. $7x + 17y - 17z + 14 = 0$ **25.** $\frac{x+1}{2} = \frac{y}{3} = \frac{z}{4}$

26. $17 \cdot \sqrt{\frac{3}{406}}; \frac{-19}{\sqrt{1218}}, \frac{29}{\sqrt{1218}}, \frac{4}{\sqrt{1218}}$ **27.** $\sqrt{3}; (0, 1, 2)$

28. $x - 2y - 2z = 3, 2x + y - 2z + 3 = 0$

29. $\left|\frac{d - d'}{\sqrt{a^2 + b^2 + c^2}}\right|$

30. $4x - 4y + 7z + 38 = 0, 4x - 4y + 7z - 34 = 0$

31. $2(-2 + \sqrt{11})x + (5 + \sqrt{11})y - (1 + 3\sqrt{11})z = 14,$
$2(-2 - \sqrt{11})x + (5 - \sqrt{11})y + (-1 + 3\sqrt{11})z = 14.$

32. A, C and B, C; $\left(-\frac{1}{5}, -\frac{11}{5}, -\frac{4}{5}\right)$

34. $\frac{x - y + 2z - 4}{\sqrt{3}} = \pm\frac{(4\sqrt{3} - 6)x + y - z - 1}{3\sqrt{3} - 4}$

35. (i) $11x + 19y + 31z = 18$ (ii) $2x + 7y - 5z = 21$

36. (i) $35x + 23y - 38z = 24$ (ii) $17x - 49y - 14z = 54$

37. $x + 2y + 3z = 2$ **38.** $\frac{x-2}{7} = \frac{y+1}{-11} = \frac{z-3}{1}$

40. $\frac{x + 7/11}{4} = \frac{y + 16/11}{-3} = \frac{z + 43/11}{2}$ **42.** $3 \cdot \sqrt{3/14}$

43. $\frac{x}{1} = \frac{y}{30} = \frac{z}{17}; \left(\frac{1}{7}, \frac{30}{7}, \frac{17}{7}\right)$ **44.** $10x - 12y - 9z + 60 = 0$

45. $\sin^{-1}\frac{6}{7}$ **46.** $(-5, -4, -3)$ **47.** $4x - 2y + z - 8 = 0$

48. $\frac{x+1}{2\sqrt{6} - 1} = \frac{y+1}{8 - \sqrt{6}} = \frac{z}{5 + 2\sqrt{6}}$

49. $\frac{11}{\sqrt{342}}; 10x - 29y + 16z = 0 = 13x + 82y + 55z - 109$

51. $x + y + z = 0$ **52.** $\cos^{-1}\frac{193}{5 \cdot \sqrt{1558}}$

53. $x - y + z + 3 = 0; \left(2 \pm \frac{9}{\sqrt{14}}, 4 \pm \frac{6}{\sqrt{14}}, -1 \mp \frac{3}{\sqrt{14}}\right)$

55. $36\{(x-1)^2 + (y-2)^2 + (z-3)^2\} = (2x + y - 2z - 7)^2$

56. $x^2 + y^2 + z^2 = 9$ **57.** $x^{-2} + y^{-2} + z^{-2} = 9p^{-2}$

58. $x^2 + y^2 + z^2 - 2xy - 2zy - 2zx = 1$

59. $ayz + bzx + cxy + abc = 0$

61. (b)	**62.** (c)	**63.** (a)	**64.** (b)
65. (c)	**66.** (d)	**67.** (b)	**68.** (a)
69. (d)	**70.** (b)	**71.** (d)	**72.** (a)

□

Chapter Test

Time: 60 minutes

1. (i) The equation of the locus of points equidistant from the points (1, –3, 4) and (3, 1, –2) is

A. $2x + 5y + 3z = 0$ B. $x + 2y - 3z = 0$ C. $x + 2y - 3z + 3 = 0$ D. none of these

(ii) The plane $\lambda x + (3 - 2\lambda) y - z = 0$ is a plane passing through a fixed line for $\lambda \in R$, the fixed line being

A. $\frac{x}{2} = \frac{y}{1} = \frac{z}{3}$ B. $x = 3y = -z$ C. $x = -2y = -z$ D. none of these

(iii) If the plane $x + 2y - z = 0$ is perpendicular to the plane $(\lambda + 1)x - \lambda y + 3z = 1$ then λ is

A. 0 B. 2 C. 1 D. –2

2. Find the point of intersection of the line $\frac{x-1}{2} = \frac{y}{1} = \frac{z+2}{-1}$ and the plane $2x + y - 3z - 4 = 0$. Also, find the equation of the line in which the given line is projected on the plane.

3. Find the equation of the plane passing through the line $\frac{x+1}{2} = \frac{y-1}{-1} = \frac{z+3}{1}$ and parallel to the line of intersection of the planes $x - y - 5z - 6 = 0$ and $3x + 5y + 3z = 4$.

4. Let ABC be a triangle. Prove that the planes $x = y\cos C + z\cos B$, $y = z\cos A + x\cos C$ and $z = x\cos B + y\cos A$ pass through the same line $\frac{x}{a} = \frac{y}{b} = \frac{z}{c}$, where a, b, c are the lengths of the sides of the triangle in usual notations.

5. Let P(1, 3, 5) and Q(–2, 1, 4) be two points from which perpendiculars PM and QN are drawn to the x–z plane. Find the angle that the line MN makes with the plane $x + y + z = 5$.

6. The line $\frac{x+1}{7} = \frac{y}{4} = \frac{z+2}{-2}$ cuts the plane $2x + 3y - z = 0$ at the point P. Find the measure of the least angle through which the line has to be turned about P so that the line can lie on the plane.

7. A plane is parallel to two lines whose direction ratios are (1, 0, –1) and (–1, 1, 0), and it contains the point (1, 1, 1). If the plane cuts the coordinates axes at A, B, and C respectively then find the volume of the tetrahedron OABC.

8. AB is a rod which is held such that A = (1, –2, 3) and B = (2, 3, –4). A source of light is at the origin O. Find the length of the shadow of the rod on a plane screen whose equation is $x + y + 2z = 1$.

9. Find the equation of a plane containing the line whose equations are $2x - y + z = 3$ and $3x + y + z = 5$ if the plane is at a distance $\frac{1}{\sqrt{6}}$ from the point (2, 1, –1).

□

Answers

1. (i) C (ii) A (iii) D

2. $\left(0, -\frac{1}{2}, \frac{-3}{2}\right)$; $\frac{x}{6} = \frac{y + \frac{1}{2}}{3} = \frac{z + \frac{3}{2}}{5}$

3. $5x + 3y - 7z = 19$

5. $\sin^{-1}\frac{4}{\sqrt{30}}$

6. $\sin^{-1}\left(2 \cdot \sqrt{\frac{14}{69}}\right)$

7. $\frac{9}{2}$ unit^3

8. $\frac{1}{15}\sqrt{371}$ units

9. $2x - y + z = 3$ or $62x + 29y + 19z = 105$

□

Practice Tests

1. Integrated Tests

(*Instruction: Answer each set in two hours*)

Set A

1.1 Fill in the blanks in each of the following so that the resulting statement becomes true.

(a) The number of dissimilar terms in the expansion of $(3a + 2b + c)^{12}$ is ______ and the coefficient of $a^6 b^4 c^2$ is ______ .

(b) The value of $\int \dfrac{dx}{\sqrt{1-\sqrt{x}}} =$ ______ .

(c) Let $\vec{r} = l(\vec{b} \times \vec{c}) + m(\vec{c} \times \vec{a}) + n(\vec{a} \times \vec{b})$ and $[\vec{a}\ \vec{b}\ \vec{c}] = 3$. Then $l + m + n =$ ______ in terms of $\vec{r}, \vec{a}, \vec{b}$ and $\vec{c}$.

(d) If $10\sin^3 x + 9\sin^2 x - \sin x = 20$ then the number of solutions of the equation is ______ .

1.2 Each of the following has a number of alternative answers of which one or more are correct. Choose the correct answer(s).

(a) An ordinary coin is tossed n times. If the probability of getting 7 heads is equal to that of 9 heads then the probability of getting 2 heads is

A. $\dfrac{15}{2^{13}}$ B. $\dfrac{2}{15}$ C. $\dfrac{15}{2^8}$ D. none of these

(b) If a normal to the parabola $y^2 = 4ax$ makes an angle ψ with its axis then it will cut the curve again at an angle

A. $\dfrac{\psi}{2}$ B. $\tan^{-1}(2\tan\psi)$ C. $\tan^{-1}\left(\dfrac{1}{2}\tan\psi\right)$ D. none of these

(c) If $f(x) = [\tan^2 x]$, where $[x]$ is the greatest integer function then

A. $\lim\limits_{x \to 0} f(x)$ does not exist B. $f(x)$ is continuous at $x = 0$

C. $f(x)$ is not differentiable at $x = 0$ D. $f'(0) = 1$

(d) If $a_1, a_2, a_3, \ldots, a_n$ are positive numbers in AP whose common difference is d and $s_n = \sum\limits_{k=1}^{n} a_k$ then

A. $s_n \geq n \cdot \sqrt{a_1^2 + (n-1)da_1}$ B. $s_n \geq n \cdot \sqrt[n]{a_1 \cdot a_2 \cdot \ldots \cdot a_n}$

C. $\sum\limits_{k=1}^{n} \dfrac{1}{a_k} \cdot \sqrt[n]{a_1 \cdot a_2 \cdot \ldots \cdot a_n} \geq n$ D. none of these

(e) $P_r(x_r, y_r)$; $r = 1, 2, 3, \ldots$ is a sequence of points on the curve $y^3 = x$. The tangent at P_n cuts the curve again at P_{n+1} for all $n \in N$. Then $y_1, y_2, y_3, \ldots$ are in

A. AP B. GP C. HP D. none of these

1.3 Prove, by induction, that $n! > n^{n/2}$ for all $n > 2$.

1.4 In ΔABC, $a^2 + b^2 = 5c^2$. Prove that the medians to the sides a and b are perpendicular to each other.

1.5 A chord AB of the circle $x^2 + y^2 - 2ax - 2by = 0$ subtends an angle of $\frac{\pi}{4}$ at the origin. Find the equation of the locus of the middle point of the chord AB.

1.6 Let $\Delta = \begin{vmatrix} mx^2 + cy^2 + bz^2 - 1 & (m-c)xy & (m-b)zx \\ (m-c)xy & my^2 + az^2 + cx^2 - 1 & (m-a)yz \\ (m-b)zx & (m-a)yz & mz^2 + bx^2 + ay^2 - 1 \end{vmatrix}$

Prove that $m(x^2 + y^2 + z^2) - 1$ is a factor of Δ. Also, find the remaining factor.

1.7 A horse is tethered by a rope fastened at one end A on the boundary of a circular field whose centre is O. The horse can graze over $\frac{1}{n}$th part of the field reaching the farthest point B on the boundary that he can reach. If $\angle AOB = \theta$, prove that

$$\sin\theta + (\pi - \theta)\cos\theta = \left(1 - \frac{1}{n}\right)\pi.$$

1.8 Let C be any circle with the centre $(0, \sqrt{2})$. Prove that at most two points with rational coordinates can be there on C.

1.9 A number of n digits, where $5 \le n \le 24$, is made at random with the digits 0, 1, 2, 3, 4. What is the chance that it is a number of ten digits?

1.10 C_1 is a curve passing through the point (1, 0) having the differential equation $2xy\,dy = (x^2 + y^2)dx$. Find the area in the first quadrant bounded by C_1 and the lines $x = 0$, $y = 0$ and $y = \int_0^{1/\pi^2} (\sin^{-1}x + \cos^{-1}x)^2\,dx$.

1.11 If $f''(x)$ is continuous in $[a, b]$ then prove that

$$\int_a^b xf''(x)dx = (b-a)\cdot\left[\int_\lambda^\mu f''(x)dx + \mu f''(\mu)\right], \text{ where } a < \lambda < b, a < \mu < b.$$

1.12 (a) Prove that $y^3 + 3y = x$ defines a single-valued function $y = f(x)$.

(b) $\vec{a}, \vec{b}$ and $\vec{c}$ are three noncoplanar unit vectors and $\vec{x}$ is any vector in space equally inclined with the three vectors. Prove that $\vec{x}$ in terms of $\vec{a}, \vec{b}$ and $\vec{c}$ is given by

$$\vec{x} = t\,\frac{\vec{b}\times\vec{c} + \vec{c}\times\vec{a} + \vec{a}\times\vec{b}}{[\vec{a}\ \vec{b}\ \vec{c}]}, \text{ where } t \text{ is any real number} \ne 0.$$

Set B

2.1 Fill in the blank in each of the following so that the resulting sentence becomes true.

(a) A discontinuous function $y = f(x)$ satisfying $x^2 + y^2 = 4$ is given by ______ .

(b) An ellipse has OB as the semiminor axis. F and F′ are its foci and the angle FBF′ is a right angle. The eccentricity of the ellipse is ______ .

(c) A dice is thrown 3 times and the sum of the three numbers shown is 15. Then the probability of the first throw being four, is ______ .

(d) If $\sin\theta = \frac{2}{3}, \frac{5\pi}{2} < \theta < 3\pi$ then $\sin\frac{\theta}{4} =$ ______ .

2.2 Each of the following has a number of alternative answers of which one or more are correct. Choose the correct answer(s).

(a) The number of positive roots of the equation $xe^x = 2$ falling in the interval (0, 1) is

A. one B. two C. zero D. infinite

(b) $\left(\frac{x + |x|}{2}\right)^2 + \left(\frac{x - |x|}{2}\right)^2$ is equal to

A. $x|x|$ B. $|x|^2$ C. x^2 D. none of these

(c) If $f(x) = x\left\{1 + \frac{1}{3}\sin\log x^2\right\}, x \neq 0$

$= 0, x = 0$

then the function $f(x)$ is

A. continuous and monotonic increasing everywhere

B. continuous and differentiable everywhere

C. nondifferentiable and monotonic everywhere

D. differentiable and monotonic everywhere

(d) If $x \in [0, 1]$ then the minimum value of $x^3 + x + 1$ is

A. $\frac{3}{4}$ B. 1 C. 3 D. none of these

(e) The point on the parabola $x^2 = 4ay$ which is nearest to the circle

$x^2 + y^2 - 24ax + 128a^2 = 0$, is

A. $(-4a, 4a)$ B. $(2a, a)$ C. $(4a, 4a)$ D. none of these

2.3 If $\tan\theta = m\tan\phi$, $m > 1$ then prove that $|\tan(\theta - \phi)| \leq \frac{m-1}{2\sqrt{m}}$.

2.4 Prove, by induction, that $\int_0^{2\pi} \frac{\cos(n-1)x - \cos nx}{1 - \cos x}\,dx = 2\pi$, for all $n \in N$.

2.5 For what real a, the graph of the function $f(x) = \frac{ax^2 + 2(a+1)x + 9a + 4}{x^2 - 8x + 32}$ is always below the x-axis?

2.6 Let f be a differentiable function satisfying $f(x+y) = f(x) + f(y)$ and $f'(x) = 3$ for all $x \in R$. If $f(x) = x^2 \cdot g(x)$ then find $g(x)$.

2.7 Prove that $\sum_{k=1}^{n} (n-k)\cos\frac{2k\pi}{n} = -\frac{n}{2}$, where $n \geq 3$ is an integer.

2.8 If $\vec{p}$, $\vec{q}$ and $\vec{r}$ are noncoplanar vectors, prove that

$$[\vec{q} \times \vec{r} \;\; \vec{r} \times \vec{p} \;\; \vec{p} \times \vec{q}]^{1/2} \, (\vec{a} \times \vec{b}) = \begin{vmatrix} \vec{p} & \vec{q} & \vec{r} \\ \vec{p}\cdot\vec{a} & \vec{q}\cdot\vec{a} & \vec{r}\cdot\vec{a} \\ \vec{p}\cdot\vec{b} & \vec{q}\cdot\vec{b} & \vec{r}\cdot\vec{b} \end{vmatrix}.$$

2.9 $\triangle$ABC is inscribed in a circle of radius r. If $AB = AC$ and h is the altitude from A to BC then write the perimeter P of the $\triangle$ABC as a function of h and r. Also find

$$\lim_{h \to 0} \frac{\text{ar}(\triangle ABC)}{P^3}.$$

2.10 Let z_1 and z_2 be roots of the equation $z^2 + pz + q = 0$ where p, q are complex coefficients. If the points A and B represent z_1 and z_2 respectively in the Argand plane and $OA = OB$, $\angle AOB = \alpha (\neq 0)$ where O is the origin then prove that $p^2 = 4q\cos^2\left(\frac{\alpha}{2}\right)$.

2.11 Let $f(x) = (1 + |\sin x|)^{\frac{a}{|\sin x|}}, -\frac{\pi}{6} < x < 0$

$b, x = 0$

$e^{\frac{\tan 2x}{\tan 3x}}, 0 < x < \frac{\pi}{6}.$

Determine a and b such that f is continuous at $x = 0$.

2.12 Evaluate $\int_{-\frac{1}{2}}^{1} (1 + x^2)\sqrt{x^2 + x + 1}\; dx.$

Set C

3.1 In each of the following, fill in the blank so that the sentence becomes true.

(a) The straight line joining the points (0, 3) and (5, –2) is a tangent to the curve $y = \frac{c}{x+1}$ if $c =$ ______ .

(b) The range of the function $f(x) = 4x^3 - 12x, x \in [-1, 3]$ is ______ .

(c) Two vertices of an equilateral triangle are (–1, 0) and (1, 0) and its third vertex lies above the x-axis. The equation of the circumcircle of the triangle is ______ .

(d) From a group of 10 doctors, 8 engineers and 7 lecturers, a man unknown to their status makes a selection of 8 persons. The probability that in his selection there are at least two doctors, two engineers and 1 lecturer, is ______ if no distinction is made between persons of the same profession.

3.2 In each of the following, alternative answers are given. Choose the correct answer(s).

(a) The number of solutions of the equation $12\cos^3 x - 7\cos^2 x + 4\cos x = 9$ is

A. zero B. two C. infinite D. none of these

(b) If $x^2 + px + q$ is an integer for all integral values of x then

A. p is a fraction and q is an integer B. p and q both must be integers

C. p and q both are fractions D. none of these

(c) If $\hat{a}, \hat{b}$ are unit vectors then $(\hat{a} \times \hat{b})^2 + (\hat{a} \cdot \hat{b})^2$ is equal to

A. 0 B. 1 C. –1 D. none of these

(d) The bag A contains 2 white and 3 red balls and the bag B contains 4 white and 5 red balls. One ball is drawn at random from a randomly selected bag. It is found to be red. The probability that it was drawn from the bag B is

A. $\frac{5}{14}$ B. $\frac{5}{16}$ C. $\frac{5}{18}$ D. $\frac{25}{52}$

(e) $f(z)$ is a function of the complex variable z such that the remainders are $1 + i$ and $1 - i$ when $f(z)$ is divided by $z - i$ and $z + i$ respectively. When $f(z)$ is divided by $z^2 + 1$ the remainder will be

A. $z - 1$ B. 0 C. $z + 1$ D. none of these

3.3 If $(1 + x + x^2 + x^3)^n = \sum_{r=0}^{3n} a_r x^r$ then find the value of $a_0 + a_4 + a_8 + a_{12} + \dots$.

3.4 India and Australia play five test matches, each match ending in a win or loss or draw for India. What is the largest number of different forecasts for the results in the five test matches? What is the probability of making wrong forecasts in two of the test matches?

3.5 Three coterminous edges of a rectangular parallelepiped are of lengths a, b and c. Find the shortest distance between a diagonal of the parallelepiped and an edge connecting the faces of the terminal points of the diagonal, the edge not meeting the diagonal.

3.6 Let A_r; $r = 1, 2, 3, \ldots, n$ are n points on a circle of unit radius. Prove that the sum of the squares of their mutual distances is not greater than n^2.

3.7 Circles are inscribed in the acute angle made by the pair of lines $\sqrt{3}(x^2 + y^2) - 4xy = 0$ in the first quadrant. If each circle touches the neighbouring circle and the smallest circle has the radius a then find the sum of the radii of the first n such circles.

3.8 Prove that two parabolas $y^2 = 4ax$ and $y^2 = 4c(x - b)$ cannot have a common normal other than the common axis unless $\dfrac{b}{a - c} > 2$.

3.9 A right cylindrical tower is surmounted by a right cone. From a point on the level ground, the angles of elevation of the nearest point of the top of the tower and top of the cone are 30° and 45° respectively, and from a point nearer to the tower by a distance a, they are respectively 45° and $\tan^{-1}\left(\dfrac{3}{2}\right)$. Find the height of the top of the cone and the diameter of the tower.

3.10 Prove, by induction, that $\displaystyle\int_0^{\frac{\pi}{2}} \sin^m\theta \cdot \{\log(\sin\theta)\}^n \cdot \sin 2\theta \, d\theta = (-1)^n \cdot \frac{2(n!)}{(m + 2)^{n+1}}$

where m, n are positive integers.

3.11 A curve $y = f(x)$ passes through the point P(1, 1). The normal to the curve at P is $a(y - 1) + (x - 1) = 0$. If the slope of the tangent at any point on the curve is proportional to the ordinate of the point, determine the equation of the curve. Also obtain the area bounded by the y-axis, the curve and the normal to the curve at P.

3.12 (a) Find the functions that coincide with their derivatives for all $x \in R$.

(b) Prove that the function $y = f(x)$, given parametrically by $x = \sin t$, $y = \sin kt$, satisfies the differential equation $(1 - x^2) f''(x) - x f'(x) + k^2 f(x) = 0$ where dashes denote differentiation w.r.t. x.

□

Answers

Set A

1.1 (a) 91, $27720 \times 3^6 \times 2^3$ (b) $-\frac{4}{3}(2+\sqrt{x})\sqrt{1-\sqrt{x}}+c$ (c) $\frac{1}{3}\vec{r}\cdot(\vec{a}+\vec{b}+\vec{c})$ (d) 0

1.2 (a) A (b) C (c) B (d) A, B, C (e) B **1.5** $x^2+y^2-2ax-2by+\frac{1}{2}(a^2+b^2)=0$

1.6 $bcx^4+cay^4+abz^4+b(c+a)z^2x^2+a(b+c)y^2z^2$
$+c(a+b)x^2y^2-(b+c)x^2-(c+a)y^2-(a+b)z^2+1$

1.9 $\frac{4\times 5^5}{5^{20}-1}$ **1.10** $\frac{4+\sqrt{5}}{32}+\frac{1}{8}\log\frac{\sqrt{5}+1}{2}$

Set B

2.1 (a) $f(x)=\sqrt{4-x^2}, x\ge 0; f(x)=-\sqrt{4-x^2}, x<0$ (b) $\frac{1}{\sqrt{2}}$ (c) $\frac{1}{5}$ (d) $\sqrt{\frac{2\sqrt{3}+\sqrt{5}-1}{4\sqrt{3}}}$

2.2 (a) A (b) B, C (c) A (d) B (e) C **2.5** $a<-\frac{1}{2}$

2.6 $g(x)=\frac{3}{x}, x\ne 0$ **2.9** $P=2(\sqrt{2rh}+\sqrt{2rh-h^2}), 0\le h\le 2r; \frac{1}{128r}$

2.11 $a=\frac{2}{3}, b=\sqrt[3]{e^2}$ **2.12** $\frac{67\sqrt{3}}{64}+\frac{51}{128}\log(2+\sqrt{3})$

Set C

3.1 (a) 4 (b) $[-8, 72]$ (c) $x^2+y^2-\frac{2}{\sqrt{3}}y-1=0$ (d) $\frac{3^5-1}{3^8-1}$

3.2 (a) C (b) B (c) B (d) D (e) C

3.3 4^{n-1} **3.4** $3^5, \frac{40}{243}$

3.5 $\frac{ca}{\sqrt{c^2+a^2}}$ or $\frac{ab}{\sqrt{a^2+b^2}}$ or $\frac{bc}{\sqrt{b^2+c^2}}$ **3.7** $\frac{a}{2}\cdot\frac{2\sqrt{2}-\sqrt{3}+1}{\sqrt{3}-1}\left\{\left(\frac{\sqrt{3}-1+2\sqrt{2}}{2\sqrt{2}-\sqrt{3}+1}\right)^n-1\right\}$

3.9 $3a, \frac{4\sqrt{3}-6}{\sqrt{3}-1}a$ **3.11** $y=e^{a(x-1)}, 1+\frac{1}{2a}(2e^{-a}-1)$

3.12 (a) $y=ce^x$

□

2. Objective Tests

(*Instruction : Answer each set in one hour*)

Set A

Section I

Single Correct Choice Type

This section contains 12 multiple choice questions. Each question has 4 choices (A), (B), (C) and (D) for its answer, out of which only one is correct.

1. If the sum of the first n terms of an AP is cn^2 then the sum of squares of these n terms is

A. $\frac{n(4n^2-1)c^2}{6}$ B. $\frac{n(4n^2+1)c^2}{3}$ C. $\frac{n(4n^2-1)c^2}{3}$ D. $\frac{n(4n^2+1)c^2}{6}$

2. The number of seven digit integers, with sum of the digits equal to 10 and formed by using the digits 1, 2 and 3 only, is

A. 55 B. 66 C. 77 D. 88

3. Let $z = x + iy$ be a complex number where x and y are integers. Then, the area of the rectangle whose vertices are the roots of the equation $z(\bar{z})^3 + \bar{z}z^3 = 350$, is

A. 48 B. 32 C. 40 D. 80

4. Let $z = \cos\theta + i\sin\theta$. Then, the value of $\sum_{m=1}^{15} I_m (Z^{2m-1})$ at $\theta = 2°$, is

A. $\frac{1}{\sin 2°}$ B. $\frac{1}{3\sin 2°}$ C. $\frac{1}{2\sin 2°}$ D. $\frac{1}{4\sin 2°}$

5. The locus of orthocentre of the triangle formed by the lines $(1+p)x - py + p(1+p) = 0$, $(1+q)x - qy + q(1+q) = 0$ and $y = 0$ where $p \neq q$, is

A. a hyperbola B. a parabola C. an ellipse D. a straight line

6. Tangents drawn from the point $P(1, 8)$ to the circle $x^2 + y^2 - 6x - 4y - 11 = 0$ touch the circle at the points A and B. The equation of the circumcircle of the $\triangle PAB$, is

A. $x^2 + y^2 + 4x - 6y + 19 = 0$ B. $x^2 + y^2 - 4x - 10y + 19 = 0$

C. $x^2 + y^2 - 2x + 6y - 29 = 0$ D. $x^2 + y^2 - 6x - 4y + 19 = 0$

7. The normal at a point P on the ellipse $x^2 + 4y^2 = 16$ meets the x-axis at Q. If M is the mid-point of the line segment PQ then the locus of M intersects the latus rectum of the given ellipse at the points

A. $\left(\pm\frac{3\sqrt{5}}{2}, \pm\frac{2}{7}\right)$ B. $\left(\pm\frac{3\sqrt{5}}{2}, \pm\frac{\sqrt{19}}{4}\right)$ C. $\left(\pm 2\sqrt{3}, \pm\frac{1}{7}\right)$ D. $\left(\pm 2\sqrt{3}, \pm\frac{4\sqrt{3}}{7}\right)$

8. The line passing through the extremity A of the major axis and extremity B of the minor axis of the ellipse $x^2 + 9y^2 = 9$ meets its auxiliary circle at the point M. Then, the area of the triangle with vertices at A, M and the origin O is

A. $\frac{31}{10}$ B. $\frac{29}{10}$ C. $\frac{21}{10}$ D. $\frac{27}{10}$

9. If $\vec{a}, \vec{b}, \vec{c}$ and $\vec{d}$ are unit vectors such that $(\vec{a} \times \vec{b}) \cdot (\vec{c} \times \vec{d}) = 1$ and $\vec{a} \cdot \vec{c} = \frac{1}{2}$ then

A. $\vec{a}, \vec{b}, \vec{c}$ are noncoplanar
B. $\vec{b}, \vec{c}, \vec{d}$ are noncoplanar
C. $\vec{b}, \vec{d}$ are nonparallel
D. $\vec{a}, \vec{d}$ are parallel and $\vec{b}, \vec{c}$ are parallel

10. A line with positive direction cosines passes through the points P (2, –1, 2) and makes equal angles with the coordinate axes. The line meets the plane $2x + y + z = 9$ at point Q. The length of the line segment PQ equals

A. 1 B. $\sqrt{2}$ C. $\sqrt{3}$ D. 2

11. Let $P(3, 2, 6)$ be a point in space and Q be a point on the line $\vec{r} = (\hat{i} - \hat{j} + 2\hat{k}) + \mu(-3\hat{i} + \hat{j} + 5\hat{k})$. Then, the value of μ for which the vector $\overrightarrow{PQ}$ is parallel to the plane $x - 4y + 3z = 1$, is

A. $\frac{1}{4}$ B. $-\frac{1}{4}$ C. $\frac{1}{8}$ D. $-\frac{1}{8}$

12. Let f be a nonnegative function defined on the interval [0, 1]. If $\int_0^x \sqrt{1 - \{f'(t)\}^2}\, dt = \int_0^x f(t)\, dt,\ 0 \le x \le 1$ and $f(0) = 0$ then

A. $f\left(\frac{1}{2}\right) < \frac{1}{2}$ and $f\left(\frac{1}{3}\right) > \frac{1}{3}$
B. $f\left(\frac{1}{2}\right) > \frac{1}{2}$ and $f\left(\frac{1}{3}\right) > \frac{1}{3}$
C. $f\left(\frac{1}{2}\right) < \frac{1}{2}$ and $f\left(\frac{1}{3}\right) < \frac{1}{3}$
D. $f\left(\frac{1}{2}\right) > \frac{1}{2}$ and $f\left(\frac{1}{3}\right) < \frac{1}{3}$

Section II

Multiple Correct Choice Type

This section contains 9 multiple choice questions. Each question has 4 choices (A), (B), (C) and (D) for its answer, out of which one or more is/are correct.

13. If $\frac{\sin^4 x}{2} + \frac{\cos^4 x}{3} = \frac{1}{5}$, then

A. $\tan^2 x = \frac{3}{2}$
B. $\frac{\sin^8 x}{8} + \frac{\cos^8 x}{27} = \frac{1}{125}$
C. $\tan^2 x = \frac{2}{3}$
D. $\frac{\sin^8 x}{8} + \frac{\cos^8 x}{27} = \frac{2}{125}$

14. For $0 < \theta < \frac{\pi}{2}$, the solution(s) of $\sum_{m=1}^{6} \operatorname{cosec}\left(\theta + \frac{m-1}{4}\pi\right) \cdot \operatorname{cosec}\left(\theta + \frac{m}{4}\pi\right) = 4\sqrt{2}$ is/are

A. $\frac{\pi}{4}$
B. $\frac{\pi}{6}$
C. $\frac{\pi}{12}$
D. $\frac{5\pi}{12}$

15. In a triangle ABC with fixed base BC, the vertex A moves such that $\cos B + \cos C = 4\sin^2 \frac{A}{2}$. If a, b and c denote the lengths of the sides of the triangle opposite to the angles A, B and C respectively then

A. $b + c = 4a$
B. $b + c = 2a$
C. locus of A is an ellipse
D. locus of A is a pair of straight lines

16. The tangent PT and the normal PN to the parabola $y^2 = 4ax$ at a point P on it meet its axis at points T and N respectively. The locus of the centroid of the $\Delta\ PTN$ is a parabola whose

A. vertex $= \left(\frac{2a}{3}, 0\right)$ B. directrix is $x = 0$

C. latus rectum $= \frac{2a}{3}$ D. focus $= (a, 0)$

17. An ellipse intersect the hyperbola $2x^2 - 2y^2 = 1$ orthogonally. The eccentricity of the ellipse is reciprocal of that of the hyperbola. If the axes of the ellipse are along the coordinate axes then the

A. equation of the ellipse is $x^2 + 2y^2 = 2$ B. foci of the ellipse are $(\pm 1, 0)$

C. equation of the ellipse is $x^2 + 2y^2 = 4$ D. foci of the ellipse are $(\pm\sqrt{2}, 0)$

18. Let $L = \lim\limits_{x \to 0} \dfrac{a - \sqrt{a^2 - x^2} - \dfrac{x^2}{4}}{x^4}$, $a > 0$. If L is finite then

A. $a = 2$ B. $a = 1$

C. $L = \frac{1}{64}$ D. $L = \frac{1}{32}$

19. For the function $f(x) = x \cos \frac{1}{x}$, $x \geq 1$

A. for at least one $x \in [1, \infty)$, $f(x + 2) - f(x) < 2$

B. $\lim\limits_{x \to \infty} f'(x) = 1$

C. for all $x \in [1, \infty)$, $f(x + 2) - f(x) > 2$

D. $f'(x)$ is strictly decreasing in $[1, \infty)$

20. If $I_n = \displaystyle\int_{-\pi}^{\pi} \frac{\sin nx}{(1 + \pi^x) \cdot \sin x} dx$; $n = 0, 1, 2, \ldots$ then

A. $I_n = I_{n+2}$ B. $\sum\limits_{m=1}^{10} I_{2m+1} = 10\pi$

C. $\sum\limits_{m=1}^{10} I_{2m} = 0$ D. $I_n = I_{n+1}$

21. The area of the region bounded by the curve $y = e^x$ and lines $x = 0$ and $y = e$, is

A. $e - 1$ B. $\int_1^e \log_e (e + 1 - y)\, dy$

C. $e - \int_0^1 e^x\, dx$ D. $\int_1^e \log_e y\, dy$

Set B

Section I

Single Correct Choice Type

This section contains 14 multiple choice questions. Each question has 4 choices (A), (B), (C) and (D) for its answer, out of which only one is correct.

1. Let p and q be real numbers such that $p \neq 0, p^3 \neq q$ and $p^3 \neq -q$. If α and β are nonzero complex numbers satisfying $\alpha + \beta = -p$ and $\alpha^3 + \beta^3 = q$ then a quadratic equation having $\frac{\alpha}{\beta}$ and $\frac{\beta}{\alpha}$ as its roots is

   A. $(p^3 + q)x^2 - (p^3 + 2q)x + (p^3 + q) = 0$ B. $(p^3 + q)x^2 - (p^3 - 2q)x + (p^3 + q) = 0$

   C. $(p^3 - q)x^2 - (5p^3 - 2q)x + (p^3 - q) = 0$ D. $(p^3 - q)x^2 - (5p^3 + 2q)x + (p^3 - q) = 0$

2. Let $S = \{1, 2, 3, 4\}$. The total number of unordered pairs of disjoint subsets of S is equal to

   A. 25 B. 34 C. 42 D. 41

3. For $r = 0, 1, 2, 3, \ldots, 10$ let A_r, B_r, C_r denote respectively the coefficient of x^r in the expansions of $(1 + x)^{10}$, $(1 + x)^{20}$ and $(1 + x)^{30}$. Then

$$\sum_{r=1}^{10} A_r (B_{10} B_r - C_{10} A_r) \text{ is equal to}$$

   A. $B_{10} - C_{10}$ B. $A_{10}(B_{10}^2 - C_{10} A_{10})$ C. 0 D. $C_{10} - B_{10}$

4. The number of 3×3 matrices A whose entries are either 0 or 1 and for which the system $A\begin{bmatrix} x \\ y \\ z \end{bmatrix} = \begin{bmatrix} 1 \\ 0 \\ 0 \end{bmatrix}$ has exactly two distinct solutions, is

   A. 0 B. $2^9 - 1$ C. 168 D. 2

5. If the angles A, B and C of a triangle are in an AP and if a, b and c denote the lengths of the sides opposite to A, B and C respectively then the value of the expression $\frac{a}{c}\sin 2C + \frac{c}{a}\sin 2A$ is

   A. $\frac{1}{2}$ B. $\frac{\sqrt{3}}{2}$ C. 1 D. $\sqrt{3}$

6. Let P, Q, R and S be points on the plane with position vectors $-2\hat{i} - \hat{j}, 4\hat{i}, 3\hat{i} + 3\hat{j}$ and $-3\hat{i} + 2\hat{j}$ respectively. The quadrilateral $PQRS$ must be a

   A. parallelogram, which is neither a rhombus nor a rectangle

   B. square

   C. rectangle but not a square

   D. rhombus but not a square

7. Two adjacent sides of a parallelogram $ABCD$ are given by $\vec{AB} = 2\hat{i} + 10\hat{j} + 11\hat{k}$ and $\vec{AD} = -\hat{i} + 2\hat{j} + 2\hat{k}$. The side AD is rotated by an acute angle α in the plane of the parallelogram, so that AD becomes AD'. If AD' makes a right angle with the side AB then the cosine of the angle α is given by

   A. $\frac{8}{9}$ B. $\frac{\sqrt{17}}{9}$ C. $\frac{1}{9}$ D. $\frac{4\sqrt{5}}{9}$

8. The equation of the plane containing the straight line $\frac{x}{2} = \frac{y}{3} = \frac{z}{4}$ and perpendicular to the plane containing the straight lines $\frac{x}{3} = \frac{y}{4} = \frac{z}{2}$ and $\frac{x}{4} = \frac{y}{2} = \frac{z}{3}$ is

   A. $x + 2y - 2z = 0$ B. $3x + 2y - 2z = 0$ C. $x - 2y + z = 0$ D. $5x + 2y - 4z = 0$

9. If the distance of the point $P(1, -2, 1)$ from the plane $x + 2y - 2z = \alpha$, where $\alpha > 0$, is 5 then the foot of the perpendicular from P to the plane is

A. $\left(\frac{8}{3}, \frac{4}{3}, \frac{-7}{3}\right)$ B. $\left(\frac{4}{3}, -\frac{4}{3}, \frac{1}{3}\right)$

C. $\left(\frac{1}{3}, \frac{2}{3}, \frac{10}{3}\right)$ D. $\left(\frac{2}{3}, -\frac{1}{3}, \frac{5}{2}\right)$

10. Let f, g and h be real-valued functions defined on the interval $[0, 1]$ by $f(x) = e^{x^2} + e^{-x^2}$, $g(x) = re^{x^2} + e^{-x^2}$ and $h(x) = x^2e^{x^2} + e^{-x^2}$. If a, b and c denote the absolute maximum of f, g and h on $[0, 1]$ then

A. $a = b$ and $b \neq c$ B. $c = a$ and $a \neq b$ C. $a \neq b$ and $b \neq c$ D. $a = b = c$

11. The value of $\lim\limits_{x \to 0} \frac{1}{x^3} \int_0^x \frac{t\log(1+t)}{t^4+4} dt$ is

A. 0 B. $\frac{1}{12}$ C. $\frac{1}{24}$ D. $\frac{1}{64}$

12. Let f be a real-valued function defined on the interval $(-1, 1)$ such that $e^{-x} . f(x) = 2 + \int_0^x \sqrt{t^4+1}\, dt \;\forall\; x \in (-1, 1)$. Let f^{-1} be the inverse function of f. Then $(f^{-1})'(2)$ is equal to

A. 1 B. $\frac{1}{3}$ C. $\frac{1}{2}$ D. $\frac{1}{e}$

13. Let ω be a complex cube root of unity with $\omega \neq 1$. A fair die is thrown three times. If r_1, r_2 and r_3 are the numbers obtained on the die then the probability that $\omega^{r_1} + \omega^{r_2} + \omega^{r_3} = 0$ is

A. $\frac{1}{18}$ B. $\frac{1}{9}$ C. $\frac{2}{9}$ D. $\frac{1}{36}$

14. A signal which can be green or red with probability $\frac{4}{5}$ and $\frac{1}{5}$ respectively, is received by station A and then transmitted to station B. The probability of each station receiving the signal correctly is $\frac{3}{4}$. If the signal received at station B is green then the probability that the original signal was green is

A. $\frac{3}{5}$ B. $\frac{6}{7}$ C. $\frac{20}{23}$ D. $\frac{9}{20}$

Section II

Multiple Correct Choice Type

This section contains 5 multiple choice questions. Each question has four choices (A), (B), (C) and (D) out of which one or more may be correct.

15. Let z_1 and z_2 be two distinct complex numbers and let $z = (1-t)z_1 + tz_2$ for some real number t, where $0 < t < 1$. If arg (ω) denotes the principal argument of a nonzero complex number ω then

A. $|z - z_1| + |z - z_2| = |z_1 - z_2|$ B. $\arg(z - z_1) = \arg(z - z_2)$

C. $\begin{vmatrix} z - z_1 & \bar{z} - \bar{z}_1 \\ z_2 - z_1 & \bar{z}_2 - \bar{z}_1 \end{vmatrix} = 0$ D. $\arg(z - z_1) = \arg(z_2 - z_1)$

16. Let ABC be a triangle such that $\angle ACB = \frac{\pi}{6}$ and let a, b and c denote the lengths of the sides opposite A, B and C respectively. The value(s) of x for which $a = x^2 + x + 1$, $b = x^2 - 1$ and $c = 2x + 1$ is (are)

A. $-(2+\sqrt{3})$ B. $\sqrt{3}+1$ C. $2+\sqrt{3}$ D. $4\sqrt{3}$

17. Let A and B be two distinct points on the parabola $y^2 = 4x$. If the axis of the parabola touches a circle of radius r having AB as its diameter then the slope of the line AB can be

A. $-\frac{1}{r}$ B. $\frac{1}{r}$ C. $\frac{2}{r}$ D. $-\frac{2}{r}$

18. The value(s) of $\int_0^1 \frac{x^4(1-x)^4}{1+x^2}\,dx$ is (are)

A. $\frac{22}{7}-\pi$ B. $\frac{2}{105}$ C. 0 D. $\frac{71}{15}-\frac{3\pi}{2}$

19. Let f be a real-valued function defined on the interval $(0, \infty)$ by $f(x) = \log_e x + \int_0^x \sqrt{1+\sin t}\,dt$. Which of the following statement(s) is (are) true?

A. $f''(x)$ exists for all $x \in (0, \infty)$

B. $f'(x)$ exists for all $x \in (0, \infty)$ and $f'(x)$ is continuous on $(0, \infty)$ but not differentiable on $(0, \infty)$

C. there exists $\alpha > 1$ such that $|f'(x)| < |f(x)|$ for all $x \in (\alpha, \infty)$

D. there exists $\beta > 0$ such that $|f(x)| + |f'(x)| \le \beta$ for all $x \in (0, \infty)$

□

Answers

Set A

1. C	**2.** C	**3.** A	**4.** D	**5.** D	**6.** B
7. C	**8.** D	**9.** C	**10.** C	**11.** A	**12.** C
13. B, C	**14.** C, D	**15.** B, C	**16.** A, D	**17.** A, B	**18.** A, C
19. B, C, D	**20.** A, B, C	**21.** B, C, D			

Set B

1. B	**2.** A	**3.** D	**4.** A	**5.** D	**6.** A
7. B	**8.** C	**9.** A	**10.** D	**11.** B	**12.** B
13. C	**14.** C	**15.** A, C, D	**16.** B	**17.** C, D	**18.** A
19. B, C					

□

3. Screening Tests

(*Instruction: Answer each set in one hour*)

Set A

1. If $a_1, a_2, a_3, \ldots, a_n, \ldots$ are in GP then the value of the determinant

$$\begin{vmatrix} \log a_n & \log a_{n+1} & \log a_{n+2} \\ \log a_{n+3} & \log a_{n+4} & \log a_{n+5} \\ \log a_{n+6} & \log a_{n+7} & \log a_{n+8} \end{vmatrix} \text{ is}$$

A. 0 B. 2 C. –2 D. 1

2. If $z = x + iy$ and $z^{1/3} = p + iq$ then $\left(\frac{x}{p} - \frac{y}{q}\right) / (p^2 + q^2)$ is equal to

A. 2 B. 1 C. –1 D. –2

3. If $(1 + \omega^5)^n = (1 + \omega^{13})^n$, where ω is a nonreal cube root of unity then the least positive integral value of n is

A. 2 B. 3 C. 5 D. 6

4. For all $x \in R$, $x^2 + 2ax + 10 - 3a > 0$. The interval in which a lies is

A. $a < -5$ B. $-5 < a < 2$ C. $a > 5$ D. $2 < a < 5$

5. If $A = \begin{bmatrix} \alpha & 2 \\ 2 & \alpha \end{bmatrix}$ and $|A^3| = 125$ then the value of α is

A. ±1 B. ±2 C. ±3 D. ±5

6. If a, b, c and d are positive and $abcd = 1$ then the minimum value of $(1 + a)(1 + b)(1 + c)(1 + d)$ is

A. 4 B. 16 C. 8 D. 32

7. The term independent of x in the expansion of $\left(x^2 + 1 + \frac{1}{x^2}\right)^7$ is

A. 43 B. 253 C. 392 D. 393

8. The number of arrangements of the letters of the word GARDEN with the vowels in alphabetical order is

A. 120 B. 240 C. 360 D. 480

9. If $f: R \to S$, defined by $f(x) = \sin x - \sqrt{3}\cos x + 1$ is an onto function then the interval S is

A. [–1, 1] B. [–1, 3] C. [–1, 2] D. [0, 1]

10. If $f(x) = \sin x + \cos x$ and $g(x) = x^2 - 1$ then $g\{f(x)\}$ is invertible in the domain

A. $[0, \pi/2]$ B. $[-\pi/4, \pi/4]$ C. $[-\pi/2, \pi/2]$ D. $[0, \pi]$

11. If $\lim\limits_{x \to \infty} \left(1 + \frac{a}{x} + \frac{b}{x^2}\right)^{2x} = e^2$ then the values of a and b are

A. $a \in R,\ b \in R$ B. $a = 1,\ b \in R$ C. $a \in R,\ b = 2$ D. $a = 1, b = 2$

12. Let $g(x) = \frac{f(x)}{x+1}$, where $f(x)$ is differentiable in [0, 5] such that $f(0) = 4$, $f(5) = -1$. There exists $c \in (0, 5)$ such that $g'(c)$ is

A. $-\frac{1}{6}$ B. $\frac{1}{6}$ C. $\frac{-5}{6}$ D. -1

13. If $\int \frac{\sin x}{\sin (x - \alpha)} dx = Ax + B \log \sin (x - \alpha) + C$ then the value of (A, B) is

A. $(\sin \alpha, \cos \alpha)$ B. $(\cos \alpha, \sin \alpha)$ C. $(-\sin \alpha, \cos \alpha)$ D. $(-\cos \alpha, \sin \alpha)$

14. The value of $\int_{-2}^{3} |1 - x^2| \, dx$ is

A. $\frac{28}{3}$ B. $\frac{14}{3}$ C. $\frac{7}{3}$ D. $\frac{1}{3}$

15. If $f(x) = \frac{e^x}{1+e^x}$, $I_1 = \int_{f(-a)}^{f(a)} x \cdot g\{x(1-x)\} \, dx$ and $I_2 = \int_{f(-a)}^{f(a)} g\{x(1-x)\} \, dx$ then the value of $\frac{I_2}{I_1}$ is

A. 2 B. -3 C. -1 D. 1

16. The general solution of the differential equation $y \, dx + (x + x^2 y) \, dy = 0$ is

A. $-\frac{1}{xy} = c$ B. $\frac{-1}{xy} + \log y = c$ C. $\frac{1}{xy} + \log y = c$ D. $\log y = cx$

17. If θ and ϕ are acute angles and $\sin \theta = \frac{1}{2}$, $\cos \phi = \frac{1}{3}$ then the value of $\theta + \phi$ belongs to the interval

A. $\left(\frac{\pi}{3}, \frac{\pi}{2}\right]$ B. $\left(\frac{\pi}{2}, \frac{2\pi}{3}\right]$ C. $\left(\frac{2\pi}{3}, \frac{5\pi}{6}\right]$ D. $\left(\frac{5\pi}{6}, \pi\right]$

18. The value of x for which $\sin \{\cot^{-1}(1+x)\} = \cos (\tan^{-1} x)$, is

A. $\frac{1}{2}$ B. 1 C. 0 D. $\frac{-1}{2}$

19. The sides of a triangle are $\sin \alpha$, $\cos \alpha$ and $\sqrt{1 + \sin \alpha \cdot \cos \alpha}$, where α is an acute angle. The greatest angle of the triangle is

A. $\frac{\pi}{3}$ B. $\frac{\pi}{2}$ C. $\frac{2\pi}{3}$ D. $\frac{5\pi}{6}$

20. The greatest distance between two points on the curve $4x^2 + 9y^2 + 24x - 6y + 33 = 0$ is

A. 2 B. 1 C. $\frac{2}{3}$ D. $\frac{4}{3}$

21. If two of the feet of normals from a point to the parabola $(y+1)^2 = 4x$ be (1, 1) and (4, 3) then the foot of the third normal to the parabola from the point is

A. $(0, -1)$ B. $(9, 5)$ C. $(9, -7)$ D. $\left(\frac{9}{4}, 2\right)$

22. The line $2x + \sqrt{6}y = 2$ touches the hyperbola $x^2 - 2y^2 = 4$. The point of contact is

A. $(-2, \sqrt{6})$ B. $(-5, 2\sqrt{6})$ C. $\left(\frac{1}{2}, \frac{1}{\sqrt{6}}\right)$ D. $(4, -\sqrt{6})$

23. The centre of the circle of radius 1 touching the parabola $x^2 = y$ externally at (1, 1) is

A. $\left(\frac{\sqrt{5}-2}{\sqrt{5}}, \frac{\sqrt{5}+1}{\sqrt{5}}\right)$ B. $\left(\frac{\sqrt{5}-2}{\sqrt{5}}, \frac{\sqrt{5}-1}{\sqrt{5}}\right)$ C. $\left(\frac{2+\sqrt{5}}{\sqrt{5}}, \frac{\sqrt{5}-1}{\sqrt{5}}\right)$ D. none of these

24. Lines $\frac{x-1}{2} = \frac{y+1}{3} = \frac{z-1}{4}$ and $\frac{x-3}{1} = \frac{y-k}{2} = \frac{z}{1}$ intersect. Then k equals

A. 3/2 B. 9/2 C. $-2/9$ D. $-3/2$

25. Let $\vec{a}$, $\vec{b}$ and $\vec{c}$ be three nonzero vectors such that no two of these are collinear. If the vector $\vec{a}+2\vec{b}$ is collinear with $\vec{c}$ and $\vec{b}+3\vec{c}$ is collinear with $\vec{a}$ then $\vec{a}+2\vec{b}+6\vec{c}$ equals (λ being a scalar $\neq 0$)

A. $\lambda\vec{a}$ B. $\lambda\vec{b}$ C. $\lambda\vec{c}$ D. $\vec{0}$

26. The unit vector which is orthogonal to the vector $\vec{a}=3\vec{i}+2\vec{j}+6\vec{k}$ and is coplanar with the vectors $\vec{b}=2\vec{i}+\vec{j}+\vec{k}$ and $\vec{c}=\vec{i}-\vec{j}+\vec{k}$, is

A. $\frac{1}{\sqrt{41}}(2\vec{i}-6\vec{j}+\vec{k})$ B. $\frac{2\vec{i}-3\vec{j}}{\sqrt{13}}$

C. $\frac{3\vec{j}-\vec{k}}{\sqrt{10}}$ D. $\frac{4\vec{i}+3\vec{j}-3\vec{k}}{\sqrt{14}}$

27. The probability that A speaks truth is $\frac{4}{5}$ while that for B is $\frac{3}{4}$. The probability that they contradict each other when asked to speak on a fact, is

A. $\frac{3}{20}$ B. $\frac{1}{5}$ C. $\frac{7}{20}$ D. $\frac{4}{5}$

28. If three distinct numbers are chosen at random from the first 100 natural numbers then the probability that all three of them are divisible by 2 and 3 is

A. $\frac{4}{25}$ B. $\frac{4}{35}$ C. $\frac{4}{33}$ D. $\frac{4}{1155}$

Set B

1. If $4x^4+9y^4=64$ then the maximum value of x^2+y^2 is

A. $\frac{4}{\sqrt{3}}$ B. $\frac{4}{3}\sqrt{13}$ C. $\frac{32}{3}$ D. $\frac{32}{13}$

2. If $\log_{30}3=a$ and $\log_{30}5=b$ then $\log_{30}8$ is equal to

A. $1+a+b$ B. $3(1-a-b)$ C. $3(1+a+b)$ D. $a+b$

3. $a\in R$, $b\in R$ and the equation $ax^2+bx+1=0$ cannot have distinct real solutions then the minimum value of $a+b$ is

A. –3 B. –2 C. –1 D. 0

4. An infinite GP has the first term x and the sum 5. Then x belongs to the interval

A. $(-\infty,-10)$ B. $(-10,0)$ C. $(0,10)$ D. $(10,+\infty)$

5. Let $A=\begin{bmatrix}1&-1&1\\2&1&-3\\1&1&1\end{bmatrix}$ and $10B=\begin{bmatrix}4&2&2\\-5&0&\alpha\\1&-2&3\end{bmatrix}$. If B is the inverse of the matrix A then α is

A. –2 B. –1 C. 2 D. 5

6. The number of ways of distributing 8 identical balls in 3 distinct boxes so that none of the boxes is empty, is

A. 5 B. 21 C. 3^8 D. ${}^8C_3$

7. The coefficients of the middle term in the binomial expansions of $(1+\alpha x)^4$ and $(1-\alpha x)^6$ are equal when expanded in powers of x if α equals

A. –5/3 B. 10/3 C. –3/10 D. 3/5

8. If ${}^{n-1}C_r=(k^2-3)\cdot{}^nC_{r+1}$ then k belongs to

A. $(-\infty,-2)$ B. $(2,+\infty)$ C. $[-\sqrt{3},\sqrt{3}]$ D. $(\sqrt{3},2]$

9. The graph of the function $y=f(x)$ is symmetrical about the line $x=2$. Then

A. $f(x+2)=f(x-2)$ B. $f(2+x)=f(2-x)$

C. $f(x)=f(-x)$ D. $f(x)=-f(-x)$

10. The sides of a triangle are in the ratio $1 : \sqrt{3} : 2$. Then angles of the triangle are in the ratio

A. 1 : 3 : 5 B. 2 : 3 : 4 C. 3 : 2 : 1 D. 1 : 2 : 3

11. The value of $\cos^{-1} \sin \cos^{-1}\left(\sin \frac{10\pi}{7}\right)$ is equal to

A. $\frac{10\pi}{7}$ B. $\frac{3\pi}{7}$ C. $\frac{4\pi}{7}$ D. $\frac{13\pi}{14}$

12. $f(x)$ is a differentiable function which is increasing. Then

$\lim_{x \to 0} \frac{f(x^2) - f(x)}{f(x) - f(0)}$ equals

A. 1 B. 0 C. –1 D. 2

13. If $f(x) = x^3 + bx^2 + cx + d$ and $0 < b^2 < c$ then in $R, f(x)$

A. is increasing B. has local maximum C. is decreasing D. is bounded

14. If $f(x) = x^{\alpha} \log x$ and $f(0) = 0$ then the value of α for which Rolle's theorem can be applied in [0, 1] is

A. –2 B. –1 C. 0 D. $\frac{1}{2}$

15. If $f(x)$ is differentiable and $\int_0^{t^2} x f(x)\, dx = \frac{2}{5} t^5$ then $f\left(\frac{4}{25}\right)$ equals

A. 2/5 B. –5/2 C. 1 D. 5/2

16. Let $f(x) = \frac{1 - \tan x}{4x - \pi}$, $x \neq \frac{\pi}{4}$, $x \in [0, \pi/2]$. If $f(x)$ is continuous in $[0, \pi/2]$ then $f(\pi/4)$ is

A. 1 B. $\frac{1}{2}$ C. $-\frac{1}{2}$ D. –1

17. If $2f(x) \cdot f(y) = f(x + y) + f(x - y)$ for all $x, y \in R$ and $f(0) \neq 0$ then

A. $f(2) + f(-2) = 0$ B. $f(3) + f(-3) = 0$ C. $f'(2) + f'(-2) = 0$ D. $f'(3) = f'(-3)$

18. The area enclosed by the curves $y = ax^2$ and $x = ay^2$, $a > 0$ is 1 unit^2. Then a is

A. $\frac{1}{\sqrt{3}}$ B. $\frac{1}{2}$ C. 1 D. $\frac{1}{3}$

19. $\lim_{n \to \infty} \sum_{r=1}^{n} \frac{1}{n} \cdot e^{r/n}$ is equal to

A. e B. $e - 1$ C. $1 - e$ D. $1 + e$

20. The angle between tangents from the point (1, 4) to the parabola $y^2 = 4x$ is

A. $\pi/6$ B. $\pi/4$ C. $\pi/3$ D. $\pi/2$

21. The length of the diameter of the rectangular hyperbola $xy = 4$ which is inclined at an angle $\pi/4$ with the transverse axis, is

A. $2\sqrt{2}$ B. 2 C. 4 D. $4\sqrt{2}$

22. In an ellipse, AB and CD are the major and minor axes respectively. If CA = CD then the eccentricity of the ellipse is

A. 2/3 B. $\sqrt{2/3}$ C. 1/2 D. $1/\sqrt{2}$

23. A variable circle passes through the fixed point $A(p, q)$ and touches the x-axis. The locus of the other end of the diameter through A is

A. $(x - p)^2 = 4qy$ B. $(x - q)^2 = 4py$ C. $(y - p)^2 = 4qx$ D. $(y - q)^2 = 4px$

24. If the straight lines $x = 1 + s, y = -3 - \lambda s, z = 1 + \lambda s$ and $x = \frac{t}{2}, y = t, z = 2 + t$ with parameters s and t respectively are coplanar then λ equals

A. –2 B. $-\frac{1}{2}$ C. $\frac{1}{2}$ D. 0

25. A line makes the same angle θ with each of the x- and z-axes. If the angle β which it makes with the y-axis, is such that $\sin^2\beta = 3\sin^2\theta$ then $\cos^2\theta$ equals

A. 2/3 B. 1/5 C. 3/5 D. 2/5

26. Let $\vec{u}, \vec{v}, \vec{w}$ be such that $|\vec{u}| = 1$, $|\vec{v}| = 2$ and $|\vec{w}| = 3$. If the projection of $\vec{v}$ along $\vec{u}$ is equal to that of $\vec{w}$ along $\vec{u}$, and $\vec{v}, \vec{w}$ are perpendicular to each other then $|\vec{u} - \vec{v} + \vec{w}|$ equals

A. 2 B. $\sqrt{7}$ C. $\sqrt{14}$ D. 14

27. Let $\vec{a}, \vec{b}$ and $\vec{c}$ be nonzero vectors such that $(\vec{a} \times \vec{b}) \times \vec{c} = \frac{1}{3}|\vec{b}||\vec{c}|\vec{a}$. If θ is the acute angle between the vectors $\vec{b}$ and $\vec{c}$ then $\sin\theta$ equals

A. 1/3 B. $\sqrt{2}/3$ C. 2/3 D. $2\sqrt{2}/3$

28. Let $\vec{u}$ be a unit vector and $\vec{v} = 2\vec{i} + \vec{j} - \vec{k}$, $\vec{w} = \vec{i} + 3\vec{k}$. The maximum possible value of the scalar triple product $[\vec{u}\ \vec{v}\ \vec{w}]$ is

A. –1 B. $\sqrt{10} + \sqrt{6}$ C. $\sqrt{59}$ D. $\sqrt{60}$

□

Answers

Set A

1. A	**2.** D	**3.** B	**4.** B	**5.** C	**6.** B	**7.** D
8. C	**9.** B	**10.** B	**11.** B	**12.** C	**13.** B	**14.** A
15. A	**16.** B	**17.** B	**18.** D	**19.** C	**20.** A	**21.** C
22. D	**23.** C	**24.** B	**25.** D	**26.** C	**27.** C	**28.** D

Set B

1. B	**2.** B	**3.** C	**4.** C	**5.** D	**6.** B	**7.** C
8. D	**9.** B	**10.** D	**11.** B	**12.** C	**13.** A	**14.** D
15. A	**16.** C	**17.** C	**18.** A	**19.** B	**20.** C	**21.** D
22. B	**23.** A	**24.** C	**25.** C	**26.** C	**27.** A	**28.** C

□

4. Main Tests

*(**Instruction:** Answer each set in two hours)*

Set A

1. Solve the equation in $[0, \pi]$: $[\cos x + \sin x] = [\cos x] + [\sin x]$ where $[y]$ is the greatest integer less than or equal to y.

2. If the equation $ax^2 + bx + 2 = 0$, where $a \in R, b \in R$, cannot have real roots, find the greatest value of $b^2 - a^2$.

3. An equilateral $\triangle ABC$ has its vertex A at the point (1, 1, 1) and the other vertices lie on the line $\frac{x-1}{1} = \frac{y-2}{3} = \frac{z}{2}$. Find the vertices B and C.

4. If z and w be two complex numbers satisfying $\bar{z} + i\,\bar{w} = 0$ and $\arg(zw) = \pi$ then find $\arg z$.

5. Find the domain and range of the function $f(x) = {}^{7-x}C_{x-3}$ where ${}^nC_r$ is the number of selections of n different things taking r at a time.

6. Let $\pi < \alpha - \beta < 3\pi$. If $\sin\alpha + \sin\beta = \frac{-21}{65}$ and $\cos\alpha + \cos\beta = \frac{-27}{65}$ then find the value of $\cos\frac{\alpha - \beta}{2}$.

7. Let the function $f(x)$ be such that its second order derivative $f''(x) = 6(x - 1)$. The graph of $y = f(x)$ passes through the point (2, 1) and at that point the tangent to the graph has the equation $y = 3x - 5$. Find the function.

8. $\vec{a}, \vec{b}, \vec{c}$ and $\vec{d}$ are distinct vectors for which $\vec{a} \times \vec{b} = \vec{c} \times \vec{d}$ and $\vec{a} \times \vec{c} = \vec{b} \times \vec{d}$. Prove that $\vec{a} \cdot \vec{b} + \vec{c} \cdot \vec{d} \neq \vec{a} \cdot \vec{c} + \vec{b} \cdot \vec{d}$.

9. Prove that $\int_0^1 e^{x^2}\,dx + \int_1^e \sqrt{\log x}\,dx = e$.

10. If A and B are two independent events and C is the event that exactly one of A and B occurs then prove that

$$P(A \cup B) \cdot P(A' \cap B') \le P(C).$$

11. Let $\vec{a} \cdot \vec{b} = 0$. Find the unit vector inclined to each of the vectors $\vec{a}$ and $\vec{b}$ by an angle $\pi/3$.

12. Find the area of the region S bounded by the circle $x^2 + y^2 - 4x = 0$ and the lines $|x - 2| = 1$. Also, find the range of values of λ for which the point $(\lambda + 1, \lambda - 1)$ belongs to the region S.

13. If z_1, z_2 and z_3 are complex numbers, prove that

$$|z_1| + |z_2| + |z_3| \le |z_1 + z_2 - z_3| + |z_2 + z_3 - z_1| + |z_1 + z_2 - z_3|.$$

14. From a point on the axis of the parabola $x^2 = 4ay$, the angle between two normals to the parabola, other than the axis, is $\frac{2\pi}{3}$. Find the point.

15. The joined equation of two planes is $2x^2 + x(y - z) - (y - z)^2 = 0$. Prove that the two planes are perpendicular to each other and also find the intercept made on the line $\frac{x-3}{-1} = \frac{y+1}{2} = \frac{z+4}{5}$ by the two planes.

16. A parabola has its vertex at the centre of the ellipse $7x^2 + 16y^2 = 112$ and the focus of the parabola is that focus of the ellipse whose x-coordinate is positive. Find the equations of the common tangents to the parabola and the ellipse.

17. Let
$$f(x) = \begin{cases} b \sin^{-1} \frac{1}{2}(x + c), & -\frac{1}{2} < x < 0 \\ \frac{1}{2}, & x = 0 \\ \frac{e^{ax/2} - 1}{x}, & 0 < x < \frac{1}{2}. \end{cases}$$

If $f(x)$ is differentiable at $x = 0$ and $|c| < 2$ then find a and prove that $64b^2 = 4 - c^2$.

18. Let $M = \begin{bmatrix} a_1 & b_1 & c_1 \\ a_2 & b_2 & c_2 \\ a_3 & b_3 & c_3 \end{bmatrix}$ and $M^T M = I =$ unit matrix. If $|M| = 1$ then prove that $|M - I| = 0$.

19. Evaluate: $\displaystyle\int_{-\pi/3}^{\pi/3} \frac{\pi + 4x^3}{2 - \cos\left(|x| + \frac{\pi}{3}\right)} dx$.

20. If $s_n = \sum_{r=0}^{n} \frac{1}{{}^nC_r}$ and $t_n = \sum_{r=0}^{n} \frac{r}{{}^nC_r}$ then prove that $t_n = \frac{n}{2} s_n$.

Set B

1. If $|z + 1| + |z - 1| < 4$ then prove that $|iz + 4 - 3i| < 7$.

2. C and D are the end points of the minor axis of an ellipse. If the normal at P, a point other than C or D, passes through C or D then prove that the eccentricity of the ellipse belongs to the interval $\left[\frac{1}{\sqrt{2}}, 1\right)$.

3. Integrate $\int \frac{\sqrt{x}}{1 + x^3} dx$.

4. If $a > 0$, $b > 0$ and $c > 0$ then prove that $\{(1 + a)(1 + b)(1 + c)\}^7 > 7^7 a^4 b^4 c^4$.

5. Prove that $(1 - x^2)\cos x + 2x \sin x \geq 1$ where $0 \leq x \leq \pi$.

6. In a ΔABC let $p - q = \frac{1}{3} r$, where $p = \Sigma \sin A$, $q = \Sigma \sin 2A$ and $r = \Sigma \sin 3A$. Prove that the triangle is equilateral.

7. Prove that the locus of middle points of chords of the parabola $y^2 = 4ax$ whose lengths are equal to $4a$, is

$$(y^2 - 4ax)(y^2 + 4a^2) + 16a^4 = 0.$$

8. a, b and c are three integers for which the equation $f(x) \equiv ax^2 + bx + c = 0$ has an irrational root. Prove that $|f(p/q)| \geq \frac{1}{q^2}$, where p and q are integers, and $q \neq 0$.

9. Find the number of ways in which 16 flowers, of which 4 are distinct flowers, can be set on a garland.

10. Let $f(x) = \sqrt{(1 - \cos x) + \sqrt{(1 - \cos x) + \sqrt{(1 - \cos x) + \ldots \text{ to } \infty}}}$

and $\phi(x) = \sqrt{x^2 + \sqrt{x^2 + \sqrt{x^2 + \ldots \text{ to } \infty}}}$. Find $\lim\limits_{x \to 0} \dfrac{1 - f(x)}{1 - \phi(x)}$.

11. Let $\phi(x) = f(x) + f(1 - x)$ and $f''(x) < 0$ for $x \in [0, 1]$. Prove that $\phi(x)$ is increasing in $\left(0, \dfrac{1}{2}\right)$.

12. Find the possible real values of (x, y) satisfying $\sin x + \cos x = \dfrac{y^2 - 2y + 4}{y^2 + 2y + 4}$.

13. Prove that the lines $\dfrac{x}{1} = \dfrac{y+1}{3} = \dfrac{z-3}{0}$, $\dfrac{x+2}{3} = \dfrac{y-4}{-2} = \dfrac{z}{3}$ and $\dfrac{x-2}{-1} = \dfrac{y-1}{1} = \dfrac{z+1}{4}$ are noncoplanar but they are concurrent.

14. Two coins are tossed together. If they show the same side, three dice are thrown. But if they show opposite sides, two dice are thrown. What is the probability of the dice thrown to show a sum 7 ?

15. The ellipse $\dfrac{x^2}{a^2} + \dfrac{y^2}{b^2} = 1$ is cut by the circle whose centre is the same as that of the ellipse and the diameter is the mean of the major and minor axes of the ellipse. Prove that their points of intersection are vertices of a rectangle. Also, find the rectangular hyperbola through these points for which transverse axis is along the major axis of the ellipse.

16. $\vec{a}$ and $\vec{b}$ are two given nonzero orthogonal vectors, and $\vec{r}$ is a unit vector such that $\vec{r} \cdot \vec{a} = \vec{r} \cdot \vec{b} = [\vec{r} \cdot \vec{a} \cdot \vec{b}]$. Find $\vec{r}$.

17. Prove that the number of integral points (i.e., points with both coordinates integers) on a circle of the family of circles with the common centre $(\sqrt{5}, 0)$ is at most two. Also find the form of the radii of circles with two integral points on them.

18. If $I_m = \displaystyle\int_0^{2a} x^m \cdot \sqrt{2ax - x^2}\, dx$, prove that $I_m = \dfrac{2m+1}{m+2} a I_{m-1}$.

19. A curve passes through (2, 0) and the slope of tangent at the point P(x, y) equals $\dfrac{(x+1)^2 + y - 3}{x+1}$. Find the equation of the curve, and the area enclosed by the curve and the x-axis in the fourth quadrant.

20. Let the roots of $x^3 + px + q = 0$ be α, β and γ. Form the equation whose roots are $(\alpha - \beta)^2, (\beta - \gamma)^2$ and $(\gamma - \alpha)^2$. If α, β and γ are real then prove that $p < 0$ and $4|p|^3 \geq 27q^2$.

□

Answers

Set A

1. $0, \frac{\pi}{2}, \pi$ **2.** 16 **3.** $\left(\frac{8}{7}, \frac{17}{7}, \frac{2}{7}\right), \left(\frac{5}{7}, \frac{8}{7}, \frac{-4}{7}\right)$ **4.** $\frac{3\pi}{4}$ **5.** domain = {3, 4, 5}, range = {1, 3}

6. $-\frac{3}{\sqrt{130}}$ **7.** $(x-1)^3$ **11.** $\frac{1}{2|\vec{a}||\vec{b}|}\{|\vec{b}|\vec{a} + |\vec{a}|\vec{b} \pm \sqrt{2}\,\vec{a}\times\vec{b}\}$

12. $\left(\frac{2\pi}{3}+\sqrt{3}\right)$ unit^2, [0, 2] **14.** $(0, 5a), \left(0, \frac{7a}{3}\right)$ **15.** $\frac{9}{2}\sqrt{30}$

16. $y = \frac{3}{4}x + 4, y = -\frac{3}{4}x - 4$ **17.** 1 **19.** $\frac{4\pi}{\sqrt{3}}\tan^{-1}\frac{1}{2}$

Set B

3. $\frac{2}{3}\tan^{-1}(x\sqrt{x}) + c$ **9.** 1365 **10.** $\frac{1}{2}$

12. $(x, y) = \left(2n\pi + \frac{\pi}{4} \pm \cos^{-1}\frac{k}{\sqrt{2}}, \frac{1+k\pm\sqrt{(k-3)(1-3k)}}{1-k}\right)$; where $k \in \left[\frac{1}{3}, \sqrt{2}\right] - \{1\}$

and $\left(2n\pi + \frac{\pi}{4} \pm \frac{\pi}{4}, 0\right)$

14. $\frac{17}{144}$ **15.** $x^2 - y^2 = \frac{(a-b)(a^2+4ab+b^2)}{4(a+b)}$

16. $\frac{\pm 1}{\sqrt{\frac{1}{\vec{a}^2}+\frac{1}{\vec{b}^2}+\frac{1}{(\vec{a}\times\vec{b})^2}}}\left\{\frac{1}{\vec{a}^2}\vec{a}+\frac{1}{\vec{b}^2}\vec{b}+\frac{1}{(\vec{a}\times\vec{b})^2}(\vec{a}\times\vec{b})\right\}$

17. $r = \sqrt{(m-\sqrt{5})^2+n^2}$; where m, n are integers **19.** $y = x^2 - 2x, \frac{4}{3}$ unit^2

20. $x^3 + 6px^2 + 9p^2y + 4p^3 + 27q^2 = 0$

□

Miscellaneous Exercises

Miscellaneous Exercises

Make pairs by taking from column I and column II which correspond.

1. Column I — **Column II**

(i) $\sin(\pi[x])$ — (a) differentiable everywhere

(ii) $\sin\{\pi(x-[x])\}$ — (b) differentiable nowhere

(c) nondifferentiable at $|x| = 1$

2. Column I — **Column II**

(i) $x + \sin x$ — (a) increasing

(ii) $\sec x$ — (b) decreasing

(c) neither increasing nor decreasing

Fill in the blanks so that the resulting sentences become true.

3. If $u \sin t + v \cos t = 5$ and $u \cos t - v \sin t = 7$ then $\ddot{u}v - u\ddot{v} =$ ______ where dots denote differentiation w.r.t. t.

4. If g is the inverse function of f and $f'(x) = \sin x$ then $g'(x) =$ ______ .

5. Let $f(x) = x + 1$ and $\phi(x) = x - 2$. Then $|f(x) + \phi(x)| = |f(x)| + |\phi(x)|$ implies $x \in$ ______ .

6. One diagonal of a square is the portion of the line $7x + 5y = 35$ intercepted between the axes. Then the extremities of the other diagonal are ______ and ______ .

7. Let $f(x) = 2\sin^2\beta + 4\cos(x+\beta) \cdot \sin x \cdot \sin\beta + \cos 2(x+\beta)$.

Then the value of $\{f(\alpha)\}^2 + \left\{f\left(\frac{\pi}{4} - \alpha\right)\right\}^2 =$ ______ .

8. A positive integral power of the imaginary quantity i is taken at random. The probability of the value being real is ______ .

9. $\lim\limits_{n \to \infty} \dfrac{\sqrt[3]{n^2} \cdot \sin(n!)}{n+1} =$ ______ .

10. $\displaystyle\int \frac{\sqrt[3]{x + x^3}}{x^4}\, dx =$ ______ .

11. $\displaystyle\int_{-0.5}^{0.5} \cos x \cdot \log \frac{1+x}{1-x}\, dx =$ ______ .

12. If z is a complex number such that $z\bar{z} = 1$ and $\text{amp}\,\dfrac{z}{\bar{z}} = \dfrac{\pi}{2}$ then $z =$ ______ .

13. If $4\sin^{-1}x + \cos^{-1}x = \pi$ then $x =$ ______ .

14. If $f'(x) > g'(x)$ where $f(x) = \frac{1}{2} \cdot 5^{2x+1}$ and $g(x) = 5^x + 4x \log 5$ then $x \in$ ______ .

15. $\dfrac{d(x^{\sin^{-1}x})}{d(\cos^{-1}x)} =$ ______ .

16. If $x^2 - 4x - 2\pi \le \sin^{-1}(\sin 5)$ then x belongs to the interval ______ .

17. $\lim\limits_{n \to \infty} \dfrac{a^n}{n!}$ is equal to ______ .

18. If $f(x) = x^2 - 4$ and $g(x) = f(x) + 5$ such that $f\{g(x)\} > 0$ then the set of possible values of x is ______ .

19. The differential coefficient of 2^x w.r.t. x^2 is ______ .

20. Let $\vec{a}$, $\vec{b}$ and $\vec{c}$ be three nonzero and noncoplanar vectors and $\vec{p}$, $\vec{q}$ and $\vec{r}$ be three vectors given by

$\vec{p} = \vec{a} + \vec{b} - 2\vec{c}$, $\vec{q} = 3\vec{a} - 2\vec{b} + \vec{c}$ and $\vec{r} = \vec{a} - 4\vec{b} + 2\vec{c}$.

If the volume of the parallelepiped determined by $\vec{a}$, $\vec{b}$ and $\vec{c}$ is V_1 and that of the parallelepiped determined by $\vec{p}$, $\vec{q}$ and $\vec{r}$ is V_2 then

$V_2 : V_1 =$ ______ .

21. If $3\sin\theta + 4\cos\theta = \dfrac{5(2a-3)}{2-3a}$ has a real solution for θ then the set of possible real values of a is ______ .

22. $\lim\limits_{x \to 0} \dfrac{\sin x^a}{(\sin x)^b} =$ ______ $(a > 0, b > 0)$.

23. Let A = (–2, 0) and B = (2, 0). The range of real values of a for which the line segment AB subtends an obtuse angle at C = $(a, a+1)$, is ______ .

24. The family of circles touching the positive axes of reference has the differential equation whose order = ______ and degree = ______ .

25. Let $p(x)$ be a polynomial of degree $n \geq 3$. It leaves the remainders -2, -5 and 10 when divided by $x+1$, $x-2$ and $x+3$ respectively. The remainder, when $p(x)$ is divided by $(x+1)(x-2)(x+3)$, will be ______ .

26. $\lim_{h \to 0} \frac{2}{h} \int_2^{2+h} e^{-x^2}\, dx =$ ______ .

27. The equation of the parabola which passes through the points (0, 0) and $(b, 0)$, and has the vertex at $\left(\frac{b}{2}, h\right)$, axis of symmetry being perpendicular to the x-axis, is ______ .

28. If $|\vec{a}+\vec{b}| = |\vec{a}| + |\vec{b}|$ and $\vec{a}, \vec{b}$ are not zero vectors then $\vec{a} \times \vec{b} =$ ______ .

29. $\lim_{x \to 1+} 3^{-3^{\frac{1}{1-x}}} =$ ______ .

30. The equations of the lines perpendicular to the line $x+y=3$ from which the circle $x^2+y^2-2x-3=0$ cuts off intercepts each of length 2, are ______ and ______ .

31. The number of values of x such that $0 \leq x \leq 2\pi$ and

$$\sin^3 x \cdot \cos x + \sin^2 x \cdot \cos^2 x + \sin x \cdot \cos^3 x = 1,$$

is ______ .

32. In a certain test a_i students gave wrong answers to at least i questions where $i = 1, 2, 3, \ldots, k$. No student gave more than k wrong answers. The total number of wrong answers given is ______ .

33. The domain of the function $f(x) = \log_2 \sin^{-1}(\log_e x)$ is ______ .

34. ABCD is a cyclic quadrilateral in which $AD = 2AB$ and $CD = 2BC$. If $\angle BAD = \frac{2\pi}{3}$ then ar(quadrilateral) : ar(circle) = ______ .

35. The set of values of the parameter a for which a real θ satisfies the equation

$$2a \cos\frac{\theta}{2}\left(\cos\frac{\theta}{2} + \sin\frac{\theta}{2}\right) - \left(\cos\frac{\theta}{2} - \sin\frac{\theta}{2}\right)^2 = 0$$

is given by the interval ______ .

36. A natural number less than or equal to 200 is written down at random. The probability of the number being a perfect square is ______ .

37. A box contains 100 tickets numbered 1, 2, 3, ... , 100. Two tickets are chosen at random. It is given that the maximum number on the two chosen tickets is not more than 10. The minimum number on them is 2 with probability ______ .

Choose the correct answer(s) from the given alternative options.

38. If $\vec{a}, \vec{b}$ are nonzero vectors inclined at an angle $\frac{\pi}{4}$ then $\{(\vec{a}+\vec{b}) \times (\vec{a}-\vec{b})\}^2$ is equal to

A. $|\vec{a}|^2\, |\vec{b}|^2$ B. $|\vec{a}|^2 - |\vec{b}|^2$

C. $2|\vec{a}|^2\, |\vec{b}|^2$ D. none of these

39. If $f(x) = \sin\left(\frac{\pi}{5}+x\right) - \sin\left(\frac{\pi}{5}-x\right) - \sin\left(\frac{2\pi}{5}+x\right) + \sin\left(\frac{2\pi}{5}-x\right)$ then the range of the function is the interval

A. $[-1, 1]$ B. $\left[-\frac{1}{2}, \frac{1}{2}\right]$

C. $[0, 1]$ D. none of these

40. A purse contains 4 copper coins, 3 silver coins and a second purse contains 6 copper coins and 2 silver coins. A coin is taken at random from the purses. The probability that it is a copper coin is

A. $\frac{4}{7}$ B. $\frac{3}{4}$ C. $\frac{3}{7}$ D. $\frac{37}{56}$

41. $\lim_{n \to \infty} \frac{(-2)^n + 3^n}{(-2)^{n+1} + 3^{n+1}}$ is equal to

A. -1 B. 1 C. $\frac{1}{3}$ D. $\frac{3}{2}$ E. $\frac{2}{3}$

42. If $f(n+1) = \frac{1}{2}\left\{f(n) + \frac{9}{f(n)}\right\}$, $n \in N$ and $f(n) > 0$ for all $n \in N$ then $\lim_{n \to \infty} f(n)$ is equal to

A. 3 B. -3

C. $\frac{1}{2}$ D. none of these

43. Let $f(x) = \frac{1}{2}a_0 + (a_1 \cos x + a_2 \cos 2x + \ldots + a_n \cos nx) + (b_1 \sin x + b_2 \sin 2x + \ldots + b_n \sin nx)$.

Then $\int_{-\pi}^{\pi} f(x) \cdot \cos kx\, dx$, $(k \leq n)$ is equal to

A. a_k B. b_k C. πa_k D. πb_k

44. If $\vec{x}, \vec{y}, \vec{a}, \vec{b}$ and $\vec{c}$ are nonzero vectors such that $[\vec{x}\ \vec{y}\ \vec{a}] = 0$, $[\vec{x}\ \vec{y}\ \vec{b}] = 0$, and $[\vec{a}\ \vec{b}\ \vec{c}] = 0$ then the value of $[\vec{x}\ \vec{y}\ \vec{c}]$ is equal to

A. $\vec{a} \cdot \vec{b}$ B. 0

C. $(\vec{a}+\vec{b}) \cdot \vec{c}$ D. none of these

45. A boy can throw a ball using any one of his hands. The probability that he throws the ball beyond 50 m is $\frac{1}{4}$ when he throws with the left hand and it is p when he throws with the right hand. He throws the ball using any one of his hands. If the probability that the ball will fall beyond 50 m is $\frac{3}{8}$ then p is equal to

A. $\frac{1}{8}$ B. $\frac{3}{4}$

C. $\frac{1}{2}$ D. none of these

46. The equation $ax^4 + 3x^2 - (a+1) = 0$ will have real and unequal roots if

A. $a < 0$ B. $a > -1$

C. $-1 < a < 0$ D. none of these

47. If the point $(2a, a+1)$ belongs to the angle between the pair of lines $2x^2 - 5xy + 2y^2 - 2x - 2y - 4 = 0$ containing the origin then

A. $a > -\frac{1}{3}$ B. $a < \frac{1}{3}$

C. $-\frac{1}{3} < a < \frac{1}{3}$ D. none of these

48. The joined equation of two altitudes of an equilateral triangle is $x^2 - 3y^2 - 4x + 6\sqrt{3}y - 5 = 0$. The third altitude has the equation

A. $x + 2 = 0$ B. $y = \sqrt{3}$

C. $x = 2$ D. none of these

49. A has n coins and B has $n+1$ coins. They toss their coins simultaneously. If p be the probability that B will have more heads than A then

A. $p < \frac{1}{2}$ B. $p = \frac{1}{2}$

C. $p > \frac{2}{3}$ D. none of these

50. One bag contains m white and n black balls. Another bag contains n white and m black balls. The chance of drawing a white ball from a bag chosen at random is p. Now all the balls of the two bags are put in another bag. The probability of drawing a white ball from the third bag is p'. Then

A. $p' > p$ B. $p + p' = 1$

C. $p = p'$ D. none of these

51. If $2\cos^{-1}x = \sin^{-1}(2x\sqrt{1-x^2})$, $x > 0$ then the solution set of x is

A. $\left[\frac{1}{\sqrt{2}}, 1\right]$ B. $\left\{\frac{1}{\sqrt{2}}\right\}$

C. $\{1\}$ D. $\left\{\frac{1}{\sqrt{2}}, 1\right\}$

52. If $f(x) = px^2 + \frac{1}{2}(2p^2 - 81)x - 12$ has a maximum at $x = \frac{9}{4}$ then p is equal to

A. $\frac{9}{2}$ B. -9

C. $\frac{9}{2}$ or -9 D. none of these

53. Which of the following functions are continuous in the interval $(0, \pi)$?

A. $f(x) = \tan x$

B. $f(x) = \int_0^x t\sin\frac{1}{t}\,dt$

C. $f(x) = 1,\ 0 < x \le \frac{3\pi}{4}$
$\quad 2\sin\frac{2x}{9},\ \frac{3\pi}{4} < x < \pi$

D. $f(x) = x\sin x,\ 0 < x \le \frac{\pi}{2}$
$\quad \frac{\pi}{2}\sin(\pi + x),\ \frac{\pi}{2} < x < \pi$

54. The number of solutions of the equation $|\sin x| = \sin x + 2$ is

A. 1 B. infinite

C. 0 D. none of these

55. If $f(x) = 3x^2 + 12x - 1,\ -1 \le x \le 2$
$\quad 37 - x,\ 2 < x \le 3$

then which of the following is (are) true?

A. $f(x)$ is increasing in $[-1, 2]$

B. $f(x)$ is continuous

C. $f'(2)$ does not exist

D. $f(x)$ has the maximum value at $x = 2$

56. In an one-day international cricket match between England and Australia if England bat first, the chance of England winning is $\frac{3}{5}$ while if they field first, their chance of winning is $\frac{2}{7}$. What is the probability that England will win the match if the side winning the toss opts for batting first and the match ends in victory of a team?

A. $\frac{31}{35}$ B. $\frac{11}{35}$

C. $\frac{31}{70}$ D. none of these

57. If $\vec{a} \times \vec{b} + \vec{b} \times \vec{c} + \vec{c} \times \vec{a} = 0$ for the nonzero vectors $\vec{a}$, $\vec{b}$ and $\vec{c}$ then

A. $\vec{a}, \vec{b}, \vec{c}$ are coplanar B. $\vec{a} \times \vec{b} \cdot \vec{c} = 0$

C. $\vec{a} \times \vec{b} \perp \vec{c}$ D. $(\vec{a} \times \vec{b}) \times \vec{c} = \vec{0}$

58. If $\vec{a}, \vec{b}$ and $\vec{d}$ are coplanar and $[\vec{a}\ \vec{b}\ \vec{c}] = 0$ then

A. $\vec{a} \times \vec{b} \parallel \vec{c} \times \vec{d}$ B. $\vec{a} \times \vec{b} \perp \vec{c} \times \vec{d}$

C. $\vec{a}, \vec{b}, \vec{c}$ and $\vec{d}$ are coplanar

D. none of these

59. If $\vec{a}, \vec{b}$ are nonzero vectors then

$\frac{1}{\vec{b}^2}\{(\vec{a} \cdot \vec{b})\vec{b} + \vec{b} \times (\vec{a} \times \vec{b})\}$ is equal to

A. $\vec{b}$ B. $\vec{a}$

C. $\vec{a} \times \vec{b}$ D. none of these

60. The equation $ax^2 + by^2 + cx + dy + k = 0$ represents a parabola if

A. $a = 0$ B. $b = 0$

C. $a = 0, d = 0$ D. $b = 0, d = 0$

61. The function $f(x)$ is defined as follows:

$$f(x) = \frac{1 - \cos(1 - \cos x)}{x^4}, \quad x \neq 0$$

$$a, x = 0.$$

If $f(x)$ is continuous everywhere then a is equal to

A. $\frac{1}{2}$ B. $\frac{1}{4}$ C. $\frac{1}{8}$ D. $\frac{1}{16}$

62. If $\sum_{r=1}^{n} a_r = k$ and $\sum_{r=1}^{n-1} a_r a_{r+1} = S$ where $a_r \geq 0$ for $r = 1, 2, 3, \ldots, n$ then

A. $S < \frac{k}{2}$ B. $S > \frac{k^2}{2}$ C. $S \leq \frac{k^2}{4}$ D. $S > \frac{k}{4}$

63. $[\vec{a}\ \vec{b}\ \vec{i}\,]\vec{i} + [\vec{a}\ \vec{b}\ \vec{j}\,]\vec{j} + [\vec{a}\ \vec{b}\ \vec{k}\,]\vec{k}$ is equal to

A. $\vec{a} \times \vec{b}$ B. $\vec{a} + \vec{b}$ C. $\vec{a} - \vec{b}$ D. $\vec{b} \times \vec{a}$

64. Let $a^2 + a + 1$ be divisible by 3. If a is divided by 3 the remainder will be

A. 2 B. 1

C. 0 D. none of these

65. The equation $ax^2 + by^2 + cx + dy + k = 0$ represents a hyperbola if

A. $a > 0, b < 0$ B. $a > 0, b > 0$

C. $a < 0, b > 0$ D. $a < 0, b < 0$

66. The tangent to the parabola $y^2 = 4ax$ at the point (x_1, y_1) cuts the axis of the parabola at P. The abscissa of the point P is

A. y_1 B. $-x_1$

C. x_1 D. none of these

67. Of the three lines $x + \sqrt{3}y = 0$, $x + y = 1$ and $x - \sqrt{3}y = 0$, two are equations of two altitudes of an equilateral triangle. The centroid of the equilateral triangle is

A. (0, 0) B. $\left(\frac{\sqrt{3}}{\sqrt{3}-1}, -\frac{1}{\sqrt{3}-1}\right)$

C. $\left(\frac{\sqrt{3}}{\sqrt{3}+1}, \frac{1}{\sqrt{3}+1}\right)$ D. none of these

68. If x and y are positive real numbers and m, n are any positive integers then $\frac{x^n y^m}{(1 + x^{2n})(1 + y^{2m})}$ will be

A. greater than $\frac{1}{4}$ B. less than $\frac{1}{4}$

C. less than or equal to $\frac{1}{4}$ D. none of these

69. If p, q and r are real numbers then

A. $\max\{p, q\} < \max\{p, q, r\}$

B. $\min\{p, q\} = \frac{1}{2}(p + q - |p - q|)$

C. $\max\{p, q\} < \min\{p, q, r\}$

D. none of these

70. The equation $\frac{x^2}{6 - \lambda} + \frac{y^2}{2 - \lambda} = 1$ represents a hyperbola if

A. $\lambda < 2$ B. $\lambda > 6$

C. $2 < \lambda < 6$ D. none of these

71. The probability of choosing an integer r from the set $\{1, 2, 3, \ldots, 2n\}$ is inversely proportional to $r^3 (1 \leq r \leq 2n)$. The probability that the chosen number from the set is odd, is p. Then

A. $p = \frac{1}{2}$ B. $p > \frac{1}{2}$

C. $p < \frac{1}{2}$ D. none of these

72. If $(1 - x^3)^n = \sum_{r=0} a_r x^r (1 - x)^{3n - 2r}$ then find a_r, where $n \in N$.

73. 16 persons are occupying seats round a table. If every person refuses to work with any one of his neighbours, in how many ways can a committee of 6 be made with those persons under the chairmanship of a particular person?

74. Let A = (6, 5), B = (2, –3) and C = (–2, 1) be the vertices of a triangle. Find the point P in the interior of the triangle such that ΔPBC is an equilateral triangle.

75. A, B, C and D are four consecutive vertices of a regular septagon (polygon of seven sides). If the radius of the circumcircle be unity, prove that $AD + AC - AB = \sqrt{7}$.

76. Prove that $b\bar{z} + \bar{b}z = c$, $b \neq 0$, $c \in R$ and b is a complex constant, represents a straight line in the Argand plane. If z_1 is the reflection of the point z_2 by the above line then prove that $\bar{b}z_1 + b\bar{z}_2 = c$.

77. If $y = \frac{1}{2^{n-1}} \cos(n \text{ arc cos } x)$ then prove that
$$(1 - x^2)\frac{d^2y}{dx^2} - x\frac{dy}{dx} + n^2y = 0.$$

78. Show that $f(x) = x + \cos x - a$ is monotonic increasing in R. Hence, deduce that $x + \cos x = a$ has no positive root if $a < 1$ and has only one positive root if $a > 1$.

79. Boys of a group decide to buy a transistor costing Rs 170 or more but less than or equal to Rs 195 by contributing equally. If 2 of the boys back out, each of the remaining boys will have to contribute an extra rupee to buy the transistor. How many boys are there in the group?

80. Evaluate $\int \frac{(x - \sin x)^{\frac{3}{2}}}{\sqrt{x}} \cdot \left\{\frac{6x^2 \sin^2 \frac{x}{2}}{x - \sin x} + 3x\right\} dx.$

81. Three circles of radii a, b and c touch one another externally and the tangents at their points of contact meet in a point P. Prove that the distance of P from either of the points of contact is $\sqrt{\frac{abc}{a+b+c}}$.

82. Evaluate $\lim\limits_{x \to \infty} \left(\sin \frac{1}{x} + \cos \frac{1}{x}\right)^x$.

83. If $\prod\limits_{n=1}^{n} x_n$ denotes $x_1 \cdot x_2 \cdot x_3 \cdot \ldots \cdot x_n$ then prove, by induction, that
$$\left\{\prod_{n=1}^{n} f_n(x)\right\}' = \sum_{i=1}^{n} \{f_1(x) \cdot f_2(x) \cdot \ldots \cdot f_i'(x) \cdot \ldots \cdot f_n(x)\}$$
where dash denotes derivative w.r.t. x.

84. Let P and Q be two points on the ellipse $x^2 + 4y^2 = 16$ whose eccentric angles are $\frac{\pi}{4}$ and $\frac{3\pi}{4}$ respectively. The tangent at P and the normal at Q cut each other at R and the normal at Q cuts the ellipse again at M. Find the area of the triangle PRM.

85. Solve for $\vec{r}$: $\vec{r} \times \vec{b} = \vec{c} \times \vec{b}$,
$\vec{r} \cdot \vec{a} = 0$ and $\vec{a} \cdot \vec{b} \neq 0$.

86. If $f(t)$ is a function of t satisfying $f'(t) = \frac{G}{100V} - kf(t)$ where G, V and k are independent of t then find $\lim\limits_{t \to \infty} f(t)$.

87. Assuming that the angle of incidence is equal to the angle of reflection, prove that a ray of light emanating from
 (i) a focus of an elliptical mirror passes through the other focus after reflecting from the mirror.
 (ii) the focus of a parabolic mirror is reflected parallel to the axis.

88. Locate the region in the Argand plane determined by $z^2 + \bar{z}^2 + 2|z|^2 < 8i(\bar{z} - z)$.

89. If $f(x)$ is a polynomial function of the nth degree, prove that
$$\int e^x f(x)dx = e^x[f(x) - f'(x) + f''(x) - f^{(3)}(x) + \ldots + (-1)^n f^{(n)}(x)]$$
where $f^{(n)}(x)$ denotes $\frac{d^n f}{dx^n}$.

90. A bag contains n white and m black balls. If two balls are drawn one by one without replacement then the chance of getting both the balls of the same colour is p. If the two drawings are made with replacement then the chance of both balls being of the same colour is q. Prove that $p < q$.

91. Solve $\sin x + \cos x = 1 + \sin x \cdot \cos x$.

92. If h_1, h_2, h_3, k_1, k_2 and k_3 are six non-negative real numbers and the algebraic sum of lengths of the perpendiculars drawn from points $(h_1, 4)$, $(h_2, 5)$, $(h_3, -3)$, $(2, k_1)$, $(3, k_2)$ and $(7, k_3)$ on a variable line through the point $(2, 1)$ is always zero then find the values of the six numbers.

93. Three concurrent lines OA, OB and OC are produced to D, E and F respectively. Prove, by using vectors, that the intersections of the lines AB and DE, BC and EF, and CA and FD are collinear.

94. If $(1 + x)^n = c_0 + c_1x + c_2x^2 + \ldots + c_nx^n$ then prove that
$$\frac{n!}{x(x+1)(x+2)(x+3)\ldots(x+n)} = \frac{c_0}{x} - \frac{c_1}{x+1} + \frac{c_2}{x+2} - \ldots + (-1)^n \frac{c_n}{x+n}, x > 0.$$

95. From a point, common tangents are drawn to the circle $2(x^2 + y^2) = a^2$ and the parabola $y^2 = 4ax$. Find the area of the quadrilateral formed by the common tangents, the chord of contact of the tangents to the circle and that to the parabola.

96. Find the sum
$$\sin\theta + 2\sin 2\theta + 3\sin 3\theta + \ldots + n\sin n\theta.$$

97. Show that the values of θ satisfying the equation
$$\sin 2\theta \cdot \cos^2(\alpha - \beta) - \sin 2\alpha \cdot \cos^2(\beta + \theta) - \sin 2\beta \cdot \sin^2(\alpha + \theta) = 0$$
are given by $n\pi + \alpha$ and $\left(n\pi + \frac{\pi}{2}\right) - \beta$ where $n \in Z$.

98. Integrate $\int \frac{dx}{\sqrt[4]{(x-1)^3 \cdot (x+2)^5}}$.

99. If $f(x) = x, \quad x < 1$

$x^2, \quad 1 \le x \le 4$

$2^x, \quad x > 4$

then determine the inverse function of $f(x)$ and find the domain of the inverse function.

100. Prove that $f(x) = \left(2 - \sin \frac{1}{x}\right) \cdot |x|, x \ne 0$

$0, x = 0$

is continuous and it has a minimum at $x = 0$.

101. Show that $\frac{x}{1+x} < \log(1+x) < x$ for $x > 0$.

102. A bag contains a number of balls of which some are white. A ball is drawn repeatedly by replacement after each draw. If the probability of drawing a white ball in a single trial be p, prove that the number of white balls that is most probable in n trials is equal to the greatest integer in $p(n+1)$.

103. Prove that each member of the family of ellipses $\frac{x^2}{t^2} + \frac{y^2}{t^2 - c^2} = 1, t^2 > c^2$ cuts each member of the family of hyperbolas $\frac{x^2}{t^2} - \frac{y^2}{c^2 - t^2} = 1, c^2 > t^2$ at right angles.

104. Two sides of a triangle have the joined equation $x^2 - 2xy - 3y^2 + 8y - 4 = 0$. The third side, which is variable, always passes through the point $(-5, -1)$. Find the range of values of the slope of the third side so that the origin is an interior point of the triangle.

105. Prove, by induction, that

$$\sum_{r=1}^{n} \sin r\theta = \sin \frac{n\theta}{2} \cdot \operatorname{cosec} \frac{\theta}{2} \cdot \sin \frac{n+1}{2} \theta.$$

Use the result to find $\sum_{r=1}^{n} r$.

106. Find the set of positive values of the parameter a, for each of which the function

$f(x) = \sin 2x - 8(a+1)\sin x - (4a^2 + 8a - 14)x$

is monotonic increasing in R and has no critical point.

107. Let S be a square of unit area. Consider any quadrilateral which has one vertex on each side of S. If a, b, c and d denote the lengths of the sides of the quadrilateral, prove that the AM of a^2, b^2, c^2 and d^2 lies between $\frac{1}{2}$ and 1.

108. Give the most general solution of x and y if $2\sin x + \cos y = 2$.

109. Give a rough sketch of the function

$\phi(x) = f(x-1) + f(x+1)$ where

$f(x) = 1 - |x|, |x| \le 1$

$0, |x| > 1.$

110. Prove that $202^{303} > 303^{202}$, without using table.

111. Evaluate $\int \frac{\sqrt{x+1}}{\sqrt{x-2} + \sqrt{3}} dx, 2 < x < 5.$

112. Solve for x:

$\tan(x+\alpha) \cdot \tan(x+\beta) + \tan(x+\beta) \cdot \tan(x+\gamma) + \tan(x+\gamma) \cdot \tan(x+\alpha) = 1.$

113. Prove that two parabolas having the same focus and their axes in opposite directions, cut at right angles.

114. If a, b, c and d are positive real numbers such that $a+b+c+d = 1$ then prove that

$ab + bc + cd + da \le \frac{1}{4}$.

115. A vertical lamp-post of height h stands at a point on the boundary of a circular field. A man of height a is running round the boundary. Prove that the end of the shadow of the man will also travel on a circle. Find the ratio of the radii of the two circles.

116. Sanjoy and Sanjeev throw a dice. Sanjoy throws first and then Sanjeev throws. Sanjeev wins if he throws a higher number than Sanjoy. When Sanjoy throws 3, what is the chance that Sanjeev will win? What is the probability that Sanjoy will throw 3 and lose?

117. If $f(x)$ is differentiable and $f(xy) = f(x) + f(y)$ for all $x, y \in R$ then prove that $f(x) = c\log x + k$ where c, k are constants.

118. Classify the loci represented by $\frac{x^2}{a^2+\lambda} + \frac{y^2}{b^2+\lambda} = 1$ $(a > b)$ for different real values of λ. Show that each ellipse or hyperbola represented by the equation has the same foci.

119. For every real root α of the equation

$x^2 + 2(m-2)x + m^2 = 5$ in x if $|\alpha| \le 2$

then find the set of possible values of m.

120. For a fixed positive integer r, prove by induction or otherwise that

$(r+1)(r+2)(r+3) \ldots (r+n) \ge \{(r+1)(r+n)\}^{n/2}$

Use this result to prove that $300! > 100^{300}$

121. Find the equations of straight lines which pass through the intersection of the lines $x - 2y - 5 = 0$ and $7x + y - 50 = 0$, and divide the circumference of the circle $x^2 + y^2 = 100$ into two arcs whose lengths are in the ratio 3 : 1.

122. The circle $x^2 + y^2 = 1$ cuts the x-axis at P and Q. Another circle with the centre at Q and radius variable intersects the first circle at R above the x-axis and the line segment PQ at S. Find the maximum area of the ΔQSR.

123. Show that

$$\left|\cos\theta \cdot \left(\sin\theta + \sqrt{\sin^2\theta + \sin^2\alpha}\right)\right| \le \sqrt{1 + \sin^2\alpha}$$

124. Show that in a ΔABC,

$$\Sigma \sin 2\lambda A = (-1)^{\lambda+1} \cdot 4\sin\lambda A \sin\lambda B \sin\lambda C, \lambda \in N$$

125. Establish the identity

$$(ay - bx)^2 + (bz - cy)^2 + (cx - az)^2$$
$$= (a^2 + b^2 + c^2)(x^2 + y^2 + z^2) - (ax + by + cz)^2.$$

Hence or otherwise prove that the minimum value of $x^2 + y^2 + z^2$ is $\dfrac{k^2}{a^2 + b^2 + c^2}$ if $ax + by + cz = k$.

126. Prove that the function $f(x) = (1 + x^2) \cdot \phi(x)$ is discontinuous but its inverse is a continuous function where

$$\phi(x) = \begin{cases} 1, & x > 0 \\ 0, & x = 0 \\ -1, & x < 0. \end{cases}$$

127. Prove that $f(x) = \begin{cases} \dfrac{x}{2} + x^2 \sin\dfrac{1}{x}, & x \ne 0 \\ 0, & x = 0 \end{cases}$

is not monotonic in any interval containing the origin.

128. An urn contains some 50-paisa coins and 1-rupee coins. The number of 1-rupee coins is one less than that of 50-paisa coins. Five coins are drawn at random from the urn. The chance of getting one 50-paisa coin more than 1-rupee coins is p while the chance of getting one 1-rupee coin more than 50-paisa coins is q. If $p - q = \dfrac{25}{231}$, find the number of 1-rupee coins in the urn.

129. Evaluate $\displaystyle\int_2^3 \frac{x^2 + 1}{(x^2 - 1)\sqrt{x^4 + 1}}\, dx.$

130. The equation of an altitude of an equilateral triangle is $\sqrt{3}x + y = 2\sqrt{3}$ and one vertex is $(3, \sqrt{3})$. Find the other vertices and the orthocentre of the triangle.

131. $A_1A_2A_3 \ldots$ is a regular polygon whose each side is a and each internal angle is α. Prove that

$$A_1A_{n+1} = a\sin\frac{n(\pi - \alpha)}{2} \cdot \operatorname{cosec}\frac{\alpha}{2}.$$

132. Show that the equation $\sin(x + \alpha) = a\sin 2x + b$ has four roots whose sum is equal to $(2n - 1)\pi$ where $n \in \mathbb{Z}$.

133. Find $a \in R$ such that the equation

$$|a^3 + a^2x| + |a^2x + 1| = 1 - a^3$$

has four or more integral solutions of x.

134. Prove that $f(x) = \dfrac{\sin x}{x}$ is monotonic decreasing in $\left(0, \dfrac{\pi}{2}\right]$. Hence prove that $\dfrac{2x}{\pi} < \sin x < x$ for $x \in \left(0, \dfrac{\pi}{2}\right)$.

135. Let $f(x) = x^2 - x + 1, x \ge \dfrac{1}{2}$

and $g(x) = \dfrac{1}{2} + \sqrt{x - \dfrac{3}{4}}, \; x \ge \dfrac{3}{4}$.

Show that the functions $f(x)$ and $g(x)$ are inverse to each other. Hence, solve the equation

$$x^2 - x + 1 = \frac{1}{2} + \sqrt{x - \frac{3}{4}}.$$

136. Solve $x^2 + \left(\dfrac{x}{x+1}\right)^2 = 3.$

137. Solve $(\cos 2x - \cos^4 x)\cot 3x + \dfrac{\sin 5x - \sin x}{8\sin 3x} = 0.$

138. Let the mirror image of the point A(5, 6) with respect to the line $2x + 3y = 15$ be B. Find the equation of the circle described on AB as diameter. AC is any chord of the circle meeting the x-axis at D such that $AD = 10 \cdot DC$. How many such chords are there?

139. A circle passes through three points A, B and C with the line segment AC as its diameter. A line through A intersects the chord BC at a point D inside the circle. If $\angle DAB = \alpha$, $\angle CAB = \beta$ and the distance between the point A and the middle point of the line segment DC is d, prove that the area of the circle is

$$\frac{\pi d^2 \cos^2\alpha}{\cos^2\alpha + \cos^2\beta + 2\cos\alpha\cos\beta\cos(\beta - \alpha)}.$$

140. Examine the continuity of the function $f(x)$ in R where

$$f(x) = \begin{cases} \cos\dfrac{\pi x}{2}, & |x| \le 1 \\ |x - 1|, & |x| > 1. \end{cases}$$

141. Evaluate $\displaystyle\lim_{x \to \infty} x^2 \sin\left(\log\sqrt{\cos\frac{\pi}{x}}\right).$

142. A line segment AB of fixed length l moves with its extremities on two fixed intersecting perpendicular lines. P is a point on AB such that $AP : PB = k : 1$. Prove that the locus of P is an ellipse whose area is $\dfrac{\pi k l^2}{(k+1)^2}$.

143. Given $\vec{a} = \frac{1}{7}(2\vec{i} + 3\vec{j} + 6\vec{k})$, $\vec{b} = \frac{1}{7}(3\vec{i} - 6\vec{j} + 2\vec{k})$ and $\vec{c} = \frac{1}{7}(6\vec{i} + 2\vec{j} - 3\vec{k})$. Find a unit vector $\vec{x}$ making an angle of $\frac{\pi}{3}$ with $\vec{b} + \vec{c}$ such that $[\vec{x}\ \vec{a}\ \vec{c}] = 0$.

144. Evaluate $\int \frac{dt}{t - \sqrt{1 - t^2}}$.

145. Mr Kumar along with his wife, parents and four children of which two are boys and two are girls, put up at a rest house with four bedrooms for a night. There are beds for two persons in each room. If either a couple or two persons of the same sex sleep in a room, what is the chance that Mr and Mrs Kumar will sleep together?

146. If $a_1 + a_2 \sin x + a_3 \cos x + a_4 \sin 2x + a_5 \cos 2x = 0$ is an identity in x then prove that

$$(a_1, a_2, a_3, a_4, a_5) = (0, 0, 0, 0, 0).$$

147. Solve $\tan x^2 = \cot 5x$.

148. If $A_1, A_2, A_3, \ldots, A_{2n+1}$ are the vertices of a regular polygon of $(2n + 1)$ sides then prove that

$$A_1A_2 + A_1A_4 + \ldots + A_1A_{2n} = A_1A_3 + A_1A_5 + \ldots + A_1A_{2n+1}.$$

149. Vectors $\vec{x}, \vec{y}$ and $\vec{z}$ are each of magnitude $\sqrt{2}$ and make angles of $\frac{\pi}{3}$ with each other. If

$$\vec{x} \times (\vec{y} \times \vec{z}) = \vec{a},\ \vec{y} \times (\vec{z} \times \vec{x}) = \vec{b}$$

then find $\vec{x}, \vec{y}$ and $\vec{z}$ in terms of $\vec{a}, \vec{b}$ and $\vec{c}$ when (i) $\vec{x} + \vec{y} = \vec{c}$ (ii) $\vec{x} \times \vec{y} = \vec{c}$.

150. If $A + B + C = \pi$ and

$$\tan \frac{1}{4}(B + C - A) \cdot \tan \frac{1}{4}(C + A - B) \times \tan \frac{1}{4}(A + B - C) = 1$$

then prove that the sum of the cosines of the three angles A, B and C is equal to the cosine of the sum of the angles.

151. Prove that the function $f(x) = x^3 + 3x^2 + x + a$, a being a constant, is monotonic decreasing in $[\alpha, \beta]$ where $x = \alpha$ and $x = \beta$ are two successive points of extremum of the function.

152. Find $a > 0$ for which the area of a curvilinear trapezoid bounded by the curve $y = \frac{1}{6}x + x^{-2}$ and the lines whose joined equation is $y(x^2 - 3ax + 2a^2) = 0$, takes the least value.

153. If $\{\sin(\alpha - \beta) + \cos(\alpha + 2\beta)\sin \beta\}^2 = 4\cos \alpha \cdot \sin \beta \cdot \sin(\alpha + \beta)$ and $\tan \alpha = 3\tan \beta$ then find the general solution for β.

154. Find all $f(x)$ satisfying the relation

$$\{f(x)\}^2 + 4f(x) \cdot f'(x) + \{f'(x)\}^2 = 0.$$

155. A number of five distinct digits is written down at random using the digits 1, 2, 3, 4, 5 and 6. What is the chance that the number is divisible by 3 but not by 12?

156. If each side of a regular polygon of n sides subtend an angle α at the centre of the polygon and each exterior angle of the polygon is β then prove that

$$\cos \alpha + \cos(\alpha + \beta) + \cos(\alpha + 2\beta) + \ldots + \cos(\alpha + \overline{n - 1}\,\beta) = 0.$$

157. Let $\hat{a}, \hat{b}$ and $\hat{c}$ be unit vectors such that $\hat{a} \times \hat{b} = \hat{c}$ and $\hat{a} \cdot \hat{b} = 0$. Also, $\vec{x}$ is any vector such that

$$[\vec{x}\ \hat{b}\ \hat{c}] = 3,\ [\vec{x}\ \hat{c}\ \hat{a}] = 4 \text{ and } [\vec{x}\ \hat{a}\ \hat{b}] = 2.$$

Find $\vec{x}$ in terms of $\hat{a}, \hat{b}$ and $\hat{c}$.

158. Find the area of the portion of the circle $x^2 + y^2 = 5$ that falls between the pair of lines

$$x^2 - 4y^2 + 6x + 9 = 0.$$

159. Define two functions f and g which are non-differentiable at $x = c \in$ the domain of their definitions if $(f + g)'(c)$ and $(fg)'(c)$ exist.

160. Find the area of the region in the Argand plane satisfying $|z - 2| + |z + 2| \le 8$.

161. Let $y = f(x)$ be a curve. Lines are drawn through any point (x, y) on the curve parallel to the axes of reference. If the tangent to the curve at (x, y) divides the rectangle formed by the axes of reference and the lines drawn in two parts, the ratio of whose areas is $3:1$ then prove that the curve must be a parabola.

162. Let $S_n = \sin \theta + \sin 2\theta + \sin 3\theta + \ldots + \sin n\theta$.

Prove that $\lim_{n \to \infty} \frac{S_1 + S_2 + S_3 + \ldots + S_n}{n} = \frac{1}{2}\cot \frac{\theta}{2}$.

163. ABCD is a convex quadrilateral in which $AB^2 + CD^2 = BC^2 + AD^2$. Prove by vectors that diagonals AC and BD are orthogonal to each other.

164. Let $I_n = \int x \operatorname{cosec}^n x\, dx$, $n \ge 3$. Prove that

$$I_n = \frac{-1}{n - 1}\operatorname{cosec}^{n-2}x \cdot \left(x\cot x + \frac{1}{n - 2}\right) + \frac{n - 2}{n - 1} I_{n-2}.$$

165. Let $f(x) = \begin{vmatrix} x + \alpha & x + 1 & x + 1 \\ x + 2 & x + \beta & x + 1 \\ x + 2 & x + 2 & x + \gamma \end{vmatrix}$.

Prove that $2g(1) - g(2) = f(0)$ where

$$g(x) = (\alpha - x)(\beta - x)(\gamma - x).$$

166. Find α, β and γ if the ellipse

$$4x^2 + y^2 + \alpha x + \beta y + \gamma = 0$$

touches the x-axis at the origin and passes through the point $(-1, 1)$.

167. If $\vec{a}$, $\vec{b}$ and $\vec{c}$ are noncoplanar vectors then prove that

$$[\vec{a}\ \vec{b}\ \vec{c}]^2 = \begin{vmatrix} \vec{a}\cdot\vec{a} & \vec{b}\cdot\vec{a} & \vec{c}\cdot\vec{a} \\ \vec{a}\cdot\vec{b} & \vec{b}\cdot\vec{b} & \vec{c}\cdot\vec{b} \\ \vec{a}\cdot\vec{c} & \vec{b}\cdot\vec{c} & \vec{c}\cdot\vec{c} \end{vmatrix}.$$

168. Find the point on the parabola $y^2 = 8x$ which is at the minimum distance from the circle $x^2 + y^2 + 12y + 35 = 0$. Also find this distance.

169. A is a fixed point on the circumference of a circle of radius r. Let AY be the perpendicular from A to the tangent to the circle at any point P on the circle. Find the greatest area of the ΔAPY.

170. ABCD is a rectangle and M is any point in the plane of the rectangle. Prove that $\overrightarrow{MA}\cdot\overrightarrow{MC} = \overrightarrow{MB}\cdot\overrightarrow{MD}$.

171. Prove, by induction or otherwise, that

$$\int_{-1}^{1} (1-x^2)^n\,dx = \frac{2^{2n+1}(n!)^2}{(2n+1)!}$$

where $n = 0, 1, 2, 3, \ldots$.

172. Solve $\dfrac{\sin^3\frac{x}{2} - \cos^3\frac{x}{2}}{2+\sin x} = \dfrac{\cos x}{3}$.

173. The equation of a family of ellipses is $\dfrac{x^2}{4} + \dfrac{y^2}{\lambda^2} = 1$ where λ is the parameter. Prove that the tangents at corresponding ends of their latus recta pass through a fixed point.

174. A bag contains 6 balls. Three drawings are made in succession and the ball on each occasion is found to be white. What is the probability of the ball in the 4th drawing to be white if the ball is not replaced after each drawing?

175. The equation of a pair of lines is

$$36x^2 - 63xy + 20y^2 + 54x - 17y - 10 = 0.$$

If a circle touches one of the lines at $(-3, -1)$ and the other line at some point then find the equation of the circle.

176. Prove that complex numbers z satisfying $\alpha\bar{z} + \bar{\alpha}z = 0$ lie on a straight line through the origin. If $z_1, z_2, z_3, \ldots, z_n$ are points on the same side of the line then prove that $z_1 + z_2 + z_3 + \ldots + z_n \neq 0$ and $\dfrac{1}{z_1}, \dfrac{1}{z_2}, \dfrac{1}{z_3}, \ldots, \dfrac{1}{z_n}$ are points on the same side of the straight line $\bar{\alpha}\bar{z} + \alpha z = 0$.

177. Let $\vec{a} = \vec{i} + 2\vec{j} + 3\vec{k}$, $\vec{b} = 2\vec{i} + 3\vec{j}$, $\vec{c} = 2\vec{i} + \vec{j} - \vec{k}$ and $\vec{d} = \vec{i} - \vec{j} + 4\vec{k}$.

Find a unit vector along the line of intersection of the two planes, one containing $\vec{a}$, $\vec{b}$ and the other $\vec{c}$, $\vec{d}$.

178. Let $z = a(1+i) + i, a \in R$. If z satisfies $|z - 2i| > 2$ and $|z - 4i| < 4$ then find the set of possible values of a.

179. Examine the type of the second degree curve whose equation is

$$5(x^2 + y^2) - 6xy + 2x - 14y + 9 = 0.$$

Find the points on the curve farthest from the line $x - y + 1 = 0$.

180. A point moves in a plane such that the sum of the squares of its distances from the four vertices of a square of side a is 4 times the square of the length of a side of the square. Find the area of the region enclosed by the locus of the point outside the square.

181. Solve for $x, y, z \in R$:

$$6\cdot 2^x + 2y - 3\sin^{-1}z = 5$$
$$2^{x+1} - y - \sin^{-1}z = -5$$
$$10\cdot 2^x - y + \sin^{-1}z = 3(2a+1)$$

indicating the set of values of $a \in R$ for which the solutions are possible.

182. Let $f(x)$ be a polynomial function. It satisfies the equation $2 + f(x)\cdot f(y) = f(x) + f(y) + f(xy)$ for all x and y. If $f(2) = 5$, find $f\{f(2)\}$.

183. Let S, S′, be the foci of an ellipse whose eccentricity is e. P is a variable point on the ellipse. Prove that the locus of the incentre of the ΔPSS′ is an ellipse of the eccentricity $\sqrt{\dfrac{2e}{1+e}}$.

184. If $n \in N$ and

$$n - \frac{n(n^2-1)}{2!} + \frac{n(n^2-1)(n^2-4)}{2!\cdot 3!} - \frac{n(n^2-1)(n^2-4)(n^2-9)}{3!\cdot 4!} + \ldots$$

$= s_1$ when n is even

s_2 when n is odd

then prove that $s_1 + s_2 = 0$.

185. Evaluate $\displaystyle\int \frac{dx}{\sqrt{e^{5x}}\cdot\sqrt[4]{(e^{2x} + e^{-2x})^3}}$.

186. Determine the equation of the ellipse $\dfrac{x^2}{a^2} + \dfrac{y^2}{b^2} = 1$ such that it has the least area but contains the circle $(x-1)^2 + y^2 = 1$.

187. Prove that

$$[\vec{a}\times\vec{x}\ \ \vec{b}\times\vec{y}\ \ \vec{c}\times\vec{z}] + [\vec{a}\times\vec{y}\ \ \vec{b}\times\vec{z}\ \ \vec{c}\times\vec{x}] + [\vec{a}\times\vec{z}\ \ \vec{b}\times\vec{x}\ \ \vec{c}\times\vec{y}] = 0.$$

188. Let $f(n)$ be a function of $n \in N$, representing the number of proper divisors of $7(60)^n$. Find $\sum_{n=1}^{m} f(n)$.

189. If $a_n = {}^{2n}C_n \cdot \frac{1}{2^{2n}} \cdot \sqrt{2n+1}$ then prove, by induction, that the sequence $a_1, a_2, a_3, \ldots, a_n, \ldots$ is monotonic decreasing.

190. If $p = a^2 - bc$, $q = b^2 - ca$ and $r = c^2 - ab$ then prove that $p^3 + q^3 + r^3 - 3pqr$ is a perfect square.

191. If p is an integral multiple of 4 and lies between the coefficients of x^{10} and x^7 in the expansion of $\left(\frac{1}{x} + x^2\right)^{11}$ then find the sum of all the possible values of p.

192. Let $f(x)$ be a continuous function such that $f(x+4) = f(x+2) - f(x)$. Prove that $\int_{\lambda}^{\lambda+12} f(x)\,dx$ is independent of λ.

193. Show, by induction, that

$$\frac{d^n}{dt^n} \int_{-t}^{t} f(x+t)dx = 2^n \cdot f^{n-1}(2t),\ n \in N$$

where $f^{n-1}(x)$ denotes the $(n-1)$th derivative of $f(x)$ w.r.t. x.

194. Let the solution-set of $|x| \le \sqrt{x}, x \in R$ be the interval I. The interval I is subdivided in $(2n+1)$ equal parts and the middle part is removed. Each of the remaining parts is again subdivided in $(2n+1)$ equal parts and the middle part is removed in each case. The process is repeated 10 times. What is the total length of the interval removed?

195. If z is a nonzero complex number satisfying $|z - i| \le 3$ and $|z - 1| \le 2$ then prove that the value of $|2(z+i) - 5|$ cannot exceed 10.

196. Prove that the chords of a hyperbola that touch the conjugate hyperbola, are bisected at the point of contact.

197. If $ax^2 - bx + c = 0$ where $a, b, c \in N$ has distinct real roots lying between 0 and 1 then prove that $abc \ge 25$.

198. If three real normals can be drawn to the parabola $y^2 = 4ax$ from the point (a^3, a) then prove that $|a| > \sqrt{2}$.

199. In how many ways can 25 identical things be distributed among five persons if

(i) each gets odd number of things

(ii) each gets at least one thing but not more than eleven?

200. If p be the length of perpendicular from the origin O to the tangent to the curve C at any point P(x, y) where $OP = r$ then the functional relation between p and r is called the pedal equation of the curve C. On the basis of this definition, find the pedal equation of the ellipse $\frac{x^2}{a^2} + \frac{y^2}{b^2} = 1$.

201. If the sum and the product of three positive numbers x, y, z be respectively 8 and 18 then prove that $p \in \left[144, \frac{8^4}{27}\right]$ where $p = (8-x)(8-y)(8-z)$.

202. If $a_r; r = 1, 2, 3, 4$ are four distinct real numbers then prove that $4x^3 - 3\Sigma a_1 \cdot x^2 + 2\Sigma a_1 a_2 \cdot x - \Sigma a_1 a_2 a_3 = 0$ has three real roots lying in (α, β) where

$\alpha = \min\{a_r;\ r = 1, 2, 3, 4\}$ and

$\beta = \max\{a_r;\ r = 1, 2, 3, 4\}$.

203. In how many different ways can 20 one-rupee coins be given to 4 persons so that no two get equal amount and each gets at least one rupee?

204. Let $f(x) = \cos x + \sin x$ and

$$g(x) = \min\{f(t),\ 0 \le t \le x\},\ 0 \le x \le \pi$$
$$\cos x,\ x > \pi.$$

Define $g(x)$ in terms of trigonometrical or algebraic functions of x and discuss its continuity in $[0, +\infty)$.

205. If $(1+x)^n = \sum_{r=0}^{m} a_r x^r$, $n \in N$ then prove that

(i) $\sum_{r=0}^{n} (a_r + a_{r+1})^2 = \frac{(2n+2)!}{((n+1)!)^2} - 1$

(ii) $\sum_{0 \le r < s \le n} (a_r + a_s)^2 = 4^n + (n-1)\frac{(2n)!}{(n!)^2}$.

206. Solve $\frac{dy}{dx} + \frac{y}{x} = \frac{1}{(1 + \log x + \log y)^2}$.

207. Let S be the region in the x-y plane which consists of the set of points given by

$$\{(x, y) \mid x^2 + y^2 - 6(|x| + |y|) + 9 \le 0,$$
$$|x| + |y| \le 3\}.$$

Find the area of S, using integration.

208. The tangent at any point P of a curve C meets the x-axis at Q whose abscissa is positive and $OP = OQ$, O being the origin. Show that C belongs to a family of parabolas having a common axis.

209. Let a, b, c be real numbers such that $a + 2b + c = 4$. Find max $(ab + bc + ca)$.

210. The probability of selecting an integer r from the set $\{1, 2, 3, \ldots, 2n\}$ is proportional to $\log r$. An integer is selected at random and it is found to be an even integer. What is the probability of the number to be the smallest even integer?

211. Prove that the points A, B, C, D whose position vectors are respectively $3\vec{i} + 4\vec{j}, 3\vec{i} - 4\vec{k}, 5\vec{k}$ and $\lambda\vec{j} + (\lambda + 5)\vec{k}$ are vertices of a plane quadrilateral for all nonzero $\lambda \in R$. Find λ for which it is a cyclic quadrilateral.

212. Prove that $\int_0^\infty \left[\frac{n}{e^x}\right] dx = \log\left(\frac{n^{n-1}}{(n-1)!}\right)$ where $n \in N$ and [ ] denotes the greatest integer function.

213. Evaluate $\int \frac{\left(\text{cosec}\,\frac{x}{2} - \sec\frac{x}{2}\right)^2}{2 + \cos 2x}\,dx.$

214. The normal to the ellipse $\frac{x^2}{a^2} + \frac{y^2}{b^2} = 1$ at any point meets the axes at G and G′. Find the equation of the locus of the middle point of GG′.

215. The vertex of a parabola and the centre of a rectangular hyperbola are the same and the axis of the parabola is along the transverse axis of the rectangular hyperbola. Prove that the locus of the point whose chord of contact w.r.t. the parabola touches the rectangular hyperbola is an ellipse with the same centre as the rectangular hyperbola.

216. An elliptical field has the equation of its boundary $x^2 + 3y^2 = 3$ with A at an end of its major axis. A tower stands vertically at A and from the points B, C on the boundary the angles of elevation of the top of the tower are found to be $\frac{\pi}{4}$ each. If A, B, C are on the same side of the minor axis and BC subtends a right angle at the centre of the ellipse then find the height of the tower.

217. Let O and I be the circumcentre and incentre respectively of the ΔABC. Prove that $OI^2 = R^2 - 2Rr$ where R, r are the circumradius and the inradius respectively. Hence or otherwise prove that the circumradius will be $(\sqrt{2} + 1)$ times of the inradius if O is on the incircle.

218. A right angle is divided in three parts α, β and γ. Prove that for all possible divisions

$$\tan\alpha + \tan\beta + \tan\gamma > 1 + \tan\alpha \cdot \tan\beta \cdot \tan\gamma.$$

219. If in the triangle ABC, $\cos A \cdot \cos B \cdot \cos C = \frac{1}{16}$, prove that there exist exactly two isosceles triangles which are dissimilar.

220. If $(1 + x + x^2 + \ldots + x^9)^4 \cdot (x + x^2 + x^3 + \ldots + x^9) = \sum_{i=1}^{45} a_i x^i$ then find the value of $a_2 + a_6 + a_{10} + \ldots + a_{42}$.

221. How many five-digit numbers can be made such that the sum of the digits leaves the remainder 1 when divided by 3?

222. Let $x^2 + 3y^2 = 3$ be the equation of an ellipse in the x-y plane. A and B are two points whose position vectors are $-\sqrt{3}\,\vec{i}$ and $-\sqrt{3}\,\vec{i} + 2\vec{k}$. Find the position vector of a point P on the ellipse such that $\angle APB = \frac{\pi}{4}$.

223. Find $\vec{x}$ if $\vec{x} \times \vec{a} = \vec{b}$ and $\vec{a}, \vec{b}$ are two given vectors perpendicular to each other. Is the vector $\vec{x}$ unique?

224. If $ax^2 + 2hxy + by^2 = 1$, prove that the maximum and minimum values of $x^2 + y^2$ are given by the values of r^2 satisfying the relation

$$\left(a - \frac{1}{r^2}\right)\left(b - \frac{1}{r^2}\right) = h^2.$$

225. Evaluate: $\int \frac{dx}{(\sin x + 3\cos x)(\cos x + 3\sin x)}$.

226. Use calculus to find the area bounded by the circle $x^2 + y^2 - 2x - 2y - 23 = 0$ and the pair of lines $x^2 + 2xy + y^2 - 7x - 7y + 12 = 0$.

227. Let $f_i(x)$; $i = 1, 2, 3$ be continuous in $[0, a]$ such that $f_1(a - x) = f_1(x), f_2(a - x) + f_2(x) = 0$ and $f_3(x) + 2f_3(a - x) = 3$.

Show that $\int_0^a f_1(x) \cdot f_2(x) \cdot f_3(x)dx = 0$.

228. Solve: $x\cos\frac{y}{x} \cdot (y\,dx + x\,dy) = y\sin\frac{y}{x} \cdot (x\,dy - y\,dx)$.

229. If $f(a) = \int_0^b F(a, x)\,dx$ where b is independent of a, it is known that $\frac{df(a)}{da} = \int_0^b \frac{\partial F(a, x)}{\partial a}\,dx$, $\frac{\partial F}{\partial a}$ denoting differential coefficient of F w.r.t. a when x is treated as a constant.

On the basis of this, prove that

$$\int_0^{\pi/2} \log(\sin^2\theta + k^2\cos^2\theta)\,d\theta = \pi\log\frac{1 + k}{2}, k \geq 0.$$

230. If $\vec{a}, \vec{b}, \vec{c}$ are three given unit vectors forming a right-handed triad and $\vec{p}$ is a constant vector such that $(\vec{r}-\vec{p})\cdot\vec{b}=0$ and $(\vec{r}-\vec{p})\cdot\vec{c}=0$ then prove that $\vec{r}$ is the position vector of any point on a fixed line.

231. If $\vec{a}, \vec{b}, \vec{c}, \vec{d}$ are four vectors of which no three are coplanar then find a vector $\vec{p}$ along the line of intersection of the planes of $\vec{a}, \vec{b}$ and $\vec{c}, \vec{d}$. Also write $\vec{p}$ as a linear combination $\lambda\vec{a}+\mu\vec{b}$ and find λ, μ.

232. Let (V, ABC) be a tetrahedron in which the edges VA, VB, VC are mutually perpendicular and N is the foot of the perpendicular from the vertex V to the plane ABC. Prove vectorially that N is the orthocentre of the ΔABC.

233. Let the equations of a circle and an ellipse be $x^2+y^2=r^2$ and $\frac{x^2}{a^2}+\frac{y^2}{b^2}=1$ where $r>a$. If the chord of contact of tangents from P to the circle has its pole with respect to the ellipse on the circle then prove that the locus of P is a concentric ellipse. Find its eccentricity if the eccentricity of the given ellipse is e.

234. The eccentricity of an ellipse concentric with a circle is e. Tangents from any point on the circle to the ellipse cut the circle again at A and B. The tangents at A and B to the circle cut at P. Prove that the locus of P is an ellipse.

235. In how many ways can some apples from 45 apples be distributed among 5 persons so that none gets more than 9 apples and the total number of apples given is a multiple of 3?

236. Show that

$${}^nC_0\cdot{}^nC_0-{}^{n+1}C_1\cdot{}^nC_1+{}^{n+2}C_2\cdot{}^nC_2-\ldots=(-1)^n.$$

237. Solve the inequation

$\cos x+\cos 3x+\sqrt{3}\sin x>0.$

238. Two words, one of four letters and the other of six letters, are made at random with the letters AABBCCDEFG. What is the probability of both the words to have at least two identical letters?

239. A number of five digits is written down at random. What is the probability that the sum of the digits used leaves the remainder 2 when divided by 4?

240. If $f(x)=\sin x+\int_{-\pi/2}^{\pi/2}(\sin x+t\cos x)f(t)\,dt$ then find the maximum and minimum values of $f(x)$.

241. If the equation $ax^2+bx+c=0$ where a, b, c are sides of a triangle and the equation $x^2+2\lambda x+\lambda^2+1=0$, $\lambda\in R$ have a common root then find the set of possible values of λ.

242. In a ΔABC, $AB=6$ cm and $AC=4$ cm. The bisector of $\angle BAC$ cuts BC at D. Prove that the length of AD is less than $2\sqrt{6}$ cm for all possible measures of angle A.

243. 10 pairs of gloves are kept in a drawer. 10 gloves are taken out at random from the drawer. Find the probability of the selection to contain at least one pair but not more than 4 pairs.

244. Let $m\in N$ and $n\in N$ such that the number of proper divisors of n is $(p-2)$ where p is a prime number. Prove that $n-m^2=4$ does not hold for any pair (n, m).

245. Prove that the number 111...1, there being k ones where k is a composite number, is a composite number which is equal to the sum of a GP of positive integers.

246. Find $x\in R$ which satisfies the equation

$2\log_2\log_2 x+\log_{1/2}\log_2(2\sqrt{2}x)=1.$

247. Let $\alpha+i\beta$ be a root of the equation $x^3+qx+r=0$; $\alpha\in R$, $\beta\in R$, $q\in R$, $r\in R$. Find a real cubic equation independent of α and β whose one root is 2α.

248. For complex numbers z and w, prove that $|z|^2w-|w|^2z=z-w$ if and only if $z=w$ or $\bar{z}w=1$.

249. Let $n\in N$. Prove that

$$\sum_{k=0}^{m}\frac{{}^{2n-k}C_k}{{}^{2n-k}C_n}\cdot\frac{2n-4k+1}{2n-2k+1}\cdot 2^{n-2k}=\frac{{}^nC_m}{{}^{2n-2m}C_{n-m}}\cdot 2^{n-2m}$$

for each non-negative integer $m\le n$.

250. Prove, by mathematical induction or otherwise, that p divides ${}^{mp}C_r$ where $m\in N$, p is a prime number and p does not divide the positive integer r.

251. Let $f(x)$ be a real-valued function satisfying the following.

(i) $f(0)=2, f(1)=1$

(ii) $f(x)$ has a minimum at $x=\frac{5}{2}$.

(iii) for all $x\in R$,

$$f'(x)=\begin{vmatrix}2ax & 2ax-1 & 2ax+b+1\\ b & b+1 & -1\\ 2(ax+b) & 2ax+2b+1 & 2ax+b\end{vmatrix}$$

where a, b are constants.

Determine a, b and $f(x)$.

252. Prove that $\int_0^1\tan^{-1}\frac{1}{1-x+x^2}dx=2\int_0^1\tan^{-1}x\,dx.$

Hence or otherwise, find the value of

$$\int_0^1\tan^{-1}(1-x+x^2)\,dx.$$

253. Let $f(x)=a_0+\sum_{k=1}^{20}a_k x^k$ and $\alpha=\cos\frac{2\pi}{7}+i\sin\frac{2\pi}{7}$.

Find the value of $\sum_{k=0}^{6}f(\alpha^k x)$ independent of α.

254.

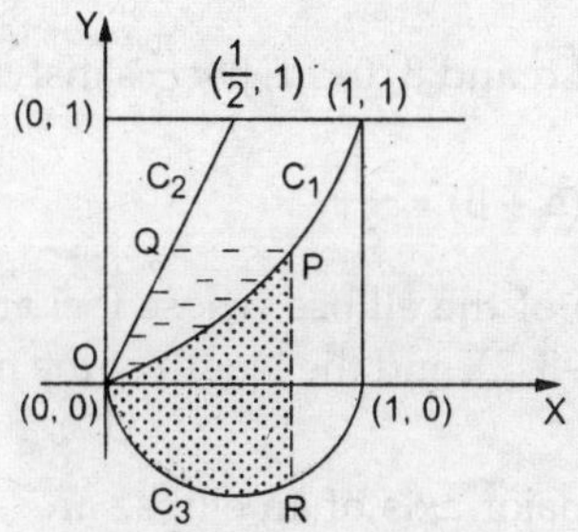

In the above figure C_1, C_2 are the graphs of the functions $f_1(x) = x^2$ and $f_2(x) = 2x$, $0 \le x \le 1$ respectively. C_3 is the graph of the function $y = f_3(x)$, $0 \le x \le 1$, $f_3(0) = 0 = f_3(1)$. P is any point on C_1 and lines through the point P parallel to the axes cut C_2 and C_3 at Q and R respectively. If ar(OPQO) = ar(OPRO) for all positions of P on C_1 then determine the function $f_3(x)$.

255. Integrate: $\int \frac{x^3 + 3x + 2}{(x^2+1)^2 \cdot (x+1)} dx$

256. Evaluate: $\int_0^{\pi} \frac{e^{\cos x}}{e^{\cos x} + e^{-\cos x}} dx$

257. Let $f(x)$ be a continuous function given by

$$f(x) = \begin{cases} 2x, & |x| \le 1 \\ x^2 + ax + b, & |x| > 1. \end{cases}$$

Find the area of the region in the third quadrant bounded by the curves $x + 2y^2 = 0$ and $y = f(x)$ lying on that side of the line $8x + 1 = 0$ which is opposite to the origin.

258. If $f(x) = \frac{x^2 - x}{x^2 + 2x}$ then find the domain and the range of $f: R \to R$. Show that f is one-one. Also, find the function $\frac{df^{-1}(x)}{dx}$ and its domain.

259. Find the constants a, b, c such that the function $f(x)$ defined below is continuous at $x = 0$.

$$f(x) = \begin{cases} (1 + ax)^{1/x}, & x < 0 \\ b, & x = 0 \\ \dfrac{(x+c)^{1/3} - 1}{(x+1)^{1/2} - 1}, & x > 0 \end{cases}$$

260. Evaluate $\int_0^{\pi/6} \frac{\sqrt{3\cos 2x - 1}}{\cos x} dx$.

261. Find the area of the region lying inside the circle $x^2 + (y-1)^2 = 1$ and outside the curve $c^2x^2 + y^2 = c^2$ where $c = \sqrt{2} - 1$.

262. Let $\vec{u}$ and $\vec{v}$ be unit vectors. If $\vec{w}$ is a vector such that $\vec{w} + (\vec{w} \times \vec{u}) = \vec{v}$ then prove that

$$|(\vec{u} \times \vec{v}) \cdot \vec{w}| \le \frac{1}{2}$$

and that the equality holds if and only if $\vec{u} \perp \vec{v}$.

263. For any two vectors $\vec{u}$ and $\vec{v}$, prove that

$$(1 + |\vec{u}|^2)(1 + |\vec{v}|^2) = (1 - \vec{u} \cdot \vec{v})^2 + |\vec{u} + \vec{v} + \vec{u} \times \vec{v}|^2.$$

264. Prove vectorially that the point of intersection of the diagonals of any trapezium lies on the line passing through the middle points of the parallel sides.

265. An arc AC of a circle subtends a right angle at the centre O. The point B divides the arc in the ratio 1 : 2. If $\vec{OA} = \vec{a}$ and $\vec{OB} = \vec{b}$ then find $\vec{OC}$ in terms of $\vec{a}$ and $\vec{b}$.

266. If $\vec{a}, \vec{b}, \vec{c}$ are noncoplanar vectors and $\vec{d}$ is a unit vector then find the value of

$$|(\vec{a} \cdot \vec{d})\vec{b} \times \vec{c} + (\vec{b} \cdot \vec{d})\vec{c} \times \vec{a} + (\vec{c} \cdot \vec{d})\vec{a} \times \vec{b}|$$

independent of $\vec{d}$.

267. Find the coordinates of all the points on the ellipse $\frac{x^2}{a^2} + \frac{y^2}{b^2} = 1$ for which the area of the ΔPON is the maximum where O is the origin and N is the foot of the perpendicular from O to the tangent at P.

268. A curve passing through the point (1, 1) has the property that the perpendicular distance of the origin from the normal at any point P of the curve is equal to the distance of P from the x-axis. Determine the equation of the curve.

269. Solve: $(x^2 + 4y^2 + 4xy)\, dy = (x + 2y + 1)\, dx$.

270. Find the normals to the ellipse $\frac{x^2}{9} + \frac{y^2}{4} = 1$ which are farthest from its centre.

271. Sum the series:

$$\sum_{n=0}^{\infty} \frac{1}{\lfloor n} \left[\sum_{k=0}^{n} (k+1) \int_0^1 2^{-(k+1)x} dx \right]$$

272. Consider the family of circles $x^2 + y^2 = r^2$, $2 < r < 5$. If in the first quadrant, the common tangent to a circle of this family and the ellipse $\frac{x^2}{25} + \frac{y^2}{4} = 1$ meets the axes at A and B then find the equation of the locus of middle point of AB.

273. Let T_1, T_2 be two tangents from (–2, 0) to the circle C whose equation is $x^2 + y^2 = 1$. Determine the circles touching C and having T_1, T_2 as their pair of tangents from (–2, 0). Further, find the equations of all possible common tangents to these circles when taken two at a time.

274. The equations of two equal sides AB and AC of an isosceles triangle ABC are $x + y = 5$ and $7x - y = 3$ respectively. Find the equations of the side BC if ar(ΔABC) = 5 unit^2.

275. Find the equation of the largest circle with centre (1, 0) that can be inscribed in the ellipse

$x^2 + 4y^2 = 16$.

276. Prove that the circle $x^2 + y^2 - 6y + 4 = 0$ and the parabola $y^2 = x$ touch. Find the common tangent at the point of contact.

277. C_1 and C_2 are two concentric circles, the radius of C_2 being twice that of C_1. From a point P on C_2, tangents PA and PB are drawn to C_1 touching at A and B. Prove that the centroid of the ΔPAB lies on the circle C_1.

278. The angle between a pair of tangents drawn from a point P to the parabola $y^2 = 4ax$ is $\pi/4$. Show that the locus of the point P is a hyperbola.

279. Eight players $P_1, P_2, \ldots, P_8$ play a knock-out tournament. It is known that whenever the players P_i and P_j play, the player P_i will win if $i < j$. Assuming that the players are paired at random in each round, what is the probability that the player P_4 reaches the final?

280. Three players A, B and C toss a coin cyclically in that order till a head shows. Let p be the probability that the coins show a head. Let α, β, γ be respectively the probabilities that A, B, C gets the first head. Prove that $\beta = (1 - p)\alpha$. Also determine α, β and γ in terms of p.

281. Let ABC be a triangle having O and I as its circumcentre and incentre respectively. Show that the triangle BIO is a right-angled triangle if and only if a, b and c are in AP.

282. The exradii of the ΔABC are in HP. If its area is 24 cm^2 and its perimeter is 24 cm, find the lengths of the sides.

283. A bird flies in a circle on a horizontal plane. An observer stands at a point on the ground. Suppose that 60° and 30° are the maximum and the minimum angles of elevation of the bird and they occur when the bird is at P and Q respectively on its path. Let θ be the angle of elevation of the bird when it is at a point on the arc of the circle exactly midway between P and Q. Find the numerical value of $\tan^2\theta$. (Assume that the observer is not inside the vertical projection of the path of the bird.)

284. A vertical tree stands at a point A on a bank of a canal. The angle of elevation of its top from a point B on the other bank of the canal and directly opposite to A is 60°. The angle of elevation of the top from another point C is 30°. If A, B and C are on the same horizontal plane, $\angle ABC = 120°$ and $BC = 20$ m then find the height of the tree and the width of the canal.

285. Find the values of α and β, $0 < \alpha, \beta < \frac{\pi}{2}$ satisfying $\cos\alpha \cdot \cos\beta \cdot \cos(\alpha + \beta) = \frac{-1}{8}$.

286. Find the equation of the ellipse whose foci are the points (1, 2) and (–3, 2), and the length of the minor axis is $4\sqrt{3}$.

287. The ends of the major axis of an ellipse are (–2, 4) and (2, 1). If it passes through the point (1, 3) then obtain the eccentricity and the latus rectum of the ellipse.

288. If in the ΔABC, c is the longest side and $a^2 + b^2 = 2Rc$, where R is the circumradius of the triangle then prove that ΔABC is right angled.

289. ABC is a right-angled triangle in which $\angle A = \frac{\pi}{2}$ and the distance of BC from A is h. If P, Q are points of trisection of BC and $\angle PAQ = \alpha$ then prove that $\tan\alpha = \frac{3h}{2a}$.

290. Find the number of real solutions of
$$2^{|x|} \cdot |1 - |x|| = 1.$$

291. ABC and PQR are two triangles in the same plane such that the perpendiculars from A, B and C to the sides QR, RP and PQ respectively are concurrent. Prove that the perpendiculars from P, Q and R to BC, CA and AB respectively are also concurrent.

292. An unbiased die with faces 1, 2, 3, 4, 5, 6 is thrown n times and the list of n numbers shown is noted down. What is the probability that among the numbers 1, 2, 3, 4, 5, 6 exactly three numbers appear in the list?

293. Prove that $\sqrt{4n+1} < \sqrt{n} + \sqrt{n+1} < \sqrt{4n+2}$ where $n \in N$. Hence or otherwise prove that $[\sqrt{n} + \sqrt{n+1}] = [\sqrt{4n+1}]$ where $[x]$ denotes the greatest integer $\le x$.

294. If $\phi(n) = \sum_{r=1}^{n} \frac{1}{r}$

and $\psi(n) = \frac{n+1}{2} - \sum_{r=1}^{n-2} \frac{r}{(n-r+1)(n-r)}$

then prove that $\phi(n) = \psi(n)$ for all $n \in N$.

295. Find the smallest positive values of x, y satisfying $x - y = \frac{\pi}{4}$ and $\cot x + \cot y = 2$.

296. Solve:
$$\log_{3/4}\log_8(x^2+7) + \log_{1/2}\log_{1/4}(x^2+7)^{-1} + 2 = 0.$$

297. Prove that
$$\left(\sqrt{^nC_1} + \sqrt{^nC_2} + \ldots + \sqrt{^nC_n}\right)^2 \le n \cdot (2^n - 1).$$

298. The sides of a triangle touch a parabola and two of its vertices lie on a parabola having its axis parallel to that of the first parabola. Prove that the locus of the third vertex of the triangle is also a parabola.

299. In the expansion of $\left(\dfrac{1+x-x^2}{(1+x^2)^{-1}}\right)^{10}$ in powers of x, prove that the sum of the coefficients of odd powers of x is zero. Also, find the sum of the coefficients of the even powers of x.

300. Prove analytically that the middle point of the line segment joining the orthocentre and circumcentre of a triangle is at a constant distance from the feet of the altitudes of the triangle.

301. Let $A = \{(x, y) \mid x^2 + y^2 = 3,\ x \in R^+, y \in R^+\}$ be a set of points in the x-y plane and P = (3, 4) be a given point. If the distance $d(Q, R)$ be defined as

$$\frac{\sqrt{(x_1-x_2)^2+(y_1-y_2)^2}}{1+\sqrt{(x_1-x_2)^2+(y_1-y_2)^2}}$$

where $Q = (x_1, y_1)$, $R = (x_2, y_2)$

then find the point of the set A which is nearest to the point P in the new sense of distance.

302. If $\vec{a}, \vec{b}, \vec{c}$ are any three vectors in space and $\vec{A} = p\vec{a} + q\vec{b} + r\vec{c}$, $\vec{B} = q\vec{a} + r\vec{b} + p\vec{c}$, $\vec{C} = r\vec{a} + p\vec{b} + q\vec{c}$ are three vectors of which no two are collinear then prove that $\vec{A}, \vec{B}, \vec{C}$ are coplanar if and only if they are represented by the sides of a triangle.

303. Let $x, y, z \in R$ and $(x+y+z)^3 = x^3 + y^3 + z^3$. Prove that $(x+y+z)^n = x^n + y^n + z^n$ where n is an odd positive integer.

304. Let f and g be two functions such that $f'(x) = g(x)$ and $g'(x) = f(x)$. If $f(0) = 2$, $g(0) = 1$ then find $f(2)$.

305. Find the number read by the last three digits of 17^{256}.

306. In how many different ways can 40 passengers board four specific big compartments of a train so that the number of people in a compartment is at least 5 and the number of people in the compartment are in AP? (Sitting arrangement in a compartment is not taken into account.)

307. Show that there is no rational number x satisfying the equation $x^2 - [x] = 4$ where $[x]$ = the greatest integer $\le x$. Also find the roots of the equation.

308. Let a, b be the roots of the equation $x^2 - kx + k = 0$, $k \in R$. If $a^2 + b^2$ is the minimum then find the roots of the equation.

309. Two parabolas have the same focus (3, –28). Their directrices are the x-axis and y-axis respectively. Find the slope of their common chord.

310. Let $g(x) = x^3 - x$ and $f(x)$ be a polynomial of degree ≤ 100. If $f(x)$ and $g(x)$ have no common factor and $\dfrac{d^{100}}{dx^{100}}\left\{\dfrac{f(x)}{g(x)}\right\} = \dfrac{p(x)}{q(x)}$ then find the degrees of the polynomials $p(x)$ and $q(x)$.

311. Prove that $f(x)$ is differentiable at $x = \alpha$ if and only if there is a function $g(x)$ which is continuous at $x = \alpha$ and satisfies $f(x) - f(\alpha) = (x - \alpha)\cdot g(x)$ for all $x \in R$.

312. In a Δ ABC, $a^2 + b^2 = 5c^2$. Prove that two of its medians cut at a right angle.

313. Let E_1 be an ellipse. E_2 is another ellipse of the same eccentricity as that of E_1 and the foci of E_2 coincides with the ends of the minor axis of E_1. If E_2 passes through the ends of the major axis of E_1, find the eccentricity of E_2.

314. Let $f(x) \equiv n_0x^4 + n_1x^3 + n_2x^2 + n_3x + n_4 = 0$, where $n_i \in Z$ (the set of integers). If $f(x) = 0$ has two distinct irrational roots whose product is also irrational then prove that $\left|f\left(\dfrac{p}{q}\right)\right| \ge \dfrac{1}{q^4}$ where p and q are integers.

315. Let $x^2 - px + q = 0$ and $x^2 - qx + p = 0$ both have unequal integral roots where $p \in N$, $q \in N$. Prove that the possible number of solutions of the ordered pair (p, q) is 2. Find them.

316. If a_i s are non-negative such that $\sum_{i=1}^{18} a_i = 27$ then prove that the greatest value of $\sum_{i=1}^{16} a_i a_{i+1} \cdot a_{i+2}$ is 729.

317. The envelope of a family of straight lines is the curve which touches each member of the family of lines. Prove that the envelope of the family of lines

$$(\cos\theta + \sin\theta)x + (\cos\theta - \sin\theta)y + 2\sin\theta - \cos\theta - 4 = 0,$$

where θ is the parameter, is a fixed circle. Find its centre and radius.

318. For what range of values of $a \in R - \{0\}$ does the equation $a\cos 2\theta - 3\cos\theta + a + 1 = 0$ admit only real values of θ? Also, solve the equation for the greatest possible value of a.

319. Prove analytically that the minimum value of $(y_1 - y_2)^2 + (\sqrt{4 - y_1^2} - y_2 - 6)^2$ where $y_1, y_2 \in R$, is $3\sqrt{2} - 2$.

320. Find the $\displaystyle\lim_{x\to 0} \frac{\sqrt{x+2} - \sqrt{2}}{\sqrt[3]{1-x} - \sqrt{1+x}}$.

321. If the equation of three planes be $\vec{r}\cdot\vec{a} = \lambda$, $\vec{r}\cdot\vec{b} = \mu$ and $\vec{r}\cdot\vec{c} = \nu$ where $\vec{a}$, $\vec{b}$ and $\vec{c}$ are noncoplanar then prove that the point of intersection of the three planes has the position vector

$$\frac{\nu\, \vec{a} \times \vec{b} + \lambda\, \vec{b} \times \vec{c} + \mu\, \vec{c} \times \vec{a}}{[\vec{a}\ \vec{b}\ \vec{c}\,]}.$$

322. Let the parabolas $y^2 = 4a(x-h)$ and $x^2 = 4a(y-k)$ always touch each other where a is a constant and h, k are variables. Prove that the point of contact always lies on the curve $xy = 4a^2$.

323. Prove that $\sum_{r=1}^{2n} (r)!$ is not divisible by $\sum_{r=1}^{n} (2r)!$ for any $n \in N$.

324. Let $2^{a_1} + 3^{a_2} + 5^{a_3} + 7^{a_4} + 9^{a_5}$ be divisible by 4 where a_1, a_2, a_3, a_4, a_5 are digits. Find the largest possible number of values of $(a_1, a_2, a_3, a_4, a_5)$.

325. Let ABC be an acute-angled triangle whose incentre and centroid are respectively I and G. AI, BI and CI cuts the sides of the triangle at P, Q and R respectively. If p_1, p_2 and p_3 are the lengths of the altitudes through A, B and C respectively and G lies on PQ then prove by vector method that $p_1 + p_2 = p_3$.

326. Let each of the urns $U_1, U_2, U_3, \ldots$ contain m white and n black balls. A ball is selected at random from U_1 and put in U_2. Then a ball is selected at random from U_2 and put in U_3 and so on. If p_K denotes the probability of drawing a white ball from U_K in the process then prove that

$$p_{K+1} = \frac{m+1}{m+n+1} p_K + \frac{m}{m+n+1}(1 - p_K).$$

Hence, prove by induction that $p_K = \frac{m}{m+n}$ for all $K \in N$.

327. If λ is a positive proper fraction then prove that $e^{\lambda\pi + (1-\lambda)e} < \lambda e^{\pi} + (1-\lambda)e^{e}$.

328. Let $f(x) = [\tan x]$ and $g(x) = \sec x$. Evaluate

$$\int_0^{\pi/4} [\sin x + \{[f(x) + g(x)]\}] dx$$

where $[x]$ = the greatest integer $\leq x$.

329. Prove by induction that $\sum_{m=0}^{n} {}^{2n+1}C_{2m+1} \cdot 2^{3m}$ is not divisible by 5 for all $n \in N$.

330. The sides a, b, c of the ΔABC are in GP whose common ratio is $\frac{2}{3}$ and the circumradius of the triangle is $6 \cdot \sqrt{\frac{7}{209}}$. Find the longest side of the triangle.

331. Let the vertices of the ΔABC represent the complex numbers z_1, z_2, z_3 respectively and their modulii be equal. Prove that the circumcentre of the triangle represents the complex number

$$z_1 \sin 2A + z_2 \sin 2B + z_3 \sin 2C.$$

332. Evaluate $\int_e^{e^2} \left\{\frac{1}{(\log x)^2} + \log(\log x)\right\} dx.$

333. Solve the equation:

$$(b^2x + a^2x^4y^2)\frac{dy}{dx} + 2a^2x^3y^3 + b^2y = 0.$$

334. Let $x^2 + y^2 = 5$ be the equation of a circle and P(3, 3) be a point. If Q(2, 1) be another point then find the equation of the line QR such that PQ and QR are equally inclined with the circle.

335. Let $g(x) \cdot (x^2 - x - 2) = h(x) \cdot (x^2 + x - 2)$ where $g(x)$ and $h(x)$ are polynomials. Prove that $g(x) \cdot h(x) + x^4 - 5x^2 + 4 = 0$ has exactly four real roots which are integers.

336. Let A = (1,3) and B(x, y) be a variable point such that $|x| \leq 2$ and $AB \leq 4$. Find the area of the region in the x-y plane in which B can lie.

337. In a ΔABC if $\cot\frac{A}{2}$, $\cot\frac{B}{2}$, $\cot\frac{C}{2}$ are in AP then prove that $\tan\frac{B}{2}$ is the maximum when the triangle is equilateral.

338. Let $2x^2 + y^2 - 3xy = 0$ be the equation of a pair of tangents drawn from the origin O to a circle of radius 3 with centre in the first quadrant. If A is one of the points of contact, find the length of OA.

339. Let $f(x), x \geq 0$, be a non-negative continuous function and let $F(x) = \int_0^x f(t)dt, x \geq 0$. If for some $c > 0, f(x) \leq c\,F(x)$ for all $x \geq 0$ then show that $f(x) = 0$ for all $x \geq 0$.

340. Let $b \neq 0$ and for $j = 0, 1, 2, \ldots, n$ let S_j be the area of the region bouned by the y-axis and the curve $xe^{ay} = \sin by$, $\frac{j\pi}{b} \leq y \leq \frac{(j+1)\pi}{b}$. Show that $S_0, S_1, \ldots, S_n$ are in GP. Also, find their sum for $a = -1$ and $b = \pi$.

341. Let P be a point on the ellipse $\frac{x^2}{a^2} + \frac{y^2}{b^2} = 1, 0 < b < a$. Let the line parallel to the y-axis passing through P meet the circle $x^2 + y^2 = a^2$ at the point Q such that P and Q are on the same side of the x-axis. If $r > 0$, $s > 0$, find the locus of the point R on PQ such that $PR : RQ = r : s$ as P varies over the ellipse.

342. A hemispherical tank of radius 2 metres is initially full of water and has an outlet of 12 cm^2 cross-sectional area at the bottom. The outlet is opened at some instant. The flow through the outlet is according to the law $v(t) = 0.6 \times \sqrt{2gh(t)}$ where $v(t)$ and $h(t)$ are respectively the velocity of flow through the outlet and the height of water level above the outlet at time t, and g is the acceleration due to gravity. Find the time it takes to empty the tank.

343. Let $f(x) = \sqrt{x^2 + 3x + 4}$ be a rational-valued function of the rational variable x. Then, find the domain and the range of the function.

344. If t_1 and t_2 are the lengths of the tangents from the point $(-1, -4)$ to the parabola $y^2 = 8x$ then prove that $|t_1 - t_2| = 4\sqrt{3}$.

345. If I_n is the area of the n-sided regular polygon inscribed in a circle of unit radius and C_n is the area of the n-sided regular polygon circumscribing the given circle, prove that

$$\frac{I_n}{C_n} = \left(1 + \sqrt{1 - \left(\frac{2I_n}{n}\right)^2}\right) \cdot \frac{1}{2}.$$

346. Find the image of the point P(2, 1, 6) w.r.t. the plane passing through the points (2, 1, 0), (5, 0, 1) and (4, 1, 1).

347. Find the value of r such that the area enclosed by the tangents from P(6, 8) to the circle $x^2 + y^2 = r^2$ and the chord of contact of the tangents is the maximum.

348. A right circular cone with radius r and height h contains a liquid which evaporates at a rate proportional to its surface area in contact with air (constant of proportion $= k > 0$). Find the time after which the cone is empty.

349. If $\vec{u}, \vec{v}, \vec{w}$ are three noncoplanar unit vectors and α, β, γ are the angles between $\vec{u}$ and $\vec{v}$, $\vec{v}$ and $\vec{w}$, $\vec{w}$ and $\vec{u}$ respectively and $\vec{x}, \vec{y}, \vec{z}$ are unit vectors along the bisectors of the angles α, β, γ respectively, prove that

$$[\vec{x} \times \vec{y} \;\; \vec{y} \times \vec{z} \;\; \vec{z} \times \vec{x}] = \frac{1}{16} \sec^2 \frac{\alpha}{2} \cdot \sec^2 \frac{\beta}{2} \cdot \sec^2 \frac{\gamma}{2} \cdot [\vec{u} \; \vec{v} \; \vec{w}]^2.$$

350. If $P(1) = 0$ and $P'(x) > P(x)$ for all $x \geq 1$ then prove that $P(x) > 0$ for all $x > 1$.

351. Normals are drawn from the point P with slopes m_1, m_2, m_3 to the parabola $y^2 = 4x$. If the locus of P with $m_1 m_2 = k$ is a part of the parabola itself then find k.

352. A family of ellipses has the same focus and the same corresponding directrix. Prove that the locus of an end of the minor axis of the ellipses is a parabola.

353. If α, β, γ are the roots of the equation $x^3 + 1 = x(x + 5)$, prove that the roots are all real and find the value of $[\alpha] + [\beta] + [\gamma]$ where $[y]$ = the greatest integer less than or equal to y.

354. Find a point on the curve $x^2 + 2y^2 = 6$ whose distance from the line $x + y = 7$ is the minimum.

355. If f is an odd function from $[-2a, 2a]$ to R such that $f(x) = f(2a - x)$ for $x \in [a, 2a]$ and $f'(a - 0) = 0$ then find the value of $f'(-a - 0)$.

356. A is targetting B while B and C are targetting A. The probabilities of hitting the target by A, B and C are $\frac{2}{3}, \frac{1}{2}$ and $\frac{1}{3}$ respectively. If A is hit then find the probability that B hits the target while C misses it.

357. A ray of light passing through the point (4, 1, 1) is incident on the plane $x - y + z = 2$ at the point $(1, -1, 0)$. Find the equation of the line along which the reflected ray travels.

358. If the equation $x^3 + kx^2 + k^2x = 7$ has at least one positive root whose integral part is 1 then find the range of real values of k.

359. Let $f : [0, 4] \to R$ be a differentiable function. Show that

(a) $\{f(4)\}^2 - \{f(0)\}^2 = 8f'(a) \cdot f(b)$ for $a, b \in [0, 4]$

(b) $\int_0^4 f(t)\, dt = 2\{\alpha f(\alpha^2) + \beta f(\beta^2)\}$ for all α, β such that $0 < \alpha, \beta < 2$.

360. If $\vec{x} + \vec{y} + \vec{z} = \vec{a}$, $\vec{x} \times \vec{y} = \vec{b}$, $\vec{y} \times \vec{z} = \vec{c}$, $\vec{a} \cdot \vec{x} = \vec{a} \cdot \vec{y} = 1$ and $|\vec{a}| = \sqrt{3}$ then find the vectors $\vec{x}, \vec{y}, \vec{z}$ in terms of the vectors $\vec{a}, \vec{b}, \vec{c}$.

361. If $f(x)$ is a continuous function of $x \in R$ and $f(x) + f(x + 1) = 1$ for $0 \leq x \leq 1$ then find $\int_0^2 f(x)\, dx$.

362. If $f : R \to [0, \infty)$ be a function such that $f(x - 1) + f(x + 1) = \sqrt{3} f(x)$ then prove that $f(x + 12) = f(x)$.

363. A straight line passing through the origin moves in space such that it always makes equal angles θ with the y- and z-axes. Prove that θ always lies between $\frac{\pi}{4}$ and $\frac{3\pi}{4}$.

364. Prove that $\sum_{r=0}^{n} \frac{2r + 1}{2r + 2} \cdot {}^nC_r = \frac{1}{n + 1}\left\{n \cdot 2^n + \frac{1}{2}\right\}$.

365. A rabbit runs up the y-axis with a constant speed u, starting from the origin. A dog starts from the point $(c, 0)$ at the same time with a constant speed v always directed towards the rabbit. Find the equation of the path of the rabbit.

366. Let α, β be the roots of the equation $ax^2 + (a-1)x + 1 = 0$, $a \in R - \{0\}$ such that $x_1\alpha - y_1\beta = y_1 - x_1$ where (x_1, y_1) is a point on the circle $x^2 + y^2 = 1$. Prove that $a \notin (3 - 2\sqrt{2}, 3 + 2\sqrt{2})$.

367. Prove that the lines $\frac{x-2}{2} = \frac{y-3}{3} = \frac{z+1}{-1}$ and $\frac{x-3}{3} = \frac{y+5}{-5} = \frac{z-1}{1}$ are intersecting. Find the equation of the line through the point of intersection of the lines and lying in their plane such that the line is perpendicular to the line $\frac{x}{1} = \frac{y}{1} = \frac{z}{4}$.

368. The plane π passes through the point (1, 1, 1) and the line $\frac{x}{1} = \frac{y}{2} = \frac{z}{3}$ is perpendicular to the plane. Find the equations of the projection of the line $\frac{x-1}{1} = \frac{y+1}{2} = \frac{z-3}{-4}$ in the plane π.

369. If $p(x) = 51x^{101} - 2323x^{100} - 45x + 1035$, use Rolle's theorem to prove that at least one root of $p(x) = 0$ lies in the interval $(45^{1/100}, 46)$.

370. Let $A = \begin{bmatrix} a & 0 & 1 \\ 1 & c & b \\ 1 & d & b \end{bmatrix}$, $B = \begin{bmatrix} a & 1 & 1 \\ 0 & d & c \\ f & g & h \end{bmatrix}$, $U = \begin{bmatrix} f \\ g \\ h \end{bmatrix}$ and $V = \begin{bmatrix} a^2 \\ 0 \\ 0 \end{bmatrix}$ where $ab \neq 1$. If there is a vector matrix X such that $AX = U$ has infinitely many solutions then prove that $BX = V$ cannot have a unique solution. If $afd \neq 0$ then prove that $BX = V$ has no solution.

371. It is known that $\sum_{n=1}^{\infty} \frac{1}{n^2} = \frac{\pi^2}{6}$. Hence or otherwise evaluate $\int_0^1 \frac{\log x}{1-x^2}\, dx$.

372. Let $L_1 \equiv bx \cos\phi + ay \sin\phi - ab = 0$ and
$L_2 \equiv ax \sin\phi - by\cos\phi - (a^2 - b^2) \cdot \sin\phi \cdot \cos\phi = 0$
form a family of pairs of straight lines, ϕ being the parameter. Find the equation of the locus of the vertices of the pairs of straight lines.

373. Let $f(x) = ax^4 + bx^3 + cx^2 + dx + \lambda$ and $\frac{f(1)}{1} = \frac{f(2)}{2} = \frac{f(3)}{3}$.
If $\lim_{x \to \infty} \frac{f(x)}{x^4} = 1$ then find $f(i) + f(\bar{i})$ where $i = \sqrt{-1}$.

374. Let ABCD be a cyclic quadrilateral in which $AB = p$, $BC = q$, $CD = r$ and $DA = s$. Show that $\frac{AC}{BD} = \frac{ps + qr}{pq + rs}$.

375. A circle touches the line $2x + 3y + 1 = 0$ at the point $(1, -1)$ and is orthogonal to the circle which has the line segment having end points $(0, -1)$ and $(-2, 3)$ as a diameter. Find the equation of the first circle.

376. ABCD is a square of side 8 cm. An ellipse is drawn touching the sides of the square, the axes being along the diagonals. Find the lengths of the axes if its eccentricity is $\frac{1}{2}$.

377. Evaluate $\int \frac{\sin^{3/2}x + \cos^{3/2}x}{\sqrt{\sin^3 x \cdot \cos^3 x \cdot \cos(x+\alpha)}}\, dx$.

378. Prove that in a ΔABC,
$a\cos^2\frac{A}{2} + b\cos^2\frac{B}{2} + c\cos^2\frac{C}{2} \leq \frac{3}{4}(a+b+c)$.

379. If $a > 0, b > 0$ and the cubic equation $x^3 + ax^2 + bx + 2 = 0$ has all its roots real then prove that $a + b \geq 3(\sqrt[3]{2} + \sqrt[3]{4})$.

380. T is a parallelepiped whose one face has the vertices A, B, C and D. The face just above it has corresponding vertices A′, B′, C′ and D′. T is now compressed to S with the face ABCD remaining the same and A′, B′, C′ and D′ are shifted to A″, B″, C″ and D″ respectively in S The volume of the parallelepiped S is 90% of T. Prove that the locus of A″ is a plane.

381. S is the region bounded by the lines $|y| = \sqrt{3}$ and the curves $x^2 + y^2 = 4$ where $1 \leq |x| \leq 2$. Find the possible values of a for which the point $P(2a-1, 3-2a)$ belongs to the interior of S. Also, find the area of S.

382. Let $A = (-1, 2, 4)$ and $B = (1, 1, 2)$. The point A is shifted towards B by a distance 4 to reach the point A′. Next A′ is shifted parallel to the line $2x = 3y = z$ by a distance 4 so that the point has all the coordinates positive in the new position A''. Find the coordinates of A''.

383. Evaluate $\int_0^{\sqrt{5}} (\log_{x+\sqrt{x^2+4}} 2)^{-1}\, dx$.

384. If $f: [-1, 1] \to R$ and $f'(0) = \lim_{n \to \infty} nf\left(\frac{1}{n}\right)$ and $f(0) = 0$, find the value of $\lim_{n \to \infty} \left\{\frac{2}{\pi}(n+1)\cos^{-1}\left(\frac{1}{n}\right) - n\right\}$.
Given that $0 < \left|\lim_{n \to \infty} \cos^{-1}\left(\frac{1}{n}\right)\right| < \frac{\pi}{2}$.

385. If $f(x) = \int_{x^2/16}^{x^2} \frac{\cos x \cdot \cos\sqrt{\theta}}{1 + \sin^2\sqrt{\theta}}\, d\theta$ then find $f'(\pi)$.

386. At any point P on the parabola $y^2 - 2y - 4x + 5 = 0$, a tangent is drawn which meets the directrix at Q. Find the locus of R which divides QP externally in the ratio 1 : 2.

387. The y-coordinates of three points on the parabola $y^2 = 4x$ are 4, –3 and –1. Prove that the normals to the parabola at these points are concurrent and find the point of concurrency.

388. If $a_i \geq 0;\ i = 1, 2, 3, \ldots, n$ and $\sum_{i=1}^{n} a_i = 2m$ then find the greatest possible value of $\sum_{i=1}^{n} a_i \cdot a_{i+1}$.

389. Let $f(n)$ = number of ways in which $n \in N$ can be expressed as sum of 1s and 2s, the order of 1s and 2s being taken into account.

Find the value of $f(f(5))$.

390. Evaluate $\int \sin x \cdot \operatorname{cosec} 4x\, dx$ where $\cos x > \frac{1}{\sqrt{2}}$.

391. Given a triangle, a similar triangle ABC is drawn where A is a fixed point and B is any point on a given line. Find the locus of C.

392. Prove that the four vertices of a quadrilateral, the combined equation of whose sides is $|x^2 - y^2| - 2(|x + y| + |x - y|) + 4 = 0$, are concyclic. Also, find the equation of the circle whose director circle passes through the vertices of the quadrilateral.

393. If $|\theta| \leq \frac{\pi}{2}$ and $x \in R - \left\{1, -\frac{1}{3}\right\}$ then find the set of values of θ for which $2\sin\theta = \frac{1 - 2x + 5x^2}{3x^2 - 2x - 1}$.

394. Find the equation of the tangent to the curve $y = f(x)$ at the point (1, 2) where $f(x)$ is a function of x such that $|f(x_1) - f(x_2)| < (x_1 - x_2)^2$ for all $x_1, x_2 \in R$.

395. Find the locus of the middle point of the chord of contact of tangents of any point of the hyperbola $4x^2 - 9y^2 = 36$ with respect to the circle $x^2 + y^2 = 9$.

396. Find the equation of the common tangent to the circle $x^2 + y^2 = 16$ and the ellipse $4x^2 + 25y^2 = 100$ cutting off positive intercepts on the axes of reference. Also, find the intercept on the common tangent between the coordinate axes.

397. Find the equation of the curves for which the length of tangent at any point is 1.

398. Let $f(x)$ be a polynomial of the second degree such that

$$\begin{bmatrix} 4a^2 & 4a & 1 \\ 4b^2 & 4b & 1 \\ 4c^2 & 4c & 1 \end{bmatrix} \begin{bmatrix} f(-1) \\ f(1) \\ f(2) \end{bmatrix} = 3 \begin{bmatrix} a(a+1) \\ b(b+1) \\ c(c+1) \end{bmatrix}.$$

Find $f(x)$. Let V be the point on the curve $y = f(x)$ whose ordinate have the greatest value and A be the point of intersect of the curve with the negative side of the x-axis. The chord AB of the curve subtends a right angle at V. Find the area enclosed by the curve and the chord AB.

399. A cricket player played in n matches of a series scoring $\frac{1}{4}(n + 1)(2^{n+1} - n - 2)$ runs in all. Runs scored by him in the rth match was $r \cdot 2^{n+1-r}, 1 \leq r \leq n$. Find the number of matches of the series in which he played and the total runs scored by him in the series.

400. A ray of light coming from A is incident on a line mirror at the point B and after reflection travels along BC. The unit vectors along the rays AB and BC are $\hat{v}$ and $\hat{w}$ respectively. If the unit vector along the internal bisector BD of $\angle ABC$ be $\hat{u}$ then express $\hat{w}$ in terms of $\hat{u}$ and $\hat{v}$.

401. A square circumscribes the circle $|z - 1| = \sqrt{2}$. If one of the vertices represents the complex number $2 + i\sqrt{3}$ then find the complex numbers represented by the other vertices.

402. Evaluate

$$\int_0^{\pi} e^{|\cos x|} \left\{2\sin\left(\frac{1}{2}\cos x\right) + 3\cos\left(\frac{1}{2}\cos x\right)\right\} \sin x\, dx.$$

403. Let f and g be two functions for which

$$f(x - y) = f(x)\, g(y) - f(y) \cdot g(x)$$
$$g(x - y) = g(x)\, g(y) + f(x) \cdot g(y) \text{ for all } x, y \in R.$$

If $f'(0^+)$ exists, prove that $f(x)$ is differentiable at $x = 0$. Also, find $g'(0)$.

404. Of the four concyclic points on the parabola $y^2 = 4x$, three are the feet of normals drawn to the parabola from the point (2, 3). Find the fourth point.

Answers

1. (i) ↔ (*a*), (ii) ↔ (*c*)

2. (i) ↔ (*a*), (ii) ↔ (*c*)

3. 74

4. $\frac{1}{\sqrt{1 - (k - x)^2}}$

5. $(-\infty, -1] \cup [2, +\infty)$

6. (–1, 1), (6, 6)

7. 1

8. $\frac{1}{2}$

9. 0

10. $-\frac{3}{8}\left(1 + \frac{1}{x^2}\right)^{\frac{4}{3}} + c$

11. 0

12. $\pm\frac{1}{\sqrt{2}}(1 + i)$

13. $\frac{1}{2}$

14. $(0, +\infty)$

15. $-x^{\sin^{-1}x} \cdot \log x - \frac{\sqrt{1 - x^2}}{x} \cdot \sin^{-1}x \cdot x^{\sin^{-1}x}$

16. $[-1, 5]$

17. 0

18. $(-\infty, -1) \cup (1, \infty)$

19. $2^{x-1} \cdot \frac{\log 2}{x}$

20. 15 : 1

21. $(-\infty, -1] \cup [1, +\infty)$

22. 0 or 1 or ∞

23. $\left(\frac{-1 - \sqrt{7}}{2}, \frac{-1 + \sqrt{7}}{2}\right)$

24. 1, 2, respectively **25.** $x^2 - 2x - 5$

26. $2.e^{-4}$ **27.** $by = 4x(b - x)$

28. $\vec{0}$ **29.** 1

30. $x - y = 1 \pm \sqrt{6}$ **31.** 0

32. $a_1 + a_2 + a_3 + \ldots + a_k$ **33.** $(1, e]$ **34.** $5\sqrt{3} : 7\pi$

35. $R - (-4, 0)$ **36.** $\frac{7}{100}$ **37.** $\frac{4}{5}$.

38. C **39.** A **40.** D **41.** C

42. A **43.** C **44.** B **45.** C

46. C **47.** A **48.** C **49.** B

50. B, C **51.** A **52.** B **53.** B, C

54. B **55.** A, B, C, D **56.** C

57. A, B, C **58.** A, C **59.** B **60.** A, B, C

61. C **62.** C **63.** A **64.** B

65. A, C **66.** B **67.** A **68.** C

69. B **70.** C **71.** B **72.** ${}^nC_r \cdot 3^r$

73. 126 **74.** $(2\sqrt{3}, 2\sqrt{3} - 1)$ **79.** 20

80. $2x^{3/2} \cdot (x - \sin x)^{3/2} + c$ **82.** e **84.** $20\frac{4}{51}$

85. $\vec{r} = \vec{c} - \frac{\vec{a} \cdot \vec{c}}{\vec{a} \cdot \vec{b}} \vec{b}$ **86.** $\frac{G}{100kV}$

88. interior of the parabola $x^2 = 4y$

91. $2n\pi, 2n\pi + \frac{\pi}{2}$ where $n \in Z$

92. each equal to 0 **95.** $\frac{15}{4}a^2$

96. $\frac{1}{4}\sin n\theta \cdot \operatorname{cosec}^2\frac{\theta}{2} - \frac{n}{2}\cos\left(n + \frac{1}{2}\right)\theta \cdot \operatorname{cosec}\frac{\theta}{2}$

98. $\frac{4}{3}\left(\frac{x-1}{x+2}\right)^{\frac{1}{4}} + c$

99. $x = y, y < 1;\ x = \sqrt{y},\ 1 \le y \le 16;$
$x = \log_2 y, y > 16;$ domain $= R$

104. $m \in \left(-1, \frac{1}{5}\right)$ **105.** $\frac{1}{2}n(n+1)$

106. $(0, \sqrt{6} - 2)$

108. $x = n\pi + (-1)^n \sin^{-1}\left(1 - \frac{t}{2}\right)$,
$y = 2m\pi \pm \cos^{-1} t$
where $n, m \in Z$ and $0 \le t \le 1$

111. $\sqrt{(x-2)(x+1)} + 9\log(\sqrt{x-2} + \sqrt{x+1})$
$+ 6\sqrt{2}\log\frac{\sqrt{3} - \sqrt{x-2} + \sqrt{2(x+1)}}{\sqrt{x-2} + \sqrt{3}}$
$- 2\sqrt{3} \cdot \sqrt{x+1} + c$

112. $\frac{1}{3}\left(n\pi + \frac{\pi}{2} - \alpha - \beta - \gamma\right),\ n \in Z$

115. $h : (h - a)$ **116.** $\frac{1}{2}, \frac{1}{12}$ respectively

118. ellipse if $\lambda > -b^2$, hyperbola if $b^2 < -\lambda < a^2$,
no locus if $\lambda < -a^2$

119. $\left[\sqrt{13} - 2, \frac{9}{4}\right]$

121. $x - 2y - 5 = 0, 7x + y - 50 = 0$ **122.** $\frac{4}{3\sqrt{3}}$

128. 5 **129.** $\frac{1}{\sqrt{2}}\log\frac{4(4 + \sqrt{34})}{3(3 + \sqrt{41})}$

130. $(0, 0), (3, -\sqrt{3})$ and $(2, 0)$; $(0, 0), (0, 2\sqrt{3})$ and $(1, \sqrt{3})$

133. $(-\infty, -3] \cup \left[-\frac{1}{\sqrt{3}}, \frac{1}{2}\right]$ **135.** 1

136. $\frac{1 \pm \sqrt{5}}{2}, \frac{-3 \pm \sqrt{3}i}{2}$

137. $\frac{n\pi}{3} + \frac{\pi}{6}, n\pi + \frac{\pi}{4}$ where $n \in Z$

138. $x^2 + y^2 - 6x - 6y + 5 = 0$, two

140. continuous in $R - \{-1\}$ **141.** $\frac{-\pi^2}{4}$

143. $\frac{1}{7\sqrt{2}}(8\vec{i} + 5\vec{j} + 3\vec{k}),\ \frac{1}{7\sqrt{2}}(4\vec{i} - \vec{j} - 9\vec{k})$

144. $\frac{1}{2}\log(t - \sqrt{1 - t^2}) - \frac{1}{2}\sin^{-1} t + c$ **145.** $\frac{1}{37}$

147. $\frac{1}{2}\{-5 \pm \sqrt{25 + (4n+2)\pi}, n \ge -2$ and $n \in Z$

149. (i) $\vec{x} = \frac{1}{2}(\vec{c} - \vec{a} - \vec{b}),\ \vec{y} = \frac{1}{2}(\vec{a} + \vec{b} + \vec{c})$,
$\vec{z} = \frac{1}{2}(\vec{b} + \vec{c} - \vec{a})$

(ii) $\vec{x} = \vec{z} - \vec{b},\ \vec{y} = \vec{z} + \vec{a}$ where
$\vec{z} = \frac{(\vec{a} + \vec{b}) \times (\vec{c} - \vec{a} \times \vec{b})}{(\vec{a} + \vec{b})^2}$

152. 1 **153.** $n\pi, 2n\pi \pm \cos^{-1}\frac{\pm 1}{2\sqrt{2}}$

154. $f(x) = c_1 e^{(\sqrt{3} - 2)x}, c_2 e^{-(\sqrt{3} + 2)x}$

155. $\frac{1}{4}$ **157.** $3\hat{a} + 4\hat{b} + 2\hat{c}$

158. $5\left(\pi - \sin^{-1}\frac{24}{25}\right) + \frac{24}{5}$

159. $f(x) = -x, x < 0$ and $g(x) = x, x < 0$
$0, x = 0$ $\quad 0, x = 0$
$x, x > 0$ $\quad -x, x > 0$

160. $8\sqrt{3}\pi$ **166.** $\alpha = 0, \beta = -5, \gamma = 0$

168. $(2, -4); 2\sqrt{2}-1$

169. $\frac{3\sqrt{3}}{8}r^2$

172. $2n\pi+\frac{\pi}{2}, n \in Z$

174. $\frac{4}{5}$

175. $3(x^2+y^2)+84x-82y+140=0$,

$\left(x+\frac{24}{11}\right)^2+\left(y+\frac{23}{11}\right)^2=\left(\frac{15}{11}\right)^2$

177. $\frac{1}{\sqrt{622}}(9\vec{i}+10\vec{j}-21\vec{k})$

178. $\left(\frac{3-\sqrt{23}}{2}, \frac{1-\sqrt{7}}{2}\right)\cup\left(\frac{1+\sqrt{7}}{2}, \frac{3+\sqrt{23}}{2}\right)$

179. $\left(\frac{3}{2}, \frac{3}{2}\right)$ and $\left(\frac{1}{2}, \frac{5}{2}\right)$

180. $\frac{\pi-2}{2}a^2$

181. $x=\log_2(a+1)-1, y=4, z=\sin(a+2)$ where

$a \in \left(-1, \frac{\pi}{2}-2\right]$

182. 26

185. $-(1+e^{-4x})^{1/4}+c$

186. $2x^2+6y^2=9$

188. $\frac{1}{3}m(m+1)(3m^2+13m+17)$

191. 13068

194. $1-\left(\frac{2n}{2n+1}\right)^{10}$

199. (i) 1001 (ii) 7051

200. $r^2+\frac{a^2b^2}{p^2}=a^2+b^2$

203. 552

204. $g(x)=1, 0\le x<\frac{\pi}{4}$

$\sqrt{2}\cos\left(x-\frac{\pi}{4}\right), \frac{\pi}{4}\le x\le\pi$

$\cos x, x>\pi$

which is continuous everywhere except at $x=\frac{\pi}{4}$

206. $x^2+c=2xy\{(1+\log x+\log y)^2-2(\log x+\log y)\}$

207. $9(\pi-2)$

209. 4

210. $\frac{1}{n+\log_2(n!)}$

211. -5

213. $-\frac{4}{3}\cot x+\frac{8}{3\sqrt{3}}\tan^{-1}\left(\frac{\tan x}{\sqrt{3}}\right)+\frac{4}{3}\log\left(\cot\frac{x}{2}\right)$

$+\frac{4\sqrt{2}}{3}\tan^{-1}(\sqrt{2}\cos x)+c$

214. $4(a^2x^2+b^2y^2)=(a^2-b^2)^2$

216. $\sqrt{\frac{3}{2}}$ units

220. 22500

221. 30000

222. $\pm\vec{j}$

223. $\lambda\vec{a}+\frac{\vec{a}\times\vec{b}}{\vec{a}^2}$,

where λ is an arbitrary scalar, not unique

225. $-\frac{1}{8}\log\frac{\tan x+3}{3\tan x+1}+c$

226. $25\left(\sin^{-1}\frac{\sqrt{2}}{5}-\sin^{-1}\frac{1}{5\sqrt{2}}\right)+2\left(\sqrt{\frac{23}{2}}-\frac{7}{4}\right)$

228. $xy=c\sec\frac{y}{x}$

231. $\pm(\vec{a}\times\vec{b})\times(\vec{c}\times\vec{d})$; $\lambda=\mp[\vec{b}\ \vec{c}\ \vec{d}]$, $\mu=\pm[\vec{a}\ \vec{c}\ \vec{d}]$

233. $e\sqrt{2-e^2}$

235. 33333

237. $\bigcup_{n\in Z}\left(2n\pi-\frac{\pi}{6}, 2n\pi+\frac{5\pi}{6}\right)$

238. $\frac{23}{70}$

239. $\frac{1}{4}$

240. $\frac{\sqrt{5}}{3}, -\frac{\sqrt{5}}{3}$ respectively

241. (0, 2)

243. $\frac{4170}{4199}$

246. 8

247. $x^3+qx-r=0$

251. $a=\frac{1}{4}, b=\frac{-5}{4}, f(x)=2+\frac{x}{4}(x-5)$

252. $\log_e 2$

253. $7(a_0+a_7x^7+a_{14}x^{14})$

254. $x^2(x-1)$

255. $\frac{x}{1+x^2}+\frac{3}{2}\tan^{-1}x-\frac{1}{2}\log(x+1)+\frac{1}{4}\log(x^2+1)+c$

256. $\frac{\pi}{2}$

257. $1\frac{65}{192}$

258. domain of $f=R-\{0,-2\}$, range of $f=R-\left\{1, \frac{-1}{2}\right\}$,

$\frac{df^{-1}(x)}{dx}=3(1-x)^{-2}$ and its domain $=R-\left\{1, -\frac{1}{2}\right\}$

259. $a=\log_e\frac{2}{3}, b=\frac{2}{3}, c=1$

260. $\sqrt{\frac{2}{3}}\pi-2\sin^{-1}\sqrt{\frac{2}{3}}$

261. $\pi+\frac{1}{2\sqrt{2}}(2-\pi)$

265. $-\sqrt{3}\vec{a}+2\vec{b}$

266. $|[\vec{a}\ \vec{b}\ \vec{c}]|$

267. four points which are $\left(\frac{\pm a^2}{\sqrt{a^2+b^2}}, \frac{\pm b^2}{\sqrt{a^2+b^2}}\right)$

268. $x^2+y^2-2x=0$

269. $2y=\log\{1+(x+2y+1)^2\}+c$

270. $\sqrt{3}x-\sqrt{2}y=\pm\sqrt{5}, \sqrt{3}x+\sqrt{2}y=\pm\sqrt{5}$

271. $\frac{1}{2}(2e+\sqrt{e})\cdot\log_2 e$

272. $\frac{25}{x^2}+\frac{4}{y^2}=4$

273. $C_1: 9(x^2+y^2)+24x+15=0$;

$C_2: x^2+y^2-8x+7=0$.

The common tangents to C, C_1 are

$$y = \pm\frac{1}{\sqrt{3}}(x+2), \quad x+1=0.$$

The common tangents to C, C_2 are

$$y = \pm\frac{1}{\sqrt{3}}(x+2), \quad x=1.$$

The common tangents to C_1, C_2 are

$$y = \pm\frac{1}{\sqrt{3}}(x+2),\ 5x+4 = \pm\sqrt{39}\cdot y.$$

274. $3x+y-2=0, 3x+y-12=0,$
$x-3y=-21, x-3y=-1$

275. $3(x^2+y^2)-6x-8=0$

276. $2y = x+1$ **279.** $\frac{4}{35}$

280. $\alpha = \frac{1}{p^2-3p+3}$, $\beta = \frac{1-p}{p^2-3p+3}$, $\gamma = \frac{(1-p)^2}{p^2-3p+3}$

282. 6 cm, 8 cm, 10 cm **283.** 0.6

284. height $= \frac{5}{4}(\sqrt{3}+3\sqrt{11})$ m, width $= \frac{5}{4}(1+\sqrt{33})$ m

285. $\alpha = \beta = \frac{\pi}{3}$ **286.** $3x^2+4y^2+6x-16y-29=0$

287. $e = \sqrt{\frac{5}{6}}$, latus rectum $= \frac{5}{6}$

290. three **292.** $\frac{10(3^{n-1}-2^n+1)}{6^{n-1}}$

295. $x = \frac{5\pi}{12}, y = \frac{\pi}{6}$ **296.** ± 3

299. 2^{10} **301.** $\left(\frac{3\sqrt{3}}{5}, \frac{4\sqrt{3}}{5}\right)$ **304.** $\frac{1}{2}(3e^2+e^{-2})$

305. 681 **306.** $\frac{40!}{(10!)^4} + \frac{40!}{7!\,9!\,11!\,13!}\times 4!$

307. $-\sqrt{2}, \sqrt{6}$ **308.** $\frac{1\pm\sqrt{3}i}{2}$ **309.** -1

310. $6\cdot 2^{99}-101, 6\cdot 2^{99}$ respectively

313. $\sqrt{\frac{3-\sqrt{5}}{2}}$ **315.** (5, 6), (6, 5)

317. $\left(-\frac{1}{2}, \frac{3}{2}\right), 2\sqrt{2}$

318. $(-\infty, -2)\cup\left[1, \frac{9}{8}\right], 2n\pi \pm \cos^{-1}\frac{2}{3}$ **320.** $\frac{-3}{5\sqrt{2}}$

324. 72×5^4 **328.** $\frac{\pi}{4}$ **330.** $\frac{7}{4}$

332. $e - \left(\frac{1}{2}-\log 2\right)e^2$ **333.** $x^2y = \frac{b^2}{a^2xy}+c$

334. $2x+11y=15$

336. $\sqrt{15}+3\sqrt{7}+16\sin^{-1}\frac{\sqrt{7}+3\sqrt{15}}{16}$ **338.** $9+3\sqrt{10}$

340. $\frac{\pi(e+1)(e^{n+1}-1)}{(1+\pi^2)(e-1)}$ **341.** $\frac{x^2}{a^2}+\frac{y^2}{\left(\frac{ar+bs}{r+s}\right)^2}=1$

342. $\frac{14\pi}{27\sqrt{g}}\cdot 10^5$

343. dom $f = \{x \mid x = \frac{4-p^2}{2p-3}, p\neq\frac{3}{2}$ and p is rational$\}$

range $f = \{y \mid y = \frac{p^2-3p+4}{2p-3}, p\neq\frac{3}{2}$ and p is rational$\}$

346. (6, 5, –2) **347.** 5 **348.** h/k **351.** 2

353. 0 **354.** (2, 1) **355.** 0 **356.** 0.5

357. $\frac{x-1}{5}=\frac{y+1}{10}=\frac{z}{-1}$

358. $\left[-3, -\frac{1+\sqrt{2}}{\sqrt{2}}\right)\cup\left(\frac{1-\sqrt{2}}{\sqrt{2}}, 2\right]$

360. $\vec{x} = \frac{1}{3}(\vec{a}+2\vec{a}\times\vec{b}-\vec{c}\times\vec{a}), \vec{y} = \frac{1}{3}(\vec{a}-\vec{a}\times\vec{b}-\vec{c}\times\vec{a}),$
$\vec{z} = \frac{1}{3}(\vec{a}-\vec{a}\times\vec{b}+2\vec{c}\times\vec{a})$

361. 1

365. $2y = \frac{c^a}{1-a}\cdot x^{1-a} - \frac{c^{-a}}{1+a}\cdot x^{1+a} - \frac{2ac}{1-a^2}$,
where $a = \frac{u}{v}$

367. $\frac{x}{1}=\frac{y}{11}=\frac{z}{-3}$ **368.** $\frac{7x-6}{3}=\frac{7y+9}{6}=\frac{7z-18}{-5}$

371. $-\frac{\pi^2}{8}$ **372.** $\frac{x^2}{a^2}+\frac{y^2}{b^2}=1$

373. –20 **375.** $x^2+y^2-5x-\frac{5}{2}y+\frac{1}{2}=0$

376. major axis $= 16\cdot\sqrt{\frac{2}{7}}$, minor axis $= 8\cdot\sqrt{\frac{6}{7}}$

377. $c-2\left\{\sqrt{\frac{\cot\alpha-\tan x}{\sin\alpha}}+\sqrt{\frac{\cot x-\tan\alpha}{\cos\alpha}}\right\}$

381. $\frac{1}{2}(3-\sqrt{3}) < a < \frac{3}{2}, 2\sqrt{3}+\frac{8}{3}\pi$ **382.** $\left(\frac{71}{21}, \frac{38}{21}, \frac{100}{21}\right)$

383. $\sqrt{5}\log_2(3+\sqrt{5})-\log_2 e$ **384.** $1-\frac{2}{\pi}$

385. $2\pi+\frac{\pi}{12\sqrt{2}}$ **386.** $(y-1)^2\cdot(x+1)+4=0$

387. $\left(\frac{21}{4}, \frac{3}{2}\right)$ **388.** m^2 **389.** 34

390. $\frac{1}{2\sqrt{2}}\log\frac{1+\sqrt{2}\sin x}{\sqrt{\cos 2x}} - \frac{1}{4}\log\tan\left(\frac{\pi}{4}+\frac{x}{2}\right)+c$

391. $4x^2 \cdot \left[\frac{p^2}{r^2}\{x^2 + (y+h)^2\} - h^2\right]$

$$= \left[\frac{p^2 - q^2}{r^2}\{x^2 + (y+h)^2\} + x^2 + y^2 - h^2\right]^2$$

where A = $(0, -h)$ and B is on the x-axis

392. $x^2 + y^2 = 2$

393. $\left[-\frac{\pi}{2}, -\frac{\pi}{10}\right] \cup \left[\frac{3\pi}{10}, \frac{\pi}{2}\right]$

394. $y = 2$

395. $4x^2 - 9y^2 = \frac{4}{9}(x^2 + y^2)^2$

396. $2x + \sqrt{3}y = 4\sqrt{7}, \frac{14}{\sqrt{3}}$

397. $\sqrt{1 - y^2} + \log \frac{1 - \sqrt{1 - y^2}}{|y|} = \pm x + c$

398. $1 - \frac{1}{4}x^2, 41\frac{2}{3}$

399. 7, 494

400. $\hat{w} = \hat{v} - 2(\hat{u} \cdot \hat{v})\hat{u}$

401. $-\sqrt{3}i, (1 - \sqrt{3}) + i, (1 + \sqrt{3}) - i$

402. $\frac{24}{5}\left[e\left\{\cos\frac{1}{2} + \frac{1}{2}\sin\frac{1}{2}\right\} - 1\right]$

403. 0

404. (0, 0)

□

Some New Pattern Problems

Comprehension/Paragraph Type

[In this category of problems the purpose is to assess the capability of the students to understand a *new concept* or a *new situation* by reading a passage and apply themselves to answer related questions on them which are done mainly by their knowledge and working techniques that they have already done in previous classes.]

Example *(with a new concept)*

- The curve $y = f(x)$ passes through the origin O and the tangent to the curve at the origin is the x-axis. The tangent at any point $P(x, y)$ on the curve makes an angle ψ with the x-axis and the arc length OP is s. Then the relation between ψ and s will be called the intrinsic equation of the curve.

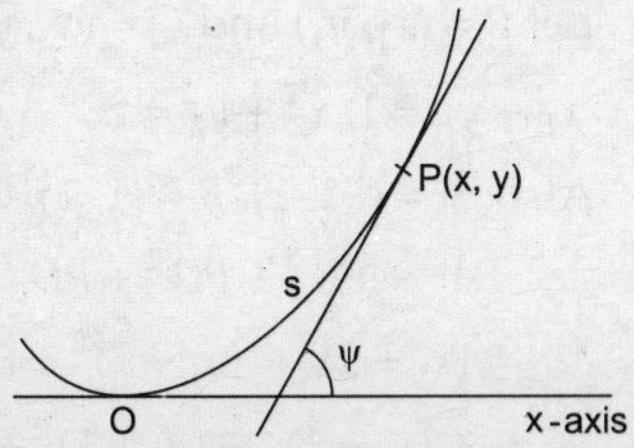

Also, the length l of the arc AB on the curve $y = f(x)$, where $A = (x_1, y_1)$ and $B = (x_2, y_2)$ is given by

$$l = \int_{x = x_1}^{x_2} ds \quad \text{where } (ds)^2 = (dx)^2 + (dy)^2 \text{ and so}$$

$$\frac{ds}{dx} = \sqrt{1 + \left(\frac{dy}{dx}\right)^2}$$

1. The distance between the points (0, 1) and (1, 2) along the curve $y = x^2 + 1$ is

A. $\frac{\sqrt{5}}{2} + \frac{1}{4}\log(2 + \sqrt{5})$ B. $\frac{1}{4}\log(2 + \sqrt{5})$

C. $\sqrt{5} + \frac{1}{2}(2 + \sqrt{5})$ D. $\sqrt{5}$

2. In the parabola $y^2 = 4x$ the ends of the double ordinate through the focus are P and Q. Let O be the vertex. Then the length of the arc POQ of the parabola is

A. $\sqrt{2} + \log(\sqrt{2} + 1)$ B. $2\sqrt{2} + \log(3 + 2\sqrt{2})$

C. $2\log(\sqrt{2} + 1)$ D. $\log(\sqrt{2} + 1)$

3. The intrinsic equation of the curve $x = a(\theta + \sin\theta)$, $y = a(1 - \cos\theta)$ is

A. $s = 4a\tan\psi$ B. $s = a\cos\psi$

C. $s = a\psi$ D. $s = 4a\sin\psi$

Solution

1. Both points are on the curve. So, the required distance is the arc length of the curve $y = x^2 + 1$ from the point (0, 1) to (1, 2).

$\therefore$ the required distance

$$= \int_{x=0}^{1} ds = \int_0^1 \frac{ds}{dx}\,dx = \int_0^1 \sqrt{1 + \left(\frac{dy}{dx}\right)^2}\,dx$$

$$= \int_0^1 \sqrt{1 + (2x)^2}\,dx$$

$$= \left[\frac{1}{2}\left\{\frac{2x}{2}\sqrt{4x^2 + 1} + \frac{1}{2}\log(2x + \sqrt{4x^2 + 1})\right\}\right]_0^1$$

$$= \frac{\sqrt{5}}{2} + \frac{1}{4}\log(2 + \sqrt{5}).$$

So, the correct answer is A.

2. Here $y^2 = 4x \Rightarrow \frac{dy}{dx} = \frac{2}{y}$.

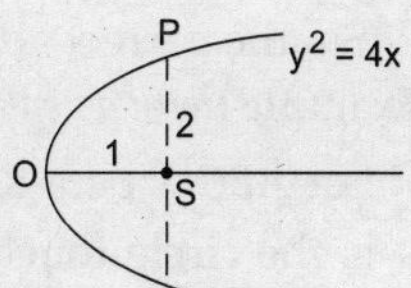

So, arc $POQ = 2 \times$ arc $OP = 2\int_0^1 ds$

$$\therefore \quad \text{arc } POQ = 2\int_0^1 \sqrt{1 + \left(\frac{dy}{dx}\right)^2}\,dx$$

$$= 2\int_0^1 \sqrt{1 + \frac{4}{y^2}}\,dx$$

$$= 2\int_0^1 \sqrt{1 + \frac{4}{4x}}\,dx = 2\int_0^1 \sqrt{\frac{x + 1}{x}}\,dx$$

$$= 2\int_0^{\pi/4} \sqrt{\frac{\sec^2\theta}{\tan^2\theta}} \cdot 2\tan\theta \cdot \sec^2\theta\,d\theta$$

(putting $x = \tan^2\theta$)

$$= 4\int_0^{\pi/4} \sec^3\theta\,d\theta.$$

$$\text{Now, } I = \int_0^{\pi/4} \sec^3\theta \, d\theta = (\sec\theta \cdot \tan\theta)_0^{\pi/4}$$

$$- \int_0^{\pi/4} \tan\theta \cdot \sec\theta \cdot \tan\theta \cdot d\theta$$

$$= \sqrt{2} - \int_0^{\pi/4} (\sec^3\theta - \sec\theta) \, d\theta$$

$$= \sqrt{2} - I + [\log(\sec\theta + \tan\theta)]_0^{\pi/4}$$

$$\therefore \quad 2I = \sqrt{2} + \log(\sqrt{2} + 1).$$

So, arc $POQ = 4I = 2\sqrt{2} + \log(3 + 2\sqrt{2})$.

So, the correct answer is B.

3. At (0, 0), $\theta = 0$ and $\dfrac{dy}{dx} = \dfrac{dy/d\theta}{dx/d\theta} = \dfrac{a\sin\theta}{a(1+\cos\theta)} = \tan\dfrac{\theta}{2}$.

So, at the origin $\dfrac{dy}{dx} = 0$ and the tangent at the origin is the x-axis.

As $\dfrac{dy}{dx} = \tan\psi = \tan\dfrac{\theta}{2}$, we have $\psi = \dfrac{\theta}{2}$.

$$\text{Now } s = \int_0^{\theta} ds = \int_0^{\theta} \sqrt{1 + \left(\frac{dy}{dx}\right)^2} \, dx$$

$$= \int_0^{\theta} \sqrt{1 + \tan^2\frac{\theta}{2}} \cdot \frac{dx}{d\theta} d\theta$$

$$= \int_0^{\theta} \sec\frac{\theta}{2} \cdot a(1+\cos\theta) \, d\theta$$

$$= \int_0^{\theta} 2a \sec\frac{\theta}{2} \cdot \cos^2\frac{\theta}{2} \, d\theta = 2a \int_0^{\theta} \cos\frac{\theta}{2} \, d\theta$$

$$= 2a \cdot \left[2\sin\frac{\theta}{2}\right]_0^{\theta} = 4a\sin\frac{\theta}{2} = 4a\sin\psi.$$

So, the correct answer is D.

Example *(with a new situation)*

- Let $ABCD$ be a square of sides of length 2 units. C_1 is the circle touching all the sides of the square and C_2 is the circle circumscribing the square. l is a line through A.

4. If P is a point on C_1 and Q is a point on C_2 then $\dfrac{PA^2 + PB^2 + PC^2 + PD^2}{QA^2 + QB^2 + QC^2 + QD^2}$ equals

A. $\dfrac{3}{4}$ B. $\dfrac{5}{4}$

C. 1 D. $\dfrac{1}{2}$

5. A circle touches the line l and the circle C_1 externally such that both the circles are on the same side of the line. The locus of the centre of the circle is

A. part of a straight line B. parabola

C. ellipse D. hyperbola

6. A line m through A is drawn parallel to BD. A point S moves such that its distances from BD and the vertex A are equal. If the locus of S cuts the line m at T_2 and T_3, and AC at T_1 then ar($\Delta\, T_1T_2T_3$) in square units is

A. $\dfrac{1}{2}$ B. $\dfrac{2}{3}$

C. 1 D. 2

Solution

4. The equation of C_1 is $x^2 + y^2 = 1$ and that of C_2 is $x^2 + y^2 = 2$.

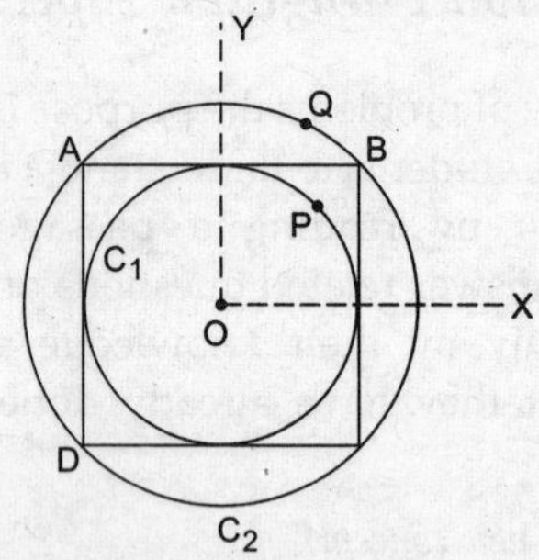

Let $P = (x_1, y_1)$ and $Q = (x_2, y_2)$. Then,

$x_1^2 + y_1^2 = 1,\ x_2^2 + y_2^2 = 2$.

Also, $A = (-1, 1),\ B = (1, 1),\ C = (1, -1),\ D = (-1, -1)$.

$$\therefore \quad PA^2 + PB^2 + PC^2 + PD^2$$

$$= (x_1+1)^2 + (y_1-1)^2 + (x_1-1)^2 + (y_1-1)^2$$

$$+ (x_1-1)^2 + (y_1+1)^2 + (x_1+1)^2 + (y_1+1)^2$$

$$= 2\{(x_1+1)^2 + (x_1-1)^2\} + 2\{(y_1-1)^2 + (y_1+1)^2\}$$

$$= 4(x_1^2+1) + 4(y_1^2+1) = 4(x_1^2+y_1^2) + 8 = 4 + 8 = 12.$$

Similarly, $QA^2 + QB^2 + QC^2 + QD^2$

$$= 4(x_2^2 + y_2^2) + 8 = 4 \times 2 + 8 = 16.$$

$\therefore \quad$ value $= \dfrac{12}{16} = \dfrac{3}{4}$. So, the correct answer is A.

5. The circle C always touches the line l and the fixed circle C_1, whose radius is 1. Take a line $p \parallel l$ at a distance 1 from l not cutting C, whose centre is G. Clearly, the distance of G from the centre O of C_1 is equal to the distance of G from the line p. Thus, the centre G of the circle C moves keeping equal distances from the fixed point O and the fixed line p. Hence, the locus is a parabola whose focus is O and the directrix is the line p. So, the correct answer is B.

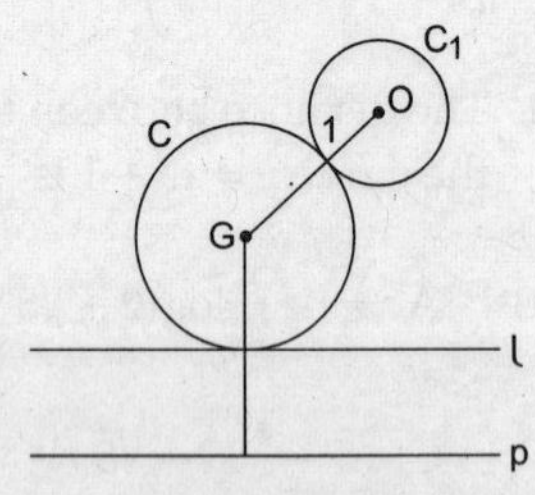

6. $AT_1 = T_1O = \dfrac{1}{2}AO = \dfrac{1}{4}AC = \dfrac{1}{4} \cdot 2\sqrt{2} = \dfrac{1}{\sqrt{2}}$.

Clearly, the locus of S is a parabola whose focus is A and BD is the directrix. So, OA is the line of axis and T_2T_3 is the latus rectum.

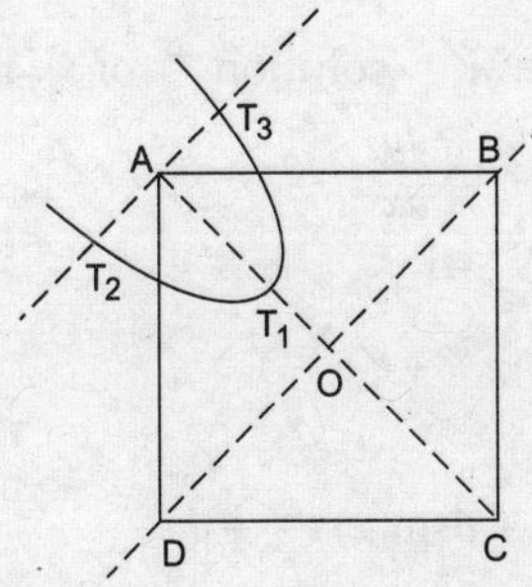

$\therefore \quad T_2T_3 = 4 \times AT_1 = 4 \times \frac{1}{\sqrt{2}} = 2\sqrt{2}$

$\therefore \quad \text{ar}(\Delta T_1T_2T_3) = \frac{1}{2} \cdot T_2T_3 \cdot AT_1 = \frac{1}{2} \cdot 2\sqrt{2} \cdot \frac{1}{\sqrt{2}} = 1.$

So, the correct answer is C.

Problems for Exercise

- The envelope of a family of curves $f(x, y, c) = 0$ is a curve whose equation is obtained by eliminating the parameter c from

$$f(x, y, c) = 0 \text{ and } \frac{\partial f}{\partial c} = 0,$$

where $\frac{\partial f}{\partial c}$ is the differential coefficient of f with respect to c, treating x and y as constants.

Moreover, the envelope of the family of normals to a curve is known as the envolute of the curve.

. The envelope of the family of straight lines whose sum of intercepts on the axes is 4 is

A. $(x + y)^2 - 8(x - y) + 16 = 0$

B. $(x - y)^2 - 8(x + y) + 16 = 0$

C. $(x - y)^2 = 4(x + y)$

D. $(x + y)^2 = 4(x - y)$

3. The envelope of the family of tangents to the curve $y^2 = x$ is

A. $x + y^2 = 0$ B. $x^2 = y$

C. $x^2 + y = 0$ D. $y^2 - x = 0$

9. The evolute of the curve $x^2 = 4y$ is

A. $27x^2 = 4(y - 2)^3$ B. $4y^2 = 27(x - 2)^3$

C. $27x^2 = 4y^3$ D. none of these

- The arc rate of turning of tangent to a curve at a point is called the curvature of the curve at that point. The angle that the tangent to the curve at the point (x, y) makes with the positive direction of the x-axis is denoted by ψ. Thus, curvature $= \frac{d\psi}{ds}$, where s is the arc distance of the point (x, y) from a fixed point on the curve.

Also, for a curve, $(ds)^2 = (dx)^2 + (dy)^2$ and the reciprocal of the curvature at a point is known as the radius of curvature.

10. The radius of curvature of the curve $y = f(x)$ at (x, y) is equal to

A. $\dfrac{\frac{d^2y}{dx^2}}{\left\{1 + \left(\frac{dy}{dx}\right)^2\right\}^{3/2}}$ B. $\dfrac{\sqrt{1 + \left(\frac{dy}{dx}\right)^2}}{\frac{d^2y}{dx^2}}$

C. $\dfrac{\left\{1 + \left(\frac{dy}{dx}\right)^2\right\}^{3/2}}{\frac{d^2y}{dx^2}}$ D. $\dfrac{\frac{d^2y}{dx^2}}{\sqrt{1 + \left(\frac{dy}{dx}\right)^2}}$

11. The radius of curvature of the curve $y^2 = 4x$ at the point (1, 2) is

A. $4\sqrt{2}$ B. $2\sqrt{2}$

C. $\frac{1}{\sqrt{2}}$ D. $\sqrt{2}$

12. The radius of curvature of $9x^2 + 4y^2 = 36$ is the greatest at the point

A. (0, 2) B. (2, 0)

C. (0, 3) D. $\left(\sqrt{3}, \frac{3}{2}\right)$

- Suppose we define definite integral using the formula $\int_a^b f(x)\,dx = \frac{b-a}{2}\{f(a) + f(b)\}$. For more accurate result, we have $\int_a^b f(x)\,dx = \frac{b-a}{4}\{f(a) + f(b) + 2f(c)\}$ when $c = \frac{a+b}{2}$. Also, let

$$F(c) = \frac{c-a}{2}\{f(a) + f(c)\} + \frac{b-c}{2}\{f(b) + f(c)\} \text{ when } c \in (a, b).$$

13. $\int_0^{\pi/2} \sin x\,dx$ equals

A. $\frac{\pi}{8}(1 + \sqrt{2})$ B. $\frac{\pi}{4}(1 + \sqrt{2})$

C. $\frac{\pi}{8\sqrt{2}}$ D. $\frac{\pi}{4\sqrt{2}}$

14. $\int_a^b f(x)\,dx$ is equal to

A. $F\left(\frac{a-b}{2}\right)$ B. $F\left(\frac{a+b}{2}\right)$

C. $F\left(\frac{a-b}{4}\right)$ D. $F\left(\frac{a+b}{2}\right)$

15. If $f''(x) < 0\ \forall x \in (a, b)$ and $(c, f(c))$ is a point lying on the curve $y = f(x)$, where $a < c < b$ and for that value of c, $F(c)$ has a maximum then $f'(c)$ equals

A. $\dfrac{f(b)-f(a)}{b-a}$ B. $\dfrac{2}{b-a}\{f(b)-f(a)\}$

C. $\dfrac{2f(b)-f(a)}{2b-a}$ D. 0

- Two one-parameter families of curves are called orthogonal trajectories of one another if each member of a family cuts all the members of the other family at right angles.

It is found that the orthogonal trajectories of the family of curves $f\left(x, y, \dfrac{dy}{dx}\right) = 0$ is the family of curves $f\left(x, y, -\dfrac{dx}{dy}\right) = 0.$

16. The family of parabolas with a common vertex at the origin whose foci are on the x-axis and the directrices are parallel to the y-axis can have the equation (a being a parameter)

A. $y^2 = 3ax$ B. $x^2 = 4ay$

C. $y^2 = 4a(x+2)$ D. $(y-1)^2 = 4ax$

17. The orthogonal trajectory of the family of parabolas $y^2 = 4ax$ is

A. $x^2 + y^2 = c^2$ B. $x^2 + 2y^2 = c^2$

C. $2x^2 + y^2 = c^2$ D. $y^2 - x^2 = c^2$

18. The curve passing through the point (1, 2) that cuts each member of the family of parabolas $y^2 = 4ax$ orthogonally is

A. $2x^2 + y^2 = 6$ B. $x^2 + y^2 = 5$

C. $x^2 + 2y^2 = 9$ D. $y^2 - x^2 = 3$

- The general solution of a differential equation of the form

$$\frac{d^ny}{dx^n} + a_1\frac{d^{n-1}y}{dx^{n-1}} + a_2\frac{d^{n-2}y}{dx^{n-2}} + \ldots + a_ny = 0$$

(a_is being constants)

is given by the following rules.

(i) If the roots of the corresponding auxiliary equation

$$D^n + a_1D^{n-1} + a_2D^{n-2} + \ldots + a_n = 0$$

in D has unequal real roots $\alpha_1, \alpha_2, \alpha_3, \ldots, \alpha_n$ then the general solution will be

$y = c_1e^{\alpha_1 x} + c_2e^{\alpha_2 x} + \ldots + c_ne^{\alpha_n x}$, where c_is are arbitrary constants.

(ii) If the roots of the corresponding auxiliary equation has two equal roots (say, $\alpha_1 = \alpha_2$) and the rest are unequal real roots then the general solution will be

$$y = (c_1x + c_2)\,e^{\alpha_1 x} + c_3e^{\alpha_3 x} + \ldots + c_ne^{\alpha_n x}.$$

(iii) If two roots of the corresponding auxiliary equation are $\alpha_1 = \alpha + i\beta$, $\alpha_2 = \alpha - i\beta$ and the rest are real and unequal then the general solution will be

$$y = (A\cos \beta x + B\sin \beta x)\,e^{\alpha x} + c_3e^{\alpha_3 x} + \ldots + c_ne^{\alpha_n x}.$$

19. The general solution of the equation $\dfrac{d^3y}{dx^3} - 7\dfrac{d^2y}{dx^2} + 16\dfrac{dy}{dx} - 12y = 0$ is

A. $c_1e^{2x} + c_2e^{-2x} + c_3e^{-3x}$

B. $(c_1x + c_2)\,e^{2x} + c_3e^{3x}$

C. $(c_1x + c_2)\,e^{-3x} + c_3e^{2x}$

D. $(A\cos x + B\sin x)\,e^{2x} + c_3e^{3x}$

20. $y = (c_1\cos x + c_2\sin x)\,e^{-x} + c_3e^{x}$ is the general solution of the equation

A. $\dfrac{d^3y}{dx^3} - \dfrac{d^2y}{dx^2} + 2y = 0$ B. $\dfrac{d^3y}{dx^3} - y = 0$

C. $\dfrac{d^3y}{dx^3} + \dfrac{d^2y}{dx^2} - 2y = 0$ D. $\dfrac{d^2y}{dx^2} - y = 0$

21. The general solution of the equation $\dfrac{d^3y}{dx^3} + y = 0$ is

A. $ce^{-x} + e^{x/2}\left(a\cos\dfrac{\sqrt{3}}{2}x + b\sin\dfrac{\sqrt{3}}{2}x\right)$

B. $c_1e^{\sqrt{3}/2\,x} + c_2e^{x/2} + c_3e^{-x}$

C. $c_1e^{-x} + c_2e^{-\omega x} + c_3e^{-\omega^2 x}$, where $\omega = \dfrac{-1+\sqrt{3}i}{2}$

D. none of these

- Let $A = \begin{bmatrix} 1 & 0 & 0 \\ 2 & 1 & 0 \\ 3 & 2 & 1 \end{bmatrix}$ and U_1, U_2, U_3 are three column matrices such that

$$AU_1 = \begin{bmatrix} 1 \\ 0 \\ 0 \end{bmatrix},\ AU_2 = \begin{bmatrix} 2 \\ 3 \\ 0 \end{bmatrix} \text{ and } AU_3 = \begin{bmatrix} 2 \\ 3 \\ 1 \end{bmatrix}.$$

U is a 3×3 matrix whose first column is the column in U_1, the second column is the column in U_2 and the third column is the column in U_3. On the basis of this, answer the following questions.

22. The value of $|U|$ is

A. 3 B. –3

C. $\dfrac{3}{2}$ D. 2

23. The sum of the elements of U^{-1} is

A. 0 B. –1

C. 1 D. 3

24. The sum of the elements of the product $[3\ 2\ 0]\cdot U \cdot \begin{bmatrix} 3 \\ 2 \\ 0 \end{bmatrix}$ is

A. $\dfrac{5}{2}$ B. $\dfrac{3}{2}$

C. 4 D. 5

- r is called the rank of the matrix A if there exists at least one nonzero minor of order r and every minor of order $r+1$ of the matrix equals zero. Again if A is a square matrix and I is the unit matrix of the same order then the equation $|A - xI| = 0$ is called the characteristic equation and the roots of the characteristic equation are called eigenvalues of the matrix A.

Let $A = \begin{bmatrix} 8 & -6 & 2 \\ -6 & 7 & -4 \\ 2 & -4 & 3 \end{bmatrix}$

25. The rank of A is
A. 3 B. 2
C. 1 D. none of these

26. The characteristic equation of A is
A. $x^3 - 18x^2 + 45x = 0$ B. $x^3 - 6x^2 + 9x - 4 = 0$
C. $x^3 + 18x^2 - 45x = 0$ D. $x^3 + 6x^2 - 9x + 4 = 0$

27. One of the eigenvalues of A is
A. 1 B. 2
C. 3 D. 5

- It is known that

$$3205 = 3 \times 10^3 + 2 \times 10^2 + 0 \times 10^1 + 5 \times 10^0.$$

This method of representing a number is in the denary scale (scale of 10) of notation and it is a denary number. Similarly, in a scale of r,

$$3205 = 3 \times r^3 + 2 \times r^2 + 0 \times r^1 + 5 \times r^0.$$

More generally,

$$a_n a_{n-1} a_{n-2} \ldots a_0 = a_n r^n + a_{n-1} r^{n-1} + \ldots + a_1 r + a_0,$$

where $a_0, a_1, \ldots, a_n$ are digits whose values range from 0 to $r - 1$. Thus, in the scale of 7, the digits will be 0, 1, 2, 3, 4, 5, 6.

Conversion of a denary number to the scale of 5

5	268	
5	53	... 3
5	10	... 3
	2	... 0

$\therefore$ 268 = 2033 in the scale of 5

Addition of numbers in the scale of 5

34 $\quad 34 = 3 \times 5 + 4$
+ 22 $\quad 22 = 2 \times 5 + 2$
111 $\quad$ sum $= 5 \times 5 + 6 = 1 \times 5^2 + 1 \times 5 + 1$

Multiplication of numbers in the scale of 5

23 $\quad 3 \times 4 = 12 = 2 \times 5 + 2$
× 4 $\quad 2 \times 4 + 2 = 10 = 2 \times 5 + 0$
202

On the basis of the above discussions, answer the following.

28. The denary number 43125 in the scale of 6 will be represented by
A. 353135 B. 531353
C. 515313 D. 55453

29. The sum of the numbers 2053 and 412 in the scale of seven is
A. 2465 B. 3020
C. 3004 D. none of these

30. The product of the numbers 5623 and 6 in the scale of eight is
A. 41672 B. 33738
C. 42562 D. 45262

- Z is the set of integers and N is the set of natural numbers. If $b \in Z, c \in Z$ and $n \in N$ are such that when b and c are divided by n, we get the same remainder then b and c are said to be congruent with respect to modulo n. Symbolically it is written as $b \equiv c \pmod n$

31. Which of the following is true?
A. $4 \equiv 7 \pmod 5$ B. $118 \equiv 18 \pmod 5$
C. $110 \equiv 93 \pmod{12}$ D. $63 \equiv 9 \pmod{11}$

32. Let $a \equiv b \pmod n$, $a' \equiv b' \pmod n$ and $d, m \in N - \{1\}$. Then which of the following need not be true?
A. $a + a' \equiv b + b' \pmod n$ B. $aa' \equiv bb' \pmod n$
C. $a^m \equiv b^m \pmod n$ D. $\frac{a}{d} \equiv \frac{b}{d} \pmod n$

33. If $5x \equiv 2 \pmod 7$ then
A. $x \not\equiv 6 \pmod 9$ B. $x \equiv 6 \pmod 7$
C. $x \equiv 7 \pmod 6$ D. $x \equiv 2 \pmod 7$

- The quadratic equation $ax^2 + bx + c = 0$ will have two roots equal to zero if $b = 0, c = 0$. Also, roots of $cx^2 + bx + a = 0$ are reciprocal to the roots of $ax^2 + bx + c = 0$.

A straight line $y = mx + c$ will be an asymptote to the curve $f(x, y) = 0$ if the line touches the curve at infinity. As a result, $f(x, mx + c) = 0$ should have two infinite roots.

34. The value of λ for which the equation $\lambda^2 - 4\lambda + 3 + (\lambda^2 + \lambda - 2)x + 6x^2 - 5x^3 = 0$ will have two roots equal to zero is
A. 3 B. −2
C. 0 D. 1

35. The value of (λ, μ) for which $(\lambda - \mu)x^3 - (\lambda - 2)x^2 - 3x + 7 = 0$ will have two infinite roots is
A. (0, 0) B. (2, 2)
C. (1, 1) D. (2, 0)

36. The equations of asymptotes to the curve $x^2 + 4xy + 3y^2 + 4x - 3y + 1 = 0$ are
A. $2x + 2y = 7$ and $x + 3y + 15 = 0$
B. $x + y - 7 = 0$ and $x + 3y + 15 = 0$
C. $2(x + y) - 7 = 0$ and $2(x + 3y) + 15 = 0$
D. $x + y = 7$ and $2(x + 3y) + 15 = 0$

- According to Fermat's theorem, $n^p - n = M(p)$ = a multiple of p when p is a prime number and $n \in N$. It is also known that the highest index of power of a prime number p which divides $n\,!$ is equal to $\left[\frac{n}{p}\right] + \left[\frac{n}{p^2}\right] + \left[\frac{n}{p^3}\right] + \ldots$, where $[x]$ = greatest integer less than or equal to x.

37. If $n(>2)$ is a prime number which is coprime with a, b and $a + b$ then $a^{n-2} \cdot b - a^{n-3} \cdot b^2 + a^{n-4} \cdot b^3 - \ldots + ab^{n-2}$ is equal to

A. $M(n) - 1$ B. $M(n) + 1$
C. $M(n)$ D. none of these

38. $17^{22} - 1$ is a multiple of

A. 16 B. 44
C. 46 D. 27

39. The highest power of 7 which divides 1000 ! is

A. 7^{164} B. 7^{162}
C. 7^{167} D. 7^{142}

- If $\alpha_1, \alpha_2, \alpha_3, \ldots, \alpha_n$ are in an AP whose common difference is λ then we have

$$\sum_{i=1}^{n} \sin \alpha_i = \frac{\sin \frac{n\lambda}{2}}{\sin \frac{\lambda}{2}} \sin \frac{\alpha_1 + \alpha_n}{2}.$$

A quadrant $A_0 A_{90}$ of a unit circle is divided into 90 equal parts by inserting the points $A_1, A_2, A_3, \ldots, A_{89}$ on the arc of the circle. Let $\cot 0.5° = a$.

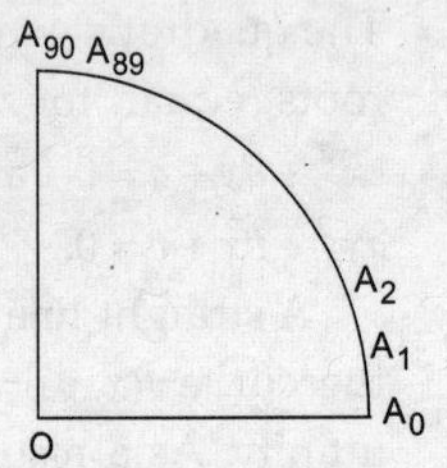

40. $\sum_{i=1}^{90} \sin \angle A_0 OA_i$ is equal to

A. $\frac{a+1}{\sqrt{2}}$ B. $\frac{a-1}{2\sqrt{2}}$
C. $\frac{a+1}{2a}$ D. $\frac{a+1}{2}$

41. $\sum_{i=1}^{90} \text{ar}(\Delta A_0OA_i)$ is equal to

A. $\frac{a+1}{4}$ B. $\frac{a+1}{2\sqrt{2}}$
C. $\frac{a-1}{2\sqrt{2}}$ D. $\frac{a+1}{4a}$

42. $\sum_{i=1}^{90} \sqrt{1 + \cos 2\angle A_0OA_i}$ is equal to

A. $\frac{a+1}{2\sqrt{2}}$ B. $\frac{a-1}{\sqrt{2}}$
C. $\frac{a-1}{\sqrt{2a}}$ D. $\frac{a-1}{2}$

- It is known that $\frac{x^2}{a^2} - \frac{y^2}{b^2} = -1$ is the equation of the conjugate hyperbola of the hyperbola $\frac{x^2}{a^2} - \frac{y^2}{b^2} = 1$. Also, the pair of lines $\frac{x^2}{a^2} - \frac{y^2}{b^2} = 0$ are the asymptotes of the hyperbola $\frac{x^2}{a^2} - \frac{y^2}{b^2} = 1$.

Let the equation of a hyperbola be $x^2 - 3y^2 - 2x - 8 = 0$.

43. The equation of its conjugate hyperbola is

A. $x^2 - 3y^2 - 2x + 8 = 0$ B. $3x^2 - y^2 - 2y - 8 = 0$
C. $x^2 - 3y^2 - 2x + 10 = 0$ D. $x^2 - 3y^2 = -8$

44. The equation of the asymptotes of the hyperbola is

A. $x^2 - 3y^2 - 2x = 0$ B. $x^2 - 3y^2 = 0$
C. $3x^2 - y^2 - 2y = 0$ D. $x^2 - 3y^2 - 2x + 1 = 0$

45. If $(a - 1, 2)$ is a point interior to the hyperbola but exterior to its conjugate hyperbola then the number of positive integral values of a is

A. 3 B. 2
C. 1 D. 4

- A second-degree curve inscribing the ΔABC, where the sides AB, BC and CA have the equations $L_1 = 0$, $L_2 = 0$ and $L_3 = 0$ respectively, has the equation $L_1L_2 + \lambda L_2L_3 + \mu L_3L_1 = 0$, where λ and μ are parameters. Let the equations of AB, BC and CA be $y = 2x$, $y + 2x = 0$ and $x = 1$ respectively.

46. The circumcentre of the ΔABC is

A. $\left(\frac{5}{4}, \frac{5}{8}\right)$ B. $\left(\frac{5}{2}, 0\right)$
C. $\left(0, \frac{5}{4}\right)$ D. $\left(\frac{5}{4}, -\frac{5}{8}\right)$

47. The parabola circumscribing ΔABC and passing through the point (4, 4) has the focus

A. (0, 1) B. (4, 0)
C. (1, 0) D. (0, 4)

48. The parabola circumscribing ΔABC and passing through the point (4, 4) has the latus rectum

A. 4 B. 1
C. 16 D. 8

- Two diameters AB and CD of an ellipse are called conjugate diameters if chords parallel to one of them are bisected by the other diameter. It is found that the difference of eccentric angles of A and C is $\frac{\pi}{2}$. Let the equation of the ellipse be $9x^2 + 16y^2 = 144$.

49. If $A = \left(2, \frac{3\sqrt{3}}{2}\right)$ then coordinates of C are

A. $\left(-2, \frac{3\sqrt{3}}{2}\right)$ B. $\left(\sqrt{2} - \sqrt{6}, \frac{3(\sqrt{3}+1)}{2\sqrt{2}}\right)$

C. $\left(-2\sqrt{3}, \frac{3}{2}\right)$ D. $(0, 3)$

50. If the equation of the diameter AB is $x = y$ then the equation of the conjugate diameter CD will be

A. $9x + 16y = 0$ B. $x + y = 0$

C. $16x + 9y = 0$ D. $9x + 16y = 7$

51. The area of the quadrilateral formed by the tangents at the ends of conjugate diameters AB and CD, where $A = \left(\frac{12}{5}, \frac{12}{5}\right)$, is

A. 96 B. 24

C. 12 D. 48

- A ray of light emancipating from the point source $P(1, -3, 2)$ and travelling parallel to the line $\frac{x-2}{1} = \frac{y}{3} = \frac{z+1}{2}$ is incident on the plane $x + y - 3z = 0$ at the point Q. After reflecting from the plane the ray travels along the line QR. It is also known that the incident ray, reflected ray and the normal to the plane at the point of incidence are in the same plane.

52. The point Q is

A. $(3, 15, 6)$ B. $(3, 6, 3)$

C. $(-3, -6, -3)$ D. $(-3, -15, -6)$

53. The equations of the line QR are

A. $\frac{x-12}{15} = \frac{y-22}{37} = \frac{z-4}{10}$

B. $\frac{x-3}{3} = \frac{y-15}{7} = \frac{z-6}{2}$

C. $\frac{x-3}{15} = \frac{y-6}{37} = \frac{z-3}{10}$

D. $\frac{x+3}{3} = \frac{y+6}{7} = \frac{z+3}{2}$

54. The equation of the plane PQR is

A. $5x + 2y - z + 3 = 0$ B. $11x - 5y + 2z = 30$

C. $5x - y - z = 6$ D. $x - y + z = 6$

- The sphere whose centre $= (\alpha, \beta, \gamma)$ and radius $= a$, has the equation $(x-\alpha)^2 + (y-\beta)^2 + (z-\gamma)^2 = a^2$. Again, the intersection of a sphere by a plane is a circle.

55. The distance of the centre of the sphere $x^2 + y^2 + z^2 - 2x - 4y = 0$ from the origin is

A. 5 B. $\sqrt{5}$

C. $2\sqrt{5}$ D. $\frac{5}{2}$

56. The radius of the circle of intersection of the sphere $x^2 + y^2 + z^2 = 9$ by the plane $3x + 4y + 5z = 5$ is

A. $\sqrt{\frac{17}{2}}$ B. 3

C. $\sqrt{34}$ D. $\frac{1}{\sqrt{2}}$

57. The area of the circle on the sphere $x^2 + y^2 + z^2 = 25$ whose centre is $(1, 1, 1)$ is

A. 13π B. 3π

C. 22π D. 25π

- There are n urns each containing $n + 1$ balls such that the ith urn contains i white balls and $(n + 1 - i)$ red balls. Let u_i be the event of selecting ith urn and w denote the event of getting a white ball.

58. If $P(u_i) \propto i$, where $i = 1, 2, 3, \ldots, n$ then $\lim_{n \to \infty} P(w)$ is equal to

A. $\frac{2}{3}$ B. $\frac{3}{4}$

C. $\frac{1}{4}$ D. 1

59. If $P(u_i) = c$, a constant then $P\left(\frac{u_n}{w}\right)$ is equal to

A. $\frac{1}{n+1}$ B. $\frac{2}{n+1}$

C. $\frac{n}{n+1}$ D. $\frac{1}{2}$

60. If n is even and E denotes the event of choosing even-numbered urn and u_is are equiprobable then the value of $P\left(\frac{w}{E}\right)$ is

A. $\frac{1}{n+1}$ B. $\frac{n}{n+1}$

C. $\frac{n+2}{2(n+1)}$ D. $\frac{n+2}{2n+1}$

- If any point of the closed interval $[\alpha, \beta]$ is a sample point and any point of the closed interval $[a, b] \subseteq [\alpha, \beta]$ be a favourable point for the event E then the probability of the event E is defined by $P(E) = \frac{\text{length of } [a, b]}{\text{length of } [\alpha, \beta]}$.

61. The set of values of $p \in R$ for which $x^2 + px + \frac{1}{4}(p+2) \geq 0$ for all $x \in R$ is

A. $(-2, 1)$ B. $(2, +\infty)$

C. $[-1, 2]$ D. $[1, 2]$

62. The set of values of $p \in R$ for which the equation $x^2 + px + \frac{1}{4}(p+2) = 0$ will have real roots is

A. $[2, +\infty)$ B. $(2, +\infty)$
C. $(-\infty, 2)$ D. $R - (-1, 2)$

63. If p is chosen at random from the interval [0, 6] then the probability that the roots of the equation $x^2 + px + \frac{1}{4}(p + 2) = 0$ will be real is

A. $\frac{3}{5}$ B. $\frac{1}{2}$
C. $\frac{3}{4}$ D. $\frac{2}{3}$

• In a series of n independent trials for an event of constant probability p, the most probable number r of successes is given by $(n + 1)p - 1 < r < (n + 1)p$. Hence, the most probable number of successes is the integral part of $(n + 1)p$. But if $(n + 1)p$ is an integer, the chance of r successes is equal to that of $r + 1$ successes. Both r and $r + 1$ are most probable numbers of successes.

A bag contains 2 white balls and 1 black ball. A ball is drawn at random and returned to the bag.

64. The experiment is done 10 times. The probability that a white ball is drawn exactly 5 times is

A. $\frac{10!}{5!\,5!}\left(\frac{2}{3}\right)^5$ B. $\frac{10!}{5!\,5!}\left(\frac{1}{3}\right)^5$
C. $\frac{10!}{(5!)^2}\left(\frac{2}{9}\right)^5$ D. $\frac{10!}{5!}\left(\frac{2}{9}\right)^5$

65. The experiment is done 5 times. The probability that a white ball is drawn most of the time is

A. $\frac{8}{81}$ B. $9 \cdot \left(\frac{2}{3}\right)^6$
C. $\left(\frac{2}{3}\right)^6$ D. $\frac{8}{27}$

66. The experiment is done 10 times. The most probable number of successes is

A. 7 B. 6
C. 8 D. 5

• Let A, B, C be three sets of complex numbers as given below.

$A = \{z : I_m(z) \geq 1\}$, $B = \{z : |z - 2 - i| = 3\}$,
$C = \{z : \text{Re}((1 - i)z) = \sqrt{2}\}$

67. The number of elements in the set $A \cap B \cap C$ is

A. 0 B. 1
C. 2 D. ∞

68. Let z be any point in $A \cap B \cap C$. Then, $|z + 1 - i|^2 + |z - 5 - i|^2$ lies between

A. 25 and 29 B. 30 and 34
C. 35 and 39 D. 40 and 44

69. Let z be any point in $A \cap B \cap C$ and let ω be any point satisfying $|\omega - 2 - i| < 3$. Then, $|z| - |\omega| + 3$ lies between

A. –6 and 3 B. –3 and 6
C. –6 and 6 D. –3 and 9

• Consider the function $f : R \to R$ defined by $f(x) = \frac{x^2 - ax + 1}{x^2 + ax + 1}$, $a \in (0, 2)$.

70. Which of the following is true?

A. $(2 + a)^2 \cdot f''(1) + (2 - a)^2 \cdot f''(-1) = 0$
B. $(2 - a)^2 \cdot f''(1) - (2 + a)^2 f''(-1) = 0$
C. $f'(1) f'(-1) = (2 - a)^2$
D. $f'(1) f'(-1) = -(2 + a)^2$

71. Which of the following is true?

A. $f(x)$ is decreasing in (–1, 1) and has a local minimum at $x = 1$
B. $f(x)$ is increasing in (–1, 1) and has a local maximum at $x = 1$
C. $f(x)$ is increasing in (–1, 1) but has neither a local maximum nor a local minimum at $x = 1$
D. $f(x)$ is decreasing in (–1, 1) but has neither a local maximum nor a local minimum at $x = 1$

72. Let $g(x) = \int_0^{e^x} \frac{f'(t)}{1 + t^2}\, dt$.

Which of the following is true?

A. $g'(x)$ is positive on $(-\infty, 0)$ and negative on $(0, \infty)$
B. $g'(x)$ is negative on $(-\infty, 0)$ and positive on $(0, \infty)$
C. $g'(x)$ is changes sign on both $(-\infty, 0)$ and $(0, \infty)$
D. $g'(x)$ does not change sign on $(-\infty, \infty)$

• Consider the lines $L_1 : \frac{x + 1}{3} = \frac{y + 2}{1} = \frac{z + 1}{2}$

$L_2 : \frac{x - 2}{1} = \frac{y + 2}{2} = \frac{z - 3}{3}$

73. The unit vector perpendicular to both L_1 and L_2 is

A. $\frac{1}{\sqrt{99}}(-\hat{i} + 7\hat{j} + 7\hat{k})$ B. $\frac{1}{5\sqrt{3}}(-\hat{i} - 7\hat{j} + 5\hat{k})$
C. $\frac{1}{5\sqrt{3}}(-\hat{i} + 7\hat{j} + 5\hat{k})$ D. $\frac{1}{\sqrt{99}}(7\hat{i} - 7\hat{j} - \hat{k})$

74. The shortest distance between L_1 and L_2 is

A. 0 B. $\frac{17}{\sqrt{3}}$
C. $\frac{41}{5\sqrt{3}}$ D. $\frac{17}{5\sqrt{3}}$

75. The distance of the point (1, 1, 1) from the plane passing through the point (–1, –2, –1) and whose normal is perpendicular to both the lines L_1 and L_2 is

A. $\frac{2}{\sqrt{75}}$ B. $\frac{7}{\sqrt{75}}$

C. $\frac{13}{\sqrt{75}}$ D. $\frac{23}{\sqrt{75}}$

- Consider the functions defined implicitly by the equation $y^3 - 3y + x = 0$ on various intervals in the real line.

  If $x \in (-\infty, -2) \cup (2, \infty)$, the equation implicitly defines a unique real valued differentiable function $y = f(x)$.

  If $x \in (-2, 2)$ the equation implicitly defines a unique real valued differentiable function $y = g(x)$ satisfying $g(0) = 0$.

76. If $f(-10\sqrt{2}) = 2\sqrt{2}$ then $f''(-10\sqrt{2}) =$

A. $\frac{4\sqrt{2}}{7^3 \cdot 3^2}$ B. $-\frac{4\sqrt{2}}{7^3 \cdot 3^2}$

C. $\frac{4\sqrt{2}}{7^3 \cdot 3}$ D. $-\frac{4\sqrt{2}}{7^3 \cdot 3}$

77. The area of the region bounded by the curve $y = f(x)$, x-axis, and the lines $x = a$ and $x = b$, where $-\infty < a < b < -2$, is

A. $\int_a^b \frac{x}{3\{(f(x))^2 - 1\}} dx + b\,f(b) - a\,f(a)$

B. $-\int_a^b \frac{x}{3\{(f(x))^2 - 1\}} dx + b\,f(b) - a\,f(a)$

C. $\int_a^b \frac{x}{3\{(f(x))^2 - 1\}} dx - b\,f(b) + a\,f(a)$

D. $-\int_a^b \frac{x}{3\{(f(x))^2 - 1\}} dx - b\,f(b) + a\,f(a)$

78. $\int_{-1}^{1} g'(x)dx =$

A. $2g(-1)$ B. 0

C. $-2g(1)$ D. $2g(1)$

- A circle C of radius 1 is inscribed in an equilateral triangle PQR. The points of contact of C with the sides PQ, QR, RP are D, E, F respectively. The line PQ is given by the equation $\sqrt{3}x + y - 6 = 0$ and the point D is $\left(\frac{3\sqrt{3}}{2}, \frac{3}{2}\right)$. Further, it is given that the origin and the centre of C are on the same side of the line PQ.

79. The equation of the circle C is

A. $(x - 2\sqrt{3})^2 + (y - 1)^2 = 1$

B. $(x - 2\sqrt{3})^2 + (y + \frac{1}{2})^2 = 1$

C. $(x - \sqrt{3})^2 + (y + 1)^2 = 1$

D. $(x - \sqrt{3})^2 + (y - 1)^2 = 1$

80. Points E and F are given by

A. $\left(\frac{\sqrt{3}}{2}, \frac{3}{2}\right), (\sqrt{3}, 0)$ B. $\left(\frac{\sqrt{3}}{2}, \frac{1}{2}\right), (\sqrt{3}, 0)$

C. $\left(\frac{\sqrt{3}}{2}, \frac{3}{2}\right), \left(\frac{\sqrt{3}}{2}, \frac{1}{2}\right)$ D. $\left(\frac{3}{2}, \frac{\sqrt{3}}{2}\right), \left(\frac{\sqrt{3}}{2}, \frac{1}{2}\right)$

81. Equations of the side QR, RP are

A. $y = \frac{2}{\sqrt{3}}x + 1,\ y = -\frac{2}{\sqrt{3}}x - 1$

B. $y = \frac{x}{\sqrt{3}},\ y = 0$

C. $y = \frac{\sqrt{3}}{2}x + 1,\ y = -\frac{\sqrt{3}}{2}x - 1$

D. $y = \sqrt{3}x,\ y = 0$

- Let $\mathcal{A}$ be the set of all 3×3 symmetric matrices all of whose elements are either 0 or 1. Five of the elements are 1 and 4 are 0.

82. The number of matrices in $\mathcal{A}$ is

A. 12 B. 6

C. 9 D. 3

83. The number of matrices A in $\mathcal{A}$ for which the system of linear equations $A\begin{bmatrix} x \\ y \\ z \end{bmatrix} = \begin{bmatrix} 1 \\ 0 \\ 0 \end{bmatrix}$ has a unique solution, is

A. less than 4

B. at least 4 but less than 7

C. at leat 7 but less than 10

D. at least 10

84. The number of matrices A in $\mathcal{A}$ for which the system of linear equations $A\begin{bmatrix} x \\ y \\ z \end{bmatrix} = \begin{bmatrix} 1 \\ 0 \\ 0 \end{bmatrix}$ is inconsistent, is

A. 0 B. more than 2

C. 2 D. 1

- A fair die is tossed repeatedly until a six is obtained. Let X denote the number of tosses required.

85. The probability that $X = 3$ equals

A. $\frac{25}{216}$ B. $\frac{25}{36}$

C. $\frac{5}{36}$ D. $\frac{125}{216}$

86. The probability that $X \geq 3$ equals

A. $\frac{125}{216}$ B. $\frac{25}{216}$

C. $\frac{5}{36}$ D. $\frac{25}{36}$

87. The conditional probability that $X \geq 6$ given $X > 3$ equals

A. $\frac{125}{216}$ B. $\frac{25}{216}$

C. $\frac{5}{36}$ D. $\frac{25}{36}$

- The circle $x^2+y^2-8x=0$ and hyperbola $\frac{x^2}{9}-\frac{y^2}{4}=1$ intersect at the points A and B.

88. The equation of a common tangent with positive slope to the circle as well as to the hyperbola is

A. $2x-\sqrt{5}y-20=0$ B. $2x-\sqrt{5}y+4=0$

C. $3x-4y+8=0$ D. $4x-3y+4=0$

89. The equation of the circle with AB as its diameter is

A. $x^2+y^2-12x+24=0$ B. $x^2+y^2+12x+24=0$

C. $x^2+y^2+24x-12=0$ D. $x^2+y^2-24x-12=0$

- Consider the polynomial $f(x)=1+2x+3x^2+4x^3$. Let s be the sum of all distinct real roots of $f(x)=0$ and let $|s|=t$.

90. The real number s lies in the interval

A. $\left(-\frac{1}{4},0\right)$ B. $\left(-11,-\frac{3}{4}\right)$

C. $\left(-\frac{3}{4},-\frac{1}{2}\right)$ D. $\left(0,\frac{1}{4}\right)$

91. The area bounded by the curve $y=f(x)$ and the lines $x=0$, $y=0$ and $x=t$, lies in the interval

A. $\left(\frac{3}{4},3\right)$ B. $\left(\frac{21}{64},\frac{11}{16}\right)$

C. $(9, 10)$ D. $\left(0,\frac{21}{64}\right)$

92. The function $f'(x)$ is

A. increasing in $\left(-t,-\frac{1}{4}\right)$ and decreasing in $\left(-\frac{1}{4},t\right)$

B. decreasing in $\left(-t,-\frac{1}{4}\right)$ and increasing in $\left(-\frac{1}{4},t\right)$

C. increasing in $(-t, t)$

D. decreasing in $(-t, t)$

- Tangents are drawn from the point $P(3,4)$ to the ellipse $\frac{x^2}{9}+\frac{y^2}{4}=1$ touching the ellipse at the points A and B.

93. The coordinates of A and B are

A. $(3, 0)$ and $(0, 2)$

B. $\left(-\frac{8}{5},\frac{2\sqrt{161}}{15}\right)$ and $\left(-\frac{9}{5},\frac{8}{5}\right)$

C. $\left(-\frac{8}{5},\frac{2\sqrt{161}}{15}\right)$ and $(0, 2)$

D. $(3, 0)$ and $\left(-\frac{9}{5},\frac{8}{5}\right)$

94. The orthocenter of the triangle PAB is

A. $\left(5,\frac{8}{7}\right)$ B. $\left(\frac{7}{5},\frac{25}{8}\right)$

C. $\left(\frac{11}{5},\frac{8}{5}\right)$ D. $\left(\frac{8}{25},\frac{7}{5}\right)$

95. The equation of the locus of the point whose distances from the point P and the line AB are equal, is

A. $9x^2+y^2-6xy-54x-62y+241=0$

B. $x^2+9y^2+6xy-54x+62y-241=0$

C. $9x^2+9y^2-6xy-54x-62y-241=0$

D. $x^2+y^2-2xy+27x+31y-120=0$

Matching and Fill in/Completion

Problems for Exercise

Match the following (more than one may match with the same).

96. (A) $\int_0^{\pi/2}(\sin x)^{\cos x}\cdot\{\cos x\cot x-\log(\sin x)^{\sin x}\}dx$ (p) 1

(B) Area bounded by $-4y^2=x$ and $x-1=-5y^2$ (q) 0

(C) Cosine of the angle of intersection of curves $y=3^{x-1}\log x$ and $y=x^x-1$ is (r) $6\log 2$

(D) $\lim\limits_{x\to\infty}\frac{\left(\int_0^x e^{x^2}dx\right)^2}{\int_0^x e^{2x^2}dx}$ is equal to (s) $\frac{4}{3}$

97. Normals are drawn at points P, Q and R lying on the parabola $y^2=4x$ and they intersect at $(3, 0)$. Then,

(A) Area of ΔPQR (p) 2

(B) Circumradius of ΔPQR (q) $\frac{5}{2}$

(C) Centroid of ΔPQR (r) $\left(\frac{5}{2},0\right)$

(D) Circumcentre of ΔPQR (s) $\left(\frac{2}{3},0\right)$

98. (A) Two rays in the first quadrant are $x+y=|a|$ and $ax-y=1$ which intersect each other if $a\in(a_0,\infty)$. The value of a_0 is (p) 2

(B) The point (α,β,γ) lies on the plane $x+y+z=2$. If $\vec{a}=\alpha\vec{i}+\beta\vec{j}+\gamma\vec{k}$ and $\vec{k}\times(\vec{k}\times\vec{a})=0$ then γ is (q) $\frac{4}{3}$

(C) $\left|\int_0^1(1-y^2)\,dy\right|+\left|\int_1^0(y^2-1)\,dy\right|$ equals (r) $\left|\int_0^1\sqrt{1-x}\,dx\right|+\left|\int_{-1}^0\sqrt{1+x}\,dx\right|$

(D) If in a ΔABC, $\sin A\sin B\sin C+\cos A\cos B=1$ then the value of $\sin C$ is (s) 1

99. (A) If $\sum_{n=1}^{\infty}\tan^{-1}\frac{1}{2n^2}=t$ then $\tan t$ equals (p) 0

(B) If in the ΔABC the sides a, b, c are in AP and $\cos\theta_1 = \frac{a}{b+c}$, $\cos\theta_2 = \frac{b}{a+c}$ and $\cos\theta_3 = \frac{c}{a+b}$ then $\tan^2\frac{\theta_1}{2} + \tan^2\frac{\theta_3}{2}$ equals (q) 1

(C) A line is perpendicular to $x + 2y + 2z = 0$ and passes through (0, 1, 0). The perpendicular distance of the line from the origin is (r) $\frac{\sqrt{5}}{3}$

(D) $\int_0^1 e^{x^2} dx + \int_1^e (-1 + \sqrt{\log x})\, dx$ has the value (s) $\frac{2}{3}$

Complete the following statements

100. If the roots of the equation $x^2 - 10cx - 11d = 0$ are a, b and those of $x^2 - 10ax - 11b = 0$ are c, d then the positive value of $a + b + c + d$ is (a, b, c and d being distinct numbers)

101. If $a_n = \frac{3}{4} - \left(\frac{3}{4}\right)^2 + \left(\frac{3}{4}\right)^3 - \ldots + (-1)^{n-1} \cdot \left(\frac{3}{4}\right)^n$ and $b_n = 1 - a_n$ then the smallest natural number n_0, such that $b_n > a_n \forall n > n_0$, is

102. The value of $5050 \times \frac{\int_0^1 (1 - x^{50})^{100}}{\int_0^1 (1 - x^{50})^{101}\, dx}$ is

103. If $f(x)$ is a twice differentiable function such that $f(a) = 0$, $f(b) = 2$, $f(c) = -1$, $f(d) = 2$ and $f(e) = 0$, where $a < b < c < d < e$, then the minimum number of zeros of $g(x) = \{f'(x)\}^2 + f''(x) \cdot f(x)$ in the interval $[a, e]$ is

Assertion–Reason Type Questions

- [Each question contains STATEMENT-1 (Assertion) and STATEMENT-2 (Reason). Each question has 4 choices out of which only one is correct.]

104. Suppose four distinct positive numbers a_1, a_2, a_3, a_4 are in GP. Let $b_1 = a_1$, $b_2 = b_1 + a_2$, $b_3 = b_2 + a_3$ and $b_4 = b_3 + a_4$.

STATEMENT-1: The numbers b_1, b_2, b_3, b_4 are neither in AP nor in GP.

and

STATEMENT-2: The numbers b_1, b_2, b_3, b_4 are in HP.

A. Statement-1 is True, Statement-2 is True; Statement-2 is a correct explanation for Statement-1

B. Statement-1 is True, Statement-2 is True; Statement-2 is *Not* a correct explanation for Statement-1

C. Statement-1 is True, Statement-2 is False

D. Statement-1 is False, Statement-2 is True

Solution

Here $b_1 = a_1$, $b_2 = a_1 + a_2 = a_1(1 + r)$, $b_3 = a_1(1 + r) + a_1r^2 = a_1(1 + r + r^2)$, $b_4 = a_1(1 + r + r^2) + ar^3 = a_1(1 + r + r^2 + r^3)$, r being the common ratio of the GP.

Clearly b_1, b_2, b_3, b_4 are neither in AP nor in GP nor in HP.

$\therefore$ STATEMENT-1 is True but STATEMENT-2 is False. So, C is correct.

105. STATEMENT-1: The curve $y = -\frac{x^2}{2} + x + 1$ is symmetric with respect to the line $x = 1$.

because

STATEMENT-2: A parabola is symmetric about its axis.

A. Statement-1 is True, Statement-2 is True; Statement-2 is a correct explanation for Statement-1

B. Statement-1 is True, Statement-2 is True; Statement-2 is *Not* a correct explanation for Statement-1

C. Statement-1 is True, Statement-2 is False

D. Statement-1 is False, Statement-2 is True

Solution

Here $y = -\frac{x^2}{2} + x + 1$ or $y = -\frac{1}{2}(x^2 - 2x + 1) + \frac{3}{2}$

or $(x - 1)^2 = 2\left(\frac{3}{2} - y\right)$. Putting $x - 1 = Y$, $\frac{3}{2} - y = X$

the equation becomes $Y^2 = 2X$ which is a parabola whose axis is $Y = 0$, i.e., $x = 1$.

We know that a parabola is symmetrical about its axis, i.e., STATEMENT-2 is correct. So, the STATEMENT-1 is also true. Hence, A is correct.

Problems for Exercise

106. Consider the planes $3x - 6y - 2z = 15$ and $2x + y - 2z = 5$.

STATEMENT-1: The parametric equations of the line of intersection of the given planes are $x = 3 + 14t$, $y = 1 + 2t$, $z = 15t$.

because

STATEMENT-2: The vector $14\vec{i} + 2\vec{j} + 15\vec{k}$ is parallel to the line of intersection of given planes.

A. Statement-1 is True, Statement-2 is True; Statement-2 is a correct explanation for Statement-1

B. Statement-1 is True, Statement-2 is True; Statement-2 is *Not* a correct explanation for Statement-1

C. Statement-1 is True, Statement-2 is False

D. Statement-1 is False, Statement-2 is True

107. Let the vector $\vec{PQ}, \vec{QR}, \vec{RS}, \vec{ST}, \vec{TU}$ and $\vec{UP}$ represent the sides of a regular hexagon.

STATEMENT-1: $\vec{PQ} \times (\vec{RS} + \vec{ST}) \neq \vec{0}$.

<u>because</u>

STATEMENT-2: $\vec{PQ} \times \vec{RS} = \vec{0}$ and $\vec{PQ} \times \vec{ST} \neq \vec{0}$.

A. Statement-1 is True, Statement-2 is True; Statement-2 is a correct explanation for Statement-1

B. Statement-1 is True, Statement-2 is True; Statement-2 is *Not* a correct explanation for Statement-1

C. Statement-1 is True, Statement-2 is False

D. Statement-1 is False, Statement-2 is True

108. Tangents are drawn from the point (17, 7) to the circle $x^2 + y^2 = 169$.

STATEMENT-1: The tangents are mutually perpendicular.

<u>because</u>

STATEMENT-2: The locus of the points from which mutually perpendicular tangents can be drawn to the given circle is $x^2 + y^2 = 338$.

A. Statement-1 is True, Statement-2 is True; Statement-2 is a correct explanation for Statement-1

B. Statement-1 is True, Statement-2 is True; Statement-2 is *Not* a correct explanation for Statement-1

C. Statement-1 is True, Statement-2 is False

D. Statement-1 is False, Statement-2 is True

109. Lines $L_1 \equiv y - x = 0$ and $L_2 \equiv 2x + y = 0$ intersect the line $L_3 : y + 2 = 0$ at the points P and Q respectively. The bisector of the acute angle between L_1 and L_2 intersects L_3 at R.

STATEMENT-1: The ratio $PR : RQ = 2\sqrt{2} : \sqrt{5}$.

<u>because</u>

STATEMENT-2: In any triangle, bisector of an angle divides the triangle into two similar triangles.

A. Statement-1 is True, Statement-2 is True; Statement-2 is a correct explanation for Statement-1

B. Statement-1 is True, Statement-2 is True; Statement-2 is *Not* a correct explanation for Statement-1

C. Statement-1 is True, Statement-2 is False

D. Statement-1 is False, Statement-2 is True

110. The curve C is the locus of a point whose sum of distances from the points $S(\sqrt{3}, 0)$ and $S'(-\sqrt{3}, 0)$ is 4.

STATEMENT-1: The curve C cuts off intercept $2\sqrt{3}$ from the line $2y - 1 = 0$.

<u>because</u>

STATEMENT-2: The equation of the curve C is $\frac{x^2}{4} + 2y^2 = \frac{5}{4}$.

A. Statement-1 is True, Statement-2 is True; Statement-2 is a correct explanation for Statement-1

B. Statement-1 is True, Statement-2 is True; Statement-2 is *Not* a correct explanation for Statement-1

C. Statement-1 is True, Statement-2 is False

D. Statement-1 is False, Statement-2 is True

111. Let $F(x)$ be an indefinite integral of $\sin^2 x$.

STATEMENT-1: The function $F(x)$ satisfies $F(x + \pi) = F(x)$ for all real x.

<u>because</u>

STATEMENT-2: $\sin^2(x + \pi) = \sin^2 x$ for all real x.

A. Statement-1 is True, Statement-2 is True; Statement-2 is a correct explanation for Statement-1

B. Statement-1 is True, Statement-2 is True; Statement-2 is *Not* a correct explanation for Statement-1

C. Statement-1 is True, Statement-2 is False

D. Statement-1 is False, Statement-2 is True

112. Let $f(x) = 2 + \cos x$ for all real x.

STATEMENT-1: For each real t, there exists a point c in $[t, t + \pi]$ such that $f'(c) = 0$.

<u>because</u>

STATEMENT-2: $f(t) = f(t + 2\pi)$ for each real t.

A. Statement-1 is True, Statement-2 is True; Statement-2 is a correct explanation for Statement-1

B. Statement-1 is True, Statement-2 is True; Statement-2 is *Not* a correct explanation for Statement-1

C. Statement-1 is True, Statement-2 is False

D. Statement-1 is False, Statement-2 is True

113. STATEMENT-1: $1 \cdot 3 \cdot 5 \cdot \ldots \cdot (2n - 1) > n^n, n \in N$.

<u>because</u>

STATEMENT-2: The sum of the first n natural numbers is equal to n^2.

A. Statement-1 is True, Statement-2 is True; Statement-2 is a correct explanation for Statement-1

B. Statement-1 is True, Statement-2 is True; Statement-2 is *Not* a correct explanation for Statement-1

C. Statement-1 is True, Statement-2 is False

D. Statement-1 is False, Statement-2 is True

114. Let $(1 + x)^{36} = a_0 + a_1 x + a_2 x^2 + \ldots + a_{36} x^{36}$.

STATEMENT-1: $a_0 + a_3 + a_6 + \ldots + a_{36} = \frac{2}{3}(2^{35} + 1)$

<u>*because*</u>

STATEMENT-2: $a_0 + a_1 + a_2 + \dots + a_{36} = 2^{36}$ and
$a_0 + a_2 + a_4 + \dots + a_{36} = 2^{35}$

A. Statement-1 is True, Statement-2 is True; Statement-2 is a correct explanation for Statement-1

B. Statement-1 is True, Statement-2 is True; Statement-2 is *Not* a correct explanation for Statement-1

C. Statement-1 is True, Statement-2 is False

D. Statement-1 is False, Statement-2 is True

115. Let $z_1 = r_1 e^{i\theta_1}$ and $z_2 = r_2 e^{i\theta_2}$, where $r_1 > 1, r_2 > 1$.
STATEMENT-1: $\left|1 - \bar{z}_1 z_2\right| > \left|z_1 - z_2\right|$.
<u>*because*</u>
STATEMENT-2: $\left|z_1 - z_2\right| < \left|z_1\right| + \left|z_2\right|$.

A. Statement-1 is True, Statement-2 is True; Statement-2 is a correct explanation for Statement-1

B. Statement-1 is True, Statement-2 is True; Statement-2 is *Not* a correct explanation for Statement-1

C. Statement-1 is True, Statement-2 is False

D. Statement-1 is False, Statement-2 is True

116. Given in ΔABC, $a : b : c = \cos A : \cos B : \cos C$.
STATEMENT-1: ΔABC is equilateral.
<u>*because*</u>
STATEMENT-2: $\cos A = \dfrac{b^2 + c^2 - a^2}{2bc}$,
$\cos B = \dfrac{c^2 + a^2 - b^2}{2ca}$, $\cos C = \dfrac{a^2 + b^2 - c^2}{2ab}$.

A. Statement-1 is True, Statement-2 is True; Statement-2 is a correct explanation for Statement-1

B. Statement-1 is True, Statement-2 is True; Statement-2 is *Not* a correct explanation for Statement-1

C. Statement-1 is True, Statement-2 is False

D. Statement-1 is False, Statement-2 is True

117. Let $H_1, H_2, H_3, \dots, H_n$ be mutually exclusive and exhaustive events with $P(H_i) > 0$; $i = 1, 2, 3, \dots, n$. Let E be any other event with $0 < P(E) < 1$.
STATEMENT-1: $P(H_i / E) > P(E/H_i) \cdot P(H_i)$ for $i = 1, 2, 3, \dots, n$.
<u>*because*</u>
STATEMENT-2: $\sum_{i=1}^{n} P(H_i) = 1$.

A. Statement-1 is True, Statement-2 is True; Statement-2 is a correct explanation for Statement-1

B. Statement-1 is True, Statement-2 is True; Statement-2 is *Not* a correct explanation for Statement-1

C. Statement-1 is True, Statement-2 is False

D. Statement-1 is False, Statement-2 is True

118. Let A and B are two independent events.
STATEMENT-1: If $P(A) = 0.3$ and $P(A \cup \overline{B}) = 0.8$ then $P(B)$ is $\frac{2}{7}$.
<u>*because*</u>
STATEMENT-2: $P(\overline{E}) = 1 - P(E)$, where E is any event.

A. Statement-1 is True, Statement-2 is True; Statement-2 is a correct explanation for Statement-1

B. Statement-1 is True, Statement-2 is True; Statement-2 is *Not* a correct explanation for Statement-1

C. Statement-1 is True, Statement-2 is False

D. Statement-1 is False, Statement-2 is True

119. The equations of two straight lines are
$$\frac{x-1}{2} = \frac{y+3}{1} = \frac{z-2}{-3} \text{ and } \frac{x-2}{1} = \frac{y-1}{-3} = \frac{z+3}{2}.$$
STATEMENT-1: The given lines are coplanar.
<u>*because*</u>
STATEMENT-2: The equations
$$2r - s = 1$$
$$r + 3s = 4$$
$$3r + 2s = 5 \text{ are consistent.}$$

A. Statement-1 is True, Statement-2 is True; Statement-2 is a correct explanation for Statement-1

B. Statement-1 is True, Statement-2 is True; Statement-2 is *Not* a correct explanation for Statement-1

C. Statement-1 is True, Statement-2 is False

D. Statement-1 is False, Statement-2 is True

120. Let $0 < \alpha < \beta < \frac{\pi}{2}$.
STATEMENT-1: $\dfrac{\tan \beta}{\tan \alpha} > \dfrac{\alpha}{\beta}$.
<u>*because*</u>
STATEMENT-2: If $f''(x) > 0$, $x_1 > x_2$ then $f(x_1) > f(x_2)$.

A. Statement-1 is True, Statement-2 is True; Statement-2 is a correct explanation for Statement-1

B. Statement-1 is True, Statement-2 is True; Statement-2 is *Not* a correct explanation for Statement-1

C. Statement-1 is True, Statement-2 is False

D. Statement-1 is False, Statement-2 is True

121. Let $I_n = \int_0^{\pi/4} \tan^n x\, dx$, where $n \in N$.

STATEMENT-1: $\int_0^{\pi/4} \tan^4 x\, dx = \frac{3\pi - 8}{12}$.

because

STATEMENT-2: $I_n + I_{n-2} = \frac{1}{n-1}$.

A. Statement-1 is True, Statement-2 is True; Statement-2 is a correct explanation for Statement-1

B. Statement-1 is True, Statement-2 is True; Statement-2 is *Not* a correct explanation for Statement-1

C. Statement-1 is True, Statement-2 is False

D. Statement-1 is False, Statement-2 is True

122. STATEMENT-1: The cubic equation $4x^3 - 15x^2 + 14x - 5 = 0$ has a root in the interval (2, 3).

because

STATEMENT-2: If $f(x) = 0$ is a polynomial equation which has two real roots α, β $(\alpha < \beta)$ then $f'(x) = 0$ will have a root γ such that $\alpha < \gamma < \beta$.

A. Statement-1 is True, Statement-2 is True; Statement-2 is a correct explanation for Statement-1

B. Statement-1 is True, Statement-2 is True; Statement-2 is *Not* a correct explanation for Statement-1

C. Statement-1 is True, Statement-2 is False

D. Statement-1 is False, Statement-2 is True

123. Consider $L_1 : 2x + 3y + p - 3 = 0$

$L_2 : 2x + 3y + p + 3 = 0,$

where p is a real number and

$C : x^2 + y^2 + 6x - 10y + 30 = 0.$

STATEMENT-1: If the line L_1 is a chord of the circle C then the line L_2 is not always a diameter of the circle C.

and

STATEMENT-2: If the line L_1 is a diameter of the circle C then the line L_2 is not a chord of the circle C.

A. Statement-1 is True, Statement-2 is True; Statement-2 is a correct explanation for Statement-1

B. Statement-1 is True, Statement-2 is True; Statement-2 is *Not* a correct explanation for Statement-1

C. Statement-1 is True, Statement-2 is False

D. Statement-1 is False, Statement-2 is True

124. Let a, b, c, p, q be real numbers. Suppose α, β are the roots of the equation $x^2 + 2px + q = 0$ and $\alpha, \frac{1}{\beta}$ are the roots of the equation $ax^2 + 2bx + c = 0$, where $\beta^2 \notin \{-1, 0, 1\}$.

STATEMENT-1: $(p^2 - q)(b^2 - ac) \geq 0$.

and

STATEMENT-2: $b \neq pa$ or $c \neq qa$.

A. Statement-1 is True, Statement-2 is True; Statement-2 is a correct explanation for Statement-1

B. Statement-1 is True, Statement-2 is True; Statement-2 is *Not* a correct explanation for Statement-1

C. Statement-1 is True, Statement-2 is False

D. Statement-1 is False, Statement-2 is True

125. Let a solution $y = y(x)$ of the differential equation

$$x\sqrt{x^2 - 1}\, dy - y\sqrt{y^2 - 1}\, dx = 0$$

satisfy $y(2) = \frac{2}{\sqrt{3}}$.

STATEMENT-1: $y(x) = \sec\left(\sec^{-1} x - \frac{\pi}{6}\right)$.

and

STATEMENT-2: $y(x)$ is given by $\frac{1}{y} = \frac{2\sqrt{3}}{x} - \sqrt{1 - \frac{1}{x^2}}$.

A. Statement-1 is True, Statement-2 is True; Statement-2 is a correct explanation for Statement-1

B. Statement-1 is True, Statement-2 is True; Statement-2 is *Not* a correct explanation for Statement-1

C. Statement-1 is True, Statement-2 is False

D. Statement-1 is False, Statement-2 is True

126. Consider three planes $P_1 : x - y + z = 1$

$P_2 : x + y - z = -1$

$P_3 : x - 3y + 3z = 2$

Let L_1, L_2, L_3 be the lines of intersection of the planes P_2 and P_3, P_3 and P_1, and P_1 and P_2 respectively.

STATEMENT-1: At least two of the lines L_1, L_2 and L_3 are non-parallel.

and

STATEMENT-2: The three planes do not have a common point.

A. Statement-1 is True, Statement-2 is True; Statement-2 is a correct explanation for Statement-1

B. Statement-1 is True, Statement-2 is True; Statement-2 is *Not* a correct explanation for Statement-1

C. Statement-1 is True, Statement-2 is False

D. Statement-1 is False, Statement-2 is True

127. Consider the system of equations

$$x - 2y + 3z = -1$$
$$-x + y - 2z = k$$
$$x - 3y + 4z = 1$$

STATEMENT-1: The system of equations has no solution for $k \neq 3$.

and

STATEMENT-2: The determinant $\begin{vmatrix} 1 & 3 & -1 \\ -1 & -2 & k \\ 1 & 4 & 1 \end{vmatrix} \neq 0$, for $k \neq 3$.

A. Statement-1 is True, Statement-2 is True; Statement-2 is a correct explanation for Statement-1
B. Statement-1 is True, Statement-2 is True; Statement-2 is *Not* a correct explanation for Statement-1
C. Statement-1 is True, Statement-2 is False
D. Statement-1 is False, Statement-2 is True

128. Consider the system of equations

$ax + by = 0$, $cx + dy = 0$, where $a, b, c, d \in \{0, 1\}$.

STATEMENT-1: The probability that the system of equations has a unique solution is $\frac{3}{8}$.

and

STATEMENT-2: The probability that the system of equations has a solution is 1.

A. Statement-1 is True, Statement-2 is True; Statement-2 is a correct explanation for Statement-1
B. Statement-1 is True, Statement-2 is True; Statement-2 is *Not* a correct explanation for Statement-1
C. Statement-1 is True, Statement-2 is False
D. Statement-1 is False, Statement-2 is True

129. Let f and g be real-valued functions defined on interval $(-1, 1)$ such that $g''(x)$ is continuous, $g(0) \neq 0$, $g'(0) = 0$, $g''(0) \neq 0$ and $f(x) = g(x) \cdot \sin x$

STATEMENT-1:

$$\lim_{x \to 0} \{g(x) \cdot \cot x - g(0) \cdot \operatorname{cosec} x\} = f''(0).$$

and

STATEMENT-2: $f'(0) = g(0)$.

A. Statement-1 is True, Statement-2 is True; Statement-2 is a correct explanation for Statement-1
B. Statement-1 is True, Statement-2 is True; Statement-2 is *Not* a correct explanation for Statement-1
C. Statement-1 is True, Statement-2 is False
D. Statement-1 is False, Statement-2 is True

Matrix-Match Type

Statements (A, B, C, D) in **Column I** have to be matched with statements (p, q, r, s) in **Column II.** The answers to these questions have to be appropriately bubbled as illustrated in the following example.

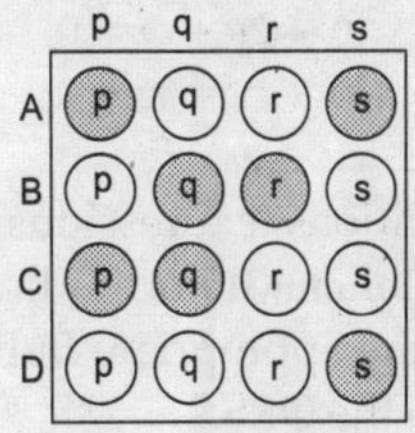

If the correct matches are A–p, A–s, B–q, B–r, C–p, C–q and D–s then the correctly bubbled 4 × 4 matrix should be as given above.

130. Match the integrals in **Column I** with the values in **Column II** and indicate your answer by darkening the appropriate bubbles in the 4 × 4 matrix.

Column I	Column II
(A) $\int_{-1}^{1} \frac{dx}{1+x^2}$	(p) $\frac{1}{2}\log\left(\frac{2}{3}\right)$
(B) $\int_{0}^{1} \frac{dx}{\sqrt{1-x^2}}$	(q) $2\log\left(\frac{2}{3}\right)$
(C) $\int_{2}^{3} \frac{dx}{1-x^2}$	(r) $\frac{\pi}{3}$
(D) $\int_{1}^{2} \frac{dx}{x\sqrt{x^2-1}}$	(s) $\frac{\pi}{2}$

131. In the following $[x]$ denotes the greatest integer less than for equal to x.

Match the functions in **Column I** with the properties in **Column II** and indicate your answer by darkening the appropriate bubles in the 4 × 4 matrix.

Column I	Column II
(A) $x\lvert x\rvert$	(p) continuous in $(-1, 1)$
(B) $\sqrt{\lvert x\rvert}$	(q) differentiable in $(-1, 1)$
(C) $x + \lvert x\rvert$	(r) strictly increasing in $(-1, 1)$
(D) $\lvert x-1\rvert + \lvert x+1\rvert$	(s) not differentiable at least at one point in $(-1, 1)$

132. Match the statement in **Column I** with the properties in **Column II** and indicate your answer by darkening the appropriate bubbles in the 4 × 4 matrix.

Column I	Column II
(A) Two intersecting circles	(p) have a common tangent
(B) Two mutually external circles	(q) have a common normal

(C)	Two circles, one strictly inside the other	(r)	do not have a common tangent
(D)	Two branches of a hyperbola	(s)	do not have a common normal

133. Consider the following linear equations

$$ax + by + cz = 0,$$
$$bx + cy + az = 0,$$
$$cx + ay + bz = 0.$$

Match the conditions/expressions in **Column I** with statements in **Column II** and indicate your answer by darkening the appropriate bubbles in the 4×4 matrix given in the ORS.

Column I	Column II
(A) $a + b + c \neq 0$ and $a^2 + b^2 + c^2 = ab + bc + ca$	(p) the equations represent planes meeting only at a single point.
(B) $a + b + c = 0$ and $a^2 + b^2 + c^2 \neq ab + bc + ca$	(q) the equations represent the line $x = y = z$.
(C) $a + b + c \neq 0$ and $a^2 + b^2 + c^2 \neq ab + bc + ca$	(r) the equations represent identical planes.
(D) $a + b + c = 0$ and $a^2 + b^2 + c^2 = ab + bc + ca$	(s) the equations represent the whole of the three dimensional space.

134. Let (x, y) be such that

$$\sin^{-1}(ax) + \cos^{-1}(y) + \cos^{-1}(bxy) = \frac{\pi}{2}.$$

Match the statements in **Column I** with statements in **Column II** and indicate your answer by darkening appropriate bubbles in the 4×4 matrix.

Column I	Column II
(A) If $a = 1$ and $b = 0$ then (x, y)	(p) lies on the circle $x^2 + y^2 = 1$
(B) If $a = 1$ and $b = 1$ then (x, y)	(q) lies on $(x^2 - 1)(y^2 - 1) = 0$
(C) If $a = 1$ and $b = 2$ then (x, y)	(r) lies on $y = x$
(D) If $a = 2$ and $b = 2$ then (x, y)	(s) lies on $(4x^2 - 1)(y^2 - 1) = 0$

135. Let $f(x) = \dfrac{x^2 - 6x + 5}{x^2 - 5x + 6}$.

Match the expression/statements in **Column I** with expression/statements in **Column II** and indicate your answer by darkening the appropriate bubbles in the 4×4 matrix.

Column I	Column II
(A) If $-1 < x < 1$ then $f(x)$ satisfies	(p) $0 < f(x) < 1$
(B) If $1 < x < 2$ then $f(x)$ satisfies	(q) $f(x) < 0$
(C) If $3 < x < 5$ then $f(x)$ satisfies	(r) $f(x) > 0$
(D) If $x > 5$ then $f(x)$ satisfies	(s) $f(x) < 1$

136. Consider the lines given by

$$L_1 : x + 3y - 5 = 0,$$
$$L_2 : 3x - ky - 1 = 0,$$
$$L_3 : 5x + 2y - 12 = 0.$$

Column I	Column II
(A) L_1, L_2, L_3 are concurrent, if	(p) $k = -9$
(B) One of L_1, L_2, L_3 is parallel to at least one of the other two, if	(q) $k = -6/5$
(C) L_1, L_2, L_3 form a triangle, if	(r) $k = 5/6$
(D) L_1, L_2, L_3 do not form a triangle, if	(s) $k = 5$

137. Consider all possible permutations of letter of the word ENDEANOEL.

Column I	Column II
(A) The number of permutations containing the word ENDEA is	(p) 5!
(B) The number of permutations in which the letter E occurs in the first and the last positions is	(q) 2 (5!)
(C) The number of permutations in which none of the letters D, L, N occurs in the last five position	(r) 7 (5!)
(D) The number of permutations in which the letters A, E, O occur only in odd positions is	(s) 21 (5!)

138.

Column I	Column II
(A) The minimum positive value of $\dfrac{x^2 + 2x + 4}{x + 2}$, $x \in R$ is	(p) 0
(B) Let A and B be 3×3 matrices of real numbers, where A is symmetric, B is skew-symmetric and $(A + B)(A - B) = (A - B)(A + B)$. If $(AB)^T = (-1)^K AB$, where $(AB)^T$ is the transpose of AB then the possible values of k are	(q) 1
(C) Let $a = \log_3 \log_3 2$. An integer k satisfying $1 < 2^{-k + 3^{-a}} < 2$, must be less than	(r) 2
(D) If $\sin \theta = \cos \varphi$, then the possible values of $\dfrac{1}{\pi}\left(\theta \pm \varphi - \dfrac{\pi}{2}\right)$ are	(s) 3

139. Match the statements/expressions given in **Column I** with the values given in **Column II.**

Column I	Column II
(A) Root (s) of the equation $2\sin^2\theta + \sin^2 2\theta = 2$	(p) $\pi/6$
(B) Points of discontinuity of	(q) $\pi/4$

$f(x) = \left[\frac{6x}{\pi}\right] \cos \left[\frac{3x}{\pi}\right]$, where

$[y]$ = greatest integer less than or equal to y

(C) The volume of the parallelopiped with edges represented by the vectors $\vec{i} + \vec{j}$, $\vec{i} + 2\vec{j}$ and $\vec{i} + \vec{j} + \pi\vec{k}$ — (r) $\pi/3$ (s) $\pi/2$

(D) The angle between vectors $\vec{a}$ and $\vec{b}$; where $\vec{a}, \vec{b}$ and $\vec{c}$ are unit vectors satisfying $\vec{a} + \vec{b} + \sqrt{3}\,\vec{c} = \vec{0}$ — (t) π

140. Match the statements/experessions given in **Column I** with the values given in **Column II.**

Column I	Column II
(A) The number of solutions of the equation $xe^{\sin x} - \cos x = 0$ in the interval $\left(0, \frac{\pi}{2}\right)$	(p) 1
(B) The value(s) of k for which the planes $kx + 4y + z = 0$, $4x + ky + 2z = 0$ and $2x + 2y + z = 0$ intersect in a straight line	(q) 2
(C) The value(s) of k for which $\lvert x+1\rvert + \lvert x-2\rvert + \lvert x+1\rvert + \lvert x+2\rvert = 4k$ has integral solution(s)	(r) 3
(D) If $y' = y + 1$ and $y(0) = 1$ then value(s) of $y(\log_e 2)$	(s) 4
	(t) 5

141. Match the statements/expressions in **Column I** with the open intervals in **Column II.**

Column I	Column II
(A) The interval contained in the domain of definition of nonzero solutions of the differential equation $(x-3)^2 \cdot y' + y = 0$	(p) $\left(-\frac{\pi}{2}, \frac{\pi}{2}\right)$
	(q) $\left(0, \frac{\pi}{2}\right)$
(B) The interval containing the value of the interval $\int_1^5 (x-1)(x-2)(x-3)(x-4)(x-5)\,dx$	(r) $\left(\frac{\pi}{8}, \frac{5\pi}{4}\right)$
(C) The interval in which at least one of the points of local maximum of $\cos^2 x + \sin x$ lies	(s) $\left(0, \frac{\pi}{8}\right)$
(D) The interval in which $\tan^{-1}(\sin x + \cos x)$ is increasing	(t) $(-\pi, \pi)$

142. Match the conic sections in **Column I** with the statements/expressions in **Column II.**

Column I	Column II
(A) Circle	(p) The locus of the point (h, k) for which the line $hx + ky = 1$ touches the circle $x^2 + y^2 = 4$
(B) Parabola	(q) Points z in the complex plane satisfying $\lvert z+2\rvert - \lvert z-2\rvert = \pm 3$
	(r) Points of the conic have parametric representation $x = \sqrt{3}\left(\frac{1-t^2}{1+t^2}\right), y = \frac{2t}{1+t^2}$
(C) Ellipse	(s) The eccentricity of the conic lies in the interval $1 \le x < \infty$
(D) Hyperbola	(t) Points z in the complex plane satisfying $\text{Re}\,(z+1)^2 = \lvert z\rvert^2 + 1$

143. Match the statements in **Column I** with those in **Column II.**

[**Note:** Here z takes values in the complex plane and Im(z) and Re(z) denote the imaginary part and real part of z respectively.]

Column I	Column II
(A) The set of points z satisfying $\lvert z - i\lvert z\rvert\rvert = \lvert z + i\lvert z\rvert\rvert$ is contained in or equal to	(p) an ellipse with eccentricity $\frac{4}{5}$
	(q) the set of points z satisfying Im(z) = 0
(B) The set of points z satisfying $\lvert z+4\rvert + \lvert z-4\rvert = 10$ is contained in or equal to	(r) the set of points z satisfying $\lvert \text{Im}(z)\rvert \le 1$
(C) If $\lvert\omega\rvert = 2$ then the set of points $z = \omega - \frac{1}{\omega}$ is contained in or equal to	(s) the set of points z satisfying $\lvert\text{Re}(z)\rvert \le 2$
(D) If $\lvert\omega\rvert = 1$ then the set of points $z = \omega + \frac{1}{\omega}$ is contained in or equal to	(t) the set of points z satisfying $\lvert z\rvert \le 3$

144. Match the statements in **Column I** with the values in **Column II.**

Column I	**Column II**
(A) A line from the origin meets the lines $\frac{x-2}{1}=\frac{y-1}{-2}=\frac{z+1}{1}$ and $\frac{x-\frac{8}{3}}{2}=\frac{y+3}{-1}=\frac{z-1}{1}$ at P and Q respectively. If the length $PQ = d$ then d^2 is	(p) –4
	(q) 0
(B) The values of x satisfying $\tan^{-1}(x+3)-\tan^{-1}(x-3) = \sin^{-1}\left(\frac{3}{5}\right)$ are	(r) 4
(C) Nonzero vectors $\vec{a}, \vec{b}$ and $\vec{c}$ satisfy $\vec{a}\cdot\vec{b}=0$, $(\vec{b}-\vec{a})\cdot(\vec{b}+\vec{c})=0$ and $2\lvert\vec{b}+\vec{c}\rvert=\lvert\vec{b}-\vec{a}\rvert$. If $\vec{a}=\mu\vec{b}+4\vec{c}$ then the possible values of μ are	(s) 5
(D) Let f be the function on $[-\pi, \pi]$ given by $f(0)=9$ and $f(x)=\frac{\sin\frac{9}{2}x}{\sin\frac{1}{2}x}$ for $x \neq 0$. The value of $\frac{2}{\pi}\int_{-\pi}^{\pi} f(x)\,dx$ is	(t) 6

Integer Answer Type

The answer to each of the questions is a single digit integer, ranging from 0 to 9. The answers to these questions have to be appropriately bubbled. If the correct answers to questions X, Y, Z (say) are 3, 0, 6 respectively then the correct darking of bubbles will look like the following.

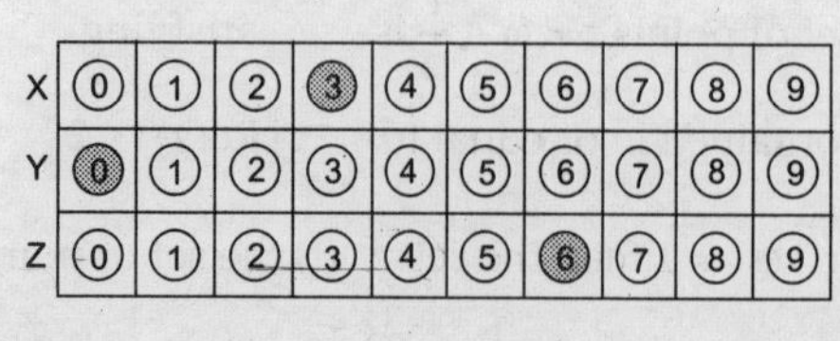

145. Let (x, y, z) be points with integral coordinates satisfying the system of homogeneous equations

$$3x - y - z = 0,$$
$$-3x + z = 0,$$
$$-3x + 2y + z = 0.$$

Then the number of such points for which $x^2 + y^2 + z^2 \leq 100$ is

146. Let ABC and ABC' be two non-congruent triangles with sides $AB = 4$, $AC = AC' = 2\sqrt{2}$ and angle $B = 30°$. The absolute value of the difference between the areas of these triangles is

147. The centres of two circles C_1 and C_2 each of unit radius are at a distance of 6 units from each other. Let P be the midpoint of the line segment joining the centres of C_1 and C_2, and C be a circle touching circles C_1 and C_2 extrenally. If a common tangent to C_1 and C passing through P is also a common tangent to C_2 and C then the radius of the circle C is

148. Let $p(x)$ be a polynomial of degree 4 having extremum at $x = 1, 2$ and $\lim_{x\to 0}\left(1+\frac{p(x)}{x^2}\right) = 2$. Then, the value of $p(2)$ is

149. The maximum value of the function $f(x) = 2x^3 - 15x^2 + 36x - 48$ on the set $A = \{x \mid x^2 + 20 \leq 9x\}$ is

150. If the function $f(x) = x^3 + e^{x/2}$ and $g(x) = f^{-1}(x)$ then the value of $g'(1)$ is

151. Let $f: R \to R$ be a continuous function which satisfies $f(x) = \int_0^x f(t)\,dt$. Then the value of $f(\log_e 5)$ is

152. The smallest value of k for which both the roots of the equation $x^2 - 8kx + 16(k^2 - k + 1) = 0$ are real, distinct and have values at least 4, is

153. Let $w = \cos\frac{2\pi}{3} + i\sin\frac{2\pi}{3}$. Then, the number of distinct complex number z satisfying

$$\begin{vmatrix} z+1 & w & w^2 \\ w & z+w^2 & 1 \\ w^2 & 1 & z+w \end{vmatrix} = 0 \text{ is equal to}$$

154. Let S_k, $k = 1, 2, 3, \ldots, 100$ denote the sum of the infinite geometric series whose first term is $\frac{k-1}{\lfloor k}$ and the common ratio is $\frac{1}{k}$. Then, the value of

$$\frac{100^2}{\lfloor 100} + \sum_{k=2}^{100} \lvert (k^2 - 3k + 1)S_k \rvert \text{ is}$$

155. The number of all possible values of θ, where $0 < \theta < \pi$, for which the system of equations

$$(y + z)\cos 3\theta = (xyz)\sin 3\theta$$
$$x\sin 3\theta = \frac{2\cos 3\theta}{y} + \frac{2\sin 3\theta}{z}$$
$$(xyz)\sin 3\theta = (y + 2z)\cos 3\theta + y\sin 3\theta$$

have a solution (x_0, y_0, z_0) with $y_0 z_0 \neq 0$ is

156. The number of values of θ in the interval $\left(-\frac{\pi}{2}, \frac{\pi}{2}\right)$ such that $\theta \neq \frac{1}{5} n\pi$, where $n = 0, \pm 1, \pm 2$ and $\tan \theta = \cot 5\theta$ as well as $\sin 2\theta = \cos 4\theta$ is

157. The maximum value of the expression

$$\frac{1}{\sin^2\theta + 3\sin\theta \cdot \cos\theta + 5\cos^2\theta} \text{ is}$$

158. The line $2x + y = 1$ is tangent to the hyperbola $\frac{x^2}{a^2} - \frac{y^2}{b^2} = 1$. If this line passes throught the point of intersection of the nearest directrix and the x-axis then the eccentricity of the hyperbola is

159. If $\vec{a}$ and $\vec{b}$ are vectors in space given by $\vec{a} = \frac{\hat{i} - 2\hat{j}}{\sqrt{5}}$ and $\vec{b} = \frac{2\hat{i} + \hat{j} + 3\hat{k}}{\sqrt{14}}$ then the value of

$(2\vec{a} + \vec{b}) \cdot \{(\vec{a} \times \vec{b}) \times (\vec{a} - 2\vec{b})\}$ is

160. If the distance between the plane $Ax - 2y + z = d$ and the plane containing the lines $\frac{x-1}{2} = \frac{y-2}{3} = \frac{z-3}{4}$ and $\frac{x-2}{3} = \frac{y-3}{4} = \frac{z-4}{5}$ is $\sqrt{6}$ then $|d|$ is

161. Let f be a real-valued differentiable function on R, the set of real numbers, such that $f(1) = 1$. If the y-intercept of the tangent at any point $P(x, y)$ on the curve $y = f(x)$ is equal to the cube of the abscissa of P then the value of $f(-3)$ is equal to

162. For any real number x, let $[x]$ = largest integer less than or equal to x. Let f be a real valued function defined on the interval $[-10, 10]$ by

$$f(x) = \begin{cases} x - [x] & \text{if}[x] \text{ is odd} \\ 1 + [x] - x & \text{if}[x] \text{ is even} \end{cases}$$

Then, the value of $\frac{\pi^2}{10} \int_{-10}^{10} f(x) \cos \pi x \, dx$ is

163. Let $a_1, a_2, a_3, \ldots, a_{11}$ be real numbers satisfying $a_1 = 15, 27 - 2a_2 > 0$ and $a_k = 2a_{k-1} - a_{k-2}$ for $k = 3, 4, \ldots, 11$. If $\frac{a_1^2 + a_2^2 + \ldots + a_{11}^2}{11} = 90$ then the value of $\frac{a_1 + a_2 + \ldots + a_{11}}{11}$ is equal to

164. Two parallel chords of a circle of radius 2 are at distance $\sqrt{3} + 1$ apart. If the chords subtend at the centre, angles of magnitude $\frac{\pi}{k}$ and $\frac{2\pi}{k}$, where $k > 0$ then the value of $[k]$, the largest integer less than or equal to k, is

165. Consider a ΔABC. Suppose $BC = 6$, $CA = 10$ and the area of the triangle is $15\sqrt{3}$. If $\angle ACB > 90°$ and r denotes, the radius of the incircle of the triangle then r^2 is equal to

166. Let k be a positive real number and

$$A = \begin{bmatrix} 2k-1 & 2\sqrt{k} & 2\sqrt{k} \\ 2\sqrt{k} & 1 & -2k \\ -2\sqrt{k} & 2k & -1 \end{bmatrix} \text{ and } B = \begin{bmatrix} 0 & 2k-1 & \sqrt{k} \\ 1-2k & 0 & 2\sqrt{k} \\ -\sqrt{k} & -2\sqrt{k} & 0 \end{bmatrix}.$$

If det (adj A) + det (adj B) = 10^6 then $[k]$ is equal to

[**Note:** adj M denotes the adjoint of the matrix M and $[k]$ denotes the largest integer $\leq k$.]

167. Let f be a function defined on R, the set of real numbers, such that $f'(x) = 2010(x-2009)(x-2010)^2(x-2011)^3(x-2012)^4$ for all $x \in R$. If g is a function defined on R with values in $(0, \infty)$ such that $f(x) = \log_e g(x)$ for all $x \in R$ then the number of points in R at which g has a local maximum is

Answers

7. B	8. D	9. A	10. C
11. A	12. B	13. A	14. B
15. A	16. A	17. C	18. A
19. B	20. C	21. A	22. A
23. A	24. D	25. B	26. A
27. C	28. B	29. A	30. C
31. B	32. D	33. B	34. D
35. B	36. C	37. B	38. C
39. A	40. D	41. A	42. B
43. C	44. D	45. B	46. B
47. C	48. A	49. C	50. A
51. D	52. D	53. A	54. B
55. B	56. A	57. C	58. A

59. B	60. C	61. C	62. D
63. D	64. C	65. B	66. A
67. B	68. C	69. D	70. A
71. A	72. B	73. B	74. D
75. C	76. B	77. A	78. D
79. D	80. A	81. D	82. A
83. B	84. B	85. A	86. D
87. D	88. B	89. A	90. C
91. A	92. B	93. D	94. C
95. A			

96. (A) $\leftrightarrow$ (p) (B) $\leftrightarrow$ (s) (C) $\leftrightarrow$ (p) (D) $\leftrightarrow$ (q)

97. (A) $\leftrightarrow$ (p) (B) $\leftrightarrow$ (q) (C) $\leftrightarrow$ (s) (D) $\leftrightarrow$ (r)

98. (A) $\leftrightarrow$ (s) (B) $\leftrightarrow$ (p) (C) $\leftrightarrow$ (q, r) (D) $\leftrightarrow$ (s)

99. (A) $\leftrightarrow$ (q) (B) $\leftrightarrow$ (s) (C) $\leftrightarrow$ (r) (D) $\leftrightarrow$ (q)

100. 1210 In matrix form:

0	1	2	3	4	5	6	7	8	9
○	●	○	○	○	○	○	○	○	○
○	○	●	○	○	○	○	○	○	○
○	●	○	○	○	○	○	○	○	○
●	○	○	○	○	○	○	○	○	○

101. 5 In matrix form:

0	1	2	3	4	5	6	7	8	9
●	○	○	○	○	○	○	○	○	○
●	○	○	○	○	○	○	○	○	○
●	○	○	○	○	○	○	○	○	○
○	○	○	○	○	●	○	○	○	○

102. 5051 In matrix form:

0	1	2	3	4	5	6	7	8	9
○	○	○	○	○	●	○	○	○	○
●	○	○	○	○	○	○	○	○	○
○	○	○	○	○	●	○	○	○	○
○	●	○	○	○	○	○	○	○	○

103. 6 In matrix form:

0	1	2	3	4	5	6	7	8	9
●	○	○	○	○	○	○	○	○	○
●	○	○	○	○	○	○	○	○	○
●	○	○	○	○	○	○	○	○	○
○	○	○	○	○	○	●	○	○	○

106. D	**107.** C	**108.** A	**109.** C
110. C	**111.** D	**112.** A	**113.** D
114. B	**115.** B	**116.** A	**117.** B
118. A	**119.** A	**120.** C	**121.** A
122. A	**123.** C	**124.** B	**125.** C
126. D	**127.** A	**128.** B	**129.** B

130.

	p	q	r	s
A	○	○	○	●
B	○	○	○	●
C	●	○	○	○
D	○	○	●	○

(i.e., $A \to s, B \to s,$
$C \to p, D \to r$)

131.

	p	q	r	s
A	●	●	●	○
B	●	○	○	●
C	○	○	●	●
D	●	●	○	○

(i.e., $A \to p, A \to q, A \to r,$
$B \to p, B \to s, C \to r,$
$C \to s, D \to p, D \to q$)

132.

	p	q	r	s
A	●	●	○	○
B	●	●	○	○
C	○	●	●	○
D	○	●	●	○

(i.e., $A \to p, A \to q,$
$B \to p, B \to q,$
$C \to q, C \to r,$
$D \to q, D \to r$)

133.

	p	q	r	s
A	○	○	●	○
B	○	●	○	○
C	●	○	○	○
D	○	○	○	●

(i.e., $A \to r, B \to q,$
$C \to p, D \to s$)

134.

	p	q	r	s
A	●	○	○	○
B	○	●	○	○
C	●	○	○	○
D	○	○	○	●

(i.e., $A \to p, B \to q,$
$C \to p, D \to s$)

135.

	p	q	r	s
A	●	○	○	○
B	○	●	○	○
C	○	●	○	○
D	●	○	○	○

(i.e., $A \to p, B \to q,$
$C \to q, D \to p$)

136.

	p	q	r	s
A	○	○	○	●
B	●	●	○	○
C	○	○	●	○
D	●	●	○	●

137.

	p	q	r	s
A	●	○	○	○
B	○	○	○	●
C	○	●	○	○
D	○	●	○	○

138.

	p	q	r	s
A	○	○	●	○
B	○	●	○	●
C	○	○	●	●
D	●	○	●	○

139.

	p	q	r	s	t
A	○	●	○	●	○
B	●	○	●	●	●
C	○	○	○	○	●
D	○	○	●	○	○

140.

	p	q	r	s	t
A	●	○	○	○	○
B	○	●	○	○	○
C	○	●	●	●	●
D	○	○	●	○	○

141.

	p	q	r	s	t
A	●	●	○	●	○
B	●	○	○	○	●
C	●	●	●	○	●
D	○	○	○	●	○

142.

	p	q	r	s	t
A	●	○	○	○	○
B	○	○	○	●	●
C	○	○	●	○	○
D	○	●	○	●	○

143.

	p	q	r	s	t
A	○	●	●	○	○
B	●	○	○	○	○
C	●	○	○	●	●
D	○	●	●	●	●

144.

	p	q	r	s	t
A	○	○	○	○	●
B	●	○	●	○	○
C	○	●	○	○	○
D	○	○	●	○	○

145	146	147	148	149	150	151	152
0	0	0	●0	0	0	●0	0
1	1	1	1	1	1	1	1
2	2	2	2	2	●2	2	●2
3	3	3	3	3	3	3	3
4	●4	4	4	4	4	4	4
5	5	5	5	5	5	5	5
6	6	6	6	6	6	6	6
●7	7	7	7	●7	7	7	7
8	8	●8	8	8	8	8	8
9	9	9	9	9	9	9	9

153	154	155	156	157	158	159	160
0	0	0	0	0	0	0	0
●1	1	1	1	1	1	1	1
2	2	2	2	●2	●2	2	2
3	●3	●3	●3	3	3	3	3
4	4	4	4	4	4	4	4
5	5	5	5	5	5	●5	5
6	6	6	6	6	6	6	●6
7	7	7	7	7	7	7	7
8	8	8	8	8	8	8	8
9	9	9	9	9	9	9	9

161	162	163	164	165	166	167
0	0	●0	0	0	0	0
1	1	1	1	1	1	●1
2	2	2	2	2	2	2
3	3	3	●3	●3	3	3
4	●4	4	4	4	●4	4
5	5	5	5	5	5	5
6	6	6	6	6	6	6
7	7	7	7	7	7	7
8	8	8	8	8	8	8
●9	9	9	9	9	9	9

□

Recent IIT Questions

SET 1

Section I

Single Correct Choice Type

This section contains 7 multiple-choice questions. Each question has 4 choices (a), (b), (c) and (d) for its answer, out of which *only one* is correct.

1. Let (x_0, y_0) be the solution of the following equations

$$(2x)^{\ln 2} = (3y)^{\ln 3}$$
$$3^{\ln x} = 2^{\ln y}.$$

Then x_0 is

(a) $\frac{1}{6}$ (b) $\frac{1}{3}$

(c) $\frac{1}{2}$ (d) 6

2. Let $P = \{\theta : \sin\theta - \cos\theta = \sqrt{2}\cos\theta\}$ and $Q = \{\theta : \sin\theta + \cos\theta = \sqrt{2}\sin\theta\}$ be two sets. Then

(a) $P \subset Q$ and $Q - P \neq \phi$ (b) $Q \not\subset P$

(c) $P \not\subset Q$ (d) $P = Q$

3. Let $\vec{a} = \hat{i} + \hat{j} + \hat{k}$, $\vec{b} = \hat{i} - \hat{j} + \hat{k}$ and $\vec{c} = \hat{i} - \hat{j} - \hat{k}$. A vector $\vec{v}$ in the plane of $\vec{a}$ and $\vec{b}$, whose projection on $\vec{c}$ is $\frac{1}{\sqrt{3}}$, is given by

(a) $\hat{i} - 3\hat{j} + 3\hat{k}$ (b) $-3\hat{i} - 3\hat{j} + \hat{k}$

(c) $3\hat{i} - \hat{j} + 3\hat{k}$ (d) $\hat{i} + 3\hat{j} - 3\hat{k}$

4. Let α and β be the roots of $x^2 - 6x - 2 = 0$, $\alpha > \beta$. If $a_n = \alpha^n - \beta^n$, $n \geq 1$ then the value of $\frac{a_{10} - 2a_8}{2a_9}$ is equal to

(a) 1 (b) 2

(c) 3 (d) 4

5. A straight line L through the point $(3, -2)$ is inclined at an angle $60°$ to the line $\sqrt{3}x + y = 1$. If L also intersects the x-axis then the equation of the line L is

(a) $y + \sqrt{3}x + 2 - 3\sqrt{3} = 0$ (b) $y - \sqrt{3}x + 2 + 3\sqrt{3} = 0$

(c) $\sqrt{3}y - x + 3 + 2\sqrt{3} = 0$ (d) $\sqrt{3}y + x - 3 + 2\sqrt{3} = 0$

6. The value of $\int_{\sqrt{\ln 2}}^{\sqrt{\ln 3}} \frac{x\sin x^2}{\sin x^2 + \sin(\ln 6 - x^2)}\,dx$ is

(a) $\frac{1}{4}\ln\frac{3}{2}$ (b) $\frac{1}{2}\ln\frac{3}{2}$

(c) $\ln\frac{3}{2}$ (d) $\frac{1}{6}\ln\frac{3}{2}$

7. Let the straight line $x = b$ divide the area enclosed by $y = (1 - x)^2$, $y = 0$ and $x = 0$ into two parts $R_1(0 \leq x \leq b)$ and $R_2(b \leq x \leq 1)$ such that $R_1 - R_2 = \frac{1}{4}$. Then b equals

(a) $\frac{3}{4}$ (b) $\frac{1}{2}$

(c) $\frac{1}{3}$ (d) $\frac{1}{4}$

Section II

Multiple Correct Choice Type

This section contains 4 multiple-choice questions. Each question has four choices (a), (b), (c) and (d) out of which ***one* or *more*** may be correct.

8. The vector(s) which is/are coplanar with vectors $\hat{i} + \hat{j} + 2\hat{k}$ and $\hat{i} + 2\hat{j} + \hat{k}$, and perpendicular to the vector $\hat{i}+\hat{j}+\hat{k}$ is/are

(a) $\hat{j}-\hat{k}$ (b) $-\hat{i}+\hat{j}$

(c) $\hat{i}-\hat{j}$ (d) $-\hat{j}+\hat{k}$

9. Let the eccentricity of the hyperbola $\frac{x^2}{a^2}-\frac{y^2}{b^2}=1$ be reciprocal to that of the ellipse $x^2+4y^2=4$. If the hyperbola passes through a focus of the ellipse then

(a) the hyperbola has the equation $\frac{x^2}{3}-\frac{y^2}{2}=1$

(b) a focus of the hyperbola is (2, 0)

(c) the eccentricity of the hyperbola is $\sqrt{5/3}$

(d) the hyperbola has the equation $x^2-3y^2=3$

10. Let M and N be two 3×3 non-singular skew symmetric matrices such that $MN = NM$. If P^T denotes the transpose of P then $M^2N^2(M^TN)^{-1}(MN^{-1})^T$ is equal to

(a) M^2 (b) $-N^2$

(c) $-M^2$ (d) MN

11. Let $f: R\to R$ be a function such that $f(x+y)=f(x)+f(y)$, $\forall\ x, y\in R$. If $f(x)$ is differentiable at $x=0$ then

(a) $f(x)$ is differentiable only in a finite interval containing zero

(b) $f(x)$ is continuous $\forall\ x\in R$

(c) $f'(x)$ is constant $\forall\ x\in R$

(d) $f(x)$ is differentiable except at finitely many points

Section III

Paragraph Type

This section contains 2 paragraphs. Based upon one of paragraphs 2 multiple-choice questions and based on the other paragraph 3 multiple-choice questions are to be answered. Each of these questions has four choices (a), (b), (c) and (d) out of which *only one* is correct.

Paragraph for Question Nos. 12 to 13

Let U_1 and U_2 be two urns such that U_1 contains 3 white and 2 red balls and U_2 contains only one white ball. A fair coin is tossed. If head appears then 1 ball is drawn from U_1 and put in U_2. However, if tail appears then 2 balls are drawn at random from U_1 and put in U_2. Now 1 ball is drawn at random from U_2.

12. The probability of the drawn ball from U_2 being white is

(a) $\frac{13}{30}$ (b) $\frac{23}{30}$

(c) $\frac{19}{30}$ (d) $\frac{11}{30}$

13. Given that the drawn ball from U_2 is white, the probability that head appeared on the coin is

(a) $\frac{17}{23}$ (b) $\frac{11}{23}$

(c) $\frac{15}{23}$ (d) $\frac{12}{23}$

Paragraph for Question Nos. 14 to 16

Let a, b and c be three real numbers satisfying

$$[a\ b\ c]\begin{bmatrix}1 & 9 & 7\\ 8 & 2 & 7\\ 7 & 3 & 7\end{bmatrix}=[0\ 0\ 0]\ \dots (E)$$

14. If the point $P(a, b, c)$, with reference to (E), lies on the plane $2x+y+z=1$ then the value of $7a+b+c$ is

(a) 0 (b) 12

(c) 7 (d) 6

15. Let ω be a solution of $x^3-1=0$ with $\text{Im}(\omega)>0$. If $a=2$ with b and c satisfying (E) then the value of $\frac{3}{\omega^a}+\frac{1}{\omega^b}+\frac{3}{\omega^c}$ is equal to

(a) −2 (b) 2

(c) 3 (d) −3

16. Let $b=6$, with a and c satisfying (E). If α and β are the roots of the quadratic equation $ax^2+bx+c=0$ then

$$\sum_{n=0}^{\infty}\left(\frac{1}{\alpha}+\frac{1}{\beta}\right)^n \text{ is equal to}$$

(a) 6 (b) 7

(c) $\frac{6}{7}$ (d) ∞

Section IV

Integer Answer Type

This section contains 7 questions. The answer to each question is a single digit integer. The bubble corresponding to the correct answer is to be darkened in the ORS.

17. Consider the parabola $y^2 = 8x$. Let Δ_1 be the area of the triangle formed by the end points of its latus rectum and the point $P\left(\frac{1}{2}, 2\right)$ on the parabola and Δ_2 be the area of the triangle formed by drawing tangents at P and the end points of the latus rectum. Then $\frac{\Delta_1}{\Delta_2}$ is

18. Let $f(\theta) = \sin\left\{\tan^{-1}\left(\frac{\sin\theta}{\sqrt{\cos 2\theta}}\right)\right\}$, where $-\frac{\pi}{4} < \theta < \frac{\pi}{4}$.

Then the value of $\frac{d}{d(\tan\theta)} f(\theta)$ is

19. Let $f: [1, \infty) \to [2, \infty)$ be a differentiable function such that $f(1) = 2$. If $6\int_1^x f(t)dt = 3x(f(x) - x^3$ for all $x \geq 1$ then the value of $f(2)$ is

20. The positive integral value of $n > 3$ satisfying the equation

$$\frac{1}{\sin\frac{\pi}{n}} = \frac{1}{\sin\frac{2\pi}{n}} + \frac{1}{\sin\frac{3\pi}{n}} \text{ is}$$

21. Let $a_1, a_2, a_3, \ldots, a_{100}$ be in AP with $a_1 = 3$ and $S_p = \sum_{i=1}^{p} a_i$, $1 \leq p \leq 100$. For any integer n with $1 \leq n \leq 20$, let $m = 5n$. If $\frac{S_m}{S_n}$ does not depend on n then a_2 is

22. The minimum value of the sum of the real numbers $a^{-5}, a^{-4}, 3a^{-3}, 1, a^8$ and a^{10} with $a > 0$ is

23. If z is a complex number satisfying $|z - 3 - 2i| \leq 2$ then the minimum value of $|2z - 6 + 5i|$ is

Answers

1. c	**2.** d	**3.** c	**4.** c
5. b	**6.** a	**7.** b	**8.** a, d
9. b, d	**10.** c	**11.** b, c	**12.** b
13. d	**14.** d	**15.** a	**16.** b

17.	0	1	**2**	3	4	5	6	7	8	9
18.	0	**1**	2	3	4	5	6	7	8	9
19.	0	1	2	3	4	5	**6**	7	8	9
20.	0	1	2	3	4	5	6	**7**	8	9
21.	0	1	2	3	4	5	6	7	8	**9**
22.	0	1	2	3	4	5	6	7	**8**	9
23.	0	1	2	3	4	**5**	6	7	8	9

Hints and Solutions to Selected Questions

1. $(2x)^{\log 2} = (3y)^{\log 3}$

$\Rightarrow \log 2 \cdot (\log 2 + \log x) = \log 3 \cdot (\log 3 + \log y)$

or $(\log 3)^2 - (\log 2)^2 = \log 2 \cdot \log x - \log 3 \cdot \log y \quad \ldots(1)$

$3^{\log x} = 2^{\log y} \Rightarrow \log x \cdot \log 3 = \log y \cdot \log 2 \quad \ldots(2)$

From (1) and (2),

$$\log y = \frac{-(\log 3)^2 + (\log 2)^2 + \log 2 \cdot \log x}{\log 3}$$

or $\frac{\log x \cdot \log 3}{\log 2} = \frac{\log 2 \cdot \log x - (\log 3)^2 + (\log 2)^2}{\log 3}$

or $\log x \{(\log 3)^2 - (\log 2)^2\} = \log 2\{-(\log 3)^2 + (\log 2)^2\}$

$\therefore \log x = -\log 2 = \log\frac{1}{2} \Rightarrow x = \frac{1}{2}$.

2. $\sin\theta - \cos\theta = \sqrt{2}\cos\theta$ or $\sin\theta = (\sqrt{2}+1)\cos\theta$

$\therefore\quad \tan\theta = \sqrt{2}+1 \left(\because\ \theta \neq 0, \dfrac{\pi}{2}\right)$

$\sin\theta + \cos\theta = \sqrt{2}\sin\theta$ or $\cos\theta = (\sqrt{2}-1)\sin\theta$

$\therefore\quad \tan\theta = \dfrac{1}{\sqrt{2}-1} = \dfrac{\sqrt{2}+1}{2-1} = \sqrt{2}+1.$

$\therefore$ the two sets P and Q are equal.

3. Let $\vec{v} = x\vec{a} + y\vec{b}$

$= x(\hat{i}+\hat{j}+\hat{k}) + y(\hat{i}-\hat{j}+\hat{k}).$

Here, $\vec{v}\cdot\dfrac{\vec{c}}{|\vec{c}|} = \dfrac{1}{\sqrt{3}}$

$\therefore\quad \dfrac{\hat{i}-\hat{j}-\hat{k}}{\sqrt{3}}\cdot\{(x+y)\hat{i} + (x-y)\hat{j} + (x+y)\hat{k}\} = \dfrac{1}{\sqrt{3}}$

or $\quad x+y-(x-y)-(x+y) = 1$

$\therefore\quad y - x = 1$

$\therefore\quad \vec{v} = x(\hat{i}+\hat{j}+\hat{k}) + (x+1)(\hat{i}-\hat{j}+\hat{k})$

$= (2x+1)\hat{i} - \hat{j} + (2x+1)\hat{k}$

Taking $x = 1$, $\vec{v} = 3\hat{i} - \hat{j} + 3\hat{k}$.

4. Here, $\alpha^2 - 6\alpha - 2 = 0 \ \therefore\ \alpha^{10} - 6\alpha^9 - 2\alpha^8 = 0$ ... (1)

$\beta^2 - 6\beta - 2 = 0 \ \therefore\ \beta^{10} - 6\beta^9 - 2\beta^8 = 0$ ... (2)

From (1) – (2),

$$\alpha^{10} - \beta^{10} - 6(\alpha^9 - \beta^9) - 2(\alpha^8 - \beta^8) = 0$$

$\therefore\quad a_{10} - 6a_9 - 2a_8 = 0$

$\therefore\quad \dfrac{a_{10} - 2a_8}{2a_9} = \dfrac{6a_9}{2a_9} = 3.$

5. Let L be $y + 2 = m(x-3)$, $x \neq 0$.

Given, $\left|\dfrac{m-(-\sqrt{3})}{1+m(-\sqrt{3})}\right| = \sqrt{3};$

$\therefore\ m + \sqrt{3} = \pm\sqrt{3}(1-\sqrt{3}m)$

or $\ m + \sqrt{3} = \pm(\sqrt{3} + 3m)$

$\therefore\ m = 0, \sqrt{3}.$

But $m \neq 0$.

$\therefore$ the equation of L is

$y + 2 = \sqrt{3}(x-3) \Rightarrow y - \sqrt{3}x + 2 + 3\sqrt{3} = 0.$

6. Putting $x^2 = z$, $I = \displaystyle\int_{\log 2}^{\log 3} \frac{\frac{1}{2}\sin z}{\sin z + \sin(\log 6 - z)}\,dz$

$= \dfrac{1}{2}\displaystyle\int_{\log 2}^{\log 3} \frac{\sin z}{\sin z + \sin(\log 6 - z)}\,dz$

Putting $\log 6 - z = t$,

$I = \dfrac{1}{2}\displaystyle\int_{\log 3}^{\log 2} \frac{\sin(\log 6 - t)}{\sin(\log 6 - t) + \sin t}(-dt)$

Adding, $2I = \dfrac{1}{2}\displaystyle\int_{\log 2}^{\log 3} \frac{\sin\theta + \sin(\log 6 - \theta)}{\sin\theta + \sin(\log 6 - \theta)}\,d\theta$

$= \dfrac{1}{2}\displaystyle\int_{\log 2}^{\log 3} d\theta = \dfrac{1}{2}(\log 3 - \log 2)$

$\therefore\quad I = \dfrac{1}{4}\cdot\log\dfrac{3}{2}.$

7. $R_1 = \displaystyle\int_0^b y\,dx = \int_0^b (1-x)^2\,dx$

$R_2 = \displaystyle\int_b^1 y\,dx = \int_b^1 (1-x)^2 dx$

$\therefore\quad R_1 - R_2 = \displaystyle\int_0^b (1-x)^2\,dx - \int_b^1 (1-x)^2 dx$

$= \left[-\dfrac{(1-x)^3}{3}\right]_0^b - \left[-\dfrac{(1-x)^3}{3}\right]_b^1$

$= -\dfrac{(1-b)^3}{3} + \dfrac{1}{3} - \dfrac{(1-b)^3}{3}$

$\therefore\quad \dfrac{1}{4} = -\dfrac{2}{3}(1-b)^3 + \dfrac{1}{3};$

$\therefore\quad \dfrac{2}{3}(1-b)^3 = \dfrac{1}{3} - \dfrac{1}{4} = \dfrac{1}{12};$

$\therefore\quad (1-b)^3 = \dfrac{1}{8};$

$\therefore\quad 1 - b = \dfrac{1}{2};$

$\therefore\quad b = \dfrac{1}{2}.$

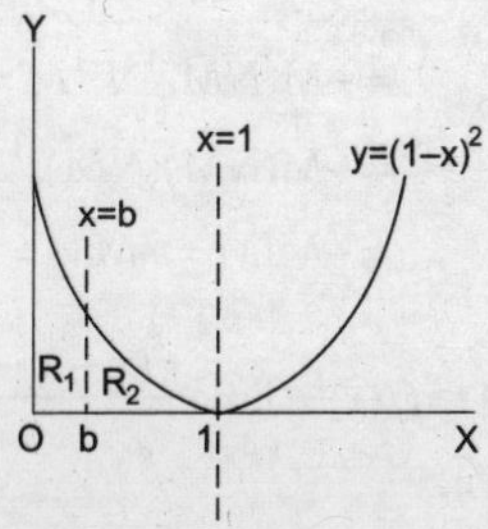

8. Here, $\vec{r} = x(\hat{i}+\hat{j}+2\hat{k}) + y(\hat{i}+2\hat{j}+\hat{k})$

$\therefore\quad \vec{r}\cdot(\hat{i}+\hat{j}+\hat{k}) = 0 \Rightarrow x(1+1+2) + y(1+2+1) = 0$

$\therefore\quad y = -x.$ So, $\vec{r} = x(\hat{i}+\hat{j}+2\hat{k}) - x(\hat{i}+2\hat{j}+\hat{k})$

$= x(-\hat{j}+\hat{k}).$

Taking $x = \pm 1$, $\vec{r} = \pm(-\hat{j}+\hat{k})$.

9. For the ellipse, $x^2 + 4y^2 = 4$ or $\dfrac{x^2}{2^2} + \dfrac{y^2}{1^2} = 1.$

$\therefore\quad 1^2 = 2^2\cdot(1-e^2);$

$\therefore\quad 1 - e^2 = \dfrac{1}{4};$

$\therefore\quad e = \sqrt{\dfrac{3}{4}} = \dfrac{\sqrt{3}}{2}.$

$\therefore$ the eccentricity of the hyperbola $= \dfrac{1}{\sqrt{3}/2} = \dfrac{2}{\sqrt{3}}.$

The foci of the ellipse are $\left(\pm 2\cdot\dfrac{\sqrt{3}}{2}, 0\right)$,

i.e., $(\pm\sqrt{3}, 0)$.

The hyperbola passes through a focus of the ellipse

$\Rightarrow\ \dfrac{3}{a^2} = 1 \Rightarrow a^2 = 3.$

$\therefore\quad b^2 = a^2(e^2 - 1) \Rightarrow b^2 = 3\left\{\left(\dfrac{2}{\sqrt{3}}\right)^2 - 1\right\}$

$$\Rightarrow b^2 = 3\left(\frac{4}{3} - 1\right) = 1.$$

$\therefore$ the equation of the hyperbola is $\frac{x^2}{3} - \frac{y^2}{1} = 1$,

i.e., $x^2 - 3y^2 = 3$.

A focus of the hyperbola is

$$(ae, 0) = \left(\sqrt{3} \cdot \frac{2}{\sqrt{3}}, 0\right) = (2, 0).$$

10. $M^2N^2(M^TN)^{-1}(MN^{-1})^T = M^2NNN^{-1}(M^T)^{-1}(N^{-1})^TM^T$

$[\because (AB)^{-1} = B^{-1}A^{-1}, (AB)^T = B^TA^T]$

$= M^2N(-M)^{-1}(N^T)^{-1}(-M)$

$[\because (A^T)' = (A')^T, AA^{-1} = I, M^T = -M,$ being skew symmetric$]$

$= M^2NM^{-1}(-1)^2(-N)^{-1}M$

$[\because N^T = -N,$ being skew symmetric$]$

$= -M^2NM^{-1}N^{-1}M = -M(MN)(NM)^{-1}M$

$= -M(NM)(NM)^{-1}M \quad [\because MN = NM,$ given$]$

$= -MIM = -MM = -M^2$.

11. $f'(x) = \lim\limits_{h \to 0} \frac{f(x+h) - f(x)}{h} = \lim\limits_{h \to 0} \frac{f(h)}{h}$

$[\because f(x+h) = f(x) + f(h),$ given$]$

$= \lim\limits_{h \to 0} \frac{f(0+h) - f(0)}{h}$

$[\because f(x+0) = f(x) + f(0) \Rightarrow f(0) = 0]$

$= f'(0) =$ constant.

$\therefore f(x) = f'(0) \cdot x + c; \quad \therefore f(0) = c = 0.$

$\therefore f(x) = f'(0) \cdot x.$

12. By the theorem of total probability, required prob.

= prob. of head appearing {prob. of drawing white from U_1 × prob. of drawing white from 2 whites from U_2 + prob. of drawing red from U_1 × prob. of drawing white from 1 white from U_2}

+ prob. of tail appearing {prob. of drawing 2 whites from U_1 × prob. of drawing white from 3 whites from U_2 + prob. of drawing 2 reds from U_1 × prob. of drawing white from 1 white and 2 reds from U_2 + prob. of drawing 1 white and 1 red from U_1 × prob. of drawing white from 2 whites and 1 red from U_2}

$$= \frac{1}{2}\left\{\frac{3}{5} \times 1 + \frac{2}{5} \times \frac{1}{2}\right\}$$

$$+ \frac{1}{2}\left\{\frac{^3C_2}{^5C_2} \times 1 + \frac{^2C_2}{^5C_2} \times \frac{1}{3} + \frac{^3C_1 \cdot {^2C_1}}{^5C_2} \times \frac{^2C_1}{^3C_1}\right\}$$

$$= \frac{2}{5} + \frac{1}{2}\left\{\frac{3}{10} + \frac{1}{10} \times \frac{1}{3} + \frac{6}{10} \times \frac{2}{3}\right\} = \frac{2}{5} + \frac{11}{30} = \frac{23}{30}.$$

13. By Bayes's theorem on inverse probability, required probability

$$= \frac{\{\text{prob. of head appearing (prob. of drawing white from } U_1 \times \text{prob. of drawing white from 2 whites from } U_2 + \text{prob. of drawing red from } U_1 \times \text{prob. of drawing white from 1 white from } U_2)\}}{\text{total probability}}$$

$$= \frac{\frac{1}{2}\left\{\frac{3}{5} \times 1 + \frac{2}{5} \times \frac{1}{2}\right\}}{\frac{23}{30}} \text{ (see Q.N. 12)}$$

$$= \frac{12}{23}.$$

14. Here $[a + 8b + 7c \quad 9a + 2b + 3c \quad 7a + 7b + 7c] = [0 \ 0 \ 0]$

$$\Rightarrow \left.\begin{array}{r} a + 8b + 7c = 0 \\ 9a + 2b + 3c = 0 \\ a + b + c = 0 \end{array}\right\} \Rightarrow \frac{a}{1} = \frac{b}{6} = \frac{c}{-7} = t \text{ (say)}.$$

$P(a, b, c)$ is on the plane

$2x + y + z = 1 \Rightarrow 2t + 6t - 7t = 1 \Rightarrow t = 1.$

$\therefore \quad a = 1, b = 6, c = -7;$

$\therefore \quad 7a + b + c = 7 + 6 - 7 = 6.$

15. $x = 1, \frac{-1 \pm \sqrt{3}i}{2}$;

so, $\omega = \frac{-1 + \sqrt{3}i}{2}$ and $\omega^3 = 1, 1 + \omega + \omega^2 = 0$

As in Q.N. 14, $\frac{a}{1} = \frac{b}{6} = \frac{c}{-7}$; so, $\frac{2}{1} = \frac{b}{6} = \frac{c}{-7}$.

$\therefore \quad b = 12, c = -14$ when $a = 2$.

So, required value $= \frac{3}{\omega^2} + \frac{1}{\omega^{12}} + \frac{3}{\omega^{-14}}$

$$= \frac{3\omega^3}{\omega^2} + \frac{1}{(\omega^3)^4} + \frac{3}{\omega^{-14} \cdot \omega^{15}}$$

$$= 3\omega + 1 + \frac{3}{\omega} = 3\omega + \frac{3\omega^3}{\omega} + 1$$

$$= 3(\omega + \omega^2) + 1 = 3(-1) + 1 = -2.$$

16. As in Q.N. 14, $\frac{a}{1} = \frac{b}{6} = \frac{c}{-7}$. So, $\frac{a}{1} = \frac{6}{6} = \frac{c}{-7}$

$\therefore a = 1, c = -7.$

$\therefore \alpha, \beta$ are roots of $x^2 + 6x - 7 = 0$

or $(x + 7)(x - 1) = 0; \therefore \alpha, \beta = -7, 1.$

$$\therefore \sum_{n=0}^{\infty}\left(\frac{1}{\alpha} + \frac{1}{\beta}\right)^n = \sum_{n=0}^{\infty}\left(\frac{1}{-7} + 1\right)^n = \sum_{n=0}^{\infty}\left(\frac{6}{7}\right)^n$$

$$= 1 + \frac{6}{7} + \left(\frac{6}{7}\right)^2 + \dots \text{ to } \infty = \frac{1}{1 - \frac{6}{7}} = 7.$$

17. The parabola is $y^2 = 8x$

or $y^2 = 4 \times 2 \times x \quad \therefore a = 2$.

$\therefore$ end points of the latus rectum are

$$(a, \pm 2a) = (2, \pm 4).$$

$\therefore \quad \Delta_1 = \text{ar}(\Delta PLM) = \frac{1}{2} \cdot \left(2 - \frac{1}{2}\right) \cdot 8 = 6.$

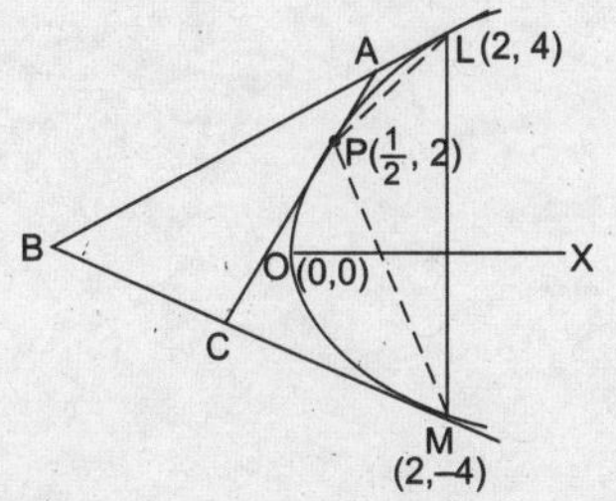

Tangents at $L(2, 4)$, $M(2, -4)$ are

$y = x + 2$ and $-y = x + 2$.

Tangent at $P\left(\frac{1}{2}, 2\right)$ is $y = 2x + 1$.

$\therefore \quad B = (0, -2), A = (1, 3)$ and $C = (-1, -1)$

$$\therefore \quad \Delta_2 = \text{ar}(\Delta BCA) = \left| \frac{1}{2} \begin{vmatrix} 0 & -2 & 1 \\ -1 & -1 & 1 \\ 1 & 3 & 1 \end{vmatrix} \right| = 3.$$

$$\therefore \quad \frac{\Delta_1}{\Delta_2} = \frac{6}{3} = 2.$$

18. Let $\tan^{-1}\left(\frac{\sin\theta}{\sqrt{\cos 2\theta}}\right) = \lambda$. Then $\tan\lambda = \frac{\sin\theta}{\sqrt{\cos 2\theta}}$.

$$\therefore \quad f(\theta) = \sin\lambda = \frac{\sin\theta}{\sqrt{\sin^2\theta + (\sqrt{\cos 2\theta})^2}}$$

$$= \frac{\sin\theta}{\sqrt{\sin^2\theta + \cos^2\theta - \sin^2\theta}} = \frac{\sin\theta}{\cos\theta} = \tan\theta.$$

$$\therefore \quad \frac{d}{d(\tan\theta)} f(\theta) = \frac{d}{d(\tan\theta)} \tan\theta = 1.$$

19. Differentiating w.r.t. x, $6f(x) = 3f(x) + 3xf'(x) - 3x^2$

or $xf'(x) - f(x) = x^2$ or $\frac{dy}{dx} - \frac{1}{x}y = x$

where $y = f(x)$

$\therefore$ integrating factor $= e^{\int -\frac{1}{x}dx} = e^{-\log x} = e^{\log x^{-1}} = x^{-1}$

$\therefore \quad \frac{d}{dx}\{y \cdot x^{-1}\} = x \cdot x^{-1} = 1; \quad \therefore \quad \frac{y}{x} = x + c.$

From the question, when $x = 1$, $y = f(1) = 2$.

So, $2 = 1 + c$; $\quad \therefore c = 1$.

$\therefore \quad \frac{y}{x} = x + 1$ or $y = f(x) = x^2 + x$

$\therefore \quad f(2) = 2^2 + 2 = 6$.

20. Let $\frac{\pi}{n} = \theta$.

Then the equation is $\frac{1}{\sin\theta} = \frac{1}{\sin 2\theta} + \frac{1}{\sin 3\theta}$

or $\frac{\sin 3\theta - \sin\theta}{\sin\theta \cdot \sin 3\theta} = \frac{1}{\sin 2\theta}$

or $\frac{2\cos 2\theta \cdot \sin\theta}{\sin\theta \cdot \sin 3\theta} = \frac{1}{\sin 2\theta}$

$\therefore \quad \sin 3\theta = 2\cos 2\theta \cdot \sin 2\theta = \sin 4\theta$;

$\therefore \quad 3\theta + 4\theta = \pi; \quad \therefore \quad 7 \cdot \frac{\pi}{n} = \pi \Rightarrow n = 7.$

21. $S_p = \frac{p}{2}\{2 \times 3 + (p-1)d\}$.

$$\text{Then } \frac{S_m}{S_n} = \frac{S_{5n}}{S_n} = \frac{\frac{5n}{2}\{6 + (5n-1)d\}}{\frac{n}{2}\{6 + (n-1)d\}}$$

$$\therefore \quad \frac{S_m}{S_n} = \frac{5(6 - d + 5nd)}{6 - d + nd}.$$

This will be independent of n if $6 - d = 0 \Rightarrow d = 6$.

$\therefore \quad a_2 = a_1 + d = 3 + 6 = 9$.

{$d = 6$ can also be obtained by equating the values of $\frac{S_m}{S_n}$ when $n = 1, 2$.}

22. AM $\geq$ GM for positive numbers.

$$\text{So, } \frac{a^{-5} + a^{-4} + a^{-3} + a^{-3} + a^{-3} + 1 + a^8 + a^{10}}{8}$$

$$\geq \sqrt[8]{a^{-5} \cdot a^{-4} \cdot a^{-3} \cdot a^{-3} \cdot a^{-3} \cdot 1 \cdot a^8 \cdot a^{10}} = \sqrt[8]{1 \cdot a^0} = 1$$

$\therefore \quad a^{-5} + a^{-4} + 3a^{-3} + 1 + a^8 + a^{10} \geq 8$.

23. $|2z - 6 + 5i| = 2\left|z - \left(3 - \frac{5}{2}i\right)\right|$

$= 2 \times \left[\text{distance between } z \text{ and the point } \left(3, -\frac{5}{2}\right)\right]$

$|z - 3 - 2i| \leq 2 \Rightarrow |z - (3 + 2i)| \leq 2$

$\Rightarrow z$ is a point on or inside a circle whose radius is 2 and the centre is (3, 2).

$\therefore$ minimum value of $|2z - 6 + 5i| = 2 \times \frac{5}{2} = 5$.

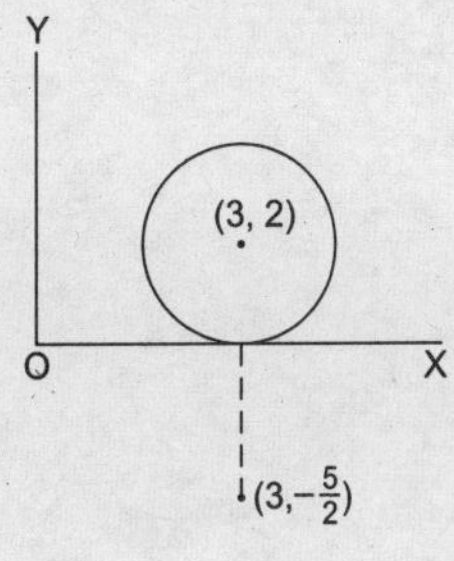